EVOLUTION OF THE INSECTS

Insects are the most diverse group of organisms to appear in the 3-billion-year history of life on Earth, and the most ecologically dominant animals on land. This book chronicles, for the first time, the complete evolutionary history of insects: their living diversity, relationships, and 400 million years of fossils. Whereas other volumes have focused on either living species *or* fossils, this is the first comprehensive synthesis of *all* aspects of insect evolution. Current estimates of phylogeny are used to interpret the 400-million-year fossil record of insects, their extinctions, and radiations. Introductory sections include the living species, diversity of insects, methods of reconstructing evolutionary relationships, basic insect structure, and the diverse modes of insect fossilization and major fossil deposits. Major sections cover the relationships and evolution of each order of hexapod. The book also chronicles major episodes in the evolutionary history of insects: their modest beginnings in the Devonian, the origin of wings hundreds of millions of years before pterosaurs and birds, the impact that mass extinctions and the explosive radiation of angiosperms had on insects, and how insects evolved the most complex societies in nature.

Evolution of the Insects is beautifully illustrated with more than 900 photo- and electron micrographs, drawings, diagrams, and field photographs, many in full color and virtually all original. The book will appeal to anyone engaged with insect diversity: professional entomologists and students, insect and fossil collectors, and naturalists.

David Grimaldi has traveled in 40 countries on 6 continents collecting and studying recent species of insects and conducting fossil excavations. He is the author of *Amber: Window to the Past* and is Curator of Invertebrate Zoology at New York's American Museum of Natural History, as well as an adjunct professor at Cornell University, Columbia University, and the City University of New York.

Michael S. Engel has visited numerous countries for entomological and paleontological studies, focusing most of his field work in Central Asia, Asia Minor, and the Western Hemisphere. In addition to his positions as Associate Professor in the Department of Ecology and Evolutionary Biology and Associate Curator in the Division of Entomology of the Natural History Museum at the University of Kansas, he is a Research Associate of the American Museum of Natural History and a Fellow of the Linnean Society of London.

David Grimaldi and Michael S. Engel have collectively published more than 250 scientific articles and monographs on the relationships and fossil record of insects, including 10 articles in the journals *Science, Nature,* and *Proceedings of the National Academy of Sciences.*

Evolution of the Insects

David Grimaldi
American Museum of Natural History

Michael S. Engel
University of Kansas

CAMBRIDGE
UNIVERSITY PRESS

CAMBRIDGE UNIVERSITY PRESS
Cambridge, New York, Melbourne, Madrid, Cape Town, Singapore, São Paulo

Cambridge University Press
40 West 20th Street, New York, NY 10011-4211, USA

www.cambridge.org
Information on this title: www.cambridge.org/9780521821490

First published 2005
Reprinted 2006

Printed in Hong Kong

A catalog record for this publication is available from the British Library.

Library of Congress Cataloging in Publication Data

Grimaldi, David A.
Evolution of the insects / David Grimaldi, Michael S. Engel.
p. cm.
Includes bibliographical references and index.
ISBN 0-521-82149-5 (alk. paper)
1. Insects – Evolution. I. Engel, Michael S. II. Title.

QL468.7.G75 2004
595.7′138 – dc22 2004054605

ISBN-13 978-0-521-82149-0 hardback
ISBN-10 0-521-82149-5 hardback

An orthopteran of the extinct family Elcanidae in 120 MYO limestone from Brazil's Santana Formation. AMNH; length of elcanid (including antennae) 98 mm(3.8 in.).

For the entomophiles,
winged and larval

CONTENTS

PREFACE

Writing a book on a subject as vast as the evolution of the most diverse lineage of organisms had one simple justification for us: it was needed. Having taught Insect Diversity and Insect Systematics at the City University of New York, Columbia University, Cornell University, and the University of Kansas, we became acutely aware of a gaping hole in entomology. No volume integrates the unprecedented diversity of living and extinct insects, particularly within the evolutionary framework of phylogeny. Some excellent texts, popular books, and field guides cover insect identification, structure, and living diversity, as well as physiology, behavior, and general biology, of which *The Insects of Australia* (Naumann, 1991a) is perhaps the best example. For our lectures to students we thus found ourselves pulling an extremely scattered literature together. Instead of trudging through the insect families – interesting as they are – we found that students were fascinated by an approach of folding Recent insect diversity into one large context of phylogeny, biogeography, ecology, and the fossil record. The big picture engaged them. After four years of intensive literature research and writing, study and imaging of important museum specimens, and thousands of figures, we like to think we've succeeded in our goal.

Our approach to the volume was tempered by our own experience and interests with fossil insects. Entomologists typically ignore fossils, and since we too work on speciose groups of living insects, we have always been intrigued by the dismissiveness of most entomologists. Why ignore such illuminating parts of evolutionary history? We hope that this book will reveal to our colleagues the significance, and even esthetics, of insect fossils. There are several comprehensive treatments of the insect fossil record, particularly the hexapod section of the *Treatise on Invertebrate Paleontology* (Carpenter, 1992) and the more recent *History of Insects* by Rasnitsyn and Quicke (2002). But these volumes are devoted entirely to fossil insects, so something more inclusive, and accessible, was needed.

Compiling a book like this is humbling, not only because of the scope of the subject, but also because discoveries and new work reported every month in paleontology and insect systematics continually revise the field. As this book was nearing completion, for example, two large projects were launched. One of these is the U.S. National Science Foundation's *Tree of Life* project, which seeks to examine the phylogeny of major groups of organisms using all existing data and vast new morphological and DNA data. The other is the Dresden conference on insect phylogeny, which met for the first time in 2003 (e.g., Klass, 2003), and which is intended to meet every few years. Like the insects themselves, our understanding is thus evolving. As more genes become sequenced for hundreds more species of insects, for example, phylogenetic hypotheses will be revised, or at least discussed. But, thirty years ago a book like this would have been very different and much slimmer. Our knowledge of insect relationships has advanced tremendously over this period of time, and dozens of spectacular fossil deposits of insects have been discovered. Tomorrow's discoveries will reinforce, revise, and entirely redefine our present knowledge, but one needs to start somewhere. The optimal moment is always elusive. We hope that thirty years from now – indeed, twenty – much of what we present here will not fall far from the mark. Should we be so fortunate, new editions of this volume will attempt to keep abreast of developments.

Working at the American Museum of Natural History has also given us a keen appreciation for appealing to the nascent naturalist and scientist, not only to the landed professional. We were very deliberate in developing a volume that would be visually engaging to insect and fossil collectors, general naturalists, botanists, and other biologists, as well as to student and professional entomologists. Although we tried to avoid the thick jargon of entomology and systematics, it was not entirely avoidable (some of the jargon is useful), and we hope our colleagues will understand this was done deliberately to make the subject more digestible. The nearly 1,000 images were also included to make the book more engaging. Should the images and captions whet the reader's appetite, a healthy meal of text is also available.

A volume like this would not have been possible without the assistance of authoritative colleagues around the world, who kindly reviewed large chunks of text. These authorities include: chapter 1: Lee Herman, Valerie Schawaroch, and Craig Gibbs; chapter 2: Derek Briggs (fossilization) and Vladimir Blagoderov (deposits); chapter 4: Ismael A. Hinojosa-Díaz; chapter 5: Michael Ohl; chapter 7: Daniel J. Bennett, George W. Byers, and Kumar and Valerie Krishna (termites); chapter 8: Niels P. Kristensen, Lance Durden (Phthiraptera), Bruce Heming (Thysanoptera), Penny Gullan (Sternorrhyncha), and Nils Møller Andersen (Heteroptera); chapter 9: Michael Ohl; chapter 10: Lee Herman, Caroline Chaboo, Jim Liebherr, Marc Branham, and Jeyaraney Kathirithamby; chapter 11: Ricardo Ayala and Charles D. Michener; chapter 12: George Byers (Mecoptera), Robert Lewis (Siphonaptera), and Dalton Amorim and Vladimir Blagoderov (Diptera); chapter 13: Niels Kristensen (entire), David Wagner and Eric Quinter (Lepidoptera), and Phil DeVries (butterflies); chapter 14: Dalton S. Amorim, Amy Berkov, Peter Cranston (entire), and William L. Crepet (angiosperms). Charles D. Michener and Molly G. Rightmyer generously reviewed various sections. We take all responsibility for the final version of the book since, in a few cases, we felt compelled to disagree with reviewers.

Numerous colleagues and institutions loaned images: Alex Rasnitsyn; Bryn Mawr College; Carl Rettenmeyer; Carlos Brandão; Caroline Chaboo; Carsten Brauckmann; Deutsche Entomologische Institut; Thomas D. Seeley; Dieter Waloszek; Enrique Peñalver; Helmut Sturm; Holger H. Dathe; Horst and Ulrike Aspöck; Jim Davis; Librarians of the American Museum of Natural History, particularly Mary DeJong; Liz Brozius; Michael Dolan; Nick Fraser; Bibliothèque Centrale of the Museum National d'Histoire Naturelle; Ray Swanson; *Science News*; Scott Elias; University of Massachusetts, Amherst; Wilfried Wichard; and Xavier Martínez-Delclòs. In this regard, we are particulary grateful for being able to use the portraits of important entomologists provided by George W. Byers and the many beautiful images of living insects and of entomologists provided by our colleagues Phil DeVries, Janice Edgerly-Rooks, Valerie Giles, Steve Marshall, Cristina Sandoval, Ray Swanson, and Alex Wild.

We are also grateful to the many individuals who assisted us in our museum travels to examine important specimens, particularly Peter Jell (Queensland Museum), Robert Jones (Australian Museum), Phil Perkins (Harvard University), Alexandr P. Rasnitsyn (Paleontological Institute, Moscow), Andrew J. Ross (Natural History Museum, London), and Tim White (Yale University). Many people provided loans and gifts of specimens: Dalton Amorim, David Wagner, Jeff Cumming, Jens von Holt, Jeyaraney Kathirithamby, Keith Luzzi, Ken Christiansen, Klaus-Dieter Klass, Lance Durden, Mike Picker, Penny Gullan, Robert Lewis, Roy Larimer, Susan Hendrickson, and the late Jake Brodzinsky. We are extremely grateful for the hard work and generosity of the New Jersey amber collectors, particularly Keith Luzzi, Paul Nascimbene and the late Steve Swolensky. Roy Larimer and Keith Luzzi have been extremely generous with their time and resources in the field, and they are two of the finest field workers we know.

Particularly generous were Dr. Herbert Axelrod and Dott. Ettore Morone. Dr. Axelrod donated the collection of Santana Formation fossils to the American Museum of Natural History (AMNH), and provided generous funding over the years in support of research on this collection. The senior author often visited Ettore for the study of his magnificent collection of Dominican amber, and he was as accommodating and gracious a host as one could ever have. Images of beautiful specimens from these two collections grace the volume throughout.

Our work over the years has been funded by various sources: the National Geographic Society; Sigma Xi, the Scientific Research Society; the U.S. National Science Foundation; Kansas Technology Enterprise Corporation-Kansas NSF EPSCoR (KAN29503); the late Henry G. Walter, former trustee of the AMNH; Henry and Meryl Silverstein; and donations in memory of Steve Swolensky.

Last, and hardly least, Mr. Robert G. Goelet, Chairman Emeritus and Trustee of the American Museum of Natural History, has been steadfast in his generosity toward the wonderful collection of amber fossils at the AMNH, for funding fieldwork, and for funding Michael S. Engel as a research scientist at the AMNH for two enjoyable years. Mr. Goelet also generously donated funds to help defray the cost of publication for this book to make it more available, indeed, possible. We hope this volume is a pleasant reminder of your former teacher, the late Professor Frank Carpenter.

Production of this book would not have been possible without the skilled assistance of four AMNH staff. Paul Nascimbene (Collections Specialist) has been a dedicated and diligent preparator of thousands of amber and rock fossil specimens; and Simone Sheridan (Curatorial Assistant) meticulously attended to databases, references, and specimen preparation. Tam Nguyen (Senior Scientific Assistant) and Steve Thurston (Graphic Artist) produced most of the images. Tam did all the SEMs and many of the photomicrographs. Steve rendered cladograms and other diagrams, and both he and Tam composed many of the plates. The thousands of images for the book would have been impossible without the use of fine optics, lighting, and digital photography available from Infinity, Inc., and MicrOptics, Inc., all made possible by the expertise of Roy Larimer. Words fail to express the complete extent of our gratitude to these people.

The gestation of this book took longer than we expected, so we deeply appreciate the patience of our students and colleagues while we cloistered ourselves. The patience and support of the editors at Cambridge University Press, Kirk Jensen, Shari Chappell, and Pauline Ireland, are also appreciated. We are especially grateful to Ward Cooper, former Acquisitions Editor at Cambridge, for his initial interest in this work, his enthusiasm for the project, and his calming influence. Camilla Knapp was a Production Editor *par exellence*, who should never have to endure a work of this size and complexity again. This volume could not have come to fruition without her skill and experience.

Everyone is a product of their past, and to a large extent this volume reflects several influential teachers of ours, whose tutelage and support we will always fondly remember: Thomas Eisner, Charles Henry, John Jaenike, James Liebherr, Charles Michener, Quentin Wheeler, and the late George C. Eickwort and William L. Brown, Jr. Lastly, without the stalwart patience and support of our families and loved ones, we could not possibly have waded through this work: Karen, Rebecca, Emily, Nicholas, and Dominick; Jeffrey, Elisabeth, Donna, and A. Gayle. They quietly endured our absences and steadily encouraged us. They understand.

COMMONLY USED ABBREVIATIONS

MY	million years
MYA	million years ago
MYO	million years old
BP	base pairs (of DNA)
DNA	deoxyribonucleic acid
RNA	ribonucleic acid
YA	years ago

Time Periods

T	Tertiary
K	Cretaceous
J	Jurassic
Tr	Triassic
P	Permian
C	Carboniferous

1 Diversity and Evolution

INTRODUCTION

Evolution begets diversity, and insects are the most diverse organisms in the history of life, so insects should provide profound insight into evolution. By most measures of evolutionary success, insects are unmatched: the longevity of their lineage, their species numbers, the diversity of their adaptations, their biomass, and their ecological impact. The challenge is to reconstruct that existence and explain the unprecedented success of insects, knowing that just the veneer of a 400 MY sphere of insect existence has been peeled away.

Age. Insects have been in existence for at least 400 MY, and if they were winged for this amount of time (as evidence suggests), insects arguably arose in the Late Silurian about 420 MYA. That would make them among the earliest land animals. The only other terrestrial organisms of such antiquity are a few other arthropods, such as millipede-like arthropleuridans and scorpion-like arachnids, and some plants. But age alone does not define success. Various living species belong to lineages that are hundreds of millions of years old, like horsetails (*Equisetum*), ginkgo, horseshoe "crabs" (*Limulus*), and the New Zealand tuatara (Rhynchocephalia), all of which, and many more species, are vestiges of past diversity. The living coelacanth (*Latimeria*), as another example, is the sole survivor of a 380 MYO lineage, and the very synonym for "relict." Not so for the insects. While there are some very significant extinct insect lineages, such as the beaked Palaeodictyopterida, most modern insect orders appeared by 250 MYA, and many living insect families even extend to the Cretaceous about 120 MYA. Some living insect families, in fact, like staphylinid beetles and belostomatid water bugs, appeared in the Late Triassic approximately 230 MYA. By comparison, 120 MYA only the earliest and most primitive therian mammals had appeared, and not until 60 MY later did modern orders of mammals appear. Perhaps the most recited example of evolutionary persistence concerns 300 million years of cockroaches, but this also brings up a very important aspect about fossils, which is their proper interpretation.

Fossil "roachoids" from 320 MYA to 150 MYA were actually early, primitive relatives of living roaches that retained a large, external ovipositor and other primitive features of insects (though they did have a shield-like pronotum and forewings similar to modern roaches). To interpret roachoids or any other fossil properly, indeed the origin and extinction of whole lineages, it is crucial to understand phylogenetic relationships. The incompleteness of fossils in space, time, and structure imposes challenges to understanding them, which is why most entomologists have avoided studying fossil insects, even beautifully preserved ones. Fortunately, there has never been more attention paid to the phylogenetic relationships of insects than at present (Kristensen, 1975, 1991, 1999a; Boudreaux, 1979; Hennig, 1981; Klass, 2003), including research based on DNA sequences (Whiting *et al.*, 1997; Wheeler *et al.*, 2001; Whiting, 2002), so an interpretive scaffolding exists and is being actively built. Entomologists are beguiled by the intricacy of living insects, their DNA, chemistry, behavior, and morphological detail, as the electron micrographs throughout this book partly reveal. But, ignoring fossils relegates us to a small fraction of all insects that have ever existed and seriously compromises our understanding of insect evolution.

Fossils provide unique data on the ages of lineages, on radiations, and on extinctions (Figure 1.1). Social bees, for example, occur today throughout the world's tropics. However, based on diverse fossils in amber from the Baltic region – an area today devoid of native advanced social bees aside from the western honey bee, *Apis mellifera* – they were unexpectedly diverse in the Eocene 40–45 MYA (Engel, 2001a,b). Ants and termites existed for 50–100 MY before they became diverse and abundant (Grimaldi and Agosti, 2000; Dlussky and Rasnitsyn, 2003), indicating that sociality per se is insufficient for ecological dominance (rather, highly advanced societies in huge colonies make *certain* ants and termites ecologically dominant today). Tsetse flies (Glossinidae)

1.1. A fossil plant hopper of the living family Issidae, in Miocene amber from the Dominican Republic. Fossils are the only direct evidence of extinct life so they contribute unique insight into reconstructing evolutionary history. M3445; wingspan 8 mm; Photo: R. Larimer.

occurred in Europe and North America in the Oligocene and latest Eocene, 30–40 MYA, far outside their range in Africa today. Giant odonatopterans – griffenflies – cruised the Permian skies, their size possibly enabled by the high oxygen content of the atmospheres at the time (Dudley, 2000). When fossils provide insights like these, the greatest sin of omission arguably is avoidance of the fossil record, despite the challenges to studying fossils. Such avoidance is certainly not for a shortage of insect fossils.

The insect fossil record is surprisingly diverse and far more extensive than most entomologists and paleontologists realize. Hundreds of deposits on all continents harbor fossil insects (Rasnitsyn and Quicke, 2002; Chapter 2). Also, the manner in which insects have become fossilized exceed that of probably all other organisms except plants (Chapter 2). Insects are commonly preserved as compressions in rock (particularly their wings), but they are also preserved as exquisite three-dimensional replicas in carbon, phosphate, pyrite, and silica; as original cuticular remains from Pleistocene and Holocene tar pits, bogs, and mammalian mummies; as remains of their galleries and nests; and as inclusions in chert, onyx, gypsum, and of course amber. Insects are the most diverse and abundant fossils in ambers around the world (Grimaldi, 1996), though fossil resin records only the last third of insect evolutionary history. More recent exploration of fossilized plants has revealed a wealth of insect feeding damage (Scott, 1991; Scott *et al.*, 1991; Labandeira, 1998), including specialized relationships between insects and plants.

Fortunately, the voluminous and scattered primary literature on fossil insects is now summarized in several compendia. The treatise by Carpenter (1992) is a catalogue of fossil insect genera described up to 1983, illustrated with reproduced drawings of the type species for many genera. Since 1983 about 500 families and 1,000 genera have been added to the insect fossil record. Carpenter's treatise is nicely complemented by the volume by Rasnitsyn and Quicke (2002), since the latter reviews major fossil insect deposits, insects in ancient ecosystems, and the fossil record and relationships within orders, particularly of extinct families. The volume by Rasnitsyn and Quicke, though, uses names of insect groups from Laicharting (1781), which no one else uses or even

1.2. A common halictine bee, visiting a flower in Vancouver, Canada. Flowering plants, and therefore much of terrestrial life, depend in large part on insect pollinators. Nearly half of all living insects directly interact with plants. Photo: R. Swanson.

1.3. The diversity of life shown as proportions of named species.

recognizes, and their systems of relationships (based almost entirely on fossil evidence) often conflict with phylogenies based on expansive evidence from living insects. Short reviews of the fossil record of insects include Wootton (1981, for Paleozoic insects only), Carpenter and Burnham (1985, now rather dated), Kukalová-Peck (1991), Ross and Jarzembowski (1993), Willmann (1997, 2003), Labandeira (1999, 2001), and Grimaldi (2001, 2003a). The volume by Hennig (1981) attempted to synthesize the geological record of insects with relationships of living insects, but the evidence he drew from was very limited compared to what is now known. We have adopted Hennig's approach here, drawing fossils into the fold of the spectacular Recent diversity of insects, but in a much more comprehensive treatment and based on original study of many fossils.

Species and Adaptive Diversity. The daunting number of Recent species of insects is well known to naturalists (Figures 1.2 and 1.3). Though there are nearly one million described (named) species, the total number of insects is believed to be between 2.5 million and 10 million, perhaps around 5 million species. In an age of such technological sophistication and achievement, it is remarkable that there is an error range for estimates of insect species in the millions. Despite this fundamental problem, without a doubt the diversity of any other group of organisms has never been more than a fraction of that of insects. The enduring question, of course, is: Why? The arthropod design of an exoskeleton with repetitive segments and appendages preadapted insects for terrestrial existence, and wings further refined this design by vastly improving mobility, dispersal, and escape. Judging just from Recent species, though, a more recent innovation in insect evolution spurred their success, which is holometabolous development. Just four orders today, Coleoptera, Diptera, Hymenoptera, and Lepidoptera account for approximately 80% of all insects, and these have a larva, or "complete" metamorphosis. It is uncertain, though, why a larval stage is so advantageous, as we discuss later. Two lineages within the holometabolan "big four" contain the two largest lineages of plant-feeding animals: the Lepidoptera (150,000 species) and phytophagan beetles (100,000 species). In each of these two lineages, almost all species feed on angiosperms, and many are restricted to particular species or genera of angiosperms. Indications are that these and other insect groups (indeed, nearly half of all insects) have co-radiated with the angiosperms beginning 130 MYA, but exactly how host plant specialization promotes speciation still needs to be resolved.

Another measure of diversity besides number of species is the variety of structures and behaviors that adapt insects to environmental challenges. The most obvious of these is wings. Insects are one of only four lineages of animals that had or have powered flight, the others being (in order of appearance) pterosaurs, birds, and bats. Insects evolved flight just once (based on the apparent common ancestry of all winged insects, or pterygotes), at least 100 MY before pterosaurs and perhaps 170 MY before them if *Rhyniognatha* (Figure 5.8) was actually winged. A time traveler going into the mid-Carboniferous to the mid-Triassic, 330–240 MYA, would have seen only insects in the air. Insects indeed. During the Permian, giants like *Meganeuropsis permiana* had a 27 inch (70 cm) wingspan and were the apex of aerial predators. Today, the flight of most insects outperforms that of birds and bats in energetic efficiency, wing beat frequency, and agility, though not speed. Birds and bats are the major vertebrate predators of Recent insects, but they clearly didn't wrest the air from insects; insects may have even spurred the evolution of flight in early insectivorous ancestors of these vertebrates. As birds and bats improved their abilities in flight, insects evolved an arsenal of defenses against them. No group of animals, for example, matches the camouflage and mimicry seen in insects (e.g., Figures 7.24 to 7.27, 13.62, 13.77, 13.87). Night-flying insects repeatedly evolved hearing organs sensitive to the ultrasonic calls of bats so they divebomb or fly in loops to escape an approaching bat. Myriad day-flying insects have evolved warning, or aposematic, coloration either to advertise their venomous or toxic defenses or to mimic such species (e.g., Figures 13.88, 13.90). No group of animals possesses the chemical repertoire of insects from pheromones to toxic defensive secretions (Eisner, 2003). Only plants are as diverse in their chemical defenses, and in many cases phytophagous insects sequester host plant toxins for their own use.

Our time traveler to 330–240 MYA would also have noticed no chorusing frogs or song birds, not even dinosaurs. Other than the occasional squawk or grunt of a labyrinthodont or other early tetrapod, animal sounds would have been largely from singing insects. Fossilized wings of orthopterans are preserved complete with stridulatory structures, and in one case were used to reconstruct the song (Rust *et al.*, 1999). One can only imagine that Triassic Titanoptera (Figure 7.43) had a deep, resonant song, like a bullfrog. By the Jurassic the familiar nocturnal trill of crickets filled the air.

Sociality is perhaps the most striking and sophisticated innovation by insects (Wilson, 1971). Only one mammal (the naked mole rat of Africa) has advanced sociality, a behavior involving closely related individuals of different generations living together and specialized for particular tasks, particularly reproduction. Otherwise, sociality is entirely an arthropod innovation that occurs in groups as diverse as mantis shrimps and some spiders (Choe and Crespi, 1997) but that has evolved approximately 20 times in insects (Chapter 11; Table 11.7). The colonies of some attine (leaf cutter) ants, army and driver ants, and termitid termites contain millions of individuals housed in labyrinthine nests – the most elaborate constructions in nature. Such large colonies usually have extreme specialization: major and minor workers, soldiers, a queen replete with huge ovaries to produce thousands of eggs per year, and expendable males. No societies, including those of humans, have such efficiency.

To some extent adaptive diversity is both the cause and the effect of species diversity, but it also seems to be an intrinsic aspect of insect design, with refinements building on the basic design. Having six legs allows for the front pair to become raptorial or fossorial without losing the ability to walk. Wings facilitate mobility, but when the fore pair is hardened as in Heteroptera and Coleoptera, they protect the flight pair and abdomen when the insect is wedged in tight spaces and burrowing into substrates. An impervious exoskeleton guards against injury and desiccation on land but also protects insects from their own toxic secretions (Blum, 1981).

Ecological Dominance. In terms of biomass and their interactions with other terrestrial organisms, insects are the most important group of terrestrial animals. Remove all vertebrates from earth, by contrast, and ecosystems would function flawlessly (particularly if humans were among them). Insects, moreover, have invaded virtually every niche except the benthic zone, including ocean shores and in one instance (the water strider *Halobates*) the open ocean. On land, though, insects reign.

Angiosperms are the defining terrestrial life form, but even these have co-radiated with the insects. Approximately 85% of the 250,000 species of angiosperms are pollinated by insects, and the inspiring diversity of flowers, in fact, is due in large part to insects lured to them (Figure 1.2). Thousands of generalized insect species visit and feed from flowers today, so similar liaisons in the Early Cretaceous must have spurred the diversification of angiosperms, and fossils indicate that specialized insect pollinators evolved quickly after angiosperms appeared. When bees evolved about 120 MYA, and later radiated eventually to form the current fauna of 20,000 species, the world truly blossomed. Bees are extremely efficient foragers and pollinators, and without doubt these insects alone are the most important agents of pollination.

The impact of insects, as plant-feeding organisms (phytophages), eclipses that of all other animals, the most impressive testament being crop pests. No other group of organisms affects agriculture and forestry as much as insects. A few of the more devastating ones include the boll weevil (*Anthonomus grandis*), Colorado potato beetle (*Leptinotarsa decemlineata*), and Mediterranean fruit fly (*Ceratitis capitata*), which alone inflict annual damage amounting to hundreds of millions of dollars, and for which tons of insecticides are broadcast. Migratory locusts (*Schistocerca*) form swarms of biblical proportions – billions of individuals covering several thousand square kilometers – and because they have indiscriminate diets, their swarms denude entire landscapes. Bark beetles (Scolytidae) and gypsy moths (*Lymantria*) can destroy or denude entire forests. In all, the cumulative effect of approximately 400,000 species of plant-feeding insects must be staggering. It has been estimated, in fact, that every species of plant has at least one species of insect that feeds on it, and probably all plants have many more than this (some host dozens of insect species). Even on the savannas of eastern Africa, renowned for the vast herds of ungulates, insects like orthopterans, beetles, caterpillars, and termites consume more cellulose than all mammalian herbivores combined. The array of plant chemical defenses is arguably attributed to the herbivory of insects, two groups that have been waging an arms race for 350 MY or more.

Insect vectors of pandemic diseases have probably affected humans more than any other eukaryotic animals. Tens of millions of people have died throughout historical times as a result of just six major insect-borne diseases: epidemic typhus (a spirochete carried by *Pediculus* lice), Chagas's disease (a trypanosome carried by triatomine bugs), sleeping sickness (another trypanosome, carried by *Glossina* tsetse), and the three big ones, malaria (*Plasmodium* carried by *Anopheles* mosquitoes), yellow fever (a virus carried by *Aedes* mosquitoes), and plague (a bacterium carried by *Xenopsyllus* and *Pulex* fleas). Two mutations in humans, sickle cell anemia and the delta-32 gene, are actually genetic adaptations to millennia of selection by malaria and plague, respectively. While these microbes are the immediate agent of selection, their mosquito and flea vectors are the only metazoans known to have affected the evolution of humans. Given the scale with which humans have been affected, blood-feeding insects have obviously had an immense effect on natural populations of various land vertebrates.

While earthworms are absolutely essential for soil-building (humification), certain insect detritivores, particularly termites (Isoptera), play a role that earthworms can't. Termites comprise an estimated 10% of all animal biomass in the tropics; one virtually cannot kick into a rotting log in a tropical forest without having termites spill out. In tropical regions they consume an estimated 50–100% of the dead wood in forests, as well as dead grasses, humus, fungi, and herbivore dung, and so are absolutely essential in mineralization of plant biomass. The huge termite mounds on the savannas of Africa, South America, and Australia are chimneys for the waste gases from the huge underground nests. A large nest has the respiratory capacity of a cow, and it has even been estimated that termites contribute 2–5% of the annual global atmospheric methane. The amount of soil that is moved by these insects is prodigious: one geological formation in eastern Africa, formed between 10,000 and 100,000 years ago by the living mound-building species *Macrotermes falciger*, consists of 44 million cubic meters of soil (Crossley, 1986). Some ants vie with the excavation abilities of these termites, particularly leaf-cutter (attine) ants. Unrelated *Pogonomyrmex* ants, which form modest-sized colonies of approximately 5,000 individuals, excavate sand that is more than 100 times the weight of the colony in just 4 days (Tschinkel, 2001). Since the biomass of ants in the world's tropical river basins is estimated to be up to four times that of vertebrates, their impact on humification and mineralization, as well as the predation of other arthropods is likewise prodigious. But perhaps no other fact speaks to the ecological significance of ants as this: More than 2,000 species in 50 families of arthropods mimic ants, hundreds of plant species in 40 families have evolved specialized structures for housing ant colonies, and thousands of hemipteran species engage in intimate protective alliances with ants in exchange for honeydew. Ants have had a pervasive effect on the evolution of other insects and are clearly keystone consumers in the tropics.

Because insects have been so destructive to agriculture and human health, less informed people gladly imagine a world devoid of insects. But *if ants, bees, and termites alone were removed from the earth, terrestrial life would probably collapse.* Most angiosperms would die, the ensuing plant wreckage would molder and ferment for lack of termites, soil depleted of nutrients would barely be able to sustain the remaining plants; erosion would choke waterways with silt.

Vast tropical forests of the Amazon, Orinoco, Congo, and other river basins would die off, and the earth's atmosphere and oceans would become toxic.

Without a doubt, the ecological significance of insects, their diversity, and the longevity of the insect lineage makes this the most successful group of organisms in earth's history, and a subject completely worthy of our understanding.

SPECIES: THEIR NATURE AND NUMBER

To understand evolution and its history, it is essential to understand what is a species. The concept of species is so entrenched in biology that it should be very easy to define or describe, but it has meant different things to different biologists. Species (singular and plural) have generated a great deal of discussion (perhaps too much), but it is important to review it briefly here because the hallmark of insects is that there are more species of them than any other group of organisms. Without question, species comprise a real unit – the fundamental unit of nature (Wilson, 1992) – and not a category defined at somewhat of an arbitrary level, like genera and families. Fortunately, we can draw on several intensively studied insects to illustrate the empirical nature of species.

Species have been recognized well before Linnaeus, who erected this as a formal category for classification ("species" means "kind" in Latin). In the first half of the twentieth century, the New Synthesis in evolutionary biology was preoccupied with variation and its significance in evolutionary change. One of its architects, Ernst Mayr, reacted strongly to the traditional systematic concept of species. To Mayr (1942, 1963), the concept of species up to that point was *typological*, wherein systematists grouped individual organisms into a species if they all conformed to a particular standard or ideal. Mayr, as a bird systematist, was familiar with the constant variation within species that sometimes confounded interpretations of species' boundaries. Most systematists dismissed the variation as trivial, but to Mayr and other evolutionists the variation was highly significant. Mayr's definition of species, the *biological species concept*, was "a group of actually or potentially interbreeding populations, which are reproductively isolated from other such groups." In other words, if two individuals mate and produce offspring, they're the same species, because they share the same gene pool. There were difficulties with this concept. First, "potentially" was an unfortunate adverb to use. Many closely related species can be forced to breed in the laboratory, zoo, or barnyard, but they produce infertile offspring or hybrids, like mules, but hybrids of some species are fertile. It was argued, in response, that individuals within a species would only breed *naturally*, but, again, such hybrids also occur, like the "red wolf" of the southern United States, which is a wolf-coyote hybrid. Also, what about parthenogenetic organisms, including bacteria, all bdelloid rotifers, many insects, and even some vertebrates, all of which are easily classifiable as species on the basis of morphology and DNA? Or fossils? Individuals separated by thousands of generations may belong to the same species, but they are hardly reproductively compatible. Lastly, the daily work of systematists is deciphering species from preserved specimens, so breeding experiments are just too impractical, and yet great progress has been made in deciphering species. In fact, Mayr (1942, 1963) used these traditional systematic studies with their "typological" concepts quite successfully in formulating the biological species concept.

Another major criticism leveled against the biological species concept is that it defines species on the basis of the process by which they arise: Species are formed when an isolated population or group of individuals becomes reproductively isolated from other populations. Defining species as reproductively isolated (or interbreeding) groups of individuals is thus circular. In response, some systematists defined species using different criteria, leading to evolutionary (Simpson, 1944; Wiley, 1978), phylogenetic (Wiley, 1978; De Queiroz and Donoghue, 1988; Cracraft, 1989; Wheeler and Meier, 2000), and other concepts of species (reviewed in Futuyma, 1998). The first two of these are actually not very different, and they also accommodate the process by which systematists work. A reasonable consensus of the evolutionary and phylogenetic definitions of a species is that it is *a discrete group of individual organisms that can be diagnosed, or defined on the basis of certain specialized features, and that had a common ancestor and unique evolutionary history*. The species could be defined on the basis of any feature of its genotype or phenotype, including morphology and behavior. Strict adherence to this definition, however, is not without *its* problems. First, how can a "unique evolutionary history" actually be observed? It can only be inferred, based on the strength of the evidence defining the species, like the morphological characters or the DNA sequences. If the sole criterion for circumscribing species is that they be discrete groups of individuals, then some variants could be called different species, like the color morphs of many butterflies or castes of an ant colony. A few phylogeneticists might not have any problem calling color morphs of a butterfly as different species, but we actually know that the morphs differ by just one or a few genes that affect coloration, and in all other respects they are identical.

In reality, systematists have been using a phylogenetic and evolutionary species concept all along. They assess variation and then lump individuals on the basis of consistent similarities. It is very reassuring that the results of this practice have largely agreed with results based on the biological species concept. This is well revealed by the study of two genera of insects, *Drosophila* fruitflies and *Apis* honey bees. Years of scrutiny of each of these two genera – their morphology, genetics, behavior, ecology, and hybrids – have provided probably more empirical evidence on the nature of species than have any other kind of organisms.

DROSOPHILA

That stupid little saprophyte.
–William Morton Wheeler, on *Drosophila melanogaster*

Drosophila fruitflies may not have the behavioral repertoire of ants that so fascinated the famous entomologist W. M. Wheeler, but *Drosophila* has revolutionized biology more than any other organism. Contrary to popular belief, *Drosophila* does not naturally live in little vials. There are approximately 1,000 species in the genus, which breed in a great variety of plants and other substrates. Some species are highly polyphagous and have followed humans around the globe, the so-called tramp or garbage species. The laboratory fruitfly, *Drosophila melanogaster*, is one such tramp species. It was originally used by T. H. Morgan and his "fly group" at Columbia University for probing the elements of heredity and the behavior of chromosomes (see Sturtevant, 1965; Kohler, 1994). Because its genetics became so well known, *D. melanogaster* has been and is still used in all sorts of laboratory research, from cell biology, to physiology, behavior, and ecology (Lachaise *et al.*, 1988; Ashburner, 1989), making it, arguably, the best known eukaryotic organism. To better understand *D. melanogaster*, there has been intensive comparison of this species to its three closest relatives: *D. simulans*, which is a polyphagous African species introduced around the world; *D. mauritiana*, endemic to the islands of Mauritius and Rodriguez in the Indian Ocean; and *D. sechellia*, endemic to the Seychelles Islands, also in the Indian Ocean. The ancestral distribution of *D. melanogaster* is

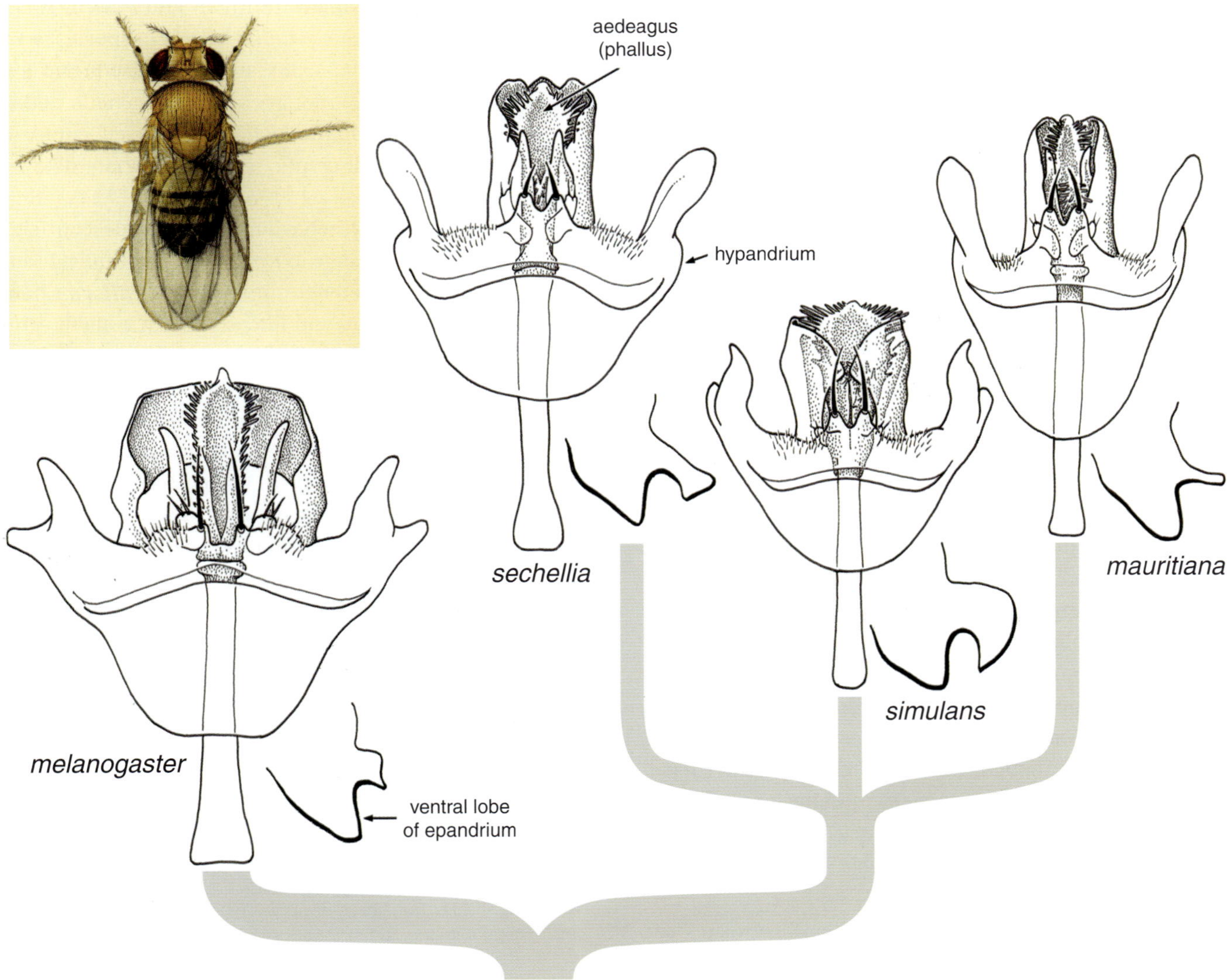

1.4. Relationships among closely related species in the *Drosophila melanogaster* complex, differences being best reflected in the male genitalia (shown here). Relationships based on Hey and Kliman (1993) and Kliman *et al.* (2000).

believed to be central Africa. Collectively, these species comprise the *melanogaster* complex of species.

Individuals of the *melanogaster* complex are consistently separated and grouped on the basis of male and female genitalia (Figure 1.4), mating behavior (Cowling and Burnet, 1981; Cobb *et al.*, 1986), chromosomes (Ashburner and Lemeunier, 1976; Lemeunier and Ashburner, 1976), DNA sequences (Hey and Kliman, 1993; Kliman and Hey, 1993; Kliman *et al.*, 2000; Schawaroch, 2002), and other features, including larval diet. For example, even though *D. simulans* and *D. melanogaster* breed in a great variety of decaying fruits, *D. sechellia* is very specialized and breeds naturally only in fruits of *Morinda citrifolia* (Rubiaceae), which contain toxins that the other species can't tolerate. *Drosophila simulans, D. sechellia,* and *D. mauritiana* are most closely related, based on DNA sequences (Kliman *et al.*, 2000), their *homosequential* polytene chromosomes (there are no distinguishing inversions), and fertile F_1 hybrid females (F_1 males are sterile). In a comprehensive study of 14 genes and nearly 40 strains of these species (Hey and Kliman, 1993; Kliman and Hey, 1993; Kliman *et al.*, 2000), all or most strains of these species are grouped according to traditional separation using morphology and chromosomes. Interestingly, though, a few strains of *D. simulans* grouped with *D. sechellia* or *D. mauritiana*, but groupings varied depending on the gene.

Apparently, *D. sechellia* and *D. mauritiana* evolved nearly contemporaneously as peripheral, isolated populations of *D. simulans*. This has fundamental implications for systematics because in this case a living species is considered ancestral and not a simple two-branched divergence from an extinct common ancestor. In a mainstream phylogenetic view, at least some strains of *D. simulans* would not belong to that species, because they make *D. simulans* a *paraphyletic* taxon (basically everything left over after *D. mauritiana* and *D. sechellia* were extracted). Yet, *D. simulans* has distinctive (diagnosable) and consistent differences with other species in the complex. Also, a typical assumption in phylogenetic analyses is that divergence is bifurcating, or two-branched, even though traditional models of speciation allow for the simultaneous origin of species. Traditionally, it has been thought that isolated populations on the periphery of the range of an ancestral species can diverge into species, the old "Reisenkreiss" model of speciation, which may actually be the case for *D. simulans, D. mauritiana,* and *D. sechellia.* Most importantly, though, when all the evidence is considered in total, from DNA sequences to behavior, individual flies in the *melanogaster* complex are consistently categorized into discrete groups of individuals, which can be done even on the basis of morphology alone.

Hybrids in the *melanogaster* complex have also been intensively studied, and the genetics of hybrid sterility are known to be controlled by at least five genes on the X chromosome (Coyne and Charlesworth, 1986; Wu *et al.*, 1993), and probably many more loci overall (Wu and Palopoli, 1994). Interestingly, it has been estimated on the basis of molecular clock estimates (Kliman and Hey, 1993; Kliman *et al.*, 2000) that *D. sechellia* and *D. mauritiana* diverged from *D. simulans* merely 420,000 and 260,000 years ago, respectively.

A few other examples in *Drosophila* show more of a continuum of groupings or divergence among individuals, perhaps the best studied being in the *Drosophila willistoni* species group. The *willistoni* group consists of 25 Neotropical species, six of which are "sibling" (cryptic) species, and among these six there are 12 "semispecies" and "subspecies," most of them in *Drosophila paulistorum* (reviewed by Ehrman and Powell, 1982).[1] The semispecies of *paulistorum* are morphologically indistinguishable so far as is known (one is never sure that very subtle features are being overlooked), and were first identified on the basis of chromosomal inversions. They also have distinct male courtship songs (Kessler, 1962; Ritchie and Gleason, 1995), and the hybrids of most crosses produce sterile males (Ehrman and Powell, 1982). DNA sequences of some *paulistorum* semispecies were examined (Gleason *et al.*, 1998), and these also group discretely. Thus, under evolutionary and phylogenetic definitions of species, *Drosophila paulistorum* itself could be considered a complex of cryptic species, but more data are needed to address this.

Interestingly, mating behavior (usually male courtship behavior) appears to diverge in *Drosophila* more quickly and prior to noticeable differences in morphology (e.g., Chang and Miller, 1978; Gleason and Ritchie, 1998; Grimaldi *et al.*, 1992), and this appears to be the case as well in many insects (Henry, 1994). It is known that just a few amino acid changes in a protein can dramatically affect, for example, an important component of *Drosophila* courtship song, the pulse interval (coded by the *period* gene; Wheeler *et al.*, 1991). Most morphological characters, by contrast, such as merely the shape of a lobe on the male terminalia of *Drosophila* (Coyne *et al.*, 1991), are highly polygenic. Divergence in mating behavior probably leads to further divergence (Liou and Price, 1994), which is eventually expressed morpologically.

[1] Sibling species and semispecies are categories devised largely by drosophilists and can be ambiguous terms. Sibling species are morphologically very similar or even identical, but the word "sibling" implies a close relationship, much like "sister group" in phylogenetics (which we discuss later). In fact, there are six sibling species in the *willistoni* group, some of which are closest relatives. Thus, we prefer the term "cryptic species" to simply mean morphologically indistinguishable or very subtly different species. The terms "semispecies" and "incipient species" imply these are not quite species, but are perhaps in the process of becoming species. But, because we can't predict the future, simply calling them populations and forms also adequately conveys their nature.

APIS

I hate myself, I hate clover, and I hate bees!
–Charles R. Darwin, in letter to J. Lubbock (3 September 1862)

The western honey bee, *Apis mellifera*, has perhaps received more intensive study than any animal except *Drosophila melanogaster*, white mice, and humans. Like horses, dogs, and other domesticated animals, a cultural bond was forged between humans and honey bees from the earliest civilizations, and *A. mellifera* has even been woven into mythology and religions (Ransome, 1937; Crane, 1983, 1999). Honey bees are eusocial, living in perennial colonies within nests constructed principally of wax from the sternal glands of worker bees. The genus is native to the Old World (with the exception of the Australian Region and Pacific islands) but has been globally distributed by humans. There is, in fact, scarcely a vegetated place on earth where *Apis* is not found. While the pollination of honey bees is not always as efficient as that of wild bees (Buchmann and Nabhan, 1996), apiculture is a multibillion dollar industry, and the demand for honey alone makes it highly unlikely that *Apis* will be commercially displaced by native pollinators anytime soon.

Unlike *Drosophila*, with about 1,000 species, honey bees in the genus *Apis* have just seven currently recognized species (Engel, 1999e) (Figure 1.5), although some distinctive Asian populations are frequently elevated to specific status (e.g., Sakagami *et al.*, 1980; McEvoy and Underwood, 1988; Otis, 1991, 1996). This lack of species diversity, however, has not hindered systematists from classifying the extensive variation in honey bees. While drosophilists cite their sibling species and semispecies, apidologists refer to *subspecies* or *races*. Indeed, perhaps more scientific names (species, subspecies, and races) have been proposed for *Apis mellifera* than for any other organism, 90 to be precise (Engel, 1999e). Despite the effort concentrated on species of *Apis*, the recognition of natural groupings in the genus has been confusing. Numerous attempts to classify the variation in *Apis* have resulted in the recognition of from four to 24 species at any one time (e.g., Gerstäcker, 1862, 1863; Smith, 1865; Ashmead, 1904; Buttel-Reepen, 1906; Enderlein, 1906; Skorikov, 1929; Maa, 1953; Ruttner, 1988; Engel, 1999e).

Species of *Apis*, particularly *A. mellifera* and *A. cerana*, are widely distributed (even without the aid of humans), and they have a striking range of variation across their various habitats (Ruttner, 1988). The most noticeable variation is in coloration, but it also includes subtle morphological differences like the size and shape of cells in the wings. These variants were alternatively treated as species or subspecies in the past because they corresponded to geographical regions and climatic zones. As the New Synthesis began to influence apidologists, morphometric analyses (mostly of wing venation) were used to segregate individuals into "morphoclusters." Backed by the appearance of statistical rigor, these morphoclusters were then united into newly defined subspecies and species (Ruttner, 1988), and these studies became the norm for segregating honey bees into what were believed to be natural groups. Contradictions between the morphoclusters and numerous biological traits and molecular data were increasingly found (Hepburn and Radloff, 1998; Hepburn *et al.*, 2001), and large regions of hybridization further blurred the traditional distinctions of these forms. Subtle morphometrics of wing venation have proven to be of little systematic value.

Like most groups of insects, species of the genus *Apis* can be distinguished on the basis of differences in male genitalic structure to varying degrees (Ruttner, 1988; Koeniger *et al.*, 1991) and other morphological details of adults and even larvae (Ruttner, 1988; Engel, 1999e) (Figure 1.5). These differences are largely congruent with ecological, behavioral, chemical, and molecular features, and they serve to define most of the honey bee species, regardless of the preferred species concept. Adoption of the biological species concept, however, sent generations of apidologists into apiaries and fields seeking mating differences in honey bee populations that might be congruent with the traditional morphoclusters (i.e., subspecies). Differences potentially restrictive to gene flow were considered enough evidence to warrant species status for isolated subspecies. For example, the timing and location of mating flights is important in *Apis* biology because this is when virgin queens meet drones, with synchronization being critical for the two sexes to meet. Temporal segregation of drone flight times and spatial differentiation of drone congregation areas has therefore been used as evidence of reproductive isolation, and the separation of species in the absence of morphological features (e.g., Underwood, 1990; Hadisoesilo and Otis, 1996; Koeniger *et al.*, 1996). These behavioral differences are indeed significant because they likely represent incipient isolation, the first step in speciation. Such traits, however, are difficult to use for defining species. Even though forms can be segregated from each other at their point of contact, drone flight time varies considerably over its entire distribution within a species. On this basis, traits for species recognition are only applicable to one or a few locales and do not diagnose the species as a whole. It is difficult, if not impossible, to distinguish the species in its entirety from its peripherally distinct morphs. This is a common problem because the Biological Species Concept (BSC) is testable in regions of contact only. The BSC is not amenable to complete testing because some allopatric populations, such as the distinct island populations of giant honey bees (*Apis dorsata*), do not come into geographical contact. Most accounts ignore the historical relationships of the species and their populations and fail to think in terms of defining individual species on a global scale. In other words, how is it that we define *A. cerana* or *A. dorsata* across the

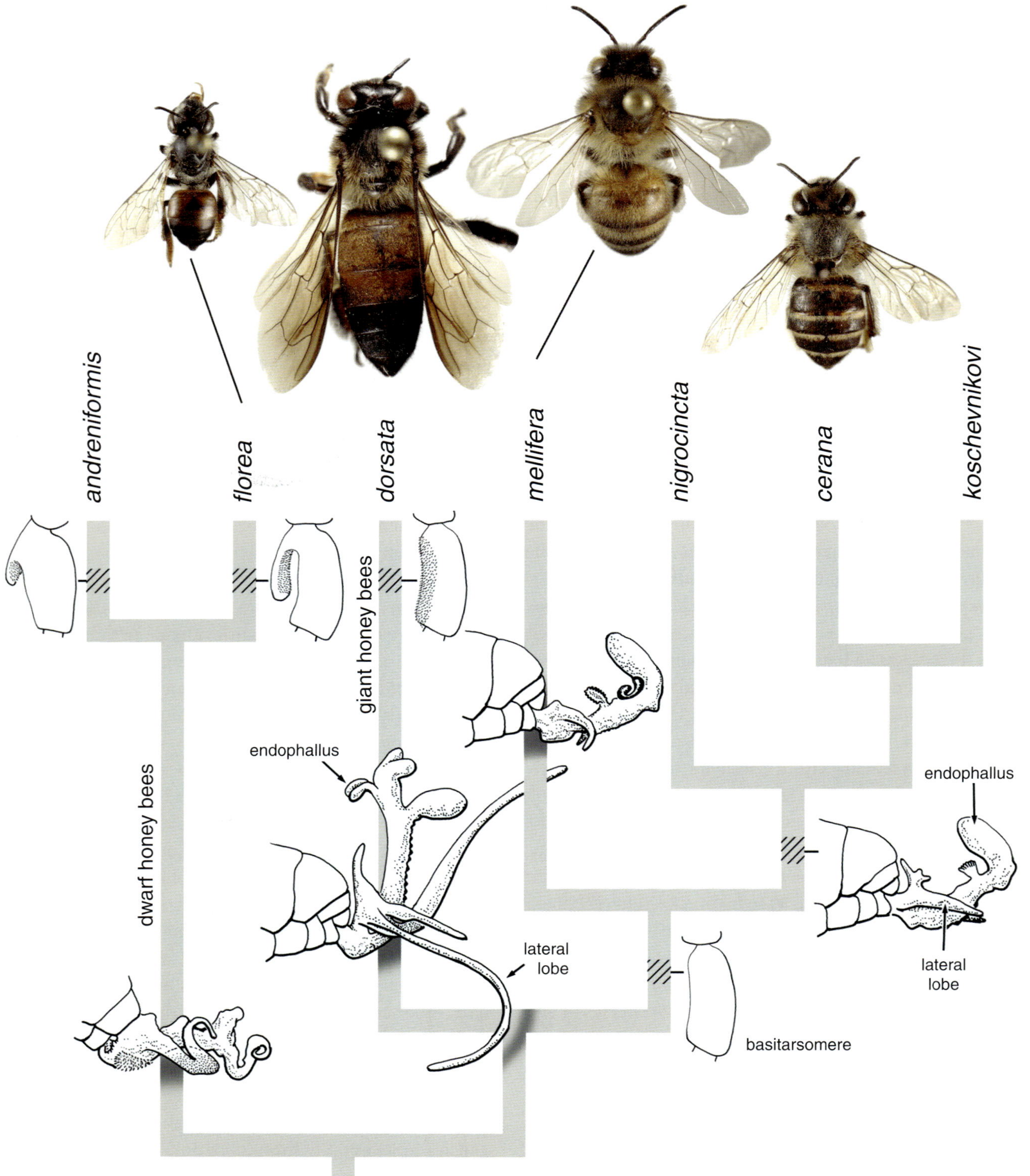

1.5. Relationships among species of Recent honey bees, genus *Apis*, showing important variations in tarsomeres and male genitalia. Relationships from Engel and Schultz (1997).

entirety of their ranges, distinct from regional morphotypes or ethotypes, and that may be reproductively isolated at fine geographical scales?

Perhaps the most dramatic development of variation is seen in the Cape honey bee, *Apis mellifera capensis*. This subspecies is facultatively parthenogenetic and a social parasite on colonies of other honey bee subspecies. While *A. mellifera capensis* is still reproductively compatible with other subspecies of *A. mellifera*, gene flow is asymmetrical and the Cape bee dominates during introgressions (Johannsmeier,

1983; Hepburn and Radloff, 1998). This may be a rare example of incipient speciation. Similar cases, but not involving the evolution of parasitic behavior, occur in the widely distributed *A. cerana* and *A. dorsata*, in which great variation is related to local differences in habitat such as elevation. *Apis cerana nuluensis* is often considered specifically distinct because it is found only in the mountains of Sabah above 1800 m, with mating flights temporally separated from the overall *A. cerana* population occurring at lower elevations (Otis, 1996). Workers forage together, and aside from remarkably variable differences in coloration correlated with latitude or elevation, there are no derived traits to support species status of *A. cerana nuluensis*. These morphs are all derivatives of the larger, ancestral *A. cerana*, thereby leaving the mother species paraphyletic if the isolates themselves are recognized as species (e.g., Tanaka *et al.*, 2001).

Species of *Apis*, however, unlike higher taxonomic levels, are almost never monophyletic. Indeed, based on DNA evidence, *Apis nigrocincta* (which lacks fixed, morphologically diagnostic traits) is a derivative of *A. cerana* (Smith *et al.*, 2000), much the way *Drosophila sechellia* and *D. mauritiana* appear to be derivatives of *D. simulans*. Differences in nest architecture occur between *A. nigrocincta* and *A. cerana* across their ranges (Hadisoesilo and Otis, 1998), which are congruent with differences in drone flights and morphometric clusters (Hadisoesilo *et al.*, 1995; Hadisoesilo and Otis, 1996). *Apis koschevnikovi* of Malaysia, Indonesia, and Borneo is reproductively isolated and genetically and morphologically distinct from other *Apis* species. This species is perhaps a more ancient example of peripheral specialization, being restricted to wet primary forests (Otis, 1996), while having derived from an ancestral *cerana*-group stock and becoming secondarily sympatric with *A. cerana*.

Teasing out subtle details that distinguish cryptic insect species is not an academic exercise, but it is a practical necessity in cases involving vectors of serious diseases or major crop pests. Controlling the diseases carried by cryptic species in the *Anopheles gambiae* complex or the *Simulium damnosum* complex, for example, was completely confounded until the species were accurately defined, and slight differences in their biology were deciphered. Also, if most individuals can be grouped into discrete and diagnosable species, as in *Drosophila* and *Apis* (why should other insects be different?), this would have profound implications for evolutionary biology and systematics. Traditionally, species are believed to have formed gradually, through the steady accretion of small genetic changes, called *phyletic gradualism*. Studies on *Drosophila* have found good correlation between genetic distance and degree of reproductive isolation (Ayala *et al.*, 1974; Coyne and Orr, 1989), supporting the view of phyletic gradualism. But if this were the standard mode for speciation, one would expect many examples of intermediates, individuals with features of *D. melanogaster*, *D. simulans*, or other species, or at least many species with great ranges of variation. The extensive genetic and phenotypic evidence from *Drosophila* and *Apis* indicates that there exist discrete groupings of individuals – species – though in some cases to define the groups this may require extensive data on mtDNA, courtship songs, swarming behavior, and other evidence. Perhaps species actually are "typological," contrary to Mayr (1942). Moreover, discrete groups would suggest that the time for the formation of a species is quick relative to its entire lifespan, which is consistent with the concept of *punctuated equilibrium*, but this is an area that still needs considerable exploration.

A great deal more, in fact, could be discussed about the exact nature of species, but to explore 400 MY in the evolution of insects, we need to consider how many species of them presently exist.

HOW MANY SPECIES OF INSECTS?

Scientists know far more about (and spend vastly more money studying) the systematics of stars than the systematics of earthly organisms. Consequently, they have as good a knowledge of the number of atoms in the universe – an unimaginable abstraction – as they do of the number of species of plants and animals.

–Robert May, 1992

Numerus specierum in entomologia fere infinitus et nisi in ordinen redigantur, chaos semper erit entomologia. [The number of species in entomology is almost infinite, and if they are not brought in order entomology will always be in chaos.]

–J. C. Fabricius, 1778, *Philosophia entomologica* VI, section VI, para. 3 [translation by Tuxen, 1967a]

I have heard it stated upon good authority that 40,000 species of insects are already known, as preserved in collections. How great, then, must be the number existing in this whole globe!

–W. Kirby and W. Spence, 1826

Insects are so diverse that their numbers are impressive even in the most parochial of places. Cockroaches, of course, are expected in New York City dwellings, but a quick entomological survey of a typical apartment can yield 20 or more species of arthropods (Volk, 1995). New species of midges, an ant, and various other insects are known throughout the eastern United States, some of which even occur in New York's Central Park, the most visited green space on earth. A new dwarf genus of arrupine millipedes, in fact, was discovered in Central Park in 2000. It is clearly introduced, probably from eastern Asia or western North America, but so far the genus is known only from Central Park (Foddai *et al.*, 2003). In a forgotten study done in the 1920s, Frank Lutz of the American Museum of Natural History surveyed the insect species in a typical one-acre yard in the suburbs of northern New Jersey: he found 1,250 species. That was before we had refined concepts of species among the myriad tiny acalyptrate flies, parasitoid wasps, and staphylinoid beetles, so the number is probably at least 1,500 species. Another surprise about these kinds of studies is that there have been very few intensive surveys of the insects or terrestrial arthropods of natural areas (e.g., Proctor, 1946; Woodley and Hilburn, 1994), even though that type of study is so important to estimating how many species of insects exist.

One million species is commonly recited for the diversity of named living insects, but even this figure is ambiguous. Estimates range from 750,000 (Wilson, 1992) to approximately 1.4 million (Hammond, 1992), but the number appears close to **925,000 named species** based on recent figures for the "big four" orders (Hymenoptera, Lepidoptera, Coleoptera, and Diptera) (Gaston, 1991; Resh and Cardé, 2003) (Table 1.1; Figure 1.6). Diptera is the only major group of insects where the world species have been catalogued within the last few decades, and it will be necessary for similar catalogues to be produced before accurate tallies of all described species are made. Proper species catalogues require tedious checking and verifying of old literature (names, dates, types, etc.) so it has attracted little effort, even though these are the very scaffold for other work in systematics.

TABLE 1.1. Numbers of Described Species of Extant Hexapods

Wingless Orders:	Species
Entognatha:	
Protura	600
Collembola	9,000
Diplura	1,000
Archaeognatha	500
Zygentoma	400
Paleopterous Orders:	
Ephemeroptera	3,100
Odonata	5,500
Polyneopterous Orders:	
Grylloblattodea + Mantophasmatodea	41
Phasmatodea	3,000
Orthoptera	20,000
Dermaptera	2,000
Embiodea	500
Plecoptera	2,000
Zoraptera	32
Dictyoptera:	
Blattodea	4,000
Mantodea	1,800
Isoptera	2,900
Paraneoptera:	
Psocoptera	4,400
Phthiraptera	4,900
Thysanoptera	5,000
Hemiptera	90,000
Holometabola:	
Neuropterida	6,500
Coleoptera	350,000
Strepsiptera	550
Mecoptera	600
Siphonaptera	2,500
Diptera	120,000
Hymenoptera	125,000
Trichoptera	11,000
Lepidoptera	150,000
Approximate Total	**926,400**

What has engendered most of the discussion about insect diversity, though, are the estimates of total numbers of insect species, described and mostly undescribed. These estimates differ wildly, from approximately 2 million species (Hodkinson and Casson, 1991), to 8.5 million (Stork, 1988, 1996; Hammond, 1992) to 30 million or more (Erwin, 1982, 1983a). Other recent estimates place the number at approximately 5 million insect species globally (Gaston, 1991), which is within a much earlier estimate (Brues *et al.*, 1954) of 3.75 million to 7.5 million species. In a time when advances in technology allow measurements of drifting continents (an average of 2.5 cm per year), the mean diameter of the earth (7,913 miles), or the mass of an electron (9.1×10^{-28} grams), one would expect more precision on species numbers. The discrepancies lie in how the estimates are made.

Erwin's (1982) estimate of 30 million species of insects is widely criticized, but in all fairness it was the first study to bring attention to the nebulous problem of total numbers of insect species. This work also exposed a whole new biota in the canopies of tropical forests (Erwin, 1983a,b, 1990), and led to similar studies by others in forests of southeast Asia (e.g., Allison *et al.*, 1993; Stork, 1987, 1991, 1997) and elsewhere (reviewed by Basset, 2001). The basic technique for all these studies uses a fog of insecticide that is blasted into the canopy, which degrades quickly, and the insects that rain down into basins are then collected, preserved, and sorted later back in the laboratory. The original study by Erwin (1982) extracted arthropods out of the canopies of trees in Panama, and one particular tree, *Luehea seemannii*, was used to extrapolate total diversity. From multiple individuals of this tree Erwin found, among hundreds of beetle species, 163 species occurring only on this tree, presumably restricted to it. By his calculations, because there are approximately 50,000 species of tropical trees, the number of beetle species living in tropical forest canopies would be 8,150,000. Because beetles comprise approximately 40% of all terrestrial arthropods, the number of tropical forest arthropods is likely to be 20 million species. But the canopy is only part of the fauna, so Erwin estimated that species on the ground comprise approximately half the number of canopy species, which is how the estimate of 30 million species total of *tropical arthropods* was made.

The estimates were critiqued on the basis of unrealistic

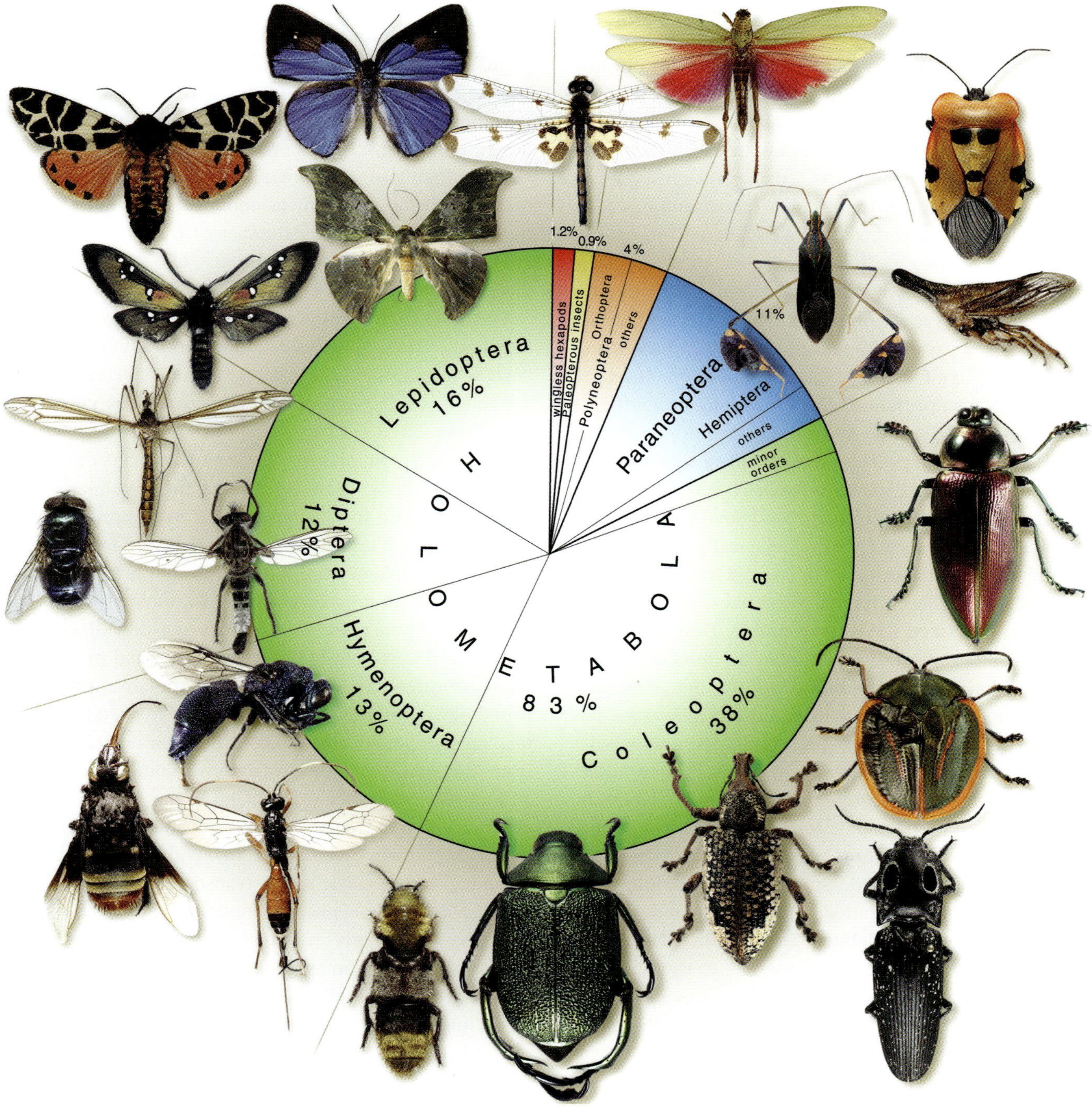

1.6. The diversity of Recent hexapods as proportions of named species.

assumptions, particularly the high proportions of species specialized to particular species of trees (May, 1988; Stork, 1988). These assumptions have dramatic effects on the estimates. Another reason for Erwin's high estimates may be that his studies were in the neotropics, which is the most diverse biotic realm for many insect groups, like carabid beetles and weevils. Indeed, his samples of beetles were consistently more diverse than canopy samples from the Old World tropics (Erwin, 1997). It also needs to be remembered that not all groups of terrestrial arthropods are most speciose in the tropics: sawflies, cynipid and ichneumonid wasps, spiders, and bees are as diverse in temperate and xeric regions as they are in the tropics, or even more so.

A completely different approach was used by Gaston (1991): survey systematists instead of forests. Systematists have at their disposal large collections in the groups of their expertise, among which lurk large numbers of undescribed species. By surveying hundreds of systematists, a reasonable estimate can be made of the proportions of undescribed species, but this approach, too, has its problems.

First, systematists' collections may or may not accurately reflect true diversity, depending on the thoroughness of the field collecting techniques used to amass the collection. Significant new diversity is usually not discovered until a curious entomologist discovers that hundreds of new species of *Aulacigaster* or odiniid flies, for example, can be found by sweeping up and down the trunks of dying rain forest trees, or taken in fogged samples from tropical forest canopies. Also, the diversity of some groups can be completely unexpected and may turn out to be "bottomless pits" of species. The largely neotropical fruit fly genus, *Cladochaeta*, was one such instance: from 13 species originally known, 105 new species were discovered (Grimaldi and Nguyen, 1999), which is now known to still be a fraction of the probable actual diversity. The worldwide staphylinid genus *Pseudopsis* contained six species prior to 1975. Later it was discovered that one widespread "species" of *Pseudopsis* was actually a complex of 24 species, and another 21 species have also been found (nearly seven times the pre-1973 diversity) (Herman, 1975). Intensive surveying in the forests of La Selva Biological Station in Costa Rica and surrounding areas has uncovered hundreds of new species of tiny gracillariid moths, where only a few had previously been described (D. Wagner and D. Davis, unpubl.). Their total neotropical diversity must be immense. These studies are based just on morphological differences, so if detailed genetic and behavioral comparisons were made, it is likely that even more (cryptic) species would be uncovered.

Another way in which a survey of systematists may miscalculate diversity is that accurate estimates usually cannot be made in lieu of a monographic revision. Revisionary studies pull all the available material together, hundreds to thousands of dissections are made, specimens are carefully compared, distributions are plotted, new species are described, and known species are redefined. They usually take years to finish, and it would not be unreasonable to estimate that fewer than 5% of all described insect species have been treated in a modern revisionary monograph. Yet, these give us accounts that have the least tarnish. In the recent 794-page revision of the huge ant genus, *Pheidole*, which took about 20 years to produce, the New World fauna was more than doubled, from 287 species to 624 species (Wilson, 2003). It was estimated that, of the 900 world species now known in the genus, perhaps 1,500 actually exist. Revisions of this scope are exceptional, however. Some genera of truly daunting size may never be revised, like *Lasioglossum* bees, *Megaselia* scuttleflies, and *Agrilus* staphylinid beetles, with between 2,000 and 3,000 named species in each. Given that such large genera account for a considerable proportion of the known species, a lack of monographic revisions on them seriously biases our estimates of diversity.

Monographic revisions are also necessary in documenting *synonyms*, which are different names referring to the same species. Early taxonomic descriptions of insects were often imprecise, or even vague, so to determine the identity of some species accurately it is necessary to examine *type specimens*, which is standard protocol for revisions. Types of some species may not exist, and of the many that do exist some have never been reexamined. This may lead to erroneous descriptions of species as new. *Drosophila melanogaster*, for example, being so widespread, was given five other names between 1853 and 1862 by taxonomists ignorant of the fact that Meigen already named the species in 1830. Other species are far more taxonomically notorious, like *Apis mellifera*, as discussed previously. Recent records of the numbers of synonyms (based on figures from monographs) are unexpectedly large: The number of synonyms in insects recognized each year is 25–30% the number of new species (Gaston, 1991). On this basis, estimates of new species need to be scaled down.

Lastly, and perhaps most significantly, it is very difficult for even expert systematists to predict the numbers of rare species, and rare species now appear to comprise a major proportion of all insects, as they probably do for all organisms. In a sample of beetles fogged from forest canopies in Borneo (why are they always beetles?), most of the species (499, or 58%) were represented by only one individual, and an additional 133 species (15%) were represented by only two individuals (Stork, 1996); just six very common species comprised one quarter of all the individuals. In a study of fruit-feeding nymphalid butterflies in Ecuador, species known from only one specimen ("singletons") comprised between 14 and 30% of all species collected in the traps, and these continued to trickle in after 11,861 specimens and 5 years of the study (DeVries and Walla, 2001). Without very thorough sampling, rare species, which realistically comprise about half of a fauna, would never be found.

So, how many species of insects, total, exist? We believe the estimate of Gaston (1991), about **5 million species total**, is most accurate, despite the inherent biases of those methods. Thus, only about 20% of the global insect fauna is probably known and named, so clearly a great deal of basic exploration is needed. This is not merely an academic exercise because knowing the true numbers of species is crucial for wise stewardship of earth's biodiversity (Wilson, 1992). Also, deciphering the evolution of insects would be a perversion without knowledge of the end product, and any one discovery in the Recent fauna can have a dramatic impact on our understanding of evolutionary history, an impact as profound as the discovery of an important fossil. The discovery in 2002, for example, of living African species in the new insect order Mantophasmatodea allowed much better interpretation of fossils of these insects, just like coelacanths did for fish. In fact, now knowing the close relationship of Mantophasmatodea to another small order, Grylloblattodea (which are in the Northern Hemisphere) helps to unravel the relict nature of a lineage that was quite diverse in the

Mesozoic. The probability is quite high that, among perhaps 4 million more insect species, another mantophasmatodean or even *Drosophila melanogaster* will emerge, so shouldn't scientists "map" them as well as they do stars, or even better?

It is even more sobering to estimate the number of all insect species that have ever lived, if this can be credibly done. We know that huge radiations of insects that feed on angiosperms, like lepidopterans and phytophagan beetles, barely existed before the Cretaceous, but we also know other groups today are vestiges of an extremely diverse past, like Paleozoic odonatopterans and Mesozoic mecopterans, grylloblattodeans, and many others. Some, like the Paleodictyopterida, left no survivors at all. Also accounting for modest beginnings 400 MYA, and that the lifespan of an average insect species is conceivably around 5 MY, a reasonable "guesstimate" would be that **100 million** insect species have ever lived. This may be an underestimate, but it still reflects the magnitude of the challenge in reconstructing the evolutionary history of insects.

RECONSTRUCTING EVOLUTIONARY HISTORY

It has often been said that theory of evolution is the most unifying concept in biology. Indeed, every aspect of an organism – its mating display, the mode of photosynthesis, a mutation in a gene – can be explained from an evolutionary perspective. Evolution is only very rarely observed in human time, say, over decades (e.g., Grant, 1999; Grant and Grant, 2002), with millennia being a more typical time scale. Thus, reconstructing biological history relies on the fossil record and comparisons among living, or Recent, species. But even for groups with extensive fossil records, like foraminiferans, the record is never complete, and most groups leave a very spotty fossil record. Moreover, no fossil is complete; they never have behavior (they may leaves traces of it), and even ones in amber probably don't even have DNA. So systematists, who seek the relationships among species, living and extinct, can improve their interpretation of fossils by also studying the virtually limitless features of Recent species. How, exactly, are extinct species related to living ones? In what ways have lineages changed?

But, first, why even reconstruct phylogeny? Besides identifying lineages of organisms, we can record the success and demise of those lineages, and perhaps even provide explanations for these outcomes. Phylogenies also allow the interpretation of evolutionary patterns. Another salient scientific advantage to understanding phylogeny, though, is that it allows predictions. Armed with knowledge about the closest relatives, accurate predictions can be made about any species, and so phylogenies explain very well. There are many ways to classify organisms – things good to eat, bad to eat, things that sting – but a phylogenetic classification, one that reflects and summarizes phylogeny, is the most useful and a goal of modern systematists.

To understand modern ideas in phylogenetic reconstruction, such as the concept of homology, knowing the history of the ideas helps us understand the logical construction of systematic and evolutionary theory. Evolutionary thinking was born of a need to classify and name organisms, and thus we must reach back into the history of classification to find its origins. Also, the history of thought is not a steady progression of ideas. There are, actually, many false starts, dead ends, reversals and changes in direction. In actuality, history shapes the way we think and we are merely a part of history. In other words, the present state of science today is not a final product, and our concepts of insects evolve, just like the insects themselves, and hopefully the concepts become refined as well.

SYSTEMATICS AND EVOLUTION

Prior to the establishment of what we consider the Linnean System, groupings and classifications of organisms were not based on their evolutionary relationships to one another. Folk taxonomy, or common names, dominated the world. Although this type of naming had great local practicality, the difficulties with such systems were that most species did not have a name (e.g., most insects lack common names). Thus, names were applied only to the most commonly encountered organisms, or ones most useful to know about. Moreover, the names varied greatly with region and were therefore only locally applicable. As a result, a village could adopt a new name at any time, and its meaning was lost to other villages. The classification was derived from tradition and sometimes included few actual attributes of the biological world; in fact, fanciful creatures such as unicorns and basilisks were classified alongside flies and horses. Lastly, all languages were naturally included (e.g., *bee* versus *pchel* [ПЧЕЛЬ]!); there was no standardization. To bring together the knowledge of all humanity, it required a polyglot. Thus was born the need for a formal and universal taxonomy, or development of scientific names. Such a system was advantageous in that it would be universally applicable to all organisms and useful in all countries and cultures. It would abide by a standardized set of pragmatic rules (*nomenclature*) and empirical evidence to ensure its stability. The system would recognize only natural groups of organisms; mythical beasts and illogical groupings would be abandoned. The now extinct languages of Latin or ancient Greek were adopted so as to avoid the pitfalls of any nationalism, and early on these classical languages were the communication of academic scholarship. However, a formalized system did not appear overnight. There is a long history of the development of taxonomy, nomenclature, and systematics, for which we provide only a brief outline.

The Greeks

Today we herald Darwin as the architect of evolutionary thought, and rightly so; however, the first proponent of evolutionary ideas was an ancient Greek. Anaximander, living during the sixth century B.C., developed the idea that living creatures were derived from water, specifically that terrestrial animals, including humans, formed directly from fish. His ideas are simplistic and somewhat Lamarckian by today's standard; nevertheless, it was a prescient idea concerning the diversity of life.

Well after Anaximander was Aristotle (384 B.C.–322 B.C.), the famous student of Plato (himself a student of Socrates) and the father of empirical thought. Aristotle is best known as the father of logic, logical argumentation via syllogisms, and epistemology (the theory of knowledge), but he also wrote compendia summarizing all human knowledge and, as a result, was the first careful observer of plant and animal life. He summarized his biological information in a series of books, most notably his *Historia Animalia*, which ultimately discussed 520 species of animals. Another contribution of Aristotle was his attempt to explain things by common or general terms ("primacy of the universal"), which enabled him to look for the overarching patterns in nature. Aristotle attempted to explain organisms in terms of similarities in function and behavior (analogous to the methods of the much later writer Cuvier). Although he was a great proponent of knowledge based on direct observations from nature, he still was a creature of his era and included in his volumes numerous "facts" that must have been acquired as rumors from travelers abroad. Overall, however, his volumes would be the basis for scientific inquiry for over 1,000 years to come since scientific thought grew relatively little after Aristotle. For example, many of his followers and various Roman scientists mostly attempted to expand upon his work. One of them was the Roman scholar Pliny the Elder (23?–79 A.D.), who, interestingly, died while trying to examine Vesuvius' eruption during the destruction of Pompeii, and whose encyclopedia on natural history was used well into the 1600s.

One last Greek worth mentioning was Porphyry (233–304 A.D.) who developed what would become known as the *Tree of Porphyry* (Figure 1.7). Porphyry's tree represented a hierarchical, dichotomous system of everything and was based on the Greek's fascination with reconciling opposites in nature. Although quaint from a modern perspective, Porphyry's ideas formed the basis of our dichotomous identification keys and, in general, our ideas for hierarchical classification, or groups nested within larger groups.

1.7. A fourteenth-century representation of the Tree of Porphyry (note the Latin phrase *arbor Porphiriana* at the top of the image), the original tree of which was the earliest depiction of a dichotomous tree. Along one side of the dichotomous divisions, represented as leaves sprouting from the image of the man, are *corporea, animatum, sensible, rationale,* and *mortale,* while the branches of the right side depict the opposite. Beneath the figure are the names of Plato, Homo [man] and Socrates. Photo: Gordon MS 92, Bryn Mawr College Library.

Tragedy Falls

During the Fall of the Roman Empire beginning in 378 A.D. and culminating in the destruction of Rome in 410 A.D., the world turned to the writings of St. Augustine, particularly his book, *The City of God* (413–414 A.D.). Therein he taught individuals to ignore nature and that all knowledge was to be found by focusing solely on the afterlife. Centuries of careful and original observation of the natural world virtually disappeared in tradition and as written records, except for what was copied by scribes in monasteries and known from a few surviving ancient texts. The reigning authorities of the time forbade inquiry into the natural world, and the Dark Ages began. Fortunately, some books survived the Roman collapse, and volumes like Martianus Cappella's *Seven Liberal Arts* (the precursor from which we get today's Colleges of Liberal Arts and Sciences) would form the foundation of thought for centuries (Burke, 1985). During the Dark Ages there were some modifications of systematic thought. Porphyry's tree of all knowledge was modified and expanded into the *Scala Naturae*, or the *Chain of Being*. The Scala Naturae represented a permanent, unchanging hierarchy, which was immutable to reflect Divine perfection. The chain included the creator and heavenly angels down through the natural world to the lowest levels of creation. It would become the dominant idea for organizing nature until the eighteenth century.

Resurgence of Scientific Inquiry and Enlightenment

During the eleventh-century crusades of El Cid and his mercenaries against the fiefdoms in southern Spain, at that time under Arab control, a wealth of information was rediscovered.

The Arabs had preserved and studied copies of most ancient Greek texts, with ideas now expanded by Arab scholars or adopted from cultures further East, such as India. The invading crusaders were overwhelmed by the luxury of life in southern Spain, which was in part the result of the preservation of such knowledge. Over the course of two centuries, monks transcribed the Arabic texts into Latin so that scholars throughout Europe could easily read them (Burke, 1985). It was at this time that Arabic words like "algebra" and "zero," entered the lexicon of western Europe. As this rediscovered knowledge flowed across Europe, particularly Aristotle's ideas on logic, the absolute authority of the Church began to be questioned. New observations of nature were being made, and direct investigation of the natural world was once again cultured. The new fervor was fueled by technological achievements like Guttenberg's moveable type, such that the dissemination of knowledge no longer required years of labor by monkish scribes. The development of the microscope in the 1600s allowed levels of scrutiny never before imagined. Indeed, some of the earliest scholars at this time began to focus on insects and produced magnificent tomes. Some of the most significant were Jan Swammerdam's (1637–1680) *Historia Insectorum Generalis* (published posthumously in 1685; Figures 1.8, 1.9) and later John Ray's (1627–1705) *Historia Insectorum* (also published posthumously in 1710; Figure 1.10). John Ray's study would, in fact, set the stage for future developments in the classification of insects. This new-found desire to investigate nature, coupled with the various new technologies, opened a new age of exploration. Explorers set sail from western Europe to map the world, bringing back with them specimens and stories from the furthest points on the globe. The founder of modern nomenclature entered the scene during the later part of this era.

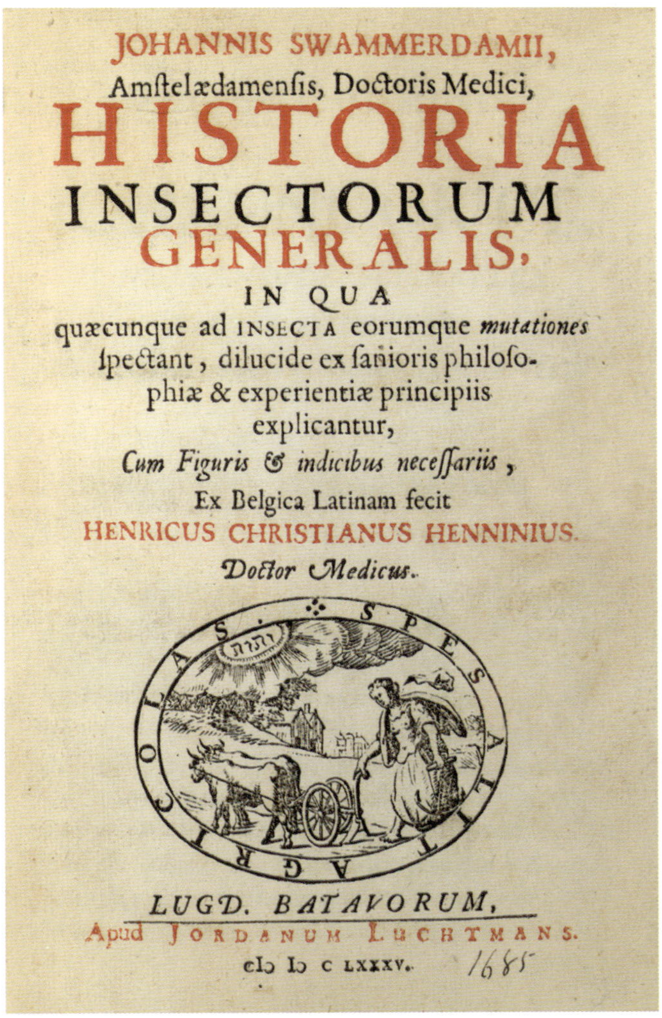
JOHANNIS SWAMMERDAMII,
Amſtelædamenſis, Doctoris Medici,
HISTORIA
INSECTORUM
GENERALIS,
IN QUA
quæcunque ad INSECTA eorumque *mutationes*
ſpectant, dilucide ex ſanioris philoſophiæ & experientiæ principiis explicantur,
Cum Figuris & indicibus neceſſariis,
Ex Belgica Latinam fecit
HENRICUS CHRISTIANUS HENNINIUS.
Doctor Medicus.
SPES ALIT AGRICOLAS
LUGD. BATAVORUM,
Apud JORDANUM LUCHTMANS.
CIↃ IↃ C LXXXV. 1685

1.8. The title page of Jan Swammerdam's *Historia Insectorum Generalis* (1685). Swammerdam was a masterful anatomist and one of the earliest scholars to use microscopy to study insects. Photo: American Museum of Natural History (AMNH) Library.

Karl Linnaeus and Beyond

Karl Linnaeus (1707–1778) (Figure 1.11) was a Swedish botanist who took the science of systematics on its next greatest leap. It is ironic that the father of biological nomenclature should have such confusion surrounding his own name. He was actually born "Linnaeus," and it is a common misconception that Linnaeus is a latinization of his "real" name "Linné." He did not acquire the ennobled name Linné until late in life, at which time he became Carl von Linné (Blunt, 2001).

Linnaeus did not operate under a model of evolution. However, he did note that nature was roughly hierarchical and thus placed his classification into a hierarchy, or sets of categories, the *Linnean Hierarchy*. He also, as alluded to, was the first to consistently employ a *binomial system* so as to condense information because previously the description of organisms involved lengthy paragraphs of Latin (botanists retain a vestige of this and still publish brief diagnoses of new taxa in Latin). Looking for general patterns, as advocated by Aristotle, Linnaeus distilled generalized features into a genus and the most salient distinguishing feature of the individual kinds, or species, into a single epithet (the *specific epithet*). The specific epithet was then followed by the more standard, lengthy description. However, any given organism would be readily and easily referred to by its binomial composed of a genus and species, like *Apis mellifera*. With these components, Linnaeus built a classification of all plants and animals. Thus, he was the first systematist to categorize the entire biological world as he understood it into a hierarchical, binomial system (Figure 1.12).

As biologists continued to investigate the world around them and produce classifications, they noted that the characters of organisms sometimes suggested different hierarchies. In the earliest years (e.g., around the time of, and shortly after, Linnaeus), the debate centered around identifying a single character or suite of characters, like the reproductive parts of flowers, that, owing to its various biological properties,

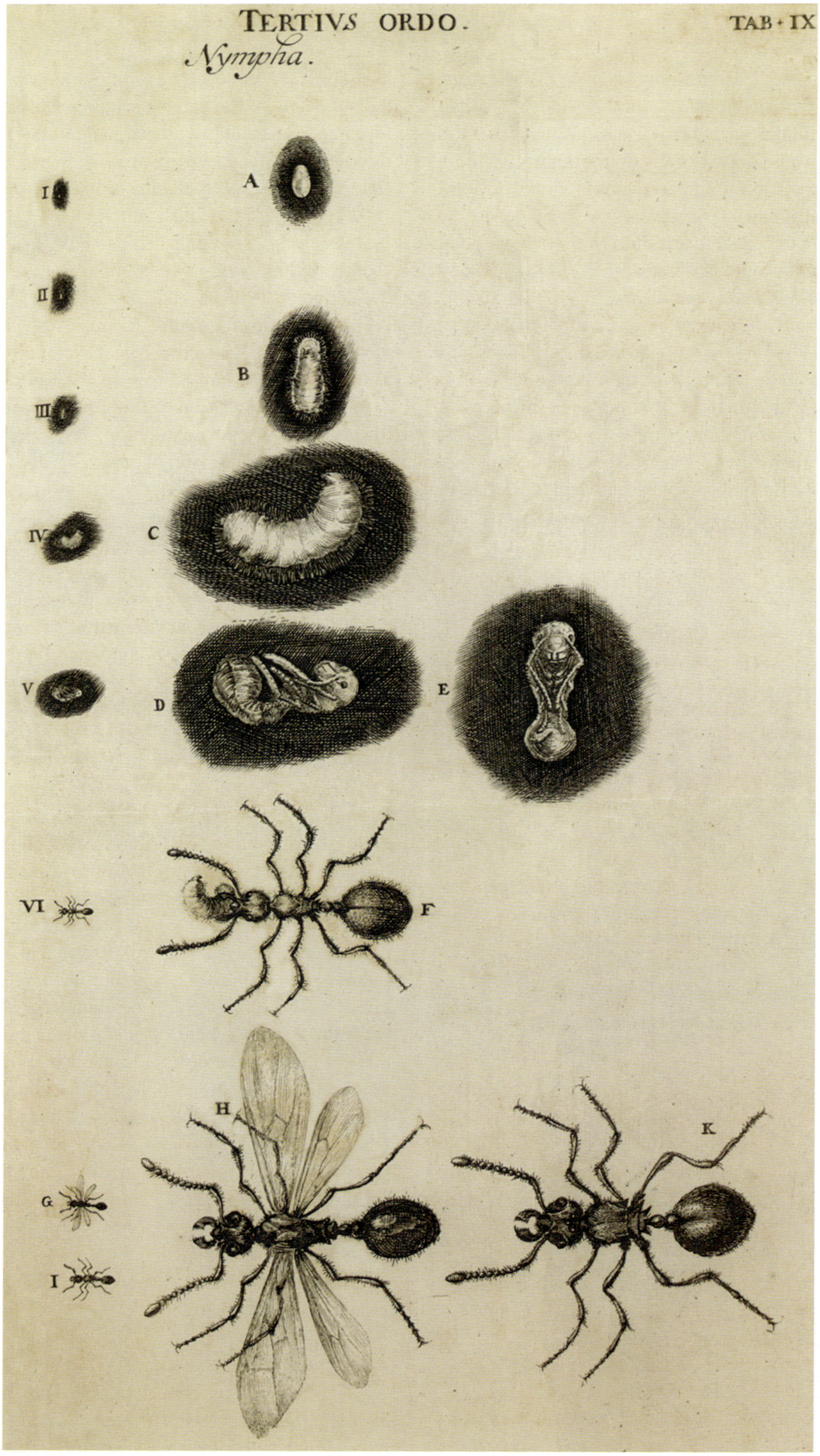

1.9. A plate from Swammerdam's *Historia Insectorum Generalis* (1685) depicting the metamorphosis of an ant. Photo: AMNH Library.

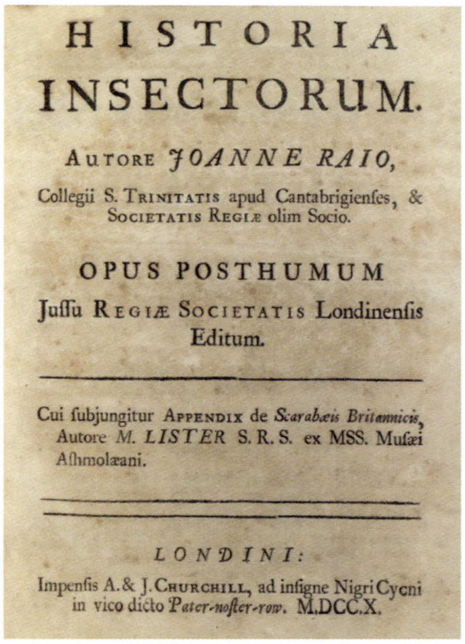
HISTORIA
INSECTORUM.
Autore JOANNE RAIO,
Collegii S. Trinitatis apud Cantabrigienses, & Societatis Regiæ olim Socio.
OPUS POSTHUMUM
Jussu Regiæ Societatis Londinensis Editum.
Cui subjungitur Appendix de Scarabæis Britannicis, Autore M. LISTER S. R. S. ex MSS. Musæi Ashmolæani.
LONDINI:
Impensis A. & J. Churchill, ad insigne Nigri Cycni in vico dicto Pater-noster-row. M.DCC.X.

1.10. John Ray's *Historia Insectorum* (1710) was an influential work not only for summarizing entomological knowledge of its day but also for taxonomic science in general. Photo: AMNH Library.

would produce a natural classification. For Linnaeus, the orders of insects should be defined on the basis of their wing number and a bit of their structure – hence names like Aptera, Diptera, Hymenoptera, and Neuroptera. To his student, J. C. Fabricius, feeding was more important because it provided the sustenance of life; therefore, mouthparts took precedence over the wings. Hardly known is that both Linnaeus and Fabricius relied on the work of Maria Sibylla Merian (1647–1717) for their descriptions and classification of many insects from tropical South America. She studied engraving and painting in Germany and was inspired by the intricacy of insects. Funded by Dutch scientists to study the insects from that colony, she separated from her husband and moved to Suriname with her daughters for two years. It was there that she produced her masterpiece, *The Metamorphosis of the Insects of Suriname* (1705) (e.g., Figure 1.13).

Fabricius (1745–1808) (Figure 1.14), in fact, was far more observant of insects than was his mentor Linnaeus, who was primarily a botanist. As Fabricius maintained, mouthparts are indeed complex structures that reflect phylogeny, as we elaborate upon throughout this book. Fabricius described 9,776 species of insects (Linnaeus merely about 3,000), and he published a major reference in entomology, *Philosophia Entomologica* (1778). Most importantly, he recognized that in classifying insects, as for any organisms, at least some groupings should be natural combinations of species: "those whose nourishment and biology are the same, must then belong to the same genus" [Fabricius, 1790, as translated by Tuxen (1967a)]. Besides mouthparts, Fabricius even predicted that genitalia, which are complex in male insects, would provide many important characters, but was himself limited to the

1.11. The Swedish botanist Karl Linnaeus (1707–78), founder of our modern system of binomial nomenclature. Photo: AMNH Library.

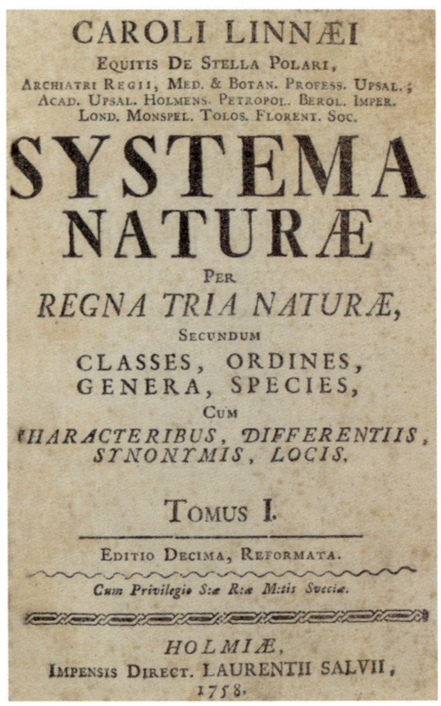
CAROLI LINNÆI
Equitis De Stella Polari,
Archiatri Regii, Med. & Botan. Profess. Upsal.; Acad. Upsal. Holmens. Petropol. Berol. Imper. Lond. Monspel. Tolos. Florent. Soc.
SYSTEMA
NATURÆ
Per
REGNA TRIA NATURÆ,
Secundum
CLASSES, ORDINES, GENERA, SPECIES,
Cum
CHARACTERIBUS, DIFFERENTIIS, SYNONYMIS, LOCIS.
Tomus I.
Editio Decima, Reformata.
Cum Privilegio S:æ R:æ M:tis Sveciæ.
HOLMIÆ,
Impensis Direct. LAURENTII SALVII,
1758.

1.12. Opening page of the tenth edition of Linnaeus' *Systema Naturae* (1758), the starting point of zoological nomenclature. Photo: AMNH Library.

1.13. A plate from Maria Sibylla Merian's *Metamorphosis Insectorum Surinamensium* (1705). Merian's beautiful and detailed works had a strong influence on Linnaeus' later treatment of insects. Photo: AMNH.

1.14. Johann C. Fabricius (1745–1808), student of Linnaeus and the first specialist on entomology. Photo: Deutsche Entomologische Institut.

use of a hand lens. Fabricius' thinking even predated evolutionism: "die nach und nach in Arten übergehende [sic] festen Abänderungen" (p. 24) ["the stable varieties which little by little change into species," as translated by Tuxen (1967a)]. Justifiably, Fabricius is considered the original insect taxonomist, whose study of insects far eclipsed that of his renowned mentor, and even Linnaeus later on accomodated Fabricius' classification of insects into his own.

Another famous entomologist of this era was Pierre André Latreille (1762–1833) (Figure 1.15), who was even called the "foremost entomologist" by luminaries such as Fabricius (Geoffroy Saint-Hilaire *et al.*, 1833; Dupuis, 1974), with whom he was a regular correspondent. Latreille received a formal education and attended seminary, eventually becoming a priest. However, during the formative years of the French Revolution he failed to take the newly instigated civic oath for priests and was therefore condemned for execution and imprisoned. While in prison, Latreille identified a new species of beetle, *Necrobia ruficollis* and, with the aid of two fellow naturalists, was able to secure his release as an entomologist (perhaps the only time the discovery of a new species saved someone's life!). Latreille relinquished his priesthood and, through a series of teaching positions, arrived at the Museum National d'Histoire Naturelle in Paris, eventually receiving a professorship there at the age of 68. Although Latreille was quite prolific and produced numerous fine volumes on the classification of arthropods, he is most noted for his *Précis des Caractères Génériques des Insectes* (Latreille, 1796). In this work he attempted a natural classification of the Arthropoda, delimiting within each order for the first time what we would today call families, although he did not formally name them until subsequent publications.

The pursuit of a natural classification during the late 1700s and early 1800s eventually developed into evolutionary theory, methods of phylogenetic reconstruction, and modern predictive classifications that allow us to explore the diversification of life and evolution of biological phenomena. The debates at the time included questions such as, *What could be the origin of such apparent "natural affinities"?* and *What made groups natural?* Ideas varied from patterns in nature reflecting the thoughts of a divine creator, to the other extreme with no pattern, all of which was a figment of human imagination (order imposed on an otherwise chaotic world). Other scholars believed that nature was harmonic and fell into mathematical sets, the most famous being the Quinarians whose system set nature into groups of five. Naturalists increasingly found that biological traits (*characters*) of organisms formed hierarchical groups and that these groups did not correspond to harmonic numbers and were not arbitrary. As Darwin (1859) himself noted: "From the most remote period in the history of the world organic beings have been found to resemble each other in descending degrees, so that they can be classed in groups under groups. This classification is not arbitrary like the grouping of the stars in constellations." Another question also plagued naturalists, *What was the origin of species? Where did species come from?* The answer was simple: from God. For insects, this is best seen in

1.15. The famous French entomologist, Pierre André Latreille (1762–1833). During the French Revolution, Latreille was scheduled to be executed but was spared after he discovered a new species of beetle in his prison cell. Photo: © Bibliothèque Centrale MNHN Paris 2003.

1.16. A plate from Johann Scheuchzer's *Physique Sacrée, ou Histoire Naturelle de la Bible* (1732), depicting a divine creation of insects (cf. Figure 1.9). Photo: Carl A. Kroch Library, Cornell University.

1.17. Baron Georges Cuvier (1769–1832), the great French morphologist who was the first to carefully document that fossils were the remains of extinct animals. Photo: AMNH Library.

the volume *Physique Sacrée, ou Histoire Naturelle de la Bible* of the Swiss naturalist, Johann Jacob Scheuchzer (1672–1733). There Scheuchzer depicted creation for many groups, including terrestrial arthropods, but of course he only gleaned the surface (Figure 1.16).

Numerous luminaries contributed to the debates about classification. Georges L. Leclerc, Comte de Buffon (1707–1788) published a 44-volume series on natural history. Influenced by Sir Isaac Newton and the concept of physical laws, Buffon worked toward the production of a *classification of natural classes*, which was based on functional morphology, and he was not interested in the systematic methods of his contemporary Linnaeus. In systematic theory, however, two contemporaries at the newly founded Museum National d'Histoire Naturelle in Paris (formed from the collections of Buffon) were to make the greatest contributions, and their debates set the stage for some of the most critical ideas in evolutionary thinking. Georges L. C. F. D. Cuvier (1769–1832) (Figure 1.17), who was eventually made a Baron, decided that the characters that formed natural groups were *adaptive*. Cuvier in large part continued the tradition of Buffon and worked toward a classification based on functional laws. He also decided that fossils were truly the remains of *extinct* organisms, although this troubled him because he believed that all species were created at 9 A.M. on the 26th of October 4004 B.C. (as calculated by the theologian Bishop Ussher: Ussher, 1650). Could God change his mind? The notion of the Biblical Flood, however, could easily account for the loss of such animals, and this idea would be carried forward by numerous scientists including William Buckland, who had a significant influence on the young Charles Darwin. Also in the Museum National d'Histoire Naturelle was Etienne Geoffroy St. Hilaire (1772–1844), a gentleman who did the most to develop the concept of homology (although he employed the antonym "analogy" for what we today call "homology") and Jean Baptiste P. A. M. de Lamarck (1744–1829), who developed an alternative explanation. Lamarck considered that God created a few forms, which then transformed into the various kinds we see. This explanation was certainly a precursor to Darwin's theory of evolution; however, Lamarck failed to develop a plausible mechanism by which such transformations might form.

Geology and Evolution Step In

While the French debated homologies and the Scandinavians devised classifications, the British naturalists and geologists, such as William Smith, were making remarkable discoveries about the Earth and its diversity. Among them, Sir Charles Lyell (1797–1875) (Figure 1.18) produced a major synthesis that led to new concepts of the earth. Lyell, building upon his

1.18. Sir Charles Lyell (1797–1875), whose extensive work on the long, slow accretion of geological forces had a major impact on Darwin's concept of evolutionary time. Photo: AMNH Library.

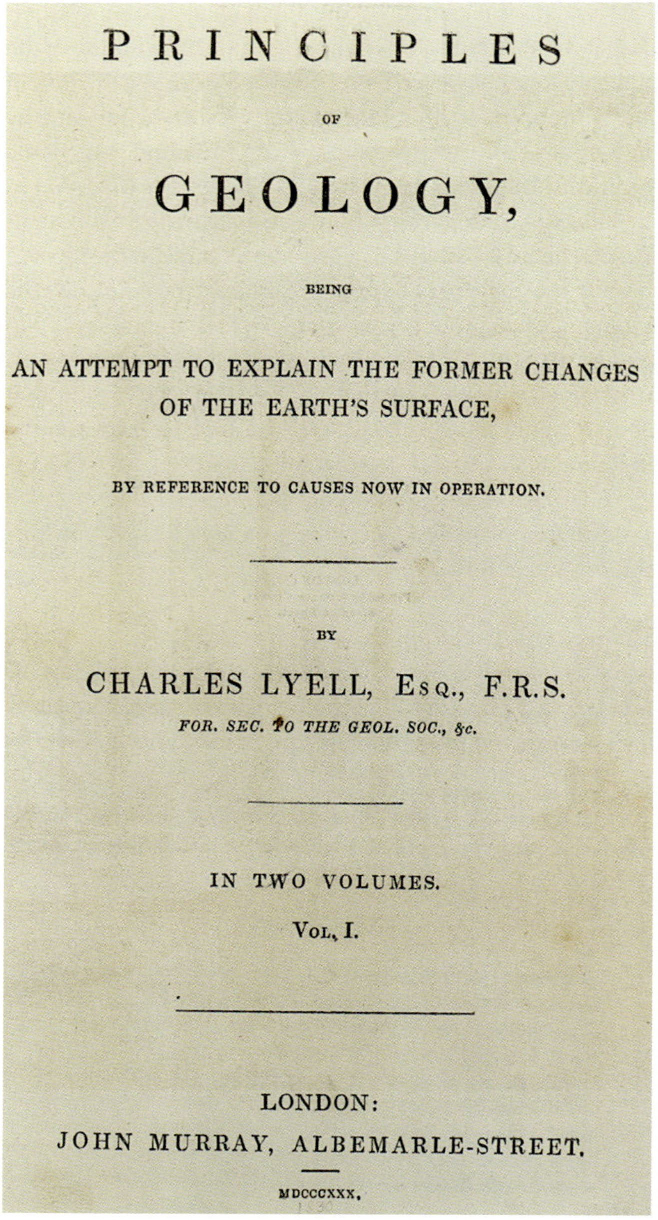
PRINCIPLES

OF

GEOLOGY,

BEING

AN ATTEMPT TO EXPLAIN THE FORMER CHANGES
OF THE EARTH'S SURFACE,

BY REFERENCE TO CAUSES NOW IN OPERATION.

BY

CHARLES LYELL, Esq., F.R.S.

FOR. SEC. TO THE GEOL. SOC., &c.

IN TWO VOLUMES.

VOL. I.

LONDON:
JOHN MURRAY, ALBEMARLE-STREET.

MDCCCXXX.

1.19. The title page of volume one of Lyell's *Principles of Geology* (1830), the work that revolutionized geological thought. Photo: AMNH Library.

extensive travels and historical accounts from around the world, pieced together geological observations and united them with the discoveries of his predecessors and contemporaries. Lyell united ideas on present-day mechanisms such as the accumulation of sediments and erosion, with geological patterns such as continuity of stratigraphic layers and index fossils. As such, he provided the logic for an ancient earth. The publication in 1830 of Lyell's three-volume *Principles of Geology* (Figure 1.19), revealed that the earth was ancient, and that massive formations accumulated slowly over time by forces still acting today (called *uniformitarianism*). As is well known, this work had a great impact on another British scientist whose theory of natural selection revolutionized biology.

Darwin and Wallace

Charles R. Darwin (1809–1882) (Figure 1.20) and Alfred Russel Wallace (1823–1913) (Figure 1.21) were naturalists, the systematists of the day, and they synthesized their knowledge of nature into ideas on the minute, everyday increments that accumulate over geological time to produce the diversity of organisms. They noted that forms of life were interrelated in a seemingly hierarchical fashion, and that if the hierarchy of relationships was spread out over geological time it formed a branching tree. If the earth is as old as Lyell suggested, the continuum of life would be over millions of years (today we understand it to be at least 3.8 *billion* years old). Darwin noted various slow processes that create variations in organisms, such as animal breeding that produces good, bad, or even exotic traits, and how some variations are better adapted to survive than others. Thus, Darwin revisited the question of his time: *What is the origin of species*? (Figure 1.22). His answer, which was that species come from ancestral species that changed through time to produce new

1.20. Charles R. Darwin (1809–82), the architect of modern evolutionary thought and, among other things, a talented field naturalist. Photo: AMNH Library.

species, was not particularly original. His explanation for the mechanism of evolutionary change was, however, entirely original: It is the result of natural selection. A natural by-product of Darwin and Wallace's mechanism was that, by acting over great expanses of geological time, evolutionary change via natural selection would produce a hierarchy of life. Without altering the practice of systematics, Darwin and Wallace revolutionized the theory behind it. Pre-evolutionary systematists had extensive evidence that evolution had occurred. Darwin and Wallace extracted patterns from systematics and natural history and simply *added* the evolutionary interpretation to it. This theory and mechanism could then also explain patterns seen in the fossil record, variations among species around the world, the distribution of related organisms, the similarities seen in embryology, etc. Evolution and the mechanism of natural selection further explained the hierarchical nature of life.

Darwin's influence on classification was strictly theoretical. His work had little effect on the day-to-day practice of systematics: "Systematists will be able to pursue their labours as at present" (Darwin, 1859: p. 405). Indeed, nothing needed to be changed because the findings of systematists created evolutionary theory in the first place. The theory of evolutionary change by means of natural selection merely provided an explanation, a theoretical scaffolding, for the variation that systematists studied. Darwin himself concluded that all natural classifications produced by systematists are genealogical, and, if reconstructed carefully, they are *evolutionary classifications*. In fact, though, the theory of natural selection explained anagenetic change, or the evolution of particular characters, not the origin or formation of new species. This would become a major issue in the twentieth century, in which entomologists had a substantial impact.

1.21. Alfred R. Wallace (1823–1913), adventurer, naturalist, and prolific insect collector. Wallace was coauthor with Darwin on the original paper proposing evolution by means of natural selection. Photo: AMNH Library.

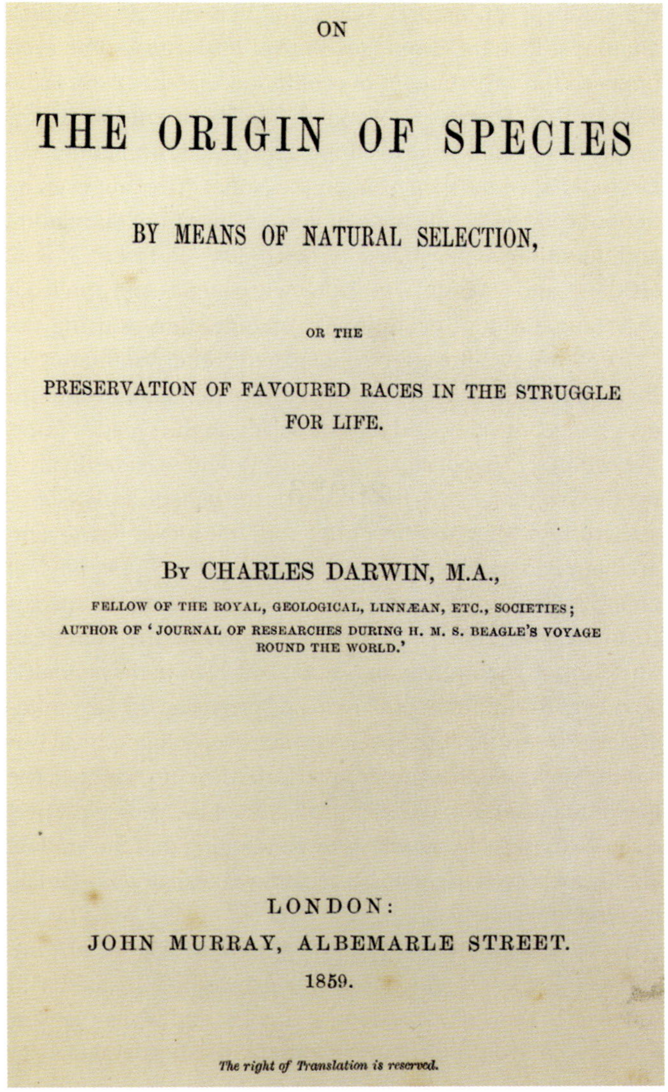

ON

THE ORIGIN OF SPECIES

BY MEANS OF NATURAL SELECTION,

OR THE

PRESERVATION OF FAVOURED RACES IN THE STRUGGLE FOR LIFE.

By CHARLES DARWIN, M.A.,

FELLOW OF THE ROYAL, GEOLOGICAL, LINNÆAN, ETC., SOCIETIES;
AUTHOR OF 'JOURNAL OF RESEARCHES DURING H. M. S. BEAGLE'S VOYAGE ROUND THE WORLD.'

LONDON:
JOHN MURRAY, ALBEMARLE STREET.
1859.

The right of Translation is reserved.

1.22. The title page of Darwin's *On the Origin of Species* (1859). The first edition of Darwin's book contains a single figure – a diagram of a phylogenetic tree. Photo: AMNH Library.

After Darwin

Many biologists adopted evolutionism after the publication of *The Origin of Species*, but some influential biologists resisted the concept of natural selection, like the great British anatomist Richard Owen (see Rupke, 1994), and some even opposed the very thought of evolution (Agassiz, 1896). From 1905 to 1920 Thomas H. Morgan's "fly lab" at Columbia University discovered scores of important genetic phenomena in *Drosophila*, which won Morgan the Nobel Prize, but even he denounced the significance of natural selection. Population and quantitative geneticists in the 1920s – R. Fisher, J. B. S. Haldane, and S. Wright – took the new genetics and applied it to explanations of evolutionary change, thus starting the "New Synthesis," the marriage of genetics and systematics. To them, mutation was the source of raw material for evolutionary change, the clay, and natural selection the creative force, the sculptor. Theodosius Dobzhansky, who was originally a beetle systematist, was impressed by the genetic and systematic work of Morgan's insightful student, Alfred Sturtevant, and applied studies of wild *Drosophila* to questions of genetic variation in nature and the formation of species (Dobzhansky, 1937). A bird systematist at the American Museum of Natural History, Ernst Mayr, mined systematic studies in the tradition of Darwin and proposed the very influential idea of allopatry: Species evolve from geographical isolation (Mayr, 1942). Another American Museum scientist, the paleontologist George G. Simpson, added the geological perspective to the New Synthesis by concluding that the gradual change observed in the fossil record was explained by small, incremental genetic changes (a concept challenged 30 years later by Eldredge and Gould's *punctuated equilibrium*). Between the 1940s and 1960s, biology was preoccupied with the mechanisms of evolutionary change, and while the New Synthesis relied on the work of systematists, systematics per se was marginalized. Ironically, during the foment of the New Synthesis, the brilliant geneticist Alfred Sturtevant was quietly working away on the systematics of drosophilid flies. The methods he used in his classification of the genus *Drosophila* (Sturtevant, 1942) actually presaged the theoretical and empirical work of another highly influential insect systematist, Willi Hennig. Sturtevant's paper was overlooked, perhaps because it was published the same year as Mayr's highly influential *Systematics and the Origin of Species*.

In the 1950s and 1960s scholars began to address these deficiences, particularly attempting to get at an objective means for reconstructing relationships among taxa. One response was the development of numerical taxonomy, or phenetics. Developed principally at the University of Kansas by the entomologists Robert Sokal (Figure 1.23) and Charles D. Michener and by the microbiologist Peter Sneath from Britain, phenetics attempted to develop a method of grouping organisms based on their overall similarity (e.g., Michener and Sokal, 1957; Sokal and Sneath, 1963; Sneath and Sokal, 1973). The goal was to make classifications logical, repeatable, and methodological, in an attempt to remove as much subjectivity as possible. It also required users to record

1.23. Three prominent entomologists and major architects of three philosophies of systematic thinking – from left to right, Willi Hennig, fly systematist and founder of phylogenetics; Robert Usinger, bug systematist and proponent of evolutionary taxonomy; and Robert Sokal, fly population geneticist and cofounder of phenetics. Photo: G. W. Byers, University of Kansas Natural History Museum.

as many differences as could be measured among species, which was another advantage over previous methods of classification. Rather than presuming that one or a few characters have particular evolutionary importance, phenetics attempted to obtain a holistic picture of diverse attributes of taxa. One recited advantage of the approach was that it offered an explicit and testable methodology, but explicitness and testability per se (criteria later espoused by cladists, below) are the most minimal of scientific criteria. An absurd hypothesis ("the moon is made of green cheese") can be very testable. Phenetics had other problems too. First, the choice of specimens and characters to measure imposed a substantial bias, particularly since different systematists considered different characters to be important for study, and a change in a few characters could yield dramatically different results. Also, there was an overall loss of information, such that characters were summarized in a similarity matrix, so the significance of individual characters was hidden in the statistical wash (but this problem is again resurfacing in some molecular analyses). Similarly, different analytical methods gave radically different results, so the classifications turned out not to be stable. Additionally, the predictive value of the phenetic classifications was low, and thus obscured evolutionary patterns, which is because phenetics measures dissimilarity, not relationships. In a phenetic scheme, an insect species that lost its wings might be classified in a genus or family completely different from a closely related winged species. The overarching problem with the use of phenetics is that biologists are interested in phylogeny and require classifications that reflect natural, genealogical relationships. Rather than estimate phylogenetic relationships, the method measured phenetic *similarity*, failing to distinguish between evolutionary similarity (the result of descent) from convergence and that from parallelism. Unfortunately, even today, some phenetic analyses (masked under the name of distance methods) are erroneously used to obtain what are believed to be genealogical relationships. Until DNA sequencing became available, bacteriologists relied heavily on phenetic methods because their organisms lack numerous useful morphological features. Other biologists, however, largely abandoned phenetic methods or never accepted them because such classifications consciously ignored phylogeny. Various authors tinkered with the method in an attempt to make it more representative of phylogeny, but the results were not persuasive. As a result of these problems, pheneticists adopted a nominalist philosophy in which only individuals exist and all other groups are artifacts of the human mind. Thus, the classification of plants and animals could be treated the same as inanimate objects in this system. Now, species are classified as lineages, groups that are closely related by common descent.

Roughly contemporaneous with the development of phenetics, another response to the NeoDarwinists was being pursued, but this one was more fruitful. Willi Hennig (1913–76) (Figures 1.23, 4.21), a German fly systematist, operated under the principle that phylogeny can be reconstructed and that all classifications should be based on the reconstructed pattern of genealogical descent, thus echoing Darwin's original call for evolutionary, or natural, classifications. Hennig provided a rigorous method for analysis of phylogenetic relationships, called *cladistics* (Hennig, 1965b, 1966). The basic idea was that a truly phylogenetic system of classification would be the most useful as a general reference system for biology. If a classification is to reflect phylogeny exactly and have explanatory power, then all taxa classified must be strictly *monophyletic*. *Paraphyletic* and *polyphyletic* groups are not accepted as taxa suitable for ranking in the classification (see also Farris, 1974, for characterization of these terms). Monophyletic groups were natural because they contained the common ancestor and all the descendent species. Paraphyletic groups contained the common ancestor and some, but not all, of its descendent species. The most problematic groupings are polyphyletic, which are those that contain some of the descendants of a common ancestor but not the common ancestor itself (Figure 1.24).

These distinctions serve to highlight another of Hennig's major contributions, the concept that among the mosaic of an organism's characters only those that are derived ("specialized," "modified") and shared with other species are informative of phylogeny. Hennig noted that, for the purposes of phylogenetic reconstruction, there are actually two kinds of character similarity. Apomorphies are similarities that arose in a most recent common ancestor, or a recently evolved ("advanced") feature that appears only in a group of closely related species. Plesiomorphies are similarities that arose in a distant common ancestor, or "primitive" feature. Use of "advanced" and "primitive" is generally discouraged because these terms imply a scale of perfection or adaptation, while plesiomorphic features are merely apomorphic features deeper in phylogeny. Wings are a plesiomorphy of butterflies, but an apomorphy of pterygote insects. In other words, it is important to note that "plesiomorphy" and "apomorphy" are relative terms.

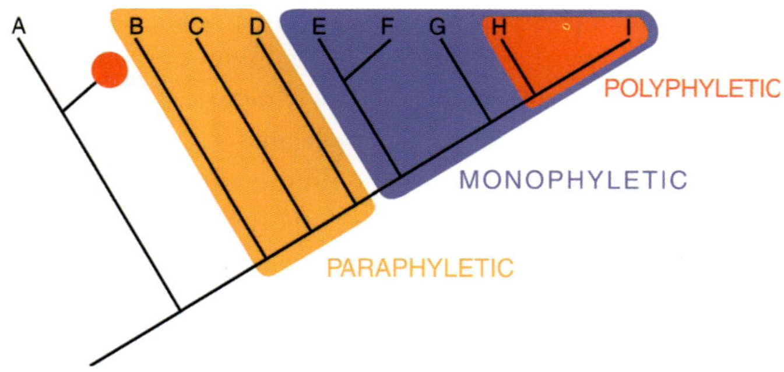

1.24. Three kinds of groupings – monophyletic, paraphyletic, and polyphyletic. Natural classifications attempt to categorize and name only monophyletic groups.

Thus, the critical aspect of recognizing monophyletic groups was to base them on shared apomorphies (called *synapomorphies*) and not on shared plesiomorphies (called *symplesiomorphies*). Grouping taxa based on plesiomorphies would only produce paraphyletic groups. A group is monophyletic if it is characterized by the possession of *synapomorphies*, derived characters shared by the members of the group. For example, wings are a synapomorphic trait that unites the winged insects into a monophyletic group called Pterygota. Taxa that are paraphyletic are based on *plesiomorphies*. In the same example, before the evolutionary novelty of wings arose, insects were primitively wingless. Thus, the absence of wings is plesiomorphic for insects and the apterygotes (wingless insects) are paraphyletic, as some apterygote taxa are actually more closely related to Pterygota than they are to other wingless species. Lastly, polyphyletic groups are frequently based upon convergent characters that are only superficially similar but not of a common evolutionary derivation. The sucking insects, for example, are a polyphyletic assemblage, the development of such mouthparts having arisen several times and in different configurations. Hennig thereby provided us with the framework to define characters and then to polarize them into apomorphic and plesiomorphic states. As already discussed, the critical aspect of recognizing monophyletic groups was to base them on shared apomorphies and not on shared plesiomorphies. The presence of an apomorphy in a single species and in no other species is not informative of relationships because it groups the taxon with no other lineage. These are called *autapomorphies*.

However, how does one know whether a character is

apomorphic or plesiomorphic? This is the problem of determining *character polarity*, and various methods have been proposed to determine this, the most widely used being *outgroup comparison*. Lineages closely related to the one under study but sharing traits through a more ancient common ancestor (the *outgroup*) share plesiomorphies; those traits that are unique within the *ingroup* are *apomorphies*. Other, less often employed or defunct methods involve criteria based on ontogeny or development, frequency of characters, and paleontology. The ontogenetic criterion uses developmental evidence, which sometimes works but is rarely practical. The frequency criterion states that the most common character state is plesiomorphic. This can be very misleading because groups that recently radiated will have the most species with derived features and only a few species will remain with plesiomorphic features. The paleontological criterion maintains that the characters found in the oldest fossils are plesiomorphic. This can also be misleading because fossil taxa often have some very highly modified traits, as we show throughout this book.

Cladistic analyses are all about congruence. Each character is allowed to influence the final set of relationships, unlike in phenetics where the effects of individual characters are obscured and irretrievable. We then need to find the overall hierarchical pattern implied by the various similarities. Before we elaborate on finer aspects of the methodology, however, we must step back for a moment and consider what these similarities really are.

Homology

Early in the history of comparative biology, scientists recognized a general similarity among some organisms. Common features could be recognized in an overall *body plan*, but scientists could not explain why. Organisms seemed to be made of similar parts organized into a similar body plan with simple modifications at various points that created diversity, a phenomenon that required explanation. One of the most famous examples of the early recognition of this similarity in body plans is by Belon (1555), who published a book on comparative natural history in which he noted various topological similarities in the skeletons of a human and a bird. He identified, based on relative positions and connections, "identical" bones in the common tetrapod body plan.

Various naturalists, particularly vertebrate anatomists, played with this idea, and it formed the basis for all comparative biology. The next major development in the concept of homology was not until the early 1800s, when Etienne Geoffroy Saint-Hilaire (1818) at the Paris Museum wrote his *Philosophie Anatomique* (Figure 1.25). Geoffroy used what he called the *principle of connections*. Based on the relative connections of structures in the overall body plan, he identified similar structures even if they had been slightly or dramatically modified between two organisms. He went so far as to apply his method throughout the vertebrates and developed in his book the idea that the inner ear bones were in fact bones modified from part of the jaw in lower vertebrates, a revolutionary thought at the time. Many people ridiculed him (particularly his prestigious adversary Cuvier), but an origin of the vertebrate ear bones from jaw bones is entirely accepted today. This principle of connections allowed for the recognition of homologies, *common structures in a common organismal plan* (although Geoffroy himself used the term "analogy" for what we today call "homology").

1.25. The great French anatomist and natural philosopher, Etienne Geoffroy Saint-Hilaire (1772–1844). Geoffroy laid the foundation for recognizing homologous traits, although he personally employed the term "analogy" for what we today call homology. Photo: © Bibliothèque Centrale MNHN Paris 2003.

Richard Owen (1866) expanded on Geoffroy's concept and noted that some structures might appear superficially similar but that, based on the principle of connections, they were indeed not identical in the common body plan. He therefore called these "deceptive" characters *analogies*. He, being a bit of a fan of Plato, more fully developed the idea of an *archetype*, that there was some "ideal" form and that all structures in organisms were modifications of God's ideal. Comparing any organism to the archetypal ideal would allow for a meaningful comparison between the degenerate forms observed on Earth. Owen, like his predecessors, however, still explained homologies as evidence of the hand of God, which is one reason why he and Darwin clashed.

Darwin (1859) once again provided the critical synthesis

recognizing this observational phenomenon of topological similarities (i.e., homologies) among organisms and how these formed the basis for defining hierarchical groups. Thus, if the observed hierarchy was explained as a product of evolutionary descent, *then* the observed homologies shared among organisms must be shared because of their *ancestry*. In other words, organisms share traits because their common ancestor had them. Darwin's contribution was simply a matter of interpretation and explanation, but it was critical and fundamental.

Darwin was quick to point out a logical separation for the observation of homology versus its explanation via ancestry, just as for the hierarchy of life and its explanation by descent. His vehement follower Huxley, however, unified homology and ancestry such that, in a single term, we had the observation and its explanation. Today, we tend to forget the observational basis of homology and focus simply on the fact that it is *a trait shared between two organisms because they inherited it through their most recent common ancestor*. The observational phenomenon and the explanation of that observation are often combined and confused, but there is indeed such a distinction within the single concept of homology, which is critical for keeping evolutionary theory from becoming logically circular. Although we *interpret* homology as a shared trait inherited from a common ancestor, we do not *observe* common ancestors. We need criteria for recognizing homology or rules that allow us to hypothesize why two or more species are closest relatives (Brady, 1985, 1994).

The modern definition of homology is basically an equation with synapomorphy. This unification is fine as long as one keeps in mind the logical separation between the observation of homology and the demonstration of ancestry, the practice and the conclusion. To distinguish observational homology from that equal to synapomorphy, some authors have employed the term "primary homology" for the former and "secondary homology" for the latter (e.g., de Pinna, 1991). *Primary homologies* are observational homologies not yet tested by a cladistic analysis, while *secondary homologies* are synapomorphies (i.e., those tested by a cladistic analysis and interpreted to be the product of ancestry).

Many people have gone about characterizing criteria for observational homology (i.e., primary homologies). The most extensive work was that of Remane (1952), who provided three "criteria of similarity" that can be used to identify homologous features.

1. *Criterion of Position.* A structure in different species that has a similar position relative to other landmarks in the body plan is likely to be homologous (this is essentially the *topological identity* criterion of Geoffroy Saint-Hilaire). This is also the main criterion (and perhaps the most important) for molecular homology because identifying a base-pair substitution depends on what base-pairs flank it. In morphology, too, this is a fundamental criterion. For example, the criterion of position allows one to readily identify that stalked eyes are not homologous across various families of flies. An important landmark for these topological comparisons is the position of the antennae (Figure 1.26). In the Diopsidae, for example, the antennae are positioned at the apex of the stalks and near the compound eyes; in other families, such as various tephritoid flies, they are normally positioned near the midline of the face. The position of the antennae is the clue that the "stalk" is composed of paraocular integument in tephritoids but also of the frons and face itself in Diopsidae (Figure 1.26). The fossorial legs of many insects is another excellent example where topological identity allows for the discernment of homologous versus nonhomologous traits (Figure 1.27).
2. *Criterion of "Special Similarity."* Complex structures that agree in their details are more likely to be homologous. An excellent example concerns the reduced, club-like wings in flies (Diptera) and the small order Strepsiptera. Some molecular evidence suggests that these orders are closely related (Whiting *et al.*, 1997; Wheeler *et al.*, 2001), but the halteres of flies are the hind wings (Figure 12.23), and those of Strepsiptera are forewings (Figure 10.79). Whiting and Wheeler (1994) were so convinced by the "special similarity," essentially shape, of the halteres, that they proposed a developmental switch of the mid- and hind thoracic segments in the two orders (see Chapter 10). For them, the criterion of "special similarity" overrode the criterion of position.
3. *Criterion of Continuation.* Structures in two or more taxa may be quite dissimilar, but if other taxa can be found that have intermediate forms so that dissimilar extremes can be linked, then the homology may be more readily recognized. For example, the proboscis of glossatan moths is usually very long and coiled and is formed from extremely long galeae (Figure 13.19). It is possible that the galeae could not be recognized if early intermediates between glossatan and mandibulate moths were not known.

Modifications of these criteria have been adapted for the identification of behavioral and molecular homologies as well (e.g., Baerands, 1958; Wenzel, 1992; Greene, 1994; Hillis, 1994; Miller and Wenzel, 1995). Interestingly, these two types of characters are somewhat related by the means in which homologies are identified.

Molecular data principally consist of identifying the sequence of amino acids (for proteins) or nucleotides (for DNA) from a particular enzyme or gene. This produces a linear sequence of point-by-point identifications. In order to compare two or more sequences, they must first be aligned so that corresponding points in the sequences are paired. Although this sounds trivial on the surface, it is, in fact, often

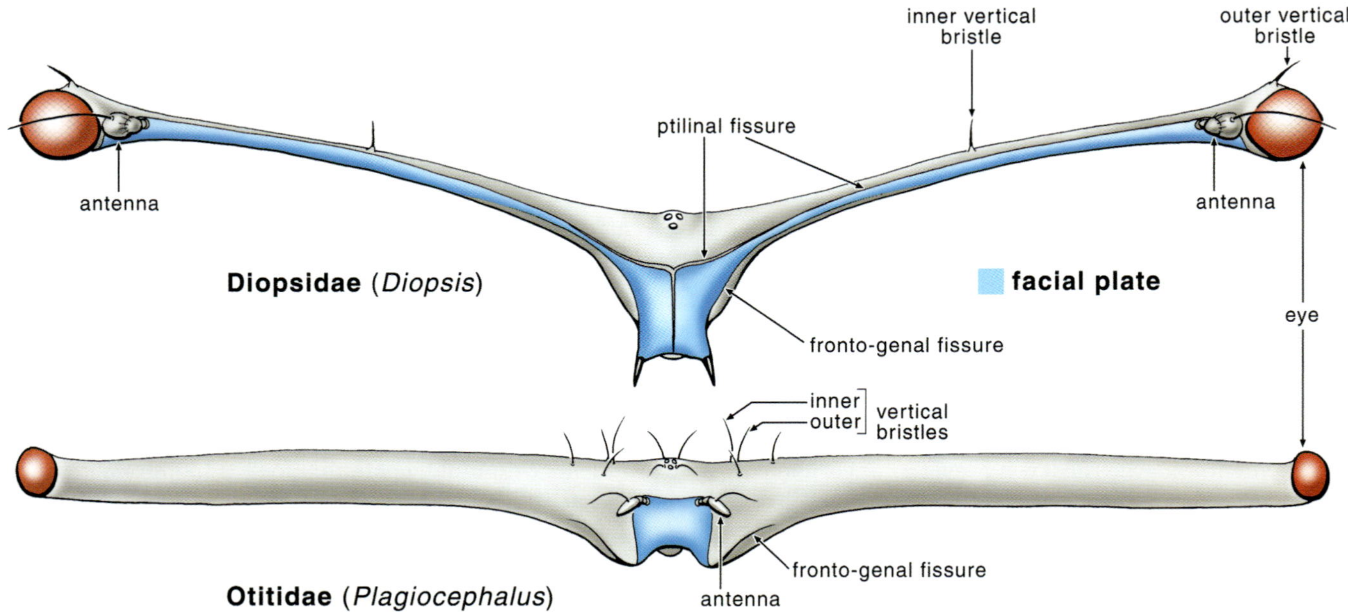

1.26. An example of homologizing structures using the criteria of size, position, and "special similarity": the heads of stalk-eyed flies in different families. Diopsid antennae are near the end of the stalks; this and the boundaries (sutures) between the plate-like sclerites indicate that the face of diopsids is expanded. In the other flies, the frontoorbital plates on the inside margin of each eye are expanded, and antennae remain close together on the face. Not to the same scale.

difficult and nonintuitive. Deletions or insertions into the sequence of one or more taxa create linear sequences of unequal lengths. Thus, to make the sequences align, gaps indicating where the deletion or insertion took place may have to be hypothesized so that the remaining amino acids or nucleotides are aligned. Gaps, however, are not observed; they are hypothesized ad hoc because a strand of DNA does not possess a physical gap, just a lost section of nucleotides. For regions of DNA that code for proteins, the codon structure provides a nice frame of reference for aligning sequences. However, noncoding regions can be problematic, with numerous insertions and deletions. Once again, this is a critical first step for molecular analysis because it is essentially the identification of primary homologies. Each nucleotide position is a character in the analysis; consequently, alternative alignments, which compare nonidentical positions across the same strands, may arrive at wildly different sets of relationships, skewing our interpretation of evolutionary history. Whereas in morphology the last two of Remane's three criteria can be called upon to refine our hypothesis of observed homology, molecular data rely almost solely on topological position.

The recognition of behavioral homologies can be analogous to the identification of molecular homology because it is often possible for a behavioral repertoire to be taken apart into a sequence of acts. For instance, the mating displays of some flies are ritualized performances of various acts; the male may have to wave his wings, move around the female, and tap her, in a specific manner and order before she will allow mating. Thus, like molecular data, a linear sequence of acts must be aligned when comparing two or more species. Furthermore, like molecular data, hypothesized gaps may be required. For example, in the mating ritual, one species of a genus may have lost the wing-flipping motion seen in all the other species and, instead, skip directly to the final act of tapping. In this example, a gap would need to be inserted to align properly the behaviors observed. In contrast to molecular sequences, however, the last two criteria for homology can also be applied to behavioral repertoires. Behaviors need not always appear as a repertoire and can extend beyond the organism itself. Several insects have diagnostic behaviors that result in the alteration of the environment (e.g., feeding damage) or the production of some lasting structure (e.g., nests). In some instances, different lineages produce diagnostic types of environmental changes that can frequently be preserved through geological time and provide us with early records of behavioral novelties. Leaf-cutter bees produce a diagnostic form of leaf damage often seen in Tertiary fossil leaves (e.g., Wappler and Engel, 2003). Nests are also commonly preserved, leaving behind evidence of their ancient architects. Fossilized behaviors provide important insights into the stages of evolution. Even though many people *a priori* believe that behavior is too labile to be informative for phylogeny, it is in fact no different from other types of character data. It is best never to assume *a priori* that a feature will not be phylogenetically informative because only a cladistic analysis can determine this. Indeed, some behaviors are highly variable and do not explain relationships, but in the same way that some morphological traits or gene regions can also be too labile or variable. In studies comparing the use of behavioral characters in cladistic analyses, behaviors were found to be just as informative as other kinds of data (e.g., Wenzel, 1992; de Queiroz and Wimberger, 1993; Wimberger and de Queiroz, 1996).

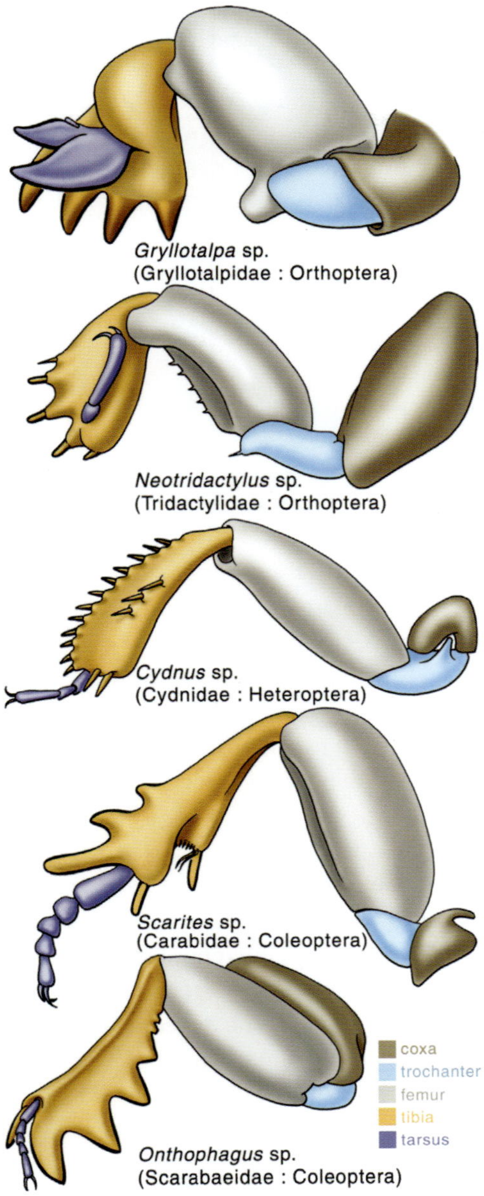

1.27. A classic example of convergence among insects: fossorial forelegs. Forelegs adapted for digging have evolved repeatedly in insects, but in each case the forelegs have been modified in different ways. Not to the same scale.

Unlike morphology or behavior, with their seemingly endless variety of comparable forms, the criteria for analyzing molecular data are more limited. Though extensive stretches of DNA sequences are easily gathered, each character is confined to only four possible states: A, T, C, and G. The language of molecular systematics is short of words. This limits the kinds of change that may take place. If multiple changes have occurred at the same position, the probability of arriving at the same nucleotide by chance alone and not by descent increases. Such multiple "hits" at a nucleotide position essentially erase the historical information that was once stored in the DNA sequence and can be difficult to discern. Because DNA sequences rely solely on Remane's first criterion of homology, multiple "hits" often lead to erroneous homologies. With no recourse to the criterion of "special similarity" (after all, an adenosine is an adenosine), nonhomologous nucleotides in identical positions can mislead in an analysis by uniting two or more species that, in actuality, are not closely related at all. The more multiple hits across the entire sequence, the more random the final inference of phylogeny becomes. This problem is avoided to some extent by sequencing genes that have appropriate substitution rates. Genes with high substitution rates ("fast evolving" genes) are used for species-level relationships; genes with low rates ("slow evolving") are used for more ancient divergences. Specialized analyses have also been designed in attempts to confront and solve this difficulty.

In reconstructing relationships, molecular data presently draw from a more limited portion of the genome than the information derived from morphology relies upon. Most morphological structures in insects are highly polygenic, so even the most cursory morphological data set may represent the cumulative effect of hundreds to even thousands of genes from across the genome. For example, the shapes of several male structures of drosophilid fruitflies, each of which comprises one or two systematic characters, are controlled by between four and ten genes (Templeton, 1977; Val, 1977; Coyne, 1983). Presently, the largest molecular studies may provide data from about eight genes, usually fewer, and reflect a much more restricted part of the genome. This is particularly true given that most studies, regardless of taxon, focus on a relatively small suite of genes for phylogenetic inference (Table 1.2). Moreover, the phenotype not only reflects genotype, but it is also an emergent phenomenon, comprised of the interaction of genotype and environment. Naturally, as sequencing technology and computational methods improve, this setback should be overcome, but there will always be a need for studying morphological characters, particularly when interpreting fossils.

Phylogenetic Analyses

At its simplest, reconstructing phylogeny boils down to a congruence test among topological identities, that is, of either morphological, behavioral, or molecular homologies. The observed homologies are analyzed cladistically – divided into apomorphies and plesiomorphies to form a hierarchical pattern, a *cladogram*. The cladogram is a type of very general evolutionary tree that indicates only relative relationships, not ancestor-descendant relationships. A cladogram calibrated with the fossil record and the geological time scale is considered a *phylogeny* (Smith, 1994). After this has been completed, the pattern of change of the individual characters can be interpreted. Homologies that support monophyletic groups are interpreted as synapomorphies. Alternatively, homologies that appear independently in different places on the phylogeny are interpreted as *homoplastic* (= analogies). In other words, the trait is incongruent with the hierarchical

TABLE 1.2. Most Frequently Employed Genes in Zoological Molecular Phylogenetics

Gene	Abbreviation	Genome	Av. Size (bp)
Cytochrome b	Cytb	Mitochondrial	1,140
Cytochrome oxidase I	COI	Mitochondrial	1,546
Cytochrome oxidase II	COII	Mitochondrial	684
Dopa decarboxylase	DDC	Nuclear	1,422
Elongation factor 1α	EF1α	Nuclear	1,374
NADH dehydrogenase (subunit 1)	ND1	Mitochondrial	971
Phosphoenolpyruvate carboxykinase	PepCK	Nuclear	1,641
rDNA subunit 12-svedbergs	12S	Mitochondrial	819
rDNA subunit 16-svedbergs	16S	Mitochondrial	1,630
rDNA subunit 18-svedbergs	18S	Nuclear	1,824
rDNA subunit 28-svedbergs	28S	Nuclear	3,720
RNA polymerase II	Pol II	Nuclear	4,360

Other increasingly employed genes: various tRNAs, cytochrome c, wingless (Wg), rhodopsin, etc.

pattern suggested by the overall body of evidence; it is *convergent*.

There are an intimidating number of methods for reconstructing phylogeny. This is almost universally done with computers today because of the vast amount of data available. However, different programs have been written to solve particular problems of analysis, so many have been proposed simply to analyze specific varieties of data (e.g., protein, morphological, DNA sequences) or to analyze these data under differing assumptions of how evolution must operate. Many programs are quite easy to use (or *misuse*) and have led some investigators to employ a "hit-and-run" approach to phylogeny reconstruction. Regardless of methodology, traditional complications for phylogeny reconstruction, such as hybridization, introgression, and lateral gene transfer, remain as problems. Today, analyzing large numbers of taxa and DNA sequences has led to even more algorithms for inferring phylogeny.

But again, at their simplest, these programs seek some degree of agreement between the characters being studied, referred to as *congruence*. But how is congruence measured?

Parsimony

Parsimony criteria are used as tree-building methods and certainly have been among the most widely used methods to reconstruct evolutionary history. One of the most basic tenets of this method is that the most preferred hypothesis of evolutionary relationships will be the one that requires the fewest number of evolutionary changes, or steps. Such hypotheses also represent the most congruent arrangement of all of the characters in an analysis. They seek the greatest support for homology and the least support for homoplasy. Generalized parsimony operates throughout science and is based on the philosophical principle of Occam's Razor, which simply states that "all things being equal, the simplest explanation is the best explanation." In other words, the simplest means of explaining the observations is better than any explanation requiring numerous ad hoc hypotheses. However, there are other forms of parsimony, such as Wagner parsimony, Dollo parsimony, and Camin-Sokal parsimony, which incorporated *a priori* models of character change, like reversals being more likely to occur relative to independent, convergent origins.

Maximum Likelihood

The idea of incorporating evolutionary models as well as a desire to impose statistical inference led to the development of likelihood methods for phylogeny reconstruction. This first became important with the increase in biochemical and molecular data and the limited number of character states (i.e., the four nucleotides) and the aforementioned difficulties. In particular, circumstances were identified in which parsimony failed to find the "correct" evolutionary history (Felsenstein, 1973, 1978, 1981, 1983). When evolutionary rates in different lineages are extremely dissimilar and evolutionary change is large (i.e., the rates of mutation are high), then the probability of arriving at the same nucleotide not by descent but by chance becomes high. This has been referred to as the Felsenstein Zone.

Maximum likelihood estimation starts with an aligned set of sequences that are constrained onto a starting hypothesis of phylogeny, and that are frequently derived from a simple parsimony analysis. The tree is arbitrarily rooted, and the likelihood of particular sites is computed as the sum of the probabilities of every possible reconstruction at ancestral states given a particular model of nucleotide substitution. The likelihood of the entire tree is calculated as a product of the likelihood of each individual site. The objective is to end up with the simplest model to explain evolution. Then, the "best" likelihood settings are used to run the phylogenetic analysis, which employs differential weighting of characters as specified by the employed models. The method strives to

TABLE 1.3. Most Frequently Employed Ranks in Zoological Nomenclature

Kingdom	Animalia
Phylum	Arthropoda
Subphylum	Mandibulata
Superclass	Panhexapoda
Epiclass	Hexapoda
Class	Insecta (Ectognatha)
Subclass	Dicondylia
Superorder	Hymenopterida
Order	Hymenoptera
Suborder	Apocrita
Superfamily (-oidea)	Apoidea
Family (-idae)	Halictidae
Subfamily (-inae)	Halictinae
Tribe (-ini)	Augochlorini
Genus	*Augochlora*
Species	*nigrocyanea* Cockerell

There are no standardized terminations in zoology for names above the rank of superfamily (ICZN, 1999).

discover the tree or set of trees that gives the highest probability of a data set being derived from it, and *not* the probability of a tree being derived from a data set (which is a common misconception of likelihood methods).

Despite the apparent appeal of maximum likelihood methods, they are not without problems. Likelihood methods require an *a priori* probabilistic model of evolutionary process. Statistical methods of phylogeny inference require a hypothesis of the processes of evolutionary change. With their small, finite number of possible character states, DNA sequences lend themselves well to such methodology. There is widespread acceptance of probabilistic models of evolution for nucleotide sequence data. According to Swofford *et al.* (1996), parsimony methods ignore information about branch length, whereas maximum likelihood methods, which assume that repeated changes are more likely on longer branches, do not ignore length. The most common forms of models for evolutionary change are those pertaining to base-pair frequencies in DNA/RNA; transition:transversion ratios; probabilities of amino acid substitutions in proteins (i.e., synonymous versus nonsynonymous codons); and the shape of complex macromolecules (e.g., DNA/RNA loops).

While some attempts have been made to apply likelihood methods to morphological data, probabilistic models of morphological change are considered to have little basis. What biological justification can be made for suggesting a particular probabilistic model of anatomical evolution, such as from the large hind wings of Mecoptera to the halteres of flies? Moreover, computationally, the calculation of probabilities is extremely time consuming, though computing ability improves every year. Some scientists still consider maximum likelihood advantageous because as sample sizes increase toward infinity and the models become more realistic, the results are believed to converge on some idea of the "truth" [and interestingly, appear to converge on a simple parsimony analysis (Goloboff, 2003)].

A recent advance in model-based phylogenetic reconstruction has been Bayesian analysis (e.g., Huelsenbeck *et al.*, 2001). Rather than relying on *a priori* probabilities for estimating the most likely tree topology, Bayesian methods employ *a posteriori* probabilities for models of nucleotide evolution. Use of posterior probabilities eases the use of complex, ideally more realistic, models of molecular evolution and allows vastly larger data sets than was feasible under strict likelihood estimation.

TAXONOMY, NOMENCLATURE, AND CLASSIFICATION

If the names are lost the knowledge also disappears.
–J. C. Fabricius, 1778, *Philosophia Entomologica*

The first part of knowledge is getting the names right.
–Chinese proverb

Nomina si nescis, perit et cognitio rerum. [Who knows not the names, knows not the subject.][2]
–Linnaeus, 1773, *Critica Botanica*

The first reaction upon encountering a physical object is to recall its name or to name it. Every human culture names the world around it, developing folk classifications for the biotic and abiotic realms. The human mind operates by circumscribing the features of relevant objects, often subconsciously, and then applying a name to refer to them. Folk taxonomies

[2] Adapted by Linnaeus from Isidore of Seville's (ca. A.D. 560–636: Patron Saint of Students) phrase, *Nisi enim nomen scieris, cognitio rerum perit*, in his *Origines seu Etymologiae: Liber I*.

were of great importance; they served to distinguish the edible from the poisonous or the cuddly from the threatening. This was standardized and codified to form the formal taxonomy of binomial nomenclature in biology. A *taxon* (plural *taxa*) refers to a group of organisms at any rank (e.g., *Pelecinus* and *Musca* are taxa at the rank of genus). The principal ranks in an ascending series are species, genus, family, order, class, phylum, and kingdom (although numerous other intercalated ranks and informal categories are also frequently used: Table 1.3). *Classification* is the codification of the results of systematic studies, using taxonomic principles.

All accumulated information of a species is tied to a scientific name, a name that serves as the link between what has been learned in the past and what we today add to the body of knowledge. For instance, how do we know that unique data on nesting biology of the honey bee referred to by Gerstäcker (1863) is the same species to which we refer today? What if Gerstäcker misidentified the species? What if he was actually mixing data from two species under a single reference? Unlike other sciences, the history of biological nomenclature is of paramount importance and the correct application of scientific names ensures that these names will be stable. Names provide for a means of universal communication and as a reference database for information storage, retrieval, documentation, and use. For this purpose, taxonomy is governed by rules of *nomenclature*, a precise system used by systematists that concerns itself with pragmatic methods for naming taxa.

The rules of nomenclature used to assign names to taxa are a strictly utilitarian, pragmatic part of taxonomy. The rules of nomenclature have nothing to do with theories about relationships (phylogenetic work determines those), species concepts, or logical consistency (i.e., parsimony). In fact, the rules of nomenclature are *not* intended to impair or interfere with a taxonomist's judgment about how taxa should be circumscribed or which taxa should be named. The rules of nomenclature are simply intended to specify *how* taxa should be named. In instances where more than one name has been proposed, they specify which name should be used or has priority. The rules of nomenclature are codified, legal documents; *International Commissions* deliberate upon any changes in the rules and pass judgment on petitions for exceptions to the rules in special circumstances.

Rules of nomenclature have not always existed. Even in Linnaeus' time, there was no regulation for how to name things formally. However, as more and more species were discovered and scientists expanded biological disciplines (taxonomic and otherwise), chaos and confusion began to develop as different names were applied to the same species or identical names were given to radically different taxa. In the absence of such rules, the scientific literature was becoming utterly confusing, and in some instances meaningless, because there was no stability to the use of names. The benefits of having a formal taxonomy were being lost. To ensure the continuity and conveyance of biological information, rules were established, and the International Commissions were formed. There are presently three independent rules of nomenclature: Botanical, Zoological, and Bacteriological. Naturally, the names of insects are governed by the *International Code of Zoological Nomenclature* (ICZN, 1999).

Scientific names have a codified format. Each animal has the binomial name (genus and species), the name of its author, and the year it was established: for example, *Apis mellifera* Linnaeus, 1758, or Genus *Zorotypus* Silvestri, 1912, or Family Megachilidae Latreille, 1802. The author name indicates who proposed that part of the name and in what year. In the aforementioned examples, Linnaeus proposed the species name *mellifera* in 1758 as a taxon in the genus *Apis*; Silvestri proposed the generic name *Zorotypus* in 1912; and Latreille proposed the family name Megachilidae in 1802. The author's name is always capitalized (sometimes abbreviated for famous individuals, e.g., Linnaeus is often abbreviated as "L.") and never underlined or italicized. This allows us to keep track of the name and to recognize when someone has proposed an identical name for a different biological taxon that would lead to confusion. Such identical names are called *homonyms*. For example, Eversmann in 1852 established the name *Bombus modestus* for a species of bumble bee living in Eurasia (the name was thus, *Bombus modestus* Eversmann, 1852). Numerous other bumble bee species were also established in *Bombus*, and it was later discovered that in 1861 Smith had described a completely different species from Mexico which he had also called *Bombus modestus*, unaware that an identical name already existed. (Indeed, a third *B. modestus* was proposed by Cresson in 1863!) These names are homonyms. Eversmann's name is older and is thus called a *senior homonym* while Smith's name is the *junior homonym*. These species have distinct biologies and distributions, and these data are referenced by their names. However, it is confusing that they have identical names, so how do we objectively decide which one to call *B. modestus* and which should be named something else? This decision has nothing to do with the biological distinctions between the species but merely how we will reference and access the data associated with each. The nomenclatural *Principle of Priority* states that the oldest available name is the valid name for the taxon. Thus, in the example of *B. modestus*, the name established by Smith had to be replaced to avoid confusion with an already existing, identical name. This correction was made by Dalla Torre (1890), and the name of the second species is now *B. trinominatus*. This system of priority simultaneously provides a simple, objective basis for deciding among competing, different names for the same species. Any names applied to the same species that are proposed subsequent to the first available name are *junior synonyms*. Priority avoids more subjective aspects like the quality of the original description because these are open to extreme differences of interpretation as to what makes a "good" versus "bad" original description.

Every species name occurs in a *combination*, specifically in combination with some generic name (hence *binomial nomenclature*). As systematics continues to refine our understanding of relationships among organisms, the taxonomic placement of species is often changed. Thus, although nomenclature seeks to stabilize taxonomy, it necessarily recognizes that taxonomy is, by its nature, dynamic. Species are often moved from one genus to another (i.e., their combination is altered from their original assignment). This poses the difficulty of historically tracing the data when a given species name has shifted. For example, the metallic green sweat bee *Andrena metallica* Fabricius, 1797, was moved to the genus *Augochloropsis*. In zoological nomenclature, the simple system of placing parentheses around the author's name indicates a change in combination; for example, "*Andrena metallica* Fabricius, 1793" becomes "*Augochloropsis metallica* (Fabricius, 1793)." Taxonomic catalogues are the repositories for such data: tracking the usage of names since their inception and unifying them with the body of scientific literature tied to the taxon. Such compendia are the most fundamental works on any group of organisms (e.g., Herman, 2001) and the first reference sought at the opening of any line of inquiry. With a group as diverse as insects, taxonomic catalogues are essential but lag far behind our needs.

Tracking the names is only a small part of the problem, which is becoming increasingly manageable with databases. While we can readily ascertain what Fabricius wrote, how do we know exactly what he was looking at and that it is conspecific with what we may today be studying in the field? For that, we must examine Fabricius' type specimens. As in all arenas of biology (e.g., ecology, physiology), taxonomy has its system of vouchers. However, unlike other fields, the nomenclatural system of vouchering is widely misunderstood and erroneously cited as an explanation for taxonomy being an outdated, Victorian pursuit. The confusion ironically centers on the *name* of the vouchers used in nomenclature. The concept of *typification*, or the application of *nomenclatural types*, is confused with the older concept of Platonic types or archetypes. These concepts were at one time inseparable, but the two are no longer related. Earlier in the history of taxonomy it was believed that a single specimen was selected to represent each species and that this individual best approximated the *eidos* or ideal form of the taxon. Today, the nomenclatural *type* serves the purpose of vouchering biological data and determining species identity when published descriptions are inadequate.

There is a suite of rules in nomenclature for the designation of a type specimen, and types consist of two general kinds: name-bearing (or primary) types and non-name-bearing (or secondary) types. Primary types include *holotypes*, *lectotypes*, and *neotypes*, while all other types (e.g., *paratypes*) are secondary types. A *holotype* is a unique, name-bearing type specimen designated by the original author. The holotype is the single individual of a species that serves as the voucher for a given species name. All taxa conspecific with the holotype must use the name associated with that holotype. A holotype can be designated only by the original author and in the publication in which that author established the name. It is the original author's voucher for his/her species. *Paratypes* are additional specimens that were examined and designated by the original author in the original publication as being likely to be conspecific with the holotype and are frequently represented by the degrees of variation known at that time. While this distinction between the holotype and paratypes may appear on the surface to be no different than the older concept of archetypes, the selection of a single specimen to hold the name is strictly pragmatic. Should future studies discover that the original author confused two or more species in the original publication, to which should we apply the proposed name and to which should we give another name? As we discussed earlier about the more frequent discovery of cryptic species, determining the name of a species frequently relies on examining types. What the original author believed to be a range of variation in a single species may now be understood to be discrete sets of variation for two species. The set of individuals that are conspecific with the holotype gets the original name, along with any paratypes or newly discovered material that is similarly conspecific. The other paratypes and newly discovered specimens become the type series for a different species. Recall that early in the formalization of taxonomy there were no rules of nomenclature; as such, early authors frequently did not designate types. However, often their original series of specimens can be located in museums. In order to stabilize the current and future application of those names, a type is subsequently selected from their original series of specimens. While the holotype indicates that the original author made the selection of the name-bearing type, the *lectotype* is reserved for those name-bearing types designated by later authors studying the original series of early scientific names. The lectotype is identical to a holotype except that the original author did not distinguish that specimen from his original series of specimens. Similarly, holotypes may be lost or destroyed; many types have been destroyed during European wars. In such instances, a new unique individual is selected to serve as the name-bearing type for the species; these are called *neotypes*.

Higher groups also have types: Genera have *type species*, and families have *type genera*. These vouchers similarly serve to stabilize the usage of these names in the same manner that they stabilize the usage of species names. If a genus is broken up into two or more genera as the result of a phylogenetic study, the name of the original composite genus goes to the new group containing its type species, and new names are required for the other sets of species.

Numerous other rules of nomenclature exist, all designed to form the database of life's diversity. More details on the historical development of zoological nomenclature are provided by Melville (1995), while the complete set of rules can be found in the most recent edition of the International Code of Zoological Nomenclature (ICZN, 1999).

The current developments in information technology are creating a revolution in systematics and taxonomy. Now it is possible to access, link, and synthesize data for individual species on scales never before imagined. However, at the same time taxonomy is under attack. Technological arguments have been put forward that taxonomy is lingering in the past, failing to discover taxa at a rate keeping up with human-induced extinctions. Such arguments include a variety of misplaced attacks. "*Taxonomy is lost in ancient Latin and Greek.*" What would it matter if it were in French, Spanish, Chinese, or Farsi? By using ancient forms of Latin and Greek, taxonomy avoids the vagaries of nationalism. "*Taxonomy should automate and use numbers in place of names for easy use by computers.*" Computers serve us – not the other way around. A computer cares not whether we refer to an organism as *Drosophila melanogaster* or as taxon 2789.63; the machine will read both the same. More importantly, there is meaning in names. Ideally, the author of a species constructed the name to be descriptive; referring to a distinguishing feature or location of the species. Some names are even poetic; for example, the fossil butterfly *Prodryas persephone*, so exquisitely preserved in rock, is named for the Greek goddess Persephone, the beautiful daughter of Zeus who was abducted by Hades to be his queen of the Underworld. Computers can handle data organized by name the same as by number, but human language and cognition simply make words more recognizable and understandable. Indeed, we all have numbers assigned to us (e.g., phone, social security, passport numbers), yet we consider our traditional names to best reflect our identity and so we retain them. Names describe; they give identity.

PALEONTOLOGY

Before phylogenetic systematics, or cladistics, became mainstream in systematics, it was typical for systematists to view the fossil record as the only source for reconstructing evolutionary history. Indeed, fossils are the only *direct* source of information on past life. But, a scheme of relationships is required to reveal patterns, including the fundamental pattern of change through time, or evolution (Eldredge and Cracraft, 1980). Without this context, fossils are just the remains of extinct organisms and tell us very little about evolution. Also, even the best preserved fossils are always seriously incomplete, so the virtually limitless source of characters in Recent species usually provides an important scaffold with which to interpret fossils. For groups that are entirely extinct and ancient, though, comparisons to Recent species may be very limited.

With such an overwhelming diversity of extant species of insects, the study of fossil insects may seem quaint, particularly because they have a notoriously incomplete fossil record. Actually, as we discuss later, the insect fossil record is far more diverse than commonly assumed, and just as in other kinds of organisms it affords remarkable insight into their evolution. Fossils uniquely provide information on basically four aspects of evolution:

1. Documentation of extinct species and lineages and patterns among them
2. The actual and estimated ages of lineages
3. Phylogeny
4. Biogeography

Each deserves some commentary. The books by Kemp (1999) and Smith (1994) also provide excellent and readable overviews of the systematic study and evolutionary significance of fossils.

1. Documentation of Extinct Species and Lineages and Patterns Among Them. This aspect may seem self-evident, but there are many popular misconceptions about extinct taxa and extinctions, perhaps best illustrated by the bird-dinosaur debate. Evidence strongly favors the interpretation that birds are recently evolved saurischian dinosaurs, specifically that they are raptors closely related to dromeosaurs and are highly specialized for powered flight (Chiappe, 1995). Thus, dinosaurs didn't go extinct; the non-avian ones did. Understanding relationships is crucial to documenting extinctions. Vague concepts of relationships have, conversely, led to misconceptions about the *stasis* of lineages. Perhaps the best example of this is the popular notion that cockroaches are ancient, having persisted unchanged for 300 MY. This notion is based on the abundance of insects in the Carboniferous with roach-like wings and pronota, but as we discuss later, these insects were not true roaches because they primitively possessed a long ovipositor, among other features (Figures). Paleozoic "roachoids" gave rise to modern roaches, mantises, and termites (the Dictyoptera), a situation analogous to that of dinosaurs and birds.

These examples bring up a crucial point regarding the naming and classification of fossils, namely that early fossils usually lack some of the synapomorphies seen in modern relatives. Thus, to accommodate early fossils into a group, the defining features may need to be less restrictive. This is the situation with the most basal ants, which are sphecomyrmines from the Cretaceous (Grimaldi *et al.*, 1997; Grimaldi and Agosti, 2000). Sphecomyrmines were, as most ants living today are, stinging Hymenoptera with a metapleural gland, a petiole, worker morphology (and thus were probably highly social), but they lacked the long antennal scape seen in modern ants. Some very early fossils may be just too plesiomorphic to accommodate into the definitions of modern groups, like the Paleozoic roachoids. Paleoentomology is, unfortunately, rife with the procrustean practice of squeezing such early and very basal taxa into taxonomic categories that have much more restrictive definitions. Equally problematic is the opposite approach, of describing a new genus and family for every fossil of significant age.

A thorough understanding of the fossil record and relationships in a group can also provide unique evolutionary insights. One such example regards polytomies in cladograms. Systematists strive to fully resolve a cladogram into dichotomous branches and often search for characters, sometimes extremely nuanced ones, in an effort to do so. Polytomies, in fact, may actually be a common pattern in evolution (Hoelzer and Melnick, 1994), of which the fossil record provides an independent test. Mammals, for example, have one of the best fossil records of terrestrial animals, so there is excellent documentation that they explosively radiated in the Paleocene (e.g., Simpson, 1945; Alroy, 1999). Some molecular data has provided interesting support for a lineage of African orders (Stanhope *et al.*, 1998; van Dijk *et al.*, 2001), and there is paleontological data for the relationships and origins of whales (Gingerich *et al.*, 2001; Thewissen *et al.*, 2001), but the relationships among placental mammal orders largely remain ambiguous (Shoshani and McKenna, 1998; Liu and Miyamoto, 1999; Novacek, 1999; Miyamoto *et al.*, 2000). The relationships of mammal orders may be difficult to resolve because they originated so close in time. Echinoderms, which likewise have an excellent fossil record, show a similar pattern (Bodenbender and Fisher, 2001). *The more similar in age the groups are, the more data will be required to reconstruct the chronology of their divergence.* The relationships within speciose insect groups like ditrysian Lepidoptera and schizophoran flies may be difficult to unravel for this reason because they appear to have had spectacular radiations in the early Cenozoic, but more exploration is required to determine this.

Evolutionary radiations are repeatedly discussed throughout this book. Simply put, a radiation (R) refers to the rapid diversification of a group, where $R = n/t$, and where n is the number of lineages and t is time. The higher the value of R, the greater the radiation. Salient aspects in the history of most taxa concern radiations: the radiation of vascular plants; animal phyla in the Cambrian; winged insects in the Carboniferous; angiosperms, phytophagan beetles, and Lepidoptera in the Cretaceous; long-tongue bees, higher termites, and orders of mammals in the Cenozoic, to name a few. Fossils are the ultimate source of evidence for the timing of such radiations.

Taxic analysis is a popular approach for examining trends in past diversity. This approach simply plots taxonomic diversity – usually number of genera, families, or orders – over time. It has been used extensively in the study of fossil insects (Wootton, 1990; Labandeira and Sepkoski, 1993; Jarzembowski, 1995a; Jarzembowski and Ross, 1996; Labandeira, 1997; Dmitriev and Ponomarenko, 2002). Even though it is convenient, the approach unfortunately has certain fundamental problems (reviewed by Smith, 1994). First, taxonomic categories are not equivalent evolutionary units: A genus of butterflies is hardly equivalent (say, evolved at the same time) to a genus of cupedid beetles. Second, taxonomic categories are also not equivalent in terms of numbers of species or ecological significance. There are genera with thousands of species, and others with only one species, so comparing just genera or families may misrepresent the actual diversity. Also, if geological trends in diversity are used to understand ecological or evolutionary trends, taxic analyses may underestimate the significance of groups. It is unrealistic to consider any family of psocopterans, for example, as an equivalent unit to ants, the family Formicidae. Third, particular deposits and taxonomists can have a profound effect on diversity curves. For example, the Late Jurassic spike in insect diversity seen in the preceding studies is entirely the result of the vast deposit at Karatau, which has been studied for nearly 70 years, and by some taxonomists rather notorious for their prolific names, particularly B. B. Rohdendorf. Likewise, the spike in Eocene diversity is largely the result of the vast deposits of Baltic amber. Lastly, taxic analyses do not accurately measure evolutionary patterns (Smith, 1994) and can seriously underestimate, for example, mass extinctions. If insufficient attention is paid to the monophyly of taxa being plotted, this will lead to spurious results. For example, Gall *et al.* (1998) dispute the view (e.g., Labandeira and Sepkoski, 1993) that the end-Permian extinctions had a major impact on insect diversity. That is because Gall *et al.* (1998) use several paraphyletic orders, like Protorthoptera, some portions of which extended into the Triassic or later. Later, we discuss the trends in insect diversity over time that these taxic analyses have proposed and compare them with trends based on phylogeny.

2. The Actual and Estimated Ages of Lineages. Knowing the age of a group allows inference on *rates* of evolution, be it speciation or the rate of nucleotide substitutions, and the *chronology* of lineages. As we just discussed, it is crucial when discussing ages or origins of groups in the fossil record that these be defined with characters. It is well known that the age of the earliest known fossil of a group is a minimum age because we are never sure that older ones existed. By studying fossils in the context of phylogeny, however, powerful predictions can be made about *absolute ages* of lineages, even when there is a major gap in the fossil record (Smith, 1994). This is done by examining the correlation of the

positions of fossils on a cladogram and their stratigraphic positions or ages, the so-called *stratigraphic-clade rank correlation*. Norell and Novacek (1992) found an excellent such correlation for horses (Equidae), but a poor one for primates, which reflects the fact that the primate fossil record is not nearly as complete as that for horses, but which may also indicate problems with the known phylogeny of primates. This approach requires that cladograms be done well (i.e., they are stable to the addition of new characters or from different types of analytical procedures). Even so, though, an excellent correlation will probably be rare for most groups, including insects, for several reasons. One, there can be great differences in the life span of lineages within a group, and fossils come from thin, random slices during that life span. This is less of a concern with groups that have excellent fossil records, like foraminiferans and horses. Second, there can be a considerable range in error on the estimated ages of the fossils.

Two examples from insects, involving small midges preserved in amber, show reasonable stratigraphic-clade rank correlations, largely because their stratigraphic sampling is better than in most insects, and the excellent preservation allows accurate phylogenetic placement of the fossils.

The Lygistorrhinidae is a small Recent family of gracile fungus gnats, most of which have a long, thin proboscis (Figure 1.28). A recent study reported diverse species in Cretaceous ambers, which are phylogenetically basal to the Cenozoic species (Grimaldi and Blagoderov, 2002; Blagoderov and Grimaldi, 2004). The excellent correlation between stratigraphic and cladistic ranks of genera (Figure 1.28) allows several inferences. One, a proboscis probably evolved in the Early Tertiary (perhaps in the Paleocene), since the basal, Cretaceous genera do not have a proboscis, and the Eocene genus in Baltic amber (*Palaeognoriste*) had a short proboscis. Second, the family, *as defined by certain features of reduced wing veins* (not the specialized proboscis), probably originated in the earliest Cretaceous. Another excellent correlation was found in the family of tiny biting midges, the Ceratopogonidae (Borkent, 2000a). These midges are among the most abundant and diverse insects preserved in amber through the Cretaceous and Tertiary. Moreover, they have been exhaustively described, as we discuss later, and the relationships of living and fossil genera have been well explored (Borkent, 1995, 1996, 2000a). The most basal genera, living and extinct, are found in the oldest ambers from Austria and Lebanon and were probably all blood feeding. Lineages of intermediate cladistic rank first occurred in the mid- to Late Cretaceous, and most were probably predators of other midges; the most derived lineage is found exclusively in Cenozoic ambers. On this basis, we can predict that ceratopogonids perhaps originated in the Late Jurassic and fed on blood, and that the insectivorous ones evolved in the later part of the Early Cretaceous, perhaps 110–120 MYA.

The ages of lineages are also estimated using molecular methods, or *molecular clocks*. In this approach the ages of lineages are estimated based on the amount of nucleic acid or amino acid divergence, which assumes a steady (clock-like) rate of mutation. Molecular dating still requires fossils for calibrating a time scale, and the phylogenetic position of the fossils must be accurately known. While nontranscribed parts of genes may be selectively neutral (i.e., introns) and therefore change at a steady rate, this assumption is unrealistic for protein-coding regions. Indeed, there can be intense selection on particular genes and their protein products, a good example of which is the remarkable similarity in sequences and tertiary structure of hemoglobins from such disparate organisms as nitrogen-fixing bacteria, chironomid midge larvae, and vertebrates. Also, rates of genetic change can vary greatly among lineages and among genes (Britten, 1986; Li, 1997), just as do any morphological characters. Moreover, genes must be selected for divergence within an appropriate time frame, such as in the Late Paleozoic (300–250 MYA) if one is going to estimate the ages of most insect orders. One way to circumvent the problem of rate heterogeneity is to sequence proteins or genes from multiple genes so that the different rates will average out.

There is a consistent and troublesome result from molecular dating: the overestimation of age (Conway Morris, 2000; Rodriguez-Trelles *et al.*, 2002). Unfortunately, the molecular studies that hypothesize older estimates are routinely interpreted to mean that the early fossil record is deficient. One example of molecular overestimation is an age of metazoans extending to one billion years ago (Wray *et al.*, 1996; Bromham *et al.*, 1998; Heckman *et al.*, 2001), which is nearly twice the age known from the admittedly spare fossil record of early metazoans. Another such estimate is the origin of angiosperms 200–210 MYA (Sanderson, 1997), more than 80 MY earlier than the earliest definitive fossil evidence of angiosperms. Because fossil angiosperm pollen is recognizable and well preserved, lack of it in the Jurassic (along with any macrofossil angiosperms) is not simply a sampling artifact. Dating using molecular clocks has fundamental methodological problems (Rodriguez-Trelles *et al.*, 2002), one of which is that estimates are commonly made in lieu of statistical measures of error, and this leads to an illusion of precision of such estimates (Graur and Martin, 2004).

Now, Bayesian methods are in vogue for molecular age estimation (Huelsenbeck *et al.*, 2000; Kishino *et al.*, 2001). This technique, described previously, assumes that rates of genetic change vary across lineages based on a theory of probabilities, but it still relies on fossils as calibration points and on large numbers of molecular sequences. Bayesian methods have provided estimates of age or divergence times that are more consistent with those based on fossils (e.g., a 550 MYO divergence of metazoan phyla [Aris-Brosou and Yang, 2002]), but some Bayesian estimates are still implausibly old

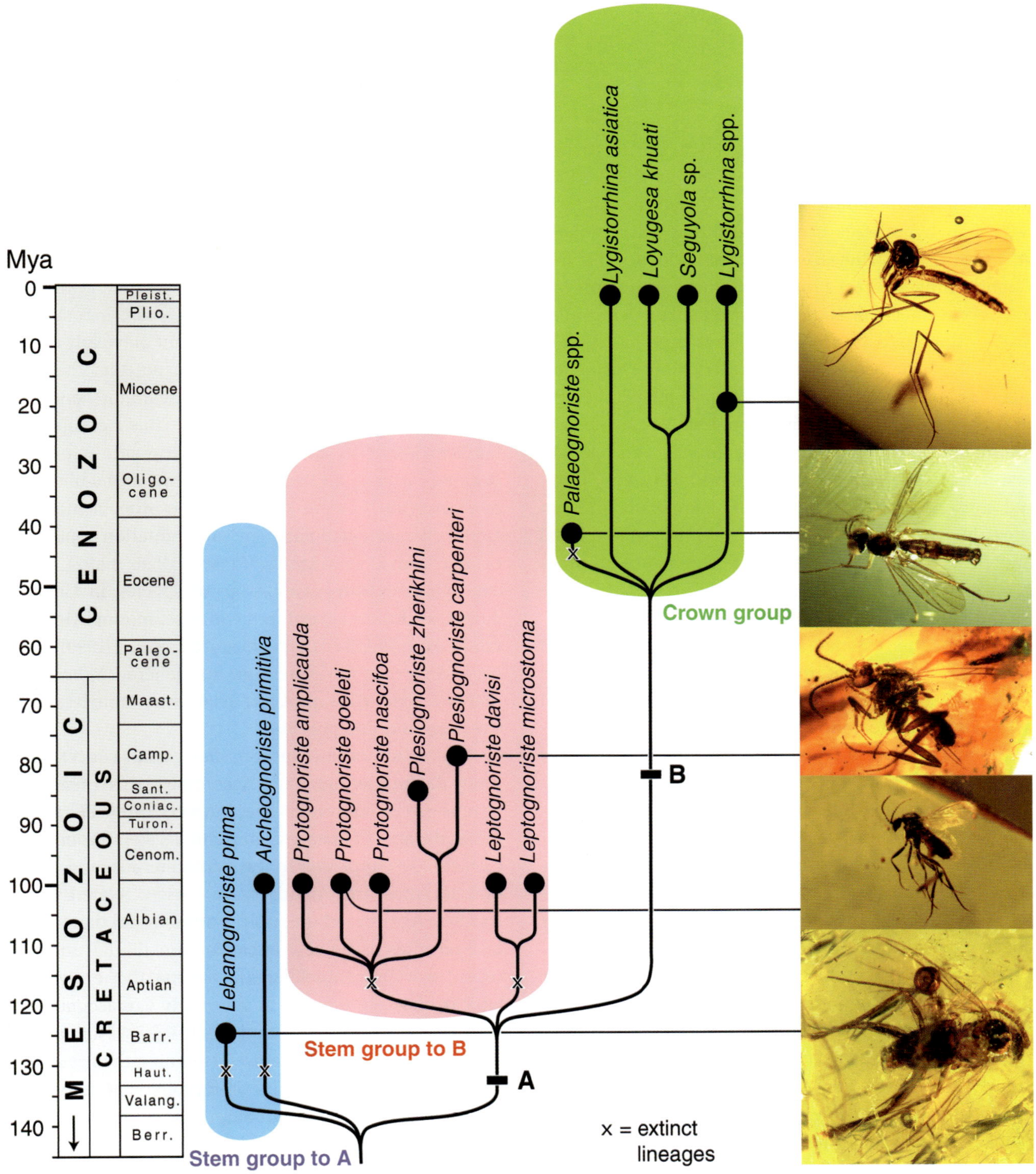

1.28. Phylogeny of the fungus gnat family Lygistorrhinidae (Diptera: Sciaroidea), showing the concepts of a paraphyletic stem group and a monophyletic crown group. This example also shows a good relationship between geological age and position in a phylogeny: The oldest and most generalized groups are basal in the phylogeny; the youngest and most specialized ones branch off last. Based on Grimaldi and Blagoderov (2002) and Blagoderov and Grimaldi (2004).

(e.g., Thorne and Kishino, 2002). Throughout this book we refer to various studies where molecular dating has been used, and we compared those estimates to ones based strictly on the phylogenetic analysis of fossils.

3. Phylogeny. The significance of fossils in reconstructing phylogenetic relationships is controversial. Paterson (1981) reviewed evidence that indicated fossils actually have little to no effect on hypotheses of relationships of Recent taxa. Gauthier *et al.* (1988), however, presented evidence that fossils significantly affect a phylogeny where extinction has extensively pruned the earliest lineages, like coelacanths, horseshoe crabs, or early amniotes. This is largely an issue of the numbers of characters and the preservation. The incompleteness of fossils will always compromise digital phylogenetic analyses of them and their Recent relatives, specifically by producing many cladograms of equally short length ("most parsimonious" cladograms), and bushy, very unresolved consensus trees. This is a very significant problem because, in some situations, fossils could belong to a basal lineage by virtue of missing so many characters (in other words, most derived characters were not preserved). Some researchers exclude highly fragmentary fossils from a cladistic analysis and then estimate its phylogenetic position based on the few preserved features, though the validity of this approach is debated. As expected, fossils that are very well preserved contribute more significantly to phylogenetic hypotheses. The diverse fauna of bees in Baltic amber, for example, provides even more compelling evidence for a single origin of social corbiculate bees than does evidence based just on Recent species (Engel, 2001a,b).

But, there is no magic formula, and each situation needs to be assessed individually. A scrap of a fossil may have preserved a unique and very significant character that makes its phylogenetic placement certain. *Rhyniognatha*, for example, is merely a pair of mandibles in Devonian-aged chert, but their structure is unique among arthropods to the winged insects (Engel and Grimaldi, 2004a). This character would make the age of pterygotes approximately 75 MY older than previously known. Isolated mandibles preserved in much younger matrix, say Miocene amber, wouldn't warrant a second look. Generally, older fossils are more fragmentary but are also tolerated more easily.

An aspect of fossils that is not controversial is that they often provide a chronology for the appearance of synapomorphies. This is not restricted to fossils, but the earliest fossils of a group usually reveal such a chronology that no Recent species can; recall the previously mentioned example for *Sphecomyrma* and Recent ants. Also, fossils can have unique combinations of characters unlike any Recent species. For example, the tiny brachyceran fly in Early Cretaceous amber, *Chimeromyia* (Figures 12.92, 12.93), has an unpredicted combination of features shared with empidoids and Cyclorrhapha, though this fly does not seriously affect phylogenetic analyses of these groups (Grimaldi and Cumming, 1999).

Lastly, and perhaps the most significant phylogenetic aspect of fossils, is that *stem groups* can be found, and these may be very significant evolutionarily. The problem with stem groups is that they are paraphyletic assemblages of basal taxa (Figure 1.28). Though they possess some of the features of their Recent relatives (the *crown group*), they are not defined by any feature that is apomorphic to the crown group. The common view is that this may be caused by the incompleteness of fossils, such that a synapomorphic feature wasn't preserved. This case is very likely for seriously incomplete specimens, like isolated wings, but the nearly complete preservation of specimens in amber indicates that extinct stem groups are real. *Sphecomyrma*, again for example, is a basal assemblage of Cretaceous species that lacks specialized features found in all other ants (Grimaldi *et al.*, 1997). Because virtually all Recent characters can be observed in *Sphecomyrma*, its stem-group nature is probably not an artifact of character sampling. The same situation applies to the extinct "genus" *Prioriphora* (Figures 12.96, 12.97), which is also a stem-group assemblage of species in Cretaceous amber, in this case basal to the diverse, monophyletic living family of phorid flies (Grimaldi and Cumming, 1999). It seems very likely that these stem-group "genera," *Sphecomyrma* and *Prioriphora*, are *ancestral* to large monophyletic groups, the living ants and phorids.

Another important aspect of stem groups concerns sister-group dating (Hennig, 1981). According to this concept, sister groups are of equivalent age, but very often this overestimates the ages of lineages. For example, stem-group roaches ("roachoids") appeared in the Carboniferous, and we know that termites and mantises are closely related to living roaches. Ages based on sister groups would date mantises and termites as equally ancient, even though their fossil record provides compelling evidence that they evolved much later in the Late Jurassic to very Early Cretaceous. Are we missing 150 MY of termite and roach fossils? No, because in this situation we would be overestimating the ages of mantises and termites as a result of improperly identifying their sister group. While roaches comprise a *living* sister group to termites and mantises, their *immediate* sister groups are particular Mesozoic lineages of stem-group roachoids. Thus, the study of fossils can avoid a serious pitfall of dating lineages based on living sister groups.

4. Biogeography. *Ecological* biogeography is concerned with ecological processes, like dispersal and colonization ability, population size, and other parameters used to describe the distribution of species. *Historical* biogeography, which is germane to fossils, deals with historical factors that have contributed to the present-day distributions of taxa: continental drift, climate change, extinction, and the like. One popular

school of historical biogeography is *vicariance biogeography*, which attempts to explain separated or vicariant distributions of taxa based on their relationships and the geological relationships of the areas they inhabit (Platnick, 1976; Nelson and Platnick, 1981; Humphries and Parenti, 1986). Vicariance biogeography is most commonly practiced by neontologists, who seek to discover vicariant patterns among living species and, as such, perhaps an historical mechanism for distributions, like the formation of a seaway or a mountain range. As we discuss in depth toward the end of this book, perhaps the most famous vicariant distribution is that of *austral disjunction*. This pattern is commonly repeated in diverse organisms, where closely related taxa are widely separated in the southern temperate regions of Chile and Argentina, New Zealand, Australia, and sometimes Tasmania, New Caledonia, and southern Africa. This distribution is almost always interpreted as a remnant of Cretaceous-aged continental drift, the breakup of Gondwana. But, fossils of austral groups from the Northern Hemisphere sometimes indicate that they had nearly global distributions (Grimaldi, 1992; Eskov, 2002 see Chapter 14 in this volume), so fossils can potentially disprove vicariance hypotheses. Conversely, the age of a fossil can provide supporting evidence that a group is old enough to have been affected by a particular geological event.

2 Fossil Insects

INSECT FOSSILIZATION

Too often insect fossils are thought to be merely impressions of wings scattered in slabs of rock. Wings are indeed a common insect fossil. They do not readily decay or digest, which is why birds and spiders typically leave the wings after devouring the rest of an insect. Fortunately, wing veins are also a veritable road map to the identification and phylogeny of insects, much the way crowns and pits are on the teeth of mammals. Most isolated insect wings, in fact, are readily identifiable to the family level. Also, the generally small size of terrestrial arthropods, along with their durable cuticle, have allowed them to be preserved in many more ways than have fossil vertebrates, and certainly in many more ways than just as compressed wings.

Terrestrial vertebrates are almost always preserved just as bony remains (or inorganic casts thereof), the original bone usually having been replaced by the mineral apatite [$Ca_5(PO_4)_3(F,Cl,OH)$]. Occasionally, mummified or frozen vertebrates are found, but their age is usually no more than several thousand years. Fossils of insects, in contrast, are preserved as organic compressions and inorganic impressions; as three-dimensional, permineralized, and charcoalified replicas; and as inclusions in amber and even within some minerals (Schopf, 1975). There is also abundant fossil evidence for the behavior of extinct insects, including feeding damage on fossil vegetation and in wood, fecal pellets, and nests in fossil soils. Dinosaur behavior, by contrast, is recorded mostly as footprints and coprolites.

The common denominator among most deposits of fossil insects and terrestrial plants is the lake environment. Those insects that became preserved were either living in the fossil lake (*autochthonous*) or carried into it from surrounding habitats by winds, stream currents, or their own flight (*allochthonous*). Drowning and dying insects not eaten by fish and other predators settle to the bottom, where they may be preserved in the lake's sediments (*lacustrine*) under appropriate conditions. Even amber, or fossil resin from trees, requires a watery environment that is lacustrine or brackish in order to be preserved. Without protection in anoxic sediments, amber would gradually disintegrate; it is never found buried in fossil soils. Various factors contribute greatly to what kinds of insects become preserved and how well, if indeed at all, including lake depth, temperature, and alkalinity; type of sediments; whether the lake was surrounded by forest or vast and featureless salt pans; and if it was choked in anoxia or highly oxygenated.

There are some major exceptions to the lacustrine theme of fossil insects, the most famous being the Late Jurassic limestones from Solnhofen and Eichstätt, Germany, which are marine (Barthel *et al.*, 1990). These deposits are famous for pterosaurs and the earliest bird, *Archaeopteryx*. The limestones were formed by a very fine mud of calcite that settled within stagnant, hypersaline bays isolated from inland seas. Most organisms in these limestones, including rare insects, were preserved intact, sometimes with feathers and outlines of soft wing membranes, indicating that there was very little decay. The insects, however, are like casts or molds, having relief but little detail (Figures 2.1, 6.43). In some cases iron oxides precipitated around wing veins, revealing better detail.

Lagerstätten are exceptional fossil deposits. *Konservat Lagerstätten* are deposits in which organisms, including soft-bodied ones, are exceptionally well preserved in any one of a variety of ways. *Konzentrat Lagerstätten* are ones that preserve an exceptional assemblage of individuals, from one to many species. Use of these terms often depends on the scale of diversity. For vertebrates and a great variety of marine invertebrates, Solnhofen limestones can be considered a Lagerstätte. For plants and insects, however, Solnhofen is not particularly remarkable. The finest terrestrial Lagerstätten are lacustrine, and these have provided most information on the evolutionary history of insects.

Taphonomy, or the study of how organisms fossilize, is essential for understanding differential preservation of organisms in the fossil record, and can help explain extinctions and the ecology of extinct organisms and communities (Allison and Briggs, 1991; Donovan, 1991; Briggs, 1995;

2.1. A dragonfly from the famous Jurassic limestone quarries of Solnhofen, Germany. The wing venation is clearly visible, which is exceptional preservation for insects from this deposit. Private Collection; wingspan 93 mm.

Martin, 1999). One aspect of taphonomy is *biostratinomy*, or those environmental factors that affect which species become trapped and fossilized. The vegetation surrounding a water body, for example, dictates which kinds of insects waft into the water. Deep, cold, still, and anoxic waters will usually preserve specimens better than shallow, warm, oxygenated waters where dead organisms will decay faster and also become fragmented by currents and scavengers. *Diagenesis* refers to changes that take place after burial. Very deep sediments and faulting, for example, can dramatically transform a fossil with extreme pressure, and even obliterate it.

TYPES OF PRESERVATION

Compressions and Impressions

Compressions and Impressions are the most extensive types of insect fossils, occurring in rocks from the Carboniferous to Recent. Impressions are like a cast or mold of a fossil insect, showing its form and even some relief (like pleating in the wings) but usually little or no color from the cuticle (Figure 2.2). Compressions preserve remains of the cuticle, so color also distinguishes structures (Figure 2.3). In exceptional situations microscopic features such as microtrichia on sclerites and wing membranes are even visible, but preservation of this scale also requires a matrix of exceptionally fine grain, such as in *micritic* muds and volcanic tuffs. Because arthropod sclerites are held together by membranes, which readily decompose, many fossil arthropods are known only by isolated sclerites. Far more desirable are complete fossils.

Just as bones and teeth are to the vertebrate fossil record, sclerites and wings are the enduring record in arthropod evolution. Sclerites in some fossil arthropods may appear remarkably preserved, but rarely is the molecular composition intact; that is, rarely are the sclerites actually cuticle. Insect cuticle is composed of a readily denatured protein component, and chitin, which is a pleated polymeric sheet of mucopolysaccharides. Chitin is crosslinked by other molecules to the protein, which makes it particularly durable. By chemically analyzing a spectrum of insect remains from the Holocene to the Carboniferous, it has been possible to determine how

2.2. A roach forewing from the Cretaceous of India, with the intricate pleating still preserved. Most insect fossils occur only as isolated wings because these structures are so durable. Fortunately, wing venation is very informative about the identity and relationships of insects. Museum of Comparative Zoology, Harvard University (MCZ) 2008-4012; length 24 mm.

2.3. A dragonfly wing from the mid-Miocene of Oregon, with the dark coloration still preserved. MCZ 4895a; length 38 mm.

and where insect cuticle preserves (Stankiewicz *et al.*, 1997a,b, 1998a,b; Briggs *et al.*, 1998a,b; Briggs, 1999).

The chitinous portion of fossil insect cuticle preserves much longer than does protein, as expected. Even in relatively recent fossils, like those in bog peats or from tar pits, 2,000 to 40,000 years old, protein is highly degraded or virtually absent, but much of the chitin is still present. Age, however, has probably less effect on preservation of chitin than does the conditions under which the cuticle was preserved and the nature (i.e., the thickness) of the cuticle itself. Insects that settled into anoxic levels of a lake decompose more slowly than those that settled into oxygenated levels, and so preserved better. Chitin also preserves better in freshwater sediments than in marine sediments. Cuticle that is thick and heavily sclerotized preserves far better than thin, poorly sclerotized cuticle, which explains why roach tegmina and beetle elytra are so widespread in the insect fossil record. But chitin has definite limits to its lifespan. The most ancient chitin known thus far is from the elytra of 25 MYO weevils preserved in ancient lake sediments of Westerwald, Germany. Older insect fossils (and younger ones that are less well preserved) have sclerites composed largely of just aliphatic and aromatic hydrocarbons, probably a product of the polymerization of lipids that coat arthropod cuticles (*epicuticular waxes*) and lipids that are contained within the body. As the chitin in the insect cuticle polymerized, it was chemically transformed, leaving a layer visually indistinguishable in some cases to arthropod cuticle. Fragmentary arthropods and isolated sclerites from the Devonian, for example, appear strikingly like modern cuticle (e.g., Shear *et al.*, 1984; Subias and Arillo, 2002) (Figure 3.24), and have even been interpreted by some as *original* cuticle. Though the composition of these Devonian remains has not yet been determined, it is almost certainly completely modified.

Detailed looks at compression fossils sometimes reveal astonishing preservation. Perhaps the most interesting is the preservation of spores and pollen from carbonized gut remains (Krassilov and Rasnitsyn, 1982; Rasnitsyn and Krassilov, 1996a,b). Spores and pollen are among the most persistent, and pervasive, biological structures in terrestrial sediments, which makes them so useful for dating. Pollen grains are even durable enough to be voided in the feces of bees, flower flies, and other pollen-grazing insects, so it is not surprising that they have endured within long-extinct insects. Spores and pollen have been recovered from the guts of generalized Permian Hypoperlida and Grylloblattida (sensu lato), and from xyelid sawflies from the Early Cretaceous. Amounts and content of the meals indicated the insects actually foraged on the spores and pollen, and thus probably had a significant impact on pollination. Detailed examination of fossil insect guts is rarely done, so systematic study of them using scanning electron microscopy may reveal many more examples like these.

There is no relationship between detail of preservation, mode of preservation, and age. Insects from the Triassic of Virginia (ca. 220 MYO), for example, are entirely two-dimensional silvery films on a very fine-grained, black shale. In this situation, the body of the insect was compressed into metamorphosed carbon, and this acted as a template for the precipitation of aluminosilicates in clays (D. Briggs, pers. comm.), which is what gives them the silvery appearance (Figures 8.69, 10.16, 10.26). But even microtrichia less than 1 μm thick are routinely preserved in these fossils, which is detail few Cenozoic deposits show. Similar preservation occurs in a deposit of Early Cretaceous insects from South Korea (Engel *et al.*, 2002, unpublished data). One of the most celebrated Lagerstätte is Grube Messel, a mid-Eocene deposit of oil shales southeast of Frankfurt, Germany (Schaal and Ziegler, 1992). It was originally a 30–40 m deep, anoxic lake periodically choked by algal blooms. It preserved a great variety of completely intact organisms, including insects, and some beetles even retained iridescence.

Concretions

These are stones with a fossil at the core whose chemical composition differs from that of the surrounding matrix, usually formed as a result of mineral precipitation from decaying organisms (Sellés-Martinez, 1996). The most significant deposit consists of various localities of the Late Carboniferous Francis Creek Shale of the Carbondale Formation at Mazon Creek, Illinois, which are composed of shales and coal seams yielding oblong concretions. Within most concretions is a mold of an animal (Figures 2.4, 3.7) and sometimes a plant that is usually marine in origin. Rare insects were preserved when their bodies were transported to flooded coastal swamp-forests and then rapidly buried in anoxic sediments. This is the most diverse deposit of early winged insects, and many of the specimens are preserved complete. Carbonates in the sediments were replaced by siderite ($FeCO_3$), assisted by decay from anaerobic bacteria. This process caused the formation of a hard rind of mud around the insect that metamorphosed into the stones. Insects within the concretions have relief, though they are not completely three-dimensional nor do they have significant detail. The diversity

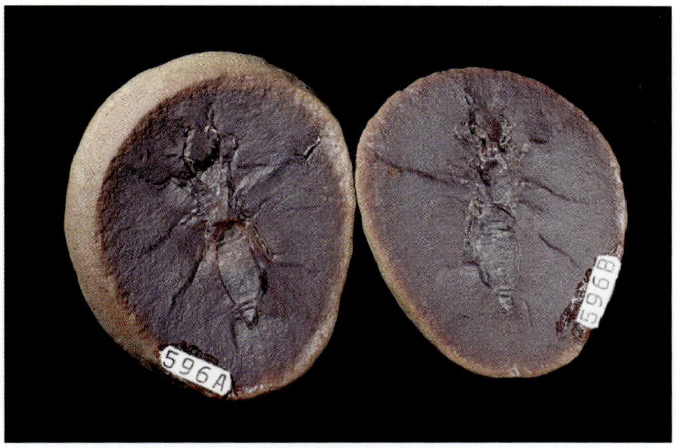

2.4. An arachnid preserved in an ironstone concretion from the Upper Carboniferous at Mazon Creek, Illinois. Mazon Creek arthropods are often articulated and have relief, but the detail of preservation varies greatly. Yale University Peabody Museum of Natural History (YPM) 66–596; body length 27 mm.

and taphonomy of Mazon Creek fossils were reviewed by Baird *et al.* (1986) and Shabica and Hay (1997). Another deposit of insect-bearing concretions is from the Early Miocene of Izarra, northern Spain (Barrón *et al.*, 2002). Fragmentary plants and diverse terrestrial arthropods in the Izarra concretions are replicated in calcite, many of which are allochthonous.

Mineral Replication

When an insect is partly or wholly replaced by minerals, usually completely articulated and with three-dimensional fidelity, it is *replicated.* This is also called *petrifaction,* as in "petrified" wood. Insects preserved this way are often, but not always, preserved as concretions, or within nodules of minerals that formed around the insect as its nucleus. Such deposits generally form where the sediments and water are laden with minerals, and where there is also quick mineralization of the carcass by coats of bacteria (Seilacher *et al.*, 1985; Allison, 1988b). *Permineralization* is a form of replication that, strictly speaking, is a result of microbial decay (Briggs, 2003).

The most significant replicated arthropods are the so-called Orsten (nodule)–preserved animals from the Paleozoic, some 550–500 MYA, which are composed of phosphate (Figure 3.4). They come from the Early Ordovician of Sweden and Newfoundland, the Late Cambrian of Poland and Sweden, the Middle Cambrian of Australia, and the Lower Cambrian of Shropshire, England (Müller and Walossek, 1986, 1991; Hinz, 1987; Andres, 1989; Roy and Fåhraeus, 1989; Walossek and Szaniawski, 1991; Müller and Hinz, 1992; Müller and Hinz-Schallreuter, 1993; Walossek *et al.*, 1993, 1994). These arthropods were tiny (<2 mm), benthic marine species, many of them the larval stages of stem-group arthropods and other phyla, including Pantopoda (Chelicerata), Crustacea, Tardigrada, and even Pentastomida. Pentastomida are parasitic, probably highly modified crustaceans, presently known only from the respiratory tracts of terrestrial vertebrates. The phylum Tardigrada (Figure 3.4), discussed later, comprises minute, membranous animals that are probably a living sister group to the arthropods. Though the Paleozoic Orsten deposits were formed 150 MY before the origin of insects, these fossils provide unique information on the early evolution of arthopods and closely related phyla. Insects were preserved in similar fashion, some even also by phosphates, but much later.

Perhaps the most famous and extensive deposit of replicated insects comes from the Barstow Formation (Miocene, 18 MYA), in the Calico Mountains of southern California in the Mojave Desert (Palmer, 1957). The deposit was originally a shallow, highly alkaline lake, which preserved insects living in it as well as ones that wafted into the water from surrounding areas. By dissolving calcareous nodules with acids, minute arthropods can be extracted, preserved like minute glass sculptures (Figures 2.5 to 2.7). Hairs and fine appendages are preserved, and even internal organs like the brain and digestive tract are known for some specimens. Most of the specimens are composed virtually entirely of celestite ($SrSO_4$), quartz (SiO_4), or both, with trace quantities of other minerals.

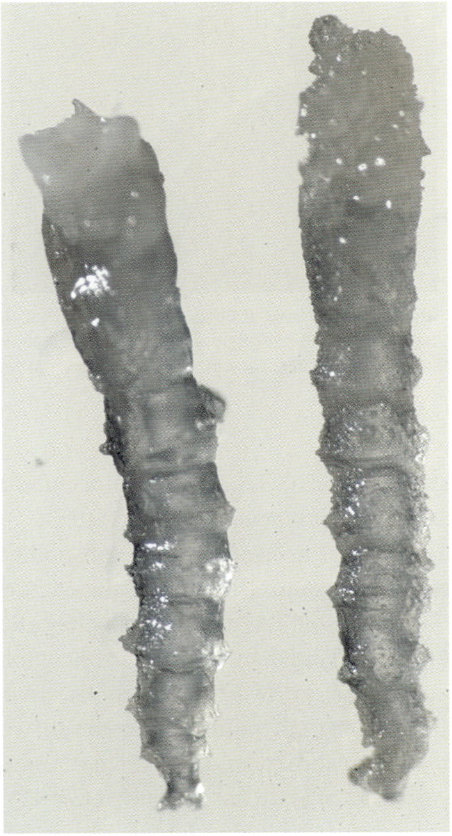

2.5. Pupae of ceratopogonid midges replicated in silica, from the Miocene-aged Barstow Formation of California. AMNH; body length 2.8 mm.

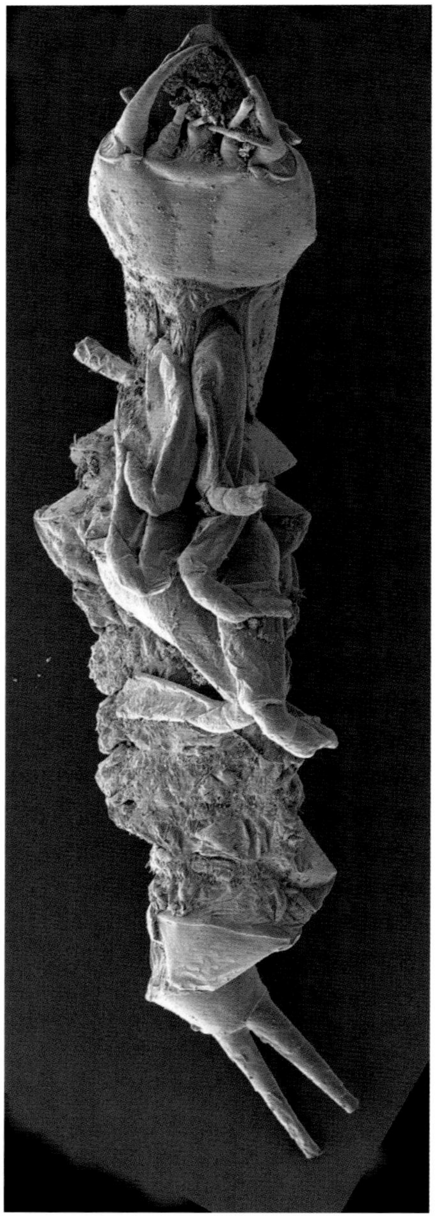

2.6. An early instar beetle larva of the aquatic family Dytiscidae, from the Barstow Formation. Like the midge pupae (Figure 2.5), it too is replicated in silica. Scanning electron micrograph. AMNH; body length 4.7 mm.

Replicas in gypsum ($CaSO_3 \cdot 2H_2O$) also occur, but are rare (Palmer, 1957; Doberenz *et al.*, 1966; Park, 1995).

Miocene (20 MYO) arthropods and soft-bodied invertebrates from Mfwangano Island in Lake Victoria, Kenya (Leakey, 1952) are also three-dimensional; they are, however, composed of calcite and do not occur in concretions. Among the specimens are an earthworm, a caterpillar, arachnids, various insects, roach nymphs, and even a colony of weaver ants, *Oecophylla* (Wilson and Taylor, 1964) (Figure 2.8). This is one of only several known fossil ant colonies. Over 350 ants were found in the colony, including larvae, pupae, and adults. Excellent preservation allowed measurement of the size distribution of major and minor workers, which is remarkably

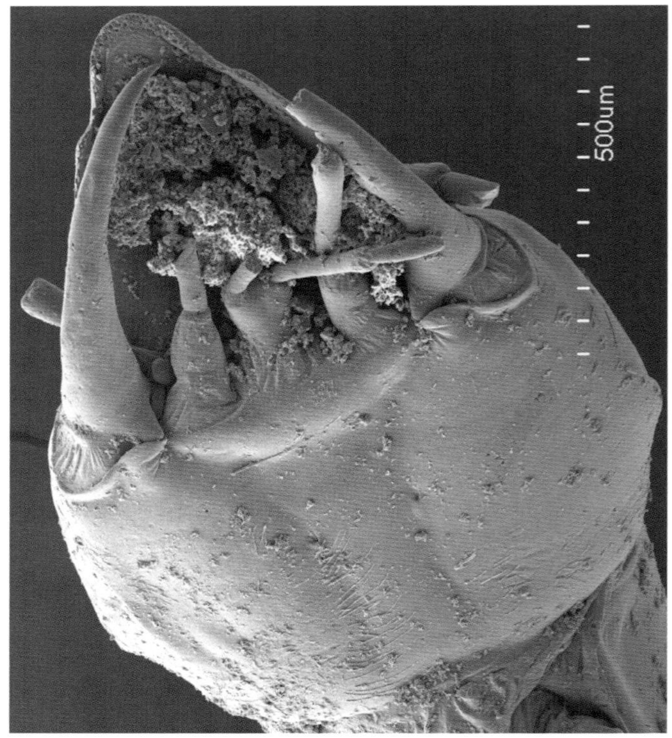

2.7. Head of the beetle larva in Figure 2.6, showing the detail of preservation under scanning electron micrography.

similar to living species of the genus. Another deposit of three-dimensional insects not preserved within concretions occurs in Oligo-Miocene (25 MYO) limestone from Riversleigh, Queensland, northern Australia (Duncan and Briggs, 1996; Duncan *et al.*, 1998). Rare insects replicated in calcium phosphate [$Ca_5(PO_4)_3$] are found here among abundant and diverse vertebrates. Internal tissues decayed, but microscopic structure of cuticle and ommatidia are preserved in fine detail. Similar phosphatized, three-dimensional insects occur from the Eocene of Quercy, France (Handschin, 1944) and the Oligocene of Ronheim, Germany (Hellmund and Hellmund, 1996).

The huge Cretaceous Lagerstätte from Ceará, northeastern Brazil, has preserved insects, plants, and various other organisms with consistent detail and relief (Martill, 1988; Grimaldi, 1990a) (Figure 2.9). Insects occur in fine-grained limestone of the Crato Member, Santana Formation, which was an evaporating, shallow, alkali lake approximately 120 MYA. Small fish (*Dastilbe*) died en masse in the lake, probably as it evaporated and the minerals concentrated, as did aquatic insects and diverse terrestrial ones that wafted into the water from amongst vegetation that surrounded the lake. The organisms were preserved quickly and with very little disturbance; almost all are completely articulated and show little sign of decay (e.g., Figure 8.48). Unlike many three-dimensional fossil insects preserved by phosphates, these are preserved as goethite (FeO(OH)) – a form of rust – though portions of many also have calcite in the body cavity. Internal tissues replicated in goethite have preserved even

2.8. Phosphatized replicas of a roach nymph (left) and a weaver ant (*Oecophylla*) pupa, from the Miocene of Kenya. Natural History Museum, London (NHM); body length of pupa 5.5 mm.

myofibrils in the muscles (Grimaldi, 2003a) (Figure 2.10), similar to the scale of preservation known for diverse fish from the Romauldo Member of the Santana Formation (Martill, 1988, 1990). The gut contents of some insect specimens are preserved, some even with pollen. Significant difference between the density of the insect replica and the softer limestone matrix has allowed high-resolution CT scanning (HRCT), providing images of the complete insect from any view, even of structures concealed deep in matrix (Figure 2.9).

Similar to the Santana Formation insects, though less extraordinary, are insects from the Late Eocene (ca. 40 MYA) Bembridge Marls of the Isle of Wight, southern England (McCobb *et al.*, 1998). A diverse insect fauna (220 species in 12 orders) was preserved in dense mudstone produced by the sediments of a brackish lagoon or estuary. Preservation is a spectrum from isolated wings to three-dimensional, fully articulated specimens (Figure 2.11). Some three-dimensional specimens are mere voids inside (Figure 2.12); others have remnants of soft tissues, even myofibrils of the muscles, replicated in calcite.

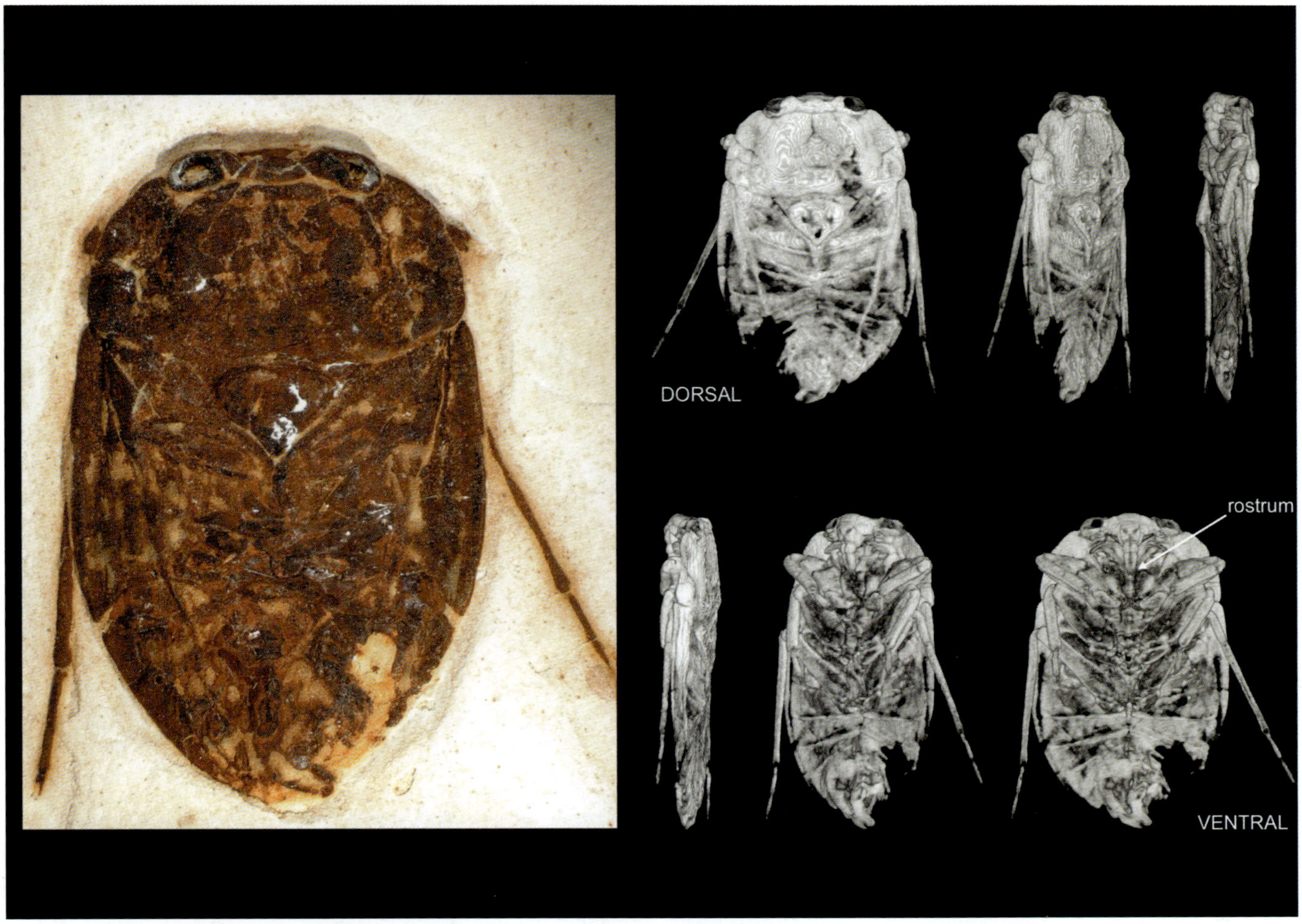

2.9. A predatory, aquatic water bug, family Belostomatidae, from the Early Cretaceous Santana (Crato) Formation of Brazil. The bug is replicated in iron hydroxide and phosphates and lies in a matrix of soft limestone. This allows high-resolution CT scan images (right), which reveal hidden details on the ventral surface of the bug. AMNH; body length 15 mm.

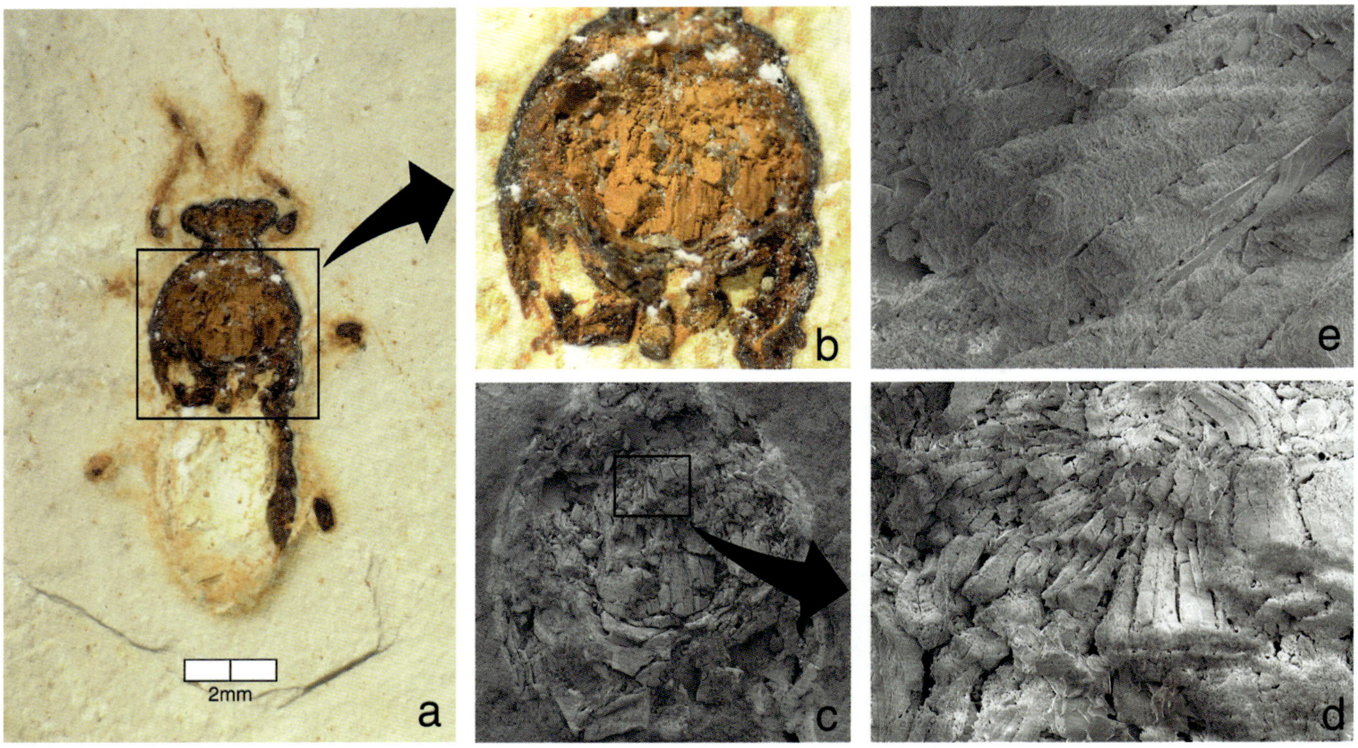

2.10. A plant bug (Heteroptera) in limestone from the Early Cretaceous Santana Formation, showing the preservation of muscle tissue. Scanning electron micrographs reveal bundles of muscle fibers and even their striations. AMNH; body length 6.5 mm. a: whole insect. b, c: thorax with exposed muscles (b, light Photomicrograph; c, scanning electronmicrograph). d, e: details of muscle fibers.

2.11. A beetle preserved with complete relief in Late Eocene–Early Oligocene clays from the Bembridge Marls of Isle of Wight, UK. Because of the detailed, three-dimensional preservation of insects from this deposit, one entomologist has called the Bembridge clay "opaque amber." NHM In. 26121; body length 6 mm.

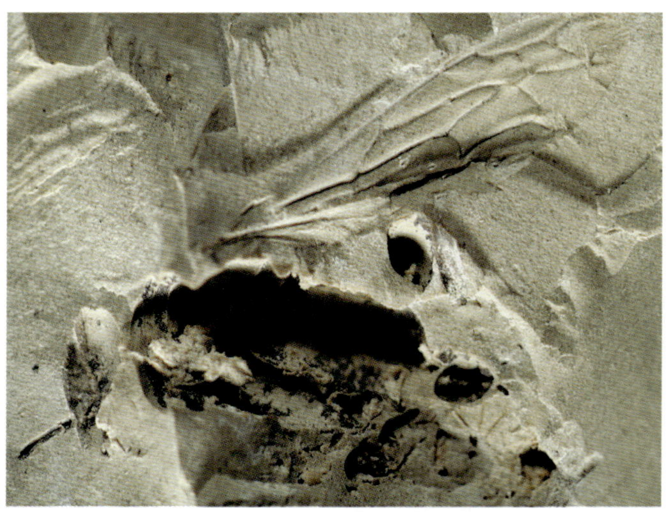

2.12. A wasp, *Sceliphron brevior*, in Bembridge Marl clay, with the hollow body cavity exposed. NHM In. 17472; wing length 9 mm.

2.13. "Gold bugs," pyritized beetle remains preserved in the Eocene London Clay. NHM; length of largest 5.1 mm.

Exceptional situations are insects replicated by the metallic mineral pyrite ("fool's gold") and its less stable form, marcasite (both FeS_2). These generally form in highly reducing environments where sulphur is abundant, such as in clays and other fine sediments from brackish or marine waters that have stagnated with decaying organisms. Fresh waters are usually sulphate-poor, so pyritization is rare in lacustrine sediments. *Pyritization* is common in marine sites, such as Beecher's Trilobite beds of New York state, United States (Late Ordovician), and the Early Devonian Hunsrückschiefer (Hunsrück slate) of western Germany (reviewed by Martin, 1999). The former is a depauperate fauna of trilobites; the latter has preserved diverse echinoderms, and molluscs, even worms and other soft-bodied invertebrates, many in virtually complete relief. External and internal soft tissues of various marine invertebrates, even an octopus, are remarkably replicated in calcite, apatite, and pyrite from the Jurassic (ca. 160 MYA) of La Voulte-sur-Rhône, southern France (Wilby *et al.*, 1996). The uppermost Paleocene and lowermost Eocene London Clay is one of the rare Tertiary marine deposits in which insects are preserved, in this case also by pyritization. In it are scattered small (<7 mm), woodland insects, particularly beetles and even larvae (Rundle and Cooper, 1971; Allison, 1988a; Jarzembowski, 1992) (Figures 2.13, 2.14). Pyritized insects occur sporadically in amber (e.g., Baroni-Urbani and Graeser, 1987; Grimaldi *et al.*, 2000a). Water laden with sulphurous minerals can seep into fine cracks in the amber and replace the insects' tissues when the pyrite crystallizes. Even though insects preserved this way may be in turbid amber, they can be readily imaged using high-resolution radiographs (Schlüter and Stürmer, 1984).

Encapsulation

Related to permineralized insects from evaporate deposits is a rare form of preservation wherein minerals form around the insect, sometimes a single crystal (Tillyard, 1922a; Schlüter and Kohring, 2001; Schlüter, 2003a). In the latter two studies, dragonfly nymphs were reported entombed in gypsum crystals from the Late Miocene (ca. 5.2 MYA) of northern Italy. These were formed during the Messinian Crisis, a period of such extreme aridity that the Mediterranean Sea virtually evaporated. The nymphs are preserved in the crystals as three-dimensional or compressed inclusions with hollow bodies.

The scientifically remarkable biota preserved in the Rhynie Chert, from the Old Red Sandstone in the Devonian of Scotland has a similar preservation, though organisms are not encapsulated within single crystals. Chert is microcrystalline silica, SiO_2, formed in a volcanic process. The chert is translucent and, if pieces are trimmed thinly enough, three-dimensional fragments of tiny organisms can be observed (Figure 5.8). These were trapped in ancient shallow pools created by hot springs and then rapidly silicified (Trewin, 1989). The Rhynie Chert has preserved some of the finest examples of earliest terrestrial life. Similar encapsulation is known from Tertiary onyx (Figure 2.15).

Charcoalified (Fusainized) Remains

Recently discovered are three-dimensional insect remains, buried in Cretaceous clays and lignitic peats amidst abundant fossil plant material (Grimaldi *et al.*, 2000a). The remains were found by paleobotanists prospecting for flowers, cones, and other structures that had been fusainized, or rendered to charcoal by ancient forest fires (e.g., Friis *et al.*,

2.14. A group of pyritized larvae, presumably in a gall or wood cavity, from the Eocene London Clay. NHM In. 64736; diameter of gall 11 mm.

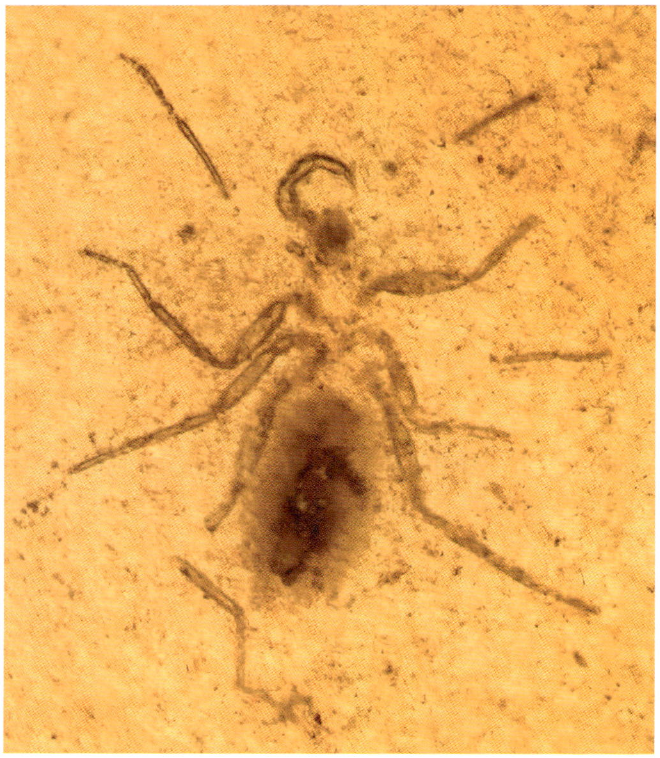

2.15. Arachnid (*Calcitro fisheri*) preserved in Miocene-aged onyx from Arizona, which is a form of silica. YPM 17380; length 3.2 mm.

1999; Herendeen *et al.*, 1999). Exposed vegetation burned, and some of it rendered to ash; however, plant fragments and insects buried beneath and within leaf litter and humus (thus sealed somewhat from oxygen) were replicated in carbon with perfect fidelity. Their cuticles, even cell walls, were preserved in exquisite microscopic detail. Diverse flowers preserved this way have revolutionized understanding of angiosperm evolution. Unfortunately, fusainized insects, while preserved in great detail, are extensively disarticulated and represented mostly by heavily sclerotized structures like heads, and isolated mandibles, legs, and elytra.

Trace Fossils (Ichnofossils)

Some groups of insects don't readily fossilize, particularly soft-bodied ones, but remains of their activities persist. These largely involve structures on plants, like larval feeding mines on leaves, chew and puncture marks on leaves and stems, galls, and galleries in wood. Plants are abundant in the terrestrial fossil record, and because insects are the predominant group of herbivores, these trace fossils provide a unique and direct record of past plant-insect associations and remarkable persistence of some associations. Also involved are larval cases and burrows in soil (Figures 2.16 to 2.18). Fortunately, the insects that made trace fossils can frequently be identified (at least to order, sometimes to family) by the geometry of the burrows, galleries, and feeding traces, as well as by the type of substrate or plant in which they occur. *Frass*, or fecal pellets, also has been used in identification. Zherikhin (2003) reviewed types of insect trace fossils and proposed a nomenclature for them based on functional types. We prefer to not use that nomenclature because descriptive names of the traces are most recognizable.

Fossil Burrows and Nests. These occur in fossil soils, or *paleosols*, and are usually formed by beetles, wasps, and bees, sometimes by ants and termites. In all cases the burrows are not mere excavations, but rather constructions consisting of corridors with brood cells that are provisioned with food for developing larvae. The architecture of some can be impressive. Genise *et al.* (2000) thoroughly reviewed all known 58 examples of fossil insect nests/burrows. Some types of burrows are abundant and widespread, such as *Coprinisphaera* (a form genus, or generic name used for a trace fossil). Though the actual beetle is unknown, *Coprinisphaera* burrows were made by scarabaeine dung beetles, modern species of which provision brood cells with balls of dung for development of their larvae. Nests of modern dung beetle species are usually distinctive (Halffter and Edmonds, 1982). *Coprinisphaera* lived from the Paleocene to the Pleistocene and was widespread from Antarctica to Ecuador, eastern Africa and Asia. Occurrence of these nests is believed to coincide with the evolution of ecosystems with abundant mammalian herbivores (Retallack, 1984), like savannas and pampas, since scarabaeines specialize on the dung of these animals. Scarabs even dined much earlier on dinosaur dung, as tunnels have been found in coprolites of herbivorous dinosaurs from the Late Cretaceous of Montana (Chin and

2.16. Subterranean bee cells replicated in calcite, from the Oligocene of Badwater, Wyoming. AMNH; length of largest cell 15 mm.

2.17. Subterranean nest of a bee replicated in sandstone from the Paleocene Ascencio de Palacio Formation of Uruguay. AMNH; greatest length 49 mm.

2.18. Portion of nest galleries of *Macrotermes* termites, replicated in phosphates, from the Pliocene of northern Tanzania. AMNH; greatest width 55 mm.

2.19. A caddisfly case constructed of sand grains, from the Early Cretaceous of Mongolia. AMNH; length 8.3 mm.

Gill, 1996). The fossil burrows provide a better temporal and spatial record of scarabs than do body fossils.

Nests of bees are another common type of fossil insect nest (Figure 2.17). Like scarabs, subterranean excavations of modern halictids are structurally diverse (Sakagami and Michener, 1962). *Celliforma* is a widespread form genus of halictid nests, occurring from the Late Cretaceous to the Pliocene, from North and South America, Europe, and Africa. Those from the Dakota Formation of Arizona (Elliott and Nations, 1998), if indeed bees, are among the oldest known evidence of bees. Not all the makers of fossil nests can be identified, though, even those nests with distinctive architecture.

Larval Cases. Various larval insects construct cases from gathered materials or feces, which they carry with them and in which they can retreat for protection. Some small moths produce larval cases from bits of their food woven together with silk, such as the "clothes" and scavenging moths (Tineidae), and "bag worms" (Psychidae) (Figures 13.34, 13.35, 13.37). Chrysomelid ("leaf") beetles also produce cases, such as many "tortoise" beetles (Cassidinae), which excrete strands of feces molded into a structure like a bird's nest, held over the larva. Tiny cryptocephaline beetles feed amongst decaying vegetation on the forest floor, dragging themselves around in a small sac of particulate debris (Figure 10.65). Because these cases are composed of materials that readily decay, and the species are terrestrial, fossils of these preserve only in exceptional cirumstances.

Cases of larval caddisflies, or Trichoptera, however, are diverse and locally abundant in the fossil record (Figure 2.19). The larvae dwell on the bottom of lakes, ponds, and rivers and build cases from durable materials, like sand grains, minute pebbles, and bits of vegetation (Figures 13.2, 13.5). Moreover, the shape and composition of the cases are distinctive to most modern genera (Wiggins, 1977), allowing the separation of species and even identification of some fossil cases to genus and family. The oldest Trichoptera cases occur in the mid- to Late Jurassic, 160–145 MYA (Sukatsheva, 1982, 1999; Ivanov and Sukatsheva, 2002); none are known from the Triassic, even though stem-group Trichoptera (Necrotaulidae) are as old as the mid-Triassic (living families are no older than the mid-Jurassic). Caddisfly cases are diverse from the Cretaceous of Asia, with some 9 form genera and 200 ichnospecies described. The most peculiar fossil caddis cases are in the ichnogenus *Piscindusia*, which are constructed of minute fish scales and bones (Jarzembowski, 1995b).

Leaf/Stem Feeding Damage. Traces of chewing and punctures on fossil plants made by feeding arthropods are the most abundant and diverse kind of insect trace fossil. Evidence of this, including galls and wood/stem boring, was reviewed by Scott and Taylor (1983), Taylor and Scott (1983), Scott (1991), Scott *et al.* (1992), Stephenson and Scott (1992), and Labandeira (1998). The evidence is based mostly on leaf damage, particularly marginal and surface feeding, and mining. Virtually all groups of insects with chewing mandibles that feed on plants either leave chew marks along the leaf margins or rasp small holes in the surface. Feeding traces are sometimes distinctive, such as the large, circular pieces chewed from the margins of leaves by attine (fungus growing) ants (Figure 11.58) and megachiline (leaf-cutter) bees. Insects that feed continuously along the margin, like sawflies and caterpillars, leave indistinctive traces. Some insects *skeletonize* the leaf, eating just the epidermis among the veins and leaving the veins.

The earliest evidence of marginal and surface feeding on leaves involves both the Carboniferous seed fern *Neuropteris* and *Glossopteris*, small to large woody plants that were widespread throughout Gondwana from the Permian to early Mesozoic (Scott *et al.*, 1992; Labandeira *et al.*, 1998). The most comprehensive study involves herbivory on Gigantopteridaceae from the Early Permian of Texas (Beck and Labandeira, 1998). Gigantopterids were plants of enigmatic relationship, having large, spreading leaves. Those from the Texas Permian had various kinds of feeding marks, and, remarkably, showed extensive herbivory: up to 83% of the leaves and some 4.4% of the leaf area for some species of the plants (Beck and Labandeira, 1998). By 300 MYA herbivory was a routine lifestyle for terrestrial arthropods.

Mines. These blotches or meandering tunnels between the epidermal layers of a leaf are the pathways within which a larva feeds on the mesophyll layer. When the larva molts, the mine enlarges; at the end of the mine is a pupation chamber and exit hole. The size, shape, and path of the mine, the kind of leaf it is on, and even how the frass is deposited within the mine help determine the identity of the miner. Leaf mining is exclusively holometabolous, caused by the larvae of some Coleoptera, some Diptera (especially Agromyzidae), some Hymenoptera (especially Pergidae and Argidae sawflies), and especially by 10 major, basal families of "micromoths." Major reviews of leaf-mining insects are by Needham *et al.* (1928) and Hering (1951). Putative leaf mines are reported from the Carboniferous (Scott *et al.*, 1992), which is doubtful, and possible mines exist from the Permian (Beck and Labandeira, 1998). Definitive mines first appear on the leaves of Triassic conifers and pteridosperms, but their identity is unknown. Like galls, leaf mines become diverse and abundant with the radiation of angiosperms in the Cretaceous and Tertiary (Figures 2.20, 13.32) (e.g., Rozefelds, 1988) and the evolution of the major mining groups of insects.

Traces of plant feeding by insects have documented apparent persistence of some plant-insect relationships over tens of millions of years. For example, there has been intimate association between various living genera of basal, leaf-mining families of moths on trees like *Quercus* (oaks) and *Populus* (poplars) for approximately 20 million years (Opler, 1973), and on *Cedrela* (Meliaceae) for approximately 40 million years (Hickey and Hodges, 1975). Leaf mines from the mid-Cretaceous Dakota Formation (100 MYO) have been identified to *living* genera of Nepticulidae and Gracillariidae

2.20. An exceptionally well-preserved leaf from the Eocene of Anglesea, Victoria, Australia, with the mine of a larval moth. Museum Victoria (VM) 180365; length of leaf 64 mm.

moths (Labandeira *et al.*, 1994), which have been used to infer basal radiations of the Lepidoptera deep into the Early and mid-Jurassic. Other evidence of host plants use has been found on ginger (Zingerberaceae) leaves from the Upper Cretaceous of Wyoming (Wilf *et al.*, 2000). Distinctive, parallel chew marks on the surface are indistinguishable from those made by modern hispine beetles on *Heliconia* plants: an association that has persisted for some 70 million years.

Galls. These are excessive growths of plant tissue on stems, leaves, cones, and flowers, created by a feeding insect (immature or adult) or an ovipositing female. Substances in the saliva or oviposition fluids cause plant tissue to grow around the developing insects, encapsulating them. Gall formers are scattered throughout arthropods, including some mites (Acari), some aphids and scale insects (Sternorrhyncha: Aphidoidea, Coccoidea), a few leafhoppers, some lacebugs (Auchenorrhyncha; Heteroptera: Tingidae), some thrips (Thysanoptera: Phlaeothripidae), a few Lepidoptera, various Diptera (some Tephritidae, Agromyzidae, and many Cecidomyidae), and various Hymenoptera (some sawflies and many Cynipidae). The largest groups of gall formers are several derived groups of gall midges (Cecidomyidae) and the gall wasps (Cynipidae: Cynipinae). The biology and diversity of galls has been extensively reviewed (Felt, 1940; Mani, 1964; Ananthakrishnan, 1984b; Meyer, 1987; Gagné, 1989, 1994; Williams, 1994). Fossil galls have been reviewed by Larew (1986, 1992), Scott *et al.* (1994), and Whittlake (1981). The size and shape of the gall, its location on a plant, and the type of plant the gall is found on are usually necessary to identify the galler. In an exceptional situation the inhabitants were preserved: galls on the seeds of a *Sequoia* from the Miocene of Germany contained larval and pupal Cecidomyidae (Möhn, 1960).

The oldest galls are well-described structures from the petioles of 300 MYO *Psaronius* tree fern fronds (Late Carboniferous) of Illinois (Labandeira and Phillips, 1996a). The galls are elliptical structures with a central cavity about 3.5 × 0.50 cm, showing wound tissue and an exit hole and filled with cylindrical fecal pellets. Size frequency of the pellets grouped in four classes, the size of each class differing from adjacent classes by a factor of 1.3, which is similar to the size differences between instars of terrestrial arthropods. Thus, the inhabitant was a developing arthropod that spent at least four instars in the gall. By process of elimination, the gall maker was identified as the earliest known holometabolous insect (Labandeira and Phillips, 1996a), because more basal galling insects, hemipteroids, feed on plant vascular fluids and produce liquid feces. No Holometabola are known from the Carboniferous, the earliest records being body fossils of Neuropterida, Coleoptera, and Mecoptera from the Early Permian (*Srokalarva*, from the Carboniferous of Mazon Creek, was proposed as the earliest insect larva [Kukalová-Peck, 1991] but is now believed to be a myriapod). Even though the Carboniferous gall and frass therein are much larger than ones made by living mites, it is quite possible that mites produced this gall. Some mites, including living ones, can be quite large (5 mm or more in size); wood-boring mites are known from the Carboniferous (Figure 2.21); and various Paleozoic arachnids were quite large.

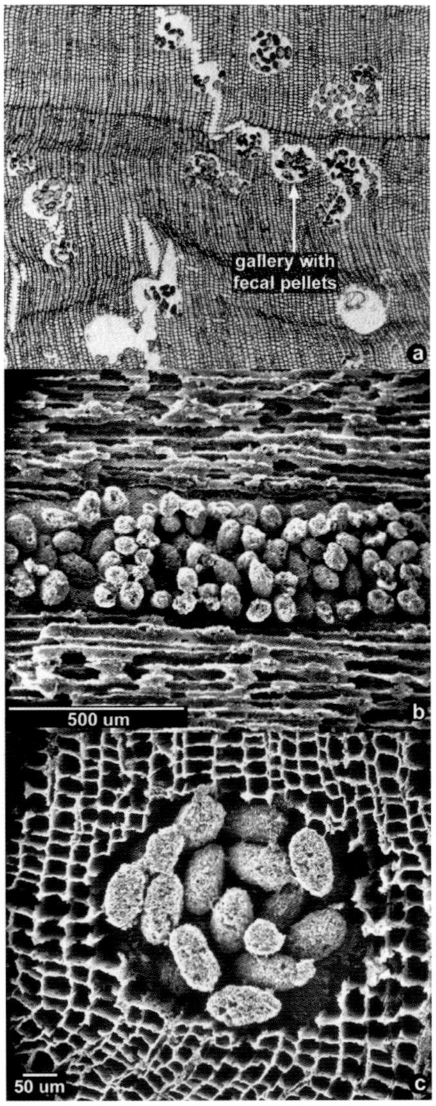

2.21. Scanning electron micrographs of galleries in fossilized *Premnoxylon* wood from the Carboniferous. The galleries were probably formed by mites, and they are filled with fossilized fecal pellets (frass). These are among the earliest remains of plant feeding. Photos: T. & E. Taylor, University of Kansas.

Insect galls are encountered much more in the fossil record with the radiation of angiosperms, approximately 100 MYA, with diverse galls found on angiosperm leaves from the Cretaceous and Cenozoic (e.g., Figure 2.22). Miocene galls on oak leaves from western North America reveal that for a minimum of 20 million years certain kinds of cynipid gall wasps have been intimately associated with trees in the genus *Quercus* (Waggoner and Poteet, 1996; Waggoner, 1999).

2.22. Gall on the petiole of a *Populus* leaf, from the Miocene of Oeningen, Germany. This gall is very similar to those made by some pemphigine aphids on poplar leaves today. MCZ 16735; diameter of gall 5.5 mm.

Borings and Galleries in Wood. Thick, highly lignified stems and branches of plants (wood) preserve well in the fossil record and have also preserved a unique record of the workings of extinct arthropods. An early paper on these trace fossils was by Brues (1936). Labandeira *et al.* (1997) summarized the fossil record of wood borings.

Some arthropods merely excavate the wood and feed upon fungus that grows in the galleries or upon the nutritious cambial layer. These include most wood-boring beetles, some sawflies (Hymenoptera), and some Lepidoptera. A few groups of insects actually eat the wood, particularly termites and some closely related roaches. Beetles are the most diverse, though termites are the most ecologically significant group that excavates wood. The earliest records of wood borings were probably produced by mites (Acari) from the Carboniferous (Cichan and Taylor, 1982; Rothwell and Scott, 1983; Scott and Taylor, 1983; Rex and Galtier, 1986; Labandeira *et al.*, 1991). Among these are exquisitely preserved borings in the outer layers of the wood of *Premnoxylon*, having chambers filled with frass (Cichan and Taylor, 1982) (Figure 2.21). The earliest *possible* beetle borings are in Permian glossopterids. The earliest *definitive* beetle borings are from the Triassic of Europe and Arizona, and later in the Mesozoic beetle borings become more diverse with the radiation of various families of beetles (reviewed in Labandeira *et al.*, 2001) (Figure 2.23). An interesting report concerns engraved galleries of scolytid (bark) beetles in Eocene wood, ca. 45 MYO (Labandeira *et al.*, 2001). Body fossils of Scolytidae are abundant and diverse in the Tertiary, and their borings even occur in the Cretaceous (though not the body fossils) (Figure 13.32). The Eocene engraving is of the extant beetle genus *Dendroctonus* in larch (*Larix*) wood from Axel Heiburg Island in the high arctic.

Termite borings are found in the Cretaceous and Tertiary, which reflects their early history. Many of them are preserved three-dimensionally in silicified wood, the galleries filled with frass (e.g., Rogers, 1938; Rozefelds and de Baar, 1991) (Figure 2.24). The earliest termite workings are from the Late Cretaceous Javelina Formation of western Texas (Rohr *et al.*, 1986) (Figure 7.84). They were attributed to Kalotermitidae, since the preserved wood (from *Diospyros*) appeared to be sound, and kalotermitids today excavate sound, often dry, wood. This is the oldest social insect nest. The only other kalotermitids from the Cretaceous occur in mid-Cretaceous amber from Myanmar.

Triassic fossil burrows attributed to termites and bees

2.23. Petrified wood studded with beetle bore holes, from the Eocene of central Queensland, Australia. Queensland Museum (QM) F. 14679; length of piece 38 mm.

2.24. Mineralized termite frass in wood from the Eocene of Queensland, Australia. QM ML511A; length of pellet ca. 2 mm.

(Hasiotis and Dubiel, 1995) illustrate how some ichnofossils are overly interpreted. Small burrows in wood from trees of the Late Triassic Petrified Forest (ca. 220 MYA) of Arizona resemble those of termites. The order Isoptera, however, is known no earlier than Early Cretaceous, approximately 130 MYA, and it is highly unlikely that body fossils of termites are missing from nearly 100 million years of the fossil record. Moreover, all Cretaceous termites belong to basal families, indicating that the order is probably no older than Late Jurassic in age. Bees almost certainly originated in the mid- to Late Cretaceous (Engel, 2001b), and the earliest aculeate (stinging) Hymenoptera occur in the Late Jurassic and Early Cretaceous, 150–140 MYA. Triassic bees are likewise inconceivable, and it is most likely that the insect burrows from the Petrified Forest are from beetles.

Amber Inclusions

Now dim, now bright, trapped in its amber tear
A bee seems sealed in its own nectar clear;
For a life of endless toil, most fitting pay
Surely a bee would wish to die this way!
–Martial's Epigrams, ca. 89 A.D. (translation by Valerie Krishna)

Amber, sometimes called resinite, is ancient tree resin. Vast deposits from the Baltic region have been collected for charms, amulets, and objets d'art for at least 13 millennia – making Baltic amber the original precious substance. When the resin was sticky, insects and other small organisms became mired on the surface and were gradually engulfed by the flowing resin, preserving them in finer fidelity than perhaps any other kind of fossil. The use of amber in jewelry and decorative objects, and an intrigue with inclusions have created a special popular appeal of the substance (Grimaldi, 1996).

Fossil resins are scattered throughout earth's surface in deposits from the Carboniferous to the Holocene. Unusual filaments of fossil resin are known from vessels of Carboniferous tree ferns, but true amber first appears in the Triassic (ca. 235 MYA). Amber is usually secreted externally by the plant into oozing masses, believed to be a way trees seal an injury from invading insects and fungi. Resins are complex mixtures of terpenoids and other biomolecules (Mills *et al.*, 1984), with over 100 individual compounds identified from 7 botanically distinct amber deposits alone (Grimalt *et al.*, 1987). Resins (and therefore amber) from different species of tree (and even populations of some species) have unique chemical profiles, usually characterized using pyrolysis-gas chromatography/mass spectroscopy. This feature is commonly exploited to help identify the type(s) of trees that produced an amber deposit, though associating the amber with plant macrofossils is equally important for identification (e.g., Shedrinsky *et al.*, 1991; Anderson and LePage, 1995; Grimaldi *et al.*, 2000a,b). The recent book by Langenheim (2003) comprehensively treats Recent and fossil resins.

Almost all of the amber produced in the Mesozoic, and much of the Cenozoic amber, is from conifers. Rare exceptions involve a few cases of Cretaceous angiosperm amber (Langenheim, 1969, 2003; Grimaldi *et al.*, 2000b). Some large Cenozoic deposits, though, were formed from broad-leaved (dicot) trees, such as Dominican and Mexican amber (*Hymenaea*: Leguminosae), and Arkansas and Borneo amber, possibly formed from Dipterocarpaceae. Resin production is widespread among conifers but sporadic among angiosperms. The largest deposit of amber in the world occurs on the southern coasts of the Baltic Region and was formed by a conifer, probably a pine (Pinaceae) (Langenheim, 1995). The oldest deposits that contain insects are from the Early Cretaceous of Austria (Schlee, 1984); England, Lebanon, and Jordan (Azar, 2000); and Choshi, Japan (Fujiyama, 1994), 140–120 MYA. The taphonomy of fossil insect assemblages in amber, and world amber deposits, has been comprehensively reviewed by Martínez-Delclòs *et al.* (2004).

Because there is such a spectrum in ages of fossil resin, from several hundred years old (generally referred to as *copal*) to millions, it is often confusing as to when buried resin becomes amber. Resin begins to polymerize and crosslink almost immediately after exuding and will continue to do so for centuries and millennia in the ground, thus rendering amber much more inert than modern resins. But, the degree to which these processes occur depends on many factors: the molecular composition of the original resin, exposure to ultraviolet light, age, and the amount of geothermal energy imposed by overlying sediments and faulting. Dominican amber (approximately 20 MYO), for example, is far more inert and reacts less with solvents than many, much older Cretaceous ambers, largely because of its composition.

One proposal has suggested a solution to the ambiguity of amber and resin based entirely on the practical criterion of ^{14}C dating (Anderson, 1996). Organic materials up to 40,000

2.25. A small menagerie of arthropods in Oligocene amber from southern Mexico. Other forms of fossilization can also preserve aggregations, but no form preserves minute and delicate organisms with the fidelity of amber. AMNH A231; greatest length of piece 31 mm.

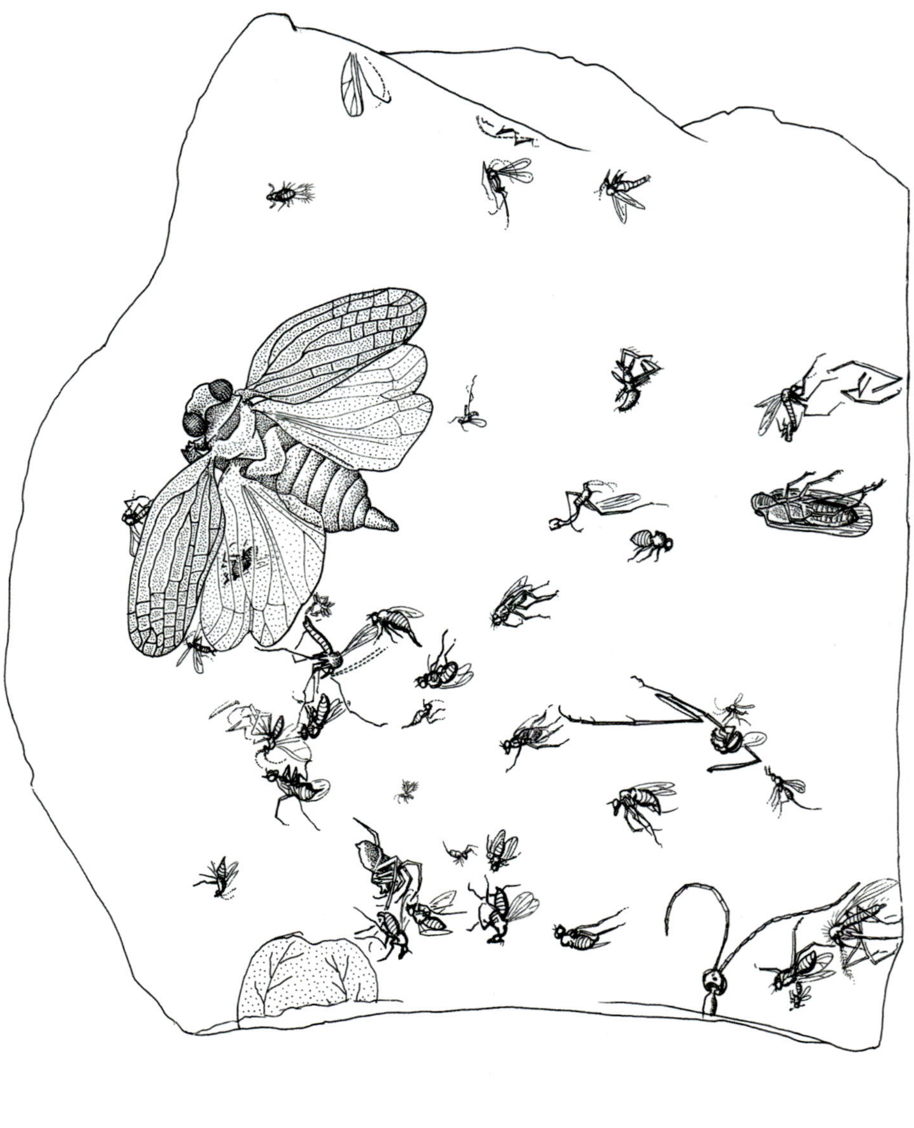

2.26. A map of the menagerie in amber in Figure 2.25. There are 36 individual arthropods, including insects and spiders, belonging to four insect orders, ten families, and approximately 13 species. Such pieces are small slices of ancient ecosystems.

years old, including amber, can be reliably dated using the technique. In this proposal, material 250 years old or less is *modern* or *recent resin* or *copal*; that between 250 and 5,000 years old is *ancient resin*; resin 5,000 to 40,000 years old is *subfossil resin*; and material older than 40,000 years is *amber, fossil resin*, or *resinite*. While this is an arbitrary classification, it is at least a very practical way to distinguish between modern and ancient resins (<40,000 years old) and amber (>40,000 years old). In general, too, copals are lighter in color than amber, they readily melt, and the surface forms a system of fine cracks (or crazing) over several years. In the ^{14}C scheme, material from Madagascar; East Africa; Colombia; Mizunami, Japan; and eastern portions of the Dominican Republic (Burleigh and Whalley, 1983) are recent to ancient resins, not amber.

Forests bleed large quantities of resin as a result of damage by storms, fires, and outbreaks of wood-boring insects. Any one or several of these factors contributed to the prolific amounts of amber produced by ancient forests from the Baltic region, the Dominican Republic and Mexico, and several Cretaceous sites (the most significant deposits are discussed individually later). Productive deposits like Baltic and Dominican amber have captured a great variety of organisms (Figures 2.25 to 2.29): leaves, flowers, portions of vines and stems, fungi, myriad arachnids and insects, swarms and mating pairs of insects, hosts with parasites, even scorpions, small lizards, and frogs. So diverse are these amber Lagerstätte that detailed reconstructions have been made of their paleoenvironments (Brues, 1933b; Larsson, 1978; Grimaldi, 1996). A significant proportion of the biodiversity in amber comprises tiny arthropods a millimeter or less in size, like mites; cecidomyiid and ceratopogonid midges; scelionids and mymarid and mymarommatid fairy "flies" (all Hymenoptera); and minute ptiliid ("featherwing") beetles. Indeed, the smallest arthropod fossils are known from amber, and because of the exquisite preservation, minute structures can be observed at the micron scale (e.g., Figures 10.83, 10.84). Many of the insects in amber died on the surface of a resin flow and then were sealed when more resin flowed over them. If the amber is split along the flow line, exposing a cast

2.28. A cecidomyiid gall midge caught in Dominican amber while laying its eggs. Some insects reflexively exude eggs as they are dying, which probably happened here. Minute scenes like this are commonly preserved in amber. AMNH DR14–704; length of midge 1.8 mm.

2.27. Cicadellid leafhoppers in Miocene Dominican amber, captured while mating. AMNH DR15-5; total length (both) 6.2 mm.

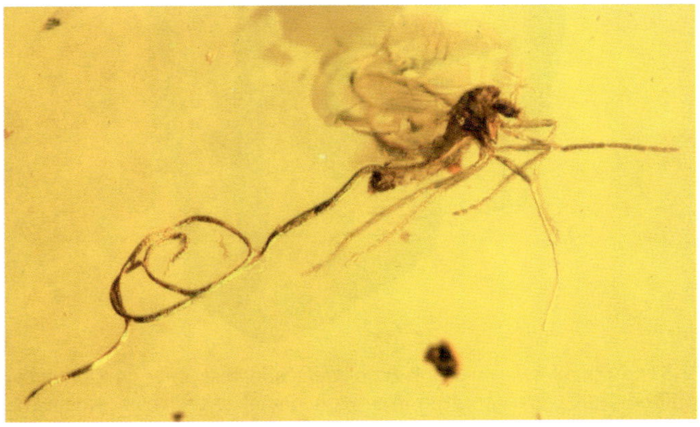

2.29. A chironomid midge with two parasitic mermithid nematodes bursting from its abdomen, preserved in mid-Cretaceous amber from Burma. AMNH Bu320; length of midge 1.2 mm.

2.30. A milichiid fly in Dominican amber with the vivid red pigment of its eyes preserved. Amber rarely preserves such color on insects. AMNH DR14–1316; body length 2.6 mm.

of the insect, scanning electron micrography can reveal extremely fine external details. Color patterns are frequently preserved, and in some cases the original vivid color remains (Figure 2.30).

Microscopic-scale preservation of internal soft tissues within amber-encased insects was known as early as 1903, but electron microscopy has revealed unexpected, lifelike fidelity (Mierzejewski, 1976; Kohring, 1998), even subcellular structure (Henwood, 1992a,b; Grimaldi *et al.*, 1994). The degree of preservation varies greatly, largely as a result of the chemistry of the resin and how quickly the organism was embedded. Small organisms in Baltic amber, for example, commonly have a milky coating, which is actually a froth of microscopic bubbles, probably formed by gases exuding from the decaying body cavity. Organs and tissues of such insects are usually poorly preserved, or the body cavity is merely a void. Insects well preserved in amber, including from the Cretaceous, often have the organs intact, with virtually none of the shrinkage typically seen in specimens that have been merely dehydrated. Muscles lie in their original positions, and the banding is vivid, with the digestive tract, tracheae, even the nerves, brain (Figures 2.33, 2.34), and symbiotic microbes preserved. For example, modern, wood-boring platypodid and other beetles have small pockets under the cuticle ("mycangia") specialized for harboring symbiotic "ambrosia" fungus, which they inoculate into their galleries and then feed upon. The mycangia of platypodids in Dominican amber were likewise found to contain mycelia and spores of the fungus (Grimaldi *et al.*, 1994).

Even more impressive is the cellular and subcellular preservation, including muscle cells and neurons, myofibrils and sarcomeres, membranes, and mitochondria (Henwood, 1992a,b; Grimaldi *et al.*, 1994). Most recently, symbiotic bacteria attached to the membrane of symbiotic protists have been observed in hind gut tissue of a termite in

2.31. A small swarm of 11 *Proplebeia* stingless bees in Dominican amber. They were captured in a fresh runnel of the resin, which was then engulfed by successive flows that formed layers like a stalactite. AMNH DR14–1054; length of amber 43 mm.

2.32. A *Proplebeia* stingless bee in Dominican amber preserved with balls of resin on its hind legs. Stingless (meliponine) bees harvest resin to construct their nests and frequently become trapped, which is why *Proplebeia* is so common in Dominican amber. Their abundance in the amber has made this bee a favored subject for studies of tissue and molecular preservation. AMNH; length of bee 2.9 mm.

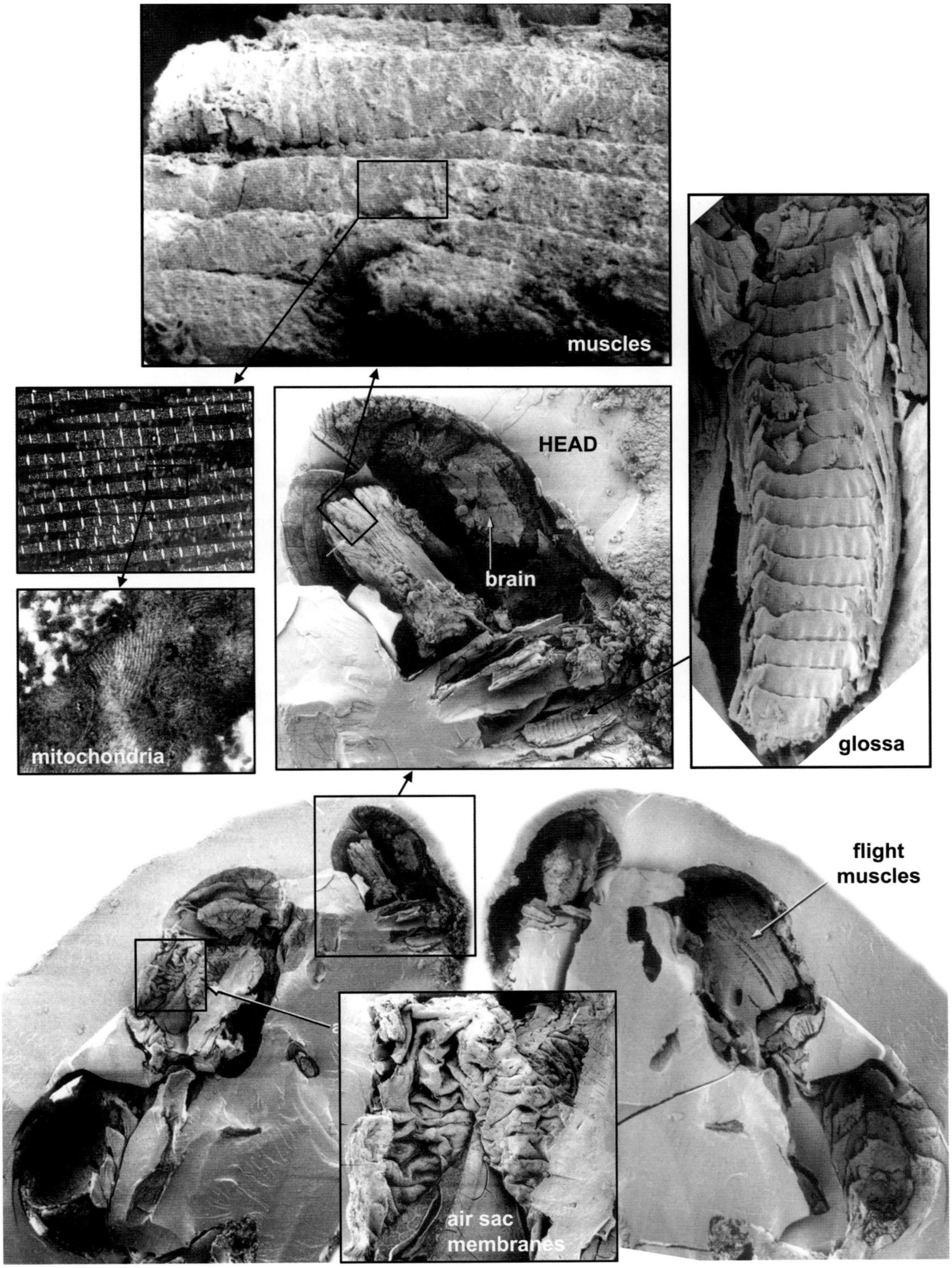

2.33. A *Proplebeia* bee exhumed from Dominican amber and imaged with an electron microscope. The amber was carefully split in half, exposing the body cavity and internal organs and tissues. Under high magnification (10,000X) even the banding of muscle myofibrils and the folded cristae of mitochondria are preserved. Reports in the early 1990s of DNA sequences from amber fossils now seem to have been a result of contaminants. AMNH.

2.34. Muscle fibers from a beetle preserved in 125 MYO amber from Lebanon, preserved with the tube-like tracheoles that carry oxygen and carbon dioxide to and from tissues. Preservation is not necessarily better in younger ambers but mostly depends on the composition of the resin. AMNH.

Dominican amber (Wier *et al.*, 2002). The unparalleled fine structure of insects embalmed in amber inspired studies on their molecular preservation, and no molecule is of greater evolutionary significance than the one that is the basis of inheritance, DNA.

Does Amber Preserve Ancient DNA? Changes in DNA inherited through generations eventually become part of the blueprint of a lineage, so by reconstructing the sequence of changes we can retrace the evolutionary history and relationships of diverse organisms. Unfortunately, estimates of the times and extent of genetic divergence have traditionally been relegated to comparisons among living species. When polymerase chain reaction (PCR) was developed for amplifying minute quantities of DNA, this led to an avid search for DNA that was truly ancient, from fossils millions of years old. Insects in amber played a leading role in that search.

Interest in DNA from fossils seriously began with study of compression fossil leaves in Miocene sediments from Clarkia, Idaho, a site in which insects are also beautifully preserved (discussed later). Dense mudstones at Clarkia sealed the fossils from destruction by oxygen for nearly 18 million years. When unearthed, some leaves are a vivid green, or autumn yellow, like the day they fell, but within minutes they oxidize to a blackish brown. Cellular and even subcellular preservation of Clarkia plants, including chloroplasts, has been well documented (Smiley *et al.*, 1975; Niklas *et al.*, 1985). The first report of DNA millions of years old was a 770-bp fragment of the chloroplast gene rbcL from leaves of an extinct tree at Clarkia, *Magnolia latahensis* (Golenberg *et al.*, 1990). This gene is of standard use in the molecular systematics of plants, and sequences of it were found to differ by 17 bp (or 2.2% sequence divergence) between the fossil and a closely related living species. The pitfall of PCR techniques is that contaminant DNA is commonly amplified, even under scrupulously clean conditions, which led some to criticize the study (Golenberg, 1991). Soon after this came a report on DNA from another Clarkia fossil plant, this time a bald cypress (*Taxodium*) (Soltis *et al.*, 1992). Using the same gene, they found 11-bp changes among the 1,320 they sequenced, representing a divergence of less than 1%. These results soon led to popular accounts on the scientific implications of truly ancient DNA (Gould, 1992). For the first time, it seemed, genetic divergence over millions of years could be directly measured.

While work on Clarkia fossils was being discussed, several other labs focused attention on insect fossils in Miocene amber (20 MYO) from the Dominican Republic, renowned for their preservation. Two papers were published within a month of each other, one on the common stingless bee in Dominican amber, *Proplebeia dominicana* (Cano *et al.*, 1992) (Figures 2.31 to 2.33), the other on the relict termite, *Mastotermes electrodominicus* (DeSalle *et al.*, 1992) (Figure 7.81). Both came on the heels of the best-selling novel, *Jurassic Park*, wherein dinosaurs are resurrected from cloned DNA extracted from the blood meals of mosquitoes preserved in amber. Amber had never been so popular.

DNA of the fossil bees' 18S rRNA gene was 7% divergent from several living species in the genus *Plebeia*. The fossil termite sequences were divergent by 10% from the sole living species of *Mastotermes* (*darwiniensis*) in the 16S rRNA gene, an unexpectedly large amount. Phylogenetic analysis, though, provided compelling evidence for a termite identity of the fossil DNA. Soon after the publication of these reports, reviewed and unreviewed reports of DNA being extracted from *Hymenaea* leaves and chrysomelid beetles in Dominican amber appeared, and then a major critique of all reports on ancient DNA was published (Lindahl, 1993). Of all biomolecules, DNA is perhaps the most labile (Eglinton and Logan, 1991). It is particularly susceptible to destruction by hydrolysis and oxidation, and even spontaneously decays under ideal conditions. No natural space is completely devoid of water and oxygen, including amber, and preservation of the molecule over millions of years appeared implausible.

Despite Lindahl's caution, a paper was published two months later in the same journal (*Nature*), reporting DNA from a weevil preserved in Lebanese amber, some 125 million years old (Cano *et al.*, 1993). Publication of the paper on the same day that the film version of *Jurassic Park* publicly debuted (10 June 1993) was not coincidental, and this propelled the popularity of amber even more. The unique weevil specimen had been splintered open, its tissue extracted, and 541 bp of the 18S rRNA gene were sequenced and compared to other insects, including weevils. Similar to the Dominican amber fossils, the DNA of the Lebanese amber weevil was 7% divergent from the living nemonychid weevil that was sequenced, and phylogenetic analysis again indicated authenticity.

The most convincing test for the authenticity of the ancient DNA is reproducibility. Two independent attempts failed to replicate the extraction of DNA from *Proplebeia* bees in Dominican amber (Austin *et al.*, 1997; Walden and Robertson, 1997). Even attempts to extract DNA from other kinds of insects in fossil resins failed, including platypodid beetles (Howland and Hewitt, 1994) and phorid flies (Austin *et al.*, 1997) in Dominican amber, as well as bees in much younger East African copal (Austin *et al.*, 1997). In fact, these attempts consistently found contaminant DNA. By the time a report was published on the revival of bacterial spores from Dominican amber (Cano and Borucki, 1995), it was justifiably regarded with widespread skepticism. If the contamination of PCR products was notoriously difficult to control, how much more would it be for ubiquitous *Bacillus* bacteria? Also, if DNA of insects in amber is at best highly fragmented, why should even highly resistant bacterial endospores endure so long, their genomes perfectly intact?

Attempts to replicate extractions of DNA from the Lebanese amber weevil and Domincan amber *Mastotermes* have not been made. The weevil was a unique specimen, and destructive sampling of it also generated controversy concerning the study of unique and rare amber specimens. Reanalysis of the published sequences from that specimen, though, indicates that they are probably contaminants from another beetle (Gutiérrez and Marín, 1998). A similar fate perhaps awaits the putative DNA sequences of the extinct *Mastotermes*. Studies made on the racemization of amino acids in amber fossilized tissues, though, support the possibility that DNA is preserved by amber (Bada *et al.*, 1999). Racemization is the formation of equal proportions of D and L enantiomers of a molecule, and certain amino acids racemize over steady rates. Apparently, the extent of amino acid racemization in tissues of insects preserved in amber is very similar to that of modern species, suggesting that DNA could be similarly preserved. But, amino acids are particularly durable molecules (Savage *et al.*, 1990; Bada, 1991; Kemp, 2002). Also, the cuticles from *Proplebeia* bees in Dominican amber contain no trace of chitin or protein (Stankiewicz *et al.*, 1998c). If a molecule as durable as chitin is completely degraded in amber, it is highly likely that DNA will also be degraded. In fact, it is highly unlikely that any DNA is preserved in ancient fossils of any sort.

Despite controversy and serious suspicion over DNA from amber fossils, these studies brought closer attention to the remarkable preservative qualities of amber, which have a fidelity that is far greater and more consistent than any other kind of fossil.

Pleistocene and Holocene Traps

The accumulation of insects in sediments that formed during the Quaternary (1.7 MYA to present) provides unique insight on climate change and the duration of species, a subject treated thoroughly by Elias (1994). Most Quaternary remains occur in peats from mature successional stages of bogs, which are the edges. Here, the thickly sclerotized, durable elytra, pronota, and heads of beetles predominate. Fortunately, the gross structure and microsculpturing of beetle sclerites allow detailed matches with modern species. Scudder (1877, and various papers thereafter) was among the first to carefully study Quaternary insects, of which he was mainly preoccupied with deposits from Scarborough, Ontario. He described 50 beetle species from Scarborough, all but two presumed to be extinct. It was not until the work of Carl Lindroth (1948), a coleopterist, that Quaternary insects were revealed to be generally extant, not extinct. Russell Coope, in Britain, systematically challenged the dogma that Pleistocene insects were largely extinct species like mammals. He essentially developed the study of Quaternary insects and was the first to document contractions in the distributions of modern insect species, some of which are dramatic. For example, fossils of the scarab *Aphodius holderei* and the staphylinid *Tachinus caelatus* occur in Britain, but these today are found in the Himalayas and Mongolia, respectively. Their present day distribution is a relict vestige of a time when cooler climates embraced most of Europe. Because dozens, even hundreds, of insect species can occur in a Quaternary site, these deposits provide abundant evidence of past climates along with fossil pollen and leaves.

Besides beetles, heavily sclerotized remains of other insects are also preserved, such as ant heads, oribatid mites, and the larval cases of caddisflies. The sclerotized head capsules of otherwise soft-bodied midge larvae are extremely abundant in lake sediments. Intricate structures on the head capsules allow species identification of these remains, some hundreds of thousands of years old. The tarpits of La Brea, California, are famous for the impressive mammals that were trapped and preserved there, but insects were also victims (Miller, 1983) (Figure 2.35). Mammoths and remains of other Pleistocene mammals frozen in tundra permafrost occassionally yield parasites (Dubinin, 1948; Grunin, 1973) (Figure 2.36), as do human mummies (Figure 2.37). Perhaps the most intriguing Quaternary fossils come from pack rat middens of the American southwest, sheltered amidst rock overhangs and caves. As the name indicates, pack rats stock their nest with gathered materials, like pebbles, cactus spines, and vertebrate bones. These materials form a conglomerate mass with feces and urine – the midden – from

2.35. Remains of predatory dytiscid beetles from the Pleistocene tar pits at La Brea, near Los Angeles, California. The beetles may have been living in water that pooled on top of the tar pits. AMNH; length of middle beetle 23 mm.

generations of occupation of the nest, and this attracted scavenging insects, many of them beetles.

When Is It a Fossil?

Well-preserved remains of living insect species in the Quaternary, and even lifelike preservation of extinct species in amber millions of years old, forces the question as to when remains are considered fossils. After one million years? After the original remains have been replaced by minerals? For some, a fossil is the remains of a species that has become naturally extinct, but for insects it isn't always certain if a species is extinct. Several species of scarab beetles from the La Brea tarpits, for example, are unknown among the living fauna (Miller, 1983). It is plausible they became extinct with the sloths, mammoths, and other mammalian fauna on whose dung they depended. Or, the species may persist, say in remote areas of Mexico. A similar situation pertains to a species of ptinid beetle found in 12,000- and 30,000-year-old pack rat middens, but unknown from anywhere else (Spilman, 1976). A more practical and perhaps equally valid definition is that a fossil is the remains or workings of any species, living or extinct, that have been *naturally* preserved for several thousand years or more.

DATING AND AGES

There is considerable confusion among nonpaleontologists as to how fossils are dated. Usually, the layers or strata in which the fossils occur are dated, but some fossils can be directly dated using isotope methods, and for yet other kinds of fossils the taxon itself reflects a geological stage or period. The classical geological time scale, with periods, epochs, and ages, codifies major biotic episodes in earth history. When eighteenth- and early nineteenth-century geologists

2.36. A botfly puparium (*Cobboldia*) from the stomach of a mammoth found in Siberia. Paleontological Institute Moscow (PIN) Q-TA-1/1; length 19 mm.

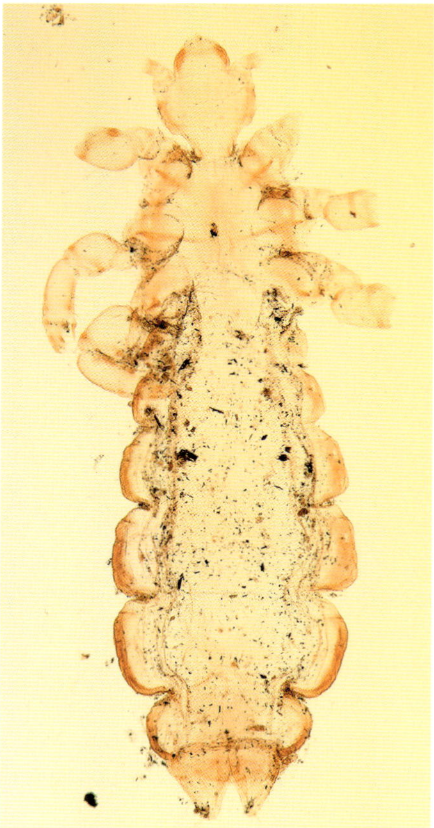

2.37. The human louse, *Pediculus humanus*, found on a 1,000-year-old human mummy from Peru (cf. Figure 8.7). AMNH.

2.38. A beautifully preserved mayfly in Pleistocene clays from North America. AMNH; wing spread 25 mm.

described the stratification of rock layers and their marine animal fossils, their intentions were to define a succession of extinct faunas, and work started small. The Devonian Epoch, for example, is named for a sequence of rocks described in 1839 from Devonshire, England; the Cambrian and Silurian, from sequences in Wales, named after ancient tribes from there. Correlating sequences of fossils in various geological columns (*stratigraphic correlation*) on a global scale transformed a provincial system into one that is as universal to geology as the Periodical Table of the Elements is to physics and chemistry, and the Linnean Hierarchy is to biology.

Correlation requires fossil taxa that are widespread, common, and diverse and that have fairly long histories (usually a million years or more). Most fossilization has involved invertebrates from marine continental shelves that had hard parts, and certain of these groups have become particularly important in defining epochs and ages. Graptolites (an extinct group of marine, colonial hemichordates), for example, are very diagnostic of ages within the Paleozoic, ammonites (extinct nautilus-like animals) within the Mesozoic, and planktonic Foraminifera (minute, shelled protists) and pollen for the Cenozoic. Typical index fossils for terrestrial sediments are pollen and spores. By describing stretches of similarity, change, and gaps, the early stratigraphers were essentially defining periods of biotic stasis and extinction events. Indeed, the boundaries between periods, epochs, and ages largely reflect abrupt events of extinction and biotic turnover. Geological strata continue to be studied using fossil correlation, but assigning absolute ages to strata ultimately requires physical methods.

The most commonly used physical dating method uses isotopes, wherein the proportion of isotopes of an element is measured. Because isotopes of an element are formed at a steady rate (the *decay constant* or *half-life*), the amount of isotope reflects the age of the substance. Elements have vastly different decay constants, which is why ^{235}U (uranium)-^{207}Pb (lead) is used for Paleozoic ages; ^{40}K (potassium)-^{40}Ar (argon) is used for Mesozoic and Cenozoic strata; and ^{14}C (carbon) is used for ages of 40,000 years or less. Another constraint to isotope dating involves the composition of the strata or fossil. The minerals used in isotope dating are most common in igneous and metamorphic rocks, but most terrestrial fossils occur in sedimentary rocks. Dating fossiliferous rocks thus requires overlaid or intrusive igneous or volcanic ashes that contain datable crystals like zircons.

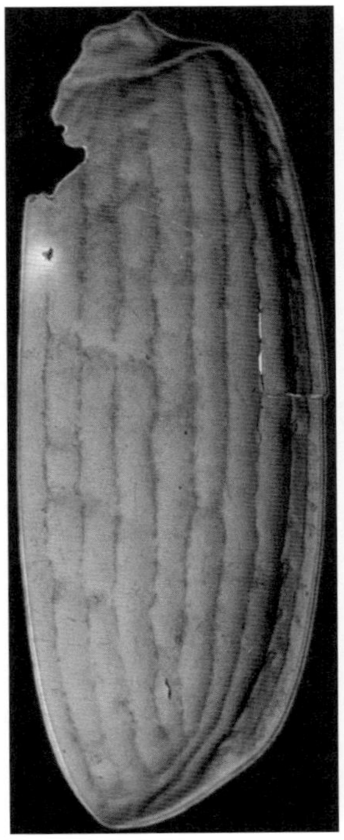

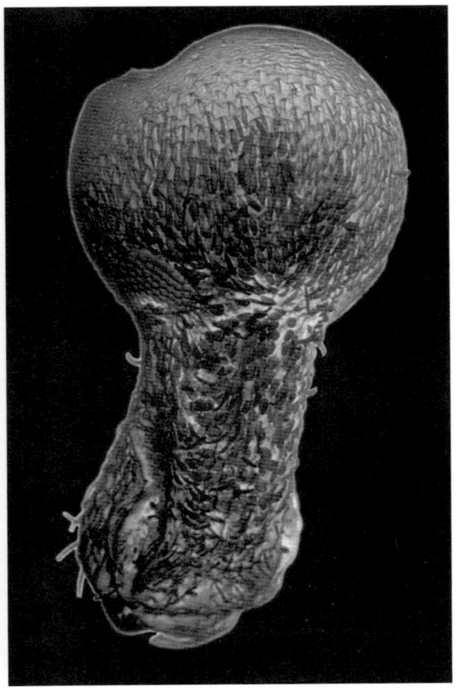

2.39 (Left). Hardened forewing, or elytron, of a carabid beetle from Wisconsin-aged (Holocene) bog deposits in Alaska. Photo: Scott Elias.

2.40 (Right). Head of a Holocene weevil from bog deposits in Alaska. Photo: Scott Elias.

PERIOD	EPOCH	AGE	BEGINNING
HOLOCENE			0.01 MYA
PLEISTOCENE			1.7
TERTIARY	PLIOCENE	Piacenzian	
		Zanclean	5.2
	MIOCENE	Messinian	7.0
		Tortorian	11.0
		Seravallian	14.8
		Langhian	16.4
		Burdigalian	21.0
		Aquitanian	23.6
	OLIGOCENE	Chattian	29.0
		Rupelian	34.5
	EOCENE	Priabonian	37.8
		Bartonian	41.7
		Lutetian	49.5
		Ypresian	55.6
	PALEOCENE	Thanetian	59.0
		Selandian	61.0
		Danian	65.0
MESOZOIC	CRETACEOUS	Maastrichtian	73.0
		Campanian	83.5
		Santonian	86.0
		Coniacian	89.0
		Turonian	92.0
		Cenomanian	98.0
		Albian	112
		Aptian	122
		Barremian	129
		Hauterivian	133
		Valanginian	138
		Berriasian	145
	JURASSIC	Tithonian	151
		Kimmeridgian	154
		Oxfordian	158
		Callovian	162
		Bathonian	167
		Bajocian	175
		Aalenian	179
		Toarcian	188
		Pliensbachian	195
		Sinemurian	202
		Hettangian	207
	TRIASSIC	Rhaetian	209
		Norian	222
		Carnian	231
		Ladinian	236
		Anisian	241
		Olenekian	245
		Induan	247
←PALEOZOIC	PERMIAN	Tatarian	252
		Ufimian-Kazanian	256
		Kungurian	260
		Artinskian	269
		Sakmarian	282
		Asselian	290
	CARBONIFEROUS	Gzelian	296
		Kasimovian	303
		Moscovian	311
		Bashkirian	323
		Serpukhovian	327
		Visean	342
		Tournaisian	358
	DEVONIAN	Famennian	364
		Frasnian	370
		Givetian	380
		Eifelian	391
		Emsian	400
		Pragian	412
		Lockhovian	414

2.41. The geological time scale, with the system of periods used in this book. Based on Palmer and Geissman (1999) and others.

Dating with carbon-14 (^{14}C) obviously requires an organic material.

Geomagnetic Polarity Timescale (GPTS), another physical dating technique, measures the orientation of iron oxide crystals aligned to the earth's magnetic field when the rock in which the crystals occur was formed. Earth's magnetic polarity is constantly reversing over periods of less than a million years. Regardless of the location on earth, crystals from contemporaneous paleomagnetic reversals all have the same orientation, and these reversals must be calibrated using radiometric techniques. GPTS is useful for most of the history of winged insects, to approximately 300 MYA (Late Carboniferous), but the error range in dating is far smaller for strata younger than half this age (Jurassic-Cretaceous boundary), which can be as little as several thousand years.

Depending on the time frame and types of fossils, fossil correlation has proved to be a very reliable dating method for estimating ages, particularly when the fossils are correlated with a column dated with physical methods. Physical dating methods, however, have revolutionized estimates of absolute ages (Figure 2.41 – the system used in this book). Over the past century, physical dating, for example, indicates an age of the earth that is more than 100 times Lord Kelvin's estimate of 40 million years.

MAJOR FOSSIL INSECT DEPOSITS

The geological or fossil record of insects is often dismissed by paleontologists and even by entomologists as too incomplete. In the words of one Oxford evolutionary biologist, the fossil record is "corrupted." Such adjectives gloss over the fact that hundreds of deposits of fossil insects occur on all seven continents, from the Devonian to the Holocene. The deposits vary greatly in preservation and diversity but collectively form a geological record that is, while not the envy of paleobotanists and vertebrate paleontologists, actually more impressive than for most groups of terrestrial animals. In some aspects the insect fossil record is unique or virtually so, such as the myriad life forms embalmed in amber and preserved as three-dimensional mineralized replicas. The mostly tiny insects preserved these ways together with larger insects preserved in sediments provide a complementary and vivid fossil record.

Our review here is not complete; it does not discuss all fossil deposits known to have yielded insects. Instead, we have focused discussion on the largest and most diverse deposits, ones that have yielded particularly significant finds, or those that are from poorly represented corners of the globe. An overview of fossil insect deposits was by Hennig (1981), which is rather out of date and general and which has now been replaced by the review in Rasnitsyn and Quicke (2002), itself incomplete but the most comprehensive to date. Schlüter (1990, 2003b) reviewed insect deposits from Gondwana, or the southern continents plus India. Evenhuis (1994) provided an extensive list of Mesozoic and Tertiary deposits in which flies are preserved (thus, no Paleozoic localities), but there was very little discussion, and the ages of some deposits require updating, which we have done here. The encyclopedic references on fossil insects – Rohdendorf (1962, 1991) and Carpenter (1992) – provided no overview at all on the various deposits.

Most fossil insect deposits occur in the Northern Hemisphere, which may be attributable simply to centuries more exploration and study in regions where paleontology developed, namely Europe and then North America. Arid regions of uplift, like the American west, Patagonia, Mongolia, and parts of Australia, harbor many fossil formations where sparse vegetation and erosion exposes fossil beds. Formations overgrown with thick rain forest are only occasionally exposed by mudslides or eroded river banks, which partly explains the paucity of deposits from tropical countries.

PALEOZOIC

The Paleozoic Era saw the most dramatic biotic changes on earth. At the opening of the era (i.e., the beginning of the Cambrian, 543 MYA) the biotic world experienced a literal explosion of diversity, all of it marine. During the "Cambrian Explosion" came the development of major animal body plans and the rapid proliferation of complex life (e.g., Conway Morris, 1979, 1989). By the end of the Paleozoic, life had invaded land – a littoral explosion of diversity, first by the plants, and shortly followed by arthropods and then other animals. Terrestrial ecosystems came into existence during the Paleozoic, which not only affected the surface of the earth but the atmosphere as well. For the story of insects, however, the latter half of the Paleozoic interests us. Land plants made their first appearance in the Ordovician (490–443 MYA), and terrestrial arthropods migrated into these miniature plant communities during the later Silurian (443–414 MYA), but the Silurian fauna as it is generally understood consisted entirely of primitive arachnids and myriapods (e.g., Jeram *et al.*, 1990). Very soon thereafter were the first hexapods, which were undoubtedly already present by the end of this epoch (e.g., Engel and Grimaldi, 2004a). The Early Devonian (ca. 410 MYA) heralded the first hexapods (at least the first preserved as fossils) and it was not long thereafter that insects dominated earth, becoming the first to fly and then rapidly proliferating during the Carboniferous and Permian. Owing to constant tectonic change of the earth since the Paleozoic, however, there are not as many terrestrial deposits with insects as there are from later eras.

Devonian (414–358 MYA)

The earliest evidence of hexapods, the group of arthropods to which insects belong, comes from a few fragmentary remains of apterygote lineages from the Devonian. Although plants had long since colonized the land, this invasion was slow, and in the Early Devonian most plants were confined to moist, lowland environments or still consisted of mats growing on the surface or edges of pools. These primitive vascular plants were not complex and consisted of relatively simple shoots that generally reached no higher than a meter. The entire biotic world literally resided at knee-height or below, but Devonian plants rapidly diversified and took over an essentially empty landscape. During the Late Pragian through Givetian (412–370 MYA), arborescence evolved, and by the end of this time period the structural support provided by woodiness allowed plants to dramatically increase their physical size. The first forests developed at this time, and plants began to venture further from the ecologically restricted moist, riparian environment. By the end of the Devonian, medium-sized to giant tree ferns would appear (e.g., cladoxylopsids, lycopsids) as well as archaeopterid and aneurophyte progymnosperm trees. Some Late Devonian archaeopterids actually exceeded 30 m in height. Near the end of the Devonian (during the Famennian, ca. 364 MYA), the first seed plants appeared. As can be imagined, such a considerable explosion of plant life on land had a dramatic impact on the world's environment. In fact, the impact on CO_2 levels was significant, resulting in a precipitous decline between the Late Devonian and Early Carboniferous (Berner, 1997; Alego *et al.*, 2001), and this even caused a brief episode of continental glaciation at the end of the Devonian (Caputo, 1985). The origin and spread of forests also transformed the soil, particularly acidification of the soil and slow weathering processes that retarded the transport of sediments.

During this dramatically changing world insects appeared and spread. The earliest hexapods are known from the Early Devonian, a "miniature" terrestrial world, and their rise would follow that of the plants that formed their microcosm.

Rhynie Chert. The first and perhaps the most famous Devonian hexapod is *Rhyniella praecursor* Hirst and Maulik (1926), a springtail from the Rhynie chert of Scotland (Figure 3.31). The age of the hot spring chert from Rhynie, Scotland, has been assigned to the Lockhovian-Pragian (as a maximal age) from spore evidence, while radiometric dating has indicated a slightly younger age of Emsian (Trewin, 1994; Rice *et al.*, 1995: ca. 396 MYA). The Rhynie paleoenvironment was likely one of a marsh or swamp (Tasch, 1957). Rhynie chert is exceptional among Paleozoic deposits for the detail of the remains. Most fossils are preserved as inclusions, three-dimensionally embedded in the microcrystalline, translucent chert. Somewhat mimicking the quality of preservation known from considerably younger amber inclusions, specimens entombed in Rhynie chert preserve fine microscopic details such the structure of setae and cuticle. The fauna mostly includes arachnids, such as trigonotarbids and mites, but also contains remains of crustaceans, eurypterids, and centipedes (Shear *et al.*, 1987, 1998; Anderson and Trewin, 2003) and a pair of mandibles – described by Tillyard as *Rhyniognatha hirsti* in 1928b, but recently reported as the earliest definitive insect (Engel and Grimaldi, 2004a) (Figure 5.8). Although the identity of the hexapod remains from Rhynie was challenged by Crowson (1985), who believed the remarkably modern *Rhyniella* to be a later contaminant, the recovery of additional specimens have established that springtails were definitively present in the Early Devonian environment of Scotland (Scourfield, 1940a,b; Whalley and Jarzembowski, 1981). Even with the excellent preservation of Rhynie chert, the higher-level assignment of *Rhyniella* has been difficult but appears to be within the modern family Neanuridae (Massoud, 1967) or more likely Isotomidae (Greenslade and Whalley, 1986).

Gaspé Bay, Canada. In eastern Québec on the northern and southern shores of Gaspé Bay are outcrops representing a terrestrial environment in the Battery Point Formation. Spore and brachiopod assemblages date the formation from the beginning of the Middle Devonian (near the Emsian-Eifelian boundary, ca. 390 MYA) (Boucot *et al.*, 1967; Richardson and MacGregor, 1986). The flora of the Gaspé fossil beds is perhaps one of the most thoroughly studied Devonian botanical assemblages. The deposits are understood to represent a relatively tropical fluvial and delta-plain environment, with freshwater marshes and lacustrine deltas (Lawrence and Williams, 1987; Hotton *et al.*, 2001). From macerated material taken from the Gaspé fossil beds, two fragments of a bristletail head and thorax were recovered (Labandeira *et al.*, 1988). These remains represent the oldest record of insects in North America, although Jeram *et al.* (1990) suspected them to be recent contaminants. The Gaspé fragments preserve only primitive features of the Archaeognatha and thereby of the Insecta as a whole and accordingly may represent an extinct lineage basal to Recent apterygotes. However, not enough evidence presently exists to make a more conclusive assignment.

Gilboa, New York. In upstate New York near the town of Gilboa rests a layer of mudstone famous for its fossils from the Middle Devonian (Givetian). The fossils are extracted from the mudstone matrix by macerating the material in an acid wash and then sorting the minute pieces of cuticle from the dissolved rock. From this site many fragmentary remains of early arthropods have been recovered. They are mostly chelicerates and myriapods (Figure 3.24) but also include tantalizing pieces of cuticle with scales, indicative of Archaeognatha or Zygentoma (Shear *et al.*, 1984, 1987).

Together these three deposits represent the earliest evi-

2.42. An protodonatan nymph preserved in an ironstone concretion from the famous Upper Carboniferous deposits at Mazon Creek in northcentral Illinois. FM PE30272; length of concretion 51 mm.

dence of terrestrial hexapods. Interestingly, all these sites occur within 10° of the Devonian equator. Potential sites that deserve attention in the future, but that to date have revealed no insects, are from the Falkland Islands and southern Greenland.

Carboniferous (358–289 MYA)

The Carboniferous consists of two major periods, the Mississippian (358–324 MYA) and the Pennsylvanian (323–289 MYA). By the Early Carboniferous, vast coal swamps had developed, and large arborescent plants dominated the landscape. Land plants had finally taken hold, and forests covered many regions of the world. Insects also radiated and had become diverse in the Carboniferous. Unfortunately, from the Early Carboniferous (i.e., the Mississippian) no outcrops have as of yet yielded insects. Thus, there is a considerable gap in our knowledge from this critical time period. Essentially, the fossil record terminates with the few fragmentary fossils in the Early and Late-Middle Devonian and resumes with a diverse fauna in the earliest Pennsylvanian. The recovery and characterization of the Mississippian insect fauna is one of the greatest challenges and discoveries awaiting insect paleontology. While in the Devonian, insects were flightless, terrestrial creatures, by the time their record resumes in the Late Carboniferous, there are numerous winged forms representing most of the major superordinal lineages, albeit as distinctly plesiomorphic forms. Thus, the Early Carboniferous, and perhaps the Devonian (Engel and Grimaldi, 2004a), witnessed the incredible radiation of winged insects.

By the Late Carboniferous several deposits are found throughout the world: the central United States as far east as Pennsylvania, southern England, western Europe, Moravia, Brazil, and Argentina (e.g., Sellards, 1904; Handlirsch, 1906b; Bolton, 1916; Pruvost, 1927; Carpenter, 1933, 1940, 1963c; 1970; Laurentiaux, 1952; Laurentiaux and Laurentiaux-Vieira, 1980; Durden, 1984, 1988; Pinto, 1986, 1990; Nelson and Tidwell, 1987; Shear *et al.*, 1992), but three deposits, in particular, have revealed the most significant material in terms of diversity.

Commentry and Montceau-les-Mines, France. Some of the earliest discovered and most famous Carboniferous deposits are the fossil beds of Commentry in Allier, France, and Montceau-les-Mines of Central Massif, France. Both deposits are of the same geological formation (dating from the lower Bashkirian) but represent rather different faunas (Burnham, 1981). It is from these deposits that the earliest glimpses of the giant insects (e.g., griffenflies = Protodonata) were obtained (Brongniart, 1884, 1893). The Coal Measures of the Commentry Basin include a remarkable diversity of early insects (Brongniart, 1878, 1884, 1885a,b, 1893; Bolton, 1917; Carpenter, 1943b, 1951, 1961, 1963a,b, 1964a; Kukalová, 1969a,b, 1970), particularly of the extinct lineage Palaeodictyopterida as well as "protorthopterans" (ancestral polyneopterous insects). Specimens are preserved in fine-grain sandstone at Commentry and in ironstone nodules at Montceau-les-Mines; both preserve remarkable detail of both wing and body characters.

Mazon Creek, Illinois. Among the most famous invertebrate fossils in North America are those preserved in ironstone concretions of the Carbondale Formation near Mazon Creek, Illinois (Shabica and Hay, 1997) (Figures 2.4, 2.42, 2.43, 3.7). The material is scattered in old coal strip mines and predominantly includes rather large and robust specimens (Nitecki, 1979). The deposits are of Upper Carboniferous age (ca. 300 MYA) and represent a coastal region during that time period (Baird, 1997). Numerous accounts have been written on the insect fauna preserved at Mazon Creek (e.g., Handlirsch, 1911; Richardson, 1956; Carpenter and Richardson, 1968, 1971; Carpenter, 1997; Kukalová-Peck, 1997).

Hagen-Vorhalle, Germany. The rich deposits in the former brickyard quarry of Hagen-Vorhalle in the Ruhr area of Germany are of lowermost Bashkirian age and are therefore tantalizingly on the borderline with the Early Carboniferous (Brauckmann *et al.*, 1994: ca. 315 MYA). Many specimens beautifully preserve body structures in addition to wing venation, making them critical for a broader understanding of early insect evolution. To date nearly two dozen species are recognized from five orders (Brauckmann and Koch, 1982, 1994; Brauckmann, 1984, 1986, 1988, 1989, 1991; Kukalová-Peck and Brauckmann, 1990; Brauckmann *et al.*, 2003) (e.g., Figures 6.28, 6.29, 7.1).

2.43. Prospecting for Carboniferous nodules at Mazon Creek, late 19th century. This formation yields the redeposited remains of coal swamps, including insects. Mazon Creek fossils include a remarkable array of terrestrial arthropods, particularly arachnids, myriapods, and insects. Photo: The Field Museum, Chicago; negative GEO 85145.

Permian (290–248 MYA)

The world of the Permian saw a steady decline in the intensely hyperoxic and tropical global climate of the Carboniferous. Even though the opening of this period was similar to that of the Carboniferous world, its close would be marked by the most traumatic cataclysm earth ever experienced (Kaiho *et al.*, 2001). Like the Carboniferous, relatively few deposits provide significant glimpses into the insect fauna at this time. In fact, there are five principal deposits – two from the Early Permian, three from the Late Permian – which have generally shaped our knowledge of Permian insects (discussed later). Other, lesser-studied deposits are from Colorado (Lesquereux, 1882; Scudder, 1890b), Texas (Carpenter, 1962), New Mexico (Kukalová-Peck and Peck, 1976), Brazil (Rosler *et al.*, 1981; Pinto and Pinto, 1981), Argentina (Pinto and Mendes, 2002), France (Gand *et al.*, 1997; Nel *et al.*, 1999; Béthoux *et al.*, 2001, 2002a,b, 2003), Rhodesia (Zeuner, 1955), Germany (Hörnschemeyer, 1999), Democratic Republic of the Congo (Pruvost, 1934), South Africa (Pinto and Ornellas, 1978; Geertsema and van der

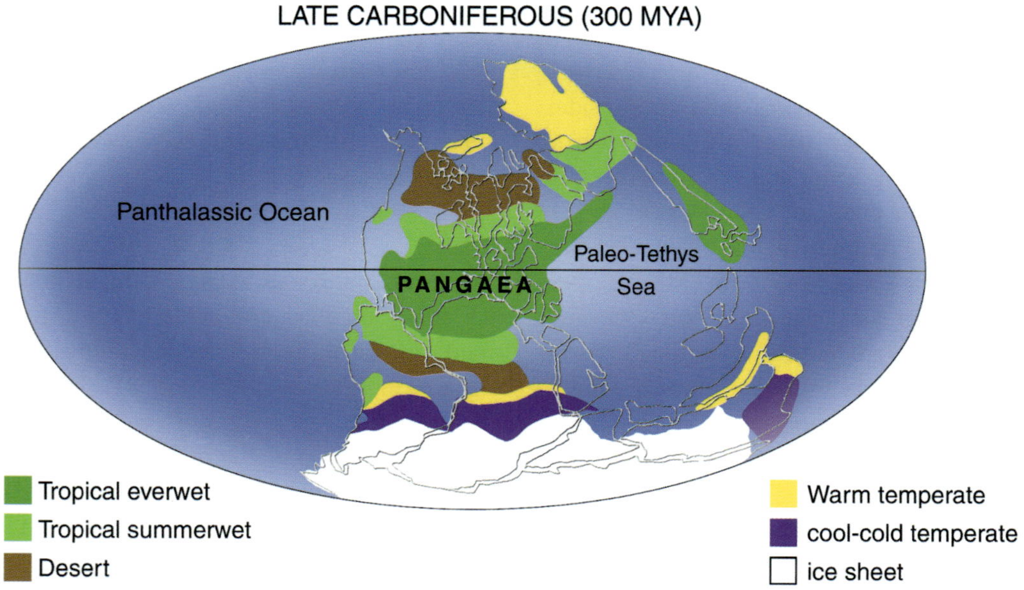

2.44. Continental configurations and climate during the Late Carboniferous. All paleomaps based on Scotese's *Paleomap* website and from Willis and McElwain (2002).

2.45. Frank Carpenter at the famous Permian insect locality in Elmo, Kansas in the late 1920s. First studied by R. J. Tillyard, the Permian insects at Elmo (e.g., Figures 6.9, 6.25) were then studied by Carpenter for nearly 60 years. Photo: Liz Brozius, Kansas Geological Survey.

2.46. The hills at Elmo, Kansas today, where there outcrops the world's most prolific and diverse deposits of Permian insects. These are found in the Early Permian Wellington Formation, which has yielded insects also at Midco, Oklahoma. The largest known insect, *Meganeuropsis permianum*, is from the Wellington Formation. Photo: M. S. Engel.

Heever, 1996), Kazakhstan (Vilesov and Novokshonov, 1994), Mongolia (Gorokhov, 1992), China (Lin, 1978; Lin and Han, 1985; Lin and Liang, 1988), and India (Srivastava, 1988).

Elmo, Kansas, and Midco, Oklahoma. Perhaps the most productive deposits in the world with Permian insects are those from the Wellington Formation of central Kansas (Figures 2.45, 2.46) and northeastern Oklahoma in the United States. These small lenses of fossils are approximately 267 MYO (Artinskian) and occur in at least three limestone layers clustered at the bottom of the series. The stratigraphy is poorly worked out, with the most significant reports being those of Dunbar (1924), Raasch (1940), Tasch and Zimmerman (1959, 1962), and Tasch (1962, 1963). The lenses with insects originated as lakes – some freshwater – while others were apparently playas. The entire region was coastal, with nearby bodies of water that freshened as the sea regressed. The insect fauna was extensively studied in the past by individuals such as Sellards, Tillyard, and most impressively by Carpenter. To date over 15,500 specimens have been amassed from Elmo, representing 150 species from 17 orders (Engel, 1998c; Beckemeyer, 2000) (no tally has yet been made for the sites in Oklahoma). Although work on these fossils essentially ceased in 1998, a great deal remains to be completed both in the basic descriptive and taxonomic work as well as more synthetic studies of the evolutionary-phylogenetic implications of these taxa. The largest insect ever, *Meganeuropsis permiana* (*M. americana* is a junior synonym), is known from the Wellington Formation.

Oboro, Czech Republic. The lacustrine mudstone of the Boskovice Furrow in Moravia, Czech Republic, is of either Early Artinskian (Kukalová-Peck and Willmann, 1990) or Sakmarian (Zajic, 2000) in age. Regardless, a diversity of Lower Permian insects has been recovered and described from these deposits (e.g., Kukalová, 1955, 1960, 1963, 1964, 1965, 1969c; Kukalová-Peck, 1975; Kukalová-Peck and Willmann, 1990; Carpenter and Kukalová, 1964), representing one of the most important Paleozoic insect localities in the world.

European Russia. Historically, the most extensively studied Permian locality is that of Tshekarda. The mudstones of the Koshelevka Formation were originally believed to be of earliest Permian (even latest Carboniferous) age, but have subsequently proven to derive from the later half of the Permian (Kungurian) (Ponomaryova *et al.*, 1998). The deposits are exposed along the Sylva River in the Urals of Russia. Soyana is another, albeit slightly younger (Kazanian), significant Upper Permian locality. Along the Soyana River in the northern Urals of Russia (Arkhangelsk region), the Iva-Gora limestones continue to yield new specimens.

Belmont, Australia. The fine-grained chert of the Newcastle Coal Measures near Belmont, Australia, are of Tatarian age (Kristensen and Wilson, 1986) and contain a limited, seemingly biased, group of insects heavy in primitive paraneopterans and mecopterids. The fauna has been most recently studied by Knight (1950), Evans (1947, 1958) and Riek (1953, 1968, 1971).

Natal, South Africa. In the Beaufort Series in southern Africa, principally near Natal, South Africa, along the Moori River, insects representing the Upper Permian (Tatarian) can be found in relative abundance. The stratigraphy of the Beaufort Series has been examined by Botha and Linstrom (1978). The fauna has not been extensively studied but material has been described by Riek (1973), van Dijk (1997), Geertsema and van Dijk (1999), and van Dijk and Geertsema (1999).

The Paleozoic was terminated by an event where approximately 85% of marine taxa and 70% of terrestrial taxa became extinct (Valentine *et al.*, 1978; Raup, 1979; Sepkoski, 1989; Erwin, 1993, 1994; Bentop and Tooitchett, 2003). The End Permian Event (EPE) is shrouded in mystery, much of the geological evidence for its cause obscured by the action of tectonics and time. Various theories account for the cause and the possible duration of the change, and it is likely that a combination of factors contributed to the extinctions. The EPE appears to have been confined to a period of approximately one million years (Bowring *et al.*, 1998) – most taxonomic extinctions occurred relatively quickly in both the marine and terrestrial environments (e.g., Erwin, 1993; Retallack, 1995; Eshet *et al.*, 1995; Rampino and Adler, 1998; Jin *et al.*, 2000). During the EPE there was significant geological turmoil: Extensive volcanism and basalt flows occurred, the supercontinent of Pangaea formed around 300 MYA, much of the continental shelf was lost, and the oceans began to regress (Holser and Magaritz, 1987; Wignall and Hallam, 1992, 1993, 1996; Wignall and Twitchett, 1996; Wignall *et al.*, 1996). Like the more famous Cretaceous-Tertiary boundary impact, the EPE has also been attributed at times to experiencing a massive extraterrestrial impact, as now seems to be the case for most major extinction events (e.g., Becker *et al.*, 2001; Kaiho *et al.*, 2001). Whatever its source, the Paleozoic-Mesozoic transition marks the single most pervasive extinction event for life, and for the insects as well. While insects suffered little when tetrapods met their demise 65 million years ago at the end of the Cretaceous, the Permian-Triassic event dealt everything a heavy blow.

Although a seemingly abrupt episode of upheaval on the planet, some dramatic changes were certainly more progressive and were rewriting the composition of our planet regardless of these catastrophes. Atmospheric concentrations changed significantly during this period, with oxygen levels dropping dramatically and continuously across the Permian from their previous hyperoxic state in the Carboniferous. Had no "event," whatever it may have been, taken place at the end of the Permian, significant changes in the flora and fauna must have already been in the process of shaping. For example, the giant insects present in early periods could not have continued to diffuse oxygen to their body core and would likely have become extinct naturally. Despite the End Permian Event and the devastating toll it weighed upon life, greater things were yet to come.

MESOZOIC

The middle episode in the history of life begins with the earliest Triassic, 247 MYA, and marks the end of the ancient, Paleozoic realm. There is perhaps no more dramatic transition in biotas than that between the Paleozoic and Mesozoic. Little is known of the effects of this extinction for the terrestrial biota during this time because the stratigraphic sampling for most terrestrial groups is grossly incomplete compared to that of marine invertebrates with durable, calcified shells. Preliminary indications from some groups, though, like vascular plants, suggest that there was a gradual replacement of floras from the late Permian to Triassic, not a cataclysmic extinction (Knoll, 1984). Overall, the fossil record clearly shows that the Mesozoic is a period of modernization for terrestrial life, including the insects.

Triassic (247–208 MYA)

The Triassic was dramatically different than the Permian, with a mean global temperature (mgt) of near 22°C, compared

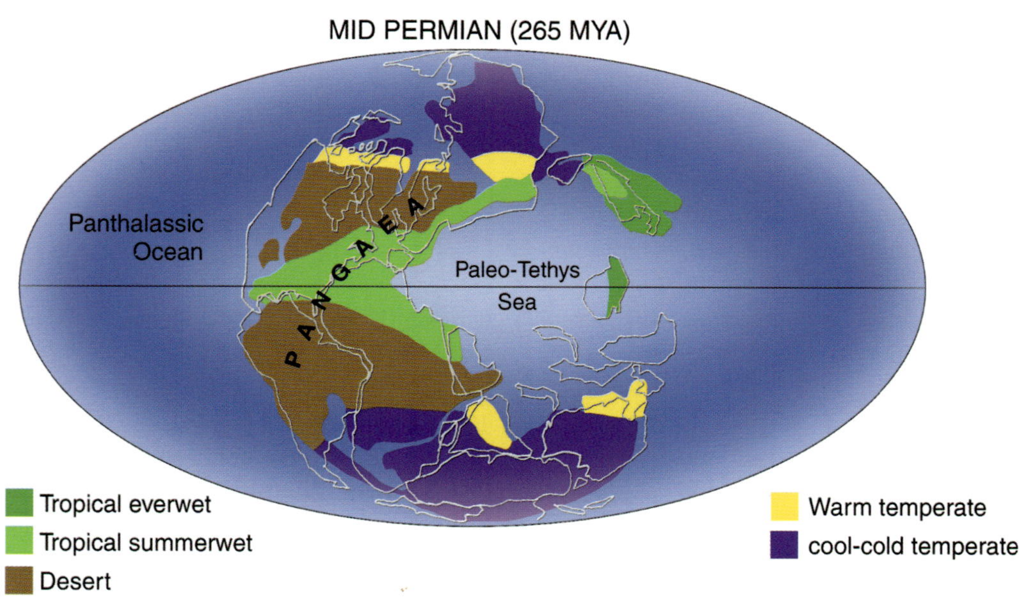

2.47. Continental configurations and climate during the Permian.

to the 12–15°C mgt in the first half of the Permian. No ice occurred at either pole, and floras changed from archaic lycopsids, ferns, cordaites, and pteridosperms, to radiations of cycads, ginkgos, conifers, and the angiosperm-like Bennettitales. Because insects are so intimately associated with plants, this floristic change probably affected the change in insect faunas in the Triassic. Though the effects of a Permo-Triassic extinction on insects is debated (which we discuss elsewhere), the first modern families like Tipulidae (Diptera), Staphylinidae (Coleoptera), Belostomatidae and Naucoridae (Heteroptera), and Xyelidae (Hymenoptera), to name a few, undoubtedly appeared during the Triassic. A review of Triassic life was presented by Lucas (1999). With the exception of the Bugarikhta Formation of central Siberia (which is latest Permian to earliest Triassic), very few insect deposits are known from the Early Triassic, ca. 247–241 MYA. Most deposits, in fact, are from the Carnian or later, 231–207 MYA.

Europe. Triassic insects have been known from Europe longer than in any other region, and the one most studied has been the Bundsandstein from Bavaria and Thuringia, Germany (Anisian to Carnian). The most significant European Triassic deposit, though, is the Grès à Voltzia from the mid-Triassic (Anisian: 240 MYA) of the Vosges mountains in France (Gall, 1996; Marchal-Papier, 1998). Diverse insects, arachnids, myriapods, marine worms, bivalves, crustaceans, fish, and plants occur in finely laminated clay and siltstone deposited in a shallow, brackish environment (Gall, 1971, 1985). Over 5,000 specimens, representing some 200 species and 11 orders are known (Marchal-Papier, 1998), though 40% of the individuals are roaches. Papers have been published on the Orthoptera (Marchal-Papier *et al.*, 2000), roaches (Papier *et al.*, 1994; Papier and Grauvogel-Stamm, 1995), and the oldest mygalomorph spider (Selden and Gall, 1992). Various deposits occur throughout the Keuper Basin in western Europe, which is generally Norian (222–209 MYO) in age. Near Bergamo in northern Italy, the lower Rhaetian (209 MYO) Argilliti di Riva di Solto Formation has yielded diverse odonates and some Coleoptera and Orthoptera (Whalley, 1986b; Bechly, 1997). These deposits have also yielded exquisitely preserved pterosaurs, which are the earliest in the fossil record. In southern Switzerland and northern Italy the Ladinian-aged Meride limestone (234 MYO) has yielded a few insects (Krzeminski and Lombardo, 2001), but this deposit is best known for the diverse vertebrates. The Triassic in Britain has yielded insects from Rhaetian-aged (209 MYO) deposits from Stensham (Hereford-and-Worcester) and Forthampton (Gloucestershire) (Popov *et al.*, 1994; Krzeminski and Jarzembowski, 1999).

Asia. The largest Triassic deposits are probably those of central Asia, in the regions of Kazakhstan, Uzbekistan, and Kyrgyzstan; vast collections from which reside in the Paleontological Institute in Moscow. Issyk-Kul', a 225 MYO lake bed in the Tien Shan mountains, has yielded over 3,000 specimens, from which B. B. Rohdendorf described 53 species of Diptera alone, though these need serious revision. This deposit has recently been reevaluated as being Early Jurassic. An extremely large and diverse deposit for insects is the Madygen Formation, from the Ladinian-Carnian (236–220 MYA) of the Fergana Valley of Uzbekistan, Kyrgyzstan, and Tajikistan (Martynova, 1958; Rohdendorft, 1961, 1962; Bekker-Migdisova, 1962; Sharov, 1968; Ponomarenko, 1969, 1977b; Papier and Nel, 2001) (Figure 2.48). Ponomarenko (1995) mentioned that the Madygen Formation yielded 15,000 insect specimens. Other central Asian deposits are the Maltsevo Formation of the Kuznetsk Basin in Siberia (Early Triassic), the Tolgoy Formation of western Kazakhstan (Norian to Rhaetian, ca. 210 MYA), and the Protopivskya Formation of southern Ukraine (Carnian in age). From the Asian far east Triassic insects are known from Hon Gay and Ke Bao Island in Vietnam, from Japan (Fujiyama, 1973, 1991), and from various localities in China. The China localities included Szechuan and Guizhous Provinces, the Tongchuan Formation in Shaanxi Province (Ladinian: 235 MYA), the Beishan Formation in Jilin Province, and the Shangu Formation in Hebei Province (these latter two are Rhaetian: 208 MYO) (Lin, 1982, 1986). Triassic deposits containing insects are actually fairly common throughout the Far East, but the remains consist largely of beetle elytra and roach tegmina. In the Japanese deposits, Fujiyama (1991) reported some 6,000 insect specimens recovered from the Momonoki Formation (Carnian: 225 MYO) at the Ominé Coal Field in Miné, Yamaguchi, Japan. Over half of these are isolated tegmina and elytra of roaches and beetles, about 20% are Auchenorrhyncha, and 10 orders comprise the remaining specimens.

2.48. The hills of Fergana Valley near the confluence of Uzbekistan, Kyrgyzstan and Tajikistan, which have yielded fossiliferous outcrops of the Triassic-aged Madygen Formation. This formation has prolifically yielded insects. Photo: Paleontological Institute, Moscow (PIN).

North America. Until recently the remains of Triassic insects from this continent were sparse and scattered. Earliest reports were of borings and galleries (Walker, 1938) in the wood of the conifer, *Araucarioxylon arizonicum*, the tree that

2.49. Fine-grained shales of the Late Triassic Cow Branch Formation, exposed here in the Solite Quarries near Martinsville, Virginia. This deposit has preserved the oldest definitive aquatic insect fauna, along with myriad other arthropods, preserved as two-dimensional, silvery images on a black shale (Figures 8.69, 10.26). Photo: Virginia Museum of Natural History (VMNH).

largely comprises the Petrified Forest National Monument in Arizona (Chinle Formation: Carnian, 231–222 MYA). More recent reports, of similar galleries, have reported these as nests of bees and termites, for which the Triassic is far too early. These galleries are almost certainly from beetles. Lucas (1999) mentioned rare and poorly preserved insects in the Bluewater Creek Formation of New Mexico, and a poorly preserved staphylinid beetle was reported from the Norian (220 MYA) of northern Virginia (Gore, 1988). Without doubt, the most significant deposit of North American Triassic insects is in Cascade, Virginia, on the Virginia–North Carolina border (Figure 2.49). Here there are exposures of the upper part of the Cow Branch Formation, which is a series of extremely fine-grained shales showing cyclical changes in sedimentation attributed to Van Houten cycles in climate (Olsen, 1986). These cycles are controlled by 21,000-year cycles of the precession of the equinoxes. The Cow Branch Formation is part of a series of Triassic- and Jurassic-aged rift basins from eastern North America called the Newark Supergroup (Olsen *et al.*, 1978). In a rich, fossiliferous quarry at Cascade more than 30 such cycles are known, one of which has yielded diverse insects preserved in great detail as silvery two-dimensional films. Some 11 orders, 30 families, and perhaps 60 species are presently known (Fraser *et al.*, 1996; Fraser and Grimaldi, 2003), but more excavation is still needed. The oldest Staphylinidae (Figures 10.26, 10.27) and aquatic insect fauna (e.g., Figure 8.69) are from the Cow Branch Formation.

South America. Triassic insects from South America were first known to occur in the Rhaetian-aged Potrerillos and Los Rastros Formations (uppermost Triassic: 209–207 MYA) of Mendoza and Los Rastros Provinces, Argentina, which also extend into southern Brazil (Wieland, 1925, 1926). Pinto and Purper (1978) described stoneflies (Plecoptera) from this formation, and an odonate was described by Carpenter (1960). Other insects were treated by Martins-Neto and Gallego (1999), and overall diversity was reviewed by Gallego and Martins-Neto (1999). This fauna was diverse, including odonatoids, plecopterans, miomopterans, grylloblattodeans, orthopterans, auchenorrhynchans, glosselytrodeans, and various undetermined species. Apparently, the Argentinian deposits are very similar to those of Australia's Triassic Ipswich Series (Martins-Neto *et al.*, 2003) and, no doubt, reflect a time when these continents were connected via Antarctica. Anderson and Anderson (1993) mentioned the Los Rastros deposit as being Carnian, based on paleobotanical evidence, though Martins-Neto and Gallego (1999) indicated a slightly older, Ladinian-Carnian age. A small deposit is also known from Rio Grande del Sul and Santa Catarina in Brazil (Pinto, 1956; Pinto and Ornellas, 1974). Most recently, Martins-Neto *et al.* (2003) reviewed the South American Triassic deposits.

Africa. Several vast deposits occur in southern Africa, including the Stormberg Series from the uppermost Permo-Triassic Karroo suite of Lesotho and Transvaal (Anderson and Anderson, 1993). Because this series straddles the P-Tr boundary, it is quite important for assessing the impact of the Permian extinctions on terrestrial arthropods. In Cape Province there are deposits at Birds River near Mount Fletcher, and the Molteno Beds (Carnian: 231 MYA), probably the richest gondwanan Triassic insect site (Zeuner, 1961; Riek, 1974a, 1976a,b). Riek (1974a, 1976a) largely described 32 species of insects in 22 families and 11 orders, based just on 70 specimens from the Molteno Formation. Anderson and Anderson (1993) mentioned that there is vast insect diversity based on new collections from the Molteno Formation, with some 335 recognizable species in 18 orders, based on 2,056 specimens (Anderson and Anderson, 1993; Anderson *et al.*, 1996). They estimated, however, using a Poisson distribution of species abundance, that there may actually be 7,740 species of insects preserved in the Molteno Formation! In lieu of published results it is difficult to assess the accuracy of the 335 recognizable species, which probably has a dramatic effect on the estimate. An estimate of 7,740 insect species seems extremely excessive, as few places on earth today probably harbor this kind of diversity (even the richest tropical forests), and the fossilized diversity of a region is always a fraction of the actual diversity. Molteno insects are preserved mostly as isolated wings. Oddly, there is an absence of Diptera in this formation, which are often among the most abundant orders in Mesozoic deposits, so there may be something peculiar about the taphonomy of the Molteno.

Australia. Diverse Triassic insects are preserved near Sydney, New South Wales, in mid-Triassic sandstones (Anisian: 240 MYA) (Riek, 1954) and in Late Triassic–Early Jurassic shales (Etheridge and Olliff, 1890). The major Triassic deposits are near Mt. Crosby in Queensland, northern Australia, at Dinmore and Denmark Hill (the Ipswich Series: Carnian) (Tillyard and Dunstan, 1916; Tillyard, 1917b, 1925; Tindale, 1945; Riek, 1955; Evans, 1956; Rozefelds and Sobbe, 1987). A small deposit occurs near Perth. Cockroaches, beetles, and auchenorrhynchans (Figure 8.42) predominate in the Australian Triassic, but there is also a unique diversity of early mecopteroids. Other orders include Neuroptera (Figure 9.15), Coleoptera (Figure 10.7), Odonata, Orthoptera, Phasmatodea, rare Hymenoptera (Figure 11.4), and the extinct order Titanoptera. In fact, some of the most spectacular insect fossils are the patterned wings of large, presumably predatory Titanoptera from the Australian Triassic (Figure 7.43). Smaller deposits of similar Triassic age occur in Tasmania (Riek, 1962).

Jurassic (207–146 MYA)

The Jurassic was a 62 MY period in the middle of the Mesozoic (207–146 MYA), which is now probably the poorest sampled period for fossil insects because the great bulk of Jurassic insect deposits are Palearctic. Gondwanan Jurassic insects are extremely sparse and require significant exploration.

Abundant evidence indicates that a meteoritic impact on earth at the Triassic–Jurassic boundary caused substantial extinction of marine and terrestrial organisms. For vertebrates, this had a profound effect because the extinction of labyrinthodonts and other archaic reptiles by the end of the Triassic, some 207 MYA, apparently allowed the ecological release and radiation of the dinosaurs in the Jurassic (Olsen *et al.*, 2002). Dinosaurs became much more diverse and larger in size less than a million years into the Jurassic. There seems to have been little differentiation, though, between insect faunas of the Late Triassic and early Jurassic. Significant floristic changes of the Jurassic include radiations of the cycads, ginkgos, bennettitaleans, and especially the conifers, with modern families of the last group appearing, like the Pinaceae, Taxodiaceae, and Podocarpaceae. In the Early Jurassic there was no ice, and the poles were cool temperate, all other regions having been warm temperate to tropical (Figure 2.50). By the Late Jurassic this changed, though, and limited ice formed at the poles, and climates were slightly cooler and more seasonal, no doubt a result of extensive rifting and continental separation. By the Late Jurassic, about 155 MYA, the supercontinent of Pangaea separated into Laurasia and Gondwana. Most Jurassic insects belong to extinct families or to stem groups of basal Recent families.

Europe. As would be expected by the history of paleontology, the earliest studied Jurassic insects are from Europe, particularly

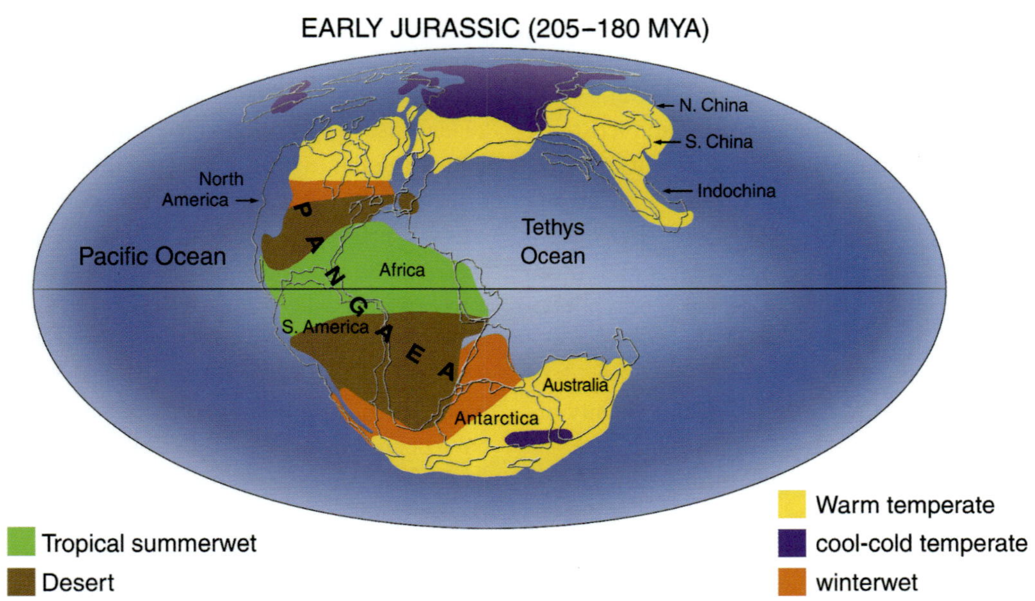

2.50. Continental configurations and climate during the Early Jurassic.

Germany and Britain (e.g., Brodie, 1845). Southern Britain has extensive deposits from the Lias (the first 27 MY of the Jurassic), which vary in age within this interval. Deposits from Gloucestershire, Warwickshire, and Worcestershire are slightly older (by about 20 MY) than deposits from Dumbleton, Aldertone, and other places. The best-studied deposit is from the Sinemurian-aged outcrops (ca. 200 MYO) from the cliffs at Charmouth, Dorset, which has been monographed by Whalley (1985). That study was based on just 400 specimens, which had been collected over many years since the insects are rare and widely scattered in the deposits. The Dorset Jurassic insects are preserved in a marine or deltaic deposit of calcareous mudstone and were allochthonous. Beetle elytra predominate (40% of all insects), with Orthoptera second in abundance (22%) among the 11 orders and 66 species known thus far. Perhaps the most significant find from Dorset is the oldest lepidopteran, discussed later.

In continental Europe, Lias insect deposits are widely distributed through Germany, Switzerland, and Luxembourg, which have been reviewed in Ansorge (1996, 2003b). A small deposit is known from the Early Jurassic (ca. 205 MYA) of Odrowaz, near Kielce, in central Poland. The most significant of the continental Europe sites are the deposits at Dobbertin (Mecklenberg), Schandelah (Saxony), and Grimmen (Vorpommern), the last of which has been monographed by Ansorge (1996). Ansorge's monograph is a model of systematic paleoentomology because it is based on careful observations, detailed documentation, and the reexamination of old types by Geinitz (1883), Handlirsch (1906a,b, 1907, 1908, 1939), and Bode (1953). The Grimmen deposits are marine clays from the Toarcian (150 MYA), with the insects preserved in carbonate concretions. As of 1996, 1,200 specimens representing 91 species were known, most of them isolated wings of small insects (<5 mm wing length), and allochthonous in origin. Like the Dorset Jurassic insects Orthoptera and Coleoptera were abundant at Grimmen, but at Grimmen Diptera are the most abundant insects, comprising 23% of all insects; Auchenorrhyncha are also an abundant order (Ansorge, 2003b).

Fossils from the rich deposits at Solnhofen and Eichstätt, Germany, have provided some of the first views of Jurassic life, which has been nicely reviewed by Barthel *et al.* (1990) and Frickhinger (1994). These deposits have been made famous by the six skeletal specimens of the oldest and most basal bird, *Archaeopteryx lithographica*, which is one of the premier examples of transitional forms in the entire fossil record (in this case between raptor dinosaurs and true birds). The fossils from these localities include a great diversity of vertebrates (bony fish, sharks and rays, turtles, ichthyosaurs, plesiosaurs, crocodilians, lizards, beautifully preserved pterosaurs, and a dinosaur), plants (seed ferns, ginkgos, conifers), marine invertebrates (mollusks, horseshoe "crabs" [Xiphosura], jellyfish, corals, squids, ammonites, various crustaceans, and echinoderms), and insects. They are preserved in very fine-grained, layered limestone, *Plattenkalke*, which has been quarried for millennia, even by the Romans. The fossils were preserved in micritic mud of calcite that settled to the bottoms of isolated, anoxic, and highly saline lagoons. The terrestrial organisms wafted in or flew from surrounding land. Unlike the vertebrates and crustaceans, the Solnhofen insects are not particularly well preserved, though they are usually complete (Figures 2.1, 6.43). Among the 12 orders and at least 50 genera of insects, significant examples include an impressive diversity of dragonflies, and the large insects *Chresmoda* and *Kalligramma*. The classification of *Chresmoda*, discussed later, has been controversial, entirely as a result of the typically poor preservation of details in Solnhofen insects. Anton Handlirsch provided many of the early descriptions of Solnhofen insects, and other, later studies on these insects include Carpenter (1932), Kuhn (1961), Ponomarenko (1985), Tischlinger (2001), and various papers on assorted taxa.

Asia. Jurassic deposits from Asia are extensive, and have been reviewed by Rasnitsyn (1985), Hong (1998), and by Eskov (2002), so we are providing a superficial overview here (Table 2.1). Eskov (2002) mapped some 45 Asian and Eurasian Jurassic insect localities, most of which lie in central Asia and China. The Eurasian and central Asian collections alone comprise approximately 50,000 Jurassic insect specimens from 20 major localities, housed in the Paleontological Institute in Moscow. These span the Early to latest Jurassic and thus provide a unique and nearly continuous fossil record of insect life from approximately 200 to 150 MYA. The most significant of all the deposits is the famous Karatau deposit (Figure 2.51), which is one of the truly great insect Lagerstatten. Without Karatau, our knowledge of Jurassic insects would be far more incomplete. Karatau is comprised of various outcrops in the Karatau range of mountains in southern Kazakhstan, which is a spur of the Tien Shan mountains. The main fossiliferous locality for insects is near the village of Aulie (formerly called Mikhailovka), from which 20,000 insect specimens alone were collected. The age is Kimmeridgian to Oxfordian, Late Jurassic (ca. 152–158 MYO). Insects are preserved in dark grey shales as isolated wings or entire specimens, usually in a detail so fine that even fine setae can be discerned on tiny specimens a few millimeters long (Figures 7.68, 9.7, 9.28, 10.12, 11.16, 12.4 to 12.6). The deposit is lacustrine and has preserved diverse plants (Doludenko and Orlovskaya, 1976), including diverse bennettitaleans, cycads, and conifers. The insects from Karatau have been intensively studied, as monographed in Rohdendorf (1968) and numerous subsequent papers. These include 19 orders and several thousand species. Coleoptera comprise approximately half of all insects (55%), then Diptera (14%), Blattodea (10%), Heteroptera (6.6%), other Hemiptera (3.3%), Orthoptera (2.2%), Raphidioptera (2.2%), Neuroptera (1.8%),

TABLE 2.1. Asian and Eurasian Jurassic Insect Deposits

Formation	Localities	Country/Region	Age	Notes	Reference
Jiaoshang	Hunan Province	China	Early		Hong, 1983
Shiti	Guanxi Province	China	Early	20 spp.	Lin, 1986
Beipiao	Liaoning, Hebei	China	Early	Very diverse	
Hansan	Anhui Province	China	Early/mid		Lin, 1986
Haifanggou	Liaoning Province	China	mid		
Jiulongshan	Beipiao, Hebei	China	mid	Very diverse	Ren, 1995
Sanjianfangzi	Xingjian Province	China	mid		
Houcheng	Hebei, Beijing	China	Late		
Dabeigou	Hebei Province	China	latest		
Phra Wihan	Phra, Nan Province	Thailand	mid?	Few	Heggemann *et al.*, 1990
Osinovskya	Chernyi Etap	East Siberia	Early		Rasnitsyn 1985
Unnamed	South Fergana	Kyrgystan	Early	Diverse	Martynov, 1925b, c
Dzhil	Sogyuty	Kyrgystan	Early	Diverse	
Cheremkhovo	Ust-Baley	Irkutsk Region	Early		Rasnitsyn, 1985
Sogul	South Fergana	Kyrgystan	Early/mid		
Karabastau	Karatau Range	Kazakhstan	Late	Exceptional	Rohdendorf,1968
Zhargalant	Altai Mountains	Mongolia	Early		Rasnitsyn, 1985
Unnamed	Uver-Khangay	Mongolia	mid	1,000 collected	
Togo-Khongor	Bayan-Khongor	Mongolia	mid/Late	5,000 collected	Sinitza, 1993
Shar-Teg	Gobi-Altai	Mongolia	Late	200 families	Ponomarenko,1998
Ulughey	S. Gobi Aymag	Mongolia	Latest	1,300 collected	Ibid.
Ulan-Ereg	Dund-Gobi	Mongolia	Latest	2,000 collected	Ibid.
Ichetuy	Transbaikalia	East Siberia	Early/mid	Diverse	Rasnitsyn, 1985
Itat	Yenissey River	Central Siberia	mid		Rasnitsyn, 1985
Cheremkhora	Angara: Iya River	Siberia	mid		Rasnitsyn, 1985
Bada	Mogzon Depres.	East Siberia	Late		Rasnitsyn, 1985
Uda	Transbaikalia	Central Siberia	Late	1,500 collected	Rasnitsyn, 1985
Glushkovo	Transbaikalia	Central Siberia	Latest/K		Rasnitsyn, 1988a

and various other orders with 1% or less. Among the more impressive insects are diverse odonates, roaches with long ovipositors (which is one of the latest occurrences of this feature in the fossil record), beautiful Kalligrammatidae (Figure 9.25) and Raphidioptera (Neuropterida) (Figure 9.7), diverse Brachycera, and some early Apocrita (Figures 11.10, 11.16). Karatau is a testament to how views of insect evolution can be dramatically affected by the discovery of just one exceptional fossil deposit.

North America. Though no Jurassic insect deposits were indicated from eastern North America in the map by Eskov (2002), the earliest such records from this continent were from the Early Jurassic of Massachusetts (Hitchcock, 1858). Sparse insects occur in lacustrine sediments of the ancient rift lakes of the Newark Supergroup, which is nicely reviewed and discussed by Huber *et al.* (2003). These include beetle elytra, a roach, abundant larvae (*Mormolucoides*), and various undetermined fragments. One type of beetle is *Holcoptera*, which is a dytiscid with distinctive patterning on the elytra and occurs from the Late Triassic to the Early Cretaceous. Assorted localities from western North America include two mid- and one Late Jurassic site. The Late Jurassic

2.51. A small outcrop of the Late Jurassic-aged Karabastau Formation at Karatau, central Kazakhstan. The Karabastau Formation is the world's most prolific source of Jurassic insects. Seventy years of study of the insects from Karatau by specialists at the Paleontological Institute in Moscow have revealed most of what we know about Jurassic insect life. Photo: Paleontological Institute, Moscow.

(Kimmeridgian: 152 MYA) Morrison Formation has preserved rare caddisfly cases. The Toldito Formation (Callovian: 160 MYA) of northern New Mexico has preserved nymphs of two species of predatory nepomorphan water bugs (Polhemus, 2000). The Sundance Formation of northern Wyoming and southern Montana (also Callovian) is probably the most diverse Jurassic deposit in North America for insects, albeit they are poorly preserved. They include about 15 species of aquatic Hemiptera (nepomorphs), Coleoptera (Dytiscidae, including *Holcoptera*), and rare trichopteran cases (Santiago-Blay *et al.*, 2001).

Antarctica. Jurassic insects from Antarctica are of great biogeographic interest because they should help reveal the nature of nonglaciated Antarctica, at a time when it was joined to the other southern continents. Isolated insect specimens have been recovered from strata of undetermined Jurassic ages at Mount Flora, Grahamland (Zeuner, 1959), and two sites in southern Victoria Land (Carapace Nunatak and Beardmore Glacier area) (Carpenter, 1969; Tasch, 1973, 1987).

The only other gondwanan Jurassic insects are found in a deposit from the Early Jurassic (Kotá Formation) of Andhra Pradesh in central India (e.g., Rao and Shah, 1959) (India was connected to Africa and Antarctica in the Jurassic), scattered occurrences from southern South America, and one locality each in Africa and New Zealand. Insects from the Jurassic of India are the most diverse yet known from Gondwana, and they include Auchenorrhyncha, Blattodea, Coleoptera, Diptera, Ephemeroptera, Heteroptera, Hymenoptera, and Neuropterida (Tasch, 1987; Mostovski and Jarzembowski, 2000).

Cretaceous (145–65 MYA)

Until about 30 years ago, the Cretaceous (145–65 MYA) was one of the poorest known geological periods for insects; now it is one of the best known. At least 25 major deposits of Cretaceous insects occur around the world (and at least as many less significant ones), but the Northern Hemisphere (with 97 Cretaceous deposits) has nearly four times as many known deposits as the Southern Hemisphere (23 deposits: Eskov, 2002). The Cretaceous is of exceptional biological significance for five main reasons, which we will discuss in greater detail later in this book.

1. This is the period when the origin and radiation of the angiosperms took place (approximately Hauterivian/Barremian to Turonian: 135–90 MYA). Angiosperms are the predominant life form on land, the diversity of which defines biomes from Arctic tundra to tropical forests. Because insects are intimately associated with plants, especially as pollinators and phytophages of angiosperms, the radiation of angiosperms appears to be both a cause for and an effect of contemporaneous radiations of insects.
2. The Cretaceous is when most of the Recent families of insects first appeared, many of which are probably associated with the angiosperm radiations, though not all. There are large groups, though, that largely radiated in the Cenozoic, like the families of ditrysian Lepidoptera and schizophoran Diptera. It is during the Cretaceous that there first appeared ants, termites, and vespid wasps – the three main groups of insects with advanced sociality.
3. The fragmentation of most of the land masses into present-day continents (though not their modern configurations) occurred in the Cretaceous. By the Early Cretaceous Pangaea had already separated into the northern (Laurasia) and southern continents (Gondwana), and further fragmentation of Gondwana, took place in the later part of the Early Cretaceous, approximately 115–110 MYA. This fragmentation dramatically affected global climate; during the Early Cretaceous it was very hot (particularly in the dry interiors of equatorial regions) but became more temperate and even seasonal in the Late Cretaceous.
4. The very end of the Cretaceous is marked by the most famous mass extinction event in earth history (though hardly the largest), since this is when the nonavian dinosaurs, ammonites, and rudist bivalves and some other marine life became extinct. It has been thoroughly established that a large meteorite crashed near the Yucatan Peninsula at or just before 65 MYA, and that it had global effects. What is not perfectly established is the effect of this catastrophe on the extinctions of dinosaurs and other organisms because many of these groups were in decline before 65 MYA. The end-Cretaceous, or K/T, extinctions appear to have had minor or only regional effects on insects, though there is some controversy about this.
5. During the Cretaceous, amber appeared in abundance as a remarkable preservation medium. Nodules of fossilized resin, or amber, are known since the Triassic, but they are small and scattered until the Cretaceous. For reasons that are not entirely clear, large quantities and globules of amber appear in the Early Cretaceous, ca. 130 MYA, formations of which then continue through the Tertiary. There are nine major Cretaceous deposits of amber that yield diverse insect inclusions. These deposits may be attributed to the origin and spread of certain resin-producing conifers in the Cretaceous. The family Araucariaceae (*kauri*, *Agathis*, etc.) is routinely implicated as the source of all or most Cretaceous ambers by a few workers, but for only a few deposits (e.g., Alava amber: Alonso *et al.*, 2000) is the evidence compelling. Taxodiaceae (redwoods and cedars) (Grimaldi *et al.*, 2000a,b), the extinct family Cheirolepidiaceae (Azar, 2000), and other families probably produced many Cretaceous ambers. Alternatively, and perhaps additionally, large

TABLE 2.2. Major Deposits of Cretaceous Insects

Formation(s)	Amber (A) / Compression (C)	Localities	Country/Region	Age[a] (MYA)	Diversity	Reference
Europe						
Lulworth	C	Purbeck Group	Southern England	140	Very diverse	Ross and Jarzembowski, 1996
Durlston	C	Purbeck Group	Southern England	140	Very diverse	Ibid.
Various	C	Wealden Group	Southern England	130	Diverse	Jarzembowski, 1984
Wessex+Vectis	A	Wealden Group	Isle of Wight	130	Modest	Nicholas *et al.*, 1993
Not named	A	Gröling	Austria	130?	Modest	
Montsec	C	Lleida Province	NE Spain	130?	Diverse	Martínez-Delclòs, 1991
Las Hoyas	C	Cuénca Province	Eastern Spain	125	Diverse	Martínez-Delclòs, 1991
Nograro	A	Álava	Northern Spain	115	Very diverse	Alonso *et al.*, 2000
Not named	A	Sarthe	France	100	Modest	Schlüter, 1978
Not named	A	Charente-Maritime	Western France	110	Diverse	Néraudeau *et al.*, 2003
Eurasia, Middle East, Asia						
Zaza	C	Baissa	Central Siberia	135	Exceptional	Zherikhin *et al.*, 1999
Turga	C	Semyon	Central Siberia	130?	Diverse	Zherikhin, 1978
Turga	C	Turga	Central Siberia	140	Diverse	
Byankino	C	Bolboy	Central Siberia	140?	Modest	Rasnitsyn, 1985
Emanra	C	Khetana River	Russian Far East	105	Diverse	
Arkagala	C	Arkagala	Russian Far East	95	Modest	Zherikhin, 1978
Olsk	C	Obestchayustchy	Madagan	95	Diverse	
Dolgan	A	Taimyr Peninsula	Northern Siberia	95	Diverse	Zherikhin and Eskov, 1999
Begichev	A	Taimyr Peninsula	Northern Siberia	110–95	Modest	Ibid.
Kheta	A	Taimyr Peninsula	Northern Siberia	85	Diverse	Ibid.
Dolgan-Kheta	A	Taimyr Peninsula	Northern Siberia	85–95	Diverse	Ibid.
Timmerdyakh	A	Yakutia	Eastern Siberia	95	Modest	Ibid.
Kempendyay	C	Yakutia	Eastern Siberia	140?	Diverse	Sinitshenkova, 1976
Tsagan-Tsab	C	East Gobi Aymag	Mongolia	140	Diverse	Rasnitsyn *et al.*, 1998
Khotont	C	Ara-Khangay	Mongolia	140	Diverse	Ibid.
Gurvan-Eren	C	Khovd	Mongolia	130	??	Rasnitsyn, 1986
Khurilt	C	Bon-Tsagan	Mongolia	125	Exceptional	Zherikhin, 1978
Shavarshavan	A	Caucasus Mtns.	Armenia	88	Modest	Zherikhin and Eskov, 1999
Agdzhakend	A	Caucasus Mtns	Azerbaidzhan	98	Modest	Ibid.
Not named	C	Kzyl-Zhar	Kazakhstan	90	??	Zherikhin, 1978
Various	A	Various locales	Lebanon	135–120	Exceptional	Azar, 2000
Aarda-Subeihi	A	Zerqa River	Jordan	110	Modest	Bandel *et al.*, 1997
Tayasir	C		Israel	130	Modest	Dobruskina *et al.*, 1997
Ora	C		Israel	90	Modest	Dobruskina *et al.*, 1997
Yixian	C	Liaoning, Hebei	NE China	130?	Exceptional	Ren, 1995
Laiyang	C	Shandong	NE China	130?	Very diverse	Zhang, 1985, 1989
Laoqun	C	Zhejiang	China	130?	Diverse	Lin, 1980
Lushangfeng	C	Near Beijing	China	125	???	Ren, 1995
Iwaki	C	Honshu Island	Japan	85	Modest	Schlee, 1990
Choshi	A	Honshu Island	Japan	125	Modest	Fujiyama, 1994
Kuji	A	Honshu Island	Japan	85	Modest	Schlee, 1990
Undetermined	A	Kachin Province	Myanmar	100?	Exceptional	Grimaldi *et al.*, 2002
North America						
Redmond	C	Labrador	Eastern Canada	100?	Modest	
Magothy	A	New Jersey	NE U.S.	90	Exceptional	Grimaldi *et al.*, 2000a
Foremost	A	Manitoba and Alberta	Western Canada	80	Exceptional	McAlpine and Martin, 1969

(*continued*)

TABLE 2.2. *(Continued)*

Formation(s)	Amber (A) / Compression (C)	Localities	Country/Region	Age[a] (MYA)	Diversity	Reference
South America						
Santana	C	Ceará	NE Brazil	120	Exceptional	Grimaldi, 1990a
Africa	C					
		Orapa	Botswana	90	Very diverse	Rayner *et al.*, 1998
Australia						
Koonwarra	C	Victoria	Australia	120	Very diverse	Jell and Duncan, 1986

[a] Ages are approximate.

formations of Cretaceous amber were formed when certain wood-boring insects radiated because it is well known that certain trees today produce copious resin in response to insect attacks. Since most species of insects are 3–4 mm in length or less, exquisite preservation of the myriad smaller species in amber has vastly improved our understanding of insect evolution. Also extensive formations of layered limestone were deposited during much of the Early Cretaceous, which resulted in exceptional insect Lagerstätten in Brazil, Spain, and elsewhere.

Significant Cretaceous deposits for insects are summarized in Table 2.2, with major deposits reviewed below.

Europe. The main European deposits of compression/impression-fossilized insects are from the Early Cretaceous of Britain (the Purbeck and Wealden groups) and Spain (Montsec and Las Hoyas), and for Cretaceous amber fossils the main deposits are from northern Spain and France.

The Purbeck Group of deposits from southern Britain is of exceptional significance because it is the only major, very diverse assemblage of insects of known earliest Cretaceous age (Berriasian, 145–138 MYO). The Purbeck is stratigraphically well constrained (Allen and Wimbledon, 1991), the paleoclimate is well characterized (fresh and brackish water lagoons with surrounding hinterlands of Mediterranean climate and flora [Allen, 1998]), and the fossil insects are well explored (Ross and Jarzembowski, 1996; Coram and Jarzembowski, 2002; Coram, 2003). As of 2003 there were 200 named insect species for the entire Purbeck, representing some 17 orders, with over 70 species from Wiltshire alone (Ross and Jarzembowski, 1996). The great percentage of insects are isolated wings and other disarticulated remains, principally Coleoptera elytra, then Hemiptera and Diptera. An autochthonous, brackish water aquatic insect fauna is indicated by the taxonomic composition, lithology, and somewhat depauperate nature of the fauna. Approximately 700 "morphospecies" have actually been collected from the Purbeck (Coram and Jarzembowski, 2002), and these authors even made estimates using abundance-diversity curves that 1,400 species may be preserved in the Purbeck. This high estimate should be taken with caution, though because their curves showed no obvious asymptote, and thus little basis for extrapolation. Also, careful study of one group of Purbeck insects, the roaches, indicates that there is actually one-third the number of described species of these insects (A. Ross, pers. comm.).

The Wealden is an extensive series of Early Cretaceous outcrops of mud- and siltstones from southern Britain (Figure 2.52) and limestones in Belgium and Germany (the famous iguanodons from Bernissart in Belgium are from the Wealden). In Britain the 'Weald Clay Group' is divided into an Upper (early Barremian: ca. 128 MYO) and a Lower Weald Clay (Late Hauterivian: ca. 130 MYO). The British Wealden yields diverse remains of vertebrates (including the occasional dinosaur fragment), plants, and disarticulated insects. Unlike the Purbeck, aquatic insects are uncommmon. Stratigraphy

2.52. Outcrops of the Early Cretaceous Wealden strata in England. The Wealden group was the first intensively studied assemblage of Cretaceous insects, and the various outcrops are also well dated and so insects from here are an important source of comparison for other Cretaceous deposits. Photo: Natural History Museum, London (NHM).

2.53. Fossiliferous limestone outcrops containing diverse insects from the Early Cretaceous of Las Hoyas, Spain. Photo: X. Martínez-Delclòs.

has been discussed by Jarzembowski (1977, 1984, 1987, 1991, Worssam (1978), Ross and Cook (1995), and Cook and Ross (1996),) has discussed the insects.

The finely grained, laminated lithographic limestones of Montsec (Lleida Province) and Las Hoyas (Cuénca Province) (Figures 2.53, 2.54) yield one of the most significant Cretaceous deposits of insects in Europe besides the Wealden and Purbeck. Las Hoyas is approximately Barremian in age (130 MYO) (Whalley and Jarzembowski, 1985; Martínez-Delclòs, 1989, 1991); Montsec has often been thought to be late Berriasian to early Valanginian (about 140 MYO), but some investigators attribute an early Barremian age (129–125 MYO) to this deposit. Both deposits have yielded a total of 13 orders and nearly 50 families of insects (reviewed in Peñalver *et al.*, 1999). Among these are numerous aquatic insects, such as Ephemeroptera nymphs and Belostomatidae, and abundant larvae of stratiomyid flies. Insects of particular significance are diverse odonates, Hemiptera, early aculeate wasps and weevils, large kalligrammatid lacewings, early alate termites (*Meiatermes*: Figure 7.82), and the oldest known worker termite.

The main European amber deposits of Cretaceous age have been discovered relatively recently. The first European deposit to be seriously studied is from Cenomanian-aged strata of the Paris Basin in western France (Schlüter, 1978, 1983). More recently, amber of late Albian and early Cenomanian ages (100–95 MYO) has been found in Charente-Maritime in southwest France, which is more abundant and has more diverse inclusions (Néradeau *et al.*, 2003). The French Cretaceous amber is very similar in age and composition to amber from Álava, in the Sierra de Cantabria mountains of northern Spain, about 30 km south of the town of Vitoria-Gasteíz (Alonso *et al.*, 2000). Álava amber is late Aptian to mid-Albian in age (115–120 MYO), and its chemical composition and association with fossil pollen indicates an araucarian source. Of nearly 2,000 insect inclusions in Álava amber, 13 hexapod orders are known, and 50% of all inclusions are Diptera, followed by Hymenoptera (28% – almost all parasitoids). These proportions are similar to those found in the French amber, though amber from Charente-Maritime has also preserved early ants, a mole cricket, scorpion remains, and other very rare inclusions. The French and Spanish Cretaceous ambers are additionally similar in that they are turbid, resulting from a suspension of fine bubbles and organic particles. The amber must be carefully trimmed close to the surface of the inclusion for optimal observation.

2.54. Deposits of similar age as Las Hoyas, from La Cabrua, Spain. Photo: X. Martínez-Delclòs.

Asia. Cretaceous insect deposits abound in Transbaikalia, the region of Siberia that is west of Lake Baikal, the world's largest freshwater lake (it even has an endemic species of seal). Some of these deposits are summarized in Table 2.2, and the most exceptional one is on the Vitim Plateau near a small tributary of the Vitim River, called Baissa Creek (reviewed by Zherikhin *et al.*, 1999) (Figure 2.55). Five expeditions of Russian paleontologists to "Baissa" between 1959 and 2000 have uncovered nearly 20,000 insect specimens from strata of approximately Hauterivian age (ca. 135 MYO). It is the only fossil insect locality in the world where nearly all of the pterygote orders of insects are preserved, the exceptions being Embiodea and Zoraptera. Nearly 300 species of insects have been described from Baissa thus far, and an estimated 700–1,000 species (and 200 families) are thought to exist (the estimate of 7,000 species [Vrsansky, 1999] is extremely excessive). Even an apparent louse (Phthiraptera) is preserved in this deposit, which we discuss later. Aphids comprise one

2.55. Outcrops of the Early Cretaceous deposits at Baissa, central Siberia, seen on the far shore. Compression fossil insects are extremely diverse in the Baissa deposits and are preserved with microscopic detail. Photo: Paleontological Institute, Moscow.

third of all the terrestrial insects from Baissa, which must reflect the luxuriant vegetation that is known to have surrounded ancient Lake Baissa, including dense conifer forests. Fossils from Baissa are extremely well preserved (Figures 6.10, 8.12); some Coleoptera and Heteroptea are preserved with relief (though most insects are flattened), and the cuticular remains have even preserved some sensilla. Bon-Tsagan in central Mongolia is the other major deposit of Cretaceous insects from Eurasia, deposits being approximately Aptian (120 MYO) in age. Some 10,000 fossil insects have been collected by Russian paleoentomologists from this site (Zherikhin, 1978).

Siberian amber derives from outcrops of various ages on the Taimyr Peninsula in northern Sibera, the most productive of them being Yantardakh ("amber mountain"), which is Santonian (ca. 85 MYO) in age. It has yielded fossiliferous amber containing some 3,000 inclusions, and is most abundant in chironomid midges and aphids. Yantardakh is located on the Maimecha River in the eastern part of the peninsula, and smaller outcrops of the same formation are known from the Kheta River. Older Siberian amber occurs likewise in the eastern part of the Taimyr, but in the Khatanga River basin (Albian to early Cenomanian, 110–95 MYO) inclusions are not abundant. In the western part of the Taimyr Peninsula at Nizhnayaya Agapa River (called just "Agapa") are fairly rich deposits of Cenomanian-aged (95 MYO) amber.

Middle East. Early Cretaceous amber occurs in numerous outcrops from Egypt, to Israel, Lebanon (Figure 2.56), and Jordan, the so-called "Levantine amber belt." Only amber from Lebanon has yielded insects in significant quantity and preservation, and a minor number of poorly preserved inclusions occur in Jordanian amber (Bandel *et al.*, 1997). The Jordanian amber is approximately 10–15 MY younger than the Lebanese amber, though there is significant variation in the ages of the latter depending on the formations and outcrops (see Azar, 2000). Lebanese amber is arguably the most scientifically significant amber in the world because it is the oldest amber that yields a great diversity of organismal inclusions. Ages of outcrops vary from latest Jurassic (though none of these contain insects) to upper Aptian (ca. 115 MYA). Most of the outcrops yielding insect inclusions, though, come from the upper part of the Neocomian, approximately Barremian (125 MYO). Lebanese amber was first seriously explored by Dieter Schlee (formerly of the Natural History Museum in Stuttgart), based on excavations from near Jezzine (Schlee and Dietrich, 1970). He discovered early bird feathers and a significant diversity of insects, some of which he and Willi Hennig had studied (e.g., Hennig, 1970; Schlee, 1970). Subsequent collections made by Aftim Acra of the American University in Beirut (also near Jezzine) and by Dany Azar of the Museum National d'Histoire Naturelle in Paris (from many other outcrops) have uncovered a trove of insect and other inclusions. Diptera comprise approximately half of all insect inclusions (most of these chironomids and ceratopogonids) (Figures 12.28, 12.30, 12.47), then Hymenoptera (mostly parasitoids: 6–11%). Among the more significant aspects of the insect fauna are very interesting aculeate wasps, brachyceran flies, beetles (Figures 10.6, 10.39, 10.59), early termites, and Lepidoptera (Figure 13.21). Various papers have been written on some of these inclusions, many of which we cite elsewhere in this book, but a great deal more research is needed. Lebanese amber is very fractured and brittle, so it requires special embedding techniques in order to trim the amber for observation of inclusions.

Far East Asia. Rich amber deposits from Burma (presently Myanmar) (Figure 2.57) have been known to be a source of material for carvings in Peking (Beijing) for several millennia,

2.56. Strata like this one from the Early Cretaceous of Lebanon are the world's oldest source of insects in amber. Amber from the Early Cretaceous occurs throughout the Middle East but only the material from Jordan and especially Lebanon has yielded insects.

and one collection of approximately 1,200 inclusions was made in the turn of the 20th century and housed at the Natural History Museum in London (Ross and York, 2000). An excellent history of its exploitation has been written (Zherikhin and Ross, 2000). Until about 1997, Burmese amber was thought to have been abandoned or depleted, but a larger collection of 3,500 inclusions in this material has recently been made (Grimaldi *et al.*, 2002). Burmese amber was originally believed to be Miocene to Eocene in age, but the study of insect inclusions indicate it is clearly Cretaceous (Rasnitsyn and Ross, 2000). In fact, comparison of Burmese amber insects to those from amber deposits with better dating indicate a Cenomanian age of this amber, approximately 95 MYO (Grimaldi *et al.*, 2002), which is corroborated from modest data based on pollen and an ammonite (Cruikshank and Ko, 2003 [these authors suggest a slightly older, late Albian to early Cenomanian age, 100–105 MYO]). Burmese amber is the richest Cretaceous amber deposit in the world. Among the rare inclusions are an onycophoran (Figure 3.3), primitive ants (Figure 11.70), the only Mesozoic Embiodea (Figure 7.13) and Zoraptera (Figures 7.15, 7.16), the oldest Strepsiptera (Figures 10.85, 10.86), a very primitive mosquito, and several archaic taxa from the earlier Mesozoic (Mesoraphidiidae [Raphidioptera], Protopsyllidiidae [Hemiptera: Figure 8.21], and Pseudopolycentropodidae [Mecopterida: Figure 12.3]). There are also very diverse Coleoptera and Diptera, and 27 hexapod orders in total, with approximately 130 families now known, representing some 300 species or more. Clearly, future exploration of these deposits will uncover some exciting discoveries.

The Yixian and Laiyang Formations yield the most abundant compression fossil Cretaceous insects in the Far East (reviewed by Ren, 1995; Lin, 1998). The insects are extremely diverse and beautifully preserved in light lacustrine and volcanic shales (Figure 12.2), but their significance is unfortunately overshadowed by the equally spectacular vertebrate finds from this area. The most exciting vertebrates include a beautiful specimen of *Jeholodens*, a stem-group triconodont mammal, and an unexpected diversity of early feathered dromeosaurs that have provided unique insight into the early evolution of birds. For insects, the Yixian and Laiyang insects include a vast diversity in most orders, including spectacular Neuropterida, Odonata, Mecoptera, and Hymenoptera. Unfortunately, the ages have been seriously confused. The Yixian Formation was originally promoted as latest Jurassic in age, so when beautiful angiosperm plants (*Archaefructus*) were discovered from the deposits, they were announced as the first Jurassic and earliest angiosperms (Sun *et al.*, 1998). Now it is generally agreed that the Yixian Formation is Early Cretaceous, probably Hauterivian to Barremian in age, approximately 130 MYO, which would place it nearly contemporaneous with the Wealden and Baissa. The Cretaceous dating is based on microfossils as well as isotopes (Barrett, 2000).

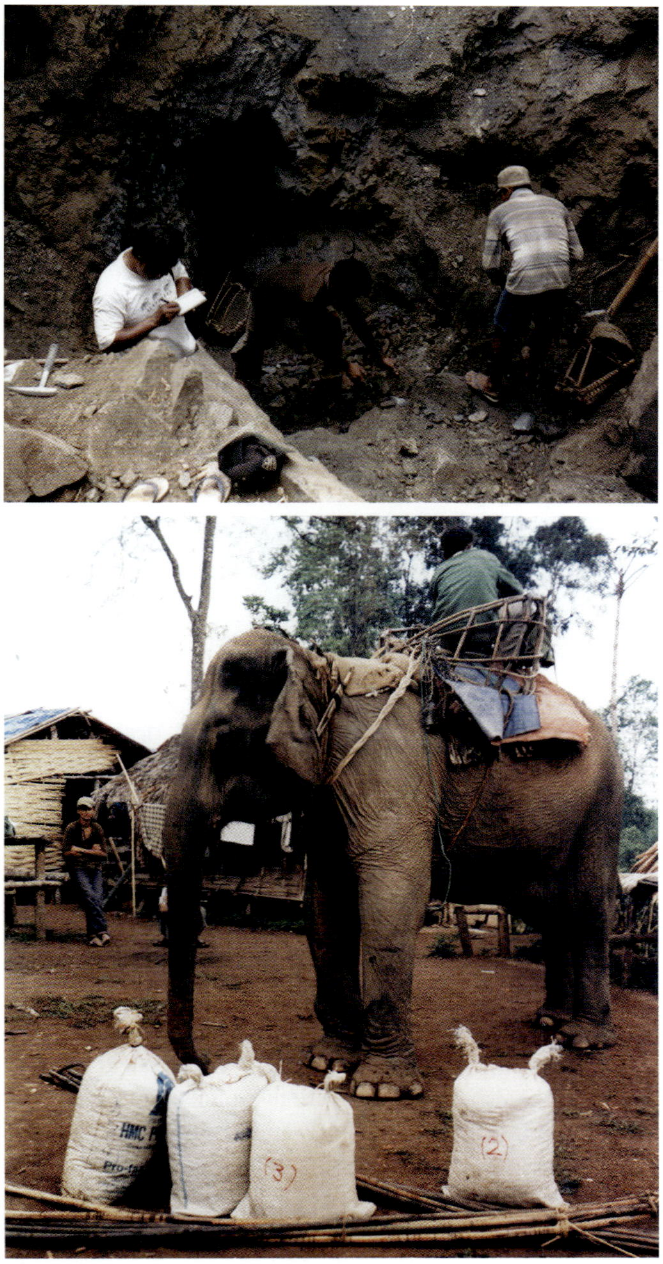

2.57. Excavations (above) of amber from the mid-Cretaceous of northern Burma, and transportation of the amber (below) in sacs to be loaded onto the elephant. Photos: Jim Davis, Leeward Capital.

This is a prime example as to how erroneous dating can seriously affect interpretations of evolution.

North America. The only truly diverse deposits of Cretaceous insects from this continent are of amber. The world's first major deposit of Cretaceous amber to be seriously studied, in fact, is from western Canada (Carpenter *et al.*, 1937; McAlpine and Martin, 1969). McAlpine and Martin (1969) listed nearly 40 localities of Cretaceous amber from western North America (including Alaska), but the main deposits came from two localities: Cedar Lake, Manitoba, and Medicine Hat, Alberta. Since then, another major site has been

2.58. Excavations of amber from the Late Cretaceous Magothy Formation of New Jersey. Virtually all amber deposits occur in sediments rich in black, fossilized peat and lignite, shown piled here. Lignites and peat are the remains of vegetation that were buried with amber in lagoons, estuaries, and along coastlines. Photo: K. Luzzi.

discovered, from Grassy Lake, Alberta (e.g., Pike, 1994, 1995). Most of the Canadian amber derives from the Foremost Formation, of Campanian age (ca. 75 MYO). In contrast to most other Cretaceous amber, Canadian amber has excellent clarity, though abundant flows require that pieces be trimmed to best observe the inclusions (Figure 10.84). The insect fauna in this amber is also distinctive, with nymphal aphids being the most abundant inclusions (one third to two thirds of all insects), with Diptera a close second (nematocerans are abundant in virtually all ambers). Also unusual is the abundance of mites, which comprise about a quarter of all inclusions. As of approximately the year 2000, some 150 species of hexapods in 17 orders and nearly 80 families were known from Canadian amber. This amber is believed to have been formed from an extinct species of tree in the Taxodiaceae.

2.59. An excavation of Cretaceous amber in New Jersey. Cretaceous amber occurs in the Atlantic Coastal Plain of the eastern United States from Massachusetts to Georgia, with particularly rich deposits like this one known from Staten Island and central New Jersey. Photo: K. Luzzi.

Surrounded by urban sprawl in central New Jersey is the other diverse Cretaceous deposit of insects for North America (Figures 2.58, 2.59). Scattered pieces of amber had been collected in Cretaceous clay pits in Staten Island and central New Jersey for more than a century, but the discovery of the first Cretaceous ant, *Sphecomyrma* (Figures 11.62, 11.63) brought serious attention to this amber (Wilson *et al.*, 1967). Subsequent study in the late 1980s and early 1990s uncovered rich, localized deposits from the Turonian-aged Magothy Formation (ca. 90 MYO) in Middlesex County, New Jersey (Grimaldi *et al.*, 2000a,b). Chemistry of the amber, and its match to amber preserved in wood and cone scales of fossil conifers, indicates that it was also formed by a taxodiaceous tree, or possibly by an early pine (Pinaceae). New Jersey amber is significantly more diverse than Canadian amber, which may reflect its lower (and more tropical) paleolaltitude and climate. It contains 19 orders and approximately 120 families and 250–300 species of hexapods (Grimaldi *et al.*, 2000a). Diptera comprised 34% of all inclusions, Hymenoptera 24%, Hemiptera 13% (most of these coccoids), and Coleoptera 8%. Among the more significant finds were the oldest fossil mushrooms, a tardigrade (Figure 3.5), early ants (including the only known Cretaceous formicine) (Figures 11.66 to 11.68), and an unexpected diversity of scale insects, Neuropterida (Figures 9.9, 9.33, 9.36, 9.37), and Lepidoptera (Figures 13.22, 13.28).

South America. The only major Cretaceous deposit of insects from South America – and arguably the largest and most diverse of the approximately 25 known gondwanan localities of insects – is from the Aptian-aged Crato Member of the Santana Formation in Ceará, northeastern Brazil (Grimaldi, 1990a). The deposit is a classic nearshore *Plattenkalke* deposit (Figure 2.60), except that the insects and other organisms are preserved as completely articulated permineralized replicas in remarkable relief and with microscopic detail. Even the fine structure of muscle tissues, like myofibrils, is well preserved (Figure 2.10). The Santana Formation has preserved a diverse autochthonous fauna of aquatic insects, chiefly Ephemeroptera adults (Figure 6.12) and nymphs (Figures 6.11, 6.13), nepomorphan waterbugs (Figure 2.9), and some odonates (Figures 6.39, 6.44 to 6.46), along with abundant roaches (Figures 7.70, 7.72), orthopterans (see cover), diverse terrestrial Hemiptera (Figures 8.47, 8.48, 8.79, 8.80), diverse Neuropterida, and other terrestrial groups from nearby vegetated areas. Thus far, approximately 300 species in 18 orders and approximately 100 families are known

2.60. The Tatajuba quarry in Ceará, Brazil, containing finely laminated limestones of the Early Cretaceous Santana (Crato) Formation. The Santana Formation is probably the richest source of Cretaceous insects from the Southern Hemisphere; the insects are remarkably preserved as detailed, mineralized replicas (e.g., Figures 2.9, 2.10, and many others in this book). Photo: J. Maisey.

(Grimaldi, 1990a; Martins-Neto, 1999). This number will clearly change when taxonomic specialists assess the voluminous descriptions by Martins-Neto. Among the more significant aspects of this insect fauna are the the oldest known Thysanura (Figure 5.6); the earliest blattid roaches with egg cases (oothecae) (Figure 7.72); a superb early mantis (*Santanmantis*) (Figures 7.97, 7.98); the only Southern Hemisphere snakeflies (Raphidioptera) (Figure 9.8); and a remarkable "long-tongued" brachyceran (*Cratomyia*) (Figure 14.15), which is one of the earliest specialized pollinators.

Africa and Australia. As the Santana Formation is to South America, so are Orapa to Africa and Koonwarra to Australia. Orapa was formed from the eruption of a kimberlite pipe and

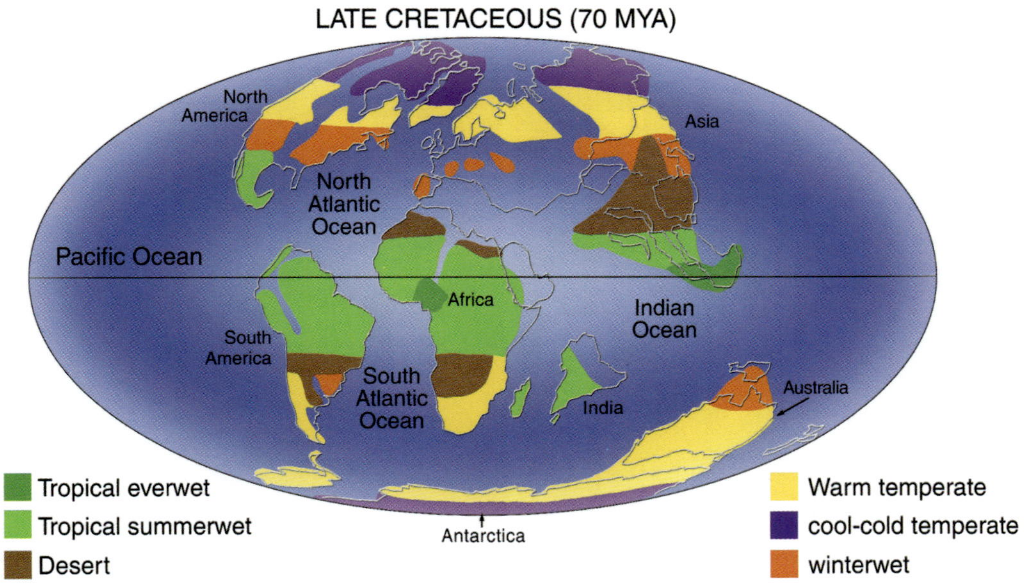

2.61. Continental configurations and climate during the Late Cretaceous. For additional reconstructions during the Cretaceous of the Southern Hemisphere, see Figure 14.27.

then sedimentary filling of a crater lake in Botswana during the Turonian-Coniacian (95–87 MYA). Preserved in the fine-grained shales there is a significant diversity of plants, arachnids, and insects in eight orders (McKay and Rayner, 1986; Rayner *et al.*, 1998). Of approximately 3,000 insects collected, reports exist thus far only for some Diptera (Waters, 1989a,b) and carabid beetles. Koonwarra is a shallow lacustrine deposit formed probably during the Aptian (120 MYA) in South Gippsland, Victoria, Australia, thus contemporaneous with the Santana Formation. The Koonwarra deposit was nicely documented in a monograph (Jell and Duncan, 1986), which shows a significant diversity of soft-bodied invertebrates, along with 12 orders of insects. Immature Epheme-roptera and Diptera are most abundant, but Hemiptera, Coleoptera, and Diptera are most diverse. Among the more significant insects are damselfly (zygopteran) nymphs, plecopteran nymphs, mesoveliid (Figure 8.65) and gelastocorid bugs (Heteroptera), diverse adult and larval Coleoptera, and abundant blackfly larvae (Simuliidae). There is also an exceptional specimen of an ectoparasite, *Tarwinia*, which appears to be closely related to modern fleas (Siphonaptera) (Figures 12.19, 12.20).

CENOZOIC

The Cenozoic, from 65 MYA to the present, is when modern insect faunas became refined. Many Recent insect families appeared in the Cretaceous, and some even appeared in the Jurassic or Triassic, but several diverse lineages of insects radiated in the Cenozoic: the "higher" mantises, termites, and scale insects (Mantoidea, Termitidae, and Neococcoidea, respectively); many ectoparasitic groups (fleas, lice, batflies); the schizophoran flies; bees and ants; and large lineages of phytophagous insects like the ditrysian Lepidoptera and phytophagan beetles. The Cenozoic is also the geological era for which we have the best fossil record for insects and all life; younger fossil deposits have been least destroyed by constant subduction, faulting, erosion, and other earth processes. Dramatic geological processes occurred during the Cenozoic that had great influences on biotic diversity and evolution. Uplift of some of the largest and highest mountain ranges (Himalayas, Alps, and Rockies) occurred during the Cenozoic, which then created deserts and arid grasslands in their rain shadows. Between 60 and 50 MYA was one of the warmest periods in earth history, and now tropical groups ranged nearly worldwide. In the Oligocene the continents reached their present positions (Figure 2.62), and this had tremendous impact on global climates and thus distributions of terrestrial organisms. For example, about 30 MYA the Drake Passage – the ocean passage between Australia, Antarctica, and South America – opened up, allowing the circulation of ocean currents around Antarctica. This allowed the cooling and glaciation of Antarctica, though it was not until the Pliocene about 5 MYA that the southernmost landmass became fully glaciated. Seeing Antarctica today it is difficult to imagine that only 10 MYA lush forests harbored a biota there similar to what is found today in New Zealand and Patagonian South America. Land bridges during the Pliocene and Pleistocene connected Europe, eastern Asia, and North America, and the biotas of North America and South America began to mix when the isthmus of Panama connected about 3 MYA. All these events, and more, are beautifully documented in the insect fossil record.

For North America, Cenozoic deposits with insects are entirely restricted to areas west of the Appalachian Mountains, most even being within or west of the Rocky Moun-

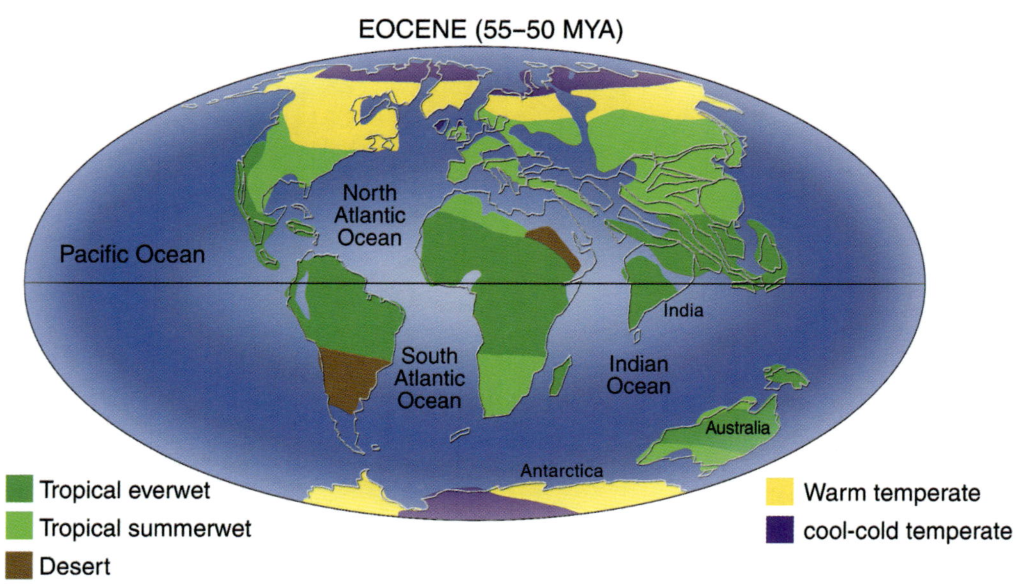

2.62. Early Cenozoic (Eocene) continental configurations and climates.

tains. The magnificent monograph by Scudder (1890a) on the North American Cenozoic insects is still an important reference. South America is sparse for Cenozoic deposits, probably because little prospecting has been done. Europe and Asia contain several very impressive deposits.

Nearctic and Palearctic

Paleocene (65–55 MYA). The Paleocene is a very poorly known period in the geological record of insects. Though major lineages of insects, like families, were largely unaffected by the mass extinctions at the end of the Cretaceous, we have very few details about how insect faunas responded to this cataclysm (see, for example, Labandeira *et al.*, 2002). North America was in the direct wake of the ejecta from the giant meteorite that fell at Chixculub, Mexico 65 MYA, so Paleocene insect faunas from North America would be particularly interesting.

The most significant Paleocene insect deposit in North America is from the Paskapoo Formation of Red Deer River, near Blackfalds, Alberta (Mitchell and Wighton, 1979; Wighton, 1982), a region that was probably protected within the shadows of the Rocky Mountains when the meteorite hit. Insects from the Paskapoo Formation are preserved as detailed, carbonized compressions in fine-grained, calcitic limestone. The insects were largely autochthonous, with a preponderance of aquatic insect larvae (including many beetles); 8–9 orders and about 20 families are reported.

In Eurasia, the Paleocene is slightly better represented than in North America, albeit it is still underrepresented. The two most significant deposits with insects are those of the lacustrine shales of Menat in France (Gaudant, 1979; Olliveier, 1985), and the marine diatomites of the "Mo Clay" in the Fur Formation (Thanetian) of Denmark (Willmann, 1990c; Thomsen and Schack-Pedersen, 1997; Rust, 1999) (e.g., Figure 2.63). Perhaps one of the most significant discoveries from the Paleocene of Europe has been the discovery of silky lacewings (Andersen, 2001), a lineage today known only from southern Africa, southeast Asia, and Australia. Among the most interesting fossils of the Eurasian Paleocene are those preserved in amber from Sakhalin Island in the Russian Far East (Zherikhin, 1978). The amber has been found in the Due Formation and is of approximately Late Paleocene (Thanetian) age. The major collection of Sakhalin amber is located in the Paleontological Institute, Russian Academy of Sciences, Moscow.

Eocene (55–38 MYA). Climatically, the Eocene is the most dramatic period in the Tertiary. Changes during this time had profound impact on the global distributions of insects. During the Early Eocene no ice occurred on earth, even at the poles, and tropical organisms ranged to the highest latitudes. Lemurs and crocodiles roamed among forests where arctic tundra is today. By the end of the Eocene and the early part of the Oligocene, the glaciation of Antarctica had begun. The Eocene is particularly well represented in North America, along with the Miocene. The farthest east Eocene insects occur in North America is Kentucky; all other sites are restricted to western North America.

GREEN RIVER FORMATION. This is one of the largest fossil lake systems in the world; it is some 65,000 km^2 in area and 600 m thick, and is also among the most prolific sources of compression fossils in the world, including insects. An informative review of the Green River Formation was provided by Grande (1984), which focused mostly on the fishes. The formation was formed by three paleolakes in what are now eastern Utah, southwestern Wyoming, and western Colorado: "Fossil Lake" (the smallest, Early Eocene), "Lake Gosiute" (Early to mid-Eocene), and "Lake Uinta" (the largest, Late Paleocene to Late Eocene). "Green River" is perhaps best known for the diverse and beautifully articulated fish, some birds, reptiles (including a boid snake), and the oldest known bat, *Icaronycteris index*. Most of the fossil insects derived from the U-2 (or "Ray-domed") and the U-4 (or "Bonanza") localities, both in the Parachute Creek Member of Lake Uinta. Vertebrates that are so abundant and diverse elsewhere in the Green River Formation are scarce in these localities, but plants are diverse and well studied (MacGinitie, 1969). Thus, the paleofloral context of the insect fauna is well known. The flora indicates that the paleoenvironment was warm temperate to subtropical, represented by species that are a curious

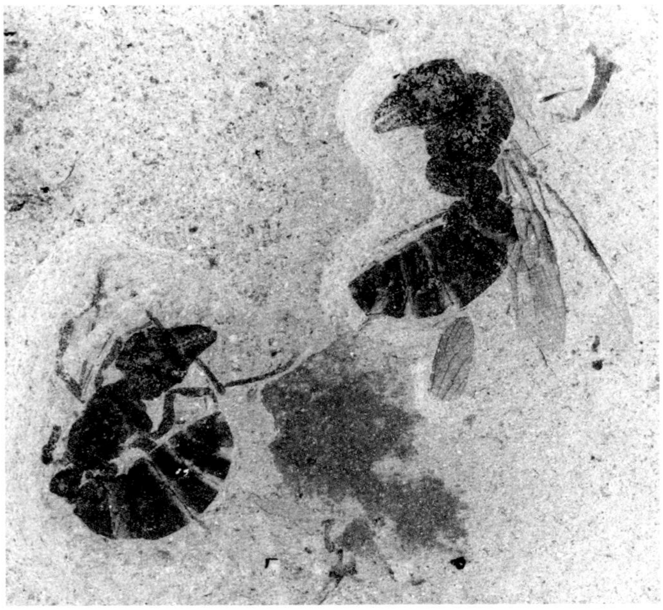

2.63. The Mo-Clay of Denmark is one of the few diverse deposits of Paleocene insects in the world, and these giant ants are distinctive to the deposit. Assessing the impact of the Cretaceous-Tertiary extinctions on insects will require discovery of more Paleocene deposits. Photo: Zoological Museum, University of Copenhagen.

2.64. A bibionid (March fly) from the Eocene of Washington state. Bibionids are extremely common and diverse in Cenozoic compression deposits of the Northern Hemisphere. University of Washington Burke Museum (UWBM); length 5.1 mm.

mixture of tropical and temperate, Asian and North American. Approximately 14 orders, and nearly 100 families and 300 species of aquatic and terrestrial insects are recorded from the Green River, among the most famous of which are several very rare butterflies. T. D. A. Cockerell described many of the Green River insects. Large historical collections of Green River insects are in Harvard's Museum of Comparative Zoology; another large collection is at the Smithsonian Institution.

OKANONGAN HIGHLANDS. Eocene outcrops in western British Columbia and northern Washington state have produced diverse plants and insects (e.g., Figure 2.64), preserved in mudstone, in shales, and in volcanic tuffs formed during periods of extensive volcanism. Most of the deposits belong to the Early to mid-Eocene Allenby Formation (48–50 MYO); the one from Republic, Washington, is in the contemporaneous Klondike Mountain Formation. Insect fossils from this region were known since the time of Scudder (1890a), but the deposits are still being explored for insects (Wilson, 1977, 1978; Lewis, 1992; Douglas and Stockey, 1998; Archibald and Mathewes, 2000) and plants (Wolfe and Wehr, 1987; Wehr, 1996). The paleoflora from the Klondike Mountain Formation, in fact, is the richest one known for the Eocene of western North America (approximately 250 species and 23 families of angiosperms alone), even more diverse than the Green River Formation. Paleoclimate of the Okanongan Eocene was more temperate (12–13°C mean annual temperature) than the Green River (15–21° C), and is a particularly early record of montane conifer forests. The insects have not been studied as long as those from the Green River Formation, but systematic study of insects from Quilchena, British Columbia, for example has revealed at least 13 orders and 40 families thus far (Archibald and Mathewes, 2000).

CLAIBORNE GROUP. Two thick clay formations in this mid-Eocene unit of the Gulf Coastal Plain have yielded significant fossil insects: the Claiborne Formation of Arkansas and the Tallahatta Formation in Mississippi (the Holly Springs Formation may also belong here). The Tallahatta Formation contains diverse compressions of leaves, flowers, seeds, fruits, and various terrestrial and freshwater aquatic arthropods, including six orders of insects. Among them are abundant caddisfly (Trichoptera) cases (Johnston, 1993, 1999). In very large clay pits in Malvern County, Arkansas, are deep exposures of the Claiborne Formation, within which is abundant fossiliferous amber (Saunders *et al.*, 1974). The amber occurs as large (1–5 cm), rounded lumps with a thick weathered rind. The translucent (rarely transparent) interiors of the amber contain an array of arthropod families typical for Tertiary amber; even several flowers have been found. Quality of preservation is not very good, no doubt as a result of the chemistry of the original resin. The amber was not formed by a conifer, but probably from a dipterocarp-like tree. This is the most productive of only several known Tertiary deposits of fossiliferous amber in North America. Despite its significance it has not been fully exploited.

FLORISSANT FORMATION. The first major fossil insect deposit to be studied in North America is the now famous Florissant fossil beds (Figure 2.65), or Florissant Formation, a deposit from the latest Eocene to lowermost Oligocene (Evanoff *et al.*, 2001), approximately 100 km south of Denver, Colorado, 9,000 ft (3,000 m) in elevation. Prolific study of Florissant began as early as 1873 by S. H. Scudder (1878a, 1889, 1890a) and then by T. D. A. Cockerell, A. L. Melander, and many others. Nearly 200 families and 1,100 species of insects, and 140 species of plants, are known from the Florissant beds, though the taxonomy of some groups (particularly the beetles) require restudy (Meyer, 2002). Ancient Lake Florissant was formed around 38 MYA by volcanic mudflows that dammed a river valley. One to two million years later, repeated eruptions produced thick ash that blanketed the area, preserving organisms on fine-grained shales in exqui-

2.65. Historical photo, c. 1890s, of the outcrops excavated by Samuel Scudder, at Florissant, Colorado. These outcrops are now known to have been formed in the latest Eocene to earliest Oligocene, 38 MYA. The Florissant Formation has been among the most prolific of all fossil insect deposits, perhaps second only to the deposits of Baltic amber. Photo: H. Meyer, U.S. National Park Service.

site two-dimensional detail, including minute setae, eye facets, and scales on the wings of Lepidoptera. As one early entomologist put it, ancient Florissant was an "insectan Pompeii." The area was verdant and forested; some of the plants have relatives found only in southeast Asia today, like *Koelreuteria* (rain tree) and *Ailanthus* (tree of heaven). Just like the evolutionary history of horses, fossils throughout western North America reveal that *Ailanthus* was once a native of North America, but then it became extinct on that continent. It is now back, but as an introduced weed that grows amidst concrete in northeastern U.S. cities. North American "castaways" among Florissant insects are many, the most famous including spoon-winged lacewings (*Marquettia americana* [Nemopteridae: Figure 9.18], the genus also occurs in the Ruby River beds), and tsetse flies (Glossinidae: Figure 12.106). Nemopteridae are now found only in Australia, South America, southern Eurasia, and Africa, but the most diverse are found in Africa. Tsetse flies from Florissant were twice the size of extant species, which occur exclusively in Africa. Butterflies are exceptionally rare in all the fossil deposits in the world, but Florissant has yielded 12 species – far more than any other place (e.g., Figures 4.12, 13.65, 13.66). The unique specimen of *Prodryas persephone* (Figure 13.66) was shown to Frank Carpenter as a boy, who was so impressed with the exquisitely preserved wing veins and color patterns that it sparked an interest that developed into a 70-year career in paleoentomology. Much of the Florissant Formation outcrops became officially protected in 1969, as the Florissant Fossil Beds National Monument (*http://www.nps.gov/flfo/paleopage.htm*). Major collections of Florissant insects are at the Museum of Comparative Zoology, Harvard University; University of Colorado; and the American Museum of Natural History, New York.

Assorted other Eocene deposits have yielded sparse remains of insects, the most significant being the Wilcox Group. The Wilcox Group is comprised of the Wilcox Formation, exposures of which have yielded Coleoptera and leaf-mined and galled leaves in Arkansas and Kentucky. Some place the Holly Springs Formation in this Group (alternatively in the Claiborne Group). Exposures of the Holly Springs Formation have yielded Coleoptera in Arkansas, and five orders of insects from several localities in Tennessee. Among the more significant Holly Springs Formation insects are a giant ant, *Eoponera berryi* (wingspan of 57 mm [2.4 in.]), and the termite *Blattotermes*, which belongs to the relict family Mastotermitidae, extant in Australia.

Across Eurasia, deposits with insects abound; however, most of these localities preserve relatively few fossils, with low diversity and unimpressive preservation. Several of the world's most prolific deposits of Eocene insects, though, are from northern Europe. Additional sites of significance include Monte Bolca, Italy (Lutetian); Pesciara di Bolca (Lutetian); and Menat, France (Ypresian).

MESSEL AND ECKFELD, GERMANY. The central European fossil Lagerstätten Eckfeld and Messel have been of special interest to paleontologists because their diversity includes almost everything from lithified bacteria to articulated mammals with gut contents and soft tissue preservation (e.g., Schaal and Ziegler, 1992; Lutz and Neuffer, 2001). Specimens are preserved in a fine oilshale with little or no turbation. The diversity of insects at both localities is high, with most specimens preserving coloration and minute structural details (Lutz, 1990, 1992; Tröster, 1992, 1993). Eckfeld represents the Middle Lutetian, with an established age of ca. 44.3 MYO (Franzen, 1993; Mertz *et al.*, 2000). Messel is slightly older, marking the lowermost Lutetian (ca. 49 MYO). Not surprisingly, several taxa occurring in the Messel and Eckfeld deposits are similar to those found in the somewhat contemporaneous Baltic amber (e.g., Wappler and Engel, 2003).

The crater of the Eckfeld Maar near Manderscheid, Eifel, Germany, originally had a diameter of 900 m and a depth of about 170 m. The depth of the maar lake initially exceeded 110 m and might have reached 150 m (Pirrung, 1992, 1998; Fischer, 1999; Pirrung *et al.*, 2001). Anoxic, alkaline conditions, and a raised content of electrolytes help to explain the perfect preservation of both the lamination of the oilshale and of the fossils (Wilde *et al.*, 1993; Mingram, 1998). To date more than 30,000 fossils have been excavated, all of which document a highly diverse terrestrial flora and fauna (e.g., Schaal and Ziegler, 1992; Neuffer *et al.*, 1996; Lutz *et al.*, 1998; Wilde and Frankenhäuser, 1998; Lutz and Neuffer, 2001; Wappler and Engel, 2003).

Messel is located on the eastern shoulder of the northern Rhine rift valley. The Messel Pit today has a diameter of 700–1000 m. Despite the fact that Messel's perfectly preserved fossils have been studied for nearly a century (e.g., Ludwig, 1877; Lutz, 1990; Tröster, 1991, 1992, 1993, 1994; Hörnschemeyer, 1994), its formation was a matter of considerable debate until recently (Schaal and Ziegler, 1992; Pirrung, 1998; Liebig and Gruber, 2000). New geophysical data (Harms, 2002) demonstrate that the Messel locality was a maar-like Eckfeld (Lutz *et al.*, 2000; Pirrung *et al.*, 2001), as had previously been proposed by Pirrung (1998). However, with an initial diameter of about 1,500 m, the Messel maar was certainly considerably larger than Eckfeld.

BALTIC AMBER. The most productive and extensive amber deposits in the world stretches across northern Europe in what is known as the *blau Erde* ("Blue Earth"). Pockets of amber can be found from Denmark and Sweden into Lithuania and south into Germany and Poland, although the greatest concentration can be found on the Samland Peninsula. The formation containing the amber runs approximately 45 m below the surface and approximately 5 m below sea level. Amber frequently erodes from the shores of the Baltic and can be carried as far as the East Anglian coast. Stratigraphic studies of the blaue Erde indicate it to be middle

Eocene (ca. 44 MYO) in age (Kosmowska-Ceranowicz and Müller, 1985; Kosmowska-Ceranowicz, 1987), a dating congruent with K-Ar radiometric measures (Ritzkowski, 1997). The Albertus Universität in Königsberg (present day Kaliningrad), situated near the base of the Samland Peninsula, once held the most significant and largest collections of Baltic amber, with inclusions numbering near 100,000 at one time. Most of this material was, unfortunately, lost or destroyed during World War II, although a surviving portion is preserved in the Institut und Museum für Geologie und Paläontologie in Göttingen. The Saxonian amber (also known as Bitterfeld amber) is actually contemporaneous and chemically identical to Baltic amber. Although now located in Miocene deposits, Saxonian amber has been shown to be Baltic amber that had eroded and been redeposited in Miocene sediments (Weitschat, 1997). Ukrainian amber, of slightly younger age, was perhaps also part of the Baltic amber forest, although at an extreme of the distribution, having a slightly different faunal composition. The same can likely be said for the Paris Basin amber (Oise, France) of slightly older age (Ypresian) (Plöeg *et al.*, 1998).

Baltic amber has historically been considered the resin of an extinct species in the pine family (Pinaceae). Even though some authors have argued against this hypothesis based on chemical analyses and favor an araucarian origin (e.g., Langenheim, 1969), a pinaceous origin for Baltic amber has gained additional support. Among the thousands of inclusions in Baltic amber are numerous cones and needles of pines (e.g., Weitschat and Wichard, 1998; Wichard *et al.*, 2002) as well as wood fragments with microstructural details indicative of Pinaceae (Pielinska, 1997; Turkin, 1997). No araucariaceous inclusions have ever been found in Baltic amber. Moreover, it has been discovered that both extant and extinct species of the pine genus *Pseudolarix* produce succinic acid; indeed, *Pseudolarix* resin is chemically similar to Baltic amber (Anderson and LePage, 1995). Today *Pseudolarix* occurs in Asia but included at least the Arctic during the Eocene. Numerous Baltic amber fossils have relationships to taxa today occurring in southeast Asia or sub-Saharan Africa (Larsson, 1978; Engel, 2001a). Indeed, Baltic amber was perhaps produced by taxa related to *Pseudolarix* or even by an extinct species of this genus. Preservation of the insects in Baltic amber generally does not have the fidelity seen in some other ambers, such as Dominican amber. Insects in Baltic amber commonly have a milky coating (e.g., Figure 10.88), which is a microscopic froth that exuded from the bodies during decomposition.

Many tons of Baltic amber have been gathered by humans for nearly 13 centuries, but systematic study of the fossils did not begin until the 1700s (Sendel, 1742). Several thousand species of insects alone have been described from Baltic amber, making this *the* most prolific source of fossil insects (Figures 11.84 to 11.86, 11.93, 12.40, 12.62, 12.73, 12.103, 14.37). Major summaries of the fauna include Larsson (1978), Weitschat and Wichard (1998), and Wichard *et al.* (2002).

Oligocene (38–23.6 MYA)

CREEDE AND ANTERO FORMATIONS. Approximately contemporaneous with Florissant are two other Colorado formations also formed by lake deposits and volcanic ash. Outcrops of the Creede Formation are approximately 100 mi (160 km) northeast of Florissant, near Pike's Peak. The paleoflora was dominated by conifers, not by angiosperms as at Florissant. It is quite productive, with some five orders and eight families represented among 2,000 specimens housed in the Museum of Comparative Zoology, Harvard. Outcrops of the Antero Formation in South Park, Colorado, have been less productive, though it has preserved an *in situ* aquatic fauna dominated by Heteroptera and Diptera. Some of the insect species in the Creede and Antero Formations also occur in the Florissant Formation (Carpenter *et al.*, 1938; Durden, 1966).

KISHENEHN BASIN. Outcrops of the Kishenehn Basin of northwestern Montana and southeastern British Columbia contain abundant plants, terrestrial and aquatic gastropods, fishes, mammals, and 6 orders and more than 14 families of terrestrial insects (Constenius *et al.*, 1989). The Kishenehn deposits were dated by K/Ar and fission track methods as approximately 30 MYO (mid-Oligocene).

RENOVA FORMATION, MONTANA. Outcrops of the Late Oligocene–Early Miocene Renova Formation near Alder, Ruby River Basin (Passamari Member), and Canyon Ferry Reservoir, both in southwestern Montana, have yielded 13 orders and over 30 families of insects. The Ruby River Basin insects were discovered by Zuidema (1950), although Lewis (e.g., 1971, 1973, 1978) published many papers on assorted insects from here. Among them are rare bee flies (Bombyliidae), eumastacid grasshoppers, and sialid "alder flies" (Megaloptera). The Canyon Ferry outcrops were only recently discovered (CoBabe *et al.*, 2002). The Ruby River Basin deposits are typical lacustrine sediments, though they are called *paper shales* because they are more finely laminated than most such sediments. The Canyon Ferry deposits are lacustrine and volcanic tuffs, dominated by abundant corixids ("water boatmen"), but most of the diversity comprises families that feed on terrestrial plants. A satyrid butterfly wing is a notable record from Canyon Ferry. Both deposits have preserved an impressive record of insects now extinct in this region, or even for North America and the Western Hemisphere. For example, in the Ruby River Basin there is a nemopterid lace wing, and a stalk-eyed fly (Diopsidae). Diopsidae today occur in Africa and southeast Asia, with two species of the basal genus *Sphyracephala* living in eastern North America, but the Ruby River basin diopsid is more closely related to Old World species. Hodotermitid termites occur in both deposits, though today they are found in western North America and

Eurasia. As at Florissant, Canyon Ferry has preserved an osmylid lacewing in the subfamily Kempyniinae, a group that today is austral. Also at Canyon Ferry are giant hornets in the genus *Vespa*, whose natural distribution today is Eurasian.

BEMBRIDGE MARLS. These are highly fossiliferous limestones from the northern portion of the Isle of Wight, United Kingdom, formed near the Eocene–Oligocene boundary. The limestones are very finely grained and preserved insects in microscopic detail but also with complete relief (Figure 2.11). The body cavities of many insects are preserved (Figure 2.12), in some cases with muscles still preserved (Figure 13.65). Thousands of insects have been collected from this deposit; most of them are stored in the Natural History Museum in London. Among the more significant finds are several beautifully preserved butterflies (Figure 13.65), and termites belonging to the presently Australian genus *Mastotermes* (Figure 7.80). An important reference on the deposit is by Jarzembowski (1980), but much work remains on understanding this paleofauna.

AIX-EN-PROVENCE. The Late Oligocene gypsum marls that occur in Aix-en-Provence, France, contain abundant insects and plants. Though productive and also known and studied much longer than the Bembridge Marls, these deposits are not quite as diverse for insects as this other deposit. Most of the insects were described by Théobald (1937) but are in need of revision before meaningful comparisons can be made to other Oligocene sites. Among the more significant fossil insects from Aix are six species of butterflies, which are exceedingly rare as fossils (Table 13.3).

ROTT. The most famous Oligocene deposit in Europe occurs in Rott, Germany near Bonn, which are also historically among the longest studied insect compression fossils in Europe. Insects from this deposit are preserved with very fine detail (e.g., Figure 11.87), partly because the matrix consists of extremely fine-grained paper shales. Age of the deposit is slightly ambiguous, being either latest Oligocene (Chattian) or earliest Miocene (Aquitanian). A very large collection of Rott fossil insects is at the Natural History Museum of Los Angeles in California, which is the collection of Georg Statz, who published several large papers on this fauna between 1936 and 1950.

Miocene (23.6–5.2 MYA)

OENINGEN AND RADOBOJ. Perhaps the richest Miocene insect deposits in Europe are from Oeningen in Switzerland and Radoboj in Croatia. Fossils from these localities were extensively monographed by Oswald Heer (Heer, 1849), and relatively little work has been done on this deposit since. As a result, much of the fauna is in need of modern revision. Insects from Oeningen are of Late Miocene (Messinian) age and are preserved as compressions in freshwater limestone. Insects from Radoboj are Early Miocene (Burdigalian) in age and are preserved similar to those at Oeningen. Among the more significant insects from Radoboj are three species of very rare fossil butterflies (Table 13.3). Classic reconstructions of the biota and climate of Switzerland during the Miocene were made by Heer (1865), based principally on the Oeningen fossils. The Early Miocene Randecker Maar in Germany has received recent attention (e.g., Schweigert and Bechly, 2001), although the insect fauna is much smaller than that of Oeningen or Radoboj.

RUBIELOS DE MORA, SPAIN. This deposit is particularly interesting and has been monographed recently (Peñalver Mollá, 1998). The deposit is located in Teruel, in the Rubielos de Mora Basin, which formed during the Early Miocene (ca. 20 MYA) from the deposits of a meromictic lake. Diversity of the insects is not exceptional; some seven orders of insects are preserved. But, the preservation is remarkable (Figure 2.66). The matrix is a fine-grained, light oil shale, against which the dark, shiny cuticle of the insects stands out. Resolution of preservation is remarkable, including the fringe of setae around the margin of thrips' wings, and even the microtrichia on the wing. Adults and soft-bodied larvae are preserved.

SHANWANG FORMATION. This formation of diatomites occurs near Linqu, in Shandong Province, China, from which approximately 400 species of insects in 84 families have been recorded (e.g., Zhang, 1989; Zhang *et al.*, 1994).

LATAH FORMATION. Numerous and prolific outcrops of this mid-Miocene (ca. 18 MYA) formation occur in western Idaho and eastern Washington state, including compressions of entire insects and occasional insect mines and galls among diverse fossil plants (e.g., Carpenter *et al.*, 1931; Lewis, 1969, 1985). Preservation varies greatly, that of outcrops near Spokane, Washington and Juliaetta, Idaho, being mostly carbonized wings. Among the more significant fossils are *Bombus proavus* and rhinotermitid termites that are closely

2.66. A beautifully preserved aphid from the Miocene of Rubielos de Mora, Spain, showing the long, slender cornicles and venation. MPZ-96/18. Photo: Enrique de Peñalver.

related to species from southeast Asia (Emerson, 1971). Perhaps the most famous of the outcrops is an unpretentious roadcut at Clarkia, Idaho, which yields beautifully preserved plants and completely articulated insects. Cellular-level preservation of these fossils led to early efforts in the extraction of ancient DNA. At Clarkia alone some 9 orders and 30 families of insects have been identified.

SAVAGE CANYON FORMATION (STEWART VALLEY). This is probably the most complete paleocommunity known for the Cenozoic of North America. It occurs in southwestern Nevada, and K/Ar dating indicates an age of 16–10.5 MYO (mid- to Late Miocene). The hot, arid environment in the region today contrasts with the paleoenvironment, which was cool, wet, and covered in mixed conifer-deciduous forest. Fossils from Stewart Valley include 50 families of mammals, 30 families of plants, and terrestrial and freshwater molluscs; fine-grained shales contain articulated fish skeletons, bird feathers, and complete leaves and insects. Even small, delicate insects like midges are preserved, intact with fine setae and often with color patterns. Ten orders and 50 families of insects are known, 50% of the individuals being Diptera, with Hymenoptera second in abundance due mostly to ants (Schorn *et al.*, 1989).

BARSTOW FORMATION (CALICO MOUNTAINS). This is one of the most distinctive fossil insect deposits in the world. The Barstow Formation occurs approximately 100 miles northeast of Los Angeles, California, in the Mojave Desert near the town of Yermo. While fossiliferous outcrops of the formation occur at Mt. Pinos and in the Frazier Mountains, there are three beds in the upper part of the Barstow Formation unique to the nearby Calico Mountains that yield insects some 13–14 MYO. Insects occur within 5- to 60-cm-sized nodules, one or two of which are extracted from each nodule by digesting it with acids. The insect remains resemble microscopic glass sculptures (Figures 2.5 to 2.7). Original work on the deposit was by Palmer (1957), and there has been a recent paleoecological study of it (Park and Downing, 2001). The fauna was mostly preserved in situ in a drying "alkali" or "soda"/"salt" lake, similar to lakes found today in the western United States (Mono Lake, California, or Big Soda Lake, Nevada), and throughout the Middle East and eastern Africa. This paleoenvironment accounts for the remarkable preservation and impoverished autochthonous fauna. The fauna is dominated by fairy shrimp (*Anostraca*), dytiscid beetles, and immature stages of *Dasyhelea* midges (Ceratopogonidae), much as one would find in alkali lakes today. Rare arthropods include mites, spiders, thrips, psyllids, leafhoppers, heteropterans, and even a caterpillar, which apparently wafted into the lake. W. D. Pierce, whose taxonomy is notorious, described diverse species, genera, and even some new families from this deposit, the last of which is extremely odd for such relatively young fossils (e.g., Pierce, 1963, 1966; Pierce and Gibron, 1962).

ALASKA. Two fairly diverse deposits of insects occur in the Late Miocene of Alaska, formed during a period of considerable biogeographic importance: formation of the Bering Land Bridge. One deposit is uppermost Miocene (5.7 MYO), from the northern part of the Seward Peninsula (Hopkins *et al.*, 1971). The other is slightly older (6.7 MYO), from Suntrana, central Alaska (Grimaldi and Triplehorn, unpubl.).

Pliocene (5.2–1.7 MYA) and Pleistocene (1.7 MYA–10,000 YA)

Evolutionary significance of the Pliocene and Pleistocene for insects largely concerns the origins of modern species and sweeping changes in their distributions in the more recent past.

North American Pliocene insects are scarce, known from only four modest deposits. These occur in Alaska (Matthews, 1970), California (Squires, 1979), Nevada (Sleeper, 1968), and Texas (Carpenter, 1957). Remains of beetle elytra, chironomid head capsules, and other durable parts of insects preserved in Pleistocene lake beds and bogs are often identifiable to species. As a result, changes in the distribution of extant species can be tracked. Just as elephants (i.e., mammoths), lions, cheetahs, camels, and other "African" megafauna became extinct from North America, there have been dramatic changes among Pleistocene insects. Pleistocene and Holocene insects are well represented throughout the world, which is presented in detail by Elias (1994). One fascinating Pleistocene deposit is the La Brea tarpits outside of Los Angeles, which is famous not only for the impressive preservation and diversity of vertebrates but also for its diverse insects (Miller, 1983).

As in North America, Pliocene deposits of insects are uncommon and known from only scattered deposits in Greenland (Bennike and Bocher, 1990; Heie, 1995; Bocher, 1995, 1997), Italy (Pedroni, 1999, 2002), Sicily (Kohring and Schlüter, 1989; Schlüter and Kohring, 1990), France (Balazuc, 1989; Nel, 1987, 1988a,b, 1991b), Germany (Weidner, 1979; Harz, 1980; Schlüter, 1982; Rietschel, 1983; Grabenhorst, 1985, 1991; Kohring and Schlüter, 1993; Briggs *et al.*, 1998b; Brauckmann *et al.*, 2001), Turkey (Nel, 1988a,b), Georgia (Kabakov, 1988), Japan (Hayashi, 1999, 2000, 2001a,b; Hayashi and Shiyake, 2002; Mori, 2001), Chad (Duringer *et al.*, 2000a,b; Schuster *et al.*, 2000), Malawi (Crossley, 1984), Tanzania (Ritchie, 1987; Sands, 1987; Kaiser, 2000), and Antarctica (Ashworth *et al.*, 1997; Ashworth and Kuschel, 2003).

Central America and the Caribbean

Central America and the Caribbean – so-called nuclear America – is largely devoid of fossil insect deposits, with two very dramatic exceptions: rich deposits of Oligocene and Miocene amber from Chiapas, Mexico, and the Dominican Republic (Figure 2.67), respectively. The amber from both deposits was formed by extinct tree species in the genus

2.67. Hillside excavation of Miocene amber at the Palo Alto mine, northern Dominican Republic, ca. 1995. Preservation in Dominican amber is perhaps the finest of all ambers. Photo: R. Larimer.

Hymenaea, which today comprises canopy species found throughout lowland rain to deciduous dry forests of the neotropics (one species, *H. verrucosa*, occurs in eastern Africa). The living species exude copious resin, as the extinct ones did. Botanical source of the amber is confirmed by chemistry and abundant inclusions of sepals, flowers, and leaves (Langenheim, 1966; Hueber and Langenheim, 1986).

Fossils in Dominican amber are more renowned, even though Mexican amber has been studied scientifically longer (Hurd *et al.*, 1962; various papers in *University of California Publications in Entomology*, Volume 31 [1963], volume 63 [1971]: see Engel, 2004a). This may be attributable to more effective commercial exploitation of Dominican amber, more productive deposits, or both. Inclusions in Mexican amber are usually slightly to obviously compressed; those in Dominican amber are often perfect. Indeed, the preservation of organisms in Dominican amber are preserved with a finer and more consistent fidelity – externally and internally – than is any other amber deposit in the world (e.g., Grimaldi *et al.*, 1994). Arthropods in Dominican amber are arguably the most beautiful such fossils in the world and the most diverse Miocene fauna of insects known. Unfortunately, despite its significance, the age of Dominican amber was confused, with unsubstantiated but popularized claims of Eocene age. It is now known to be definitively younger (Grimaldi, 1994b); specifically mid-Miocene, approximately 17–20 MYO (Iturralde-Vinent and MacPhee, 1996). Age of Mexican amber appears to be Late Oligocene, based on foraminiferans. The most significant collection of Mexican amber is at the University of California Museum of Paleontology, Berkeley, assembled by Hurd and others in the 1950s.

The first serious work on Dominican amber inclusions began with Dieter Schlee at the Staatlichen Museum für Naturkunde in Stuttgart, Germany. He assembled an impressive collection (Schlee, 1980, 1984, 1986, 1990), much of it on permanent display. The Morone Collection in Turin, Italy, is the finest one, containing many superb specimens of rare and impressive organisms (e.g., Grimaldi, 1996) (e.g., Figures 3.16, 9.24, 10.43, 13.69). The American Museum of Natural History (New York), Smithsonian Institution (Washington, D.C.), and Natural History Museum (London) have collections amounting to approximately 20,000 pieces containing 30,000 inclusions, mostly smaller insects and arachnids. Published and unpublished work on these collections indicates that over 400 families and 1,500 species of insects exist in Dominican amber. Many of the same families and even genera of insects occur in Mexican and Dominican amber, with a great variety of other impressive inclusions found in the latter. Dominican amber contains, for example, diverse plants (especially flowers, some 30 families); solpugids, scorpions, and mature amblypygids (Figure 3.16); feathers; *Anolis* and *Sphaerodactylus* lizards; *Eleutherodactylus* frogs;

and even the partial remains of a mammal (Grimaldi, 1996; MacPhee and Grimaldi, 1996).

The paleobiota in both deposits is distinctly tropical. Most species are related to ones presently living in Central or South America, though there have been impressive extinctions. For example, *Mastotermes* termites (Figure 7.81) and a variety of other insects in Dominican and Mexican amber belong to groups now found only in Australia, Africa, or southeast Asia. Caribbean landmasses had a complex history of drift, submergence, and land bridges (Iturralde-Vinent and MacPhee, 1996), so understanding the evolution of its biota is challenging. Dominican and Mexican amber contributes unique insight on this subject, as well as on the origins of modern tropical ecosystems.

South America

For more than 60 million years South America was isolated from its gondwanan neighbors, its biota evolving in what paleontologist G. G. Simpson called "a splendid isolation." Approximately 3.5 million years ago a profound event occurred that integrated faunas from North and South America: the Panamanian Land Bridge closed. This "Great American Interchange" and the earlier Cenozoic history of South America's fauna have largely been unraveled by the study of fossil mammals (e.g., Simpson, 1948). Unlike fossil mammals, Cenozoic insects from South America are scarce, with only four Cenozoic formations yielding significant numbers of insects, plus several smaller formations. Petrulevicius and Martins-Neto (2000) catalogued 73 named Cenozoic insects in 11 orders, not including diverse fossil nests and burrows from southern South America.

The first intensively studied formation is the Margas Verdes Formation from Sunchal, Jujuy Province, Argentina. Petrulevicius and Martins-Neto (2000) refer to this as the Maíz Gordo Formation, Late Paleocene (ca. 60 MYO). T. D. A. Cockerell published various papers on this formation (e.g., Cockerell, 1936), in which he described diverse beetles (of putative Carabidae and Curculionidae), as well as Orthoptera, Dermaptera, and Auchenorrhyncha, based on elytra and tegmina.

Probably the most diverse and significant formation is the Tremembé Formation (Oligocene), from Taubaté Basin, São Paul state, Brazil. Thus far, it contains six orders, the most significant being several rare Lepidoptera. These include two butterflies, *Archaeolycorea ferreirai* (Nymphalidae: Danainae) and *Neorinella garciae* (Satyrinae), and a noctuid moth, *Philodarchia cigana*. Other smaller Cenozoic deposits are the Pirassununga Formation (Oligocene) of São Paulo, Brazil, and the Fonseca Formation of Minais Gerais, Brazil. The latter has yielded the mastotermitid termite *Spargotermes limai*. A. E. Emerson, who described the termite, attributed the Fonseca Formation to the Eocene, but Petrulevicius and Martins-Neto (2000) indentify it as Oligocene. The Ventana Formation (Paleocene-Eocene) of Pichileufú, Río Negro Province, has yielded several ants. The Early Eocene "Tufolitos Laguna del Hunco" deposit of Chubut, patagonian Argentina is a caldera lake deposit preserved with leaves, insects, and caddisfly cases. The cases are composed of bits of plant material, not sand grains or pebbles (Genise and Petrulevicius, 2001).

A fascinating formation is the Palacio Member of the Ascencio Formation (latest Cretaceous–earliest Paleocene), Uruguay, which contains abundant and diverse burrows and nests in paleosols. The ichnofossils were first studied by Frenguelli (1939), now mostly by Genise (1999; Genise and Bown, 1994; Genise and Laza, 1998). The burrows were formed by scarabeids and by some unidentified insects; most significantly there appear to be burrows from perhaps 10 species of bees. Identification of particular groups of bees (e.g., Halictinae) is based on nest architecture, which can be distinctive for certain groups. Bee fossils are scarce, and these would be among the oldest ones known.

Africa and Australia

While paleoentomological work in Africa and Australia, albeit limited, has focused principally on the Paleozoic and Mesozoic, hardly any attention has been paid to the Tertiary of these regions. The result is that our knowledge of past insect diversity is heavily lopsided, with most information derived from the northern continents. Given the diversity of potential Tertiary sites throughout Australia and Africa, there is a tremendous amount of work and potentially valuable discoveries to be made. Although some insect sites are already known, only a couple have been explored to any extent, and they have proven to be rich in material.

After the separation of the southern continents, Australia experienced a long and isolated history, made famous by its unique flora and vertebrate fauna, and the insect fauna as well (e.g., CSIRO, 1991). Perhaps the best-known Tertiary site are the Paleocene insects of the Redbank Plain Formation and Eocene Dinmore Formation in southeast Queensland (e.g., Riek, 1952; Rix, 1999) as well as other Early Tertiary deposits in the same region (e.g., Duncan *et al.*, 1998). Those from Riversleigh are three-dimensionally preserved and are remarkable for their level of detail (e.g., Duncan and Briggs, 1996), but have still not been significantly explored for their systematic and evolutionary implications.

In Africa, records are even more spotty, most Tertiary records of insects being trace fossils, such as beetle borings in Pliocene-Pleistocene mammal bones, or termite mounds of similar age (Crossley, 1984). Records of preserved insects do exist, such as some presumably Early Paleocene tenebrionids from Namaqualand, but they are exceedingly few. Both Africa and Australia represent relatively untapped frontiers for paleoentomology.

3 Arthropods and the Origin of Insects

Multicellular life arose in the Precambrian Period. While many animal phyla appear to have originated near the end of the Precambrian (e.g., putative annelids), the first diverse assemblages of animals is not known until the Cambrian, the so-called Cambrian Explosion (Conway Morris, 1979, 1989, 1993, 1998, 2000, 2003). We recognize these lineages as phyla because of the dramatic and fundamental differences in body organization (i.e., each phylum represents a basic *groundplan* or *bauplan* for animal design). Among the diversity of groundplans for building an animal, the phylum Arthropoda, which has achieved a level of evolutionary success unrivaled in evolutionary history, is clearly more dominant on Earth relative to all others. The arthropodan groundplan is the most commonly encountered form of life, having radiated into more species and into more habitats than any other lineage. It is also an ancient phylum, being well represented in the Cambrian faunas by an already impressive diversity, the most famous of which are the familiar trilobites.

Traditionally the arthropods have been considered to belong to a larger grouping of phyla called the Articulata, or the segmented animals (e.g., Cuvier, 1817; Haeckel, 1866; Hatschek, 1878; Snodgrass, 1938; Lauterbach, 1972; Rouse and Fauchauld, 1997; Wägele *et al.*, 1999; Nielsen, 2001; Wägele and Misof, 2001; Scholtz, 2002), and, aside from Arthropoda, to consist of the Annelida, Tardigrada, and Onychophora. The latter two phyla comprise, along with Arthropoda, the Panarthropoda (= Haemopoda of Cavalier-Smith, 1998; = Lobopodia of Snodgrass, 1938, although this name is often restricted to Onychophora today) (Figure 3.1). The annelid theory for the ancestry of the panarthropods has been widely recognized (e.g., Brusca and Brusca, 1990) but is increasingly perceived as incorrect. Most recent analyses of morphological and molecular data consider the Annelida distantly related to Panarthropoda (e.g., Aguinaldo *et al.*, 1997; Giribet and Ribera, 1998; Zrzavý *et al.*, 1998a, 2001; Zrzavý, 2001, 2003). Annelid worms are alternatively believed to be related to the Mollusca and Sipuncula, while panarthropods are allied to a series of phyla that are characterized by the absence of locomotory cilia and the presence of a trilayered cuticle (consisting of multilayered epicuticle, exocuticle, and chitinous endocuticle), which molts as they grow via ecydsteroid-induced cycles. This larger clade of moulting animals, referred to as Ecdysozoa (Aguinaldo *et al.*, 1997), consists of the Panarthropoda sister to the Nematoda, Nematomorpha, Kinorhyncha, Priapula, and Loricifera (e.g., Zrzavý, 2003) (Table 3.1). The Ecdysozoa itself is apparently sister to the enigmatic and little understood phylum Gastrotricha (e.g., Zrzavý, 2003), a small group of microscopic worms inhabiting aquatic habitats.

The Pentastomida, a phylum formerly included in the "Articulata," includes enigmatic parasitic worms (about 95 species) that live in the lungs or nasal passageways of various vertebrates (Storch, 1993). Like the panarthropods, pentastomids have a nonchitinous cuticle that is periodically molted

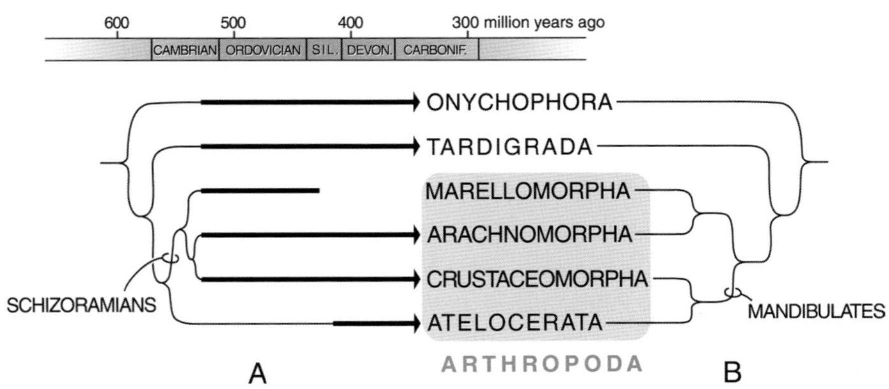

3.1. Phylogeny of panarthropod phyla, the Onychophora, Tardigrada, and Arthropoda; and relationships among the four subphyla of arthropods under the schizoramian (A) and mandibulate (B) hypotheses.

TABLE 3.1. Hierarchical Classification of Ecdysozoa

—ECDYSOZOA—
Phylum Gastrotricha (gastrotrichs)
Introverta
Nematoida
Phylum Nematoda (roundworms)
Phylum Nematomorpha (hairworms)
Cephalorhyncha (= Scalidophora)
Phylum Priapulida (priapulans)
Scalidorhyncha
Phylum Kinorhyncha (mud dragons)
Phylum Loricifera (loriciferans)
Panarthropoda
Phylum Onychophora (velvet worms)
Tritocerebra
Phylum Tardigrada (water bears)
Phylum **Arthropoda**

during growth, and the principle body cavity is a hemocoel. However, the nervous system is Crustacean-like, the larvae are nearly identical to Crustacean larvae, and sperm structure and embryogenesis are identical to crustaceans (Wingstrand, 1972; Riley *et al.*, 1978; Storch and Jamieson, 1992). Molecular data have also supported a crustacean origin of the pentastomids (Abele *et al.*, 1989). Indeed, it is now hypothesized that pentastomids are highly derived Crustaceans, perhaps near the Maxillopoda (Martin and Davis, 2001), although limited paleontological evidence tends to favor the placement of the group sister to Arthropoda proper (e.g., Walossek and Müller, 1994, 1998; Walossek *et al.*, 1994; Zrzavý, 2001; Waloszek, 2003).

Despite the controversy surrounding the larger placement of Panarthropoda, extensive evidence indicates that the group is monophyletic.

ONYCHOPHORA: THE VELVET WORMS

The phylum Onychophora, or "velvet worms," consists of approximately 100 species of legged worms that have a pantropical distribution. Living Onychophora are classified into two families, Peripatidae and Peripatopsidae, and are now entirely terrestrial, generally living amongst moist leaf litter in forests (Figure 3.2). Species tend to be predatory, spraying a proteinaceous "glue" from the oral papillae that ensnares their victims, which include snails, worms, and small arthropods. The body is elongate, with a weak cuticle that is finely annulate (*pseudosegmentation*) and beset with dermal papillae. The true segments have distinct, unjointed lobopods, which on one segment of the head form sensory structures superficially similar to the antennae of arthropods. The phylum has characteristic peribuccal and large oral papillae, and even has a tracheal system like that of the myriapods and hexapods within Arthropoda.

3.2. A modern velvet worm, *Peripatus* sp. (Peripatidae), from Panama. Onychophora is a small phylum of approximately 100 living species closely related to arthropods. Photo: P. J. DeVries.

Interest in the Onychophora largely stems from their apparent phylogenetic position among major lineages of Panarthropoda, specifically as basal to the Tardigrada + Arthropoda lineage (e.g., Zrzavý *et al.*, 1998b; Nielsen, 2001). The phylum unites primitive features of typical "worms" (e.g., Nematoda and Nematomorpha) with those of other panarthropods. Numerous fossils from the Cambrian have been allied with the Onychophora (e.g., Dzik and Krumbiegel, 1989; Ramsköld and Hou, 1991; Hou and Bergström, 1995), including several enigmatic forms from the Middle Cambrian Burgess Shale such as *Hallucigenia* and *Aysheaia* (Ramsköld and Hou, 1991). These Paleozoic velvet worms are traditionally placed in their own class, Xenusia, and likely are a paraphyletic stem group to modern Onychophora (class Euonychophora), or are stem group lobopodians. Xenusians, unlike modern members of the phylum, were entirely marine and had a terminal (vs. ventral) mouth apparently lacking oral papillae. The earliest Euonychophora (i.e., terrestrial and with a ventral mouth), is known from the Upper Carboniferous (Thompson and Jones, 1980); however, the next record of the phylum is not until the mid-Cretaceous Burmese amber (Grimaldi *et al.*, 2002), a vacuum of nearly 200 million years. This gap is probably due to the fact that the soft bodies of onychophorans very rarely preserve in sediments. While the Carboniferous *Helenodora* is considered basal within Euonychophora (Thompson and Jones, 1980), the Cretaceous amber *Cretoperipatus burmiticus* (Figure 3.3), is remarkably modern and even belongs to the living family Peripatidae.

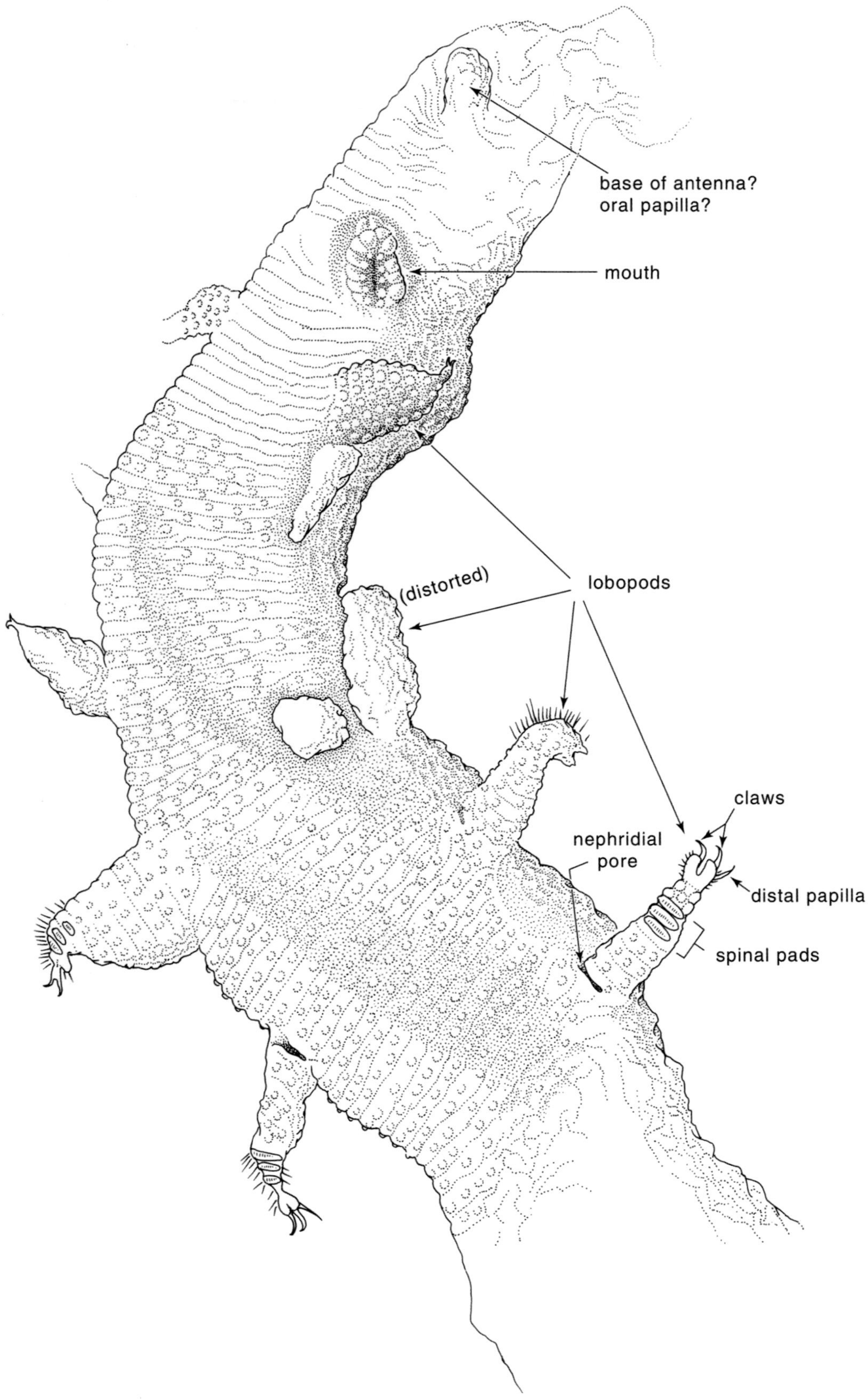

3.3. The oldest velvet worm in amber; *Cretoperipatus burmiticus* (Peripatidae) in 100 MYO Cretaceous amber from Myanmar. Onychophorans date from the Cambrian, but this is the only known Mesozoic member of the phylum. AMNH Bu218; preserved length 5 mm; from Grimaldi *et al.* (2002).

TARDIGRADA: THE WATER BEARS

Tardigrades are a small phylum of 840 species of minute animals (generally 200–500 μm in length) that live in moss, lichens, leaf litter, and freshwater or even marine habitats. Species feed on mycelia, algae, plant cells, rotifers, nematodes, and even other tardigrades. They are segmented, possess paired, clawed legs, and molt. Based on these traits, other morphological features, and DNA sequences, tardigrades have been placed as the closest, extant relatives of arthropods (Dewell and Dewell, 1996, 1998; Garey *et al.*, 1996; Giribet *et al.*, 1996; Yeo-Moon and Kim, 1996; Nielsen, 2001). General works on the phylum include Greven (1980), Ramazzotti and Maucci (1983), Nelson and Higgins (1990), Dewell *et al.* (1993), and Kinchin (1994), while Garey *et al.* (1999) have provided the most recent cladistic analysis of the group. Defining features of the group include the structure of the eyes; the presence of a nerve between the lateral protocerebral lobes and the ganglion of the first pair of walking legs; the modification of the anterior claws into stylets; and the absence of a heart and metanephridia. The occurrence of "Malpighian tubules" in some tardigrades is convergent with those seen in arthropods.

The best-known feature of the phylum is the ability of some species to endure extreme conditions in a dormant, or cryptobiotic, state: years, probably even decades, of complete desiccation (Baumann, 1927); temperatures well above boiling point and near absolute zero (Rahm, 1921, 1924, 1925); intensities of X-rays that are more than 100-fold the lethal dose for mammals (May *et al.*, 1964); and pressures more than six times that known in the deepest oceanic trenches (Seki and Toyoshima, 1998). For these reasons, tardigrades inhabit some of the harshest regions on earth. Six species live in mosses and lichens in eastern Antarctica (Miller *et al.*, 1996). Most species are widely distributed, if not cosmopolitan. The ability of tardigrades to "encyst," to become highly resistant to extreme environmental conditions, has likely been a principal factor in their distribution. Once encysted, tardigrades can easily be carried by wind or in soil carried by other organisms. Eggs are similarly hardy and may also be easily distributed.

The minute size and membranous integument of tardigrades makes their fossilization by mineralization or compression highly unlikely or undetectable, although tardigrade-like fossils have been described from mid-Cambrian deposits in Siberia (Müller *et al.*, 1995), phosphatized in complete relief and with microscopic detail (Figure 3.4). These specimens differ from living tardigrades by having three pairs of legs rather than four (although homologues of these may be present in one of the fossils figured by Waloszek, 2003), a simplified head morphology, and no posterior head appendages (lateral cirri and clavae: although Waloszek, 2003, considers fine sensorial structures of the fossils to correspond to these traits among living tardigrades). The Cambrian fossils probably represent a stem group to the living Tardigrada (Walossek and Müller, 1998).

3.4. Earliest fossils of the phylum Tardigrada, from the Cambrian of northern Europe. They are exquisitely preserved as phosphatized replicas. Tardigrades are the closest relatives of arthopods. Scanning electron micrographs; photos: D. Waloszek.

Besides the Cambrian phosphatized tardigrades, the only other fossils are several rare specimens in Cretaceous amber. The oldest of these is *Milnesium swolenskyi* in New Jersey amber (Figure 3.5); detailed preservation indicates that the structure of its claws and mouthparts are virtually indistinguishable from the living cosmopolitan species *M. tardigradum* (Bertolani and Grimaldi, 2000). The other amber fossil tardigrades are two specimens in amber from western Canada (Cooper, 1964), 15–20 million years younger than *M. swolenskyi*. The best preserved Canadian amber specimen was described in its own genus and family, *Beorn*

3.5. *Milnesium swolenskyi* (Milnesiidae), a tardigrade in 90 MYO Cretaceous amber from New Jersey. Tardigrades are remarkably durable animals that can persist in dormancy for such extended periods of time (called cryptobiosis) as to challenge concepts on the longevity of individuals. This species is barely distinguishable from a widespread living species. AMNH NJ796; length 0.85 mm.

leggi (Beornidae), but it bears a resemblance to several genera in the contemporary family Hipsibiidae (R. Bertolani, pers. comm.). The existence of a recently derived tardigrade lineage in the mid-Cretaceous is consistent with origins of the phylum during the "Cambrian explosions" (Gould, 1989), although such morphological stasis, or bradytely, is extraordinary.

Extreme bradytely (Simpson, 1944; Eldredge and Stanley, 1984) is well known, albeit rare, in evolution. Perhaps the most famous examples from the animal fossil record that show little or virtually no morphological change over millions of years are horseshoe "crabs" (Chelicerata: Xiphosura: *Limulus*), and the coelacanth (*Latimeria*). The living Atlantic horseshoe crab, *Limulus polyphemus*, is very similar to a species from the Upper Cretaceous (c. 70 MYO), *L. coffini* (Fisher, 1984). The only living coelacanth, *Latimeria chalumnae*, is the sole survivor of the Actinistia fishes, which thrived from the Devonian to the Upper Cretaceous, 380–79 MYO (Forey, 1984). The tadpole shrimp, *Triops cancriformis* (Crustacea: Branchiopoda), is another, less well known example. This "oldest known living animal species" (Tasch, 1969; Schram, 1986; Futuyma, 1998) is indistinguishable from 180-MYO Jurassic fossils. Bradytely in *Triops* tadpole shrimp and *Milnesium* tardigrades may be attributable to their remarkable ability to become dormant. Living *T. cancriformis* inhabit nonsaline ponds that are often ephemeral. When the water evaporates, desiccated eggs can remain viable in the sediment for nearly a decade, and withstand temperatures of >90°C for short periods of time. Unlike tardigrades, though, adult *Triops* cannot enter into such dramatic dormancy, nor can they endure the extremes that tardigrades can. Cryptobiosis probably acts as a general adaptation to various environmental conditions, freeing these organisms from developing suites of morphological and behavioral adaptations, and thus slowing the rate of morphological change. Cryptobiotic tardigrades, in fact, are probably the most durable animals.

ARTHROPODA: THE JOINTED ANIMALS

Over three quarters of all species on earth belong to the Arthropoda. Arthropods have become ubiquitous in every habitat on our planet except for the extreme poles. Nearly everyone can intuitively recognize an arthropod, and they have almost universally been recognized as a natural group for centuries. Even Linnaeus (1758), who was a botanist, was able to recognize arthropods as a group. His Kingdom Animalia was divided into several groups of vertebrates (Pisces, Reptilia, Aves, Mammalia) and two classes of animals without backbones: Insecta and Vermes. Linnaeus' "Class Insecta" corresponds to what we now call Phylum Arthropoda.

The arthropods are defined by numerous features (e.g., Lankester, 1904; Snodgrass, 1938; Boudreaux, 1979; Weygoldt, 1986; Brusca and Brusca, 1990). Some of these features are external and internal body segmentation with regional specialization, or *tagmosis*; a hardened exoskeleton composed of cuticle that is hardened through calcification (mineral deposition) or by sclerotization (protein cross-linking); an exoskeleton composed of articulated plates; body segments that primitively bear paired, articulated appendages (and hence the name Arthropoda, meaning "jointed foot"); frequently paired compound eyes and some median simple eyes; coelom reduced to portions of reproductive tract and excretory system (the main body cavity is an open *hemocoel*); an open circulatory system with dorsal, ostiate heart; a complete digestive tract; a ventral nerve cord; stepwise growth via molting (which, as we have seen, is not unique to Arthropoda); and muscles striated and arranged in isolated segmental bands and generally in opposing pairs of flexor and extensor muscles. Perhaps one of the most important features of arthropods is their organization into *tagma* (plural *tagmata*), or sets of segments specialized into functional units. *Tagmosis* has allowed arthropods to diversify their overall body design. For example, the pattern of tagmosis is used, in conjunction with other traits, to identify major arthropod groups.

Despite this impressive array of traits, the monophyly of arthropods has been questioned. Tiegs (1947) considered that the arthropods were actually an artificial combination of two unrelated groups: the Myriapoda, Hexapoda, and Onychophora (the "Uniramia") and the Trilobita, Crustacea, and Chelicerata ("TCC" of Cisne, 1974). Tiegs posited that these two groups originated independently from annelid-like ancestors, converging on "arthropod" traits. This hypothesis was later expanded to consider the three TCC lineages as each being independently derived, expanding the polyphyly to four separate origins (Tiegs and Manton, 1958; Manton, 1964, 1966, 1972, 1973, 1979; Anderson, 1973, 1979; Willmer, 1990; Fryer, 1996, 1998). The principal notion behind the Tiegs and Manton hypothesis of independent origins of arthropods is that the various arthropod lineages could not be considered relatives if the putative ancestor of them all possessed anatomical structures that were, hypothetically, nonfunctional (particularly appendicular structures). Alternatively, if character states observed among the arthropod lineages could not be immediately derived from other characters already existing in modern taxa, then these authors did not believe common ancestry could be supported. In their scenario, the use of the limb base (gnathobase) for grinding food in Trilobita, Chelicerata, and Crustacea was fundamentally different from the use of the apex of an appendage in the other lineages (composite in Myriapoda and Hexapoda). All other characters supporting Arthropoda were ignored along with the possibility that mandibular structures had simply diverged in favor of a functional scenario of appendage evolution. Such a concept of "functionalism" is not valid in phylogenetic reconstruction (Kristensen, 1975; Ax, 1984; Weygoldt, 1986). No rigorous study of panarthropod relationships based on molecular or morphological data has been able to convincingly establish arthropod polyphyly. Indeed, every modern study has strengthened the concept of a monophyletic Arthropoda (Field *et al.*, 1988; Turbeville *et al.*, 1991; Wheeler *et al.*, 1993a; Giribet *et al.*, 1996; Giribet and Ribera, 1998; Giribet and Wheeler, 1999; Nielsen, 2001; Regier and Shultz, 2001a).

The complete phylogeny and evolution of Arthropoda is outside the scope of this work and would fill volumes alone. We have provided here only a brief outline of the major lineages of the phylum so as to place the insects in a greater context and for understanding their origin (Table 3.2). The arthropods consist of at least four major lineages (considered subphyla): Marellomorpha, Arachnomorpha, Crustaceomorpha, and Atelocerata (Hexapoda and Myriapoda). Neontologists and paleontologists differ considerably on their interpretation of the relationships among these groups, mostly concerning the position of the Crustaceomorpha as either sister to Atelocerata (the Mandibulata, supported by most neontologists) or to Arachnomorpha, along with Marellomorpha (the Schizoramia, supported by most paleontologists) (Figure 3.1). Molecular biologists have come in on both sides of the issue and, while providing additional important data, have not generally swayed the conclusion overwhelmingly to one hypothesis or the other. Furthermore, among those authors who favor the Mandibulata hypothesis, there is a schism concerning the monophyly of the Atelocerata. Within Mandibulata the hexapods are either allied to the Myriapoda (the traditional Atelocerata) or to the Crustacea (the Pancrustacea hypothesis). Despite these points of contention, some major groups are generally accepted.

MARELLOMORPHA: THE LACE CRABS

The extinct subphylum Marellomorpha consists of several enigmatic marine fossils from the Cambrian to Devonian, the most famous of which is *Marella* (the "lace crab") from the Burgess Shale (Wolcott, 1912), and is perhaps the most common nontrilobite arthropod in these deposits. Other genera include *Mimetaster* and *Vachonisia*. The lineage is supported as monophyletic based on four traits (Stürmer and Bergström, 1976; Wills *et al.*, 1998): number of head appendages, large number of trunk somites, regular decrease in the length of the trunk appendages, and the division of the trunk endopods into five podomeres. Marellomorphs may be the sister group to Arachnomorpha (see discussion that follows) or basal within a larger grouping called Schizoramia, which also consists of Arachnomorpha but as the sister group of Crustaceomorpha (see debate concerning Schizoramia monophyly, later in this chapter).

ARACHNOMORPHA: TRILOBITES, ARACHNIDS, AND RELATIVES

The subphylum Arachnomorpha is a diverse lineage consisting of those arthropods with the anus in a ventral position in the penultimate somite of the trunk, development of a styliform terminal projection, presence of trunk gut diverticulae, a marginal rim on the cephalic shield, the fusion of four (or more) appendages into the head, and the presence of six podites in the inner rami of the appendages (Wills *et al.*, 1998), although each of these traits is secondarily modified or lost in various lineages of the arachnomorphs. The extensive extinction of basal Arachnomorpha has perhaps been one of the greatest obstacles to studies attempting to study arthropod phylogeny based solely on the modern fauna. Two principal lineages comprise this subphylum: the trilobites (Trilobita) and the cheliceriformes (chelicerates and their extinct relatives). The trilobites are most closely related to the Cheliceriformes, a large assemblage containing, among other groups, the familiar arachnids (spiders, mites, scorpions *et al.*) and horseshoe crabs.

TABLE 3.2. Hierarchical Classification of Phylum Arthropoda

—PHYLUM ARTHROPODA—

- Subphylum †Marellomorpha
- Subphylum Arachnomorpha
  - Infraphylum †Trilobita (trilobites)
  - Infraphylum Cheliceriformes
    - Superclass †Sidneyiida
    - Superclass †Emeraldellida
    - Superclass †Sanctacarida
    - Superclass Chelicerata
      - Epiclass Pycnogonida (sea spiders)
      - Epiclass †Aglaspidida
      - Epiclass Euchelicerata
        - Class Xiphosura (horseshoe crabs)
        - Class †Eurypterida (sea scorpions)
        - Class Arachnida
          - Subclass Micrura
            - Order Palpigradi
            - Order †Haptopoda
            - Order †Trigonotarbida
            - Order Araneae (spiders)
            - Order Amblypygida (whip scorpions)
            - Order Uropygida (vinegaroons)
            - Order Schizomida
            - Order Ricinulei (ricinuleids)
            - Order Acari (mites, ticks)
          - Subclass Dromopoda
            - Order †Phalangiotarbida
            - Order Opiliones (harvestmen)
            - Order Scorpiones (scorpions)
            - Order Pseudoscorpionida
            - Order Solfugida (sun scorpions)
- Subphylum Mandibulata
  - Infraphylum Crustaceomorpha
    - †*Martinssonia et al.*
    - Superclass Crustacea
      - Epiclass †Phosphatocopida
      - Epiclass Eucrustacea
        - Class Branchiopoda
          - Subclass Sarsostraca
          - Subclass Phyllopoda
        - Class Remipedia
        - Class Cephalocarida
        - Class Maxillopoda
          - Subclass Thecostraca[a]
          - Subclass †Ascothoracida
          - Subclass †Orstenocarida
          - Subclass Tantulocarida
          - Subclass Branchiura (fish "lice")
          - Subclass Pentastomida
          - Subclass †Skaracarida
          - Subclass Mystacocarida
          - Subclass Copepoda
        - Class Ostracoda
          - Subclass Myodocopa
          - Subclass Podocopa
        - Class Malacostraca (crabs, isopods, etc.)
          - Subclass †Nahecarida
          - Subclass Phyllocarida
          - Subclass Hoplocarida
          - Subclass Eumalacostraca
  - Infraphylum Atelocerata (= Tracheata)
    - Superclass Myriapoda
        - Class Chilopoda (centipedes)
        - Progoneata
        - Class Symphyla
      - Epiclass Dignatha
        - Class Pauropoda
        - Class Diplopoda (millipedes)
          - Subclass Pselaphognatha (polyxenids)
          - Subclass †Arthropleuridea[b]
          - Subclass Chilognatha
    - Superclass Panhexapoda
        - †*Devonohexapodus et al.*
      - Epiclass Hexapoda
        - Class Entognatha
        - **Class Insecta** (Ectognatha)

[a] Includes Cirripedia (barnacles).
[b] Includes orders Arthropleurida, Eoarthropleurida, and Microdecemplicida.

Trilobita

Aside from the lumbering relics and casts of dinosaurs, perhaps the most famous lineage of fossilized organisms is that of the trilobites (Figure 3.6). These rather ovoid, marine creatures have fascinated both professional and amateur paleontologists for centuries. The group is well documented in the fossil record and was present from the Cambrian until the Permian, having become extinct during the crisis marking the end of the Paleozoic. The group was most abundant during the Cambrian and Ordovician periods, apparently experiencing declines through the later Paleozoic. Trilobites were probably benthic feeders, although a few may have been predatory. Some bore elaborate ornamentations, perhaps to prevent other marine animals from preying upon them. True

3.6. Trilobite from the Devonian of Morocco. Trilobites were the most abundant and diverse marine arthropods in the Paleozoic (there are nearly 4,000 species known). They succumbed to extinction in the Permian. Length 26 mm.

trilobites are monophyletic and are supported by the rounded terminal segment (which bears the anus), the structure of the eye, and a unique tagmosis of the pygidium composed of a series of fused segments (Fortey and Whittington, 1989; Ramsköld and Edgecombe, 1991; Wills *et al.*, 1998; Edgecombe and Ramsköld 1999; Fortey, 2001).

Cheliceriformes

Basal cheliceriformes such as Aglaspidida and Chasmataspida show an intuitive primitive similarity to the Trilobites but also resemble the early chelicerates such as xiphosurans (horseshoe crabs). Along with the marellomorphs, these lineages were at one time considered as a group, called Trilobitomorpha, that has since been recognized to be artificial (e.g., Wills *et al.*, 1998). Little is known of the extinct Aglaspidida and Chasmataspida and the best understood lineage is that of the Chelicerata. These groups are primitively similar to the xiphosurans but lack traits such as chelicerae (Briggs *et al.*, 1979), so they are likely to be stem-group Chelicerata.

Chelicerata. Included within the chelicerates are the sea "spiders," horseshoe crabs, arachnids, and their extinct relatives. Chelicerates are united by the presence of visible ecdysial lines and the loss of inner rami on the trunk appendages (Wills *et al.*, 1998). However, the most prominent, defining feature of the chelicerates is the presence of chelicerae, a trait from which they derive their name. Chelicerae are modified appendages of the first body segment. The chelicerae serve a variety of roles, principally in feeding, and form the familiar "fangs" of lineages such as the spiders. The presumed basal lineage of the Chelicerata is the Pycnogonida, or sea spiders, a group difficult to place phylogenetically and sometimes excluded from the chelicerates. The pycnogonids (also known as Pantopoda, mostly applied for the clade of living species only) are considered basal to a clade consisting of the "true" chelicerates (Euchelicerata): Xiphosura, Eurypterida, and Arachnida. There are around 1,000 living species known of pycnogonids, most of which are predators although a few feed on algae. The opisthosoma of the pycnogonids is dramatically reduced, and there is a short proboscis preceding the chelate segment of the prosoma. Pycnogonids are recorded from as far back as the Cambrian, and most paleontological work on the group has been undertaken by Hedgepeth (1955a,b), Bergstrom *et al.* (1980), and Waloszek and Dunlop (2002). Euchelicerates are united by the presence of six pairs of prosomal appendages (including the chelicerae), a 12-segmented opisthosoma, and the presence of a post-anal telson (Selden and Dunlop, 1998).

XIPHOSURA. The horseshoe crabs are one of the classic examples of evolutionary stasis and "living fossils." Xiphosurans are the only living lineage of marine euchelicerates, the five extant species being the sole survivors of a once greater radiation. Modern species of *Limulus* are remarkably similar to Paleozoic fossils and attest not only to the longevity of this group but also to the success of their design, having survived several cataclysmic extinctions throughout evolutionary history. The head, or prosoma, is covered by a large dorsal shield, with lateral compound eyes, while the opisthosoma is similarly covered by a large shield, although primitively segmented in one order (the paraphyletic "Synziphosurina"). Numerous fossil genera are recorded for Xiphosura, dating from as far back as the Ordovician. Ordovician through Devonian xiphosurans had a segmented opisthosoma and are likely a stem group to true Xiphosura (Anderson and Selden, 1997). The Xiphosura is considered to be the sister group to all other Euchelicerata (Boudreaux, 1979; Paulus, 1979; Weygoldt and Paulus, 1979; Weygoldt, 1980; Wills *et al.*, 1995; Selden and Dunlop, 1998).

EURYPTERIDA. The extinct "sea scorpions" were large, amphibious chelicerates that superficially looked like elongate xiphosurans with a segmented opisthosoma; indeed, they were once classified with them into an artificial group called Merostomata (e.g., Woodward, 1865). The eurypterids were among the first arthropods to venture onto land, although they remained principally marine. The legs were frequently modified into paddles for swimming, while the chelicerae could at times be dramatically altered into elongate grasping "arms," useful for capturing prey during aquatic chases. The telson of the body was distinctly flattened and formed a terminal spine or paddle. The eurypterids could be quite large, exceeding two meters in total body length, and were likely terrifying predators in coastal waters. Eurypterids could easily have preyed upon early vertebrate lineages (no wonder the vertebrates quickly moved onto land!). The group persisted from the Ordovician until the Permian and had a described diversity of about 300 species. Recent classificatory treatments of the Eurypterida include the works of Plotnick (1983) and Tollerton (1989).

ARACHNIDA. The arachnids are entirely terrestrial chelicerates (except for water mites, which secondarily returned to an aquatic lifestyle), and they ventured on to land sometime during the Silurian, perhaps the Ordovician. The group is almost universally predatory or parasitic, principally victimizing other arthropods, like insects. With over 80,000 described species, the arachnids are certainly the most successful lineage of the cheliceriformes. In terms of numbers of species, the spiders (Araneae) and the mites (Acari) dominate the Arachnida; they are also the most diverse ecologically. The arachnids consist of numerous orders: Opiliones (the harvestmen), Scorpiones (scorpions), Pseudoscorpionida (pseudoscorpions), Solifugida (wind "scorpions"), Palpigradi (palpigrades), Araneae (spiders), Amblypygida (whip "spiders"), Uropygida (vinegaroons), Schizomida (schizomids), Ricinulei (ricinuleids), and Acari (mites and ticks), in addition to a few extinct orders known from the Paleozoic, Phalagiotarbida, Haptopoda, and Trigonotarbida (Figure 3.7). Arachnid phylogeny has been most recently treated by Selden and Dunlop (1988), Shultz (1989, 1990), Dunlop (1999), and Dunlop and Webster (1999).

The scorpions consist of approximately 1,900 species and are among the most ancient arachnids, today occurring in tropical and warm temperate areas worldwide. Modern species range from 8 mm to 21 cm in length and live in everything from xeric to tropical habitats. Scorpions tend to be nocturnal, remaining concealed during the day in crevices or under stones. Prey is captured with their large, chelate pedipalps and is usually stunned with venom from a sting at the end of a narrow, five-segmented, tail-like metasoma. The anterior segments of the opisthosoma ("mesosoma") are relative broad and flattened, bearing the four pairs of walking legs. Scorpions exhibit maternal care, and the young, which are born live, are often carried on the back of the adult for several instars (Figure 3.8). Interestingly, the integument of scorpions will fluoresce under ultraviolet light and the use of blacklights is a standard method of collection. In fossilized forms where remains of the integument are preserved, some fluorescence may occur even after hundreds of millions of years. Fossils of the order are known from as far back as the early Silurian (e.g., Størmer, 1977) and could reach nearly a meter in length. Even though scorpions today are, like all arachnids, terrestrial, some forms from the Silurian-Carboniferous were aquatic (e.g., Rolfe and Beckett, 1984), with terrestrial species first appearing in the Devonian (Selden and Jeram, 1989; Walossek *et al.*, 1990). The earliest true scorpions are *Proscorpius osborni* from the Silurian (Figure 3.9). Mesozoic fossils are restricted to a single Triassic

3.7. Trigonotarbids (here: *Architarbus rotundatus*, from the Carboniferous of Illinois) were a diverse group of Paleozoic, terrestrial arachinds that superficially resembled large mites. Note the segmentation on the opisthosoma. YPM 00185; length of opisthosoma approx. 15 mm.

3.8. An *Opishacanthus* scorpion from Panama with its young on its back. Photo: P. J. DeVries.

3.9 (left). The earliest scorpion, *Proscorpius osborni*, from the Silurian of New York. AMNH; length 38 mm.

3.10 (right). An Early Cretaceous scorpion from Brazil's Santana Formation, approximately 120 MYO. Morone Collection; 43 mm. Photo: R. Larimer.

record (Gall, 1971) and several in the Cretaceous (e.g., Campos, 1986; Ross, 1998; Lourenço, 2001, 2002; Grimaldi *et al.*, 2002) (Figure 3.10). Scorpions are also represented in Tertiary resins (e.g., Lourenço and Weitschat, 2001; Weitschat and Wichard, 2002). Scorpion biology has been reviewed by Polis (1990), the world species cataloged by Fet *et al.* (2000), and a phylogenetic treatment provided by Stockwell (1989) and Prendini (2001).

Most spiders (Araneae) are immediately recognizable. There are about 35,000 species of living spiders, making them the most diverse of all arachnids in terms of described species. The order is ubiquitous and includes taxa with a wide range of biologies. All species are predatory; however, their biology ranges from solitary to group hunters, or even commensals and cleptoparasites. Some species even live in large, social colonies. Certainly the principal factor in the success of spiders is their silk, which is used for prey capture as well as to construct elaborate retreats and protective cases for their eggs. The body of spiders is composed of a well-divided prosoma (sometimes called the *cephalothorax*) and opisthosoma, the former bearing the walking legs and mouthparts, the latter region being robust and bearing on its ventral surface near the apex a set of spinnerets. The spinnerets are perhaps the hallmark trait of spiders, which was how a very fragmentary Devonian fossil was identified (Shear *et al.*, 1989). The order is divided into two suborders: the Mesothelae and Opisthothelae, the latter being further divided into the infraorders Mygalomorphae and Araneomorphae (Coddington and Levi, 1991). Mesothele spiders are the most primitive, living members of the order and are generally large. The suborder is immediately notable for having a segmented opisthosoma; all other spiders have the segments indistinguishably fused. Mygalomorph spiders include the familiar tarantulas, trap-door spiders, and other large, hairy taxa that do not spin aerial webs (Figure 3.11) and are largely tropical but that are also well known in xeric habitats. The silk is used to construct burrows, either in the soil or in wood, and generally extends from the opening for some distance and is used as an extension of the spider's sensory area, detecting prey that walks across the mat of silk. Araneomorph spiders include all other lineages, from the common garden spider, to crab (Figure 3.12) and ground spiders. Many, but certainly not all, araneomorph species spin orb webs – the familiar nets used for sieving the air for prey (Figure 3.13). The oldest evidence of spiders is *Attercopus fimbriunguis* (Shear *et al.*, 1987;

3.11. Tarantulas and other mygalomorph spiders are hairy, massive spiders, many of which spin trip lines along the ground to detect prey passing by their burrows. Photo: Valerie Giles.

3.12. A crab spider consuming a moth. Some spiders are sit-and-wait predators, such as this thomisid. It is cryptic among the blossoms, allowing it to ambush wary pollinating insects. Photo: Valerie Giles.

Selden *et al.*, 1991), considered the sister group to all other Araneae, and an unnamed spinneret (Shear *et al.*, 1989), both from the Devonian of New York. The earliest representatives of Araneae proper are Carboniferous representatives of the Mesothelae (Selden, 1996). The first opisthothele spiders are known from the Triassic and are of the Mygalomorphae (Selden and Gall, 1992), while araneomorphs are first known from the Jurassic (Eskov, 1984; Eskov and Golovatch, 1986). Spiders were numerous in the Cretaceous and Tertiary (e.g., Wunderlich, 1986; Selden, 1990, 2001, 2002; Johnston, 1993), particularly in ambers from throughout these periods (e.g., Wunderlich, 1988, 2000; Eskov, 1992; Eskov and Wunderlich, 1994; Penney, 2000, 2001, 2002). Foelix (1982) provided the most detailed account of spider biology, while Coddington and Levi (1991) summarized the higher classification of the order.

The pseudoscorpions are minute, predatory arachnids with large, chelate pedipalps. There are about 2,500 species, found in leaf litter and moss or under stones or bark. Many species are phoretic on other arthropods, grasping with their pedipalps to "hitch" rides. Species exhibit subsocial behavior with an extended brood care and even build brood chambers with silk extruded from glands in their chelicerae. Unlike other arachnids, the pseudoscorpions and solfugids, their nearest relatives, lack a patellar segment in the leg (Shultz, 1990). The Pseudoscorpionida is well represented in Tertiary (e.g., Schawaller, 1982; Weitschat and Wichard, 2002) and Cretaceous ambers (e.g., Schawaller, 1991; Azar, 2000; Judson, 2000; Grimaldi *et al.*, 2002). Remains of pseudoscorpions have been recovered from the Devonian of New York (Shear *et al.*, 1989; Schawaller *et al.*, 1991). Pseudoscorpion biology is reviewed by Weygoldt (1969).

Wind scorpions (Solfugida) are moderate-sized (7–70 mm) arachnids, which are remarkably swift runners that chase down prey (Figure 3.14). There are about 900 species known from xeric regions of the world except for Australia. Perhaps the most notable feature of the solfugids are their enormous, stout, chelate chelicerae. These tremendous "jaws" allow solfugids to shred their prey, which for the larger species can also include small vertebrates. The pedipalps are rather stout relative to the walking legs and serve a tactile function. The oldest representative of the order is *Protosolpuga carbonaria* (Petrunkevitch, 1913; Selden and Shear, 1996); otherwise, solfugids are unknown until the Early Cretaceous (Selden, 1996) and Miocene Dominican amber.

The harvestmen (Opiliones) are spider-like arachnids, noted for their rather short, robust bodies and long, thin legs. The approximately 5,000 species occur throughout the world and range in size from less than a millimeter up to about 23 mm in body length, although their leg span can be several times this length. Species are omnivorous, but, unlike other arachnids, they digest solid food (the other orders pre-orally digest their prey and then consume the dissolved fluids). The prosoma is broadly fused to the opsithosoma, giving them the appearance of having a single body tagma. Fossils of

3.13. A black widow (*Latrodectus*: Theridiidae) perched in her web, in the Dominican Republic. The genus is renowned for the potency of its venom and cannibalism of the males by females. Photo: D. Grimaldi.

3.14. A wind scorpion, or solpugid, in southern Texas, with a roach in its chelicerae. These are swift ground predators. Photo: P. J. DeVries.

3.15. A whip scorpion, or amblypygid, in Panama. They are flat and live under large rocks, on the walls of caves, on tree trunks, and under loose bark. Photo: P. J. DeVries.

harvestmen are known from the Early Carboniferous (e.g., Petrunkevitch, 1913; Wood *et al.*, 1985), but their diversity in the fossil record is best documented from Cretaceous and Cenozoic deposits (e.g., Jell and Duncan, 1986; Weitschat and Wichard, 1998, 2002).

The palpigrades are minute (less than 3 mm long), soft-bodied, soil- or humus-dwelling arachnids with a modern diversity of about 125 species. They have a long, jointed flagellum at the apex of the opsithosoma and superficially resemble minute schizomids (discussed later). The only fossil palpigrade is *Palaeokoenenia mordax* from the Pliocene of Arizona (Rowland and Sissom, 1980).

3.16. A rare amblypygid in Miocene amber from the Dominican Republic. Morone Collection, M0699; body length 9 mm.

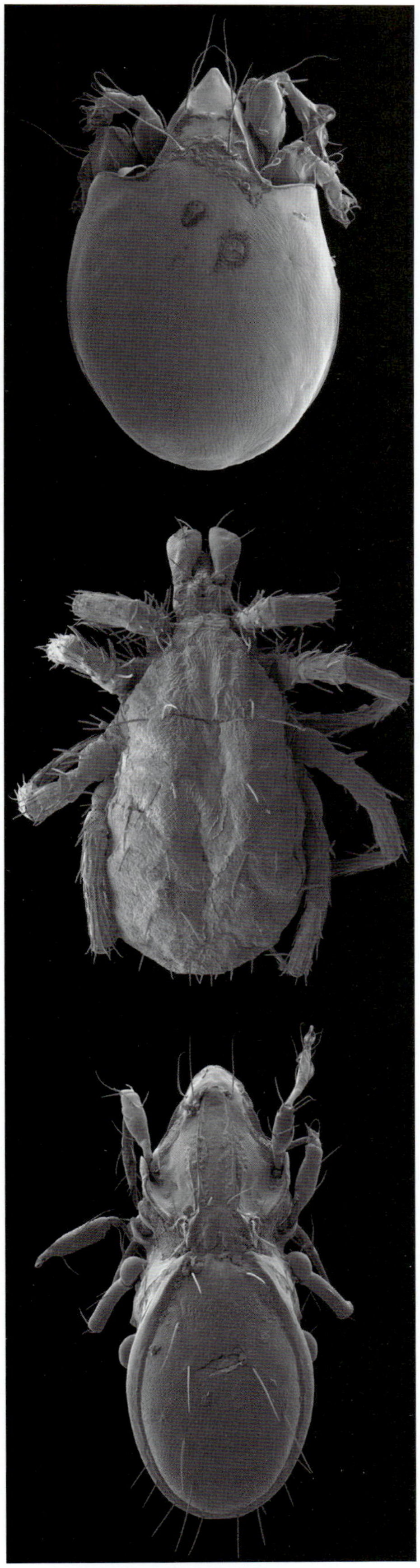

3.17. Representative mites (Acari). Mites are the most diverse lineage of arachnids, of which there are vast numbers of undescribed species. Scanning electron micrographs; not to same scale.

3.18. A mite in mid-Cretaceous amber from Myanmar. Mites are relatively common and diverse in fossiliferous ambers, but they are essentially unstudied. AMNH.

Amblypygida, or whip spiders, are moderate to large (15–47 mm) arachnids with a flattened, rather circular body and long, thin legs held close to the substrate (Figure 3.15). Species live in stone crevices, caves, and hollow trees; under loose bark; and in leaf litter, principally in tropical environments. The front pair of legs are particularly elongate and are not used in locomotion but instead are used as "antennae," which they sway back and forth to detect prey. The pedipalps are enlarged and beset with numerous, stout spines that allow them to snare prey easily, which are then consumed using the chelicerae. Approximately 80 species are recognized today. Definitive fossil whip spiders are known from the Late Carboniferous of North America and Europe (Dunlop, 1994) but are mostly represented in Tertiary deposits (e.g., Schawaller, 1979) (Figure 3.16). However, fragments of a putative amblypygid have been recovered from the Devonian of New York (Shear *et al.*, 1984). Weygoldt (1996) has provided the most comprehensive treatment of the order.

The Uropygida, commonly referred to as vinegaroons, are superficially similar to scorpions because of their large,

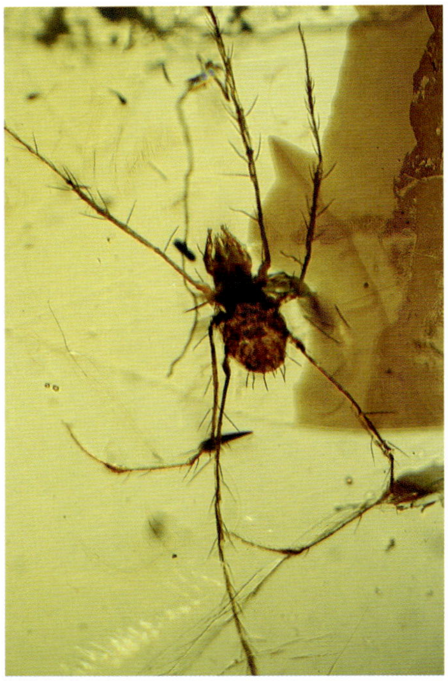

3.19. A mite in Early Cretaceous Lebanese amber. AMNH.

3.20. A spined mite in amber from Myanmar. AMNH Bu342; length 1.0 mm.

chelate pedipalps and their defense posture of raising their opsithosoma. Their common name refers to the spray of acetic acid (essentially vinegar), which they disperse from pygidial glands when disturbed. Species live in subterranean burrows and hunt small arthropods. Vinegaroons are known from as early as the mid-Carboniferous of Europe (Brauckmann and Koch, 1983) as well as the Early Cretaceous of Brazil (Dunlop, 1998) but are otherwise unknown from the fossil record.

The order Schizomida is overall rather similar to the vinegaroons but they are smaller (1.5–15 mm), and indeed are essentially "miniaturized" uropygids (e.g., Shultz, 1990; Selden and Dunlop, 1998). The oldest fossils are a single species from the Oligocene of China (Lin *et al.*, 1988) and three from the Pliocene of Arizona (Petrunkevitch, 1945).

Ricinulei are small, blind, tick-like arachnids that occur in leaf litter and caves in the equatorial tropics of the Americas and Africa. The most remarkable trait for the order is the presence of a "hood" (*cucullus*), which hinges to the front of the prosoma and effectively covers the chelicerae. The fossil ricinuleids were revised by Selden (1992) who revealed a dramatic diversity in the Carboniferous, significantly greater than that today, but otherwise fossils of the order are unknown.

Second in diversity for the numbers of described (named) species of arachnids are the mites and ticks (Acari), with approximately 30,000 species known (Figure 3.17). The number of mite species will eventually far exceed that of spiders owing to the remarkable number of undescribed species from virtually every habitat. Most mites are minute and, like the spiders, have taken over a dramatic range of environments, including 5,000 species that are aquatic. Mites are particularly abundant in soil and organic debris (such as the forest floor), where the number of individuals can easily outnumber all other arthropods. Many mites are ectoparasitic on both vertebrate (e.g., ticks) and invertebrate hosts, in some cases co-evolving with their hosts. In addition, the order includes scavengers and the only herbivorous arachnids, some of which can be quite damaging to crops. Like the ricinuleids (their closest, extant relatives) the Acari have a

3.21. An argasid ("soft") tick, *Carios jerseyi*, in Late Cretaceous amber from New Jersey. It probably fed on birds or feathered dinosaurs. AMNH NJ8; length 520 μm.

distinct *gnathosoma* that bears the chelicerae and pedipalps. Despite the assertions of van der Hammen (1972, 1989), mites are considered to be monophyletic (e.g., Shultz, 1990) and of two basic lineages – the Anactinotrichida and the Actinotrichida, the latter including the ticks (Ixodida). Anactinotrichid mites are known from as early as the Rhynie chert of Scotland (Hirst, 1923) and other Devonian sites (Norton *et al.*, 1988, 2002; Kethley *et al.*, 1989). Mites are also common as fossils in both Cretaceous and Tertiary ambers (e.g., Sellnick, 1931; Azar, 2000; Rasnitsyn and Ross, 2000; Grimaldi *et al.*, 2002; Weitschat and Wichard, 2002) (Figures 3.18, 3.19, 3.20), and ticks have also been found as far back as the Cretaceous (Klompen and Grimaldi, 2001; Grimaldi *et al.*, 2002) (Figure 3.21). General references on the biology, ecology, and evolution of Acari include Krantz (1970), Woolley (1988), Schuster and Murphy (1991), Evans (1992), Houck (1994), and Walter and Proctor (1999).

CRUSTACEOMORPHA

This, almost entirely marine, subphylum is perhaps the insectan analogue for the oceans. Whereas insects have become vitally influential in terrestrial ecosystems, so the crustaceans have become in the oceans. Some have moved onto land (e.g., Isopoda), and others into freshwater (e.g., crayfish, some copepods). The group includes the familiar, living Crustacea as well as numerous extinct taxa that are considered to be stem groups to either constituent lineages of crustaceans or to the Crustacea as a whole. There are approximately 50,000 living crustacean species, which range in size from miniscule (less than a millimeter) to enormous (355 cm). Six extant classes are recognized in the Crustacea (Schram, 1986; Martin and Davis, 2001): Branchiopoda (water fleas, brine, tadpole shrimp), Remipedia (remipedes), Cephalocarida (cephalocarids), Malacostraca (crabs, lobsters, isopods, crayfish, shrimp), Ostracoda (seed shrimp, ostracods), and Maxillopoda (barnacles, branchiurans, pentastomids, copepods), but these are not all monophyletic (e.g., Schram and Hof, 1998). They are immediately recognizable for their five pairs of head appendages: one set of mandibles, two sets of maxillae, and two pairs of antennae. Most species belong to the Malacostraca (which includes the familiar amphipods, isopods, and crabs) and are benthic creatures.

Like the insects, their biology is incredibly diverse with species ranging from detritivorous to predatory to parasitic, and from solitary to social. A complete treatment of the diversity of both form and biology in the Crustacea is beyond the scope of this volume. Major accounts include Abele (1982), Schram (1983a, 1986), Gore and Heck (1986), Bauer and Martin (1991), Jones and Depledge (1997), Schram and Hof (1998), and Martin and Davis (2001).

MANDIBULATA

Mandibulata Versus Schizoramia

The placement of Crustaceomorpha has been of considerable contention, with the primary schism lying between paleontologists who believe the group to be sister to the Arachnomorpha, and the neontologists who support a mandibulate arthropod group (i.e., Crustacea allied to the Hexapoda and Myriapoda) (Figure 3.1). In most recent analyses, albeit ones that did not include several critical fossils, the Mandibulata is supported as monophyletic (e.g., Scholtz *et al.*, 1998; Bitsch, 2001; Giribet *et al.*, 2001; Fanenbruck *et al.*, 2004) and will therefore be considered as the working hypothesis for the discussion herein.

Atelocerata Versus Pancrustacea

Within the Mandibulata we are faced with a similar problem concerning the relationships of lineages. As discussed, the mandibulate arthropods include the Crustacea, the Myriapoda, and the Hexapoda. Traditionally, myriapods and hexapods have been considered sister groups (e.g., Snodgrass, 1938) and together called either Tracheata or, more widely, Atelocerata (Figure 3.22). Familiar traits defining the Atelocerata include the loss of the second antennal pair, presence of a tentorium (internal head skeleton), a respiratory system involving a system of fine tubules or *tracheae*, and Malpighian tubules. However, based mostly on recent molecular studies, the monophyly of Atelocerata has been challenged, and an alternative relationship between Crustacea and Hexapoda put forward (e.g., Zrzavý and Stys, 1997). This alternative grouping, called Pancrustacea (= Tetraconata), has gained considerable support among molecular and developmental biologists (e.g., Zrzavý *et al.*, 1998b). The Pancrustacea is supported by some molecular analyses (e.g., Field *et al.*, 1988; Turbeville *et al.*, 1991; Ballard *et al.*, 1992; Boore *et al.*, 1995, 1998; Friedrich and Tautz, 1995; Giribet *et al.*, 1996; Giribet and Ribera, 1998), while morphological traits are not outwardly apparent and are poorly understood across a variety of taxa: for example, suppression of distal mandibular segments (Popadic *et al.*, 1996, 1998; Deutsch, 2001), neurogenic pattern-formation processes (Whitington *et al.*, 1991; Osorio *et al.*, 1995; Dohle, 1998, 2001), and ultrastructure of the compound eye (Paulus, 1979; Osorio and Bacon, 1994; Osorio *et al.*, 1995; Dohle, 1998, 2001). This grouping also indicates that several complex morphological features, specifically the tentorium, tracheae, and Malpighian tubules, were independently evolved. An extensive study of both morphological and molecular data by Edgecombe *et al.* (2000), however, supported Atelocerata monophyly. For the time being Atelocerata will be adopted pending the accumulation of more evidence to the contrary.

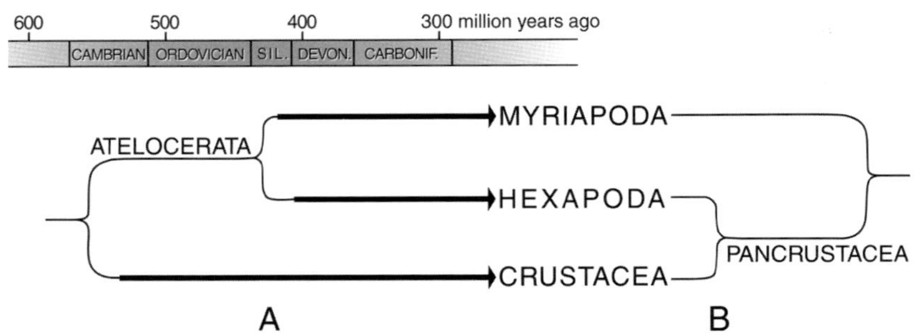

3.22. Phylogeny of Mandibulata showing alternative relationships of the Atelocerata (A) and Pancrustacea (B).

Myriapoda

The Myriapoda has not been widely supported as a natural group (although see Zrzavý *et al.*, 1998b). Indeed, several studies indicate that the centipedes (Chilopoda) and symphylans are basal (although not themselves related), while the Dignatha (Pauropoda and Diplopoda [millipedes]) comprise a sister group to the Hexapoda (e.g., Wheeler, 1998; Kraus, 2001). The symphylans are sometimes classified with the Dignatha into a larger group called the Progoneata. Other views on the phylogeny of Myriapoda are presented by Wheeler *et al.* (1993a), Kraus and Kraus (1994), Borucki (1996), Kraus (1998, 2001), Ax (1999), and Regier and Shultz (2001b). Overall there is little consensus on myriapod phylogeny, and we have adopted for the time being the conservative, traditional view of relationships (Figure 3.23). The biology of myriapods is summarized by Camatini (1979).

The centipedes (Chilopoda) are, along with the millipedes (Diplopoda), the best-known group of Myriapoda. Centipedes range in size from minute to gigantic and are active, terrestrial predators. Species occur in numerous habitats but are particularly abundant in the tropics, with a diversity of approximately 2,600 species worldwide. The body of centipedes is typically somewhat compressed dorsoventrally, with the first maxillae expanded at their base and forming a ventral cover for the other mouthparts. The number of leg pairs varies from 15 to 177 and centipedes are capable of quick movement, although not owing to so many legs but instead to the fact that the body generally is suspended below the attachments to the legs, allowing them to step over the legs of preceding segments during movement. Centipedes are principally nocturnal and are carnivorous except for species of Geophilomorpha, which are omnivorous and prefer a diet of plant tissue. The appendages of the first trunk segment are developed into maxillipeds, or more commonly "forcipules," and are used to poison prey. The bases of the maxillipeds fuse to form a lower, shovel-like base for the head. Other defining features of the centipedes include the composition of the stemmata of the eyes and the complete loss of median eyes. Centipede phylogeny has been elaborated upon by Shear and Bonamo (1988), Borucki (1996), Prunescu (1996), Shultz and Regier (1997), Edgecombe *et al.* (1999), Giribet *et al.* (1999), Kraus (2001), and Regier and Shultz (2001b). Fossils centipedes are among the earliest terrestrial arthropods known (Jeram *et al.*, 1990; Shear *et al.*, 1998) (e.g., Figure 3.24).

Symphyla are small centipede-like animals with 15–22 segments, 12 pairs of legs, and, like the centipedes, long antennae (Figure 3.25). Development, like the Pauropoda (below) is anamorphic; juveniles hatch with 6–7 pairs of legs and progressively add appendages until the full compliment is achieved. Defining traits for symphylans include the unpaired genital opening near the anterior end of the trunk, loss of eyes, a pair of spiracles on the sides of the head, second maxillae fused to form a labium-like structure (analogous to the labium of Hexapoda), spermatheca formed as pockets positioned in the mouth, and terminal spinnerets. They occur in moist soil, in decaying wood, in moss, and under stones, and species are herbivorous. Only two fossil records are known for the Symphyla: one in Baltic amber (Bachofen-Echt, 1949) and one in Dominican amber.

The Pauropoda are minute, infrequently encountered myriapods living in moist leaf litter, in the soil, or under stones or bark (Figure 3.26). There are approximately 500 species principally occurring in tropical or warm temperate regions. Juveniles begin their life as hexapods (only three pairs of legs) but add pairs as they mature, ultimately reaching a total of nine pairs on the anterior trunk segments. The heads of pauropods tend to be relatively small, possessing characteristically branched antennae, and they lack eyes. The posterior head segment lacks appendages and is separated from the remainder to form a circular collar (*collum*), similar

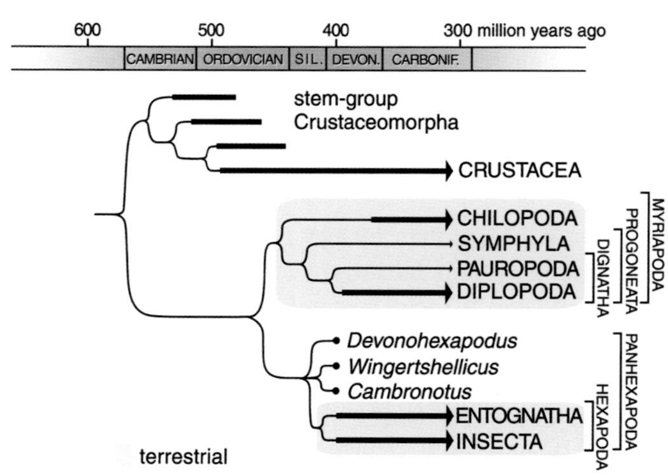

3.23. Phylogeny of Mandibulata, indicating relationships among the major lineages of Atelocerata.

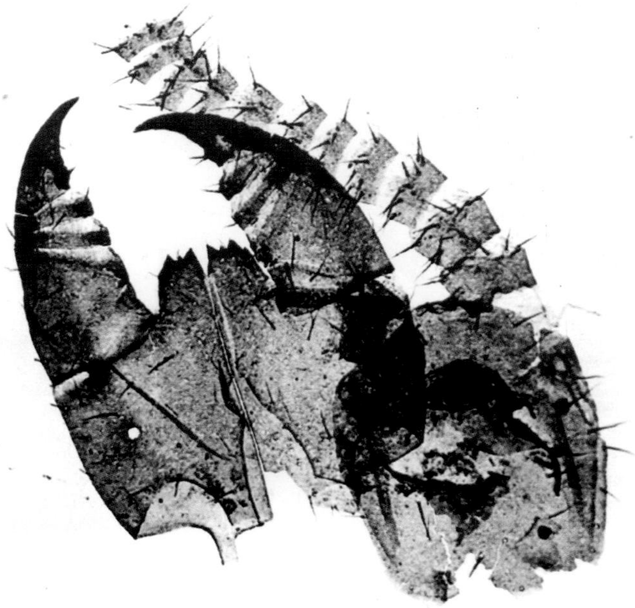

3.24. Fragments of a mid-Devonian centipede from New York. Arthropods are the earliest known land animals. AMNH 411-7-AR97; length 1.5 mm; from Shear and Bonamo (1988).

3.25. Scanning electron micrograph of a Recent symphylan. Length 2.1 mm.

to the millipedes, although in the latter vestiges of appendages are usually present. The trunk is composed of 11 segments. As in the millipedes, each tergal plate covers two body segments (*diplosegments*); however, the legs are not doubly paired as in most Diplopoda. Aside from antennal structure, the defining features of the class include the reduction and fusion of the second maxillae and the occurrence of an eversible vesicle on the first trunk segment. Little is known about pauropod biology or phylogeny. The sole fossil record of a pauropod is *Eopauropus balticus* in mid-Eocene Baltic amber (Scheller and Wunderlich, 2001), a species that is quite modern in appearance. The pauropods are likely quite ancient, but their small body size, delicate bodies, and habitat preference precludes fossilization.

Millipedes are generally herbivorous or detritivorous and, like the Pauropoda, tend to live in leaf litter, in the soil, or beneath stones, logs, and bark. They are also the most diverse of all myriapods, with around 10,000 species known worldwide. Individuals are relatively common and occur worldwide. They range in size from 2 mm to an incredible 28 cm (some fossil diplopods of the Arthropleurida exceeded 1.8 meters!) (e.g., Kraus and Brauckmann, 2003). The trunk is composed of diplosegments in most species and bears two sets of legs on each, except for the anterior three trunk segments, which possess a single pair of legs each. The number of legs varies widely with a maximum record of 350 pairs (700 legs!) in *Illacme plenipes*. Although not equipped with poisonous "fangs" like the centipedes, millipedes produce cyanogenic compounds from repugnatorial glands to protect themselves from predators. Fossils of millipedes extend back at least to the early Devonian and are known from numerous localities from that time period until the present day (e.g., Shivarudrappa, 1977; Dzik, 1981; Shear, 1981; Hannibal, 1984; Donovan and Veltkamp, 1994; Duncan *et al.*, 1998; Schneider and Werneburg, 1998; Grimaldi *et al.*, 2002). Millipede phylogeny has been studied by Enghoff (1984, 2000), Regier and Shultz (2001b), and Sierwald *et al.* (2003). Hopkin and Read (1992) have summarized millipede biology.

THE INVASION OF LAND

Insects are principally terrestrial organisms. Indeed, despite frequent colonization of freshwaters by mayflies, dragonflies, diving beetles, predatory water bugs, various midges, and other groups, they are all terrestrial organisms by original design. The transition to land took place in the ancestor of insects and their closest relatives, the Entognatha. The freshwater life-histories of immature mayflies (Ephemeroptera), dragonflies and damselflies (Odonata), and many other insects evolved later. The occurrence of marine, stem-group Hexapoda suggests that the invasion of land occurred independently in the Myriapoda and Hexapoda (Figure 3.23).

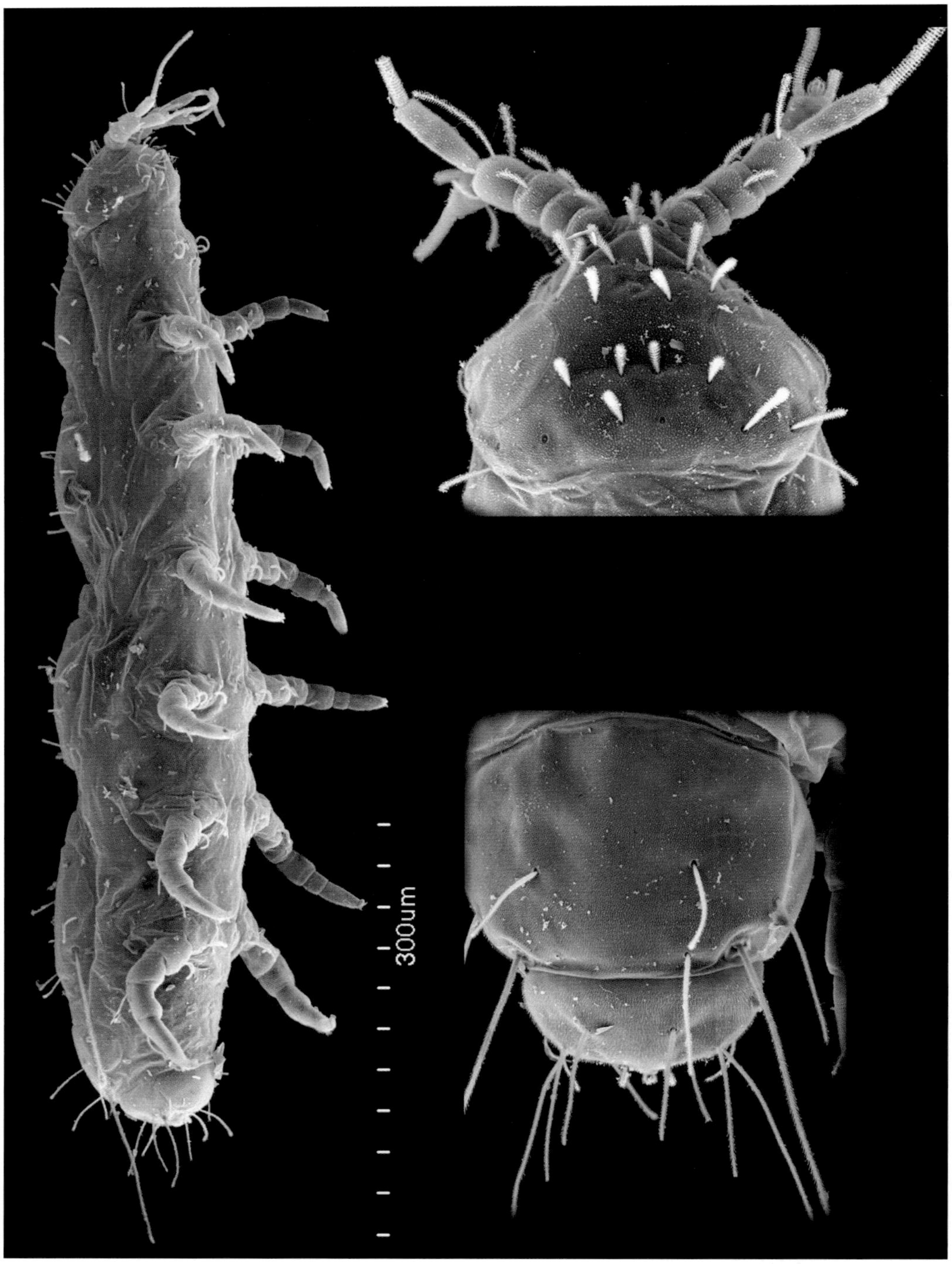

3.26. A representative pauropod, showing the eyeless head (above, right), the distinctively branched antennae, plumose setae, and the terminal trunk segments (below, right). Not to same scale, body length 1.1 mm.

Terrestrialization also occurred independently in the Crustacea (Isopoda), Cheliceriformes (Chelicerata), Tardigrada, and Onychophora (Euonychophora). So when did all of these groups depart from the waters and first explore the terrestrial biosphere? The earliest assemblages of terrestrial arthropod fossils are from the Late Silurian (Jeram *et al.*, 1990). However, fossilized trackways of arthropods on land are known from the Early to mid-Ordovician (Sharpe, 1932; Johnson *et al.*, 1994; MacNaughton *et al.*, 2002), tens of millions of years earlier. These fossilized tracks are not of insects but instead appear to be early cheliceriforms. Indeed, the earliest evidence of insects is from the Early Devonian. Early terrestrial tracks document the presence of various arthropod lineages and support the view that insects themselves originated in a terrestrial environment. It is interesting to note that the arthropods comprised the earliest known terrestrial animals. The fact that these early land animals were all predatory indicates that the selective pressure for terrestrial living was perhaps not an herbivorous diet of land plants. Instead, early, amphibious arthropods may have ventured on to land as part of their reproductive cycle; they may have sought temporary refuge on land from predators lurking in coastal waters or to feed upon worms and other animals feeding on microbial and algal mats growing at the water's edge.

HEXAPODA: THE SIX-LEGGED ARTHROPODS

The epiclass Hexapoda consists of the entognathous hexapods and the true insects (Table 3.3). The group is supported by the fusion of the second maxillae to form a labium (convergent with Symphyla); the loss of an articulating endite on the mandible; fixation, at least primitively, of the number of abdominal segments to 11; and loss of jointed abdominal appendages (Kristensen, 1991). Early, nonterrestrial "hexapodous" arthropods are known from the earliest Devonian of Germany (Bartels, 1995; Briggs and Bartels, 2001; Haas *et al.*, 2003). These fossils lack the true hexapod condition of 11 abdominal segments (or less) and loss of appendages on the tenth abdominal segment. Instead, these fascinating marine organisms have loosely differentiated thoracic and abdominal tagma, with numerous abdominal segments bearing appendages (Figure 3.27). Like true Hexapoda, they show a reduction to a single pair of antennae (although preservation in some of these fossils is a bit ambiguous), as well as well-developed appendages on the "thoracic" segments in comparison to the trunk. Rather than include such stem-group marine forms into the Hexapoda, we prefer to consider them as members of superclass **Panhexapoda**, a larger clade containing Hexapoda (i.e., Entognatha and Insecta), and these stem-lineage, hexapodous, marine organisms (Figure 3.23; Table 3.2). Certainly exploration of the Devonian and latest Silurian, both terrestrial and marine, will give us our most profound insights into the origination and differentiation of the hexapods.

The first major dichotomy among hexapods is the division into Entognatha and Ectognatha, the latter more widely known as the Insecta. These two divisions were recognized as early as 1888 by Grassi, but their defining features were best established by Hennig (1953, 1969, 1981).

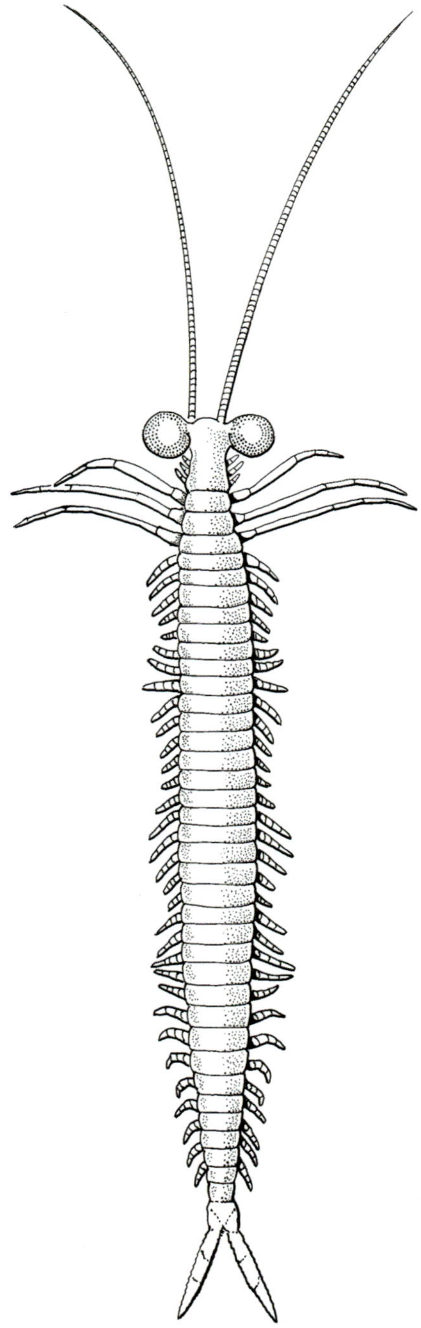

3.27. A reconstruction of the Early Devonian marine panhexapod, *Devonohexapodus bocksbergensis*. *Devonohexapodus* and other marine panhexapods are stem groups to terrestrial hexapods (Entognatha and Insecta). Redrawn from Haas *et al.* (2003).

TABLE 3.3. Hierarchical Classification of Epiclass Hexapoda

Epiclass HEXAPODA
Class Entognatha
Order Diplura
Ellipura
Order Protura
Order Collembola
Class Insecta (= Ectognatha)

ENTOGNATHA: PROTURA, COLLEMBOLA, AND DIPLURA

Three orders (each sometimes given the rank of class) are included in this group: the Collembola (springtails), the Protura (proturans), and the Diplura (diplurans). All have generally edaphic lifestyles and, except for Collembola, are not widely encountered. As implied by the name, the principal feature of this group is the development of *entognathy*, in which the mouthpart appendages are recessed within a *gnathal pouch* on the head capsule. More primitive lineages (e.g., Symphyla, Diplopoda) as well as other hexapods all have ectognathous mouthparts. Entognathy is also unique in that during embryogenesis lateral folds of the head (called *plica orales*) form over the buds of the mouthparts (Tuxen, 1959). The plica orales grow downward to fuse with the base of the labium (at the postmentum). The result is the formation of the gnathal pouch that entirely encloses the mandibles and maxillae. The jaws lie essentially horizontal, and the mandibles are long and narrow, with their monocondylic (i.e., singly articulated) bases near the back of the head and often sunken into the posterior wall of the gnathal pouch. Throughout hexapods the maxillae are protrusible and retractible, owing to the articulations between the cardo and stipes, and entognaths have the derived feature of protrusible/retractible mandibles as well. This movement is achieved by a set of dorsal muscles that originate on the head capsule (Tuxen, 1959). As can be imagined, the tentorium has been radically and uniquely rearranged so as to accommodate this distinctive style of mouthparts. Without question, entognathan mouthparts are highly specialized and derived and partly define the monophyly of Entognatha. Other defining features of the group include the reduced or completely absent compound eyes (although this may be convergent owing to similar edaphic lifestyles), reduced Malpighian tubules, and elongate, saclike ovarioles (except Japygidae, which is convergently more similar to Insecta).

Within the Entognatha the Collembola appear to be more closely related to the Protura and are sometimes together referred to as Class Ellipura (Börner, 1910). Defining features of the Ellipura include the absence of cerci and the presence of simple papillae in place of Malpighian tubules, paired ovarioles developed as elongate sacs, and a linea ventralis (Tuxen, 1958, 1959). The *linea ventralis* is a longitudinal groove that runs along the middle of the ventral part of the body, which has lateral crests and extends from the opening of the labial glands caudad onto the neck membrane in Protura and to the preabdominal tube in Collembola. They are further characterized by unsegmented (i.e., monomeric) tarsi and simple claws.

Although the Diplura are similarly entognathous, this order has at times been placed as sister to the Insecta, thereby leaving Entognatha paraphyletic (e.g., Kukalová-Peck, 1987, 1991; Koch, 1997; Kraus, 1998). A relationship between Diplura and Insecta (i.e., the "Euinsecta"), however, is only weakly supported, and recent morphological and molecular studies have further recognized a monophyletic Entognatha (e.g., Bitsch and Bitsch, 1998, 2000; Carapelli *et al.*, 1998; Frati *et al.*, 1998; Wheeler *et al.*, 2001; D'Haese, 2002). Those who support a Diplura + Insecta relationship base their argument on the divided ovarioles, epimorphic development, and paired claws common to both groups. Furthermore, such authors cite slight differences in the development of entognathy in Diplura relative to Ellipura, whereby the plica orales extend to the labium but remain differentiated from it by a longitudinal sulcus (e.g., Ikeda and Machida, 1998). Frequently a small sclerite, called the *admentum*, forms between the prementum and the plica orales, a further difference between Ellipura and Diplura. However, such differences likely merely represent unique features in diplurans as an elaboration upon the standard entognathous condition and not independent derivations of entognathy.

Fertilization is indirect in the Entognatha, and as such the genitalic structure is impressively simple, consisting externally merely of gonopores used for either depositing a spermatophore or receiving one. This feature has not been considered of general phylogenetic importance, though the simplified gonopore may represent a further defining feature of this group. Perhaps the most interesting aspect of the Devonian marine panhexapods comes from the structure of the genital appendages described for *Devonohexapodus bocksbergensis* (Haas *et al.*, 2003). A pair of abdominal segments near the apex of this tagma are similar to primitive ectognathan genital segments, showing apparent gonapophyses; thus they are similar to a primitive ovipositor that in Insecta appears on the eighth and ninth abdominal segments. If such structures truly existed in basal lineages of Panhexapoda rather than being derived at the origin of Hexapoda or Insecta, then the formation of a rudimentary ovipositor is phylogenetically more primitive than once believed. Furthermore, it suggests that the complete loss of genital appendages in the Entognatha is a secondary reduction and, thus, is a derived trait uniting these lineages rather than being a vestige from a more distant ancestry.

Protura

The Protura are rarely encountered, minute hexapods that are overall rather simple in their morphology (Figure 3.28). Approximately 500 species of Protura are distributed across all zoogeographic regions. Tuxen (1963) explored the relationships among the genera recognized at that time and monographed the world species the following year (Tuxen, 1964). Numerous species have been subsequently added (e.g., Tuxen, 1967b), and a new revision of the order is needed. More recent cladistic studies have focused on the suprageneric classification of proturans (e.g., Yin, 1983, 1984), but have not been widely followed; see also Francois (2003). Perhaps the premier feature of the order is that, while hexapodous, the proturans are *functionally* tetrapods. The anterior legs are directed forward and are not used in locomotion; they are instead lifted above the ground to function as sensory appendages and are, in fact, covered with sensory structures. The types and distributions of these sensory sensillae are of taxonomic importance in the group. Other defining features of the Protura include 12 abdominal segments; no antennae; rudimentary appendages on the first three

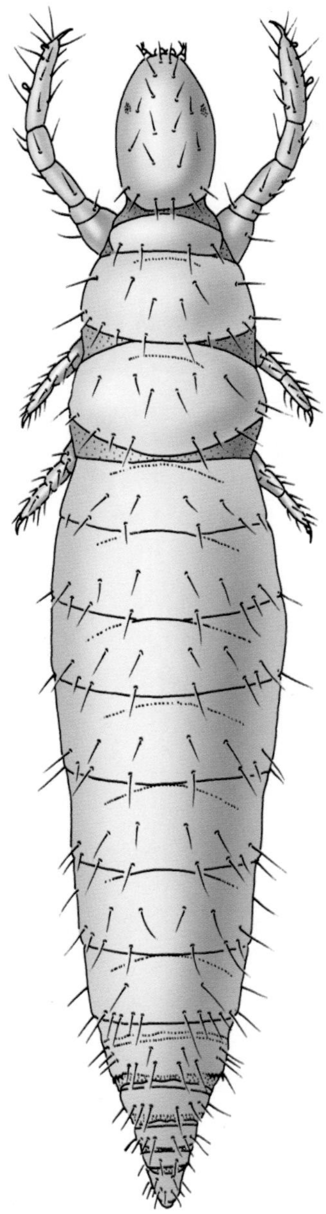

3.28. A proturan. The forelegs are not used in walking but are modified into sense organs, functioning like antennae.

abdominal segments; eversible vesicles at the apices of the abdominal appendages; the gonopore positioned on the eleventh abdominal segment; a transverse sclerite (sometimes considered vestigial and fused gonocoxae) in the genital chamber; a pair of lateral, genital plates (sometimes considered parameres) in the male; reduced deutocerebrum of the brain; partial fusion of the ganglia in the ventral nerve cord; and no peritrophic membrane in the gut. These are highly modified arthropods. As in Collembola the cerci are lacking, but proturans do not molt after sexual maturity (springtails, diplurans, and primitive insects do). In addition, developing Protura add segments between molts, starting with nine abdominal segments and progressing to the full compliment of 12 as in myriapods and possibly retained from an ateloceratan ancestor.

The order is currently divided into two superfamilies: the Eosentomoidea and the Acerentomoidea, each with two families (although see Yin, 1983, 1984, for an alternative familial classification). The superfamilies differ from each other by the occurrence of spiracles, tracheae, and a striate band on the eighth abdominal segment. Eosentomoidea lack the striate band, while possessing tracheae and spiracles; Acerentomoidea have just the opposite. The features defining Acerentomoidea are notable, derived characters and justify its recognition as a monophyletic group. However, the eosentomoids are based solely on the absence of acerentomoid synapomorphies and are certainly paraphyletic (although Yin, 1983, 1984, suggests the complete opposite).

Proturans occur in moss, rotting wood, soil, and leaf litter, where they are believed to feed on mycorrhizal fungi. The biology of proturans is poorly understood, and fossils of the order are entirely lacking. Indeed, their minute, soft bodies would not easily fossilize. Amber preservation would be ideal, but proturans are not arboreal and would therefore not readily encounter resin.

Collembola: The Springtails

The most familiar of all the entognaths are understandably the springtails (Figures 3.29, 3.30), which are also the most diverse and commonly encountered lineage of Entognatha with about 6,000 species. Collembola live in diverse habitats worldwide; from caves, to alongside fresh or marine waters, to soil and decomposing vegetation. Most species feed on fungal matter, decomposing debris, and fecal material of other invertebrates or will prey on microorganisms. A few species feed on fresh plant material. The most recent major account of the Collembola is that of Hopkin (1997) and the phylogenetic studies of Lee *et al.* (1995) and D'Haese (2002). Christiansen and Bellinger (1998) have treated the North American fauna; Greenslade (1994), the Australian fauna; and Mari-Mutt and Bellinger (1990), the Neotropical fauna.

The order is universally supported as monophyletic and is easily characterized by the reduction of the abdomen to six segments (although owing to partial fusion it sometimes appears to have even fewer). They also have short, typically four-segmented antennae (the fourth segment is sometimes subsegmented); thoracic sterna divided into lateral basisternites by the linea ventralis; legs with tibiae and monomeric tarsi fused to form a tibiotarsus; a pair of eversible vesicles at the apex of a ventral tube (called the *collophore*) on the first abdominal segment (Figure 3.29); and the location of the gonopore on the fifth abdominal segment. Despite this impressive suite of derived traits, the hallmark character of the Collembola is their "spring." On the third and fourth abdominal segments are interlocking structures that form a spring mechanism, allowing the springtails to propel themselves into the air. Although hardly a form of controlled flight, the spring is an effective means of escaping predation. The

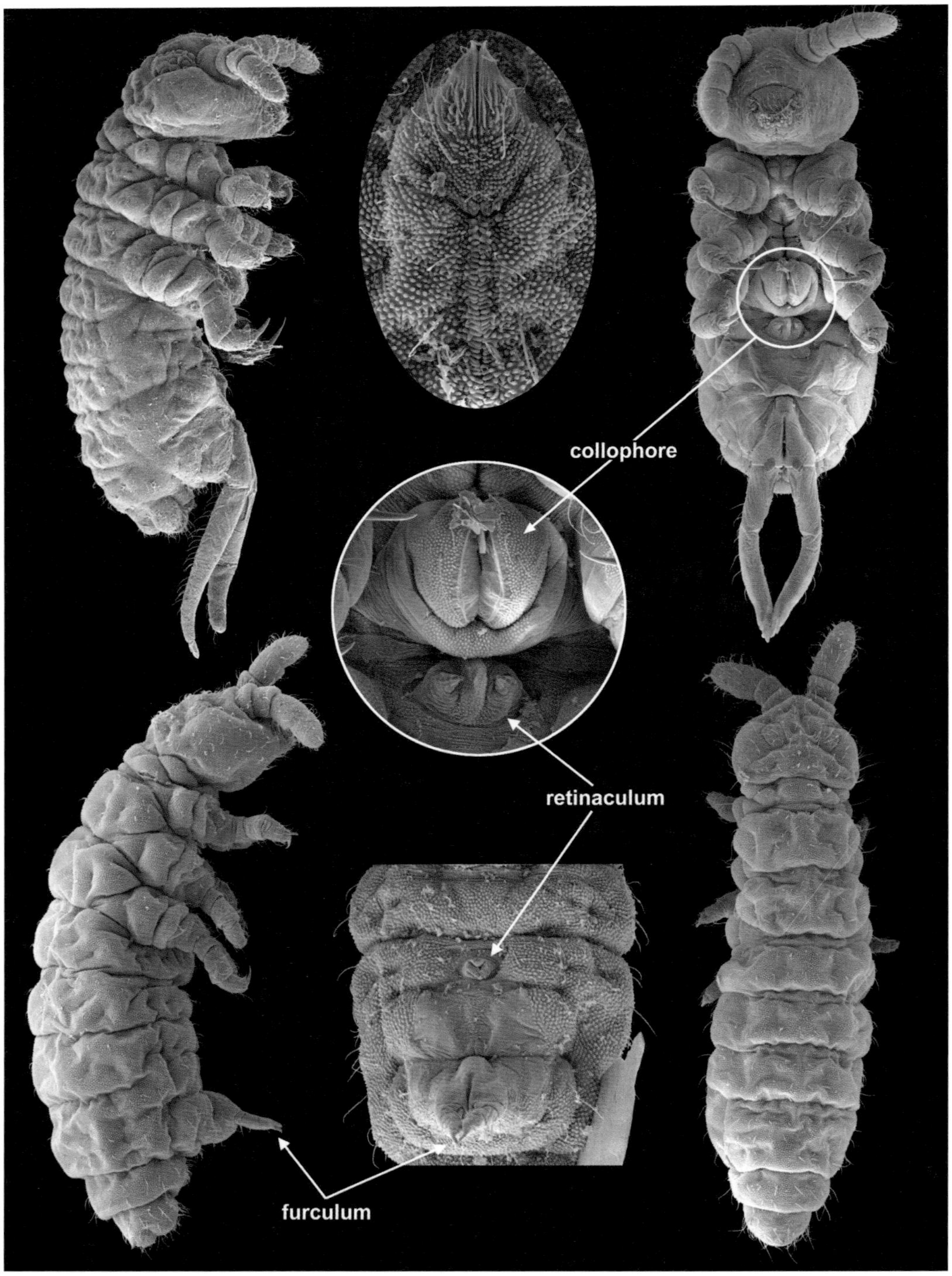

3.29. Scanning electron micrographs depicting typical features of springtails (Collembola) based on species of Poduridae. In the center at top is the opening to the gnathal pouch, in which the mouthparts reside (as in all Entognatha). The other central images depict defining features of the order: the collophore and the "spring," the latter formed of the furculum and the retinaculum. Not to same scale.

actual moving portion of the spring is the *furculum*, formed from fused abdominal appendages on the fourth abdominal segment (Figure 3.29). The furculum is ventrally located and has a broad base called the *manubrium* that bears paired, frequently elongate, finger-like processes at its apex called the *dens* (themselves sometimes bearing small processes at their own apices called *mucrones*). The furculum can recline into a small "lock," the *retinaculum*, which is located on the third abdominal sternum. Some lineages have lost the spring. Other features of the order include the presence of compound eyes (although these are lost in some families) and the absence of tracheae except in the suborder Symphypleona, which have a single pair of spiracles in the collar and a rudimentary tracheal system.

Springtails are presently classified into three suborders (Arthropleona, Neelipleona, and Symphypleona) (Figure 3.30),

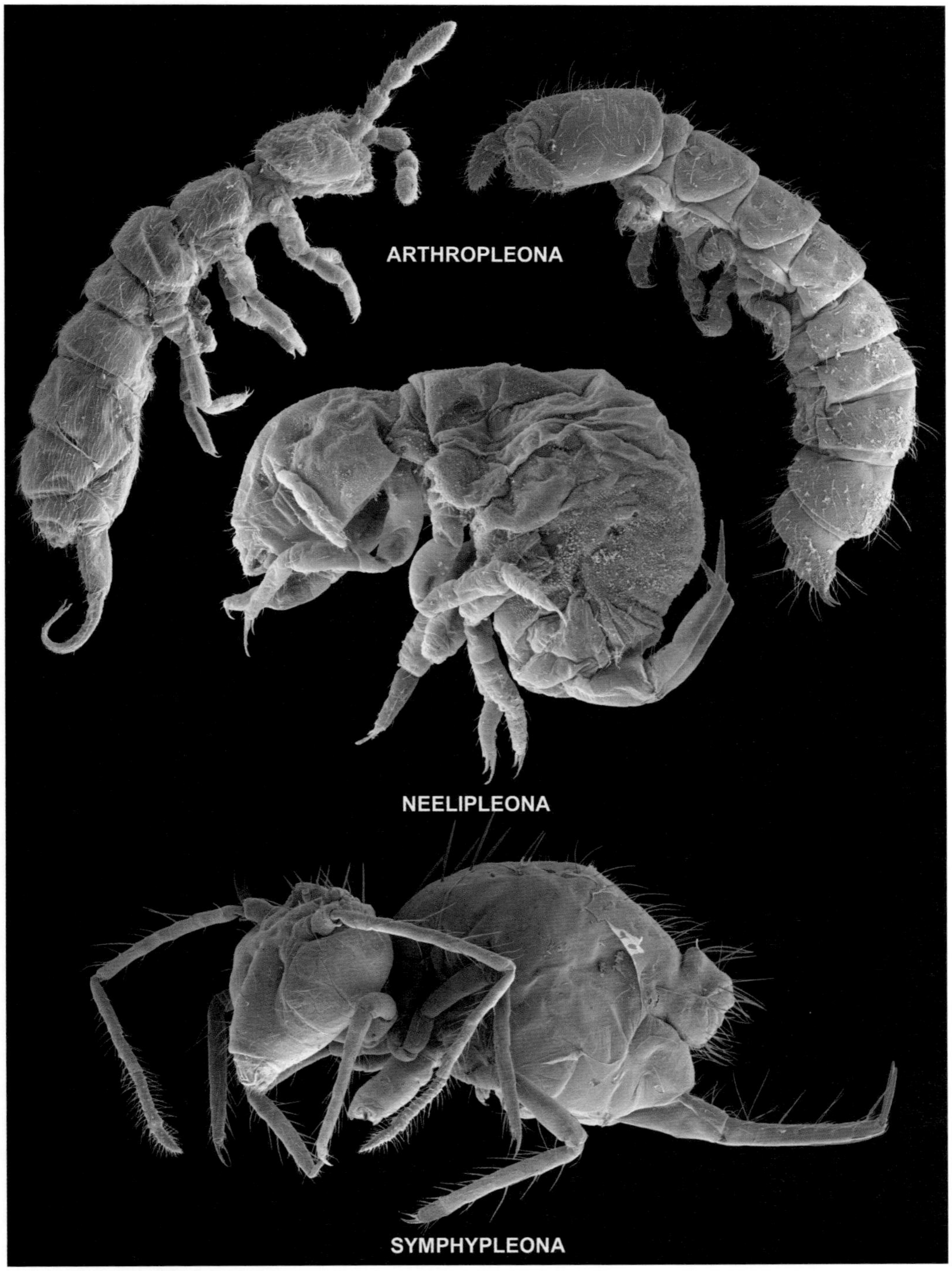

3.30. Representative springtails (Collembola). Scanning electron micrographs; not to same scale.

one of which is definitively paraphyletic. The Arthropleona is a paraphyletic assemblage of families from which the Symphypleona and Neelipleona are derived, the latter itself likely derived from among the symphypleones (perhaps allied to the Sminthuridae). Arthropleona have primitively elongate bodies with relatively complete abdominal segmentation and the mouth typically opening anterior to the ocelli (i.e., the heads are *prognathous*, or with the mouthparts held forward). By constrast, the Neelipleona and Symphypleona (comprising the Neopleona) have globular bodies with the first four abdominal segments fused, sometimes also fused with the meso- and metathoracic segments. In addition, the mouth typically opens ventral to the ocelli (i.e., the heads are hypognathous). Neelipleona is poorly understood and consists of about 25 minute species in a single family (Neelidae), which are blind and live in caves or in the soil. The neelipleones differ from the Symphypleona, from which they are certainly derived, by the short antennae (being shorter than

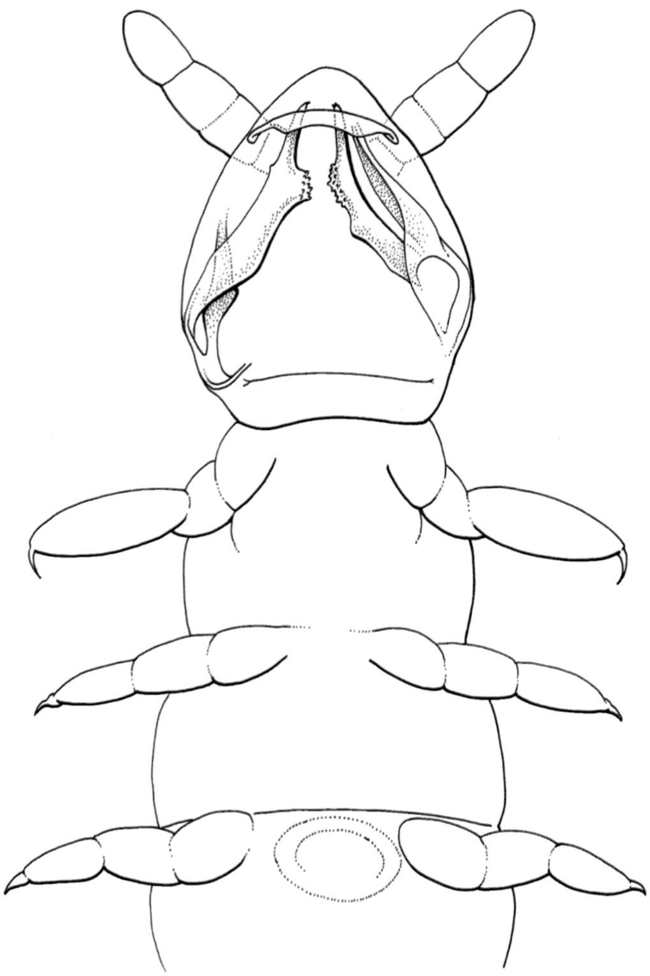

3.31. Reconstruction of *Rhyniella praecursor*, the earliest fossil of Entognatha, from the Early Devonian chert of Rhynie, Scotland. The entognathous mouthparts are well preserved, and the remains of a collophore and furculum indicate it was a collembolan.

the head), absence of ocelli, absence of bothriotrichia, and presence of sensory regions on the abdomen. Conversely, the symphypleones have long antennae, ocelli, bothriotrichia, while lacking the sensory regions on the abdomen, all primitive traits relative to neelids.

By stark contrast to Protura and Diplura, the springtails have an extensive fossil record. Indeed, one of the oldest hexapods is a springtail. *Rhyniella praecursor* from the Early Devonian (Pragian) Rhynie Chert of Scotland is a rather typical collembolan (Hirst and Maulik, 1926; Tillyard, 1928b; Scourfield, 1940a,b; Massoud, 1967; Whalley and Jarzembowski, 1981; Greenslade and Whalley, 1986) (Figure 3.31). Although at one time placed in its own family (e.g., Paclt, 1956), it has since been recognized as being most similar to the arthropleone family Isotomidae (perhaps the most basal of all collembolan families) (Greenslade and Whalley, 1986). Unfortunately, there is a gap in the fossil record of the order of nearly 300 MY. The next oldest springtails are in ambers from the Cretaceous (Christiansen and Pike, 2002a,b; Simon-Benito *et al.*, 2002) and are also quite common in Cenozoic ambers (Christiansen, 1971; Mari-Mutt, 1983; Lawrence, 1985), which are mostly represented by Arthropleona but also include Symphypleona. Neelipleona are unknown in the fossil record. While several fossil species have been described, the phylogenetic implications of these taxa have not yet been explored.

Diplura

The diplurans consist of two groups of rather divergent lineages: the suborders Campodeomorpha and Japygomorpha. The campodeomorphs (Figure 3.32) have multisegmented cerci and a movable, mandibular prostheca, which is a process near the molar surface developed either as a sclerite or fringe. The japygomorphs lack the mandibular prostheca and have unsegmented, forcipate cerci, similar to the cercal forceps of earwigs (Figure 3.33). Species of both lineages live in soil, rotting wood, or leaf litter but otherwise differ in their biology. The campodeomorphs are generally not aggressive and are mostly herbivorous. Japygomorphs are fiercely predatory, principally victimizing small insects and other invertebrates, subduing them by grasping them with their maxillae or their impressive cercal forceps. The genus *Heterojapyx* is particularly interesting because some species behave like antlions, burying themselves head-down into the soil with only the apices of the forceps extending above the ground. Once an unsuspecting insect approaches, the *Heterojapyx* seizes the prey with its forceps, emerges from the soil, and consumes its victim.

Like all entognaths, diplurans have external fertilization. Females deposit eggs in small clumps within rotting wood, vegetation, or cracks in the soil surface. Interestingly, diplurans can be subsocial, with females guarding their eggs and immatures for several molts, just as in many earwigs. However, this maternal devotion can sometimes lead to unfortunate consequences as japygomorphs are at times cannabilistic, with the young devouring their mother when they grow. Development, in contrast to other entognaths, is *epimorphic*, with relatively little change in postembryonic stages aside from the number of antennal segments or alternations in chaetotaxy. However, like all primitive hexapods, molting continues after adulthood, with up to 30 molts recorded for some *Campodea*.

Dipluran monophyly has not been robustly supported in the past but has been consistently recovered by rigorous and recent studies of basal hexapods (e.g., Bitsch and Bitsch, 2000). Aside from the unique form of entognathy previously discussed, all Diplura have a monocondylic articulation between the trochanter and femur and between the femur and tibia (these articulations are dicondylic in almost all other Hexapoda). Additional features of the order include the absence of eyes (both ocelli and compound eyes) and the presence of panoistic ovarioles (except campodeomorphs,

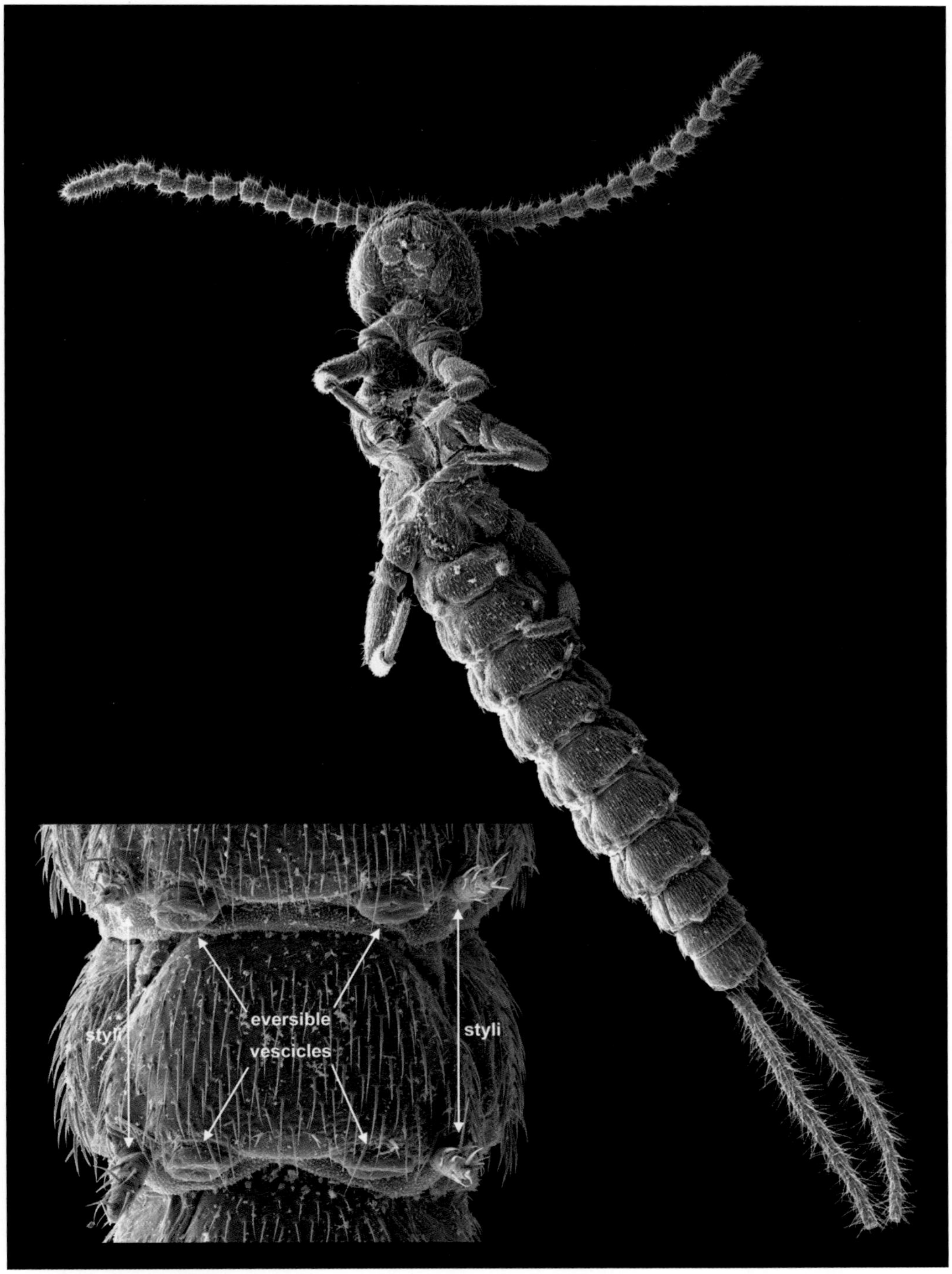

3.32. A campodeid dipluran (Entognatha), with the sternal styli and eversible vesicles indicated. Scanning electron micrograph, length 2.2 mm.

which are more like that of Ellipura), similar to primitive insects. Molecular studies concentrating on basal hexapods (e.g., Carapelli *et al.*, 1998; Frati *et al.*, 1998) have also found a monophyletic Diplura (in addition to a monophyletic Entognatha). The presence of paired pretarsal claws in Diplura is tantalizingly similar to the same condition seen in insects. Similarly, most diplurans have the gonopore recessed into a pouch between the eighth and ninth abdominal segments (the same position of the insectan gonopore), although some taxa have it developed between the seventh and eighth abdominal segments. The antennae are long, moniliform, and multisegmented. Major classifications of the order have been provided by Pagés (1997), although his system has proven to be rather unstable (e.g., Bitsch and Bitsch, 2000), and the more conservative classifications of Paclt (1957) and Condé and Pagés (1991), which are more widely employed.

The fossil record of diplurans is exceptionally poor given that they presumably evolved in the Early Devonian judging

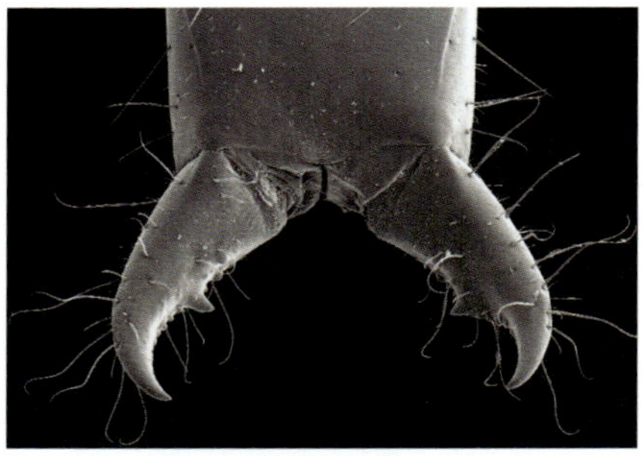

3.33. The cercal forceps of a japygid dipluran are similar to those in earwigs and serve a similar purpose: subduing prey. Scanning electron micrograph.

3.34. A campodeid dipluran in Early Miocene Dominican amber. Fossil diplurans are extremely rare. M2232; length 2.6 mm.

from their phylogenetic position. The only records of campodeomorphs are from the Tertiary, with a few specimens in Baltic (Middle Eocene) and Dominican (Early Miocene) ambers (Figure 3.34). Japygomorphs are also known from the Tertiary (in Dominican amber and Pliocene deposits of Arizona) but records also extend into the Mesozoic, albeit based on few specimens. Typical japygomorphs have been described from the Lower Cretaceous Santana deposits of Brazil (Bechly, 2001; Wilson and Martill, 2001). These represent the oldest, definitive members of the order but are remarkably similar to modern taxa (also indicative of an ancient origin for Diplura). The most controversial fossil is *Testajapyx thomasi* from the Upper Carboniferous of Mazon Creek. This fossil has been described as having well-developed compound eyes, relatively externalized mouthparts, a series of reduced abdominal appendages (leglets), among other enigmatic traits (Kukalová-Peck, 1987). However, the preservation of the fossil is quite poor, it has not been reexamined by additional entomologists to evaluate these very unusual features, and as such it cannot be conclusively considered a dipluran (e.g., Bitsch, 1994; Kristensen, 1995).

4 The Insects

Certainly the most famous and successful lineage of arthropods is the Insecta. Insects dominate our world and have silently witnessed the rise of vertebrates, the fall of nonavian dinosaurs, the proliferation of mammals, and the rapid evolution and industrialization of humans. Despite dramatic changes on earth, they have not only persisted but, by ever imaginable criterion of success, excelled.

In order to understand their modern diversity and history, we still examine them firsthand with microscopes and the naked eye. To compare species separated by eons, entomologists must make detailed comparisons of common structural designs. Thus, in order to discuss insect evolution, we first need a framework for understanding their basic structure. The comparative study of anatomical structure is *morphology*.

MORPHOLOGY OF INSECTS

What makes things baffling is their degree of complexity,
not their sheer size;
a star is simpler than an insect.
–Martin Rees, 1999 (*Scientific American*)

Morphology is perhaps the oldest of the biological sciences, but it is still of paramount importance. It is through morphology that our minds first encounter anything in the world. We perceive the size of an object, its color and texture, and its shape. It is the way in which we first come to *know* a thing, and it provides the foundation for other subsequent inquiries, like behavior and molecular biology.

There are two basic kinds of questions in the study of morphology: (1) What does it do? (a question of biomechanics or *functional morphology*), and (2) Where did it come from? (a question of its evolutionary history and origin, or *comparative morphology*). Structure is a result both of its function and its particular evolutionary history. One premise of functional morphology is that structures have a purpose and that physics can be used to describe their performance and function. It is a powerful tool for attempting to explain convergent features that are correlated with certain behaviors or other features.

Evolutionary or comparative morphology is principally practiced in the broader field of systematics. Even though biomechanics is a powerful tool, it cannot explain everything we observe. All organisms have a unique history that is reflected by modifications to existing structures. This imposes historical and developmental constraints on what forms a particular structure may ultimately acquire. There is, therefore, an interplay between function and history. One goal of evolutionary morphology is to explain unique, never-to-be-repeated historical events (e.g., the origin of wings) and to examine the origins of structure. As such, it is not always exactly testable by experimentation, but it does utilize the *comparative method* and is critically founded on identifying homologous structures. The comparative method is not unique to biology; it is also used, for example, in geology and astronomy.

It is critical to remember, particularly when considering fossils, that organisms are *mosaics* of primitive and derived traits. That is to say, numerous homologous features can be identified, but after congruence tests many will prove to be primitively retained features and others evolutionary derived features indicative of relationship. Even the most ancient fossil will represent some combination of these attributes, and the derived novelties may be ones that were unsuccessful and not repeated in evolutionary history.

In order to understand the remainder of this volume, it is imperative to have a working knowledge of the general construction, or *bauplan*, of the insects.

GENERAL STRUCTURE

Like other segmented animals, insects are composed of a series of repeated units. Ancestrally these units, called metameres, were identical and self-contained. Specialization of metameres allowed for versatility in the overall design of the body. The segments of the insect body are organized into three major tagmata: the head, the thorax, and the abdomen.

BOX 4.1.

Dorsal

Dorsum: the entire upper portion of a segment or of the whole insect.
Tergum (pl. *terga*): the dorsal exoskeletal plate or plates of a segment (also called *notum*; pl. *nota*).
Tergite: a subdivision of the tergum.

Lateral

Pleural Area: the lateral portions of a segment or of the whole insect.
Pleuron (pl. *pleura*): the lateral exoskeletal plate or plates of a segment.
Pleurite: a sclerotized subdivision of the pleuron.

Ventral

Venter: the entire undersurface of a segment or of the whole insect.
Sternum (pl. *sterna*): the ventral exoskeletal plate or plates of a segment.
Sternite: a subdivision of the sternum.

The thorax consists of three metameres, the abdomen consists primitively of 11 metameres (the eleventh metamere is lost in almost all holometabolous insects), while the head consists of an uncertain number. Even though these tagmata are dramatically modified in some groups, typically the function of the head is for sensory input and feeding, the thorax is for locomotion, and the abdomen is for visceral functions, mating, and various modes of sensory input.

Insects are, naturally, arthropods; therefore, they are encased by a chitinized cuticle. The cuticle provides protection, support, and locomotion; prevents water loss via a wax layer; provides a site for waste product deposition; protects from ultraviolet radiation; serves communication functions (inter- and intraspecifically) via hydrocarbons on the surface or through coloration, etc. The cuticle is divided into a series of distinct plates or *sclerites*. The boundaries of sclerites may or may not correspond to boundaries between body segments (frequently they *do not*). Differing terms are given to sclerites located on different parts of the body (Box 4.1).

Another important feature of the integument is the externally visible lines used to subdivide the sclerites. *Sutures* are intersegmental lines, demarcating two primitive metameres, now fused (e.g., postoccipital suture). This term is almost universally misapplied by insect systematists, biologists, and others for any line on the integument of the insect. It is important to be precise, however, because this will greatly affect our interpretations as to where particular structures originally derived. A *sulcus* (pl. *sulci*) is a line marking where an internal ridge is formed for muscle attachment (e.g., epistomal sulcus). The internal ridge associated with the sulcus is called a *costa* (sometimes the costa forms a large, internal plate called a *phragma*). An external ridge (i.e., an evagination, rather than invagination) is called a *carina* or crest. Costae and carinae are used either to strengthen the cuticle or to serve as a site for muscle attachment. An *ecdysial cleavage line* is a point of weakness in the cuticle where it splits at molting (e.g., epicranial "suture," or, more correctly, epicranial line). Such lines are never indicated internally by a costa or phragma. *Apodemes* (synonymous with *apophysis*) are invaginations of the cuticle that form a rod or bar; frequently the external sign of such a structure is a *pit*. Not all apodemes have externally visible pits to betray their existence. For example, abdominal terga and sterna typically have long, lateral processes to which muscles attach, and these are apodemes; mandibles are another example since they have large apodemes for the attachment of strong mandibular muscles but no pit.

Studying insect segmentation is more complicated than simply seeking suture lines. Except for the most primitive lineages, most insects show some degree of *secondary segmentation* in the thorax and abdomen (Figure 4.1). Secondary segmentation is often most evident in the abdomen. Terga overlap each other posteriorly, such that the posterior border

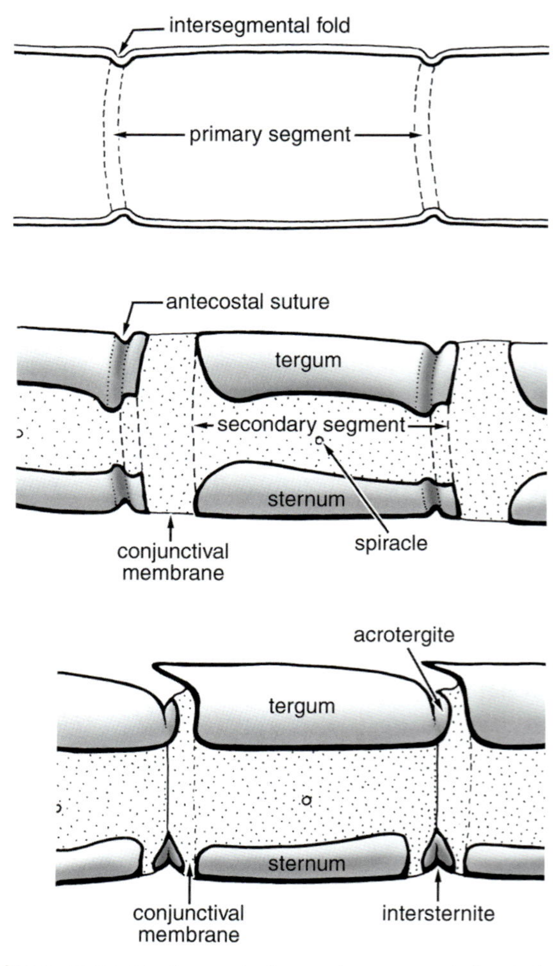

4.1. Stages in the development of secondary segmentation in insects.

of tergum overlaps the anterior border of the tergum behind it. Along the anterior edge of each tergum is a costa called the *antecosta* (and the associated suture is the *antecostal suture*; as the name implies, this is a true intersegmental suture). The antecosta, however, is frequently not exactly on the leading edge of the tergum and the sclerotized strip anterior to it is the *acrotergite*. Immediately anterior to the acrotergite is a membranous region (*conjuctiva*) marking the region between adjacent terga. By comparing the structure of the terga and sterna as well as the associated internal organs across higher arthropods and throughout the insects, we see that the antecosta marks the true intersegmental boundary and the acrotergite belongs to the preceding body metamere. Thus, the visible membranous regions between terga are a form of secondary segmentation. The primary segmentation is only apparent internally by the arrangement of costae and muscles. Primary segmentation is retained in Onychophora, Tardigrada, and some other Arthropoda as well as in embryonic forms of insects. Expansion of the acrotergite in the thorax of many insects leads to the development of a *postnotum*. Primitively (or in embryonic forms) each notum has a suture separating it from others and the associated costa is the attachment site for internal muscles. The simplest form of secondary segmentation is where the posterior portion of the notum separates and becomes the acrotergite of the following notum. The next most complicated form of secondary segmentation is that in which the antecostae become developed into phragmata and some thoracic acrotergites expand to form distinct postnota. The ultimate level of complexity is when the postnotal plates and associated phragmata become separated by secondary membranes and appear dissociated upon superficial examination.

In the sterna of pterygote insects, the overlapping pattern in the thorax is just the opposite of the tergal pattern, but this pattern returns to normal in the abdomen (i.e., the posterior border of the sternum is concealed underneath the anterior margin of the immediately following sternum). The sternum of the third thoracic segment, however, overlaps on both ends the preceding and following sterna, and abdominal sterna then return to normal overlapping orientation. Sometimes the sterna are fragmented, reduced, or lost entirely . . . thus confusion reigns! The sternum is subdivided into a *eusternum* and a *spinasternum*. The eusternum can be further subdivided into a *basisternum* (anteriorly) and a *sternellum* (posteriorly), marked externally by a transverse median line, the *sternacostal sulcus*. The eusternum bears two large *sternal apophyses* visible externally as the apophyseal pits. These, along with the spina (see discussion that follows) serve as the attachment sites for ventral thoracic muscles. These can either fuse or be attached via muscles to pleural apophyses. On thoracic sterna, the costa marking the intersegmental line is reduced to a spine, called the *spina*. The spinasternum is the sclerite that bears the *spina*; it marks the true intersegmental boundary. When coxae are close to the ventral midline, the sterna tend to be reduced. The sternal apophyses may fuse to form a *furca* (internally) and leave externally only the apophyseal pit. The eusternum may be reduced to form a tiny *furcasternum* (= to the portion sometimes called the sternellum) or an internal *cryptosternum* indicated externally by a single median, longitudinal line called the *discrimen*.

The advantage of all this complexity created by secondary segmentation is that it provides a series of landmarks for associating structures with their original, ancestral segments (i.e., metameres), important for careful comparative morphology.

THE HEAD

Cephalization has occurred in all arthropods, where one tagma is always present at the anterior end of the body and minimally has the functions of food intake, sensory perception, and neural integration and command. Most of the head is occupied by musculature, and the brain takes up little space. The details of cephalization in the different arthropod lineages, however, differ and can be used to define major arthropod lineages. For any given lineage, we can ask, "How many segments are present in the head?", "How many pairs of appendages on these segments are modified into other structures, such as mouthparts?", and "How distinct is the head from the rest of the body?"

The head of insects has three general positions in relation to the mouthparts. *Hypognathous* heads are those in which the mouthparts are oriented ventrally (e.g., Orthoptera); this type of head is the primitive condition for insects. *Prognathous* heads are those in which the mouthparts are oriented anteriorly (e.g., many Coleoptera, Raphidioptera); those structures that typically face anterior in a hypognathous insect are directed dorsally in a species with prognathy. The last type of gross head morphology is essentially an elaboration of the aforementioned condition of hypognathy. *Opisthognathous* heads are those in which the mouthparts are in a ventral position but are oriented to the rear of the insect (e.g., Blattodea, Hemiptera).

The lateral surfaces of the head capsule are called the *parietal regions*. The *gena* is that portion of the parietal region above the *subgenal sulcus*. The *vertex* is essentially the top of the head, above the ocelli and between the compound eyes (Figure 4.2).

The head is etched by a series of lines, most of which are not true sutures (as frequently referred to) but are instead sulci marking internal costae. Because costae are typically employed for functional purposes of strengthening the head capsule, it is perhaps not surprising that as the functional requirements of the insects vary, so do the position and shape of these sulci. The sole exception among extant insects is the

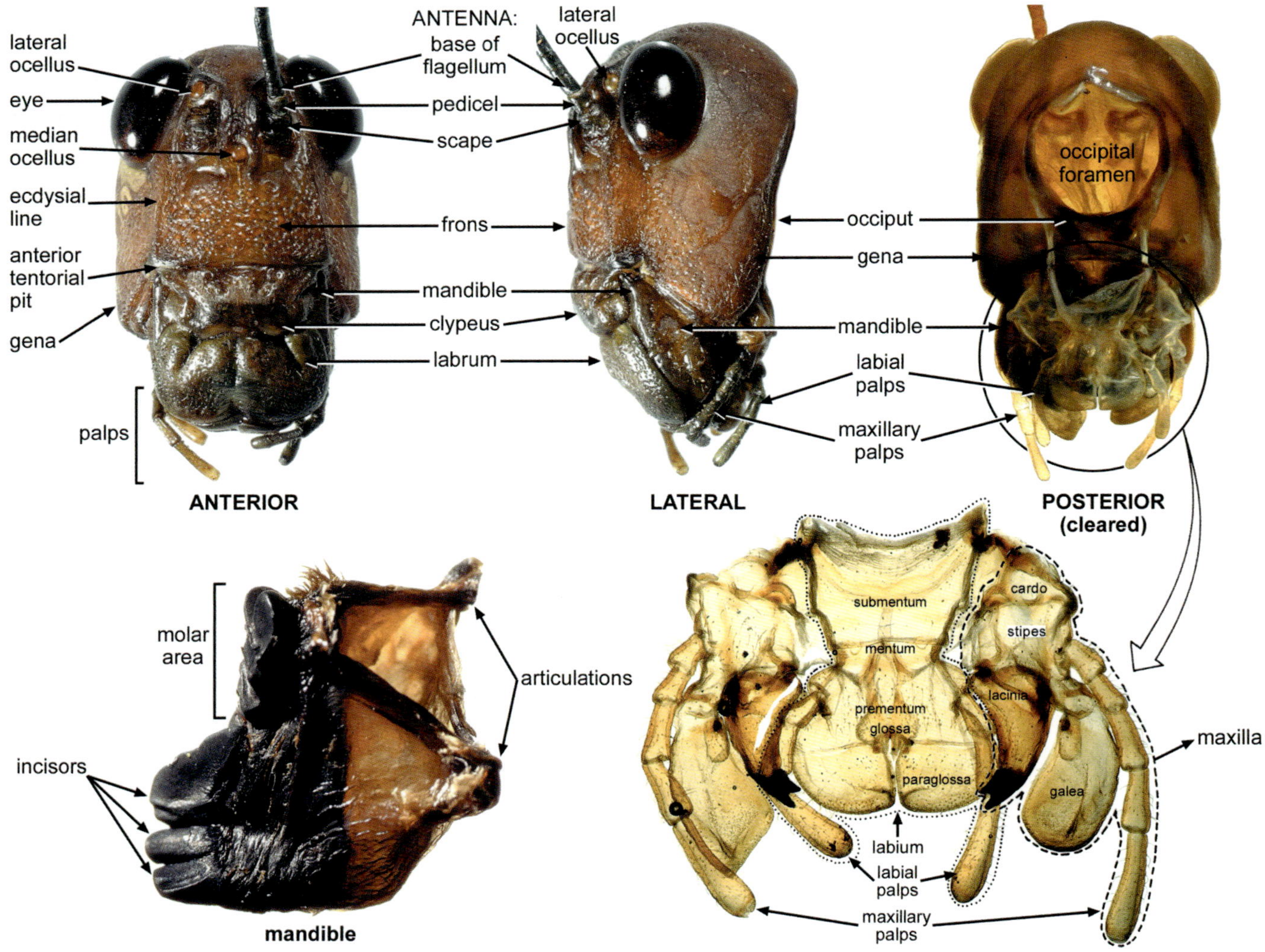

4.2. Basic head and mouthpart morphology of insects, illustrated with a grasshopper.

postoccipital suture. This line marks the boundary between the labial and maxillary segments of the head and is evident from between the labium and maxillae, demarcating behind it a very narrow region called the *postocciput*. The *occipital sulcus* is just anterior to the postoccipital suture and defines, between these two lines, the *occiput*. Frequently, the postoccipital suture is faint or indistinct, and the bulk of the back of the head is composed of the occiput. The *frontal line* delineates a region on the front of the head called the *frons* and connects to the *coronal line* that runs dorsally toward the vertex. The coronal line is most often not present in adults but is easily seen in most larvae. The frontal and coronal lines (together referred to as the *epicranial lines*) are ecdysial cleavage lines, despite being referred to by many entomologists as sulci or sutures, and are where the integument breaks during molting. Circumscribing the head on its inner surface, near where the mouthparts articulate, is a continuous *costa*. Externally, this costa is visible by a line that is given different names on different regions of the head. The utility of this complex naming system is that it is immediately clear whether one is referring to the front, side, or back of the head, but it does obscure the fact that the different names refer to a single continuous line and internal ridge. The *epistomal* (or frontoclypeal) *sulcus* is that portion of the line that marks the boundary between frons and clypeus on the anterior surface of the head. The *clypeus* is that region of the head below the epistomal sulcus and anteriorly articulates with the movable *labrum* (Figure 4.2). The *subgenal sulcus* is that portion of the line on the side of the head, defining a region called the subgena between the subgenal sulcus and the mouthparts. The subgenal sulcus is often subdivided into two parts: the *pleurostomal sulcus* above the mandible (the sclerite below this is the *pleurostoma*) and the *hypostomal sulcus* behind the mandible (the sclerite below this is the *hypostoma*). Some insects, such as crickets (Orthoptera: Ensifera) have a sulcus between the bottom of the compound eye and the epistomal-subgenal sulcus. This is the *subocular sulcus*. There are also sulci around the sockets of the compound eyes and the antennae; these are the *circumocular sulci* and the *circumantennal sulci*.

The head capsule contains an internal skeleton called the *tentorium*, which not only provides strength to the head but also provides some sites for the attachment of muscles that insert on mouthparts. On the external surface of the head, the existence of the tentorium is evidenced by *anterior* and *posterior tentorial pits*. Within the head the tentorium is typically an H-shaped structure with anterior and posterior arms on each side; it is joined in the middle by a *tentorial bridge* but generally positioned toward the back of the head. Frequently, the dorsal arms of the tentorium extend toward the upper surfaces of the head capsule. The anterior tentorial pit is located in the epistomal/subgenal sulcus, while the posterior tentorial pit is located in the postoccipital suture. The *postgena* is that region posterior and ventral to the gena and above the subgenal sulcus. In some prognathous insects the region just proximal to the posterior tentorial pits is expanded anteriorly from the neck region to form a median sclerite called the *gula* (e.g., Coleoptera).

Mouthparts

The mouthparts principally consist of three pairs of appendages, each pair corresponding to a fused segment. These are (from anterior to posterior): *mandibles*, *maxillae*, and *labium* (Figure 4.2). As already discussed, there are two major divisions of the mouthpart appendages within the Hexapoda: *entognathous* and *ectognathous*. In Collembola, Protura, and Diplura the mouthparts are recessed into a pocket within the head capsule, a condition called *entognathy*. These mouthparts can be extruded during feeding and withdrawn at other times. The Insecta (also known as Ectognatha) consist of all remaining hexapod orders, and their mouthparts are not retracted into the head capsule and are dramatically modified in various lineages. The groundplan morphology for the major structures is outlined next under their respective head segments.

Head Segmentation

Most insects have the following head appendages: antennae, mandibles, maxillae, and a labium. The last three are widely accepted as serially homologous with the thoracic legs and thereby attest to at least four ancestral segments, or metameres, comprising the insect head capsule. The principal criteria for identifying a metamere (proposed by Rempel, 1975) include a pair of appendages, a pair of apodemes corresponding to the appendages, a neuromere (i.e., a ganglion associated with the ancestral segment), and a pair of mesodermal somites (coelomic sacs during development, which also correspond to the ancestral body cavity of the original segment). Using these criteria, insects appear to have six or perhaps seven fused segments in the head, although the debates continue to rage. For now there are six purported head segments.

The Preantennal Segment. This segment is a bit counterintuitive because there is no obvious pair of appendages anterior to the antennae. In fact, some authors dismiss the idea of a preantennal segment entirely (e.g., Snodgrass, 1935: despite the fact that he himself cites some embryological studies that identify "a pair of evanescent appendage-like lobes in the embryo . . . lying anterior to the antennae"). The appendages of the preantennal segment (argued by Rempel, 1975) are the fused components that form the *labrum*. The labrum in insects is an articulated sclerite anterior to the clypeus. The labrum is clearly associated with coelomic sacs in development and forms embryologically from the fusion of two appendage-like outgrowths of the first segment (e.g., in *Pieris*, *Tenebrio*, *Lytta*, some Phasmatodea, and almost any other insect that has been studied). Thus, it appears that the labrum is the remnant of an ancestral appendage.

The Antennal Segment. The origin of antennae has been of some debate and has been interpreted in two ways: (1) as a pair of modified appendages serially homologous with mouthparts and thoracic legs (favored by Rempel and most other workers) or (2) as sensory structures associated with the presegmental part of the body, at least analogous, if not homologous, with tentacles on the heads of certain worms (an interpretation favored by Snodgrass). Modern developmental and phylogenetic evidence favors the first hypothesis and antennae meet all four of the criteria for a true segment discussed earlier. Antennae arise as lateral outgrowths of the segment, the antennal buds are provided with large coelomic cavities, the antennae have their own neuromere (which becomes the deutocerebrum of the brain), and apodemal invaginations are also present. The antennae are composed of three principal units (from base to apex): *scape*, *pedicel*, and *flagellum* (itself subdivided into *flagellomeres*). In primitive hexapods (i.e., Collembola and Diplura; the Protura have lost their antennae), there are intrinsic muscles to each flagellomere; in all other groups the scape and pedicel only have musculature. The scape is inserted within a membranous antennal socket and articulates via a single point called the *antennifer*. The *pedicel* is typically a small joint in insects. In adult insects the pedicel contains the *Johnston's organ* (a *chordotonal organ*, or specialized group of cells designed to detect deformations of the cuticle). The *flagellum* is variable in length and shape and frequently divided into annuli or flagellomeres, which are joined to each other by membranes so that it is overall quite flexible. Minute sensory structures (*sensilla*) are most abundant on the flagellum, and nerves traversing the flagellum are entirely sensory, not motor neurons.

The Intercalary Segment. There is now conclusive evidence for a segment between the antennae and the mandible of insects (believed to be the homologue of the second antennal segment of Crustacea). This segment is evident as a coelomic cavity in development and is associated with the tritocerebrum, although in the later stages of development the tritocerebrum becomes secondarily associated with the labrum.

The Mandibles. The mandibles of hexapods principally vary in their articulation with the head capsule. The point at which a mandible articulates with the head is called a *condyle*. The condyle is a ball-like structure that fits into a socket, called the *acetabulum*, to form the articulation. Two types of mandibles are known. *Monocondylic* mandibles articulate via a single, dorsal condyle and therefore make a rotary motion around this point (Figure 5.9). The condyle is on the mandible and the acetabulum on the head capsule. *Dicondylic* mandibles articulate via two condyles (one anterior condyle and one posterior condyle) (Figures 4.2, 5.9). Thus, the mandible can only move in a single plane of motion. The posterior condyle is homologous with the dorsal condyle of monocondylic mandibles; at the posterior point of articulation, the ancestral condyle is on the mandible and the acetabulum on the head capsule; the anterior articulation point is the novel feature of the dicondylic hexapods, and the morphology is reversed – condyle is on the head capsule and acetabulum is on the mandible.

The Maxillae. The maxilla has five basic components: *cardo, stipes, galea, lacinia,* and *maxillary palpus* (Figure 4.2). The homology of mouthparts with leg structures is most obvious in the maxillae and is identical in insects with generalized chewing mouthparts. The cardo has a monocondylic articulation with the head, identical to that of a mandible.

The Labium. The labium is essentially a fused set of maxillae and homologous with the "second maxillae" of the Myriapoda and Crustacea (Box 4.2). Interestingly, in embryonic development the labium starts off as separate limb buds (like the mandibles and maxillae), which then fuse. The main elements are *postmentum, prementum, paraglossa, glossa,* and *labial palpus* (Figure 4.2). The postmentum is typically subdivided into two sclerites: *mentum* (anterior) and *submentum* (posterior). The *ligula* is a general term referring to the glossa and paraglossae; these are often fused into a single structure in holometabolous insects. Anterior to the labium is a lobe that forms the posterior wall of the mouth (or *gnathal*) region called the *hypopharynx*. A small cavity is formed between the posterior wall of the hypopharynx and the anterior surface of the labium called the *salivarium*. The labial salivary glands open into the salivarium. Posterior to the labrum is a lobe called the *epipharynx*. Between the hypopharynx and epipharynx is a cavity called the *cibarium*.

Some authors hypothesized that there is an additional segment in the head associated with the compound eyes, and that the eyes are modified appendages on a hypothetical metamere called the *ocular segment* (see discussion under eyes later in this chapter). For now we retain the more traditional number of head segments pending further investigation.

BOX 4.2. Serial Homology Between Maxillary and Labial Structures

Labium		*Maxilla*
Postmentum	=	Cardo
Prementum	=	Stipes
Paraglossa	=	Galea
Glossa	=	Lacinia
Labial palpus	=	Maxillary palpus

Eyes

Compound eyes are present throughout arthropods, although they are absent in Recent Arachnida, Protura, and Diplura, as well as some other more minor lineages, like Symphyla. Compound eyes were, however, present in some fossil Arachnida (e.g., Paleozoic scorpions, Trigonotarbida) and trilobites and may have even been present in Paleozoic, xenusian Onychophora (Dzik, 2003). The basic unit of an insect compound eye is the *ommatidium*. Eyes range from a single ommatidium up to 28,000 ommatidia. For example, a collembolan has two ommatidia, while a dragonfly can have around 28,000. Each ommatidium is composed of multiple cells and divided into two functional units: the dioptric apparatus and the receptor apparatus. The *dioptric apparatus* is functionally the lens and gathers and focuses light on the receptor apparatus. The receptor apparatus receives the focused light from the dioptric apparatus and translates it into receptor potentials that are sent via neural axons to the optic lobe. In addition to the compound eyes, there are three *ocelli* (simple eyes) present in most adult insects as well as larvae of Holometabola. Ocelli are composed of a transparent cuticular cornea and cannot form images but are highly sensitive to low light intensity. Simple eyes in holometabolous larvae (e.g., caterpillars) are called *stemmata* and are structurally similar to ocelli, although they are structurally identical to ommatidia in some mecopteran larvae.

Some researchers believe the eyes are highly modified appendages and thereby evidence for yet another ancestral segment in the head. This theory has some developmental evidence (e.g., Schmidt-Ott *et al.*, 1994, 1995; Scholtz, 1995; Rogers and Kaufman, 1996; Queinnec, 2001) and supports the original hypothesis of this segment (the "ocular segment") by the Russian paleontologist A. Sharov (1966).

Thus, under this theory, the various segments of the insectan head, from anterior to posterior, would be labral, perhaps ocular (perhaps present across all arthropods but reduced or lost several times independently), antennal, intercalary, mandibular, maxillary, and labial.

THE THORAX

The thorax is the middle tagma of insects and is the main unit for locomotion because it bears the legs and, in pterygotes, the wings. The thorax primitively consists of three metameres although the first abdominal segment is closely associated with the thorax in Pterygota and is completely fused to it in the Apocrita (Hymenoptera). Each thoracic segment has one pair of legs. Anteriorly, there is a membranous region where the head attaches to the thorax. This is the *cervix* (Latin, meaning "neck"). This neck region may not be a true intersegmental boundary, and the region as a whole may be composite in origin. The evidence for the composite origin of the neck derives from the fact that there is no antecosta on the pronotum, the dorsolongitudinal muscles go directly from the antecosta of the mesothorax (i.e., from the first phragma) to the back of the head (to the postoccipital ridge), and the ventral longitudinal muscles go directly from the sternal apophysis of the prothorax to the posterior arm of the tentorium. Several cervical sclerites form a fulcrum on which the head rotates. The head is protruded when the muscles attaching to these sclerites contract. When the muscles relax, the head is pulled back to the thorax.

The dorsal structure of the thorax is very similar to abdominal terga (i.e., with typical secondary segmentation, see preceding discussion). In apterygotes and nymphal pterygotes the terga do not overlap each other; however, the terga overlap in adult pterygota. In winged insects the terga of the winged segments typically divide into a *postnotum* (which bears the phragma) and the *alinotum* (which bears the wing sclerites). In several lineages the *alinotum* becomes divided by a *scuto-scutellar sulcus* (or completely divided, typically called a suture but more appropriately called a fissure) and forms an anterior *scutum* and a posterior *scutellum*. In the mesothorax this separates the *mesoscutum* from the *scutellum* (more appropriately called the *mesoscutellum*) and in the metathorax it is the metascutum and metascutellum (these modifications are only rarely present in the metathorax). This separation allows for specific changes in the thoracic structure for flight.

The pleuron is the side of the thorax and is where the legs join the body (Figure 4.3). These sclerites are the least like the abdomen of any sclerites and have been interpreted in dramatically different ways by different authors. The most robust theory is the *subcoxal theory*, wherein the the pleura are composed of the subcoxa of the appendages. In this theory the subcoxa was perhaps primitively a podite (the basic units of a jointed leg) that became incorporated as the lateral wall of the body. After incorporation into the body wall, the subcoxa was primitively divided into three sclerites: the *anapleurite*, the *coxopleurite* (also called the *katapleurite*), and the *sternopleurite*. The former two are obvious in groups like Plecoptera. The sternopleurite is perhaps fused into the lateral portions of the eusternum (there is no distinct sternopleurite in Hexapoda, although it does occur in Chilopoda). The coxopleurite (katapleurite) is involved in the articulation with the coxa of the leg (the other articulation is hypothesized to have been on the sternopleurite), and when present in Arthropoda, it is there; otherwise in Hexapoda there is primitively no ventral articulation. Instead, a secondary articulation is present. The anapleurite and coxopleurite typically fuse to form the pleural wall of the body. In winged insects a pleural wing process is formed dorsally to make up the other part of the fulcrum in the wing articulation (see the discussion of Pterygota and wings later in this chapter). A pleural sulcus is formed running from the pleural wing process to the coxal articulation (at the *pleural coxal process*). The pleural sulcus corresponds to an interior ridge to strengthen the pleuron during the contraction of the flight muscles. The pleural sulcus divides the pleuron into anterior and posterior regions. The area anterior to the pleural sulcus is the *episternum*; the area posterior to the pleural sulcus is the *epimeron* (Figure 4.3). Thus the two main pleural sclerites (recall that the sternopleurite is perhaps fused into the eusternum in Hexapoda or lost altogether) can be divided into an *anepisternum*/*anepimeron* (from the anapleurite) and a *katepisternum*/*katepimeron* (from the coxopleurite/katapleurite). Sometimes these regions can be further subdivided, like a distinct anterior portion of the episternum, called the *preëpisternum*. Typically the pleural sclerites are completely fused making any distinction between the anapleurite and coxopleurite impossible.

The mesothoracic and metathoracic pleura possess *spiracles* (also present laterally on the abdomen), which are openings for respiration. The spiracle is situated in a sclerite called the peritreme and attaches internally to *tracheae*, the main branches of the tracheal system. Sometimes the spiracle first opens into a small chamber, called the *atrium*, which can bear on its sides a series of folds or spines that serve dual functions – preventing the entrance of foreign particles and simultaneously catching water vapor before gases depart from the body. The tracheae are invaginations of the exoskeleton and are shed during molting.

Legs

The legs articulate directly with the pleural sclerites and are composed of a series of segments, or podites. The hexapod leg consists of six podites: *coxa*, *trochanter*, *femur*, *tibia*, *tarsus*, and *pretarsus* (Figure 4.4). The basalmost podite is the *coxa*. The coxa of the leg primitively has a single (monocondylic) articulation with the pleuron; however, dicondylic articulations have evolved in many insect lineages. The secondary articulation is either with the sternum (perhaps the sternopleurite?) or the *trochantin*, which is visible in generalized Pterygota but lost or fused in most higher lineages. The trochantin is a precoxal sclerite perhaps derived from the

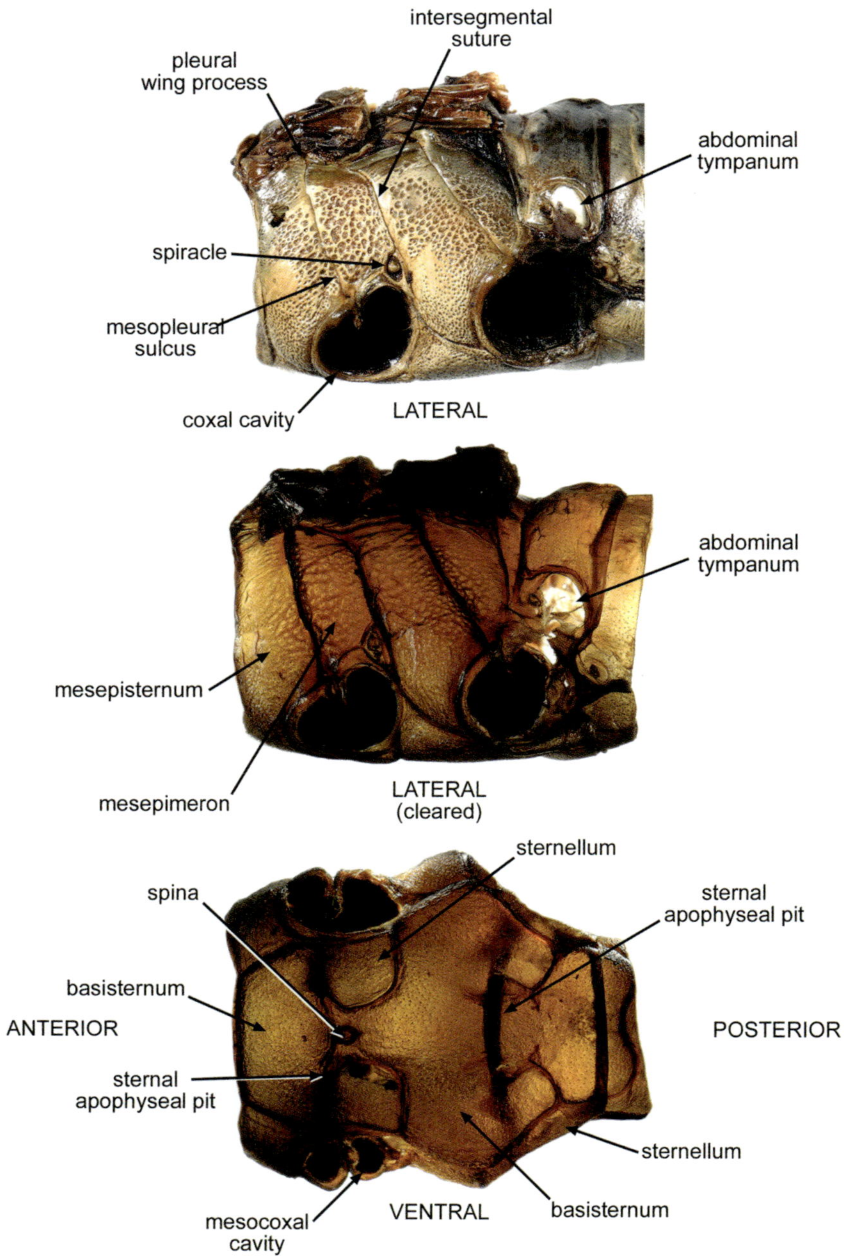

4.3. Basic thoracic structure of a grasshopper.

trochanteral articulation. As noted, muscles operating the coxa are extrinsic muscles, while those operating the other podites are typically *intrinsic muscles* (where origins and insertions are *within* the appendage). Flexion and extension of the podites is accomplished by antagonistic sets of muscles (flexors and extensors, also called depressors and levators). *Flexors* are used to flex a section of an appendage; flexors bend an appendage by moving portions of it toward the body (somewhat synonymous with *depressors*, which lower an appendage). *Extensors* are used to extend a section of an appendage; extensors straighten an appendage by moving portions of it away from the body (somewhat synonymous with *levators*, which raise an appendage). The *trochanter* is the second podite of the leg (the basalmost segment of the *telopodite*) and is generally rigidly attached to the base of the femur so that the articulation no longer functions; however, it is sometimes completely fused to the femur. There are two trochanters in Odonata, the first and second trochanters. A reductor muscle originates at the base of the trochanter and inserts on the femur. In the Odonata the second trochanter is a true trochanter because the reductor attaches to the base of the second trochanter. In some lineages a second trochanter (called the *trochantellus* in some groups of Hymenoptera) is developed from the base of the femur and is, therefore, not a homologue of the second trochanter seen in Odonata or other arthropod lineages. The reductor muscles attach internally at the base of the true trochanter and attach to the base of the trochantellus, so the trochantellus is merely the base of the femur demarcated by an outwardly visible sulcus. The femur is typically the largest podite of the leg and contains muscles that originate near its base and insert on the tibia (the tibial extensors above and the tibial flexors below: sometimes called the *tibial levators* and the *tibial depressors*, respectively). Highly developed tibial levators occur in several "jumping" groups such as grasshoppers. It also contains the origin of the pretarsal depressor (= pretarsal flexor). The tibia is typically a slender podite and the second largest (and often the longest) in the leg, with its basal end slightly bent toward the apex of the femur so

anterior portion of the primitive coxopleurite (i.e., the katepisternum) and forms the secondary articulation with the coxa along its anterior margin. The coxa is divided into proximal and distal portions by the *basicostal sulcus*, which is an external indication of the internal *basicosta* (a specialized costa) and the insertion point for *extrinsic muscles* (ones where the origins lie outside the appendage) that move the coxa. The portion of the coxa proximal to the basicostal sulcus is the *basicoxite*. The posterior part of the basicoxite is frequently developed into a large lobe called the *meron* (e.g., Neuroptera, Mecoptera, Trichoptera, Lepidoptera; but fused with the pleural wall in many Diptera). A *coxal sulcus* is often present and runs from the base of the coxa to the anterior

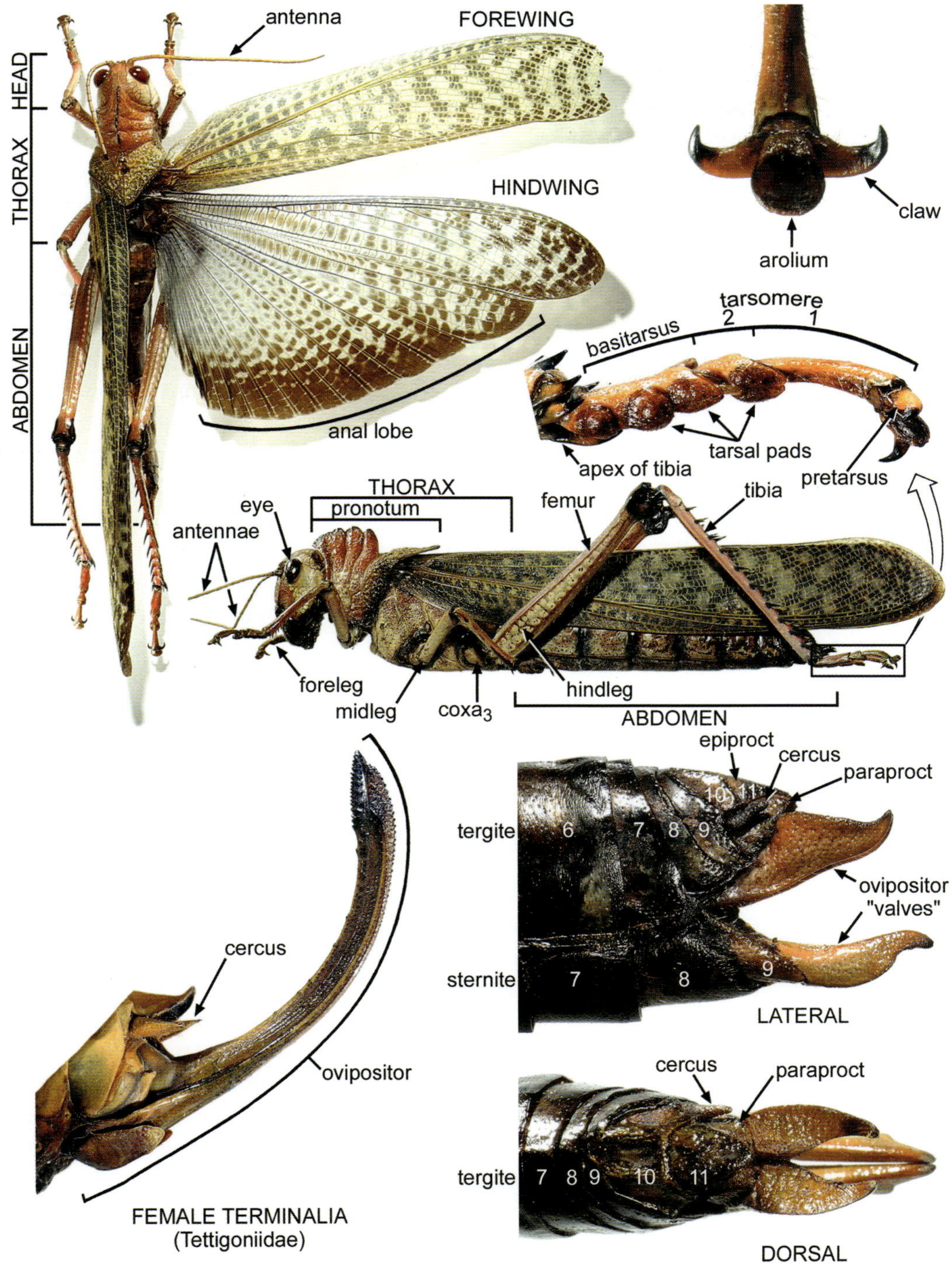

4.4. Basic external morphology of insects, based on orthopterans. The large ovipositor is from a katydid (Tettigoniidae); all other parts are from the acridid grasshopper shown in full. Not to the same scale.

that it forms the major joint of the hexapod leg. This bent head allows the tibia, when depressed (i.e., flexed) to be closely appressed to the undersurface of the femur, which is particularly important in jumping insects.

The tarsus is primitively one-segmented but is frequently subdivided into two to five subunits called *tarsomeres* (in Protura, some Collembola, and most larvae it retains the primitive condition of only one unit) (Figure 4.4). The tarsomeres are typically movable, but the tarsus never has muscles intrinsic to itself and is operated entirely by muscles that originate in the tibia and insert on the base of the tarsus, the tarsal levators, and tarsal depressors. The basal tarsomere is typically enlarged and called the *basitarsus*. Externally the tarsomeres sometimes possess small pads called *tarsal pulvilli* or *euplantulae*. The *pretarsus* is the apicalmost podite of the leg and, despite its minute size, is quite complicated. The pretarsus consists of lateral *claws* (*ungues*, which are erroneously and frequently called the "tarsal claws") and a median *arolium* (rarely entirely sclerotized) (Figure 4.4). The claws are attached via membranes to the *unguifer*, a small dorsal process of the last tarsomere. There can sometimes be minute, lateral sclerites near the base of the claws called *auxiliae*. Ventrally the pretarsus consists of a sclerite called the *unguitractor*, which is typically partially invaginated into the apex of the last tarsomere. The unguitractor can be further subdivided at times or have a distal sclerite called a *planta*. The pretarsus is moved via a long tendon that originates in the tibia and femur and inserts on the unguitractor plate of the pretarsus. This tendon is sometimes called the retractor of the claws. This muscle is a depressor (= flexor); there is no levator (= extensor) for the pretarsus, and elevation is done entirely by touching the substrate. The tendon is attached to the unguitractor plate of the pretarsus by a thin apodeme that extends back through the tarsus and tibia, and the actual flexor is in the base of the tibia or the femur.

Joints are formed by regions of membranous cuticle called *arthrodial membranes* formed between adjacent podites. The joints make movement between podites possible, while the type of articulation controls what kind of movement is possible. The main coxal articulation is situated at the ventral terminus of the pleural sulcus. Dicondylic coxae come in two forms; the first has a secondary articulation formed anteriorly with the trochantin (the trochantin is lost in many higher orders), while another secondary articulation is formed ventrally with the eusternum. All other joints in adult insects are dicondylic within the appendage. Larvae of holometabolous insects, however, frequently have monocondylic articulations even though the adults are dicondylic. Articulations are typically formed of an anterior and posterior point of articulation except at the trochanteral-femoral articulation, where (if present) it is sometimes composed of dorsal and ventral points of articulation.

Wings

While it is understood how vertebrate wings evolved from forelimbs, homologues and origins of insect wings have been confusing and controversial. Although often depicted in general texts as relatively simple structures with a few veins running through them, the insect wing is a structure of daunting complexity. Here, we provide just a basic account. Functional wings are present only in adult insects, the mayflies (Ephemeroptera) being the only insects where the last nymphal instar (the *subimago*) primitively possess functional wings. Wings begin to develop in earlier instars, either externally or internally, but do not become functional until after the final molt in all other pterygotes. Insect wings occur exclusively on the middle (i.e., mesothoracic) and metathoracic segments, together called the *pterothorax*, and they articulate to the body via a series of sclerites called *pteralia*. While numerous bones and muscles shape the foil of the vertebrate wing during flight, an insect wing is actively operated only at its base in the same way a lever hinges on a fulcrum. In the wing itself there are no muscles that allow the insect deliberate control over the movements of the wing beyond its base. *This is not to indicate that insect flight is as simple as flapping up and down*. Indeed, insects are capable of a greater range of movements than are the wings of vertebrates. Insects, particularly those that hover, are the most acrobatic fliers. These movements are a result of various muscles attaching at the wing base that pull on the pteralia, as well as how veins, folds, and flexion lines buttress and fold the wing.

As noted, however, the insect wing is similar to a lever acting on a fulcrum. As such, the structure of the pterothorax is also critical to wing movement. The wing extends between the dorsal plate of the insect thorax (the notum) and the side of the thorax (the pleuron), and from these are processes that function in the wing's articulation – the anterior and posterior notal wing processes, and the pleural wing process. The pleural wing process forms the fulcrum for the wing while two notal processes push down against the base of the wing on either end. Two additional plates, situated in membrane on either side of the pleural wing process and called the epipleurites, are the *basalare* and *subalare*. These plates provide insertion points for muscles that control the tilt of the wing during flight and assist in the downstroke of flight. In the Neoptera, the basalare also serves to extend the wing from its folded position over the abdomen.

The wing itself is formed of two epidermal layers, an upper layer and a ventral layer, which grow out from the body and fuse together. They are living structures, complete with hemolymph, tracheae, and nerves. Cavities form during this development, called *lacunae*, and form channels through which tracheae move along with some nerves. At eclosion to an adult, most of the epidermal cells die and form a cuticular wing membrane with cavities, called *veins*. This is an extremely durable structure that fossilizes much more readily than any other part of the body. The veins provide some structural support to form a more-or-less stable wing foil.

The veins are perhaps the most notable feature of the insect wing and a rich source of characters for understanding the evolutionary relationships of numerous groups and most insect fossils. They also provide information on flight biomechanics. The insect wing is primitively fluted, like a Japanese fan, with the veins alternating between concave and convex (typically denoted as "+" for convex and "−" for concave). This corrugation provides strength to the wing as it experiences various flight stresses. Convex veins sit on an elevated ridge, while concave ones lie in a trough or depression. Several systems have been proposed for naming veins; they are based

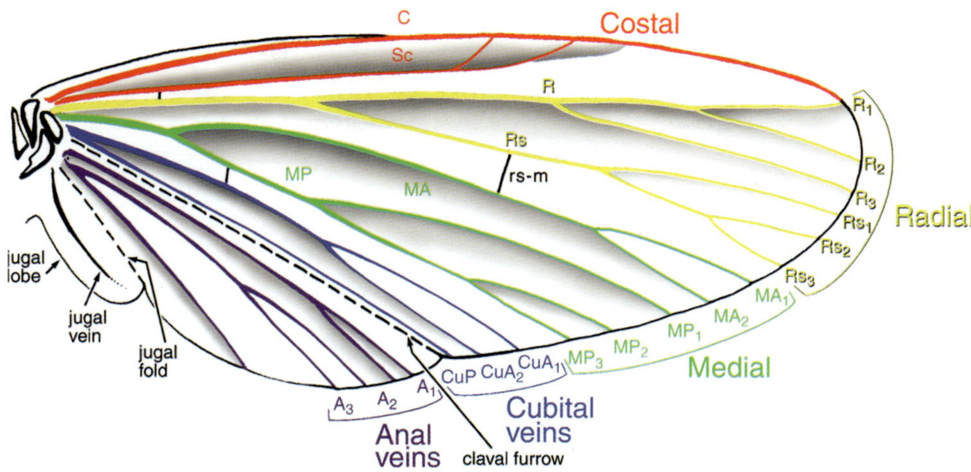

4.5. A generalized wing, indicating major vein systems and the terminology used in this book.

on various modifications of the Comstock-Needham system (e.g., Comstock and Needham, 1898, 1899; Kukalová-Peck, 1991). Here, we adopted the system that is most similar to that espoused by Wootton (1979) (Figure 4.5). Major longitudinal veins typically have major branches, each given names, and are indicated by uppercase letters with their branches indicated by subscripts. *Crossveins* are small, secondary veins running between the longitudinal veins and are generally indicated by lowercase letters; a hyphen separates the anterior-posterior longitudinal veins that they connect (e.g., sc-r = a crossvein from the subcosta to radius). The *archedictyon* is an irregular network of many short crossveins between the longitudinal veins and is believed to be the primitive condition for winged insects. The major, longitudinal vein systems in insects follow (refer also to the section on pteralia later in this chapter).

Costa (C+): this vein is usually on or just behind the anterior wing margin (if there is a small membranous area anterior to C, then it is typically called the *precostal area*). The costa can meet at its base a small sclerite called the *humeral plate* (see the discussion of pteralia later in this chapter).

Subcosta (Sc−): this vein sometimes has two branches and contacts the first axillary sclerite at the wing base.

Radius (R): the radius branches into two components, the true radius (R+) and the *radial sector* (Rs−). The base of the radius contacts the second axillary sclerite. Although it is a major longitudinal vein, Rs− does not extend directly to a sclerite at the base of the wing and instead originates directly as a branch from the radius; it is therefore considered part of the radial system. Rs− forks a variable number of times into veins R_{2-5}, sometimes called Rs_{1-4}. The original stem of Rs is concave, but its branches can sometimes alternate between convex and concave, though typically they are all concave.

Media (M): the base of the media contacts the distal end of the medial plate. Typically the media has two major branches (which will also fork) called the *media anterior* (MA+) and the *media posterior* (MP−). MA+ typically forks to form two branches (M_{1-2} or MA_{1-2}) as does MP− (M_{3-4} or MP_{1-2}). MP− can have more branches at times.

Cubitus (Cu): this is typically a three-branched vein that contacts the distal medial plate. Like the median, there is an *anterior cubitus* (CuA+) and a *posterior cubitus* (CuP−). CuA+ typically branches.

Anal veins (A+): the number of anal veins (called vannal by Snodgrass and earlier authors) is variable, and the veins are usually unbranched.

As noted, these veins contact pteralic sclerites that form the articulation at the base of the wing. The principal components of the pteralia are the axillary sclerites, tegula, humeral plate, and in Neoptera the medial plates. Typically there are three axillary sclerites that are positioned in different membranes of the wing (Figure 4.6). The *first axillary sclerite* lies in the dorsal membrane and articulates at its base with the anterior notal wing process and distally with vein Sc and the second axillary sclerite. The *second axillary sclerite* runs in both dorsal and ventral membranes and articulates ventrally with the pleural wing process and distally with vein R. It also attaches to the third axillary sclerite. The *third axillary sclerite* lies in the dorsal membrane and articulates with the posterior notal wing process and the anal veins. In the Hymenoptera and Orthoptera there is sometimes a fourth axillary sclerite (perhaps derived posteriorly from the third) that lies between the third axillary sclerite and the posterior notal wing process. The third axillary sclerite is Y-shaped with a flexor muscle inserting in the crutch of the Y, the other end of the muscle originating on the inner surface of the pleuron. When this muscle flexes, the wings fold over the abdomen

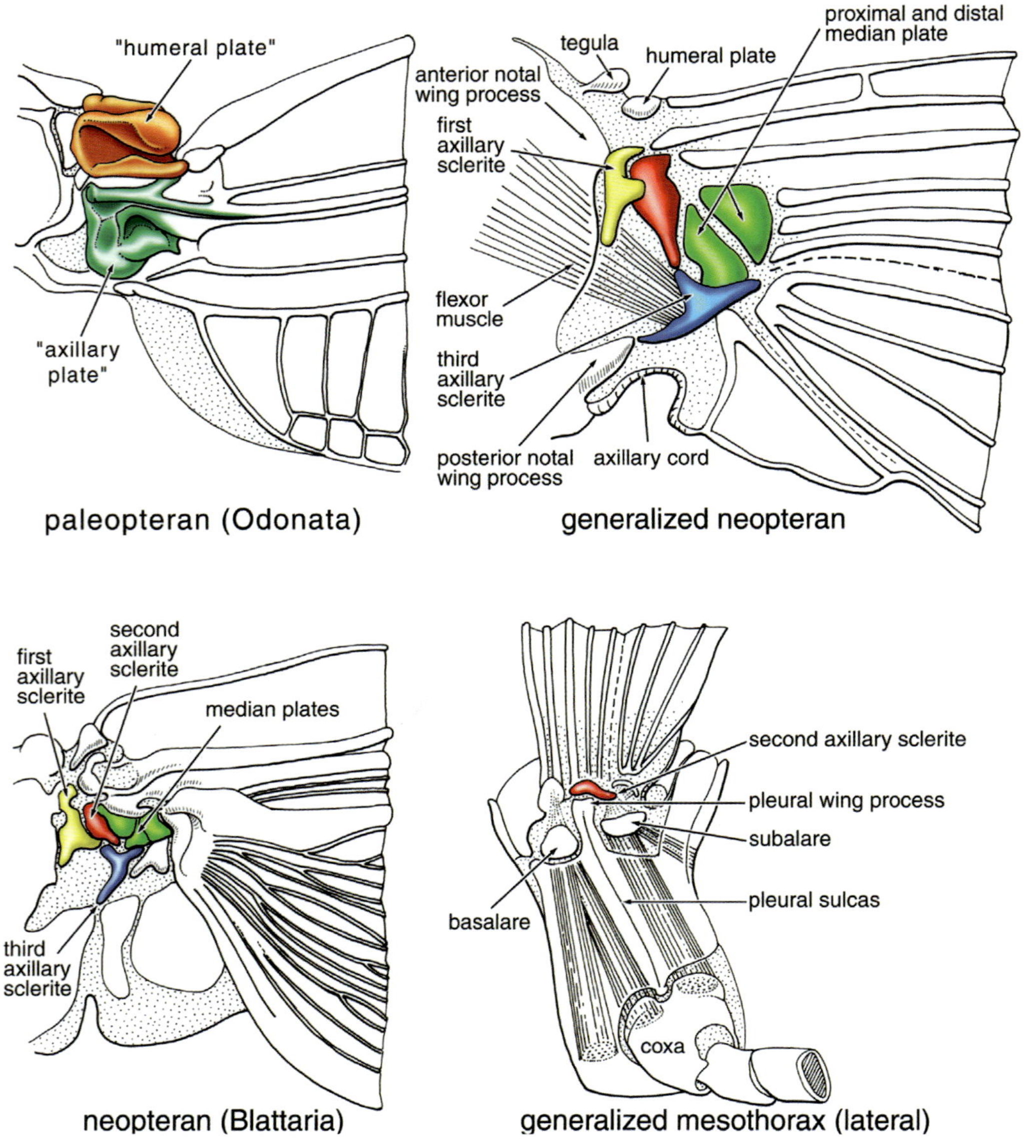

4.6. Major types of wing articulation in Ephemeroptera, Odonata, and Neoptera.

during rest, which is what defines the neopterous condition. The *medial plate* is perhaps derived from the third axillary sclerite. There are most frequently two medial plates, but sometimes there is only one. From the distal medial plate, veins M and Cu arise. The distal and proximal medial plates are separated by an oblique line over which the fold occurs during the flexion of the wings over the abdomen. The distal plate may be reduced in size or only a slightly more sclerotized region of membrane (leading to the condition of only one medial plate). The small *humeral plate* articulates with the base of vein C. The *tegula* is frequently developed to cover much of the wing base. When it is reduced in size, it lies anterior in the membrane between the humeral plate and the thoracic notum. Although tiny and often overlooked by even the most trained of entomologists, these plates orchestrate the complex suite of wing movements during flight.

In the primitively winged insects (e.g., Ephemeroptera, Odonatoptera) the arrangement is somewhat different. The Ephemeroptera wing base is exceedingly complex and very similar to other pterygote insects, but they have the axillary sclerites less defined and there are no medial plates (Figure 4.6). Odonata, alternatively, have only two larger articular plates, the humeral and axillary plates, the result of the fusion of several smaller sclerites; they are hinged together with the thoracic notum (Figure 4.6).

The *cells* of the wing are the regions between veins and crossveins. There are numerous naming systems for the cells, none of which are consistent across orders (we indicate these in the systematic sections of the book). One of the more common naming systems for cells uses the name of the vein that marks the anterior margin of the cell, though numerous other names are in use, like marginal, submarginal, discal, and discoidal. Only two terms are universal in entomology when referring to wing cells: *Closed cells* are those that are bordered on all sides by veins, while *open cells* are those where part of the cell is not delineated by a vein because it is at the edge of the wing.

To provide weight and strength to the leading edge of the wing, there is frequently a region of either expanded anterior veins (veinal composition differs by group) or highly pigmented and slightly sclerotized cuticle in an anterior cell to form a *pterostigma*. The pterostigma prevents the wing from fluttering during changes in the stroke of the wing by adding weight to the leading edge.

The most ingenious structures on the insect wing, however, are its folds. Recall that the insect wing does not have muscles within it to control the fine adjustments to its shape that are necessary during flight. The alterations must come about through the structure of the wing itself. The wing is beset with a series of *flexion lines* that respond to physical stresses and allow, in a very controlled fashion, the wing to partially collapse or fold upon itself, principally to generate vortices when developing lift. Some of these points of weakness are common to most insects. For example, the *median flexion line* is typically an oblique line of flexion running between Rs and MA and the *nodal line* is a transverse line of flexion, typically dividing a stiffer proximal region from a more deformable distal wing.

Because insects are dramatically varied, so are their flight mechanics. Hovering insects, for example, have veins near the wing tip coalesced to keep it rigid. As such, different groups require their wings to undergo different changes in shape so as to achieve the type of flight necessary for that particular lineage. *Alar fenestra* is a generalized term for any of various, shorter lines of flexion in various places on the wing, often seen as areas of weakness running across veins.

THE ABDOMEN

The abdomen is specialized for "visceral" functions of digestion, excretion, respiration, gametogenesis, and copulation. Suprisingly, the abdomen is the least modified tagma from the arthropod groundplan. It is a flexible structure that can swell with a meal, production of eggs, etc. The complete absence of legs on the abdomen is a derived feature of hexapods, since Myriapoda, Trilobita, and others possess numerous postcephalic locomotory appendages. Movements of the abdomen generally function to circulate air through the tracheal system, and hemolymph through the hemocoel. Primitively, the abdomen is composed of 11 metameres, plus one postmetameric segment, which is homologous to the *pygidium* of annelids and sometimes called the *telson* or *periproct*. Abdominal segments 8 and 9 form the genital segments, while abdominal segments 1–7 are pregenital and provide the clearest examples of metameric segmentation in insects. Abdominal segment 1 is often modified for articulation with the thorax, and abdominal segment 10 (when present) is always reduced in size. Abdominal segment 11 may be represented only by a dorsal lobe, called the *epiproct*, and two lateroventral lobes, called the *paraprocts*, as seen in Orthoptera. Only in adult Protura and embryos of many hemimetabolous insects is the full complement of 11 complete abdominal segments easily discernable, all other insects show reduction in at least one segment, and frequently more. The telson is sometimes lost altogether or reduced to a circumanal membrane. Odonata, for example, have three small sclerites surrounding the anus that are likely remnants of the telson. Abdominal segmentation can also differ with instar. Protura, for example, hatch with eight segments and add three segments in successive molts; they are *anamorphic*.

The basic components of the insect abdomen are the terga (dorsal plates) and the sterna (ventral plates). The tergum consists of basic secondary segmentation, that is, with terga possessing an acrotergite and antecosta, and with longitudinal musculature running between antecostae. There is no further division in the groundplan of the abdominal terga, so there are no postnota. The sterna typically are organized like the terga. In Archaeognatha the ventral surface of the pregenital abdominal segments each consist of three sclerites, where the small, median, subtriangular plate is hypothesized to be the true sternum. The large, lateral sclerites bear median *eversible vesicles* and lateral movable appendages called *styli* (Figure 4.7); the sclerites are hypothesized to be homologous with coxae of ancestral legs. Thus, these sclerites are typically called *coxopodites*, and they are even provided with musculature arising on the tergum like the coxae of thoracic legs. Under this interpretation the styli are homologous with thoracic legs, but they have also been hypothesized to merely be epipodites and homologous to gills in Crustacea and aquatic insects. Eversible vesicles are present in apterygotes and myriapods but are lost in pterygotes except for the independent acquisition of a peculiar median vesicle present in Grylloblattodea. They are used to absorb water and are operated mostly by hemolymph pressure, although some have weak extrinsic muscles. Some insects may have lateral plates called *pleurites* and while in some cases these are like the pleura of the thorax, most are lateral fragments of the terga or sterna. Spiracles are typically present on segments 1–8, but there may be fewer segments and spiracles. In the groundplan condition, the spiracles are located in the pleural membrane between the terga and sterna, but may be incorporated into the terga or sterna.

The *cerci* are appendages on segment 11 (Figure 4.4), typically with a sensory function, although these are developed into forceps in Dermaptera and some Diplura and serve a grasping function. Since the eleventh segment is typically lost, fused, or highly reduced, they typically appear to articulate with the tenth tergum or other segments. Typically, the

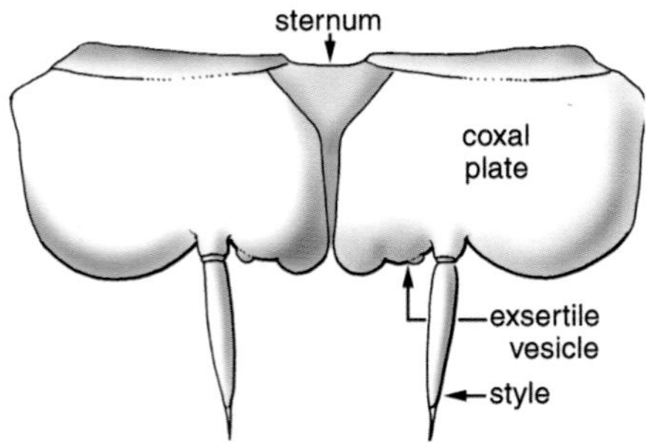

4.7. A generalized abdominal sternum of a bristletail. Apterygote insects retain the basic hexapod feature of styli and eversible vesicles that are primitive for all other insects.

cerci are implanted in membrane between the epiproct and the lateral paraprocts. The paraprocts sometimes retain a ventral membranous connection that can be developed into a small, subanal lobe called the *hypoproct*. Cerci are operated either by muscles originating on the tenth tergum or the epiproct, never on the paraprocts. In some primitive apterygotes the cerci retain an apparent coxopodite. Males of Embiodea and a few Orthoptera have asymmetrical cerci that also function in copulation.

The telson is highly reduced, and in groups like Odonata it is present only as three circumanal sclerites called periprocts; a few other primitive insects also have a circumanal membrane, which might also be the periproct. The telson is otherwise unknown except in embryonic forms but is primitively present in Protura. It is a postmetameric segment and therefore never has appendages.

The most fundamental function of the insect abdomen is for reproduction. Eggs can be laid and matured singly or in prodigious quantities, the current world record in a single day being 86,400 eggs by queens of the termite *Odontotermes obesus*. The male genitalic system is internally composed of testes, vasa deferentia, seminal vesicles, accessory glands, and an ejaculatory duct that is primitively located on the ninth abdominal segment. Females have ovaries, lateral oviducts, a common oviduct, accessory glands and a spermatheca, and a genital chamber (*bursa copulatrix*), which is primitively located between the eighth and ninth abdominal sterna, but the position may vary in different groups of insects. The ovaries of insects occur in three types, the morphology and organization of which is determined by nurse cells. In the *panoistic* (*pan* = "all"; *oon* = "egg") condition all oogonia except stem-line oogonia are eventually transformed to oocytes, and no nurse cells are present. In the *meroistic* (*mero* = "part"; *oon* = "egg") condition oogonia divide to form two types of cells, oocytes and nurse cells. Among meroistic ovarioles two additional forms are known: *polytrophic*, where nurse cells travel down the ovariole with the oocytes, and *telotrophic*, where nurse cells are retained within the germarium and are connected to oocytes passing down the ovariole via long cytoplasmic filaments. It would appear, however, that panoistic ovaries are primitive for Hexapoda, with a unique development of meroistic ovaries in Ephemeroptera that might not be homologous with meroism in Neoptera.

Insects also have a series of sclerotic structures used in coupling and egg deposition. In entognathous hexapods there are no such structures, and mating occurs via external fertilization with a spermatophore, there are no complex external genitalic structures, and eggs are deposited on the ground directly out of the common oviduct. For true insects, that is, ectognathous hexapods, the condition is quite different and that in Archaeognatha, basalmost living order of insects, is considered close to the groundplan for all insects and most useful as an example. Also in Archaeognatha male and female genitalia are the most similar, and the construction of these structures is very simple. Furthermore, the genital segments share many structures in common with abdominal segments 1–7. As mentioned previously, bristletails have large coxopodites with a small, triangular sternum. Each coxopodite has a lateral stylus and a median eversible vesicle. In both male and female Machilidae abdominal segments 8 and 9, termed the *first* and *second genital segments*, are basically alike and similar to segments 1–7. The coxopodites with styli are consistently present; these sclerites are called *gonocoxae* and the styli are referred to as *gonostyli*. The gonocoxae are also called *valvifers* in some systems of genitalic nomenclature. The gonocoxae of the eighth segment are called the *first gonocoxae*, and those of the ninth are the *second gonocoxae* (likewise for the gonostyli), also known as *first* and *second valvifers*. Where eversible vesicles would occur, the genital segments have long, narrow, annulated structures called *gonapophyses*, which are also known as *valvulae*. The gonapophyses are typically longer and more conspicuous in females where they function, albeit somewhat inefficiently in Machilidae, as ovipositor valves (Figure 4.8). In male bristletails the gonapophyses do not act as an intromittent organ; sperm transfer is indirect, similar to the situation in entognaths; and some males lack the first pair of gonapophyses. The male produces a spermatophore and silken threads leading to it which he then induces the female to follow, to the spermatophore.

The bristletail male has a single, median penis, which is a membranous tube situated in the intersegmental membrane between the ninth and tenth abdominal segments (Figure 4.9). The valvulae in the male are heavily sclerotized and fit together into a common shaft. The second gonapophyses, those from segment 9, are rotated 180° and fused dorsally. A tongue-and-groove joint connects the second and first gonapophyses. Muscles originating on the gonocoxa insert on the base of gonapophyses, with muscles originating on terga inserted on gonocoxae. The ovipositor valvulae are not joined together at a common base; each is attached to its own gonocoxa, and each gonocoxa articulates dorsally to its

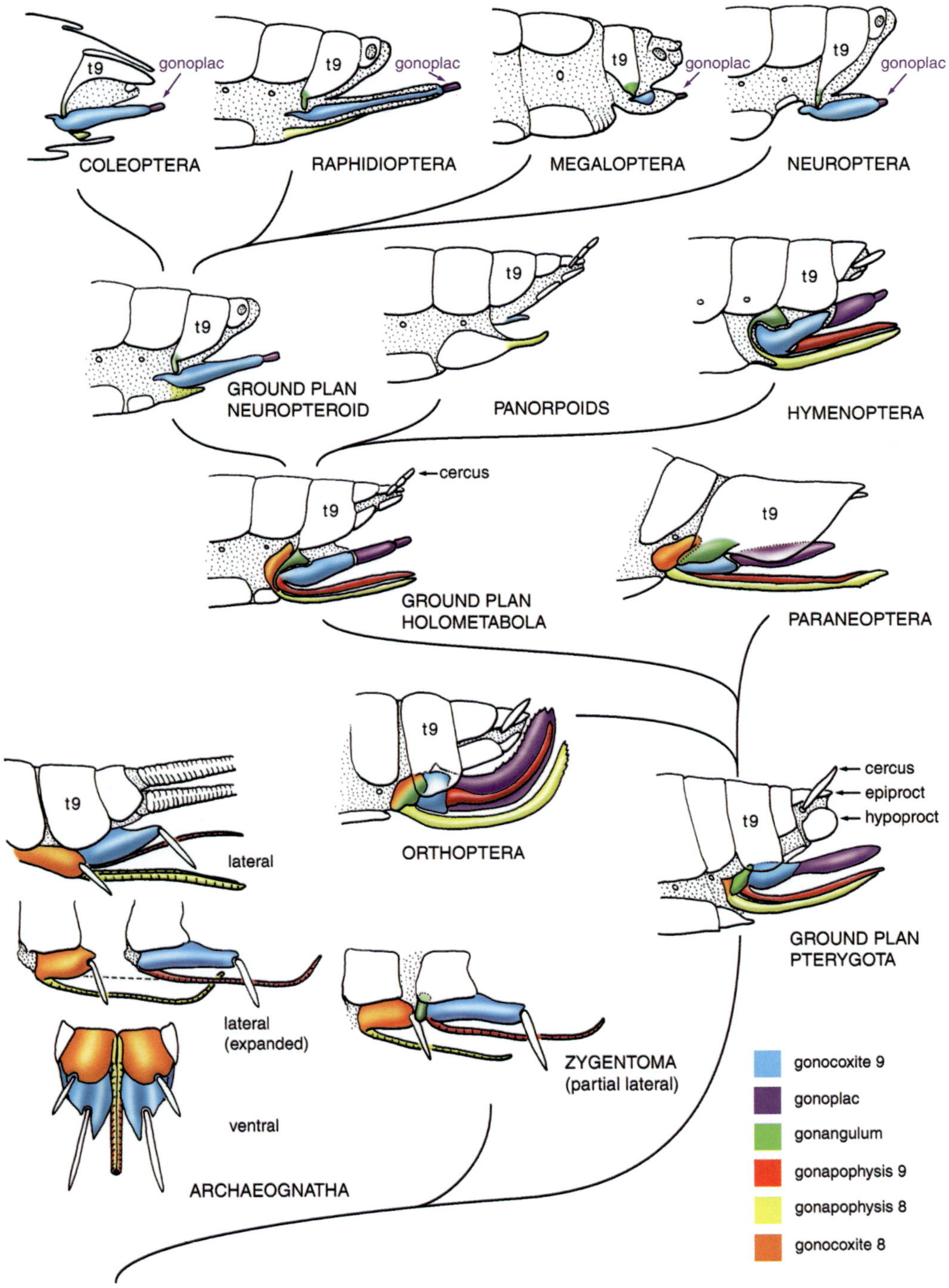

4.8. Evolution of the insect ovipositor, showing homologous structures. Based on Snodgrass (1935) and Mickoleit (1973).

corresponding tergum. In the female these structures together form the *ovipositor* (Figure 4.8). The female genital opening, or *gonopore*, is located between the eighth and ninth abdominal sterna. Many insects, however, have lost the ovipositor. For example, in Ephemeroptera each ovary opens to the outside via its own oviduct; there is no median, common oviduct opening via a single median, gonopore. Instead there are paired, lateral gonopores, each opening into conjuctival membrane behind the seventh sternum, and the male genitalia are similarly paired. Similarly, in Dermaptera the lateral oviducts unite into a very short, median oviduct, with the gonopore opening immediately behind the seventh sternum. Similarly, some groups have a structure that secondarily functions as an ovipositor but is modified from nonappendicular structures like sternites, called an *oviscapt*. The distal segments of the abdomen are tapered and capable of being retracted like a collapsible telescope. The distal parts may even be heavily sclerotized for piercing the substrate.

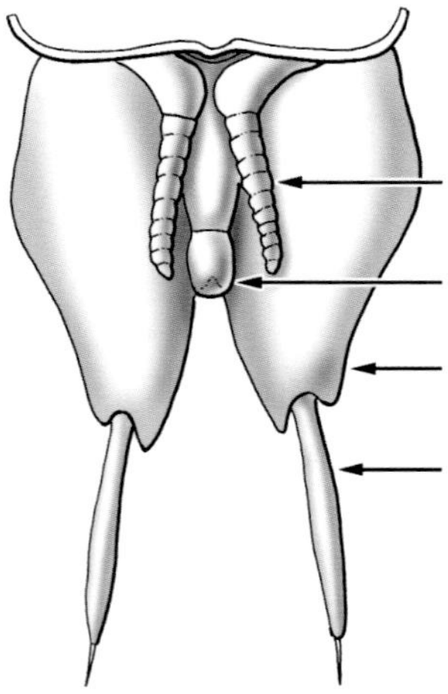

4.9. Generalized male genitalic structures of a bristletail. Male genitalia of basal insects are relatively simple; in pterygotes the diversity and complexity of genitalic structure becomes bewildering.

Some subsequent modifications also occur in the ovipositor of insects. The *gonangulum* is a derived structure occurring in the Zygentoma and most Pterygota, which is a triangular sclerite that improves the mechanical efficiency of the ovipositor. Recall that each gonocoxa had an articulation with the tergum. The gonangulum forms a third articulation that strengthens the base of the ovipositor. In its generalized form it attaches to the base of the first gonapophysis (first valvula) and articulates with the second gonocoxa and ninth tergum. It appears as a distinct, triangular sclerite in Lepismatidae and some other groups, like Hymenoptera, but is fused with the gonocoxae in most other groups though frequently still evident by lines of fusion. In instances where the fusion is so complete that the sutures are gone, the typical articulations with the second gonocoxa, with the ninth tergum, and the base of the first valvulae indicate its presence.

The ovipositor is further modified in the Pterygota (Figure 4.8). First, the first gonostylus is apparently lost. Second, a new structure appears called the *gonoplac*, or the *third valvula*. This is on the apical margin of the second gonocoxa and is similar to that of the second gonostyli but develops differently from the true styli. These arise as outgrowths of the second gonocoxae, with the gonostyli at their apices. Gonoplacs sometimes form part of the ovipositor shaft but more often they form a sheath that surrounds the ovipositor shaft. Third, the inner surface of the gonapophyses are lined with scales or ridges. These are arranged so as to produce spines that point backward and prevent the egg from sliding back up the shaft of the ovipositor. Lastly, the gonocoxae of the eighth tergum tend to shift forward and form a second articulation with the ninth tergum.

While the male genitalia in Archaeognatha and Zygentoma do not differ greatly from the groundplan provided for female Machilidae (Figure 4.9), they are fundamentally different in all other insects. In males the external genitalic structures serve two principal functions: They are the primary organs for delivery of sperm to the female, and they are also used for seizing and holding the female during mating. The primitive condition for Hexapoda is indirect fertilization, and external genitalic structures are absent altogether or present but not functioning as intromittant organs, such as in Machilidae. Fertilization is internal; *transfer* is external. This is in contrast to marine Arthropoda, such as in *Limulus*, where fertilization is usually external. In Archaeognatha and Zygentoma the external male genitalia are identical to the female's except that the gonapophyses are smaller and frequently missing the first pair. In addition to the typical elements just mentioned (i.e., gonocoxae, gonostyli, and gonapophyses), there is a median, membranous tube that arises from between the second gonocoxae (second valvifers) and the second gonapophyses (second valvulae). The *ejaculatory duct* runs through this tube and opens to the outside of the body via the gonopore. This median tube is the penis, or *aedeagus*. Primitively the position of the penis is between segments 9 and 10, and there are differences of opinion as to whether it arose as an outgrowth of the ninth or tenth abdominal segment.

Copulation is the direct transfer of sperm from the male into the female's gonopore or some other opening for sperm reception. This method of reproduction has evolved independently in numerous arthropod groups and related lineages. Within Hexapoda, the direct transfer of sperm is an additional defining feature for the winged insects, Pterygota. Some forms of copulation (i.e., direct fertilization) can be brutal. For example, in some species of Onychophora (admittedly not insects) the male deposits the spermatophore anywhere on the outside of the female's body. The cuticle "dissolves" beneath the spermatophore and the sperm swim inside. Within the insects *traumatic insemination* (e.g., in bedbugs: Cimicidae) whereby the male uses his penis to puncture the body wall of the female at a notch on the fourth sternum beneath which is a special structure called the copulatory tube. Sperm are released into the Organ of Berlese and from there they swim through the hemolymph to the female reproductive system.

Pterygotes transfer sperm directly into the female. Male genitalic structures form on the ninth abdominal segment. The aedeagus is often weakly sclerotized or entirely membranous and is flanked by *parameres*, which typically form clasping structures. The parameres bear at their apices gonostyli although in many groups these are given alternate names because the homology with true gonostyli is not always evident. Basally the parameres are articulated with the *phallobase*, or *gonobase*. Invaginated within the aedeagus is typically an *endophallus*, and through which the ejacultory duct runs. The endophallus is often everted, turning inside out, so that the gonopore is extruded. The opening of the endophallus is called the *phallotreme*. Various orders possess a plethora of additional sclerites or paraphyses.

The origin of male structures in Pterygota is perplexing. They are fundamentally different from anything seen in the wingless insects and have at times been considered merely further elaborations on the original structures of appendicular origin, or alternatively de novo structures. As such, there are two suites of terminologies for the basic components of male genitalia in Pterygota depending on whether or not they are consider to be of appendicular derivation. Many sclerites are unique to particular orders or families; consequently, a mountain of terms exist for the unusual additions.

The Complexity and Diversity of Male Genitalia

To the entomological neophyte, entomologists appear to have a fetish; they are unnaturally preoccupied with male genitalia in insects. Male genitalia in insects are actually the richest source of morphological characters, and species of insects are almost universally diagnosed on the basis of the sclerotized male genitalic appendages. These can be exceedingly complex structures, and probably as a result of this they differ among closely related species more than any other structures on the body, though it is quite clear that these structures evolve more quickly than do other parts of the body. By genitalia we refer to the *penis*, which is the organ of sperm delivery and usually intromission (male apterygotes, for example, deposit a spermatophore, which the female picks up), as well as to the associated pairs of *appendages that flank the penis*, like the gonocoxae, gonostyli, and parameres. These latter structures are generally assumed to have a clasping function in insects, but the mechanics of insect mating is an obscure topic that has been barely studied. We are not referring to the soft internal organs, like testes.

A bewildering complexity of male genitalia is not limited to insects but occurs in diverse animals that have internal fertilization, or where copulation takes place. Examples of copulating animal groups where their diverse male genitalia are routinely used for systematics include the intromittent organs of various worms, the claspers of particular fish (like guppies and sharks), the hemipenes of snakes, and the bacula of certain mammals. The baculum is the penis bone, which occurs in five mammal orders [Primates (except humans), Rodents, Insectivores, Carnivores, and Chiroptera (bats)]. Insect male genitalia, though, are in their own special category of exuberance. So diverse is their morphology that homologies of male genitalia among orders is not always possible to determine, and a volume of specialized vocabulary is necessary just to name the structures (Tuxen, 1970). Female genitalia, in contrast, are relatively simple and differ little among closely related species. The most significant traditional explanation for this is that male genitalia provide cues to a female that her mate is the appropriate species, the so-called lock and key, or hand-in-glove, hypothesis. Eberhard (1985), however, provided a detailed and comprehensive thesis that challenges this view. He accumulated compelling evidence to show that male genitalia are a result of sexual selection, just like more overt, sexually dimorphic structures like antlers, manes, and breeding plumage. In their own subtle way, male genitalia also "court" the female, and whether or not she "chooses" to actually copulate or even remain in tandem may depend on how she is titillated.

Eberhard's (1985) theory explains several interesting phenomena. First, in animal groups where fertilization is external, the male genitalia are monotonous in structure, such as in most fish, where the males simply broadcast their sperm over the laid eggs. In hexapods, the primitively wingless entognaths (springtails and their relatives), bristletails, and silverfish reproduce via spermatophores. In these groups these structures are packets of sperm on a stalk that the female picks up (there is no copulation), and male genitalia are either quite simple or only subtly different among species. Second, in those groups of animals where the male doesn't use the penis for insemination, but instead uses some other structure, the "secondary" genitalia are wildly diverse. In all these situations the male transfers the sperm from his genitalia to the secondary genitalia before copulation. Perhaps the best known example of this are the pedipalps of male spiders (a group in which species are almost never described in lieu of a male specimen), but other examples include the chelicerae of solifugids and certain mites, and the secondary male genitalia of damselflies and dragonflies. Third, in groups where there is elaborate courtship or sexually dimorphic features (usually these occur together), the male genitalia are often (but not always) relatively uniform. This suggests a tradeoff between overt and copulatory courtship. It is difficult to assess the complete complexity of courtship, though, because pheromones and inaudible stridulation or other sound production may also be involved.

Without doubt, insect male genitalia are evolving extremely rapidly, which is amply illustrated by the hundreds of species of parasitic fruitflies in the genus *Cladochaeta* (family Drosophilidae) (Figure 4.10). We draw on this group as an example not only because we know it well (Grimaldi and Nguyen, 1999) but also because externally the species are monotonous, yellowish little flies that have an exceptional complexity and diversity of male genitalia. Some species have paraphyses (lobes that flank the penis, and that may or may not be homologues with parameres!) that are like spatulas, prongs, or corkscrews, and the surstyli (the claspers) are similarly diverse. Moreover, the female genitalia of *Cladochaeta* were also carefully examined (which is generally not done in insect systematics, or at least for flies), and the female genitalia consistently differ among species, only much more subtly than for males. This is one of many possible

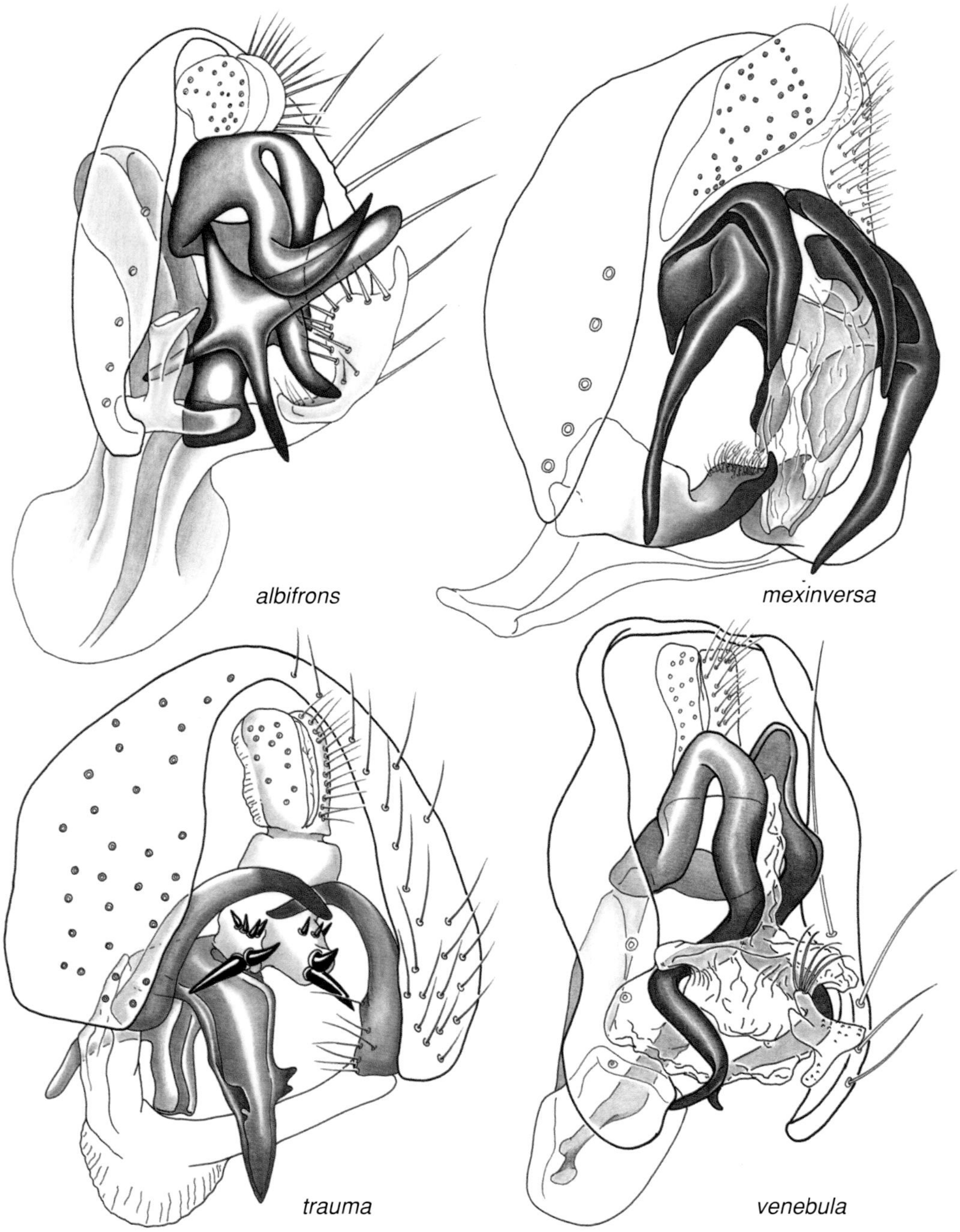

4.10. An exuberance of male genitalia in select species of the large fruit fly genus *Cladochaeta* (Drosophilidae). As for many groups of insects, closely related species differ most on the basis of male genitalia. Insect male genitalia are elaborate for developmental and functional reasons: They are segmented appendages that also perform subtle, internal "courtship." From Grimaldi and Nguyen (1999).

examples that illustrate the fundamental difference in complexity of male and female genitalia in insects. In many pterygotes, female genitalia are sternal in origin: They are derived from plain, basic plates. Male genitalia in all insects, by contrast, are derived from arthropod-style pairs of appendages, with their typical segmented construction. With such a design, one pair of appendages may become grotesquely modified, but the other pair may assume some of the original function. Likewise for segments within an appendage, the gonocoxite may evolve a huge lobe, but the gonostylus will retain its original clasping function. The repetition of a segmented structure allows innovation without entirely compromising the original function.

The theory that sexual selection accounts for the dramatic diversity and complexity of insect male genitalia makes great sense, but it does not entirely account for these remarkable structures. The basic design of insect male genitalia predisposes them to be more developmentally, structurally, and functionally complex than female genitalia. Homologies matter too.

DEFINING FEATURES OF THE INSECTS

What then defines an insect? Certainly six legs is what most individuals think of; however, entognaths as well as stem-group panhexapods have or had six legs. The insects are without a doubt a monophyletic group, universally supported by morphological and molecular features. The defining features of the Insecta include the following:

- Loss of musculature in the antenna beyond the scape
- The presence of a chordotonal organ in the antennal pedicel (the Johnston's Organ)
- The development of the posterior tentorium into a transverse bar
- The loss of articulations between the coxae and the sterna
- The subsegmentation of the tarsus into units called tarsomeres
- The articulation of the pretarsal claws with the apicalmost tarsomere rather than the pretarsal base
- The presence in females of an ovipositor formed by gonapophyses on the eighth and ninth abdominal segments (although this trait may be more primitive since progenitors of these structures were apparently present in marine panhexapods)
- The presence, at least primitively, of a long terminal filament on the dorsum of the eleventh abdominal segment (Kristensen, 1991).

As we will see through this discourse, most of these traits may be secondarily modified or reduced in one or more insect lineages. For example, the ovipositor is vestigial or independently lost in several groups, like the orders Zoraptera, Phthiraptera, and Coleoptera and the orders of Panorpida.

With this concept of an insect in mind, we can begin to unravel their evolutionary history. As we will see, their history as we know it begins around 410 MYA, in the alien world of the middle Paleozoic. Before considering the various lineages, we provide here a brief outline of the major groups of insects and the history of studies that contributed to our present understanding of their relationships.

RELATIONSHIPS AMONG THE INSECT ORDERS

A BRIEF HISTORY OF WORK

The history of entomology, although an engaging topic, has been more thoroughly covered elsewhere (Essig, 1931; Smith, 1973). The following is an account of those post-Darwinian authors who have contributed most significantly to our knowledge of insect phylogeny and the fossil record.

Even though early authors did consider the various affinities among major groups, like the quinarians of the early nineteenth century, it was not until Darwin provided his theory of evolution that systematists really tackled the phylogeny of insects in a major way. The first individual to depict a phylogeny for the insects was Ernst Haeckel (1834–1919), although he did not take into account fossil forms in any critical manner. Indeed, many of the early advances in paleoentomology had not yet taken place while he was developing his system (Haeckel, 1866, 1890, 1909) (Figure 4.11). Haeckel's treatment was, moreover, superficial as he was more interested in the higher phylogeny of all life (and later in various racist ideas on the evolution of humans). Alpheus Hyatt (1838–1902) and Jennie M. Arms (1852–1937: later Jennie M. Sheldon) provided another early evolutionary tree of insects in a small guide for teaching entomology (Hyatt and Arms, 1890). Even though they did not explicitly include fossil taxa, their phylogeny did abstractly depict the surface of the globe with the living lineages arising from the disc of the Earth and progenitors of the orders coalescing beneath the disc. Examples of other authors giving early genealogical arrangements of the orders were Schoch (1884), Brauer (1885), Emery (1886), Packard (1869, 1886), Sharp (1895, 1898), Comstock (1888), and Comstock and Comstock (1895).

In the late 1800s, two authors, Samuel H. Scudder (1837–1911) and Charles Brongniart (1859–99), dramatically expanded the study of fossil insects and, as such, discussions of insect relationships and evolution. Although previous students of fossil insects such as Oswald Heer, Ernst F. Germar, P. B. Brodie, and Christoph G. A. Giebel had cursorily described taxa from several deposits, none provided a significant context in which to understand their finds. Scudder worked on deposits of various ages ranging from the Late Carboniferous to the Pleistocene of North America (Figure 4.12), while Brongniart studied the Late Carboniferous fauna of Commentry, France. These authors brought for the first time the numerous, enigmatic Paleozoic lineages to the forefront of entomology. They were the first to highlight the importance of fossil forms for understanding insect evolution (Brongniart, 1885a, 1893; Scudder, 1885, 1886). Their monographs served as the foundation of paleoentomology that would be expanded upon by Anton Handlirsch (1865–1935) (Figure 4.13). Handlirsch can truly be considered the architect of paleoentomology and provided the first critical phylogenetic study of both living and fossil insects (Handlirsch, 1903, 1904). In a monumental work, *Die Fossilen Insekten*, of nearly 1,500 pages, Handlirsch synthesized all that was known of fossil insects and placed them into a phylogenetic classification with the Recent fauna (Handlirsch, 1906b, 1907, 1908). Handlirsch (1925, 1937, 1939) later slightly refined his system to reflect changes in his thinking during the intervening decades.

Although not as widely recognized as Handlirsch, Karl Börner (1880–1953) published papers on the phylogeny and

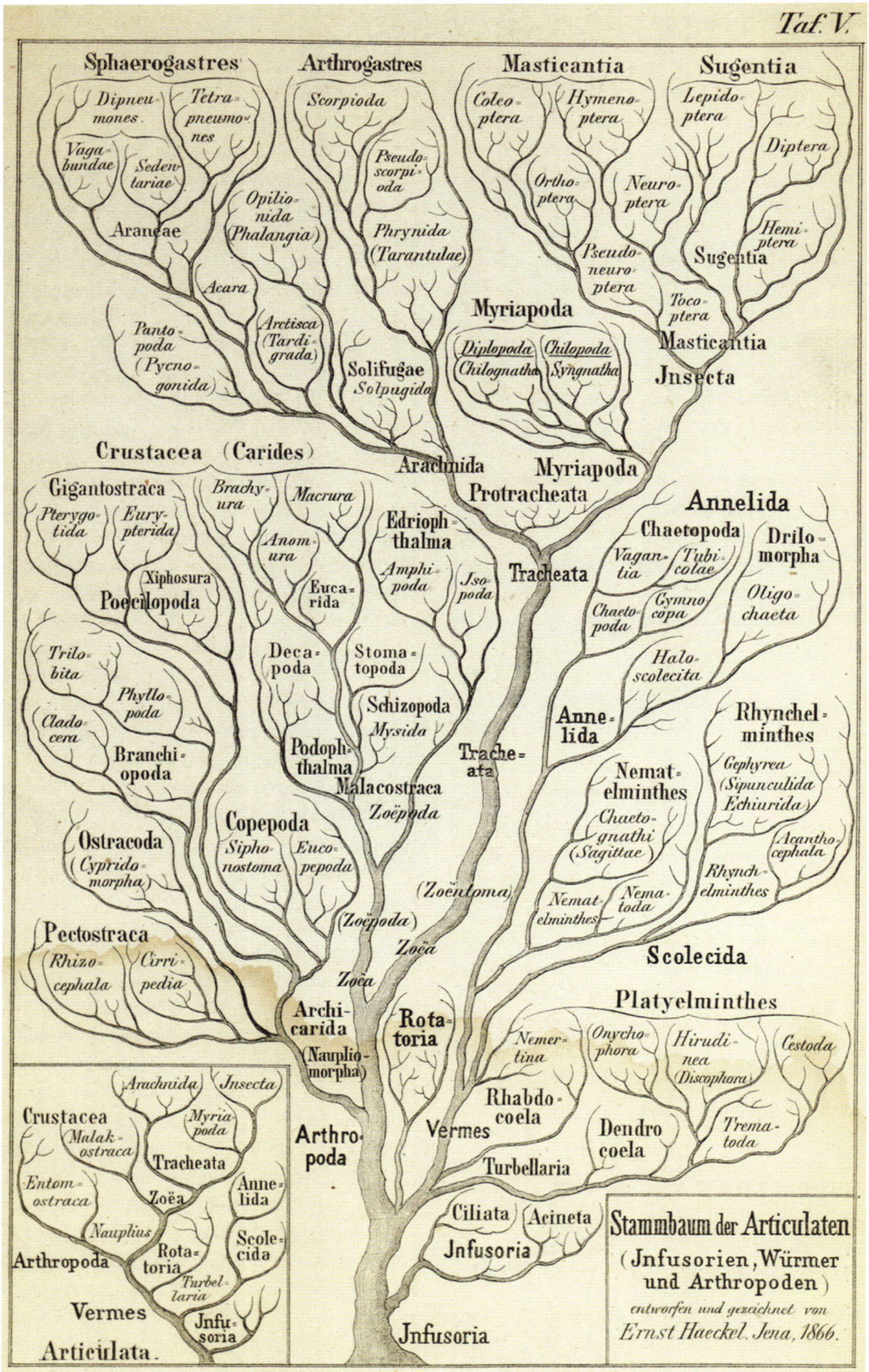

4.11. The first phylogenetic diagram depicting relationships among the insects and other arthropods. Insects were divided into Masticantia and Sugentia, for the masticating (chewing) and sucking insects, respectively, which is an artificial system. From Haeckel (1866).

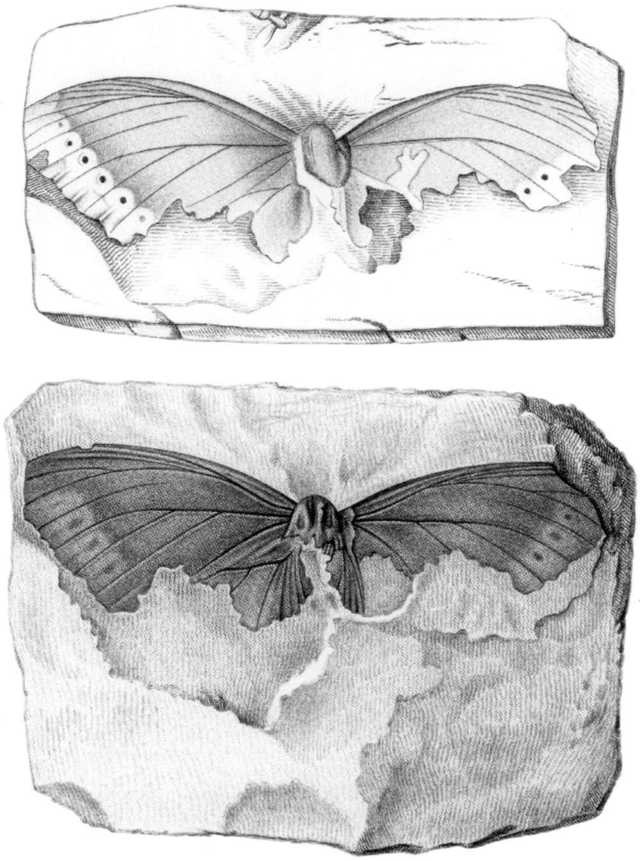

4.12. Fossil butterfly, *Mylothrites pluto*, from Florissant, Colorado, from the first major work on fossil butterflies (Scudder, 1875). Samuel H. Scudder (1837–1911) was the first entomologist to specialize on fossil insects. He published sumptuous monographs on orthopterans and North American butterflies, as well as fossils. Photo: AMNH Library.

4.13. Anton Handlirsch (1865–1935), Director of the Naturhistorisches Museum in Vienna and the first major paleoentomologist. Photo: Deutsche Entomologische Institut.

classification of Recent insects at about the same time that the former was putting forth his major contributions to the subject. It can be easily said that some of Börner's concepts on insect ordinal relationships were ahead of his time. He was a careful comparative anatomist and made such distinctions as separating the silverfish from the bristletails and positing a closer relationship between Odonata and Neoptera based on the loss of a subimaginal molt (Börner, 1904, 1909).

Two other prominent paleoentomologists who contributed significantly to the ordinal classification of insects began producing papers around the time of Handlirsch's magnum opus: Robin J. Tillyard (1881–1937: Dunbar, 1937) (Figure 4.14) of Australia and Andreas V. Martynov (1879–1938: Carpenter, 1938b) of Russia (Figure 4.15). Both of them, like Handlirsch, avidly studied the living insect fauna. Martynov, who was early interested in caddisflies and crustaceans, eventually took up the study of the great fossil insect deposits of the newly formed Soviet Union. Because Martynov was a keen comparative morphologist of living insects, he was adept at interpreting prolific fossils from diverse sites such as Karatau and Soyana. Martynov developed a major classification of all insects based on his studies on fossils, in which he recognized the classical division of the winged insects into paleopterous versus neopterous forms among other supraordinal groupings (Figure 4.16), and which was simultaneously recognized by G. C. Crampton (discussed later). At the same time, Tillyard was developing his own ideas on insect evolution based on studies of the Paleozoic deposits of Australia and Kansas. In particular, Tillyard was the first individual to truly recognize the significance of the Lower Permian fossils recovered from Elmo, Kansas, which he monographed. It is little known that Tillyard was also a devotee of the occult, who believed that mysticism could be used to understand the lives of long vanished organisms. It has even been recounted that he once visited the Museum of Comparative Zoology with the idea of having a local Bostonian mystic bring back to life the huge Permian griffenfly specimens preserved there so that he could observe

4.14. Entomologist and paleontologist Robin John Tillyard (1881–1937), who was the first major worker on Permian insects. He sought a mystic to reveal the deep past so that he could see the giant Permian griffenflies in flight, but he was also a careful and thoughtful scientist. Photo: *Science News*.

their behavior (e.g., Evans and Evans, 1970). Certainly other paleoentomologists were also at work at this time, such as the prolific Theodore D. A. Cockerell (1866–1948: Weber, 1965), but none were as influential on paleoentomology or neoentomology.

While Tillyard and Martynov toiled away on Paleozoic deposits, several neontologists were at work attempting to construct a framework of insect evolution based on the comparative morphology of living forms. Guy C. Crampton (1881–1951: Mallis, 1971) (Figure 4.17) was a gifted, albeit obsessive, morphologist who filled his tiny apartment with vials of specimens and mountains of papers for his studies on the anatomy and phylogeny of insects. He published extensively during the early 20th century on the phylogeny and classification of insect orders (e.g., Crampton, 1924, 1928, 1931, 1938). His work was focused on the modern fauna, but he did consider the fossil record as it had been documented by Handlirsch, Tillyard, and others. Crampton's detailed studies of numerous invertebrates mirrored those of the more widely recognized Robert E. Snodgrass (1875–1962) (Figure 4.18) who wrote the definitive work, even to this date, on the morphology of insects (Snodgrass, 1935). While Snodgrass contributed perhaps more than any other person to our understanding of the anatomy of insects, his studies principally focused on the larger picture of arthropods among other invertebrates, the position of hexapods, and the deciphering of homologous structures across insect orders. Thus, Snodgrass provided a wealth of data for interpreting insect evolution but he did not provide a phylogenetic synthesis of the orders. Other researchers on the modern fauna worth mentioning include Anton Krausse (Krausse, 1906; Krausse and Wolff, 1919), Charles T. Brues and Axel L. Melander (Brues and Melander, 1915, 1932), Karl L. Escherich (Escherich, 1914: building upon Prell, 1912); John B. Smith

4.15. The Russian entomologist and paleontologist Andreas V. Martynov (1879–1938). His phylogeny of insect orders (Figure 4.16) was ahead of its time, and he proposed several major lineages now recognized (cf. Figure 4.24). From Carpenter (1938b).

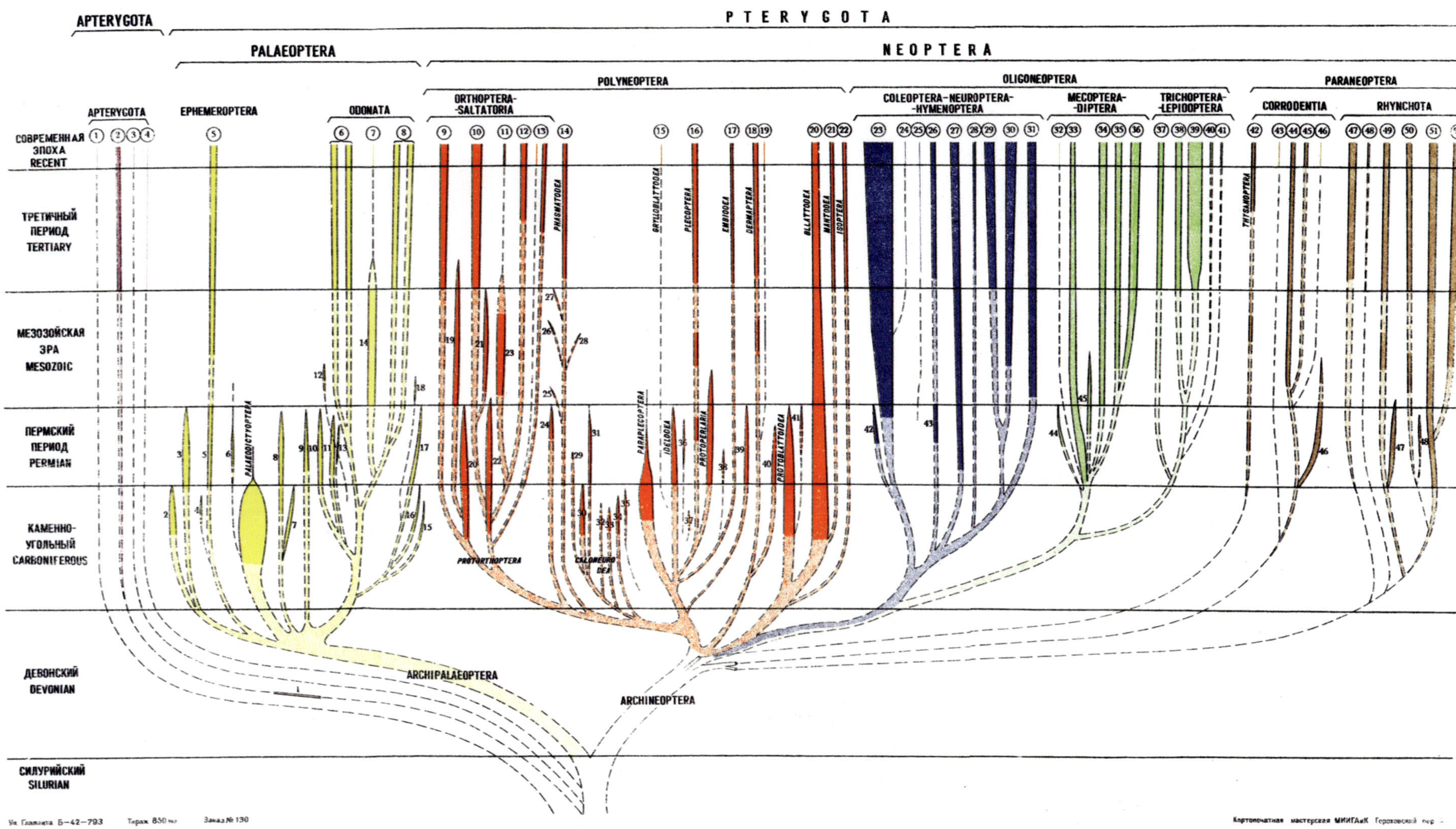

4-16. Phylogenetic relationships among insect lineages as proposed by Martynov (1938).

4.17. The distinguished, albeit eccentric, insect morphologist Guy Chester Crampton (1881–1951) of the University of Massachusetts. Crampton's devotion to his subject was total: He surrounded himself in his small apartment with thousands of vials of specimens. Photo: Special Collections and Archives, W. E. B. DuBois Library, University of Massachusetts, Amherst.

4.18. Three giants of entomological science; from left to right, Howard E. Hinton (1913–77), noted insect biologist; Robert E. Snodgrass (1875–1962), the most famous comparative arthropod anatomist; and Sir Vincent B. Wigglesworth (1899–1994), the prominent insect physiologist. Photo taken approximately 1956. Photo: G. W. Byers, University of Kansas Natural History Museum (UKNHM).

4.19. Frank M. Carpenter (1902–94), curator of fossil insects at Harvard's Museum of Comparative Zoology for 60 years and the most influential paleoentomologist of his generation. Photo taken 1956. Photo: G. W. Byers, UKNHM.

(1897a,b), Franz Klapálek (1904, 1905); and Charles Woodworth (1906, 1907, 1930).

Picking up the reins of Tillyard and Martynov were the paleoentomologists Frank M. Carpenter (1902–94: Furth, 1994) of the United States (Figure 4.19) and Boris B. Rohdendorf (1904–77) of the Soviet Union (Figure 4.20). While Carpenter focused on the Paleozoic fauna and conservatively approached the assignment of taxa to orders, Rhodendorf prolifically described taxa, naming numerous new groups. In contrast to his predecessor Martynov, Rohdendorf produced volumes of cursory descriptions, which the current generation of Russian paleoentomologists is extensively revising. Carpenter's methodical and careful studies gained him tremendous insight into the early evolution and diversification of insects. Perhaps his greatest contributions were the synthesis of his findings on the North American Paleozoic insect fauna with those taxa described from Carboniferous and Permian deposits from elsewhere in the world. Carpenter's conservatism led him to reduce the number of extinct orders, from approximately 50 to nine. He used as minimal criteria evidence from venation and mouthparts, but also recognized large paraphyletic groups, like Protorthoptera, and was too much of a taxonomic lumper.

Meanwhile, Ryuichi Matsuda (1920–86: Ando, 1988) revived the tradition of Crampton and Snodgrass and prepared several volumes on the comparative morphology of insects, either refining previous observations or providing alternative views for those expressed by earlier authors (Matsuda, 1965, 1970, 1976). Like his predecessors, Matsuda did not provide an explicit phylogenetic outline of the insect orders, but he did contribute considerably to the understanding of homologies across the orders.

Certainly the most influential insect systematist was Willi Hennig (1913–76: Schlee, 1978) (Figures 1.23, 4.21). Aside from developing cladistic methodology, Hennig was also the first to apply this critical method of analysis to the phylogeny of insects. Hennig's major accounts in 1953 and 1969 constitute even to this day some of the greatest advances in insect phylogeny (Figure 4.22). These works were updated posthumously in an extensively annotated and edited volume (Hennig, 1981). Hennig was simultaneously a paleontologist and liberally used information from fossil insects to shape his hypotheses of primary homology, character polarity, and evolution.

4.20. Boris B. Rohdendorf (1904–77), a prolific student of Martynov whose work is now being extensively revised by a new generation of Russian paleoentomologists. Photo taken 1964. Photo: G. W. Byers, UKNHM.

4.21. Willi Hennig (1913–76), founder of phylogenetic systematics and renowned entomologist. Hennig also produced very important work in paleoentomology and biogeography. Photo: G. W. Byers, UKNHM.

Perhaps the single most important paper in systematic entomology is the 1975 review by Niels P. Kristensen (Figure 4.23), who revised and built upon Hennig's work. Kristensen's (1975) review and updated accounts (Kristensen, 1978a, 1981, 1989a, 1991, 1995, 1999a) have provided our modern concepts for ordinal relationships of Recent insects. Shortly after the appearance of Kristensen's classic paper, H. Bruce Boudreaux, a mite specialist from Louisiana, provided his own interpretation of arthropod phylogeny, with a particular emphasis on insect relationships (Boudreaux, 1979). His compendium is another important source of characters regarding insect ordinal relationships, although his classification is not entirely supported by more recent analyses.

The number of workers interested in insect phylogeny has proliferated greatly in the last couple of decades, most of them focusing on molecular data or individual Recent orders of hexapods. Despite the fervor generated by new molecular advances, morphology and paleontology are as relevant as ever. Indeed, their importance is perhaps even more acutely felt today than they have been in preceding years, and synergistic studies are more and more the norm. One might even say that paleoentomology is experiencing a renaissance with numerous new students entering the field. Today the most prominent paleoentomologists actively researching the evolution of insect orders include Jarmila Kukalová-Peck, Alexandr P. Rasnitsyn, André Nel, and Rainer Willmann. The study of fossil insects and insect phylogeny is perhaps more active than ever before, but the two are rarely married. That is why our approach here was to fold extinct insects into a framework of relationships with Recent species.

4.22. (facing page) Relationships among major insect lineages as proposed by Hennig (1953). This phylogenetic hypothesis was the first one produced for insect orders using Hennig's phylogenetic methodology. Photo: Deutsche Entomologische Institut.

A ROADMAP TO THE PHYLOGENY OF INSECTS

The remainder of this book concerns itself with the major groups of insects, with accounts of their relationships, biology, and evolution. We have attempted to outline the relationships among the Recent and extinct insect orders as we believe are best supported by current morphological, molecular, and paleontological evidence. Figure 4.24 is a phylogeny of orders and the principal superordinal groups employed throughout this volume, with the classification summarized in Table 4.1.

4.23. Niels P. Kristensen, an architect of modern insect phylogenetics and authority on Lepidoptera. This photo was taken one year after his influential 1975 paper on insect phylogeny. Photo: G. W. Byers, UKNHM.

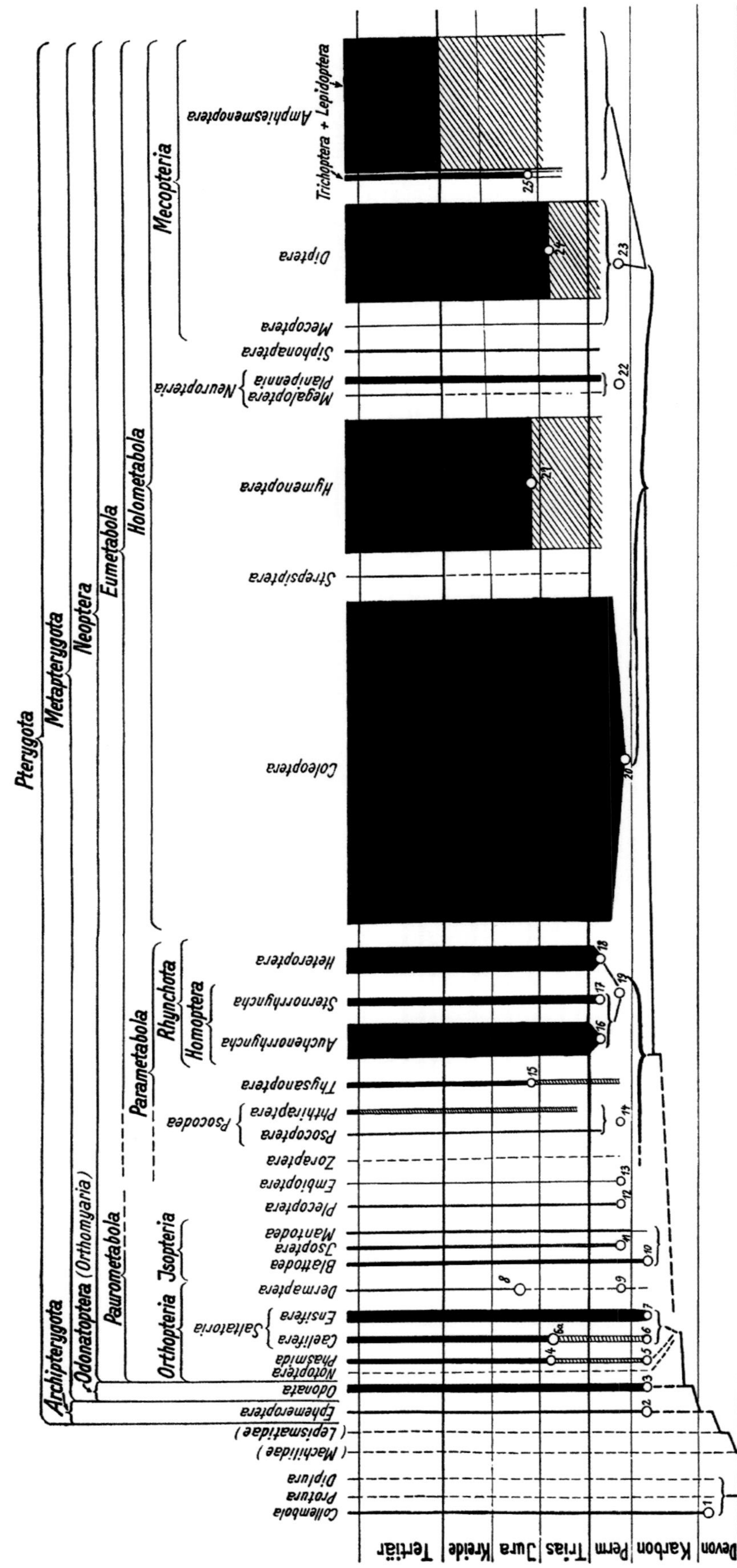

Pterygota
Metapterygota
Neoptera
Eumetabola
Holometabola
Archipterygota
Odonatoptera (Orthomyaria)
Paurometabola
Parametabola
Orthopteria
Isopteria
Rhynchota
Homoptera
Mecopteria
Collembola
Protura
Diplura
(Machilidae)
(Lepismatidae)
Ephemeroptera
Odonata
Notoptera
Phasmida
Caelifera
Ensifera
Saltatoria
Dermaptera
Blattodea
Isoptera
Mantodea
Plecoptera
Embioptera
Zoraptera
Psocoptera
Phthiraptera
Psocodea
Thysanoptera
Auchenorrhyncha
Sternorrhyncha
Heteroptera
Coleoptera
Strepsiptera
Hymenoptera
Megaloptera
Planipennia
Neuropteria
Siphonaptera
Mecoptera
Diptera
Trichoptera + Lepidoptera
Amphiesmenoptera
Tertiär
Kreide
Jura
Trias
Perm
Karbon
Devon

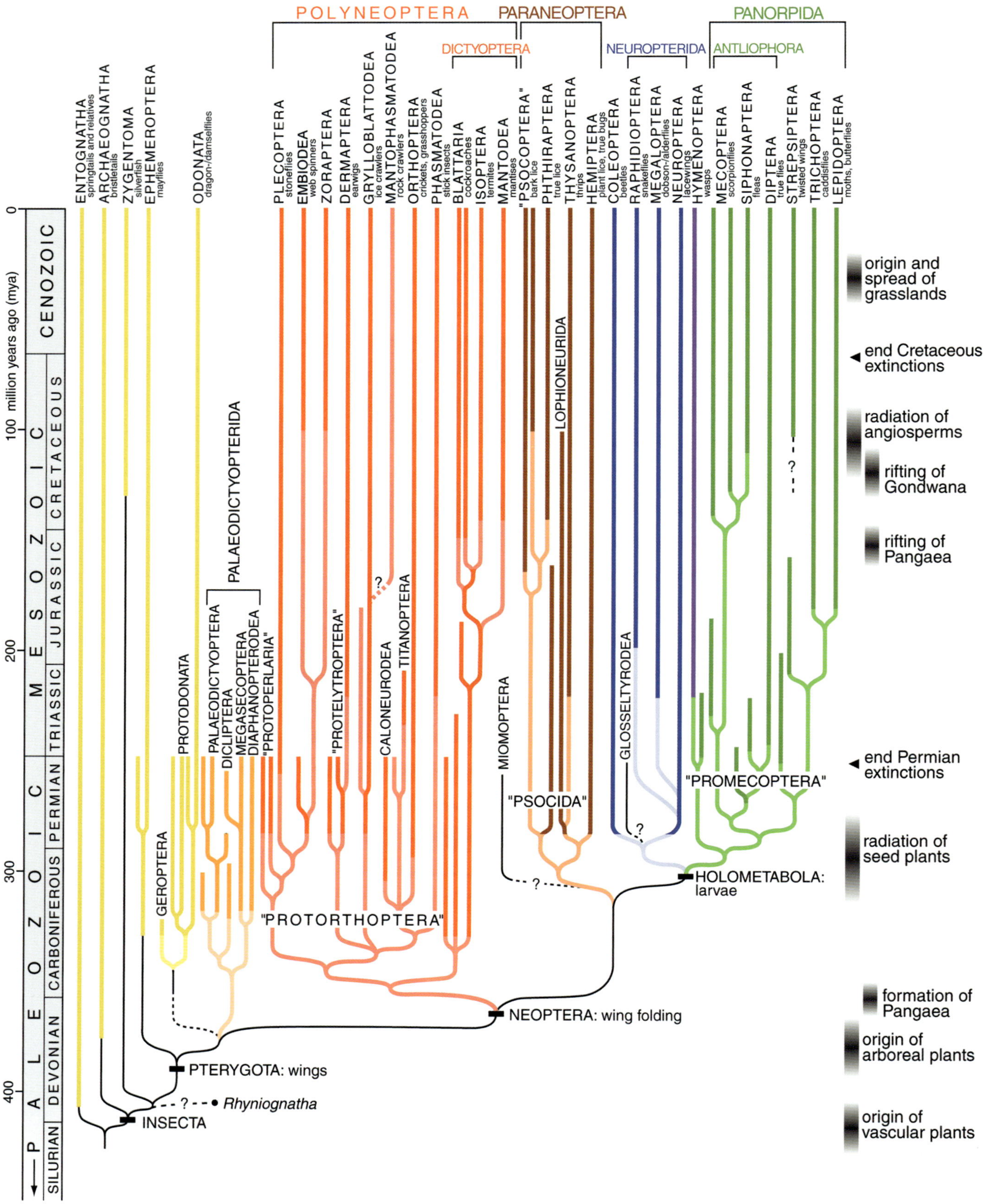

4.24. The phylogeny of living and extinct insect orders of insects used in this book, based on various sources (see text). Colors denote most major lineages; darker colors indicate the known extent of fossils.

TABLE 4.1. Hierarchical Classification of Class Insecta

—Class INSECTA (= Ectognatha)—

- Order Archaeognatha
- Dicondylia
  - Order Zygentoma
  - Pterygota
    - Order Ephemeroptera
    - Metapterygota
      - †Palaeodictyopterida
        - Order †Palaeodictyoptera
        - Order †Megasecoptera
        - Order †Dicliptera
        - Order †Diaphanopterodea
      - Odonatoptera
        - Order †Geroptera
        - Holodonata
          - Order †Protodonata
          - Order Odonata
      - Neoptera
        - †Paoliidae (†Protoptera)
        - Polyneoptera
          - "†Protorthoptera" (*stem-group, polyphyletic*)
          - Anartioptera
            - Order Dermaptera
            - Order Grylloblattodea
            - Order Mantophasmatodea
            - Superorder Plecopterida
              - Order Plecoptera
              - Order Embiodea
              - Order Zoraptera
            - Superorder Orthopterida
              - Order Phasmatodea
              - Order †Caloneurodea
              - Order †Titanoptera
              - Order Orthoptera
          - "†Blattodea" (*stem-group Dictyoptera*)
            - Superorder Dictyoptera
              - Order Mantodea
              - Order "Blattaria"
              - Order Isoptera
- *Incertae sedis* (Paraneoptera, Holometabola)
  - Order †Miomoptera
  - Order †Glosselytrodea

- Eumetabola
  - Paraneoptera
    - Superorder Psocodea
      - Order "Psocoptera"
      - Order Phthiraptera
    - Superorder Condylognatha
      - Order Thysanoptera
      - Order Hemiptera
  - Holometabola (= Endopterygota)
    - Order Coleoptera
    - Superorder Neuropterida
      - Order Raphidioptera
      - Order Megaloptera
      - Order Neuroptera
    - Superorder Hymenopterida
      - Order Hymenoptera
    - Superorder Panorpida
      - Antliophora
        - Order "Mecoptera"
        - Order Siphonaptera
        - Order Strepsiptera
        - Order Diptera
      - Amphiesmenoptera
        - Order Trichoptera
        - Order Lepidoptera

5 Earliest Insects

ARCHAEOGNATHA: THE BRISTLETAILS

Bristletails, or Archaeognatha (= Microcoryphia), are the most primitive of living insects, having persisted since at least the mid-Devonian. These cryptic, somewhat cylindrical insects occur under loose bark or stones (Figure 5.1). Except for a rare few, bristletails are nocturnal and typically hide in crevices during the day. About 500 species are known worldwide and live in diverse habitats, including elevations as high as 4,800 meters (15,749 feet) in the Himalayas. The typical diet of bristletails is composed of algae and lichens, which they glean using their monocondylic mandibles as picks. While they are not predatory, some species will scavenge remains of arthropods and some will even eat their own exuviae. Major references for the Archaeognatha include accounts by Paclt (1956), Sturm (1984, 1994a,b, 1995a,b, 1997, 2003a), Kaplin (1985), Mendes (1990, 2002), Sturm and Bach de Roca (1993), Larink (1997a,b), Bitsch and Nel (1999), and Sturm and Machida (2001).

Particularly noteworthy is the mating behavior of Archaeognatha, which have three principal modes of sperm transfer. Species do not technically copulate, and sperm transfer is indirect even though fertilization is internalized. Males transfer to the female droplets of sperm or spermatophores. Perhaps as a result, the genitalia of archaeognaths are simple and males and females differ relatively little in the appendicular structures of the genital segments. Couples remain in close contact during mating and have distinctive courtship behaviors. In many machilids the males use a "carrier-thread" (Sturm, 1952, 1955). The male taps the female with his maxillary palpi and once she becomes receptive he secretes a thread of silk. The silken line is drawn out, to which the male attaches droplets of sperm. The female is prevented from moving forward by the male and while the female is facing away from the thread, the sperm droplets are transferred to her ovipositor and eventually into her gonopore. In the "firebrat," *Petrobius*, the sperm droplets are set directly onto the female ovipositor and in Meinertellidae the male creates stalked spermatophores that are retrieved by the female (Sturm and Machida, 2001). The ovipositor is used to lay eggs in deep crevices or in holes actually dug by the ovipositor.

The Archaeognatha consists of only four families and approximately 500 species: Meinertellidae, Machilidae, Triassomachilidae, and Dasyleptidae (the last two are extinct). The Meinertellidae occur predominantly in the Southern Hemisphere (Sturm, 1984). The Machilidae, considered by many authors to possess more primitive features than the Meinertellidae (e.g., Sturm and Machida, 2001), are mostly found in the Northern Hemisphere, although a few machilines can be found below the equator in Africa and Asia.

Together with silverfish (Zygentoma, which are discussed later), the bristletails comprise the only surviving orders of primitively wingless insects and were at one time considered a single group, Thysanura (e.g., Remington, 1954), as well as often being united with the Entognatha into a larger grouping called the Apterygota, neither of which are monophyletic. Defining features of the Archaeognatha are the large compound eyes that meet at the top of the head, and the well-developed ocelli, both presumably adaptations for nocturnal living. Other features include "jumping," which is actually a sudden flexion of the abdomen that propels the insect into the air. Also, the meso- and metapleura consist of a single sclerite with large pleural apodemes. Primitively, the archaeognaths have monocondylic mandibles, the head skeleton is composed of paired anterior and posterior plates, "styli" (exopodites?) are usually on the mid and hind coxae, and, unlike abdominal styli, these lack musculature. They also have long, seven-segmented maxillary palpi (even longer than the legs); a terminal filament; eversible vesicles; and abdominal styli (Figure 4.7). Like the Zygentoma, the integument of archaeognaths is generally covered with scales that sometimes form patterns.

Even though the putative archaeognaths are among the oldest fossil insects on record, the overall record of the order is still sparse enough that it has provided little insight into the internal evolution of the order. Among macerated material taken from the mid-Devonian Gaspé fossil beds of Quebec

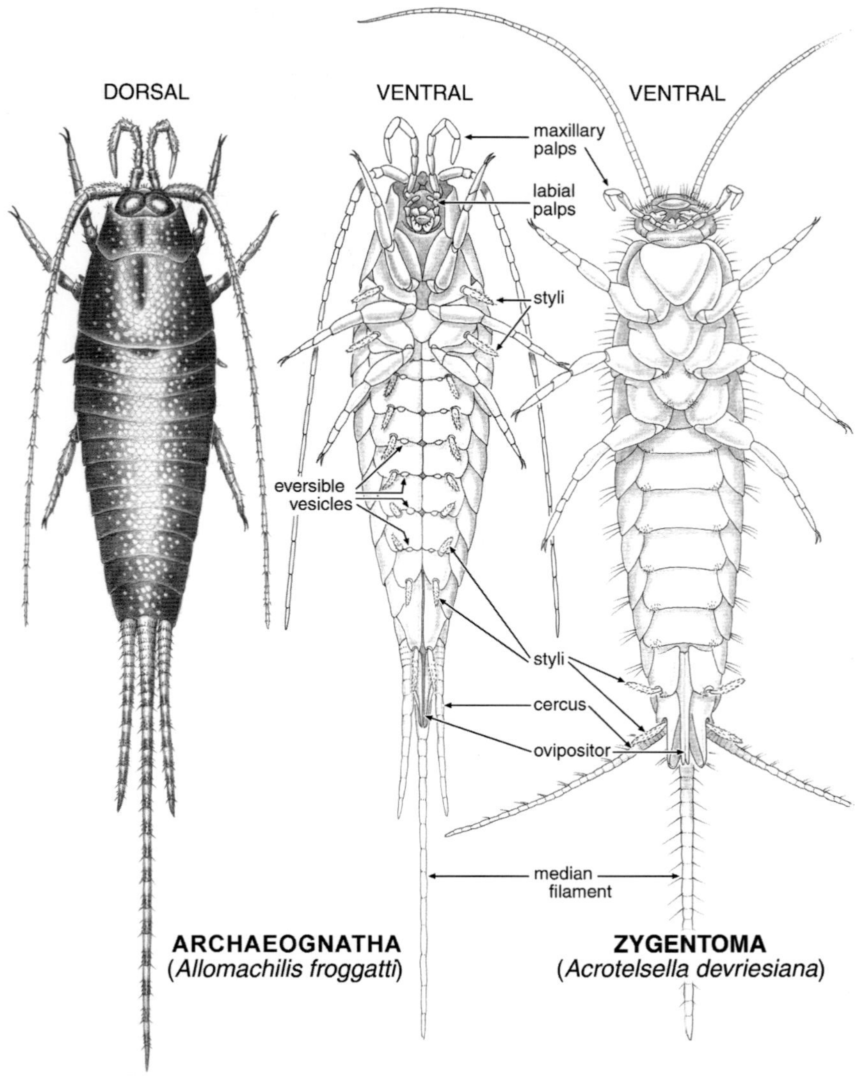

5.1. Representative basal, wingless insects: a bristletail (Archaeognatha) and a silverfish (Zygentoma). These are modern species belonging to groups that evolved at least as early as the Devonian, 400 MYA. Redrawn from *Insects of Australia*.

were two fragments of a bristletail, a head and thorax, representing the oldest record of insects in North America (Labandeira *et al.*, 1988). The Gaspé fragments, however, preserve only primitive features of the Archaeognatha, and no more conclusive assignment can be made concerning them.

Several species from the Carboniferous and Permian are placed in the extinct genus *Dasyleptus* (Dasyleptidae) (Figures 5.2, 5.3). These enigmatic Paleozoic fossils were previously considered to represent an extinct order of monocondylic ectognaths called Monura, closely related to the Archaeognatha (Sharov, 1957). This position was strongly supported by a reconstruction of *Dasyleptus* (e.g., Kukalová-Peck, 1987, 1997); however, this intrepretation has been challenged (e.g., Kaplin, 1985; Bitsch and Nel, 1999; Rasnitsyn, 1999), and, instead, individuals of *Dasyleptus* appear to be typical juvenile silverfish, albeit larger. Rasnitsyn (1999) has recently reviewed the family Dasyleptidae.

The earliest Mesozoic fossils of the order are *Triassomachilis uralensis* (Triassomachilidae) from the Upper Triassic of Russia (Sharov, 1948). *Triassomachilis* has at times been excluded from the Archaeognatha (e.g., Kukalová-Peck, 1991; Sinitshenkova, 2000c). The fossils are poorly preserved but do exhibit the dorsal, thoracic hump typical of Archaeognatha as well as annulated, abdominal styli and a terminal filament in addition to the long cerci. However, Sharov's renderings of the fossils show relatively small compound eyes that do not meet on the vertex, as well as short maxillary palpi, which are primitive differences from all other Archaeognatha including the much older Dasyleptidae. Paclt (1972) considered the differences of *Triassomachilis* and other bristletails to be artifacts of preservation misinterpreted by Sharov. The original material should be newly studied to determine the affinities of *Triassomachilis* and the validity of a family Triassomachilidae.

5.2. *Dasyleptus brongniarti* (Dasyleptidae) from the Permian of Russia was formerly placed in an extinct order, Monura, but is today considered an immature bristletail and a sister group to all other Archaeognatha. PIN 1197; length 10.5 mm.

5.3. *Dasyleptus sharovi* from the Early Permian of Elmo, Kansas. MCZ; length 11 mm.

All other fossil bristletails occur in Cretaceous or Tertiary ambers. The earliest is a meinertellid in Early Cretaceous amber from Lebanon, a species typical of the *Machiloides* group of genera and attesting to the antiquity of this family (Sturm and Machida, 2001). Other Cretaceous amber archaeognaths are known from Burma and New Jersey (Grimaldi *et al.*, 2000a, 2002). Fossil machilids are also known from Baltic amber, while meinertellids are known from Dominican amber (e.g., Silvestri, 1912; Sturm and Machida, 2001). Although described as an archaeognath, the Pliocene fossil *Onychomachilis fischeri* (Pierce, 1951) is actually a silverfish, perhaps near the Nicoletiidae (Sturm and Machida, 2001).

DICONDYLIA

The traditional taxon Thysanura, or Apterygota, has been recognized as an unnatural group for decades because silverfish (Zygentoma) are more closely related to the winged insects (Pterygota) than to the bristletails (e.g., Snodgrass, 1935). In his monumental work on the phylogeny and classification of insects, Hennig (1953, 1969, 1981) proposed the name Dicondylia for the group uniting the silverfish with the winged insects. Recent molecular studies have also supported the Dicondylia as a lineage (e.g., Wheeler *et al.*, 2001). The hallmark character of this group is the presence of a novel, secondary articulation to the mandible (i.e., the *dicondylic mandible*) (Figure 5.9). This second articulation results in the movement of the mandible being roughly confined to a single plane of motion rather than the rotating motion possible in Archaeognatha. It is homologous to the monocondylic joint in Archaeognatha and Entognatha, since the condyle for the new point of articulation is located on the head capsule with the acetabulum on the mandible itself. Another significant character defining Dicondylia is the development of the gonangulum in the ovipositor base (Figure 4.8). The gonangulum represents a basal differentiation of the second gonocoxa to form a sclerite with three points of articulation – to the ninth abdominal tergum, the first gonapophysis, and the second gonocoxa. Other features of dicondylic insects are the reduced maxillary palpi (primitively five-segmented, but this is also observed within Entognatha, the Ellipura in particular), and the development of tracheal commissures and connectives in the abdomen (Kristensen, 1981). Thus, despite the great overall similarity of bristletails and silverfish, important but subtle features indicate that silverfish are actually more closely related to winged insects.

ZYGENTOMA: THE SILVERFISH

Silverfish are similar to bristletails in gestalt but are more flattened; they lack the distinctive hump of the latter group and, therefore, do not jump. Like bristletails, these primitively wingless insects have a long terminal filament between the cerci and a surface covering of scales (which are lost in Nicoletiidae and Maindroniidae) (Figures 5.1, 5.4). The biology of the order is superficially similar to that described for the Archaeognatha, including the occurrence of indirect sperm transfer via a spermatophore; however, Zygentoma are mostly diurnal and omnivorous with the notable exception of the family Nicoletiidae, which is principally subterranean and vegetarian. Though they are not capable of jumping, silverfish are very agile and run swiftly, as anyone who has chased one across a kitchen counter or sink knows. In contrast to the Archaeognatha, the compound eyes are reduced or absent in Zygentoma, and most families except Lepidotrichidae lack ocelli entirely. Defining features of the silverfish are not immediately apparent, though the enlargement and modification of the distalmost palpal segment of the labium and the dorsoventrally flattened and enlarged coxae may be significant. The enigmatic

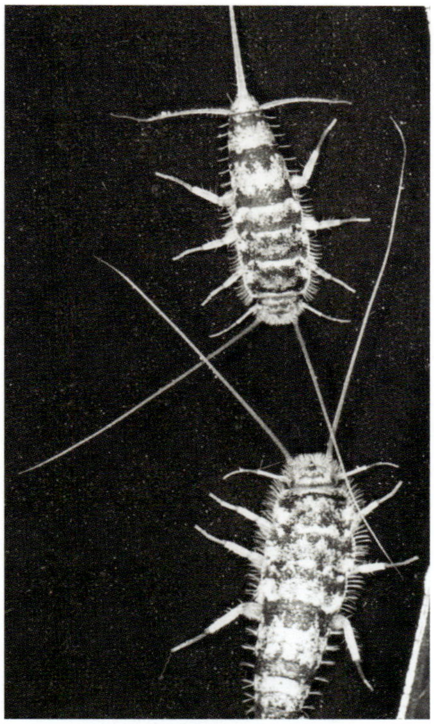

5.4. A male and female of the common European silverfish, *Thermobia domestica* (Lepismatidae), as they initiate their courtship. Photo: H. Sturm.

Lepidotrichidae are at times excluded from the order. Otherwise, most differences between Zygentoma and Archaeognatha lie in those traits the former shares with winged insects. Major recent references for the Zygentoma include Mendes (1991, 1994, 2002), Larink (1997a,b), and Sturm (1997, 2003b).

The order consists of five Recent families in two groups: Lepidotrichidae, and the families Lepismatidae, Nicoletiidae, Ateluridae, and Maindroniidae. The apparently primitive family Lepidotrichidae today is represented by a single modern species living in northern California, *Tricholepidion gertschi* (Wygodzinsky, 1961) (Figure 5.5), but the family was originally described from a fossil species, *Lepidothrix pilifera*, in mid-Eocene Baltic amber. This family, however, may not be monophyletic, since *Lepidothrix* may be more closely related to the remainder of the Zygentoma (Euzygentoma). *Tricholepidion* possesses distinct ocelli while *Lepidothrix* and Euzygentoma lack ocelli. Regardless of potential paraphyly of Lepidotrichidae, the family possesses a number of putative primitive features relative to Euzygentoma, including the large abdominal sterna and large number of abdominal styli and eversible vesicles. However, eversible vesicles are only absent in Lepismatidae and Maindroniidae, and this character is shared (primitively?) with Nicoletiidae and Ateluridae. Indeed, *Tricholepidion* and Nicoletiidae share a unique modification of sensillar structures on the terminal filament (Wygodzinsky, 1961) and the former shares sperm conjugation with Lepismatidae (Wingstrand, 1973; Kristensen, 1997), whereby individual sperm cells pair in the vas deferens of the testes. Evidence for sperm conjugation in Nicoletiidae (Jamieson, 1987) is apparently not conclusive (Dallai *et al.*, 2001). It is possible that Lepidotrichidae is the sister group to all other Zygentoma, with Nicoletiidae and Ateluridae being sister to a clade consisting of Lepismatidae and Maindroniidae. Interestingly, *Tricholepidion* has five-segmented tarsi, like the presumed primitive condition for Pterygota. Archaeognatha and the remaining silverfish families have two- or three-segmented tarsi, conditions typically interpreted as derived, independent reductions. Alternatively,

5.5. The relict silverfish, *Tricholepidion gertschi* (Lepidotrichidae), from California. The only other member of the family is found in Baltic amber, which together may comprise the sister group to the remainder of the order Zygentoma. Photo: H. Sturm.

5.6. A silverfish from Brazil's Early Cretaceous Santana Formation. Although silverfish undoubtedly are of Devonian age, this relatively modern-looking species is among the earliest records of the order Zygentoma, leaving a presumed 280 MY gap in their earliest fossil record. AMNH; body length 14 mm.

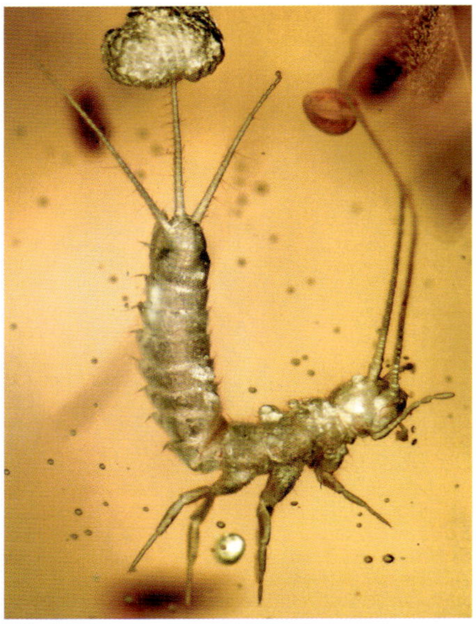

5.7. An immature silverfish (Zygentoma) in Miocene Dominican amber. Morone Collection; length 4 mm.

but an unlikely scenario, pentamerous tarsi developed independently in *Tricholepidion* and Pterygota, or these are sister groups (e.g., Wheeler *et al.*, 2001).

The Euzygentoma have been defined by the reduction of the abdominal sterna (although this also occurs in Archaeognatha), the reduced number of abdominal styli, and the absence of ocelli. Reduction in the number of tarsomeres may also represent a defining feature of this lineage. The Maindroniidae is a rare family of three species restricted from Asia Minor and Chile. Little is known or understood of these presumed relics, but they may be derivatives of the Lepismatidae (e.g., Remington, 1954). The Lepismatidae is the largest family of the order, with over 200 species worldwide and as such are the group most frequently encountered. The families Nicoletiidae and Ateluridae are close relatives, both lacking eyes and at times having been included in a single family (e.g., Remington, 1954; Paclt, 1963, 1967). The Nicoletiidae are, like the lepismatids, cosmopolitan in distribution. Interestingly, some nicoletiids can reproduce parthenogenetically. Ateluridae are inquilines that live in subterranean ant and termite nests, though a few lepismatids are also inquilines.

Despite the apparent antiquity of Zygentoma as the sister group to all other dicondylic insects, there is a huge gap in their early fossil record. We would expect to find fossil Zygentoma as early as the Devonian, but their fossil record is almost entirely restricted to Cretaceous and Tertiary resins (e.g., Mendes, 1997, 1998; Sturm and Mendes, 1998), although nice compressions of Lower Cretaceous silverfish have been recovered from the Sanatana Formation of Brazil (e.g., Sturm, 1998: Figures 5.6, 5.7). Cuticular fragments from the Devonian of Gilboa, New York, may represent a species of Zygentoma (Shear *et al.*, 1984), but a conclusive assignment remains impossible to make. Similarly, *Carbotriplura kukalovae* from the Late Carboniferous of the Czech Republic is likely a silverfish (Kluge, 1996), and although its assignment to Zygentoma is tentative, its placement in a separate suborder of wingless insects is unjustified. The Pliocene fossil *Onychomachilis fischeri* was described as a bristletail (Pierce, 1951) but is actually a silverfish (Sturm and Machida, 2001).

RHYNIOGNATHA

Rhyniella, the collembolan in Early Devonian Rhynie chert, has long been heralded as the oldest hexapod, while fragmentary remains from the same chert had been mostly forgotten. In the original paper announcing the discovery of *Rhyniella*, Hirst and Maulik (1926) also reported a pair of mandibles preserved with largely unidentifiable tissues surrounding them (Figure 5.8). Later, Tillyard (1928b) formally described the mandibular elements and named them, *Rhyniognatha hirsti*. Tillyard was the first to note that they were insect-*like*, but he did not place them formally into any group. Indeed, subsequent authors relied on Tillyard's illustration and followed his intrepretation that the mandibles were "suggestive" of an insect (e.g., Hennig, 1981). Most authors, however, considered *Rhyniognatha* as too fragmentary to make any determination, and Carpenter (1992) even excluded it from his monumental treatise on insect paleontology.

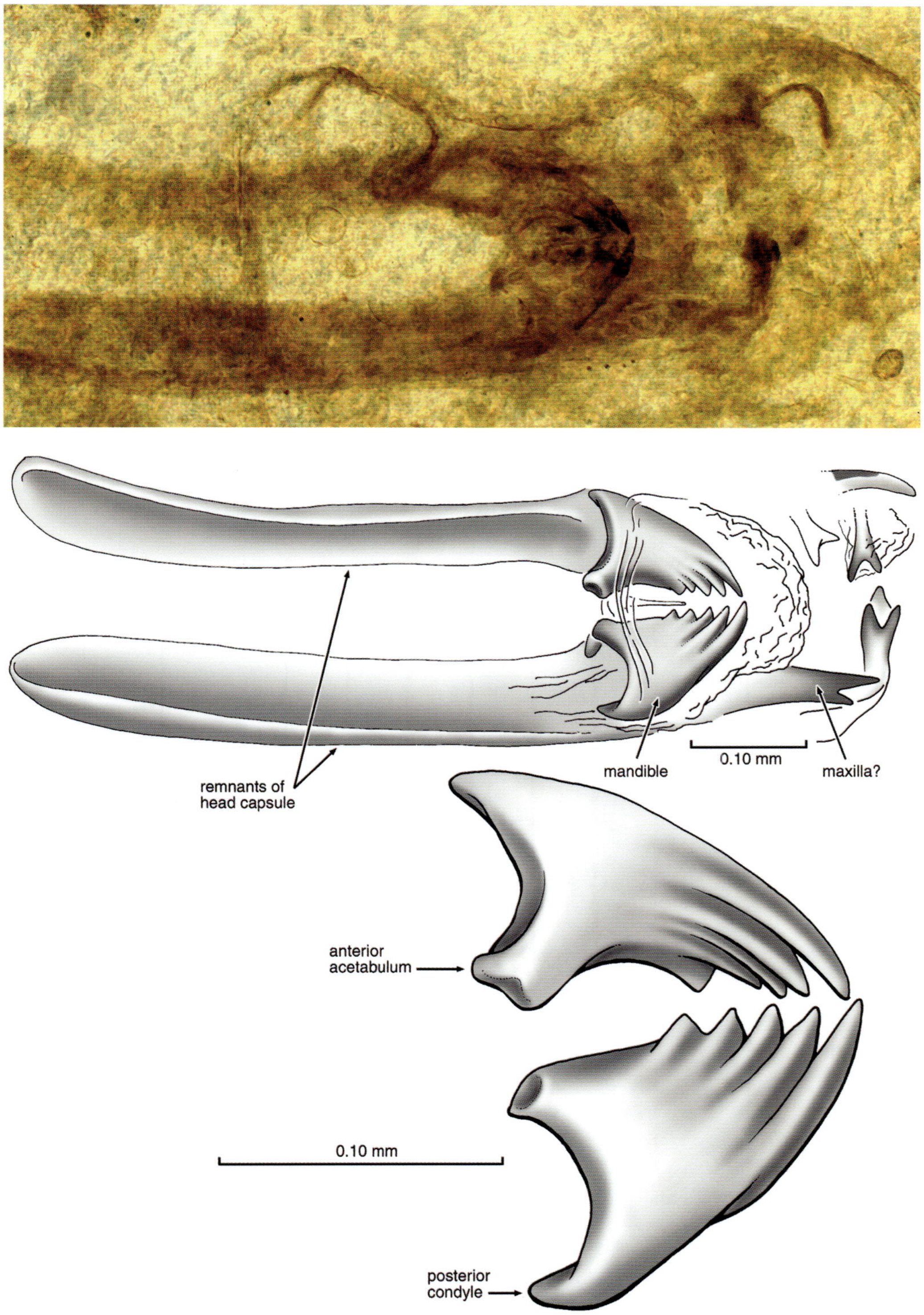

5.8. The oldest insect, *Rhyniognatha hirsti*, from the Early Devonian chert near Rhynie, Scotland. Only portions of the head are preserved, but the dicondylic mandibles indicate it was an insect; their triangular shapes indicate it may even have been a winged insect. NHML In. 38234.

In a recent study of the unique holotype of *Rhyniognatha* (Engel and Grimaldi, 2004a), the presence of an anterior acetabulum and posterior condyle on the mandibles was confirmed, conclusively demonstrating that the mandibles are dicondylic (Figures 5.8, 5.9). Furthermore, the mandibles are short and triangular, a morphology characteristic of a subset group among the pterygote insects, the Metapterygota, implying that *Rhyniognatha* possessed wings. This would place the origin of wings at least 80 million years earlier than previous fossil evidence allowed, and interestingly agrees with a recent molecular study that estimated insects originated in the Early Silurian and neopterous insects in the

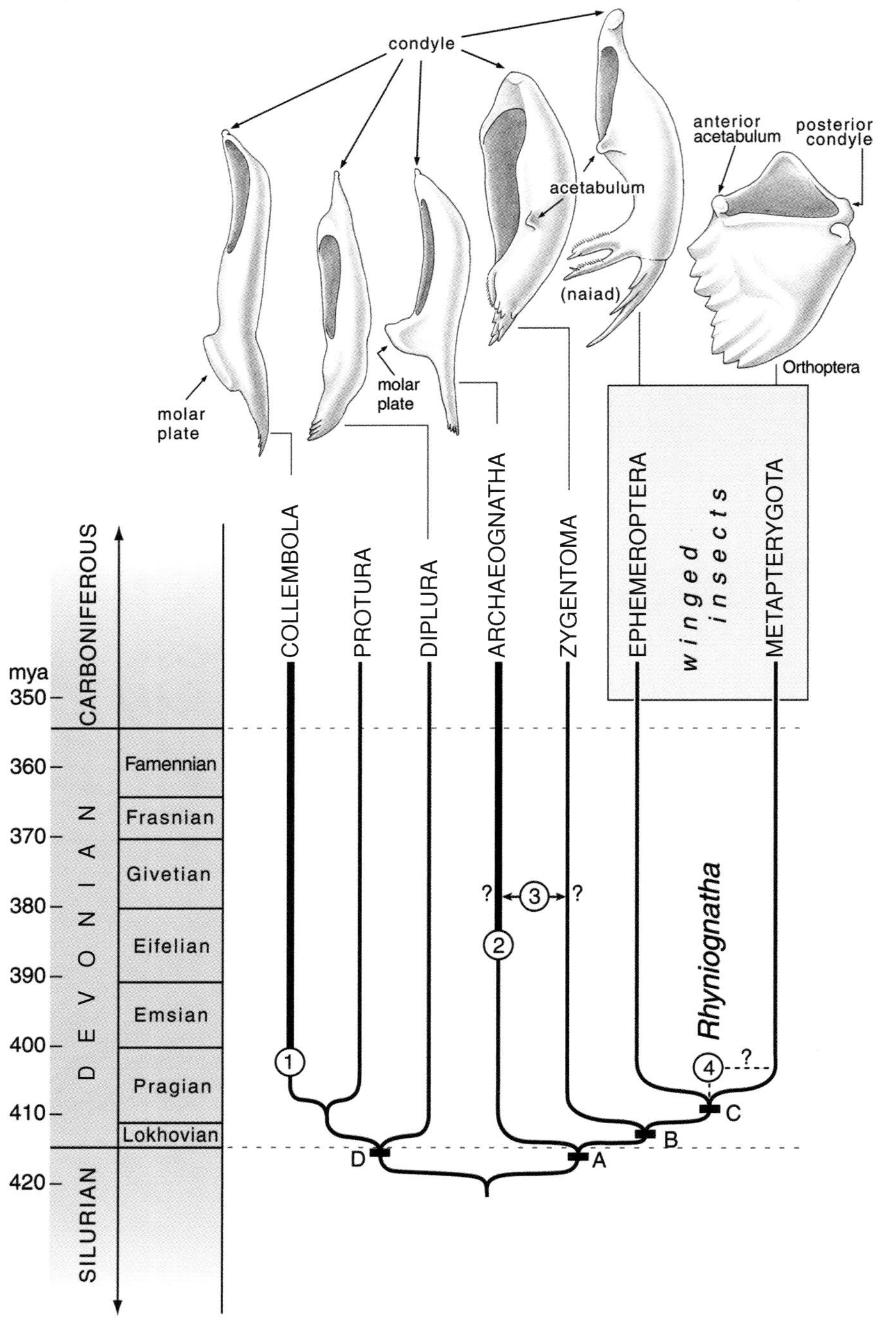

5.9. Phylogeny of basal insect lineages indicating the position of *Rhyniognatha*, the oldest insect, as based on the structure of mandibles. Fossils (numbers): 1. *Rhyniella praecursor*; 2,3. undescribed; 4. *Rhyniognatha hirsti*. Characters (letters): A. insectan (see text), B. dicondylic mandibles, C. wings, D. entognathous mouthparts.

mid-Devonian (Gaunt and Miles, 2002) (see also the *Origin of Wings* in Chapter 6). A Devonian origin of wings could only be conclusively proven with fossilized wings from that period.

It is impossible to say to what order *Rhyniognatha* might have belonged, or if it belonged to an unknown, extinct lineage of primitive insects. All than can be said is that *Rhyniognatha* is the oldest insect, and that it was more derived than bristletails and silverfish, and probably more than Ephemeroptera. Regardless, *Rhyniognatha*'s occurrence in the Early Devonian indicates that insects likely originated in the latest Silurian and were among the earliest of terrestrial faunas (Engel and Grimaldi, 2004a). *Rhyniognatha* also reflects the serious need for intensive exploration of insects from the Devonian and Early Carboniferous.

6 Insects Take to the Skies

PTERYGOTA, WINGS, AND FLIGHT

Flight is usually considered to be the relatively recent acquisition of wings in vertebrates such as pterosaurs, birds, and bats. In fact, insects were the first organisms to have developed powered flight and took to the skies at least 90 MY prior to the earliest winged vertebrates, perhaps even 170 MY earlier (e.g., Engel and Grimaldi, 2004a). They are also the only group of invertebrates to have acquired powered flight (Figure 6.1). The most obvious effect of wings is on the organism's ability to disperse. A flying insect can readily exploit new spaces, and should the local environment become unfavorable, it can more effectively seek better habitats. Similarly, when faced with a predator or other threat, wings allow for a quick retreat. While the "springs" of springtails allow quick escape, there is little or no control over directionality, and a collembolan might find itself in a worse situation after leaping. Flight also enhances locating a mate, allowing once remote, inbred populations to experience a new influx of genes, thus increasing panmixis and genetic variability. Wings as a form of locomotion were clearly the first major morphological innovation of insects, but have been refined through time. Wings even serve functions in addition to flight. Just as the extinct reptile *Dimetrodon* is presumed to have done with its great dorsal fan, many insects use their wings for thermoregulation, acquiring heat from sunlight to recover from the torpor of cold nights. But this thermoregulation is generally related to flight because flight muscles must reach a critical temperature to function. In some insects, powerful flight muscles vigorously contract while the wings are held motionless, and this quickly generates the heat needed for flight. Wings also can provide passive and active protection, the way folded elytra of beetles protect the abdomen, or leathery forewings of membracid treehoppers are flailed against attacking wasps. Some mantises, katydids, and stick insects are efficiently camouflaged because their forewings are remarkably leaf-like (Figure 7.26); in other groups the wings have gaudy patterns to advertise toxicity or sex. Wings also function in auditory communication, for which Orthoptera are best known but hardly the only order of insects to use these structures for sound.

Wings and the refinement of flight have arguably comprised the most critical morphological innovation in the success of insects, and it is quite possible that those insects with complete metamorphosis would not have been so successful if flight did not precede this type of development. With the advent of wings, neural capabilites were expanded to control not just flight but also sensory and integrative neural systems so that the insect could cope with a vaster, three-dimensional environment. Indeed, some of the most "intelligent" insects (i.e., most capable of learning), and those with the most acute vision and olfaction, seem to be predators and pollinators that are active fliers.

Defining features of the Pterygota include the loss of eversible vesicles, the presence of a transverse stipital muscle, the fusion of the pleural apophyses with the sternal apophyses (strengthening the thorax during flight deformations), the formation of a pleural sulcus to strengthen the pterothoracic walls, two coxal proprioreceptor organs, a corporotentorium, sperm transfer through copulation (rather than via external spermatophores), and, of course, two pairs of wings (Kristensen, 1991). Wings are not merely modified limbs because the limbs homologous with those in apterygotes are still present in pterygotes. There is, unfortunately, no readily identifiable structure that can easily account for the appearance of wings, and debates over the origin of insects wings have raged for over a century. These twofold arguments highlight the dual nature of a question like, "What is the origin of insect wings?" This seemingly simple query actually consists of two components: (1) From what morphological elements are insect wings composed? (i.e., the homology question); and (2) For what purpose were wings, or winglike structures, first employed? That is, what conditions spurred the origin of wings? To answer these questions, we must first consider how wings function.

6.1. A paperwasp takes off from its nest in Ecuador. Insects were the first organisms to fly, they evolved various flight designs, and have the most maneuvered flight of all animals. Photo: R. Swanson.

INSECT WINGS

Wing Function

Detailed reviews of insect flight mechanics are provided by Wootton (1992), Brodsky (1994), Grodnitsky (1999), Dudley (2000), and Alexander (2002), with only the more salient points elaborated here. The complex system of membrane, veins, flexion lines, and overall shape provides a strong but lightweight, flexible structure that can change shape in a controlled (but entirely passive!) way as it moves through air. To achieve flight, all flying animals must produce lift and thrust. Lift is the force that raises the insect off the ground, while thrust is the force moving the insect either forward or backward. The wings form what is called an *air foil*. This is owing to a slight convex curvature to the overall wing surface with a concave or flat ventral surface. The degree of curvature is the wing's camber: A low camber is weakly convex on the top, while a high camber is strongly convex on top. As air moves over the surface of the wing, it moves slightly faster over the convex surface than it does the ventral, concave surface. This generates an area of lower pressure on the upper surface of the wing (i.e., Bernoulli's principle) creating a force that lifts the wing, and thereby the remainder of the insect, into the air. The air speed and camber of the wing are critical for determining the amount of lift that is created. While the overall body of the wing is a passive actor in flight, muscles pulling on the pteralic plates and epipleurites at the base of the wing alter its tilt in the air stream. As a result, insects fly by maneuvering the wings in a convoluted figure-eight motion where the costal edge leads, not by merely flapping the wings up-and-down as is typically supposed. By tilting the leading edge of the wing downward, an insect can alter its *angle of attack* relative to the air stream. The angle of attack is the change in the position of the wing owing to a forward or backward tilt, created by pulling the leading edge downward (*pronation* mostly caused by pulling on the basalare) or upward (*supination* mostly caused by pulling on the subalare), respectively. Changing the angle of attack by tilting the wing forward is equivalent to altering the camber of the wing by simulating a more strongly curved surface. Thus, additional lift is generated for flight. Thrust, on the other hand, is generated by the pushing movement of the wing against the air mass. Flight thus proceeds by dipping the wing forward (i.e., pronating) from its highest point until the wing has reached the bottom of its downstroke. During the relatively slow downstroke, the wing is also being moved forward (called promotion), thereby generating most of the lift required for flight as well as some thrust. Once reaching the trough of the downstroke, the wing is strongly tilted backward (i.e., supinated) such that the leading edge is brought upward as the wing begins its upstroke. Simultaneously the wing is shifted slightly to the rear (called remotion), thereby cutting across the path of its downstroke (hence the figure-eight motion) before reaching the peak of its upstroke and repeating the process. By comparison to the downstroke, the upstroke is relatively fast so as to minimize the loss of lift. During all of this gyrating, portions of the wing foil may fold along their lines of flexion, frequently generating vortices of air and additional lift or thrust.

Numerous modifications of this generalized pattern occur among insects, all associated with the peculiarities of flight among orders, families, or species. Highly maneuvered flight is made possible by synchronizing the two pairs of wings, and many orders have developed mechanisms for linking the wings in flight (e.g., Hymenoptera, Lepidoptera), or even by virtually dispensing with one pair of wings (e.g., Diptera). The Odonata are noteworthy exceptions because the forewings and hind wings are out of synchrony. The forewing generates vortices that are captured by the hind wing in hovering flight (see the section on Odonata for more details).

The powerhouse of insect flight, alluded to before, involves the *indirect flight muscles*. These consume almost all of the available space in the pterygote thorax and do not pull directly on the wing for generating the up- and downstroke of flight (hence their name as *indirect*). Instead, the muscles are attached such that contractions deform the overall shape of the entire thorax, causing the notum and pleuron to push on the base of the wing and move it up and down. The upstroke is generated by a series of dorsoventral muscles that pull down on the notum during a contraction. The notal wing processes thereby press downward on the leading and posterior edges of the wing base and cause the wing to move upward on the pleural wing process, which provides a pivot point from below. The downstroke is generated by the dorsolongitudinal muscles running lengthwise through the thorax. A contraction of these muscles causes the notum to buckle upward and moves the notal wing process inward. Elastic forces stored in the reinforced walls of the thorax thereby pull the wings upward. Thus, in insect flight the muscles are actually deforming the overall shape of the thorax and not pulling directly on the wings themselves. Despite the apparent absurdity of this design, it is remarkably efficient and powerful. The only exception to this design is once again found in the Odonata. Odonates fly with direct flight muscles, the name of which is self-explanatory regarding their operation. The flight muscles of dragonflies and damselflies are oriented dorsoventrally and are connected above to expanded plates, which themselves attach via tendons to the subalare, basalare, humeral plate, and axillary plate (refer back to the discussion on odonatoid pteralia). The upstroke of odonate flight is caused by contraction of these muscles, which insert just inside of the pleural wing process. The downstroke is produced by muscles inserting lateral to the pleural wing process.

Neuronal control of the wingbeat is not the same for all insects. In insects that beat their wings relatively slowly, each muscle contraction is stimulated by a nerve impulse. This obviously sets an upper limit to how fast a wing can move owing to the recovery time for the nerve to build an action potential. Such muscles are called *synchronous flight muscles* because one muscle contraction is associated with each nerve impulse. However, some insects beat their wings far more rapidly than a nerve impulse can be conducted, the record being around 1,000 cycles per second, with more typical species ranging from 400 to 600 cycles per second. Flight muscles that contract this rapidly are obviously not dependent on a single nerve transmission, but they function by fibrillation. In fact, the nerve impulse in such insects is instead a signal for beating to begin or to cease. Such muscular systems are called *asynchronous flight muscles* and are the most metabolically active tissues in nature. Insects with asynchronous muscles tend to have fewer muscles, although each is quite massive.

Most insect flight operates under what is generally called steady-state aerodynamics. For insects that are large enough, the physics of flight (although not the action) is fundamentally the same as in other flying animals. This simply means that the forces of lift and thrust are generated from a steady stream of continuous air flowing across the wing surface. Differences in the direction and velocity of air flow create the differences in thrust, lift, and drag. Incredible flight patterns can be generated under such physical conditions. However, insect flight continues to amaze researchers as they delve further into the complexities of model systems or examine the diversity of flight across species. Indeed, some insects employ non-steady-state aerodynamic principles. This is particularly true in minute insects, for which air is a viscous medium. Moving through air for such species is equivalent to a human swimming through a vat of melted chocolate! The best-studied system is that of the encyrtid wasp, *Encarsia*. This wasp employs a "clap-and-fling" flight mechanism in which the wings are moved up and down and forcefully clap above and below the body. In this system lift is not generated by air moving across the wing's surface but instead by vortices that swirl around the long axis and tips of the wings. These vortices create the lift necessary for flight.

Paleopterous Versus Neopterous

A single origin of winged insects is now generally undisputed (although see Lemche, 1940; Manton, 1977; La Greca, 1980; Matsuda, 1981), but relationships among its basal members are far from settled. Numerous arguments have been made for relationships among the four main branches of the Pterygota, namely the Ephemeroptera (mayflies), the Palaeodictyopterida (an extinct superorder of haustellate insects), the Odonatoptera (a superorder containing the dragonflies, damseflies, and their extinct relatives), and the Neoptera, which comprises all other winged insects.

Martynov (1925a) was the first author to describe the two major differences in the construction of insect wings, although this division was independently noted by Crampton (1924) at about the same time. They noted that most insects were capable of flexing the wings over the abdomen during rest. This consisted of a flexor muscle pulling on the third axillary sclerite, which in turn helped to collapse or fold the posterior part of the wing, thereby pulling the entire structure over the abdomen (Figure 4.6). The adaptive significance of this feature presumably lies in better exploitation of the environment while storing and protecting the wings, specifically the invasion of tight spaces such as under bark, in soil, among fallen leaves, and even through water. This opened the way for other major wing modifications, such as the development of the forewings into protective covers, like the elytra of beetles and hemelytra of true bugs. If true, it is remarkable to think that a microscopic muscle attached to a microscopic sclerite contributed to the great success of insects. Martynov noted that, in contrast to the neopterous lineages, a few groups were incapable of such movement, and the wings were therefore restricted to being held outstretched either at the insect's side or above the body. In this latter condition, although the wings can be brought together above the body, they cannot be twisted or flexed, such that the wing surface would become parallel to the abdomen; instead, the wing membrane remains perpendicular to the body's long axis. Martynov aptly termed this the *paleopterous* condition and believed it to be primitive relative to the neopterous insects. Martynov thus initiated a debate that continues today. He proposed two major groups of winged insects, which he named Palaeoptera, for the Ephemeroptera and Odonatoptera, and Neoptera, for all other winged insects.

The Neoptera has been universally supported as a

monophyletic group based on the development of wing flexion via unique musculature attachments to the third axillary sclerite; the development of a median plate in the wing articulation, which is divided such that it can fold during wing flexion; the simplification of vein R, which does not branch from the extreme wing base; and the development of the gonoplac. In addition, studies based on DNA sequences have repeatedly recovered the Neoptera as a natural group (e.g., Wheeler *et al.*, 2001).

The "Palaeoptera," on the other hand, presents quite a confusing tale and has been of legitimate contention. Some authors have argued that the palaeopterous condition is itself a derived condition, either respective to neoptery or relative to some unknown, presumably more simplistic, wing design mechanism that has since become extinct. Under these scenarios the Palaeoptera should be recognized as a monophyletic lineage and some authors have cited the bristle-like antennae of Odonata and Ephemeroptera, the aquatic lifestyle of their immatures, and the formation of intercalary veins, among other traits, as further evidence for the monophyly of this group (e.g., Hennig, 1981; Kukalová-Peck, 1983, 1985, 1987, 1991, 1992, 1997). Indeed, limited studies from 18S and 28S rDNA sequences have also supported the Palaeoptera (Hovmöller *et al.*, 2002) but require expansion in taxon sampling before conclusive decisions can be made (e.g., inclusion of Zygentoma as an outgroup for character polarity). Other recent analyses of the same data have failed to support a monophyletic Palaeoptera (Ogden and Whiting, 2003). Alternatively, the Palaeoptera has been considered paraphyletic, but even here there is a difference of opinion. Kristensen (1975) highlighted the primitive attributes of mayflies and considered Ephemeroptera to be the sister group to Odonata + Neoptera (a position earlier heralded by Hennig (1953). In fact, Börner (1904) had already considered Odonata and Neoptera to represent a natural group and had even given them the name Metapterygota. The Metapterygota was defined by several attributes, the most notable being the absence of a subimago (i.e., the loss of molts after the mature, winged form). Ephemeroptera are the only insects to molt when they have wings, which is probably a vestige of the primitive condition seen in apterygotes, which have indefinite numbers of molts. Additional traits defining the Metapterygota are the absence of the caudal filament (which is present in Archaeognatha, Zygentoma, and Ephemeroptera and considered part of the hexapod groundplan), and the fixation of the anterior mandibular articulation. Boudreaux (1979) reversed this hypothesis by placing Odonata basal to Ephemeroptera + Neoptera (as had Lemche, 1940). Recent and extensive morphological and molecular work has continued to support Palaeoptera paraphyly and the monophyly of Metapterygota, which we believe to be the more compelling hypothesis. Thus, we must seek glimpses of traits possibly characteristic of the first fliers among the paleopterous orders.

Whence and Whither Wings?

How, when, and why insect wings originated is one of the most perplexing conundrums in evolution. While wings have been repeatedly lost among insects, it is essentially certain that insect wings evolved only once in spite of their stunning structural diversity. Indeed, a monophyletic Pterygota has been supported by every serious study of insect relationships. Arguments in support of a single origin for insect wings include, first, the basic structure of wing veins, which can be homologized across insect orders. Although minor arguments continue over points of detail, particularly in the number of branches for specific vein systems, the overlying pattern is consistent across all orders. Additionally, with the exception of Odonata, the thoracic musculature operating the wings can be homologized across all orders. Furthermore, the wings are always composed of membranous cuticle supported by veins, are always present on the same thoracic segments, and are associated with the same suite of thoracic modifications, such as the development of notal and pleural wing processes and the formation of pteralia. Wings are also congruent with other features that define features of the Pterygota not associated with flight, such as the formation of the gonoplac in the genitalia and evidence from DNA sequences.

With structures as complex as insect wings, it is not surprising that numerous theories have been proposed to explain the morphological and functional origin of wings. The plethora of ideas can be distilled into two current but contrasting theories (Figure 6.2).

Paranotal Lobes. The development of wings from fixed extensions of the thoracic terga, called *paranotal lobes*, is a traditional theory championed by classic workers such as Snodgrass (1935), and later by Hamilton (1971) and Quartau (1986). Under this theory paranotal lobes provided early insects with the ability to glide, and eventually, with the acquisition of an articulation at their base, these were used for controlling the aerial descent of the insect from perches on tall plants. The presence of paranotal lobes on the prothoracic segments of some Paleozoic insects (Figures 6.17, 6.21, 6.24, 7.9), complete with vein patterns similar to miniature wings, has been heralded as critical evidence for this theory. Indeed, such fossils have at times been referred to as "six-winged" insects! The prothoracic paranotal lobes, however, were not articulated in any of the known fossils so far as anyone can discern. Among extant insects, silverfish possess distinct paranotal lobes that can be used to control their descent while falling (e.g., Hasenfuss, 2002).

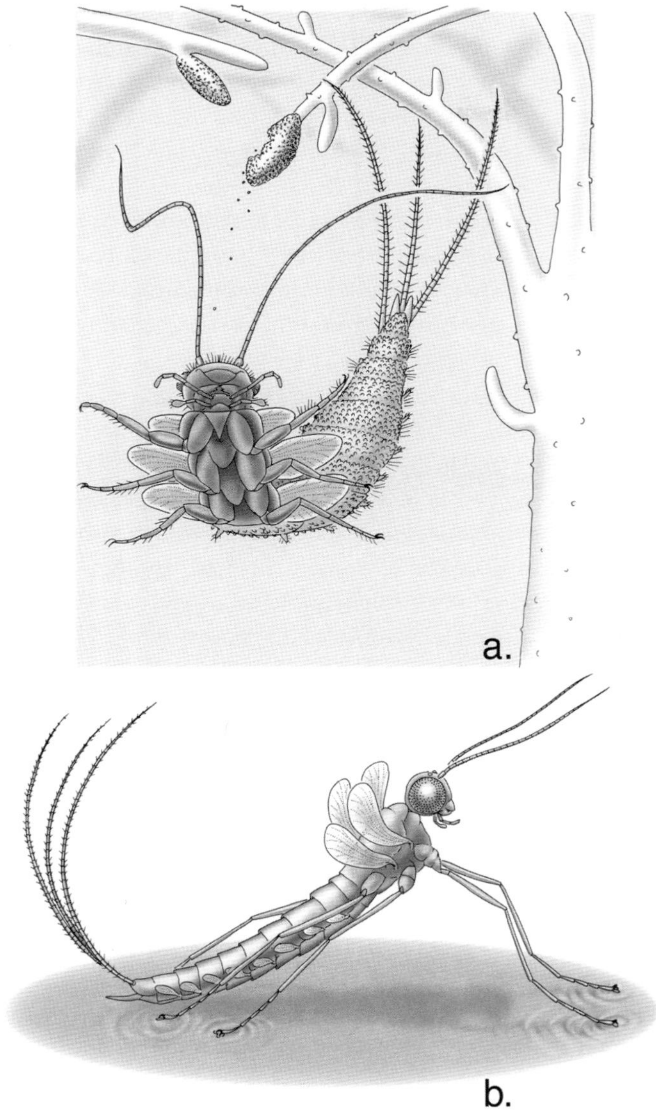

6.2. Alternative hypotheses on the origin of wings from either pronotal lobes (a) or from modified gills (b).

Exite or Gill Theory. The "gill theory" for wing origins hypothesizes that wings are serially homologous with the movable abdominal gills on aquatic naiads of mayflies (Kukalová-Peck, 1991). In this theory, the gill itself is a modified exite of a hypothetical, basal leg podite called the *epicoxa*. The epicoxa is considered to form the junction between the pleuron and the thoracic dorsum. The articulating exite of this leg segment then became modified into gills and into wings. Indeed, the gills of immature mayflies superficially resemble tiny wings, and their pattern of tracheation tantalizingly mimics that of wing venation. Non-homologous exites are present on the coxae and abdominal segments of some apterygote insects, but these cannot be considered evidence for an exite origin of wings because exites in these lineages are different appendicular podites. Recent developmental evidence (e.g., Averof and Cohen, 1997) has supported the exite theory by examining the expression of homologues in crustaceans for two wing-related genes in insects. The pattern of expression was associated with basal leg appendages and to some extent the lateral extremities of the thoracic dorsum.

An interesting idea not yet fully explored is a hybrid between the paranotal and exite theories. As noted, the wing is composed of two epidermal layers. It could be hypothesized that the dorsal epidermal layer is an evolutionary derivative of the tergum (i.e., a partial paranotal lobe), while the lower layer is derived from the pleuron (i.e., from tissues ancestrally associated with the basal segment of an appendage). There is presently no evidence, however, supporting such an idea.

Depending on the theory of wing origin, the exite or paranotal, insects evolved wings, respectively, either as sails at the water surface or as parachutes at the tips of branches. The exite theory was recently endorsed by observations that wings are used by some aquatic, basal groups of insects to sail along the surface ("skimming") (Marden and Kramer, 1994; Marden, 2003). In this scenario, gill-like exites or gill covers (Kukalová-Peck, 1978, 1991; Averhof and Cohen, 1997) expanded to catch surface breezes and carry the newly emerged adult insect along, which then evolved into structures capable of powered flight. This theory has been critiqued on several bases (e.g., Will, 1995; Rasnitsyn, 2003), some of which follow.

1. While skimming is widespread in adult Plecoptera (Marden, 2003), it is very sporadic or even rare in more basal pterygotes, the Ephemeroptera and Odonata.
2. The most basal insects (apterygotes), the Zygentoma and Archaeognatha, are fully terrestrial, suggesting that the ancestral pterygotes were also terrestrial.
3. A terrestrial origin of pterygotes is further supported by fossil evidence, which indicates that freshwater aquatic insects did not appear until the Triassic (Zherikhin *in* Rasnitsyn and Quicke, 2002), nearly 100 MY after the earliest known winged insects. Indeed, the phylogeny of insect orders (Figure 4.24), and the different adaptations of nymphs in various aquatic insect orders (e.g., locations and structure of gills), indicate entirely independent colonizations of water by these orders.
4. Contrary to claims, mutational evidence does not provide evidence of homology, and, in fact, structures as complex as wings clearly have a highly polygenic basis. Also, if gills are homologous to wings, as this theory proposes, one would expect some structural similarity between the two, such as in the branching patterns of the gills and wing veins. Other than the gill covers and their dichotomous branching seen in some mayflies, there is little evidence for this similarity.
5. Many of the earliest winged insects from the mid-

Carboniferous had large body size and broad wings, which would have precluded skimming. Also, no fossilized aquatic nymphs of these insects are known, despite the great bias toward insect fossilization in lacustrine sediments.

When Did Wings Originate?

Wings originated sometime prior to the Late Carboniferous, but the fossil record presently lacks conclusive evidence about the evolution of flight. The presumed insect "wing" mentioned by Schram (1983b) from the Early Carboniferous of Scotland is actually a crustacean. Surprisingly, some of the earliest occurrences of winged taxa are those of the Paoliidae from the Early Bashkirian (= Namurian) of Germany along with Palaeodicytoptera, Megasecoptera, and Geroptera, as well as some putative wings of primitive Paraneoptera (e.g., Brauckmann *et al.*, 1996). Although many of these are paleopterous taxa, paoliids were, however, already truly neopterous (i.e., true Neoptera, not "neopterous" paleopterans like the Diaphanopterodea). During the Late Carboniferous a diversity of wing morphologies and pterygote lineages were already established (e.g., the aforementioned putative paraneopteran wings). Thus, the origin of wings took place earlier and is often hypothesized to have occurred during the Late Devonian or Early Carboniferous. Discovery of metapterygotan features in the fragmentary fossil *Rhyniognatha hirsti* from the Early Devonian (Pragian) chert of the Old Red Sandstone, Scotland, strongly suggests that the age of winged insects may extend earlier by perhaps 80 MY (Engel and Grimaldi, 2004a). As already noted, the possibility that *Rhyniognatha* was a pterygote agrees with recent molecular estimates for the age of pteryogtes (Gaunt and Miles, 2002). Such an ancient age for the origin of wings better accounts for the diverse pterygotes found at the opening of the Late Carboniferous. Similarly, this potentially ancient origin of wings has implications for understanding how flight might have evolved.

One hypothesis concerning the origin of wings indicates that metabolically expensive insect flight evolved in the hyperoxic atmospheres of the Carboniferous and Permian (e.g., Dudley, 2000). However, the oxygen content of the atmosphere during the Devonian was far less than during those two periods (Berner and Canfield, 1989). Protowings, or paranotal lobes, were perhaps used for controlled gliding, similar to that of modern Zygentoma (Hasenfuss, 2002). This transitional form of insect flight is analogous to those seen with *Archaeopteryx* and modern birds (Chiappe, 1995) and with the flying lemurs and bats (Novacek, 1992). Gliding, however, requires a perch. There is abundant evidence that various Paleozoic insects grazed on the nutritious spores from sporangia (Edwards *et al.*, 1995; Rasnitsyn and Krassilov, 1996a,b; Afonin, 2000), which in many Devonian plants were produced at branch tips (e.g., Gensel and Andrews, 1984; Alego *et al.*, 2001). *Rhyniognatha*, and likely early insect fliers, occurred as tracheophytes had evolved into shrubby plants approximately one meter tall. Thus, the potential occurrence of wings in *Rhyniognatha* more closely matches the chronology for the evolution of plant arborescence (Engel and Grimaldi, 2004a).

Certainly numerous discoveries remain to be made pertaining to wing origins and pterygote diversification, particularly from the Devonian and Early Carboniferous. An insect equivalent of an *Archaeopteryx* remains elusive but certainly existed.

EPHEMEROPTERA: THE MAYFLIES

Mayflies hold the distinguished position of being the most basal, extant lineage of winged insects. Though Recent species comprise one recently evolved lineage, these provide a glimpse of some features of the earliest winged insects (Figure 6.3). Mayflies, loved by fish and fisherman, support an industry of "fly tying" that mimicks the appropriate "hatch" of these and other aquatic insects. Mayfly nymphs are the proteinaceous base of the freshwater food chain (Allan, 1995), along with some other aquatic insect groups. Immature mayflies are entirely aquatic and respire through a series of lateral abdominal gills (Figure 6.4). They are sensitive to impurities and particularly dissolved oxygen levels and are reliable indicators of water quality (e.g., Hubbard and Peters, 1978), but they are also critical for filtering particulate matter from streams (e.g., Merrit and Cummins, 1978; Ward, 1992), so they have considerable significance for aquatic ecosystems and human economies. Defining features of the order include the vestigial to absent mouthparts of adults; aristate antennae (convergent with Odonata); a distinctive costal brace at the base of the forewing; greatly reduced, if not entirely absent, hind wings that nearly lack an anal region; and long forelegs and paired penes in males. The order is the only one among pterygotes to primitively retain a caudal, median filament and a subimaginal molt. Major references to mayflies include Needham *et al.* (1935), Burks (1953), Peters and Peters (1970), Edmunds (1972), Edmunds *et al.* (1976), Hubbard and Peters (1978), Flannagan and Marshall (1980), Campbell (1987), Hubbard (1990), Alba-Tercedor and Sanchez-Ortega (1991), Corkum and Ciborowski (1995), and Domínguez (2001).

As the name suggests, adult mayflies have ephemeral lives. Because they don't feed, none live more than a few days, and many live for just a few hours (Figure 6.2). The sole purpose of adult mayflies is to reproduce and disperse. Owing to this exceedingly short lifespan, it is not surprising that the emergence of adult male and female mayflies must be tightly synchronized, which result in mass emergences at certain times of the year. In some places, such emergences can easily

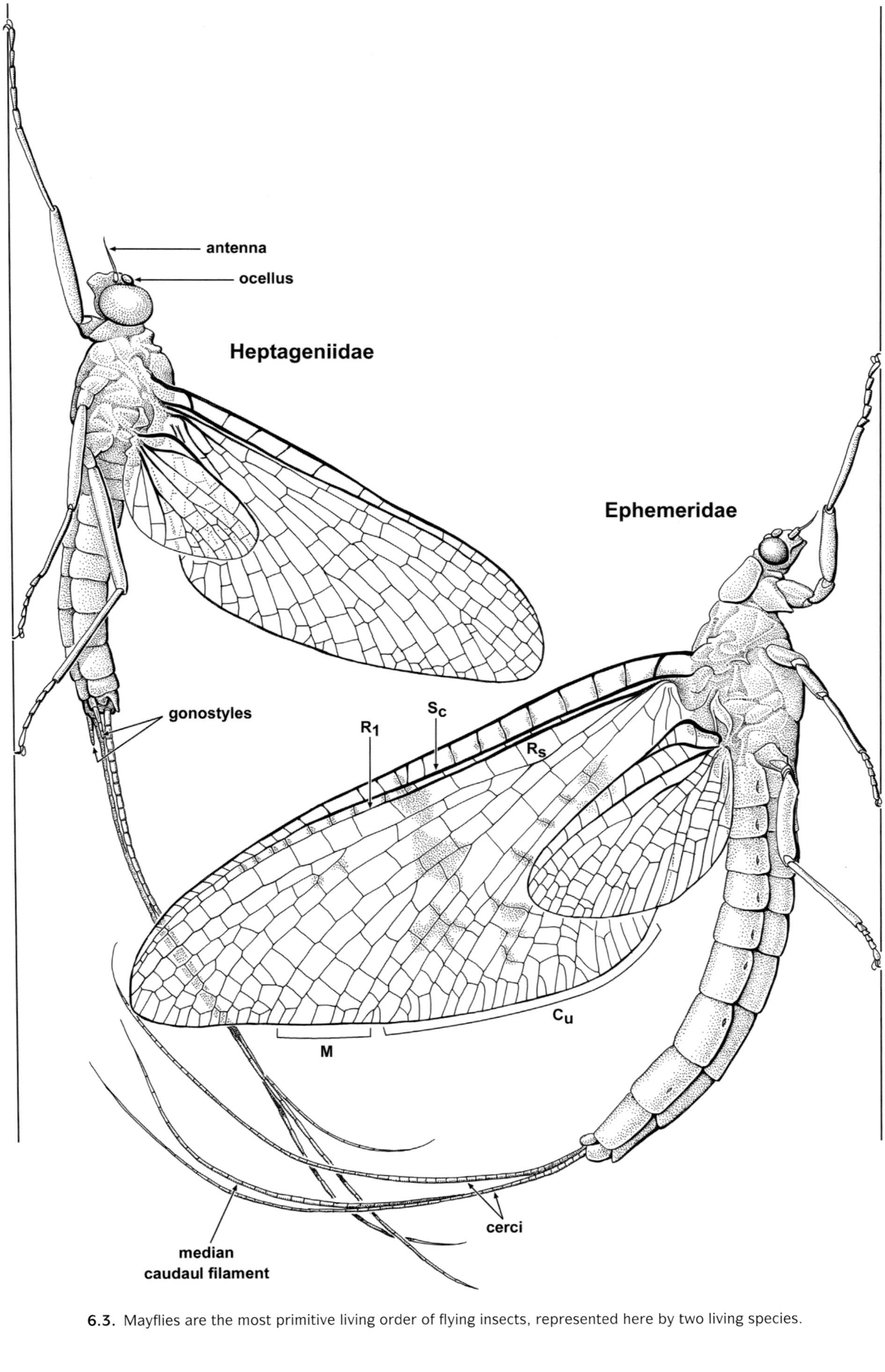

6.3. Mayflies are the most primitive living order of flying insects, represented here by two living species.

6.4. A mayfly naiad (nymph) of *Hexagenia* (above), which has tusks protruding from its mandibles that are used for burrowing. A more typical mayfly naiad is below. Like the adults, immature mayflies have three terminal filaments, a groundplan feature of insects. Length 23 mm (above), 8 mm (below).

number into the tens of millions of individuals and form clouds of dense swarms. Within a day of swarming, their corpses can pile up on roadways or in towns where they had been attracted by lights. In temperate regions adults emerge in warm seasons such as late spring or early summer, usually during the evening hours. In the tropics they swarm earlier in the day or even in the morning. Typically swarms are timed to occur during a single evening or over a couple of days, although they can persist for months, slowly dribbling out individuals that quickly seek a partner before passing away. Once a male and female locate each other, copulation takes place in flight. The male grabs the female with his long, outstretched forelegs and then clasps her abdomen with his genital forceps (which are modified gonostyli). The paired penes (a defining feature of the order) are inserted into the female gonopores located between the seventh and eighth abdominal sterna. Most females oviposit directly into the water either by inserting the apex of the abdomen into the water, or they may "air drop" packages of eggs. A few species carefully place their eggs on aquatic substrates such as stones.

Immature, wingless mayflies, or naiads, are immediately notable for their three caudal filaments and lateral abdominal gills (Figure 6.4). Species occur in a diversity of aquatic systems, from small, cold streams to slow, murky rivers, and lakes, occupying littoral to benthic zones. Naiads are capable of "swimming" but typically adhere close to the substrate and overall are relatively flattened. Swimming occurs by ventral-dorsal undulations of the body and terminal filaments, in stark contrast to the unique lateral undulations of the abdomen of stonefly nymphs. Naiads are herbivorous or detritivorous, typically scraping the substrate for algae or diatoms, though a few are carnivorous, such as some heptageniids (e.g., McCafferty and Provonsha, 1986). Naiads of the Ephemeroidea and some of the Leptophlebioidea have developed elaborate "tusks," which are actually mandibular processes that point forward (Figure 6.4). The tusks and sometimes modified forelegs are used for digging burrows or for feeding (Keltner and McCafferty, 1986; McCafferty and Bae, 1992; Bae and McCafferty, 1995).

Ephemeroptera are the only pterygote insects with a winged subimago; loss of this molt occurs in all other pterygote orders. Emergence is actually done as the subimago, later molting into true, reproductively mature adults. In some species mating takes place in the subimago instar, which probably are males mating with females that will soon emerge as imagos.

Recent phylogenetic and classificatory treatments of the order are by McCafferty and Edmunds (1979), Landa and Soldán (1985), Kluge (1989, 1998, 2000, 2004), McCafferty (1990, 1991, 1997a), and Tomka and Elpers (1991). Different classifications are advocated by various authors, and there is presently no stability in the application of suborders and infraorders within the Ephemeroptera. Numerous cladistic studies have been undertaken on families and superfamilies of mayflies (e.g., Bae and McCafferty, 1995; Wang and McCafferty, 1995; Wang *et al.*, 1997; McCafferty and Wang, 2000), and a complete classification of the order is perhaps imminent. The traditional division into Schistonota and Pannota (McCafferty and Edmunds, 1979) is no longer followed because the latter is derived from the former (e.g., McCafferty, 1991). Presently three suborders are recognized: Furcatergalia (including the former Pannota), Carapacea, and Pisciforma (including the Setisura and Branchitergalia) (Figure 6.5).

Mayflies are relatively well represented in the fossil record, and, not surprisingly given their phylogenetic position, there are some ancient fossils of them. The earliest definitive mayflies stem from the earliest Permian, although some enigmatic fossils from the Carboniferous are likely close relatives to these. Among the most curious creatures representing early

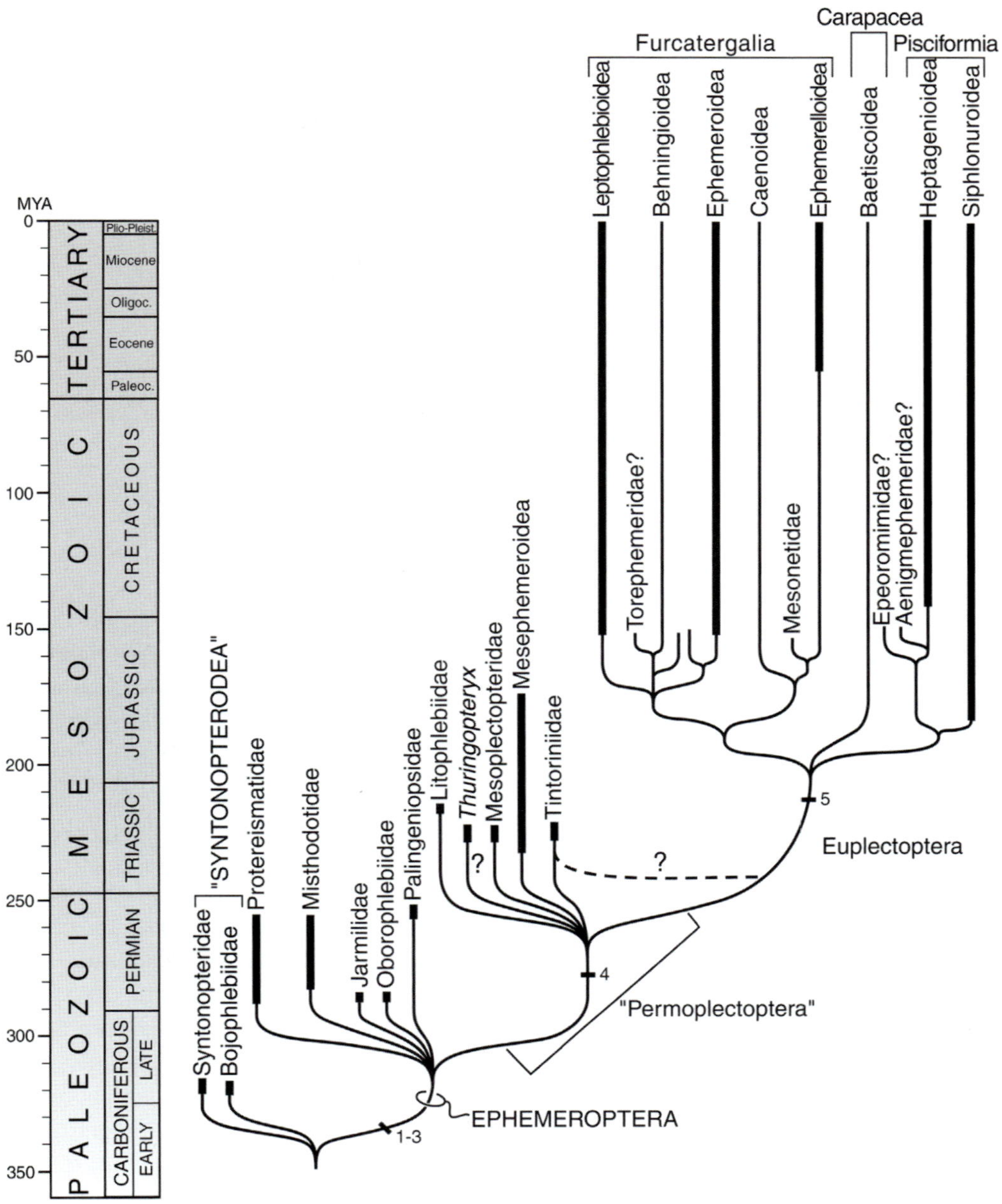

6.5. Phylogeny of the mayflies, order Ephemeroptera, with significant characters indicated (Table 6.1). Modified from Willmann (1999), McCafferty (1997), McCafferty and Wang (2000), and Kluge (1998, 2000).

TABLE 6.1. Significant Characters in Ephemeroptera Phylogeny[a]

1. Costal brace at base of forewing
2. Anal region in hind wing reduced
3. Aquatic naiads
4. Crossvenation more irregular and sparse; hind wings (where known) somewhat shortened
5. Heteronomous wings (hind wing reduced); male forelegs elongate; tarsi reduced; male terminalia paired; adult mouthparts reduced (unknown in most fossil forms)

[a] Numbers correspond to those on phylogeny, Figure 6.5.

mayflies or stem-group mayflies is *Lithoneura lameerei* (Syntonopteridae) from the Late Carboniferous of Mazon Creek, Illinois (Figure 6.6). The fossil was originally described as a member of the extinct order Palaeodictyoptera (Carpenter, 1938a, 1987, 1992), but was later assigned to the Ephemeroptera by several other authors (e.g., Edmunds and Traver, 1954; Edmunds, 1972; Wootton, 1981) or placed in its own order/suborder, Syntonopterodea (e.g., Laurentiaux, 1953). Hubbard and Kukalová-Peck (1980) and Kukalová-Peck (1985) were the first to report the putative presence of a haustellate beak, swollen clypeus, englarged compound eyes, and elongate antennae in *Lithoneura*, which would corroborate the palaeodictyopteran hypothesis. Willmann (1999), however, countered the reconstruction of

6.6. The Late Carboniferous *Lithoneura lameerei* (Syntonopteridae) from Mazon Creek has been a controversial fossil concerning the phylogeny of early winged insects. *Lithoneura* and other "Syntonopterodea" are now understood to be stem-group Ephemeroptera. MCZ 4537; greatest width 65 mm.

6.7. An early mayfly from the Early Permian of Elmo, Kansas, *Protereisma permianum* (Protereismatidae). These early mayflies, while sharing some traits with modern Ephemeroptera, had homonomous wings. YPM; forewing length 18 mm.

Lithoneura, indicating that certain traits are not preserved in the fossil and that it indeed shares a derived formation of a triad between Rs, M, and Cu near the wing base with the Ephemeroptera. However, the fore- and hind wings of *Lithoneura* were very similar to each other (they were *homonomous*), slight differences being that the hind wing possessed a large anal fan and that the forewing lacked a costal brace (Willmann, 1999). Thus, *Lithoneura* cannot be considered a true mayfly. Instead, *Lithoneura* is representative of a stem group that eventually gave rise to the true Ephemeroptera.

Another enigmatic fossil is *Triplosoba pulchella* (Triplosobidae) from the Late Carboniferous of Commentry, France (Brongniart, 1893). Like *Lithoneura*, this fossil has weighed considerably on the mind of systematists and generated wide opinions concerning its identity, although most have agreed that it is somehow related to the mayflies (e.g., Handlirsch, 1906b; Lameere, 1917; Martynov, 1923; Tillyard, 1932; Carpenter, 1963a; Rasnitsyn and Quicke, 2002). The body preserves three caudal appendages and is therefore excluded from the Metapterygota, which includes the Odonatoptera, Palaeodictyopterida, and Neoptera. Willmann (1999), however, noted that *Triplosoba*, unlike *Lithoneura* and Ephemeroptera, has the stems of R and Rs fused near the wing base, a feature considered by him to be a derived trait of Metapterygota and *Triplosoba*. The Triplosobidae may represent a lineage sister to the Metapterygota, the group that consists of all other winged insects.

Various Permian–Jurassic families of Ephemeroptera are believed to be stem groups to modern Ephemeroptera, such as the extinct families Protereismatidae, Bojophlebiidae, Misthodotidae, Mesoplectopteridae, and Mesephemeridae (and perhaps also Jarmilidae and Oboriphlebiidae, but these are known only from naiads); indeed, these families possess the distinctive costal brace in the forewing and have a reduced anal region in the hind wing, similar to modern mayflies (Figures 6.7, 6.8, 6.9). However, unlike modern Ephemeroptera, these families possessed homonomous wings and apparently had well-developed mouthparts in adults. These families are often placed in a separate suborder, the "Permoplectoptera" (e.g., Tillyard, 1932; Hubbard, 1990; Kluge, 1998), but this group is likely paraphyletic and a stem group to the modern suborders of Ephemeroptera (the latter is sometimes referred to as Euplectoptera or Euephemeroptera: e.g., Kluge, 2000, 2004). These stem-group taxa persisted throughout the Triassic (e.g., Lin, 1986; Via and Calzada, 1987; Sinitshenkova, 2000c; Krzeminski and Lombardo, 2001) and, on the basis of a few fossils of Mesephemeridae, into the Late Jurassic, disappearing from the record at that time.

Two fragmentary Triassic fossils, at times associated with the Ephemeroptera, deserve particular mention. The family Litophlebiidae was originally described as a mayfly from the Triassic of South Africa (Riek, 1976b; Hubbard and Riek, 1978). The family was moved into the Megasecoptera by Hubbard and Kukalová-Peck (1980), where it would

6.8. Forewing detail of *Protereisma permianum* (Protereismatidae) from the Early Permian of Kansas. Note the costal brace near the base of the wing, which is a defining feature of Ephemeroptera. MCZ 34056; forewing length 20 mm.

6.9. A naiad (nymph) of *Protereisma americanum* (Protereismatidae) from the Early Permian of Kansas. The wingpads of many Paleozoic naiads protruded from the body, unlike modern insects. MCZ 80356; wingpad length 6.5 mm.

6.10. Like early relatives of dragonflies, some extinct mayflies reached impressive proportions, such as this naiad of *Ephemeropsis melanurus* from the Early Cretaceous of Baissa. PIN 3064/3313; length 37 mm (1.5 in.).

potentially represent a post-Permian persistence of this latter, palaeodictyopterid order. As considered by Rasnitsyn and Quicke (2002), the fossil does not permit assignment to Megasecoptera and instead is an ephemeropteran proper, like other Triassic mayflies. Similarly, the Triassic fossil *Thuringopteryx* from Germany was described as a dragonfly (Kuhn, 1937), moved into the Orthoptera (Zeuner, 1939), and then into the Palaeodictyoptera (Bechly, 1997). Like, *Litophlebia*, the assignment of *Thuringopteryx* to a palaeodictyopterid order would suggest a post-Permian survival of this otherwise Paleozoic superorder and would be an extremely significant discovery. However, the fossil is based solely on fragments of a hind wing that possesses no derived features of any of these lineages and instead more closely resembles a mayfly hind wing (e.g., Rasnitsyn and Quicke, 2002). Indeed, a definitive placement of *Thuringopteryx* is not possible until more complete material is discovered, but an ephemeropteran assignment presently seems more justified than any other ordinal placement.

Ephemeroptera with fore- and hind wings very different in structure (*heteronomous*) are first known from

6.11. A mayfly naiad, *Protoligoneura limai*, from the Early Cretaceous of Brazil's Santana Formation. Mayfly naiads are abundant in this deposit. AMNH; length 9 mm (excluding terminal filaments).

6.12. An adult mayfly from the Santana Formation of Brazil. AMNH; body length 16 mm (excluding terminal filaments).

6.13. Although superficially resembling a myriapod, this is actually an extremely unusual, laterally compressed mayfly naiad with stubby gills that appear like legs. It is preserved in 120 MYO limestone from Brazil. AMNH; length 18.5 mm.

the Early Jurassic, and several modern families were already present by the end of this period. It would appear that a lineage consisting of heteronomous Ephemeroptera came into existence sometime in the Triassic and were diverse by the end of the Jurassic. Certainly in Jurassic and Early Cretaceous deposits mayfly naiads are not uncommon, and some, like *Ephemeropsis melanurus* (Figure 6.10), could reach remarkable proportions, with a forewing length of nearly 40 mm (1.7 inches) and wingspans near 90 mm (3.5 inches). By the Cretaceous mayflies are more abundant and can be extremely numerous regionally, although not very diverse (e.g., Tshernova and Sinitshenkova, 1974; Sinitshenkova, 1975, 1976, 1986; Jell and Duncan, 1986; McCafferty, 1990; Lin and Huang, 2001) (Figure 6.12). Naiads can be so abundant in some Early Cretaceous lacustrine deposits that they are used for stratigraphic dating (Figures 6.11, 6.13). Mayflies are rare in amber but have been discovered in several Cretaceous (e.g., Tshernova, 1971; Kluge, 1993, 1997; McCafferty, 1997b; Peters and Peters, 2000; Sinitshenkova, 2000a,b; Grimaldi *et al.*, 2002) (Figure 6.14) and Tertiary (e.g., Demoulin, 1954a, 1955, 1956, 1965, 1968, 1970a,b; Hong, 1979; McCafferty, 1987; Staniczek and Bechly, 2002) resins (Figure 6.15). Numerous fossils of a diversity of families are known as compression fossils from throughout the Tertiary (e.g. Lewis, 1978; McCafferty and Sinitshenkova, 1983; Fujiyama, 1985; Zhang, 1989; Richter and Krebs, 1999; Sinitshenkova, 1999; Masselot and Nel, 1999), and, although not providing insight into the higher-level relationships within the order, these do provide a perspective on Cenozoic taphonomy, biogeography, and evolution of mayflies.

METAPTERYGOTA

The defining features of this group were mentioned earlier but are repeated here for clarity: loss of the subimaginal molt; loss of the median, caudal filament; fusion into an arch of the anterior and posterior tracheal trunks in the wings and legs;

6.14. A beautiful adult male mayfly in mid-Cretaceous amber from New Jersey. AMNH NJ1018; body length 6 mm.

6.15. A mayfly, family Baetidae, in Miocene amber from the Dominican Republic. Many mayflies have very small hind wings, and in baetids the hind wings are often extremely reduced. Morone Collection M3351. Photo: R. Larimer.

posterior tracheation of the pterthoracic legs; and fixation of the anterior mandibular articulation (although present in Ephemeroptera, it is loose and can act as a slider). The extinct superorder Palaeodictyopterida is included here because they lost the caudal filament.

The Odonatoptera have at times been united with the Ephemeroptera owing to the aristate antennae (hence the name Subulicornia for this group). However, as noted by Grimaldi (2001), the homology between the "aristate" condition seen in these two lineages is suspect. The overall morphology of the "arista" suggests that they are not homologous (Figure 6.16). In the Ephemeroptera the flagellum consists of a series of short, nearly indistinguishable flagellomeres, appearing almost as annulations rather than distinct units. By contrast, the flagellum of Odonata consists of 2–4 distinct flagellomeres, each longer than wide and with well-defined articulations between them. The basalmost flagellomere in odonates is noticeably stouter and more elongate. Moreover, representatives of stem-group lineages of both Odonatoptera (e.g., *Namurotypus*) and Ephemeroptera (e.g., *Protereisma*) have relatively long flagella, indicating that the reduction to an aristate condition occurred independently in both groups.

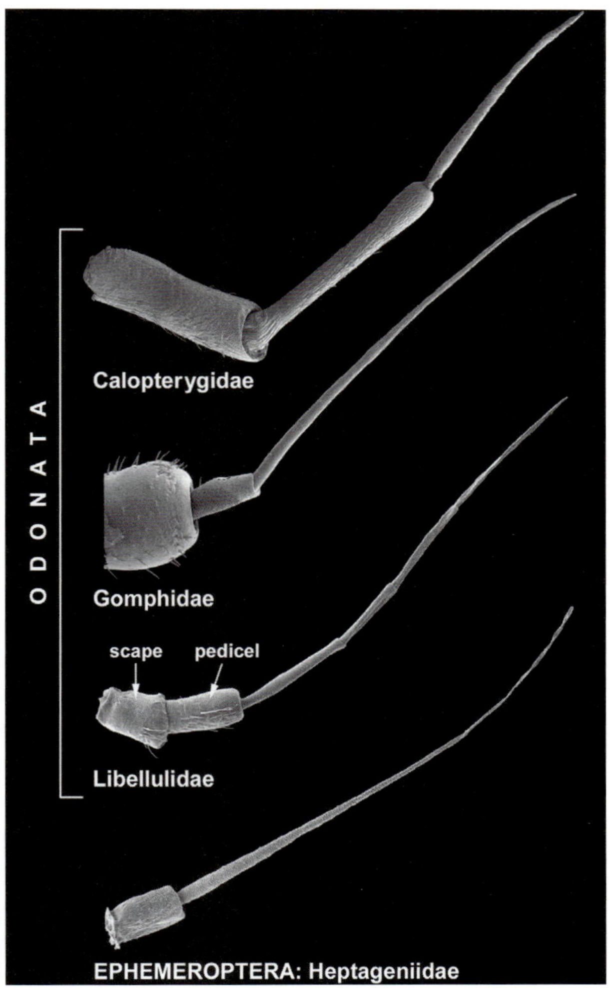

6.16. The aristate antenna of Ephemeroptera and Odonata is frequently cited as a trait uniting the two orders. However, detailed structure of the antenna differs in the two groups, and early fossils of each group had long antennae, indicating that these antennae evolved convergently. Scanning electron micrographs; not to same scale.

6.17. Reconstruction of a fairly typical paleodictyopteridan, *Stenodictya lobata* (order Paleodictyoptera), from the Late Carboniferous of Commentry, France. The mouthparts in this superorder formed a sucking beak, which was apparently used for feeding on plants. Paleodictyopteridans were the dominant insects of the Paleozoic.

PALAEODICTYOPTERIDA: EXTINCT BEAKED INSECTS

The palaeodictyopterids were fascinating, sometimes enormous insects that comprised about 50% of the known Paleozoic insect species. They radiated into a diversity of forms and presumably niches that, after their extinction at the end of the Permian, appeared to have been filled by the new Mesozoic insect fauna. The Palaeodictyopterida apparently comprised mostly herbivorous insects from

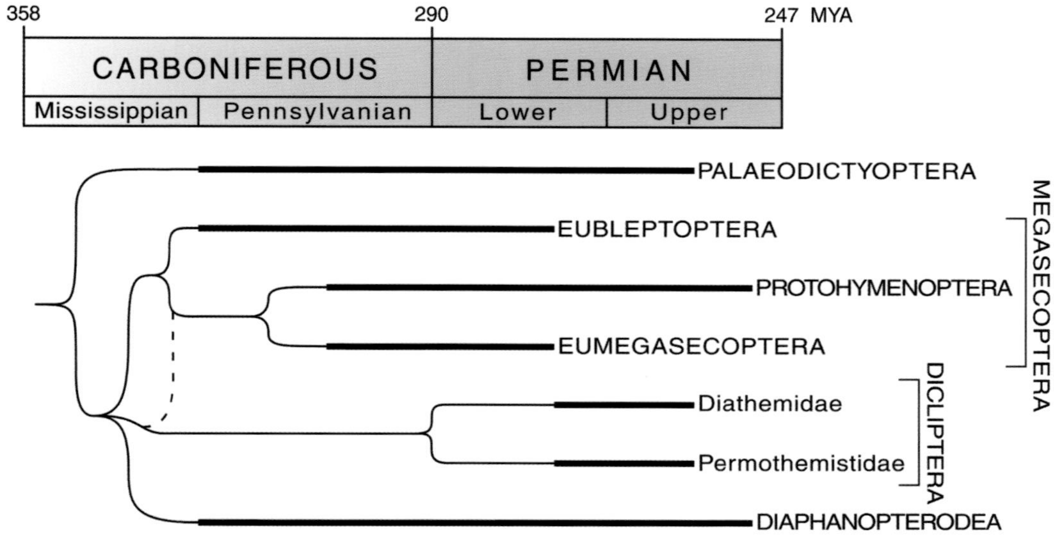

6.18. Phylogeny of the extinct superorder Palaeodictyopterida. This is the only major lineage of insects to have become extinct, which apparently was at the end of the Permian.

the mid-Carboniferous to the Late Permian, having vanished probably at the End Permian Event about 250 million years ago.

Many palaeodictyopterids (Palaeodictyoptera and Eubleptoptera) were remarkable for their prothoracic paranotal lobes, complete with venation resembling that of the flight wings, albeit reduced (Figure 6.17). The lobes of some species were quite large, which provided much of the basis for notions of "six-winged" insects, although the lobes were apparently never articulated as functional wings. Another defining feature of the superorder was the development of a hypognathous head with haustellate mouthparts or beak, with five stylets, and an enlarged postclypeus (indicative of a cibarial pump for sucking fluids). The "beak" was long in most lineages but a few had shorter, stouter proboscides, perhaps used for puncturing tougher plant tissues. Their mouthparts were designed for piercing-sucking and resembled those of the living order Hemiptera. Other features of the palaeodictyopterids were the long, filamentous, multisegmented cerci typically covered with dense setae. The legs were simple but slightly short for such stout bodies, primitively with five-segmented tarsi but reduced to three segments in Diaphanopterodea. The antennae were usually long and filiform but could also be considerably shortened, as in some Megasecoptera. The ovipositor was short, stout, and equipped with many serrations in most taxa, suggesting that they oviposited into plant tissues.

Nymphs of palaeodictyopterids were all apparently terrestrial (e.g., Carpenter and Richardson, 1968), with wing pads held in an oblique, lateral position instead of close over the thorax and abdomen. Given that terrestrialization is primitive for insects as a whole, this observation furthers the notion that the aquatic naiads of Odonatoptera and Ephemeroptera were independently developed, particularly if Palaeodictyopterida is basal to an Odonatoptera + Neoptera clade (this would also be further evidence that aristate antennae are convergent in Odonata and Ephemeroptera).

Handlirsch (1908, 1925) believed the palaeodictyopterids to be the stem lineage from which all other winged insects derived. He argued originally that the winged insects, via the Palaeodictyoptera, evolved directly from Trilobita and that all other insects were derived from individual branches within Palaeodictyopterida (e.g., Megasecoptera gave rise to what we now called the Panorpida). This concept has not held up to scrutiny, and as we have seen the palaeodictyopterids are not basal within insect phylogeny but rather were a specialized lineage that was also the sole superordinal complex to have become entirely extinct. All other superorders have at least a few survivors today. They were also the first major lineage of herbivorous insects.

The Palaeodictyopterida is divided into groups (Figure 6.18), whose relationships are not entirely understood, although the order Palaeodictyoptera retains more primitive traits and may be paraphyletic to all other members of the superorder. Similarly, the Eubleptoptera of the Megasecoptera may be paraphyletic to the remaining taxa, Eumegasecoptera, Protohymenoptera, Dicliptera, and Diaphanopterodea. Extensive cladistic work within the Palaeodictyopterida is needed, and we have made no attempt to resolve relationships within these more or less well-defined groups. Most work on the group has consisted of the descriptions of taxa, the documentation of their character combinations, and detailed accounts of their stratigraphic occurrence.

PALAEODICTYOPTERA

The Palaeodictyoptera is perhaps paraphyletic to all other palaeodictyopterid "orders," and, pending a study of their phylogeny, they should be segregated into natural groups. Many "ordinal" names already exist for paleodictyopterans (e.g., Eugereonoptera of Crampton, various names of Handlirsch), which may need to be resurrected. The order can be characterized by large paranotal lobes, many of which resemble small wings, a well-developed archedictyon in the wing, and typically a broader, roughly triangular hind wing. All of these are primitive traits and the Palaeodictyoptera lacked traits that were derived relative to other palaeodictyopterid lineages.

The wings of Palaeodictyoptera were frequently boldly patterned (e.g., *Dunbaria fascipennis*: Figure 6.19) and may have been used for communication between conspecifics or in disruptive, startle responses when attacked by predators. Palaeodictyopterans could be enormous, achieving wingspans of around 550 mm (22 inches) (e.g., *Mazothairos*) (Figures 6.20, 6.21)! Indeed, the order included some of the largest insects on record, second only to the giants of the Protodonata. A walk through a Late Carboniferous or Early Permian forest would encounter many palaeodictyopterans, and one can imagine startling a perched spilapterid or *Stenodictya* that was feeding on a plant cone or leaf, flashing its wings in response.

Specimens of Palaeodictyoptera (e.g., *Delitzschala bitterfeldensis*) are among the earliest records of pterygote insects, having been recovered from near the Mississippian–Pennsylvanian boundary in Germany and in North America (Nelson and Tidwell, 1987; Brauckmann *et al.*, 1996). The order is recorded from most of the known Paleozoic deposits and from a diversity of genera and families.

DICLIPTERA

This group includes the Diathemoptera and Permothemistida; however, it was referred to as Martynov's Archodonata in Kluge's (2000) treatment of palaeopterous insects. We prefer the more descriptive name Dicliptera, in reference to the vestigial or absent hind wings. Dicliptera are notable among palaeodictyopterids for an absence of paranotal lobes and archedictyon, presence of a distinct pterostigma (as a heavily infuscated region, and not formed of fused or expanded veins as in Hymenoptera), reduction

6.19. *Dunbaria fascipennis* from the Early Permian of Elmo, Kansas. The wings of some species of paleodictyopterans had striking patterns. YPM 1002a; wing spread 35 mm.

6.20. Some palaeodictyopterans reached gigantic proportions. An example is *Moravia grandis* from the Early Permian of Midco, Oklahoma, which is preserved as a small portion of the hind wing here. MCZ 8647; preserved length 73 mm.

6.21. The large paleodictyopteran *Lithomantis carbonarius* from the Late Carboniferous of Scotland. Although faint in this specimen, the large paranotal lobes of the first thoracic segment are visible and give the impression of a "six-winged" insect. NHM I.8118; preserved width 89 mm.

or complete loss of the hind wings, and reduction of crossveins to a single rs-m vein. The Diathemoptera (e.g., Diathemidae) were clearly more primitive owing to the presence of vestigial hind wings, while permothemistids had lost the hind wings entirely. The reduced crossvenation is likely a derived feature uniting the group with the Megasecoptera. The absence of paranotal lobes and reduced venation could be characters uniting the Dicliptera with the Protohymenoptera and Eumegasecoptera in the Megasecoptera. In the latter two groups the wings are "costalized" (veins C, Sc, and R are tightly brought together along the anterior margin of the wing) and petiolate, while the Dicliptera have broader wings and are not costalized. Dicliptera are presently recorded only from the Permian of Eurasia but likely extended into the Late Carboniferous. Kansasiidae is a possible diclipteran (near Diathemidae) from the Early Permian of North America (Tillyard, 1937a; Demoulin, 1954b).

MEGASECOPTERA

Megasecoptera in the sense used by Carpenter (1992) are possibly paraphyletic, with the Eubleptoptera being more primitive. The order is difficult to define as distinct from the Palaeodictyoptera when the Eubleptoptera is included, and Carpenter (1992) suggested that Megasecoptera might be merged with Palaeodictyoptera. Such a decision, however, would only further cloud relationships within this lineage, and we have, conservatively, retained the principal lineages pending phylogenetic study. Overall, megasecopterans were relatively smaller than individuals of Palaeodictyoptera. Unlike palaeodictyopterans, an archedictyon was only rarely

6.22. *Pseudohymen* (Megasecoptera) from the Early Permian of Tshekarda in the Ural Mountains of central Russia. PIN 1700/4153.

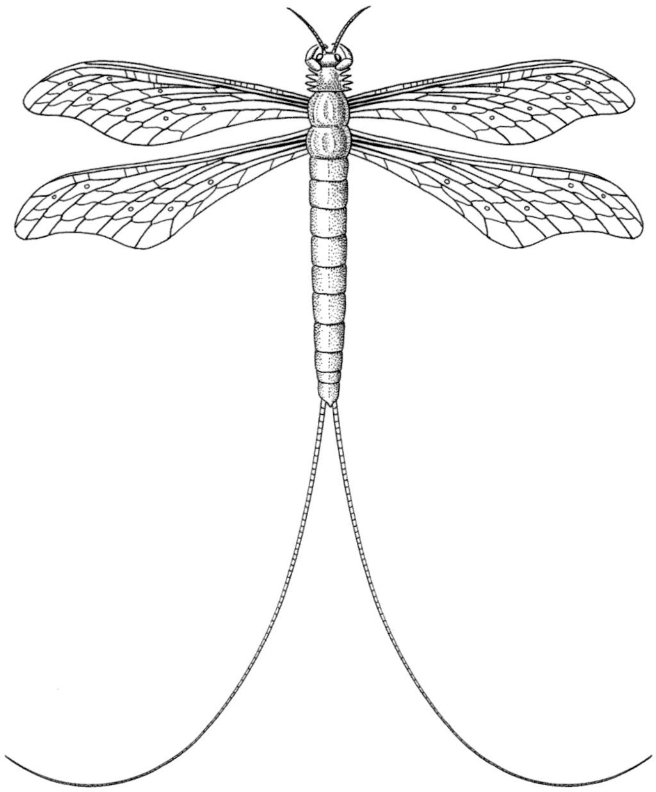

6.23. Reconstruction of *Mischoptera nigra* (Megasecoptera) from the Late Carboniferous of Commentry, France.

present and occurred only in what appear to have been the most primitive families. Megasecoptera had distinct but variable numbers and arrangements of crossveins. Veins Sc and R were positioned close together, and indeed in subgroups of the Megasecoptera there was costalization of the wing margin. Overall, the wings of Megasecoptera could be quite spectacular (Figure 6.22). Like the Palaeodictyoptera, some were patterned, but the patterns were never as bold as those in palaeodictyopterans. However, the petiolate wings were distinctive and probably reflect a unique flight among Paleozoic insects, possibly even hovering (Wootton and Kukalová-Peck, 2000). In addition to the petiolate base, some taxa had distinctly falcate wings and a pronotum studded with spines, making them appear almost fanciful (e.g., *Mischoptera*) (Figure 6.23). The Megasecoptera have been divided into two groups: those families with distinct paranotal lobes (Eubleptoptera) and those lacking the lobes (Eumegasecoptera and Protohymenoptera), though this is probably an unnatural grouping.

Nymphs are known only of the Mischopteridae and Brodiidae, which had wing pads that protruded from the body and with a remarkably well-developed venation (Carpenter and Richardson, 1968). Unlike the wing pads of modern insects, the wing pads of Megasecoptera were apparently only joined to the thorax at the point of articulation and were held free at the sides of the body like adult wings. As is generally true for the Palaeodictyopterida as a whole, nymphs lacked gills or any other modifications indicative of an aquatic lifestyle.

The Eubleptoptera (a.k.a., Eubleptidodea) were primitive megasecopterans primitively retaining paranotal lobes, three anal veins, a normal costal space, and numerous crossveins in the wings (frequently with an archedictyon), all features reminiscent of Palaeodictyoptera, which is where Carpenter (1992) had placed them. However, there were distinctly fewer crossveins compared to all other Palaeodictyoptera, similar to that of other Megasecoptera. Eubleptoid families (e.g., Eubleptidae, Namurodiaphidae, Anchineuridae, Engisopteridae, Sphecocorydaloididae, and "Xenopteridae") are recorded from the Pennsylvanian of Europe (Carpenter, 1963a; Kukalová-Peck and Brauckmann, 1990), North America (Handlirsch, 1906a), and Argentina (Pinto, 1986, 1994) through the Early Permian of Oboro (Kukalová-Peck, 1975). The group is perhaps a stem group to all other Megasecoptera.

The remaining two lineages of Megasecoptera (Eumegasecoptera and Protohymenoptera) are united by the loss of paranotal lobes and the elongate wings (petiolate in many taxa); possession of a single, pectinate anal vein; costalization of the wing (crowding of veins C, Sc, and R); and further reduction in the number of crossveins. The two groups differed principally in extent of the costal space in the crowded anterior wing margin. While the Eumegasecoptera retained a distinct costal space (e.g., Mischopteridae, Corydaloididae, Sphecopteridae, Vorkutiidae, Carbonopteridae, Moravohymenidae), the Protohymenoptera almost entirely lacked a costal space (e.g., Ancopteridae, Aspidothoracidae, Aspidohymenidae, Bardohymenidae, Brodiidae, Brodiopteridae, Caulopteridae, Hanidae, Protohymenidae, Scytohymenidae). Both lineages are known from the Pennsylvanian and Permian of Europe and North America, although Protohymenoptera is also known from the Late Permian of South Africa.

DIAPHANOPTERODEA

The Diaphanopterodea, also known as the Paramegasecoptera, were relatively rare insects of moderate to large size and are known from some of the earliest deposits containing winged insects (e.g., Kukalová-Peck and Braukmann, 1990; Kukalová-Peck, 1992; Brauckmann *et al.*, 1996). Defining features of the order include a reduction of the tarsus to three tarsomeres (trimerous), complete loss of the archedictyon (perhaps a character uniting Dicliptera, Protohymenoptera, Eumegasecoptera, and some Eubleptoptera?), the simple MA vein, and flexion of the wings. Indeed, perhaps the single most fascinating aspect of the Diaphanopterodea is the convergence on neoptery, or the ability to flex the wings over the abdomen during rest (Tillyard, 1936; Carpenter, 1947; Carpenter and Richardson, 1971; Sharov, 1973; Kukalová-Peck,

1974). The venational similarity with Megasecoptera (in which many species of diaphanopterodeans were at one time placed) and their possession of a sucking beak clearly indicate their place in the Palaeodictyopterida, and not at all with Neoptera. Furthermore, Neoptera possess a third axillary sclerite responsible for the flexion of the wings, and Diaphanopterodea apparently lacked this sclerite (based on the limited preservation), having developed neoptery via a novel means. In the Palaeodictyopterida the wing base was apparently composed of a series of large, articular plates, so flexion of the wings over the abdomen in Diaphanopterodea was enabled by the reduction of the various articular plates along with the formation of a basal fold line (Kukalová-Peck, 1974).

PALEOZOIC HERBIVORY

The Palaeodictyopterida were the first major group of herbivorous insects. We tend to think of the impact of herbivorous orthopterans, phytophagan beetles, plant bugs, and caterpillars on modern ecosystems, but during the Paleozoic the palaeodictyopterids were among the primary herbivores. They probably caused much of the plant-tissue damage during the Permian and Late Carboniferous (perhaps even into the Early Carboniferous, although body fossils are lacking), and in fact feeding scars can be found readily on Paleozoic plants. In a comprehensive study of Permian plants, 83% of the leaves showed evidence of insect herbivory, and 4.5% of the leaf area was consumed by insect herbivores (Beck and Labandeira, 1998). Though all the insects that caused the damage are not definitively known, Palaeodictyopterida were certainly among them, as indicated by holes made with beaks.

Long before definitive evidence for winged insects, mites and other terrestrial arthropods probably exploited plants. Indeed, the damage observed in some Devonian plants is identical to that made by modern phytophagous mites. Early Devonian rhyniophytes and trimerophytes had herbivore damage, and they contained arthropod coprolites (presumably of myriapods and mites) that consisted of spores (e.g., Edwards, 1966; Banks, 1981; Trant and Gensel, 1985; Banks and Colthart, 1993) (e.g., Figure 2.21). Evidence of pierced plant tissues is also known from this time period (Kevan *et al.*, 1975; Labandeira and Phillips, 1996b); perhaps the piercings were made by mites and springtails, the latter group of which is also known from this time period (e.g., *Rhyniella*). However, the frequency of herbivory was perhaps minor as terrestrial arthropod communities were dominated by predators, particularly among the arachnids. Basal insects (e.g., archaeognaths and zygentomans) were probably mostly detritivorous (although silverfish certainly also consumed spores or pollen) and probably presented little threat to the plants that radiated across the barren landscapes of the Early Devonian.

As we have already discussed, the rise of forests in the Devonian may have led to the development of flight and it is then, among the pterygotes, that plants finally felt the force of insects, most impressively from Palaeodictyopterida during the Paleozoic. The Odonatoptera, like their modern counterparts, were almost certainly predatory and played little part in these nascent plant-insect interactions.

Contemporaneous with the Palaeodictyopterida were numerous families of "Protorthoptera," which were clearly herbivorous. Unlike the paleodictopterids, protorthopterans had chewing mouthparts and must have fed on external foliage (e.g., Müller, 1982; Scott and Taylor, 1983; Beck and Labandeira, 1998; Labandeira and Beall, 1990; Obordo *et al.*, 1994; Castro, 1997), or they consumed spores and pollen (e.g., Rasnitsyn, 1977c; Scott and Taylor, 1983; Rasnitsyn and Krassilov, 1996a,b; Krassilov and Rasnitsyn, 1997).

Palaeodictyopterids, however, could exploit tissues within the plant through the then novel means of piercing and sucking. To what extent the paleodictyopterid beak could take in fluids or was relegated to feeding on nutritious internal tissues is unknown (bear in mind, too, that the vascular systems of many Permian and Carboniferous plants were not as developed as plants are today). Damage from piercing mouthparts has been found on various Paleozoic seed ferns (e.g., Medullosales, Cordaitales), tree ferns (Marattiales), and lycophytes (Lepidodendrales) (e.g., reviewed by Labandeira, 1998), all presumably the activity of palaeodictyopterids owing to their physical proportions and trace morphology (e.g., Labandeira and Phillips, 1996b). Some Palaeodictyoptera (e.g., Eugeronidae, Homiopteridae) had particularly long, possibly flexible beaks reaching nearly 32 mm (1.3 in.) in length. Such insects could presumably insert their stylets into the inner tissues and extract phloem and xylem. The base of the beak was broad and resulted in a conical surface scar, indicative of palaeodictyopterid feeding. Similar piercing scars are also seen in fossilized seeds, in which the insect bored its beak through the protective layers and into the embryonic tissues (Sharov, 1973). Feeding punctures into plant stems are documented from the mid-Pennsylvanian (Scott and Taylor, 1983; Taylor and Scott, 1983). When palaeodictyopterids became extinct at the end of the Paleozoic, diverse and efficient new insect herbivores took their place.

ODONATOPTERA: DRAGONFLIES AND EARLY RELATIVES

The Odonatoptera are the most recognizable of the primitive pterygotes. The familiar dragonflies are conspicuous day-flying insects common to most parts of the world. They are also among the most ancient of winged insects, and with the Ephemeroptera and Palaeodictyopterida comprise the former "Palaeoptera" – an unnatural grouping that has been abandoned (see earlier discussion in this chapter).

Three orders are included in this group: the Geroptera, Protodonata, and Odonata, which are collectively defined by a reduction of the anal region of the wings; a distinctive form of bracing where there is a "kink" in CuP where it meets the anterior anal vein; and two articular plates at the wing base (Riek and Kukalová-Peck, 1984; Bechly, 1996; Bechly *et al.*, 2001). The fusion of several axillary sclerites forming two large plates in the wing articulation is distinctive to Odontatoptera and is not found in any other group of flying insects. In Odonata, the points of fusion between the original, smaller plates are indistinguishable; in Geroptera and Protodonata suture lines (and purportedly some membrane, but this is based on compression fossils) still exist that demarcate the original, joining sclerites.

It is difficult to surmise when the Odonatoptera originated. The basal Geroptera and Protodonata occurred as early as the Late Carboniferous and, in fact, the records of the Geroptera are among the oldest of any winged insects (e.g., Brauckmann *et al.*, 1996). Thus, the age of the last common ancestor of Odonatoptera is speculative, but may have been in the Early Carboniferous or even the latest part of the Devonian.

GEROPTERA

The Geroptera comprise a single family of primitive odonatopteroids, Eugeropteridae, from the early Late Carboniferous (Early Bashkirian = Namurian) of Argentina (Riek and Kukalová-Peck, 1984). While a quick glance at the primitive wings of geropterans reveals little affinity to anything one might identify as a dragonfly or damselfly, finer study indicates a shared, albeit distant, ancestry between the groups (as noted earlier). Superficially, the order more closely resembles the Palaeodictyopterida and, like some palaeodictyopterids, geropterans had pronotal lobes but lacked an archedictyon (Figure 6.24). Very little is known of Geroptera and the monophyly of the few species in the order is entirely speculative. Within the Odonatoptera the pronotal lobes are derived, and so these may serve as a defining feature of the Geroptera, though they are convergent with similar structures in unrelated Paleozoic orders and apparently occurred in Erasipteridae (a primitive protodonate family that might be best included in Geroptera). Alternatively, paranotal lobes may be the groundplan design of all pterygotes, retained in basal members of various orders but lost independently in most species in each lineage.

It is entirely unknown whether or not the Geroptera, like Odonata, had aquatic nymphs. The ancestor of Odonata certainly had an aquatic nymph, and some protodonatans may have had as well, but it is not known whether this mode of life evolved in the common ancestor of Protodonata + Odonata, or was even a feature of all Odonatoptera. It would further be very significant to know if Geroptera had terrestrial immatures because it would cast light on whether aquatic living in Odonatoptera and Ephemeroptera were independently derived, as current evidence suggests. Other basal metapterygotans, the Palaeodictyopterida, were certainly terrestrial and tentatively suggest that Odonatoptera evolved aquatic nymphs independently of Ephemeroptera.

HOLODONATA: PROTODONATA AND ODONATA

The Holodonata includes two orders that better approximate what most people know as dragonflies and damselflies. The wings had a characteristically long, slender, "odonatoid" appearance. In fact, most of the defining features of this

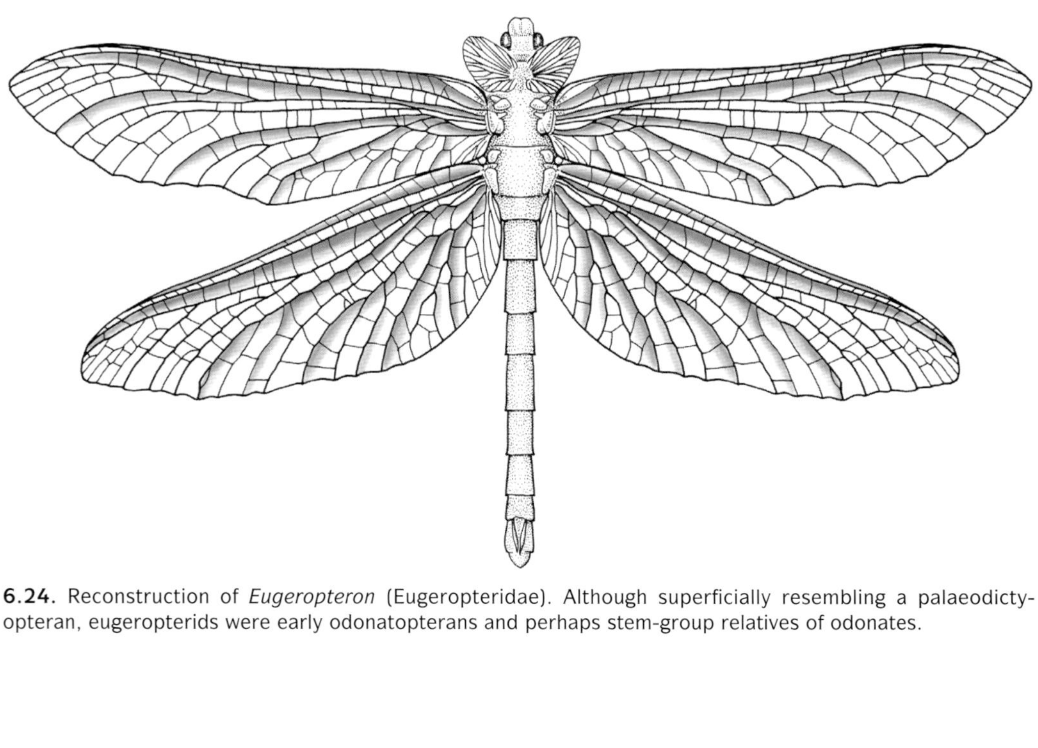

6.24. Reconstruction of *Eugeropteron* (Eugeropteridae). Although superficially resembling a palaeodictyopteran, eugeropterids were early odonatopterans and perhaps stem-group relatives of odonates.

6.25. Wing of *Megatypus schucherti* (Meganeuridae: Protodonata) from the Early Permian of Elmo, Kansas, shown here at life size. The meganeurid "griffenflies" were early relatives of modern odonates and included the largest known insect, which was *Meganeuropsis permiana* with a wingspan of approximately 640 mm (26 in.). YPM 1021; length 160 mm.

group occur in the structure of the wings. Unlike the Geroptera, holodonotans have a large, proximal, hornlike sclerite on the posterior articular plate of the wing articulation, and there is fusion of various veins. For example, MA is fused with RP, while the stem of M at the wing base is vestigial (in Protodonata and some basal Odonata), or it is fused entirely with the Cu stem (in most Odonata). Vein MP originates from the combined stem of veins M + Cu, rather than as part of a stem of M as in Geroptera, and the area between veins MA and MP is expanded and filled with intercalary, longitudinal veins. An interesting feature of the Holodonata is that the thorax is slanted posteriad. The slanting is weak in the few Permian Protodonata that had some pleural sclerites preserved. In Odonata the slanting is much more dramatic, particularly in damselflies where the wings extend over the abdomen when held together at rest, similar to the neopterous condition but where the wings are actually folded flat over the abdomen.

PROTODONATA: THE GRIFFENFLIES

Although frequently called giant dragonflies, the Protodonata cannot truly be considered dragonflies. This is a Paleozoic stem group to the true Odonata, the dragonflies and damselflies. The name "griffenflies" more aptly highlights this distant relationship, rather than the name "giant dragonflies," which implies a much closer affinity. As the cladogram shows, dragonflies (Epiprocta) are distantly removed from the Protodonata and radiated after most protodonatan lineages became extinct.

While "dragonflies" is a misnomer, "giant" is not. Among the Protodonata insects attained grandiose proportions (Figures 6.25, 6.26, 6.27). The largest insect to have ever lived was *Meganeuropsis permiana*, from the Early Permian of Elmo, Kansas, and Midco, Oklahoma (it is also known by the synonymic name *M. americana*) (Carpenter, 1939, 1947). This magnificent griffenfly attained wingspans of approximately 710 mm (28 inches), which dwarfs the largest odonates found today. Protodonatans were almost certainly predaceous, as all nymphal and adult odonates are today. Most fossils of

6.26. Wing of *Tupus gracilis* (Meganeuridae: Protodonata) from the Early Permian of Midco, Oklahoma. MCZ 4818; length 145 mm.

6.27. Wing of *Arctotypus sinuatus* (Meganeuridae: Protodonata) from the Permian of Russia. PIN 3353/87; length 120 mm.

these insects consist only of wings, but among the few preserved body parts are large, toothed mandibles, enormous compound eyes, and stout legs with spines, thrust forward in a similar manner to Odonata – all indicative of their being aerial predators. It is intriguing to imagine how these insects flew, perhaps streaking through Paleozoic swamps and forests, landing on unsuspecting animals like a bird of prey. At their prodigious size, they must have preyed on virtually all other insects and even small vertebrates.

Although immature griffenflies are yet unknown, the close relationship of the group to Odonata might suggest that their naiads were aquatic, but there is no direct evidence. It would be particularly fascinating to know if griffenfly naiads possessed the labial "mask" characteristic of odonate naiads (Figure 6.33), which is the prey capture device. Some of the larger naiads of Recent Odonata can even capture small vertebrates such as fish or tadpoles, so given the size of *Meganeuropsis* (its naiad must have been up to 18 inches in length), it must have been a formidable predator. It would also have been impressive to see such a naiad eclose into an adult, its huge, soft gossamer wings gradually expanding before taking its first flight. Evidence for aquatic protodonate naiads, however, like those of Paleozoic Ephemeroptera, is equivocal (Wootton, 1988).

The famous French paleontologist Charles Brongniart (1893) brought us our first image of these giants and coined their scientific name. Brongniart's dissertation was a study of the insects from the Carboniferous coal measures of Commentry, France, and among the fossils he discovered was the first griffenfly, *Meganura monyi*, which was the largest insect known until the discovery of *Meganeuropsis* by Carpenter approximately 50 years later. He even published in his dissertation a life-sized, fold-out reconstruction of the insect, although at the time only the wings were known. Unfortunately, even today little is known of Protodonata, much remains educated speculation, and even some is pure myth, such as the existence of extinct dragonflies with 2 m (6 ft) wingspans. Most specimens are preserved as wing fragments only, a few as virtually complete wings, and even fewer with some body structures (Figure 6.28).

The wings of protodonatans, unlike those of the Odonata, lacked the distinctive pterostigma or a nodus (formed by the abrupt termination of Sc into a transverse nodal crossvein near wing midpoint), among other typical odonate features (Figure 6.29). Although frequently believed to be an unnatural (paraphyletic) group, the exclusion of Geroptera from Protodonata, as well as a few basal odonate suborders, makes the traditional families Meganeuridae and Paralogidae apparently monophyletic. In a recent study of odonatoid relationships (Rehn, 2003) Protodonata were redefined in a more restricted sense, based on a small lobe on the outside edge of the costal axalare (a portion of the anterior axillary plate at the wing base) and intercalary longitudinal veins between IR_1 and RP_2. The evidence is not substantial, so it is still possible that griffenflies are paraphyletic, stem-group odonatoids.

The Protodonata ruled the Paleozoic skies from the Late Carboniferous until the Late Permian, disappearing from the

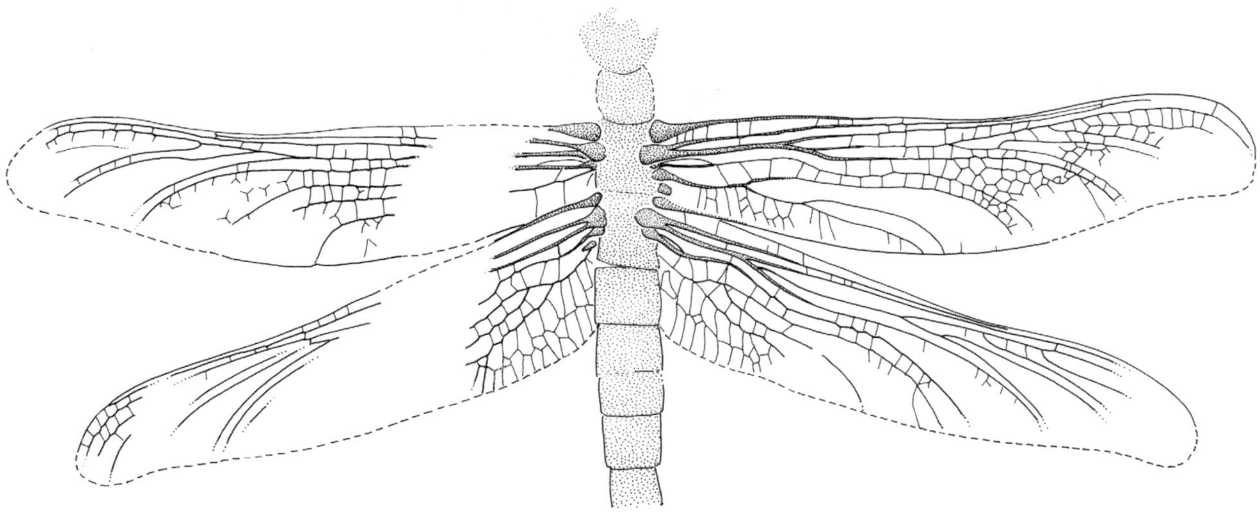

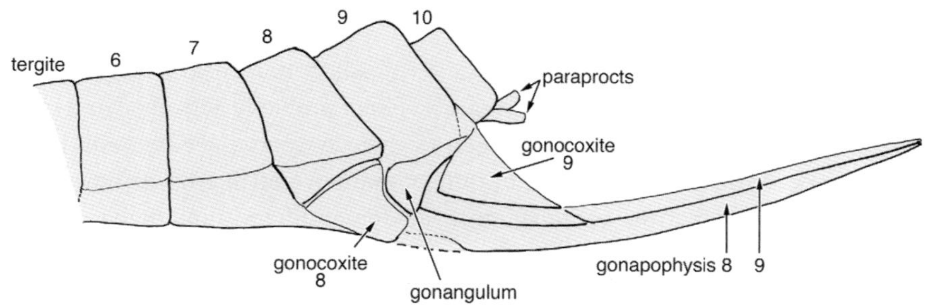

6.28. The early odonatopteran, *Erasipteroides valentini* from the Late Carboniferous of Hagen-Vorhalle in Germany. Although odonates today do not have ovipositors, stem-group taxa such as *Erasipteroides* had well-developed, primitive ovipositors. Redrawn from Bechly *et al*. (2001).

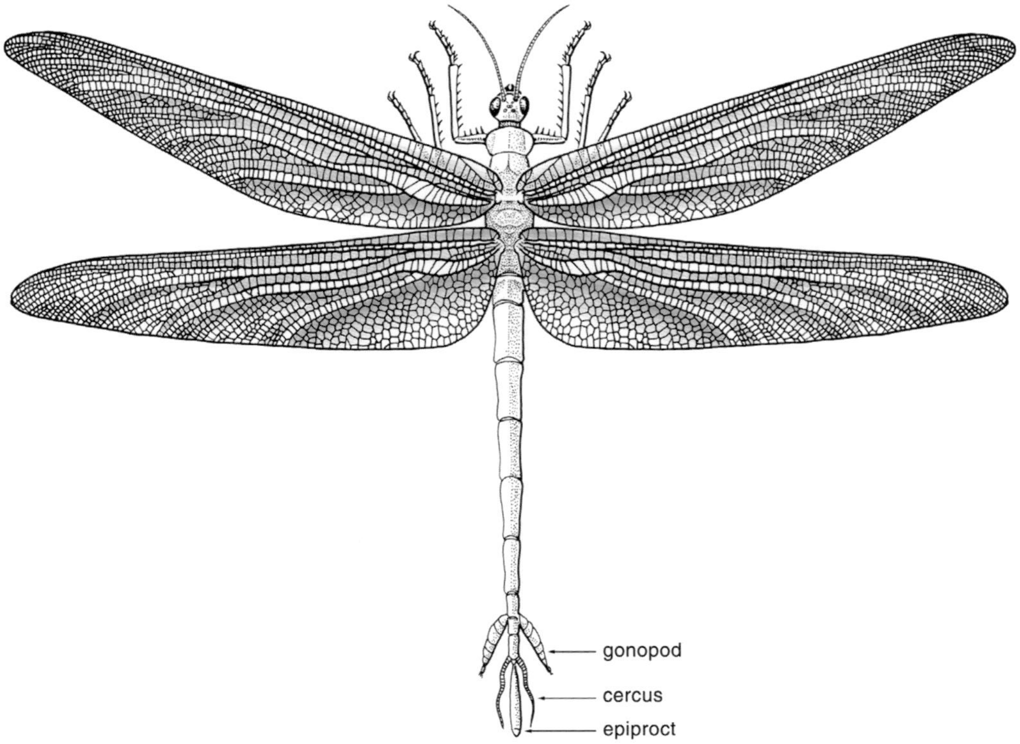

6.29. A reconstruction of one of the large, extinct odonatopterans, *Namurotypus sippeli*, from the Carboniferous of Hagen-Vorhalle. Based on Bechly *et al*. (2001).

record after the End Permian Event (ca. 247 MYA). While the Protodonata vanished, the Odonatoptera as a whole persisted through this catastrophe, including true Odonata from the Permian. The flourishing of Odonata in the Mesozoic may be a result of the demise of the protodonatans.

Gigantism

Griffenflies were not the only giants during the Paleozoic. Enormous mayflies, myriapods, scorpions, palaeodictyopterids, and others were all contemporaries of these aerial juggernauts (Kraus, 1974; Hünicken, 1980; Briggs, 1985; Shear and Kukalová-Peck, 1990; Kraus and Braukmann, 2003). But, it is a very common misconception that all Paleozoic insects were giants when in fact most species were only a few centimeters or less in size, not unlike the situation today. Also, gigantism occurred in primitive amphibians during the Carboniferous (Carroll, 1988), and of course in some lineages of nonavian dinosaurs. The development of gigantism and its disappearance is an intriguing evolutionary and mechanistic question. The repeated evolution of unusually large size can be a feature of the lineage (e.g., sauropod dinosaurs), but in general there must also be some environmental factors conducive to gigantism, such as defense against predators (Vermeij, 1987; Shear and Kukalová-Peck, 1990). Another explanation concerns changes in the atmospheric concentration of gases (specifically oxygen) during the Late Paleozoic and Early Mesozoic (Graham *et al.*, 1995), which has been discussed mostly in terms of insects because of their manner of respiration.

Insects breathe through a tubular system of *tracheae*, which are connected to the outside of the animal by minute, valved openings (spiracles). Air moves through the insect's tracheae and it is the passive diffusion of oxygen that allows the insect to respire. Insects can enhance the movement of airflow by contracting small "bellows" located at various points in the tracheal system or by expanding and contracting their abdomen, but overall it is the simple physics of diffusion that allows them to breathe. This action immediately imposes constraints on the body size of an insect because it becomes increasingly difficult to get oxygen to the interior of a larger animal: the greater the mass, the disproportionately more tracheae are required to reach the deepest muscles, and respiration becomes very inefficient. Increased partial pressures of oxygen in the atmosphere have the effect of allowing the gas to diffuse further through the network of fine tubes. Thus, as the atmosphere becomes hyperoxic the upper limits of arthropod size may have increased (Graham *et al.*, 1995). Indeed, several authors have hypothesized that giant insects would have occurred during episodes of increased oxygen concentration (Rutten, 1966; Schidlowski, 1971; Tappan, 1974; Budyko *et al.*, 1987; Dudley, 1998). Interestingly, the Late Paleozoic, when these giants existed, was a period of high oxygen concentrations.

During the Devonian, plants invaded land and rapidly proliferated. This expanded flora produced large volumes of oxygen as a photosynthetic byproduct, and concentrations continued to increase until reaching a peak during the Late Carboniferous (Berner and Canfield, 1989; Graham *et al.*, 1995; Dudley, 1998, 2000). Although the peak was in the Carboniferous, what might be considered a hyperoxic atmosphere first came about during the mid- to Late Devonian when oxygen concentrations began to exceed today's levels. During the very end of the Paleozoic, oxygen concentrations began to decline, and indeed concentrations went steadily from their Carboniferous peak to well below today's level across the Permian. Thus, the decline of giant insects may not have been a result of the fateful End Permian Event but instead a factor of physics. This is highlighted by the brief reappearance of giant mayflies in the Hexagenitidae in the Cretaceous, when hyperoxic conditions were reached once again. It must also be emphasized that the Paleozoic giants were probably actively flying insects. The structure of meganeurid wings and wing veins indicate that these insects had a maneuvered flight, which would have been too metabolically demanding without high oxygen levels.

Caution is required, though, when interpreting the gigantic sizes of some extinct insects. First, our basis of comparison is only the Recent – a geologically instant slice of time. The 400-MYO fossil record of insects draws from a collective insect fauna that was many orders of magnitude more diverse than what exists today, so we may just be more likely to encounter rare giants by surveying the fossil record. This is especially true given the preservational bias toward larger insects as compressions in rocks. Also, while we tend to think that the giant insects have vanished, some still persist. Damselflies of the family Pseudostigmatidae can reach wingspans similar to those of *Megatypus* griffenflies (Protodonata: Meganeuridae), and there exist beetles today that are 4–6 inches long, walking sticks about 12 inches long, katydids (Tettigoniidae) with 8-inch wingspans, and some lepidopterans (e.g., the "white witch," *Thysania agripinna* [Noctuidae]) can have a wingspan up to 10 inches. No insects, though, ever matched the size of some of the meganeurid griffenflies.

ORDER ODONATA: THE DRAGONFLIES AND DAMSELFLIES

Odonata are "bird-watcher's" insects (Figure 6.30). The aerial displays and complex behaviors of the order invite even casual observers, and it is little wonder that ornithologists make excellent odonatologists. Species are almost entirely diurnal and have acute vision and an active, powerful, and maneuvered flight. Because of their popularity and diversity (approximately 6,000 species), dragonflies and damselflies have received a great deal of attention from professionals and laymen alike. Overviews of odonate biology and taxonomy are by Allen *et al.* (1984, 1985), Bechly (1996), Steinmann (1997a,b),

6.30. A dragonfly rests before taking flight in White Rock, Canada. Photo: R. Swanson.

with major regional treatments by Westfall and May (1996) and Needham *et al.* (2000) for North America, Askew (1988) for parts of Europe, Needham (1930) for parts of Asia, Watson *et al.* (1991) for Australia, and Pinhey (1951, 1961) for Africa. Classic references to the Odonata are by Tillyard (1917a), Corbet (1999), and Silsby (2001). The order has undergone numerous recent classificatory rearrangements based on phylogenetic studies (e.g., Fraser, 1954, 1957; Pfau, 1971, 1986, 1991; Carle, 1982; Trueman, 1996; Lohmann, 1996; Bechly, 1996), but the cladogram presented herein is based on the excellent study of adult and immature morphology by Rehn (2003) (Figure 6.31).

The order is defined by numerous traits discussed by Rehn (2003) as well as others. They include the development of a pterostigma, the formation of a nodus (albeit somewhat weak in some of the extinct, basal suborders), a complete absence of vein CuP beyond its attachment to CuA at the wing base, the presence of an arculus (Figure 6.32), the reduction of the thoracic terga, a mespisternum nearly touching the wings, direct flight muscles that power the wings out of phase with each other, a prehensile labial mask in naiads (Figure 6.33), a bristle-like antennal flagellum (convergent with Ephemeroptera), and pronounced skewness of the thorax. The degree of thoracic skew is variable within the order; in fact, the extremely oblique thorax of damselflies is a defining feature for that suborder (Needham and Anthony, 1903; Rehn, 2003). Although entirely paleopterous (i.e., unable to flex the wings posteriorly so that they fold *over* the abdomen during rest, usually flatly), damselflies bring their wings together during rest but over the abdomen (Figure 6.35). In damselflies the dorsal surface of the thorax nearly faces to the rear such that the wing apices are directed to the tail end.

Perhaps the most remarkable trait for the order is, however, the suite of modified male copulatory structures (Figure 6.34). The male terminalia have evolved into grasping appendages, while the actual copulatory organs are distant, on the ventral surface of abdominal segments 2 and 3. Males still produce sperm and emit sperm from a gonopore on the ninth abdominal segment at the tip of the abdomen, but the sperm must be transferred to the secondary genitalia before copulation is initiated. As such, reproduction in Odonata is far from simple. It begins when a female enters the territory of a male. Territorial males occur in two types: perchers and fliers. Fliers incessantly patrol a particular habitat, whereas perchers make regular, short exploratory flights from a fixed, local position, defending their territories from conspecifics and at times from males of other species. Conspecific males and females first recognize one another by their flight behavior, followed by coloration and overall body shape. Sometime during this watch a male will transfer sperm from his gonopore at the tip of the abdomen to his secondary genitalia underneath abdominal segment 2. Males will then grasp females as quickly as a positive identification can be made, generally in flight. The male will then attempt to align himself in tandem (*tandem-linkage*), grasping the female behind her head (in Epiprocta) or on the prothorax (in Zygoptera) with his terminalic claspers. The male may hold the female in this manner for several minutes to several hours, until she initiates copulation. Copulation takes place when the female extends her abdomen underneath her and forward, and her genitalia interlock with the male secondary genitalia, thereby transferring the sperm. Such a mating couple forms a characteristic *copulation wheel* (Figure 6.35).

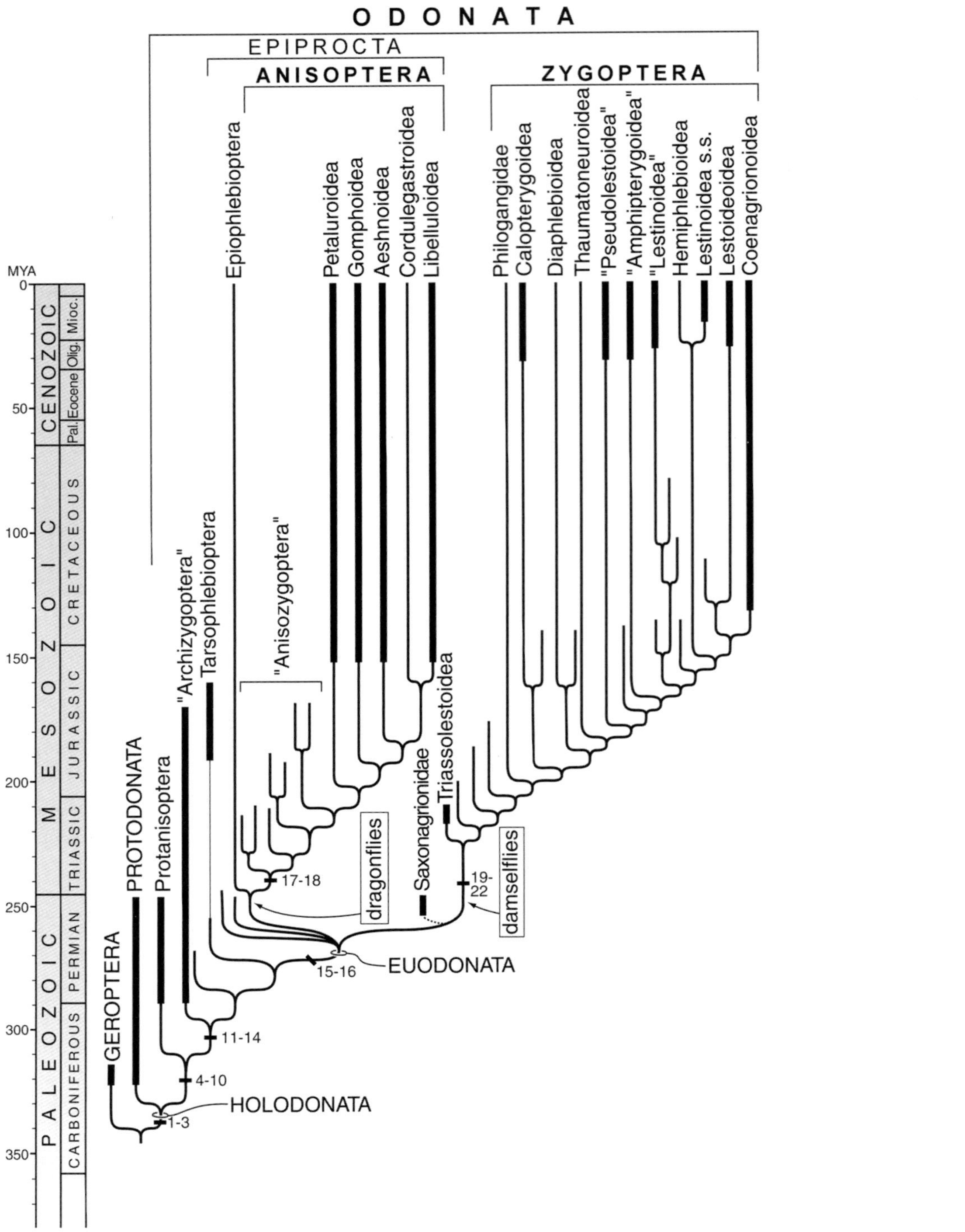

6.31. Phylogeny of the Odonatoptera (living odonates and their extinct relatives), with significant characters indicated (Table 6.2). Modified after Rehn (2003).

Interestingly, the male copulatory organ is designed not only to insert sperm but also to remove it. Females are not monogamous and sperm competition between competing suitors can be intense. The penal structures of males are modified in different lineages to scoop any competitors' sperm out of the female's bursa copulatrix before depositing his own. Alternatively, some males will simply pack competitors' sperm tightly into the female such that his own will be accessible during the fertilization of eggs prior to oviposition. This strategy is quite successful because sperm from the last male are more likely to encounter the egg as it passes the fertilization pore during oviposition. Males of some species further ensure the success of their progeny by forcibly guarding females after copulation and until the female has deposited her eggs. This guarding behavior can even take the form of essentially holding onto the female until she has oviposited, or even forcing eggs out (e.g., Libellulidae) by using his abdomen to thrust the female's abdomen into the water. It is

TABLE 6.2. Significant Characters in Odonatoptera Phylogeny[a]

1. Proximal process on posterior articular plate
2. MA fused with RP, base of M vestigial or fused to stem of Cu
3. Thorax slanting (slightly in Protodonata, more so in most Odonata)
4. Formation of nodus (albeit incipient in basal suborders)
5. CuP absent beyond attachment to CuA
6. Presence of arculus
7. Reduction of thoracic terga
8. Mesepisterna nearly touching dorsally
9. Prehensile labial mask in naiad
10. Antennal flagellum reduced ("aristate")
11. More completely formed nodus
12. Formation of true pterostigma
13. Loss of secondary branches in anal vein system
14. Fusion of posterior anal vein with posterior wing margin
15. Formation of posterior arculus at RP-MA separation
16. Subdiscoidal vein present
17. Frons bulbous
18. Epiprocts enlarged to form appendage of male claspers
19. Head transversely elongate
20. Absence of epiprocts in adults
21. Thorax strongly oblique
22. Three caudal gills in naiads

[a] Numbers correspond to those on phylogeny, Figure 6.31.

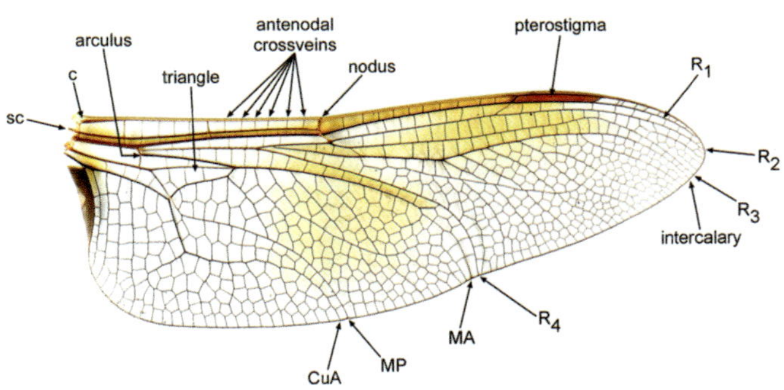

6.32. Representative odonate fore wing.

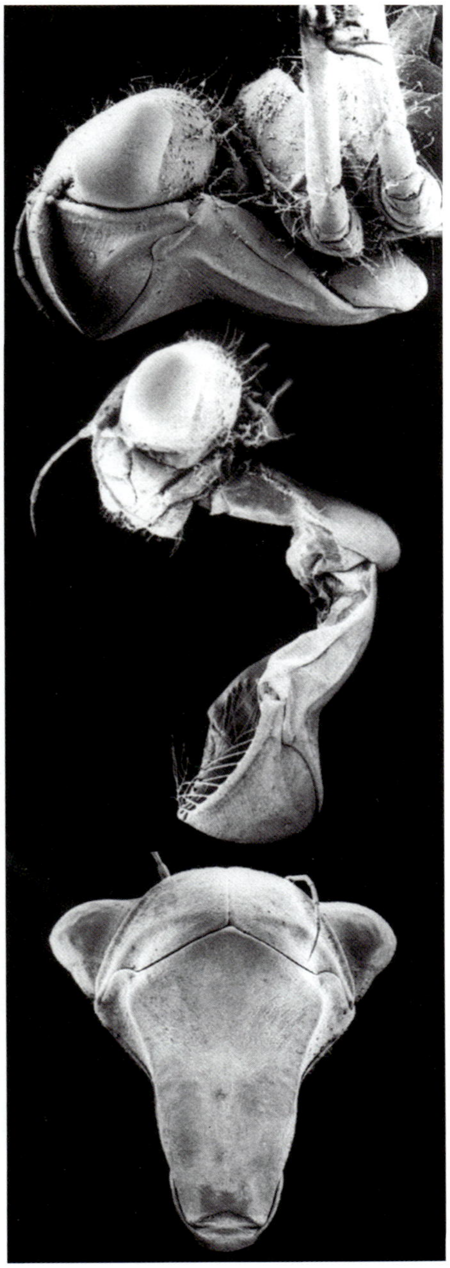

6.33. Odonate naiads have a prehensile labial mask for capturing prey. The labium unfolds, spine-like palps impale the prey, and the labium quickly folds back to the mouth with the prey. Scanning electron micrographs. Photos: W. Wichard.

unclear at what point in the evolution of Odonata (or Odonatoptera) this remarkable system of mating first came about.

Females deposit eggs directly into the water or in vegetation near fresh or brackish waters; some species prefer phytotelmata (e.g., water captured in plants, such as in the tanks of bromeliads). Development proceeds quickly through the earliest instars (total number ranging from 11 to 13 instars), which rely mostly on nutrition from stored yolk while the young refine their predatory behavior. After the second or third instar, the naiads become more aggressive and adept at prey capture. Odonates use a *prehensile mask*, or modified labium, for prey capture (Figure 6.33). Prey are generally detected visually and captured by a rapid extension of the mask. Spine-like palpi at the apex of the labial mask impale the prey, and the folding labium draws the prey into the mandibles and mouth. Depending on the size of the species, prey ranges from small invertebrates (typically arthropods) to larval fish and amphibians. Aquatic respiration is achieved through the integument, supplemented in Epiprocta by a rectal chamber lined with gill pads or by long, external anal gills in Zygoptera. Eventually the naiad crawls from the water onto nearby vegetation to molt to the adult (Figure 6.36).

Hunting is also well developed in adults that, unlike mayflies, continue to feed and are, in fact, voracious aerial predators. Hunting behavior is slightly different between major lineages and once again breaks down into the fliers versus

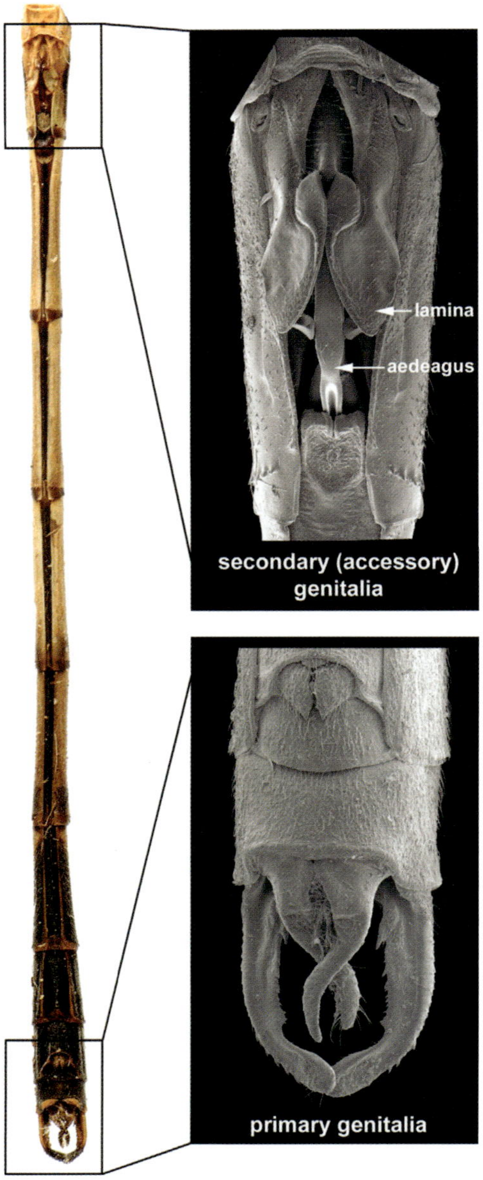

6.34. The primary and secondary male genitalia of a damselfly (Zygoptera). Before mating the male will transfer sperm from the primary genitalia to his secondary genitalia. Photo- and scanning electron micrographs.

6.35. A damselfly mating wheel. The male (above) uses his primary genitalic claspers to grasp the female (below) by the neck and the female becomes impregnated by coupling with his secondary genitalia. Photo: S. Marshall.

6.36. The exuvium of a dragonfly naiad. Before emergence the naiad climbs up on a stem; even the chitinous tracheae (the light filaments here) are shed. Photo: V. Giles.

the perchers. The slanted thorax proves to be advantagous for prey capture. The oblique thorax, while pushing the wing bases backward, thrusts the legs forward. The legs are elongate and beset with numerous spines that together can form a basket of sorts. This allows individuals to grasp and control a victim effectively while aloft and to snag midges and other small insects in flight. Fliers, not surprisingly, capture and consume their prey while in flight. Perchers, which include most damselflies as well as most Gomphidae, Petaluridae, and Libellulidae, capture prey and then return to their roost to feast. Some damselflies pluck small arthropods off stems, and the long, tropical pseudostigmatines hover quite nicely while plucking insects out of spider webs. All types have visual acuity that is remarkable for arthropods. In fact, Odonata have the largest compound eyes (these occupy nearly the entire head), which also possess the most facets of all insect eyes.

Subordinal Relationships and Early History

While the odonatopteran lineage is old, dragonflies (Epiprocta) and damselflies (Zygoptera) in the strict sense are not all that more ancient than many other insect lineages (see discussion that follows). Odonata in its broader sense stems well into the Paleozoic, based on suborders that are basal to a monophyletic lineage consisting of the Zygoptera and Epiprocta. The Odonata as a whole have had a remarkably successful geological history, and although still quite diverse today, they were equally diverse, if not slightly more so, in the past.

Suborder Protanisoptera. This group includes the extinct families Ditaxineuridae and Permaeschnidae (Carpenter, 1931, 1992) from the Early to Late Permian, which disappeared by the End Permian Event. Species have been recovered from deposits ranging from the central United States to Russia and Australia, and the suborder was likely cosmopolitan in distribution during its day. Unfortunately, protanisopterans are known only from their wings; the bodies and naiads remain to be discovered. They had relatively broad wings, nonpetiolate wings; the forewing being more slender than the hind wing with its broader base. The distinctive odonatan nodus was present but poorly formed, and the nodal crossvein was not completely developed (though the wing margin possessed a notch where such a nodal point would later appear). Similarly, the arculus was not entirely formed yet, and the "pterostigma" was merely a diffuse precursor crossed by the radial vein. Protanisoptera is currently held as the basalmost group of Odonata, all other members being united by a more complete formation of the nodus (albeit still somewhat incipient in Archizygoptera); a formation of a true pterostigma; a loss of secondary branches in the anal vein system, leaving just the anal brace; and a fusion of the posterior anal vein with the posterior wing margin (see also Rehn, 2003).

Suborder Archizygoptera. Basal relationships among early odonate families are contentious and continue to fluctuate in studies. We have therefore taken a conservative position herein and included the Protozygoptera within the Archizygoptera (as was done by Carpenter, 1992). The Archizygoptera are the first suborder to truly take on a definite odonate appearance. The group is perhaps paraphyletic with respect to all other Odonata, and resurrection of Protozygoptera and other putative suborders may be warranted as future studies continue to resolve relationships among families, some of the best know of which are the extinct families Kennedyidae, Permagrionidae, Protomyrmeleontidae (not to be confused with antlions of the Neuroptera!), Permolestidae, Permepallagidae, and Batkeniidae. Archizygopterans occurred throughout Europe, today's northern and central Asia, North America, Australia, and the Falkland Islands from the Permian through the Jurassic. Overall most species were relatively small and had petiolate wings resembling those of damselflies (hence the name Protozygoptera). Unlike the Protanisoptera, Protodonata, and Geroptera, the Archizygoptera lacked a precostal area (similar to other Odonata); however, they had an incipient nodus (more closely approximating that of other Odonata) and a completely developed pterostigma, although the arculus remained undeveloped in most taxa.

Suborder Triadophlebiomorpha. This group derives from the Triassic of central Asia and has at times been confused as a post-Paleozoic member of the Protodonata, owing to the loss of the pterostigma (e.g., Grauvogel and Laurentiaux, 1952). Moreover, based on incomplete specimens, it was believed that the nodus was similarly absent, once again a primitive feature of Protodonata. However, the nodus was actually present in these insects, and an arculus was similarly developed. The wings, like Archizygoptera, tended to be petiolate although with crowding and fusion of longitudinal veins at the wing base. A complete formation of the nodus, including aligned nodal crossveins, and the formation of the discoidal cell unite Triadophlebiomorpha with all other Odonata. The group contains the Triassic families Triadophlebiidae, Triadotypidae, Mitophlebiidae, Zygophlebiidae, and Xamenophlebiidae.

Suborder Tarsophlebioptera. The extinct family Tarsophlebiidae has the distinction of being the sister group to the remainder of the modern Odonata. Although presently known only from the Jurassic (Figure 6.37), the tarsophlebiids likely occurred as early as the Permian owing to their phylogenetic position. *Tarsophlebia* and its relatives appear to be the sister group to Zygoptera + Epiprocta based on a costal triangle in the wing and vein Sc turning sharply perpendicular to the long axis of the wing before meeting the costa (i.e., the completed formation of the nodus), among other traits discussed by Bechly (1996) and Rehn (2003).

The remaining two suborders are what most students of entomology truly think of as Odonata. The damselflies (Zygoptera) and dragonflies (Epiprocta) are supported as monophyletic in almost every study of odonate relationships (see earlier references). Defining features of the group include the development of the posterior arculus at the point of separation between veins RP and MA and the presence of a subdiscoidal crossvein.

Suborder Zygoptera. The damselflies are monophyletic and defined by the broad head, with the compound eyes separated by more than their own width; an absence of epiprocts in adults; a strongly oblique thorax; and the presence of three caudal gills in naiads. Damselflies are also characterized by similar, petiolate fore- and hind wings (Figure 6.38). The group occurs throughout the world, with species in the tropics reaching particularly large sizes (e.g., up to 6 inches in length and 8-inch wingspans in some pseudostigmatines).

6.37. A primitive odonate, *Turanophlebia,* from the famous Jurassic limestone of Solnhofen, Germany. Tarsophlebioptera were primitive odonates, and although they looked like dragonflies (Epiprocta), they were actually relatives of both Epiprocta-Zygoptera together. NHM In. 46336; forewing length 38 mm.

The earliest definitive Zygoptera are of the family Triassolestidae from the Triassic of South America, Australia, and Central Asia (Pritykina, 1981; Carpenter, 1992). Although called damselflies, the Late Permian Saxonagrionidae were actually more basal than the Zygoptera + Epiprocta lineage (Nel *et al.*, 1999). Damselflies are abundant in Cretaceous and Tertiary deposits (Figure 6.39), although those in fossiliferous resins are rare (Figure 6.40).

Suborder Epiprocta. The suborder Epiprocta was established by Lohmann (1996) for the former suborders Anisoptera and "Anisozygoptera." Epiproctans are what most individuals consider true dragonflies (Figures 6.41, 6.42). These insects are robust with huge compound eyes that touch (or nearly so) and nearly occupy the entire head, and their wings are held out to the sides at rest. Defining features of the suborder include the large, bulbous frons; the enlargement of the epiprocts to form an inferior appendage as part of male terminalic grasping organ; and the development of a rectal chamber in the naiad lined with rectal pads for respiration. Additional features based on wing venation are discussed by Rehn (2003).

INFRAORDER EPIOPHLEBIOPTERA. The infraorder Epiophlebioptera of Epiprocta contains only the family Epio-

6.38. Representative Recent damselflies. To the same scale.

6.39. An Early Cretaceous damselfly, *Eoprotoneura hyperstigma*, from the Santana Formation in Brazil. AMNH 44203; forewing length 18 mm.

6.40. A relatively modern damselfly in Miocene amber from the Dominican Republic. Morone Collection, M0219.

phlebiidae. The family consists of only two species, *Epiophlebia superstes* from Japan and *E. laidlawi* from the eastern Himalayan region (Nepal). Epiophlebiids are "living fossils," which combine physical attributes of both Zygoptera and Anisoptera (Asahina, 1954). Overall the body of *Epiophlebia* resembles that of any other dragonfly. However, like the damselflies, the wings are petiolate, and the fore- and hind wings are relatively similar in overall shape. Both species occur in mountainous areas and breed in fast-flowing waters (possibly even in high-altitude waterfalls in the case of *E. laidlawi*). Epiophlebiids, unlike other Epiprocta, have functional ovipositors (similar to Zygoptera and more basal extinct lineages), and even though the naiads also have a rectal chamber lined with gills, they are apparently not capable of the "jet" propulsion so characteristic of Anisoptera. Epiophlebiids are not known from the fossil record, but their position as the living sister group to Anisoptera suggests that the lineage is at least as old as the Triassic (based on early "anisozygopteran" fossils).

INFRAORDER ANISOPTERA. The infraorder Anisoptera is used here in an expanded sense to include the large part of the former "Anisozygoptera." Anisozygoptera is a paraphyletic stem group to Anisoptera, and, despite some recent contentions (e.g., Carpenter, 1992), it does not include the living, relict family Epiophlebiidae. Instead, the numerous fossil Anisozygoptera are a stem-group assemblage leading to the Anisoptera (Rehn, 2003), which is now considered an infraorder of Epiprocta (Lohmann, 1996). The Anisozygoptera, excluding Epiophlebiidae, is linked with Anisoptera based on a hind wing that is broader than that of the forewing and which has a different venation, as well as a costal nodal "kink" (a small extension of the costal vein along the nodal crossvein).

Defining features of the Anisoptera in its traditional sense (here called Eteoanisoptera) are the following: a vestigial ovipositor; the ability of the naiad to propel itself using water pressure from its rectal chamber ("jet" propulsion); a pterostigmal brace vein (a supporting vein under the inner edge of the pterostigma); a distinctive anal loop (a series of distinctive cells demarcated in the anal region of the wing); a small, frequently darkened area of veinless membrane at the base of the wings near the anal region (often homologized with the jugum, though it appears to be secondarily evolved); a secondary "CuP" vein that is associated with an expanded

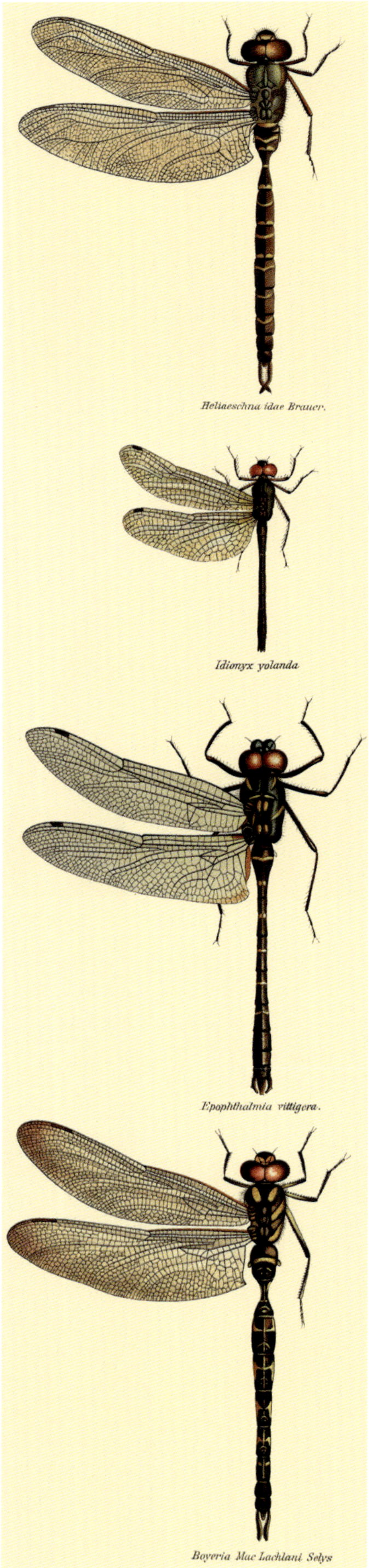

6.41 (left). Representative Recent dragonflies. From *Genera Insectorum*.

6.42 (right). Representative Recent dragonflies. From *Genera Insectorum*.

6.43. A stunning Jurassic dragonfly, *Libellulium longialata*, from Solnhofen, Germany. NHM In. 28201; wingspan 140 mm.

6.44. Not all extinct odonatoids were giants; the smallest odonate that ever lived was *Parahemiphlebia mickoleiti* from the Early Cretaceous of Brazil. It is shown at approximately life size in the box. AMNH; wingspan 18.5 mm.

anal region in the hind wing; and division of the discoidal cell into two triangular cells in the fore and hind wings. The Anisoptera includes the eteoansiopteran superfamilies Aeshnoidea, Libelluloidea, Petaluroidea, Gomphoidea, and Cordulegastroidea; extinct anisozygopteran lineages include families such as Archithemistidae, Asiopteridae, Euthemistidae, Heterophlebiidae, Isophlebiidae, Karatawiidae, Liassophlebiidae, Oreopteridae, Progonophlebiidae, and Turanothemistidae. Relationships among these extinct families is uncertain, although Lohmann (1996), Bechly (1996), and Rehn (2003) all indicate them to be paraphyletic to Eteoanisoptera. The earliest fossils are from the Triassic, with definitive Eteoanisoptera in the Early Jurassic (e.g., Liassogomphidae of western Europe). Fossils of dragonflies are abundant (e.g., Figures 6.43, 6.44), including those of the aquatic naiads (e.g., Figures 6.45, 6.46), and are rich in characters that allow deciphering their relationships to modern odonate lineages.

6.45. An impressive naiad of the dragonfly *Pseudomacromia sensibilis* (Macromiidae) from the Early Cretaceous of Brazil. The long legs of this naiad suggest that it climbed amongst submerged vegetation to ambush prey. The long antennae may have served to detect prey in dense growth. AMNH 44205; length 21.3 mm.

6.46. A dragonfly naiad in Early Cretaceous limestone from the Santana Formation in Brazil. AMNH; length 21 mm.

7 Polyneoptera

NEOPTERA

This group of winged insects appeared in the earliest Late Carboniferous (early Bashkirian) and subsequently radiated into every imaginable terrestrial and freshwater niche. Features of the Neoptera (discussed earlier, and briefly reiterated here) include wing flexion via special muscles attached to the third axillary sclerite (Figure 4.6), the formation of a median plate in the wing base, the radial vein never forking from the wing base, and the development of a gonoplac (i.e., the third "valvula").

Why should this group have been so successful among the flying insects? The ability to flex the wings over the abdomen when at rest is much more significant than it might at first appear, and it is a remarkable quirk of nature that a few tiny muscles attached to a minute sclerite should be one of the main reasons for the great success of insects. Wings are vital means of dispersal and thus require protection; they need to be stored when not in use and to minimize damage while the insect is moving amidst leaves, under bark or rocks, or in other tight spaces. The wings themselves can also serve to protect the abdomen, which is the function of leathery forewings, or tegmina, in roaches and some orthopteridans, and the entirely sclerotized earwig tegmina and beetle elytra. The ability to adeptly control the wings when not aerial was certainly a major innovation among the flying insects, as is shown by the fact that when fossil neopterans appeared, they quickly outnumbered paleopterous insects.

Some of the earliest neopteran insects include members of the Carboniferous family Paoliidae (Figure 7.1). Paoliids were rare, large insects that have at times been placed in their own order, Protoptera (e.g., Sharov, 1966). The group is characterized by numerous primitive features, and they actually lacked any derived traits (at least none observable in preserved specimens). This family is very likely a stem group to all other Neoptera. Ten genera and 12 species are presently recorded from the early Pennsylvanian of Europe and North America (e.g., Smith, 1871; Handlirsch, 1906a,b, 1919; Laurentiaux, 1950; Kukalová, 1958; Brauckmann, 1984, 1991; Laurentiaux-Vieira and Laurentiaux, 1986; Maples, 1989). The antennae were long and filiform, and the legs were unmodified, slender, and had five-segmented tarsi. Unfortunately, the paoliids are poorly understood, with few body characters known; fortunately, however, the wings are preserved. These insects had relatively broad, homonomous wings with rich crossvenation that formed an archedictyon, and the hind wing lacked an anal fan. The absence of an anal fan in most non-neopteran lineages suggests that this is a primitive trait for Neoptera and that this feature may indeed support the monophyly of Polyneoptera (with several, independent reversals therein). Paranotal lobes were not present in paoliids, and the occurrence of these structures among some "Protorthoptera" families, particularly those allied to the Plecoptera, may be a derived trait for those lineages as well (and convergent with the paranotal lobes in Palaeodictyoptera and Geroptera). Alternatively, but a less parsimonious explanation, is that paranotal lobes are primitive for all pterygotes (and homologous to the "protolobes" of Zygentoma), in which case the loss of such lobes would have occurred independently in the Ephemeroptera lineage (including *Lithoneura*), in Holodonata, in Dicliptera + Diaphanopterodea + Eumegasecoptera + Protohymenoptera, in Paoliidae, and in Eumetabola and frequently among polyneopterans.

Neoptera is generally divided into three major lineages: Polyneoptera, Paraneoptera, and Holometabola, the latter two being sister groups and each definitively monophyletic, while strong support for a polyneopteran group is elusive.

The Polyneoptera, under one concept or another, have traditionally gone by the name of Paurometabola (exclusive of several orders), Orthopteroidea, or simply the orthopteroid insects. This is a disparate group of generalized and specialized unlikely relatives, and indeed, the group is ill-defined and may not be natural. Some superordinal groups appear well supported, particularly the Plecopterida, Orthopterida, and Dictyoptera (discussed later), while the orders Dermaptera, Grylloblattodea, and Mantophasmatodea are

7.1. *Kemperala* (Paoliidae) from the mid-Carboniferous of Germany. Paoliids, which occurred from the mid- to Late Carboniferous, were among the earliest known winged insects, but they also folded their wings over their backs and may have been the earliest neopterans. Photo: C. Brauckmann.

difficult to place into or near any one of these lineages. Even though our understanding of relationships among hexapods is congealing, the ancient origins and relationships among early polyneopterans has been difficult to interpret. The polyneopterans may represent the first major radiation of neopteran insects (note here that we say "neopterans," not "neopterous," because neopterous insects appeared convergently within the superorder Palaeodictyopterida, i.e., the Diaphanopterodea). The polyneopteran radiation began in the Paleozoic; however, the origination of what we recognize as the modern polyneopterous orders did not occur until later as a second, post-Permian radiation.

Polyneopterans stem from the late Paleozoic and are most widely known by the heterogeneous assemblage of Paleozoic families united into the "Protorthoptera." At present, the "Protorthoptera" are a cloud of genera and families, among which all other orders of Polyneoptera have arisen. Thus, "protorthopterans" have been a receptacle for any Paleozoic or Early Mesozoic polyneopterous insect not readily assigned to one of the modern orders; it is a polyphyletic conglomeration of unrelated families that retained traits primitive not only for Polyneoptera but in many cases for Neoptera as a whole. The recognition of this problem is not new; even Tillyard (1928d) recognized that the "Protorthoptera" were an "*omnium gatherum*." The most fruitful work on these extinct insects will clarify phylogenetic relationships for the families and genera of "Protorthoptera." Basal Neoptera diversified during the Late Paleozoic, and soon after the end of the Permian, the Polyneoptera appear to have coalesced into those clades we recognize today. Thus, most of the polyneopterous orders we are familiar with today are Early Mesozoic in origin. Once the true relationships of many of the protorthopteran lineages are elucidated, it will be necessary to resurrect many of Handlirsch's and Tillyard's orders, such as Cnemidolestodea. Interesting groups are already recognizable from "Protorthoptera," which provide insight into the earliest differentiation of higher groups within the insects, but relationships remain veiled by unnatural, classificatory edifices (e.g., Rasnitsyn and Quicke, 2002).

WHAT ARE POLYNEOPTERA?

Fundamental to our understanding of relationships within Polyneoptera is the question surrounding their monophyly. The defining feature for the group is the expansion of the anal region in the hind wing by the addition of numerous anal veins (Figure 7.2), apparently secondarily reduced in Zoraptera + Embiodea and unknown for the apterous orders Grylloblattodea and Mantophasmatodea. Interestingly, recent molecular studies have also supported the Polyneoptera to some extent (e.g., Wheeler *et al.*, 2001). Additional traits uniting polyneopterans may also be present in the neuroanatomy and other internal organs (e.g., Ali and Darling, 1998; Pass, 2000).

Reductions of various structures commonly obscure homologies in the Polyneoptera, but fortunately fossils help to clear some of the confusion. For example, three-segmented (*trimerous*) tarsi occur in some Orthoptera, one lineage of Phasmatodea (*Timema*), extant Dermaptera, extant Plecoptera, and Embiodea. By examining solely Recent species one might unite these groups on this distinctive trait (e.g., Grimaldi, 2001). However, fossils of stem-group Plecoptera and stem-group Dermaptera all primitively retained five-segmented tarsi while also having features shared with each of their crown groups. In other words, the reduction to three-segmented tarsi has occurred independently *within* each of these orders. It is imperative that a paleontological perspective be applied when attempting to resolve the relationships of highly modified survivors of an ancient radiation, in this instance, ones that were Early Mesozoic or latest Paleozoic in origin (Gauthier *et al.*, 1989).

For the moment, three groups are readily definable within the Polyneoptera. These are the superorders Plecopterida (stoneflies, webspinners, and zorapterans), Orthopterida (walking sticks, crickets, grasshoppers, wetas, and their relatives), and Dictyoptera (roaches, mantises, and termites) (Figure 7.3). Dictyopteran monophyly and internal relationships are elaborated upon later. The earwigs (Dermaptera), African rock crawlers (Mantophasmatodea), and ice crawlers (Grylloblattodea) remain difficult to place within Polyneoptera. Grylloblattodea and Mantophasmatodea are probably basal orthopteridans, but the loss of wings in both of

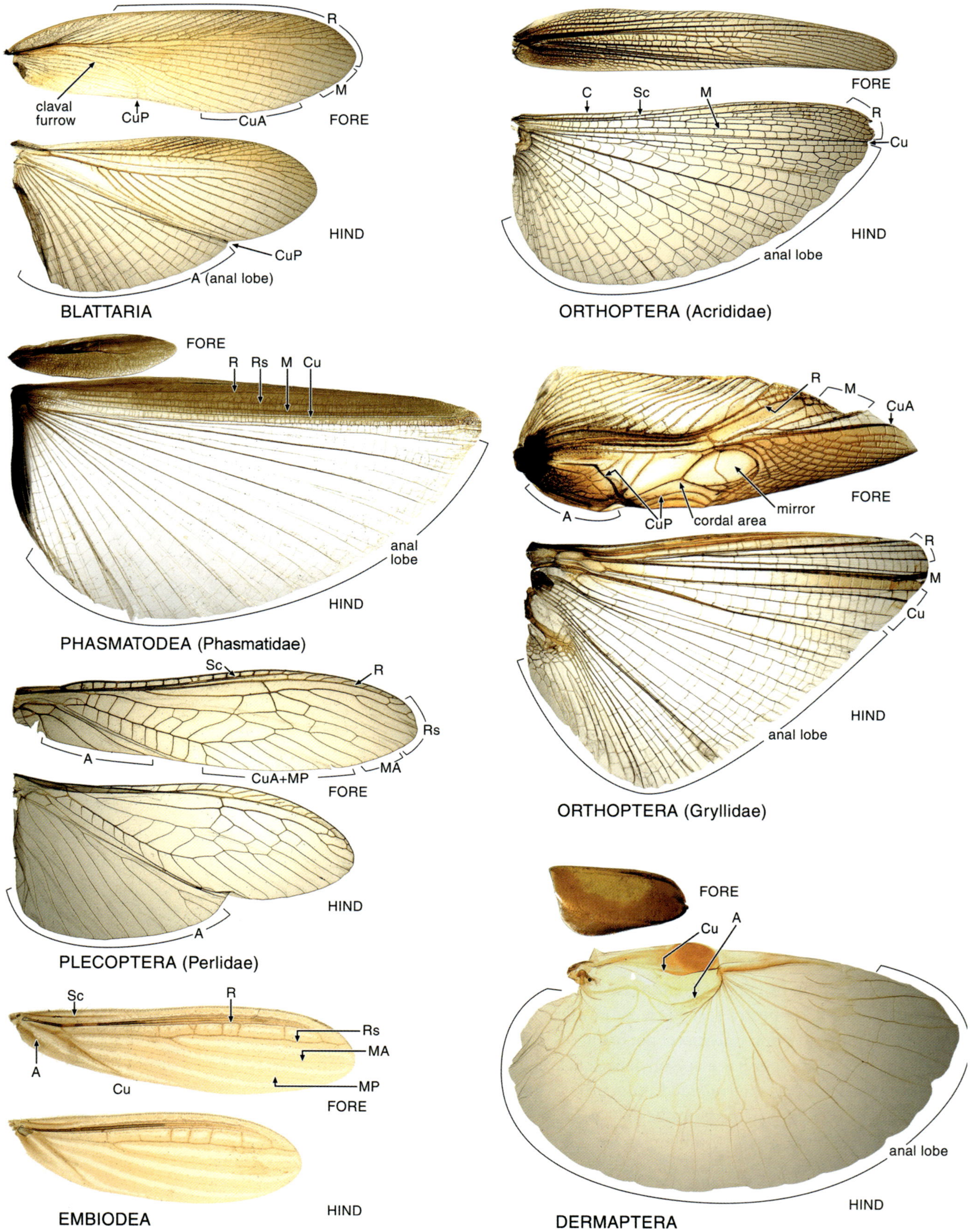

7.2. Wings of representative living polyneopterous insects, showing the diversity of wing venation. Not to same scale.

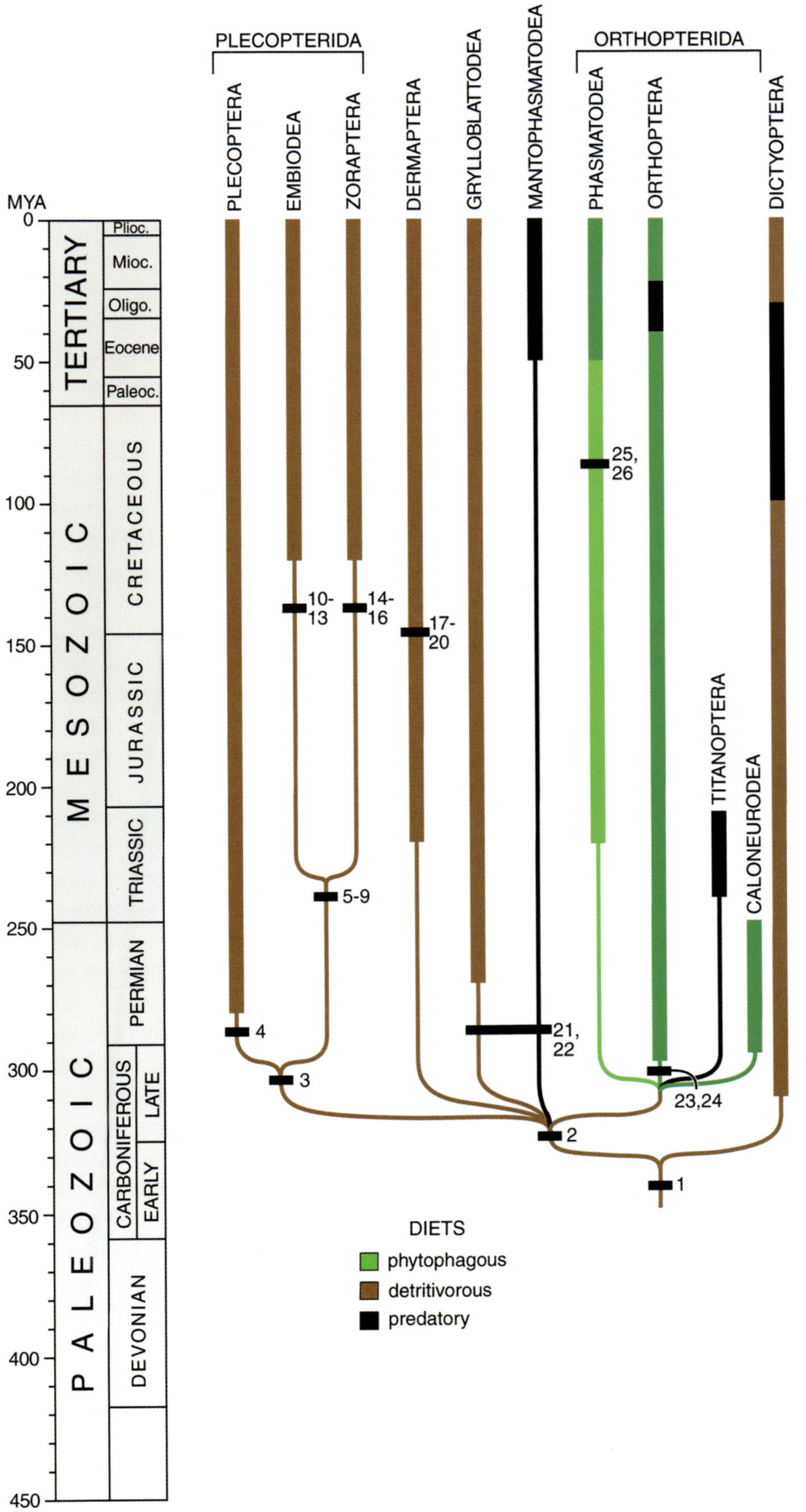

7.3. Phylogeny and the evolution of diets of polyneopterous insects, with significant characters indicated (see Table 7.1). The thick lines indicate the known extent of the fossil record. From various sources (see text).

TABLE 7.1. Significant Characters in Polyneopteran Phylogeny[a]

1. Hind wing with expansive, fanlike anal lobe
2. Head prognathous
3. Ovipositor lost
4. Nymphs aquatic, with filamentous tracheal gills on thorax/abdomen, or both
5. Anal lobe of hind wing lost
6. Cercus reduced to one to two segments
7. Wings dehiscent
8. Hind femora enlarged, with distinctive musculature
9. Communal behavior
10. Secrete silk with enlarged fore basitarsus having numerous silk glands
11. Wing veins lie in blood sinuses that inflate
12. Distinctive, reduced wing venation
13. Asymmetrical male terminalia
14. Highly reduced, distinctive venation
15. Wingless, blind morphs; winged, eyed morphs
16. Hind femur with thick spines
17. Ovipositor highly reduced
18. Forewings tegminous, short (hemeltyra)
19. Hind wings with unique folding mechanism, very large anal fan
20. Cerci modified into sclerotized, unsegmented forceps
21. Ocelli lost
22. Apterous (all individuals, not as morphs) + molecular characters
23. Pronotum large, with cryptopleuron
24. Saltatorial hind legs, large hind femur
25. Prothoracic defensive glands
26. Male with vomer (tenth sternite, used in mating)

[a] Numbers correspond to those on phylogeny, Figure 7.3.

these groups, perhaps in the common ancestor of both, obscures their relationship. Similarly, the placement of earwigs is contentious, and rightly so, with much of the debate centering on an extinct group known as the Protelytroptera. Molecular studies suggest an affinity of earwigs with the Dictyoptera (e.g., Wheeler *et al.*, 2001) but the only morphological support for this is the enlargement of the coxae; the overlapping sternites referred to by Hennig (1981) is likely plesiomorphic.

PLECOPTERIDA

Three disparate relatives within the Polyneoptera are the very generalized stoneflies and the highly specialized webspinners and zorapterans. The placement of these three orders in the grander scheme of the insects has been troublesome for many years. The crux of the problem is that each is without clear affinity to any other order and each is a recent remnant of an otherwise ancient lineage. As alluded to before, it is exactly for such groups that paleontological data are hypothesized to be most critical for resolving relationships (Gauthier *et al.*, 1989). The plecopterid orders are united principally by numerous reductions from the polyneopteran groundplan. A prognathous head capsule is a potential synapomorphy for the group, but more likely this feature is a more inclusive trait uniting Plecopterida to Orthopterida (hypognathous in some Orthoptera). Plecopterids also share the reduction of the ovipositor and suppression of male styli. Ovipositor reduction is not homologous to the distinctive reduction noted for Dictyoptera. Furthermore, Early Mesozoic and Paleozoic stem-group Dictyoptera (i.e., blattodean "roachoids") had well-developed ovipositors (Figures 7.65–7.67). Instead, in Plecoptera, Embiodea, and Zoraptera the ovipositor is not merely reduced as it is in Dictyoptera (Figure 7.62) but actually entirely lost. Correspondingly, putative stem-group plecopterids from the Paleozoic appear to have had well-developed ovipositors. Alternatively, the reduction in Plecoptera and Embiodea + Zoraptera may be independently derived (discussed further later in this section), thereby eroding the use of this trait to support the more inclusive clade Plecopterida. Other features are not universally understood across plecopterid orders and therefore have limited significance so far, the most notable being the median, ventral holes (*ostia*) in the dorsal blood vessel (presently unknown in Zoraptera).

Zwick (1973, 2000) and Hennig (1981) both dismissed a close relationship of Plecoptera and Embiodea, stressing the many remarkably primitive neopteran traits in stoneflies, and they even suggested that Plecoptera were the living sister group to the remainder of Neoptera. It is ironic that Hennig should have argued on such grounds since plesiomorphic traits are not indicative of relationship. Instead, the available evidence suggests that stoneflies are in the Polyneoptera on the basis of the enlarged anal fan in the hind wing as well as sperm and ovarian ultrastructure (e.g., Büning, 1998; Fausto *et al.*, 2001). Hennig (1981) noted that the argument for Polyneoptera monophyly based on the anal fan was weak because Sharov (1966) derived the Paraneoptera and Holometabola from among the polyneopterous orders (i.e., Polyneoptera was paraphyletic to all other Neoptera in his system). This actually reflects more on a confused phylogenetic interpretation by Sharov than on the homology of the polyneopteran anal fan.

The controversial placement of stoneflies as the living sister group to Paraneoptera + Holometabola, also known as the Planoneoptera hypothesis (Ross, 1955; Hamilton, 1972a,b), was based upon the structure of the mesotrochantin, which is a slender, straplike structure in the latter groups (although highly modified through much of the Holometabola). Interestingly, this same trait may support the Plecopterida as we view them. In fact, Zoraptera and Embiodea both possess similar mesotrochantins while all three differ from the Holometabolan condition, in which the mesotrochantin is rigidly fused at its base to the episternum and is retained merely as a small projection articulating with the mesocoxa. It may be significant that a slender "eumetabolan-like" mesotrochantin that is separated from the episternum by membrane or a sulcus is present in Plecoptera, Embiodea, and Zoraptera. Dermaptera are the only other polyneopterans that have a similar trochantin.

Zoraptera can be excluded from the Paraneoptera based on their primitive retention of cerci, the presence of a first abdominal sternum, and the broad "orthopteroid" lacinia. Zoraptera have also been placed as a living sister group to Paraneoptera or as basal Eumetabola (i.e., Paraneoptera + Holometabola). Characters supporting their Eumetabola relationship include the development of a *jugal bar* (Hamilton, 1972a), which has been oddly assigned to the zorapterans at times (e.g., Wheeler *et al.*, 2001) when it is actually absent. Alternative hypotheses are entirely unfounded, specifically the theory that tubercles on the furcasternum of Plecoptera approximate or transform into the tricondylic articulation seen in Holometabola (Adams, 1958; Rasnitsyn, 1998; but see Engel and Grimaldi, 2000, for Zoraptera). Evolution is replete with examples where structures have been co-opted for novel functions, such as some vertebrate mandibular bones incorporated into the inner ear. However, in each of these instances, there is significant evidence not only for the basic homology but also for the criterion of continuation and what all other characters indicate.

Assuming plecopterid monophyly for the moment, Plecoptera are basal relative to Embiodea + Zoraptera. Plecoptera retain the polyneopteran anal fan, entirely lost in Embiodea + Zoraptera who both exhibit dehiscent, narrow, paddle-shaped wings without any anal region. Furthermore, the webspinners and zorapterans share asymmetrical male genitalia, apterous morphs, reduced cerci (relative to Plecoptera), tarsomere reduction (likely not homologous to tarsal reduction in stoneflies, as evidenced by stem-group plecopterans), loss of gonostyli, gregarious and cryptic life histories, and large hind femora with overdeveloped tibial depressors (in sharp contrast to the developed tibial levators of other orders) (Engel and Grimaldi, 2000).

ORTHOPTERIDA

Perhaps allied to Plecopterida by the prognathous head (secondarily hypognathous in Orthoptera), fusion of the premental lobes, and neuroanatomical traits (Ali and Darling, 1998) is the Orthopterida. This group in its strict sense is relatively free of controversy and comprises the two living orders Orthoptera and Phasmatodea and the two extinct orders Titanoptera and Caloneurodea. Orthopterids have the second valvulae reduced, concomitant development of the gonoplac (i.e., the ill-named *third valvula*) as the functional ovipositor, and an enlarged precostal field in the forewing. Attempts to incorporate other orders into the Orthopterida, such as Grylloblattodea or Embiodea, are not strongly supported at present. Embiodea has at times been argued to be the living sister group of the walking-sticks owing to the presence of an operculum on the egg as well as certain mouthpart muscles (Kristensen, 1975; Tilgner, 2000). Despite these similarities, they must be interpreted as convergence against the larger body of evidence showing that Embiodea is within Plecopterida and sister to Zoraptera (see preceding discussion), and that Phasmatodea is closely related to Orthoptera (Kristensen, 1975; Flook and Rowell, 1998; Wheeler *et al.*, 2001). Orthopterida is perhaps related to Grylloblattodea and Mantophasmatodea as these orders exhibit a similar development of the gonoplac over the second valvulae, although to a much lesser degree. The **Polyorthoptera** incorporates the Orthopterida, Grylloblattodea, Mantophasmatodea, and perhaps the Dermaptera.

Possibly included in this group is the enigmatic fossil family Chresmodidae. Chresmodids contained a single known genus and four species (Engel and Grimaldi, in prep.) that superficially resembled water striders (Hemiptera: Gerridae) with robust bodies and extremely long legs. Like water striders, chresmodids were apparently aquatic and may have skated across water surfaces, or perhaps treaded over vegetation exposed at the water surface. Chresmodids are known from the Late Jurassic of Solnhofen, Germany, and the Early Cretaceous of Spain (Montsec) (Figure 7.4), China, and Brazil (Santana Formation) (Figure 7.5). The difficulty of placing *Chresmoda* is highlighted by its classification over the last 164 years – at times being considered related to or in Gerridae (Hemiptera), Mantodea, Paraplecoptera, and Phasmatodea. Handlirsch (1906b) was the first to move the chresmodids into the Phasmatodea, and Martynov (1928) later erected a separate suborder, Chresmododea, believing them to be stem-group phasmatodeans (a position also held by Sharov, 1968). Popov (1980), highlighting the fact that chresmodids were aquatic and resembled gerrids (like Oppenheim, 1888, before him), considered the family as belonging to the Gerromorpha (Hemiptera). However, Ponomarenko (1985) noted the presence of cerci and pentamerous tarsi, traits that excluded the Chresmodidae from the Paraneoptera, and

7.4. *Chresmoda aquatica* (Chresmodidae) from the Early Cretaceous of Spain. Chresmodidae resembled modern water striders, but these giant, apparently semiaquatic insects were relatives of Orthoptera and Phasmatodea. Photo: X. Martínez-Delclòs.

7.5. Chresmodidae from the Early Cretaceous of Brazil, which is the first Western Hemisphere species of this enigmatic, Mesozoic lineage of insects. AMNH; body length 37 mm.

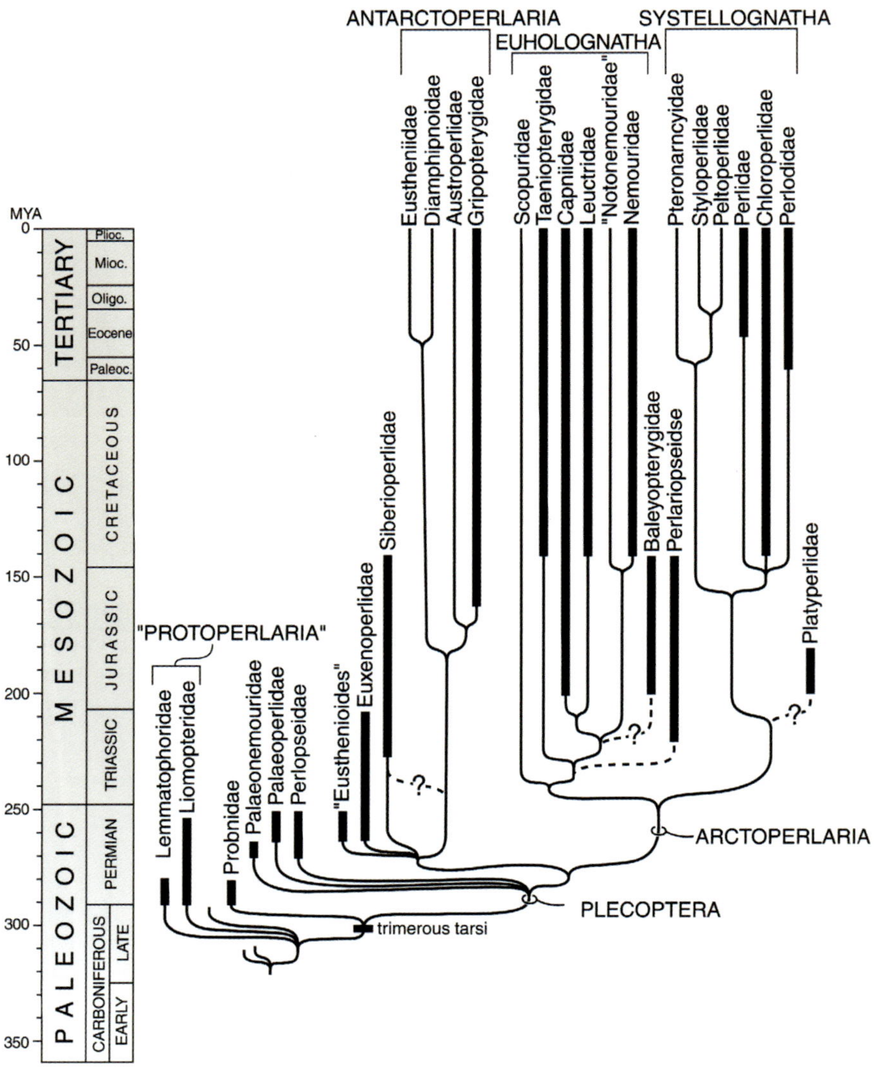

7.6. Phylogeny of Plecoptera. Thick lines indicate the extent of known fossils. Based on Zwick (1998).

alternatively placed the family in the extinct Paraplecoptera (which are stem-group polyneopterans). Rasnitsyn (in Rasnitsyn and Quicke, 2002) considered chresmodids as of uncertain affinity among polyneopterans, dismissing any relationship with Phasmatodea.

PLECOPTERA: THE STONEFLIES

Stoneflies are unique among basal neopterans for their aquatic life history, the nymphs of which live entirely in freshwater. Nymphs of a few species are terrestrial, but they live in very wet or moist habitats. Owing to the unique development of strong abdominal muscles, nymphs are capable of swimming via lateral undulations of the body, much like fish (Zwick, 1980, 2000). Species are principally detritivores or omnivorous; few species are strictly carnivorous. Many nymphs are grazers, gleaning the surfaces of stones for algae. The approximately 2,000 species are segregated into 16 modern families, which divide themselves into Laurasian and Gondwanan lineages (Zwick, 1973, 2000) (Figure 7.6). Major works on the order include those by Claassen (1940), Illies (1966), and Zwick (1973, 2000).

Adult stoneflies are primitive-looking neopterans (Figure 7.7), which has contributed to the notion that the order is basal among neopterous insects. As discussed earlier, however, the Plecoptera possess some presumed apomorphies with other polyneopterous groups. Additionally, an aquatic life history for stonefly nymphs (Figure 7.8) is often considered to be primitive owing to aquatic naiads in Ephemeroptera and Odonata, when in fact it may partly define monophyly of Plecoptera. One hypothesis of Plecoptera as nested within Polyneoptera (e.g., Ali and Darling, 1998; Wheeler *et al.*, 2001) suggests a transition to aquatic living in this group, which is of significant interest. Various aspects of stonefly biology have been equated with the groundplan condition for all Neoptera, and some studies have focused on stoneflies as models for understanding the

7.7. Representative adults of Recent Plecoptera. Not to the same scale.

7.8. A Plecoptera nymph.

origins of flight within insects (e.g., Marden and Kramer, 1994, 1995; Thomas *et al.*, 2000). Certainly, *the different gill types alone in Plecoptera, Ephemeroptera, and Odonata suggest independent origins of aquatic lives.* Furthermore, the placement of Plecoptera within Polyneoptera erodes the notion that stoneflies can be considered early fliers even if aquatic naiads are primitive.

Some Plecoptera oviposit during flight, merely dropping masses of eggs from the air into the water or skimming over the water surface and "washing" eggs off of their venter. The ovipositor is vestigial and is cited as a defining specialization for the order (Willmann, 1997; Zwick, 2000), but here it is considered as a more inclusive trait uniting Plecoptera with Embiodea and Zoraptera. A few species with secondarily developed ovipositors (e.g., Notonemouridae) oviposit into wet crevices. Other, less visible, synapomorphies for the order include the lopped gonads, where the anterior apices of the left and right ovaries and testes are fused in the middle (Zwick, 1973), and the presence of an accessory circulatory organ ("cercal heart") (Pass, 1987; Zwick, 2000). Modern species of the order have three-segmented tarsi, but this is not homologous with the trimerous condition seen in Embiodea, Dermaptera, Timematodea, and some other orders since Paleozoic stem-group Plecoptera had five-segmented tarsi.

The most recent cladistic classification of Plecoptera is by Zwick (2000), which found support for the two traditional suborders or lineages, Antarctoperlaria and Arctoperlaria (Figure 7.6). Limited cladistic analyses have suggested paraphyly of Antarctoperlaria, by placing Gripopterygoidea (an antarctoperlarian) basal within Arctoperlaria, but these studies require corroboration (Nelson, 1984: reanalyzed by Will, 1995). As alluded to in their names, the two suborders are divided between hemispheres; Antarctoperlaria in the Southern Hemisphere and Arctoperlaria in the Northern. Colonization of the Southern Hemisphere has occurred in some derived genera of Arctoperlaria – the family Perlidae into South America and the Notonemouridae into Australia, South America, and Africa (Zwick, 2000). The suborders are defined unfortunately by characters that are unknown from fossils or unlikely to ever be preserved as fossils, such as muscles. Antarctoperlaria is supported by the presence of a unique sternal depressor muscle of the fore trochanter and concomitant absence of the typical tergal depressor muscle of this structure as well as the presence of floriform-chloride osmoregulatory cells. Arctoperlaria is defined by "drumming," a behavior of males advertising to females. The male taps on the substrate with the tip of his abdomen, on which he typically has special modifications (called the "hammer") for producing the sound (Zwick, 1973, 1980).

The oldest stoneflies (*stoneflies in stone!*) are a few groups from the Early Permian, but these likely represent stem-group families paraphyletic to all other Plecoptera. Indeed, families such as Lemmatophoridae had terrestrial nymphs

(Figure 7.9) and, along with others such as Liomopteridae and Probnidae, appear allied to the Plecoptera while plesiomorphically retaining five-segmented tarsi. At times such families have been grouped into an extinct order, Paraplecoptera or Protoperlaria (e.g., Tillyard, 1928d; Martynov, 1938); however, recognition of such a taxon is premature and is at present defined entirely on the absence of modern stonefly apomorphies. It is possible that paranotal lobes in some of these lineages comprise a defining specialized feature of an early branch of stoneflies (e.g., Lemmatophoridae + Liomopteridae). Like other polyneopterans, the record of the order does not truly develop until the Mesozoic, from which numerous fossils are known, including early representatives of modern families (e.g., Notonemouridae). Permian records for the modern families Eustheniidae and Taeniopterygidae deserve critical evaluation, particularly because these are not basal families for the order (Zwick, 2000). The biogeographical separation of the suborders corresponds to the breakup of Pangea into Laurasia and Gondwanaland during the Late Jurassic, and these groups likely became differentiated during this time period.

7.9. Although the immatures of Early Permian Lemmatophoridae were apparently not aquatic, species such as *Lemmatophora typa* were primitive relatives of modern stoneflies. YPM 5115; length 8.5 mm.

EMBIODEA[1]: THE WEBSPINNERS

Webspinners are gregarious insects occurring principally in the tropics, though some extend into warm temperate regions (Figure 7.10). There are approximately 360 described species in approximately nine families (Ross, 1970, 2001, 2003a,b; Szumik, 2004), although the higher classification of the order is presently under investigation, and these figures will undoubtedly change dramatically in the very near future. The foremost features of the order are intimately associated with their biology. Species live in "galleries" produced by silk from glands inside the enlarged fore basitarsi (Figure 7.11). Silk is spun during all nymphal instars and by adults. The silk is extruded through specialized, hollow setae (Figure 7.11) that internally connect to unique glands of ectodermal origin. The tarsi are three-segmented but, as discussed earlier, the trimerous condition is perhaps not homologous with that seen in Plecoptera and is certainly not homologous with that in Dermaptera or Timematodea.

Another distinctly embiodean trait is the desclerotization of the longitudinal wing veins and the development of *blood sinuses* in the wings, which are hollows through which hemolymph is pumped. When hemolymph is withdrawn into the body, the wings become flexible and collapse upon themselves, even flipping over the thorax and head during backward movement through the narrow tunnels and chambers of the gallery. When flight is necessary, the wings are made more rigid by filling the sinuses with hemolymph. Females are apterous, while males may possess dehiscent wings.

7.10. A webspinner in its web. Photo: J. Edgerly.

[1] **Lively Wings?** Various names abound for this order, the most common of which is Embiidina. Embiidina was favored by authors over Embioptera since the latter was less than truly descriptive. In Greek Embioptera means "lively wing," even though webspinners are anything but spectacular fliers, although it might have been meant as a reference to how males flip their wings over and back. Regardless, several ordinal names are not very meaningful as currently constructed (e.g., Psocoptera, which combines a reference to their feeding habits and their wings!). Although ordinal names are not regulated in zoölogical taxonomy, family-group names are and the suffix *–ina* is a standard termination for the rank of subtribe. Thus, the name Embiidina is misleading, suggesting a group within the family Embiidae. Thus, we have adopted the name Embiodea, which avoids the difficulties cited for both of the former names.

7.11. Scanning electron micrographs of representative webspinners (Embiodea). Silk is extruded from the tips of specialized, hollow setae on the enlarged foretarsus, which houses the silk glands.

Other traits for Embiodea include a prognathous head, closed ventrally by a gula between the submentum and occipital foramen; absence of ocelli; presence of a dorsal, paraglossa flexor muscle in the mouthparts (Rähle, 1970: not well surveyed across the order but also in Phasmatodea); and those traits discussed previously in relation to the overall placement of the order within Plecopterida.

Webspinners are communal, with females taking close care of their offspring (Edgerly, 1987a, 1988). Females may share a composite gallery but participate only in rearing their own young (Ross, 2000b), and, indeed, communal behavior is facultative (Edgerly, 1987b, 1994). Such societies are similar to those of zorapterans, although communal living is obligate within Zoraptera given that individuals separated from colonies do not survive.

The basal family for the order is Clothodidae (Davis, 1939b, 1940; Ross, 1970, 1987; Szumik, 1996; Szumik *et al.*, 2003). Clothodids are restricted to the South American tropics and include the "giants" of the Embiodea (Ross, 1987). Unlike the "higher" webspinners, Clothodidae have

7.12. A webspinner in Miocene amber from the Dominican Republic. Morone Collection, M3473; length 7.5mm Photo: R. Larimer.

symmetrical, unlobed male terminalia and more complete and sclerotized wing venation (Ross, 1987, 2000a). Relationships among the "higher" families are remarkably confused, and the largest family, Embiidae, is paraphyletic if not polyphyletic (Szumik, 1996; Szumik *et al.*, 2003). Comprehensive work on relationships and classification is desperately needed.

Webspinner fossils are rare. Tertiary fossils derive almost exclusively from mid-Eocene Baltic or Early Miocene Dominican amber (Ross, 1956; Szumik, 1994, 1998) (Figure 7.12), but a single compression fossil is also known from the latest Eocene of Florissant, Colorado (Cockerell, 1908b; Ross, 1984). The Miocene fossil described by Hong and Wang (1987) as a clothodid webspinner, *Clothonopsis miocenica*, is actually a bibionid fly (Zhang, 1993)! All the Tertiary fossils belong to the families Teratembiidae or Anisembiidae or the polyphyletic family Embiidae. There are only two pre-Cenozoic records for the order, both in mid-Cretaceous amber from Myanmar (Cockerell, 1919; Davis, 1939a; Grimaldi *et al.*, 2002; Engel and Grimaldi, unpubl.) (Figure 7.13). Both Cretaceous species are typical webspinners but represent extinct families, perhaps close to Australembiidae.

Earlier authors attributed several Permian or Early Mesozoic fossils to Embiodea. For instance, Tillyard (1937b) proposed the suborder Protembiaria for what he believed to be the earliest representatives of the webspinners, *Protembia permiana* (Protembiidae) from the Lower Permian deposits of Elmo, Kansas. Carpenter (1950), however, demonstrated that these Permian fossils were not webspinners. Similarly, Martynova (1958) proposed an extinct suborder, Sheimiodea, for a Late Permian fossil from Russia that she believed to be a basal webspinner. Again Carpenter (1976: see also Ross, 2000a) intervened, showing that this fossil, like *Protembia*, preserved no character indicative of Embiodea.

Most recently, a putative webspinner was reported by Kukalová-Peck (1991) from the Permian of Russia, and another was depicted by Rasnitsyn (in Rasnitsyn and Quicke, 2002) from the Jurassic of Karatau. The figure of Kukalova-Peck's specimen is consistent in its overall shape to that of

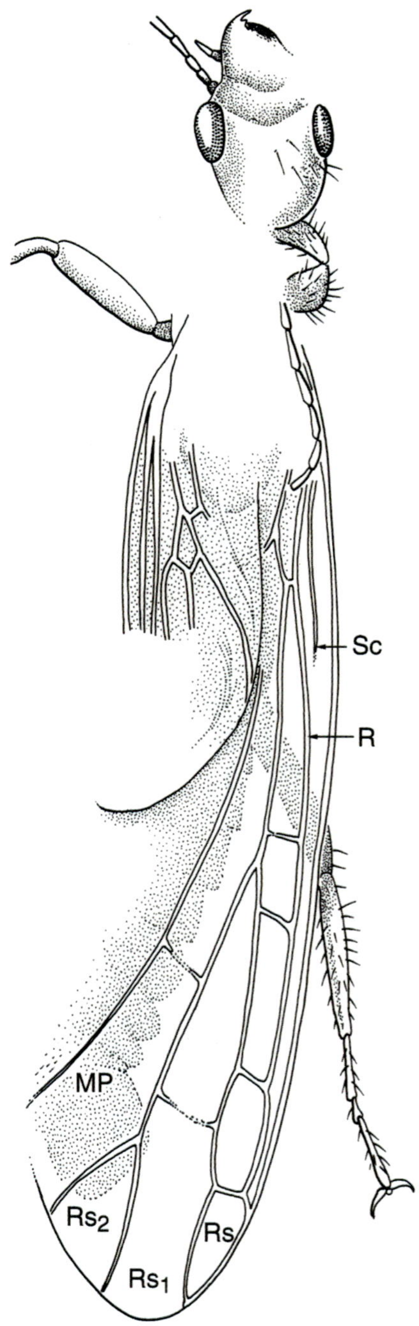

7.13. The earliest definitive fossil webspinners (order Embiodea) are two species in mid-Cretaceous amber from Myanmar, one shown here. Neither of them is particularly primitive, indicating that webspinners existed much earlier than the Cretaceous. AMNH Bu227; length 5.5 mm.

Embiodea, such as the homonomous wings with narrow bases (though venation is barely depicted) and apparently asymmetrical male genitalia. The latter are not unique to the order (e.g., Grylloblattodea, Zoraptera, Timematodea), and it is not clear whether asymmetry is merely the result of imperfect preservation. Despite its name, *Permembia* was not considered a relative of Embiodea but instead of the Psocoptera (Tillyard, 1928c, 1937b). Permian webspinners are not impossible, but they seem unlikely.

The Jurassic specimen is more consistent with the presumed Mesozoic age of the order, but again an absence of defining features prevents definitive assignment of the Karatau fossil. In fact, the fossil possesses a large anal fan in the hind wing, unmodified tarsi, and unmodified hind femora, so it is clearly not a true webspinner. Of those fossils definitively assigned to Embiodea, all are derived compared to the Clothodidae. Thus, basal divergence in the order must be at least prior to the appearance of *Burmitembia* and others in the latest Albian to Cenomanian (Engel and Grimaldi, unpubl.), perhaps extending to the Early Jurassic. Pre-Cretaceous webspinners will be difficult to recover owing to the poor preservation of their soft bodies and wings in rocks, and as of yet insect-bearing ambers prior to the Early Cretaceous are unknown.

ZORAPTERA: THE ZORAPTERANS

Zorapterans are minute insects, ca. 3 mm long (0.12 in.), superficially resembling barklice (Psocoptera) (Figure 7.14). They live gregariously under the bark of decaying logs or within termite nests, where they principally feed on fungal hyphae, nematodes, or minute arthropods like mites and Collembola (Engel, 2003a, 2004b). Despite being physically obscure, they have been of considerable evolutionary interest regarding relationships. The zorapterans presently comprise 32 modern species, all classified into a single genus (*Zorotypus*) and family (Zorotypidae), distributed pantropically. Attempts to divide extant *Zorotypus* into multiple genera (e.g., Kukalová-Peck and Peck, 1993; Chao and Chen, 2000) have rendered *Zorotypus* paraphyletic, creating an unnatural classification, so the traditional system is retained (Engel and Grimaldi, 2000, 2002; Engel, 2003e). Where known, each species has two adult morphs – an eyed, winged form (i.e., alates) and an eyeless, apterous one. Zoraptera monophyly is well established based on the peculiar wing venation (Figure 7.15); two-segmented tarsi (with the basal segment greatly reduced; the more elongate second segment probably results from the fusion of two segments); peculiar mating via a "mating hook" (even evident in Cretaceous fossils); unsegmented cerci (except in one apomorphic Miocene species); stout metafemoral spines; and moniliform, nine-segmented antennae. Furthermore, they possess peculiar behavioral traits that help to define the group (Valentine, 1986).

Many authors contend that zorapterans are poor fliers and therefore have limited dispersal abilities (e.g., Huang, 1980; Kukalová-Peck and Peck, 1993). Indeed, zorapteran wings are not well-adapted for flight; however, species live in relatively ephemeral, subcortical habitats. Such habitats suggest that individuals are quite capable of dispersal and also consistent with the dimorphism within species. During the general life of a zorapteran colony, blind, wingless morphs predominate. As the population grows, resources become limited either owing to the natural decomposition of the logs in which they reside or through the consumption of local nutrients by the larger numbers of individuals. Such crowding or nutrient deficiencies trigger the production of fully eyed alates capable of dispersing to new nesting sites; females of these winged morphs probably mate prior to dispersal, thereby accounting for the relatively low abundance of alate males. Once arriving at a new log, individuals shed their wings, the way termites, ants, and some male webspinners do. Deälated individuals can often be found in young colonies. Experimental evidence lends credence to this scenario because both habitat quality and crowding can lead to the production of alates (Choe, 1992). Furthermore, the distributions of various species are increasingly understood to cover large geographic ranges, suggesting some dispersal capabilities (e.g., Engel, 2001d). Little emphasis should be paid to absences in distributions, but despite the intensive efforts of Australian insect surveys, no zorapteran is yet known from the mainland of Australia. A single species has been discovered on Christmas Island (New, 1995), politically an Australian territory but geographically and biologically part of Indonesia and a region where zorapterans are already known to occur. If indeed the order does truly occur in the Australian region, then they would be expected in tropical Queensland or New Caledonia, areas that are typical components of old, relict distributions and particularly those affected by continental vicariance (Engel and Grimaldi, 2002).

Perhaps more than any other polyneopteran order, Zoraptera have puzzled entomologists for decades and fueled considerable debate regarding their relationships to other orders. At one time or another Zoraptera has been considered to be the living sister group to Isoptera (Caudell, 1918; Crampton, 1920; Weidner, 1969, 1970; Boudreaux, 1979), to Isoptera + Blattaria (Silvestri, 1913), Paraneoptera (Hennig, 1953, 1969, 1981; Kristensen, 1975), Embiodea (Minet and Bourgoin, 1986; Engel and Grimaldi, 2000, 2002; Grimaldi, 2001; Engel, 2003a,e), Holometabola (Rasnitsyn, 1998), Dermaptera (Carpenter and Wheeler, 1999), Dermaptera + Dictyoptera (Kukalová-Peck and Peck, 1993),

7.14. The minute zorapterans, such as *Zorotypus hubbardi* (Zorotypidae) shown here are dimorphic within each species; colonies predominantly have individuals that are blind and wingless. Zorapterans superficially resemble barklice or termites, but have distinctive mouthparts and enlarged hind femora with stout spines. Phylogenetic position of the order has been controversial, but they appear closely related to webspinners. Scanning electron micrographs; length 2.5 mm.

7.15. The most primitive known zorapteran, *Xenozorotypus burmiticus* (Zorotypidae), in mid-Cretaceous amber from Myanmar. AMNH Bu-182; length 1.9 mm; from Engel and Grimaldi (2002).

basal within Thysanoptera (Karny, 1922), or Psocoptera (Karny, 1932); or unresolved among Orthoptera, Phasmatodea, and Embiodea (Kukalová-Peck, 1991). The only other order that has such confusing relationships is the Strepsiptera. Despite the confusion, Zoraptera have been demonstrated to belong to the Polyneoptera (Boudreaux, 1979; Carpenter and Wheeler, 1999; Engel and Grimaldi, 2000) within which a close relationship to the Embiodea is best supported.

Unfortunately, the current geological record of Zoraptera, like Embiodea, is extremely sparse. Until recently, the only fossil Zoraptera known were two species in Miocene amber from the Dominican Republic (Engel and Grimaldi, 2000). Not surprisingly, these species are remarkably modern in appearance, with only *Zorotypus goeleti* possessing any notably primitive features (i.e., two-segmented cerci). The order is known from only one other fossil deposit. Four species occur in mid-Cretaceous amber from Myanmar, three of which belong to the modern genus *Zorotypus* (Engel and Grimaldi, 2002) (Figure 7.16). One species, *Xenozorotypus burmiticus* (Figure 7.15), primitively retains an additional vein in the hind wing, but it is like a modern zorapteran in all other respects, indicating that it is probably sister to all other Zoraptera. Highlighting their essentially modern character, Cretaceous zorapterans occurred in two morphs, with fossils known as both alates and apterous, blind morphs.

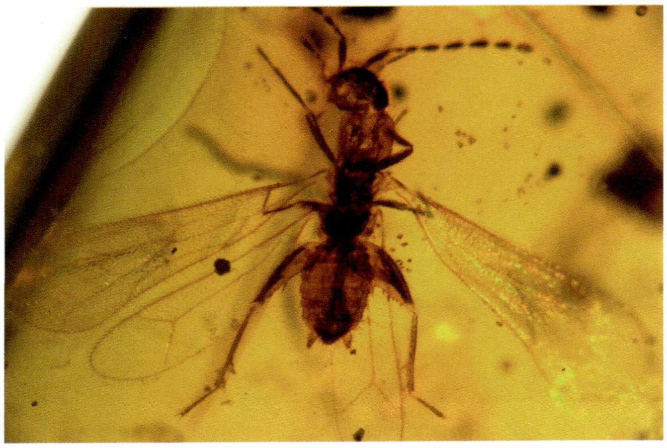

7.16. Although 100 MYO, *Zorotypus nascimbenei* (Zorotypidae) and several other zorapterans in Burmese amber are amazingly similar to modern species, attesting to the antiquity of the group. AMNH Bu341; length 1.5 mm; from Engel and Grimaldi (2002).

ORTHOPTERA: THE CRICKETS, KATYDIDS, GRASSHOPPERS, WETAS, AND KIN

The poetry of Earth is never dead:
When all the birds are faint with the hot sun,
And hide in cooling trees, a voice will run
From hedge to hedge about the new-mown mead;
That is the Grasshopper's – he takes the lead
In summer luxury, – he has never done
With his delights; for when tired out with fun
He rests at ease beneath some pleasant weed.
The poetry of Earth is ceasing never:
On a lone winter evening, when the frost
Has wrought a silence, from the stove there shrills
The Cricket's song, in warmth increasing ever,
And seems to one in drowsiness half lost,
The Grasshopper's among some grassy hills.

–*On the Grasshopper and Cricket*, John Keats (1795–1821), 30 December 1816

Most polyneopterous lineages consist of a few thousand species (or much less!), but Orthoptera is the only polyneopteran order with any sizeable diversity, having around 22,500 described species. The order has attracted the attention of humans since antiquity, developing into a "love-hate" relationship. On the one hand, orthopterans have been a source of wonderment for their diversity and soothing songs. Images of raphidophorine cave crickets, for example, have even been drawn onto the walls of caves in southern France by Paleolithic people, along with large mammals. Larger orthopterans are eaten by some indigenous peoples. In China, crickets are kept in tiny bamboo cages as pets and signs of good luck, as they have for millennia, and the chirping of woodland and field crickets is considered more enjoyable than that of a songbird (Laufer, 1927). Conversely, swarms of grasshoppers have been scourges to agriculture.

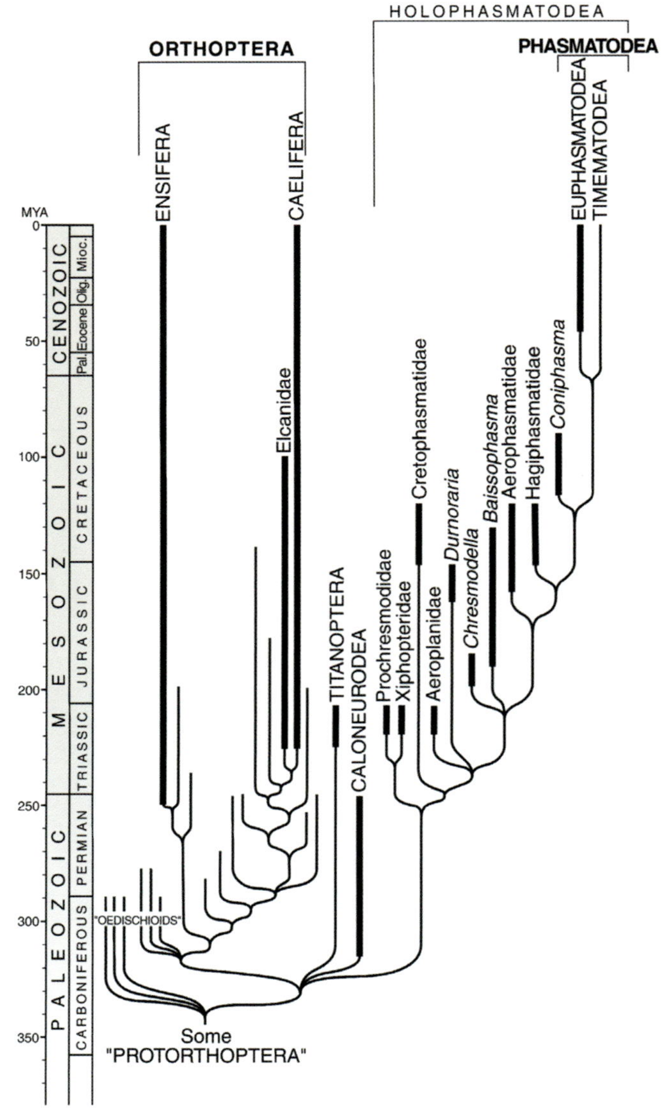

7.18. Phylogeny of the Orthopterida. Modified from Béthoux and Nel (2002); Phasmatodea after Willmann (2003).

7.17. Aposematic grasshopper nymphs in Costa Rica improve their individual chemical defense by clustering. Some orthopterans, like pygomorphids, defend themselves very effectively with toxic secretions. Photo: D. Grimaldi.

Ancient texts have repeated references to pestilence and plague brought by these insects, perhaps the most famous being those from the Bible: "All thy trees and fruit of thy land shall the locust consume" (Deuteronomy 28:42, King James Version). The order is, indeed, principally phytophagous but carnivorous, predatory species do exist. Major references to the Orthoptera include Uvarov (1928, 1966), Chopard (1938), Otte (1981, 1984, 1994), Gwynne and Morris (1983), Gangwere *et al.* (1997), Field (2001), Gwynne (2001), and Béthoux and Nel (2002).

Orthoptera have traditionally been divided into two major lineages, presently recognized as suborders, Ensifera and Caelifera (Figure 7.18). Although these have at times been elevated to ordinal status (e.g., Kevan, 1977, 1986; Vickery and Kevan, 1985), this division is based on superficial, phenetic differences between the two suborders. The monophyly of Orthoptera and of extant members in the suborders is

7.19. A female katydid consuming a nuptial meal: a large spermatophore left by a recent mate. The large packets of sperm are far larger than is needed for fertilization and appear to have evolved as one means of courtship. Photo: P. J. DeVries.

strongly founded and widely supported by both morphological and molecular data (Boudreaux, 1979; Hennig, 1981; Kuperus and Chapco, 1996; Flook and Rowell, 1997, 1998; Rowell and Flook, 1998; Flook *et al.*, 1999; Maekawa *et al.*, 1999; Wheeler *et al.*, 2001). A *cryptopleuron*, developed from the lateral extension of the pronotum over the pleural sclerites and desclerotization of the latter, is typical of Orthoptera, though this feature is lost in Proscopiidae (Caelifera). Another famous orthopteran apomorphy is the possession of saltatorial (i.e., jumping) hind legs, with straightening of the femur-tibia articulation for maximal leg extension, and a thick femur packed with muscles. Additional defining features of the order are the hind tibia with paired, longitudinal rows of teeth or spines on the dorsal surface; a horizontal division of the prothoracic spiracle; wings inclined over the abdomen during rest; and a reversal in the orientation of nymphal wing pads during later instars (Kristensen, 1991). Similarly, although monophyly of Ensifera (crickets, katydids, wetas, and their relatives: Figure 7.19) has been doubted, cladistic studies have consistently recovered them as a natural group, supported principally by the long, flagellate antennae (e.g., Flook and Rowell, 1997, 1998), but this group might eventually prove to be plesiomorphic when compared to Paleozoic orthopterans. The distinctive protibial auditory organs of ensiferans are believed to have evolved twice within the suborder and therefore do not define the entire group (Gwynne, 1995). Ragge (1977) and Gwynne (1995) both support two monophyletic branches within Ensifera, the Tettigoniida and Gryllida. Caelifera (grasshoppers, locusts, and their relatives) is similarly monophyletic and is presently divided into eight superfamilies (Flook *et al.*, 2000) united by the reduced antennae and complete reduction of the ovipositor to only two valvular pairs. The absence of prothoracic auditory organs (when present the tympana are abdominal) is sometimes cited as a defining feature of Caelifera, but this is certainly plesiomorphic.

Orthopterans are the most "vocal" of all orders, with calling behavior playing a major role in the biology and evolution of the order. Indeed, behavioral differences in mating calls are critical for the recognition of many species that can differ very little morphologically, so it is not uncommon for new orthopteran species to be described not only by their anatomy but also by their songs. Males regularly chorus on warm evenings for females. Sound is produced either by rubbing a specialized area of the wing against a corresponding area on the other, overlapping forewing (Ensifera) or by scraping the legs against stiff edges of the forewings (Caelifera). Scrapers or files are used to create the rasping sounds (Figure 7.20), these being amplified by specialized membranes of the wings referred to as "mirrors" (Figure 7.21). Many factors can affect the sounds produced, such as the number of ridges or teeth on the files, the size and density of these teeth, the position of the mirror relative to the scraper, the size of the mirror, and the rate at which the file and scraper are rubbed. The manner of stridulation is very diverse in assorted lineages of both suborders. For example, crickets stridulate by scissoring their shortened, leathery forewings together, typically the right wing is rubbed across the left wing. Perhaps the most remarkable form of stridulation in the order does not involve the wings at all. Cylindrachetids, a relict family of the Tridactyloidea (Caelifera), stridulate by rubbing their mandibles together, obviously independently derived from other modes of sound production. Another remarkable means of producing and altering the song is found in the gryllotalpids (Caelifera) in which individuals build cone-shaped "amphitheatres" at the opening of their subterranean tunnels, which amplify their calls.

Concomitant with sound production is, of course, the ability to hear the songs. Ensiferans typically produce long-distance, airborne sounds and detect these with tibial tympana (Figure 7.22). Some species produce substrate-borne drumming for close-range communication. For example, Stenopelmatidae were long believed to be "silent" despite the fact that they possess tibial ears; they do not sing but drum (e.g., Weissman, 2001). The tibial ears of ensiferans, when present, are located on the fore-tibia, facing forward, and their separation from one another (i.e., when the legs are spread apart) improves the ability of the insect to determine the directionality of the sound.

The pulse rate of songs is temperature-dependent, with the rate increasing with temperature. Songs are used in vari-

7.20. The microscopic file on the forewing of a *Gryllus* cricket, which when scraped against a small knob on the other forewing produces the familiar stridulating trill of Gryllidae. Orthopterans sing to attract mates and advertise territories, and the songs are almost always species-specific. Scanning electron micrograph.

7.21. The circular, drum-like *mirror* is easily seen on this cricket forewing from the Early Cretaceous of Brazil. The mirror produces sound by amplifying stridulation from the file, which lies very close to the mirror. AMNH; wing length 9 mm.

ous contexts, and most species produce entire suites of context-dependent as well as species-specific songs. "Calling songs" attract mates while "courtship songs," of low frequency (presumably to avoid drawing the attention of competing males), lure the female into copulation. Crickets and some other species also have "fighting songs," which are used for display and ritualized fighting. The diversity of orthopteran songs can be attributed to sexual selection, and some studies have documented how natural selection keeps songs from becoming too elaborate.

Males sing to females but also advertise themselves to eavesdropping parasitoids and predators. Pheromones are usually very receptor-specific, and few predators have evolved the ability to track such chemical cues, but sounds are very conspicuous. As male-male competition becomes more intense, structures of combat and display become more exaggerated, but they are kept in check by predation and parasitoidism. For example, singing male crickets attract ormiine tachinid flies, a worldwide group of parasitoids that lay their larva on or near singing males. The fly larva burrows into the cricket, feeds internally, and eventually kills the host. These flies possess a unique ear that is specialized for hearing ensiferan calls. Portions of the song that are most attractive to females are also the ones most attractive to the flies (Cade,

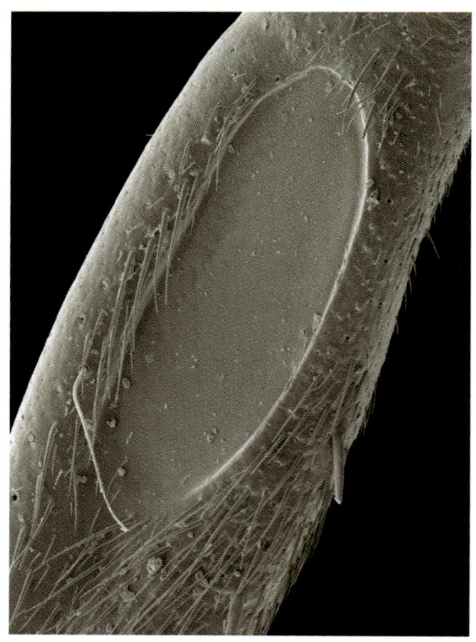

7.22. Foretibial tympanum of a cricket, which serves as an ear drum. Scanning electron micrograph.

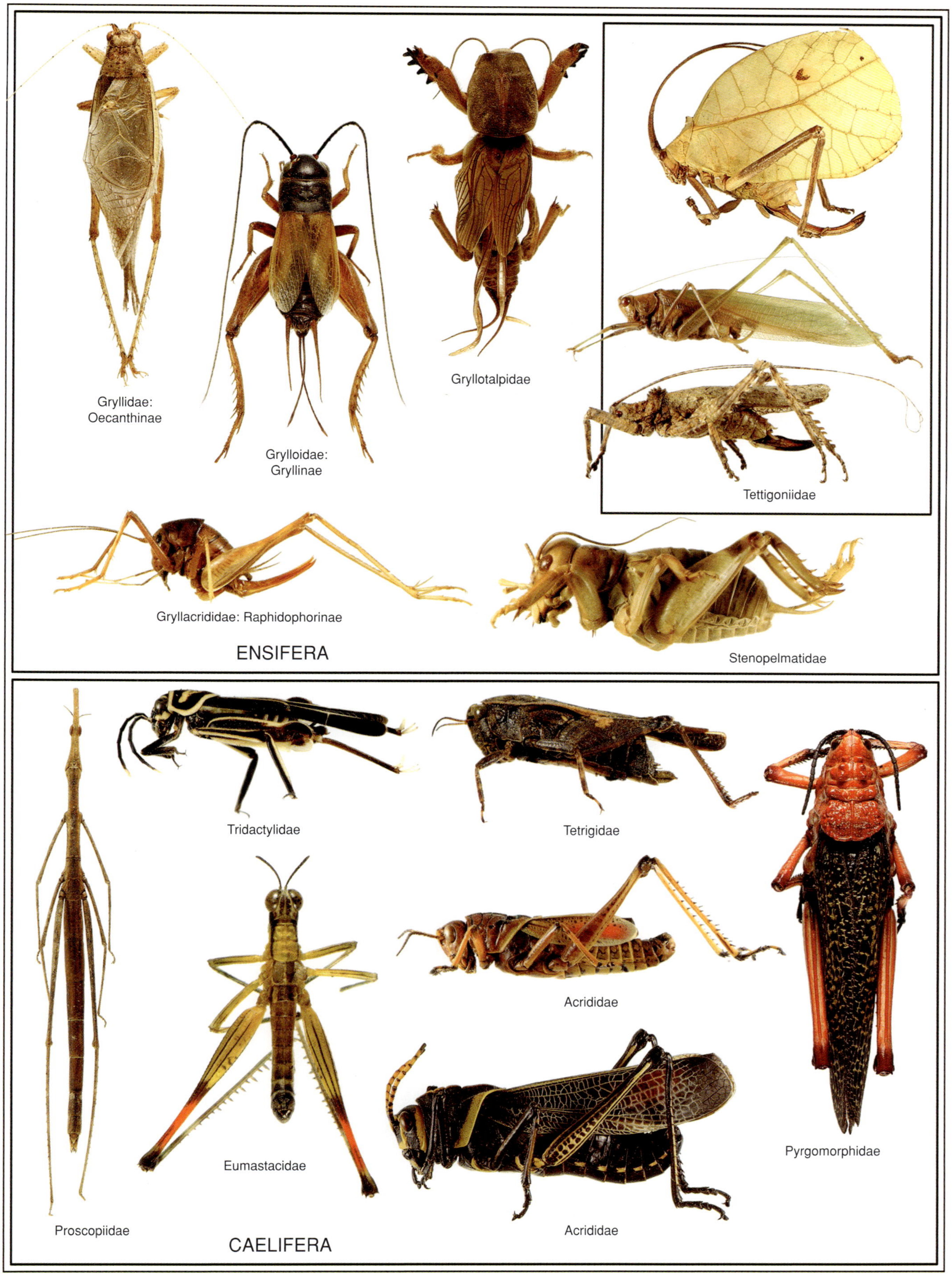

7.23. Representative Recent Orthoptera. Not to the same scale.

7.24. Katydid (Tettigoniidae) mimicking lichens on a tree trunk in Costa Rica. Orthopteridans, including stick insects, have some of the best camouflage in nature. Photo: P. J. DeVries.

1975; Burk, 1982; Wagner, 1996; Allen, 1998). This system is analogous to predation of frog-eating bats on the Tungara frog, *Physalaemus pustulosus*. The male choruses atop a rock to advertise his song. The sexiest part of the song, however, also attracts the frog-eating bat, which snags the frog from its perch. Another such system exists between katydids and bats. Not all insectivorous bats are echolocators. In neotropical forests some bats are silent gleaners, snagging katydids from their calling perches at the tips of branches. Tropical forest katydids possess an odd cycle of momentary bursts of calls interspersed by minutes of silence – apparently an adaptation against bat predation. Katydids in clearings and in Old World forests (which do not have such bat gleaners) have songs that are more continuous.

Today, orthopteran songs are best appreciated at night in a tropical forest. A cacophony of squeaks, clicks, stutters, and high-pitched whines mixed with the calls of frogs nearly drown out any other noise. Triassic forests, however, were devoid of birds, and other than the occasional groans and squeaks of tetrapods, most song probably derived from the resonant clacking of titanopterans, backed by trills of haglids and the chirping of early gryllids. The structure of the wing can allow some prediction as to the nature of songs produced by extinct ensiferans, but few studies have been undertaken. The most interesting to date is that of Rust *et al.* (1999) on the Paleocene tettigoniid, *Pseudotettigonia amoena*, from Denmark. There is a close relationship between the size of the area of the mirror and the frequency of the sound in Recent Tettigoniidae, just as small drums generate higher frequency sounds and a bass kettle drums produces a low sound. Among tettigoniids, *Pseudotettigonia* was a baritone.

Besides their songs, Orthoptera are also well known as pests, a distinction based on just a few species. Though most orthopterans are generalist phytophages, most families are of no particular concern to agriculture, but the most devastating species are in the family Acrididae (Caelifera). Some acridids – the plague locusts – have solitary and gregarious phases, the latter of which form enormous "clouds" that number into the tens to hundreds of millions of individuals. Such species are extreme generalist feeders; they don't specialize on crop-plant species, but crops simply offer concentrated food resources and are thus natural targets. Today the most problematic plague species are the desert locust, *Schistocerca gregaria*, and the migratory locust, *Locusta migratoria*, particularly in northern Africa. Some swarms of the former species have been calculated to number near

7.25. Another camouflaged katydid, which mimics moss in Central America. Photo: P. J. DeVries.

7.26. Leaf katydid camouflaged among dead leaves on the floor of a Panamanian forest. The wings of pseudophylline katydids, like this *Mimetica*, are remarkably leaflike, complete with leaf veins, splotches, and even chew marks. Photo: P. J. DeVries.

50 billion individuals and to cover thousands of square miles. The most devastating swarm in recent history was a 1949 plague in California that resulted in 3,000 square miles of destruction. Eventually the feeding frenzies created by swarming locusts result in the denuding of the foliage; individual locusts then begin to starve, leaving locust corpses strewn across the landscape – the single most impressive example of the extent of insect biomass. Swarms can be carried by the wind and swept out to sea, or even dumped onto glacier surfaces, leaving frozen locusts preserved in the ice. Various other orthopterans are also crop pests, but none instill the same level of fear as locust swarms. It is little wonder why early civilizations were so fearful of locusts and envisioned them as a plague imposed by a wrathful god.

Like the stick insects, many Orthoptera are cryptic in coloration and body structure, resembling foliage or bark substrate (Figures 7.23 to 7.25, 7.27, 7.28). Some of the most remarkable are the Pseudophyllinae, which resemble dried leaves complete with leaf veins, splotches, and chew marks (Figure 7.26). This crypsis is so effective that even seasoned entomologists miss these katydids sitting in plain sight. Aposematic coloration is also common throughout the order. Some of the most peculiar morphologies are found among wingless species that have specialized for living in the soil or in caves. For example, the famous cooloola monster (Stenopelmatidae: Cooloolinae) was hesitatingly placed in Orthoptera when first discovered, owing to its odd anatomical construction even for an ensiferan (Rentz, 1980). Another extreme example is the Myrmecophilidae (ant crickets), which are small, wingless, microptic, and flattened, adapting them well to a life as inquilines in the nests of ants.

Of all polyneopterous orders, the Orthoptera, perhaps not surprisingly, has the most extensive geological record, second only to that of the roaches and roachoids (Dictyoptera, "Blattodea"). However, Paleozoic Orthoptera, historically placed in Ensifera (e.g., Oedischiidae in Ensifera: Carpenter, 1992), are perhaps a stem group that gave rise independently to the two suborders. Families of this stem group ("oedis-

7.27. A katydid suspended from a branch in Panama, encrusted with lichen-like markings. Another katydid, which has just molted, is behind it. Photo: P. J. DeVries.

7.28. A long grasshopper, camouflaged among palm fronds. Photo: P. J. DeVries.

chioids") persisted well into the Late Mesozoic. By the end of the Permian and the Early Triassic, there is the first definitive evidence of the suborders Ensifera and Caelifera. Béthoux *et al.* (2002) have recently described an ensiferan from the Late Permian of France that may be a sister group to all other members of the suborder, while the earliest definitive Caelifera come from the Triassic of Asia, Australia, and Europe (Tillyard, 1922b; Sharov, 1968; Jarzembowski, 1999) as well as the Early Jurassic of England (Whalley, 1985; Zessin, 1983). Early Permian families such as Permoraphidiidae and Permelcanidae have at times been considered to be ensiferans (e.g., Kukalová-Peck, 1991) but may be stem-group Caelifera, with the extinct family Elcanidae (cover, Figure 7.29) closest to true caeliferans (Béthoux and Nel, 2002). A complete synthesis of fossil Orthoptera, or Orthopterida, has yet to be achieved, the most extensive treatments being those of Zeuner (1939) and Sharov (1968).

ENSIFERA

The Ensifera consists of about 10,000 species in 10 families. The phylogeny of Ensifera has been most recently investigated by Gwynne (1995) and Desutter-Grandcolas (2003). Among ensiferans are some clear relicts, such as members of the Stenopelmatoidea and Hagloidea. As alluded to earlier, Ensifera is the older of the two suborders, with putative members recorded from the very end of the Permian (Béthoux *et al.*, 2002), and the lineage was certainly established during the Triassic. Numerous ensiferan families are represented in the Mesozoic (Figures 7.29 to 7.32), and even putative gryllids are documented from the Triassic along with the presently relict haglids. Tettigoniidae are documented from as long ago as the Early Cretaceous. Other modern families such as Stenopelmatidae (Gryllacridinae) and Gryllotalpidae are known from the Early Tertiary. Among the various extinct lineages of Ensifera, perhaps the most distinctive are the Phasmomimidae from the Late Jurassic. Phasmomimids had long wings and may have been related to the Haglidae.

The superfamily **Stenopelmatoidea** consists of four cricket-like families (Stenopelmatidae, Rhaphidophoridae, Schizodactylidae, Anostostomatidae), principally occurring

7.29. One of the last occurrences of Elcanidae is in 120 MYO limestone from the Cretaceous of Brazil. Elcanidae were diverse and abundant from the Late Triassic to the mid-Cretaceous, having gone extinct presumably during the Late Cretaceous. Although they had long, ensiferan-like antennae, they were probably more closely related to grasshoppers and locusts. AMNH; body length 14 mm.

7.30. Cricket (Gryllidae) from the Early Cretaceous of Brazil. The earliest crickets occur in the Late Triassic; by the time of the Santana Formation in Brazil 120 MYA, they were diverse and abundant. AMNH; body length 19 mm.

in New Zealand and Australia. Some stenopelmatoids have an imposing habitus and include the Jerusalem crickets, wetas, cave crickets, and king crickets. Most species are flightless and nocturnal, scavenging for arthropod remains, but they can be omnivorous or even predatory. The former family Gryllacrididae is now considered a subfamily of the Stenopelmatidae, as are the Schizodactylidae in some classifications (e.g., Gorochov, 2001). Gryllacridids are best known for the raphidophorines, the so-called camel crickets, which are named for the short, hump-back body but which also have extremely long antennae that they constantly wave back and forth. The large genus *Ceuthophilus* contains hundreds of species, which are common inhabitants of caves, basements, and rocky areas.

Tettigonioidea consists of the familiar katydids and bush crickets and, with about 6,000 species, is the most diverse lineage of Ensifera. Presently, the superfamily consists of only the nominate family and several subfamilies that are sometimes given family rank. Tettigoniids are large insects occurring throughout the world. While many are of little economic importance, just as many species can cause extensive damage to crops and reach pest levels, a few even swarm similar to locusts. Perhaps the most widely known of such pests is the Mormon cricket, *Anabrus simplex*. Many tettigoniids are arboreal or live in bushes (hence the name bush crickets) and are remarkably cryptic, typically mimicking leaves (Figure 7.26), but also mimicking lichens (Figures 7.24, 7.27) and mosses (Figure 7.25).

The superfamily **Hagloidea** contains a single modern family with two subfamilies that seem to intermingle traits of the Tettigonioidea and Grylloidea. Hagloids were much more diverse in the past, extending at least into the Triassic, apparently diminishing in diversity through the Cretaceous.

Grylloidea contains the familiar crickets (Gryllidae) as well as the mole crickets (Gryllotalpidae) and ant crickets (Myrmecophilidae). Each of these families have a worldwide distribution, but only the gryllids are of any significant diversity. Gryllotalpids and myrmeophilids together amount to

7.31. Another cricket from the Cretaceous of Brazil. AMNH; body length 21 mm.

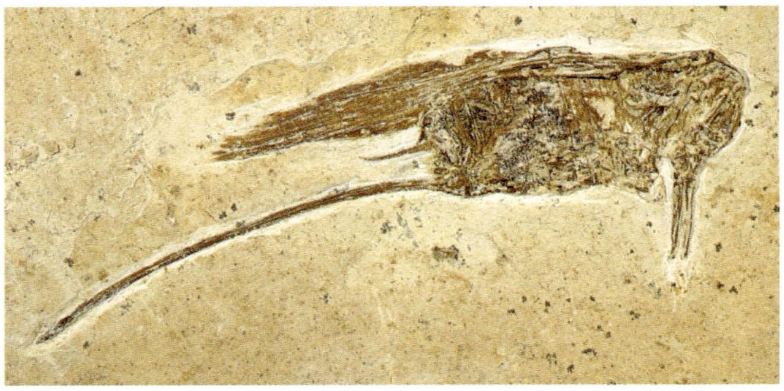

7.32. An ensiferan with a long, thin ovipositor from the Early Cretaceous of Brazil. AMNH; length 42 mm.

slightly more than 100 species, while Gryllidae accounts for about 570 species. Crickets are the most famous of all ensiferans, particularly for their songs. While crickets are familiar to everyone, the other grylloid families are a bit more unusual. Mole crickets have remarkably molelike fossorial forelegs (Figure 1.26), modified for digging their way through soil and sand in a system of tunnels just like moles. These forelegs are also used for excavating the roots of various plants on which individuals feed. Myrmecophilids are small, apterous, scale-covered inquilines of ants, a few of which are associated with termites.

CAELIFERA

The suborder Caelifera consists of about 11,000 species in 20 families, depending on the classification followed. The phylogeny of Caelifera has been repeatedly investigated using molecular data (e.g., Flook and Rowell, 1997, 1998; Rowell and Flook, 1998; Flook *et al.*, 1999, 2000), while similar studies based on morphology have lagged behind. The most extreme classification of a subset of Caelifera was that of Dirsh (1975) who went so far as to break the acridomorphs into a series of separate orders! Permian families such as Permelcanidae are stem groups to the Elcanidae (Triassic through Cretaceous), which itself is either paraphyletic or a sister group to the Caelifera despite having elongate, ensiferan-like antennae (Béthoux and Nel, 2002). The earliest definitive caeliferans are those from the Triassic, represented by several extinct families such as Locustopseidae and Locustavidae, which may prove to be stem groups to some of the modern superfamilies. Caelifera were diverse by the end of the Triassic and have been very significant phytophages in ecosystems ever since. Modern families are first documented in the Cretaceous, such as Eumastacidae, Tetrigidae, and Tridactylidae, although putative tetrigids and tridactylids have been described from the Early Jurassic. The familiar Acrididae is first definitively recorded from the Eocene. Lin (1980) has described a putative acridid from the Early Cretaceous of China, but this is perhaps a member of the extinct family Locustopseidae. Some of the Tertiary records of fossil acridids also document gregarious associations (e.g., Arillo and Ortuño, 1997). Most caeliferans are preserved as compression fossils, but smaller species are often captured in amber, particularly nymphs (Figure 7.33).

The most diverse and familiar superfamily of caeliferans is the **Acridoidea**, with about 8,000 species and comprising the grasshoppers and locusts. Aside from the nominate family Acrididae the superfamily includes the Recent families Lathiceridae, Lentulidae, Ommexechidae, Pamphagidae, Pamphagodidae, Pauliniidae, Romaleidae, and Tristiridae. Lathicerids are confined to xeric regions of southern Africa, while pamphagids and pamphagodids are principally found in Africa although the latter family also ranges into the Palearctic and southwestern Asia. Tristirids are confined to the Andean region; they are cryptically colored and blend in with small stones or leaf litter. Ommexechidae are widely distributed in South America, principally in dry, sandy areas. Like ommexechids, pauliniids are also widely distributed in South America, although the two genera of the family are nocturnal, and individuals can swim through or skate across the surface of water. Pauliniids feed on aquatic vegetation and have even been employed in control programs for water hyacinth. The Lentulidae is an African family of wingless acridoids that primarily occur on bushes and oviposit into relatively dry soil. The two major families of the Acridoidea are the Romaleidae and Acrididae; the former is principally found in the Americas, while the latter is cosmopolitan in distribution. These are the grasshoppers most individuals are familiar with, although the acridids also include the infamous locusts.

Four superfamilies, the **Pyrgomorphoidea**, **Tanaoceroidea**, **Pneumoroidea**, and **Tetrigoidea**, consist of a single family each. Pyrgomorphidae is most diverse in the Old World

7.33. A grasshopper nymph (family Eumastacidae) in Miocene Dominican amber, caught while poised to leap. Caeliferans are rare in amber because most of them are ground dwelling; they can usually extract themselves from the sticky resin using powerful hind legs. Morone Collection; 7 mm. Photo: R. Larimer.

tropics, particularly in the afrotropics and Madagascar. The pyrgomorphids are represented in the Western Hemisphere only in tropical Mexico. Some pyrgomorphids are toxic, even deadly, and these generally have very aposematic coloration. Tanaocerids comprise only a few species of nocturnal caeliferans occurring in the deserts of Mexico and western North America. Pneumoridae contains about 20 species confined to sub-Saharan Africa. The family Tetrigidae is sometimes broken into a series of families. Unlike the aforementioned families, the tetrigids are moderately diverse with about 1,000 species distributed throughout the world and include the pygmy grasshoppers. Tetrigids are generally small (1 cm or less in length) and are very cryptic among the brown leaves of the forest floor.

The families Eumastacidae and Proscopiidae are the only members of the superfamily **Eumastacoidea**. Both families have long, angular heads and tend to be wingless, the latter exclusively so. The South American proscopiids are commonly called the false stick insects owing to their cryptic, sticklike body structure, and they are often confused with species of the Phasmatodea. Eumastacids are distributed in most areas around the world.

The superfamily **Trigonopterygoidea** consists of two enigmatic families, the Asian Trigonopterygidae and the Mexican Xyronotidae, the latter family formerly included in the Tanaoceridae. Neither family is very diverse, the xyronotids in particular accounting for only two species in a single genus.

7.34. Pygmy mole cricket (Tridactylidae) from the Early Cretaceous of Brazil. AMNH; length 9 mm.

The superfamily **Tridactyloidea** contains three families, the closely related Tridactylidae and Rhipipterygidae and the relict Cylindrachetidae. These insects superficially resemble true crickets and mole crickets, and they also tend to be small and frequently gregarious. The tridactylids and rhipipterygids are most closely related, both occurring in the tropics or subtropics except for a few tridactylids that extend into temperate habitats. The family Cylindrachetidae consists of nine peculiar species commonly known as sand gropers. Cylindrachetids are confined to the Southern Hemisphere, living in subterranean galleries in southern South America (Patagonia), Australia, and New Guinea. These insects are extremely reduced, having minute eyes, no vestiges of wings, and a soft, fleshy abdomen (Figure 14.31). The mid and hind legs are very short, and the forelegs are remarkably convergent with those of gryllotalpids. Because they do not live above ground, their dispersal ability must be very limited. That fact, and their disjunct austral distribution, suggests that they may have been affected by the drifting of gondwanan continents in the Cretaceous, which is an age that agrees with that of their close relatives, the tridactylids and rhipipterygids (Figure 7.34).

PHASMATODEA[2]: THE STICK AND LEAF INSECTS

Like many Orthoptera, the stick and leaf insects are remarkable mimics of the stems, twigs, and leaves (Figures 7.35 to 7.37) on which they feed. Some species are very large; in fact, the longest extant insect is a stick insect, *Pharnacia kirbyi* from Borneo, which can be up to 555 mm (22 inches) in length. The order has more than 3,000 species distributed in temperate and tropical habitats. Bedford (1978), Mazzini and Scali (1987), Brock (1999), and Bradler (2003) have most recently summarized the available biological information on the order. Phasmatodean monophyly is clear owing to the unique development of anterior dorsolateral defensive glands on the prothorax, which provides one of their few means of defense, but can be very effective. Additionally, males of Phasmatodea have a *vomer*, a sclerite of the tenth

[2] **Phasmida Versus Phasmatodea.** Much confusion surrounds the formation of scientific names, which is not surprising. Most authors are not familiar with the intricacies of Greek and Latin. These two names naturally originate from *phasma* (Greek, meaning "spirit"), which is a neuter, third declension noun with an augmented stem such that its combining form is actually *phasmatos*. This is why when family-group names are formed from such words, the family name is slightly different from the genus upon which it was based (e.g., *Phasma* forms Phasmatidae). Therefore, the correct formation of any familial or ordinal name from *Phasma* results in Phasmatida or Phasmatodea. The alternative names of Phasmida and Phasmodea are improperly formed in Greek.

7.35. A large living species of walking stick, *Eurycnema cercata* from Australia (order Phasmatodea), shown here life size. The longest insect is *Pharnacia kirbyi* from Borneo, which can reach lengths up to 55 cm (22 in.). Body length 20 cm.

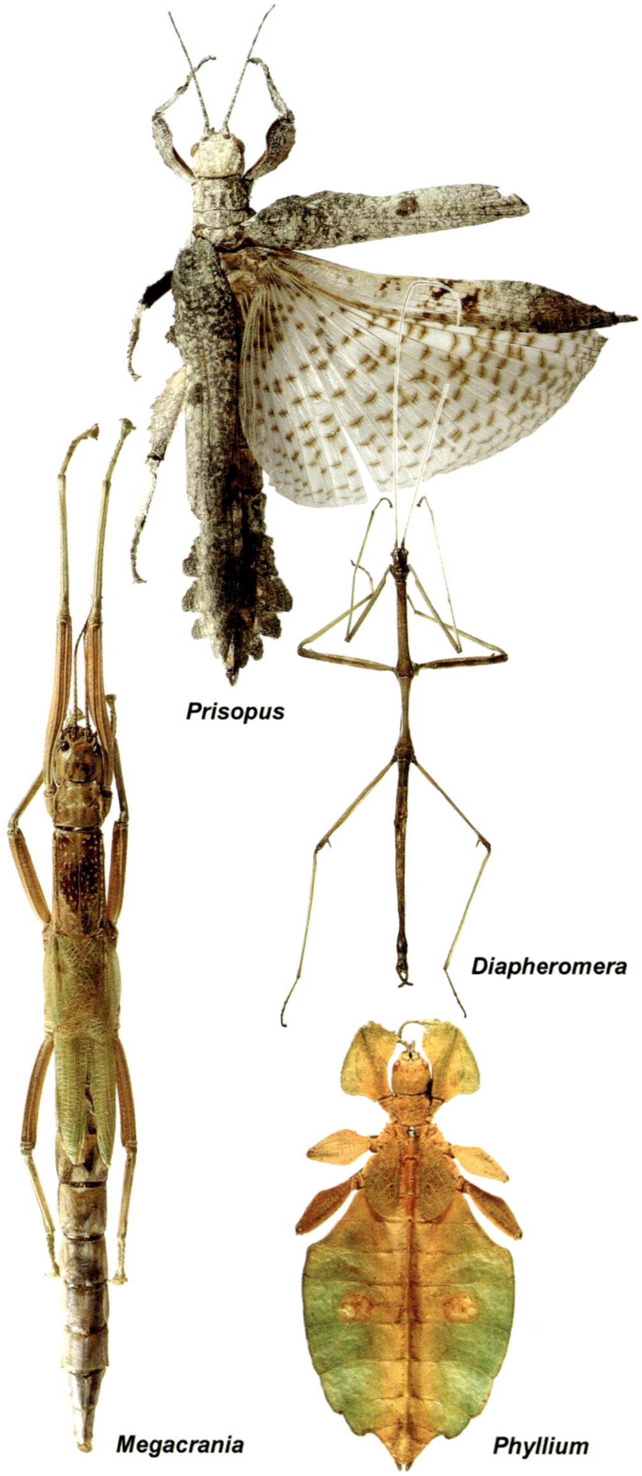

7.36. Representative living stick insects. Not all Phasmatodea are sticklike. To the same scale.

sternum that permits males to clasp the female during copulation (Sinéty, 1901; Pantel, 1915; Snodgrass, 1937) but that has only recently been established as part of the stick-insect groundplan (Bradler, 1999; Tilgner *et al.*, 1999). Other, principally internal, features of the order are discussed by Kristensen (1975, 1991) and Tilgner *et al.* (1999). Lastly, the egg of phasmatodeans have an operculum, a lidlike section of the oocyte, and this trait, in combination with a unique structure of the micropyle (Sellick, 1998), supports the order's monophyly.

Aptery occurs throughout the order, but in all species that retain forewings these are relatively reduced. The forewings,

7.37. A walking stick moving across a log in an Ecuadorian forest. Most phasmatodeans live amongst the vegetation they feed on, remaining motionless or even swaying with branches to better camouflage themselves. Photo: R. Swanson.

when present, are tegminous and abbreviated. The hind wings are often altered so as to either fold tightly against the long, thin body, or modified to resemble leaves, like the forewings in such species. Cryptic coloration, elongation of the body and legs, or, alternatively, broadening and flattening of the body to resemble leaves are other important adaptations. Stick insects also "sway," or gently rock backward and foreward, resembling rustling leaves or branches in a breeze. Phasmatodeans also protect themselves by feeding at night and remaining virtually motionless during the day. If disturbed or attacked, most stick insects become cataleptic, falling from their perch and laying motionless for hours. Some species will attempt to stand their ground, armed with sharp, thorny spines, or emitting noxious secretions, even regurgitating their gut contents. The secretions are sprayed from exocrine glands opening at the anterolateral corners of the prothorax, and at least some include chemicals such as quinoline and can cause blindness, although the chemistry of most are unknown.

Stick insects apparently take little care in egg deposition. The ovipositor is reduced relative to other orthopterids, and female phasmatodeans tend to scatter eggs from upon high. Some species deposit eggs within the soil or cement them to plants, but typically, without changing position, a stick insect will fling eggs from the tip of the abdomen so that they fall amidst leaf litter. To protect eggs, crypsis in Phasmatodea extends even to this early stage, with a bewildering array of egg modifications across the order (Figure 7.38). Many eggs resemble seeds and, like seeds, may not hatch for years and are resistant to various forms of damage. A *capitulum* is developed in some lineages that flings their eggs to the ground. This knoblike process on the anterior end of the egg resembles the elaiosomes of some seeds. Elaiosomes are lipid-rich processes of seeds that attract ants, who then collect and disperse the seeds. Amazingly, ants readily collect these eggs and in removing them from the leaf litter to their nests, young stick insects have the protective confines of the subterranean colony and thereby avoid parasitism and predation. Eventually, the egg opens from the operculum and the nymph emerges. It resembles the adult except in the smaller number of antennal segments and rudimentary wings and genitalia. A study of crypsis from a phylogenetic perspective in adult and immature Phasmatodea will be of great significance, and already egg morphology is proving important for Phasmatodea classification (Sellick, 1997a,b,c, 1998).

Phylogenetic work within Phasmatodea has been essentially lacking, and the classification has changed little since Günther's (1953) treatment of the order. Bradley and Galil (1977) and Kevan (1977, 1982) had dramatically different classifications, neither of which are entirely well founded. Kristensen (1975) highlighted the segregation of the relict genus *Timema* (Timematidae) from other Phasmatodea, and the importance of this group is recognized today by its placement into a separate suborder, Timematodea, versus all other families (i.e., the Euphasmatodea). The 21 species of *Timema* occur in western North America, principally in California (Vickery, 1993; Sandoval and Vickery, 1996, 1998; Vickery and Sandoval, 1997, 1998, 1999, 2001) (Figure 7.39), but were assigned to the paraphyletic "Areolatae" of earlier classifications (e.g., Bradley and Galil, 1977). Timematidae is one of the only groups of stick insects to have been studied phylogenetically, principally to investigate the evolution of parthenogenesis, which appears to have arisen at least five times within the family (Sandoval *et al.*, 1998; Crespi and Sandoval, 2000; Law and Crespi, 2002) (many other stick insects are also parthenogenetic). *Timema*, like all organisms, is a mosaic of derived and primitive traits, and its monophyly is supported by three-segmented tarsi, development of a mesal lobe on the right cercus, and egg-laying behavior, in which females ingest soil and then coat the eggs with this material (Tilgner *et al.*, 1999). The genus is excluded from Euphasmatodea by the primitive features of a molar lobe on the mandible, separation of the prothoracic ana- and coxopleurites, and retention of prothoracic sternal apophyses (Kristensen, 1975). Unfortunately, there is no fossil record for Timematodea.

The grouping within Euphasmatodea of two infraorders (traditionally considered suborders), namely the Areolatae and Anareolatae, is unnatural, and neither group is monophyletic (Bradler, 1999, 2003). These groups were established on the presence or absence of the *area apicalis*, a sharply defined region near the apex of the mid- and hind tibiae (Redtenbacher, 1906; Bradley and Galil, 1977). The presence of this structure in *Timema* suggests that it is primitive for the order and thus the Areolatae is paraphyletic. In fact, some Areolatae and some Anareolatae share the derived feature of a gula; consequently, it is likely that the area apicalis was

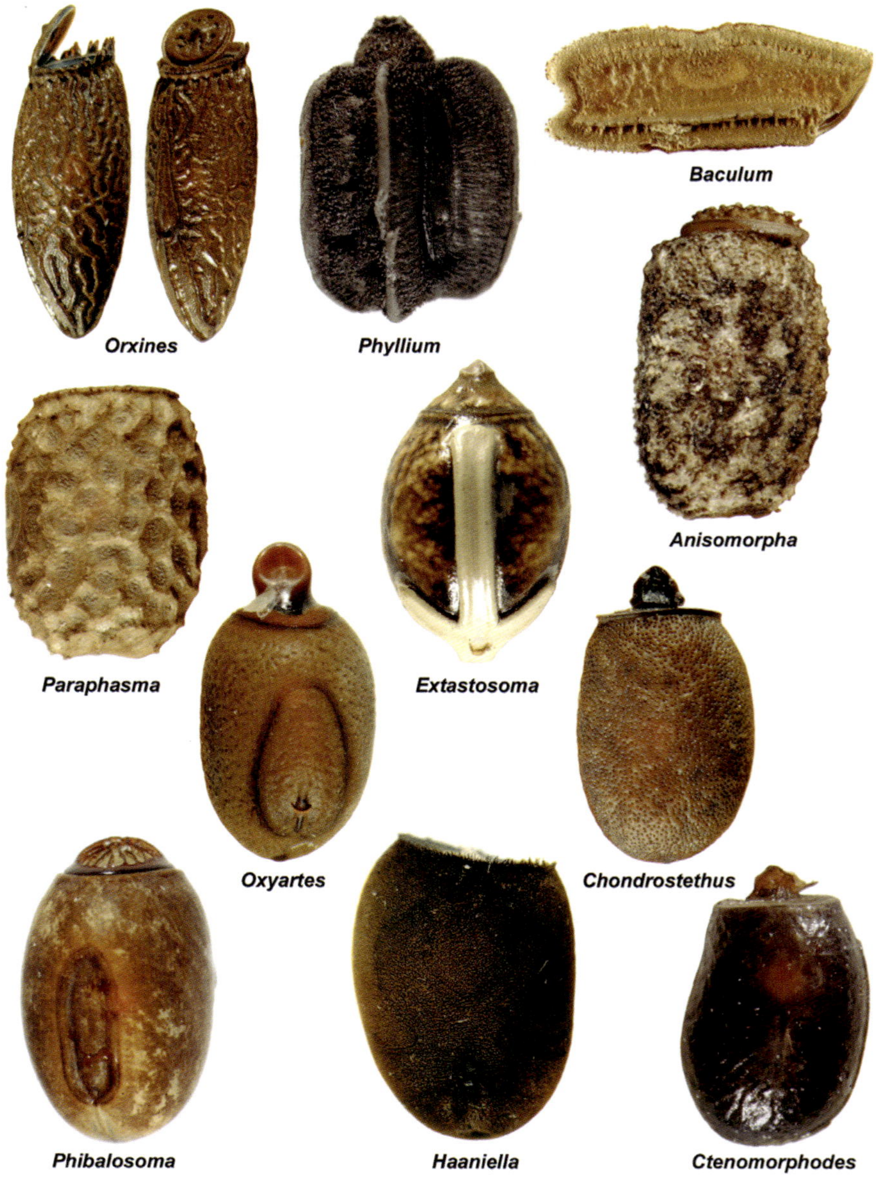

7.38. Photomicrographs of representative stick insect eggs. The diverse and elaborate eggs are dropped from the plant and sometimes brought back to the nests of seed-gathering ants, where they are protected. Specimens: University of Georgia.

reduced twice within the order, rendering Anareolatae polyphyletic (Bradler, 1999, 2003). The most recent work on the phylogeny of Euphasmatodea arrived at a novel set of relationships based on molecular data, suggesting that, under the most parsimonious reconstruction, wings, once lost, were reacquired several times independently (Whiting *et al.*, 2003). This is not to say that wings in many Phasmatodea are novel structures, unrelated to wings in other Pterygota. Indeed, the basic structure of the wings is identical to other lineages, complete with the typical venation, etc. Instead, it appears that the genes controlling expression of the wings were suppressed early in stick-insect evolution, becoming reactivated several times independently throughout the order's history. However, Trueman *et al.* (2004) noted that since wings are commonly lost in diverse insect lineages, this better accounts for phasmatodean wing evolution, rather than hypothesizing an ancestral loss and multiple reacquisitions. Indeed, stem-group Phasmatodea were fully winged (Willmann, 2003), lending support to the multiple-loss hypothesis. Without a fossil record for Timematodea and a more extensive record for basal Euphasmatodea, however, the paleontological data cannot presently resolve the debate (see following discussion), so additional paleontological study is needed.

Sharov (1968), Carpenter (1992), Rasnitsyn and Quicke (2002), and Willmann (2003) considered the stick insects as having an extensive geological record extending back to the Permian or Triassic (Figure 7.18). In sharp contrast, Tilgner

7.39. *Timema*, the most basal lineage of Phasmatodea. The genus occurs in western North America. Photo: C. Sandoval.

7.40. A nymph of the Eocene phasmatodean *Pseudoperla* in Baltic amber. AMNH; body length 13 mm.

(2000) concluded that these fossils possessed no defining features of Phasmatodea. Although the development of subparallel veins in the forewing has been considered a phasmatodean trait (e.g., Rasnitsyn and Quicke, 2002), there are both Ensifera (e.g., Proparagryllacrididae) and Caelifera that also possess such characters, and putative stick insects such as *Xiphopterum* approximate caeliferan families such as Locustopseidae. The Mesozoic fossils are perhaps stem-group Phasmatodea as demonstrated by Willmann (2003). Tertiary stick insects are rare and only a few species are documented from Baltic and Dominican ambers (Figures 7.40, 7.41).

As discussed before, eggs of stick insects are diverse in structure and are very hardy; some eggs even occur in the fossil record. Definitive eggs of Euphasmatodea are known from as old as the mid-Cretaceous of Myanmar (Rasnitsyn and Ross, 2000) as well as from Eocene deposits of North America (Sellick, 1995) and in Dominican amber (Tilgner, 2000).

TITANOPTERA: THE TITANIC CRAWLERS

Among the most impressive Orthopterida are the giant titanopterans (Figure 7.42). These insects, known only from the Triassic of Australia and central Asia (Tillyard, 1925; Riek, 1954; Sharov, 1968; Jell and Lambkin, 1993), could reach 400 mm (15.75 inches) in wingspan (Sharov, 1968). Although the order was first recognized under the name Mesotitanoptera by Crampton (1928), it was not accepted by authors until the more extensive monograph of Sharov (1968). Unlike Orthoptera, titanopterans had five-segmented tarsi, cursorial (running) legs not capable of jumping, and wings that were held flat over the abdomen during rest. The forelegs appeared to be raptorial, being ventrally armed with stout spines on the femora and tibiae. The forewings of many species possessed large stridulatory structures, so these were clearly very vocal animals. Because the wings were held over the abdomen during rest, sound was perhaps produced by rubbing the stridulatory files/scrapers of the tegminous forewings together, typical of modern Ensifera. Given the preserved details of the stridulatory structures, it is likely that

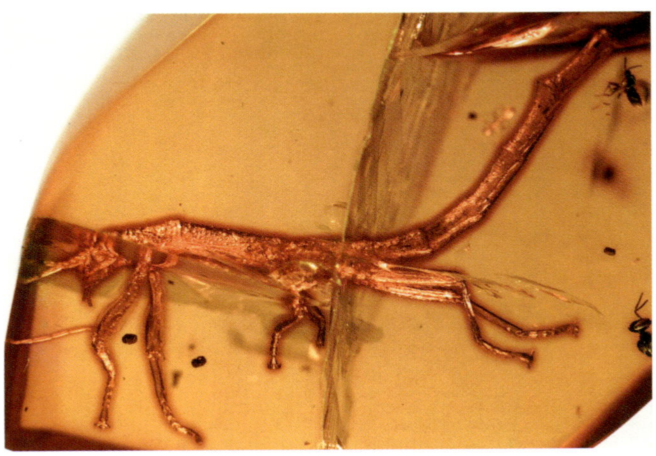

7.41. A Miocene phasmatid in amber from the Dominican Republic. Morone Collection; length 48 mm. Photo: R. Larimer.

7.42. Wing of *Clatrotitan andersoni*, from the Triassic of Australia. Titanopterans were giant, raptorial orthopterids from the Triassic. Large stridulatory structures on the patterned forewings show that they were vocal like their orthopteran relatives; in fact, they were probably the baritones of Mesozoic insects. Australian Museum (AM) F.36274; length 139 mm.

these would have had relatively deep calls, probably a resonant call like a bullfrog. The pronotum extended laterally over the pleura and hypognathous head capsule, which are features shared with Orthoptera, and it is possible that, like other orthopterid orders, Titanoptera is derived from within a paraphyletic assemblage of Paleozoic orthopterids, their closest relatives being the spectacular Permian Geraridae (Figure 7.43) (Gorokhov, 2001). Certainly much remains to be discovered concerning these giants. Although presently known only from deposits in present-day Australia and Asia, the consolidation of continents during the Triassic into Pangea implies that they will likely be discovered in strata of similar age from Africa, North America, and South America. The absence of Titanoptera from Jurassic deposits in Europe and Eurasia indicates that these fascinating insects were narrowly restricted to the Triassic, which may be a consequence of Titanoptera being a crown group to the Paleozoic Geraridae.

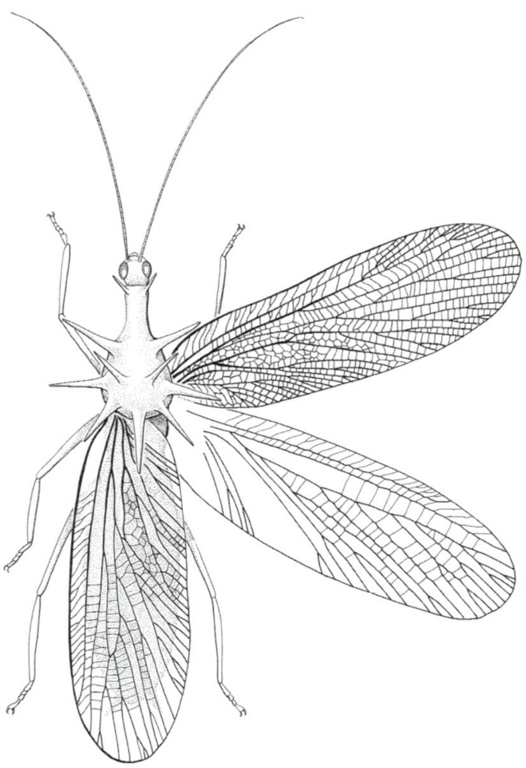

7.43. Reconstruction of *Gerarus danielsi* (Geraridae), from the Late Carboniferous of Mazon Creek, Illinois. Gerarids were Paleozoic orthopterids and a stem group to the Titanoptera. Redrawn from Burnham (1983).

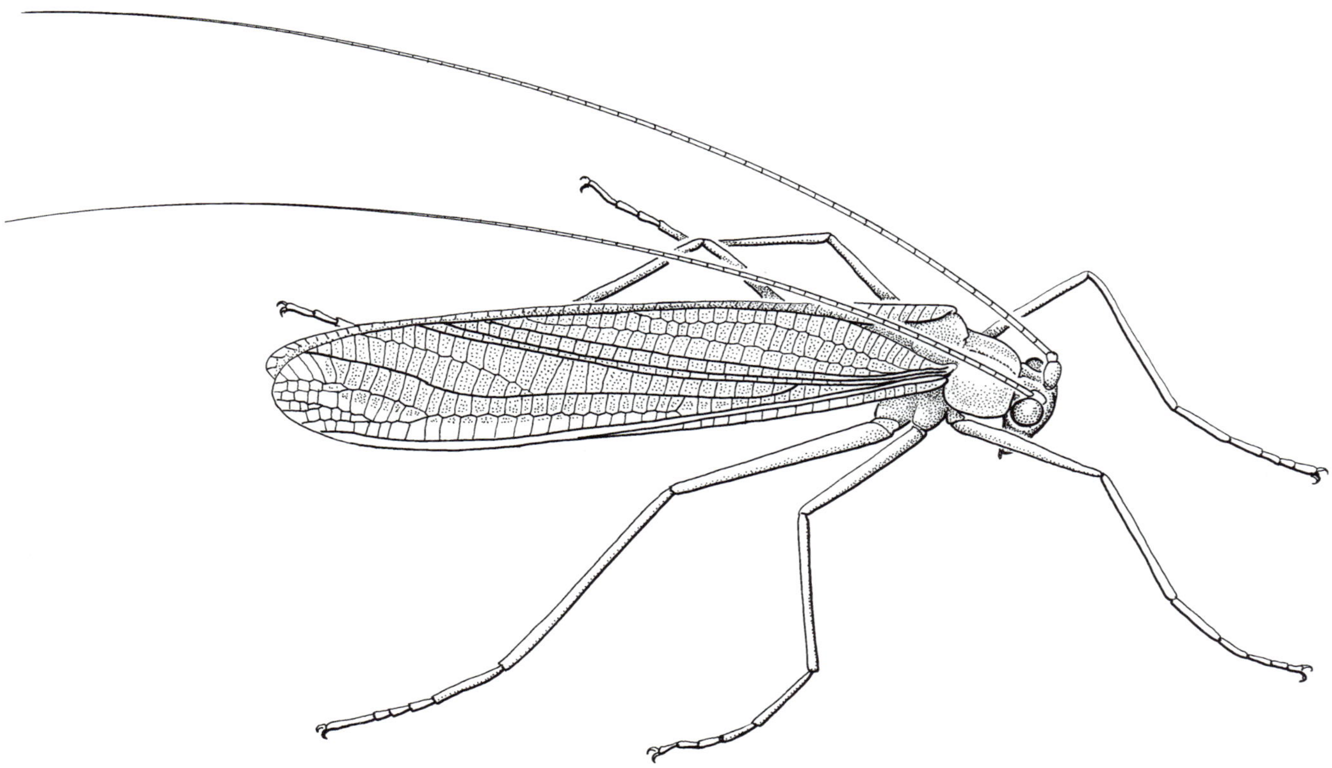

7.44. The caloneurodeans, such as *Paleuthygramma tenuicornis* (Paleuthygrammatidae) from the Permian of Russia, were enigmatic relatives of early Orthoptera and Phasmatodea that became extinct probably at the end of the Permian. Length 24 mm; redrawn from Carpenter (1992).

CALONEURODEA: THE CALONEURODEANS

Little is known of this extinct Paleozoic order of polyneopteran insects (Figure 7.44). The order is noteworthy for the secondary loss of the anal area in the hind wing, the unbranched and nearly parallel (or fused in a few groups) veins CuA and CuP in both the fore- and hind wings, and, where known, unsegmented cerci. Other features of the order include the strongly convex and concave wing veins; five-segmented tarsi; long, multisegmented antennae; and fore- and hind wings having a similar shape, venation, and texture. Presently there are nine families, many with a single genus, and the systematics of the group requires considerable revision. Nothing is known of caloneurodean biology aside from the fact that they were terrestrial, apparently primitively resembling cursorial orthopterids. The caloneurodeans are known only from the Late Carboniferous and Permian.

DERMAPTERA: THE EARWIGS

Almost anyone can immediately recognize an adult earwig with its stout, terminal forceps (Figure 7.45) and the short, tegminous forewings. Less conspicuous are the three-segmented tarsi; very distinctive, greatly expanded hind wing vannus (Figure 7.50), with a dramatically reduced remigium and unique folding mechanism; a prognathous head (lacking a gula); absence of ocelli; a subgenital plate formed by an enlarged seventh sternum in females; and a vestigial ovipositor. The uniquely expanded anal fan of the hind wing may eventually prove to be independently derived from that of other polyneopterans, in which case Dermaptera would be classified elsewhere. Earwigs are overflowing with unique features and the chronology of these is even apparent in the fossil record.

Dermaptera are distributed globally except Antarctica and the extreme Arctic, but most of the nearly 1,900 described species occur in tropical to warm-temperate habitats (e.g., Steinmann, 1986, 1989a,b,c, 1993) (Figure 7.46). Earwigs typically live in riparian habitats, in crevices, in leaf litter, or under the bark of trees. Most species are nocturnal and omnivorous; only a very few species are strictly herbivorous or carnivorous (Chopard, 1938). The cercal forceps are used to capture prey and are employed in mating and in folding the hind wings under the tegmina (Kleinow, 1966). Female earwigs exhibit extended maternal care over the eggs and early instar nymphs, carefully cleaning them to protect from invasive fungi (Herter, 1943; Rentz and Kevan, 1991). After two molts, however, nymphs must fend for themselves, or they will be eaten by the mother.

Earwigs have a long geological history. The oldest fossils to date are tegmina from the Late Triassic–Early Jurassic of England and Australia (e.g., Jarzembowski, 1999). As already

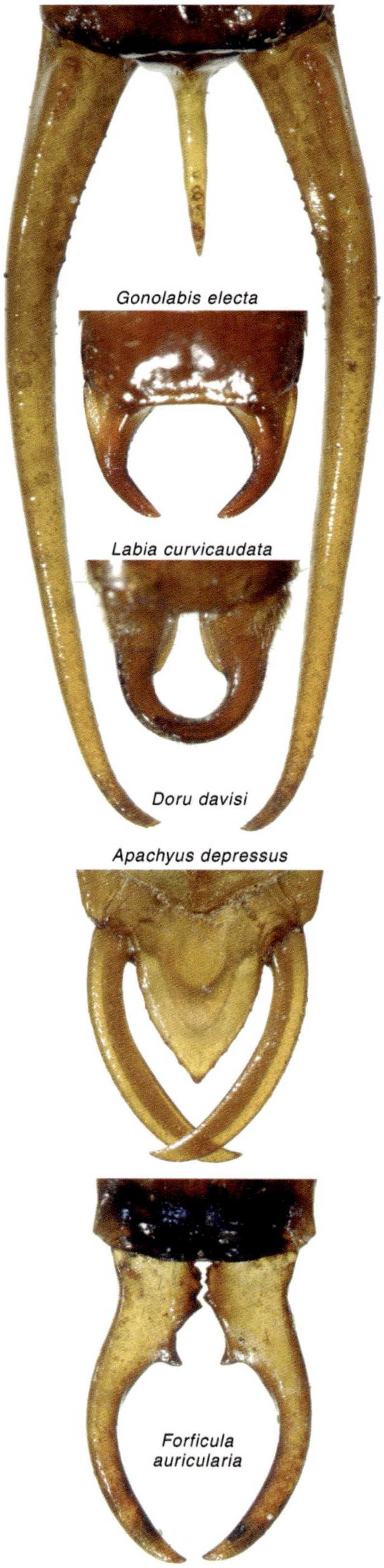

7.45. The modification of cerci into forceps is the most recognizable trait of the earwigs (order Dermaptera). The forceps are used to defend, to capture prey, and to assist in folding the fanlike wings under the short tegmina. Not to the same scale.

discussed and what should be of little surprise, some traits believed by neontologists to be diagnostic for an earwig do not apply to the earliest earwigs. Basal, extinct members of the order all had five-segmented tarsi, well-developed ovipositors, veined tegmina, and long, multisegmented cerci. Traditionally, primitive earwigs from the Late Jurassic and Early Cretaceous were classified in the suborder Archidermaptera (e.g., Bei-Bienko, 1936; Vishniakova, 1980; Carpenter, 1992), a paraphyletic stem group to modern Dermaptera. Archidermaptera in a restricted sense contains the Jurassic–Cretaceous families Protodiplatyidae, Turanoviidae, and Dermapteridae, which comprise the basalmost lineage of the order (Willmann, 1990a; Haas and Kukalová-Peck, 2001; Engel, 2003c) (Figure 7.47). Archidermapterans are a sister group to the Pandermaptera, which comprise two further suborders: Eodermaptera, for the Jurassic-Cretaceous families Semenoviolidae and Turanodermatidae, and the Neodermaptera, which contains all the modern lineages. Eodermapterans share with Neodermaptera the derived development of unsegmented, forcipate cerci but primitively retain venation in their tegmina, presence of ocelli, and pentamerous tarsi. The Neodermaptera have three-segmented tarsi, no ocelli, and lost venation in their tegmina. They first appear in the Early Cretaceous (e.g., Popham, 1990; Engel *et al.*, 2002) but may have originated in the latest Jurassic since there is a putative, undescribed pygidicranoid from the Jurassic of central Asia (Rasnitsyn and Quicke, 2002). Certainly, definitive neodermapterans (Figure 7.48) and recognizable pygidicranids are known by the mid-Cretaceous (Grimaldi *et al.*, 2002; Engel and Grimaldi, 2004c) (Figure 7.49). Tertiary earwigs, mostly of the Forficulidae, are preserved in several deposits, and a relatively unexplored diversity of species is known in Baltic (Burr, 1911), French (Nel *et al.*, 2002c); and Dominican ambers (Figure 7.51).

The internal phylogeny and classification of Neodermaptera has been in constant flux, with dramatically different arrangements of families and superfamilies by contemporaneous authors. Recent phylogenetic work has begun to shed light on relationships within the suborder (e.g., Haas, 1995; Haas and Kukalová-Peck, 2001; Colgan *et al.*, 2003) (Figure 7.51). Within Neodermaptera, the infraorder Protodermaptera, including the superfamilies Pygidicranoidea and Karschielloidea, is basal but perhaps paraphyletic (e.g., Haas, 1995; Haas and Kukalová-Peck, 2001). Unlike all other earwigs, the protodermapterans have ventral cervical sclerites of equal size, carinae on the femora, and a segmented pygidium. Almost all Cretaceous records of Dermaptera are protodermapterans. The infraorder Epidermaptera, including all other Neodermaptera, is characterized by the enlargement of the posterior ventral cervical sclerite, rounded femora, and fusion of the three pygidial sclerites (Popham, 1985).

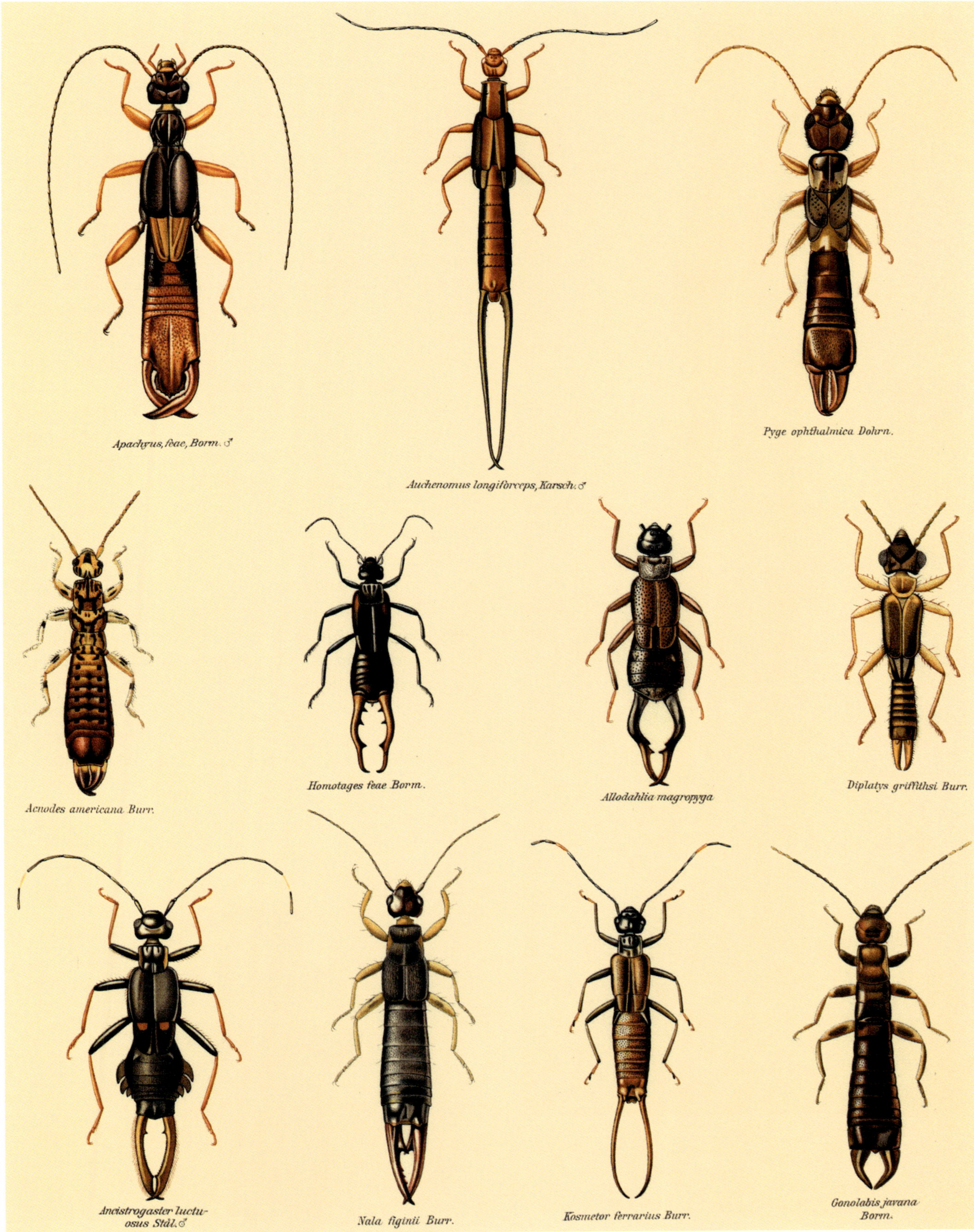

7.46. Representative Recent earwigs. Assembled from *Genera Insectorum*.

Although two additional suborders (or infraorders) have been traditionally recognized for the parasitic earwigs, both are actually derived Epidermaptera. The families Hemimeridae and Arixeniidae are ectoparasites with a suite of *paedomorphic* characters (i.e., the retention of nymphal traits in the adult). The Hemimeridae is perhaps closely related to Apachyidae (Klass, 2001b), while Arixeniidae is related to, if not derived from within, Spongiphoridae (Popham, 1985). Hemimerids live on murid rodents in Africa. Two genera are recognized: *Hemimerus*, which consists of nine species living on *Cricetomys* rats, and *Areomerus*, which consists of two species living on *Beamys* rats (Rehn and Rehn, 1936; Nakata and Maa, 1974). The five species of arixeniids live in the roosts of *Cheiromeles* bats (Molossidae) in southeast Asia, where they feed on secretions from the skin or on other insects invading the roost (Medway, 1958; Nakata and Maa, 1974). Hemimerids were at one time considered a distinct order, called Diploglossata, Dermodermaptera, or Hemimerina (e.g., Verhoeff, 1902; Brues and Melander, 1915; Popham, 1961). As with many ectoparasitic insects, fossil Hemimeridae and Arixeniidae have yet to be discovered, and the families are probably no older than the mid-Tertiary.

Dermaptera are believed by some to stem from the Permian order Protelytroptera owing to tegminous forewings and the large, uniquely formed anal fan (Kukalová-Peck, 1991; Haas and Kukalová-Peck, 2001). Some families have erroneously been placed in Protelytroptera, specifically the Cretaceous Umenocoleidae, which is actually a highly specialized lineage of roaches (Figures 7.70, 7.71). The possibility exists that Permian Protelytroptera were earwig progenitors, but definitive evidence is still required to establish this.

7.47. An archidermapteran, *Microdiplatys campodeiformis* (Protodiplatyidae), from the Late Jurassic of Karatau in Kazakhstan. Archidermaptera lacked the cercal forceps and primitively retained some veins in the forewings, though the tegminous forewings were short as in modern earwigs. PIN 2904/441; length, excluding cerci, 10 mm.

7.48. (right) A neodermapteran earwig from the Early Cretaceous of Brazil. AMNH; length 5 mm.

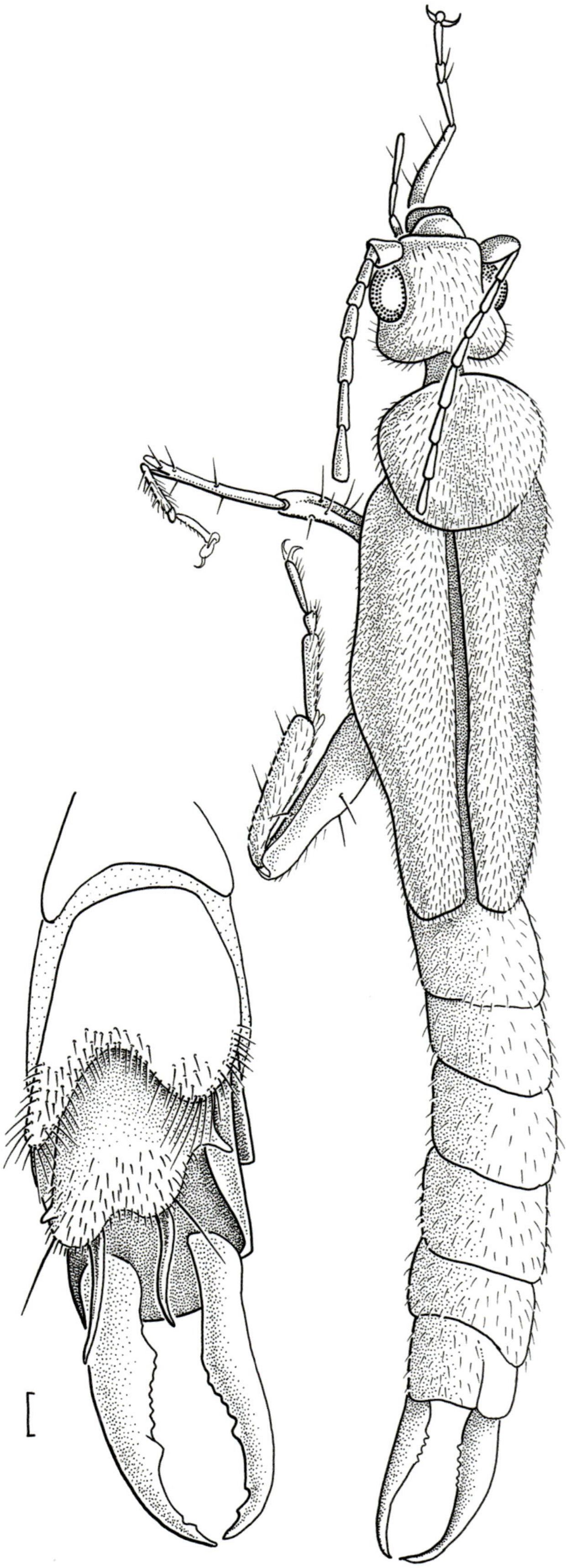

7.49. The earliest pygidicranoid earwig, *Burmapygia resinata* (Pygidicranidae), in mid-Cretaceous amber from Myanmar. Pygidicranids are among the most primitive of living earwigs. AMNH Bu274; length 6 mm; from Engel and Grimaldi (2004c).

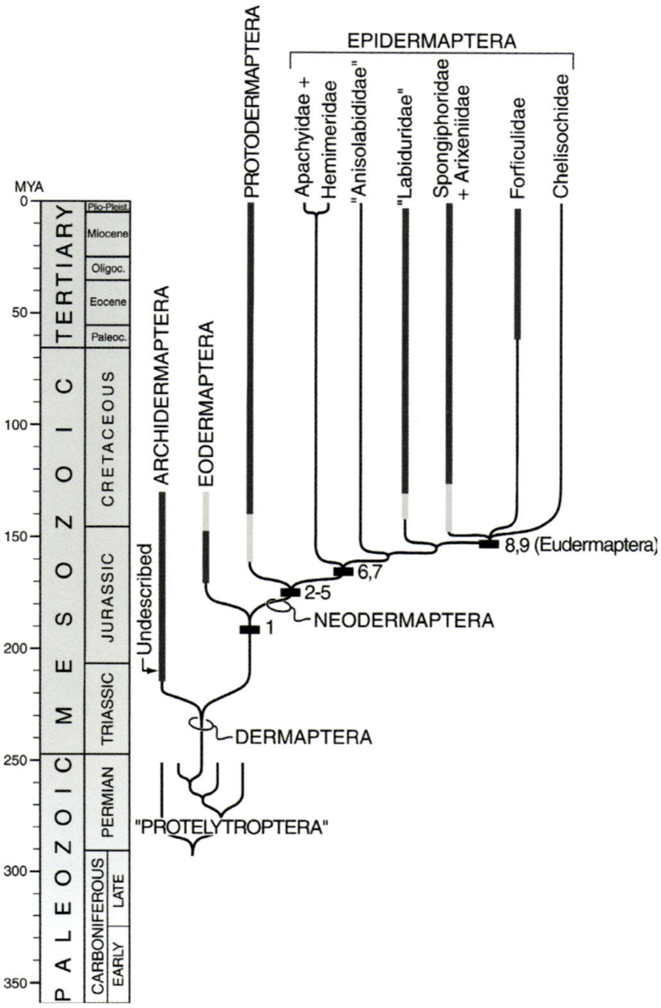

7.50. Phylogenetic relationships among major lineages of earwigs. Significant characters indicated in Table 7.2. Relationships based on Haas and Kukalová-Peck (2001). Thick dark lines are the known extent of fossils, lighter thick lines indicate fossils possibly belonging to those groups.

TABLE 7.2. Significant Characters in Dermaptera Phylogeny[a]

1. Cerci unsegmented, forcipate
2. Trimerous tarsi
3. Ocelli lost
4. Tegminal veins lost
5. Ovipositor reduced
6. Posterior, ventral cervical sclerite enlarged
7. Three pygidial subsegments fused
8. Reduction to single penal lobe and single virga
9. Expanded regions of anal and intercalary veins distinctly separated

[a] Numbers correspond to those on phylogeny, Figure 7.50.

7.51. A forficulid earwig in Miocene amber from the Dominican Republic. The hind wings of earwigs are remarkably distinctive, with most of the veins fused at the base and a large anal fan comprising most of the wing. Morone Collection, M3400; body length 6 mm. Photo: R. Larimer.

GRYLLOBLATTODEA: THE ICE CRAWLERS

Grylloblattodea (or Notoptera) represent an interesting, relict lineage today confined entirely to the Northern Hemisphere. In a class renowned for its overwhelming diversity, the Grylloblattodea, along with Zoraptera and Mantophasmatodea, hold the distinction of being the least diverse of insect orders. Today there are 26 species classified into five genera within a single family – *Grylloblatta* from the northwestern United States and southwestern Canada; *Grylloblattella* and *Grylloblattina* from the Russian Far East; *Galloisiana* from Japan, Korea, China, and Russia; and *Namkungia* from Korea but which probably makes *Galloisiana* paraphyletic (Storozhenko, 1997, 1998; Storozhenko and Park, 2002). These soft-bodied, wingless insects are typically found in leaf litter or under stones in cold temperate forests, often at higher elevations, although some blind Asian species have been discovered in caves (Namkung, 1982; Nagashima, 1990). Species are active at cold temperatures, and several studies have indicated optimal temperatures around 1–4°C (Mills and Pepper, 1937; Henson, 1957) (Figure 7.52). Although ice crawlers prefer low temperatures, they are not impervious to freezing. Individuals of *Grylloblatta* can be killed by ice formation within the body at around −8°C (Morrissey and Edwards, 1979). During winter months when night temperatures drop below freezing, ice crawlers likely survive under the snow-pack and near the soil where temperatures may deviate little from freezing (Atchison, 1979). These insects are omnivorous and typically scavenge dead arthropods but rely on plant material when frozen carcasses become scarce (Pritchard and Scholefield, 1978; Nagashima *et al.*, 1982).

Modern Grylloblattodea have numerous defining features, such as a median, eversible sac on the first abdominal sternum; the loss of ocelli (also in Mantophasmatodea and perhaps shared through common ancestry in these two lineages); and the asymmetrical male genitalia. Unique among all hexapods is the presence of a spina on the metathoracic sternum. Although difficult to place in the greater scheme of insect phylogeny, ice crawlers are probably basal members of Orthopterida and were also thought to be "living fossils" even by their discoverer (Walker, 1914, 1937). Wings are absent in modern grylloblattids and other characteristics are generally primitive: the five-segmented tarsi; long, multisegmented cerci; and an ovipositor composed of three stout pairs of "valvulae" intermediate to the orthopterid ovipositor. The third pair of valvulae are, in fact, the gonoplacs. In Grylloblattodea the gonoplac is greatly developed and incorporated into the ovipositor, representing an intermediate stage among orthopterids where the function of the reduced, second gonapophyses is assumed by the gonoplac. Gonoplac development in ovipositor construction is a typical orthopterid feature.

The presence of enlarged coxae, which is typical of basal Dictyoptera, has at times been used to support a relationship

7.52. A rock crawler, *Grylloblatta* (Grylloblattodea). The order today consists of 26 species in the northern parts of the Holarctic Region. They require cool to cold temperatures. Photo: Alex Wild.

of Grylloblattodea, along with Dermaptera and Mantophasmatodea, with this group. This placement has also been supported to some degree by molecular analyses (e.g., Maekawa *et al.*, 1999; Wheeler *et al.*, 2001). However, if this trait is primitive, as suspected by Hennig (1981), then it has little bearing on the placement of these orders. Alternatively, the analyses of Kamp (1973: later reanalyzed by Kuperus and Chapco, 1996) weakly supported a Grylloblattodea + Dermaptera relationship but placed these orders as relatives of the Orthopterida. Ultrastructure of the spermatozoa and embryological development also support a relationship to Orthoptera (Ando and Nagashima, 1982; Baccetti, 1982).

Generally believed to have an extensive geological history (e.g., Storozhenko, 1997, 1998; Vršanský *et al.*, 2001; Rasnitsyn and Quicke, 2002), the Grylloblattodea has essentially taken on the role of "Protorthoptera" in some modern systems of insect classification. In such systems, most protorthopteran families have been transferred to the Grylloblattodea. As previously discussed, many of these Paleozoic and Early Mesozoic families are not related, and any discussion of past "grylloblattodean" diversity based on these is misleading. In fact, this is highlighted by the phylogenetic outlines of such systems, where Grylloblattodea is depicted as giving rise to all other Polyneoptera (e.g., Storozhenko, 1997, 1998; Rasnitsyn and Quicke, 2002). Thus, quite in opposition to other authors, we presently believe the ice crawlers to have been of modest diversity in the beginning as well as today. Some taxa in the fossil record appear to share with modern Grylloblattodea a similar ovipositor construction and may truly represent stem-group ice crawlers. Interestingly, these Jurassic and Permian families, like Blattogryllidae (Jurassic), Megakhosaridae (Jurassic), and Tillyardembiidae (Permian), possessed fully formed wings (Figures 7.53, 7.54). The wing venation of these fossils has not been critically explored for derived features potentially uniting Grylloblattodea with other orders. Unlike Orthopterida, however, the precostal field was hardly developed in these fossils, but they did possess enlarged anal fans in the hind wing, supporting the polyneopteran position of the order. If Grylloblattodea and Mantophasmatodea are living sister groups, then the loss of wings might be a defining feature of the combined lineage, and the Mesozoic fossils may represent a stem group to both (further suggesting that the two orders should be united; see discussion

7.53. Although at one time considered an ally of the webspinners, *Tillyardembia antennaeplana* (Tillyardembiidae), from the Permian of Russia, is now recognized as an early relative of the Grylloblattodea and perhaps Mantophasmatodea. PIN 1700/1177; length 22 mm.

7.54. *Blattogryllus karatavicus* (Blattogryllidae) from the Late Jurassic of Karatau in Kazakhstan, a stem-group grylloblattodean. Modern Grylloblattodea and their close relatives Mantophasmatodea are wingless, but stem groups to the lineage were fully winged and possessed the anal fan typical of polyneopterans. PIN 2554/227; length 28 mm.

later in this chapter). Other fossil lineages sometimes placed in the Grylloblattodea are related to other orders, such as Lemmatophoridae, which is more closely related to Plecoptera. Limited available evidence suggests that Grylloblattodea diversity has changed little through geological time, although the loss of wings in the Cretaceous or Early Tertiary represents a significant morphological modification in Recent ice crawlers.

MANTOPHASMATODEA: THE AFRICAN ROCK CRAWLERS

The African rock crawlers are the most recently discovered order of insects (Klass *et al.*, 2002) (Figures 7.55, 7.56), and, not surprisingly, they are already stirring debate. Little information has accumulated or permeated into the literature, and thus any discussion for the moment must remain tentative. Modern Mantophasmatodea occur in xeric, rocky habitats in southern Africa. These insects are apparently aggressive carnivores, pouncing on prey and grasping their victims with the fore- and mid-legs. Rock crawlers tend to be nocturnal, feeding on unsuspecting moths, silverfish, and roaches, but take most small arthropod prey they can catch and subdue. During daylight hours rock crawlers hide among stones and the bases of plants, particularly clumps of grass or spiny shrubs in South Africa's Succuluent Karoo (e.g., Walker, 2003). The 15 species are segregated into three families, although these should perhaps be downgraded to subfamilies of a single family owing to the homogenous habitus of all members of the group. The principal papers for Mantophasmatodea are Zompro (2001), Klass *et al.* (2002, 2003a,b), Zompro *et al.* (2002, 2003), and Engel and Grimaldi (2004b).

The group is monophyletic, defined by the combination of the following traits: a loss of ocelli (although, as mentioned, perhaps a trait of Grylloblattodea + Mantophasmatodea); a hypognathous head; loss of the epistomal sulcus, and the unique subgenal sulcus that loops from the posterior mandibular articulation to the anterior tentorial pit and then back to the anterior mandibular articulation; a loss of wings (perhaps also shared with Recent Grylloblattodea); an enlarged pretarsal arolium with a series of long setae; a vomer-like process articulating on the apical margin of the tenth sternum; and unsegmented cerci (modified for clasping in males) (Klass *et al.*, 2002, 2003a). Typical for a polyneopteran, few traits clearly unite the group with any other order.

Shortly after the description of Mantophasmatodea, Tilgner (2002) highlighted that the group might represent a derived lineage of Caelifera, perhaps near the Proscopiidae. This hypothesis was based on the observation that the cryptopleuron, diagnostic for Orthoptera, has been secondarily lost in Proscopiidae and that the apparently five-segmented tarsi of Mantophasmatodea are, in fact, *synsclerotic* (i.e., united to form a single, compound structure) and resemble some "trimerous" Caelifera. Tilgner believed that there was little to truly differentiate Mantophasmatodea from such Orthoptera. The hypognathous head, which is similar to that of many Orthoptera, lends credence to this observation. However, as Klass (2002) indicated, derivation from within Orthoptera is unlikely owing to the absence of defining orthopteran characters, such as the development of a cryptopleuron and jumping hind legs. Furthermore, the synsclerotic conditions in Orthoptera and Mantophasmatodea are not homologous. In Orthoptera, the basal three tarsomeres are entirely fused to form a single subsegment of the podite. Conversely, in Mantophasmatodea the three basal segments are still differentiated by distinct, dorsal grooves (Klass *et al.*, 2002, 2003a; Klass, 2002). The possibility, although very unlikely, does exist that Mantophasmatodea may prove to be derived Orthoptera, with several secondary reversals to primitive traits, just as Phasmatodea or Titanoptera may similarly prove to be derived from the Caelifera. None of these seem likely, and it is far more likely that Mantophasmatodea are the living sister group to Grylloblattodea.

The gonoplac is short in Mantophasmatodea but is sclerotized and more developed than the second valvulae, as in Orthopterida, and this structure possibly acts as the functional ovipositor. This would further support a placement of

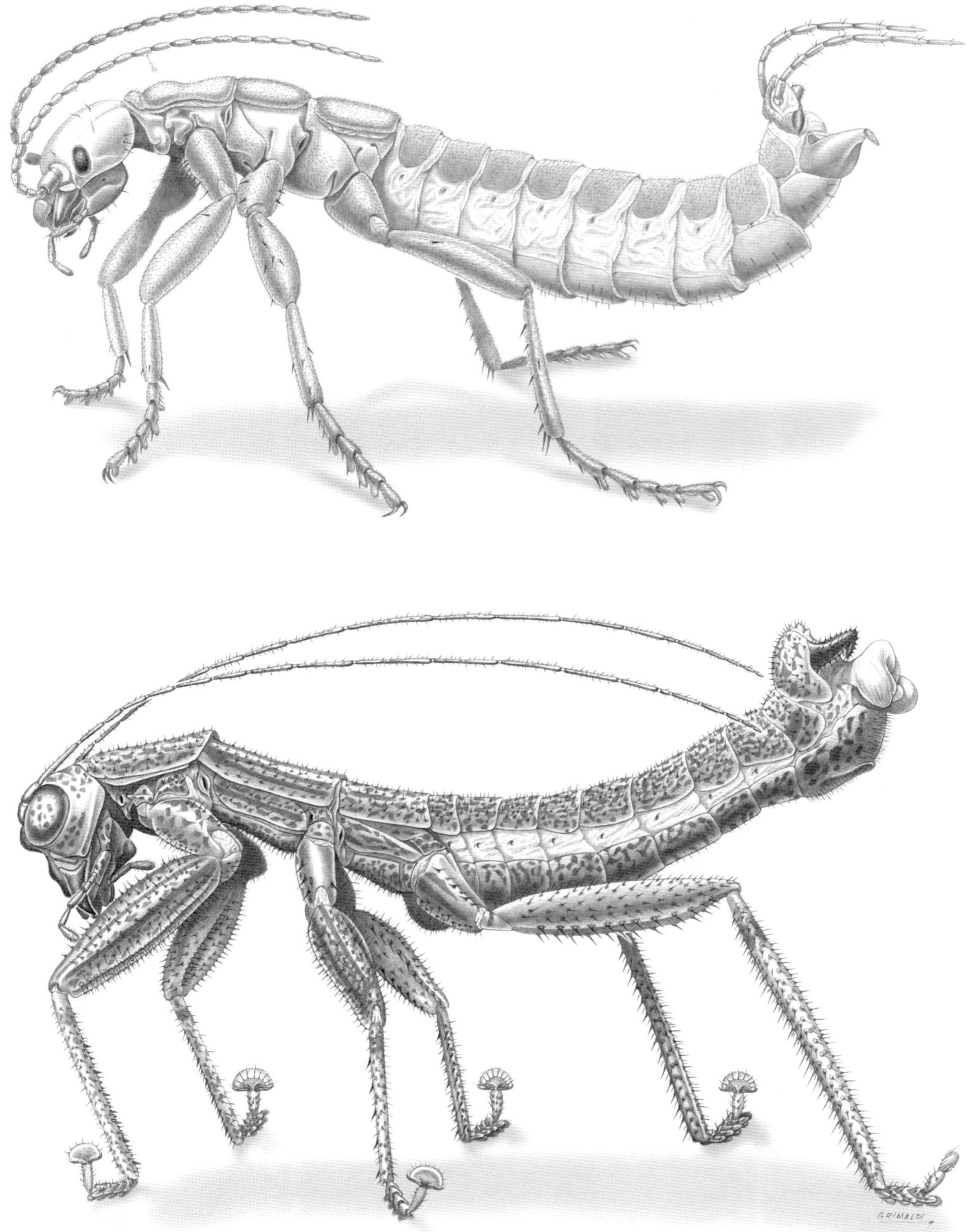

7.55. The closely related and relict orders Grylloblattodea (above: *Grylloblatta washoa*) and Mantophasmatodea (below: *Karoophasma bieolouwensis*). The former is today a Northern Hemisphere group, while the latter lives in southern Africa. To the same scale, above: length 10.5 mm.

7.56. A male *Lobophasma redelinghuysensis* (Austrophasmatidae), of the recently described order Mantophasmatodea, from the Western Cape Province's fynbos in South Africa. Photo: M. D. Picker.

Mantophasmatodea near Orthopterida. Mantophasmatodea possess a vomer-like process on the apex of the tenth sternum and therefore somewhat resemble Phasmatodea. However, Klass *et al.* (2002, 2003a) dismissed this feature as homologous with the vomer of Phasmatodea owing to the articulation of this sclerite along its *posterior* margin to the tenth sternum, versus the anterior margin in Phasmatodea. Alternatively, it could be interpreted that this inversion is merely a unique alteration of the trait in Mantophasmatodea. Despite the superficial similarity of many Mantophasmatodea to *Timema*, the former lack a micropylar plate in eggs, although they do possess a circular ridge reminiscent of an operculum (Klass *et al.*, 2002), and the diets of *Timema* and mantophasmatodeans are completely different. At present there is no justification for a placement of Mantophasmatodea within Orthopterida, particularly not near Phasmatodea.

Interestingly, like the coelacanth, this group was known as a fossil before living species were known. The mid-Eocene Baltic amber genus *Raptophasma* was identified for years as an enigmatic insect perhaps allied to walking sticks (e.g., Arillo *et al.*, 1997), but it was not formally described until recently (Zompro, 2001). A second Baltic amber genus, *Adicophasma*, was identified and described as being more closely allied to the modern species than to *Raptophasma* because it had stout spines on the legs and body typical of some Recent species (Engel and Grimaldi, 2004b) (Figure 7.57). These Tertiary fossils possessed some of the distinctive traits of the order, such as the peculiar track of the epistomal sulcus and the very large, setose arolium, as well as the fused basal tarsomeres. However, these fossils lack the reduction of the compound eyes seen in living Mantophasmatodea.

Biogeographically, a restriction of Mantophasmatodea to sub-Saharan Africa (Klass *et al.*, 2002, 2003a; Picker *et al.*, 2002) is a tantalizing gondwanan juxtaposition to the Grylloblattodea, itself confined to Laurasia. If a Mantophasmatodea + Grylloblattodea relationship is ever conclusively demonstrated, then this would be highly significant and sim-

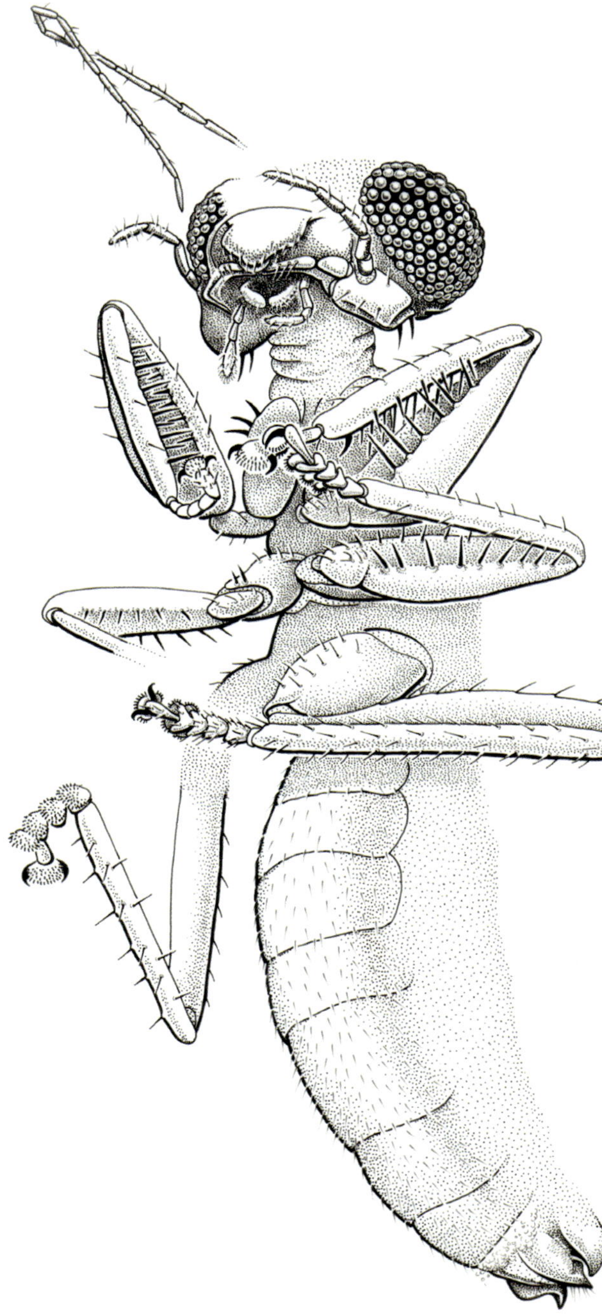

7.57. Mantophasmatodeans today are restricted to sub-Saharan Africa; at least as recently as the Eocene they were more widespread, as shown by *Adicophasma spinosa* in Baltic amber. AMNH; length 4.1 mm; from Engel and Grimaldi (2004b).

7.58. *Phylloblatta gallica* (Phylloblattidae), from the Late Carboniferous of Commentry, France. Paleozoic insects closely resembling modern roaches were diverse and abundant; they were not true roaches, however, but rather stem-group dictyopterans, or "roachoids." NHM In. 7296; body length 40 mm.

7.59. Phylloblattid "roachoid" from the Late Triassic of New South Wales, Australia. AMF38257; longest length 62 mm.

ilar to the apparent Laurasian–Gondwanan split in the Plecoptera. Under such an hypothesis the loss of wings and ocelli would be shared features that evolved in an immediate common ancestor of both orders, and Mesozoic "Grylloblattodea" fossils could be stem groups to both lineages, presumably with these traits appearing sometime in the Cretaceous or earliest Tertiary.

DICTYOPTERA

Few insect lineages have species as disparate as those in the Dictyoptera, comprising the predatory mantises, saprophagous roaches, and the highly social, cellulose-feeding termites. A close relationship of these orders would seem implausible were it not for distinctive structures in the male and female reproductive systems, the proventriculus, and evidence from DNA sequences, as well as several relict, transitional species. Roaches are commonly believed to be ancient insects evolving since the Carboniferous, though in fact fossils of modern families are no older than Cretaceous – an age on a par with the other two orders. Dictyoptera, in fact, is probably relatively recent, extending to the Jurassic, but for which there is currently very little evidence.

A popular belief in Paleozoic roaches (e.g., Guthrie and Tindal, 1968) is understandable because abundant Carboniferous fossils possessed many of the features of modern roaches, including the tegminous forewings and large, shield-like pronotum (Figures 7.58, 7.59). However, Paleozoic "roachoids" differed from modern roaches in several key respects, most significantly by possession of a large external ovipositor – a very primitive trait appearing before insects even evolved flight. The common ancestor of the lineage that includes the modern families of roaches, termites, and mantises had a highly reduced ovipositor, as all species have today. This ancestor probably derived from one group of the Paleozoic roachoids, perhaps sometime in the Jurassic (Grimaldi, 1997b). Names have been proposed to distinguish these groups: Order *Blattaria* for the modern families of roaches; *Dictyoptera* for the orders Blattaria, Isoptera, and Mantodea and Paleozoic roachoids; and *Blattodea* or *Blattoptera* for the paraphyletic assemblage of Paleozoic roachoids (Hennig, 1981; Grimaldi, 1997b).

DICTYOPTERAN RELATIONSHIPS

Understanding relationships in Dictyoptera is essential for understanding key evolutionary events, such as the origin of eusocial termites or specialized, predatory mantises. Unfortunately, virtually every possible set of relationships among the three orders has been proposed, which is partly the result of different analyses and kinds of data but is more fundamentally caused by a common (and erroneous) assumption that all three orders are monophyletic. Nalepa and Bandi (2000), Thorne *et al.* (2000), and Deitz *et al.* (2003) provided the most recent summaries of the various hypotheses and their proponents. If the Isoptera are actually closely related to particular roaches, as current evidence indicates, the Blattaria would be a paraphyletic order, or, conversely, termites should be classified as a highly specialized group of roaches (Figure 7.60).

The seminal work on the comparative morphology of Dictyoptera was by McKittrick (1964). Kristensen (1975) provided the first rigorous hypothesis of dictyopteran relationships, resorting to more structures than did McKittrick and organizing the characters cladistically. Thorne and Carpenter (1992) cladistically analyzed morphological and biological characters of the three orders gleaned from published works, but recoding the polarity of several characters in their study was found to alter relationships substantially (Kristensen, 1995; Klass, 1997; Deitz *et al.*, 2003). Considerable effort has been spent on examining the relationships of these insects, such as the following: Grandcolas (1994, 1996), using primarily roach male genitalia; Kambhampati (1996), using a minimal sampling of 440 bp of the 12S rRNA gene; and Grandcolas and D'Haese (2001), reanalyzing DNA with morphological data. Lo *et al.* (2000) sequenced approximately 2,300 bp of three genes (18S rDNA, COII, and EG cDNA). Finally, Klass (1997, 1997/8, 1998, 2001a) has done detailed, comprehensive studies of the proventriculus (see also Miller and Fisk, 1971), male genitalia, and the vestigial ovipositor. These are very significant structures for understanding dictyopteran evolution, which deserve special commentary. The paper by Deitz *et al.* (2003) is based on much of Klass's morphological work.

The proventriculus is a gizzard-like structure in the foregut of various insects. Its walls are muscular and lined with spines or teeth, which grind food. In most Dictyoptera the proventriculus has six internal, longitudinal folds, or *plicae*, each possessing sclerites with teeth (Figure 7.61). In roaches, these plicae are bilaterally symmetrical. Imagine in a cross section of the foregut that plica 1 is at the 12:00 position, then continuing clockwise, plica 4 is at the 6:00 position and plica 6 at the 10:00 position. In roaches, plicae 2 and 6 are identical, as are plicae 3 and 5. Plicae 1 and 4 each are unique. *Cryptocercus* has only slight bilateral symmetry, and in termites the symmetry has become entirely radial (all plicae are identical). In mantises the proventriculus is bilaterally symmetrical, but the teeth are much smaller than in roaches.

TABLE 7.3. Significant Characters in Dictyopteran Phylogeny[a]

1. Pronotum large, shield-like
2. Forewing tegminous; with arched, groove-like claval suture
3. Ovipositor reduced: small and external, or highly vestigial and largely/mostly internal
4. Ovipositor extremely reduced, valvulae mostly internal (slightly protruding from vestibulum in Mantodea); eggs deposited in an ootheca formed from secretions of colleterial glands
5. Tentorium perforated
6. Male genitalia asymmetrical
7. Ovipositor vestigial, valvulae entirely internal
8. Median ocellus lost
9. Ootheca formed within vestibulum (this feature lost in all but one species of termite)
10. Fat bodies with specialized cells (mycetocytes) harboring symbiotic bacteria, which are transferred to offspring via ovaries
11. Proventriculus with hexaradial arrangement of folds and teeth (vs. bilateral symmetry)
12. Feed on wood or other lignocellulose plant material
13. Digestion of lignocellulose via mutualistic hindgut protists (*Trichonympha, Leptospironympha, Oxymonas*)
14. Sociality: living in colonies with extended parental care, or eusocial
15. Isopteran features (see Figure 7.88, Table 7.4)
16. Oothecal rotation within vestibulum
17. Ovoviviparity/viviparity: oothecae retained in uterus, birth of live young
18. Predatory, forelegs raptorial with large apical spur or spine on tibia
19. "Pseudovein" present: an oblique, veinlike structure near the basal forks of M, CuA
20. Fore femur armed with thick, stiff spines
21. Claval furrow on forewing reduced
22. Fore femur with "discoidal spines" on ventral surface near proximal end
23. Metathoracic hearing organ
24. Pronotum long

[a] Numbers correspond to those on phylogeny, Figure 7.60.

Male roaches and mantises have genitalia hung to the left, and are so asymmetrical that large phallomeres – highly elaborate ventral appendages of abdominal segment IX – on the left side have barely recognizable corresponding parts on the vestigial right side. Roach phallomeres have various sclerites, folds, lobes, and spines, and these have challenged morphologists seeking to identify homologous parts, even leading to seemingly prosaic disputes (e.g., Grandcolas, 1999, versus Klass, 2001a). Phallomeres in Isoptera are almost impossible to identify because the male genitalia are reduced to minute knobs, virtually lost in all but a few termites like *Stolotermes*.

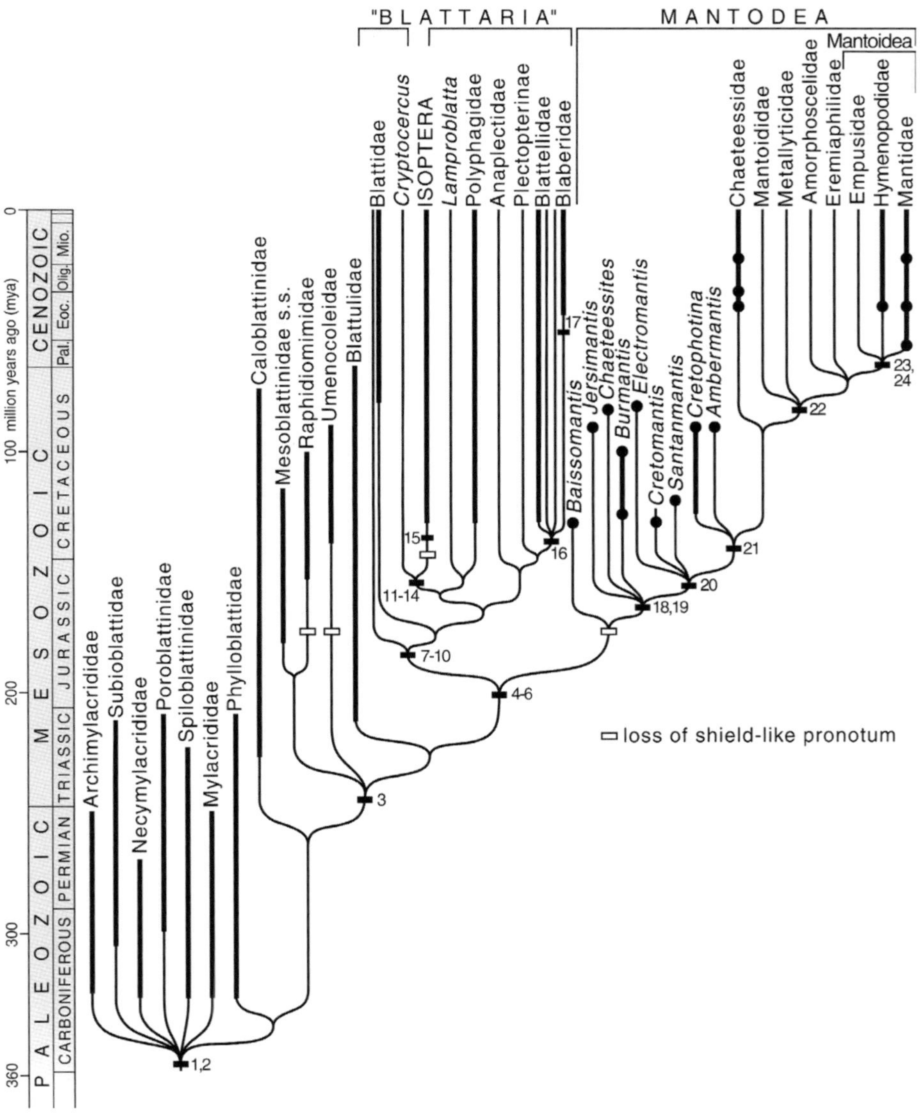

7.60. A phylogeny of Dictyoptera, including roaches and the positions of mantises and termites. See Table 7.3 for characters (numbers). Relationships among stem-group Dictyoptera ("roachoids") based on Vršanský *et al*. (2002); those of Recent roach families are from Klass (1995) and Klass and Meier (2000); those of basal mantises are from Grimaldi (2003b).

The mechanics of how termites couple and copulate is unknown.

The ovipositor in Dictyoptera conforms to the basic structure of the insect ovipositor (Scudder, 1961), only it is much reduced (Figure 7.62). It belongs to abdominal segments VIII and IX and is comprised of paired appendages. These include two pairs of *gonocoxae* ("*valvifers*") on segments VIII and IX, so called because of apparent serial homology with the coxae of the legs; two pairs of *gonapophyses* (first and second valves or valvifers), one each on segments VIII and IX; a pair of *gonoplacs* on segment IX (third valvifers); and a pair of *gonangula* also on segment IX. The longest appendages appear in Dictyoptera gonapophyses VIII and the gonoplacs, the others often being rudimentary knobs. Mantises have the best-developed ovipositor, which is slightly protruding; roaches and the termite *Mastotermes* have an ovipositor entirely concealed within an internal pouch, the *vestibulum*, and suspended from its roof. All other termites have an ovipositor that is virtually lost.

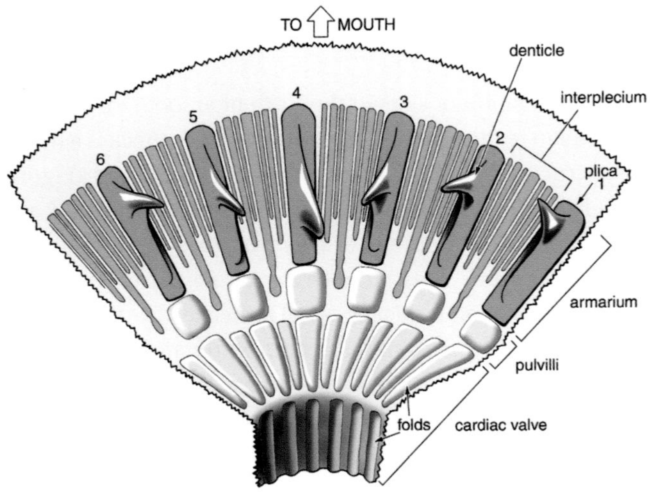

7.61. Schematic diagram of generalized dictyopteran proventriculus ("gizzard"), based mostly on roaches. The structure has been split lengthwise and unfolded. Redrawn from Klass (1998).

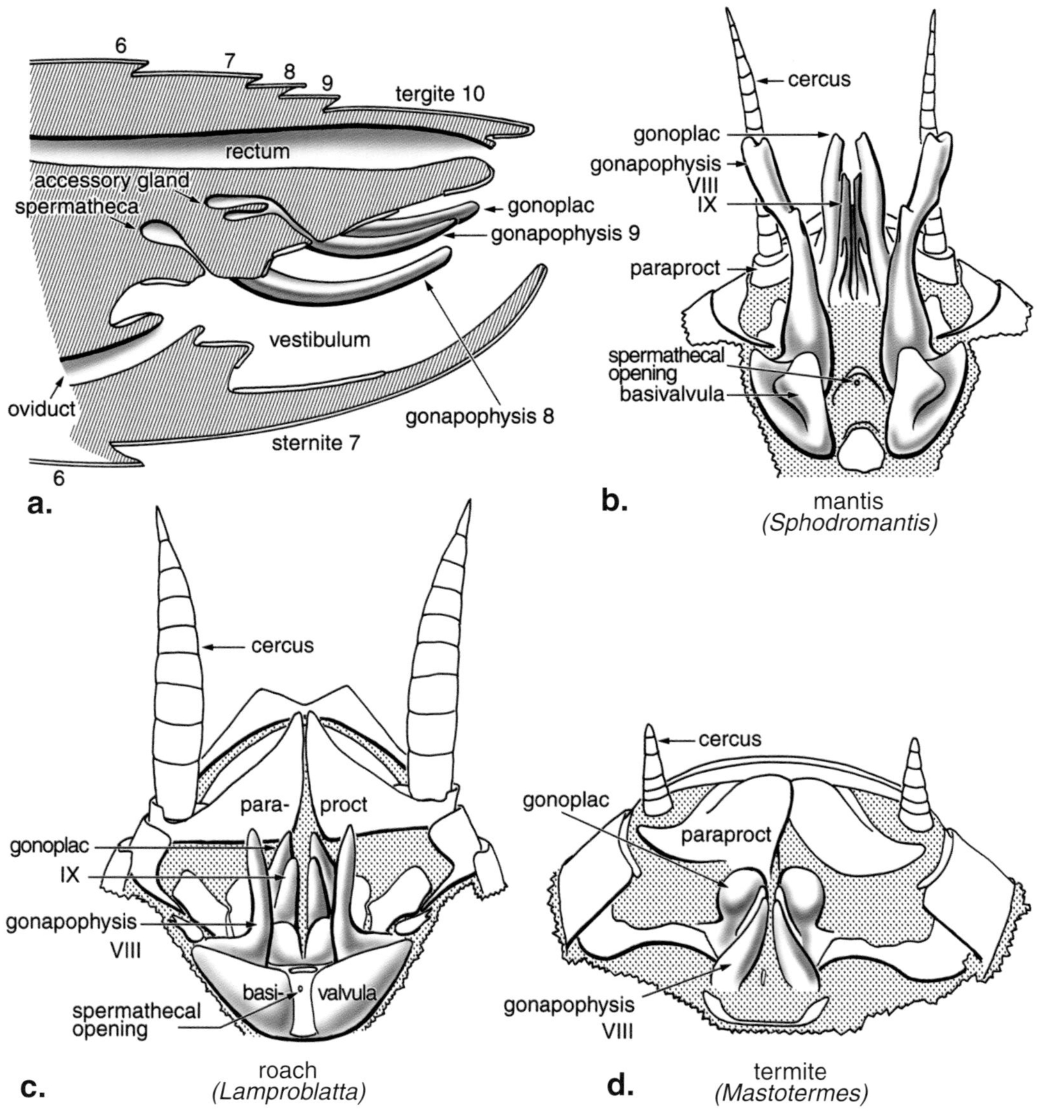

7.62. The reduced ovipositor in Recent Dictyoptera. (a) Lateral view through a midsection of the abdominal apex of a generalized roach. (b–d) Three representative Dictyoptera, showing parts of the ovipositor attached to the dorsal wall of the vestibulum (the terminal sternites have been removed). Redrawn from Klass (1997).

The deposition of eggs in the form of an *ootheca* is also a groundplan feature of the Dictyoptera. The ootheca is a hardened structure in which rows of eggs are encased (Figure 7.63). It is produced by all mantises, most roaches (except species that are *viviparous* or *ovoviviparous*), and the most primitive termite, *Mastotermes darwiniensis*, though it is extremely vestigial in this termite. Oothecae clearly function to protect the eggs from desiccation, predators, and parasitoids. The structure of the ootheca in roaches and mantises is considerably different, though in both groups they are formed as the eggs pass into the vestibulum and are coated with secretions from the *accessory* or *colleterial glands* (in mantises the coating is applied externally). Mantises lay a soft, foamy ootheca around a plant stem or twig, or against a stone or tree trunk, which hardens into a consistency like styrofoam. In roaches the ootheca is smooth and beanlike, with a highly sclerotized covering. The structure of roach and mantis oothecae varies among taxa, a result of molding by the ovipositor valves. Because roach-type oothecae could not pass through a long, narrow ovipositor, and since they require manipulation by the tiny ovipositor valves, oothecae were clearly not laid by Paleozoic roachoids with long ovipositors, despite the occasional report of putative oothecae from the Paleozoic (Brown, 1957).

BLATTARIA: THE ROACHES

Roaches are mostly denizens of wet, tropical forests, cryptically dwelling under stones and bark and in logs, emerging under darkness. Various species have become *troglobites*; a few have become conspicuous, diurnal insects; but they are all extremely polyphagous, feeding on decaying and fresh leaves, fruits, fungi, rotten wood, even bird droppings, guano, and dung. Pest species are easily cultured, which, with their

7.63. Egg pods, or oothecae, of a roach (protruding from the abdomen) and of two mantises, which is one of the defining features of living Dictyoptera. All species of termites, except for the most basal one, have lost the oothecal trait; they lay their eggs singly. Not to the same scale.

generally large size, makes roaches highly desirable for experimental research. Many species are pests in the tropics, but in northern cities *Periplaneta americana*, *Naupheta cinerea*, and *Blatella germanica* have become pervasive occupants. The large, flightless, Madagascar species, *Gromphadorhina portentosa*, is a popular pet, delighting children when it hisses by forcing air through its spiracles. The biology of roaches has been reviewed by Guthrie and Tindal (1968), Cornwell (1968), and Roth (1991), the last of whom also reviewed the taxonomy of the Blattaria.

To those only aware of the pest species, roaches are surprisingly diverse (Figure 7.64), with some 4,000 described species in 460 genera, most of them tropical. The family-level classification, unfortunately, is unsettled. The number of families that are typically recognized centers around six, but there is considerable dispute about the taxonomic rank and composition of each and even the monophyly of some. Roth (1991) presented a traditional classification, based largely on the work of McKittrick (1964).

The families of roaches traditionally recognized are the Blaberidae, Blattidae, Blattellidae, Cryptocercidae, Nocticolidae, and Polyphagidae (Roth, 1991). The Anaplectidae (a name based on the distinctive network of hind-wing veins) is a small family not mentioned by Roth (1991). The Blattidae is the largest family, with approximately 525 species, including all-too-familiar pest species of *Periplaneta*. Females in Blaberidae possess a brood sac, wherein the ootheca or loose eggs are retained until the eggs hatch (ovoviviparity). The blaberid *Diploptera punctata*, widely distributed in the IndoPacific, is truly *viviparous*, because the embryo is nourished through the uterine walls.

The phylogeny of roach families was explored by McKittrick (1964), Grandcolas (1994, 1996), Kambhampati (1996), and Klass (1997, 1998, 2001a). Kambhampati's study is based on a minimal sampling of taxa and only a portion of one gene. The more comprehensive studies by Grandcolas and Klass relied heavily on the morphology of male and female genitalia, which Klass supplemented with observations on the musculature of the genital structures. Hypotheses by both morphologists agree on most aspects of relationships: Blattidae ((Polyphagidae + Cryptocercidae) (Anaplectidae (Blattellidae (Blaberidae)))). While Klass and Grandcolas agree that the basalmost divergence is between the Blattidae and all other roaches, the family is also considered to be polyphyletic (Klass, 1997, 1998). Also, the Blattellidae is probably paraphyletic with respect to the Blaberidae. The most intriguing roaches are in the genus *Cryptocercus*, which is the subject of greatest disparity between the phylogenies by Grandcolas and Klass. Figure 7.60 summarizes relationships of roach families and particular genera.

Fossil Record

Paleozoic "roachoids" were among the most abundant animals in the extensive coal swamps of the Carboniferous (e.g., Figure 7.65). Their roachlike features included a large, discoid pronotum concealing most of the head; large, flattened, splayed coxae for running; a flattened body with tegminous forewings; and, in some, forewings with a distinctive, strongly curved CuP vein, or claval furrow. But, they primitively possessed a long, external ovipositor, a feature that persisted up to the Late Jurassic (e.g., Vishniakova, 1968) (Figure 7.66). Generally three pairs of valvulae are sometimes discernable (Sellards, 1904; Vishniakova, 1968), and what has been interpreted as the first pair of valvulae (gonapophyses 8) are the longest appendages (Figure 7.67), which are also the longest ones in modern Dictyoptera. Presumably, these roachoids used their long ovipositor to insert eggs into soil and crevices within rotting wood and humus. The taxonomy of Paleozoic roachoids is a confusing array of names, many based on fragmentary

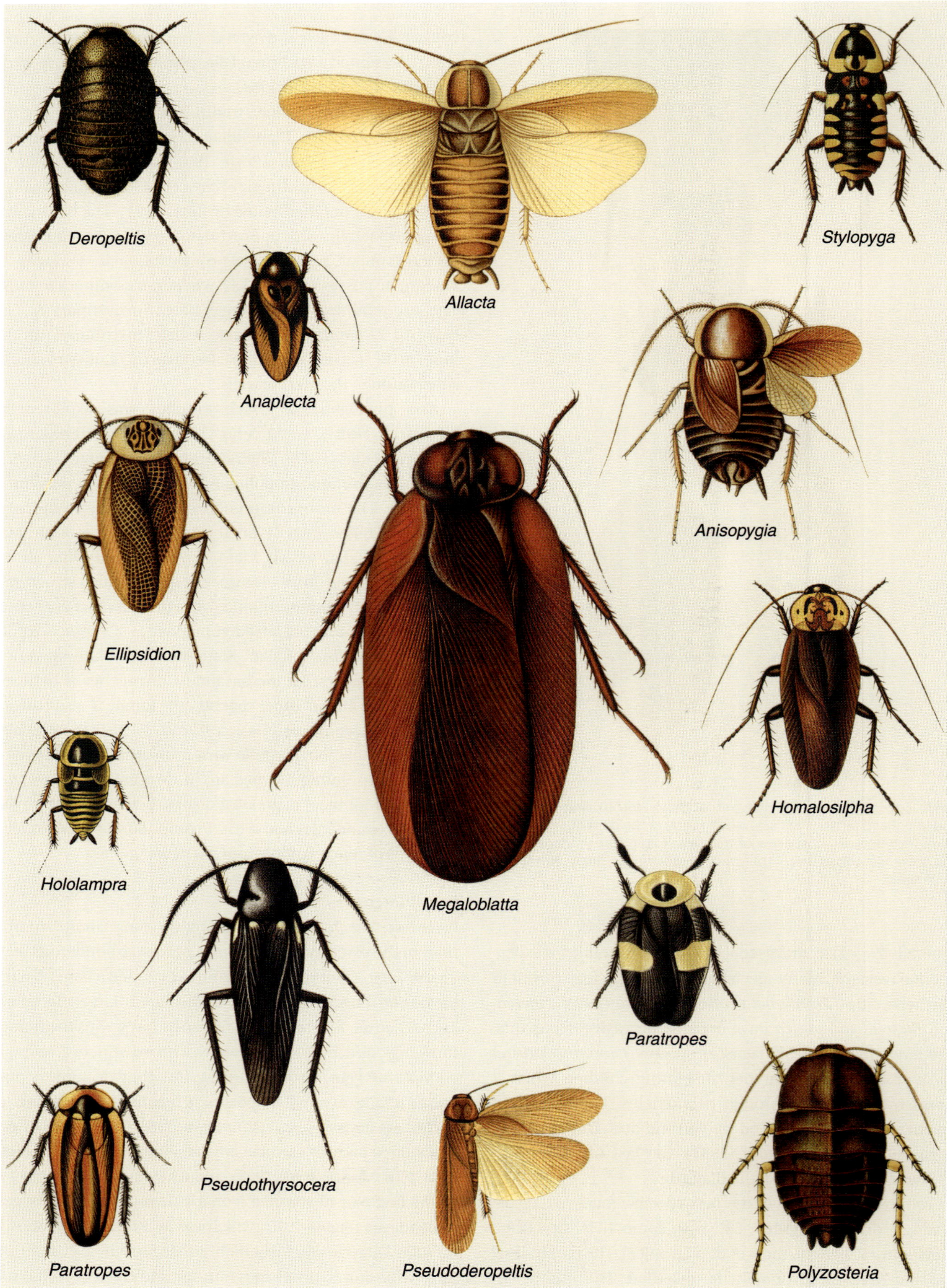

7.64. Representative Recent roaches. Compiled from *Genera Insectorum* (Shelford 1908–1910, fasc. 55, 73, 74, 109).

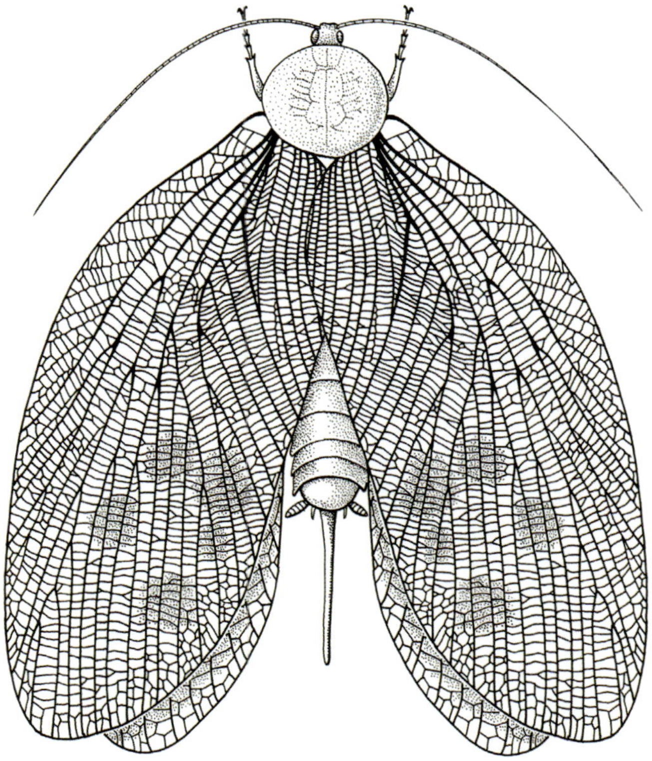

7.65. Reconstruction of a stem-group "roachoid," *Manoblatta* Archimylacrididae *bertrandi*, from the mid-Carboniferous of France, 310 MYA. Redrawn from Laurentiaux *et al.* (1979); forewing length 4.5 cm.

7.66. *Karataublatta longicaudata*, from the Late Jurassic of Kazakhstan. Note the long ovipositor. This is one of the last appearances in the fossil record of roachoids with such long ovipositors. PIN 2066/774; body length 32 mm.

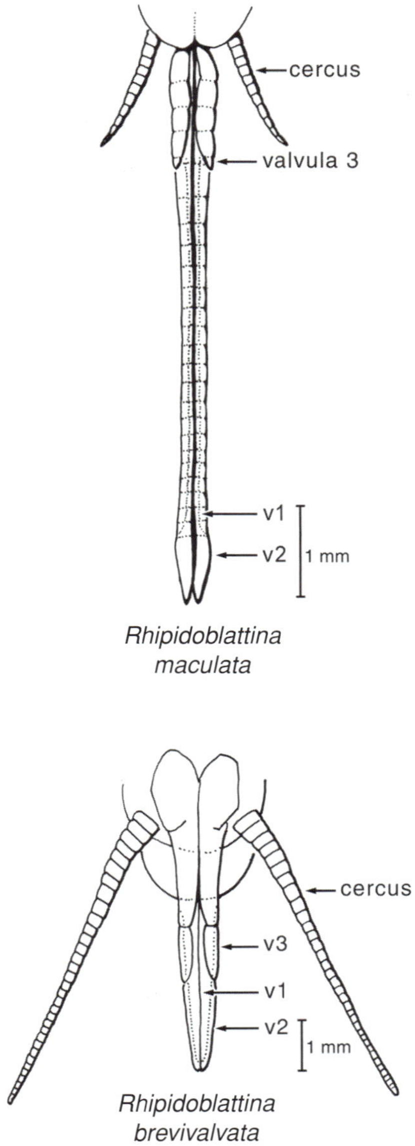

7.67. Ovipositors of two stem-group roachoids from the Late Jurassic of Kazakhstan. Reconstructions redrawn from Vishniakova (1968).

specimens, to which Schneider (1983, 1984) has brought some order. The recognition of the three Paleozoic families Archimylacrididae, Necymylacrididae, and Mylacrididae (Vršanský *et al.*, 2002) is largely based on the classification of Schneider (1983). First appearances occurred in the early part of the Late Carboniferous, (approximately 320 MYA) to the Late Permian (255 MYA). Most of these Paleozoic roachoids had broad tegmina with intricate, dichotomous venation that was very similar to the pinnules of *Neuropteris* tree ferns. Fossils of the two, in fact, are commonly confused, which led one entomologist to suggest that perhaps Paleozoic roachoids camouflaged themselves in foliage.

Three other families of roachoids replaced these Paleozoic families in the early Mesozoic, one of which, the Phylloblattidae, is considered the sister group to all living Dictyoptera as

7.68. A slender raphidiomimid with patterned wings, *Rhipidoblattina katavica*, from the Late Jurassic of Kazakhstan. Raphidiomimidae were predatory roachoids from the Jurassic and Cretaceous. PIN 2066/441; body length (including ovipositor) 28 mm.

well as six extinct families that existed later in the Mesozoic and became extinct presumably by the end of the Cretaceous. We tend to think of early fossils as very generalized, primitive organisms, but two groups of Mesozoic roachoids were highly specialized. One, the Raphidiomimidae (Late Jurassic to mid-Cretaceous), appears to have been predatory. These insects had a long, prognathous head, long palps, a narrow pronotum with the head entirely exposed; long, slender legs and wings; and the fore-femur and -tibia had two rows of sharp, stiff spines. Moreover, the fore legs appear to have been held forward, not splayed out to the side, and thus they were probably convergent with mantises. Many species from the Late Jurassic of Karatau had colorful wing patterns (Visniakova, 1968) (Figure 7.68), and the family is also now known from the Early Cretaceous Yixian Formation of China and the mid-Cretaceous Burmese amber (Grimaldi and Ross, 2004) (Figure 7.69).

Another interesting and highly specialized Mesozoic family of roachoids is the Umenocoleidae. These occur entirely within the Cretaceous, and are known from the Early Cretaceous of Siberia (Baissa), China (Yixian), Brazil (Santana) (Figure 7.70), Lebanon (in amber), and in amber from the Late Cretaceous of New Jersey (Figure 7.71). Umenocoleidae are the roach equivalents of beetles. Like some living polyphagids, their tegmina were heavily sclerotized, with a venation that was highly reduced and even lost in some species. Umenocoleidae also had a small, narrow pronotum, which exposed most of the head, and the head itself was quite broad and the eyes distantly separated. They possessed a short ovipositor, as did Raphidiomimidae, so unlike Vršanský *et al.* (2002) we consider these families basal to Recent Dictyoptera. Great reduction of the ovipositor in living Dictyoptera may be a synapomorphy for this group, but homologizing

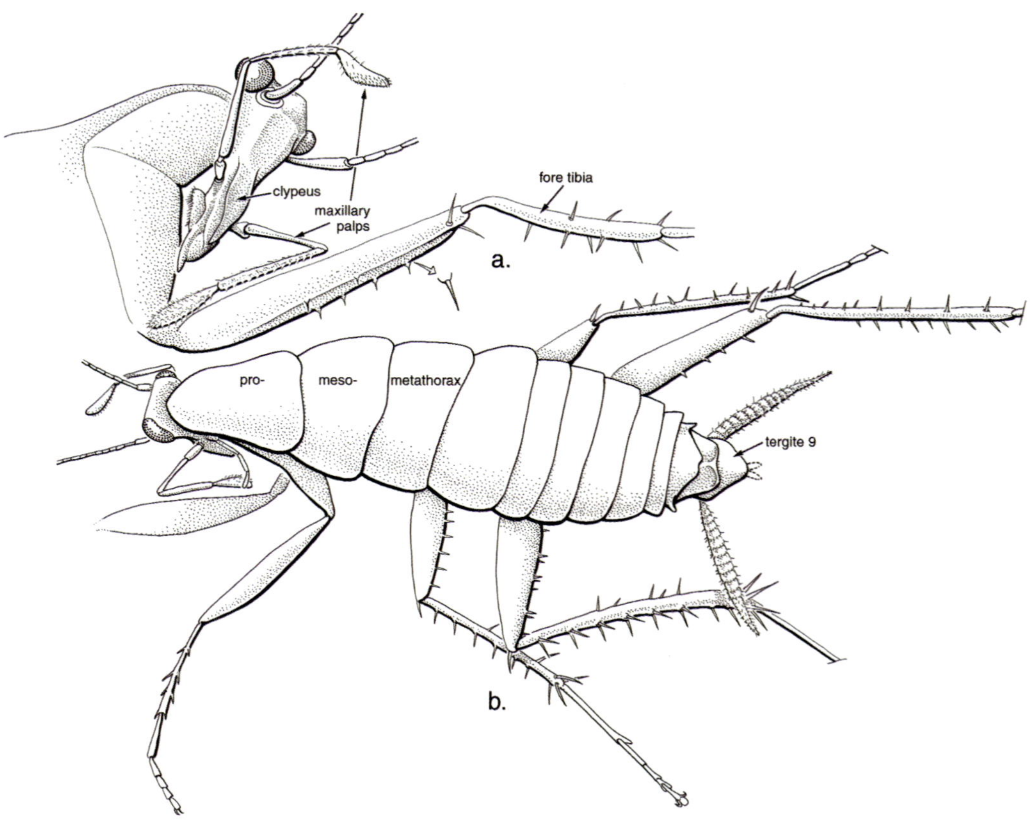

7.69. A nymphal raphidiomimid, *Raphidiomimula*, in mid-Cretaceous amber from Burma. NHM In. 20150; body length 10.5 mm.

7.70. *Ponopterix axelrodi* (Umenocoleidae) from the Cretaceous of Brazil, ca. 120 MYO. Umenocoleid roaches are known from the Late Jurassic to Cretaceous, though a putative living species exists. AMNH; body length 4.9 mm.

7.72. A female roach of the Recent family Blattellidae, from the Early Cretaceous of Brazil, preserved with an ootheca still lodged in her terminalia. AMNH; body length (excluding ootheca) 16 mm.

highly reduced structures must always be done with caution because very little remains on which to base this reduction homology decision. It is quite possible that vestigial ovipositors were independently acquired among several lineages of Recent Dictyoptera. Contrasting with Carboniferous roachoids, fossils of the extant roach families are no older than Early Cretaceous (reviewed in Ross, 2001; Vršanský *et al.*, 2002). The oldest polyphagids are known from the Early Cretaceous of Spain, England, and Asia (Martínez-Delclòs, 1993; Vršanský, 1999; Ross, 2001; Vršanský and Ansorge, 2001). The oldest Blattellidae occur in the Early Cretaceous of England and Asia (Vršanský, 1997; Ross, 2001), during which time they were diverse. They were abundant during the time of the Santana Formation in Brazil, approximately 120 MYA, though they were not particularly diverse there (A. Ross, pers. comm.). Some of these Santana blattellids even had oothecae still lodged in their terminalia (Figure 7.72), which confirms this mode of reproduction in early species of the family. Other than isolated oothecae in Burmese amber, of unknown roach origin, these are the earliest definitive oothecae. The oldest Blattidae is *Stantoniella* from the Late Cretaceous of North America (Bekker-Migdisova, 1962). The oldest Blaberidae are known only from nymphs in Eocene Baltic amber (Shelford, 1910), and *Cryptocercus* has no fossil record. A reasonable hypothesis is that the group that includes the modern families of roaches and all termites probably evolved in the Jurassic.

7.71. Beetle-like umenocoleid roach, in amber from the mid-Cretaceous of New Jersey. The tegmina and pronotum of this particular species are hairy. AMNH NJ749; body length 4.1 mm.

The Relict Wood Roach, *Cryptocercus*

In forests of the southern Appalachian and Cascade Mountains of North America and of southern China and Korea are several species of wingless, long-lived roaches that feed their way into soft, rotten logs, where they nest and raise their offspring. Details of the behavior and life history of *Cryptocercus punctulatus* (Figure 7.73) have been provided by Seelinger and Seelinger (1983), Nalepa (1984), and Nalepa and Bandi (2000). On average, a pair of parents and about 20 *C. punctulatus* offspring inhabit the galleries in a log, and they remain together for at least three years, the nymphs maturing in approximately six years. The nymphs – small, pale, and with highly reduced eyes, like termites – feed on liquids exuded from the anus (*proctodeal trophallaxis*) of an adult for approximately their first year (Figure 7.74). This behavior allows them to acquire mutualistic protists that reside in the hindgut and are required for digestion of the wood. Unique

7.73. The wood roach, *Cryptocercus* sp. (Cryptocercidae), dorsal (above) and ventral. *Cryptocercus*, though wingless and with other specialized features, is probably the living sister group to the termites. Length 14.5 mm.

7.74. Social behavior in the relict wood roach, *Cryptocercus*: early instar nymphs engaging in proctodeal trophallaxis with an adult. This is how nymphs obtain from their nestmates the symbiotic protists required for the digestion of wood. Drawn from photo in Seelinger and Seelinger (1983).

for cockroaches, the young nymphs also groom the older nymphs and adults, spending up to 20% of their time on this activity. Besides their aptery, these roaches are further specialized in possessing glands in the membrane between the abdominal tergites. The genus name is based on the small cerci enclosed between the last large tergite and sternite. Similar species of *Cryptocercus* inhabit the Pacific Northwest of North America and eastern Asia.

Cryptocercus has traditionally been believed to comprise three species with a highly disjunct distribution: *Cryptocercus primarius* from southcentral China, *C. relictus* from eastern Russia, and *C. punctulatus* from the southern Appalachians and Pacific Northwest of North America. Despite attention to *Cryptocercus* as a possible "missing link" between roaches and termites, it wasn't until 1996 that *C. "punctulatus"* was found to actually be several morphocryptic species (Kambhampati *et al.*, 1996b; Nalepa *et al.*, 1997; Burnside *et al.*, 1999). Not surprisingly, the widely separated populations from the east and west were found to be different species, with *C. clevelandi* described for the species from the Pacific Northwest. Unexpectedly, the eastern populations were found to comprise three additional morphocryptic species besides the original *C. clevelandi*. Separation of the species is based on sequences from five genes and karyotypes, the heterogametic diploid number varying from 37 to 47 (Kambhampati *et al.*, 1996b), and subtle morphological differences.

Recent discovery of four closely related species of *Cryptocercus* in the southern Appalachians is not too surprising. This region of North America is well known for harboring numerous species of animals with poor vagility, such as plethodontid salamanders, freshwater mussels, millipedes, and centipedes. Speciation in these groups and others is promoted by the wet, mild temperate climate and the many mountains and valleys that isolate populations. The three other Appalachian species of *Cryptocercus* are *C. garciai* (from northwest Georgia), *C. darwini* (from Tennessee, Alabama, and North Carolina), and *C. wrighti* (from North Carolina and Tennessee), though it will be important to confirm these species with anatomical features. *Cryptocercus punctulatus* is now known only from Virginia and West Virginia. Even more recently, a new *Cryptocercus* has been discovered in Korea (Grandcolas *et al.*, 2001), and it would be surprising if yet additional species are *not* found in mountainous regions of southern China.

Cryptocercus has exceptional biological significance because of its apparent close relationship to termites, a concept first proposed by Cleveland *et al.* (1934) based on his exquisite and comprehensive studies of the mutualistic protists found in the hindguts of this roach and basal termites. This view was promoted by Wilson (1971) but dismissed by Thorne and Carpenter (1992) and has become a subject of significant dispute. *Cryptocercus* is often placed in the monogeneric family Cryptocercidae (i.e., McKittrick, 1964), but Grandcolas (1996, and later) concluded that the genus is in the Polyphagidae. Polyphagids comprise a family of some 200 species, many of them inhabitants of deserts, scrub, and dry forests. Grandcolas found *Cryptocercus*, in fact, to be deeply embedded within the Polyphagidae, closely related to *Therea*, a genus from Indian dry forests. Studies by Klass, on the other hand, uphold the traditional relationship of *Cryptocercus* to termites, though together these two groups comprise a sister group to the Polyphagidae + *Lamproblatta*, so *Cryptocercus* is not distant from polyphagids. Grandcolas's studies have better taxon sampling, but Klass's morphology is more detailed and provides more evidence (see, for example, Klass, 2001a). Also, Grandcolas's results on *Cryptocercus*-termite relationships are biased by his use of termites as an outgroup, as are the results of Kambhampati (1995, 1996). Interestingly, in a reanalysis of morphological data with Kambhampati's molecular data (Grandcolas and D'Haese, 2001), one of the few consistent results was a grouping of *Cryptocercus* with termites, which contradicts Grandcolas's own views. Lastly, and quite significant, is the study by Lo *et al.* (2000), which is the most comprehensive molecular study yet on these insects. It provides strong support for a close relationship between *Cryptocercus* and termites.

Cryptocercus shares with two families of lower termites several genera of distinctive protists, though some studies indicate a potential for transfaunation of the protists, or inoculation between unrelated roaches and termites. Despite claims that "it is difficult to support [a *Cryptocercus*-termite relationship] based on any character except gut fauna" (Thorne and Carpenter, 1992: 255), there is abundant and compelling evidence for their close relationship, including the distinctive proventriculus discussed previously (Klass, 1998) as well as dentition of the mandibles (Ahmad, 1950). Oddly, an obvious feature that has never been discussed is the antenna. Antennae that are truly moniliform, with spherical segments like a string of beads, are actually rare in insects but occur in *Cryptocercus*, termites, and several other roaches (especially those that excavate galleries or burrows). Detailed structure of the segments indicates homology of *Cryptocercus* and termite antennae.

Perhaps most significant are monogamy, extended biparental care, allogrooming, and proctodeal trophallaxis, which are also features uniquely shared between *Cryptocercus* and termites. Some unrelated panesthiine roaches have habits and structures similar to *Cryptocercus*. Some species of *Panesthia*, for example, excavate galleries in wood where they live gregariously, and they shed their wings (though not by a basal suture as in termites); they even have a reduced cercus and moniliform antennae (though detailed structure indicates these are convergently derived). Moreover, these large roaches digest cellulose in wood via mutualistic intestinal amoebae (but not flagellates as in *Cryptocercus* and termites). *Geoscaphius* roaches in Australia even excavate extensive underground burrows, where the nymphs are tended and adults provision them with leaves from the surface. Though a remarkable convergence between *Cryptocercus* and termites is possible, all the available evidence indicates their resemblance is the result of common ancestry.

Estimating an age of *Cryptocercus* based strictly on cladistic "sister-group dating" would place the genus as the same age as the order Isoptera, perhaps 140–150 MYO. However, *Cryptocercus* is clearly highly modified and is thus itself a specialized descendant of the roach-termite ancestor. Clark *et al.* (2000) applied an interesting method of dating the genus and certain species, based on molecular divergence of a bacterium symbiotic in *Cryptocercus* which is *Blattabacterium cuenoti*. They used what is believed to be a universal substitution rate of 0.5–1% sequence divergence per 50 million years in the 16S rRNA gene of bacteria. Their estimates place the divergence of Asian–North American species of *Cryptocercus* at 70–115 MYA (Late Cretaceous), the divergence of eastern and western North American species at 88–53 MYA (Late Cretaceous–Paleocene), and the divergence of the eastern North American species 13–38 MYA. Even without a fossil record for *Cryptocercus*, these ages and their hypothesized Jurassic origin of the genus are excessively old, probably a result of molecular clock estimates.

Grandcolas (1999) hypothesized much younger ages for *Cryptocercus*, with which we agree, although for different reasons. A disjunct distribution comprising the southeastern United States and eastern Asia is a common biogeographic pattern, involving over 100 genera of vascular plants, such as *Magnolia*, *Liriodendron* (tulip tree), *Hamamelis* and *Liquidambar* (sweet gum trees), many oaks (Fagaceae), *Pachysandra*, and *Symplocarpus* (skunk cabbage) (Little, 1983; Wu, 1983; Ying, 1983). This distribution of closely related disjunct species conforms to areas affected by Pleistocene glaciations. Moreover, fossils vividly document formerly more widespread Holarctic distributions for many of these and other plants (Axelrod, 1983; Davis, 1983; Hsü, 1983). The present distribution of *Cryptocercus* is probably a result of Pleistocene and/or Pliocene land bridges connecting Asia and North America. That *Cryptocercus* is an apparent *living* sister group to a lineage that appeared in the Early Cretaceous or earlier does not mean that the genus is this old, or even Early Tertiary in age. *Cryptocercus* could be the sole surviving taxon of an entire lineage of Cryptocercidae that is now largely extinct. It is reasonable, in fact, to envision a winged, colonial, wood-feeding cryptocercid in the Late Jurassic or earliest Cretaceous, ancestral to insects that gave rise to modern *Cryptocercus* and to the grand architects among insects, the termites.

CITIZEN ROACH: ISOPTERA (TERMITES)

Termites are essentially highly social, morphologically reduced roaches, whose digestion of cellulose via symbiotic microbionts has allowed them to invade a vast, largely unexploited niche among insects. Though their diversity is modest compared to some orders (there are approximately 2,900 described species), termites are probably the most ecologically important group of insects besides the bees, and termites even have a global impact on geochemical cycles. Termites, like Mantodea, Lepidoptera, and a few other orders, are also among the most recently evolved major insect orders, with the earliest fossils appearing in the Early Cretaceous. This situation raises questions as to how wood, leaf litter, humus, herbivore dung, and other abundant plant matter was processed prior to the Cretaceous (e.g., Raymond *et al.*, 2000).

Like honey bees, some vespid wasps, all ants, and a few other arthropods, termites have advanced eusociality, with individuals behaviorally specialized for particular tasks in the colony. Only termites and ants have morphologically highly specialized reproductives, workers, and soldiers. Unlike social hymenopterans, termite soldiers and workers belong to both sexes. Also, termites are diploid, versus haplodiploid in hymenopterans, so workers are not any more closely related to each other than they would be to their own offspring. Lacy (1980) hypothesized that termite parents and sibs have higher relatedness to offspring of the same sex than to those of a different sex, as a result of chromosomal sex-linked translocations found in termites. However, such chromosomal behavior confers only slightly higher relatedness, not the degree found in haplodipolid systems. Certain ecological conditions and intrinsic biological features of termites other than sex determination must have contributed to the evolution of eusociality in this group.

Also unlike the eusocial hymenopterans, individual termites have the ability to transform into other castes. The colony is bathed in a milieu of pheromones, and changes in their concentration regulate the production of hormones that induce particular castes. Optimal proportions of caste members are maintained by the colony, and workers may even destroy surplus individuals. The "primary" reproductives are fully sclerotized and winged (with only the basal scales of the dehisced wings remaining), while the "secondary," "supplementary," or "neotenic" reproductives are less sclerotized and lack wings or fully developed ones. Workers and soldiers have vestigial eyes and gonads and lack wings. They specialize in foraging, nursing, and feeding soldiers and reproductives, as well as in nest construction and repair, and they have few specialized features. The families Termopsidae and Kalotermitidae lack true worker castes, with colony tasks instead being done by immatures (*pseudergates*). Loss of the worker caste no doubt is related to the fact that they nest in the wood they consume, resulting in a loss of persistent pressure for specialized individuals that forage. Soldier termites, like ants, often have huge heads occupied by powerful mandibular muscles. In many Nasutitermitinae, the mandibles are vestigial, and the head is modified into a peculiar bulb that sprays noxious secretions through a fine pore at the tip of the nozzle (Figure 7.75). When the wall of a termite nest is damaged, soldiers immediately crowd at the edge of the hole, snapping their mandibles or spraying, while workers frantically wall it up.

Reproduction takes place in nuptial flights once or several times each year. Various external cues synchronize colonies of particular species, and collectively the swarming termites can cloud acres of grasslands and savannas in some regions. After alighting the males and females shed their wings along a weak suture at the base of the wing and begin excavating a nest. As the first-born nymphs care for the queen, she lays more eggs and becomes cloistered within a royal chamber. Here, with an abdomen eventually engorged to grotesque proportions (*physogastry*), she can produce up to 2,000–3,000 eggs per day (Figure 7.76). Some queens have been known to live for a decade, and plausibly have produced more than five million offspring in a lifetime – the most fecund insects known. The colonies of some Termitidae have been known to reach to three million individuals, which comprise the largest societies in nature along with army and driver ants (Ecitoninae and Dorylinae) and leaf-cutter (attine) ants. Colonies are known to persist for decades; one *Nasutitermes* colony lived more than a century. Social and nest structure varies greatly and has been reviewed by Noirot (1970), Wilson (1971), Shellman-Reeve (1997) Thorne (1997), Higashi *et al.* (2000), Noirot and Darlington (2000), Roisin (2000), and Thorne and Traniello (2003).

Symbiotic Microbiota of the Hindgut

Cellulose, including its common form lignocellulose, is perhaps the most ubiquitous biomolecule on land. Unfortunately, it is also relatively inert, requiring cellulases and other enzymes secreted mostly by microbes for its degradation. Consequently, few higher organisms have evolved a diet that exploits this superabundant resource. The "lower" termites have broken through the barrier as a result of mutualistic protists in their hindguts that metabolize cellulose. The higher termites, family Termitidae, instead harbor symbiotic

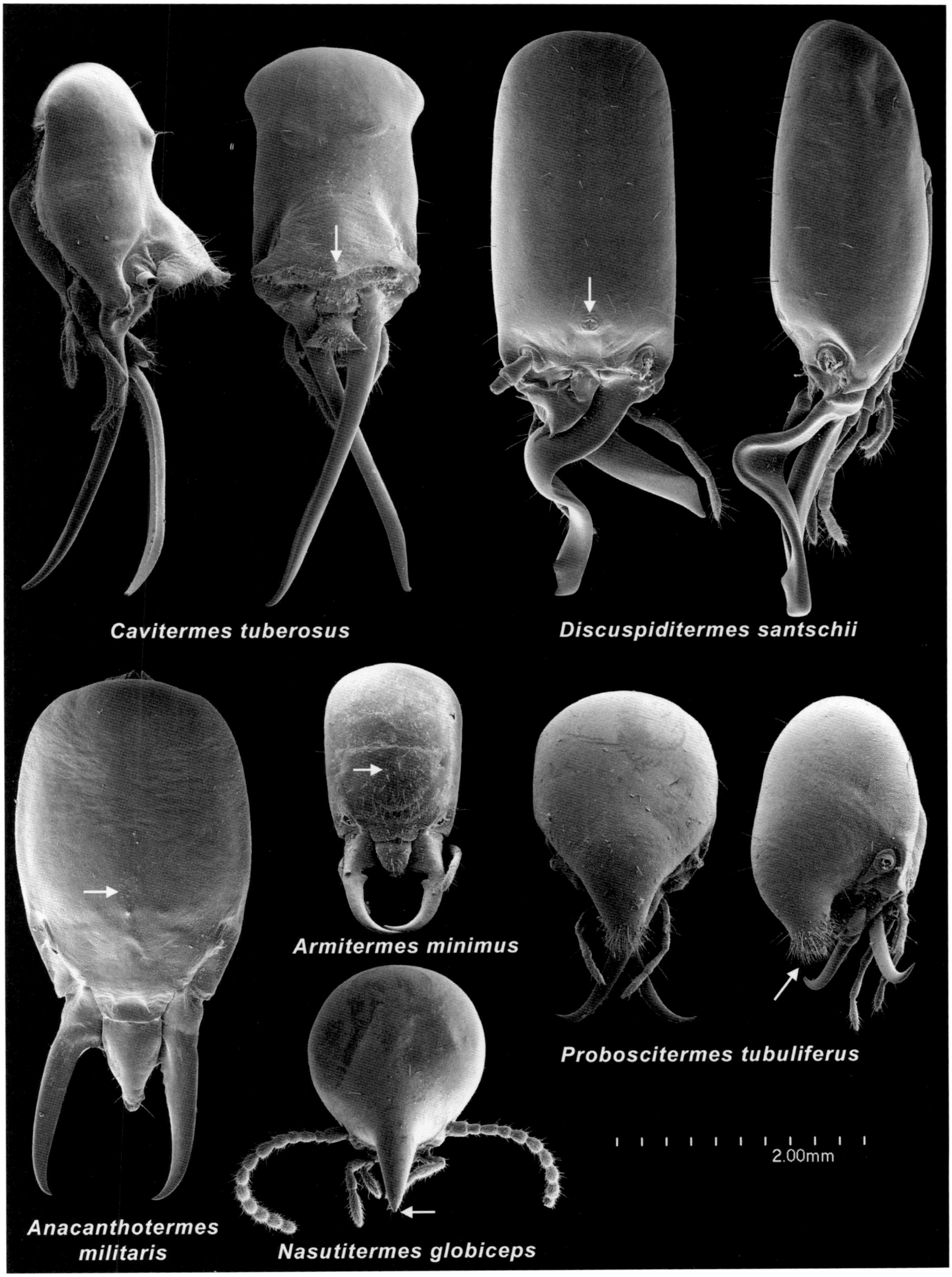

7.75. Heads of soldiers from select genera of Termitidae, showing varied development of the fontanelle (arrows) and mandibles. Scanning electron micrographs.

7.76. A bloated, physogastric queen termite (*Nasutitermes* sp.: Termitidae) being tended by her minions in Costa Rica. Photo: Carl Rettenmeyer, Connecticut State Museum of Natural History.

bacteria that assist in the metabolism of cellulose. Important reviews of termite intestinal microbiota are by Honigberg (1970), Breznak and Brune (1994), Breznak (2000), and Inoue *et al.* (2000), and there is a very useful catalogue of protists and their termite hosts by Yamin (1979).

Defaunation studies, in which the microbiota is expelled by feeding antibiotics to the termites, show that the lower termites cannot live without the protists packed into their hindgut. Joseph Leidy, the great early American naturalist, described the exodus of these protists from the hindgut of a dissected termite "like citizens spilling from a crowded meeting hall." Indeed, one study indicates the protists can comprise up to one fifth the mass of a termite. When an insect molts, the cuticle that lines the hindgut (the *intima*) is shed with the external cuticle, and in termites the protists are shed as well. As a result, newly molted individuals are devoid of protists, as are newly hatched nymphs. The protists cannot form resistant cysts that endure in the termite frass (they can exist only in strictly anaerobic environments), so a termite cannot inoculate itself with protists by coprophagy. Instead, termites must engage in *proctodeal trophallaxis*, or feeding on the anal secretions of a nest mate. In the Termitidae proctodeal trophallaxis is lost.

The lower termites harbor protists of the families Oxymonadidae, Devescovinidae, Calonymphidae, Trichonymphidae, and Trichomonadidae, which, except for the first belong to the very basal lineage of eurkaryotes, the Parabasalea. Parabasaleans lack mitochondria, have 70S-sized ribosomes like bacteria, and have protein-coding genes homologous to mitochondrial genomes, and so are considered evolutionary precursors of the eukaryotic cell. Why termites have evolved obligate mutualisms with these primordial organisms is unclear, though protists are not the only microbes inhabiting the hindgut of lower termites. Methanogens – minute, methane-producing archaebacteria – and spirochetes – tightly coiled eubacteria – are also abundant. Spirochetes, in fact, are more diverse in termites than virtually any other place on earth. The gut of the well-studied pest species, *Reticulitermes flavipes*, was recently found to contain at least 21 new species of spirochetes in the genus *Treponema*, to which the species that causes syphilis in humans also belongs. Like small fuel cells, the spirochetes derive energy from the metabolism of hydrogen, which is an abundant byproduct in the termites' gut. Moreover, some spirochetes blanket the surface of protists and propel the protist like undulating cilia. The huge protist *Mixotricha paradoxa*, found only in the most basal termite, *Mastotermes darwiniensis*, has incorporated spirochetes and other bacteria into itself – a chimeric organism of five genomes! From approximately 5% of all termite species sampled thus far, we already know 450 species of protists and myriad bacteria (Dolan, 2001). Obscure as it might seem, the hindgut of lower termites is an expanse of primordial diversity and evolution, awaiting exploration.

Several genera of termite protists have attracted particular attention because they are shared with an apparent sister group to termites, the roach genus *Cryptocercus*, discussed earlier in this chapter. These protists are the complex hypermastigotes *Leptospironympha* (found also in *Stolotermes* termites) and *Trichonympha* (Figure 7.77) (found also in Hodotermitidae, Termopsidae, and Kalotermitidae), as well as some oxymonads (i.e., *Oxymonas*). Many species of protists are specific to certain species of termites and *Cryptocercus*. Oddly, the most basal termite, *Mastotermes darwiniensis*, does not have these protist genera but rather an array of unique protists, like *Koruga*, *Deltotrichonympha*, and *Mixotricha*.

There is significant debate as to whether *Cryptocercus* shares these distinctive protists with some termites purely as a result of inheritance by common ancestry. Early experiments successfully inoculated the intestinal microbiota from *Cryptocercus* into termites whose guts had been sterilized. *Cryptocercus* and some termites also share rotten logs and will feed on the fresh remains of each other (Thorne, 1990, 1991), indicating the potential for one species to inoculate another in the recent or distant past. This seems an unlikely

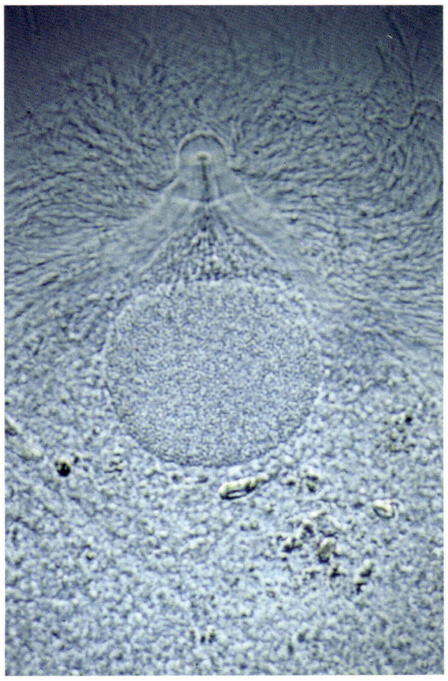

7.77. *Trichonympha*, a flagellate protozoan that lives in the hindgut of *Cryptocercus* roaches and lower termites and that metabolizes cellulose. These insects presumably inherited *Trichonympha* from their common ancestor. Photo: M. Dolan.

scenario, though, since *Cryptocercus* shares protists with three termite families, and it is highly unlikely for the roach to have acquired them independently three times (Nalepa, 1991). Also, some panesthiine roaches burrow into wood, but none of them are known to harbor any of these protists. Flagellate protists occur in the hindgut of the wood-eating Brazilian roach, *Parasphaeria boleiriana* (Blaberidae: Zetoborinae) (Pellens *et al.*, 2002), but proctodeal trophallaxis was not observed in this species nor is it known if these protists even metabolize cellulose. In fact, the ultrastructure of the protists, which is essential to determining their identity, has not been studied, but it is most likely that the protists are not the same as ones in *Cryptocercus* and lower termites. Since the transfaunation studies were done, additional evidence has made the case for a sister-group relationship of *Cryptocercus* and termites, thus bolstering the case for inheritance of the microbionts through common descent. In the most recent study, the DNA of roaches, *Mastotermes*, and their *Blattabacterium* symbionts appear to form parallel phylogenies and further indicate a close relationship between *Cryptocercus* and termites (Clark *et al.*, 2000; Lo *et al.*, 2003). Very likely, the ancestral termite had a diverse complement of protists, similar to what is found in *Cryptocercus*, and various termite lineages lost some kinds of protists and acquired or evolved others.

Ecological Significance of Termites

Exploitation of lignocellulose as a food by large eusocial colonies seems to be the best explanation for the ecological dominance of termites. Tropical ecosystems, comprising approximately 20% of earth's land surface, harbor about 70% of all termite biomass, and termites are virtually absent above or below 45°N and 45°S latitudes (Eggleton, 2000). One study estimated that termites comprise 10% of all animal biomass in the tropics. Arguably, there is more termite biomass than any other order of insects. Important reviews of termite ecology are by Wood (1978), Wood and Sands (1978), and Bignell and Eggleton (2000).

The main effects of termites in ecosystem functioning are in carbon *mineralization* and in *humification* (soil formation and enrichment). They have even been implicated as important global sources for the greenhouse gases methane (CH_4) and carbon dioxide (CO_2). An original report (Zimmerman *et al.*, 1982) estimated that as much as 40% of the annual global methane is emitted by termites. More comprehensive reports, however, have taken into account great variation in methane production among regions and termite feeding groups (Khalil *et al.*, 1990; Brauman *et al.*, 1992; Martius *et al.*, 1993; Sanderson, 1996; Sugiomoto *et al.*, 2000). For example, termites that consume soil ("soil feeders," who actually extract nutrition from the humus fraction) produce approximately ten times the methane that wood feeders produce. These more recent reports estimate that termites overall emit approximately 2–5% of annual global methane. Though a great deal more methane is anthropogenic (produced by rice paddies, livestock, and gas and coal mining), this is still an impressive amount considering that it is produced by insects approximately a centimeter in length that are largely condensed within one fifth of the earth's land surface.

Consumption of plant matter by tropical termites can be prodigious. Though estimates have been more difficult to make for rainforests, termites are reported to consume up to 40% of all dead wood in Australian sclerophyll forests and up to 100% in similar forests in Africa. In tropical forests, tunnels constructed of frass, soil, and salivary secretions anastomose around tree trunks and throughout the forest floor, converging on subterranean colonies or arboreal carton nests. Foraging termites course through these galleries unseen, and most rotting logs teem with workers from colonies whose nests are many meters away. Termites have even greater impact on carbon mineralization in grassland or savanna ecosystems, consuming up to 20% of the available plant biomass – more than mammalian herbivores in some situations (Dettling, 1988), though fires are probably the most significant mineralizer of plant carbon. Most forms of lignocellulose are consumed, including wood, dry grass indigestible to ungulates, and even dung. On the African savanna, beneath virtually every drying pile of elephant dung is a swarm of termites extracting the finely milled plant material.

Also in Africa, the savannas of Australia, and other tropical regions, there are edifices up to eight meters (25 ft) high, constructed by *Nasutitermes*, *Bellicositermes*, and *Macrotermes*

7.78. *Macrotermes* mound on a savanna in Kenya, eastern Africa, circa 1925. Termites have a profound effect on soils and the cycling of carbon in tropical ecosystems. Photo: AMNH Library Archives.

(Figure 7.78) Like mound nests constructed by other Termitidae, the termites actually reside in convoluted galleries beneath the ground, not in the tower. These towers are essentially chimneys that facilitate the diffusion of CO_2 out of the labyrinth chambers of the nest, and O_2 into it. The amount of respiration in the largest nests is actually equivalent to that of a cow, most of it coming from the symbiotic fungus that carpets the chambers. With the vents raised high into the air, away from the still, boundary layer of air near the ground, the wind aids in the diffusion of gases. Termite towers are largely constructed of earth and testify to the amount of soil moved by these organisms, the most vivid example of this being some subfossil nests from eastern Africa (Crossley, 1986). On the western shore of Lake Malawi is a geological formation comprised of a 5-m-thick layer of clays and sands spread over 8,800 km^2. Between 10,000 and 100,000 years ago some 44 million cubic meters of deep soil were transported to the surface by *Macrotermes falciger*, which formed this layer, and the termites continue this work as residents of the area today. We tend to assess ecological impacts in terms of years, but effects like these over millennia justifiably earn termites the role of civil engineers among the insects. The collective impact of termites over 140 million years of their evolution, but particularly the last 40–50 million years when the Termitidae evolved, must be immense.

Living and Fossil Diversity

The Isoptera are the only major order of insect without an extinct family, though this may simply reflect little phylogenetic work on early termite fossils. There are seven families of termites, six if the system of Emerson (1955) and Krishna (1970) are used, where the Termopsidae are considered a subfamily of the Hodotermitidae (as Termopsinae).

The fossil record of termites was reviewed by Nel and Paicheler (1993) and Thorne *et al.* (2000). The taxonomy of termites was reviewed by Krishna (1970) and Kambhampati and Eggleton (2000), and a database of the world species is nearly complete (Krishna and Engel, in prep.).

The so-called lower termites comprise the basal families Mastotermitidae, Hodotermitidae, Termopsidae, and Kalotermitidae, as well as the Rhinotermitidae. The last family is the sister group to the large, recently evolved family Termitidae. With the exception of the hodotermitids, most of the basal families construct nests in their food substrate, which is moist or dry and even sound wood.

Family Mastotermitidae. This family is the undisputed sister group to the rest of the termites. The sole living member is *Mastotermes darwiniensis*, native to nonforested regions of tropical, northern Australia and introduced into southern New Guinea (Figure 7.79). It retains a suite of primitive, roach-like features, including large size, relatively large pronotum, large anal lobe on the hind wing, full complement of wing veins, and legs with five tarsomeres, among other features. In its fat bodies it even retains specialized cells, *mycetocytes*, that harbor a symbiotic bacteria (*Blattabacterium*) found only in roaches. Moreover, it lays a pod of up to 24 eggs, neatly arranged in two rows. The pod has a covering that is much thinner than the heavily sclerotized case seen in cockroach oothecae. As in roaches, *Mastotermes*' egg case is molded in the female's vestibulum and the covering is secreted by the accessory (colleterial) glands, so *Mastotermes*' egg pod is clearly a vestigial ootheca (Nalepa and Lenz, 2000). All other termites lay single eggs.

Fossils likewise indicate that *Mastotermes darwiniensis* is an evolutionary relict, the family and even the genus once having a nearly global distribution (Emerson, 1965). The genus itself is known as wing impressions from the Miocene through the Eocene of Croatia, England, France, Germany, and Spain, with two species preserved in amber from Mexico (*M. electromexicus*) and the Dominican Republic (*M. electrodominicus*) (Figures 7.80, 7.81). The species in amber, preserved as entire specimens for all castes, are morphologically very similar to the living species (Krishna and Emerson, 1983;

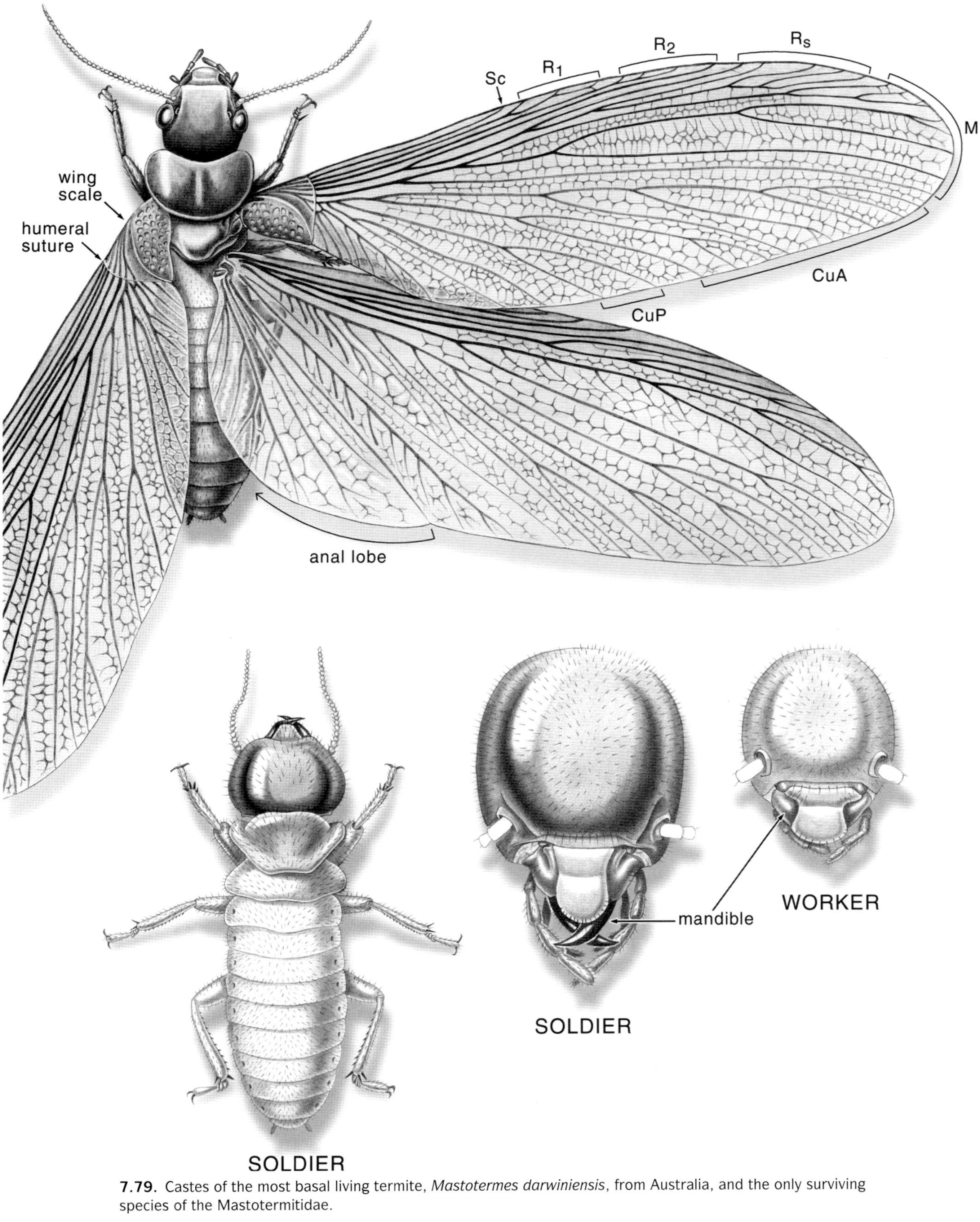

7.79. Castes of the most basal living termite, *Mastotermes darwiniensis*, from Australia, and the only surviving species of the Mastotermitidae.

7.80. *Mastotermes anglicus*, from the Bembridge Marls of England, ca. 38 MYO. Mastotermitids were nearly global in the early Tertiary. NHM In 24571; body length 34 mm.

Krishna and Grimaldi, 1991). Three extinct genera are known from the Tertiary: *Blattotermes* (Eocene of Australia and Tennessee, Oligocene of France), *Miotermes* (Miocene of Yugoslavia and Germany), and *Spargotermes* (Eocene of Brazil, formerly believed to be much younger [Mio-Pliocene]). Oddly, though occurring in Europe during the Eocene, Mastotermitidae is unknown from the vast deposits of Baltic amber. More importantly, only two records of Mastotermitidae are known from the Cretaceous, the placement of both equivocal: "*Mastotermes*" *sarthensis*, based on poorly preserved specimens in mid-Cretaceous amber from France, and *Valditermes*, an extinct genus known from wing impressions from the Lower Cretaceous of England and Upper Cretaceous of Mongolia. There is an undescribed termite from the Cretaceous of Brazil that may also be a mastotermitid. Features clearly diagnostic for the Mastotermitidae on each of these are obscure, so the attribution of *M. sarthensis* to *Mastotermes* is particularly dubious. Given the basal position of Mastotermitidae in termite phylogeny, many more fossils in this genus would be expected from the Cretaceous. Instead, only hodotermitids and termopsids were diverse in the Cretaceous, unless some of the lesser-preserved compression fossils are misidentified mastotermitids.

Despite its appearance, *Mastotermes darwiniensis* behaves like anything but a "living fossil." It does not merely reside in the logs it consumes like most other lower termites; rather, it constructs nests and extensive foraging galleries, and the colonies can be huge. Some "colonies" in Queensland have been recorded to harbor a million individuals in massive underground nests formed in rotten logs, though these are probably a fusion of numerous colonies resulting from precocious dispersal in this species. *Mastotermes darwiniensis* is the most destructive pest in northern Australia, consuming virtually every form of cellulose from wood to paper, even creosote-soaked utility poles, and the rubber from underground utility cables and discarded tires. Such a diet allows *Mastotermes* to outcompete all other termites in northern Australia, so if the feeding habits of extinct relatives were as notoriously polyphagous as the living species, the question arises: Why did *Mastotermes* become extinct from most of its distribution?

There is evidence that extinct *Mastotermes* was highly polyphagous like the living species, based on exceptional preservation of hindgut microbiota in a Miocene *Mastotermes*. Hindgut tissue from *M. electrodominicus* in Dominican amber contained wood fragments, some spirochetes, bacteria, and remnants of protists that are recognizable from the living species (Wier *et al.*, 2002), an indication that at least some extinct relatives had a similar diet. Perhaps abiotic factors like climatic change caused the nearly global extirpation of the Mastotermitidae.

Family Hodotermitidae. This is a small family of 19 living species in three genera (*Anacanthotermes*, *Hodotermes*, and *Microhodotermes*), which are ground-nesters in arid regions of Africa and Eurasia. Their common name, "harvester" termites, derives from their rather specialized diet among the lower termites of dry grasses. The Termopsidae have often been placed as a subfamily within the hodotermitids, partly because they both lack ocelli, though this may be convergence.

Like Mastotermitidae, the Hodotermitidae were once more diverse and very widely distributed, most of the fossils known only as wing impressions. The extinct genus *Ulmeriella* has been considered intermediate between Mastotermitidae and the Hodotermitidae, on the basis of mandible dentition, eye structure, segmentation of cerci (where it has been preserved), and the distinctive branching of Rs wing veins. As such, it would be the most basal genus of the family. *Ulmeriella*, though, has been found only in the Tertiary, based on 11 species from the Pliocene to Oligocene of Japan, Europe, and the United States. Emerson (1968) reviewed species of *Ulmeriella* in detail, to which Nel and Paicheler (1993) added subsequent specimens, species, and

a

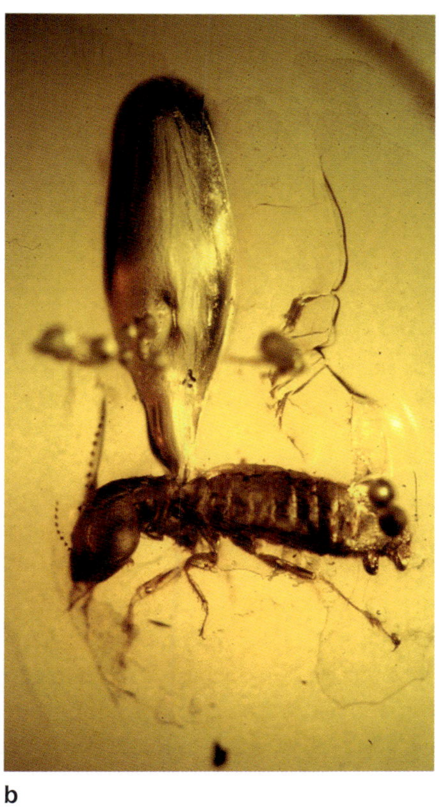

b

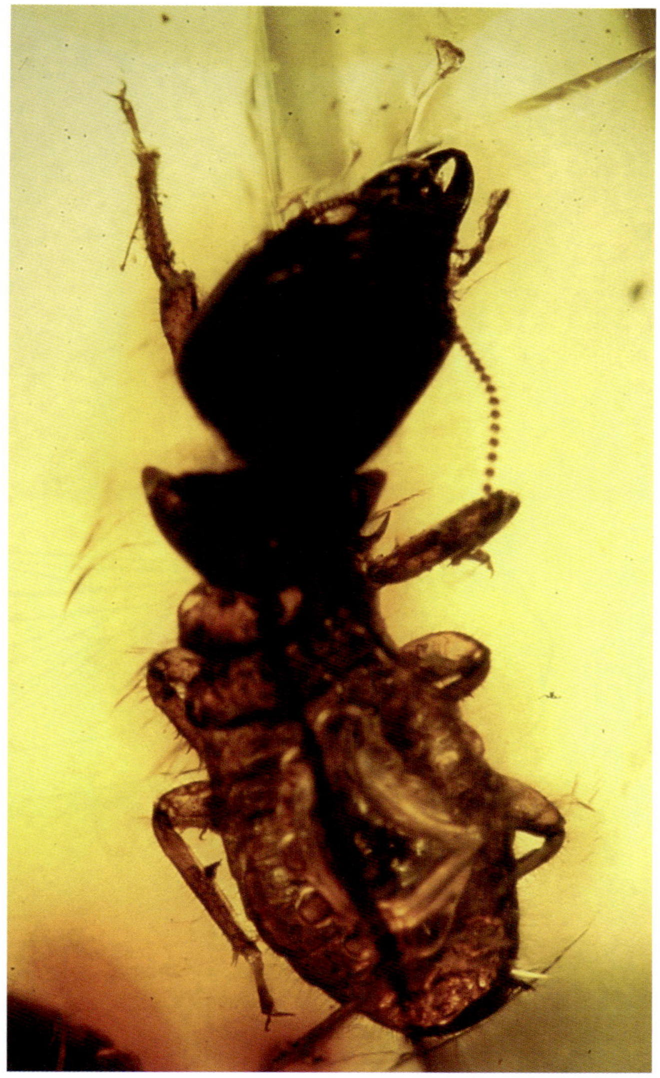

c

7.81. An alate (a), worker (b), and soldier (c) of *Mastotermes electrodominicus*, in Miocene amber from the Dominican Republic. The species is extremely similar to the one living species of the genus in Australia. AMNH; length of alate 31 mm.

7.82. Alate of one of the oldest known termites, *Meiatermes bertrani*, from the Early Cretaceous of Spain. LC-807-IEI. Photo: X. Martínez-Delclòs.

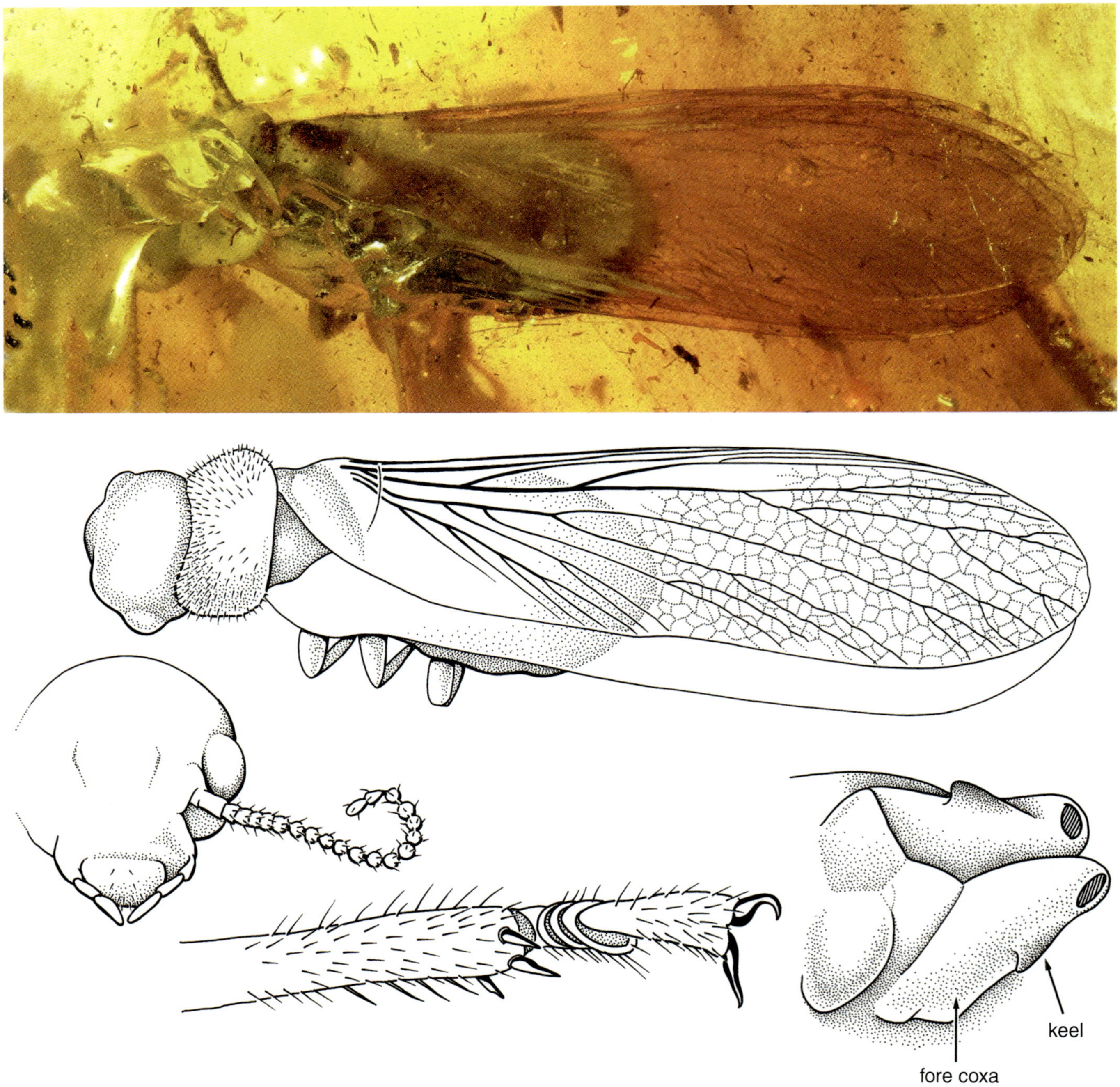

7.83. An early hodotermitid termite, *Carinatermes nascimbeni*, in 90 MYO New Jersey amber, showing the front of the head, the primitively keeled forecoxae, and the tarsus. AMNH NJ124; forewing length 10.2 mm.

taxonomic changes. All the other genera are from the Cretaceous and are likely to be more basal members of the family.

Meiatermes is the oldest termite genus, known from the Lower Cretaceous of Spain (*M. bertrani*) and Brazil (*M. araripina*) (Figure 7.82). A worker specimen is also known from this deposit, perhaps of *M. bertrani*, making this the oldest record of insect castes in the fossil record, and establishing the eusociality of termites approximately 130 MYA (Martínez-Delclòs and Martinell, 1995). It remains the only worker termite known from the Cretaceous; all others are alates. Besides *M. araripina* there are additional, undescribed hodotermitids from the Santana Formation of Brazil, perhaps the most diverse fauna of termites known from the Cretaceous. Three poorly diagnosed genera are reported from the Lower Cretaceous of China. One species, *Luteitermes prisca*, occurs in mid-Cretaceous amber from France; there are, however, several better preserved fossils also in Cretaceous amber. *Carinatermes nascimbeni* is known from a completely preserved specimen in mid-Cretaceous amber from New Jersey (Figure 7.83); it has peculiar keels on the forecoxae, as in *Mastotermes*, which is almost certainly a primitive character for termites. An undescribed hodotermitid-like termite is also known from the Early Cretaceous shales of Baissa in central Siberia.

Grasses did not appear until the very late Cretaceous and very early Tertiary, and grasslands did not flourish until the Miocene with the earliest appearance of C_4 grasses, which dominate today (Jacobs *et al.*, 1999). The Cretaceous hodotermitids must have foraged on other herbaceous and probably even woody vascular plants.

Family Termopsidae. A highly disjunct distribution, as well as the fossil record, indicates that this family is likewise relict today. The 20 living species in five genera are distributed in western North America (*Zootermopsis*), eastern Eurasia (*Hodotermopsis*), and central Eurasia (*Archotermopsis*). Two genera, sometimes placed in the Hodotermitidae, are extremely southern: *Porotermes* (one species each in southeast Australia, Tasmania, Chile, and southern Africa) and *Stolotemes* (three species in Australia, one in Tasmania, two in New Zealand, and one in southern Africa). This bipolar and particularly disjunct austral range may reflect a Cretaceous history ravaged by extinction, leaving a distribution fragmented within temperate biomes. Most recently, study of *Stolotermes* suggests that this genus may actually be more closely related to the Kalotermitidae, Rhinotermitidae, and Termitidae, than it is to Termopsidae (Deitz *et al.*, 2003).

All four of the Cretaceous genera have been assigned to the extinct subfamily Cretatermitinae (three doubtful genera from China, one from North America); the Termopsinae includes living and extinct genera known from the Eocene to the present. The extinct termopsine genera are *Parotermes*, from the Eocene of Colorado; *Paleotermopsis*, from the Oligocene of France; and *Termopsis*, in Eocene Baltic amber and Miocene and Oligocene rocks. A species of *Archotermopsis* in Baltic amber is interesting since the one living, closely related species of the genus occurs in the Himalayan region; the other two living genera are known from the Miocene of Japan (*Hodotermopsis*) and possibly the Eocene of Colorado (*Zootermopsis*). The fossil record of this family is, thus, entirely Laurasian, but it is unclear if this distribution is merely the result of huge gaps in our sampling of Southern Hemisphere fossils.

Family Kalotermitidae. The "dry wood" termites are a circumtropical group of 21 genera and 417 living species, and are the only diverse family of termites to have been monographed (Krishna, 1961). Fossils were reviewed by Emerson (1969), with new records and taxonomic changes provided by Nel and Paicheler (1993). With the important exception of "*Kalotermes*" *swinhoei* in Burmese amber, all fossils are known only from the Paleocene to Pliocene of most continents. Placement of "*Kalotermes*" *swinhoei* in this living genus is uncertain, but it does nearly double the age of the Kalotermitidae. Very interesting fossil remains of putative kalotermitids are a nest excavated in a section of *Diospyros* wood from the Late Cretaceous Javelina Formation of western Texas (Rohr *et al.*, 1986) (Figure 7.84). Galleries in the center of the well-preserved wood, which even possesses growth rings, are filled with hexagonal frass pellets, suggestive of the workings of kalotermitids. The nest is quite small, occupying a diameter of only about 3.5 cm (1.37 in.). A kalotermitid supposedly occurs in Lebanese amber (André Nel, pers. comm. to DG), which would be the oldest one known.

Family Rhinotermitidae. This is a group approximately the size of the Kalotermitidae, with 15 genera and 368 species. The monotypic family Serritermitidae, which lives within the walls of *Cornitermes* nests in Brazil, appears to be the sister group to the Rhinotermitidae and Termitidae (Emerson and Krishna, 1975). Soldiers in these three families possess a fontanelle, which is a large pore or a group of fine ones through which defensive secretions are sprayed. The fontanelle is normally in the middle of the head, but lies at the tip of a snout in nasute soldiers of the Termitidae.

Some rhinotermitids, such as *Reticulitermes*, are important pests in northern temperate regions, where they generally feed on rotting wood. Only six genera are known as fossils (two of them extinct), with most fossil species preserved in amber (Emerson, 1971; Nel and Paicheler, 1993). *Parastylotermes*, known from Eocene Baltic amber and Miocene shales and concretions of western North America, appears closely related to the living southeast Asian genus *Stylotermes*. Recent discovery of a rhinotermitid in Burmese amber doubles the known age of the family and compresses the early diversification of termite families (excluding the Termitidae) into the Early Cretaceous (Krishna and Grimaldi, 2003). *Archeorhinotermes rossi* in Burmese amber (Figure 7.85) has a left mandible unlike any known termite, being quite long with exceptionally long, sharp teeth. How such teeth could masticate wood or other plant material is perplexing, and perhaps reflects a unique diet. Despite this discovery, Emerson's (1971) hypothesis that many modern rhinotermitid genera probably originated in the Upper Cretaceous is still unlikely because *Archeorhinotermes* has many primitive features and appears to be the basal genus of the family, though also too specialized to be ancestral for Rhinotermitidae.

Family Termitidae. These are the "higher" termites, comprising over 80% of all termite species, with 2,020 species in 237 genera, the great proportion of them tropical. The nests of Termitidae are the most elaborate and intricate dwellings constructed in nature (Figure 7.86). Wood is a minor part of the diet of this family, but where it is consumed it is usually permeated and decayed by fungal mycelia. Most species feed on soil humus, leaf litter, grass, dung, and/or fungi. Corresponding to the diversity in diet is the structure of the

7.84. Cross section of a nest of Kalotermitidae termites in mineralized *Diospyros* wood, from the Late Cretaceous of Texas, showing details of cavities filled with frass. This is the oldest evidence of the family. Kalotermitidae distinctively form small colonies in sound wood. MCZ; largest diameter 47 mm.

digestive system, which has been important in defining certain clades (e.g., Noirot, 1995, 2001; Bitsch and Noirot, 2002.

The family is now divided into six subfamilies, most of whose relationships have been discussed by Krishna (1970), Miura *et al.* (1998), and Donovan *et al.* (2000). Miura's study is based on a very limited sampling of species using DNA sequences, and Donovan *et al.* were unable to show the monophyly of all subfamilies save the Nasutitermitinae. Despite the morphological distinctness of nasutitermitines, Bitsch and Noirot (2002) were not able to define the group as monophyletic on the basis of gut structure, and the mandibulate nasutitermitines have even been recently classified into a separate subfamily, the Syntermitinae (Engel and Krishna, 2004). Most of these studies agree, however, with the following relationships: The subfamily Macrotermitinae is the sister group to the other four subfamilies, then the Foraminitermitinae (Engel and Krishna, 2004), the Apicotermitinae, and lastly the Termitinae and Nasutitermitinae *sensu lato* are sister groups. *Sphaerotermes* was recently removed from the Macrotermitinae and placed in its own subfamily, the Sphaerotermitinae (Engel and Krishna, 2004). With such a speciose group, it may be some while before relationships are thoroughly and confidently worked out.

Macrotermitinae are Old World mound builders or ground dwellers, many of them cultivating the symbiotic basidiomycete fungus, *Termitomyces*. Like unrelated symbiotic basidiomycetes cultivated by attine (leaf-cutter) ants, *Termitomyces* occurs exclusively in the colonies of these termites. The fungus belongs to the large family Tricholomataceae, which is best known for the various little mushrooms that cover rotting logs, though *Termitomyces* rarely matures into fruiting bodies. The fungus is cultivated in the nest on intricate combs of fecal material, and the termites consume nodules and mycelia of the fungus, providing nitrogen and a more readily digestible form of cellulose. Some of the termites, in fact, appear not to be able to subsist without the

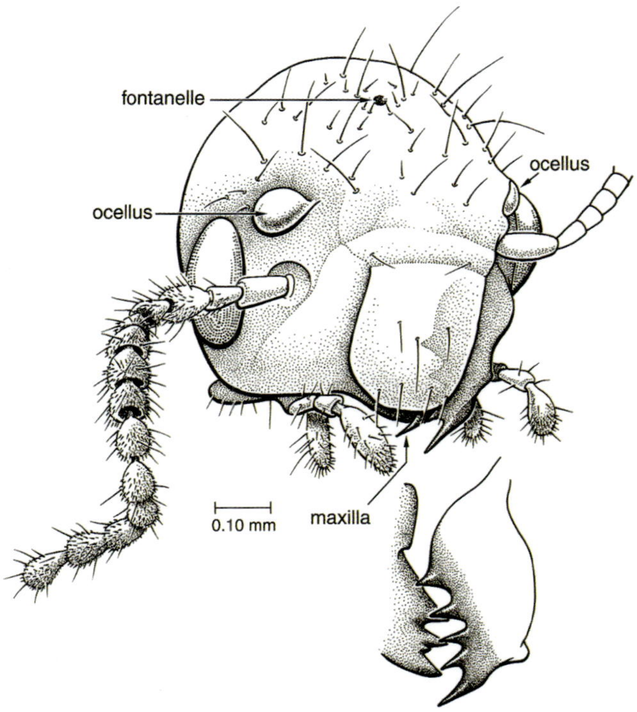

7.85. Head and mandibles of the oldest rhinotermitid termite, *Archeorhinotermes rossi*, in mid-Cretaceous amber from Myanmar. NHM In. 20160; length of forewing 4.4 mm; from Krishna and Grimaldi (2003).

fungus. Thus, cultivation of the fungus has replaced proctodeal trophallaxis, but individual termites are still dependent on the colony for their nutrition. Despite claims that a one-to-one relationship exists between species of termites and species of *Termitomyces* fungus, recent evidence corroborates the view that *species* of the fungus tend to be associated with particular *genera* of macrotermitine termites (Sands, 1969; Rouland-Lefevre *et al.*, 2002).

Perhaps the most impressive aspect of the Termitidae is the diversity of soldier morphology, particularly the mandibles. In many genera the largest soldiers have a massive head that dwarfs the rest of their body, sporting crushing mandibles or long scimitar-like blades that function as shears (Figure 7.75). The most derived genera of Nasutitermitinae have very specialized soldiers, the *nasutes*, with mandibles reduced to functionless stubs, and the head distended into a snout with the fontanelle at the tip. These soldiers douse intruders with a noxious and gluey spray composed of terpenes and other compounds.

Termitidae is a very young group, the oldest fossils of which occur in Eocene amber from the Baltic region, and the oldest nasutes occur in Miocene amber from the Dominican Republic (Figure 7.87).

Phylogeny

There are six major studies on the phylogeny of termite families thus far, three of them based on morphology (Krishna, 1970; Emerson and Krishna, 1975; Donovan *et al.*, 2000), and three based on DNA sequences. The first molecular hypothesis of relationships was by Kambhampati *et al.* (1996a), who used a modest sampling of ten exemplar species and genera and one gene (16S rRNA, 420 bp). Hodotermitidae were not sampled in that study. A second study (Kambhampati and Eggleton, 2000) had slightly better sampling (20 species in 17 genera, with Hodotermitidae included), but it too used a single gene (NADH 5 dehydrogenase, 426 bp). Appearing at the same time as this second study were two other, more comprehensive studies. Thompson *et al.* (2000) sequenced two genes (a rRNA and COII, 1366 bp) and 18 termite genera. The morphological study by Donovan *et al.* (2000) used 197 morphological and biological characters for 49 exemplar genera, but there was relatively low support for many nodes and for the entire consensus tree, probably as a result of coding of

7.86. An arboreal nest of a species of *Nasutitermes* in Peru. Note the meandering galleries on the trunk below the nest.

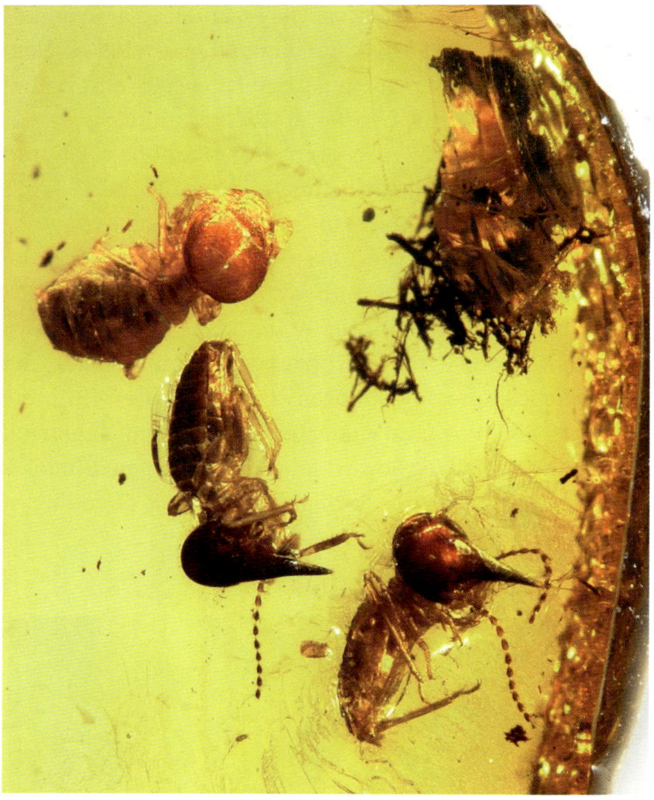

7.87. Two nasute soldiers and a worker of *Nasutitermes electrodominicus* in Miocene amber from the Dominican Republic. AMNH DR14-589; body lengths ca. 2.2 mm.

various characters. Also, characters from the alates were not used in that study, and the taxonomic sampling among the lower termites was relatively modest so the study is not particularly useful for interpreting earliest termites. Nonetheless, it is the most comprehensive study of termite phylogeny so far and largely agrees with molecular results based on four genes. Important areas of agreement and difference among the more recent phylogenies follow.

- Mastotermitidae is undoubtedly the sister group to all other termites; then the Hodotermitidae is the next most basal group.
- The sequence of relationships among the more recently derived families is almost consistently Kalotermitidae [Serritermitidae (Rhinotermitidae + Termitidae)].
- The position of Termopsidae is most uncertain. Kambhampati and Eggleton (2000) indicated it was the sister group to Serritermitidae + Rhinotermitidae + Termitidae, which is highly doubtful. Thompson *et al.* (2000) indicated it was the sister group to Hodotermitidae, which is possible given the traditional view of this grouping (e.g., Krishna, 1970). Other studies, however, indicate that Termopsidae is the sister group to Kalotermitidae plus more derived families (Kambhampati *et al.*, 1996a; Donovan *et al.*, 2000).

Fortunately, this relatively stable framework of family relationships allows sound interpretation of the fossil record and termite evolution (Figure 7.88). Significant evolutionary trends are the following: Colony size varies from small in basal termites (*Mastotermes* being a notable exception) to

TABLE 7.4. Significant Characters in Termite Phylogeny[a]

1. Eusociality: castes with alate reproductives, wingless soldiers and usually workers (vs. solitary or colonial lifestyle, no castes)
2. Wings dehiscent, shed along humeral or basal suture after nuptial flight (vs. not dehiscent)
3. Pod of eggs, the ootheca, highly reduced or eggs laid singly (vs. eggs laid within a heavily sclerotized casing, the ootheca)
4. Male genitalia highly reduced and symmetrical (vs. well developed and asymmetrical)
5. Pronotum highly reduced in size (vs. large, generally concealing head)
6. Ootheca lost, eggs laid singly (vs. ootheca present, albeit vestigial)
7. Tarsi with four or fewer segments (vs. five)
8. Hind wing without an anal lobe (vs. with an anal lobe)
9. Mycetocytes lost (vs. present)
10. Ocelli absent (vs. all three or just lateral ocelli present)
11. Specialized worker individuals not present (vs. present)
12. Soldier with unpigmented eyes (vs. pigmented)
13. Head with central, pore-like defensive gland, the fontanelle (vs. fontanelle absent)
14. Hindgut without mutualistic protists, with bacteria only (vs. protists and bacteria)
15. Soldier with fontanelle at tip of projection (vs. fontanelle not projecting)

[a] Numbers correspond to those on phylogeny, Figure 7.88.

TABLE 7.5. Significant Fossil Termites[a]

a. *Valditermes brenanae*
b. Undescribed, Santana Formation.
c. ?*Mastotermes sarthensis*
d. *Blattotermes*, *Spargotermes*
e. *Blattotermes*
f. *Mastotermes*
g. *Mastotermes*
h. Undescribed, Baissa (Siberia)
i. *Meiatermes bertrani*
j. Various genera, China
k. *Meiatermes araripena*
l. *Valditermes acutipennis*
m. *Luteitermes prisca*
n. *Carinatermes nascimbeni*
o. *Cretatermes carpenteri*
p. ?*Kalotermes swinhoei*
q. Kalotermitid nest, Javelina Formation, Texas
r. *Archeorhinotermes rossi*

[a] Letters refer to ones on phylogeny, Figure 7.88.

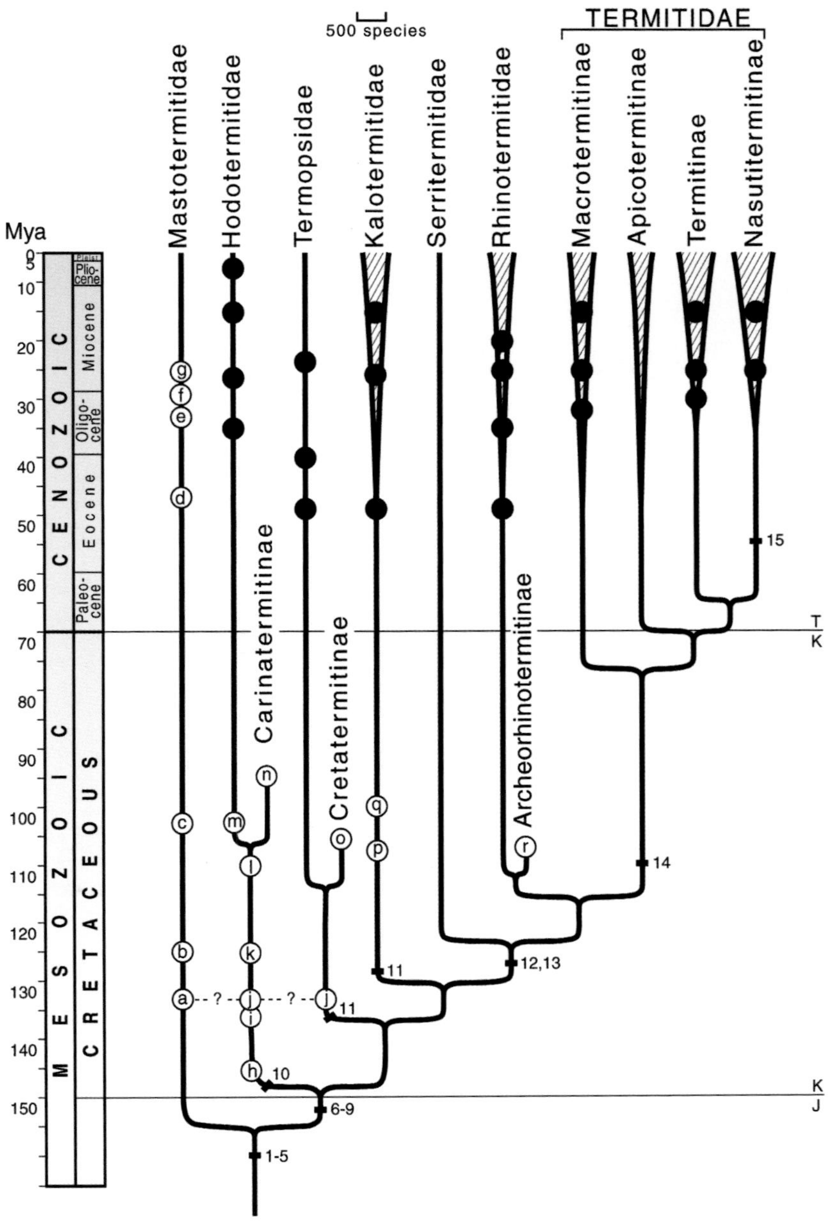

7.88. Phylogeny of termite families and major subfamilies of the Termitidae. Significant Cretaceous fossils are indicated by circled letters, as are Tertiary Mastotermitidae; all other Tertiary records are indicated by dots. See Table 7.4 for characters (numbers) and Table 7.5 for fossils (circled letters). Relationships of fossils were summarized in Thorne *et al.* (2000) for most fossils; relationships among families based on Donovan *et al.* (2000), Kambhampati and Eggleton (2000), and Thompson *et al.* (2000).

huge in the Termitidae; nest construction generally varies from simple excavations to complex structures; the developmental potential of individuals in particular castes varies from great to very little plasticity; and morphology varies from generalized to reduced, as in wing venation, eye structure, and number of tarsomeres. Also, the three most basal families are relict, each with diverse extinct species distributed over larger areas than presently occur.

Fossil Record and Origins

Some primal termite knocked on wood
And tasted it and found it good
And that is why your Cousin May
Fell through the parlor floor today.

–Ogden Nash

There is a strong correspondence between relationships of termite families and the chronology of fossils, with the most basal families appearing first in geological time. Mastotermitidae is anomalous because, save for two ambiguous records, it is virtually absent during the Cretaceous, a time when the family should have been at least as diverse and abundant as it was in the Early Tertiary. The basal families Hodotermitidae and Termopsidae comprise virtually all the fossils in the Cretaceous. Three records of the phylogenetically intermediate families Kalotermitidae and Rhinotermitidae are known from the mid- to Late Cretaceous, two in Burmese amber. Very importantly, the rhinotermitid and kalotermitid in Burmese amber are among the most basal species known for those families, so their appearance in the Cretaceous fits into the general trend of the families: The radiation of termite families is condensed within a 50-million-year period in the Early Cretaceous. Termitidae first appeared in the Eocene and didn't become diverse until the Oligocene and Miocene; they appear to presently be in the heyday of their evolution.

7.89. A female mantis from Costa Rica guarding her ootheca. Photo: P. J. DeVries.

The pattern of termite phylogeny and their fossil record indicates that nascent stages of termite evolution were in the Upper Jurassic, doubtfully earlier. Jurassic ancestors may have eluded paleontologists for any number of reasons: the scarcity of Jurassic deposits outside Eurasia, highly localized populations, general rarity, or a cryptic lifestyle that precluded fossilization.

The ancestral termite consumed wood, within which it also dwelled in social groups. Though the evidence is somewhat controversial, the close relationship of termites to *Cryptocercus* roaches looks likely. *Cryptocercus* itself is too specialized to be ancestral to termites, but a more generalized, extinct cryptocercid probably appeared very much like a basal termite. As in *Cryptocercus*, the use of domiciles by the ancestral termite predisposed the lineage to evolve extended parental care, which eventually led to overlapping generations and then eusociality – a trend repeated in unrelated lineages of insects (see Chapter 11). A groundplan dictyopteran diet of decayed plant material, as is found in modern roaches, probably also facilitated the evolution of termites.

Emerson (1961) summarized termite evolution as "regressive," or the progressive development of many characters of reduction: loss of pigmenation, smaller eyes and pronotum, fewer tarsomeres, extremely reduced genitalia, and the like. Nalepa and Bandi (2000) extended this concept to one of *paedomorphosis*, the retention of juvenile roach-like traits in the adult termite. Though roaches secrete some cellulases for the digestion of plant diets, they too have a diverse anerobic intestinal microbiota that facilitates digestion, and young roaches are more dependent on this microbiota than are older ones. This is probably the main reason why coprophagy is prevalent in early instar roach nymphs: The gut needs to be inoculated (Nalepa and Bandi, 2000). Interestingly, too, among roaches allogrooming of nestmates occurs uniquely in *Cryptocercus* so far as is known, and only in nymphs. A keystone trait of termites, proctodeal trophallaxis, clearly evolved to inoculate newly hatched and molted nestmates with the symbiotic protists they require – an exaggerated nymphal behavior reflecting their roach ancestry.

THE PREDATORY ROACHOIDS: MANTODEA (MANTISES)

Though mantises comprise the smallest order of the Dictyoptera, most of the 2,300 species of Mantodea are the most impressive in the group. Their diverse structures (Figure 7.90) and fascinating habits are largely related to their hallmark lifestyle, predation. Coloration and body structures exquisitely camouflage them on vegetation (Figure 7.89), adapting them for ambush and avoidance of larger predators. Their keen binocular vision allows accurate strikes with large raptorial forelegs. The biology of mantises has been reviewed by Chopard (1949), Beier (1968), and Prete *et al.* (1999). Oddly, the phylogeny of the group has attracted virtually no attention, and their systematics remains largely taxonomic (Beier, 1968; Roy, 1999), especially on a regional basis (Vickery and

7.90. Diverse Recent mantises. From plates in *Genera Insectorum* (fascicles 119, 144, 177, 196).

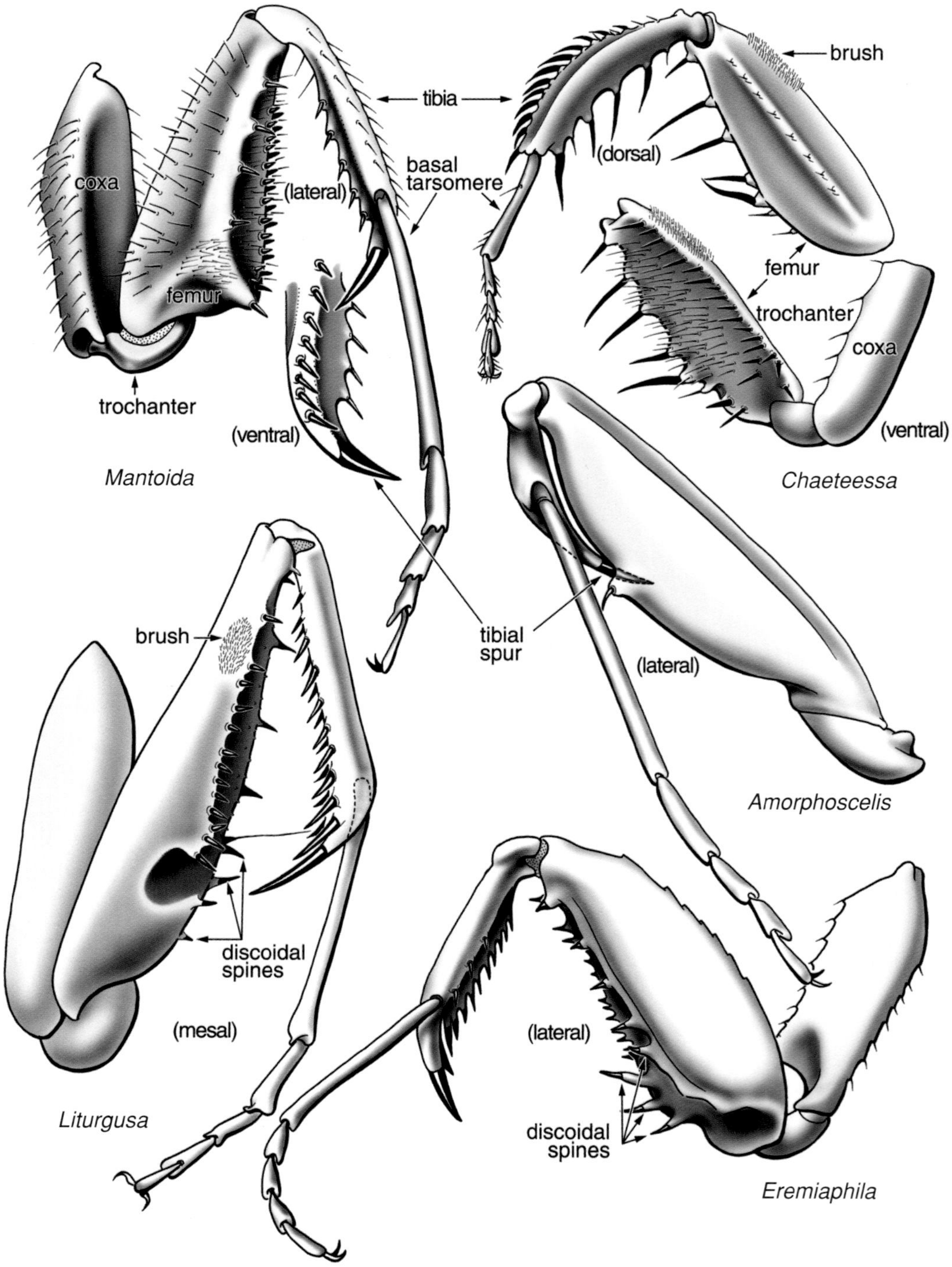

7.91. Forelegs of various Recent basal mantises and a "higher" mantis (*Liturgusa*).

Kevan, 1983; Balderson, 1991; Wang, 1993; Terra, 1995; Kaltenbach, 1996). The classification in general use has been that of Beier (1968), which is an improvement over Victorian-era classifications but is still based largely on rather superficial features of color, pronotal shape, and foreleg spination. The classification has most recently been updated by Ehrmann (2002), in a wonderful monograph cataloguing the world species, and diagnosing the 434 genera. This is the single most important reference on the order.

All mantises are predatory and have the forelegs distinctly modified for grasping prey (Figure 7.91), and some of the larger species even prey on small lizards and frogs. At rest the fore-tibiae and tarsi are folded tightly against the femora, and both forelegs are folded close to each other and against the underside of the body (hence the common name "praying" mantis; "*mantis*" itself is Greek for diviner, seer, or prophet). The femur is long and thick, with rows of spines on the ventral surface that enmesh with spines on the ventral surface of the tibia, a jawlike trap for prey. Variation in the number, arrangement, and shapes of spines is taxonomically very significant (Figure 7.91). Reach is extended by the long forecoxae, which pivot at both ends, and by the connection of the forelegs to the anterior end of a long prothorax. Basal families of mantises have a small prothorax, but in the "higher" mantises the prothorax length is 2 to 10 times its width, even up to 20 times its width in the African species *Leptocola stanleyana*. The mid and hind legs, which do the walking, are distinctively long and slender for Dictyoptera. These also serve in lunging, which the mantis often does when it attacks prey outside the normal strike distance. The effectiveness of raptorial forelegs is illustrated by the repeated origin of the feature in various predatory insects, such as mantispid and rhachiberothine neuropterans; emesine, phymatine, and nepid heteropterans; *Ochthera* (Ephydridae), hemerodromiine (Empididae), and extinct microphorine Diptera; and extinct Raphidiomimidae (Blattaria) and Titanoptera.

The eye of mantises is large and bulging (*exophthalmic*), with a large frontal field and *fovea*, the latter a spot where facets have the sharpest resolution. Mantises move their unusually mobile head and prothorax to fixate on an object in the foveal field and to judge its distance accurately. This is also why they frequently sway from side to side. Like stalk-eyed flies, mantises have exceptional binocular vision among insects (Kral, 1999). They are also fastidious, cleaning their eyes and head after a meal with a patch of fine, feathery hairs on the inside surface of each fore femur (Figure 7.92), in the most feline of manners.

The effectiveness of mantises as predators is due to their vision, raptorial forelegs, quickness of strike, and camouflage. While everyone is familiar with green mantises – remarkably camouflaged against virtually any foliage – there is a stunning array of stem, grass, leaf, and flower mimics in the order (Figures 7.90, 7.93, 7.94). Their ability at crypsis is rivaled only by Phasmida and some of the katydids such as the pseudophyllines (Tettigoniidae). "Leaf species" mimic live green leaves or dead brown ones, and they usually have flat, leaflike ("foliate") extensions of the head, pronotum, tegmina, or even the legs. The tegmina, moreover, can have a lacy venation typical of dicot leaves. Not surprisingly, mantises in scrub and grassland habitats mimic sticks and grass stems: The body is very long and thin, appendages are held close to the body, and the head may even be held in the prognathous instead of the usual orthognathous position. These mantises will frequently hold the body against the stem, as many phasmids do. The most impressive mimics are those of flowers. These forms, exemplified by the African genera *Pseudocreobotra* and *Chlidonoptera* and the Asian genus *Hymenopus*, hold the abdomen and thorax erect, and the body is usually covered with frills and extensions. They have elaborate coloration, boldly patterned in colors highly unusual for insects, like pink, lavender, and pure white. With

7.92. Scanning electron micrograph of the brush on the inner surface of the forefemur of a mantis. Mantises wipe the brush over their eyes and mouthparts, particularly after a meal and in the most feline of manners.

7.93. A nymphal mantis from Costa Rica that mimics a plant tendril. Photo: P. J. DeVries.

7.94. A white version of the orchid-mimic mantis, *Hymenopus coronatus*, from Viet Nam. These mantises will perch at the tip of a branch and grab pollinating insects that visit them. Photo: D. Grimaldi.

7.95. An aggregation of young hymenopodid mantis nymphs in Ecuador, shown clinging to the ootheca from which they just emerged (above) and moving along the ground en masse (below). The color, body shape, and movements mimic army ants (*Eciton*) (cf. Figure 11.60) and help protect the vulnerable young nymphs. Photo: P. J. DeVries.

typical mantis reflexes, they snag agile hovering insects like bees and flower flies attracted to the apparent flower.

Many mantises show remarkable color and shape polymorphisms among and within individuals. Individuals reared in bright light and low humidity (as in a tropical dry season), for example, become brownish when they molt, or green when reared in lower light and higher humidity. Like chameleons, some of the flower mantises can actually change color to match the color of the flowers, though it generally takes several days to do so. First and second instar nymphs of many Hymenopodidae mimic ants in shape, color, and behavior (ants are a pervasive model among many insects) (Figure 7.95). Older instars, though, lose this habitus.

The infamous "black widow" behavior of female mantises involves copulating with him as she devours him, usually headfirst, which was vividly described more than a century ago (Howard, 1886; Fabré, 1897). One belief is that cannibalism is unnatural (e.g., Balderson, 1991), observations of it based primarily on caged situations where mantises are forced into close proximity. Indeed, mantises are thinly dispersed in nature. Nymphs usually scatter after hatching, and adult males probably seek females through widely dispersed pheromones, so individuals rarely meet except to mate. But, in roughly three quarters of the mantis species whose mating has been observed, up to 50% of the natural prey items of some females are males (Maxwell, 1999). Some people have construed this mating to be "sacrificial" on the part of the male; others believe it is adaptive because a meal of male provides her with abundant protein required for egg production, and ultimately the production of his offspring. This is excessively adaptationist, though, because it would always be to the male's advantage to mate more than once. Sexual cannibalism in fact appears not to be a standard part of mating, because courting males are usually quite wary of the female, carefully maneuvering toward her when she turns her head away, or ambushing her from behind. As in almost all insects the females are generally larger than the males, sometimes considerably so (male *Hymenopus* and other Asian Hymenopodinae, for example, are half the size of females, though this is extreme). This size difference makes the male mantis particularly vulnerable, plus the fact that mantises by nature are solitary, as well as voracious and indiscriminate predators.

Two aspects of mantis mating, though, appear to be adaptations resulting from sexual cannibalism. One, which has never been discussed, is the position of the male on top of the female. Except for some of the Holometabola, this is an unusual mating position in insects and unique in the Dictyoptera. Cockroaches and termites mate tail-to-tail, facing opposite directions. Though female mantises are known to grab males who are on top of them, mating tail-to-tail would put the male within easier reach. Second, decapitated males show a perplexing capacity for copulation. Famous experiments by Roeder (1963, 1967) on male mantises revealed that severing the ventral nerve cord at any location (including decapitation) elicited the stereotypical array of mating movements: bending of the abdomen, distension of genitalia, coupling, and copious insemination. Though decapitated roaches also elicit some mating movements, these are very incomplete compared to mantises. Indeed, copulation of decapitated male mantises can last up to 20 hours! Roeder (1963) hypothesized that cutting the ventral nerve cord blocks nerve impulses that normally inhibit mating reflexes. These studies probably led to the popular but

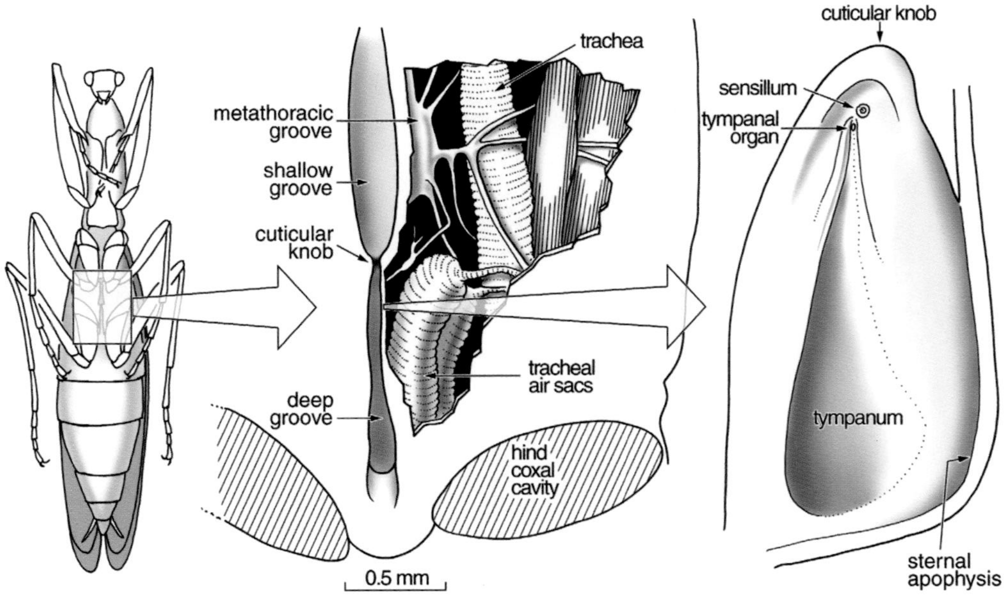

7.96. The metathoracic, "cyclopean" ear of higher mantises. It is sensitive to ultrasonic calls of bats and presumably evolved to protect flying mantises. Redrawn from Yager and Hoy (1987).

erroneous thoughts that female mantises can only mate with decapitated males, even though male mantises are quite adept at mating with their heads on. The frequency of sexual cannibalism may have selected for strong copulatory reflexes, even by dismembered males.

Because mantises are highly visual animals it has always been assumed they were deaf. With exception of the phylogenetically basal species, mantises have a unique ear in the insects, located on the metathorax between the sternites and just anterior to the hind coxae (Figure 7.96). Its gross and ultrastructural morphology, neural activity, behavioral function, homology, and even phylogenetic distribution has been presented in detail (Yager and Hoy, 1986, 1987; Yager *et al.*, 1990; Yager, 1992, 1999b).

The opening to the mantis ear is a groove, the anterior end with a pair of knobs and the flat inner walls of the groove facing each other. The tympanum is a drop-shaped area on each wall, with the tympanal organ – which transduces the sound waves into neural impulses – at the narrow tip. Sound waves are amplified by ligaments and an inflated air sac attached to the inside surface of the tympanum. Mantises are tone deaf and auditory cyclops: They cannot discriminate frequency or direction. Their hearing, though, is acute between 25–50 kHz, the region of ultrasound frequently used by echolocating bats. Though commonly believed to be diurnal insects, mantises will fly at night, especially males who are in search of female pheromone plumes. A flying mantis that hears a bat call undergoes a typical maneuver, including stretching out its forelegs, rolling the head to one side, curling the abdomen, and divebombing. Field tests indicate mantises are highly effective at evading bats. Moreover, there is a direct relationship between wing development and the occurrence and development of ears. An ear is absent or vestigial in young nymphs and in species that are brachypterous or apterous, in which case these are usually females. Thus, the mantis ear is an organ that is probably pleiotropically co-evolving with flying mantises as an adaptation against bat predation.

The unique cyclopean ear of mantises is also one of the best phylogenetic characters in the order. It occurs in the "higher" mantises, or the Mantoidea (Empusidae, Hymenopodidae, Mantidae), also defined partly on the basis of an elongate pronotum. It is absent in the basal families Chaeteessidae, Eremiaphilidae, Mantoididae, Amorphoscelidae, and Metallyticidae, though Yager (1999b) suggested the possibility of loss of the ear in the last two families. All the basal species dwell on the ground or tree trunks, and absence of the ear may be ecologically related to these habits. Alternatively, what may be a unique origin of the ear in the Mantoidea is perhaps related to the relatively young age of this group, an age approximately contemporaneous with the origin of insectivorous bats in the Late Paleocene to Early Eocene.

Early Fossils and Evolution

Prior to 1990, Cretaceous mantises were unknown, even though their history was thought by some to extend into the Paleozoic. The first major report on Mesozoic mantises was by Gratshev and Zherikhin (1993), which described and named 11 Cretaceous mantises from Eurasia (and one from the Oligocene of Siberia). More recently, Grimaldi (2003b) reported additional Cretaceous species along with a phylogeny of the Cretaceous genera and living families of mantises, which revealed an evolutionary history very similar to that of termites. Fossils occur throughout the Cretaceous, all

7.97. An early and very basal mantis, *Santanmantis axelrodi*, in 120 MYO Cretaceous limestone from Brazil. Staatlichen Museum für Naturkunde, Stuttgart (SMNS) 112.

of which are very primitive, and it is even possible that *Regiata scutra* from the Lias of Dorset, England, is a very primitive mantis or at least a stem-group mantodean (A. Ross, pers. comm., 2004). It was originally described in the Haglidae (Whalley, 1985: p. 131–2). If *Regiata* is indeed mantodean, this would be the oldest dictyopteran belonging to a modern order by some 50 MY. The most diverse group of the order, the superfamily Mantoidea (comprising 90% of all living species), radiated in the Early to mid-Tertiary like the Termitidae.

Cretaceous mantises are preserved as compressions and mineralized replicas in rock and as inclusions in amber. Most of the compressions are isolated wings, which of course provide no information on foreleg spination, pronotal shape, and other important features of the body. Most of the specimens in amber are nymphs, which lack wings and thus are difficult to compare with the compression fossils, though they do exquisitely preserve foreleg spination and other details. All Cretaceous mantis fossils lack some specialized features of Recent mantises. For example, the extinct genera

7.98. Reconstruction of the Early Cretaceous *Santanmantis axelrodi*, based on a series of eight specimens and HRCT scans. Forewing length 12.5 mm.

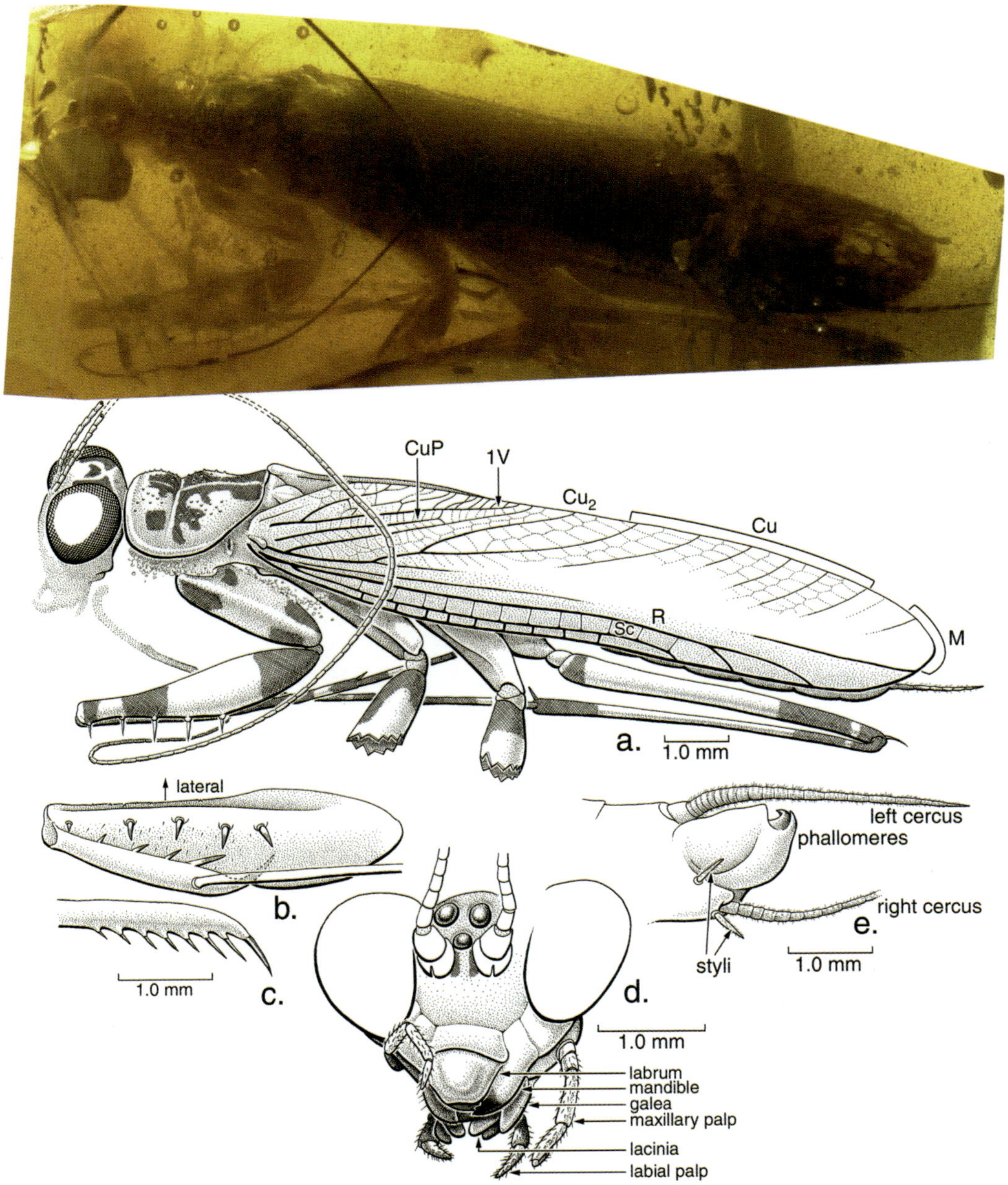

7.99. *Ambermantis wozniaki*, a completely preserved, basal adult mantis in a piece of turbid Cretaceous amber from New Jersey, showing details of the head, foreleg, and terminalia. AMNH NJ 1085; body length 15 mm; from Grimaldi (2003).

Chaeteessites and *Jersimantis* in amber from Siberia and from Burma and New Jersey (respectively) lack the brush and thick spines that occur on the fore-femora of Recent species of mantises, nor do they have an elongate pronotum. The best preserved specimens are two species preserved as complete adults: *Santanmantis axelrodi* from the Santana Formation of Brazil (Figure 7.97) and *Ambermantis wozniaki* in New Jersey amber. Because there is a series of *Santanmantis* specimens and some were even imaged using high-resolution CT scans, it has been possible to reconstruct most of this animal (Figure 7.98). *Santanmantis* is superficially similar to the basal Recent mantises *Chaeteessa* and *Mantoida*, but it lacks some of their derived wing veins and leg spines. The fossil also possesses an unusual, very long "pseudovein," which is an oblique veinlike structure that occurs in all Recent mantises but is always restricted to the basal forks of the M and CuA veins. *Ambermantis* is slightly more derived than *Santanmantis*, which may reflect its slightly younger age (by approximately 30 MY). *Ambermantis* has more derived features of Recent mantises, but lacks, for example, the discoidal spines seen in almost all Recent mantises. This fossil is peculiar for the extremely long, stiltlike mid and hind legs (Figure 7.99).

The most basal living families are the monogeneric families Chaeteessidae (*Chaeteessa*), Mantoididae (*Mantoida*), and Metallyticidae (*Metallyticus*). The first two live in tropical forests of Central and South America; the last one is a brightly colored, iridescent mantis (hence the name) from southeast Asian tropical forests. They are all ground predators or stalk their prey on tree trunks – habitats reminiscent of their roach ancestors. Despite claims of Chaeteessidae in the Cretaceous (Gratshev and Zherikhin, 1993), definitive fossil chaeteessids do not occur until the Early Oligocene (*Lithophotina*, from Florissant) and in Miocene amber from the Dominican Republic, though *Arverineura* from the Paleocene of France may be a chaeteessid (Nel and Roy, 1996). *Chaeteessa* has a unique arrangement of thick, dark foreleg spines, virtually forming a basket (Figure 7.91) and primitively lacks a feature found in all other living mantises. The fore-tibia in all other living mantises (and even in the Cretaceous *Electromantis* and *Ambermantis*) has a large spine at the apex of an extension of the tibia, the "claw." A particularly significant fossil is a fragment of a forewing, also from the Paleocene of France, *Prochaerododis enigmaticus* (Nel and Roy, 1996). Despite such an incomplete specimen, the intricately reticulate venation is very similar to that of living leaf-mimic Mantidae. Other fossil Mantidae occur in Baltic, Mexican, and Dominican ambers (e.g., Figure 7.100).

The correspondence between the phylogeny of mantises and chronology of fossils indicates an origin of the order in the Jurassic, but the most significant radiations occurred only in the Tertiary. These Tertiary radiations involved the superfamily Mantoidea, comprising 90% of all living species in three families, the Empusidae, Hymenopodidae, and Mantidae (some subfamilies of the last are sometimes given family status). If the Mantoidea uniquely possess the cyclopean ear, which is an adaptation to bat predation, then this group may have originated around the time of the earliest and most basal microchiropteran (insectivorous) bats in the Early Eocene to possibly the Late Paleocene (Simmons and Geisler, 1998). Such a date corresponds with the fossil record. While the origin and spread of grasslands contributed to the Tertiary radiation of the Termitidae, there is apparently no similar paleoenvironmental explanation for the relatively recent radiation of mantises into the exuberance of forms seen today.

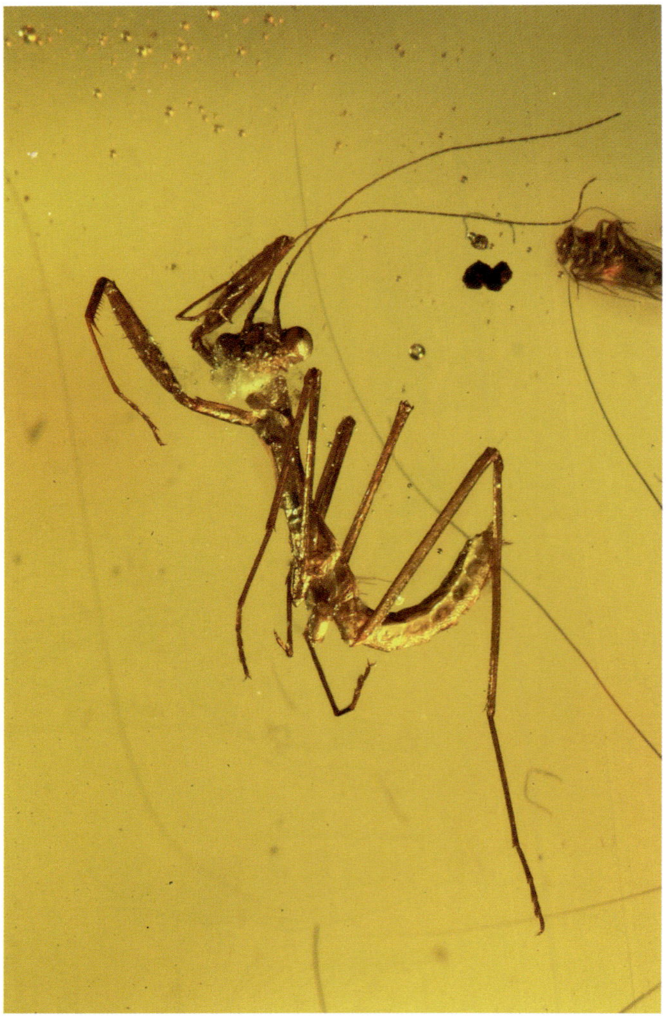

7.100. A nymph of the family Mantidae in Miocene amber from the Dominican Republic. AMNH; body length 4.5 mm.

AGES OF THE DICTYOPTERA

Earlier, entomologists promoted views of Dictyoptera as being ancient, no doubt influenced by the existence of Carboniferous roachoids. For example, termites have been hypothesized to have evolved in the Triassic (Emerson and Krishna, 1975; Carpenter, 1992), and the mantises as early as the Permian (Carpenter, 1992). Even Hennig (1981), who understood the paraphyletic nature of Paleozoic roachoids, admitted the possibility of Paleozoic Mantodea. But, these authorities were writing at a time when no Cretaceous fossils had been found for mantises, and only two Cretaceous fossils of termites were known. The current evidence indicates that the Recent orders of Dictyoptera evolved more recently (Grimaldi, 1997b, 2003b; Thorne *et al.*, 2000), since there is a consistent pattern of basal families of mantises and termites occurring in the Cretaceous and more derived families in the Tertiary. This indicates a Jurassic origin of these orders, greatest radiations of which occurred in the Tertiary.

8 The Paraneopteran Orders

The evolutionary history of the Paraneoptera – the bark lice, true lice, thrips, and hemipterans – is a history beautifully reflected in structure and function of their mouthparts. There is a general trend from the most generalized "picking" mouthparts of Psocoptera with standard insect mandibles, to the probing and puncturing mouthparts of thrips and anopluran lice, and the distinctive piercing-sucking rostrum or beak of the Hemiptera. Their mouthparts also reflect diverse feeding habits (Figures 8.1, 8.2, Table 8.1). Basal paraneopterans – psocopterans and some basal thrips – are microbial surface feeders. Thysanoptera and Hemiptera independently evolved a diet of plant fluids, but ancestral heteropterans were, like basal living families, predatory insects that suction hemolymph and liquified tissues out of their prey. Diets then reverted to the ancestral habit of phytophagy among "higher" heteropterans. Blood feeding arose independently between the ectoparasitic lice and disparate, small groups of Heteroptera. Thus, there are multiple origins of sucking mouthparts and a liquid diet in the Paraneoptera (Figure 8.1).

There is little question about the monophyly of the Paraneoptera, based on morphological and molecular evidence. Some features, like reduction in the number of tarsomeres (to three or fewer) and loss of cerci, occur throughout insects. Other features, though, are not so homoplasious and reliably reflect monophyly, such as abdominal trichobothria, loss or great reduction of the labial palps, structure of the laciniae in the more basal taxa, and the large postclypeus and cibarial muscles, among others. Paraneopterans also have a structure for coupling the fore- and hind wings in flight that is rather conservative in design. In Psocopterans, many Sternorrhyncha, and even basal Heteroptera, a small clasp formed from clumps of stout microtrichia on the hind edge of the forewing attaches to a clasp on the fore edge of the hind wing. Another interesting feature defining the Paraneoptera, but one hardly mentioned, is that the antennal flagellomeres have fine annuli. This is consistently seen throughout the group but lost in Heteroptera and obscured in many Auchenorrhyncha because of their reduced antenna. Also, many paraneopterans fold their wings rooflike at rest over the abdomen, but thrips and Heteroptera fold them flat over the abdomen, which probably relates to the structure of axillary sclerites and other minute structures at the base of the wing (i.e., Yoshizawa and Saigusa, 2001).

Relationships among paraneopteran orders have been discussed by Seeger (1975, 1979), Kristensen (1975, 1991), Hennig (1981), Wheeler *et al.* (2001), and most recently by Yoshizawa and Saigusa (2001). These studies generally agree on the monophyly of the order Hemiptera and most of its suborders and a close relationship of the true lice (order Phthiraptera) with the most basal group, the "bark lice" (Psocoptera), which comprise the Psocodea. One major issue is the position of thrips (order Thysanoptera), which either comprise the sister group to Psocodea or to Hemiptera. We prefer the latter of these, in which case the thrips plus Hemiptera comprise a group called the Condylognatha. The other major issue is the monophyly and relationships of the leaf hoppers, plant hopper, and tree hoppers (the Auchenorrhyncha).

PSOCOPTERA: THE BARK LICE

Small and inconspicuous, the approximately 4,400 described species of psocopteran "book lice" and bark lice probably are a fraction of the actual diversity. Many live in concealed spaces, microscopic structures are required for their identification, and there is virtually no medical or agricultural significance of the group. Thus, they are poorly studied. Relationships are also poorly explored, even though the order may be paraphyletic with respect to the true lice, much the way certain scorpionflies (Mecoptera) appear closely related to the fleas. There is little doubt about a monophyletic Psocodea (psocopterans + lice), and an intriguing hypothesis even links the true lice with a particular family of Psocoptera – possibly a rare example of transitional forms between free-living insects and highly specialized parasites. Moreover, the Psocodea are the living sister group to the rest of the Paraneoptera, and a very basal, extinct group of psocopterans has

TABLE 8.1. Characters Defining Paraneopteran Relationships[a]

Paraneoptera:
1. Legs with three or fewer tarsomeres
2. Cerci lost
3. Ganglia in abdomen fused into one large mass
4. Winged forms usually with abdominal trichobothria
5. Labial palps reduced or lost
6. Asymmetrical mandibles
7. Maxillary laciniae slender, long, detached from stipes; independently movable
8. Postclypeus large, with large cibarial dilator muscles

Psocodea:
9. Simplified ovipositor
10. Rupturing mechanism at the base of antennnal flagellum
11. Postclypeus bulbous, protruding
12. Cibarium and hypopharynx specialized for water vapor absorption
13. Cardo lost

Psocoptera:
14. Wings held rooflike over abdomen
15. Forewings with "areola postica" cell
16. Pearman's organ on hind coxa
17. Egg with thin, unsculptured chorion; aeropyles and micropyles lost

Liposcelidae + Phthiraptera:
18. Reduction in wings: brachypterous or apterous
19. Body flattened, including dorsoventral flattening of head
20. Hind femora enlarged
21. Fusion of meso- and metanotum
22. Loss of abdominal spiracles one and two
23. Reduction or loss of labial palpi
24. Prognathous head
25. Eyes reduced or lost

Phthiraptera:
26. Ectoparasitic on warm-blooded vertebrates
27. Apterous
28. Eyes highly reduced (only several facets), usually completely lost
29. Head with very limited movement
30. Dorsal tentorial arms lost
31. Reduction or loss of lacinial stylets
32. Reduction or loss of maxillae
33. Antennae reduced to three flagellomeres
34. Ocelli lost
35. Egg with hydropile and operculum, cemented to host hair or feather
36. Fusion of third thoracic ganglion with abdominal ganglion
37. Three (versus four) nymphal instars
38. Loss of metathoracic spiracle

Amblyceran lice:
39. First flagellomere pedunculate
40. Antennae concealed in fossae

All Other Lice:
41. Connective tissue occluding occipital foramen ("obturaculum")
42. Spiracular glands
43. Occipital apodeme extending into thorax
44. Saucer-shaped antennal sensilla ("pore organs")

Ischnoceran Lice:
45. Number of sensilla coelonica and basiconica on antennae are reduced (to two)
46. Small rhombic sclerite separated from pronotum

Rhyncophthirina + Anopluran Lice:
47. Pretarsus with only one claw
48. Pronotum and forecoxae fused, do not articulate
49. Loss of cervical sclerites
50. Virtual fusion of head and thorax
51. Extreme prognathy
52. Antennal muscles attach to dorsum of head, not to tentorium
53. Loss of anterior tentorial pits
54. Posterior tentorial pits are reduced
55. Loss of lacinia

Rhyncophthirina:
56. Elongate rostrum, at apex of which are the mandibles
57. Mandibles rotated 180°

Anopluran Lice:
58. All thoracic segments fused
59. Meso- and metathoracic terga reduced
60. Complete loss of tentorium
61. Hypopharynx and labium developed into piercing stylets

Condylognatha:
62. Opisthognathous head
63. Expanded hypopharyngeal apodemes
64. Unicondylar mandibular and lacinial stylets
65. Narrowed labrum
66. Dorsal shift of anterior tentorial pits

Thysanoptera:
67. Wings narrow, with fringe of long setae, venation reduced
68. Claws reduced in adult
69. Pretarsus with eversible vesicle
70. Piercing lacinial stylets
71. Left mandibular stylet piercing, right one lost; mouthcone asymmetrical

Hemiptera:
72. Maxillary and labial palps lost
73. Mouthparts developed into suctorial beak, with two pairs of mandibular and maxillary stylets lying in a long, grooved labium

Sternorrhyncha:
74. Absence of vannus, vannal fold in hind wing
75. Labium originates from prosternum

Auchenorrhyncha:
76. Complex tymbal acoustic system
77. Aristate antennal flagellum

Coleorrhyncha + Heteroptera:
78. Wing coupling mechanism
79. Thoracic scent glands
80. Gula developed between occipital foramen and mouthparts

Coleorrhyncha:
81. Body highly flattened, with areolate wings

Heteroptera:
82. Reduced tentorium
83. Nymphal dorsal scent gland on tergites three through seven

[a] Numbers correspond to numbers on phylogeny, Figure 8.1.

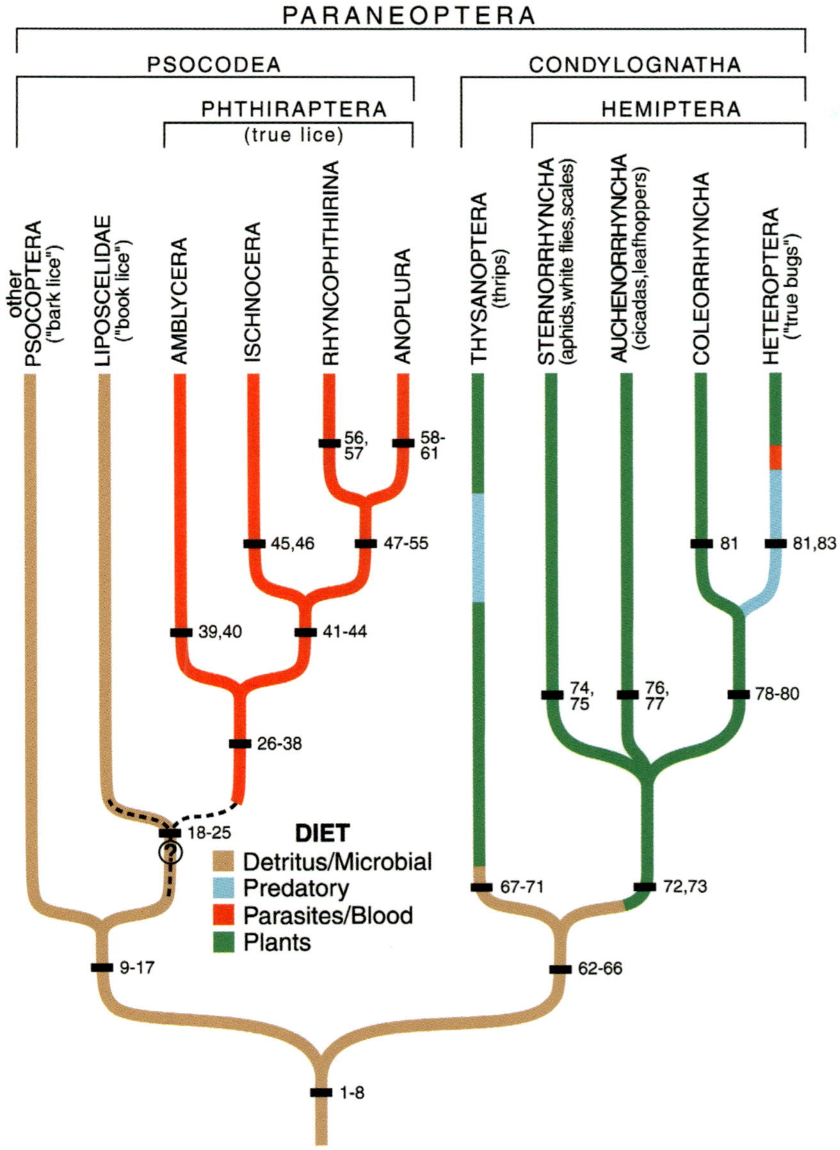

8.1. Relationships among Recent orders and suborders of the Paraneoptera, with their diets. Significant characters indicated in Table 8.1. Based on Lyal (1985) for Psocodea, Kristensen (1975, 1991) and others for orders, and Wheeler *et al*. (1993b) for Hemiptera.

even been proposed as a sister group to the thrips. General accounts on the biology, morphology, and taxonomy of the group are by Badonnel (1951), Smithers (1967, 1972, 1991), Weidner (1972), Günther (1974, the taxonomy of central European species), New (1974, 1987), and Mockford (1987, 1993, North American species). There is a recent world catalogue of the species (Lienhard and Smithers, 2002).

Psocopterans are distinctive, though evidence for monophyly of the order is modest. They often have bulging eyes, they usually have a bulging postclypeus, and in winged species and forms they always have a small prothorax and large pterothorax with the wings held rooflike over the abdomen when at rest. A small knob or brush near the forewing pterostigma hooks onto the fore margin of the hind wing when at rest, and in flight a small hook on vein CuP couples the forewing to the hind wing. The forewings have vein CuA with a fork near the margin ("areola postica"). Most of these wing features are similar to that of some Hemiptera. The Sternorrhyncha and Auchenorrhyncha also hold the wings rooflike over the abdomen, with some Psyllidae and Aleyrodidae also having the areola postica. Some Sternorrhyncha and even basal Heteroptera have similar wing coupling structures (Figure 8.3).

Morphology of the psocodean head and mouthparts is presented by Badonnel (1934) and Symmons (1952), and there are several very distinctive microscopic features. In the hypopharynx are minute sclerites (*sitophore sclerites*), which appear mechanically involved in a pharyngeal structure specialized for absorbing moisture from atmospheric water vapor (Rudolf, 1982) (Figure 8.2). The ability to absorb water vapor occurs in various insects, but psocodeans are the only ones to do so in the adult stage, and their efficiency at this surpasses all other insects, being capable of swelling with water to several times their original mass. The true lice, Phthiraptera, have a similar but even more modified hypopharynx, and are likewise efficient at absorbing water

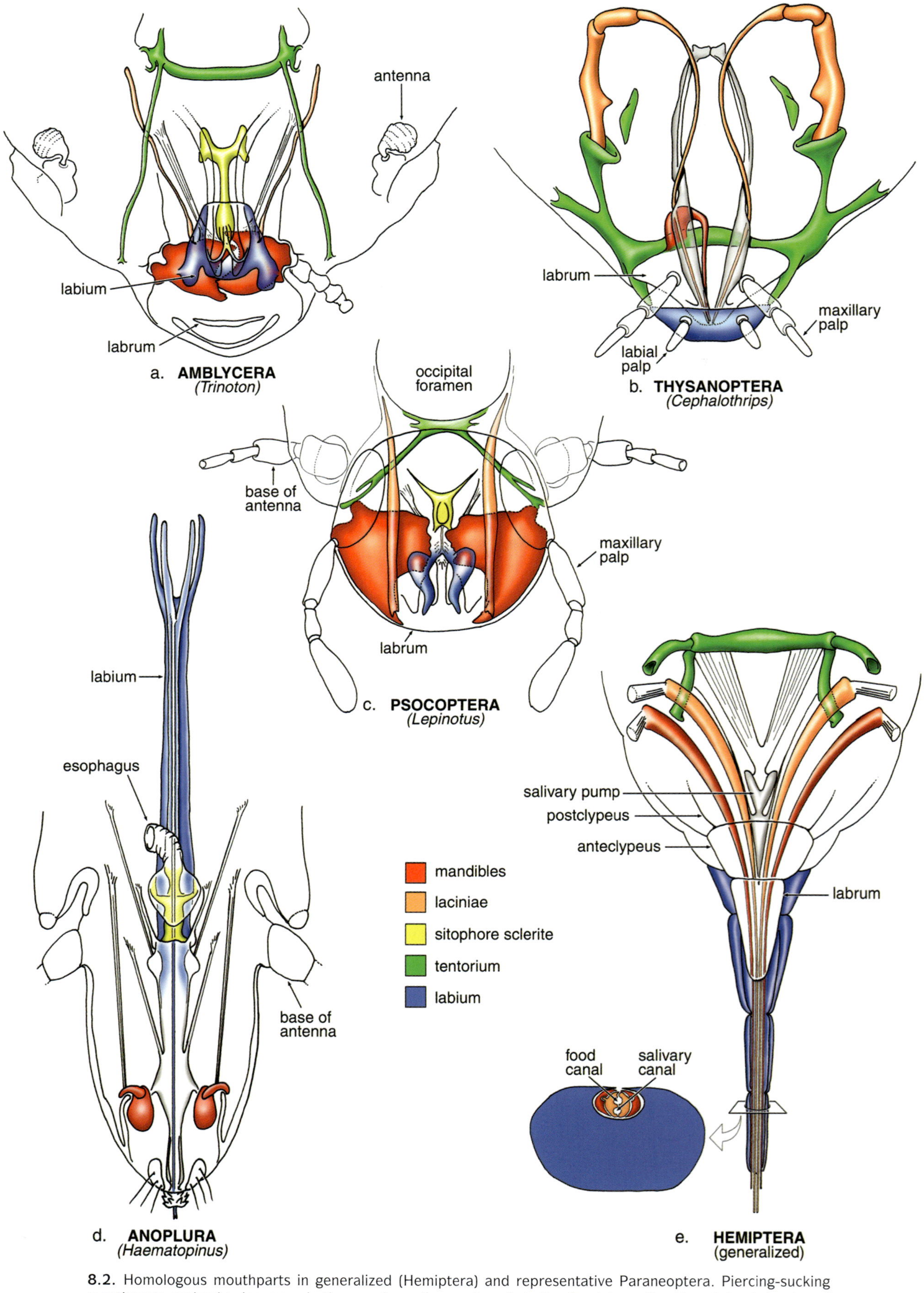

8.2. Homologous mouthparts in generalized (Hemiptera) and representative Paraneoptera. Piercing-sucking mouthparts evolved twice, once in the anopluran lice, and again in the Condylognatha, though best developed in the Hemiptera.

8.3. A psocopteran, *Pseudocaecilius* sp., showing features typical of Psocoptera (a, c), or Paraneoptera (b). (a) Wing-folding mechanism; (b) wing-coupling mechanism; (c) Pearman's organ. Scanning electron micrographs; body length 1.3 mm.

vapor (Rudolf, 1983). This unique feature may in fact have preadapted lice to invading the desiccating environment of feathers and fur, made even more parched by frequent dust baths by their hosts.

The hind coxa of most psocopterans has a Pearman's organ (Figure 8.3c), which is a small rough dome with an adjacent tympanum that is supposedly stridulatory. Also, basal flagellomeres of the antenna have rings of weakened cuticle (the "antennal rupturing mechanism") that allows the antenna to detach (Seeger, 1975), but this structure too, like the pharyngeal sclerites, occurs in the true lice. Psocopteran eggs have a thin, unsculptured chorion, with the aeropyles and micropyles lost, but the eggs of few species are known. Even the mouthparts, with a pair of long, fine, freely movable maxillary laciniae, are primitive compared to the mouthpart structure of the rest of the Paraneoptera, the so-called Condylognatha (thrips + hemipterans). No structures were found in the base of the forewing that are distinct to the Psocoptera, though structures unique to other paraneopteran orders were found (Yoshizawa and Saigusa, 2001). Based on the cibarium and hypopharynx structures, and the flagellar rupturing mechanism, bark lice and true lice are closely related, which is supported by molecular data (Wheeler *et al.*, 2001; Yoshizawa and Johnson, 2003).

Psocopterans feed by using the toothed laciniae to scrape spores, fungal hyphae, lichens, algae, and films of yeast and bacteria from the surfaces of living and decaying plants. Most live in cryptic and concealed places, including amongst leaf litter and compost, under decaying bark and stones, in galleries of wood-boring insects, in caves, and in nests of paper wasps and bees, termites, ants, and vertebrates. Not surprisingly, many species are brachypterous or apterous, even polymorphic for winged and wingless forms. Some species are brightly colored, such as the tropical Caeciliidae found on forest leaves. Some species have elaborate mating behavior, including courtship that involves drumming the abdomen against the substrate, a habit also seen in stoneflies and lacewings. Particularly unusual mating is seen in *Phlotodes australis*, which performs the ultimate male display: standing on their heads (New, 1987).

Biology of tropical and subtropical Archipsocidae is interesting since some species construct sheets of webbing against tree trunks or branches much like Embiidina, only the psocopterans secrete silk from labial glands, not the forelegs. Parthenogenesis, which is found sporadically in the order (Mockford, 1971b), also occurs in this family, as does even viviparity. Given that domicile use and close genetic relatedness appear to be preconditions for social behavior, it would be very interesting to see if some archipsocids have a social system similar to that of some other paraneopterans, specifically certain galling thrips and aphids. New (1987) mentioned that both parents and offspring produce the webbing, and the latter remain with the parents, but no detailed study has yet been made of the behavior of archipsocids.

The classification of three suborders and approximately 50 families of living psocopterans in general use today (e.g., Smithers, 1991) is based primarily on Badonnel (1951), itself largely based on that of Pearman (1936) and Roesler (1944). The former of these last two studies was the first to use microscopic structures of these insects seriously. Indeed, besides wing venation, color, and vestiture, microscopic structures of the genitalia, tarsal claws, laciniae, and hypopharynx provide many characters. Among the three suborders, only the Psocomorpha is particularly diverse with 37 families. Relationships among these families have recently been explored using morphological (Yoshizawa, 2002) and molecular (Johnson and Mockford, 2003; Yoshizawa and Johnson, 2003) evidence. The other two suborders, Trogiomorpha and Troctomorpha, have five and eight small to moderately sized families, respectively. Smithers's (1972) monograph is a very useful compendium of the families and subfamilies of the world, including fossils, but his phylogenetic trees were artificial. Dichotomous branches in those trees were defined by the presence and absence of characters, the result being that some "groups" are not defined by any derived character. Relationships among living psocopterans, let alone Mesozoic and putative Paleozoic ones, are only beginning to be explored (Yoshizawa, 2002; Johnson and Mockford, 2003), though an hypothesis by Lyal (1985b) on a psocopteran sister group to the true lice proposes a very intriguing phylogenetic relationship.

Lyal (1985b) postulated that the psocopteran family Liposcelidae (Figure 8.4) is a living sister group to the true lice. Liposcelidae comprise a global family of six genera and approximately 100 described species, most in the genus *Liposcelis*. Species of *Liposcelis*, or book lice, are notoriously known as the minute, pale insects scuttling between damp papers, cardboard, and pages of books, where they feed on mold and glue. Their feeding can also damage pinned insect specimens. The natural habitats of liposcelids typically include tight, concealed spaces under decaying bark and stones and in compost and amongst leaf litter, with several species of *Liposcelis* commonly occurring in ant nests (*L. formicaria*, *L. myrmecophila*, and *L. prenolepidis*) (Broadhead, 1950; Lienhard, 1990). Some are even known to feed on insect eggs (Williams, 1972). Liposcelidae have also been found in mammal and bird nests feeding on nest debris (Rapp, 1961; Wlodarczyk, 1963, Mockford, 1971a; New, 1972; Baz, 1990; Lienhard, 1990), even amongst the feathers and fur of some birds and mammals (Pearman, 1960; Mockford, 1967; Badonnel, 1969). The cosmopolitan, parthenogenetic species, *Liposcelis bostrychophila*, for example, has been found among the fur of Asian rats and African tree mice and in the nests of cliff swallows in Nebraska, weaverbirds in Africa, and the small African primate *Galago*, though it has a

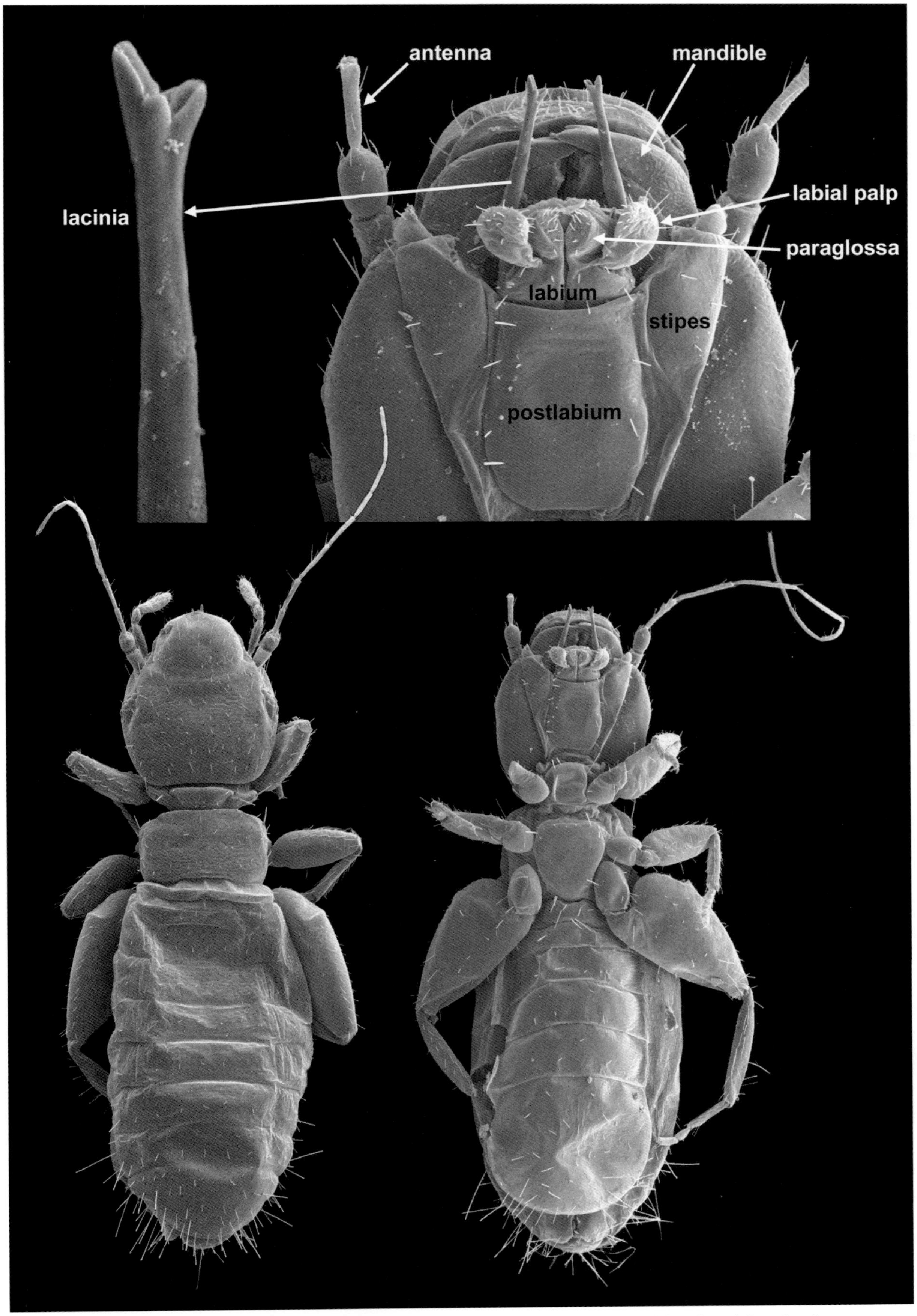

8.4. The psocopteran *Liposcelis* sp., showing mouthpart structures in ventral view (above), and dorsal and ventral views of the body. The prognathous head and enlarged hind femora seen here are some of the features why Liposcelidae are considered the closest living relatives of true, parasitic lice. Scanning electron micrographs; body length 1.2 mm.

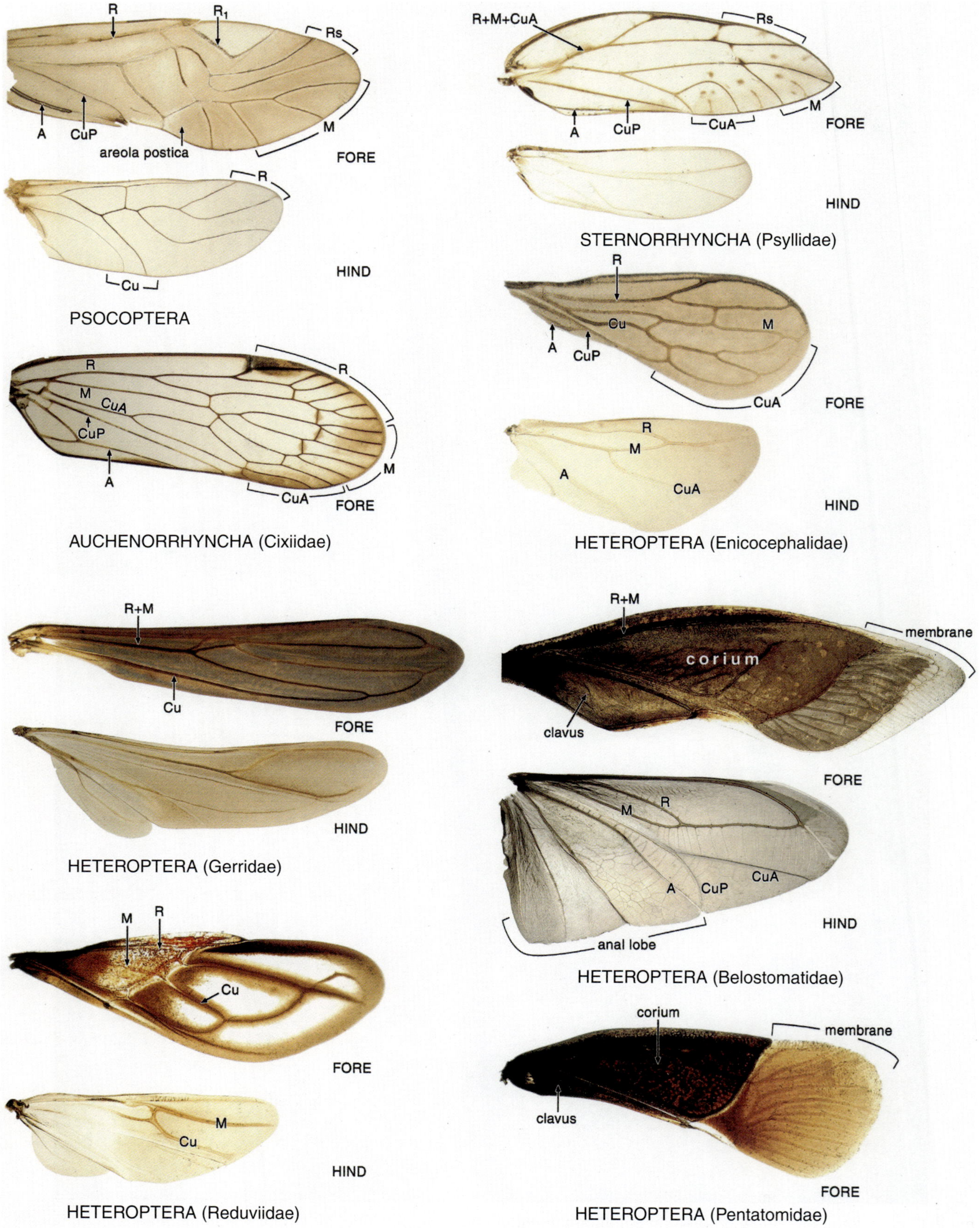

8.5. Wings of assorted Recent Paraneoptera, with homologous veins labeled. Not to the same scale.

8.6. Recent families of Psocoptera preserved in Cretaceous amber. The oldest known Liposcelidae, in Burmese amber, ca. 100 MYO (above); Prionoglariidae in Lebanese amber, ca. 125 MYO (below). There are only five known species of prionoglariids today; they inhabit caves in Eurasia. Body lengths: 1.1 mm (liposcelid) (AMNH Bu1449), 1.3 mm (prionoglariid) (AMNH L-AE46).

diet that ranges far outside nests. In an interesting study by Broadhead and Hobby (1944), they reared *L. bostrychophila* on various natural and synthetic foods, which predictably had a significant effect on survivorship, feundity, and longevity. Unexpectedly, some individuals lived for 268 days! Any predilection that Liposcelidae have for the abodes and sometimes the bodies of warm-blooded vertebrates is clearly preadapted by their feeding and dwelling habits and their body structure.

When wings are present in Liposcelidae, they are characteristically short, have vestigial venation (cf. Figure 8.5), and are held flat over the abdomen instead of rooflike (*Liposcelis* itself is apterous). Liposcelidae, and other Psocoptera like some Pachytrochidae, share with true lice a suite of reduced structures apparently correlated with wing loss. These include the loss of ocelli, great reduction of eyes, reduction of the pterothorax (including fusion of the meso- and metanotum), reduction or loss of the ctenidiobothria on the tarsi and trichobothria on the genitalia, and even shortening of the antennae. Features shared between Liposcelidae and true lice that appear *not* to be convergently correlated with wing loss are the dorsoventrally flattened bodies, a prognathous head with loss or great reduction of epicranial sutures, significant reduction of labial palpi, loss of abdominal spiracles 1 and 2, and enlarged hind femora. Given the possibility for significant convergence in body structure, a close relationship between Liposcelidae and true lice is ripe for testing with molecular data. Just recently, sequences from the 12S and 16S rDNA genes confirmed the close relationship of these two groups (Yoshizawa and Johnson, 2003). The only known Cretaceous liposcelid, in Burmese amber (100–110 MYO) (Figure 8.6), implies that the family may have first appeared in the Early Cretaceous or Late Jurassic, which is consistent with what is thought to be a Late Mesozoic origin of the true lice.

Psocoptera is often regarded as an ancient group deriving at least from the Permian (Smithers, 1972, Carpenter, 1992; Kukalová-Peck, 1991; Rasnitsyn, in Rasnitsyn and Quicke, 2002), but Paleozoic fossils attributed to them actually have an array of features considerably primitive to living psocopterans. These Paleozoic fossils include the "Permopsocida," a paraphyletic group comprised of the Psocidiidae (including the Dichenotomidae) and the Permopsocidae, from the Early to mid-Permian (ca. 260–265 MYA) of the Czech Republic, the Ural Mountains, and Kansas. They commonly represent 5% of all insect specimens from these deposits, so were quite abundant. Permopsocidans, however, had the hind wing similar in size to (versus smaller than) the forewing, had a long subcostal vein, a poorly developed pterostigma, and vein M with four (versus three) branches. Where preserved, small cerci are present (versus absent in living Paraneoptera), and the legs had between four and five tarsomeres (versus three or fewer in living Paraneoptera) – all plesiomorphic features. The only feature linking these Paleozoic fossils with Psocoptera is the areola postica, which, as mentioned earlier, is not unique to Psocoptera (Figure 8.5). Hennig (1981), in fact, maintained that the Paleozoic permopsocidans could not even clearly be linked to Psocoptera and may in fact be stem-group Paraneoptera, with which we agree. Another paraneopteran stem group that resembles Psocoptera is the Archipsyllidae, from the Jurassic and Cretaceous. Oddly, these primitive paraneopterans are very rare in the early Mesozoic, in contrast to their earlier abundance.

8.7. The infamous louse, *Pediculus humanus*, clinging to head hair (ventrally, left; dorsally, right). For nearly two centuries this species was thought to be two separate species or subspecies, *Pediculus humanus capitus*, living on the head; and *P. h. humanus*, on the body, but they are not genetically distinct. Scanning electron micrographs; body length 1.05 mm.

The oldest records of living groups of Psocoptera are wings of putative Amphientomidae and Psocidae from the Late Jurassic of Karatau (ca. 152 MYO). The first definitive records of living families appear in Cretaceous ambers from the following localities: Lebanon (Prionoglariidae [Figure 8.6], Sphaeropsocidae, and others), Myanmar (Liposcelidae [Figure 8.6], Pachytrochidae, Psyllipsocidae, Trogiidaae, and others), New Jersey (diverse: Grimaldi *et al.*, 2000a, unpublished), Siberia (Amphientomidae, Ellipsocidae, Lachesillidae, Psocidae, Psyllipsocidae, Sphaeropsocidae, and Trogiidae: Vishniakova,

8.8. Frontal view of the human louse, *Pediculus humanus*. Like other anoplurans, the mouthparts are miniscule, eyes absent, and the legs modified for grasping hair shafts between the tibial spines and the stout tarsus with large claws. Scanning electron micrograph.

1975), and Spain (Ephemeriidae: Baz and Ortuño, 2001). Only the Siberian amber psocopterans have been mostly described; however, the Spanish amber ones have been partially described; virtually all those in the other Cretaceous ambers still require just family-level identification. Eventually, study of these fossils will yield unique insight into the origins of modern psocopteran groups, but not until there is a rigorous phylogenetic understanding of Recent and extinct psocopteran families. Psocoptera are diverse in Tertiary ambers from the Baltic region (Eocene) and the Dominican Republic (Miocene), and these species belong to living genera or are very closely related to living genera.

PHTHIRAPTERA: THE TRUE LICE

The invasion of fur and feathers spawned a radiation of psocodeans that was certainly very successful, as the nearly 5,000 named species of lice exceed the known species diversity of the free-living Psocoptera (Durden and Musser, 1994; Price *et al.*, 2003). All Phthiraptera are ectoparasitic on birds and mammals, and their extreme modification leaves no doubt about a common ancestry for the group. All lice are apterous and dorsoventrally flattened; they not only have completely or virtually lost their eyes but also have antennae reduced to between three and five segments, highly modified mouthparts, and distinct oviposition (Figures 8.7, 8.8). Louse eggs ("nits") are generally large relative to the adult body, are few in number (generally 1–10 laid per day), and are cemented to the shaft of a hair or feather barbule with a secretion from an accessory gland (Figure 8.9). The eggs have a well-formed cap (operculum), and some are highly sculptured and ornamented. Among all arthropods ectoparasitic on vertebrates, lice are the only ones except mites to spend the entire "lice cycle" on their hosts. Polyctenidae bugs, which are closely related to bed bugs, come close to this, but they may actually lay their eggs off their bat hosts. As a result of this extreme dependence, host specificity and cospeciation with hosts is better seen in lice than in any other parasitic insects.

Phthiraptera are much better studied than Psocoptera because of their medical and agricultural significance. Infestations commonly occur on poultry and livestock, but lice are sporadically vectors of significant human disease, with several notable exceptions. The human body louse, *Pediculus humanus*, is the notorious vector of three human epidemics. One of these is Trench Fever (caused by *Bartonella quintana*), which was epidemic among soldiers in World Wars I and II and is now an emerging infection among inner city homeless in several cities. Another sporadic, louse-borne epidemic is relapsing fever, which is caused by *Borrelia recurrentis* (*Borrelia* is the genus of spirochetes that also causes Lyme disease). The third, and probably worst, is Epidemic Typhus (caused by *Rickettsia prowazekii*), which has erupted in Burundi but was particularly devastating centuries ago when louse infestations were thick. Thomas á Becket, the medieval Archbishop of Canterbury, heaped on cloaks in his cold cathedral, rarely (if ever) bathing during cold months. When he lay dying, stabbed by his conspirators, "waves" of lice were described as coming from his body. Epidemic typhus killed millions of people in Europe in past centuries, its spirochete spreading when infected feces of the louse are scratched into the feeding wounds. Off their host, lice can live at most several days, so transmission of lice involves direct contact of hosts, such as among nestmates. Lice have occasionally been described as phoretic on winged, bloodsucking insects like mosquitoes and hippoboscid flies, but with the exception of some bird lice this is rare.

8.9. Louse eggs cemented to a shaft of hair. Egg length: 0.8 mm.

Depending on the group of louse, lice feed on the keratin in hair or feathers, oily secretions, or blood. Most groups of birds are parasitized, and among mammals the only ones not parasitized are the monotremes (platypus, echidna), anteaters, armadillos, cetaceans (whales, porpoises), sirenians (sea cows), and bats. Even pinnipeds (seals) have lice (Echinophthiriidae) living amongst the dense short fur. *Lepidophthirus macrorhini*, for example, is known to live on elephant seals for the several months these mammals spend at sea. Cetaceans and sirenians, though, have extremely sparse hair and never come ashore, so absence of lice in these two orders is understandable. It could be argued that monotremes are an ancient group that never acquired lice, though these basal mammals appear to be no older than

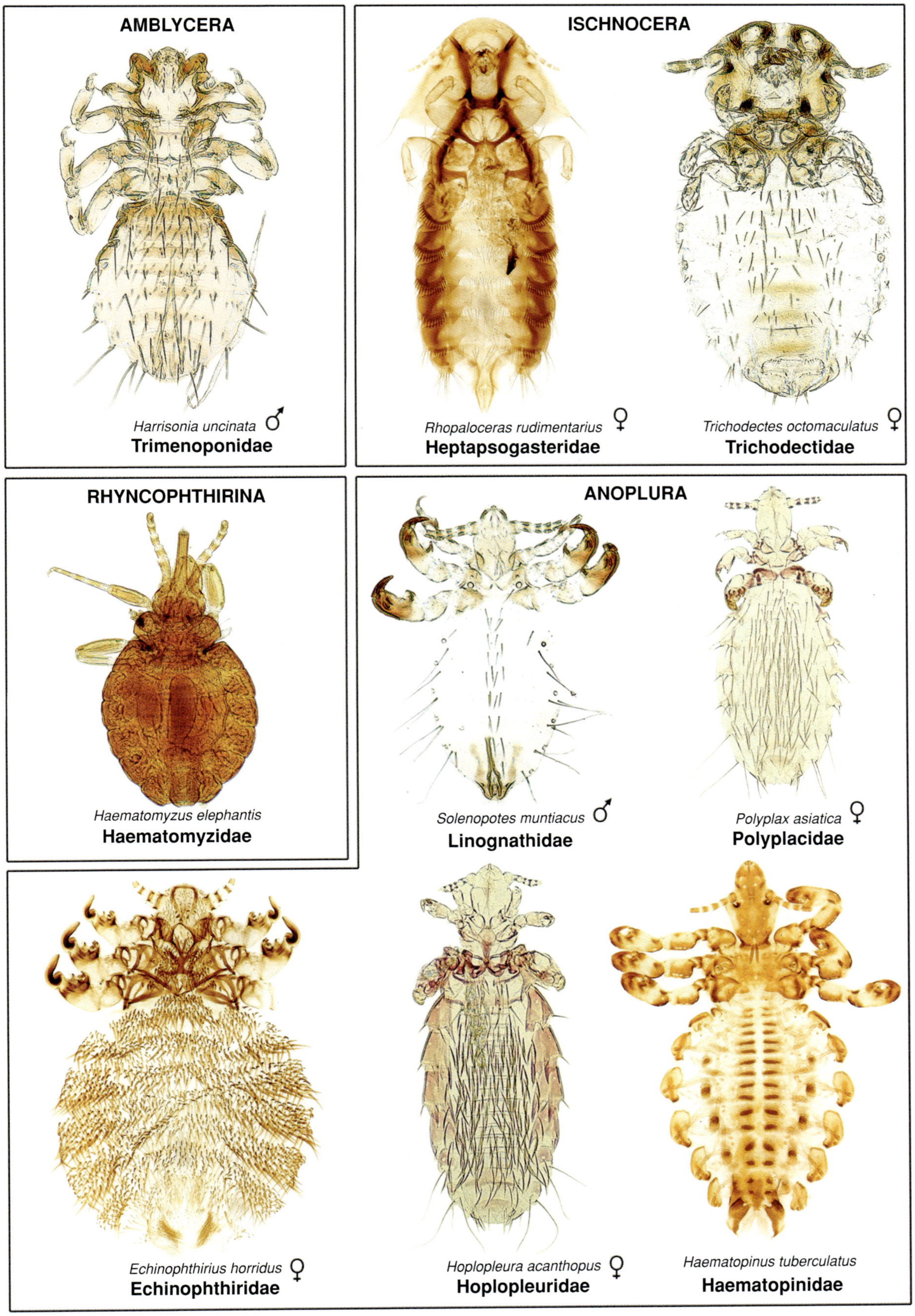

8.10. Representative lice, order Phthiraptera, from the four suborders and exemplar families. Specimens: L. Durden, Georgia Southern University; not to same scale.

birds. The absence of lice on anteaters, armadillos (which actually have fur), and especially bats is an enigma. Bats are a veritable sink for ectoparasites, harboring a unique and wide array of highly specialized ectoparasitic insects and mites (see Chapter 14). Besides species specificity of lice and their hosts, there can be extreme site specificity as well. For example, *Piagetiella* lice live inside the pouches of pelicans and cormorants, and then there is the infamous dietary preference of the human lice, *Pediculus humanus* (the head and body louse, Figures 8.7, 8.8) and *Pthirus pubis (*the pubic louse, Figure 8.11).

Phthiraptera have traditionally been classified into two separate suborders, the biting lice, or Mallophaga, and the sucking lice, or Anoplura. Though this classification has been supported by some (Kim and Ludwig, 1978b, 1982; Kim, 1985a–c), only the Anoplura is a monophyletic group. Biting lice are now classified into three suborders: the Amblycera, Ischnocera, and Rhyncophthirina (Figure 8.10). This is based largely on earlier morphological work, and more recently by the phylogenetic study of Lyal (1985b), a scheme supported by molecular analyses (Johnson and Whiting, 2002; Barker *et al.,* 2003). Another molecular study (Cruikshank *et al.*, 2001) did not resolve monophyly or relationships of most suborders, even of the highly modified Anoplura, probably because only minimal DNA sequences were used (347 bp of the EF-1 alpha gene). Taxon sampling in the molecular study by Johnson and Whiting (2002) was minimal (only 21 genera), versus 70 genera (Cruikshank *et al.*, 2001), though both studies sampled all suborders. The former study also had nearly an order of magnitude more sequences: 2,130 bp from one mitochondrial and two nuclear genes (COI, EF-1 alpha, and 18S).

The suborder **Amblycera** is the most basal, and antennal and other characters indicate it is clearly monophyletic (Cruikshank *et al.*, 2001, Marshall, 2003). A major monograph on the group is by Clay (1970), and Marshall (2003) presented phylogenetic relationships of genera within and among four of the seven families. Three families with approximately 1,200 species are parasites of birds: Laemobothriidae (on rails, storks, and hawks), Menoponidae (on various groups), and Ricinidae (on hummingbirds and various small passerines). Four families, comprising only about 170 species, are parasites on mammals in South America and Australia, the most interesting aspect being that marsupials are parasitized by Boopiidae in Australia and by Trimenoponidae in South and Central America. Two other Neotropical families, Abrocomophagidae and Gyropidae, feed on caviomorph rodents (guinea pigs and relatives), as do some Trimenoponidae. It is tempting to assume that, because Amblycera are basal lice, the ancestral host of lice was a basal mammalian group like marsupials (e.g., Vanzolini and Guimarães, 1955; Lyal, 1987), but this appears unlikely since boopiids are not the most basal amblyceran lice (Marshall, 2003).

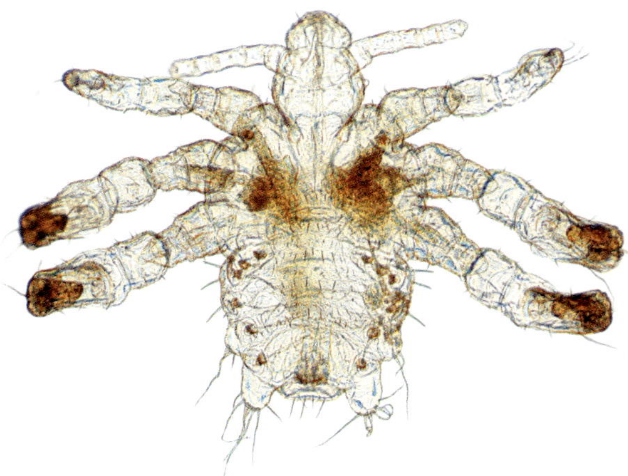

8.11. The human louse, *Pthirus pubis*. The most closely related species of *P. pubis* and *Pediculus humanus* are found on the great apes.

The suborder **Rhyncophthirina** is comprised merely of three species in the genus *Haematomyzus,* though they are very intriguing ones. *Haematomyzus elephantis* is a parasite of the African and Asian elephants, and two species live on wild pigs in eastern Africa: *H. hopkinsi* on the wart hog (*Phacochoerus aethiopicus*) and *H. porci* on the bush pig (*Ptoamochoerus aethiopicus*) (Ferris, 1931; Emerson and Price, 1988). All have an elongate rostrum, the tip of which bears mandibles that are rotated 180° from the normal position. Despite the highly modified structure of Rhyncophthirina (Weber, 1969), it is almost certainly a sister group to the Anoplura.

The suborder **Anoplura** has extensive modification of the head and mouthparts (Hirsch, 1986), and all 550 known species (in 15 families) feed on mammals (Table 8.2). Major works include the monographs by Ferris (1951) and a world catalogue of the species that also summarizes hosts (Durden and Musser, 1994). The only attempt at a phylogenetic hypothesis of relationships is by Kim and Ludwig (1978a,b). Unfortunately that study is based on only 22 morphological characters, several of which have dubious character states, such as arbitrary divisions among multistate characters like body size and hairiness. Interpretations of host-anopluran cospeciation (Kim, 1985b,c) are thus still difficult to discern. Ungulates and squirrels (the latter in the family Sciuridae) are particularly heavily parasitized by anoplurans, and several small or even monotypic families of anoplurans parasitize phylogenetically isolated mammalian groups. These include Hybophthiridae on aardvarks (order Tubulidentata), Neolinognathidae on elephant shrews (Order Macroscelidea), and Hamophthiriidae on "flying lemurs" (order Dermoptera) (Table 8.2). Using rigorous analyses of many more anopluran characters, it would be very interesting to see what patterns of host use might emerge. Particularly interesting are the anoplurans parasitizing humans and apes. Humans have domesticated animals for at least 10,000 years but have only occasionally

TABLE 8.2. Hosts of Anopluran Lice

Louse Family	Hosts
Echinophthiriidae	Pinnipedia (seals, walruses) Carnivora: Mustelidae: *Lutra* (otters)
Enderleinellidae	Rodentia: Sciuridae (squirrels)
Haematopinidae	Artiodactyla: Bovidae, Cervidae, Suidae Perissodactyla: Equidae
Hamophthiriidae	Dermoptera: *Cynocephalus* ("flying lemur")
Hoplopleuridae	Insectivora, Lagomorpha, Primates, Rodentia
Hybophthiridae	Tubulidentata: *Orycteropus* (aardvark)
Linognathidae	Artiodactyla: Bovidae, Cervidae, Giraffidae Carnivora: Canidae Uranotheria: *Procavia* (hyraxes)
Microthoraciidae	Artiodactyla: Camelidae
Neolinognathidae	Macroscelididae (elephant shrews)
Pecaroecidae	Artiodactyla: Tayasuidae (peccaries)
Pedicinidae	Primates: Cercopithecidae (Old World monkeys)
Pediculidae	Primates: Cebidae (New World monkeys), Hylobatidae (gibbons), Pongidae (apes), humans
Pthiridae	Primates: Pongidae, humans
Ratemiidae	Perissodactyla: Equidae

acquired lice from them, such as *Heterodoxus longitarsus* from dogs. Humans, instead, have retained lice in the genera *Pthirus* and *Pediculus*, which are shared with other higher primates, particularly the apes (family Pongidae). The Late Miocene, when hominids and pongids diverged, presumably saw the origins of the infamous association of lice and men.

Ischnocera is the largest suborder and most problematic regarding monophyly. Fortunately, several excellent studies have clarified relationships (Lyal, 1985a; Smith, 2000, 2001). In fact, this group is now phylogenetically the best understood one for the Phthiraptera. Approximately 2,700 species feed on birds and 380 species are on mammals. Morphological support for the monophyly of this suborder is weak (Lyal, 1985a), though there is additional evidence for this from 18S gene sequences (Johnson and Whiting, 2002). Limited taxon sampling from the molecular study, however, may be insufficient to address adequately the question of common origin for all species of this diverse suborder. The main problem in systematics of the Ischnocera has been the family Philopteridae, which contains 70% of the species of the suborder and appears extensively paraphyletic (Smith, 2001). The taxonomy of that family will need extensive revision, and fortunately the excellent world catalogue of biting lice will guide this effort (Price *et al.*, 2003). The family Trichodectidae, species of which feed on various eutherian mammals, is monophyletic and appears to be the basalmost family of the suborder. Heptapsogasteridae and Goniodidae are sister groups (Smith, 2000, 2001) and feed on birds. The only ischnocerans to feed on mammals besides Trichodectidae are members of the Trichophilopteridae, a monotypic family feeding on lemurs and indriids.

FOSSILS AND AGES

Because lice are almost always restricted to their host, fossilization depends on preservation of the host. Bones are the usual fossil remains of terrestrial vertebrates, so lice have virtually no fossil record save for some exceptional examples. Putative louse eggs have been reported on strands of hair in Baltic amber (Voigt, 1952), but this has very little systematic value. Cuticular remains of a microscopic arthropod from the Triassic of India (Kumar and Kumar, 2001) cannot be phthirapteran but appear to be that of an oribatid mite. Lice have been reported from frozen Pleistocene ground squirrels in Russia (Dubinin, 1948), and from human mummies from New Zealand, Peru, and Egypt, some 1,000 to 4,000 years old (Figure 2.37). Incredibly, lice on Pleistocene mammals seem never to have been systematically recovered, even though there are abundant frozen remains and some intriguing questions. Were mastodons and mammoths, for example, hosts to extinct species of Rhyncophthirina? The only definitive body fossil of a louse of significant age is *Megamenopon rasnitsyni*, a compression-fossilized amblyceran of the family Menoponidae from Eocene oil shales at Eckfeld, Germany. Not only it is very well preserved, but feather barbules preserved in its gut definitively prove it was a chewing louse (Wappler *et al.*, 2004).

A fascinating exception to the paltry fossil record may be *Saurodectes vrsanskyi*, a putative louse preserved in shales from the Zaza Formation (Early Cretaceous, ca. 130 MYO) of Baissa, central Siberia (Rasnitsyn and Zherikhin, 1999) (Figure 8.12). Its placement in the Phthiraptera was largely based on a process of elimination of other orders, particularly since *Saurodectes* has several features highly unusual for lice. Its eyes are anomalously large, and the insect itself is larger (17 mm [0.67 in.] long) than any living louse. The prothorax is fairly long, with the mid- and hind legs (and meso- and metathorax) in the middle of the body, distant from the forelegs (bases of the legs are usually close together in extant lice). There appear to be none of the large, stiff setae often seen in extant lice, and the claws are unusually small. A pair of thick spines behind each forecoxa is unique. The oddest feature, though, is a pair of large appendages jutting from the sides of the head. These were interpreted as huge trabeculae, processes on the side of the head in some extant lice that are typically quite small though sometimes large. These structures are certainly not antennae, as there is no evidence of segmentation.

The possibility that *Saurodectes* was a larval or nymphal stage of a free-living order was dismissed by the original authors, and indeed it possesses features of an ectoparasite.

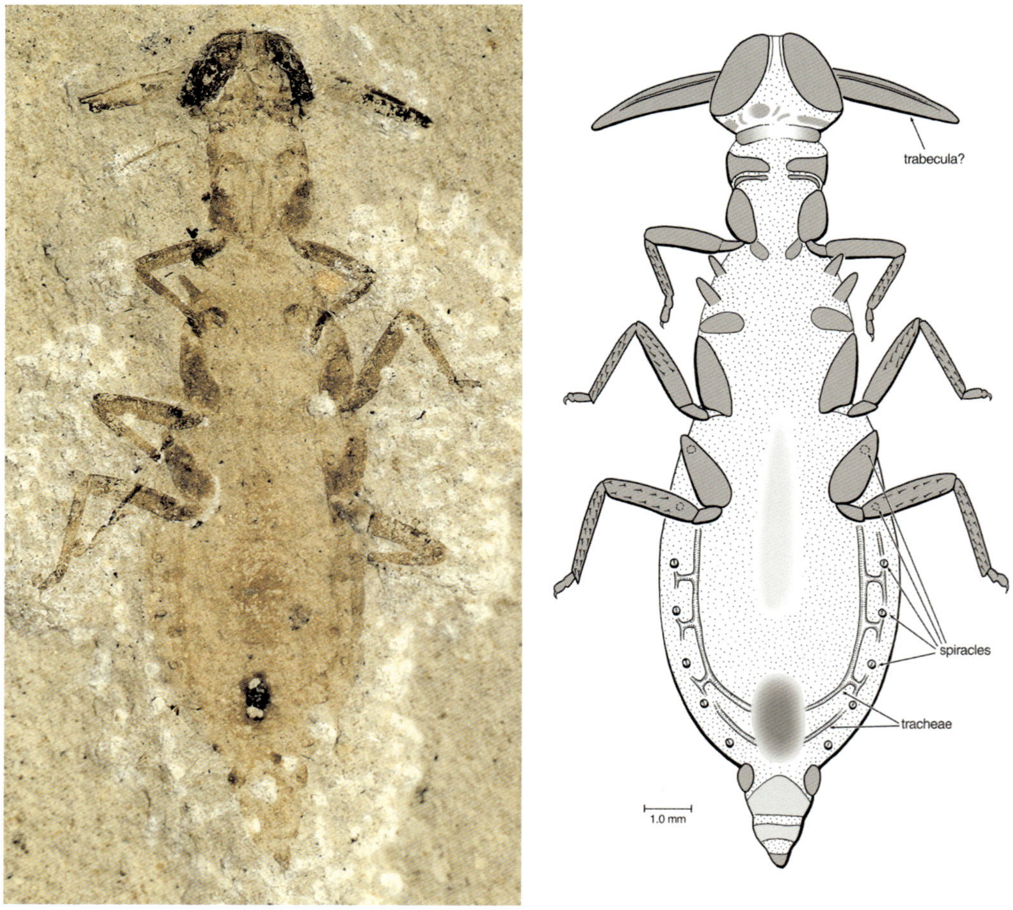

8.12. A large Mesozoic louse? *Saurodectes vrsanskyi*, photo and reconstruction, from the Early Cretaceous of Baissa, Siberia (ca. 140 MYO). This peculiar, apterous insect shares some features with Phthiraptera. PIN; body length 17 mm.

Coxae are widely separated, indicating it had sprawling legs, so it was probably a very flat insect. Short legs and a large, thick abdomen indicate that it was probably not very mobile. Tarsomeres reduced to one or two per leg is similar to that of lice, and the smaller prothorax is characteristic of the Psocodea. Ventrally (the only surface preserved), sclerites are highly reduced, and most of the insect, particularly the abdomen, is membranous. Some ectoparasites are heavily sclerotized, but all have extensive membranous areas to expand the body after feeding or to accommodate eggs. Lastly, the abdomen has thick tracheal trunks running along each side of the abdomen, and large spiracles, similar to those of lice.

The original suggestion that *Saurodectes* is an ischnoceran louse (Rasnitsyn and Zherikhin, 1999) has no basis, though the suggestion it parasitized mammals is intriguing. *Saurodectes* has only one claw on each leg, which is a feature found only on mammal lice. Rasnitsyn and Zherikhin (1999) suggested that *Saurodectes* parasitized pterosaurs because some pterosaurs appear to have had hairlike vestiture. Indeed, it appears to be too large to have parasitized Cretaceous mammals, which were all small. There is a rough correlation between body size of Recent lice and their hosts, so this suggests that *Saurodectes* had a huge host. Enigmatic as it is, *Saurodectes* is a likely ectoparasite, plausibly with phthirapteran affinities.

Because lice are virtually absent in the fossil record, the only alternative is estimating an age of the Phthiraptera using phylogenies and the much better fossil records of their vertebrate hosts. Dating using ages of hosts is a useful approach, however, only if the hosts and parasites have parallel phylogenies. Various lines of evidence indicate that host colonization, or host shifts, of lice are considerably less frequent than is speciation of the lice with their hosts, since in situations where it would be most expected it is actually rare. For example, seabirds usually roost on shoreline cliffs in dense colonies of mixed species, but their lice are mostly highly specific to species of the birds. Transmission of lice from prey to predator would also intuitively seem highly likely. Yet, raptors and mammalian carnivores do not have diverse faunas of unrelated lice derived from their prey. In the Trichodectidae, for example, there is a major clade of carnivore lice and one of ungulate lice (Lyal, 1985a), but with no apparent mixing of the two. Lastly, particularly persuasive examples involve the lice of cleptoparasitic birds and their hosts (Marshall, 1981). Cuckoos (*Cuculus:* Cuculiformes) and cowbirds (*Molothrus:* Passeriformes) are well known for laying their eggs in the nests of other birds, yet the cleptoparasites retain their own distinctive lice, not the lice of their foster parents.

Recent work on the phylogenetic relationships of Ischnocera reveals varying degrees of correspondence between the phylogenies of hosts and lice (Lyal, 1985a, 1987; Smith, 2000, 2001). Host colonization clearly does occur, but it appears to be rare in the Trichodectidae. For example, among the "carnivore clade" of these lice, obvious colonizations include *Lorisicola* onto lorises (primates) and *Geomydoecus* on pocket gophers. The lice of pocket gophers, in fact, have extensively speciated with their hosts (see discussion that follows). Host use patterns among bird ischnocerans are much more complicated. Some lice are restricted to particular groups of birds, like Heptapsogasteridae on tinamous and *Austrogoniodes* on penguins, among others. Others, like the *Philopterus* complex of genera, parasitize diverse bird orders (Smith, 2001), though it is quite likely that, when studied, relationships within such louse complexes may reveal more host use patterns. And still others, like the columbiforms (pigeons and doves) have had at least three colonizations of ischnoceran lice.

Even if there has been perfect or near-perfect cospeciation of lice with their hosts, various factors will obscure this pattern, one being the effects of incomplete sampling. It is estimated that the known Australian fauna of lice is a mere tenth of the actual fauna (Calaby and Murray, 1991), and that half to one third of all Anoplura species are presently known. In this regard, studies on cospeciation between lice and hosts may have outpaced their data. Also, knowing relationships among birds and mammals is necessary to examine host use patterns, but these are still being deciphered and are hardly without controversy (Shoshani and McKenna, 1998; Liu *et al.*, 1999; Cracraft, 2001; van Dijk *et al.*, 2001; Thewissen *et al.*, 2001; Barker *et al.*, 2002).

What is probably most significant, though, is extinction of louse and host. Simpson (1945) documented that 54% of the families and 67% of the genera of all mammals are extinct. Far more fossil mammals have been described in the past 60 years than have living ones (McKenna and Bell, 1997), so the known proportions of extinct mammals have become significantly greater. Given the host specificity of lice, it is most likely that extinction has affected them as much as, and perhaps even more than, their mammalian hosts, and bird lice are probably no different. Barker (1994) even proposed that extinction of lice on their hosts may be widespread because of competitive exclusion. This seems unlikely because evidence for louse competition is ambiguous and contradicted by other evidence. Even so, it is remarkable that patterns of host use are so discernable among living lice. Because of pervasive host extinction, close phylogenetic correspondence between lice and their hosts will probably be most obvious at the species level, particularly in those groups whose hosts have recently speciated.

When, then, did the Phthiraptera evolve? Lyal (1985b) indicated there is little reason to assume lice are any older than birds and therian (marsupial + placental) mammals. Birds are closely related to dromaeosaurs, from which they diverged perhaps in the mid-Jurassic (Chiappe, 1995). *Archaeopteryx*, the earliest bird from the Late Jurassic of Solnhofen (ca. 152 MYO), had feathers similar to modern flight feathers. At least some Cretaceous dromaeosaurs also had feathers (though none known for flight), so it is quite possible that they too were parasitized by lice. For mammals, the basal lineages of triconodonts and multituberculates extend to the Triassic (McKenna and Bell, 1997; Rougier and Novacek, 1998; Novacek, 1999). Tantalizing scraps of evidence indicate that some Cretaceous multituberculates had hair, so it is possible that ones in the Jurassic may have too. Phthirapterans probably appeared in the Early Cretaceous to Late Jurassic, 140–150 MYA.

Cospeciation and Coevolution

The ecological relationships among some unrelated organisms can be so intimate that evolution of one group transforms the other. This is most prevalent in obligate symbionts like parasites. Among terrestrial organisms, insects provide particularly diverse and specialized examples of coevolution and cospeciation.

Coevolution results from selection pressures exerted by one species that cause direct, genetically based change in another species, and vice versa (Futuyma, 1998). This is usually revealed in behavioral or morphological traits, most of which can safely be assumed to have a substantial genetic basis. Plants with toxic chemical defenses commonly have insect herbivores physiologically adapted to feed exclusively on them. Caterpillars of heliconiine butterflies, for example, feed on particular species of toxic *Passiflora* vines. Some *Passiflora*, in turn, have developed tiny structures that mimic *Heliconius* eggs and thus deter oviposition. Coevolution is thus reciprocal, an evolutionary "arms race" where adaptation and counteradaptation have escalated into highly specialized features.

Cospeciation is where the formation of one species, say a host, induces the formation of another, unrelated but ecologically dependent species, such as a parasite. Cospeciation (also called *parallel cladogenesis*) is identified by comparing a cladogram of the hosts with that of their symbionts (Figure 8.13). If the cladograms have very similar or identical paths, the two groups may have cospeciated. Coevolution involves just anagenetic change, but cospeciation involves cladogenesis and may involve anagenetic change as well.

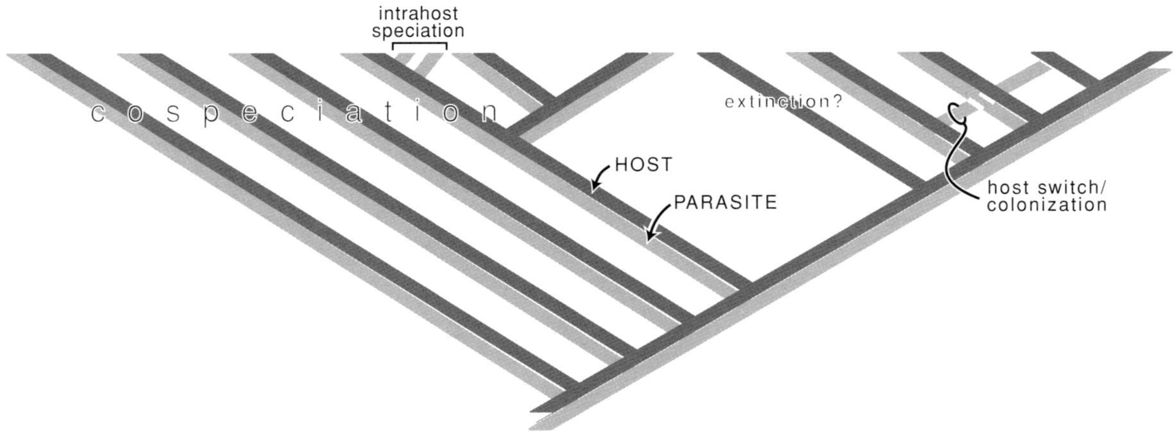

8.13. Congruence and incongruence between cladograms of hosts and parasites, and their interpretation. Absence of a parasite from a host is ambiguous; it could mean extinction of the parasite, incomplete sampling, or lack of colonization.

One particular study involves fungus-growing ants (tribe Attini) and the fungus they cultivate (Chapela *et al.*, 1994). There are about 200 described species of attines, all of which depend on their symbiotic fungus for nourishment. Foundress queens of the more derived genera, in fact, carry an inoculum of the fungus with them during their nuptial flight for their new nest. The fungus is found only in the ant nests, and those fungi in the nests of the "higher" attines are further specialized by having minute nodules on the hyphae, or gongylidia, which comprise a significant part of the ant diet. A phylogeny of the fungi based on the 28S rDNA gene (Chapela *et al.*, 1994) and of the ants based on larval morphology (Schultz and Meier, 1995) revealed important aspects of this symbiosis. The basal attine genus *Apterostigma* cultivates a fungus in the Tricholomataceae, and all the other attines cultivate Lepiotaceae. Despite the fact that fungal cultivation evolved only once in ants, there was widespread host colonization, and the only cospeciation-like pattern was found among species of the more recently evolved genera *Trachymyrmex* and *Atta*. Given the dispersal abilities of microscopic fungal spores, perhaps such widespread host colonization should be expected.

Cladograms of hosts and their symbionts are never perfectly congruent, and this can be statistically measured by several means (Brooks and O'Grady, 1989; Page, 1994). One such method is comparing the actual cladograms against a random assortment of topologies in order to test if the match significantly deviates from random. It is fundamentally important, though, that the final cladograms be stable, or not seriously affected by additional data or analyses, otherwise match and mismatch will be spurious. How the cladograms match to each other is the result of several "sorting events": cospeciation, host colonization/shift, intrahost speciation, extinction, or any combination of these.

Because parasites are so exquisitely adapted to life on or in their host, their vagility and colonization ability is often extremely compromised, which is why parasites present ideal models for cospeciation (Brooks and McLennan, 1993). Among insects, no group is more specialized for parasitism than are lice (Phthiraptera) because they are probably the only insect parasites to spend their entire life cycle, from egg to adult, on one host. All other ectoparasites, by comparison, have free-living stages, such as the larvae of fleas. This, plus the fact that species of lice are typically restricted to one species of host, has made them unique subjects regarding cospeciation. Despite Barker's (1994) claim that cospeciation is "not [a] prevailing pattern in the Phthiraptera," the few detailed studies done thus far actually indicate that cospeciation of lice with their hosts may be typical.

A study of various biting lice on diverse sea birds (Procellariiformes and Sphenisciformes) shows a close fit between cladograms of the birds and lice (Paterson *et al.*, 2000). Depending on the analytical methods, there were between one and four speciation events on the same host, one host shift, and nine cospeciation events (Figure 8.14). Another study of bird lice involved 13 species and subspecies of the genus *Dennyus* (Amblycera: Menoponidae) parasitic on 12 species and subspecies of swiftlets (Collocalliinae) from the Indopacific (Page *et al.*, 1998). Cladograms of lice and swiftlets were constructed using 505 bp of the mitochondrial gene cytochrome B. Relationships among one group of lice (the *distinctus* species group) were difficult to resolve, but a comparison of the other lineages to a cladogram of their hosts was largely congruent.

The most intensively studied and well-known example of louse-host cospeciation involves the parasites of pocket gophers, which comprise 40 species in five genera of solitary, fossorial rodents in the family Geomyidae. The family occurs throughout western North America, through parts of Central America, and into northern Colombia. Host associations and taxonomy of the lice have been intensively studied for nearly 30 years (e.g., Price, 1975; Hellenthal and Price, 1991, among

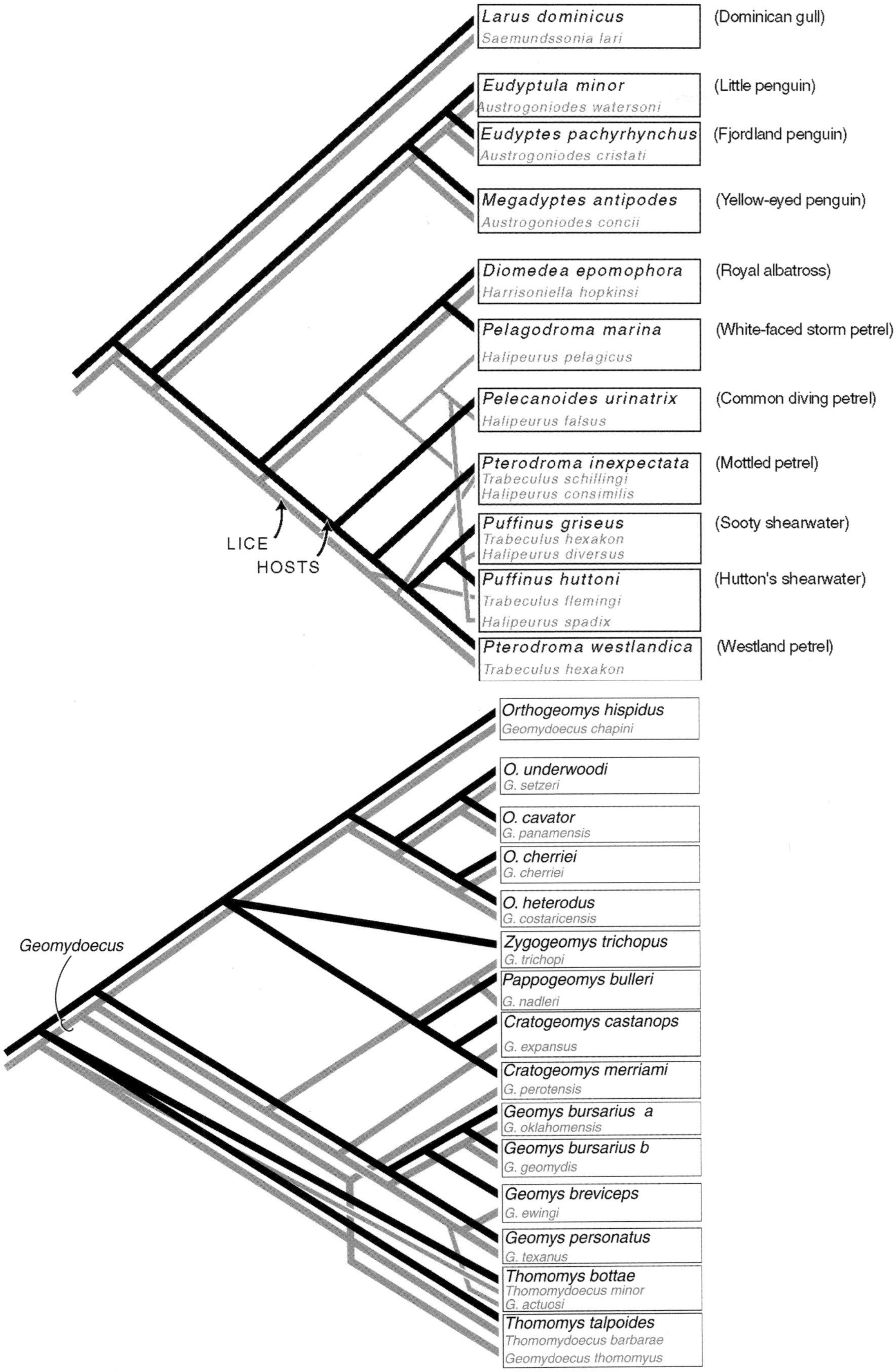

8.14. Relationships among various seabirds (above), pocket gophers (Rodentia: Geomyidae) (below), and their louse parasites. Some louse lineages track the host lineages perfectly, but other "sorting events" occur in both groups. From Paterson *et al*. (2000), seabirds; and Hafner *et al*. (1994) and Hafner and Page (1995), pocket gophers.

many other papers), so sampling of hosts and parasites is about as thorough as is known among any diverse group of lice. The lice are trichodectids in two genera, *Geomydoecus* and *Thomomydoecus*, containing approximately 122 species and subspecies. The first phylogenetic study on this system sampled 14 species of gophers and 17 species of lice from among the major groups of each, with 379 bp of the mtCOI gene sequenced for both groups (Hafner *et al.*, 1994). Though this is minimal taxon and character sampling, there was excellent congruence among host and louse cladograms. Of the estimated 20 "sorting" events, there were five intrahost speciation events, 10 cospeciation events, and only one host switch (Hafner and Page, 1995) (Figure 8.14).

Relationships among all 122 species of *Geomydoecus* and *Thomomydoecus* have been studied morphologically (Page *et al.*, 1995), but relationships of all the gopher species unfortunately have not yet been thoroughly studied. Monophyletic groups of the lice do generally correspond to groups like genera and subgenera of gophers, but different lineages of lice on the same gophers also hint at a complex history between hosts and lice. It may be impractical to sequence all species of the lice, but a study of relationships among all species of gophers is clearly needed. Interestingly, when experimental gophers are fumigated and then inoculated with foreign *Geomydoecus* lice, lice that live on closely related gophers thrive well, but those from distantly related ones do not (Reed and Haffner, 1997). Thus, even though dispersal is probably the major impediment to colonizing a new host, intrinsic features of the hosts probably prevent shifts to dramatically different hosts.

Among all of these DNA studies on lice, the rates of nucleotide substitutions were much higher in the lice than in their hosts, indicating that the lice are genetically diverging much faster than their hosts, even though morphological differences among closely related species of lice is notoriously subtle. In fact, lice appear to have 100 times or more faster rates of mitochondrial DNA change than occurs in any other organisms (Johnson *et al.*, 2003). Though it would be useful to confirm this observation with nuclear genes, lice probably are genetically diverging much faster than their hosts. This has been attributed to the much shorter generation times of lice (Hafner *et al.*, 1994), but the rates are even higher than in other short-lived insects like *Drosophila* (Page *et al.*, 1998). Rapid genetic divergence in lice possibly is the result of small effective population sizes, which, with isolation and poor dispersal, would result in pronounced *founder effects*. Populations of pocket gophers, for example, are highly fragmented and genetically differentiated (Patton and Smith, 1994), so their lice should be considerably more so. Even sea birds, some of which migrate thousands of miles, commonly have fidelity to roosting sites on certain islands and cliffs, and this leads to even greater isolation of the lice anchored to the birds. Probably the best example of genetic divergence among lice involves a global study of the human louse, Pediculus *humanus* (Leo *et al.*, 2002). Despite the unparalled dispersion and mobility of its host, *P. humanus* has some geographically restricted mtDNA haplotypes and appears considerably more genetically differentiated than its host. Rapid change in mtDNA is not restricted to lice, however, but also occurs in the sister group to lice, the psocopteran family Liposcelidae (Yoshizawa and Johnson, 2003). Thus, the ectoparasitic lifestyle of lice is not the sole factor for the unusual genetic evolution of lice.

FRINGE WINGS: THYSANOPTERA (THRIPS)

Known best as merely plant pests, some of the 5,500 species of Thysanoptera have life histories and habits that are unexpectedly diverse and sophisticated, such as advanced social behavior. Thrips (singular and plural) are also among the most recognizable insects, albeit tiny, and their mouthpart structure suggests a close relationship to the sucking insects, the Hemiptera. Thrips biology has been reviewed by Ananthakrishnan (1984a), Lewis (1973), and Mound and Heming (1991).

Monophyly of the Thysanoptera has never been questioned because thrips host a suite of uniquely derived features. The most obvious one is the wing, which is fringed with long, fine setae (Figures 8.15, 8.16) and is thus the basis for the ordinal name ("fringe wing"). In the family Phlaeothripidae these setae are fixed in position, but in other thrips the setae have a socketed base and are cocked into position for flight or rest by preening the wings with the legs and long, flexible abdomen. The fringe increases the functional surface area of the wing, which is a feature that has repeatedly evolved among insects that are minute, such as in "fairy flies" (Mymaridae: Hymenoptera) and featherwing beetles (Ptiliidae), which, at 0.2–0.3 mm in length, are the smallest known adult insects. But the fringe is retained even in large thrips 8–10 mm in length, probably because the narrow, strap-shaped wings typical of all thrips have a surface area that is insufficient for flight (why thrips simply haven't evolved broader wings is unclear). All winged Recent thrips have highly reduced venation, and the family Phlaeothripidae has even lost virtually all venation (they do retain a short, basal spur of vein). Polymorphism for wing size is fairly common within many thrips; these species have *micropterous* (small-winged), *macropterous* (fully winged), and sometimes even *apterous* (wingless) individuals.

Another distinctive thrips feature concerns the tarsi (Figure 8.15). Like many paraneopterans, the tarsi are one- or two-segmented, but thrips also possess an eversible bladder on the pretarsus and the claws are tiny in adults. Heming

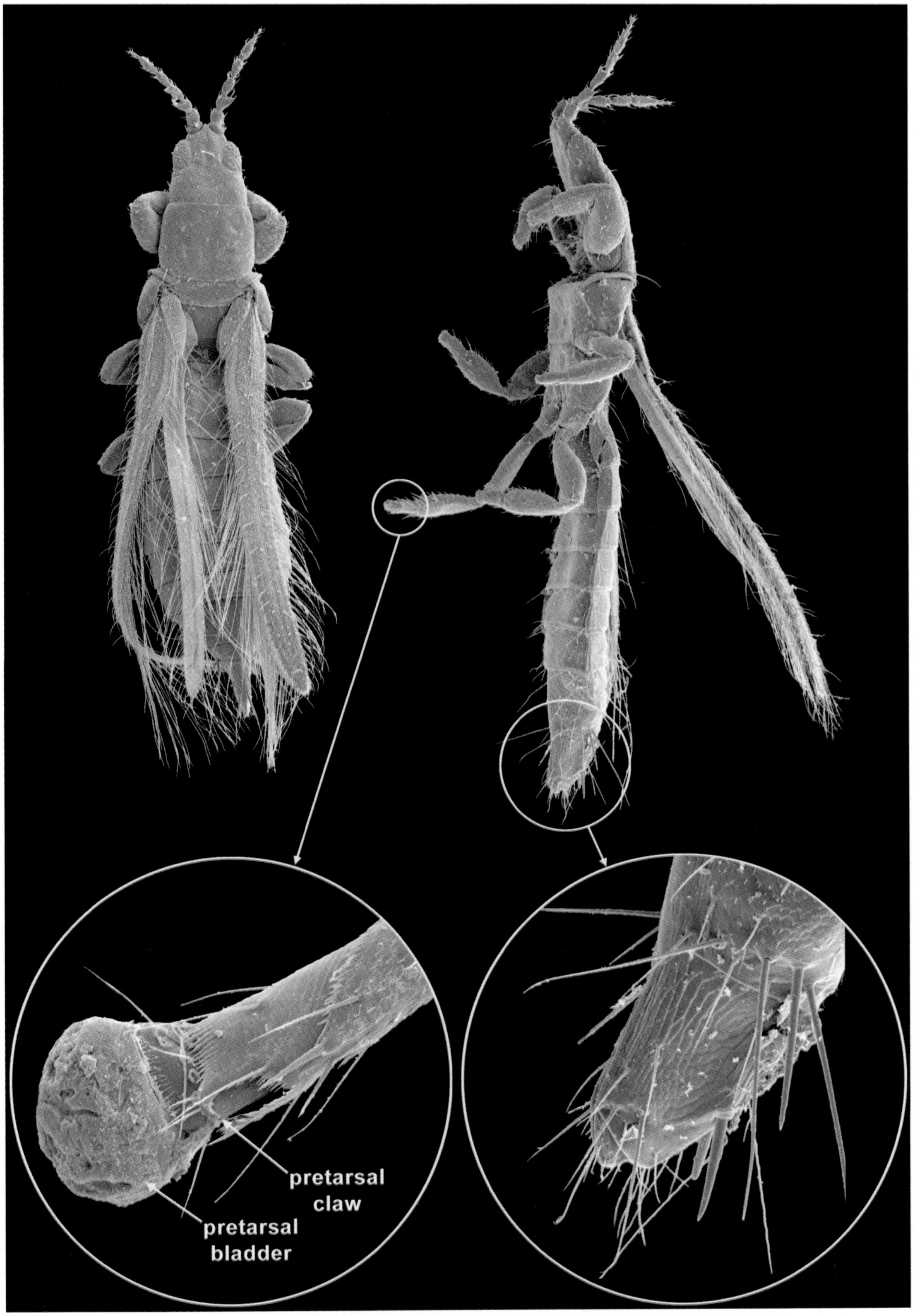

8.15. The barley thrips, *Limothrips denticornis*, showing details of the ovipositor, and the tarsal bladder typical of the order. Body length: 1.2 mm.

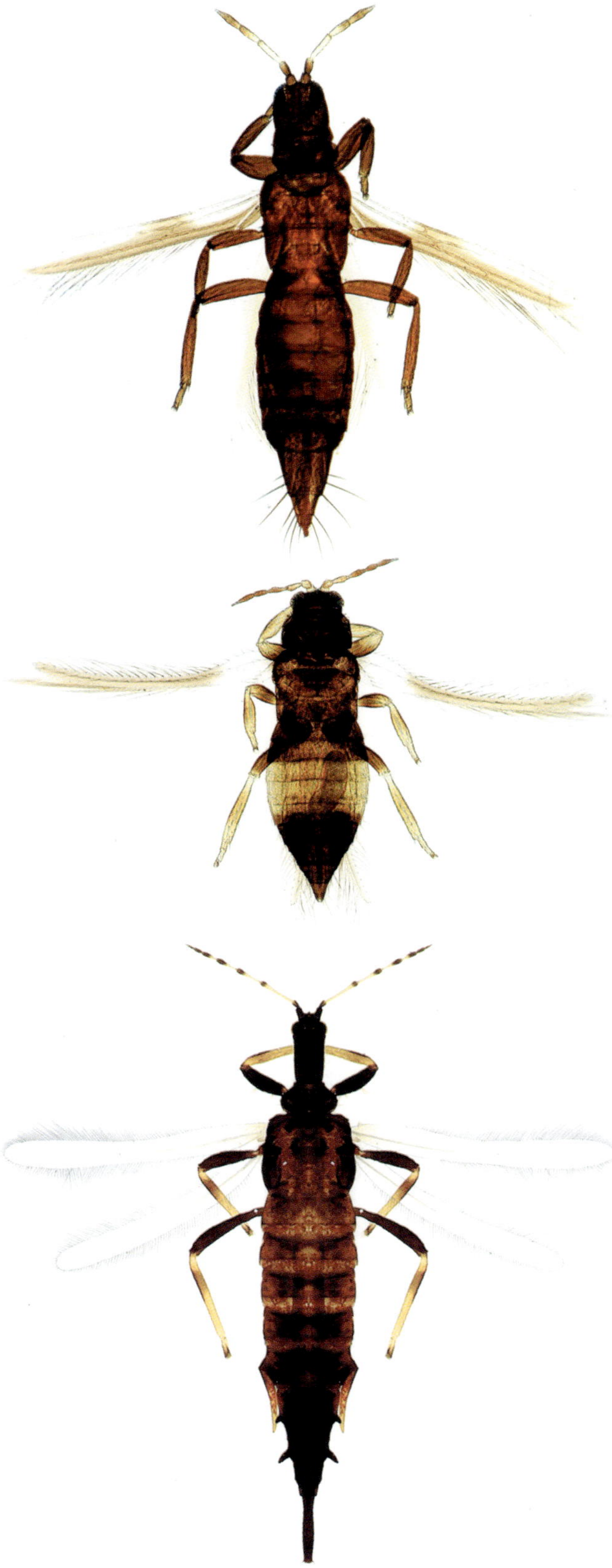

8.16. Representative Recent thrips.

(1971, 1972) described the structure of the thysanopteran pretarsus in detail, and how they function. The bladder, which is the arolium, inflates with hemolymph and retracts when certain tendons and muscles are relaxed. This structure allows thrips to adhere on slippery plant surfaces because it is coated with secretion from specialized tibial glands. Thrips also have very unusual development, similar to that of white flies (Aleyrodidae) and male scale insects (Coccoidea). Thrips instars I and II are the "larva," an active, miniature, wingless version of the adult, typical of development in nonholometabolan insects (Heming, 1975, 1991; Moritz, 1991). At the end of instar II, some thrips spin a cocoon. Instars III and IV (and in Phlaeothripidae, instar V) are quiescent, the first one or two being called the "prepupa" and the last one the "pupa." The pupa has fewer antennal segments or lacks them entirely, has nonfunctional mouthparts, and develops wing buds; additionally, the internal organs reorganize (Moritz, 1997), similar to the metamorphosis of Holometabola. Aleyrodoids have only one quiescent instar, but male coccoids have two as in thrips. This distinctive development may have been independently derived in thrips and these basal hemipterans, or it may actually be an ancestral feature of the Condylgnatha.

Thrips mouthparts are particularly significant for understanding their relationships within Paraneoptera. Though thrips have piercing-sucking mouthparts, they are less specialized for sucking fluids than are those of hemipterans. The mouthparts comprise a "mouthcone," consisting of the labrum, labium, a pair of maxillary stipites and laciniae (the latter called maxillary "stylets"), and another stylet that is a slender left mandible (Reyne, 1927; Mickoleit, 1963; Mound, 1971; Heming, 1978, 1993) (Figure 8.2). The right mandible is lost, or virtually so, though vestiges of it are seen in early embryos before they degenerate later in development (Heming, 1980). As a result of "left-handed" mandibles, the mouthcone in thrips is asymmetrical, particularly so for the more derived species. Thrips have a pair each of maxillary and labial palps, with the latter reduced as is typical in the lower paraneopterans; both pairs are lost in all hemipterans. A feeding thrips punctures a surface with the mandibular stylet, then the maxillary stylets are inserted and usually probed back and forth (Chisholm and Lewis, 1984). The maxillary stylets are fused on their protraction into a thin tube by a tongue-and-groove structure along their inner margins and by interlocking fingers at the tip. Fluids, chloroplasts, and particles of plants or prey are pumped into the fine tube through a small hole near the tip, which looks like the end of a sewing needle. Unlike many hemipterans, the stylets of most thrips are too short to tap into the vascular system of the plant, but in thrips feeding on Australian *Casuarina* trees the stylets are so long they are coiled within the head capsule. Uncoiled, they reach half the length of the body.

FEEDING HABITS

Most species feed on plant tissues, but various thrips feed on pollen and fungal mycelia and spores; however, some are predacious. Predacious species are not particularly specialized and feed on minute animals that are innocuous or easily subdued, such as mites, nematodes, insect eggs, scale insects, and other thrips. A bizarre feeding habit is in a species of *Taeniothrips* that milks secretions from the abdominal glands of lycaenid caterpillars (Downey, 1965). There is even an ectoparasitic species, the South American heterothripid *Aulacothrips dictyotus* (Heterothripidae), whose host is an aethalionid leaf hopper (Izzo *et al.*, 2002). Immatures of this species occur under the wings and wing pads of the host, and these thrips even have some specialized features of ectoparasites, namely reduction of antennae and a recessed groove on the abdomen into which the folded wings rest. No observations were described of the diet, but presumably the thrips feeds on hemolymph.

Mycophagy evolved several times in thrips, but because species in the basal family Merothripidae feed on fungal mycelia, the habit is considered ancestral (e.g., Mound, 2003; Heming, 1993). Actually, mycophagy is widespread in thrips and evolved various times. Species in the very derived subfamily Idolothripinae (Phlaeothripinae) have particularly wide feeding tubes, through which whole fungal spores can pass and then be ground up in a specialized proventriculus. For unknown reasons, mycophagous thrips commonly have biphasic male allometry. In these species large or major (*oedymerous*) males sport stout forelegs and spines; small or minor (*gynaecoid*) males are little different from females. These morphs are probably related to aggregations and group behavior commonly found in mycophagous thrips.

Pollen feeding seems to have evolved at least three times in thrips (Heming, 1993). Feeding on pollen poses challenges for such minute insects because pollen is generally 25–40 μm in diameter, but the core of thrips feeding tubes is generally 1–2 μm in diameter, sometimes 5–10 μm. Most pollen-feeding thrips, therefore, puncture the tough coat and drain individual grains (Kirk, 1984). Some species are so numerous in flowers that they can be significant pollinators, such as of huge dipterocarp trees in southeast Asia (Ashton, 1988), and of various other flowering plants (Lewis, 1973; Ananthakrishna, 1984a).

Insects that form galls are usually *monophagous*, or specific to a species of plant, and, in the case of thrips, the plants are almost always woody and persistent (a "predictable" resource in ecological terms). Otherwise, monophagy is rare in thrips; *oligophagy* (feeding on several related species or genera) and *polyphagy* are common. Thrips secluded within their gall continue to feed upon the nutritive lining of the gall. The great majority of galling thrips are Phlaeothripidae in the Old World tropics, including Australia. In Australia, galls are common and diverse on *Acacia*, *Casuarina*, *Geijera*, and *Ficus*, and appear to be adaptations of the insects to the hot, arid outback. Thrips galls are usually created by the feeding activity of numerous individuals, which causes a young, developing leaf to tightly furl or blister. The galls are also made by glueing leaves or phyllodes together with sticky anal secretions, so thrips galls are commonly open, with individuals freely passing in and out. This also allows individuals of other species to enter the gall, some of which are merely inquilines, while others are cleptoparasites.

Work on the Australian fauna has revealed a suite of species that occupy the galls of other species. Some thrips will occupy abandoned galls; others move in with resident species but cause them little or no harm, like *Adventathrips inquilinus* inhabiting the galls of *Dunatothrips* on Australian acacia (Morris *et al.*, 2000). Generally, though, intruding thrips are invaders, cleptoparasites that usurp the gall and sometimes kill the residents in the process, for which many have effective weapons. *Xaniothrips*, for example, invades Acacia galls by wielding a flexible abdomen armed with spines (Mound and Morris, 1999). *Phallothrips* invades *Iotatubothrips* galls on *Casuarina*, initially with weakly armed macropterae that wall themselves off from their hosts within the same gall. These macropterae produce a brood of armed apterae that then attack the residents (Mound *et al.*, 1998). In the blistering silence of an afternoon in the Australian outback, miniature dramas unfold within these galls.

SOCIAL BEHAVIOR

A spectrum of behavior occurs in thrips from solitary to gregarious, colonial, subsocial, to eusocial – a sophistication unexpected for insects commonly assumed to spend their life merely sucking plant juices. *Anactinothrips gustaviae*, for example, is a 5-mm (0.2-in.)-long thrips that feeds on the lichens growing on tree trunks in Panama. The thrips form communal "bivouac" sites of up to 200 individuals, where adults guard eggs and larvae. They even lay chemical trails to coordinate their group foraging (Kiester and Strates, 1984). Others cooperate in building a domicile, defending it, and even cooperatively breeding, such as *Lichanothrips* and *Carcinothrips* on acacias in Australia. *Carcinothrips* is unusual because it has huge, chelate forelegs, which it uses for folding leaves to create its domicile.

Truly advanced social behavior, where individuals are morphologically and behaviorally specialized into castes, is found in the sister genera *Oncothrips* and *Kladothrips* (Phlaeothripidae) (Crespi and Mound, 1997; Morris *et al.*, 1999). Six species in these genera form galls on Acacia trees in Australia. A foundress female initiates the colony, sometimes with a male; multiple foundresses do not occur as they do in some wasps. The first brood contains soldiers, which are individuals with reduced wings and stout forelegs bearing thick spines. They defend the colony against intruders, and often sacrifice themselves when confronting cleptoparasites.

Though unexpected, such advanced social behavior in these thrips makes sense in the light of their biology. First, these thrips live in domiciles, wherein offspring can be tended for long periods of time. Wherever advanced sociality has evolved in animals, this appears to have been a universal prerequisite (see Chapter 11). Second, like Hymenoptera, thrips are haplodiploid, or have a genetic sex-determining mechanism where females hatch from fertilized, diploid eggs, and males hatch from unfertilized, haploid eggs (Heming, 1995). This mechanism confers close relatedness among female sibs, which facilitates the evolution of a seemingly altruistic, nonreproductive, female worker caste. In the Australian Acacia thrips, though, both sexes are soldiers, which may have evolved because of the inordinately high relatedness among *all* colony members (Chapman *et al.*, 2000). In *Kladothrips harpophyllae*, in fact, relatedness is virtually clonal: Each individual measured was genetically indistinguishable. Such high relatedness is a result of brother-sister incest, matings in this case taking place between male and female soldiers from the same gall, and between male and female macropterae.

DIVERSITY AND RELATIONSHIPS

Thrips have traditionally been classified into two suborders, the Terebrantia and Tubulifera, the latter containing only the large family Phlaeothripidae (Mound and Heming, 1991). A main feature separating them is the ovipositor, which is well developed and saw-like in most terebrants, but developed into a chute in tubuliferans. The tubuliferan chute consists of vestiges of sternite eight and their anterior valvulae, and by a tubular segment ten (Heming, 1970). Bhatti (1988) provided an excellent review of tubuliferan characters, including the ovipositor. For virtually each of the 37 characters he discussed, though, the terebrantian condition was primitive. On this basis, the Phlaeothripidae is clearly monophyletic, and the Terebrantia are almost certainly paraphyletic. An ovipositor with fully developed valvulae, for example, is a primitive feature found in the other paraneopterans (Psocoptera and Hemiptera) as well as basal pterygotes.

Traditionally there are nine families, and the Merothripidae has been proposed as the most basal family (Mound and O'Neill, 1974; Mound *et al.*, 1980). This is a small but virtually worldwide family of 15 species in three genera, with most species in the neotropics and North America. Merothripids feed on fungal mycelia amongst dried leaves on the forest floor, under bark of dead and decaying trees, and on polypore fungi. This is considered the ancestral diet for the order.

Five "families" of thrips are likewise small, but the limits and definitions of two of these are currently vague. The Uzelothripidae is monospecific, with only *Uzelothrips scabrosus*, which is known from dead wood and leaves in southern Brazil and southeast Asia. "Adiheterothripidae" consists of three flower-feeding genera from the eastern Mediterranean, India, and western North America. Heterothripidae has three genera from the Western Hemisphere, and "Fauriellidae" consists of four genera from Europe, southern Africa, and California. Until recently, the Melanothripidae was considered a subfamily of Aeolothripidae.

The Aeolothripidae (250 species, 27 genera) was traditionally believed to be the most basal family of the order because these thrips have the broadest wings with fewest veins lost. Their wings are usually bicolored, and the bodies often have bold colors, which has apparently preadapted them to mimic the constricted body shape of ants. *Mymarothrips* and some *Franklinothrips* even palpate antennae and move quickly like ants, quite unlike the typical, slow thrips. These genera are also predatory, though most aeolothripids feed in flowers; none are known to feed on leaf tissue of plants. Aeolothripidae are also the only thrips family with a relict austral distribution, involving genera like *Dorythrips* and *Geolothrips*, each having species in southern South America and Australia; *Cranothrips* in Australia and southern Africa; and *Cycadothrips* breeding in male cones of *Macrozamia* cycads in Australia. The distributions of such small, easily transported insects are difficult to determine, but these genera suggest that Aeolothripidae is a particularly ancient family.

The two families Phlaeothripidae and Thripidae comprise 95% of the species in the order. Thripidae (2,500 species, 260 genera) are among the most specialized flower-feeding species. Adult Dendrothripinae have a metasternum with an internal, forked strut (*endofurca*), to which are attached powerful muscles connected to the hind legs, allowing them to spring like fleas and flea beetles. Phlaeothripidae is the largest and most diverse family (with 3,200 species), with species that form galls; feed on leaves, flowers, on fungal spores; and are predatory. Besides having segment ten of the adult extended into a tube, the wings have lost the veins and overlap at rest, these are bare of microtrichia, and setae of the marginal fringe do not have socketed bases. Advanced sociality occurs in this family.

Several studies have hypothesized a sister-group relationship of the Phlaeothripidae with the rest of the thrips (Bhatti, 1988; Crespi *et al.*, 1996). Phenetically, the most morphologically modified group would be quantitatively most divergent. Though Bhatti (1988) did not phylogenetically analyze taxa and characters, his discussion and conclusion is essentially phenetic because it emphasizes differences between the highly derived Phlaeothripidae and all other thrips. His ranking, in fact, of the two traditional thrips suborders into two orders exaggerates these taxa even more. The DNA sequence study by Crespi *et al.* (1996) reported well-supported monophyly of Tubulifera and Terebrantia, but that study had minimal character and taxon sampling, using 1050 bp from two genes (COI, and 18S rDNA), from eight species in four families. The preferred tree from their combined analysis did not define a monophyletic Thysanoptera, which is actually without any doubt.

The few morphological studies done thus far are equally equivocal about thrips relationships. The study by Mound *et al.* (1980) presented results that were contradictory, with Phlaeothripidae either as the sister group to all other thrips or as a derived sister group to the panchaetothripine Thripidae, the latter being the preferred alternative to us. Since that study dealt with family-level taxa, polymorphism among genera complicated the analyses, and most characters had multiple states. Some of the results also contradict those in a later study by Mound and Marullo (1998), such as a relationship between Fauriellidae and Heterothripidae that was thought to be close in the first study. In fact, Mound and Marullo found the Fauriellidae and Adiheterothripidae to be extensively paraphyletic, which appears to indeed be the case (Mound, 2003). Thrips phylogeny has most recently been addressed with another molecular study (Morris and Mound, 2003: in Klass, 2003), which sequenced just 600 bp of the 18S rDNA gene, but which greatly expanded taxon sampling from earlier studies to 52 species from most of the families. Not surprisingly, Tubulifera was monophyletic and not the sister group to all other thrips, and there was extensive paraphyly at the base of the terebrantians, even a paraphyletic Aeolothripidae. Clearly, more sequences and genes will need to be made.

Despite disagreements there is a coarse but consistent pattern of relationships: Aeolothripidae, Merothripidae, and Uzelothripidae are basal thrips; Thripidae and Phlaeothripidae actually appear to be the most derived families; and the remaining families and genera are intermediate (Figure 8.17). A basically similar scheme was originally proposed by Schliephake and Klimt (1979). Basal thrips have the following primitive features: slight asymmetry in mouthparts; a fully formed tentorium; nine-segmented antennae (versus fewer segments); two longitudinal veins in the forewing, with some crossveins (versus none); three-segmented maxillary palps (versus two-segmented); two tarsomeres (versus one); and a valvular ovipositor, though it is sometimes reduced as in some Merothripidae.

FOSSILS AND ORIGINS

The closest apparent relative to the thrips is Lophioneurida, a group that occurred from the Permian to the Cretaceous (Figure 8.18). Lophioneurida may have affinities with stem-group paraneopterans like the Permopsocida, but they also possessed several thrips features (Vishniakova, 1981). These features are nearly symmetrial forks of veins Rs and M, a similar branching pattern of the bases of these veins, and narrow (but not straplike) wings, and some apparently had asymmetrical mouthcones with stylet-like parts. Russian paleoentomologists classify the approximately 10 genera and 20 species of Lophioneurida in Thysanoptera (their "Thripida": Zherikhin, in Rasnitsyn and Quicke, 2002), but Lophioneurida lack the wing fringe, venation, and pretarsal bladders seen in true thrips and are actually a paraphyletic stem group to thrips. Extremely basal thrips in the Late Triassic and Jurassic indicate an origin of Thysanoptera from lophioneurid ancestors sometime in the Early Triassic, at most latest Permian. Lophioneurida actually extended into the Cretaceous, the latest occurrences being in Late Cretaceous ambers from Siberia and Myanmar.

The oldest definitive thrips, *Triassothrips virginicus* and *Kazachothrips triassicus* from the Triassic of Virginia and Kazakhstan, respectively, are approximately 220 MYO (Grimaldi *et al.*, 2004a) (Figure 8.17). These are stout bodied and have wings that are broader and with a venation more complete than any living thrips, quite similar to that of *Karataothrips jurassica,* also from Kazakhstan but some 70 MY younger. All of these have a wing fringe shorter than in any living species, and a venation intermediate between living Aeolothripidae and the extinct order Lophioneurida. In fact, *Zoropsocus stanleyi* from the Permian of Australia has a venation remarkably similar to these thrips, intermediate between these true thrips and other Lophioneurida. A Cretaceous aeolothripid, *Cretothrips antiquus* (Figure 8.19) in 90 MYO amber from New Jersey, is similar to relict Recent genus *Cycadothrips* of Australia, which feeds on cycad spores.

The first occurrence in the fossil record of modern groups of thrips is in the Cretaceous. Not suprisingly, these diminutive fossils are preserved in amber. Diverse thrips occur in Cretaceous amber from western Canada, Myanmar, New Jersey, Siberia, and northern Spain (80–110 MYO), but the oldest from the Cretaceous are in Lebanese amber (ca. 125 MYO). Zur Strassen (1973) described five new families based only on seven specimens in Lebanese amber, which he classified in the Heterothripoidea, a group to which belongs only one small living family, the Heterothripidae. Unfortunately, this taxonomy of Lebanese amber species is not phylogenetic. Species in the Rhetinothripidae, for example, possess the very plesiomorphic feature of 15–16 antennal segments (versus 6–9 in extant thrips). Like Psocoptera in Cretaceous ambers, eventual study of many more Lebanese amber thrips and ones in other Cretaceous ambers will be exceptionally important for understanding the origins and evolution of modern families.

Baltic amber (42 MYO) has an extremely diverse thrips fauna, prolific descriptions of which have been provided most recently by Schliephake (1990, 1993, 1997, 1999, 2000, 2001). Virtually all Baltic amber thrips belong to modern families, and the earliest phlaeothripids, which are also very diverse, appear then. Despite a major gap between approxi-

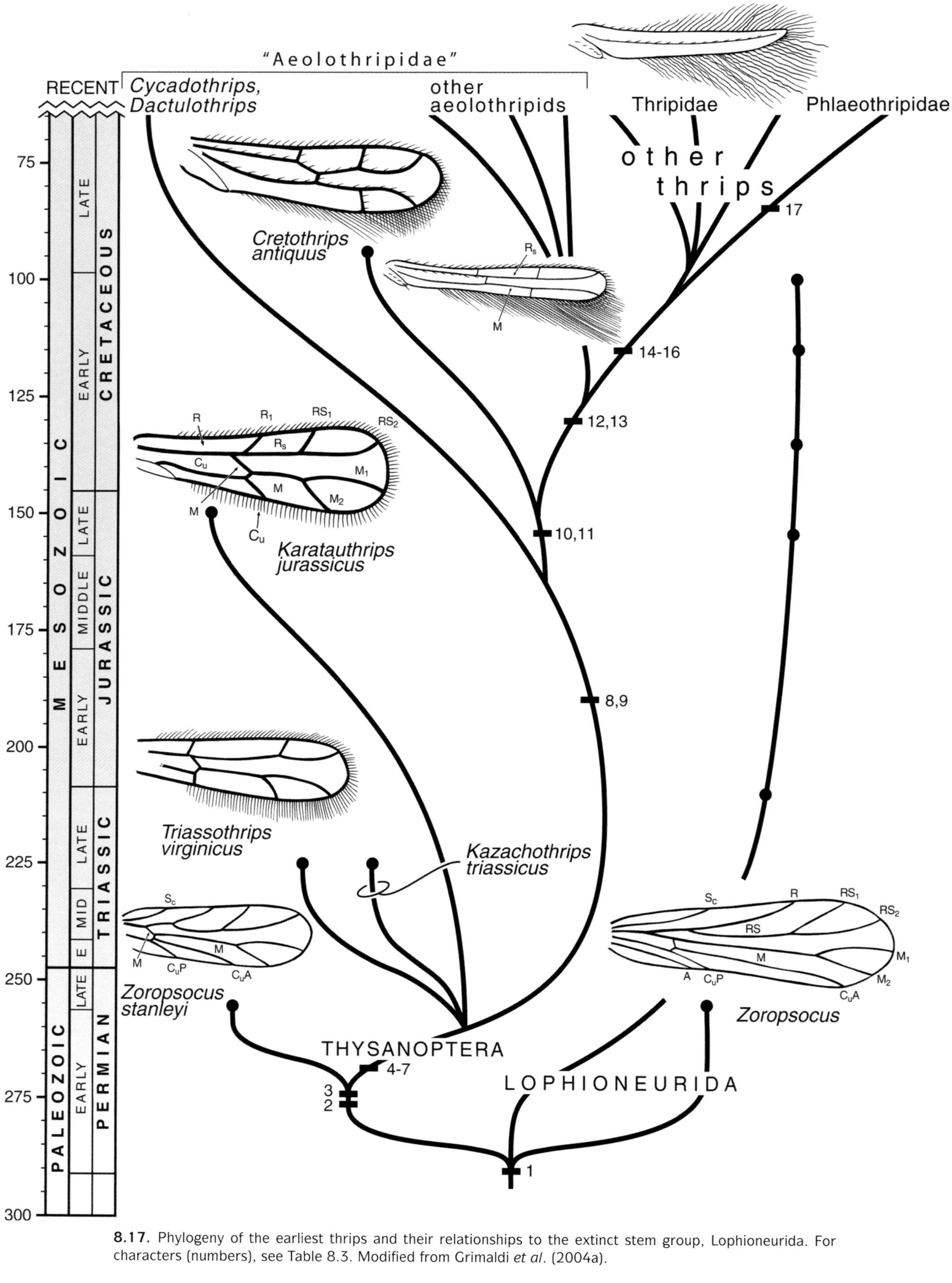

8.17. Phylogeny of the earliest thrips and their relationships to the extinct stem group, Lophioneurida. For characters (numbers), see Table 8.3. Modified from Grimaldi *et al*. (2004a).

TABLE 8.3. Characters Defining Relationships of Basal, Mesozoic Thrips (Thysanoptera)[a]

1. Veins Rs and M parallel or nearly so, with opposing forks of Rs_1-Rs_2 and M_1-M_2
2. Base of vein M shortened
3. Veins R_1, RS_1, CuA, and CuP shortened
4. All veins except the bases of Rs and M are short and braced with the wing margin
5. Wing has a fringe of long, fine setae
6. Vein Sc is lost
7. Vein CuP is lost
8. Base of vein M is very short, and perpendicular to veins R and stem of M
9. Base of vein M significantly distal to level of fork of R_1-Rs (versus very near or even proximal to fork)
10. Apical branch of vein Cu is lost
11. Crossvein m-cu is lost
12. R and M veins are linear
13. Veins Rs_1 and M and crossveins are very short, and perpendicular to veins R and M
14. Apex of wing acute, not rounded
15. Vein M_2 lost
16. Vein Rs_1 lost
17. All veins virtually lost

[a] Numbers correspond to those on phylogeny, Figure 8.17.

8.18. Reconstruction of a lophioneurid, *Zoropsocus itschetuensis*, from the Early to mid-Jurassic of Siberia. Original reconstruction based on Becker-Migdisova (1962); body length 2.3 mm (exclusive of wings).

8.19. *Cretothrips antiquus*, in 90 MYO amber from New Jersey. The thrips is remarkably primitive and very similar to the living genus *Cycadothrips* from Australia. AMNH NJ432b; body length 1.0 mm.

mately 70–45 MYA, indications are that the most diverse group, the Phlaeothripidae, radiated in the Cenozoic or latest Cretaceous. Like many other orders of phytophagous insects, most thysanopteran radiation was probably tied to the radiation of angiosperms.

THE SUCKING INSECTS: HEMIPTERA

The myriad aphids, scale insects, tree hoppers, leaf hoppers, plant hoppers, and predatory and plant bugs comprise a clearly defined monophyletic group based on the loss of maxillary and labial palps and the unique structure of the rostrum or "sucking beak." There has been unnecessary confusion over ordinal-level names for the group, specifically the names Hemiptera, Homoptera, and Heteroptera, which is still promoted by some popular textbooks (i.e., Borror *et al.*, 1989). "Homoptera" is an abandoned name formerly used for all the hemipterans except the Heteroptera, or true bugs.

Homopterans are a paraphyletic group, now divided into three monophyletic orders/suborders: Sternorrhyncha (white flies, scale insects, aphids, and jumping plant lice), Auchenorrhyncha (leaf hoppers, tree hoppers, plant hoppers, cicadas), and the relict group Coleorrhyncha.

Hemiptera is the largest group of non-holometabolous insects, their diversity probably related to angiosperm radiations. Virtually all Sternorrhyncha and Auchenorrhyncha feed on plants. Basal Heteroptera are predatory, with some large and recently evolved groups having reverted to ancestral phytophagy. Their efficiency as plant feeders is a consequence of mouthpart structure (Figure 8.2), which also has preadapted some for predation. Hemipteran mouthparts consist of two pairs of long, fine *stylets* – the mandibles and maxillae – lying in a grooved, gutter-like labium. Bases of the mandibles and maxillae articulate within the head, such that the stylets can be slid against each other, back and forth, when penetrating host tissue. Large muscles attached to the ceiling of the *cibarium* create strong suction, so that fluids can be drawn or injected through the fine *food* and *salivary canals* that lie between the maxillary stylets. Saliva is pumped down one canal, and liquid food is sucked up the other. Secretions from large salivary and accessory glands flush the fine vessels, facilitate penetration of the proboscis, and form a *stylet sheath*, or feeding tube on or within the plant. In predatory Heteroptera, copious saliva pumped into the prey digests it in its own body, and the brew is sucked out. The digestive system also is well adapted for a fluid diet, with the anterior gut in contact with or even encapsulated by the hind gut. This, the *alimentary filter system*, facilitates the absorption of water and sugars into the hind gut by plant feeders, which is rapidly excreted (as *honeydew*); the amino acids, nitrogen, and other nutrients in the sap are sequestered. Because of the poor nutritional quality of plant fluids, some nutrients are supplied to hemipterans by endosymbiotic bacteria that reside in specialized cells, the *mycetocytes* or *bacteriocytes*.

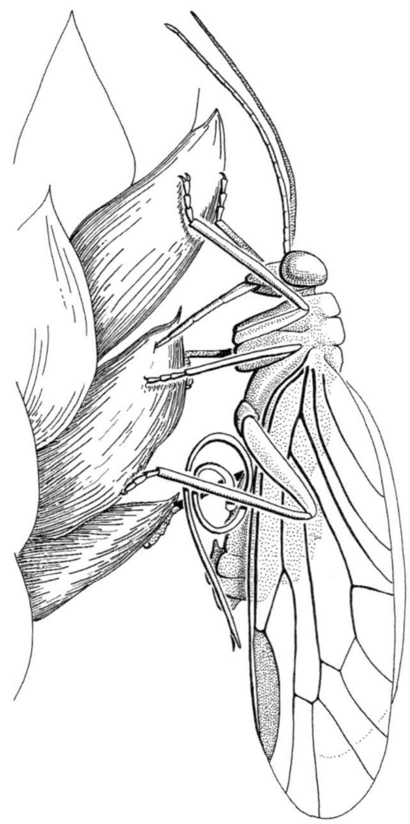

8.20. Reconstruction of a Permian species of the extinct group archescytinidae, considered to be the most primitive Hemiptera. Redrawn from Rasnitsyn and Quicke (2002).

Early History and Hemipteran Suborders

The most basal heteropterans clearly occurred in the Permian, as definitive members of extant hemipteran suborders occurred by the Triassic. Indeed, there is a considerable diversity of Permian hemipterans, but their relationships are obscure. The Archescytinidae (Figure 8.20), occurring from the Early Permian to the Late Triassic, are believed to be the most basal hemipterans (Shcherbakov and Popov, in Rasnitsyn and Quicke, 2002), though these authors also included the group in a cladogram at the base of the Sternorrhyncha. A rostrum is well preserved in some of the fossils, so there is no question about Archescytinidae being hemipterans, even though venation is the only basis for judging their relationships within the order. Martynov (1938) and Bekker-Migdisova (1962) included Archescytinidae within or very near the Sternorrhyncha; Carpenter (1931) and J.W. Evans (1963) indicated the group belonged to neither Sternorrhyncha nor Auchenorrhyncha. Archescytinidae are unusual in that some have a thin, coiled ovipositor suspended beneath the abdomen and then projected backwards (Figure 8.20), which they may have used to insert eggs into plants.

Four suborders of Hemiptera are now usually accepted, with the Sternorrhyncha, Coleorrhyncha, and Heteroptera clearly monophyletic, and the last two certainly closely related. A few studies have suggested that the Auchenorrhyncha is paraphyletic with respect to the Heteroptera (e.g., Bourgoin *et al.*, 1997), but the morphological evidence seems to favor monophyly of this suborder. For Sternorrhyncha, two of the more significant defining features concern the structure of the proboscis, and loss of the vannus and vannal fold in the hind wing (where the hind wing is present; in some, like male coccoids, this wing is vestigial). Sternorrhynchans are extremely opisthognathous, such that the labium originates from or close to the prosternum. Sternorrhynchans also have a one- or two-segmented tarsus, though tarsal reduction commonly occurs among diverse insects. Defining characters of the other groups are discussed later. The major question concerns the relationships and monophyly of the

Auchenorrhyncha because the Sternorrhyncha are almost always considered as the living sister group to all other Hemiptera (e.g., Wheeler *et al.*, 1993), though others consider relationships of these two suborders unresolved (e.g., Hennig, 1981).

STERNORRHYNCHA: APHIDS, WHITEFLIES, PLANT LICE, AND SCALE INSECTS

Some insects have evolved sophisticated mechanisms of escape and dispersal, but for most sternorrhynchans life is unmoving, spent entirely on a leaf, stem, or virtually in one spot. Females or certain life stages of most sternorrhynchans are sedentary, or even sessile, an evolutionary result of their mouthpart structure and feeding habits. Sternorrhynchans have fine, hairlike feeding stylets, coiled at rest in the head capsule; when extended, they snake through a plant's cells to tap into phloem vessels, to a distance often longer than the body of the insect. (Phloem transports nutrients from leaves to the rest of the plant and lies adjacent to xylem but deeper toward the core of the stem.) Stylets efficiently tap the plant, but are not quickly withdrawn, so the resultant sedentary life, exposed on plants, makes the insects particularly conspicuous to parasites and predators. A plant infested with aphids or scale insects is the insect equivalent of the Serengeti Plains, with these insects playing the part of the vast herds of grazing ungulates, and the predators played by myriad parasitoid wasps, certain larval Diptera, larval and adult Neuroptera, beetles, and even some carnivorous caterpillars. A diverse array of adaptations, though, defends sternorrhynchans from complete assault.

The winged and sedentary forms of many sternorrhynchans secrete wax from specialized glands and pores, which has the dual function of preventing desiccation and reducing detection and injury by predators. Some secrete hard, external shells, or live within galls that gradually grow around the growing insects. The most fascinating defenses, however, involve how these insects have co-opted the defenses of pugnacious ants, which tend and even herd some species in order to harvest the honeydew (reviewed in Way, 1963; Hölldobler and Wilson, 1990; Gullan, 1997). Lastly, the growth of sternorrhynchan populations can be astonishingly rapid because many are parthenogenetic and viviparous, which is why seriously destructive outbreaks occur in some crops. Populations of virtually all organisms cycle, with those of predators lagging behind the prey (Ricklefs, 1991), so exploding prey populations can swamp the effects of predators and parasites for a period until predator numbers eventually catch up and limit prey population size.

Sternorrhyncha are typically classified into four groups: Psylloidea ("plant lice"), Aleyrodoidea ("white flies"), Aphidoidea (aphids), and Coccoidea (scales, including mealy bugs). Most schemes group the first two as closest relatives (Psyllomorpha) and then group the latter two (Aphidomorpha) (e.g., Börner, 1934; Carver *et al.*, 1991; Schlee, 1969a–d). Psyllomorpha have adults with jumping hind legs and eggs with short peduncles, the latter structure of which may serve to absorb water from the plant. Schlee (1969 a,b) discussed at least ten synapomorphies of these two groups. Aphidomorpha have winged morphs with a composite vein in the forewing (comprised of R and the bases of M and Cu), a reduction or even loss of male genitalia, and polymorphic individuals. Basal coccoids, like some of the larger margarodids, have venation that is easily homologized with that of aphids (Schlee, 1970). Aleyrodoidea, though, share some characters with aphids and coccoids: antennae are reduced to six or fewer segments (versus ten, though some basal coccoids like female Margarodidae have between 10 and 11 segments); they have sedentary or sessile nymphs; venation is extremely reduced (these last two features are in the Coccoidea); and the rostrum has four segments where segment two telescopes into segment one. All but the last of these are plausibly convergent features of reduction.

The Protopsyllidiidae appears to be an extinct sister group to Sternorrhyncha (Grimaldi, 2003c), even though it is traditionally believed to be closely related to the Psyllomorpha or even within it (Evans, 1956; Hennig, 1981; Shcherbakov and Popov, 2002). The family has an extensive record, with at least 30 genera and 50 species of compression fossils from the Late Permian to the mid-Cretaceous. Complete preservation of a peculiar, highly modified genus, *Postopsyllidium* (Figure 8.21), in Cretaceous amber reveals features that are primitive for all Recent Sternorrhyncha. These features include tarsomere (three versus two) and flagellomere numbers (ten versus eight and generally fewer), an auchenorrhynchous type of ovipositor; and bristly wing veins plus ventrally-inserted antennae that are typical of fulgoroid plant hoppers (Grimaldi, 2003c). Like Sternorrhyncha, though, *Postopsyllidium* has a similar rostrum, loss of a vannal fold in the hind wing, and fine annulations on the tibiae. Protopsyllidiidae clearly evolved in parallel with Sternorrhyncha since species in both groups coexisted in the Late Mesozoic.

The **Psylloidea** are the most morphologically and biologically generalized sternorrhynchans (Figure 8.22), generally classified into six families. Nymphs are typical hemipterans, not highly reduced as in most other sternorrhynchans; venation is the least reduced; and even the mouth stylets are exposed basally in nymphs, where they loop at the base of the labium. It would not be at all suprising if Psylloidea were the living sister group to the rest of the Sternorrhyncha, particularly since Aleyrodoidea share some striking features with aphids and coccoids. Triozidae and some Psyllidae produce galls; honeydew excreta from nymphs in the Spondylaspidinae produce hardened structures in the form of shells or baskets, called "lerps," that are specific to the taxon. Fossils of the group are rare, the earliest of which are the Liadopsyllidae and Malmopsyllidae from the Early Jurassic to mid-Cretaceous of Eurasia. Psylloids are rare in Cretaceous and even in Tertiary deposits.

Aleyrodoidea is comprised only of the family Aleyrodidae. Most species are tropical and feed almost entirely on angiosperms, especially woody species. Adults are usually covered with a mealy, flocculent wax, which is secreted by pairs of large glands on several abdominal sternites and spread over the body using combs of hairs on the hind legs. Like female coccoids the first instar is mobile, with subsequent instars having atrophied appendages and settling to a sedentary life. Nymphs have a unique structure at the tip of the abdomen, the vasiform orifice, for expelling honeydew. Because surfaces coated with honeydew often grow a layer of sooty mold, it is important for nymphal sternorrhynchans not to become polluted. Tergite eight in nymphs has a deep fold in the middle into which the anus opens; droplets formed in the orifice are flicked away with a tiny flap, the lingula. Parthenogenesis is common in aleyrodids, and females commonly lay eggs in a circle by ovipositing while pivoting on her feeding spot. Like coccoids and thrips, aleyrodoids have quiescent "prepupal" and "pupal" instars, which are the penultimate and final instars. Their minute size precludes their being fossilized in sediments with reasonable preservation, though Shcherbakov and Popov (in Rasnitysn and Quicke, 2002) and Whalley (1985) figured compressions of very plausible aleyrodoid pupae from the Late Jurassic and Early Cretaceous, respectively. A putative aleyrodoid pupa from the Permian of South Africa (Kukalová-Peck, 1991) is doubtful. Otherwise, the earliest aleyrodoids, *Heidea cretacica*, *Bernaea neocomica*, and some undescribed species and genera (Figure 8.23) are well preserved in Early Cretaceous amber from Lebanon (Schlee, 1970). These and other aleyrodoids in this amber are stem-group species to Tertiary and Recent Aleyrodoidea. A definitive pupa occurs in Burmese amber (Figure 8.24), and various rare adults occur in other Cretaceous, and Tertiary, ambers, as yet unstudied.

Aphids, or **Aphidoidea**, comprise approximately 5,000 species and are most diverse in temperate regions of the

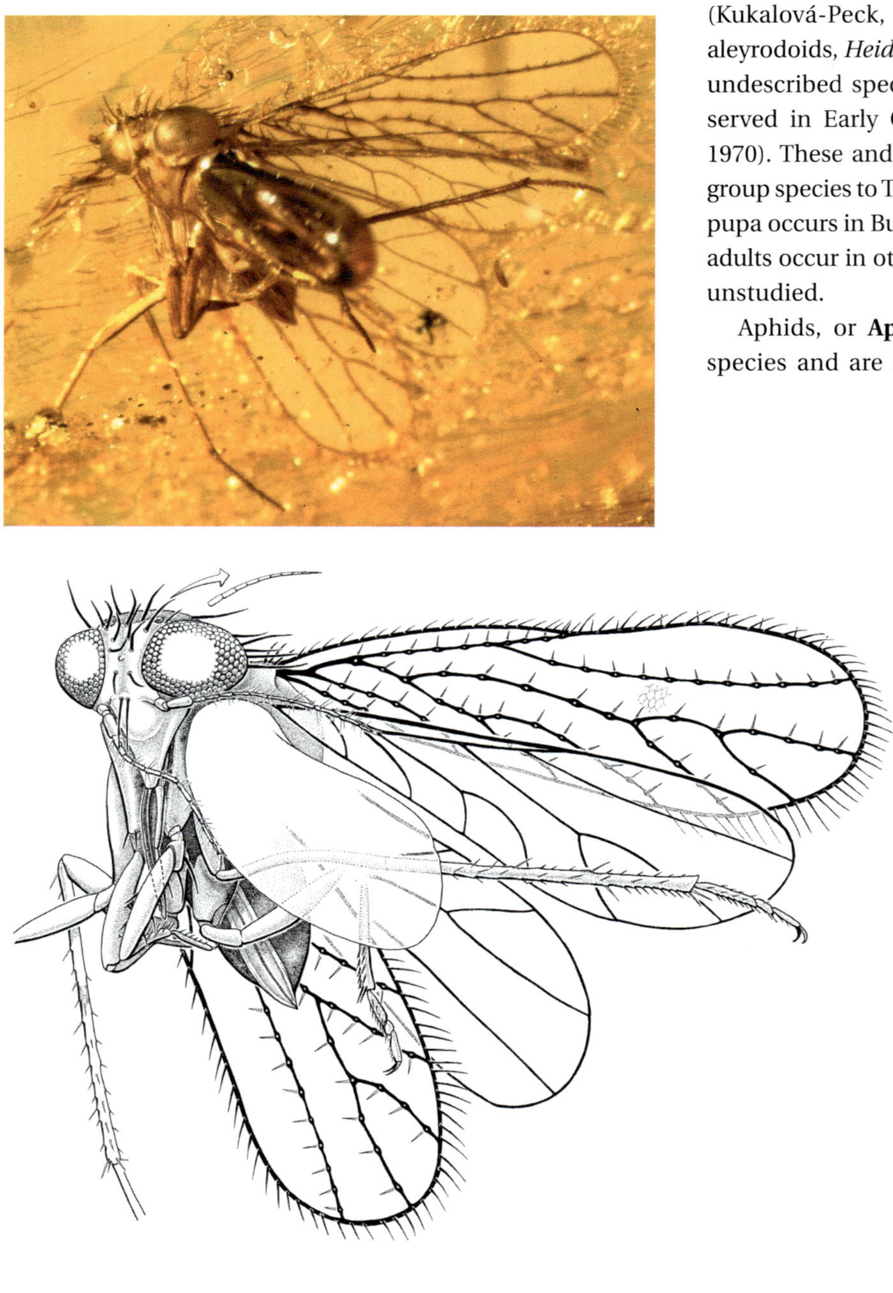

8.21. Photomicrograph and drawing of *Postopsyllidium emilyae* in 90 MYO New Jersey amber. This unusual hemipteran belongs to the extinct sister group to the Sternorrhyncha, the Protopsyllidiidae. This family was most diverse in the early Mesozoic and is a rare instance of a long-extinct group preserved in amber. AMNH NJ623; forewing length 1.5 mm.

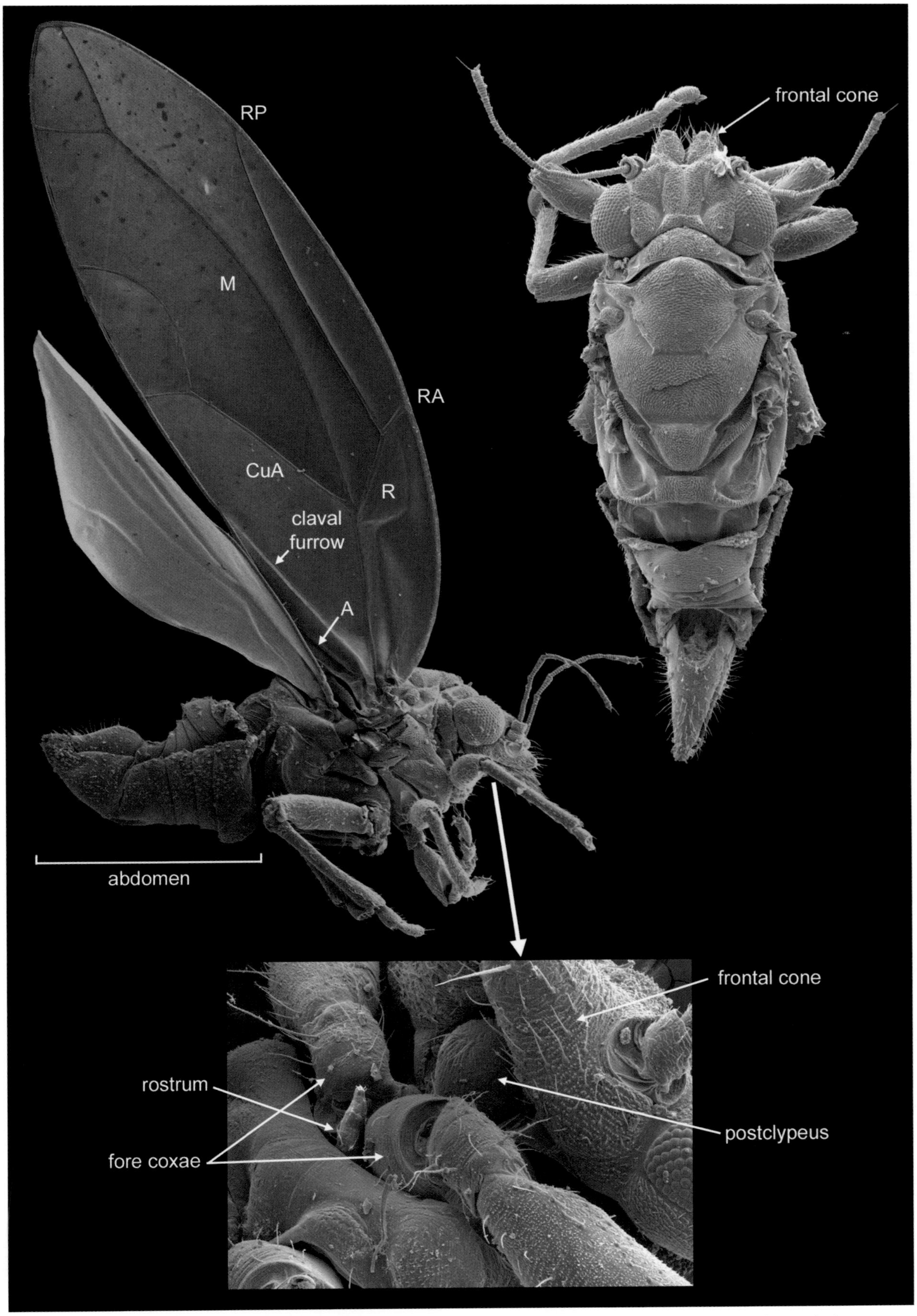

8.22. *Trioza* sp., a typical psyllid, with dorsal and lateral views of the body, and showing the area beneath the head and thorax where the tip of the rostrum protrudes between the fore coxae. Scanning electron micrograph; body length 1.6 mm.

8.23. A whitefly, or aleyrodoid, in 125 MYO amber from Lebanon. Like the genera *Bernaea* and *Heidea* also in Lebanese amber, this genus has primitive features indicating they are all stem groups to Recent Aleyrodoidea. AMNH JG 250/2; body length 1.33 mm.

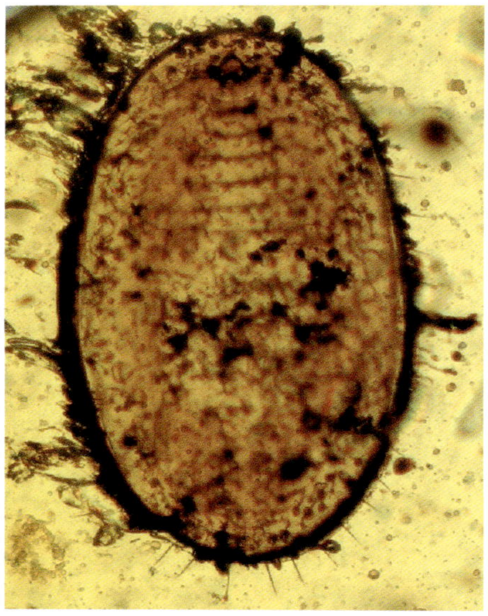

8.24. The highly reduced pupa of an aleyrodoid, in Burmese amber, ca. 100 MYO. AMNH Bu961; body length 0.91 mm.

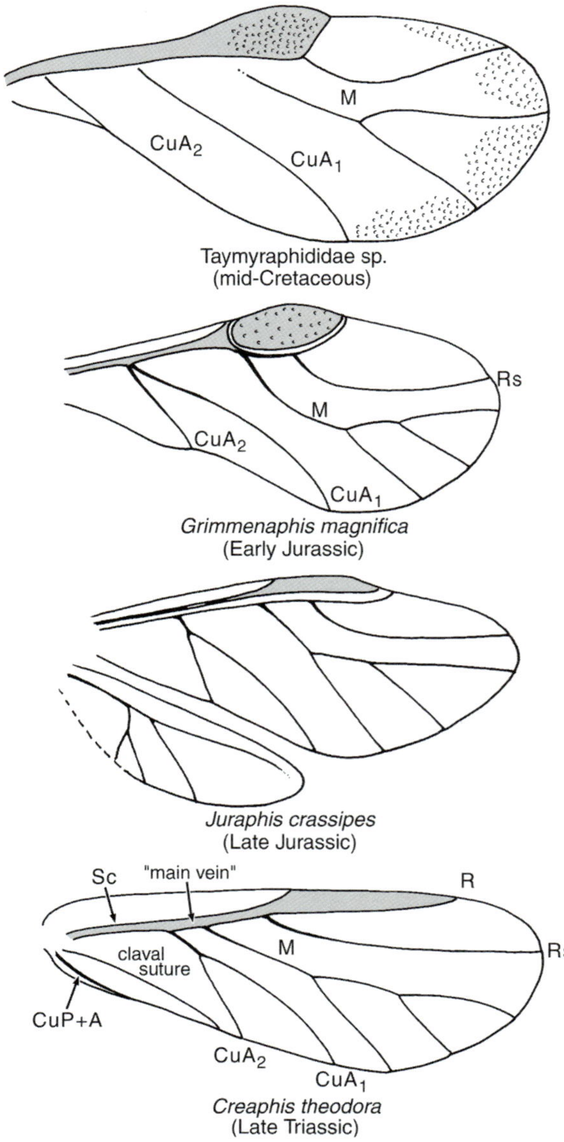

8.25. Wings of primitive, extinct aphidoids (*Juraphis*, *Grimmenaphis*, Taymyraphididae) and an early, stem-group aphidomorphan (*Creaphis*).

Northern Hemisphere. Subfamilies of the large family Aphididae are sometimes given family rank (e.g., Heie, 1981, 1987), but the common classification we use here employs just the three families Adelgidae, Phylloxeridae, and Aphididae sensu lato. Like coccoids, species taxonomy of aphids is advanced because of the agricultural significance of these insects. Winged (alate) forms have distinctive venation, such that even isolated wings compressed in rock are easily recognized. The venation is comprised of a thick longitudinal vein near the costal edge (a fusion of the bases of R, M, and Cu), which terminates in a large stigma and from which basally diverge M and Cu veins (Figure 8.25). Life cycles are complex and, like some thrips, sophisticated social behavior has been discovered relatively recently.

Many aphids have morphs that are alate and apterous, sexual and parthenogenetic, and egg-laying (*oviparous*, the females being *oviparae*) or *viviparous* (*viviparae*); these occur in various combinations, and the generations alternate. In the autumn, the typical cycle involves sexual oviparae, which produce fertilized eggs that overwinter, which then hatch in spring and later summer generations into winged and wingless parthenogenetic viviparae. To complicate the cycles further, about 10% of the species alternate on disparate hosts (*heteroecy*), the primary host being taxonomically conserved and woody, used during autumn, winter, and spring; secondary hosts are various summer herbaceous plants. The development of stages is affected by seasonal changes, crowding, and the health of the host plant. A complex life cycle like heteroecy perhaps evolved originally from seasonal morphs on a single host plant. The morph early in the year evolved to maximize fecundity on the spring flush, when predator populations are also most sparse; late summer morphs evolved to maximize dispersal, when hosts are being depleted and predators building on the earlier hosts.

Adelgidae is a small Holarctic family of eight genera and about 50 species that may be the sister group of all other aphids, though some molecular evidence indicates that this position may be held by Phylloxeridae (Cook *et al.*, 2002). The closest living relatives of Adelgidae are usually considered to be the Phylloxeridae (Heie, 1981, 1987; Carver *et al.*, 1991), but adelgids primitively retain the ovipositor that all other living aphids have lost. Both families have lost vein Rs and have only one M vein (which is typically forked in aphidoids). Adelgids are highly host specific on conifers. Also, both the Adelgidae and Phylloxeridae have (primitively) only oviparous females, and like adelgids the Phylloxeridae are a small (ca. 70 species) Holarctic group. Phylloxerids are specific feeders on various Junglandaceae (hickories and ashes), Fagaceae (oaks and beeches), and Vitaceae (grapes). In fact, the grape phylloxera, *Dakulosphaira vitifolii*, which is a native of North America, nearly devastated European grapes in the 1870s and 1880s when it was introduced.

The Aphididae sensu lato have parthenogenetic viviparae and sexual oviparae and are well known for their *cornicles* or *siphunculi*, a pair of spigot-like structures on or near tergite five (Figures 8.26, 8.28). Contrary to popular belief, the cornicles do not secrete honeydew but rather droplets of liquid containing alarm pheromones. Aphids nearby detect the chemical alarm and usually drop off the plant, and tending ants may also rush to the rescue. A droplet of honeydew is secreted through the anus near a modified apex, the cauda (tergite ten), and either flicked away or held aloft for an eager ant.

Unlike Phylloxeridae and Adelgidae, Aphididae harbor symbiotic *Buchnera* bacteria, which are closely related to *E. coli*. Aphids fare poorly without the bacteria, and these bacteria are known only from aphidids. Besides providing nutrients that are in limited supply, *Buchnera* also appears to kill the larvae of parasitoid wasps developing within the host aphid (Oliver *et al.*, 2003). Molecular cladograms of the bacteria apparently correspond well with a cladogram of hosts based on DNA sequences (Munson *et al.*, 1991; Moran *et al.*, 1993; Brynnel *et al.*, 1998; Baumann *et al.*, 1999; von Dohlen and Moran, 2000; Martinez-Torres *et al.*, 2001), suggesting intimate cospeciation of the bacteria and aphids. These cladograms do not correspond, however, with the morphological scheme by Heie (1981, 1987). For example, DNA sequences of aphids and their *Buchnera* bacteria indicate that Lachninae may be a basal group in Aphididae (von Dohlan and Moran, 2000; Martinez-Torres *et al.*, 2001), not recently derived near the subfamily Aphidinae according to Heie.

Around the same time that social thrips were discovered, aphids with advanced sociality were also discovered (Aoki, 1982). Behavior of the social species is described in many papers by Aoki (e.g., 1977, 1978) and reviewed (Foster and Northcott, 1994; Stern and Foster,1997). First-instar nymphs in these societies serve as soldiers, which either do not molt and remain as soldiers or molt into normal individuals. Soldiers are aggressive, and, depending on the species, they are adorned with horns or raptorial fore- and/or mid-legs; some even use the proboscis for impaling intruders. About 50 species of social aphids occur in six of the nine tribes of Hormaphidinae and Pemphiginae, so at least six independent origins of the soldier caste appear to have evolved. Some species have dimorphic soldiers, but this is related to host alternation. Soldiers of *Pseudoregma bambucicola* on the primary host, for example, are barely different from normal individuals; secondary host soldiers have horns and thick forelegs. Aphids in these two subfamilies produce galls, with genetically identical individuals (clones) living inside. Soldiers are also housekeepers in the galls, disposing of wax-wrapped droplets of honeydew. Interestingly, species with soldiers have particularly persistent galls, some lasting up to two years. Thus, preconditions were well established in certain aphids for the evolution of advanced sociality: high relatedness, and a long-lasting domicile in which siblings can be well defended.

Many fossil aphids have been described, most of them in a series of papers by Heie and colleagues (Heie, 1987, 1996; Heie and Pike, 1992; Heie and Wegierek, 1998; Heie and Azar, 2000), in Canadian amber (Richards, 1966), Siberian amber (Kononova 1975, 1976), and New Jersey amber (Wegierek, 2000). Aphids are diverse and abundant in some ambers from Cretaceous and younger deposits; older fossils are mostly isolated wings in rock. Stem-group Aphidoidea appeared in the Triassic, 220–210 MYA, the best preserved of which is *Creaphis theodora*, from the Triassic of Kirghizstan (Shcherbakov and Wegierek, 1991) (Figure 8.25). This and another Triassic fossil (a very fragmentary one from Australia) have venation that is

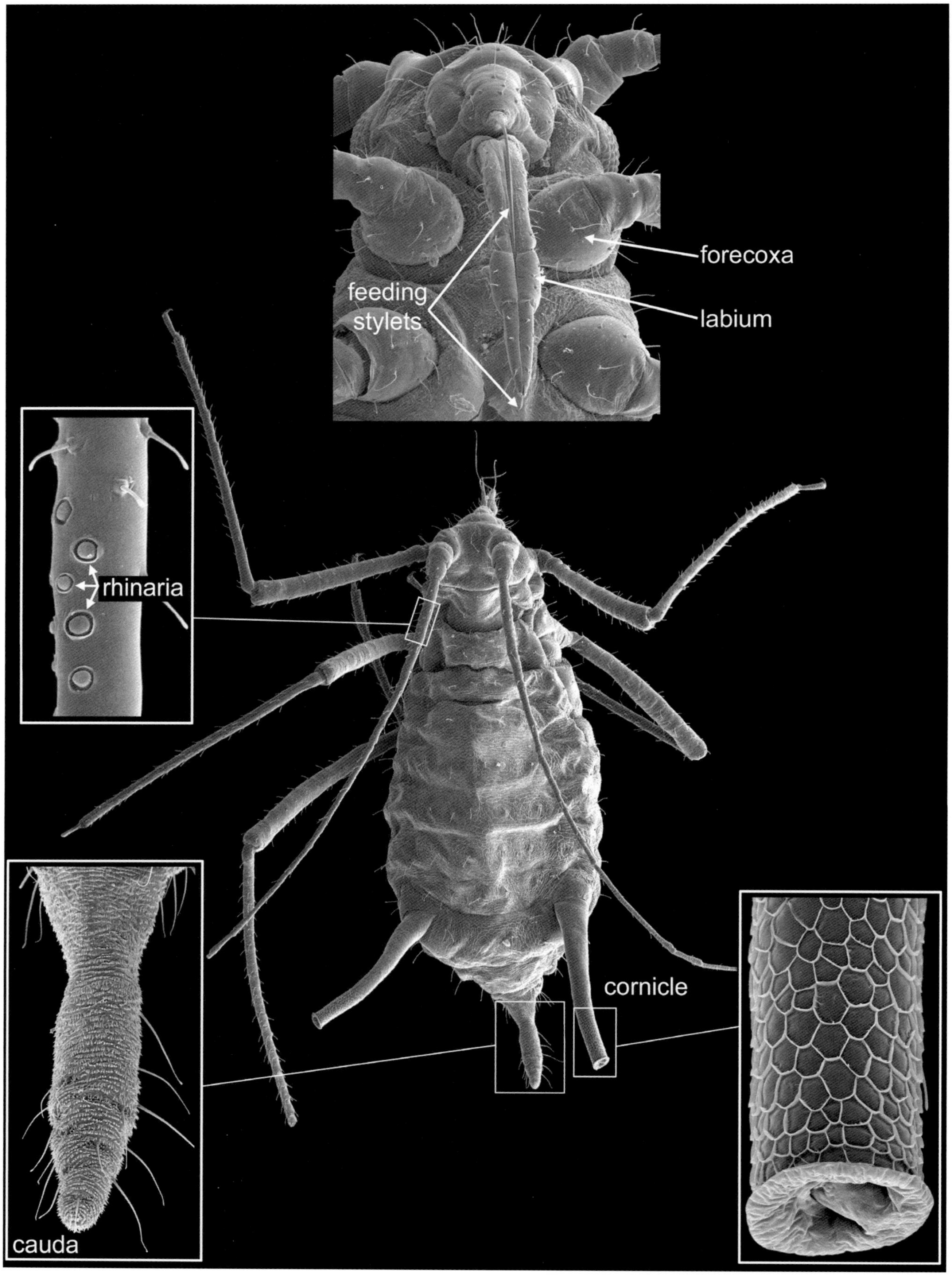

8.26. Scanning electron micrograph of a typical apterous aphid (Aphididae), showing various aphid features. Body length 1.8 mm.

primitive to all other aphids by their very long stem of Cu, and long straight Rs and M veins. Shcherbakov and Popov (in Rasnitsyn and Quicke, 2002) even proposed that the Permian families Boreoscytidae and Pincombeidae are related to aphidoids. Indeed, true aphid venation is readily derived from the venation of some genera in these families (e.g., *Archescytinopsis* and *Pincombea*) but not from other genera. These two genera are extremely primitive for Aphidoidea, having Rs forked, CuA with a very long stem and short forks, and a distinct (though small) clavus.

True aphidoids first appear in the Early to Late Jurassic of Eurasia, 170–150 MYA, specifically *Genaphis, Grimmenaphis, Juraphis,* and *Tinaphis* (Figure 8.25). Like living aphids, they have a large pterostigma, a sharply curved Rs vein, and deeply forked, two-branched Cu veins. However, they are primitive to all living aphids in the following features: Rs is long, usually near half the length of the wing (versus one third or shorter); M is three-branched (which occurs in some Recent species, but is usually just two-branched); Cu is forked with a short stem at the base or branches of Cu meet at the basal vein (this occurs in some Phylloxeridae, but not in Aphididae); and, where known (e.g., *Juraphis*), Cu in the hind wing is forked. The Jurassic species, and certain Cretaceous species with this primitive venation, are clearly the most basal Aphidoidea.

In Cretaceous ambers from the northernmost deposits of Siberia and western Canada, aphids comprise 5–10% and 35–40% of the total number of inclusions, respectively. Thus, they were abundant and probably feeding on the conifers that produced the amber. These two deposits have six and seven extinct "families," respectively, of the total ten that are known from the Cretaceous (Kononova, 1975, 1976, 1977). In Cretaceous amber from Lebanon, New Jersey, and Burma, aphids are quite rare, with aleyrodoids and coccoids being abundant instead (Azar, 2000; Grimaldi *et al.*, 2000a, 2002) (e.g., Figure 8.27). This likely reflects the more seasonal paleoclimate of the Siberian and Canadian deposits, to which aphids would have been better adapted. Collectively, approximately 65 aphidoid species in eight extinct and two extant "families" occur in the Cretaceous, though relationships and taxonomic ranks are obscure. Some, like Mesozoicaphididae, Tajmyraphididae, and Elektraphididae, are considered closely related to the basal living family Adelgidae. Others, like Canadaphididae and Cretamyzidae, are considered closely related to the Aphididae (e.g., Heie and Pike, 1992; Heie and Azar, 2000). A putative aphidine, *Aphidocallis caudatus*, occurs in Siberian amber; Pemphiginae and Drepanosiphinae occur in Siberian amber, and the latter in Canadian amber as well. Many of the Cretaceous aphids had extremely long proboscides, up to twice the length of the body, as do some basal living Drepanosiphinae and Pemphiginae. This striking feature is believed to be related to feeding on hosts with thick, rough bark, as in conifers.

8.27. Alate aphid in 100 MYO Burmese amber, in the extinct family Taymyraphididae. AMNH; wing spread 2.76 mm.

The only Tertiary group of aphids that is extinct is the Elektraphididae, which is diverse and abundant in Baltic amber (Eocene, 42 MYO) but also occurs as early as 85 MYA in Siberian amber. Otherwise, the Tertiary aphids belong to modern families and subfamilies of Aphididae (Figures 2.66, 8.28). The

8.28. Apterous aphid in 20 MYO Dominican amber, captured while its cornicles were exuding alarm pheromones. AMNH DR14–629; body length 0.95 mm.

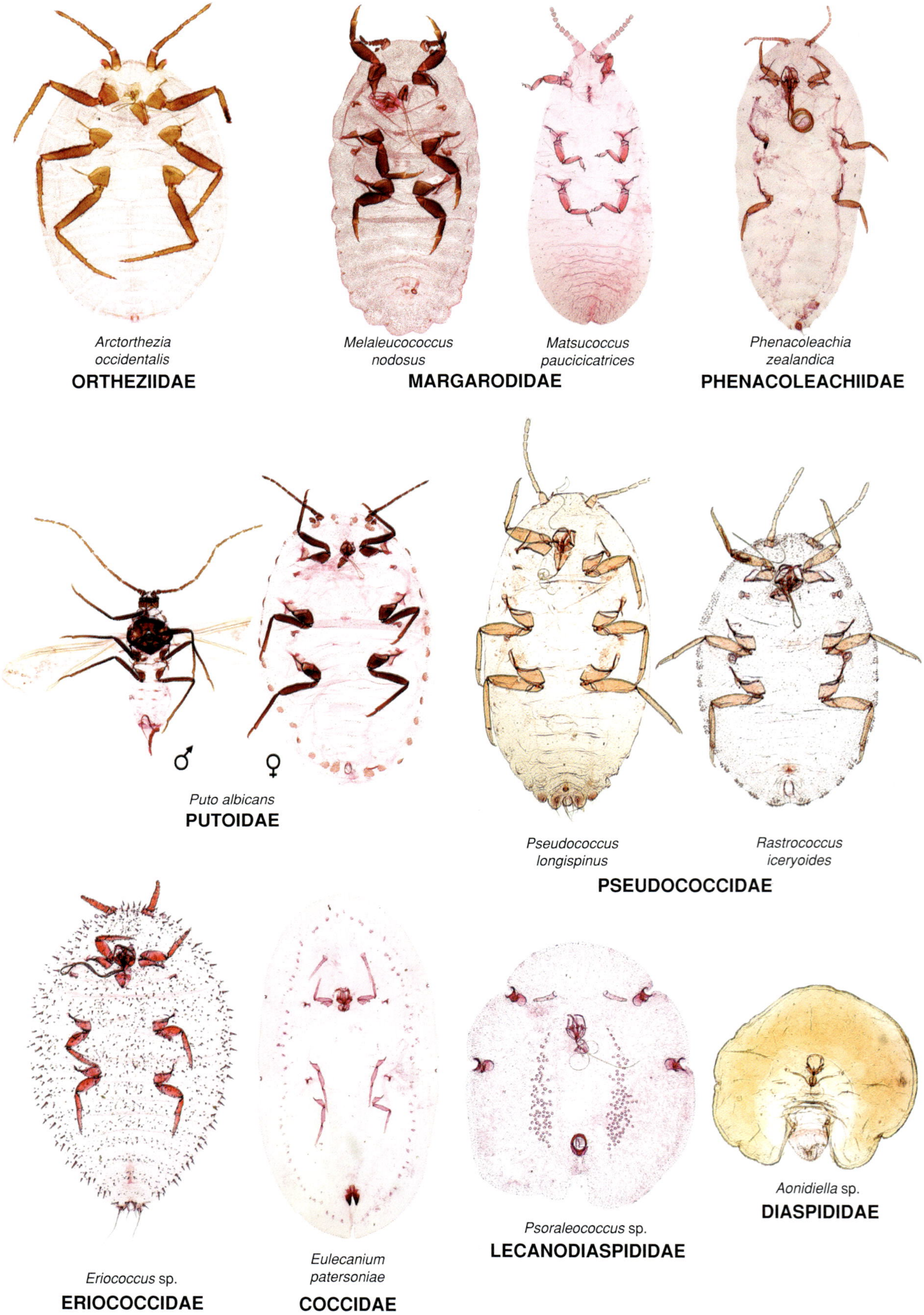

8.29. Female Coccoidea (and the male of *Puto*) of representative families. Not to scale. Specimens: P. Gullan, University of California, Davis.

Elektraphididae appear to be a basal family, near Adelgidae, and like that family the Baltic amber species probably also fed on pines, in this case the extinct *Pinites succinifer*. Dramatic change in the Tertiary aphids has been attributed to floristic changes, particularly the radiation of Compositae and grasses (Heie, 1996). The most pressing need for a thorough understanding of aphid evolution is a comprehensive phylogenetic study of living and extinct families and subfamilies.

The **Coccoidea**, or scale insects, have taken the "sit-and-suck" lifestyle to the extreme, but with 7,700 species this mode of life has obviously been evolutionarily very successful. Several coccoids have been intimately associated with human culture for thousands of years. Species of *Dactylopius* (Dactylopidae) are a source of scarlet cochineal dye, and *Kerria* (especially *K. lacca*: Kerriidae) are the source of natural lacquer. The copious honeydew from *Trabutina mannipara* (Pseudococcidae), which can encrust whole branches, is believed to be the source of the manna that sustained the ancient Israelites during years of traveling the desert. Since many coccoids are serious agricultural pests there has been voluminous species-level taxonomy on the group for some regions (e.g., Ferris, 1937–1955; McKenzie, 1967).

Extreme sexual dimorphism leaves no question that the Coccoidea is monophyletic. Females are highly neotenic, larviform insects that have neither eyes nor wings and, in most species, they have highly reduced antennae and legs (Figure 8.30), which have been entirely lost in the recently evolved families Halimococcidae and Diaspididae. Females in the phylogenetically basal families have functional legs, but these are rudimentary in the recently evolved families (Figure 8.29). Otherwise, the first-instar *crawler* is the dispersal stage. Males also disperse since in most species this sex has wings, but the hind pair is reduced, most often to tiny knobs (the *hamulohalteres*). Adult males also have brief lives because they completely lack mouthparts and so do not feed. Not surprisingly, with such sedentary females, parthenogenesis is common, but more than seven chromosomal systems of parthenogenesis occur in coccoids (Nur, 1980), making this probably the most genetically complex group of animals (e.g., Normark, 2003). The taxonomy of the group is based almost entirely on the females, which usually employs microscopic features like pores, ducts, pits, sensilla, and setae.

In standard classifications of Coccoidea, there are 19–20 Recent families, traditionally grouped into archaeococcoids and neococcoids. In contrast to their living sister group, the aphids, there has actually been significant phylogenetic work using coccoid morphology, as well as DNA sequences (Koteja, 1974; Miller, 1984; Gullan and Sjaarda, 2001; Cook *et al.*, 2002; Hodgson and Miller, 2002). Archaeococcoids are almost certainly a paraphyletic assemblage of the most basal families: Margarodidae sensu lato, Ortheziidae, Carayonemidae, and most recently the families Phenacoleachiidae and Putoidae (Figure 8.31). The Phenacoleachiidae is a relict group comprised of just two species of *Phenacoleachia*, from New Zealand, Auckland, and the Campbell Islands, one species of which feeds on the relict southern beeches, *Nothofagus*. Though *Phenacoleachia* has been shifted among three families, in one very important respect these appear to be the most basal coccoids: They have a four-segmented labium. All other coccoids have three or usually fewer segments (Koteja, 1974). Another basal archaeococcoid group, the small family Putoidae, was formerly placed in the large family Pseudococcidae but appears to be phylogenetically more basal (Cook *et al.*, 2002) The Margarodidae includes the largest coccoids (a few are more than 1 cm [0.3 in.] long), and the biology of some is very interesting. The female of both *Callipappus* and *Platycoelostoma*, for example, retracts the apex of the

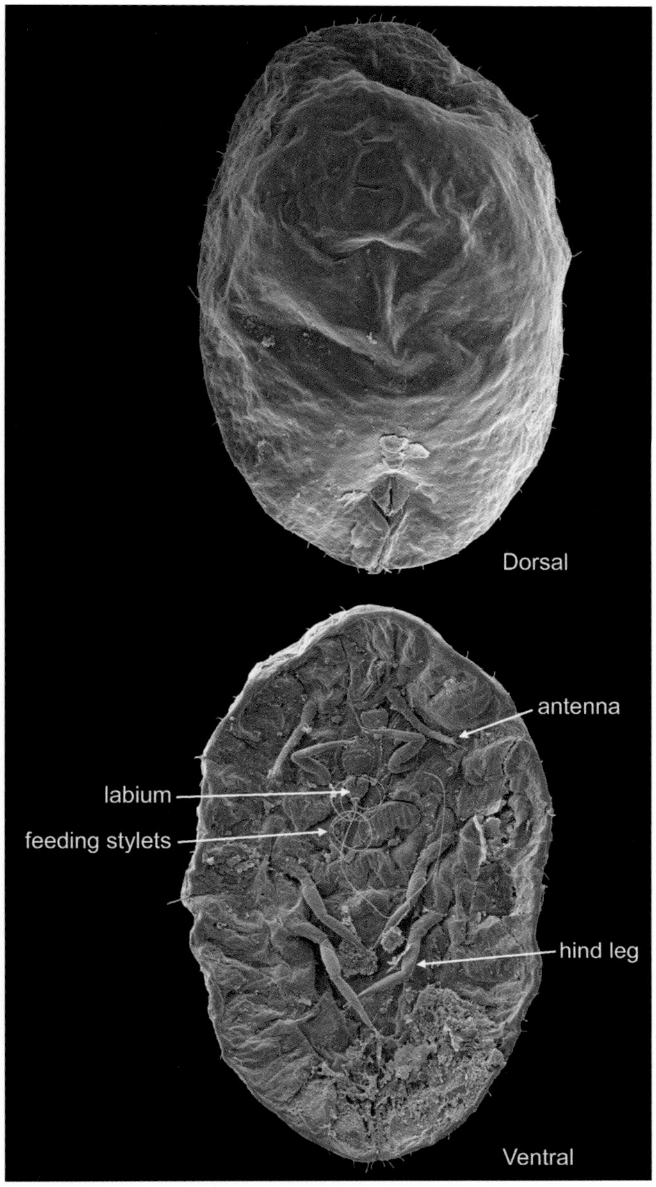

8.30. A female soft scale (family Coccidae), showing the extreme reduction of appendages and the hairlike feeding stylets. Above, dorsal view; below, ventral view. Scanning electron micrographs; body length 1.3 mm.

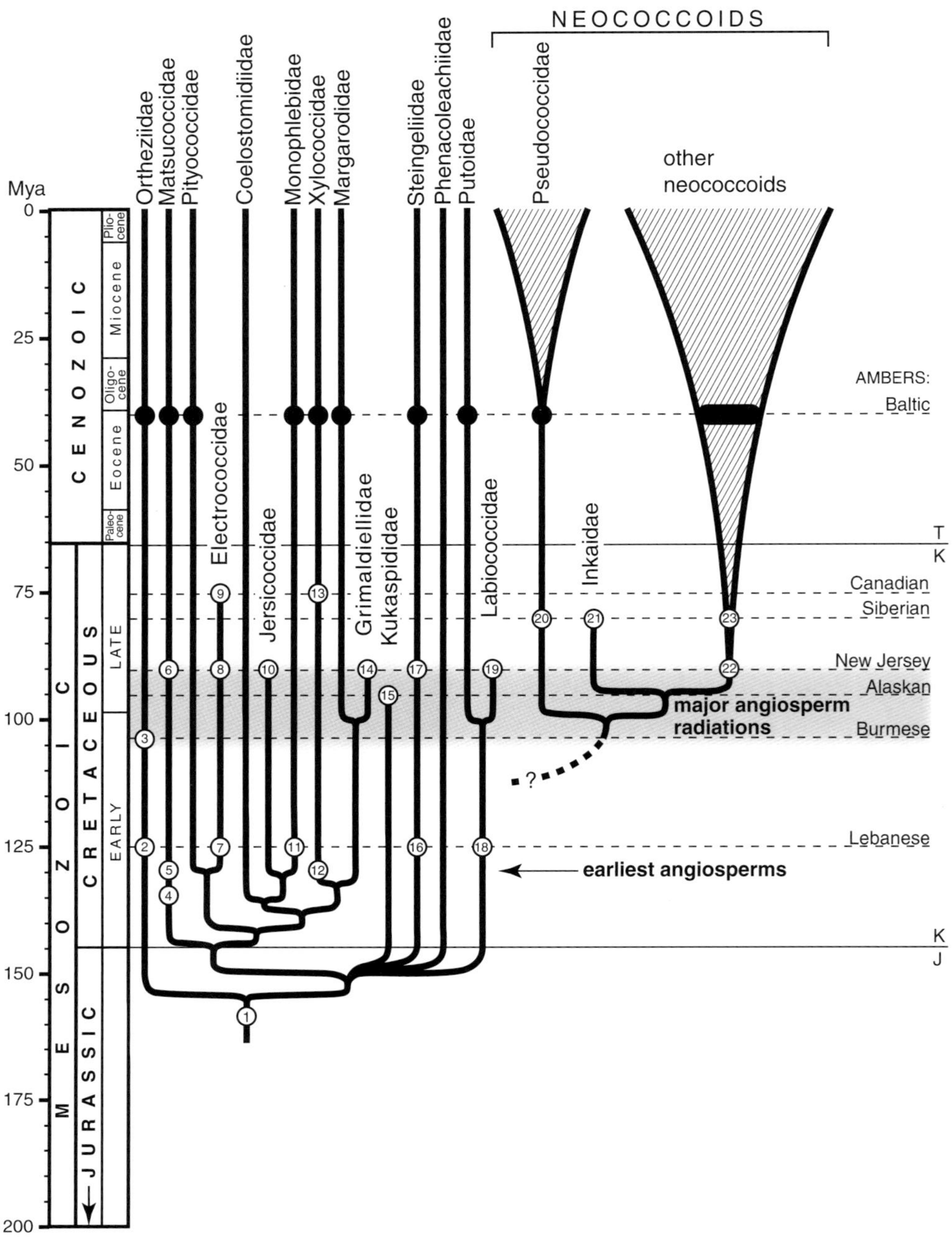

8.31. Phylogeny of basal Coccoidea. Only Mesozoic fossils are numbered and listed, and only Baltic amber fossils are indicated for the Tertiary. The most diverse lineage, the neococcoids, appear to be a product of the angiosperm radiations. See Table 8.4 for numbered fossils. Relationships based on Gullan and Sjaarda (2001) and Cook *et al.* (2002); placement of fossils from Koteja (2000a, b; 2001) and others.

abdomen into her body to form a *marsupium*, in which she then broods the eggs. The Margarodidae may not be monophyletic (Gullan and Sjaarda, 2001), in which case Koteja's classification of recognizing the subfamilies of margarodids as families is probably the best system.

The neococcoids are certainly monophyletic, defined by obvious derived features such as the loss of abdominal spiracles in the females and the male eyes being just individual facets (Figure 8.31). In fact, there is a continuum among coccoids, from males having fully compound eyes in the most basal coccoid families to their having a row of facets in the intermediate families to their having just a pair of facets on each side of the head in the neococcoids. Neococcoids also have a unique chromosomal system of sex determination, or paternal genome elimination (PGE). The youngest embryos of neococcoids are diploid and have no sex chromosomes, so the paternal chromosomes have become heterochromatic, or are lost, in the early stages of male development (Nur, 1980; Normark, 2003). The large family Pseudococcidae, commonly called "mealybugs" for the flocculent, mealy wax coating the body, appears to be the sister group to the rest of the neococcoids, based on male structure and DNA sequences

TABLE 8.4. Mesozoic Coccoidea[a,b]

1. Undescribed, late Jurassic compression
2. Undescribed Ortheziidae: Lebanese amber
3. Undescribed Ortheziidae: Burmese amber
4. *Eomatsucoccus sukachevae*, *E. popovi*: compressions, Transbaikal, Russia
5. *Eomatsucoccus andrewi*: compression, Wealden, England
6. *Eomatsucoccus casei*: New Jersey amber
7. Undescribed Electrococcidae: Lebanese amber
8. *Turonicococcus beardsleyi*, *T. grimaldii*: New Jersey amber
9. *Electrococcus canadensis*: Canadian amber
10. *Jersicoccus kurthi*: New Jersey amber
11. Undescribed Monophlebiidae: Lebanese amber
12. *Baissococcus victoria*: compression, Baissa: Transbaikal, Russia
13. Undescribed xylococcid: Canadian amber
14. *Grimaldiella resinophila*, *G. gregaria*: New Jersey amber
15. *Kukaspis usingeri*: Alaskan amber
16. Undescribed Steingeliidae or near: Lebanese amber
17. *Steingelia cretacea*: New Jersey amber
18. Undescribed Putoidae : Lebanese amber
19. *Labiococcus joosti*, *Solicococcus nascimbenei*: New Jersey amber
20. Undescribed Pseudococcidae: Siberian amber
21. *Inka minuta*: Siberian amber
22. *Keithia luzzii* (in paraphyletic Eriococcidae): New Jersey amber
23. Undescribed "Eriococcidae": Siberian amber

[a] Species and their classifications taken from the extensive work of Jan Koteja (2000a,b, and other papers).
[b] Numbers correspond to those on phylogeny, Figure 8.31.

(Cook *et al.*, 2002). Many intimate symbioses with ants have evolved in this family. Monophyly of another large neoccoid family, the Eriococcidae or "felt scales," is disputed (e.g., Cook *et al.*, 2002; Hodgson and Miller, 2002). The two largest families of coccoids are neococcoids, which are the Coccidae, or "soft scales" (1,000 species), and the Diaspididae, or "armored scales" (1,700 species). Armored scales secrete a waxy material that hardens into a protective covering, usually with their exuviae attached.

Fossil coccoids are rarely found in rocks, probably because the female is sedentary and the males are ephemeral and minute, with a wingspan of generally less than 2 mm. Putative coccoids from the Permian and Triassic (*Permaleurodes*, *Mesococcus*) are small, wingless, presumably nymphal insects, and virtually no characters are preserved to allow a convincing placement. The oldest definitive coccoids in rock are in the extant family (or subfamily of Margarodidae) Matsucoccidae, from the Early Cretaceous of England and central Siberia 130–140 MYA (Koteja, 1999). Patterns of fine, pinnate grooves on the wing that are so distinctive to the Matsucoccidae are well preserved even in compression fossils. Despite their absence from the Jurassic, it is virtually certain that true Coccoidea occurred then, but unless they had similar distinctive features, it is unlikely they would be recognized. Just as minute, delicate midges are commonly trapped in amber, so are male coccoids. Females and crawlers are only occasionally preserved as such (Figures 8.32, 8.33). Unfortunately, with the exception of a few excellent studies on particular families (e.g., Ghauri, 1962; Afifi, 1968; Koteja, 1986; but see also Hodgson, 2002), male coccoids are not as well surveyed taxonomically as are the females. Fortunately, though, coccoids fossilized in amber have been studied almost entirely by Jan Koteja, a morphologist who prepares and studies the amber fossils with the same finesse he applies to slide-mounted living species (e.g., Koteja, 1984, 1990, 1996, 2000a,b, 2001). In fact, coccoids have one of the best fossil records among insects for the past 120 MY, entirely because of their diversity in amber.

Most Cretaceous coccoids belong to extinct and living basal families ("archaeococcoids") (Figure 8.31); these have male eyes either entirely compound or each reduced to at most a lateral row of facets. Among the few Cretaceous neococcoids is *Inka minuta* (Inkaidae) in 80 MYO Siberian amber, which has each eye further reduced to merely a pair of facets, typical of male neococcoids. Another Cretaceous neococcoid is *Keithia luzzii* (Eriococcidae), known from a crawler in 90 MYO

8.32. Female ortheziid scale insect in Burmese amber, ca. 100 MYO. Strands of exuded wax are still attached to the abdomen. AMNH Bu962; body length 2.50 mm.

8.33. Males of basal scale insect families in Burmese (left) and Lebanese amber (right); note the complete compound eyes, which are reduced in more recently evolved scales. Body lengths: 1.71 mm (AMNH Bu1416) (left); 2.32 mm (AMNH LAE-93, right).

New Jersey amber. The exact phylogenetic position of each of these among neococcoids is difficult to assess. The first diverse "modern" fauna of coccoids occurs in Eocene Baltic amber, which contains abundant matsucoccids but also the earliest known Coccidae, Diaspididae, and Pseudococcidae (summarized by Koteja, 2000b). The trend of basal coccoids in Cretaceous amber and diverse neococcoids by the Eocene indicates that the major lineages of Coccoidea were in existence by the Early Cretaceous, and the entire group probably originated in the Jurassic. Some coccidologists insist that Coccoidea originated in the Triassic, a view that is based on the close relationship of this group to aphids. As previously discussed, though, a *stem-group* aphidomorphan is known from the Late Triassic, and true aphids are not known until the Jurassic. So, if sister-group dating is to be applied to the Coccoidea, their age would have to be Jurassic. Radiations of the most diverse group, the neococcoids, occurred at least 100 MY after the earliest coccoids presumably appeared. These events are probably related to the evolution of two other groups of organisms intimately associated with coccoids: ants and flowering plants.

Sternorrhyncha and Ants

Many hemipterans in the Sternorrhyncha and Auchenorrhyncha excrete sugary honeydew, but it is produced in prodigious amounts in Sternorrhyncha, often several times the body weight of the insect per day. Honeydew is eagerly collected by ants (Figure 8.34), which in many cases is solicited from the hemipterans using rapid strokes of their antennae, and the bug then readily delivers (Buckley, 1987). Though the amino acid and nitrogen needs of ants must usually be supplemented with insect prey, honeydew can provide the bulk of the nutritional needs for a colony. It has been estimated, for example, that a large colony of *Formica* ants can harvest 500 kg of honeydew in a year – far more honey than a colony of bees produces. Where ants are few or absent, whole new interactions have evolved between sternorrhynchans and other organisms. In New Zealand, where the cool, wet forests host few native ants, sooty beech scales (*Ultracoelostoma*) can reach densities of 200 scales per square foot of tree trunk. The honeydew encrusts whole branches and trunks and is an essential diet for some native birds.

Ants seem preadapted for "milking" hemipterans since trophallaxis, or liquid food exchange, is an intricate part of

8.34. Ants gathering honeydew from aphids and coccoids. Ants pugnaciously guard their sources of honeydew, thus providing one of the best defenses for these vulnerable hemipterans. Some predators slip through the defenses though: a predatory fly larva feeds unnoticed on an aphid (right). Left and right: tending aphids; middle: mealybugs tended by *Camponotus* ants. Photos: P. J. DeVries and V. Giles.

their social life, made possible by the structure of their proventriculus. Sedentary hemipterans are extremely vulnerable, so the main benefit to them is protection against a veritable onslaught of predators and parasitoids by ants. Even the pugnacious and efficient defenses of ants, though, are occasionally broken. The larva of a North American green lacewing (*Chrysopa slossonae:* Chrysopidae), for example, are insect wolves in sheep's clothing: They cover themselves with the fluffy, white wax of their aphid prey, and so go undetected by the ants that "herd" the aphids (Eisner *et al.*, 1978). Sternorrhyncha also benefit from tending ants by the improved hygiene as well as protection of the plant on which they feed, since ants typically dispatch other insects feeding on the same plant as their herds. Though hemipterans have been described as the wards of ants (Hölldobler and Wilson, 1990), another view has the ants as extortionists, giving protection (indeed, refraining from killing) after being appeased with payment. Their payment is merely excrement, but it is sufficient to maintain a widespread and very successful symbiotic relationship. Where scale insects have evolved effective defenses of their own, they don't require ants. Armored scales (family Diaspididae), for example, protect themselves with a tough coat, and they don't even have an anus to exrete honeydew. Species of *Dactylopius* scales (Dactylopiidae), or cocchineal insects, produce approximately 2% of their body weight in carminic acid, a potent and offensive quinone that has been used for millennia by humans as a brilliant red dye. The dye efficiently repels generalist predators like ants, but the predatory caterpillars of a pyralid moth, *Laetilia coccidovora,* feed with impunity on the cochineal scales, and even use the dye in defense by regurgitating it on intruders (Eisner *et al.*, 1980). No defense, it seems, is perfect.

Many studies on ant-hemipteran symbiosis have focused on aphids (reviewed by Hölldobler and Wilson, 1990), which may reflect a temperate region bias, but nonetheless have revealed how the tending ants can dramatically manipulate their "herds." An extreme example among aphids involves *Lasius neoniger* ants tending *Aphis maidiradicis:* The ants overwinter the aphids in their colonies and then deliver them back to the host plants in the spring. Even more specialized symbioses, however, are found between ants and coccoids, particularly Pseudococcidae. In Java, *Hippeococcus* mealybugs that are disturbed climb onto the *Hypoclinea* ants tending them. In peninsular Malaysia *Hypoclinea cuspidatus* ants tend *Malaicoccus formicarii*, and the mealybug is known to occur only with this ant. The mealybugs feed on the developing shoots of various plants, to which they are transported by the ants. The ants are actually nomadic, forming a nest with the bodies of workers where they bivouac, like army ants, with their brood and mealybug "herds" cloistered within. When the mealybugs require better feeding sites, an entourage of the entire colony and their mealybugs may move to greener pastures dozens of meters away.

Most recently, a comprehensive study by Dill *et al.* (2002) has found that herding mealybugs is an important lifestyle for *Dolichoderus* ants in southeast Asia. In their study, 12 species of these ants herded 36 species of mealybugs (all in one tribe, the Allomyrmococcini) on 57 families of forest plants. Moreover, the association is species specific, and the ants derive most of their nutrition from the honeydew. Like *Hypoclinea*, the *Dolichoderus* ants transport their mealybug herds to young growth, shelter them in the nest, and relocate their colony where the mealybugs feed best.

Myrmecophytes are plants that harbor colonies of ants in specialized structures, or *domatia*. These are important components of tropical forests, as at least 100 genera in 39 families and hundreds of species of plants have domatia (Hölldobler and Wilson, 1990). The ants receive a domicile and sometimes nutrition from extrafloral nectaries or from specialized, proteinaceous packets, or *food bodies*, growing on the host plant. The plants receive valuable nitrogen from the colony's waste and protection from insect herbivores – with one notable exception: hemipterans. The inner walls of many domatia are lined with coccoids. In the neotropics, for example, the walls of *Azteca* ant nests within the chambers of *Cecropia* trees are carpeted with pseudococcids (Longino, 1991), and the walls of *Crematogaster* nests living in *Macaranga* euphorbs in southeast Asia are similarly lined with coccid soft scales (Heckroth *et al.*, 1998). In fact, domatia may have evolved from sheltered feeding sites used by tended hemipterans, so the ants may not be protecting the plants per se but rather their hemipteran herds (Benson, 1985; Gullan, 1997).

Another remarkable example of symbiosis involves the pantropical ant genus *Acropyga*, the members of which tend root-feeding mealybugs (Pseudococcidae) in vast underground colonies. These ants are so dependent on the mealybug honeydew that swarming virgin queens carry them in their mandibles to inoculate their new nests, not releasing the mealybug even during copulation (e.g., La Polla *et al.*, 2002). Incredibly, *Acropyga* queens carrying mealybugs have been captured in Dominican amber (Johnson *et al.*, 2001) (Figure 8.35). Two examples are known of groups of worker ants captured in amber with hemipterans they presumably were tending, one in Dominican amber (Johnson *et al.*, 2001) and another in Baltic amber (Wheeler, 1915).

Thus, there is direct evidence that ant-hemipteran symbiosis has existed for at least 40 MY, and that highly specialized, obligate symbiosis has been around for at least 20 MY, but the alliance between these two kinds of insects probably began in the Late Cretaceous. Most ants tending hemipterans are in the Dolichoderinae, Formicinae, and Myrmicinae. Dolichoderines are known from Canadian amber (ca. 75–80 MYO) (Wilson, 1985b), and there is a formicine in New Jersey amber (ca. 90 MYO)(Grimaldi and Agosti, 2000) and coccoids were diverse and numerous when these ambers were produced. The exploitation of hemipterans by ants in these subfamilies probably contributed to the explosive radiation of both of these groups in the early Tertiary.

Sternorrhynchan Evolution and the Angiosperm Radiations

A major radiation of ants in the early Tertiary was probably affected partly by sternorrhynchans, of which the most diverse groups – the neococcoids and Aphididae sensu lato – also seemed to have radiated around the same time. Sternorrhynchan evolution itself was certainly fueled by the Late Cretaceous explosion of angiosperms (e.g., Heie, 1996; von Dohlan and Moran, 2000). Among living sternorrhynchans, feeding on conifers is concentrated among basal taxa, such as the adelgid and lachnine aphids (the latter of which may be a basal subfamily of Aphididae), *Neophyllaphis*, and various genera in Margarodidae s.l. *Neophyllaphis* is interesting because it retains some very primitive features for aphids (like mating males and females throughout all seasons; and winged, mating females), and it has a relict distribution with its conifer hosts, *Podocarpus* and *Araucaria*. Moreover, the genus *Aniferella* in Cretaceous amber is very similar to this living genus. The abundance of basal groups of aphids and coccoids in coniferous ambers indicates that many early sternorrhynchans fed on conifers: Canadaphidae, Cretomyzidae, Elektraphididae, and Tajmyraphididae (Aphidoidea), and various archaeococcoids (Koteja, 2000a,b). There is even direct evidence that some Cretaceous coccoids fed on conifers (Grimaldi *et al.*, 2000a). Host switching from conifers to angiosperms in the Late Cretaceous is believed to have spurred the radiation of aphids that took place later in the Tertiary, beginning about 40 MYA, probably a result of the radiation of composites and grasses (Heie, 1996; von Dohlen and Moran, 2000). Aphididae, for example, comprise only 4% of the species from the Early Tertiary, but more than 50% in the modern fauna.

Koteja (e.g., 1985; and Gullan and Kosztarab, 1997) reviewed several lines of evidence as to why ancestral coccoids were not conifer feeders, but saprophages living amongst leaf litter feeding on fungal hyphae on phytophages of roots. While ancestral root feeding (on conifers!) is plausible, mycophagy, which is found in a few archaeococcoids, is almost certainly recently evolved since all other sternorrhynchans feed on plants, including the phylogenetically basal

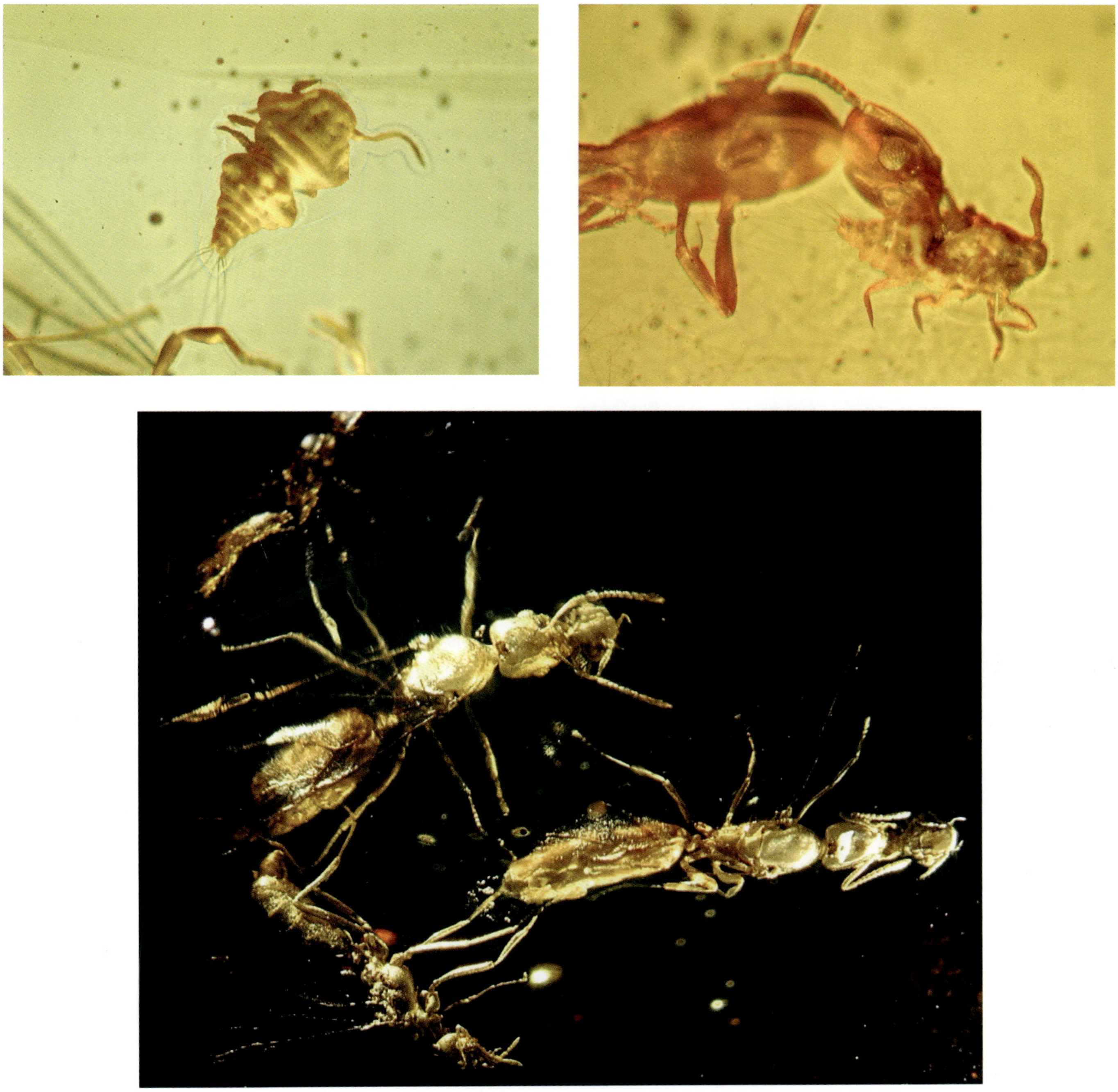

8.35. An ancient and intimate symbiosis: Pseudococcid scale insects (*Electromyrmococcus*) in Miocene Dominican amber, with their *Acropyga* ant hosts. These ants require pseudococcid honeydew for feeding the colony, and when a young queen flies to begin a new colony, she carries with her one of these hemipterans, to start the new "herd." Above, left: a mealybug released by a queen ant when they were immersed in the resin; above, right: head of a queen ant with a mealybug in her mandibles; bottom: a small swarm of four queens, two of them with mealybugs. All AMNH; coccid lengths ca. 0.75 mm.

psylloids. Regardless of whether their ancestors fed on mycelia or conifer roots, the chronology and phylogeny of coccoids from the Early Cretaceous through Miocene is compelling evidence that coccoids also explosively radiated in the Tertiary, another impact of the angiosperm radiations. If ancestors of each of the four major groups of Sternorrhyncha fed on conifers, each group must have colonized angiosperms independently, and those events probably took place in the Late Cretaceous.

AUCHENORRHYNCHA: THE CICADAS, PLANT HOPPERS, AND TREE HOPPERS

Like Sternorrhyncha, virtually all of the approximately 25,000 described species of auchenorrhynchans feed on plant vascular fluids, with a few feeding on fungi. In fact, the great majority of auchenorrhynchans feed on angiosperms. Despite several unique morphological characters, monophyly of the group has been debated. Monophyly of Auchenorrhyncha

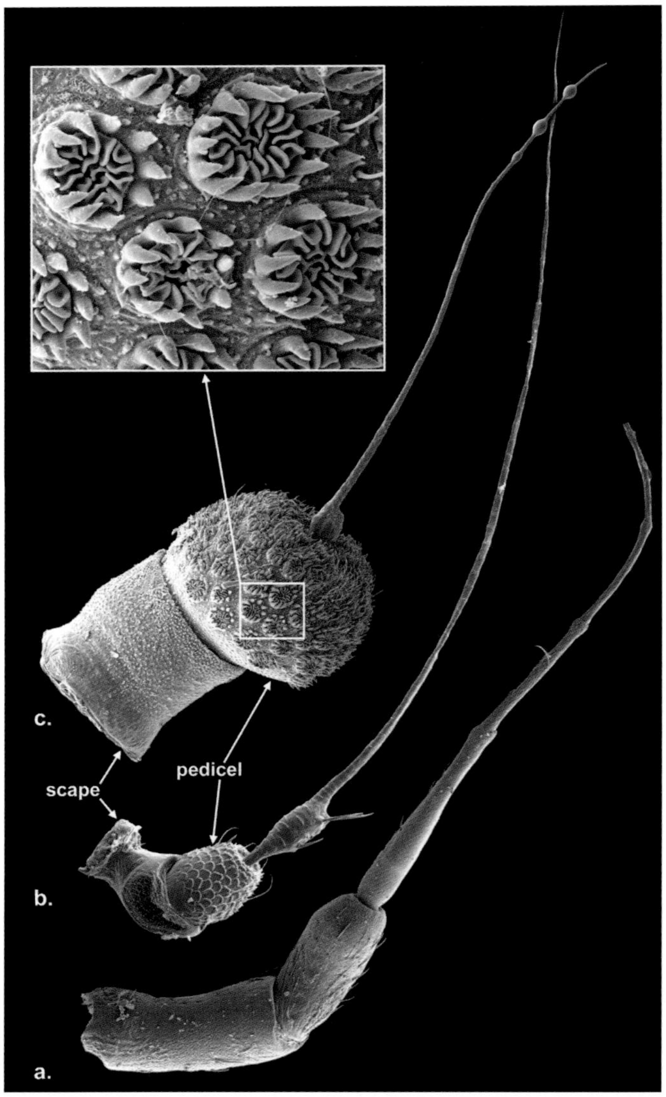

8.36. A defining feature of the Auchenorrhyncha: the aristate antenna. Fulgoromorphs have an enlarged pedicel with numerous sensillar plaque organs, shown in detail. (a) Cicadidae, (b) Membracidae, (c) Fulgoridae. Scanning electron micrographs.

has been supported by Kristensen (1975, 1991), Hennig (1981), and Carver *et al.* (1991). Hamilton (1981, 1996), however, maintained that the Sternorrhyncha was sister group to one lineage of Auchenorrhyncha (the Cicadomorpha). Some DNA and morphological studies indicate a close relationship between the other main lineage of auchenorrhynchans, the Fulgoromorpha, and the Heteroptera (Bourgoin *et al.*, 1997). We feel that the evidence for a monophyletic Auchenorrhyncha is most compelling.

Auchenorrhyncha have a distinctive, aristate antenna, where the basal segments are large but the apical segments are reduced to a short, thin, tapered arista (Figure 8.36). They have a large pronotum that folds down along the sides and a very large clypeus that houses a large cibarium used in siphoning plant fluids. The structure of the wing base is also unique (Yoshizawa and Saigusa, 2001), as is the complex tymbal acoustic system. On each side of tergites one or two is a tymbal and tympanum, the former composed of thin plates that vibrate (Ossiannilsson, 1949; Claridge, 1985). Most auchenorrhynchans are inaudible, but cicadas are very loud because the tymbals are large and elaborate, though sound in both groups is used for communication among individuals and even for defense (e.g., Cocroft, 1999).

The suborder Auchenorrhyncha is traditionally divided into two groups, each apparently monophyletic. The infraorder **Cicadomorpha** (Figure 8.37) comprises the cicadas (Cicadoidea), frog hoppers and spittlebugs (Cercopoidea), and leaf hoppers and tree hoppers (Cicadelloidea or Membracoidea). In these the tegula at the base of the wing is absent; the wings (or at least the hind pair) have an ambient vein that runs parallel to and near the margin; and the gut has a filter chamber that contains the Malpighian tubules. The infraorder **Fulgoromorpha** (Figure 8.38) comprises all others, defined by carinae running down the face; immobile hind coxae; ocelli and bases of the antennae situated below the eyes; and a large antennal pedicel with numerous sensillar plaque organs on the first flagellomere.

Three DNA sequencing studies concluded that Fulgoromorpha is more closely related to the Heteroptera than to Cicadomorpha (von Dohlen and Moran, 1995; Bourgoin *et al.*, 1997; and the study reported by Campbell *et al.*, 1994, 1995, and Sorensen *et al.*, 1995). Unfortunately, character and taxon sampling in all of these studies was minimal, comprising no more than ten exemplar species and families, and using only one gene (18S rDNA). Moreover, in each study the Fulgoromorpha + Heteroptera relationship was either unstable or support for it was barely more than that for a monophyletic Auchenorrhyncha.

In the paper by Sorensen *et al.* (1995), it was assumed that the relationships based on their sequence data was "superior to, and definitely more objective than . . . morphology" (p. 43). They then applied this assumption to dismissing the unique morphological features of auchenorrhynchans. For example, by emphasizing the differences in antennal structure and tymbals between fulgoromorphs and cicadomorphs, this ignores the fundamental similarity of these structures in the two groups, which is phenetic thinking. Sorensen *et al.* (1995) also appealed to a unique fossil specimen, *Megaleurodes megocellata,* for their arugment. It was placed tentatively in the Aleyrodoidea (?Boreoscytidae) by Hamilton (1990) because it apparently has non-aristate antennae, albeit only partially preserved. Sorensen *et al.* (1995) suggested that this specimen belonged to the Jurassic family Fulgoridiidae, which purportedly has flagellate antennae, and thus appears to be intermediate between fulgoromorphs and Heteroptera. Shcherbakov (1996), however, found that the long rostrum in fulgoridiids was misinterpreted as an antenna by the original describer, and that this group is actually closely related to or even belongs in the Cixiidae! Prior to citing features of critical

8.37. Representative families of Cicadomorpha (Auchenorrhyncha). Not to the same scale.

8.38. Representative families of Fulgoromorpha (Auchenorrhyncha). Not to the same scale.

8.39. A newly emerged adult cicada. Photo: P. J. DeVries.

fossils in a controversial topic, it is best if the fossils are actually reexamined. The characters proposed by Sorensen *et al.* (1995) for a Fulgoromorpha + Heteroptera relationship are likewise weak. In light of the compelling morphological evidence listed earlier, and the ambiguous DNA studies, we prefer to regard the Auchenorrhyncha as monophyletic.

Auchenorrhychans appear to have first evolved in the Permian, which may be too ancient for genes like 18S to probe. Unfortunately, too, interpretations of Permian and early Mesozoic auchenorrhynchan fossils and evolution are disparate (cf., J. W. Evans 1963; Hamilton, 1996; Shcherbakov, 1996; Shcherbakov and Popov, 2002), revealing just how clouded our understanding of the topic is. What is most needed is a rigorous morphological and molecular study on the relationships of living Auchenorrhyncha at the family level. From this study one could identify structures that reliably define families and groups of families and reexamine important fossils for these structures. Though very useful, J. W. Evans's (1963, 1964) and Dworakowska's (1988) studies of wing venation need to be supplemented with other characters, which then need to undergo considered cladistic analyses with other characters. In lieu of that, the best we can presently do is to identify the earliest records of extant families and family groups.

The three superfamilies of the infraorder Cicadomorpha are the Cicadoidea (cicadas), Cercopoidea (frog hoppers, spittle bugs), and Cicadelloidea (leaf hoppers and tree hoppers, also called Membracoidea). Cicadas are probably the most familiar; their piercing drone is synonymous with the heat of temperate summers and the tropics. They call from treetops to attract mates, in choruses specific to the species, but they also produce a loud shrill as a startle response to predators, which anyone who has tried picking up a grounded adult cicada will know. Nymphs have large, curved, chelate forelegs, used for moving through soil as they feed from

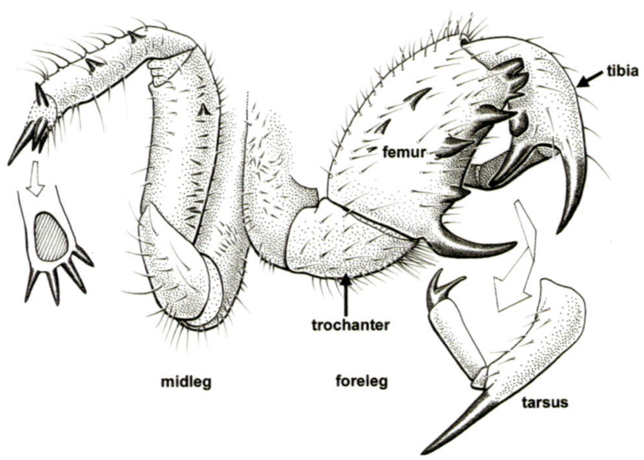

8.40. Fossorial forelegs of cicada nymphs, from a Recent species of *Magicicada* (Cicadidae) (above) and from an undetermined Cretaceous species in New Jersey amber (below). Fossil: AMNH NJ300.

the roots of trees. Nymphs are most commonly encountered as brown husks – the exuviae of last (fifth) instar nymphs – clinging to tree trunks (Figure 8.39). The distinctive forelegs of cicada nymphs were apparently lacking in Cretaceous Cicadaprosbolidae (Hamilton, 1990) (presumably they were not fossorial), though a cicada nymph in mid-Cretaceous amber from New Jersey has this distinctive foreleg structure (Figure 8.40). The distinctive calls of cicadas are produced by a pair of tymbals at the base of the abdomen, discussed later (Figure 8.41).

Tettigarctidae is a relict family of hairy cicadas with only two species (both in the genus *Tettigarcta*), now restricted to southeastern Australia and Tasmania and morphologically primitive to all other cicadas. "Cicadaprosbolidae" from the Jurassic and Cretaceous of South America, Europe, and Asia have been classified as early tettigarctids (Shcherbakov, 1996). *Eotettigarcta*, from the Eocene of Scotland, indicates that the family was formerly quite widespread. Dunstaniidae were large, hairy Triassic cicadas derived from the Paleontinidae (Figure 8.42), which was a Permian–Jurassic stem group of the Cicadoidea. The earliest Cicadoidea, or true cicada, is *Liassocicada*, from the Early Jurassic of England. The large living family Cicadidae is worldwide and comprises 1,200 species, earliest definitive records of which begin much younger, in the Paleocene (Figure 8.43). In Cicadidae there is a pair of tymbals and tympana in the male, one on each side of abdominal tergite one. The tymbals are thin, chitinous, parallel plates recessed in a chamber that is covered by a flap grown out of tergite two. Sound is produced when the plates rapidly vibrate from rapid contractions of muscles attached to them; it is amplified by large tracheal air sacs that lie against the walls of the tymbal chamber. A tympanum, which hears the calls of other cicadas, lies beneath the tymbal and is covered by a flaplike outgrowth of the metathoracic sternite, called the operculum (Figure 8.41). Tettigarctidae primitively lack the air sacs and tympana, and both sexes have tymbals.

The other renowned aspect of cicadas are the eastern North American species that are "periodical," comprising three species of *Magicicada*. These have the longest time to maturity of any insect. When they emerge en masse, their abundance can be impressive: 1.5 million (or 1.5 tons) per acre in one study based in Illinois. Not all individuals of a species emerge in the same year, but their emergence is staggered into 13- and 17-year "broods." Each brood has a unique distribution and year of emergence. Because there is no reproduction among the broods, these insects have attracted evolutionary biologists interested in gene flow and speciation (Simon, 1988). The 13- and 17-year cohorts of species are barely separable morphologically; they readily hybridize in the lab and have identical calls, but are genetically distinct.

The nymphs of Cercopoidea, or spittle bugs, secrete a frothy or gelatinous mass in which they reside while feeding from the plants (Figure 8.44); they emerge from the spittle as adults. The mass is produced by excreta and glandular secretions from the abdomen, into which air may be bubbled from the spiracles. This is probably an adaptation against parasitoidism and desiccation. Two families in this group, the Aphrophoridae and Clastopteridae, are often classified in the large family Cercopidae. Species in the small African and Australasian family Macherotidae have the distinctive behavior of secreting a calcareous material that hardens and is used for the gradual construction of a tube. The nymph lives in the tube, immersed in its spittle. The earliest Cercopoidea are Procercopoidae (Early Jurassic to Late Cretaceous); the extinct family Cercopionidae is known from Early Cretaceous limestone of Brazil (110–115 MYO), and true Cercopidae/Clastopteridae are known only from the Tertiary (Figure 8.45).

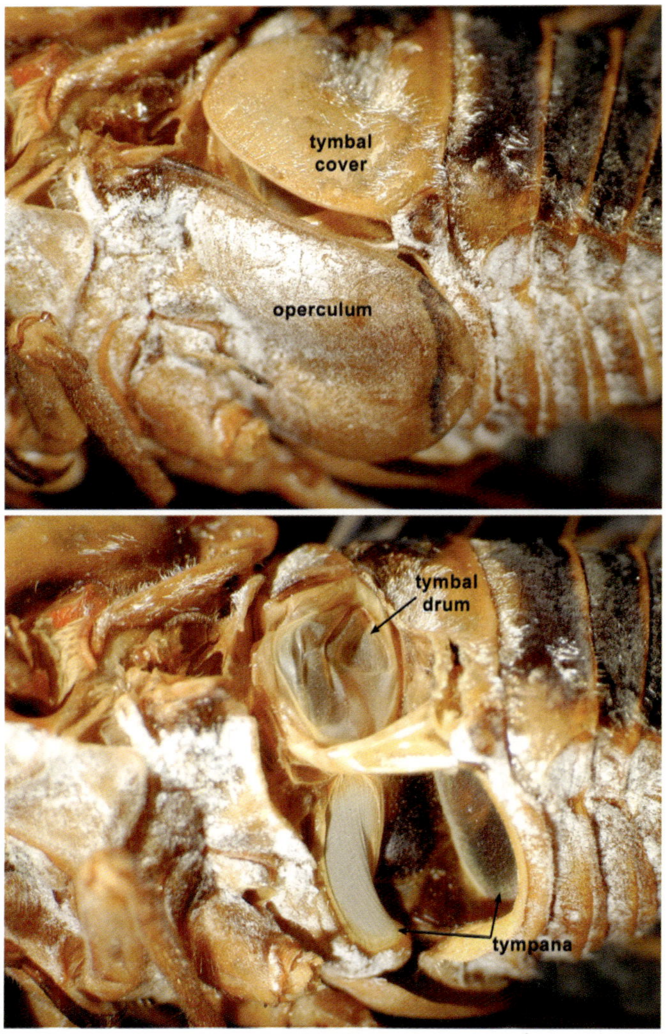

8.41. Acoustic system of a cicada, with the opercula intact (above) and removed showing the tympana (below). Beneath the tymbal drum are a series of parallel plates, the tymbals, which vibrate to produce the familiar sound of cicadas.

The Cicadelloidea are probably the best studied auchenorrhynchans, for their species taxonomy, behavior, and relationships. These comprise the largest family of auchenorrhynchans, the Cicadellidae (leaf hoppers), with some 10,000

8.42. *Fletcheriana triassica*, from the Late Triassic of Brookvale, Australia. The extinct (Permian-Jurassic) family Paleontinidae were large, hairy, stem-group cicadoids that may have fed on ginkgos. Some color patterns are still present on the wings. AM F.25234; total length 80 mm.

8.43. A Cicadidae, *Platypedia primigenia*, from the Late Eocene of Florissant, Colorado. NHM In19232; body length 24 mm.

8.45. Undetermined Cercopidae, from the Late Eocene of Florissant, Colorado. NHM; wing length 27 mm.

species, and five families in the membracid group (3,500 species). Cicadellidae are well known for their stripes and spots of vivid colors, which contrast with membracids, who mimic parts of the stems on which they feed. Relationships among families have been presented by Deitz and Dietrich (1993) and Deitrich and Dietz (1993). Hamilton (1983) proposed that three subfamilies of Cicadellidae (Jassinae, Ledrinae, and Ulopinae) are more closely related to Membracidae than to other, true cicadellids. Relationships within the fam-

8.46. Membracid nymphs tended by a vespid wasp for their honeydew. Photo: P. J. DeVries.

8.44. A newly emerged adult froghopper (Cercopidae), in its mass of spittle. Photo: S. Marshall, University of Guelph.

8.47. Membracidae in Miocene Dominican amber. Unlike cicadas, membracids appear to have evolved relatively recently, in the latter half of the Cenozoic. Morone Collection M0584. Photo: R. Larimer.

8.48. Cicadellidae, from the Lower Cretaceous of Brazil, ca. 120 MYA. This formation has yielded the earliest and most diverse Mesozoic fossils of this large extant family. AMNH SA45250; wing spread 13 mm.

8.50. *Carpopodus difficilis*, from the Early Cretaceous of Brazil, which belongs to the extinct fulgoroid family Lalacidae. AMNH 43604; body length 17 mm.

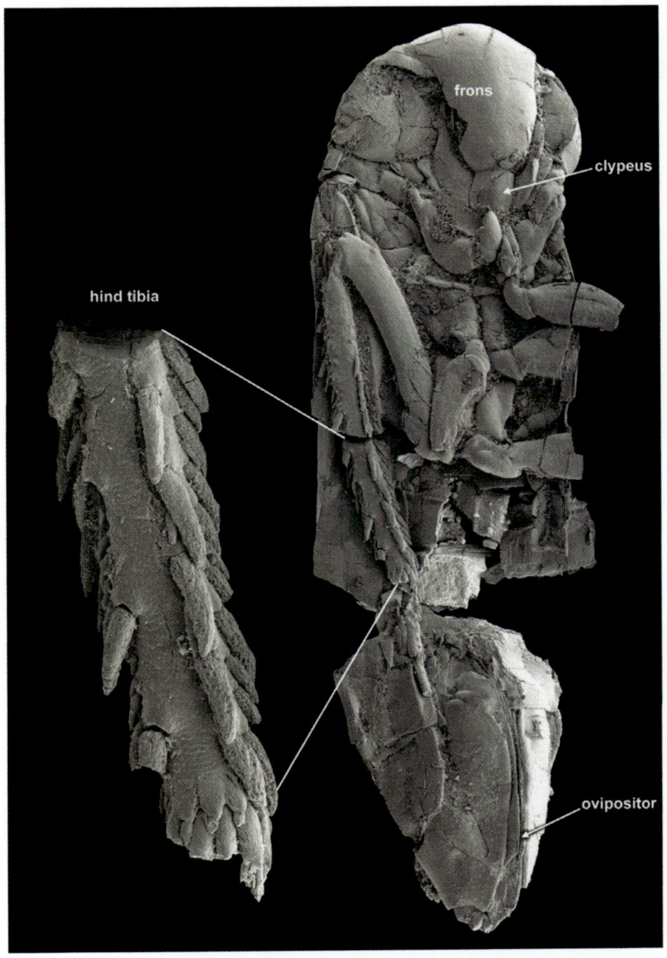

8.49. Scanning electron micrograph of a Cretaceous cicadellid extracted from Santana Formation limestone. AMNH; body length 2.6 mm.

8.51. Undetermined nymphal fulgoroid in 100 MYO Burmese amber, with an extremely long rostrum jutting out between the legs and abdomen. AMNH; length 4.2 mm.

8.52. A fulgoroid in Miocene amber from the Dominican Republic. AMNH DR15-4; wing length 5.13 mm.

ily Membracidae have been explored by Dietrich *et al.* (2001). Species of the family Membracidae are best known for their elaborate pronota, many of which have evolved into outrageously bizarre and diverse structures: keels, protuberances, thorns, and even hanging ornaments. Nymphs are usually gregarious and are often tended by ants and even wasps (Figure 8.46); some species have parental care. Membracidae appear to be a relatively young group, as all fossils are Tertiary, most of them in amber (Figure 8.47). The earliest Cicadelloidea are the extinct families Karajassidae and Jascopidae (mid-Jurassic to Early Cretaceous). The earliest Cicadellidae are diverse (three genera, ten species) from the Early Cretaceous of Brazil (110–115 MYO) (Hamilton, 1990) (Figures 8.48, 8.49).

The infraorder Fulgoromorpha contains just the superfamily Fulgoroidea, comprised of 20 families (reviewed in Carver *et al.*, 1991) and approximately 8,000 described species. Nymphs commonly produce long strands of fluffy wax from the abdomen, as in many Sternorrhyncha, and the heads of many are produced into a snout. Though the small, relict family Tettigometridae appears to be the most basal one among fulgoroids (Asche, 1988; Bourgoin *et al.*, 1997), the oldest fulgoroids are the Cixiidae. Cixiidae are known from the Jurassic, with a definitive one occuring in Lebanese amber. The closely related family Lalacidae was extremely diverse in the Early Cretaceous (Hamilton, 1990) (e.g., Figure 8.50), and yet other families are represented in Cretaceous ambers, particularly from Burma (Figure 8.51). Most other fulgoroids, though, have a much younger fossil record, in the Tertiary (Figure 8.52).

COLEORRHYNCHA

This is a small group of obscure, cryptic insects that have attracted considerable attention as the closest relatives of the large, diverse suborder Heteroptera. There is only one living family, the Peloridiidae, with 13 genera and 25 described species. J. W. Evans (1981) reviewed their taxonomy and biology; Popov and Shcherbakov (1996), their relationships and fossil record; and Schlee (1969a), their relationships to other Hemiptera.

Peloridiids live amongst wet moss and leaf litter in temperate *Nothofagus* forests of southern South America, Tasmania, eastern Australia, New Zealand, New Caledonia, and Lord Howe Island; they are a classic example of austral disjunction. Even though peloridiids are relicts they are hardly generalized. Specialized features include a very flattened, broad body with eyes widely separated and on short

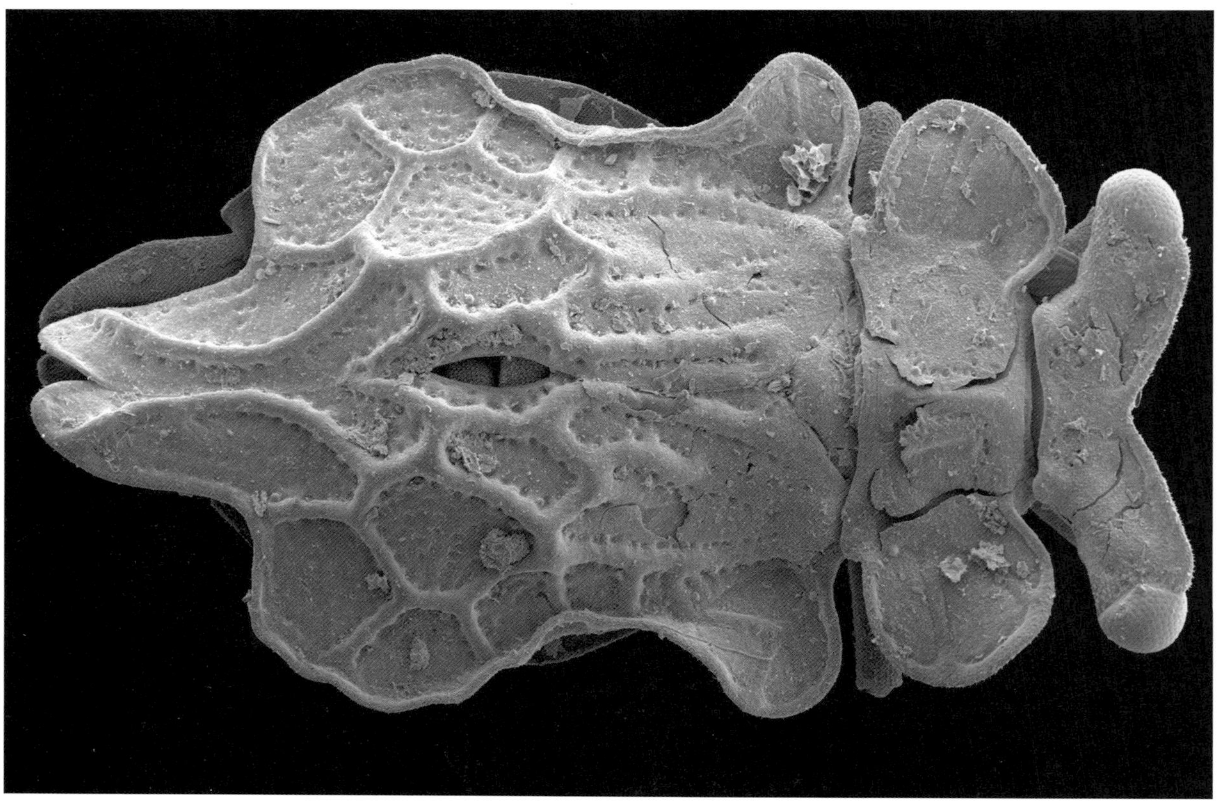

8.53. A living species of *Peloridora* (Coleorrhyncha: Peloridiidae) from Chile. Peloridiids inhabit cool, wet *Nothofagus* forests of the southern landmasses (except for Africa) and are considered a premier example of relicts from the Cretaceous when Gondwana drifted apart. The flaking is a dried layer of slimy "skin" typical of these insects. Scanning electron micrograph; body length 3.7 mm.

peduncles; forewings with a netlike arrangement of veins forming many cells ("areolae"); and a pronotum expanded into paranotal lobes (Figure 8.53). The tegminous forewings are translucent, the rest of the body is brownish-green and is covered with a slimy layer, which, with their slow movements, make the insects very cryptic amidst wet moss.

Schlee (1969a) outlined evidence for the close relationship of Peloridiidae to Heteroptera, but this was disputed by Cobben (1978) and later by Popov and Shcherbakov (1996). The critiques of the latter authors are not very convincing, and in fact both of the only DNA sequencing studies that included peloridiids found them to be the sister group to the Heteroptera (Wheeler *et al.*, 1993b; Campbell *et al.*, 1995). These odd bugs share the following features with Heteroptera: clavate (though short, three-segmented) antennae; a flattened abdomen with spiracles on the ventral surface; overlapping forewings lying flat against the abdomen; very similar wing-coupling devices (e.g., D'Urso, 1993); a gula, or ventral sclerite on the head, which is responsible for a prognathous condition; thoracic scent glands in adults; and various structures of the terminalia. There is, thus, little question that the peloridiids are a relict, living sister group to the very diverse suborder Heteroptera, which is why they have been accorded their own suborder among hemipterans, the Coleorrhyncha (so called because of the tegminous forewings). Schlee (1969a) proposed the group name Heteropteroidea to include both of these suborders.

Despite the phylogenetic and biogeographic interest in Peloridiidae, the only attempt at examining relationships has been by Popov and Shcherbakov (1996). They used 20, mostly multistate characters, apparently derived from the literature, and the methods of analysis were ambiguous. They proposed three lineages: one in New Zealand and New Caledonia (A), another in Australia and Tasmania (B), and a South American one (C), with B + C being closely related. These relationships and distributions led the authors to conclude that there was a Cretaceous origin of the family, and that the major lineages of the family diverged when the landmasses on which they lived drifted apart. A detailed, comprehensive study of the morphology and phylogenetics of all species of peloridiids would be exceedingly interesting.

Fossils of the group are sparse but very interesting and support a Cretaceous origin for the Recent family. The only definitive Coleorrhyncha from the Mesozoic are the extinct genera and families *Karabasia* (Karabasiidae) and *Hoploridium* (Haploridiidae), from the latest Jurassic and Early Cretaceous of Eurasia. These have the broad, flat body of Peloridiidae, with overlapping wings, paranotal lobes, and widely separated eyes (Figure 8.54). However, the wing venation is not as modified as in peloridiids, so these are stem-group Coleorrhyncha. It would not be surprising if true peloridiids

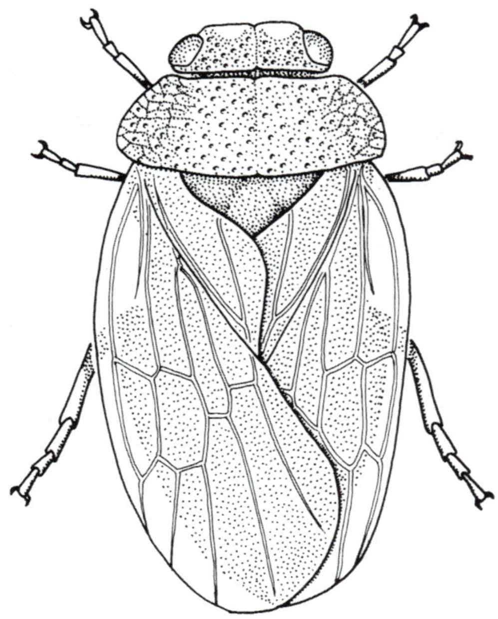

8.54. Reconstruction of *Karabasia evansi*, an extinct hemipteran related to Peloridiidae, from the Late Jurassic of Kazachstan, ca. 155 MYO. Body length 3.5 mm.

8.55. The extinct family Progonocimicidae is an early relative of peloridiids. Here is a species from the Early Cretaceous of Brazil, AMNH SA45253; body length 7.5 mm.

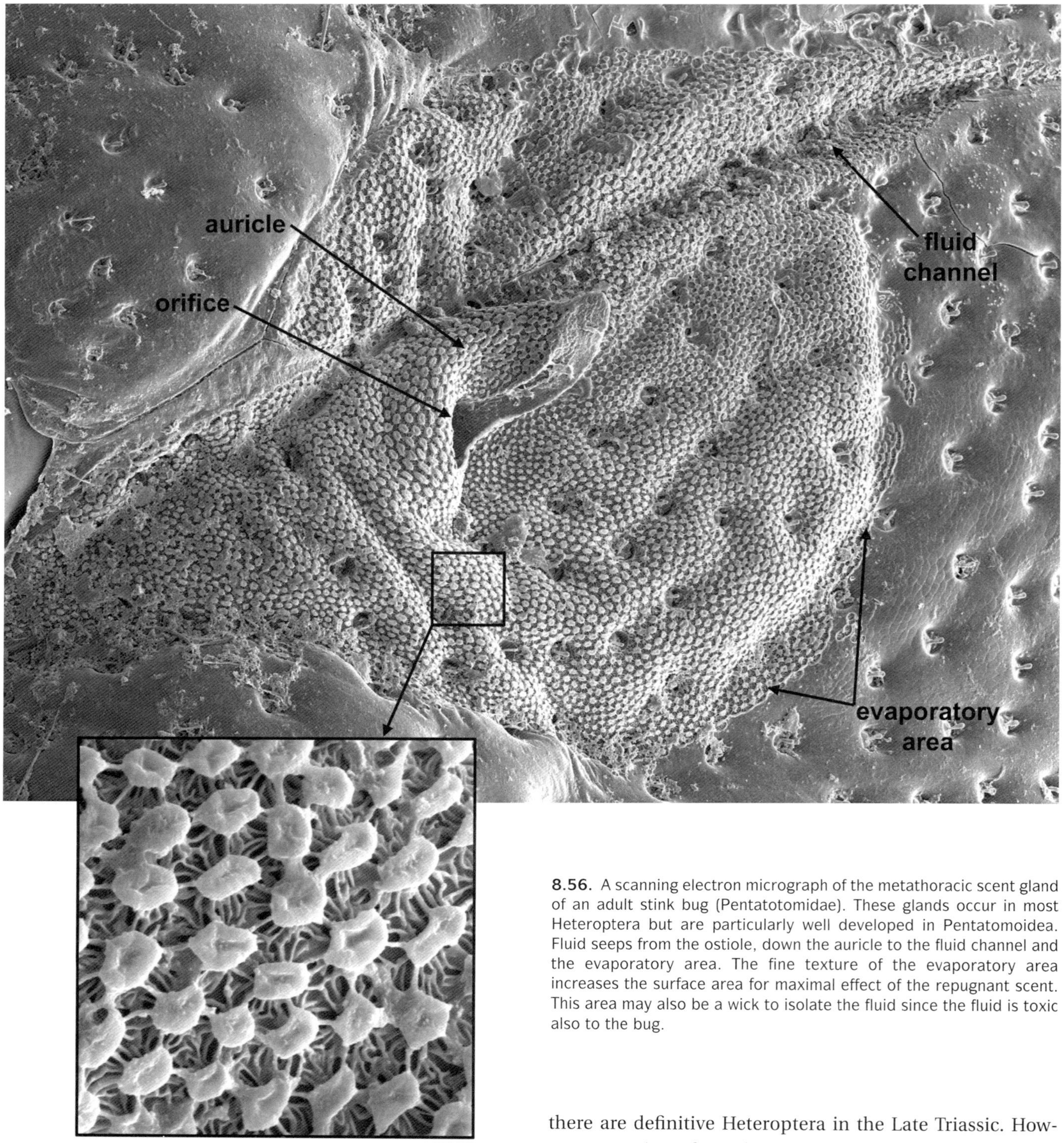

8.56. A scanning electron micrograph of the metathoracic scent gland of an adult stink bug (Pentatotomidae). These glands occur in most Heteroptera but are particularly well developed in Pentatomoidea. Fluid seeps from the ostiole, down the auricle to the fluid channel and the evaporatory area. The fine texture of the evaporatory area increases the surface area for maximal effect of the repugnant scent. This area may also be a wick to isolate the fluid since the fluid is toxic also to the bug.

eventually surface in Eocene Baltic amber, which has preserved an array of presently austral taxa. Popov and Shcherbakov (1996) further link with the extinct and Recent Coleorrhyncha the Progonocimicidae (Figure 8.55), a family from the Late Permian to the Early Cretaceous. Stem-group coleorrhynchans must extend to the Early Triassic, when this group presumably diverged from the Heteroptera, because there are definitive Heteroptera in the Late Triassic. However, venation (the only real source of characters) of Progonocimicidae is only slightly differentiated from basal Auchenorrhyncha, so these are clearly stem-group coleorrhynchans.

HETEROPTERA: THE "TRUE BUGS"

The Heteroptera is one of those rare groups that have attracted considerable attention and effort by serious entomologists. This diverse suborder is defined by scent glands

8.57. Enicocephalidae in Early Cretaceous amber from Lebanon, 125 MYO, are the earliest fossils of the group. As the basalmost heteropterans, enicocephalids would be expected to occur as early as the Triassic. AMNH; body length 2.3 mm (AMNH LAE3-11).

arrayed on tergites three through seven in nymphs and by a reduced tentorium, but they are most recognizable for the hemelytrous forewings, which have a *corium* on the anterior half and a membranous distal half. This unique wing structure is a feature only of the more recently evolved groups of Heteroptera (Figure 8.5). Thoracic scent glands serve a defensive function and are barely developed in aquatic and semiaquatic bugs, but potent in plant feeding forms (Figure 8.56). The earliest heteropteran is generally acknowledged to be *Paraknightia magnifica*, from the Late Permian of Australia. It has large paranotal lobes, a long ovipositor, and a forewing with a distinct clavus but a venation that is more reduced than occurs in the basal living heteropterans. Unfortunately, a head is not preserved. Shcherbakov and Popov (2002), who consider the scytinopteroids as an extinct sister group to the Heteroptera, placed this fossil in this group. Excellent overviews of the biology and taxonomy of the suborder include Poisson (1951), Miller (1956), Carver *et al.* (1991), and Schuh and Slater (1995).

The Heteroptera is generally divided into seven monophyletic infraorders (which have a *–morpha* ending), relationships of which have been discussed by Wheeler *et al.* (1993a). Most of the morphological characters in that study derive from the superb morphological work of René Cobben (1968, 1978), and much more comprehensive molecular studies are currently under way. Considerably different from these hypotheses of relationships is that of Shcherbakov and Popov (in Rasnitsyn and Quicke, 2002), who interpret heteropteran phylogeny largely from fossils. They, for example, placed Enicocephalidae among dipsocoromorphs, instead of at the base of the Heteroptera, and have Gerromorpha derived from the Leptopodomorpha. Given the stability and support for the cladogram by Schuh, though, that system is adopted here.

The basalmost infraorder, **Enicocephalomorpha** has only the family Enicocephalidae, though this family has been split into the Aenictopecheidae and Enicocephalidae s.s. by Štys (e.g., 1995, in Schuh and Slater, 1995). This division is really unnecessary, as both families obviously form a monophyletic group, defined in part by the short, chelate foretarsi armed with small spines, and the head being divided into anterior and posterior halves by a constriction (Figure 8.58). Aenictopecheinae retains more primitive features than does the nominal subfamily, like the head being only slightly constricted, the wing coupling device being remarkably similar to that of Peloridiidae, and various features of the genitalia.

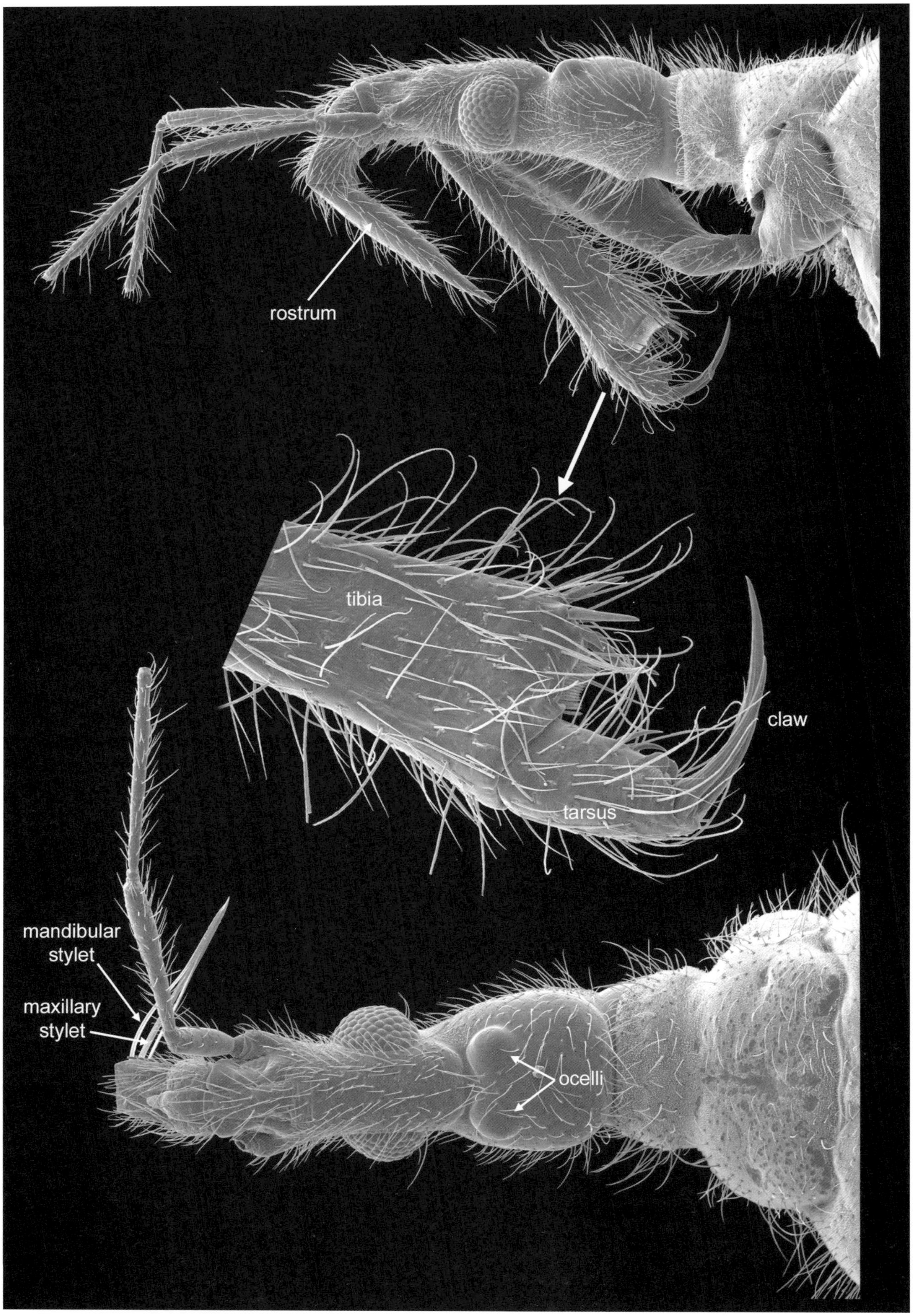

8.58. An enicocephalid, *Systelloderes* sp. from Costa Rica, showing the distinctive head in lateral and dorsal view, and the grasping forelegs used to capture small prey. Enicocephalidae are the living sister group to the rest of the Heteroptera. Scanning electron micrograph.

Indeed, as diagnosed by Štys, Aenictopechinae is probably paraphyletic with respect to Enicocephalinae. Approximately 450 species occur in the family, all of which are predatory and have cryptic lifestyles of living under stones, logs, and bark or in leaf litter. Most are dull brown, but some *Enicocephalus* and Megenicophalini are bright red. Occasionally they are encountered as mating swarms (with the sexes often mixed, not just males), a habit from which the common name "gnat bugs" derives. A major work on the group is by Wygodzinsky and Schmidt (1991) on the New World fauna. As a living sister group to the rest of the Heteroptera, enicocephalid-like fossils would be expected from the Jurassic and Triassic, but the earliest ones known are in amber from the Early Cretaceous of Lebanon (Grimaldi *et al.*, 1993; Azar *et al.*, 1998) (Figure 8.57) and mid-Cretaceous of Myanmar (Cockerell, 1917; Štys, 1969; Grimaldi *et al.*, 2002). If they were preserved only as wings in sediments, they should still be recognizable for the distinctive, albeit primitive, venation.

Dipsocoromorpha (Figure 8.59) is the smallest infraorder, with approximately 200 described species, and this group also has the smallest adult bugs, most merely 0.5–4.0 mm in length. There are five obscure families, the Schizopteridae being the only one with over 100 species. These are ground- and litter-dwelling bugs, which are usually flat and broad and have an antenna that is long and whiplike, often with long hairs. Like Enicocephalidae, the venation is relatively complete and, in fact, not very divergent from Peloridiidae. Shcherbakov and Popov (2002) mentioned that the Early Jurassic Cuneocoridae were "probably" Dipsocoroidea, though limited preservation makes this difficult to assess. Dipsocoromorphs would be expected even in the Triassic, and in fact a possible fossil of the group is known from Late Triassic shales of Virginia (Figure 8.60). Otherwise, the oldest definitive fossils of these bugs occur in Early Cretaceous amber from Lebanon and in mid-Cretaceous amber from Myanmar (Figure 8.61).

The **Gerromorpha** contain what are probably the most recognizable heteropterans, the water striders or pond skaters (family Gerridae), though there are seven other families, most of them small (less than 4 mm), obscure insects. All 1,860 species are predaceous and associated with watery environments, many of them walking or skating over the surface of water. These habits have allowed at least 14 colonizations by gerromorphs (Andersen, 1998) of marine habitats – a biome that insects have rarely invaded. Few groups of Heteroptera are as well understood as the gerromorphs, largely as a result of the work of Nils Møller Andersen, who has synthesized their behavior, morphology, biogeography, systematics (e.g., Figure 8.62), and fossil record (e.g., Andersen, 1982, 1998; and many other references cited therein). The group is clearly monophyletic, with numerous distinctive features, among which are the following: adults with three pairs of head trichobothria inserted into deep pits; mandibular levers; pretarsi with dorsal and ventral arolia, the ventral one often highly modified with hydrofuge hairs; and a female genital tract that has an intricate "gynatrial complex." They also have a fine, hydrophobic pile that covers the head, thorax, and some of the abdomen and that also occurs in some Nepomorpha and many Leptopodomorpha.

Three families of Gerromorpha – the Macroveliidae, Paraphrynoveliidae, and Hermatobatidae – have only between two and eight species each, the last family (called "coral treaders") remarkable for lodging themselves in pockets of air amongst coral at high tides and feeding during low tides. Mesoveliidae, Hebridae, and Hydrometridae have between 20 and 160 species; the hydrometrids, commonly called "water measurers" or "water treaders," are best known for their slow, deliberate strides over the water surface using long, stiltlike legs. The largest families are the closely related Veliidae (840 spp.) and Gerridae (670 spp.). Veliidae, or "riffle bugs," are generally smaller than gerrids, most of them living at the water's edge and in riffle areas, though some are intertidal or terrestrial, and others even live in phytotelmata such as the water in tree holes and in bromeliads. Veliids have the most diverse pretarsal structure in Gerromorpha, the most distinctive feature being the ventral arolium of the mid tarsus, which is developed into a "swimming fan" and most developed in *Rhagovelia* (Figure 8.64). The Gerridae are built for an entirely aquatic life, with long, thin mid and hind legs (which are the ones used for locomotion) attached to large coxae; the coxae are actually connected to the sides of the thorax. They can propel themselves against very swift currents, but they can't walk on land. Water striders of the subfamily Eotrechinae, particularly the Asian genera *Eotrechus* and *Onychotrechus*, reverted to a life on wet soils and seepages, and they jump. Prey can be detected via surface ripples, which is also how individuals purportedly communicate. Among the more remarkable genera are *Rheumatobates*, some males of which have prehensile antennae, and both sexes have prehensile hind legs. *Halobates* is the only pelagic insect.

Given the habits of gerromorphans, fossils of these insects would be expected to be abundant in lacustrine sediments the way nepomorphans are, but they aren't. In some deposits they are fairly diverse, such as in the Paleo-Eocene Mo-Clay of Denmark (55 MYO), which has fossilized three species each of Gerridae and Hydrometridae, and a species of Macroveliidae (Andersen, 1998), but this deposit is exceptional in this regard. Gerromorphans are particularly scarce in Mesozoic deposits, from which only four, possibly five, species are known in total. Two of these Mesozoic fossils belong to the apparently basal family of the group, Mesoveliidae: *Duncanovelia extensa* (Koonwarra, Australia, ca. 120 MYO) (Figure 8.65), and *Karanabis kiritshenkoi* (Karatau, Kazachstan, ca.

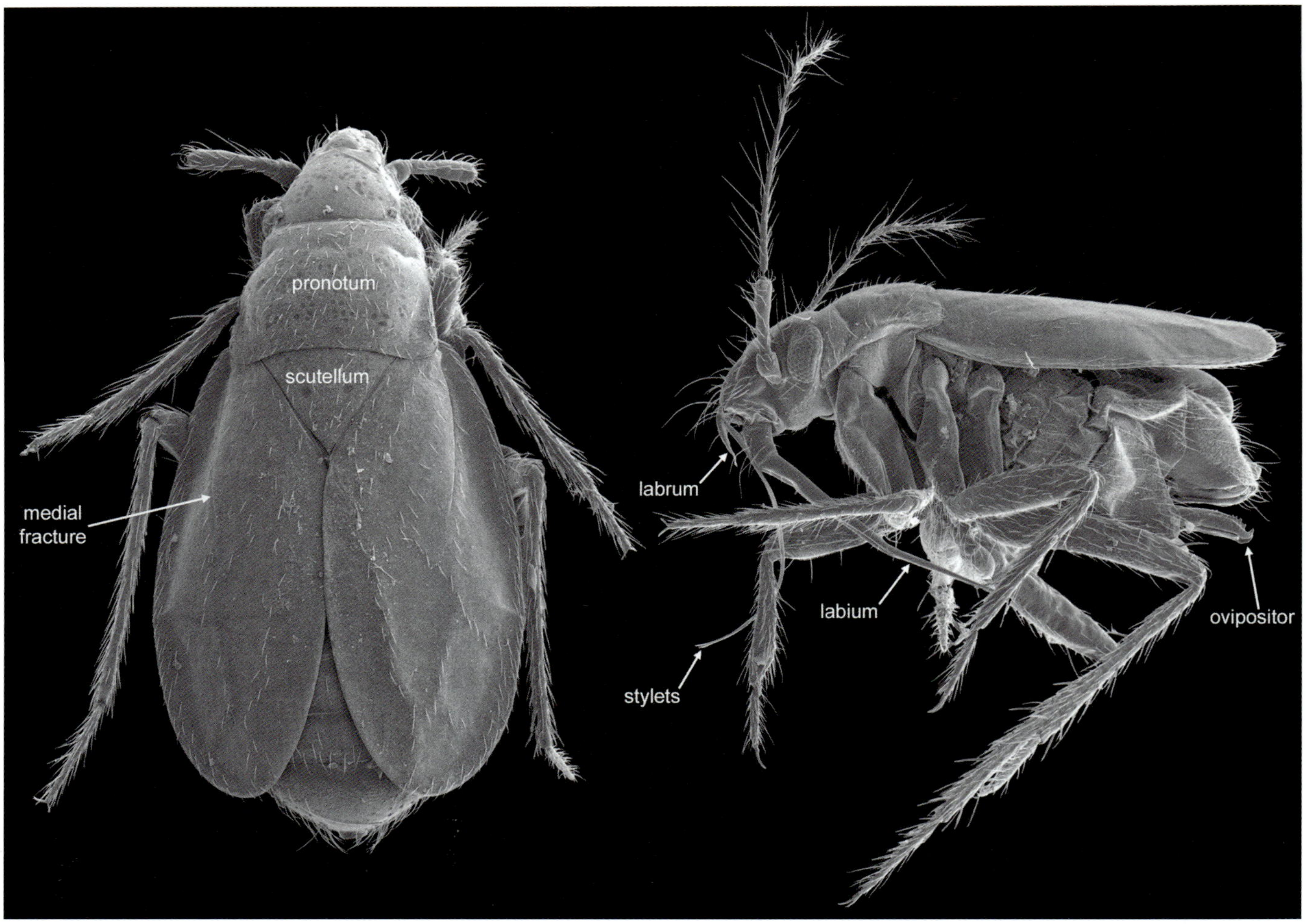

8.59. *Ceratocombus* (Ceratocombidae) from Australia, belonging to another basal group of the Heteroptera, the Dipsocoromorpha. In basal heteropterans, the forewings are not hemelytrous, or differentiated into areas of corium and membrane. The stylets have separated from the labial trough in this specimen. Scanning electron micrograph; body length 1.50 mm.

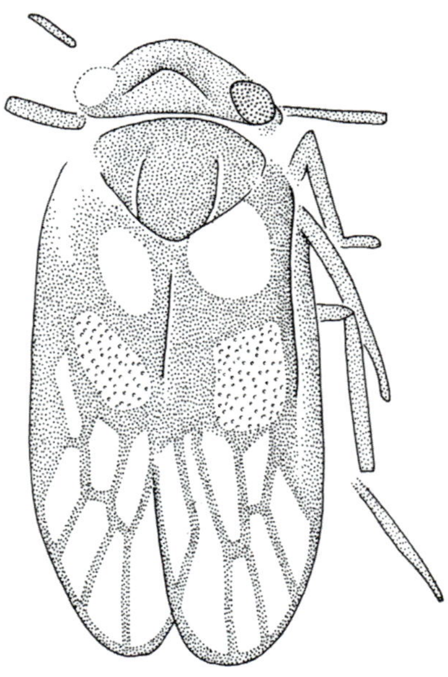

8.60. A fossil of a possible Dipsocoromorpha, from the Late Triassic (ca. 220 MYO) of Virginia, perhaps the earliest known fossil of the group. VMNH 736; body length 4 mm.

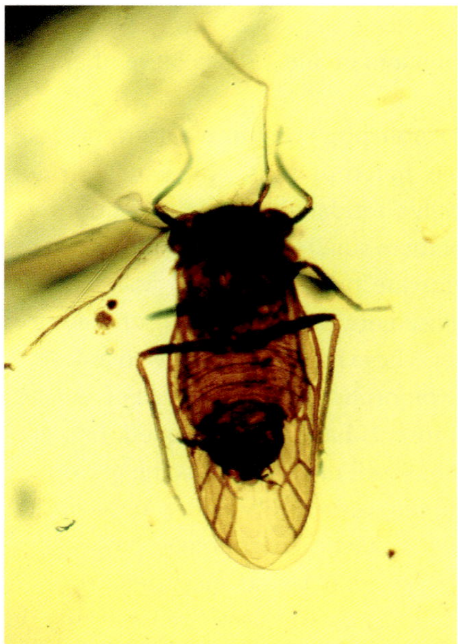

8.61. Schizopteridae (Heteroptera), in mid-Cretaceous Burmese amber. AMNH Bu720; body length 1.33 mm.

152 MYO). The latter fossil was not mentioned by Andersen (1998), but it needs to be reexamined to confirm its identity. Two others are Hydrometridae: *Cretaceometra brasiliensis* (Santana Formation, Brazil, ca. 120 MYO) and *Carinametra burmensis* (Burmese amber, ca. 100 MYO [Andersen and Grimaldi, 2001]) (Figure 8.66). *Carinametra* is a basal member of the family, but *Cretaceometra* is a hydrometrine, suggesting a Late Jurassic age for this particular family. A putative veliid also occurs in the Koonwarra deposit, though this needs restudy. If Gerromorpha is indeed basal to Nepomorpha, then we should expect fossils of them in the Triassic. Suggestions are that the Recent families Mesoveliidae and Hydrometridae are Late Jurassic in age, which is consistent with an older origin of the infraorder, but the lack of earlier Mesozoic fossils is an enigma. The existence of Paleocene and Eocene gerrines (e.g., Figure 8.67) indicates an origin of the Gerridae (and probably the Veliidae) in the Cretaceous, but probably no earlier because these are the most recently evolved families of the infraorder. *Halobates ruffoi*, from the mid-Eocene of Verona, indicates that the minimum age for insect colonization of the sea is at least 45 MYO.

The **Nepomorpha** contains the only truly aquatic bugs – the only insect group besides some beetles that spend their entire lives under water, including the adult stage. Some nepomorphs are also the top predators among Heteroptera, in particular the Belostomatidae and Nepidae, which can reach 10 cm (4 in.) in length. They are visual predators, tracking prey with large, bulging eyes. The larger species of belostomatids even feed on small fish and amphibians. An intriguing behavior of belostomatids is that males of some species guard the eggs or carry them on their back (Figure 8.68). Other distinctive features are tiny antennae concealed beneath the eyes; a short, stout labium; the mid and hind legs fringed with stiff hairs for swimming; and short, grasping forelegs. Relationships of the 11 families (containing approximately 2,000 species) have been studied by Rieger (1976) and Mahner (1993). Although these authors disagree on the placement of four families, some groupings appear certain. The broad, flat Belostomatidae ("water bugs") and gracile Nepidae ("water scorpions") are clearly closely related (comprising about 380 species), as both have abdominal tergite eight with a pair of respiratory appendages. These structures siphon air from the surface into an airspace beneath the wings, which is then "inhaled" through abdominal spiracles. In belostomatids these appendages are short *airstraps*. Nepidae have long breathing tubes, the tips of which protrude just above the surface of the water while the insect is suspended on submerged vegetation, waiting in ambush. It is quite likely that nepids are merely highly modified belostomatids.

The two small riparian (shore-dwelling) families, the Ochteridae (55 species) and Gelastocoridae (100 species), are also closely related. Gelastocoridae ("toad bugs") are flat, squat, and warty; Ochteridae are the most morphologically generalized nepomorphs. Another group of closely related families comprises the Notonectidae ("backswimmers": 350 species) and two small families, of which the Pleidae (40 species) are also known to swim ventral side up. Relationships of four families are uncertain, including the Naucoridae (400 species), which resemble small belostomatids, and Corixidae (560 species). The Corixidae ("water boatmen") are distinctive for their short forelegs, the tarsus of which is scoop-shaped with a brush of hairs that they use for gleaning the algae on which they feed. Although corixids will also feed on small, soft-bodied animals like midge larvae, they are the only group of Heteroptera outside the Cimicomorpha and Pentatomorpha that is not exclusively predatory.

In lakes from the mid-Jurassic through the mid-Cretaceous, 180–100 MYA, nepomorphs were abundant and diverse worldwide (Figure 8.62), and they have left a fossil record that is better than any other group of Heteroptera. That record has been reviewed by Popov (1971) and Carpenter (1992) and in Rasnitsyn and Quicke (2002). In fact, the oldest definitive Heteroptera are nepomorphs from the Late Triassic, including Belostomatidae (Figure 8.69), Naucoridae (and extinct relatives, Triassocoridae), Notonectidae, and Ochteroidea. These bugs are very abundant in some Cretaceous *Plattenkalke* deposits (Figure 8.70).

The **Leptopodomorpha** is a small infraorder of "shore" bugs, which have attracted inordinate attention relative to their modest diversity of 300 species. Various species are intertidal; others live near streams and lakes. There are four families, three of which have 13, 5, and 1 species each. The family Saldidae contains 270 species; their name derives from their ability to jump. Jurassic Archegocimicidae (e.g., *Saldonia*) and Cretaceous Enicocorinae are putative leptopodomorphans (Shcherbakov and Popov, 2002). Otherwise, definitive fossils appear late, in Miocene amber from the Dominican Republic (Figure 8.71) and Mexico. These amber fossils are of the family Leptopodidae, which has only two rare Recent species from northwestern South America and 11 Old World species.

The **Cimicomorpha** and Pentatomorpha comprise 90% of all Heteroptera species, probably owing to the evolution of plant feeding. Close relationship of the two groups is based largely on the loss of the arolia on the pretarsi, even in first instar nymphs. These groups correspond to the old concept of "Geocorisae," or land bugs. Parental care has evolved repeatedly among these bugs, particularly in pentatomoids (stink bugs and allies) (Figure 8.78), and in Tingidae. Such care may simply involve brooding of the eggs by the female to covering young nymphs beneath the body and actively

TABLE 8.5. Significant Records of Mesozoic and Several Early Tertiary Heteroptera[a]

1. *Enicocephalinus acragrimaldii* and undescribed enicocephalids: Lebanese amber
2. *Disphaerocephalus constrictus* Cockerell; *Paenicotechys fossilis* (Cock.): Burmese amber
3. Possible dipsocoromorphan: Virginia (USA)
4. Undescribed Dipsocoridae: Lebanese amber
5. Undescribed Dipsocoridae: Burmese amber
6. *Karanabis kiritshenkoi*: Karatau (Kazachstan)
7. *Duncanovelia extensa*: Koowarra (Australia)
8. *Daniavelia morsensis*: Mo-Clay (Denmark)
9. *Cretaceometra brasiliensis*: Santana Formation (Brazil)
10. *Carinametra burmensis*: Burmese amber
11. Possible Veliidae: Koonwarra
12. *Paleogerris* spp.: Mo-Clay (Denmark)
13. Undescribed Belostomatidae: Virginia (USA)
14. *Lethonectes, Tarsabedus*: Dorset (England)
15. *Nettelstedtia*: Nettelstedt (Germany)
16. *Mesonepa, Scarabeides, Stygeonepa*: Solnhofen (Germany)
17. Undescribed Belostomatidae: Purbeck, Dorset (England)
18. *Lethopterus*: Bon-Tsagan (Mongolia)
19. *Iberonepa*: Las Hoyas (Spain)
20. Undescribed Belostomatidae: Santana Formation (Brazil)
21. *Propreocoris*: Dorset (England)
22. *Lufengnacta*: Yipinglang Formation (China)
23. *Liassocorixa*: Dorset (England)
24. *Haenbea, Bakharia*: Mongolia
25. *Archaeocorixa, Gazimuria, Karataviella, Shurabella*: Karatau (Kazachstan)
26. *Baissocorixa, Diapherinus, Mesosigara*: Baissa (Russia)
27. *Bumbacorixa, Corixonecta, Corixalia, Cristocorixa*: Mongolia
28. *Rhomboidella*: Argentina
29. *Mesosigara*: Russia
30. Undescribed Naucoridae: Virginia (USA)
31. Undescribed Naucoridae: New Mexico (USA)
32. *Aidium, Angaronecta, Heleonaucoris, Liadonaucoris, Nectodes, Nectonaucoris, Scaphocoris*: Karatau (Kazachstan)
33. *Aphlebocoris, Apopnus, Scaphodemopsis*: Solnhofen (Germany)
34. *Cantaeronecta*: La Cantera Formation (Argentina)
35. Undescribed Naucoridae: Santana Form. (Brazil)
36. *Nepidium*: England
37. *Asionecta*: Karatau (Kazachstan)
38. *Notonectites, Pelonecta*: Solnhofen (Germany)
39. *Clypostemma*: Baissa (Russia)
40. *Britannicola*: Dorset (England)
41. *Saldonia*: Karatau (Kazachstan)
42. Various Archegocimicidae: Early Cretaceous of Siberia, China, Mongolia
43. *Pterocimex*: Dorset (England)
44. *Lygaenocoris, Mesopentacoris, Miridoides, Scutellifer*: Karatau (Kazachstan)
45. *Cymocoris, Coriopachys, Monocoris* (Miridae), *Golmonia, Sinaldocader* (Tingidae), "*Aradus*" (Aradidae): Bon-Tsagan (Mongolia)
46. Undescribed Cimicomorpha: Burmese amber
47. *Vianagramma* (Tingidae): New Jersey amber
48. Undescribed Cydnidae: Dobbertin, Mecklenberg (Germany)
49. Undescribed Cydnidae: Kirghistan
50. Undescribed Cydnidae: Karatau (Kazachstan)
51. Six "genera" by Handlirsch: Solnhofen (Germany)
52. *Clavicoris, Cretacoris*: Mongolia
53. *Latiscutella, Pricecoris*, undescribed spp.: Santana Formation (Brazil)
54. Aradidae, Piesmatidae, Lygaeoidea: Burmese amber

[a] Numbers correspond to those on phylogeny, Figure 8.62.

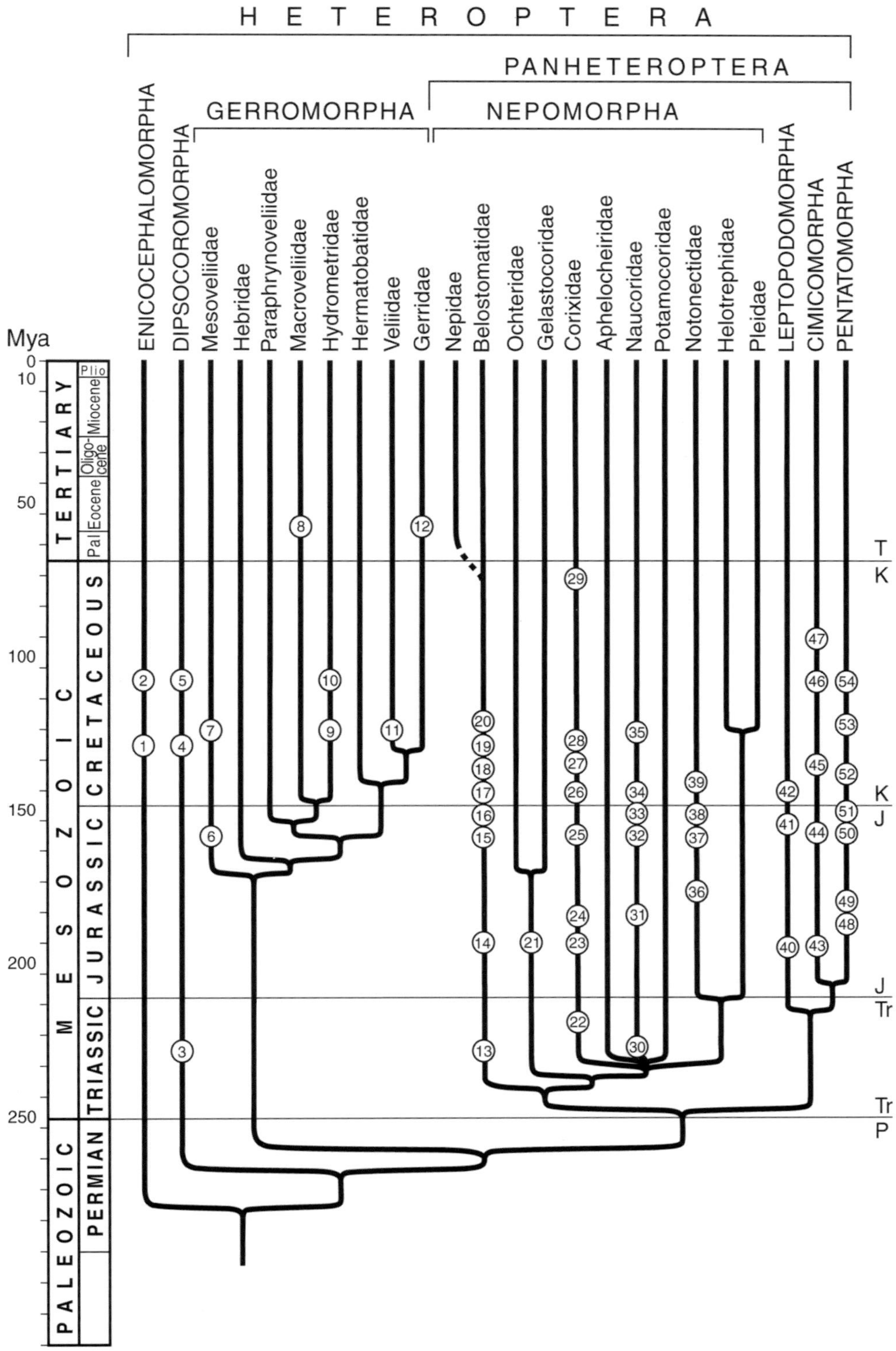

8.62. Phylogeny of the Heteroptera, showing significant Mesozoic fossil records and a few Cenozoic records. Families are shown for only the aquatic bugs (Nepomorpha), and the semiaquatic bugs, Gerromorpha (water striders and relatives). Nepomorphs have an excellent fossil record but the Mesozoic record for the gerromorphs is surprisingly sparse. See Table 8.5 for a list of the fossils. Relationships based on Wheeler *et al.* (1993) for infraorders, Andersen (1998) for Gerromorpha, and Rieger (1976) and Mahner (1993) for Nepomorpha.

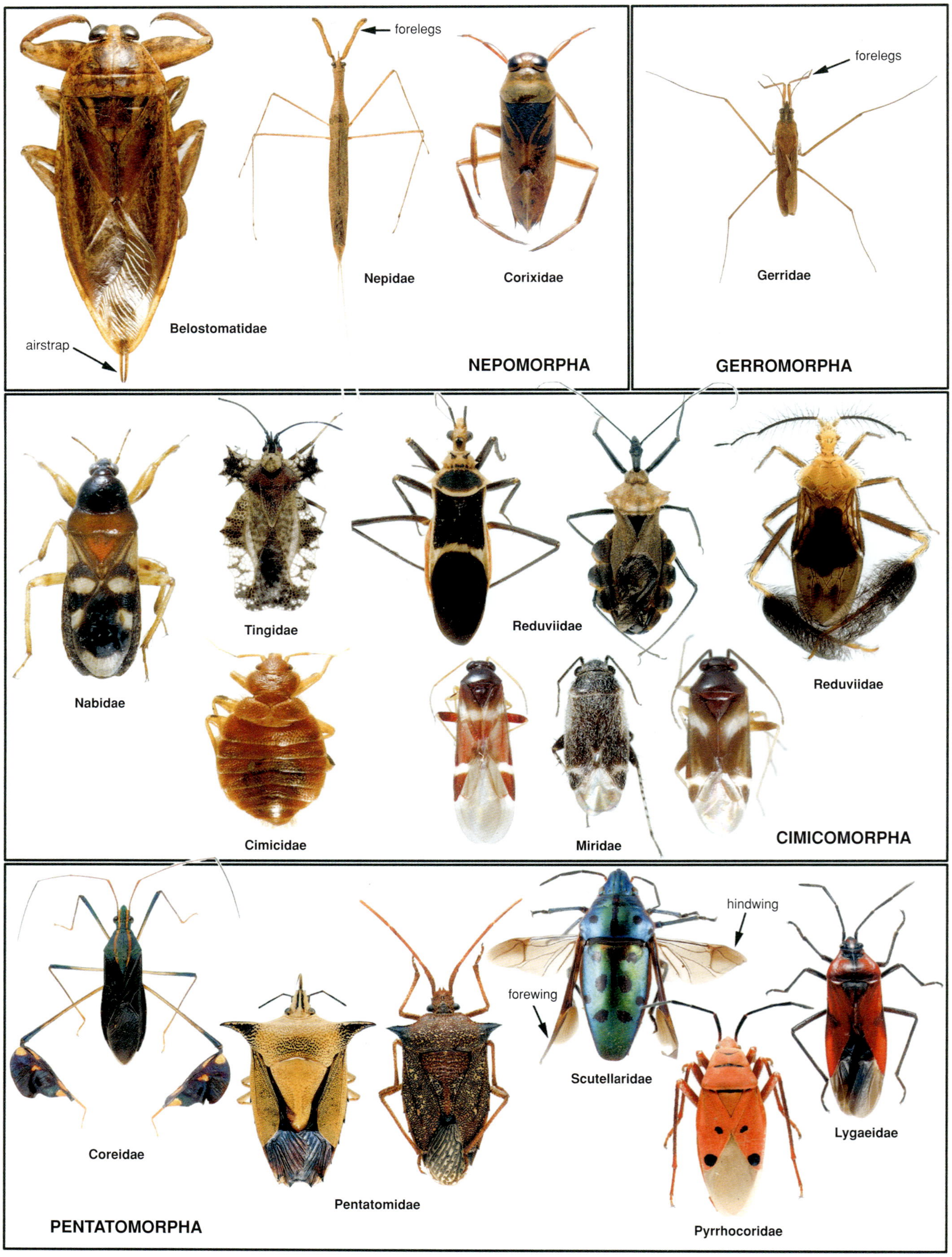

8.63. Representative living Heteroptera. Not to the same scale.

8.64. Pretarsus of *Rhagovelia*, showing the unfolded, intricate swimming fan that allows the bug to skate along the water's surface. Scanning electron micrograph.

8.65. *Duncanovelia extensa* (Mesoveliidae), from the Early Cretaceous of Victoria, Australia. VM P27044; body length 5 mm.

defending them against predators and parasitoids. Interestingly, extended parental care in Heteroptera and in some other insects, like cassidine chrysomelids, appears to be related to rooflike extensions of the wings, under which nymphs take cover, like a mother hen with chicks. Another interesting aspect of these bugs is the repeated evolution of ant mimics, or *myrmecomorphy*. Myrmecomorphy occurs within seven families and has evolved probably no less than 20 times. Resemblance to ants in some of the mimics seems merely superficial, but their quick, furtive movements and palpating antennae make them astonishingly antlike. Some of the best ant mimics are the heterogastrine lygaeids, various mirids and nymphal alydids, whose bodies are very antlike in form, even when dead.

The Cimicomorpha comprises 16 families, two of which are the largest families of Heteroptera, the Miridae and Reduviidae (e.g., Figure 8.72). Cimicomorphs don't store sperm in the spermatheca, which is vestigial; this situation is a precursor to the bizarre hemocoelic insemination that evolved later in this group among the bed bugs and their close relatives. An intriguing aspect of cimicomorphan natural history involves various species that are inquilines in spider webs. Though spider webs are designed to snare insects, some emesine reduviids, nabids (e.g., *Arachnocoris*), Miridae (e.g., *Ranzovius*), and all Plokiophilidae easily walk on the webs, where they feed on the remains of prey, a behavior that has even been captured in Cretaceous amber (Figure 8.73). This habit probably evolved as a result of the gracile body and slender legs of most cimicomorphs, a generalist predatory diet, and some peculiar structure of the pretarsus. Relationships among the families of Cimicomorpha have been discussed by Schuh and Štys (1991). The Reduviidae appear to be a basal family in the infraorder, with two distinctive superfamilies, the Miroidea and Cimicoidea.

Some of the large Reduviidae, like Harpactorinae, are the terrestrial equivalent of belostomatids, being sturdy predators with powerful "bites." Reduviids (ca. 6,500 species) are extremely diverse, ranging from delicate, stilt-legged emesines (Wygodzinsky, 1966), to tanklike harpactorines. Most have a distinctive neck, a curved labium, and a filelike stridulatory structure on the prosternum against which the tip of the labium is rubbed. Phymatinae are generally cryptic and have short, stout raptorial forelegs, which they use when ambushing prey, like small mantises. Apiomerini apply resin to the foretarsi to better grasp prey, a behavior that has even been preserved in Miocene amber. All reduviids are predatory, with exception of the blood-feeding triatomines (111 species; Lent and Wygodzinsky, 1979). Triatomines live in nests and lairs, and like bed bugs emerge at night to feed from their hosts. *Rhodnius*, a blood feeder, is commonly used in physiological research. Some species are vectors of Chagas's disease in the Western Hemisphere, which results when the bug's bite is contaminated with the flagellate protozoan *Try-*

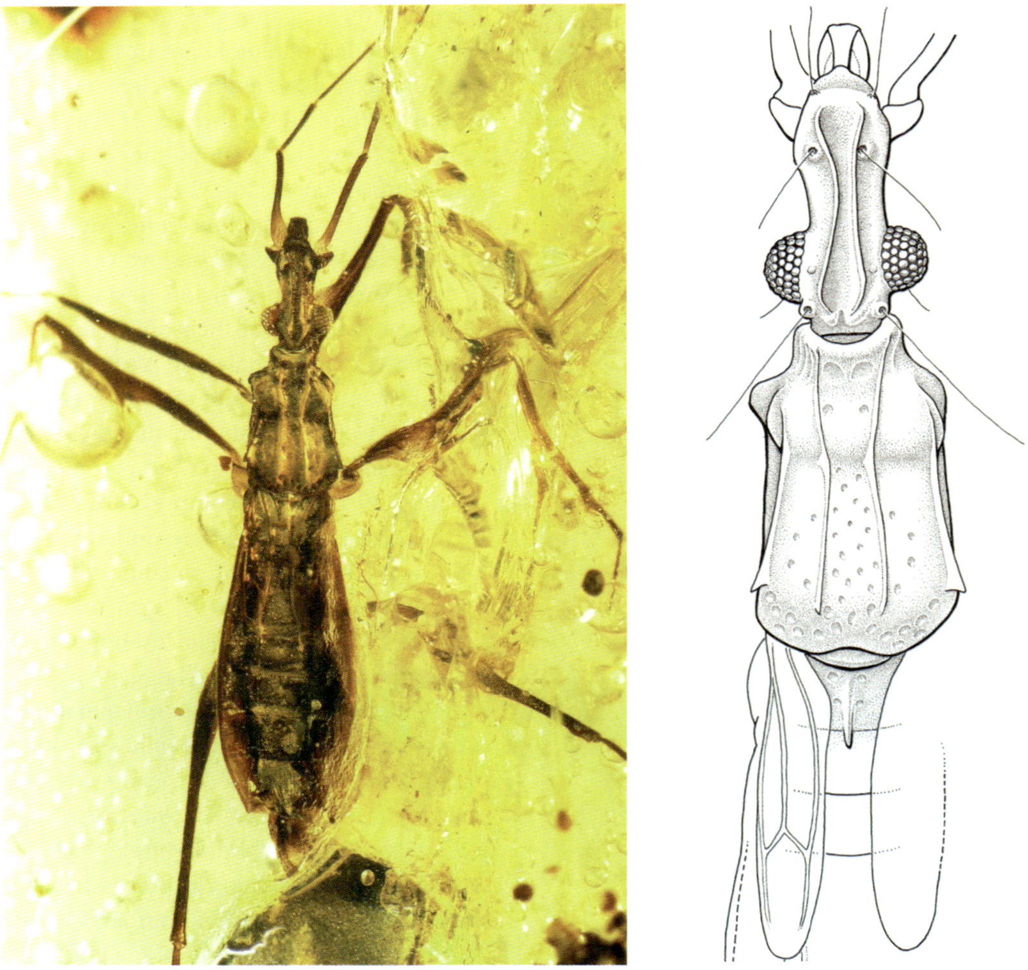

8.66. An early, very primitive water measurer (Hydrometridae), *Carinametra burmensis*, in Burmese amber, with detail of its head and thorax. AMNH Bu1098; body length 4 mm.

8.67. An Eocene waterstrider, *Limnoporus* (Gerridae), from Driftwood Creek, British Columbia. Water striders are the most recently evolved gerromorphans. MCZ 75a; body length 14 mm.

8.68. A male belostomatid guarding eggs laid on a twig in Costa Rica. In some belostomatids, the female lays her eggs on the male's back, which are protected there and aerated while he swims around. Photo: P. J. DeVries.

8.70. A belostomatid from the Early Cretaceous of Brazil. Belostomatidae have the best fossil record of all Heteroptera. AMNH SFHT13; body length 14 mm.

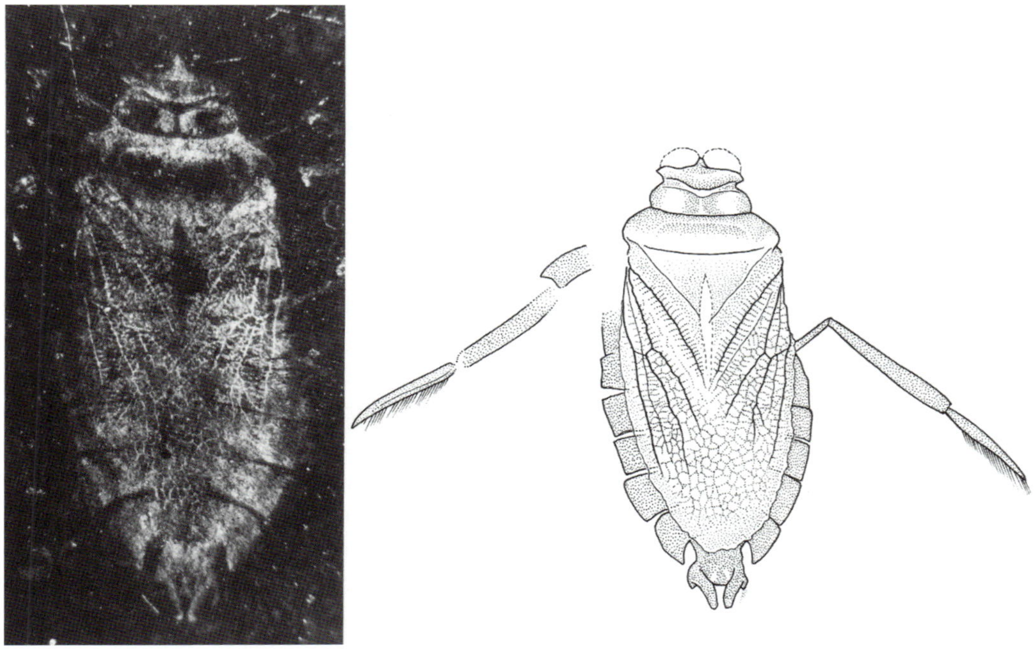

8.69. Earliest fossils of the predatory water bugs, Belostomatidae, from the Late Triassic of Virginia. This deposit has preserved the earliest diverse fauna of freshwater insects. VMNH; body length 12 mm.

panosoma cruzi via the bug's feces. Other species of *Trypanosoma* cause sleeping sickness in Africa but are transmitted by tsetse flies (Glossinidae). Perhaps the most unusual reduviids are holoptilines, which possess a tuft of fine hairs on abdominal sternite three that acts as a wick for an underlying gland. In at least one case, in *Ptilocerus ochraceus* from Java, the bug waits by an ant nest or column and lures ants, which gnaw at the tuft. They become intoxicated by the secretions, whereupon the bug injects its proboscis into the staggering ant and drains it. In the original description (Jacobson, 1911) piles of ant husks were found at the base of a tree where dozens of the bugs stationed themselves. Similar behavior has been captured in amber.

With 10,000 species, Miridae ("plant bugs") is the largest family of Heteroptera, world catalogues of which have been prepared by Carvalho and completely updated by R. T. Schuh (1995). Pretarsal structures are diverse and particularly important in mirid classification. When disturbed, mirids generally fall off their host plant, but before hitting the ground they catch themselves on foliage below using a very peculiar feature: an eversible, sticky rectum. Most species are phytophagous, in which case they are generally specific to a genus or family of host plants. Their biology has been extensively reviewed by Wheeler (2001). Predatory species occur in all the subfamilies, and if Isometopinae (which feed on scale insects) are indeed the most basal mirids (Schuh and Slater, 1995), predation may be the ancestral habit of feeding for the family. A. G. Wheeler (2001) suggested that phytophagy arose at least five times among mirid subfamilies, and many more times within subfamilies, though this needs to be tested with a phylogeny of the entire family. Closely related to the Miridae are the Tingidae (2,000 species) and Thaumastocoridae, or "palm bugs" (16 species, some feeding on palms); both these groups are phytophagous. Tingidae, or lace bugs, are well known for the intricately beautiful forewings, though the basal species have merely punctate forewings (Figure 8.73). Interestingly, most of these basal tingids live with ants and feed on roots within their nests.

8.71. Fossil shorebug (*Leptosalda* sp.: Leptopodidae: Leptopodomorpha) in Miocene amber from the Dominican Republic. It is very similar to a rare living species, *Saldolepta kistnerorum*, from northwestern South America. AMNH DR14-58; body length 1.78 mm.

In the superfamily **Cimicoidea** the male punctures a membrane in the female's abdomen with the phallus, rather than copulating vaginally. Primitively, the phallus may puncture vaginal walls (as in Nabidae), but the sperm migrate through hemolymph to the ovaries, not through oviducts. In the more derived situations the phallus is actually inserted into intersegmental membranes, or into a membranous "copulatory tube" atop the female abdomen. The extreme condition is found in the closely related Anthocoridae and bed bugs (Cimicidae sensu lato), which have a sharp, hooked-shaped left paramere – literally a hypodermic penis (Figure 8.74). Their courtship is extremely brief, and their mating aptly termed "traumatic insemination." It is quite likely that the small family Polyctenidae (30 species) are merely highly modified Cimicidae (70 species), much the way that nycteribiid bat flies are merely highly modified streblids. Polyctenidae feed on the blood of just bats; Cimicidae feed from bats, certain birds (swifts and swallows), and a few other vertebrates. *Cimex lectularius* is the notorious bed bug. Both of these "families" have highly reduced eyes and wings, are dorsoventrally flattened, and are covered with dense, stiff hairs. Polyctenidae are further modified by being completely blind and viviparous and by having ctenidia; in fact, when Polyctenidae were first known, they were even thought to be batflies or lice. Usinger's (1966) monograph on the Cimicidae is an extremely thorough account of the species in the family. An evolutionary precursor to obligate hematophagy in Cimicidae may be the habit of *Lyctocoris* (Lyctocoridae), which are generally predatory cimicoids, but some of them feed on the blood of ground-nesting mammals.

The **Pentatomorpha** comprise approximately 12,000 almost wholly plant-feeding species, best known for the stink bugs. Relationships among the families have been analyzed by Leston *et al.* (1954) and Henry (1997), the latter of which we are adopting here. Among the more significant structures for their classification are abdominal trichobothria, though the most basal family, Aradidae, lacks these. Aradidae (1,800 species) are commonly known as bark bugs for their habit of living on or under the bark of dead and dying trees. Most are flat, dark brown, and rugose, resembling bark (Figure 8.75). The bizarre, small family Termitaphididae (10 species) may actually be just highly modified aradids, which have lost their

8.72. An assassin bug (family Reduviidae) which has ambushed a halictid bee at a flower. Photo: V. Giles.

eyes and wings and have become extremely flat and oval like coccoids (Figure 8.76). These are inquilines in termite nests. Both groups tap fluids from fungal mycelia using fine, hair-like stylets that are coiled inside the head at rest. Species in the small family Phloeidae resemble aradids (convergently) but feed on plant vascular fluids.

The superfamily Pentatomoidea comprises nine families including stink bugs and their allies, many of them readily distinguished by the large scutellum. In two families, the Scutellaridae (450 species) and Plataspidae (30 species), the scutellum has become so large that it covers the abdomen like the elytra of beetles. Some scutellerids (450 species) have technicolor iridescence; a few southeast Asian plataspids feed on phloem, secrete honeydew, and are tended by ants. The Cydnidae, or "burrowing bugs" (600 species), have thick spines on the front of the head and on their stout, fossorial forelegs. Pentatomidae is a large family (4,100 species) renowned for the effectiveness of their scent glands. All are phytophagous with exception of the Asopinae, which are predatory (Figure 8.77). Among the five families of Coreoidea, only the Coreidae (1,800 species) has significant diversity. Coreids include some of the largest terrestrial Heteroptera (e.g., *Thasus* spp.), and many have swollen hind femora and broad flanges on the hind tibiae. All species feed on plants, with complicated mixtures of specialist and generalist feeders (Schaefer and Mitchell, 1983).

Many Pyrrhocoroidea and Lygaeoidea feed on the nutritious, highly caloric endosperm of seeds. The family Lygaeidae as traditionally recognized is extensively paraphyletic, which Henry's (1997) classification has rectified. He raised 11 lygaeid subfamilies to family rank, leaving only the Ischnorhycninae, Lygaeinae, and Orsillinae within Lygaeidae (sensu stricto). Most Geocoridae are predators, and Cleradinae (Rhyparochromidae) are vertebrate blood feeders in southeastern Asia. The common milkweed bug, *Oncopeltus*, is the best known hemipteran because it is used extensively in laboratory research.

The classification and phylogenetic study of Cimicomorpha and Pentatomorpha bugs depends on microscopic features like trichobothria and the structure of pretarsi, glands, and genitalia. As such, it is exceedingly difficult – in most cases impossible – to assign compression fossil Heteroptera to a family with complete certainty. It is far easier to identify compression fossils of groups that have obvious structures, such as in Tingidae, certain pentatomoids, and Coreidae. It is reasonable to assume that these two infraorders evolved in

8.73. Cimicomorphan bugs in Cretaceous amber. Top: possibly in the family Plokiophilidae, in Canadian amber (75 MYO). Plokiophilids walk on spider webs and feed on the trapped insects; this specimen was trapped while still attached to several strands of silk. Middle and bottom: early and primitive lacebugs in New Jersey amber (Tingoidea). Top: body length 1.40 mm. (CAS 37). Bottom: body length 2.5 mm (AMNH NJ624).

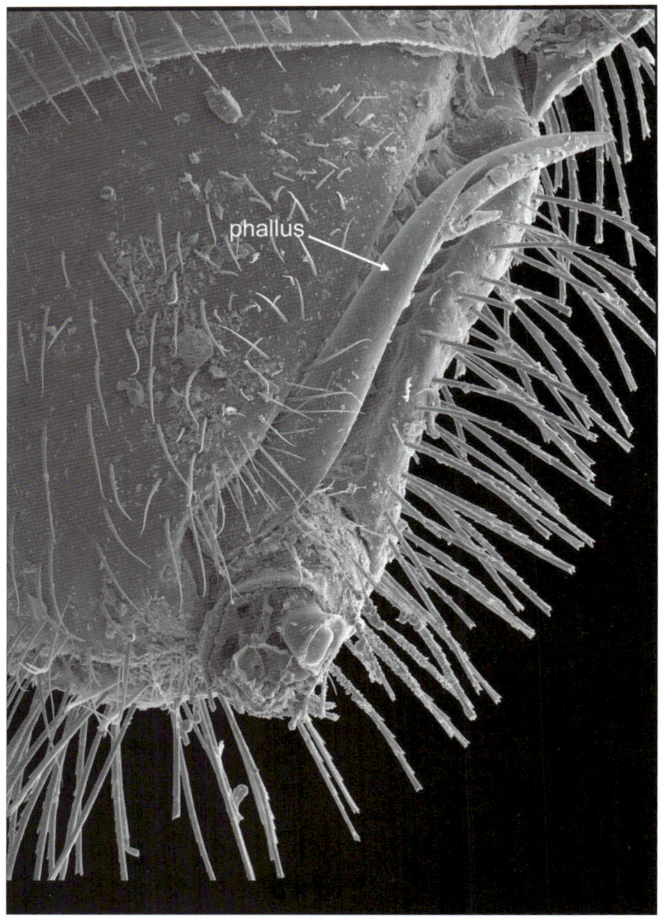

8.74. Phallus of the bed bug, *Cimex* (Cimicidae). In the cimicoid bugs the male injects sperm into the female through her abdomen using his hypodermic-like phallus. Scanning electron micrograph.

8.75. An early bark bug (Aradidae), *Archearadus burmensis*, in Cretaceous Burmese amber. AMNH Bu167; body length 5.1 mm.

8.76. A Miocene termite bug, *Termitaradus* (Termitaphididae), in Dominican amber. These wingless and eyeless bugs are closely related to Aradidae, possibly even highly modified members thereof. They live in termite nests where they seem to feed on fungi. AMNH DR14-425; body length 3.6 mm.

the Jurassic, as did at least their *basal* families, but it will probably be impossible to confirm the identity of putative Miridae, Nabidae, and Velocipedidae reported from Jurassic rocks (see Popov *et al.*, 1994). Late Jurassic fossils assigned to the Cydnidae have a more convincing basis, being based on a distinctive scutellum and leg structure. Other Late Mesozoic compression fossils assigned to modern families are the Anthocoridae, Coreidae, Reduviidae, and Tingidae, from several Early Cretaceous deposits in Eurasia, 140–120 MYA. With the exception of the tingid, the family identities of these are difficult to confirm, so most of the earliest definitive records of cimicomorphs and pentatomorphs are in Cretaceous ambers. These include the cimicomorphans Lasiochilidae, Miridae, Thaumastocoridae, and Tingidae in 90 MYO New Jersey amber (Golub and Popov, 2000, 2003, and unpubl.), a plokiophilid in 75 MYO Canadian amber (Figure 8.73), and the pentatomorphans Aradidae, Lygaeoidea, and Piesmatidae in 100 MYO amber from Myanmar (Heiss and Grimaldi, 2001; Grimaldi *et al.*, 2002). A putative cimicoid in Burmese amber is broad, flat, and pilose, but is primitive to all known cimicids by the long rostrum, complete wings, and larger eyes (Grimaldi *et al.*, 2002). An aradid preserved in 90 MYO rocks from Siberia illustrates the difficulty of interpreting the rock fossils. It was assigned to the living genus *Aradus*, which is highly doubtful even though it superficially resembles the genus. Another Cretaceous aradid, completely preserved in Burmese amber (Figure 8.75), is a very primitive member of the family, as based on the primitive presence of pulvilli and other microscopic features.

While the more basal families of Cimicomorpha and Pentatomorpha probably originated in the Late Jurassic, the more recently evolved families in the Cimicoidea, Pentatomoidea, Coreoidea, Lygaeoidea, and Pyrrhocoroidea probably are no older than Cretaceous. Indeed, most of the diversification of these groups probably occurred in the Late Cretaceous and the Tertiary. Some, like the Cimicidae (including Polyctenidae), probably did not evolve much earlier

8.77. An asopine stink bug (Pentatomidae) feeding on a membracid tree hopper. Most pentatomids are plant feeders. Photo: V. Giles.

8.78. A pentatomid female guarding her brood in Ecuador. Parental care has repeatedly evolved in exposed, plant-feeding insects with broadly rimmed bodies, like lace bugs, cassidine leaf beetles, and pentatomids. Photo: P. J. DeVries.

than the Eocene, when bats first appeared and before swallows and swifts even evolved.

Dating the radiations of these two heteropteran infraorders is important to consider because most of the Heteroptera are phytophagous species from these two groups. Diverse lineages of phytophagous Heteroptera evolved at least three times: once in Pentatomorpha and twice or more in the Miroidea. This is clearly a recently derived habit for the suborder; nevertheless, it has spawned 80% of all heteropteran species. The Cretaceous Piesmatidae, Tingidae, and Thaumastocoridae are all primitive members of these phytophagous families. This supports the view that heteropteran diversity today is very much a result of the Cretaceous angiosperm radiations.

8.79. An Early Cretaceous cimicomorphan bug from Brazil. AMNH SA45265; body length 5 mm.

8.80. A pentatomorphan bug from the Early Cretaceous of Brazil. AMNH SA45246 (ventral view); body length 13 mm.

9 The Holometabola

Holometabola are those insects with so-called complete metamorphosis, versus insects that are hemimetabolous (incomplete metamorphosis), or (as in the case of apterygotes) insects that display virtually no changes from immature stages to adult (*ametabolous*). Metamorphosis itself refers to a dramatic change in an organism from one developmental stage to another, which occurs in some amphibians and various invertebrates, including many marine and terrestrial arthropods. In *hemimetabolous* insects, which include the paleopterans, polyneopterans, and paraneopterans, which are groups we have already discussed in detail, the wing buds increase in size with each nymphal instar (e.g., Figure 9.1). *Holometabolous* insects have a soft-bodied, wingless, morphologically reduced *larval* stage and a mostly quiescent *pupal* stage. Adults of all insects are defined as those individuals that are reproductively mature, even in species where the mature female is *larviform* or *paedogenic*, retaining some or most larval features. Examples of paedogenic female holometabolans include most Strepsiptera (Figure 10.80), some Diptera (e.g., *Miastor* [Cecidomyiidae]), and some Coleoptera (e.g., *Micromalthus*, a few Ripiphoridae), among others.

The larva is the truly defining feature of Holometabola. This immature form is generally a slender, soft-bodied insect with three short pairs of thoracic legs or none (sometimes they also have abdominal *prolegs*), and no external wing buds. Primitively they have a sclerotized head capsule with mandibles, minute palps and antennae, and reduced eyes or usually just isolated ommatidia separated into *stemmata*. During metamorphosis, which takes place as a pupa, much or all of the larval cuticle is replaced with adult cuticle. This new cuticle derives from *imaginal discs* (sometimes also called by the German term, *anlage*), including the legs, eyes, genitalia, and wings. Imaginal discs are pockets of specialized larval epidermal cells (*histoblasts*) that lie under the cuticle; the histoblasts proliferate and differentiate during the late larval and pupal stages. An internal development of wings is why Holometabola are also sometimes called Endopterygota (e.g., Kristensen, 1999a).

The holometabolous lineages account for the stunning diversity of insects, since Holometabola comprise nearly 85% of all insects, and in this regard the feature is usually considered one of the two most important innovations of insects along with wings. There is no doubt that Holometabola is a monophyletic lineage (e.g., Kristensen, 1975, 1999a; Hennig, 1981; Whiting *et al.*, 1997; Whiting, 2002), though a similar "pupal" stage has evolved independently in whiteflies (Aleyrodoidea: Sternorrhyncha) and thrips (Thysanoptera), wherein late instar nymphs are reduced, nonfeeding, and quiescent. These are entirely convergent with holometabolan pupae, though, based on relationships and the fact that these paraneopterans do not have a larval stage. The most recent and comprehensive presentation of insect development is the comprehensive treatment by Heming (2002). As we discuss later, the adaptive advantage of the insect larva is not entirely clear, and larvae may in fact provide multiple improvements over nymphal development.

PROBLEMATIC FOSSIL ORDERS

Two extinct "orders" are enigmatic and have uncertain position, though these have generally been affiliated with the Holometabola. The Paleozoic Miomoptera have been considered to be stem-group Holometabola even though their venation could also be interpreted as primitive relative to Paraneoptera. The other order, the Glosselytrodea from the Paleozoic and Early Mesozoic, is perhaps a composite group, apparently consisting of stem-group Neuropterida and families of uncertain phylogenetic affinity.

MIOMOPTERA

The Miomoptera were generally minute insects, with fore- and hind wings similar in shape and venation (Figure 9.2). The hind wings lacked an anal lobe, and both sets of wings lacked crossveins. Little is understood of miomopterans because they are known almost entirely as wing fragments.

9.1. Nymphal stages and the adult of a typical hemimetabolous insect, a barklouse (Psocoptera: family Lepidopsocidae). Holometabolan metamorphosis involves larval and pupal stages. Scanning electron micrographs.

Where body fragments are known, species were apparently mandibulate and had conspicuously thick antennae, with 15–20 antennomeres; tarsi were four-segmented; and the cerci were short. The position of Miomoptera is extremely unclear but they have been treated by most authors as stem-group Holometabola (e.g., Rasnitsyn and Quicke, 2002). They have even at times been considered as a stem group to the Hymenoptera (e.g., Rasnitsyn, 1980). Martynov (1938) originally asserted that Miomoptera were allied to the Plecoptera, but, like the Holometabola hypothesis, there is little evidence to support this notion. Miomoptera are herein considered as stem-group Paraneoptera (Figure 4.24) but may eventually be considered allies of the Psocodea or Condylognatha.

There are presently only three families included in Miomoptera. Various authors have placed several other families in the order, but many of these had large anal fans in the hind wing and numerous crossveins and appear to be more polyneopterous in character. The polyneopterous families included by earlier authors have all been removed to the "Protorthoptera" or to Neoptera incertae sedis. The order is known only from the Late Carboniferous and Permian, having apparently gone extinct during the End Permian Event. All three families are known from the Late Carboniferous but the Metropatoridae are presently unknown from the Permian, while the other two families are also known from the Permian.

GLOSSELYTRODEA

The glosselytrodeans as presently constituted (Carpenter, 1992) may be an unnatural group, consisting on the one hand of the Early Permian Permoberothidae and on the other hand as an assemblage of Late Permian and Triassic-Jurassic families (Archoglossopteridae, Glosselytridae, Glossopteridae, Jurinidae, Polycytellidae, and Uskatelytridae). The Permoberothidae are stem-group Neuropterida (e.g., Carpenter, 1943a). Permoberothidae should perhaps be referred to as **Protoneuroptera** (Figure 9.2), and the Glosselytrodea should be restricted to the aforementioned suite of families. The straight CuP vein marking off a distinct anal region with crowded anal veins and the relatively straight MA vein running beside the anterior of MP is reminiscent of some Jurinidae, which was the justification for unifying Permoberothidae with other Glosselytrodea. The tegminous forewings of Glosselytrodea s.str. have a superficial resemblance with protocoleopteran elytra, perhaps also furthering the notion that these families were "neuropteroid."

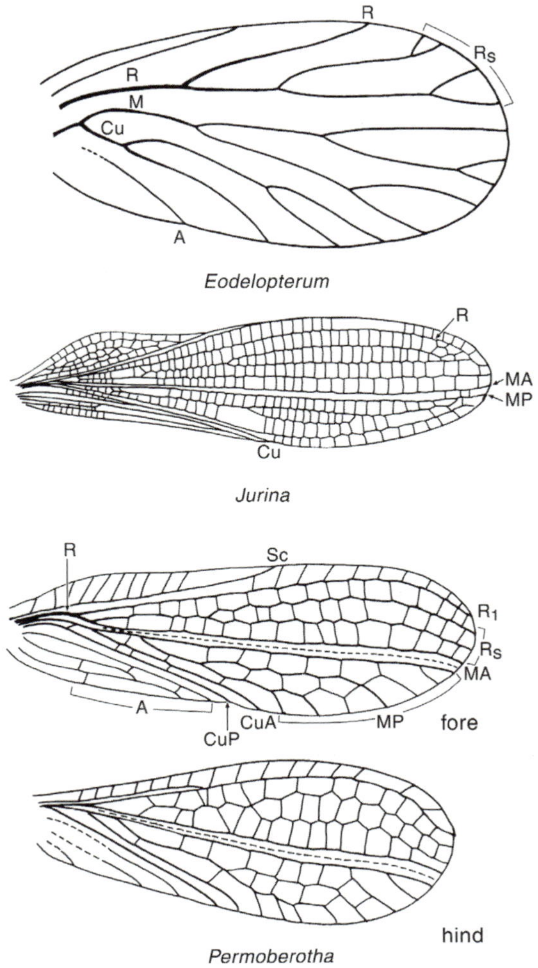

9.2. Wings of the problematic orders Miomoptera and "Glosselytrodea," which may have had close affinities to the Holometabola. From top to bottom, *Eodelopterum priscum* (Archaemiopteridae) from the Late Carboniferous of Germany, *Jurina marginata* (Jurinidae) from the Permian of Central Asia, and *Permoberotha villosa* (Permoberothidae) from the Early Permian of Kansas.

The Glosselytrodea s.str. had very distinctive wings (Figure 9.2). Most were relatively small insects, with wings subequal in shape and size. The forewing possessed an expanded precostal area (similar to Orthopterida, and where this set of families may properly belong) and formed a prominent bulge at the wing base. The hind wing lacked a precostal area, the anal region was well demarcated by a deep claval furrow in the forewing, and vein CuP was straight and the anal veins were clustered together. The wings possessed numerous crossveins, which is apparently the basis for the belief that they were neuropteroid, even though dense crossvenation is by no means confined to Neuropterida. From the few fossils with body fragments, glosselytrodeans were apparently hypognathous, with slender legs and short, multi- or unsegmented cerci. Typical for holometabolous insects, the radial and medial systems are fused at the wing base, and thus it may indeed be that this group is allied to the Neuropterida along with Permoberothidae (e.g., Zalessky, 1932; Carpenter, 1964b). However, Béthoux *et al.* (2001) have argued that Glosselytrodea s.str. are relatives of the Orthoptera. The phylogenetic position of glosselytrodeans remains contentious, and considerable work remains to be undertaken before Glosselytrodea s.str. can be considered to be clearly polyneopterans.

THE ORIGINS OF COMPLETE METAMORPHOSIS

Given the great success of holometabolans, the major question is: How did they evolve? Just like wings and any other feature of adaptive significance, there is a spectrum of specialization in holometaboly, perhaps the most important aspect involving the imaginal discs. In the more basal holometabolans, such as neuropterids, many beetles, mecopterans, nematocerous flies, and symphytan wasps, some of the larval epidermis is retained in the adult. Imaginal discs in these groups develop in the last larval instar. In cyclorrhaphan Diptera, Lepidoptera, and apocritan Hymenoptera, all or virtually all of the adult cuticle is formed from imaginal discs, and the imaginal discs begin developing in the embryo, which is a more evolved form of holometaboly. Most of the work on imaginal discs has been done on *Drosophila melanogaster* since disc development can be manipulated not just experimentally but genetically as well. This fly has a series of imaginal disc pairs that develop into the adult head (the labial, humeral, clypeolabral, and antennal-eye discs), the thorax (the fore-, mid, and hind legs, wing, and halter discs), and abdomen (the genitalic discs and four small groups of histoblasts) (Figure 9.3). In the case of appendages, as these develop they expand and either fold on itself or telescope to fit within the confines of the imaginal disc pocket; they then unfold and expand upon pupation and further upon eclosion. Each disc has compartments of predetermined regions that develop into particular structures. The extensive work on *Drosophila* has revealed a cascade of genes controlling the development of the basic body plan to the structure of appendages, which has been reviewed by Heming (2002).

Two theories basically account for how larvae evolved. One is that holometabolous larvae and hemimetabolous nymphs are "homologous" life history stages, and the pupa developed from an intermediate form as holometabolans evolved larvae that increasingly differed from adults (Hinton, 1955; Sehnal *et al.*, 1996). The other hypothesis, which we feel is best supported, is that *the holometabolan larva is a protracted version of the hemimetabolous pronymph* (Truman and Riddiford, 1999). This is actually an old idea that began with the great early entomologist A. Berlese (1913), who hypothesized that that larvae were essentially extended, free-living embryos, but this idea has been extensively expanded and revised based on recent embryology and endocrinology (Truman and Riddiford, 1999). The *pronymph* is an obscure

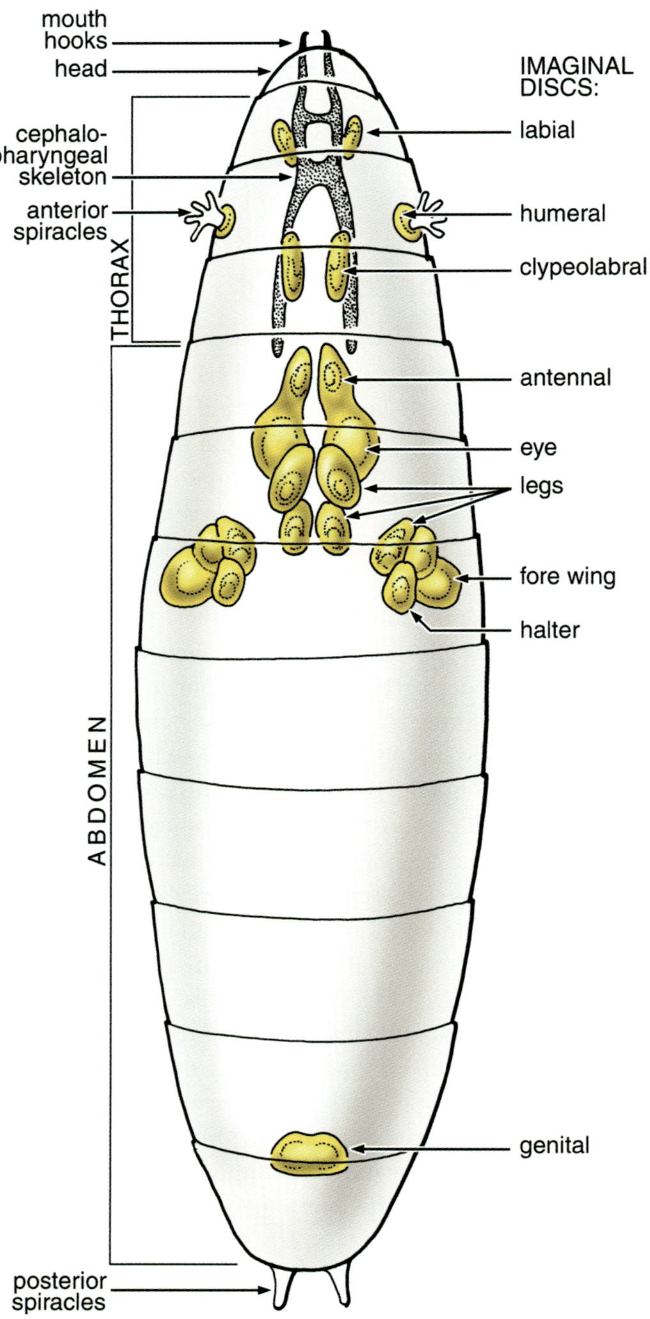

9.3. Schematic diagram of a fruitfly larva (*Drosophila* sp. [Drosophilidae]), showing the position of its imaginal discs, which are pockets of specialized ectoderm that will form adult structures. Imaginal discs are unique to holometabolous insects.

stage between the hatching and first instar nymph of hemimetabolous insects. It is usually very brief, lasting from 3–4 days in apterygotes to just a few hours after hatching in basal pterygotes (in which most of this stage is spent in the egg, or entirely in the egg in the case of Paraneoptera). Pronymphs have shorter appendages than first instar nymphs and do not feed (except in Ephemeroptera). Evidence that the holometabolan larva is actually a pronymph is considerable and includes the following similarities between the two:

- The first-instar cuticle of both are secreted around the same time late in embryonic life.
- Pronymphs and first instar larvae have similar unsclerotized cuticles, including the mandibles (which is why a specialized "egg burster" structure has developed in many embryonic insects).
- Both lack wing buds.
- They have similar, reduced nervous systems.
- Both have precociously high levels of juvenile hormone (JH), which in larvae leads to an earlier maturation and suppression of embryonic tissues. Larvae, in addition, maintain high levels of JH.

Indeed, the primitive condition in holometabolan larvae is development of the imaginal discs at the end of larval growth, which corresponds to the time when early instar hemimetabolous nymphs sprout minute wing pads. This exciting hypothesis provides new insight into how the insect larva evolved, but *why* should it have evolved? Why was a heterochronic shift in development, specifically neoteny, so successful?

Traditional, adaptationist explanations for the origin of the insect larva is that this kind of development allows adults and immatures to have different diets. Indeed, the larvae of most holometabolans, and virtually all of the basal lineages, live in particulate and liquid substrates, or in plants, either as saprophages, leaf miners, wood borers, and endoparasitoids, whereas adults are free-living. The internal development of wings, thus, is considered adaptive to protect larvae while in the confines of a substrate. Notable exceptions to this larval habit are the larvae of symphytan wasps, many of which are external plant feeders, and larvae and adults of Neuropterida and adephagan beetles, both stages of which generally are free-living and have the same kinds of prey. In some holometabolous insects the adult (generally the female) is larviform in structure and habits, such as *Micromalthus* and some phengodid beetles, female Strepsiptera, and assorted flies like *Miastor* (Cecidomyiidae) and *Ascodipteron* (Hippoboscidae). An avoidance of competition between adults and larvae is often invoked as a selective force for the evolution of disparate diets, but in fact competition for food limitation is not well documented in natural populations of insects (e.g., Speight *et al.*, 1999; but see Grimaldi and Jaenike, 1984). Alternatively, larvae could simply exploit abundant resources for which the adults were not optimally designed. It would also seem that the physical, not just dietary, separation of larvae and adults would spare the less mobile and vulnerable immatures from the same predators and parasites that attack the adults.

Another, though not alternative, traditional explanation for the holometabolous larva is that the life span of immatures is shortened, and they are thus exposed to less predation and parasitism. In fact, holometabolans have the shortest life cycles in insects. *Drosophila*, like many cyclorrhaphan flies, develops from egg to adult in 10 days or less, and some leiodid beetles complete larval development in an astonishing period of two days. But, holometabolans also have some of the longest life spans, and the larvae of many wood-boring beetles are known to last for several years. What is probably of greater significance in this regard is that *holometabolous larvae have more effective control of development.* They can prematurely pupate in situations of extreme crowding or food shortage and readily diapause under adverse abiotic conditions. Diapause, like hibernation, is a period of arrested development and metabolism that occurs under periods of environmental stress, like northern winters or tropical dry seasons. It is innate but also environmentally induced (Tauber *et al.*, 1986; Denlinger, 2002). Very likely it is a host of explanations like these that accounts for the adaptive success of the insect larva.

Kristensen (1999a) cautioned about inferring the causes of radiations directly on any one diagnostic feature of a group. Indeed, for virtually any key innovation in insects the original form appears to have been modified and improved upon in the more recently evolved lineages, and the evolution of the larva appears to be no different. *Subsequent refinement* of larval development probably led to holometabolan success, much as the features of neopteran wing folding and flight led to the great success of winged insects. For example, some holometabolan orders have only very modest species diversity (<1,000 species) and the extant basal lineages in each of the "big four" (Coleoptera, Diptera, Hymenoptera, and Lepidoptera) are not particularly diverse. The question, according to Kristensen (1999a), is not "Why are there so many Holometabola?" but really "Why are there so many apocritan Hymenoptera, cyclorrhaphan Diptera, staphylinoid and phytophagan beetles, and ditrysian lepidopterans?" These are presently the truly diverse groups. As we discuss later, abundant fossil evidence indicates that mecopterans and archostematan beetles, for example, were also quite diverse in the early Mesozoic. Such groups may have been eclipsed by more recently evolved relatives having more refined adaptations, larval and otherwise.

Holometabola presumably evolved in the Late Carboniferous, but there is no unequivocal evidence for this. Two significant reports of putative Holometabola from the Carboniferous concern *Srokalarva* (Kukalová-Peck, 1991, 1997) and some feeding traces (Labandeira and Phillips, 1996a). *Srokalarva* is an elongate animal with approximately eight pairs of legs from the Late Carboniferous of Mazon Creek, Illinois, but we agree with Willmann (1997) that it is not at all certain that this specimen is a larva, and it may in fact be a myriapod. It requires independent examination from various entomologists. The feeding traces comprise frass pellets in the gall of a Carboniferous tree fern petiole. Earlier we discussed why it is too ambiguous to attribute the gall to Holometabola, which may in fact be from large Paleozoic mites. The basal divergence in Holometabola comprised two lineages: Coleoptera + Neuropterida and Hymenoptera + Panorpoid orders (Kristensen, 1975, 1991, 1999a; Hennig, 1981), though the 18S rDNA data of Whiting (2002) doesn't agree. We maintain that the morphological evidence is more compelling. Given the paucity of external diagnostic features for adult Holometabola and the fact that early insect fossils are mostly of wings, it may be a long time before fossils shed light on the earliest history of the Holometabola.

ON WINGS OF LACE: NEUROPTERIDA

Ah, the romance of the Neuroptera.
–W. L. Brown, Jr.

The only relict lineage of Holometabola besides the mecopteroids are the three orders of Neuropterida. Together these three orders – Neuroptera, Megaloptera, and Raphidioptera – comprise the unlikely sister group of the beetles. Superficially, Coleoptera and Neuropterida share apparently little in common. However, the evolutionary relationship of the two is clear in the details, both sharing derived traits of the ovipositor and wing articulation (Mickoleit, 1973; Achtelig, 1978; Hörnschemeyer, 1998, 2002). Molecular studies have also repeatedly recovered a Coleoptera-Neuropterida clade (e.g., Wheeler *et al.*, 2001).

Defining features of the Neuropterida include the fusion of the gonoplacs in the ovipositor, a medially divided metapostnotum, the first abdominal tergum having a caudally bifid longitudinal sulcus, and a proventriculus with an unpaired diverticulum (Achtelig, 1975; Kristensen, 1991). While neuropterid monophyly is unquestioned, relationships among the three orders have been difficult to resolve. Achtelig (1967) considered the Megaloptera to be paraphyletic with respect to the Raphidioptera, and thus demoted the latter to a family within the former. Achtelig and Kristensen (1973) later revised this work, concluding that Raphidioptera was the living sister group to Megaloptera, rather than derived from within, and that together they were the sister group of the Neuroptera. Later studies of the abdominal base and female genitalia further supported this arrangement (e.g., Achtelig, 1975, 1976, 1978). Boudreaux (1979), however, argued that Megaloptera was more closely related to Neuroptera based on the fusion of veins Sc and R near the wing apex, long stipites in the adult mouthparts, reduction of the gonoplac, and absence of a pigmented pterostigma on the wing. Kristensen (1981) convincingly argued against the use of these characters, and Achtelig (1981) and Kristensen (1991) further suggested that the presence of a gula, telotrophic ovarioles, unique grooming behavior, and various internal modifications of the thorax

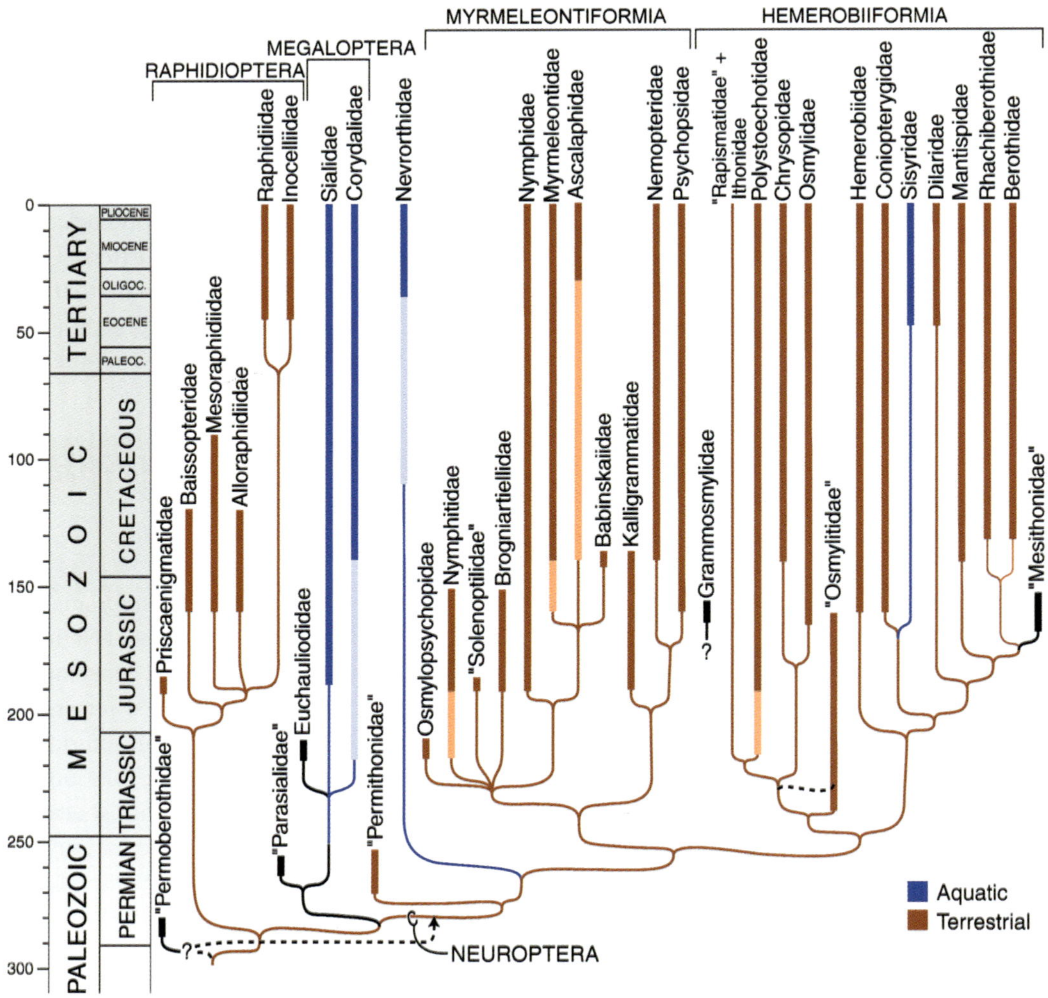

9.4. Phylogenetic relationships among families of Neuropterida (Raphidioptera, Megaloptera, and Neuroptera). Relationships of orders and extant families after U. Aspöck *et al.* (2001). Lighter colors are inferred fossil ranges.

and abdomen were traits uniting the Megaloptera and Raphidioptera.

The most recent and comprehensive study on relationships within the Neuropterida is that of U. Aspöck *et al.* (2001) (Figure 9.4). Their cladistic analysis has revived Boudreaux's hypothesis of a basal position for the Raphidioptera relative to the Megaloptera and Neuroptera. These authors have argued that the latter two orders are united by the presence of aquatic larvae (present, however, in only two families of Neuroptera) and the integration of the larval cardines into the head capsule and in the structure of the male genitalia. While the latter characters are indeed suggestive, the significance of aquatic larvae is unclear. An apparently basal lineage of Neuroptera, Nevrorthiformia, is aquatic, which could indicate that aquatic immatures are a defining feature of Megaloptera-Neuroptera monophyly. In such case, the terrestrial larvae of most neuropterans would be secondarily derived, but an aquatic ancestor of Megaloptera and Neuroptera is ambiguous. It is equally logical and more intuitive to assume that aquatic larvae arose independently in the Megaloptera and in Nevrorthiformia, and that terrestrial larvae are primitive for both lineages. Many of the characters purported to unite Megaloptera and Raphidioptera were not included in the study of U. Aspöck *et al.* (2001) and so an expanded cladistic analysis might reveal if a Megaloptera–Neuroptera clade is still recovered. The exarate (i.e., mobile) pupa of Megaloptera and Raphidioptera is a feature of these orders that is more primitive than all other holometabolans.

9.5. A female *Mongoloraphidia eklipes* (Raphidiidae) cleaning her foreleg in Uzbekistan. Snakeflies are relict today, and Central Asia harbors more modern species of Raphidioptera than any other region. Photo: A. Schumacher, courtesy H. and U. Aspöck.

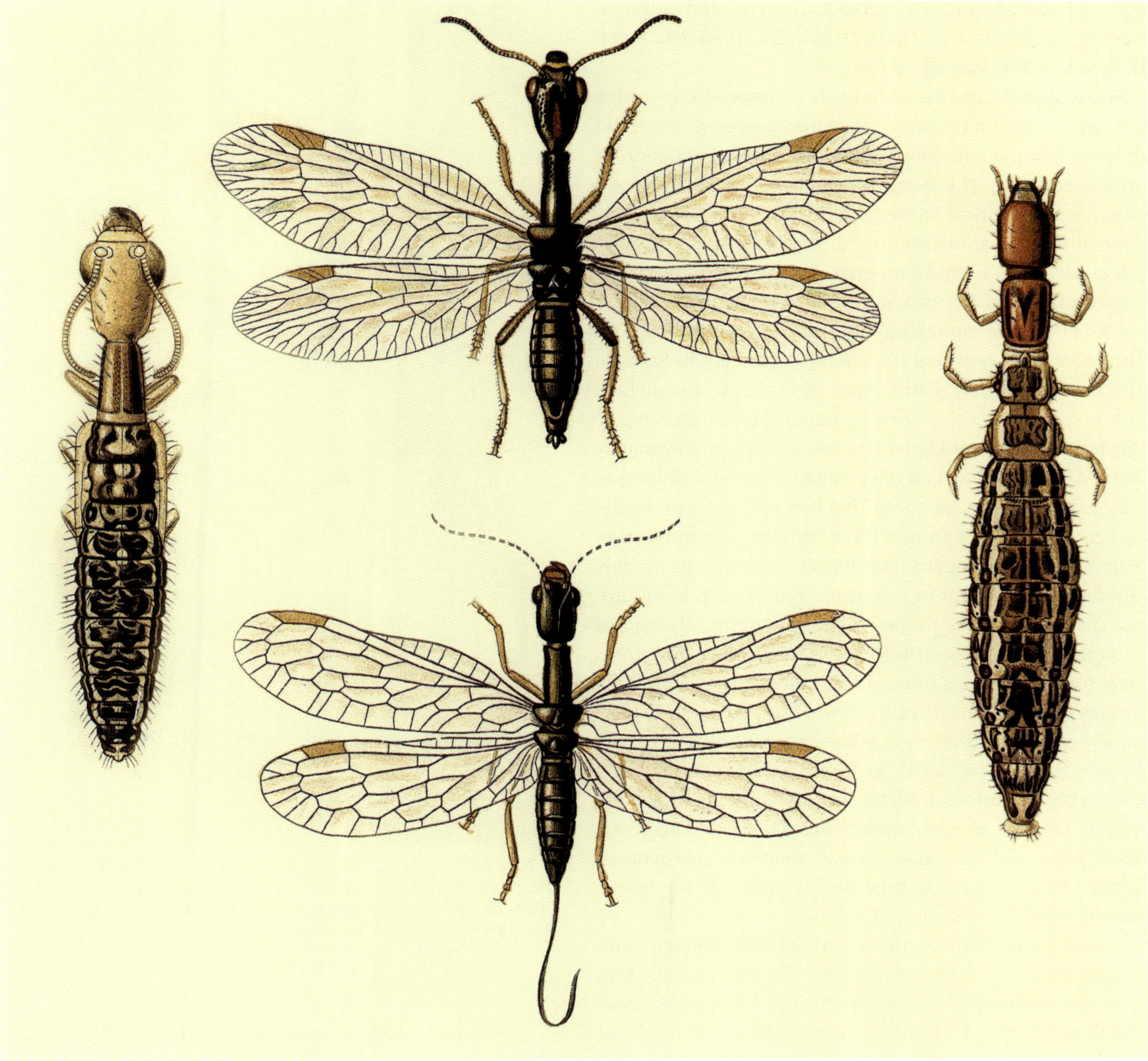

9.6. Representative Raphidioptera: a pupa (left), larva (right), an adult male raphidiid (above), and an adult female inocelliid (below). Raphidiopterans have the most primitive development among all holometabolans, and their pupae are in fact active. Compiled from *Genera Insectorum*.

The Neuropterida are an ancient group since representatives of putative Megaloptera and Neuroptera occur as early as the Late Permian, such as the family Permoberothidae, which are perhaps stem-group Neuropterida (see preceding discussion of Glosselytrodea).

RAPHIDIOPTERA: THE SNAKEFLIES

Although the number of known snakefly species has grown significantly during the past forty years, the order remains the least diverse lineage of Holometabola. They are also the only endopterygote order to have been completely monographed to the species level, which is made possible by the fact that their extant diversity consists of just nearly 220 species (H. Aspöck *et al.*, 1991; with numerous small supplements). Snakeflies are immediately recognizable owing to the elongate prothorax (i.e., the famous "snake neck"), a stout and elongate ovipositor in females and typical wing venation (Figures 9.5, 9.6). Despite this characteristic habitus, diagnostic features of the order are not immediately evident, but perhaps include the bilobed third tarsomere (not to be confused with the bilobed *fourth* tarsomere of Sialidae), the termination of vein Sc into the anterior wing margin, and the absence of the pretarsal arolium (Kristensen, 1991). The biology and biogeography of the order has been recently summarized by H. Aspöck *et al.* (1991), H. Aspöck (1998, 2002), and U. Aspöck and H. Aspöck (2003a).

Modern snakeflies live exclusively on trees, also requiring at least some period of cold, near-freezing temperatures in order to complete development (either before pupation or prior to eclosion). This feature of their biology has, in combination with historical factors (e.g., H. Aspöck, 2000), largely defined the global distribution of the order today. Species principally occur in cold temperate regions, extending southward into Central America, northernmost Africa, and southern Asia only in mountainous areas above 1,100 m (3,609 ft). The order is today absent from South America, sub-Saharan Africa, and the Australian region. Raphidioptera is divided into two families, the assuredly natural Inocelliidae (ca. 20 species) and the Raphidiidae (ca. 190 species), unfortunately defined mostly by the lack of inocelliid traits. Inocelliidae, as their name suggests, lack ocelli, but they also lack a pterostigmal crossvein. Aside from a basal, extinct subfamily, other defining traits of modern inocelliids include the posteriorly tapering head (except in Electrinocelliinae) and basal mp-cua crossvein oblique to the longitudinal veins (orthogonal in Raphidiidae and Electrinocelliinae). The long ovipositor is ideal for placing eggs under bark, and the larvae of many species develop subcortically. However, some raphidiid larvae live in the detritus or soil at the base of shrubs or in rock crevices (H. Aspöck, 2002). Larvae are predatory on the eggs and larvae of various insects, as well as adults of minute arthropods such as mites, spiders, springtails, barklice, Sternorrhyncha, and Auchenorrhyncha. Adults are also generalist predators, consuming prey similar to that of the larvae, and the pupae are exarate and active.

While the Raphidioptera are undoubtedly ancient, convincing records of their occurrence in the Paleozoic are now known to be erroneous, leaving as the oldest definitive snakeflies those from the Early Jurassic (Engel, 2002a). All Paleozoic snakefly families (i.e., Fatjanopteridae, Letopalopteridae, Permoraphidiidae, and Sojanoraphidiidae) have been transferred to other orders in the Polyneoptera. Although primitive with respect to the extant Raphidiidae and Inocelliidae, the earliest snakeflies from the Jurassic already possessed features typical for the order. While no direct evidence currently supports a Paleozoic origin of Raphidioptera, as believed by earlier authors (e.g., Carpenter, 1992), indirect evidence from the apparent geological ages of related lineages (i.e., Neuroptera and Megaloptera) implies that snakeflies derive at least from the Triassic and, with stem groups perhaps extended into the Paleozoic. Continued paleontological work will undoubtedly refine these estimates as more material becomes available from Triassic deposits. Based on currently available evidence, crown-group snakeflies likely originated and diversified in the earliest Mesozoic after the Permo-Triassic boundary.

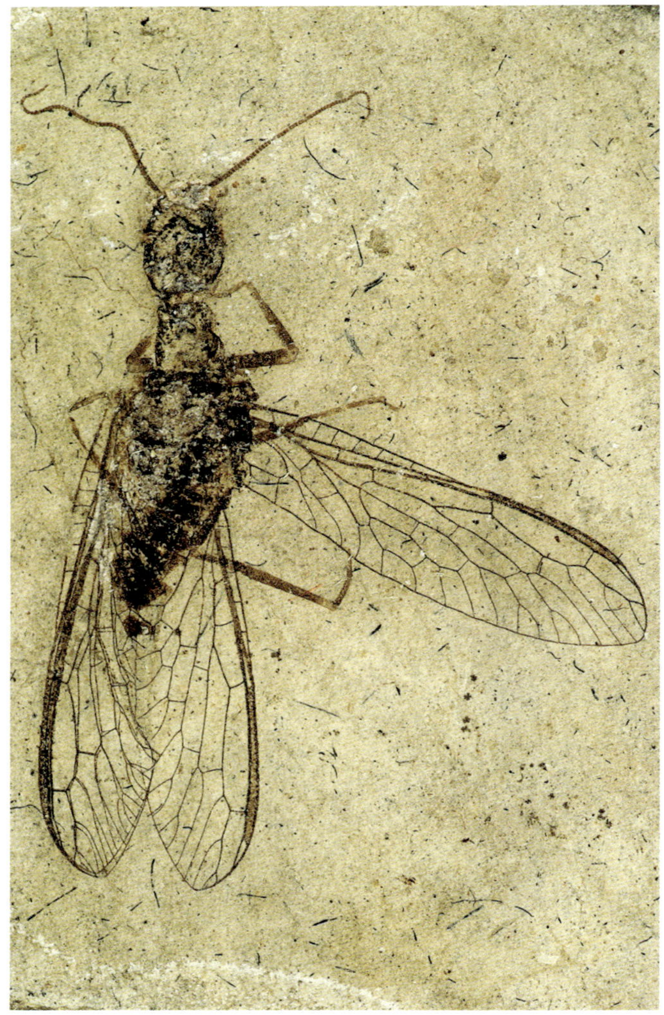

9.7. *Mesoraphidia pterostigmalis* (Mesoraphidiidae) from the Late Jurassic of Karatau in Kazakhstan. Numerous genera and species of Mesozoic snakeflies indicate that the extant fauna is a vestige of past diversity. PIN 2511/89; forewing length 13 mm.

Mesozoic snakeflies are segregated into four extinct families (Engel, 2002a) and were very diverse in terms of species, thus reinforcing the view of modern Raphidioptera as relict. The Early Jurassic Priscaenigmatidae were likely stem-group Raphidioptera, primitive in most traits of the order although the hind wing is typical for snakeflies. Priscaenigmatidae has been relegated to a basal suborder, Priscaenigmatomorpha; all other fossils (exhibiting the typical raphidiopteran venation in the forewing) have been placed in the Raphidiomorpha. Mesozoic snakeflies of the latter suborder (families Baissopteridae, Alloraphidiidae, and Mesoraphidiidae) are mostly known from the Late Jurassic or Early Cretaceous of Asia and South America (e.g., Martynov, 1925b; Oswald, 1990; Ren, 1997; Engel, 2002a) (Figures 9.7, 9.8), with some others from the Jurassic of Europe and Cretaceous of North America (e.g., Whalley, 1985; Willmann, 1994a). To date only two raphidiomorph snakeflies have been discovered in Cretaceous ambers, *Nanoraphidia electroburmica* from Myanmar

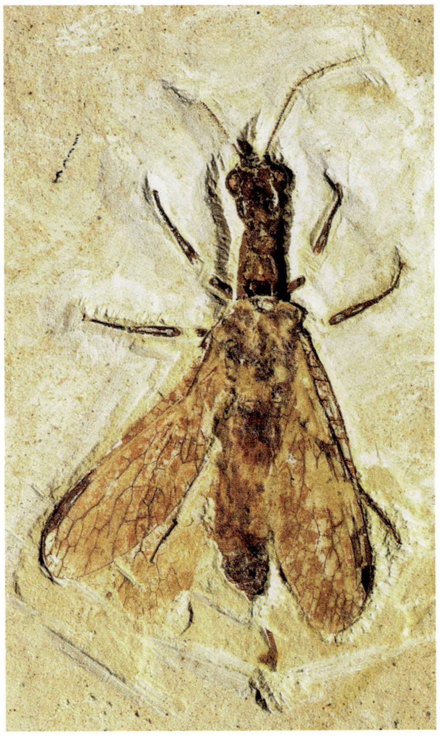

9.8. Although today snakeflies are extinct in the Southern Hemisphere, they were globally distributed in the Mesozoic, as evidenced by species such as *Baissoptera brasiliensis* (Baissopteridae) from the Early Cretaceous of Brazil, shown here. AMNH 43287; length 19.5 mm.

and *Mesoraphidia luzzii* from New Jersey (Grimaldi, 2000a; Engel, 2002a) (Figure 9.9). Unfortunately, the families into which these Jurassic-Cretaceous snakeflies are classified are undoubtedly not natural (e.g., Willmann, 1994a; Engel, 2002a). The baissopterids appear to represent a basal grade owing to the dense crossvenation, and the long wings of the "alloraphidiids" render the Mesoraphidiidae paraphyletic.

9.9. *Mesoraphidia luzzii* (Mesoraphidiidae) in Late Cretaceous amber from New Jersey. This is the last known occurrence of this Mesozoic family and one of the most completely preserved Mesozoic snakeflies. AMNH NJ-435; length 9.2 mm.

9.10. *Raphidia funerata* (Raphidiidae) from the Eocene-Oligocene boundary of Florissant, Colorado. This is the most completely preserved Tertiary snakefly from North America. NHML In 26922; length 16.5 mm.

Considerable work remains to elucidate the relationships among Mesozoic snakefly genera conclusively. Tertiary snakeflies are known as compressions from the Eocene-Oligocene boundary of Florissant, Colorado (e.g., Engel, 2002a, 2003d) (Figure 9.10), the Miocene of France (Nel, 1993), and the Oligocene of British Columbia (Handlirsch, 1910), as well as inclusions in Baltic amber (Carpenter, 1956; Engel, 1995c; Weitschat and Wichard, 2002; U. Aspöck and H. Aspöck, 2004).

Even though modern snakeflies are not found in humid, tropical environments, it is clear that the order has not always been so restricted. The Cenomanian forests responsible for producing Burmese amber and in which *Nanoraphidia* occurred were perhaps the most tropical of all the Cretaceous amber localities (Grimaldi *et al.*, 2002). This is also true, although to a much lesser degree, of the amber-producing forests of the middle Eocene in northern Europe and where snakeflies are also known to have occurred (Carpenter, 1956; Engel, 1995c; Weitschat and Wichard, 2002; U. Aspöck and H. Aspöck, 2004). Thus, tropical or subtropical snakeflies persisted at least into the early Tertiary but were perhaps relict and eventually extinguished by the climatic changes marking the Eocene-Oligocene transition. It is also abundantly clear that the order was globally distributed in the past (e.g., species from the Early Cretaceous of South America, Figure 9.8) and has undergone significant extinction. The broader distribution and climatic range of extinct snakefly lineages are likely linked. Overall the Cretaceous was much warmer than the Cenozoic (particularly so following precipitous drops in the Paleogene) and climatic changes perhaps drove many Raphidioptera lineages near extinction. As hypothesized by H. Aspöck (1998, 2000), the decline of the order was perhaps accentuated by the extraterrestrial impact and concomitant climatic disruption at the close of the Cretaceous. Those snakefly lineages not already adapted for colder climates in the early Tertiary would have been quickly extinguished by the Eocene-Oligocene transition. Significant extinction across the order has left the cold-adapted lineage of the Raphidiidae and Inocelliidae free to diversify, albeit within modest limits, in the temperate habitats of the Northern Hemisphere.

MEGALOPTERA: THE ALDERFLIES AND DOBSONFLIES

The Megaloptera are impressive insects, frequently attaining large sizes and, in males of the well-known dobsonfly *Corydalus*, having enormous, sickle-shaped mandibles. In many ways, the Megaloptera are rather primitive, the adults generally resembling large lacewings but retaining a broad anal region in the hind wing, and the aquatic larvae lack the sucking modifications typical of Neuroptera and possess lateral gills. Overall it is difficult to identify derived traits on which to define the order; indeed, Megaloptera has been considered by some to be paraphyletic respective to Raphidioptera (e.g., Achtelig, 1967; Afzelius and Dallai, 1988). As discussed previously for the Neuropterida as a whole, it is highly plausible that the aquatic lifestyle of megalopteran larvae is independently derived from that of the Neuroptera, implying that this life history trait along with the lateral gills are derived traits for the order. The argument that aquatic larvae and lateral gills are primitive features of Holometabola is not supported when comparing Megaloptera with basalmost families of the various basal holometabolous orders (i.e., Coleoptera, Raphidioptera, Hymenoptera, most basal mecopteroids). Despite the lack of compelling evidence for megalopteran monophyly, existence of the group has some support from molecular studies (e.g., Wheeler *et al.*, 2001). A world revision of Megaloptera was published by Weele (1910), but it is significantly dated and of little use for identification. The supraspecific classification has most recently been reviewed by New and Theischinger (1993), who have also, along with the more recent treatments by U. Aspöck and H. Aspöck (2003b), summarized the available information on megalopteran biology. Contreras-Ramos (1999) has catalogued the neotropical species, and H. Aspöck *et al.* (2001) have catalogued the European fauna, while Theischinger and Houston (1988) have catalogued the Australian taxa. The Megaloptera are divided into two extant families – the dobsonflies or hellgrammites, family Corydalidae; and the alderflies, family Sialidae.

Sialidae are typically small (with wingspans less than 30 mm), have the fourth tarsomere bilobed, and lack ocelli (Figure 9.11). The larvae have seven pairs of lateral gills and a terminal abdominal filament but lack prolegs on the distal abdominal segment. Sialids are perhaps further defined by the partial desclerotization of vein MA (Ansorge, 2001). Sialids occur throughout the world, with most species in the Holarctic and belonging to the nominate genus *Sialis*. Larvae occur in freshwater ranging from small ponds and streams to lakes or to large, slow-flowing rivers, frequenting areas of dead vegetation. The immatures are generalist predators of invertebrates, feeding on prey such as annelid worms, mollusks, or other aquatic insects. The larvae crawl onto the shore and pupate in a chamber dug into the soil. Adults are relatively short-lived and feed on pollen. Whiting (1994) investigated the phylogenetic relationships of North American *Sialis* species, but cladistic studies are otherwise lacking for the approximately 70 species of the family. Six to eight genera are sometimes recognized in the family (e.g., New and Theischinger, 1993), but several of these genera likely render *Sialis* paraphyletic and should be reevaluated.

Corydalids are large, robust animals, with wingspans over 30 mm, not uncommonly extending up to 175 mm (Figure 9.12). In contrast to the sialids, corydalids primitively

9.11. An alderfly, *Sialis* sp. (Sialidae: Megaloptera). Length 19 mm.

9.12. A male dobsonfly, *Corydalus* sp. (Corydalidae). Females have mandibles of normal size, and the function of the huge male mandibles remains unknown. Length 82 mm.

retain three ocelli and have the fourth tarsomere simple. Dobsonfly larvae can be differentiated from sialids by the possession of eight lateral gill pairs and prolegs on the distal abdominal segment but lack a terminal filament. The approximately 200 species of corydalids are segregated into two subfamilies and 25 genera – the Corydalinae and the Chauliodinae (sometimes called fishflies). Corydalines occur in the New World, southern Africa, and Asia, while chauliodines are in North America, Chile, Australia, New Zealand, Madagascar, southern Africa, and southeast Asia. The classification of the subfamily Corydalinae was reviewed by Glorioso (1981). While larval chauliodines are found in a diversity of freshwater habitats, the corydalines typically frequent streams. Immatures of both lineages are generalist predators of small invertebrates, but they may sometimes scavenge. Like sialids, pupation occurs in the nearby soil, and adults are short-lived, living 1–2 weeks.

The earliest putative Megaloptera are fossils from the Late Permian of Russia, assigned to the extinct family Parasialidae (e.g., Ponomarenko, 1977a, 2000). Despite their name, parasialids were perhaps stem-group Megaloptera, sister to all other members of the order (e.g., Ansorge, 2001). The only other extinct family presently recognized is Euchauliodidae from the Triassic of South Africa (Riek, 1974a), although it has been suggested that *Euchauliodes* is a polyneopteran allied to the Grylloblattodea (Ansorge, 2001). The families Permosialidae and Tychtodelopteridae, formerly included in the Megaloptera, were transferred to the Miomoptera and Mecoptera, respectively (Rasnitsyn and Quicke, 2002). Numerous larval fossils ranging from the Late Permian through the Cretaceous have been described as Megaloptera, principally owing to lateral abdominal structures on them that might be gills. These fossils have been mostly attributed to the "corydalid-type" (i.e., lacking the terminal filament of sialids; e.g., Ponomarenko, 2002a), but no clearly derived trait truly unites these with Corydalidae. Definitive fossils of corydalid adults are entirely restricted to Baltic amber (Weitschat and Wichard, 2002). Fossil Sialidae were, until recently, confined to the Tertiary, although the family, along with Corydalidae, is assuredly of Mesozoic origin. Most species have been discovered as amber inclusions, principally in Baltic amber (Wichard, 1997; Wichard and Engel, unpubl.), although a single species is known in Eocene French amber (Nel *et al.*, 2002a) and in Dominican amber (Engel and Grimaldi, in prep.). Compression fossils of sialids are also recorded from the Paleocene of Australia (Lambkin, 1992), Oligocene of Turkey (Nel, 1988a), Miocene of France (Nel, 1988a, 1991b), and Pliocene of Germany (Illies, 1967) and as subfossils from the Late Holocene of Germany and other Quaternary sediments (Frey, 1964; Fehler, 1999). Ansorge (2001) demonstrated that the Early Jurassic "mecopteran" *Dobbertinia reticulata* is a sialid. However, *Dobbertinia*, known only from wing fragments, is generally primitive relative to modern Sialidae but does exhibit the partially desclerotized MA vein typical of the family.

NEUROPTERA: THE LACEWINGS, ANTLIONS, AND RELATIVES

The Neuroptera are archaic members of the Holometabola and may be less diverse today than in the past. The order consists of elegant insects, notable for their diaphanous wings laced with intricate venation. Defining features of the Neuroptera include the association of the ninth gonocoxites with the gonarcus and the peculiar sucking-mandibulate mouthparts of the larvae (U. Aspöck *et al.*, 2001). The larval mouthparts are unique in that the maxillae and the mandibles form a sucking tube. Adults and larvae are mostly predaceous, particularly on small, soft-bodied insects such as Sternorrhyncha. Despite being considered archaic among holometabolous insects, the earliest definitive evidence of Neuroptera is the family Permithonidae in the Late Permian of Eurasia. The family is known only from wing fragments and is difficult to assign as it principally preserves primitive features, so it may just represent stem-group neuropterans. Major accounts of the order include the recent phylogenetic studies of U. Aspöck *et al.* (2001) and U. Aspöck (2002). New (1986, 1989) and U. Aspöck and H. Aspöck (2003c) have provided major summaries of the biology of Neuroptera, while H. Aspöck *et al.* (1980, 2001) have catalogued the European fauna, Penny *et al.* (1997) the Nearctic fauna north of Mexico, and New (1996) the Australian taxa.

Nevrorthiformia

The basalmost lineage of the Neuroptera is the suborder Nevrorthiformia, which consists of the single family Nevrorthidae with eleven species in three genera: one in the Mediterranean region, one in eastern Asia, and the third in Australia. Nevrorthids are relatively small lacewings, approximately 12 mm in wingspan, which were formerly classified as a subfamily in the Sisyridae (spongillaflies). Like the sisyrids, larval nevrorthids are aquatic (Figure 9.13), and the adults have some superficial similarities, hence the former unification of the two groups. Nakahara (1958) was the first to demonstrate that the biology of immature nevrorthids was different from that of sisyrids and subsequently Zwick (1967) demonstrated that the two groups were not related, suggesting an isolated position and relict nature of Nevrorthidae. Aside from a few, brief accounts, little is known of nevrorthid biology (e.g., Zwick, 1967; New, 1978; Malicky, 1984). Larvae tend to inhabit streams with fast-moving water, occurring in vegetation and under rocks where they perhaps prey on other aquatic insects. Although nevrorthids most likely pupate on land, Malicky (1984) and Wichard *et al.* (2002) indicated that *Nevrorthus* is capable of pupating in

9.13. The aquatic larva of *Nevrorthus fallax* (Nevrorthidae), a relict group of primitive Neuroptera. Most larval neuropteridans are predatory, as are the adults. Length 12 mm. Photo: P. Duelli, courtesy of H. and U. Aspöck.

freshwater. The nevrorthids were catalogued by Monserrat (1977), with two additions by H. Aspöck *et al.* (1977) and U. Aspöck and H. Aspöck (1983). Nevrorthids are, unfortunately, poorly documented in the fossil record, and some of the few existing records may be misidentifications with Sisyridae. The most definitive records of fossil nevrorthids are adults and larvae of the genus *Rophalis* in Baltic amber (Nel and Jarzembowski, 1997; Weitschat and Wichard, 2002). Grimaldi *et al.* (2002) recorded a larva preserved in mid-Cretaceous amber of Myanmar that may be a nevrorthid, thereby representing the oldest record of the suborder. Nevrorthiformia is excluded from a clade consisting of the remaining two suborders – the Myrmeleontiformia and the Hemerobiiformia. The latter two lineages are united by the reduction of the larval gula (eventually lost in the higher hemerobiiforms) and paired male ectoprocts (of which there are several secondary reversals) (U. Aspöck *et al.*, 2001). Nevrorthidae retain the primitive condition of a large, well-developed gula on the larval head capsule as well as a single, dorsally fused, crescent-shaped, male ectoproct.

Myrmeleontiformia

The Myrmeleontiformia contain some of the largest and most spectacular living members of the Neuroptera (e.g., Figure 9.14, except Polystoechotidae), as well as some of the more impressive of the Mesozoic lacewings (e.g., Figure 9.15). The suborder consists of five extant families; the Psychopsidae (silky lacewings), Nemopteridae (spoon- or thread-winged lacewings), Nymphidae (split-footed lacewings), Myrmeleontidae (antlions), and Ascalaphidae (owlflies). By comparison to the Hemerobiiformia, the myrmeleontiform families are relatively homogeneous. Defining features of the suborder are found in the larval head capsule (the prementum is modified and is segment-like, resembling a labial palp segment) and adult genitalia (U. Aspöck *et al.*, 2001). Most families of the suborder exhibit southern Old World distributions. For example, species of Nymphidae are almost exclusively found in Australia and New Guinea, while Psychopsidae are today confined to southern Africa, Asia, and Australia. Similarly, the Nemopteridae are most abundant in tropical and arid regions of Africa, Asia Minor, and Australia, with relatively weak representation in southern Europe and the New World. Only the families Mrymeleontidae and Ascalaphidae, which are close relatives, have relatively diverse species outside of these regions.

The silky lacewings (family Psychopsidae, ca. 26 species) are nocturnal, with spectacularly patterned, hairy wings – hence their common name, silky lacewings. Adults live for a few months, but the arboreal larvae can live for several years under or within deep crevices of bark, where they are generalist predators. As will be repeatedly seen in many Neuroptera families, psychopsids had a wider distribution during the Mesozoic and Tertiary. Fossil psychopsids or stem-group psychopsids are recognizable from the distinctive venation and are known from as early as the Late Triassic of Australia and Africa, Jurassic-Cretaceous of Asia, North America, and South America, and the Eocene of Europe and North America (Martins-Neto, 2000; Grimaldi, 2000a; Andersen, 2001). The phylogeny and classification of the family was treated by Oswald (1993a).

Nemopterids are among the most visually striking of all lacewings (Figures 9.14, 9.16). The characteristic, thin, elongate, or sometimes spoon- or leaf-shaped hind wings are immediately diagnostic of the family, which today consists of approximately 150 species in principally arid habitats. Two subfamilies are recognized based largely on differences in hind-wing shape (Mansell, 1986, 1992). The Crocinae (thread-winged lacewings) tend to be smaller in size than other nemopterids and have filiform hind wings. Nemopterinae (spoon-winged lacewings), on the other hand, are usually larger and have the elongate hind wings apically dilated. The hind wings may also bear a bulla, or patch of specialized setae, which has been believed to function as a sex organ (Tjeder, 1974). Crocines are also noteworthy for the larvae,

9.14. Select Recent Neuroptera. Note the superficial similarity between the primitive hemerobiiform, Polystoechotidae, and other families that are Myrmeleontiformia. Not to same scale.

9.15. *Petropsychops superba* (Osmylopsychopsidae) from the Triassic of Denmark Hill, Australia. UQC2135–6; length 29 mm.

appearing fanciful with their robust heads and bodies separated by a peculiar long neck (Figure 9.17). Larvae of nemopterines are more squat and primitive in most traits relative to Crocinae. Adults feed on pollen (e.g., Stange and Williner, 1981; Picker, 1984), while the larvae are generalist predators. Crocine larvae occur in caves, under stones, or in crevices (e.g., Mansell, 1980, 1981), while those of Nemopterinae are found in the sand or soil of more open areas and behave like antlions. Nemopterines burrow into the sand head-first and, upon capturing prey, consume their victims beneath the surface. The earliest fossil nemopterid is from the Early Cretaceous of South America (Martins-Neto, 2000). Today nemopterids do not exist in North America, but fossils

9.16. *Nemoptera bipennis* (Nemopteridae) from Spain. The extremely long hind wings are distinctive to many species of this family. Photo: P. Duelli, courtesy H. and U. Aspöck.

9.17. The mature larva of *Laurhervasia setacea* (Nemopteridae) from Namibia, showing the remarkable neck typical of the family. Photo: P. Duelli, courtesy H. and U. Aspöck.

from the Eocene-Oligocene boundary of Florissant, Colorado, document the occurrence of the family in this region during the Tertiary (Cockerell, 1907a; Pierce and Kirkby, 1959) (Figure 9.18). Like many families of insects with presently southern distributions, nemopterids were widespread and then retreated from the Nearctic and likely from the northern Palaearctic as well, probably during the climatic shift of the Eocene-Oligocene transition.

The 27 or so species of Nymphidae were historically treated as two separate families: Myiodactylidae and Nymphidae s.str. Little is known of nymphid biology, but where known the larvae prey on termites or on caterpillars (Gallard, 1935; New, 1986). Like other neuropteran families, the larvae will cover themselves with debris presumably as camouflage. Immatures of *Nymphes* occur in leaf litter or under logs. By contrast, larvae of *Myiodactylus* and *Osmylops* are apparently arboreal, do not cover themselves with debris, and are relatively flattened; the latter two traits are perhaps factors of their living under or within crags of bark. Adults are sometimes rather large, with wingspans approximately

9.18. Although nemopterids do not occur today in North America, *Marquettia americana* (Nemopteridae) from the Florissant Formation in Colorado shows that they were present there as well during the Early Tertiary. NHM In.8428; wingspan 64 mm.

80 mm in some species. While some species can have relatively broad wings, many are more elongate and therefore resemble to some degree adult antlions (Myrmeleontidae), although, unlike the latter family, nymphids have long, filiform antennae. The phylogeny of the family was investigated by New (1984). Fossil nymphids are known from as far back as the Jurassic of Eurasia and are remarkably similar to modern genera (Carpenter, 1929; Panfilov, 1980). The only other published records are fossils in Baltic amber (Krüger, 1923; MacLeod, 1970). The fossil records of the family also show a more extensive distribution in the past, in contrast to the modern isolation of the family in New Zealand and Australia today.

The antlions (Myrmeleontidae) have received more attention from biologists than perhaps any other family of the Neuroptera. The fascinating pit traps dug by some myrmeleontid larvae have stimulated considerable research into the biology of the group. The family is also the most diverse group of neuropterans, with around 2,000 species throughout the world but concentrated in arid, sandy regions. The higher classification of the family is in flux, and a comprehensive cladistic study is still needed. Stange and Miller (1990) provided the most recent account of the family, based on characters of the larvae, which like most neuropterans are predaceous. The immatures of most antlion groups do not dig the characteristic pits with which we are all familiar. In these lineages, the larvae live in detritus, on bark, under rocks, or underneath the soil surface and become covered with lichens that blend them with their surroundings, particularly since they remain motionless for long durations before suitable prey passes by. Genera with pit-building larvae, which actually constitute a relatively small proportion of the family, construct their distinctive pits by scuttling backward into a tighter spiral while displacing sand to the sides, forming a small cone-shaped depression. They stop when the critical angle of pit slope is reached, or that point where the addition of a grain or two of sand causes the side to slump in. The larva lies in wait at the bottom of the bit, covered in sand. An ant or other prey may stumble into the pit, where it is quickly dispatched, but if it teeters at the edge, the larva will flick sand grains up at the rim of the pit to dislodge the prey. The condition of the soil can dramatically affect the construction and morphology of the pit (e.g., Reddy and Kotikal, 1986; Lucas, 1989). Indeed, the effects of the soil alter the stability of the sloping sides of the pit, which must be near the point of collapse in order to ensnare passing prey. Finer grain soils are apparently preferred by antlions because these more readily dislodge prey, rather than for building a steeper

9.19. An Early Cretaceous antlion, one of various Neuroptera from Brazil's Santana Formation. AMNH; body length 32 mm.

pit (Botz *et al.*, 2003). The larvae of some species can move readily through the sand and may even seek out the immatures of other antlions underneath the sand (Stange, 1970). Myrmeleontids are documented from the Early Cretaceous (perhaps latest Jurassic) of China (Ren and Guo, 1996), Brazil (Martins-Neto, 2000) (Figure 9.19), and Israel (Dobruskina *et al.*, 1997). *Palaeoleon ferrogeneticus* from Labrador has been traditionally classified as a myrmeleontid (Rice, 1969), but some authors prefer to recognize multiple antlion families for the Cretaceous taxa (e.g., Dobruskina *et al.*, 1997; Martins-Neto, 2000) and place this fossil in its own lineage, the Palaeoleontidae. Whalley (1980) reported an incomplete, unnamed specimen of a putative myrmeleontid in Lebanese amber, but the only definitive myrmeleontid in amber is a species of the

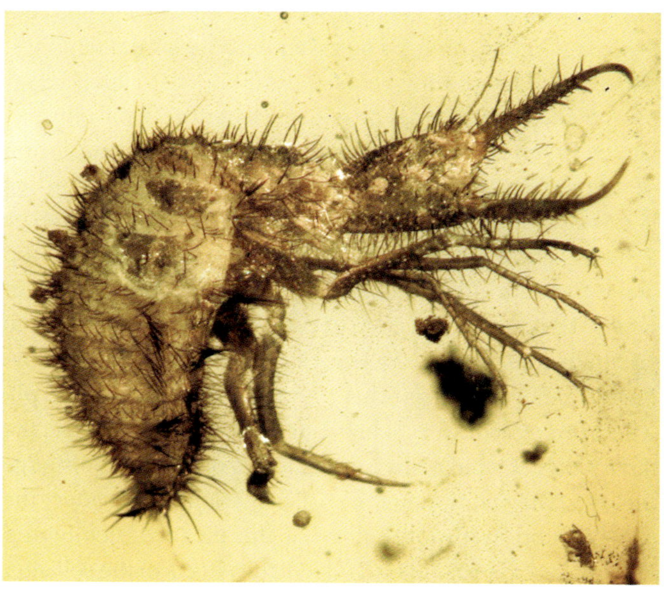

9.20. A mature larva of the antlion *Porrerus* (Myrmeleontidae) in Early Miocene Dominican amber. Not all larval antlions construct pits in the ground; some live amongst leaf litter or on tree trunks, which is probably how this one became captured in resin. Morone Collection, M1282; length 4.5 mm.

9.21. Adults of *Cordulecerus* owlflies (Ascalaphidae), roosting in Panama. Photo: P. J. DeVries.

living genus *Porrerus* in Miocene Dominican amber (Engel and Grimaldi, in prep.), which is known from both adults and larvae (Figure 9.20). The only other Tertiary record of the family is *Dendroleon septemmontanus* (Statz, 1936a).

The owlflies, family Ascalaphidae, are close relatives of the antlions and comprise approximately 430 species. Adults have long, clubbed antennae that readily distinguish them, among other traits, from myrmeleontids (Figure 9.21). The family is most diverse in both xeric and mountainous regions of the subtropics or tropics. Owlflies are frequently large, and some groups (e.g., Haplogleniinae) appear like hybrids of other orders of insects, even superficially similar to dragonflies. Like dragonflies, adults are swift aerial predators, and some species even roost. Female ascalaphids lay eggs in clusters on twigs or grass stems. The first instar larvae aggregate at twig apices into a defensive ring (Henry, 1972). The larvae eventually disperse to live solitarily in the litter or on trees where they are generalist predators, frequently covering themselves with soil or debris to camouflage themselves (Henry, 1977) (Figure 9.22). The larvae inject their victims with poison, which serves to paralyze the prey while the owlfly begins feeding (Henry, 1977). The geological record of Ascalaphidae is exceedingly sparse with four definitive ascalaphids from Tertiary compression fossils in Europe and the Russian Far East (e.g., Nel, 1991b; Makarkin, 1998), one species known only from the larva in Baltic amber (MacLeod, 1970), and one known from both the adult and the larva in Dominican amber (Engel and Grimaldi, in prep.) (Figures

9.22. A larval owlfly (Ascalaphidae) in Panama, showing the large, impaling jaws typical of most larval myrmeleontiform Neuroptera. The jaws are comprised of the mandibles and maxillae. Photo: P. J. DeVries.

9.23, 9.24). A fossil from the Early Cretaceous of China has been dubiously assigned to the family but lacks definitive traits of Ascalaphidae (Ren, 1995).

Myrmeleontiforms are relatively diverse in the fossil record. Aside from the various records of modern lineages in deposits ranging from the Jurassic through Pliocene, several extinct families have been described from the Mesozoic. Many of these extinct families appear to represent stem groups to the Nymphidae-Myrmeleontidae-Ascalaphidae, to the Psychopsidae-Nemopteridae, or to both. Unfortunately most of these families are documented strictly as fore- or hind wing compressions, and we know nothing of their body morphology. However, this does not diminish their significance nor their esthetic appeal. Furthermore, some of these were among the largest neuropterans known. Perhaps the most famous is the Eurasian family Kalligrammatidae, which were large insects with spectacularly patterned wings, and among the few fossil neuropterans with their bodies preserved (Figure 9.25). Kalligrammatids were the "butterflies" of the Jurassic. Their densely setose and patterned wings and bodies and their long palpi gave them the superficial appearance of large moths or butterflies, despite occurring 60–90 MY before the first papilionoid fluttered. Kalligrammatids were apparently rather diverse as well, with numerous genera and species occurring in the Jurassic and earliest Cretaceous faunas of Europe and Central Asia. The large, broad wings were sometimes marked with distinct patterns, most commonly with eyespots similar to those found on large moths and fulgoroid plant hoppers, who flash them when startled, presumably to mimic vertebrate eyes. It is tantalizing to speculate how the eyespots of kalligrammatids served a similar function of startle defense. It is equally interesting as to what peculiar Jurassic vertebrates kalligrammatids might have been mimicking, perhaps some small, insectivorous reptile or even an archaeopterygid bird? The kalligrammatids, with

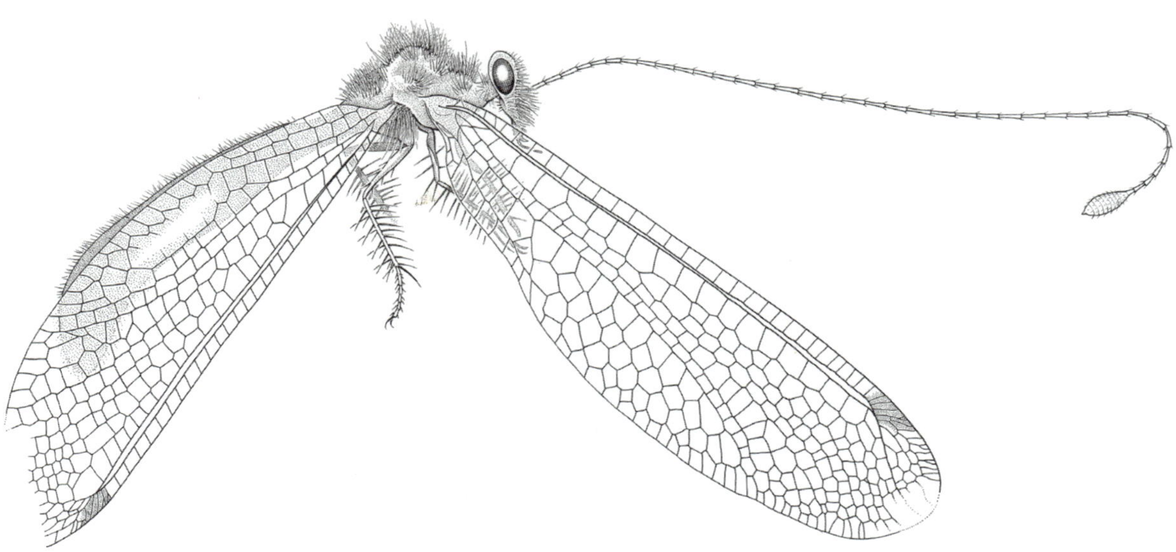

9.23. An adult *Ululodes* (Ascalaphidae) in Early Miocene Dominican amber. AMNH; length 31 mm.

9.24. An *Ululodes* larva in Miocene Dominican amber. Morone Collection, M3550; total length 8 mm.

their *vena triplica* (apical fusion of Sc, R, and Rs in the hind wing), are assuredly related to the Psychopsidae-Nemopteridae, but there is no evidence for uniting them with Psychopsidae (e.g., Andersen, 2001), as this would render the latter family paraphyletic. The family Brongniartiellidae may also be related to this lineage (as asserted by Andersen, 2001), but pending more extensive evidence (e.g., the vein triplica is not completely formed in many taxa), we have retained it in an unresolved position, perhaps allied to the nymphids. Several other families have been described, some of which may prove to be stem-group Nymphidae (e.g., Nymphitidae), or stem groups to the entire Nymphidae-Myrmeleontidae-Ascalaphidae lineage.

Hemerobiiformia

The Hemerobiiformia is the most heterogeneous group of neuropteridans, and its monophyly is the most controversial. The suborder is defined on the basis of larval traits, in particular the elongate cardines, a modified neck, and the formation of the "maxillary head" (U. Aspöck *et al.*, 2001), in which the underside of the head is almost entirely composed of maxillary sclerites. As presently defined, the suborder includes the families Polystoechotidae (giant lacewings), Ithonidae (moth lacewings), Osmylidae (osmylid lacewings), Chrysopidae (green lacewings), Hemerobiidae (brown lacewings), Coniopterygidae (dustywings), Sisyridae (spongillaflies), Dilaridae (pleasing lacewings), Mantispidae (mantispid lacewings), Rhachiberothidae (thorny lacewings), and Berothidae (beaded lacewings). Although relationships among the families of the smaller Myrmeleontiformia have been rather stable, our understanding of hemerobiiform phylogeny has changed dramatically over the years and is still

9.25. The giant lacewing *Meioneurites spectabilis* (Kalligrammatidae) from the Late Jurassic of Karatau in Kazakhstan. Kalligrammatids are strikingly convergent with butterflies and may even have been dayflying. They lived during the Jurassic and Cretaceous. PIN 2784/1069.

in a state of flux. The most recent and comprehensive treatment is that of U. Aspöck *et al.* (2001). Hemerobiiforms did not appear until the Mesozoic, and the only Triassic records are those of the Osmylitidae, which may be a stem group to the clade consisting of Polystoechotidae, Osmylidae, and related families. Alternatively, osmylitids may represent a stem group to the entire suborder as it retains primitive features of most groups. Unfortunately, osmylitids are known only from wings ranging in age from the Triassic through the Late Jurassic. The only other extinct family recognized in Hemerobiiformia is the Mesithonidae, which are either stem-group Berothidae (perhaps even a synonym of Berothidae) or a stem group to the Berothidae-Rhachiberothidae.

The giant lacewings, family Polystoechotidae, comprise three or four extant and highly disjunct species restricted to the New World: *Polystoechotes* in Chile and North America (Figure 9.14), *Platystoechotes* in California, and *Fontecilla* in Chile (Carpenter, 1940; Oswald, 1998a). Polystoechotids are large lacewings, with wingspans of 40–75 mm in modern species. Essentially nothing is known of polystoechotid biology, and the family is clearly relict today. Polystoechotids are generally primitive, sharing several plesiomorphic features with other relatively basal families of Hemerobiiformia, such as Ithonidae and Osmylidae. Indeed, owing to the numerous primitive features of wing venation in these families, it has been difficult to assign fossils to particular families, and several have been assigned to either Osmylidae or Polystoechotidae. However, polystoechotid or polystoechotid-like fossils, easily distinguishable from Osmylidae, are known from several Mesozoic deposits (Lambkin, 1988; Ren *et al.*, 2002). That these stem-group polystoechotids are found in the Jurassic and early Cretaceous of Asia, Australia, and Europe demonstrates that the lineage was once widespread, has experienced significant extinction, and remains in just the Nearctic and Chilean regions today.

Like the Polystoechotidae, the Recent distribution of the family Ithonidae is also dramatically disjunct, with two genera in North America (Penny, 1996), three in Australia (Riek, 1974b), and one in southeast Asia (Barnard, 1981). The latter genus, *Rapisma*, has been historically classified in its own family, the Rapismatidae. *Rapsima* consists of about 20 species in montane areas of southeast Asia (Barnard, 1981; Barnard and New, 1985) but is clearly related to the more typical ithonids as is evidenced by the discovery of *Adamsiana* in Honduras (Penny, 1996). "Rapismatids" are merely specialized, montane Ithonidae. Ithonids are robust lacewings with wingspans around 60 mm, whose biology has not been elucidated. Where observed, ithonids oviposit in loose soil or sand, and larvae appear to be subterranean (New, 1986). Larvae tend to be long-lived (2–3 years) and associated with decaying bark. Ithonid larvae may actually feed on decaying plant tissues, perhaps even on decaying bark (Gallard, 1932). Interestingly, ithonids undergo mass emergences, although many fall prey to ants or birds (Tillyard, 1922c; Riek, 1974b), with relatively few adults surviving. Ithonids are presently unknown in the fossil record.

Osmylids are generally primitive lacewings and have tentatively been proposed as the sister group to the more familiar Chrysopidae. The family is most diverse in the Old World, represented by only two (of a total of eight) subfamilies in South America and are presently unknown in the Nearctic. There is speculation that some osymlids have aquatic larvae because adults prefer moist habitats near water sources. However, there is no evidence as yet for aquatic larvae; instead, they appear to live simply in damp, terrestrial habitats (Figure 9.26). Osmylids, or stem-group osmylids, abound in the fossil record, extending from the Late Jurassic through the late Tertiary (e.g., Lambkin, 1988).

Chrysopids are the most intensely studied of all Neuroptera with the antlions (Myrmeleontidae) (Figure 9.27). The Chrysopidae is also the family most individuals envision when they think of Neuroptera owing to the abundance of many species. The "green" lacewings, with approximately 1,500 species, are distributed worldwide and have been the

9.26. A larva of *Osmylus fulvicephalus* (Osmylidae). Photo: P. Duelli, courtesy of H. and U. Aspöck.

9.27. A green lacewing, *Chrysopa* (Chrysopidae), feeding on nectar in Vancouver, Canada. Adults of many myrmeleontiforms supplement their carnivorous diet with nectar and pollen. Photo: R. Swanson.

subject of several phylogenetic studies (Schlüter, 1984; Brooks and Barnard, 1990; Brooks, 1997). Many species are indeed green, as their common name suggests, but occur in other colors as well, like brown, yellow, and red. Chrysopid larvae are active, generalist predators, and most species are arboreal (Figure 9.28). Some species rapidly cover themselves with debris (held in place by long, curved setae), which may include remains of their prey (Eisner *et al.*, 1978; Eisner and Silberglied, 1988). Adults are commonly predators, although some are specialized to feed on honeydew and harbor symbiotic yeasts, presumably to facilitate digestion. Some adult green lacewings have "ears" located in the base of the radial vein of the forewings, representing the smallest tympanal organs known. The organ functions similarly to those of noctuid moths and is used to detect bat signals, allowing the lacewing to avoid becoming prey itself. Lacewings also communicate via substrate-borne vibrations that are detected by hearing organs in the legs. This vibrational communication functions in courtship and is species-specific, representing a useful character for differentiating morphologically cryptic taxa. Chrysopids have been extensively employed in biological control programs, particularly for the control of aphid and coccid populations (e.g., New, 1975, 1999, 2002; McEwen *et al.*, 2001). Although numerous chrysopid fossils have been described or reported from the Early Jurassic through the Miocene (e.g., Carpenter, 1935; Statz, 1936a; Adams, 1967; Panfilov, 1980; Ansorge and Schlüter, 1990; Makarkin, 1991, 1994, 1998; Willmann and Brooks, 1991; Nel and Henrotay, 1994; Ren and Guo, 1996) (Figure 9.29), only four have been

9.28. *Chrysoperla plorabunda* (Chrysopidae) in California feeding on a nymphal aphid. Length 8 mm. Photo: P. Duelli, courtesy of H. and U. Aspöck.

9.29. An extinct *Chrysopa* (Chrysopidae) from the Eocene of British Columbia, Canada. MCZ.

discovered in amber – one in Baltic, three in Dominican (Weitschat and Wichard, 2002; Engel and Grimaldi, in prep.) (Figures 9.30, 9.31).

The brown lacewings, family Hemerobiidae, comprise approximately 560 species of typically small lacewings superficially resembling chrysopids. Hemerobiids occur throughout the world and are generalist predators, although some adults feed on honeydew as in Chrysopidae (Stelzl, 1990). The phylogeny and classification of the family have been investigated by Oswald (1993b, c, 1994, 1996). The earliest records of hemerobiids are fossils of *Promegalomus* from the Late Jurassic of Kazakhstan (originally attributed to their own family) and *Mesohemerobius* from the Early Cretaceous of China (Oswald, 1993b), but the family generally is scarce during the Mesozoic. Tertiary hemerobiids are more abundant, being known in Baltic (e.g., Krüger, 1923; Weitschat and Wichard, 2002) and Dominican ambers (Oswald, 1999), as well as some compressions from the Eocene of Denmark (Henriksen, 1922), Eocene of Germany (Illies, 1941), Oligocene of England (Jarzembowski, 1980), and British Columbia (Scudder, 1878). The phylogenetic implications of the fossil hemerobiids have not been critically explored.

The dustywings, family Coniopterygidae, appear as the least likely members of the Neuroptera. These minute insects (ca. 450 species), typically with wingspans less than 7 mm, are easily recognizable among all other neuropterous families (Figures 9.32, 9.33). Aside from their Lilliputian size, their reduced and characteristic wing venation and dense covering by a waxy or mealy secretion (and hence the name dustywings) are immediately diagnostic. Indeed, the dramatic difference from other Neuroptera led to the general belief that they had differentiated from all other lineages of the order early in geological history, perhaps representing the sister group to all other Neuroptera except Ithonidae (e.g., Withycombe, 1924; Meinander, 1972) and arising in the Paleozoic (e.g., Schlüter, 1986). It is now understood that the dustywings are merely a highly derived lineage of hemerobiiform Neuroptera (e.g., U. Aspöck *et al.*, 2001), and their reduced venation is most likely a factor of their size. Interestingly, these, the tiniest members of the order, have attracted considerable attention from biologists, principally owing to the potential role of dustywings in controlling mite populations (e.g., Fleschner, 1950; Fleschner and Ricker, 1953; Putman

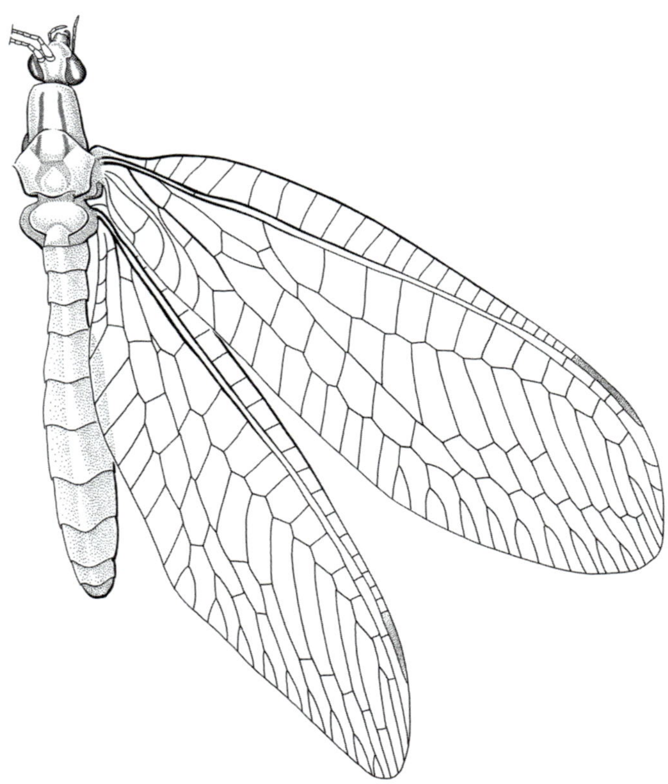

9.30. A Miocene *Chrysopa* (Chrysopidae) in Dominican amber. M1221; length 12.1 mm.

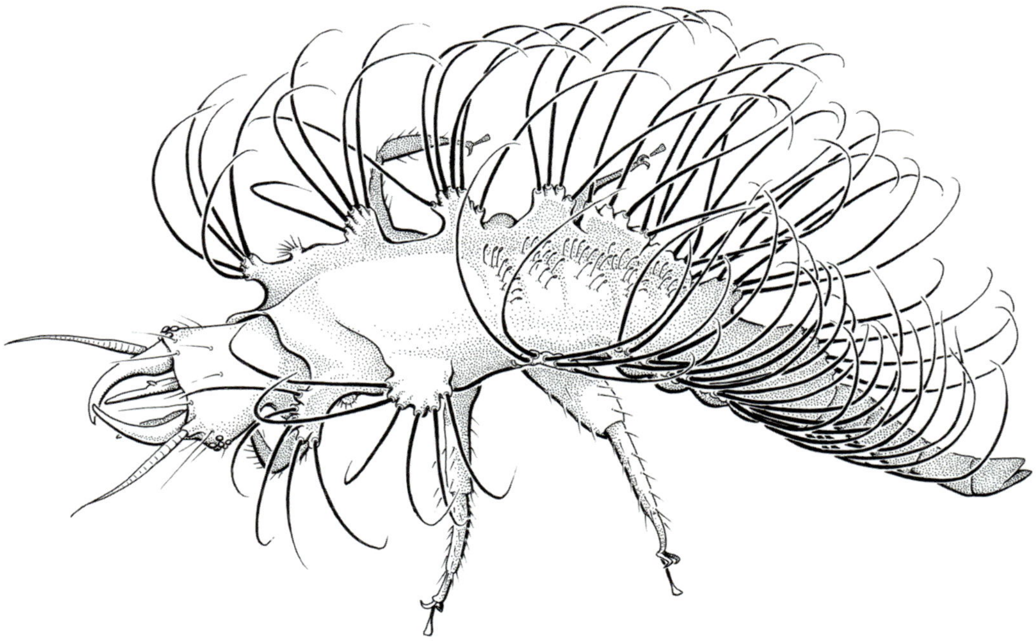

9.31. An early instar larval lacewing (Chrysopidae), in Miocene amber from the Dominican Republic. Some lacewing larvae collect debris in the long setae on their back, which then camouflauges them. AMNH; length 1.3 mm.

and Herne, 1966; New, 1999). Although adults can be predatory, they apparently feed more frequently on plant tissues. The larvae, however, are aggressive predators of small arthropods such as mites and Sternorrhyncha. The family has received considerable attention systematically, and the world fauna has been revised (Meinander, 1972), and numerous taxa have been since added. Despite some phylogenetic hypotheses for dustywing lineages (Meinander, 1972, 1979), no cladistic study of the family has been undertaken, and several genera and tribes are doubtfully monophyletic. Indeed, even the monophyly of the subfamilies is of some question. The earliest evidence of dustywings is *Juraconiopteryx zherichini* from the Late Jurassic of Kazakhstan (Meinander, 1975), apparently referable to Aleuropteryginae. The only other Jurassic "dustywing," *Archiconiopteryx*, was discovered to be a hemipteran (Ansorge, 1996). All other definitive dustywing fossils are preserved in amber except for a compression from the Oligocene of France (Nel, 1991b). Primitive species of both subfamilies are recorded from Cretaceous (Meinander, 1975; Whalley, 1980; Azar *et al.*, 2000; Grimaldi, 2000a; Engel, 2002e, 2004d) (Figures 9.32, 9.33) and Tertiary ambers (e.g., Meinander, 1998; Dobosz and Krzemiński, 2000; Engel and Grimaldi, in prep.), although most Mesozoic species are of the Aleuropteryginae and primitively retain three-branched medial veins (unknown in the modern fauna). Unfortunately, the aforementioned difficulties concerning the monophyly of various higher dustywing taxa compromise meaningful discussions of their biogeography. Thus, apparent disjuncts such as the occurrence of *Spiloconis* in Dominican amber (Meinander, 1998; Engel and Grimaldi, in prep.) may be the result of paraphyletic, taxonomic definitions rather than localized extinction.

The cosmopolitan Sisyridae, with about 60 modern species, are remarkable neuropterans whose larvae are specialized predators of freshwater sponges. Adult females oviposit on vegetation or other materials overhanging streams or ponds, and the newly hatched larvae drop immediately into the water. The larvae actively seek out sponges and then settle to feed by piercing the cells with their elongate mouthparts. The larvae leave the water and spin a loose, silken cocoon in the surrounding vegetation or crevices where pupation occurs. The adults are omnivorous, feeding on pollen, fungi, and other plant materials, in addition to small arthropods such as aphids, mites, or even eggs of other

9.32. A mid-Cretaceous dustywing, *Glaesoconis baliopteryx* (Coniopterygidae), in amber from Myanmar. Coniopteryids are the smallest neuropterans. AMNH Bu198; length 1.3 mm.

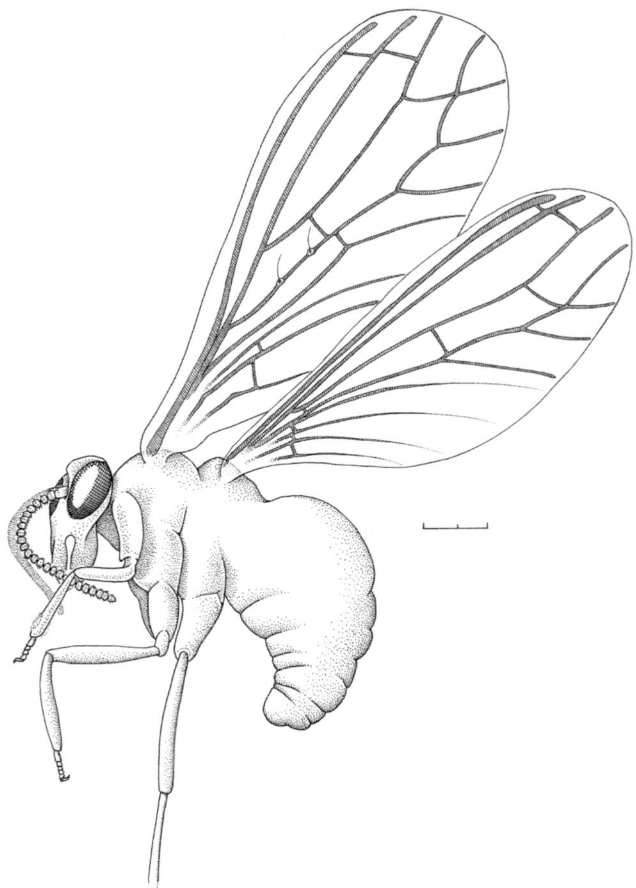

9.33. *Apoglaesoconis ackermani* (Coniopterygidae) in Late Cretaceous amber from New Jersey. AMNH NJ-294; length 1.2 mm.

insects (Tjeder, 1944; Parfin and Gurney, 1956; Kokubu and Duelli, 1983). Although described as the earliest sisyrid, *Cratosisyrops* from the Early Cretaceous Santana Formation in Brazil (Martins-Neto, 1997) preserves no diagnostic features of the family and must be considered as incertae sedis. Definitive fossil spongillaflies are known in Eocene ambers from France and the Baltic (Nel and Jarzembowski, 1997; Nel *et al.*, 2002b). A putative sisyrid has been described from an Eocene wing fragment from the Isle of Wight (Jarzembowski, 1980).

As noted, relationships within the Hemerobiiformia are contentious and the focus of continued investigation, but one clade is consistently recovered within the suborder. The "dilarid clade" comprises the families Dilaridae, Mantispidae, Rhachiberothidae, and Berothidae and is easily the most strongly supported lineage among the hemerobiiforms, based on genitalic and larval characters (e.g., U. Aspöck *et al.*, 2001). The most notable of these characters are the flattened head capsule and the long postmentum. Relationships within this clade are also quite stable, with the Dilaridae being the living sister group to all other families. The sole controversy in the dilarid clade is the position of Rhachiberothidae, which has been considered as basal to either Mantispidae or to Berothidae (Willmann, 1990b, 1994b; U. Aspöck and Mansell, 1994) and has accordingly been elevated or demoted in rank (from family to subfamily) by many authors. Despite the controversy, most evidence supports this family as the closest living relative to Berothidae (U. Aspöck, 1992, 1995, 2002; U. Aspöck and Mansell, 1994; U. Aspöck *et al.*, 2001).

The pleasing lacewings, family Dilaridae, are distinctive lacewings in which males have pectinate antennae and females have long ovipositors. Almost nothing is known of dilarid biology; most information has been derived from a few observations of *Nallachius americanus*. In this species, the larvae occur under bark of dead *Quercus* or *Liriodendron* trees and are apparently predators of soft-bodied, wood-dwelling arthropods (Gurney, 1947; MacLeod and Spiegler, 1961). Larvae of *Dilar septentrionalis*, however, occur in the soil (Ghilarov, 1962; Oswald, 1998b). The 69 or so species of the family are segregated into two subfamilies, Dilarinae and Nallachiinae (Oswald, 1998b; Mirmoaydi and Yassayie, 1999; Oswald and Schiff, 2001), the former occurring in the Palaearctic and Oriental regions and the latter principally in the New World, with a species in Vietnam and southern Africa each. Given the abundance of Early Cretaceous records for other families in the dilarid clade as well as the basal phylogenetic position of pleasing lacewings within this lineage, the dilarids are presumably rather ancient, perhaps relict today. Unfortunately, while the dilarids may have diverged from close relatives sometime in the Jurassic, the sole record of the family comes from the Tertiary: *Cascadilar eocenicus*, in Baltic amber (Engel, 1999d). Interestingly, the species, with its doubly pectinate basal flagellomere, may be related to some southeast Asian species of *Dilar*, in which case *C. eocenicus* should be transferred to *Dilar*. As has been noted repeatedly, a biogeographical connection between modern southeast Asian and southern African taxa and fossils in Baltic amber is common (e.g., Larsson, 1978; Weitschat and Wichard, 2002).

The raptorial Mantispidae are cosmopolitan in distribution but typically more diverse and abundant in the tropics. Mantispids are immediately recognizable for their mantis-like appearance, with stout, spiny forelegs, elongate "necks," and well-developed eyes (Figure 9.34). The family consists of highly specialized parasites of spiders or social Hymenoptera. Many mantispids are mimics of aculeate Hymenoptera, particularly polybiine wasps (Vespidae), presumably escaping some predation pressures because it is doubtful that they prey on vespids. Lambkin (1986) has undertaken the most recent cladistic study of the family, recognizing four subfamilies – Symphrasinae, Drepanicinae, Calomantispinae, and Mantispinae – the latter two being sister groups and symphrasines basal to them all. Symphrasines, like other neuropterans, are generalist predators, feeding on the soft-bodied immatures of various other

9.34. A living mantispid. Some tropical species are remarkable mimics of vespid wasps.

insects such as larval Diptera, Coleoptera, and Lepidoptera (Redborg, 1998). Calomantispines are also generalist predators, feeding on sedentary, soft-bodied arthropods, but the biology of Drepanicinae unfortunately remains unknown. This rather typical neuropteran biology does not, however, apply to the Mantispinae. Mantispines are obligate predators of spider egg cases, and some larvae are ectoparasitoids, feeding on the hemolymph of adult spiders before production of the spider's egg case (Redborg, 1998). Given the abundance of spiders in the fossil record, particularly in fossilized resins, it is remarkable how few fossil mantispids are known. This may reflect a young age for the Mantispinae, perhaps having radiated during the Tertiary, but this relationship

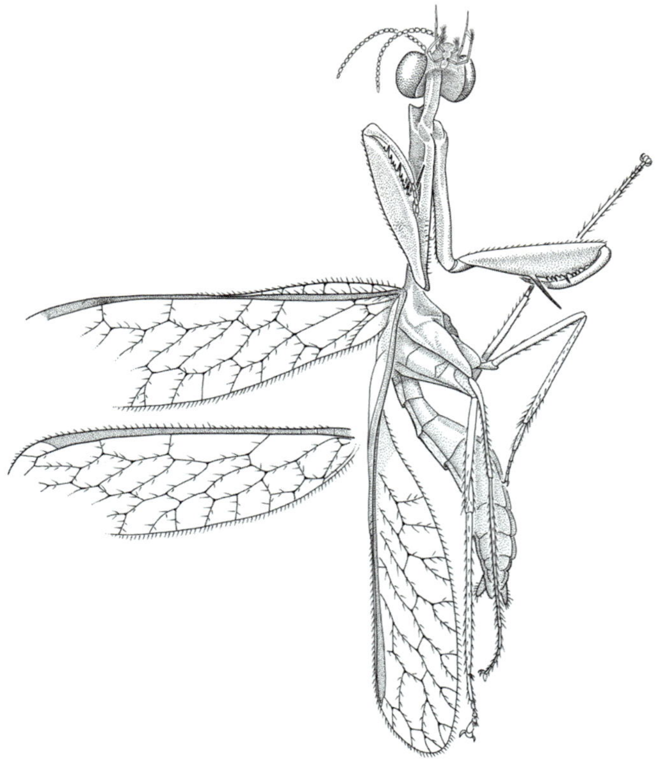

9.35. A rare *Mantispa* in Oligo-Miocene amber from Chiapas, Mexico. AMNH CH-MP5; length 10.4 mm.

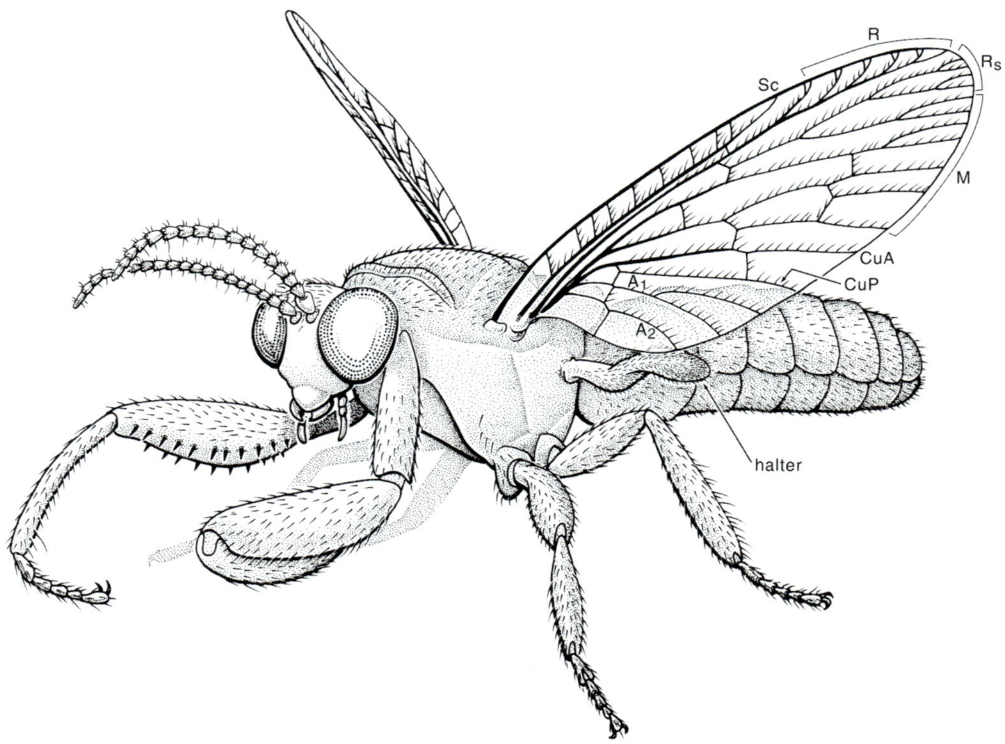

9.36. Reconstruction of the unusual neuropteran *Mantispidiptera henryi*, the genus of which is only known in mid-Cretaceous amber from New Jersey. Forewing venation is highly reduced; and the hind wings are dipteran-like halteres. The genus is apparently closely related to Mantispidae. Body length 3.7 mm; redrawn from Grimaldi (2000).

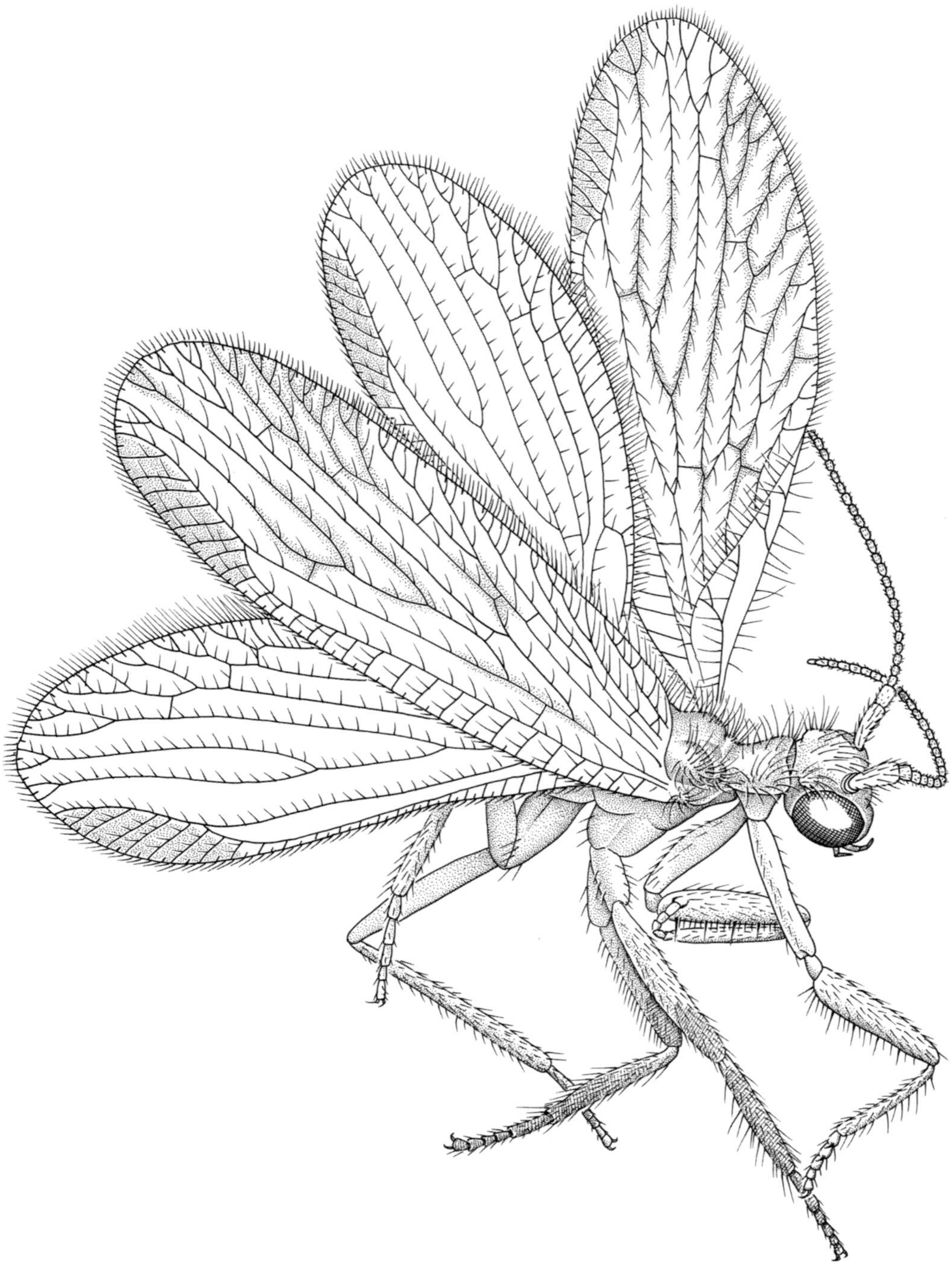

9.37. *Rhachibermissa splendida* (Rhachiberothidae), in mid-Cretaceous amber from New Jersey. While thorny lacewings are today relict in Africa, the family was once widespread. AMNH NJ-615; length 3 mm.

remains speculative as of yet. Indeed, fossils of Mantispinae are not known until the Oligocene and Miocene (Nel, 1988b; Engel and Grimaldi, in prep.) (Figure 9.35). Certainly Cretaceous mantispids are known, but they either belong to the extinct subfamily Mesomantispinae or are putatively members of the basal Drepanicinae (Makarkin, 1990, 1996). The oldest record of the mantispids is *Promantispa similis* from the Late Jurassic of Kazakhstan, although likely a mantispid subfamilial position is uncertain because it preserves mostly primitive traits (Lambkin, 1986). The Baltic amber *Whalfera venatrix*, originally described as a mantispid (Whalley, 1983), has been tentatively transferred to the Rhachiberothidae (see discussion that follows). One of the more remarkable of the extinct Mesozoic mantispids is the enigmatic *Mantispidiptera* (Figure 9.36), named for the reduction of the hind wings to halteres, convergent with those in true flies. *Mantispidiptera* is known only from two species in Late Cretaceous amber from New Jersey (Grimaldi, 2000a).

The family Rhachiberothidae has been of considerable controversy and has been considered at different times during its history as a subfamily of Berothidae (Tjeder, 1959) or even of Mantispidae (Willmann, 1990b, 1994b). The "thorny lacewings" generally resemble other berothids except in the raptorial forelegs, which are convergent with those of the mantispids. While thorny lacewings are today confined to sub-Saharan Africa (U. Aspöck and Mansell, 1994; U. Aspöck and H. Aspöck, 1997), they were clearly distributed throughout the world in the Cretaceous and perhaps into the Tertiary as well, disappearing from other regions apparently during the Eocene-Oligocene transition. Fossil rhachiberothids have been discovered in Cretaceous ambers from Lebanon (Whalley, 1980), France (Schlüter, 1978), New Jersey (Grimaldi, 2000a) (Figure 9.37), and Myanmar (Engel, 2004c, unpubl. data). In addition, *Whalfera venatrix* in Baltic amber (Whalley, 1983; Engel, 2004c), which has at times been placed in Mantispidae (Whalley, 1983), is more likely a rhachiberothid (e.g., U. Aspöck and Mansell, 1994; Engel, 2004c), representing the most recent fossil record of the family and documenting their occurrence outside of Africa during the early Tertiary.

The berothids, or beaded lacewings, comprise about 100 species of superficially hemerobiid-like neuropterans. Berothids occur throughout the world, although all but one subfamily has a rather restricted distribution. While Berothinae occur in most regions of the world, with a particular diversity in Australia and Africa, the Cyrenoberothinae are confined to Chile and southern Africa; the Nosybinae, to Africa; the Nyrminae, to Anatolia; the Protobiellinae, to Australia and New Zealand; and the Trichomatinae, to Australia (U. Aspöck and Nemeschkal, 1998). The biology of berothids is rather primitive for Hemerobiiformia. The larvae are generally predaceous, those of *Lomamyia* preying on termites (Gurney, 1947; Tauber and Tauber, 1968), and occur in detritus or dead logs. Indeed, larvae may use chemical baits to lure in their prey (Johnson and Hagen, 1981). Unfortunately, the biology of the family has been very poorly documented. The classification and phylogeny of the family has been investigated by MacLeod (1967), MacLeod and Adams (1967), U. Aspöck (1986), U. Aspöck and H. Aspöck (1988), and U. Aspöck and Nemeschkal (1998). The earliest definitive berothids are those from the Early Cretaceous, although the Jurassic mesithonids are possibly primitive Berothidae. Cretaceous berothids have been recovered in Lebanese (Whalley, 1980), Burmese (Grimaldi *et al.*, 2002), New Jersey (Grimaldi, 2000a), and Canadian ambers (Klimaszewski and Kevan, 1986). Tertiary berothids are presently known only from Baltic amber (Krüger, 1923; Weitschat and Wichard, 2002), and none yet have been found in Dominican or Mexican ambers.

10 Coleoptera and Strepsiptera

There is a story, perhaps apocryphal, of the distinguished British biologist, J. B. S. Haldane, who found himself in the company of theologians. On being asked what one could conclude as to the nature of The Creator from a study of His Creation, Haldane is said to have answered, "An inordinate fondness for beetles."

–G. E. Hutchinson, 1959

Familiar as garden grubs, tree borers, and flour beetles among many other pests, Coleoptera have also been a source of inspiration to centuries of collectors enamored with their diversity (Figures 10.1, 10.2, 10.63), and for good reason. Coleoptera comprises 350,000 named species, which is the largest order of organisms and 40% of all insects. One in every four kinds of animal is a beetle.

The diversity of Coleoptera is probably related to a suite of features that adapt the adults for dwelling in concealed, tight spaces, while yet retaining the ability to fly. These include the following defining adult features of the order:

- Elytra, which are heavily sclerotized forewings that cover and protect the abdomen and hind wings. In some beetles the elytra flap in flight, but they do so very slowly and so contribute very little to flight. The meeting edges of elytra generally fit together with a tongue-and-groove structure. In all Recent beetles the venation has been lost on elytra, but it was retained in some Mesozoic fossils. The rows of punctures or striae seen in most modern species may be vestiges of these veins. Related to the loss of forewings in flight is also a loss of eight thoracic muscles (Beutel and Haas, 2000) and a reduced mesothorax, but in other holometabolans it is usually the metathorax that is reduced.
- Hind wings that are folded lengthwise and crosswise, and tucked under the elytra, where they are protected. The hind wings also have reduced venation and complex mechanisms of folding. Tension created by contractions of the direct flight muscles causes wings to fold along lines of weakness, though folding may be assisted by abdominal movements in some groups. When the muscles relax and the elytra lift, wings naturally unfold.
- Prothorax, including the pronotum, which freely articulates with the rest of the thorax (pterothorax) and which is always large and shieldlike. Moreover, the pterothoracic segments are strongly joined to each other and to the base of the abdomen. This joining and the large pronotum affords further mechanical protection to the beetle, while the pro-pterothoracic joint allows flexibility.
- Heavily sclerotized abdominal sternites and less sclerotized tergites, which are related to the dorsal protection provided by the elytra. In groups where the elytra have been shortened, like Staphylinidae, the abdominal tergites are heavily sclerotized.
- Ocelli lost (rarely redeveloped, and never as all three at once), and cerci absent.

All these features are related to the cryptic lifestyle of beetles. They serve to protect the beetles' wings and body from damage while wedged in tight galleries under rocks or bark, or for life in particulate substrates like soil and sand as well as in water. This suite of features also protects Coleoptera from infection by microbes and fungi. Invasion of these niches not only affords access to diets few other adult holometabolans exploit, but it also allows adult beetles to avoid direct exposure to stressful conditions. If freezing weather sets in, for example, beetles can take refuge deep in crevices or under rocks and logs. Also, the elytra, which cover the abdominal and some thoracic spiracles, controls the amount of transpiration and thus water loss. Indeed, beetles are the most diverse insects in the driest places on earth, infesting even bins of flour in human settlements in the Sahara Desert.

Other features of beetles include a generally flattened body, but many species have evolved a thick body, particularly ones that are exposed feeders on plants, dung, or carrion. Beetles also typically have abdominal sternite one reduced, but this may be related to the strengthened joint between the pterothorax and the abdomen. The genitalia of beetles are largely internal in both sexes, and the ovipositor is highly reduced as a result. This feature is also probably related to the general body structure of beetles because delicate genitalia can be protected internally when the beetle is wedged into tight spaces or in soil. Mouthparts and leg structure are generalized in most beetles, but overall the body plan

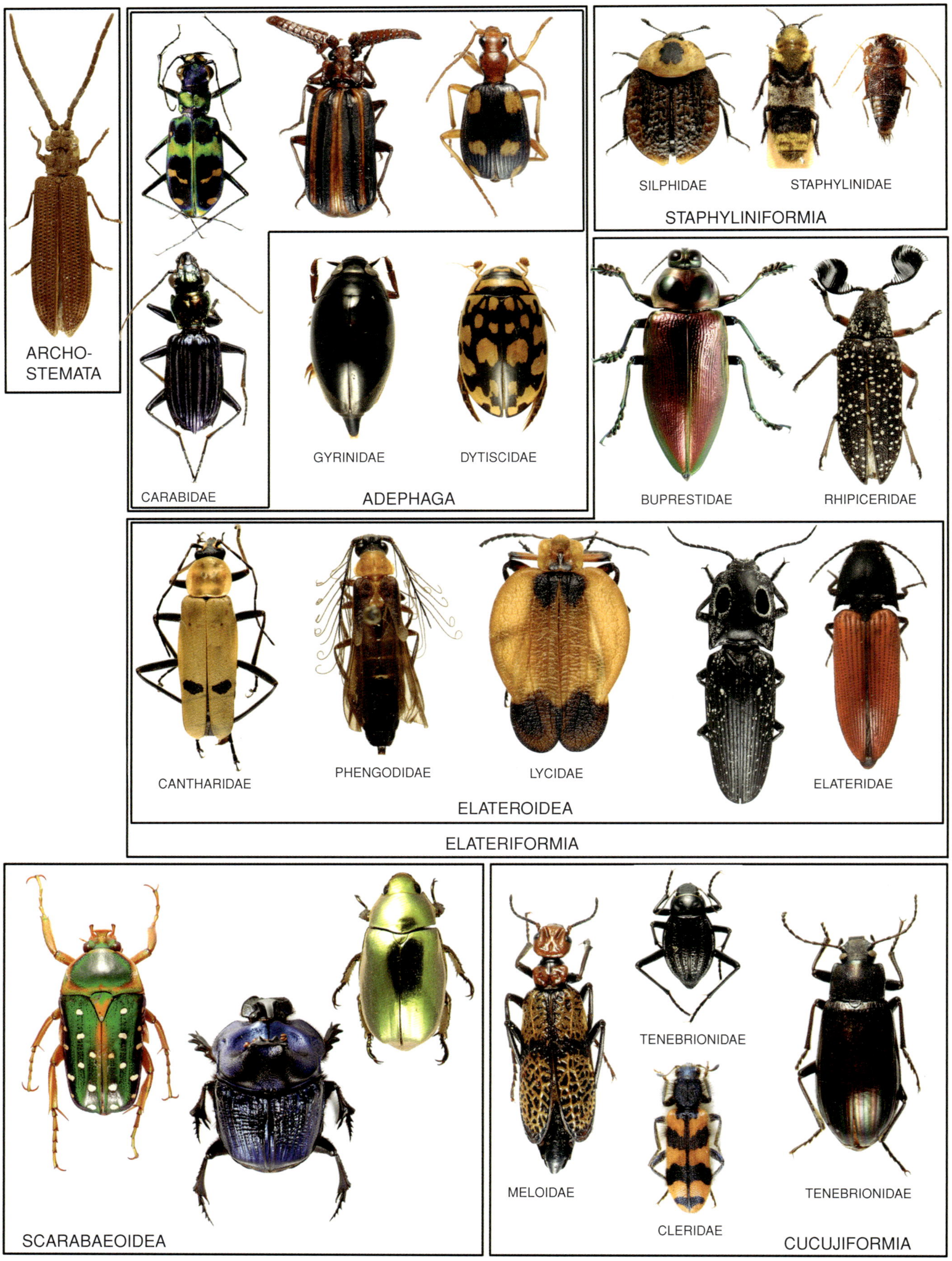

10.1. Assorted representatives of the most diverse order of organisms, the beetles, exclusive of phytophagan beetles (weevils, long-horned beetles, leaf beetles: see Figure 10.63). Not to the same scale.

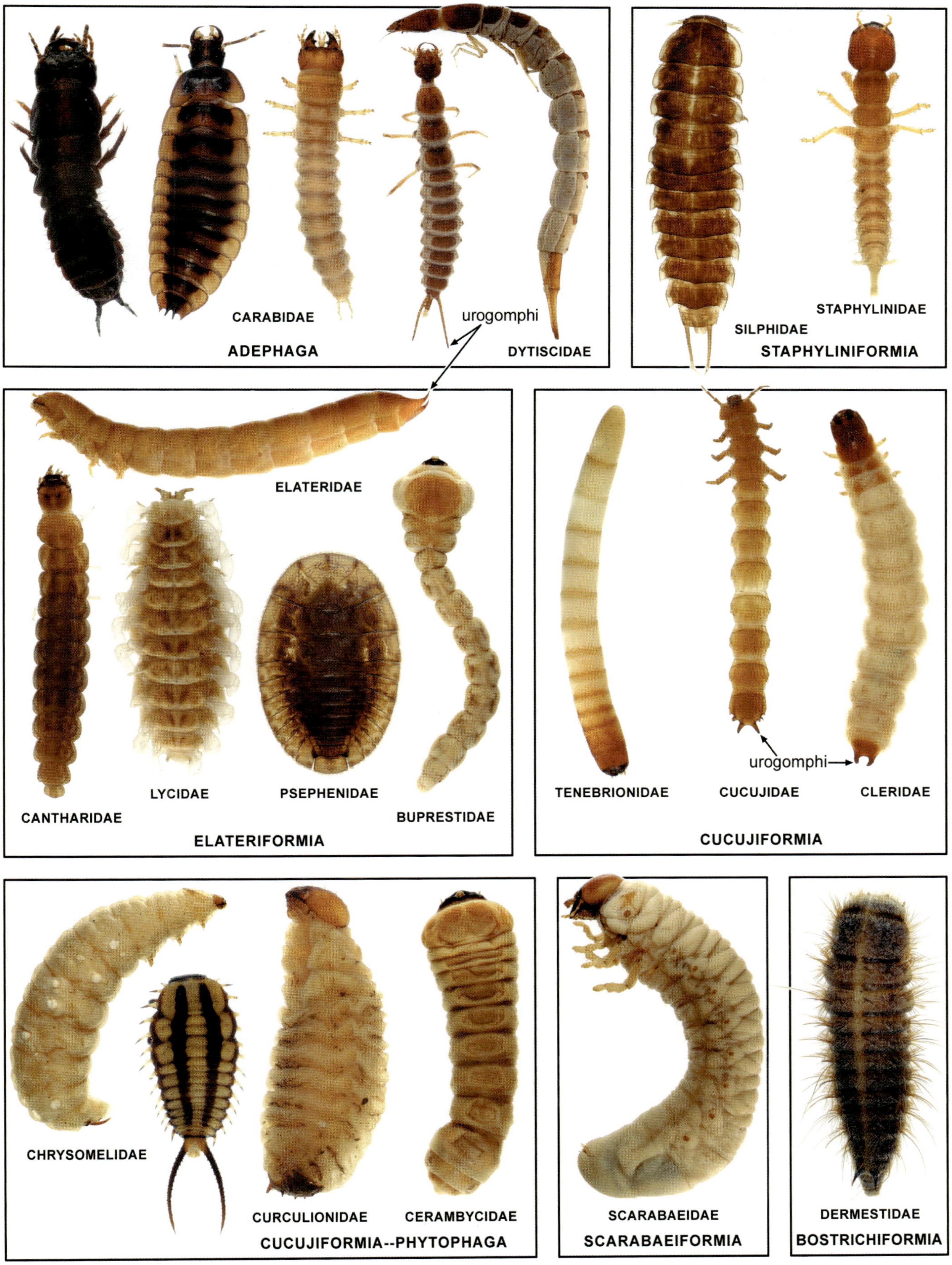

10.2. Beetle larvae from representative families. Not to the same scale.

of beetles comprises a suite of features related to the development of elytra.

There are four major groups, or suborders, of beetles. These include the following:

- Archostemata, which comprises approximately 35 Recent species and is consistently indicated as the most basal lineage in all studies on the relationships of beetles (Crowson, 1981; Lawrence and Newton, 1995; Maddison *et al.*, 1999; Beutel and Haas, 2000; Shull *et al.*, 2001; Caterino *et al.*, 2002; Vogler and Caterino, *in* Klass, 2003). These specialized wood borers have a venation that is least vestigial among all Recent beetles.
- Myxophaga, which is an obscure group of approximately 65 species of tiny, specialized aquatic and semi-aquatic beetles. These are considered as either a sister group to Adephaga + Polyphaga (Maddison *et al.*, 1999; Caterino *et al.*, 2002; Vogler and Caterino, *in* Klass, 2003), or to Polyphaga (Beutel and Haas, 2000; Shull *et al.*, 2001). Some studies have even indicated that myxophagans are highly modified relatives of basal polyphagan beetles (Beutel *et al.*, 1999).
- Adephaga, which comprises approximately 10% of all beetle species. These include ground and aquatic beetles in 12 Recent and five extinct families, which are mostly predatory as adults and larvae.
- Polyphaga, which is the group that includes 90% of all beetle species and accounts for the great diversity of the order. As the name indicates these have extremely diverse diets.

The suborders have been recognized for a long time, though relationships among them have only recently been scrutinized. Despite the scrutiny there is significant disagreement, as we mentioned earlier and will discuss in more detail later. For relationships among suborders (Figure 10.3), we have adopted the scheme discussed by Crowson (1981) and Beutel and Haas (2000) because it is based on diverse morphological features from adults and larvae, and these characters are also easy to evaluate. The molecular studies (Maddison *et al.*, 1999; Shull *et al.*, 2001; Caterino *et al.*, 2002) are based on very limited sequences from just the 18S rDNA gene, which has been shown to give misleading results in various studies (see, for example, the discussion in Caterino *et al.*, 2002).

A great deal has been written on beetles, but most of it is taxonomic and descriptive. Among the more synthetic reviews is the book by Roy Crowson (1981), who is acknowledged as having been one of the world's most knowledgeable coleopterists. Lawrence and Britton (1991) provided a very condensed, detailed, and authoritative account that emphasizes the Australian fauna, though most of the families occur in other regions as well. Similar reviews, of larvae and world families, are by another world authority on beetles, John Lawrence (1982, 1991; Lawrence *et al.*, 2000a,b). The classification of beetle families (Lawrence and Newton, 1995) is another important reference. Brief overviews of Coleoptera include ones by Lawrence and Newton (1982) and Liebherr and McHugh (2003).

TABLE 10.1. Signficant Characters in the Phylogeny of Coleoptera[a]

1. Forewings tegminous, with reduced venation
2. Elytra present: forewings heavily sclerotized, with venation reduced (and various other features, see text)
3. Adults with scales on elytra (lost in some Archostemata)
4. Labrum fused to head capsule
5. Larvae with urogomphi: pair of appendages on abdominal tergite nine (lost in various groups)
6. Loss of eight pterothoracic muscles
7. Hind coxae immovable
8. Abdominal sternite two divided by hind coxae
9. Pygidial defense glands present
10. Larvae prognathous; labium and clypeus fused; mouthparts modified for liquid feeding
11. Adult legs with fringes of stiff hairs for swimming
12. Larvae flattened, with sensory hairs, spiracular gills, pupation in the last larval exuvium
13. Apex of hind wing folded, with specialized hind wing folding
14. Cryptonephridic Malpighian tubules

[a] Numbers correspond to those on phylogeny, Figure 10.3.

EARLY FOSSILS AND OVERVIEW OF PAST DIVERSITY

Besides having a cryptic lifestyle and the myriad adaptations related to this, another feature attributed to the evolutionary success of beetles is their age. Beetles diverged from their common ancestors with Neuropterida and perhaps Glosselytrodea (Permobesothidae) probably in the Early Permian, particularly since there is as yet no definitive evidence for a Carboniferous holometabolan. The oldest stem-group coleopteroid is from the Early Permian, 280 MYA, making this lineage one of the earliest holometabolans in the fossil record. True Coleoptera did not appear, however, until the Triassic 230 MYA. The stunning diversity of Coleoptera appears to have been achieved by the Late Jurassic some 155–160 MYA, when even diverse weevils and other derived groups existed. The fossil record of Coleoptera has been reviewed or discussed by Ponomarenko in Arnol'di *et al.* (1977 [1991 English translation]), Crowson (1981), Carpenter (1992), Kukalová-Peck (1991), and Ponomarenko (2002b). Ponomarenko has been the most prolific worker of fossil Coleoptera, having based his studies on the vast Eurasian deposits of the former Soviet Union. We give more details on fossils for each suborder and infraorder later in this chapter.

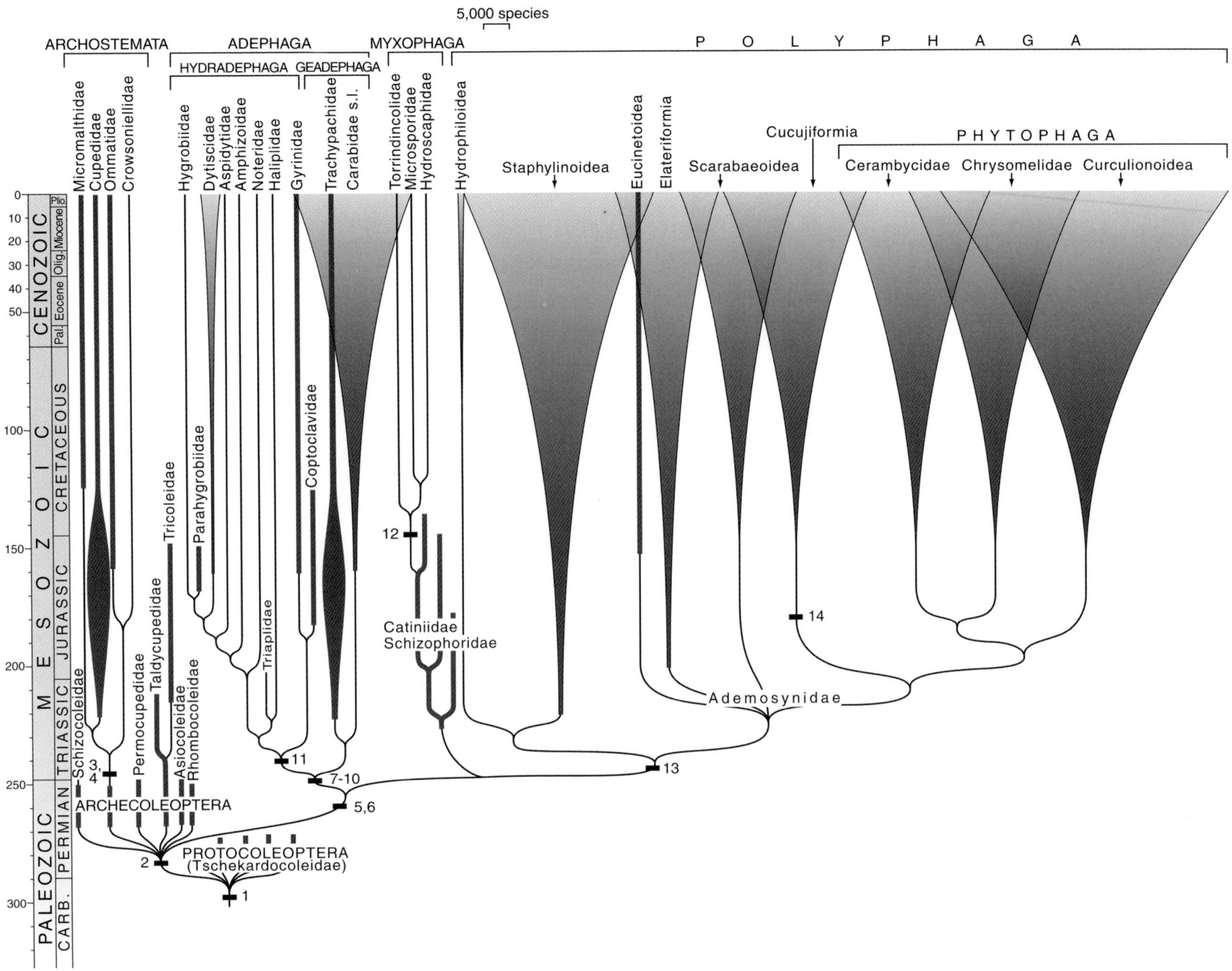

10.3. A phylogeny of the major living and extinct lineages of Coleoptera, with estimates of past and living diversity. The thick lines indicate the known occurrence in the fossil record. Significant characters are listed in Table 10.1. Compiled from various sources (see text). "Cucujiformia" here includes Lymexylidae, Cucujoidea, and Tenebrionoidea.

Unfortunately, the fossil record of beetles has been very difficult to interpret since, with the modification of beetle forewings into veinless elytra, "the main source of characters present in nearly all other fossil insects is eliminated" (Carpenter, 1992: 281). Thus, the identification of their pre-Cretaceous (i.e., pre-amber) fossils, all of which are preserved in rock, relies on the remains of coxae, sutures, tarsal segmentation, and the general form of the body and elytra. Most fossil beetles, in fact, are isolated elytra in rocks, and hundreds of species have been described on this basis. As a result, their classification is considerably less than clear. In some cases the preservation of distinctive features allows better interpretation of fossils, like the rostrum of weevils, short elytra of staphylinids, and the elytral sculpturing of cupedoids.

The distinctive pattern of sculpted pits on the elytra of cupedoids has allowed interpretation of the earliest obvious relatives of beetles. These are fossils in the probably paraphyletic family Tshekardocoleidae, which consists of 12 genera from the Early Permian (Artinskian-Kungurian, ca. 270 MYA) of Chekarda in the Ural Mountains and the Czech Republic (Figure 10.4). These "beetloids" were exceptionally primitive and to define them as beetles really stretches the point, which is why Crowson regarded them as Protocoleoptera and is a view that we adopt here. Preserved features that were primitive to Recent Coleoptera include 13 flagellomeres (versus 11 or less), small forecoxae with widely separated bases, and the structure of the forewings. Protocoleopteran forewings were apparently coriaceous and leathery, not fully sclerotized as in many modern beetles, narrow and apically pointed, and extended well beyond the tip of the abdomen. Most importantly, these wings had nearly a full complement of wing veins, including veins Sc, R, Rs (with several branches), M, CuA, and CuP. They had a system of dense, square cells recessed between the veins and crossveins – a precursor to the rows of square punctures seen in Recent cupedoids. One well-preserved genus at least, *Moravocoleus*, even had a short, pointed, external ovipositor (Kukalová, 1969c), indicating that adult protocoleopterans were free-living as are many Recent cupedoids, though larval protocoleopterans may have been wood borers. Adults were prognathous also like Recent cupedoids and had distinctive paranotal lobes, which are structures that gradually decreased in size as cupedoids evolved into the Mesozoic. Protocoleopterans were rare among all insects from Early Permian deposits, but it is still odd that none of these insects was found in the most prolific outcrops of this time from Elmo, Kansas, and Midco, Oklahoma, until recently, which is a protocoleopteran forewing from Midco (Lubkin and Engel, in press).

In the Late Permian, 260–255 MYA, there appeared five other families of very basal, cupedoid-like insects, which Crowson (1981) classified as Archecoleoptera: Asiocoleidae, Permocupedidae, Rhombocoleidae, Schizocoleidae, and Taldycupedidae (Figure 10.4). These also had the rows of square punctures, but there was slight reduction in the venation, especially of Rs. Though they too were rare, these insects

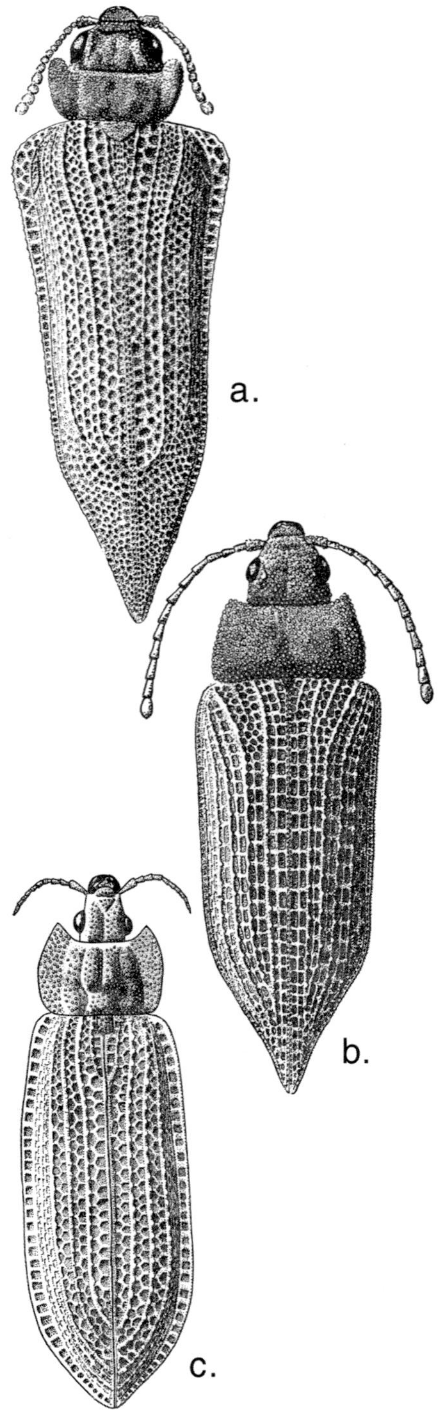

10.4. Primitive protocoleopterans and archostematan beetles from the Permian and Early Mesozoic. (a) *Sylvacoleus sharovi* (Permian); (b) *Permocupoides sojanensis* (Permian); (c) *Notocupoides triassicus* (Triassic). From Ponomarenko (1969).

had spread throughout Pangaea, and are known from the Permian of Africa and Australia, as well as Europe and Asia. As the Permian drew to an end, archecoleopterans became less rare. One of these families (Taldycupidae) persisted into the Triassic, and another family (Tricoleidae) occurred from the Triassic to the Jurassic.

The Early Triassic is unfortunately barely represented for fossil insects. However, by the Late Triassic, 240–220 MYA, 250 known species and 20 families had evolved, including true beetles with hardened, veinless elytra (Ponomarenko, 2002b). Moreover, beetles were now becoming common, representing some 20% of the individual insects from the deposits. Triassic families included some of the oldest families of Recent insects, including Cupedidae and Ommatidae, and four families of polyphagans. Triassic polyphagans include putative Hydrophilidae and Artematopidae, probable Elateridae, and definitive Staphylinidae. The earliest Adephaga also appeared in the Triassic, which were Trachypachidae (close relatives of carabids) and water beetles (Hydradephaga). A record of Triassic weevils (Curculionoidea) is highly unlikely based on reexamination of the fossil.

The position of three extinct families – the Ademosynidae, Schizophoridae, and Catiniidae – which first appeared in the Triassic, are ambiguous. Though these were originally described as archostematans, they lack elytral veins and square cells, and some were very broad and had expansive paranotal lobes. Crowson (1981) interpreted Ademosynidae as stem-group Polyphaga, as did Lawrence (1999). The latter author, in fact, pointed out that *Ademosyne* appears quite similar in habitus to the basal polyphagan family Scirtidae. Crowson (1981) further maintained that Catiniidae were stem-group Myxophaga, a position that Ponomarenko (2002a) felt also included Schizophoridae. These three families became extinct in the Jurassic and Cretaceous.

Beetle diversity further mushroomed in the Jurassic, with some 35 families and 600 species known, but this was not accompanied by the dramatic transition seen between Permian and Triassic faunas. Unfortunately, the Jurassic record is regionally biased to Europe and central Asia, though it preserves an important episode that occurred in the Late Jurassic: a dramatic diversification of terrestrial beetles. Individual beetles in the great Late Jurassic deposit of Karatau are comprised of 80% polyphagans (Ponomarenko, 2002a), which is almost equivalent to their modern diversity and abundance. Beetle faunas in the Early Cretaceous were more similar to ones from the Jurassic than to younger ones in the Late Cretaceous, which is probably the result of the angiosperm radiations. By the Late Cretaceous, approximately 100 to 65 MYA, there occurred radiations of speciose beetle groups that presently feed on angiosperms, and this greatly contributed to the spectacular diversity of beetles seen today.

ARCHOSTEMATA

Abundant cladistic and fossil evidence indicates that archostematans are the basal suborder of beetles. Despite this position, they are specialized wood borers, especially the lightly sclerotized and generally legless larvae, and one species – *Micromalthus debilis* – has what is arguably the most complex life cycle in insects. Relationships among the four families have been discussed by Beutel and Hörnschemeyer (2002) and reviewed by Lawrence (1999). Their more distinctive features include a fusion of the adult labrum to the head capsule, which is considered related to wood boring. Also, many species have fine scales on the elytra, which Beutel and Hörnschemeyer (2002) have interpreted as being lost in *Micromalthus* and *Crowsoniella*. These authors have also discussed several derived features of the larvae that reveal the monophyly of this basal lineage. Lastly, archostematans have a distinctive way of tucking the wings under the elytra, wherein the tips are rolled, not folded.

There are four small families of Recent Archostemata, possibly five if the monotypic Sikhotealinidae from the Russian Far East is included, which has just one rare species that needs further examination. The largest family of archostematans is the Cupedidae, which is nearly worldwide and includes 9 genera and approximately 30 Recent species. These are generally quite flat and have the elytra sculpted into rows of square punctures. This distinctive sculpturing has allowed identification of myriad cupedoid fossils from the Mesozoic. Similar to Cupedidae are the presently austral Ommatidae, which consists of one genus (*Tetraphalerus*) from South America and another (*Omma*) from Australia (Lawrence, 1999). Some *Omma* are quite colorful and even have maxillary and labial combs specialized for feeding on the pollen of flowers, on which they are commonly found. Ommatidae were more widespread in the Mesozoic; there are fossils of the family from the Jurassic of Europe and Asia and the Cretaceous of Asia. The family Crowsoniellidae is monotypic, with the one species, *Crowsoniella relicta*, known only from Italy. Lastly, the family Micromalthidae is also presently monotypic, but there is some evidence that *Micromalthus debilis* may actually consist of several cryptic species.

Micromalthus (Figure 10.5) is one of the most intriguing insects. These are tiny beetles apparently native to North America, and perhaps Central America, with larvae and adults feeding on or living within decaying wood, particularly oaks, but they will also infest hemlock, acacias, and eucalypts. Long considered one of the rarest and most enigmatic beetles, sporadic infestations from imported timber have occurred in distant parts of the world, including in deep diamond mines in South Africa and even vaults of a Wall Street bank in New York lined with oak timbers. *Micromalthus* is highly specialized in structure and lifestyle. Morphologically, the adult

10.5. The enigmatic archostematan beetle, *Micromalthus debilis*, which probably has the most complex life cycle known in insects. These larvae are the triungulin type. Note the vestigial venation on the short elytra of the adults (left, above). Scanning electron micrographs; body length 1.1 mm (adult), to same scale.

prothorax lacks sutures, and the adults are soft-bodied and the elytra shortened. The elytra do not bear the scales and distinctive rows of punctures seen in Cupedidae and Ommatidae, but they do retain some vestigial venation (Figure 10.5). The life cycle of *Micromalthus* is perplexingly complex (reviewed by Pollock and Normark, 2002). The species usually reproduces as a larviform female that bears live young parthenogenetically. These young include an active, triungulin-like first instar larva with long legs and setae (Figure 10.5), and legless "cerambycoid" second through fourth instars. There is also a stage where arrhenotokous females produce a single, large egg that hatches into a curculionoid-type larva (thick, C-shaped, and legless), which develops into a haploid male that devours its mother. As one can imagine, this stage is rare, otherwise a population would no longer be able to reproduce. There is, lastly, a stage where an amphitokous female (parthenogenetically producing males and females) is capable of both of these modes of ovoviviparous and oviparous reproduction. Pollock and Normark (2002) have proposed the intriguing hypothesis that this bizarre lifecyle evolved from the effects of maternally inherited bacteria that help the beetles digest wood. If the bacteria are not passed to males or are senescent there, the male must inoculate itself by being a cannibal. This makes a great deal of sense since infections by *Wolbachia* and other spirochetes are known to have profound effects on mortality of males in various insects. It is quite possible, though, that the variation in life cycles is merely the result of incomplete infection by microbes and is not necessarily adaptive.

The relationships among Recent families of Archostemata were discussed by Beutel and Hörnschemeyer (2002), who hypothesized that Ommatidae and *Crowsoniella* are most closely related, as are Cupedidae and *Micromalthus*. All major studies have concluded that the suborder is monophyletic. Recent species are, however, small twigs on the entire evolutionary tree of Archostemata because there is a

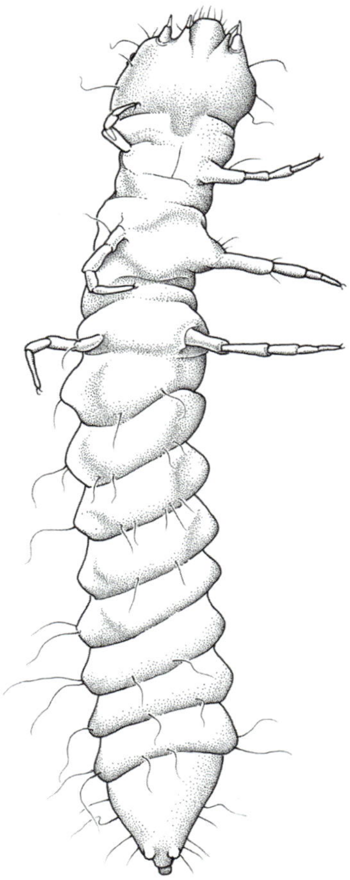

10.6. Larval Micromalthidae preserved in Early Cretaceous amber from Lebanon, ca. 125 MYO, the oldest record of the family. Length 0.85 mm.

much more diverse fossil fauna of cupedoids than living ones. As one would expect, fossils of Micromalthidae occur only in amber because of their minute size. The triungulin form of larva is preserved in Eocene Baltic amber, Oligocene amber from Mexico (Rozen, 1971), and Early Cretaceous amber from Lebanon (Figure 10.6). Thus, amber fossil micromalthids must have been feeding on the amber trees, which include a dicot (Mexican amber) and conifers (Baltic and Lebanese amber).

Not including the Protocoleoptera and archecoleopterans from the Permian, true Archostemata first appeared in the Late Triassic. Twenty genera of Cupedidae and two genera of Ommatidae are known, all from the Mesozoic (Figures 10.7 to 10.9) (e.g., Ponomarenko, 1963). The ommatids included *Omamima* from the Jurassic of Europe and Asia and the Cretaceous of Asia and *Tetraphalerites*, a genus apparently closely related to the Recent genus *Tetraphalerus*. Cupedoids dominated the Triassic faunas, representing some 30% of the beetle species and individuals in Eurasian and Australian deposits, though cupedoids were less common in the American Triassic, where polyphagans were more diverse. Dominance of cupedoids in the European and Asian deposits continued into the Early Jurassic; by the Late Jurassic (Karatau: Kazachstan), these beetles comprised 10% of the individuals and slightly more than 10% of the species – still far more diverse than today but clearly eclipsed by polyphagans.

10.7. (Far left). Elytron of an early cupedoid beetle (family Schizocoleidae?), from the Triassic of Queensland, Australia. The distinctive elytra of cupedoids allows interpretation of the earliest beetle fossils. QM C1528; length 7.1 mm.

10.8. (Left). Elytron of another early cupedoid beetle (family Cupedidae), from the Triassic of Queensland, Australia. QM C1474; length 4.0 mm.

10.9. An entire cupedid beetle, preserved in Early Cretaceous limestone from Las Hoyas, Spain. Cupedoid beetles were diverse through the Early Mesozoic and waned by the Cretaceous. Photo: X. Martínez-Delclòs.

ADEPHAGA

The largely predaceous ground and water beetles comprise approximately 10% of all beetles, or some 45,000 species. Most of these are Carabidae (sensu lato), which is one of the largest families of insects with 40,000 species. The suborder Adephaga had traditionally been classified into two groups, the Hydradephaga (water beetles) and Geadephaga (ground beetles) (e.g., Crowson, 1960; Beutel, 1998). This system fell out of favor but recently has been supported by several DNA sequence studies (Shull *et al.*, 2001; Ribera *et al.*, 2002a). Caution, however, must be used when interpreting these studies because they are based on just one gene (18S rDNA). While these two adephagan groups need better definition with morphological characters, the monophyly of the Adephaga itself is on terra firma. Morphological characters include immobile hind coxae that are raised only slightly from the plane of the thorax and abdomen, abdominal sternite two divided by the hind coxae (Figure 10.10), several features of the mouthparts (see Beutel and Haas, 2000), and specialized (*pygidial*) glands at the apex of the abdomen that secrete defensive compounds. The larvae of Adephaga are more like those of the coleopteran sister group (Neuropterida) than any of the other suborders, being active predators with well-developed legs having six podites, their mouthparts are prognathous, and the mandibles are even generally modified for feeding on liquids. Adephagans have an excellent rock fossil record, no doubt because their aquatic and riparian habits facilitated preservation in ancient lake sediments.

There are nine families of Recent Adephaga, depending on how one classifies certain groups within Carabidae. The most phylogenetically interesting family is Trachypachidae, a family of just six Recent species from the Holarctic Region (*Trachypachus*) and temperate Chile and Argentina (*Systolosoma*). These are small, carabid-like beetles that live in riparian to dry habitats and have been considered as being either the sister group to the Hydradephaga (e.g., Bell, 1966), or to the Geadephaga, or are a separate, third lineage (e.g., Crowson, 1960). Molecular evidence places Trachypachidae as the sister group to the Geadephaga, though evidence for this is not particularly strong (Maddison *et al.*, 1999). Like water beetles, Trachypachidae have the primitive type of adephagan hind coxae, which do not rock (as in Carabidae). It is quite likely that this family is closest to the ancestral adephagan, particularly since Trachypachidae are known from the Triassic. Trachypachidae, in fact, were far more diverse than they are today, with some seven genera and 20 species alone known from the Late Triassic to Early Cretaceous of Asia and Europe (Ponomarenko, 1977b) (Figure 10.11). This rich fossil record reveals how extinction created the relict, bipolar distribution of trachypachids seen today.

The Hydradephaga, as the name implies, live in water, and beetles in all but two small families are adept at swimming on or in water using coordinated strokes of legs outfitted with thick fringes of setae. Their bodies are generally very streamlined, with the head, thorax, and elytra confluent. These are the only insects besides nepomorphan bugs that spend virtually

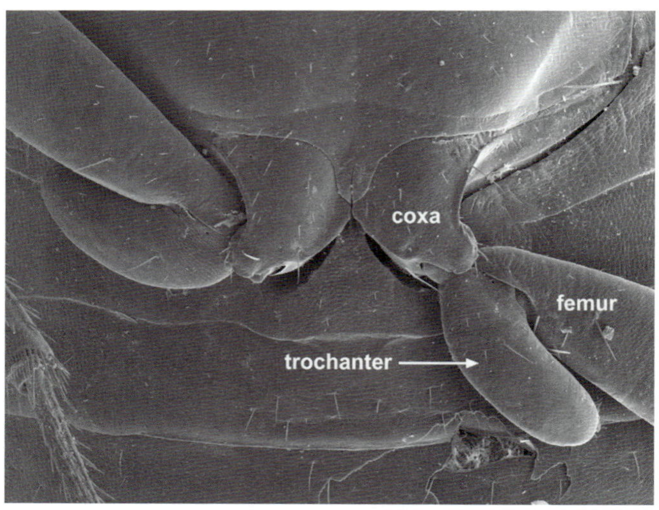

10.10. A distinguishing feature of adephagan beetles: immovable hindcoxae virtually fused to the thorax, and the first sternite interrupted by the hindcoxae. Scanning electron micrograph of carabid beetle.

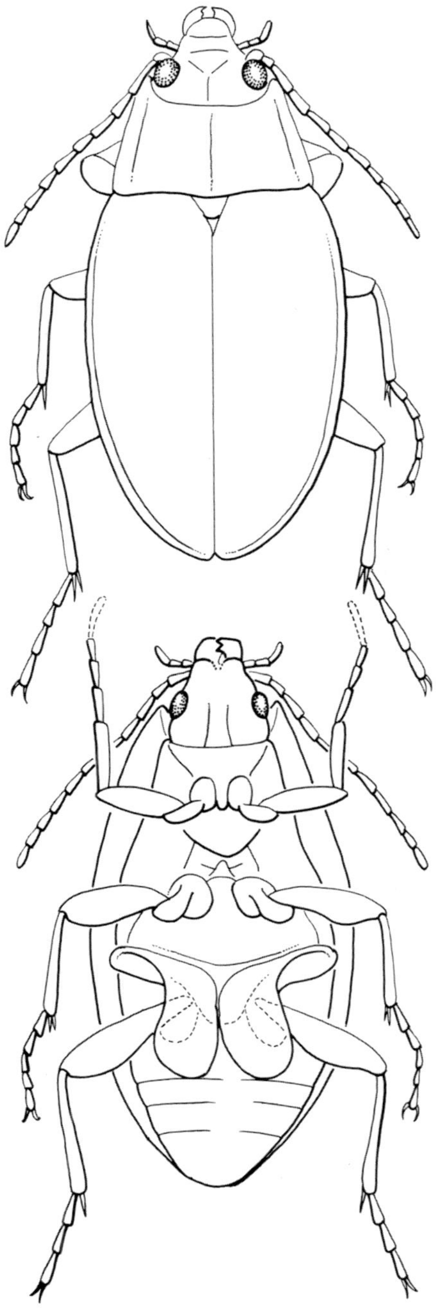

10.11. Reconstruction of the trachypachid beetle *Psacodromeus guttata*, from the Late Jurassic of Karatau. Trachypachidae are small predatory ground beetles that are relict today but that were diverse in the Mesozoic. They are either the closest relatives of the large family Carabidae, or to the predatory water beetles, the Hydradephaga. Body length 10 mm; redrawn from Ponomarenko (1977b).

their entire life cycle in water, the exception being the beetle pupal stage, which usually is spent in shoreline sediments. Most species are active predators as adults and larvae, though the Haliplidae (cosmopolitan, with 220 species) feed \on algae and stoneworts (Charophyceae). Species of the Noteridae ("burrowing beetles," 270 species) may be detritivorous as larvae, though this is not entirely clear. The small, disjunct families Amphizoidae (3 species in China, 3 in North America) and Aspidytidae (1 species in South Africa, 1 in China) do not swim. The latter of these is the most recently discovered family of Hydradephaga (Ribera *et al.*, 2002b). Another small family is the Hygrobiidae (four species from Australia, one in China, one from western Europe and northern Africa), which is distinctive for the stridulatory mechanism on the inner surface of the elytra and apex of the abdomen.

The largest hydradephagan family is the Dytiscidae ("diving" beetles), with 3,500–4,000 species, phylogeny of which has been treated by Miller (2001). These are swift swimmers with large hind legs (including coxae), and the males of some species have the foretarsi modified into suction cups for grasping the smooth elytra of the female during mating. Larval dytiscids are also active swimmers with fringed legs. They have a vestigial mouth but feed by injecting enzymes into their prey through fine channels in the mandibles and then sucking up the liquefied remains. Dytiscids colonize diverse aquatic habitats, depending on the species, from saline soda lakes to the water that pools within forest canopy bromeliads. The most basal family of hydradephagans according to Ribera *et al.* (2002a) and Shull *et al.* (2001) comprises the whirligig beetles, Gyrinidae (with approximately 1,000 species). Their name derives from the distinctive habit of rapid, constant swimming in loops on the surface of the water. They usually occur in dense aggregations but rarely collide, which may be attributable to their distinctive vision: Their eyes are divided into dorsal and ventral halves, for aerial and aquatic views. They also have compact antennae recessed in grooves, each of which has a sensitive Johnston's Organ that can detect surface ripples. The front legs of gyrinids are long and raptorial.

Hydradephagans occur as early as the Late Triassic of Eurasia (approximately Carnian to Norian, 230–220 MYA). These include the extinct and obscure family Triaplidae, which Ponomarenko (1977b) placed near the Haliplidae, and Colymbothetidae, an unplaced family. The extinct family Coptoclavidae (Jurassic to Early Cretaceous) had impressive larvae with broad, paddle-shaped mid and hind tarsi, and raptorial forelegs (Figure 10.12). Adults of these beetles apparently had long, raptorial forelegs and divided eyes (Figure 10.13), which indicates they are closely related to Gyrinidae, possibly even stem-group gyrinids. Other Gyrinidae (*Anagyrinus, Angarogyrus*) are recorded from the Late Jurassic of Germany and Kazakhstan and three other genera from the Cretaceous. Two other Jurassic families, Liadytidae and Parahygrobiidae, are known only on the basis of larvae, which had fringed swimming legs (Figure 10.14). Dytiscidae are actually rather rare in the Mesozoic. A putative Jurassic dytiscid is *Angaragabus*, which is known just as a larva that lacked fringes of leg hairs. Thus, it appears that by approximately 150 MYA most major lineages of hydradephagans had appeared. Diverse hydradephagans occur through-

10.12. Larva of the extinct predatory beetle, *Coptoclava longipoda* (Coptoclavidae), from the Early Cretaceous of Baissa, Siberia. It swam using paddle-shaped mid and hind legs. PIN 4210/525; body length (exluding cerci) 16 mm.

out the Tertiary, as compressions and even in amber (e.g., Figure 10.15).

Exclusive of the Trachypachidae, the Geadephaga is comprised almost entirely of the huge, diverse family Carabidae, particularly with the Paussinae, Cicindelinae, and Rhysodinae as part of this family (Liebherr and Will, 1998; Maddison *et al.*, 1999). Carabidae have attracted a great deal of serious attention in behavior, systematics, and particularly biogeography (e.g., Jeannel, 1942; Darlington, 1943, 1965; Lindroth, 1969, 1992; Erwin *et al.*, 1979; Ball, 1985; Liebherr and Zimmerman, 1998; Ball *et al.*, 1998; Larochelle and Larivière, 2003). Interest in their biogeography is due not only to their diversity, but also to the fact that they have invaded most terrestrial habitats from tundra to all sorts of forest, steppe, caves, and even the intertidal zone. Moreover, flightlessness has repeatedly evolved, particularly in species restricted to isolated islands and montane habitats, and this promotes isolation, speciation, and narrow endemism.

Though these beetles are best known as generalized ground predators, they are in fact extremely diverse ecologically (Thiele, 1977). Some Harpalinae feed on seeds that they also cache in burrows. Among the more specialized groups are the Paussinae and Pseudomorphinae, which are inquilines in ant nests. The former possess glandular trichomes that must produce secretions for appeasing their hosts, and they also have highly specialized antennae. Larval

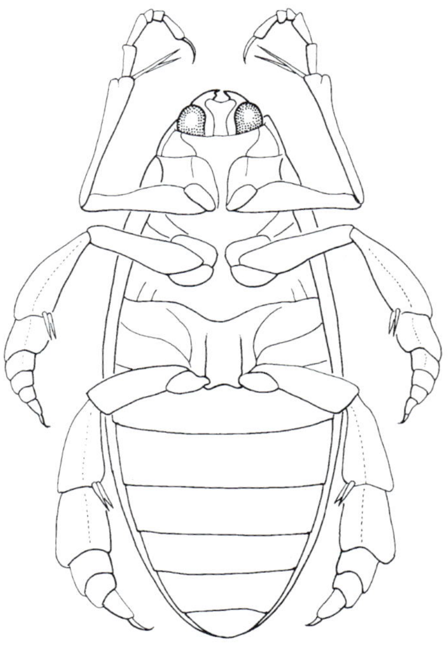

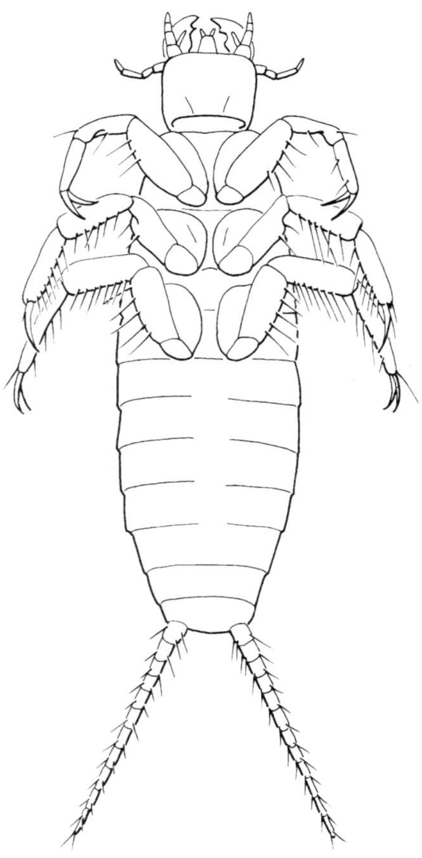

10.13. (Right). Reconstruction of the adult of *Coptoclava longipoda* (ventral view). The family Coptoclavidae lived from the Early Jurassic to the Early Cretaceous and is closely related to the Gyrinidae and Dytiscidae. Redrawn from Ponomarenko (1977).

10.14. (Far right). Reconstruction of the predatory hydradephagan larva, *Parahygrobia natans*, from the Late Jurassic of Siberia. Body length (excluding cerci) 6 mm; redrawn from Ponomarenko (1977b).

10.15. A predaceous diving beetle of the living genus *Copelatus* (family Dytiscidae) in Miocene amber from the Dominican Republic. Some species of this genus live in the pools of water that accumulate in epiphytic bromeliads, which may be why this aquatic beetle was caught in tree resin. AMNH DR15–141; length 4.7 mm.

Rhysodinae are highly modified for feeding on slime mold plasmodia in rotten wood. Some Lebiinae and Brachinini even have ectoparasitic larvae. Most carabids are dietary opportunists, but predatory species can be remarkably specialized and have distinct preferences, such as for snails. Snail eaters typically have sickle-shaped mandibles and a long, narrow pronotum for reaching into the shell and plucking out the occupant. *Notiophilus biguttatus* has such excellent vision and reflexes that it can nab springing Collembola. The familiar tiger beetles, subfamily Cicindelinae, comprise a major lineage of mostly diurnal, brightly colored carabids (e.g., Trautner and Geigenmüller, 1987). These chase down their prey in open ground and are believed to be the fastest running insects. Studies have been done on predatory carabids that show they consume their body weight or more each day.

One feature of carabids that may have contributed to their remarkable success is the sophisticated development of the pygidial glands, which is an efficient defense system. The most famous examples of this are the bombardier beetles in the Paussini, Metriini, but particularly Brachinini, which spray a hot, caustic secretion at attackers. The glands empty hydroquinone, peroxide, and enzymes into a sclerotized "firing chamber," which produces an explosive discharge of water, oxygen, and quinones (Schildknecht *et al.*, 1968; Eisner *et al.*, 1977). There is an arsenal of other compounds used for defense by carabids, including organic acids (formic, methacrylic, isobutyric), phenols, and aldehydes (reviewed by Moore, 1979; Will *et al.*, 2000), depending on the lineage. Interestingly, the more recently evolved groups of carabids defend themselves with stronger irritants like hydrocarbons and formic acid, and they are primarily tropical (Will *et al.*, 2000). It is quite likely that ants, which are the dominant animal group in tropical lowlands of the world, were the main selective pressure for the potency of tropical carabids.

Carabid systematics has a tradition of monography that puts the taxonomy of subfamilies and tribes on firm ground. Relationships among the major lineages, however, have been very difficult to unravel, and there has been significant disagreement among morphological studies as well as between these and the molecular studies that are now beginning to be done. For example, Paussinae is considered a basal lineage on the basis of the female reproductive system (Liebherr and Will, 1998) and the lack of an antennal cleaner on the foretibia, among other features. Evidence from the 18S rDNA

10.16. A very early fossil of predatory ground beetles, family Carabidae, from the Late Triassic of Virginia. Carabidae are an ecologically diverse family of approximately 40,000 living species. VMNH; length 4.9 mm.

10.17. A beautifully preserved, unusual caraboid beetle preserved in Early Cretaceous limestone from Brazil, approximately 120 MYO. AMNH SA46250; length 6.2 mm.

10.18. A Miocene species in Dominican amber of the unusual and highly specialized carabid subfamily Rhysodinae. These beetles are found in decaying wood, presumably feeding on slime mold plasmodia. AMNH DR10–2065; length 4.1 mm.

gene, however, suggests that this group is considerably more derived (Maddison *et al.*, 1999). Rhysodinae and Cicindelinae have likewise been considered to be structurally and ecologically very disparate lineages at the base of carabid phylogeny (Beutel and Haas, 1996; Kavanaugh, 1998; Liebherr and Will, 1998), but the molecular evidence groups these two lineages as closely related and near the Harpalinae. Either there will need to be far more sampling of taxa and characters in further studies of carabid phylogeny, or the unresolved relationships found in some studies (e.g., Liebherr and Will, 1998; Maddison *et al.*, 1999) may actually reflect an explosive radiation of the major lineages. The earliest carabids are diverse compression fossils from the Triassic of Virginia (Figure 10.16) and six genera from the Late Jurassic of Kazachstan and Germany, approximately 155–160 MYO. By the Cretaceous there were diverse subfamilies, all of them preserved as compressions (e.g., Figure 10.17), but carabids are diverse in Tertiary ambers as well as in Tertiary rocks. Among the more significant carabid groups preserved in Tertiary amber are Lebiinae, Paussinae, and Rhysodinae (Figure 10.18).

MYXOPHAGA

Myxophaga is a small, obscure suborder of 65 Recent species and five families. Adults and larvae are minute (generally no more than 2 mm in body length) (Figure 10.19), which are highly specialized for an aquatic and semi-aquatic life living amongst sand grains and other particles grazing on films of green and blue-green algae. They are most commonly found at the edges of streams and rivers, in the splash zones of rapids and waterfalls, and in seepages. Adults respire via a *plastron*, which is a layer of air held close to the body and over spiracles by a dense pile of fine hydrophobic hairs. Both stages are also highly flattened, no doubt an adaptation for living in moving water. The suborder is clearly monophyletic partly because of these features, as well as the fact that pupation occurs in the last larval exuvium, like cyclorrhaphan flies (though any aquatic adaptiveness of this feature is unknown) (Beutel *et al.*, 1999; Beutel and Haas, 2000). Interestingly, these beetles also tuck their wings under the elytra using a rolling mechanism, not unlike that in Archostemata. Lastly, the larvae have tracheal gills. The families Microsporidae and Hydroscaphiidae have larvae with balloon-like tracheal gills, which may be a highly modified form of the segmented tracheal gills seen in larval Torridincolidae. Beutel *et al.* (1999) hypothesized a close relationship between Microsporidae and Hydroscaphiidae, partly on the basis of the peculiar larval gills but also because of the unusual larval mouthparts, which are partially hidden in a pouch. They were unable to include in their study the small family Lepiceridae

10.19. Beetles of the obscure suborder Myxophaga. (a) *Hydroscapha natans* (Hydroscaphidae); (b) *Torrindicola rhodesica* (Torrindicolidae). Myxophaga are minute aquatic beetles that appear to be the closest relatives of Polyphaga. Scanning electron micrographs, Photos: R. Beutel.

(= Cyathoceridae), which consists of two Central American species.

What is most intriguing about the study by Beutel *et al.* (1999), but which was not addressed there or in a subsequent study (Beutel and Haas, 2000), is that Myxophaga may be embedded within Polyphaga. Beutel *et al.* (1999) hypothesized that Myxophaga may, in fact, be most closely related to the basal polyphagan family Hydraenidae, and another polyphagan family, the Leiodidae, may be related to this group as well. Hydraenidae and Leiodidae are usually classified as very basal members of the polyphagan infraorder Staphyliniformia, but curiously the first of these families has habits very similar to those of myxophagans. Hydraenids feed on algae in aquatic and semi-aquatic habitats, and the adults breathe via a plastron on their ventral surface. The larvae of one genus, *Tympanogaster*, even have short spiracular gills. Thus, it should probably be seriously considered that Myxophaga are highly specialized, basal polyphagans, but clearly this will require further work to confirm. Leiodidae are much more generalized and probably are entirely terrestrial, and both adults and larvae feed on fungal and slime mold spores. Myxophagans have no fossil record, probably because they are so small and obscure, and their habits would preclude fossilization in the only medium that would sufficiently preserve them, which is amber.

POLYPHAGA

The great diversity of Coleoptera is essentially attributable to the suborder Polyphaga, which contains 90% of beetle diversity, or approximately 315,000 species. This group is, as the name implies, the most diverse ecologically, though the radiations of certain groups on angiosperms accounts for a great deal of the species. Polyphaga feed as larvae and adults on plants, fungi, slime molds, dung, and other insects; in wood; and as larval parasitoids. They are also the most diverse group in terms of structure, including size. These include minute featherwing beetles (Ptiliidae), which are among the smallest adult insects (down to 0.25 mm body length), to heavy African Goliath scarabs and the South American cerambycid *Titanus giganteus* (the latter up to 20 cm [8 in.] long), which are several million times the size of ptiliids. Monophyly of the Polyphaga is partly defined by the prothorax, wherein the propleuron is fused with the trochantin and entirely concealed, the so-called *cryptopleuron*. The suborder is typically classified into infraorders or "series" (with a *-formia* ending) (Crowson, 1985; Lawrence, 1991), which is a system used here.

Staphyliniformia

Staphyliniformia is a large group that has some aquatic species but that is mostly terrestrial and edaphic, living amongst soil, humus, leaves, and decaying wood. Monophyly of the infraorder is defined by a spring mechanism for folding and unfolding the hind wings (in Staphylinidae this is also accomplished using the flexible abdomen) and by a basal

10.20. A staphylinoid beetle of the family Scydmaenidae. Like most staphylinoids, these largely live in leaf litter and decaying wood, and are predatory. Scanning electron micrograph; length 1.8 mm.

articulation of the urogomphi. Many of the groups have active, campodeiform larvae that are predaceous, and adults are commonly predaceous as well. One of the superfamilies, the Hydrophiloidea, includes the Histeridae, some of which are very flattened beetles with large mandibles that prey on other insects as larvae and adults, and also the family Hydrophilidae. Hydrophilidae, as implied by the name, are mostly aquatic but also live in decaying vegetation, dung, and similar substrates. Also, the adults are generally omnivorous, not predaceous like the larvae.

The seven families in the Staphylinoidea comprise a huge group of species, classifications of which have been presented by Lawrence and Newton (1982, 1995), Naomi (1985), and Newton and Thayer (1988), though that of Naomi is generally not used. This group of beetles is arguably the most ecologically diverse in Coleoptera because it includes saprophages, phytophages, highly specialized inquilines of social insects, predators, spore feeders, fungivores, pollenivores, and even vertebrate ectoparasites. Ectoparasitic beetles are highly unusual, but for some reason they have evolved much more often in staphylinoids than any other group of beetles. Based on the proportions of newly described species of staphylinoids, this large group may eventually exceed the diversity of the Phytophaga, which includes the leaf beetles and weevils. Phytophagans are mostly exposed plant feeders, but staphylinoids are generally small insects living in cryptic microhabitats that have not been surveyed nearly as well.

Most staphylinoids have a pair of short elytra, though the elytra are not abbreviated in Hydraenidae, Leiodidae, many Silphidae, and some Staphylinidae, like many species in the subfamily Pselaphinae. Hydraenidae, as mentioned earlier, is an unusual small family that may be closely related to the Myxophaga (Beutel *et al.*, 1999). Scydmaenidae (Figures 10.20, 10.21), which have flat larvae somewhat resembling woodlice (Isopoda), have habits similar to those of saprophagous Leiodidae. Leiodidae are generally black, rounded beetles whose adults and larvae feed on fungal hyphae and spores, the fruiting bodies of slime "molds," and even carrion. Some Leiodidae are specialized inquilines in the nests of ants, termites, and bees. Scotocryptines, for example, live in the nests of stingless (meliponine) bees, where they probably feed on the fecal material that accumulates there (Roubik and Wheeler, 1982). Some scotocryptines,

10.21. A Cretaceous Scydmaenidae in amber from northern Burma, approximately 100 MYO. AMNH Bu113; length 0.6 mm.

10.22. Staphylinoid beetle of the family Ptiliidae, commonly called featherwing beetles for the long fringe on the margin of the hind wings (seen here protruding from under the elytra). They feed principally on fungal mycelia and spores. This is one of the larger species; others are among the tiniest adult insects, barely 0.3 mm long. Scanning electron micrograph; length 0.8 mm.

in fact, are even phoretic on the bees, grabbing on to them to hitch a flight to puddling areas, where the beetles can find another host bee from a different hive. Even more specialized leiodids are ectoparasites of mammals. Though they don't generally feed on blood, these have some or most of the features that comprise an ectoparasitic syndrome (e.g., Marshall, 1981), such as highly reduced eyes and wings, a flattened body, as well as short, thick legs for grasping to the pelage of the host. A European species of *Catopidius* lives on rabbits, and six species of Platypsyllinae live on rodents in the Holarctic Region. *Platypsyllus castor* is an obligate parasite of the beaver in North America, and a European species of *Silphopsyllus* is parasitic on shrews. Other staphylinoid ectoparasites include approximately 65 neotropical species of staphylinids in the tribe Amblyopinini, which live in the fur of rodents.

The largest staphylinoids are the Silphidae, or carrion beetles, which can be up to 4 cm (1.5 in.) in length. Actually, only some of these breed in carrion; others are predatory and a few are even phytophagous. These beetles are best known because of the extended parental care that is known in *Nicrophorus* (Eggert and Müller, 1997) and best studied in the European species *N. vespillo* (Pukowski, 1933) and several North America species, especially *N. tomentosus* (Milne and Milne, 1976; Pellissier Scott, 1994, 1996, 1997). A mated male and female or several unrelated females (depending on the species) excavate soil from underneath the carcass of a small animal and cover it. The female, and sometimes the male, molds the carrion into a bolus with a depression, and she lays eggs away from the carrion so that they are not fouled. The newly hatched larvae make their way to a depression in the bolus and are fed regurgitate by the adults, like nestling birds, and as they mature they feed on the carrion themselves. The adults stridulate by rubbing the abdomen against the inside of the elytron, which purportedly can even attract the larvae. The carrion is kept from putrifying by the adults, who apply secretions to it and constantly glean the surface of the bolus. Cooperative and communal burying by *Nicrophorus* appears to be related to selection via intense competition with other beetles and carrion-breeding flies (Pellissier-Scott, 1994). More larvae can be raised by burying the larval breeding sites before flies can oviposit. Dung-breeding insects similarly compete, which also accounts for the dung-burying behavior of scarabs.

The smallest staphylinoids are in the Ptiliidae (e.g., Figure 10.22), which as mentioned earlier also have some of the smallest adult insects (e.g., *Nanosella fungi*: 0.25 mm body length). Associated with microscopic size are narrow, strap-like wings with a fringe of long, socketed setae. As in other minute insects like thrips and mymarid wasps, the fringe increases the surface area of the minute wing, and because the setae can be folded against each other, the wing can be folded away when at rest. Many ptiliids feed on fungal spores and hyphae in decaying vegetation, and some live in the minute pores of bracket fungi feeding on the spores that are produced there. A study of the reproduction of ptiliids has revealed at least one functional constraint to the lower limits of body size in adult insects (Dybas and Dybas, 1981). Ptiliids are so miniaturized that the male of the smaller species produces a single sperm and the female a single egg at a time, and when mated, the tail of the sperm protrudes from the female! Some other species of ptiliids are parthenogenetic (Dybas, 1966). Lawrence (1991) also includes in this family the limulodids, which have traditionally been placed in their own family. Limulodine ptiliids are much larger than ptiliines and have a characteristic drop-shaped body (Figure 10.23), which is a common feature of insects that are obligate symbionts of ants and termites.

The Staphylinidae (including the Scaphidiinae and Pselaphinae) is a vast group of generally small, gracile beetles with a flexible abdomen (Figures 10.24, 10.25), which worm their way amidst the interstices of decaying leaves and humus. Though the greatest diversity of the family is tropical, staphylinids are also the dominant beetle family in temperate forests, comprising half or more of beetle individuals and one quarter or more of the beetle biomass. This family is studied by a devoted group of specialists (most of them European), which is a good thing because it is arguably the largest family of all organisms. The recent, 4,200-page catalogue of world species (exclusive of the subfamilies Aleocharinae, Paederinae, Pselaphinae, and Scaphidiinae) treats the nomenclature, citations, and significant biological references for 23,500 valid species names (Herman, 2001). In total, Staphylinidae probably comprises some 47,000 *described* species, but the total number is probably several times this total. Without a comprehensive opus like this catalogue, taxonomy on such a huge group would be virtually intractable because references are scattered among thousands of papers. This work also is probably the single largest project done by one entomologist, a fitting achievement considering the size of the group.

10.23. A ptiliid beetle of the subfamily Limulodinae, preserved in 20 MYO amber from the Dominican Republic. Many limulodines live in ant and termite nests, feeding on colony detritus. The drop-shaped body is a feature that has repeatedly evolved among various insects that are social insect inquilines. It presumably helps protect them, should they be recognized by workers. AMNH DR14–619; length 1.5 mm.

A staphylinid is the oldest definitive polyphagan beetle, which is comprised of a series of specimens of a Triassic species from the Carnian (ca. 225–230 MYA) of Virginia (Fraser *et al.*, 1996) (Figures 10.26, 10.27). This beetle is slender and fusiform, has a dense covering of short hairs, and short, truncate elytra that expose six abdominal segments, which is so typical of the family Staphylinidae. In Late Jurassic deposits from Karatau (152 MYO), staphylinids were diverse (Tichhomerova, 1968), though the affinities of compression fossils are unclear since diagnostic features of Recent subfamilies depend on features too microscopic to be preserved in rock (e.g., Figure 10.28). An even greater diversity of Mesozoic staphylinids is collectively preserved in amber from the Early Cretaceous of Lebanon, the mid-Cretaceous of New Jersey (e.g., Gusarov, 2000), and a particularly diverse fauna from the mid-Cretaceous of Burma (Figure 10.29). When all of these specimens are studied in detail, they will be extremely important to understanding the origins of Recent subfamilies and tribes. It already appears that Staphylinidae originated quite early, perhaps as much as 240 MYA, and by at least 100 MY later some Recent subfamilies had appeared.

Classification of the family is comprised of a system of 32 subfamilies and many tribes, only a few of which can be mentioned here. The subfamily Paederinae is best known by the type genus *Paederus*, some species of which are aposematic and secrete a potent toxin, paederin (Pavan, 1963, 1982). A minute droplet of paederin causes a persistent, oozing lesion on human skin. Paederin is not only the most complex beetle toxin known (Dettner, 1987) but also one of the most complex non-proteinaceous substances known in nature. Though most staphylinids are predatory as adults and larvae, many of them will scavenge, and overall they have diverse diets. Some feed on carrion (e.g., *Omalium, Atheta, Proteinus*), *Aleochara* are larval ectoparasitoids of fly pupae, gyrophaenines feed on spores in mushrooms, and adults and larvae of the large genus *Bledius* burrow up to 40 cm (15 in.) in sand along the shores of rivers, lakes, and seas, where they graze on interstitial algae and diatoms using brushy mouthparts (Herman, 1986). Among all staphylinids, though, certainly the most highly

10.24. A typical member of the huge family Staphylinidae. This is one of the largest and most ecologically diverse of all insect families. Scanning electron micrograph; body length 4.2 mm.

10.25. A staphylinid in Miocene Dominican amber. Morone Collection, M1425; length 5.1 mm.

10.26. The oldest known staphylinid beetle, from the Late Triassic of Virigina, approximately 220 MYO. Staphylinidae is the oldest definitive polyphagan family. VMNH; length 2.1 mm.

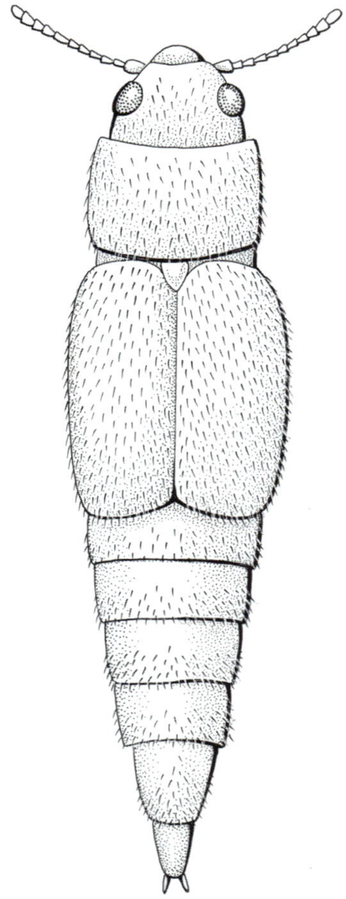

10.27. Reconstruction of the Triassic staphylinid from Virginia, based on a series of specimens. Staphylinidae is one of few living families of insects known to extend back to the Triassic.

10.28. A staphylinid in Early Cretaceous limestone from Brazil, 120 MYO. AMNH SA43295; length 11.4 mm.

10.29. A staphylinoid in mid-Cretaceous amber from northern Burma, 100 MYO. AMNH Bu1464; length 1.0 mm.

10.30. A larval staphylinid captured in Miocene Dominican amber while clinging to a coprolite, probably that of a snail. Many staphylinids are predatory as both adults and larvae, and some are attracted to dung, where they feed on fly larvae. Morone Collection, M2226; length 2.3 mm.

modified are those that are specialized inquilines of social insects. The adults of some of these are virtually unrecognizable as staphylinids.

Ectosymbionts of Social Insects. Many insects, collectively called *inquilines* or *ectosymbionts*, are specialized for living within the colonies of ants (these are *myrmecophiles*), termites (*termitophiles*), bees (*melittophiles*), and vespids (*sphecophiles*) (Wilson, 1971). Casual visitors usually lack defenses and disguises for penetrating the sanctum. Among the 32 families of social insect ectosymbionts in insects, the habit has evolved more often in beetles than any other order, and in beetles most of these are staphylinoids (reviewed in Wilson, 1971). Consequently, many beetle ectosymbionts are probably caused by the predominance of beetles in edaphic microhabitats, where they would commonly encounter ants and termites. Most of the ectosymbionts are merely scavengers feeding on the middens, which are ignored by the ants or termites. Some are predators, ectoparasitoids, and cleptoparasites, particularly if there is larval brood and provision for them, and all these are very well disguised chemically and sometimes also structurally. Other species of ectosymbionts are regarded as members of the colony that may even be fed and protected. The transition to specialized ectosymbionts appears to have most commonly been from saprophagy, and less so from a predatory diet.

Among social insects the ants and termites have a much larger fauna of ectoparasitoids associated with them than do bees and vespids. This difference probably is attributable to the fact that the ants and termites generally have their colonies in the ground, where ectosymbionts encounter them more often, whereas most social bees and vespids nest above ground. Also, ants and termites have larger colonies and construct larger nests than most social bees and vespids. Indeed, those ants and termites that form the largest colonies, like driver and army ants (Dorylinae, Ecitoninae), leaf-cutter ants (Attini), and higher termites (Termitidae), have the most diverse faunas of ectosymbionts (Wilson, 1971). This range is not just the result of the greater amount of resources in the colony, like huge middens, but also the persistence of these huge colonies. Occupied nests of some termitids and attines are known to have existed for nearly a century, and such longevity promotes the evolution of specialized ectosymbionts, much the way caves in which bats have roosted for millennia have been a sink for specialized bat parasites.

There is a host of features that adapt ectosymbionts for life among social insects (Wilson, 1971). Many, for example, are lighter in color than free-living relatives, and a flat, drop-shaped body is commonly found in such disparate groups as atelurid silverfish, *Thaumatoxena* phorid flies, and cephaloplectine and other staphylinid beetles. Presumably the drop-shaped body helps prevent ant mandibles from getting a purchase on the insect, should it be recognized as an intruder. Many beetles secrete appeasement substances. *Termitella* staphylinids, for example, live in the nest of *Nasutitermes* termites, and a huge pair of glands at the apex of the abdomen secretes droplets that the termites eagerly consume. Different glands are used by some myrmecophilous staphylinids, like *Atemeles,* to calm and appease workers, and yet other glands on other beetles will even entice worker ants to bring the beetle back to the nest. *Trichomes* are fluffs of fine hairs that serve as wicks for disseminating scents in various beetles; these attract worker ants. The most structurally peculiar ectosymbiont beetles are various aleocharine staphylinids whose abdomens have been grotesquely swollen and physogastric, some of which have protruding lobes, and they are generally held curled over the thorax and head. These peculiar beetles have been interpreted as physical mimics of their hosts, which is unlikely because vision of most of their ant and termite hosts is rudimentary. More likely, the swollen abdomens produce secretions that appease or mimic the host nestmates. Perhaps most remarkable are the arthropods that mimic the action of a nestmate in soliciting liquid food through trophallaxis. Lomechusine staphylinids, for example, tap on the labium of their host ants with their antennae and tarsi, and the replete ant delivers a droplet of meal.

Scarabaeiformia (Superfamily Scarabaeoidea)

Scarabaeiformia is a large group of approximately 31,000 species, comprising the scarabs, dung beetles, grubs, chafers, and various horned beetles. Monophyly of the infraorder or superfamily is quite clear, based on the following features (among others): (1) burrowing adaptations, including large forecoxae and tibiae with series of teeth; (2) hind wings with reduced venation and a spring mechanism for folding them; (3) adult antennae with a club, or a *lamellate* apex having a series of flattened extensions; (4) grublike, or *scarabaeiform*,

10.31. A grub, or scarab larva, preserved in Miocene Dominican amber, with the characteristic C-shaped body and well-developed head capsule and legs. Morone Collection, M1216; length 10 mm.

larvae that are thick and C-shaped, with well-developed legs but no urogomphi (Figure 10.31). There are 13 families in the group, three of them rather small (Belohinidae, Ceratocanthidae, Hybosoridae).

Several families of scarabaeoids stridulate (usually as adults), which is used for communication in their burrows and nests. Lucanidae, which are best known for the males with large mandibles, have stridulatory structures on the mid coxa and hind trochanter. Passalidae ("bess beetles") are plug-shaped, shiny black beetles that live in galleries in rotten wood as larvae and adults. Adults stridulate by rubbing a rough surface on the hind wing over another rough surface on the abdomen. These beetles are like the wood-eating roach *Cryptocercus* because there is extended parental care (Schuster and Schuster, 1985; Halffter, 1997). Trogids, or "hide beetles," are brown, rugose beetles that feed on very dried remains of carcasses, often burrowing beneath it. These stridulate by rubbing rough areas on the abdomen against the inside of the elytra. Locations of the stridulatory structures indicate independent origin of the structures in the three families.

Scarabaeidae is the largest family of the group with 28,000 species, which is classified into a system of subfamilies. Some are exceptionally colorful and beautiful (like many Cetoniinae and some Rutelinae), and one Central American species, *Plusiotes resplendens,* is well known for its unique color, which has an appearance of being entirely gold-plated (Figure 10.1). Melolonthinae are the mostly brown "chafers," the grub pests that infest agricultural and ornamental plants. Perhaps the most famous are the dung beetles, which are many species in the Aphodiinae and Scarabaeinae. The adults of these subfamilies have specialized mouthparts with a brushy labrum and reduced mandibles for feeding on fluids.

The dung-rolling scarabs have intrigued people at least since the ancient Egyptians, who believed these insects were responsible for the cycle of day and night and so were a common symbol of reincarnation. We now know that dung rolling is done simply so that a female scarab can protect her breeding site. The biology of these insects has been amply reviewed (Halffter and Edmonds, 1982; Hanski and Cambefort, 1991). The female, sometimes with the assistance of the male, extracts a chunk of dung generally larger than the beetles themselves, and rolls it by pushing it backwards with her hind legs and away from the frenzy of the dung heap, sometimes for many meters, around obstacles and up slopes – the Sisypheans among insects. A tunnel is then excavated in which the ball is interred, the female remains with it while periodically laying eggs on the dung, and the larvae feed on the dung. Others excavate tunnels beneath the dung pile itself. This habit no doubt protects the cached dung from desiccation and exposure of the larvae to predators, parasitoids, and competing larvae of other insects, like flies and other beetles.

Dung beetles occur throughout the world, in tropical forests to deserts, but they are most diverse in tropical grasslands where the ungulate fauna is diverse, particularly Africa. Wherever they occur, though, competition for fresh dung is fierce, and dung beetles are among the first insects to arrive at a fresh heap, sometimes by the thousands (Ridsdill-Smith, 1991). In the Australian outback the largest native herbivores are the kangaroos and wallabies, which are very efficient at conserving water, and they produce dry, hard dung, which quickly and thoroughly dries in the arid climate. The competiton for fresh dung is so intense here that some species of *Onthophagus* (a huge, worldwide genus) have evolved prehensile tarsi, which are used for clinging to the hair around the anus of wallabies. There they hang, awaiting a fresh delivery from their life-giving portal, and when it is delivered they fall to the ground with it and promptly bury it. It takes little imagination to wonder what may be the next evolutionary step.

As the cattle industry flourished in Australia, native bush flies (*Musca vetustissima*), which breed in all sorts of dung, became particularly oppressive. Original efforts to control the cow pats and thus the flies using native dung beetles failed because these scarabs were adapted for handling the tidy, dry dung of marsupials. Twenty-six species of scarabs were established in Australia from around the world, and these have been very effective in burying the cow pats and controlling bush flies, which attests to the abilities of these beetles to process waste.

Among the heaviest insects are African scarabs in the genus *Goliathus*, larvae of which breed in decaying wood and the adults feed on fruits and sap. This genus doesn't have the impressive length of *Titanus* and *Megasoma*, but it does have more girth than any insect. A live specimen of *Goliathus* was left with entomologists at the American Museum of Natural History in late 1958, which was found among a shipment of bananas. It survived for seven months and was even allowed to fly around the hallways, occasionally crashing into closed windows. When it died, the beetle gave scarabs a further distinction: the only insect to have an obituary and photograph, in this case in the July 9, 1959 edition of *The New York Times.*

Horns and Dimorphism. Another major feature of scarab biology concerns the repeated origin of male horns (Figure 10.32). Horned beetles occur in various families but are most prevalent in the Scarabaeoidea (Arrow, 1951). One to three horns can each occur on the prothorax, head, mandibles, and clypeus. The males of the rotund Geotrupidae usually have short pronotal horns and one or two large horns sticking up from the head. Scarabaeids in the subfamily Dynastinae are renowned for their spectacular, long horns (see the world monograph by Enrődi, 1985). Studies on the behavior of horned beetles and development of the horns have revealed evolutionary and developmental mechanisms as to how such exotic structures evolve. First, the male usually has horns (females very rarely do so, and they are always small), indicating that these are products of sexual selection, or intense competition among males for mating. In many scarabaeoids, males fight to control access to breeding sites and to females (Eberhard, 1980). Those species with large horns use the horns directly in combat. By flexing the head, a male can separate the pronotal and cephalic horns to grasp another male like pincers and then throw him off the plant.

Second, male dimorphism usually occurs in horned beetles: Large males have very large horns, and small males have diminutive horns (Emlen, 2000). Such *bimodal* or *biphasic allometry* occurs throughout insects that have elaborate structures, including the heads and mandibles of ants having soldiers and major and minor workers, and flies where the male has eyes at the tips of long stalks (Emlen and Nijhout, 2000). At some intermediate, boundary size, larger individuals have a disproportionately larger structure, but the structure is disproportionately small below this size. The biphasic allometry relates to adult body size, which for holometabolous insects is determined during larval growth. Larger larvae become larger adults with larger horns. Appendages such as horns, eyestalks, and mandibles develop from imaginal discs late in larval development (usually just before pupation), and beyond a threshold body size the epidermal cells in imaginal discs proliferate disproportionately more, all of which is mediated by levels of juvenile hormones. The great variation among closely related species in horn size, shape, and location is determined by the number, location, and JH sensitivity of imaginal disc cells. As for any

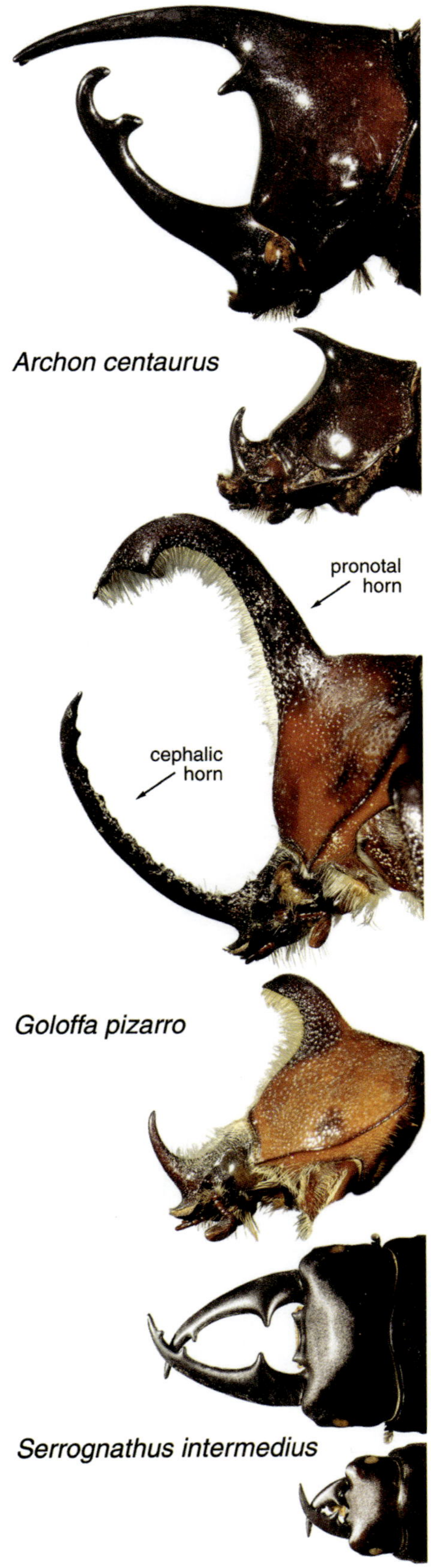

10.32. Male scarabaeoid beetles showing dimorphism in mandibles (below: Lucanidae) and pronotal and cephalic horns (above: Dynastinae). The mandibles and horns are used in combat to grasp and throw male opponents vying for females. Dimorphism typically occurs in insects with elaborate structures, including worker and soldier ants and stalk-eyed flies. Not to the same scale.

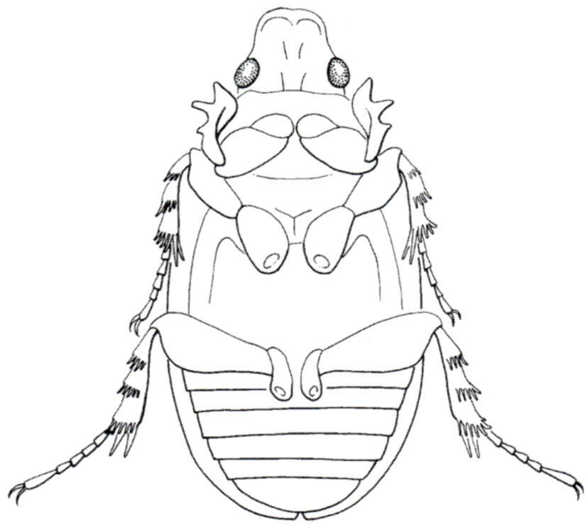

10.33. Scarabs appeared in the fossil record in the Late Jurassic and were diverse by the Cretaceous. *Holcorobeus* was a Cretaceous genus that was global in distribution. This is a reconstruction of *Holcorobeus vittatus*, from the Early Cretaceous of Baissa in Siberia. Length 12 mm; redrawn from Ponomarenko (1977b).

10.34. Scarab preserved in 120 MYO limestone from the Santana Formation of Brazil, which preserves diverse early scarabs. The broad, shovel-like front of the head is easily seen. AMNH SA46253; length 14.8 mm.

elaborate structure, though, there are upper limits to its development. These include functional constraints (the largest horns are awkward), as well as pleiotropic effects. Beetles with larger head horns, for example, have smaller eyes, and those with larger pronotal horns have smaller wings (Emlen, 2000; Emlen and Nijhout, 2000), so structures of fundamental importance can be seriously compromised in individuals with the most elaborate horns, mandibles, or eyestalks.

The most recent analysis of relationships in Scarabaeoidea is the morphological study by Browne and Scholtz (1998, 1999). Superimposed on this scheme was a summary of the fossil record of scarabs (Krell, 2000). Krell catalogued and critically assessed the 189 plausible and definitive fossil species of scarabs, omitting 27 species that lacked diagnostic features and whose identities were ambiguous. Most of these are body fossils and some are trace fossils, which are the remains of distinctive scarabaeid burrows. Early Jurassic records of body fossils attributed to scarabs lack sufficient detail, though it is possible that primitive scarabs existed by this time. Certainly, abundant vegetation and dung existed at this time to support scarabs, the latter from the large herbivorous dinosaurs. In fact, fossilized dung from Late Cretaceous herbivorous dinosaurs in Montana contains scarab burrows (Chin and Gill, 1996), though this does not indicate the habit necessarily existed much earlier. The oldest apparent scarab is *Holcorobeus nigrimontanus*, from the Late Jurassic of Karatau, Kazachstan (ca. 152 MYO), which had the typical burrowing legs of scarabs. By the Early Cretaceous, 130 MYA, diverse genera in some five families and subfamilies existed, many of them preserved in the rich Lagerstätte from ancient Lake Baissa in Siberia (Figure 10.33). Diverse scarabaeoids are also preserved in the Early Cretaceous Santana Formation of Brazil (Figure 10.34), but these still require study. By the Oligocene, fossils of almost all scarabaeoid families and subfamilies existed. The "form genus" *Coprinisphaera* consists of various fossilized burrows of scarabaeine burrows (reviewed by Genise *et al.*, 2000). These were abundant and scattered nearly globally from the Paleocene to the Pleistocene, suggesting that the evolution of dung beetles was relatively late for Scarabaeoidea.

Elateriformia

Elateriformia is a heterogeneous infraorder of six superfamilies (five if the cantharoids are placed within the Elateroidea). Larvae in this group are generally long-lived compared to the adults, and the adults either have a complex system that locks the pro- and mesothoracic segments together (like the elateroids), or they are aposematic and defend themselves chemically (like the cantharoids). Adults also commonly have pectinate or flabellate antennae. Larval life histories vary greatly, from ectoparasitoids of cicada nymphs (Rhipiceridae: Dascilloidea) to free-living aquatic larvae.

The small superfamily Scirtoidea contains four families:

10.35. Beetle of possibly the family Dryopidae, from the Santana Formation of Brazil, 120 MYO. Dryopidae larvae are aquatic. AMNH SA43296; length 8.0 mm.

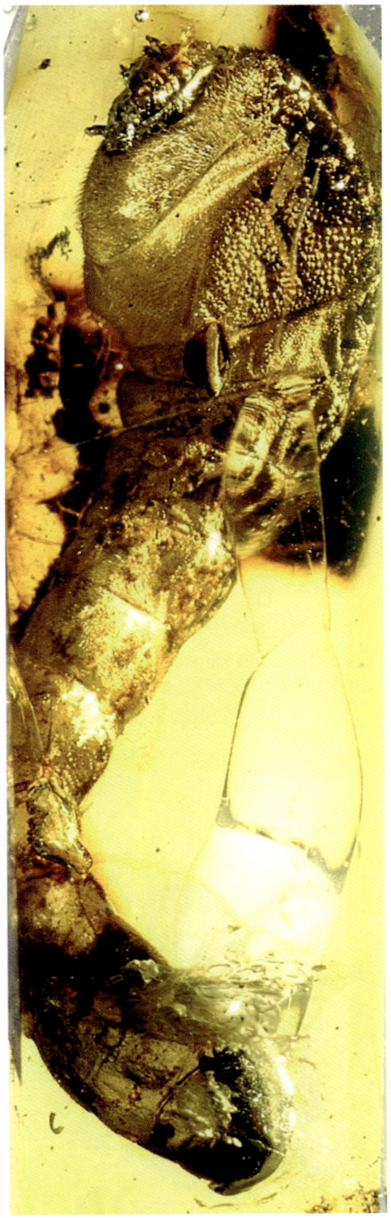

10.36. Larva of a buprestid ("jewel") beetle, in 90 MYO amber from central New Jersey. Buprestid larvae with a swollen thorax like this one are typically miners under the bark of living trees, so this beetle was probably living in the tree that produced the amber. Length 11.0 mm.

Scirtidae, Eucinetidae, Declinidae, and Clambidae. These have a hypognathous head that fits tightly against the prothorax. Scirtidae (Helodidae) are considered very primitive polyphagans because the larvae have features typical of basal holometabolans: They are flattened, active, and aquatic or semi-aquatic; they feed by filtering particulate detritus from submerged surfaces (using combs of hairs on the maxillae and hypopharynx); and they have long, multiarticulate antennae. Eucinetidae and Clambidae appear to be closely related, and these are terrestrial and feed on fungal spores and slime "molds."

The Byrrhoidea contains most of the families whose larvae live in wet environments, which is a departure for Polyphaga. This superfamily contains 12 families according to the classification of Lawrence and Newton (1995), and these are either terrestrial (Byrrhidae, Callirhipidae, Dryopidae [Figure 10.35], most Ptilodactylidae), semi-aquatic (Limnichidae, Heteroceridae) or fully aquatic as larvae (Psephenidae) or as larvae and adults (many Elmidae). Psephenidae ("water pennies") have very flat, disk-shaped larvae (Figure 10.2) that graze algae off the surface of rocks in fast currents; the larvae of Elmidae have anal gills and the adults breathe via bubbles on the body.

The Buprestoidea contains just the family Buprestidae, or jewel beetles, renowned for the many species with brilliant metallic colors that are avidly collected. These beetles are very streamlined and bullet-shaped; the body regions are confluent with the head recessed into the thorax. The larvae are distinctive, being soft-bodied and with an expanded thorax. These mine the phloem under the bark of trees, but others bore into stems and even mine leaves. The earliest body fossils attributed to Buprestidae are from the Late Jurassic of Karatau (two Triassic records are erroneous), and there is a remarkable larval specimen in mid-Cretaceous amber from New Jersey, 90 MYO (Figure 10.36). Some fossil buprestids in the Eocene oil shales from Messel, Germany have even retained their metallic sheen, and other compression fossils of them are diverse throughout the Tertiary of Europe and North America (e.g., Figure 10.37).

The superfamily Elateroidea includes 15 families, including the cantharoids, larvae of which have mouthparts adapted for fluid feeding, either as predators, saprophages,

10.37. *Ancylocheira*, an adult buprestid beetle from the Miocene of Oeningen, Germany. NHM In.59502; length 27 mm.

or phytophages. They feed using extraoral digestion, by essentially regurgitating enzymes onto their meal and sucking up the brew. Three of the families (Elateridae, Eucnemidae, Throscidae) possess the well-known clicking mechanism, which is a small, hard tongue on the prosternum that snaps into a groove in the mesosternum and suddenly flexes the body. A click beetle on its back seems unable to right itself without clicking, which is one possible function of this structure. Another function is probably antipredatory because these smooth, seed-shaped beetles are difficult to hold when they click, particularly because the clicking mechanism produces a powerful flexion of the body. Elateroids have a distinctive slender body shape and a prothorax with backward-pointing spines on the posterior corners, which have made it much easier to interpret compression fossils. The larvae of many bore into rotting wood permeated with fungal rot, though the larvae of some feed on other wood-boring larvae or the roots of plants (like "wireworms"). Elateroids have a rich fossil record that is well studied (Dolin, 1975, 1976; Muona, 1993). Diverse Elateridae are known from the Late Jurassic (160–150 MYA), from which some 30 genera have been described, most of them from the rich deposit at Karatau. Diverse adults occur in various Cretaceous deposits (including amber) (Figures 10.38, 10.39), though most of

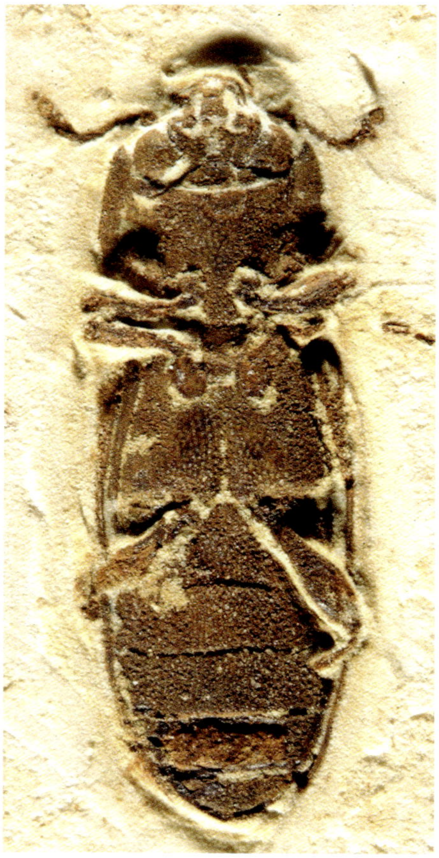

10.38. An Early Cretaceous click beetle (family Elateridae) from the Santana Formation of Brazil, ventral view. The sternal process that is involved in sudden flexion of the body is clearly visible. This behavior may be a defense against predators and a way to upright the insect when on its back. AMNH SA46545; length 6.8 mm.

10.39. Elaterid click beetle in Early Cretaceous amber from Lebanon, 125 MYO. The first occurrence of elateroid beetles is in the Early Jurassic, and they become increasingly abundant and diverse through the later Mesozoic. AMNH LAE-126; length 3.4 mm.

10.40. Beetles of the elateroid family Eucnemidae in 90 MYO amber from central New Jersey. Like the larvae of many modern eucnemids and other elateroids do today, these beetles were probably breeding in decaying wood. AMNH NJ-263; length 2.9 mm.

10.41. A beetle of the elateroid family Throscidae, preserved in Miocene Dominican amber. Bubbles exuded from glandular pores on the elytra, probably delivering a defensive secretion. AMNH DR8–394; length 2.3 mm.

these have not been studied. In the Eocene Baltic amber there are 22 genera and 45 species of eucnemids and throscids alone (Muona, 1993), most of them belonging to Recent genera. The diverse Elateridae preserved in other Tertiary ambers have unfortunately not been studied, but the trend is clear: Elateroids probably originated in the Early to mid-Jurassic, proliferated rapidly by the Late Jurassic, and many Recent genera were established by the Early Tertiary (Figures 10.40, 10.41).

The cantharoid beetles are distinctive for their elytra, which in many species are broad, flat, leathery soft, and boldly marked in yellow, orange, and black to advertise the distastefulness of the beetles. These include the familiar Lampyridae, or "lightning bugs" ("fireflies"), the phengodid glow-worms, the soldier beetles (family Cantharidae), the lacewing beetles (Lycidae) (Figure 10.42), and eight other, small families (e.g., Figure 10.43), all of which are usually classified within the Elateroidea (e.g., Lawrence and Newton, 1995). The distinctive, soft elytra probably evolved because the adults are generally exposed feeders, including predators and even those that feed on pollen. Many larvae are predatory, though some are saprophagous, like Lycidae. The most fascinating aspect of cantharoid biology, without question, is bioluminescence.

Bioluminescence is the production of light via specialized metabolic processes and usually specialized cells in various organisms. It occurs in bacteria, fungi, some plants, and 11 phyla of animals. In insects it occurs sporadically, including Diptera (e.g., some keroplatine fungus gnat larvae), but bioluminescence appears most frequently in Coleoptera (Lloyd, 1983). Reports of bioluminescent Collembola need confirmation. Among all bioluminescent insects the mechanism of light production involves a luciferin in the presence of oxygen, the enzyme luciferase, and the cellular energy source ATP (adenosine triphosphate). The reaction of these produces oxyluciferin, CO_2 and light (Wood, 1995). Because ATP is ubiquitous among virtually all organisms, this reaction has been used in early and distant space probes searching for extraterrestrial life. Planetary soil is scooped into the chamber of a remote probe that contains luciferin and luciferase, which would glow if life forms like bacteria were present; the light would be easily detected by sensors, and this information would be transmitted. More recently, luciferin and luciferase have been used in molecular probes to help locate and quantify metabolic activity, and some organisms have even had these genes transgenically inserted for such purposes. This example illustrates how technology benefits from basic knowledge of nature.

Among beetles, bioluminescence occurs in some Elateridae and in many cantharoids. There is a report of a bioluminescent staphylinid from Brazil, but because many staphylinids feed on fungi (including bioluminescent fungi), this needs to be confirmed. Bioluminescent elaterids are in

10.42. Lycid beetle in Miocene Dominican amber. Most extant Lycidae are aposematically and boldly colored black with red, yellow, or orange. The color patterns are still preserved on this specimen. AMNH DR10–698; length 2.8 mm (excluding antennae).

the Pyrophorinae, which include the well-known *Pyrophrous* click beetles that have luminous eye-spots on their pronotum and a photic organ under the thorax near the abdomen. When cruising through the dark understory of a tropical forest at night, they leave a luminescent green or orange trail. Less well known, but even more impressive, is the larva of *Pyrearinus termitilluminans* from northeastern Brazil. Larvae of these click beetles inhabit niches in the high termite mounds that dot the Brazilian cerrado, and their glowing attracts insects that they prey upon. The nighttime displays of these larvae have been likened to skyscrapers in a miniature city. The pyrotechnic masters among insects, though, are the cantharoids.

There have been various approaches to understanding the evolution of bioluminescence in cantharoid beetles, from adaptationist (Lloyd, 1983) to phylogenetic (Branham and Wenzel, 2001, 2003). In cantharoids at least, the light-producing photic organs are composed of highly specialized fat body cells backed by a reflective layer of uric acid crystals and lying underneath translucent cuticle. Understanding the evolution of photic organs has become possible with recent knowledge of relationships among the cantharoid families (Branham and Wenzel, 2001). The small basal families of cantharoids, Plastoceridae and Drilidae (from Eurasia and Africa), are not bioluminescent, and sexes communicate with each other pheromonally. In the next most basal family, the Omalisidae (from southern Europe), it is just the larvae that are luminescent, and this probably represents the original or ancestral condition for a lineage that comprises all the bioluminescent cantharoids. Some cantharoids later evolved bioluminescence in the adults as well, in an array of photic organs generally on the apex of the abdomen (e.g., Figure 10.44). Indeed, in all the adult cantharoid beetles having adult photic organs, where their larvae are also known these are luminescent. The larval photic organs generally comprise a pair of luminous spots per segment, and their original function appears to have been aposematic. The photic organs in lampyrid larvae appear to have homologues in adults, though adult lampyrids have evolved novel photic organs as well.

Luminescence appears to have originally evolved once in most cantharoids, including the familiar Lampyridae and their apparent living sister group, the small family Rhagophthalmidae (Branham and Wenzel, 2001). These are the only two families of cantharoids where the adult females use abdominal photic organs for courtship communication. Bioluminescence was then lost in a middle grade of taxa including the the small Holarctic family Omethidae and in the cosmopolitan families Lycidae and Cantharidae (the latter of these appears to also be a paraphyletic grade)

10.43. Bizarre larva of the cantharoid beetle family Brachypsectridae, preserved in 20 MYO Dominican amber. These predatory larvae are found on tree trunks – the beetle equivalents of larval ascalaphid owlflies (cf. Figure 9.22). Morone Collection, M2198; length 5 mm.

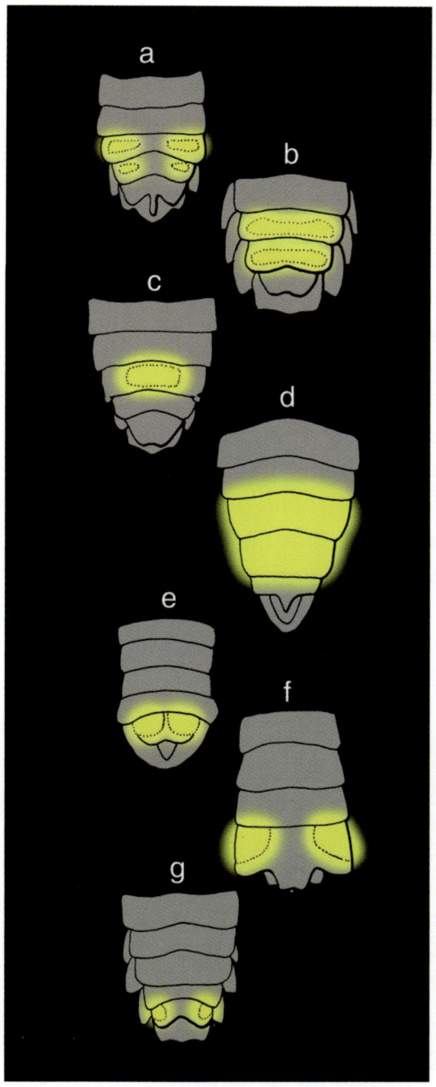

10.44. Positions of the photic organs on the abdominal sternites of various lampyrid beetles, or fire flies. (a) *Cratomorphus*, (b) *Pyrocoelia*, (c) *Erythrolychnia*, (d) *Bicellonycha*, (e) *Robopus*, (f) *Pteroptyx*, (g) *Pleotomus*. Besides position and sizes of the photic organs, the flash duration, frequency, and habitat location of the beetle communicate species identity. It is hypothesized that cantharoid bioluminescence originally evolved as aposematic warnings in larvae, which then developed into a system of adult communication in Lampyridae. Redrawn from Branham and Wenzel (2003).

(Branham and Wenzel, 2001). Both lycids and cantharids are well-known beetles because they are conspicuous feeders commonly seen at flowers, many species of which are involved in mimicry complexes. Thus, it is plausible that chemical defense and warning colors in these families have taken over the original role of larval bioluminescent aposematism seen in other cantharoids. Luminescence then appears to have redeveloped in the recently evolved family Phengodidae (Branham and Wenzel, 2001). Phengodids are best known for the larval and neotenic female glow-worms and for the large pectinate antennae of the adult males. Such antennae suggest that these beetles communicate with pheromones, and indeed adult bioluminescence in phengodids is probably aposematic, not a form of sexual communication. In fact, only two genera of phengodids have photic organs in both males and females, which are *Pseudophengodes* and the curious genus *Phrixothrix*. Besides having the typical pairs of luminescent spots, larvae of *Phrixothrix* ("railroad worms") also have a pair of photic organs on the head that produce a ruby-red light. These are voracious predators and the "head lamps" are believed to function like night-vision goggles in locating prey, particularly because most insects are insensitive to red wavelengths.

Bioluminescence is best studied in Lampyridae because this family has the most elaborate development of photic organs and communication systems. Adults in the basal lineages of lampyrids actually do not have photic organs, but where adult signaling has evolved it generally appears in both sexes, indicating it clearly evolved in adults as a form of sexual communication (Branham and Wenzel, 2003). Indeed, communication by light appears to have replaced pheromonal communication in most lampyrids. Besides differences in the location of the adult photic organs (on either one or more of sternites 6, 7, 8 [Figure 10.44]), and their sizes and shapes, lampyrid light signals are generally species-specific also by the duration of the flashes, their frequency, and the location of the flasher (i.e., in trees or on the ground). The most intriguing signalers involve aggressive mimicry and synchronous flashing. In some species of "femme fatale" *Photuris* lampyrids, females mimic the signals of other species to draw in unsuspecting males, which they devour (Figure 10.45). Some *Photuris* females apparently can even mimic the signals of as many as five species. Other species have impressive displays where males and females flash their

10.45. A *Photuris* femme fatale firefly, consuming a male *Photinus* firefly that was drawn to her beacon. *Photuris* females can mimic the flashes of *Photinus* females. Photo: Marc Branham.

10.46. Two Lampyridae fireflies engulfed in resin while mating, which then hardened into amber 20 MYA in the Dominican Republic. Morone Collection, M1244.

photic organs in synchrony, though the adaptive significance of this is not understood.

Cantharoids appear to be quite young, at least judging from the fossil record. The oldest ones known are Cantharidae and Lycidae in Eocene Baltic amber, and younger ones occur in Miocene Dominican amber. Compression fossils of Lampyridae have been reported from the Miocene of Germany and Eocene–Oligocene boundary of Florissant, Colorado, but the identity of these is difficult to confirm. Diverse lampryids are preserved in Miocene Dominican amber (e.g., Figure 10.46), including rare larvae.

Bostrichiformia

This is a monophyletic assemblage of seven families (Lawrence and Britton, 1991; Lawrence and Newton, 1995), many of which have a modified excretory system that is *cryptonephridic*. This is a system wherein the Malpighian tubules (which open into the rectum and normally float free in the hemolymph) are fused to the hindgut and rectum. This allows for resorption of water and is an adaptation for living in very dry habitats. Among the best known families are Bostrichidae and Anobiidae, the so-called "powderpost" and "deathwatch" beetles, respectively. These beetles bore into recently felled or dried dead wood and gradually render the core to powder (Bostrichidae), or riddle the wood with extensive tunnels or galleries (Anobiidae). Anobiids get their common name, deathwatch beetles, from the habit of rapping their heads against the walls of the galleries, which is a way for males and females to communicate with and find each other. One can imagine a silent Victorian parlor with people sitting around the coffin of a recently departed loved one, and the only sound heard being that of faint clicking. According to legend, the beetles click the hour of death. This family also includes the Ptininae (formerly placed in its own family), which are rotund little beetles with long, scurrying legs that

10.47. A Cretaceous bostrychoid beetle, in 90 MYO amber from central New Jersey. Bostrychid and anobiid beetles today live in dry, dead wood. AMNH NJ1264; length 1.8 mm.

10.48. Larva of the family Dermestidae, preserved in 100 MYO amber from Burma. Dermestids scavenge dried remains of animals. This is the oldest record of the family. AMNH Bu231; length 0.71 mm.

resemble spiders. Ptinines feed more on dried plant and animal remains and are common occupants of granaries, nests, and some are even ectosymbionts of ants. Another, notorious bostrichiform family is Dermestidae, the "carpet" beetles. Larvae of dermestids feed on keratin (like that in woolen carpets) and the very dried, proteinaceous remains of carcasses. Adults of dermestids are usually covered in a vestiture of fine scales, and the larvae have long hairs that easily detach and supposedly entangle arthropod attackers. Larvae of the larger *Dermestes* are employed in museum "bone rooms" for cleaning the defleshed, dried bones of vertebrate specimens; meanwhile, much smaller *Anthrenus* insidiously consume skins and pinned insects that have been improperly stored. The only definitive Mesozoic bostrychoids are several adults and a few larvae preserved in amber from New Jersey and Burma, 90–100 MYA (Figures 10.47, 10.48), and other bostrichiforms are known in Tertiary amber.

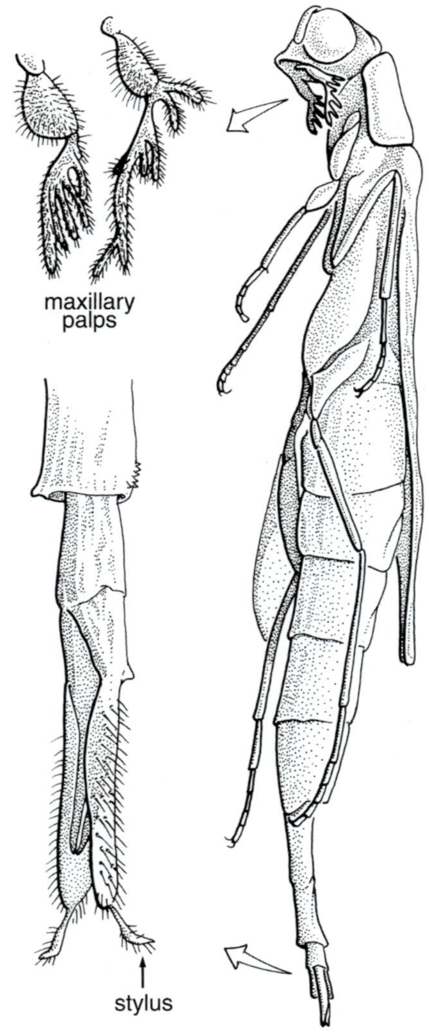

10.49. Oldest known lymexylid beetle, preserved in 100 MYO Burmese amber. The beetle is a very primitive member of the living genus *Atractocerus*. AMNH; length 9.5 mm.

Cucujiformia

The infraorder Cucujiformia is a very large assemblage of 82 families, or nearly half of the approximately 170 families of Recent Coleoptera. Because this group includes the very speciose phytophagan beetles (chrysomelids, cerambycids, and weevils), Cucujiformia contains the largest proportions of beetle species. Monophyly is defined in part by a unique cryptonephridic excretory system. There are six superfamilies, which are all quite different except that most have a distinct predilection for fungi, foliage, or wood. The most unusual cucujiforms are the ship timber beetles, family Lymexylidae, which are placed in their own isolated superfamily and which have been monographed (Wheeler, 1986). These beetles are long and soft, with a very long abdomen, large eyes, and soft elytra; in two of the seven genera, the elytra are short. The larvae are also long and slender, resembling worms. The males of some have very elaborate, pectinate palps, which must have chemosensory functions. The larvae bore into wood, and they feed on a fungus that lines the walls, much the way ambrosia beetles do (Scolytinae, Platypodinae: Curculionidae). The oldest lymexylid is an adult in 100 MYO amber from Burma, which is a very primitive species of *Atractocerus* (Figure 10.49), indicating that by the mid-Cretaceous at least some major lineages of lymexylids had appeared. Otherwise, the fossil record of lymexylids is restricted to adult and larval *Atractocerus* preserved in Miocene Dominican amber (Figure 10.50). Cleroidea includes beetles whose adults are usually exposed predators with hairy, bold-colored bodies, and the larvae are mostly predators of wood-boring insects. Adults of Clerinae and Melyridae, though, feed on pollen.

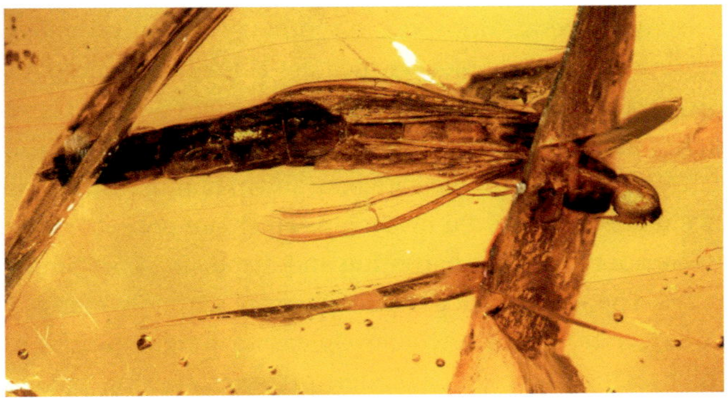

10.50. Adult timber beetle, family Lymexylidae, preserved in Miocene Dominican amber. These beetles bore into wood and inoculate the galleries with an ascomycete fungus that their larvae feed upon. Larvae are also preserved in this amber. AMNH; length 29 mm.

The Cucujoidea comprises 31 families of the Cucujiformia, and most of these are small families (<1,000 species) of obscure, minute brown beetles. These are to Coleoptera what the acalyptrates are to Diptera and the chalcidoids are

10.51. Mating lady bug beetles (family Coccinellidae), the female of which is also feeding on aphids. Some of the aphids have secreted droplets of alarm pheromone at the tip of the cornicles. Coccinellidae are unusual for cucujoid beetles since they live in exposed places and are aposematically colored, and many of them are predaceous. Photo: V. Giles.

10.52. Coccinellid beetle preserved in Miocene amber from the Dominican Republic. The color patterning is still preserved. Morone Collection, M1781; length 5.1 mm.

to Hymenoptera. Moreover, mycophagy and myxophagy (feeding on slime "molds," which are actually colonial protists) is a predominant diet of the larvae and adults. These beetles will feed either on the exposed sporophores or graze on hyphae and spores beneath bark and in humus. Cucujoids have repeatedly evolved mycangia, which are pockets in the cuticle specialized for harboring and dispersing host fungal spores. The most attractive cucujoids are certainly many of the larger tropical Erotylidae ("pleasing fungus beetles"), which can be orange, lavender, or other striking colors. Nitidulidae are well known as the small, shiny beetles that congregate around fermenting fruits, sap fluxes, and similar substrates, in which the larvae also breed.

Coccinellidae, the "ladybird" beetles or "ladybugs," are the most familiar cucujoids, and their habits of preying on small, soft-bodied insects like aphids and coccoids (Figure 10.51) is considerably divergent with the habits of other cucujoids.

10.53. Coccinellid beetle larva in Dominican amber. Morone Collection, M3021.

Epilachine coccinellids are actually phytophagous as adults and larvae. Many coccinellids obviously advertise themselves with spots and red/yellow colors with black, and they are indeed chemically defended. Coccinellids bleed pungent secretions (alkylpyrazines) from the joints between the tibia and femur to advertise their toxicity, and they further defend themselves with toxic alkaloids in their hemolymph called

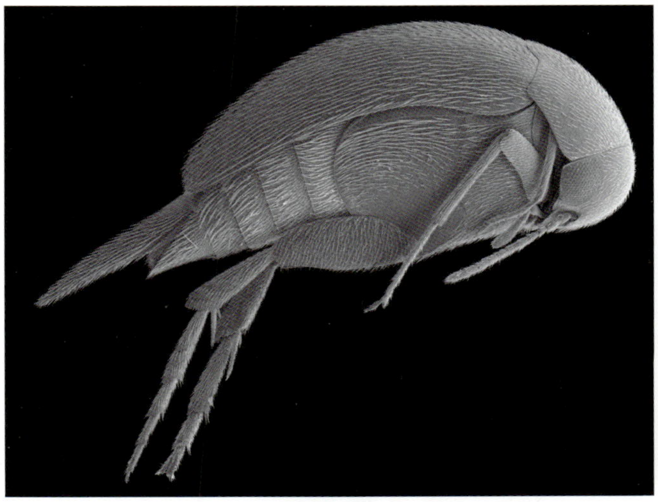

10.54. Tenebrionoid beetle of the family Mordellidae, species of which have a distinctive, drop-shaped, hump-backed body. They are usually found on flowers (where they feed on pollen) and tumble off when disturbed. Scanning electron micrograph.

10.56. A deeply grooved beetle of the large, diverse family Tenebrionidae, in Dominican amber. Tenebrionids have invaded dry habitats around the world (among other niches), where they scavenge dried plant debris and fungus. AMNH DR14–1027; length 4.4 mm.

10.55. A mordellid beetle in mid-Cretaceous amber from New Jersey, 90 MYO. AMNH NJ879; length 2.4 mm.

coccinellines. Fossil coccinellids occur only in Tertiary amber (Figures 10.52, 10.53). Perhaps the most unusual cucujoid diets are of some Cucujidae and Bothrideridae. Most species of these families are larval and adult predators of wood-boring insects, but some have become larval ectoparasitoids.

Tenebrionoidea, with some 30 families, are generally larger cucujiforms that have much more diverse life histories than cucujoids. Included here are the "tumbling flower beetles," family Mordellidae, so named for the flower-feeding adults that typically fall off the plant when startled. These have a distinctive, drop-shaped, curved body with a pointed tail end (Figures 10.54, 10.55), which has made the interpretation of *Praemordella* in Late Jurassic rocks from Karatau a very plausible member of the family. If indeed a mordellid, *Praemordella* would be among the earliest plausible cucujiforms. Tenebrionidae (Figure 10.56) is the largest family of the Cucujiformia (including here Alleculidae as a subfamily). These are generally scavengers on dried plant remains, but which also feed on lichens, fungi, and decaying wood. Certain tenebrionids are occupants of the driest places on earth and have fascinating adaptations for acquiring water. Species in the parched Namib Desert of Africa allow water to condense on their bodies at dawn, and by raising the abdomen the condensation trickles along specialized grooves on the body and collects as droplets near the mouth. "Flour beetles" are the species of *Tribolium* that spend their entire life cycle in this desiccated substance, acquiring water from vapor in the air. Because *Tribolium* is so easily reared, it is used extensively in research on genetics, physiology, and population biology (Sokoloff, 1972/1974/1977). The southern African species *Parastizopus armaticeps* has been well studied because it has sophisticated parental behavior (Rasa, 1998, 1999). These beetles live in dry, barren areas and are monogamous; females carefully choose large males during nocturnal displays following heavy rains. The male and female then excavate a burrow into sand, in which eggs are laid and tended. The female forages for scarce plant detritus at night, which she brings back to the burrow for feeding her offspring. Males continually excavate the burrow and assist in defense. The offspring are cared for through their entire larval and pupal stages and for approximately the first two weeks of their adult lives, after which the family disperses. Relationships of Tenebrionidae have been considered by Doyen and Tschinkel (1982).

Three families of tenebrionoids are best known for their

synthesis of a potent toxin, cantharidin, which are Meloidae ("blister beetles"), Oedemeridae ("false blister beetles"), and Pyrochroidae. Cantharidin is a monoterpene anhydride that occurs in hemolymph or particular exocrine glands and is secreted through pores in intersegmental membranes when the beetles are harmed. Cantharidin is also a potent drug, acting principally as a vasodilator. In fact, some meloids had been used many years ago as an aphrodisiac called "Spanish Fly," making cantharidin (in minute quantities) the original, natural "Viagra" (Prischam and Sheppard, 2002). It apparently has prolonged and painful effects, and is lethal in sufficient quantities. These beetles are typically aposematic, and meloids themselves commonly have soft, metallic blue or green bodies that exude copious yellow secretions. In Meloidae only the males and larvae produce cantharidin (McCormick and Carrell, 1997). In Pyrochroidae and Meloidae, cantharidin is transferred from the male to the female during mating, and she uses it to defend eggs (Eisner *et al.*, 1996). Meloidae are also interesting because the first instar larvae are bristly and active, and attach themselves to bees that they will parasitize for the remainder of their larval life. A similar habit occurs in another tenebrionoid family, Rhipiphoridae. These are peculiar, slender beetles with soft, pubescent bodies, which have a range of morphology from those that have nearly complete elytra and little sexual dimorphism to those where males have huge flabellate antennae and minute elytra. The females of some rhipiphorids lay their eggs on flowers, as do many meloids, where the hatching triungula encounter their hosts. Females of some species (in the Rhipidiinae) are even larviform, lacking wings and having reduced mouthparts. This syndrome of features and parasitoid lifestyle led early entomologists to believe that these beetles were closely related to the enigmatic twisted-winged parasites, order Strepsiptera. Indeed, the convergence between these two groups is absolutely remarkable. The oldest definitive rhiphiphorids occur in mid-Cretaceous Burmese amber (Grimaldi *et al.*, 2002); modern genera appear in Tertiary ambers (e.g., Figure 10.57).

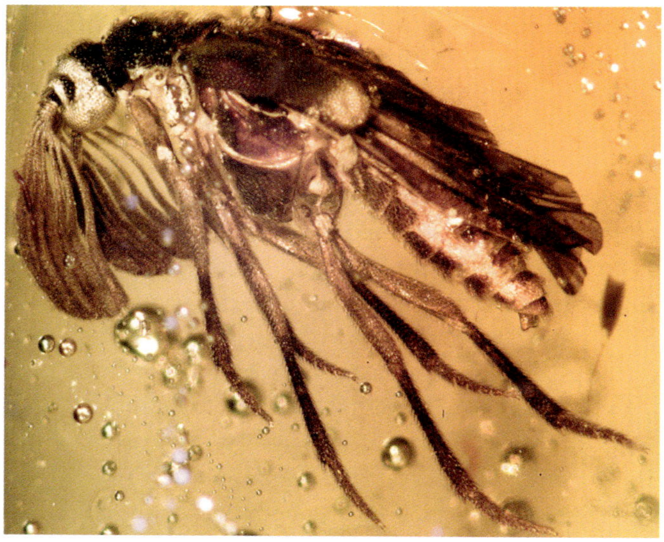

10.57. Beetle of the family Rhipiphoridae in Miocene Dominican amber, showing the distinctive flabellate antenna with long branches. Larval rhipiphorids are parasitoids on roaches, wasps, and various wood-boring insects. The biology, immature stages, and adult morphology of these beetles are remarkably convergent with that of the order Strepsiptera. Morone Collection, M1177; body length 5.6 mm.

10.58. A Cretaceous tenebrionoid beetle of the family Prostomidae, in 100 MYO amber from Burma. Prostomids live in very decayed wood. AMNH Bu1422; length 4.8 mm.

10.59. A Cretaceous tenebrionoid beetle of the family Aderidae, in 125 MYO amber from Lebanon. This is one of the earliest definitive records of the superfamily Tenebrionoidea. AMNH LAE3-10; length 1.6 mm.

10.60. A Cretaceous cucujoid beetle in 90 MYO amber from New Jersey; its red color is still preserved. AMNH NJ885; length 2.2 mm.

10.61. A predatory larva of the cucujoid family Discolomidae, preserved in Dominican amber. AMNH DR10–2009; length 2.6 mm.

10.62. Cucujid beetle in Dominican amber. The bases of the antennae are large and pointed, resembling mandibles. AMNH DR14–975; length 1.8 mm.

Identification of cucujiform beetles, particulary smaller species, critically depends on microscopic features that generally aren't preserved in fossils compressed in rocks. Indeed, Mesozoic records of Nitidulidae (e.g., *Nitidulina* from the Late Jurassic of Karatau) and Tenebrionidae (*Ulomites*, from the Late Triassic of Queensland) simply can't be assessed, though the family placement of the latter one is almost certainly incorrect. Earliest records of many cucujiforms, particularly cucujoids, are several families preserved in Cretaceous amber (Figures 10.58 to 10.60). These include Acanthocnemidae, Cerophytidae, Cryptophagidae, Lathridiidae, Nitidulidae, and Scraptiidae preserved in 80 MYO Siberian amber (Zherikhin and Sukatsheva, 1973), but without doubt the most diverse assemblage of well-preserved, Mesozoic cucujiforms is in 100 MYO Burmese amber. Besides the families found in Siberian amber, additional Recent families in Burmese amber include Aderidae (e.g. Figure 10.59), Anthicidae, Ciidae, Cleridae, Colydiidae, Cucujidae, Melandryidae, Melyridae, Mordellidae, Oedemeridae, Rhipiphoridae, and Salpingidae (Rasnitsyn and Ross, 2000; Grimaldi *et al.*, 2002). Most of these still require detailed study. Ancestral Cucujiformia probably appeared in the Jurassic, and most Recent families had appeared by the mid- to Late Cretaceous, 100–80 MYA. By the Eocene, as preserved in Baltic amber, many genera of cucujiforms that are Recent or closely related to them had evolved (Abdullah, 1964; Weitschat and Wichard, 1998). Additional, and much more recently derived taxa occur in Miocene Dominican amber as both adults and larvae (e.g., Figures 10.61, 10.62).

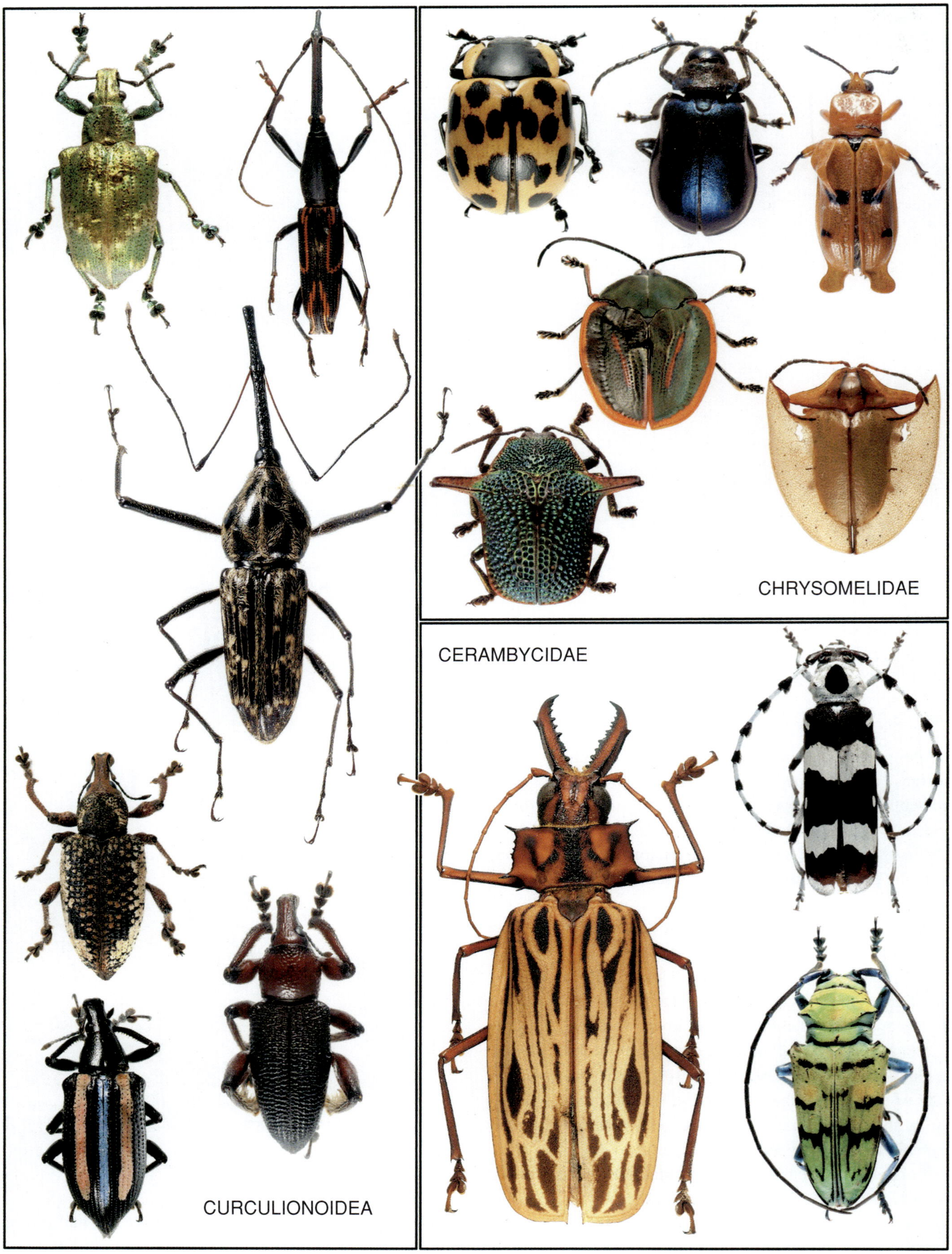

10.63. Representatives of phytophagan beetles. This group of beetles comprises one of the largest lineages of plant-feeding animals, second only to the Lepidoptera. Phytophagan beetles appear to have radiated in the latter half of the Cretaceous and in the Tertiary, probably in response to the angiosperm radiations. Not to the same scale.

Phytophagan Beetles. This is the largest group of Cucujiformia, and it includes the leaf beetles (family Chrysomelidae) and long-horned beetles (family Cerambycidae), which are placed in a superfamily Chrysomeloidea, and the weevils (superfamily Curculionoidea) (Phytophaga is a name sometimes used just for the Chrysomeloidea) (Figure 10.63). Phytophagans comprise the second largest lineage of plant-feeding animals after the Lepidoptera because virtually all species of these beetles are phytophagous, and more than 99% of them feed on angiosperms. As a result, the group has been the subject of studies on the coradiations of insects and angiosperms. Monophyly of the phytophagans is based on the structure of the tarsi, which appear to be four-segmented because segment four is concealed between two tarsomeres. The reduction of the male aedeagus and structure of the internal male reproductive organs further define monophyly of phytophagan beetles (Lawrence and Britton, 1991).

The family Cerambycidae is comprised of approximately 35,000 Recent species in 4,000 genera. These are generally medium-sized to huge beetles with cylindrical to flattened bodies and antennae that are often longer than the body. The antennae are typically flexed backward and held flat over the thorax and abdomen. Adults are active and feed on leaves or bark, as well as pollen; larvae generally mine the phloem of trees or bore into the heartwood. They seem to prefer freshly injured or felled trees, and some neotropical species girdle and even completely chew off small branches like miniature beavers. Because adults are active and exposed, and feed on flowers, many species are aposematic and part of mimicry complexes with wasps or toxic insects like lycid beetles. The earliest apparent fossil appears to be *Cerambycomima* from the Late Jurassic of Karatau, ca. 152 MYO, which is usually classified as a chrysomelid (specifically as an aulacosceline). However, the long antennae and flagellomeres are very typical of cerambycids. There are essentially no Cretaceous cerambycids, making their fossil record basically entirely Tertiary. Indeed, diverse cerambycids occur in the Eocene Baltic amber, the Eocene-Oligocene shales from Florissant, Colorado, and in Miocene amber from the Dominican Republic (e.g., Figure 10.64).

Chrysomelidae is a family of similar size, with approximately 38,000 described Recent species and 2,500 genera (Seeno and Wilcox, 1982), which are extremely diverse in size and shape. There is a devoted following of specialists on the family, and the taxonomy and general biology of chrysomelids have been amply reviewed (Jolivet and Hawkeswood, 1995; Jolivet and Cox, 1996; Jolivet, 1997). Until recently, though, classifications have been based on limited numbers of characters, the result being great disparity among the earlier classifications (reviewed by Suzuki, 1996; Schmitt, 1996). The most recent work involves phylogenetic study of adult and larval morphology (Reid, 1995, 2000), as well as some preliminary molecular studies.

10.64. Cerambycid beetle in Miocene Dominican amber. The larvae of most cerambycids feed within wood on the bark, phloem, or heartwood. Morone Collection, M3413; length 10.2 mm.

One of the principal subfamilies of Chrysomelidae includes Megalopinae, which is sometimes placed in its own family. This group is consistently considered basal in Chrysomelidae and, interestingly, has host plant relationships that are arguably ancient. *Palophagus* in Australia, for example, feeds on *Araucaria* pollen, and some *Donacia* feed

10.65. Larva of a cryptocephaline leaf beetle (family Chrysomelidae) with its case, preserved in Dominican amber. Larvae feed on dead leaves on the forest floor and protect themselves with a case constructed of particles of feces and humus. Most chrysomelids feed on foliage, including the adults of cryptocephalines. Morone Collection, M1223; length 1.8 mm.

10.66. Flea beetle (ventral view) of the subfamily Alticinae (family Chrysomelidae). The name derives from their ability to spring into the air, aided by muscles in the stout hind femora. Feeding alticines typically perforate leaves. Scanning electron micrograph.

on basal angiosperms in the Nymphaceae (water lilies). Sagrinae are most diverse in Australia and are distinctive for their size and brilliant colors, as well as the spiny hind legs built like nutcrackers that grasp and impale attackers. Cryptocephalinae are small species, larvae of which have a diet unique for chrysomelids: dead leaves. The larvae are further distinctive in that they carry a case composed of fecal and debris particles, fossils of which are even preserved in Miocene amber (Figure 10.65). There is something about a diet of detritus that facilitates the evolution of a larval case, since the habit is also found in larvae of psychid and tineid moths and trichopterans. Two closely related subfamilies, the Galerucinae and Alticinae, comprise the largest lineage within chrysomelids. The alticines, or "flea beetles," are well known for their enlarged, jumping hind legs (Figure 10.66). Larvae of Bruchinae feed on seeds, which for many toxic angiosperms is the most toxic portion of the plant. In at least one case, bruchine larvae actually derive nutrition from the toxins. *Caryedes brasiliensis* feeds within the seeds of the neotropical legume *Dioclea megacarpa*, which contains L-canavanine. This compound is a potent neurotoxin and analog of the amino acid L-arginine, but the larvae metabolize it as a major source of nitrogen (Rosenthal *et al.*, 1982).

The subfamily Chrysomelinae includes many external feeders that defend themselves with a potpourri of toxins (Dettner, 1987). The paropsines, for example, secrete hydrogen cyanide; *Chrysomela* pupae retain a larval cuticle laced with aldehydes (adults even rub themselves in the cuticle soon after emergence); and *Chrysolina* contains cardenolides, which are a potent defense by many insects including monarch butterflies. "Tortoise beetles" (Cassidini) are sometimes placed in their own subfamily, but these are actually a recently evolved lineage within Hispinae (Crowson, 1981; Chaboo, 2004), all of which comprise approximately 6,000 species. Many hispines are leaf miners, some with flat larvae that graze the epithelium within rolled leaves. Cassidines are broad, with an expansive pronotum and elytra, and some will protect the larvae until they pupate (Figure 10.67), herding them away from predators and confronting parasitoids. Some larval cassidines protect themselves with elaborate shields constructed of larval exuviae and feces.

The fossil record of Chrysomelidae has been reviewed by Carpenter (1992) and Santiago-Blay (1994). The latter author is unfortunately mistaken about an origin of the phytophagan beetles near the Permo-Triassic boundary, which is based on neither phylogenetic evidence nor credible fossils. Indeed, a putative chrysomelid from the Triassic is implausible and cannot be evaluated, and the identities of three compression-fossil genera (*Protoscelis*, *Protosceloides*, and *Pseudomegamerus*) from the Late Jurassic of Karatau are likewise impossible to assess (let alone their placement in the Recent subfamily Aulacoscelidinae). Records of Cretaceous chrysomelids are oddly sparse and include just three records: an unstudied beetle in 100 MYO amber from Burma, a sagrine in 72 MYO amber from Canada, and 65 MYO feeding traces of an

10.67. Adult *Acromis* tortoise beetle (subfamily Cassidinae, family Chrysomelidae) tending its larval offspring in Panama. Adults guard the larvae until they pupate, fending off parasitoids, ants, and various predators. Photo: P. J. DeVries.

10.68. Chrysomelid in Dominican amber, with its color patterns still preserved. AMNH DR15–558; length 4.8 mm.

10.69. Another chrysomelid in Dominican amber. Chrysomelid beetles are replete with defensive secretions, many of which are sequestered from poisonous host plants. AMNH DR10–811; length 3.6 mm.

apparent hispine. Leaves of a fossil ginger, *Zingerberopsis* (Zingiberaceae), from the latest Cretaceous and Eocene of North Dakota have traces very similar to those made by two Recent tribes of hispines that feed today on *Heliconia* and related plants (Wilf *et al.*, 2000). Assuming that certain hispines were the only insects to leave such marks and using a prior phylogeny (Hsiao and Windsor, 1999), Wilf *et al.* extrapolated from this a Late Cretaceous divergence of most chrysomelid subfamilies and radiation of the family with the angiosperms. Despite the paucity of Cretaceous chrysomelids, this hypothesis makes great sense, so it will be very interesting to see if it is supported by future discoveries of fossils and by rigorous phylogenetic work. The phylogeny used in that study is based just on one gene (18S rDNA) and very limited taxa, and the relationships of subfamilies and tribes in it differ considerably from cladograms based on comprehensive morphological evidence (Reid, 1999; Chaboo, 2004). Chrysomelid fossils are diverse in the Tertiary (Figures 10.68, 10.69).

The Curculionoidea, or weevils, comprises 11 families, one of which, Curculionidae, has the distinction of being one of the largest families of organisms with 44,000 Recent, described species. Adults of most species have the distinctive snout or *rostrum*, at the end of which is a small pair of mandibles, and other mouthparts have undergone some reduction and fusion (Figures 10.70, 10.71). Weevils, thus, are biting and chewing insects, not sucking insects as they have sometimes been popularly portrayed. Adults further have a clubbed antenna with a long scape that lies in a scrobe, or a groove on the side of the rostrum; they have a rigid body; and the larger species that feed externally often have a thick, almost impenetrable cuticle. Larvae are apodous, though the larvae of the more basal groups have small legs.

Nemonychidae is often regarded as the most primitive weevil family (Kuschel, 1995; Marvaldi and Morrone, 2000;

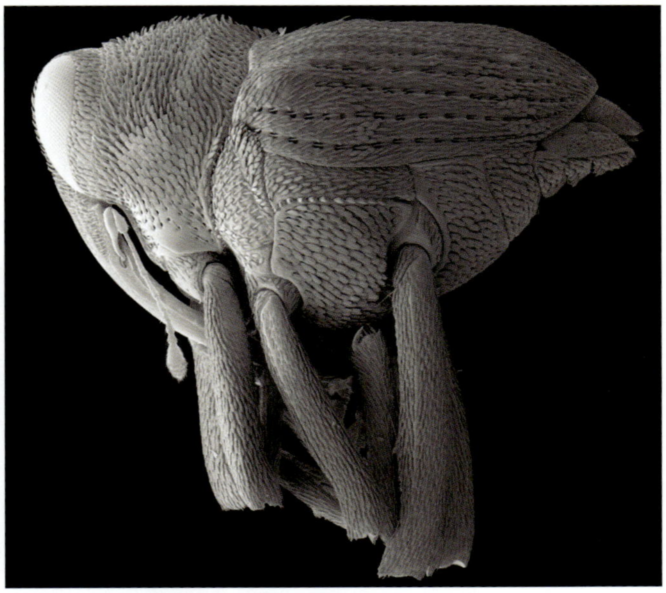

10.70. A weevil of the subfamily Conoderinae. The adults and larvae of most weevils feed on living plants, but many in this subfamily live beneath the loose bark of rotten logs, probably feeding on fungal hyphae. Weevils belong to the superfamily Curculionoidea, containing some eight families. Curculionidae has more described species than any other insect family (44,000), though Staphylinidae may eventually become a larger family. Scanning electron micrograph.

10.71. A weevil of the family Apionidae, many of which feed on or within legume pods. Scanning electron micrograph.

Marvaldi *et al.*, 2002). These feed as adults and larvae on the pollen of araucarian conifers in the Austral Region (south temperate South America, Australia, New Zealand, and surrounding areas). Many other weevil lineages bore through and feed in wood as adults and larvae, including some Anthribidae, Belidae, most Brentidae, and the cossonine, cryptorhynchine, and scolytine Curculionidae. Scolytinae sensu lato are well known as bark and ambrosia beetles, for the distinctive, radiating galleries they leave as larvae and adults while feeding under bark. Platypodinae (Figure 10.72) are sometimes separated into their own family, but they are actually recently evolved curculionid weevils, and platypodines even appear to be a lineage within scolytines (Farrell *et al.*, 2001). Some of these are serious timber pests of conifers, such as *Ips* on Ponderosa pine (*Pinus ponderosae*). The ecological impact of these weevils is much greater than appears because, as acres of trees succumb to the beetles, they become foci for forest fires. The devastating wildfires in southern California in 2003 were essentially a result of bark beetles killing so many trees. There evolved in scolytines at

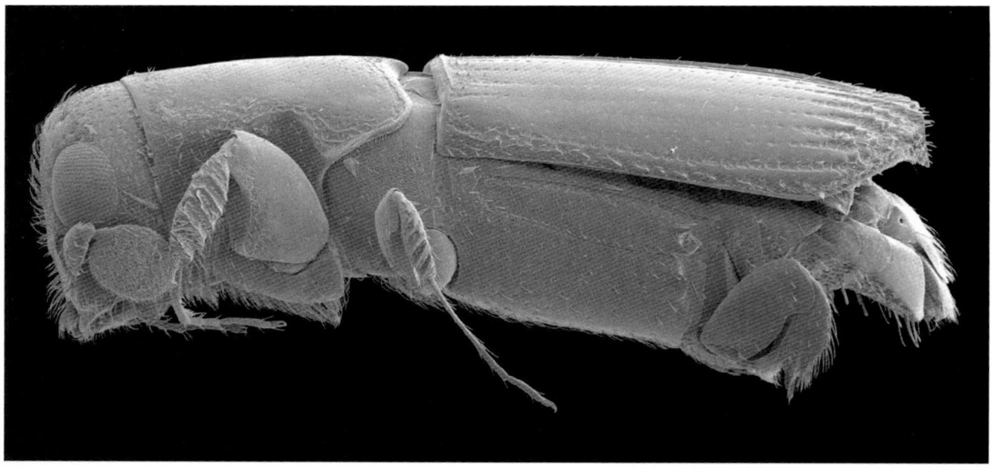

10.72. A weevil of the subfamily Platypodinae (family Curculionidae). These are closely related to scolytine weevils, all of which bore into wood and can cause serious damage to forests. These weevils have lost the snout and are also one of the most recently evolved lineages of weevils. Scanning electron micrograph.

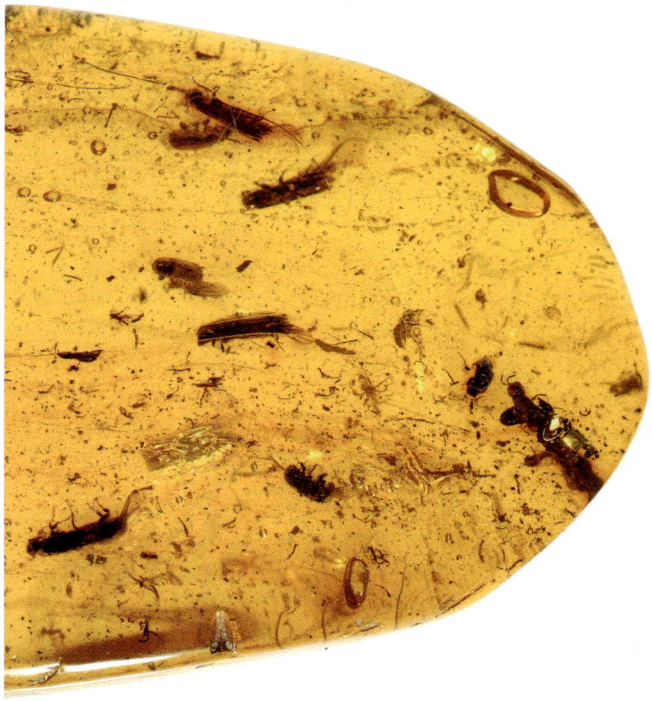

10.73. Platypodine and scolytine weevils in Miocene Dominican amber. These beetles are common and diverse in this amber, and their feeding damage may even have caused the prolific resin production of the Dominican amber tree. The particles are wood fragments the beetles pushed out of their galleries in the wood. Detail shows a scolytine and platypodine up close. AMNH DR10–1632; length of beetle 1.4 mm.

least seven times a diet of ambrosia fungus, which grows in the wood and is even transmitted by some of these beetles in specialized pockets called *mycangia*. The fungus is an ascomycete that carpets the walls of their galleries and the larvae and adults graze on this. The fungus can kill the tree, and it is often effective in plugging the resin canals in wood so that the beetles can burrow and feed without being engulfed, though some amber fossils reveal the effectiveness of resin in flushing out the beetles (Figure 10.73). In one of the most recently evolved lineages of scolytines, the Xyleborini, there has evolved – incredibly – advanced social behavior (Kent and Simpson, 1992; Kirkendall *et al.*, 1997). Interestingly, these beetles are haplodiploid, which is a sex-determining mechanism known to promote cooperative living among relatives. But of the approximately 1,500 xyleborines, only one is known thus far to be social – *Australoplatypus incompertus*, which breeds in *Eucalyptus* in Australia.

A few curculionids are actually aquatic, particularly adult and larval Erirhininae; adults breathe with a ventral plastron. The Australian genus *Tentegia* is convergent with dung beetles: adults roll balls of wallaby dung under stones and logs and lay eggs on them for the larvae to develop. Most weevils, though, feed on angiosperm flowers and leaves, and some of the more conspicuous external feeders are beautiful and jewel-like. The most notorious weevil is *Anthonomus grandis*, the boll weevil, which virtually devastated the cotton industry in the southern United States in the early 1900s. The adults feed on the flower buds and fruits of cotton (*Gossypium*), wherein the larvae also develop.

The fossil record of Curculionoidea is considerably better than that of the other phytophagans and most recently reviewed by Gratshev and Zherikhin (2003). This may just reflect the fact that a rostrum is easy to recognize in compression fossils, or it may actually reflect an early predominance of weevils. The earliest fossils ascribed to the Curculionoidea are from the Triassic, specifically the Obrieniidae from the mid- to Late Triassic of Eurasia and Australia, some of which had snouts of modest size (Zherikhin and Gratshev, 1993). Examination of these beetles by authoritative coleopterists indicates that they are not weevils (C. H. Lyal and R. Oberpreiler, pers. comm.. 2000; J. F. Lawrence in Reid, 2000). In fact, based on the structure of the metepisternum, the Obrienidae may even be archostematans (Gratshev and Zherikhin, 2003). Thus, Triassic Curculionoidea have the same status as Triassic Chrysomeloidea: a relationship that is highly improbable and presently impossible to verify.

10.74. An early weevil from the Late Jurassic of Karatau, Kazakhstan. This exceptional deposit has preserved the earliest record of definitive and diverse weevils. PIN 2784/1451; length (including snout) 6 mm.

10.75. Weevil from the Early Cretaceous Santana Formation of Brazil. Most weevils in the Early to mid-Cretaceous belong to basal groups. AMNH SA73629; length 3.7 mm.

The earliest definitive Curculionoidea are from the Late Jurassic deposit of Karatau, Kazakhstan, about 152 MYO (e.g., Arnol'di *et al.*, 1977; Gratshev and Zherikhin, 1995) (Figure 10.74). This deposit contains some 46 described species and 17 genera as of 2003, most of them in the extinct and possible stem-group family Eobelidae, though several of these genera were later transferred to the Nemonychidae, Belidae and a family consisting of the isolated genus *Car* and close relatives (Caridae). Thus, weevils may have originated in the mid- to Early Jurassic, 175–200 MYA, but additional fossils are needed to test this theory. Cretaceous weevils occur in major compression Lagerstätte from Baissa, Siberia, and the Santana Formation of Brazil (Figure 10.75), including the living families Belidae, Brentidae, Eccoptarthridae, Nemonychidae, and the earliest members of the family Curculionidae. A nemonychid exists in Lebanese amber (oddly no fossils of the family occur after that), and a few unstudied weevils occur in mid-Cretaceous amber from Burma and New Jersey, including adults and larvae (none are known in Siberian or Canadian amber). Two weevils in 90 MYO amber from New Jersey are an attelabid, and a genus closely related to the primitive living Australian genus *Car* (Gratshev and Zherikhin, 2000) (Figure 10.76). Other extinct relatives of *Car* occur in Eocene Baltic amber. In fact, this deposit houses a significant diversity of genera closely related to Recent ones from South America, Asia, and Australia (Kuschel, 1992).

10.76. *Cretocar luzzii*, a weevil of the family Caridae in 90 MYO amber from New Jersey. Caridae is a small, relict weevil family today that was most diverse in the Late Jurassic and Early Cretaceous. Carid weevils feed on cedars (Cupressaceae). AMNH NJ493; length 1.1 mm.

Because scolytines and platypodines leave such distinctive galleries in wood, fossilized wood with their traces has been identified from the mid-Cretaceous of Siberia (Figure 10.77) and the Eocene of the Canadian arctic. Oligocene amber from southern Mexico and Miocene amber from the Dominican Republic and Sicily contain numerous Recent genera and an abundance of platypodines and scolytines, and body fossils of the living genus *Scolytus* are described from the Miocene of France (Petrov and Zherikhin, 2000). It would appear that scolytines and platypodines probably originated in the latest Cretaceous but radiated in the Tertiary, which agrees with some estimates based on molecular dating (Farrell *et al.*, 2001).

10.77. Beetle galleries in fossilized wood from the mid-Cretaceous (100 MYO) of northern Siberia, attributed to scolytine weevils. PIN 3308/12; width 72 mm.

Given that 99% of the phytophagan beetles feed on angiosperms, these insects clearly radiated with or at least diversified as a result of the angiosperms. That was the subject of one commonly cited study (Farrell, 1998), which examined relationships and patterns of host use in Chrysomelidae, Cerambycidae, and Curculionoidea. It concluded that basal Recent lineages of each of these feed on cycads and conifers, and huge radiations in each group evolved independently as a result of feeding on angiosperms. We believe this makes a great deal of sense. But, as for the Wilf *et al.* (2000) paper, this conclusion needs to be tested with more data and rigorous analyses. The study by Farrell (1998) was based just on the one commonly used gene 18S rDNA for 0.1% of the species in Phytophaga, and the morphological data has been critiqued (see Reid, 2000: 838). The study has also been critiqued on the basis of accepting the Triassic genus *Obrienia* as a weevil and some of the putative chrysomelids from the Jurassic as belonging to a Recent subfamily, although later origins of these groups actually provide further support for Cretaceous radiations of these beetles.

Beetles are diverse probably because they are the only flying insects besides ants to have thoroughly invaded cryptic niches on the ground. The statement that coleopteran diversity was "enabled by the rise of the flowering plants" (Farrell 1998: 558) does not account for two very significant groups, Adephaga and the Staphylinoidea. Based on the proportions of undescribed species being discovered in Staphylinoidea alone, it is very likely that the size of this group will eclipse all phytophagan beetles. Angiosperms certainly contributed to the spectacular diversity of beetles, but these insects were well on their way at least 100 million years before angiosperms came on the scene.

STREPSIPTERA: THE ENIGMATIC ORDER

The relationships of the "twisted-winged" parasites, order Strepsiptera, which comprise approximately 600 named species, have perplexed entomologists more than any other group. Extreme morphological specialization of the larvae, adults, and even their genome has confounded efforts to identify the order most closely related to Strepsiptera. In fact, Strepsiptera have historically been considered to be close relatives of Lepidoptera, Diptera, or Hymenoptera, but traditionally they are believed to be most closely related to Coleoptera or even within this group (e.g., Crowson, 1981). Suggestions have even been made that the evidence is not entirely convincing that these insects are holometabolans (Kristensen, 1981, 1999a). Before reviewing the most compelling hypotheses about relationships, a review of the diversity and unusual biology of the order is necessary, which we have largely extracted from Kathirithamby (1989, 1991), Kinzelbach (1971, 1978), and various other references mentioned in the discussion.

10.78. A well-preserved weevil in Miocene amber from the Dominican Republic. Morone Collection, M0771; length 9 mm.

Every species, stage, and sex of strepsipterans shows dramatic modification. The male is the winged, dispersal stage; females are apterous, larviform, and viviparous and permanently reside within their host in all but the most basal family. Adult males are immediately recognizable for the flabellate antennae (with the reduced number of eight or fewer segments); eyes with large, bulging facets like raspberries (these are very sensitive to motion [Pix *et al.*, 2000]); highly reduced or lost labial palps, and maxillary palps reduced to two or one segment; hind coxae fused to the metathorax; and many features of the thorax (Figure 10.79). The forewings of male strepsipterans are reduced to clublike halteres, and the expansive hind wings are used in flight. Hind wings retain only longitudinal veins (no crossveins), and even these are mostly incomplete. Interestingly, strepsipteran halteres function very similar to those in flies (Pix *et al.*, 1993), with sensilla sensing forces and relaying impulses to control movements of the head and abdomen. Male strepsipterans are agile fliers, even capable of hovering. Most species are based just on the basis of the males because these are more readily captured by conventional means using nets and Malaise traps, and they have more features for comparisons. Males apparently live only for a few hours.

In all species except those in the primitive family Mengenillidae, females are legless and completely wingless and lack external genitalia, antennae, eyes, and a rectum (Figure 10.80). They remain within their hosts for virtually their entire lives (beginning when they entered as a first instar larva), except for the mengenillids, whose females emerge from the host just before pupariation and are then free-living. Mengenillid females have short antennae, rudimentary compound eyes, and legs. All other females mate and bear live young while encased in their host, usually in its abdomen. The sclerotized head end (cephalotheca) of larviform females typically protrudes slightly from between host tergites (Figure 10.81), and glands on this region disperse pheromones to attract males. The male fertilizes her in situ, copulating through a brood-canal opening on the cephalothorax.

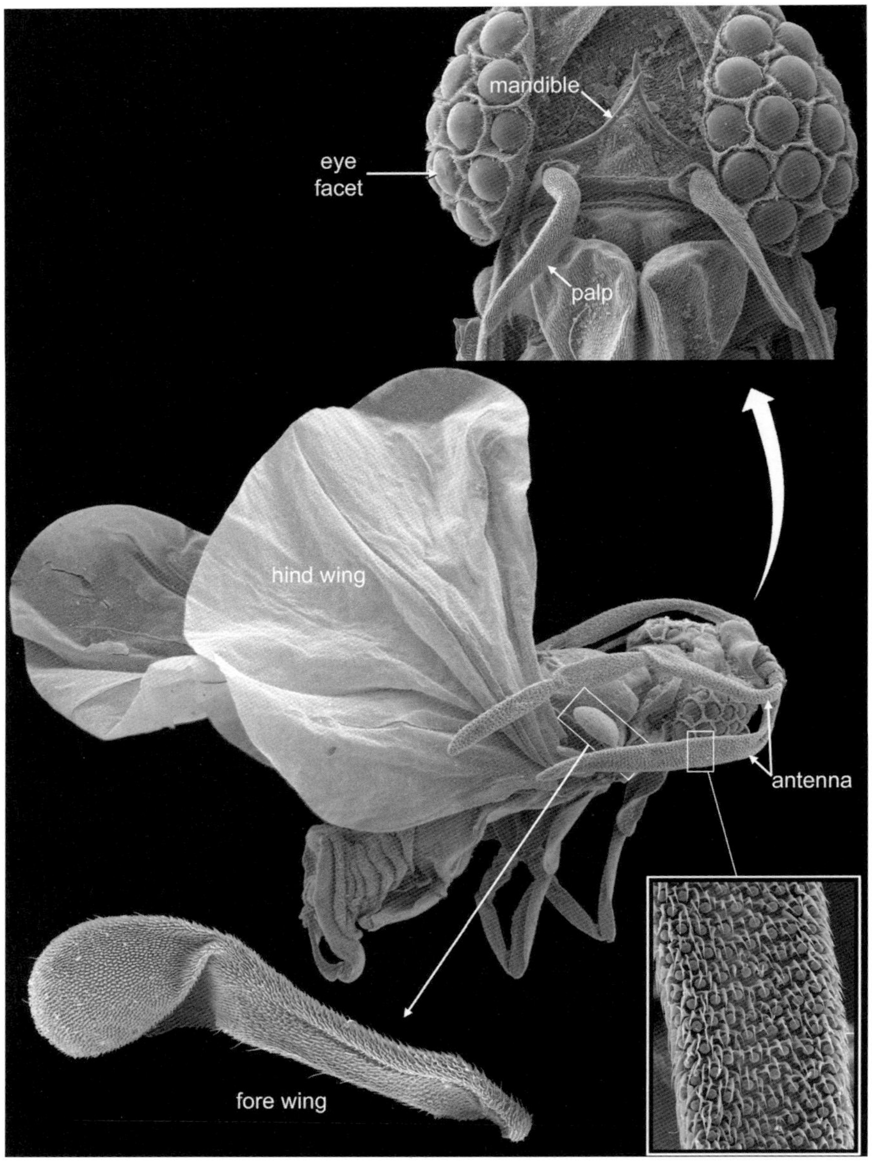

10.79. An adult male strepsipteran (Myrmecolacidae), showing the highly specialized eyes, mouthparts, antennae, and wings in this order. Identifying the closest relatives of Strepsiptera has perplexed entomologists more than any living order of insects. Scanning electron micrograph; body length 1.4 mm.

Strepsipteran reproduction is unique among insects because, where known, the ovaries autolyze in late instars and the oocytes spill into the hemocoele. Larvae develop within the hemocoele and exit via the brood-canal opening. Females are typically very fecund, giving birth from 1,000 to 750,000 larvae. First instar larvae are minute (approximately 0.1 mm), bristly, and legged (Figure 10.82), and actively search for hosts as in the triungula of endoparasitoid insects like Meloidae, Rhiphiphoridae (Coleoptera), and other insects. Morphology of the first instars has been treated by Pohl (2002). They have distinctive, long terminal setae that are apparently used for springing up to 1 cm. When they find a host, they attach to it, and enzymes allow it to penetrate the host's integument. This mode of entry is probably related to a unique mechanism among insect endoparasitoids of escaping the host's internal defenses. Many host insects attack endoparasitoids by encapsulating them with specialized cells (insects do not have immune systems), among other mechanisms of internal defense. In the case of Strepsiptera, the larva appears to masquerade as the host's own self (Kathirithamby *et al.*, 2003). As the first instar "digests" its way into the host's cuticle, the epidermis separates from the basalmost layer of cuticle (endocuticle), a sac of epidermis forms around the larva, and this pinches off to entirely envelop the larva. The host, thus, can't recognize the intruder. The larva then molts to a legless second instar.

This method of larval defense may have adapted strepsipterans to invade hosts in 34 families and 7 orders, which, given only 600 species, is a remarkably diverse array. Endoparasitoids usually have a much narrower range of hosts because they need to evolve specialized defenses against the host's internal defense mechanisms. The orders most parasitized by strepsipterans are auchenorrhynchans (such as Cicadellidae, Membracidae, and various fulgoroids), as well as aculeate Hymenoptera. Individual species of strepsipterans can have hosts in multiple genera and even several

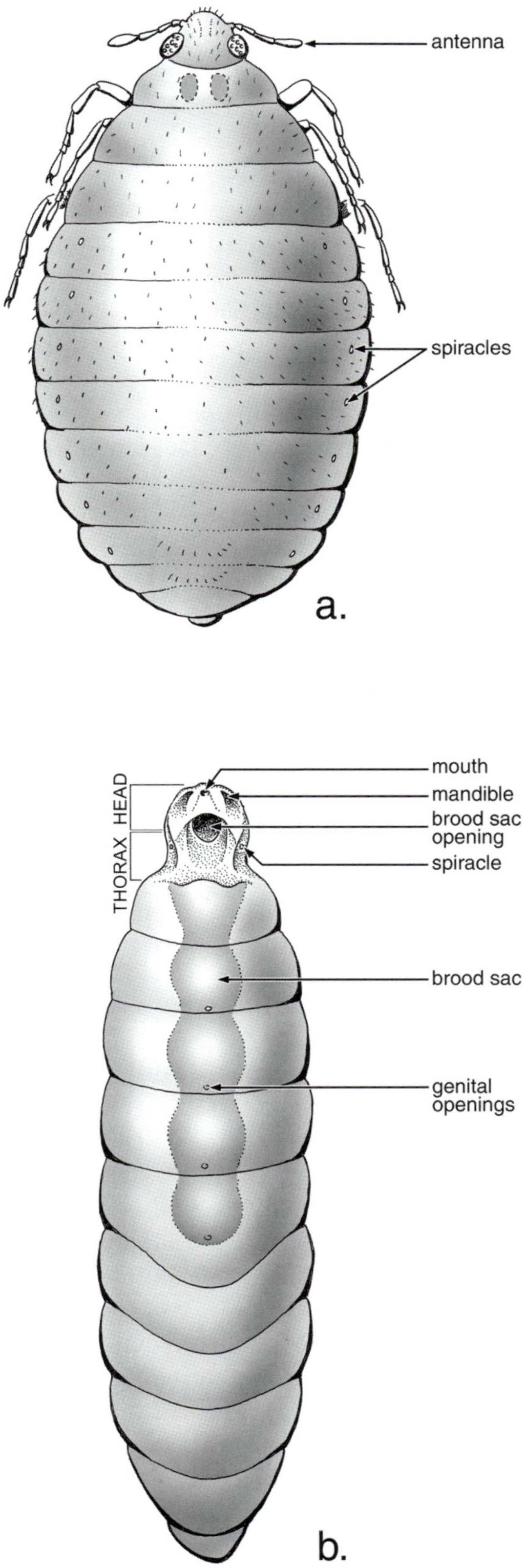

10.80. Mature female strepsipterans. (a) *Eoxenos* (Mengenillidae), (b) *Coriophagus* (Halictophagidae). Mengenillidae is the basalmost living family of Strepsiptera, and the adult female is free-living. In all other living strepsipterans, the larviform female (b) remains within the host. Lengths: (a) 4.5 mm, (b) 8.5 mm; redrawn from Kathirithamby (1991).

10.81. A female strepsipteran, with her cephalothorax protruding from between the abdominal tergites of her vespid wasp host.

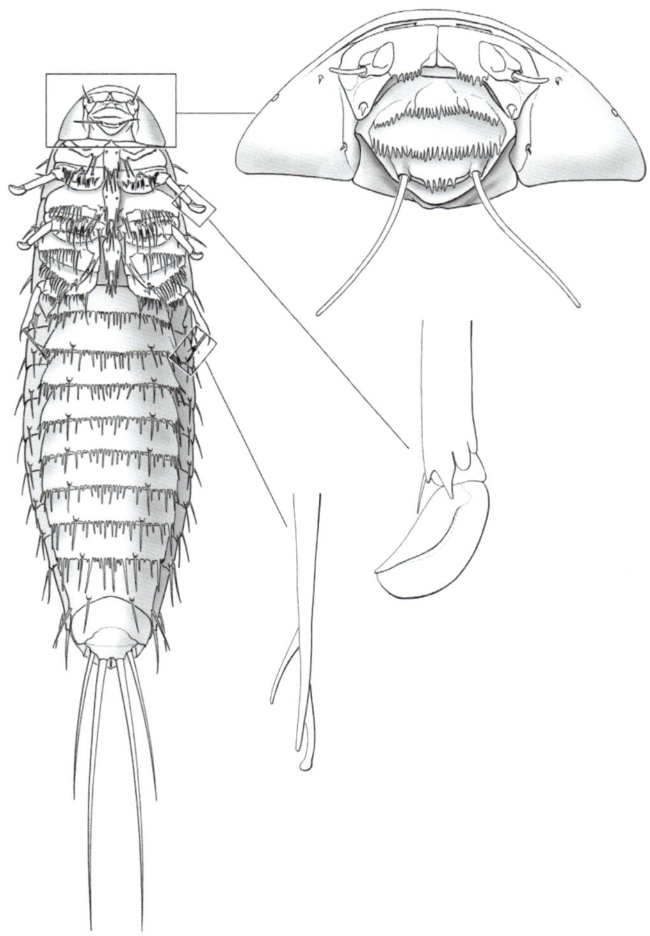

10.82. First instar (triungulun) of *Eoxenos laboulbeni* (Mengenillidae: Strepsiptera), ventral view, with details of the head and the unusual tarsi. Redrawn from Pohl (2002).

families, though the ones attacking wasps tend to be more specific. Larvae and females usually occupy the abdomen of the host, and they often do not kill the host but usually alter its fitness, including neutering it. The mature male larva pupariates within the tanned exuvium of the last larval instar, just as in cyclorrhaphan flies.

DIVERSITY

There are eight Recent families and one extinct family (Mengeidae, in Baltic amber). Mengeidae and the Recent family Mengenillidae are clearly the most basal families (though very doubtfully a monophyletic group, contrary to Kinzelbach, 1990). This is based on male morphology for both families, which is less reduced than in the other families, and the morphology and free-living habits of female mengenillids. Mengenillids occur on all vegetated continents except North and South America, and they are parasitoids of Zygentoma. Most other Strepsiptera are grouped into the suborder Stylopidia except for the subfamily Triozocerinae (Corioxenidae) (Pohl, 2002). Stylopidia are defined on the basis of first instar larval characters (Pohl, 2002), adults with four or fewer tarsomeres, having claws that are highly vestigial to absent, and various features of the face. The largest and one of the most recently evolved families is the Stylopidae, which parasitize aculeates. Closely related to these (sensu Kinzelbach, 1971, 1990; Pohl, 2002) is the family Myrmecolacidae, of which the females parasitize ants but the males parasitize polyneopterans (Orthoptera, Mantodea)! Such disparate sexual difference in host use is unique among insects, and until recently this has completely confounded efforts to match males and females of the same species. Only recently have the male and female of the same species of a myrmecolacid been definitely associated, using DNA techniques (e.g., Halbert *et al.*, 2001). The male of *Caenocholax fenyesi* parasitizes a *Dolichoderus* ant, and the female parasitizes a cricket (Kathirithamby and Johnston, 2003). Relationships among families and some subfamilies is based thus far on the morphology of males (Kinzelbach, 1971; 1990; Pohl *et al.*, 2004) and first-instar larvae (Pohl, 2002). These schemes largely agree and reveal a gross trend in host use: The most primitive family parasitizes Zygentoma, the intermediate families parasitize mostly polyneopterans, and the most recently evolved families mostly parasitize aculeate Hymenoptera.

RELATIONSHIPS TO OTHER ORDERS

Traditionally, the Strepsiptera have usually been considered to be most closely related to the Coleoptera (e.g., Kinzelbach, 1971; 1990; Kristensen, 1975, 1981; Kathirithamby, 1989, 1991; Kukalová-Peck and Lawrence, 1993), or even to lie within the Coleoptera (Crowson, 1981). In Crowson's most recent account (1981), in fact, he placed Strepsiptera close to the unusual cucujoid beetle family Lymexylidae, based on the structure of the mentum, thorax, and the reduced, soft elytra of these beetles. Male Strepsiptera have an absolutely remarkable resemblance to another family of beetles, the Rhipiphoridae, which also have flabellate antennae, large eyes, short and soft elytra, and parasitoid larvae with first instar triungula (Figure 10.83). It has been proposed that Strepsiptera may not even be Holometabola because the larvae have rudimentary compound eyes, vestigial wing buds appearing at the beginning of the second instar of male larvae, first instars with a well-developed eleventh abdominal segment, and there is no change in the internal organs from larva to adult.

Perhaps the most intriguing, and certainly the most debated, hypothesis of relationships is that Strepsiptera are closely related to Diptera. Studies on the 18S rDNA gene support a close relationship between these two orders (Whiting and Wheeler, 1994; Chalwatzis *et al.*, 1995, 1996; Whiting *et al.*, 1997; Whiting, 1998). On this basis, morphological structures of both orders were re-interpreted so that they might better reflect this relationship, including an hypothesis that the halteres of both orders are homologous but that the thoracic segments switched positions as a result of a homeotic gene (Whiting and Wheeler, 1994; Whiting, 1998). To christen this new group, the name "Halteria" was proposed (Whiting and Wheeler, 1994; Whiting *et al.*, 1997).

Evidence for a monophyletic "Halteria," however, is ambiguous.

- It is based essentially on one gene (18S rDNA). Additional studies of 5.8S and 28S rDNA were unable to decipher relationships, though they did confirm that Strepsiptera are holometabolans (Hwang *et al.*, 1998). That study further indicated that the rDNA of Strepsiptera and Diptera convergently accrued large amounts of nucleotide substitutions.
- The high rDNA substitution rates in Diptera and Strepsiptera probably led to an artificial linking of the two orders, the so-called long-branch attraction (Carmean and Crespi, 1995; Huelsenbeck, 1997, 1998, 2001). This feature affects parsimony, maximum-likelihood, and Bayesian methods of sequence analysis.
- The axillary sclerites and muscles of strepsipteran flight wings lack derived features of Diptera and even the larger group to which Diptera belongs, the Antliophora (Hörnschmeyer, 2002).
- Strepsiptera lack certain specialized features of the antliophoran pterothorax (Kristensen, 1995).
- Putative synapomorphies of Halteria are largely reductions and losses, homologies of which are always difficult to evaluate. These include, for example, a one-segmented labial palp in Strepsiptera, but in Diptera the labial palpi are fused into a fleshy labellum at the end of the proboscis.

Also, the loss of an ovipositor is a feature shared with all Panorpida, not just Diptera.

- Paired claws of larval Mengenillidae is a primitive feature for the panorpoids.
- The *engrailed* gene has an intron that is distinctive to the panorpoid orders but that is absent in Coleoptera, Orthoptera, Strepsiptera, and other arthropods (Rokas *et al.*, 1999).
- The structure of the strepsipteran halter differs from that of Diptera, indicating that they are convergent. Mengenillid halters retain vestiges of venation that no flies retain, and the number, position, arrangement, and structure of sensilla at the base of the halter differs between the two groups (cf. Figures 10.79, 12.23).
- Most Strepsiptera have a pair of slender, blade-like, cruciate mandibles, which has been considered synapomorphic (Whiting, 1998) with an apparent groundplan feature for the Antliophora, specifically the Mecoptera (Mickoleit, 1971). The mandibles of the earliest and most primitive fossil strepsipterans, however, lack these features, indicating that the groundplan of the strepsipteran mandible was much more generalized.

Resolution of the ambiguous relationships will probably depend on the study of additional genes and gene regions, detailed study of all stages of Mengenillidae (especially larvae), and discovery of additional Cretaceous fossils in amber. It would not be surprising, though, if this order was found to be a highly modified lineage derived from some extinct, archaic holometabolans, such as stem-group neuropteroids or panorpidans.

FOSSILS

As is expected, virtually all fossil strepsipterans are preserved in amber because most species are 2 mm or less in length. Only recently have the first Cretaceous triungula been discovered, which are preserved in 100 MYO Burmese and 75 MYO Canadian amber (Figures 10.83, 10.84) along with an adult male in Burmese amber. Even amber has limits to its preservation, so the microscopic size of the larvae and their even finer features are difficult to observe. The male in Burmese amber, however, is exceptionally primitive (Figures 10.85, 10.86). It possesses the typical array of strepsipteran features and in fact closely resembles a mengenillid or mengeid, but it possesses several features that are strikingly primitive even for those two basal families. The eye facets are not particularly large, nor are they separated by strips of plush, fine pile as occurs in modern species and in *Mengea tertiaria* from Baltic amber (formerly the most primitive known strepsipteran; Figure 10.87). Also, the fore trochanters are not entirely fused to the femora, and the pterostigma is well defined and not diffuse. Most importantly, the mandibles

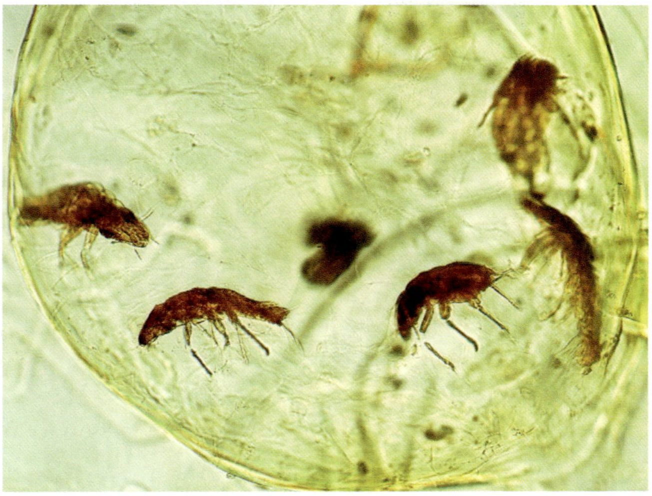

10.83. Group of six triungula, in mid-Cretaceous amber from Burma probably of the beetle family Rhipiphoridae, which are similar to Strepsiptera. AMNH; body length 0.1 mm.

10.84. Possible Strepsipteran triungulun, in amber from the Late Cretaceous (ca. 75 MYO) of western Canada. CNC CAS651; length 0.1 mm.

10.85. The oldest male strepsipteran, in Cretaceous Burmese amber. It is one of the two most primitive known strepsipterans. AMNH Bu1558; length 1.4 mm.

10.87. *Mengea tertiaria* in Eocene Baltic amber, a basal strepsipteran. AMNH; length 2 mm.

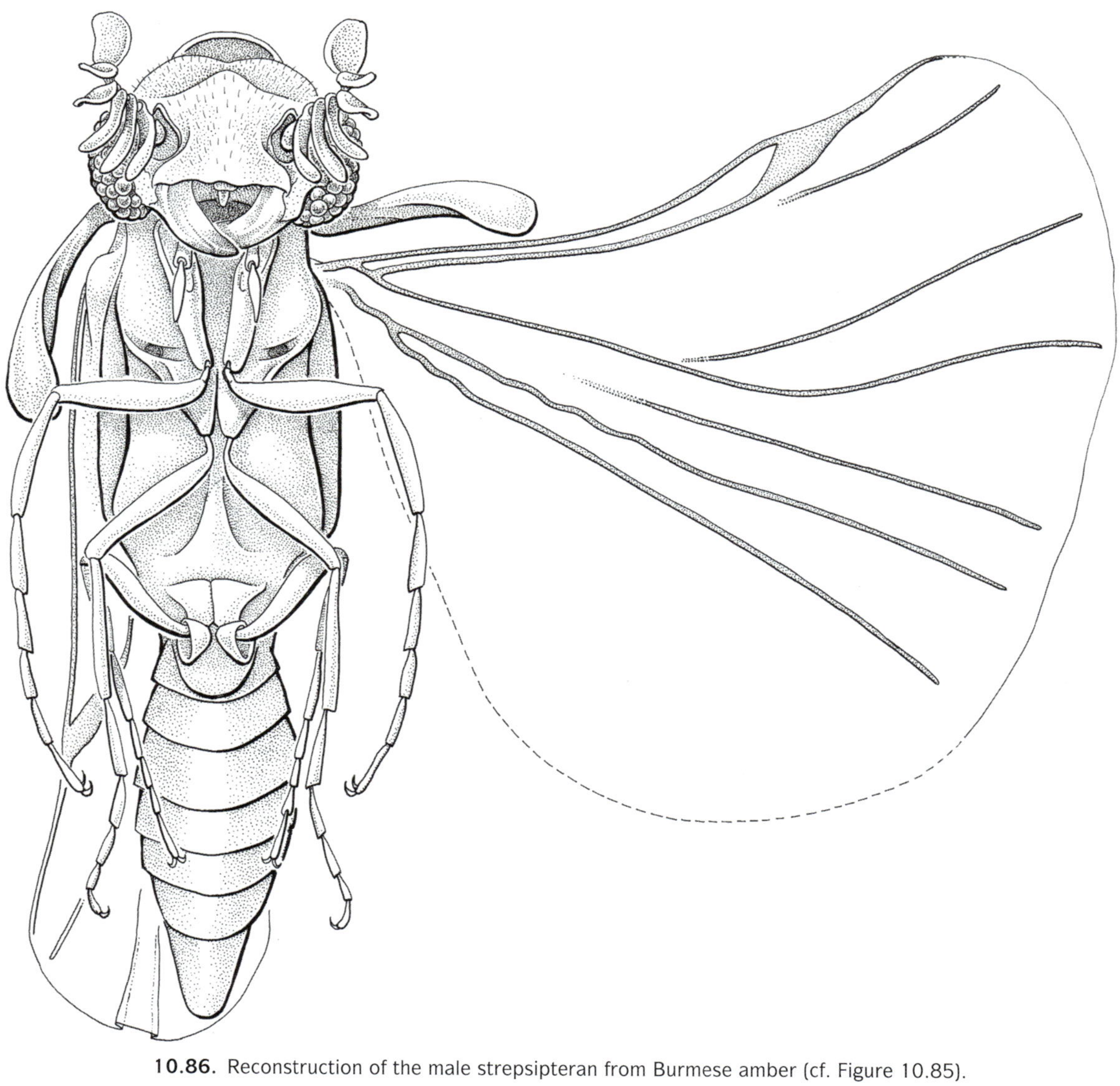

10.86. Reconstruction of the male strepsipteran from Burmese amber (cf. Figure 10.85).

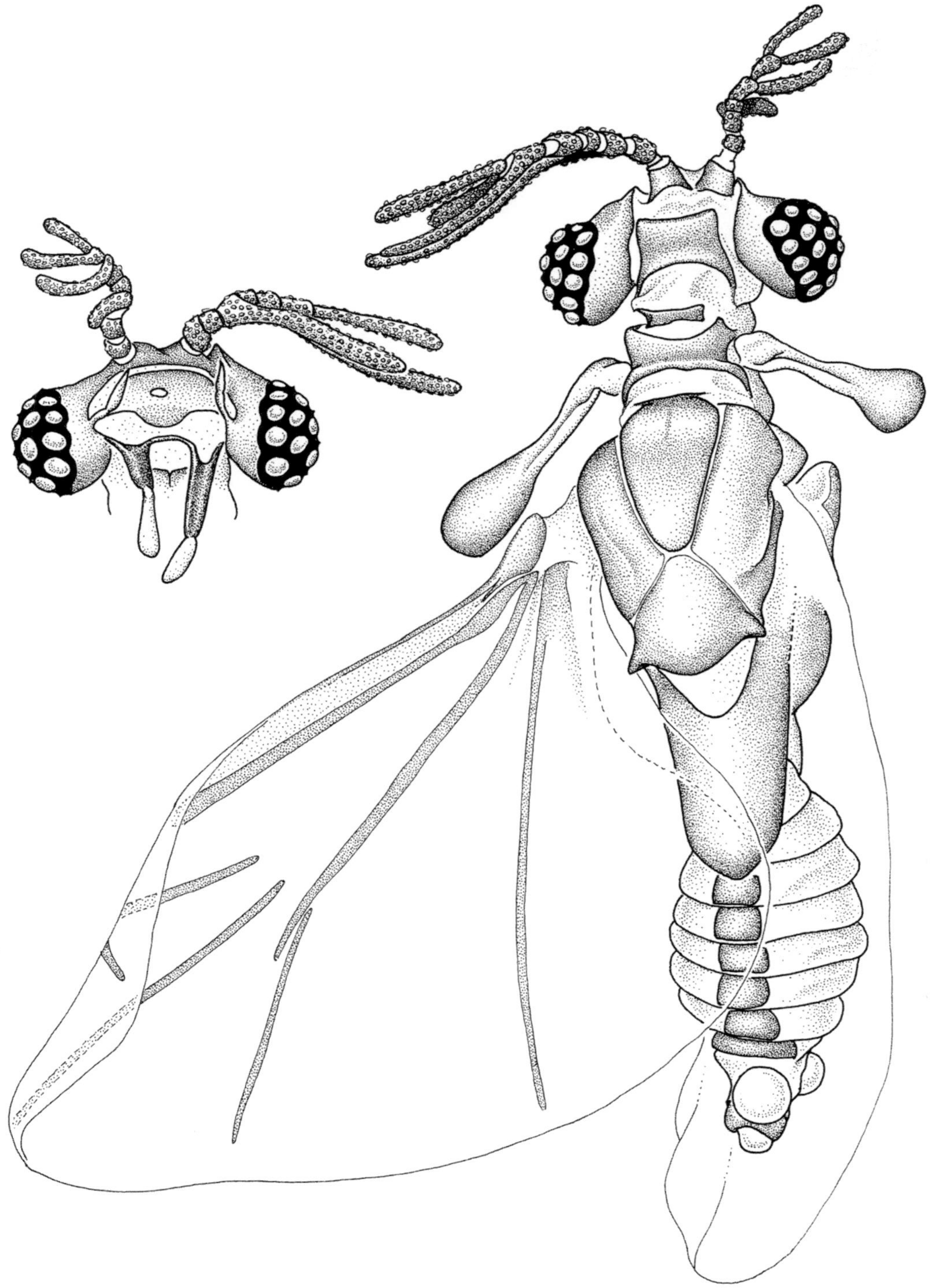

10.88. *Bohartilla kinzelbachi* (Bohartillidae) in Miocene Dominican amber. It is very similar to the living species *B. megalognatha*. AMNH DR10–6; length 1.4 mm.

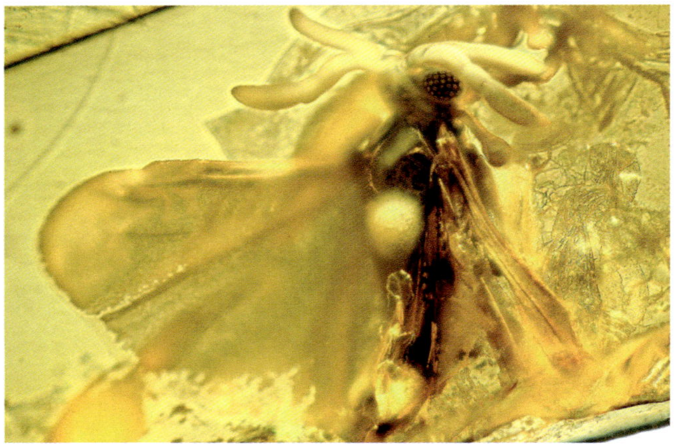

10.89. *Stichotrema weitschati* (Myrmecolacidae), in Baltic amber, which belongs to a more recently evolved lineage of Strepsiptera, the Stylopidia. AMNH.

of the Cretaceous fossil are broad and generalized, not narrow, blade-like, and crossing each other as in all other strepsipterans. The male of another very primitive strepsipteran was recently discovered in Eocene Baltic amber (Pohl *et al.*, in press). Among its most primitive features are the large size (>7 mm body length), eight antennal segments, short meta-postnotum, and heavily sclerotized abdominal tergites (in modern strepsipterans the sternites are more heavily sclerotized, as in beetles). While these two primitive fossils don't reveal ordinal relationships of Strepsiptera, they better define what early strepsipterans were like.

Diverse Strepsiptera occur in 42–45 MYO Baltic amber (Ulrich, 1927, 1943; Kulika, 1978, 1979, 2001; Kinzelbach and Pohl, 1994; Pohl and Kinzelbach, 1995, 2001), and there are two compression fossils from the Eocene of Germany (Kinzelbach and Lutz, 1985; Lutz, 1990). One of the compression fossils is a minute larva from Halle-an-der-Salle, *Pseudococcites eocaenicus*, which apparently bears a gross resemblance to myrmecolacids of the living genus *Stichotrema*. The other compression is of two male puparia (also apparently of *Stichotrema*) preserved with their *Camponotus* ant host from the famous oil shales of Grube Messel.

Besides *Mengea tertiaria* and *M. mengei*, most other strepsipterans in Baltic amber are myrmecolacids: two species of *Stichotrema* (Kinzelbach and Pohl, 1994; Pohl and Kinzelbach, 1995) (Figure 10.89), three species of the extinct genus *Palaeomyrmecolax* (Kulika, 2001), and a female preserved while parasitizing an ant (Pohl and Kinzelbach, 2001). The only fossil of the Stylopidae is *Jantarostylops kinzelbachi*, in Baltic amber (Kulika, 2001). Miocene Dominican amber – half the age of Baltic amber – has preserved species very similar to living species (in some cases they are virtually indistinguishable) (Kinzelbach, 1979, 1983; Kathirithamby and Grimaldi, 1993). Most of these are myrmecolacids (*Caenocholax, Myrmecolax, Stichotrema*), but Bohartillidae (*Bohartilla;* Figure 10.88), and Elenchidae (*Protelencholax*) are also included.

The chronological sampling of strepsipteran fossils is very limited, but there is a basic trend, from an essentially modern fauna belonging to recently evolved families in the Miocene, to an older fauna in the Eocene that contains some archaic taxa (e.g., *Mengea*), to a very primitive genus in the mid-Cretaceous. Additional Cretaceous strepsipterans will likely be very revealing.

But, for the point of wisdom,
I would choose to
Know the mind that stirs
Between the wings of
Bees and building wasps.

–George Eliot, *The Spanish Gypsy*

11 Hymenoptera: Ants, Bees, and Other Wasps

The order Hymenoptera comprises one of the four "hyperdiverse" insect lineages; the others – Diptera, Lepidoptera, and, of course, Coleoptera – are also holometabolous. Among holometabolans, Hymenoptera is perhaps the most difficult to place in a phylogenetic framework, excepting the enigmatic twisted-wings, order Strepsiptera. Hymenoptera are morphologically isolated among orders of Holometabola, consisting of a complex mixture of primitive traits and numerous autapomorphies, leaving little evidence to which group they are most closely related. Present evidence indicates that the Holometabola can be organized into two major lineages: the Coleoptera + Neuropterida and the Panorpida. It is to the Panorpida that the Hymenoptera appear to be related, owing to the reduction into a single claw on the larval leg, the presence of labial silk glands in the larva, and a sclerotized sitophore plate in the cibarium of the adult mouthparts (Königsmann, 1976; Kristensen, 1981, 1991, 1995, 1999a), as well as significant reduction of the prothorax. Molecular studies have also supported this position (e.g., Whiting *et al.*, 1997; Wheeler *et al.*, 2001). The sister-group relationship between the Hymenoptera and the panorpid orders is intriguing and might, on the surface, suggest that the order should stem well into the Permian because numerous Paleozoic mecopteroids (stem-group Antliophora and Panorpida) are known from this period.

The earliest definitive Hymenoptera, which are easily recognizable by the distinctive wing venation, are from the Triassic of Australia, Central Asia, and Africa (Riek, 1955; Rasnitsyn, 1964, 1969; Schlüter, 2000). Are the Hymenoptera more ancient than presumed, owing to the antiquity of their sister group? They likely are not. As we shall discuss later, numerous fossilized "mecopteroid" wings from the Permian exhibit only primitive traits for the Hymenoptera + Panorpida complex, and defining features of Panorpida cannot be determined for many of these taxa. It is, therefore, possible that some of these Paleozoic groups are in fact stem-group lineages to the Hymenoptera + Panorpida, with the Hymenoptera and panorpoid orders as we know them today having arisen from a paraphyletic cloud of Paleozoic taxa at various times between the Late Permian and Early Triassic. Thus, unlike some of the basal holometabolan orders, the Hymenoptera have a relatively recent origin, first appearing in the Late Triassic. Since the Triassic, the Hymenoptera have truly come into their own, having radiated extensively in the Jurassic, again in the Cretaceous, and again (within certain family-level lineages) during the Tertiary. The hymenopteran bauplan, in both structure and function, has been tremendously successful.

While the beetles today boast the largest number of species among all orders, Hymenoptera may eventually rival or even surpass the diversity of coleopterans (Kristensen, 1999a; Grissell, 1999). Already it is known that in well-sampled temperate regions, Hymenoptera outnumber Coleoptera (e.g., Gaston, 1991), and the order rivals the beetles in a few well-studied tropical ecosystems (e.g., Noyes, 1989; Stork, 1991; Hanson and Gauld, 1995). Today there are approximately 125,000 named species of Hymenoptera, but even the most conservative estimates on the total diversity of the order approximate figures between 600,000 and 1,200,0000 species (Gaston, 1991; Grissell, 1999; Austin and Dowton, 2000b). The sobering aspect of these estimates is that 60–88% of the order remains undescribed, perhaps 96% if Stork's radical estimate of 2.5 million species is adopted (Stork, 1996). This vast diversity is largely hidden among the relatively poorly explored microhymenopterans (e.g., Proctotrupoidea, Chalcidoidea, Platygastroidea) and where vast new faunas of species are continually discovered. However, even among the more physically "robust" lineages, like Ichneumonoidea, the number of species is rising rapidly as more comprehensive sampling and monographs are undertaken of megadiverse regions, such as Malaysia, Indonesia, Ecuador, Peru, and Zaire. Major treatments covering both the systematics and biology of the Hymenoptera include Evans and West-Eberhard (1970), Iwata (1972, 1976), Spradberry (1973b), Krombein *et al.* (1979), Gauld and Bolton (1988), Schedl (1991), Goulet and Huber (1993), LaSalle and Gauld (1993), Godfray (1994), Hanson and Gauld (1995), Quicke (1997), Austin and Dowton (2000a), and O'Neill (2001). Although

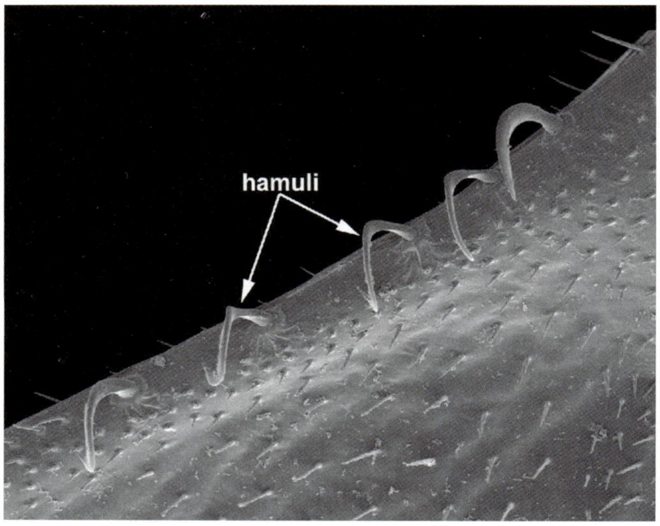

11.1. Hamuli on the hind wing of a wasp, which is a defining feature of the Hymenoptera. These microscopic hooks link to the posterior margin of the forewing and couple the wings in flight. Hymenoptera effectively fly like flies, with functionally one pair of wings. Scanning electron micrograph.

certainly dated, the book by Malyshev (1966: translated 1968) is a remarkable source for information on the biology of hymenopterans.

As mentioned before, the Hymenoptera possess numerous primitive traits combined with several unique features. Notable primitive features in the groundplan of Hymenoptera include the mandibulate mouthparts and the generalized ovipositor of the "lepismatoid" type (i.e., the typical arrangement for Dicondylia with a gonangulum), although the first gonocoxite in the order has been entirely lost and is wholly replaced by the gonangulum (and comments about a first valvifer or gonocoxa in the order are erroneous and actually refer to the gonangulum) (Figure 4.8). Hymenoptera are defined by the following specialized features: relatively small hind wings that are linked to the forewings by a series of minute hooks (*hamuli*) on the leading edge and that grasp the posterior edge of the forewing (Figure 11.1); the anal veins of the forewing not reaching the posterior wing margin; a modification of the protibial spur into an "antenna cleaner"; articulation of the profurcal apophyses to the propleura; the presence, at least among basal hymenopterans, of *cenchri* (paired oval structures on the metascutum, which can engage the forewing undersides in symphytans); and the presence of volsellae in the male genitalia (other morphological traits are outlined by Vilhelmsen, 2001; Schulmeister, 2003b). Perhaps one of the more interesting, derived traits of the Hymenoptera, and one that plays an important role in kin selection throughout the social lineages, is a haplodiploid sex determination system, though this also occurs sporadically throughout other groups of insects. In all members of the order, females have two sets of chromosomes (i.e., are diploid), being the usual product of the union of two gametes. Males, however, are produced from unfertilized eggs and as such have only a single compliment of the genome (i.e., are haploid). Diploid males do at times occur, particularly among the social aculeates, but typically they do not survive to maturity, but instead die as larvae.

Hymenoptera are well known as *parasitoids* (Figure 11.2), a mode of life in which they have succeeded and far surpassed all other insect parasites. As such, the group is of paramount importance in natural ecosystems, but it has also made them a focus for biological control agents in integrated pest management (e.g., Waage and Greathead, 1986). Parasitoids, unlike parasites, develop from nutrients extracted from a single host organism either externally (*ectoparasitoids*) or internally (*endoparasitoids*), and they kill the host as a direct or indirect result (a parasite, while inflicting minimal to severe ill effects, does not kill its host). The host, however, remains alive for the larger part of the parasitoid's period of feeding. This mode of life is remarkably efficient, and parasitoid wasps are so diverse that they exploit most groups of terrestrial arthropods. Some parasitoid wasps are *hyperparasitoids*, in which one parasitoid attacks another

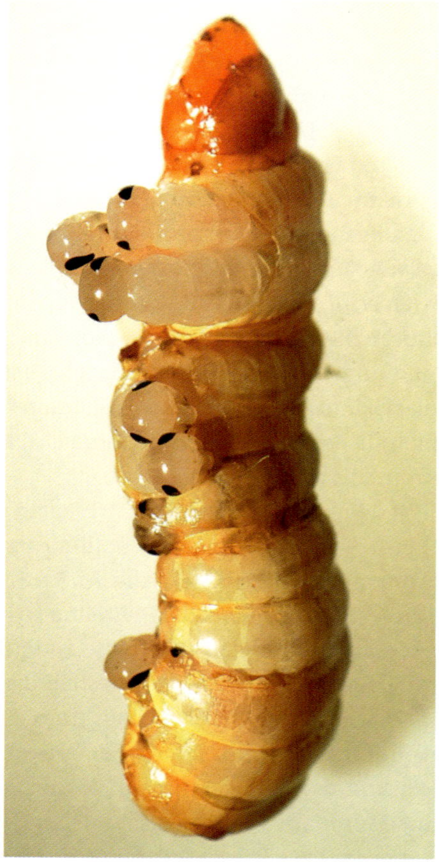

11.2. Parasitoid wasps emerging from their beetle larva host. The order Hymenoptera contains the greatest diversity and numbers of insect parasitoids, which are largely responsible for regulating populations of pest insects and so are used extensively in biocontrol.

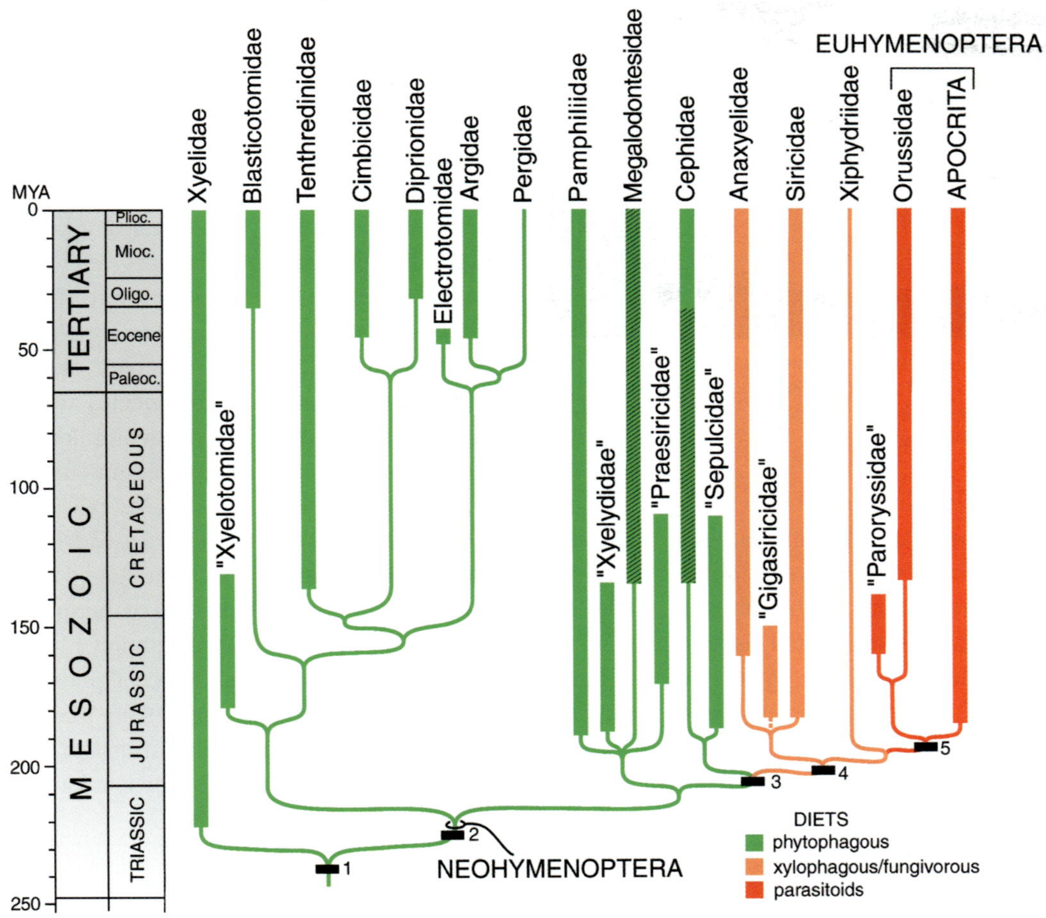

11.3. Phylogenetic relationships of the sawflies and wood wasps (symphytan Hymenoptera). Numbers refer to some significant characters (see Table 11.1). Based on Vilhelmsen (2001) and Schulmeister (2003b), with fossils added. Crosshatching indicates putative ranges which require more careful authentification.

TABLE 11.1. Significant Characters in the Phylogeny of Basal Hymenoptera[a]

1. Hamuli; protibial spur with velum; haplodiploid
2. Forewing Rs not furcate apically, Sc fused to R; Sc absent in hind wing
3. Reduction of posterior protibial spur ("Unicalcarida")
4. Absence of larval eyes
5. Parasitoidism; thoracic legs of larva lost

[a] Numbers correspond to those on phylogeny, Figure 11.3.

parasitoid that is already on or within a primary host. Under some circumstances such wasps can be a detriment, effectively killing the primary parasitoid being used to control a pest.

The Hymenoptera are presently divided into two suborders: the demonstrably paraphyletic "Symphyta" (sawflies and wood wasps) and the Apocrita ("true" wasps or parasitic wasps). Symphytans are, simply put, anything that is not an apocritan. Although this situation is certainly understood, the phylogeny and classification of Hymenoptera is presently the subject of intense study, and revisions to this long dated system are expected in due course. Rather than adopt a novel subordinal classification in advance of such studies, we shall refer to these imperfect subdivisions but shall not employ "Symphyta" as a formal suborder; instead, we refer informally to a grade of symphytan superfamilies and use the adjectival form of the name.

The symphytan superfamilies are the basal grade of Hymenoptera (Figure 11.3). By contrast to the Apocrita, symphytan families are generally most diverse in cool, temperate regions and are primitively phytophagous. Interestingly, the oldest lineage of Hymenoptera is one that is still extant and one of about eight oldest insect families that extend to the Triassic, approximately 230 MYA. The primitive family Xyelidae (sole family of the **Xyeloidea**) is well supported as the sister group of all other Hymenoptera (e.g., Rasnitsyn, 1988a; Vilhelmsen, 2001; Schulmeister *et al.*, 2002; Schulmeister, 2003b) and is also a family that is relatively well documented in the fossil record. The earliest records of Hymenoptera are of putative xyelids from the Late Triassic of Australia (Riek,

11.4. The earliest Hymenoptera are some primitive xyelid species from the Triassic, such as this forewing of *Archexyela* (Xyelidae) from Mt. Crosby, Australia. Xyelidae is the sister group to all other Hymenoptera. QMF.44154; length 10 mm.

1955) (Figure 11.4), central Asia (Rasnitsyn, 1964, 1969), and southern Africa (Schlüter, 2000). However, although modern Xyelidae are monophyletic (e.g., Schulmeister, 2003b), no convincing traits indicate that the family is natural when all the Triassic taxa are included (a few Triassic archexyelines, such as *Madygenius*, apparently possessed the derived, enlarged first flagellomere of true xyelids), and the family including *all* Archexyelinae may be paraphyletic. On the whole, Xyelidae has a relatively well-documented history through the later Mesozoic (Figure 11.5) and even into the Tertiary. As alluded to, xyelids (**Archihymenoptera**) are the sister group to all other Hymenoptera (**Neohymenoptera**), the latter being united by the derived fusion of the forewing subcosta (Sc) into the radial vein (R), forewing Rs not branched apically, absence of a subcosta in the hind wing, fusion of cubital (Cu) and medial (M) stems proximally in the hind wing, presence of postspiracular sclerites in the mesothorax, presence of well-developed cervical apodemes, and loss of metapleuro-mesosternal musculature (Vilhelmsen, 2001; Schulmeister, 2003b). Hinton (1971) also indicated that xyelids were unique in the primitive possession of dectiticous pupae; however, as noted by Vilhelmsen (2001), this trait needs to be more extensively surveyed across the symphytan families. Xyelids have a Holarctic distribution, with a particularly "rich" fauna in North America. During the Mesozoic they were apparently more widely distributed, occurring in distinctly warmer climates than are presently preferred by species of the family. Females oviposit in staminate cones of *Pinus* (Pinaceae) and larvae feed on the developing sporophylls, although larvae of some genera live in developing buds or young shoots of pines and firs (*Abies*: Pinaceae), and those of *Macroxyela* feed on leaves of deciduous elm (Ulmaceae) or walnut (Juglandaceae) trees. Once the larva completes feeding, it drops to the ground where it constructs a brood cell in the soil, typically overwintering as a prepupa and pupating in the spring.

When most people encounter a sawfly it is usually a member of the **Tenthredinoidea**, the nominate family in particular. The superfamily Tenthredinoidea is the most diverse lineage of sawflies today (around 7,000 species), with at least six disparate families: Blasticotomidae, Argidae, Pergidae, Diprionidae, Cimbicidae, and Tenthredinidae. Tenthredinoids are defined by the deeply curved pronotum, with lateral concavities for reception of the prothoracic spiracle, the position of crossvein 2r-rs distal of 2rs-m, and numerous internal structural arrangements (e.g., Vilhelmsen, 2001; Schulmeister, 2003b). Larvae of tenthridinoids principally feed on the leaf tissue of a wide variety of plants ranging from horsetails and ferns to a diversity of gymnosperm and angiosperm hosts, and some species are even leaf miners. Within the superfamily the Blasticotomidae is the most primitive, with only nine species feeding on ferns in temperate Eurasia (Gussakovskij, 1935; Smith, 1978). The families Argidae and Pergidae are close relatives, with 800 and 400 species, respectively, and are linked by the shared fusion of the pronotum and mesepisternum ventral to the spiracle, among other more minute details of morphology (Vilhelmsen, 2001; Schulmeister, 2003b). Argids occur pantropically, with few temperate species (Benson, 1938; Malaise, 1941), while pergids are almost entirely confined to Central and South America and Australia, with merely a few species extending

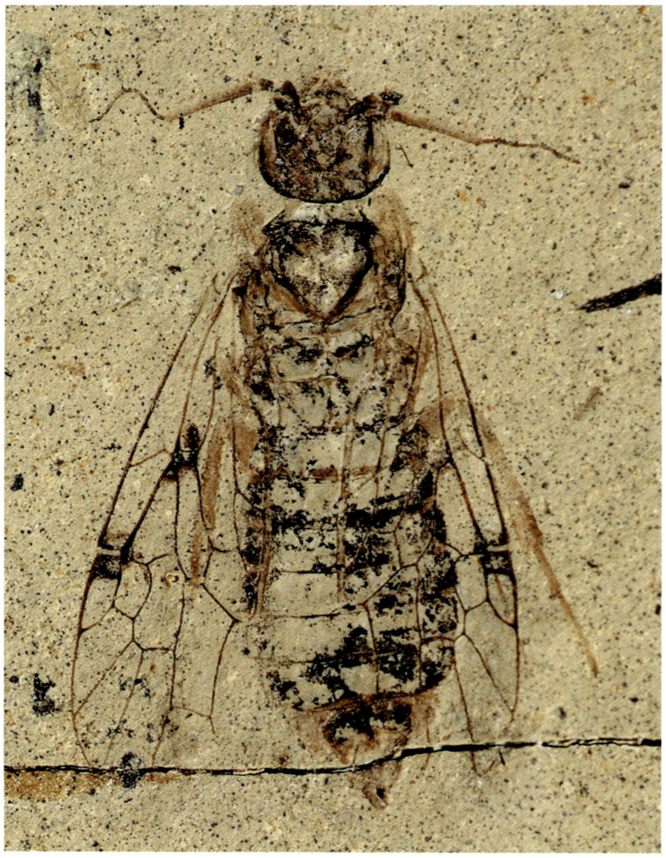

11.5. *Angarixyela vitimica* (Xyelidae) from the Early Cretaceous of central Asia. Xyelids are a small Holarctic family today (with most species in North America), and they were apparently more diverse in the Mesozoic. PIN 3064/1919; length 10 mm.

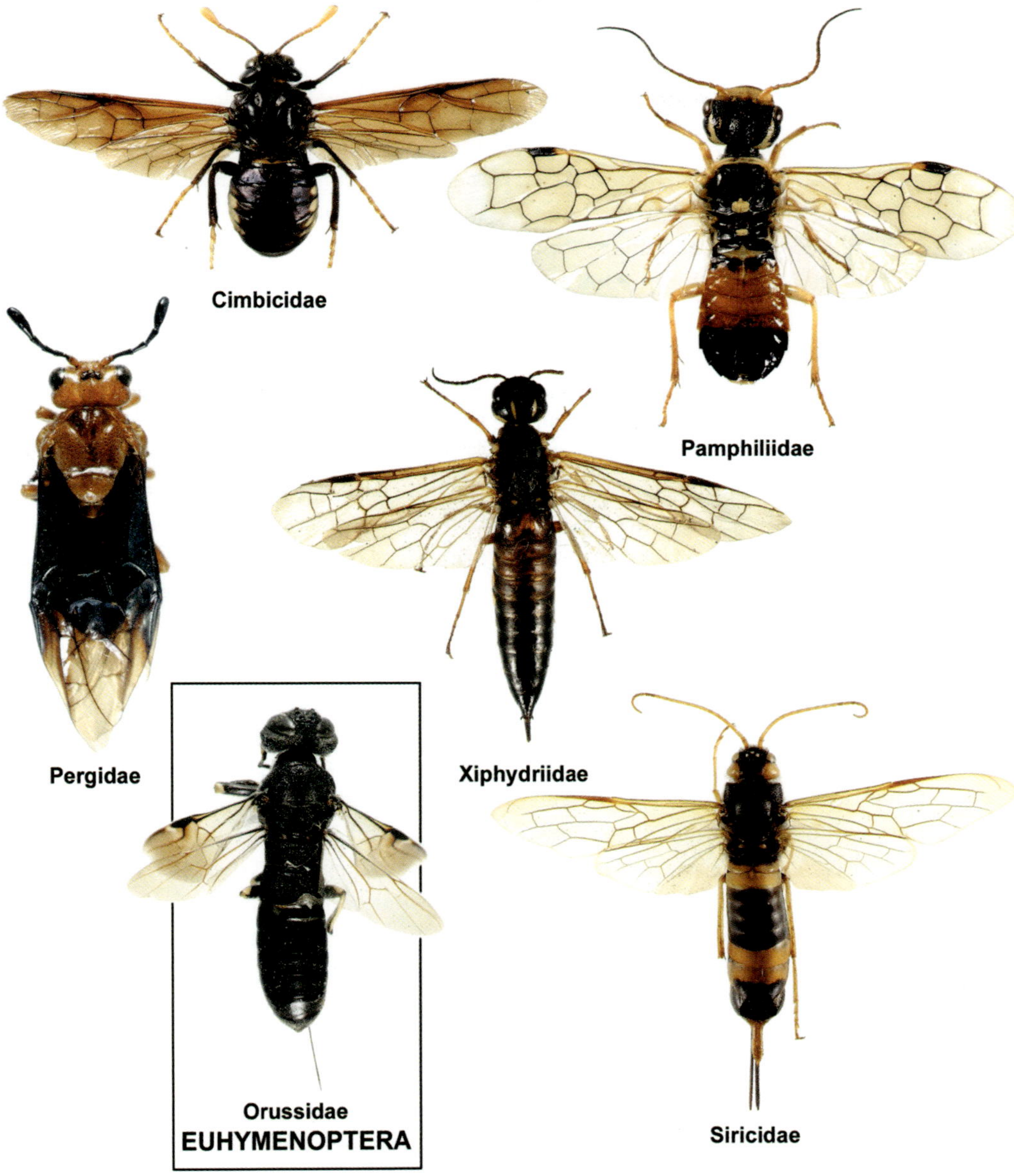

11.6. Representatives of Recent symphytan Hymenoptera. Not to the same scale.

into North America (Benson, 1938; Riek, 1970a; Smith, 1990) (Figure 11.6). The families Diprionidae and Cimbicidae, like the blasticotomids, are both relatively small families, with about 90 and 130 species, respectively. Cimbicids are robust insects, among the largest of all Hymenoptera, and they are easily recognized by their clubbed antennae (Figure 11.6). While cimbicids can be found around the world, diprionids are entirely confined to the Northern Hemisphere. The Tenthredinidae is the largest family of the superfamily, and, indeed, tenthredinids account for the majority of all symphytans, numbering about 6,000 species mostly in temperate environments of the Northern Hemisphere.

Tenthredinoids have a rather good fossil record. The earliest records of the superfamily are those of the Xyelotomidae, an extinct, stem-group lineage from which all other tenthredinoids arose. Xyelotomids are known from the Late Jurassic and Early Cretaceous of Eurasia (e.g., Rasnitsyn, 1969, 1977a; Rasnitsyn *et al.*, 1998; Rasnitsyn and Ansorge, 2000a). Putative tenthredinids are recorded from the earliest Cretaceous, but they may be stem-group taxa (e.g., Zhang, 1985). Compressions of blasticotomids, diprionids, argids, and cimbicids are known from the Eocene-Oligocene of Florissant, Colorado (e.g., Brues, 1908; Rohwer, 1908c; Zhelochovtzev and Rasnitsyn, 1972; Lewis *et al.*, 1990; Riou, 1992), while diprionids and argids are also known as amber inclusions from the Baltic region and Dominican Republic (e.g., Bachofen-Echt, 1949; Smith and Poinar, 1992). A fossil tenthredinoid larva in Baltic amber was described in its own family, Electrotomidae, and was believed to be closely allied to the argids and pergids (Rasnitsyn, 1977b). Pergidae are presently unknown as fossils. As one might imagine, the Tenthredinidae are abundant in the fossil record, albeit mostly from the Tertiary; in fact they are more heavily represented than any other group of symphytans. Deposits from North America and Europe are richly supplied with tenthredinids (e.g., Brues, 1908; Rowher, 1908b,c; Zhelochovtzev and Rasnitsyn, 1972; Carpenter, 1992), and their diversity reflects the large number of modern species. Fossilized leaf mines from the Neogene have been attributed to the Tenthredinidae (e.g., Straus, 1977; Givulescu, 1984).

Two families, Megalodontesidae (called Megalodontidae in the older literature: Springate, 1994; ICZN, 1996) and Pamphiliidae, together comprise the superfamily **Pamphilioidea**, which today are found in temperate Eurasia and North America. The approximately 40 species of megalodontesids feed on herbaceous plants, while pamphiliids (Figure 11.6), which number about 200 species, roll leaves to form tubes or spin silken chambers in which they feed. Some pamphiliid species are gregarious, with their larvae spinning large communal webs in the same manner of the tent caterpillars. The classification of the group has been examined by Gussakovskij (1935), Benson (1945); and van Achterberg and van Aartsen (1986). Pamphilioidea is first recorded from the Early Jurassic by putative Pamphiliidae as well as two extinct families, Xyelydidae and Parapamphiliidae, the latter two perhaps forming a grade to at least the Megalodontesidae, if not actually representing a stem group of the entire superfamily (e.g., Rasnitsyn, 1968, 1969, 1977a, 1983b). Definitive megalodontesids are not known as fossils, but pamphiliids are well documented in the Tertiary (e.g., Zhelochovtzev and Rasnitsyn, 1972; Rasnitsyn, 1983b). All fossil pamphilioids, like their modern counterparts, have been found in the Northern Hemisphere.

The superfamily **Cephoidea** contains a single, living family that is generally referred to as the stem sawflies. Cephids, especially Cephini, bore in stems of various grasses (Gramineae) or in stems of herbaceous Rosaceae (e.g., Hartigiini). The biology has been elucidated by Holmes (1954) and Hanson (1986). The approximately 100 species of stem sawflies are mostly found in Eurasia, with a single species in Madagascar and a small diversity in North and South America (Muche, 1981; Smith, 1988). The position of Cephidae has had a rather interesting history as some authors once believed them to be the sister group to the Apocrita, owing to the convergent loss of cenchri and the formation of an incipient "waist" in the abdomen (e.g., Königsmann, 1977; van Achterberg and van Aartsen, 1986). Exploration of a wider array of characters (e.g., Gibson, 1985; Vilhelmsen, 1996, 2001), however, has not supported such a notion. Cephoids are known from as far back as the early Jurassic, with those from the Jurassic presently being placed in the extinct family Sepulcidae, a stem group to Cephidae (Rasnitsyn, 1988b). Additional fossils of definitive stem sawflies occur in the Tertiary of the Northern Hemisphere (e.g., Konow, 1897; Cockerell, 1913; Weitschat and Wichard, 2002). Cephoidea are united to the remainder of Hymenoptera most visibly by the reduction of the inner protibial spur, leaving a single spur (Schulmeister *et al.*, 2002); although overlooked by these authors, this character is apomorphically reversed in Ceraphronoidea, which have two protibial spurs. Numerous other characters, which are however mostly internal, also support this grouping (e.g., Vilhelmsen, 2001; Schulmeister, 2003b).

Three superfamilies (Siricoidea, Xyphidrioidea, and Orussoidea) constitute the remainder of the symphytan lineages, and together are more closely related to Apocrita (e.g., Vilhelmsen, 2001; Schulmeister *et al.*, 2002; Schulmeister, 2003b). This group represents an important biological transition in the order, having experienced a shift from feeding principally on the foliage of plants to boring and feeding in wood. Thus, the Siricoidea and Xyphidrioidea are commonly known as wood wasps (Figure 11.6). Although once considered a group in and of themselves, they are actually a grade leading to the Orussidae + Apocrita. The wood wasps have a fascinating biology, having developed a symbiotic relationship with ascomycete or basidiomycete fungi that are a necessary part of the larval diet (Francke-Grossmann, 1939; Parkin, 1942; Stilwell, 1965, 1966; Spradberry, 1973a). In fact, the fungal infection of the wood that is required by the larvae is transmitted by the adult wasps themselves. Adult females have a pair of abdominal mycangia (sacs in the cuticle for carrying fungal spores) that open at the base of the ovipositor. The adult female also has hypopleural organs that nourish the fungi while in transport. Spores are injected into the tree at the same time as the egg is deposited so that the fungus develops in the wood with the larva. Interestingly, *Xeris spectrum* lacks the mycangia and exclusively oviposits in trees that have already been infected by other siricids. The superfamily **Siricoidea** is basal in this lineage (e.g., Schulmeister *et al.*, 2002; Schulmeister, 2003b), with two families extant – Siricidae (also known as horntails) and Anaxyelidae. Although well represented in the fossil record, Anaxyelidae is today confined to a single species, *Syntexis libocedrii*, from western North America. *Syntexis libocedrii* attacks recently burnt cedar and juniper trees (Middlekauff, 1974). Although today confined to the Nearctic, anaxyelids are documented from Late Jurassic and Early Cretaceous deposits as geographically separated as Brazil, England, Spain, and Central Asia (e.g., Rasnitsyn, 1969; Darling and Sharkey, 1990; Rasnitsyn *et al.*, 1998; Rasnitsyn and Martínez-Delclòs, 2000; Rasnitsyn and Ansorge, 2000b) (Figure. 11.7). Siricidae has nearly 100 species that prefer but do not confine themselves to dead or nearly dead Pinaceae (subfamily Siricinae) or hardwood angiosperm trees (subfamily Tremicinae) that are more susceptible to fungal infection. Horntails are found in temperate forests of the Northern Hemisphere, the

11.7. *Prosyntexis gouleti*, a primitive wood wasp of the living family Anaxyelidae from the Early Cretaceous of Brazil, 120 MYO. The wood wasps and horntails are xylophagous-fungivorous relatives of the parasitoid Euhymenoptera. AMNH 43270; length 12.8 mm.

world fauna having most recently been reviewed by Benson (1943) and Smith (1978), although several regional treatments do exist (e.g., Smith and Schiff, 2002). Like the anaxyelids, siricids have an ancient history extending back to the Early Jurassic, although records for horntails are much more extensive than the former family (e.g., Brues, 1926; Rasnitsyn, 1969, 1980; Königsmann, 1977; Fidalgo and Smith, 1987; Nel, 1988c, 1991a; Gromov *et al.*, 1993; Wedmann, 1998). The identity of the Tertiary "horntail" *Lithoserix williamsi* (Brown, 1986) has been questioned and is perhaps an aculeate (Nel, 1991a). Borings in Oligocene wood from Poland have been attributed to the activity of siricids (Rajchel and Uchman, 1998). The Jurassic family Gigasiricidae is likely a stem group to all other Siricoidea.

The superfamily **Xiphydrioidea** consists of a single family of about 80 species occurring in most regions of the world except Africa. Xiphydriids are defined by the elongate propleuron creating a defined "neck" and elongate cervical sclerites (Schedl, 1991) (Figure 11.6). Females oviposit into fresh timber of deciduous trees in the families Aceraceae, Betulaceae, Salicaceae, and Ulmaceae. The larvae tunnel through the wood, pupating in a chamber near the sapwood (Chrystal and Skinner, 1932). As described for the Siricoidea, xyphidriids are dependent on a symbiotic fungus in their tunnels, the wasps presumably consuming the mycelia as well as portions of infected wood (Deyrup, 1984). Benson (1954) revised the world genera and Smith (1978) catalogued the species. Xiphydriids are unknown as fossils. Xiphydriids are noteworthy as they are the closest relatives of the Euhymenoptera (= Vespina), the parasitic hymenopterans. Notable derived traits of both xiphydriids and Euhymenoptera include the loss of occipital sulci, the presence of a parapsidal signa on the mesoscutum (an anteroventral mark or line), the mesopostnotum dorsally covered by the metanotum, the elongations of the mesofurcal apophyses, the absence of hind wing tegulae, and the absence of metathoracic trochantins (Vilhelmsen, 2001; Schulmeister, 2003b).

THE EUHYMENOPTERA AND PARASITISM

Great fleas have little fleas upon their back to bite 'em,
And little fleas have lesser fleas and so ad infinitum.
–Augustus de Morgan

That which is not good for the beehive
cannot be good for the bees.
–Marcus Aurelius

Without question, the major feature of hymenopteran evolution was the development of parasitoidism. More so than any other insect lineage, the Hymenoptera perfected the parasitoid mode of life and represents the single, largest radiation of parasitic arthropods. As we have already discussed, the basal lineages of the order are phytophagous, with more recently evolved xylophagous and fungivorous groups. Among the xylophages of the symphytan superfamilies arose the small but phylogenetically very important family **Orussidae** (Figure 11.6). Orussids share numerous primitive morphological structures with other symphtan families – most notably the broad attachment of the thorax to the abdomen. However, unlike other symphytans, orussids have long, slender ovipositors and are external parasitoids (*ectoparasitoids*) of wood-boring beetles and, interestingly, siricoid wood wasps (Burke, 1917; Gourlay, 1951; Rawlings, 1957; Powell and Turner, 1975; Nuttall, 1980; Vilhelmsen, 2003). The parasitic mode of life is one of many traits uniting the orussids with the suborder Apocrita, and indeed some of the unusual traits of the family associated with its parasitic lifestyle led Rohwer and Cushman (1917) to create a separate suborder for them, Idiogastra. Indeed, the name Idiogastra should perhaps be resurrected for Orussidae, as a suborder sister to Apocrita. The 100 or so species of orussids occur throughout the world in temperate and tropical habitats, although the greatest diversity of species is in Africa and Australia. Aside from parasitism, orussids share numerous other anatomical features with the apocritan wasps and perhaps represent one of the most robust sister-group relationships among the Hymenoptera (Gibson, 1985; Johnson,

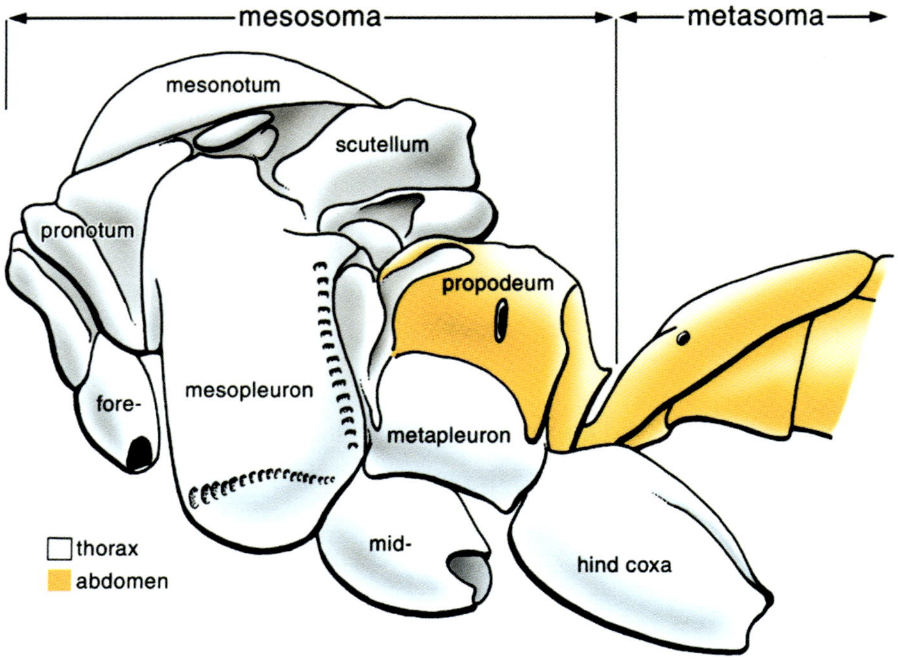

11.8. A major feature in hymenopteran evolution was the formation of the "wasp waist," a defining feature of the Apocrita. It results from fusion of the first abdominal segment, or *propodeum*, to the thorax. The result is a combined thorax + propodeum (the *mesosoma*) and the remaining abdominal segments (the *metasoma*).

1988; Whitfield *et al.*, 1989; Basibuyuk and Quicke, 1995; Vilhelmsen, 1996, 1997, 2001, 2003; Vilhelmsen *et al.*, 2001; Schulmeister *et al.*, 2002; Schulmeister, 2003a, 2003b). Orussids share with the Apocrita several traits, the most notable of which include the loss of thoracic legs in larvae, the reduction of the larval antenna to a single segment, the accommodation of the adult mesocoxae into metepisternal depressions, and several reductions in hind wing venation (a complete list of synapomorphies is provided by Vilhelmsen, 2001; Schulmeister, 2003b). The Orussidae and Apocrita together form the **Euhymenoptera** (Figure 11.3), often called Vespina, but this latter name is confusing owing to the usage of the suffix *-ina* for subtribes in zoölogical nomenclature and should not be employed. Definitive orussids are, unfortunately, exceptionally rare in the fossil record, which consists of only two specimens. Interestingly, both records are from Cretaceous ambers. *Mesorussus taimyrensis* was described from Late Cretaceous amber of Siberia (Rasnitsyn, 1977a), while *Minyorussus luzzii* was discovered in Late Cretaceous amber from New Jersey (Basibuyuk *et al.*, 2000a). Although described as an orussid, the Miocene fossil, *Sinicorussus luzhongensis* (Lin, 1982), has subsequently proved to be an ichneumonid (Zhang *et al.*, 1994). The extinct family Paroryssidae is likely paraphyletic and representative of stem-group orussids, thereby extending the orussoid lineage into the Late Jurassic of Central Asia (Martynov, 1925c; Rasnitsyn, 1968, 1969, 1980). Given that definitive apocritan wasps are known from the Early Jurassic, the orussoids are certainly of similar age. Taken together, these records indicate that the origin of parasitism in the Hymenoptera occurred sometime in the earliest Jurassic, approximately 200–205 MYA.

While the shift to parasitism was perhaps the most significant event in hymenopteran evolution, parasitism did not immediately lead to a rapid radiation of the order. Instead, further refinements of the parasitoid system were developed, and these refinements led to their diversity, traits that we use to characterize the suborder **Apocrita** (the "true" wasps or parasitic wasps). The hallmark trait of the Apocrita is the development of the "wasp waist," the result of a constriction between the first and second abdominal segments (Figure 11.8). In apocritans the first abdominal segment has become incorporated into the thorax as the *propodeum*, and the remainder of the abdomen has a narrow connection to this hybrid region. To avoid confusion in identifying tagma, hymenopterists employ the terms *mesosoma* (thorax + first abdominal segment) and *metasoma* (remaining abdominal segments) when referring to these body regions (ant specialists call these the *trunk* or *alitrunk* and the *gaster*, respectively). Thus, the first metasomal segment of wasps is the second abdominal segment. The formation of the "wasp waist" allows for greater maneuverability in controlling a long ovipositor. Ectoparasitism on wood-boring insects, as described for Orussidae, is certainly the primitive habit for Apocrita. Indeed, basal lineages of Apocrita (e.g., Stephanoidea, Megalyroidea) are also ectoparasitoids of wood-boring beetles or symphytans, with subsequent shifts to alternative hosts. Relationships within Apocrita are, however, debated and of considerable controversy (e.g., Königsmann, 1978a; Rasnitsyn, 1988a; Whitfield, 1992a, 1998; Dowton and Austin, 1994, 2001; Ronquist *et al.*, 1999; Sharkey and Roy, 2002). Rather than attempt to support one of the numerous, unstable topologies for apocritan phylogeny, we have provided a relatively conservative outline of superfamilial relationships (Figure 11.9).

The **Stephanoidea** have one of the more prominent posi-

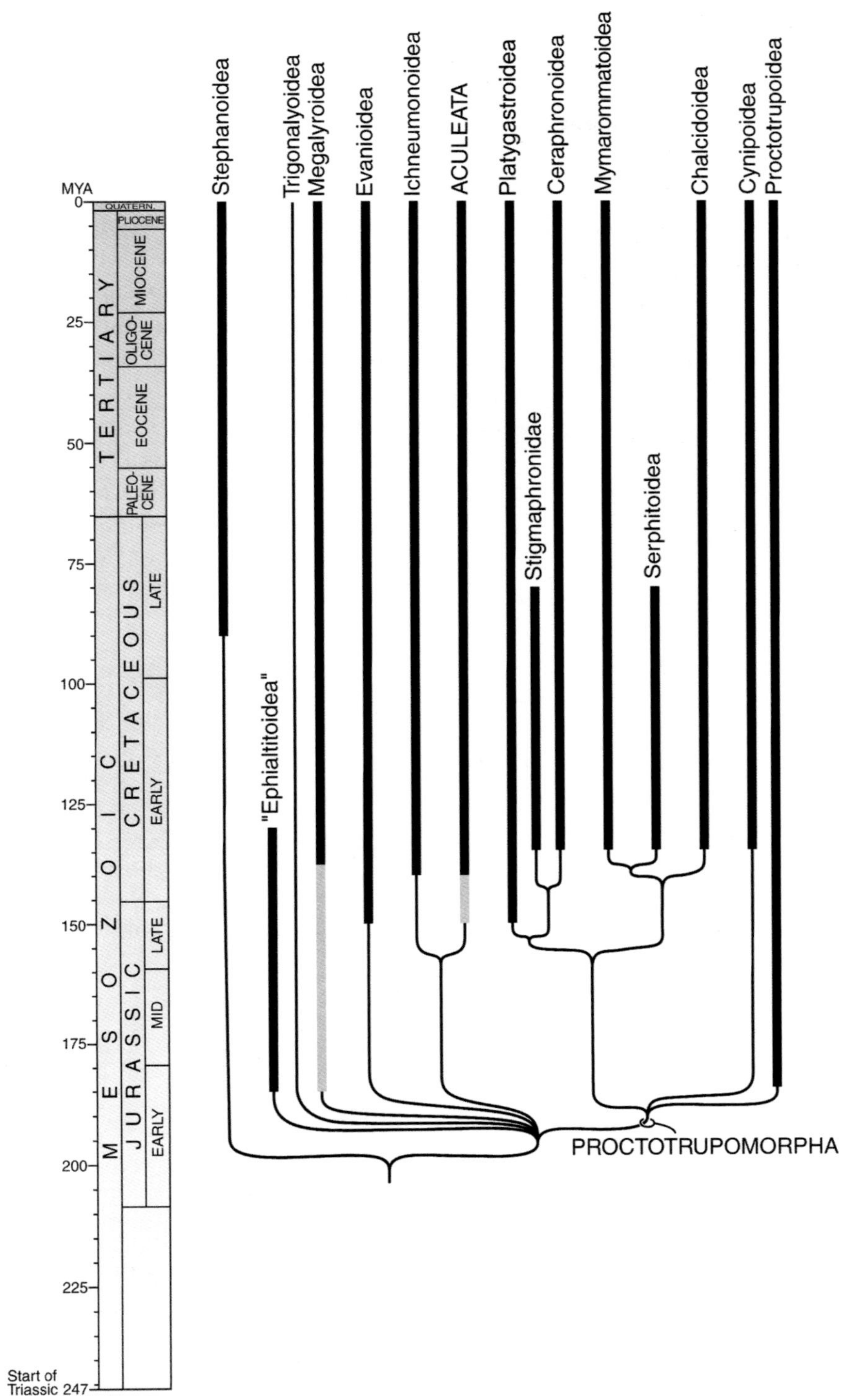

11.9. Phylogeny of the Apocrita. The thick lines indicate the known extent of fossils. Based on various sources (see text).

tions in apocritan phylogeny, being the putative living sister group to all other Apocrita (e.g., Whitfield, 1992a, 1998; Downton and Austin, 1994; Vilhelmsen, 1996, 1997), although in some analyses stephanids group with Megalyroidea and Trigonalyoidea, or other families (e.g., Ronquist *et al.*, 1999; Dowton and Austin, 2001). The superfamily consists of a single Recent family of principally pantropical distribution, although some species do range into regions outside of the tropical zone (Enderlein, 1905; Elliott, 1922; Orfila, 1949; Townes, 1949a; Benoit, 1984; Aguiar, 1998, 2001; van Achterberg, 2002; Aguiar and Johnson, 2003). Species are parasitoids of wood-boring beetle larvae of the families Buprestidae and Cerambycidae as well as of siricid wood wasps (Figure 11.12). Although we know little about stephanid biology, it appears to be similar to that of the Orussidae. Based on this unique position in apocritan phylogeny, it

11.10. *Stephanogaster magna* (Ephialtitidae) from the Late Jurassic of Karatau in Kazakhstan. It was a representative of a primitive apocritan superfamily, Ephialtitoidea, which flourished in the Jurassic and Early Cretaceous. Long ovipositors in ephialtoids indicate that they were probably parasitoids of wood-boring larvae. PIN 2784/1205; body length (excluding ovipositor), 15 mm.

would seem that stephanids are ancient, and they may very well be. However, stephanids have an exceedingly poor fossil record, one that extends only slightly into the Mesozoic. The earliest, definitive stephanid is *Archaeostephanus corae* from the Late Cretaceous amber of New Jersey (Engel and Grimaldi, 2004d), which is apparently a member of the primitive subfamily Schlettereriinae. Three other genera of fossil stephanids are known – two from Baltic amber and one preserved as a compression from the Eocene-Oligocene boundary of Florissant, Colorado (Cockerell, 1906; Brues, 1933a; Aguiar and Janzen, 1999; Engel and Grimaldi, 2004d). The Tertiary stephanids all appear to belong to the Stephaninae (Engel and Grimaldi, 2004d).

The extinct superfamily **Ephialtitoidea** has at times been considered to consist of stem-group stephanoids (e.g., Rasnitsyn, 1975; Königsmann, 1978a) or as a stem group to all other Apocrita (Rasnitsyn, 2002) (Figures 11.10, 11.11). However, based on the little available evidence, it would appear that they are a stem group to Megalyroidea or other basal Apocrita. The lineage is known from the Early Jurassic through the Early Cretaceous (Rasnitsyn, 1975, 1980, 2002). Ephialtitoids were apparently cosmopolitan in distribution, occurring in Asia (e.g., Rasnitsyn, 1975; Zhang *et al.*, 2002a; Rasnitsyn *et al.*, 2003), South America (e.g., Darling and Sharkey, 1990), and Europe (e.g., Zessin, 1985; Rasnitsyn and Martínez-Delclòs, 2000). Although assuredly parasitoids, their biology is uncertain. Given their apparent phylogenetic position and the fact that apocritans (particularly ones with long ovipositors) are primitively ectoparasitoids of wood-boring insects, it is logical to believe that ephialtitoids were parasitoids of wood-boring insects. For the moment, however, such a deduction remains speculative.

Other apparently basal Hymenoptera are the spectacular wasps of the **Megalyroidea**. Megalyroids are today essentially pantropical and austral in distribution, with their greatest diversity in the Old World and only a few species reaching

11.11. *Cratephialtites kourios* (Ephialtitidae), from the Early Cretaceous of Brazil, which is one of the last known occurrences of the superfamily. AMNH 46321; length 10 mm (excluding ovipositor).

north of the equator (Shaw, 1990, 2003). Like other primitive apocritans, megalyrids appear principally to victimize wood-boring beetles such as cerambycids (e.g., Froggatt, 1906; Hacker, 1913, 1915; Fahringer, 1928; Rodd, 1951), and, like stephanids, they are ectoparasitoids. However, at least one species, *Megalyra troglodytes*, is a parasitoid of a spheciform wasp (Naumann, 1983, 1987); *Megalyra*, however, is not basal within the family and this host association is not indicative of the primitive condition for Megalyridae. Compared to Stephanidae, the Megalyridae have relatively good representation as fossils, and were probably more diverse and common than they are today. Megalyridae are known from three deposits of Cretaceous amber – Siberian amber (Rasnitsyn, 1977a), Burmese amber, and New Jersey amber (Engel and Grimaldi, unpubl.). Another interesting Cretaceous amber fossil is *Maimetsha* from Siberian amber (Rasnitsyn, 1975). *Maimetsha* preserves some notably primitive features relative to all other megalyrids such as the greater number of antennal segments. *Maimetsha* has been at times considered an intermediate form between typical megalyrids and the minute wasps of the Ceraphronoidea, inclusive of the Cretaceous family Stigmaphronidae (e.g., Rasnitsyn, 1988a, 2002). Unfortunately, the only known specimen was accidentally destroyed, making it difficult to evaluate the position of this fossil. From the original description, however, it would appear that *Maimetsha* is not a ceraphronoid but instead belongs among the megalyrids (Shaw, 1988, 1990; Engel and Grimaldi, unpubl.). Two megalyrids are described from Baltic amber (Brues, 1923b, 1933a), however, and several other species are known to occur in these deposits (Engel, unpubl.). The family is otherwise unknown from the Cenozoic, even among the many prolific rock fossil deposits. Together these records account for the definitive history of the family, demonstrating that the family once had a much wider distribution and has experienced significant northern extinctions. A series of compressions ranging from the Early Jurassic through the Early Cretaceous (e.g., *Cleistogaster*, *Mesaulacinus*), however, has been putatively assigned to the Megalyroidea (Rasnitsyn, 1975) but exhibit few conclusive traits to consider them as megalyrid or stem-group megalyroids (Engel and Grimaldi, unpubl.).

The **Trigonalyoidea** are yet another enigmatic lineage of apparently basal apocritans and are sometimes considered to be close relatives of the Megalyroidea (Figure 11.12). Aside from several derived features of their biology, the trigonalyoids have primitive wing venation, tarsal plantulae, and mouthparts (particularly the paraglossae), which led some authors in the past to believe that they are basal within Apocrita (e.g., Malyshev, 1968; Naumann, 1991b), and Börner (1919) even proposed a separate division for them, the Archiglossata. The superfamily consists of a single cosmopolitan family of about 120 species (e.g., Schulz, 1907; Bischoff, 1938; Weinstein and Austin, 1991; Carmean and Kimsey, 1998). Unlike the groups discussed thus far, trigonalyids are *endoparasitoids*, developing inside of their host. However, this is quite an understatement. The trigonalyids are among the most remarkable of all endoparasitoids because they reach their hosts only after passing through a separate intermediate host, a mode of parasitism otherwise known only in the tachinid flies (Askew, 1971). Trigonalyids are essentially *hyperparasitoids*, with the female ovipositing on foliage that is consumed by a primary host (typically a caterpillar or sawfly), the egg then hatches inside of the primary host and the first instar larva punctures the host's gut wall to enter the hemocoel. If a primary parasitoid (e.g., ichneumonid larva) is already present inside of the primary host's body, then the trigonalyid immature proceeds to attack the parasitoid, using it as its secondary host. Even more complex are those cases where the caterpillar hosting the trigonalyid is captured by a vespid wasp. The caterpillar is, like all vespid prey, cut up and brought back to the nest to feed to the developing wasp larvae. The trigonalyid is consumed again, this time via the consumption of its primary host's carcass by the wasp larva. Once inside this secondary host, the trigonalyid resumes its development by penetrating the wasp's gut and feeding in its hemocoel. The scant information on the biology of these remarkably derived parasitoids is reviewed by Clausen (1940), Carmean (1991), and Weinstein and Austin (1991).

The **Evanioidea** comprise the first group of moderately diverse apocritans in hymenopteran phylogeny. Three families are known from the modern fauna. Among them the family Aulacidae retains the largest number of primitive features both in biology and morphology, but it is certainly not basal based on other traits. As is similar to the primitive condition for Apocrita, aulacids are parasitoids of wood-boring beetles (Buprestidae and Cerambycidae) and wood wasps (Xiphydriidae) (Carlson, 1979). Aulacidae today consist of approximately 150 species with an overall cosmopolitan distribution. Often overlooked, the Evaniidae and Gasteruptiidae, in contrast to aulacids, are less so parasitoids and perhaps more accurately considered larval predators, having reached this condition independently. Evaniids are "parasites" of cockroach oothecae, which is why it is not uncommon to find these wasps inside the windows of New York City apartments (Figures 11.12, 11.13). Females oviposit into the ootheca and the larva consumes the cockroach eggs (e.g., Brown, 1973). Gasteruptiids are larval predators of wasps and solitary bees (Crosskey, 1962; Malyshev, 1964; Houston, 1984a,b, 1987; Jennings and Austin, 1994a,b, 1997a,b, 2002; Engel, 1995a) (Figure 11.12). The egg is deposited either in or near a "host" brood cell. The gasteruptiid larva consumes rather than parasitizes the "host" larva and then proceeds to ingest the provisions originally intended for the victim. Like the Aulacidae, evaniids and gasteruptiids are cosmopolitan in distribution; however, they are notably more diverse with approximately 500

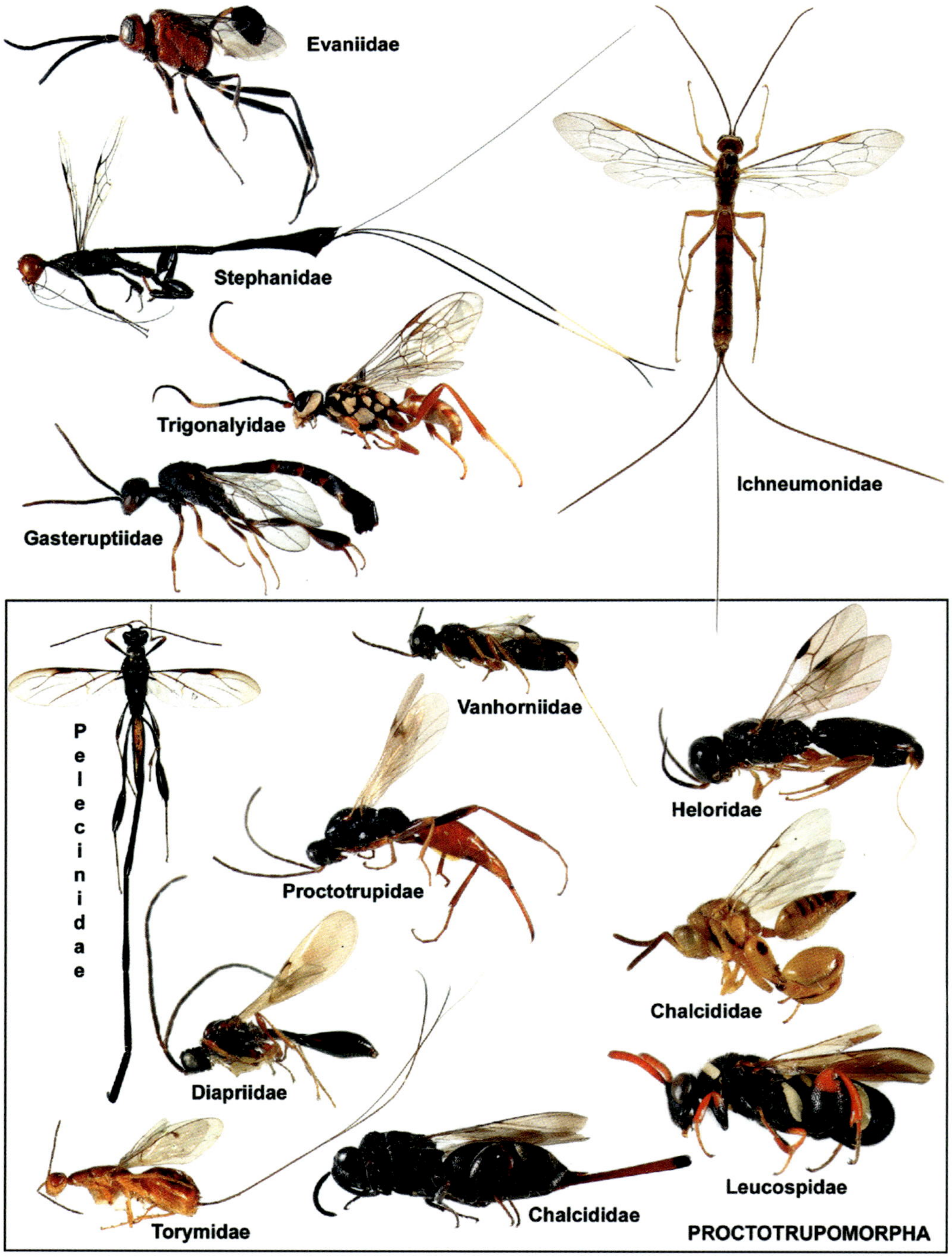

11.12. Representative Recent parasitoid apocritan wasps. Not to the same scale.

species each, with a greater abundance of species in tropical environments.

The most distinctive trait shared by the evanioid families is the connection of the metasoma high up on the propodeum. Rather than articulating in a ventral position, near the hind coxae, the metasoma attaches dorsally and well apart from the coxal bases, sometimes near the propodeal base. Such an articulation is sometimes found in other lineages of Apocrita (e.g., Ichneumonoidea, Cynipoidea, Chalcidoidea), but these are generally different from the evanioid condition (Naumann, 1991). However, such convergence has led some authors to also consider the Evaniidae as unrelated to the Aulacidae and Gasteruptiidae (e.g., Townes, 1949b, 1950; Crosskey, 1951, 1962; Carlson, 1979), believing the condition to be convergent between the two lineages. Molecular studies, alternatively, support the view that the evanioids are a natural group (e.g., Dowton and Austin, 1994, 2001) as do some morphological analyses (e.g., Ronquist *et al.*, 1999). Gauld and Bolton (1988) considered the loss of functional spiracles on all metasomal segments

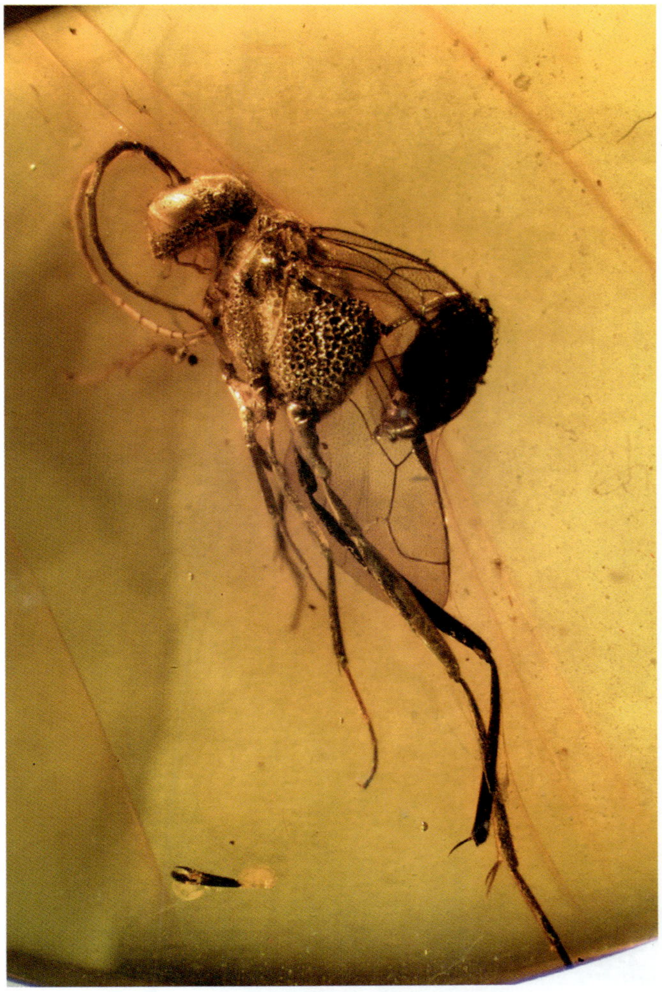

11.13. An ensign wasp (Evaniidae) in Early Miocene Dominican amber. Evaniidae are parasitoids of roach oothecae. Morone Collection, M0140.

except the eighth to be another derived feature of the superfamily.

Earliest evidence of evanioids comes from compression fossils of the extinct family Praeaulacidae. The Praeaulacidae were widely distributed in the Jurassic and Early Cretaceous of Australia and Asia (e.g., Rasnitsyn, 1972, 1990; Jell and Duncan, 1986). The family shares with other evanioids the dorsal propodeal articulation with the metasoma but is otherwise primitive in all features (e.g., the long pronotum), perhaps representing a stem group to the remainder of the superfamily. The remainder of the superfamily can be segregated into two lineages: the Evaniiformes, including only the Evaniidae, and the Aulaciformes, consisting of the modern families Evaniidae and Gasteruptiidae (inclusive of Kojutellitinae) as well as the Cretaceous family Baissidae. The Evaniiformes previously consisted of several families, but cladistic work on evanioids (e.g., Basibuyuk *et al.*, 2002) suggests that these divisions are not warranted. Evaniiformes are defined by the elongate head and petiolate first metasomal segment with a stout, short gaster and short ovipositor (apparently reversed in the fossil *Andrenelia*), while Aulaciformes are defined by the loss of the jugal lobe in the hind wing, fusion of the first and second metasomal terga, sexually dimorphic antennal segmentation, elongate propleuron, and rigid attachment of the pronotum and mesepisternum (Naumann, 1991; Mason, 1993). Quicke *et al.* (1994) also identified some shared features in ovipositor construction of the Aulaciformes.

The monogeneric family Andreneliidae from the Early Cretaceous of Spain (Rasnitsyn and Martínez-Delclòs, 2000) consistently groups among primitive evaniid genera, and its presumed primitive features, such as the more elongate mestasoma, did not exclude it from the family in the most conservative analyses by Basibuyuk *et al.* (2002). The family is therefore considered a synonym of Evaniidae. Similarly, the former family Cretevaniidae from the Early to Late Cretaceous of Asia and Europe (Rastnitsyn, 1975; Rasnitsyn *et al.*, 1998; Zhang and Zhang, 2000) nests well within Evaniidae and should not be recognized, as was also the conclusion of Basibuyuk *et al.* (2002). Aside from the compression fossils, evaniids are also well documented as inclusions in Cretaceous resins from Lebanon (Basibuyuk *et al.*, 2002; Engel and Grimaldi, unpubl.), Myanmar (Basibuyuk *et al.*, 2000c; Engel and Grimaldi, unpubl.), and New Jersey (Basibuyuk *et al.*, 2000b; Engel and Grimaldi, unpubl.). Tertiary evaniids are not uncommon (e.g., Brues, 1933a), and despite some having been placed in extinct genera, these most likely belong to modern groups. Among the Aulaciformes, the Baissidae (= Manlayinae) is presumably sister to a group comprised of the Aulacidae + Gasteruptiidae, which apparently share with these families sexually dimorphic antennal segmentation (Basibuyuk *et al.*, 2002) but are primitive in other traits for the lineage and superficially resemble aulacids. Baissids are known only from the Early Cretaceous of Europe and Asia (e.g., Rasnitsyn, 1975, 1980; Rasnitsyn *et al.*, 1998). Kotujellitinae is a basal, extinct subfamily of the Gasteruptiidae from the Cretaceous of Asia (Rasnitsyn, 1975, 1991). Like modern Gasteruptiidae, the notauli (a pair of posteriorly converging lines on the mesoscutum) fuse posteriorly before reaching the scuto-scutellar sulcus, and the articulation of the metasoma is near the propodeal base. Kotujellitines, however, possessed a rs-m crossvein and lacked the enlarged metatibiae, which are clearly primitive features, and so the subfamily may eventually prove to be paraphyletic.

Most of the order's small parasitoids, or "microhymenopterans," are placed in a group called the **Proctotrupomorpha**. Given the instability of hypotheses on apocritan phylogeny, it is noteworthy that the Proctotrupomorpha is perhaps the most robustly supported group in the suborder. This is a diverse lineage, and when all is said and done, it may reveal itself to be the most species-rich lineage of the Hymenoptera (Figure 11.14). Presently the group includes

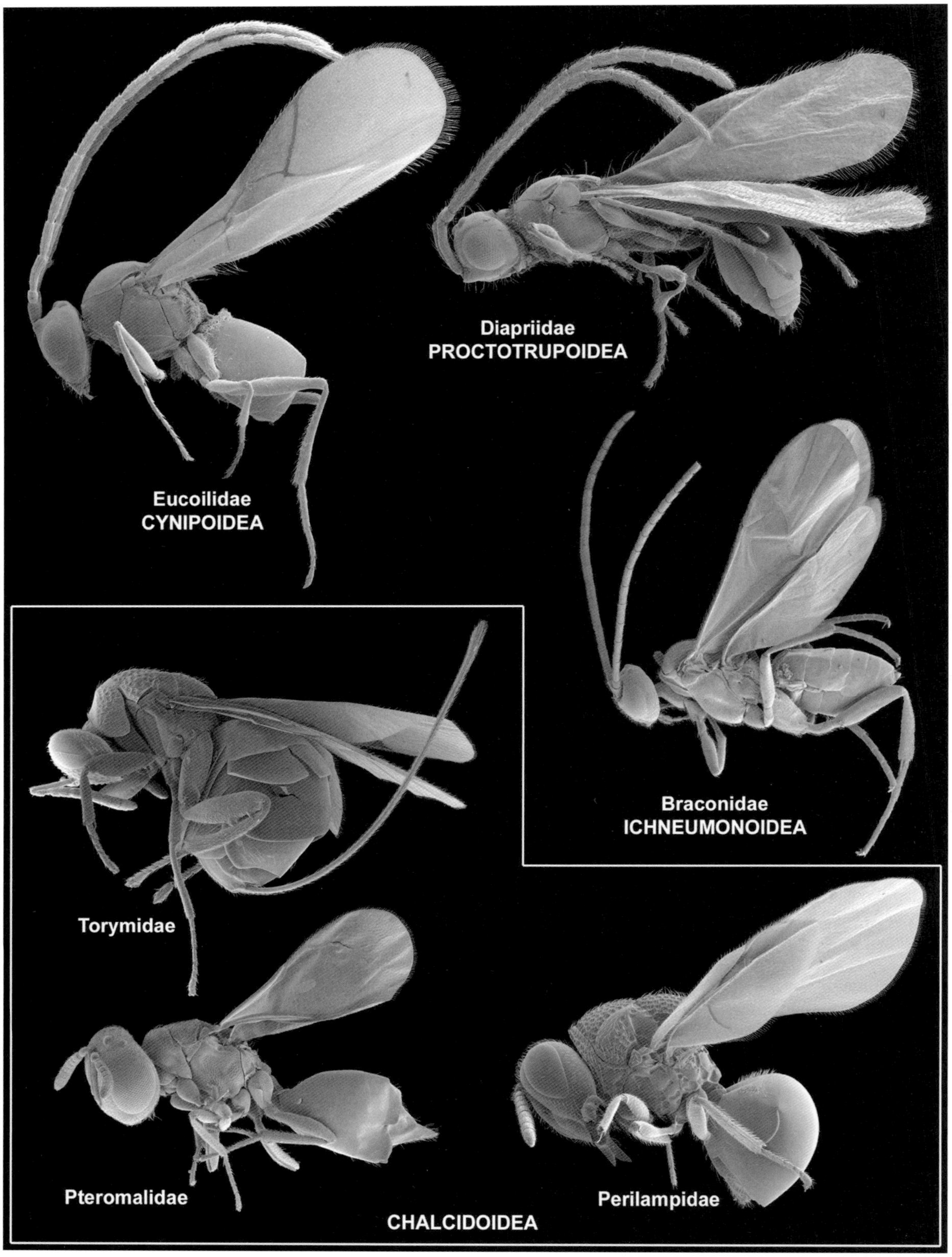

11.14. Representative "microhymenopteran" Apocrita. Scanning electron micrographs; not to the same scale, all 2 mm or less in length.

seven superfamilies, although one, Proctotrupoidea, is most assuredly paraphyletic with respect to other proctotrupomorphs. Defining features of the proctotrupomorphs include the lower tentorial bridge of the head intervening between the postgenae (replacing the postgenal bridge) to reach the occipital foramen and the loss of the medial scutal sulcus (Rasnitsyn, 1988a). Although the superfamily Ceraphronoidea is sometimes excluded from the infraorder (e.g., Rasnitsyn, 1988a), the group shares many credible apomorphies with other proctotrupomorphs (e.g., Basibuyuk and Quicke, 1997; Ronquist *et al.*, 1999).

The basal "superfamily" of the Proctotrupomorpha is the **Proctotrupoidea**, which is demonstrably an unnatural group consisting of a series of families from which all the other proctotrupomorph superfamilies are derived. In particular, many analyses indicate that families such as Diapriidae are more closely related to Cynipoidea and their relatives, than they are to other proctotrupoids. Because a thorough understanding of higher-level relationships in Hymenoptera is elusive, we have for the moment taken the conservative stance of retaining this superfamily, albeit with the understanding that in the future it must be segregated into meaningful natural entities. At present the Proctotrupoidea consists of eleven extant families – the diverse Diapriidae and Proctutrupidae (Figures 11.14, 11.15), and the much smaller Austroniidae, Heloridae, Maamingidae, Monomachidae, Pelecinidae, Peradeniidae, Proctorenyxidae, Roproniidae, and Vanhorniidae. Biologies of the nearly 3,000 species, most of which are diapriids, are sparsely known (Masner, 1993a). Where records do exist, all are endoparasitoids of larval beetles, flies, green lacewings, or symphytans. Most species are small to minute except for the giants of the family Pelecinidae, which can reach to almost 60 mm in total length owing to their very long metasoma, which is remarkably dextrous and used to probe the soil for its hosts. Interestingly, some of the oldest records of the Apocrita are purported proctotrupomorphs from the Early Jurassic of Asia (e.g., Rasnitsyn, 1983a), and proctotrupoid fossils can be found in most deposits through the present day (e.g., Kozlov, 1968; Rasnitsyn, 1975; Rasnitsyn, Jell and Duncan, 1986; Darling and Sharkey, 1990; Rasnitsyn *et al.*, 1998; Martínez-Delclòs, 2000) (Figures 11.15, 11.16). Reflective of this age, it is perhaps not surprising that many families are relict. The family Peradeniidae, for example, consists of two Recent species restricted to Australia (Naumann and Masner, 1985), but it has been recently discovered in Baltic amber (Johnson *et al.*, 2001). Similarly, the Pelecinidae, which today is confined to the New World (Johnson and Musetti, 1999), is also known from Baltic and Sakhalin amber (Johnson, 1998; Engel, 2002d), and putative stem-group pelecinids are known from Asia (e.g., Kozlov, 1974; Zhang *et al.*, 2002b).

Cynipoidea is a relatively small group of apocritans with about 3,000 presently recognized Recent species (Ronquist,

11.15. A diapriid wasp in Early Miocene Dominican amber, which is very similar to modern species occurring in the West Indies and Central America. Morone Collection, M0063; length 1.2 mm.

11.16. Proctotrupomorph wasps are among the first apocritans found in the fossil record. This is a primitive proctotrupoid, *Mesoserphus karatavicus* (Mesoserphidae), from the Late Jurassic of Karatau in Kazakhstan. PIN 2554/1281; length 7 mm (excluding ovipositor).

1999), but Nordlander (1984) has estimated that it may grow to as many as 20,000 species. The superfamily is one of the few among Apocrita that includes parasitoids as well as secondarily phytophagous species (i.e., the gall wasps: Cynipidae). Basal lineages of the superfamily are generally parasitoids of wood-boring or cone-boring insect larvae – Austrocynipidae on oecophorid moth larvae, Ibaliidae on siricid larvae, and Liopteridae on beetle larvae of the families Buprestidae, Cerambycidae, and Curculionidae (Ronquist, 1995b, 1999). The Figitidae, sister group to the gall wasps, are parasitoids of a diversity of hosts, mostly of fly larvae but also cynipids (e.g., Ronquist and Nieves-Aldrey, 2001), hemerobiids, chrysopids, and as hyperparastoids of braconid and chalcidoid parasitoids of aphids and psyllids (Ronquist, 1995b, 1999). The Cynipidae, alternatively, is entirely phytophagous, and these produce characteristic galls in angiosperm tissue, including phytophagous inquilines of other cynipids (e.g., Ronquist, 1994). These are perhaps best known for the characteristic round galls they form as larvae on the leaves of various oaks. Indeed, the oaks (family Fagaceae) comprise a major host plant family for Cynipidae. The superfamily is defined by the combination of the first and second metasomal sterna broadly joined or fused, the laterally compressed metasoma, the unique course of the forewing median vein, and the absence of the forewing costa and antennal radicle (which is a basal, constricted area of the scape), among other traits (Ronquist, 1999). The classification and phylogeny of the superfamily has been summarized by Ronquist (1999), Ronquist and Nieves-Aldrey (2001), and Vårdal *et al.* (2003).

Relationships among the families, basal relationships within several families, and various fossils, all suggest Jurassic divergences for the principal lineages of Cynipoidea (e.g., Ronquist, 1995a, 1999; Nordlander *et al.*, 1996; Liu, 1998a,b, 2001; Liu, Grimaldi, and Engel, unpubl.). Indeed, the cynipoids as a whole likely stem from the Jurassic, but fossil evidence for the earliest evolution of the superfamily is lacking. Cynipoid fossils are rare, but a variety of parasitoid lineages were already well established in the Cretaceous (e.g., Kinsey, 1937; Kovalev, 1994, 1995, 1996; Liu, Grimaldi, and Engel, unpubl.) (Figure 11.17), supporting the hypothesis of a Jurassic age. Several extinct families have been described from the Cretaceous of Asia (Kovalev, 1994, 1995, 1996). Ronquist (1999) suggested that the Gerocynipidae and Rasnicynipidae were stem groups to Cynipidae and that the Palaeocynipidae were actually basal within Figitidae. Archaeocynipidae from the Early Cretaceous of Asia (Rasnitsyn and Kovalev, 1988) apparently do not belong to the Cynipoidea (Ronquist, 1999), but may be a stem group to Cynipoidea between Diapriidae and cynipoids (e.g., Ronquist *et al.*, 1999). The Rasnicynipidae is actually a member of Figitidae (Liu, Grimaldi, and Engel, unpubl.). The parasitoid families are also represented in Tertiary deposits (e.g., Kinsey, 1919; Kovalev, 1994, 1995). All Cretaceous cynipoids are of parasitoid families, the sole exceptions being Gerocynipidae and a fossil from New Jersey amber that may be a true cynipid. As an apparently close relative of Cynipidae, Gerocynipidae may have been gallers, and certainly Fagaceae occurred in the mid-Cretaceous, but such biological associations are entirely conjectural at present (Liu, Grimaldi, and Engel, unpubl.). *Aulacidea succinea* in Baltic amber, although incorrectly placed as to genus (Liu, Grimaldi, and Engel, unpubl.), is the oldest fossil that can be unquestionably placed in Cynipidae and is therefore the oldest gall wasp, presumably having induced galls on Rosaceae like possible modern relatives. Cockerell (1921) described two slightly younger fossils (Oligocene) as *Rhodites vetus* and *Andricus vectensis*. Ronquist (1999) suggested that the former might well be correctly placed in the tribe Diplolepidini, which make galls on *Rosa*, but the placement of *A. vectensis* may be misplaced in Cynipini, which produce galls on oaks and beeches (Fagaceae). Fossilized galls that can definitively be assigned to Cynipidae are all fairly recent, coming from the late Tertiary (e.g., Straus, 1977; Scott *et al.*, 1994; Dieguez *et al.*, 1996; Waggoner and Poteet, 1996; Waggoner, 1999) or even from the late Pleistocene (Larew, 1987). This is far younger than the host groups (e.g., Fagales in the mid-Cretaceous [Figure 14.5]). Some galls from the mid- to Late Cretaceous have been tentatively attributed to Cynipidae (e.g., Hickey and Doyle, 1977; Scott *et al.*, 1994), but the insects that formed these are uncertain.

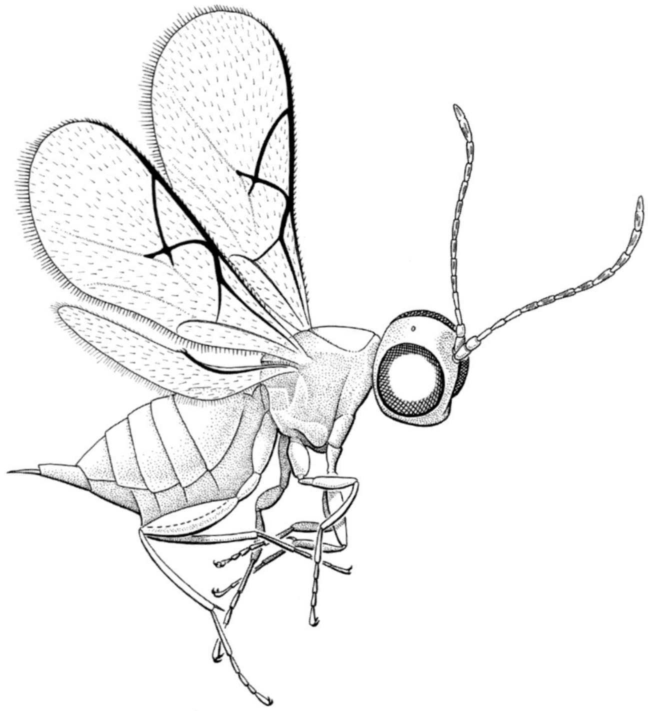

11.17. An early relative of modern gall wasps, in mid-Cretaceous amber from New Jersey, which was probably a parasitoid. Gall-forming cynipids apparently did not evolve until the Tertiary. AMNH; length 1.1 mm.

11.18. A chalcidoid wasp in Costa Rica inspects eggs of its host prior to injecting its own eggs. Photo: P. J. DeVries.

The superfamily **Chalcidoidea** represents one of the spectacular radiations of parasitoid Hymenoptera, second in number of species only to the Ichneumonoidea. Presently there are nearly 20,000 described chalcidoids (e.g., Noyes, 1990), but the bulk of the diversity remains undescribed. Overall, chalcidoids are small to minute, typically less than 5 mm in length (Figure 11.18), although some relative giants do exist in the families Chalcididae and Leucospidae. Species are endo- and ectoparasitoids on a wide range of hosts and also include a diverse array of hyperparasitoids. Parasitic flies attack a wide range of hosts and have even been argued to have more diverse hosts than any other insect parasitoids (e.g., Eggleton and Belshaw, 1992), but actually the greatest breadth of hosts is in the chalcidoids (Gordh, 1979; Grissell and Schauff, 1997). Hosts include an array of insect orders, but also mites, spiders, pseudoscorpions, and even nematodes. Not all chalcidoids are parasitoids, however. The family Agaonidae is secondarily phytophagous, and species of the subfamily Agaoninae, in particular, are important pollinators of figs (*Ficus*: Moraceae) (Figure 11.19). Interestingly, and perhaps not surprisingly, the Agaonidae is not a natural group (e.g., Rasplus *et al.*, 1998), and phytophagy may have arisen several times in Chalcidoidea. Among the pollinating fig wasps (Agaoninae), there has been careful study on the pattern of coevolution between the wasps and the particular figs that they visit, demonstrating at least partial cocladogenesis (e.g., Herre *et al.*, 1996; Machado *et al.*, 1996, 2001; Kerdelhue *et al.*, 1999; Lopez-Vaamonde *et al.*, 2001; Weiblen, 2001, 2002; Weiblen and Bush, 2002). Major overviews of chalcidoid biology are by Clausen (1940), Askew (1971), Bendel-Janssen (1977), Gordh (1979), Gauld and Bolton (1988), and Gibson *et al.* (1997). Chalcidoids are principally known from a diversity of Cretaceous (e.g., Yoshimoto,

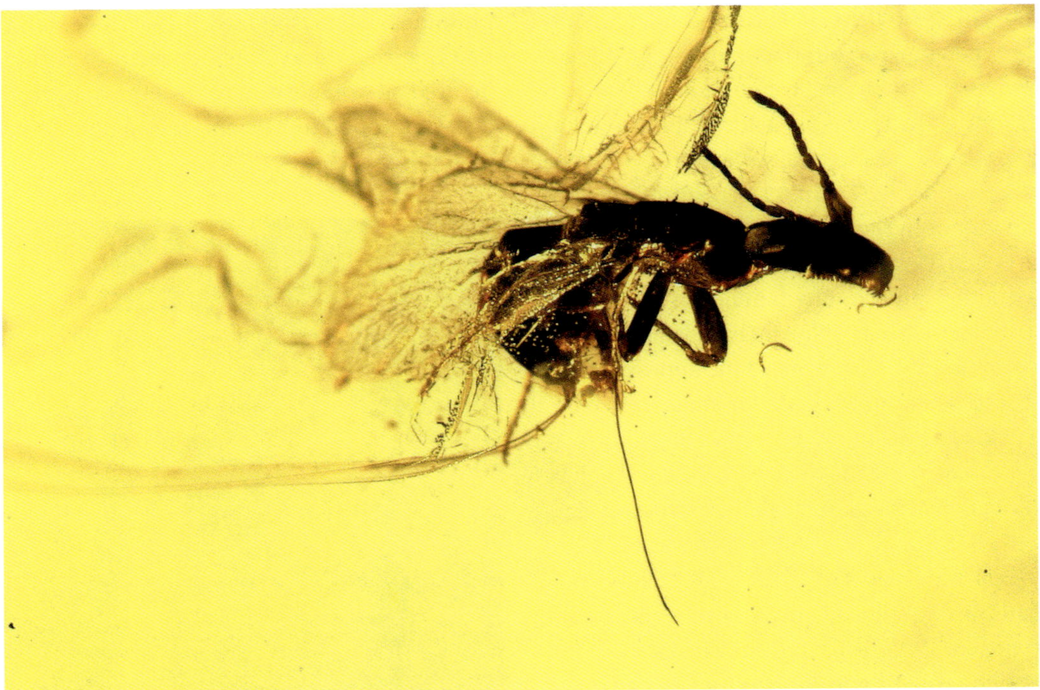

11.19. Fig wasps (Agaonidae) are obligate, specialist pollinators of figs (*Ficus*: Moraceae). Agaonidae in Dominican amber indicate that fig trees were present in the Miocene forests of Hispaniola. This fossil wasp was preserved with pollen grains (probably of *Ficus*) and minute, possibly parasitic, nematodes. AMNH DR-16-576; length 1.4 mm.

11.20. Chalcidoids are rare in Cretaceous ambers but become more diverse and common in Tertiary ambers, such as this wasp of the family Chalcididae in Dominican amber. Morone Collection, M2537; length 3 mm. Photo: R. Larimer.

1975) and Cenozoic ambers (e.g., Brues, 1923a; Doutt, 1973a,b; Grissell, 1980; Skalski, 1988a; Darling, 1996; Weitschat and Wichard, 2002; Engel, 2002c; Simutnik, 2002; (Figures 11.19, 11.20), which is expected given the typically small body size of most species, although some compressions do exist (e.g., Lewis *et al.*, 1990; Peñalver and Martínez-Delclòs, 2000).

Defining features of the Chalcidoidea include the presence and position of the *prepectus*, a triangular sclerite separating the base of the forewing from the pronotum (Gibson, 1986). The prepectus can be reduced in some chalcidoid lineages and thus be difficult to discern. The mesothoracic spiracle is situated between the pronotum and the lateral margin of the mesoscutum, rather than the pronotum and the mesopleuron (Gibson, 1986). Lastly, the presence of long, longitudinal sensilla raised above the antennal surface are distinctive for chalcidoids (Gibson, 1986) (Figure 11.21). Further discussion of chalcidoid monophyly can be found in Gibson (1986) and Gibson *et al.* (1999). Despite the confidence in chalcidoid monophyly, the relationships among families in the Chalcidoidea and even the monophyly of the included families is of considerable debate. Indeed, only a few of the approximately 20 families appear to be natural (e.g., Mymaridae, Leucospidae, Tanaostigmatidae). Recent studies on chalcidoid relationships have illuminated the massive work required before a complete picture of chalcidoid phylogeny is achieved (e.g., Heraty *et al.*, 1997; Rasplus *et al.*, 1998; Gibson *et al.*, 1999; Campbell *et al.*, 2000; Gauthier *et al.*, 2000).

Mymarommatoidea contains a single family and extant genus, *Palaeomymar*, of tiny wasps (typically less than 1 mm in length), of which we know little. Nothing is known of mymarommatid biology, although about nine species are known from around the world and an additional ten species have been found in Cenozoic and Cretaceous ambers (e.g., Huber, 1986; Gibson, 1993; Gibson *et al.*, 1999; Fursov *et al.*, 2002; Engel and Grimaldi, unpubl.) (Figure 11.22). *Palaeomymar* (under the synonym *Mymaromma*) was early on considered allied with the Mymaridae (which is in the Chalcidoidea) owing to the superficial similarity with paddle-like forewings with long marginal fringe of setae. Among Apocrita the mymarommatids are notable for their two-segmented petiole, reticulate forewing membrane, and reduced and forked hind wing (which are haltere-like and consist of a single vein). Although the function is entirely unknown, mymarommatids also have a peculiar bellows-like structure on the back of the head, with the posterior surface separated by a pleated membrane that is capable of expanding and contracting. If these tiny wasps are egg parasitoids, which is likely, the head bellows may be a mechanism for bursting from the host egg. Additional traits of the Mymarommatoidea are discussed by Gibson (1986) and Gibson *et al.* (1999). The extinct **Serphitoidea**, also with a single family, is known only from Cretaceous ambers (Kozlov and Rasnitsyn, 1979; Engel and Grimaldi, unpubl.) (Figures 11.23, 11.24) and is likely a close relative of Mymarommatoidea because both superfamilies have the distinctive two-segmented petiole. Indeed, it has been suggested that they should be combined into one superfamily, the Serphitoidea (Nikolskaya, 1978; Kozlov and Rasnitsyn, 1979). Unlike Mymarommatoidea, Serphitidae lacked the bellowed head and retained hind wings, and their forewings had venation. Serphitoidea also lacked the reticulate forewing surface and marginal fringe of long setae. Serphitids had a characteristic, enlarged pterostigma, a notable apomorphy among chalcidomorph superfamilies (i.e., Serphitoidea, Mymarommatoidea, Chalcidoidea) – assuming that Diapriidae and other derived proctotrupoid families can be used as suitable outgroups for comparison. The Serphitidae were apparently diverse throughout the Cretaceous, with

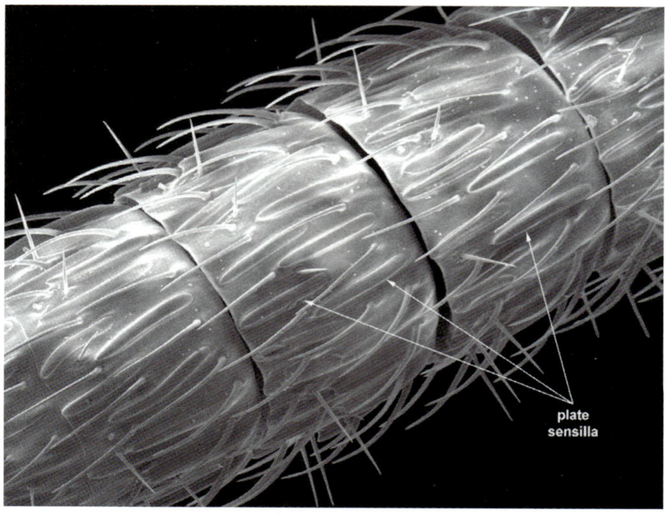

11.21. Raised, longitudinal sensilla of the antennal flagellum is one of the defining features of wasps in the Chalcidoidea. Scanning electron micrograph.

11.22. A fossil Mymarommatidae, in mid-Cretaceous amber from New Jersey. While the family still exists, it was formerly more diverse. These are minute wasps whose entire fossil record is from amber. The biology of the few living Mymarommatidae is unknown, but it is likely that they and their extinct relatives, the Serphitidae, are and were parasitoids of insect eggs. AMNH; length 0.8 mm.

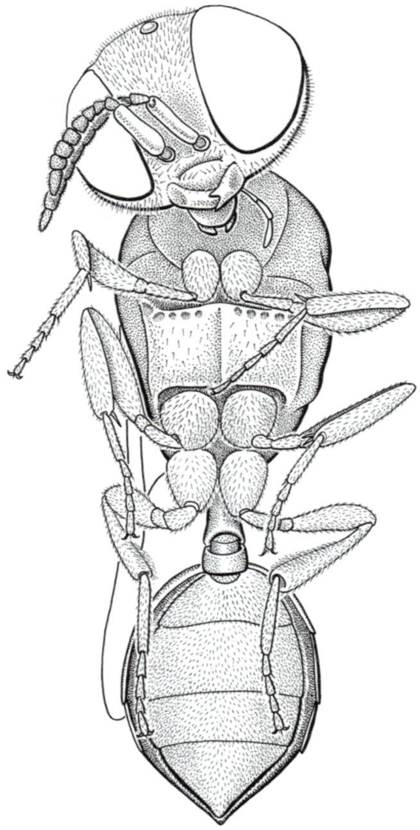

11.23. A wasp of the extinct family Serphitidae, in mid-Cretaceous amber from New Jersey, showing the petiole of the metasoma that is so distinctive to this family and the Mymarommatidae. Serphitids are known only from Cretaceous ambers 125–75 MYA. Some species had impressive mandibles with large, clawlike teeth. AMNH; length 1.7 mm.

several species and genera known from deposits ranging in time from the Early Cretaceous to the end of the Mesozoic as well as in space – from Siberia, New Jersey, Canada, and Myanmar (Brues, 1937; Kozlov and Rasnitsyn, 1979; Engel and Grimaldi, unpubl.).

The superfamily **Platygastroidea** is sometimes confused with Chalcidoidea but these are actually quite different wasps. Traditionally, the two families of Platygastroidea – Platygastridae and Scelionidae – were placed in the Proctotrupoidea. Defining features of the platygastroids are the unique telescoping ovipositor mechanism (Austin, 1983; Austin and Field, 1997) and the occurrence of unique basiconic sensilla on the ventral surface of the first flagellomeres of the antennal clavus, which appear to serve a secretory function (Bin and Vinson, 1986; Masner, 1993b). Scelionids are principally endoparasitoids of insect and spider eggs, which accounts for their minute body size (typically 1–3 mm in length). Hosts range from the eggs of spiders to those of crickets, mantises, webspinners, moths, beetles, and even floating gerrid eggs (e.g., Allen, 1972; Eberhard, 1975; Masner, 1978; Spence, 1986). Most platygastrid larvae begin as egg parasitoids, although in contrast to the scelionids, they do not complete development until the host has nearly reached the adult stage. Some platygastrids are more similar biologically to scelionids, being egg parasitoids of such insects as

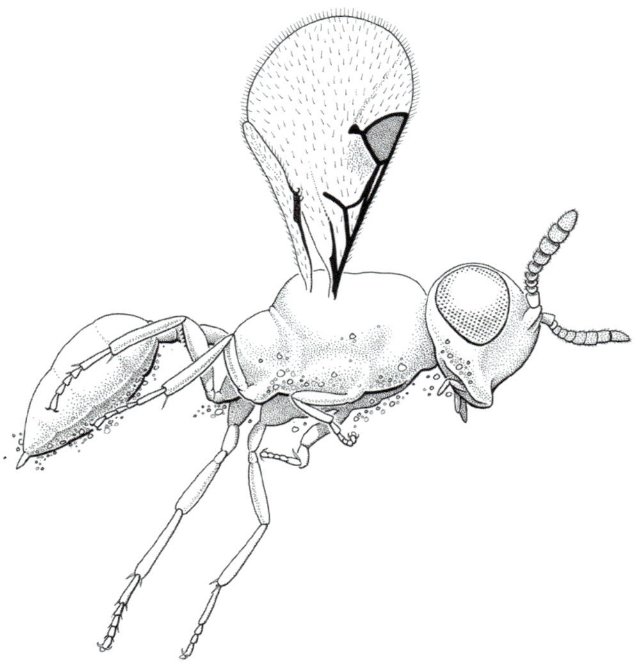

11.24. Another serphitid in mid-Cretaceous amber from New Jersey, showing the wing venation. AMNH; length 1.8 mm.

11.25. Wasps of the superfamily Platygastroidea experienced their greatest diversity during the Cretaceous. The fossil record of these minute wasps is almost exclusively preserved in fossil resins, such as this species in Early Cretaceous amber from Lebanon. AMNH; length 1.8 mm.

11.26. Another platygastroid wasp in Lebanese amber. AMNH; length 1.9 mm.

beetles and sternorrhynchans, or of early nymphal instars of sternorrhynchans (e.g., on Coccoidea), and completing their development within a single life-stage of the host. *Superparasitism* – the occurrence of more than one parasite in a single host – is uncommon in both families but not unheard of despite the minute size of the host (e.g., Clancy, 1944). Teratocytes released by the scelionid egg at hatching destroy other parasitoid eggs and discourage gregarious associations (e.g., Strand, 1986). Presently there are slightly more than 4,000 characterized Recent species, but this is believed to be much less than half of their true diversity, most species being within the Scelionidae (Masner, 1993b). Owing to their minute size, fossils of platygastroids are, not surprisingly, virtually restricted to amber. Indeed, fossil platygastroids comprise one of the most diverse and abundant families of insects in Cretaceous ambers (e.g., Figures 11.25, 11.26). The Cretaceous fossils, though virtually unstudied, show that Scelionidae may have experienced their greatest diversity during the Cretaceous, dwindling during the Cenozoic. The superfamily is also well documented in Baltic amber (e.g., Brues, 1940). Interestingly, the Mexican amber species *Palaeogryon musesebecki* is also known among the modern fauna of southern Mexico (Masner, 1969). This species challenges ideas surrounding the ages of species and stasis within a lineage, but it is not unique for Mexican amber (e.g., Engel, 2004a). Obviously, records like these invite further scrutiny.

The **Ceraphronoidea** consists of two families – Ceraphronidae and Megaspilidae – which are minute parasitoid wasps with two protibial spurs. Both families have a cosmopolitan distribution, with ceraphronids accounting for about 360 species and megaspilids, about 450 species. The superfamily is defined by, in addition to the two protibial spurs, the fusion of the costa and subcosta-radial (Sc + R) veins into a bar running along the anterior wing margin to the pterostigma and the enlarged second metasomal segment (Masner and Dessart, 1967; Masner, 1993c). Biological information on ceraphronoids is scant. Where known, the Ceraphronidae are endoparasitoids of cecidomyiids, thrips, caterpillars, coniopterygids, chrysopids, caddisfly pupae, or puparia of higher flies (Priesner, 1936; Bakke, 1955; Parnell, 1963; Chiu *et al.*, 1981; Oatman, 1985; Gilkesen *et al.*, 1993; Luhman *et al.*, 1999). There is also some evidence that they may be parasitoids of spheciform wasp larvae (e.g., Melo and Campos, 1993). Species can also be hyperparasitoids on microgastrine braconids (e.g., Lever, 1933; Gauld and Bolton, 1988). Megaspilids are ectoparasitoids on a diversity of hosts – Coccidae, Boreidae, Hemerobiidae, Chrysopidae, Coniopterygidae, Cecidomyiidae, Syrphidae, Chloropidae, Chamaemyiidae, Muscidae – or are hyperparasitoids (e.g., Cooper and Dessart, 1975; Walker and Cameron, 1981; Dessart, 1985; Polaszek and Chinwada, 2000). Definitive fossil ceraphronoids are known only from Cretaceous and Cenozoic ambers (e.g., Alekseyev and Rasnitsyn, 1981; Szabó and Oehlke, 1986).

The Cretaceous family Stigmaphronidae (Figure. 11.27) should either be incorporated into the Ceraphronoidea or considered a distinct superfamily sister to the ceraphronoids. Stigmaphronids, like their modern relatives, were apparently cosmopolitan in distribution. Species are known solely from Cretaceous ambers of Siberia, New Jersey, Alaska, Lebanon, and Myanmar (Muesebeck, 1963; Rasnitsyn, 1975, 1991; Engel and Grimaldi, unpubl.). Like modern Ceraphronoidea, C and Sc + R are united, although in some species a faint impression in the "costal bar" is visible, as perhaps a remnant

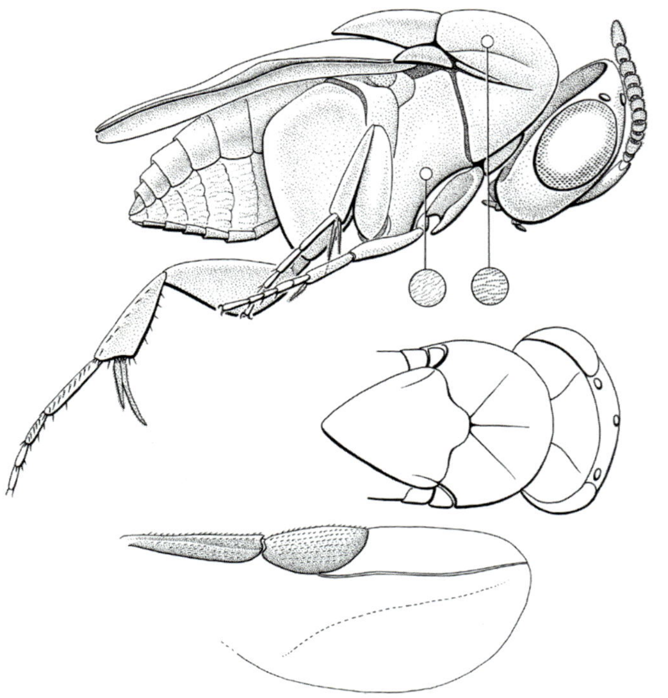

11.27. A minute stigmaphronid wasp in amber from New Jersey; this extinct family is known only in Cretaceous ambers, like Serphitidae. Stigmaphronids were perhaps insect egg parasitoids and were close relatives of the living superfamily Ceraphronoidea. AMNH; length 1.3 mm.

of the fusion. Unlike modern Ceraphronoidea, however, primitive species possessed a single protibial spur and the second metasomal segment was not enlarged. The body was extremely compact, and most species had a greatly expanded scutellum and enlarged hind coxae that were laterally compressed, superficially resembling in these respects species of Elasmini (Chalcidoidea: Eulophidae, *sensu* Gauthier *et al.*, 2000).

Aside from the Proctotrupomorpha and the sundry, principally basal apocritan superfamilies, two groups remain, the Ichneumonoidea and the Aculeata. The **Ichneumonoidea** are among the most favored biological control agents. Wasps in this diverse complex, with an estimated 100,000 species in two families (Braconidae and Ichneumonidae), are remarkable for their array of hosts and modes of parasitoidism. Basal lineages of both families tend to be ectoparasitoids of wood-boring beetles, caterpillars, and wood wasps (as noted, an undoubtedly primitive habit for Apocrita) (Figures. 11.12, 11.28, 11.29, 11.30), but species are also endoparasitoids on a wide range of arthropod hosts, having switched from ecto- to endoparasitoidism multiple times (e.g., Shaw, 1983; Gauld, 1988; Whitfield, 1992b). The success of ichneumonoids may partly be attributable to their complex venoms, which manipulate the host's physiology and development for sustaining the parasitoid larva (e.g., Shaw, 1981; Piek, 1986; Coudron, 1990). In addition, the developing larva further affects the host's physiology and behavior. Perhaps the most remarkable of these is the pimpline ichneumonid *Hymenoepimecis argyraphaga*, whose larva is an ectoparasitoid on the tetragnathid spider *Plesiometa argyra* (Eberhard, 2000a,b). The spider continues with a relatively normal life even after having been temporarily paralyzed by the adult ichneumonid during oviposition. The parasitoid larva feeds regularly from the host but begins to exert a physiological control over the spider toward the end of the larva's development, resulting in the spider constructing a unique form of web, one designed for the parasitoid to build its cocoon after the spider is killed.

Endoparasitoids have a slightly more complex life because these wasps must counter the "immune" systems of the hosts in which they reside. A solution to this obstacle has most successfully been achieved by members of the Ichneumonoidea, and apparently independently and convergently in the Ichneumonidae and Braconidae, respectively. Certain endoparasitoid wasps in each of these lineages have symbiotic viruses, which may be the only known beneficial viral-eukaryote interaction (Whitfield, 1990), and which are critical in overcoming the host's defenses. The most completely understood of these systems are the polydnaviruses (pronounced "poly-D-N-A-viruses"). Polydnaviruses are entirely integrated into the wasp's chromosomal DNA and are therefore transmitted vertically (from parent to offspring). As would be suspected from such means of transmission, there is relatively tight cospeciation between the wasp lineages and the polydnaviruses they host (Whitfield, 2000). In the wasp's ovaries the polydnavirus DNA is packaged into nucleocapsids consisting of several different, circular, double-stranded

11.28. A female of *Megarhyssa* (Ichneumonidae) ovipositing through bark into a wood-boring larva. Photo: S. Marshall.

11.29. An ichneumonid larva feeds from the opisthosoma of the giant orbweaver spider, *Nephila clavipes*, in Panama. Some larval ichneumonids even alter the behavior of their host to the benefit of the parasitoid. Photo: D. Grimaldi.

11.30. Braconid larvae living off their caterpillar host. Photo: S. Marshall.

DNA elements (hence their name, *poly* and *DNA*) and are exported (injected) into the wasp's host insect – typically caterpillars. The polydnaviruses do not replicate within the host (Stoltz, 1993), but the virus does express several genes that compromise the host's immune response (Edson *et al.*, 1981; Lovallo *et al.*, 2002). Some of these genes are not of viral origin; indeed, it appears that the polydnaviruses have incorporated some of the wasp's own venom genes (originally part of the venom), that have an enhanced expression (Webb and Summers, 1990; Webb and Cui, 1998). Thus, the target of the virus' attack is a species other than the one in which it resides, that is, not the virus' host (the wasp), but the wasp's host. These viruses may represent complex symbioses between viruses and wasps for which the viral source population is as of yet unknown; that is, it is not clear from what group of viruses these originated. Alternatively, and less likely, they may not be viruses at all, instead representing an intricate genetic development by the wasps themselves for manipulating their hosts (e.g., Whitfield, 1990). Although both families of Ichneumonoidea have incorporated polydnaviruses, the two classes of viruses are quite distinct and are referred to as ichnoviruses (in Ichneumonidae) and bracoviruses (in Braconidae) for the respective types. Regardless of their type, however, the effect is the same. The cascade of physiological effects created by the expression of the polydnavirus genes results in the survival of the wasp larva, to the extreme detriment of the caterpillar host. Within at least the Braconidae, this association is certainly representative of a single evolutionary event (Whitfield, 1997), while the number of origins in Ichneumonidae has not been critically evaluated. Dating the origin of this association is rife with difficulty.

Even though ichneumonoids are well documented in the fossil record, at least through the Early Cretaceous (e.g., Rasnitsyn, 1975, 1983c; Rasnitsyn and Sharkey, 1988), it is, of course, impossible to determine the presence of polydnaviruses in such compression remains. Given that remains of protists and bacteria can be found within tissues of insects encapsulated in amber (Wier *et al.*, 2002), it may be possible to visualize viruses. However, this would surely involve destruction of unique or very rare wasp fossils and would not be advisable. Calibrating a molecular phylogeny of braconids with fossil records, Whitfield (2002) was able to arrive at an age of approximately 74 MYO for the microgastroid lineage housing the polydnaviruses. Given the apparently tight vertical transmission of the polydnaviruses in their hosts and their fixation across species in this complex, it seems that this association appeared sometime in the Campanian, persisting to the present day.

Ichneumonoids are generally considered to be the living sister group to the Aculeata, principally owing to the presence of valvilli in the ovipositor (Oeser, 1961) and a narrow propodeal foramen (the point where the metasoma articulates to the mesosoma) with dorsolateral apodemes called propodeal "teeth" (e.g., Rasnitsyn, 1988a; Sharkey and Wahl, 1992). Ichneumonoidea itself is defined by the following features: a unique articulation of the first and second metasomal segments by dorsolateral condyles on the hind margin of the first tergum and the anterior margin of the second (Mason, 1987); the division of the first metasomal sternum into a strongly sclerotized anterior sclerite and a comparatively weakly sclerotized posterior region (Mason, 1981, 1987); the fusion of the prepectus (a pleural sclerite between the pronotum and mesepisternum) to the posterolateral margin of the pronotum with the spiracle immediately dorsal to the prepectus (Gibson, 1985); the absence of the 2rs-m crossvein in the forewing (Rasnitsyn, 1980); and the positioning of the costa and radius in the forewing adjacent to each other such that the costal cell is effectively absent (Sharkey and Wahl, 1992; Quicke *et al.*, 1999).

The first ichneumonoids are known as compression fossils from the Early Cretaceous of Asia and Australia. The extinct family Praeichneumonidae from the Early Cretaceous of Mongolia (Rasnitsyn, 1983c; Rasnitsyn and Sharkey, 1988) is difficult to place among the superfamily. The costal cell is narrowed in *Praeichneumon* but is distinctly present, and the antenna with more than 15 antennal segments is presumably primitive. While certainly suggestive of an ichneumonoid, the fossil preserves only primitive features of the superfamily and cannot, therefore, be conclusively placed within the Ichneumonoidea (Sharkey and Wahl, 1992; Quicke *et al.*, 1999). It is certainly possible that *Praeichneumon* represents a stem group to Ichneumonoidea, but even this assignment is conjectural. The first definitive Ichneumonoidea are fossils of the extinct family Eoichneumonidae from the Early Cretaceous of Asia as well as from Australia (Jell and Duncan, 1986; Rasnitsyn and Sharkey, 1988), along with an enigmatic, primitive genus *Tanychora* from Early Cretaceous of Asia (Townes, 1973). While certainly exhibiting some of the defining traits of the superfamily, *Tanychora* does not share any derived features with recognizable clades in the Ichneumonoidea (Sharkey and Wahl, 1992). Eoichneumonids, on the other hand, may be related to the Braconidae (Sharkey and Wahl, 1992) with the former actually representing a natural group rather than a grade (Sharkey and Wahl, 1992; Quicke *et al.*, 1999). The earliest records of the modern families Braconidae and Ichneumonidae are also from the Lower Cretaceous of Mongolia (Rasnitsyn, 1983c). In the younger Cretaceous amber deposits, braconids are best represented (Figure 11.31), although some ichneumonids are also known (e.g., Brues, 1937; Townes, 1973; Basibuyuk *et al.*, 1999). Certainly ichneumonoids are abundant as fossils throughout the Tertiary as both amber inclusions as well as compression fossils (e.g., Brues, 1910, 1923a, 1933a, 1939; van Achterberg, 1982, 2001; Tobias, 1987; Lewis and Luhman, 1988; Kasparyan, 1994, 2001; Weitschat and Wichard, 2000).

11.31. A braconid wasp in mid-Cretaceous amber from New Jersey. AMNH; length 2.2 mm.

ACULEATA

The **Aculeata** (ants, bees, and other stinging wasps) certainly contain the most famous Hymenoptera, which is what most people think about when we imagine wasps, particularly stinging wasps. Aculeates are woven into the fabric of human culture, and bees alone occupy a symbolic role in themes of creation, fertility, and divine power. Aculeata is easily defined: Rather than the ovipositor serving for the deposition of eggs, it is a *sting*, an offensive and defensive weapon used for injecting venom. Eggs are instead deposited through an opening at the base of the sting, rather than traveling through the channel formed by the ovipositor components. The sting itself is a sharp, stiff pair of gonapophyses, the base of which has a reservoir of venom called the bulb (Figure 11.32).

Without doubt, the sting evolved to subdue prey that is sometimes much larger than the wasp itself, since basal lineages of aculeates kill or paralyze prey to feed to their larvae. In the social aculeates as well as bees and other apoid flower foragers, the sting is also or exclusively used for defense. The effectiveness of the aculeate sting is perhaps best reflected by the thousands of species of insects that mimic the wasps with which they coexist: These make memorable models. In one of the more distinctive entomological papers, aculeate stings were rated on a "pain scale" from 0 (no pain) and 1 (slight pain) to 4 (traumatic pain) (Starr, 1985). Honey bees and bumble bees (*Apis*, *Bombus*) ranked 2 (simply "painful"); and certain paper wasps (*Polistes*: Vespidae), velvet "ants" (Mutillidae: *Dasymutilla*), and Australian bulldog ants (*Myrmecia*) ranked 3 ("sharply and seriously painful"). The aculeates packing the greatest punch are the large tarantula hawk wasps (Pompilidae) of the genus *Pepsis*,

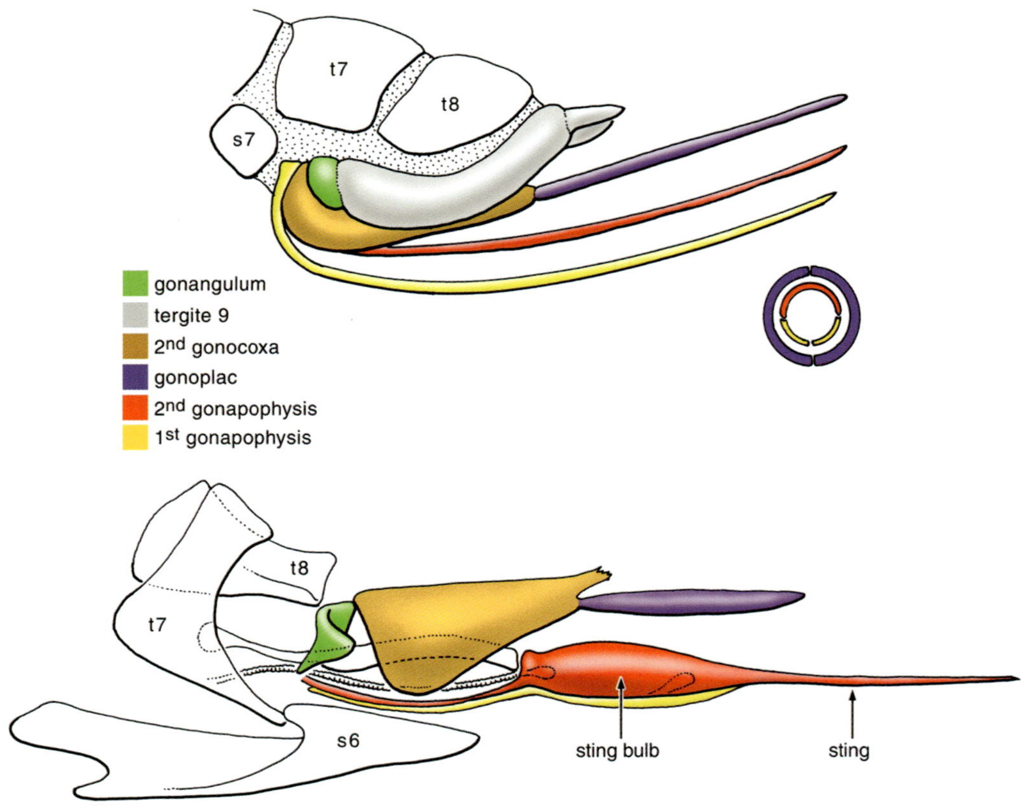

11.32. Above: a typical apocritan ovipositor; below: a generalized sting (after Rightmyer, 2004). A defining feature of the Aculeata is the sting, which is an ovipositor modified to inject venom but where eggs pass through an opening at the base of the sting.

and the giant neotropical "bullet" or "bala" ant, *Paraponera clavata*, the latter so named for the feel of its sting. A sting's effect is based on its size and the depth it penetrates, how much venom is delivered, and the nature of the venom.

The earliest evidence of aculeates is a series of compression fossils from the latest Jurassic of Central Asia, all placed in the family Bethylonymidae. Bethylonymids are putatively sister to all other Aculeata on the basis of their generally bethylid-like appearance, with 13 antennomeres and very short sting (e.g., Rasnitsyn, 1975, 1988a, 2002), although none of these can be taken as conclusive for placement with aculeates. Bethylonymids are certainly suggestive of early Aculeata and do indeed occur in a period one would predict for stem-group aculeates. Hopefully additional material will be discovered in the Early Cretaceous ambers where their characters can be studied with higher fidelity.

The remainder of the Aculeata, entirely restricted to the Cretaceous or younger deposits, is segregated into three superfamilies: Chrysidoidea, Vespoidea, and Apoidea (Figure 11.33). Primitive aculeates such as the Chrysidoidea and some members of Vespoidea and spheciform Apoidea are ectoparasitoids, endoparasitoidism being rare, although some groups (e.g., dryinids) seem to cross the boundaries of both lifestyles. Basal among the true aculeates are the families of **Chrysidoidea**, accounting for the smallest portion of species. Chrysidoids are excluded from Vespoidea + Apoidea (**Euaculeata**) by the primitive retention of identical antennomere counts in both sexes (vespoids and apoids have sexually dimorphic antennae, with 13 and 12 antennomeres in male and females, respectively) and the female seventh metasomal tergum exposed and evenly sclerotized (vespoids and apoids have the seventh metasomal tergum of females hidden and largely desclerotized). The superfamily is defined by the inflated femora of females, the anteriorly narrowed first metasomal tergum that is fused to the sternum, the reduction of Cu_2 in the forewing, and by some internal traits (Brothers and Carpenter, 1993). Relationships among chrysidoid families were investigated by Carpenter (1986, 1999), Brothers (1999), and Brothers and Carpenter (1993).

The family Plumariidae constitutes the basalmost branch of the chrysidoid tree (Figure 11.33) and is therefore of considerable interest. However, nothing is known of their biology. There are only five genera and fewer than 20 species of plumariids, all of which occur in South America (Ecuador to Argentina and Chile) and southern Africa (Evans, 1966c; Nagy, 1973; Brothers, 1974; Day, 1977; Roig-Alsina, 1994). There are no definitive fossil records of the Plumariidae, and a putative "plumariid" from the Late Cretaceous New Jersey amber may be sister to all other Chrysidoidea (Brothers, unpubl.).

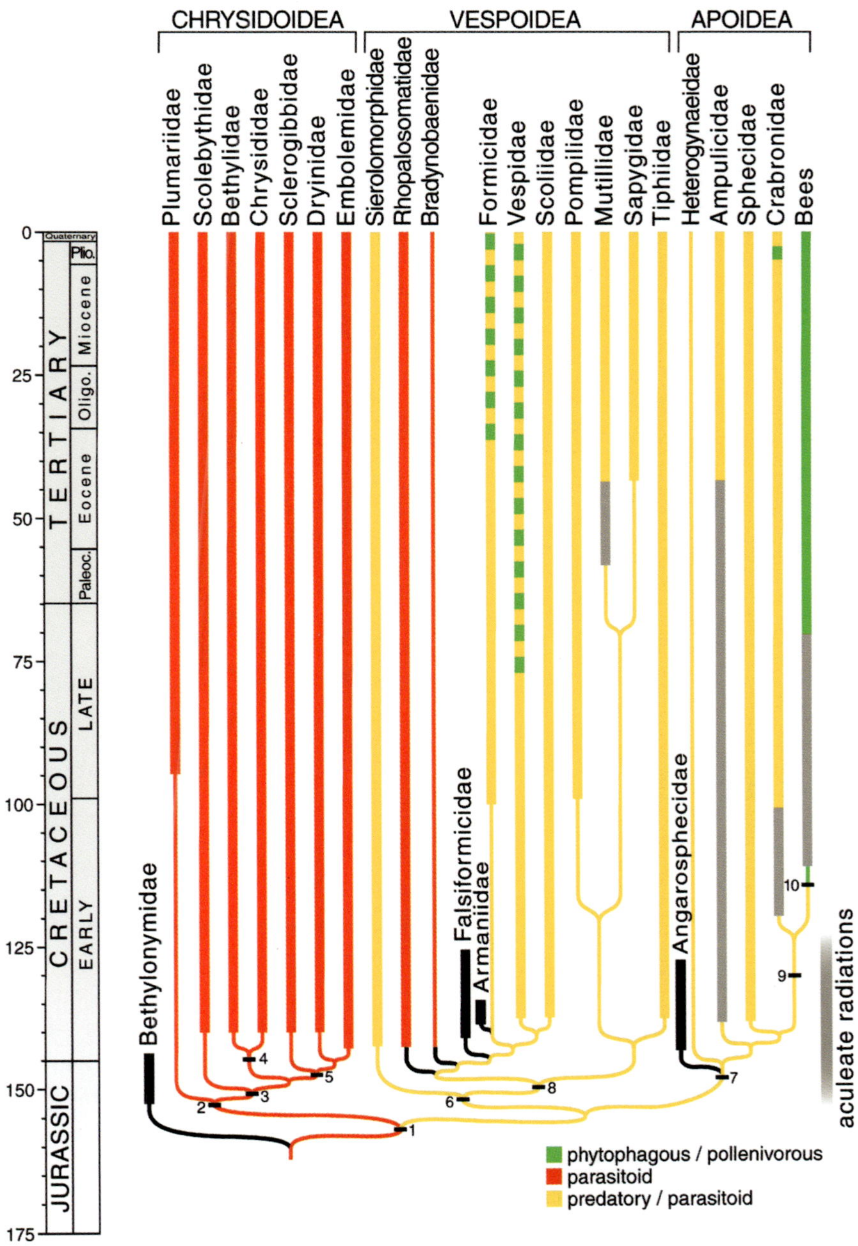

11.33. Phylogeny of the Aculeata with significant characters indicated (see Table 11.2). Thick lines indicate the known extent of fossils. Relationships based on Brothers and Carpenter (1993), with fossils added.

TABLE 11.2. Significant Characters for Aculeata Phylogeny[a]

1. Sting definitively present
2. Femora inflated; first metasomal tergum narrowed and fused with sternum; Cu2 reduced
3. Forewing with seven (frequently fewer) closed cells
4. Metapostnotum shortened and hidden
5. Hind wing with anal veins absent; ovipositor furcula absent
6. Prepectus reduced
7. Pronotal lobe present; metapostnotum forming "propodeal triangle"
8. Metapostnotum partially invaginated; hind wing jugal lobe reduced
9. Sixth metasomal tergum of female with pygidial plate (secondarily reduced in various bee lineages); mesocoxal carina present
10. Branched setae; larva feeding on pollen mixed with floral oils and nectar

[a] Numbers correspond to those on phylogeny, Figure 11.33.

Another poorly understood family of primitive chrysidoids are the five modern species of Scolebythidae (Evans, 1963b). The living species occur in South and Central America, South Africa, Madagascar, Australia, and Fiji (Evans, 1963b; Nagy, 1975; Beaver, 2002; Fernandez *et al.*, 2002), although the species from Fiji is likely synonymous with *Ycaploca evansi* from Africa and is perhaps a recent introduction like the Australian records (Naumann, 1990). Only scant biological information is available for scolebythids, but species conform to the primitive apocritan syndrome of being ectoparasitoids of wood-boring beetles of the families Cerambycidae and Anobiidae (Brothers, 1981; Melo, 2000). Based on the living species, authors have considered the scolebythids to be a classic example of an austral disjunct distribution. However, there are more fossil species of scolebythids than there are of living species, and these are found in resins dating from the Early Cretaceous to Early Miocene across the Northern Hemisphere (e.g., Prentice *et al.*, 1996; Brothers and Janzen, 1999; Lacau *et al.*, 2000; Engel and Grimaldi, unpubl.) (Figures 11.34, 11.35). Interestingly, *Dominibythus*, known from South America and in Dominican amber (Prentice *et al.*, 1996; Azevedo, 1999), has proved to be a synonym of the Baltic amber genus *Pristapenesia* (Brothers and Janzen, 1999).

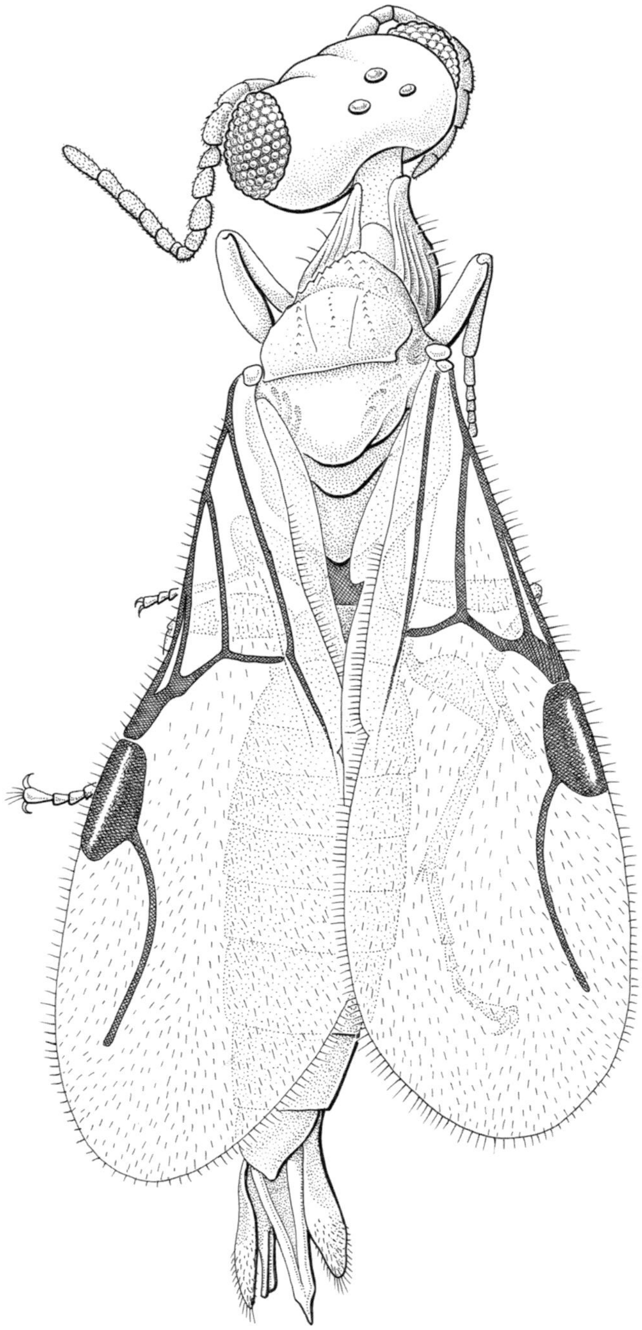

11.34. Restricted to a few species in the Southern Hemisphere today, wasps of the basal aculeate family Scolebythidae were diverse in the past and occurred throughout the world. This scolebythid is in Early Cretaceous amber from Lebanon. AMNH.

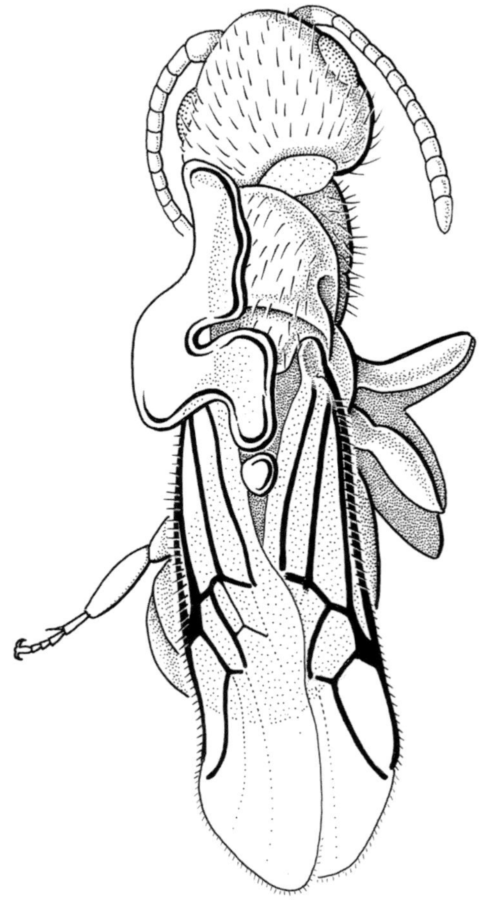

11.35. A northern scolebythid wasp in mid-Cretaceous amber from New Jersey. AMNH.

The closely allied families Bethylidae and Chrysididae represent the largest diversity of chrysidoid wasps, with about 2,200 and 3,000 species, respectively (Figure 11.41). Both families are of cosmopolitan distribution but are particularly abundant in the tropics. Bethylids are parasitoids of lepidopteran and coleopteran larvae (Clausen, 1940; Nagy, 1969; Doutt, 1973c), while chrysidids are either parasitoids on a wide range of hosts – Lepidoptera and Tenthredinoidea (Diprionidae, Tenthredinidae) larvae and Phasmatodea eggs – or are cleptoparasites of vespids and bees. The most comprehensive treatments of Bethylidae are those of Kieffer (1914) and Evans (1964, 1978), while Argaman's (1988a) classification is entirely unstable. The chrysidids have been comprehensively covered by Kimsey and Bohart (1990), and their biology has been reviewed by these authors as well as treated in works by Piel (1933), Gauss (1964), Dahlsten (1961, 1967), Carrillo and Caltagirone (1970), Krombein (1983), and Yamada (1987). Both families are present as fossils in Cretaceous and Tertiary resins (Cockerell, 1917; Brues, 1923a, 1933a, 1939; Evans, 1969, 1973a; Krombein, 1986; Engel and Grimaldi, unpubl.) as well as compressions from North America (Cockerell, 1907c; Rohwer, 1909).

Sclerogibbidae are enigmatic, obligate ectoparasitoids of webspinners (Embiodea) and presently consist of only 14 species found throughout the world (Callan, 1939; Yokoyama and Tsuneyoshi, 1958; Argaman, 1988b, 1993; Ross, 2000b, Penteado-Dias and van Achterberg, 2002), though many other species likely exist. Currently there are only two records of sclerogibbids as fossils, a single individual in Early Cretaceous Lebanese amber (Engel and Grimaldi, unpubl.) and a record of an undescribed specimen from Dominican amber (Rasnitsyn, 2002). Interestingly, the oldest Embiodea are from the slightly younger Burmese amber (e.g., Grimaldi *et al.*, 2002; Engel and Grimaldi, unpubl.), although the Lebanese amber sclerogibbid implies that embiodeans were also present at that time, in lieu of actual fossils of them from that time.

The families Dryinidae and Embolemidae are close relatives within the superfamily. Wasps of the family Dryinidae are cosmopolitan in distribution and comprise approximately 1,400 species (Figure 11.36). Dryinid larvae victimize nymphal and adult Auchenorrhyncha. Adult females are frequently apterous and superficially antlike, most (except Aphelopinae) possessing a distinctive chelate protarsus, with combs, used for grasping the host. Although there are some interesting variations in dryinid biology, females generally sting the host and oviposit between overlapping segmental sclerites. The developing larva eventually forms a *thylacium* (formed of exuviae from previous instars), which extends from the host's abdomen, and dryinids effectively have stages that are endo- and ectoparasitoids. Fossilized leafhoppers with attached thylacia have even been preserved in amber (Figure 11.37). The last instar emerges from the expired host and pupates on the foliage or in the soil before eclosion (Jervis, 1980a,b; Giri and Freytag, 1986). The world fauna has been treated by Olmi (1984), with several supplements to that classification since that time (e.g., Olmi, 1995, 1999, 2002). Aside from bethylids and chrysidids, the Dryinidae have the next most extensive fossil record for the superfamily, one that is entirely confined to fossiliferous resins. The earliest dryinids are in Lebanese amber and interestingly of the most basal subfamily, Aphelopinae, and putatively in the living genus *Aphelopus* (Olmi, 1998). The next oldest fossils are those in mid-Cretaceous Burmese amber (Engel, 2003b, unpubl.) and the Late Cretaceous resins of Siberia and Canada (Ponomarenko, 1975, 1981). A diversity of dryinids representing most of the subfamilies is known from Tertiary ambers (Olmi and Bechly, 2001).

Biological data on Embolemidae is scarce, but their development and hosts appear similar to those of dryinids (Bridwell, 1958; Wharton, 1989). Some species have been found in ant nests (Donisthorpe, 1927; Hirashima and Yamagishi, 1975) or in small mammal burrows (Heim de Balsac, 1935), and the ant records are perhaps explained by the occurrence of some auchenorrhynchans as symbionts with ants. The species of the family have been revised by Olmi (1996) and van Achterberg and van Kats (2000). Interestingly, the earliest fossil records for the superfamily are a series of putative Embolemidae from the earliest Cretaceous of Baissa in Central Asia (Rasnitsyn, 1996). Other fossil embolemids are from the slightly younger Cretaceous deposits of Mongolia (Rasnitsyn, 1996). Otherwise fossils of embolemids are only known in resins from Myanmar (Rasnitsyn and Ross, 2000), the Baltic, and the Dominican Republic (Olmi, 1995; Rasnitsyn, 1996).

11.36. A female wasp of the family Dryinidae in Miocene Dominican amber. Females of this family use their distinctive, long forelegs and chelate foretarsi to grasp their fulgoroid hosts while they inject an egg. Morone Collection, M1404; body length 4.8 mm.

11.37. A dryinid larva in its sac-like thylacium in Miocene Dominican amber, still attached to its fulgoroid host. AMNH DR-16–305; length of nymph 4.2 mm.

The **Vespoidea** consists of ten modern and two extinct families of distinctive wasps, which, although a natural group, are defined principally by traits that are either subsequently modified within the superfamily or appear secondarily in other families (Figure 11.41). The superfamily has been studied most extensively by Brothers and Carpenter (1993) and Brothers (1999), the studies from which the relationships presented herein are drawn. These studies define Vespoidea by the combination of a reduced prepectus (also present in Chrysididae) and sparse setae on the hypopharynx (a feature reversed in Rhopalosomatidae and Pompilidae; Brothers and Carpenter, 1993).

11.38. The earliest tiphiiform wasp, *Architiphia rasnitsyni* (Tiphiidae) from the Early Cretaceous of Brazil, ca. 120 MYO. AMNH 43265; length 16 mm.

The family Sierolomorphidae is the most basal one in Vespoidea. It consists of a single genus from central Asia, North America (Panama through Canada), and Hawaii (Evans, 1961; Nagy, 1971; Miller, 1986). Unfortunately, nothing is known of *Sierolomorpha* biology. Various Cretaceous amber wasps that have numerous primitive traits superficially resemble sierolomorphids, and so these have been placed in this family (Rasnitsyn, 2000), although they may actually represent stem-group vespoids.

The family Tiphiidae is a relatively diverse group of cosmopolitan parasitoid, aculeate wasps ranging in size from 3 to 25 mm (Figure 11.41). The approximately 2,000 known species are segregated into seven subfamilies (Kimsey, 1991), of which the biology for two – Anthoboscinae and Brachycistidinae – is entirely unknown but are likely ectoparasitoids of larval scarabs. The other subfamilies are also ectoparasitoids of grounddwelling larval beetles (Scarabaeidae, Carabidae: Cicindelinae, and perhaps Curculionidae) except the southern Australian Diamminae, which victimizes mole crickets (Clausen, 1940; Pate, 1947a; Lloyd, 1952; Burdick and Wasbauer, 1959; Krombein, 1968; Wilson and Farish, 1973; Rivers *et al.*, 1979; Knisley *et al.*, 1989). Interestingly, male tiphiids are co-opted as pollinators by many orchid species (Stoutamire, 1983; Peakall, 1990; Handel and Peakall, 1993; Alcock, 2000). The orchids mimic the shape of female wasps as well as their sex pheromones, so the male wasps attempt to copulate with the flowers. While the male wasp is busy the orchid places a pollinium on him, with which he unwittingly departs. There has apparently been enough fidelity between the orchids and the wasps that some cospeciation between the wasps and flowers has taken place (Mant *et al.*, 2002). Tiphiid fossils are uncommon but species have been recovered from the Early Cretaceous of Brazil (Darling and Sharkey, 1990) (Figure 11.38) and the Oligocene-Miocene of Asia (Rasnitsyn, 1986; Zhang *et al.*, 1994).

Wasps of the family Pompilidae are famous as the spider wasps and tarantula "hawks" (Figure 11.39). The family is relatively diverse, with about 5,000 species distributed throughout the world, particularly in tropical or xeric regions. While the size of spider wasps ranges dramatically, many can be quite large, around 50 mm in total length and striking, at times with deep metallic blue bodies and orange wings (obviously aposematic with their powerful stings) (Figure 11.41). Species are ectoparasitoids of spiders (hence their name), and adult females hunt down and paralyze spiders, including large tarantulas. While some spider wasps seek out exposed spiders in the environment, they are not shy and will readily enter a spider's burrow to meet the host on its own turf (e.g, Richards and Hamm, 1939; Evans and Yoshimoto, 1962). Head-to-head encounters in the spider's burrow are common, with the wasp rushing in and stinging the spider in the central ganglion. After a female has subdued her victim, she provisions a simple subterranean burrow with a spider and a single egg, which she places on the host's opisthosoma.

11.39. Spider wasps (Pompilidae) hunt various spiders, including tarantulas, and their stings are renowned for being among the most painful. Photo: S. Marshall.

Sometimes this is done within the spider's own burrow (such as those that attack spiders of the family Ctenizidae), or within a separate burrow prepared by the wasp herself. While most prefer subterranean burrows, others such as *Auplopus*, construct mud nests to house their victim and young. Pompilids attack a wide range of spiders, ranging from giant tarantulas to wolf spiders to the smaller ground spiders and even web spinners (Shimizu, 1994; Wasbauer, 1995; O'Neill, 2001), sometimes exhibiting a diversity of hosts within a single species (e.g., Evans and Yoshimoto, 1962). Most remarkably, some species of the tribe Ageniellini have communal behavior, co-operating in mud nest construction and defense, although still provisioning each brood chamber with its own spider and egg (e.g., Wcislo *et al.*, 1988; Evans and Shimizu, 1996). Relationships among spider wasps have been most thoroughly explored by Shimizu (1994). Pompilids are documented in Baltic amber (e.g., Meuner, 1917; Larsson, 1978; Weitschat and Wichard, 2002) and as compressions from the Eocene-Oligocene (e.g., Heer, 1849; Cockerell, 1906, 1908b, 1941; Rohwer, 1909; Statz, 1936b, 1938; Théobald, 1937; Arillo, 2001). The oldest, definitive spider wasp is an undescribed species in mid-Cretaceous amber from Myanmar (Grimaldi *et al*, 2002; Engel and Grimaldi, unpubl.) (Figure 11.40). Given the abundance of spiders in fossilized resins, it is surprising that fossilized pompilids are not encountered more frequently. Despite its name, the Cretaceous genus *Pompilopterous*, originally described as a pompilid (Rasnitsyn, 1975), is actually a spheciform wasp (Rasnitsyn *et al.*, 1998).

The family Sapygidae is a group of robust wasps that are widely distributed everywhere except the Australian region. Species are solitary and are either ectoparasitoids of larval eumenine wasps (subfamily Fedtschenkiinae) or cleptoparasites of solitary bees of the Megachilinae and Xylocopini (subfamily Sapyginae) (Pate, 1947b). The biology of *Sapyga* has been investigated by Torchio (1972), who observed females depositing eggs into completed brood cells of *Megachile*. The *Sapyga* larvae compete with each other until a single cleptoparasite remains, which then proceeds to dispatch the bee egg before consuming the pollen and nectar provisioned for the immature bee. Although these wasps are poorly understood biologically, they are of considerable importance since at least *Sapyga* can reach levels detrimental to *Megachile rotundata*, which is used for alfalfa pollination (Torchio, 1979). The only reference to a fossil sapygid is a specimen purportedly of *Sapyga* in Baltic amber (Spahr, 1987).

Wasps of the family Mutillidae are frequently called velvet ants owing to the superficial resemblance of the fuzzy, wingless females to some true ants (Formicidae) (Figure 11.41). There are around 8,000 species of mutillids occurring predominantly in xeric habitats around the world, but also in tropical and temperate regions. Mutillids are ectoparasitoids of the terrestrial immatures of various other insects, particularly bees and solitary wasps but also flies, limacodid moths, beetles, and cockroaches. Males and females are difficult to associate owing to frequent dimorphism in color patterns, pubescence, and aptery in females. This has resulted in a confusing classification of mutillids, with males and females of a single species at times being confused and placed in sep-

11.40. The earliest known fossil spider wasp (Pompilidae) in mid-Cretaceous amber from Myanmar. AMNH Bu051; length 5 mm.

11.41. Representative Recent aculeate wasps. Wingless specimens are females. Not to the same scale.

arate genera. Females are infamous for their remarkably painful sting, which has led to their being known as "cow killers" in some regions. The biology of the family has been reviewed by Mickel (1928), Clausen (1940), Ferguson (1962), Brothers (1972, 1978, 1989, 1995a), and O'Neill (2001). Relationships within the Mutillidae have been comprehensively treated by Brothers (1975, 1999) and Lelej and Nemkov (1997). Mutillids are rare as fossils, the oldest being those in Baltic amber (Bischoff, 1916; Lelej, 1996), although specimens are also known from Dominican amber (Manley and Poinar, 1991, 1999; Brothers, 2003).

The Rhopalosomatidae is another small family (ca. 40 species) of poorly understood wasps, occurring throughout the world (Townes, 1977). The little available information on the family indicates that they are ectoparasitoids of crickets (Gurney, 1953). Apparently the adult female attaches an egg behind the hind coxa of the cricket, and the larva feeds on its abdomen. Eventually the larva departs from the host to construct a burrow in the soil within which it spins a cocoon (Gurney, 1953). Only two rhopalosomatid fossils have been recovered – *Mesorhopalosoma* from the Early Cretaceous of Brazil (this deposit contains abundant cricket fossils; Darling and Sharkey, 1990) and *Propalosoma* from the Eocene of Washington (Dlussky and Rasnitsyn, 1999). Nel (1991a) suggested that

Lithoserix, originally described as a siricid wood wasp, may be a rhopalosomatid, although this identification seems doubtful.

Bradynobaenidae are, like Sapygidae, widespread but absent from the Australian region, with about 150 mostly tropical species. Bradynobaenids have rather extreme sexual dimorphism, and the association of sexes is difficult without capturing individuals during mating. These wasps are presumably ectoparasitoids of other arthropods, little of their biology has been studied aside from documenting that some are nocturnal. Brothers (1995b) reported that larvae of *Typhoctes* were retrieved from an immature arachnid of the order Solifugida. The most extensive works on the family have been by Mickel (1967), Brothers (1975, 1995b), Genise (1986), and Pagliano (2002). No fossils are recorded for the family.

The Falsiformicidae is one of the few extinct families in aculeates and is known only from the Early Cretaceous. The family is perhaps a relative of the group consisting of the vespids, scoliids, armaniids, and ants (Formicidae) (Rasnitsyn, 1975, 2002). Species are known only from Cretaceous ambers and are poorly understood.

About 300 species from around the world comprise the Scoliidae. Like other vespoid families, the biology of these surprisingly conspicuous wasps is little understood. Where known, scoliids are ectoparasitoids of larval ground-burrowing beetles (Scarabaeidae) (e.g., Clausen, 1940; King and Saunders, 1984; Piek, 1986; Lai, 1988). The sole exception is the recently discovered parasitoidism of the spheciform wasp, *Trypoxylon politum*, by some species of *Scolia* (Molumby, 1995). Adult females are well equipped with spines to dig in the soil and seek out larval beetles, which they paralyze with their sting before depositing an egg. The immature wasps feed upon the host, eventually devouring the beetle except for its skin. The wasp spins a cocoon from which it will exit through a characteristic cap at the anterior end (Clausen, 1940). The most primitive extant scoliids are those of the genus *Proscolia* (Proscoliinae), from the eastern Mediterranean and Asia Minor (Day *et al.*, 1981). Fossils, attributed to a basal subfamily Archaeoscoliinae, have been recovered from the Early Cretaceous of Europe, Brazil, and Asia (Rasnitsyn, 1993; Rasnitsyn and Martínez-Delclòs, 1999, 2000; Zhang *et al.*, 2002c), as well as from the Eocene-Oligocene boundary of North America (Rasnitsyn, 1993).

The Vespidae is second only to the ants (Formicidae) among vespoid aculeates in terms of the attention paid to them. These conspicuous, frequently social wasps are cosmopolitan in distribution and comprise approximately 4,500 species. Earlier classifications frequently divided the vespids into three separate families, but today they are united into a single family with six extant subfamilies (Carpenter, 1982, 1991) and a single extinct subfamily (Carpenter and Rasnitsyn, 1990). The basal subfamilies – Euparagiinae, Masarinae, and Eumeninae – are solitary, while the primitively eusocial Stenogastrinae are sister to the highly eusocial Polistinae and Vespinae. The basalmost, living subfamily Euparagiinae is restricted to the xeric regions of the southwestern United States and Mexico, but included at least Africa and Central Asia during the Cretaceous (Carpenter and Rasnitsyn, 1990; Brothers, 1992). Euparagiines construct simple burrows in the soil and are predatory on larval beetles. The Masarinae are the famous pollen wasps and, although of cosmopolitan distribution, these are principally diverse in the Southern Hemisphere. Masarines construct relatively simple burrows, which, like bees, they provision with pollen and nectar, and indeed they appear to be significant pollinators (Gess, 1996; O'Neill, 2001). Eumeninae are solitary or subsocial wasps that either nest in the soil or in stems or construct often elaborate mud nests that can resemble artistic clay pots (hence the common name potter wasps). Nests are provisioned with the carcasses of larval beetles, lepidopterans, or symphytans (e.g., Iwata, 1953; Evans, 1973b; Krombein, 1978; Cowan, 1991). The eusocial subfamilies comprise one lineage, with

11.42. A female of *Mischocyttarus* (Vespidae) tending her nest in Ecuador. Note an egg in the bottom of each cell. Photo: P. J. DeVries.

11.43. Species of the nocturnal vespids *Apoica* form impressive colonies in the New World tropics, such as this *Apoica pallens* nest in Ecuador. Photo: P. J. DeVries.

the Stenogastrinae, or hover wasps, basal to these. Stenogastrines are found only in the Indo-Pacific tropics. These are primitively eusocial and construct either mud or paper nests in concealed locations of relatively constant temperature and humidity. The wasps are predators and provision the brood with masticated remains of prey, including spiders plucked from their webs. The biology of stenogastrines has been presented by many authors (e.g., Yoshikawa *et al.*, 1969; Spradberry, 1975; West-Eberhard, 1978; Hansell *et al.*, 1982; Pardi and Turillazzi, 1985; Turillazzi, 1987, 1989, 1990a,b, 1991, 1996). Most studies on vespids, however, have focused on the highly eusocial species of the Polistinae (paper wasps) and Vespinae (hornets, yellow jackets; see also discussion of eusociality later in the chapter) (Figures 11.42 to 11.44), the former of which is by far the more diverse with about 800 species worldwide, compared to about a tenth of that in Vespinae, which is confined to the Holarctic and Oriental regions. These too provision the nest with masticated prey. The biology of polistines and vespines are thoroughly covered by authors such as Spradberry (1973b), Matsuura and Yamane (1990), Ross and Mathews (1991), and Gadagkar (2001).

The nests of vespids are elaborate and quite distinctive not only for particular species but for whole lineages, corresponding with the phylogeny of various groups (e.g., Turillazzi, 1990b; Ohgushi *et al.*, 1990; Wenzel, 1991, 1993; Yamane, 1993; Figure 11.45). Keys have even been developed to identify the nests of some groups (e.g., Wenzel, 1998; Turillazzi, 1999), which attests to the diagnostic value of the architecture.

Vespids from the Cretaceous are rare and are of the solitary or subsocial subfamilies (Carpenter and Rasnitsyn, 1990; Brothers, 1992; Carpenter, 2000); nevertheless, a polistine nest has been recovered from the Late Cretaceous of Utah (Brown, 1941; Wenzel, 1990), indicating that social forms were already present at that time. Diverse Tertiary fossils are recorded, mostly as compressions from the Eocene through Miocene (e.g., Burnham, 1978), although species of Vespinae and Polistinae are known from Baltic and Dominican ambers (Cockerell, 1910; Carpenter and Grimaldi, 1997) (Figure 11.46). Masarines are entirely unknown in the fossil record.

11.44. A paperwasp (Vespidae) tending its nest in Ecuador. Photo: R. Swanson.

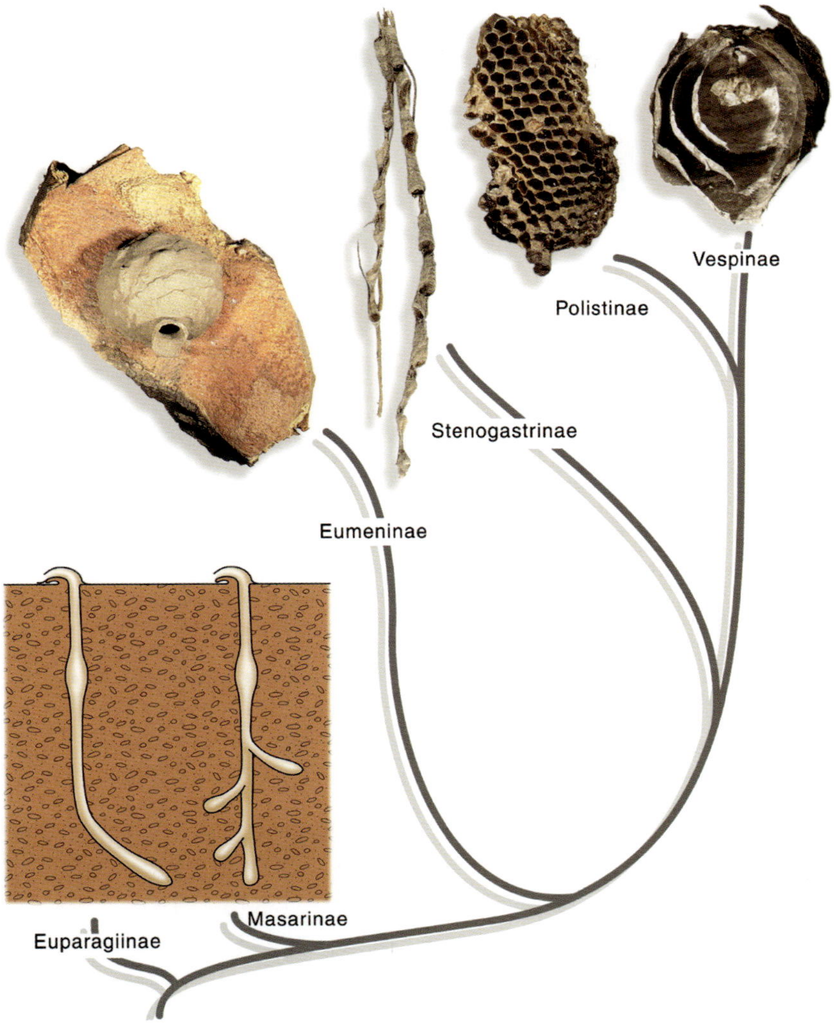

11.45. Representative vespid nests and the relationships of subfamilies. Not to the same scale.

11.46. A rare vespid fossil, of the living genus *Agelaia*, in Miocene Dominican amber. Morone Collection, M1217, length 14 mm.

THE ANTS

The Ant has made himself [sic] illustrious
Through constant industry industrious.
So what?
Would you be calm and placid
If you were filled with formic acid?
–Ogden Nash

All of the approximately 9,500 named species of ants, or family Formicidae, are highly social vespoid wasps, which live in societies having overlapping adult generations, where nestmates cooperate in the care of the brood, and where just a few colony members reproduce. Ants are arguably among the most ecologically important group of insects along with bees (pollinators) and termites (detritivores); ants, however, are primary consumers. They are most diverse in the great forested tropical river basins of the world, where their abundance can be spectacular. It has been estimated, for example, that ants in the Amazon Basin comprise four times the biomass of all vertebrates combined. Indeed, the activity one initially senses upon entering a tropical forest is the constant milling of ants. They can be prodigious excavators: A colony of just 5,000 Florida harvester ants, each one merely 3 mm long, was observed to move 44 lb (20 kg) of sand in 5 days, through approximately 35 feet (10 m) of tunnels (Tschinkel, 2001). The collective impact that ants have had on biological communities is perhaps best reflected in the thousands of plant and animal species that have specialized features associated with ants. Ants, for example, are the models for over 2,000 species of arthropods in 54 families that mimic ant body shape, color, and texture, and even parody their movements (McIver and Stonedahl, 1993) (e.g., Figure 11.47). These are *myrmecomorphs*, which are to be distinguished from the thousands of other arthropods (*myrmecophiles*) that have infiltrated ant colonies as parasites, inquilines, or predators but that do not necessarily resemble their host. Additionally, hundreds of species of plants in nearly 40 families (*myrmecophytes*) house ant colonies in specialized structures called *domatia*. Thus, ants have had an absolutely profound and pervasive influence on terrestrial communities. Their origins approximately 120 MYA appear to have been very modest, which is well documented in Cretaceous ambers from three continents; their rise to dominance 70 MYA is even better documented in Tertiary amber and rock compressions.

Such fascinating biology has inspired numerous serious entomologists to publish comprehensive treatises on ants. We have drawn on these works for this very brief review. The first truly comprehensive work (for the time) was by W. M. Wheeler (1910), followed much later by Wilson (1971) for social insects in general. The three most important references were published in the 1990s: the sumptuous treatise by Hölldobler and Wilson (1990), the only book on insects to be

11.47. An ant-mimicking cerambycid beetle and *Camponotus* ant in Miocene Dominican amber. The influence of ants on terrestrial life is partly reflected by the thousands of arthropods that have evolved *myrmecomorphy*. Ant mimics appear in amber after the radiation of ants in the early Tertiary. AMNH; length of beetle 8 mm.

11.48. Major and minor workers of *Camponotus* (carpenter) ants tending their larval brood. Photo: P. J. DeVries.

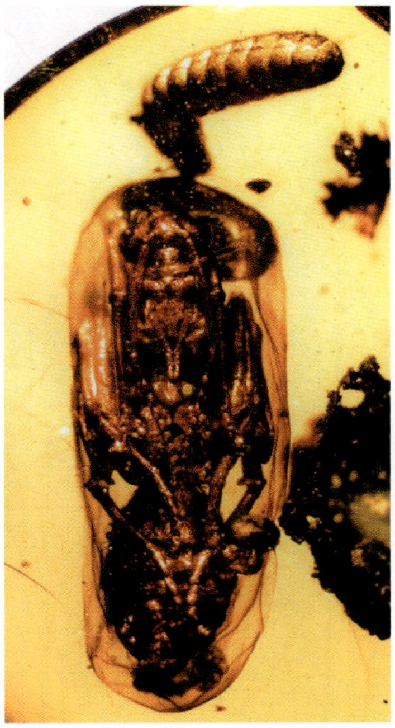

11.49. (left). Ant larva and pupa in Dominican amber. The cocoon became transparent in the resin, revealing the pupa within. Not all ants produce a cocoon. Morone Collection, M3380; length 3.6 mm.

11.50. Basic ant structure using *Zacryptocerus* (subfamily Myrmicinae). Formicologists use the terms "trunk" and "gaster" for the mesosoma and metasoma, respectively. The metapleural gland is a unique feature of ants. The anterior tergites of the gaster of *Zacryptocerus* are fused into a solid plate, and the flat head has a very specialized use (see Figure 11.51). Scanning electron micrograph; body length 2.9 mm.

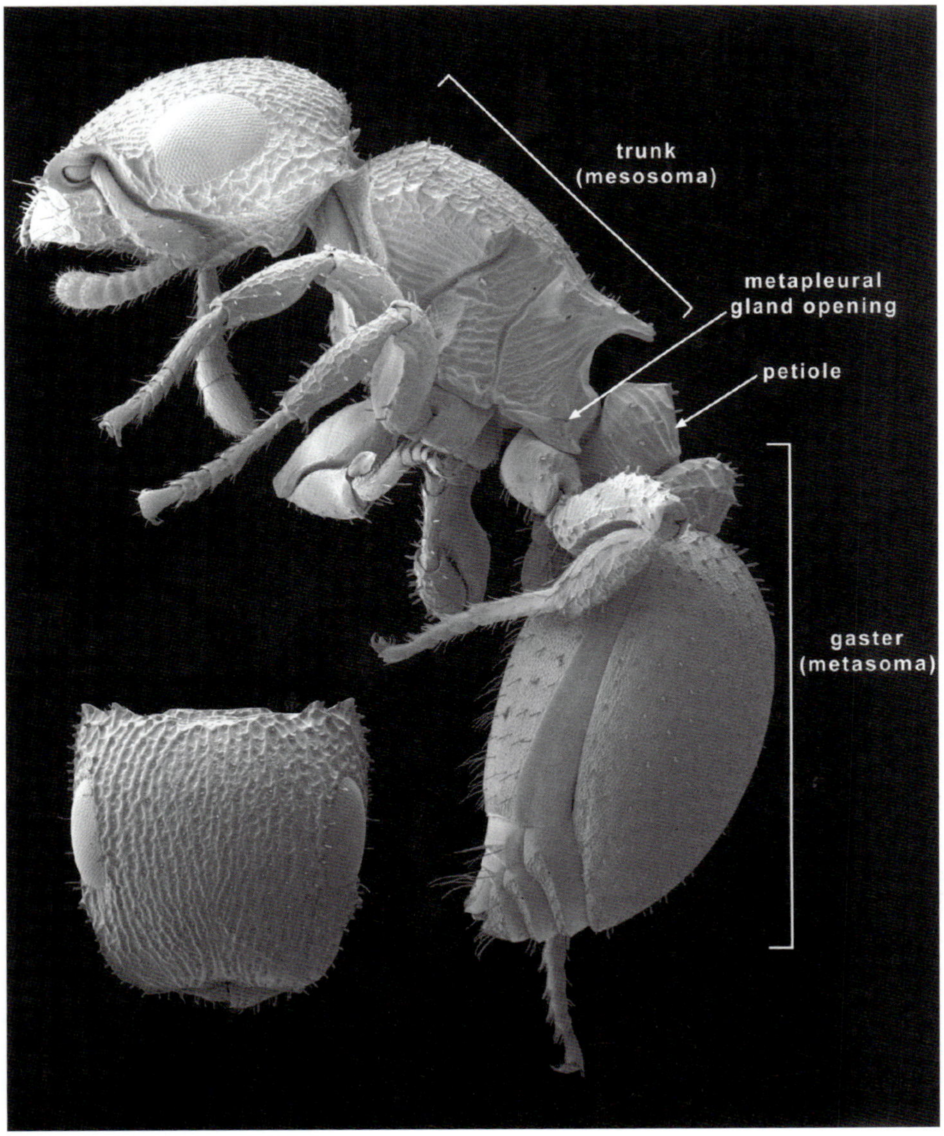

awarded a Pulitzer Prize; the guide to the world genera by Bolton (1994), beautifully illustrated with scanning electron micrographs; and the catalogue of world species (Bolton, 1995). No other diverse family of insects except Apidae and Drosophilidae has been studied so intensively.

An ant colony generally contains one or more queens, the immatures or brood (eggs, larvae, pupae), and the workers (all of which are females, and usually sisters, which is why Ogden Nash's poem is incorrect) (Figures 11.48 to 11.51). Like all hymenopterans, they have haplodiploid sex determination, where males develop from unfertilized eggs and are haploid, and workers develop from fertilized eggs and are diploid. Because male sperm is genetically homogeneous, daughter ants having the same mother and father are more closely related than any of these daughters would be to their own offspring. This, kin selection is generally agreed to be a major force for the evolution of advanced sociality in ants and other aculeate wasps because workers maximize their overall genetic (or *inclusive*) fitness by caring for sisters. It is very significant that new queens generally mate just once, and can store the sperm sometimes for as long as a decade, gradually apportioning it egg by egg. Having the same father ensures close genetic relatedness among workers. In accounting for the evolution of eusociality in ants, however, one cannot ignore a widespread behavioral feature of vespoid wasps: use of a domicile for the immatures that is provisioned with food. This has probably been a pivotal evolutionary step to eusociality in ants, bees, and paper wasps because many nonsocial aculeates provision simple burrows. Thrips, after all, are also haplodiploid, but social behavior has evolved in one small lineage of these that live in galls, so genetics is a necessary though not sufficient condition for the evolution of sociality.

11.51. A colony of the New World ant genus *Zacryptocerus*, in a hollow stem. Use of a persistent domicile was a crucial factor in the evolution of advanced sociality in ants and other insects. The opening of this nest is a small hole at the top. When the nest is disturbed, a *Zacryptocerus* worker plugs the entrance with its broad, flat head, even though in this case the rest of the nest has been sliced away. Photo: P. J. DeVries.

Probably as a result of these genetic constraints, a typical view of ants has them as finely oiled automatons instinctively and altruistically serving the greater needs of the colony. In fact, there can be significant conflict among nest mates. In the more primitive ants, which have small colonies, dominance hierarchies occur among workers, and some even (parthenogenetically) produce their own sons until the queen intercedes. There is a dramatic spectrum in the sophistication of ant societies, from small colonies with little differentiation among nestmates to huge colonies with multiple castes. Another common misconception is that colonies have only one queen. Actually, the majority of species probably have two or more queens (called *polygyny*), and there can be dominance hierarchies among these individuals too. There are queenless species in five genera of ponerines, where reproduction is done by workers. In some of these species, the mother parthenogenetically produces diploid females. There are also workerless species, which are actually social parasites given that a queen inserts herself into the colony of another species and her brood is cared for by that colony. Slave-making species, such as the species in *Polyergus*, are well known. This is done either by an intruding queen killing the resident queen and then having the resident workers care for her brood or by directing the workers to assault an entire colony, kill the workers and queen, and steal the brood to rear as their own. In any case, the concept of a colony as a superorganism (Wilson, 1971) is an apt one because all nest mates collectively function far more effectively than they would separately, regardless of the reason. The efficiency of the colony is essentially a result of the division of labor (particularly castes among workers) and highly effective chemical communication.

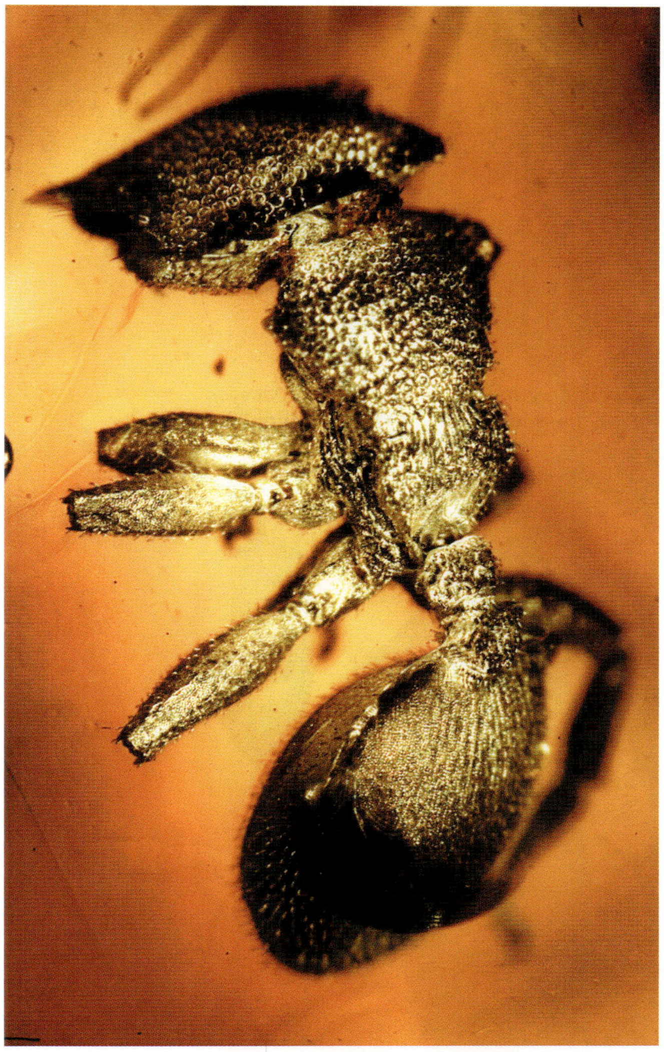

11.52. *Zacryptocerus* in Miocene amber from the Dominican Republic. The fossil record of ants is largely revealed in amber, particularly the earliest ants in the Cretaceous. Morone collection, M0082; length 2.7 mm.

Castes are a hierarchy of social classes, defined by their functions in the society. In most species of ants, the worker castes are largely discerned by behavioral differences (they are *monomorphic*). Others have evolved castes that are also distinguished morphologically, and the differences can be remarkable in terms of both size and specialized structures among the workers (Figure 11.53). The largest workers are generally *soldiers,* which are built for defense. In the African driver ants (*Dorylus*), for example, the smallest workers can be 1% of the size of the largest soldiers. Associated with this extreme disparity in size is allometry, or the disproportional growth of certain structures. In the neotropical army ants, *Eciton,* for example, soldiers have huge hanging jaws like ice tongs, and a massive head housing the jaw muscles, whereas other workers have heads typical of other ants. Interestingly, many of the species with the most dramatic caste polymorphism have some of the largest colonies, with up to several million individuals, like *Eciton, Dorylus,* the leaf-cutter ants (*Atta*), and the seed harvesters (*Pheidole*), and these are also among the most dominant ants in various biological communities. Coordinating so many individuals is only possible through an efficient system of communication, which in ants in essentially based on the transfer of chemicals from an array of exocrine and other glands by grooming and trophallaxis (some species also stridulate). Ants, in fact, have more kinds of glands than any other group of Hymenoptera, perhaps more than any other family of insects. Phermones secreted by the exocrine glands include ones for trail construction, recruitment to food sources, alarm, and recognition of worker nest mates and queens, among others. Efficient communication has allowed ants to mobilize thousands of workers quickly for defense and foraging, and this is what makes them such formidable competitors.

There are 16 subfamilies of ants, 7 of which have modest generic and species diversity (although see the recent reclassification by Bolton, 2003). The largest subfamilies and some of their more significant genera are the following.

- *Dolichoderinae.* Members of this subfamily have lost the sting and have Pavan's gland, which is a source of trail pheromone under sternite seven in this group. Many Dolichoderinae are arboreal. The subfamily includes such genera as *Azteca, Dolichoderus, Iridomyrmex,* and *Tapinoma. Azteca* is a neotropical genus, many species of which live within the internodes of hollow stems of *Cecropia* plants. *Tapinoma* is a worldwide genus of tiny ants with extremely rapid movements and an insidious way of invading houses in the tropics.
- *Formicinae.* This subfamily has also lost the sting, so formicines defend themselves by spraying formic acid through an acidopore at the tip of the gaster. The acidopore is generally ringed with fine hairs. These include *Formica, Lasius,* and other familiar ants of northern temperate regions, as well as the slave makers (*Polyergus*) and the carpenter ants (*Camponotus*). It also includes the weaver ants (*Oecophylla*: Asia, Africa, Indo-Australia), which construct nests by pulling leaves together (Figure 11.54) and sewing the edges with silk from larvae. *Cataglyphis* are long-legged desert ants, which can scurry rapidly over hot sand during midday, orienting by the sun (pheromone trails are not very effective in such circumstances). *Myrmecocystus* are the New World honey pot ants. Some workers in a *Myrmecocystus* colony hang from the roof of the nest and continually feed on honeydew brought by other workers. Their gasters distend with honeydew, which sustains the colony during dry spells.
- *Myrmicinae.* This subfamily is the most diverse subfamily, ecologically and in terms of species numbers. It includes such ants as the worldwide genus *Crematogaster*; fire ants (*Solenopsis*), which have become notorious pests; *Pogonomyrmex* harvester ants, which have a potent sting venom;

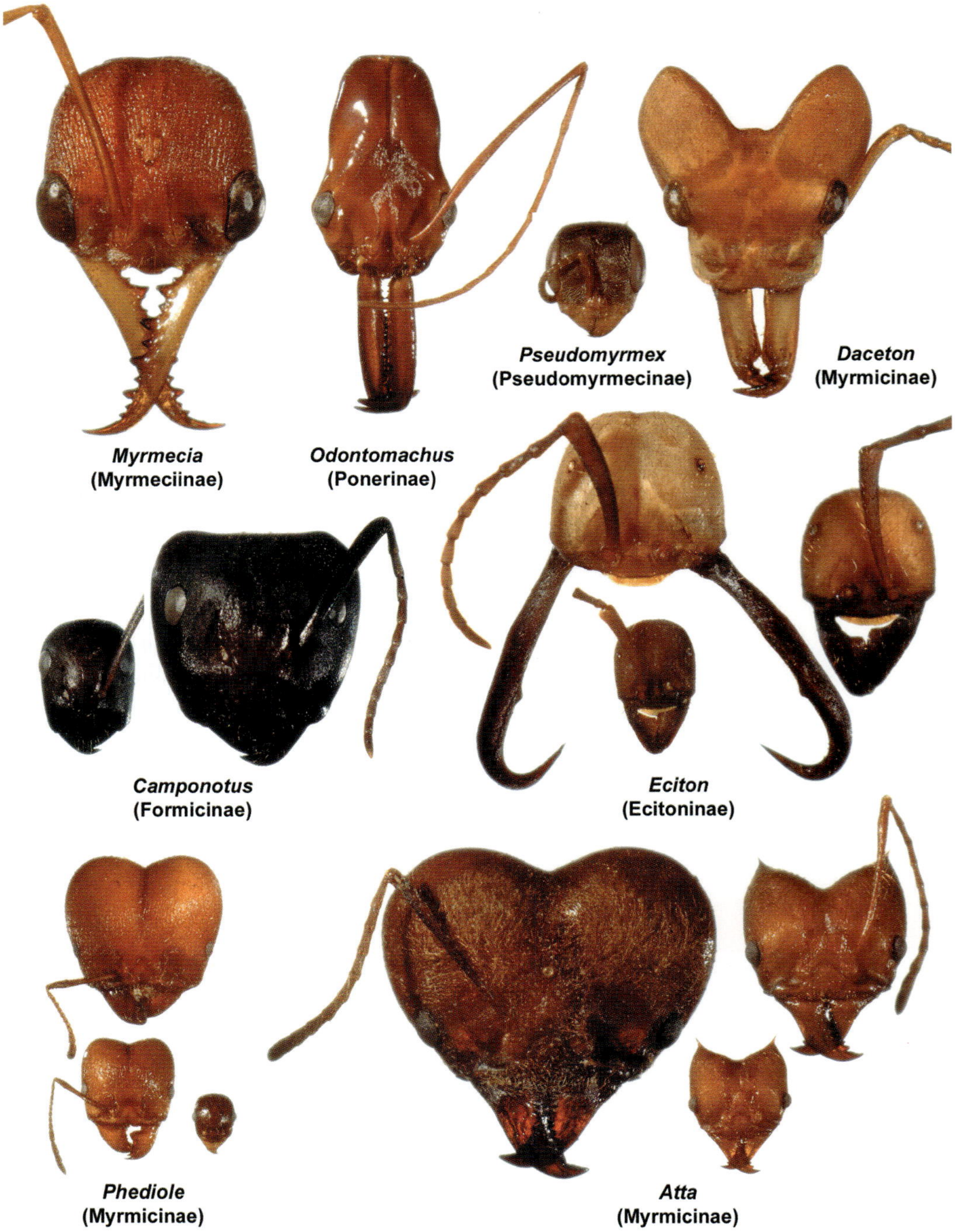

11.53. Heads of representative ant workers and soldiers. Extreme polymorphism among workers (minors, majors, soldiers) generally occurs in ants that form huge colonies. Ants in the top row have monomorphic workers. To the same scale.

and various genera of Dacetonini. Dacetonines generally have heart-shaped heads and walk around with gaping, gaff-shaped mandibles (e.g., Figure 11.55). When fine hairs between the mandibles touch a prey item, these trigger the mandibles to snap shut, which is why some can snag a collembolan as it begins to spring. Perhaps the best known myrmicines, though, are ones in the neotropical tribe Attini, or leaf-cutters and fungus growers (Figure 11.56). These forage on vegetation, which they chew and apply to the walls of their nests. A fungus grows on the walls and then is eaten. Leaf-cutters have an obligate symbiotic relationship with fungi in the families Tricholomataceae and Lepiotaceae. The more recently evolved lineages of attines use just the lepiotaceous fungus and there is even significant cospeciation among the fungi and their ant hosts (Chapela *et al.*, 1994). Leaf-cutter colonies can persist for a century, and when they senesce the fungus will sprout an immense mushroom to disperse spores

11.54. Weaver ants (*Oecophylla*: subfamily Formicinae) in Brunei. These ants fashion nests by bending leaves together and suturing the edges with silk exuded from the larvae. Photo: P. J. DeVries.

11.55. The neotropical ant *Daceton* (subfamily Myrmicinae). Myrmecines are structurally and behaviorally the most diverse subfamily of ants. Many ants in the tribe Dacetonini typically walk around with long mandibles gaping. When fine hairs between the mandibles are touched, the mandibles snap shut. Photo: P. J. DeVries.

11.56. Leaf-cutter ants, genus *Atta* (subfamily Myrmicinae). Fragments of leaves and flowers are used to cultivate fungus in labyrinthine nests underground, and the ants feed on the fungus. The tribe Attini occurs entirely in the tropics and warm temperate regions of the Western Hemisphere. Here two minor workers ride a leaf fragment carried by a major worker and guard the major against parasitoid phorid flies. Photo: P. J. DeVries.

11.57. Giant mushrooms of the subterranean fungus cultivated by *Atta* leaf-cutter ants. When an *Atta* colony senesces, the spore-producing stage of the fungus emerges. These are rarely seen because the colonies can persist for decades. In photo: P. J. DeVries; photo by D. Grimaldi.

11.58. Leaf-gathering behavior of attine ants is at least 20 MYO, based on evidence from fossils like this *Acromyrmex* ant preserved in a piece of Dominican amber with chewed leaf fragments. The evolution of attines, beginning perhaps 35 MYA, profoundly affected forest ecosystems of the New World tropics. AMNH.

(Figure 11.57). Attines have dramatic impacts on tropical forests, not only in terms of folivory, but also through aeration of soil (their underground nests are huge) and the succession of grasslands to forests.

- *Ponerinae.* This "subfamily" appears to be a basal grade of ants, many of them with a constriction between the first and second gaster segment, but they are best known for the powerful stings. An intimate encounter with the large neotropical ant *Paraponera clavata* is memorable.

Despite all of the attention lavished on the biology and species taxonomy of ants, there has been limited work on the relationships among subfamilies. Original work on this was by W. L. Brown (1954), but this has been replaced by more recent phylogenetic hypotheses (Baroni-Urbani *et al.*, 1992; Ward, 1994; Grimaldi *et al.*, 1997; Grimaldi and Agosti, 2000; Ward and Brady, 2003). There is significant ambiguity about the relationships of some genera, but areas of agreement generally include the following (see Figure 11.59).

- Ponerinae is a paraphyletic grade, many of which are in fact basal, particularly the Amblyoponini and Paraponerini (Ward and Brady, 2003).
- There is a bush of unresolved relationships among many subfamilies.
- The bulldog ants of Australia (Myrmeciinae), which includes *Myrmecia* and the primitive genus *Nothomyrmecia*, comprise one of the basal lineages in this bush, along with the large-eyed arboreal ants in the Pseudomyrmecinae.
- The large subfamilies Formicinae and Dolichoderinae are closely related.
- The Old World driver ants (Dorylinae) and the New World army ants (Ecitoninae) are closely related (Grimaldi *et al.*, 1997; Brady, 2003). These are nomadic, carrying their brood to temporary nesting sites, and thus form huge swarms and raiding columns (Figures 11.61, 11.62). The eyes in both groups are extremely vestigial or the ants are completely blind.
- The extinct, Cretaceous subfamily Sphecomyrminae is the most primitive group of true ants (Wilson *et al.*, 1967; Hölldobler and Wilson, 1990; Bolton, 1994; Grimaldi *et al.*, 1997; Grimaldi and Agosti, 2000; Ward and Brady, 2003).

There has been some controversy, none serious, that the Cretaceous sphecomyrmines were not true ants because they lacked the long scape that is seen in modern groups of ants

TABLE 11.3. Early Fossil Ants[a]

1. Sphecomyrminae (undescribed): Burmese amber
2. *Sphecomyrma freyi* (Figures 11.60, 11.63), *Sphecomyrma* sp. (undescribed): New Jersey amber
3. *Cretomyrma* spp., *Dlusskyidris zherichini* (Sphecomyrminae): Siberian amber
4. *Sphecomyrma canadensis*: Canadian amber
5. *Gerontoformica cretacica* (French amber), *Haidomyrmex cerberus* (Burmese amber) (Figure 11.69)
6. *Dolichoformica helferi*: French amber
7. Myrmeciinae? (undescribed): Burmese amber
8. *Kyromyrma neffi* (Formicinae): New Jersey amber (Figure 11.66, 11.68)
9. *Brownimecia clavata* ("Ponerinae"): New Jersey amber (Figure 11.67)
10. *Canapone dentata* ("Ponerinae"): Canadian amber
11. *Eotapinoma macalpini* (Dolichoderinae): Canadian amber
12. *Protopone* spp. ("Ponerinae"): Sakhalin amber
13. *Chimaeromyrma* sp. (Formicinae): Sakhalin amber
14. *Aneuretellus* sp. (Aneuretinae): Sakhalin amber
15. *Eotapinoma zherichinius* (Dolichoderinae): Sakhalin amber

[a] Numbers correspond to those circled on phylogeny, Figure 11.59.

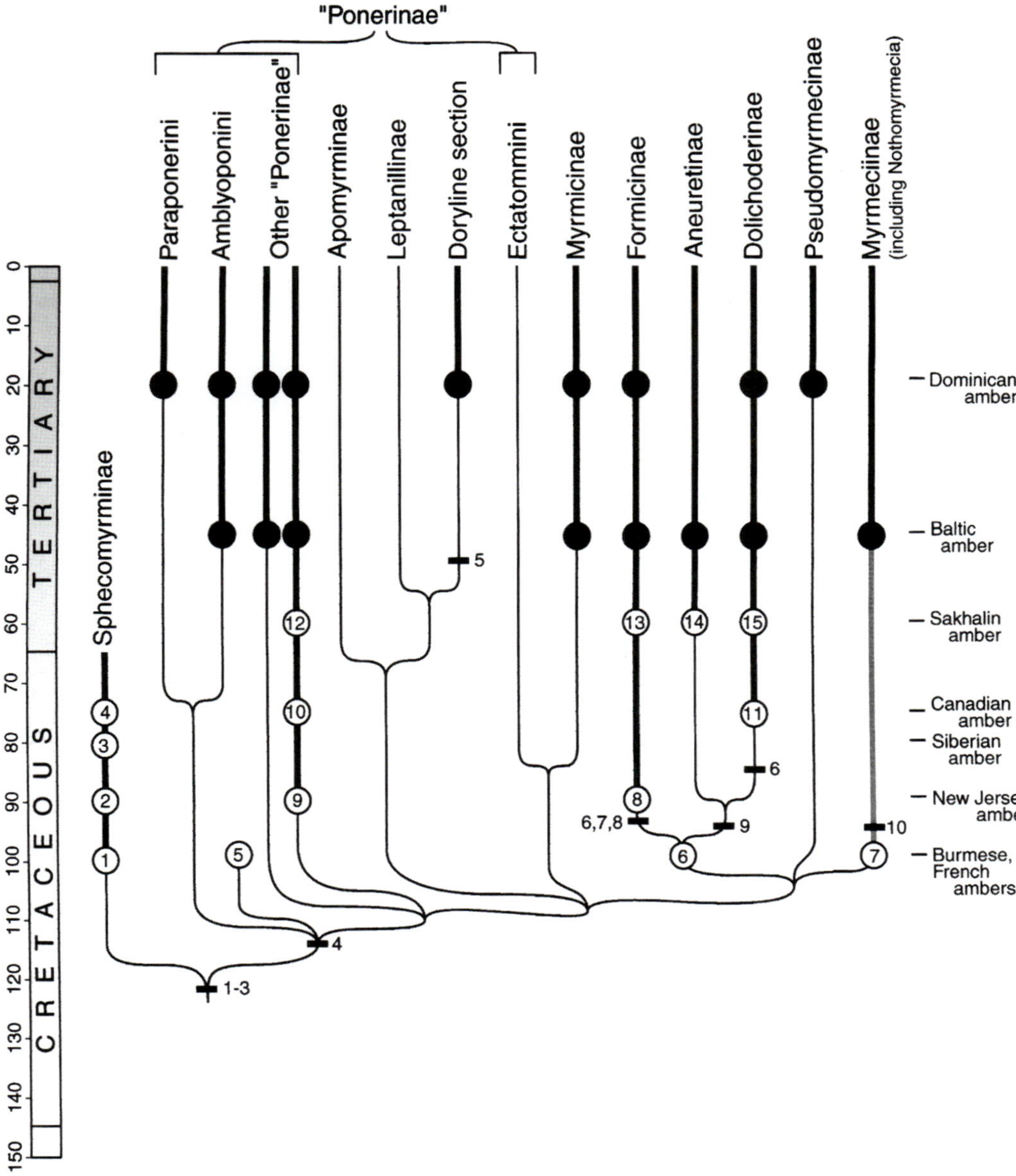

11.59. Phylogeny of most major lineages of ants (subfamilies and some tribes), based on several studies (see text). Circled numbers refer to Cretaceous and Paleocene fossils (see Table 11.3); other numbers refer to significant characters (see Table 11.4). Ants appear to have originated around 120 MYA.

TABLE 11.4. Significant Characters in Ant Phylogeny[a]

1. Metapleural gland
2. Eusociality (wingless female castes)
3. Petiole
4. Elongate scape
5. Legionary behavior (army and driver ants)
6. Sting lost
7. Acidopore present; spray formic acid
8. Pygidial gland lost
9. Loss of Pavan's gland (between sternites six and seven, produces trail phermones)
10. Numerous characters (Myrmeciinae): see Ward and Brady, 2003.

[a] Numbers correspond to those on phylogeny, Figure 11.59.

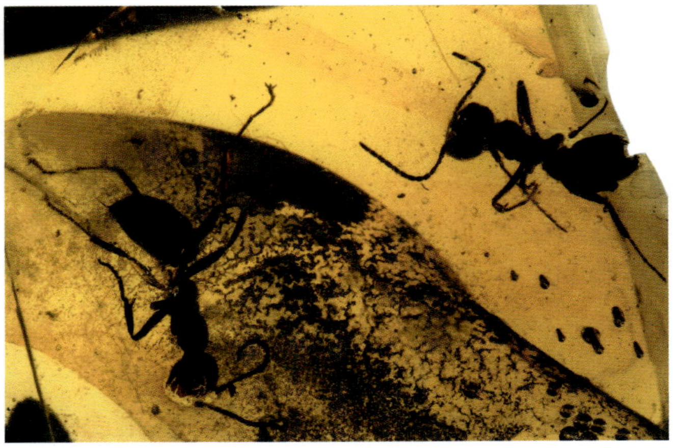

11.60. Two workers of *Sphecomyrma freyi* in a piece of mid-Cretaceous amber from central New Jersey. Sphecomyrmines are the most primitive known ants, and these specimens suggest they foraged socially. AMNH NJ943; length of ant 4 mm.

11.61. Soliders of army ants carrying larval brood. Ecitonines do not nest in permanent structures but are nomadic and form bivouac sites with their bodies, with the queen and brood cloistered within. Photo: Carl Rettenmeyer, Connecticut State Museum of Natural History.

11.62. Soldier and workers in a raiding column of the South American army ants, *Eciton*. Photo: P. J. DeVries.

and that gives most ants the characteristic elbowed (or *geniculate*) antenna. But, like the early fossils of all taxa, sphecomyrmines possessed some, not all, of the features of living ants (Figures 11.60, 11.63). Features they shared with modern ants include a petiole between the trunk and gaster, the unique ant feature of a metapleural gland, and they were apparently social. The metapleural gland opens on the propodeum in ants and secretes a substance thought to aid in disinfecting the nest. It is clearly preserved in specimens in amber. The body proportions of wingless individuals preserved in Cretaceous amber are similar to those of modern worker ants (Wilson, 1987), and even the preservation of *Sphecomyrma freyi* in New Jersey amber suggests social behavior. Cretaceous ants were remarkably rare, always comprising far less than 1% of all insects from any one deposit. Thus, the probability that two wingless individuals would be caught in one piece (e.g., Figure 11.60) is very remote based on chance alone, unless of course they were workers that were foraging socially.

Apparently more primitive than sphecomyrmines is the extinct family Armaniidae (Figure 11.64), which is classified as an extinct subfamily of the Formicidae by Dlussky (1983). These had a broad petiole, which would make them more primitive than even sphecomyrmines, but because armaniids are known only as compressions (from the Early to mid-Cretaceous), it is impossible to verify their position using more reliable, microscopic characters. With little question, sphecomyrmines represent the most basal lineage of true ants and were perhaps even the ancestral ants. Earliest occurrence of sphecomyrmines is in Burmese amber (ca. 100

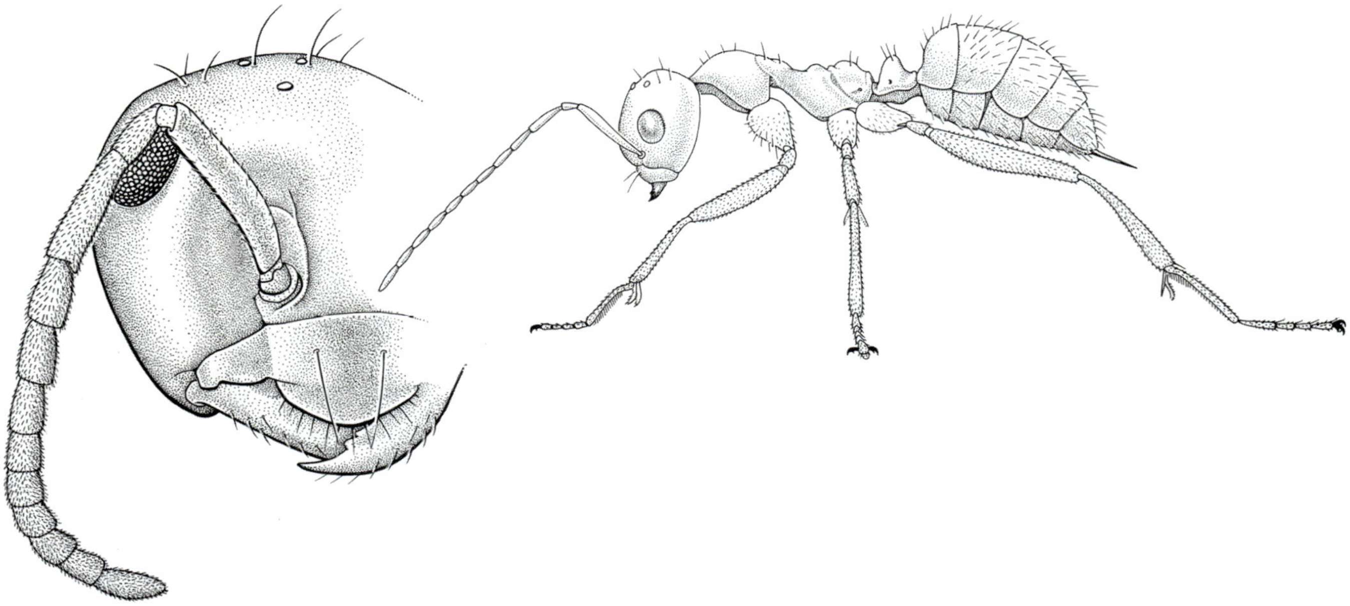

11.63. *Sphecomyrma freyi*, reconstruction and detail of front of head.

11.64. *Armania robusta* (Armaniidae), an early close relative of true ants. Armaniidae are known only from the mid-Cretaceous of central Asia. The petiole is well preserved in this specimen, but is considerably wider than occurs in true ants. PIN 3901/155.

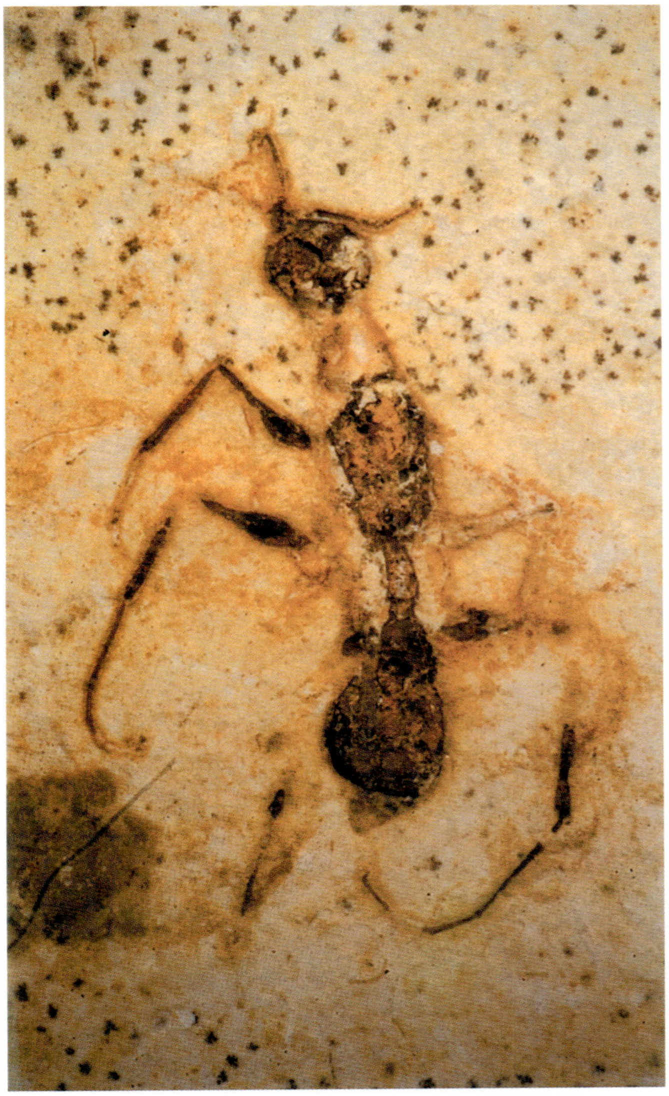

11.65. A putative early ant, *Cariridris bipetiolata*, from the Early Cretaceous Santana Formation of Brazil. Appearance is suggestive of the Myrmeciinae, but preservation does not allow observation of critical details of the mandibles, antenna, and metapleural gland, and it has even been attributed to the Ampulicidae (Apoidea) as well. Length 5.2 mm. Photo: R. Brandão.

MYO) (Engel and Grimaldi, in press), and they also occur in Cretaceous amber from New Jersey, Siberia, and western Canada (the latter approximately 72–75 MYO). But sphecomyrmines are not the only ants known from the Cretaceous; others include extinct genera belonging to the modern subfamilies Formicinae, Dolichoderinae, Myrmicinae, and "Ponerinae." Cretaceous ants have been described or reviewed in nine significant papers (Dlussky, 1987 [Siberian amber]; Dlussky, 1999, and Wilson, 1985b [Canadian amber]; Wilson *et al.*, 1967, Grimaldi *et al.*, 1997, Grimaldi and Agosti, 2000, and Engel and Grimaldi, in press [New Jersey amber; Figures 11.66 to 11.68]; Dlussky, 1996, and Engel and Grimaldi, in press [Burmese amber]; and Nel *et al.*, 2004, [French amber]). The earliest "ponerine" is *Brownimecia clavata* in New Jersey amber (Figures 11.66, 11.67); the earliest formicine is *Kyromyrma neffi*, also in New Jersey amber (Figure 11.68); and the earliest dolichoderine is *Eotapinoma macalpinei* in Canadian amber. Others are not so easily classified. *Dolichoformica helferi*, in French Cenomanian amber (about the same age as Burmese amber) has been classified as a stem-group species of the formicine-dolichoderine lineage, and the relationships of *Gerontoformica cretacica* (also in French amber) are obscure (Nel *et al.*, 2004). Two species in Burmese amber likewise have obscure relationships, and they are also extremely unusual. *Haidomyrmex cerberus* has very peculiar, sickle-shaped mandibles and a small clypeus with a brush of fine and scale-like hairs (Figure 11.69). Another Burmese amber ant may be a primitive myrmeciine or belong to one of the 'ponerine' lineages (Engel and Grimaldi, in press) (Figure 11.70). Most recently, Dlussky *et al.* (2004) have recorded compressions of the first Cretaceous ants – in the subfamilies Ponerinae and Myrmicinae – from the Southern Hemisphere (Orapa, Botswana).

11.66. Early ant in 90-MYO amber from central New Jersey, *Brownimecia clavata*, belonging to the basal poneroid grade of ants.

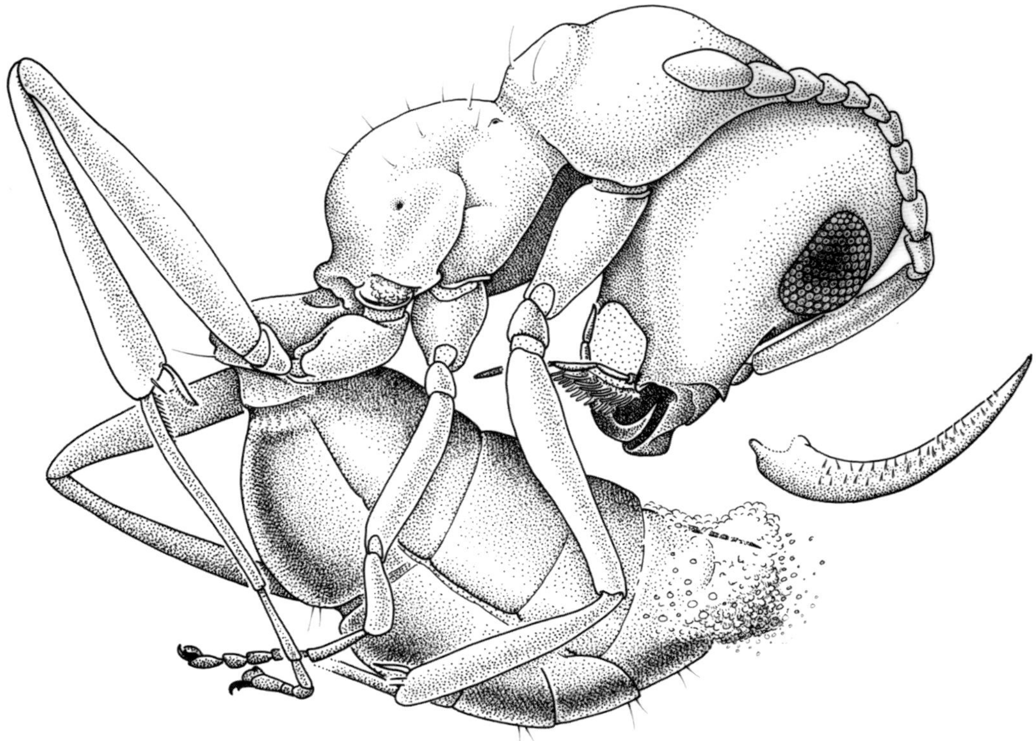

11.67. *Brownimecia clavata* in New Jersey amber, with a detail of its unusual, toothless mandible. The constriction between gaster segments one and two is a feature of myrmeciines and the poneroid grade of ants. An extruded sting is partially visible through froth. AMNH NJ667; length 3.4 mm; from Grimaldi *et al.* (1997).

In total, the Cretaceous record of ants indicates they were very rare and primitive (Grimaldi and Agosti, 2000; Dlussky and Rasnitsyn, 2002), and that major lineages like formicines, dolichoderines, and 'ponerines' were established between 90 and 75 MYA. The apparent sister group to Formicidae, which is the family Armaniidae, has a fossil record that extends to approximately 125 MYA, but monophyly of that extinct family is doubtful so it doesn't help much in dating. Using the same reasoning as we do for the bees, the lower limits on the age of ants can be constrained by the collective fossil record of their vespoid relatives, which in this case extends to approximately 130 MYA. Because ants are among the most recently evolved vespoids based on phylogenetic studies, it is reasonable to estimate *the origin of the Formicidae as no earlier than 120 MYA*. A significant contradiction to this is a recent estimate of the age of the doryline lineage, the army and driver ants (Brady, 2003). This lineage is considered very recently derived in most phylogenies (e.g., Baroni-Urbani *et al.*, 1992); it may possibly have evolved in the early Tertiary (Grimaldi *et al.*, 1997; Grimaldi and Agosti, 2000). This is actually difficult to estimate based on the virtual lack of fossils for this group. Brady (2003), however, has estimated an age of the doryline lineage itself as mid-Cretaceous, approximately 110 MYO, with the more basal ant lineages diverging deep into the Cretaceous (140 MYO). In our view these estimates are far too old and probably a result of methods of molecular dating, and they also contradict the fossil evidence. The Cretaceous age of drift between South America and Africa is not much use in dating the divergence of the dorylines and ecitonines, because fossil ants indicate that some groups were very widespread at one time. It would be ideal to test these hypotheses with the discovery of fossil army/driver ants.

The collective fossil record of ants indicates that it was not until 70–80 MY after their origin about 120 MYA that ants evolved into the dominant organisms we know today.

The fossil record of Tertiary ants is excellent, and it documents an explosive radiation around 50 MYA. Little is actually known about the Paleocene fauna, as is the case for all fossil insects, but two significant papers include Dlussky (1988: Sakhalin amber) and Rust and Andersen (1999: Fur Formation of Denmark). The Eocene record is superb, and major papers for this period include the following: Carpenter (1930) on Florissant, Colorado; Dlussky and Rasnitsyn (2003) on Green River, Wyoming; and Wheeler (1915) and Dlussky (1997) on the rich Baltic amber fauna. As for many other groups preserved in Baltic amber, there have been some impressive extinctions, perhaps the most notable being the extinction of myrmeciines, which are now relegated to Australia but which also included southern South America (*Ameghinoia* and *Polanskiella*) and the Baltic region (*Prionomyrmex*) (Ward and Brady, 2003) during the Eocene. There

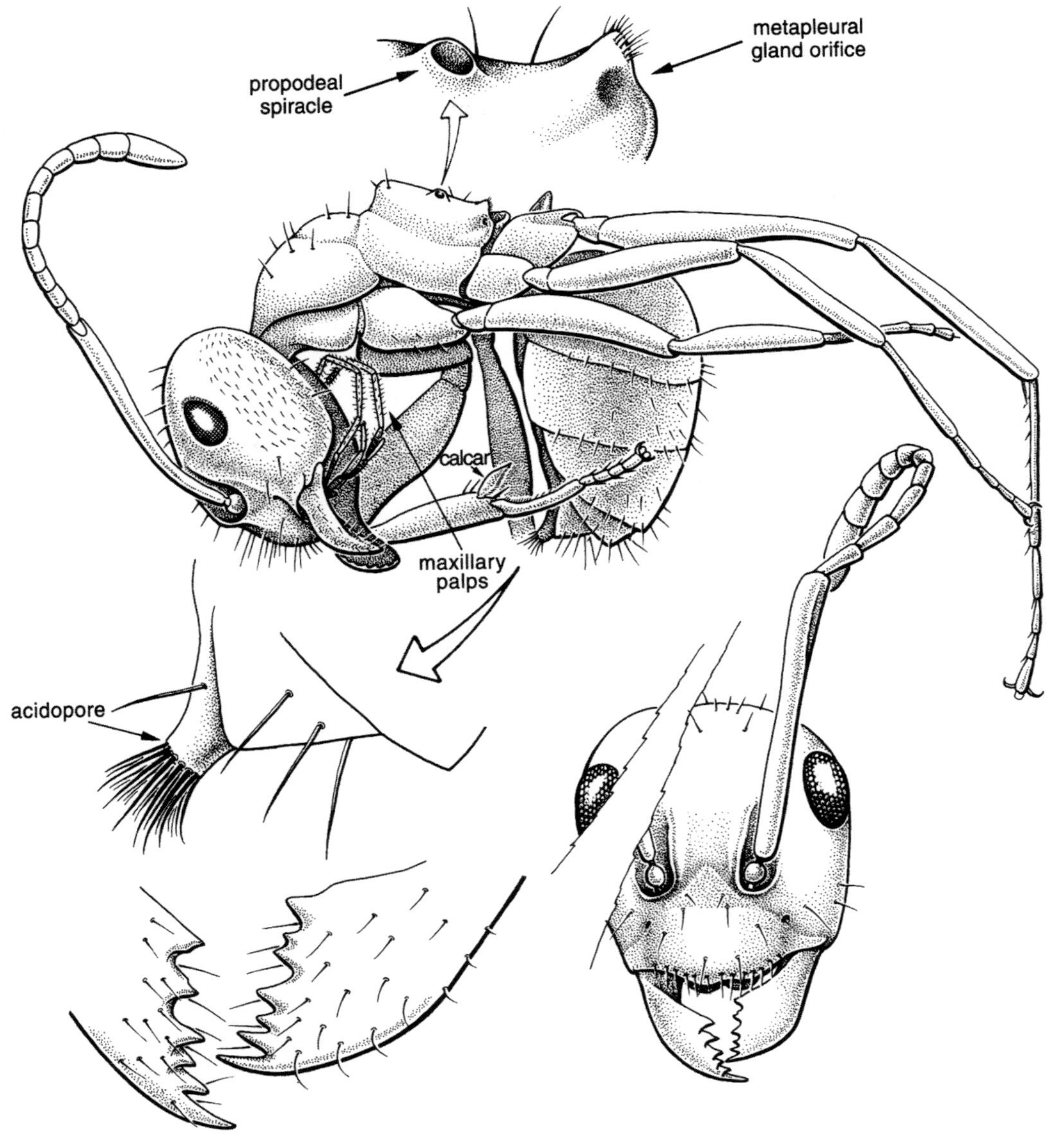

11.68. *Kyromyrma neffi* (Formicinae) in New Jersey amber, with details of the mandibles, metapleural gland opening, and acidopore. Formicines have lost a sting and defend themselves by spraying formic acid through the acidopore. *Kyrmomyra* is an unspecialized formicine, but presence of the subfamily 90 MYA indicates divergence of major ant lineages approximately 100 MYA or more. AMNH NJ1029; length 2.2 mm; from Grimaldi and Agosti (2000).

are numerous other examples. By the early Oligocene, approximately 35 MYA, there is a dramatic spike in the abundance of ants in compression and fossil deposits, which has been attributed to the rise of diverse ant genera that today have huge colonies, like *Azteca*, *Camponotus*, *Dolichoderus*, and *Formica*. By the time the large amber deposits from southern Mexico and Hispaniola were formed, 25–20 MYA, ant faunas were essentially modern.

The bees and spheciform wasps together form the **Apoidea** (e.g., Brothers, 1975; Königsmann, 1978b; Brothers and Carpenter, 1993). The superfamily as a whole comprises approximately 30,000 described species, most of them bees. The superfamily is defined by the posterolateral expansion of the pronotum into a distinct lobe (*pronotal lobe*), the midventral fusion of the prepectus, and the posteromedial expansion of the metapostnotum ("*propodeal triangle*"). Historically, the bees have been classified into five or six families, while the spheciforms, sometimes called "digger wasps," were all lumped into the family "Sphecidae." Sphecids in this sense are undoubtedly not a natural group and, in fact, form a grade of lineages of which the bees comprise one recently evolved lineage (Lomholdt, 1982; Alexander, 1992; Brothers, 1999). Recognizing this situation, the former Sphecidae has been divided into a series of families: Heterogynaidae, Ampulicidae, Sphecidae, and Crabronidae, with the latter representing the living sister group to the bees (Prentice,

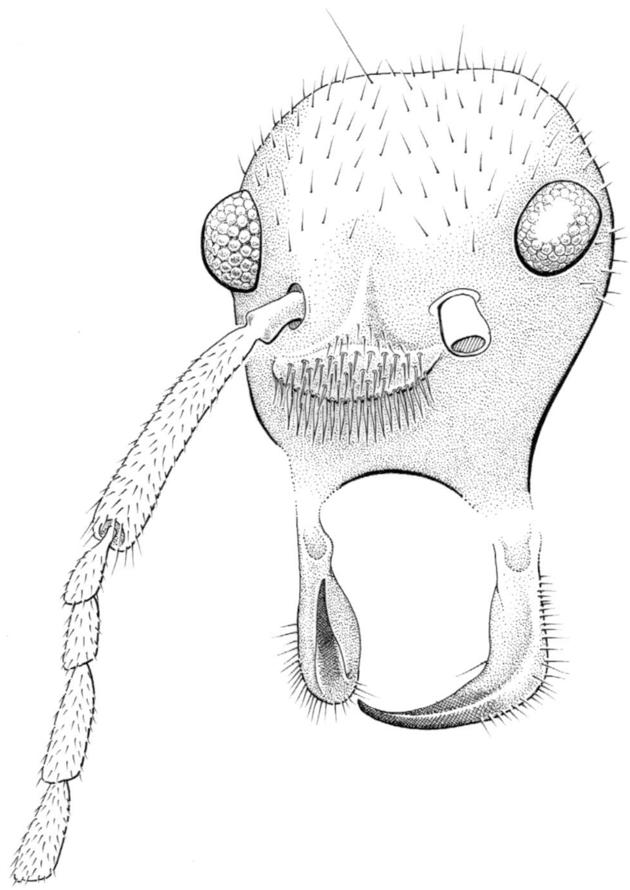

11.69. Head of the unusual Cretaceous ant, *Haidomyrmex cerberus*, in mid-Cretaceous Burmese amber. The clypeus has a unique brush of stiff setae, but the toothless, sickle-shaped mandibles, small eyes, and a pronotal "neck" strongly suggest it may be related to *Brownimecia*. NHM 20182; head width 0.3 mm.

1998; Melo, 1999). Bohart and Menke (1976) provided the most comprehensive account of the sphecoid grade, but their classification is clearly dated. Species of Ampulicidae, Crabronidae, and Sphecidae are all known from the Cretaceous (e.g., the "ant" described by Brandão *et al.*, 1989, from the Early Cretaceous of Brazil may be an ampulicid: Rasnitsyn, 2002) (Figure 11.65), while the relict Heterogynaidae is presently unknown from the fossil record. While ampulicids are recorded from the Early Cretaceous, definitive sphecids and crabronids are not known until the mid-Cretaceous (e.g., Antropov, 2000). A single extinct family is recognized – Angarosphecidae, which occurs in Early Cretaceous deposits throughout the world (e.g., Rasnitsyn, 1975, 1980, 2000; Rasnitsyn *et al.*, 1999; Rasnitsyn and Ansorge, 2000b; Rasnitsyn and Martínez-Delclòs, 2000; Darling and Sharkey, 1990) (Figure 11.71). Angarosphecids are primitive in most traits relative to other spheciforms, and the family itself may not be a natural group. During the Cenozoic, a diversity of spheciforms are known as both amber inclusions and compression fossils and are relatively modern in appearance (e.g., Cockerell, 1906, 1909b, 1910; Rohwer, 1908a; Pulawski and Rasnitsyn, 1980; Sorg, 1986; Menke and Rasnitsyn, 1987; Nemkov, 1988, 1990; Antropov and Pulawski, 1989, 1996; Budrys, 1993; Antropov, 1995. Cladistic studies of spheciforms that extensively incorporate fossils have not yet been completed so it may be some while before this fossil record is fully appreciated. Phylogenetic relationships based strictly on modern representatives indicate that Heterogynaidae is basal for living families (angarosphecids are perhaps basal within Apoidea), followed by Ampulicidae, then by Sphecidae, and lastly by Crabronidae as the sister group to the bees (Alexander, 1992; Prentice, 1998; Melo, 1999) (Figure 11.33).

Spheciforms are predatory wasps, provisioning their nests with the paralyzed bodies of small arthropods (see the various prey records summarized by Bohart and Menke, 1976). While the larvae are certainly carnivorous, adults, like many aculeates, feed on nectar and possibly pollen. Adult females typically construct in the ground relatively simple burrows that terminate in one or more brood cells. The brood cell is filled with prey insects that were paralyzed by stings, then an egg was deposited on the prey mass, and the cell was sealed.

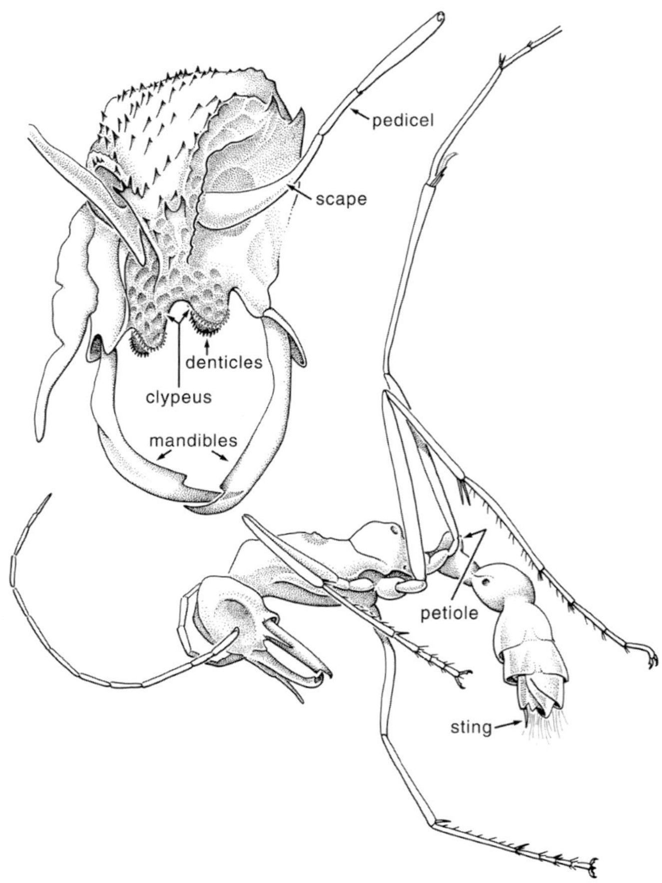

11.70. Another early ant in Burmese amber, approximately 100 MYO. It was a fairly large, gracile ant with long legs and a gastral constriction, bearing some resemblance to Myrmeciinae and the "poneroid grade" of ants. Though distorted by compression, it clearly had an unusual, sculptured head, bilobed clypeus, and long mandibles. Ants in Burmese and French amber are the oldest definitive ants. AMNH Bu0014; length 8.1 mm.

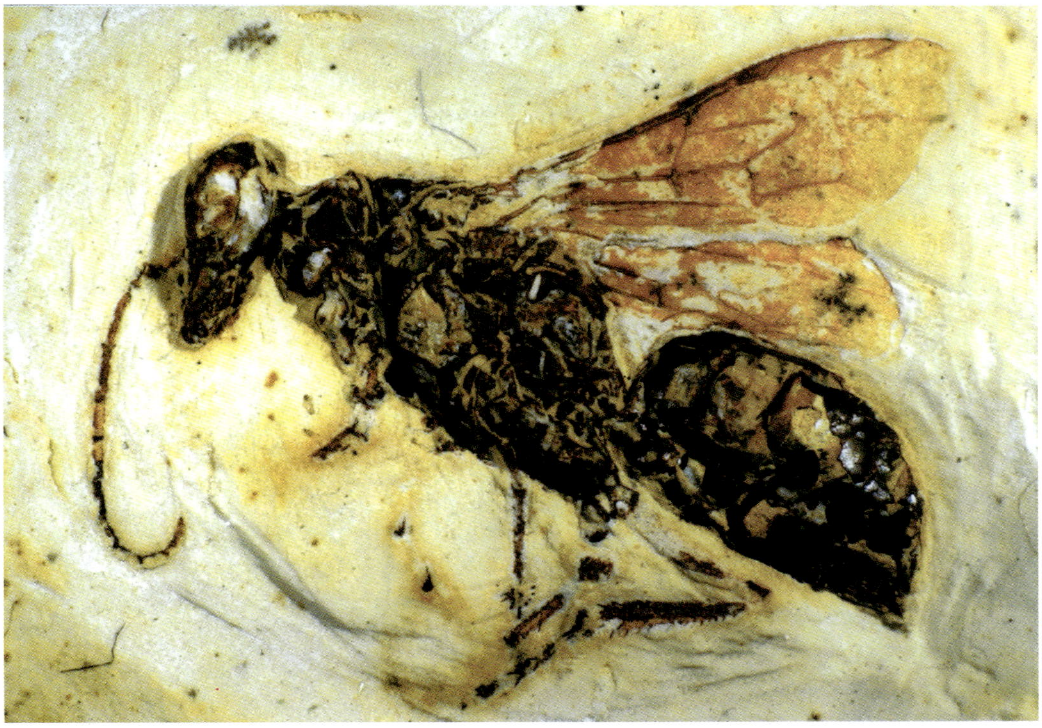

11.71. The earliest apoid wasps were primitive species like *Angarosphex magnus* (Angarosphecidae) from the Early Cretaceous of Brazil's Santana Formation. AMNH; length 8 mm.

The egg hatches and the larva proceeds to feed on the provisioned prey. Some species construct exposed nests, building mud chambers on exposed surfaces (e.g., the side of homes). Twig nesting spheciforms generally do not construct their burrows, rather co-opting a preexisting cavity or abandoned burrow. The biology of spheciforms is most interestingly reviewed by Evans (1962, 1963a, 1966a,b), Evans and West-Eberhard (1970), and O'Neill (2001). In a very few groups, which are derived among the spheciforms, there are exceptions to this general pattern of biology. Some species are secondarily ectoparasitoids, the adult female temporarily paralyzing the prey before ovipositing on the host (e.g., Gwynne and Evans, 1975; Iwata, 1976; Gess, 1984; Castner and Fowler, 1987; Hudson *et al.*, 1988). The wasp larva feeds on the live host for several instars before entering the host's body cavity and feeding as an endoparasitoid until the host expires, then spinning a cocoon and pupating therein. Such ectoparasitoids occur in the Ampulicidae, Sphecidae (e.g., *Chlorion*), and Crabronidae (e.g., *Larropsis*). The Nyssonini and the genus *Stizoides* are cleptoparasites (i.e., cuckoos) of their close relatives (Evans, 1966b). Lastly, the Sri Lankan wasp *Krombeinictus nordenae* (Crabronidae) feeds its offspring pollen and nectar, similar to the bees (Krombein and Norden, 1997a,b). The biology of spheciforms is most thoroughly covered by Evans and O'Neill (1988) and O'Neill (2001). Spheciforms occur throughout the world, although some interesting distributions are known for some lineages. For example, the relict family Heterogynaidae is known only from Botswana, Madagascar, and the eastern Mediterranean (Day, 1984; Argaman, 1986). Perhaps one of the most interesting biogeographic distributions among spheciforms is that of the species of *Clitemnestra* that occur in Australia and Chile, a classic disjunct. It will be interesting if species of this genus are eventually discovered in Tertiary ambers from the Northern Hemisphere, or if Cretaceous fossils of their early relatives are found.

11.72. *Agapostemon virescens* (Halictidae) visiting flowers in Vancouver, Canada. Bees are the most important group of pollinators. Most halictines are generalists, but some lineages of bees specialize on particular groups of angiosperms. Photo: R. Swanson.

THE BEES (ANTHOPHILA[1])

Now what delight can greater be
Then secrets for to knowe,
Of Sacred Bees, the Muses Birds,
All of which this booke doth shew.
And if commodity thou crave,
Learne here no little gaine
Of their most sweet and sov'raigne fruits,
With no great cost or paine.
If pleasure then, or profit may
To read induce thy minde;
In this smale treatise choice of both,
Good Reader, thou shalt find.
–Charles Butler, 1609

To make a prairie it takes a clover and one bee,
One clover, and a bee,
And reverie.
–Emily Dickinson, *Prairie*

Bees are among the most recognized of all insects, albeit for only a few species – honey bees (*Apis*) and bumble bees (*Bombus*). Indeed, these familiar bees account for only a tiny fraction of the true diversity of bees, most of which are solitary and typically overlooked by the layman. Whether we generally recognize them or not, the bees – all of the nearly 20,000 species – are intimately intertwined with our lives. Bees are the life and breath of any flowering area: gardens, orchards, crops, blooming waysides, deserts, prairies, plains, and forests, including vast areas in the great tropical river basins. While humans measure themselves in terms of technological achievements, our lives still depend upon these quiet agriculturists for the pollination of our crops, and honey remains one of the few foods not yet synthetically produced. Bees underpin and sustain global ecosystems through their role as the most significant pollinators of flowering plants (Figure 11.72); a service that began nearly 120 MYA (although the earliest bees probably were not as ecologically significant as Recent ones) and without which earth would wither. For these reasons the bees, and the common western honey bee in particular, have earned a sacred place in our societies as the most famous of all arthropods; they are inextricably woven into mythology, religions, and culture (e.g., Ransome, 1937; Crane, 1983, 1999).

[1] The name for bees and their relatives is a source of some confusion. Classically the spheciform wasps were placed in the superfamily Sphecoidea and as sister to the bees, superfamily Apoidea. With the increased realization that the sphecoids were paraphyletic to the bees, the two were naturally united. The family-group name based on *Apis* (Apoidea) predates the family-group name based on *Sphex* (Sphecoidea) and, applying the *Principle of Priority*, the former becomes the valid name for the combined group. Thus, "Apoidea" refers to the bees and the grade of spheciform families previously called sphecids. There is, therefore, no formal name for the group comprising the families of bees. The informal name Apiformes has been at times used, but we prefer to use the name, Anthophila, which predates Apiformes by more than a century and was widely used well into the twentieth century. Anthophila is also a far more descriptive name, meaning "flower loving."

Mountains of literature exist for the bees, even when excluding the vast libraries dedicated to apiculture of the western honey bee, *Apis mellifera*. Major works concerning the systematics and biology of bees include Friese (1923), Bischoff (1927), Grütte (1935), Malyshev (1935), Michener (1944, 1965, 1974, 1979, 2000), Frisch (1967), Bohart (1970), Iwata (1976), Seeley (1985, 1995),Winston (1987), O'Toole and Raw (1991), Barth (1991), and Radchenko and Pesenko (1994), among many others. Major classifications have been proposed by Kirby (1802), Latreille (1802a,b), Lepeletier de Saint Fargeau (1835, 1841), Schenck (1861, 1869), Thomson (1872), Schmiedeknecht (1882), Friese (1895), Ashmead (1899), Robertson (1904), Börner (1919), Michener (1944, 1965, 2000), Sustera (1958), and Warncke (1977). Other significant papers concerning themselves with floral relationships are by Thorp (1979), Wcislo and Cane (1996), and Westerkamp (1996). Michener (2000) provides additional references to most aspects of bee biology and systematics. Dalla Torre (1896), although greatly dated, remains the sole world catalogue for the Anthophila.

As already discussed, the bees are a derived, natural group of the spheciform wasps and as such are vegetarian wasps. The monophyletic origin of the bees has almost never been questioned (e.g., Müller, 1872; Michener, 1944, 2000; Brothers, 1975, 1999; Lomholdt, 1982; Alexander, 1992; Brothers and Carpenter, 1993; Alexander and Michener, 1995; Prentice, 1998; Melo, 1999; Engel, 2001a). Since the time of Aristotle (and likely even much earlier!), there has been relatively little difficulty recognizing bees, even when species are remarkably wasplike. Certainly the pollenivorous habits of bees are immediately characteristic, although some wasps (e.g., Masarinae and *Krombeinictus*) also provision their brood with pollen and nectar, and three highly derived bees scavenge from carcasses (Camargo and Roubik, 1991). Anatomical characteristics of bees (Figure 11.73) include the presence of branched or plumose setae, the subantennal sulci internally connected to fanlike sheets of the tentorium, presence of a mesotibial comb, expanded metabasitarsus (broader than following tarsomeres), absence of a metatibial strigil, crossvein cu-a in the hind wing generally shorter than second abscissa of M + Cu, and the division of the seventh metasomal tergum of females into lateral hemitergites (see also Michener, 2000; Engel, 2001a). The phylogeny of bees has been investigated by several authors, the most recent of which include Michener and Greenberg (1980), Roig-Alsina and Michener (1993), Alexander and Michener (1995), and Engel (2001a). The six extant families of bees are divided into two informal groups – the short-tongued and long-tongued bees.

While the long-tongued bees (families Megachilidae [Figure 11.74] and Apidae) are united by several traits, most notably the long and flattened labial palpi, the short-tongued bees are a grade, with the family Melittidae being the closest living relatives to the long-tongued families. Colletidae (with

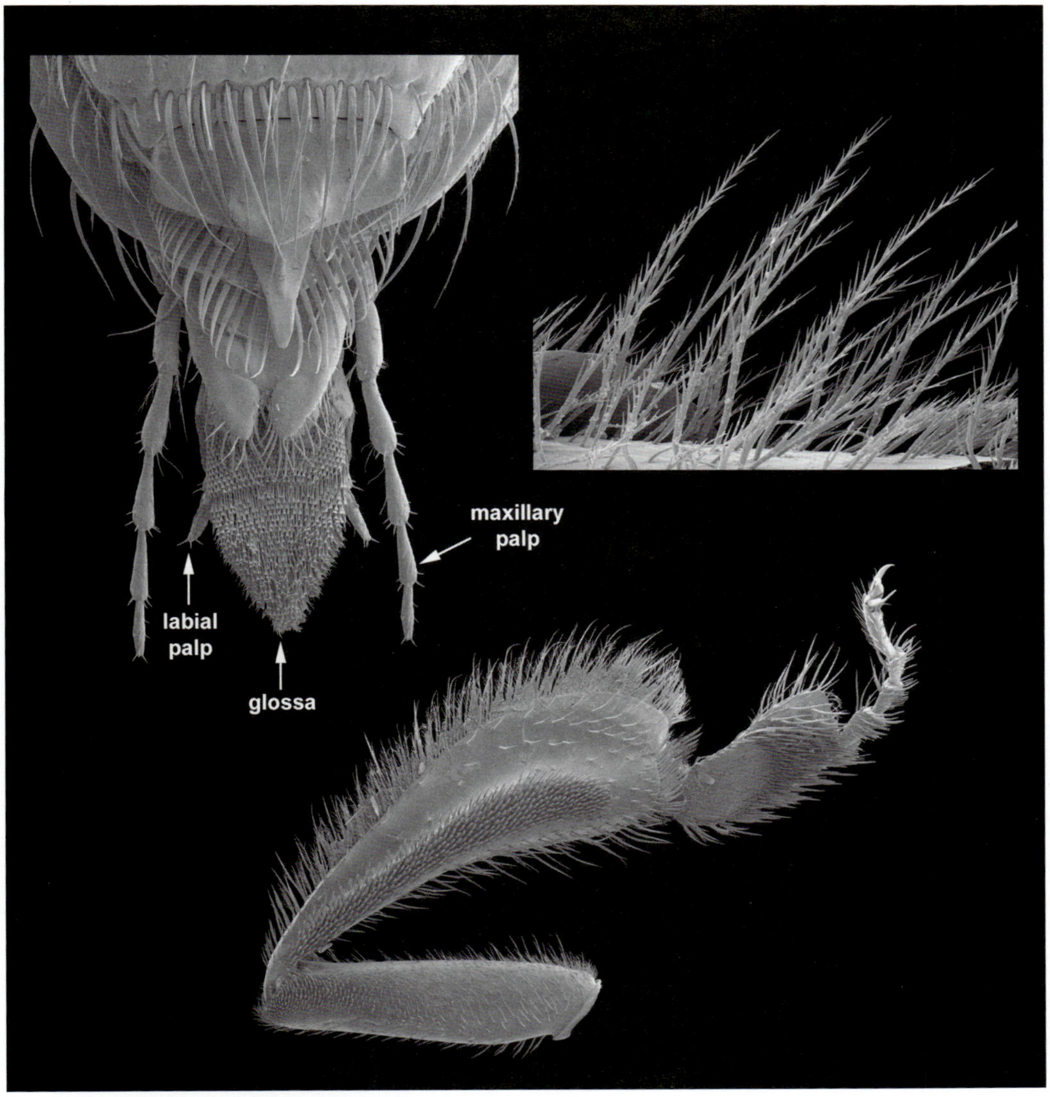

11.73. Significant traits of bees, including the "tongue" (glossa), plumose hairs, and scopa of the hind leg. These parts are from a stingless bee (Meliponinae). Scanning electron micrographs; not to the same scale.

11.74. A mating pair of *Osmia* bees (Megachilidae) in White Rock, Canada. Photo courtesy R. Swanson.

its synonym Stenotritidae) is the most basal family of bees, primitively retaining the short, bifid glossa found among the spheciform wasps. Interestingly, many of the putatively primitive subfamilies within many bee families have their greatest diversity in or are confined to the Southern Hemisphere, for example the Paracolletinae (Colletidae), Euherbstiinae (Andrenidae), Meganomiinae (Melittidae), and Fideliinae (Megachilidae). Certainly other lineages occur also in the Northern Hemisphere, but typically basal bee lineages are southern in distribution. In fact, some of these subfamilies are clearly relict, like Fideliinae, which were likely more widely distributed in the past and have undergone significant extinction (Engel, 2002b). Many bees are oligolectic, who prefer the pollen of a few related flower species, although some lineages such as Halictinae are predominantly polylectic. Aside from the collection of pollen, other forms of flower specialization have occurred numerous times throughout the bees. A derived group of the genus *Megachile* (Megachilidae)

11.75. The orchid bee *Eulaema meriana* (Apidae) gathering oils from *Antherium* in Costa Rica. Photo: P. J. DeVries.

comprise the familiar leaf-cutter bees, which cut semicircular swaths from petals or leaves and then use these pieces in nest construction. Some anthidiines (also of Megachilidae) shave plant trichomes to similarly be used in building their nests. Numerous bees are equipped with combs designed for the collection of oils (e.g., Redivivini in the Melittidae), which they mix with the pollen provisions. Perhaps the most remarkable are the orchid bees (Euglossini) (e.g., Figure 11.75). Male euglossines collect oils from one or more particular orchid species, which they then transfer to modified metatibial glands to produce an attractant for the females. Nest design is also varied throughout the bees. Primitively, bees excavated burrows in the soil, but numerous lineages have independently transitioned to using hollowed twigs, or constructing their nests in hard wood (Figure 11.76). Others build nests of cemented mud or stones on exposed surfaces. Lastly, the waxen nests of the social honey bees, bumble bees, and their relatives are well-known to most individuals (Figures 11.77 to 11.80), particularly the vertical combs with hexagonal cells characteristic of *Apis*.

Interestingly, while bees are perhaps most widely known for their social behavior, the majority of species are solitary, the social forms accounting for barely a few percent of the total world diversity. However, the social systems of certain bees have achieved levels of complexity that can easily rival those of ants and termites. It is, therefore, of little wonder that the intrinsic interest in social organisms has focused so acutely on bees, and honey bees (*Apis*) in particular. The social bees are considered further in our discussion of eusociality, later in this chapter.

As equally fascinating as sociality are the bees that are cleptoparasitic (i.e., *cuckoo bees*). Like sociality, cleptoparasitism

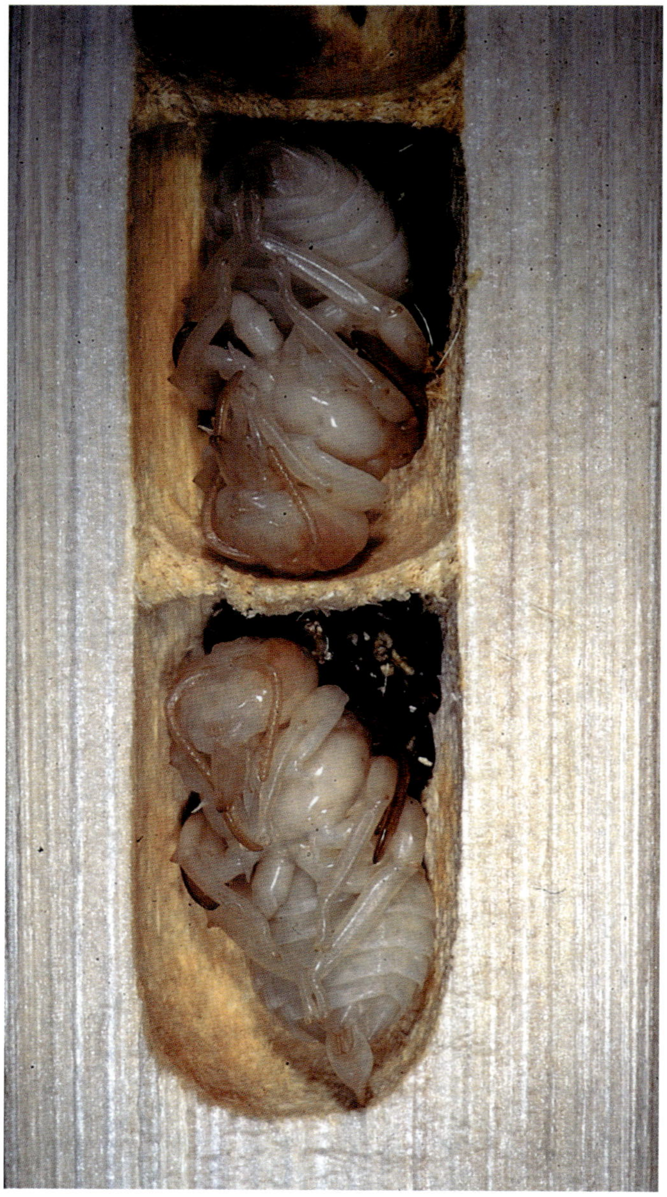

11.76. Pupae of the giant carpenter bee, *Xylocopa virginica* (Apidae) in New York. Carpenter bees, as their name suggests, excavate nests in stems or wood, including the wooden eaves of houses. Photo: V. Giles.

11.77. Brood comb of a socially advanced stingless bee of the genus *Plebeia* (Apidae) in Panama. Photo: P. J. DeVries.

11.78. (left). Honey pots for storing nest resources in a nest of *Plebeia* (Apidae) stingless bees in Panama. Photo: P. J. DeVries.

11.79. Primitive honey bees (tribe Apini) build exposed nests. Giant honey bees, *Apis dorsata* (Apidae), build giant nests beneath overhangs of cliffs, like this one in Nepal. Photo: B. Underwood.

11.80. Workers of the dwarf honey bee, *Apis florea* (Apidae), salvaging wax from a comb. Photo: T. D. Seeley.

originated multiples times in the bees (Figure 11.82), but in stark contrast is the fact that many cleptoparasitic lineages have radiated into hundreds of species, such as the Nomadinae. Parasitism in Hymenoptera is a predominant trend, and this is true within bees as well, with cleptoparasitism having arisen perhaps 27 times among the bees (Rozen, 2000). Cleptoparasitic bees generally have lost, or have at least greatly reduced, their pollen-collecting apparati and have a more wasplike appearance, often with a coarse integument and various projections and prominences for defending themselves. Parasitic bees are still pollenivorous, but instead of the adult constructing and provisioning her own nest, females steal into the nest of a host bee and deposit an egg in the brood cell before making her escape.

There are various modes of cleptoparasitism in bees (Rozen, 1991, 2003). In some the parasite deposits her egg in a completed brood cell (i.e., provisioned and with a host bee larva present). The female parasite either must dispatch the host bee larva before oviposition or, as is done more frequently, leave this task to her own offspring. In such lineages, the earlier larval instars (typically the first instar) of the parasite are greatly modified, having structures specialized for dispatching the host bee larva. Such murderous larvae are referred to as *hospicidal* (Rozen, 1989, 1991), and they are generally first instars, though some cleptoparasites such as *Kelita* have hospicidal second and third instars (e.g., Ehrenfeld and Rozen, 1977). Later stages of the parasite tend to be typical for bees and designed for a cloistered existence consuming pollen, upon which the parasite is now free to focus. In other cleptoparasitic bee lineages (e.g., Nomadinae), the adult parasite places her egg in a brood cell that is still in the process of being provisioned by the host bee. This immediately poses the difficulty that the host bee will detect the parasite's egg. Such bees typically insert the egg into the lining or wall of the brood cell. The egg itself is modified to minimize any exposed surfaces, and the external surface mimics the texture of the cell wall (e.g., Rozen, 2003; Rozen and Özbek, 2003). Once the brood cell is provisioned and the host has deposited her egg, the cell is sealed, and the two immatures, host and parasite, are left to their own interaction, usually to the detriment of the host larva. There is evidence, however, that the host's larva is not entirely without defensive capabilities (e.g., Rozen, 1984). When cleptoparasitism evolved in bees is entirely conjectural. *Protomelecta brevipennis* from the Eocene-Oligocene of Florissant, Colorado, is the sole fossil of a parasitic lineage (in this case, Melectini) (Cockerell, 1908a); however, this putative assignment to a parasitic group has yet to be confirmed. Based on the phylogenetic position of several of the cleptoparasitic lineages (Figure 11.82), however, one would expect at least some cleptoparasitic bees to have been present in the Late Cretaceous. It would appear that it did not take long for some bees to take advantage of their industrious relatives once the group had evolved. Interestingly, cleptoparasitism has also evolved among the spheciforms, but apparently only twice and with relatively few species (Bohart and Menke, 1976; Ohl, 1997, 1999; Ohl and Linde, 2003).

The geological history of bees is frustratingly obscure. Fossils of bees are uncommon to rare. However, by unifying the sparse record of fossil bees and bee traces with the more complete history of spheciforms, it is possible to obtain a defined picture of early bee evolution. The almost complete lack of fossils from the Cretaceous certainly limits any discussion. The oldest fossil of a bee is *Cretotrigona prisca* from the Late Cretaceous of eastern North America (Michener and Grimaldi, 1988a; Engel, 2000a) (Figure 11.83). Although *Cretotrigona* is, quite unfortunately, a singular fossil, it reveals much owing to its derived position in a derived family (Apidae), implying that the grossest level of bee diversification (i.e., the differentiation of families, subfamilies, and many tribes) had already occurred by the Late Cretaceous (Figure 11.82). Earlier evidence of bee activity is suggested by presumed halictine nests of Cenomanian age from Arizona, ca. 95 MYO (Elliott and Nations, 1998), which again indicate that bees are older than this epoch, perhaps as old as late in the Early Cretaceous to early in the mid-Cretaceous, about 120 MYA (Engel, 1996, 2000a, 2001a). Also, the disjunct, austral distributions of basal bees would suggest a Cretaceous age, prior to the complete fragmentation of Gondwana, but fossils indicate that such austral taxa were not always restricted to the southern land masses. There are mid-Eocene and younger leaves with damage typical of leaf-cutter bees, which are derived species of the genus *Megachile* (e.g., Berry, 1916, 1931; Brooks, 1955; Lewis, 1994; Wappler and Engel, 2003), and this suggests a much older age of Megachilidae. The derived position of the leaf-cutters, not only among all bees, but also within Megachilidae itself, implies that significant divergence had occurred earlier in bee history (Figure 11.82).

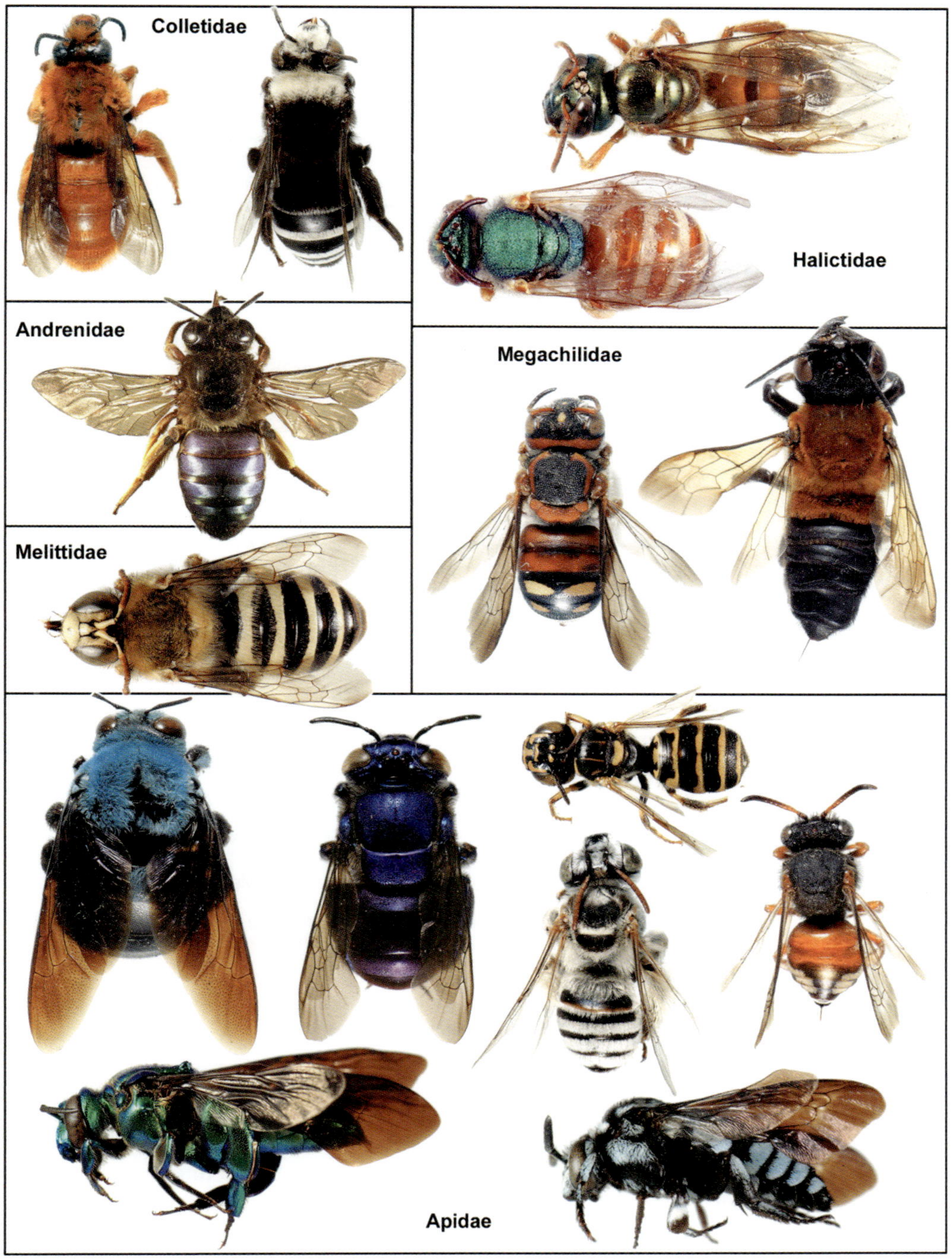

11.81. A diversity of Recent bees. Note the large ocelli of one of the halictids, which is the nocturnal bee *Megalopta genalis*. Not to the same scale.

While the Cretaceous records provide an upper boundary of bee diversification, the origin and initial radiation is actually relatively simple to pinpoint. As already discussed, bees are derived from within the spheciforms and as such must be younger than this grade. The first spheciforms appear in the Early Cretaceous but are primitive relative to all other spheciforms (angarosphecids and a putative ampulicid) and are certainly not of the Crabronidae. Sphecidae and Crabronidae are first definitively recorded from the Albian-Cenomanian boundary and then on the basis of a diversity of forms. Thus, the Crabronidae had already appeared by approximately 110 MYA and likely as long ago as 115 – 120 MYA. *It would, therefore, appear that the crabronid-bee divergence took place approximately 120 MYA.* Pre-Cretaceous accounts of bees are all based on misidentifications (see Engel, 2001a). Differences between bees and crabronids can at times be subtle, and it may be difficult to discern early bee fossils, particularly in compressions where minute structural details are not

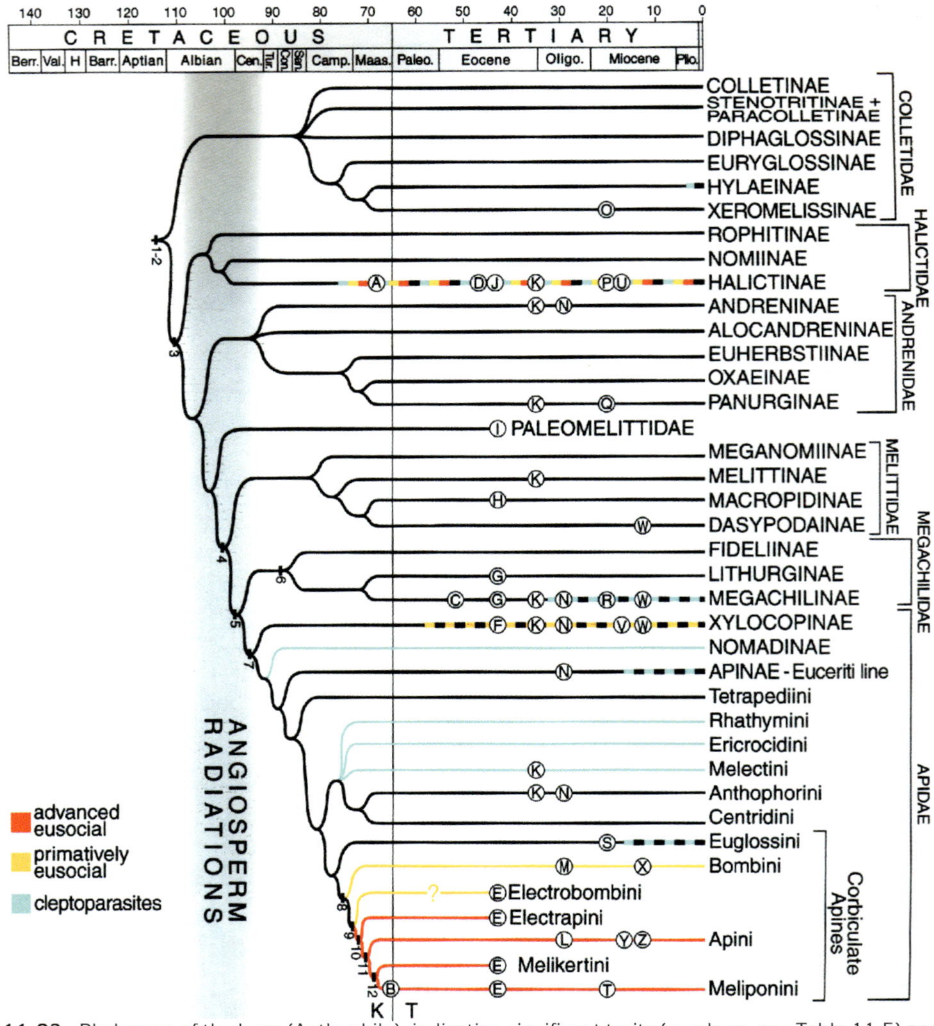

11.82. Phylogeny of the bees (Anthophila), indicating significant traits (numbers, see Table 11.5) and significant fossils (letters, see Table 11.6). Relationships based on Engel (2001a).

always clearly preserved. Thus, it may prove difficult to distinguish the earliest bee fossils from Cretaceous crabronid-like specimens.

The "phylogeny-based" age of the bees is intuitively pleasing because it agrees perfectly with major evolutionary events for plants. While flowering plants certainly arose in the earliest Cretaceous (see Chapter 14), they did not experience significant increases in diversity for some time following their origin. Paleobotanical evidence indicates the most dramatic radiation in angiosperm diversity between 110 and 90 MYA (Lidgard and Crane, 1988; Crane and Lidgard, 1990), a period that agrees with our estimated origin of bees. This does not imply that bees were responsible for the radiation of angiosperms because the association between small, generalized insects and early flowering plants probably existed for millions of years prior to bees. Early bees likely took advantage of the resources offered by early flowers to myriad insects. Given the occurrence of *Cretotrigona* by the Maastrichtian, however, it is clear that the bees did not take long to perfect their association, rapidly diversifying in the Cretaceous, certainly adding additional fuel to the angiosperm rise to dominance by the end of the Mesozoic. The episode of early bee evolution coincides with the appearance of numerous, derived floral structures associated with insect pollination (Crepet, 1996). Indeed, by the Turonian, approximately 90 MYA, flower types had evolved that today are tightly associated with bee pollination (Crepet and Nixon, 1998). Recent Bayesian estimates based on molecular data have been unable to refine divergence dates for bee evolution (Danforth *et al.*, 2004), the date ranges falling within those based on paleontological data (e.g., Engel, 1996, 2001a). The question of bee origins was further discussed by Engel (2001a).

While the initial diversification of bees took place in the mid-Cretaceous, subsequent diversification into some of the Recent subfamilies and tribes appears to have occurred during the Tertiary, particularly following the Eocene-Oligocene transition approximately 34 MYA. This is where many neontologists become confused. It does not necessarily follow that the age of a lineage is identical to the age of the group (tribe, genus, etc.) that we are familiar with today. Indeed, the

TABLE 11.5. Significant Characters for Bee (Anthophila) Phylogeny[a]

1. Feeding on pollen mixed with floral oils or nectar
2. Some body setae branched or plumose and hind basitarsus broader than following tarsomeres; see other traits in Michener, 2000
3. Glossal apex tapered
4. Submentum V-shaped, mesocoxa fully exposed, hind wing jugal lobe shortened
5. "Long-tongued" condition: first two labial palpal segments elongate, flattened, and sheath-like; see also Michener and Greenberg, 1980
6. Pollen carried in metasomal scopa
7. Clypeus laterally bent posteriorly; five pairs of ostia in dorsal vessel of metasoma; four or more ovarioles per ovary
8. Hind tibia modified into corbicula
9. Anterior condyle partly covered by clypeus; eusocial; see also Engel, 2001a
10. Arolia present; single or no metatibial spur; alar papillae absent
11. Outer mandibular grooves lost; see also Engel, 2001a
12. Strigil without anterior velum; reduced distal hamuli

[a] Numbers correspond to those on phylogeny, Figure 11.82.

TABLE 11.6. Significant Fossil Bees[a]

A. Various Late Cretaceous halictine nests (some putative nests have also been reported from the Cenomanian, which would potentially extend this lineage significantly)
B. *Cretotrigona prisca*
C. Various *Megachile* – cut leaves (e.g., Wappler and Engel, 2003).
D. ?*Halictus savenyei*
E. Numerous Baltic amber corbiculate apines (Engel, 2001a)
F. *Boreallodape* spp.
G. *Protolithurgus ditomeus*
H. *Eomacropis glaesaria*
I. *Paleomelitta nigripennis*
J. *Electrolictus antiquus*
K. Florissant Formation bees (Zeuner and Manning, 1976; Engel, 2001c, 2002f)
L. *Apis henshawi*
M. Various European Oligocene *Bombus* spp.
N. Various European non-apid bees
O. *Chilicola* in Dominican amber (Engel, 1999c).
P. *Oligochlora*, *Eickwortapis*, *Augochlora leptoloba*, *Neocorynura electra*
Q. *Heterosarus eickworti*
R. *Megachile* (*Chalicodomopsis*) *glaesaria*
S. *Euglossa moronei*
T. *Proplebeia* spp.
U. *Halictus*
V. Miocene *Xylocopa* spp. from Germany
W. Various Miocene bees from Shandong, China
X. *Bombus vetustus*
Y. *Apis armbrusteri*
Z. Undescribed, middle Miocene of Japan

[a] Letters correspond to those on phylogeny, Figure 11.82.

sequence of synapomorphies in a lineage may be spread across eons, and extinction can cull much of the evidence. For example, orchid bees (Euglossini) are certainly sister to all other corbiculate apines (Schultz *et al.*, 1999, 2001; Engel, 2001a,b; Noll, 2002) (Figure 11.82) and are a *lineage* that differentiated from the common ancestor of other corbiculate bees sometime during the Late Cretaceous, but it does not follow that Euglossini originated in the Cretaceous. Indeed, in this example euglossines, as we know them, probably are Tertiary in origin and acquired their specialized features sometime after the Paleocene. Stem-group euglossines from the Early Tertiary may only be remotely recognizable as being of that a euglossine-like lineage (perhaps exhibiting only one of the myriad of traits we know of for euglossines: e.g., the jugal comb at the base of the hind wing). Evolution does not bring fully formed groups into existence at the moment of cladogenesis; if it did, we would have little difficulty in reconstructing phylogeny even at the most ancient levels of life, and fossil "problematica" would not exist.

As already discussed, Paleocene deposits are not abundant, and it is, therefore, of little wonder that bees from this epoch are unknown. By the middle of the Eocene epoch, the period from which most bee lineages are first documented as fossils, representatives of the principal families and subfamilies are present, generally represented by taxa that are not basal for their respective lineages (e.g., Engel, 2001a) (e.g., Figures 11.84 to 11.86). Early Tertiary bees are almost exclusively known from Holarctic deposits (e.g., Engel, 1998a,b, 2001a,c, 2002f; Engel and Archibald, 2003; Wappler and Engel, 2003), making the exploration for fossil bees in southern deposits more pressing. The most diverse deposit of fossil bees is the 42 MYO Baltic amber. The most productive compression deposit is from Florissant, Colorado, but the nature of preservation renders positive identification for the majority of individuals difficult, if not impossible. Although bee fossils tend to be rare, the Dominican amber stingless bee, *Proplebeia dominicana* (Figure 11.89), is extremely common, clearly because these were captured while harvesting resin to construct nests (Figure 11.90). Other species of bees do occur in this amber (e.g., colletids, halictids, an andrenid, a megachilid, other apids), but each of these other species are known from a single individual or a rare few (e.g., Engel, 1995b, 1997, 1999a,b,c, 2000b; Engel and Rightmyer, 2000; Rozen, 1996; Camargo et al., 2000). The bees from the Oligocene and younger deposits are generally modern in appearance (Figure 11.91), providing little insight into relationships among bee lineages, but they do provide information on distributional changes through time. For example, today orchid bees (Euglossini), stingless bees (Meliponini), and species of *Neocorynura* (Halictidae) do not occur on

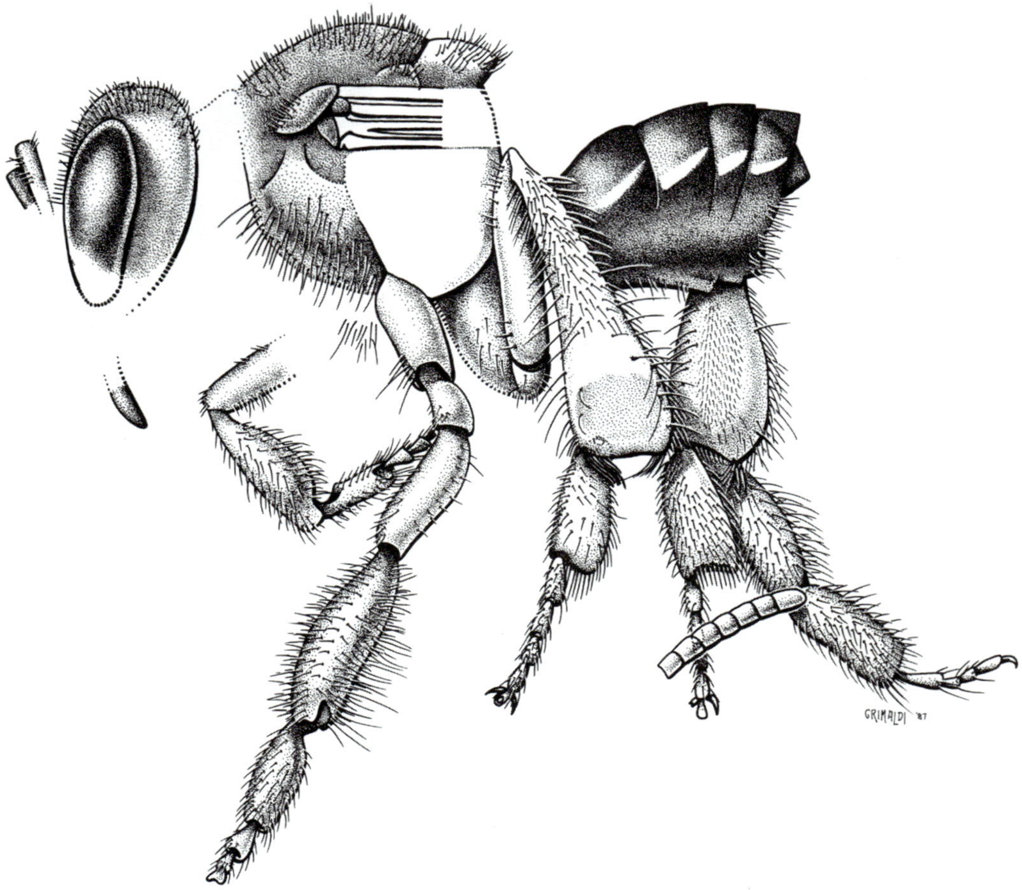

11.83. The oldest fossil bee, *Cretotrigona prisca* (Apidae), in amber from probably the latest Cretaceous of New Jersey, ca. 65 MYA. The oldest bee is also a social bee, attesting to the antiquity of sociality in the corbiculate apines. AMNH C88720; length 5.2 mm; from Michener and Grimaldi (1988a) and Engel (2000a).

11.84. An extinct relative of stingless bees and honey bees, *Protobombus indecisus* (Apidae), in Eocene Baltic amber. *Apis mellifera* is the only honey bee native to Europe today, but Baltic amber records an unexpectedly rich former diversity there of eusocial honey bees. Climatic cooling at the end of the Eocene and into the Oligocene probably contributed to their virtual extinction in Europe. AMNH B-JH98; length 8.3 mm.

11.85. The advanced social bee *Melikertes stilbonotus* (Apidae) in mid-Eocene Baltic amber is an extinct relative of the stingless bees (tribe Meliponini) that are found today throughout the world's tropics. AMNH B-JH82; length 3.7 mm.

11.86. *Ctenoplectrella viridiceps* (Megachilidae) in mid-Eocene Baltic amber, an extinct relative of some Old World mason bees. AMNH B-JH95; length 6.9 mm.

11.87. The primitive honey bee, *Apis henshawi* (Apidae) from the Oligocene of Rott, Germany. MCZ 6316; length 14 mm.

11.88. *Oligochlora eickworti* (Halictidae) in Miocene amber from the Dominican Republic. The bee fauna of Dominican amber, although harboring a few groups not found today in the West Indies, is remarkably modern and attests to the antiquity of halictids. Morone Collection, M0684; length 5.8 mm.

Hispaniola, although extinct members of these groups are documented from Dominican amber. Mexican amber, of roughly contemporaneous age, offers a mainland counterpart to Dominican amber, similarly documenting some taxa that today have more restricted ranges, such as the stingless bee genus *Nogueirapis* (Wille, 1959; Engel, 2004a).

In summary, the major episodes in the evolution of the Hymenoptera are:

- Origin and divergence from stem-group Hymenoptera + Panorpida sometime in the mid- to Late Triassic, about 230 MYA;
- Origin of the Euhymenoptera and parasitoidism likely near the Triassic-Jurassic boundary, about 210 MYA, with rapid radiation of Apocrita in the earliest Jurassic about 195 MYA, as evidenced by ephialtitoids and the earliest proctotrupomorphs;
- Origin of the stinging Hymenoptera (Aculeata) in the latest Jurassic, about 155 MYA, and radiation of many aculeate families in the earliest Cretaceous, about 140 MYA;
- Origin of the ants and bees in the mid-Cretaceous and during the rise of angiosperms, about 125–120 MYA, but ants not radiating extensively until later in the Tertiary;
- Origin of eusociality during at least the mid- (for ants and wasps) and Late (for bees) Cretaceous.

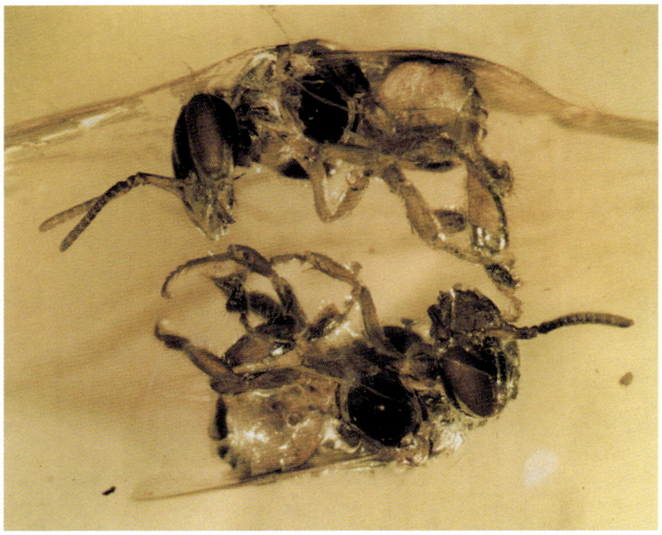

11.89. Two workers of the stingless bee *Proplebeia dominicana* (Apidae: Meliponinae) in Dominican amber. No stingless bees occur in Hispaniola today. Morone Collection, M0236; length 2.6 mm.

11.90. A small swarm of *Proplebeia dominicana* (Apidae) in Miocene Dominican amber. The species is common in Dominican amber because foragers became mired while collecting resin to construct their nests, a behavior that meliponines have today. AMNH DR-14-460.

EVOLUTION OF INSECT SOCIALITY

Of all adaptations in insects, behavioral and morphological, sociality is certainly among the most significant. Social insects are competitively far superior to other insects, and they dominate various terrestrial ecosystems, particularly tropical forests. Virtually all advanced societies in nature are those of arthropods, but there is also a great spectrum in the development of sociality. The simplest insect societies are extended brood care or gregarious associations among individuals of the same generation. This *subsocial* behavior occurs in a great diversity of animals, including such arthropods as scorpions, mites, opiliones, spiders, and numerous insects in the orders Coleoptera, Embiodea, Dermaptera, Heteroptera, and Zoraptera (Eickwort, 1981). *Communal* behavior entails the sharing of a common nest structure by a cohort of usually related individuals of the same generation, but these individuals care for their own offspring. Such aggregations are common among bees (e.g., Andrenidae, Halictidae) and in some bark beetles, larval symphytan wasps, and caterpillars (e.g., tent caterpillars). Communal living offers individuals a more effective group defense against predators and parasites, and members may possess complex communication systems to warn other nest members. This differs from the communication seen in subsocial insects, in which the transfer of information is just between mother and offspring.

Among all forms of societies, however, *eusociality* is the most sophisticated. Eusocial societies are those in which overlapping generations collectively care for the brood, which is produced by a single female or small set of related females, the remaining individuals being either sterile or reproductively inactive (Wilson, 1971). The related condition *semisocial* refers to those societies that lack the overlap of generations. Among eusocial animals there are varying degrees of complexity in the development of the sterile worker caste. In highly eusocial societies the workers are morphologically differentiated from the reproductive caste and usually never reproduce, such as in many termites and ants. In primitively eusocial societies there is a gradation in anatomy between queen and workers (such as in abdomen or head size), or there is no perceptible difference at all as in bumble bees, many sweat bee societies, and primitive ants like amblyoponines and myrmeciines. In these primitively eusocial insects, workers may reproduce when circumstances allow.

Some degree of advanced social (semisocial or eusocial) behavior is found in arthropods as disparate as snapping shrimp (Duffey, 1996), spiders (Vollrath, 1986), and, of course, a diversity of insects (Wilson, 1971). Advanced social behavior appears to have evolved about 20 times in insects alone (Table 11.7), and social insects have developed the most sophisticated societies in nature. These include termites, ants, some bees, a few spheciform wasps (*Microstigmus*), many vespid wasps, and some thrips and aphids. Outside of arthropods, eusociality occurs only in the naked mole rat of Africa, *Heterocephalus glaber* (Rodentia: Bathyergidae); interestingly, mole rat castes have been found to be anatomically differentiated, albeit weakly (O'Riain *et al.*, 2000). At various times in history, humans developed eusocial societies, when eunuchs guarded royal harems, but despite the oppressive size and physical complexity of human societies in general, they still lack the individual specialization and sheer efficiency seen in arthropod societies.

Two traits are required for the evolution of an advanced society: living in a nest or other persistent domicile, and close genetic relatedness. The nest provides a discrete enclosure where individuals meet and store food, and, most importantly, it allows for the cooperative protection and extended care of the brood. If it is a particularly sturdy structure, this will allow multiple generations to occupy it. The construction of a nest or cooption of a preexisting structure for a domicile is a feature common to all eusocial organisms, which in fact appears to have phylogenetically preceded social behavior. For example, social aphids in the Pemphiginae and Hormaphidinae, and *Cladothrips* and *Oncothrips* Thysanoptera, belong to lineages that construct persistent galls, but most species in these lineages are not social. Social Hymenoptera are all aculeates, and many of these construct burrows or nests that they provision with food for their larvae (Evans, 1964; Evans and West-Eberhard, 1970).

The other major catalyst for advanced social behavior is close relatedness among the colony members. Indeed, in most advanced insect societies the colony members are closely related. In the social aphids, they are clones of each other because they reproduce parthenogenetically. All Hymenoptera are haplodiploid, which is a genetic sex-determining mechanism where males are haploid and females diploid, conferring high relatedness among female sibs (0.75, instead of the typical 0.50 seen in diploid species). This genetic system is considered the main reason for the repeated evolution of eusociality in Hymenoptera. Because workers are more closely related to each other than to any offspring they would produce, their "inclusive fitness" is maximized by tending their sibs (Hamilton, 1964). Kin selection and haplodiploidy probably explains why eusociality has evolved more often and involves more species in Hymenoptera than in any other insect order. Supporting facts that seem almost too facile for the kin selection theory of social evolution involve the social thrips and *Austroplatypalpus* bark beetles, which belong to groups that are also haplodiploid! The only major exception to the kin selection theory involves termites, where colony members are diploid and sibs are not significantly more closely related to each other than they would be to their own offspring. A chromosomal sex-determining mechanism has been implicated in making termite sibs slightly more closely related to each other than they would be to their own offspring (Lacy, 1980), but the traditional explanation is that termite sociality originated and is maintained by the physiological need of nest mates to be inoculated with symbiotic microbes. Kin selection is clearly a driving force in the evolution of societies, but this force needs an appropriate ecological context: a domicile.

How and when did advanced insect societies begin? As we discussed in several previous sections, eusociality clearly first evolved in the common ancestor of all termites in the Early Cretaceous, perhaps 140 MYA, and in the ancestor of all ants in the mid-Cretaceous perhaps 115–120 MYA, and there is no

11.91. A bee of the living genus *Halictus* (Halictidae) from Rubielos de Mora in Spain. The fossil record of halictid bees was, until recently, believed to be sparse, but diverse fossil species are now known. MPZ-98/423; length 6.5 mm. Specimen courtesy E. Peñalver.

11.92. A colony of *Plebeia* bees (Apidae), an advanced social stingless bee, in Panama. Workers surround the nest entrance, ready to protect the colony with irritating bites. Photo: P. J. DeVries.

fossil evidence that social insects existed before then. The earliest known ants are fossilized in mid-Cretaceous amber from Burma, France, and New Jersey (Wilson *et al.*, 1967; Grimaldi *et al.*, 1997, 2002; Engel and Grimaldi, in press), and the earliest termites are compression fossils from approximately 130 MYA (see discussion of Isoptera). The worker-like structure of some sphecomyrmine ants from the Cretaceous indicates that the most primitive known ants were social (Wilson, 1987; Grimaldi *et al.*, 1997), but the body proportions of Cretaceous Armaniidae (close relatives to ants) suggest they were not social (Wilson, 1987). Termites presumably diverged from a common ancestor with wood-eating cryptocercid roaches in the earliest Cretaceous, ca. 140–145 MYA. In the bees and vespid wasps, however, there are many species intermediate between solitary and social, and sociality has repeatedly evolved and been lost, so these insects provide special insight into the evolution of sociality. Species of halictine bees, in particular, range from communal to semisocial to eusocial, and *within* some species they can even vary from solitary to eusocial, which is why they have been scrutinized in studies of social evolution (e.g., Michener, 1974, 1990). The origins of eusociality are more obscure in corbiculate bees than in halictines, despite the fact that this group has received more attention than any other social insects besides ants, particularly the honey bees.

The corbiculate bees consist of four modern tribes – the highly eusocial honey bees (Apini: *Apis*) and stingless bees (Meliponini), the primitively eusocial bumble bees (Bombini: *Bombus*), and the solitary or more frequently communal orchid bees (Euglossini). Unlike halictines, corbiculate apines have no living species that represent intermediate stages in social behavior. Interestingly, the most enduring hypothesis of apine social evolution was made in *The Origin of Species*, in which it was noted that stingless bees were intermediate between honey bees and bumble bees in social organization (Darwin, 1859). Darwin lacked information at the time on orchid bees and so he didn't discuss them, but he implied a single origin of eusocial behavior in the ancestor of the three eusocial corbiculate bee tribes, and then a single origin of advanced eusociality in the common ancestor of stingless and honey bees. Darwin's hypothesis was corroborated by subsequent studies (Cockerell, 1908c; Michener, 1944; Schultz *et al.*, 1999, 2001). Some recent studies have hypothesized different relationships for the corbiculates (e.g., Winston and Michener, 1977; Cameron, 1993; Mardulyn and Cameron, 1999; Koulianous *et al.*, 1999), but these have not held up to scrutiny against larger, more comprehensive studies based on morphological, molecular, and behavioral data (e.g., Prentice, 1991; Roig-Alsina and Michener, 1993; Chavarría and Carpenter, 1994; Schultz *et al.*, 1999, 2001; Ascher *et al.*, 2001; Serrão, 2001; Hart and Ratnieks, 2002; Noll, 2002).

Paleontological data also weigh into Darwin's hypothesis for the single origin of eusociality in corbiculate bees. Compared to the early fossil record of other eusocial insects, corbiculates have a relatively good fossil record, which indicates that from the Oligocene to the present they were typical members of modern tribes (Cockerell, 1931; Wille, 1959; Wille and Chandler, 1964; Rasnitsyn and Michener, 1991; Engel, 1998a,b; Riou, 1999; Camargo *et al.*, 2000). During the Eocene, however, there were diverse corbiculate bees in not only the living tribes (e.g., *Liotrigonopsis* stingless bees in Baltic amber), but also in tribes with traits intermediate between modern ones (Engel, 1998b; 2001a,b; Wappler and Engel, 2003). Moreover, many of the Eocene bees were morphologically workers (e.g., *Electrapis*, *Melikertes*, *Melissites* [e.g., Figure 11.93], *Protobombus*, *Roussyana*, *Succinapis*, *Thaumastobombus*) – a feature only of highly eusocial insects. Early fossils thus corroborate a single origin of eusociality in these bees by revealing transitional forms. Eusocial corbiculate bees actually evolved in the Late Cretaceous (the oldest fossil, *Cretotrigona prisca*, is apparently from the latest Cretaceous) (Michener and Grimaldi, 1988a,b; Engel, 2000a), and these bees appear – unexpectedly – to have peaked in their distribution and diversity in the Eocene. Because corbiculates have been considered one of the most highly and recently evolved lineages of bees, they have traditionally been thought to presently be in nascent stages of their evolution. Their decline may be a result of the cooling of the earth since the Eocene, since these bees are essentially tropical, or less competitive corbiculates may have succumbed to recently evolved superbees like *Apis* (Engel, 2001a,b).

11.93. Although no stingless bees occur in Europe today, during the Eocene true meliponines occurred there, as did such extinct relatives as this worker of *Melissites trigona* (Apidae) in Baltic amber. Eocene fossils of corbiculate bees have proven critical for reconstructing the phylogeny and evolution of advanced social bees. AMNH B-JH102; length 4.2 mm.

The situation for the Vespidae is perfectly analogous to that of the corbiculate bees. The eusocial vespids comprise

TABLE 11.7. Eusocial Animals

Taxa	No. Origins	Domiciles	Nonreproductive Caste	Sex Mechanism
Vertebrata				
Naked mole rat[a]	1	Extensive burrows	Workers	Diploid
Crustacea				
Synalphaeus snapping shrimp	3	Inside marine sponges	Workers? Soldiers	?
Araneae[a]				
Various species (e.g., *Anelosimus*)	>2	Silk galleries	Workers?	?
Thysanoptera				
Phlaeothripine thrips (*Oncothrips*, *Kladothrips*)	>2	Acacia galls	Workers? Soldiers	Haplodiploidy
Hemiptera				
Aphidoidea aphids (some Pemphigidae, Hormaphididae)	6	Galls	Soldiers	Thelytoky
Isoptera				
All termites	1	Various	Workers Soldiers	Diploid (sex linkage)
Coleoptera				
Platypodidae *Austroplatypus incompterus*[a]	1	Subcortical galleries	Workers	Diploid
Hymenoptera				
Vespidae	1	Paper nests	Workers	Haplodiploidy
Ants				
All Formicidae	1	Various	Workers Soldiers	Haplodiploidy
Crabronidae				
Microstigmus[a]	1	Paper nests	Workers	Haplodiploidy
Anthophila (bees)				
Halictinae[a]	>4[b]	Ground/twig cavities/rotten logs	Workers	Haplodiploidy
Xylocopinae	1	Stem cavities	Workers	Haplodiploidy
Apinae	1	Waxy/resin combs	Workers	Haplodiploidy

[a] Primitively eusocial; caste differences are slight.
[b] In addition to origins of eusociality, there have been several reversions from a eusocial ancestry to solitary behavior.

one lineage within which the primitively eusocial stenogastrines are basal. Limited molecular studies have suggested multiple origins of eusociality in these wasps (Schmitz and Moritz, 1998), but as for the bees such studies have not held up to scrutiny (Carpenter, 2003). Eusociality also appears to be ancient in vespids, since there is a fossil polistine nest from the Late Cretaceous, and so eusociality probably evolved in these wasps around the time it did in ants, in the mid-Cretaceous.

Besides the unexpected extinction of Eocene corbiculate bees, fossils have provided other insight into the evolution of social insects. Specifically, all evidence indicates that termites and ants were rare and primitive in the Cretaceous, representing usually far less than 1% of all insect specimens. These insects then exploded in biomass and diversity in the Eocene, possibly slightly earlier, and have maintained ecological dominance ever since. It seems that their late dominance is attributable to the radiations of *particular* termites and ants, namely those that form huge colonies, like the Termitidae, and in ants the Formicinae, Dolichoderinae, Dorylinae, and the leaf-cutting attines. The earliest fossil evidence for termites and ants indicates that they were social even then, but for the first 90 MY of termite existence and 70 MY of ant existence they were very minor players in ecosystems. Most likely, Cretaceous ants and termites lived in very small colonies, and the dominance of these groups today is a result of the evolution of huge colonies in the Tertiary.

12 Antliophora: Scorpionflies, Fleas, and True Flies

PANORPIDA

While beetles comprise essentially one-half of holometabolan diversity, Panorpida and Hymenoptera comprise the other half. Panorpida is a group consisting of the orders Mecoptera (scorpionflies), Siphonaptera (fleas), Diptera (true flies), Trichoptera (caddis flies), and Lepidoptera (moths and butterflies) (Figure 12.1; Table 12.1), and possibly the bizarre, twisted-winged parasites in the small order Strepsiptera. The sister group to Panorpida appears to be Hymenoptera. The name Panorpida, or panorpoid orders, began with Hinton (1958) and has been used occasionally (e.g., Kristensen, 1989a); the names Mecopteroidea and Mecopterida have also been used for this group (Kristensen, 1975, 1991; Willmann, 1989; Whiting *et al.*, 1997). We are using the name Panorpida here since Mecopterida will be used in a more restrictive sense (Figure 12.1). The close relationship of Hymenoptera to the Panorpida is based on two larval characters (leg with one claw; silk produced by labial glands), the 18S and 28S rDNA sequences (Whiting *et al.*, 1997), and the adult prothorax, which is reduced to a size half or less that of the meso- or metathoracic segments. Monophyly of the Panorpida themselves is based on several rather obscure morphological features, as well as the overt reduction and even loss of the ovipositor (Mickoleit, 1973). Hymenoptera have retained a complete ovipositor. Obscure features of Panorpida include a muscle connecting the pleura and the first axillary sclerite at the base of the wing; several structures of the larval maxilla and labium, such as loss of labial palp muscles; and basal fusion of veins CuP and A_1 in the hind wing (Kristensen, 1975, 1991). The monophyly of Panorpida is also supported by molecular data (Whiting *et al.*, 1997).

The last common ancestor of Diptera, Mecoptera, and Trichoptera was certainly in the Permian, when stem groups to these three orders existed (Figure 12.1). The order Lepidoptera is much younger. Hymenoptera evolved in the Triassic since the most basal taxa first appear about 220 MYA; therefore, this order is younger than the Panorpida. As the sister group to the Panorpida, Hymenoptera must have diverged from stem-group panorpoids in the early Triassic.

It is difficult to say what the lifestyle was of the ancestral Panorpida. The larva of the most basal Recent mecopteran (Nannochoristidae) is aquatic, as are virtually all Trichoptera larvae. The larvae of most basal Diptera (Tipuloidea, other nematocerans) live in wet soils or the sediments of swamps, lakes, and rivers. If the ancestral panorpoid larva was aquatic, then terrestrial larvae independently evolved in Lepidoptera, Mecoptera, and Diptera.

ANTLIOPHORA: THE SCORPIONFLIES, TRUE FLIES, AND FLEAS

The Antliophora is comprised of the Diptera and mecopterids, the latter being the Mecoptera + Siphonaptera. Data from DNA sequences (Whiting *et al.*, 1997) suggest placement of the enigmatic order Strepsiptera in this group, which is controversial and was discussed previously (chapter 10). With or without Strepsiptera, molecular support for a monophyletic Antliophora is strong. The group was originally defined by the presence of a sperm pump in the male, a structure that aids the transfer of sperm during copulation ("antlia" is Greek for pump). Early groups of insects, such as apterygotes and basal neopterans, transfer sperm in a packaged spermatophore; sperm is liquid in virtually all Antliophora. However, a sperm pump does not seem to be in the antliophoran ground plan but is independently derived in Diptera and mecopteroids since its structure differs in the two groups. It does not occur in the basal mecopteroids Nannochoristidae and Boreidae (Willmann, 1989), and its presence and structure is difficult

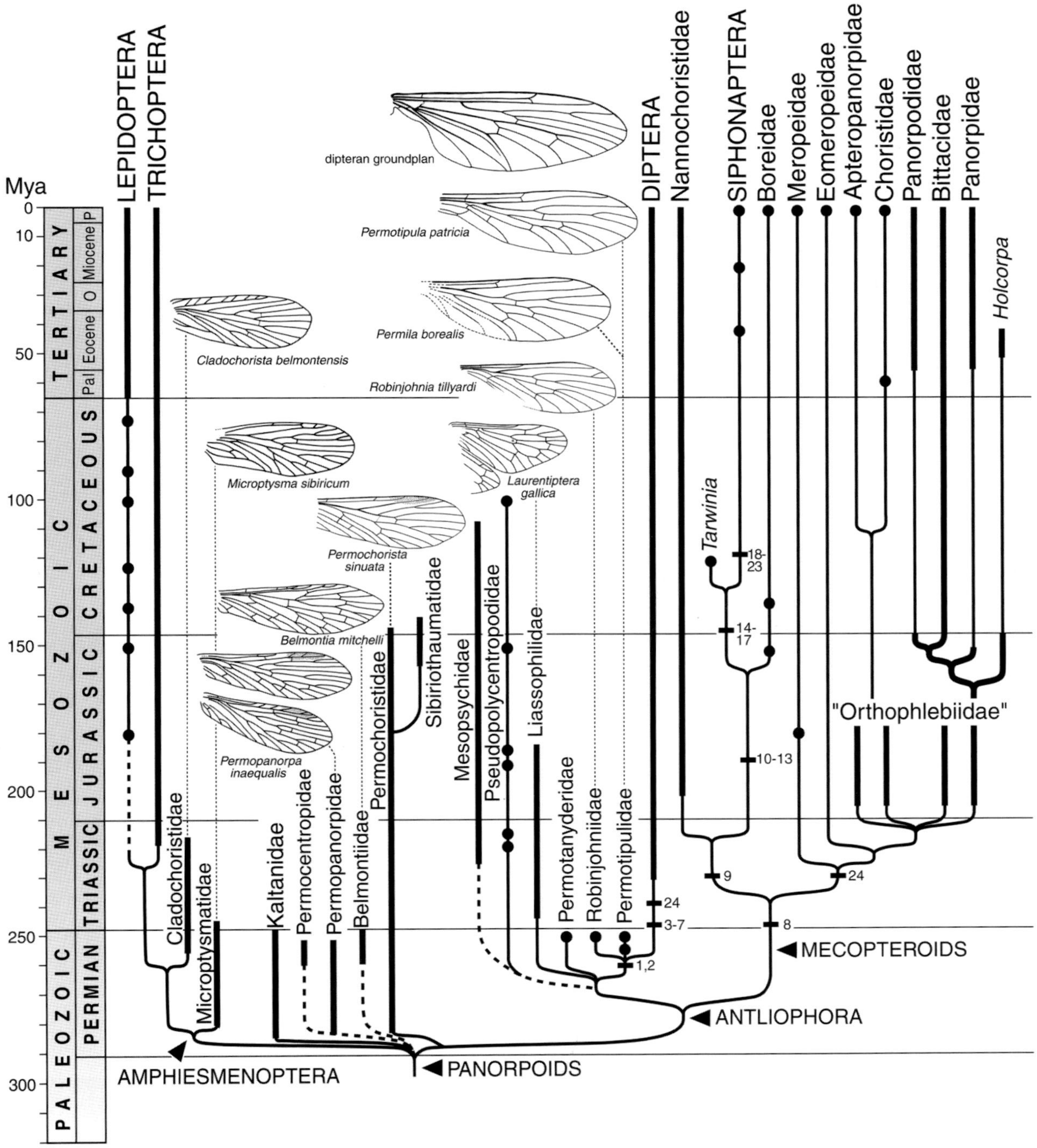

12.1. Phylogeny of the Panorpida, which includes the scorpionflies, fleas, true flies, caddisflies, and lepidopterans. Relationships of the fossils are based on Willmann (1989), Novokshonov (2002), and others; most relationships of mecopteroids are based on Willmann (1989) and Whiting (2001). Significant characters are denoted by numbers for some lineages, which are listed in Table 12.1. Isolated or particularly significant fossils are denoted by dots; diverse fossils and entire fossil ranges are denoted by thick lines.

to determine for the Siphonaptera because the male terminalia of fleas are so complex. Other structures also define the group (Kristensen, 1975, 1991); many of them are on the mouthparts: dagger-like adult mandibles with the anterior articulation reduced (mandibles are lost in most Diptera), one endite lobe on the maxilla and none on the labium, the labial palp reduced to two segments, loss of certain labial muscles, and the posterior notal wing process has a pleural muscle inserted on it. Basal Diptera, like the craneflies and relatives, superficially resemble mecopterans, with their long wings and long, dangling legs. It is now widely agreed that the mecopterans are paraphyletic with respect to the fleas because one particular group of scorpionflies seems closely related to fleas.

TABLE 12.1. Significant Characters in Antliophoran Phylogeny

1. Forewing with radial branching and number of crossveins reduced
2. Forewing with anal region and venation reduced
3. Reduction of the pro- and metathorax
4. Hind wings reduced to halteres
5. Labial palps fused into a labellum
6. Mandibles reduced (present in females of bloodsucking species)
7. Male abdominal spiracle eight lost
8. Sclerotized spicules (acanthae) in the proventriculus
9. Molecular characters; ovary structure (Simiczyjew, 2002)
10. Wings reduced or absent
11. Pupae form a silken cocoon
12. Jumping ability
13. Structure of reproductive system (Biliński *et al.*, 1998)
14. Lateral compression of the body
15. Winglessness; great reduction of the thorax
16. Antennal flagellum compact
17. Pygidium with patch of sensilla (the sensilium)
18. Eyes lost or virtually so
19. Ocelli lost
20. Antennae reduced, with nine flagellar segments, recessed into scrobe
21. Mouthparts suctorial, blood feeding
22. Hind legs with large hind coxae and femora, pads of resilin
23. Larva is legless, with long, stiff setae and pair of anal struts
24. Sperm pump (convergently evolved)

[a] Numbers correspond to those on phylogeny, Figure 12.1.

MECOPTERIDA: MECOPTERANS AND SIPHONAPTERA

Like Neuroptera, the Recent species of scorpionflies appear to be a vestige of a long, diverse evolutionary history, which is beautifully reflected in their anatomy, habits, fossils, and distributions. Mecoptera are among the most morphologically generalized Holometabola, the adults generally having a host of primitive features: three large ocelli; mandibulate mouthparts; both pairs of palps; long, filiform antennae; and little differentiation of the meso- and metathoracic segments, with each of the thoracic segments having a well-defined scutum, scutellum, and postnotum. Also, their wing venation is relatively complete and intricate, and the shapes of fore- and hind wings are very similar; they have five tarsomeres, an 11-segmented abdomen, and little or no fusion of ganglia in the ventral nerve cord. Their larvae are generally eruciform, with three pairs of well-developed, three-segmented thoracic legs, and the pupae are exarate and decticous, with all appendages including mandibles fully formed and adult-like. Even their habits are primitive, as larvae and adults are generally scavengers. Moreover, the distributions of some families are dramatically disjunct, past distributions having been widely fragmented by extinction. Lastly, at the family and generic levels, fossil diversity is at least three times the Recent diversity, an amount of extinction unseen in any other group of Holometabola.

EARLY HISTORY

There is a bewildering array of Permian and early Mesozoic forms of mecopteroids, most of course known only as wings, and so their relationships are obscure and even confusing. Not surprisingly, the earliest mecopteroids represent paraphyletic stem groups to the Recent mecopterans (Figure 12.1, Table 12.1). Nevertheless, because Diptera diverged from a stem-group mecopteroid in the Permian, there is considerable interest in deciphering this early history. With nearly 400 fossil species in the order Mecoptera, sorting out the extinct diversity would be virtually impossible if not for

12.2. Early Cretaceous mecopteroids from the Yixian Formation of China, belonging to the paraphyletic Mesozoic "family" Orthophlebiidae. Above, forewing length 22 mm; below, forewing 15 mm.

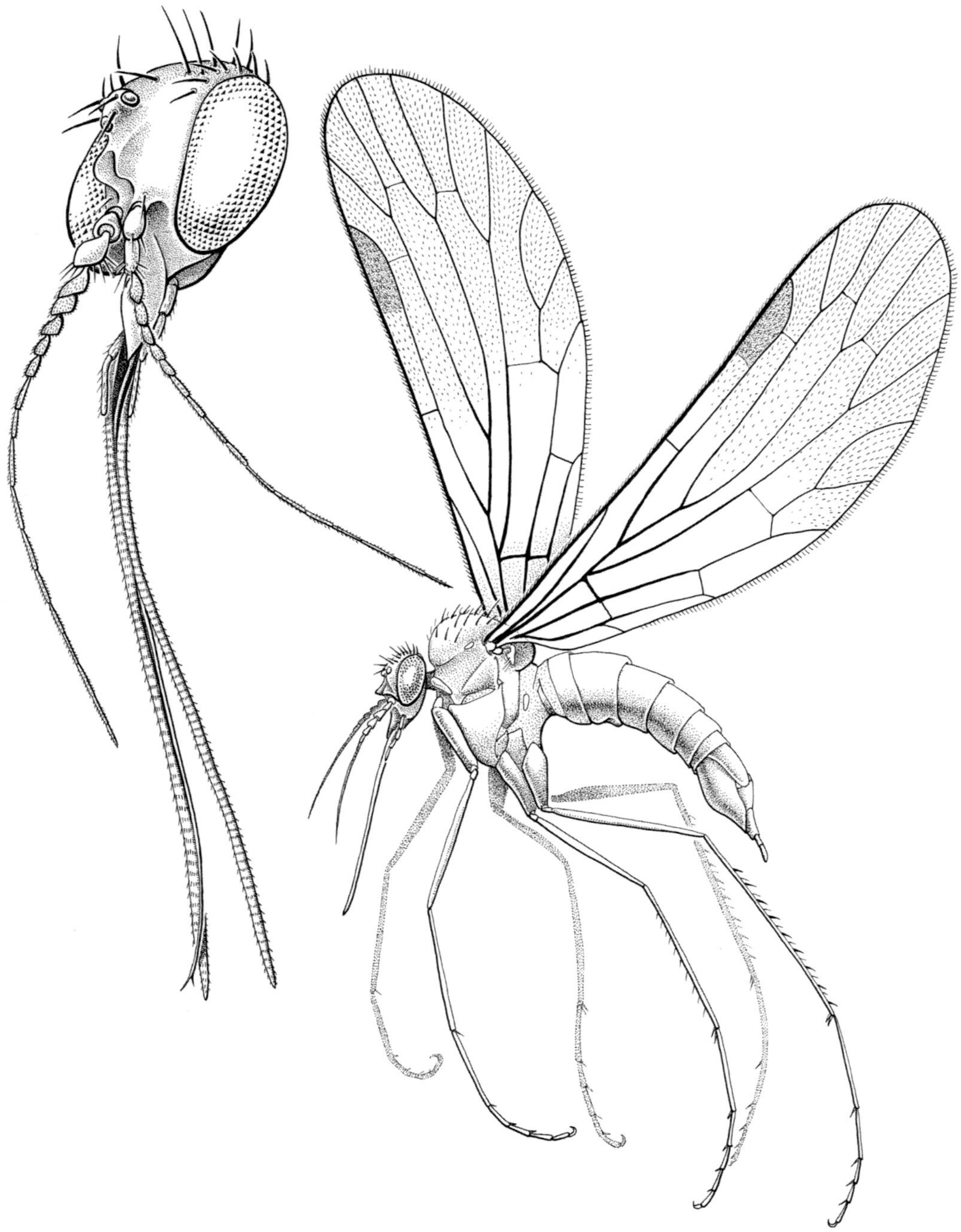

12.3. A mosquito-like scorpionfly, family Pseudopolycentropodidae (reconstructed) from mid-Cretaceous amber of Burma. Similar to flies, the hind wings have been reduced to small lobes. The proboscis does not appear to have been for feeding on blood. Body length 3.0 mm.

the catalogues of fossil mecopterans by Willmann (1978) and Carpenter (1992). Still, relationships are ambiguous. Willmann (1978), for example, recognized 25 extinct families; Novokshonov (in Rasnitsyn and Quicke, 2002), 19 families; and Carpenter (1992), only 16. To add to the confusion, at least the Permo-Triassic family Permochoristidae and the Mesozoic family Orthophlebiidae are paraphyletic, the former supposedly ancestral to basal Recent families of mecopterans (Novokshonov, 1998, 2002) and the latter ancestral to the more recently evolved families (Willmann, 1989; Novokshonov, 2002; Figure 12.2). Four genera alone in these two families – *Mesochorista, Petromantis, Mesopanorpa,* and *Orthophlebia* – occur from the Permian through the Jurassic of Eurasia and Australia, and they contain one third of all fossil mecopteran species. Novokshonov (1997) monographed fossil mecopteroids but emphasized just Eurasian taxa. This work was summarized elsewhere (Novokshonov, 1998, 2002). Some groupings he proposed are similar to those presented in the monograph by Willmann (1989), but others are wildly different. For example, Novokshonov (2002) placed the relict

Recent family Meropeidae amidst a distinctive, extinct lineage of three mecopteroid families, for which there is actually little basis. Thus, we have basically followed Willmann (1989) for the relationships of fossil mecopteroids and stem-group Antliophora (Figure 12.1).

Mecopterids were very abundant and diverse in the Late Permian, Triassic, and Jurassic, and by the Cretaceous their numbers waned considerably. The earliest putative mecopteroids are the Permian families Kaltanidae, Permopanorpidae, and somewhat later, Permocentropidae, Belmontidae and several other families (Figure 12.1). Willmann (1989) considered these to be stem group panorpoids, basal to Amphiesmenoptera as well as Antliophora.

Perhaps the most significant species from the Permian comprise a group of three, possibly four, families that appear to be the closest relatives of the flies, order Diptera. These are the Permotipulidae, Permotanyderidae, Robinjohniidae, and perhaps some of the Permian species of Permochoristidae, such as *Prochoristella* (Figure 12.1). The Mesozoic family Liassophilidae is probably also closely related to these. According to Shcherbakov *et al.* (1995), Robinjohniidae is the closest relative of the true flies; Willmann (1989) considered Permotipulidae (*Permotipula* and *Permila*) to be the closest dipteran relative. Indeed, both of these genera are close relatives to the flies, though it is difficult to determine which may be closer.

Another early mecopteroid group of particular interest comprises three families from the Early to Late Mesozoic: Mesopsychidae, Aneuretopsychidae, and Pseudopolycentropodidae. These do not appear to be true mecopterans but rather stem-group mecopteroids showing remarkable parallel evolution with the Diptera. They had reduced venation, and the Pseudopolycentropodidae also had small hind wings, which led early workers to consider that some of these mecopteroids (and a few others) were the closest relatives or even ancestors of Diptera. Even more remarkable were the long, stylet-like proboscides in *Aneuretopsyche* and some of the younger species of the family. A species of *Pseudopolycentropus* from the Upper Jurassic of Kazakhstan had a long, stylet-like proboscis, but two similar species in mid-Cretaceous amber from Myanmar were even more remarkable. Species in Burmese amber had a long, fine, rigid proboscis – very similar to that of a mosquito – and hind wings reduced to small knobs, even smaller than the halteres of Diptera (Grimaldi *et al.*, in press) (Figure 12.3). Female terminalia were very primitive and similar to those of the basal Recent mecopteran, *Nannochorista*, and thus confirm Willmann's (1989) hypothesis that Pseudopolycentropodidae are basal mecopteroids. Both Burmese amber pseudopolycentropodid species even lacked the small nygmata on the wing that virtually all mecopterans have. Oddly, the labial palps were highly reduced or lost, and the antenna evolved an arista, in which the apical flagellomeres were very fine and whiplike. In the Holometabola, such an antenna is seen only in the higher (brachyceran) flies and in one Cretaceous family of nematocerous flies, the Archizelmiridae (Grimaldi *et al.*, 2003a). Wing, thorax, and antennal structure suggest these mecopteroids were agile fliers. They may have just been predatory or scavenging, like mecopterans today, but such a long, fine proboscis usually is found in insects that feed on two diets: the nectar and pollen of flowers, or vertebrate blood. The pseudopolycentropodid proboscis probably was not capable of piercing skin because it didn't have fine teeth at the tip of the stylets. Moreover, blood and nectar feeding are not mutually exclusive since female mosquitoes and other culicomorphan flies feed on blood but both sexes (particularly males) commonly feed on nectar.

12.4. Detail of the specimen in Figure 12.5 showing a ventral view of the body of *Saurophthirus longipes*. A huge clypeus is visible, with small mouthparts at the tip.

Among the most intriguing fossils of all insects must be two apterous, apparently ectoparasitic, mecopteroids from the Late Mesozoic of Eurasia. *Saurophthirus longipes*, from the Early Cretaceous Zaza Formation of Baissa, central Siberia, is extremely well preserved in part and counterpart and was a large insect (12 mm [0.5 in.] body length), with very

12.5. A sprawling mecopteroid from the Late Jurassic of Kazakhstan, *Saurophthirus longipes*. Compact antennae; the small, wingless thorax; and apparently suctorial mouthparts suggest it was an ectoparasite, probably on a large vertebrate. Coloration was even well preserved: The body and tarsi were reddish, and the coxae and femora were dark. PIN 3064/1898; body length 12 mm.

long, slender legs (Figures 12.4, 12.5). Even the coloration is apparent: the body, tibiae, and tarsi were reddish, and the bases of the legs were dark brown or black. It was clearly an apterous insect, the adult having well-preserved (and unusual) terminalia and a small, wingless thorax. An ectoparasitic lifestyle is further suggested by the membranous, perhaps distensible abdomen, aptery, and head and leg structure. Legs were long but splayed, and the very large claws suggest they were for grasping. Similar proportions of body and legs occur in some ectoparasitic batflies. Details of the head are very well preserved, revealing a large clypeus and a short, slender beak, so clearly it was suctorial. Antennae were multiarticulate but short and with compact segments, as in ectoparasites. Lastly, the body was covered with rows of short, stiff spicules that pointed backward and that probably functioned like the specialized combs of bristles (*ctenidia*) that help anchor many Recent ectoparasites to the hairs and feathers of their hosts. Ponomarenko (1976) proposed, by process of elimination, that *Saurophthirus* is a mecopteroid that was probably closely related to fleas. We agree with the mecopteroid placement, based on tarsal and antennal structure, but none of its features indicate a special relationship with fleas. Ponomarenko (1976) further proposed that *Saurophthirus* was a pterosaur ectoparasite because some pterosaurs are known to have had hairlike vestiture. Given its age, it is also quite possible for *Saurophthirus* to have parasitized early birds or feathered dinosaurs, though mammal hosts are less likely because at the time they were relatively small compared to the insect.

The other remarkable fossil is *Strashila incredibilis*, from

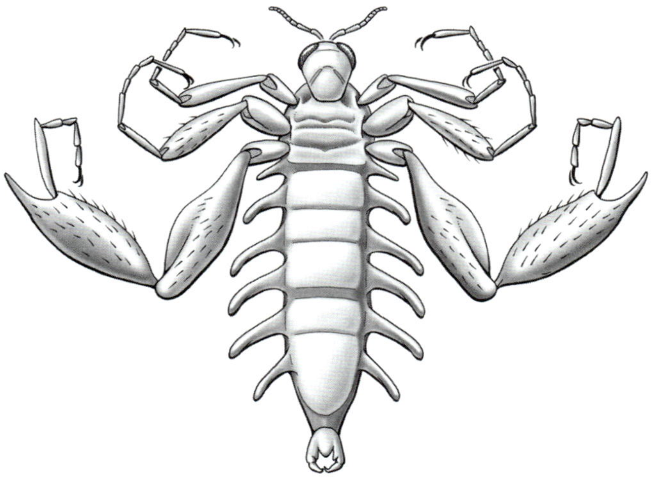

12.6. Another bizarre, apparently ectoparasitic mecopteroid, *Strashila incredibilis*, reconstructed from the part and counterpart specimen from the Late Jurassic of eastern Siberia. The huge hind legs were probably used for grasping onto its host. The lobes on the abdomen are an enigma. PIN 3084/60; body length 6 mm.

the Late Jurassic Bada Formation of Khilok, Russia (Figure 12.6). It was about half the size of *Saurophthirus*, and similarly had a reduced, wingless thorax and compact antennae. The unique specimen was clearly an adult because the male terminalia are very well preserved; in fact, these had an antliophoran structure. Mouthparts are not evident, though a suctorial beak was figured and described in the original description by Rasnitsyn (1992). Most striking, though, are the legs and appendages on the abdomen. Six pairs of fleshy lobes projected from the sides of the abdomen (presumably formed from the pleural membranes), a pair per segment. It is possible that these were more elaborate versions of the eversible abdominal glands presently found in higher mecopterans. Similar structures are also known in a few Recent species of empidid flies that have long, annulate lobes projecting from the abdomen, like the tentacles of an alien occupant. The fore- and mid legs of *Strashila* were slender and fairly long, and the hind legs were huge. The swollen hind femur and huge tibia, as well as the folded position of the tarsi, indicate this insect could tightly grasp its host. Widely separated coxae indicate the insect had sprawling legs. In fact, the leg structure of *Strashila* suggests it probably scuttled easily over its host but may not have been able to walk on the ground.

Without wings, the exact relationships of *Strashila* and *Saurophthirus* among mecopteroids will remain speculation. Yet, these and the mosquito-like *Pseudopolycentropus* provide rare glimpses of past mecopteroid faunas with species that were diverse, and bizarre in some cases.

RECENT DIVERSITY AND RELATIONSHIPS

Though there are only about 600 Recent species of mecopterans, they have attracted an inordinate amount of attention among entomologists. Significant publications include the following: Byers (1987), a review of immature stages; Byers (1991), a review emphasizing the Australian fauna; Grassé (1951) and Kaltenbach (1978), comprehensive reviews of the taxonomy, morphology, and biology; Byers and Thornhill (1983), a brief review of biology; and Willmann (1989), a major monograph on relationships of Recent and fossil families. A website gives a checklist of the world species up to 1991 (*http://www.calacademy.org/research/ entomology/mecoptera*).

Despite their primitive nature, the Recent mecopterids (including Siphonaptera) appear to be a cohesive and monophyletic group with some interesting specializations. Mecopterids have been defined (Kristensen, 1975; Mickoleit, 1971) by a medially divided first abdominal sternite, the female genital chamber formed by an invagination behind sternum nine, fusion of the clypeus and labrum (and the resulting loss of labral muscles), and the loss of certain other mouthpart muscles. Many of them have a rostrum, formed by elongation of the clypeus + labrum, mandibles, hypopharynx, and other mouthparts (Figure 12.24); predatory species even have long, tong-shaped mandibles. Mecopterids also have a distinctive proventriculus lined with spicule-like teeth, or *acanthae*. Mecopteran acanthae are entirely cuticular, and each is formed from a single epithelial cell (Richards and Richards, 1969). There is a great diversity of acanthae and similar proventricular spines in insects, reflecting their multiple origins (Boudreaux, 1980). Though Boudreaux (1980) suggested that acanthae existed in all antliophorans, but were lost in all flies, the uniform proventricular structure in mecopterids actually supports the view that fleas are highly modified mecopteroids. Acanthae probably never occurred in flies because they diverged very early in the evolution of antliophorans. Another common feature of mecopterans is a bulbous male genital capsule, with complex genitalia and large, clasping gonocoxites. This structure is held curled over the abdomen in the Panorpidae, superficially resembling the tail of scorpions (and hence the common name of the mecopterans). Small, sclerotized spots on the wing – nygmata – are found in various basal

Holometabola, but in mecopterans two of them characteristically occur near the basal fork of R_S and another one slightly distad. Wings have been lost or extremely reduced in mecopteroids at least six times, including once for the fleas.

Relationships among the nine Recent families of mecopteroids have been analyzed by Willmann (1981, 1983, 1987, 1989), using fossils and morphological features, and by Whiting (2001), using sequences from four genes (18S, 28S, EF-1alpha, and COII) (Figure 12.1). Willmann did not discuss the position of fleas among mecopteroids, which Whiting did address. Much of the morphology is also based on the pioneering work of Mickoleit (1975, 1976, 1978) and Willmann (1989). Major areas of agreement between the Willmann and Whiting studies include the following: basal positions of the Nannochoristidae and Boreidae, a close relationship between Panorpidae and Panorpodidae, and the four remaining families intermediate between these. The major difference between the two hypotheses concerned the family Bittacidae, which either is closely related to the Panorpodidae (Whiting, 2001) or is the most basal family after the Nannochoristidae (Willmann, 1989). Despite an impressive amount of DNA sequences, only the monophyly of families was well supported in the molecular study, and the relationships among families had considerably weaker support. Nonetheless, Panorpidae and Panorpodidae may indeed be closely related because they have unusual, eversible glands on the male abdomen that disperse phermones for attracting mates. In Panorpidae these glands are within the genital capsule; in bittacids they are between abdominal segments 6–7 and 7–8. The fossil record appears to support a very basal position of the Bittacidae, since some Triassic species have been attributed to this family. Willmann (1989), however, classified these fossils as stem-group Neorthophlebiidae, and true bittacids apparently did not appear until the Late Jurassic. The Late Jurassic is still about 100 MY older than the oldest fossil panorpids and panorpodids, but this discrepancy may be resolved by the "Orthophlebiidae." This is a diverse paraphyletic fossil group, some Jurassic taxa of which apparently gave rise to the Panorpidae and Panorpodidae (Willmann, 1989; Novokshonov, 2002). Thus, ages of fossil bittacids and panorpids may not be too contradictory.

Nearly 95% of the Recent world species are concentrated in just two families, the Panorpidae (380 species) and Bittacidae (180 species), and most of these are in just three genera, *Bittacus*, *Panorpa*, and *Neopanorpa*. Panorpidae is a Laurasian group, occurring primarily in temperate forests of the Holarctic Region and IndoChina. Bittacidae is primarily Gondwanan, being particularly diverse in Australia, South America, and comprising the entire mecopteran fauna of Africa (approximately 50 species of *Bittacus*). If these two families are indeed closely related this distribution would be bipolar (similar to that in the stoneflies), and it would indicate a Jurassic age for these mecopterans since this is when the supercontinent Pangaea drifted apart into Laurasia and Gondwana. The larvae of Panorpidae uniquely possess large eyes, comprised of 25–30 closely packed stemmata, whereas all other mecopterans have seven or fewer stemmata. The larval eye in Panorpidae is certainly recently evolved, since in more basal families it is simpler, but interestingly the structure of the panorpid larval eye is very similar to that of the true, adult compound eye (Bierbrodt, 1942). Bittacid larvae have pairs of protuberances on each segment that bear setae (*scoli*) on which the larva fashions particles of soil cemented with feces, camouflaging itself amidst soil and humus.

The bulbous and curled male terminalia in Panorpidae reaches an absurd extreme in some Recent and fossil genera, notably the southeast Asian genus *Leptopanorpa* and the Upper Eocene genus *Holcorpa* from Florissant (the latter sometimes placed in its own family) (Figure 12.8a). In these genera the bulbous genitalia are on a stalk formed from abdominal segments seven and eight that are nearly the length of a leg (their mating positions must be interesting). Bittacidae are distinctive for the one large claw on each pretarsus, which folds against the toothed, penultimate tarsomere and is used for grasping prey as well as fine stems and leaf edges from which they suspend themselves (hence their common name, "hangingflies"). While panorpids may occasionally suspend themselves, bittacid legs are so modified that the insect cannot stand on surfaces or walk. Bittacids and panorpids are commonly seen fluttering amidst ferns and other understory plants in damp forests, sometimes carrying a dead insect on which the insect will feed. They feed while suspended beneath a plant, and males may offer a meal to females as an invitation to copulation – a behavior (called *nuptial feeding*) that has been thoroughly studied by Thornhill (reviewed in Byers and Thornhill, 1983; Figure 12.7). A ritualized form of the behavior in Panorpidae involves trophallaxis, where the female feeds on a small pillar of gelatinous salivary secretions offered by the male. Male *Panorpa*, in fact, have huge salivary glands that occupy most of the body cavity. Females even choose among males based on the quality of the offered meal, which is a form of sexual selection and a behavior very similar to that of some male empidine dance flies (family Empididae). Fossil Panorpidae indicate that this family had a recent origin, since the only fossils known are approximately ten species of *Panorpa* from the Oligocene and Eocene of Europe and North America (Figure 12.8), though ancestral panorpids may eventually be recognized amongst Cretaceous orthophlebiids (Willmann, 1989; Novokshonov, 2002).

Closely related to the Panorpidae is the small family Panorpodidae. The distribution of this family is relict, with two species of *Brachypanorpa* in the southern Appalachian Mountains and two species in the wet, coastal regions of the Pacific northwest of North America, plus five species of *Panorpodes* in Japan and Korea. A very similar distribution

12.7. Nuptial feeding in bittacid scorpionflies from western Australia. Photo: S. Marshall.

occurs in the relict wood roach, *Cryptocercus* (discussed earlier), nymphomyiid flies, and other groups and may be a consequence of fairly recent events like Pleistocene glaciation. During the Eocene, Panorpodidae also occurred in Europe as evidenced by two fossil *Panorpodes* in Baltic amber.

Species in the small family Apteropanorpidae (with just two species of *Apteropanorpa* from Tasmania) are, as the name implies, wingless insects. They are superficially similar to the micropterous Boreidae. *Apteropanorpa* may be closely related to another Australian family, the Choristidae (Whiting, 2001), which makes biogeographic sense. There are ten species of Choristidae in three genera, all occurring in southeast Australia, the area closest to Tasmania. Male choristids are distinctive for the maxillary and labial palps, which are large, curled, and sclerotized, resembling tusks, though it's unknown how these are used. Nuptial trophallaxis has been observed in some choristids, so it is plausible that the male palps are prehensile, used perhaps for holding the female palps during the transfer of regurgitate. Cretaceous and Eocene fossils indicate that the Choristidae have been in Australia since these times, and fossils of them aren't known from elsewhere.

The families Eomeropidae (monospecific: *Notiothauma reedi*, from Chile) and Meropeidae have very similar, broad wings, which have an intricate venation and are held flat over the abdomen. Meropeidae differ in having male genitalia that have large forceps, but exactly how these function is probably unknown since meropeids are very rarely found. This family has just two species, *Merope tuber* from eastern North America, and *Austromerope poultoni* from western Australia – an exceptionally disjunct and relict distribution. Fossil Meropeidae are known from the mid-Jurassic of Siberia, and putative Eomeropidae from the Triassic. Definitive Eomeropidae (*Eomerope*) are known from the Miocene to Paleocene of Russia and North America, which indicate formerly more widespread distributions. Whiting (2001) did not include *Notiothauma reedi* in his study, in which the Meropeidae was indicated as the sister group to much of the Mecoptera. Willmann (1989) hypothesized that *Notiothauma* is more closely related to higher mecopterans than to Meropeidae, and he was more equivocal about the position of the Meropeidae. It would seem that the two families are actually closely related, though basal to most mecopterans.

12.8. Tertiary fossil Panorpidae. (a) Male *Holcorpa maculosa* from Florissant; its genitalia are at the end of a very long, folded stalk. (b) *Panorpa* in Baltic amber. *Holcorpa*: AMNH 01887; forewing 19.5 mm. *Panorpa*: AMNH; body length 19 mm.

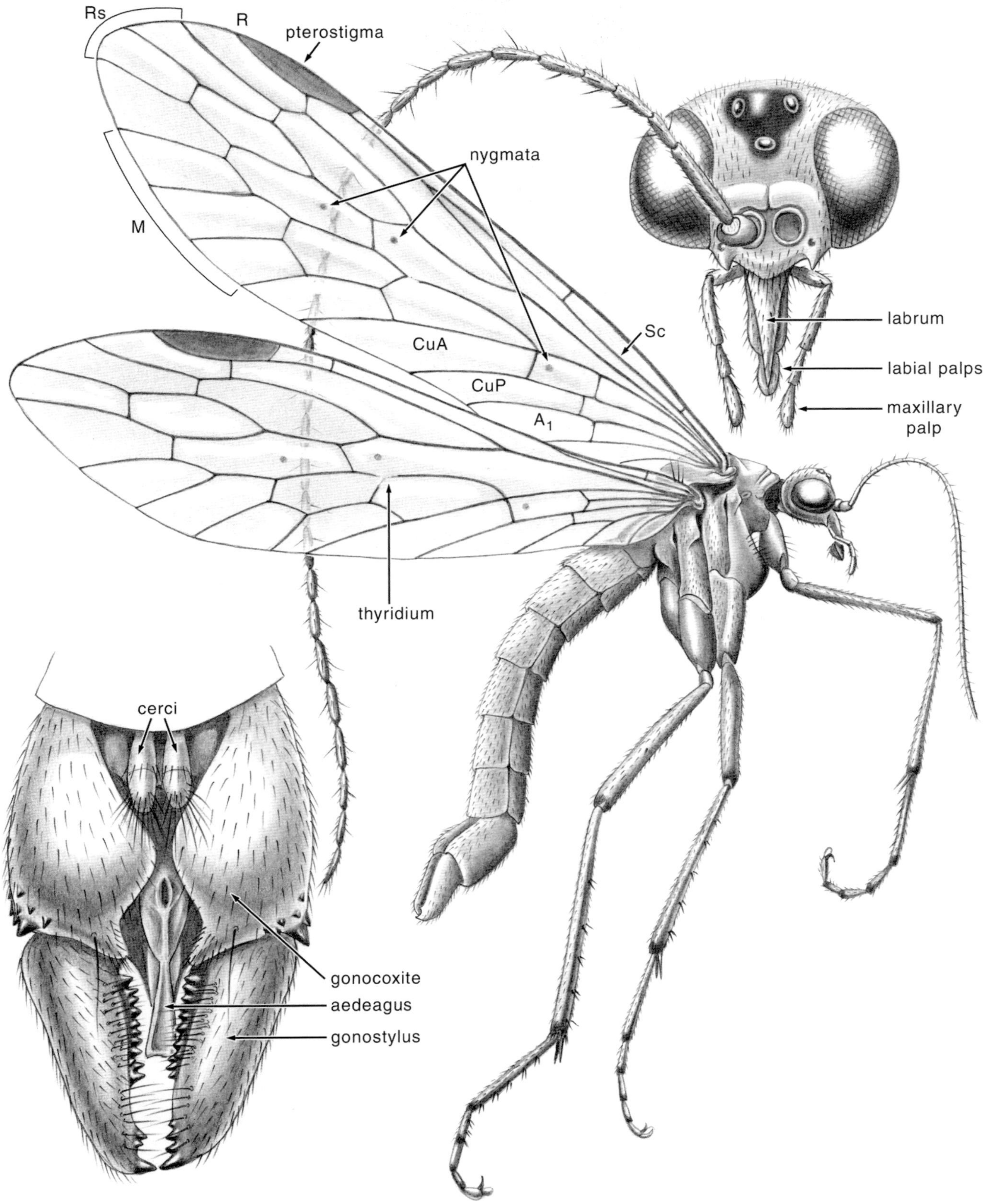

12.9. A living species of the relict mecopteran, *Nannochorista neotropica* from Chile, with details of the head and male genitalia. Nannochoristidae is considered to be the living sister group to a lineage comprised of the Boreidae and fleas, or possibly the most basal of all living mecopteroids. Body length 3.6 mm.

Without doubt the most basal mecopterans are in the family Nannochoristidae, whose phylogenetic position is in agreement between the studies of Willmann (1987, 1989) and Whiting (2001), though in the latter study this family is allied with the Boreidae as the most basal lineage. Nannochoristidae truly deserve the title of "living fossils" because they are evolutionary relicts and both adults and larvae appear morphologically intermediate between mecopterans and dipterans (Figure 12.9). Wood and Borkent (1989), in fact, proposed on the basis of larval characters that nannochoristids are the living sister group to the Diptera, though most evidence indicates nannochoristids are indeed basal mecopteroids, so their resemblance to flies is based largely on plesiomorphic features. There are eight species of nannochoristids, five of them in Tasmania, southeast Australia (a distribution like that of *Apteropanorpa* and Choristidae) and New Zealand, and three in the temperate regions of southern South America (Byers, 1989; Kristensen, 1989b). The monotypic genus *Microchorista* from New Zealand was synonymized with *Nannochorista* by Kristensen (1989b). The larva of *Nannochorista philpotti* is aquatic, long, and slender, and (like Diptera) it lacks the prolegs commonly seen in other mecopterans. Nannochoristids have a long fossil record, though the placement of *Nannochoristella* from the Permian in this family is almost certainly incorrect. Definitive nannochoristids appear in the Early Jurassic and were quite abundant by the mid-Jurassic and Early Cretaceous in Siberia and China (Figure 12.10), so these mecopterans too had a widespread distribution. If the Nannochoristidae is closely related to the Boreidae (Whiting, 2001), this would be a third relict, bipolar distribution among mecopteroids.

The last Recent family of mecopteroids for discussion, the Boreidae, is the most intriguing. Molecular studies (Whiting *et al.*, 1997; Whiting, 2001) indicate that this family is the closest relative of the fleas, order Siphonaptera – a situation reminiscent of the proposed relationship between liposcelid psocopterans and the true lice, order Phthiraptera. Boreidae comprise approximately 24 described species in three genera from the Holarctic Region: *Boreus* (Holarctic: 21 species), *Hesperoboreus* (western North America: 2 species), and *Caurinus* (western Oregon: 1 species) (Svensson, 1972; Penny, 1977; Russell, 1979). Their life histories and biology were reviewed by Cooper (1972, 1974) and Penny (1977). Their larvae are stout and slightly curled, with only the thoracic legs present in *Boreus* and *Hesperoboreus* (the larvae are *scarabaeiform*), but which are vestigial in larval *Caurinus*. They and the adults apparently feed on mosses, in which they also form pupal cells. Adults are stout-bodied insects with long legs and vestigial wings; they walk over the snow on sunny days in late winter and early spring (thus their traditional common name, "snow fleas," which is apparently also a prescient one given the new views on their relationships) (Figures 12.11, 12.12). Adults have an elongate rostrum like Panorpidae, formed from the clypeus and labrum, genae, and maxillo-labium. The wings in females are reduced to small oval pads; in males they are modified into narrow, hooked straps, with rows of teeth on the inner edges that are actually used to grasp the female over his back during mating (Figure 12.12).

Caurinus is particularly interesting because it has some remarkable specializations as well as features primitive to all other boreids (Russell, 1979, 1982). Its unique features are absence of hind wings, the first abdominal tergite fused to the thorax, abdominal segments 2–6 partially fused and heavily sclerotized (which give it an armored appearance),

12.10. Nannochoristid scorpionfly from the Early Cretaceous Yixian Formation of China. The family today is relict, living in southern South America, New Zealand, Tasmania, and southeastern Australia. Body length 9.5 mm.

12.11. Male *Boreus* (Boreidae) walking on the surface of snow. They are active on warmer, sunny days in late winter and early spring. Boreids are the closest living relatives of fleas. Photo: S. Marshall.

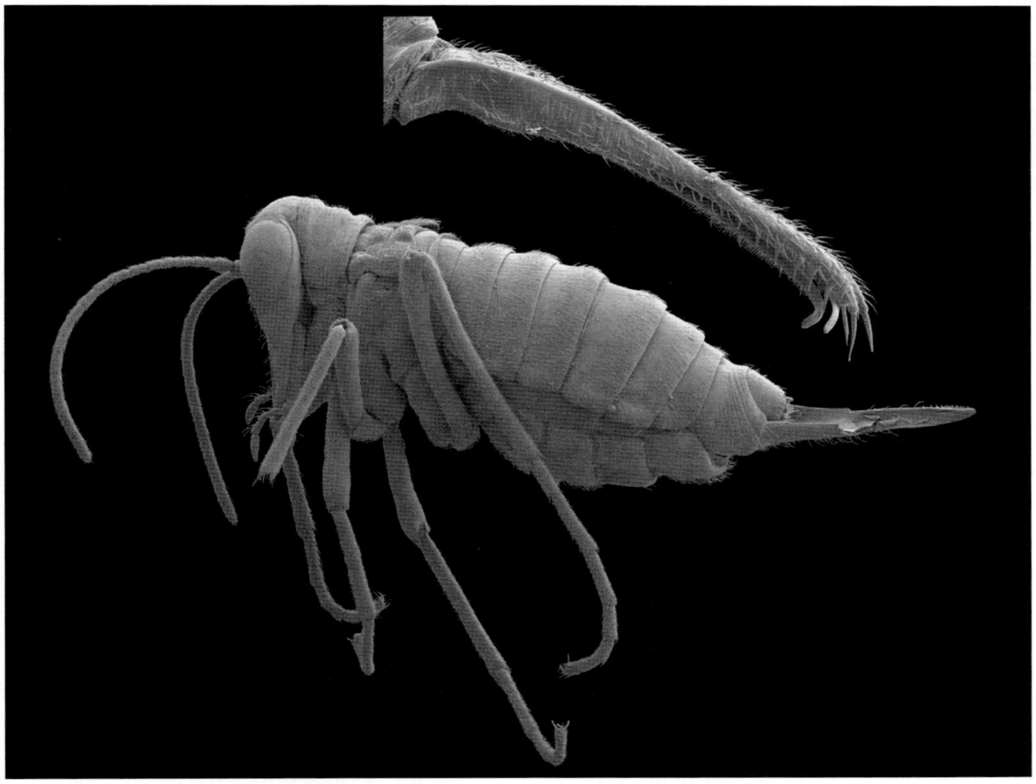

12.12. Female *Boreus*, with a detail of a male wing. Spines on the male wings help to grasp the female over his back during mating. The dorsal valve of the ovipositor was lost in this specimen. Scanning electron micrograph; body length 2.6 mm.

and legless larvae. Primitive features are a short rostrum, short mandibles with large teeth, presence of a clypeolabral suture (which is absent in other mecopteroids), females with free (not fused) cerci, and larvae with a pair of large eyes.

The consistent grouping of boreids with fleas on the basis of DNA sequences has forced a reexamination of anatomical and other phenotypic features that would support this grouping, and they are significant albeit traditionally overlooked. Among the more obscure features are multiple sex chromosomes in both groups (Bayreuther and Brauning, 1971), sperm ultrastructure (an axoneme coiled around the mitochondrion), and the structure and function of the female reproductive system (Biliński *et al.*, 1998). Boreids and fleas have panoistic ovaries (these have no nurse cells), the initial stages of oogenesis are completed postembryonically, there are multiple nucleoli in the oocytes, and the late, previtellogenic ooplasm is separated into two regions. *Nannochorista* shares these features (Simiczyjew, 2002), which supports the close relationship of this family with the boreids and fleas proposed by Whiting (2001). Other obscure features that support this relationship are an absence of extrinsic labral muscles, sexual dimorphism in the ventral nerve cord (males having more ganglia than females), and the structure of the proventriculus and spiny acanthae within (Richards and Richards, 1969). Overt characters that have been neglected are ocelli that are highly reduced or lost (which occur in various insects), elongation of the labrum and maxillolabium, and some features of larvae (those of *Caurinus* have vestigial thoracic legs, and flea larvae are completely legless). Interestingly, though the ability to produce silk is widespread among panorpoids, among mecopteroids only fleas and boreids form a silken pupal cocoon. Larval silk production also appears in the Lepidoptera and Trichoptera, some basal Diptera, and more sporadically in the Hymenoptera. Lastly, and surprisingly overlooked, is the ability of boreids and fleas to jump. While the mechanism and behavior of the flea's jump is well studied, it has not been for the boreids. Even more suprising is that both boreids and fleas will often feign death after their leaps (Cooper, 1974; Penny, 1977; Marshall, 1981).

If there were early fossil boreids with fully formed wings, they probably wouldn't be recognizable, which makes perplexing Novokshonov's view (in Rasnitsyn and Quicke, 2002) that the Boreidae were derived from the extinct Permochoristidae. This interpretation could only be based on venation. There are two very intriguing, wingless fossil insects attributed to the Boreidae (Sukatscheva and Rasnitsyn, 1992), but both of them are wingless: *Palaeoboreus zherikhini*, from the Late Jurassic of Mongolia, and *P. baissicus*, from the Early Cretaceous of central Siberia, ca. 150–140 MYA. *Palaeoboreus zherikhini* is better preserved (Figure 12.13), and is clearly an

12.13. An early boreid mecopteran, *Palaeoboreus zherichini*, from the Late Jurassic of Mongolia. The female terminalia are well preserved. PIN 3688/1218a; body length 3 mm.

apterous adult insect with a small thorax and well-preserved female terminalia. Small eyes and what appears to be a complete lack of wings are actually more advanced features of the fossils than occur in Recent boreids, but the structure of the terminalia is very similar to that of Boreidae. Specifically, the female had a pair of long, sclerotized eighth gonocoxosternites and pointed apical cerci. *Palaeoboreus* is primitive to Recent boreids in not having a long rostrum; it is extremely significant because it verifies a Mesozoic age for the sister group to the fleas.

THE FLEAS

The Order Siphonaptera consists of wingless, laterally flattened ectoparasites that feed on the blood of birds and mammals, which are renowned for their jumping ability (Figure 12.14; Table 12.2). The lineage is considerably more successful evolutionarily than their closest relatives; with 2,500 validly named species, their Recent diversity is almost five times that of the mecopterans. They have also evolved far more recently than have most mecopterans, no doubt a consequence of the radiations of mammals in the Paleocene and Eocene. Lewis (1998) recently summarized the taxonomy of fleas, which currently consists of 15 recognized families. Species-level taxonomy has been complicated by a plethora of trivial subspecies names, much like the traditional taxonomy of their hosts but now essentially ignored. In some cases the use of subspecies became absurd, such as in *Ctenophthalmus agyrtes* with 23 subspecies! Despite this problem, species taxonomy of the Siphonaptera is among the most thorough of any insect order, partly as a result of the monumental, multivolume monograph by G. H. E. Hopkins and Miriam Rothschild (1953–71), and others on the famous Rothschild collection of fleas at the Natural History Museum in London. The significance of fleas as vectors of disease also contributed to this body of taxonomy, particularly since certain fleas transmit a disease that has been among the most devastating in human history, which is plague.

Plague is caused by a bacterium, *Yersinia pestis*, which is normally harbored in rodents with little to disastrous effect, and transmitted by fleas. Plague can wipe out, for example, whole colonies of prairie dogs. Toward the mid-1300's, poor crops in Europe drove rural people into cities, resulting in excessive urban crowding amidst the typical poor hygiene and sanitation. Populations of the brown rat, *Rattus rattus*, likewise swelled. Feeding off these rats are several fleas in the large genus *Xenopsylla* (Pulicidae), three of which are now known to transmit plague, though *X. cheopis* is a particularly efficient vector and most often implicated in plague. The

TABLE 12.2. Families of Fleas[a]

Family	Species	Distribution	Main Hosts
Ancistropsyllidae	3	Asia	Artiodactyla
Ceratophyllidae	403	Worldwide, Holarctic	Various mammals (mostly rodents), birds
Chimaeropsyllidae	17	Africa	Rodentia, Insectivora
Coptopsyllidae	19	Palearctic	Rodentia (gerbils)
Ctenophthalmidae	548	Worldwide, Holarctic	Rodentia, Insectivora, Marsupialia
Hystricopsyllidae	36	Holarctic, neotropical	Rodentia, Insectivora
Ischnopsyllidae	122	Worldwide	Chiroptera (bats)
Leptopsyllidae	230	Worldwide	Rodentia, Lagomorpha, birds
Malacopsyllidae	2	Neotropical	Edentata: Dasypodidae (armadillos)
Pulicidae	182	Worldwide	Various mammals
Pygiopsyllidae	166	Eastern Hemipshere (+ neotropical)	Rodentia, marsupials, some birds
Rhopalopsyllidae	122	Western Hemisphere	Various mammals
Stephanocircidae	51	Southern South America, Australia	Marsupials, Rodentia
Vermipsyllidae	3	Holarctic	Carnivores, pikas, ungulates
Xiphiopsyllidae	8	Africa	Rodentia

[a] From Lewis, 1998.

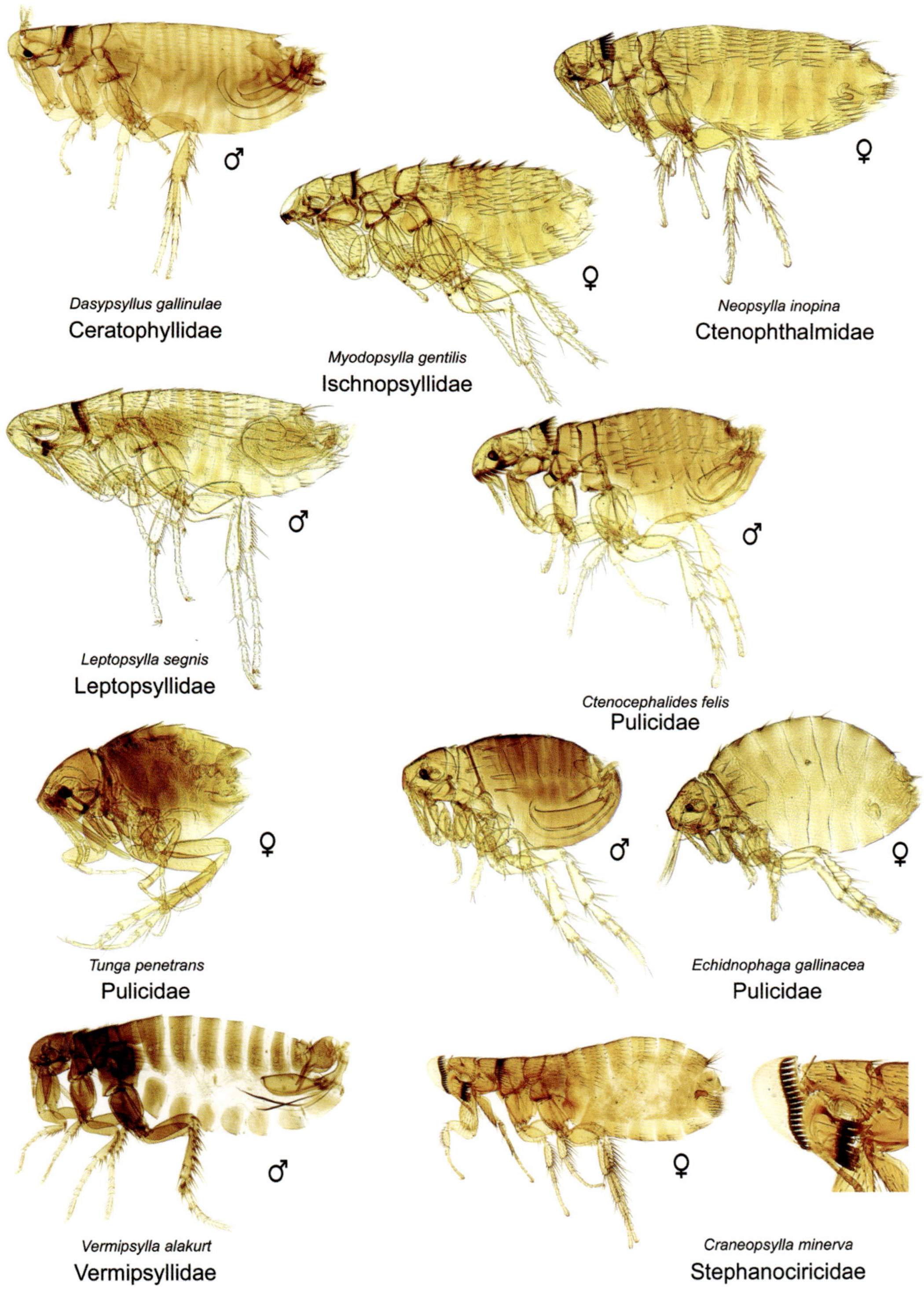

12.14. Representative fleas, order Siphonaptera (not to the same scale). Specimens: R. Lewis, Iowa State University.

common pest *Pulex irritans* may also have been responsible for these notorious pandemics. When rats die of plague, indeed when any host dies, their fleas usually turn to the next most available host. In 1348, plague erupted in central Europe (the "Black Death"), killing so many people that it would take nearly two centuries for their populations to be restored. It subsided after several years, lingered among the rat populations, and again erupted at various times throughout Europe. In 1665, for example, it surfaced in London, killing 65,000 people. It was not until 1914 that the British-Indian Plague Commission discovered the mechanism of transmission (Hirst, 1953; Ziegler, 1969; Busvine, 1976). When plague is transmitted by fleas it causes blackish swellings where bites were made ("bubos," hence *bubonic* plague), but a person infected with plague can transmit the bacterium via aerosols from coughing (*pneumonic* plague). Other flea-borne diseases, like endemic typhus (*Rickettsia typhi* – also transmitted by *X. cheopis*), and various worm parasites, are minor by comparison.

The extreme modification of fleas leaves no doubt about their monophyly, but this has also obscured traditional study of their relationships based on morphology. Adults are heavily sclerotized, with the typical complement of ectoparasite traits: loss of ocelli, virtual or complete loss of eyes, reduction of antennae, wings lost and the thorax highly reduced, ctenidia frequently (though not always) developed, stiff hairs over the body that point backward, large claws, specialized mouthparts, and a flattened body (Figure 12.15). In fact, fleas are the only insect ectoparasites where the body is laterally flattened; all others are flattened dorsoventrally. The antennae are short and stout, with nine compressed flagellomeres, and lie in deep grooves (*scrobes*). Another distinctive feature of Siphonaptera is a patch of sensilla on the pygidium (probably tergite 9), the *sensilium*, but the real hallmark feature is their jumping ability. Fleas have large, powerful hind legs with large coxae, but which only partly contribute to the leap. A pair of minute structures near the base of the hind legs contains pads of resilin, a highly elastic protein that can be greatly compressed, and when released its immediate expansion puts the spring into a flea's leap.

Fleas feed by piercing the skin and sucking blood. The piercing is done with the stylet-like labrum (usually identified as the "epipharynx," but see Michelsen [1996/7]) and a pair of serrated maxillary laciniae, all of which are held together by labial palps (Figure 12.15). Snodgrass (1946) presented a definitive study on the morphology of flea mouthparts, but the identities of some structures were revised recently based on musculature of the structures (Michelsen, 1996). Besides the labrum, other significant revisions to the study of Snodgrass are the lingua (= "hypopharynx"), stipes (= "lacinial lever"), first segment of the maxillary palp (= "maxillary lobe"), the prementum (= "postmentum"), and the first segment of the labial palp (= "prementum"). When a flea feeds, it injects saliva, which acts as anticoagulant, and blood is sucked up using powerful cibarial muscles. Unlike some hematophagous insects, like biting flies, both sexes of fleas, not just the female, feed on blood. Water is extracted via rectal papillae, and undigested blood is excreted as blackish droplets that dry into fine granules; these granules accumulate in the hair of the host and fall to the floor of the nest or den. In fleas, the spiny acanthae lining the proventriculus presumably help prevent the backflow of blood into the esophagus. It is one of the more remarkable quirks of nature that this microscopic structure contributed to the spectacular carnage of Medieval Europe and Elizabethan London.

Fleas feeding on a host infected with plague will have the *Yersinia* bacteria multiply in their midgut, which may then cause partial or complete blockage of the proventriculus. Fleas without a blocked proventriculus can easily transmit plague to a host via their mouthparts, but blocked fleas tend to regurgitate blood when they bite, infecting the bite even more. Moreover, fleas with blocked proventriculi starve, so they bite more often. Another reason why some fleas are so effective in spreading plague is that they can be indiscriminate feeders.

Fleas are largely mammal ectoparasites, since only about 6% of the species feed on birds (reviewed by Marshall, 1981; Lewis, 1998), and approximately 75% of all flea species feed on rodents alone (Table 12.2). Their prevalence on rodents is not just a result of the fact that rodents are the most diverse mammals (they comprise 41% of all mammal species), since by this line of reasoning bats (order Chiroptera, 21% of all mammals) would be a strong second. In fact, the proportions of flea species on various mammals are approximately 8% on insectivores, 5% on marsupials, 5% on bats, 3% on lagomorphs, 3% on carnivores, and 1% on all others (monotremes, edentates, Pholidota, Hyracoidea, and artiodactyls). Fleas do not infest as many orders of mammals as do lice because they do not occur on flying lemurs (Dermoptera), elephants (Proboscidea), or primates. Humans are merely surrogate hosts for the fleas of their domesticated animals. Like lice, fleas do not occur on wholly aquatic mammals, like whales (Cetacea) or sea cows/manatees (Sirenia), and no fleas are further known from seals (Pinnepedia, which do come to shore) though seals do harbor lice. Oddly, no lice are known to parasitize bats, but in fleas the family Ischnopsyllidae is restricted to bat hosts, primarily ones with large, perennial colonies. While some fleas are known from only one host, others are known to use diverse hosts. *Ctenocephalides felis*, the "cat" flea, for example, has been recorded feeding on hosts in eight orders of mammals, and even on lizards. A bird flea, *Ceratophyllus gallinae*, is known to feed on over 100 species of birds, 75 in Britain alone. While there are patterns of host use among some genera and tribes of fleas, patterns are not as obvious as they are in many lice. Flea polyphagy is partly the result of their life histories.

The female deposits eggs generally either directly on the floor of the nest or den or on the host while in the nest. The rabbit flea, *Euhoplopsyllus glacialis*, is exceptional since its eggs are laid directly on the fur of its host, though larvae live in the den. Flea eggs have a sticky coating, to which adhere fine particles of debris. Larvae feed on fecal remains of the host, exfoliated skin, the granules of dried blood excreted by adults, and some are predatory and even cannibalistic. Larval stages are generally poorly known, the morphology and diversity of which have been reviewed by Ebel (1991) and Pilgrim (1991). There are three larval instars, the first having

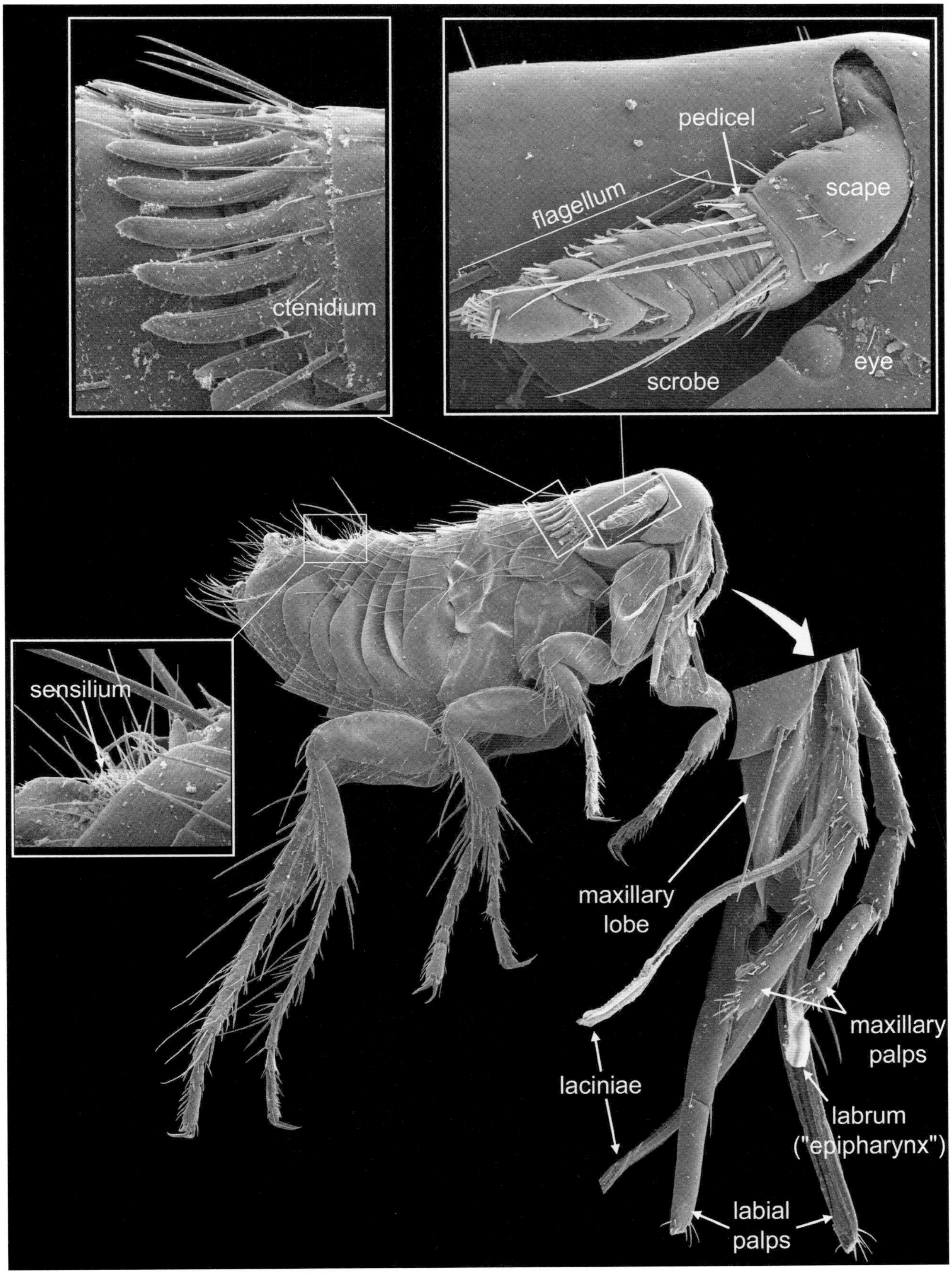

12.15. A typical adult flea, showing the main siphonapteran features. Body length 1.9 mm.

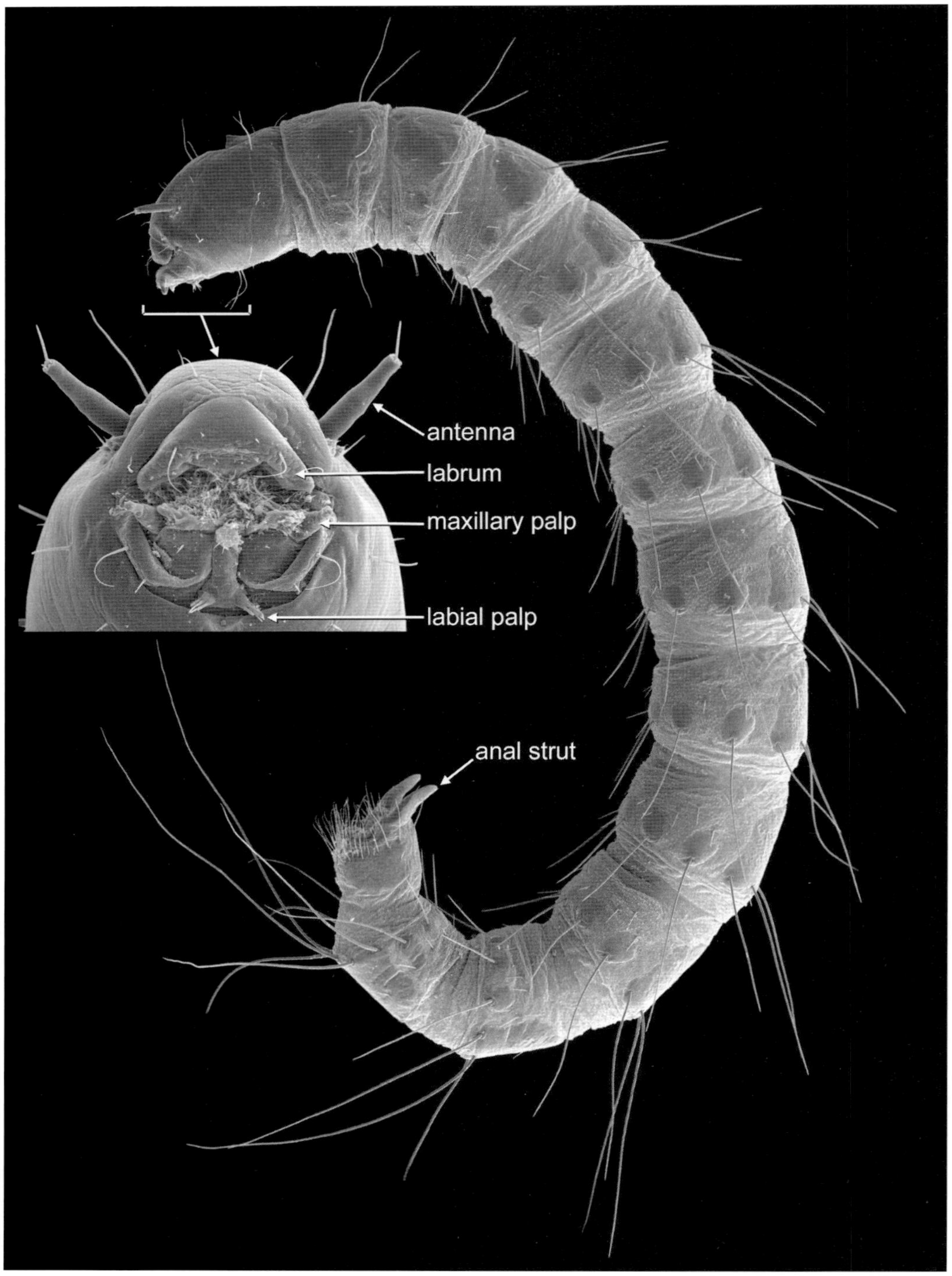

12.16. Typical legless larva of a flea, family undetermined. The larva was found in a swallow nest in California. Body length 3.1 mm.

an egg burster, or blunt spine, on the head capsule. As the name implies, this helps the larva emerge from the egg. All instars are legless and slender, with a well-developed head capsule bearing the full complement of mouthparts and other structures, except that stemmata are lost. Among the more distinctive features is that the body is covered with slender, stiff, backward-projecting setae (Figure 12.16), the structure and distribution of which is taxonomically important, just as for the hairs of caterpillars. A mature larva will spin a loose silken cocoon in which it pupates. Remarkably, the new adult may remain in this cocoon for over a year, until stimulated by CO_2 or other sign of a nearby host. This is why a house vacant for months can become suddenly infested with hungry fleas. Moreover, unfed adults can live for several months. The ability of adults to remain quiescent, and for newly emerged and starved adults to feed indiscriminately, may be adaptations of fleas for inhabiting winter or seasonal dens of mammals or the breeding nests of birds.

There are a few interesting exceptions to this general life history of fleas. Some involve *neosomy*, or a stage in which new cuticle is formed during an active stadium. In insects, this occurs only in a few fleas and in the bizarre streblid fly *Ascodipteron*. In these neosomic insects the duration of immature stages is reduced, and in some the larvae may not even feed at all. Neosomy in fleas occurs in the Tunginae, Malacopsyllidae, and Vermipsyllidae and is particularly well developed in *Tunga*, the chigoe fleas (Figure 12.14). There are ten species of *Tunga*, occurring in southeast Asia, Africa, and in the Western Hemisphere (mostly the tropics), of which *Tunga penetrans* is a particularly notorious pest of swine and humans. A female *T. penetrans* burrows into skin between toes, under toenails or in some other area, where she will then feed. The only exposed part is the tip of the abdomen: the anus, gonopore, and apical spiracles. In approximately a week her abdomen swells to nearly 1,000 times her normal mass and she exudes thousands of minute eggs, which fall to the ground. Embedded chigoes often become infected and painful. A lifestyle that is equally bizarre is that of *Uropsylla tasmanica*, which is an ectoparasite of dasyurid marsupials in Tasmania and southern Australia. The eggs are attached to hair, larvae burrow into skin, and adults live permanently on the hosts. Except for the pupa, which develops on the ground, *Uropsylla* is certainly the louse among fleas.

Relationships and Origins

Families of fleas are traditionally classified into superfamilies, although there has been little phylogenetic work on the order. Medvedev (1994, 1998) and Whiting (2001) are the only studies to deal with family-level relationships in the order, the former of which is based on just 19 species in eight families but significant numbers of DNA sequences from four genes. Medvedev's classification is little used. The monographic study by Cheetham (1988) is a detailed comparative study of the male genitalia in just the families of Pulicoidea, so the relationships of most families are not discussed. Not surprisingly, relationships found by Whiting (2001) are rather different than are traditionally recognized, and among the most intriguing results is that Stephanocircidae was the most basal family – a relationship prophesized by Hennig (1981: 407). Stephanocircidae are morphologically very specialized, with a head that is so reduced it looks as if comprised entirely of the genal comb (Figure 12.14), from which the name "helmet fleas" is derived. The family is comprised of just 51 species (Lewis, 1998), the subfamily Craneopsyllinae of which parasitizes rodents and some insectivores and marsupials in Chile, Argentina, and other southern regions of South America. The subfamily Stephanocircinae has eight species, parasitizing marsupials in Australia. This is the only flea group with a distribution that is so obviously disjunct in the austral region, so if it is indeed the most basal family its distribution may reflect a Cretaceous age, when Gondwana drifted apart to form the southern continents. Alternatively, given that there was widespread extinction from the northern continents of Cretaceous marsupials (see Chapter 14), a group today concentrated in Australia, such an interpretation may be naive.

Like lice, the ages and origins of fleas must be inferred almost wholly from the relationships of Recent species because the fossil record is so sparse. In fact, the fossil record of true fleas is comprised of approximately a dozen specimens in Eocene Baltic amber and Miocene Dominican amber (Figures 12.17, 12.18). All the Baltic amber fleas are species in the genus *Palaeopsylla* (Ctenophthalmidae): *P. klebsiana*, *P. dissimilis*, and *P. baltica* (Peus, 1968; Beaucorneau and Wunderlich, 2001). There are 52 Recent species of *Palaeopsylla*, which parasitize small insectivores in the Palearctic Region. All the eight or so flea specimens in Dominican amber are Rhopalopsyllidae (Figure 12.18), with one exception: a pulicid specimen. Rhopallopsyllidae is comprised of 127 described species, all from the Western Hemisphere, which parasitize primarily rodents, some birds, and the genus *Rhopalopsyllus* is on a broad range of mammal hosts. The pulicid in Dominican amber belongs to the genus *Pulex*, which is comprised of six Recent species parasitizing a variety of mammals (including man) and even a bird. Interestingly, the fossil, *Pulex larimerius*, was preserved in a piece of amber also containing very rare strands of hair, confirming that it fed on mammals (Lewis and Grimaldi, 1997) (Figure 12.17). Clearly, all these fossil fleas were captured where the fossil resin was produced: on trees. Despite vast deposits of compression fossil insects from the Tertiary, and the fact that most fleas are not arboreal, it is quite surprising that no fleas are known from Tertiary rocks. Nonetheless, the fine preservation of the amber fleas allows definitive classification of these extinct species, which further indicates that by the mid-Tertiary modern species-groups of fleas had evolved. This suggests at least a Cretaceous age of Siphonaptera.

The only credible Cretaceous fossils of Siphonaptera to be reported are in the Early Cretaceous Koonwarra sediments of Victoria, southern Australia (Jell and Duncan, 1986). There is little basis for the claims that *Strashila* and *Saurophthirus*, from the Late Jurassic and Early Cretaceous of Eurasia, respectively, are fleas. As discussed earlier, these are probable mecopteroid ectoparasites of obscure relationships. The Koonwarra specimens are *Tarwinia australis* (Figures 12.19, 12.20), *Niwratia elongata*, and two undetermined specimens ("indet. 1, 2") (Jell and Duncan, 1986). Specimen "indet. 1" is too fragmentary to identify, and *Niwratia* is very suggestive of an early flea, but unfortunately no diagnostic features are preserved. For example, it had a large abdomen and small thorax, and it doesn't appear to have merely been a nymph because terminalia are well preserved. Moreover, eyes appear reduced, and a pair of appendages protruded from the head, which are possibly mouthparts but difficult to discern

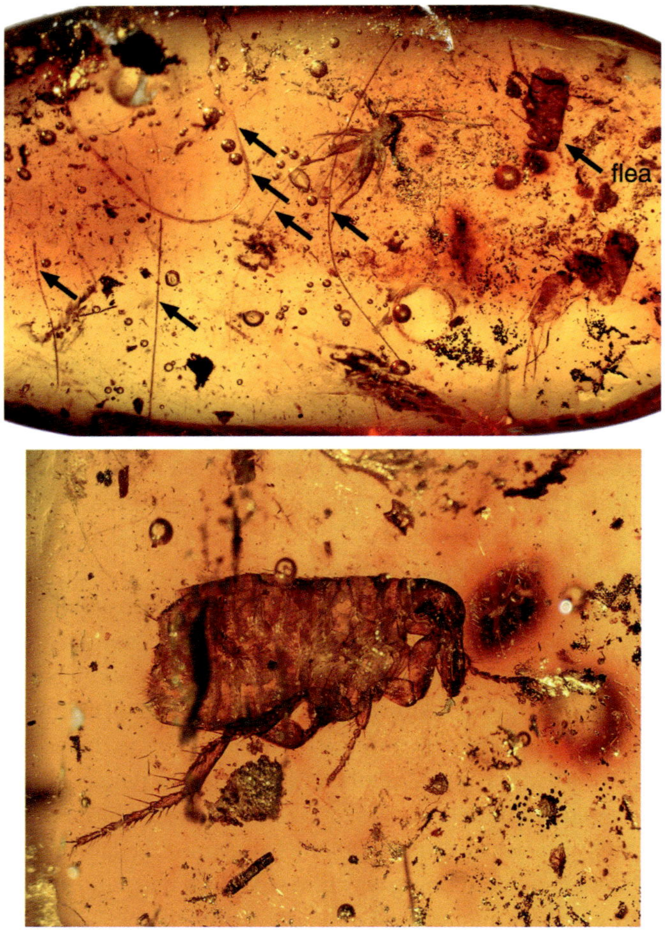

12.17. A Miocene flea, *Pulex larimerius*. It is preserved in a piece of Dominican amber with five strands of hair (arrows, above), indicating that it was a mammal parasite. *Pulex* fleas today are ectoparasites of swine and humans. AMNH DR-14–1140; body length 2.26 mm.

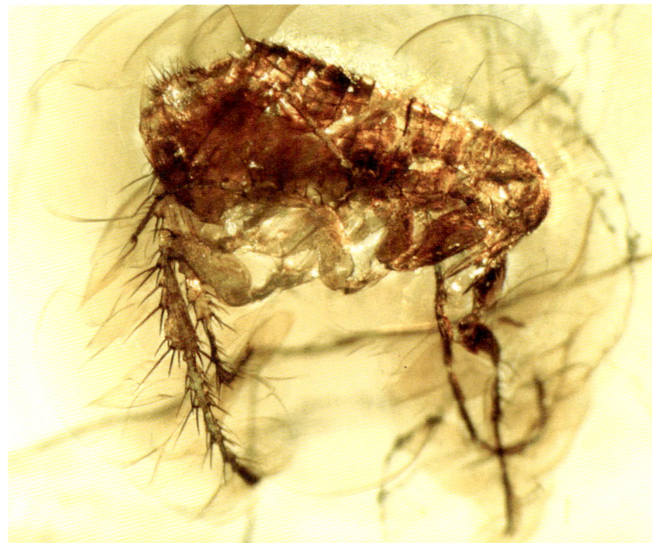

12.18. Another Miocene flea in Dominican amber, family Rhopalopsyllidae. Modern species of rhopalopsyllids feed primarily on rodents. Morone Collection, M-3147; body length 2.6 mm.

because of the preservation. It did not have the long, stiff setae typical of fleas, large hind legs or coxae, nor even any structures like ctenidia that would suggest it was ectoparasitic. Far more intriguing and beautifully preserved, though, is *Tarwinia* (Figures 12.19, 12.20), which we believe is the only clear and early close relative of fleas from the Mesozoic.

When *Tarwinia* was first reported as a flea (Riek, 1970b), this was critiqued by many (e.g., Hennig, 1981). It was displayed by Riek at the 1970 International Congress of Entomology in Washington, D.C., where various flea specialists examined it. The general agreement among them was that it was not a flea (R. Lewis, pers. comm. to DG, 2003). Admittedly, it is a *stem-group* siphonapteran, based on various features that are primitive for Recent Siphonaptera. *Tarwinia* has a flagellum with 15 segments (versus 9 in true fleas); long, slender legs, even for the hind pair; and hind coxae of moderate size, not huge as in modern fleas. Oddly, the fore coxae are quite large. Unfortunately, mouthparts are not preserved, and if the stiff, sclerotized stylets typical of fleas were present, they probably would have been preserved. *Tarwinia* did have, though, a pointed clypeus, suggesting that it had small, perhaps puncturing, mouthparts. Like fleas, *Tarwinia* possessed flagellomeres that were compact (though not as compact as in true fleas), large claws, a small wingless thorax, and a large abdomen that certainly appears to have been laterally flattened. Most convincing, though, are the terminalia: *Tarwinia* possessed a sensilium, though it was larger than in Recent fleas (Figure 12.20). This insect was clearly ectoparasitic because it had ctenidia that uniquely occurred on the fore- and hind tibiae but that no Recent fleas have. The leg ctenidia of *Tarwinia* were formed from stout, sharp setae on the margins of these tibiae, though never figured in detail (Figure 12.20) (some Recent fleas have rows of extra bristles on the mid and hind tibia, but never ctenidia on these segments). The "pulicid indet. 2" specimen described by Jell and Duncan (1986) likewise had tibial ctenidia, but on the mid tibiae. Its overall structure is strongly suggestive of Siphonaptera, including what looks like a projecting proboscis. However, its preservation is more incomplete than in *Tarwinia*. Hennig (1981: 321, 407) was pessimistic that "it is unlikely that it will ever be possible to recognize any fossils as belonging to the stem-group of the fleas." *Tarwinia*, when actually examined, refutes that claim.

Was *Tarwinia* a relict in its own time, or was it among the earliest fleas? Perhaps the only evidence for answering this lies in the fact that fleas are essentially mammal parasites, parasitizing marsupials and (unlike lice) even monotremes. Flea parasitism of birds is generally acknowledged to be the result of multiple, relatively recent colonizations from mammals. History of the earliest mammals, the primitive triconodonts and multituberculates, extended into the Jurassic and latest Triassic; the earliest monotremes occur from the mid-Cretaceous; and the earliest placental mammals apparently

12.19. Part and counterpart of *Tarwinia australis*, from the Early Cretaceous of Koonwarra, Australia. It possesses some, but not all, features of fleas and was an early close relative of true fleas. VMP 26202a,b; body length 7 mm.

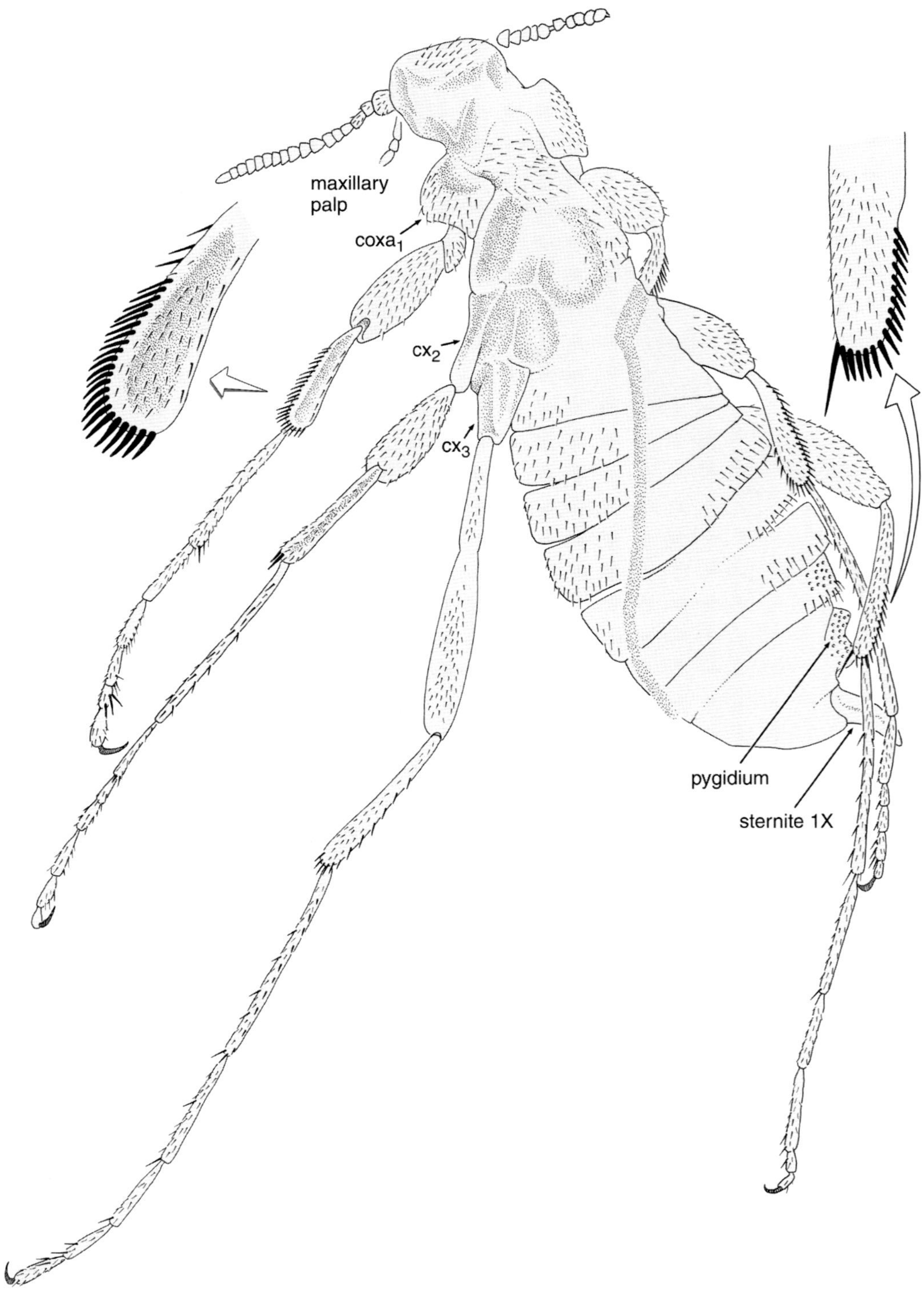

12.20. *Tarwinia australis*, showing details of important structures. It had a laterally flattened body and a sensilium on the pygidium, indicating its close relationship to fleas. Ctenidia indicate it was ectoparasitic, but their locations on the tibiae are unique for Siphonaptera.

diverged in the later parts of the Early Cretaceous (McKenna and Bell, 1997; also reviewed in Novacek, 1999). Without direct evidence, it is still probably safe to assume that Jurassic and Early Cretaceous triconodonts and multituberculates had hair, and it is even more reasonable to assume that the earliest therians (marsupials + placentals) had hair. It could be argued, based just on the monotreme-therian fossil record, that fleas originated in the Early Cretaceous. *Paleoboreus*, from the latest Jurassic and Early Cretaceous, appears closely related to the Boreidae, a Recent mecopteran family that is possibly the living sister group to the fleas. *Paleoboreus* and *Tarwinia* are tantalizing glimpses that suggest an Early Cretaceous divergence of boreids and Siphonaptera. Even if fleas did originate 120 million or more years ago, the radiation of the modern orders of placental mammals much later in the Paleocene (Shoshani and McKenna, 1998; Liu and Miyamoto, 1999; Novacek, 1999; Thewissen *et al.*, 2001) clearly led to a quantum leap in the evolution of fleas.

EVOLUTION OF ECTOPARASITES AND BLOOD FEEDERS OF VERTEBRATES

Though blood-feeding insects derive their nutrition from other animals, they are generally not considered *ectoparasitic* unless they are specialized for living on the host for at least part of their life (Marshall, 1981; Kim and Adler, 1985). That is, ectoparasites will typically remain on the host even when not feeding. Ectoparasites are not necessarily blood feeders; some feed on secretions, or the keratin in hair, feathers, and skin. A species of ectoparasitic insect may have any number of features in the full suite of adaptations, which are the following: wings lost or highly reduced (aptery, brachyptery), with a concomitant reduction of the pterothorax; eyes reduced or lost, as are the ocelli; antennae reduced in size and usually protected within grooves; the body flattened and generally covered with dense, fine hairs, or stiff hairs that point rearward; and tarsi or combs of thick setae (ctenidia) specialized for anchoring the insect within the pelage of the host. Mouthparts of ectoparasites can be generalized, chewing mandibles, but in blood feeding (*hematophagous*) insects the mouthparts are usually modified for puncturing, the most obvious being the proboscides of mosquitoes (Figures 12.24, 12.36), phlebotomine sandflies, and tabanid flies, as well as anopluran lice. Since blood feeders need to puncture the skin, these sometimes have a series of fine, sharp teeth or serrations on the hypodermic structures of the mouthparts (Figure 12.46). In blood feeders the hypopharynx and mandibles are modified into a tube through which blood can be drawn, using suction from an enlarged, muscular cibarial pump. Blood feeders also usually have large salivary glands, since saliva is necessary to prevent coagulation of the blood while feeding and clotting within the fine vessels of the mouthparts. Blood may be high in protein and many amino acids, but not all, so many blood feeders supplement their diet via specialized structures in the digestive tract that harbor symbiotic bacteria (*mycetomes*); these bacteria thrive off the ingested blood and secrete additional nutrients for the insect. Blood feeding has been reviewed by Balashov (1999) and Lehane (1991).

The proteins that are required for the growth and development of eggs are sparsely scattered in nature. They are thinly spread in plants (more occurs in seeds), they can be gleaned as pollen and films of yeast and bacteria, but proteins are most concentrated in the form of other animals. While there are many predatory insects, feeding on vertebrates either as an ectoparasite or a hematophage actually occurs in a minute fraction of all insects (16,000 species, or $<0.01\%$). But, this lifestyle evolved approximately 30 times in insects, indicating it has been evolutionarily very significant (Table 12.3). The lice and fleas contain three quarters of all ectoparasitic insects (representing just two origins of this lifestyle), and five other orders have sporadic occurrence of hematophagy and ectoparasitism. The facts that even some beetles have evolved into ectoparasites and that a moth sucks blood are vivid reflections of the evolutionary plasticity of insects.

As Waage (1979) pointed out, there appear to be two major routes that foster the evolution of feeding on vertebrates. One involved being a predator. Blood feeding triatomine and cimicid bugs, for example, evolved from predatory heteropterans, and *Symphoromyia* and other hematophagous rhagionid flies (Figure 12.68) evolved from predatory ones. In at least one case, the opposite has evolved: Ancestral blood feeding in ceratopogonid midges gave rise to insectivory in more recently evolved lineages (Borkent, 2000a). Culicomorphan midges (including ceratopogonids), overall, have more blood feeders than any other group, and the habit probably evolved just once from a predatory ancestor and was then lost independently in five families (Figure 12.25). The other route to hematophagy and parasitism involved a close but benign association with the host, such as living in the nest or breeding in feces. The larvae of various cyclorraphan fly families, for example, breed in the dung or nest of the animals that the adults feed from, like *Carnus* (Carnidae), *Stomoxys* (stable flies), *Haematobia* (horn flies) (both Muscidae), and the peculiar bat flies *Mormotomyia* (Mormotomyiidae) and *Mystacinobia* (Calliphoridae). As discussed earlier, lice may have evolved from a psocopteran ancestor like Liposcelidae, which are known to forage within nests and even among feathers. The "host association route" appears to have been more pervasive in evolution, though in some cases the inquiline ancestor also appeared to have been predatory, so the two routes are hardly exclusive to each other.

Most insects that feed on vertebrates feed on homeothermic, or warm-blooded vertebrates, the birds and mammals. This is largely because the two major groups of ectoparasites, lice and fleas, are adapted for living in pelage. Indeed, no insect ectoparasites are known from amphibians or reptiles, even though diverse mites and ticks parasitize reptiles. Blood feeders, though, are less restricted, and the free-living hematophagous flies are known to take meals from various reptiles as well as homeotherms (Figure 12.35). Corethrellid and some other midges, in fact, even specialize on frogs. There is evidence that some Culicomorpha in the Triassic (like *Aenne*: Chironomidae) had biting mouthparts. By the Jurassic there were blood-feeding sandflies, blackflies, horseflies, and perhaps mosquitoes, as well as some bizarre ectoparasitic mecopteroids (Figures 12.4 to 12.6) (Kalugina, 1991; Lukashevich and Mostovski, 2003). Blood-feeding cyclorrhaphan flies, like tsetse, are known from the Late

TABLE 12.3. Insects of the Recent Fauna That Are Ectoparasitic (in bold) and Bloodfeeding (in red)

Subgroups	No. Spp.	Feeding Sex		Feeding Type	Diet	Hosts
			DERMAPTERA			
Hemimeridae	11	M, F		Chewing	Skin, secretions	Rodents
Arixenia	5	M, F		Chewing	Secretions	Bats
			PHTHIRAPTERA			
Amblycera	900	M, F		Chewing	Keratin[a]	Birds, mammals
Ischnocera	1,800	M, F		Chewing	Keratin	Birds, mammals
Rhyncophthirina	3	M, F		Chewing	Keratin	Mammals
Anoplura	500	M, F		Sucking	Blood	Mammals
			HETEROPTERA			
Reduviidae: Triatominae	155	M, F		Sucking	Blood	Birds, mammals
Lygaeoidea: Cleradini	50	M, F		Sucking	Blood	Birds, mammals
Cimicidae	90	M, F		Sucking	Blood	Birds, mammals
Polyctenidae	32	M, F		Sucking	Blood	Bats
			COLEOPTERA			
Leiodidae: *Catopidius*	1	??		Chewing	??	Rabbits, hares
Leptinidae	6	??		Chewing	??	Rodents: beavers, rats
Platypsyllidae	2	??		Chewing	??	Beavers, insectivores
Staphylinidae: Amblyopinini	60	??		Chewing	??	Rodents
Languriidae: *Loberopsyllus*	2	??		Chewing	??	Rodents
Scarabaeidae: *Uroxys, Trichillium*	3	??		Chewing	??	Sloths (Edentata)
			SIPHONAPTERA			
All	2,500	M, F		Sucking	Blood	Mammals, some birds
			DIPTERA			
Psychodidae: Sycorinae	100	F		Sucking	Blood	Various vertebrates
Psychodidae: Phlebotominae	700	F		Sucking	Blood	Various vertebrates
Culicidae	3,000	F		Sucking	Blood	Various vertebrates
Corethrellidae		F		Pool[b]	Blood	Frogs
basal Chironomidae	10	F		Pool[b]	?Blood	??
basal Ceratopogonidae	1,000	F		Sucking	Blood	Various vertebrates
Simuliidae	1,500	F		Pool[b]	Blood	Birds, mammals
some Rhagionidae	50	??		Pool[b]	Blood	Birds, mammals
Tabanidae	3,500	F		Pool[b]	Blood	Various vertebrates
Heleomyzidae: *Chiropteromyza*	1	M, F		Sponging	??	Bats
Mormotomyidae	1	??		Sponging	??	Bats
Carnidae: *Carnus*	4	M, F		Pool[b]	Blood	Birds
some Muscidae	50	M, F		Pool[b]	Blood	Mammals
Glossinidae	25	M, F		Pool[b]	Blood	Mammals
Pupipara	400	M, F		Pool[b]	Blood	Birds, mammals (esp. bats)
Calliphoridae:						
some larvae	50	M, F		Pool[b]	Blood	Mostly birds
Mystacinobia	1	M, F		Sponging	??	Bats
			LEPIDOPTERA			
Noctuidae: *Calyptra*	1	M		Sucking	Blood	Mammals
Pyralidae: *Bradypophila*	3	M, F?		Sucking	Secretions?	Sloths (Edentata)

[a] keratin can refer to hair, feathers, or skin.
[b] Pool feeding involves puncturing the skin and feeding on the blood that wells beneath or on the surface.

Eocene, but like batflies they certainly did not evolve before the beginning of the Tertiary. As discussed previously, Phthiraptera may have originated in the Late Jurassic to Early Cretaceous, and the fleas sometime in the Cretaceous. The very isolated other instances of hematophagy and ectoparasitism indicate that these are all quite recently evolved late into the Tertiary. All the available evidence indicates that the original blood-feeding insects, and the plague of dinosaurs, were the Diptera.

DIPTERA: THE TRUE FLIES

Busy, curious, thirsty fly,
Drink with me, and drink as I;
Freely welcome to my cup,
Could'st thou sip, and sip it up;
Make the most of life you may,
Life is short and wears away.
Just alike, both mine and thine,
Hasten quick to their decline;
Thine's a summer, mine no more,
Though repeated to threescore;
Threescore summers when they're gone,
Will appear as short as one.

–William Oldys, *On a Fly Drinking Out of a Cup of Ale*

God in His wisdom
Made the fly
And then forgot
To tell us why

–Ogden Nash

The most ecologically diverse group of insects is certainly the true flies, or Diptera (e.g., Figures 12.21, 12.22). The menu of fly diets comprises nearly the complete insect smorgasbord: blood feeders, endo- and ectoparasites of vertebrates, gall makers, larval and adult predators, leaf miners, parasitoids, pollinators, saprophages, and wood borers. As diverse as flies are, none to our knowledge are social. The approximately 120,000 species of Recent flies are virtually entirely catalogued for the major zoogeographic regions (Stone *et al.*, 1965; Papavero, 1966–?; Delfinado and Hardy, 1977; Crosskey, 1980; Soós, 1984–?; Evenhuis, 1989; and a World Diptera Database underway), making this the only one of the "big four" orders to have virtually all the described species globally inventoried. All the named fossil species are even catalogued (Evenhuis, 1994). The true diversity of flies, in fact, probably vies with that of the Coleoptera because there are much higher proportions of undescribed Diptera (Gaston, 1991). Very few people collect and study flies entirely for pleasure as they do butterflies and some beetles, but the order has been studied more intensively by biologists than any other.

Certainly the most serious reason for studying flies is their significance as vectors of disease. In fact, the diseases transmitted by mosquitoes, blackflies, tsetse, and others presently affect more people than all other arthropod-borne diseases combined. Like plague, diseases such as malaria, yellow fever, leshmaniasis, and sleeping sickness, to name just a few, have profoundly impacted human history, altering the course of wars and human demographics. Indeed, flies are the only multicellular organisms known to have affected human evolution besides fleas.

There are intrinsic reasons too for studying flies, apart from human destruction. Arguably the best understood eukaryotic organism is a fecund little fly, *Drosophila melanogaster*, originally from Africa and now introduced around the world (Figure 12.86). Study of it in the early twentieth century propelled genetics and led to the discovery of such fundamental processes as chromosomal sex linkage, crossing over, inversions, duplications, translocations, and gene mapping (Sturtevant, 1965; Redei, 1974; Kohler, 1994). *Drosophila melanogaster* was the first complex eukaryote to have its genome sequenced (it has 13,600 genes [Adams *et al.*, 2000]), but even more importantly, sophisticated understanding of its genetic loci, made possible by so many mutant strains, makes the fly unique for *functional* genetics and development (Lindsley and Zimm, 1992; see also Flybase: *http://flybase.bio.indiana.edu*). Early interest in *Drosophila* was partly the result of the very large polytene chromosomes in its larval salivary glands, the distinctive banding patterns of which facilitated gene mapping. Actually, the salivary glands of many larval flies have polytene chromosomes, and study of them in flies like the midges *Chironomus* and *Rhynchosciara* led to profound insights on the structure of chromosomes and sites of transcription. The blowfly *Phormia* has been a paradigm organism in sensory physiology (Dethier, 1976).

The study of fly diversity and evolution has benefited from some brilliant minds, most notably Willi Hennig, who we discussed earlier in this book. Though he was known primarily for his formulation of phylogenetic systematics, he was foremost a dipterist. His study of dipteran wing venation, genitalic morphology, larvae, biogeography, and fossils has laid a foundation for the order that far exceeds the efforts of anyone else (e.g., Hennig, 1936, 1948, 1950a,b, 1952, 1954, 1958, 1965a,b, 1970, 1971, 1973). In biogeography, the work of Lars Brundin on the systematics and biogeography of chironomid midges from southern temperate ("transantarctic") regions was a landmark study that partly inspired the concept of vicariance biogeography (Brundin, 1966). Fly systematics has empirically progressed from major taxonomic works, such as the series *Die Fliegen der Palearktischen Region*, *Manual of Palaearctic Diptera*, and *Flies of the Nearctic Region*. One of the most comprehensive and lavishly illustrated insect references, *Manual of Nearctic Diptera* (McAlpine et al., 1981; 1987; McAlpine and Wood, 1989) has become the standard for morphological study. Other useful references are Séguy (1951) and Colless and McAlpine (1991).

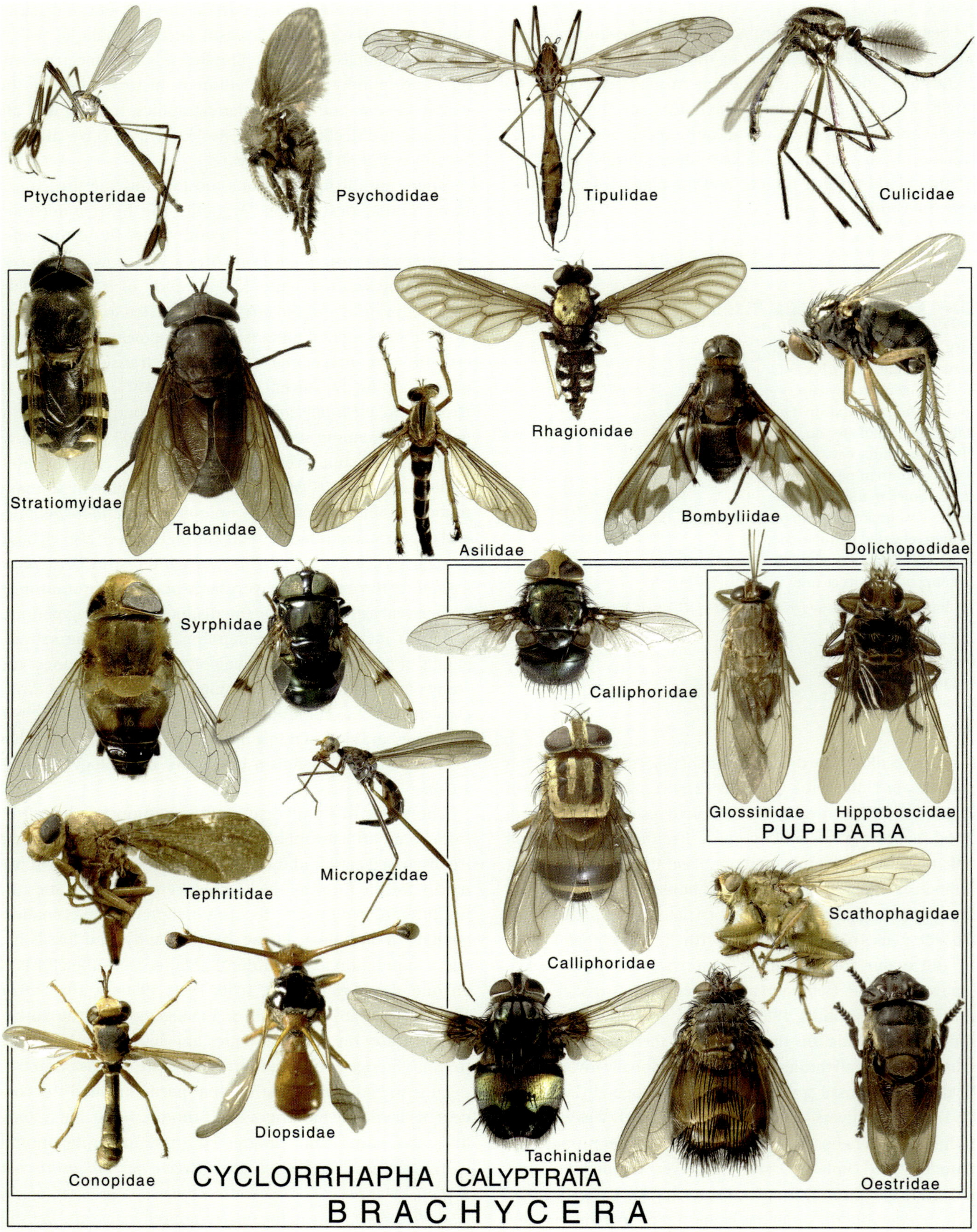

12.21. Adult flies of representative major lineages and families. Not to the same scale.

12.22. Larval flies of representative families and major lineages. Not to the same scale.

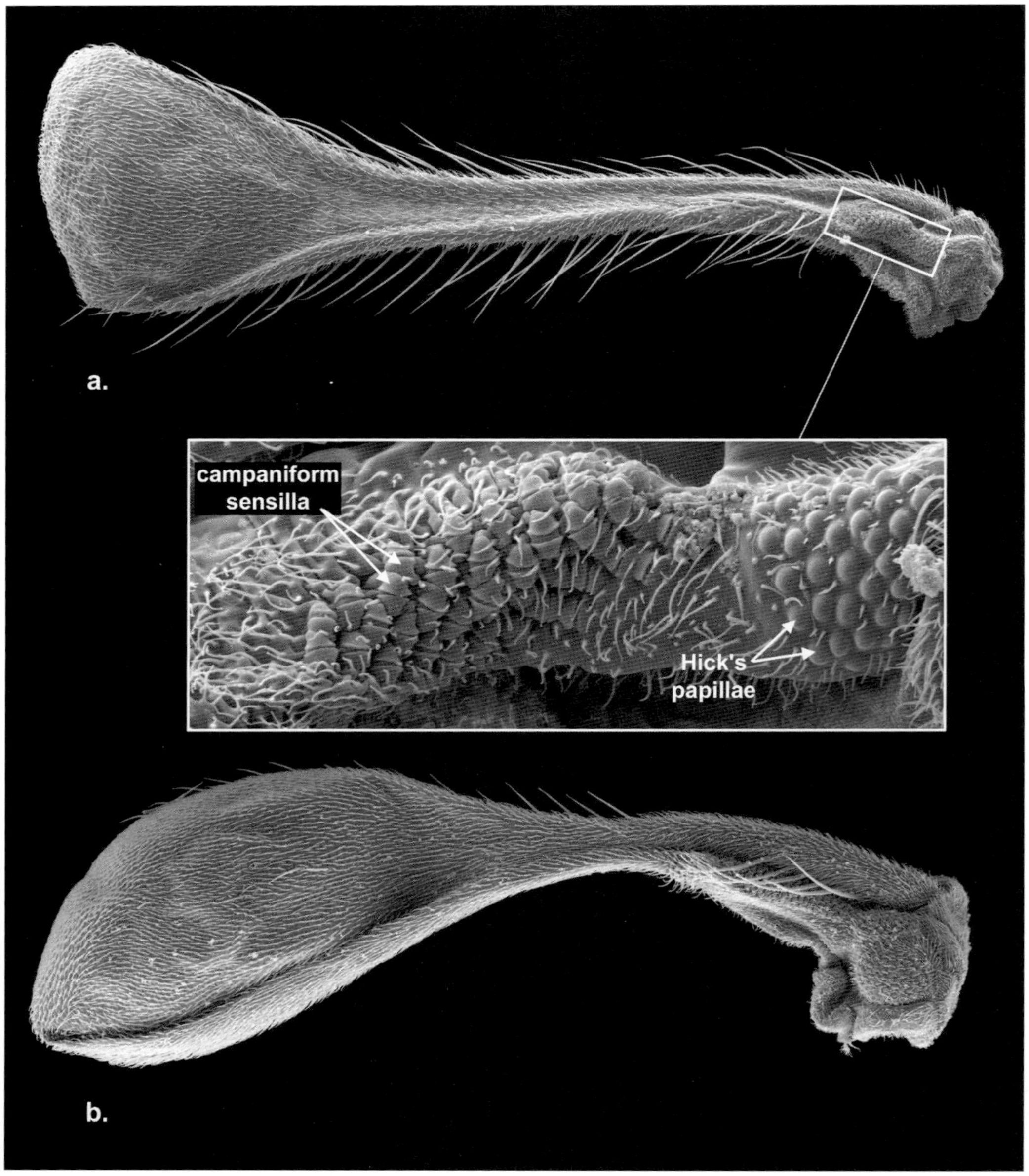

12.23. A dipteran feature: the hind wing reduced into a clublike halter. Rotation of the halteres stabilizes the fly in flight, like a gyroscope. Distinctive rows of sensilla at the base sense movements of the halter. (a) *Limonia* (Tipuloidea), (b) *Sylvicola* (Anisopodidae). Scanning electron micrographs.

Unfortunately, major treatments for tropical regions have not yet been done. Though the comments on evolution and relationships are outdated, Oldroyd's (1964) *The Natural History of Flies* is an engaging account of these diverse insects.

Everyone can recognize most flies, with their distinctive pair of hind wings reduced to clublike *halteres* (Figure 12.23), but not all flies are easily recognized. Some groups of flies have become so anatomically modified that they are barely recognizable as insects. Early naturalists, for example, described some as mites. Even the most bizarre adult flies can be recognized, though, after the larvae are known because larvae are rarely so dramatically modified. Halteres, actually, are not unique to flies. Situations where one of the pair of wings (usually the hind pair) is reduced to buds or small clubs have evolved at least seven times in insects. These involve, besides flies, some mayflies (e.g., Baetidae), male Coccoidea, some Derbidae, the forewings of male Strepsiptera, and the Cretaceous insects *Mantispidiptera* (Neuroptera) (Grimaldi, 2000a) and *Pseudopolycentropus* (Mecopteroidea) (Grimaldi *et al.*, 2004b) (e.g., Figure 12.3). In most cases it is known to be associated with improved, maneuvered flight. In flies, the halteres gyrate and stabilize the insect, partly allowing it to hover and control roll and pitch. Many basal flies, like midges, form mating swarms wherein the males flit back and forth, into which females fly for mating, so halteres and swarming probably evolved together early in the history of flies. Because the forewings are powering the flight of flies, the mesothorax is very

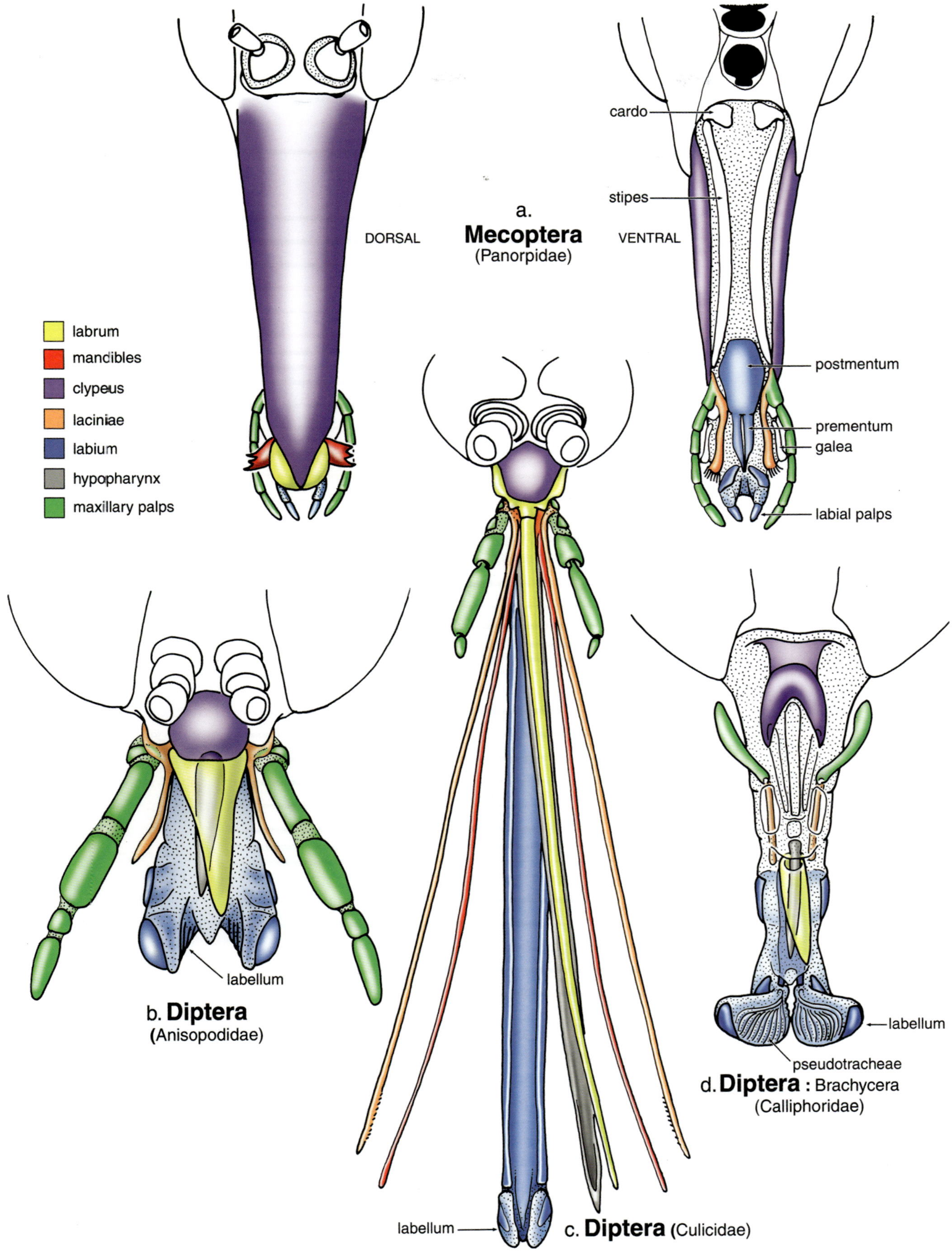

12.24. Mouthparts and their homologies in select Antliophora.

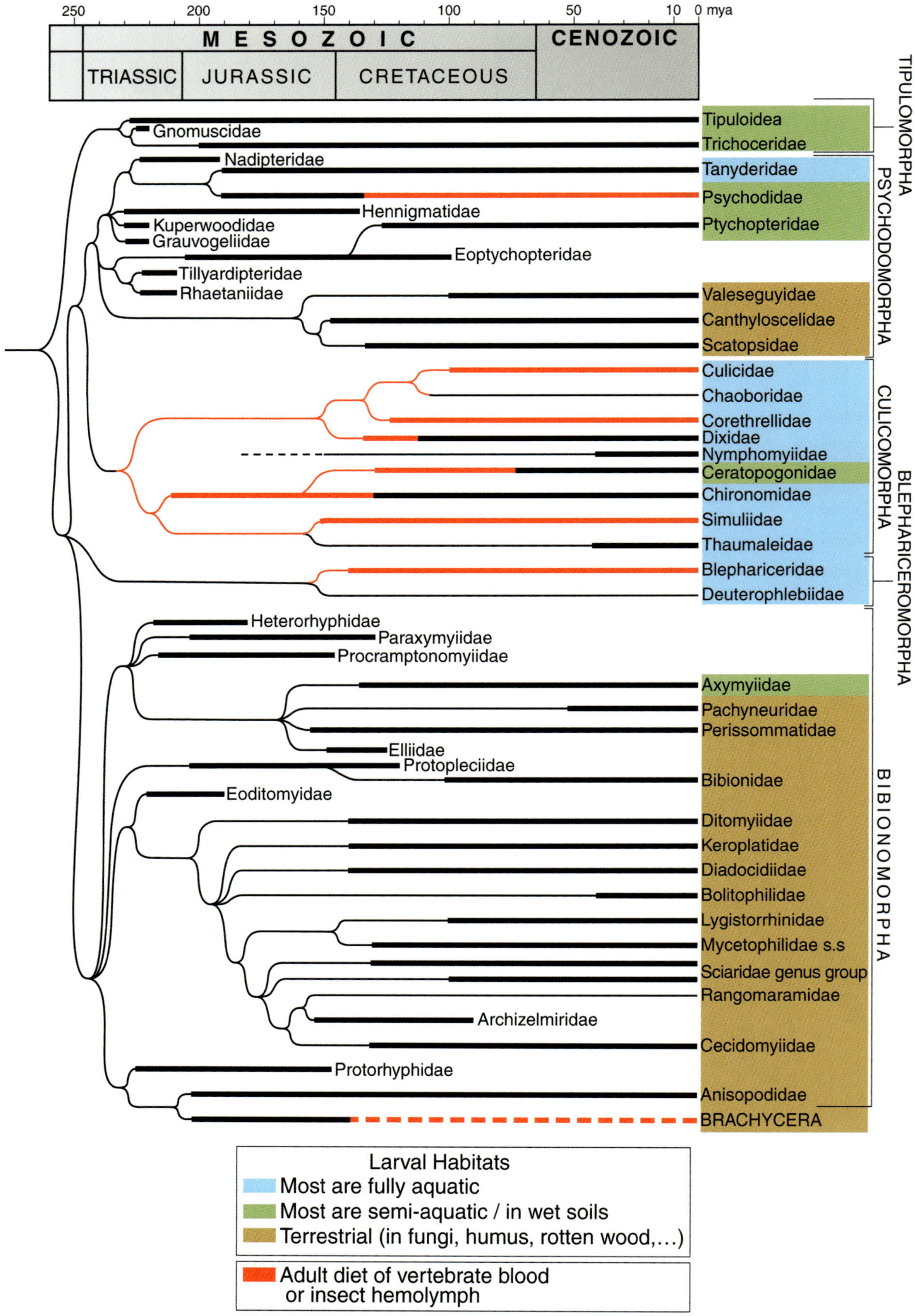

12.25. Phylogeny of the basal (nematocerous) Diptera, indicating larval habitats and lineages with blood-feeding adults. Thick lines indicate the extent of the known fossil record for that lineage. True, two-winged flies probably evolved in the earliest Triassic, about 240 MYA, and by the Late Triassic, 210 MYA, most major infraorders had appeared. From various sources (see text).

large, and the pronotum and metanotum are quite small. Another major innovation in flies concerns the mouthparts (Figure 12.24).

Flies do not have chewing mandibles; they feed on liquids. The mandibles in most flies are vestigial or lost, but where present they are best developed in the females of blood-sucking groups, functioning as lancets. Associated with fluid feeding in Diptera are the labial palpi at the apex of the labium, which are fused into a pair of fleshy lobes, the labella (Figure 12.24). In the higher flies, the Brachycera, the labella are broad and have a system of fine canals, or *pseudotracheae*, radiating from the oral cavity. These aid in mopping up and perhaps filtering fluid food. Where the food is solid, such as a grain of sugar, a fly must first apply saliva to dissolve and then ingest it.

Of the four major holometabolan orders, only the Coleoptera are similar in age to the earliest Diptera in the mid-Triassic; Hymenoptera are slightly younger (Late Triassic); and Lepidoptera is one of the youngest insect orders (Jurassic). The earliest close relatives of flies are three families of mecopteroids from the Late Permian of Australia and Eurasia, 260–250 MYA. These include Permotipulidae (*Permotipula* and *Permila*) (Willmann, 1989), Robinjohniidae (*Robinjohnia*) (Shcherbakov *et al.*, 1995), and somewhat less closely related, the Permotanyderidae (Figure 12.1). The first two are known only from isolated wings (presumably the forewings), and the latter two are known from entire specimens. *Robinjohnia* had four wings of approximately equal size; the hind wings of *Choristotanyderus nana* (Permotipulidae) were half the size of the forewings, and its mesothorax was large. In all these genera the venation was reduced relative to other mecopteroids and close to the hypothetical groundplan venation of Diptera (Hennig, 1973; Willmann, 1989) (Figure 12.1). The earliest Diptera appeared in the Triassic (Krzemiński and Krzemińska, 2003), with *Grauvogelia arzvilleriana* from the mid-Triassic Grés-a-Volzia Formation of France (ca. 240 MYA) being the oldest (Figures 12.25, 12.26) (Krzemiński *et al.*, 1994). By the Late Triassic most major infraorders of Diptera appeared, including Bibionomorpha, Culicomorpha, Psychodomorpha, and Tipulomorpha (e.g., Figure 12.25). These come from the Late Triassic of Australia (Mt. Crosby and Blackstone Formations), Britain (Lilstock Formation: Krzemínski and Krzemińska, 2003), Kazakhstan (Tolgoi and Madygen Formations: Shcherbakov *et al.*, 1995), and Virginia USA (Cow Branch Formation: Krzemiński, 1992; Fraser *et al.*, 1996). Oddly, no Diptera are yet known from the prolific Triassic outcrops of the Molteno Formation of southern Africa. Given that early close relatives, or stem groups, of Diptera occurred in the Late Permian, and four extant infraorders appeared in the Late Triassic, it is reasonable to estimate that the origin of true, two-winged flies took place in the earliest Triassic about 245 MYA.

Interpretation of early fossil flies has been hampered by a poor understanding of the basal relationships of Recent flies. Indeed, the relationships among nematocerous flies, based so far mostly on the structure of adults and larvae of Recent species, are ambiguous and controversial (Hennig, 1973; Wood and Borkent, 1989; Oosterbroek and Courtney, 1995). The family composition of some infraorders is in significant disagreement, and the relationships among the infraorders are obscure. Traditionally, flies were classified into two suborders, Nematocera ("thread horned," for the long antennae), and Brachycera ("short horned"). It is now widely acknowledged that the Nematocera are paraphyletic with respect to the Brachycera, which is why throughout this book we refer to them as nematocerous or nematoceran flies. Oddly, no

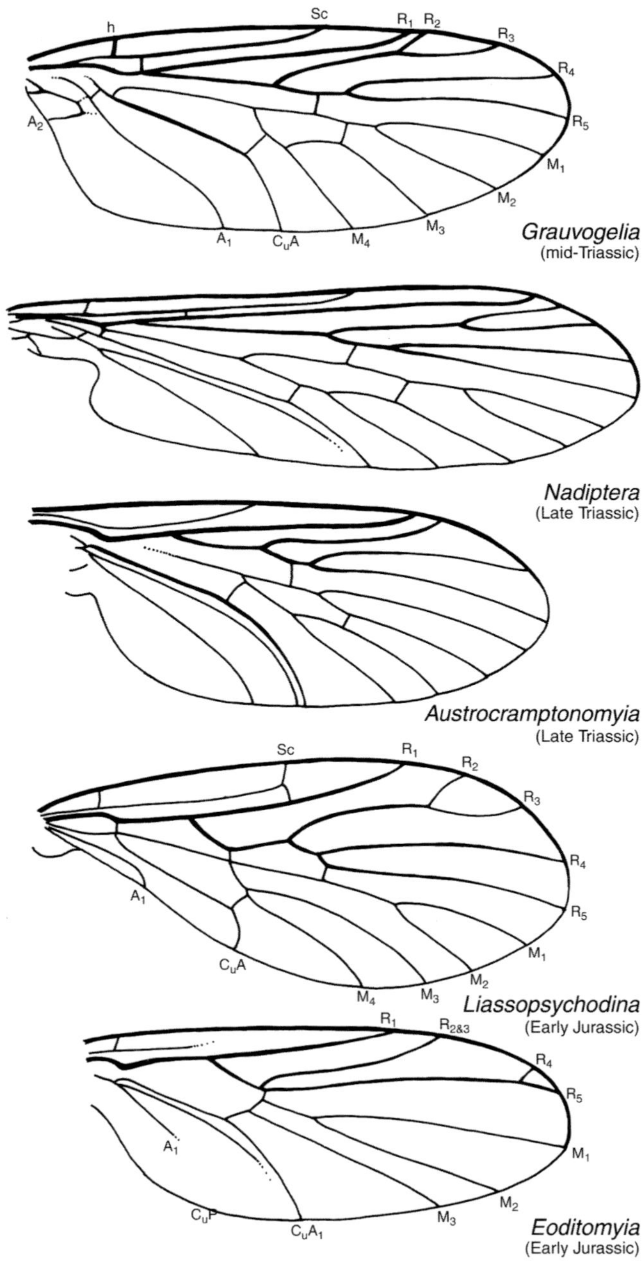

12.26. Wings and venation of Early Mesozoic flies in three infraorders. Not to the same scale.

major DNA sequencing study of relationships among nematoceran families has yet been done. If nematoceran infraorders evolved very close in time, as the Triassic fossil evidence suggests, resolving their basal relationships may be very difficult, perhaps intractable (Figure 12.25).

The infraorder **Tipulomorpha** is comprised of the family Tipulidae sensu lato (craneflies) and perhaps the Trichoceridae and is traditionally considered the most basal group of Diptera, or the sister group to all other flies (Hennig, 1973). The family Tipulidae is split into three families in European classifications: the generally small and delicate Limoniidae (11,000 Recent species); the phytophagous Cylindrotomidae (70 species); and the Tipulidae sensu stricto (4,000 species), or Tipulinae of other classifications. *Pedicia* is a particularly striking tipuline, with a 2-inch-long (5-cm-long) body, long banded wings, and stalklike legs spanning 4 inches (10 cm). The Tipulidae s.l. is the largest family of Diptera, incredibly most species of which were described by one person, Charles P. Alexander (1889–1981), who was a professor at the University of Massachusetts. Having described approximately 11,000 species (one species per day for his career), Alexander was the most prolific of all taxonomists. Tipulid larvae are quite distinctive, with their partially sclerotized, retractable head capsule, and most generally live in semiliquid media like mud and wet soils and are saprophagous. Some, like the larvae of pediciines and hexatomines, are predatory. Neotropical *Geranomyia* may have the most bizarre life cycle, its larva living suspended in a clear gelatinous globule, which it probably leaves periodically in order to forage on leaf microflora (Grimaldi and Young, 1992). *Chionea* is a Holarctic genus of stout-bodied, apterous craneflies (Byers, 1983), which are most commonly seen walking over the snow on warm winter days, like the mecopteran *Boreus*. The Trichoceridae, or "winter craneflies" (100 Recent species), are likewise seen in early Spring and late Fall, usually swarming over rotting stumps and under rock ledges. Some dipterists include the trichocerids in the Psychodomorpha. They have a bipolar distribution, with genera occurring in cool temperate areas of the Holarctic Region and southern South America, the Australian Region, and subantarctic islands.

The relationships of the Tipulomorpha to other flies are not entirely clear. They have been considered as being the sister group to all other flies (Hennig, 1973; Wood and Borkent, 1989; Shcherbakov *et al.*, 1995), but also as close relatives of the Brachycera along with the woodgnats, family Anisopodidae (Oosterbroek and Courtney, 1995). We think the former of these is more likely, based partly on work like that of Michelsen (1996) on musculature of the neck region of flies. Late Triassic tipulomorphs, like Limoniinae and the extinct Vladipteridae, indicate the group is ancient, but no more so than other infraorders. Fossil Limoniinae were diverse, with over 300 species known from the Jurassic through the Cenozoic; they are even preserved in Cretaceous ambers (Figure

12.27. Fossil craneflies (Tipuloidea: Limoniidae). Above, in Early Cretaceous limestone from Brazil. Below, in Early Cretaceous amber from Lebanon. Tipuloids are traditionally considered to be the basalmost living flies. AMNH (limestone); 2.9 mm body length. AMNH LAE152 (amber); wing length 2.13 mm.

12.27). Tipulines and Cylindrotomiines, in contrast, are entirely Cenozoic, suggesting that these radiated quite recently. Trichoceridae fossils are sparse, the earliest of which occur in the Early to mid-Jurassic of Eurasia, and are not seen in the geological record again until the Eocene, in Baltic amber.

Psychodomorpha is a group of considerable contention. We have adopted a classification here that includes the moth flies (Psychodidae) and their apparent closest relatives, the phantom craneflies (Ptychopteridae) and Tanyderidae, as well as the scatopsids and their relatives. Relationships have been discussed by Hennig (1973), Wood and Borkent (1989), Amorim (1994, 2000), and Oosterbroek and Courtney (1995). Psychodomorpha probably had a very early origin and isolated phylogenetic position. Most attention has been given to the family Psychodidae because of their diversity (some 3,000 species) and medical significance. The most commonly seen mothflies are the small, hairy *Psychoda* with broad wings (hence their common name) (Figure 12.21), which rest on the walls of bathrooms, garages, and basements – wherever they can emerge from open drains where they were breeding.

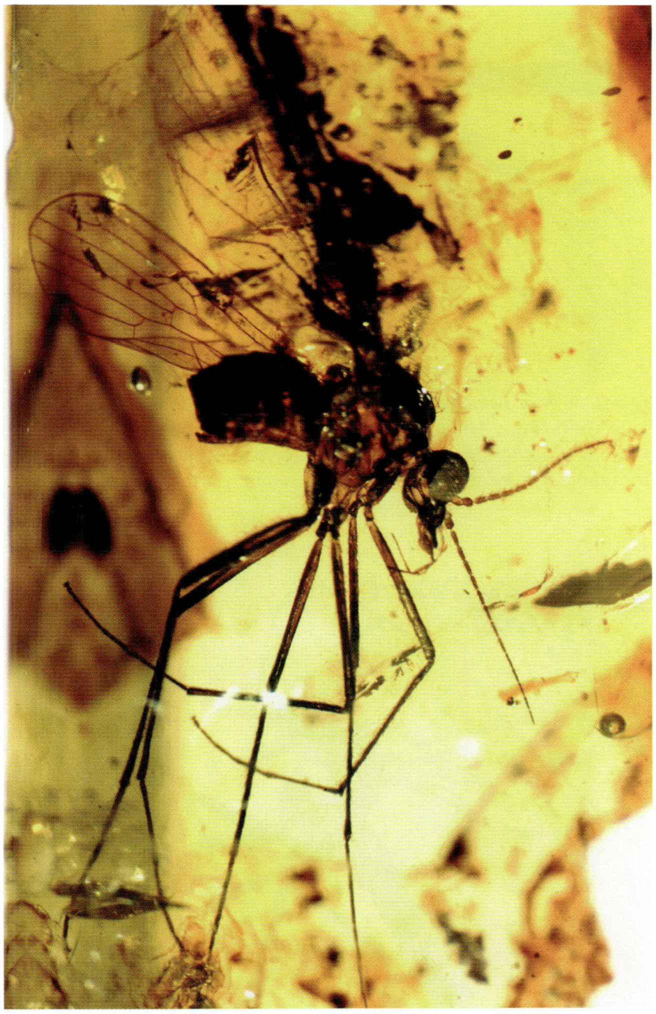

12.28. Tanyderidae in Lebanese amber. This is a relict living family closely related to the Psychodidae. AMNH LAE3-16; wing length 2.6 mm.

Species of greatest significance, though, are the phlebotomines (aptly named "vein lovers" in Greek) – tiny, delicate midges with an irritating bite.

In the Old World the major genus of psychodid blood feeders is *Phlebotomus*, and in the New World tropics it is *Lutzomyia*, which are actually very similar. The sycorine psychodids also feed on blood, but the phlebotomines transmit diseases. Phlebotomines transmit several viruses, reptilian malarias, and bartonellosis or Carrion's disease in Andean South America, the last of which is caused by a bacterium, *Bartonella bacilliformis*. The dreaded phlebotomine disease, though, is leshmaniasis, caused by various species of trypanosome protozoans in the genus *Leshmania* (Lewis, 1973). Different species of phlebotomines carry different leshmanias, some of which are lethal and others merely disfiguring. The worst involves *Leshmania brasiliensis* in South America, a "cutaneous" leshmaniasis that causes leprosy-like lesions on the nose, pharynx, cheeks, and other areas, and *L. donovani*. The latter causes a virulent "visceral" leshmaniasis in southeast Asia, Africa, and parts of South America (sometimes called kala-azar), which results in high fevers, emaciation, and death in untreated individuals. Several outbreaks over the past 150 years in parts of eastern India of kala-azar – their "black death" – killed more than a quarter of the population at times and wiped out entire villages. Various mammals are the normal reservoirs for human leshmanias, and the larvae breed in sandy soils, from which phlebotomines get the name "sandflies."

Putative close relatives of psychodids are the small Recent families Ptychopteridae and Tanyderidae, larvae of which are semiaquatic and aquatic, respectively. Tanyderidae (only about 45 Recent species) have traditionally been viewed as the most primitive living flies, though their structure and fossil record are not particularly more primitive than, say, tipuloids (Figure 12.28). Probably the best known ptychopterid (60 Recent species) is *Bittacomorpha*, which has black and white banded legs and swollen basitarsomeres (Figure 12.21) and holds its legs outstretched when flying along in a bobbing flight. The behavior and coloration make it difficult to focus on the fly while it is in flight. Both families have early relatives from the Early Jurassic, and even the Late Triassic for the Ptychopteridae. Mesozoic ptychopterids are generally grouped into a paraphyletic family, the Eoptychopteridae (Lukashevich *et al.*, 1998), species of which show progressively closer relationship to the Recent species with their decreasing age. The last known "eoptychopterids," in fact, occur in Cretaceous amber from Lebanon and Myanmar (Figure 12.29), and possess (among other features) the hallmark structure of ptychopterids, the *prehalter*, which is a lobe at the base of the halter. Shcherbakov *et al.* (1995) included three extinct families in the Psychodomorpha, all occurring as early as the Late Triassic (of Australia, Britain, and Kazachstan): the Nadipteridae ("Rhaetaniidae" is very similar to this family; Figure 12.26), Hennigmatidae, and Tillyardipteridae. The first two are considered to be quite closely related to Ptychopteridae.

Fossil Psychodidae are particularly interesting because early forms show connections to the Tanyderidae; also, the family is diverse in Cretaceous ambers. Several very primitive psychodids from the Early Jurassic of Germany (ca. 180 MYA) appear to be close relatives to the Tanyderidae, specifically *Nannotanyderus* and *Tanypyscha* (Ansorge, 1994). *Liassopsychodina*, a primitive psychodid from the German Lias also occurs in the Early Cretaceous of Lebanon, where it is completely preserved in amber (Figure 12.30). Lebanese amber likewise contains the earliest phlebotomines – *Phlebotomites* and *Lebanophlebotomus* – which are very similar to Recent *Phlebotomus* and reflect the antiquity of this subfamily and their blood-feeding habits (Hennig, 1971; Azar *et al.*, 1999) (Figure 12.31). The more derived subfamilies Psychodinae and Sycoracinae first appeared in Late Cretaceous amber from New Jersey and western Canada, respectively, and were generically diverse by the time the Baltic amber was formed in the

12.29. Eoptychopteridae (*Leptychoptera*) in Lebanese amber. Eoptychopterids existed from the Early Jurassic to the mid-Cretaceous and comprised a paraphyletic stem group to the living family Ptychopteridae. Note the curled, grasping tarsi. AMNH LAE250; body length 2.9 mm.

12.30. A very primitive moth fly, *Psychotanyderus* (Psychodidae), in Early Cretaceous Lebanese amber. The genus appears intermediate between Tanyderidae and Psychodidae, earliest fossils of which are from the Early Jurassic. AMNH LAE96; wing length 1.23 mm.

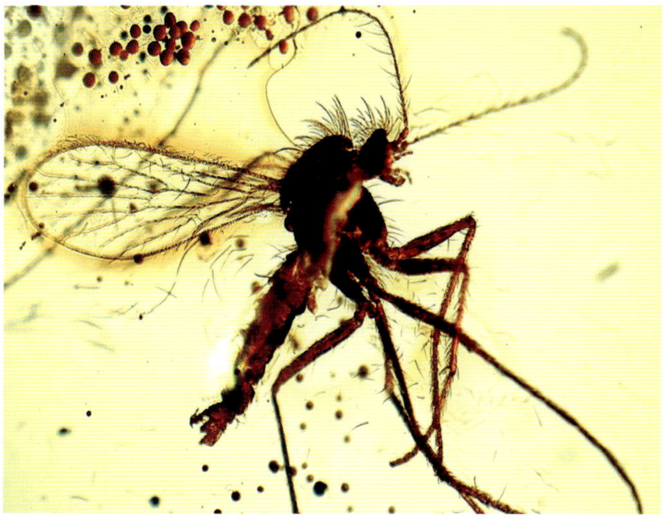

12.31. Primitive male sandfly, *Libanophlebotomus lutfallahi* (Psychodidae: Phlebotominae) in Lebanese amber. Female Phlebotominae feed on vertebrate blood, and some modern species transmit leshmaniasis (kala azar). The earliest known species of the subfamily occur in Lebanese amber. AMNH LAE2; body length 1.43 mm.

Eocene. Because microbial symbionts are preserved in some insects fossilized in amber (Wier *et al.*, 2002), it would be interesting to see – providing that expendable specimens were available – if the Early Cretaceous phlebotomines harbored *Leshmania* or some similar protozoans.

Also included in this infraorder are the scatopsoids, including the families Scatopsidae (black scavenger gnats) (Figure 12.32), Canthyloscelidae, and the unusual genus *Valeseguya* (Figure 12.33), their phylogeny of which has been discussed by Amorim (1994, 2000). Scatopsids can occur in tremendous numbers around foul substrates in which the larvae are breeding. Scatopsidae and the small, closely related family Canthyloscelidae (s.l.) occurred first in the Jurassic, and scatopsids are fairly common in Cretaceous ambers. Lastly, there is a group of three peculiar scatopsoid species, known from one Recent species of *Valeseguya* in Australia (Colless, 1990), another in Miocene Dominican amber, and a Cretaceous species in Burmese amber (Grimaldi, 1991; Amorim and Grimaldi, in prep.), so they were formerly more diverse and widespread. Besides highly modified

12.32. Mating pair of black scavenger gnats, family Scatopsidae, in Miocene Dominican amber. Morone Collection, M-2711; body length 1.9 mm.

12.33. *Valeseguya disjuncta* (Scatopsoidea) in Dominican amber. Its closest living relative is a very rare species from southern Australia. AMNH DR11653; wing length 5.0 mm.

venation, these flies have very reduced mouthparts, and the female has a long, filamentous ovipositor. Their very unusual morphology made their former placement (in Anisopodidae) ambiguous.

Blephariceromorpha includes delicate midges whose larvae live in cold, swift streams. The bizarre neotenic midges, the Nymphomyiidae, have traditionally been placed in this group largely on the basis of having larval abdominal prolegs (Wood and Borkent, 1989; Courtney, 1991; Oosterbroek and Courtney, 1995), but they are now believed to be culicomorphs, as is discussed later. Just two small families exist in the Recent fauna. The Blephariceridae includes 300 Recent species, females of which feed on soft-bodied insects like mayflies and other midges. They feed by grasping prey with large claws that fold against the tarsi, lacerating it with sharp, serrated mandibles and hypopharynx, and then siphoning out body fluids. Males do not have functional mandibles. Adults of the Deuterophlebiidae, comprised of 14 Recent species from western North America and central and eastern Asia (Courtney, 1990, 1994), have vestigial mouthparts and live for only a day or so. Males, in fact, spend their entire brief lives on the wing, never resting even during mating. There is no doubt about the close relationship of these flies, as the larvae of both families are flattened and have abdominal prolegs with ventral suction disks (Figure 12.22). This body structure allows the larvae to cling to smooth stones in fast currents while they graze on films of algae. Emergence from the submerged pupa is very quick, and the wings are immediately functional, which may be related to the distinctive fine folds on the wing membrane of these flies. As would be expected from their habits, these flies are rarely fossilized. The Late Jurassic genus *Ansorgius* is considered a sister group to the Blephariceridae (Shcherbakov *et al.*, 1995; Lukashevich and Shcherbakov, 1997) and had fuller venation than in any Recent species. *Ansorgius*, moreover, had large claws with folding tarsi and short antennae, as in Blephariceridae. *Megathon zwicki*, from the mid-Cretaceous of Mongolia, had a venation very similar to that of the Recent genus *Bibiocephala*. Lastly, there is a beautifully preserved male in mid-Cretaceous amber from Myanmar (Figure 12.34). Fossil blephariceromorphs are too rare, unfortunately, to determine if they are as old as other dipteran infraorders.

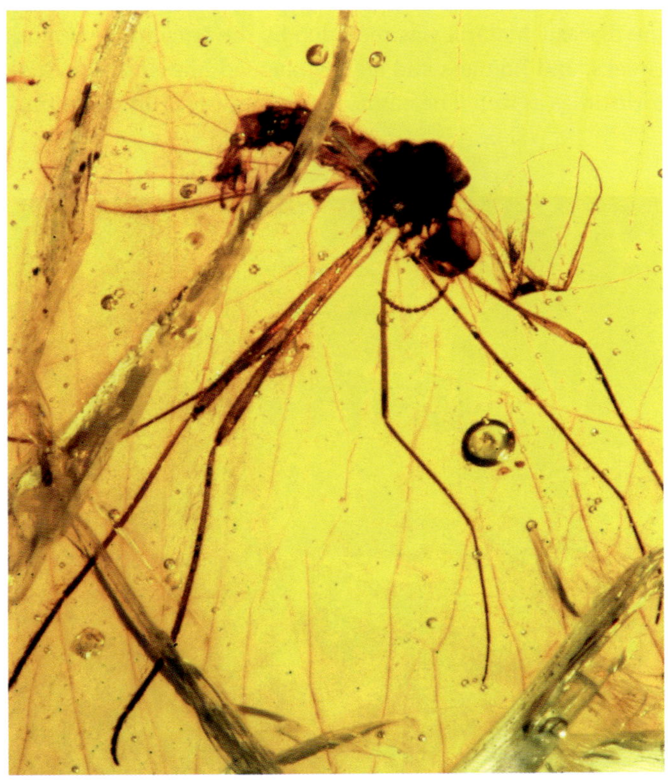

12.34. Fossil net-winged midge (Blephariceridae) in amber from the mid-Cretaceous of Burma. Larval blephariceridss (Figure 12.22) live in swift, cold streams. Fossils of the family are extremely rare. AMNH Bu310; body length 3.8 mm.

The **Culicomorpha** are flies of singular distinction. These are the biting midges, mosquitoes, blackflies, and their relatives, which have affected humans more than any other insects. Malaria is the primary scourge, but viruses, bacteria, nematodes, and other protozoans are also transmitted by culicomorphans to humans, primarily in tropical regions. More than 300 viruses are transmitted by mosquitoes and ceratopogonids alone (e.g., Mellor *et al.*, 2000), the most serious being epidemics of dengue ("breakbone fever"), yellow fever, and various encephalatides. Notable nematode diseases are onchocerciasis ("river blindness") and elephantiasis. The former of these is caused by *Onchocerca volvulus*, transmitted primarily by cryptic species of the *Simulium damnosum* complex of blackflies. Filariae accumulate in various parts of the body, including the retina, which then blind the host. Elephantiasis is the grotesque enlargement of extremities resulting from local blockage of the lymphatic system by infections of filariae, generally from the worm *Wuchneria bancrofti*, which is transmitted by mosquitoes. The most devastating of the culicomorphan diseases is malaria.

Malaria is caused by a protozoan parasite, *Plasmodium*, four species of which infect humans, and all of them are transmitted by *Anopheles* mosquitoes. The most serious of these is *Plasmodium falciparum*, transmitted by species of the *Anopheles gambiae* complex. It is estimated that 200 million people are infected with malaria, most of them in sub-Saharan Africa where a million children die each year from the disease. Malaria was recorded by the ancient Egyptians, Greeks, and Romans and clearly plagued humans in prehistoric times as well. The name, in fact, derives from "malairia," in the belief that the swamps surrounding ancient Rome generated "bad air." It wasn't known until 1898 that mosquitoes transmitted the disease. Millennia of severe selection pressure actually have caused humans to genetically evolve some resistance to it in the form of *sickle cell anemia*, making this only the second instance where humans have evolved in response to insects (the other instance involves lice that transmit plague, of which the delta-32 gene in humans confers resistance). Sickle cell anemia is an inherited disorder of the red blood cells, in which the cells become crescent-shaped because of a recessive mutation of hemoglobin, called hemoglobin S. In homozygous individuals the anemia is debilitating, but heterozygous individuals have mild *sickle cell trait*. The heterozygous form actually confers some resistance to the plasmodial parasite of malaria, which is why this trait has evolved among people where malaria has been epidemic for thousands of years, such as in Africa, the Mediterranean region, the Middle East, and the Indian subcontinent. Malaria was introduced to the New World. The efficiency of some culicomorphs as vectors is attributable to how the adults feed (Figure 12.35), but the manner of larval feeding is why these midges can be so abundant and thus particularly problematic.

12.35. The tables are turned: a fly (*Culex territans*: Culicidae) feeding on a frog. Mosquitoes feed on diverse vertebrates, also including reptiles, birds, and mammals, though many mosquitoes have distinct diets. Photo: S. Marshall.

With the exception of a few transitions to semi-aquatic habitats, many larval culicomorphan midges are fully aquatic and feed by filtering or gleaning plankton, algae, or protozoans from the water and from surfaces. Some, like chaoborids and a few groups of mosquitoes, have predatory larvae. The larvae of many chironomids ("blood worms") live and feed in benthic ooze, respiring by means of concentrated hemoglobins that give them the color of blood. The larvae of blackflies, family Simuliidae, anchor themselves to rocks in streams and feed by filtering out flowing particles (Figure 12.39); they can become so abundant as to carpet whole rocks. Larval mosquitoes, or "wrigglers," are notorious for being able to breed in stagnant water, where they feed by seining out protozoans and other microbiota (Figure 12.36). The larvae of *Wyeomyia* mosquitoes, in fact, thrive in the soup within pitcher plants amongst decaying insects; wrigglers of other species cram the pools among the moss hummocks on arctic tundra, and some species even thrive in the brackish water of salt marshes. Such larval lifestyles make culicomorphans as ubiquitous as water.

Adult culicomorphans that feed on vertebrate blood are the females of most Culicidae, Simuliidae, and Corethrellidae, basal genera of the Ceratopogonidae, and a few primitive chironomids (reviewed in Borkent, 1996). Though it has been suggested that the evolution of culicomorphan blood feeding has occurred multiple times (i.e., Pawlowski *et al.*, 1997), it is likely that it evolved only once in the ancestral culicomorphan and was lost several times. This is based on phylogeny as well as on the basic uniform structure of the adult mouthparts. The females of biting species have bladelike

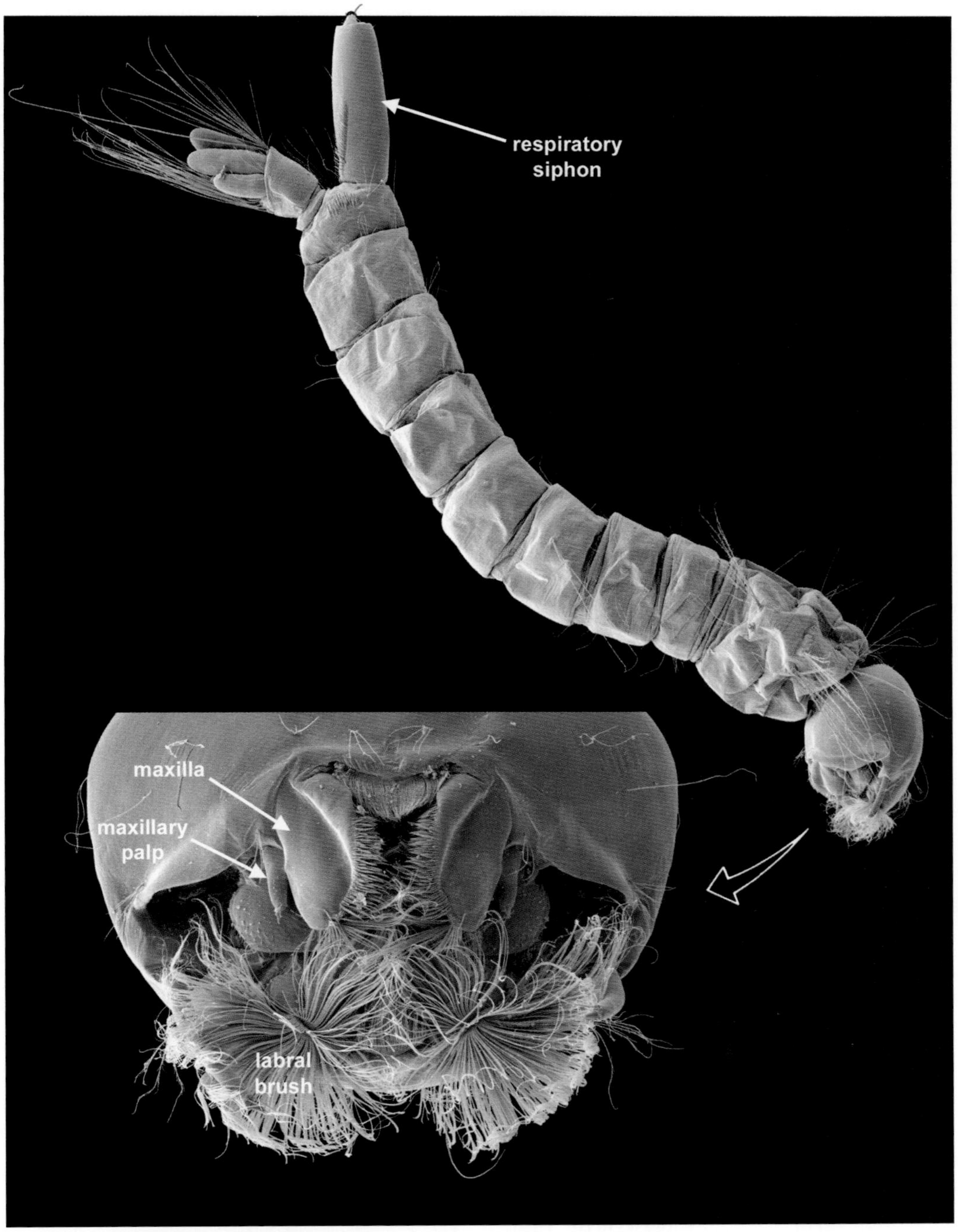

12.36. Larval mosquito (*Aedes*) in its breathing position near the water surface. The larva breathes by sticking the spiracle at the tip of the respiratory siphon through the water surface, and the suspended wriggler feeds by seining microbiota from the water with its labral brushes. Scanning electron micrograph; body length 3.2 mm.

mandibles and maxillary laciniae, and generally a rigid labium; at least the first two of these structures have finely serrate edges for puncturing the skin, and sometimes the labrum and hypopharynx do as well (Figures 12.24c, 12.37). Lengths of the mouthparts, though, vary dramatically. Mosquitoes actually siphon blood; blackflies feed from blood that wells to the surface, but so far as is known all the biting culicomorphans have copious salivary secretions that prevent coagulation from closing the wound or clogging the fine mouthparts. As a result, microbes multiplying in the large salivary glands or even the foregut are easily injected into the host. In species where the males have mandibles, these structures are generally poorly developed or at least not functional because many of them feed just on nectar or honeydew. How, and when, these unparalleled vectors evolved is intriguing, and, fortunately, some families of them have an excellent fossil record.

The family composition of the Culicomorpha is rather firmly established. However, only some of the relationships among families are known with certainty, even though these are the best studied nematoceran flies. Studies where the relationships of culicomorphan families have been

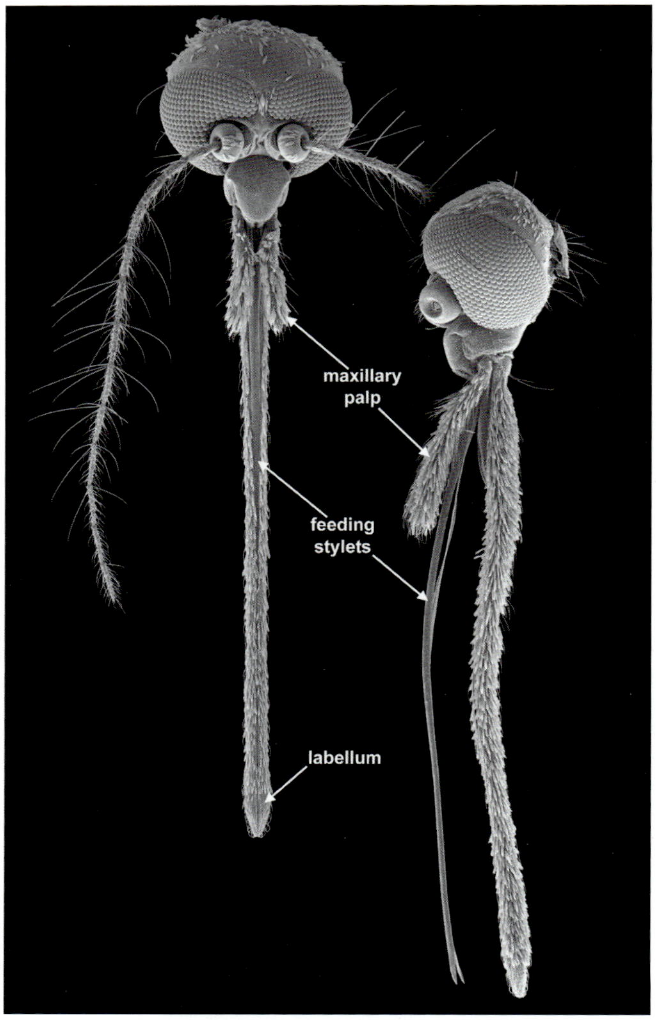

12.37. Mosquito head (genus *Aedes*), revealing the piercing feeding stylets. The stylets are comprised of the labrum and a pair each of mandibles and laciniae (Figure 12.24), which rest in the labial trough (the long, scaled appendage). Like most hematophagous flies, only the females feed on blood. Scanning electron micrograph; head length 1.3 mm.

treated are based on morphological (Wood and Borkent, 1989; Oosterbroek and Courtney, 1995; Saether, 2000) and molecular evidence (Miller *et al.*, 1997; Pawlowski *et al.*, 1997; Aransay *et al.*, 2000; Moulton, 2000). The only consistent grouping is of mosquitoes (Culicidae) with the phantom midges (Chaoboridae) and these with the Corethrellidae, all comprising the "culicoids" (Figure 12.25). The sister group to the culicoids is ambiguous, and for which chironomids (Miller *et al.*, 1997), ceratopogonids (Pawlowski *et al.*, 1997), or all other culicomorphs (Aransay *et al.*, 2000) have been proposed on the basis of molecular data. The morphological evidence for Dixidae being the closest relatives of culicoids is actually quite good (Wood and Borkent, 1989; Oosterbroek and Courtney, 1995; Saether, 2000). Dixidae is a small family (100 Recent species) of delicate, long-legged midges with distinctive venation (Figure 12.38), the larvae of which have habits similar to those of the other culicoids.

Though several of the molecular studies do not closely group the Ceratopogonidae and Chironomidae, the morphological evidence for this relationship appears sound. In fact, the two basal lineages of these families, the intertidal chironomid *Telmatogeton* (Saether, 2000) and the beach-dwelling ceratopogonid *Leptoconops* (Borkent, 2000a), have distinctive long cerci in females. This feature was probably in the common ancestor of the two families, but then lost in the largest lineage of each family. Both families are large (for Ceratopogonidae 5,500 species [Borkent and Wirth, 1997]; for Chironomidae about 5,000 species), but the ceratopogonids are the notorious blood feeders. Actually, only the basal four genera of ceratopogonids feed on blood: *Austroconops*, *Leptoconops*, *Culicoides*, and *Forcipomyia* (*Lasiohelea*). More recently evolved groups are predatory on other small insects; and *Atrichopogon* and some *Forcipomyia* even pierce the wing veins of large insects like mantises, dragonflies, stick insects, and butterflies and feed on their hemolymph (Lane, 1984). Virtually all adult chironomids have vestigial mandibles, or none at all, but the primitive Recent genera *Archaeochlus* and *Austrochlus* have biting mandibles (Cranston *et al.*, 2002), which appear to be a vestige of the biting ancestors of these families. *Aenne*, the earliest and most primitive chironomid, known only as Mesozoic fossils, had biting mandibles (Cranston, unpubl.).

The blackflies, family Simuliidae (1,700 Recent species), have been proposed as the sister group to the Dixidae in two molecular studies (Miller *et al.*, 1997; Aransay *et al.*, 2000), which seems unlikely on the basis of morphology. Rather, Simuliidae appear closely related to the small family Thaumaleidae (70 Recent species), which has a bipolar distribution in the Holarctic Region and southern temperate regions. Adults in both of these families are quite similar, being unusually stout-bodied midges with short, compact antennae and similar wing venation. Like simuliids (e.g., Figure 12.39), larval thaumaleids attach themselves to wet rocks, but

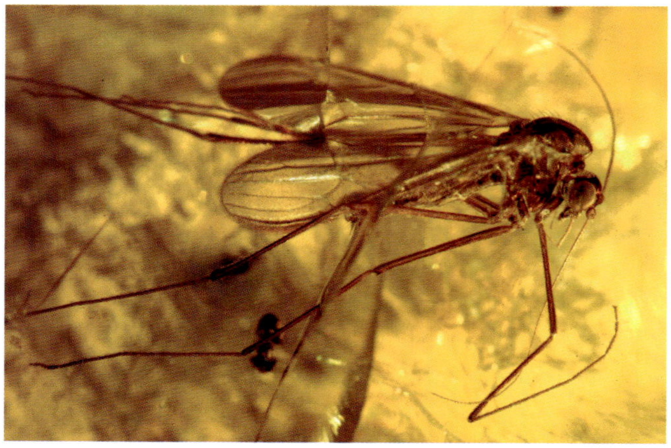

12.38. Midge of the culicomorphan family Dixidae (genus *Paradixa*) in Eocene Baltic amber. This family is a primitive living relative of mosquitoes. AMNH; body length 2.2 mm.

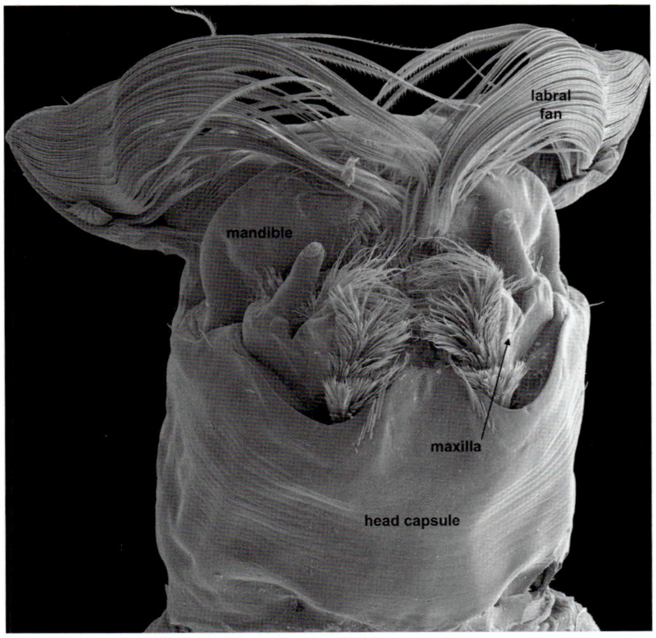

12.39. Head of a larval blackfly (Simuliidae: *Cnesia*), showing the large labral fans. These are used to seine particulate food from the water flowing past the anchored larva. Scanning electron micrograph.

generally ones that are vertical and not entirely submerged. Adult feeding in thaumaleids is unknown, so these have probably lost the blood-feeding habit.

Lastly, the peculiar small, Holarctic family Nymphomyiidae has recently been placed in the Culicomorpha, though of uncertain affinities because they are so highly reduced and neotenic. Adults are very tiny and slender, with a vestigial tentorium, digestive tract, mouthparts, and antennae; their wings, which are also shed after the mating flight, are narrow straps with highly reduced venation and a long marginal fringe like that of thrips; and they retain various larval features. Formerly, nymphomyiids were placed in the Blephariceromorpha, but their larval mouthparts are very similar to those of the Culicomorpha. In fact, it would not be surprising if nymphomyiids were highly modified members of another family, the way *Oreadomyia* is in the Chironomidae, or *Baeonotus* is for cecidomyiid gall midges. Both of these genera are wingless and neotenic. Courtney (1994) provided a detailed review of the taxonomy, morphology, and biology of this interesting family. That there are any fossil nymphomyids is remarkable; nevertheless, there is one – *Nymphomyia succina* in Baltic amber (Wagner *et al.*, 2000) (Figure 12.40). It is very similar to the Recent species, all of which occur in eastern North America and eastern Asia, so the distribution of the family included Europe in the Eocene as well.

The excellent fossil record of Culicomorpha is largely the result of the abundance of chironomids and chaoborids in lake sediments and the preservation of various families in amber. The oldest culicomorphan, and the only known Triassic specimen, is *Aenne triassica* from the Late Triassic of England, ca. 210 MYO (Krzeminski and Jarzembowski, 1999). Though it is based just on the distal half of the wing, distinctive branching of the small r-r crossvein near the apex, and the base of Rs and M, shows it is very similar to Jurassic species of this genus from Germany (Ansorge, 1999), and even to an undescribed species of the genus from the Early Cretaceous amber of Lebanon. The amber specimens confirm that these are indeed chironomids, and not stem-group chironomoids with a primitive venation. In fact, the *Aenne*-like species in amber had biting mandibles. Otherwise, the Jurassic record indicates this is the period when most Recent families of Culicomorpha diverged. Besides abundant Chironomidae, Jurassic records include Dixidae (*Syndixa* and Eucorethrinae: Early to mid-Jurassic of Kazakhstan and Siberia [Lukashevich, 1996]), Thaumaleidae (*Mesothaumalea*: Late Jurassic to Early Cretaceous of Siberia), Simuliidae, and the extinct family Rhaetomyiidae. Rhaetomyiidae are considered a stem group to the culicoids on the basis of venation (Ansorge, 1996, 1999), or as chaoborids (Lukashevich, 1996), and these were among the most abundant Diptera in the Jurassic and Early Cretaceous, particularly as pupae. The earliest definitive Simuliidae are *Kovalevimyia lacrimosa*, an adult from the Late Jurassic of Siberia, and *Simulimima grandis*, a pupa from the mid-Jurassic of Siberia (Mesozoic Simuliidae are reviewed in Currie and Grimaldi [2000]). *Kovalevimyia* appears to belong to the basal tribe Prosimuli-

12.40. A very rare fossil of the unusual family Nymphomyiidae (*Nymphomyia succina*) in Eocene Baltic amber. Larvae of these minute, delicate midges live in clean, cold streams; adults are highly modified. The wings are long and slender (opaquely preserved here) and fringed with long hairs. AMNH Ba-JH9; body length 1.3 mm.

12.41. Reconstruction of an early blackfly, *Archicnephia ornithoraptor*, in 90 MYO amber from New Jersey. Structure of the tarsal claws suggests that it fed on birds or feathered dinosaurs. AMNH NJ 115; length 1.0 mm.

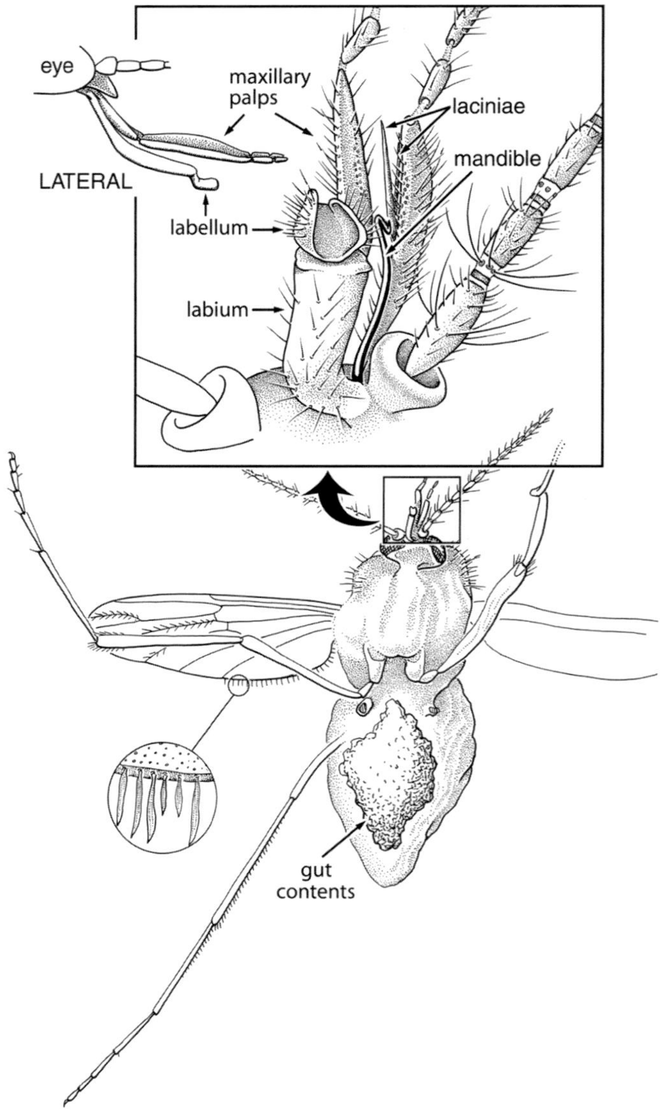

12.42. The oldest and most primitive mosquito, in mid-Cretaceous amber from Burma. AMNH Bu32.

ini. Early Cretaceous simuliids are *Baisomyia incognita*, from Baissa, central Siberia, and larvae from the Early Cretaceous of Koonwarra, Australia. Adults of *Archicnephia ornithoraptor* are the only simuliids in Cretaceous amber (from New Jersey, ca. 90 MYO) (Figure 12.41), and are the most basal members of the Recent tribe Simuliini. Structure of their claws indicates this fly fed on birds or closely related feathered dinosaurs (Currie and Grimaldi, 2000).

Culicomorpha in Cretaceous and Cenozoic ambers have left a unique fossil record. The only Mesozoic mosquitoes, in fact, are two specimens in amber from western Canada (Pike, 1995) and northern Myanmar (Grimaldi *et al.*, 2002) (Figure 12.42), and the only Mesozoic Corethrellidae is in Lebanese amber (Szadziewski, 1996). The Canadian amber mosquito, *Paleoculicis minutus* (ca. 75 MYO) is supposedly closely related to the Culicinae. If so, this would suggest an origin of the Culicidae in perhaps the Early Cretaceous. A culicoid in older (100 MYO) Burmese amber primitively lacks the scales on the wing veins or body that all mosquitoes have, but it has a proboscis much longer than in any chaoborid as well as four-segmented palps like mosquitoes (other Culicoidea have five segments). Thus, it seems to be a rare intermediate in the fossil record. Higher level relationships in Culicidae have only recently been addressed on the basis of morphology (Harbach and Kitching, 1998) and DNA sequences (Mitchell *et al.*, 2002). Traditionally, three subfamilies of Culicidae are recognized: Anophelinae, Culicinae, and Toxorhynchitinae. The last of these is monogeneric, and the adults of *Toxorhynchites* can be strikingly large and iridiscent; their larvae are predaceous, even on other mosquito larvae. Recent studies show that *Toxorhynchites* is actually a highly evolved culicine.

The two most diverse and abundant families, Chironomidae and Ceratopogonidae, generally comprise 10% or more of all insect inclusions in the major deposits of insect-bearing ambers from the Cretaceous (e.g., Figure 12.45). This is probably because most amber-producing trees lived close to brackish or fresh water, and most midges in these families swarm. Unfortunately, the Chironomidae in these ambers are virtually unstudied. Study of the Ceratopogonidae in Cretaceous ambers, though, is a model for the study of insect evolution.

Unlike Chironomidae, there have been prolific descrip-

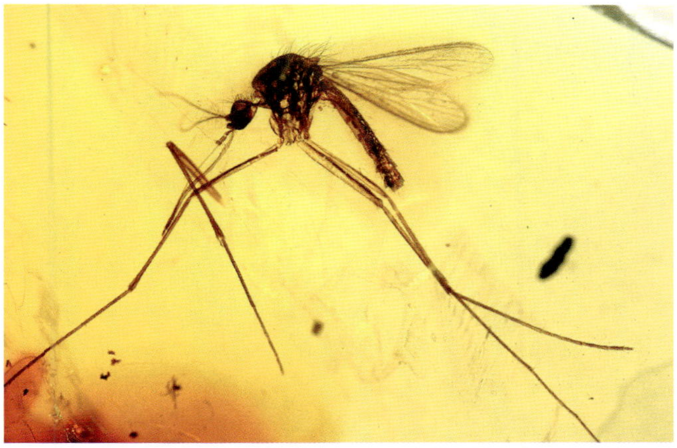

12.43. Female culicine mosquito in Miocene Dominican amber. Mosquitoes probably evolved in the Early Cretaceous, and modern genera of them appeared by the Eocene. AMNH DR11642; body length (excluding proboscis) 2.9 mm.

12.44. A male *Trichoprosopon* mosquito from Ecuador, guarding a raft of eggs just laid by his mate. Such guarding is unusual for mosquitoes; usually the egg raft is abandoned after oviposition. Photo: P. J. DeVries.

tions of the Ceratopogonidae fossilized in Cretaceous ambers (Szadziewski and Schlüter, 1992; Szadziewski and Arillo, 1994; Borkent, 1995, 1996, 1997, 2000a,b, 2001; Szadziewski, 1996, 2000), and in Tertiary amber (e.g., Szadziewski, 1988, 1990, 1993; Szadziewski and Grogan, 1998a,b). Moreover, relationships among living and fossil genera reveal an interesting pattern. The most basal living genera, *Leptoconops* and *Austroconops* (which are blood feeders), appeared first in Early Cretaceous Lebanese amber (Figures 12.46, 12.47), along with six extinct genera. Genera and tribes of intermediate phylogenetic position (Borkent, 2000a; Beckenbach and Borkent, 2003) appear in mid- and Late Cretaceous ambers, many of which are insect feeders and some of which are vertebrate blood feeders (*Forcipomyia, Culicoides*) (Figure 12.48). The most recently derived lineage first appeared in the Eocene Baltic amber. This lineage contains Recent species whose females are known to suck the hemolymph of the males with whom they are mating – the ultimate nuptial gift. Without doubt, vertebrate blood feeding occurred in the ancestral ceratopogonid, which is a habit inherited from the ancestor of probably all culicomorphans. Moreover, the remarkable correlation between chronology and phylogenetic position of the fossils indicates that Ceratopogonidae originated probably in the Late Jurassic but not much earlier, probably from a stem-group chironomoid. The combined fossil record of culicomorphans indicates that biting midges in general probably plagued land vertebrates beginning in the Late Triassic, 220 MYA.

Bibionomorpha is the most complex infraorder of nematocerous flies. It may be a paraphyletic group since one family, Anisopodidae, has been proposed as the closest living relative to the diverse suborder Brachycera (Oosterbroek and Courtney, 1995). The morphological features studied by Michelsen (1996), which were sclerites and musculature of the neck region, indicate no special affinity of any particular family of Bibionomorpha to the Brachycera, though that work does provide strong evidence for a close relationship between these two large groups. Michelsen (1996), in fact, proposed the name *Neodiptera* to comprise the Bibionomorpha + Brachycera. Blaschke-Berthold (1994) discussed relationships within the infraorder based on detailed morphology.

12.45. A male midge (family Chironomidae) in 125 MYO amber from Lebanon. Males of many culicomorphan flies have brushy antennae, which are used for sensing the location of females. Chironomids are very abundant in Cretaceous ambers and have a fossil record that begins in the Triassic. AMNH LAE105; length 2.5 mm.

12.46. *Austroconops macmillani* (Ceratopogonidae), from Australia. This is one of the two most basal and oldest genera of biting midges. Scanning electron micrograph; body length 0.95 mm.

Besides various adult structures, close relationship of the Bibionomorpha to the Brachycera is also indicated by the larvae, both of which are essentially terrestrial. All other nematocerans are basically aquatic or semi-aquatic. Larvae of bibionomorphans worm their way through humus, leaf mold, decaying vegetation, and rotten wood and under decaying bark, where they feed on fungal mycelia or directly on fungal fruiting bodies such as mushrooms. The only major group of bibionomorphans to have evolved into a completely different niche are the cecidomyiids, or gall midges, whose larvae feed within galls on plants (some are predatory), but even the basal lineages of cecidomyiids retain the ancestral mycophagous habit.

There are 13 Recent families of bibionomorphans and 12 extinct ones. Relationships among the Recent families has been discussed by Amorim (1993). Recent work on sciaroid fungus gnats provides good justification for classifying five "subfamilies" of Mycetophilidae as families within the Sciaroidea. Interpreting their fossil diversity has been complicated by Rohdendorf's taxonomy of Jurassic and Cretaceous forms (Rohdendorf, 1964), which is now being revised by various authorities. Three of the Recent families of Bibionomorpha are relict and small, with generally fewer than ten species each and all having evolved since the Jurassic. These include Axymyiidae (Holarctic), Perissomatidae (Australia and southern South America), and Pachyneuridae (Holarctic). The last of these families, including the Recent genus *Cramptonomyia*, is the only surviving member of an old and diverse group that occurred from the Triassic to the Early Cretaceous. This group included Procramptonomyiidae (Figure 12.49) and Elliidae. By the mid-Cretaceous, their diversity greatly diminished, with Elliidae probably being the stem group to the Recent species of Pachyneuridae. Though they are a basal family of Bibionomorpha, Bibionidae ("March flies") are very rare as Mesozoic fossils; they then become some of the most common insects in Tertiary lake deposits (Figure 2.64). Over 300 fossil bibionids, in fact, have been described from the Tertiary (reviewed in Evenhuis, 1994). This pattern is probably because the family "Protopleciidae" is a paraphyletic stem group to the Bibionidae.

12.47. *Austroconops fossilis*, in Early Cretaceous Lebanese amber. The two basal genera of ceratopogonids, *Austroconops* and *Leptoconops*, are the only living genera of the family that occur so early in the Cretaceous. AMNH LAE29; body length 0.78 mm.

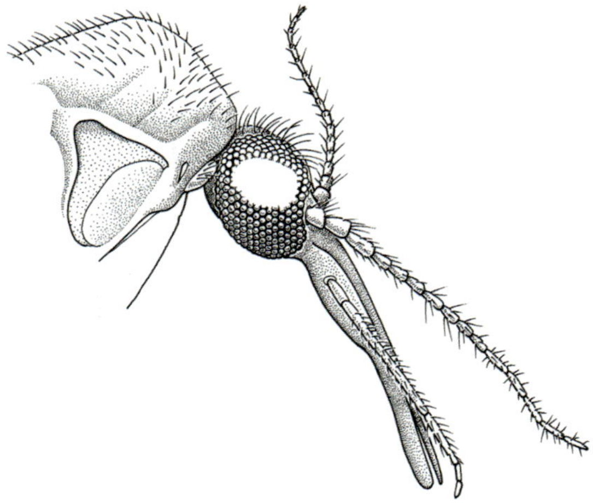

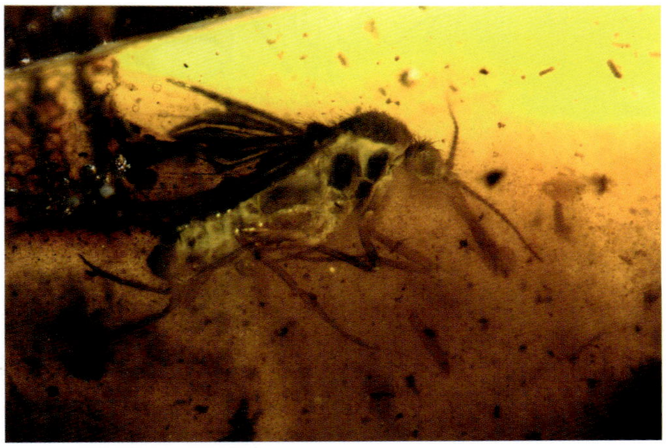

12.48. The mosquito of ceratopogonid midges: *Culicoides yoosti* in mid-Cretaceous amber from New Jersey. Its proboscis, shown above, is much longer than any other species in the Ceratopogonidae, which was used for feeding on blood and perhaps for also probing flowers. AMNH NJ1112; body length (excluding proboscis) 2.2 mm.

12.49. A Jurassic procramptonomyiid fly from Kazakhstan, the family of which is an extinct close relative of other bibionomorphans. PIN 2239/2104; length 7.2 mm.

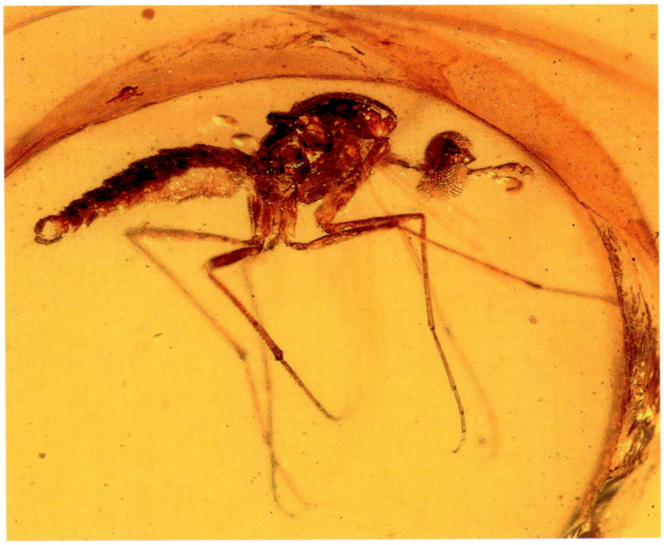

12.50. *Burmazelmira aristica* in mid-Cretaceous Burmese amber, one of the last occurrences of Archizelmiridae in the fossil record. The antenna of this fungus gnat (Figure 12.51) was similar to that of Brachycera in having a compact base and bristle-like flagellum. AMNH Bu178; body length 2.5 mm.

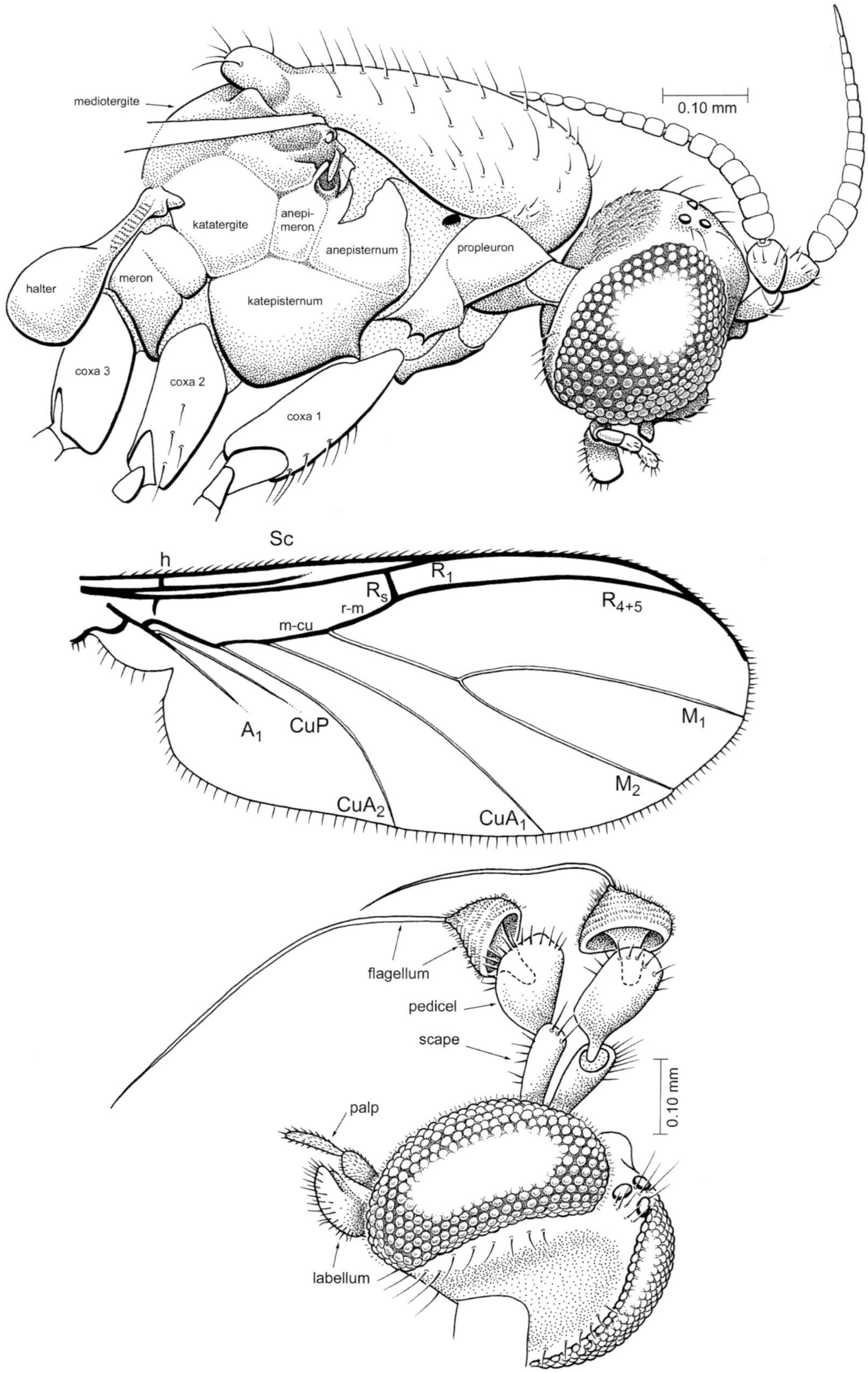

12.51. Archizelmirids from Cretaceous amber: thorax and head with wing of *Zelmiarcha lebanensis* (above), in Lebanese amber; head of *Burmazelmira aristica* (below), showing the unusual antenna. Archizelmirids are extinct relatives of black fungus gnats (family Sciaridae) and gall midges (family Cecidomyiidae).

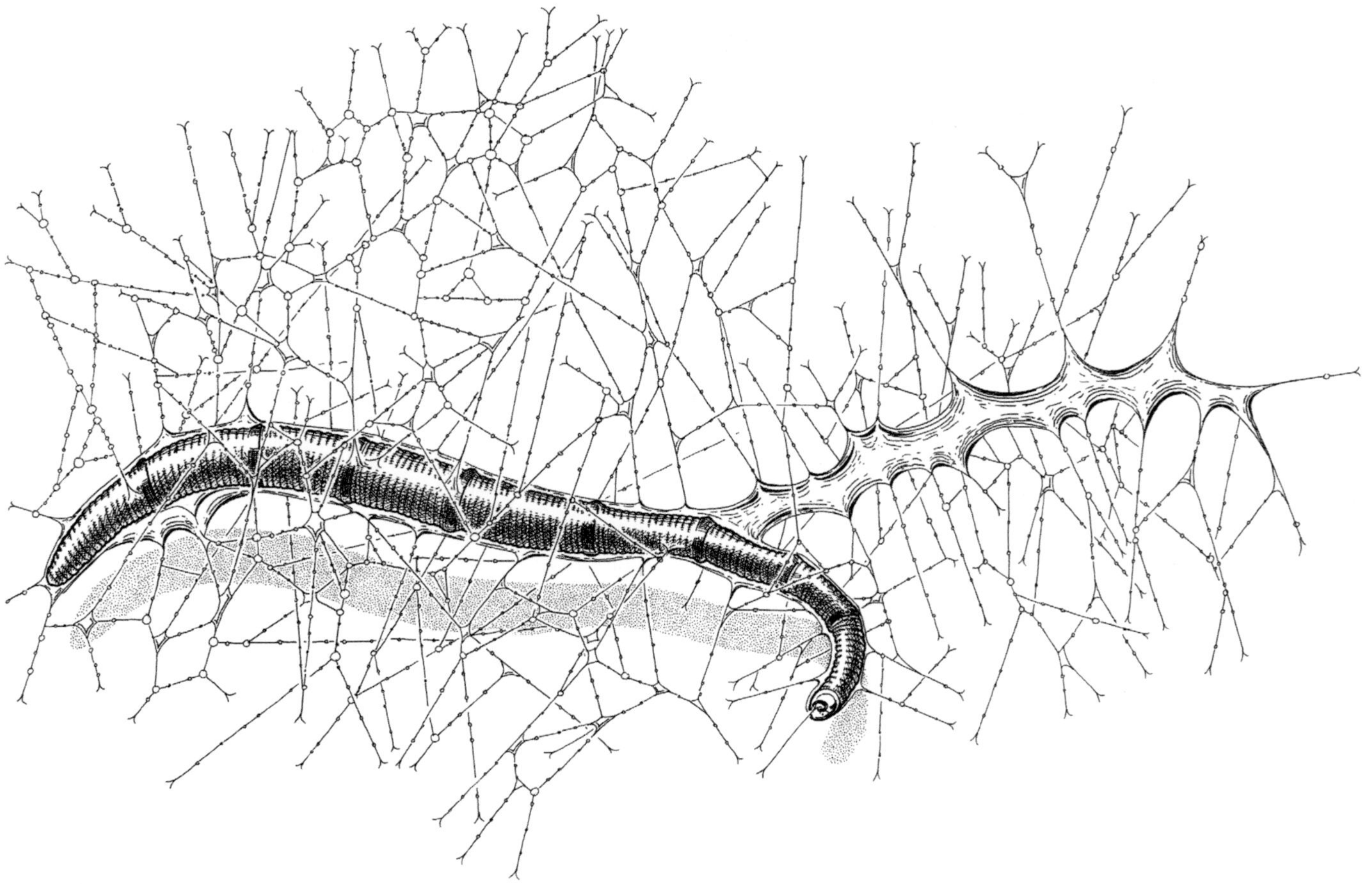

12.52. Larva of a North American species of *Asindulum* fungus gnat (Keroplatidae) in its web. The sticky web is produced from mucus and salivary secretions. The web collects fungal spores and snares minute animals, on which the larva feeds. Length of larva 6.6 mm.

Protopleciids were diverse from the Early Jurassic to the mid-Cretaceous.

The Sciaroidea, comprising the "fungus gnats" and "gall midges," is by far the most heterogeneous group in the Bibionomorpha in terms of life histories, but they are a definitive monophyletic lineage. Included in this group are six extinct families from the Late Triassic to Early Cretaceous, almost all from Eurasia (V. G. Kovalev, 1990; O. V. Kovalev, 1996; Blagoderov, 1993; Grimaldi *et al.*, 2003). Some extinct families appear to be stem groups to Recent families, like Eoditomyidae and the Recent family Ditomyidae. Another family, the Archizelmiridae, is in a group of basal sciaroid genera, the so-called "*Heterotricha* group" (Chandler, 2002), which have a distinctive, primitive venation and whose relationships are proving difficult to unravel. Archizelmiridae are interesting because they also occur in Cretaceous amber, and a species in Burmese amber is the only nematocerous fly with the antennal segments differentiated into an arista – a feature seen only in the Brachycera (Grimaldi *et al.*, 2003) (Figures 12.50, 12.51). The recently described family, Rangomaramidae (five Recent species from New Zealand) (Jaschoff and Didham, 2002) appears very similar to the Archizelmiridae. It is during the Cretaceous, in fact, that most Recent families of Sciaroidea first appeared: Bolitophilidae, Diadocidiidae, Lygistorrhinidae, Mycetophilidae (sensu stricto), and Cecidomyiidae (Blagoderov and Grimaldi, 2004). Keroplatidae first appeared in the Late Jurassic (Blagoderov, 1993). The earliest fossils of the large, homogeneous family Sciaridae (ca. 3,000 species) occur in the Late Jurassic; they are scarce in the Cretaceous but become some of the most abundant inclusions in Tertiary ambers. There are thousands of undescribed sciarid species in the Recent fauna.

Mycetophilidae (sensu stricto: 3,500 species) radiated in the Cretaceous (Figure 12.53), and by the Eocene they became very abundant and diverse inhabitants of moist temperate forests, as they are today. Their relationships have been treated by Soli (1997). The Keroplatidae, which have been monographed by Matile (1990), certainly have some of the most peculiar sciaroids. Many larval keroplatids construct mucous webs and tunnels under rock overhangs or logs, where they graze on fungi or feed on small snared arthropods (Figure 12.52). Species of the New Zealand "glowworms," *Arachnocampa*, have taken the habit one step further. They suspend themselves from the ceilings of caves by mucous strands, and luminescence from the larva attracts small midges, which are snared in the strands and devoured

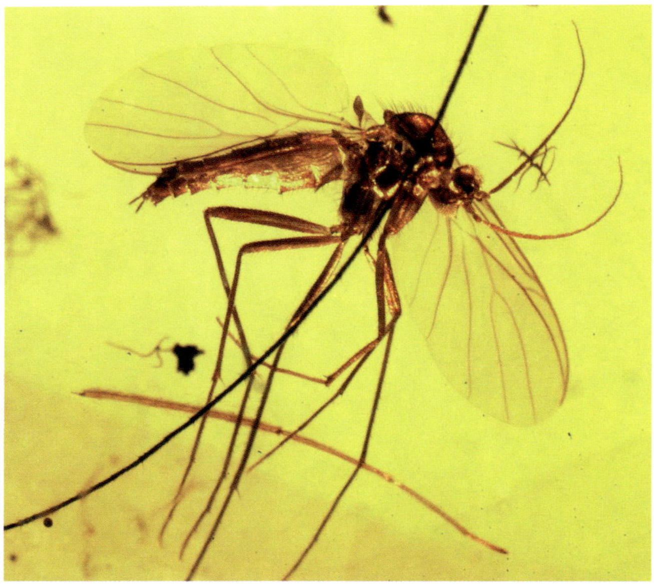

12.53. Fossil fungus gnat (Keroplatidae: Macrocerinae), in amber from the mid-Cretaceous of Burma. AMNH Bu314; body length 2.8 mm.

by the larva. Luminescence is generated by swollen ends of the Malphighian tubules, which are embedded in tracheae that reflect the light through the thin, translucent cuticle. The "light show" from the ceilings of Waitomo caves in New Zealand is a spectacle, which supports $5 million annually in tourism revenue. Even more bizarre is *Planivora* from New Zealand and southern South America, the larva of which is a parasite of land planarians. Land planarians can be several centimeters long and leave slime trails like slugs and keroplatid larvae, so conceivably this anomalous lifestyle evolved from predaceous keroplatine larvae following planarian trails and then invading the planarian.

The gall midges, family Cecidomyiidae (5,000 Recent species), have left a rich fossil record. Because they are small, frail midges, they are particularly common and diverse in ambers from the Tertiary and Cretaceous, much like the Ceratopogonidae (Figure 12.54). Unlike the Ceratopogonidae, though, most of these fossil cecidomyiids are unstudied. Only six species have been described from Cretaceous ambers, of Canada and Spain only (Gagné, 1977; Arillo and Nel, 2000). The only other Cretaceous cecid is *Catotricha mesozoica*, in Late Jurassic-Early Cretaceous rock, and it is very similar to the primitive living genus *Catotricha* (it is uncertain, though, whether it belongs in this Recent genus). Cecidomyiidae are an ideal insect group for studying the effects of the angiosperm radiations. The great proportions of living species are gall formers on angiosperms, a habit that apparently evolved once in the largest and most recently evolved subfamily, Cecidomyiinae (Gagné, 1986; Harris, 1994). Taxonomy, ecology, and host relations of Cecidomyiidae have been nicely reviewed by Gagné (1989, 1994). Cecid larvae induce gall formation by applying salivary secretions to leaf tissues, and this causes the tissues to proliferate and grow around the larva. Many species are monophagous, particularly those on perennial plants like trees, and they form characteristic galls. Though the gall midges from the canopies of tropical forests are unexplored, the actual numbers of

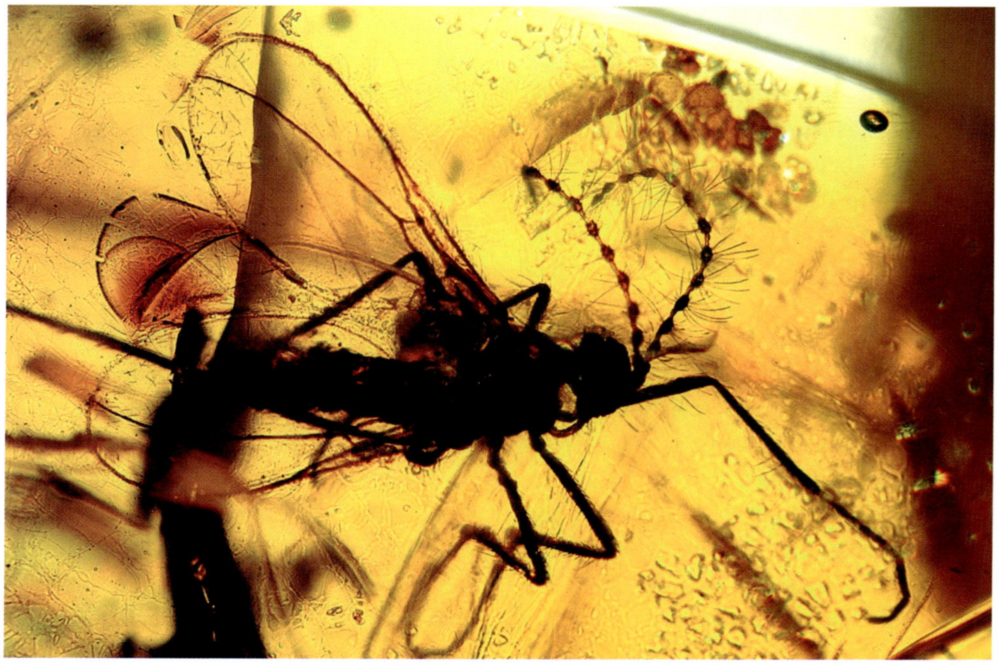

12.54. Gall midge (Cecidomyiidae) in Early Cretaceous Lebanese amber. Larvae of primitive cecidomyiids feed on fungi, as this one probably did. Gall-forming cecidomyiids did not evolve until the Tertiary. Antennal flagellomeres that are like a string of beads are distinctive to this family. AMNH JG 79/10; body length 1.3 mm.

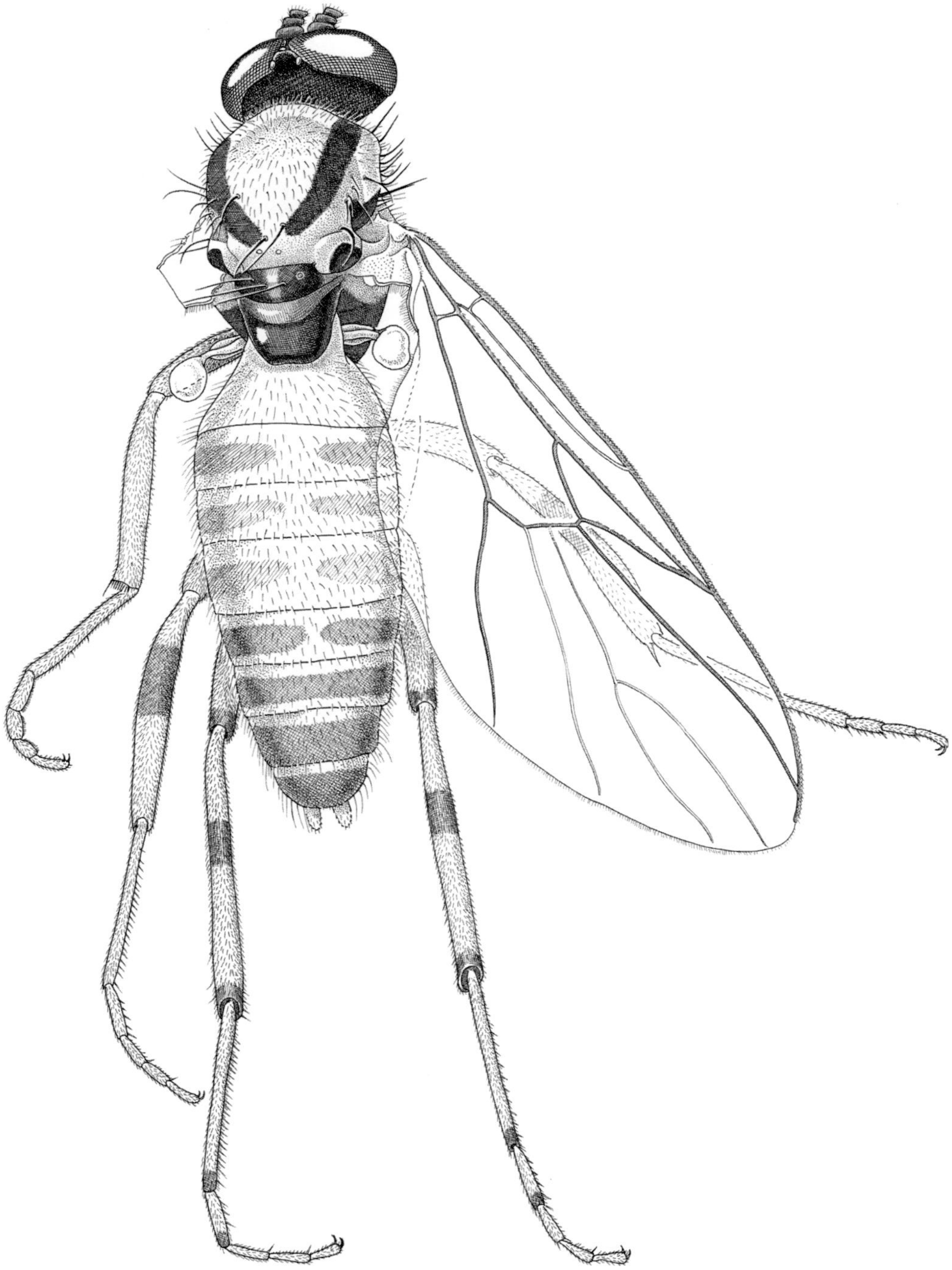

12.55. *Mesochria neotropica* (Anisopodidae) in Dominican amber. AMNH DR5–15; body length 2.8 mm.

plant-feeding species is likely to be truly vast, with thousands, perhaps tens of thousands, of new species. Thus far, most or all of the cecids in Cretaceous ambers appear to belong to the basal, mycophagous subfamilies Catotrichinae, Lestremiinae, and Porricondylinae, but these specimens require detailed study. Also, though diverse galls are known in the fossil record, no Cretaceous galls positively attributed to Cecidomyiidae are as yet known. It would be fascinating to see if there is a phylogenetic trend from the earliest fossil cecids in Lebanese amber to much younger ones in Dominican and Mexican amber, and if these trends correspond to major episodes in the evolution of plants, particularly the radiation of angiosperms.

The Anisopodidae is a small Recent family (120 species, 7 genera), but it is phylogenetically the most intriguing of all nematocerous flies because this group may be the living sister group to the Brachycera, based on larval and adult features (Woodley, 1989; Oosterbroek and Courtney, 1995; Shcherbakov *et al.*, 1995). The subfamilies of Anisopodidae (Olbiogastrinae, Mycetobiinae, and Anisopodinae) (Figures 12.55, 12.56) are sometimes classified as families (e.g., Amorim and Tozoni, 1994), though this is usually not adopted

12.56. *Olbiogaster perezi* (Anisopodidae), in Miocene amber from the Dominican Republic. Anisopodidae have an extensive fossil record beginning in the Late Triassic. AMNH DR11–657; body length 4.4 mm.

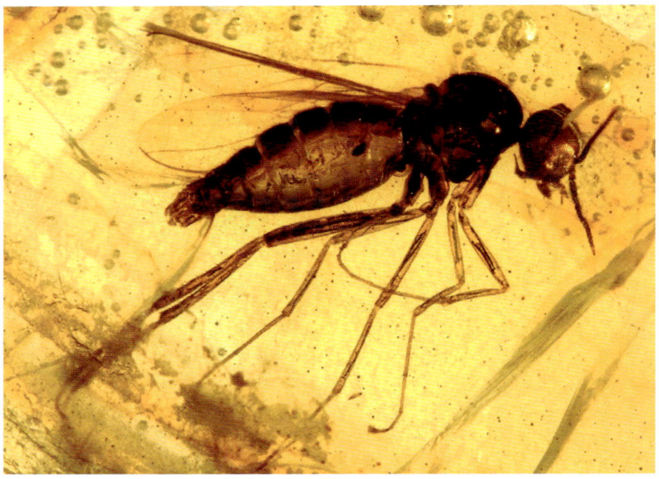

12.57. An extinct genus of Anisopodidae, *Mesorhyphus*, in Cretaceous amber from New Jersey. Other species of *Mesorhyphus* are known only as compressions in rocks from the Jurassic and Early Cretaceous. Anisopodidae have particular phylogenetic significance among nematocerous flies because they appear to be the living sister group to the Brachycera. AMNH NJ153c; body length 4.5 mm.

(Michelsen, 1999). The extinct family Protorhyphidae, from the Late Triassic to the Late Cretaceous (mostly of Eurasia), appears to be a stem group to the Anisopodidae. The Anisopodidae itself appears to be a vestige of its former diversity, including nearly 10 Mesozoic genera. One of these genera, *Mesorhyphus*, was known only from the Jurassic and Early Cretaceous of Eurasia, but completely preserved specimens now known in Cretaceous amber (Figure 12.57) will help to establish relationships of at least one extinct group. Mycetobiines occur in ambers from the Early Cretaceous to the Miocene and have changed very little in 125 MY. *Crosaphis*, from the Late Triassic of Australia, has venation that is a primitive version of the mycetobiines, so at least stem-group Anisopodidae extended to about 220 MYA. In the Early Jurassic there appeared the first definitive fossils of the most ecologically diverse group of Diptera, the Brachycera.

THE BRACHYCERA

The Brachycera is a clearly defined lineage comprised of the more familiar quick, stout-bodied flies, like the horseflies, houseflies, and fruitflies, within which there is a far greater ecological diversity than among nematocerans. It is also an extremely successful group, having evolved for the last 200 million years into approximately 80,000 Recent species in approximately 95 families. Brachyceran relationships have been better explored than have the relationships among families of nematocerans (Hennig, 1971; Griffiths, 1972; McAlpine, 1989; Woodley, 1989; Sinclair *et al.*, 1994; Stuckenberg, 2001; Yeates, 2001). Thus, paraphyletic groups from traditional classifications, like the "Orthorrhapha," "Aschiza," and "Acalyptrata," have now been abandoned because those groupings were based on primitive characters.

Monophyly of the Brachycera is based on molecular data and many reliable morphological features of the larval head capsule, adult head appendages, and venation. In Brachycera, the posterior portion of the larval head capsule is extended into the thorax, which is also mostly desclerotized except for several rodlike apodemes (Figure 12.58) (Cook, 1949). The mouthparts are held parallel to each other and move up, out, and down, not scissoring against each other horizontally in the primitive, nematoceran fashion. This feature alone probably led to a major innovation in Diptera larvae: predation. The larvae of almost all basal Brachycera are no longer saprophagous, but feed on other insects and soft-bodied invertebrates that they subdue by slashing with the *mouth hooks*, and some have become parasitoids. Each mouth hook is comprised of two blades, one of which may be derived from the maxilla and the other is clearly a mandible. Salivary ducts in the larval mandibles of some families have even been modified for the delivery of poison. Larvae of the more recently evolved Cyclorrhapha, affectionately called "maggots," have a head that is dramatically modified: It is entirely invaginated into the thorax, and all portions except the mouth hooks and an internal truss (the *cephalopharyngeal skeleton*) are unsclerotized. The mouth hook in Cyclorrhapha is comprised of only one part. Cyclorrhaphans reverted to the larval saprophagous lifestyle by innovating: They live in their food. These larvae work by repeatedly raking into the substrate with the mouth hooks while secreting copious saliva, and shoveling the digested soup into the

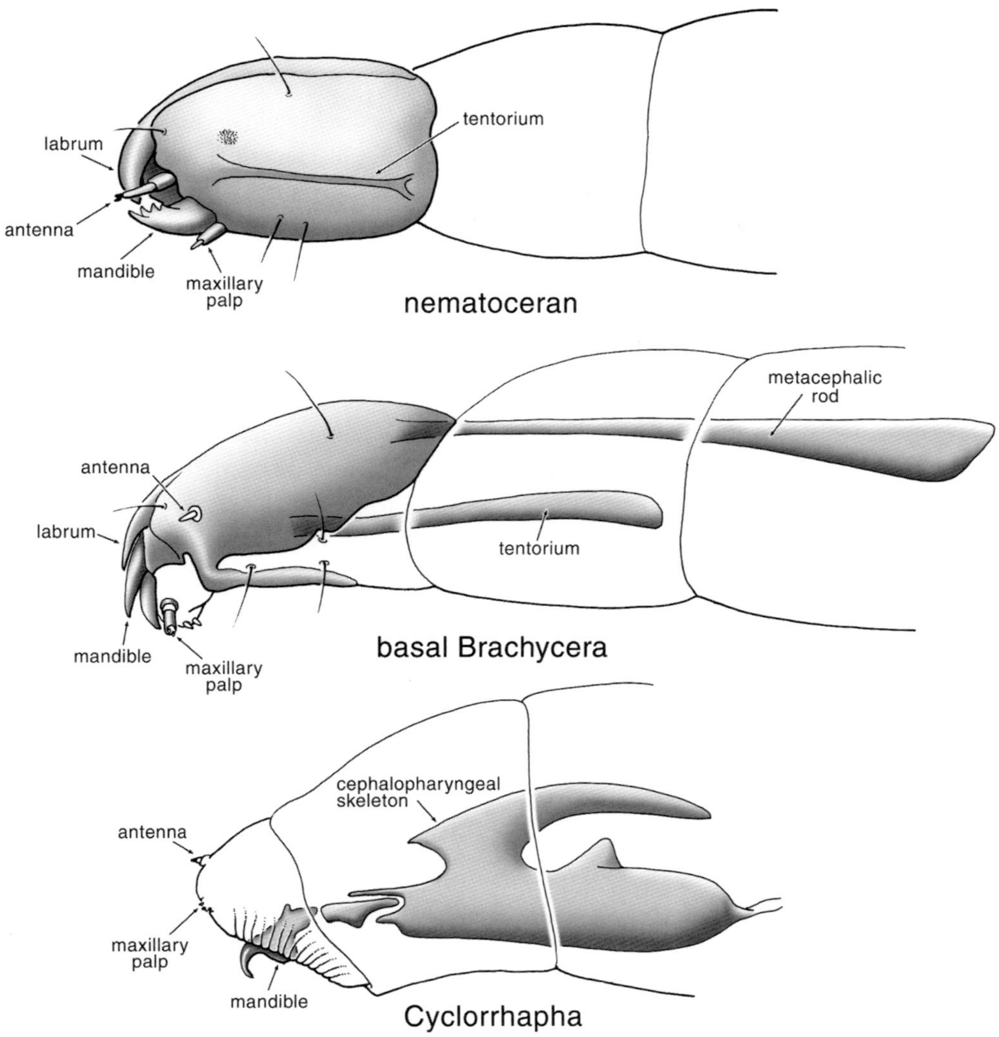

12.58. Schematic diagram of larval heads in flies, showing reduction and invagination of the head capsule in the Brachycera, particularly for the Cyclorrhapha.

mouth cavity. The development of cyclorrhaphan immatures is thus not confined to feeding in aquatic ooze like nematocerans but is based on exploiting a smorgasbord from flesh to fruit.

The Brachycera get their name (Greek, for "short horned") from the compacted antennae, comprised primitively of eight flagellomeres but most often reduced to four or fewer segments. The basal flagellomeres are typically unaffected, but the apical ones are reduced to a slender *style* or hairlike *arista* (Figure 12.59) that partly serves as a mechanoreceptor for the swift, maneuvered flight of brachycerans (Stuckenberg, 1999). Indeed, some brachycerans, particularly those that visit flowers, are among the most adept hoverers. The generally compact body of brachycerans is also related to their maneuvered flight, and for streamlining they generally hold the midlegs forward and against the thorax while flying instead of holding them out. This is made possible by the greater mobility of their coxae (Frantsevich and Gladun, 2002).

Adult Brachycera have maxillary palps reduced to two segments or (in Eremoneura) to one. Because palps largely function for taste, this reduction is probably related to the development of a specialized labellum in Brachycera. All flies have the labial palps fused at the end of the proboscis into a labellum, but in Brachycera this structure is further modified into a broad pad with radiating gutters of *pseudotracheae* that lead to the oral cavity (Figure 12.60c). It is particularly well developed in Cyclorrhapha but is reduced in flower feeding and predatory species. The labellar pseudotracheae greatly improve the efficiency of mopping up liquid food and are arrayed with rows of chemosensilla for tasting (Dethier, 1976). The evolution of this structure has also been implicated as to how Brachycera power their expensive flight by exploiting an important source of energy, sugary hemipteran honeydew (Downes and Dahlem, 1987). Indeed, the radiation of prolific honeydew producers, like the coccoids and aphidoids, was contemporaneous with that of the Brachycera in the mid-Jurassic through the Cretaceous and Tertiary. The adults of Recent, basal Brachycera, however, are largely

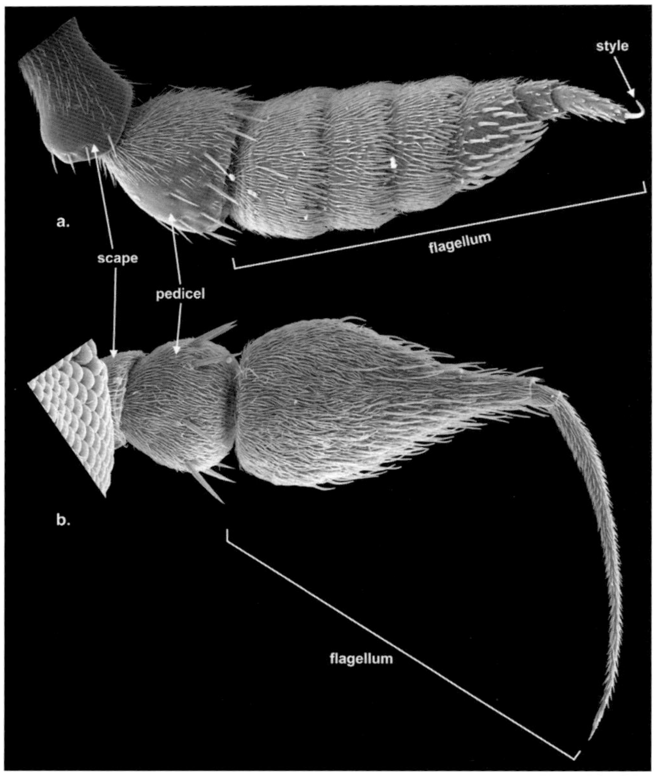

12.59. The antenna of two flies, showing consolidation of basal segments typical of Brachycera. In lower Brachycera (a, Stratiomyidae) the basal 5–7 segments are consolidated. In more recently evolved Brachycera (e.g., b, *Microphor*), the basal segments are consolidated or reduced to a single large segment. The apical style or arista is probably the apical-most flagellomere. Scanning electron micrograph.

flower feeders, so feeding on honeydew is widespread mostly in the cyclorrhaphan flies.

Adult Diptera are also unique among Antliophora by possessing a *crop*, or a large diverticulum off the esophagus for holding recently ingested liquids. With the mopping labellum and crop, a fly can ingest a large meal in minutes. Cyclorrhaphan flies have further evolved a *cardia*, a swollen structure joining the fore- and mid guts that secretes the *peritrophic membrane* (King, 1991). This membrane is a fine sleeve of cuticle that lines the lumen of the mid and hind gut, separating it from direct contact with food. The membrane may be involved in water absorption or for protecting the fly from infection by ingested microbes, or both. Many adult Cyclorrhapha feed on the surfaces of necroses where their larvae breed, and this liquid is laden with microbes that could invade their body.

Brachycera are also defined partly by a venational feature, in which veins A and CuA converge closely or even meet before the wing margin. This feature is extremely significant because it has been used for interpreting several Triassic wings as being the earliest fossil Brachycera (Krzemiński, 1992; Krzemiński and Krzemińska, 2003). Like most venational characters, these veins are not perfectly consistent; there are some nematocerans (e.g., Axymyidae) where these veins converge and various Brachycera where the veins do not meet at all. Though stem-group Brachycera from the Late Triassic probably existed, it would be best to base this on additional veins and particularly on the preservation of other structures like antennae. The oldest apparent Brachycera are from the Early Jurassic, though, again, some of these identifications are based just on the convergent CuA_2 and A_1 veins. Early Jurassic Brachycera include Rhagionidae (a Recent family of diffuse meaning at present), the extinct family Oligophrynidae, and putative, undescribed Nemestrinidae (Ansorge and Krzemiński, 1994; Ansorge, 1996; Krzemiński and Ansorge, 2000).

Like nematocerans, families of Brachycera are currently classified into infraorders (which have a *–morpha* ending), relationships and fossils of which are summarized in the phylogeny in Figure 12.61. The larvae of basal, or "orthorrhaphous," Brachycera are active predators of other insect larvae or soft-bodied invertebrates that live in rotting wood, under decaying bark, or in soil, sand, or streams. The most basal infraorders of Brachycera are the **Xylophagomorpha** and **Stratiomyomorpha**. The former is a small group comprised of the families Xylophagidae (120 species; Figure 12.62) and Pantophthalmidae (20 species). The latter family has the distinction of being the largest flies; these have large heads and broad and flat abdomens; they can be up to 2.2 inches (5.5 cm) long. Stratiomyomorpha is comprised of the small family Xylomyidae (110 species) and the large family Stratiomyidae (2,000 species). This lineage is defined by several peculiar larval features: The skin is encrusted with calcium carbonate, and the puparium forms within the last (third) instar, which is a feature convergent with Cyclorrhapha. Stratiomyidae get their common name "soldier flies" from the bold patterns of yellow and black or similar contrasting colors – a feature repeatedly evolved in flower-visiting insects. Stratiomyid larvae are preserved in limestone from the Early Cretaceous of Spain, but the most primitive known species of the family is a diminutive species in 75 MYO amber from Canada, *Cretaceogaster pygmaeus* (Figure 12.63).

Among Brachycera whose larvae are predators, the most distinctive are certainly the so-called worm-lions, which are the larvae of the small family Vermileonidae (with 30 species). Like larval antlions (Neuroptera: Myrmeleontidae), these larvae construct pits in sandy soil into which prey tumbles and then is dispatched at the bottom. The main exceptions to the larval predatory habit are the Nemestrinidae (tangle-veined flies: 260 species), Acroceridae (spider parasites: 700 species), and the large family Bombyliidae (beeflies). The larvae of Bombyliidae are external parasitoids of mostly holometabolous larvae, which they eventually kill by very gradually consuming the body contents (Figure 12.64), so this is not dramatically different from the predatory habit in the other families of basal Brachycera. Larvae of Acroceridae and Nemestrinidae, however, are unique in being internal

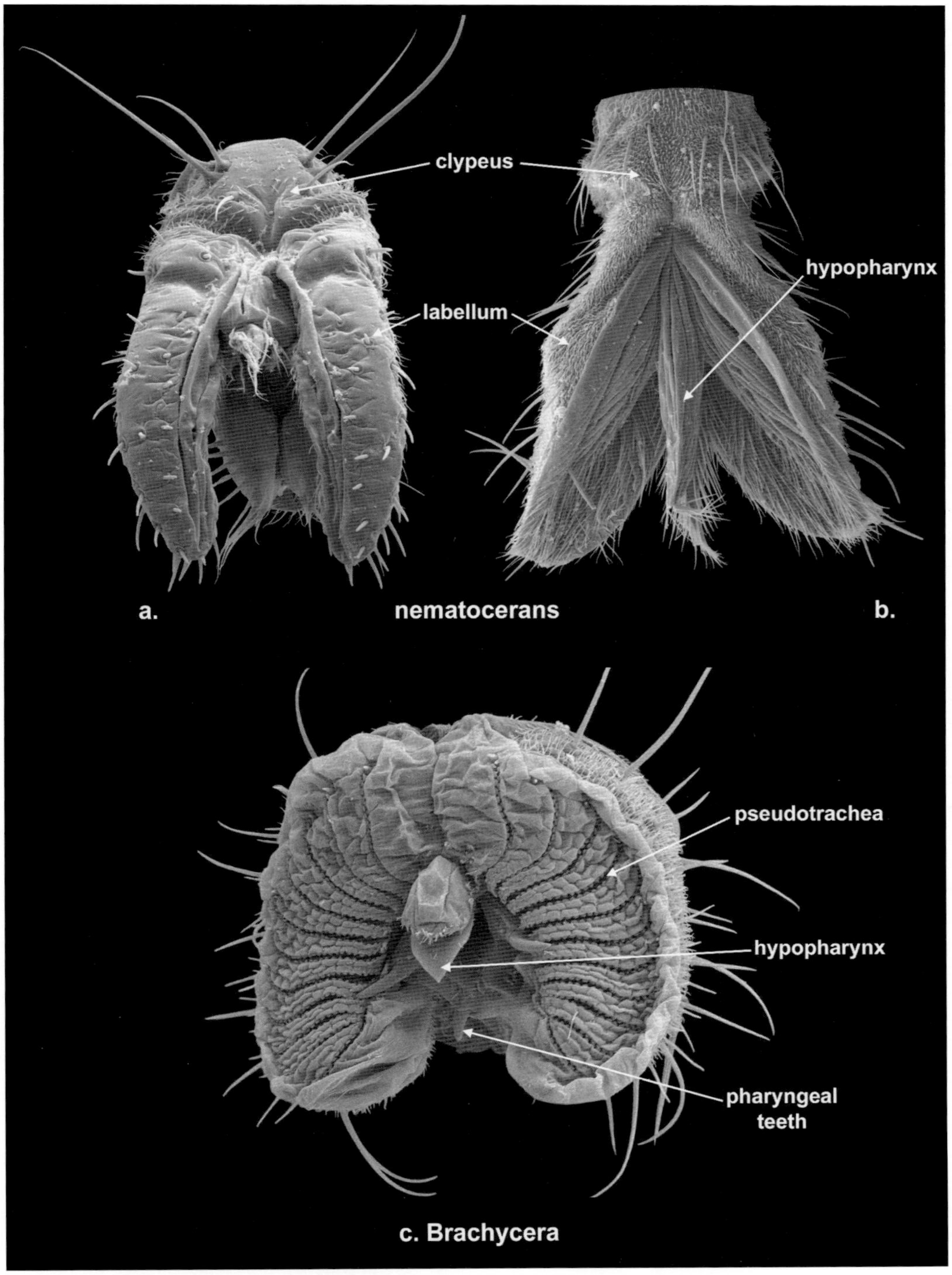

12.60. The dipteran labellum. Fine, gutter-like pseudotracheae are absent in nematocerans but a defining feature of Brachycera. (a) *Limonia* (Tipuloidea); (b) *Sylvicola* (Anisopodidae); (c) *Microphor* (Dolichopodidae). Scanning electron micrographs.

parasitoids (*endoparasitoids*), which is why these two families are often considered closely related even though the adults are significantly different. Acroceridae are restricted to feeding on spiders; the few records of nemestrinid parasitoidism include immature grasshoppers and scarab beetle larvae. The first instar larva in these three families is a *planidium*, a slender, bristled larva that actively searches for a host. Because planidia are so distinctive, it had been thought that these three families were closely related, but most of the evidence (from adult morphology) indicates that Bombyliidae belong to a different group, the Asiloidea. The Bombyliidae is a large, heterogeneous family containing some 5,000 species, from minute mythicomyiines (e.g., Figure 12.75) to large, beelike anthracines. Mythicomyiines are sometimes placed in a separate family (e.g., Evenhuis, 1994, 2002a) because they are the sister group to all other beeflies (Yeates, 1994).

Adult "orthorrhaphous" Brachycera commonly feed on

TABLE 12.4. Mesozoic Records of Basal ("Orthorrhaphous") Brachycera

1. *Ganeopteromyia calypso*: Karatau (Mostovski, 1999)
2. Rhagionemestriidae: Karatau
3. Rhagionemestriidae: Montsec, Spain
4. *Sinonemestrius*: Yixian Formation, Purbeck, England
4a. Oligophrynidae: Dorset, England; Issyk-Kul, central Asia (Ansorge and Krzeminski, 1994)
5. ?*Xylomya shcherbakovi*: Karatau (Mostovski, 1999)
6. *Archisolva cupressa*: Upper Jurassic, China
7. Xylomyidae, undescribed: Lebanese amber (Grimaldi *et al*., in prep.)
8. *Cratomyia*: Santana Formation, Brazil (Mazzarollo and Amorim, 2000)
9. Stratiomyidae, undescribed: Karatau (Mostovski, 1999)
10. *Montsecia martinezdelclosi*: Spain; undescribed larvae: Baissa, Siberia (Mostovski, 1999)
11. Stratiomyidae, undescribed: Ola'a Formation, Mongolia (Mostovski, 1999); undescribed, partial: New Jersey amber (Grimaldi and Cumming, 1999)
12. *Cretaceogaster pygmaeus*: Canadian amber
13. *Paleobrachyceron willmanni*, *Grimmyia baltica*, *Protobrachyeron* sp.: Toarcian, Germany (Ansorge, 1996)
14. *Ija problematica*: mid-Jurassic, Siberia
15. Various "Rhagionidae" (*Jurabrachyceron*, *Paleobrachyceron*, etc.): mid-Jurassic, Siberia
16. Various "Rhagionidae" (*Archirhagio*, *Paleobolbomyia*, etc.): Karatau
17. Various "Rhagionidae": Late Jurassic, China
18. *Mongolomyia latitarsis*: Early Cretaceous, Mongolia
19. *Ptiolinites* spp.: Purbeck, England (Mostovski, 1999); *Ptiolinites cretaceous*: western Mongolia
20. Various "Rhagionidae" (*Leborhagio*, *Mesobolbomyia*, *Paleochrysopilus*): Lebanese amber (Grimaldi and Cumming, 1999)
21. *Ptiolinites almuthae*: Montsech, Spain (Mostovski, 1999)
22. *Ptiolinites oudatchinae*: Baissa, Siberia (Mostovski, 1999)
23. *Atherimorpha festuca*: Koonwarra, Australia (Jell and Duncan, 1986); undescribed, Santana Formation (Grimaldi, 1990a)
24. Undescribed, Spanish amber
25. Undescribed, Burmese amber
26. *Jersambromyia borodini*, undescribed "Rhagionidae": New Jersey amber (Grimaldi and Cumming, 1999)
27. *Zarzia zherikhini*: Siberian amber
28. Undescribed "Rhagionidae": Canadian amber (Grimaldi *et al*., in prep.)
29. Undescribed Tabanidae: New Jersey amber (Grimaldi *et al*., in prep.)
30. Tabanidae: ?Pangoniinae: *Palaepangonius, Eopangonius*: Yixian Formation (Ren, 1998a,b); undescribed, Las Hoyas, Spain (Mostovski, 1999)
31. *Batgana*, *Shulmas*: Karatau, Kazakhstan
32. Undescribed Nemestrinidae: Grimmen, Germany (Ansorge and Mostovski, 2000)
33. Various Nemestrinidae (6 genera, 14 species): Karatau (Ansorge and Mostovski, 2002); *Prohirmoneura jurassica*: Eichstätt, Germany
34. *Florinemestrius, Protonemestrius, Sinonemestrius*: Yixian Form., China (Ren, 1998a,b), Wealden (Jarzembowski and Mostovski, 1999)
35. *Hirmoneura* spp., *Iberomusca*: Montsec, Spain (Mostovski and Martínez-Delclòs, 2000)
36. *Prosoeca saxea*: Cuenca, Spain (Mostovski and Martínez-Delclòs, 2000)
37. *Hirmoneura elenae*, ?*Nemestrius incertus*: Baissa, Siberia (Mostovski and Martínez-Delclòs, 2000)
38. *Iberomosca ponomarenkoi*: Bon-Tsagaan, Mongolia (Mostovski and Martínez-Delclòs, 2000)
39. *Archocyrtus gibbosus*: Karatau
40. Acroceridae, undescribed: Burmese amber (Grimaldi *et al*., in prep.)
41. *Apystomima zaitzevi*: Karatau (Mostovski, 1999)
42. *Hilarimorphites* sp., undescribed: Burmese amber (Grimaldi *et al*., in prep.)
43. *Hilarimorphites* (4 species): New Jersey amber (Grimaldi and Cumming, 1999)
44. *Paleoplatypygus zaitzevi*: mid-Jurassic, Siberia
45. *Procyrtosia sukatshevae*: Siberian amber
46. *Psilocephala electrella*: Burmese amber
47. *Proratites simplex*: New Jersey amber (Grimaldi and Cumming, 1999)
48. Putative Scenopinidae, undescribed: Karatau (Mostovski, 1999)
49. Apsilocephalidae: *Burmapsilocephala cockerelli*, Burmese amber (Mostovski, 1999)
50. Putative Therevidae, undescribed: Baissa, Siberia (Mostovski, 1999)
51. *Araripogon axelrodi*: Santana Formation, Brazil
52. Putative Asilidae, undescribed: Karatau (Mostovski, 1999)
53. Putative Asilidae, undescribed: Baissa, Siberia (Mostovski, 1999)
54. Asilidae, undescribed: New Jersey amber (Grimaldi and Cumming, 1999)
55. *Alleremonomus*: Yixian Formation, China (Ren, 1995)
56. *Dissup, Eremochaetosoma*: Late Cretaceous, Mongolia
57. *Eremochaetus, Pareremochaetus*: Karatau
58. Putative Apioceridae, undescribed: Baissa (Mostovski, 1999)
59. Putative Apioceridae, undescribed: Bon-Tsagaan, Mongolia (Mostovski, 1999)

[a] Numbers correspond to those circled on phylogeny, Figure 12.61.[b]
[b] More recent references (i.e., which do not appear in Evenhuis, 1993) are given.

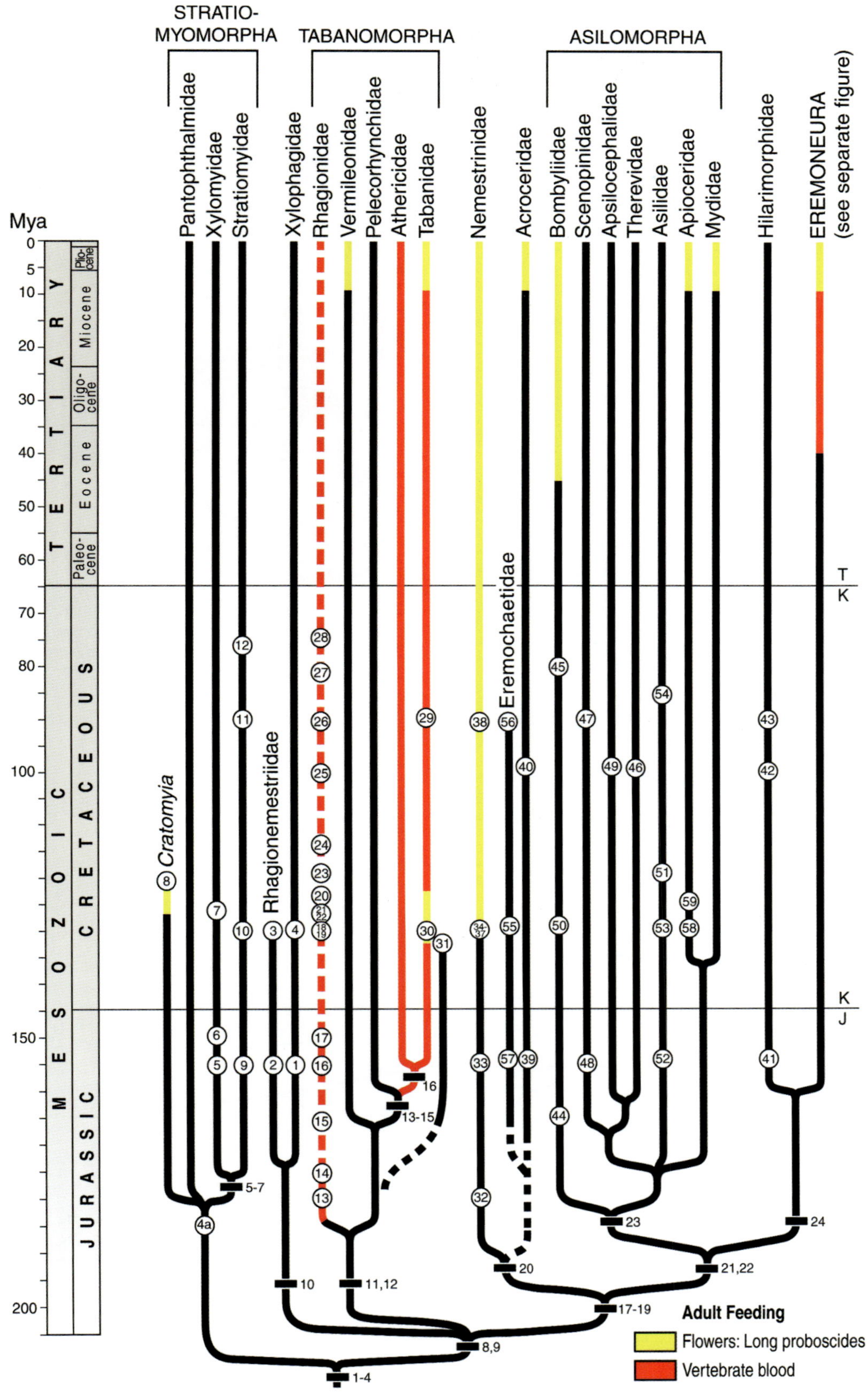

12.61. Phylogeny of basal brachyceran flies and the evolution of two major diets. Flower feeding is widespread among these flies, but just those that are specialized flower feeders (with long proboscides) are indicated. Rhagionidae are a polyphyletic group, but they are shown here as one family for convenience only. Some, but not all, rhagionids feed on blood. The numbers in circles refer to significant fossils (listed in Table 12.4); the other numbers refer to significant characters (listed in Table 12.5). Relationships among families based on Woodley (1989), Yeates (2001), and others.

TABLE 12.5. Significant Characters in the Phylogeny of Basal Brachycera[a]

1. Antennal flagellum with eight or fewer flagellomeres
2. Maxillary palps with one to two segments, labellum with pseudotracheae
3. Veins CuA_2-A_1 convergent
4. Larval mandibles working in a vertical plane
5. Pupation in last larval (instar three) cuticle
6. Larval cuticle with nodules of calcium carbonate
7. Loss of tibial spurs on front leg
8. Larvae predatory (secondarily saprophagous)
9. Larval mouthparts with slashing distal hook, salivary channel
10. Larval head capsule elongate, strongly sclerotized
11. Female cercus with basal segment having small lobe
12. Larval head retracts into thorax
13. Clypeus bulbous
14. Larval mandible with brush
15. Larval mandible with poison canal
16. Female cercus one-segmented
17. Antennal flagellum with four or fewer segments
18. Female cercus one-segmented (convergent with Tabanidae + Athericidae)
19. All apical tibial spurs lost
20. Larvae are endoparasitoids
21. Pretarsus with bristle-like empodium
22. Female terminalia with spine-bearing acanthophorites
23. Larva with posterior spiracle on penultimate segment
24. Vein M_3 lost

[a] Numbers correspond to those uncircled in phylogeny, Figure 12.61.

12.62. Xylophagidae in Eocene Baltic amber. Xylophagidae are basal Brachycera whose fossil record extends to the Jurassic. Larvae (Figure 12.22) live in decaying wood and are predaceous. AMNH; wing length 12 mm.

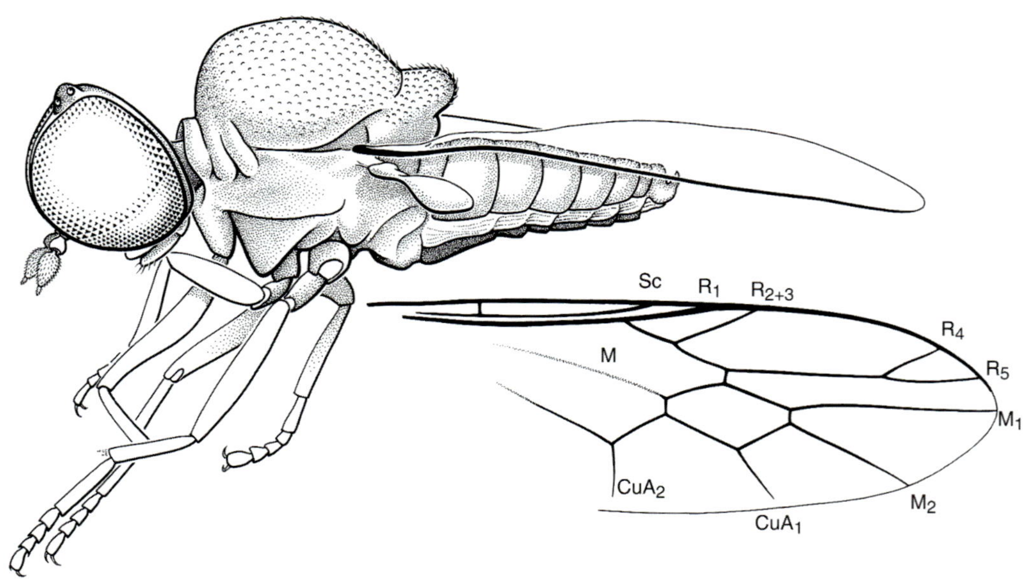

12.63. An early stratiomyid, *Cretaceogaster pygmaeus*, in Late Cretaceous Canadian amber. This is the most primitive known stratiomyid. MCZ 6572; body length 2.2 mm.

12.64. Larval beefly (Bombyliidae: *Anthrax*) gradually consuming its host; a larval carpenter bee (*Xylocopa*) in its nest cell. A healthy bee larva resides in the next cell. Larval bombyliids, nemestrinids, and acrocerids are parasitoids. Photo: V. Giles.

12.65. A beefly, *Bombylius*, feeding from a flower. Bombyliidae are a large family of anthophilous Brachycera, and many species in this and several other families of lower Brachycera have evolved long proboscides specialized for feeding from deep flowers (Figures 14.14 to 14.17). Nemestrinidae from the Early Cretaceous, in fact, appear to be the earliest specialized pollinators. Photo: S. Marshall.

nectar and pollen from flowers, though some have become specialized predators or blood feeders. Flower feeding has been recorded in nearly all these families, and in seven of them there has evolved a long slender proboscis specialized for probing flowers (Figure 14.16) (Szucsich and Krenn, 2000, 2002). Such proboscides generally have a reduced labellum at the tip with just a few pseudotracheae per lobe. In none of the families do all species have long proboscides, but even those flies with short proboscides will commonly feed from open flowers. In a few families, like the Acroceridae, there are some species that even have vestigial mouthparts. Bombyliids are probably the most commonly seen flower feeders, as *Bombylius* frequents spring wildflowers in north temperate regions (Vogel, 1954; Grimaldi, 1988) (Figure 12.65). Most of the specialized flower feeders, interestingly, live in xeric regions such as deserts or areas with Mediterranean-type flora and climate, like California, Chile, and South Africa. For example, "long-tongued" species of Vermileonidae occur in South Africa (e.g., Stuckenberg, 2000); certain pangoniine tabanids, like *Scaptia* and *Philoliche*, occur in southern South America and southern Africa, respectively; and *Eulonchus* and *Lasia* (Acroceridae) are an important pollinator guild in the southwestern United States (Grant and Grant, 1965). In fact, the insect with the longest proboscis relative to its body length is *Moegistorrhynchus longirostrus* (Nemestrinidae) from southern Africa, having a proboscis nearly 10 cm (4 inches) in length, or 10 times the length of its body (Figure 14.17)! It and several other long-tongued flies are important pollinators of herbs with tubular flowers in southern Africa (Goldblatt and Manning, 2000). Species that are flower feeding and have a long proboscis generally also have dense, fine setae covering the body, and the wing veins near the apex of the wing curve upward or even coalesce. The furry coat probably serves in thermoregulation, which is crucial in regulating the temperature of flight muscles in actively flying insects (Heinrich, 1979). The curved wing veins, which even coalesce is some species, must serve in strengthening the apex of the wing while these flies hover at flowers (Grimaldi, 1999).

Oddly, adults of the large family Asilidae (with 5,000 species) are strictly predators of other insects (Figure 12.66), and no species has an adult diet that is transitional between this family and other asiloids. Asiloidea includes the large families Bombyliidae, Asilidae, and Therevidae, as well as the small families Apioceridae, Mydidae, and Scenopinidae (some species in the first two of the small families are anthophiles with long proboscides). A new family of asiloids

12.66. Robberfly (family Asilidae) with its prey, a membracid tree hopper. The fly's sharp, sclerotized proboscis penetrated the armored treehopper through intersegmental membranes. Photo: P. J. DeVries.

12.67. An early robberfly (Asilidae), *Araripogon axelrodi*, in Early Cretaceous limestone from Brazil. Asilids probably originated in the Jurassic and evolved into a large family of 5,000 modern species. It is the only family of lower Brachycera with adults that are exclusively predatory. AMNH SA43517; body length 10 mm.

was recently discovered from central Chile, Ocoidae, for the sole species *Ocoa chilensis* (Yeates *et al.*, 2003). Asilids generally perch at the tip of a twig or other spot that affords a view of passing insects, which they ambush with a quick, darting flight. Robberflies grasp their prey with spiny legs and large claws and then puncture the prey with the lance-like proboscis; the prey is often carried back to the perch and slowly siphoned. The sclerotized, jutting proboscis is formed from a fused prementum and labella, within which is a fine, sharp hypopharynx that injects a toxic saliva and sucks out the liquified contents. A beard of stiff facial hairs, the *mystax*, protects the fly's head while it impales its flaling victim. These are voracious insects, capable of dispatching insects larger than themselves, even stinging wasps. In fact, some species mimic aculeate wasps, not as protection against predators but to camouflage themselves from their prey. *Laphria* and *Mallophorina*, for example, are excellent mimics of bumble bees (Bombini), which are some of their preferred prey (Figure 13.92). The taxonomy of world Asilidae has been reviewed by Hull (1962), but is in serious need of revision. The earliest known Asilidae are from the Late Jurassic and Early Cretaceous (e.g., Figure 12.67).

A diet of vertebrate blood has evolved at least three times in the lower Brachycera but is restricted to **Tabanomorpha**. Family-level relationships of this infraorder have been studied by Woodley (1989), and by Wiegmann *et al.* (2000b) based on the 28S rDNA gene. The largest family of this group is the Tabanidae (2,500 species), the well-known "horseflies," "deerflies," or "greenheads." Their larvae generally breed in swampy muck, and the adults can occur in oppressive numbers. The bite of tabanids is painful because it involves slicing the skin with serrated, blade-like mandibles and their maxillae, and sponging up with their labella the blood that wells to the surface. Some species are known to transmit diseases, but they are not nearly as important a group of vectors as are the culicomorph biting midges. The small, closely related families Pelecorhynchidae (50 species) and Athericidae (100 species) superficially resemble tabanids, but adults of only the latter are known to suck blood. Rhagionidae is a basal family of Tabanomorpha with a very confused classification. As traditionally classified (e.g., Nagatomi, 1982), the family is polyphyletic, and recent phylogenetic work indicates three lineages of this "family" are interspersed throughout the tabanomorphs. One of these is a basal lineage including *Chrysopilus, Symphoromyia,* and *Spania,* the so-called Spaniidae (Stuckenberg, 2001). The unusual genus *Bolbomyia* has an isolated position, and true Rhagionidae in a restricted sense should include *Rhagio* and a few other genera that are closely related to the tabanids. Blood feeding occurs in some Spaniidae (Figure 12.68), but most of these flies and some Rhagionidae sensu stricto are predacious. In fact, the ancestral diet of tabanomorphan flies was probably a predatory one, from which blood feeding clearly arose.

Overall, the ancestral diet of adult Brachycera appears to have been the nectar and pollen of flowers, since even the most basal and structurally generalized Brachycera in the Xylophagidae and Stratiomyomorpha commonly feed from flowers. As such, it is perplexing as to what may have been the adult food of early Brachycera. All basal lineages of Brachycera, or infraorders, appeared by the mid- to Late Jurassic, at least 30–40 million years before angiosperms first appeared. Reproductive structures of gnetaleans and other Jurassic plants may have provided food for these flies, but it is uncertain if

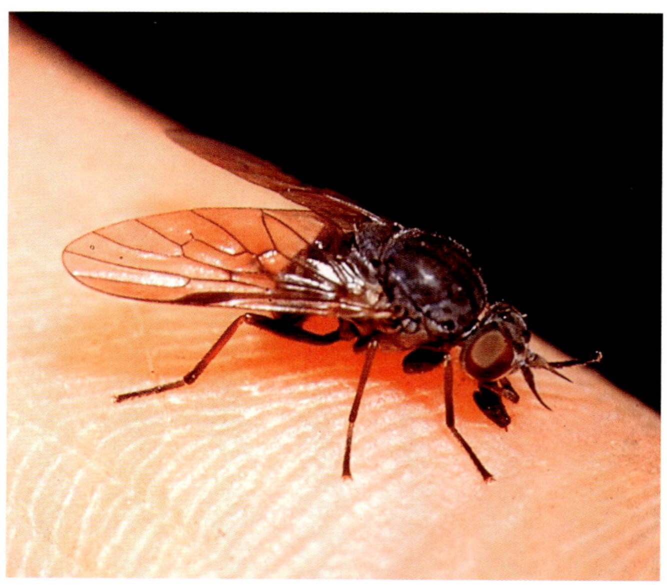

12.68. Rhagionidae in the process of puncturing skin for feeding on blood. Only some rhagionids are blood feeders; others are predaceous. The notorious blood-feeding horseflies and deerflies (family Tabanidae) and their closest relatives (Athericidae) probably evolved from a rhagionid-like ancestor. Photo: S. Marshall.

these plants were abundant enough to sustain the evolution of such a diverse group. The suggestion by Downes and Dahlem (1987) is very plausible; they suggested that early brachyerans foraged on droplets of hemipteran honeydew scattered over the surfaces of leaves, though in Recent flies such habits are only common in Cyclorrhapha.

The fossil record of flies, including basal Brachycera, has been catalogued by Evenhuis (1994) up to 1993, and more recent discoveries include reports based on Cretaceous and Jurassic flies from Europe and Asia (e.g., Ansorge, 1996; Mostovski, 1999; Mostovski and Martínez-Delclòs, 2000; myriad reports are summarized in Figure 12.61, Table 12.4). These include compressions from the diverse deposits in China, Mongolia, Kazachstan, central Siberia, Germany, England, and Spain. A few are known from the Jurassic of India and Early Cretaceous deposits in southern Africa, Australia, and Brazil. There is no obvious trend between phylogenetic position and the chronology of fossils for basal Brachycera, which is probably attributable to very incomplete sampling of early brachyceran fossils and understanding of their relationships. Perhaps the only such trend is that Asiloidea appear slightly later in the fossil record than other basal Brachycera, and this may reflect a close relationship of asiloids to the recently evolved Eremoneura, a group comprising the empidoids and Cyclorrhapha.

Though Brachycera clearly evolved well before the radiation of angiosperms beginning approximately 130 million years ago, it is difficult to separate the evolution of these two groups based on the feeding habits of Recent basal Brachycera as well as our knowledge of fossils. Indeed, the earliest fossil insects with specialized adaptations for flower feeding are Nemestrinidae, which include *Prosoeca*-like nemestrinids from the Early Cretaceous of Spain and *Florinemestrius* from the Yixian Formation (Ren, 1998a,b; Mostovski and Martínez-Delclòs, 2000) (Figure 14.14). Another Early Cretaceous brachyceran with a long proboscis, which is beautifully preserved, is *Cratomyia macrorrhyncha*, from the Santana Formation of Brazil, about 120 MYA (Figure 14.15). Because it belongs to a group of Recent flower feeders, the Stratiomyomorpha (Mazzarolo and Amorim, 2000), its proboscis almost certainly was used from probing flowers. No Jurassic flies, nemestrinids or otherwise, are known to have had a long proboscis. Some tabanomorph flies from the Early Cretaceous Yixian Formation of China had a jutting proboscis, which were ascribed to the pangoniine tabanids (Ren, 1998a,b). These have also been implicated as early pollinators, though pangoniines today with similar proboscides can also be blood feeders (Grimaldi, 1999). Other than these, though, flies with an elongate proboscis oddly do not appear again until the Tertiary. In fact, the largest family of pollinating flies, the Bombyliidae, did not become diverse until the Tertiary, when some 50 species and 30 genera of fossils appeared (Evenhuis, 2002b) (Figures 12.74, 12.75, 14.19). The proliferation of flowers beginning approximately 120 MYA and the gradual warming and drying of Cretaceous climates probably spurred the radiation of much of the basal Brachycera. Independent evidence from biogeography likewise indicates divergence of some basal Brachycera in the Cretaceous. Disjunct patterns of Southern Hemisphere mydids and apiocerids, for example, correspond to the fragmentation of Gondwana that occurred between 120 and 100 MYA (Yeates and Irwin, 1996).

12.69. A large, 35 MYO horsefly (Tabanidae), from Florissant, Colorado. The long proboscis is very well preserved. AMNH; body length 21 mm.

The radiation of angiosperms and the appearance of a dry Cretaceous climate does not account, though, for the tabanomorphs, which have adults that are predatory or blood feeding (if feeding at all) and larvae that generally breed in wet soils or mud. There are virtually no fossils of the small families of Tabanomorpha, and even the Tabanidae appeared to have evolved quite recently in the Early and mid-Cretaceous (Figures 12.70, 12.71). Fossils attributed to the polyphyletic family Rhagionidae, however, are among the earliest and most diverse "orthorrhaphous" brachycerans in the fossil record. Understanding the relationships of the fossil rhagionids has

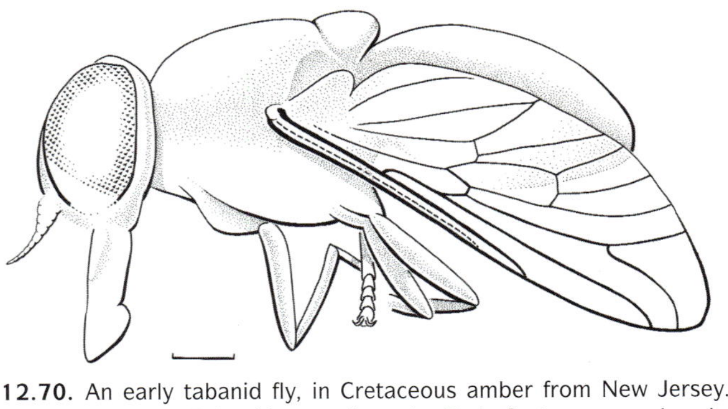

12.70. An early tabanid fly, in Cretaceous amber from New Jersey. The oldest known Tabanidae are from the Early Cretaceous and probably plagued dinosaurs. AMNH; body length 9 mm.

12.71. By the time of the Miocene Dominican amber, Tabanidae had evolved into modern genera. Morone Collection, M-0790; body length 9 mm.

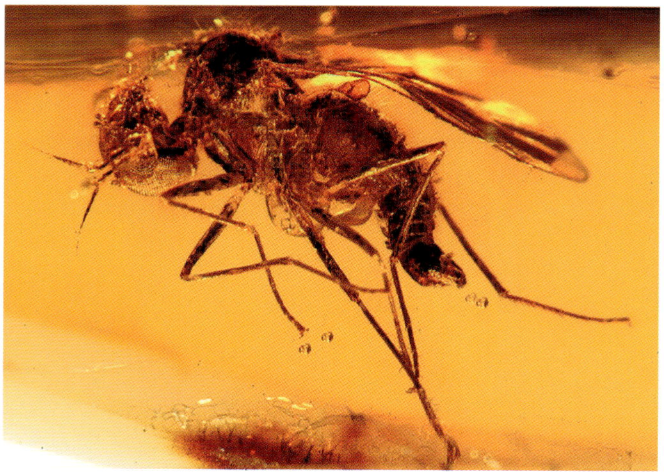

12.72. Rhagionidae preserved in mid-Cretaceous amber from Burma. Rhagionidae has been a problematic family and is now known to be polyphyletic. Fossils attributed to the family are among the most common and diverse Brachycera in the Jurassic and Cretaceous. AMNH Bu128; body length 2.15 mm.

12.73. Rhagionidae in Eocene Baltic amber. It has a long, thin, stiff proboscis that probably was used for feeding on blood. AMNH Ba JH31; body length 3.5 mm.

12.74. Early Tertiary beefly, *Alepidophora peali* (family Bombyliidae), from the latest Eocene of Florissant. Bombyliidae appear to have radiated largely in the Tertiary. YPM 102; body length 12 mm.

12.75. A minute beefly, *Glabellula brunnifrons*, in Dominican amber. It belongs to the primitive subfamily Mythicomyiinae, the fossil record of which extends to the Jurassic. AMNH; body length 1.6 mm.

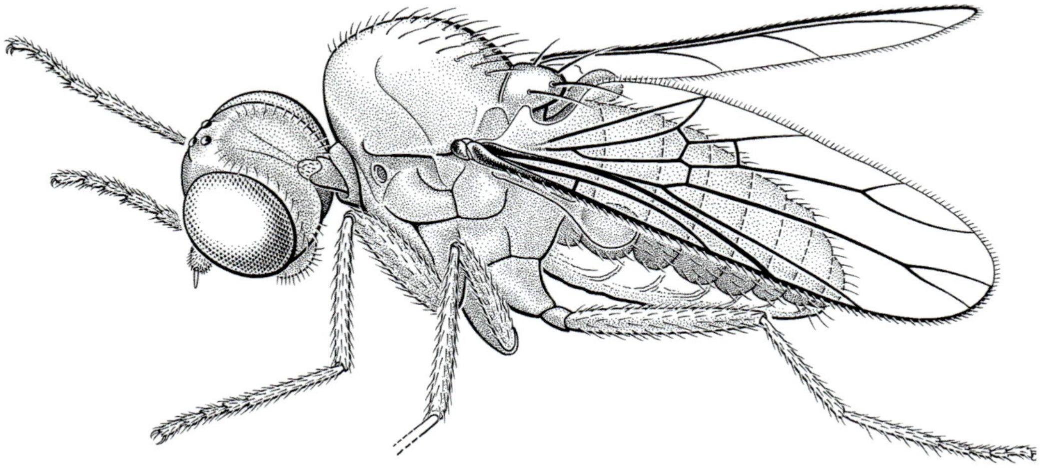

12.76. *Hilarimorphites yeatsi* (Hilarimorphidae), in 90 MYO New Jersey amber. This is a small Holarctic family today, and possibly relict; the family may also be the living sister group to the large lineage Eremoneura. AMNH NJ60; body length 1.3 mm.

been compromised by the generally poor understanding of the Recent species, until recently (Stuckenberg, 2001; Hibbs, 2004). Some of the Cretaceous species appear to have had blood-sucking or predatory proboscides (Grimaldi and Cumming, 1999) (Figures 12.72, 12.73), so it is reasonable to assume that the groundplan diet of this infraorder was predatory or one that was based on the blood of vertebrates of the time.

Basal Brachycera are generally large flies, some 5–10 mm (1.2–2.5 in.) in length, and often much larger. Eremoneura, on the other hand, are generally small flies, and this fortunately has resulted in a diverse fossil record from the Cretaceous and later that is preserved in amber (Grimaldi and Cumming, 1999). The few basal Brachycera in Cretaceous amber are considerably smaller than most compression fossils of these flies. Cretaceous amber has provided a better early record of eremoneuran flies than exists for the largely compression fossilized brachycerans, in both diversity and quality of preservation.

The **Eremoneura** are the empidoids and Cyclorrhapha, a group that is defined by overt anatomical features (Figure 12.78; Tables 12.6, 12.7), by obscure features of male terminalia, and from molecular evidence (Collins and Wiegmann, 2002a,b). Empidoids include the "dance flies" and the even more familiar "long-legged" flies (Dolichopodidae), which appeared approximately 40 MY before the earliest Cyclorrhapha. Even in the Jurassic, though, empidoids were scarce, and only three, possibly four, records exist. These include *Protoreogeton* from the mid-Jurassic of Mongolia, and *Protempis* and *Homalocnemis* from the Late Jurassic of Karatau (Mostovski, 1999; Figure 12.77). The wing venation of these genera is so primitive that they cannot be placed into any Recent empidoid lineage. An even older empidoid, from the upper Lias (185 MYO) of Grimmen, Germany, has been mentioned but not described (Mostovski, 1999). The life histories of empidoids and cyclorrhaphans are dramatically different, the former being adult and larval predators, and the latter being saprophages, a dichotomy that probably appeared when they diverged in the latest Jurassic or earliest Cretaceous.

Relationships of the **Empidoidea** have been discussed most thoroughly by Chvála (1983), Wiegmann *et al.* (1993), Cumming *et al.*, (1995), and Collins and Wiegmann (2002a,b), summarized in Figure 12.78 and Tables 12.6 and 12.7. The last study was based on sequences of two genes (28S rDNA, EF-1α); the others study adult morphology. These studies largely agree on the composition of the families in Empidoidea, but they differ on the relationships among families and the placement of a few subfamilies (mostly the Ceratomerinae and Brachystomatinae). Of most significance is that the Empidoidea actually appears not to be paraphyletic with respect to the Cyclorrhapha (as hypothesized by Chvála, 1983), but a monophyletic sister group to the Cyclorrhapha.

12.77. A very early empidoid fly, *Protempis antennata*, from the Late Jurassic of Kazakhstan. PIN; body length 4 mm.

The recent molecular evidence indicates that the small family Atelestidae is the most basal one in the Empidoidea, which may be the basis for earlier ideas based on morphology that Atelestidae and Cyclorrhapha are closely related (Chvála, 1983; Wiegmann *et al.*, 1993), since Atelestidae have probably diverged little from the eremoneuran ancestor. Several Early Cretaceous atelestids are known, including a complete specimen in Lebanese amber (Grimaldi and Cumming, 1999) (Figure 12.79).

All the phylogenetic studies also agree on a close relationship of the large family Dolichopodidae with the microphorines, which have often been classified in their own family, Microphoridae (e.g., Chvála, 1983). In fact, it appears that the microphorines are actually a basal and paraphyletic assemblage to the Dolichopodidae (Cumming and Brooks, 2002; Figure 12.78), so we are treating them as all in the same family here. Microphorines today comprise approximately 60 described species, most of which occur in xeric regions, but diverse fossils in Cretaceous ambers indicate they were more widespread and probably significantly more diverse (Grimaldi and Cumming, 1999) (Figures 12.80, 12.81). Certainly the most interesting microphorine is an extinct species in mid-Cretaceous Burmese amber. The males had large, shovel-shaped foretibiae rimmed with small scales and were marked in the center with a large spot (Figure 12.81). These tibiae may have been used for displaying to females, in much the same way as males of some "higher" dolichopodids semaphore with brushy flags on their foretarsi. The extinct microphorine may also have used the legs for capturing tiny prey or gleaning pollen from flowers. The "higher" Dolichopodidae appear largely to be a Tertiary radiation, and a

TABLE 12.6. Mesozoic Fossils of Eremoneura[a]

1. *Protoreogeton admirabilis*: Bayan-Hongor, Mongolia (Mostovski, 1999)
2. *Protempis antennata*, *Homalocnemimus abjugatus*: Karatau, Kazakhstan (Mostovski, 1999)
3. *Dianafranskia fisheri*: Purbeck, England (Coram *et al.*, 2000)
4. *Atelestites senectus*: Lebanese amber (Grimaldi and Cumming, 1999)
5. *Trichinites cretaceous*: Lebanese amber
6. *Alavesia subiasi*: Spanish amber (Waters and Arillo, 1999)
7. *Alavesia* sp., *Electrocyrtoma burmanica*: Burmese amber (Grimaldi *et al.*, in prep.)
8. *Pseudoacarterus orapaensis*: Botswana
9. *Cretoplatypalpus archaeus*: Siberian amber
10. *Archiplatypalpus cretaceus*: Siberian amber
11. *Cretoplaytpalpus americanus, Mesoplatypalpus carpenteri*: Canadian amber (Grimaldi and Cumming, 1999)
12. *Microphorites* spp.: Lebanese amber (Grimaldi and Cumming, 1999)
13. Various Microphorini spp: Spanish amber (Grimaldi *et al.*, in prep.)
14. Various Microphorini spp: Burmese amber (Grimaldi *et al.*, in prep.)
15. *Cretomicrophorus novemundus*: New Jersey amber (Grimaldi and Cumming, 1999)
16. *Cretomicrophorus rohdendorfi*: Siberian amber
17. *Archichrysotus incompletus*: New Jersey amber (Grimaldi and Cumming, 1999)
18. *Archichrysotus hennigi*, *A. minor*; *Retinitus*: Siberian amber
19. *Archichrysotus manitobus*: Canadian amber (Grimaldi and Cumming, 1999)
20. *Sympycnites primaevus*: Lebanese amber (Grimaldi and Cumming, 1999)
21. Undescribed Empididae: Purbeck, England (Coram *et al.*, 2000)
22. *Helempis* spp., *Protempis minuta*: Yixian Form., China; undescribed Empididae: Baissa, Siberia and Wealden, England (Coram *et al.*, 2000)
23. Lebanese amber: *Phaetempis* (Grimaldi and Cumming, 1999)
24. Spanish amber: undescribed Empididae (Grimaldi *et al.*, in prep.)
25. *Empis orapaensis*: Botswana; *Ecommocydromia difficilis*: French amber; *Burmitempis halteralis* and undescribed Empididae: Burmese amber (Grimaldi *et al.*, in prep.)
26. New Jersey amber: *Turonempis, Emplita, Nemedromia, Neoturonius* (Grimaldi and Cumming, 1999)
27. Undescribed Empididae: Siberian amber
28. *Nemedromia* sp., *Prolatomyia*, *Cretodromia*, *Apalocnemis canadambris*: Canadian amber (Grimaldi and Cumming, 1999)
29. *Chimeromyia* spp.: Lebanese amber (Grimaldi and Cumming, 1999)
30. *Chimeromyia* sp., *Chimeromyia*-like genus, undescribed: Spanish amber (Grimaldi *et al.*, in prep.)
31. *Chimeromyia* sp., undescribed: Burmese amber (Grimaldi *et al.*, in prep.)
32. *Lonchopterites prisca*: Lebanese amber (Grimaldi and Cumming, 1999)
33. *Lonchopteromorpha asetocella*: Lebanese amber (Grimaldi and Cumming, 1999)
34. *Lonchopteromorpha* sp., undescribed: Spanish amber (Grimaldi *et al.*, 1999)
35. *Opetiala shatalkani*: Purbeck, England (Coram *et al.*, 2000)
36. *Lithopetia hirsuta*: Yixian Formation, China
37. *Electrosania cretica*: New Jersey amber (Grimaldi and Cumming, 1999)
38. Undescribed genus: Lebanese amber (Grimaldi *et al.*, in prep.)
39. *Maritulus, Parnassos, Proplatypeza*: Baissa, Siberia (Mostovski, 1995)
40. *Lebambromyia acrai*: Lebanese amber (Grimaldi and Cumming, 1999)
41. *Proplatypeza* spp.: Bon-Tsagaan, Mongolia (Mostovski, 1995)
42. *Promittor, Oloplatypeza*: Ola'a Formation, Magadan (Mostovski, 1995)
43. Undescribed: Canadian amber (Grimaldi *et al.*, in prep.)
44. *Sinolesta*, *Paleopetia laiyangensis*: Yixian Form., China; *Hermaeomyia*: Baissa, Siberia (Mostovski, 1996)
45. *Paleopetia* spp., *Hermaeomyia* spp.: Bon-Tsagaan, Mongolia (Mostovski, 1996)
46. *Paleopetia* spp., *Hermaeomyia* spp., *Eridomyia*: Ola'a Form., Madygen (Mostovski, 1996)
47. Undescribed: New Jersey amber (Grimaldi *et al.*, in prep.)
48. *Cretonomyia pristina*: Canadian amber
49. *Archisciada lebanensis*: Lebanese amber (Grimaldi and Cumming, 1999)
50. *Archiphora pria*: New Jersey amber (Grimaldi and Cumming, 1999)
51. *Euliphora grimaldii*: Spanish amber (Arillo and Mostovski, 1999)
52. undescribed *Prioriphora* spp.: Burmese amber (Grimaldi *et al.*, in prep.)
53. *Prioriphora* spp.: New Jersey amber (Grimaldi and Cumming, 1999)
54. Various "prioriphorine" genera: Siberian amber (Mostovski, 1999)
55. *Prioriphora* spp.: Canadian amber (Brown and Pike, 1990)
56. *Metopina goeleti*: New Jersey amber
57. Undescribed syrphid: Siberian amber (Mostovski, pers. comm.)
58. *Cretophormia* [pupa]: Campanian, Canada (McAlpine, 1970)
59. Diverse Recent families: Baltic amber (Hennig, 1965a)
60. *Protanthomyia minuta*: Baltic amber (Michelsen, 2000)
61. *Glossina oligocenus*, *G. osborni* (Grimaldi, 1992)
62. *Glossina oligocenus*: Germany, Oligocene (Wedmann, 2000)

[a] Numbers correspond to those on phylogeny, Figure 12.78.

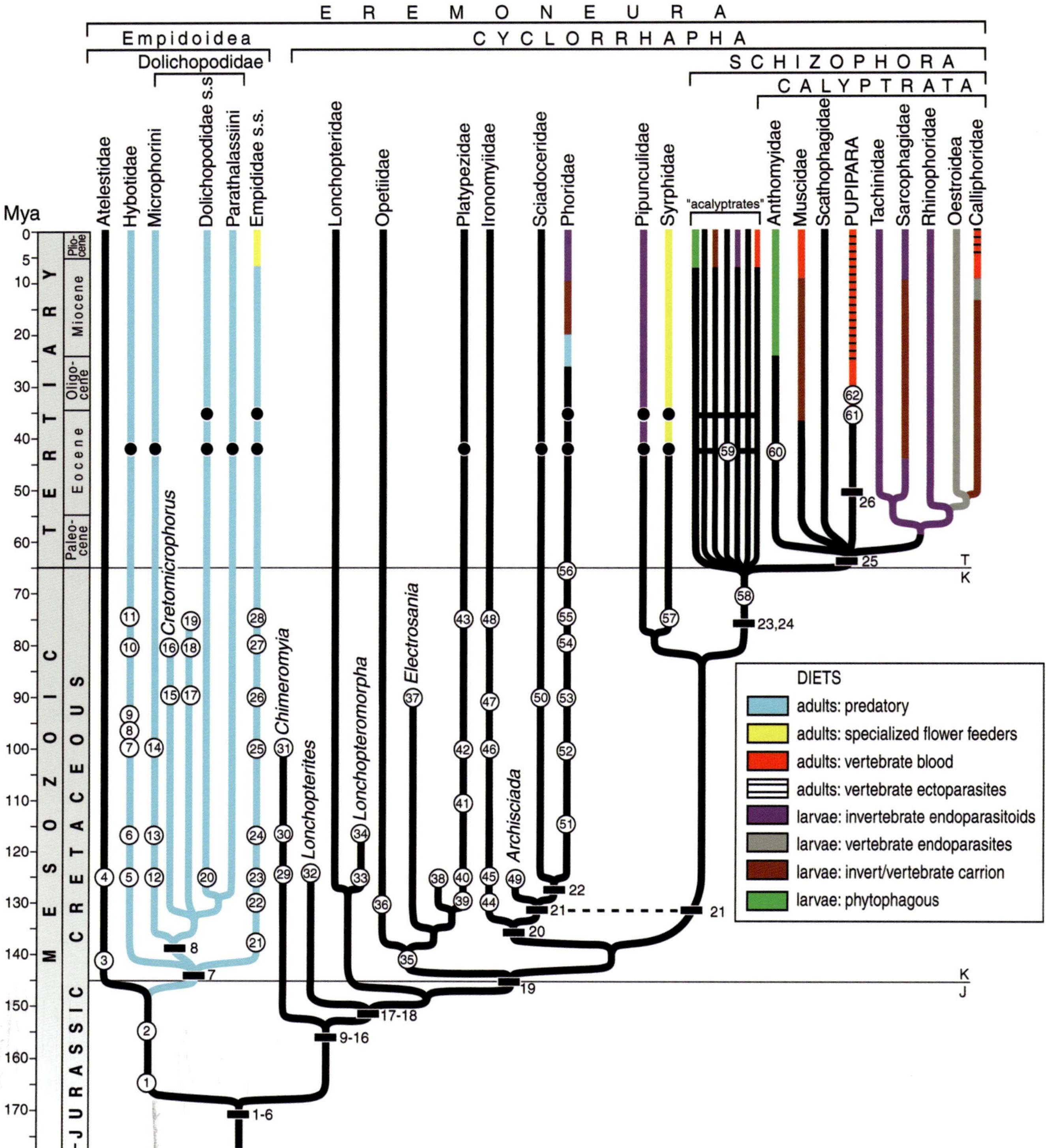

12.78. Phylogeny of the Eremoneura, showing the evolution of very diverse adult and larval diets. Circled numbers refer to significant fossils (listed in Table 12.6), and other numbers refer to significant characters (listed in Table 12.7). Only some Tertiary records are indicated. The Schizophora had one of the largest Tertiary radiations among all insects. Relationships based on Cumming *et al.* (1995), Grimaldi and Cumming (1999, in prep.), Collins and Wiegmann (2002a,b), and Pape (1992).

TABLE 12.7. Significant Characters in Eremoneuran Phylogeny[a]

1. Three larval instars
2. Larval mandible with single article
3. Adult with ocellar setae
4. Male terminalia structures (Griffiths, 1972; Cumming *et al.*, 1995)
5. Maxillary palps one-segmented
6. Wing cell *cup* closed apically or absent
7. Predatory, mouthparts with piercing labella
8. Male terminalia rotated to right
9. Face with pair or more of thick vibrissae
10. Loss of abdominal plaques in adults (Stoffolano *et al.*, 1988)
11. Embryos do not develop in amnion (Schmidt-Ott, 2000)
12. Larvae are saprophagous "maggots": maxilla lost, head desclerotized, invaginated
13. Pupariation in ecdysed cuticle of last larval instart (pharate adult)
14. Arista with two small basal segments
15. Proboscis retractile, withdrawn at rest into large oral cavity
16. Female terminalia without acanthophorite spines
17. Venation with Rs two-branched (R_{2+3} and R_{4+5} not forked)
18. Male terminalia circumverted 360°, folded under abdomen (evolved independently in Dolichopodidae)
19. Antennal pedicel with thumb-like condyle inserted into first flagellomere
20. Sc + R_1 very close or fused
21. Arista placed dorsally on flagellomere one, not at apex
22. Crossvein dm-cu shifted very basally, fork of M_1-M_2 shifted basally
23. Ptilinum present, adult emerges from puparium via circular sclerite
24. Adult with specialized structure between fore- and mid gut: cardia (King, 1991)
25. Large calypters at base of wing
26. Specialized reproduction: uterine development of larvae

[a] Numbers correspond to those uncircled on phylogeny, Figure 12.78.

12.80. A microphorine dolichopodid preserved in Lebanese amber. Microphorines appear to have been more diverse in the Cretaceous than they are today. A large mite is attached to the top of the abdomen of this fly. Azar Collection, A559a,b; length 1.3 mm.

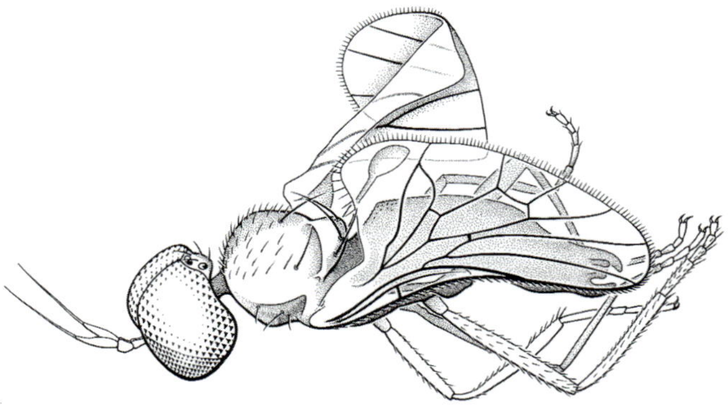

12.79. *Atelestites senectus* in Early Cretaceous Lebanese amber. Atelestidae appears to be the living sister group to the rest of the empidoid flies. Empidoids appeared in the Jurassic, but their fossil record is best preserved in amber from the Cretaceous and Tertiary. AMNH LAE72; length 1.6 mm.

12.81. A bizarre microphorine fly in Cretaceous Burmese amber, which has shovel-like tibiae. This feature probably occurred just in the males, who may have used the structures to signal to each other and to females. AMNH Bu175; body length 1.6 mm.

12.82. Swarm of male Dolichopodidae in Miocene Dominican amber. Males of many flies congregate into mating swarms, which attract females. AMNH; length of fly 1.3 mm.

12.83. A tiny empidid fly preserved in Dominican amber, caught with its cecidomyid prey. The diet of Atelestidae is unknown, otherwise most adult empidoid flies are predators; a few Empididae feed from flowers. AMNH DR14–524; body length 1.6 mm.

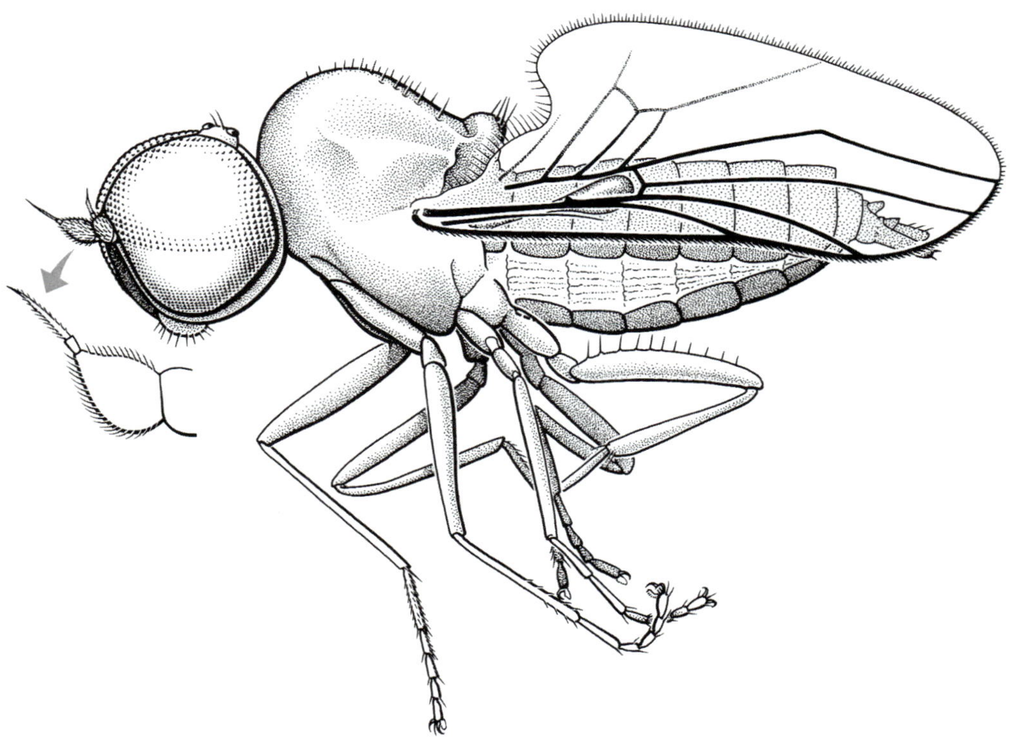

12.84. *Phaetempis lebanensis* in Early Cretaceous Lebanese amber, an early and primitive empidoid fly. AMNH LAE13; length 1.8 mm.

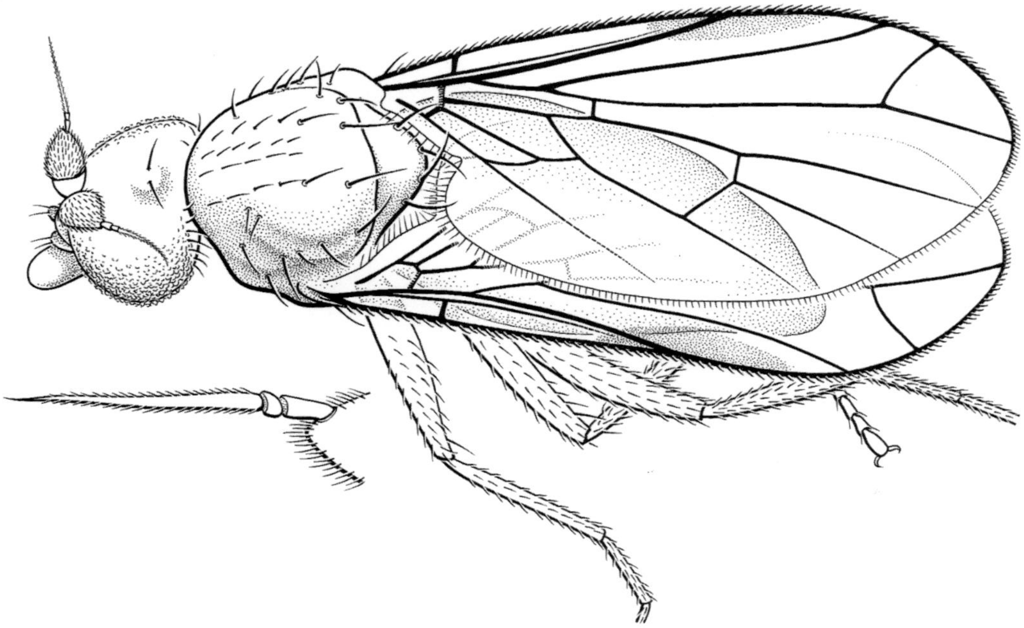

12.85. *Turonempis styx* (Empididae) in 90 MYO New Jersey amber, one of the earliest fossils of the subfamily Empidinae. AMNH NJ520; length 2.2 mm.

large one at that. There are about 6,000 described species of these flies, and given the proportions of new species, it is quite likely that 15,000 species or more actually occur. Most of the "higher" Dolichopodidae are immediately recognizable, being active, metallic green flies typically seen scuttling over leaves. They are predatory like other empidoids, but they feed by using a highly specialized labellum armed with sharp edges and sometimes spines (Figure 12.60c). Essentially all fossils of "higher" Dolichopodidae occur in the Tertiary (Figure 12.82), but by the time of Baltic amber, the "higher" Dolichopodidae were diverse and very abundant.

The family Empididae (sensu stricto) (Figure 12.83) has attracted attention because of the interesting mating habits of some groups. Many species form mating swarms, or aggregations of flying males into which females fly and become mated (hence, their common name, "dance flies"). In some species of *Empis*, *Hilara*, and *Rhamphomyia*, though, there is nuptial feeding, with each male carrying prey into the swarm and offering it to a female as a part of the courting ritual. The habit has become even more ritualized in some species of the former two genera, where the male first wraps the prey in a large silky or frothy balloon formed from salivary secretion. In yet other species, the prey is omitted altogether and just the balloon is offered (Kessel, 1955, 1959; Chvála, 1976; Cumming, 1994). Adult diets of empidids are often supplemented with nectar from flowers and extrafloral nectaries, and in a few genera the diet is exclusively nectar and pollen. Obligate flower feeders often have long proboscides, but this habit and structure almost certainly appeared relatively recently, in the Eocene or later. Empididae are diverse in Cretaceous ambers (e.g., Figures 12.84, 12.85; Grimaldi and Cumming, 1999) and they occur in slightly older Cretaceous rocks, indicating that most main lineages in the family evolved during this period.

THE CYCLORRHAPHA

Cyclorrhapha is the best defined and most diverse major lineage in Diptera, containing approximately 80 families and 65,000 species. The larvae are well-known maggots, which have a largely unsclerotized head capsule invaginated into the thorax (only the internal *cephalopharyngeal skeleton* remains of the original head capsule) (Figures 12.22, 12.86). In stark contrast to the largely predacious diet of other Brachycera larvae, cyclorrhaphan maggots are mostly saprophagous. Interestingly, some of the early cyclorrhaphans preserved in Cretaceous ambers are even larvae (Grimaldi and Cumming, 1999; Figure 12.87). Another major feature of the group is that "pupation" (actually, pupariation) occurs within the tanned cuticle of the last larval instar, which is the *puparium*. The pupa and pharate adult develop within this puparium. The adults of non-cyclorrhaphan Brachycera have small, bare patches on the sides of the abdominal tergites, the *muscle plaques*, which are external marks where muscles were attached and used by the pharate adult to wriggle through substrate to an eclosion site (Stoffolano *et al.*, 1988; Figure 12.88). The cyclorrhaphan puparium, in contrast, is immobile, as are the pupa and pharate adult within, so these muscles (and the plaques) have been lost in adult Cyclorrhapha.

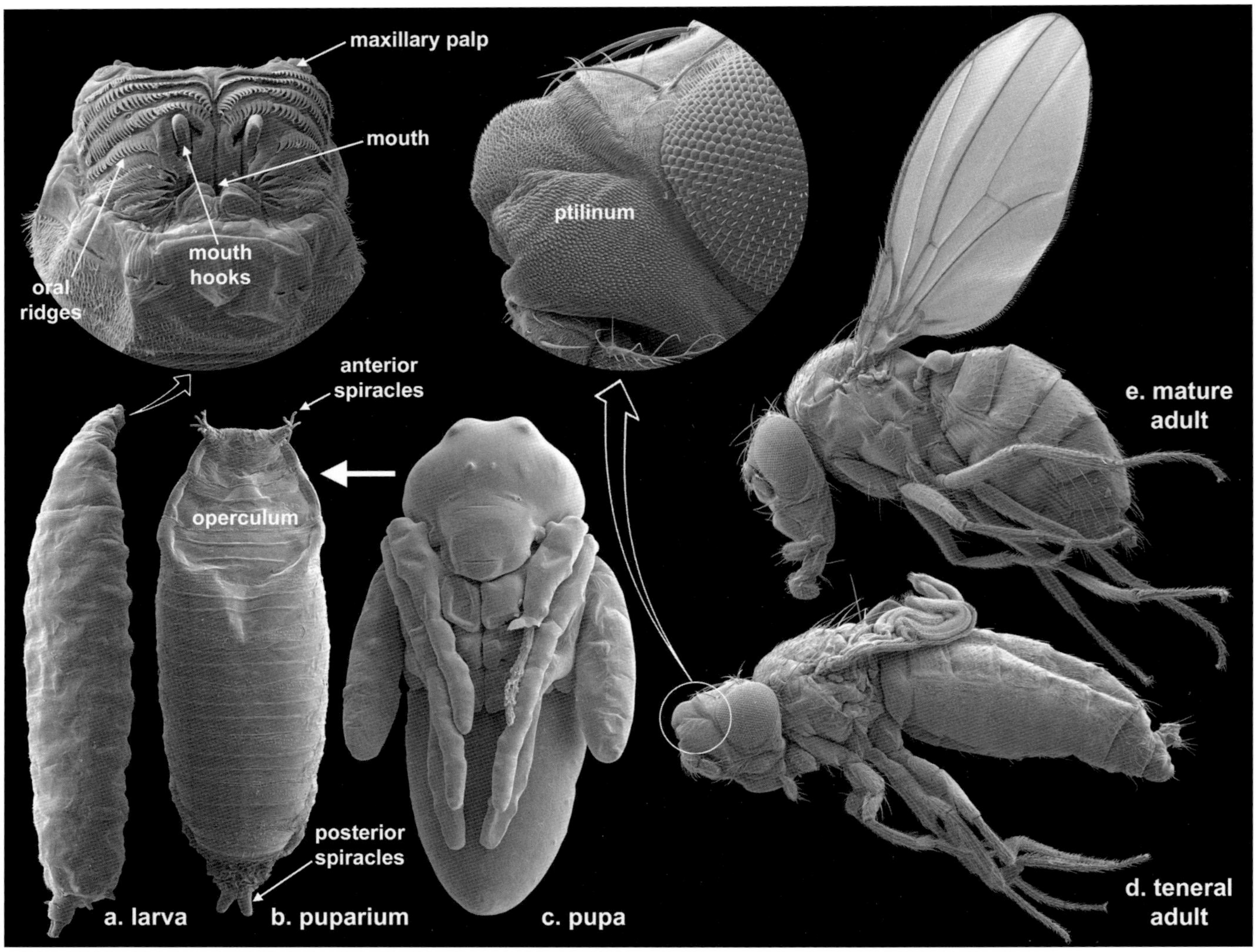

12.86. Metamorphosis of a typical cyclorrhaphan fly, the fruitfly *Drosophila melanogaster*. Cyclorrhapha form a hard puparium from the tanned cuticle of the last (third) larval instar, within which they pupate. The pupa (pharate adult) was extracted from a mature puparium. The fly emerges using its ptilinum – a sac that inflates and pushes the operculum of the puparium out; the ptilinum then withdraws into the ptilinal suture on the face. The wings quickly expand after emergence from the puparium. Scanning electron micrographs.

The most unusual feature of the Cyclorrhapha is that the male terminalia are rotated 360° (called *circumversion*), which evolved independently in "higher" Dolichopodidae. Circumversion occurs just before and just after eclosion and is revealed by the ventral nerve cord being twisted around the rectum. This unusual process may be an elaboration of what is seen in the other eremoneurans, the empidoids, wherein the male terminalia are rotated either 90° in microphorines, or rotated 180° and carried dorsally as in other empidoids (this too evolved independently, but in various nematocerans). Rotation, of course, affects the mating position. Flies without rotated terminalia generally mate tail-to-tail and face opposite each other, but in cyclorrhaphans the male is on top of the female. Why the terminalia would need to rotate into the original position probably relates to the fact that this process causes them to also be tucked under the abdomen, which presumably facilitates mating.

There are eight families of basal ("aschizous") Cyclorrhaphans, most of these being relict except for the Phoridae and Syrphidae, which are very diverse. Cyclorrhaphan relationships have been discussed in several large studies focused on homologous structures of the complex male terminalia (Griffiths, 1972; McAlpine, 1989; Cumming *et al.*, 1995; Zatwarnicki, 1996) or on DNA sequences (Collins and Wiegmann, 2002b). The most basal living cyclorrhaphans appear to be *Opetia* and the Lonchopteridae (Cumming *et al.*, 1995; Collins and Wiegmann, 2002b). *Opetia* is a genus of three Palearctic species usually classified as a separate family, Opetiidae (Figure 12.89); the Lonchopteridae is a family of approximately 40 species. *Opetia* has also been included with the Platypezidae, a family of approximately 250 species that breed mostly in fungi. Platypezids are one of the few groups of cyclorrhaphans that have retained or re-evolved the primitive habit of male swarming.

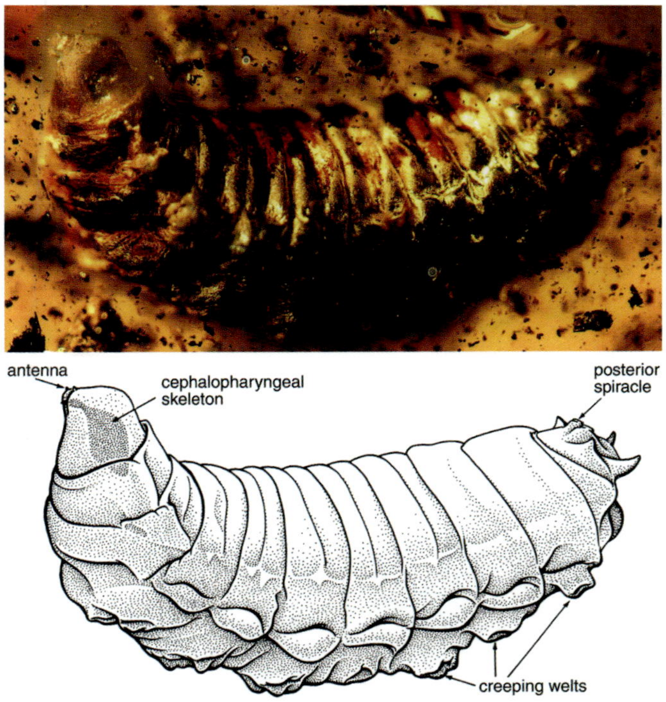

12.87. A few cyclorrhaphan larvae are preserved in Cretaceous ambers, such as this one in New Jersey amber. It is probably a primitive cyclorrhaphan like Sciadoceridae. It must have been breeding in decaying wood to be immersed in ancient resin. AMNH NJ628; length 3.9 mm.

Without question, Cyclorrhapha is a recently evolved lineage of flies. There are no Jurassic records of them, even from the prolific deposits of Karatau, though they may have originated just as the Jurassic Period eclipsed. The apparent oldest cyclorrhaphan, *Opetiala shatalkani*, is a compression fossil wing from the Purbeck of England (ca. 140 MYO), and it precedes other early cyclorrhaphans by about 10 MY. Though it was placed in the Opetiidae (Coram *et al.*, 2000), its venation is more primitive. More importantly, though, all of the Cretaceous Cyclorrhapha are of basal ("aschizous") families, and well-preserved fossils in amber reveal them to be very primitive (Grimaldi and Cumming, 1999). For example, *Lonchopterites* and *Lonchopteromorpha* in Lebanese amber resemble Recent Lonchopteridae (Figures 12.90, 12.91), but they have more primitive venation and setation. This trend is strong evidence that the Early Cretaceous witnessed the nascent evolution of the Cyclorrhapha.

Certainly the most intriguing Cretaceous cyclorrhaphan is *Chimeromyia*, a tiny fly 1–2 mm long preserved in amber from 125 to 100 MYA (Figures 12.92, 12.93). It possessed some of the features of Cyclorrhapha, like loss of muscle plaques, but not all features (like fully rotated male terminalia), hence its name. Oddly, it is not the kind of basal, cyclorrhaphan-like fly that was expected, which presumably would have been more

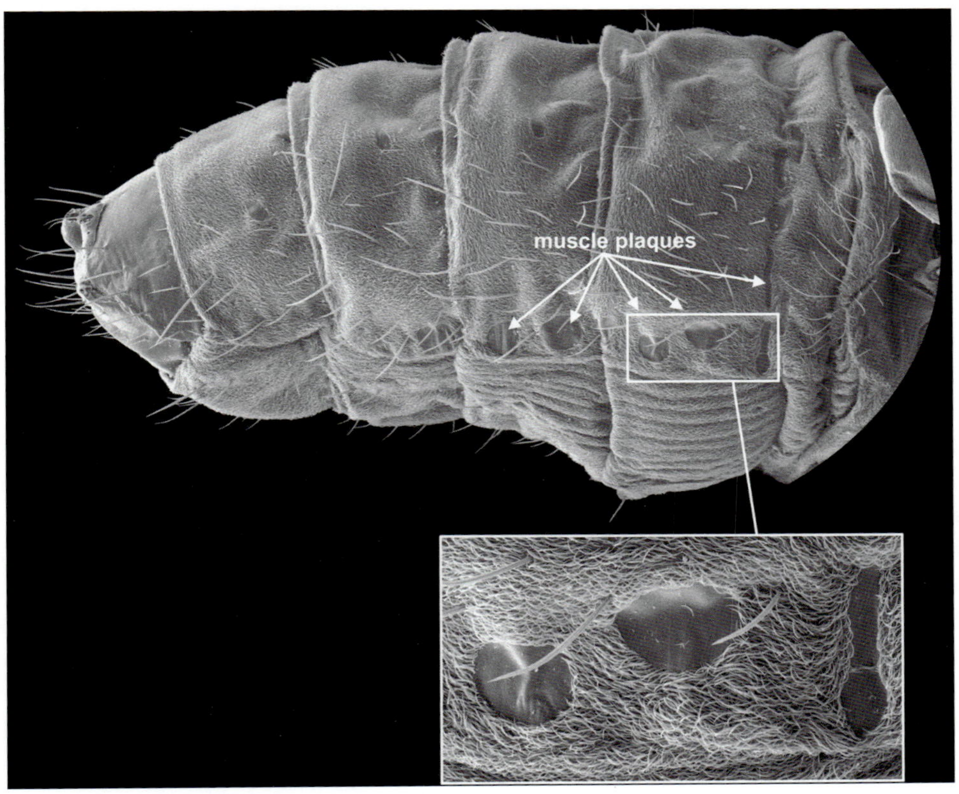

12.88. A defining feature of Cyclorrhapha: loss of muscle plaques on the adult abdomen. Basal Brachycera (such as *Microphor* here [Dolichopodidae]) have bare areas on the abdomen, which mark the sites where muscles attach internally in the pharate adult, allowing them to wriggle. Pharate adults of non-cyclorrhaphan flies are immobile. Wriggling assists the pupa in freeing itself from the substrate prior to emergence of the adult, and perhaps from being punctured by the ovipositor of parasitoid wasps. Scanning electron micrograph.

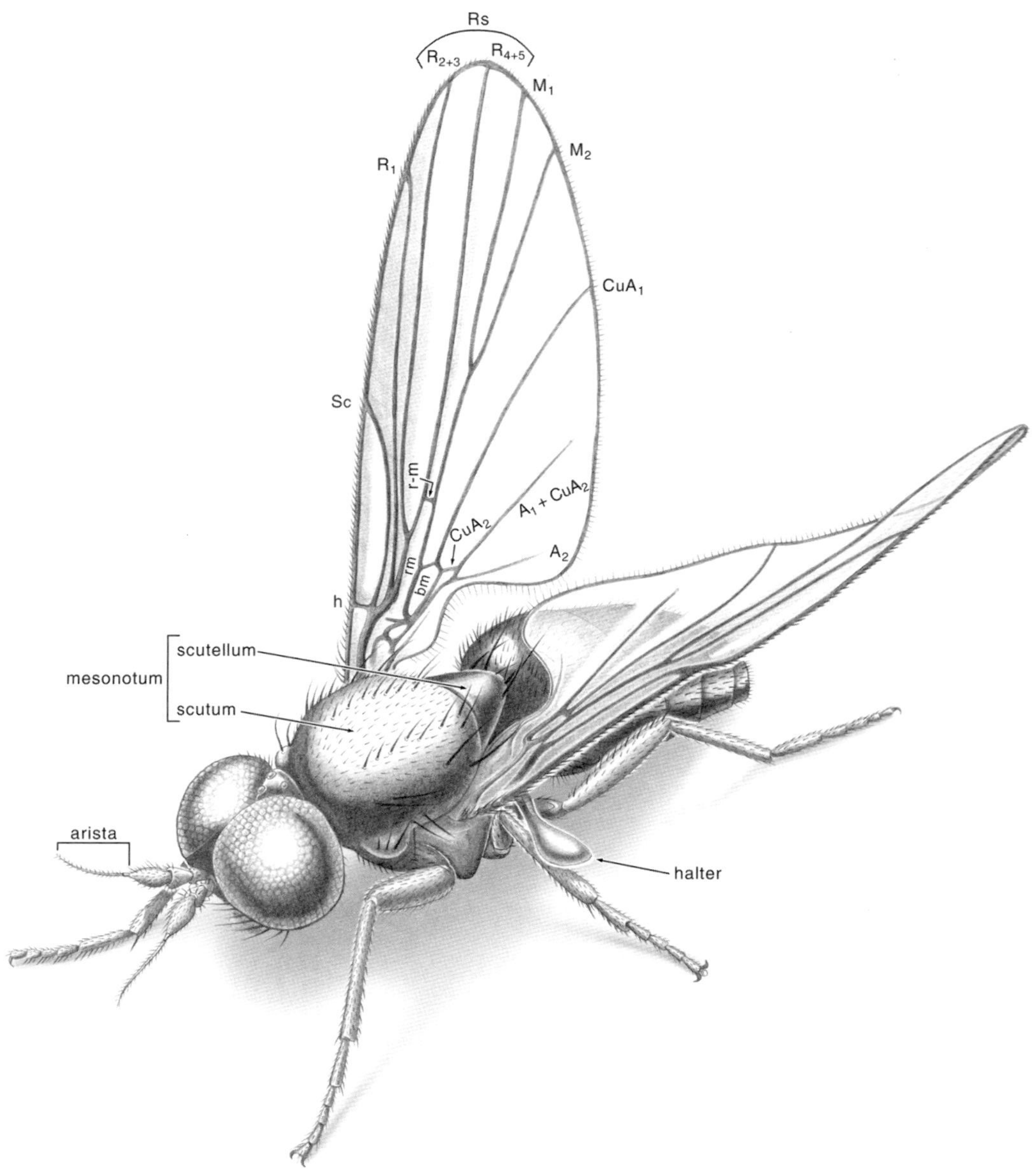

12.89. *Opetia nigra*, a primitive living cyclorrhaphan from Europe. *Opetia* is sometimes put into its own family and is thought to be the living sister group to all other Cyclorrhapha. Alternatively, it is considered a member of the Platypezidae or closely related to this family. Body length 1.9 mm.

generalized, like Atelestidae. Stem-group cyclorrhaphans occur throughout the Cretaceous (e.g., Figure 12.94).

The Phoridae is one of only two diverse families of basal cyclorrhaphans (Figure 12.95). In fact, it is perhaps the most ecologically diverse and individually numerous of all fly families. The biology of these flies was thoroughly reviewed in the book by Disney (1994) and relationships discussed by Brown (1992). There are approximately 6,000 described species, but given the rate of discovery of new species, the actual number is certainly several times this rate. "Scuttleflies" are immediately recognized for their quick, jerky way of running over surfaces, and they break all the biological rules for insects. Many are saprophagous, and *Megaselia scalaris* (the "coffin fly") has what is probably the most indiscriminate diet of all insects, having been recorded breeding in myriad things from formalin-preserved body parts to shoe polish. Other species of phorids are far more particular, including predatory, phytophagous, and even parasitoid species. For example, hundreds of species of the "decapitating flies" (genus *Apocephalus*) parasitize ants. Some specialize on the major workers of leaf-cutting (attine) ants, which they attack by landing on the leaf fragment carried by the worker and inserting an egg into the occipital foramen behind the head (minor workers frequently ride the fragments and chase some of the phorids away). The heads of major workers are huge, housing muscles for the powerful mandibles, on which the developing larval *Apocephalus* feeds. The brain of these workers must be insignificant, since parasitized individuals have been known to march along with their nestmates right up until the time its head falls off and the larva wriggles out to pupariate.

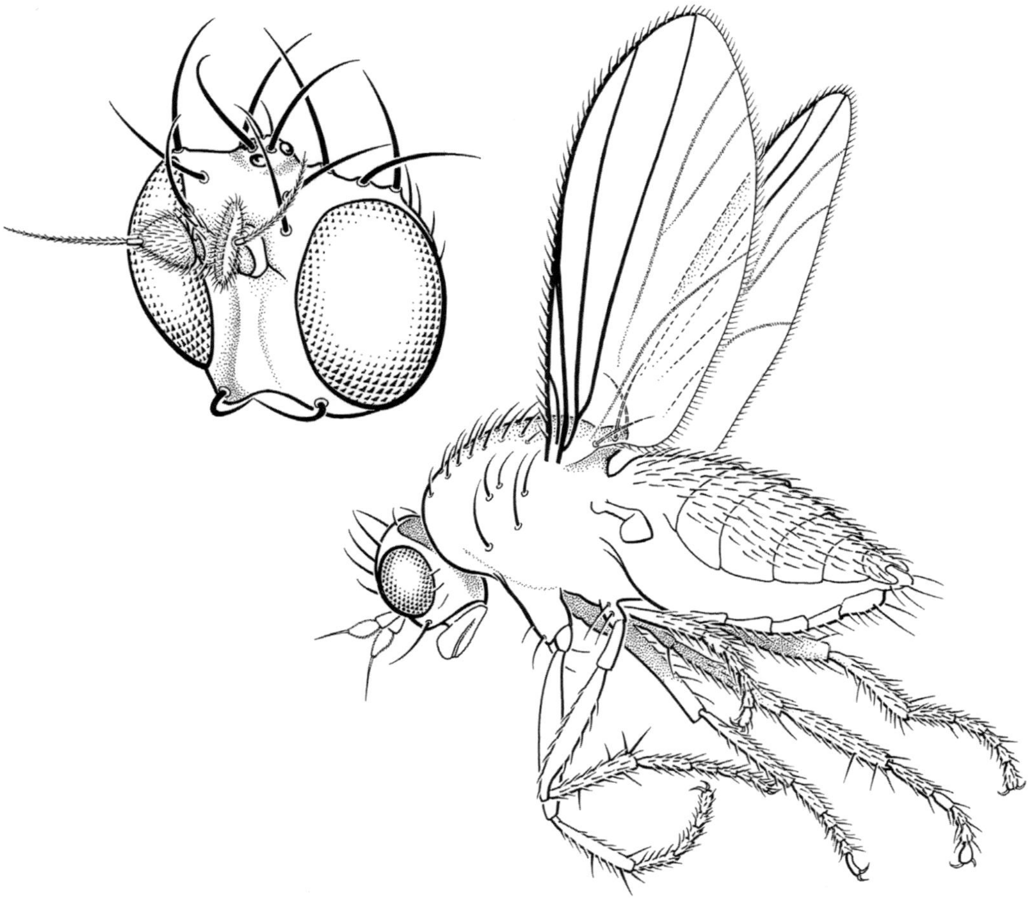

12.90. *Lonchopterites prisca* in Early Cretaceous Lebanese amber. It resembles flies in the small, living family Lonchopteridae but is more primitive. Lonchopteridae may be the most primitive living family of Cyclorrhapha. AMNH LAE79; length 1.1 mm.

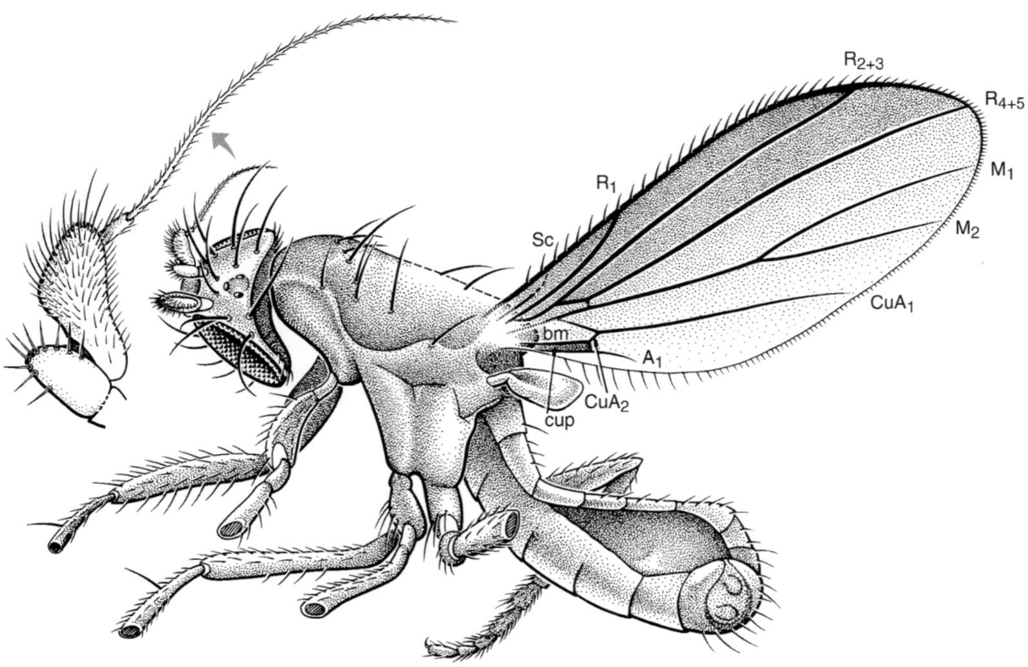

12.91. *Lonchopteromorpha asetocella* in Lebanese amber, a very early, primitive, close relative of Lonchopteridae. AMNH LAE-132; length 1.1 mm.

12.92. The most primitive cyclorrhaphan? *Chimeromyia* is known only from several tiny species preserved in amber 125–100 MYA. This one is in Burmese amber. The genus possesses features of empidoids and cyclorrhaphans. AMNH Bu1541; length 1.0 mm.

Examples of even more bizarre species are the Aenigmatiinae, many of which are parasitoids also of social insects. Though wing reduction and loss is common in phorids, some species of aenigmatiines even have females with highly reduced legs, which, with their round little bodies, gives these flies the appearance of being phoretic mites attached to their hosts.

The fossil record indicates that phorids were primitive and scarce in the Cretaceous but then explosively radiated in the Tertiary. Approximately 20 species of the stem-group genus *Prioriphora* occur exclusively in Cretaceous amber from 110 to 75 MYA (Figures 12.96, 12.97), which are primitive to all living species of phorids. There is actually no feature of *Prioriphora* or any other "genera" of these Cretaceous phorids that is derived compared to Recent phorids, so it is a premier example of a paraphyletic stem group. Indeed, *Prioriphora* is ancestral to the huge radiation of Tertiary Phoridae. Other than an anomalously old record of *Metopina goeleti* from probably the latest Cretaceous (its exact age is uncertain), the earliest significant record of diverse phorids is in the Eocene Baltic amber (Brown, 1999). By the Miocene, preserved in Dominican amber, phorids were very diverse, abundant, and similar to Recent species (Figure 12.98). Two relict families closely related to the Phoridae – Ironomyiidae and Sciadoceridae – each have a single genus and only two living species. *Ironomyia* occurs in Australia, and the two living

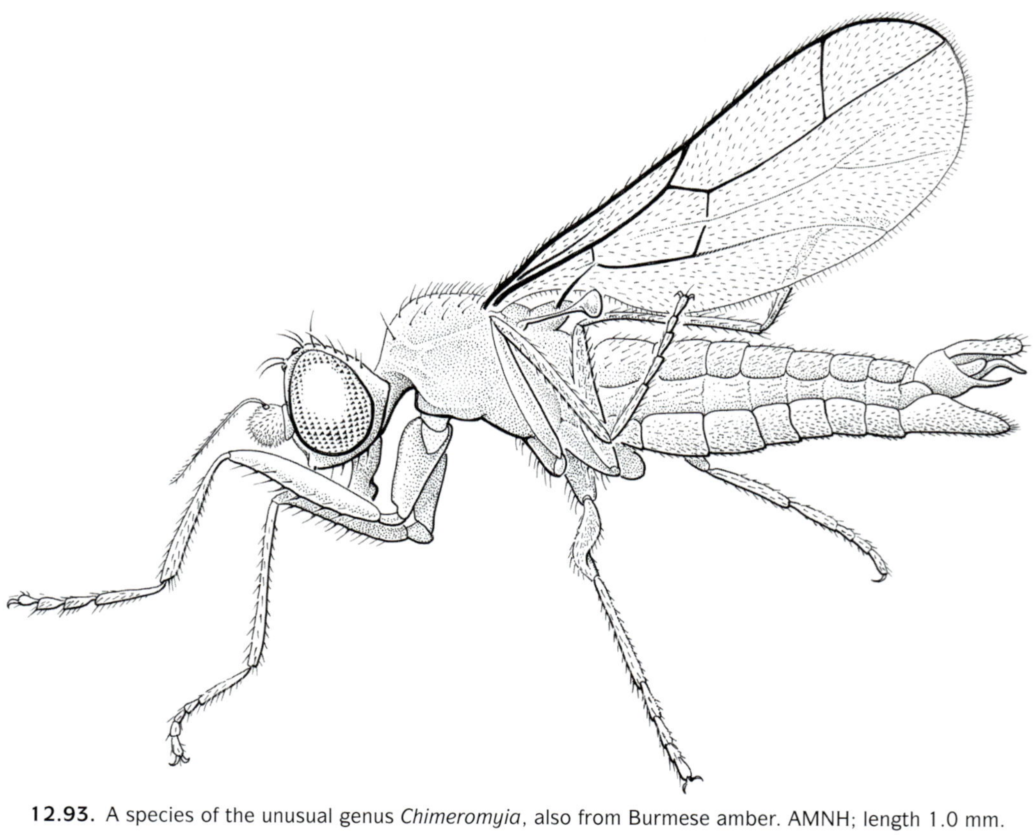

12.93. A species of the unusual genus *Chimeromyia*, also from Burmese amber. AMNH; length 1.0 mm.

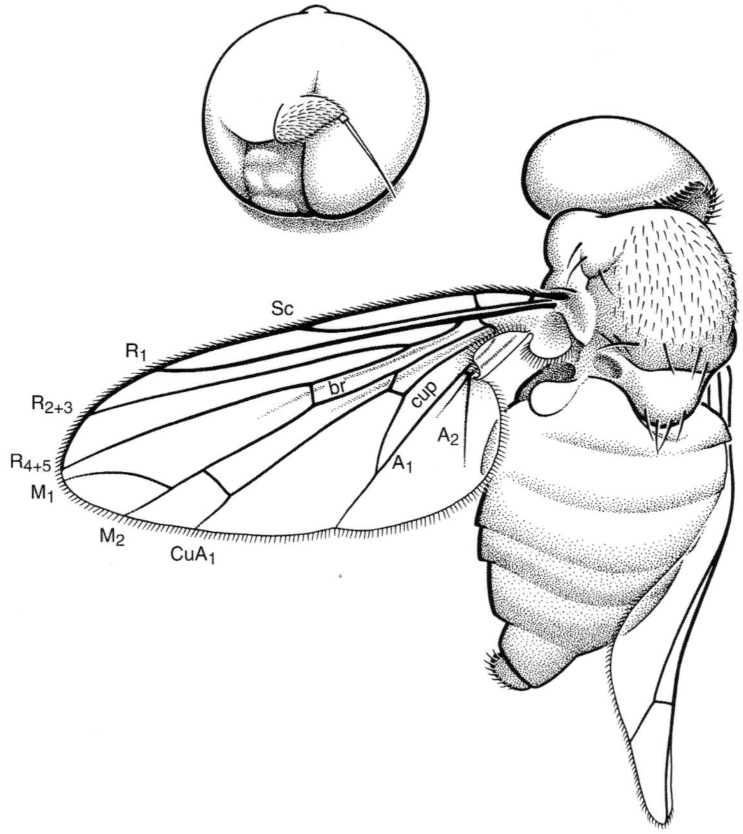

12.94. An early and very primitive platypezid fly, *Electrosania cretica*, in Cretaceous New Jersey amber, also showing the front of the head (most details are covered by a layer of froth). Primitive cyclorrhaphans like this one in Cretaceous amber indicate that the group originated in the earliest Cretaceous. AMNH NJ518a; length 3.45 mm.

species of sciadocerids are *Archiphora patagonica* from Chile and *Sciadocera rufomaculata* from Australia and New Zealand. It is tempting to interpret the austral distribution of these basal phoroids as remnants of Gondwanan drift because fossils show they definitely existed in the Cretaceous. In fact, ironomyiids appear to have been quite diverse in the Cretaceous. But fossils also indicate most of these groups occurred in the Northern Hemipshere as late as the Eocene (Figures 14.36, 14.37). Sciadocerids are sometimes classified with Phoridae as the most basal species, but this does not affect interpretation of the evolutionary history of the family.

The other major group of non-schizophoran Cyclorrhapha is the "flower" or "hover" flies, family Syrphidae. There are approximately 6,000 described species, but the numbers of new species being discovered are very modest compared to phorids despite the popularity of syrphids among amateur collectors (the only flies with apparently such appeal). Many syrphids feed amongst bees and vespid wasps on nectar and pollen, which is why many have evolved black and yellow patterns that mimic these aculeates, sometimes remarkably well (Figure 13.90). Syrphinae and Eristalinae, in fact, are probably important pollinators (Figure 12.99). Syrphids are also very acrobatic fliers and are among the most adept hovering animals. *Ornidia*, for example, is a common syrphid of neotropical forests, often seen hovering in a shaft of sunlight like a suspended blue jewel. The real ecological diversity of syrphids, though, is in the larval stages. Many eristaline larvae are saprophagous, living in wet substrates like submerged leaves, water in tree holes, or the small pools within bromeliads; about one third are predators, some even venomous. Syrphine larvae are commonly predacious on aphids; larvae of the subfamily Microdontinae feed on early instar ant larvae, their peculiar, turtle-shaped structure presumably protecting them from attack by the ants in their colony. Relationships among tribes have most recently been discussed by Ståhls *et al.* (2003).

Closely related to the syrphids are the Pipunculidae, a distinctive family of approximately 600 species, all of which have large heads comprised almost entirely of eyes. These are interesting because the larvae are endoparasitoids of auchenorrhynchans. The adults attack by grasping and holding the plant hopper in flight and then injecting an egg with a stiletto-like oviscape. Leaf hoppers and other Auchenorrhyncha are powerful leapers, so to insert an egg into one requires highly specialized means. Female dryinid wasps also parasitize leaf hoppers, which they grasp with chelate tarsi, briefly riding the host like a rodeo bull while it injects an egg. A

12.95. Living species of the large and very diverse family Phoridae: (a) male *Megaselia*, (b) female *Acontistoptera* from Costa Rica, which has bizarre, short wings fringed with long hairs. Life histories and structure of the Phoridae are probably more diverse than any other family of flies or even most insects. Scanning electron micrographs.

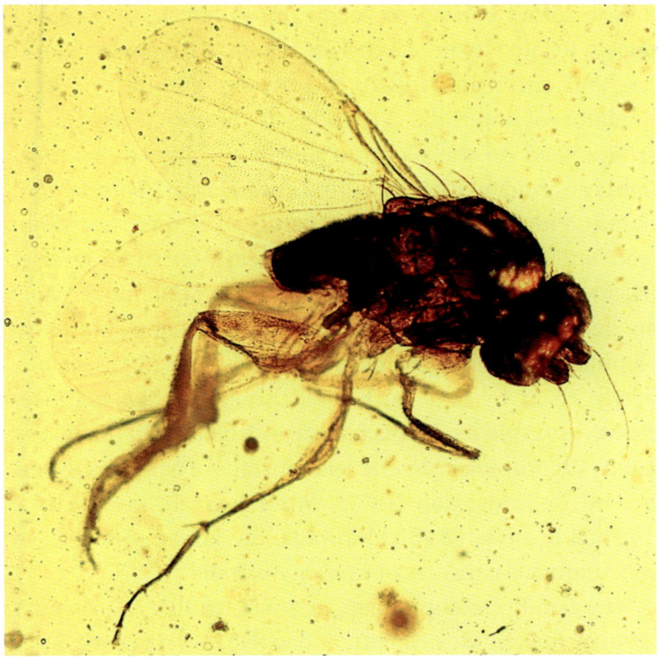

12.96. A classic paraphyletic stem group: *Prioriphora*. This "genus" of phorids is known only in Cretaceous amber, 110–75 MYO, and is primitive to all living species of phorids. This specimen is in Burmese amber. AMNH Bu1395; length 0.94 mm.

pipunculid merely holds its thrashing host aloft while it injects its eggs. Other than a fragmentary specimen of a putative syrphid in Siberian amber, all fossils of Syrphidae and Pipunculidae are Tertiary. Syrphidae, in particular, are diverse in Baltic amber (Figure 12.100), with some 40 species described (Hull, 1945, 1949), though the relationships of these need to be restudied. Like phorids, syrphids radiated in the Early Tertiary along with the Schizophora.

The **Schizophora** comprise the largest Tertiary radiation of insects along with the ditrysian lepidopterans; the Recent fauna of Schizophora comprises approximately 50,000 species. They are defined, in part, by the *ptilinum*, remnants of which are typically seen as a crescentic suture bordering the face. Invaginated within the *ptilinal fissure* is the ptilinum itself, a membranous sac that expands like an air bag to help the emerging adult rupture the puparium. Schizophora is traditionally classified into acalyptrates and calyptrates, and, as can be guessed, one of these is paraphyletic. The "acalyptrates" are myriad, small, stout-bodied flies that are almost certainly not monophyletic despite McAlpine's (1989) definition of them, which is based on primitive features. The monophyletic calyptrates, or Calyptrata, are probably derived from some group of acalyptrate.

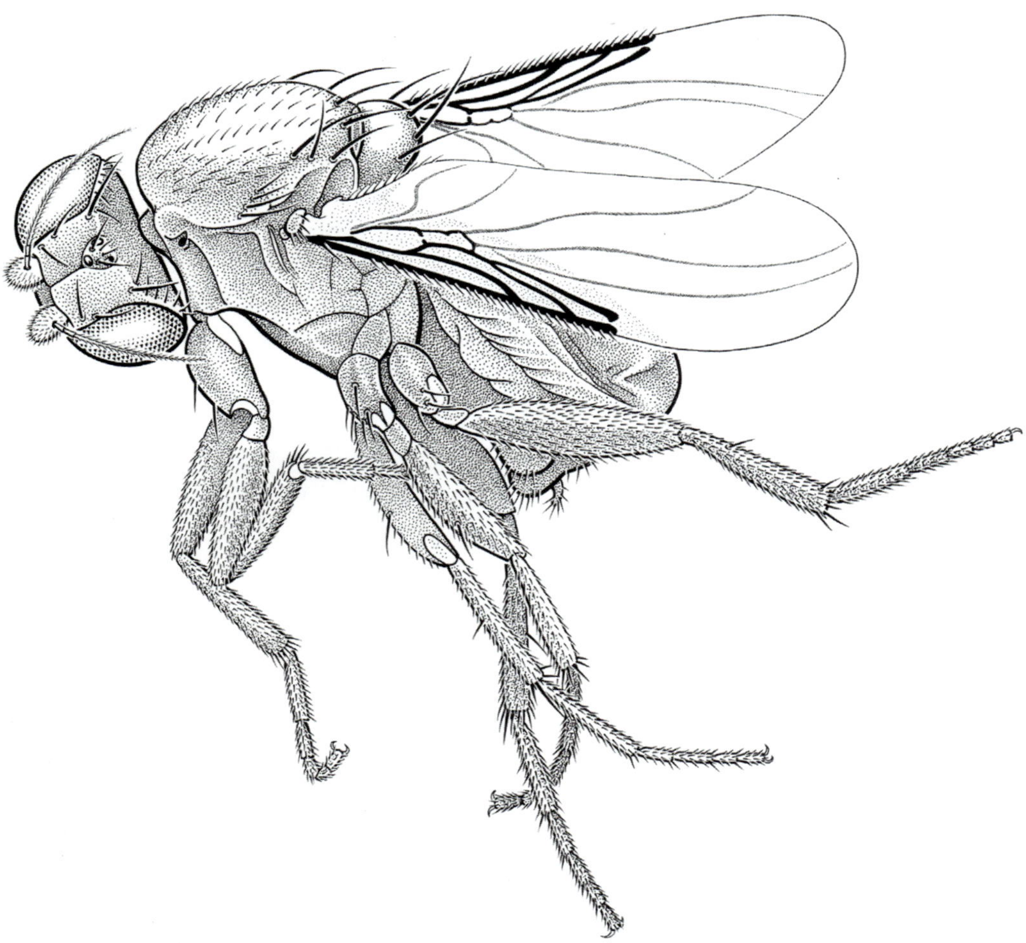

12.97. *Prioriphora casei* (Phoridae) in 90 MYO New Jersey amber. AMNH NJ230a; length 1.95 mm.

12.98. Phoridae radiated in the Tertiary. A pair, perhaps originally mating, of the living genus *Puliciphora* preserved in Miocene Dominican amber. Males of the genus are winged; females are wingless. AMNH; female length 1.1 mm.

Acalyptrates are taxonomically daunting to many entomologists because they are usually small, they have subtle differences in setae that are taxonomically important, and their male terminalia are complex but necessary for their systematics. Also, there are approximately 70 families, relationships among which have been discussed (though hardly exhausted) by Griffiths (1972) and McAlpine (1989) and in various papers by Hennig and others on particular groups of families. Ecologically, acalyptrates are exceptionally diverse; this subject is reviewed by Colless and McAlpine (1991) and in various chapters in volume two of the *Manual of Nearctic Diptera*. Most species are saprophagous, the larvae breeding in decaying foliage, fruit, fungi, necrotic cacti, sap fluxes of trees, algae, and dung. There are also leaf and stem miners, larval predators, specialized inquilines, and even endoparasitoids. Larvae of the large family Agromyzidae mine leaves and are often monophagous. Closely related to them is the small Australian family Fergusoninidae, which form galls in intimate and strict association with particular species of nematodes (Giblin *et al.*, 2003).

Acalyptrates that have endoparasitoid larvae are in the Conopidae, Pyrgotidae, and Cryptochaetidae. Conopids (Figures 12.21, 12.101) parasitize aculeate Hymenoptera, roaches, and calyptrate flies; pyrgotids parasitize adult scarab beetles; and cryptochaetids are endoparasitoids of scale insects. Scales and other sternorrhynchans are also attacked by larval acalyptrate predators in the Chamaemyiidae and by some species of Chloropidae and Drosophilidae. Other larval predators include Sciomyzidae and Pallopteridae, the latter of which feed on the larvae of wood-boring beetles and some cecidomyiid midge larvae. Sciomyzidae are particularly interesting because the larvae are aquatic predators of freshwater snails and fingernail clams (Berg and Knutson, 1978) and have even been used to control snails that are intermediate hosts of the flukes that cause schistosomiasis. Related to the sciomyzids is a small family, Dryomyzidae, one species of which, *Oedoparena glauca*, is a larval predator of intertidal barnacles in North America. Two groups feed on the blood of nestling birds: adult *Carnus* (family Carnidae) (Grimaldi, 1997a) and larval *Neottiophilum* (family Piophilidae). Female *Carnus* shed their wings and feed until the abdomen expands to nearly ten times its original size. Carnidae are closely related to the Milichiidae, which are generally saprophagous, but adults of some species are cleptoparasitic, feeding on the hemolymph of insect prey newly killed by large arthropod predators like mantises, asilids, and spiders. Cleptoparasites also occur in the diverse family Chloropidae, generic relationships of which have been treated by Andersson (1977). *Batrachomyia* has the most unusual lifestyle of all chloropids, the larvae of which are subcutaneous parasites of frogs in Australia.

The family Drosophilidae, or small fruitflies, is a microcosm of acalyptrate diversity, and with 3,500 species it is also one of the more diverse families in Cyclorrhapha (Ashburner, 1989; Grimaldi, 1990b). The family is well known for the laboratory fruitfly, *Drosophila melanogaster* (Figure 12.86), which is a typical saprophagous species that breeds in decaying fruit. The range of life histories in the family, though, is among the greatest in Diptera, yet not quite that of the Phoridae. Drosophilids pollinate and breed in flowers; they are larval inquilines in bee nests and predators of spider eggs, egg masses of black flies, and sternorrhynchans (Ashburner, 1981), and the larvae of a species in Central America even feed on the embryos of glass frogs (*Centrolenella*) (Grimaldi, 1994a). Larvae of the large genus *Cladochaeta* are parasites of spittlebug nymphs; they cling to the nymphs, rasp small holes in them, and feed on the hemolymph, though they are not known to kill their hosts (Grimaldi and Nguyen, 1999). The most peculiar larval niche of drosophilids, though, is

12.99. A flower fly, *Eristalis* (Syrphidae), feeding from a flower. Syrphids are diverse, abundant, and excellent fliers and are probably important pollinators. Photo: V. Giles.

12.100. Fossil syrphid fly in Eocene Baltic amber. This amber contains the first diverse record of early syrphids, though the family probably originated in the latest Cretaceous. AMNH BaJH43; length 4.7 mm.

occupied by three unrelated species on three distant islands, each species of which breeds in the nephric exudates of land crabs (Carson, 1974). Because various *Drosophila* can be bred in the lab, experiments are easily conducted on them, and in one case this approach was used to study how crab breeding could evolve so readily. In a little known experiment (Wallace, 1978) the common "garbage" species *Drosophila virilis* was bred in cages on pads continually soaked with human urine, to see how long it might take for the flies to develop entirely on a uric acid diet. It took three months, about six fly generations, and approximately 3 liters (1.3 gal) of urine (daintiness is not a trait of serious fly biologists). Most importantly, after 13 months of breeding on the "artificial crab," *Drosophila virilis* lost the ability to breed on standard lab medium.

The most impressive acalyptrates are ones with elaborate sexual dimorphisms (Figure 12.102), which have evolved in profusion. Particularly striking are the stalk-eyed flies, a feature where the eyes are situated at the end of long stalks, and which has evolved separately in seven families. In all families except the Diopsidae (e.g., Figure 12.103), only the males have stalked eyes, the females have normal heads, so the trait has evolved as a result of intense sexual selection. Indeed, males of dimorphic stalk-eyed flies spend inordinate effort in ritualized fighting and displays, defending sites where females get mated and then lay their eggs. In some cases the eyestalks reach absurd proportions. For example, in the neotropical species *Plagiocephalus latifrons* (Otitidae) (Figure 12.102) and the New Guinea species *Achias rothschildi* (Platystomatidae), eyestalks of the largest males are up to five times the length of the body. This doesn't seem to interfere much with their flight, and in fact with the eyes so far apart the vision of stalk-eyed flies is very binocular and has excellent depth of field. Other elaborate sexual dimorphisms in acalyptrates are "beards," tarsal "flags," "horns," and "antlers." Males of the Australasian genus *Phytalmia* have deer- or moose-like "antlers" protruding from the cheeks (Figure 12.102). They display by rising on their hind legs, pressing their antlers against each other, and pushing each other back and forth like deer in rutting season (Dodson, 1989, 1997). For some reason, most of the stalk-eyed and antlered flies are in the Tephritoidea, a group of families that includes the large family Tephritidae (with 4,000 species). Tephritids are infamous as some of the most destructive agricultural pests because their larvae riddle fresh fruit, though the larvae of others feed in stems or even form galls. Biology and relationships of the family were treated in the book by Aluja and Norrbom (2000).

The most peculiar acalyptrates are two highly specialized inquilines. One is *Mormotomyia hirsuta* (Mormotomyidae), which lives on bats in eastern Africa and on which it may be an ectoparasite (Austen, 1936). The other is the family Braulidae ("bee lice"), which contains two small genera living in the hives of *Apis* honey bees (Grimaldi and Underwood, 1986). Braulids seem to do little harm to the honey bees because the

12.101. *Physocephala* (Conopidae) mating on a clover blossom (the male is on top). These flies have jutting proboscides that they use for feeding from flowers. Like many anthophilous Diptera, they mimic stinging wasps; larvae are parasitoids. Conopidae is a very primitive family of Schizophora. Photo: D. Grimaldi.

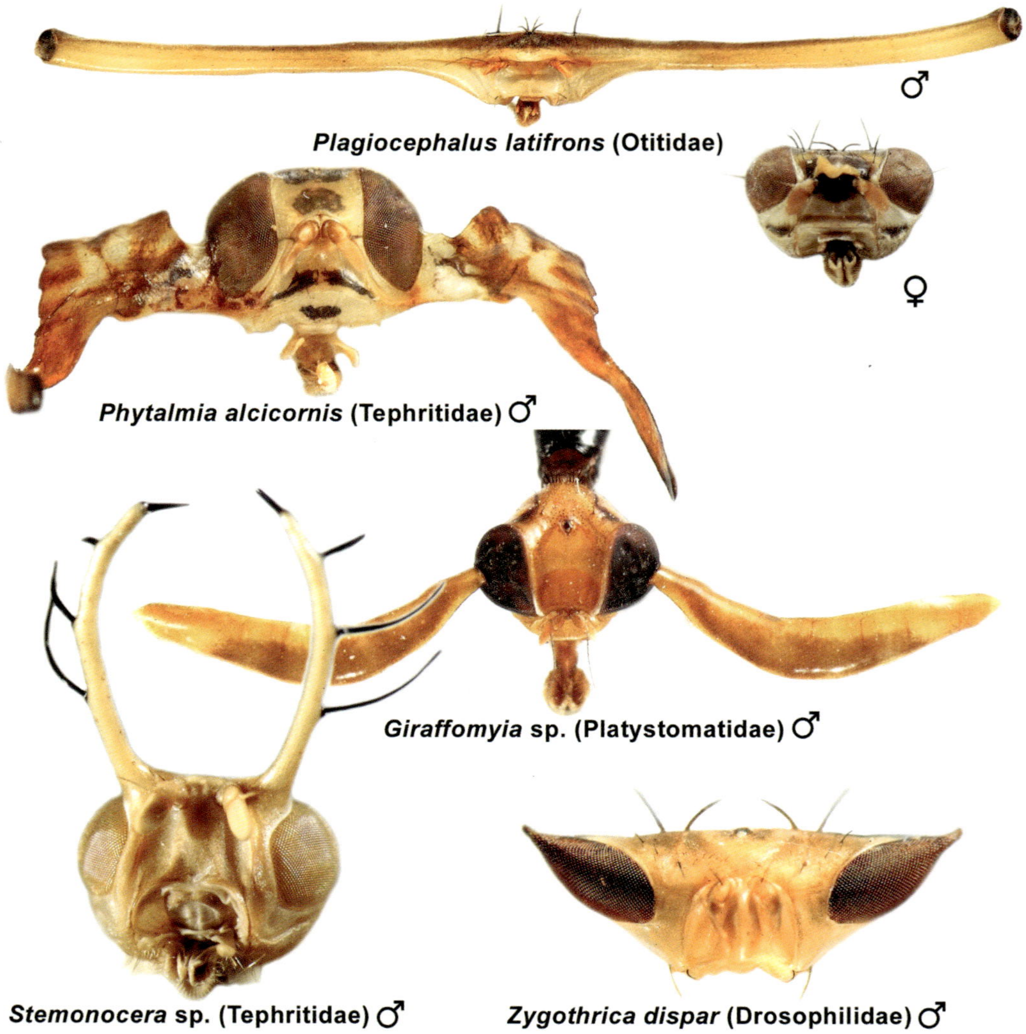

12.102. Extreme sexual dimorphisms of the head of some acalyptrate Schizophora flies. With the exception of the drosophilid, all are in closely related families in the superfamily Tephritoidea. Structures include "tusks" growing from the cheeks, stalked and pointed eyes, and facial horns. Males use these structures in displays and ritualized fighting. Not to the same scale.

larvae mine the wax combs in the hives and adults feed on regurgitate of the adults. Flies in both families have lost the eyes and ocelli, and the wings and halteres have been extremely reduced or completely lost. Braulids, moreover, have the pretarsal claws modified into combs, which they use for clinging to the branched hairs of the bees (Figure 12.104). Extreme reduction of these flies has obscured relationships, but genitalic morphology indicates these flies are closely related to several families with free-living species: *Mormotomyia* to the Heleomyzidae and Braulidae to the small family Australimyzidae (Griffiths, 1972; McAlpine, 1989).

Much more taxonomic attention has been paid to the **Calyptrata** (Figure 12.21), or calyptrate flies, because they are so important as pests of humans and livestock, and now are also used in forensics for dating decayed corpses (Smith, 1986). There are about 16,000 described species in this group. Unlike acalyptrates, calyptrates are clearly monophyletic, based on the large, flaplike lower calypter at the base of the wing, the prestomal teeth on the labellum (similar ones have evolved independently in a few acalyptrates), and other features. Griffiths (1972), McAlpine (1989), and Pape (1992) discussed relationships among the families. Adult blood feeding has arisen at least five times in two groups of calyptrates – the Muscidae (including the Fanniidae) and Pupipara. All of the hematophagous calyptrates feed by rasping skin with enlarged, sharp, scale-like prestomal teeth situated on a reduced labellum and then imbibing the blood that wells to the surface. The stable fly, *Stomoxys calcitrans*, is a well-known example, and tsetse feed in a similar way. In both of these the proboscis is long, rigid, and jutting. There are blood-feeding larvae as well, the best known among them being *Protocalliphora*, which feed on nestling birds and the dreaded Congo floor maggot, *Auchmeromyia senegalensis*, which is the leech among insects.

Most species of calyptrates have larvae that breed in dung (Scathophagidae, Muscidae, Calliphoridae) or in carcasses

12.103. An early stalk-eyed fly, *Prosphyracephala*, of the living family Diopsidae, preserved in Eocene Baltic amber. Both sexes of diopsine flies have eye stalks, though they are slightly smaller in the females of some species. AMNH; body length 5.1 mm.

(especially Calliphoridae and Sarcophagidae). Anyone who has seen an animal carcass roiling with the maggots of these flies appreciates their ability to consume. Individual females of the more common carrion-breeding species, such as *Lucilia*, can lay hundreds of eggs, and development can be completed in several weeks depending on the temperature (Smith, 1986). Over merely several weeks the carcass of an entire cow can be consumed by blowflies, making these insects the premier scavengers. Less appreciated about calyptrates, though, is that some are remarkably specialized endoparasitoids. The family best known for this habit is the Tachinidae, which, with 8,000 described species, is the largest family of Cyclorrhapha. The larvae of all Tachinidae are endoparasitoids of terrestrial arthropods, parasitizing eight orders of insects as well as some millipedes (reviewed in Wood, 1987), but the main hosts are the larvae of Lepidoptera, or caterpillars. There are a variety of ways tachinids invade their hosts, typical of different groups of Tachinidae. They either attach eggs directly onto their hosts or insert them; some have free-living, first instar *planidia* that actively seek a host; and others broadcast thousands of minute *microtype* eggs that are consumed by the host and hatch in their gut (Wood, 1987). Closely related to the Tachinidae are the Rhinophoridae, a small, mostly Old World family that are parasitoids of woodlice (Crustacea: Isopoda). Other parasitoids also occur in the Muscidae, Calliphoridae, and Sarcophagidae.

Calyptrate flies have refined the "art" of vertebrate parasitism, particularly the Oestroidea and Pupipara. The oestroids are the bots, classified into four families or into subfamilies of one family, Oestridae, and relationships of which have been studied by Pape (2001). Adults of these flies are rotund, furry flies with a quick, buzzing flight and generally vestigial mouthparts. The larvae of all oestroids are parasites of mammals, living under the skin (Cuterebrinae, Hypodermatinae), in the gut (Gasterophilinae), or in nasal and tracheal passages (Oestrinae) (Figures 12.21, 12.22). These are among the few specialized vertebrate endoparasites in the insects. Ungulates are the most common hosts, though rabbits, rodents, and even a few primates, marsupials, and the African elephant are also parasitized. They cement eggs to the hair of the host, like lice, which the host then licks and swallows, or the larvae burrow directly into the skin. Female Oestrinae flick young larvae directly into the nasal passages of their ungulate hosts.

The bot fly that regularly attacks humans is *Dermatobia hominis*, occurring in Central and South America. This fly has a novel method of attacking its hosts: it uses mosquitoes and other flies. A female *Dermatobia* catches a female mosquito and lays her eggs on it, which quickly hatch; when the mosquito feeds on a host the first-instar larvae crawl down the proboscis and burrow into the feeding site. In an obscure paper, Dunn (1930) gave a vivid personal account of bot fly parasitism. He exposed himself naked in a Panamanian forest for several hours, and mapped the sites where mosquitoes fed. He then made a log over the months that followed of the sites that developed bots, and their progress until emergence: "October 17. Last night I was awakened by a severe itching and dull burning sensation at [larval site] no. 4" (p. 331) . . . "November 17. [Larval site] No. 5 is very painful and the muscles of the leg are stiff and sore to the touch, and causing some lameness. All the lesions are oozing considerable serum and puss" (p. 337). Only genuine fascination could foster such devoted inquiry.

The **Pupipara** comprise the tsetse and the bird and bat flies. The group is clearly monophyletic, defined in part by unusual development. Females lay a single, fully mature larva that pupariates soon after emerging; it is nourished in a uterus-like structure by the secretions of specialized ("milk") glands. All adult pupiparans feed on the blood of birds or mammals, and their mouthparts are modified into a short or long, but rigid, proboscis armed with large prestomal teeth. In all of the pupiparans with wings, these are held flat and close over the abdomen. Tsetse are the most basal group of the Pupipara, the other groups being three families of highly modified ectoparasites – Hippoboscidae (on birds and mammals) and Steblidae and Nycteribiidae (on bats). Griffiths (1972) and McAlpine (1989) discussed relationships in the Pupipara, the former of whom made the persuasive argument that the nycteribiids and streblids are merely highly

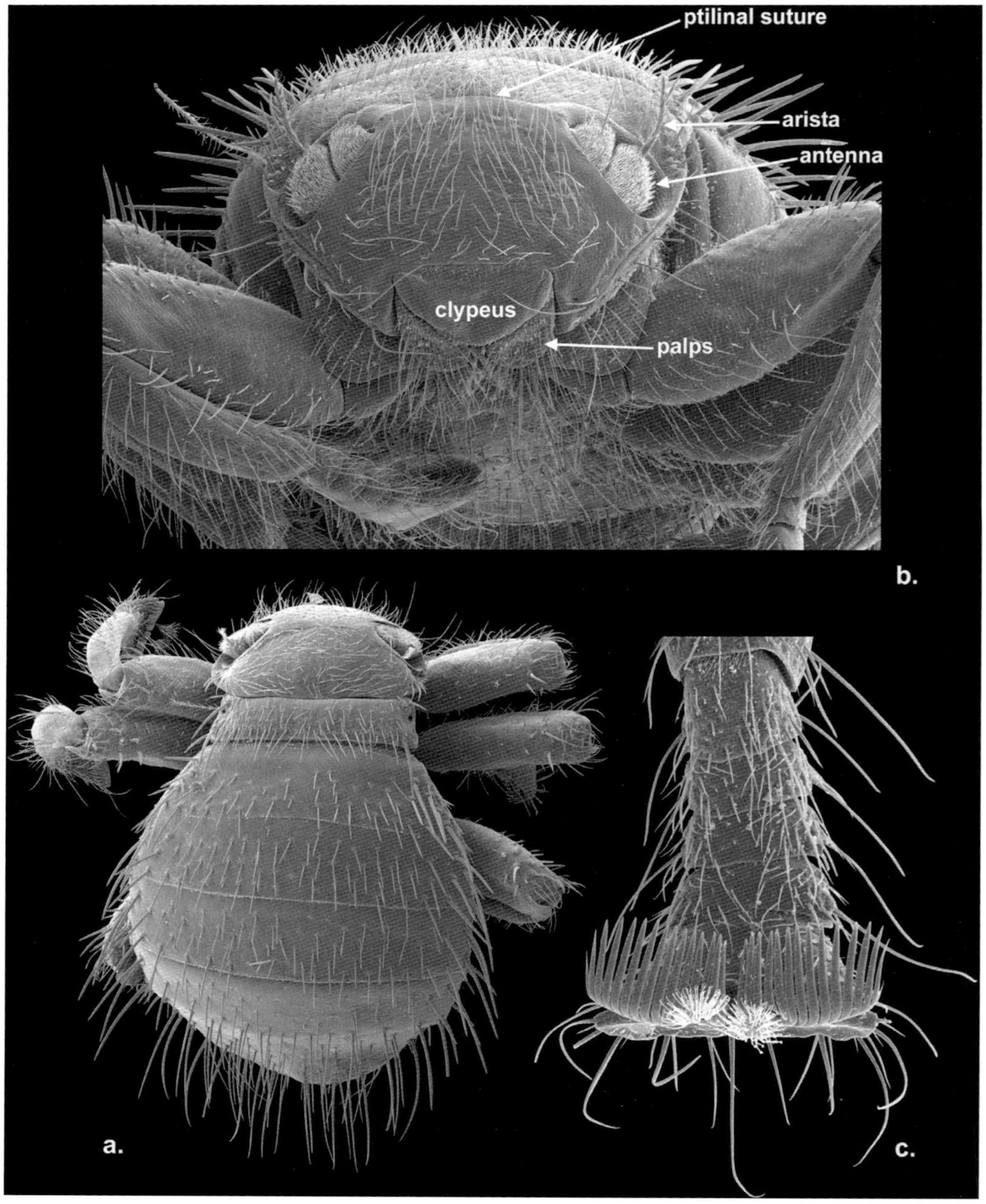

12.104. The bizarre schizophoran "bee louse," *Braula coeca* (Braulidae). Larvae and adults live in the hives of honey bees (*Apis*); adults are blind and wingless; they climb through the hairs of the bee using pretarsal combs and feed on regurgitate that they coax from worker bees. Relationships of these flies are obscure. (a) dorsal view of body, (b) front of head, (c) detail of pretarsus. Scanning electron micrograph.

modified hippoboscids. Hippoboscids themselves are flat, sprawling flies (Figure 12.21), with small wings, one quarter of which feed on mammals, the rest feed on birds. Some streblids have small wings (Figure 12.105), in others they are vestigial, and these are completely lost in all nycteribiids. Steblids and nycteribiids have long legs and pilose bodies and look like tiny spiders when they scurry over the hair of their bat hosts. Nycteribiids are primarily Old World and have the thorax extremely reduced, with a very small head that points up. Streblids are primarily New World and generally less modified, except for the bizarre *Ascodipteron*. This fly is like the tungine fleas, females of which embed themselves under the skin of the host and become bloated, egg-laying machines. In the case of female *Ascodipteron*, though, both the wings and the legs are lost. Schizophoran ectoparasites of bats have thus evolved three times: in Pupipara, in Mormotomyidae, and in a peculiar calliphorid from New Zealand, *Mystacinobia zelanica* (Holloway, 1976; Griffiths, 1982). In Chapter 14 we discuss why bats are such a magnet for specialized ectoparasites.

The tsetse of central Africa comprise 22 species of the genus *Glossina*, typically classified into three species groups

12.105. A batfly, family Streblidae. Streblids and their close relatives nycteribiids are small, flat, bristly flies that are ectoparasites of bats. Nycteribiids lack wings. Both families are recently evolved pupiparan flies that appeared in the Tertiary and are close relatives of tsetse. Scanning electron micrograph; body length 1.9 mm.

or subgenera. They resemble *Stomoxys* but are larger and hold the wings tightly folded over the abdomen. They are also rather flat, and not easily crushed when slapped, a result no doubt of tens of millions of years of selection by swatting tails. Unlike most other blood-feeding insects, both male and female tsetse feed on blood, in this case generally on suids (like warthogs and bushpig) and bovids, but also rhinoceros and hippopotamus. Certain species readily feed on humans and livestock and transmit sleeping sickness caused by *Trypanosoma brucei* and *T. gambiense*. Though sleeping sickness is not the scourge it used to be, it has caused disastrous epidemics. Between 1896 and 1906, for example, nearly one million people died of the disease in central Africa. It also destroys livestock, and so tsetse has been credited with protecting African bush and its wildlife from the encroachment of pastureland. Fossil tsetses from North America have become famous as an example of geographical extinction. Two large species, twice the size of living ones, occurred in the Late Eocene of western North America (Grimaldi, 1992) (Figure 12.106), and fossils of the family have recently been discovered in the Oligocene of Germany as well (Wedmann, 2000). Glossinidae were probably nearly worldwide 30–40 MYA.

With the exception of two ambiguous records from the latest Cretaceous, schizophoran flies are entirely Tertiary. The two apparent Cretaceous records concern a putative milichiid in amber from New Jersey (Grimaldi *et al.*, 1989) and a mineralized puparium in 70 MYO rocks from Canada (McAlpine, 1970). Identity of the milichiid is actually uncertain, and the piece of amber in which it is preserved (which also contains the oldest bee) may have been formed as early as the K/T transition, 65 MYA, or even later. Identity of the puparium, *Cretaphormia fowleri* (Figure 12.107a,b) is uncertain, particularly its attribution to the Calliphoridae or even to the Schizophora, though it is almost certainly a cyclorrhaphan. Thus, the earliest *definitive* Schizophora are preserved as compressions in Eocene deposits from western North America (e.g., Melander,

12.106. The giant extinct tsetse, *Glossina oligocena* (family Glossinidae), in 35 MYO shales from Florissant, Colorado. Now restricted to sub-Saharan Africa, tsetse occurred in North America and Europe in the Eocene to Oligocene. This species is twice the size of living species. AMNH 018839; length 21 mm.

12.107. Two puparia of *Cretophormia fowleri*, in ironstone from the Late Cretaceous of Canada. These were originally attributed to the Calliphoridae, which would be the only Cretaceous calyptrate flies. They can only be confidently attributed to Cyclorrhapha. Calyptrate flies appear to have evolved in the Tertiary. CNC 817; length 13 mm.

12.108. A very late Eocene otitid fly, *Melieria calligrapha*, from Florissant, Colorado. The wings were patterned. Eocene rocks from western North America and particularly the Baltic amber have recorded the early radiation of schizophoran flies. AMNH 026615; length 5.5 mm.

1949) and amber from the Baltic region (e.g., Hennig, 1965a, and other papers; Michelsen, 2000) (Figures 12.108 to 12.110). In fact, the Baltic amber has preserved some 32 Recent families of Schizophora, most of which have been assigned to extinct genera of acalyptrates (Hennig, 1965a), but the family relationships of some are obscure. Study of many recently excavated Baltic amber schizophorans indicate the diversity is even greater than previously thought (e.g., Weitschat and Wichard, 1998), and these specimens will be important in clarifying relationships. Oddly, only one possible schizophoran is known from the Eocene amber of Fushun, China (Hong, 1981). Numerous Oligocene and Miocene sites in Europe have yielded diverse compression fossils of Schizophora (e.g., Statz, 1940), even leaf mines of Agromyzidae (Süss, 1980). In the Miocene amber of the Dominican Republic, some 33 families of Schizophora are preserved (e.g., Figure 12.111), 10 of which are not represented in the Baltic amber. Lack of definitive Schizophora in the Cretaceous, and the primitive nature of the Baltic amber species indicate that an explosive radiation of the Schizophora began probably in the Paleocene, some 60 MYA.

Schizophora may have originated in the latest Cretaceous, but if so they must have been very scarce. The Paleocene-Eocene radiations of mammals can account for the radiations of Oestroidea and Pupipara, during which time the ditrysian Lepidoptera also radiated along with their parasites, the Tachinidae. Hypotheses by Papavero (1977) that Schizophora, and Oestroidea in particular, were affected by Cretaceous continental drift and the origins of early mam-

12.109. Early schizophoran fly from the Eocene of British Columbia, probably a member of the superfamily Tephritoidea. Archibald Collection, SBA 1122; body length 10 mm.

12.110. An extinct genus of the small living family Cypselosomatidae, in Baltic amber. Baltic amber has preserved the earliest diverse assemblage of schizophoran flies. AMNH BaJH68. length 2.6 mm.

mals are without factual basis. Given that calyptrates are essentially Tertiary in age, it is intriguing as to what insects fed on the carcasses of dinosaurs and other ponderous vertebrates of the Mesozoic. While some beetles today infest carcasses, they are not nearly as efficient as blowflies. Schizophora are by far most diverse today in the tropics, so the Eocene climate, which was nearly globally tropical, also probably contributed to their radiation.

In summary, major episodes in the evolution of Diptera follow:

- Origin and divergence from stem-group mecopteroids in the latest Permian to earliest Triassic, about 250 MYA
- Origin of Brachycera in the latest Triassic to earliest Jurassic, about 210 MYA, with origins of the major lineages of Brachycera (such as infraorders) in the Jurassic
- Origin of Cyclorrhapha in the early Cretaceous, about 145 MYA, with the evolution of basal (non-schizophoran) families in the Cretaceous
- Origin of the Schizophora in the latest Cretaceous to earliest Tertiary, about 65 MYA, but with their radiation exclusively in the Tertiary

Wiegmann *et al.* (2003) also estimated ages of these major lineages by using DNA divergence. Their estimated ages of each group are slightly older, which is typical of molecular studies, but they are still very similar to the chronology we propose.

12.111. A stilt-legged fly (family Micropezidae) in Miocene Dominican amber. The Schizophora preserved in this amber are decidedly more modern than those in Baltic amber, reflecting the younger age and tropical conditions when Dominican amber was formed. AMNH DR15–37; body length 6.5 mm.

13 Amphiesmenoptera: The Caddisflies and Lepidoptera

With very few exceptions the life histories of the orders Trichoptera (caddisflies) and Lepidoptera (moths and butterflies) are extremely different; the former have aquatic larvae, and the latter nearly always have terrestrial, plant-feeding caterpillars. Nonetheless, the close relationship of these two orders has essentially never been disputed and is supported by strong morphological (Kristensen, 1975, 1991), molecular (Wheeler *et al.*, 2001; Whiting, 2002), and paleontological evidence. Synapomorphies linking these two orders include heterogametic females; a pair of glands on sternite V (found in Trichoptera and in basal moths); dense, long setae on the wing membrane (which are modified into scales in Lepidoptera); forewing with the anal veins looping up to form a double "Y" configuration; larva with a fused hypopharynx and prelabium, which also have openings for glands that produce silk for constructing cases, webs, and cocoons; among others (Kristensen, 1984). The monophyly of each order is also firmly established, so Lepidoptera certainly did not diverge from any Recent group of Trichoptera. The best candidate for ancestors to the Recent Amphiesmenoptera is the extinct "family" Necrotauliidae (Figure 13.1), which is a paraphyletic assemblage of species that existed from the Triassic to the Cretaceous (Willmann, 1989; Ivanov and Sukatsheva, 2002). Lepidoptera and Trichoptera probably diverged from some necrotauliid ancestor in the earliest Jurassic. Slightly more distantly related are the Cladochoristidae, Microptysmatidae, Protomeropeidae, and Uraloptysmatidae, which existed even earlier, from the Permian to the Triassic.

TRICHOPTERA: THE CADDISFLIES

The Trichoptera are well known for their aquatic larvae, many of which construct cases with architectural finesse (Figures 13.2, 13.5) and as beautifully documented in the book by Wiggins (1977, 1996). The primary function of the case must be for protection against predators, primarily fish. The earliest known use of the common name "caddisfly" may be in Izaak Walton's book, *The Compleat Angler* (1653) and may derive from Old English traveling *cadice men*, who pinned bits of cloth to their coats to advertise their fabrics. A few species actually have terrestrial larvae, but even these are relegated to wet leaf litter, so many defining features of the order concern larval adaptations for an almost wholly aquatic lifestyle (Wiggins, 1977, 1996). For example, larvae are apneustic (without spiracles) and respire through a thin, permeable cuticle, some of which have filamentous abdominal gills that are simple or intricately branched (Figure 13.3). Antennae and the tentorium of larvae are reduced, though functional significance of these features is unknown. Larvae do not have prolegs on most abdominal segments, save for a pair of anal prolegs that have sclerotized hooks for anchoring the larva in its case. The order contains more than 11,500 described species (a list appears in *http://entweb.clemson .edu/database/trichopt*), making this the second largest lineage of aquatic insects after the culicomorphan flies (which have about 1,000 more species; the hydradephagan beetles are a distant third, with about 5,000 species total).

Perhaps because of their aquatic diversity Trichoptera have invaded a niche that only the larvae of a few midges (culicomorphans in fact) have also invaded, which is the marine environment. The Chathamiidae, which occur in Australia and New Zealand, are the only other Holometabola besides some chironomid midges (like *Pontomyia*) whose larvae develop in sea water, in this case in tidal pools and shallow shorelines. Salinity cannot be the sole constraint as to why more insects have not become marine because there are insects that even live in brine, like the larvae of some ephydrid flies. The oceans are packed with groups that evolved there for hundreds of millions to billions of years, which may partly explain why a terrestrial group as successful as insects has barely been able to re-invade the seas.

Though best known for the cases constructed by larvae (Figure 13.2), it is mostly just species in the suborder Integripalpia that do so. H. H. Ross (1956, 1964), Wiggins (1997), and Mackay and Wiggins (1979) categorized five *functional* groups of Trichoptera larvae, which should not be strictly interpreted as evolutionary stages. The most basic habit, and

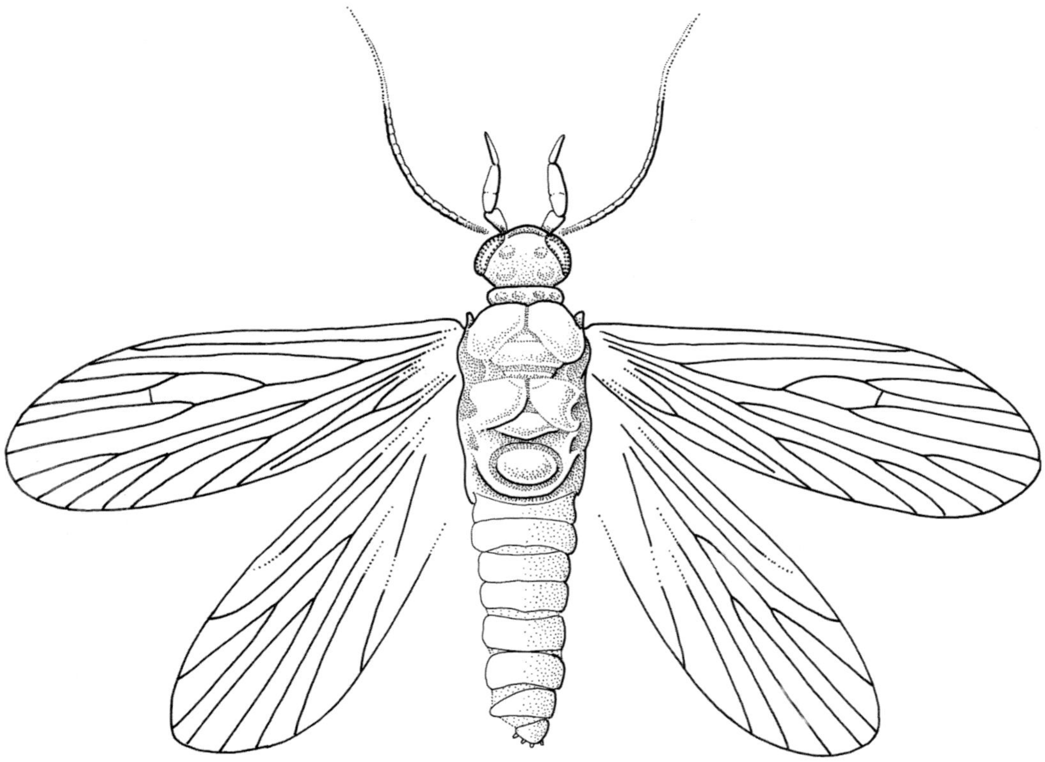

13.1. A stem-group amphiesmenopteran, *Necrotaulius tener*, from the Late Jurassic of Siberia. Lepidoptera and Trichoptera probably shared a common necrotauliid ancestor similar to this but from the Early Jurassic about 200 MYA. Reconstruction, based on Ivanov and Sukatsheva (2002). Wing length 3.7 mm.

perhaps the ancestral condition, is a free-living larva, wherein silk threads are used only as anchoring lines in currents, and silk is spun for a pupal case as in virtually all Trichoptera. This group consists mostly of predators. The ancestral larval habit was, alternatively, burrowing (Weaver and Morse, 1986). Group 2, consisting mostly of the Hydroptilidae, have the first four instars free-living, and instar 5 builds a silken purse case. Group 3 is the saddle-case makers (e.g., Glossosomatidae), which construct coarse, dome-shaped cases of irregular, small pebbles. Group 4 (the Hydropsychoidea) comprises the net spinners and retreat makers; these hide in a fixed retreat that may be used for seining food or that may have special filtering nets. Group 5 are the most familiar, the tube-case makers. Case structure is often typical of a genus or group of genera, though there also appears to be considerable convergence (Stuart and Currie, 2001, 2002). Structure depends on the shape, of course, but also on the type of building materials, which can be sand grains, tiny pebbles, small shells, or bits of vegetation. Some variation in case structure is imposed by the availability of materials, but larvae are usually deliberate in their use of materials. They actively select and even customize them by trimming fragments of leaves and stems to an appropriate size before adding to the case. "Building blocks" are fitted and arranged; then, they are sutured with silk to the growing edge of the case. The most fascinating demonstration of this was made, not by an entomologist, but by a French artist who provided captive caddisfly larvae with glass pebbles of alternating colors. The whorls of colors vividly revealed the manner of larval masonry, which is far more precise than the individual efforts of a worker ant or termite in constructing a nest. A large comparative study by Stuart (2000) used videotapes of 35 species of larvae (in 10 families) constructing their cases. She classified the cases and case-making behavior into 9 and 5 functional groupings, respectively, so several different case types are produced by the same behaviors (Stuart and Currie, 2002). Larval caddisflies build a loose, temporary case. By the time this case covers the abdomen, they begin constructing the permanent case and continue building this as the larva grows and molts. Before pupating, a mature larva will fasten its case to the substrate and seal the ends loosely with silk. The pupa then cuts its way out of its shelter, swims to the water surface, and there quickly molts to a winged adult.

With the exception of having a few distinctive structures, adult Trichoptera are rather generalized insects with long, filiform antennae and little differentiation of thoracic segments other than the smaller prothorax typical of panorpoids (Figure 13.4). The most unusual feature concerns the mouthparts, which are modified into an *haustellum* and comprised of a fused hypopharynx and prelabium; the mandibles, maxillae, and galeae are reduced. The haustellum has rows of microtrichia that form fine channels, not unlike what is seen

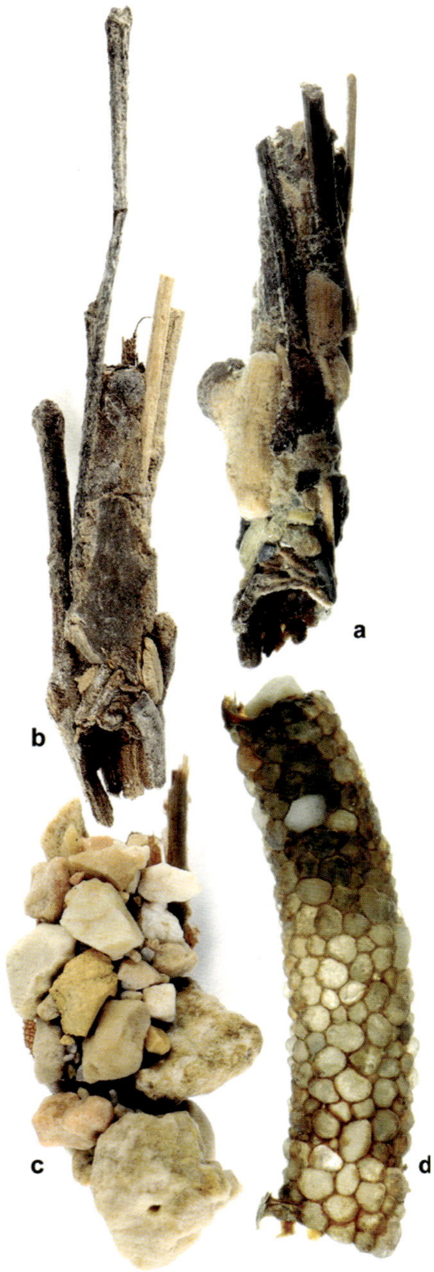

13.2. Cases of representative larval caddisflies. Larvae select and carefully apply sand grains and pebbles to the case, and even trim plant fragments. a, b, d: Limnephilidae. c: Glossosomatidae. Length of (a) 12 mm; to the same scale.

on the labella of nematocerous flies. Like flies, adult trichopterans feed on liquids, though only a few have been observed doing so. Some colorful species of Plectrotarsidae in Australia, though, have a long proboscis that they use to feed from flowers, and they even fly during the day – a striking parallel to butterflies. A relatively complete wing venation with some distinctive veins has provided a basis for interpreting some of the early fossils of the group. Fortunately, too, there is a substantial phylogenetic framework for the order.

The order is traditionally classified into two suborders, Annulipalpia and Integripalpia, which is a classification that began with Martynov (1924) and was later refined by Ross (1956), Weaver (1984), Weaver and Morse (1986), Frania and Wiggins (1997), Morse (1997), and Kjer *et al.* (2001a,b) Ivanov and Sukatsheva (2002). A third suborder, the Spicipalpia, has been proposed, but it is now known to be a paraphyletic grade of the Integripalpia. Figure 13.5 presents a phylogeny of most families, taken from Frania and Wiggins (1997) and Kjer *et al.* (2001a,b). The study by Kjer *et al.* is based on sequences of four genes and the 70 adult and larval morphological characters that were analyzed by Frania and Wiggins (1997). Relationships proposed in both studies are very similar. Integripalpia contains 25 families of primarily tube makers, and larvae are generally eruciform gatherers, scrapers, and shredders of submerged vegetation. Annulipalpia contains eight families of primarily net spinners; with generally free-living, campodeiform larvae or other larvae that filter their food using silken nets. Of the annulipalpian larvae that construct tubes, they are mostly silken, often with grains of sand woven within. There is actually a general trend in the integripalpian lineage, from free-living larvae in the most basal families (Rhyacophilidae, Hydrobiosidae) to construction in the "intermediate" families of irregular tubular cases made usually of plant fragments (sometimes coarse tiny pebbles), to tubular cases made of tightly packed ("sorted") sand grains or pebbles in the most recently evolved lineage. The most beautiful cases are of the recently evolved family Helicopsychidae, which are in the form of snail shells.

The earliest definitive trichopteran is *Liadotaulius maior* (Figure 13.6), from the latest Lias (Early Jurassic) of Germany (180–185 MYA) (Ansorge, 2002). It possessed the distinctive trichopteran feature of a sharp bend at the apex of vein CuP that was also slightly desclerotized. What were probably the males of this species had a thickened area on vein R_3, or a *corema*, thought to be used for the production and dispersing of sexual pheromones (Ansorge, 2002). Both suborders are represented by five families that were present by the mid-Jurassic. One of those families are fossils thought to be of the Recent, basal family Rhyacophilidae (Ivanov and Sukatsheva, 2002); the Philopotamidae are another Recent family of putatively Jurassic age. Suggestions are, then, that the divergence of basal Trichoptera occurred earlier in the Jurassic to the Late Triassic, no doubt from necrotauliid ancestors. No Recent integripalpian family is yet known before the Early Cretaceous, even though the origin of essentially a modern trichopteran fauna probably occurred in the Early Cretaceous to latest Jurassic (e.g., Figure 13.7). The Baltic amber fauna (Eocene) is the most diverse from the Tertiary and is essentially modern in the composition of families (Ulmer, 1912). What distinguishes the fossil record of Trichoptera from other insects is that, besides bodies, their distinctive and durable cases have also been preserved (Figures 13.8 to 13.10).

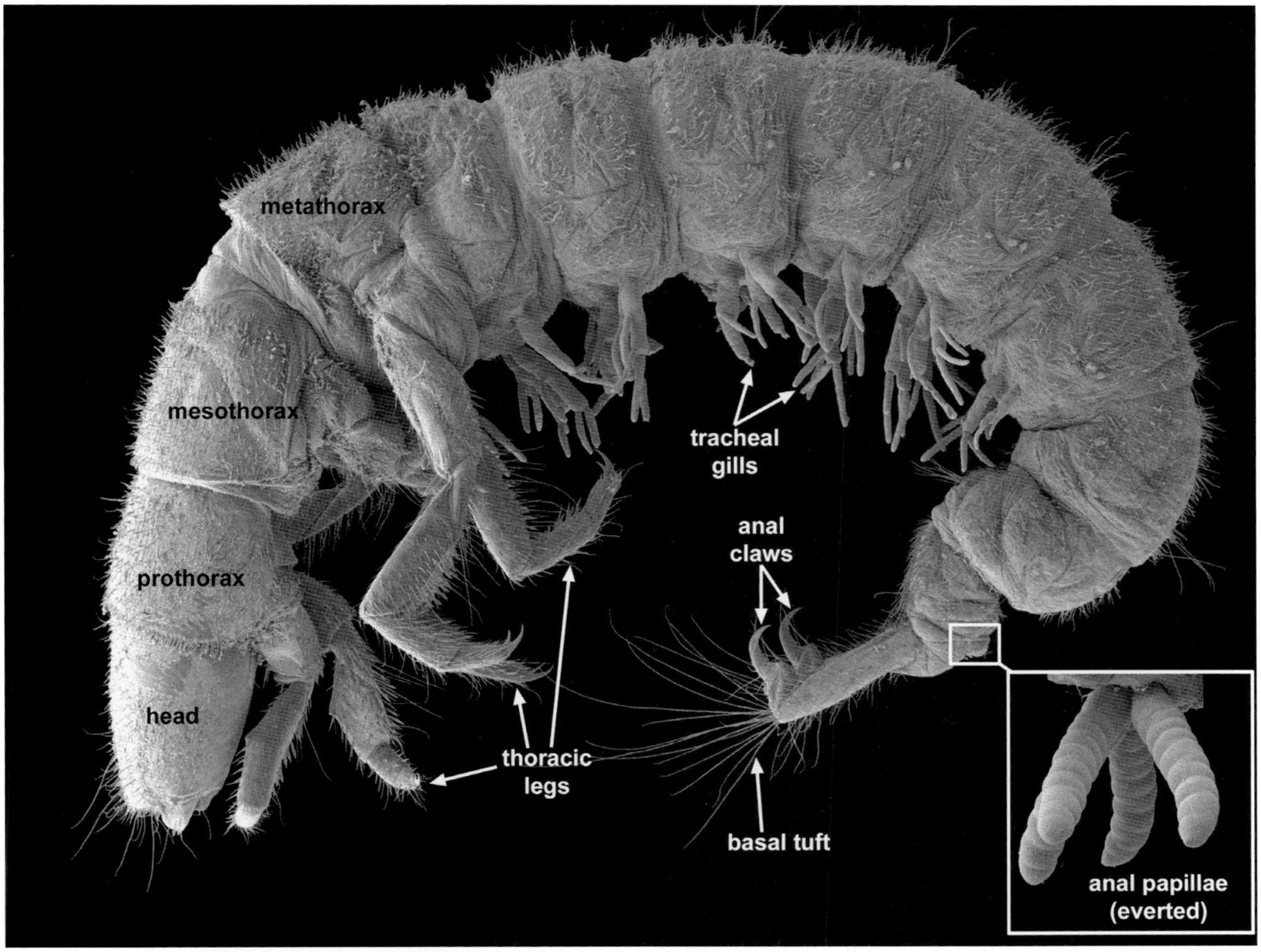

13.3. Larva of a typical caddisfly (family Hydropsychidae). The anal papillae can be retracted and everted. Anal claws help anchor the larva within its case. Length 3.2 mm.

The oldest trichopteran cases are from the Early and mid-Jurassic of Siberia and Mongolia, respectively, the former of which (*Scyphindusia*) were tubular and constructed of silk and sand grains like those of Recent Annulipalpia and basal integripalpians (Sukatsheva, 1985, 1994). The Siberian cases have been attributed to the Recent family Hydroptilidae (Ivanov and Sukatsheva, 2002), but they may belong to the extinct family Vitimotauliidae, known otherwise as compressed wings and bodies from the mid-Jurassic to the Late Cretaceous. The mid-Jurassic Mongolian cases, loosely constructed of unsorted sand grains and plant fragments, were possibly formed by another extinct family, the Baissoferidae because venation indicates that this family is an integripalpian. Fossil trichopteran cases are given taxonomic names (usually with an *-indusia* ending), but they are usually not incorporated into trichopteran classifications because they are trace fossils, or ichnofossils. *Folindusia*, for example, are fossil cases made of small bits of foliage. Indeed, the widespread convergence in Recent Trichoptera of the standard tube cases made of sand grains or bits of foliage does not allow a reliable, natural classification. By the Cretaceous, caddisfly cases were abundant and diverse. In fact, Asian Cretaceous caddisfly cases comprised 9 "form" genera and 200 "ichnospecies" (Sukatsheva, 1999). The most distinctive fossil cases were *Piscindusia*, from the Wealden (Early Cretaceous of England), composed of minute fish scales and bones (Jarzembowski, 1995b). In general, the fossil record of caddisfly cases reflects that of the adults, since the greatest Mesozoic diversity of the cases appears by the Cretaceous along with wings representing 12 Recent families. Of course, fossil cases do not record the history of all fossil Trichoptera larvae because it does not include larvae that live free or within retreats. Even delicate silken cases are occasionally preserved so, presumably, free-living larvae should be too, but there are actually very few Mesozoic larvae known. A Late Mesozoic evolution of Trichoptera also appears to be reflected in the biogeography of some Recent families.

Families of Annulipalpia and basal integripalpians ("spicipalpians") are generally widespread, but the distributions of other integripalpian families are much more provincial. Some

13.4. Adult caddisflies of representative families. Compiled from plates in Ulmer (1907).

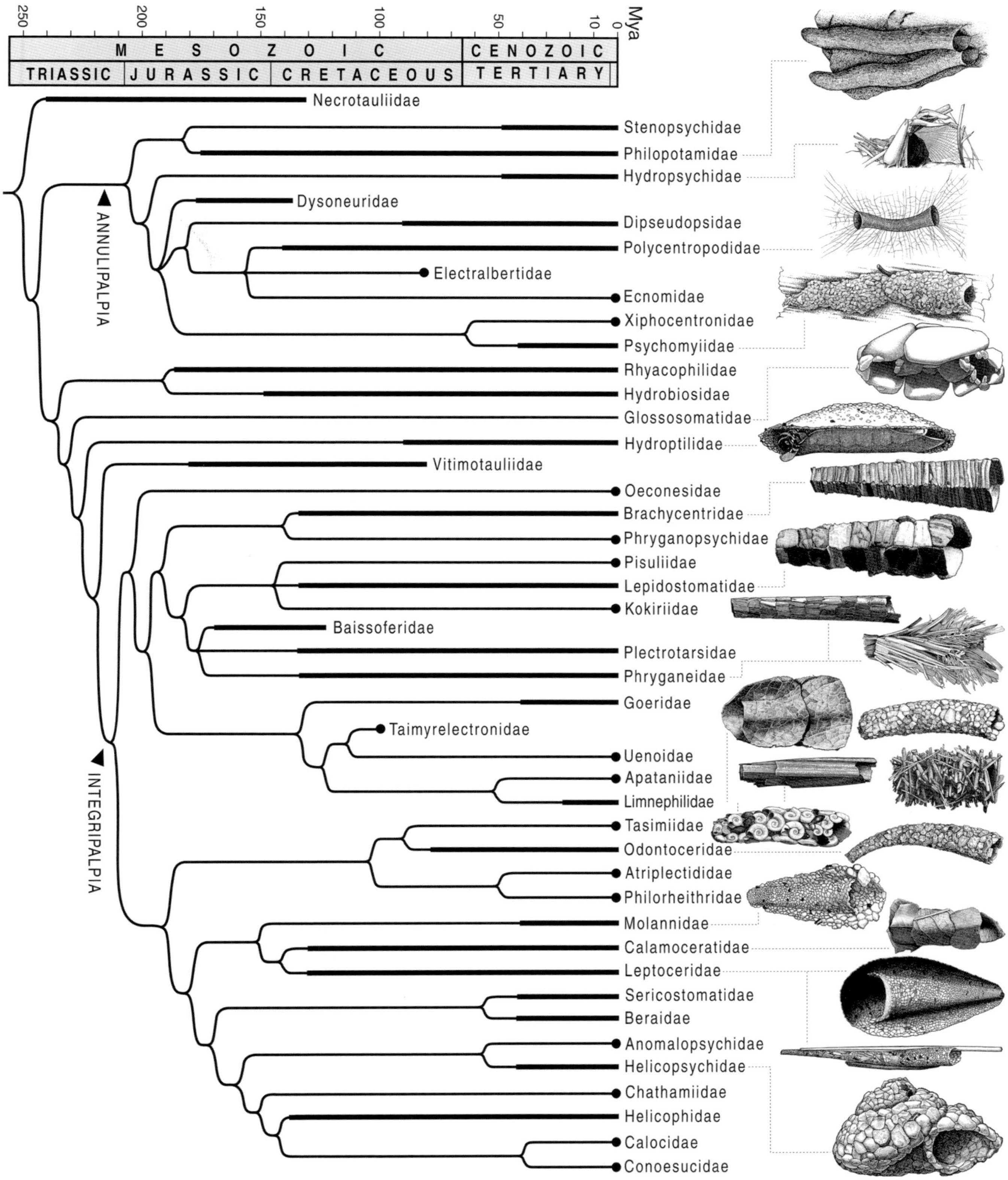

13.5. Phylogeny of caddisflies, order Trichoptera. Relationships among living families based on Kjer *et al.* (2001a,b) and Frania and Wiggins (1997); relationships of fossils based on Ivanov and Sukatsheva (2002). Six of the smaller, more obscure Recent families are not included. Figures from Wiggins (1977).

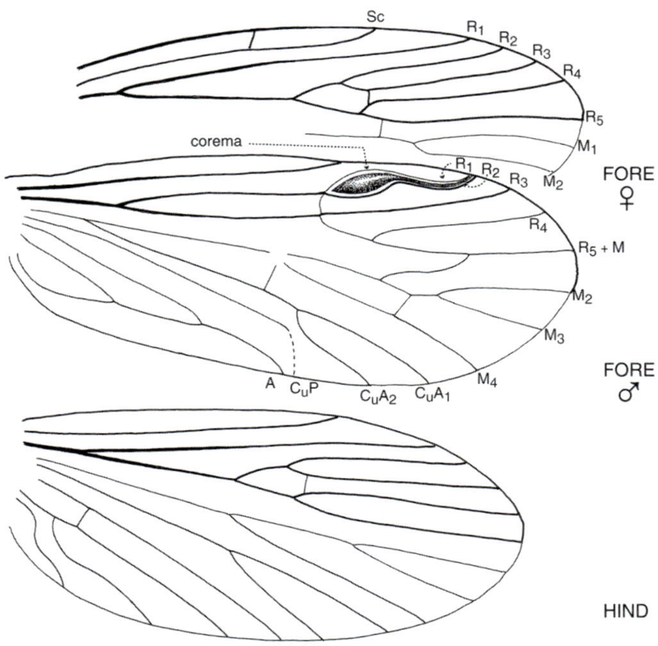

13.6. Remains of the oldest definitive caddisfly, *Liadotaulius maior*, based on sexually dimorphic wings from the Early Jurassic of Germany (ca. 180 MYO). Redrawn from Ansorge (2002).

13.7. Fossil caddisfly, in 90 MYO amber from New Jersey. AMNH NJ855; body length (without antennae) 3.4 mm.

integripalpian families are entirely Laurasian, whereas others are entirely Gondwanan, leading some to suggest that the early history of Trichoptera was affected by the fragmenting of the supercontinent of Pangaea. Such an interpretation has been made for another freshwater aquatic group, the Plecoptera, but the distribution of that order into Northern and Southern Hemisphere taxa is quite striking. According to the present phylogeny (Figure 13.5), there are several groups of closely related integripalpian families that show this northern-southern separation, such as the Nearctic Anomalopsychidae and four, mostly Australian, families. Another such group

13.8. Fossil caddisfly case, *Terrindusia* sp., from the Cretaceous of Eurasia. PIN 3709/106; length 10.2 mm.

13.9. *Molindusia variabilis*, from the Cretaceous of Mongolia. PIN 3787/76; length 14 mm.

13.10. Fossil caddisfly case, *Folindusia kemaensis*, from the Oligocene of Kena, Far Eastern Russia. It is composed of trimmed plant fragments. PIN 3136/166; length 40 mm.

consists of the northern families Brachycentridae, Phryganeidae, and Phryganopsychidae, the austral Kokiriidae, and the Australian Plectrotarsidae. One problem with interpreting a Pangean age to these distributions is that within each of these groups there are also some cosmopolitan families, which indicates either that some families dispersed widely, or that the provincial families were restricted by extinction (i.e., they are relict). Ages of Recent families of Trichoptera that are Cretaceous and even Jurassic makes it quite plausible that their past distributions, and to some extent their present ones, were affected by the rifting of Pangaea. In this regard, the Recent lineages of Trichoptera are considerably older than their closest living relatives, the Lepidoptera.

LEPIDOPTERA: THE MOTHS AND BUTTERFLIES

By virtue of butterflies alone, the Lepidoptera are the most familiar insects, and certainly the most enjoyed ones. Despite centuries of avid collecting, though, even butterflies – superfamily Papilionoidea – remain misunderstood. Their relationships are poorly explored, novel life histories are still being discovered, and there is actually a more impressive diversity of butterflies than commonly believed, with 14,500 described species (not including their closest relatives, the skippers). This is only about 6% of the total diversity of all Lepidoptera, the remaining being the non-papilionoids, or "moths." Because most moths are generally smaller, night-flying insects, it is only natural that gaudy, day-flying butterflies have attracted so much interest and have been an obsession of Victorian naturalists, nobility (Rothschild, 1983), poets (Boyd and Pyle, 2000), and even clowns (Bowden, 1985). No other order of insects has been so comprehensively and beautifully monographed. Among the more significant series for the taxonomy and identification of lepidopterans are the following: For the more basal, "microlepidopterans" there is *Microlepidoptera Palaearctica*, and the eight-volume series by J. F. G. Clarke (1941–1969) redescribing the microlepidopteran types of the early British lepidopterist Edward Meyrick. For macrolepidopterans, the single most important reference is the 14-volume series by Adalbert Seitz, published between 1906 and 1930, *Macrolepidoptera of the World*. Other important series are the "MONA" series (*Moths of North America*), *Die Schmetterlinge Mitteleuropas*, the *Fauna of British India*, *Lepidoptera Indica*, *Monographs on Australian Lepidoptera*, and *The Moths and Butterflies of Great Britain and Ireland*. The recent Lepidoptera section of the *Handbuch der Zoologie* (Kristensen, 1999c), is the best recent summary of systematics, as is the Lepidoptera chapter in *Insects of Australia*. Very useful books on biology include those by Common (1990) and Scoble (1995).

The Lepidoptera comprise the largest lineage of plant-feeding organisms. The other very large group of phytophagous organisms comprises the phytophagan beetles, which includes the weevils, leaf beetles, and long-horned beetles. Even more significant, biologically, is that the preponderance of lepidopteran plant hosts are angiosperms. The fossil record of these fragile insects is sparse and best represented as inclusions in amber and as larval mines in fossil leaves, but which still indicates that this order largely diversified in the Cretaceous and early Tertiary with the flowering plants. Thus, the Lepidoptera, among all orders of insects, appears to have radiated most recently. Fortunately, interpreting lepidopteran evolution is also made possible by a thorough understanding of relationships of the basal lineages.

13.11. A bird wing swallowtail butterfly, *Troides* (Papilionidae), from Malaysia. Photo: P. J. DeVries.

Another distinction of the Lepidoptera is that their basal relationships are probably the best understood of any major order of insects (Figure 13.13). Relationships among basal Diptera (nematocerans), by contrast, are still extremely controversial, and those of basal Hymenoptera and Coleoptera are only now being thoroughly explored. Such thorough understanding for Lepidoptera is based on phylogenies constructed from careful and comprehensive morphological work (Davis, 1975, 1978, 1986; Kristensen, 1978b, 1984, 1991; Kristensen and Nielsen, 1979; Nielsen, 1985; Nielsen and Kristensen, 1989, 1996). Rarely do cladograms based on morphology and DNA sequences largely agree, but DNA-based cladograms of basal Lepidoptera (Friedlander *et al.*, 1996, 2000; Wiegmann *et al.*, 2000a, 2002) have been unable to revise the earlier schemes of relationships. Stability in the hypotheses of relationships allows better insight into evolution, but interpreting fossils is still challenging because many of the phylogenetically useful structures in Lepidoptera are obscure and minute, and venation is of limited use. Such structures in adults include the mouthparts, particularly the proboscis, including its musculature and sensilla; the microscopic structure of wing and body scales; and, of course, genitalia. The structure of larvae is also extremely important, but these fossilize very rarely. Wings and other very fragmentary remains of the earliest lepidopterans from the Jurassic are, thus, very difficult to interpret. Mesozoic fossils of Lepidoptera have been reviewed and discussed by Whalley (1986a), Kristensen and Skalski (1999), and Grimaldi (1999).

MESOZOIC FOSSILS

The earliest fossils of Lepidoptera are from the Jurassic of three localities in Europe and central Asia. The oldest of these, *Archaeolepis mane*, is comprised of three tiny wings on a slab of rock from the Early Jurassic (Sinemurian, ca. 190 MYO) of Dorset, England (Whalley, 1985). When it was originally described, scanning electron micrographs revealed the wings to have scales with the characteristic array of fine parallel grooves. No other details of the scales, though, were apparent, and the poorly preserved venation could have been interpreted as also psocopteran, particularly since some psocopterans (like Lepidopsocidae) also have scaled wings and bodies. It was on this basis that *Archaeolepis* was placed as a "possible" lepidopteran (Grimaldi, 1999). Recent study of *Archaeolepis*, though, indicates the venation is actually better preserved than originally reported (D. Grimaldi and A. J. Ross, unpubl.). Though the basal branching of the R and M veins is not discernable, one wing (probably a forewing) had the characteristic short anal veins looped together (Figure 13.14). This configuration of anal veins also occurs in Trichoptera, but this character together with the wing scales indicates that *Archaeolepis* was indeed a lepidopteran.

13.12. Representative basal families and lineages and Lepidoptera. 1, *Micropterix*; 2, *Epimartyia*; 3, genus indet.; 4, *Dyseriocrania*; 5, *Aenetus*; 6, *Leto*; 7, *Heliozela*; 8, *Adela*; 9, *Incurvaria*; 10, *Daviscardia*; 11, *Amydria*; 12, *Gracillaria*; 13, genus indet.; 14, *Atteva*; 15, *Yponomeuta*; 16, *Arega*; 17, genus indet.; 18, *Depressaria*. Not to the same scale.

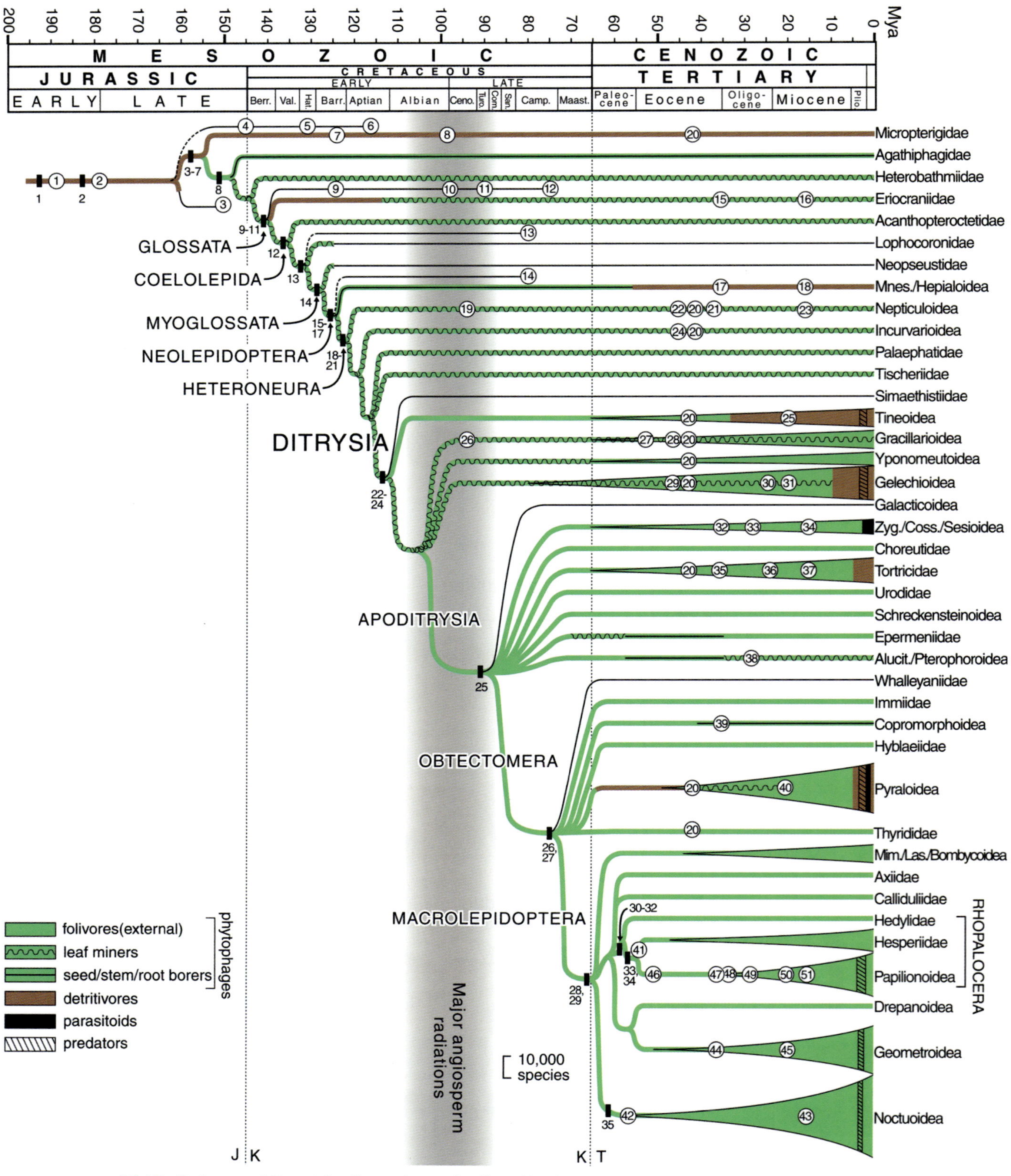

13.13. Phylogeny of Recent families and superfamilies of Lepidoptera, showing principal larval diets. Circled numbers are fossils (see Table 13.1); other numbers are significant morphological characters (see Table 13.2). Lepidoptera are the largest lineage of plant-feeding organisms, and their evolution is intimately related to the radiation of angiosperms in the Cretaceous. Some lineages have reverted to being detritivores, and others even have predatory and parasitoid caterpillars. Relationships based on Kristensen (1999c).

TABLE 13.1. Significant Lepidopteran Fossils[a]

1. *Archaeolepis mane*: Dorset, England (Whalley, 1985)
2. Eight "genera": Grimmen, Germany (Ansorge, 2002)
3. Karatau, Kazakhstan: *Auliepterix*, *Eolepidopterix*, *Karataunia*, *Palaeolepidopterix*, *Protolepis* (Kozlov, 1989; Rasnitsyn, 1983d) (not a monophyletic group)
4. *Auliepterix minima*: Khotont Somon, Mongolia; *Undopterix sukatshevae*: Unda, Siberia (Skalski, 1979)
5. *Daiopterix olgae*: Shelopugion, Transbaikal, Russia; *D. rasnitsyni*, Transbaikal (Skalski, 1984)
6. *Gracilepterix pulchra*, *Parasabatinca caldasae*, *Undopterix carirensis*: Santana Formation, Brazil (Martins-Neto and Vulcano, 1989) (4-6 do not comprise a monophyletic group)
7. *Parasabatinca aftimacrai*: Lebanese amber (Whalley, 1978)
8. *Micropteryx pervetus*: Burmese amber (Cockerell, 1919)
9. Glossatan larva: Lebanese amber (Grimaldi, 1996, 1999)
10. Glossatan adults: Burmese amber (Grimaldi *et al.*, 2002)
11. Glossatan adults: New Jersey amber (Grimaldi *et al.*, 2000a)
12. Glossatan larval head capsule: Canadian amber (Mackay, 1970)
13. Lophiocoronid-like moth: Siberian amber (Kristensen and Skalski, 1999)
14. Mnesarchaeid-like moth: Siberian amber (Skalski, 1976)
15. Eriocraniid leaf mines: Isle of Wight, Bembridge Marls, UK (Jarzembowski, 1980)
16. Eriocraniid leaf mines: Idaho (Opler, 1973)
17. Hepialidae: *Prohepialis* sp.: Isle of Wight, Bembridge Marls, UK (Jarzembowski, 1980; Robinson, 1977)
18. Hepialoidea: China (Zhang, 1989)
19. Nepticulidae leaf mines: Dakota Formation: western U.S. (Labandeira *et al.*, 1994)
20. Baltic amber: various Recent families (Skalski, 1990 [review])
21. Nepticulidae leaf mines: Victoria, Australia (Rozefelds, 1988)
22. Nepticulidae leaf mines: Republic, Washington, U.S. (Labandeira, 2002)
23. Nepticulidae leaf mines: western U.S. (Opler, 1973)
24. Incurvariid, Heliozelidae leaf mines: Republic, Washington, U.S. (Labandeira, 2002)
25. Tineidae, Acrolophidae (undescribed): Dominican Republic amber (unpubl.)
26. Gracillariidae leaf mines: Dakota Formation: western U.S. (Labandeira *et al.*, 1994)
27. Gracillariidae leaf mines: Wyoming (Hickey and Hodges, 1975)
28. Gracillariidae leaf mines: Republic, Washington U.S. (Labandeira, 2002)
29. Coelophoridae leaf mines: Republic, Washington U.S. (Labandeira, 2002)
30. Cosmopterigidae: Mexican amber (Kristensen and Skalski, 1999)
31. Oecophoridae: Dominican amber (Kristensen and Skalski, 1999)
32. *Adelopsyche* (Cossidae) and Castniidae (Sesioidea): Florissant, Colorado USA (Cockerell, 1926; Tindale 1985)
33. Zygaenidae: France (Leestmans, 1983), Germany (Naumann, 1987)
34. ?*Zygaena turolensis*: Rubielos de Mora, Teruel, Spain (Fernandez-Rubio *et al.*, 1991)
35. Tortricoidea: Florissant, Colorado (Cockerell, 1909a)
36. Tortricidae, undescribed: Mexican amber (unpubl.)
37. Tortricidae (various): Dominican amber
38. Pterophoridae: Oligocene, France (Bigot *et al.*, 1986)
39. Copromorphidae: Isle of Wight, Bembridge Marls, UK (Jarzembowski, 1980)
40. Pyraloidea, undescribed: Dominican Republic amber (unpubl.)
41. Unnamed hesperioid: Fur Formation, Denmark (Kristensen and Skalski, 1999)
42. Unnamed noctuoid: Fur Formation, Denmark (Rust, 2000)
43. *Stauropolia nekrutenkoi* (?Arctiidae): Caucasus (Skalski, 1988b)
44. *Hydriomena*: Florissant, Colorado, U.S. (Cockerell, 1922)
45. Geometridae (Ennominae), undescribed: Dominican Republic amber (unpubl.)
46. Three species of butterflies, Green River Formation, Colorado: *Praepapilio*, *Riodinella* (Durden and Rose, 1978)
47. Various butterflies, Florissant Formation, Colorado (see Table 13.3)
48. *Nymphalites obscurum*, *Lithopsyche antiqua*: Bembridge Marls, England
49. Various butterflies, Aix-en-Provence, France (see Table 13.3)
50. Riodinidae (adults, larva), Nymphalidae: Dominican amber
51. Various butterflies: Croatia, North Caucusus, Germany (see Table 13.3)

[a] Numbers correspond to those circled on phylogeny, Figure 13.13.

TABLE 13.2. Significant Characters in Lepidopteran Phylogeny[a]

LEPIDOPTERA:
1. Wings with dense covering of scales on veins and membrane
2. Wings with M veins having three branches (versus four or more)
3. Foretibia with one apical spur, or none
4. Median ocellus lost
5. Tergum one extensively desclerotized
6. Cerci lost
7. Foretibia with articulated epiphysis on inner surface

AGATHIPHAGIDAE + HETEROBATHMIIDAE + GLOSSATA:
8. Adult mouthparts with loss of paraglossae; various larval characters

GLOSSATA:
9. Nonfunctional/vestigial adult mandibles, lacinia reduced
10. Galeae long, formed into a proboscis and usually coiled at rest
11. Larva with a spinneret on apex of prelabial-hypopharyngeal lobe

COELOLEPIDA:
12. Hollow scales on wing surface

LOPHOCORONIDAE + MYOGLOSSATA:
13. Vein R (versus Rs) always unforked

MYOGLOSSATA:
14. Proboscis with intrinsic muscles (origins and insertions are within the proboscis)

NEOLEPIDOPTERA:
15. Adecticous, obtect pupae
16. Pupae with dorsal rows of spines on tergites
17. Larva with prolegs on abdominal segments 3–6 and 10, bearing crochets

HETERONEURA:
18. Female genitalia with gonopore separate from ovipore, no connection between the two tracts
19. Reduction in branching of Rs veins in hind wing
20. Wings coupled with frenulum and retinaculum
21. Loss of abdominal sternum one

DITRYSIA:
22. Connection between copulatory opening and ovipore
23. Large apodemes on abdominal sternite two
24. Muscles in proboscis with short bands, not long fibers

APODITRYSIA:
25. Apodemes of abdominal sternum two specialized, with short, large bases

OBTECTOMERA:
26. Pupal abdominal segments 1–4 immobile
27. Dorsal row of spines lost on pupal tergites

MACROLEPIDOPTERA:
28. Complete loss of CuP
29. Larval proleg crochets in a "mesoseries"

HEDYLIDAE + HESPERIIDAE + PAPILIONOIDEA:
30. Abdominal tergite one pouched
31. Eggs resting upright
32. Pupa girdled

PAPILIONOIDEA:
33. Clubbed antennae
34. Day flying

NOCTUOIDEA:
35. Metathoracic tympanal organs

[a] Numbers correspond to those uncircled on phylogeny, Figure 13.13.

The next oldest Jurassic fossils are eight monotypic "genera" originally described by Anton Handlirsch in the early 1900s from the uppermost Lias (Toarcian, ca. 180 MYO) of Dobbertin, Germany (Ansorge, 2002, 2003a). The Handlirsch material is poorly preserved, but well-preserved wings from Grimmen, Germany have intact venation (Figure 13.14), scales, and even color patterns. One of the features used to identify these early lepidopterans is a three-branched M vein, since most trichopterans (and the basal moth *Agathiphaga*) have a four-branched M. Detailed description of these fossils is still needed.

The most diverse fossils of Jurassic Lepidoptera include five genera and species from the vast deposits of Karatau, Kazakhstan (Rasnitsyn, 1983d; Kozlov, 1988, 1989), and several slightly younger deposits (e.g., Figure 13.15). The identities of these and other Mesozoic fossils were discussed by Grimaldi (1999), and recent study of the fossils confirms that these are indeed Lepidoptera, but their preservation precludes any more detailed classification of them. Indeed, lepidopteran scales make the wing and body resistant to wetting, which may be why lepidopterans are rare as compression fossils and is certainly why venation is usually obscure in the fossils. The next oldest Lepidoptera are from the Early Cretaceous, many of which have been assigned to the most basal living family of Lepidoptera, the Micropterigidae (e.g., Figure 13.17), though some Early Cretaceous lepidopterans are slightly more derived than this (e.g., Figure 13.16).

BASAL GROUPS

The Micropterigidae is one of three families in a basal grade of mandibulate moths, which have not evolved a proboscis and which retain the ancestral use of functional mandibles. They are generally small, day-flying moths superficially similar to smaller trichopterans. These three families, collectively with fewer than 250 species, provide unique and interesting glimpses as to what the Mesozoic Lepidoptera were probably like. Micropterigidae itself is the sister group to all other

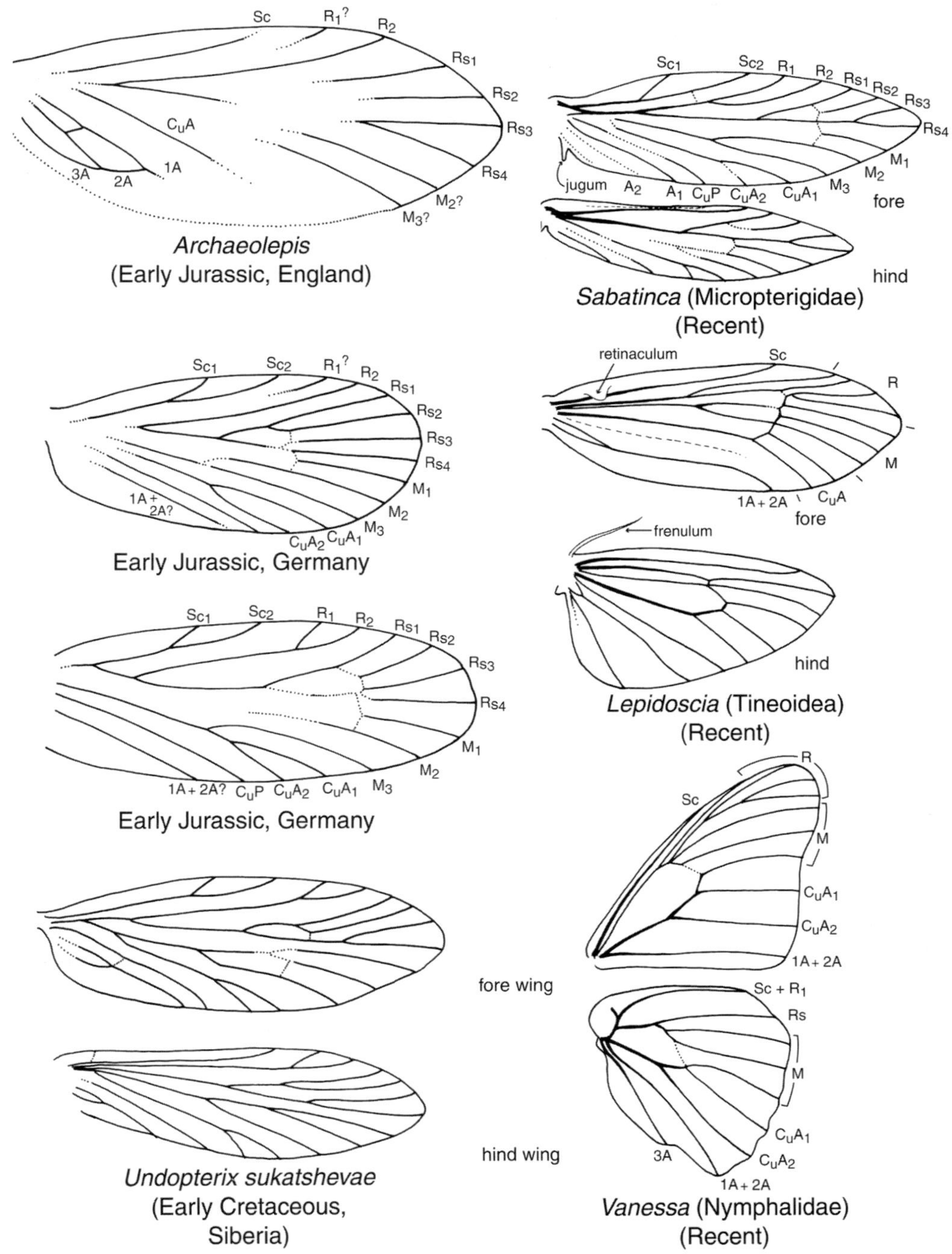

13.14. Wing venation of early fossil Lepidoptera (left column) and assorted Recent Lepidoptera (right column). The oldest Lepidoptera are wings with primitive venation from the Early Jurassic of England (*Archaeolepis*) and Germany (undescribed). *Archaeolepis* venation is revised based on new observations; the German Jurassic wings based on Ansorge (2002). Not to the same scale.

Lepidoptera (Kristensen and Nielsen, 1979, 1982, 1999b; Kozlov, 1988–90), and with 120 described species (plus 100 undescribed ones), it is also the most diverse of the mandibulate moths. Larvae have retained, like most trichopterans, what is probably the ancestral amphiesmenopteran diet, plant detritus. Larvae of micropterigids feed on fungal hyphae, liverworts, and other substrates amongst forest leaf litter, and some feed on the foliage of angiosperms. Adults feed on pollen and may even be involved in pollination of some basal angiosperms, like Winteraceae in New Caledonia. The other two mandibulate families, Agathiphagidae and Heterobathmiidae, have only a few species (two and ten, respectively), and much more specific diets. Agathiphagidae, as the name indicates, are "*Agathis* eaters," their legless larvae boring within and feeding on the seeds of relict conifers – the kauri pines – in northern Australia, Fiji, the Solomon Islands, New Hebrides, and New Caledonia.

Heterobathmiidae is the first lineage in a basal grade of

13.15. Lepidopteran, *Daiopterix olgae*, from the Glushkova Formation in central Siberia at the Jurassic-Cretaceous boundary, ca. 145 MYA. The preservation of basal lepidopterans in rocks rarely allows definitive identification to family. PIN 3063/741; length 2.3 mm.

13.16. A primitive moth from the Early Cretaceous of Brazil. It does not appear to belong to the two most basal lepidopteran families (Micropterigidae and Agathiphagidae) because it possesses a piercing oviscapt. AMNH SF46441; forewing span 6.8 mm.

nearly a dozen lineages to have leaf-mining larvae, in this case on the relict southern beeches, *Nothofagus*. These moths are distinct among more recently evolved leaf miners, though, in that their larvae (probably primitively) can actually leave and enter other leaves and are not restricted to living within one leaf. This family is also the basalmost lineage that feeds exclusively on angiosperms, and it is likely that the larval diet of Agathiphagidae represents an early evolutionary stage in Lepidoptera before angiosperms evolved. No fossils of Heterobathmiidae and Agathiphagidae are known, but two beautifully preserved micropterigids in Cretaceous amber confirm the Mesozoic existence of this family. Other Cretaceous fossils, preserved as compressions, have been classified as micropterigids (Kozlov, 1989; Martins-Neto and Vulcano, 1989; Labandeira *et al.*, 1994), but because crucial details are lacking, their identities remain obscure (e.g., Figure 13.16). A Mesozoic age of the mandibulate moths may also be reflected by their distributions. Heterobathmiidae occur in southern, temperate South America. With Agathiphagidae and some lineages of Micropterigidae (the *Sabatinca*-group of genera; *Micropterix* occurs in the Northern Hemisphere), Heterobathmiidae give the mandibulate moths one of the most obvious disjunct austral distributions in the Lepidoptera. These moths are plausibly relicts from the Cretaceous drift of Gondwana.

The large, monophyletic group comprising all other moths, the **Glossata**, are the "tongued" moths, or those in which the pair of (maxillary) galeae have become elongate, concave, and "zipped" together with rows of minute processes (*legulae*), forming a feeding tube (Krenn and Kristensen, 2000) (Figures 13.18, 13.19). Where the proboscis has become particularly distended, it is generally also coiled. Related to the development of a proboscis are adult mandibles that either are nonfunctional after emergence

13.17. A micropterigid moth in 100 MYO amber from Myanmar. AMNH Bu701; body length 2.4 mm.

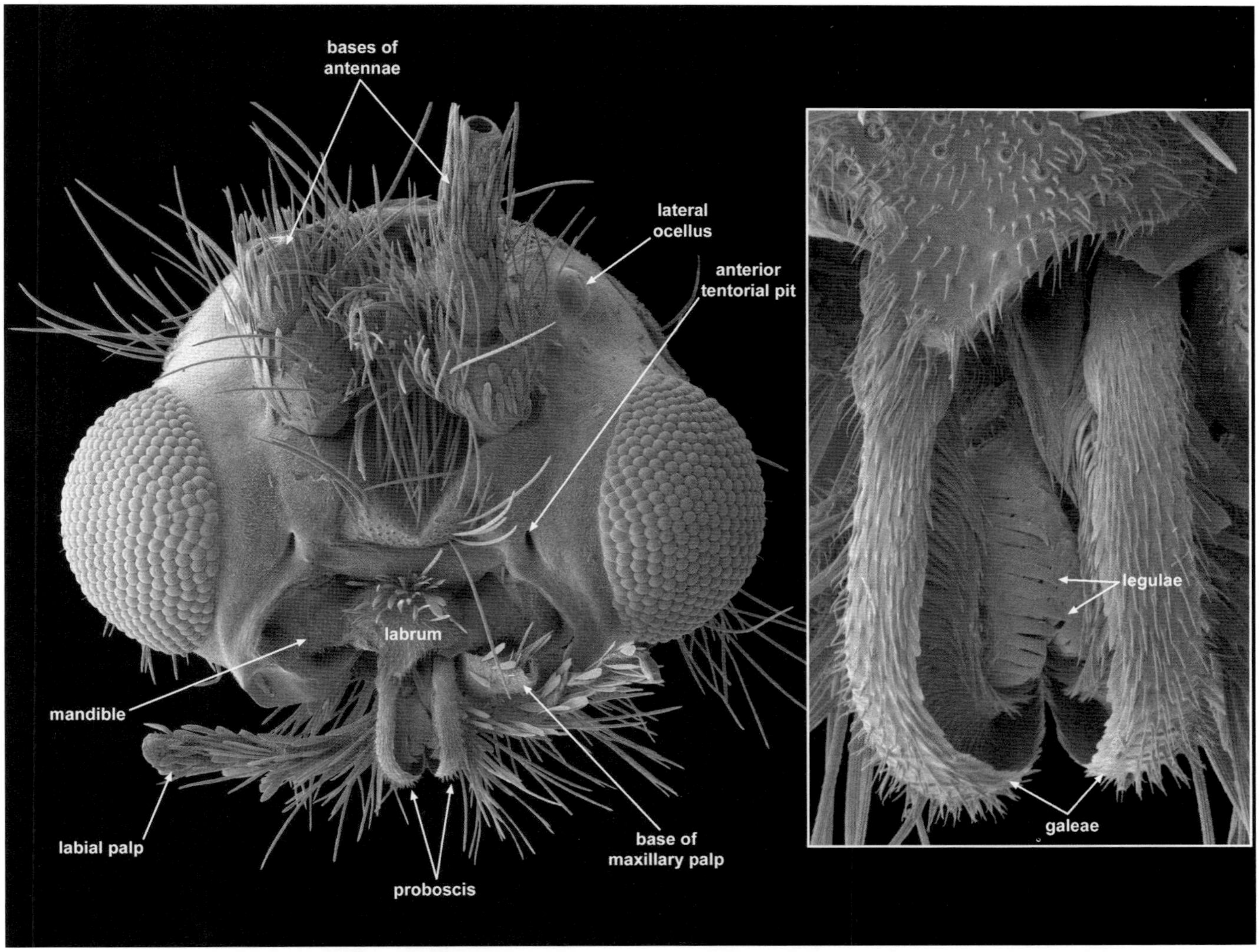

13.18. Head of a primitive glossatan (tongued) moth in the family Eriocraniidae, showing the short proboscis and its structure. Most of the scales on the head have been brushed away in order to reveal details. Scanning electron micrographs.

from the cocoon or have become vestigial and even lost in the youngest lineages along with reduction of the laciniae. Proboscis coiling is controlled by a fine filament of the elastic protein *resilin*, which lies along the dorsal surface of the proboscis (Hepburn, 1971). It was previously thought that extension and retraction of the proboscis was largely muscular, but it is now believed that extension is largely controlled by blood pressure. Hemolymph forced into the stipes at the bases of the galeae causes the galeae to unfurl quickly (Krenn, 1990), like the party toy a child blows into (usually into the ear of another child). The proboscis is further developed in most Lepidoptera (the Myoglossata) by the use of intrinsic muscles, which are muscles within the proboscis that have their origins and insertions within the proboscis itself, not just the insertions. This configuration probably gives the proboscis of myglossatans great flexibility, allowing it to wag laterally, which is clearly important for probing deep recesses of flowers.

The other hallmark feature of the Glossata is the larval spinneret, which is a minute spigot on the prelabio-hypopharyngeal lobe near the tip of the head (Figure 13.20). Secretions are exuded at the tip of this spinneret in the form of silk, which is used for constructing larval retreats and webs, for ballooning, and mostly for constructing pupal cases, or cocoons. The spinneret is how one remarkable fossil lepidopteran has been identified as the earliest definitive glossatan, which is a larva in Early Cretaceous amber from Lebanon (Figure 13.21; Grimaldi, 1996, 1999). The larva is prognathous; with its head pointing forward, the body is flattened, and it has only thoracic legs and a pair of prolegs on abdominal segment ten. Most of these features are typical of the basal lineages of leaf-mining lepidopterans, but it is plausible that this larva mined the cambium or other layers of tree bark. It is possible that a spinneret evolved before the adult proboscis, but in lieu of evidence for that it is safe to assume that the Lebanese amber larva is a true glossatan. Another

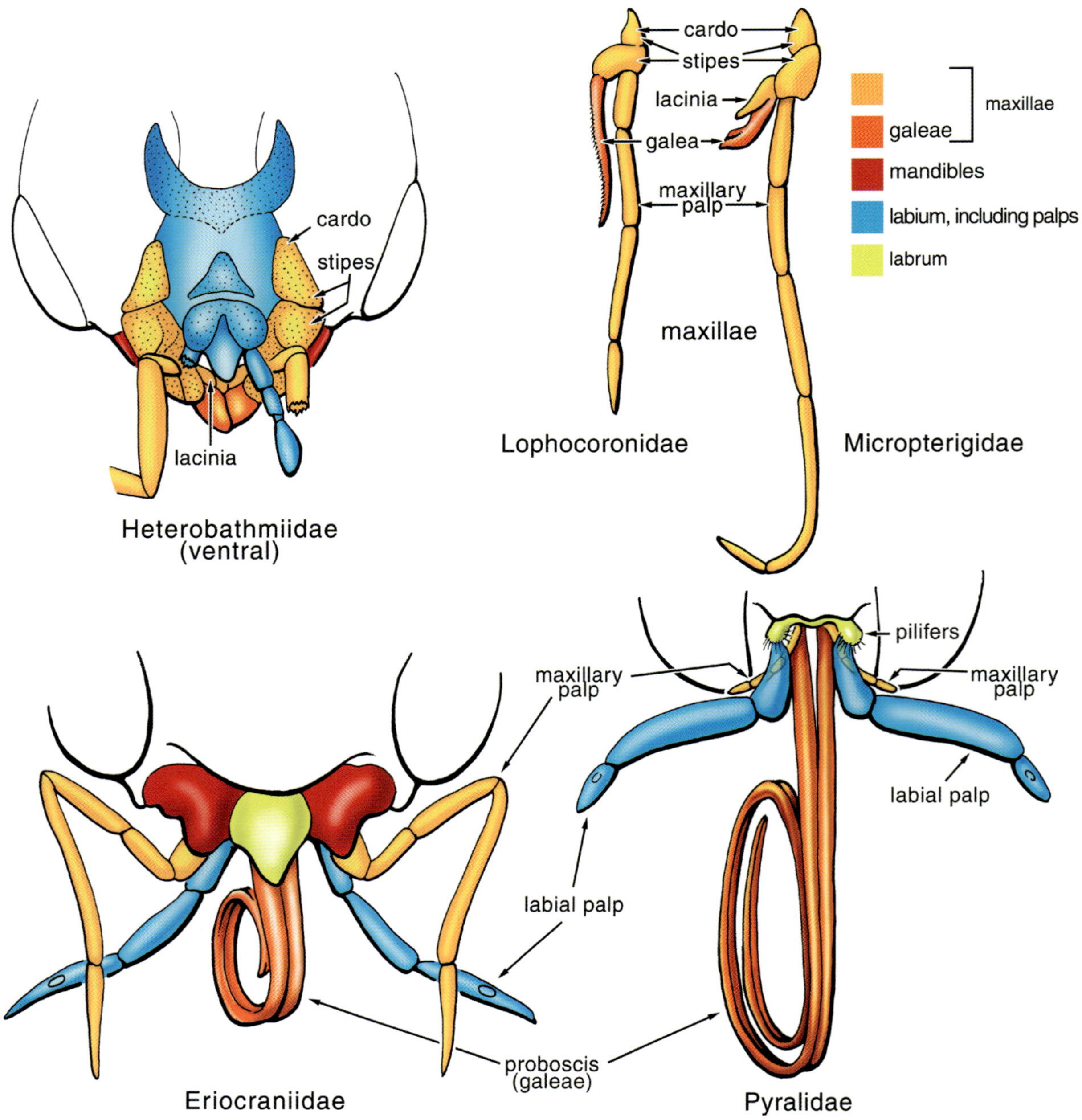

13.19. Homologous mouthpart structures in Lepidoptera. Micropterigidae and Heterobathmiidae do not have the long galeae that form the proboscis of glossatan moths; Lophocoronidae has galeae that form a short proboscis. The galeae are longer and coiled in Eriocraniidae, but much longer in more recently evolved families like Pyralidae. Redrawn from several sources.

larval glossatan is known in younger Cretaceous amber from western Canada (MacKay, 1970), but it is just a head capsule. Adult glossatan moths are also preserved in Cretaceous amber from Burma and New Jersey (Figure 13.22) and should be studied in detail so that their exact position within Lepidoptera can be determined. A moth from the Late Jurassic of Karatau, *Protolepis cuprealata*, has been identified as a glossatan (Kozlov, 1989; Kristensen and Skalski, 1999), but our restudy of it reveals that the preservation is insufficient to see the short proboscis that was originally reported. With little doubt, the evolution of the lepidopteran proboscis and an adult diet of nectar and pollen are related to the evolution of their larval host plants, the angiosperms.

The basal four families of Glossata are probably all leaf miners in angiosperms, but this can only be inferred for two of the families by the structure of the oviscapt, the egg-laying structure in Lepidoptera that is independently derived from a true, appendicular ovipositor. Though hosts of the Lophocoronidae (six Australian species) and Neopseustidae (from southeast Asia and southern South America) are unknown, these moths have a piercing oviscapt similar to other leaf miners, where the apex of the abdomen is laterally flattened,

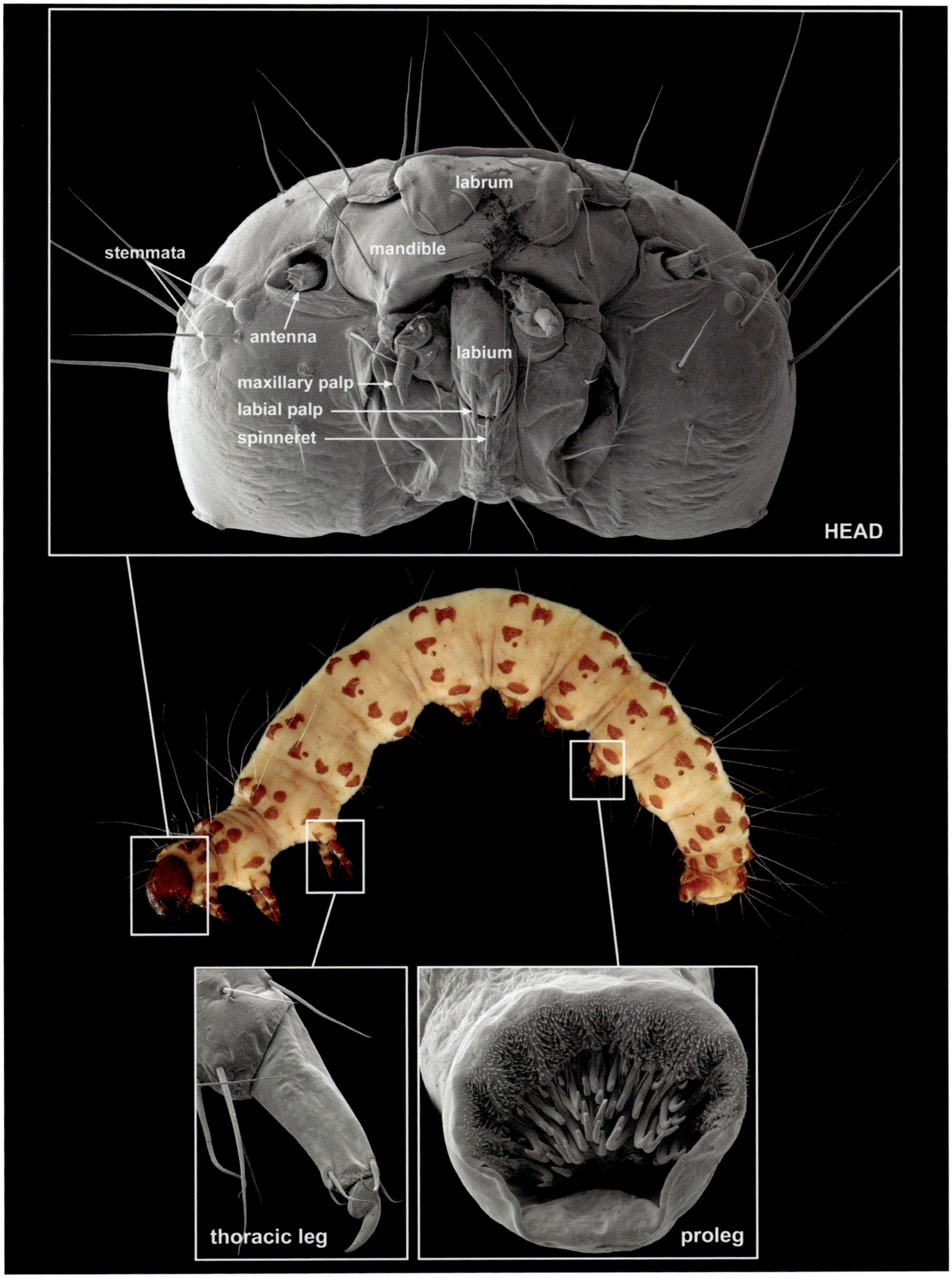

13.20. A typical glossatan caterpillar (family Pyralidae), showing details of the head capsule and the legs. Photomicrograph and scanning electron micrographs. Body length 6.3 mm.

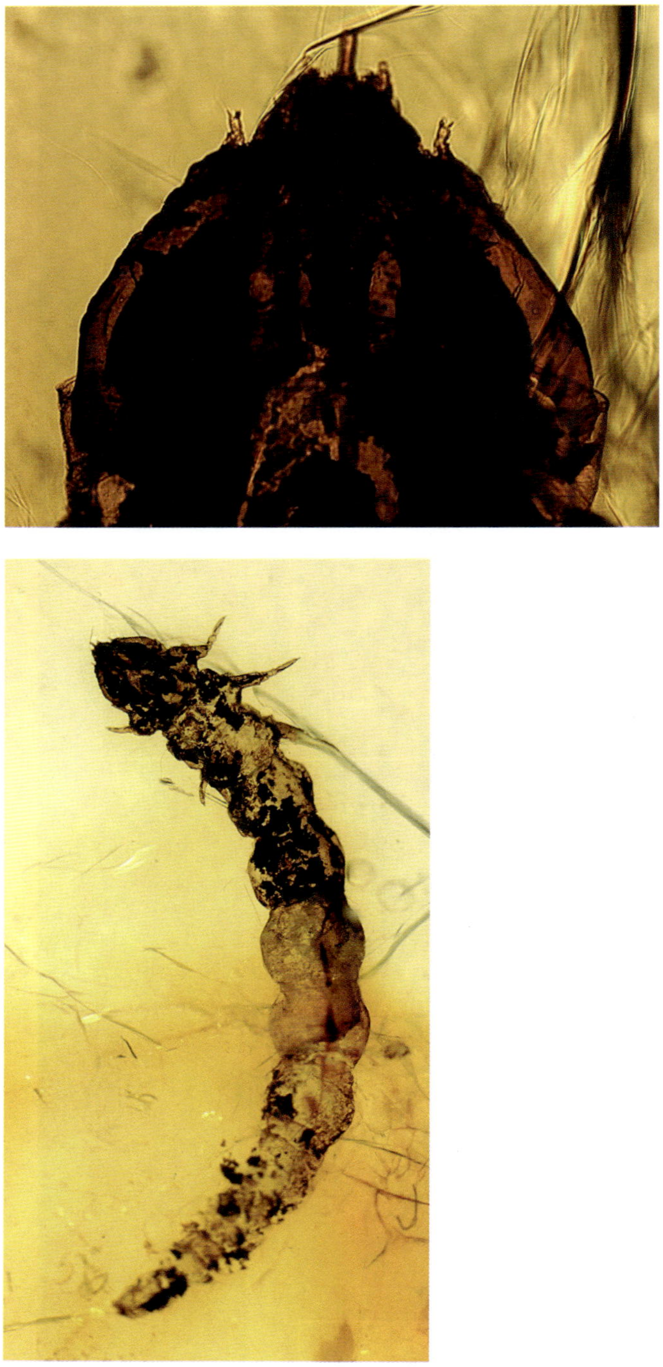

13.21. The earliest definitive glossatan moth, a caterpillar in Early Cretaceous amber from Lebanon (125 MYO). Above: a detail of the head showing the protruding spinneret at the tip. Silk is extruded from the spinneret, and this structure occurs in all Glossata. JG 19/70; body length 5.0 mm.

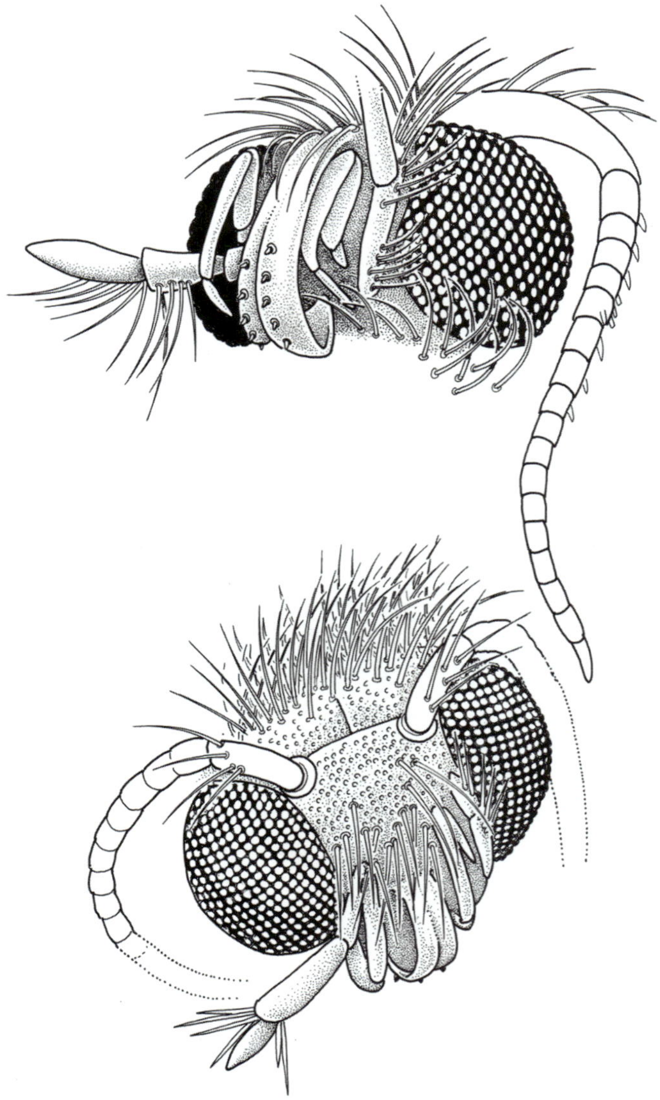

13.22. Head of an adult glossatan moth, in mid-Cretaceous amber from New Jersey, 90 MYO. Papillae on the short proboscis are clearly visible. AMNH NJ.

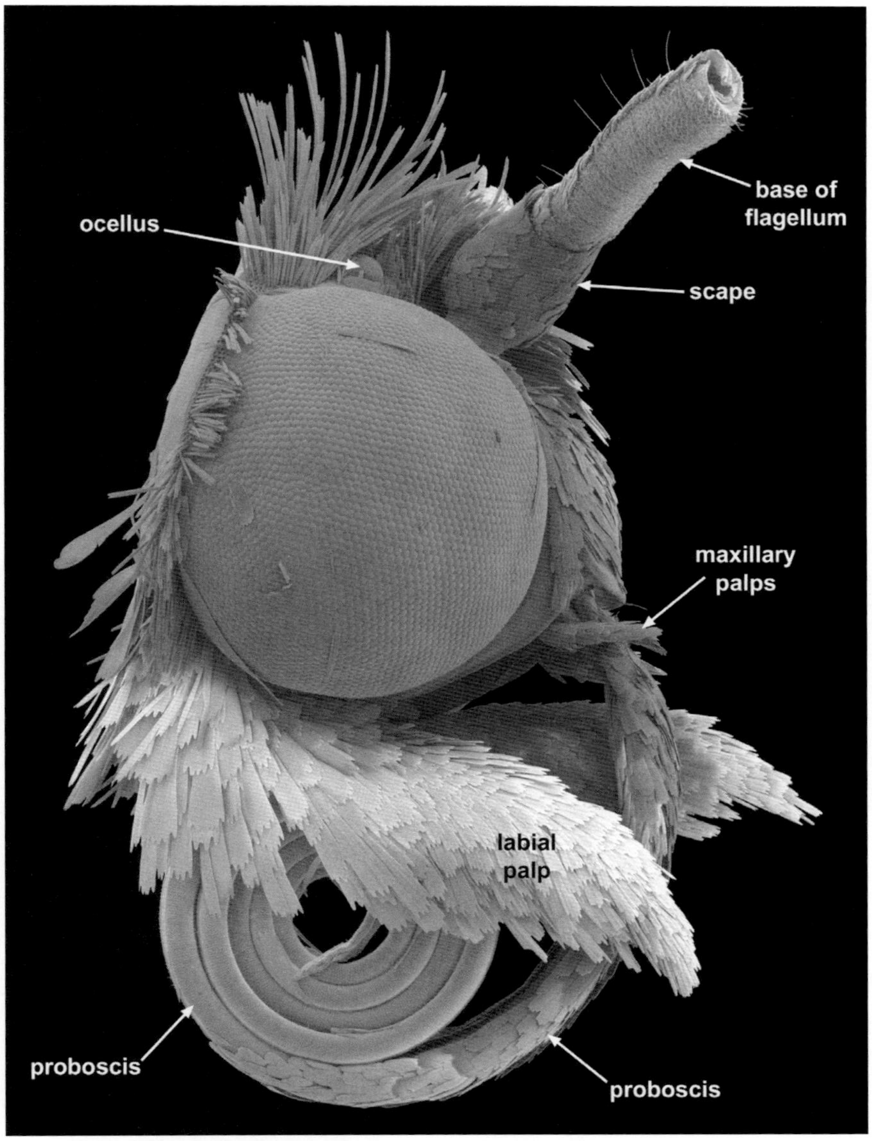

13.23. Head of a pyralid moth, showing some specialized lepidopteran head features, such as the long, coiled proboscis and reduced maxillary palps. Scanning electron micrograph.

sclerotized, and bears fine, sawlike serrations on the edges. The Eriocraniidae are miners of fagalean and rosalean trees and shrubs, with a biology very similar to that of the more basal Heterobathmiidae. It is within this grade of families that there developed further modification of the scales: They became hollow. Why scales evolved in the first place is unknown, but presumably it relates to a thermoregulatory or insulating function. Hollow scales may have evolved for this reason as well as for their aerodynamics and their role in color patterns and signaling.

Scales are flattened setae, with the fine ribs transformed into parallel ridges or vanes, and the socketed base is generally enlarged. The upper ("abwing") surface of each lepidopteran scale has the microstructure, including ridges as well as crossribs; the lower ("adwing") surface is smooth. In the most basal lepidopterans, the scales on the surface of the wing are solid and without perforations or windows (Figure 13.25), though they have apparently become convergently hollow in Agathiphagidae (Simonsen and Kristensen, 2001). In the more recently evolved lepidopterans, the **Coelolepida** ("lumen scaled"), which includes most Lepidoptera, the surface scales are hollow and have windows or perforations on the upper surface (Figures 13.26, 13.27). The scales on the margins of the wings of all Lepidoptera are hollow. The upper surface microstructure of scales provides some color, which is dependent on the type of microstructure (Ghiradella, 1984). Scattered light can impart a white color, as in pierid butterflies, or the rare Tyndall blue of certain Papilios. Diffracted light, which splits white light into its spectrum of colors, is uncommon even in butterflies. Most common is interference, where the wavelength of the reflected (visible) color depends on the distance between the repetitive grooves and other

13.24. A geometrid moth preserved in Miocene Dominican amber has its long, slender proboscis preserved in detail and still partially coiled. AMNH DR14–20; head width 1.0 mm.

microstructures of the scales. Lastly, granules of carotenoid, pteridine, and other pigments are suspended within the lumen of the scale from its upper surface and provide most of the nonreflective color (Figure 13.27). Though scales and colors are best studied in butterflies, many of even the most basal "microlepidopterans" have exquisite wing patterns with patches of golden and silvery reflective scales. In some cases, the scales are so well preserved in amber and even in compression fossils that the fine structure can be studied under high magnification (Figure 13.28) and even with a scanning electron microscope (Grimaldi *et al.*, 2000a).

Another major lineage within this basal grade, the **Neolepidoptera**, is defined by the structure of the pupae, which are adecticous (having immovable mandibles) and obtect (with appendages fused to the body, not free). Pupae of the more basal groups of neolepidopterans have spines on the tergites, which assist the wriggling pupa in freeing itself from the cocoon for adult eclosion. Other neolepidopteran features include wings with a double layer of scales (the ones in the lower layer being solid), and the larva has muscular prolegs on abdominal segments three to six and ten with an array of tiny hooks, the crochets (Figure 13.29). It is traditionally believed that the crochets evolved as an adaptation for the caterpillar to hold onto leaves, which certainly seems to be the case for the Ditrysia, but crochets perhaps originally

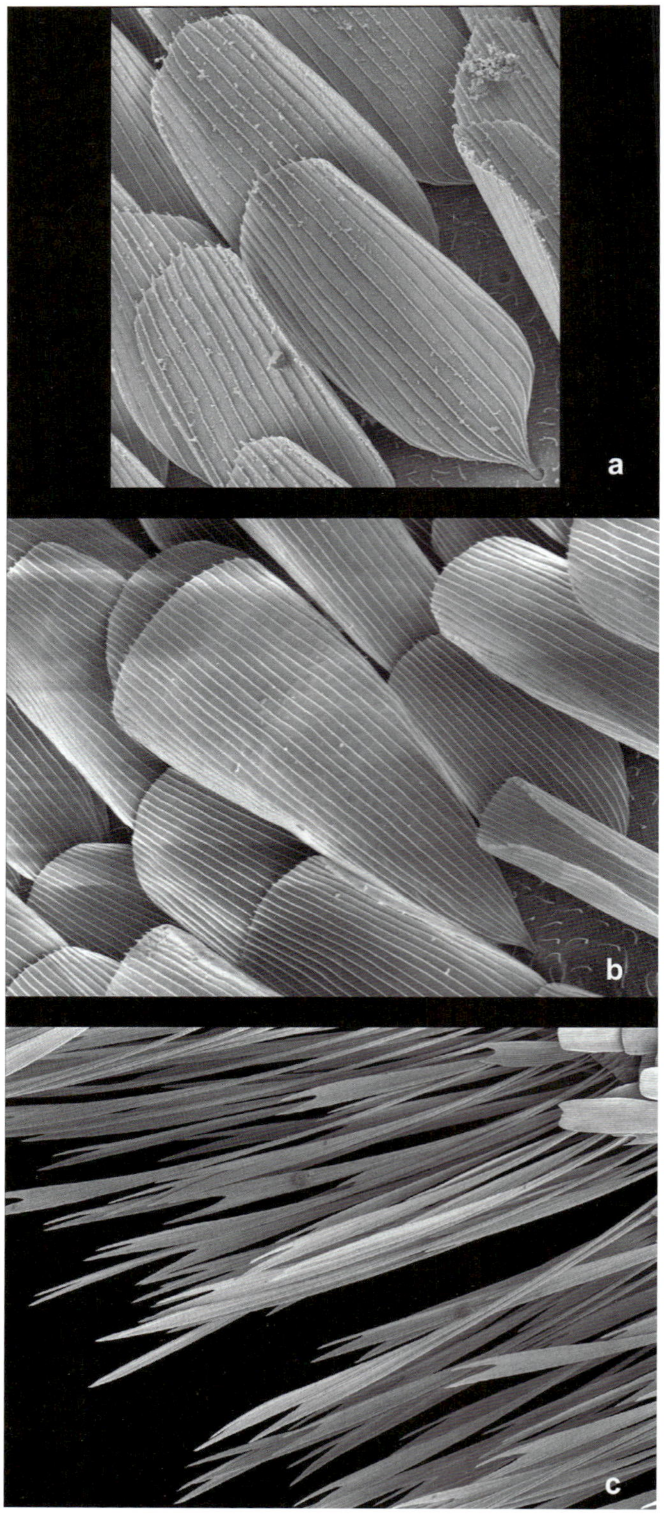

13.25. Wing scales of primitive moths, which lack perforations. Scales are flattened setae that provide reflective and diffractive coloration, and assist in thermoregulation and escape: surface scales (a, b); scales from the edge of the wing (c). a, Micropterigidae; b, c, Eriocraniidae. Scanning electron micrographs.

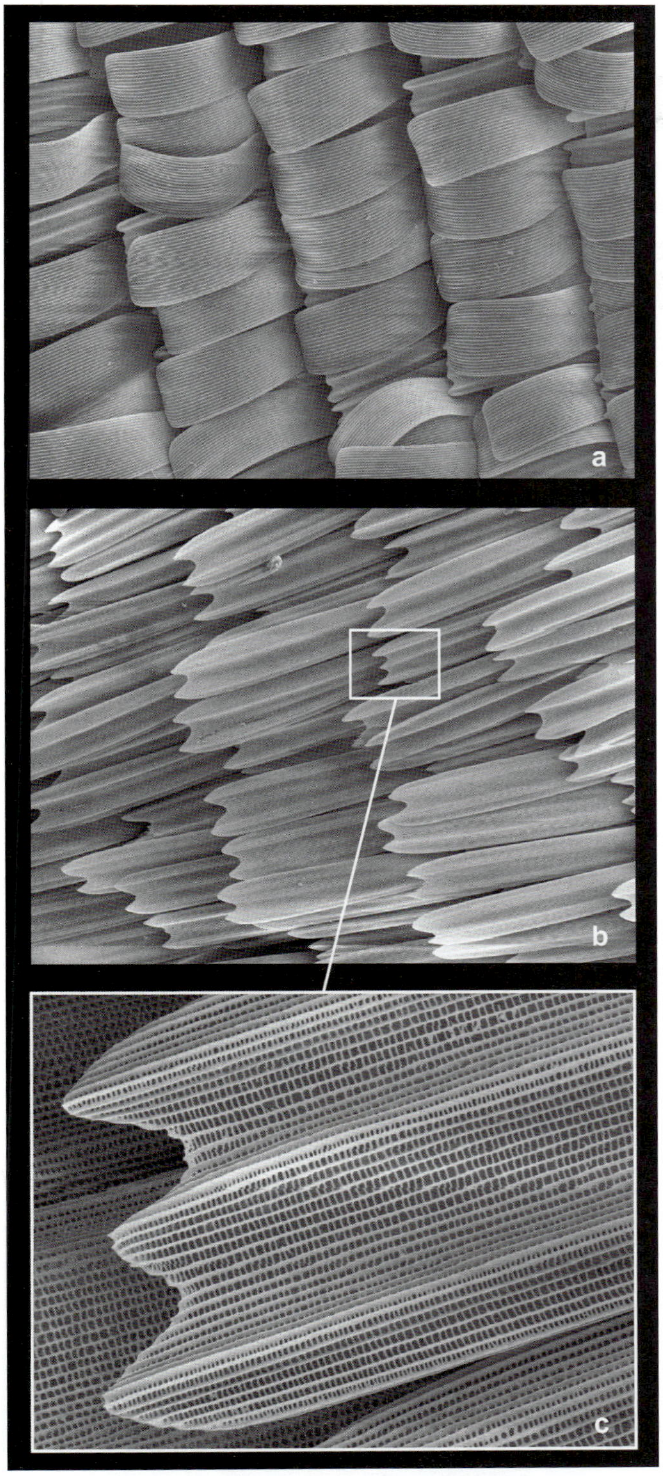

13.26. Wing scales of ditrysians, showing the perforations that diffract light into different colors. a: Uraniidae. b, c: Pyralidae. Scanning electron micrographs.

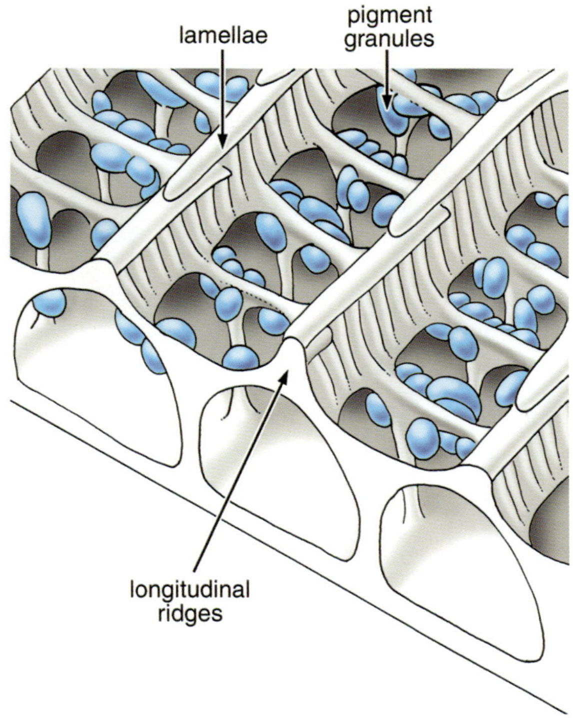

13.27. Diagram of a scale crossection, showing the trusslike structure of a perforated scale. Pigment granules also provide coloration. Redrawn from Ghiradella (1984).

allowed caterpillars to walk over the silken webbing of their retreats (Kristensen, 1997). Silk webs and retreats evolved in the Neolepidoptera.

Lastly, the neolepidopterans have a unique female reproductive system, where the copulatory opening (the "ostium bursa") is separate from the opening to the oviduct, or the duct through which the eggs pass (Figure 13.30) (Dugdale, 1974). In "monotrysian" Lepidoptera there is a common opening for copulation and oviposition, which may be external or empty into a cloaca, a cavity into which the opening for excretion also empties. This type of reproductive system occurs in all the basal ("non-ditrysian") Lepidoptera, except those that have the intermediate, exoporian anatomy. In this condition there are separate copulatory and oviposition openings, and sperm must migrate through an external fold from one pore to the other for eggs to be fertilized. An improvement on the exoporian condition occurs in the Ditrysia, where there is actually a direct internal connection (the *ductus seminalis*) between the copulatory and egg-laying tracts and separate openings for these tracts. Intuitively this seems far more efficient, which may actually be why the Ditrysia comprise most of the Lepidoptera!

Life histories and other features of the most basal group of neolepidopterans, the Mnesarchaeidae and Hepialoidea (ghost moths), deviate widely from the typical leaf-mining habit of basal "microlepidopterans." The Mnesarchaeidae is a small family (14 species) from New Zealand, the larvae of which feed on detritus in forest leaf litter. The Hepialoidea includes about 600 species worldwide, placed into five families; these can be very large moths and many of the smaller species have a vestigial proboscis. The larvae of many hepialoids, where known, construct silken galleries in leaf litter or below ground, feeding on detritus, liverworts, and fungi, and

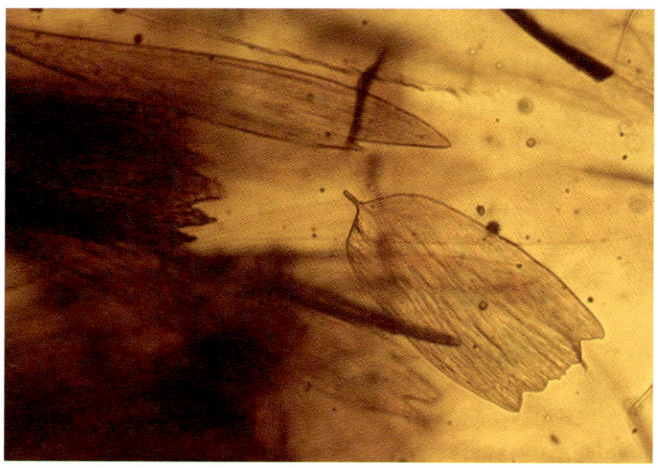

13.28. Fossilized scales, caught wafting off of a Cretaceous glossatan moth in New Jersey amber. Detailed structure of scales (below) can be studied under high magnification. AMNH NJ-638; body length 3.5 mm.

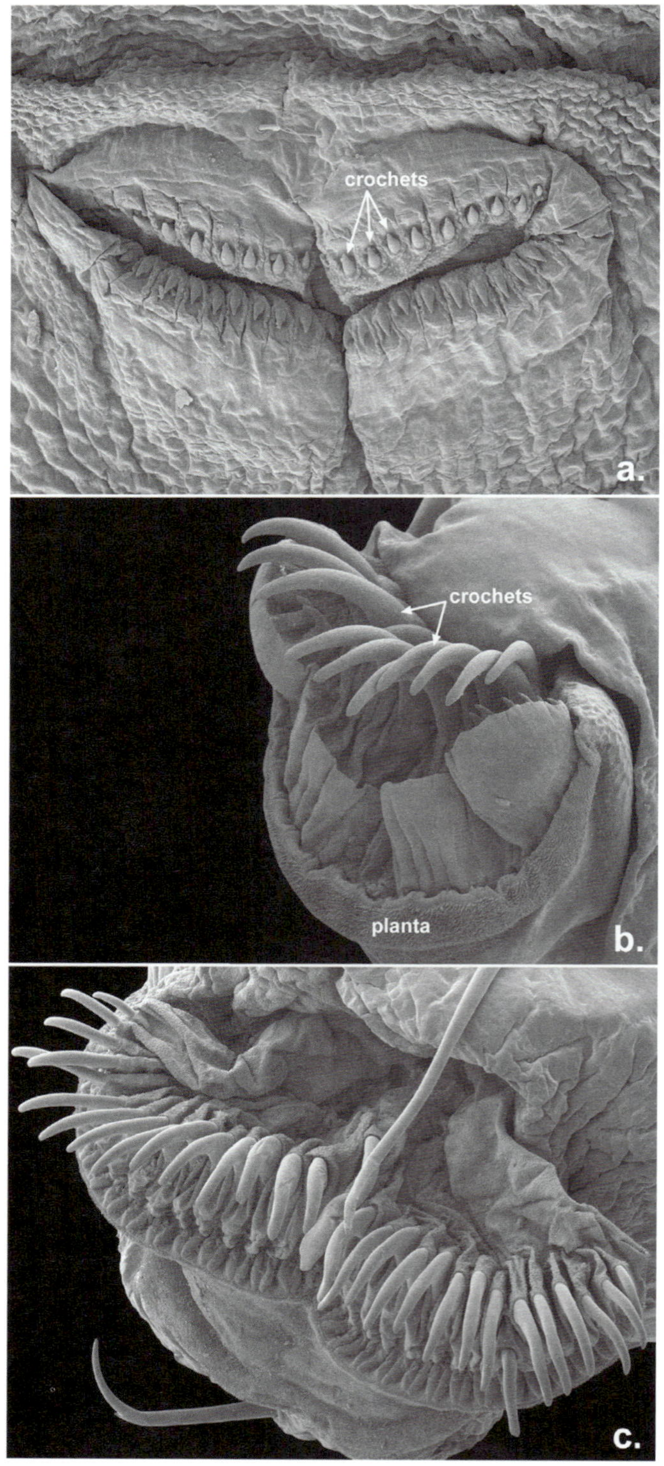

13.29. The crochets of various caterpillars, which define the group Neolepidoptera. Crochets are thought to have evolved to assist caterpillars in treading over plant surfaces or the surfaces of their silken webbings. Scanning electron micrographs.

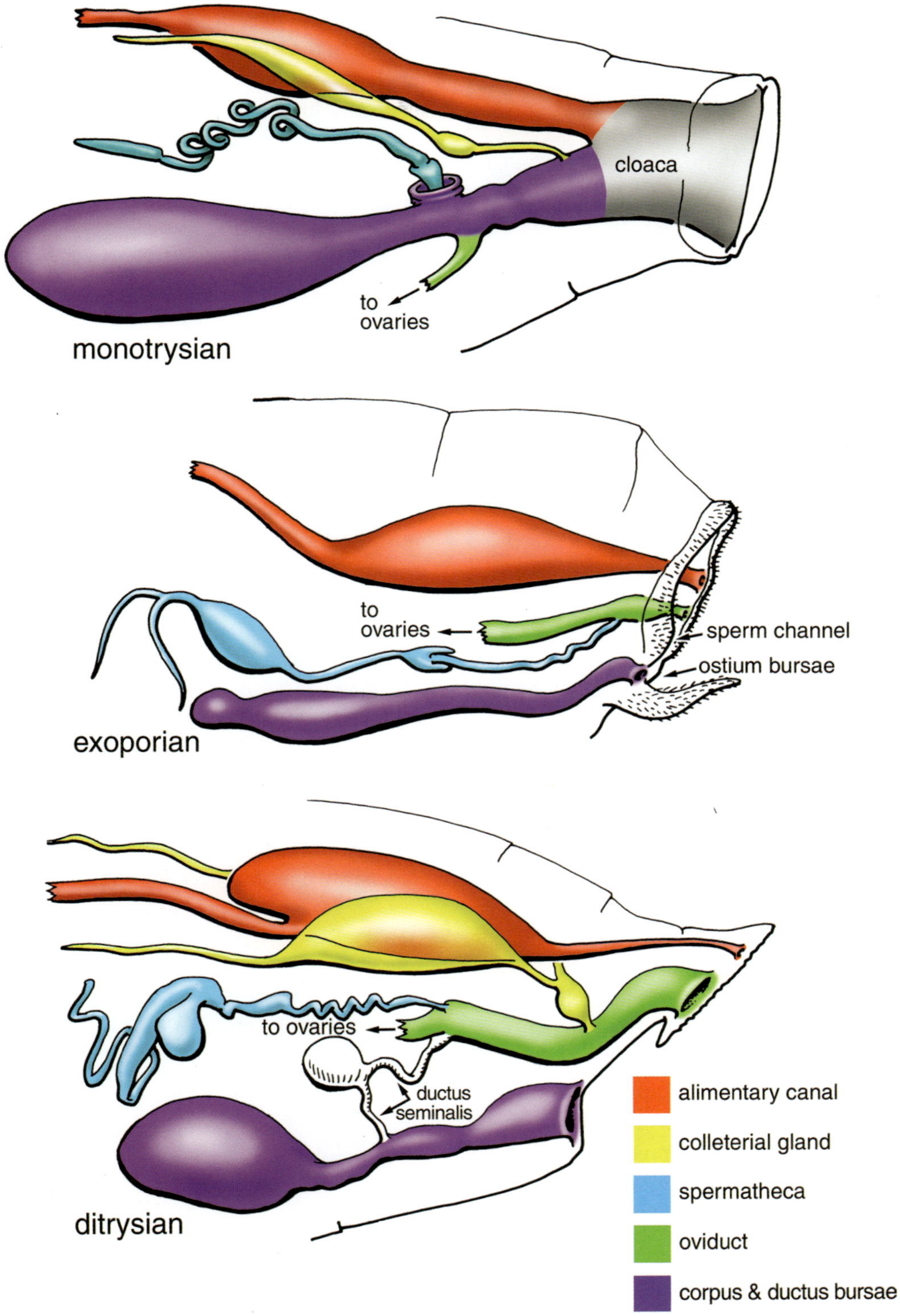

13.30. Schematic diagrams of female reproductive systems in Lepidoptera. The primitive condition is the monotrysian; the most derived system is ditrysian. Redrawn from Scoble (1992).

some bore into stems or wood. Most neolepidopterans belong to another major lineage, the **Heteroneura** ("different venation"). Among other features, heteroneurans have the pair of wings on one side coupled via a bristle-like *frenulum* at the base of the hind wing, which fits into a *retinaculum*, which is a lobe or patch of scales at the base of the forewing (Figure 13.31).

There are four major lineages (superfamilies) of basal (non-ditrysian) heteroneuran lepidopterans (Davis, 1999), most of which are leaf miners on trees (Powell *et al.*, 1999).

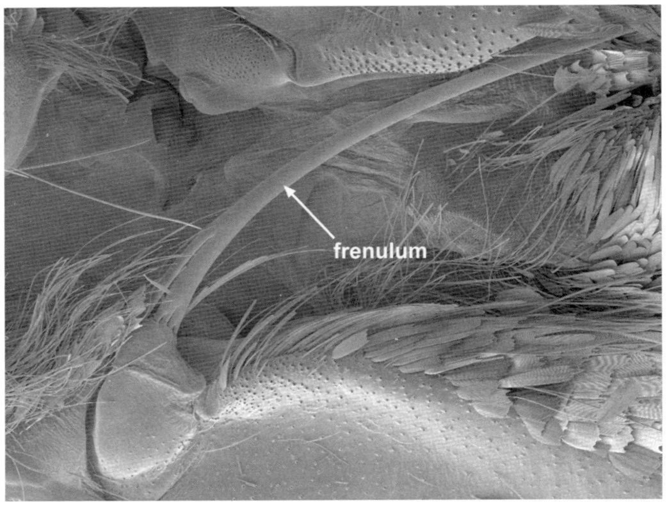

13.31. The bristle-like frenulum on the leading edge of the hind wing of a noctuid moth. It fits into a tongue, the retinaculum, on the forewing, and together these structures help keep the wings coupled in flight. This structure partly defines the Heteroneura. Many of the scales were brushed away to reveal the frenulum. Scanning electron micrograph.

The most speciose of these are the Nepticuloidea and Incurvarioidea, the former of which includes some of the smallest lepidopterans with a wing length of merely 1.5 mm (0.06 in.). Of the two families in Nepticuloidea, the Nepticulidae itself is by far most diverse, with 800 species found worldwide; Opostegidae contains about 100 species worldwide. What may be the oldest known heteroneuran is a larval leaf mine in a platanoid leaf from the mid-Cretaceous Dakota Formation of the western United States (Labandeira *et al.*, 1994), attributed to the Nepticulidae. Other fossil nepticulids include adults in Baltic amber and also leaf mines from the Eocene to Miocene of Europe, a mine from the mid-Cretaceous of Kazakhstan (Figure 13.32), and a probable adult in Late Cretaceous Siberian amber. Putative nepticulids from the Early Cretaceous of Eurasia are simply compressions in rock (Kozlov, 1989), so the systematic position of these fragmentary small moths is actually uncertain. A recently discovered family of Recent monotrysians is the Andesianidae, which contains just three species from southern Argentina and Chile (Davis and Gentili, 2003).

The Incurvarioidea includes 6 families, all of them with 100 described species or less, and larval borers in this group have very interesting biology. Larvae of the Cecidosidae produce hard, round galls in the stems of *Rhus* and *Schinus* (Anacardiaceae: the cashew and poison-ivy family) in southern South America and southern Africa. The gall falls off the plant, and the active pupa within causes the gall to roll and skip around the ground. This behavior has been interpreted as an adaptation for escaping hot ground for cool shade, but it probably also effectively prevents parasitoid wasps from getting a purchase on the gall and drilling their ovipositor into it. A similar habit has evolved in a few Tortricidae (e.g., *Cydia deshaisiana*), or the so-called Mexican jumping beans (at least in the United States, where they are sold as novelties for children). The basalmost member of the family Cecidosidae is the "bark scribbler" moth from New Zealand, whose larvae mine tree bark (Hoare and Dugdale, 2003). Besides this family, the Palaephatidae are the other incurvarioid group to have an austral distribution, in this case in southern South America and Australia. In fact, these incurvarioids are the last group in the phylogeny of Lepidoptera to show an austral distribution (Holloway and Nielsen, 1999), which may further reflect a young age of the Ditrysia in the Late Cretaceous to Tertiary, after the fragmentation of Gondwana.

Certainly the best known family of incurvarioids is the Prodoxidae, some of which are the yucca moths. The family is small (about 90 described species) and primarily North American (Davis, 1967; Davis *et al.*, 1992). The larvae are borers in seeds, fruits, or stems of diverse plant families, and most species in the subfamily Prodoxinae breed in the stems or reproductive organs of yuccas (family Agavaceae). The genera *Tegeticula* and *Parategeticula* breed in particular species of yucca and pollinate them, and their pollination by the adults has evolved into a highly specialized, obligatory relationship with the host plants. This intimate mutualism was first studied by the early agricultural entomologist C. V. Riley (1892), and has been intensively studied ever since (Powell, 1984, 1992; Baker, 1986; Pellmyr, 1992, 1999, 2003; Pellmyr and Huth, 1994; Addicott, 1996; Pellmyr and Leebens-Mack, 1999). Female *Tegeticula* and *Parategeticula* seek newly opened yucca flowers in which to oviposit, and

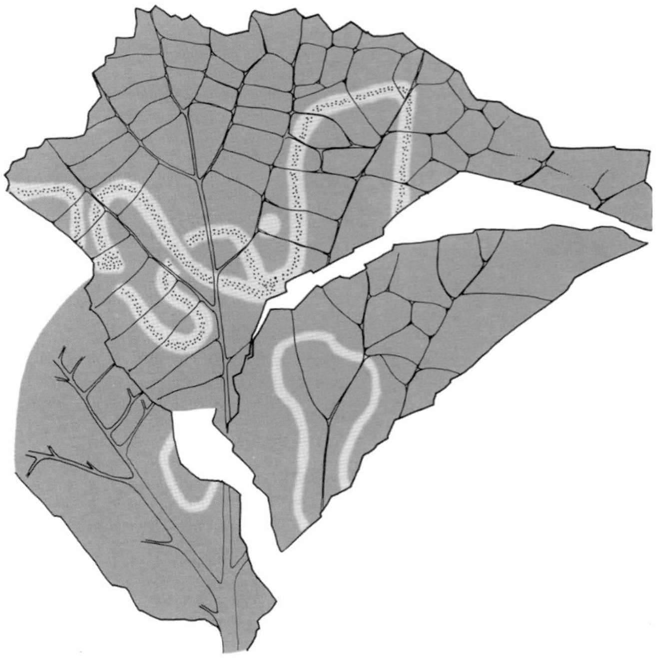

13.32. Drawing of a fossil leaf mine from the mid-Cretaceous of Kzyl-Zhar, Kazakhstan. It putatively was formed by a moth larva in the living family Nepticulidae. Paleo. Institute, Moscow.

then they may mark the site with pheromones to discourage other female moths from ovipositing there. Then she uses specialized appendages, the maxillary "tentacles," in order to insert a pollen mass collected from another flower into the style of the flower in which she just oviposited (Figure 13.33). Larvae breed in the developing ovules (*Tegeticula*) or fruit (*Parategeticula*) of *Yucca*, consuming some but not all of the embryonic plants. There are also nonpollinating prodoxines, like *Prodoxus*, which also breed in yucca flowers, thus usurping the efforts of the pollinators, as well as "cheater" species. *Tegeticula corruptrix* and *T. intermedia*, for example, evolved from pollinating ancestors. It was formerly believed that *Tegeticula* consisted of just three species, two of them monophagous and the other highly polyphagous, but the genus is now known to consist of 13 species or more (Pellmyr, 1999), most of which are monophagous and the others oligophagous.

How did such a specialized mutualism evolve? Based on relationships of prodoxines and their life histories (Nielsen and Davis, 1985; Pellmyr and Leebens-Mack, 1999), a series of stages is apparent. The first stage appears to have been larvae that are endophytic, or feeding within the developing fruits and seeds of the plants, since most members of the family do this. Along with this evolved specific oviposition preferences by the females. The next step was probably the regular, then restricted, feeding by the female on the flowers of the host plant (i.e., its pollination). This is seen, for example, in a basal prodoxid genus, *Greya*, which breeds in and pollinates saxifrages, but the effects of its pollination are swamped by bees and other efficient pollinators. A third stage would be extreme host restriction, caused by the timing of yucca flowering, by narrow preferences of the ovipositing female, or both. The fact that nonpollinating yucca moths do not restrict themselves to breeding in single species of yuccas indicates that highly specific pollination is, at least in part, a result of the very specific oviposition preference by the females. The last stage in the evolution of this obligate mutualism is the origin of specialized morphological structures, the moth mouthparts and the flower styles, which act as a lock-and-key mechanism that restricts pollination to particular pollinators. Despite expectation, there does not seem to be cospeciation between species of yuccas and species of *Tegeticula* and *Parategeticula* (Pellmyr, 2003), even though this intimate association is hypothesized to have existed since the Eocene (Pellmyr and Leebens-Mack, 1999). An Eocene age was based on molecular clock estimates calibrated with ages estimated from biogeography, so any fossil prodoxids would be of considerable interest.

Prodoxids are not the only specialized pollinators in Lepidoptera. Others include a North American pyralid moth, *Upiga virescens*, on the *Lophocereus schotti* cactus (Fleming and Holland, 1998; Holland and Fleming, 1999), and various species of the southeast Asian gracillariid moths, *Epicephala*, on *Glochidion* euphorbs (Kato *et al.*, 2003). No doubt others will be discovered. In these two examples there is a one-to-one obligate relationship between the flowers and pollinators, the flowers or the moths are morphologically specialized, and, very significantly, the larvae feed within seeds. Just like fig wasps (Agaonidae) and some other seed-breeding insects, the origin of such specific pollination is probably a product of highly restricted larval host use, which itself is largely the product of very discriminating oviposition.

13.33. Adult *Tegeticula* pollinating yucca flowers, which is one of the most specialized pollinating relationships known. Its mouthparts are uniquely built for inserting pollen into the stigmas, which is what these moths are doing. Photos: O. Pellmyr, University of Idaho.

DITRYSIA

This group is defined in part by an obscure but very significant structure: an internal passage between the copulatory tract and oviduct through which sperm can pass (Dugdale, 1974) (Figure 13.30). Ditrysia contains 98% of the species of Lepidoptera, among which occurs a major transformation of larval feeding habits. Some of the basal Ditrysia have retained ancestral habits of larval leaf mining, like the Gracillarioidea and some Gelechioidea, but many have now become external

13.34. Caterpillars of a tineoid moth, with and without the larval case. The case is composed of fecal pellets and bits of debris, held together with silk. Scanning electron micrographs; body length (without case) 2.2 mm.

feeders, feeding exposed on foliage or plant detritus. The larvae of many basal ditrysians conceal themselves, however, with webs or cases, the best example of which is in certain basal ditrysians, the tineoids.

The Tineoidea is a diverse superfamily of six families (Davis and Robinson, 1999), well known as pest species (the clothes moths) and for their larval cases – the trichopteran analogs in Lepidoptera. The superfamily is defined in part by a long, telescoping oviscapt, and the galeae on the proboscis are often reduced and only loosely "zipped" together, or even separate. Otherwise Tineoidea is a poorly defined lineage and many not even be monophyletic. Tineidae (3,000 species) is the largest group, the most ecologically significant, and also the most basal family (Robinson, 1988; Robinson and Nielsen, 1993). Larval tineids feed on plant and animal detritus and fungi, and species of *Tinea*, *Tineola*, and *Trichophaga* are notorious consumers of the keratin in wool fabrics. Other tineids feed on dung, lichens, bracket fungi, grains, feathers

13.35. Case of a tineoid moth in Miocene Dominican amber. The caterpillar is still inside. AMNH DR11–14; length 2.8 mm.

13.36. *Acrolophus* moth in Dominican amber (Tineoidea: Acrolophidae). Larvae in this family construct narrow, winding galleries through leaf litter and over tree trunks. AMNH; forewing length 9 mm.

and debris in animal nests, and dried corpses, and a few are inquilines in the nests of social insects (feeding therein, of course, on the colony's debris). The larvae of many species build portable, tube-shaped cases constructed of silk and particles of their foodstuff (Figure 13.34), or live and feed within silken galleries. Fossil tineids are abundant and diverse in Baltic (Skalski, 1979) and Dominican amber, and even their cases have been preserved (Figure 13.35), which indicates that the fossil larvae were probably feeding on or under bark. It would not be surprising if tineids or tineoids were preserved in Cretaceous amber, which would be extremely significant, but determining this must await careful study of these inclusions. The other major family of tineoids, the "bag worms" (family Psychidae, with 1,000 species), are so named because larvae carry cases made from gathered and trimmed fragments of plants. Just as in Trichoptera, the cases vary in shape and composition, being most often tubular or prismatic in shape (Figure 13.37), though even snail-shaped cases are made. Female psychids are often brachypterous or even wingless, and some are even neotenic, having retained larval features. In instances where the females are highly reduced, males mate with them in their cases. Widespread reduction of female Psychidae is probably related to the fact that there are several parthenogenetic lineages in this family because this mode of reproduction is common among insects where vagility is greatly reduced.

The Gracillarioidea have the distinction of being the largest lineage of leaf-mining Lepidoptera, and perhaps the earliest known ditrysians. The larvae are also distinctive in that they have early instars that are considerably different from later instars, such as in lacking legs and being sap feeders rather than just leaf miners. Among the four families in the superfamily, the Gracillariidae (with about 2,000 described species) are by far the most diverse, and even this is apparently a fraction of their actual diversity. In one region of Costa Rica, for example, David Wagner and Don Davis recorded hundreds of new species of gracillariids, where fewer than ten had previously been known. The oldest ditrysian to have been reported is a mid-Cretaceous leaf mine attributed to a living subfamily of Gracillariidae, the Phyllocnistinae (Labandeira *et al.*, 1994). Based partly on the derived nature of this mine and its identification in a Recent subfamily, Labandeira *et al.* (1994) estimated a Jurassic age for the Ditrysia. This estimate was also based on their acceptance of the Jurassic fossil *Protolepis cuprealata* as a glossatan, but, as explained earlier, preservation of *Protolepis* does not allow identification of it any further than lepidopteran. Unfortunately, the mines of Recent leaf-mining moths are very incompletely known (Powell, 1980). While 80–95% of the leaf mines are known for European and North American species, most of the tropical species are unknown, so the mines of perhaps only 10–20% of all described species are known.

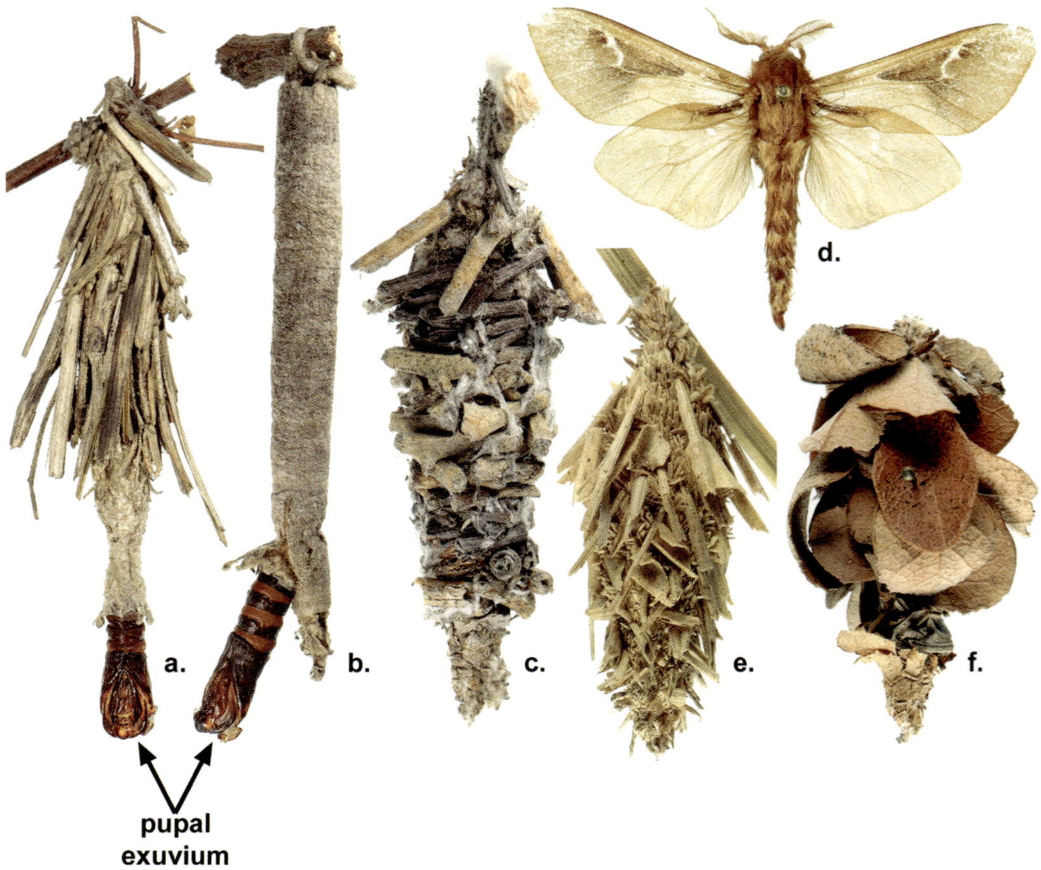

13.37. Larval cases of assorted Psychidae ("bag worms") and a representative adult. Cases are made from trimmed plant fragments sutured with silk, or just from silk, and the larva pupates within the case as well. a–d, *Oiketicus* spp.; e, *Cryptothelea*; f, genus indet. Not to the same scale.

Moreover, mines provide limited diagnostic features compared to the morphology of the insect itself. On this basis, and particularly with the vast numbers of undescribed gracillariids coming to light, it was argued (Grimaldi, 1999) that such fossilized feeding traces must be interpreted cautiously.

The Yponomeutoidea is a group of eight families that, though each comprising fewer than 500 species (usually fewer than 100), is one of the first significantly diverse lineages in lepidopteran phylogeny whose larvae feed on herbaceous plants. Previously, most larval hosts have been trees and other woody plants, including those in more basal angiosperm groups like the Fagales. Though many larval yponomeutoids bore into stems, flowers, and seeds or mine leaves and twigs, this is also the first significantly diverse group of Lepidoptera, with the gelechioids, that are external feeders on plants (Powell *et al.*, 1999). They retain ancestral larval habits of concealing themselves while feeding, like other basal lepidopterans, but usually under communal webs. *Urodus parvula* of the southeastern United States spins a delicate, lacy cocoon of golden silk, which it suspends beneath the overhang of a rock or a fallen log (Figure 13.38).

The Gelechioidea is the first mega-diverse group in the phylogeny of Lepidoptera, comprised of 15 families and approximately 16,500 described species. Family-level relationships were analyzed and discussed by Hodges (1999), including revised concepts of families and the descriptions of new families, as well as a review of the taxonomy, morphology, and biology. Like gracillariids, there are vast numbers of gelechioids, with only 10–50% of the species known, depending on the region. Most species are in the Elachistidae, Oecophoridae, and Gelechiidae, with 3,000 species or more in each family; the Xyloryctidae, Coelophoridae, and Cosmopterigidae have between 1,000 and 2,000 species each; other families have fewer than 500. It may be solely a result of their species diversity, but gelechioids are also one of the most ecologically diverse groups of Lepidoptera. Gelechioid larval diets include detritivores and phytophages, most of the latter being external feeders concealed under silk or by leaves that they tie, fold, or roll over themselves. Oecophorids in *Depressaria* and related genera feed on highly toxic umbellifers, like *Pastinaca sativa* ("cow parsnip"), and their study has contributed very interesting insight into the evolution of specialized plant feeding in insects (Berenbaum, 1981, 1983). These plants contain coumarins, compounds that vary in toxicity depending on their structure, the least potent being hydroxycoumarins and the most potent being angular furanocoumarins (AFCs). Linear furanocoumarins become more toxic by exposure to sunlight, so *Depressaria* circumvent this by feeding within the rolled, shaded leaves (some bore into stems). Plants containing hydroxycoumarins are fed upon by many polyphagous insects, but those plants with AFCs are more diverse and have diverse specialized lepidopteran feeders, including oecophorids. Beside phytophages, species in the Oecophoridae, Batrachedridae, and Coelophoridae are larval predators on sessile Sternorrhyncha, like scale insects and aphids. There are even parasitoids, like larval *Euclemensia* (Cosmopterigidae) on armored scales (Diaspididae: Coccoidea). Gelechioids are abundant and diverse in Tertiary amber (e.g., Figures 13.39, 13.40), comprising some 30% of the Lepidoptera inclusions in Baltic amber alone (Kusnezov, 1941; Skalski, 1973, 1977, 1979), and some Tertiary leaf mines have also been attributed to this family.

Most of the described Ditrysia are in the **Apoditrysia**, a group defined by the structure of the abdominal sternum. All fossil apoditrysians are Tertiary. There is a basal group of nine lineages with unresolved relationships, one of which is the Obtectomera and to which all other lepidopterans belong. This grouping is defined by the immobility of the anterior segments of pupa and the loss of tergal spines, which are functionally related because the wriggling pupa needs the spines to free itself from the cocoon or the larval gallery. Most

13.38. Pupa and adult of the southeastern U.S. species *Urodus parvula* (Urodidae). Just before the larva pupates, it spins a meshlike cocoon of golden silk, suspended from under a log. Wing spread 2 cm.

13.39. Tiny gelechioid moth in Miocene Dominican amber. Gelechioids are extremely diverse ecologically and in numbers of species, but taxonomically they are greatly understudied. The larvae of most feed while concealed under silk shelters. AMNH DR14–278; body length 1.5 mm.

13.40. Oecophoridae (Gelechioidea) in Eocene Baltic amber. The larvae of many oecophorids are detritivores. AMNH; forewing length 8.2 mm.

13.41. Assorted non-macrolepidopteran Ditrysia. 1, *Amesia*; 2, *Zygaena*; 3, *Megalopyga*; 4, *Euclea*; 5, *Archips*; 6, *Argyrotaenia*; 7, *Aulacita*; 8, *Telchin*; 9, *Melittia*; 10, *Eurrhyperia*; 11, *Syllepte*; 12, *Evergestis*; 13, *Munroessa*; 14, *Anania*. Not to the same scale.

13.42. *Zeuzera* (Cossidae) adult and pupal exuvium in the larval gallery. Many larval cossids bore into the stems and trunks of woody plants. Wingspan 29 mm.

13.43. Adult and pupa with case of *Synanthedon* (Sesiidae). Larvae in this genus can be serious pests of fruit trees, boring into the wood. Wingspan 22 mm.

of the non-obtectomeran moths are small families that have externally feeding larvae (some are miners and borers), but there are two significant lineages. One is the tortricoids, and the other is a loosely defined lineage of 12 families and some 3,000 species. These are the Zygaenoidea and the closely related superfamilies Cossoidea and Sesioidea (Edwards *et al.*, 1999). The latter two superfamilies are essentially larval stem and root borers (Figures 13.42, 13.43), and the Sesiidae in particular can be serious pests because some bore into wood or under the bark of trees. Sesiidae are best known as boldly colored, day-flying moths with narrow, often transparent wings, most of which are remarkable wasp mimics. The Castniidae, which have only about 150 species, superficially resemble butterflies (Figure 13.41), with large wings and clubbed antennae. Some Castniidae are diurnal and bold colored and mimic distasteful butterflies.

Moths in the Zygaenoidea have exceptional biology, the group of which was treated by Epstein (1996) and Epstein *et al.* (1999). The most significant fossils of these are compressed adults in rocks from the Miocene of Spain and Germany (Naumann, 1987; Fernandez-Rubio *et al.*, 1991), but the family is certainly much older than this. Many zygaenoid caterpillars are heavily defended by poisonous spines, toxins, or both. Species in the Limacodidae (with 1,000 species) include the well-known "saddle-backed" or "nettle" caterpillars, bristling with poisonous spicules (Figures 13.44, 13.45). The caterpillars of Dalceridae produce gelatinous secretions (hence the name, "gelatin" caterpillars) (Figure 13.46), which is a defense known to repel ants. Some Zygaenidae and Lacturidae are extremely toxic in all their stages because they can actually produce cyanoglycosides, not just sequester them from host plants the way many insects do. They retain these compounds even as adults, which are aposematic and visit flowers during the daytime with sluggish impunity. These moths are so toxic, in fact, that their tissues have become extremely resistant to concentrations of cyanide that easily kill other insects. Lepidopterists typically dispatch moths in jars of potassium cyanide, which kills them within seconds and thus prevents loss of scales and wing patterns, but certain Zygaenidae and Lacturidae can

13.44. "Nettle" caterpillar (Limacodidae: Zygaenoidea), a family renowned for the toxic, needle-like setae that cause a painful rash. Photo: V. Giles.

13.45. A limacodid caterpillar, *Euclea*, which feeds primarily on oaks. Photo: V. Giles.

13.46. A gelatin caterpillar (Dalceridae: Zygaenoidea), which is a neotropical family. The gelatinous coating protects dalcerid caterpillars from ants and other small predators. Photo: V. Giles.

live for half an hour in a jar of fresh cyanide. Two small families, the closely related Epipyropidae (40 circumtropical species) and Cyclotornidae (12 Australian species), have life histories that are truly bizarre for Lepidoptera: they have ectoparasitoid and ectoparasitic larvae, depending on whether they kill their hosts or not. In both families the first instars are triungula that actively seek their hosts. For Epipyropidae, hosts are auchenorrhynchans. After the larva attaches, subsequent instars remain on the host, usually concealed by an encasement of wax that is secreted by the host. Cyclotornidae are even more bizarre. First instars follow *Iridomyrmex* ants to cicadellids, the latter of which they then parasitize and leave to molt. Second instars, which are now bright and flat, are collected by the ants and brought back to the nest, where they then feed on the ant larvae and appease the adult ants with anal secretions.

The Torticidae (Figure 13.47) is a homogeneous group of approximately 7,000 described species of small to medium-sized moths (10–40 mm wingspan), usually with mottled and marbled brown colors. A recent review of the subfamilies and tribes of the world is by Horak (1999). They are defined in part by the large, flat oviscapt and are best known for the larval habits of living and feeding within rolled or tied leaves, which is why they are commonly called "leaf rollers." Actually, larval habits are significantly more diverse and include feeding in plant detritus and as borers in roots, seeds, stems, and flowers and under bark. Another distinctive feature is the eggs, which are often flattened and scale like, and sometimes "fenced in" by the ovipositing female who uses specialized scales, probably to ward off ants.

Without doubt, the most ecologically diverse group of Lepidoptera is the Pyralidae (Figure 13.48), which may even be the most speciose family of Lepidoptera, if their species diversity is ever thoroughly explored. There are 16,000 described species in the family, but it is estimated that this is just one quarter of the actual diversity (Munroe and Solis, 1999). The monophyly of Pyralidae is defined by a pair of tympanal organs on the ventrum of abdominal segment two, which face into the space between the thorax and abdomen. The relationships among subfamilies have been analyzed by Solis and Mitter (1992). The family is sometimes split into two, the Pyralidae and Crambidae, but these sub-

13.47. Tortricidae in Miocene Dominican amber. A large mite is attached to its left eye. DR8–43; body length 9.5 mm.

families are clearly closely related so it is just as well to regard them as one family. As the phorids are to the Diptera, so there is an astounding diversity of larval diets in the Pyralidae, including saprophages, phytophages, predators, inquilines of social insects and mammals, aquatic species, and predators. The larvae of basal pyralids are saprophagous on fungi, dried remains of plants and animals, algae, lichens, and some phytophagous on liverworts and ferns. Not surprisingly, some are serious pests of stored grains. Primitive saprophagy led to the suggestion (Munroe and Solis, 1999) that pyralids may have pre-dated the origin of angiosperms, which contradicts the fossil record and phylogenetic position of pyralids among Lepidoptera. As in the tineoids, a saprophagous diet is probably secondarily derived, and from this phytophagy and other diets evolved. True to the habits of the more basal lineages of Ditrysia, phytophagous pyralids are concealed feeders, living and feeding in rolled or tied leaves, under webs, and as borers; some even construct cases like tineoids. The predators feed on sternorrhynchans (the moth of which is *Laetilia coccidivora*), psychid caterpillars (*Dicymdomia julianalis*), and lymantriid and notodontid caterpillars (*Metoecis*). The larvae of various species feed on combs or detritus within the nests of social insects, such as *Galleria* wax moths (in bee hives), *Chalcoela* (in vespid paper nests), and Wurthiinae in nests of arboreal weaver ants (*Oecophylla*). The larvae of Nymphulinae are aquatic, where they feed on plants or films of algae and diatoms; some even construct cases like caddisflies. In fact, nymphulines living in fast-flowing streams usually have filamentous tracheal gills, just like many caddisflies. The most famously anomalous lepidopteran habit, though, are the pyralids whose adults are phoretic on sloths (Waage and Montgomery, 1976; Bradley, 1982). Adults in three genera – *Bradypodicola, Bradypophila, Cryptoses* – scuttle amongst the fur of neotropical tree sloths, and when the sloth descends a tree to defecate at the base of the tree, the moths momentarily leave the sloth and oviposit on the dung, where the larvae breed. It may be just the probability of sheer numbers of pyralid species, but presumably there is an intrinsic biological feature of pyralids that has spawned such diverse habits.

THE "HIGHER" DITRYSIANS: MACROLEPIDOPTERA

The Macrolepidoptera are, as the name implies, relatively larger lepidopterans (Figure 13.49), which may be a monophyletic group according to Minet (1983, 1991). These have been defined by a complete loss of vein CuP and by the crochets on the larval prolegs arranged in a "mesoseries," which is a crescentic series on the mesal (inner) side of the proleg.

13.48. Caterpillar of the huge family Pyralidae in Eocene Baltic amber. Pyralidae are ecologically the most diverse family of Lepidoptera. AMNH; length 9.8 mm.

13.49. Representative families of "macromoths," or the non-papilionoid macrolepidopterans. 1, *Psychomorpha*; 2, *Papaipema*; 3, *Noctua*; 4, *Catocala*; 5, *Lirimis*; 6, *Doa*; 7–10, genera indet.; 11, *Lymantria*; 12, *Actias*; 13, *Syssphinx*; 14, *Aglia*; 15, *Brahamaea*; 16–19, genus indet.; 20, *Lasiocampa*; 21, *Chrysiridia*; 22, genus indet.; 23, *Epides*; 24, *Tanaorrhinus*. Not to the same scale.

These lepidopterans also tend to have broad, triangular forewings and smaller, rounded hind wings, with thick to slender bodies. They comprise about 60% of all lepidopteran species. Another interesting feature of many macrolepidopterans, but not all, are hearing organs, also called tympanal organs.

Tympanal organs in lepidopterans are similar to those in most insects, being comprised of a smooth, taut membrane (tympanum) pressed against a tracheal air sac (Yager, 1999a). Sound waves cause the tympanum to vibrate, and the waves become amplified by the air sac, which are then transduced into nerve impulses by specialized receptors. Chordotonal "organs" are specialized receptors that detect distortions in a tympanum, and these always evolve where existing proprioceptors occur (these occur throughout arthropods and likewise detect distortions and pressure on cuticle). The basal moth *Micropterix*, for example, has a pair of chordotonal organs at the base of the abdomen (Kristensen, 1984), which may be an evolutionary precursor to tympanal organs that evolved much later in pyraloids and macrolepidopterans.

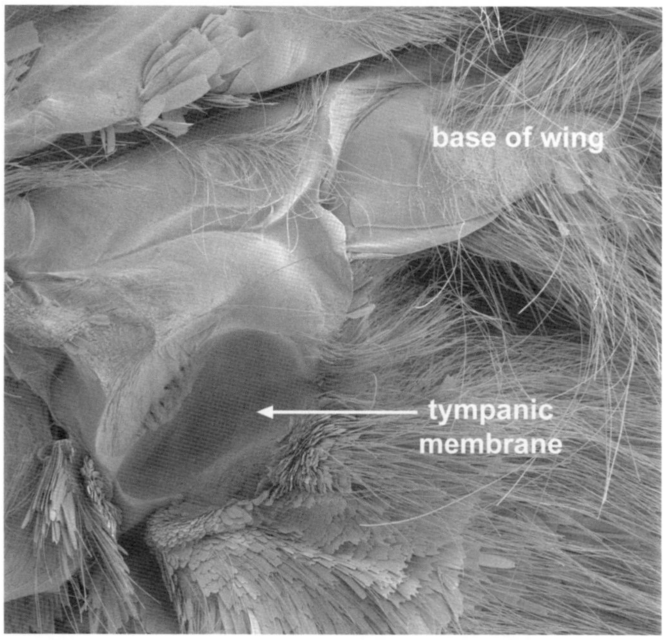

13.50. Hearing organ of a noctuid moth, showing its outer surface – the tympanum – which is at the back of the thorax. "Ears" evolved in ditrysians at least eight times and have specific locations on the body depending on the lineage, but all function to detect the ultrasonic calls of bats. Scanning electron micrograph.

So far as is presently known, tympanal organs appear to have evolved at least eight times in Lepidoptera, all in the Ditrysia. Independent origin is based on location and structure (Richards, 1933; Kennel and Eggers, 1933; Minet, 1983; Scoble, 1995). In most groups they are paired structures within abdominal segments one or two, but in some Thyrididae, Hedylidae, and Nymphalidae the tympanal organs are at the base of the wings. Abdominal tympana occur in a few tineoids, in Drepanoidea, in the related Uraniidae and Geometridae, and in Pyralidae. Finally, Noctuoidea have a unique pair of tympanal organs, on each side of the metathorax between the epimeron and the postnotum (Figure 13.50). These organs may serve several functions, possibly including courtship (Spangler, 1988), but for the nocturnal lepidopterans they are without question adaptations against bat predation.

The high-pitched squeaks of microchiropteran bats are echolocating signals used for avoiding obstacles and for finding prey in the dark, including flying insects like moths. Various insects are now known to detect bats calls, including orthopterans, mantises, and lacewings, but the ability is most widespread in moths (Roeder and Treat, 1970; Roeder, 1965; Fenton and Fullard, 1981; Fullard, 1982, 1998). In all of these insects their hearing is generally most sensitive in the range of bat calls, some 30–60 kHz, and when they hear bat sonar close by, they usually have a stereotyped avoidance behavior, which generally involves divebombing and sometimes aerial loops. Remarkably, the tympanal organs are lost or nonfunctional in species that are diurnal (like ctenuchine arctiid moths), or in the wingless sex of dimorphic species (like the females of some geometrid moths and some mantises). Arctiid moths, which are discussed later, have evolved a unique method to defend against bats: They squeak back. A sensory organ in Pyraloidea, Geometroidea, and Noctuoidea that is an adaptation against bats suggests that these higher ditrysians, like bats, evolved relatively recently, perhaps entirely within the Tertiary. The earliest bats are primitive microchiropterans from the Eocene (Simmons and Geisler, 1998) so bats plausibly originated in the Paleocene. Fossil Ditrysia are spare, but with what is also known of their relationships all of the available evidence suggests an origin of the higher Ditrysia in the earliest Tertiary to latest Cretaceous, 60–70 MYA. Later in this section, we review four major lineages of the macrolepidopterans.

A monophyletic group may be comprised of the silkworm moths (Bombycoidea), the Mimallonoidea (with 200, mostly neotropical, species), and Lasiocampidae, all defined in part by the absence (perhaps a loss) of tympanal organs (Lemaire and Minet, 1999). With the notable exception of the hawkmoths (Sphingidae), most of these moths also have adult mouthparts that are lost or highly reduced. Lasiocampidae (1,500 species worldwide) are stout-bodied, fluffy moths, with hairy caterpillars that are often urticating. *Malacosoma*, or tent caterpillars, are perhaps the most familiar lasiocampids; the larvae of some species construct extensive webs on their host plants where older instars retreat when not feeding. The Bombycoidea, as most recently classified (Lemaire and Minet, 1999), includes nine families, of which three are most significant: the saturniid moths, the sphingids, and the silkworms (Bombycidae).

13.51. The silk worm moth, *Bombyx mori* (Bombycidae), showing the adult (a), cocoon (b), and unspun silk from the cocoon (c). The species has been cultivated for at least 4,600 years, and is the only insect known entirely as a domesticated species. All to the same scale; wingspan 42 mm.

13.52. Caterpillar of *Citheronia* (Saturniidae), which feeds on leaves of trees in the families Juglandaceae (hickories) and Hamamelidaceae (sweet gums). Photos: V. Giles.

It is remarkable to think that the secretions of a caterpillar led to some of the most profound cultural events in human history. Silk produced by *Bombyx mori* has been harvested since at least the Huang-Di Dynasty, some 5,000 years ago, and was worth its weight in gold to ancient cultures of the Mediterranean. These civilizations laid the ancient Silk Road, which Marco Polo later traveled, and this opened cultural exchange between east and west. Most silk production, or sericulture, still occurs in China, but now silk is also produced in India and Brazil, with an annual global production of some 620,000 tons. Most of this silk is produced by *B. mori*, which is the only insect known entirely as a cultivated species, though closely related wild species occur in southern China. The larva feeds exclusively on mulberry (*Morus*: Moraceae) and produces a thick cocoon of silk (Figure 13.51) from labial glands that comprise one quarter the mass of the larva. The silk is harvested by dipping the cocoons in hot water, which dissolves the sticky sericin coating and allows the fibroin fiber to be unspun. Other bombycoids are also used in the production of silk. In southern China, silk from the oak silkworm, *Antheraea pernyi*, produces some 50,000 tons a year. In India, tussore and muga silks are produced by *Antheraea mylytta* and *A. assamensis*, and eria silk is produced by *Samia cynthia* (Peigler, 1993). Because *Bombyx mori* is easily raised in captivity, and is so large, its study led to major discoveries in insect physiology, such as the structure of molting hormone (ecdysone), the roles of the prothoracic gland in molting, the corpora allata in secreting juvenile hormone, several other insect hormones, and the first known insect sex pheromone. *Bombyx mori* has even been used for the transgenic production of human drugs.

The Saturniidae (1,500 species worldwide) includes some very large moths, like *Attacus*, *Coscinocera*, and *Rothschildia*, with thick bodies and expansive, lobed wings. Males of many species have plumose antennae, used to follow plumes of pheromones for locating females. The monograph by Michener (1952) remains one of the definitive taxonomic studies on this group. Saturniids are fairly typical of bombycoids, but hawk moths (sphingids) are quite different. These moths have stout, short bodies with slender, triangular forewings and are excellent fliers; many are even capable of hovering, in contrast to the slow, fluttering flight of most other bombycoids. Sphingidae contains 1,200 species, and major works on them include the huge monograph by Rothschild and Jordan (1903), the photographic atlas of type specimens by D'Abrera (1987), and the extensively annotated checklist of world species by Kitching and Cadiou (2000). While some sphingids have small or vestigial proboscides, most have long proboscides that are used to probe flowers. The proboscides of some sphingids, in fact, like *Xanthopan* and *Macroglossum*, are several times the length of the body. The most famous example of this was the early nineteenth-century discovery of the Madagascar star orchid, *Angraecum sequipedale*, which

13.53. Larva, pupa, cocoon, and adult of *Samia cynthia* (Saturniidae) from eastern North America. The larva feeds primarily on *Ailanthus*; the cocoon is constructed of leaves tied together with thick silk. Not to the same scale.

has a nectar spur up to 30 cm (12 in.) deep. Darwin predicted the pollinator would be an insect with a tongue of equal length, perhaps a sphingid, and some 40 years later *Xanthopan morganii praedicta* was described; its immense proboscis fit that role. Flowers that are pollinated by sphingids have a distinctive suite of features, or "pollinator syndrome," being white, opening at night, and very fragrant. A "hummingbird hawkmoth," which are mostly macroglossines, hovers in front of a flower, extends the long proboscis, and siphons up copious nectar. Like hummingbirds they may even trap line, or visit the same flowers in the same sequence and time each day or night. Though the function is unknown, some macroglossines have a unique auditory organ (which is not a tympanum) at the base of the pilifers (Roeder and Treat, 1970). It has been suggested that a long proboscis in sphingids helps keep a hovering hawkmoth out of the reach of predatory spiders and mantises lurking at flowers, so perhaps the pilifer ear plays a role in detecting these ambush predators. Not all sphingids feed from flowers, the most notorious being *Acherontia atropos*, the Death's Head hawkmoth, adults of which produce high-pitched sounds presumably to appease bees when raiding honey from their hive. Most sphingid larvae are camouflaged the color of their green host plants (Figure 13.54), and some of these have large eye spots on the thorax that presumably startle potential predators. The caterpillars of one sphingid genus, though, are some of the best mimics in the animal world: They mimic small snakes (Figure 13.55).

Janzen (1984) discussed the very different life history strategies of sphingids and saturniids. Sphingidae are strong fliers that repeatedly feed, are long lived, lay single eggs on scattered host plants, and are usually restricted in their host use. Saturniidae, by contrast, are poor fliers, do not feed as adults, are short lived, and lay groups of eggs on one host plant, and the caterpillars are often extremely polyphagous.

There are two megadiverse lineages of the Macrolepidoptera, which together comprise approximately half of all lepidopteran species. These are the Geometroidea and the Noctuoidea. The **Geometroidea** includes two smaller families and the very diverse family of inchworm moths, the Geometridae (Figure 13.56), the last with some 21,000 described species (Minet and Scoble, 1999; Scoble, 1999). The small family Sematuridae (40 species from the Southern Hemisphere) is probably the most basal group of geometroids because they lack the distinctive tympanal organs at the base of the abdomen seen in the Geometridae and Uraniidae. Uraniidae is best known for the diurnal species of *Alcides* (in the Australasian Region), *Chrysiridia* (in Madagascar and eastern Africa), and *Urania* (in the neotropics), which

13.54. A typical sphingid caterpillar, well camouflaged in green. The common name for sphingid caterpillars, "hornworm," is based on the dorsal horn on abdominal segment eight, though this feature is found in other Bombycoidea. Some species have large spots on the thorax; when disturbed the thorax swells and the spots resemble eyes of a small vertebrate. Photo: P. J. DeVries.

13.55. A very atypical sphingid caterpillar, the South American *Hemeroplanes*. The larva is suspended beneath a stem at rest (top). When disturbed (bottom), the thoracic legs are retracted, the thorax swells to reveal eyes and patterns that resemble scales and a mouth, and the larva even strikes. Photo: P. J. DeVries.

have velvety black, swallow-tailed wings with brilliant iridescence (as do some sematurids). Most uraniids (i.e., Epipleminae) are quite dull, and the family is defined in part by sexually dimorphic tympanal organs. True inchworm moths, Geometridae, have more modified tympanal organs that open ventrolaterally into deep cavities at the base of the abdomen (Cook and Scoble, 1992). These are lost in females that are brachypterous or apterous, the wing reduction of which occurs, inexplicably, more often in the Geometridae than any other group of lepidopterans (Sattler, 1991) (Figure 13.57). Geometrids are also biologically distinctive in that their larval hosts are mostly trees, which is a marked departure from the herbaceous host plants of many other macrolepidopterans. Most inchworm caterpillars have prolegs only on segments 6 and 10 of the abdomen, and when they move they have the very familiar "looping" movement. The green or mottled brown caterpillars are usually extremely well disguised on the leaves or stems of their host plants, adopting a typical posture when resting. In Hawaiian species of the large worldwide genus *Eupithecia*, the caterpillar rests motionless with the front of the body raised above the stem, and its long, sharp thoracic legs held outward, like tongs. A small insect or spider that passes close by is then captured by the quick plunge of the caterpillar, which then devours it. All other *Eupithecia* feed on flowers, seeds, and leaves. Most fossil geometrids are in amber (Figures 13.58 to 13.60).

13.56. Inch worms, or larvae of Geometridae, including two cryptic forms (middle, below). Below: *Synchlora*, camouflaged with bits of leaves and flower petals applied to its back. Photos: V. Giles, P. J. DeVries.

13.57. Life cycle of the unusual geometrid moth *Alsophila pometaria*, the female of which is flightless. This species is a serious defoliator of hardwood trees in North America, particularly apple (*Malus*) and elm (*Ulmus*). Not to the same scale.

The Noctuoidea is the most speciose lineage of Lepidoptera, with approximately 42,000 species, though the Pyraloidea may vie with this total when their species are better explored. There is essentially no question that the Noctuoidea is monophyletic because it is defined by the unique tympanal organs on the thorax. The superfamily includes two small families, the Oenosandridae (with three Australian species that feed on eucalypts), and the Neotropical Doidae, which feed on euphorbs and are closely related to the Notodontidae (Miller, 1991). Notodontidae are also called "prominent moths," for the prominent toothlike tuft of scales on the inner margin of the forewing of some species; these scales are quite noticeable in resting adults. "Prominents" comprise approximately 3,000 species, including the neotropical Dioptinae (Miller, 1991), which were formerly considered a separate family. Dioptinae are colorful and diurnal (e.g., Figure 13.93), which may be related to the fact that their larvae also feed mostly on toxic herbaceous plants, whereas most notodontid caterpillars (Figures 13.61, 13.62) feed on trees and are dull-colored nocturnal adults. The Lymantriidae (2,700 species), or "tussock" moths, includes one of the most

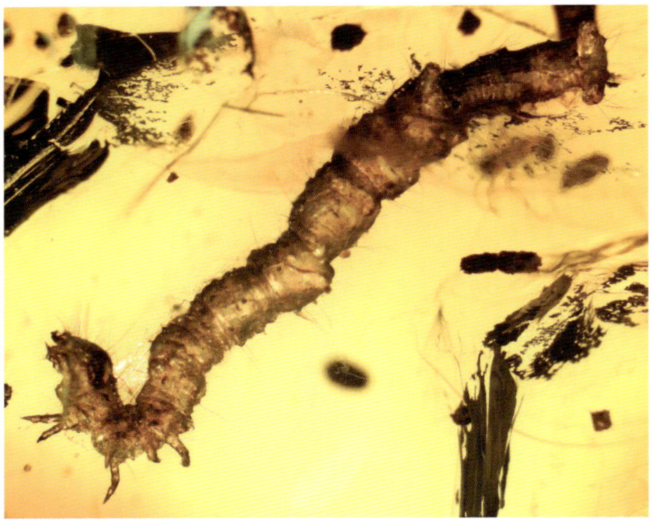

13.58. Fossil inch worm caterpillar (Geometridae) in Dominican amber. Morone Collection, M0428; length 11 mm.

13.59. Fossil inch worm moth (Geometridae: subfamily Ennominae), in Miocene amber from the Dominican Republic. AMNH; greatest length across wings 38 mm.

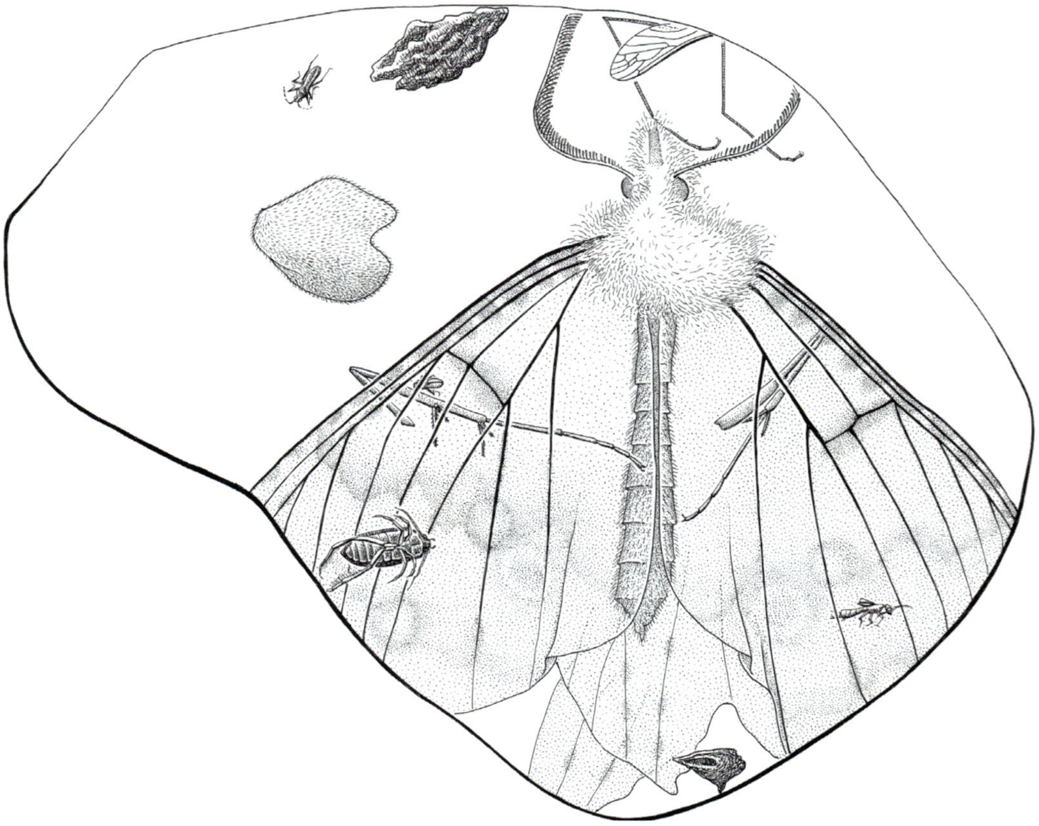

13.60. Fossil inch worm moth (Figure 13.59), mapping some of the organismal inclusions. This is the largest lepidopteran in amber. Some color patterns are preserved, and mold was growing over the thorax.

13.61. Caterpillar of a prominent moth (Notodontidae: Noctuoidea). Photos: V. Giles.

13.62. Caterpillar of *Schizura* (Notodontidae), concealed against the margin of the chewed leaf by its body color and outline. When disturbed, glands behind the head capsule can spray formic acid up to 8 inches away. Photo: V. Giles.

destructive of all lepidopterans, the gypsy moth (*Lymantria dispar*). It was introduced from Europe to North America in 1869 and has been the most serious defoliator of hardwood trees in the eastern United States ever since.

The Arctiidae, aptly named "tiger moths" for the bold, colorful wings, comprise 11,000 species, most of them neotropical. The caterpillars (Figure 13.63) of many feed on toxic herbaceous plants and sequester toxins like pyrrolizidine alkaloids and cardenolides, so the colorful adults may be aposematic. In some cases the toxins are used as pheromones, not just in defense. One of the most interesting aspects of arctiids concerns their tymbal organs, which produce pulses of high-frequency sounds (Dunning, 1968; Fullard *et al.*, 1979; Fullard, 1982). It has been suggested that the sounds are used in courtship, but they also appear to be used in defense against bats. One defensive role of the sounds may be a sort of auditory aposematism because these nocturnal moths need to advertise themselves not just visually. Alternatively, the sounds are believed to confuse bats by sending a signal that appears as an immediate reflection of the echolocating signal, indicating to the bat that an obstacle is close by.

In contrast to the colorful Arctiiidae are most of the Noctuidae, adults of which have delicate and subtle earthen patterns. These are almost certainly not a monophyletic group (Weller *et al.*, 1994; Speidel *et al.*, 1996). In fact, the Arctiidae and Lymantriidae are probably closely related, derived noctuids. There are approximately 38,000 species of noctuids as traditionally classified. The traditional classification is based on two groups, the trifines and quadrifines, which is based itself on the branching of the cubital veins in the hind wing. As can be supposed, in a classification based on a dichotomous feature like this, only one of the groups is monophyletic, and in this case it is the trifines (Kitching, 1984; Kitching and Rawlins, 1999; Mitchell *et al.*, 2000). Though taxon and gene sampling will need to be expanded, already the molecular-based relationships are revealing extensive paraphyly of the quadrifines (Mitchell *et al.*, 2000). The plusiines appear to be the sister group to the quadrifines, which includes as derived lineages the lymantriids and arctiids.

13.63. Caterpillar of a tiger moth (Arctiidae: Noctuoidea), some of which are commonly known as woolybears. Photo: V. Giles.

Thus, eventually either some of the noctuid subfamilies will need to be raised to family rank, or lymantriids and arctiids will need to be classified as subfamilies of noctuids. As indicated by Miller (1991) on the basis of detailed morphology, the notodontids appear to be at the base of the noctuoids. Most noctuids are exposed feeders, but many are concealed feeders like the agriculturally notorious cutworms (such as *Agrotis* and various other genera), which feed on stems near the ground; heliothines like the corn earworm bore into flowers and fruits. Exceptional biology includes several species in southeast Asia that suck blood as adults, which evolved from fruit-piercing ancestors (Bänziger, 1982), and the larval predators in Acontiinae that prey on scale insects or the remains of insects captured in spider webs.

BUTTERFLIES AND THEIR RELATIVES (RHOPALOCERA)

Centuries of philatelic effort have been lavished on butterflies, which more than any other group of organisms have been icons of natural beauty. A preoccupation with wing color patterns and variants within species, and the fascinating life histories of butterflies has made the natural history of this group extremely well explored. Their higher-level relationships, however, have been long neglected and are only now being seriously examined. Five families of true butterflies are traditionally recognized (Figure 13.64), classified in the superfamily Papilionoidea, and these are closely related to the skippers (Hesperioidea). The closest relative to skippers and butterflies is the small family Hedylidae, the "moth butterflies." The group name Rhopalocera has usually been applied just to the Papilionoidea or to these plus the Hesperioidea, but we are following Scoble (1995) in also including the small family Hedylidae.

Though Rhopalocera comprise only about 15% of all macrolepidopteran species, they actually have a better fossil record than any other group of macrolepidopteran. There are approximately 50 fossil specimens known, comprising some 33 named species in all living families (Table 13.3), not including very young (Pliocene and Pleistocene) fossils and obvious forgeries (e.g., d'Abrera, 2001: 65). Most of these are compression fossils, but several are inclusions in amber, and all the fossils are from the Cenozoic. The oldest rhopaloceran is an undescribed "apparent hesperioid" from the Late Paleocene Fur Formation (56–57 MYO) of Denmark (Kristensen and Skalski, 1999). The Fur Formation specimen has wings folded over each other, so the venation is not easily discerned, and head, legs, and mouthparts are obscured. The next oldest rhopalocerans are from the mid-Eocene (50 MYO) Green River shales of Colorado (Durden and Rose, 1978), but these are not particularly well preserved. Slightly younger still are several beautifully preserved specimens from Britain's Bembridge Marls (e.g., Figure 13.65) and from Florissant, Colorado. Nearly one third of the named fossil butterflies are from the fine-grained paper shales of the Florissant Formation in Colorado (Figures 13.66, 13.67), which is uppermost Eocene (36 MYO) (reviewed by Emmel *et al.*, 1992). Some of the Florissant butterflies are exquisitely preserved, with even color patterns intact. The most famous of these is *Prodryas persephone* (Figure 13.66), which lives up to its namesake, Persephone, the beautiful daughter of Zeus who was abducted by Hades to be his queen of the Underworld. Unfortunately, there still is a serious need for a systematic revision of these fossils, which will not be easy as they are widely scattered in museum and private collections. Of course, the finest preserved fossil rhopalocerans are in amber, but these are found thus far exclusively in Miocene Dominican amber (Figures 13.68, 13.69), reports of butterflies in Eocene Baltic amber notwithstanding.

Although none of the fossil rhopalocerans have affected our views of relationships in this group, they do indicate, for example, that some groups of butterflies have undergone dramatic extinction in geographic ranges, and the fossils contribute unique information on the ages of Rhopalocera. Even with the vicissitudes of the lepidopteran fossil record (especially major gaps in the Paleocene and Late Cretaceous), fossil rhopalocerans confirm the cladistic results that butterflies are a recently evolved lineage of Lepidoptera, despite estimates they evolved deep into the Cretaceous (e.g., Hall *et al.*, 2004). Butterflies and their close relatives may have originated in the early Paleocene, but even if they originated in the latest Cretaceous, 65–70 MYA, which is plausible (Vane-Wright, 2004), without question they largely evolved and entirely radiated in the early to mid Tertiary, 60–15 MYA. Indeed, the popular view of two lepidopteran groups, moths and butterflies, makes it difficult to reinforce the fact that butterflies are merely a very recent lineage of gaudy, day-flying moths, a situation analogous to dinosaurs and birds. A glimpse of early stages of butterfly evolution may be afforded by what appear to be the most basal rhopalocerans, the Hedylidae.

The Hedylidae is a small neotropical family (35 species) comprised entirely of the genus *Macrosoma*, which have traditionally been classified as geometrids. They are brown or gray-brown as adults (Figure 13.64); their larvae feed on a variety of shrubs. Compelling morphological evidence presented by Scoble (1986), Scoble and Aiello (1990), Minet (1991), and DeJong *et al.* (1996) indicate this family is the closest relative either to the true butterflies (Papilionoidea) or to this group plus the skippers (Hesperioidea). Although the relationship was disputed (Weintraub and Miller, 1987), the rhopaloceran position of the hedylids was also confirmed with DNA sequences, using a modest sampling of macrolepidopterans (Weller and Pashley, 1995). Hedylids have slender bodies and long wings like nymphalid and papilionid butterflies, and adults fold the forelegs under the thorax when

13.64. Assorted Rhopalocera, including the closest relatives to the Papilionoidea (true butterflies), the Hedylidae and Hesperiidae. 1, *Baronia brevicornis*; 2, *Bhutanitis* sp.; 3, *Parnassius bremeri*; 4, *Dismorphia amphiona*; 5, *Pieris beckerii*; 6, *Elbella etna*; 7, *Pyrrhopyge creon*; 8, *Thecloxurina fassli*; 9, *Arcas ducalis*; 10, *Macrosoma hedylaria*; 11, *Blepolensis catherinae*; 12, *Morpho cypris*; 13, *Agrias beatifica*; 14, *Heliconius elevatus*; 15, *Helicopis cupido*; 16, *Rhetus arcius*; 17, *Lyropteryx apollonia*. Approximately to the same scale.

TABLE 13.3. Pre-Pliocene Fossil Rhopalocera (Butterflies and Close Relatives)

Name	Super-/Family	Deposit	Age	Ref.
PALEOCENE				
Undescribed	Hesperioidea	Fur Form., Denmark	Late Paleocene	1
EOCENE				
Praepapilio colorado	*Papilionidae*	*Green R., Colorado*	Early Eocene	2
Praepapilio gracilis	Papilionidae	Green R., Colorado	Early Eocene	2
Riodinella nympha	?Riodinidae	Green R., Colorado	Early Eocene	2
Undescribed	?Nymphalidae	Green R., Colorado	Early Eocene	2
Prodryas persephone	Nymphalidae/-inae	Florissant, Colorado	Late Eocene	3
Vanessa amerindica	Nymphalidae	Florissant, Colorado	Late Eocene	4
Prolibythea vagabunda	Nymph.: Libytheinae	Florissant, Colorado	Late Eocene	5
Barbarothea florissanti	Nymph.: Libytheinae	Florissant, Colorado	Late Eocene	6
Lithodryas styx	Nymphalidae/-inae	Florissant, Colorado	Late Eocene	5
Chloripe wilmattae	Nymphalidae/-inae	Florissant, Colorado	Late Eocene	7
Jupitellia charon	Nymphalidae/-inae	Florissant, Colorado	Late Eocene	5
Nymphalites scudderi	Nymphalidae/-inae	Florissant, Colorado	Late Eocene	8
Apanthesis leuce	Nymphalidae/-inae	Florissant, Colorado	Late Eocene	5
Stolopsyche libytheoides	Pieridae	Florissant, Colorado	Late Eocene	5
Oligodonta florissantensis	Pieridae	Florissant, Colorado	Late Eocene	9
Nymphalites obscurum	Nymphalidae	Bembridge Marls, UK	Late Eocene	5, 10
Lithopysche antiqua	?Lycaenidae	Bembridge Marls, UK	Late Eocene	10
OLIGOCENE				
Thaites ruminiana	Papilion.: Parnassiinae	Aix-en-Provence, France	Mid-Oligocene	11
Archaeolycorea ferreirai	Nymph.: ?Danainae	Tremembé Form, Brazil	Mid-Oligocene	12
Neorinella garciae	Nymph.: ?Satyrinae	Tremembé Form, Brazil	Mid-Oligocene	12
Coliates proserpina	Pieridae	Aix-en-Provence, France	Mid-Oligocene	11
Lethites reynesii	Nymphalidae	Aix-en-Provence, France	Mid-Oligocene	11
Neorinopis sepulta	Nymphalidae	Aix-en-Provence, France	Mid-Oligocene	11
Lethe(?) corbieri	Nymph.: Satyrinae	Céreste, France	Late Oligocene	13
Pseudoneorina coulleti	Nymph.: Satyrinae	Céreste, France	Late Oligocene	14
Undescribed	Nymph.: Satyrinae	Canyon Ferry, Montana	Mid-Oligocene	23
Aquisextana irenaei	Lycaenidae	Aix-en-Provence, France	Mid-Oligocene	15
Pamphilites abdita	Hesperiidae	Aix-en-Provence, France	Mid-Oligocene	11
MIOCENE				
Vanessa karagancia	Nymphalidae	N. Caucusus	Mid-Miocene	16
Luehdorfitis bosniaskii	Papil.: Parnassiinae	Gabbro, Tuscany, Italy	Miocene	17, 18
Mylothrites pluto	Pieridae?	Radoboj, Croatia	Early Miocene	19, 11
Pontia freyeri	Pieridae	Radoboj, Croatia	Early Miocene	19, 11
Eugonia atava	Nymphalidae	Radoboj, Croatia	Early Miocene	20, 11
Voltinia dramba	Riodinidae	Dom. Republic (amber)	Early Miocene	21, 22
Undescribed	Nymphalidae	Dom. Republic (amber)	Early Miocene	here
Thanatites vetula	?? (poor)	Rott, Germany	Miocene	11

1 – Kristensen and Skalski (1999); 2 – Durden and Rose (1978); 3 – Scudder, 1875; 4 – Miller and Brown (1989); 5 – Scudder (1889); 6 – Scudder (1892); 7 – Cockerell (1907b); 8 – Beutenmüller and Cockerell (1908); 9 – Brown (1976); 10 – Jarzembowski (1980); 11 – Scudder (1875); 12 – Martins-Neto, 1993; 13 – Nel *et al.* (1993); 14 – Nel and Descimon (1984); 15 – Théobald (1937); 16 – Nekrutenko (1965); 17 – Krüger (1878); 18 – Bryk (1934); 19 – Heer (1849); 20 – Hübner (1819); 21 – Grimaldi (1996); 22 – Hall *et al.*, 2004; 23 – Cobabe *et al.*, 2002.

walking, which are reduced in males, as in nymphalid butterflies. But, hedylids retain a habit that is largely nocturnal, and they do not have the clubbed antennae found in skippers and butterflies. Interestingly, clubbed antennae appear in various unrelated groups of Lepidoptera that are diurnal, but the functional significance of this correlation is yet unknown.

The skippers, Hesperiidae (3,500 species), are mostly diurnal and are generally drab brown; they get their name from their erratic, quick flight. Some, like the pyrrhopygines, are very colorful. Megathymines are unusual in that their caterpillars bore into yuccas and *Agave*, whereas most other skippers fold or tie leaves together while feeding on grasses and other herbaceous monocots and some dicots (Figure 13.70). There is little doubt about the monophyly of skippers, since they are a rather homogeneous group defined by distinctive features, like antennal bases spaced far apart and the anten-

13.65. *Lithopsyche antiqua* (part and counterpart) (?Lycaenidae), from the latest Eocene to earliest Oligocene (38 MYA) Bembridge Marls of the Isle of Wight, Britain. NHM In 19984, 10369.

13.66. *Prodryas persephone* (Nymphalidae), from the latest Eocene of Florissant, Colorado. This is one of the most beautifully preserved compression butterfly fossils, which was shown by S. H. Scudder to Frank Carpenter as a boy and which inspired him to take up the study of fossil insects. Carpenter (Figures 2.45, 4.19) then spent 70 years studying fossil insects as a professor at Harvard. MCZ; forewing length 24 mm.

13.67. *Chloripe wilmattae*, from the latest Eocene of Florissant, Colorado, and preserved as the head, thorax, and forewings. MCZ 8602; left wing length 29 mm.

nal club with a small apical hook (see DeJong *et al.*, 1996). Kristensen (1976) and Weller and Pashley (1995) proposed Hesperiidae as the sister group to the Papilionoidea, though some authors propose that the Hedylidae is more closely related to the Papilionoidea. Ackery *et al.* (1999) thoroughly reviewed the family, including diagnoses of the world subfamilies.

The true butterflies, or Papilionoidea sensu stricto (ss), are a monophyletic group of approximately 14,500 species, essentially all of them diurnal (some morphine nymphalids, like *Morphopsis*, are apparently nocturnal). The more distinctive of their defining features include loss of the retinaculum and frenulum (wing coupling is achieved by the broad overlap of the fore- and hind wings); as well as a naked, exposed pupa that attaches itself to the host plant (Figure 13.72) or some protective substrate via the cremaster and a thin strap of silk, or "girdle" (though this latter feature occurs just in the swallowtails and sulphurs). Traditionally there are either four or five families, the Riodinidae often being treated as a subfamily of the Lycaenidae. Because butterfly workers have traditionally been concerned (actually, virtually obsessed) with variations in color patterns, there has been remarkably scant attention paid to the higher-level relationships of butterflies. The few studies on this topic are by Ehrlich (1958), Kristensen (1976), Scott and Wright (1990), and DeJong *et al.* (1996), and all of these studies are based on morphological characters. Ehrlich's study was the first comprehensive, exploratory study, but it used a phenetic analysis. Nonetheless, many of the morphological characters he reported are still used today. The study by DeJong *et al.* (1996) is now considered the most authoritative, and was recently reviewed by Ackery *et al.* (1999). In most of these studies the families Nymphalidae and Lycaenidae (sensu lato) are grouped together, and relationships of the Pieridae and Papilionidae are ambiguous. Either these latter two families are closely related, or one is more closely related to the nymphalid + lycaenid s.l. lineage. Although the pierids have lost an epiphysis like the nymphalids + lycaenids s.l., other evidence suggests that pierids and papilionids are actually closest relatives (Ackery *et al.*, 1999). More recent studies on higher-level relationships have focused within particular families, as discussed later in this section.

The species diversity of butterflies is essentially attributable to the Nymphalidae and Lycaenidae s.l., which comprise more than 85% of all butterfly species. Moreover, the greatest diversity of butterflies is in the Neotropical Region, with nearly 40% of the world's species (DeVries, 1987, 2001). In a gradient from 45° N latitude to the equator in the Western Hemisphere, the diversity of nymphalids and lycaenids expo-

13.68. A riodinid butterfly, *Voltinia dramba*, in Miocene amber from the Dominican Republic, 20 MYO, captured while possibly ovipositing on bromeliads.

13.69. A riodinid butterfly, *Voltinia dramba,* in Miocene Dominican amber. M0439.

nentially increases, while the numbers of pierids and papilionids change very little (DeVries, 2001).

As would be expected, publications on butterflies outweigh those on all other lepidopteran groups combined, the bulk of them being regional or faunistic treatments and field guides. Ackery (1984), updated to 1998 by Ackery *et al.* (1999), listed the major faunal works. The series by D'Abrera (1971–95) is a useful photographic atlas of the type specimens in the world's largest and most comprehensive butterfly collection, the Natural History Museum, London. For ardently antievolutionary commentary on classification, and butterfly taxonomy in particular, readers are referred to D'Abrera (2001).

The best known butterflies are the swallowtails, family Papilionidae, which comprise approximately 630 species, including the largest, most beautiful, and eagerly sought species, the birdwings (*Troides*, *Ornithoptera*). The family is clearly monophyletic, defined in part by a unique larval structure, the *osmeterium* (Figure 13.73). This is an eversible, forked lobe near the head that emits repugnant scents to repel enemies. Swallowtail caterpillars are often aposematically colored in bold, transverse stripes, which is probably related to the many toxic herbaceous plants they feed upon,

13.70. A particularly colorful caterpillar of a skipper, family Hesperiidae. Photo: V. Giles.

13.71. Hatching first instar pierid caterpillars. Immediately after emerging they begin on their first meal: the egg shell. Plant tissue has low nitrogen content, but the chitin in insect cuticle (including eggs shells) is rich in this vital element. Photo: P. J. DeVries.

13.72. A neotropical butterfly, *Archaeprepona*, newly emerged from its chrysalis or pupal case. Photo: P. J. DeVries.

13.73. Caterpillar of the swallowtail butterfly *Papilio polyxenes*. It is defending itself by regurgitating foul liquid and by everting the osmeterium on the prothorax (above the head), which emits a repugnant scent. Photo: V. Giles.

13.74. Pierid butterflies "puddling" in Belize, here feeding from wet soil laden with salts. Photo: P. J. DeVries.

and whose poisons they sequester. Only some of the papilionids have tails on the hind wings, including some of the basal ones. The most basal lineage of swallowtails, the subfamily Baroniinae, is comprised of a single, relict species from southwest Mexico, *Baronia brevicornis* (Figure 13.64) which feeds on *Acacia*. The subfamily Parnassiinae is small, with only 60 primarily Holarctic species, and caterpillars of the primitive species feed on toxic vines in the family Aristolochiaceae (other, nontoxic host plant families are fed upon by other parnassiines). Interestingly, Aristolochiaceae is the predominant host plant family of another swallowtail subfamily, the Troidinae. Parnassiines are the only butterflies to produce a cocoon, though it is unknown if this is a habit that reverted to an ancestral condition or was never lost from the ancestral habit. The subfamily Papilioninae contains most of the species (500 species); their greatest diversity is in the Old World tropics. Relationships in the family have been thoroughly explored (e.g., Hancock, 1983; Miller, 1987a,b), and recent molecular work supports the groupings proposed by Miller (1987a,b), including the sister-group relationship of *Baronia* to all other papilionids (Caterino *et al.*, 2001).

The Pieridae comprise 1,100 species of the familiar "sulphurs" and "whites" (Figure 13.74), though not all are white or light yellow. Many species in the Dismorphiinae, for example, mimic the gaudy neotropical heliconiine nymphalids (Figure 13.64). Pterin pigments in the wings of pierids impart the typical chalky white and yellow color, but these pigments are not exclusive to this group. Males of many pierids have more ultraviolet (UV)-absorbent pigments in their wings than do females, and because butterflies are very sensitive to UV wavelengths (Silberglied, 1984), there can be dramatic dimorphism in their color patterns that is invisible to humans. Notorious pests include *Pieris rapae* and *P. brassicae*, whose caterpillars feed on cultivated brassicas, and there has been a significant amount of behavioral work and sensory physiology done on host selection of these species. Many pierines, in fact, feed on the Brassicaceae, which they locate by the pungent mustard oils. The butterfly with perhaps the most anomalous life history is *Eucheira socialis*, a neotropical pierid whose larvae feed gregariously in a silken nest, the adults of which are fully winged but barely fly or not at all.

The family Lycaenidae (sensu stricto) (Figure 13.75–13.78) comprises some 5,000 species of blues, coppers, and hairstreaks, which are generally small to minute butterflies. Their classification has been treated by Eliot (1973), but is in need of great work, particularly since other authors have created more confusion than order in this family. Often included in this family are the riodinids, or metalmarks, a lineage of some 1,500 mostly neotropical species. While some authors concluded that the riodinids are essentially a very heterogeneous lineage deeply embedded within lycaenids (Kristensen, 1976; DeJong *et al.*, 1996), there are features of riodinids that apparently indicate they are a separate group from lycaenids, though they are closely related (Ackery *et al.*, 1999). DNA sequences from the *wingless* gene indicates that Lycaenidae and Riodinidae are monophyletic sister groups (Campbell *et al.*, 2000). Lycaenids are commonly abundant where their

13.75. Gregarious caterpillars, *Eumaeus godarti* (Lycaenidae), feeding on leaves of the cycad *Zamia* in Panama. Photo: P. J. DeVries.

13.76. Caterpillar of the lycaenid butterfly *Curetis regula*, from Brunei. This species is tended by ants. Photo: P. J. DeVries.

13.77. Camouflaged caterpillar of the lycaenid butterfly *Thereus*, in Ecuador. It is being tended by *Crematogaster* ants and is further protected by resembling its host plant. Nearly a third of all lycaenid and riodinid caterpillars are tended by ants. The caterpillars appease the ants with secretions and receive protection in return. Photo: P. J. DeVries.

13.78. A predatory North American lycaenid butterfly caterpillar, *Feniseca tarquinius*, feeding on mealybugs (Coccoidea) that are tended by *Camponotus* ants. The ants are ignoring the caterpillar, which is defended by dense hairs and probably by secretions that mask its identity. Photo: V. Giles.

13.79. A procession of metalmark caterpillars, *Euselasia* (Riodinidae), in Belize. They travel from their daytime resting spots at the base of the tree and feed on leaves at night. They are not tended by ants. Photo: V. Giles.

13.80. Caterpillar of *Ancyluris inca* in Panama, a non-myrmecophilous riodinid caterpillar. Riodinid caterpillars that are not tended by ants defend themselves with thick, long hairs. Photo: P. J. DeVries.

13.81. A "balloon head" riodinid caterpillar (*Theope*) tended by *Azteca* ants in Panama. Some riodinid caterpillars appease ants not only with secretions but also by stridulating. The stridulatory structures are located on top of the head, just below the "balloons." Photo: P. J. DeVries.

host plants occur, flitting about in full sunshine in the canopies of tropical forests or in open fields. Riodinids have a quick, darting flight in the understory; they are often active for a brief period of the day, and rest most of the time beneath leaves with their wings spread out. When flying, the wings of some riodinid species produce small flashes of light from reflective patches under their wings (hence the common name). The most significant work on riodinids is the book by DeVries (1997). Though it is based on the Costa Rican fauna, it is a comprehensive treatment replete with original observations on the distributions, immature stages, host plants, and behavior of many neotropical genera, including the myrmecophilous forms. The riodinids and lycaenids arguably have the most interesting larval biology of all butterfly caterpillars (e.g., Figures 13.76 to 13.82).

Myrmecophily, an intimate association with ants, is probably the most ecologically significant feature of lycaenids and riodinids, most instances of it being an appeasement of ants by caterpillars for protection, as in hemipterans (Cottrell, 1984; DeVries, 1990; Fiedler, 1991). Out of 850 life histories known for the lycaenids, 245 are myrmecophilous (Downey, 1962), and perhaps similar proportions occur in the riodinids. In fact, this habit is so widespread in both groups that it has oft been thought to have originated in the common ancestor of the two groups. The myrmecophilic adaptations in the caterpillars of each group are quite different, however, and the sister-group relationship of the two families indicates parallel evolution of this interesting behavior (DeVries, 1991, 1992, 1997; Fiedler, 1991; Pierce *et al.*, 2002). For example, both groups have a pair of eversible lobes (*tentacle organs*), which secrete alarm pheromones that excite ants when the caterpillar is being disturbed. These occur on abdominal segment 8 (A8) in lycaenids, but on thoracic segment 3 (T3) in riodinids. Riodinids have on A8, instead, a pair of *tentacle nectary organs*, which, as the name implies, secrete a honeydew-like substance. A functionally similar gland is the unpaired *dorsal nectary organ*, or *Newcomer's organ*, in the middle of A7 in lycaenids, which riodinids do not have. Moreover, riodinid "honeydew" is very rich in amino acids and sparse in sugars; that of lycaenids is very rich in sugars and less so in amino acids (DeVries, 1997). Lastly, some riodinids have *vibratory papillae* on T1, which are knobs with fine annulations that are rubbed over minute teeth on the top of the head, creating stridulating sounds that excite ants. In fact, these sounds are similar to those of the ants that pugnaciously defend the caterpillars (DeVries *et al.*, 1993).

Thus, there is an extensive repertoire of sounds, chemicals, and behavior that many lycaenid and riodinid larvae have to appease and even control ants. The most specialized myrmecophiles, though, are the lycaenids whose larvae live within ant nests, like *Maculinea*, feeding on the regurgitates of adult ants or devouring their brood. The caterpillars of Miletinae lycaenids (150 mostly Old World species) do not feed on plants at all but instead feed on hemipteran honeydew, on the hemipterans themselves, or within ant nests. The evolution of predation in ant nests by lycaenids very plausibly evolved from an interaction of ants first tending caterpillars for their secretions, then probably transporting the caterpillars back to their nest as they do with some hemipterans.

The Nymphalidae (Figures 13.83 to 13.86) is the largest family of butterflies, with one third the total diversity of butterflies (some 6,000 species), and it is certainly the most

13.82. (left). A riodinid caterpillar, *Eurybia patrona*, tended by ants in Costa Rica. The ants are feeding on secretions from the pair of tentacle organs at its hind end. Photo P. J. DeVries.

diverse in structure and patterns of wings. The caterpillars, too, are extremely diverse in structure, though unlike riodinids and lycaenids, they are almost entirely phytophagous. Nymphalids are the "brush-footed" butterflies, so named for the reduced forelegs of males and females, which are not used in walking. Actually, females in the small subfamily Libytheinae have functional forelegs, so this group appears to be the sister group to all other nymphalids, which is why this group traditionally was split off into a separate family. Otherwise, relationships among the subfamilies of this diverse family are poorly explored. Large subfamilies that may not be monophyletic are the satyrines (2,400 species) and limenitidines (1,000 species). The Morphinae include the brilliant blue morphos as well as the large, understory "owl" butterflies, *Caligo*. The genus *Morpho* itself is a taxonomic nightmare, thick with trivial names (DeVries and Penz, pers. comm.), and reflects the systematic problems encountered when a group is the focus of zealous philately. A genus of Charaxinae that has similar gordian taxonomy is *Agrias*, species of which have brilliant colors and are eagerly sought by collectors. The Heliconiinae comprise 400 species of bold-patterned butterflies, including the "passion vine" butterflies,

13.83. Mating clear wing butterflies, *Haetera macleannania* (Nymphalidae: Satyrinae), in Panama Photo: P. J. DeVries.

13.84. Caterpillars of assorted nymphalid butterflies; this family has the greatest structural diversity among butterflies. a, *Adelpha* (Panama); b, *Euptychia insolita* (Costa Rica) c, *Morpho achilles* (Ecuador) d, *Heliconius cydno* (Costa Rica) e, *Opsiphanes* (Panama) f, *Batesia hypochlora* (Ecuador) g, *Danaus plexippus* (New York) h, *Idea* (Brunei) i, *Epiphile* (Costa Rica). Photos: P. J. DeVries and V. Giles; not to the same scale.

13.85. A camouflaged *Hamadryas* butterfly (family Nymphalidae) from Costa Rica. They typically rest head down at the base of trees during the day. Called "cracker butterflies," the wings produce a low cracking noise when these butterflies engage each other. The caterpillars feed on Euphorbiaceae. Photo: V. Giles.

13.86. Not all butterflies feed on nectar. *Melinaea* butterflies (Nymphalidae) follow army ant swarms and feed on the droppings of ant birds. Photo: P. J. DeVries.

Heliconius and related genera, relationships of which have been treated by Penz (1999) and Brower and Egan (1997). Many heliconiines have a slow, fluttering flight in the forest understory. The Danainae, which have been monographed (Ackery and Vane-Wright, 1984), include the familiar milkweed butterflies. Danaines have the very intriguing behavior of adults feeding on flowers or plant exudates containing toxic pyrrolizidine alkaloids, which they use for producing pheromones or defensive secretions, as do some arctiid moths.

With their expansive wings and brilliant patterns to advertise themselves sexually and aposematically, it is hardly surprising that butterflies have repeatedly evolved, more than most other groups of insects, deceptive coloration and mimicry. Heliconiines and danaines, in fact, have been premier subjects in the study of natural deception.

MIMICRY

In its most general sense, mimicry is when one species benefits by resembling another. There are cryptic mimics, or species that resemble their environment (camouflage) and escape the notice of predators or even potential prey. The most common models for crypsis are of leaves and stems, remarkable examples being in Phasmida (stick and leaf insects), Mantodea (leaf and flower mimics), geometrid larvae (twig and flower mimics), and pseudophylline katydids (leaf mimics). Some black and white glistening caterpillars even mimic bird droppings (Figure 13.87), even developing formations of white uric acid on their body to do so. Mimicry in a more restricted sense is usually classified into two forms, Batesian and Müllerian. In Batesian mimicry, edible unrelated mimics closely resemble a sympatric model that is distasteful or even toxic. Müllerian mimicry is when unrelated sympatric species that are toxic or distasteful closely resemble each other, but the distinction between the two forms can be blurred, and Müllerian *mimicry rings* can also be "parasitized" by Batesian mimics. In both forms the mimics and

13.87. The humblest camouflage: a butterfly caterpillar mimicking a bird dropping. Photo: P. J. DeVries.

(Figure 13.89). This butterfly has five morphs, some of which mimic three species of distasteful *Amauris* milkweed butterflies (Nymphalidae: Danainae). Polymorphic mimics are common in butterflies, and the female is usually polymorphic, which is not too surprising because, even in butterflies that are not mimics, polymorphic individuals are usually females. Presumably, male color patterns are so important in species recognition and mating that males are rarely afforded the opportunity to evolve mimetic coloration. Also, females generally live longer than males, and some even live for weeks after mating and before oviposition. Thus, there may be particular selection upon females to defend themselves. Butterfly mimicry, by the way, is concentrated in groups where aposematic species are common, whose caterpillars models advertise themselves, generally with gaudy (*aposematic*) warning coloration. Aposematic coloration is not limited to insects, and commonly occurs in arachnids, myriapods, snakes (like coral snakes), frogs (Dendrobatidae), and a few other groups of poisonous animals. In all cases, from arthropods to vertebrates, the color of aposematic species always converge on patterns of black with red, orange, yellow, or white (or several of these), no doubt because of fundamental features of visual perception (Figure 13.88). The combination of black with a long wavelength color (yellow and red) is a biological road sign that contrasts against brown and green backgrounds. Mimicry, however, is most widespread in insects, excellent reviews of which are by Wickler (1968), Rettenmeyer (1970), and Joron (2003), and it is most intensively studied in butterflies (Carpenter and Ford, 1933; Papageorgis, 1975; Vane-Wright, 1981; Turner, 1984).

13.88. Anticamouflage in a caterpillar: the toxic and highly conspicuous "franjipani worm," *Pseudosphinx*. It is common throughout Central America and the Caribbean and feeds on *Plumeria* (Apocynaceae). Toxic and venomous animals – and the species that mimic them – usually evolve bold, contrasting patterns of black with yellow, orange, and red. Photo: D. Grimaldi.

Batesian Mimicry. Batesian mimicry is a density-dependent phenomenon. The effectiveness of a model is directly related to its abundance: If too abundant, predators quickly learn that the color pattern does not need to be avoided. The effectiveness of a color pattern is also a consequence of the predators because predators differ in their abilities of perception and learning (Speed and Turner, 1999). North American blue jays (*Cyanocitta cristata*), for example, are excellent learners, and they will refrain from ever eating a butterfly that made them ill, but most other birds and insectivores are less discriminating. The most famous example of Batesian mimicry involves distasteful monarch butterflies (*Danaus plexippus*) as the model, and viceroy butterflies (*Limenitis archippus*) as the mimic, though it is now known that viceroys are somewhat distasteful (which would make this Müllerian mimicry). Some Batesian mimics can be very polymorphic and mimic several different models, perhaps the most famous example being the African mocker swallowtail, *Papilio dardanus*

13.89. One of the most impressive examples of Batesian mimicry: the African mocker swallowtail, *Papilio dardanus*, showing morphs that mimic different butterfly species.

feed on toxic plants and sequester the toxins. These include the monarchs (danaine nymphalids), which feed on milkweeds (Asclepiadaceae); heliconiine nymphalids, which feed on passion vines (Passifloraceae); ithomiine nymphalids, which feed on Solanaceae and other digitalis-containing plants; and the troidine swallowtails, which feed on pipe vines (Aristolochiaceae). Mimicry is rare in lycaenids, hesperiids, riodinids, pierids, and satyrine nymphalids.

The most common insect models are aculeate wasps. Flower-visiting insects, like syrphid flies, sesiid and actiiid moths, and even some beetles, commonly mimic vespids (Figure 13.90) because these wasps are often found gleaning pollen from flowers as well as being extremely memorable models. But ants are probably the most common models among all insects because they are ubiquitous and also well defended with stings or formic acid. Ant mimicry often involves the evolution of an antlike body form and even furtive movements. Other aculeates that are often mimicked include the deep, metallic blue Pompilidae (spider hawk wasps), which sometimes have light-tipped antennae or bright orange wings. The sting of these wasps is among the most painful known, so they make effective models. Pompilid mimics include various other aculeates, as well as insects as diverse as katydids and reduviid bugs (Figure 13.91), which even mimic the quick, twitching movements of pompilids. Bees are also mimicked, such as bumblebees by syrphid and asilid flies (e.g., Figure 13.92). In the case of the predatory asilid mimics, though, their mimicry is not just for defense.

13.90. Some remarkable Batesian moth and fly mimics of vespid wasps. Stinging wasps are frequent models for flower-visiting insects.

13.91. Batesian mimics of spider hawks (Pompilidae; cf., Figures 11.39, 11.41), which are wasps that deliver an extremely painful sting. a, Katydid (Tettigoniidae: Orthoptera) in Costa Rica; b, assassin bug in Costa Rica (Reduviidae: Heteroptera). Photos: P. J. DeVries.

13.92. Batesian mimic of a bumblebee, the robber fly *Laphria*, with beetle prey. This mimicry is probably a camouflage that enables the fly to closely approach other insects, including bumblebees. Photo: S. Marshall.

Müllerian Mimicry. Müllerian mimicry is the more common of the two basic forms of mimicry, and in this situation the mimics often bear less of a resemblance than do Batesian models and mimics, presumably because the signal is so effectively reinforced. The best-studied example involves neotropical *Heliconius* butterflies (Papageorgis, 1975; Turner, 1977, 1981; Smiley, 1978; Gilbert, 1983; Mallet, 1993; Brower, 1997; Mallet and Joron, 2000; Joron *et al.*, 2001; Beltran *et al.*, 2002; Naisbit *et al.*, 2003). Some *Heliconius* have a remarkable number of color morphs, like *H. numata* with seven such morphs in an area, each an accurate mimic of various species of *Melinaea* read: Melinaea (Ithomiinae). *Heliconius erato* is intensively studied because it too has numerous color morphs, most of them in different areas of the distribution where this species contacts other species of models. In fact, where the ranges of two-color *H. erato* color morphs come in contact, the boundary can literally be just hundreds of meters wide. For an overall distribution of the species that encompasses the entire northern half of South America and much of Central America, such sharp boundaries reflect the dynamics of selection pressures on the color morphs and the speed of evolutionary change.

Classical studies on the genetics of butterfly color patterns by Clarke and Sheppard explain how mimics can evolve so quickly and effectively (Clarke and Sheppard, 1955, 1960a,b; Sheppard *et al.*, 1985). In the highly polymorphic *Heliconius numata*, one locus controls gross color patterns and seven alleles refine details of the pattern, and similar gene "clusters" occur in other *Heliconius*. One gene converts yellow *Papilio glaucus* to black, and thus its mimicry of the distasteful swallowtail model *Battus philenor*; other genes control the yellow spotting. In effect, a mutation in one locus can have a dramatic effect on wing color patterns, and other "modifier" mutations with small effects refine the resemblance (Nijhout, 1986, 1991).

13.93. Müllerian mimicry complexes of various day-flying moth families (and one butterfly) from Central and South America. To the same scale.

14 Insects Become Modern: The Cretaceous and Tertiary Periods

THE CRETACEOUS

The Cretaceous Period, 145–65 MYA, is one of the most significant geological periods for insect evolution of the seven major periods in which insects are preserved. Hexapods appeared in the Devonian; winged insects, in the Carboniferous; and the earliest members of most modern orders, in the Permian to Triassic. In the Cretaceous, however, there evolved a nascent modern biota, amidst unprecedented geological and evolutionary episodes. Because the Cretaceous is so much younger than the Paleozoic and earlier Mesozoic periods, the fossil record of this period has been erased less by faulting, erosion, and other earth processes. Thus, Cretaceous fossils have left a particularly vivid record of radiations and extinctions.

During the Cretaceous many modern families of insects first appeared, some of which appeared as early as the Jurassic or even the Triassic. Others, like the families of schizophoran flies and ditrysian lepidopterans, are much younger, having evolved almost entirely in the Tertiary. It is plausible that half or more of the Recent insect families appeared in the Cretaceous, and this appears to be related as both cause and effect to another biotic revolution of the time: the radiation of angiosperms.

There is no question that the approximately 250,000 species of flowering plants comprise *the* dominant life form on land, though their origins approximately 130–140 MYA were very modest compared to their radiations some 20–30 MY later. Without angiosperms, terrestrial existence would be far less colorful and entirely unrecognizable, which is why we define almost all biomes on the basis of angiosperms, from heathlands and savannas to tropical rain forests. Insects were the original partners of angiosperms, as pollinators and their primary "predators" (i.e., herbivores), and insects remain the most intimate and pervasive associates of angiosperms. It is generally understood that pollinating insects facilitate and even control gene flow in most flowering plants, and thus must affect speciation of the plants, but how host plant specialization could affect the speciation and diversity of insects is ambiguous and controversial, as we will soon discuss. Without question, though, the angiosperm radiations opened vast niches that insects exploited supremely well.

The earth was geologically more restless during the Cretaceous than most times in its history. There was dramatic climate change and tectonic activity, the latter of which resulted in widespread volcanism and the splitting and drifting of continents. The fragmentation of Gondwana into the present southern continents 120–100 MYA is often invoked to explain contemporary distributions of various plants and animals (including insects) that have closely related species occupying Australia, New Zealand, southern South America, and southern Africa. Ancestors of these austral relicts purportedly drifted with the continents, though some Cretaceous and Tertiary insect fossils indicate that many presently austral groups were very widespread. The massive meteoritic impact that ended the Cretaceous is well known for the extermination of non-avian dinosaurs, ammonites, and other groups, but insects appear to have been little affected by this biotic Armageddon, at least at a gross level. As a result, the Cretaceous can uniquely inform us about how insects radiated, how they endured mass extinctions, and thus why they are such remarkably successful organisms.

FLOWERING OF THE WORLD: THE ANGIOSPERM RADIATIONS

Angiosperms appear to be closely related to a particular group of "gymnosperms," specifically the living, relict group Gnetales. Gnetaleans have a fossil record that extends more than 60 MY before the earliest known angiosperms. Bennettitaleans (including the genus *Pentoxylon*) is an extinct group, which has also been proposed as the closest relatives of angiosperms. These plants resembled cycads and lived from the Triassic to the latest Cretaceous, and they had pollen-producing organs fused into a synagium; most were *dioecious* (where the different sexual organs are on separate plants); and they appear to have been insect pollinated (their reproductive structures were even called "flowers"). Bennettitales

14.1. One of the most profound biological episodes in the history of life was the radiation of flowering plants in the Cretaceous. Insects are the primary pollinators and consumers of angiosperms, and they have affected the evolution of each other more than have any other organisms. Here a small pyralid moth visits a coneflower (*Rudbeckia*) in Vermont.

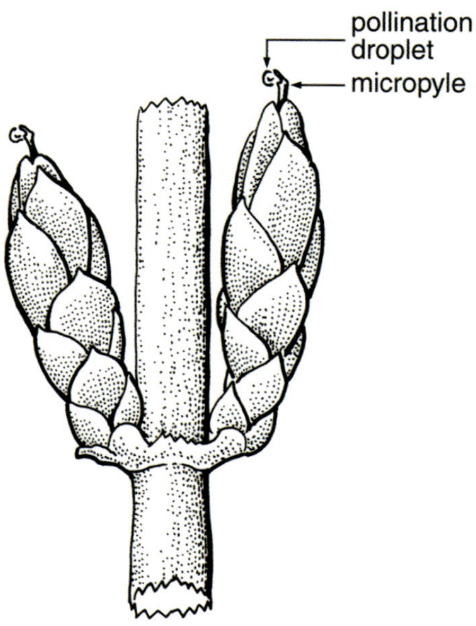

14.2. Reproductive structures of *Ephedra antisyphilitica*, a living member of the relict group Gnetales, which appear to be the closest living relatives of angiosperms, or possibly of the conifers according to recent molecular evidence. The small, sticky pollination droplets of gnetaleans not only capture wafting pollen but also attract insects and may have led to the symbiosis between pollinating insects and flowers.

appear to be most closely related to cycads (W. L. Crepet, pers. comm.). Gnetaleans may be the living sister group to the angiosperms, or they may be closely related to conifers. Molecular evidence for a close relationship between gnetaleans and conifers is compelling (Hansen *et al.*, 1999; Samigullin *et al.*, 1999; Bowe *et al.*, 2000; Chaw *et al.*, 2000), yet other studies support their close relationship with angiosperms (Doyle and Sanderson, 1997; Savolainen *et al.*, 2000). These plants actually possess several morphological features distinctive to angiosperms that seem unlikely to have evolved convergently (Loconte and Stevenson, 1990; Nixon *et al.*, 1994). There are three disparate genera of living gnetaleans. These include the cosmopolitan genus *Ephedra* with approximately 35 species (Figure 14.2), which superficially resembles horsetails (*Equisetum*), and the huge recumbant *Welwitschia mirabilis*. Both of these genera live in xeric regions, and *Welwitschia* in fact is restricted to the very dry Namib Desert in Africa. The other genus is *Gnetum* (Figure 14.3), the approximately 30 species of which are trees and vines found in the world's tropics. Fossil gnetalean pollen occurs through much of the Mesozoic, but definitive macrofossils of the group are known from the Early Cretaceous and later.

Fortunately, our current understanding of the early evolution of angiosperms has been revolutionized in the past few decades by essentially three modes of inquiry.

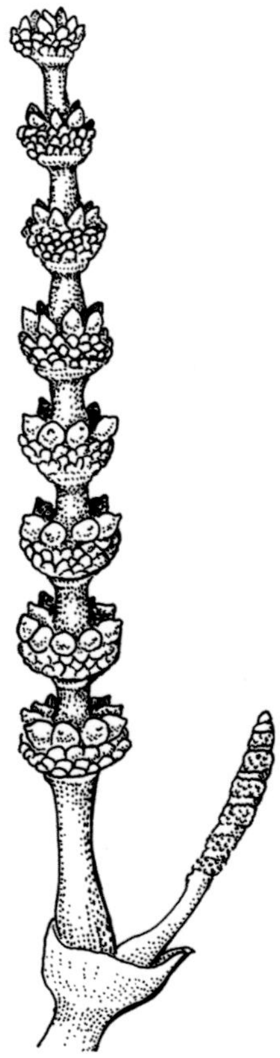

14.3. Branch with microsporangia of the living gnetalean plant, *Gnetum gnemon*.

1. The use of phylogenetics in studying the relationships of living and fossil angiosperms (Crane, 1985; Donoghue and Doyle, 1989; Doyle and Donoghue, 1992; Nixon *et al.*, 1994; Bremer, 2000; Doyle and Endress, 2000).
2. Continuous discovery of Cretaceous fossils, but particularly the remarkably preserved and diverse fusainized flowers and fruits from Europe and North America (e.g., Friis *et al.*, 1986, 1994, 1999; Drinnan *et al.*, 1990; Nixon and Crepet, 1993; Crane *et al.*, 1995; Crane and Herendeen, 1996; Crepet and Nixon, 1998; Gandolfo *et al.*, 1998) (e.g., Figure 14.4). Fusainized plant remains were produced by fires, similar to the process of charcoalification, and they are preserved with microscopic fidelity. Prior to the discovery of intact, fusainized reproductive structures the fossil record of angiosperms was largely relegated to pollen, leaves, and the occasional compressed flower. There are also a few beautifully preserved flowers in Cretaceous amber (Figures 14.5, 14.6). Important new deposits of Cretaceous compression fossils include those from the Yixian Formation of China (e.g., Sun *et al.*, 2002) and the Santana Formation of Brazil (Mohr and Friis, 2000) (Figure 14.7).
3. The application of DNA sequences to large numbers of angiosperm taxa (APG, 1998; Savolainen *et al.*, 2000; Soltis *et al.*, 2000), and particularly to the identification of the earliest "roots" in the phylogeny of living angiosperms (Matthews and Donoghue, 1999; Qiu *et al.*, 1999, 2000; Barkman *et al.*, 2000; Zanis *et al.*, 2002).

The earliest fossil evidence of angiosperms is pollen 130–140 MYO preserved in Israel, Morocco, Libya, and possibly China, and the earliest macrofossils are slightly younger leaves and flowers some 120–130 MYO. The earliest remains are sparse but become progressively more abundant and diverse into the Cretaceous, which has led to the widespread belief that angiosperms originated in the earliest Cretaceous, 130–145 MYA. This leaves a huge gap between the earliest angiosperms and earliest gnetaleans and bennettitaleans, which some recent investigators have interpreted as a fossil record deficient for angiosperms. Estimates of angiosperm origins based on molecular divergence are typically far older than those estimates based on fossils, and range from the Jurassic, 180–160 MYA (Wikström *et al.*, 2001), to even the Triassic (Li *et al.*, 1989; Kenrick, 1999). Besides the usual prob-

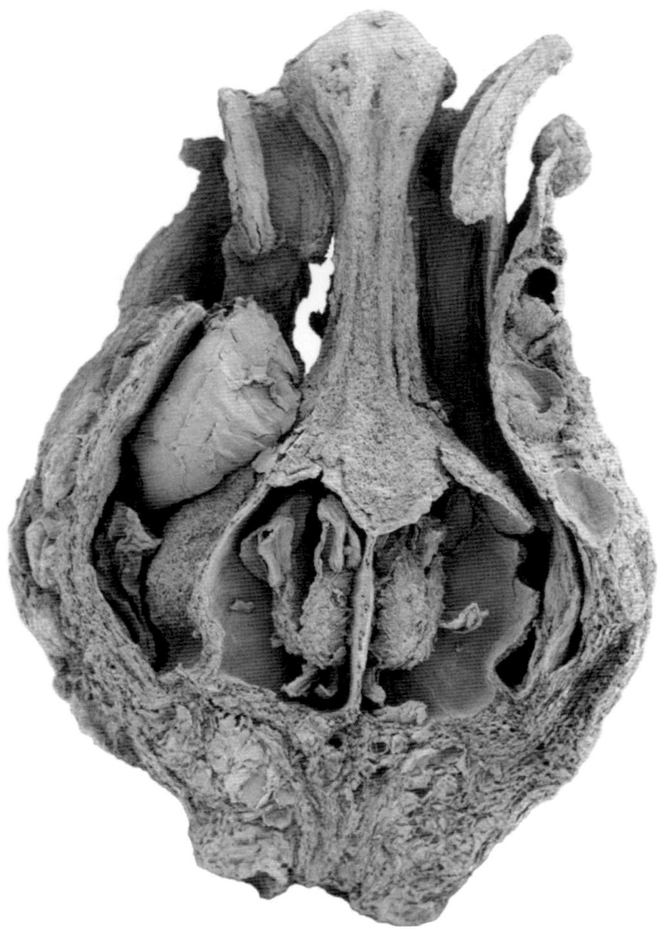

14.4. Cretaceous flower exquisitely preserved as a carbonized replica in 90 MYO clays from New Jersey. Remains of plants and animals smothered under leaf litter were apparently rendered to charcoal by ancient forest fires, or *fusainized*. The discovery of diverse fusainized flowers from the Cretaceous has revolutionized understandings of the early evolution of angiosperms. Scanning electron micrograph. Photo: W. L. Crepet.

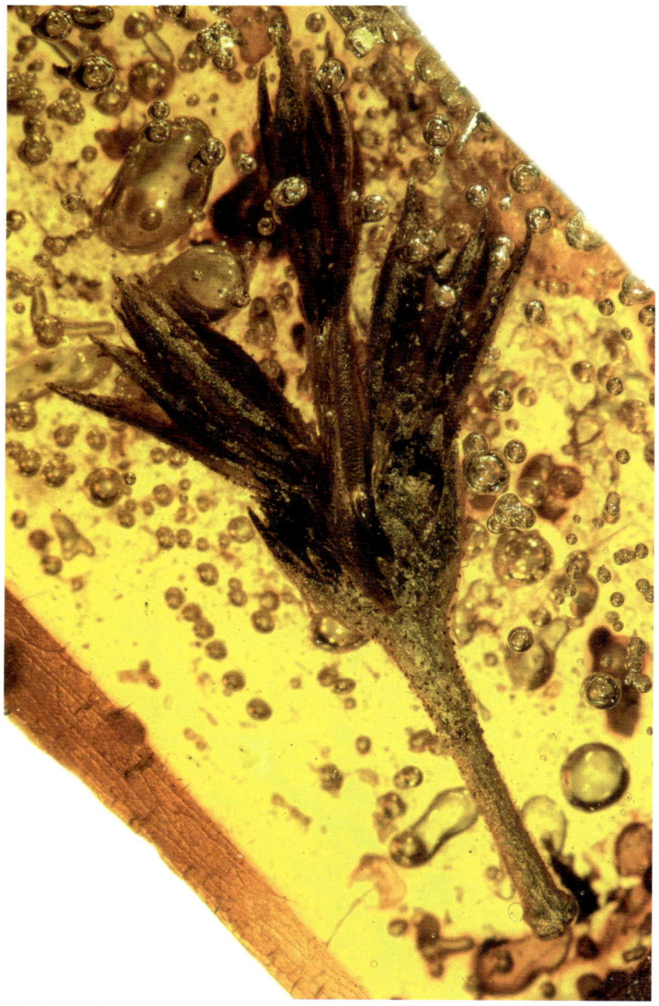

14.5. An inflorescence in 90 MYO amber from New Jersey. It is one of the earliest records of the Fagaceae, which includes the beeches and oaks. Flowers in Cretaceous amber are extremely rare, and even fewer are as detailed as this. AMNH; length 7 mm.

14.6. Another exquisitely preserved inflorescence in New Jersey amber. Length 4.5 mm.

lems with dating lineages using molecular clocks, these overestimates are probably also a result of the use of *living* sister groups for estimating divergence and rates. In cases where the basal branches of a lineage have been extensively pruned by extinction, as we suspect is the case for the base of the angiosperm tree, then this approach will be misleading (Doyle, 1998). It is very likely, in fact, that angiosperms evolved from some stem-group gnetalean in the uppermost Jurassic or earliest Cretaceous, so the great "fossil gap" may be merely illusory.

Pollen disperses very well, and its coat of sporopollenin is one of the most durable biopolymers known, so pollen has been the most important source of information about early angiosperm diversity. By the Cenomanian, approximately 95 MYA, many types of angiosperm pollen had evolved (Lupia, 1999). Using all fossil evidence, including leaves, flowers, wood, and pollen, the period of greatest angiosperm diversification was between approximately 115 and 90 MYA (Lidgard and Crane, 1988; Wing, 2000), which is the time of the *angiosperm radiations*, when the most diverse angiosperm groups – monocots and eudicots – appeared to have rapidly radiated. This timing is inferred on the basis of fossils and their relationships because the fossil record alone gives an incomplete picture. This was also a time when the diversity of more archaic groups from the earlier Mesozoic – especially the cycads and ferns – dramatically declined, though causes and effects in these trends remain purely speculative at this point. Most studies agree that the earliest angiosperms were small, weedy herbs and shrubs, based on the fact that their seeds were small (and thus probably wind dispersed), and because the wood was slender and rare

14.7. Gnetalean from the Early Cretaceous Santana Formation of Brazil. This deposit has yielded a wealth of early angiosperms and gnetaleans.

(Wheeler and Baas, 1991). By the end of the Cretaceous, large and abundant sections of fossilized wood indicate that many trees at that time were angiosperms. Cretaceous angiosperms may have proliferated first in low latitudes, where like now it was hotter, and they gradually colonized the higher latitudes where the climate was more temperate. It has been estimated, in fact, that angiosperms invaded Australia about 100–110 MYA, probably via dispersal through Antarctica (Dettman, 1990), though this interpretation conflicts with the fact that some of the most basal angiosperm lineages are austral.

The most basal angiosperms are two species of the extinct genus *Archaefructus* from the Early Cretaceous Yixian Forma-

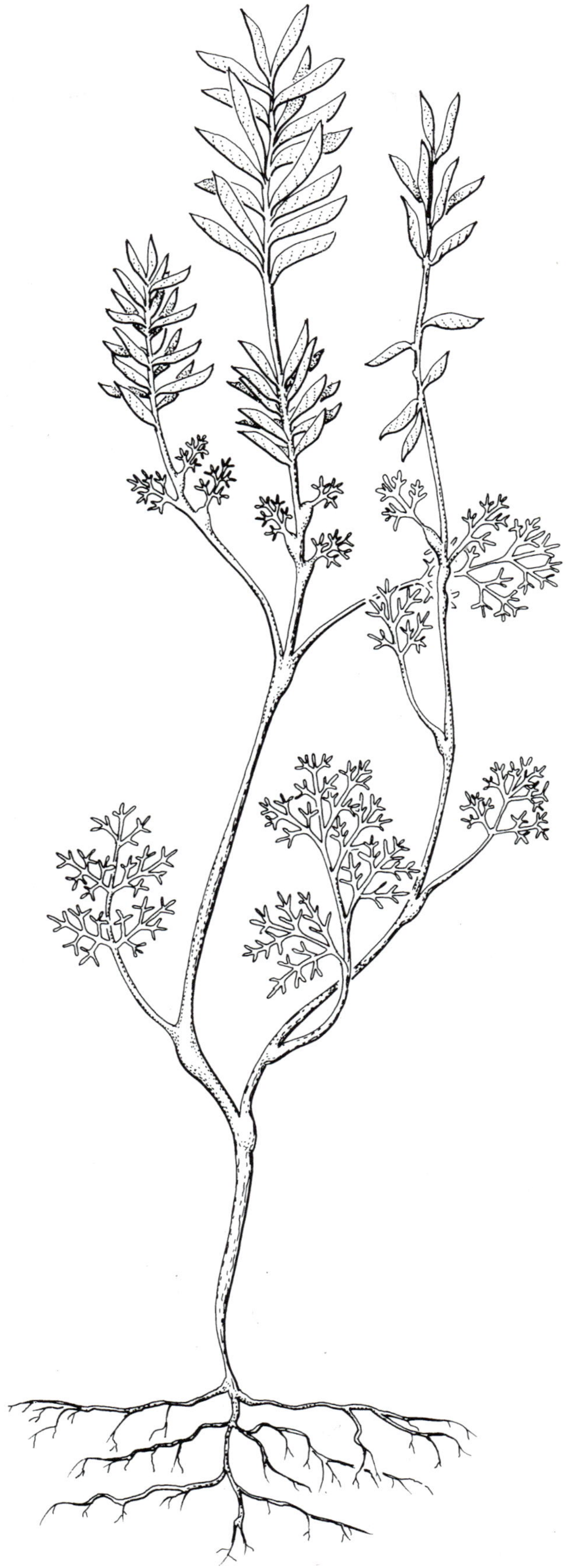

14.8. Reconstruction of the most primitive and fully preserved Cretaceous angiosperm, *Archaefructus sinensis*, from the Early Cretaceous Yixian Formation of China. The flowers are unknown, and only seed capsules were preserved on the terminal shoots, so it is difficult to infer the mode of pollination. Redrawn from Sun *et al.* (2002).

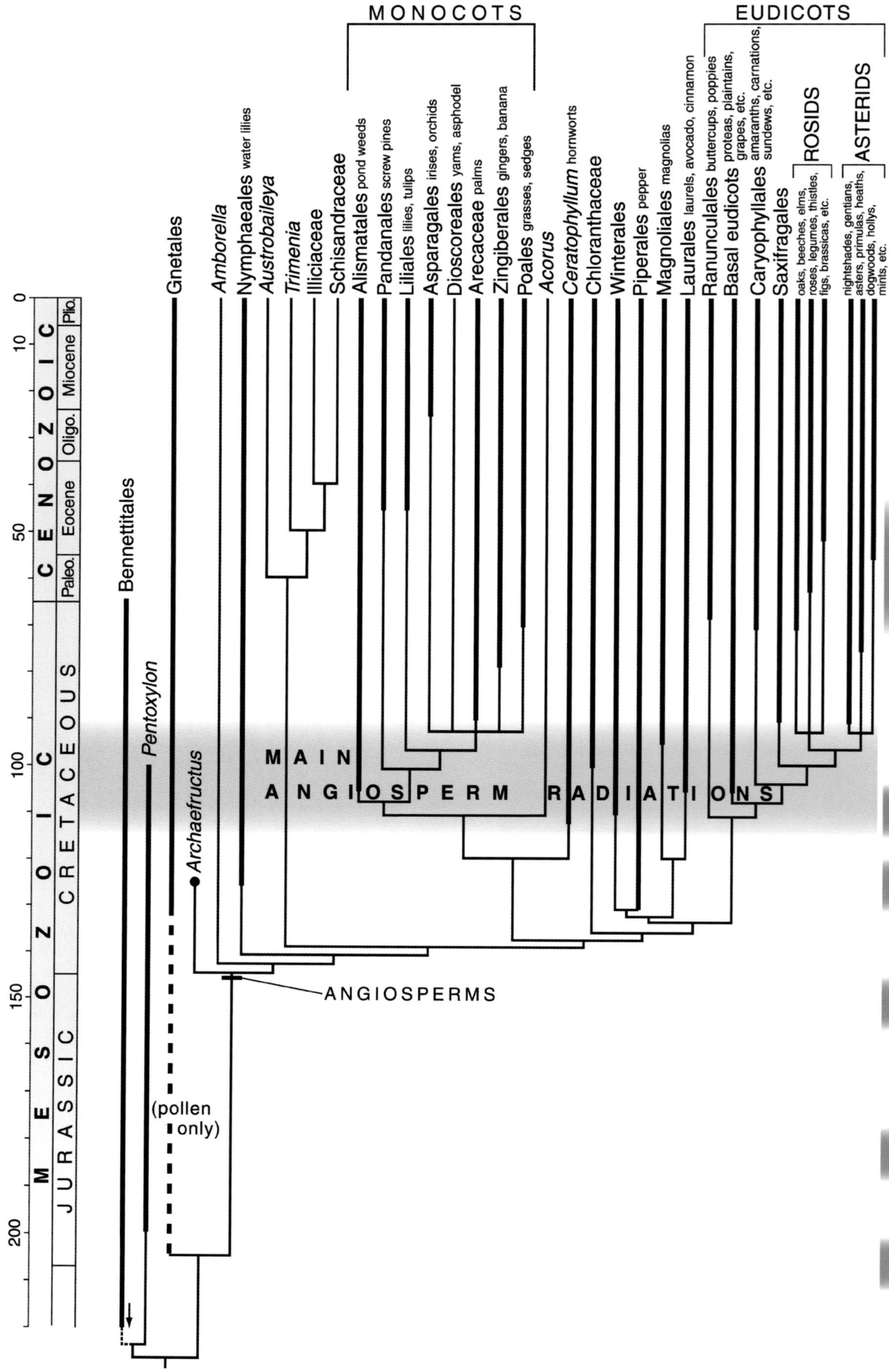

14.9. A phylogeny of angiosperms, including relationships and occurrence in the fossil record. The phylogeny is based on various sources (see text).

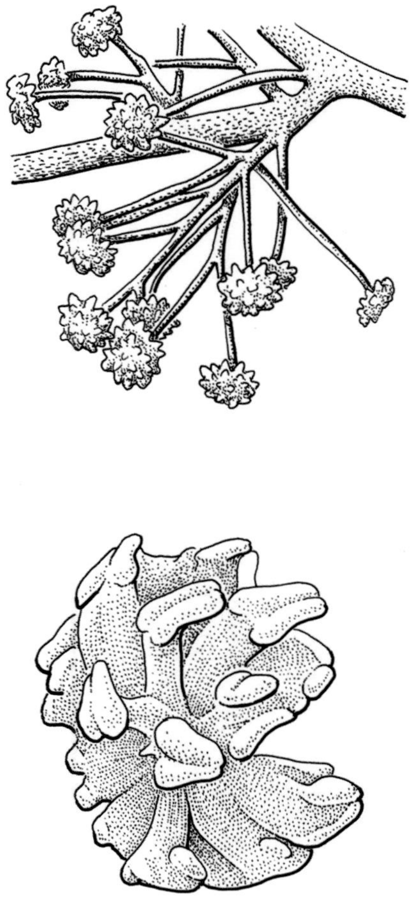

14.10. Staminate inflorescence (above) and flower (below) of the basalmost angiosperm, *Amborella trichopoda* from New Caledonia. Its mode of pollination is still unknown.

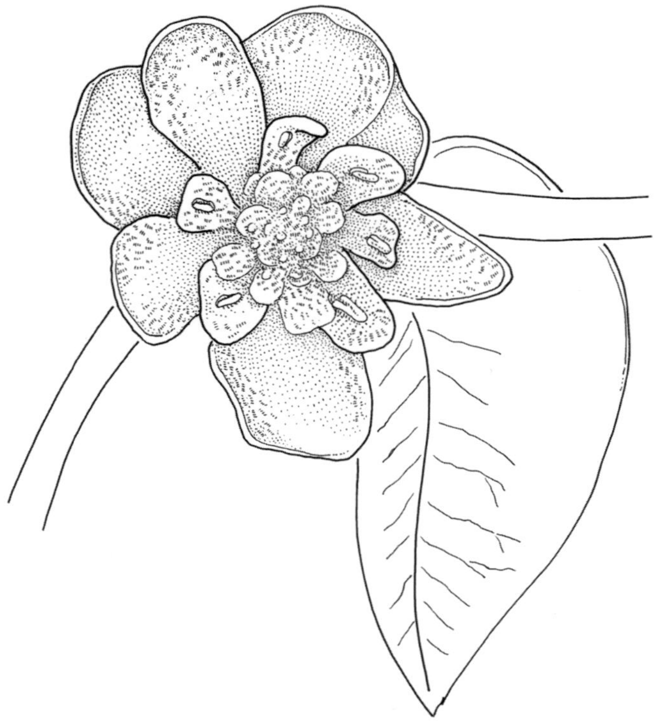

14.11. Flower of the basal angiosperm, *Austrobaileya scandens* (Austrobaileyaceae), from Australia. The flower is light green, flecked with purple, and has a putrescent odor – a pollination syndrome associated with flies.

tion of China (Sun *et al.*, 2002), which preserve the earliest known flowers. *Archaefructus* is also the most completely known Cretaceous angiosperm since abundant vegetative and reproductive remains have allowed its complete reconstruction (Figure 14.8). It was originally reported as the first Jurassic flower, but when the age of this formation was more critically examined, it appears to be approximately Barremian to Aptian in age, some 125 MYA. The basal living angiosperms comprise a grade of three lineages, the so-called ANITA group (Figure 14.9). The first of these is *Amborella*, the most basal living angiosperm, which is a small, dioecious shrub that grows in rainforests of New Caledonia, far off the northeast coast of Australia (Figure 14.10). The next is the Nymphaeales, or "water lilies," a group of about 65 species of aquatic herbs with broad, floating leaves. Fossil leaves of this group are abundant and diverse through the Cretaceous, as one might expect since the main deposits of fossil plants are lacustrine, but the flowers are considerably more rare (e.g., Krassilov and Bacchia, 2000; Friis *et al.*, 2001; Gandolfo *et al.*, 2004). Other lineages of ANITA have no known fossil record. The last lineage of ANITA, and a possible sister group to all other angiosperms, is comprised of the Illiciales (two families, the Illiciaceae and Schisandraceae, with 90 species total from southeastern Asia, southeastern United States, Cuba, Hispaniola, and the moist forests of Mexico) and *Austrobaileya*, which is an isolated genus of vines in its own family from northeastern Australia (Figure 14.11). There is then a large monophyletic group comprised of the monocots, the eudicots, and small intermediate groups. These "intermediates" include taxa that have traditionally been considered the most basal angiosperms, like magnolias, Winteraceae, and Chloranthaceae. The monocots and eudicots comprise approximately 95% of all angiosperm species.

PLANT SEX AND INSECTS: INSECT POLLINATION

What had that flower to do with being white
The wayside blue and innocent heal-all?
What brought the kindred spider to that height,
Then steered the white moth hither in the night?
What but design of darkness to appall?
If design govern a thing so small.
– Robert Frost, *Design*

Pollination is the transfer of the male gametophyte in seed plants, or pollen, to the receptive surface of the female organs, the stigma, for fertilization of the ovule. While many plants are capable of self-fertilization, only some weedy and ephemeral plants routinely do so. Some seed plants disperse

14.12. The partnership between angiosperms and pollinating insects is one of the most pervasive symbiotic relationships on earth, which probably appeared just as flowers evolved. Here, a tiny halictid bee loaded with purple pollen is lured to a flower in Kyrgyzstan. Photo: M. S. Engel.

14.13. Many insects and flowers are involved in intricate symbiotic webs. An ithomiine butterfly here is feeding on the composit *Eupatorium* in Costa Rica. *Eupatorium* is laced with toxic pyrrolizidine alkaloids, which are actually gathered by these butterflies and used for sexual displays and chemical defense. Ithomiines are frequently mimicked by other butterflies because of their toxicity. Photo: P. J. DeVries.

pollen by wind, including many conifers, the grasses, sedges, and rushes (Poaceae, Cyperaceae, and Juncaceae), and temperate species of oaks and beeches (Fagaceae), elms (Ulmaceae), birches (Betulaceae), and some willows (Salicaceae). The unparalleled success of the angiosperms, however, is very much a result of symbiotic relationships with pollinating insects that began approximately 130–140 million years ago (or earlier with gnetaleans and bennettitaleans), and evolved to include some birds and bats 100 million years later. Pollination has been studied for centuries, and there have been many comprehensive reviews of the topic (e.g., Knoll, 1956; Meeuse, 1961; Percival, 1965; Faegri and van der Pijl, 1979; Barth, 1985; Proctor *et al.*, 1996).

Pollination of flowers by animals has many advantages. Traditional views are that animal pollination allows less pollen to be wasted because there is direct transfer of the pollen among flowers of the same species. Also, gene flow is controlled by the foraging behavior of the pollinator. Because this better ensures outcrossing, animal pollination provides the advantages of sexual recombination and heterozygosity and avoids the problems of inbreeding and mutational load (Dilcher, 1995, 2001; Proctor *et al.*, 1996). The diversity of angiosperms has also been attributed to sexual isolation among populations caused by the mechanical features of flowers and the foraging behavior of pollinators, possibly even leading to the rapid formation of many angiosperms in sympatry. Such isolation is now considered an incidental consequence of allopatric speciation, though pollinators are still involved in this divergence (Grant, 1994; Johnson, 1996; Waser, 1998). Lastly, and an aspect that is rarely discussed, is that pollination by animals (particularly winged ones) allows reproduction of plants growing distant from each other. When individual plants of a species are widely and sparsely distributed, as many trees are in tropical forests, it allows more efficient exploitation of resources, and also reduces intraspecific competition and serious outbreaks of plant-feeding insects that specialize on that species. If the earliest angiosperms were widely dispersed along the margins of streams and rivers, as paleontological evidence suggests (Hickey and Doyle, 1977), then insect pollination would have been a major asset in their reproduction. The relative significance of genetic and ecological factors in pollination is a broad topic not directly relevant here, but with little question the success of angiosperms is at least partly attributed to their liaison with animals, particularly insects.

Pollinators

The main pollinators are myriad groups of insects and a few groups of birds and bats. Each group has evolved particular specializations for feeding exclusively on the nectar and sometimes the pollen of flowers. These obligate pollinators are usually adept fliers and have a long proboscis or extensible tongue for probing the recesses of flowers. Conversely, flowers have pollination "syndromes," or suites of features that adapt them for pollination by particular animals. In the Western Hemisphere the main bird pollinators are the hummingbirds (family Trochilidae) and honeycreepers (some Thraupidae); in Africa and Asia the sunbirds (Nectarinidae) assume this role, while honeyeaters (Meliphagidae) are pollinators in the IndoPacific, and the honeycreepers (Drepanidinae) in Hawaii. Bird flowers tend to be large, vivid red, and have deep corollas. Among bats the main pollinators are some Pteropinae (fruit bats) of Africa and southeast Asia, though the Macroglossinae (nectar-feeding bats) are more specialized pollinators in southeast Asia and Australia. In the Western Hemisphere, the subfamily Glossophaginae in the family of leaf-nosed bats Phyllostomidae has specialized pollinators. Bat flowers tend to be large and fleshy, open at night, and pungent, as well as have numerous exposed and "brushy" stamens. Pollinating bats occur largely in the tropics and deserts, feeding for example on the large flowers of cacti. Some small monkeys and arboreal species of rats and opposums pollinate a few species of tropical trees. The proportion of flowers pollinated by vertebrates, though, probably amounts to less than 1% of all angiosperm species.

In contrast to vertebrates, insects pollinate approximately 85% of all angiosperms, though there is great variation among biomes. At Finca La Selva biological station in Costa Rica, for example, 54 (20%) of 265 angiosperm species studied were pollinated by vertebrates, and very few were wind or water pollinated, so insects apparently pollinate some 80% of the species there (Bawa, 1974, 1990). One must bear in mind, though, it is easier to see a bird than a bee at flowers high in the canopy of a tropical tree, so this estimate may be biased toward vertebrates. In a large study on the subtropical Amami Islands south of Japan, insect pollination accounted for 93% of the pollination among the 164 angiosperm species that were studied (Kato, 2000). Other estimates of pollination, primarily from tropical forests, indicate that insects pollinate approximately 80–95% of the angiosperm species (Kress and Beach, 1994; Momose *et al.*, 1998; Frankie and Thorp, 2003). It must always be borne in mind that visiting or feeding from flowers (*anthophily*) is not necessarily pollination because pollen is not always transferred by anthophilous insects.

As one might expect, plants pollinated by insects have extremely diverse floral structures and "syndromes," and insect pollination includes some of the most specialized known mutualisms. Orchids, for example, are among the most diverse families of plants (approximately 25,000 species), and rapidly evolved within approximately the past 40 MY, perhaps as a result of their exotic modes of insect pollination (reviewed in Dressler, 1990). The flowers of some species mimic the shape of female wasps, and males unwittingly transfer pollen when they attempt to mate with the flowers, a process called *pseudocopulation*. Some South and Central American orchids secrete scents specific to each one of nearly 200 species of orchid bees (Euglossini). Male orchid bees feed from a variety of flowers but collect scents from the orchids to use for displaying to females (Dressler, 1968; Dodson *et al.*, 1969; Williams, 1978). Other orchids even mimic the appearance and odor of mushrooms and are pollinated by flies that normally congregate on mushrooms. Figs (*Ficus*: Moraceae) are large tropical trees with some of the smallest and most specialized pollinators, which are agaonid wasps (Wiebes, 1979; Janzen, 1979; Weiblen, 2002) (Figure 11.19). Many of the wasp species are restricted to a single species of fig, and there even appears to be some limited cospeciation between figs and wasps (Herre *et al.*, 1996; Lopez-Vaamonde *et al.*, 2001; Cook and Rasplus, 2003). Some flies and moths that breed in the seeds of the plants they pollinate are restricted to particular species of plants, and the yucca moths (*Tegeticula*) even have specialized mouthparts for inserting pollen into the stigma (Figure 13.33) (Després and Jaeger, 1999; Després *et al.*, 2002; Pellmyr, 2003). Flowers of Dutchman's pipe vine (*Aristolochia*), *Ceropegia* milkweeds (Asclepiadaceae), and many aroids (Araceae) lie at the bottom of leafy vessels, and various minute flies are lured into the vessels by musky or putrid odors and cannot escape because of slippery surfaces and false windows. When the vessel gradually unfurls, generally a day later, the flies are released dusted with pollen.

Fossil Evidence

How did such pervasive and diverse mutualisms between flowers and insects evolve? The answer lies in direct evidence from fossils and indirect evidence from the study of living, primitive angiosperms. The most vivid evidence of early interactions between insects and the reproductive organs of plants is in the form of pollen preserved in the guts of fossil insects. Durable pollen and spores preserve well in sediments and even in the guts of living and extinct insects (a bees' feces, for example, is largely the undigested rinds of pollen grains). The earliest records of pollen meals are from generalized polyneopterans from the Permian with gymnosperm pollen preserved in their guts (Rasnitsyn and Krassilov, 1996a,b; Krassilov and Rasnitsyn, 1999; Afonin, 2000). There also are Cretaceous records of primitive, xyelid sawflies preserved with their meal of conifer pollen (Krassilov and Rasnitsyn, 1982) and even angiosperm (i.e., *Afropollis*) pollen (Caldas *et al.*, 1989). The various reports of pollen preserved in Tertiary insects are not as significant as the Mesozoic records.

Other direct evidence for the evolution of insect pollination is the preservation in fossil flowers of specialized structures that attract favored insect partners and promote pollen delivery. A chronology of 36 specialized pollination features began approximately 120 MYA with the earliest petals and sepals; it continued with the evolution of filamentous stamens and corolla tubes and then with clawed petals, viscin threads on pollen, floral nectaries, and other features up to approximately 90 MYA (Crepet, 1996). This approach, however, underestimates the impact of generalized pollinators and pollination systems, which were probably the earliest types of alliances between flowers and insects. Nonetheless, by 90 MYA there is direct evidence that most specialized pollination systems seen today were largely intact. The complete exuberance of systems was refined in the Tertiary with the appearance of highly zygomorphic flowers like those of peas and Scrophulariaceae.

Direct evidence for the evolution of insect pollination is revealed, lastly, via the fossil insects themselves. This was comprehensively reviewed elsewhere (Grimaldi, 1999; Labandeira, 2000). The simple identification of a fossil to an insect family that presently includes pollinators is evidence for early pollination, but if specialized anthophilous structures are preserved, the inference is certain. Anthophilous adaptations usually include mouthparts that form a long proboscis, but they also include combs of setae on the mandibles or other mouthparts, for example in the flower-visiting beetle families Melyridae, Oedemeridae, and Mordellidae (Grinfel'd, 1975), though the latter type of structures does not fossilize well. Other specializations are pollen-gathering structures like scopae and pollen baskets on the legs of bees, or specialized hairs like the plumose hairs of bees or those on the legs of beeflies (Neff *et al.*, 2003). Supportive evidence for anthophily includes specialization of the wings that reflect the highly maneuvered flight of many pollinators.

Among specialized pollinating insects, a long proboscis occurs in various flies, glossatan lepidopterans, many bees

14.14. *Protonemestrius* and *Florinemestrius* (Diptera: Nemestrinidae), from the Early Cretaceous Yixian Formation of China. These and several other Cretaceous flies are the earliest specialized pollinators in the fossil record. Redrawn from Ren (1999).

14.15. *Cratomyia macrorrhyncha*, from the Santana Formation of Brazil. This beautifully preserved Early Cretaceous fly also has a long proboscis, which it probably used to probe early flowers. MZUSP DBRP 0050; length 24 mm (with proboscis).

(long-tongue bees), masarine vespids (pollen or hover wasps), and even a few meloid beetles. A long proboscis does not always denote flower feeding, though, because many flies that feed on blood also have a long proboscis, like mosquitoes, many tabanid flies, and tsetse flies (Figures 12.21, 12.37, 12.48). To complicate the matter further, some species of blood-feeding insects like mosquitoes can feed on blood (generally the females) and on nectar (generally the males). Also, even though a long proboscis is the most obvious anthophilic specialization in a fossil, in some fossils the function of a long proboscis can be ambiguous. For example, the Mesozoic scorpionfly *Pseudopolycentropus* had an impressively long, slender proboscis (Figure 12.3). The genus belongs to a group of predatory and scavenging insects (Grimaldi *et al.*, in press), but the wing structure suggests it had a highly maneuvered flight like that of anthophilous insects. If a fossil has a long proboscis along with a particular wing venation, though, it was probably an anthophilous insect. The ability to hover occurs in various insects but is best developed in anthophilous insects, particularly ones with long proboscides (Grimaldi, 1999) (Figures 12.65, 12.99, 14.17). Hovering is energetically expensive and is optimally fueled by nectar. Hovering allows an insect with a long proboscis to probe a flower without alighting (also steady the

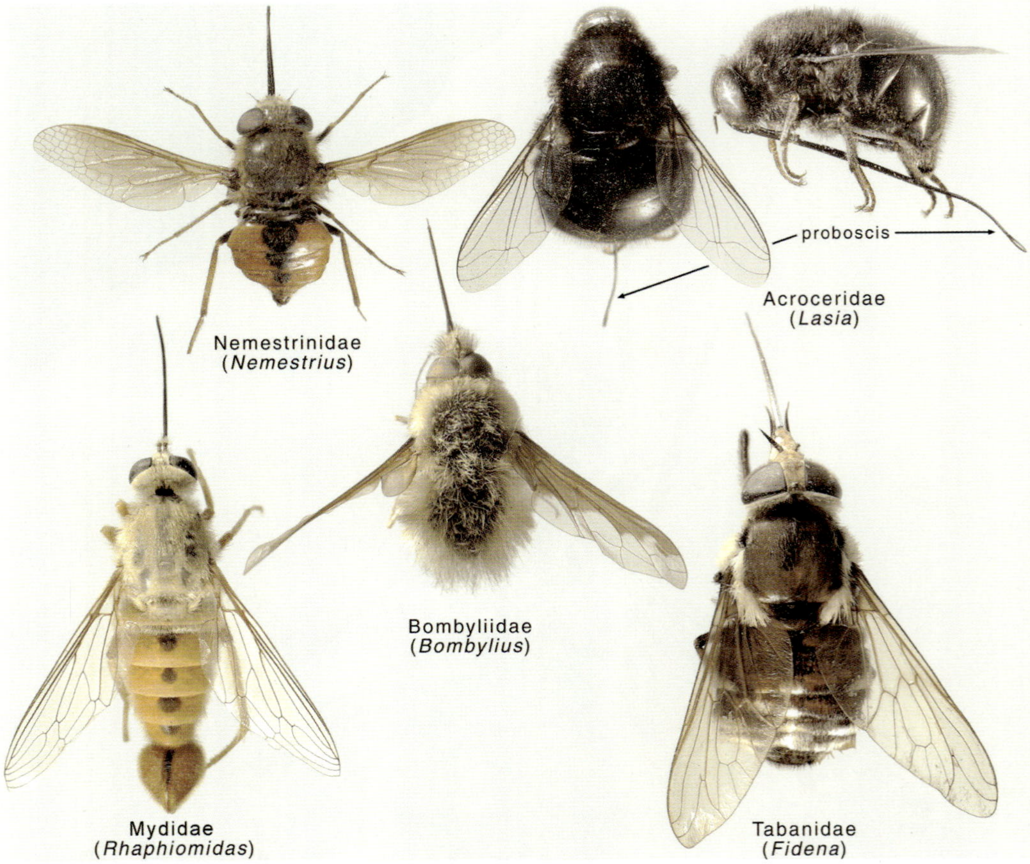

14.16. Assorted Recent species of lower Brachycera flies with long proboscides, specialized for feeding on nectar and pollen.

body so that the tip of the proboscis can be accurately aimed), and thus to visit many flowers quickly. Hovering insects have forewings that are tightly coupled to small hind wings (or, as in Diptera, the hind wings are highly reduced), and the wings are slender with a narrow tip and have veins that are upturned or coalesced near the tip. These are found in the long-tongued flies, the hover flies (Syrphidae), masarine wasps, and hovering moths. Insects that simply alight on blossoms, like butterflies, or that wrestle with the flower, like bees, lack this wing structure.

Using these features of anthophilous specialization, the earliest definitive evidence for specialized insect pollinators is found in flies (Figures 14.14 to 14.17). These are nemestrinids from the Early Cretaceous Yixian Formation of China (Ren, 1998a,b) (Figure 14.14), and *Cratomyia macrorrhyncha* from the Early Cretaceous Santana Formation of Brazil (Mazzarolla and Amorim, 2000) (Figure 14.15). They belong to groups that presently are pollinators, and they possessed long proboscides and wing venation typical of hoverers. Nemestrinidae today feed from flowers, many have long proboscides, and one species, *Moegistorrhynchus longirostris* from South Africa, has the longest proboscis of any insect proportional to its body size (Figure 14.17). *Cratomyia* belongs to the Stratiomyomorpha, a group that today are common flower visitors. Specialized pollination systems involving long-tongued flies are widespread and well known (e.g., Graenicher, 1910; Knoll, 1921; Grant and Grant, 1965; Grimaldi, 1988; Johnson and Steiner, 2000; Manning and Goldblatt, 1997; Potgieter *et al.*, 1999; Goldblatt and Manning, 2000). Invariably, the flowers these long-tongued flies visit have long, narrow corollas, but this does not mean that the Cretaceous nemestrinids and *Cratomyia* fed from deep corollas. Brachycera evolved in the Early Jurassic, and generalized Nemestrinidae with short proboscides occurred later in the Jurassic, which is well before other major pollinating groups appeared in the Cretaceous, like glossatan moths and aculeate wasps (including bees). Thus, it should almost be expected that the earliest specialized pollinators would have been brachyceran flies. Other Late Jurassic and Cretaceous insects that belong to anthophilous groups are Scenopinidae, Statiomyidae, mythicomyiine bombyliids, and diverse Empididae among the flies, and mandibulate and primitive glossatan moths, all of which were small and generalized and had no particular adaptations for flower feeding (Grimaldi and Cumming, 1999; Grimaldi, 1999). It was not until the Tertiary that diverse bees occurred (Figures 11.84 to 11.90), as well as syrphid flower flies (Figures 12.99, 12.100), long-tongued Bombyliidae and other Diptera (Figure 14.19), and long-tongued macrolepidopterans including the butterflies (Figures 13.65 to 13.69).

14.17. *Moegistorrhynchus longirostris* (Nemestrinidae) from southern Africa, which has the longest proboscis of any insect relative to its body length. The fly is an important pollinator of several flowers with deep, narrow corolla tubes, such as this *Lapeirousia anceps* (Iridaceae). Photo: J. Manning.

14.18. An *Andrena* bee wrestling with a flower. Virtually all species of the approximately 20,000 species of bees are specialized for feeding on nectar and pollen, making these insects the most important group of pollinators today and one of the most ecologically pivotal groups of all animals. Photo: R. Swanson.

14.19. A fossil beefly (Bombyliidae), *Pachysystrophus rohweri*, from Florissant, Colorado, approximately 38 MYO. It had a long proboscis, no doubt used for feeding from flowers. Fossil bombyliids appear 100 MY before this, but the family apparently did not evolve a long proboscis until the Tertiary. YPM 26175.

Other than these flies, the only Mesozoic insect that was a specialized anthophile is *Cretotrigona prisca*, in amber from New Jersey (Figure 11.83). It is a member of the recently evolved group of stingless bees (Meliponini), today found throughout the world's tropics (Michener and Grimaldi, 1988a the fossil; Engel, 2000a). Age of the fossil, unfortunately, is ambiguous, but appears to be from the latest Cretaceous (Grimaldi, 1999). Other tantalizing pieces of evidence for Cretaceous bees are nests from the Cenomanian to earliest Paleocene (Elliott and Nations, 1998; Genise and Verde, 2000; Genise *et al.*, 2002; Figure 2.17). Also, certain beautifully preserved 90 MYO flowers indicate that bees were significant pollinators at this time. Clusiaceae are typically pollinated by stingless bees today, and Cretaceous Clusiaceae flowers had specialized bee pollination structures, including resin and resin ducts (Crepet and Nixon, 1998). All of this evidence places the origin of the most important group of modern pollinators at approximately 120 MYA, no doubt evolved from generalized spheciform wasps known from that time. Virtually all of the approximately 20,000 species of bees today provision their nests with larval food made of nectar and pollen. Many forage on a particular family or even genus of plants (these are *oligolectic*), though the highly social bees (bumble bees, stingless bees, and especially *Apis*) visit diverse flowers (they are *polylectic*). Bees are intelligent insects, and an individual bee quickly learns which flowers are most rewarding, how to force itself into a flower, or even how to steal nectar by chewing a hole through the bottom of the corolla. Thus, bees are spectacularly efficient foragers (Figure 14.18), particularly the social species that can recruit workers to a productive patch of flowers, which is why they are crucial to agriculture. Of approximately 84 major crop

14.20. Craneflies (Tipulidae) feeding from a flower. Early flowers were small and structurally generalized and probably were serviced by myriad insects unspecialized for pollination, like these. Photo: S. Marshall.

plants, bees pollinate 64 (75%) of them (the others being pollinated by wind or other insects) (Buchman and Nabhan, 1996). The first diverse fauna of fossil bees is in the Eocene of Europe, which preserved an unexpected diversity of social (Engel, 2001a,b; Wappler and Engel, 2003). Bee fossils are extremely rare though, and compressed fossils from the Early Cretaceous presumably would be difficult to distinguish from related apoids.

All indications from the fossil record of plants and insects are that most of the pollinators of the earliest angiosperms were probably small, unspecialized flies (e.g., Figures 14.20, 14.21), lepidopterans, parasitoid apocritan wasps, and various beetles, which is a scenario that is bolstered by observations on the kinds of insects that visit living gnetaleans and the most basal living angiosperms.

Pollinators of Primitive Living Angiosperms

Indirect evidence for the evolution of insect pollination consists of observations on the pollination systems of seed plants, especially cycads, gnetaleans, and the most basal angiosperms. Cycads (Cycadales) are familiar plants that have a very long fossil record that begins in the Early Permian, and in the Mesozoic their diversity was impressive but today they are relict. Species of the Recent cycad family Zamiaceae are visited by beetles (particularly Curculionidae), some thrips, and flies (Norstog, 1987; Crowson, 1991; Vovides, 1991; Norstog *et al.*, 1992; Forster *et al.*, 1994; Stevenson *et al.*, 1998), and some even by such recently evolved insects as meliponine bees (DeVries, 1983; Forster *et al.*, 1994). Cycadaceae appear to be wind pollinated, as is *Ginkgo*. Of the gnetaleans where insect visitors have been observed, the insects are primarily small, generalized flies, beetles, and moths (Porsch, 1910; Pearson, 1929; Meeuse *et al.*, 1990; Evenhuis, 1994; reviewed in Friedman, 1996). Flowers of *Gnetum gnemon* from Malaysia are visited by at least 13 species of moths in the families Pyralidae and Geometridae (Kato and Inoue, 1994, 1995). Gnetaleans and other gymnosperms secrete minute, sticky "pollination droplets" on the stigmatic surface of the ovules (Figure 14.2), which serves to capture pollen but also attracts small insects that feed on the droplets. On this basis it has been proposed that nectar originally attracted insects to flowers (Lloyd and Wells, 1992), but there is direct fossil evidence that some insects dined on pollen as early as the Permian, as discussed earlier. No doubt

14.21. *Atrichopogon* midges (Ceratopogonidae) feeding from a water lily. Water lilies (Nymphaceae) are one of the most basal lineages of angiosperms and today are visited by various beetles and flies – a relationship that probably has changed little since the debut of these flowers in the Early Cretaceous. Photo: S. Marshall.

nectar (in the form of pollination droplets) and pollen were both rewards for early pollinating insects.

Insects have also been observed visiting basal living angiosperms, and though the reproductive systems of many of the most basal species (like *Amborella*) have not yet been reported, the musky odor produced by *Austrobaileya* suggests that at least this genus is visited by flies. Insects visit some Nymphaeales, which generally are beetles (e.g., Prance and Arias, 1975; Ervik and Knudsen, 2003) and small flies that are attracted to the large, fragrant flowers (Figure 14.21). Flowers in the Chloranthaceae and the Magnoliidae, like the families Winteraceae and Magnoliaceae, are largely pollinated by small flies, beetles, thrips, and primitive (e.g., micropterigid) moths (Gottsberger *et al.*, 1980; Thien, 1980; White and Thien, 1985; Bernhard t and Thien, 1987; Endress, 1987, 1990; Gottsberger, 1988; Pellmyr *et al.*, 1990; Lloyd and Wells, 1992; Proctor *et al.*, 1996; Thien *et al.*, 2000). Small, generalized insects are not restricted to basal angiosperms but are actually common visitors to many eudicots, particularly groups like umbellifers and composits (Robertson, 1928). For example, in one study in Wales, 33 of 53 flower species observed were found to have 250 species of lower apocritan (parasitoid) wasps in 15 families feeding on the nectar and sometimes the pollen (Jervis *et al.*, 1993), and this is only part of the anthophilous fauna. In a large study of the insect pollinators of the subtropical Amami Islands south of Japan, there were 610 species of insects in 12 orders found visiting 164 flower species (Kato, 2000). Most of these were Diptera (32% of the individuals), then Coleoptera (28%) and Hymenoptera (23%), though bees were the most important pollinators. For most insect species in these kinds of situations, the pollination impact of any one of them is probably barely perceptible or even negligible, but the overall impact of generalized, casual flower visitors is probably significant.

Traditional views have held that beetles were the original insect visitors and pollinators (Diels, 1916; Pijl, 1960; Gottsberger, 1977; Kevan and Baker, 1983), and that the angiosperm carpel may even have evolved as a shield against the beetles that would otherwise consume the ovules. Observations of Recent gnetaleans and basal angiosperms, however, have led to the conclusion that diverse small, generalized insects were probably the earliest pollinators in the Late Jurassic and Early Cretaceous. These must have included various nematoceran and brachyceran (e.g., empidoid) flies (Dzhambazov and Teneva, 2000), thrips (Thysanoptera), parasitoid wasps, as well as various beetles, all of which were diverse by the Early Cretaceous, 145–100 MYA. Primitive Lepidoptera, like Micropterigidae and perhaps early glossatan moths, were less diverse but also present. Thus, *diverse, small, generalized insects were on the scene when the earliest angiosperms appeared, and they probably immediately exploited flowers, then partnered with angiosperms as pollinators.* As such, insect pollination would be the primitive or groundplan condition for angiosperms. It was not until the Late Cretaceous, 100–65 MYA, that important pollinator groups achieved significant diversity, like the ditrysian lepidopterans, various brachyceran flies, and aculeate wasps. In the Early Tertiary, 65–30 MYA, there occurred radiations of butterflies and other macrolepidopterans, beeflies, syrphid flies, and – most important of all – the bees.

The pivotal role that pollinating insects played in the diversification of angiosperms has been almost universally accepted on the basis of various lines of evidence, both ecological (Robertson, 1904; Midgely and Bond, 1991; Proctor *et al.*, 1996; Dodd *et al.*, 1999; Anderson *et al.*, 2002) and paleontological (Crepet and Friis, 1987; Crepet *et al.*, 1991; Dilcher, 2001). Recently, however, this has been questioned (Gorelick, 2001). It was pointed out that many primitive seed plants are insect pollinated but never achieved the radiation that angiosperms have, and there are some angiosperm groups (like grasses) that are very diverse but wind pollinated. Gorelick's (2001) hypothesis is contradicted, first, by the fact that the diversification rates of animal-pollinated lineages of angiosperms are nearly twice the rates of abiotically pollinated lineages (Eriksson and Bremer, 1992). Second, his view assumes that the earliest pollination was as refined and efficient as in the most recently evolved situations. This is a common logical error, analogous to claiming that wings did not foster the radiation of insects because early pterygotes like Ephemeroptera and Odonata comprise only about 1% of the winged insects. Extinction actually has culled much of paleopteran diversity (as it has many cycads and probably gnetaleans), but more importantly the *refinement* of flight led to the success of later lineages of winged insects (neopterans) as a result of indirect flight muscles, rapid wing beat frequencies, axillary sclerites and wing folding, and so on. Likewise, as insect pollination systems became more specialized and flowers evolved more structural sophistication, pollen transfer became more efficient, and this had a greater effect on plant-breeding systems. The loss of insect pollination in grasses is attributed simply to the invasion of an adaptive zone (open, windy areas) that took place relatively recently, largely in the Eocene and later (Jacobs *et al.*, 1999). Likewise, the loss of wings in lice and fleas allowed the evolution of some 7,500 species of ectoparasites, and no one would claim that because these insects lack wings these structures were insignificant in the success of insects. Without question, the early partnership between angiosperms and pollinating insects and the successively specialized relationships that evolved between them were major factors in the radiation of the angiosperms. The angiosperm radiations, conversely, had a profound effect on other insects as well – ones that do not feed just from the flowers.

14.22. Insects quietly consume more plant mass than all vertebrate herbivores combined, and species of herbivorous insects vastly outnumber any other diet. Fossil evidence indicates that terrestrial arthropods have been consuming plants since at least the Devonian, for approximately 400 MY. Here, a cone-headed grasshopper peers out from behind a leaf. Photo: V. Giles.

RADIATIONS OF PHYTOPHAGOUS INSECTS

It is very likely that insects are the most significant group of herbivores, even more so than are vertebrates. Approximately 10–25% of the foliage in tropical forests is consumed by insects (Wiegert and Evans, 1967; Janzen, 1981), which doesn't include estimates of the consumption of flowers, roots, stems, wood, seeds, and fruits, all of which generally are included into what we call an herbivorous or phytophagous diet. In temperate deciduous forests, phytophagous insects comprise approximately 25% of all arthropod species, half of the arthropod biomass, and 75% of the arthropod individuals (Moran and Southwood, 1982). Even in grassland ecosystems, like savannas, which are well known for their vast herds of ungulates, some 10–75% of the herbivory is by insects, and 15–35% is by vertebrates (Sinclair, 1975; Gandar, 1982; Detling, 1988). Herbivorous insects can also have significant effects on the fecundity and survivorship of their host plants and can devastate plants by transmitting pathogens. The fungus (*Ophiostoma ulmi*) that causes Dutch elm disease, for example, is transmitted by *Scolytus* and *Hylurgopinus* bark beetles, and it nearly wiped out the American elm. Periodic outbreaks like those of migratory locusts (*Schistocercus*) and gypsy moths (*Lymantria dispar*) can denude entire landscapes. In sum, the effects of plant-feeding insects are immense and have been taking place since the Devonian some 400 MYA (Labandeira, 1998, 2002).

In the Recent fauna, phytophagous insects comprise nearly half (43%) of all insect species (Table 14.1). This is primarily the result of the following five main lineages of insects.

- *Lepidoptera.* This is the largest lineage of plant-feeding organisms. The most basal lepidopterans (Micropterigidae) are detritivorous (some are phytophagous), but all others comprise a monophyletic group of species whose caterpillars are phytophagous. There have been a few reversions to a diet of detritus, and a few groups have even become parasitoids and predators.
- *Phytophagan and other beetles.* The Phytophaga in the sense we use here includes the Chrysomelidae (leaf beetles), Curculionoidea (weevils and relatives), and Cerambycidae (long-horned beetles), which comprise what is probably a monophyletic group. Other phytophages include many scarabs and various small families of wood-boring beetles.
- *Hemiptera.* Ancestral hemipterans were clearly phytophages that siphoned plant fluids. Virtually all Sternorrhyncha and Auchenorrhyncha are phytophagous, as are

14.23. *Euselasia* (Riodinidae) caterpillars devouring a leaf en masse. Lepidoptera is the largest lineage of phytophagous animals, more than 99% of the species of which feed on angiosperms. How specialized host plant use promotes diversification remains obscure. Photo: V. Giles.

TABLE 14.1. Proportions of Phytophagous Insects in Major Groups

Group	Approximate Proportions of Phytophages (%)	Number of Phytophagous Species
Coleoptera	35	125,000
Lepidoptera	99	120,000
Hemiptera	90	82,000
Diptera	30	36,000
Orthoptera	95	19,000
Hymenoptera[a]	12	14,000
Phasmatodea	100	3,000
Thysanoptera	40	2,000
Totals		401,000

[a] Not including pollenivorous bees.

many species in the recently evolved lineages of Heteroptera, the Pentatomorpha and Cimicomorpha. Almost all the basal families of Heteroptera are predatory.

- *Orthoptera.* This group is closely related to the Phasmatodea, and the common ancestor of them both was almost certainly phytophagous. All Phasmatodea and 95% of orthopterans are phytophagous, which is a diet that apparently began in the Carboniferous and that would also make this the oldest living lineage of phytophagous organisms.
- *Hymenoptera.* The ancestral diet of Hymenoptera was plant feeding because the larvae of virtually all basal (symphytan) wasps feed on plants. Larval diets then evolved to include larval parasitoids and other habits. Secondary origins of phytophagy include larval cynipoid wasps (especially the gall-forming Cynipidae), chalcid wasps (breeding within seeds), and assorted other groups like leaf-cutter bees and leaf-cutter ants (the last of these are not strictly phytophagous but use leaves to build nests or cultivate fungus).

Among the smaller phytophagous lineages is the Thysanoptera; their ancestral diet was apparently fungus, from which phytophagy evolved several times. Phytophagy evolved multiple times in Diptera, including Cecidomyiidae (gall midges: the largest phytophagous lineage in Diptera) as well as the tephritoid fruitflies, agromyzid leaf miners, various Anthomyiidae, and many other families of Cyclorrhapha. Not included in these estimates are extinct groups, such as the Palaeodictyopterida, which was a lineage of beaked, sucking insects that apparently became extinct at the end of the Paleozoic.

The plant hosts of these Recent groups are primarily angiosperms, which may just reflect the biomass and species dominance (i.e., the availability) of angiosperms, or the generally more digestible and nutritive value of angiosperm foliage. The diversity of insect phytophages probably also reflects the general coevolutionary history of these with angiosperms. This view was disputed by Labandeira and Sepkoski (1993), who plotted the number of insect families through geological time and found no increase in their number around the time of the angiosperm radiations. As we discuss later in this chapter, such conclusions are a result of a taxic analysis. Estimating insect diversity on the basis of numbers of families makes several assumptions that appear to be unjustified: All families are monophyletic lineages and have equal diversity and ecological significance. Clearly this is not the case. Families differ dramatically in their ecological significance, and no one would suggest that Formicidae (ants) or Apidae (bees) are an equivalent unit ecologically to a family of, say, psocopterans. A phylogenetic approach, which we have used throughout this book, reveals quite a different picture. Those groups of insects where the most basal species do not feed on plants are extremely few compared to their more recently evolved phytophagous relatives (Mitter *et al.*, 1988). The most primitive moths in the family Micropterigidae, for example, comprise approximately 100 species with mostly detritivorous caterpillars, compared to 120,000 species of angiosperm-feeding caterpillars. Similar situations pertain to cecidomyiid gall midges, thrips, and various other groups, and these radiations typically involve angiosperm hosts. Moreover, the fossil record shows that during the Cretaceous there were radiations of insect lineages that, though few in numbers of families, now contain vast numbers of species: Lepidoptera, phytophagan beetles, sternorrhynchans, higher Heteroptera, various groups of Diptera, ants, and bees. With little question, the angiopserm radiations are intimately related to the radiations of major lineages of phytophagous and pollinating insects.

Divergence via Host Plants

A great deal has actually been written about the apparent coevolution between plants and phytophagous insects (e.g., Ehrlich and Raven, 1965; Strong *et al.*, 1984; Mitter *et al.*, 1991; Futuyma and Keese, 1992; McEvoy, 2002). Briefly, in the coevolutionary scenario, there are stepwise, reciprocal adaptations of plants to defend against the assault of herbivores, and insect counteradaptations to circumvent the defenses (e.g., Berenbaum and Feeny, 1981). In various studies on, for example, mustard oils in crucifers and coumarins in umbellifers (Berenbaum, 1981), the earliest defense of some plant groups were mild toxins with which most insects could contend. As herbivory increased, so did the potency of the plant toxins, to which some insects evolved resistance and even an attraction, and the plants then evolved further toxicity, and so on. This "arms race" then led to groups of plants with complex phytotoxin cocktails and a highly specialized array of herbivores, like *Heliconius* butterflies on *Passiflora* vines and danaine butterflies on milkweeds. Most evolutionary biologists regard coevolution of insects and their plant hosts as being at best merely diffuse, but phylogenetic evidence indicates that at least breaking into the "adaptive zone" of angiosperm phytophagy has greatly increased the diversity of various lineages of insects (Farrell, 1998; Farrell *et al.*, 2001).

14.24. The apple maggot fly, *Rhagoletis pomonella* (Tephritidae). Variation in this species had been attributed to races that diverged on the basis of host plant use, which was then used to explain how phytophagous insect species evolved in sympatry. Reports of insect host races have not held up to scrutiny.

While theories on the coevolution of plants and insects explain adaptive or anagenetic change, they do not account for speciation. If animal species typically form by the divergence of isolated (allopatric) populations (Mayr, 1942), how is host use involved? One theory is that insects evolve into "host races" within flight range of each other, and from these reproductively isolated populations diverge sympatrically (Bush, 1975). Conditions under which sympatric speciation can take place via host races were originally defined as the following (Bush, 1975): Mating occurs on or near the host plant, females choose where to oviposit, and the insects are very restricted in their use of host species. This initial *assortative mating* then leads to genetic divergence. For many groups of phytophagous insects, these conditions easily apply because the sexes merely rendezvous where they breed. A summary of putative examples of host races in phytophagous insects is provided elsewhere (Strong *et al.*, 1984). Jaenike (1981) proposed alternative criteria for defining host races, principally that these must be shown to differ genetically and solely on the basis of their different hosts and not by some unrelated factor.

The most famous example of host races involves the apple maggot fly, *Rhagoletis pomonella* (Figure 14.24), some individuals of which breed in apples (*Malus*) and others in hawthorn fruits (*Crataegus*), among other hosts. As originally reported, the apple and hawthorn individuals comprise genetically distinct host races (Bush, 1969, 1974). However, as pointed out by Jaenike (1981), the phenotypic differences between these "races" were not proven to be genetically based nor are the differences even consistent with each other. Thus, it is likely that these "races" are simply freely breeding individuals that sort themselves out based on individual host preferences. Similar scrutiny of other purported host races shows that none of them fulfill the criteria of host races, or else the "races" have turned out to be groups of cryptic species. For example, membracid tree hoppers of the *Enchenopa binotata* complex in North America feed on eight genera of angiopserm trees and shrubs. They are morphologically indistinguishable but behaviorally distinct, which originally led to the conclusion that these were host races (Wood, 1980; Wood and Guttman, 1982), and this was promoted as a compelling example of host races (Strong *et al.*, 1984). Now it is known that this complex comprises nine extensively sympatric species, as based on DNA sequences (Lin and Wood, 2002). It is quite likely that many so-called host races, and even highly polyphagous "species" of plant-feeding or parasitic insects are complexes of cryptic species (Claridge *et al.*, 1997; Lane, 1997; Berkov, 2002).

The fact that insects do not appear to diverge sympatrically on different hosts is probably why intimate cospeciation of phytophagous insects and their host plants (like *Phyllobrotica* beetles and *Scutellaria* plants [Farrell and Mitter, 1990]) is exceedingly unusual. Even in insects where the entire life cycle revolves around the same host plant, as for fig wasps and yucca moths, there is very limited cospeciation (Lopez-Vaamonde *et al.*, 2001; Cook and Rasplus, 2003;

Pellmyr, 2003). What mechanism then explains the remarkable diversity of phytophagous insects? It is most plausible that *specialized host use promotes allopatric divergence*. There are numerous examples of insect species where in each of them there are different but closely related host plants in distant parts of the range (reviewed in Strong *et al.*, 1984), and this is probably fairly common. In such a situation, specialized use of different hosts (that, say, evolved tolerance to local phytotoxins) may reduce or even preclude dispersal to other parts of the range, and this would limit panmixis of the entire species population. This is a subject worthy of further investigation. Despite the fact that the mechanism is obscure as to how insects diversified with angiosperms, the overall patterns are extremely clear that the angiosperm radiations had a profound impact on insects, and vice versa.

AUSTRAL ARTHROPODS: REMNANTS OF GONDWANA?

Because many living families of insects originated in the Cretaceous, and some even earlier, clearly some aspects of their evolution were shaped by Cretaceous events. Besides the angiosperm radiations, another important series of events was the breakup of the supercontinents of Pangaea, Laurasia, and Gondwana, and no area is of greater interest in this regard than is the austral region. This region includes southern South America, southern Africa, southeastern Australia, Tasmania, New Zealand, New Caledonia, and some of the smaller islands and archipelagos between Australia and South America, like Lord Howe, Norfolk, and the Fiji Islands. Before Cretaceous drifting, these landmasses were clustered around Antarctica, which at the time was fully forested but gradually glaciated until the mid- to Late Miocene, 15–10 MYA, when the last forests were buried under ice. The vegetated austral landmasses presently have several biomes, including dry steppe and some moorland, but the biome of greatest interest is the cool, wet, evergreen forests. These forests are often called southern temperate forests and are distinguished by the southern beeches (*Nothofagus*: Nothofagaceae) (Figures 14.25, 14.26) in all areas except Africa. Conifers in the primitive family Araucariaceae are also dominant trees of the area (Figure 14.28), but in southern Africa the dominant temperate forest tree is *Podocarpus*. In comparison to these forests, the "temperate" northern forests are

14.25. Southern beeches, *Nothofagus*, at Cauquenes Province in Chile. *Nothofagus* is a classic example of a group showing austral disjunction, with species occurring in the temperate regions of southern South America, New Zealand, and Australia – presumably relicts of a time when these continents were joined. Photo: L. Prendini.

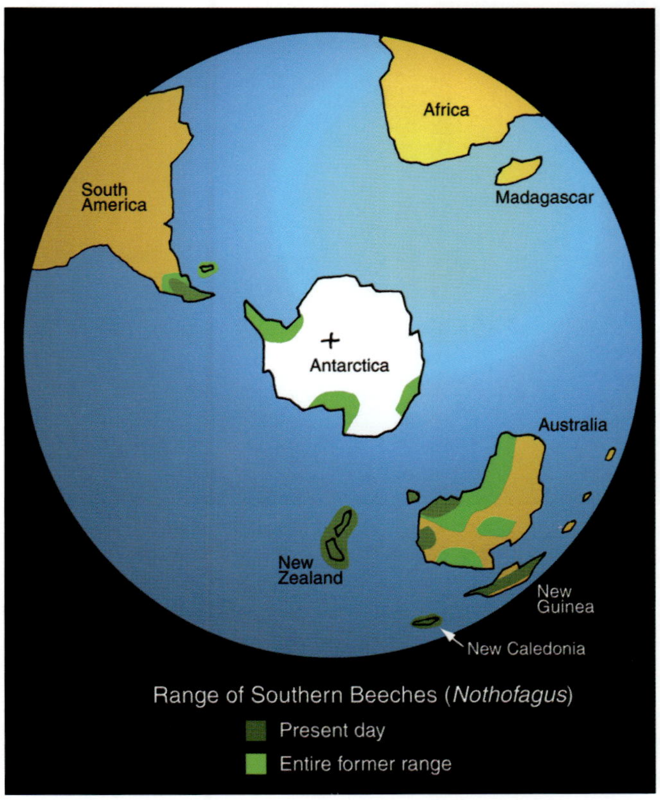

14.26. The past and present distribution of *Nothofagus*. Southern beeches had a wider distribution, including portions of Antarctica and more of Australia during warmer, wetter periods. Antarctica was a bridge among the southern continents. Adapted from Dettman *et al.* (1987).

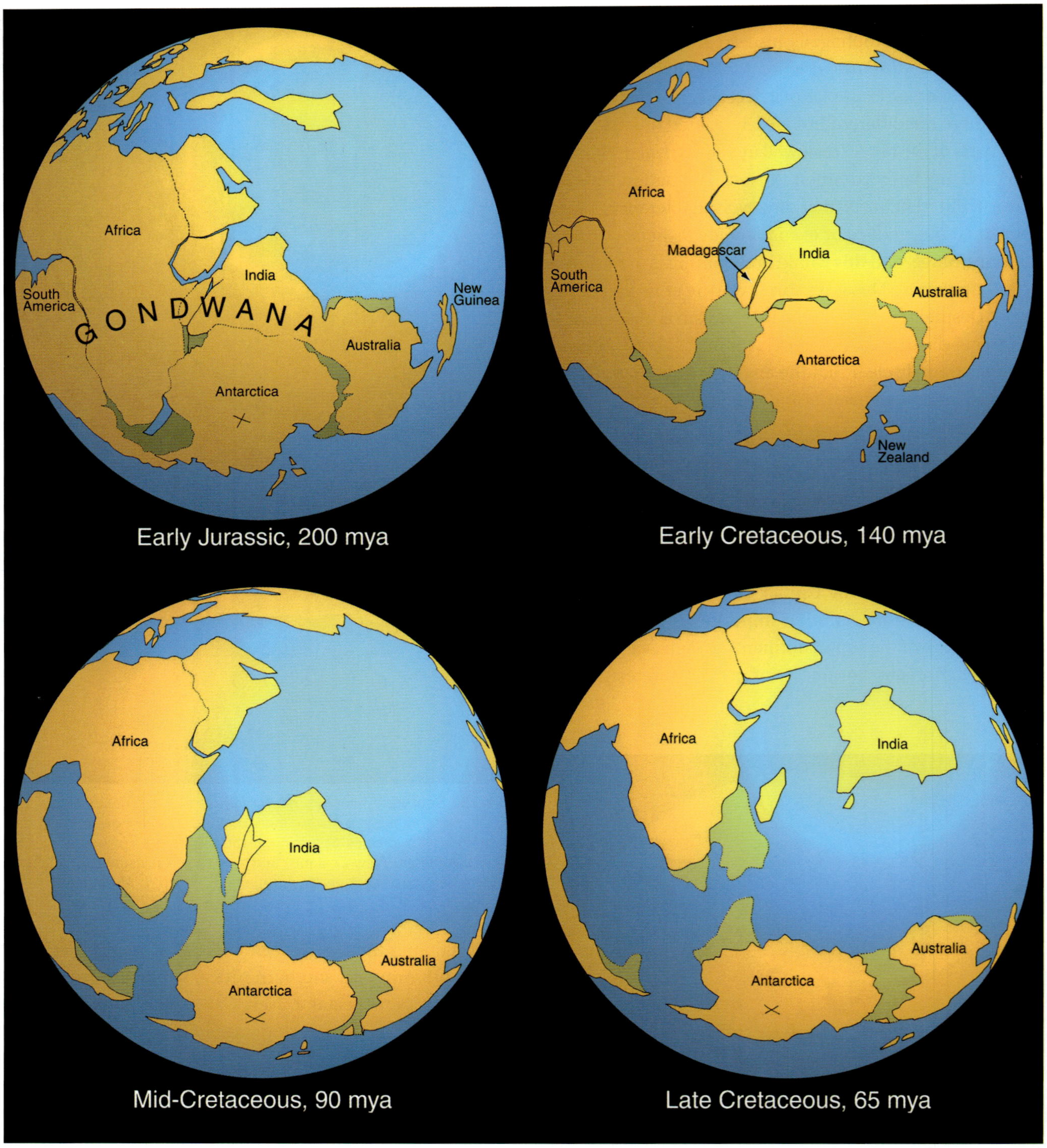

14.27. Configurations of the Southern Hemisphere during the mid- to Late Mesozoic. India has drifted more than any of the other gondwanan landmasses. From Scotese *et al.* (1988), Veevers *et al.* (1991), Wilford and Brown (1994), and de Wit: http://kartoweb.itc.nl/gondwana.

actually extremely intemperate. Northern temperate forests are deciduous because of prolonged, freezing winters and hot summers, while the southern forests rarely or never experience freezing. North America 18,000 years ago was covered with more glaciers than ever occurred in Antarctica, and so the Northern Hemisphere has experienced some of the most dramatic climatic extremes.

Among the first individuals to observe connections among the widely separated biotas in the Austral Region was Sir Joseph Hooker. In his *Flora Antarctica* (1844–1860) he documented the remarkably similar floras from Chile, Tasmania, and New Zealand, particularly the cool, wet *Nothofagus* forests. Based just on these distributions, he postulated the former existence of a continent that connected these now

14.28. The "southern pines" *Araucaria* (family Araucariaceae) in Chile. The distribution of araucarians is also austral but wider than that of *Nothofagus*. Unlike *Nothofagus*, fossils indicate that Araucariaceae was global in the past. Photo: C. Chaboo.

widely separated areas, which was a remarkably prescient idea given that it was nearly a century before continental drift was even proposed.

Alfred Wegener's theory on continental (1924) was originally met with widespread skepticism, especially by American geologists. By 1960, evidence from paleomagnetism had convinced most skeptics, along with evidence of how mid-oceanic ridges form new sea floor that spreads away from them. Older sea floor, at the margins of continental plates, is pushed beneath deep oceanic trenches and under young cordilleras like the Andes and Himalayas by the advancing new sea floor. In this process, continents drift and mountains rise. During much of the Mesozoic, all the present continents were assembled into the supercontinent Pangaea, which was relatively stable for nearly 150 MY before it began to drift apart during the Late Jurassic into Laurasia (comprised of the northern continents) and Gondwana (comprised of the southern continents and India) (Figure 14.27). Most of the separation of these two great landmasses took place in the Late Jurassic, 170–145 MYA. At approximately 180 MYA the rifting of Gondwana began, starting with that between eastern Gondwana (Africa, India, Madagascar) and western Gondwana (all other southern landmasses). A chronology of subsequent rifting and drifting follows.

160 MYA: narrow ocean opens between Africa and Antarctica + Madagascar + India + Australia
140 MYA: South America and Africa begin separation
130 MYA: India + Madagascar separate from Antarctica + Australia; the Atlantic Ocean is still a rift valley
110 MYA: narrow ocean exists between Africa and South America and between India and Australia
80 MYA: New Zealand + New Caledonia separated from the rest of western Gondwana
70 MYA: New Zealand and New Caledonia themselves split
65 MYA: India is midway between its origin and its eventual location in Asia
40 MYA: Australia and Antarctica fully separate

The rest of western Gondwana did not separate until the Oligocene, around 35 MYA, and Antarctica and South America remained conjoined until about 25 MYA. Australia separated from New Guinea and Tasmania relatively recently, anywhere from 20 to 2 MYA and were joined at various times since then. A large literature has summarized the sequence of continental rifting (e.g., Scotese, 1991; Veevers *et al.*, 1991; Wilford and Brown, 1994; and the animated reconstruction of Gondwana by de Wit: *http://kartoweb.itc.nl/gondwana*).

Almost as soon as this new geological paradigm was established in the 1960s, biogeographers used it to explain the distributions of plants and animals. Some of the earliest serious entomological studies were by René Jeannel, who applied continental drift to the distributions of beetles, particularly the small staphylinoid beetles in the family Pselaphidae and the predatory ground beetles in the large family Carabidae, which were his specialty (Jeannel, 1942, 1961). Other early studies of distributions interpreted in the context of continental drift concerned arachnids (e.g., Besch, 1964, 1969). The most celebrated example, though, involved chironomid midges (Brundin, 1965, 1966, 1967). Brundin (Figure 14.29) studied the basal subfamilies of these midges, Podonominae and Aphroteniinae, some of which are among the most primitive species of the family and even retain biting mandibles from their culicomorphan ancestors. Brundin found that closely related groups of chironomid species were widely separated on the various austral landmasses (Figure 14.30). Since that classic study, the austral region in fact has become a premier example of how geological events led to repeated and widespread vicariance, or separation, of closely related lineages. Indeed, numerous plants and animals with austral distributions appear to be remnants of gondwanan drift (Raven and Axelrod, 1974; Axelrod and Raven, 1978). Recent studies of the basal phylogeny of birds, for example, con-

14.29. Lars Brundin (1907–1995), whose work on chironomid midges from the Austral Region had a dramatic impact on biogeography. Photo: G. Byers, University of Kansas.

cluded that the major lineages of birds evolved in Gondwana (Cracraft, 2001; Ericson *et al.*, 2001; Barker *et al.*, 2002), particularly since basal groups like the large, flightless ratites (ostrich, casuary, rheas, etc.) have a presently gondwanan distribution.

Two terms are used to refer to such remnant taxa: "relicts" and "disjuncts." Relicts are species or lineages whose distributions are very restricted as a result of extinction from a wider distribution in the past. *Disjuncts* are close relatives that are widely separated, either as a result of colonization of a remote area or separation of ancestral ranges, as in continental drift. A taxon can be both relict and disjunct.

Austral Arthropods

Terrestrial arthropods are prime candidates for studying the effects of continental drift because many groups originated in the Mesozoic. However, there seems to be little correspondence between the age, dispersal abilities, and the proportions of austral disjuncts in a group. For example, even though wingless hexapods appeared in the Devonian, there are few groups of these that have austral disjunctions. Among them are some Collembola (Sminthuridae: Katianninae), Diplura (*Notocampa*), Archaeognatha (*Nesomachilis* and *Machiloides*), and Zygentoma (*Atopetalura* and *Gastrotheus*). When one considers that massive organisms like trees are still being discovered, such as the Wollemi araucarians just outside of Sydney, Australia, in 1994, documenting the distributions of cryptic soil arthropods is probably very imprecise by comparison. Ephemeroptera and Odonata, which originated in the Paleozoic and Early Mesozoic, respectively, seem to have more austral disjuncts than do the apterygotes, which may just reflect the more conspicuous nature and better surveying of these insects. Notable austral disjuncts in Ephemeroptera include Ameletopsidae, Coloburiscidae, Leptophlebiidae, Siphlonuridae, and Oniscigastridae; for Odonata these include some Neopetaliidae, Petaluridae, and certain corduliid dragonflies (subfamilies Gomphomacromiinae and Synthemistinae). Another order with Paleozoic origins and aquatic nymphs, the Plecoptera, has austral disjunctions in three families of antarctoperlarians: Austroperlidae, Eustheniidae, and Notoneumouridae (Zwick, 2000).

Some basal termites are austral relicts, most notably *Mastotermes darwiniensis* from northern Australia (Figure 7.79). Fossil *Mastotermes* and other mastotermitids from the Northern Hemisphere indicate these termites were probably global (e.g., Figures 7.80, 7.81). Only two other genera of termites have austral disjuncts, *Stolotermes* and *Porotermes*, both in the basal family Termopsidae. Both genera occur in Australia, New Zealand, and southern Africa, but unlike *Porotermes* the genus *Stolotermes* does not occur in Chile. It is unusual to have closely related species that are austral disjuncts in southern Africa and Australia but not in southern South America. There are probably two reasons for this. One, Africa separated from the rest of Gondwana earlier than did the western gondwanan continents themselves. Second, and probably most importantly, the wet temperate forests of southern Africa are extremely fragmented (most are one square kilometer in size or less, surrounded by xeric environment) and they amount to only 5,900 km^2 (less than 0. 25% the land surface of southern Africa [Low and Rebelo, 1998]). It is very plausible that many austral disjuncts in southern Africa have become extinct.

Despite the antiquity of the order Orthoptera and the Mesozoic ages of many Recent families, there are few known austral disjunctions in this group. Among these are some of the large, flightless crickets in the family Stenopelmatidae, which occur in Australia and southern Africa but not South America, as in *Stolotermes* termites. Others are certain "cave crickets" (Rhaphidophoridae) and proscopiid grasshoppers, but if ever there was a candidate group for continental vicariance it is Cylindrochaetidae (Figure 14.31). These wingless, neotenic, subterranean crickets are convergent with mole crickets (Gryllotalpidae) and surely must disperse only slightly faster than continents drift. Cylindrochaetids occur in New Guinea, Australia, and southern South America. The polyneopteran order Mantophasmatodea (Figures 7.55, 7.56) is a notable austral relict, presently restricted to southern Africa (Klass *et al.*, 2002, 2003), but some species occurred in

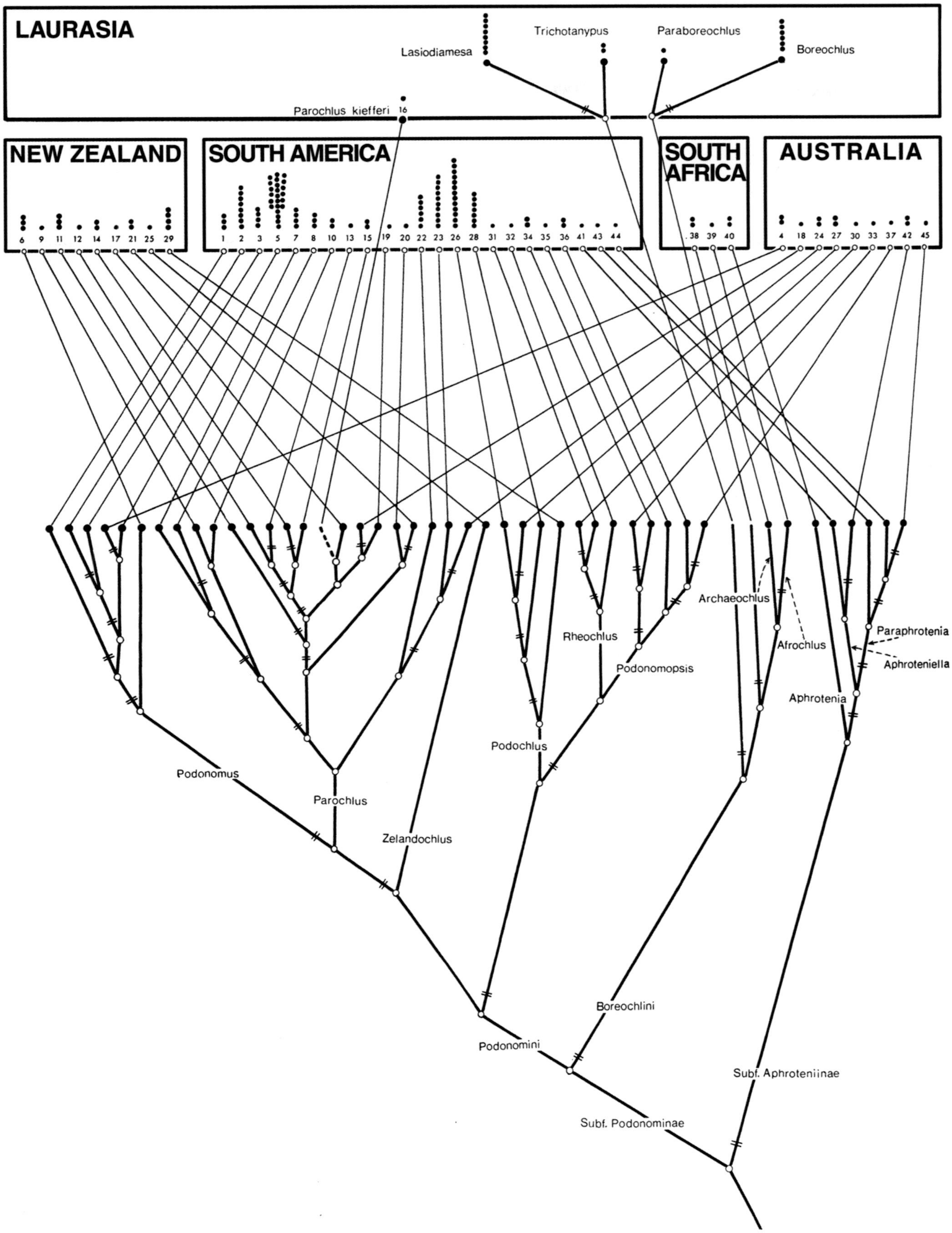

14.30. Brundin's hypothesis of relationships among basal lineages of chironomid midges and their distributions. The repeated patterns of closely related species living on separate continents are compelling evidence for a pervasive influence of drifting continents on insect distributions. Redrawn from Brundin (1966).

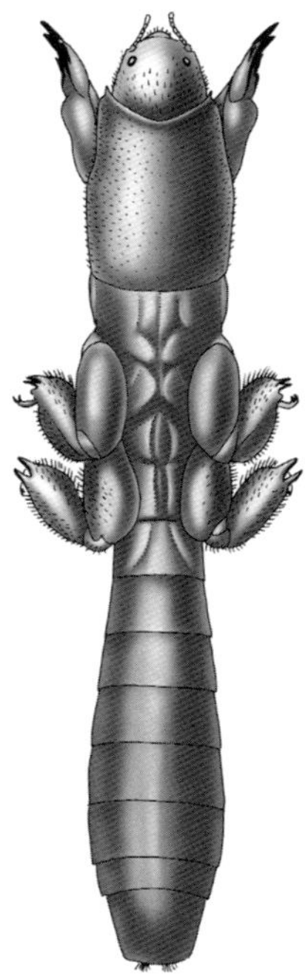

14.31. A larviform orthopteran of the family Cylindrachetidae. These orthopterans lack wings and have fossorial legs among other specializations for an entirely subterranean life (they also stridulate with their mouthparts). They do not disperse well and occur in southern Chile, Australia, and parts of New Guinea, and thus may be relicts of gondwanan drift.

the Northern Hemisphere at least during the Eocene (Figure 7.57) and may have been widespread in the Cretaceous (Engel and Grimaldi, 2004b). Mantophasmatodea is the likely sister group to the presently Laurasian order Grylloblattodea, so this group shows a bipolar distribution similar to that seen in Plecoptera. In Hemiptera, perhaps the most renowned disjunct group is Peloridiidae (Figure 8.53), which is the living sister group to the Heteroptera, and comprises 25 species living in southern South America, Australia, Tasmania, Lord Howe Island, New Caledonia, and New Zealand. Primitive peloridiomorphs occurred in the Cretaceous of South America and the Jurassic of Eurasia (Figures 8.54, 8.55), which indicates that this family is relict as well. Another austral relict hemipteran is *Tettigarcta* from Australia; it is the living sister group to all other cicadas and is also known as Eocene fossils from Scotland. Carver *et al.* (1991) mentioned 15 other hemipteran groups showing austral disjunctions. Unfortunately, phylogenetic relationships in Psocoptera and Thysanoptera are not sufficiently explored to determine fully what austral disjunctions occur, but they almost certainly exist.

Among the four major holometabolous orders, austral disjunctions occur principally among basal taxa in each and involve southern South America, New Zealand, and Australia. In Coleoptera, austral disjuncts are widespread in Adephaga and basal Polyphaga. Newton (1985) provided an excellent review of austral biogeography in staphylinoid beetles, and Darlington (1965) discussed this for carabids, as did Howden (1981) for scarabs and Kuschel (1969, 1995) for weevils. Interestingly, weevils (Curculionoidea) that show obvious austral disjunctions are also ones that feed on the southern conifers *Araucaria* and *Agathis*, such as some Nemonychidae and the genus *Araucarius* in the Cossoninae. One of the most basal families of Lepidoptera, Agathiphagidae, likewise feeds on *Agathis*, in this case as a larval borer in the seeds. Lawrence (1991) listed 12 other beetle families that have austral disjunctions, many of them Cucujoidea.

Austral disjunctions are similarly widespread in Hymenoptera. These occur principally in symphytans, like some xiphydriid and pergid sawflies, and orussids, as well as in basal apocritans like certain Gasteruptiidae (Hyptiogastrinae), Monomachidae, Diapriidae (Ambositrinae), Ichneumonidae (Labeninae), and Megalyridae. This distribution pattern is not as common in aculeate wasps and includes some Tiphiidae (Anthoboscinae), Chrysididae (Amesiginae), Pompilidae (*Epipompilus*), Scolebythidae (*Ycaploca* and *Clystopsenella*), and even a few ants and bees. Among ants, austral taxa include the primitive Australian subfamily Myrmeciinae (*Nothomyrmecia macrops* and *Myrmecia* spp.). Austral bees also include basal taxa, and disjuncts specifically include Paracolletini (including the former "family Stenotritidae") in southern South America and Australia, and bees in the tribe Fideliini from southern Africa and Chile (Engel, 2002b, 2004e).

In Diptera, further work has refined relationships and distributions of austral chironomids, including for Brundin's paradigm group, the podonomines (Cranston *et al.*, 2002; Martin *et al.*, in press). Other austral distributions in nematoceran flies include some Ceratopogonidae (*Austroconops*), Mycetophilidae (Munroe, 1974; Matile, 1990; Chandler, 2002), Psychodidae (*Nemapalpus*), and the small family Perissomatidae. In Brachycera, these occur principally among basal taxa, like Apioceridae and Mydidae (Yeates and Irwin, 1996), and some basal Bombyliidae, like the subfamily Psiloderoidinae. Eremoneuran flies with austral distributions are rare and include Sciadoceridae (South America–Australia– New Zealand), and only a few genera of empidoids. The empidoids include the basal genera *Afroclinocera* and *Homalocnemus*, which have species in the Australian region and in the south temperate areas of South America and Africa.

The austral region is not the only area to harbor relict arthropods. Notable relict insects living in the Northern Hemisphere include the silverfish *Tricholepidion* from the coastal forests of California and Oregon; the order Grylloblattodea; *Cryptocercus* roaches; *Timema* stick insects; trachypachid ground beetles; boreid scorpionflies; and *Opetia* flies, among many others that have been discussed throughout this book. But, when one considers how much larger the northern temperate forests are (21 million km^2) compared to the southern temperate forests (3.7 million km^2), and the extensive list of disjunct and relict arthropods in the austral region, the latter of these is clearly an evolutionary epicenter.

Vicariance Biogeography

Though Brundins' approach to biogeography is commonly used today, in the form of *vicariance biogeography*, it was hardly universally accepted at first. The Harvard entomologist P. J. Darlington (Figure 14.32) was an experienced naturalist who traveled throughout the world and was an ardent critic of this approach (Darlington, 1970). Ironically, he was a systematist of carabid beetles, the same group that was studied by Jeannel, an early advocate of the effects of continental drift on insect distributions. Darlington maintained that the history of animal distributions reflects climatic changes and the ebb and flow of older groups being displaced by younger, competitively superior groups into geographically and environmentally marginal areas (Darlington, 1957, 1965). According to him, ancient groups like the Carabidae, which evolved in the Late Triassic about 210 MYA and that are now widespread, were probably more diverse and abundant in the Amazon Basin at one time. As aggressive groups appeared, particularly the radiations of ants in the Tertiary, they drove many carabids from the hot tropical river basins into cool, wet montane and austral forests, where they are diverse today and where ants do not thrive as well (Brown, 1973). Indeed, ants have a dramatic effect on the distribution and abundance of all sorts of ground beetles (e.g., Kolbe, 1968, 1969). The Darlingtonian view was further developed – as the taxon cycle – by other Harvard entomologists, such as Wilson (1961, 1985a) and Brown (1973), who worked on ants. Darlington was skeptical of the significance of continental drift on insect distributions (e.g., Darlington, 1970) and so his ideas were marginalized as vicariance biogeography took hold in the 1970s and later.

Vicariance biogeography, also called *cladistic biogeography*, seeks to explain the distributions of species and higher taxa on the basis of their relationships and the geological relationships of the areas they inhabit (Nelson and Platnick, 1981; Humphries and Parenti, 1986). If the relationships of area and taxa have the same or a similar pattern, then the interpretation can be made that the distribution of the organisms was affected by geological processes that formed the areas they inhabit: the organisms are *vicariant*. If the relationships of landmasses and organisms differ, this could mean that dispersal was involved in the distributions. Brundin's (1966) study of chironomids is upheld as a model study in vicariance biogeography because the repeated patterns of midge distributions agree so well with the history of the landmasses.

Vicariance biogeography still has its critics (e.g., Briggs, 1995, 2003). First, while vicariance biogeography assumes the accepted mode of speciation via allopatry, it is routinely

14.32. Philip J. Darlington (1904–1983), the Harvard professor who maintained that dispersal, climate change, and the invasion of competitively superior groups were much more important influences on animal distributions than were drifting continents. His view is supported by fossils of austral insects from the Northern Hemisphere. Photo: G. Byers, University of Kansas.

envisioned as splitting a widespread, ancestral distribution, or vicariance. The ancestral distribution must have become widespread in the first place through dispersal. Also, the biotas of remote oceanic islands like the Hawaiian and Galapagos archipelagos (which were never in contact with continents) are the products of long distance dispersal. After Krakatau exploded in 1883, the gradual and persistent recolonization of the island has been carefully documented (Thornton, 1996). Myriad species of Recent insects that formerly had widespread distributions based on Pleistocene and Holocene remains (Elias, 1994) indicate widespread expansion and contraction of ranges. Lastly, Tertiary and Mesozoic fossils have documented extinctions of Recent groups in areas distant from where they presently occur. These include, for example, Oligocene tsetse flies in North America and Europe, and even presently austral groups that occurred in the Northern Hemisphere. Clearly, there can be sweeping changes in the distributions of various insects.

Fossils and Widespread Distributions

There is abundant evidence from fossils that many arthropod groups presently narrowly restricted to one or more areas of the austral region were once widely distributed (Eskov, 1992, 2002; Grimaldi, 1992). Much of this evidence is preserved in amber, partly because preservation with such fidelity allows accurate identifications, but also because large deposits of amber have preserved extinct insect faunas that were exceptionally diverse. Indeed, the first serious study to discuss the extinction of austral arthropods from the Northern Hemisphere was based on Baltic amber, in a comprehensive but neglected study by Ander (1942).

Ander noted that some ants in Baltic amber, for example, had their closest relatives living in Australia or southern South America, or both. *Prionomyrmex* in Baltic amber, for example, is closely related to the primitive Australian species *Nothomyrmecia macrops* (Baroni-Urbani, 2000), and both ants are closely related to the living Australian bulldog ants (*Myrmecia*) and two Eocene genera from Argentina, *Ameghinoia* and *Polanskiella* (Ward and Brady, 2003). If one looked just at the living species of this clade, it would seem their entire history was restricted to Australia. Ander also noted another relict group of Hymenoptera in Baltic amber: Megalyridae (Figure 14.33). This is a putatively basal family of Apocrita that presently occurs in the austral region and southeastern Asia, but which was present throughout the Northern Hemisphere during the Cretaceous to at least the Eocene (Rasnitsyn, 2002; Engel and Grimaldi, unpubl.). A very similar living and fossil distribution concerns the aculeate wasp families Scolebythidae (Brothers and Janzen, 1999; Engel and Grimaldi, unpubl.) (Figure 14.34), and Plumariidae (Brothers, unpubl.), as well as other parasitoid groups (e.g., Johnson *et al.*, 2001). The bee fauna preserved in Baltic amber shows widespread extinction of African and Asian groups that formerly existed in Europe, and *Protolithurgus* in Baltic amber is even closely related to basal lithurgines from southern South America (Engel, 2001a). A pattern similar to that of the bees involves the recently discovered order Mantophasmatodea, which were first discovered in Baltic amber and then as myriad living species from southern Africa (Zompro, 2001; Klass *et al.*, 2003a,b; Engel and Grimaldi, 2004b).

Another order where extinctions have been thoroughly studied is Diptera (Hennig, 1965a; Grimaldi and Cumming, 1999). One of the most basal genera of biting midges, *Austroconops*, occurs today only in western Australia, but fossil species are preserved in Cretaceous amber from Lebanon, France, and Siberia (Borkent, 2000a). Even in Brundin's own group, the podonomine chironomids, primitive species with biting mandibles are restricted today to the austral region, but there are inclusions of them in 125 MYO Lebanese amber (P. S. Cranston, unpubl.). The basal cyclorrhaphan family Ironomyiidae is restricted today to two rare species in Australia, but there are fossils in Cretaceous amber from Canada, Lebanon, and New Jersey (Grimaldi and Cumming, 1999), and also compression fossil ironomyiids from Eurasia

14.33. Parasitoid wasp of the basal apocritan family Megalyridae, preserved in Eocene Baltic amber. Cretaceous fossils indicate the family once also existed in North America and as far north as Siberia. The family today occurs in Africa, Australia, South America, and parts of southeast Asia. AMNH; length 2.2 mm.

(Mostovski, 1995). The primitive phoroid genus *Archiphora* is represented today solely by *A. patagonia* from Chile, but extinct species of the genus are preserved in 90 and 45 MY amber from New Jersey and the Baltic region, respectively (Grimaldi and Cumming, 1999) (Figures 14.36, 14.37). Stalk-eyed diopsid flies today occur in tropical regions of Africa and Asia, and tsetse in central Africa, but these flies occurred in northern Europe and in North America during the Late Eocene and Oligocene (Figures 12.103, 12.106).

The relict damselfly family Hemiphlebiidae (Zygoptera: Hemiphlebioidea) is particularly interesting because only

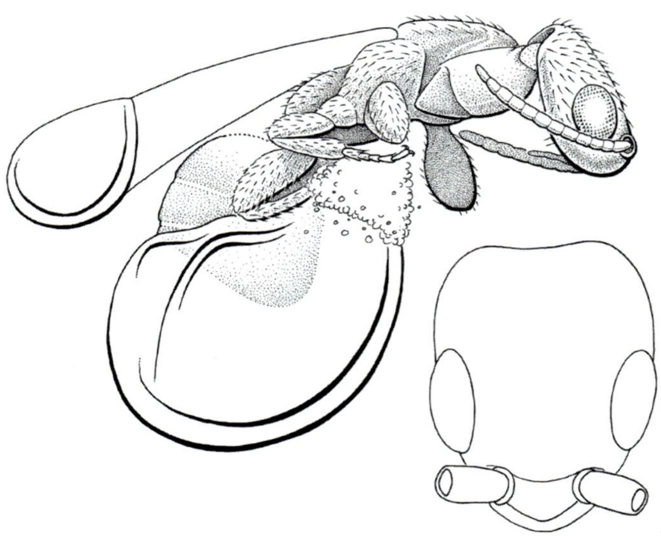

14.34. A basal aculeate wasp of the family Scolebythidae, in Cretaceous amber from New Jersey. The family presently occurs in southern portions of Africa and South America, and in Australia – a classic austral distribution. Other fossils of the family are from the Caribbean and the Baltic region.

14.35. An extinct species of the living beetle genus *Electribius* (Artematopodidae) in Eocene Baltic amber. It was first known as a fossil; then closely related species were discovered in Central America. MCZ; length 1.8 mm.

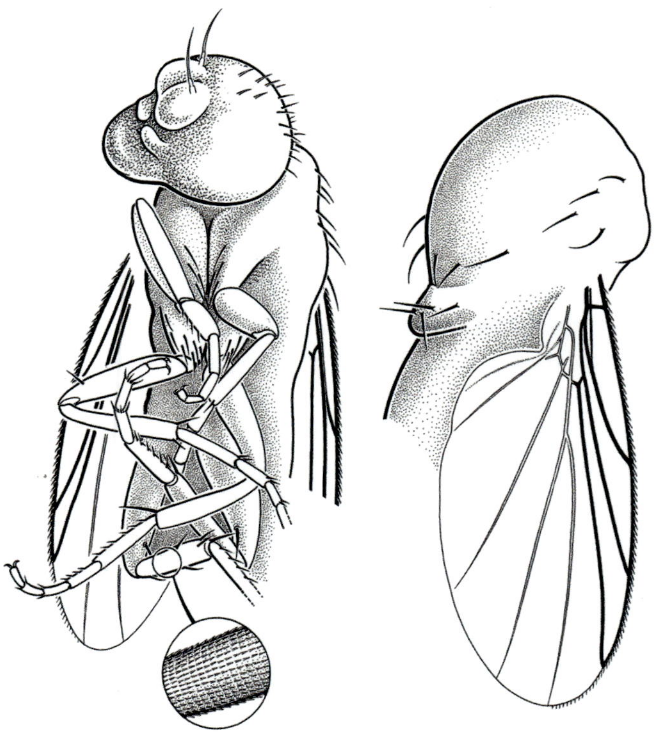

14.36. The primitive phorid fly *Archiphora pria*, in 90 MYO amber from New Jersey. It is closely related to an extinct species from the Eocene (Figure 14.37) and a living one from southern Chile. AMNH NJ 773; length 2.6 mm.

one Recent species, *Hemiphlebia mirabilis*, exists, and that is found in Victoria, Australia. There are, however, hemiphlebioids from South America (Bechly, 1996; Jarzembowski *et al.*, 1998). The family had traditionally been considered as a basal odonate or basal zygopteran (Tillyard, 1928a), but it is now understood to be relatively derived among damselflies (Rehn, 2003).

The list of extinctions is obviously too long to discuss completely, but one other superorder has revealed some rather striking examples: Neuropterida. Rhachiberothids are raptorial lacewings related to Berothidae that presently occur in southern Africa, but that existed in the Eocene of Europe and throughout the Cretaceous of the Northern Hemisphere (Grimaldi, 2000a; Engel, 2004a). Silky lacewings of the family Psychopsidae comprise only 26 Recent species in southern Africa, Australia, and portions of southeast Asia (Oswald, 1993a). Fossils, however, show that they were diverse and global from the Jurassic to the Miocene (Oswald, 1993a; Andersen, 2001; Grimaldi, unpubl.) (Figure 14.38). The spoon-winged lacewings (family Nemopteridae) are presently restricted to Australia and the Old World, but beautiful compression fossils show that they occurred in North America in the Eocene as well (Figure 9.18). Not all extinctions involve presently southern groups in the Northern Hemisphere. Raphidioptera, for example, is a small group of 206 Recent species occurring today in Europe, Asia, and western North America (Aspöck, 1998; Aspöck *et al.*, 1991, 2000), but which in the Cretaceous occurred in eastern North America and South America among other regions (Oswald, 1990; Grimaldi, 2000a; Engel, 2002a). Lastly, the relict family Polystoechotidae comprises four Recent species of large, nocturnal lacewings restricted to western North America and Chile (Oswald, 1998a). Fossil polystoechotids, however, are recorded from Jurassic and Cretaceous deposits ranging from Australia, Asia, and Europe (Lambkin, 1988; Ren *et al.*, 2002).

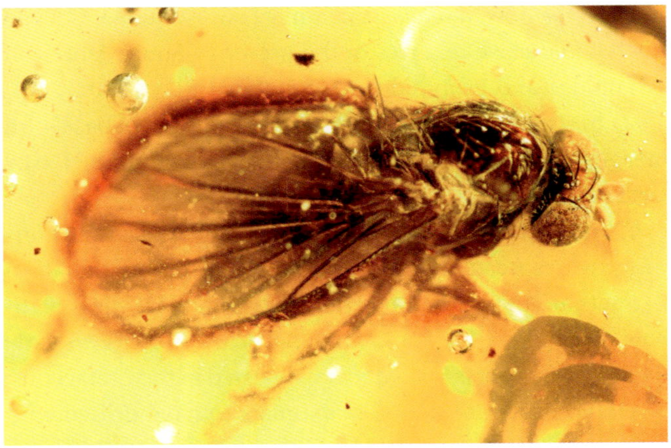

14.37. *Archiphora robusta* in Eocene Baltic amber, a close relative of the living species *Archiphora patagonica* from southern Chile. AMNH; length 2.1 mm.

14.38. Raptorial lacewing, *Rhachibermissa splendida* (Berothidae: Rhachiberothinae) in Cretaceous amber from New Jersey. Rhachiberothines today are restricted to southern Africa, but they occurred in at least Asia and North America in the Cretaceous and in Europe during the Eocene. AMNH NJ615; length 3.0 mm.

The most plausible scenario for these extinctions involves paleoclimatic change. These austral groups probably dispersed widely during periods when tropical and warm temperate conditions were essentially worldwide, such as during the Cretaceous and Eocene. They retreated south as the Northern Hemisphere cooled and north from Antarctica as it glaciated. Continents indeed wander, but their inhabitants apparently move around even more.

Former widespread distributions of austral or southern groups are not restricted to insects. There are many examples where this occurs in vertebrates as well. For example, *Palaeotis* from Messel, Germany, and other Eocene ratites indicate that this famous group of Southern Hemipshere birds occurred in the Northern Hemisphere (Mayr, 2000). Indeed, Eocene fossils of the now Australian bird family Podargidae (frog mouths) also occur at Messel and in the London Clay, as do fossils of the now Madagascan family Leptosomidae (Cuckoo rollers), and close relatives of the presently South American Cariamidae and Psophiidae (trumpeters) (Mayr, 2000; 2001, 2002a,b). Thus, Cracraft's view that the evolution of basal bird lineages was mostly or entirely gondwanan is contradicted by fossil evidence.

Mammals have a fossil record that vividly shows many formerly widespread distributions, but the group that best instructs about early gondwanan distributions is the marsupials (e.g., Cox, 1974; McKenna and Bell, 1997). As an early lineage of mammals known to have existed in the Cretaceous, we would expect ancestral marsupials to have drifted with gondwanan continents, but fossils indicate a completely different scenario: The earliest relatives of marsupials probably originated in North America. The oldest fossil marsupials are from North America, and the group became extinct there about 20 MYA (the opposum, *Didelphis virginiana*, reinvaded North America about 3 MYA). Marsupials invaded South America in the Late Cretaceous perhaps through a landmass that briefly connected the two continents, called ProtoGAARlandia, which was formed from plates of what are now the Greater Antilles and the sunken Aves Ridge. South America still retains its rich marsupial fauna. From South America they invaded Antarctica (which was forested at the time) and then Australia. In fact, the famously diverse marsupial fauna of Australia is apparently no older than Eocene, about 55 MYO. By ignoring the marsupial fossil record, it would seem that virtually the entire history of marsupials was gondwanan, with only a very recent colonization of North America. Fossils are not *sine qua non* for biogeography, but fossils provide unique insight for biogeography. Few insect groups have as rich a fossil record as that of mammals, but the mammal fossil record reminds us of the problems with interpreting past distributions solely on the basis of the Recent fauna.

Perhaps the most compelling evidence *in favor of* the effects of continental drift on distributions actually comes from other fossil evidence. Specifically, those groups of arthropods that radiated in the Cenozoic essentially show no austral distributions. Such groups include the ditrysian Lepidoptera, the "higher" termites (Termitidae), and the schizophoran flies, which have either very sparse or no appreciable Cretaceous records. This is an observation described in detail by Hennig (1966) for the Diptera. Some will claim that "absence of evidence is not evidence of absence," but when the fossil record is interpreted phylogenetically, as we presented earlier for each of these groups, their young ages are quite clear. So, *austral distributions occur only in those groups that occurred in the Cretaceous and earlier, which indicates some involvement of continental drift*, but widespread fossils indicate that drift had a minor impact on distributions compared to climatic and biotic change.

Fossil evidence of formerly widespread distributions reveals that many Recent endemic austral arthropods are not just disjuncts but relicts as well. Their presently narrow distributions seem to be a consequence of extinction and to some extent continental drift, though for the latter perhaps not as much as Brundin and others supposed. But why would relicts be so concentrated in the Austral Region when Laurasia obviously also had a Cretaceous history of drift and cool temperate environments? It is likely that ancient groups could not survive the harsh glaciation that took place in the northern continents. Also, myriad arachnids and insects took refuge in the cool, wet austral forests perhaps to escape competitively superior groups like ants, which dominate the hot tropics. In such a case, Darlington may not have been entirely wrong.

INSECTS, MASS EXTINCTIONS, AND THE K/T BOUNDARY

Mass extinctions are the extinction of diverse and unrelated lineages usually over a brief period of time in geological terms, and ultimately caused by global changes in the abiotic environment. The global changes can be of climate, ocean levels, volcanism, drifting continents, oxygen levels, an asteroid impact, or any combination of these, but the changes are catastrophic. Mass extinctions have recurred throughout earth history and classically define the boundaries between geological epochs and periods. Five major extinction events have occurred over the past 600 MY, as well as numerous smaller ones. The extinction at the very end of the Cretaceous (the K/T extinction) is the best known because it involved the non-avian dinosaurs. Some other groups were involved in the K/T extinctions, including the nautilus-like ammonites and certain marine bivalves, the most important of which were the rudists. Rudists were the principal reef builders in the Cretaceous, a niche now entirely occupied by corals. Study of the Cretaceous has transformed paleontological understanding of mass extinctions, particularly since it was the youngest major extinction event and so evidence from it is best preserved.

The new paradigm in the study of mass extinctions essentially began with Alvarez *et al.* (1980), who reported an unusual layer of the element iridium at the K/T boundary at Gubbio in the Italian Apennine Mountains, where the type section of this geological moment is exposed. Iridium and most other platinum-group elements are rare in the earth's crust, but the iridium layer at Gubbio is 30 times that normally found. Platinum-group elements are more abundant, however, in asteroids and meteorites, so Alvarez *et al.* concluded that the iridium was extraterrestrial in origin, probably from an asteroid. The search for an impact site led to a crater 180 km in diameter off Mexico's Yucatan Peninsula, near the town of Chicxulub Puerto, submerged beneath a kilometer of limestone sediments and ocean (Hildebrand *et al.*, 1995; Kyte, 1998). The crater, Chicxulub, preserves the remains of a 10-km-wide asteroid that slammed into this area 65 MYA, or just before the K/T boundary (Keller *et al.*, 2004). At the time this area was at the southern end of the Great Interior Seaway that separated eastern and western North America. The impact of the Chicxulub asteroid created massive tsunamis and spewed into the atmosphere sediments and gases with a mass several orders of magnitude greater than that of the asteroid. This wreaked havoc on the global climate, creating a nuclear winter that may have been the crushing blow to the non-avian dinosaurs, and it left other global traces. Besides the iridium layer, there were grains of shocked quartz with distinctive parallel etchings (which otherwise are found only at sites of nuclear blasts and volcanic eruptions), quartz grains that were melted into glassy spheres by the impact (called microtektites), and concentrated fern spores. The spores indicate massive wildfires because ferns are always the first colonizers of an area after a fire (soot from fires was also preserved). Such remains have now been found at the boundary of several other geological periods, indicating that periodic impacts of catastrophic asteroids have occurred throughout earth's history.

Soon after the Alvarez *et al.* (1980) study, Raup and Sepkoski (1984) published an influential article on the geological diversity of life, a plot of fossil genera over time compiled from a huge database. They quantified what had been known for some time, that some extinctions (like the End Permian Event) were much more devastating than others, but, unexpectedly, they proposed a 24 MY periodicity of extinctions. Though the periodicity is now statistically disputed, the original interpretation invoked celestial orbits as a possible cause for mass extinctions. It appeared that evolutionary change was not just gradual but also was affected by infrequent but catastrophic extraterrestrial havoc.

Even though the evidence for asteroid impacts is vivid (e.g., Kaiho *et al.,* 2001; Olsen *et al.,* 2002), the effects these impacts had on life are ambiguous and controversial. First, numerous other factors may be involved in mass extinctions, such as the effects of drifting continents, climate change, oxygen levels, massive volcanism, sweeping changes in sea levels owing to glaciation and the formation of interior marine basins (e.g., Hallam, 1989; Hallam and Wignall, 1999). Indeed, toward the end of the Cretaceous, the climate gradually cooled, and sea levels retreated. Approximately 66–67 MYA the eruptions of the Deccan Traps in west-central India occurred, which extruded half a million square kilometers of basalt and stopped right after the K/T impact. Volcanic eruptions like that of Mt. St. Helens, Mt. Punatubo, or Krakatau had global effects on climate, so one can imagine the effects on climate and ocean chemistry of an entire cordillera of volcanoes erupting simultaneously. Second, the geological and systematic conditions required for measuring biotic effects of a catastrophe are stringent. Ideally, strata containing the remains of a mass extinction event should have excellent resolution, or fine and continuous deposition before and after the extinction interval, which are conditions largely restricted to marine fossils. For example, were it not for their excellent fossil record, we would not be able to determine that the rudist bivalves actually became extinct *1.5 MY before* the K/T boundary. By this time, too, dinosaur diversity had dwindled to fewer than 10 species. Marine foraminiferans and coccoliths (which have one of the best fossil records), on the other hand, show dramatic extinctions at the K/T boundary. Also, mass extinctions must be measured against the level of background extinctions. Lastly, the sampling should ideally be global in scope because losses of taxa in a regional fossil record do not necessarily indicate overall extinction. Given that the insect fossil record is considerably less rich than for most marine groups, assessing mass extinctions in

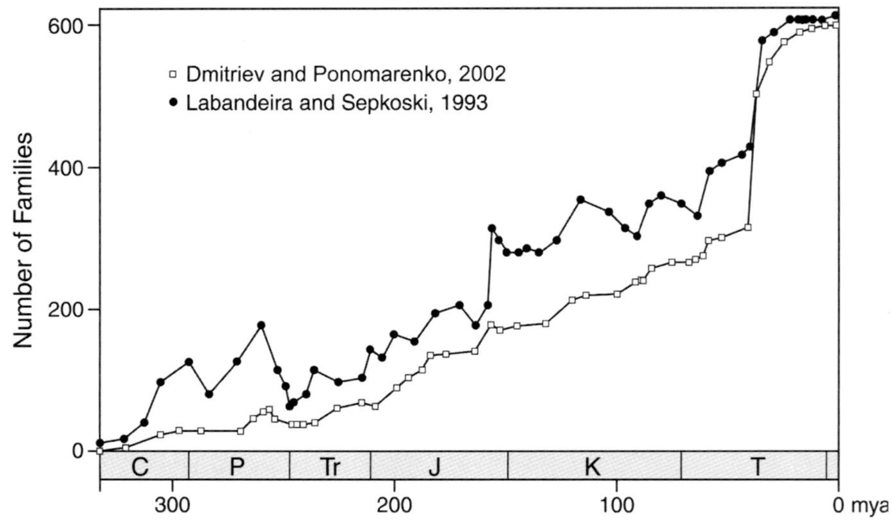

14.39. The steady increase in the numbers of insect families over time, based on two major studies. The sudden spike in the Tertiary is a result of the very diverse Baltic amber fauna. Taxic analyses like these are very susceptible to the effects of different taxonomic systems, paraphyletic groups, and a few highly prolific fossil sites. They also treat taxa as equivalent units, even though insect families vary tremendously in diversity and ecological significance.

insects is far less precise. The K/T impact event may have been a coup de grâce for groups that were waning, but it seems to have had just a minor impact on insects.

Insect Diversity: Trends and Events

The study of insect diversity over time has been based on traditional "spindle" phylogenies (e.g., Martynov, 1938; Figure 4.16; Rohdendorf, 1964; Hennig, 1981), and, more recently, on taxic analyses (Wootton, 1990; Labandeira and Sepkoski, 1993; Jarzembowski, 1995a; Jarzembowski and Ross, 1996; Labandeira, 1997; Ross *et al.*, 2000; Dmitriev and Ponomarenko, 2002; Figure 14.39). While the spindle phylogenies combine diversity and relationships, the taxic analyses (as we described earlier) are simply plots of diversity over time. Most of the insect studies deal with families as the units, though orders (Jarzembowski and Ross, 1996) have also been used, which provides the least information since taxonomic concepts of some orders can be vague, particularly with regard to early, extinct members. One study (Jarzembowski and Ross, 1996) also used genera, which should be most informative.

There are deficiencies of taxic analyses (Smith, 1994), which we have discussed earlier, some of which apply particularly well to the insect studies. First, certain deposits and taxonomists who study them can dramatically affect the diversity curve, such as the spike in Late Jurassic and mid-Tertiary diversity, owing to the vast deposits at Karatau and of Baltic amber, respectively. Conversely, the dip in Paleocene diversity (Jarzembowski and Ross, 1996) reflects the paucity of Paleocene insect deposits. Second, the use of particular classifications may have significant effects. For example, Labandeira and Sepkoski (1993) plotted 1261 insect families, but Dmitriev and Ponomarenko (2002) identified only 1049 because the latter authors considered 212 families to be synonyms of other families. A 20% difference in diversity can dramatically affect a diversity curve. Lastly, standing diversity is a result of origination, stasis, and extinction, so by just plotting total diversity (e.g., Labandeira and Sepkoski, 1993; Labandeira, 1997), it is impossible to determine if a spike is the result of higher origination rates, lower extinction rates, or both. Several studies plotted "origination" and "extinction" (Jarzembowski and Ross, 1996; Ross *et al.*, 2000), but as Dmitriev and Ponomarenko (2002) indicate, it is most appropriate to call these *first* and *last* appearances in the fossil record because origination and extinction are elusive moments in the fossil record.

What, then, are the patterns these studies show, and what might they mean? First, there was a modest to abrupt rise in family-level diversity in the Late Permian, followed by a crash or dip at the P/Tr boundary (Figure 14.39). The P/Tr reduction is clearly the end-Permian extinctions, which appear to have exterminated all or most of the Palaeodictyopterida, Caloneurodea, Miomoptera, Glosselytrodea, and primitive lineages related to modern orders. Second, there was a steady rise in diversity from the Triassic to the mid-Tertiary. The spike in Late Jurassic diversity, which is minor (Dmitriev and Ponomarenko, 2002) or large (Labandeira and Sepksoki, 1993), is probably an artifact from the large Karatau deposit. For the Cretaceous, results and interpretations differ.

Labandeira and Sepkoski (1993) reported a "slackening" in the number of insect families during the latter half of the Cretaceous – contrary to expectations that the angiosperm radiations would facilitate insect diversification. Dmitriev and Ponomarenko (2002), however, reported a steady increase of families in the Late Cretaceous through the Paleocene (Figure

14.39). This latter study and that by Ross *et al.* (2000) found that the rate of last appearances of families was elevated in the Early Cretaceous and then plateaued in the Late Cretaceous through Tertiary. This is interpreted as major faunal turnover, specifically the replacement of gymnosperm-feeding insects by new, angiosperm-feeding insects. Also, Dmitriev and Ponomarenko (2002) found a high rate of first appearances of families in the Early Cretaceous. High origination and extinction rates in the Early Cretaceous are consistent with a profound effect of the angiosperm radiations on insects, contrary to Labandeira and Sepkoski (1993). The high origination rates in the Cretaceous are caused by the radiations of large groups like the Lepidoptera, termites, brachyceran flies, polyphagan beetles, and apocritan wasps, which we have discussed throughout this book.

All the taxic analysis studies agree that the effects of the K/T extinction on insects *at the family level* are imperceptible. Recent work found significant reduction in insect herbivore diversity after the K/T boundary (Labandeira *et al.*, 2002), but this was based on a regional biota in western North America that was in the wake of the Chicxulub asteroid. Also, this study was based on feeding damage on leaves, not the actual insects. It is unknown how these herbivorous insects responded to the K/T event on a larger geographic scale, which is necessary to know when assessing extinctions. By 75 MYA (10 MY before the K/T boundary) only 2–3% of all known extinct insect families existed, and this was part of a gradual decline in the numbers of extinct families (Dmitriev and Ponomarenko, 2002). Significant examples of clearly monophyletic, Mesozoic insect families that persisted into the mid- to Late Cretaceous (extinct possibly by the K/T boundary) include the following: Elcanidae and Locustopseidae (Orthoptera; Figure 7.31, frontispiece); Aeroplanidae and Chresmodidae (Phasmatodea; Figures 7.4, 7.5); Raphidiomimidae (Blattodea; Figure 7.68); Progonocimicidae and Protopsyllidiidae (Hemiptera; Figures 8.21, 8.55); Pseudopolycentropididae (Mecoptera; Figure 12.3); Kalligrammatidae (Neuroptera; Figure 9.25); Falsiformicidae, Stigmaphronidae (Figure 11.27), and Serphitidae (Figure 11.23) (Hymenoptera); Archizelmiridae (Figure 12.50), Archisargidae, and Eremochaetidae (Diptera). Among these families are herbivores, parasitoids, and saprophages, so there is no ecological pattern to these extinctions. Most insect families in the Late Cretaceous, perhaps 95%, are the same as those found today. This is very similar to the situation in another major group of terrestrial arthropods, the spiders (Penney *et al.*, 2003). One explanation as to why insects apparently withstood the K/T extinctions so well concerns *diapause* (Whalley, 1988), the period of developmental and metabolic dormancy that is usually related to seasonal changes in climate but that can also be induced by changes in light periods and temperature. In fact, there were probably many reasons why insects were little affected by the K/T extinctions.

Although the insect fossil record lacks fine stratigraphic resolution, insects from the Late Cretaceous and Early Tertiary are compelling evidence that the K/T event had minor or just regional impact on these animals. Insects may be more resistant to natural mass extinctions than are most other animals, but the greatest challenge to their resilience is the massive environmental destruction and poisoning that is presently taking place.

THE TERTIARY

Compared to groups like mammals, Paleocene insects are poorly known. Significant deposits from this period with insect wings and body fossils are scattered among Denmark, Sakhalin Island, France, and western North America, and trace fossils occur in western North America and southern South America. A dramatic contrast to this is the Eocene fossil record, which is one of the best known geological periods for insects. Eocene deposits are widespread in the Northern Hemisphere, and include the most diverse faunas of fossil insects, such as the Baltic amber, the Green River Formation, and at the Eocene-Oligocene boundary, Florissant. A dramatic rise in insect diversity, nearly to modern levels, occurred in the Late Eocene (Labandeira and Sepkoski, 1993; Jarzembowski and Ross, 1996; Dmitriev and Ponomarenko, 2002). Great diversity is probably attributable to the fact that the earth was nearly entirely tropical and warm temperate in the Eocene as it was in the Cretaceous, and this allowed organisms to proliferate and spread as we detailed earlier for many presently tropical insects (see "Fossils and Widespread Distributions" earlier in this chapter). The high diversity of insects in the early Tertiary can be attributed to high origination and low extinction rates, but the abruptness in this rise of family diversity is almost certainly an artifact of so few Paleocene deposits and the richness of Eocene deposits.

The terrestrial world of the Early Tertiary was transformed by radiations of speciose and abundant insect groups that are primary consumers, plant feeders, parasites, and pollinators, which include the following:

- Pentatomorphan and cimicomorphan bugs, which are primarily phytophagous
- Phytophagan beetles (Cerambycidae, Chrysomelidae, and Curculionidae), which is the second most diverse lineage of phytophagous organisms after the Lepidoptera
- Ditrysian Lepidoptera, which comprises 95% of the Recent Lepidoptera species and is a group that arguably consumes more photosynthetic plant tissue than any other group of animals
- Schizophoran flies, which includes a spectrum of phytophages, saprophages, carrion breeders, endoparasitoids, and ectoparasites

- Social bees, particularly the apines, which because of their large colonies and recruitment system of foragers are particularly efficient pollinators
- Ants, particularly genera in the subfamilies Dolichoderinae, Myrmicinae, Formicinae, and Dorylinae that presently form immense colonies of one to many millions of individuals
- "Higher" termites, or the family Termitidae, which comprise 80% of all species of Isoptera, as well as form the largest termite colonies and consume the most plant matter of all termites. Like ants, termites do not become significant components (much greater than 2% of individuals) in fossil insect assemblages until the Eocene when the most recently evolved family Termitidae first appears. The explosive radiation of the Termitidae in the Lower Tertiary transformed various terrestrial landscapes, particularly with the origin of grasslands around the Oligocene and their expansion in the Miocene when the C_4 grasses proliferated (Jacobs *et al.*, 1999). The C_4 grasses have a form of photosynthesis that allows them to thrive better in more arid and seasonal environments
- Though the fossil evidence is meager to nonexistent, the radiations of fleas, lice, and smaller ectoparasitic groups like batflies almost certainly was in Tertiary, when the modern orders of mammals and largest lineages of birds (e.g., the passeriforms, or perching birds) radiated

MAMMALIAN RADIATIONS

The radiation of the Recent orders of mammals in the Paleocene and Eocene had a dramatic impact on insects, and vice versa, with both groups playing the role of the diet. A particularly dramatic example concerns the radiations of higher termites in the Early Tertiary. Some mammals like hedgehogs (Erinaceidae), tenrecs (Tenrecidae), and armadillos (Dasypodidae) readily consume ants and termites among other kinds of insects, but four mammal groups have independently evolved specialized features for obligate feeding on termites (and sometimes ants). These features include fossorial forelegs for excavating colonies; tubular snouts with long, sticky tongues to probe crevices and extract the insects; and peglike teeth. The mammals and their earliest fossil records are the following: the echidna (*Tachyglossus aculeatus*; Tachyglossidae: Miocene to present) and marsupial numbat (*Myrmecobius fasciatus*), both of Australia; the aardvark of Africa (*Orycteropus afer*; Tubulidentata: Oligocene to present); pangolins of Africa and Asia (family Manidae, Order Pholidota, with eight species: Paleocene to present); and the South and Central American anteaters and tamanduas (Figure 14.40) (Myrmecophagidae, with four species: Miocene to present) (reviewed by McKenna and Bell, 1997). This remarkably convergent evolution of mammalian termitophages was contemporaneous with the radiation of the Termitidae; it must have occurred in response to the burgeoning abundance of termites in the Early Tertiary.

14.40. Two arboreal anteaters, family Myrmecophagidae, in Panama (above, *Tamandua*; below *Cyclopes*). Despite the family name, their diet is mostly termites. There are five families of mammals that are specialized for feeding on termites and ants, which is testament to the ecological significance of these insects. Such mammals evolved between the Eocene and Miocene, which is when termites and ants achieved ecological dominance. Photos: P. J. DeVries.

The bats, Order Chiroptera, comprise another group that has affected the evolution of insects, and whose own evolution was probably dramatically affected by insects. Megachiroptera are the fruit bats, but the much more diverse and largely nocturnal microchiropterans are primarily insect feeders and arguably the most significant group of insectivores besides passerine birds (Figure 14.41). Though difficult to estimate, the huge roosts of Mexican free-tailed bats, for example, probably consume several tons of insects on each

14.41. A bat, *Eptesicus fuscus*, grabbing an insect. Echolocating bats are probably the main predators on insects and have profoundly affected the evolution of nocturnal insects. Photo: Merlin Tuttle ©, Bat Conservation International.

nightly foray. Not all microchiropterans feed on insects; some are specialized for feeding on flowers, frogs, fish, fruit, or small lizards and mammals (including other bats), and the vampires of course feed on blood. A sophisticated system of echolocation has evolved in most microchiropterans, particularly those that feed on insects, which use emissions of barely audible, high-pitched squeaks (25–50 kHz). The sound waves reflect off objects and are sensed, so the bat can "see" the location, size, and perhaps the shape of small objects in the dark. There have been at least ten independent origins among insects of tympana that are sensitive to the high frequency of bat calls, including most mantises (superfamily Mantoidea), chrysopid lacewings, some cicindelid and scarab beetles, and five major groups of moths (Hoy and Robert, 1996; Yager, 1999a). When most of these insects hear a bat, they perform evasive maneuvers, such as flying in a loop or divebombing. Some tropical katydids have also evolved calls that are a series of intermittent clicks, which make it difficult for bats to locate the source. Clearly, bats have been an important selection pressure for insects that fly or sing at night. Bats probably originated in the Paleocene but the earliest known fossils are from the Early Eocene; bats then diversified rapidly into 24 genera known during that period alone (Simmons and Geisler, 1998) (e.g., Figure 14.42), which is contemporaneous with the radiations of macrolepidopterans and mantoidean mantises. In fact, some of the Eocene bats have the stomach contents preserved, which include beetles, moths, flies, and caddisflies. The remains of the inner ear bones indicate that some of the Eocene bats were capable of at least limited echolocation (Simmons and Geisler, 1998), and this probably allowed them to locate and snatch insects off their perches at night. This behavior later evolved into aerial capture as echolocation became more refined.

Bats are also a sink for various highly specialized insect

14.42. An early microchiropteran bat from the Eocene of Messel, Germany. Bats probably originated in the Paleocene, and by the Eocene they possessed some echolocation abilities. Some of the Messel bats are preserved with the remains of insects in their stomachs. Many groups of nocturnal insects have evolved hearing organs sensitive to the ultrasound of bat calls, most species of which are in the Ditrysia Lepidoptera. Photo: R. Larimer, private collection.

ectoparasites and have more types of these than any other order of mammals or birds (Figure 14.43). These ectoparasites, discussed throughout this book, include the following: *Arixenia* (Dermaptera); Cimicidae and Polyctenidae (Heteroptera); Streblidae, Nycteribiidae, Mormotomyidae, and *Mystacinobia* (Calliphoridae) (Diptera); and Ischnopsyllidae (Siphonaptera). Oddly, bats do not have lice, which may be the result of competitive exclusion by the many other specialized ectoparasites. Bats harbor so many specialized ectoparasites probably because of their roosting behavior. These are the only mammals where the roosting sites or hibernacula are so persistent, and the larger "bat caves" are known to have been continuously inhabited for tens to hundreds of thousands of years. Under such conditions vast amounts of guano and the carcasses of dead bats accumulate, which can promote the evolution of ectoparasitism from scavenging and carrion-breeding insects. Besides bats, the Paleocene-Eocene radiation of mammals clearly had a dramatic impact on the evolution of Phthiraptera, Siphonaptera, and oestroid (bot) flies.

The transition from the Eocene to the Oligocene, approximately 35 MYA, was associated with one of the "lesser" mass extinctions, and this probably was the result of the dramatic cooling that took place. It was during the late Paleocene and Early Eocene that probably the most profound biotic events took place for the Tertiary. The climate was at its hottest and tropical regions extended nearly to both poles. Groups of insects that today are restricted to tropical and subtropical regions were widespread, as we reviewed earlier based on abundant fossil evidence. The global "hothouse" Eocene climate, though, abruptly changed to cooler, more seasonal climates in the Oligocene. As Antarctica and Australia separated, this allowed cool ocean currents to circulate freely, which had yet further dramatic impact on global climate. The Oligocene is the beginning of what is truly a modern type of insect fauna. It is probably during this period that tropical groups that formerly existed far into the north contracted into equatorial regions that have remained tropical to this day. It is also during the Oligocene that modern tropical forest ecosystems became established, as well as the great savannas, steppes, and other grasslands. Extensive glaciation began in the Late Miocene, and by the Pliocene the poles had become entirely glaciated and the earth truly polarized.

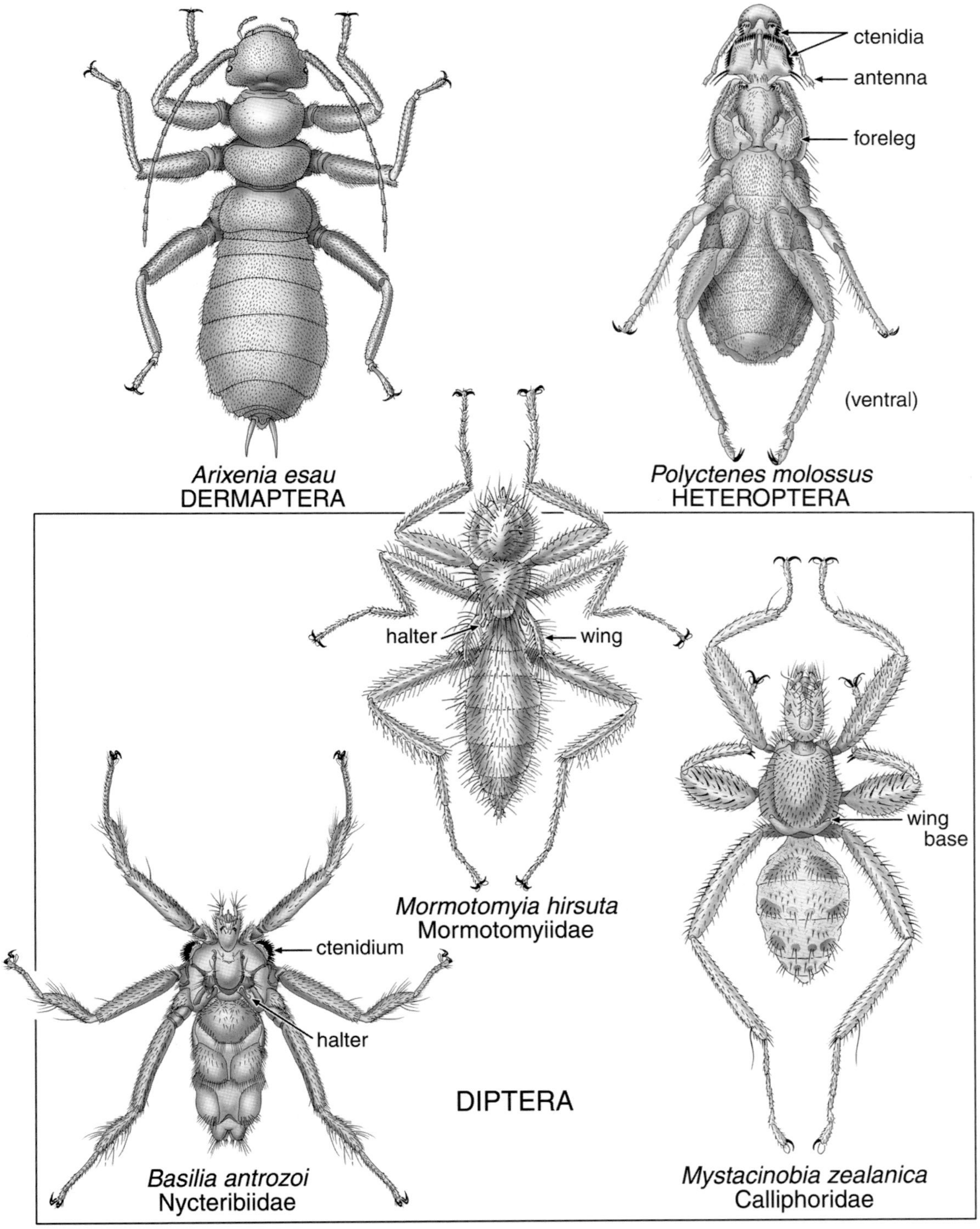

14.43. Besides being predators of insects, bats have been a sink for the evolution of specialized insect ectoparasites in four orders and eight families, some of them shown here. Caves occupied for thousands of years by roosting bats continually attract scavenging insects; this facilitated the evolution of some to become highly specialized ectoparasites. Not to the same scale.

PLEISTOCENE DISPERSAL AND SPECIES LIFE SPANS

Durable remains of sclerotized arthropod parts preserved in ancient bog peats and lake sediments provide a refined view of insects from 1. 7 MYA to several thousand years ago, during the Quaternary Period (Pleistocene and Holocene; Elias, 1994). Such remains include the head capsules of larval chironomids, cephalothoraces of spiders, chelicerae of pseudoscorpions and scorpions (even the sting), and whole mites, but the most abundant and diverse are the remains of beetles. Coleoptera are generally preserved as isolated elytra and pronota and occasionally the male genitalia, the last of which provide an excellent basis for determining species identity. In general, most of the Quaternary insects are morphologically indistinguishable from living species (Coope, 1970, 1978).

Life spans of species are crucial to understanding rates of extinction and origination, but this is probably one of the least understood aspects of evolutionary biology. For one, the fossil records of very few groups have the temporal resolution to estimate this accurately, and few fossils are sufficiently preserved to allow detailed comparison to living species. Also, it is well known that some Recent species are morphologically indistinguishable or nearly so but are genetically distinct, though morphology generally does provide a reliable means for separating closely related species. Estimates of species durations vary widely from approximately 1 to 10 MY among diverse organisms, and in a few cases they are much greater than this (Wilson, 1992). Planktonic foraminiferans, for example, which have an excellent fossil record, have species durations generally less than 10 MY, though some are known to be as great as 30 MY. This contrasts with another pelagic group, the extinct ammonites, for which the average species durations are 1.2–2 MY (such short durations are why the group is so useful for stratigraphy). Other marine groups with excellent fossil records include graptolites and echinoderms, with average species durations of 2 and 6 MY, respectively. Most reports of Recent insect species in the fossil record are from the Quaternary, but there are records of Recent species as old as 15–35 MY (Table 14.2). Some of these are based on completely preserved specimens in mid-Tertiary amber, which makes the evidence compelling. But, we suspect that an average duration of insect species is probably 3–10 MY, and that records much older than this, even of ones in amber, are the result of imprecise comparisons of fossil and living specimens.

Quaternary insects also have provided unique data on widespread changes in distributions. For example, intensive study of insects from the British Pleistocene (Coope, 1965, 1979, 1987, 1990) have revealed the former existence there of species that presently occur in the Iberian Peninsula, the Mediterranean Region, or even in Scandinavia. Perhaps the most dramatic such example involves the dung beetle *Aphodius holdereri*. This beetle was common in sediments from the last glaciation in Britain some 13,000 years ago, so preservation of myriad body parts (including genitalia) has allowed a nearly complete reconstruction (Coope, 1973). As glaciers advanced south, so did the ranges of most insects, and the insects expanded their ranges as glaciers retreated north, but this ebb and flow is not unique to insects. Mammals on the plains of Pleistocene North America, for example, resembled those of the African plains today, with cheetahs, lions, camels, proboscideans (mammoths and mastodons), and various antelope. If distributions can change so dramatically over millennia the question arises as to how static areas of endemism really are, and how much they are limited by climate rather than by geological events. There is perhaps no better reflection of the dispersal abilities of insects than in their colonization of remote oceanic islands, which have also been natural laboratories for evolution.

ISLAND FAUNAS

Some remote oceanic islands, formed in the ocean by volcanoes, are well known for their insular radiations of organisms. Perhaps the best known example is the finches on the Galapagos Islands (Lack, 1968; Grant, 1999), which were first

TABLE 14.2. Records of Long Durations of Living Species in the Insect Fossil Record

Taxa	Source	Age	Reference
Various ants and beetles	Plio-Pleistocene	2 MYO	Böcher, 1989, 1990
Micropeplus dokuchaevi (Staphylinidae)	Pliocene	5.7 MYO	Matthews, 1970
Longistigma caryae (Aphidoidea)	Greenland Miocene	15 MYO	Heie and Friedrich, 1971, 1990
Bohartilla megalognatha (Stepsiptera)	Dominican amber	20 MYO	Kathirithamby and Grimaldi, 1993
Micromalthus debilis (Coleoptera)	Mexican amber	25 MYO	Rozen, 1971
Belaphotroctes ghesquierei (Psocoptera)	Mexican amber	25 MYO	Mockford, 1972
Paleogryon muesebecki (Hymenoptera)	Mexican amber	25 MYO	Masner, 1969
Plateumaris nitida (Chrysomelidae)	Florissant	35 MYO	Askevold, 1990
Palaeomymar duisburgi (Hymenoptera)	Baltic amber	45 MYO	Doutt, 1973b

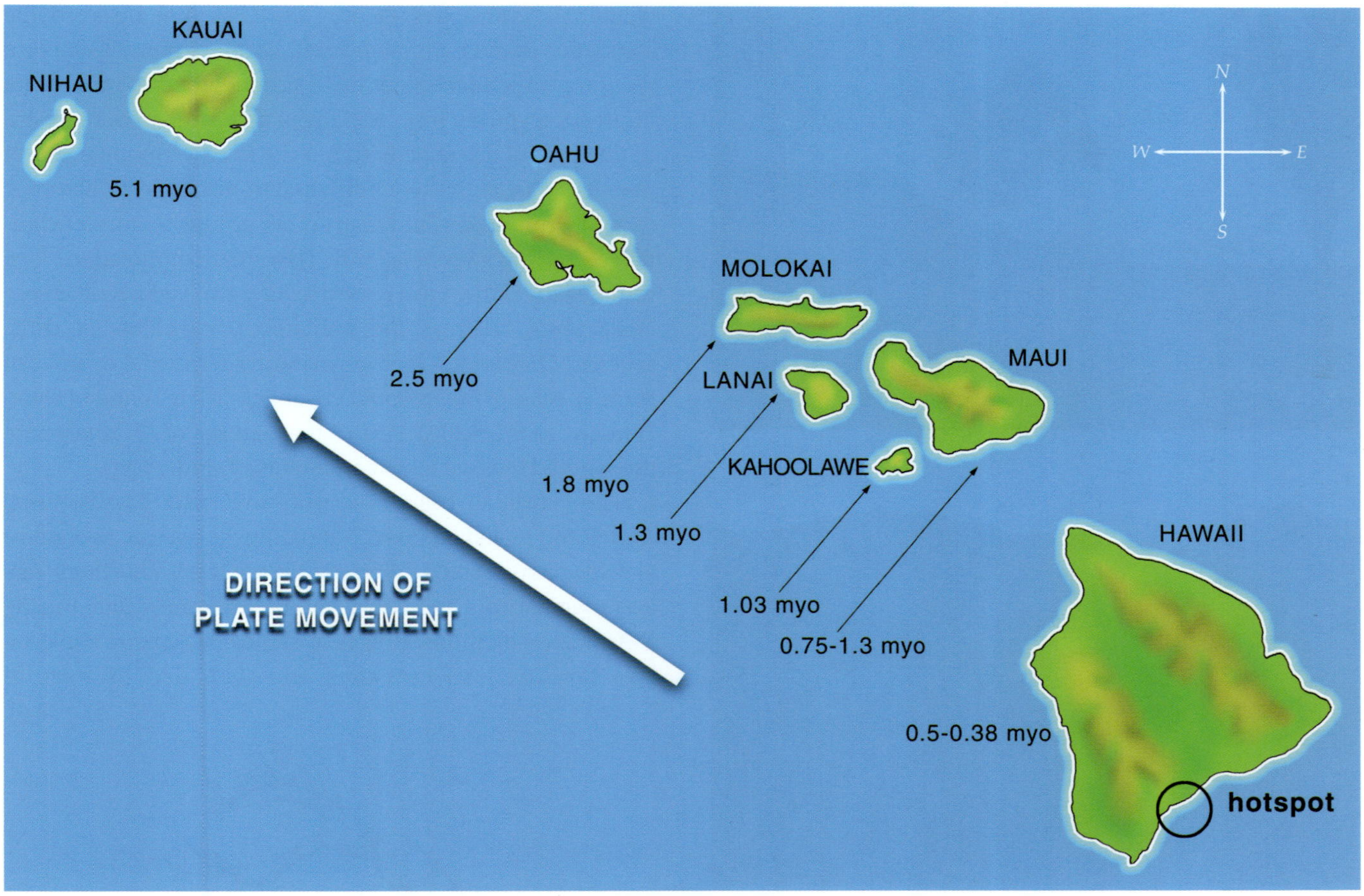

14.44. The Hawaiian Islands, frequently called a natural laboratory of evolution because no other area on earth supports so many endemic species. Thousands of species evolved in the Hawaiian Islands because of its isolation, unique geology, tropical climate, and diverse terrain. The island chain formed as the tectonic plate moved over the hotspot. The Hawaiian biota is also one of earth's most fragile and endangered.

studied by Charles Darwin and partly inspired him to formulate the theory of evolution via natural selection. The Galapagos Islands are very dry, so there has been limited speciation of insects there, but one tropical archipelago has fostered unprecedented radiations of insects and plants: the Hawaiian Islands (Figures 14.44, 14.45). These are a string of nine major islands in the middle of the Pacific Ocean, some 3,500 km from the nearest continent. On merely 16,580 km^2 (an area about the size of Connecticut or Israel) there are approximately 900 endemic species of angiosperms, 1,000 species of terrestrial snails, 2,300 species of insects, and 75 species of birds – an astounding average of 4 endemic species per square kilometer! The diversity and natural history of the Hawaiian biota has been amply reviewed (Carlquist, 1980; Wagner and Funk, 1995; and landmark papers reprinted in Kay, 1994). For terrestrial arthropods, major radiations include 250 species of endemic crickets (Otte, 1994), *Tetragnatha,* and other spiders (Gillespie and Croom, 1995; Hormiga *et al.,* 2003); 128 species of platynine carabid beetles (Liebherr and Zimmerman, 1998); and perhaps as many as 1,000 species of drosophilid flies (Hardy, 1965; Throckmorton, 1966). The radiations of drosophilid flies are particularly famous because this group appears to be the largest arthropod family on these islands, and there has been spectacular morphological diversification. Many Hawaiian drosophilids, for example, have patterned wings, and males have elaborate foretarsi and mouthparts, which are used in mating rituals (Figure 14.46). All of these appear to be monophyletic lineages, indicating that each group radiated from a single ancestral colonist. In the case of the endemic Hawaiian drosophilids, these have been routinely classified as *Drosophila,* but they actually appear to be an extremely diverse lineage within the large, cosmopolitan genus *Scaptomyza* (Throckmorton, 1966).

A combination of factors besides remoteness spawned the spectacular endemic diversity in the Hawaiian Islands. These include the fact that these islands are tropical (having some of the wettest forests on earth), and that it is an archipelago composed of high, rugged islands (up to 4,200 m high) that have been continuously produced since approximately the Paleocene 60 MYA. The Hawaiian Islands are the youngest and most easterly string of islands in a 5,800-km-long chain, which begins with the Emperor Seamount near the Aleutians (all of these islands are now submerged). The Emperor and Hawaiian Islands were formed volcanically as the Pacific

14.45. Typical of oceanic islands formed by volcanoes, the Hawaiian Islands have dramatic terrain. The lowlands (above) are hot and lush and overrun with invasive species; montane forests (shrouded in clouds in the background) are cooler and harbor more native species. The highest peaks, such as the top of Haleakala crater on Maui (below: 3040 m [10,023 ft] high) are cold and windswept. Because island insects lose some dispersal ability, mountains and valleys become formidable barriers and promote speciation.

plate moved northwesterly over a hotspot in the earth's crust, which presently is located beneath the youngest, largest, and most southeasterly island, Hawaii (Clague and Dalrymple, 1987). The oldest emergent island in the chain, Midway, is merely 3.7 m (11 ft) above sea level, and moving southeasterly the islands become progressively larger, higher, and younger. These islands have been reduced by erosion, subsidence (earth's crust buckles under the largest volcanoes), and huge land slides that disappeared into the ocean.

Insular species typically lose much of their dispersal ability, which in insects can be reduction of flight muscles or the wings themselves, from brachyptery to complete aptery. Confronted with poor dispersal ability, narrow gaps between islands become uncrossable, and deep, hot valleys separating cool mountain forests are a major barrier. Indeed, most of the speciation in the Hawaiian Islands has taken place *within* individual islands (e.g., Otte, 1994; Liebherr and Zimmerman, 1998; Hormiga *et al.*, 2003; Jordan *et al.*, 2003), and areas of endemism can be remarkably small, even confined to individual valleys or slopes (Figure 14.47). Early colonists arriving at remote islands are confronted with another situation: a plethora of unoccupied niches. Indeed, the immediate impression you get from walking in an Hawaiian rainforest is not the riot of insect activity, but the nearly complete lack of insects so distinctive to continental tropical forests, like ants, termites, butterflies, and many other groups that never capitalized on the sweepstakes of Hawaiian colonization. As a result, some of the native species have evolved peculiar specializations, such as the predatory caterpillars of three species of *Eupithecia* inchworm moths; terrestrial nymphs of some species of *Melagrion* damselflies on the island of Oahu (Jordan *et al.*, 2003); and the larvae of *Titanochaeta* drosophilid flies, which are predators of spider eggs.

The Hawaiian Islands do not have macrofossils (fossil spores and pollen do exist), but dating the ages of divergence and speciation events is perhaps uniquely possible there. Ash and magma from volcanoes can be accurately dated using K-Ar isotope methods, so ages of islands, volcanoes, and even lava flows can be measured and then tied to areas of

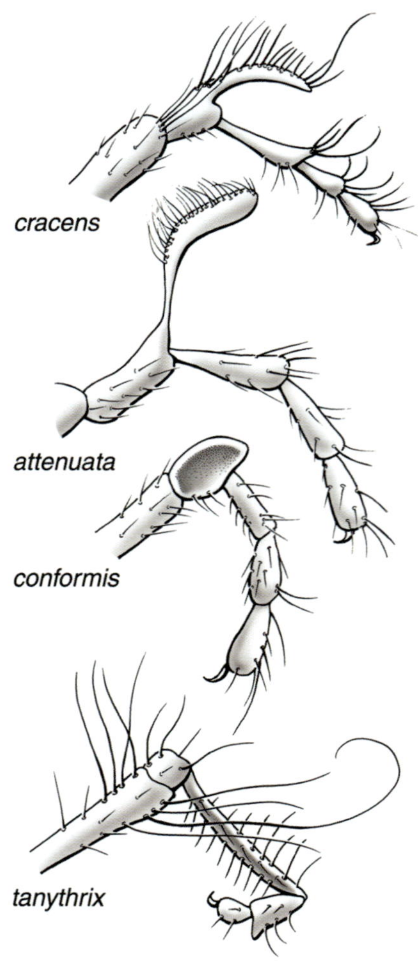

14.46. Probably the largest radiation of species in the Hawaiian Islands involves flies of the family Drosophilidae. Approximately 500 species are named and perhaps several hundred more exist. Males of many of the species have evolved elaborate structures on the forelegs that are involved in mating, some of which are shown here. Redrawn from Hardy (1965).

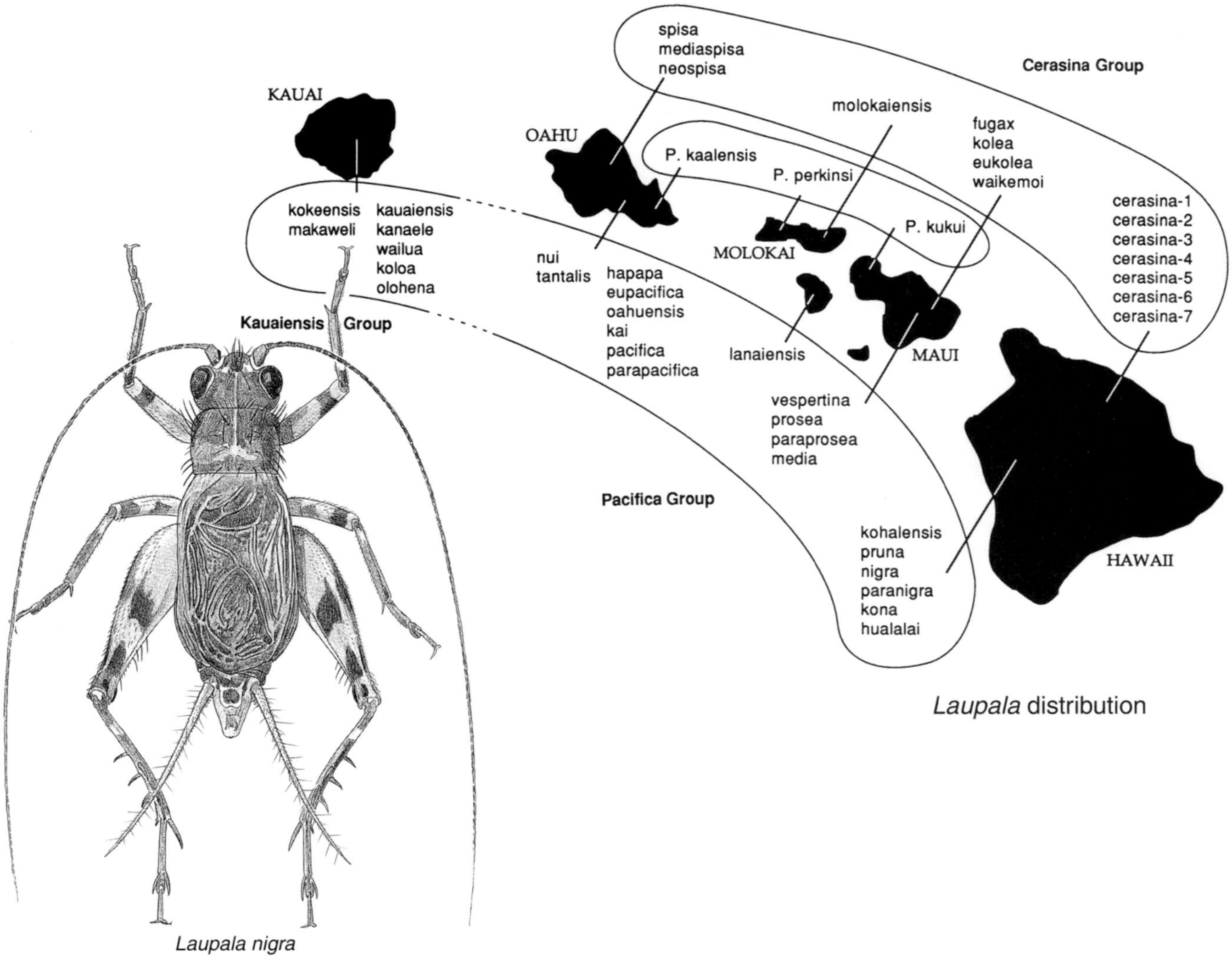

14.47 *Laupala* crickets, the genus of which has speciated into 135 Hawaiian species. The map shows the distribution of some species and how narrow the ranges can be. From Otte (1994).

endemism. The oldest "high" island (one with mountains and lush, tropical, montane forests) is Kauai, which is 5.1 MYO, and the youngest is Hawaii, merely 500,000 years old. Life in the Hawaiian Islands is certainly older than 5.1 MY because early colonists apparently arrived more than 10 MYA to islands now eroded away in the northwesterly part of the chain. Nonetheless, endemism in the current islands indicates that species and many lineages can be impressively young, having evolved as recently as 100,000–300,000 years ago (e.g., Carson, 1983a,b; Hormiga *et al.*, 2003; Jordan *et al.*, 2003). Interestingly, recent studies show that there is a repeated progression in many groups from the oldest lineages on the oldest island (Kauai) to the most recent lineages on the youngest islands (Maui and Hawaii) (Wagner and Funk, 1995; Liebherr and Zimmerman, 1998; Hormiga *et al.*, 2003). The fauna of the Hawaiian Islands may have evolved under exceptional circumstances, but it does beautifully reveal the evolutionary capacity of insects.

> All the past these mountains saw
> All the years of toil and strife
> Lives unknown that went before
> They gave us this: Our present life.
>
> –T. D. A. Cockerell, 1927

15 Epilogue

WHY SO MANY INSECT SPECIES?

This question has occupied entomologists for centuries, and some of the explanations we offer are unfortunately little different from previous ones. New knowledge of phylogeny and discoveries of fossils over the past several decades, however, have improved our understanding. Simply put: there is no single answer to the evolutionary and ecological success of insects, but is due instead to various intrinsic features of insects.

AGE

Insects are among the earliest terrestrial animals, and the recent discovery that *Rhyniognatha* from the Early Devonian was an insect (perhaps even a pterygote!) indicates that they evolved in the Silurian. As terrestrial ecosystems evolved, insects were on the scene ready to exploit the new resources. Major radiations of insects can be related in the fossil record, for example, to the evolution of vascular plants, to angiosperms, and (for ectoparasitic and blood-feeding groups) to birds and mammals.

DESIGN

Design is traditionally the main explanation for the evolutionary success of insects. No single innovation accounts for the success of insects, but rather there has been a cascade of innovations, each of which was refined later in the evolution of insects.

1. The first of these was clearly the jointed, arthropod *exoskeleton*, which provided insects with the physical protection and skeletomuscular strength to live on land. Associated with this is the tracheal system of respiration, which with an exoskeleton allowed animals with such large ratios of surface area to volume to reduce desiccation in the terrestrial environment. A protective exoskeleton also allows insects to defend themselves with potent toxins without poisoning themselves.
2. Metamerism and repetitive pairs of appendages allowed specialization of some appendages while retaining the original functions of others. Much the way gene duplication has allowed the divergence of redundant elements, *redundant appendages* led to divergent adaptations. The forelegs of various insects, for example, have become raptorial or fossorial while the mid and hind legs retained the ability for walking. Even more significant are the mouthparts. Virtually every element within the major sets of paired mouthpart appendages – the mandibles, maxillae, and labium – have been modified in some form to suit diverse diets optimally (Figures 8.2, 12.24, 13.9). Thin adult mouthparts that siphon fluids, for example, evolved in Palaeodictyopterida, Thysanoptera, Hemiptera, anopluran lice, Lepidoptera, many Diptera, bees, a few extinct Mecoptera, and even a few Coleoptera and Trichoptera. The mouthpart elements in each kind of proboscis evolved differently, illustrating the unique capacity in insects to develop different solutions to the same problem. The evolution of mouthparts probably facilitated the evolution of phytophagous diets in insects (Labandeira, 1997).
3. *Flight* greatly improves the ability to escape from predators and allows aerial reconnaissance of food and mates. Consider, for example, an early apterygote and pterygote feeding on the sporangia at the tips of a Carboniferous plant. The apterygote must climb among the branches and down the plant, the pterygote can flit among branches and plants, more easily finding sporangia and better selecting the most productive and nutritive ones. The ability to fold wings over the abdomen – neoptery – allowed insects to invade small, cryptic spaces while protecting their best means of dispersal.
4. The origin of the *larva* allowed holometabolans to accelerate the development of immatures, to facilitate diapause during periods of stress, and to exploit diets different from the adult. A separation of adult and larval diets may have

allowed the less mobile and more vulnerable immatures to invade cryptic spaces and also escape predators and parasitoids that attack the adults.

It must be stressed that, again, no one of these innovations are responsible for the success of insects. Flight probably would not have evolved if metamerism, serial appendages, and an arthropod exoskeleton did not evolve first. Neoptery then further improved wings.

CAPACITY FOR HIGH SPECIATION RATES

Evidence from the fossil record indicates that speciation rates in insects can be considerably greater than what is known in another terrestrial group of animals, the mammals. For example, two major lineages of phytophagous insects, the Lepidoptera and phytophagan beetles, comprise approximately 250,000 Recent species. Because the great proportion of these feed on angiosperms, it is reasonable to estimate that over the past 100 million years there has arisen on average two or three species every 1,000 years in these groups. This is a gross underestimate because it does not take into account the total diversity of extinct lepidopterans and phytophagans over the last 100 million years, which is impossible at present to estimate, nor the higher numbers of total Recent species (described and undescribed). Note that this is not the actual time it takes two populations to diverge into separate species (evidence from the Hawaiian fauna shows that speciation probably takes on the order of 100,000 to 300,000 years, possibly shorter), but it does reveal the remarkable rate of proliferation of insects.

Traditional explanations as to why insects are intrinsically capable of such high speciation rates have typically cited their short generation times, the reasoning being that with short generation times more mutations accrue and allow faster genetic divergence. However, the excellent fossil record of proboscideans shows that these mammals have a high speciation rate, and their generation times are on the order of about 50 years. Rather, it is the stunning *rate of increase* in insects that probably provides the genetic grist for their evolutionary mill; this is a quotient of the reproductive capacity of a species and average generation time. It has been determined, for example, that if *Drosophila melanogaster* lays 100 viable eggs per female, and that half of these produce females that do the same, the exponential rate of increase is so great (taking into account the deaths of each generation) that by the end of the twenty-fifth generation the ensuing mass of flies would be much larger than the earth itself. Obviously, there is a tremendous amount of mortality, but the actual populations of fruitflies vastly outnumber (and probably outweigh) terrestrial vertebrates, and thus generate higher genetic variation.

LOW RATES OF NATURAL EXTINCTION

Evidence for this comes primarily from the fossil record. Insects show minor to negligible effects on mass extinctions that decimated major groups of organisms, the best evidence for which is from the Cretaceous-Tertiary extinctions. The only significant example for the extinction of major monophyletic insect groups are the Palaeodictyopterida, Caloneurodea, and Miomoptera, presumably at the End Permian Event, which was a crisis that was unmatched in evolutionary history.

The combination of low extinction rates and high speciation rates, beginning well before 400 million years ago, resulted in a present day accrual of insect species that is unprecedented in the 3 billion years of life on earth. The low extinction rate itself is related to several highly adaptive designs of the insect body plan, each of which evolved in response to changes that occurred over thousands to millions of years. The environmental changes that have taken place over merely the past century, however, are proving to be the greatest evolutionary challenge to all organisms, even to ones as adaptable as insects.

THE FUTURE

People commonly think that because no species of mosquito, roach, or other pest has been eradicated, that all insects will easily thrive amidst the environmental destruction we are inflicting on our planet. There is probably no single greater misconception about insects. Some insects are indeed tenacious pests, but more than 99.99% of insect species are highly sensitive to the typical threats of pollution, pesticides, invasive species, and habitat loss (Collins and Thomas, 1991; Samways, 1994; Pullin, 1995; Deyrup, 2001). A small percentage of insects are endangered, and some have become extinct, though the number is without doubt a gross underestimate. The International Union for the Conservation of Nature (IUCN) has listed that over the past 100 years 600 plant and 491 animal species have become extinct as a result of humans, only 72 of which are insects. Forty red-listed insect species alone are from Hawaii and other remote islands, faunas of which can collapse under the effects of introduced rats, pigs, goats, mongoose, ants, and vespid wasps. Bermuda, for example, is a relatively depauperate island, and an intensive study of its fauna found 258 species of Diptera, 17 of which are probably endemics and 7 of these are possibly extinct because they have not been seen in years (Woodley and Hilburn, 1994). None of those seven species was IUCN listed. Documenting the hemorrhaging of insect biodiversity is fraught with difficulties. The large, flightless Lowd Howe Island stick insect, *Dryococelus australis*, for example, was recently discovered alive, some 80 years after it was thought to be extinct. If a creature as conspicuous as this evades notice, what about the great majority of insect species, many of which have not even been named yet? The death of the last passenger pigeon at the Cincinnati Zoo in 1914, for example, is well known, and with it were purportedly of two species of lice, *Columbicola extinctus* and *Campanulotes defectus* (these lice were later found on other species of pigeons). No one except phthirapteran specialists laments the loss of a louse, but every species has some salient aspect of biology, and the lice have taught us more about cospeciation than any other group of insect.

For many insects extinguished by humans, we do not notice their demise until well after the fact. The giant St. Helena earwig is another island casualty; however, it is known only from its remains (Figure 15.1), a dodo among insects. It succumbed probably to rats introduced to the island several centuries ago. The Antioch katydid, *Neduba extincta*, was endemic to coastal sand dunes around San Francisco, but the species only became known 40 years after the last collection of specimens.

15.1. Dodo of the Dermaptera, *Labidura herculeana*: the giant earwig (shown here actual size), known only from the remote South Atlantic island of St. Helena and last seen alive in 1965. Introduced rats and mongoose may have extirpated it. Photo: P. Naskrecki, NHM; length of entire insect 62 mm (2.5 in.).

Coastal sand dune habitats, so coveted for housing, are in fact highly endangered habitats, and three butterflies from the same area as *Neduba extincta* have followed it. One of these, the diminutive extinct butterfly *Glaucopsyche xerces*, or "Xerces blue" (Figure 15.2), is the symbol of the Xerces Society, the only organization dedicated to the conservation of invertebrates – a stark contrast to the many wealthy societies dedicated to megafauna. Another three butterflies, some tiger beetles, and other insects also endemic to coastal sand dunes of the western United States are endangered. Among them is the "Delhi Sands fly," a large, long-tongued fly in the Mydidae (cf. Figure 14.16) and an important pollinator. The proposal for its official protection caused an uproar among people who have been outraged that, of all things, *a fly* could have federal protection. Such people, of course, are completely ignorant of the spectacular diversity of flies, that some are in fact important pollinators, and that the loss of pollinators has a domino effect on rare plants (Buchmann and Nabhan, 1996).

Coastal sand dune habitats harbor many threatened species, but the situation is much worse for the coastal forests of southeastern Brazil near São Paulo. This forest contains remarkable proportions of endemic animal and plant species, including insects, and because it is tropical, the overall diversity is exceptional. Brazilian coastal forest, unfortunately, comprises less than 1% of what it was 150 years ago (Lovejoy, 1985), making it almost certain that *thousands* of insect species have gone extinct there, but we'll never know the actual numbers.

For some insects the causes of endangerment or extinction are unclear. There is probably as plentiful a supply of small mammal and bird carcasses as ever (i.e., rats and starlings), which makes the dramatic dwindling of the North American carrion beetle, *Nicrophorus americana* (Figure 15.3), so perplexing. Another carrion-feeding insect that has a similar fate, which may even be extinct, is the large, red-headed European blowfly, *Thyreophora cynophila* (Pape, 2000). Fifty years after its naming in 1794, it could no longer be found and is now considered extinct. The California condor clings to survival, but Pleistocene fossils indicate they occurred as far east as New York state, perhaps following the loss of the megafauna to indigenous and European people. Maybe *Thyreophora* is the California condor of flies.

Insects are unique for many reasons, and an additional one is that no other group of animals is the systematic target of such wholesale poisoning. The environment is laced every year with a half-million tons of insecticides in the United States alone, and the amount continues to rise as the most destructive pests evolve greater resistance (Pimentel *et al.*, 1992). The notorious chlorinated hydrocarbon DDT (banned in many countries, but not all) was developed in 1939, and by 1946 resistance of target insects was reported. Still, 50,000 tons of it was used annually in the 1950s, until it was noticed that it caused long-term endocrine problems in vertebrates (Matsumura, 1985). Now, hundreds of insecticides are in use, though most of them function in the same way (as insect neurotoxins). Despite the effort, no pest insect has been eradicated; in fact, over 500 pest species of insects and mites are resistant to at least one insecticide (McKenzie, 1996). Some species appear resistant to the complete chemical arsenal, like the green peach aphid, *Myzus persciae*; the diamondback moth, *Plutella xylostella*; the Colorado potato beetle, *Leptinotarsa decemlineata*; and the malaria mosquitoes in the *Anopheles gambiae* complex. Insecticides merely

15.2. The extinct lycaenid butterfly, *Glaucopsyche xerces* or Xerces Blue, which was endemic to coastal dune habitats of northern California. A loss of habitat through development led to its extinction. AMNH; length 13 mm.

15.3. The North America burying beetle, *Nicrophorus americanus*, was abundant at one time but now clings to survival in a few isolated populations. Its dramatic decline since the 1930s is poorly understood but may be related to pesticides. AMNH; length 21 mm.

select for individuals that have naturally occurring mutations in enzymes that detoxify the chemical, and these individuals reproduce to form whole resistant populations. The actual casualties unfortunately are the parasitoids and predators of the pest species, as well as harmless species caught in the cross fire, because they do not have the array of detoxifying enzymes typically found in phytophagous pest insects. The end result is that we will increasingly see fewer species of butterflies and wild bees but more successful strains of superbug pests, an insidious and toxic experiment of artificial selection on a global scale. Perhaps the solution is to treat insecticides and pest species the way antibiotics are encouraged to be used against resistant microbes: sparingly and intermittently.

The wholesale destruction and poisoning of earth's natural habitats is causing unprecedented extinctions, and no doubt it is an even greater spasm than the End Permian Event, which apparently had particularly dramatic effects on marine organisms. At the present rate of discovery, it will take centuries before most insect species are given a name and known only on the basis of preserved specimens and published descriptions, and that is simply too little too late. The behavior, life histories, ecological interactions, and biology of most insects in our own yards and city parks are largely unknown, let alone the millions of species in remote regions. We will never know the full extent of what we are losing.

Exploration to fill huge gaps in the fossil record of insects is another need for fully understanding insect evolution. The periods of our greatest ignorance about insects are scattered throughout the fossil record. The Paleocene is represented by very few deposits but is crucial to understanding the impact of the end-Cretaceous extinctions on insects. The Jurassic is known essentially just from Europe and Asia, so deposits from other continents would clarify pre-Cretaceous (pre-angiosperm) diversity. The Late Triassic is well represented, but the Early Triassic is essentially unrepresented for insects. Early Triassic deposits would be valuable for gauging the impact of the massive end-Permian extinctions on insects. The Early Carboniferous is completely unknown for insects, but the sudden diversity of winged insects in the earliest part of the Late Carboniferous attests to their existence well before this time, probably even into the Devonian. In fact, we consider the discovery of Early Carboniferous insects the most significant problem for insect evolution, and it may be one of the few ways – perhaps the only way – to resolve the controversy over the origin of wings. Devonian remains are extremely sparse, with fragments known from just three deposits, but this period is clearly crucial to understanding the early history of hexapods.

We can think of no more inspiring prospect than future generations exploring the millions of Recent and extinct species of insects – an essentially endless reservoir of discovery that requires very modest budgets compared to other scientific pursuits, like space travel. Unlike Recent species, however, fossils locked in the earth are under less threat since excavations for strip malls, housing developments, and hydroelectric dams can actually uncover fossils that would be otherwise unreachable (these activities can also destroy entire deposits). The most pressing problem is the protection of living diversity, not just megafauna, and if we are to provide solutions, then entire habitats and ecosystems must be protected. Only as insightful and determined stewards can we continue to appreciate the greatest evolutionary success story in the history of life on earth.

GLOSSARY

Not included in this glossary are the proper names of places, geological periods, and taxa (consult the index, or for geological periods see Figure 2.41).

abdomen the arthropod tagma posterior to the thorax, which contains most or all of the systems of digestion and reproduction.

acidopore the minute opening at the apex of the gaster (metasoma) in certain ants like Formicinae, through which is ejected defensive secretions like formic acid.

adaptation (n.), **adaptive** (adj.) a feature of an organism that is a response to environmental change and that can be behavioral, physiological, or anatomical.

aedeagus the penis in insects, or the terminal appendage involved in delivering sperm.

Afrotropical (adj.) **Afrotropical Region** (n.) the tropical region of Africa, generally sub-Saharan.

allochthonous (adj.) referring to fossils where remains were transported from and fossilized in locations distant from where the organisms lived.

allogrooming a behavior involving the licking of a conspecific, generally a nestmate.

allometry (n.), **allometric** (adj.) differential growth of structures, where one structure (such as a male feature involved in displays), grows at a rate and to proportions greater than other parts of the body.

amber fossilized resin, which for practical reasons includes resins that are more than 40,000 years old (the limits of ^{14}C dating).

ametabolous (adj.) in arthropods, involving little or no anatomical changes between instars other than an increase in size; a primitive feature occurring in apterygote insects.

amphitokous a type of parthenogenesis where male and female offspring are produced.

anagenesis (n.), **anagenetic** (adj.) adaptive change, or the evolution of individual characters (versus *cladogenesis*).

anal fan the expansive, pleated anal lobe on the hind wing of certain insects, such as in many polyneopterans.

anlage (German, n.) imaginal disc, the invaginated ectodermal tissue of immature insects that develop into appendages and other portions of the adult cuticle (Figure 9.3).

antenna the pair of long, usually multisegmented, appendages at the anterior part of the head that are involved in tactile and chemical perception.

antennomere an antennal segment, including the scape, pedicel, and flagellomeres.

anthophilous (adj.) flower loving or, specifically, flower visiting. Not all anthophilous insects are pollinators.

apodeme an internal ridge or flange of insect cuticle to which muscles attach.

apolysis the separation of the old cuticle of a molt from the new cuticle.

apomorphy (n.), **apomorphic** (adj.) a derived or specialized feature of a species or higher taxon.

aposematic (adj.) warningly colored, generally consisting of black with either yellow, orange, or red, or a combination of these. Aposematic insects are either noxious, poisonous, or mimics of these (Figure 13.88).

apterous wingless, with only minute vestiges of all four wings or no vestiges at all.

aquatic (adj.) in insects, living in fresh water (rarely sea water) as either immatures or for the entire life cycle.

arboreal (adj.) living in trees or any type of tall plant.

archedictyon a system of intricate, presumably archaic, wing venation containing numerous crossveins (e.g., Figure 6.17).

areola postica a cell on the posterior margin of the forewing of Psocoptera enclosed by veins CuA_1 and CuA_2 (Figure 8.5).

arista the shortened, bristle-like flagellum of the insect antenna, found in certain groups like Odonata, Ephemeroptera, Auchenorrhyncha, and Cyclorrhapha Diptera (Figures 6.16, 8.36, 12.59).

arolium (arolia, pl.) a fleshy median lobe on the pretarsus of insects between the claws (Figure 4.4).

arrhenotokus a type of parthenogenesis where unfertilized eggs give rise to haploid males and fertilized eggs give rise to diploid females (haplodiploidy).

asynchronous flight muscles thoracic muscles that are involved in powering the wings but contract out of synchrony with the wing movements.

Austral Region (n.), **austral** (adj.) the area that includes southeastern Australia, New Zealand, Tasmania (sometimes New Caledonia and the Fijian Islands), Chile and Argentina, and southern Africa, which have temperate to subtropical environments.

autochthonous (adj.) referring to fossils whose remains were preserved where the organisms lived.

axillary sclerites small sclerites that articulate with the base of the wing; muscles are attached to some, and the third axillary sclerite is involved in wing folding in the Neoptera (Figure 4.6).

axoneme an axial filament in the tail of sperm with an array of microtubules.

basal taxon/taxa the first lineage or lineages that branch off in a phylogeny. In strict cladistic terms, only stem groups can be basal, but basal is used in a more general sense here.

basitarsus the tarsomere of insects that is closest to the tibia.

Batesian mimicry a type of mimicry where an edible, unrelated mimic closely resembles a noxious or toxic model (Figures 13.89 to 13.92).

Bayesian analysis a statistical method of phylogenetic analysis similar to maximum likelihood but based on "posterior probabilities" for a particular model of nucleotide substitution.

beak extended mouthparts or rostrum of certain insects, such as Heteroptera, generally involved in sucking.

bimodal (biphasic) allometry a type of allometry with two discrete classes of individuals (Figure 10.32), and which have different slopes on an allometric plot.

binomial system the method of naming species using a genus name and a species name.

biogeography the study of animal and plant distributions, based on environmental factors (ecological biogeography) or evolutionary and geological factors ("historical" biogeography).

biological species concept species defined on the basis of being actually or potentially reproductively isolated groups of individuals (see also *phylogenetic species concept*).

bioluminescent the production of light by a living organism.

bipolar (biogeography) distribution of a group of related species that is widely separated in the Northern and Southern Hemispheres.

brachypterous (adj.) wings that are reduced in size.

bursa copulatrix a chamber or sac involved in copulation.

calypter the lobe at the base of and on the posterior margin of the wing in calyptrate flies.

Cambrian explosion The sudden radiation of metazoan phyla in the Cambrian Period.

cardia valve formed by the projection of the esophagus into the mid gut of Cyclorrhaphan flies.

cardo the basal part of the maxilla.

carina an external ridge on a sclerite.

caste in social insects, a group of individuals within a colony that are anatomically and/or behaviorally specialized for particular tasks, like foraging, tending the brood or queen, or defending the colony.

cenchrus (sing.), **cenchri** (pl.) a lobe on each side of the hind scutum in symphytan Hymenoptera, generally involved in holding folded wings flat over the body.

cephalopharyngeal skeleton the internal, sclerotized truss within the head and thorax of larval cyclorrhaphan flies (maggots), homologous to the head capsule of other larval insects (Figure 12.58).

cephalothorax a region of the arthropod body in which there is little or no distinction between the head and thorax, such as in mites and spiders.

cerci the pair of sensory appendages at the end of the insect abdomen (Figure 4.4).

cervical sclerites one or more pairs of small, lateral sclerites on the neck of insects.

character any aspect of a taxon that is unique to it or to a group to which it belongs.

character polarity the direction of evolution of a character as either primitive (apomorphie) or derived (plesiomorphic). It is sometimes inferred before phylogeny is reconstructed or interpreted on the basis of a phylogeny.

chitin the main biopolymer in insect cuticle, consisting of a β-pleated polysaccharide sheet.

cibarium an internal, cuticular sac in the mouthparts of insects, generally lying under the clypeus, which is involved in creating suction forces used in feeding on liquids.

circumversion the rotation of male genitalia in eremoneuran flies, generally 90°, 180°, or (in Cyclorrhapha) 360°.

cladistics (n.), **cladistic** (adj.) method of reconstructing relative evolutionary relationships (not ancestor-descendent relationships) based on shared derived characters (synapomorphies).

cladogenesis (n.), **cladogenetic** (adj.) the splitting or formation of lineages.

cladogram (n.) a diagram depicting relative evolutionary relationships based on shared, derived characters.

classification organization of species into a hierarchy of taxa, preferably based on phylogenetic relationships.

claval furrow (n.) a trough in the tegminous forewing of Blattodea in which vein CuP lies.

clavate (adj.) club-shaped.

clavus (n.) club at the end of the antenna. In the wing, the lobe at the base of the wing and on the posterior margin consisting of the anal and jugal lobes.

claw (pretarsal) the paired (sometimes singular), hooked structures at the tip of the insect pretarsus.

cleptoparasite a species that feeds off of the food or provisions of another species; a cuckoo.

clypeus a sclerite at the anterior part of the head and at the base of the mouthparts, which articulates with the labrum (Figure 4.2).

cocoon a silken sac that encases the pupa.

coevolution the interplay of two or more unrelated lineages, where one lineage affects the evolution of the other(s) or there are even reciprocal effects. Coevolution can be anagenetic or cladogenetic.

colleterial gland a gland in or near the oviduct or vestibulum that secretes a coating to encase and protect pods of eggs, such as the ootheca of roaches and mantises (Figure 7.62).

colonial (adj.) a group of related or unrelated individuals other than a mated pair that cooperate in constructing a nest and rearing offspring.

colonization in biogeography, to infiltrate new territory and establish breeding populations; in ecology/evolutionary biology, often refers to the infiltration of a parasite onto new host species.

communal related or unrelated individuals of the same generation cooperate in nest building but not in rearing offspring.

compound eye an eye of arthropods with multiple facets or ommatidia.
concretion a spheroid, mineralized nodule with a fossil as its nucleus (Figures 2.4, 2.42).
condyle an articulation or pivot point; joints of arthropod appendages can be moncondylic or dicondylic (with one or two pivot points).
consensus tree a cladogram that mimimizes the amount of conflict among characters.
convergence (n.), **convergent** (adj.) repeated origin of a derived feature that is very similar but which evolved in unrelated taxa (Figures 1.26, 1.27).
copal subfossilized to Recent resin, which for practical reasons includes resins that are less than 40,000 years old (the limits of ^{14}C dating).
coprolite fossilized feces or frass.
coriaceous (adj.) leathery, such as the forewings of Heteroptera and various polyneopterans.
cornicle the paired, spigot-like structures on the dorsum of the abdomen in aphids that secrete alarm pheromones. Synonym siphunculus (sing.), siphunculi (pl.) (Figure 8.26).
cospeciation a situation where the sequence of speciation in one lineage affects the speciation of an unrelated but ecologically intimate lineage (Figure 8.13).
costa the most anteriorly branching wing vein, which always occurs on the anterior edge of the wings and often (but not always) runs around the entire margin; abbreviated as C.
costal brace in Ephemeroptera, a thick veinlet at the base of the wing connecting vein C to R_1.
coxa the basalmost segment of the insect leg that is attached to the thorax.
crawler first instar nymphs of Coccoidea, which are a dispersive stage.
crop a diverticulum in the insect digestive system attached posteriorly to the esophagus, which stores food.
crown group a recently evolved, monophyletic lineage (Figure 1.28).
crypsis (n.), **cryptic** (adj.) hidden because of a resemblance or blending in with the substrate or background (Figures 7.24 to 7.28).
cryptic species species that are barely distinguished on the basis of anatomical features, or not at all.
cryptonephridium an excretory system in adult and larval insects where water is resorbed via Malpighian tubules fused to the rectum; found in various insects living in dry habitats, such as Neuroptera and cucujiform beetles.
cryptopleuron reduced pleuron of the thorax, occurring where the lateral part of the thorax is largely concealed by a saddle-like pronotal lobe, as in Orthoptera and polyphagan beetles.
ctenidium (sing.), **ctenidia** (pl.) a comb of stiff, often thickened, setae (Figure 12.15). Commonly found on ectoparasitic insects.
cursorial (adj.) running, generally referring to the function of legs (versus *fossorial, raptorial*).
cuticle (n.), **cuticular** (adj.) the hard external plates of arthropods, composed mostly of chitin in terrestrial arthropods.
dealate (n., adj.) an individual insect in which the wings are normally shed (e.g., a dealate termite).
decticous (adj.) mandibles that are functional in the pupal stage (versus adecticous).
dehiscent (adj.) shed, generally referring to wings that are naturally shed; occurring in certain types of insects such as termites.
deletion (DNA) the loss of a nucleotide in a sequence of DNA.
detritivore (n.), **detritivorous** (adj.) a species that feeds on decaying organic material, generally plant material.
diapause a period of arrested metabolism and development in insects, usually during adverse environmental conditions.
dimorphism (sexual) a condition where there are two morphs in a species, generally applied to differences between sexes (sexual dimorphism), such as color patterns in certain butterflies.
diploid having two sets of chromosomes.
disjunct (adj.), **disjunction** (n.) the distribution of a species or group of related species that is widely separated or remote.
divergence the splitting of lineages from a common ancestor.
DNA deoxyribonucleic acid; the molecular basis of inheritance.
domatia structure on a plant specialized for housing an ant colony.
domicile a dwelling, generally a fixed one (i.e., not a case that is carried around).
dorsal aorta/ blood vessel the main circulatory organ for insect hemolymph, occurring along the dorsal midline in the abdomen (heart) and the thorax and head (aorta). Small ostia along the sides allow for the circulation of hemolymph through the tube.
dorsoventral thoracic muscles large muscles in the thorax that have their origins and insertions on the inside of the dorsal and ventral sclerites.
ecdysone a molting hormone, found in many arthropods.
ecological biogeography study of animal and plant distributions based on environmental factors.
ectoparasite for insects, a species that lives on another animal species (generally vertebrates), derives nutrition from it, and is structurally specialized for such.
ectoproct (n.) paired plates of the tenth abdominal tergum in Neuropterida that surround the anus.
ectosymbionts usually referring to arthropods living in the nests of social insects, which derive nutrition from the midden or other source within the nest.
edaphic (adj.) of or relating to the soil, often used in reference to arthropods that live in soil or leaf litter.
elytron (sing.), **elytra** (pl.) hardened forewings, in Coleoptera.
encapsulation in paleontology, when a fossil is encased or embedded in copal, amber, or within a mineral that crystallized around the fossil (such as calcite) (Figures 2.15, 2.25).
End Permian Event an episode marking the very end of the Permian 248 MYA, when the greatest mass extinction in earth's history took place.
endoparasitoid (n., adj.) an insect that lives within the body of an unrelated species and eventually kills it.
epicoxa a hypothetical basal podite of the insect leg between the pleuron and dorsum, thought to be homologous to the gills of certain insects such as in nymphal Ephemeroptera.

epimorphosis (n.), **epimorphic** (adj.) of or pertaining to little change in development between the instars of hexapods.
epiproct the dorsal part of abdominal segment 11 in insects.
eusocial advanced sociality, where a group of related individuals from different generations cooperatively care for young, and one or a few individuals produce all of the offspring.
eversible vesicles small, paired, membranous lobes occurring on the ventral surface of the abdomen in apterygote insects, which can be everted (Figure 3.32).
exarate a type of holometabolan pupa in which the appendages are free, not fused to the rest of the body.
exite/gill theory a theory on the origin of insect wings, which postulates that wings evolved from small lobes at the base of the legs of early insects (exites), or from gills or gill covers similar to those of nymphal Ephemeroptera.
extensor muscle a muscle that extends a limb when it contracts (versus *flexor muscle*).
extinction the end of a species or a lineage.
extrinsic muscles a muscle that has its origin in one structure and its insertion in another structure (i.e., the muscles in the antennae of most insects are extrinsic, the origin lies within the head capsule).
exuvium the cast skin or cuticle of an arthropod after a molt.
facultative (adj.) taking place under some conditions but not others (versus *obligate*), such as facultative parthenogenesis, facultative parasitism.
femur the third segment of the leg from the base in insects, between the trochanter and tibia, and generally the largest segment of the leg.
filiform (adj.) slender and filamentous, where each segment is similar in size and shape; commonly used to describe the antennae of many insects.
flabellate (adj.) an appendage, usually the antenna (sometimes the palps), which has long, flat branches off one side of each of the terminal segments (Figures 10.79, 10.86).
flagellate (adj.) whiplike, generally used to describe the antennae of many insects.
flagellomere any one of the antennal segments exclusive of the basal segments (scape and pedicel).
flexion lines lines of weakness in a wing that allow flexion during flight.
flexor muscle a muscle that bends a jointed structure.
fontanelle a microscopic hole or tight cluster of holes on the anterior part of the head in Rhinotermitidae and Termitidae termites, or at the tip of the nozzle-like head in nasute soldiers (Figure 7.75).
fossil the remains or workings of an organism of reasonable age (thousands to generally millions of years old) that have been naturally preserved.
fossorial (adj.) digging, generally applied to legs specialized for digging, like those of a mole (Figure 1.27).
frass pellets of insect feces.
furculum an appendage on the ventral surface of the abdomen in Collembola responsible for springing the animal into the air (Figure 3.29).
fusainized charcoalified fossil, rendered to a (usually three-dimensional) carbonized replica by the heat from forest fires in an airless space, such as the remains of insects or plants under ancient leaf litter.
galea the outer lobe of the maxilla, which is greatly distended in most Lepidoptera to form the coiled proboscis (Figures 4.2, 13.19).
gall, plant a nodule or excessive growth on a plant caused by the workings of unrelated organisms, like fungus, nematodes, mites, or insects.
gallery meandering tunnels through a substrate, such as in wood made by wood-boring insects or in soil.
ganglion a nodule of neurons, occurring in the ventral nerve cord of arthropods, with generally one ganglion per segment or one per tagma.
gaster the metasoma (generally used just for ants), or that part of the abdomen in apocritan Hymenoptera posterior to the waist (Figure 11.50).
geniculate (adj.) appendages that are elbowed or that fold at an acute angle, such as the antennae of ants and the mouthparts of some flies.
genitalia (sing., pl.) the terminal reproductive structures in insects, which are generally sclerotized appendages that are often external. These are involved in copulation and sperm transfer in males and egg laying (i.e., the ovipositor, oviscapt) in females.
gills in insects, thin-walled cuticular structures of aquatic nymphs and larvae that absorb oxygen from water. They are generally fine, filamentous outgrowths of the tracheae.
glossa in insects, the median lobe of the labium.
gonangulum a small, triangular sclerite at the base of the true ovipositor in insects, which is attached to the base of the first pair of gonapophyses and which articulates with gonocoxa two and tergum nine (Figure 4.8).
gonapophysis paired structures of the insect ovipositor, also called the first and second valvulae (Figure 4.8).
gonarcus (n.) in Neuropterida, the paired, arch-shaped sclerites below the anus and above the aedeagus.
Gondwana (n.), **gondwanan** (adj.) the large southern continent in the Mesozoic that consisted of South America, Africa and Madagascar, India, Australia and New Zealand (and their close land masses), and Antarctica (Figure 14.27).
gonocoxa (= **gonocoxite**) the base of paired, segmented structures of male insect genitalia, which function in clasping the female terminalia during copulation (Figure 4.9).
gonoplac a paired structure in the ovipositor of winged insects, also called the third valvula, which typically forms a sheath around the ovipositor shaft (Figure 4.8).
gonostylus the segment just distal to the gonocoxa of male insect genitalia (Figure 4.9).
grade a paraphyletic group, or group of species or taxa that includes some but not all close relatives (versus clade).
gregarious (adj.) grouping of unrelated individuals, generally due to food concentrations or a mutual defense.
ground plan (bauplan) the structure of a hypothetical, idealized ancestor of a lineage.
gula a median sclerite on the ventral surface of the head in prognathous insects, lying between the submentum, the posterior margin of the head, and the genae.

halter (sing.), **halteres** (pl.) a wing reduced to small, clublike structure, occurring in Diptera (hind wings), Strepsiptera (forewings), male Coccoidea (hind wings, called hamulohalteres), and assorted other insects (Figure 12.23).

hamulus (sing.), **hamuli** (pl.) generally used for one or a row of small hooks at the base of the costal edge of the hind wing in Hymenoptera, which hook on to the posterior edge of the forewing to couple the pair of wings (Figure 11.1).

haplodiploid a type of sex-determining mechanism in insects where females are produced from fertilized diploid eggs and males are produced from unfertilized haploid eggs. Occurs in Hymenoptera, thrips, certain scolytid beetles, and in a few other insects.

haploid having a single set of chromosomes.

haustellate (adj.) referring to insect mouthparts that function by suction of liquids.

hematophagous (adj.) feeding on vertebrate blood.

hemelytra the forewings of higher Heteroptera, in which the basal half is thick and sclerotized and the apical half is membranous (Figure 8.5).

hemimetabolous (adj.) in insects, postembryonic development that has nymphal instars, or immature stages that are smaller versions of the adult but lack fully developed wings and genitalia; "incomplete" metamorphosis, exopterygotes (Figure 9.1).

hemolymph arthropod blood.

heterochrony (n.), **heterochronic** (adj.) differences among species in the timing of growth and development of a homologous structure.

heteroecy (n.) when a parasite uses different hosts in different stages of its life cycle.

heteronomous (adj.) referring to paired or serial structures that differ in size and/or shape, such as the forewings and hind wings of many insects.

historical biogeography study of animal and plant distributions based on evolutionary and geological factors.

holarctic (adj.), **Holarctic Region** (n.) temperate areas of the Northern Hemisphere, including North America, Europe, and much of Asia.

holometaboly (n.), **holometabolous** (adj.) postembryonic development in insects where there is a larval, pupal, and adult stage; "complete" metamorphosis, endopterygotes.

homology (n.), **homologous** (adj.) where a trait is possessed by two or more species that is derived from a common ancestor, with or without modification.

homonomous (adj.) referring to serial structures on an organism that are similar in size and/or shape.

homoplasy (n.), **homoplasious** (adj.) referring to a similar specialized trait in two or more species but the trait is not derived from a common ancestor; involving convergence and reversal of character change.

honeydew sugary excretion produced by various Sternorrhyncha and some Auchenorrhyncha, which is often avidly collected by ants and some other Hymenoptera.

hyperparasitoid (n., adj.) a parasitoid that lives within or off of another parasitoid.

hypognathous (adj) mouthparts situated below the head in insects (versus *prognathous, opisthognathous*).

hypopharynx a median, unpaired structure in insect mouthparts anterior to the labium (Figure 12.24).

hypoproct the ventral part of abdominal segment eleven in insects.

ichnofossil trace fossils, or the fossilized workings of animals (e.g., their burrows, nests, footprints, feeding traces), the study of which is ichnology.

imaginal discs see *anlage.*

inclusive fitness the effect on fitness (survivorship, fecundity) of a gene or genotype, measured for an individual and its relatives.

indirect flight muscles muscles that power wings but that do not connect directly to the wing base. These include dorsal longitudinal muscles whose contractions distort the thorax, causing the wings to move on their pivot and which power the downstroke.

insertion in DNA, the addition of a novel nucleotide into a DNA sequence.

instar a nymphal or larval stage between molts.

intercalary veins short veins in various paleopterous and polyneopterous insects that run parallel to longitudinal veins; they connect to the wing margin and to the longitudinal veins by short crossveins but are incomplete.

intrinsic muscle a muscle that has its origin and insertion within an appendage (versus *extrinsic muscle*).

Johnston's organ a small sensory structure on the pedicel of the antenna in many insect groups.

jugal bar one or two veinlike extensions of a putative basal sclerite near the posterior base of the wing.

jugum/jugal lobe a lobe on the posterior margin of the wing near its base, and proximal to the jugal fold and anal lobe.

juvenile hormone (JH) a hormone in insects that prevents molting.

kin selection a form of natural selection that acts on individuals based on genetic relatedness, such as workers in a polygynous colony favoring the brood most closely related to them.

K/T boundary, **K/T extinction** the very end of the Cretaceous and beginning of the Tertiary, 65 MYA (the K/T boundary), also marked by a cataclysmic meteoritic impact and the extinction of various unrelated groups of organisms.

labellum a fleshy, padlike structure at the tip of fly mouthparts, formed from the fusion of the labial palps (Figure 12.60).

labium a trough-like sclerite on the ventral surface of the mouthparts of insects (Figure 4.2).

labrum a generally triangular sclerite on the dorsal surface of the mouthparts of insects (Figure 4.2).

lacinia an inner lobe of the maxilla, which articulates with the stipes (Figure 4.2).

lacustrine (adj.) in paleontology, sediments formed by freshwater lakes, or fossils found in such sediments.

Lagerstätte (German, n.) **Lagerstätten** (pl.) a deposit with a concentration of fossils that is remarkable for the preservation (Konservat Lagerstätten), or the number of specimens and/or diversity of species (Konzentrat Lagerstätten).

larva an immature stage of holometabolous insects that is typically soft-bodied, has reduced appendages and usually imaginal discs (anlage).

larviform (adj.) similar in form to a larva, generally in reference to the adult (a form of neoteny in holometabolous insects).

larviparous (adj.) a form of viviparity where the mother gives birth to larvae.

Laurasia (n.), **laurasian** (adj.) the large northern continent that consisted of North America, Europe, and Asia, which were joined millions of years ago and then separated by continental drift.

lift the upward force that keeps flying animals airborne.

Linnean hierarchy a classification employing the classic system of Linnaeus using hierarchical taxa of formal and informal categories (Table 1.3).

long-branch attraction the artificial attraction of lineages in a phylogenetic analysis based on DNA sequences, caused by many nucleotide substitutions in each lineage.

macropterous an adult insect with fully developed wings, generally applied to individuals within a species that are polymorphic for wing development, such as in some thrips.

maggot the highly reduced larva of cyclorrhaphan flies, which have internalized remnants of the head capsule.

Malpighian tubules filamentous structures in insects that are attached between the mid gut and hindgut and that float in hemolymph space and absorb nitrogenous waste that is then excreted.

mandible a pair of mouthpart appendages that, in insects, lie anterior to the maxilla and that are usually involved in chewing.

manubrium the basal segment of the spring mechanism (furculum) of springtails (Collembola).

maxilla a segment of the insect head immediately posterior to the mandibles that bears paired appendages, including maxillary palps, galea, lacinia, and glossa plus paraglossa (Figure 4.2).

maxillary palp segmented appendage that is paired and occurs on the maxilla of insects, primitively five-segmented (Figure 4.2).

maximum likelihood a method of phylogenetic reconstruction for molecular data that relies on a priori models and probabilities of nucleotide substitution.

median caudal (or **terminal**) **filament** long, segmented, whip-like structure at the tip of the abdomen that occurs between the cerci of Zygentoma, Archaeognatha, and many nymphs and of some adults of Ephemeroptera (Figures 5.1, 6.3).

meroistic (mero = part; oon = egg) a type of ovariole in insects where the oogonia divide to form two types of cells, oocytes and nurse cells. Includes polytrophic and telotrophic ovarioles.

mesosoma the middle part of the body in apocritan wasps (the "trunk" or "alitrunk" in ants), consisting of the thorax fused to the anterior parts of the abdomen, and lying anterior to the narrow waist (Figure 11.8).

metamorphosis a developmental change between stages in the life cycle of an animal; occurs in various animals.

metapleural gland paired exocrine gland on the hind part of the trunk (the propodeum) of most ants, which secretes substances that apparently disinfect the galleries and chambers of nests (Figure 11.50).

metasoma the hind part of the body in apocritan wasps (the "gaster" in ants) lying posterior to the narrow waist, which consists of most of the abdomen (Figure 11.8).

micropterous an individual with highly reduced adult wings, generally applied to individuals within a species that are polymorphic for wing development.

mimicry resembling another species for the benefit of the mimic.

mimicry complex, mimicry ring a group of unrelated species that resemble each other and that benefit the mimics.

mine, leaf a blotch or meandering tunnel between epidermal layers of a leaf made by the feeding of certain larval insects (Figure 2.20). Structure of the mine can be distinctive to particular groups of leaf miners.

molecular clock an estimation of the rate of molecular change (e.g., nucleotide substitution), which is generally believed to vary greatly across lineages and genes.

moniliform like beads on a necklace, generally referring to the type of insect antenna where each segment is rounded and similar to the others, such as in Isoptera.

monograph a generally large paper that comprehensively treats the systematics of a taxon.

monolectic (adj.) feeding from one species of flower.

monomorphic generally used to refer to castes in social insects, where there are no anatomical differences among the non-reproductive castes.

monophagous feeding from one species of host, either as a parasite or as an herbivore.

monophyly (n.), **monophyletic** (adj.) close relationship based on a shared, common ancestor (Figure 1.24).

monotypic (adj.) a taxon that includes only one species.

morphology comparative anatomy; anatomy where structures are examined in various species and homologous parts are studied.

mouthcone usually applied to thrips, where the mouthparts are narrow and slightly protrudent, forming a small cone.

Müllerian mimicry a type of mimicry where unrelated sympatric species that are noxious or toxic closely resemble each other (Figure 13.93).

muscle plaques small bare areas on the external surface of the abdomen in non-cyclorrhaphan flies that denote sites where, in the pupa, muscles are attached to the internal wall (Figure 12.88). These muscles help the pupa to wriggle.

mycangia small pockets in the cuticle of insects specialized for harboring the spores or hyphae of symbiotic fungus; commonly found in beetles.

mycetocytes specialized internal cells in insects that harbor symbiotic bacteria, generally ones that provide essential nutrients.

mycophagy (n.), **mycophagous** (adj.) having a diet of fungus.

myrmecomorph (n.) resembling an ant.

myrmecophile (n.), **myrmecophilous/myrmecophilic** (adj.) a species (usually a terrestrial arthropod) that lives with or interacts with ants.

myxophagy (n.), **myxophagous** (adj.) having a diet of slime molds (Myxomycetes), which are colonial protists.

nasute specialized soldier of certain termites in the subfamily Nasutitermitinae, which have a head in the shape of a spraying bulb, with the fontanelle at the tip (Figures 7.75, 7.87).

nearctic (adj.), **Nearctic Region** (n.) pertaining to or specifically temperate and subtropical North America, which sometimes is meant to include northern Mexico (e.g., north of Chiapas and Oaxaca, which are tropical).

neopterous (adj.) the condition in insects where the wings fold flat over the abdomen; found in the monophyletic group Neoptera and independently evolved in the extinct group Diaphanopterodea.

neoteny (n.), **neotenic** (adj.) retaining features of juvenile stages into the adult stage (e.g., larviform insects).

neotropical (adj.), **Neotropical Region** (n.) pertaining to or specifically the tropical regions of Central and South America and the Caribbean, generally, from southern Mexico to northern Argentina and Chile, and all of the Caribbean.

nomenclature (n.), **nomenclatural** (adj.) the rules of naming taxa.

notaulus (sing.), **notauli** (pl.) converging furrows on the anterior part of the mesonotum of some insects, such as in Hymenoptera.

notum, postnotum the dorsal or upper sclerite of a segment; a portion of the notum associated with the tergum preceding it.

nucleotide the bases of DNA, adenosine (A), thymine (T), guanine (G), and cytosine (C).

nygma (sing.), **nygmata** (pl.) small, circular, thickened areas on the wings of certain insects, such as in Trichoptera and Mecoptera (Figure 12.9).

nymph immature stage of hemimetabolous insects, which is a smaller version of the adult lacking fully developed wings and genitalia.

obligate (adj.) under all conditions, such as obligate parthenogenesis and obligate parasitism (versus facultative).

occipital foramen the large hole in the back of the insect head capsule, to which the neck attaches and through which nerves, muscles, hemolymph, and esophagus pass.

occiput the part of the insect head capsule between the vertex and occipital foramen.

ocellus small, lenslike, photosensitive structures on top of the head in most adult insects; "simple" eyes (Figure 4.2).

oligolectic (adj.) feeding from the flowers of several species of related plants, such as a genus of plants.

oligophagous (adj.) feeding from several related species of hosts.

ommatidium (sing.), **ommatidia** (pl.) a facet of the insect compound eye, composed of an outer lens and a column of internal cells specialized for photoreception.

ootheca a pod of insect eggs covered with a hard covering secreted by colleterial glands in the female reproductive system; a term generally used only for certain Dictyoptera (roaches, mantises) (Figure 7.63), though similar structures occur in some beetles (some cassidine chrysomelids) and Orthoptera.

opisthognathous position of insect mouthparts where the mouthparts lie under the head capsule and point backward, as in Hemiptera.

opisthosoma the soft, globular body region behind the cephalothorax in arachnids.

Orsten fossils of remarkable three-dimensional and fine structural fidelity replicated in phosphates, generally from the Paleozoic of northern Europe.

outgroup comparison a method of determining whether a character state is derived or primitive based on comparison to a related group.

ovariole a string of oocytes in the insect ovary, highly variable in number among taxa. The tip of the ovariole has the youngest oocytes and the end of the ovariole nearest the oviduct has the oldest oocytes.

ovary internal reproductive organ of the female that occurs in pairs in insects and that produces the eggs.

oviparous (adj.), **oviparae** (n.) in aphids, referring to the stage in the annual life cycle where reproduction occurs via eggs (versus *viviparous, viviparae*).

ovipositor the egg-laying appendage of female insects that consists of pairs of appendages, such as the two pairs of gonocoxae (valvifers), and in some insects the gonapophyses (valvulae) (Figure 4.8). Not to be confused with the terminal appendages in female insects derived from sternites (the *oviscapt*).

paedomorphosis (n.), **paedomorphic** (adj.) reproduction where the mother is highly neotenic, or in insects generally where she resembles a larva or nymph.

palearctic (adj.), **Palearctic Region** (n.) pertaining to or specifically the region including Europe, the Arabian Peninsula, and Asia north of the Himalayas, Burma, Laos, and Vietnam.

paleopterous (adj.) the condition in insects where the wings primitively do not fold flat over the abdomen, but which are held out to the side or folded up over the abdomen; occurs in Ephemeroptera, Odonata, and various extinct groups of insects.

Pangaea (n.), **pangaean** (adj.) (also **Pangea**) specifically the region or pertaining to the region that was the great supercontinent when all the present continents were joined in the Mesozoic (Figure 2.47).

panoistic type of ovariole in insects that have no specialized nurse (nutritive, or trophic) cells, so yolk proteins in the oocytes are derived directly from nutrients in the hemolymph. This type occurs in Odonata, Orthoptera, Isoptera, and Siphonaptera.

pantropical (adj.) the distribution of a taxon that wholly or mostly encompasses the world tropics.

paraglossa small lobe on the labium of insect mouthparts that occurs in a pair just lateral to the ligula, which is homologous to the galea on the maxilla.

paramere paired male genitalic appendage flanking the aedeagus in insects.

paranotal lobes pair of flat, sclerotized lobes on the sides of the pronotum, often with venation, and found in various Paleozoic insects such as some Paleodictyopterida (Figures 6.2, 6.17, 6.21, 6.24, 7.9). These structures are possibly serially homologous to insect wings.

paraphyly (n.), **paraphyletic** (adj.) a group that excludes some recently evolved species that lie within it (Figure 1.24). Example: moths, which comprises all of Lepidoptera exclusive of the recently evolved butterflies.

paraproct a small pair of lobes on insect abdominal segment 11 that border the anus.

parasite an organism living off of another, unrelated species, which harms the host but generally does not kill it (versus *parasitoid*).

parasitoid an insect living off of another, unrelated insect, which eventually and generally kills the host (versus *parasite*).

parsimony (n.), **parsimonious** (adj.) a way of evaluating based on minimal assumptions, usually applied to cladograms in order to produce the fewest character changes (e.g., most parsimonious or shortest trees).

parthenogenesis (n.), **parthenogenetic** (adj.) reproduction without sex, from unfertilized eggs.

Pavan's gland a paddle-shaped exocrine gland on the seventh sternite of certain ants that produces trail pheromones.

pectinate comb-shaped, generally applied to the shape of the antenna where each flagellomere has a long branch.

penis/penes the appendage or pair of appendages in male insects involved in insemination, which delivers sperm.

peritrophic membrane a chitinous membrane that lines the midgut of the alimentary canal of insects.

petiolate with a narrowed stem, as in certain wings and the base of the metasoma of apocritan wasps.

petiole a stem, or narrow attachment base of a structure.

pharate adult the uneclosed adult of holometabolous insects within the pupal case, in which the adult cuticle has separated from the pupal cuticle.

phenetics (n.), **phenetic** (adj.) a method of classifying taxa based on overall resemblance, which usually conflicts with results based on phylogenetic or cladistic approaches.

pheromone a volatile compound dispersed to the air or laid down on substrate that is used for intraspecific communication.

photic organ an organ in certain animals specialized for light production.

phragma an internal strut of cuticle to which muscles attach.

phyletic gradualism divergence of species based on a steady accumulation of novel features.

phylogenetic species concept species defined as a discrete group of individuals identified by unique, specialized features (see also *biological species concept*).

phylogeny in the broadest sense it is a cladogram, or scheme of relationships based on shared ancestry; used in this book in a more restricted sense as a cladogram superimposed on a geological time scale (e.g., Smith, 1994).

phytophagous (adj.) feeding on plants.

planidium first instar larva of parasitoid Hymenoptera that actively seeks its host and that has long, spine-like setae. When it enters its host, it molts into a more typical, structurally reduced larva.

plastron a layer of air held close to the body in certain aquatic insects by specialized textured areas of the cuticle; it allows for respiration, either through replacement at the water surface or through diffusion of gases into the layer.

plesiomorphy (n.), **plesiomorphic** (adj.) a primitive feature, or one that evolved well before the origin of a particular taxon (e.g., wings are plesiomorphic for wasps).

polydnavirus a type of virus in parasitoid wasps that is assimilated into the wasps' DNA, inherited from the mother, and injected into the host with the wasp egg and that produces some venom-like molecules that reduce host immune defense against the parasitoid egg and larva.

polygyny (n.), **polygynous** (adj.) in social insects, where there is more than one reproductively active queen.

polylectic (adj.) feeding from a wide variety of unrelated species of flowers.

polyphagous (adj.) feeding on a wide variety of hosts.

polyphyly (n.), **polyphyletic** (adj.) an artificial group composed of two or more unrelated lineages.

polytrophic type of ovariole in insects where each oocyte has its dedicated nutritive (nurse) cells; found in Dermaptera, Phthiraptera, and most Holometabola.

postgena the area of the insect head capsule posterior to the gena, or the lateral parts of the postocciput.

postmentum basal part of the labium in insect mouthparts near the stipes (Figure 4.2).

postocciput posterior part of the insect head capsule, behind the postoccipital suture.

prehensile (adj.) grasping, used in insects usually to refer to tarsi and even the antennae of some species.

prementum part of the labium in insect mouthparts containing the stipes (Figure 4.2).

prepectus a pleural sclerite in insects below the prothoracic spiracle, which occurs in some insects like certain Hymenoptera.

prepupa in Thysanoptera and Sternorrhyncha, a quiescent stage involving one or more reduced, non-feeding nymphal instars.

prestomal teeth minute, sclerotized teeth on the labellum of certain cyclorrhaphan flies, which are used for rasping at the surface of food.

pretarsus the most distal tarsomere of the insect leg.

proboscis (sing.), **proboscides** (pl.) in insects, mouthparts that form a projecting snout.

prognathous (adj.) insect mouthparts that are situated at the very front of the insect body and project forward (vs. *hypognathous, opisthognathous*).

proleg fleshy, short legs on the abdomen of certain larval insects, such as in caterpillars (Lepidoptera), sawfly larvae, and various other insects (Figure 13.29).

pronotal lobe small, sclerotized lobes of the pronotum that are situated laterally on the insect thorax.

pronymph the brief, nonfeeding stage in hemimetabolous insects between hatching and the first instar nymph; the pronymph generally has shorter appendages. The holometabolous larva is considered to possibly be a protracted version of the pronymph.

propodeum the first abdominal segment of apocritan wasps, which is fused to the thorax and separated from the rest of the abdomen (the metasoma or gaster) by the wasp waist.

prosoma the head (versus mesosoma, metasoma in apocritan wasps).

proventriculus (n.), **proventricular** (adj.) a gizzard-like structure in the insect crop, usually lined with lobes or sclerotized teeth for straining or grinding food (Figure 7.61).

pseudotracheae in flies, fine, annulated gutters on the labellum that channel liquid food into the mouth.

pteralia minute sclerites at the base of the insect wing that articulate with each other, including the humeral plate and the axillary sclerites.

pterostigma a thickened, sclerotized area on the anterior margin of the insect wing, near the tip.

pterothorax for adults possessing wings, the meso- and metathorax combined, which are the segments possessing wings.

ptilinal fissure (= lunule) the crescentic suture above the face in cyclorrhaphan flies, which is the opening for the ptilinum (Figure 12.104).

ptilinum a membranous, saclike structure of the cyclorrhaphan fly head, which evaganinates just before the adult ecloses from the puparium in order to push the operculum open and allow the fly to escape (Figure 12.86). It invaginates soon after emergence.

pulvillus (sing.), **pulvilli** (pl.) the padlike structure(s) of the insect pretarsus.

punctuated equilibrium sudden divergence of closely related species, over a brief period of time compared to the entire lifespan of the species or lineage.

pupa the generally (but not always) quiescent life history stage between the last instar larva and the reproductively active adult in holometabolous insects.

puparium a pupal case; in cyclorrhaphan flies, pupation takes place within a puparium, which is the tanned cuticle of the third instar larva.

pygidial glands defensive exocrine glands at the tip of the abdomen in adephagan beetles.

pygidium the last abdominal tergite in insects.

radiation a rapid proliferation of species or other taxa.

raptorial (adj.) adapted for grabbing prey; in insects, generally referring to forelegs specialized for grasping prey, such as in mantises.

relict (n., adj.) a taxon or distribution that is isolated phylogenetically or geographically as a result of extinction.

remigium part of the insect wing that includes the entire surface exclusive of the anal and jugal lobes.

replication (mineral) inorganic, fossilized reproduction of an organism usually three-dimensionally and with fine structural fidelity (e.g., Orsten) (Figures 2.5 to 2.14).

riparian (adj.) living in or along the shores of rivers and streams.

rostrum in insects, beaklike mouthparts, such as in Heteroptera, weevils (Curculionidae), and some Mecoptera.

saltatorial (adj.) adapted for jumping; in insects, generally referring to the enlarged hind legs specialized for jumping, such as in Orthoptera.

saprophagous (adj.) feeding on bacteria, protists, yeasts, or other microbiota on or within decaying substrates.

scale in insects, a broad, flat form of seta with a petiolate base.

sclerite (n.), **sclerotized** (adj.) in arthropods, a hard, generally external plate bounded by sutures; a tanned, cuticular part.

scrobe a groove into which an antenna is recessed; generally occurs in fossorial and ectoparasitic insects.

scutellum the posterior section of the notum in insects (Figure 4.3).

scutum the middle part of the notum in insects (Figure 4.3).

semiaquatic living on or near water, but not in water, such as the semiaquatic water striders.

semisocial where a group of individuals of the same generation cooperate in the care of offspring, and where some (but not all) females produce the offspring.

sensillium a small dorsal sclerite at the tip the flea abdomen that bears a distinctive group of sensilla trichodea (Figures 12.15, 12.20).

sensillum (sing.), **sensilla** (pl.) microscopic structures on insects specialized for sensation.

seta (sing.), **setae** (pl.) socketed hairs in insects; minute forms are **setulae**.

sibling species Species that are very similar morphologically and in other ways, and thus considered closely related. Also, *cryptic species.*

silk a fibrous protein that is exuded into fine strands by various larval and adult terrestrial arthropods from various specialized glands, depending on the taxon. Used for constructing cases for the immatures, for domiciles, capturing prey, and/or communication.

sister group (n.), **sister-group** (adj.) the most closely related taxon, and ideally distinguished as a living sister group, or an immediate sister group (which may be living or extinct).

sitophore sclerite a minute sclerite associated with the hypopharynx of insect mouthparts, adapted for assisting in the extraction of water from air in Psocodea (Figure 8.2).

social referring to conspecific individuals that cooperate for mutual benefit.

soldier a caste in social insect colonies that is (usually structurally) specialized for defense.

species (sing., pl.) a fundamental taxonomic unit; a group of individuals that possess unique features and are reproductively compatible (see also *biological species concept* and *phylogenetic species concept*).

sperm pump an internal, sclerotized structure in male antliophoran insects (Diptera, mecopteroids, fleas) involved in ejaculation.

spermatheca an internal, generally sclerotized bulb in female insects that stores sperm.

spermatophore sperm packaged into a sac when transferred from male to female.

spine in insects, a thornlike, sclerotized structure on the cuticle that does not have a socketed base.

spiracles pairs of small, valved openings on the side of each or some postcephalic segments in certain terrestrial arthropods, including insects.

stem group an unresolved or paraphyletic assemblage of species or taxa at the base of a phylogeny or cladogram (Figure 1.28).

stemmata (pl.), **stemma** (sing.) simple larval eyes, composed of isolated facets on the head of larval insects (Figure 13.20).

sternite a ventral sclerite.

sting the ovipositor in aculeate Hymenoptera that is specialized for injecting toxin but that does not pass eggs (Figure 11.32).

stipes (sing.), **stipites** (pl.) segment of the maxilla in insect mouthparts, lying between the basal cardo and the more distal lacinia and galea (Figure 4.2).

stridulation method of producing sound by scraping a hard edge against a row of teeth or pegs, such as in many Orthoptera.

style (n.), **styli** (pl.), **stylate** (adj.) a slender appendage, used most commonly to refer to the paired ventral appendages on the abdomen of silverfish and bristletails (Zygentoma and

Archaeognatha) and to the slender, terminal flagellomeres in certain brachyceran flies.

subcortical (adj.) under bark.

subimago (n.), **subimaginal** (adj.) a winged stage in Ephemeroptera just prior to the final, reproductively mature adult.

subsocial a condition where adults (not necessarily parents) care for immature brood.

substitution (DNA) the substitution of one nucleotide (A, T, G, C) for another, which can eventually lead to a nucleotide replacement.

sulcus an external groove that marks an internal ridge or flange to which muscles attach, frequently confused with *suture*.

suture the membranous line between two metameric segments.

swarm aerial aggregation of conspecific individuals, usually for mating purposes. Males of various insects swarm, which attracts receptive females.

symbiont (n.), **symbiotic** (adj.) a species living in close association with another, unrelated species.

synapomorphy (n.), **synapomorphic** (adj.) a derived or specialized character shared among two or more species or taxa resulting from apparent or tested common ancestry.

synchronous flight muscles insect thoracic muscles that power flight and contract in synchrony with each wing beat.

systematics the science of organismal diversity, generally involving the deciphering of species, higher taxa, their phylogeny, classification, and taxonomy.

tagma, tagmosis a specialized region of metameric animals (i.e., in insects the head, thorax, abdomen; in spiders the cephalothorax and opisthosoma).

taphonomy (n.), **taphonomic** (adj.) the study of or pertaining to the process of fossilization.

tarsus the portion of the insect leg distal to the tibia, composed of one to five segments.

taxic analysis a method of comparing diversity based on numbers of taxa (e.g., genera, families), which has been criticized on the grounds that taxa of different organisms are not equivalent evolutionary units.

taxon (sing.), **taxa** (pl.) a taxonomic category, from a species to genus, family, etc.

taxonomy the science of naming species and groups of species into a classification, preferably to reflect phylogenetic relationships.

tegmina (n.), **tegminous** (adj.) leathery forewings, such as those of roaches and orthopterans.

tegulam (sing.), **tegulae** (pl.) small, scale-like sclerite at the base of the costal vein in the insect wing.

telotropic type of insect ovariole where some oocytes are nutritive and connected to other oocytes by nutritive filaments. Occurs in Heteroptera and polyphagan beetles.

temperate (adj.) in biogeography, referring to areas or zones that are generally between tropical and arctic zones, usually characterized by mild seasons and deciduous broad-leaf plants.

tentorium a system of apodemes in the insect head, to which muscles that control mouthparts and head movements are attached.

tergite dorsal sclerite.

terminalia structures on the tip of the insect abdomen, including genitalia.

terrestrial on land.

testes (incl. vas deferens) organs of the male reproductive system that produce sperm.

thelytokous a form of parthenogenesis that gives rise only to female offspring; without males.

thoracic scent glands paired, specialized exocrine glands in Heteroptera that produce a pungent, noxious secretion (Figure 8.56).

thorax (n.), **thoracic** (adj.) the middle tagma of an insect between the head and thorax.

thrust the force of airborne bodies that propels the body forward.

tibia segment of the insect leg between the femur and tarsus, generally long and slender.

trachea (sing.), **tracheae** (pl.) fine, ringed tubules that permeate the internal spaces of the hexapod body and relay oxygen and carbon dioxide directly to and from the tissues; they open to the spiracles or, in some aquatic nymphs and larvae, to the gills.

tracheole a very fine trachea, one of the last branches of a trachea.

trichobothrium (sing.), **trichobothria** (pl.) a specialized sensory structure on the cuticle of terrestrial arthropods, which consists of a long, fine seta (the trich) and a deep socket (bothrium).

trichomes in insects, tufts of fine hairs specialized for disseminating pheromones and other scents, such as in some Heteroptera and Coleoptera.

triungulin (sing.), **triungula** (pl.) the active, first instar larva of parasitoid meloid and rhipiphorid beetles; planidium.

trochantellus in Hymenoptera, a small basal segment of the femur separated from the rest of the femur by a fine groove.

trochanter small segment of the leg between the coxa and femur.

trophallaxis the direct feeding of a nestmate either orally, or through anal secretions (proctodeal trophallaxis).

trunk/alitrunk in Hymenoptera, the middle part of the body consisting of the thorax and the propodeum, generally used just for ants (Figure 11.50).

tymbals paired sound-producing organs in Auchenorrhyncha located at the base of the abdomen, consisting of papery membranes that intensely vibrate (Figure 8.41).

tympanum (sing.), **tympana** (pl.) taught membrane involved in sound detection (Figure 7.22), to which is attached a microscopic chordotonal organ that transduces the vibrations of the tympanum into nervous impulses. Tympana occur in various insects and various locations of the body.

type specimen a holotype, lectotype, neotype, paratype, allotype, or some formally designated specimen that is a taxonomic standard for a species.

urogomphus (sing.), **urogomphi** (pl.) paired dorsal processes on abdominal tergite 9 of Coleoptera (Figure 10-2).

valvifer paired appendage in the insect ovipositor, also called the gonocoxa (Figure 4.8).

valvilli minute, articulating flaps inside the first pair of gonapophyses of the ovipositor of Ichneumonoidea and Aculeata wasps.

valvula paired appendage in the insect ovipositor, also called the gonapophysis.

vannus/vannal lobe the region of the insect wing posterior/basal to the vannal fold, which extends out from the third axillary sclerite; generally the anal lobe.

vena triplica fusion of veins Sc, R, and Rs near the wing apex to form a common vein, occurring in Neuroptera.

vicariance biogeography the study of animal and plant distributions based on phylogenetic relationships and areas of endemism; a type of "historical" biogeography.

viviparity (n.), **viviparous** (adj.), **viviparae** (aphids) reproduction where a female bears live young instead of eggs.

vomer a sclerite of the tenth sternum in male Phasmatodea involved in copulation.

wasp waist the constricted, pedunculate base of the second abdominal segment of apocritan wasps, which connects the mesosoma and metasoma (Figure 11.8).

wing cells an area on the insect wing bounded by veins.

wing folds weakened areas in wing membrane and veins where the wing folds, generally during flight. Position of folds is generally taxon specific.

wing pads wing buds, or small lobes seen on the nymphs of hemimetabolous/exopterygote insects, and the pupae of certain holometabolous insects (Figure 9.1).

wing veins fine, tubular, veinlike structures in the insect wing; main branches are the costal (abbreviated C), subcosta (Sc), radius (R), median or medial (M), cubital (Cu), and anal (A) (Figure 4.5).

worker social insect caste specialized for foraging, maintenance of the nest, and queen and brood care.

xylophagous (adj.) feeding on wood.

REFERENCES

Abdullah, M. 1964. New heteromerous beetles (Coleoptera) from the Baltic amber of eastern Prussia and gum copal of Zanzibar. *Transactions of the Royal Entomological Society of London* 116: 329–46, 2 pls.

Abele, L. G., ed. 1982. *Biology of Crustacea: Volume 1: Systematics, the Fossil Record and Biogeography.* Academic Press; New York, New York; xx+318 pp.

———, W. Kim, and B. E. Felgenhauer. 1989. Molecular evidence for inclusion of the phylum Pentastomida in the Crustacea. *Molecular Biology and Evolution* 6: 685–91.

Achtelig, M. 1967. Ueber die Anatomie des Kopfes von *Raphidia flavipes* Stein und die Verwandtschaftsbeziehungen der Raphidiidae zu den Megaloptera. *Zoologische Jahrbücher, Abteilung für Anatomie und Ontogenie der Tiere* 84: 249–312.

———. 1975. Die Abdomenbasis der Neuropteroidea (Insecta, Holometabola). Ein vergleichend anatomische Untersuchung des Skeletts und der Muskulatur. *Zoomorphologie* 82: 201–42.

———. 1976. Indizien zur Monophylie der Raphidioptera und Megaloptera (Insecta, Holometabola). *Verhandlungen der Deutsche Zoologische Gesellschaft* 1976: 233.

———. 1978. Entwicklung und Morphologie der innern und äusseren weiblichen Genitalorgane der Kamelhalsfliegen (Neuropteroidea: Raphidioptera). *Entomologica Germanica* 4: 140–63.

———. 1981. Kamelhalsfliegen (Insecta, Raphidioptera) aus der Umgebung von Augsburg. *Berichte des Naturwissenschaftlichen Vereins für Schwaben* 85: 30–3.

———, and N. P. Kristensen. 1973. A re-examination of the relationships of the Raphidioptera (Insecta). *Zeitschrift für Zoologische Systematik und Evolutionsforschung* 11: 268–74.

Achterberg, C., van. 1982. The fossil species of the subfamily Blacinae described by C. T. Brues (Hym.: Braconidae). *Entomologische Berichten* 42: 91–6.

———. 2001. The first known fossil Masoninae (Hymenoptera: Braconidae) from Miocene Dominican amber. *Zoologische Mededelingen* 75: 393–6.

———. 2002. A revision of the Old World species of *Megischus* Brullé, *Stephanus* Jurine and *Pseudomegischus* gen. nov., with a key to the genera of the family Stephanidae (Hymenoptera: Stephanoidea). *Zoologische Verhandelingen* 339: 1–206.

———, and B. van Aartsen. 1986. The European Pamphiliidae (Hymenoptera: Symphyta), with special reference to the Netherlands. *Zoologische Verhandelingen* 234: 1–98.

———, and R. J. M. van Kats. 2000. Revision of the Palaearctic Embolemidae (Hymenoptera). *Zoologische Mededelingen* 74: 251–69.

Ackery, P. R. 1984. Systematic and faunistic studies on butterflies. Pp. 9–21. *In* Vane-Wright, R. I., and P. R. Ackery (eds.), *The Biology of Butterflies.* Academic Press; London, UK; xxiv+429 pp.

———, and R. I. Vane-Wright. 1984. *Milkweed Butterflies: Their Cladistics and Biology.* Cornell University Press; Ithaca, New York; ix+425 pp.

———, R. De Jong, and R. I. Vane-Wright. 1999. The Butterflies: Hedyloidea, Hesperioidea and Papilionidea. Pp. 263–300. *In* Kristensen, N. P. (ed.), *Handbuch der Zoologie: Eine Naturgeschichte der Stämme des Tierreiches: Band IV: Arthropoda: Insecta: Teilband 35: Lepidoptera, Moths and Butterflies: Volume 1: Evolution, Systematics, and Biogeography.* Walter de Gruyter; Berlin, Germany; x+491 pp.

Adams, M. D., S. E. Celniker, R. A. Holt, C. A. Evans, J. D. Gocayne, *et al.* 2000. The genome sequence of *Drosophila melanogaster. Science* 287: 2185–95. [+2 maps]

Adams, P. A. 1958. The relationship of the Protoperlaria and the Endopterygota. *Psyche* 65: 115–27.

———. 1967. A review of the Mesochrysinae and Nothochrysinae (Neuroptera: Chrysopidae). *Bulletin of the Museum of Comparative Zoology* 135: 215–38.

Addicott, J. F. 1996. Cheaters in yucca/moth mutualism. *Nature* 380: 114–15.

Afifi, S. A. 1968. Morphology and taxonomy of the adult males of the families Pseudococcidae and Eriococcidae (Homoptera: Coccoidea). *Bulletin of the British Museum (Natural History), Entomology Supplement* 13: 1–210.

Afonin, S. A. 2000. Pollen grains of the genus *Cladaitina* extracted from the gut of the Early Permian insect *Tillyardembia* (Grylloblattida). *Paleontological Journal* 34: 575–9.

Afzelius, B. A., and R. Dallai. 1988. Spermatozoa of Megaloptera and Raphidioptera (Insecta, Neuropteroidea). *Journal of Ultrastructure and Molecular Structure Research* 101: 185–91.

Agassiz, L. 1896. *Methods of Study in Natural History.* Houghton-Mifflin Company; Boston, Massachusetts; vi+319 pp.

Aguiar, A. P. 1998. Revisão do gênero *Hemistephanus* Enderlein, 1906 (Hymenoptera, Stephanidae), com considerações metodológicas. *Revista Brasileira de Entomologia* 41: 343–429.

———. 2001. Revision of the Australian Stephanidae (Hymenoptera). *Invertebrate Taxonomy* 15: 763–822.

———, and J.-W. Janzen. 1999. An overview of fossil Stephanidae (Hymenoptera), with description of two new taxa from Baltic amber, and key to species of *Electrostephanus* Brues. *Entomologica Scandinavica* 30: 443–52.

———, and N. F. Johnson. 2003. Stephanidae (Hymenoptera) of American north of Mexico. *Proceedings of the Entomological Society of Washington* 105: 467–83.

Aguinaldo, A. M., J. M. Turbeville, L. S. Linford, M. C. Rivera, J. R. Garey, R. A. Raff, and J. A. Lake. 1997. Evidence for a clade of nematodes, arthropods and other moulting animals. *Nature* 387: 489–93.

Ahmad, M. 1950. The phylogeny of termite genera based on imago-worker mandibles. *Bulletin of the American Museum of Natural History* 95: 37–86.

Alba-Tercedor, J., and A. Sanchez-Ortega. 1991. *Overview and Strategies of Ephemeroptera and Plecoptera.* Sandhill Crane Press; Gainesville, Florida; xiv+588 pp.

Alcock, J. 2000. Interactions between the sexually deceptive orchid *Spiculaea ciliata* and its wasp pollinator *Thynnoturneria* sp. (Hymenoptera: Thynninae). *Journal of Natural History* 34: 629–36.

Alego, T. J., S. E. Scheckler, and J. B. Maynard. 2001. Effects of the Middle to Late Devonian spread of vascular land plants on weathering regimes, marine biotas, and global climate. Pp. 213–36. *In* Gensel, P. G., and D. Edwards (eds.), *Plants Invade the Land: Evolutionary and Environmental Perspectives.* Columbia University Press; New York, New York; x+[1]+304 pp.

Alekseyev, V. N., and A. P. Rasnitsyn. 1981. Late Cretaceous Megaspilidae (Hymenoptera) from amber of the Taymyr. *Paleontologicheskiy Zhurnal* 1981: 127–30. [In Russian]

Alexander, B. A. 1992. An exploratory analysis of cladistic relationships within the superfamily Apoidea, with special reference to sphecid wasps (Hymenoptera). *Journal of Hymenoptera Research* 1: 25–61.

———, and C. D. Michener. 1995. Phylogenetic studies of the families of short-tongued bees (Hymenoptera: Apoidea). *University of Kansas Science Bulletin* 55: 377–424.

Alexander, D. E. 2002. *Nature's Flyers: Birds, Insects, and the Biomechanics of Flight.* Johns Hopkins University Press; Baltimore, Maryland; xix+358 pp.

Ali, D. W., and D. C. Darling. 1998. Neuroanatomy and neurochemistry: Implications for the phylogeny of the lower Neoptera. *Canadian Journal of Zoology* 76: 1628–33.

Allan, J. D. 1995. *Stream Ecology: Structure and Function of Running Waters.* Chapman and Hall; London, UK; xii+388 pp.

Allen, D., L. Davies, and P. Tobin. 1984. The dragonflies of the world: A systematic list of the extant species of Odonata. Volume 1. Zygoptera, Anisozygoptera. *Societas Internationalis Odonataologica Rapid Communications, Supplement* 3: ix+1–127.

———, L. Davies, and P. Tobin. 1985. The dragonflies of the world: A systematic list of the extant species of Odonata. Volume 2. Anisoptera. *Societas Internationalis Odonataologica Rapid Communications, Supplement* 5: xi+1–151.

Allen, D. C. 1972. Insect parasites of saddled prominent, *Heterocampa guttivitta* (Lepidoptera: Notodontidae) in the northeastern United States. *Canadian Entomologist* 104: 1609–22.

Allen, G. R. 1998. Diel calling activity and field survival of the bushcricket, *Sciarasaga quadrata* (Orthoptera: Tettigoniidae): A role for sound-locating parasitic flies? *Ethology* 104: 645–60.

Allen, P. 1998. Purbeck-Wealden (early Cretaceous) climates. *Proceedings of the Geologists' Association* 109: 197–236.

———, and W. A. Wimbledon. 1991. Correlation of NW European Purbeck-Wealden (nonmarine Lower Cretaceous) as seen from the English type-areas. *Cretaceous Research* 12: 511–26.

Allison, A., G. A. Sameulson, and S. E. Miller. 1993. Patterns of beetle species diversity in New Guinea rain forest as revealed by canopy fogging: Preliminary findings. *Selbyana* 14: 16–20.

Allison, P. A. 1988a. Taphonomy of the Eocene London Clay biota. *Palaeontology* 31: 1079–1110.

———. 1988b. Konservat-Lagerstätten: Cause and classification. *Paleobiology* 14: 331–44.

———, and D. G. E. Briggs. 1991. *Taphonomy: Releasing the Data Locked in the Fossil Record.* Plenum; New York, New York; xiv+560 pp.

Alonso, J., A. Arillo, E. Barrón, J. C. Corral, J. Grimalt, *et al.* 2000. A new fossil resin with biological inclusions in Lower Cretaceous deposits from Álava (northern Spain, Basque-Cantabrian Basin). *Journal of Paleontology* 74: 158–78.

Alroy, J. 1999. The fossil record of North American mammals: Evidence for a Paleocene evolutionary radiation. *Systematic Biology* 48: 107–18.

Aluja, M., and A. L. Norrbom, eds. 2000. *Fruit Flies (Tephritidae): Phylogeny and Evolution of Behavior.* CRC Press; Boca Raton, Florida; 944 pp.

Alvarez, L. W., W. Alvarez, F. Asaro, and H. V. Michel. 1980. Extraterrestrial cause for the Cretaceous-Tertiary extinction. *Science* 208: 1095–1108.

Amorim, D. S. 1993. A phylogenetic analysis of the basal groups of Bibionomorpha, with a critical examination of the wing venation homology. *Revista Brasiliera de Biologia* 52: 379–99.

———. 1994. A new suprageneric classification of the Scatopsidae (Diptera: Psychodomorpha). *Iheringia Serie Zoologia.* 77: 107–12.

———. 2000. A new phylogeny and phylogenetic classification for the Canthyloscelidae (Diptera: Psychodomorpha). *Canadian Journal of Zoology* 78: 1067–77.

———, and S. H. S. Tozoni. 1994. Phylogenetic and biogeographic analysis of the Anisopodoidea (Diptera, Bibionomorpha), with an area cladogram for intercontinental relationships. *Revista Brasileira de Entomologia* 38: 517–43.

Ananthakrishnan, T. N. 1984a. *Bioecology of Thrips.* Indira Publishing House; Oak Park, Michigan; iii+233 pp.

———, ed. 1984b. *Biology of Gall Insects.* Oxford and IBH Publ.; New Delhi, India; vii+362 pp.

Ander, K. 1942. Die Insektenfauna des baltischen Bernsteins nebst damit verknüpften zoogeographischen Problemen. *Lunds Universitets Årsskrift, 2 Afdeling, Medicin samt Matematiska och Naturvetenskapliga Ämnen* 38: 1–82.

Andersen, N. M. 1982. The semiaquatic bugs (Hemiptera: Gerromorpha): Phylogeny, adaptations, biogeography, and classification. *Entomonograph* 13: 1–455.

———. 1998. Water striders from the Paleogene of Denmark with a review of the fossil record and evolution of semiaquatic bugs (Hemiptera, Gerromorpha). *Biologiske Skrifter* 50: 1–157.

———, and D. Grimaldi. 2001. A fossil water measurer from the mid-Cretaceous Burmese amber (Hemiptera: Gerromorpha: Hydrometridae). *Insect Systematics and Evolution* 32: 381–92.

Andersen, S. 2001. Silky lacewings (Neuroptera: Psychopsidae) from the Eocene-Paleocene transition of Denmark with a review of the fossil record and comments on phylogeny and zoogeography. *Insect Systematics and Evolution* 32: 419–38.

Anderson, D. T. 1973. *Embryology and Phylogeny in Annelids and Arthropods.* Pergamon Press; Oxford, UK; xiv+495 pp.

———. 1979. Embryos, fate maps, and the phylogeny of arthropods. Pp. 59–105. *In* Gupta, A. P. (ed.), *Arthropod Phylogeny.* Van Nostrand Reinhold; New York, New York; xx+762 pp.

Anderson, G. J., S. D. Johnson, P. R. Neal, and G. Bernardello. 2002. Reproductive biology and plant systematics: The growth of a symbiotic association. *Taxon* 51: 637–53.

Anderson, J., and H. M. Anderson. 1993. Terrestrial flora and fauna of the Gondwana Triassic: Part 1. Occurrences. Pp. 3–25. *In* Lucas, S. G., and M. Morales (eds.), *The Nonmarine Triassic.* New Mexico Museum of Natural History and Science (Bulletin Number 3); Albuquerque, New Mexico; i+478 pp.

———, H. Anderson, P. Fatti, and H. Sichel. 1996. The Triassic explosion (?): A statistical model for extrapolating biodiversity based on the terrestrial Molteno Formation. *Paleobiology* 22: 318–28.

Anderson, K. B. 1996. The nature and fate of natural resins in the geosphere – VII. A radiocarbon (14C) age scale for description of immature and natural resins: An invitation to scientific debate. *Organic Geochemistry* 25: 251–3.

———, and B. A. LePage. 1995. Analysis of fossil resins from Axel Heiberg Island, Canadian Arctic. Pp. 170–92. *In* Anderson, K. B., and J. C. Crelling (eds.), *Amber, Resinite, and Fossil Resins.* American Chemical Society; Washington, D.C.; xvii+297 pp.

Anderson, L. I., and P. A. Selden. 1997. Opisthosomal fusion and phylogeny of Palaeozoic Xiphosura. *Lethaia* 30: 19–31.

———, and N. H. Trewin. 2003. An Early Devonian arthropod fauna from the Windyfield cherts, Aberdeenshire, Scotland. *Palaeontology* 46: 467–509.

Andersson, H. 1977. Taxonomic and phylogenetic studies on Chloropidae (Diptera) with special reference to Old World genera. *Entomologica Scandinavica, Supplement* 8: 1–200.

Ando, H. 1988. Obituary – Ryuichi Matsuda 1920–1986. *International Journal of Insect Morphology and Embryology* 17: 91–4.

———, and T. Nagashima. 1982. A preliminary note on the embryogenesis of *Galloisiana nipponensis* (Caudell *et* King). Pp. 89–95. *In* Ando, H. (ed.), *Biology of the Notoptera.* Kashiyo-Insatsu Co. Ltd.; Nagano, Japan; vi+194 pp.

Andres, D. 1989. Phosphatisierte Fossilien aus dem unteren Ordoviz von Südschweden. *Berliner Geowissenshaftliche Abhandlungen, Reihe A, Geologie und Paläortologie* 106: 9–19.

Ansorge, J. 1994. Tanyderidae and Psychodidae (Insecta: Diptera) from the Lower Jurassic of northeastern Germany. *Paläontologisches Zeitschrift* 68: 199–210.

———. 1996. Insekten as dem oberen Lias von Grimmen (Vorpommern, Norddeutschland). *Neue Paläontologische Abhandlungen* 2: 1–132.

———. 1999. *Aenne liasina* gen. *et* sp. n. – the most primitive non biting midge (Diptera: Chironomidae: Aenneinae subfam. n.) – from the Lower Jurassic of Germany. *Polskie Pismo Entomologiczne* 68: 431–43.

———. 2001. *Dobbertinia reticulata* Handlirsch, 1920 from the Lower Jurassic of Dobbertin (Mecklenburg/Germany) – the oldest representative of Sialidae (Megaloptera). *Neues Jahrbuch für Geologie und Paläontologie, Monatshefte* 2001: 553–64.

———. 2002. Revision of the "Trichoptera" described by Geinitz and Handlirsch from the Lower Toarcian of Dobbertin (Germany) based on new material. *Nova Supplementum Entomologia, Keltern* 15: 55–74.

———. 2003a. Upper Liassic Amphiesmenopterans (Trichoptera + Lepidoptera) from Germany – A review. *Acta Zoologica Cracoviensia* 46 (Supplement): 285–90.

———. 2003b. Insects from the Lower Toarcian of middle Europe and England. *Acta Zoologica Cracoviensia* 46 (Supplement): 291–310.

———, and W. Krzemiński. 1994. Oligophrynidae, a Lower Jurassic dipteran family (Diptera,

Brachycera). *Acta Zoologica Cracoviensia* 37: 115–19.

———, and M. B. Mostovski. 2000. Redescription of *Prohirmoneura jurassica* Handlirsch 1906 (Diptera: Nemestrinidae) from the Lower Tithonian lithographic limestone of Eichstätt (Bavaria). *Neues Jahrbuch für Geologie und Paläontologie, Monatshefte* 2000: 235–43.

———, and T. Schlüter. 1990. The earliest chrysopid: *Liassochrysa stigmatica* n.g., n.sp. from the Lower Jurassic of Dobbertin, Germany. *Neuroptera International* 6: 87–93.

Antropov, A. V. 1995. A new species of the genus *Trypoxylon* (Hymenoptera, Sphecidae) from Dominican amber. *Paleontologicheskiy Zhurnal* 1: 125–8. [In Russian]

———. 2000. Digger wasps (Hymenoptera, Sphecidae) in Burmese amber. *Bulletin of the Natural History Museum, London (Geology)* 56: 59–77.

———, and W. J. Pulawski. 1989. A new species of *Pison* Jurine from Baltic amber (Hymenoptera: Sphecidae). *Pan-Pacific Entomologist* 65: 312–18.

———, and W. J. Pulawski. 1996. *Pison antiquum*, a new species from Dominican amber (Hymenoptera: Sphecidae). *Journal of Hymenoptera Research* 5: 16–21.

Aoki, S. 1977. *Colophina clematis* (Homoptera, Pemphigidae), an aphid species with "soldiers". *Kontyû* 45: 276–82.

———. 1978. Two pemphigids with first instar larvae attacking predatory intruders (Homoptera, Aphidoidea). *New Entomologist* 27: 7–12.

———. 1982. Soldiers and altruistic dispersal in aphids. Pp. 154–8. *In* Breed, M. D., C. D. Michener, and H. E. Evans (eds.), *The Biology of Social Insects*. Westview Press; Boulder, Colorado; xii+419 pp.

APG [Angiosperm Phylogeny Group]. 1998. An ordinal classification for the families of flowering plants. *Annals of the Missouri Botanical Garden* 85: 531–53.

Aransay, A. M., E. Scoulica, Y. Tselentis, and P. D. Ready. 2000. Phylogenetic relationships of phlebotomine sandflies inferred from small subunit nuclear ribosomal DNA. *Insect Molecular Biology* 9: 157–68.

Archibald, S. B., and R. W. Mathewes. 2000. Early Eocene insects from Quilchena, British Columbia, and their paleoclimatic implications. *Canadian Journal of Zoology* 78: 1141–1462.

Argaman, Q. 1986. Taxonomy of Heterogynaidae (Hymenoptera: Aculeata). *Israel Journal of Entomology* 19: 7–12.

———. 1988a. A new subfamily of Bethylidae allied to Pristocerinae (Hymenoptera). *Bollettino della Società Entomolgica Italiana* 120: 139–52.

———. 1988b. Generic synopsis of Sclerogibbidae (Hymenoptera). *Annales Historico-Naturales Musei Nationalis Hungarici* 80: 177–87.

———. 1993. A taxonomic study of Sclerogibbidae especially from the circumsardinian islands. *Annali del Museo Civico di Storia Naturale "Giacomo Doria"* 89: 537–53.

Arillo, A. 2001. Presencia de la familia Pompilidae (Insecta, Hymenoptera) en el Miocene Superior de la Cuenca de La Cerdana (Lleida, NE de España). *Coloquios de Paleontologia* 52: 79–83.

———, and M. B. Mostovski. 1999. A new genus of Prioriphorinae (Diptera, Phoridae) from the Lower Cretaceous amber of Alava (Spain). *Studia Dipterologica* 6: 251–5.

———, and A. Nel. 2000. Two new fossil cecidomyiids [sic] flies from the Lower Cretaceous amber of Alava (Spain) (Diptera, Cecidomyiidae). *Bulletin de la Société Entomologique de France* 105: 285–8.

———, and V. M. Ortuño. 1997. The fossil Acrididae from the Oligocene of Izarra (Alava, Spain). The antiquity of gregarious behavior. *Geobios* 30: 231–4.

———, V. M. Ortuño, and A. Nel. 1997. Description of an enigmatic insect from Baltic amber. *Bulletin de la Société Entomologique de France* 102: 11–14.

Aris-Broscu, S., and Z. Yang. 2002. The effects of models of rate evolution on estimation of divergence dates with a special reference to the metazoan 18S rRNA phylogeny. *Systematic Biology* 51: 703–14.

Arnol'di, L. V., V. V. Zherikhin, L. M. Nikritin, and A. G. Ponomarenko. 1977. Mesozoic Coleoptera. *Trudy Paleontologischeskogo Instituta Akademii Nauk, SSSR* 161: 1–204. [In Russian: Translated in Arnol'di *et al.*, 1991]

———, V. V. Zherikhin, L. M. Nikritin, and A. G. Ponomarenko. 1991. *Mesozoic Coleoptera*. Smithsonian Institution Libraries and the United States National Science Foundation; Washington, D.C.; xii+285 pp. [English translation Arnol'di *et al.*, 1977.]

Arrow, G. H. 1951. *Horned Beetles*. W. Junk; The Hague, the Netherlands; 154 pp.

Asahina, S. 1954. *A Morphological Study of a Relic Dragonfly* Epiophlebia superstes *Selys (Odonata, Anisozygoptera)*. Japan Society for the Promotion of Science; Tokyo, Japan; iv+153 pp.

Asche, M. 1988. Preliminary thoughts on the phylogeny of Fulgoromorpha (Homoptera: Auchenorrhyncha). Pp. 47–53. *In* Vidano, C., and A. Arzone (eds.), *Proceedings of the 6th Auchenorrhyncha Meeting, Turin Italy, September 7–11, 1987*. Consiglio Nazionale delle Ricerche; Turin, Italy; 652 pp.

Ascher, J. S., B. N. Danforth, and S. Li. 2001. Phylogenetic utility of the major opsin in bees (Hymenoptera: Apoidea): A reassessment. *Molecular Phylogenetics and Evolution* 19: 76–93.

Ashburner, M. 1981. Entomophagous and other bizarre Drosophilidae. Pp. 395–429. *In* Ashburner, M., H. L. Carson, and J. N. Thompson (eds.), *The Genetics and Biology of* Drosophila [Volume 3a]. Academic Press; London, UK; 323 pp.

———. 1989. Drosophila: *A Laboratory Manual*. Cold Spring Harbor Laboratory Press; New York, New York; xliii+1331 pp.

———, and F. Lemeunier. 1976. Relationships within the *melanogaster* species subgroup of the genus *Drosophila* (*Sophophora*). 1. Inversion polymorphism in *D. melanogaster* and *D. simulans*. *Proceedings of the Royal Society of London, Series B, Biological Sciences* 193: 137–57.

Ashmead, W. H. 1899. Classification of the bees, of the superfamily Apoidea. *Transactions of the American Entomological Society* 26: 49–100.

———. 1904. Remarks on honey bees. *Proceedings of the Entomological Society of Washington* 6: 120–2.

Ashton, P. S. 1988. Dipterocarp biology as a window to the understanding of tropical forest structure. *Annual Review of Ecology and Systematics* 19: 347–70.

Ashworth, A. C., and G. Kuschel. 2003. Fossil weevils (Coleoptera: Curculionidae) from latitude 85°S Antarctica. *Palaeogeography, Palaeoclimatology, Palaeoecology* 191: 191–202.

———, D. M. Harwood, P. N. Webb, and M. C. G. Mabin. 1997. A weevil from the heart of Antarctica. *Quaternary Proceedings* 5: 15–21.

Askevold, I. S. 1990. Classification of Tertiary fossil Donaciinae of North America and their implications about evolution of Donaciinae (Coleoptera, Chrysomelidae). *Canadian Journal of Zoology* 68: 2135–45.

Askew, R. R. 1971. *Parasitic Insects*. Heinemann; London, UK; xvii+316 pp.

———. 1988. *The Dragonflies of Europe*. Harley Books; Colchester, UK; 291 pp.

Aspöck, H. 1998. Distribution and biogeography of the order Raphidioptera: Updated facts and a new hypothesis. *Acta Zoologica Fennica* 209: 33–44.

———. 2000. Die endkreidezeitliche Impakt und das Überleben der Raphidioptera. *Entomologica Basiliensia* 22: 223–33.

———. 2002. The biology of Raphidioptera: A review of present knowledge. *Acta Zoologica Academiae Scientiarum Hungaricae* 48: 35–50.

———, U. Aspöck, and H. Hölzel. 1977. *Neurorthus* [sic] *apatelios* n. sp. – eine verkannte europäische Neurorthiden-species [sic] (Neuroptera: Planipennia). *Entomologische Zeitschrift* 87: 53–7.

———, U. Aspöck, and H. Hölzel. 1980. *Die Neuropteren Europas: Eine zusammenfassende Darstellung der Systematik, Ökologie und Chorologie der Neuropteroidea (Megaloptera, Raphidioptera, Planipennia) Europas*. Goecke and Evers; Krefeld, Germany; vol. 1: 495 pp., vol. 2: 355 pp.

———, U. Aspöck, and H. Rausch. 1991. *Die Raphidiopteren der Erde: Eine monographische Darstellung der Systematik, Taxonomie, Biologie, Ökologie und Chorologie der rezenten Raphidiopteren der Erde, mit einer zusammenfassenden Übersicht der fossilen Raphidiopteren (Insecta: Neuropteroidea)*. Goecke and Evers; Krefeld, Germany; vol. 1: 730 pp., vol. 2: 550 pp.

———, H. Hölzel, and U. Aspöck. 2001. Kommentierter Katalog der Neuropterida (Insecta: Raphidioptera, Megaloptera, Neuroptera) der Westpaläarktis. *Denisia* 2: 1–606.

Aspöck, U. 1986. The present state of knowledge of the family Berothidae (Neuropteroidea: Planipennia). Pp. 87–101. *In* Gepp, J., H. Aspöck, and H. Hölzel (eds.), *Recent Research in Neuropterology: Proceedings of the Second International Symposium on Neuropterology*. Druckhaus Thalerhof; Graz, Austria; 176 pp.

———. 1992. Crucial points in the phylogeny of the Neuroptera (Insecta). Pp. 63–73. *In* Canard, M., H. Aspöck, and M. W. Mansell (eds.), *Current Research in Neuropterology: Proceedings of the Fourth International Symposium on Neuropterology*. SACCO; Toulouse, France; 414 pp.

———. 1995. Neue Hypothesen zum System der Neuropterida. *Mitteilungen der Deutschen Gesellschaft für Allgemeine und Angewandte Entomologie* 10: 633–6.

———. 2002. Phylogeny of the Neuropterida (Insecta: Holometabola). *Zoologica Scripta* 31: 51–5.

———, and H. Aspöck. 1983. Über das Vorkommen von *Neurorthus* [sic] Costa in Nordafrika (Neuropteroidea, Planipennia, Neurorthidae [sic]). *Nachrichtenblatt der Bayerischen Entomologen* 32: 48–51.

———, and H. Aspöck. 1988. Die Subfamilie Cyrenoberothinae – eine Gondwana-Element? *Mansselliberotha neuropterologorum* n. g. et n. sp. aus S.W.A./Namibia (Neuropteroidea: Planipennia: Berothidae). *Zeitschrift der Arbeitsgemeinschaft Österreichischer Entomologen* 40: 1–13.

———, and H. Aspöck. 1997. Studies on new and poorly-known Rhachiberothidae (Insecta: Neuroptera) from subsaharan Africa. *Annalen des Naturhistorischen Museums in Wien, Serie B: Botanik und Zoologie* 99: 1–20.

———, and H. Aspöck. 2003a. Ordnung Raphidioptera, Kamelhalsfliegen. Pp. 542–52. *In* Dathe, H. H. (ed.), *Lehrbuch der Speziellen Zoologie, Band I: Wirbellose Tiere. 5. Teil: Insecta.* Spektrum Akademischer Verlag; Heidelberg, Germany; xii+[1]+961 pp.

———, and H. Aspöck. 2003b. Ordnung Megaloptera, Großflügler, Schlammfliegen. Pp. 552–64. *In* Dathe, H. H. (ed.), *Lehrbuch der Speziellen Zoologie, Band I: Wirbellose Tiere. 5. Teil: Insecta.* Spektrum Akademischer Verlag; Heidelberg, Germany; xii+[1]+961 pp.

———, and H. Aspöck. 2003c. Ordnung Neuroptera (Planipennia), Netzflügler. Pp. 564–84. *In* Dathe, H. H. (ed.), *Lehrbuch der Speziellen Zoologie, Band I: Wirbellose Tiere. 5. Teil: Insecta.* Spektrum Akademischer Verlag; Heidelberg, Germany; xii+[1]+961 pp.

———, and H. Aspöck. 2004. Two significant new snakeflies from Baltic amber, with discussion on relationships on autapomorphies of the order and its included taxa (Raphidioptera). *Systematic Entomology* 29: 11–19.

———, and M. W. Mansell. 1994. A revision of the family Rhachiberothidae Tjeder, 1959, stat. n. (Neuroptera). *Systematic Entomology* 19: 181–206.

———, and H. L. Nemeschkal. 1998. A cladistic analysis of the Berothidae (Neuroptera). *Acta Zoologica Fennica* 209: 45–63.

———, J. D. Plant, and H. L. Nemeschkal. 2001. Cladistic analysis of Neuroptera and their systematic position within Neuropterida (Insecta: Holometabola: Neuropterida: Neuroptera). *Systematic Entomology* 26: 73–86.

Atchison, C. W. 1979. Winter-active subnival invertebrates in southern Canada. I–IV. *Pedologia* 19: 113–28, 153–60, 176–82.

Austen, E. E. 1936. A remarkable semi-apterous fly (Diptera) found in a cave in East Africa, and representing a new family, genus and species. *Proceedings of the Zoological Society of London* 1936: 425–31.

Austin, A. D. 1983. Morphology and mechanics of the ovipositor system of *Ceratobaeus* Ashmead (Hymenoptera: Scelionidae) and related genera. *International Journal of Insect Morphology and Embryology* 12: 139–55.

———, and M. Dowton, eds. 2000a. *Hymenoptera: Evolution, Biodiversity and Biological Control.* CSIRO; Collingwood, Australia; xi+468 pp.

———, and M. Dowton. 2000b. The Hymenoptera: An introduction. Pp. 3–7. *In* Austin, A. D., and M. Dowton (eds.), *Hymenoptera: Evolution, Biodiversity and Biological Control.* CSIRO; Collingwood, Australia; xi+468 pp.

———, and S. A. Field. 1997. The ovipositor system of scelionid and platygastrid wasps (Hymenoptera: Platygastroidea): Comparative morphology and phylogenetic implications. *Invertebrate Taxonomy* 11: 1–87.

Austin, J. J., A. J. Ross, A. B. Smith, R. A. Fortey, and R. H. Thomas. 1997. Problems of reproducibility – does geologically ancient DNA survive in amber-preserved insects? *Proceedings of the Royal Society of London, Series B, Biological Sciences* 264: 467–74.

Averof, M., and S. M. Cohen. 1997. Evolutionary origin of insect wings from ancestral gills. *Nature* 385: 627–30.

Ax, P. 1984. *Das phylogenetische System: Systematisierung der lebenden Natur aufgrund ihrer Phylogenese.* Gustav Fischer; Stuttgart, Germany; 349 pp.

———. 1999. *Multicellular Animals: The Phylogenetic System of the Metazoa* [Volume II]. Springer Verlag; Berlin, Germany; xxiv+396 pp.

Axelrod, D. I. 1983. Biogeography of oaks in the Arcto-Tertiary Province. *Annals of the Missouri Botanical Garden* 70: 629–57.

———, and P. H. Raven. 1978. Late Cretaceous and Tertiary vegetation history of Africa. Pp. 77–130. *In* Werger, M. J. A. (ed.), *Biogeography and Ecology of Southern Africa.* Junk; The Hague, the Netherlands; xv+1439 pp.

Ayala, F. J., M. L. Tracey, D. Hedgecock, and R. C. Richmond. 1974. Genetic differentiation during the speciation process in *Drosophila*. *Evolution* 28: 576–92.

Azar, D. 2000. *Les Ambres Mésozoïques du Liban.* Ph.D. Dissertation, University of Paris; Orsay, France; 202 pp.

———, G. Fleck, A. Nel, and M. Solignac. 1998. A new enicocephalid bug, *Enicocephalinus acragrimaldii* gen. nov., sp. nov., from the Lower Cretaceous amber of Lebanon (Insecta, Heteroptera, Enicocephalidae). *Estudios Museo Ciência Naturales de Alava* 14(Num. Esp. 2): 217–31.

———, A. Nel, M. Solignac, J.-C. Paicheler, and F. Bouchet. 1999. New genera and species of psychodoid flies from the Lower Cretaceous amber of Lebanon. *Palaeontology* 42: 1101–36.

———, A. Nel, and M. Solignac. 2000. A new Coniopterygidae from Lebanese amber. *Acta Geologica Hispanica* 35: 31–6.

Azevedo, C. O. 1999. A key to world species of Scolebythidae (Hymenoptera: Chrysidoidea), with description of a new species of *Dominibythus* from Brazil. *Journal of Hymenoptera Research* 8: 1–5.

Baccetti, B. 1982. The spermatozoon of Arthropoda. XXXII. *Galloisiana nipponensis* (Caudell *et* King) (Grylloblattodea). Pp. 71–8. *In* Ando, H. (ed.), *Biology of the Notoptera.* Kashiyo-Insatsu Co. Ltd.; Nagano, Japan; vi+194 pp.

Bachofen-Echt, A. 1949. *Der Bernstein und seine Einschlüsse.* Springer Verlag; Vienna, Austria; 204 pp.

Bada, J. L. 1991. Amino acid biogeochemistry. *Philosophical Transactions of the Royal Society, London (B)* 333: 349–58.

———. X. S. Wang, and H. Hamilton. 1999. Preservation of key biomolecules in the fossil record: Current knowledge and future challenges. *Philosophical Transactions of the Royal Society, London B* 354: 77–87.

Badonnel, A. 1934. Recherches sur l'anatomie des Psoques. *Bulletin Biologique de France et de Belgique, Supplement* 38: 1–241.

———. 1951. Ordre des Psocoptères (Psocoptera Psoquillae Latreille, 1810). Pp. 1301–40. *In* Grassé, P. P. (ed.), *Traité de Zoologie, Tome 10.* Masson et Cie; Paris, France; 976–1948 pp.

———. 1969. Psocoptères de l'Angola et de pays voisins, avec révisions de types africains d'Enderlein (1902) et de Ribaga (1911). *Publicações Culturais da Companhia de Diamantes de Angola* 79: 15–152.

Bae, Y. J., and W. P. McCafferty. 1995. Ephemeroptera tusks and their evolution. Pp. 377–405. *In* Corkum, L. D., and J. Ciborowski (eds.), *Current Directions in Research on Ephemeroptera.* Canadian Scholars Press, Inc.; Toronto, Canada; xiii+478 pp.

Baerands, G. P. 1958. Comparative methods and the concept of homology in the study of behavior. *Archives Néerlandaises de Zoologie* 13: 401–17.

Baird, G. C. 1997. Paleoenvironmental setting of the Mazon Creek biota. Pp. 35–51. *In* Shabica, C. W., and A. A. Hay (eds.), *Richardson's Guide to the Fossil Fauna of Mazon Creek.* Northeastern Illinois University; Chicago, Illinois; xvii+308 pp.

———, S. D. Sroka, C. W. Shabica, and G. J. Kuecher. 1986. Taphonomy of Middle Pennsylvanian Mazon Creek area fossil localities, northeast Illinois: Significance of exceptional fossil preservation in syngenetic concretions. *Palaios* 1: 271–85.

Baker, H. G. 1986. Yuccas and yucca moths – A historical commentary. *Annals of the Missouri Botanical Garden* 73: 556–64.

Bakke, A. 1955. Insects reared from spruce cones in northern Norway 1951. *Entomologisk Tidskrift* 9: 152–212.

Balashov, Yu. S. 1999. Evolution of haematophagy in insects and ticks. *Entomological Review* 79: 943–54.

Balazuc, J. 1989. Quelques insectes fossiles des diatomites de Saint-Bauzile (Ardeche). *Bulletin Mensuel de la Société Linneenne de Lyon* 58: 240–5.

Balderson, J. 1991. Mantodea. Pp. 348–56. *In* Naumann, I. D. (ed.), *The Insects of Australia: A Textbook for Students and Research Workers, Volume 1* [2nd Edition]. Cornell University Press; Ithaca, New York; xvi+[1]+542 pp.

Ball, G. E., ed. 1985. *Taxonomy, Phylogeny, and Zoogeography of Beetles and Ants: A Volume Dedicated to the Memory of Philip Jackson Darlington, Jr. (1904–1983).* W. Junk-Kluwer; Dordrecht, the Netherlands; xiii+514 pp.

———, A. Casale, and A. V. Taglianti, eds. 1998. *Phylogeny and Classification of Caraboidea (Coleoptera: Adephaga).* Museo Regionale di Science Naturali; Turin, Italy; 543 pp.

Ballard, J. W. O., G. J. Olsen, D. P. Faith, W. A. Odgers, D. M. Rowell, and P. W. Atkinson. 1992. Evidence from 12S ribosomal RNA sequences that onychophorans are modified arthropods. *Science* 258: 1345–8.

Bandel, K., R. Shinaq, and W. Weitschat. 1997. First insect inclusions from the amber of Jordan. *Mitteilungen aus dem Geologisch-Paläontologischen Institut der Universität Hamburg* 80: 213–23.

Banks, H. P. 1981. Peridermal activity (wound repair) in an Early Devonian (Emsian) trimerophyte from the Gaspé Peninsula, Canada. *Palaeobotanist* 28/29: 20–5.

———, and B. J. Colthart. 1993. Plant-animal-fungal interactions in Early Devonian trimerophytes from Gaspé, Canada. *American Journal of Botany* 80: 992–1001.

Bänziger, H. 1982. Fruit-piercing moths (Lepidoptera, Noctuidae) in Thailand: A general survey and some new perspectives. *Mitteilungen der Schweizerischen Entomologischen Gesellschaft* 55: 213–40.

Barker, F. K., G. F. Barrowclough, and J. G. Groth. 2002. A phylogenetic hypothesis for passerine birds: Taxonomic and biogeographic implications of an analysis of nuclear DNA sequence data. *Proceedings of the Royal Society of London, Series B, Biological Sciences* 269: 295–308.

Barker, S. C. 1994. Phylogeny and classification, origins, and evolution of host associations of lice. *International Journal for Parasitology* 24: 1285–91.

———, M. Whiting, K. P. Johnson, and A. Murrell. 2003. Phylogeny of the lice (Insecta: Phthiraptera) inferred from small subunit rRNA. *Zoologica Scripta* 32: 407–14.

Barkman, T. J., G. Chenery, J. R. McNeal, J. Lyons-Weiler, W. J. Ellisens, G. Moore, *et al.* 2000. Independent and combined analyses of sequences from all three genomic compartments converge on the root of flowering plant phylogeny. *Proceedings of the National Academy of Sciences, U.S.A.* 97: 13166–71.

Barnard, P. C. 1981. The Rapismatidae (Neuroptera): Montane lacewings of the oriental region. *Systematic Entomology* 6: 121–36.

———, and T. R. New. 1985. New species in the *Rapisma malayanum*-complex (Neuroptera: Rapismatidae). *Neuroptera International* 3: 165–73.

Baroni-Urbani, C. 2000. Rediscovery of the Baltic amber ant genus *Prionomyrmex* (Hymenoptera, Formicidae) and its taxonomic consequences. *Ecologae Geologica Helvetica* 93: 471–89.

———, and S. Graeser. 1987. REM-Analysen an einer pyritisierten Ameise aus Baltischen Bernstein. *Stuttgarter Beiträge zur Naturkunde, Serie B, Geologie und Paläontologie* 133: 1–16.

———, B. Bolton, and P. S. Ward. 1992. The internal phylogeny of ants (Hymenoptera: Formicidae). *Systematic Entomology* 17: 301–29.

Barrett, P. M. 2000. Evolutionary consequences of dating the Yixian Formation. *Trends in Ecology and Evolution* 15: 99–103.

Barrón, E., A. Arillo, and V. M. Ortuño. 2002. Taphonomic analysis of arthropod and plant-bearing concretions from the Lower Miocene lacustrine basin of Izarra (Álava Province, Spain). Pp. 247–56. *In* De Renzi, M., *et al.* (eds.), *Current Topics on Taphonomy and Fossilization.* Ayuntamiento de Valencia; Valencia, Spain; 544 pp.

Bartels, C. 1995. Bundenbachs fossile Schätze. *Kosmos* 95: 36–7.

Barth, F. G. 1985. *Insects and Flowers: The Biology of a Partnership.* Princeton University Press; Princeton, New Jersey; xi+297 pp.

Barthel, K. W., N. H. M. Swinburne, and S. Conway Morris. 1990. *Solnhofen: A Study in Mesozoic Palaeontology.* Cambridge University Press; Cambridge; UK; ix+236 pp.

Basibuyuk, H. H., and D. L. J. Quicke. 1995. Morphology of the antenna cleaner in the Hymenoptera with particular refernce to non-aculeate families (Insecta). *Zoologica Scripta* 24: 157–77.

———, and D. L. J. Quicke. 1997. Hamuli in the Hymenoptera (Insecta) and their phylogenetic implications. *Journal of Natural History* 31: 1563–85.

———, A. P. Rasnitsyn, K. van Achterberg, M. G. Fitton, and D. L. J. Quicke. 1999. A new, putatively primitive Cretaceous fossil braconid subfamily from New Jersey amber (Hymenoptera, Braconidae). *Zoologica Scripta* 28: 211–14.

———, D. L. J. Quicke, and A. P. Rasnitsyn. 2000a. A new genus of the Orussidae (Insecta: Hymenoptera) from Late Cretaceous New Jersey amber. Pp. 305–11. *In* Grimaldi, D. (ed.), *Studies on Fossils in Amber, with Particular Reference to the Cretaceous of New Jersey.* Backhuys Publishers; Leiden, the Netherlands; viii+498 pp.

———, M. G. Fitton, A. P. Rasnitsyn, and D. L. J. Quicke. 2000b. Two new genera of the Evaniidae (Insecta: Hymenoptera) from Late Cretaceous New Jersey amber. Pp. 313–25. *In* Grimaldi, D. (ed.), *Studies on Fossils in Amber, with Particular Reference to the Cretaceous of New Jersey.* Backhuys Publishers; Leiden, the Netherlands; viii+498 pp.

———, A. P. Rasnitsyn, M. G. Fitton, and D. L. J. Quicke. 2000c. An archaic new genus of Evaniidae (Insecta: Hymenoptera) and implications for the biology of ancestral evanioids. *Bulletin of the Natural History Museum, London (Geology)* 56: 53–8.

———, A. P. Rasnitsyn, M. G. Fitton, and D. L. J. Quicke. 2002. The limits of the family Evaniidae (Insecta: Hymenoptera) and a new genus from Lebanese amber. *Insect Systematics and Evolution* 33: 23–34.

Basset, Y. 2001. Invertebrates in the canopy of tropical rain forests: How much do we really know? *Plant Ecology* 153: 87–107.

Bauer, R. T., and J. W. Martin, eds. 1991. *Crustacean Sexual Biology.* Columbia University Press; New York, New York; ix+355 pp.

Baumann, H. 1927. Bemerkungen zur Anabiose der Tardigraden. *Zoologischer Anzeiger* 72: 1–4.

Baumann, L., P. Baumann, N. A. Moran, J. Sandstrom, and M. L. Thao. 1999. Genetic characterization of plasmids containing genes encoding enzymes of leucine biosynthesis in endosymbionts (*Buchnera*) of aphids. *Journal of Molecular Evolution* 48: 77–85.

Bawa, K. S. 1974. Breeding systems of trees species of a lowland tropical community. *Evolution* 28: 85–92.

———. 1990. Plant-pollinator interactions in tropical rain forests. *Annual Review of Ecology and Systematics* 21: 399–422.

Bayreuther, K., and S. Brauning. 1971. Die Cytogenetik der Flohe (Aphaniptera). *Chromosoma* 33: 19–29.

Baz, A. 1990. Psocoptera from weaver bird nests (Aves: Ploceidae) in Equatorial Guinea (west Africa). *Annales de la Société Entomologique de France* 26: 33–8.

———, and V. M. Ortuño. 2001. New genera and species of empheriids (Psocoptera: Empheriidae) from the Cretaceus amber of Alava, northern Spain. *Cretaceous Research* 22: 575–84.

Beaucournu, J.-C., and J. Wunderlich. 2001. A third species of *Palaeopsylla* Wagner, 1903, from Baltic amber (Siphonaptera: Ctenophthalmidae). *Entomologische Zeitschrift* 111: 296–8.

Beaver, R. A. 2002. A new species of *Ycaploca* (Hym., Scolebythidae) from Fiji. *Entomologist's Monthly Magazine* 138: 139–42.

Bechly, G. 1996. Morphologische Untersuchungen am Flügelgeäder der rezenten Libellen und deren Stammgruppenvertreter (Insecta; Pterygota; Odonata), unter besonderer Berücksichtigung Phylogenetischen Systematik und des Grundplanes der *Odonata. *Petalura, Special Volume* 2: 1–402.

———. 1997. New fossil odonates from the Upper Triassic of Italy, with a redescription of *Italophlebia gervasutti* Whalley, and a reclassification of Triassic dragonflies (Insecta: Odonata). *Rivista del Museo Civico di Scienze Naturali "Enrico Caffi", Bergamo* 19: 31–70.

———. 2001. Die faszinierende Evolution der Insekten. *Stuttgarter Beiträge zur Naturkunde, Serie C, Wissen für Alle* 49: 1–94.

———, C. Brauckmann, W. Zessin, and E. Gröning. 2001. New results concerning the morphology of the most ancient dragonflies (Insecta: Odonatoptera) from the Namurian of Hagen-Vorhalle (Germany). *Journal of Zoological Systematics and Evolutionary Research* 39: 209–26.

Beck, A. L., and C. C. Labandeira. 1998. Early Permian insect folivory on a gigantopterid-dominated riparian flora from north-central Texas. *Palaeogeography, Palaeoclimatology, Palaeoecology* 142: 139–73.

Beck, H. 1992. New view of the higher classification of the Noctuidae (Lepidoptera). *Nota Lepidopterilogica* 15: 3–28.

Beckenbach, A. T., and A. Borkent. 2003. Molecular analysis of the biting midges (Diptera: Ceratopogonidae), based on mitochondrial cytochrome oxidase subunit 2. *Molecular Phylogenetics and Evolution.* 27: 21–35.

Beckemeyer, R. J. 2000. The Permian insect fossils of Elmo, Kansas. *Kansas School Naturalist* 46: 1–15.

Becker, L., R. J. Poreda, A. G. Hunt, T. E. Bunch, and M. Rampino. 2001. Impact event at the Permian-Triassic boundary: Evidence from extraterrestrial noble gases in Fullerenes. *Science* 291: 1530–3.

Bedford, G. O. 1978. Biology and ecology of the Phasmatodea. *Annual Review of Entomology* 23: 125–49.

Bei-Bienko, G. Y. 1936. Insectes dermaptères. Pp. 1–239. *In, Payha SSSR, 5* [Fauna of the USSR,

5]. Akademii Nauk [Academy of Sciences]; Moscow, Russia; viii+240 pp. [In Russian]

Beier, M. 1968. Mantodea (Fangheuschrecken). *Handbuch der Zoologie, IV.* 2(12): 1–47.

Bekker-Migdisova, E. E. 1962. Order Blattodea. Pp. 88–111. *In* Rohdendorf, B. B. (ed.), *Fundamentals of Paleontology, Volume 9: Arthropoda: Tracheata and Chelicerata.* Akademii Nauk [Academy of Sciences]; Moscow; 561 pp. [In Russian: English translation in Rohdendorf, B. B., ed. 1991]

Bell, R. T. 1966. *Trachypachus* and the origin of the Hydradephaga. *Coleopterist's Bulletin* 20: 107–12.

Belon, P. 1555. *L'Histoire de la Nature des Oyseaux, avec Leurs Descriptions & Naïfs Portraicts Retirez du Naturel, Escrite en Sept Livres.* Guillaume Cavellat; Paris, France; 381 pp.

Beltran, M., C. D. Jiggins, V. Bull, M. Linares, *et al.* 2002. Phylogenetic discordance at the species boundary: Comparative gene genealogies among rapidly radiating Heliconius butterflies. *Molecular Biology and Evolution* 19: 2176–90.

Bendel-Janssen, M. 1977. Zur Biologie, Ökologie und Ethologie der Chalcidoidea. *Mitteilungen aus der Biologischen Bundesanstalt für Land- und Forstwirtschaft* 176: x+1–163.

Bennike, O., and J. Böcher. 1990. Forest-tundra neighbouring the North Pole: Plant and insect remains from the Plio-Pleistocene Kap Kobenhavn Formation, north Greenland. *Arctic* 43: 331–8.

Benoit, P. L. G. 1984. Stephanidae de l'Afrique occidentale et centrale (Hymenoptera). *Revue de Zoologie Africaine* 98: 215–28.

Benson, R. B. 1938. On the classification of sawflies (Hymenoptera Symphyta). *Transactions of the Royal Entomological Society, London* 87: 353–84.

———. 1943. Studies in Siricidae, especially of Europe and southern Asia (Hymenoptera, Symphyta). *Bulletin of Entomological Research* 34: 27–51.

———. 1945. Classification of the Pamphiliidae (Hymenoptera Symphyta). *Proceedings of the Royal Entomological Society of London (B)* 14: 25–53.

———. 1954. Classification of the Xiphydriidae (Hymenoptera). *Transactions of the Royal Entomological Society, London* 105: 151–62.

Benson, W. W. 1985. Amazon ant-plants. Pp. 239–66. *In* Prance, G. T., and T. E. Lovejoy (eds.), *Key Environments: Amazonia.* Pergamon Press; Oxford, UK; xiv+442 pp.

Benton, M. J., and R. J. Twitchett. 2003. How to kill (almost) all life: The end-Permian extinction event. *Trends in Ecology and Evolution* 18: 358–65.

Berenbaum, M. 1981. Patterns of furanocourmarin distribution and insect herbivory in the Umbelliferae: Plant chemistry and community structure. *Ecology* 62: 1254–66.

———. 1983. Coumarins and caterpillars: A case for coevolution. *Evolution* 37: 163–79.

———, and P. Feeny. 1981. Toxicity of angular furanocoumarins to swallowtails: Escalation in the coevolutionary arms race. *Science* 212: 927–9.

Berg, C. O., and L. Knutson. 1978. Biology and systematics of the Sciomyzidae. *Annual Review of Entomology* 23: 239–58.

Bergström, J., W. Stürmer, and G. Winter. 1980. *Palaeoisopus, Palaeopantopus,* and *Palaeothea*: Pycnogonid arthropods from the Lower Devonian Hunsrück Slate, West Germany. *Paläontologische Zeitschrift* 54: 7–54.

Berlese, A. 1913. Intorno alle metamorfosi degli insetti. *Redia* 9: 121–36.

Bernhardt, P., and L. B. Thien. 1987. Self-isolation and insect pollination in the primitive angiosperms: New evaluations of older hypotheses. *Plant Systematics and Evolution* 156: 159–76.

Berkov, A. 2002. The impact of redefined species limits in *Palame* (Coleoptera: Cerambycidae: Lamiinae: Acanthocinini) on assessments of host, seasonal, and stratum specificity. *Biological Journal of the Linnean Society* 76: 195–209.

Berner, R. A. 1997. The rise of plants and their effect on weathering and atmospheric CO_2. *Science* 276: 544–6.

———, and D. E. Canfield. 1989. A new model for atmospheric oxygen over Phanerozoic time. *American Journal of Science* 289: 333–61.

Berry, E. W. 1916. The Lower Eocene floras of southeastern North America. *United States Geological Survey Professional Papers* 91: 1–469.

———. 1931. An insect-cut leaf from the Lower Eocene. *American Journal of Science* 21: 301–4.

Bertolani, R, and D. Grimaldi. 2000. A new eutardigrade (Tardigrada: Milnesiidae) in amber from the Upper Cretaceous (Turonian) of New Jersey. Pp. 103–10. *In* Grimaldi, D. (ed.), *Studies on Fossils in Amber, with Particular Reference to the Cretaceous of New Jersey.* Backhuys Publishers; Leiden, the Netherlands; viii+498 pp.

Besch, W. 1964. Systematik und Verbreitung der südamerikanischen rheophilen Hydrachnellen. *Beiträge zur Neotropischen Fauna* 3: 77–193.

———. 1969. South American Arachnida. Pp. 723–40. *In* Fittkau, E. J., J. Illies, H. Klinge, G. H. Schwabe, and H. Sioli (eds.), *Biogeography and Ecology in South America, Volume 2.* Monographiae Biologicae. Junk; The Hague, the Netherlands; 510 pp.

Béthoux, O., and A. Nel. 2002. Venation pattern and revision of Orthoptera sensu nov. and sister groups. Phylogeny of Palaeozoic and Mesozoic Orthoptera sensu nov. *Zootaxa* 96: 1–88.

———, A. Nel, G. Gand, and J. Lapeyrie. 2001. *Surijoka lutevensis* nov. sp.: The first Glosselytrodea (Insecta) from the Upper Permian of France (Lodève Basin). *Geobios* 34: 405–13.

———, A. Nel, J. Lapeyrie, G. Gand, and J. Galtier. 2002. *Raphogla rubra* gen. n., sp. n., the oldest representative of the clade of modern Ensifera (Orthoptera: Tettigoniidea, Gryllidea). *European Journal of Entomology* 99: 111–16.

Beutel, R. G. 1998. Trachypachidae and the phylogeny of Adephaga (Coleoptera). *Bollettino del Museo Regionale di Scienze Naturali-Torino* 1998: 81–105.

———, and A. Haas. 1996. Phylogenetic analysis of larval and adult characters of Adephaga (Coleoptera) using cladistic computer programs. *Entomologica Scandinavica* 27: 197–205.

———, and F. Haas. 2000. Phylogenetic relationships of the suborders of Coleoptera (Insecta). *Cladistics* 16: 103–41.

———, and T. Hörnschemeyer. 2002. Larval morphology and phylogenetic position of *Micromalthus debilis* LeConte (Coleoptera: Micromalthidae). *Systematic Entomology* 27: 169–90.

———, D. R. Maddison, and F. Haas. 1999. Phylogenetic analysis of Myxophaga (Coleoptera) using larval characters. *Systematic Entomology* 24: 171–92.

Beutenmüller, W., and T. D. A. Cockerell. 1908. Pp. 14–19. *In* Cockerell, T. D. A., Fossil insects from Florissant, Colorado. *Bulletin of the American Museum of Natural History* 24: 59–69.

Bhatti, J. S. 1988. The orders Terebrantia and Tubulifera of the superorder Thysanoptera (Insecta). A critical appraisal. *Zoology (Journal of Pure and Applied Zoology)* 1: 167–240.

Bierbrodt, E. 1942. Der Larvenkopf von *Panorpa communis* L. und seine Verwandlung mit besonderer Berücksichtigung des Gehirns und der Augen. *Zoologische Jahrbücher, Abteilung für Anatomie und Ontogenie der Tiere* 68: 49–136.

Bignell, D. E., and P. Eggleton. 2000. Termites in ecosystems. Pp. 363–87. *In* Abe, T., D. E. Bignell, and M. Higashi (eds.), *Termites: Evolution, Sociality, Symbioses, Ecology.* Kluwer Academic Publishers; Dordrecht, the Netherlands; xxii+466 pp.

Bigot, A., A. Nel, and J. Nel. 1986. Description de la première espèce fossile connue de Ptérophore (Lepidoptera Pterophoridae). *Alexanor* 14: 283–8.

Biliński, S. M., J. Büning, and J. Simiczyjew. 1998. The ovaries of Mecoptera: Basic similarities and one exception to the rule. *Folia Histochemica et Cytobiologica* 36: 189–95.

Bin, F., and S. B. Vinson. 1986. Morphology of the antennal sex-gland in male *Trissolcus basalis* (Woll.) (Hymenoptera: Scelionidae), an egg parasitoid of the green stink bug, *Nezara viridula* (Hemiptera: Pentatomidae). *International Journal of Insect Morphology and Embryology* 15: 129–38.

Bischoff, H. 1916. Bernsteinhymenopteren. *Schriften der Physikalisch-ökonomischen Gesellschaft* 56: 141–4.

———. 1927. *Biologie der Hymenopteren; Eine Naturgeschichte der Hautflügler.* Springer; Berlin, Germany; vii+[1]+598 pp.

———. 1938. Trigonaloidae [sic]. *Hymenopterorum Catalogus* 5: 1–18.

Bitsch, C., and J. Bitsch. 1998. Internal anatomy and phylogenetic relationships among apterygote insect clades (Hexapoda). *Annales de la Société Entomologique de France* 34: 339–63.

———, and J. Bitsch. 2000. The phylogenetic interrelationships of the higher taxa of apterygote hexapods. *Zoologica Scripta* 29: 131–56.

Bitsch, J. 1994. The mophological groundplan of Hexapoda: Critical review of recent concepts. *Annales de la Société Entomologique de France* 30: 103–29.

———. 2001. The hexapod appendage: Basic structure, development and origin. *Annales de la Société Entomologique de France* 37: 175–93.

———, and A. Nel. 1999. Morphology and classification of the extinct Archaeognatha and related taxa (Hexapoda). *Annales de la Société Entomologique de France* 35: 17–29.

———, and C. Noirot. 2002. Gut characters and phylogeny of the higher termites (Isoptera: Termitidae). A cladistic analysis. *Annales de la Société Entomologique de France* 38: 201–10.

Blagoderov, V. 1993. Dipterans (Mesosciophilidae) from the Lower Cretaceous of Transbaikal. *Paleontological Journal* 27: 123–30.

———, and D. Grimaldi. 2004. Fossil Sciaroidea (Diptera) in Cretaceous ambers, exclusive of Cecidomyiidae, Sciaridae, and Keroplatidae. *American Museum Novitates*: 3433: 1–76.

Blaschke-Berthold, U. 1994. Anatomie und Phylogenie der Bibionomorpha (Insecta, Diptera). *Bonner Zoologische Monographien* 34: 1–206.

Blum, M. S. 1981. *Chemical Defenses of Arthropods*. Academic Press; New York, New York; xii+562 pp.

Blunt, W. 2001. *Linnaeus: The Compleat Naturalist*. Princeton University Press; Princeton, New Jersey; 264 pp.

Böcher, J. 1989. Boreal insect in northernmost Greenland: Palaeoentomological evidence from the Kab København Formation (Plio-Pleistocene), Peary Land. *Fauna Norvegica (B)* 36: 37–43.

———. 1990. A two-million-year-old insect fauna from north Greenland indicating boreal conditions at the Plio-Pleistocene boundary. *Proceedings of the International Conference on the Role of Polar Regions in Global Change* 2: 582–4.

———. 1995. Palaeoentomology of the Kap Kobenhavn Formation, a Plio-Pleistocene sequence in Peary Land, north Greenland. *Meddelelser om Gronland Geoscience* 33: 1–82.

———. 1997. History of the Greenland insect fauna with emphasis on living and fossil beetles. *Quaternary Proceedings* 5: 35–47.

Bode, A. 1953. Die Insektenfauna des ostniedersächsischen Oberen Lias. *Palaeontographica* 103: 1–375.

Bodenbender, B. E., and D. C. Fisher. 2001. Stratocladistic analysis of blastoid phylogeny. *Journal of Paleontology* 75: 351–69.

Bohart, G. E. 1970. *The Evolution of Parasitism Among Bees*. Utah State University (41st Honor Lecture); Logan, Utah; ii+30 pp.

Bohart, R. M., and A. S. Menke. 1976. *Sphecid Wasps of the World: A Generic Revision*. University of California Press; Berkeley, California; ix+695 pp.

Bolton, B. 1994. *Identification Guide to the Ant Genera of the World*. Harvard University Press; Cambridge, Massachusetts; 222 pp.

———. 1995. *A New General Catalogue of the Ants of the World*. Harvard University Press; Cambridge, Massachusetts; 504 pp.

———. 2003. Synopsis and classification of Formicidae. *Memoirs of the American Entomological Institute* 71: 1–370.

Bolton, H. 1916. On some insects from the British Coal Measures. *Quarterly Journal of the Geological Society of London* 72: 43–62.

———. 1917. The "Mark Stirrup" collection of fossil insects from the Coal Measures of Commentry (Allier), central France. *Memoirs and Proceedings of the Manchester Literary and Philosophical Society* 61: 1–24.

Boore, J. L., T. M. Collins, D. Stanton, L. L. Daehler, and W. M. Brown. 1995. Deducing the pattern of arthropod phylogeny from mitochondrial DNA rearrangements. *Nature* 376: 163–5.

———, D. V. Lavrov, and W. M. Brown. 1998. Gene translocation links insects and crustaceans. *Nature* 392: 667–8.

Borkent, A. 1995. *Biting midges in the Cretaceous Amber of North America (Diptera: Ceratopogonidae)*. Backhuys Publishers; Leiden, the Netherlands; 237 pp.

———. 1996. Biting midges from Upper Cretaceous New Jersey amber (Ceratopogonidae: Diptera). *American Museum Novitates* 3159: 1–29.

———. 1997. Upper and Lower Cretaceous biting midges (Ceratopogonidae: Diptera) from Hungarian and Austrian amber and the Koonwarra Fossil Bed of Australia. *Stuttgarter Beiträge zur Naturkunde, serie B, Geologie und Paläontologie* 249: 1–10.

———. 2000a. Biting midges (Ceratopogonidae: Diptera) from Lower Cretaceous Lebanese amber with a discussion of the diversity and patterns found in other ambers. Pp. 355–451. *In* Grimaldi, D. (ed.), *Studies on Fossils in Amber, with Particular Reference to the Cretaceous of New Jersey*. Backhuys Publishers; Leiden, the Netherlands; viii+498 pp.

———. 2000b. Further biting midges (Diptera: Ceratopogonidae) from Upper Cretaceous New Jersey amber. Pp. 453–72. *In* Grimaldi, D. (ed.), *Studies on Fossils in Amber, with Particular Reference to the Cretaceous of New Jersey*. Backhuys Publishers; Leiden, the Netherlands; viii+498 pp.

———. 2001. *Leptoconops* (Diptera: Ceratopogonidae), the earliest extant lineage of biting midge, discovered in 120–122 million-year-old Lebanese amber. *American Museum Novitates* 3328: 1–11.

———, and W. W. Wirth. 1997. World species of biting midges (Diptera: Ceratopogonidae). *Bulletin of the American Museum of Natural History* 233: 1–257.

Börner, C. 1904. Zur Systematik der Hexapoden. *Zoologischer Anzeiger* 27: 511–33.

———. 1909. Neue Homologien zwischen Crustaceen und Hexapoden. Die Beissmandibel der Insekten und ihre phylogenetische Bedeutung. Archi- und Metapterygota. *Zoologischer Anzeiger* 34: 100–25.

———. 1910. Die phylogenetische Bedeutung der Protura. *Biologisches Zentralblatt* 30: 636–41.

———. 1919. Stammesgeschichte der Hautflügler. *Biologisches Zentralblatt* 39: 145–86.

———. 1934. Ueber System und Stammesgeschichte der Schnabelkerfe. *Entomologische Beihefte* 1: 138–44.

Borror, D. J., C. A. Triplehorn, and N. F. Johnson. 1989. *An Introduction to the Study of Insects* [6th Edition]. Saunders College Publishing; Philadelphia, Pennsylvania; xiv+875 pp.

Borucki, H. 1996. Evolution und phylogenetisches System der Chilopoda (Mandibulata, Tracheata). *Verhandlungen des Naturwissenschaftlichen Vereins in Hamburg* 35: 5–31.

Botha, B. J. V., and W. Linstrom. 1978. A note on the stratigraphy of the Beaufort Group in north-western Natal. *Verhandelinge van die Geologiese Vereniging van Suid Afrika* 81: 35–40.

Botz, J. T., C. Loudon, J. B. Barger, J. S. Olafsen, and D. W. Steeples. 2003. Effects of slope and particle size on ant locomotion: Implications for choice of substrate by antlions. *Journal of the Kansas Entomological Society* 76: 426–35.

Boucot, A. J., L. M. Cumming, and H. Jaeger. 1967. Contributions to the age of the Gaspé Sandstone and Gaspé Limestone. *Geological Survey of Canada* Paper 67-25: 1–22.

Boudreaux, H. B. 1979. *Arthropod Phylogeny, with Special Reference to Insects*. Wiley; New York, New York; viii+320 pp.

———. 1980. Proventricular acanthae and their phylogenetic implications. *Annals of the Entomological Society of America* 73: 189–96.

Bourgoin, T., J. D. Steffan-Campbell, and B. C. Campbell. 1997. Molecular phylogeny of Fulgoromorpha (Insecta, Hemiptera, Auchenorrhyncha). The enigmatic Tettigometridae: Evolutionary affiliation and historical biogeography. *Cladistics* 13: 207–24.

Bowden, S. R. 1985. Joseph Grimaldi and the Adonis blue butterfly, *Lysandra bellargus* (Rottemburg). *Entomologist's Gazette* 36: 193–5.

Bowe, L. M., G. Coat, and C. W. DePamphilis. 2000. Phylogeny of seed plants based on all three genomic compartments: Extant gymnosperms are monophyletic and Gnetales closest relatives are conifers. *Proceedings of the National Academy of Sciences, U.S.A.* 97: 4092–7.

Bowring, S. A., D. H. Erwin, Y. G. Jin, M. W. Martin, K. Davidek, and W. Wang. 1998. U/Pb Zircon geochronology and tempo of the End-Permian mass extinction. *Science* 280: 1039–45.

Boyd, B. M., and R. M. Pyle, eds. 2000. *Nabokov's Butterflies*. Beacon Press; Boston, Massachusetts; xiv+782 pp.

Bradler, S. 1999. The vomer of *Timema* Scudder, 1895 (Insecta: Phasmatodea) and its significance for phasmatodean phylogeny. *Courier Forschungsinstitut Senckenberg* 215: 43–7.

———. 2003. Phasmatodea, Gespenstschrecken. Pp. 251–61. *In* Dathe, H. H. (ed.), *Lehrbuch der Speziellen Zoologie, Band I: Wirbellose Tiere. 5. Teil: Insecta*. Spektrum Akademischer Verlag; Heidelberg, Germany; xii+[1]+961 pp.

Bradley, J. C., and B. S. Galil. 1977. The taxonomic arrangement of the Phasmatodea with keys to subfamilies and tribes. *Proceedings of the Entomological Society of Washington* 79: 176–208.

Bradley, J. D. 1982. Two new species of moths (Lepidoptera, Pyralidae, Chrysauginae) associated with the three-toed sloth (*Bradypus* spp.) in South America. *Acta Amazonica* 12: 649–56.

Brady, R. H. 1985. On the independence of systematics. *Cladistics* 1: 113–26.

———. 1994. Pattern description, process explanation, and the history of morphological sciences. Pp. 7–31. *In* Grande, L., and O. Rieppel (eds.), *Interpreting the Hierarchy of Nature: From Systematic Patterns to Evolutionary*

Process Theories. Academic Press; San Diego, California; ix+298 pp.

Brady, S. G. 2003. Evolution of the army ant syndrome: The origin and long-term evolutionary stasis of a complex of behavioral and reproductive adaptations. *Proceedings of the National Academy of Sciences, U.S.A.* 100: 6575–9.

Brandão, C. R. F., R. G. Martins-Neto, and M. A. Vulcano. 1989. The earliest known fossil ant (first Southern Hemisphere Mesozoic record) (Hymenoptera: Formicidae: Myrmecinae). *Psyche* 96: 195–208.

Branham, M. A., and J. W. Wenzel. 2001. The evolution of bioluminescence in cantharoids (Coleoptera: Elateroidea). *The Florida Entomologist* 84: 565–86.

———, and J. W. Wenzel. 2003. The origin of photic behavior and the evolution of sexual communication in fireflies (Coleoptera: Lampyridae). *Cladistics* 19: 1–22.

Brauckmann, C. 1984. Weitere neue Insekten (Palaeodictyoptera; Protorthoptera) aus dem Namurium B von Hagen-Vorhalle. *Jahresberichte des Naturwissenschaftlichen Vereins in Wuppertal* 37: 108–15.

———. 1986. Eine neue Spilapteriden-Art aus dem Namurium B von Hagen-Vorhalle (Insecta: Palaeodictyoptera: Ober-Karbon; West Deutschland). *Dortmunder Beiträge zur Landeskunde* 20: 57–64.

———. 1988. Hagen-Vorhalle, a new important Namurian Insecta-bearing locality (Upper Carboniferous; FR Germany). *Entomologia Generalis* 14: 73–9.

———. 1991. Arachniden und Insekten aus dem Namurium von Hagen-Vorhalle (Ober-Karbon; West-Deutschland). *Veröffentlichungen aus dem Fuhlrott-Museum* 1: 1–275.

———, and L. Koch. 1982. Neue Insekten aus den Vorhalle-Schichten (oberes Namurium B) von Hagen-Vorhalle. *Dortmunder Beiträge zur Landeskunde* 16: 15–25.

———, and L. Koch. 1983. *Prothelyphonus naufragus* n. sp.: Ein neuer Geisselscorpion (Arachnida: Thelyphonida: Thelyphonidae) aus dem Namurium (unteres Ober-Karbon) von West-Deutschland. *Entomologica Generalis* 9: 63–73.

———, and L. Koch. 1994. Spinnentiere und Insekten aus dem Oberkarbon von Hagen-Vorhalle. *Fossilien* 11: 45–55.

———, B. Brauckmann, and E. Gröning. 1996. The stratigraphical position of the oldest known Pterygota (Insecta, Carboniferous, Namurian). *Annales de la Société géologique de Belgique* 117: 47–56.

———, B. Brauckmann, and E. Gröning. 2001. Anmerkungen zu den bisher beschreibenen Lepidopteren as dem Jung-Tertiär (Pliozän) von Willershausen am Harz. *Jahresberichte des Naturwissenschaftlichen Vereins in Wuppertal* 54: 31–41.

———, L. Schöllmann, and W. Sippel. 2003. Die fossilen Insekten, Spinnentiere und Eurypteriden von Hagen-Vorhalle. *Geologie und Paläontologie in Westfalen* 59: 1–89.

Brauer, F. 1885. Systematische-Zoologische Studien. *Sitzungsberichte der Königliche Akademie der Wissenschaften in Wien, Mathematische-Naturwissenschaftliche Klasse* 91: 237–431.

Brauman, A., M. D. Kane, M. Labat, and J. A. Breznak. 1992. Genesis of acetate and methane by gut bacteria of nutritionally diverse termites. *Science* 257: 1384–7.

Bremer, K. 2000. Early Cretaceous lineages of monocot flowering plants. *Proceedings of the National Academy of Sciences, U.S.A.* 97: 4707–711.

Breznak, J. A. 2000. Ecology of prokaryotic microbes in the guts of wood- and litter-feeding termites. Pp. 209–31. *In* Abe, T., D. E. Bignell, and M. Higashi (eds.), *Termites: Evolution, Sociality, Symbioses, Ecology*. Kluwer Academic Publishers; Dordrecht, the Netherlands; xxii+466 pp.

———, and A. Brune. 1994. Role of microorganisms in the digestion of lignocellulose by termites. *Annual Review of Entomology* 39: 453–87.

Bridwell, J. C. 1958. Biological notes on *Ampulicomorpha confusa* Ashmead and its fulgoroid host. *Proceedings of the Entomological Society of Washington* 60: 23–6.

Briggs, D. E. G. 1985. Gigantism in Palaeozoic arthropods. *Special Papers in Paleontology* 33: 157.

———. 1995. Experimental taphonomy. *Palaios* 10: 539–50.

———. 1999. Molecular taphonomy of animal and plant cuticles: Selective preservation and diagenesis. *Philosophical Transactions of the Royal Society, London (B)* 354: 7–17.

———, and C. Bartels. 2001. New arthropods from the Lower Devonian Hunsrück Slate (Lower Emsian, Rhenish Massif, western Germany). *Palaeontology* 44: 275–303.

———, D. L. Bruton, and H. B. Whittington. 1979. Appendages of the arthropod *Aglaspis spinifer* (Upper Cambrian, Wisconsin) and their significance. *Palaeontology* 22: 167–80.

———, R. P. Evershed, and B. A. Stankiewicz. 1998a. The molecular preservation of fossil arthropod cuticles. *Ancient Biomolecules* 2: 135–46.

———, B. A. Stankiewicz, D. Meischner, A. Bierstedt, and R. P. Evershed. 1998b. Taphonomy of arthropod cuticles from Pliocene lake sediments, Willershausen, Germany. *Palaios* 13: 386–94.

Briggs, J. C. 1995. *Global Biogeography*. Elsevier; Amsterdam, the Netherlands; xvii+454 pp.

———. 2003. Fishes and birds: Gondwana life rafts reconsidered. *Systematic Biology* 52: 548–53.

Britten, R. J. 1986. Rates of DNA sequence evolution differ between taxonomic groups. *Science* 231: 1393–8.

Broadhead, E. 1950. A revision of the genus *Liposcelis* Motschulsky with notes on the position of this genus in the order Corrodentia and on the variability of ten *Liposcelis* species. *Transactions of the Royal Entomological Society, London* 101: 335–88.

———, and B. M. Hobby. 1944. Studies on a species of *Liposcelis* (Corrodentia, Liposcelidae) occurring in stored products in Britain: Parts 1 and 2. *Entomologists' Monthly Magazine* 80: 49–59, 163–73.

Brock, P. D. 1999. The amazing world of stick and leaf-insects. *The Amateur Entomologist* 26: 1–165.

Brodie, P. B. 1845. *A History of the Fossil Insects in the Secondary Rocks of England, Accompanied by a Particular Account of the Strata in Which They Occur, and of the Circumstances Connected with Their Preservation*. J. Van Voorst; London, UK; xviii+130 pp, 11 pls.

Brodsky, A. K. 1994. *The Evolution of Insect Flight*. Oxford University Press; Oxford, UK; xiv+229 pp.

Bromham, L., A. Rambaut, R. Fortey, A. Cooper, and D. Penny. 1998. Testing the Cambrian explosion hypothesis by using a molecular dating technique. *Proceedings of the National Academy of Sciences, USA* 95: 12386–9.

Brongniart, C. 1878. Note suivante sur la découverte d'un orthoptère coureur de la famille des Phasmiens (*Protophasma dumasi*) dans les terrains supra-houillers de Commentry (Allier). *Bulletin des Seances de la Société Entomologique de France* 47: 57–60.

———. 1884. Sur un gigantesque néurorthoptère, provenant des terrains houillers de Commentry (Allier). *Comptes Rendus Hebdomadaires des Seances de l'Academie des Sciences* 98: 832–3.

———. 1885a. Les insectes fossiles des terrains primaires. Coup d'oeil rapide sur la faune entomologique des terrains paléozoïques. *Bulletin de la Société des Amis des Sciences Naturelles de Rouen* 1885: 50–68.

———. 1885b. [*Dasyleptus lucasi*]. *Bulletin des Seances de la Société Entomologique de France* 1885: ci–cii.

———. 1893. *Recherches pour servir à l'Histoire des Insectes Fossiles des Temps Primaires, Précédées d'une Étude sur la Nervation des Ailes des Insectes*. Thèse présentée à la Faculté des Sciences de Paris, no. 821; Paris, France; 495 pp., 37 pls.

Brooks, D. R., and D. A. McLennan. 1993. *Parascript: Parasites and the Language of Evolution*. Smithsonian Institution Press; Washington, D.C.; x+429 pp.

———, and R. T. O'Grady. 1989. Crocodilians and their helminth parasites: Macroevolutionary considerations. *American Zoologist* 29: 873–83.

Brooks, H. K. 1955. Healed wounds and galls on fossil leaves from the Wilcox deposits (Eocene) of western Tennessee. *Psyche* 62: 1–9.

Brooks, S. J. 1997. An overview of the current status of Chrysopidae (Neuroptera) systematics. *Deutsche Entomologische Zeitschrift* 44: 267–75.

———, and P. C. Barnard. 1990. The green lacewings of the world: A generic review (Neuroptera: Chrysopidae). *Bulletin of the British Museum of Natural History (Entomology)* 59: 117–286.

Brothers, D. J. 1972. Biology and immature stages of *Pseudomethoca f. frigida*, with notes on other species (Hymenoptera: Mutillidae). *University of Kansas Science Bulletin* 50: 1–38.

———. 1974. The genera of Plumariidae, with description of a new genus and species from Argentina (Hymenoptera: Bethyloidea). *Journal of the Entomological Society of Southern Africa* 37: 351–6.

———. 1975. Phylogeny and classification of the aculeate Hymenoptera, with special reference to Mutillidae. *University of Kansas Science Bulletin* 50: 483–648.

———. 1978. Biology and immature stages of *Myrmosula parvula* (Hymenoptera: Mutillidae).

Journal of the Kansas Entomological Society 51: 698–710.

———. 1981. Note on the biology of *Ycaploca evansi* (Hymenoptera: Scolebythidae). *Journal of the Entomological Society of Southern Africa* 44: 107–8.

———. 1989. Alternative life-history styles of mutillid wasps (Insecta: Hymenoptera). Pp. 279–91. *In* Bruton, M. N. (ed.), *Alternative Life-History Styles of Animals*. Kluwer Academic Publishers; Dordrecht, the Netherlands; xvii+616 pp.

———. 1992. The first Mesozoic Vespidae (Hymenoptera) from the Southern Hemisphere, Botswana. *Journal of Hymenoptera Research* 1: 119–24.

———. 1995a. Mutillidae. Pp. 541–8. *In* Hanson, P. E., and I. D. Gauld (eds.), *The Hymenoptera of Costa Rica*. Oxford University Press; Oxford, UK; xx+893 pp.

———. 1995b. Bradynobaenidae. Pp. 552–5. *In* Hanson, P. E., and I. D. Gauld (eds.), *The Hymenoptera of Costa Rica*. Oxford University Press; Oxford, UK; xx+893 pp.

———. 1999. Phylogeny and evolution of wasps, ants and bees (Hymenoptera, Chrysidoidea, Vespoidea and Apoidea). *Zoologica Scripta* 28: 233–49.

———. 2003. The first fossil Ephutini (Hymenoptera: Mutillidae), a new species of *Ephuta* Say from Dominican amber. *Acta Zoologica Cracoviensia* 46 (supplement): 101–7.

———, and J. M. Carpenter. 1993. Phylogeny of Aculeata: Chrysidoidea and Vespoidea (Hymenoptera). *Journal of Hymenoptera Research* 2: 227–304.

———, and J.-W. Janzen. 1999. New generic synonymy in Scolebythidae, with redescription of both sexes of *Pristapenesia primaeva* Brues from Baltic amber (Hymenoptera: Chrysidoidea). Pp. 17–26. *In* Vršanský, P. (ed.), *Proceedings of the First Palaeoentomological Conference, Moscow, 1998*. AMBA Projects; Bratislava, Slovakia; 199+[2] pp.

Brower, A.V. Z. 1997. The evolution of ecologically important characters in *Heliconius* butterflies (Lepidoptera: Nymphalidae): A cladistic review. *Zoological Journal of the Linnean Society* 119: 457–72.

———. and M. G. Egan. 1997. Cladistic analysis of *Heliconius* butterflies and relatives (Nymphalidae: Heliconiiti): A revised phylogenetic position for *Eueides* based on sequences from mtDNA and a nuclear gene. *Proceedings of the Royal Society of London, Series B, Biological Sciences* 264: 969–77.

Brown, B. V. 1992. Generic revision of Phoridae of the Nearctic Region and phylogenetic classification of Phoridae, Sciadoceridae, and Ironomyiidae (Diptera: Phoridae). *Memoirs of the Entomological Society of Canada* 164: 1–144.

———. 1999. Re-evaluation of the fossil Phoridae (Diptera). *Journal of Natural History* 33: 1561–73.

———, and E. M. Pike. 1990. Three new fossil phorid flies (Diptera: Phoridae) from Canadian Late Cretaceous amber. *Canadian Journal of Earth Sciences* 27: 845–48.

Brown, F. M. 1976. *Oligodonta florissantensis* gen. n., sp. nov. (Lepidoptera: Pieridae). *Bulletin of the Allyn Museum* 37: 1–4.

———. 1986. *Lithosirex williamsi* (Siricidae: Hymenoptera) a newly recognized fossil horntail from Florissant, Colorado. *Insecta Mundi* 1: 119–20.

Brown, R. W. 1941. The comb of a wasp nest from the upper Cretaceous of Utah. *American Journal of Science* 239: 54–6.

———. 1957. Cockroach egg case from the Eocene of Wyoming. *Journal of the Washington Academy of Sciences* 47: 340–2.

Brown, V. K. 1973. The biology and development of *Brachygaster minutus* Olivier (Hymenoptera: Evaniidae), a parasite of the oothecae of *Ectobius* spp. (Dictyoptera: Blattidae). *Journal of Natural History* 7: 665–74.

Brown, W. L., Jr. 1954. Remarks on the internal phylogeny and subfamily classification of the family Formicidae. *Insectes Sociaux* 1: 21–31.

———. 1973. A comparison of the Hylean and Congo-West African rain forest ant faunas. Pp. 161–85. *In* Meggers, B. J., E. S. Ayensu, and W. D. Duckworth (eds.), *Tropical Forest Ecosystems in Africa and South America: A Comparative Review*. Smithsonian Institution Press; Washington, D.C.; viii+350 pp.

Browne, J., and C. H. Scholtz. 1998. Evolution of the scarab hindwing articulation and wing base: A contribution toward the phylogeny of the Scarabaeidae (Scarabaeoidea: Coleoptera). *Systematic Entomology* 23: 307–26.

———, and C. H. Scholtz. 1999. A phylogeny of the families of Scarabaeoidea (Coleoptera). *Systematic Entomology* 24: 51–84.

Brues, C. T. 1908. New phytophagous Hymenoptera from the Tertiary of Florissant, Colorado. *Bulletin of the Museum of Comparative Zoology* 51: 259–76.

———. 1910. The parasitic Hymenoptera of the Tertiary of Florissant, Colorado. *Bulletin of the Museum of Comparative Zoology* 54: 1–125.

———. 1923a. Some new fossil parasitic Hymenoptera from Baltic amber. *Proceedings of the American Academy of Arts and Sciences* 58: 327–46.

———. 1923b. A fossil genus of Dinapsidae from Baltic amber (Hymenoptera). *Psyche* 30: 31–5.

———. 1926. A species of *Urocerus* from Baltic amber. *Psyche* 33: 168–9.

———. 1933a. The parasitic Hymenoptera of the Baltic amber. Part 1. *Bernstein Forschungen* 3: 4–178.

———. 1933b. Progressive change in the insect population of forests since the Early Tertiary. *American Naturalist* 67: 385–406.

———. 1936. Evidences of insect activity preserved in fossil wood. *Journal of Paleontology* 10: 637–43.

———. 1937. Superfamilies Ichneumonoidea, Serphoidea, and Chalcidoidea. Pp. 27–44. *In* Carpenter, F. M., J. W. Folsom, E. O. Essig, A. C. Kinsey, C. T. Brues, *et al.* (eds.), *Insects and Arachnids from Canadian Amber*. University of Toronto Press (University of Toronto Geological Series 40); Toronto, Canada, 62 pp.

———. 1939. New Oligocene Braconidae and Bethylidae from Baltic amber. *Annals of the Entomological Society of America* 32: 251–63.

———. 1940. Fossil parasitic Hymenoptera of the family Scelionidae from Baltic amber. *Proceedings of the American Academy of Arts and Sciences* 74: 69–90.

———, and A. L. Melander. 1915. *Key to the Families of North American Insects; An Introduction to the Classification of Insects*. Publisher by the authors; Boston, Massachusetts; vii+140 pp.

———, and A. L. Melander. 1932. Classification of insects; a key to the known families of Insects and other terrestrial arthropods. *Bulletin of the Museum of Comparative Zoology* 73: 1–672.

———, A. L. Melander, and F. M. Carpenter. 1954. Classification of insects: Keys to the living and extinct families of insects, and to the living families of other terrestrial arthropods. *Bulletin of the Museum of Comparative Zoology* 108: 1–917.

Brundin, L. 1965. On the real nature of transantarctic relationships. *Evolution* 19: 496–505.

———. 1966. Transantarctic relationships and their significance, as evidenced by chironomid midges with a monograph of the subfamilies Podonominae and Aphroteniinae and the austral Heptagyiae. *Kunglica Svenska Vetenskapsakademiens Handlingar* 11: 1–472.

———. 1967. Insects and the problem of austral disjunctive distribution. *Annual Review of Entomology* 12: 149–68.

Brusca, R. C., and G. J. Brusca. 1990. *Invertebrates*. Sinauer Associates; Sunderland, Massachusetts; xvii+[1]+922 pp.

Bryk, F. 1934. Lepidoptera: Baroniidae, Teinopalpidae, Parnassiidae, pars 1. *Das Tierreich* 64: 1–132.

Brynnel, E. U., C. G. Kurland, N. A. Moran, and G. E. Andersson. 1998. Evolutionary rates for *tuf* genes in endosymbionts of aphids. *Molecular Biology and Evolution* 15: 574–82.

Buchmann, S. L., and G. P. Nabhan. 1996. *The Forgotten Pollinators*. Island Press; Washington, D.C.; xx+292 pp.

Buckley, R. 1987. Ant-plant-homopteran interactions. *Advances in Ecological Research* 16: 53–85.

Budrys, E. 1993. Digger wasps of the subfamily Pemphredoninae (Hymenoptera, Sphecidae) from the Baltic and Taimyr amber. *Acta Entomologica Lituanica* 11: 34–56.

Budyko, M. I., A. B. Ronov, and A. L. Yanshin. 1987. *History of the Earth's Atmosphere*. Springer Verlag; Berlin, Germany; vi+139 pp.

Büning, J. 1998. The ovariole: Structure, type and phylogeny. Pp. 897–932. *In* Harrison, F. W., and M. Locke (eds.), *Microscopic Anatomy of Invertebrates* [Volume 11C]. Wiley-Liss; New York, New York; xxvi+843–1296 pp.

Burdick, D. J., and M. S. Wasbauer. 1959. Biology of *Methoca californica* Westwood (Hymenoptera: Tiphiidae). *Wasmann Journal of Biology* 17: 75–88.

Burk, T. 1982. Evolutionary significance of predation on sexually-signalling males. *Florida Entomologist* 65: 90–104.

Burke, H. E. 1917. *Oryssus* is parasitic. *Proceedings of the Entomological Society of Washington* 19: 87–9.

Burke, J. 1985. *The Day the Universe Changed*. Little, Brown, and Company; Boston, Massachusetts; 352 pp.

Burks, B. D. 1953. The mayflies, or Ephemeroptera, of Illinois. *Bulletin of the Illinois Natural History Survey* 26: 1–216.

Burleigh, R., and P. E. S. Whalley. 1983. On the relative geological ages of amber and copal. *Journal of Natural History* 17: 919–21.

Burnham, L. 1978. Survey of social insects in the fossil record. *Psyche* 85: 85–133.

———. 1981. Fossil insects from Montceau-les-Mines (France): A preliminary report. *Bulletin Trimestriel de la Société d'Histoire Naturelle et des Amis du Museum d'Autun* 100: 5–12.

Burnside, C. A., P. T. Smith, and S. Kambhampati. 1999. Three new species of the wood roach, *Cryptocercus* (Blattodea: Cryptocercidae), from the eastern United States. *Journal of the Kansas Entomological Society* 72: 361–78.

Burr, M. 1911. Dermaptera (earwigs) preserved in amber from Prussia. *Transactions of the Linnean Society of London* 11: 145–50.

Bush, G. L. 1969. Sympatric host race formation and speciation in frugivorous flies of the genus *Rhagoletis* (Diptera, Tephritidae). *Evolution* 23: 237–51.

———. 1974. The mechanism of sympatric host race formation in the true fruit flies (Tephritidae). Pp. 3–23. *In* White, M. J. D. (ed.), *Genetic Mechanisms of Speciation in Insects*. Australian and New Zealand Book Company; Sydney, Australia; 170 pp.

———. 1975. Modes of animal speciation. *Annual Review of Ecology and Systematics* 6: 339–64.

Busvine, J. R. 1976. *Insects, Hygiene, and History*. Athlone Press; London, UK; vii+262 pp.

Butler, C. 1609. *The Feminine Monarchie, or a Treatise Concerning Bees, and the Due Ordering of Them, &c.* Joseph Barnes; Oxford, U K; 216 pp.

Buttel-Reepen, H., von. 1906. Apistica. Beiträge zur Systematik, Biologie, sowie zur geschichtlichen und geographischen Verbreitung der Honigbiene (*Apis mellifica* L.), ihrer Varietäten und der überigen *Apis*-Arten. *Mitteilungen aus dem Zoologischen Museum in Berlin* 3: 117–201.

Byers, G. W. 1983. The crane fly genus *Chionea* in North America. *University of Kansas Science Bulletin* 52: 59–195.

———. 1987. Order Mecoptera. Pp. 246–52. *In* Stehr, F. W. (ed.), *Immature Insects*. Kendall-Hunt; Dubuque, Iowa; xiv+754 pp.

———. 1989. The Nannochoristidae of South America (Mecoptera). *University of Kansas Science Bulletin* 54: 25–34.

———. 1991. Mecoptera (scorpion-flies, hanging-flies). Pp. 696–704. *In* Naumann, I. D. (ed.), *The Insects of Australia: A Textbook for Students and Research Workers, Volume 2* [2nd Edition]. Cornell University Press; Ithaca, New York; [2]+543–1137 pp.

———, and R. Thornhill. 1983. Biology of the Mecoptera. *Annual Review of Entomology* 28: 203–28.

Cade, W. 1975. Acoustically orienting parasitoids: Fly phonotaxis to cricket song. *Science* 190: 1312–13.

Calaby, J. H., and M. D. Murray. 1991. Phthiraptera (lice). Pp. 421–8. *In* Naumann, I. D. (ed.), *The Insects of Australia: A Textbook for Students and Research Workers, Volume 1* [2nd Edition]. Cornell University Press; Ithaca, New York; xvi+[1]+542 pp.

Caldas, E. B., R. G. Martins-Neto, and F. P. Lima Filho. 1989. *Afropollis* sp. (pollen) no trato intestinal de *Vespa* (Hymenoptera: Apocrita: Xyelidae) no Cretaceo da Bacia do Araripe. *Atas Simposio de Geologia do Nordeste, Fortalesa* 13: 195.

Callan, E. McC. 1939. A note on the breeding of *Probethylus callani* Richards (Hymenopt., Bethylidae), an embiopteran parasite. *Proceedings of the Royal Entomological Society of London (B)* 8: 223–4.

Camargo, J. M. F., and D. W. Roubik. 1991. Systematics and bionomics of the apoid obligate necrophages: The *Trigona hypogea* group (Hymenoptera: Apidae: Meliponinae). *Biological Journal of the Linnean Society* 44: 13–39.

———, D. A. Grimaldi, and S. R. M. Pedro. 2000. The extinct fauna of stingless bees (Hymenoptera: Apidae: Meliponini) in Dominican amber: Two new species and redescription of the male of *Proplebeia dominicana* (Wille and Chandler). *American Museum Novitates* 3293: 1–24.

Camatini, M., ed. 1979. *Myriapod Biology*. Academic Press; London, UK; xviii+456 pp.

Cameron, S. A. 1993. Multiple origins of advanced eusociality in bees inferred from mitochondrial DNA sequences. *Proceedings of the National Academy of Sciences, U.S.A.* 90: 8687–91.

Campbell, B., J. Heraty, J.-Y. Rasplus, K. Chan, J. Steffan-Campbell, and C. Babcock. 2000. Molecular systematics of the Chalcidoidea using 28S-D2 rDNA. Pp. 59–73. *In* Austin, A. D., and M. Dowton (eds.), *Hymenoptera: Evolution, Biodiversity and Biological Control*. CSIRO; Collingwood, Australia; xi+468 pp.

Campbell, B. C., J. D. Steffen-Campbell, and R. J. Gill. 1994. Evolutionary origin of whiteflies (Hemiptera: Sternorrhyncha: Aleyrodidae) inferred from 18S rDNA sequences. *Insect Molecular Biology* 3: 73–89.

———, J. D. Steffen-Campbell, J. T. Sorensen, and R. J. Gill. 1995. Paraphyly of Homoptera and Auchenorrhyncha inferred from 18S rDNA nucleotide sequences. *Systematic Entomology* 20: 175–94.

Campbell, D. L., A. V. Z. Brower, and N. E. Pierce. 2000. Molecular evolution of the wingless gene and its implications for the phylogenetic placement of the butterfly family Riodinidae (Lepidoptera: Papilionoidea). *Molecular Biology and Evolution* 17: 684–96.

Campbell, I. C. 1987. *Mayflies and Stoneflies: Life Histories and Biology*. Kluwer Academic Publishers; Dordrecht, the Netherlands; ix+366 pp.

Campos, D. R. B. 1986. Primeiro registro fóssil de Scorpionoidea na Chapado do Araripe (Cretaceo Inferior), Brasil. *Anais da Academia Brasileira de Ciencias* 58: 135–7.

Cano, R. J., and M. K. Borucki. 1995. Revival and identification of bacterial spores in 25- to 40-million-year-old Dominican amber. *Science* 268: 1060–4.

———, H. N. Poinar, D. W. Roubik, and G. O. Poinar, Jr. 1992. Enzymatic amplification and nucleotide sequencing of portions of the 18S rRNA gene of the bee *Proplebeia dominicana* (Apidae: Hymenoptera) isolated from 25–40 million year old Dominican amber. *Medical Science Research* 20: 619–22.

———, H. N. Poinar, N. J. Pieniazek, A. Acra, and G. O. Poinar, Jr. 1993. Amplification and sequencing of DNA from a 120–135-million-year-old weevil. *Nature* 363: 536–8.

Caputo, M. V. 1985. Late Devonian glaciation in South America. *Palaeogeography, Palaeoclimatology, Palaeoecology* 51: 291–317.

Carapelli, A., F. Frati, F. Nardi, and C. Simon. 1998. Molecular phylogeny of apterygotan insects based on nuclear and mitochondrial genes. *Proceedings of the 5th International Symposium on Apterygota* 1998: 39.

Carle, F. L. 1982. The wing vein homologies and phylogeny of the Odonata: A continuing debate. *Societas Internationalis Odonataologica Rapid Communications* 4: 1–66.

Carlquist, S. 1980. *Hawaii: A Natural History* [2nd Edition]. Pacific Tropical Botanical Garden; Honolulu, Hawaii; vii+468 pp.

Carlson, R. W. 1979. Superfamily Evanioidea. Pp. 1109–18. *In* Krombein, K. V., P. D. Hurd, Jr., D. R. Smith, and B. D. Burks (eds.), *Catalog of Hymenoptera in American North of Mexico*. Smithsonian Institution Press; Washington, D.C.; xvi+2735 pp. [3 volumes]

Carmean, D. 1991. Biology of the Trigonalyidae (Hymenoptera), with notes on the vespine parasitoid *Bareogonalos canadensis*. *New Zealand Journal of Zoology* 18: 209–14.

———, and B. J. Crespi. 1995. Do long branches attract flies? *Nature* 373: 666.

———, and L. Kimsey. 1998. Phylogenetic revision of the parasitoid wasp family Trigonalidae [sic] (Hymenoptera). *Systematic Entomology* 23: 35–76.

Carpenter, F. M. 1929. A Jurassic neuropteran from the lithographic limestone of Bavaria. *Psyche* 36: 190–4.

———. 1930. The fossil ants of North America. *Bulletin of the Museum of Comparative Zoology* 70: 1–66.

———. 1931. The Lower Permian insects of Kansas. Part 2. The orders Palaeodictyoptera, Protodonata, and Odonata. *American Journal of Science* 21: 97–139.

———. 1932. Jurassic insects from Solnhofen in the Carnegie Museum and the Museum of Comparative Zoology. *Annals of the Carnegie Museum* 21: 97–129.

———. 1933. A new megasecopteron from the Carboniferous of Kansas. *University of Kansas Science Bulletin* 21: 365–7.

———. 1935. Tertiary insects of the family Chrysopidae. *Journal of Paleontology* 9: 259–71.

———. 1938a. Two Carboniferous insects from the vicinity of Mazon Creek, Illinois. *American Journal of Science* 36: 445–52.

———. 1938b. Andreas Vassilievitch Martynov. *Psyche* 45: 80–3.

———. 1939. The Lower Permian insects of Kansas. Part 8. Additional Megasecoptera, Protodonata, Odonata, Homoptera, Psocoptera, Protelytroptera, Plectoptera, and Protoperlaria. *Proceedings of the American Academy of Arts and Sciences* 73: 29–70.

———. 1940. A revision of the Nearctic Hemerobiidae, Berothidae, Sisyridae, Polystoechotidae and Dilaridae (Neuroptera). *Proceedings of the American Academy of Arts and Sciences* 74: 193–278.

———. 1943a. The Lower Permian insects of Kansas. Part 9. The orders Neuroptera, Raphidiodea, Caloneurodea and Protorthoptera (Probnisidae), with additional Protodonata and Megasecoptera. *Proceedings of the American Academy of Arts and Sciences* 75: 55–84.

———. 1943b. Studies on Carboniferous insects from Commentry, France. Part I. Introduction and families Protagriidae, Meganeuridae, and Campylopteridae. *Bulletin of the Geological Society of America* 54: 527–54.

———. 1947. Lower Permian insects from Oklahoma. Part 1. Introduction and the orders Megasecoptera, Protodonata, and Odonata. *Proceedings of the American Academy of Arts and Sciences* 76: 25–54.

———. 1950. The Lower Permian insects of Kansas. Part 10. The order Protorthoptera: The family Liomopteridae and its relations. *Proceedings of the American Academy of Arts and Sciences* 78: 185–219.

———. 1951. Studies on Carboniferous insects from Commentry, France, Part II. The Megasecoptera. *Journal of Paleontology* 25: 336–55.

———. 1956. The Baltic amber snake-flies (Neuroptera). *Psyche* 63: 77–81.

———. 1957. A Pliocene insect deposit in Texas. *Psyche* 64: 116.

———. 1960. A Triassic odonate from Argentina. *Psyche* 67: 71–5.

———. 1961. Studies on Carboniferous insects from Commentry, France. Part III. The Caloneurodea. *Psyche* 68: 145–53.

———. 1962. A Permian megasecopteron from Texas. *Psyche* 69: 37–41.

———. 1963a. Studies on Carboniferous insects from Commentry, France. Part IV. The genus *Triplosoba*. *Psyche* 70: 120–8.

———. 1963b. Studies on Carboniferous insects from Commentry, France. Part V. The genus *Diaphanoptera* and the order Diaphanopterodea. *Psyche* 70: 240–56.

———. 1963c. A megasecopteron from Upper Carboniferous strata of Spain. *Psyche* 70: 44–9.

———. 1964a. Studies on Carboniferous insects from Commentry, France. Part VI. The genus *Dictyoptilus* (Palaeodictyoptera). *Psyche* 71: 104–16.

———. 1964b. Lower Permian insect fauna from Elmo, Kansas. Pp. 302–3. *In* Moore, R. C. (ed.), *Paleoecological Aspects of Kansas Pennsylvanian and Permian Cyclothems*. Kansas State Geological Survey Bulletin 169.

———. 1969. Fossil insects from Antarctica. *Psyche* 76: 418–25.

———. 1970. Fossil insects from New Mexico. *Psyche* 77: 400–12.

———. 1976. The Lower Permian insects of Kansas. Part 12. Protorthoptera (continued), Neuroptera, additional Palaeodictyoptera, and families of uncertain position. *Psyche* 83: 336–76.

———. 1987. Review of the extinct family Syntonopteridae (order uncertain). *Psyche* 94: 373–88.

———. 1992. Superclass Hexapoda. Pp. 1–655. *In* Kaesler, R. L. (ed.), *Treatise on Invertebrate Paleontology, Part R, Arthropoda 3–4*. Geological Society of America; Boulder, Colorado; xxii+655 pp.

———. 1997. Insecta. Pp. 184–93. *In* Shabica, C. W., and A. A. Hay (eds.), *Richardson's Guide to the Fossil Fauna of Mazon Creek*. Northeastern Illinois University; Chicago, Illinois; xvii+308 pp.

———, and L. Burnham. 1985. The geological record of insects. *Annual Review of Earth and Planetary Sciences* 13: 297–314.

——, and J. Kukalová. 1964. The structure of the Protelytroptera, with description of a new genus from Permian strata of Moravia. *Psyche* 71: 183–97.

———, and E. S. Richardson, Jr. 1968. Megasecopterous nymphs in Pennsylvanian concretions from Illinois. *Psyche* 75: 295–309.

———, and E. S. Richardson, Jr. 1971. Additional insects in Pennsylvanian concretions from Illinois. *Psyche* 78: 268–95.

———, T. D. A. Cockerell, C. H. Kennedy, T. Snyder, and H. F. Wickham. 1931. Insects from the Miocene (Latah) of Washington. *Annals of the Entomological Society of America* 24: 307–22.

———, J. W. Folsom, E. O. Essig, A. C. Kinsey, C. T. Brues, *et al.* 1937. Insects and arachnids from Canadian amber. *University of Toronto, Geology Series* 40: 7–62.

———, T. E. Snyder, C. P. Alexander, M. T. James, and F. M. Hull. 1938. Fossil insects from the Creede Formation, Colorado. *Psyche* 45: 105–19.

Carpenter, G. D. H., and E. B. Ford. 1933. *Mimicry*. Meuthen and Company; London, UK; ix+134 pp.

Carpenter, J. M. 1982. The phylogenetic relationships and natural classification of the Vespoidea (Hymenoptera). *Systematic Entomology* 7: 11–38.

———. 1986. Cladistics of the Chrysidoidea (Hymenoptera). *Journal of the New York Entomological Society* 94: 303–30.

———. 1991. Phylogenetic relationships and the origin of social behavior in the Vespidae. Pp. 7–32. *In* Ross, K. G., and R. W. Matthews (eds.), *The Social Biology of Wasps*. Cornell University Press; Ithaca, New York; xvii+678 pp.

———. 1999. What do we know about chrysidoid (Hymenoptera) relationships? *Zoologica Scripta* 28: 215–31.

———. 2000. A vespid wasp from New Jersey Cretaceous amber. Pp. 333–7. *In* Grimaldi, D. (ed.), *Studies on Fossils in Amber, with Particular Reference to the Cretaceous of New Jersey*. Backhuys Publishers; Leiden, the Netherlands; viii+498 pp.

———. 2003. On "molecular phylogeny of Vespidae (Hymenoptera) and the evolution of sociality in wasps." *American Museum Novitates* 3389: 1–20.

———, and D. A. Grimaldi. 1997. Social wasps in amber. *American Museum Novitates* 3203: 1–7.

———, and A. P. Rasnitsyn. 1990. Mesozoic Vespidae. *Psyche* 97: 1–20.

———, and W. C. Wheeler. 1999. Cladistíca numérica, análisis simultáneo y filogenia de hexápodos. *Boletin de la Sociedad Entomológica Aragonesa* 26: 333–46.

Carrillo, J. L., and L. E. Caltagirone. 1970. Observations on the biology of *Solierella peckhami*, *S. blaisdelli* (Sphecidae) and two species of Chrysididae (Hymenoptera). *Annals of the Entomological Society of America* 63: 673–81.

Carroll, R. L. 1988. *Vertebrate Paleontology and Evolution*. Freeman; New York, New York; xiv+698 pp.

Carson, H. L. 1974. Three flies and three islands: Parallel evolution in *Drosophila*. *Proceedings of the National Academy of Sciences, U.S.A.* 71: 3517–21.

———. 1983a. Chromosomal sequences and interisland colonizations in the Hawaiian *Drosophila*. *Genetics* 103: 465–82.

———. 1983b. Evolution of *Drosophila* on the newer Hawaiian volcanoes. *Heredity* 48: 3–25.

Carver, M., G. F. Gross, and T. E. Woodward. 1991. Hemiptera (bugs, leafhoppers, cicadas, aphids, scale insects etc.). Pp. 429–509. *In* Naumann, I. D. (ed.), *The Insects of Australia: A Textbook for Students and Research Workers, Volume 1* [2nd Edition]. Cornell University Press; Ithaca, New York; xvi+[1]+542 pp.

Castillo, P. R., and G. Halffter. 1985. La estructura social de los Passalidae (Coleoptera: Lamellicornia). *Folia Entomologica Mexicana* 61: 49–72.

Castner, J. L., and H. G. Fowler. 1987. Diel patterns of *Larra bicolor* (Hymenoptera: Sphecidae) in Puerto Rico. *Journal of Entomological Science* 22: 77–83.

Castro, M. P. 1997. Huellas de actividad biológica sobre plantas del Estafaniense Superior de la Magdalena (León, España). *Revista Española de Paleontología* 12: 52–66.

Caterino, M. S., R. D. Reed, M. M. Ruo, and F. A. H. Sperling. 2001. A partitioned likelihood analysis of swallowtail butterfly phylogeny (Lepidoptera: Papilionidae). *Systematic Biology* 50: 106–27.

———, V. L. Shull, P. M. Hammond, and A. P. Vogler. 2002. Basal relationships of Coleoptera inferred from 18S rDNA sequences. *Zoological Scripta* 31: 41–9.

Caudell, A. N. 1918. *Zorotypus hubbardi*, a new species of the order Zoraptera from the United States. *Canadian Entomologist* 50: 375–81.

Cavalier-Smith, T. 1998. A revised six-kingdom system of life. *Biological Reviews* 73: 203–66.

Chaboo, C. 2004. *The systematics and biology of tortoise and leaf-mining beetles (Cassidinae s.l.: Chrysomelidae: Coleoptera)*. Ph.D. Dissertation; Cornell University; Ithaca, New York.

Chalwatzis, N., A. Baur, E. Stetzer, R. Kinzelbach, and F. K. Zimmerman. 1995. Strongly expanded 18S ribosomal-RNA genes correlated with a peculiar morphology in the insect order of Stepsiptera. *Zoology (Jena)* 98: 115–26.

———, J. Hauf, Y. van der Peer, R. Kinzelbach, and F. K. Zimmerman. 1996. 18S ribosomal-RNA genes of insects – primary structure of the genes and molecular phylogeny of the Holometabola. *Annals of the Entomological Society of America* 89: 788–803.

Chandler, P. 2002. *Heterotricha* Loew and allied genera (Diptera: Sciaroidea): Offshoots of the stem group of Mycetophilidae and/or Sciaridae? *Annals de la Societé d'Entomologique de France* 38: 101–44.

Chang, H.-C., and D. D. Miller. 1978. Courtship and mating sounds in species of the *Drosophila affinis* subgroup. *Evolution* 32: 540–50.

Chapela, I. H., S. A. Rehner, T. R. Schultz, and U. G. Mueller. 1994. Evolutionary history of the symbiosis between fungus-growing ants and their fungi. *Science* 266: 1691–4.

Chapman, T. W., B. J. Crespi, B. D. Kranz, and M. P. Schwarz. 2000. High relatedness and inbreeding at the origin of eusociality in gall-inducing thrips. *Proceedings of the National Academy of Sciences, U.S.A.* 97: 1648–50.

Chao, R.-F., and C.-S. Chen. 2000. *Formosozoros newi*, a new genus and species of Zoraptera (Insecta) from Taiwan. *Pan-Pacific Entomologist* 76: 24–7.

Chavarría, G., and J. M. Carpenter. 1994. "Total evidence" and the evolution of highly social bees. *Cladistics* 10: 229–58.

Chaw, S. M., C. L. Parkinson, Y. Cheng, T. M. Vincent, and J. D. Palmer. 2000. Seed plant phylogeny inferred from all three plant genomes: Monophyly of extant gymnosperms and origin of Gnetales from conifers. *Proceedings of the National Academy of Sciences, U.S.A.* 97: 4086–91.

Cheetham, T. B. 1988. *Male Genitalia and the Phylogeny of the Pulicoidea (Siphonaptera).* Koeltz; Koenigstein, Germany; vii+224 pp.

Chiappe, L. M. 1995. The first 85 million years of avian evolution. *Nature* 378: 349–55.

Chin, K., and B. D. Gill. 1996. Dinosaurs, dung beetles, and conifers: Participants in a Cretaceous food web. *Palaios* 11: 280–5.

Chisholm, I. F., and T. Lewis. 1984. A new look at thrips (Thysanoptera) mouthparts, their action and effects of feeding on plant tissue. *Bulletin of Entomological Research* 74: 663–75.

Chiu, S.-C., L. Y. Chou, and K. C. Chou. 1981. A preliminary survey on the natural enemies of *Kerria lacca* (Kerr) in Taiwan. *Journal of Agricultural Research of China* 30: 420–5. [In Chinese, with English summary]

Choe, J. C. 1992. Zoraptera of Panama with a review of the morphology, systematics, and biology of the order. Pp. 249–56. *In* Quintero, D., and A. Aiello (eds.), *Insects of Panama and Mesoamerica: Selected Studies.* Oxford University Press; Oxford, UK; xxii+692 pp.

———, and B. J. Crespi, eds. 1997. *The Evolution of Social Behavior in Insects and Arachnids.* Cambridge University Press; Cambridge, UK; xiii+541 pp.

Chopard, L. 1938. *La Biologie des Orthoptères.* Lechevalier; Paris, France; 541 pp.

———. 1949. Ordre des Dictyoptères. Pp. 355–407. *In* Grassé, P. P. (ed.), *Traité de Zoologie, Tome IX.* Masson et Cie; Paris, France; 1117 pp.

Christiansen, K. 1971. Notes on Miocene amber Collembola from Chiapas. *University of California Publications in Entomology* 63: 45–8.

———, and P. Bellinger. 1998. *The Collembola of North America North of the Rio Grande. A Taxonomic Analysis* [2nd Edition]. Grinnell College; Grinnell, Iowa; v+1518 pp.

———, and E. Pike. 2002a. A preliminary report on the Cretaceous Collembola. *Pedobiologia* 46: 267–73.

———, and E. Pike. 2002b. Cretaceous Collembola (Arthropoda, Hexapoda) from the Upper Cretaceous of Canada. *Cretaceous Research* 23: 165–88.

Chrystal, R. N., and E. R. Skinner. 1932. Studies in the biology of the woodwasp *Xiphydria prolongata* Geoffr. (*dromedarius* F.) and its parasite *Thalessa curvipes* Grav. *Scottish Forestry Journal* 46: 36–57.

Chvála, M. 1976. Swarming, mating, and feeding habits in Empididae (Diptera), and their significance in evolution of the family. *Acta Entomologica Bohemoslavica* 73: 353–66.

———. 1983. The Empidoidea (Diptera) of Fennoscandia and Denmark II. General Part: The families Hybotidae, Atelestidae and Microphoridae. *Fauna Entomologica Scandinavica* 12: 1–281.

Cichan, M. A., and T. N. Taylor. 1982. Wood-borings in *Premnoxylon*: Plant-animal interactions in the Carboniferous. *Palaeogeography, Palaeoclimatology, Palaeoecology* 39: 123–7.

Cisne, J. L. 1974. Trilobites and the origin of arthropods. *Science* 186: 13–18.

Claassen, P. W. 1940. A catalogue of the Plecoptera of the world. *Memoirs of the Cornell University Agricultural Experiment Station* 232: 1–235.

Clague, D. A., and G. B. Dalrymple. 1987. The Hawaiian-Emperor volcanic chain. Pp. 1–54. *In* Decker, R. W., T. L. Wright, and P. H. Stauffer (eds.), *Volcanism in Hawaii.* United States Geological Survey Professional Paper 1350; Washington, D.C.; xiii+1667 pp. [2 volumes]

Clancy, D. W. 1944. Biology of *Allotropa burrelli*, a gregarious parasite of *Pseudococcus comstocki*. *Journal of Agricultural Research* 69: 159–67.

Claridge, M. F. 1985. Acoustic signals in the Homoptera: Behavior, taxonomy, and evolution. *Annual Review of Entomology* 30: 297–317.

———, H. A. Dawah, and M. R. Wilson. 1997. Species in insect herbivores and parasitoids – sibling species, host races and biotypes. Pp. 247–72. *In* Claridge, M. F., H. A. Dawah, and M. R. Wilson (eds.), *Species: The Units of Biodiversity*. Chapman and Hall; London, UK; xvi+439 pp.

Clark, J. W., S. Hossain, C. A. Burnside, and S. Kambhampati. 2000. Coevolution between a cockroach and its bacterial endosymbiont: A biogeographical perspective. *Proceedings of the Royal Society of London, Series B, Biological Sciences* 268: 393–8.

Clarke, C. A., and P. M. Sheppard. 1955. A preliminary report on the genetics of the *machaon* group of swallowtail butterflies. *Evolution* 9: 182–201.

———, and P. M. Sheppard. 1960a. The evolution of mimicry in the butterfly *Papilio dardanus*. *Heredity* 14: 163–73.

———, and P. M. Sheppard. 1960b. Supergenes and mimicry. *Heredity* 14: 175–85.

Clarke, J. F. G. 1941–1969. *Catalogue of the Type Specimens of Microlepidoptera in the British Museum (Natural History) Described by Edward Meyrick* [7 volumes]. Trustees of the British Museum (Natural History); London, UK; 3133 pp.

Clausen, C. P. 1940. *Entomophagous Insects.* McGraw Hill; New York, New York; 688 pp.

Clay, T. 1970. The Amblycera (Phthiraptera: Insecta). *Bulletin of the British Museum (Natural History), Entomology* 25: 75–98.

Cleveland, L. R., S. K. Hall, E. P. Sanders, and J. Collier. 1934. The wood-feeding roach *Cryptocercus*, its protozoa, and the symbiosis between protozoa and roach. *Memoirs of the American Academy of Arts and Sciences* 17: 185–342.

Cobabe, E. A., K. R. Chamberlain, M. A. Ivie, and J. J. Giersch. 2002. A new insect and plant Lagerstätte from a Tertiary lake deposit along the Canyon Ferry Reservoir, southwestern Montana. *Rocky Mountain Geology* 37: 13–30.

Cobb, M., B. Burnet, and K. Connolly. 1986. The structure of courtship in the *Drosophila melanogaster* species sub-group. *Behaviour* 97: 182–212.

Cobben, R. H. 1968. *Evolutionary Trends in Heteroptera. Part 1: Eggs, Architecture of the Shell, Gross Embryology, and Eclosion.* Centre for Agricultural Publishing and Documentation; Wageningen, the Netherlands; 475 pp.

———. 1978. *Evolutionary Trends in Heteroptera. Part 2: Mouthpart Structures and Feeding Strategies.* Veeman; Wageningen, the Netherlands; 407 pp.

Cockerell, T. D. A. 1906. Fossil Hymenoptera from Florissant, Colorado. *Bulletin of the Museum of Comparative Zoology* 50: 33–58.

———. 1907a. Some old world types of insects in the Miocene of Colorado. *Science* 26: 446–7.

———. 1907b. A fossil butterfly of the genus *Chlorippe*. *Canadian Entomologist* 39: 361–3.

———. 1907c. Some fossil arthropods from Florissant, Colorado. *Bulletin of the American Museum of Natural History* 23: 605–16.

———. 1908a. Descriptions and records of bees – xix. *Annals and Magazine of Natural History, series 8*, 1: 337–44.

———. 1908b. Descriptions of Tertiary insects. II. *American Journal of Science* 25: 227–32.

———. 1908c. Descriptions and records of bees – xx. *Annals and Magazine of Natural History* 2: 323–34.

———. 1909a. A fossil tortricid moth. *Canadian Entomologist* 39: 416.

———. 1909b. Descriptions of Hymenoptera from Baltic amber. *Schriften der Physikalisch-ökonomischen Gesellschaft* 50: 1–20.

———. 1910. The fossil Crabronidae. *Entomologist* 43: 60–1.

———. 1913. Some fossil insects from Florissant, Colorado. *Proceedings of the United States National Museum* 44: 341–6.

———. 1917. Arthropods in Burmese amber. *American Journal of Science* 44: 360–8.

———. 1919. Two interesting insects in Burmese amber. *Entomologist* 52: 193–5.

———. 1921. Fossil arthropods in the British Museum, V. *Annals and Magazine of Natural History, series 9* 7: 1–25.

———. 1922. A fossil moth from Florissant, Colorado. *American Museum Novitates* 34: 1–2.

———. 1926. A new fossil moth from Florissant. *Psyche* 33: 16–17.

———. 1931. Insects from the Miocene (Latah) of Washington, II. Hymenoptera and Hemiptera. *Annals of the Entomological Society of America* 24: 309–12.

———. 1936. The fauna of the Sunchal (or Margas Verdes) Formation, northern Argentina. *American Museum Novitates* 886: 1–9.

———. 1941. Some Tertiary insects (Hymenoptera) from Colorado. *American Journal of Science* 239: 354–6.

Cocroft, R. 1999. Thornbug to thornbug: The inside story of insect song. *Natural History* 10/99: 53–57.

Coddington, J. A., and H. W. Levi. 1991. Systematics and evolution of spiders (Araneae). *Annual Review of Ecology and Systematics* 22: 565–92.

Colgan, D. J., G. Cassis, and E. Beacham. 2003. Setting the molecular phylogenetic framework for the Dermaptera. *Insect Systematics and Evolution* 34: 65–80.

COLLESS, D. H. 1990. *Valeseguya rieki*, a new genus and species of dipteran from Australia (Nematocera: Anisopodidae). *Annales de la Société Entomologique de France* 26: 351–3.

———, and D. K. MCALPINE. 1991. Diptera (flies). Pp. 717–86. *In* NAUMANN, I. D. (ed.), *The Insects of Australia: A Textbook for Students and Research Workers, Volume 2* [2nd Edition]. Cornell University Press; Ithaca, New York; [2]+543–1137 pp.

COLLINS, N. M., and J. A. THOMAS, eds. 1991. *Conservation of Insects and Their Habitats*. Academic Press; London, UK; xviii+449 pp.

COLLINS, K. P., and B. M. WIEGMANN. 2002a. Phylogenetic relationships and placement of the Empidoidea (Diptera: Brachycera) based on 28S rDNA and EF-1α sequences. *Insect Systematics and Evolution* 33: 421–44.

———, and B. M. WIEGMANN. 2002b. Phylogenetic relationships of the lower Cyclorrhapha (Diptera: Brachycera) based on 28S rDNA sequences. *Insect Systematics and Evolution* 33: 445–56.

COMMON, I. F. B. 1990. *Moths of Australia*. Melbourne University Press; Melbourne, Australia; 535 pp, 32 pl.

COMSTOCK, J. H. 1888. *An Introduction to Entomology*. Published by the Author; Ithaca, New York; iv+234 pp.

——, and A. B. COMSTOCK. 1895. *A Manual for the Study of Insects*. Comstock Publishing; Ithaca, New York; x+701 pp.

——, and J. G. NEEDHAM. 1898. The wings of insects. *American Naturalist* 32: 43–8, 81–9, 231–57, 335–40, 413–24, 561–5, 768–77, 903–11.

——, and J. G. NEEDHAM. 1899. The wings of insects. *American Naturalist* 33: 117–26, 573–82, 845–60.

CONDÉ, B., and J. PAGÉS. 1991. Diplura. Pp. 269–71. *In* NAUMANN, I. D. (ed.), *The Insects of Australia: A Textbook for Students and Research Workers, Volume 1* [2nd Edition]. Cornell University Press; Ithaca, New York; xvi+[1]+542 pp.

CONSTENIUS, K. N., M. R. DAWSON, H. G. PIERCE, R. C. WALTER, and M. V. H. WILSON. 1989. Reconnaissance paleontologic study of the Kishenehn Formation, northwestern Montana and southeastern British Columbia. Pp. 189–203. *In* FRENCH, D. E., and R. F. GRABB (eds.), *Geologic Resources of Montana, Volume I: 1989 Field Conference Guidebook* [Montana Centennial Edition]. Montana Geological Society; Billings, Montana; xvii+528 pp.

CONTRERAS-RAMOS, A. 1999. List of species of Neotropical Megaloptera (Neuropterida). *Proceedings of the Entomological Society of Washington* 101: 274–84.

CONWAY MORRIS, S. 1979. The Burgess Shale (middle Cambrian) fauna. *Annual Review of Ecology and Systematics* 10: 327–49.

———. 1989. Burgess Shale faunas and the Cambrian explosion. *Science* 246: 339–46.

———. 1993. The fossil record and the early evolution of the Metazoa. *Nature* 361: 219–25.

———. 1998. *The Crucible of Creation: The Burgess Shale and the Rise of Animals*. Oxford University Press; Oxford, UK; xxiii+242 pp.

———. 2000. The Cambrian "explosion": Slow-fuse or megatonnage? *Proceedings of the National Academy of Sciences, U.S.A.* 97: 4426–9.

———. 2003. The Cambrian "explosion" of metazoans. Pp. 13–32. *In* MÜLLER, G. B., and S. A. NEWMAN (eds.), *Origination of Organismal Form: Beyond the Gene in Developmental and Evolutionary Biology*. MIT Press; Cambridge, Massachusetts; ix+332 pp.

COOK, E., and A. J. ROSS. 1996. The stratigraphy, sedimentology and palaeontology of the Lower Weald Clay (Hauterivian) at Keymer Tileworks, West Sussex, southern England. *Proceedings of the Geologists' Association* 107: 231–9.

COOK, E. F. 1949. The evolution of the head in the larvae of the Diptera. *Microentomology* 14: 1–57.

COOK, J. M., and J.-Y. RASPLUS. 2003. Mutualists with attitude: Coevolving fig wasps and figs. *Trends in Ecology and Evolution* 18: 241–8.

COOK, L. G., P. J. GULLAN, and H. E. TRUEMAN. 2002. A preliminary phylogeny of the scale insects (Hemiptera: Sternorrhyncha: Coccoidea) based on nuclear small-subunit ribosomal DNA. *Molecular Phylogenetics and Evolution* 25: 43–52.

COOK, M. A., and M. J. SCOBLE. 1992. Tympanal organs of geometrid moths: A review of their morphology, function, and systematic importance. *Systematic Entomology* 17: 219–32.

COOPE, G. R. 1965. Fossil insect faunas from Late Quaternary deposits in Britain. *Advancement of Science* 1965: 564–75.

———. 1970. Interpretations of Quaternary insect fossils. *Annual Review of Entomology* 15: 97–120.

———. 1973. Tibetan species of dung beetles from Late Pleistocene deposits in England. *Nature* 245: 335–6.

———. 1978. Constancy of insect species versus inconstancy of Quaternary environments. Pp. 176–87. *In* MOUND, L. A., and N. WALOFF (eds.), *Diversity of Insect Faunas*. Blackwell Scientific Publications; Oxford, UK; x+204 pp.

———. 1979. Late Cenozoic fossil Coleoptera: Evolution, biogeography and ecology. *Annual Review of Ecology and Systematics* 10: 247–67.

———. 1987. The response of Late Quaternary insect communities to sudden climatic changes. Pp. 421–38. *In* GEE, J. H. R., and P. S. GILLER (eds.), *Organization of Communities: Past and Present*. Blackwell Scientific Publications; Oxford, UK; xii+576 pp.

———. 1990. The invasion of northern Europe during the Pleistocene by Mediterranean species of Coleoptera. Pp. 203–15. *In* DICASTRI, F., A. J. HANSEN, and M. DEBUSSCHE (eds.), *Biological Invasions in Europe and the Mediterranean Basin*. Kluwer Academic Publishers; Dordrecht, the Netherlands; ix+463 pp.

COOPER, K. W. 1964. The first fossil tardigrade: *Beorn leggi* Cooper, from Cretaceous amber. *Psyche* 71: 41–8.

———. 1972. A southern California *Boreus*, *B. notoperates* n. sp. I. Comparative morphology and systematics (Mecoptera: Boreidae). *Psyche* 79: 269–83.

———. 1974. Sexual biology, chromosomes, development, life histories, and parasites of *Boreus*, especially of *B. notoperates*, a southern California *Boreus*. II. (Mecoptera: Boreidae). *Psyche* 81: 84–120.

———, and P. DESSART. 1975. Adult, larva and biology of *Conostigmus quadratogenalis* Dessart & Cooper, sp. n. (Hym. Ceraphronoidea), parasite of *Boreus* (Mecoptera) in California. *Bulletin et Annales de la Société Royale Belge d'Entomologie* 111: 37–53.

CORAM, R. A. 2003. Taphonomy and ecology of Purbeck fossil insects. *Acta Zoologica Cracoviensia* 46 (Supplement): 311–18.

———, and E. A. JARZEMBOWSKI. 2002. Diversity and ecology of fossil insects in the Dorset Purbeck succession, southern England. *Special Papers in Palaeontology* 68: 257–68.

———, E. A. JARZEMBOWSKI, and M. B. MOSTOVSKI. 2000. Two rare eremoneuran flies (Diptera: Empididae and Opetiidae) from the Purbeck Limestone Group. *Paleontological Journal* 34: 370–3.

CORBET, P. S. 1999. *Dragonflies: Behaviour and Ecology of Odonata*. Harley Books; Colchester, UK; xxxii+829 pp.

CORKUM, L. D., and J. CIBOROWSKI, eds. 1995. *Current Directions in Research on Ephemeroptera*. Canadian Scholars Press; Toronto, Canada; xiii+478 pp.

CORNWELL, P. B. 1968. *The Cockroach, vol. 1. A Laboratory Insect and an Industrial Pest*. Hutchinson; London, UK; 391 pp.

COTTRELL, C. B. 1984. Aphytophagy in butterflies: Its relationship to myrmecophily. *Zoological Journal of the Linnean Society* 80: 1–57.

COUDRON, T. A. 1990. Host regulating factors associated with parasitic Hymenoptera. Pp. 41–65. *In* HEDIN, P. A. (ed.), *Naturally Occurring Pest Bioregulators*. American Chemical Society; Washington, D.C.; xii+456 pp.

COURTNEY, G. W. 1990. Cuticular morphology of larval mountain midges (Diptera: Deuterophlebiidae): Implications for the phylogenetic relationships of Nematocera. *Canadian Journal of Zoology* 68: 556–78.

———. 1991. Phylogenetic analysis of the Blephariceromorpha, with special reference to mountain midges (Diptera: Deuterophlebiidae). *Systematic Entomology* 16: 137–72.

———. 1994. Biosystematics of the Nymphomyiidae (Insecta: Diptera): Life history, morphology and phylogenetic relationships. *Smithsonian Contributions to Zoology* 550: 1–41.

COWAN, D. P. 1991. The solitary and presocial Vespidae. Pp. 33–73. *In* ROSS, K. G., and R. W. MATTHEWS (eds.), *The Social Biology of Wasps*. Cornell University Press; Ithaca, New York; xvii+678 pp.

COWLING, D. E., and B. BURNET. 1981. Courtship songs and genetic control of their acoustic characteristics in sibling species of the *Drosophila melanogaster* subgroup. *Animal Behavior* 29: 924–35.

COX, B. 1974. Vertebrate paleodistributional patterns and continental drift. *Journal of Biogeography* 1: 75–94.

COYNE, J. A. 1983. Genetic basis of differences in genital morphology among three sibling species of *Drosophila*. *Evolution* 37: 1101–18.

———, and B. CHARLESWORTH. 1986. Location of an X-linked factor causing sterility in male hybrids of *Drosophila simulans* and *D. mauritiana*. *Heredity* 57: 243–6.

———, and H. A. ORR. 1989. Patterns of speciation in *Drosophila*. *Evolution* 43: 362–81.

———, J. Rux, and J. R. David. 1991. Genetics of morphological differences and hybrid sterility between *Drosophila sechellia* and its relatives. *Genetic Research* 57: 113–22.

Cracraft, J. 1989. Speciation and its ontology: The empirical consequences of alternative species concepts for understanding patterns and processes of differentiation. Pp. 28–59. *In* Otte, D., and J. A. Endler (eds.), *Speciation and Its Consequences*. Sinauer; Sunderland, Massachusetts; xiii+679 pp.

———. 2001. Avian evolution, Gondwana biogeography and the Cretaceous-Tertiary mass extinction event. *Proceedings of the Royal Society of London, Series B, Biological Sciences* 268: 459–69.

Crampton, G. C. 1920. Some anatomical details of the remarkable winged zorapteron, *Zorotypus huddardis* [sic] Caudell, with notes on its relationships. *Proceedings of the Entomological Society of Washington* 22: 98–106.

———. 1924. The phylogeny and classification of insects. *Journal of Entomology and Zoology, Pomona College* 16: 33–47.

———. 1928. The grouping of the insect orders and their lines of descent. *Entomologist* 61: 82–5.

———. 1931. A claim for priority in dividing pterygotan insects into two sections on the basis of the position of the wings in repose, with remarks on the relationships of the insect orders. *Entomological News* 42: 130–6.

———. 1938. The interrelationships and lines of descent of living insects. *Psyche* 45: 165–81.

Crane, E. 1983. *The Archaeology of Beekeeping*. Cornell University Press; Ithaca, New York; 360 pp.

———. 1999. *The World History of Beekeeping and Honey Hunting*. Routledge; New York, New York; xxii+682 pp.

Crane, P. R. 1985. Phylogenetic analysis of seed plants and the origin of angiosperms. *Annals of the Missouri Botanical Garden* 72: 716–93.

———, and P. S. Herendeen. 1996. Cretaceous floras containing angiosperm flowers and fruits from eastern North America. *Review of Palaeobotany and Palynology* 90: 319–37.

———, and S. Lidgard. 1990. Angiosperm radiation and pattern of Cretaceous palynological diversity. Pp. 377–407. *In* Taylor, P. D., and G. P. Larwood (eds.), *Major Evolutionary Radiations*. Oxford University Press; Oxford, UK; xi+437 pp.

———, E. M. Friis, and K. R. Pedersen. 1995. The origin and early diversification of angiosperms. *Nature* 374: 27–33.

Cranston, P. S., D. H. D. Edward, and L. G. Cook. 2002. New status, species, distribution records and phylogeny for Australian mandibulate Chironomidae (Diptera). *Australian Journal of Entomology* 41: 357–66.

Crepet, W. L. 1996. Timing in the evolution of derived floral characters: Upper Cretaceous (Turonian) taxa with tricolpate and tricolpate-derived pollen. *Review of Palaeobotany and Palynology* 90: 339–359.

———, and E. M. Friis. 1987. The evolution of insect pollination in angiosperms. Pp. 181–201. *In* Friis, E. M., W. G. Chaloner, and P. R. Crane (eds.), *The Origins of Angiosperms and their Biological Consequences*. Cambridge University Press; Cambridge, UK; x+358 pp.

———, and K. C. Nixon. 1998. Fossil Clusiaceae from the Late Cretaceous (Turonian) of New Jersey and implications regarding the history of bee pollination. *American Journal of Botany* 85: 1122–33.

———, E. M. Friis, and K. C. Nixon. 1991. Fossil evidence for the evolution of biotic pollination. *Philosophical Transactions of the Royal Society of London, Series B* 333: 187–95.

Crespi, B. J., and L. A. Mound. 1997. Ecology and evolution of social behavior among Australian gall thrips and their allies. Pp. 166–80. *In* Choe, J. C., and B. J. Crespi (eds.), *The Evolution of Social Behavior in Insects and Arachnids*. Cambridge University Press; Cambridge, UK; vii+541 pp.

———, and C. P. Sandoval. 2000. Phylogenetic evidence for the evolution of ecological specialization in *Timema* walking-sticks. *Journal of Evolutionary Biology* 13: 249–62.

———, D. Carmean, L. Vawter, and C. von Dohlen. 1996. Molecular phylogenetics of Thysanoptera. *Systematic Entomology* 21: 79–87.

Cresson, E. T. 1863. List of the North American species of *Bombus* and *Apathus*. *Proceedings of the Entomological Society of Philadelphia* 2: 83–116.

Crosskey, R. W. 1951. The morphology, taxonomy, and biology of the British Evanioidea (Hymenoptera). *Transactions of the Royal Entomological Society, London* 102: 247–301.

———. 1962. The classification of the Gasteruptiidae (Hymenoptera). *Transactions of the Royal Entomological Society, London* 114: 377–402.

———, ed. 1980. *Catalogue of the Diptera of the Afrotropical Region*. British Museum (Natural History); London, UK; 1437 pp.

———. 1990. *The Natural History of Blackflies*. John Wiley and Sons; Chichester, UK; ix+711 pp.

Crossley, R. 1984. Fossil termite mounds associated with stone artefacts in Malawi, central Africa. *Palaeoecology of Africa and the Surrounding Islands* 16: 397–401.

———. 1986. Sedimentation by termites in the Malawi Rift Valley. Pp. 191–9. *In* Frostik, L. E. (ed.), *Sedimentation in the African Rifts*. Blackwell Scientific; Oxford, UK; xiii+382 pp.

Crowson, R. A. 1960. The phylogeny of Coleoptera. *Annual Review of Entomology* 5: 111–34.

———. 1981. *The Biology of the Coleoptera*. Academic Press; London, UK; xii+802 pp.

———. 1985. Comments on Insecta of the Rhynie chert. *Entomologia Generalis* 11: 57–8.

———. 1991. The relations of Coleoptera to Cycadales. Pp. 13–28. *In* Zunino, M., X. Bellés, and M. Blas (eds.), *Advances in Coleopterology*. European Association of Coleopterology; Barcelona, Spain; 323 pp.

Cruickshank, R. D., and K. Ko. 2003. Geology of an amber locality in the Hukawng Valley, northern Myanmar. *Journal of Asian Earth Sciences* 21: 441–55.

Cruikshank, R. H., K. P. Johnson, V. S. Smith, R. J. Adams, D. H. Clayton, and R. D. M. Page. 2001. Phylogenetic analysis of partial sequences of elongation factor one-alpha identifies major groups of lice (Insecta: Phthiraptera). *Molecular Phylogenetics and Evolution* 19: 202–15.

Cumming, J. M. 1994. Sexual selection and the evolution of dance fly mating systems (Diptera: Empididae: Empidinae). *Canadian Entomologist* 126: 907–20.

———, and S. E. Brooks. 2002. *Electrophorella*, a new genus of parathalassiine flies from Baltic amber, with a cladistic analysis of the Microphorinae + Dolichopodidae lineage (Diptera: Empidoidea). *Studia Dipterologica* 9: 41–54.

———, B. J. Sinclair, and D. M. Wood. 1995. Homology and phylogenetic implications of male genitalia in Diptera – Eremoneura. *Entomologica Scandinavica* 26: 121–51.

Currie, D. C., and D. Grimaldi. 2000. A new black fly (Diptera: Simuliidae) genus from mid-Cretaceous (Turonian) amber of New Jersey. Pp. 473–85. *In* Grimaldi, D. (ed.), *Studies on Fossils in Amber, with Particular Reference to the Cretaceous of New Jersey*. Backhuys Publishers; Leiden, the Netherlands; viii+498 pp.

Cuvier, G. 1817. *Le Règne Animal Distribué d'après son Organisation, pour Servir de Base à l'Histoire Naturelle des Animaux et d'Introduction à l' Anatomie Compareé* [Volume 2]. Deterville; Paris, France; 532 pp.

D'Abrera, B. 1987. *Sphingidae Mundi: Hawk Moths of the World, Based on a Checklist by Alan Hayes and the Collection He Curated in the British Museum (Natural History)*. E. W. Classey; Faringdon, UK; ix+226 pp.

———. 2001. *The Concise Atlas of Butterflies of the World*. Hill House; Melbourne, Australia; 353 pp.

Dahlsten, D. L. 1961. Life history of a pine sawfly, *Neodiprion* sp., at Willits, California. *Canadian Entomologist* 93: 182–95.

———. 1967. Preliminary life tables for the pine sawflies in the *Neodiprion fulviceps* complex. *Ecology* 48: 275–89.

Dalla Torre, K. W., von. 1890. Hymenopterologische Notizen. *Wiener Entomologische Zeitung* 9: 139.

———. 1896. *Catalogus Hymenopterorum hucusque Descriptorum Systematicus et Synonymicus. Volumen X: Apidae (Anthophila)*. Engelmann; Leipzig, Germany; 643 pp.

Dallai, R., P. Lupetti, F. Frati, F. Nardi, and B. A. Afzelius. 2001. Binucleate and biflagellate spermatozoa in *Tricholepidion gertschi* Wygodzynsky [sic] (Insecta, Zygentoma). *Tissue and Cell* 33: 606–13.

Danforth, B. N., S. G. Brady, S. D. Sipes, and A. Pearson. 2004. Single-copy nuclear genes recover Cretaceous-age divergences in bees. *Systematic Biology* 53: 309–26.

Darling, D. C. 1996. A new species of *Spalangiopelta* (Hymenoptera; Pteromalidae; Ceinae) from Dominican amber: Phylogenetic and biogeographic implications. *Journal of the Kansas Entomological Society, supplement* 69: 248–59.

———, and M. J. Sharkey. 1990. Hymenoptera. *Bulletin of the American Museum of Natural History* 195: 123–53.

Darlington, P. J., Jr. 1943. Carabidae of mountains and islands: Data on the evolution of isolated faunas, and on atrophy of wings. *Ecological Monographs* 13: 37–61.

———. 1957. *Zoogeography: The Geographical Distribution of Animals*. John Wiley and Sons; New York, New York; 675 pp.

———. 1965. *Biogeography of the Southern End of the World: Distribution and History of Far-Southern Life and Land, with an Assessment of Continental Drift.* Harvard University Press; Cambridge, Massachusetts; x+236 pp.

———. 1970. A practical criticism of Hennig-Brundin "phylogenetic systematics" and Antarctic biogeography. *Systematic Zoology* 19: 1–18.

Darwin, C. R. 1859. *On the Origin of the Species by Means of Natural Selection, or, The Preservation of Favoured Races in the Struggle for Life.* J. Murray; London, UK; ix+[1]+502 pp.

Davis, C. 1939a. Taxonomic notes on the order Embioptera. III. The genus *Burmitembia* Cockerell. *Proceedings of the Linnean Society of New South Wales* 64: 369–72.

———. 1939b. Taxonomic notes on the order Embioptera. IV. The genus *Clothoda* Enderlein. *Proceedings of the Linnean Society of New South Wales* 64: 373–80.

———. 1940. Family classification of the order Embioptera. *Annals of the Entomological Society of America* 33: 677–82.

Davis, D. R. 1967. A revision of the moths of the family Prodoxinae (Lepidoptera; Incurvariidae). *Bulletin of the United States National Museum* 255: 1–170.

———. 1975. Systematics and zoogeography of the family Neopseustida with the proposal of a new superfamily (Lepidoptera: Neopseustoidea). *Smithsonian Contributions in Zoology* 210: 1–45.

———. 1978. A revision of the North American moths of the superfamily Eriocraniodea with the proposal of a new family, Acanthopteroctectidae (Lepidoptera). *Smithsonian Contributions to Zoology* 251: 1–131.

———. 1986. A new family of monotrysian moths from austral South America (Lepidoptera: Palaephatidae), with a phylogenetic review of the Monotrysia. *Smithsonian Contributions to Zoology* 434: 1–202.

———. 1999. The monotrysian Heteroneura. Pp. 65–90. *In* Kristensen, N. P. (ed.), *Handbuch der Zoologie: Eine Naturgeschichte der Stämme des Tierreiches: Band IV: Arthropoda: Insecta: Teilband 35: Lepidoptera, Moths and Butterflies: Volume 1: Evolution, Systematics, and Biogeography.* Walter de Gruyter; Berlin, Germany; x+491 pp.

———, and P. Gentili. 2003. Andesianidae, a new family of monotrysian moths (Lepidoptera: Andesianoidea) from austral South America. *Invertebrate Systematics* 17: 15–26.

———, and G. S. Robinson. 1999. The Tineoidea and Gracillarioidea. Pp. 91–118. *In* Kristensen, N. P. (ed.), *Handbuch der Zoologie: Eine Naturgeschichte der Stämme des Tierreiches: Band IV: Arthropoda: Insecta: Teilband 35: Lepidoptera, Moths and Butterflies: Volume 1: Evolution, Systematics, and Biogeography.* Walter de Gruyter; Berlin, Germany; x+491 pp.

———, O. Pellmyr, and J. N. Thompson. 1992. Biology and systematics of *Greya* Busck and *Tetragma* new genus (Lepidoptera: Prodoxidae). *Smithsonian Contributions to Zoology* 524: 1–88.

Davis, M. B. 1983. Quaternary history of deciduous forests of eastern North America and Europe. *Annals of the Missouri Botanical Garden* 70: 550–63.

Day, M. C. 1977. A new genus of Plumariidae from southern Africa, with notes on Scolebythidae (Hymenoptera: Chrysidoidea). *Cimbebasia* 4: 171–7.

———. 1984. The enigmatic genus *Heterogyna* Nagy (Hymenoptera: Sphecidae: Heterogyninae). *Systematic Entomology* 9: 293–307.

———, G. R. Else, and D. Morgan. 1981. The most primitive Scoliidae (Hymenoptera). *Journal of Natural History* 15: 671–84.

Deitz, L. L., and C. H. Dietrich. 1993. Superfamily Membracoidea (Homoptera: Auchenorrhyncha). I. Introduction and revised classification with new family-group taxa. *Systematic Entomology* 18: 287–96.

———, C. Nalepa, and K.-D. Klass. 2003. Phylogeny of the Dictyoptera re-examined (Insecta). *Entomologische Abhandlungen* 61: 69–91.

De Jong, R. 2003. Are there butterflies with Gondwanan ancestry in the Australian region? *Invertebrate Systematics* 17: 143–56.

———, R. I. Vane-Wright, and P. R. Ackery. 1996. The higher classification of butterflies (Lepidoptera): Problems and prospects. *Entomologica Scandinavica* 27: 65–101.

Delfinado, M. D., and D. E. Hardy, eds. 1977. *A Catalogue of the Diptera of the Oriental Region* [3 volumes]. University of Hawaii Press; Honolulu, Hawaii 1931 pp.

Demoulin, G. 1954a. Essai sur quelques Ephéméroptères fossiles adultes. *Volume Jubilaire Victor van Straelen, Bruxelles* 1: 547–74.

———. 1954b. Quelques remarques sur les Archodonates. *Bulletin et Annales de la Société Royale d'Entomologie de Belgique* 90: 327–37.

———. 1955. Remarques critiques sur *Cronicus anomalus* (Pictet). Ephéméroptère de l'ambre oligocène de la Baltique. *Bulletin de l'Institut Royal des Sciences Naturelles de Belgique* 31: 1–4.

———. 1956. *Electrogenia dewalschei* n. gen. n. sp., Ephéméroptère fossile de l'ambre. *Bulletin et Annales de la Société Royale d'Entomologie de Belgique* 92: 95–100.

———. 1965. Contribution à la connaissance des Ephéméroptères de l'ambre oligocène de la Baltique. *Entomologiske Meddelelser* 34: 143–53.

———. 1968. Deuxième contribution à la connaissance des Ephéméroptères de l'ambre oligocène de la Baltique. *Deutsche Entomologische Zeitschrift* 15: 233–76.

———. 1970a. Troisème contribution à la connaissance des Ephéméroptères de l'ambre oligocène de la Baltique. *Bulletin de l'Institut Royal des Sciences Naturelles de Belgique* 46(2): 1–11.

———. 1970b. Contribution à la connaissance des Ephéméroptères du Miocène. I. *Siphurites explanatus* Cockerell. *Bulletin de l'Institut Royal des Sciences Naturelles de Belgique* 46(5): 1–4.

Denlinger, D. L. 2002. Regulation of diapause. *Annual Review of Entomology* 47: 93–122.

De Queiroz, A., and P. H. Wimberger. 1993. The usefulness of behavior for phylogeny estimation: Levels of homoplasy in behavioral and morphological characters. *Evolution* 47: 46–60.

De Queiroz, K., and M. J. Donoghue. 1988. Phylogenetic systematics and the species problem. *Cladistics* 4: 317–38.

DeSalle, R., J. Gatesy, W. Wheeler, and D. Grimaldi. 1992. DNA sequences from a fossil termite in Oligo-Miocene amber and their phylogenetic implications. *Science* 257: 1933–6.

Després, L., and N. Jaeger. 1999. Evolution of oviposition strategies and speciation in the globe-flower flies *Chiastocheta* spp. (Anthomyiidae). *Journal of Evolutionary Biology* 12: 822–31.

———, E. Pettex, V. Plaisance, and F. Pompanon. 2002. Speciation in the globeflower fly *Chiastocheta* spp. (Diptera: Anthomyiidae) in relation to host plant species, biogeography, and morphology. *Molecular Phylogenetics and Evolution* 22: 258–68.

Dessart, P. 1985. Complements a l'etude de *Dendrocerus koyamai* (Ishii, 1951) (Hymenoptera Ceraphronoidea Megaspilidae). *Bulletin et Annales de la Société Royale Belge d'Entomologie* 121: 113–17.

Desutter-Grandcolas, L. 2003. Phylogeny and the evolution of acoustic communication in extant Ensifera (Insecta, Orthoptera). *Zoologica Scripta* 32: 525–61.

Dethier, V. G. 1976. *The Hungry Fly.* Harvard University Press; Cambridge, Massachusetts; 489 pp.

Detling, J. K. 1988. Grasslands and savannas: Regulation of energy flow and nutrient cycling by herbivores. Pp. 131–48. *In* Pomeroy, L. R., and J. A. Alberts (eds.), *Concepts of Ecosystem Ecology: A Comparative Review.* Springer Verlag; Berlin, Germany; xii+384 pp.

Dettmann, M. E. 1990. The Antarctic Australian rift-valley – Late Cretaceous cradle of northeastern Australasian relicts. *Review of Palaeobotany and Palynology* 65: 131–44.

———, D. T. Pocknall, E. J. Romero, and M. C. Zamaloa. 1987. *Nothofagidites* Erdtman ex Potonié, 1960; a catalogue of species with notes on the paleogeographic distribution of *Nothofagus* Bl. (southern beech). *New Zealand Geological Survey Paleontological Bulletin* 60: 1–79.

Dettner, K. 1987. Chemosystematics and evolution of beetle chemical defenses. *Annual Review of Entomology* 32: 17–48.

Deutsch, J. S. 2001. Are Hexapoda members of the Crustacea? Evidence from comparative developmental genetics. *Annales de la Société Entomologique de France* 37: 41–9.

DeVries, P. J. 1983. *Zamia skinneri* and *Z. fairchildiana* (*Zamia*, Palmera Siempre Verde, Cycad). Pp. 349–50. *In* Janzen, D. H. (ed.), Costa Rican Natural History. University of Chicago Press; Chicago, Illinois; xi+816 pp.

———. 1987. *The Butterflies of Costa Rica and Their Natural History: Papilionidae, Pieridae, Nymphalidae.* Princeton University Press; Princeton, New Jersey; xxii+327 pp.

———. 1990. Enhancement of symbioses between butterfly caterpillars and ants by vibrational communication. *Science* 48: 1104–6.

———. 1991. Evolutionary and ecological patterns in myrmecophilous riodinid butterflies. Pp. 143–56. *In* Huxley, C. R., and D. F. Cutler (eds.), *Ant-Plant Interactions.* Oxford University Press; Oxford, UK; xviii+601 pp.

———. 1992. Singing caterpillars, ants and symbioses. *Scientific American* 267: 76–82.

———. 1997. *The Butterflies of Costa Rica and Their Natural History, Volume 2: Riodinidae.* Princeton University Press; Princeton, New Jersey; xxv+288 pp.

———. 2001. Butterflies. Pp. 559–73. *In* Levin, S. A. (ed.), *Encyclopedia of Biodiversity* [Volume 1]. Academic Press; New York, New York; xxi+943 pp.

———, and T. R. Walla. 2001. Species diversity and community structure in neotropical fruit-feeding butterflies. *Biological Journal of the Linnean Society* 74: 1–15.

———, R. B. Cocroft, and J. A. Thomas. 1993. Comparison of acoustical signals in *Maculinea* butterfly caterpillars and their obligate host *Myrmica* ants. *Biological Journal of the Linnean Society* 49: 229–38.

Dewell, R. A., and W. C. Dewell. 1996. The brain of *Echiniscus viridissimus* (Heterotardigrada): A key to understanding the phylogenetic position of tardigrades and the evolution of the arthropod head. *Zoological Journal of the Linnean Society* 116: 35–49.

———, and W. C. Dewell. 1998. The place of tardigrades in arthropod evolution. Pp. 109–23. *In* Fortey, R. A., and R. H. Thomas (eds.), *Arthropod Relationships.* Chapman and Hall; London, UK; xii+383 pp.

———, D. R. Nelson, and W. C. Dewell. 1993. Tardigrada. Pp. 115–42. *In* Harrison, F. W., and M. E. Rice (eds.), *Microscopic Anatomy of Invertebrates, Volume 12: Onychophora, Chilopoda and Lesser Protostomata.* Wiley-Liss; New York, New York; xiv+484 pp.

Deyrup, M. A. 1984. A maple wood wasp, *Xiphydria maculata*, and its insect enemies (Hymenoptera: Xiphydriidae). *Great Lakes Entomologist* 17: 17–28.

———. 2001. Endangered terrestrial invertebrates. Pp. 487–95. *In* Levin, S. A. (ed.), *Encyclopedia of Biodiversity* [Volume 2]. Academic Press; New York, New York, xxxi+826 pp.

D'Haese, C. A. 2002. Were the first springtails semi-aquatic? A phylogenetic approach by means of 28S rDNA and optimization alignment. *Proceedings of the Royal Society of London, Series B, Biological Sciences* 269: 1143–51.

Dieguez, C., J. L. Nieves-Aldrey, and E. Barron. 1996. Fossil galls (zoocecids) from the Upper Miocene of La Cerdana (Lerida, Spain). *Review of Palaeobotany and Palynology* 94: 329–43.

Diels, L. 1916. Käferblumen bei den Ranales und ihre Bedeutung für die Phylogenie der Angiospermen. *Berichte der Deutsche Botanische Gesellschaft* 34: 758–74.

Dietrich, C. H., and L. L. Deitz. 1993. Superfamily Membracoidea (Homoptera: Auchenorrhyncha). II. Cladistic analysis and conclusions. *Systematic Entomology* 18: 297–311.

———, S. H. McKamey, and L. L. Deitz. 2001. Morphology-based phylogeny of the treehopper family Membracidae (Hemiptera: Cicadomorpha: Membracoidea). *Systematic Entomology* 26: 213–39.

Dilcher, D. L. 1995. Plant reproductive strategies: Using the fossil record to unravel current issues in plant reproduction. Pp. 187–98. *In* Hoch, P. C., and A. G. Stephenson (eds.), *Experimental and Molecular Approaches to Plant Biosystematics.* Missouri Botanical Garden; St. Louis, Missouri; xxii+391 pp.

———. 2001. Toward a new synthesis: Major evolutionary trends in the angiosperm fossil record. Pp. 255–70. *In* Ayala, F. J., W. M. Fitch, and M. T. Clegg (eds.), *Variation and Evolution in Plants and Microorganisms: Towards a New Synthesis 50 Years after Stebbins.* National Academy Press; Washington, D.C.; xi+640 pp.

Dill, M., D. J. Williams, and U. Maschwitz. 2002. Herdsmen ants and their mealybug partners. *Abhandlungen der Senckenbergischen Naturforschenden Gesellschaft* 557: 1–373.

Dirsh, V. M. 1975. *Classification of the Acridomorpoid Insects.* Classey; Faringdon, UK; vii+171 pp.

Disney, R. H. L. 1994. *Scuttle flies: The Phoridae.* Chapman and Hall; London, UK; xii+467 pp.

Dlussky, G. M. 1983. A new family of Upper Cretaceous Hymenoptera – An "intermediate link" between the ants and the scolioids. *Paleontological Journal* 3: 65–78.

———. 1987. New Formicoidea (Hymenoptera) of the Upper Cretaceous. *Paleontological Journal* 1987: 146–50.

———. 1988. Ants from (Paleocene?) Sakhalin amber. *Paleontological Journal* 22: 50–60.

———. 1996. Ants (Hymenoptera: Formicidae) from Burmese amber. *Paleontological Journal* 30: 449–54.

——. 1997. Genera of ants (Hymenoptera: Formicidae) from Baltic amber. *Paleontologicheskiy Zhurnal* 6: 50–62. [In Russian]

———. 1999. New ants (Hymenoptera: Formicidae) from Canadian amber. *Paleontological Journal* 33: 409–12.

———, and A. P. Rasnitsyn. 1999. Two new species of aculeate hymenopterans (Vespidae = Hymenoptera) from the middle Eocene of the United States. *Paleontologicheskiy Zhurnal* 5: 72–5. [In Russian]

———, and A. P. Rasnitsyn. 2003. Ants (Hymenoptera: Formicidae) of Formation Green River and some other middle Eocene deposits of North America. *Russian Entomological Journal* 11: 411–36.

——, D. J. Brothers, and A. P. Rasnitsyn. 2004. The first Late Cretaceous ants (Hymenoptera: Formicidae) from southern Africa, with comments on the origin of the Myrmicinae. *Insect Systematics and Evolution* 35: 1–13.

Dmitriev, V. Yu., and A. G. Ponomarenko. 2002. Dynamics of insect taxonomic diversity. Pp. 325–31. *In* Rasnitsyn, A. P., and D. L. J. Quicke (eds.), *History of Insects.* Kluwer Academic Publishers; Dordrecht, the Netherlands; xii+517 pp.

Doberenz, A., P. Matter III, and R. W. G. Wyckoff. 1966. The microcomposition of some fossil insects of Miocene age. *Bulletin of the Southern California Academy of Sciences* 65: 229–35.

Dobosz, R., and W. Krzemiński. 2000. A new species of the Coniopterygidae (Neuroptera) from Baltic amber. *Polskie Pismo Entomologiczne* 69: 219–24.

Dobruskina, I. A., A. G. Ponomarenko, and A. P. Rasnitsyn. 1997. Fossil insects found in Israel. *Paleontologischeskiy Zhurnal* 5: 91–5. [In Russian]

Dobzhansky, T. G. 1937. *Genetics and the Origin of Species.* Columbia University Press; New York, New York; xvi+364 pp.

Dodd, M. E., J. *Silvertown*, and M. W. Chase. 1999. Phylogenetic analysis of trait evolution and species diversity variation among angiosperm families. *Evolution* 53: 732–44.

Dodson, C. H., R. L. Dressler, H. G. Hills, and R. M. Adams. 1969. Biologically active compounds in orchid fragrances. *Science* 164: 1243–9.

Dodson, G. N. 1989. The horny antics of antlered flies. *Australian Natural History* 22: 604–11.

———. 1997. Resource defense mating system in antlered flies, *Phytalmia* spp. (Diptera: Tephritidae). *Annals of the Entomological Society of America* 90: 496–504.

Dohle, W. 1998. Myriapod-insect relationships as opposed to an insect-crustacean sister group relationship. Pp. 305–15. *In* Fortey, R. A., and R. H. Thomas (eds.), *Arthropod Relationships.* Chapman and Hall; London, UK; xii+383 pp.

———. 2001. Are the insects terrestrial crustaceans? A discussion of some new facts and arguments and the proposal of the proper name 'Tetraconata' for the monophyletic unit Crustacea+Hexapoda. *Annales de la Société Entomologique de France* 37: 85–103.

Dolan, M. F. 2001. Speciation of termite gut protists: The role of bacterial symbionts. *International Microbiology* 4: 203–8.

Dolin, V. G. 1975. On the systematics of Mesozoic click beetles (Coleoptera, Elateridae). *Paleontologicheskiy Zhurnal* 1975: 51–61. [In Russian]

———. 1976. Fossil click beetles (Coleoptera, Elateridae) of the subfamilies Negastriinae and Cardophorinae from the Upper Jurassic of Karatau. *Vestnik Zoologii* 1976: 68–75. [In Russian]

Doludenko, M. P., and E. R. Orlovskaya. 1976. Jurassic floras of the Karatau Range, southern Kazakhstan. *Palaeontology* 19: 627–40.

Domínguez, E. 2001. *Trends in Research in Ephemeroptera and Plecoptera.* Kluwer Academic Publishers; Dordrecht, the Netherlands; xiii+478 pp.

Donisthorpe, H.StJ.K. 1927. *British Ants, Their Life-History and Classification.* Routledge and Sons; London, UK; xv+436 pp.

Donoghue, M. J., and J. A. Doyle. 1989. Phylogenetic analysis of angiosperms and the relationships of Hamamelidae. Pp. 17–45. *In* Crane, P. R., and S. Blackmore (eds.), *Evolution, Systematics, and Fossil History of the Hamamelidae* [Volume 1]. Clarendon Press; Oxford, UK; xii+305 pp.

Donovan, S. E., D. T. Jones, W. A. Sands, and P. Eggleton. 2000. Morphological phylogenetics of termites (Isoptera). *Biological Journal of the Linnean Society* 70: 467–513.

Donovan, S. K., ed. 1991. *The Processes of Fossilization.* Belhaven Press; London, UK; ix+303 pp.

———, and C. J. Veltkamp. 1994. Unusual preservation of Late Quaternary millipedes from Jamaica. *Lethaia* 27: 355–62.

Douglas, S. D., and R. A. Stockey. 1996. Insect fossils in middle Eocene deposits from British Columbia and Washington state:

Faunal diversity and geological range extensions. *Canadian Journal of Zoology* 74: 1140–57.

Doutt, R. L. 1973a. The genus *Polynemoidea* Girault (Hymenoptera: Mymaridae). *Pan-Pacific Entomologist* 49: 215–20.

———. 1973b. The fossil Mymaridae (Hymenoptera: Chalcidoidea). *Pan-Pacific Entomologist* 49: 221–8.

———. 1973c. Maternal care of immature progeny by parasitoids. *Annals of the Entomological Society of America* 66: 486–7.

Downes, W. L., Jr., and G. A. Dahlem. 1987. Keys to the evolution of Diptera: Role of Homoptera. *Environmental Entomology* 16: 847–54.

Downey, J. C. 1962. Myrmecophily in *Plebejus (Icaricia) icaroides*. *Entomological News* 73: 57–66.

———. 1965. Thrips utilize exudations of Lycaenidae. *Entomological News* 76: 25–7.

Dowton, M., and A. D. Austin. 1994. Molecular phylogeny of the insect order Hymenoptera: Apocritan relationships. *Proceedings of the National Academy of Sciences, U.S.A.* 91: 9911–15.

———, and A. D. Austin. 2001. Simultaneous analysis of 16S, 28S, COI and morphology in the Hymenoptera: Apocrita – evolutionary transitions among parasitic wasps. *Biological Journal of the Linnean Society* 74: 87–111.

Doyen, J. T., and W. R. Tschinkel. 1982. Phenetic and cladistic relationships among tenebrionid beetles (Coleoptera). *Systematic Entomology* 7: 127–83.

Doyle, J. A. 1998. Molecules, morphology, fossils, and the relationship of angiosperms and Gnetales. *Molecular Phylogenetics and Evolution* 9: 448–62.

———, and M. J. Donoghue. 1992. Fossils and seed plant phylogeny reanalyzed. *Brittonia* 44: 89–106.

———, and P. K. Endress. 2000. Morphological phylogenetic analysis of basal angiosperms: Comparisons and combination with molecular data. *International Journal of Plant Sciences, Supplement* 161: S121–53.

———, and M. J. Sanderson. 1997. Fossils, molecular clocks, and the age of angiosperms. *American Journal of Botany* 84: 132–42.

Dressler, R. L. 1968. Pollination by euglossine bees. *Evolution* 22: 202–10.

———. 1990. *The Orchids: Natural History and Classification*. Harvard University Press; Cambridge, Massachusetts; [6]+332 pp.

Drinnan, A. N., P. R. Crane, E. M. Friis, and K. R. Pedersen. 1990. Lauraceous flowers from the Potomac Group (mid-Cretaceous) of eastern North America. *Botanical Gazette* 151: 370–84.

Dubinin, V. B. 1948. Finds of Pleistocene lice (Anoplura) and nematodes during studies of bodies of fossil ground squirrels from the Indigirka. *Doklady Akademii Nauk, SSSR* 62: 417–20. [In Russian]

Dudley, R. 1998. Atmospheric oxygen, giant Paleozoic insects and the evolution of aerial locomotor performance. *Journal of Experimental Biology* 201: 1043–50.

———. 2000. *Biomechanics of Insect Flight: Form, Function, Evolution*. Princeton University Press; Princeton, New Jersey; xii+476 pp.

Duffey, J. E. 1996. Eusociality in a coral-reef shrimp. *Nature* 381: 512–14.

Dugdale, J. S. 1974. Female genital configuration in the classification of Lepidoptera. *New Zealand Journal of Zoology* 1: 127–46.

———, N. P. Kristensen, G. S. Robinson, and M. J. Scoble. Pp. 119–30. *In* Kristensen, N. P. (ed.), *Handbuch der Zoologie: Eine Naturgeschichte der Stämme des Tierreiches: Band IV: Arthropoda: Insecta: Teilband 35: Lepidoptera, Moths and Butterflies: Volume 1: Evolution, Systematics, and Biogeography*. Walter de Gruyter; Berlin, Germany; x+491 pp.

Dunbar, C. O. 1924. Kansas Permian insects. Part 1. The geologic occurrence and the environment of the insects. *American Journal of Science* 7: 171–203.

———. 1937. Robin John Tillyard: 1881–1937. *American Journal of Science* 33: 317–18.

Duncan, I. J., and D. E. G. Briggs. 1996. Three-dimensionally preserved insects from the Tertiary of Riversleigh, Australia. *Nature* 381: 30–1.

———, and D. E. G. Briggs, and M. Archer. 1998. Three-dimensionally mineralized insects and millipedes from the Tertiary of Riversleigh, Queensland, Australia. *Palaeontology* 41: 835–51.

Dunlop, J. A. 1994. An Upper Carboniferous amblypygid from the Writhlington Geological Nature Reserve. *Proceedings of the Geologists' Association* 105: 245–50.

———. 1998. A fossil whipscorpion from the Lower Cretaceous of Brazil. *Journal of Arachnology* 26: 291–5.

———. 1999. Poecilophysidea: A forgotten arachnid order illustrating a forgotten phylogenetic hypothesis. *British Arachnological Society Newsletter* 85: 4–6.

———, and M. Webster. 1999. Fossil evidence, terrestrialization and arachnid phylogeny. *Journal of Arachnology* 27: 86–93.

Dunn, L. H. 1930. Rearing the larvae of *Dermatobia hominis* Linn., in man. *Psyche* 37: 327–42.

Dunning, D. C. 1968. Warning sounds of moths. *Zeitschrift für Tierpsychologie* 25: 129–38.

Dupuis, C. 1974. Pierre André Latreille (1762–1833): The foremost entomologist of his time. *Annual Review of Entomology* 19: 1–13.

Durden, C. J. 1966. Oligocene lake deposits in central Colorado and a new fossil insect locality. *Journal of Paleontology* 40: 217–19.

———1984. Carboniferous and Permian entomology of western North America. Pp. 81–9. *In* Sutherland, P. K., and W. L. Manger (eds.), *Neuvième Congrès International de Stratigraphie et de Géologie du Carbonifère: Compte Rendu Volume 2: Biostratigraphy*. Southern Illinois University Press; Carbondale, Illinois; x+630 pp.

———. 1988. Hamilton insect fauna. *Kansas Geological Survey, Guidebook Series* 6: 117–24.

———, and G. G. Musser. 1994. The sucking lice (Insecta, Anoplura) of the world: A taxonomic checklist with records of mammalian hosts and geographical distributions. *Bulletin of the American Museum of Natural History* 218: 1–90.

———, and H. Rose. 1978. Butterflies from the middle Eocene: The earliest occurrence of fossil Papilionoidea (Lepidoptera). *The Pearce-Sellards Series (Texas Memorial Museum, Austin)* 29: 1–25.

Duringer, P., M. Brunet, Y. Cambefort, A. Likius, H. T. Mackaye *et al.* 2000a. First discovery of fossil dung beetle brood balls and nests in the Chadian Pliocene australopithecine levels. *Lethaia* 33: 277–84.

———, M. Brunet, Y. Cambefort, A. Beauvilain, H. T. Mackaye *et al.* 2000b. Des boules de bousiers fossiles et leurs terriers dans les sites a australopitheques du Pliocene tchadien. *Bulletin de la Société Geologique de France* 171: 259–69.

D'Urso, V. 1993. The wing coupling apparatus in *Peloridium hammoniorum* Breddin, 1897. *Spixiana* 16: 133–9.

Dworakowska, I. 1988. Main veins of the wings of Auchenorrhyncha (Insecta, Rhynchota: Hemelytrata). *Entomologische Abhandlungen* 52: 63–108.

Dybas, H. S. 1966. Evidence for parthenogenesis in the featherwing beetles, with a taxonomic review of a new genus and 8 new species (Coleoptera: Ptiliidae). *Feldiana, Zoology* 51: 11–52.

Dybas, L. K., and H. S. Dybas. 1981. Coadaptation and taxonomic differentiation of sperm and spermathecae in featherwing beetles. *Evolution* 35: 168–74.

Dzhambazov, B., and I. Teneva. 2000. Beobachtungen zum Blütenbesuch von Fliegen der Überfamilie Empidoidea (Hybotidae, Empididae). *Studia Dipterologica* 7: 553–7.

Dzik, J. 1981. An Early Triassic millipede from Siberia and its evolutionary significance. *Neues Jahrbuch für Geologie und Paläontologie, Monatshefte* 1981: 395–404.

———. 2003. Early Cambrian lobopodian sclerites and associated fossils from Kazakhstan. *Palaeontology* 46: 93–112.

———, and G. Krumbiegel. 1989. The oldest "onychophoran" Xenusion: A link connecting phyla? *Lethaia* 22: 169–81.

Ebel, R. E. 1991. Order Siphonaptera. Pp. 674–689. *In* Stehr, F. W. (ed.), *Immature Insects* [Volume 2]. Kendall-Hunt; Dubuque, Iowa; xvi+975 pp.

Eberhard, W. G. 1975. The ecology and behavior of a subsocial pentatomid bug and two scelionid wasps: Strategy and counterstrategy in a host and its parasites. *Smithsonian Contributions to Zoology* 205: 1–39.

———. 1980. Horned beetles. *Scientific American* 242: 166–82.

———. 1985. *Sexual Selection and Animal Genitalia*. Harvard University Press; Cambridge, Massachusetts; x+244 pp.

———. 2000a. Spider web manipulation by a wasp larva. *Nature* 406: 255–6.

———. 2000b. The natural history and behavior of *Hymenoepimecis argyraphaga* (Hymenoptera: Ichneumonidae) a parasitoid of *Plesiometa argyra* (Araneae: Tetragnathidae). *Journal of Hymenoptera Research* 9: 220–40.

Edgecombe, G. D., and L. Ramsköld. 1999. Relationships of Cambrian Arachnata and the systematic position of Trilobita. *Journal of Paleontology* 73: 263–87.

———, G. Giribet, and W. C. Wheeler. 1999. Phylogeny of Chilopoda: Combining 18S and 28S rRNA sequences and morphology. *Boletin de la Sociedad Entomológica Aragonesa* 26: 293–331.

———, G. D. F. Wilson, D. J. Colgan, M. R. Gray, and G. Cassis. 2000. Arthropod cladistics: Combined analysis of histone H3 and U2 snRNA sequences and morphology. *Cladistics* 16: 155–203.

Edgerly, J. S. 1987a. Maternal behavior of a webspinner (order Embiidina). *Ecological Entomology* 12: 1–11.

———. 1987b. Colony composition and some costs and benefits of facultatively communal behavior in a Trinidadian webspinner, *Clothoda urichi* (Embiidina: Clothodidae). *Annals of the Entomological Society of America* 80: 29–34.

———. 1988. Material behavior of a webspinner (order Embiidina): Mother-nymph association. *Ecological Entomology* 13: 263–72.

———. 1994. Is group living an antipredator defense in a facultatively communal wesbpinner (Embiidina: Clothodidae). *Journal of Insect Behavior* 7: 135–47.

Edmunds, G. F., Jr. 1972. Phylogenetic biogeography of mayflies. *Annals of the Missouri Botanical Garden* 62: 251–63.

———, and J. R. Traver. 1954. An outline of a reclassification of the Ephemeroptera. *Proceedings of the Entomological Society of Washington* 56: 236–40.

———, S. L. Jensen, and L. Berner. 1976. *The Mayflies of North and Central America*. University of Minnesota Press; Minneapolis, Minnesota; x+330 pp.

Edson, K. M., S. B. Vinson, D. B. Stoltz, and M. D. Summers. 1981. Virus in a parasitoid wasp: Suppression of the cellular immune response in the parasitoid's host. *Science* 211: 582–3.

Edwards, D. 1966. New insights into early land ecosystems: A glimpse of a Lilliputian world. *Review of Palaeobotany and Palynology* 90: 159–74.

———, P. A. Selden, J. B. Richardson, and L. Axe. 1995. Coprolites as evidence for plant-animal interactions in Siluro-Devonian terrestrial ecosystems. *Nature* 377: 329–31.

Edwards, E. D., P. Gentili, N. P. Kristensen, M. Horak, and E. S. Nielsen. 1999. The cossoid/sesioid assemblage. Pp. 181–98. *In* Kristensen, N. P. (ed.), *Handbuch der Zoologie: Eine Naturgeschichte der Stämme des Tierreiches: Band IV: Arthropoda: Insecta: Teilband 35: Lepidoptera, Moths and Butterflies: Volume 1: Evolution, Systematics, and Biogeography*. Walter de Gruyter; Berlin, Germany; x+491 pp.

Eggert, A.-K., and J. K. Müller. 1997. Biparental care and social evolution in burying beetles: Lessons from the larder. Pp. 216–36. *In* Choe, J. C., and B. J. Crespi (eds.), *The Evolution of Social Behavior in Insects and Arachnids*. Cambridge University Press; Cambridge, UK; vii+541 pp.

Eggleton, P. 2000. Global patterns of termite diversity. Pp. 25–51. *In* Abe, T., D. E. Bignell, and M. Higashi (eds.), *Termites: Evolution, Sociality, Symbioses, Ecology*. Kluwer Academic Publishers; Dordrecht, the Netherlands; xxii+466 pp.

———, and R. Belshaw. 1992. Insect parasitoids: An evolutionary overview. *Philosophical Transactions of the Royal Society, London (B)* 337: 1–20.

Eglinton, G., and G. A. Logan. 1991. Molecular preservation. *Philosophical Transactions of the Royal Society of London (B)* 333: 315–28.

Ehrenfeld, J., and J. G. Rozen, Jr. 1977. The cuckoo bee genus *Kelita*, its systematics, biology, and larvae. *American Museum Novitates* 2631: 1–24.

Ehrlich, P. R. 1958. The comparative morphology, phylogeny and higher classification of the butterflies (Lepidoptera: Papilionoidea). *University of Kansas Scientific Bulletin* 39: 305–70.

———, and P. H. Raven. 1965. Butterflies and plants: A study in coevolution. *Evolution* 18: 586–608.

Ehrman, L., and J. R. Powell. 1982. The *Drosophila willistoni* species group. Pp. 193–225. *In* Ashburner, M., H. L. Carson, and J. N. Thompson, Jr. (eds.), *The Genetics and Biology of* Drosophila [Volume 3b]. Academic Press; New York, New York; 362 pp.

Ehrmann, R. 2002. *Mantodea: Gottesanbeterinnen der Welt*. Natur und Tier; Munich, Germany; 519 pp. [+2 pp. inserted errata]

Eickwort, G. C. 1981. Presocial insects. Pp. 199–280. *In* Hermann, H. R. (ed.), *Social Insects* [Volume 2]. Academic Press; New York, New York; xiii+491 pp.

Eisner, T. 2003. *For Love of Insects*. Harvard University Press; Cambridge, Massachusetts; xi+448 pp.

———, and R. E. Silberglied. 1988. A chrysopid larva that cloaks itself in mealybug wax. *Psyche* 95: 15–19.

———, T. H. Jones, J. Aneshansley, W. R. Tschinkel, R. E. Silberglied, and J. Meinwald. 1977. Chemistry of defensive secretions of bombardier beetles (Brachinini, Metriini, Ozaenini, Paussini). *Journal of Insect Physiology* 23: 1383–6.

———, K. Hicks, M. Eisner, and D. S. Robson. 1978. "Wolf-in-sheep's-clothing" strategy of a predaceous insect larva. *Science* 199: 790–4.

———, S. Nowicki, M. Goetz, and J. Meinwald. 1980. Red cochineal dye (carminic acid): Its role in nature. *Science* 208: 1039–42.

———, S. R. Smedley, D. K. Young, M. Eisner, B. Roach, and J. Meinwald. 1996. Chemical basis of courtship in a beetle (*Neopyrochroa flabellata*): Cantharidin as precopulatory "enticing" agent; cantharidin as "nuptial gift." *Proceedings of the National Academy of Sciences, U.S.A.* 93: 6494–8, and 6499–6503.

Eldredge, N., and J. Cracraft. 1980. *Phylogenetic Patterns and the Evolutionary Process*. Columbia University Press; New York, New York; x+349 pp.

———, and S. M. Stanley, eds. 1984. *Living Fossils*. Springer Verlag; Berlin, Germany; xi+291 pp.

Elias, S. 1994. *Quaternary Insects and Their Environments*. Smithsonian Institution Press; Washington, D.C.; xiii+284 pp.

Eliot, J. N. 1973. The higher classification of the Lycaenidae (Lepidoptera): A tentative arrangement. *Bulletin of the British Museum of Natural History (Entomology)* 28: 371–505.

Elliott, D. K., and J. D. Nations. 1998. Bee burrows in the Late Cretaceous (Late Cenomanian) Dakota Formation, northeastern Arizona. *Ichnos* 5: 243–53.

Elliott, E. A. 1922. Monograph of the hymenopterous family Stephanidae. *Proceedings of the Zoological Society of London* 92: 705–831.

Emerson, A. E. 1955. Geographical origins and dispersions of termite genera. *Feldiana (Zoology)* 37: 465–521.

———. 1961. Vestigial characters of termites and processes of regressive evolution. *Evolution* 15: 115–31.

———. 1965. A review of the Mastotermitidae (Isoptera), including a new fossil genus from Brazil. *American Museum Novitates* 2236: 1–46.

———. 1968. A revision of the fossil genus *Ulmeriella* (Isoptera, Hodotermitidae, Hodotermitinae). *American Museum Novitates* 2332: 1–22.

———. 1969. A revision of the Tertiary fossil species of the Kalotermitidae (Isoptera). *American Museum Novitates* 2359: 1–57.

———. 1971. Tertiary fossil species of the Rhinotermitidae (Isoptera), phylogeny of genera, and reciprocal phylogeny of associated flagellata (Protozoa) and the Staphylinidae (Coleoptera). *Bulletin of the American Museum of Natural History* 146: 247–303.

———, and K. Krishna. 1975. The termite family Serritermitidae (Isoptera). *American Museum Novitates* 2570: 1–31.

Emerson, K. C., and R. D. Price. 1988. A new species of *Haematomyzus* (Mallophaga: Haematomyzidae) off the Bush Pig, *Potamochoerus porcus*, from Ethiopia, with comments on lice found on pigs. *Proceedings of the Entomological Society of Washington* 90: 338–42.

Emery, C. 1886. Ueber Phylogenie und Systematik der Insekten. *Biologisches Centralblatt* 5: 648–56.

Emlen, D. J. 2000. Integrating development with evolution: A case study with beetle horns. *BioScience* 50: 403–18.

———, and H. F. Nijhout. 2000. The development and evolution of exaggerated morphologies in insects. *Annual Review of Entomology* 45: 661–708.

Emmel, T. C., M. C. Minno, and B. A. Drummond. 1992. *Florissant Butterflies: A Guide to the Fossil and Present-Day Species of Central Colorado*. Stanford University Press; Stanford, California; vi+118 pp.

Enderlein, G. 1905. Über die Klassifikation der Stephaniden. *Zoologischer Anzeiger* 28: 473–7.

———. 1906. Neue Honigbienen und Beiträge zur Kenntnis der Verbreitung der Gattung *Apis*. *Stettiner Entomologische Zeitung* 67: 331–44.

Endress, P. K. 1987. The Chloranthaceae: reproductive structures and phylogenetic position. *Botanische Jahrbucher für Systematik, Pflazengeschichte und Pflanzengeographie* 109: 153–226.

———. 1990. Evolution of reproductive structures and functions in primitive angiosperms (Magnoliidae). *Memoirs of the New York Botanical Garden* 55: 5–34.

Endrödi, S. 1985. *The Dynastinae of the World*. W. Junk; The Hague; the Netherlands; 800 pp.

Engel, M. S. 1995a. Notes on gasteruptiid wasps (Hymenoptera: Evanioidea). *Journal of the New York Entomological Society* 103: 214–15.

———. 1995b. *Neocorynura electra*, a new fossil bee species from Dominican amber (Hymenoptera: Halictidae). *Journal of the New York Entomological Society* 103: 317–23.

———. 1995c. A new fossil snake-fly species from Baltic amber (Raphidioptera: Inocelliidae). *Psyche* 102: 187–93.

———. 1996. New augochlorine bees (Hymenoptera: Halictidae) in Dominican amber, with a brief review of fossil Halictidae. *Journal of the Kansas Entomological Society* 69: 334–45.

———. 1997. A new fossil bee from the Oligo-Miocene Dominican amber (Hymenoptera: Halictidae). *Apidologie* 28: 97–102.

———. 1998a. Fossil honey bees and evolution in the genus *Apis* (Hymenoptera: Apidae). *Apidologie* 29: 265–81.

———. 1998b. A new species of the Baltic amber bee genus *Electrapis* (Hymenoptera: Apidae). *Journal of Hymenoptera Research* 7: 94–101.

———. 1998c. *Megatypus parvus* spec. nov., a new giant dragonfly from the Lower Permian of Kansas (Protodonata: Meganeuridae). *Odonatologica* 27: 361–4.

———. 1999a. The first fossil *Euglossa* and phylogeny of the orchid bees (Hymenoptera: Apidae; Euglossini). *American Museum Novitates* 3272: 1–14.

———. 1999b. *Megachile glaesaria*, the first megachilid bee fossil from amber (Hymenoptera: Megachilidae). *American Museum Novitates* 3276: 1–13.

———. 1999c. A new xeromelissine bee in Tertiary amber of the Dominican Republic (Hymenoptera: Colletidae). *Entomologica Scandinavica* 30: 453–8.

———. 1999d. The first fossil of a pleasing lacewing (Neuroptera: Dilaridae). *Proceedings of the Entomological Society of Washington* 101: 822–6.

———. 1999e. The taxonomy of Recent and fossil honey bees (Hymenoptera: Apidae; *Apis*). *Journal of Hymenoptera Research* 8: 165–96.

———. 2000a. A new interpretation of the oldest fossil bee (Hymenoptera: Apidae). *American Museum Novitates* 3296: 1–11.

———. 2000b. Classification of the bee tribe Augochlorini (Hymenoptera: Halictidae). *Bulletin of the American Museum of Natural History* 250: 1–89.

———. 2001a. A monograph of the Baltic amber bees and evolution of the Apoidea (Hymenoptera). *Bulletin of the American Museum of Natural History* 259: 1–192.

———. 2001b. Monophyly and extensive extinction of advanced eusocial bees: Insights from an unexpected Eocene diversity. *Proceedings of the National Academy of Sciences, U.S.A.* 98: 1661–4.

———. 2001c. The first large carpenter bee from the Tertiary of North America, with a consideration of the geological history of Xylocopinae (Hymenoptera: Apidae). *Transactions of the American Entomological Society* 127: 245–54.

———. 2001d. New neotropical records for three *Zorotypus* species (Zoraptera: Zorotypidae). *Entomological News* 112: 278–80.

———. 2002a. The smallest snakefly (Raphidioptera: Mesoraphidiidae): A new species in Cretaceous amber from Myanmar, with a catalog of fossil snakeflies. *American Museum Novitates* 3363: 1–22.

———. 2002b. Phylogeny of the bee tribe Fideliini (Hymenoptera: Megachilidae), with the description of a new genus from southern Africa. *African Entomology* 10: 305–13.

———. 2002c. The first leucospid wasp from the fossil record (Hymenoptera: Leucospidae). *Journal of Natural History* 36: 435–41.

———. 2002d. The fossil pelecinid *Pelecinopteron tubuliforme* Brues in Baltic amber (Hymenoptera: Pelecinidae). *Journal of Hymenoptera Research* 11: 5–11.

———. 2002e. A new dustywing (Neuroptera: Coniopterygidae) in Turonian amber from New Jersey, with a reassessment of *Glaesoconis* in Neocomian amber from Lebanon. *Journal of the Kansas Entomological Society* 75: 38–42.

———. 2002f. Halictine bees from the Eocene-Oligocene boundary of Florissant, Colorado (Hymenoptera: Halictidae). *Neues Jahrbuch für Geologie und Paläontologie, Abhandlungen* 225: 251–73.

———. 2003a. Zoraptera. Pp. 1201–3. *In* Resh, V. H., and R. T. Cardé (eds.), *Encyclopedia of Insects*. Academic Press; San Diego, California; xxviii + [1] + 1266 pp.

———. 2003b. An anteonine wasp in Cenomanian-Albian amber from Myanmar (Hymenoptera: Dryinidae). *Journal of the Kansas Entomological Society* 76: 616–21.

———. 2003c. The earwigs of Kansas, with a key to genera North of Mexico (Insecta: Dermaptera). *Transactions of the Kansas Academy of Science* 106: 115–23.

———. 2003d. A new Eocene-Oligocene snakefly from Florissant, Colorado (Raphidioptera: Raphidiidae). *Transactions of the Kansas Academy of Science* 106: 124–8.

———. 2003e. Phylogeny of the Zoraptera. *Entomologische Abhandlungen* 61: 147–8.

———. 2004a. Arthropods in Mexican amber. Pp. 175–86. *In* Llorente, J., J. J. Morrone, O. Yáñez-Ordóñez, and I. Vargas (eds.), *Biodiversidad, Taxonomía y Biogeografía de Artrópodos de México: Hacia una Síntesis de su Conocimiento, Volumen IV*. UNAM; Mexico D.F., Mexico; viii + [ii] + 790 + [4] pp.

———. 2004b. Zoraptera. Pp. 637–40. *In* Llorente, J., J. J. Morrone, O. Yáñez-Ordóñez, and I. Vargas (eds.), *Biodiversidad, Taxonomía y Biogeografía de Artrópodos de México: Hacia una Síntesis de su Conocimiento, Volumen IV*. UNAM; Mexico D.F., Mexico; viii + [ii] + 790 + [4] pp.

———. 2004c. Thorny lacewings (Neuroptera: Rhachiberothidae) in Cretaceous amber from Myanmar. *Journal of Systematic Palaeontology* 2: 137–40.

———. 2004d. The dustywings in Cretaceous Burmese amber (Insecta: Neuroptera: Coniopterygidae). *Journal of Systematic Palaeontology* 2: 133–6.

———. 2004e. Fideliine phylogeny and classification revisited (Hymenoptera: Megachilidae). *Journal of the Kansas Entomological Society* 77: 821–36.

———, and S. B. Archibald. 2003. An Early Eocene bee (Hymenoptera: Halictidae) from Quilchena, British Columbia. *Canadian Entomologist* 135: 63–9.

———, and D. A. Grimaldi. 2000. A winged *Zorotypus* in Miocene amber from the Dominican Republic (Zoraptera: Zorotypidae), with discussion on relationships of and within the order. *Acta Geologica Hispanica* 35: 149–64.

———, and D. A. Grimaldi. 2002. The first Mesozoic Zoraptera (Insecta). *American Museum Novitates* 3362: 1–20.

———, and D. A. Grimaldi. 2004a. New light shed on the oldest insect. *Nature* 427: 627–30.

———, and D. A. Grimaldi. 2004b. A new rock crawler in Baltic amber, with comments on the order (Mantophasmatodea: Mantophasmatidae). *American Museum Novitates* 3431: 1–11.

———, and D. A. Grimaldi. 2004c. A primitive earwig in Cretaceous amber from Myanmar (Dermaptera: Pygidicranidae). *Journal of Paleontology* 78: 1018–1023.

———, and D. A. Grimaldi. 2004d. The first Mesozoic stephanid wasp (Hymenoptera: Stephanidae). *Journal of Paleontology* 78: 1192–7.

———, and D. A. Grimaldi. In press. New primitive ants in Cretaceous amber from Myanmar, New Jersey, and Canada (Hymenoptera: Formicidae). *American Museum Novitates*.

———, and K. Krishna. 2004. Family-group names for termites (Isoptera). *American Museum Novitates* 3432: 1–9.

———, and M. G. Rightmyer. 2000. A new augochlorine bee species in Tertiary amber from the Dominican Republic (Hymenoptera: Halictidae). *Apidologie* 31: 431–6.

———, and T. R. Schultz. 1997. Phylogeny and behavior in honey bees (Hymenoptera: Apidae). *Annals of the Entomological Society of America* 90: 43–53.

———, J.-D. Lim, K.-S. Baek, and L. D. Martin. 2002. An earwig from the Lower Cretaceous of Korea (Dermaptera: Forficulina). *Journal of the Kansas Entomological Society* 75: 86–90.

Enghoff, H. 1984. Phylogeny of millipedes – A cladistic analysis. *Zeitschrift für Zoologische Systematik und Evolutionsforschung* 22: 8–26.

———. 2000. Millipede phylogeny: How much do we know and what is it good for? *Fragmenta Faunistica* 43: 1–17.

Epstein, M. E. 1996. Revision and phylogeny of the limacodid-group families, with evolutionary studies on slug caterpillars (Lepidoptera: Zygaenoidea). *Smithsonian Contributions to Zoology* 582: 1–102.

———, H. Geertsema, C. M. Naumann, and G. M. Tarmann. 1999. The Zygaenoidea. Pp. 159–80. *In* Kristensen, N. P. (ed.), *Handbuch der Zoologie: Eine Naturgeschichte der Stämme des Tierreiches: Band IV: Arthropoda: Insecta: Teilband 35: Lepidoptera, Moths and Butterflies: Volume 1: Evolution, Systematics, and Biogeography*. Walter de Gruyter; Berlin, Germany; x + 491 pp.

Ericson, P. G. P., L. Christidis, A. Cooper, M. Irestedt, J. Jackson *et al.* 2001. A Gondwanan origin of passerine birds supported by DNA sequences of the endemic New Zealand wrens. *Proceedings of the Royal Society of London, Series B, Biological Sciences* 269: 235–41.

Eriksson, O., and B. Bremer. 1992. Pollination systems, dispersal modes, life forms, and diversification rates in angiosperm families. *Evolution* 46: 258–66.

Ervik, F., and J. Knudsen. 2003. Water lilies and scarabs: Faithful partners for nearly 100 million years? *Biological Journal of the Linnean Society* 80: 539–43.

Erwin, D. H. 1993. *The Great Paleozoic Crisis: Life and Death in the Permian.* Columbia University Press; New York, New York; xi+327 pp.

———. 1994. The Permo-Triassic extinction. *Nature* 367: 231–6.

Erwin, T. L. 1982. Tropical forests: Their richness in Coleoptera and other arthropod species. *Coleopterist's Bulletin* 36: 74–5.

———. 1983a. Tropical forest canopies, the last biotic frontier. *Bulletin of the Entomological Society of America* 29: 14–19.

———. 1983b. Beetles and other arthropods of the tropical forest canopies at Manaus, Brazil, sampled with insecticidal fogging techniques. Pp. 59–75. *In* Sutton, S. L., T. C. Whitmore, and A. C. Chadwick (eds.), *Tropical Rain Forests: Ecology and Management.* Blackwell; Oxford, UK; xiii+498 pp.

———. 1990. Canopy arthropod biodiversity: A chronology of sampling techniques and results. *Revista Peruviana Entomologia* 32: 71–7.

———. 1997. Biodiversity at its utmost: Tropical forest beetles. Pp. 27–40. *In* Reaka-Kudla, M. L., D. E. Wilson, and E. O. Wilson (eds.), *Biodiversity II: Understanding and Protecting Our Biological Resources.* Joseph Henry Press; Washington, D.C.; v+551 pp.

———, G. E. Ball, and D. R. Whitehead, eds. 1979. *Carabid Beetles: Their Evolution, Natural History, and Classification.* Junk; The Hague, the Netherlands; ix+635 pp.

Escherich, K. 1914. Insekten. Pp. 457–512. *In* Korschelt, E., G. Linck, F. Oltmanns, K. Schaum, H. T. Simon, M. Verworn, and E. Teichmann (eds.), *Handwörterbuch der Naturwissenschaften* [Fünfter Band: Gewürze–Kützing]. Gustav Fischer; Jena, Germany; viii+1194 pp.

Eshet, Y., M. R. Rampino, and H. Visscher. 1995. Fungal event and palynological record of ecological crisis and recovery across the Permian-Triassic boundary. *Geology* 23: 967–70.

Eskov, K. Yu. 1984. A new fossil spider family from the Jurassic of Transbaikalia (Araneae: Chelicerata). *Neues Jahrbuch für Geologie und Paläontologie, Monatshefte* 1984: 645–53.

———. 1992. Archaeid spiders from Eocene Baltic amber (Chelicerata: Araneida: Archaeidae) with remarks on the so-called 'Gondwanan' ranges of Recent taxa. *Neues Jahrbuch für Geologie und Paläontologie, Abhandlungen* 185: 311–28.

———. 2002. Geographical history of insects. Pp. 427–35. *In* Rasnitsyn, A. P., and D. L. J. Quicke (eds.), *History of Insects.* Kluwer Academic Publishers; Dordrecht, the Netherlands; xii+517 pp.

———, and S. I. Golovatch. 1986. On the origin of the trans-Pacific disjunctions. *Zoologische Jahrbücher, Abteilung für Systematik, Ökologie und Geographie der Tiere* 113: 265–85.

———, and J. Wunderlich. 1994. On the spiders from Taimyr ambers, Siberia, with the description of a new family and with general notes on the spiders from the Cretaceous resins (Arachnida: Araneae). *Beiträge zur Araneologie* 4: 95–107.

Essig, E. O. 1931. *A History of Entomology.* Macmillan; New York, New York; vii+1029 pp.

Etheridge, R., Jr., and A. S. Olliff. 1890. The Mesozoic and Tertiary insects of New South Wales. *Memoirs of the Geological Survey of New South Wales, Palaeontology* 7: 1–18.

Evanoff, E., W. C. McIntosh, and P. C. Murphey. 2001. Stratigraphic summary and ^{40}Ar/^{39}Ar geochronology of the Florissant Formation, Colorado. *Proceedings of the Denver Museum of Nature and Science* 4: 1–16.

Evans, G. O. 1992. *Principles of Acarology.* CABI Publishing; Wallingford, UK; xviii+563 pp.

Evans, H. E. 1961. A preliminary review of the Nearctic species of *Sierolomorpha* (Hymenoptera). *Brevoria* 140: 1–12.

———. 1962. The evolution of prey-carrying mechanisms in wasps. *Evolution* 16: 468–83.

———. 1963a. *Wasp Farm.* Anchor Press; Garden City, New York; 178 pp.

———. 1963b. A new family of wasps. *Psyche* 70: 7–16.

———. 1964. A synopsis of the American Bethylidae (Hymenoptera, Aculeata). *Bulletin of the Museum of Comparative Zoology* 132: 1–222.

———. 1966a. The behavior patterns of solitary wasps. *Annual Review of Entomology* 11: 123–54.

———. 1966b. *The Comparative Ethology and Evolution of the Sand Wasps.* Harvard University Press; Cambridge, Massachusetts; xvi+526 pp.

———. 1966c. Discovery of the female *Plumarius* (Hymenoptera, Plumariidae). *Psyche* 73: 229–37.

———. 1969. Three new Cretaceous aculeate wasps (Hymenoptera). *Psyche* 76: 251–61.

———. 1973a. Cretaceous aculeate wasps from Taimyr, Siberia (Hymenoptera). *Psyche* 80: 166–78.

———. 1973b. Notes on the nests of *Montezumia* (Hymenoptera, Eumenidae). *Entomological News* 84: 285–90.

———. 1978. The Bethylidae of American north of Mexico. *Memoirs of the American Entomological Institute* 27: 1–332.

———, and K. M. O'Neill. 1988. *The Natural History and Behavior of North American Beewolves.* Cornell University Press; Ithaca, New York; vii+278 pp.

———, and A. Shimizu. 1996. The evolution of nest building and communal nesting in Ageniellini (Insecta: Hymenoptera: Pompilidae). *Journal of Natural History* 30: 1633–48.

———, and M. J. West-Eberhard. 1970. *The Wasps.* University of Michigan Press; Ann Arbor, Michigan; vi+265 pp.

———, and C. M. Yoshimoto. 1962. Ecology and nesting behavior of the Pompilidae (Hymenoptera) of the northeastern United States. *Miscellaneous Publications of the Entomological Society of America* 3: 65–119.

Evans, J. W. 1947. A new fossil homopteron from Kimbles Hill, Belmont (Upper Permian). *Records of the Australian Museum* 21: 431–2.

———. 1956. Paleozoic and Mesozoic Hemiptera (Insecta). *Australian Journal of Zoology* 4: 165–258.

———. 1958. New Upper Permian Homoptera from the Belmont Beds (Homoptera: Insecta). *Records of the Australian Museum* 24: 109–14.

———. 1963. The phylogeny of the Homoptera. *Annual Review of Entomology* 8: 77–94.

———. 1981. A review of present knowledge of the family Peloridiidae and new genera and new species from New Zealand and New Caledonia (Hemiptera: Insecta). *Records of the Australian Museum* 31: 381–406.

Evans, M. A., and H. E. Evans. 1970. *William Morton Wheeler, Biologist.* Harvard University Press; Cambridge, Massachusetts; xi+[2]+363 pp.

Evenhuis, N. L., ed. 1989. Catalogue of the Diptera of the Australasian and Oceanian Regions. *Bishop Museum Special Publications* 86: 1–1155.

———. 1994. *Catalogue of the Fossil Flies of the World (Insecta: Diptera).* Backhuys Publishers; Leiden, the Netherlands; 600 pp.

———. 2002a. Catalog of the Mythicomyiidae of the world (Insecta: Diptera). *Bishop Museum Bulletin in Entomology* 10: 1–95.

———. 2002b. Review of the Tertiary microbombyliids (Diptera: Mythicomyiidae) in Baltic, Bitterfeld, and Dominican amber. *Zootaxa* 100: 1–15.

———, and D. J. Greathead. 1999. *Catalog of the Bee Flies of the World (Diptera: Bombyliidae).* Backhuys Publishers; Leiden, the Netherlands; 656 pp.

Eversmann, E. 1852. Fauna hymenopterologica Volgo-Uralensis (continuatio). Familia anthophilarium seu apidarum. *Izvêstiĭya Moskovskago Éntomologicheskago Obshchestva* 3: 3–137.

Ewing, A. W. 1989. *Arthropod Bioacoustics.* Cornell University Press; Ithaca, New York; x+260 pp.

Fabré, J. H. 1897. *Souvenirs Entomologiques: Études sur l'Instinct et les Moeurs des Insectes* [Cinquième serie]. Delagrave; Paris, France; 355 pp.

Fabricius, J. C. 1778. *Philosophia Entomologica: Sistens Scientiae Fundamenta; adiectis Definitionibus, Exemplis, Observationibus, Adumbrationibus.* Bohnii; Hamburgi et Kilonii [Hamburg and Cologne] Germany; [12]+178 pp.

———. 1790. Nova insectorum genera. *Skrifter af Naturhistorie-Selskabet Kiøbenhavn* 1: 213–28.

———. 1793. *Entomologia Systematica Emendata et Aucta. Secundum Classes, Ordines, Genera, Species adiectis Synonymis, Locis, Observationibus, Descriptionibus* [vol. 2]. Proft; Hafniae [Copenhagen], Denmark; viii+519 pp.

Faegri, K., and L. van der Pijl. 1979. *The Principles of Pollination Ecology* [3rd Edition]. Pergamon Press; Oxford, UK; xi+244 pp.

Fahringer, J. 1928. Die Megalyriden. *Archiv für Naturgeschichte, Abteilung A* 92: 98–123.

Fanenbruck, M., S. Harzsch, and J. W. Wägele. 2004. The brain of the Remipedia (Crustacea) and an alternative hypothesis on their phylogenetic relationships. *Proceedings of the National Academy of Sciences, USA* 101: 3868–73.

Farrell, B. D. 1998. "Inordinate fondness" explained: Why are there so many beetles? *Science* 281: 555–9.

———, and C. Mitter. 1990. Phylogenesis of insect/plant interactions: Have *Phyllobrotica* leaf beetles (Chrysomelidae) and the Lamiales diversified in parallel? *Evolution* 44: 1389–1403.

———, A. S. Sequeira, B. C. O'Meara, B. B. Normark, J. H. Chung, and B. H. Jordal. 2001. The evolution of agriculture in beetles (Curculionidae: Scolytinae and Platypodinae). *Evolution* 55: 2011–27.

Farris, J. S. 1974. Formal definitions of paraphyly and polyphyly. *Systematic Zoology* 23: 548–54.

Fausto, A. M., M. Belardinelli, R. Fochetti, and M. Mazzini. 2001. Comparative spermatology in Plecoptera (Insecta): An ultrastructural investigation on four species. *Arthropod Structure and Development* 30: 55–62.

Fehler, A. 1999. Erster Nachweis von fossilen Schlammfliegenlarven (Megaloptera: Sialidae) aus dem altholozänen Travertin von Bad Langensalza/Thuringen. *Aufschluss* 50: 287–90.

Felsenstein, J. 1973. Maximum likelihood and minimum-steps methods for estimating evolutionary trees from data on discrete characters. *Systematic Zoology* 22: 240–9.

———. 1978. Cases in which parsimony and compatibility methods will be positively misleading. *Systematic Zoology* 27: 401–10.

———. 1981. A likelihood approach to character weighting and what it tells us about parsimony and compatibility. *Biological Journal of the Linnean Society* 16: 183–96.

———. 1983. Parismony in systematics: Biological and statistical issues. *Annual Review of Ecology and Systematics* 14: 313–33.

Felt, E. P. 1940. *Plant Galls and Gall Makers.* Comstock Publishing; Ithaca, New York; viii+364 pp.

Fenton, M. B., and J. H. Fullard. 1981. Moth hearing and the feeding strategies of bats. *American Scientist* 69: 266–75.

Ferguson, W. E. 1962. Biological characteristics of the mutillid subgenus *Photopsis* Blake, and their systematic values (Hymenoptera). *University of California Publications in Entomology* 27: 1–91.

Fernandez, F., C. E. Sarmiento, and D. F. Campos. 2002. First records of the wasps [sic] families Megalyridae and Scolebythidae (Hymenoptera: Megalyroidea, Chrysidoidea) for northern South America. *Revista Colombiana de Entomologia* 28: 213.

Fernandez-Rubio, F., E. Peñalver, and X. Martínez-Delclòs. 1991. *Zygaena*? *turolensis*, una neuva especie de Lepidoptera Zygaenidae del Mioceno de Rubielos de Mora (Teruel): Descripción y filogenia. *Estudias Museo Cencias Naturales de Álava* 6: 77–93.

Ferris, G. F. 1931. The louse of elephants: *Haematomyzus elephantis* Piaget (Mallophaga: Haematomyzidae). *Parasitology* 23: 112–27.

———. 1937–1955. *Atlas of the Scale Insects of North America, Series I–VII.* Stanford University Press; Stanford, California; 1–8+[1824]+vii+1–278+viii+279–506+iii+203 pp.

———. 1951. The sucking lice. *Memoirs of the Pacific Coast Entomological Society* 1: 1–320.

Fet, V., W. D. Sissom, G. Lowe, and M. E. Braunwelder. 2000. *Catalog of the Scorpions of the World (1758–1998).* New York Entomological Society; New York, New York; 690 pp.

Fidalgo, P., and D. R. Smith. 1987. A fossil Siricidae (Hymenoptera) from Argentina. *Entomological News* 98: 63–6.

Fiedler, K. 1991. Systematic, evolutionary and ecological implications of myrmecophily within the Lycaenidae (Insecta: Lepidoptera: Papilionoidea). *Bonner Zoologische Monographien* 31: 1–210.

Field, K. G., G. J. Olsen, D. J. Lane, S. J. Giovannoni, M. T. Ghiselin *et al.* 1988. Molecular phylogeny of the animal kingdom. *Science* 239: 748–53.

Field, L. H., ed. 2001. *The Biology of Wetas, King Crickets and Their Allies.* CABI Publishing; Oxon, UK; xx+540 pp.

Fischer, C. 1999. Grobklastika im mitteleozänen Eckfeld Maar (Südwesteifel): Sedimentologische und petrographische Analyse. *Mainzer Naturwissenschaftliches Archiv* 37: 21–54.

Fisher, D. C. 1984. The Xiphosurida: Archetypes of bradytely? Pp. 196–213. *In* Eldredge, N., and S. M. Stanley (eds.), *Living Fossils.* Springer Verlag; Berlin, Germany; xi+291 pp.

Flanagan, J. F., and K. E. Marshall. 1980. *Advances in Ephemeroptera Biology.* Plenum Press; New York, New York; xiii+552 pp.

Fleming, T. H., and J. N. Holland. 1998. The evolution of obligate pollination mutualisms: Senita cactus and senita moth. *Oecologia* 114: 368–75.

Fleschner, C. A. 1950. Studies on searching capacity of the larvae of three predators of the citrus red mite. *Hilgardia* 20: 233–65.

———, and D. W. Ricker. 1953. Food habits of coniopterygids on citrus in southern California. *Journal of Economic Entomology* 46: 458–61.

Flook, P. K., and C. H. F. Rowell. 1997. The effectiveness of mitochondrial rRNA gene sequences for the reconstruction of the phylogeny of an insect order (Orthoptera). *Molecular Phylogenetics and Evolution* 8: 177–92.

———, and C. H. F. Rowell. 1998. Inferences about orthopteroid phylogeny and molecular evolution from small subunit nuclear ribosomal DNA sequences. *Insect Molecular Biology* 7: 163–78.

———, S. Klee, and C. H. F. Rowell. 1999. Combined molecular phylogenetic analysis of the Orthoptera (Arthropoda, Insecta) and implications for their higher systematics. *Systematic Biology* 48: 233–53.

———, S. Klee, and C. H. F. Rowell. 2000. Molecular phylogenetic analysis of the Pneumoroidea (Orthoptera, Caelifera): Molecular data resolve morphological character conflicts in the basal Acridomorpha. *Molecular Phylogenetics and Evolution* 15: 345–54.

Foddai, D., L. Bonato, L. A. Pereira, and A. Minelli. 2003. Phylogeny and systematics of the Arrupinae (Chilopoda: Geophilomorpha: Mecistocephalidae) with description of a new dwarfed species. *Journal of Natural History* 37: 1247–67.

Foelix, R. 1982. *Biology of Spiders.* Harvard University Press; Cambridge, Massachusetts; 306 pp.

Forey, P. 1984. The coelacanth as a living fossil. Pp. 166–9. *In* Eldredge, N., and S. M. Stanley (eds.), *Living Fossils.* Springer Verlag; Berlin, Germany; xi+291 pp.

Forster, P. I., P. J. Machin, L. Mound, and G. W. Wilson. 1994. Insects associated with reproductive structures of cycads in Queensland and northeast New South Wales, Australia. *Biotropica* 26: 217–22.

Fortey, R. A. 2001. Trilobite systematics: The last 75 years. *Journal of Paleontology* 75: 1141–51.

———, and H. B. Whittington. 1989. The Trilobita as a natural group. *Historical Biology* 2: 125–38.

Foster, W. A., and P. A. Northcott. 1994. Galls and the evolution of social behaviour in aphids. Pp. 161–82. *In* Williams, M. A. J. (ed.), *Plant Galls: Organisms, Interactions, Populations.* Clarendon Press; Oxford, UK; xiv+488 pp.

Francke-Grossmann, H. 1939. Über das Zusammenleben von Holzwespen (Siricinae) mit Pilzen. *Zeitschrift für Angewandte Entomologie* 25: 647–80.

Francois, J. 2003. Analyse cladistique de l'ordre Protura (Hexapoda, Protura). *Bulletin de la Société Entomologique de France* 108: 109–25.

Frania, H. E., and G. B. Wiggins. 1997. Analysis of morphological and behavioural evidence for the phylogeny and higher classification of Trichoptera (Insecta). *Royal Ontario Museum Life Sciences Series* 160: 1–67.

Frankie, G. W., and R. W. Thorp. 2003. Pollination and pollinators. Pp. 919–26. *In* Resh, V. H., and R. T. Cardé (eds.), *Encyclopedia of Insects.* Academic Press; San Diego, California; xxviii+[1]+1266 pp.

Frantsevich, L., and D. Gladun. 2002. Evolution of the middle leg basal articulations in flies (Diptera). *Acta Zoologica* 83: 125–47.

Franzen, J. L. 1993. Das biostratigraphische Alter der Fossillagerstätte Eckfelder Maar bei Manderscheid (Eifel). *Mainzer Naturwissenschaftliches Archiv* 31: 201–14.

Fraser, F. C. 1954. The origin and decent of the order Odonata based on the evidence of persistent archaic characters. *Transactions of the Royal Entomological Society of London (B)* 24: 139–46.

———. 1957. *A Reclassification of the Order Odonata.* Royal Zoological Society of New South Wales; Sydney, Australia; 134 pp.

Fraser, N. C., and D. A. Grimaldi. 2003. Late Triassic continental faunal change: New perspectives on Triassic diversity as revealed by a locality in the Danville Basin, Virginia, Newark Supergroup. Pp. 192–205. *In* LeTourneau, P. M., and P. E. Olsen (eds.), *The Great Rift Valleys of Pangea in Eastern North America* [Volume 2]. Columbia University Press; New York, New York; xi+384 pp.

———, D. A. Grimaldi, P. E. Olsen, and B. Axsmith. 1996. A Triassic Lagerstätte from eastern North America. *Nature* 380: 615–19.

Frati, F., R. Dallai, and A. Carapelli. 1998. Comparison of morphological and molecular data for the phylogeny of the Apterygota. *Proceedings of the 5th International Symposium on Apterygota* 1998: 35.

Frenguelli, J. 1939. Nidos fósiles de insectos en el Terciario del Neuquén y Rio Negro. *Notas del Museo de la Plata* 4: 379–402.

Frey, D. G. 1964. Remains of animals in quaternary lake and bog sediments and their interpretation. *Archiv für Hydrobiologie, Beihefte* 2: 1–114.

Frickhinger, K. A. 1994. *Die Fossilien von Solnhofen: Dokumentation der aus den Plattenkalken bekannten Tiere und Pflanzen.* Goldschneck Verlag; Korb, Germany; 336 pp.

Friedlander, T. P., J. C. Regier, C. Mitter, and D. L. Wagner. 1996. A nuclear gene for higher level phylogenetics: Phospoenolpyruvate carboxykinase tracks Mesozoic-aged divergences within Lepidoptera (Insecta). *Molecular Biology and Evolution* 13: 594–604.

———, J. C. Regier, C. Mitter, D. L. Wagner, and Q. Q. Fang. 2000. Evolution of heteroneuran Lepidoptera (Insecta) and the utility of dopa decarboxylase for Cretaceous-age phylogenetics. *Zoological Journal of the Linnean Society* 130: 213–34.

Friedman, W. E., ed. 1996. Biology and evolution of the Gnetales. *International Journal of Plant Sciences* 157 (6, Supplement): S1–S125.

Friedrich, M., and D. Tautz. 1995. Ribosomal DNA phylogeny of the major extant arthropod classes and evolution of myriapods. *Nature* 376: 165–7.

Friese, H. 1895. *Die Bienen Europa's (Apidae europaeae) nach ihren Gattungen, Arten und Varietäten auf vergleichend morphologisch-biologischer Grundlage* [Band 3, Theil 1]. Friedländer; Berlin, Germany; 128 pp.

———. 1923. *Die europäischen Bienen (Apidae): Das Leben und Wirken unserer Blumenwespen: Eine Darstellung der Lebensweise unserer wilden wie gesellig lebenden Bienen nach eigenen Untersuchungen für Naturfreunde, Lehrer und Zoologen*. W. Gruyter; Berlin, Germany; vi+456 pp.

Friis, E. M., P. R. Crane, and K. R. Pedersen. 1986. Floral evidence for Cretaceous chloranthoid angiosperms. *Nature* 320: 163–4.

———, K. R. Pedersen, and P. R. Crane. 1994. Angiosperm floral structures from the Early Cretaceous of Portugal. *Plant Systematics and Evolution, Supplement* 8: 31–49.

———, K. R. Pedersen, and P. R. Crane. 1999. Early angiosperm diversification: The diversity of pollen associated with angiosperm reproductive structures in early Cretaceous floras from Portugal. *Annals of the Missouri Botanical Garden* 86: 259–96.

———, K. R. Pedersen, and P. R. Crane. 2001. Fossil evidence of water lilies (Nymphaeales) in the Early Cretaceous. *Nature* 410: 357–60.

Frisch, K., von. 1967. *The Dance Language and Orientation of Bees*. Harvard University Press; Cambridge, Massachusetts; xiv+566 pp.

Froggatt, W. W. 1906. Notes on the hymenopterous genus *Megalyra* Westwood, with descriptions of new species. *Proceedings of the Linnean Society of New South Wales* 31: 399–407.

Fryer, G. 1996. Reflections on arthropod evolution. *Biological Journal of the Linnean Society* 58: 1–55.

———. 1998. A defence of arthropod polyphyly. Pp. 23–33. *In* Fortey, R. A., and R. H. Thomas (eds.), *Arthropod Relationships*. Chapman and Hall; London, UK; xii+383 pp.

Fujiyama, I. 1973. Mesozoic insect fauna of east Asia. Part 1. Introduction and Upper Triassic faunas. *Bulletin of the Natural Sciences Museum, Tokyo* 16: 331–86.

———. 1991. Late Triassic insects from Miné, Yamaguchi, Japan. Part 1. Odonata. *Bulletin of the Natural Sciences Museum, Tokyo (C)* 17: 49–56.

———. 1994. Two parasitic wasps from Aptian (Lower Cretaceous) Choshi amber. *Natural History Research* 3: 1–5.

Fullard, J. H. 1982. Echolocation assemblages and their effects on moth auditory systems. *Canadian Journal of Zoology* 60: 2572–6.

———, M. B. Fenton, and J. A. Simmons. 1979. Jamming bat echolocation: The clicks of arctiid moths. *Canadian Journal of Zoology* 57: 647–9.

Fursov, V., Y. Shirota, T. Nomiya, and K. Yamagishi. 2002. New fossil mymarommatid species, *Palaeomymar japonicum* sp. nov. (Hymenoptera: Mymarommatidae), discovered in Cretaceous amber from Japan. *Entomological Science* 5: 51–4.

Furth, D. G. 1994. Frank Morton Carpenter (1902–1994): Academic biography and list of publications. *Psyche* 101: 127–44.

Futuyma, D. J. 1998. *Evolutionary Biology* [3rd Edition]. Sinauer; Sunderland, Massachusetts; xviii+763 pp. [+ unpaginated glossary and indices]

———, and M. C. Keese. 1992. Evolution and coevolution of plants and phytophagous arthropods. Pp. 439–75. *In* Rosenthal, G. A., and M. R. Berenbaum (eds.), *Herbivores: Their Interactions with Secondary Plant Metabolites [2nd edition]: Volume 2: Ecological and Evolutionary Processes*. Academic Press; San Diego, California; xii+493 pp.

Gadagkar, R. 2001. *The Social Biology of* Ropalidia marginata: *Toward Understanding the Evolution of Eusociality*. Harvard University Press; Cambridge, Massachusetts; xiii+368 pp.

Gagné, R. J. 1977. Cecidomyiidae (Diptera) from Canadian amber. *Proceedings of the Entomological Society of Washington* 79: 57–62.

———. 1986. The transition from fungus-feeding to plant-feeding in Cecidomyiidae (Diptera). *Proceedings of the Entomological Society of Washington* 88: 381–4.

———. 1989. *The Plant-Feeding Gall Midges of North America*. Cornell University Press; Ithaca, New York; xi+356 pp.

———. 1994. *The Gall Midges of the Neotropical Region*. Cornell University Press; Ithaca, New York; xiv+352 pp.

Gaimari, S., and M. Mostovski. 2000. *Burmapsilocephala cockerelli*, a new genus and species of Asiloidea (Diptera) from Burmese amber. *Bulletin of the Natural History Museum, London (Geology Series)* 56: 43–6.

Gall, J.-C. 1971. Faunes et paysages du Grès a Voltzia du nord des Vosges: Essai paléoécologique sur le Buntsandstein supérieur. *Mémoires du Service de la Carte Géologique d'Alsace et de Lorraine* 34: 1–318.

———. 1985. Fluvial depositional environment evolving into deltaic setting with marine influences in the Buntsandstein of northern Vosges (France). Pp. 449–77. *In* Mader, D. (ed.), *Aspects of Fluvial Sedimentation in the Lower Triassic Buntsandstein of Europe*. Springer Verlag; Berlin, Germany; viii+626 pp.

———. 1996. Triassic insects of western Europe. *Paleontologia Lombarda* 5: 3–4.

———, L. Grauvogel-Stamm, A. Nel, and F. Papier. 1998. La crise biologique du Permien et la renaissance triasique. *Comptes Rendus de l'Academie des Sciences, Serie II A, Sciences de la Terre et des Planetes* 326: 1–12.

Gallard, L. 1932. Notes on the feeding habits of the brown moth-lacewing *Ithone fusca*. *Australian Naturalist* 8: 168–70.

———. 1935. Notes on the life history of the large yellow lacewing, *Nymphes myrmeleonides*. *Australian Naturalist* 9: 118–19.

Gallego, O. F., and R. G. Martins-Neto. 1999. La entomofauna mesozoica de la Argentina: Estado actual del conocimiento. *Revista de la Sociedad Entomológica Argentina* 58: 86–94.

Gand, G., J. Lapeyrie, J. Garric, A. Nel, J. Schneider, and H. Walter. 1997. Decouverte d'arthropodes et de bivalves inedits dans la Permien continental (Lodevois, France). *Comptes Rendus de l'Academie des Sciences, Serie IIA, Sciences de la Terre et des Planetes* 325: 891–8.

Gandar, M. V. 1982. The dynamics and trophic ecology of grasshoppers (Acridoidea) in a South African savanna. *Oecologia* 54: 370–8.

Gandolfo, M. A., K. C. Nixon, and W. L. Crepet. 1998. A new fossil flower from the Turonian of New Jersey: *Dressiantha bicarpellata* gen. et sp. nov. (Capparales). *American Journal of Botany* 85: 964–74.

———, K. C. Nixon, and W. L. Crepet. 2004. Cretaceous flowers of Nymphaeaceae and implications for complex insect entrapment pollination mechanisms in early angiosperms. *Proceedings of the National Academy of Sciences, USA* 101: 8056–60.

Gangwere, S. K., M. C. Muralirangan, and M. Muralirangan, eds. 1997. *The Bionomics of Grasshoppers, Katydids and their Kin*. CABI Publishing; Oxon, UK; xiii+529 pp.

Garey, J. R., M. Krotec, D. R. Nelson, and J. Brooks. 1996. Molecular analysis supports a tardigrade-arthropod association. *Invertebrate Biology* 115: 79–88.

———, D. R. Nelson, L. Y. Mackey, and J. Li. 1999. Tardigrade phylogeny: Congruency of morphological and molecular evidence. *Zoologischer Anzeiger* 238: 205–10.

Gaston, K. J. 1991. The magnitude of global insect species richness. *Conservation Biology* 5: 283–96.

Gaudant, J. 1979. Mise au point sur l'ichthyofaune paléocène de Menat (Puy-de-Dôme). *Comptes Rendus Hebdomadaires des Seances de l'Academie des Sciences, serie D: Sciences Naturelles* 288: 1461–4.

Gauld, I. D. 1988. Evolutionary patterns of host utilization by ichneumonid parasitoids (Hymenoptera: Ichneumonidae and Braconidae). *Biological Journal of the Linnean Society* 35: 351–77.

———, and B. Bolton. 1988. *The Hymenoptera*. Oxford University Press; Oxford, UK; xi+322 pp.

Gaunt, M. W., and M. A. Miles. 2002. An insect molecular clock dates the origin of the insects and accords with palaeontological and biogeographic landmarks. *Molecular Biology and Evolution* 19: 748–61.

Gauss, R. 1964. *Cleptes semiauratus* L. im Rahmen der parasitenliste von *Pristophora abietina* in Südwestdeutschland. *Zeitschrift für Angewandte Entomologie* 54: 225–32.

Gauthier, J., A. G. Kluge, and T. Rowe. 1988. Amniote phylogeny and the importance of fossils. *Cladistics* 4: 105–209.

———, A. G. Kluge, and T. Rowe. 1989. Amniote phylogeny and the importance of fossils. *Cladistics* 4: 105–209.

Gauthier, N., J. LaSalle, D. L. J. Quicke, and H. C. J. Godfray. 2000. Phylogeny of Eulophidae (Hymenoptera: Chalcidoidea), with a reclassification of Eulophinae and the recognition that Elasmidae are derived eulophids. *Systematic Entomology* 25: 521–39.

Geertsema, H., and D. E. Van Dijk. 1999. The earliest known Palaeozoic ensiferan insect from Africa, *Afroedischia oosthuizeni* gen. et sp. nov. (Orthoptera: Oedischiidae). *South African Journal of Science* 95: 229–30.

———, and J. A. Van der Heever. 1996. A new beetle, *Afrocupes firmae* gen. et sp. nov. (Permocupedidae), from the Late Palaeozoic Whitehall Formation of South Africa. *South African Journal of Science* 92: 497–9.

Geinitz, F. E. 1883. Die Flözformationen Mecklenburgs. *Archiv des Vereins der Freunde der Naturgeschichte Mecklenburg* 37: 7–151.

Genise, J. F. 1986. Las Bradynobaenidae y algunas modificaciones a la clasificacion general de Hymenoptera Aculeata. *Physis, Secciones A, B y C* 44: 39–53.

———. 1999. Fossil bee cells from the Ascencio Formation (Late Cretaceous – Early Tertiary) of Uruguay, South America. Pp. 27–32. *In* Vršanský, P. (ed.), *Proceedings of the First Palaeoentomological Conference, Moscow, 1998*. AMBA Projects; Bratislava, Slovakia; 199+[2] pp.

———, and T. M. Bown. 1994. New Miocene scarabeid and hymenopterous nests and Early Miocene (Santacrucian) paleoenvironments, Patagonian Argentina. *Ichnos* 3: 107–17.

———, and J. H. Laza. 1998. *Monesichnus ameghinoi* Roselli: A complex insect trace fossil produced by two distinct trace markers. *Ichnos* 5: 213–23.

———, and J. F. Petrulevicius. 2001. Caddisfly cases from the Early Eocene of Chubut, Patagonia, Argentina. Pp. 12–13. *In* Krzemińska, E., and W. Krzemiński (eds.), *Second International Congress on Palaeoentomology: Abstracts Volume*. Polish Academy of Sciences; Kraków, Poland; 94 pp.

———, and M. Verde. 2000. *Corimbatichnus fernandezi*: A cluster of fossil bee cells from the Late Cretaceous-Early Tertiary of Uruguay. *Ichnos* 7: 115–25.

———, M. G. Mángano, L. A. Buatois, J. H. Laza, and M. Verde. 2000. Insect trace fossil associations in paleosols: The *Coprinisphaera* ichnofacies. *Palaios* 15: 49–64.

———, J. C. Sciutto, J. H. Laza, M. G. Gonzalez, and E. S. Bellosi. 2002. Fossil bee nests, coleopteran pupal chambers and tuffaceous paleosols from the Late Cretaceous Laguna Palacios Formation, Central Patagonia (Argentina). *Palaeogeography, Palaeoclimatology, Palaeoecology* 177: 215–35.

Gensel, P. G., and H. N. Andrews, eds. 1984. *Plant Life in the Devonian*. Praeger; New York, New York; xi+380 pp.

Geoffroy Saint-Hilaire, E. 1818. *Philosophie Anatomique: Des Organes Respiratoires sous le Rapport de la Détermination et de l'Idendité de leurs Pièces Osseuses*. Méquignon-Marvis; Paris, France; xxxix+517+[1] pp., 10 pls.

———, P. L. A. Cordier, and J. V. Audouin. 1833. Discours prononcés au nom de l'Institut, du Muséum d'Histoire naturelle et de la Société entomologique de France sur la tombe de M. Latreille décédé le 6 février 1833. *Annales Sciences Naturelles* 28: 433–45.

Gerstäcker, C. E. A. 1862. *Über die geographiche Verbreitung und die Abänderungen der Honigbiene nebst Bemerkungen über die ausländischen Honigbienen der alten Welt*. Festschrift XI Wander-Versammlung deutscher Bienenwirthe; Potsdam, Germany; 75 pp.

———. 1863. On the geographical distribution and varieties of the honey-bee, with remarks upon the exotic honey-bees of the old world. *Annals and Magazine of Natural History, series 3* 11: 270–83, 333–47.

Gess, F. W. 1984. Some aspects of the ethology of *Ampulex bantuae* Gess (Hymenoptera: Sphecidae: Ampulicinae) in the eastern Cape Province of South Africa. *Annals of the Cape Province Museum (Natural History)* 16: 23–40.

Gess, S. K. 1996. *The Pollen Wasps: Ecology and Natural History of the Masarinae*. Harvard University Press; Cambridge, Massachusetts; x+[1]+340 pp.

Ghauri, M. S. K. 1962. *The Morphology and Taxonomy of Male Scale Insects (Homoptera: Coccoidea)*. Trustees of the British Museum; London, UK; vii+221 pp.

Ghilarov, M. S. 1962. The larva of *Dilar turcicus* Hag. and the position of the family Dilaridae in the order Planipennia. *Entomologicheskoe Obozrenie* 41: 402–16. [In Russian]

Ghiradella, H. 1984. Structure of iridescent lepidopteran scales: Variations on several themes. *Annals of the Entomological Society of America* 77: 637–45.

Giblin, R. M. D., R. A. Davies, K. Morris, and W. K. Thomas. 2003. Evolution of parasitism in insect-transmitted plant nematodes. *Journal of Nematology* 35: 133–41.

Gibson, G. A. P. 1985. Some pro- and mesothoracic structures important for phylogenetic analysis of Hymenoptera, with a review of terms used for the structures. *Canadian Entomologist* 117: 1395–1443.

———. 1986. Evidence for monophyly and relationships of Chalcidoidea, Mymaridae, and Mymarommatidae (Hymenoptera: Terebrantes). *Canadian Entomologist* 118: 691–728.

———. 1993. Superfamilies Mymarommatoidea and Chalcidoidea. Pp. 570–655. *In* Goulet, H., and J. T. Huber (eds.), *Hymenoptera of the World: An Indentification Guide to Families*. Agriculture Canada; Ottawa, Canada; vii+668 pp.

———, J. T. Huber, and J. B. Woolley. 1997. *Annotated Keys to the Genera of Nearctic Chalcidoidea (Hymenoptera)*. National Research Council of Canada; Ottawa, Canada; xi+794 pp.

———, J. M. Heraty, and J. B. Woolley. 1999. Phylogenetics and classification of Chalcidoidea and Mymarommatoidea – A review of current concepts (Hymenoptera, Apocrita). *Zoologica Scripta* 28: 87–124.

Gilbert, L. E. 1983. Coevolution and mimicry. Pp. 263–81. *In* Futuyma, D. J., and M. J. Slatkin (eds.), *Coevolution*. Sinauer; Sunderland, Massachusetts; x+555 pp.

Gilkesen, L. A., J. P. McLean, and P. Dessart. 1993. *Aphanogmus fulmeki* Ashmead (Hymenoptera: Ceraphronidae), a parasitoid of *Aphidoletes aphidimyza* Rondani (Diptera: Cecidomyiidae). *Canadian Entomologist* 125: 161–2.

Gillespie, R. G., and H. B. Croom. 1995. Comparison of speciation mechanisms in web-building and non-web-building groups within a lineage of spiders. Pp. 121–46. *In* Wagner, W. L., and V. A. Funk (eds.), *Hawaiian Biogeography: Evolution on a Hot Spot Archipelago*. Smithsonian Institution Press; Washington, D.C.; xvii+467 pp.

Gingerich, P. D., M. ul Haq, I. S. Zalmout, I. Hussain Khan, and M. S. Malkani. 2001. Origin of whales from early artiodactyls: Hands and feet of Eocene Protocetidae from Pakistan. *Science* 293: 2239–42.

Giri, M. K., and P. H. Freytag. 1986. Development of *Discondylus americanus* (Hymenoptera: Dryinidae). *Frustula Entomologica* 9: 215–22.

Giribet, G., and C. Ribera. 1998. The position of arthropods in the animal kingdom: A search for a reliable outgroup for internal arthropod phylogeny. *Molecular Phylogenetics and Evolution* 9: 481–8.

———, and W. C. Wheeler. 1999. The position of arthropods in the animal kingdom: Ecdysozoa, islands, trees, and the "parsimony ratchet". *Molecular Phylogenetics and Evolution* 13: 619–23.

———, S. Carranza, J. Baguna, M. Riutort, and C. Ribera. 1996. First molecular evidence for the existence of a Tardigrada+Arthropoda clade. *Molecular Biology and Evolution* 13: 76–84.

———, S. Carranza, M. Riutort, J. Baguna, and C. Ribera. 1999. Internal phylogeny of the Chilopoda (Myriapoda, Arthropoda) using complete 18S rDNA and partial 28S rDNA sequences. *Philosophical Transactions of the Royal Society, London (B)* 354: 215–22.

———, G. D. Edgecombe, and W. C. Wheeler. 2001. Arthropod phylogeny based on eight molecular loci and morphology. *Nature* 413: 157–61.

Givulescu, R. 1984. Pathological elements on fossil leaves from Chiuzbaia (galls, mines and other insect traces). *Dari de Seama ale Sedintelor Institutul de Geologie si Geofizica (Bucharest)* 68: 123–33.

Gleason, J. M., and M. G. Ritchie. 1998. Evolution of courtship song and reproductive isolation in the *Drosophila willistoni* species complex: Do sexual signals diverge the most quickly? *Evolution* 52: 1493–1500.

———, E. C. Griffith, and J. R. Powell. 1998. A molecular phylogeny of the *Drosophila willistoni* group: Conflicts between species concepts? *Evolution* 52: 1093–1103.

Glorioso, M. J. 1981. Systematics of the dobsonfly subfamily Corydalinae (Megaloptera: Corydalidae). *Systematic Entomology* 6: 253–90.

Godfray, H. C. J. 1994. *Parasitoids: Behavioral and Evolutionary Ecology*. Princeton University Press; Princeton, New Jersey; 473 pp.

Goldblatt, P., and J. C. Manning. 2000. The long-proboscid fly pollination system in southern Africa. *Annals of the Missouri Botanical Garden* 87: 146–70.

Golenberg, E. M. 1991. Amplification and analysis of Miocene plant fossil DNA. *Philosophical Transactions of the Royal Society of London (B)* 333: 419–27.

———, D. E. Giannassi, M. T. Clegg, C. J. Smiley, M. Durbin *et al.* 1990. Chloroplast DNA sequence from a Miocene *Magnolia* species. *Nature* 344: 656–8.

Goloboff, P. A. 2003. Parsimony, likelihood, and simplicity. *Cladistics* 19: 91–103.

Golub, V. B., and Yu. A. Popov. 2000. A remarkable fossil lace bug from Upper Cretaceous

New Jersey amber (Heteroptera: Tingoidea, Vianaididae), with some phylogenetic commentary. Pp. 231–9. *In* Grimaldi, D. (ed.), *Studies on Fossils in Amber, with Particular Reference to the Cretaceous of New Jersey*. Backhuys Publishers; Leiden, the Netherlands; viii+498 pp.

Gordh, G. 1979. Superfamily Chalcidoidea. Pp. 743–967. *In* Krombein, K. V., P. D. Hurd, Jr., D. R. Smith, and B. D. Burks (eds.), *Catalog of Hymenoptera in American North of Mexico*. Smithsonian Institution Press; Washington, D.C.; xvi+2735 pp. [3 volumes]

Gore, P. J. W. 1988. Paleoecology and sedimentology of a Late Triassic lake, Culpeper Basin, Virginia, U.S.A. *Palaeogeography, Palaeoclimatology, Palaeoecology* 62: 593–608.

Gore, R. H., and K. L. Heck, eds. 1986. Crustacean biogeography. *Crustacean Issues* 4: xi+1–292.

Gorelick, R. 2001. Did insect pollination cause increased seed plant diversity? *Biological Journal of the Linnean Society* 74: 407–27.

Gorochov, A. V. 2001. The higher classification, phylogeny and evolution of the superfamily Stenopelmatoidea. Pp. 3–33. *In* Field, L. H. (ed.), *The Biology of Wetas, King Crickets and Their Allies*. CABI Publishing; Wallingford, UK; xx+540 pp.

Gorokhov, A. V. 1992. New fossil Orthoptera and Phasmoptera from the Mesozoic and Cainozoic [sic] of Mongolia. *Trudy Sovmestnaya Sovetsko-Mongolskaya Paleontologischeskaya Ekspeditsiya* 41: 117–21. [In Russian]

——. 2001. On the higher classification of the Polyneoptera. *Acta Geologica Leopoldensia* 24: 11–56.

Gottsberger, G. 1977. Some aspects of beetle pollination in the evolution of flowering plants. *Plant Systematics and Evolution, supplement* 1: 211–26.

———. 1988. The reproductive biology of primitive angiosperms. *Taxon* 37: 630–43.

———, I. Silberbauer-Gottsberger, and F. Ehrendorfer. 1980. Reproductive biology in the primitive relic angiosperm *Drimys brasiliensis* (Winteraceae). *Plant Systematics and Evolution* 135: 11–39.

Gould, S. J. 1989. *Wonderful Life: The Burgess Shale and the Nature of History*. Norton; New York; New York; 347 pp.

———. 1992. Magnolias from Moscow. *Natural History* 9: 10–18.

Goulet, H., and J. T. Huber, eds. 1993. *Hymenoptera of the World: An Identification Guide to Families*. Agriculture Canada; Ottawa, Canada; vii+668 pp.

Gourley, E. S. 1951. Notes on insects associated with *Pinus radiata* in New Zealand. *Bulletin of Entomological Research* 42: 21–2.

Grabenhorst, H. 1985. Eine zweite Bremse (Tabanidae) zusammen mit ihrem Parasiten (Nematoda, Mermithoidae [sic]) aus dem Oberpliozän von Willershausen, Krs. Osterode. *Aufschluss* 36: 325–8.

———. 1991. Eine fossile Staublaus (Psocoptera) aus den oberpliozänen Ablagerungen von Willershausen, Krs. Osterode. *Aufschluss* 42: 381–3.

Graenicher, S. 1910. The bee-flies (Bombyliidae) in their relations to flowers. *Bulletin of the Wisconsin Natural History Society* 8: 91–101.

Graham, J. B., R. Dudley, N. Aguilar, and C. Gans. 1995. Implications of the late Palaeozoic oxygen pulse for physiology and evolution. *Nature* 375: 117–20.

Grandcolas, P. 1994. Phylogenetic systematics of the subfamily Polyphaginae, with the assignment of *Cryptocercus* Scudder, 1862 to this taxon (Blattaria, Blaberoidea, Polyphagidae). *Systematic Entomology* 19: 145–58.

———. 1996. The phylogeny of cockroach families, a cladistic appraisal of morpho-anatomical data. *Canadian Journal of Zoology* 74: 508–27.

———. 1999. Reconstructing the past of *Cryptocercus* (Blattaria: Polyphagidae): Phylogenetic histories and stories. *Annals of the Entomological Society of America* 92: 303–7.

———, and C. D'Haese. 2001. The phylogeny of cockroach families: Is the current molecular hypothesis robust? *Cladistics* 17: 48–55.

———, Y. C. Park, J. C. Choe, M.-D. Piulachs, X. Bellés *et al.* 2001. What does *Cryptocercus kyebangi*, n.sp. (Dictyoptera: Blattaria: Polyphagidae) from Korea reveal about *Cryptocercus* evolution? A study in morphology, molecular phylogeny and chemistry of tergal glands. *Proceedings of the Academy of Natural Sciences, Philadelphia* 151: 61–79.

Grande, L. 1984. Paleontology of the Green River Formation, with a review of the fish fauna [2nd Edition]. *Bulletin of the Geological Survey of Wyoming* 63: 1–333.

Grant, P. R. 1999. *Ecology and Evolution of Darwin's Finches* [2nd Edition]. Princeton University Press; Princeton, New Jersey; xx+492 pp.

———, and B. R. Grant. 2002. Unpredictable evolution in a 30-year study of Darwin's finches. *Science* 296: 707–11.

Grant, V. 1994. Modes and origins of mechanical and ethological isolation in angiosperms. *Proceedings of the National Academy of Sciences, U.S.A.* 91: 3–10.

———, and K. A. Grant. 1965. *Flower Pollination in the Phlox Family*. Columbia University Press; New York, New York; 180 pp.

Grassé, P. P. 1951. Ordre des Mécoptères. Pp. 71–124. *In* Grassé, P. P. (ed.), *Traité de Zoologie, Tome 10*. Masson et Cie; Paris, France; 975 pp.

Grassi, B. 1888. I progenitori dei Miriapodi e degli Insetti. VII. Anatomia comparata dei Tisanuri e considerazioni generali sull'organizzazzione degli Insetti. *Atti della Reale Accademia Nazionale dei Lincei* 4: 543–606.

Gratshev, V. G., and V. V. Zherikhin. 1993. New fossil mantids (Insecta, Mantida). *Paleontological Journal* 27: 148–65.

———, and V. V. Zherikhin. 1995. Revision of the Late Jurassic nemonychid weevil genera *Distenorrhinus* and *Procurculio* (Insecta, Coleoptera: Nemonychidae). *Paleontologicheskiy Zhurnal* 2: 83–94. [In Russian]

———, and V. V. Zherikhin. 2000. The weevils from the Late Cretaceous New Jersey amber (Coleoptera, Curculionoidea). Pp. 241–54. *In* Grimaldi, D. (ed.), *Studies on Fossils in Amber, with Particular Reference to the Cretaceous of New Jersey*. Backhuys Publishers; Leiden, the Netherlands; viii+498 pp.

———, and V. V. Zherikhin. 2003. The fossil record of weevils and related beetle families (Coleoptera, Curculionoidea). *Acta Zoologica Cracoviensia* 46 (Supplement): 129–38.

Graur, D., and W. Martin. 2004. Reading the entrails of chickens: Molecular timescales of evolution and the illusion of precision. *Trends in Genetics* 20: 80–6.

Grauvogel, L., and D. Laurentiaux. 1952. Un protodonate du Trias des Vosges. *Annales de Paleontologie* 38: 121–9.

Greene, H. W. 1994. Homology and behavioral repertoires. Pp. 370–91. *In* Hall, B. K. (ed.), *Homology: The Hierarchical Basis of Comparative Biology*. Academic Press; New York, New York; xvi+483 pp.

Greenslade, P. 1994. Collembola. Pp. 19–138. *In* Houston, W. K. K. (ed.), *Zoological Catalogue of Australia: Volume 22: Protura, Collembola, Diplura*. CSIRO; Melbourne, Australia; xi+188 pp.

———, and P. E. S. Whalley. 1986. The systematic position of *Rhyniella praecursor* Hirst & Maulik (Collembola). The earliest known hexapod. Pp. 319–23. *In* Dallai, R. (ed.), *Second International Seminar on Apterygota*. University of Siena; Siena, Italy; 334 pp.

Greven, H. 1980. Die Bartierchen, Tardigrada. *Neue Brehm Bücherei* 537: 1–101.

Griffiths, G. C. D. 1972. *The Phylogenetic Classification of Diptera Cyclorrhapha, with Special Reference to the Structure of the Male Postabdomen*. Junk; The Hague, the Netherlands; 340 pp.

———. 1982. On the systematic position of *Mystacinobia* (Diptera: Calliphoridae). *Memoirs of the Entomological Society of Washington* 10: 70–7.

Grimaldi, D. A. 1988. Bee flies and bluets: *Bombylius* (Diptera: Bombyliidae) flower-constant on the distylous species, *Hedyotis caerulea* (Rubiaceae), and the manner of foraging. *Journal of Natural History* 22: 1–10.

———, ed. 1990a. Insects from the Santana Formation, Lower Cretaceous, of Brazil. *Bulletin of the American Museum of Natural History* 195: 1–191.

———. 1990b. A phylogenetic, revised classification of genera in the Drosophilidae (Diptera). *Bulletin of the American Museum of Natural History* 197: 1–139.

———. 1991. Mycetobiine woodgnats (Diptera: Anisopodidae) from the Oligo-Miocene amber of the Dominican Republic, and Old World affinities. *American Museum Novitates* 3014: 1–24.

———. 1992. Vicariance biogeography, geographic extinctions, and North American Oligocene tsetse flies. Pp. 178–204. *In* Novacek, M. J., and Q. D. Wheeler (eds.), *Phylogeny and Extinction*. Columbia University Press; New York, New York; vi+[1]+253 pp.

———. 1994a. Description and immature stages of *Hirtodrosophila batracida* sp. n. (Diptera: Drosophilidae), a predator of frog embryos. *Entomologica Scandinavica* 25: 129–36.

———. 1994b. The age of Dominican amber. Pp. 203–17. *In* Anderson, K. B., and J. C. Crelling (eds.), *Amber, Resinite, and Fossil Resins*. American Chemical Society; Washington, D.C.; xvii+297 pp.

———. 1996. *Amber: Window to the Past*. Abrams/American Museum of Natural History; New York, New York ; 216 pp.

———. 1997a. The bird flies, genus *Carnus*: Species revision, generic relationships, and a fossil *Meoneura* in amber (Diptera: Carnidae). *American Museum Novitates* 3190: 1–30.

———. 1997b. A fossil mantis (Insecta: Mantodea) in Cretaceous amber of New Jersey, with comments on the early history of the Dictyoptera. *American Museum Novitates* 3204: 1–11.

———. 1999. The co-radiations of pollinating insects and angiosperms in the Cretaceous. *Annals of the Missouri Botanical Garden* 86: 373–406.

———. 2000a. A diverse fauna of Neuropterodea in amber from the Cretaceous of New Jersey. Pp. 259–303. *In* Grimaldi, D. (ed.), *Studies on Fossils in Amber, with Particular Reference to the Cretaceous of New Jersey.* Backhuys Publishers; Leiden, the Netherlands; viii+498 pp.

———. 2000b. Mesozoic radiations of the insects and origins of the modern fauna. *Proceedings of the XXI International Congress of Entomology, Iguassu, Brazil* 2000: xix–xxvii.

———. 2001. Insect evolutionary history from Handlirsch to Hennig, and beyond. *Journal of Paleontology* 75: 1152–60.

———. 2003a. Fossil Record. Pp. 455–63. *In* Resh, V. H., and R. T. Cardé (eds.), *Encyclopedia of Insects.* Academic Press; San Diego, California; xxviii+[1]+1266 pp.

———. 2003b. A revision of Cretaceous mantises and their relationships, including new taxa (Insecta: Dictyoptera: Mantodea). *American Museum Novitates* 3412: 1–47.

———. 2003c. First amber fossils of the extinct family Protopsyllidiidae, and their phylogenetic significance among Hemiptera. *Insect Systematics and Evolution* 34: 329–44.

———, and D. Agosti. 2000. A formicine in New Jersey Cretaceous amber (Hymenoptera: Formicidae) and early evolution of the ants. *Proceedings of the National Academy of Sciences, U.S.A.* 97: 13678–83.

———, and V. Blagoderov. 2002. A new genus of Lygistorrhinidae from Vietnam (Diptera: Sciaroidea), and phylogenetic relationships in the family. *Studia Dipterologica* 8: 43–57.

———, and J. Cumming. 1999. Brachyceran Diptera in Cretaceous ambers and Mesozoic diversification of the Eremoneura. *Bulletin of the American Museum of Natural History* 239: 1–141.

———, and J. Jaenike. 1984. Competition in natural populations of mycophagous *Drosophila. Ecology* 65: 1113–20.

———, and J. Maisey. 1990. Introduction. Pp. 5–14. *In* Grimaldi, D. A. (ed.), *Insects from the Santana Formation, Lower Cretaceous, of Brazil. Bulletin of the American Museum of Natural History* 195: 1–191.

———, and T. Nguyen. 1999. Monograph on the spittlebug flies, genus *Cladochaeta* (Diptera: Drosophilidae: Cladochaetini). *Bulletin of the American Museum of Natural History* 241: 1–326.

———, and A. J. Ross. 2004. *Raphidiomimula*, a new genus of enigmatic roach in Cretaceous amber from Myanmar (Burma) (Insecta: Blattodea: Raphidiomimidae). *Journal of Systematic Palaeontology* 2: 101–4.

———, and B. A. Underwood. 1986. *Megabraula*, a new genus for two new species of Braulidae (Diptera), and a discussion of braulid evolution. *Systematic Entomology* 11: 427–38.

———, and C. Young. 1992. Observations on the bizarre jelly mass habit of larval *Geranomyia* (Diptera: Tipulidae: Limoniinae). *Journal of the New York Entomological Society* 100: 634–7.

———, C. W. Beck, and J. J. Boon. 1989. Occurrence, chemical characteristics, and paleontology of the fossil resins from New Jersey. *American Museum Novitates* 2948: 1–28.

———, A. C. James, and J. Jaenike. 1992. Systematics and modes of reproductive isolation in the Holarctic *Drosophila testacea* species group (Diptera: Drosophilidae). *Annals of the Entomological Society of America* 85: 671–85.

———, C. Michalski, and K. Schmidt. 1993. Amber fossil Enicocephalidae (Heteroptera) from the Lower Cretaceous of Lebanon and Oligo-Miocene of the Dominican Republic, with biogeographic analysis of *Enicocephalus. American Museum Novitates* 3071: 1–30.

———, E. Bonwich, M. Delannoy, and S. Doberstein. 1994. Electron microscopic studies of mummified tissues in amber fossils. *American Museum Novitates* 3097: 1–31.

———, D. Agosti, and J. M. Carpenter. 1997. New and rediscovered primitive ants (Hymenoptera: Formicidae) in Cretaceous amber from New Jersey, and their phylogenetic relationships. *American Museum Novitates* 3208: 1–43.

———, A. Shedrinsky, and T. P. Wampler. 2000a. A remarkable deposit of fossiliferous amber from the Upper Cretaceous (Turonian) of New Jersey. Pp. 1–76. *In* Grimaldi, D. (ed.), *Studies on Fossils in Amber, with Particular Reference to the Cretaceous of New Jersey.* Backhuys Publishers; Leiden, the Netherlands; viii+498 pp.

———, J. A. Lillegraven, T. W. Wampler, D. Bookwalter, and A. Shedrinsky. 2000b. Amber from Upper Cretaceous through Paleocene strata of the Hanna Basin, Wyoming, with evidence for source and taphonomy of fossil resins. *Rocky Mountain Geology* 35: 163–204.

———, M. S. Engel, and P. C. Nascimbene. 2002. Fossiliferous Cretaceous amber from Myanmar (Burma): Its rediscovery, biotic diversity, and paleontological significance. *American Museum Novitates* 3361: 1–72.

———, D. S. Amorim, and V. Blagoderov. 2003. The Mesozoic family Archizelmiridae (Diptera: Insecta). *Journal of Paleontology* 77: 368–81.

———, A. Shmakov, and N. C. Fraser. 2004a. Mesozoic thrips and early evolution of the order Thysanoptera (Insecta). *Journal of Paleontology* 78: 941–52.

———, J.-F. Zhang, A. Rasnitsyn, and N. C. Fraser. In press. New and revised scorpionflies of the extinct family Pseudopolycentropodidae (Mecopteroidea), including two mosquito-like species in Cretaceous amber. *Insect Systematics and Evolution.*

Grimalt, J. O., B. R. T. Simoneit, P. G. Hatcher, and A. Nissenbaum. 1987. The molecular composition of ambers. *Organic Geochemistry* 13: 677–90.

Grinfel'd, E. K. 1975. Anthophily in beetles (Coleoptera) and a critical evaluation of the cantharophilous hypothesis. *Entomological Review* 54: 18–22.

Grissell, E. E. 1980. New Torymidae from Tertiary amber of the Dominican Republic and a world list of fossil torymids (Hymenoptera: Chalcidoidea). *Proceedings of the Entomological Society of Washington* 82: 252–9.

———. 1999. Hymenopteran biodiversity: Some alien notions. *American Entomologist* 45: 235–44.

———, and M. E. Schauff. 1997. Chalcidoidea. Pp. 45–116. *In* Gibson, G. A. P., J. T. Huber, and J. B. Woolley. 1997. *Annotated Keys to the Genera of Nearctic Chalcidoidea (Hymenoptera).* National Research Council of Canada; Ottawa, Canada; xi+794 pp.

Grodnitsky, D. L. 1999. *Form and Function of Insect Wings: The Evolution of Biological Structures.* Johns Hopkins University Press; Baltimore, Maryland; xiv+261 pp.

Gromov, V. V., V. Yu. Dmitriev, V. V. Zherikhin, A. G. Ponomarenko, A. P. Rasnitsyn, and I. D. Sukatsheva. 1993. Cretaceous insects from Ul'ya River basin (western Okhotsk region). *Trudy Paleontologischeskogo Instituta* 252: 5–60.

Grunin, K. Y. 1973. The first discovery of the mammoth bot-fly *Cobboldia* (*Mamontia*, subgen. n.) *russanovi* sp. n. (Diptera, Gasterophilidae). *Entomologicheskoe Obozrenie* 52: 228–33. [In Russian]

Grütte, E. 1935. Zur Abstammung der Kuckucksbienen. *Archiv für Naturgeschichte* 4: 449–534.

Gullan, P. J. 1997. Relationships with ants. Pp. 351–73. *In* Ben-Dov, Y., and C. J. Hodgson (eds.), *Soft Scale Insects: Their Biology, Natural Enemies and Control.* Elsevier; Amsterdam, the Netherlands; xxxii+790 pp.

———, and M. Kosztarab. 1997. Adaptations in scale insects. *Annual Review of Entomology* 42: 23–50.

———, and A. W. Sjaarda. 2001. Trans-Tasman *Platycoelostoma* Morrison (Hemiptera: Coccoidea: Margarodidae) on endemic Cupressaceae, and the phylogenetic history of margarodids. *Systematic Entomology* 26: 257–78.

Günther, K. 1953. Über die taxonomische Gliederung und die geographische Verbreitung der Insektenordnung der Phasmatodea. *Beiträge zur Entomologie* 3: 541–63.

———. 1974. Staubläuse, Psocoptera. *Die Tierwelt Deutschlands* 61: 1–314.

Gurney, A. B. 1947. Notes on Dilaridae and Berothidae with special reference to the immature stages of the Nearctic genera (Neuroptera). *Psyche* 54: 145–69.

———. 1953. Notes on the biology and immature stages of a cricket parasite of the genus *Rhopalosoma. Proceedings of the United States National Museum* 103: 19–34.

Gusarov, V. I. 2000. *Mesotachyporus puer*, a new genus and species of Cretaceous Tachyporinae (Coleoptera, Staphylinidae) from New Jersey amber. Pp. 255–8. *In* Grimaldi, D. (ed.), *Studies on Fossils in Amber, with Particular Reference to the Cretaceous of New Jersey.* Backhuys Publishers; Leiden, the Netherlands; viii+498 pp.

Gussakovskij, V. V. 1935. Insectes hyménoptères, I, Chalastogastra (partie 1). Pp. 1–452. *In, Payha SSSR, 2* [Fauna of the USSR, 2]. Akademii Nauk [Academy of Sciences]; Moscow, Russia; xviii+453 pp. [In Russian]

Guthrie, D. M., and A. R. Tindall. 1968. *The Biology of the Cockroach.* St. Martin's Press; New York, New York; viii+408 pp.

———, and A. R. Tindal. 1968. *The Biology of the Cockroach.* Edward Arnold; London, UK; 408 pp.

Gutiérrez, G., and A. Marín. 1998. The most ancient DNA recovered from an amber-preserved specimen may not be as ancient as it seems. *Molecular Biology and Evolution* 15: 926–9.

Gwynne, D. T. 1995. Phylogeny of the Ensifera (Orthoptera): A hypothesis supporting multiple origins of acoustical signalling, complex spermatophores and maternal care in crickets, katydids, and weta. *Journal of Orthoptera Research* 4: 203–18.

———. 2001. *Katydids and Bush-Crickets: Reproductive Behavior and Evolution of the Tettigoniidae.* Cornell University Press; Ithaca, New York; xii+[1]+317 pp.

———, and H. E. Evans. 1975. Nesting behavior of *Larropsis chilopsidis* and *L. vegeta* (Hymenoptera: Sphecidae: Larrinae). *Psyche* 82: 275–82.

———, and G. K. Morris, eds. 1983. *Orthopteran Mating Systems: Sexual Competition in a Diverse Group of Insects.* Westview Press; Boulder, Colorado; xvii+376 pp.

Haas, F. 1995. The phylogeny of the Forficulina, a suborder of the Dermaptera. *Systematic Entomology* 20: 85–98.

———, and R. G. Beutel. 2001. Wing folding and the functional morphology of the wing base in Coleoptera. *Zoology* 104: 123–41.

———, and J. Kukalová-Peck. 2001. Dermaptera hindwing structure and folding: New evidence for familial, ordinal and superordinal relationships within Neoptera (Insecta). *European Journal of Entomology* 98: 445–509.

———, D. Waloszek, and R. Hartenberger. 2003. *Devonohexapodus bocksbergensis*, a new marine hexapod from the Lower Devonian Hunsrück Slates, and the origin of Atelocerata and Hexapoda. *Organisms, Diversity and Evolution* 3: 39–54.

Hacker, H. 1913. Some field notes on Queensland insects. *Memoirs of the Queensland Museum* 2: 96–100.

———. 1915. Notes on the genus *Megachile* and some rare insects collected during 1913–1914. *Memoirs of the Queensland Museum* 3: 137–41.

Hadisoesilo, S., and G. W. Otis. 1996. Drone flight times confirm the species status of *Apis nigrocincta* Smith, 1861 to be a species distinct from *Apis cerana* F, 1793, in Sulawesi, Indonesia. *Apidologie* 27: 361–9.

———, and G. W. Otis. 1998. Differences in drone cappings of *Apis cerana* and *Apis nigrocincta*. *Journal of Apicultural Research* 37: 11–15.

———, G. W. Otis, and M. Meixner. 1995. Two distinct populations of cavity-nesting honey bees (Hymenoptera: Apidae) in South Sulawesi, Indonesia. *Journal of the Kansas Entomological Society* 68: 399–407.

Haeckel, E. 1866. *Generelle Morphologie der Organismen: Allgemeine Grundzüge der Organischen Formen-Wissenschaft, mechanische begründet durch die von Charles Darwin reformirte Descendenz-Theorie: Zweiter Band: Allgemeine Entwickelungsgeschichte der Organismen.* Georg Reimer; Berlin, Germany; clx+462 pp.

———. 1890. *Systematische Phylogenie der Wirbellosen Thiere (Invertebrata): Zweiter Theil des Entwurfs einer systematischen Stammesgeschichte.* Georg Reimer; Berlin, Germany; xviii+720 pp.

———. 1909. *Natürliche Schöpfungs-Geschichte. Gemeinverstandliche wissenschaftliche Vortrage über die Entwicklungslehre. Zweiter Teil: Allgemeine Stammesgeschichte (Phylogenie und Anthropogenie).* Georg Reimer; Berlin, Germany; [371]–832 pp.

Hafner, M. S., and R. D. M. Page. 1995. Molecular phylogenies and host-parasite cospeciation: Gophers and lice as a model system. *Philosophical Transactions of the Royal Society, London (B)* 349: 77–83.

———, P. D. Sudman, F. X. Villablanca, T. A. Spradling, J. W. Demastes, and S. A. Nadler. 1994. Disparate rates of molecular evolution in cospeciating hosts and parasites. *Science* 265: 1087–9.

Halbert, N. R., L. D. Ross, J. Kathirithamby, J. B. Woolley, R. S. Saff, and J. S. Johnston. 2001. Phylogenetic analysis as a means of species identification within Myrmecolacidae (Strepsiptera). *Tijdschrift voor Entomologie* 144: 179–86.

Halffter, G. 1997. Subsocial behavior in Scarabaeinae beetles. Pp. 237–59. *In* Choe, J. C., and B. J. Crespi (eds.), *The Evolution of Social Behavior in Insects and Arachnids.* Cambridge University Press; Cambridge, UK; vii+541 pp.

———, and W. D. Edmonds. 1982. *The Nesting Behavior of Dung Beetles (Scarabaeinae): An Ecological and Evolutive Approach.* Instituto de Ecologia, UNAM; Mexico, D.F.; 176 pp.

Hall, J. P. W., R. K. Robbins, and D. J. Harvey. 2004. Extinction and biogeography in the Caribbean: New evidence from a fossil riodinid butterfly in Dominican amber. *Proceedings of the Royal Society of London, Series B, Biological Sciences* 271: 797–801.

Hallam, A. 1989. The case for sea-level change as a dominant causal factor in mass extinctions of marine invertebrates. *Philosophical Transactions of the Royal Society, London (B)* 325: 653–5.

———, and P. B. Wignall. 1997. *Mass Extinctions and Their Aftermath.* Oxford University Press; Oxford, UK; viii+320 pp.

Hamilton, K. G. A. 1971. The insect wing. Part I. Origin and development of wings from notal lobes. *Journal of the Kansas Entomological Society* 44: 421–33.

———. 1972a. The insect wing, Part III. Venation of the orders. *Journal of the Kansas Entomological Society* 45: 145–62.

———. 1972b. The insect wing, Part IV. Venational trends and the phylogeny of the winged orders. *Journal of the Kansas Entomological Society* 45: 295–308.

———. 1981. Morphology and evolution of the rhynchotan head (Insecta: Hemiptera, Homoptera). *Canadian Entomologist* 113: 953–74.

———. 1983. Classification, morphology and phylogeny of the family Cicadellidae (Rhynchota: Homoptera). Pp. 15–37. *In* Knight, W. J. (ed.), *International Workshop on Leafhoppers and Planthoppers of Economic Importance.* Commonwealth Institute of Entomology; London, UK; 500 pp.

———. 1990. Homoptera. *Bulletin of the American Museum of Natural History* 195: 82–122.

———. 1996. Cretaceous Homoptera from Brazil: Implications for classification. Pp. 89–110. *In* Schaeffer, C. W. (ed.), *Studies on Hemipteran Phylogeny.* Entomological Society of America (Thomas Say Publications); Lanham, Maryland; iii+244 pp.

Hamilton, W. D. 1964. The genetical evolution of social behaviour. I and II. *Journal of Theoretical Biology* 7: 1–52.

Hammen, L. van der. 1972. A revised classification of the mites (Arachnidea, Acarida) with diagnoses, a key, and notes on phylogeny. *Zoologische Mededelingen* 47: 273–92.

———. 1989. *An Introduction to Comparative Arachnology.* SPB Academic Publishing; The Hague, the Netherlands; x +576 pp.

Hammond, P. M. 1992. Species inventory. Pp. 17–39. *In* Groombridge, B. (ed.), *Global Biodiversity: Status of the Earth's Living Resources.* Chapman and Hall; London, UK; xx+585 pp.

Hancock, D. L. 1983. Classification of the Papilionidae (Lepidoptera): A phylogenetic approach. *Smithersia* 2: 1–48.

Handel, S. N., and R. Peakall. 1993. Thynnine wasps discriminate among heights when seeking mates: Tests with a sexually deceptive orchid. *Oecologia* 95: 241–5.

Handlirsch, A. 1903. Zur Phylogenie der Hexapoden. *Sitzungsberichte der Königliche Akademie der Wissenschaften in Wien, Mathematische-Naturwissenschaftliche Klasse* 112: 716–38.

———. 1904. Zur Systematik der Hexapoden. *Zoologischer Anzeiger* 27: 733–59.

———. 1906a. Revision of American Paleozoic insects. *Proceedings of the United States National Museum* 29: 661–820.

———. 1906b. *Die Fossilen Insekten und die Phylogenie der rezenten Formen: Ein Handbuch für Paläontologen und Zoologen.* Engelmann; Leipzig, Germany; ix+1–640 pp.

———. 1907. *Die Fossilen Insekten und die Phylogenie der rezenten Formen: Ein Handbuch für Paläontologen und Zoologen.* Engelmann; Leipzig, Germany; 641–1120 pp.

———. 1908. *Die Fossilen Insekten und die Phylogenie der rezenten Formen: Ein Handbuch für Paläontologen und Zoologen.* Engelmann; Leipzig, Germany; 1121–1430 pp.

———. 1910. Canadian fossil insects. 5. Insects from the Tertiary lake deposits of the southern interior of British Columbia, collected by Mr. Lawrence M. Lambe, in 1906. *Contributions to Canadian Palaeontology* 2: 93–129.

———. 1911. New Paleozoic insects from the vicinity of Mazon Creek, Illinois. *American Journal of Science* 31: 297–326, 352–77.

———. 1919. Revision der palaeozoischen Insekten. *Denkschriften der Kaiserlichen Akademie der Wissenschaften, Wien, Mathematisch-Naturwissenschaftliche Klasse* 96: 511–92.

———. 1925. Phylogenie oder Stammesgeschichte. Pp. 307–76. *In* Schröder, C. (ed.), *Handbuch der Entomologie. Band III: Geschichte, Literatur, Technik, Paläontologie, Phylogenie, Systematik.* Gustav Fischer; Jena, Germany; viii+1201+[1] pp.

———. 1937. Neue Untersuchungen über die fossilen Insekten, I. *Annalen des Naturhistorischen Museums in Wien* 48: 1–140.

———. 1939. Neue Untersuchungen über die fossilen Insekten, II. *Annalen des Naturhistorischen Museums in Wien* 49: 1–240.

HANDSCHIN, E. 1944. Insekten aus den Phosphoriten des Quercy. *Schweizer Paläontologische Abhandlungen* 64: 1–23.

HANNIBAL, J. 1984. Pill millipedes from the coal age. *Bulletin of the Field Museum of Natural History* 55: 12–16.

HANSELL, M. H., C. SAMUEL, and J. I. FURTADO. 1982. *Liostenogaster flavolineata*: Social life in the small colonies of an Asian tropical wasp. Pp. 192–5. *In* BREED, M. D., C. D. MICHENER, and H. E. EVANS (eds.), *The Biology of Social Insects*. Westview Press; Boulder, Colorado; xii+419 pp.

HANSEN, A., S. HANSMANN, T. SAMIGULLIN, A. ANTONOV, and W. MARTIN. 1999. *Gnetum* and the angiosperms: Molecular evidence that their shared morphological characters are convergent, rather than homologous. *Molecular Biology and Evolution* 16: 1006–9.

HANSKI, I., and Y. CAMBEFORT, eds. 1991. *Dung Beetle Ecology*. Princeton University Press; Princeton, New Jersey; xii+481 pp.

HANSON, P. E. 1986. Biology of *Janus rufiventris* (Hymenoptera: Cephidae). *Annals of the Entomological Society of America* 79: 488–90.

———, and I. D. GAULD, eds. 1995. *The Hymenoptera of Costa Rica*. Oxford University Press; Oxford, UK; xx+893 pp.

HARBACH, R. E., and I. J. KITCHING. 1998. Phylogeny and classification of the Culicidae (Diptera). *Systematic Entomology* 23: 327–70.

HARDY, D. E. 1965. Diptera: Cyclorrhapha II, Series Schizophora, Section Acalypterae. I. Family Drosophilidae. *Insects of Hawaii* 12: 1–814.

HARMS, F.-J. 2002. Steine erzählen Geschichte(n): Ursache für die Entstehung des Messel-Sees gefunden. *Natur und Museum* 132: 1–4.

HARRIS, K. M. 1994. Gall midges (Cecidomyiidae): classification and biology. Pp. 204–11. *In* WILLIAMS, M. A. J. (ed.), *Plant Galls: Organisms, Interactions, Populations*. Clarendon Press; Oxford, UK; xiv+488 pp.

HART, A. G., and F. L. W. RATNIEKS. 2002. Task-partitioned nectar transfer in stingless bees: Work organisation in a phylogenetic context. *Ecological Entomology* 27: 163–8.

HARZ, K. 1980. Ein neuer Nachweis von *Mantis religiosa* L. im Pliozän von Willershausen. *Articulata* 1: 168.

HASENFUSS, I. 2002. A possible evolutionary pathway to insect flight starting from lepismatid organization. *Journal of Zoological Systematics and Evolutionary Research* 40: 65–81.

HASIOTIS, S. T., and R. F. DUBIEL. 1995. Termite (Insecta: Isoptera) nest ichnofossils from the Upper Triassic Chinle Formation, Petrified Forest National Park, Arizona. *Ichnos* 4: 119–30.

HATSCHEK, B. 1878. Studien über Entwicklungsgeschichte der Anneliden. *Arbeiten aus dem Zoologischen Institut der Universität Wien* 1: 57–128.

HAYASHI, M. 1999. Late Pliocene to Early Pleistocene donaciine fossils from the Uonuma Formation in Higashikubiki hills, Niigata Prefecture, Japan, with description of a new species (Coleoptera: Chrysomelidae: Donaciinae). *Bulletin of the Osaka Museum of Natural History* 53: 1–22.

———. 2000. Late Pliocene to Early Pleistocene donaciine fossils from the Uonuma Formation (Part II): A new *Donacia* from the Uonuma hills in Niigata Prefecture, Japan (Coleoptera: Chrysomelidae: Donaciinae). *Bulletin of the Osaka Museum of Natural History* 54: 31–48.

———. 2001a. Neogene donaciine fossils from Tamaniwa district, Yamagata Prefecture, Japan, with description of a new species (Coleoptera: Chrysomelidae: Donaciinae). *Bulletin of the Osaka Museum of Natural History* 55: 7–14.

———. 2001b. Fossil insects and paleoenvironments of the Plio-Pleistocene Uonuma Formation, Niigata Prefecture, Japan. *Bulletin of the Mizunami Fossil Museum* 28: 223–38. [In Japanese]

———, and S. SHIYAKE. 2002. Late Pliocene Donaciinae (Coleoptera, Chrysomelidae) from the Koka Formation, Kobiwako Group in Shiga Prefecture, Japan. *Elytra* 30: 207–13.

HECKMAN, D. S., D. M. GEISER, B. R. EIDELL, R. L. STAUFFER, *et al.* 2001. Molecular evidence for the early colonization of land by fungi and plants. *Science* 293: 1129–32.

HECKROTH, H.-P., B. FIALA, P. J. GULLAN, A. H. J. IDRIS, and U. MASCHWITZ. 1998. The soft scale (Coccidae) associates of Malaysian ant-plants. *Journal of Tropical Ecology* 14: 427–43.

HEDGEPETH, J. W. 1955a. Pycnogonida. Pp. 163–70. *In* MOORE, R. C. (ed.), *Treatise on Invertebrate Paleontology, Part P, Arthropoda 2, Chelicerata, with Sections on Pycnogonida and* Palaeoisopus. University of Kansas Press; Lawrence, Kansas; xvii+181 pp.

———. 1955b. *Palaeoisopus*. Pp. 171–3. *In* MOORE, R. C. (ed.), *Treatise on Invertebrate Paleontology, Part P, Arthropoda 2, Chelicerata, with Sections on Pycnogonida and* Palaeoisopus. University of Kansas Press; Lawrence, Kansas; xvii+181 pp.

HEER, O. 1849. Die Insektenfauna der Tertiärgebilde von Oeningen und von Radoboj in Croatien. Zweite Abtheilung: Heuschrecken, Florfliegen, Aderflügler, Schmetterlinge und Fliegen. *Denkschriften der Kaiserlichen Akademie der Wissenschaften [Wien], Mathematisch-Naturwissenschaftliche Classe* 11: 1–264.

———. 1865. *Die Urwelt der Schweiz*. Friedrich Schulthess; Zurich, Switzerland; xxix+622 pp.

HEGGEMANN, H., R. KOHRING, and T. SCHLÜTER. 1990. Fossil plants and arthropods from the Phra Wihan Formation, presumably middle Jurassic, of northern Thailand. *Alcheringa* 14: 311–16.

HEIE, O. E. 1981. Morphology and phylogeny of some Mesozoic aphids (Insecta, Hemiptera). *Entomologica Scandinavica Supplement* 15: 401–15.

———. 1987. Palaeontology and phylogeny. Pp. 367–91. *In* MINKS, A. K., and P. HARREWIJN (eds.) *Aphids: Their Biology, Natural Enemies and Control*. Elsevier [World Crop Pests, volume 2A]; Amsterdam, the Netherlands; 392 pp.

———. 1995. An aphid from the Plio-Pleistocene Kobenhavn Formation, north Greenland (Homoptera, Aphidoidea). *Entomologiske Meddelelser* 63: 17–18.

———. 1996. The evolutionary history of aphids and a hypothesis on the coevolution of aphids and plants. *Bollettino di Zoologia Agraria e di Bachicoltura* 28: 149–55.

———, and D. AZAR. 2000. Two new species of aphids found in Lebanese amber and a revision of the family Tajmyraphididae Kononova, 1975 (Hemiptera: Sternorrhyncha). *Annals of the Entomological Society of America* 93: 1222–5.

———, and W. L. FRIEDRICH. 1971. A fossil specimen of the North American hickory aphid (*Longistigma caryae* Harris) found in Tertiary deposits in Iceland. *Entomologica Scandinavica* 2: 74–80.

———, and W. L. FRIEDRICH. 1990. The hickory aphid from the Iceland Miocene. Pp. 104–7. *In* BOUCOT, A. J. (ed.), *Evolutionary Paleobiology of Behavior and Coevolution*. Elsevier; Amsterdam, the Netherlands; xxiii+723 pp.

———, and E. M. PIKE. 1992. New aphids in Cretaceous amber from Alberta (Insecta, Homoptera). *Canadian Entomologist* 124: 1027–53.

———, and P. WEGIEREK. 1998. A list of fossil aphids (Homoptera: Aphidinea). *Annals of the Upper Silesian Museum, Entomology* 8/9: 159–92.

HEIM DE BALSAC, H. 1935. Étude biologique de quelques hyménoptères du Chile. *Annales des Sciences Naturelles, Zoologie* 16: 209–356.

HEINRICH, B. 1979. *Bumblebee Economics*. Harvard University Press; Cambridge, Massachusetts; viii+245 pp.

HEISS, E., and D. GRIMALDI. 2001. *Archearadus burmensis* gen. n., sp. n., a remarkable Mesozoic Aradidae in Burmese amber (Heteroptera, Aradidae). *Carolinea* 59: 99–102.

HELLENTHAL, R. A., and R. D. PRICE. 1991. Biosystematics of the chewing lice of pocket gophers. *Annual Review of Entomology* 36: 185–203.

HELLMUND, M. and W. HELLMUND. 1996. Phosphoritisierte Insekten- und Annelidenreste aus der mitteloligozänen Karstpaltenfüllung 'Ronheim 1' bei Harburg (Bayern, Süddeutschland). *Stuttgarter Beiträge zur Naturkunde, Serie B, Geologie und Paläontologie* 241: 1–21.

HEMING, B. S. 1971. Functional morphology of the thysanopteran pretarsus. *Canadian Journal of Zoology* 49: 91–108.

——. 1972. Functional morphology of the pretarsus in larval Thysanoptera. *Canadian Journal of Zoology* 50: 751–66.

——. 1975. Antennal structure and metamorphosis in *Frankliniella fusca* (Hinds) (Thripidae) and *Haplothrips verbasci* (Osborn) (Phlaeothripidae) (Thysanoptera). *Canadian Journal of Zoology* 51: 1211–1234.

——. 1978. Structure and function of the mouthparts in larvae of *Haplothrips verbasci* (Osborn) (Thysanoptera, Tubulifera, Phlaeothripidae). *Journal of Morphology* 156: 1–38.

——. 1980. Development of the mouthparts in embryos of *Haplothrips verbasci* (Osborn) (Insecta, Thysanoptera, Phlaeothripidae). *Journal of Morphology* 164: 235–63.

——. 1991. Order Thysanoptera. Pp. 1–21. *In* STEHR, F. W. (ed.), *Immature Insects* [Volume 2]. Kendall-Hunt; Dubuque, Iowa; xvi+975 pp.

——. 1993. Structure, function, ontogeny, and evolution of feeding in thrips (Thysanoptera). Pp. 3–41. *In* SCHAEFER, C. W., and R. A. B. LESCHEN (eds.), *Functional Morphology of Insect Feeding*. Entomological Society of America (Thomas Say Publications); Lanham, Maryland; ii+162 pp.

——. 1995. History of the germ line in male and female thrips. *NATO Advanced Science Institutes Series, Series A, Life Sciences* 276: 505–35.
——. 2002. *Insect Development and Evolution.* Cornell University Press; Ithaca, New York; xv+444 pp.
Hennig, W. 1936. Beiträge zur Kenntnis des Kopulationsapparates der cyclorrhaphen Dipteren. *Zeitschrift für Morphologie und Ökologie der Tiere* 31: 328–70.
———. 1948. *Die Larvenformen der Dipteren. I. Teil.* Akademie-Verlag; Berlin, Germany; 186 pp.
———. 1950a. *Die Larvenformen der Dipteren. II. Teil.* Akademie-Verlag; Berlin, Germany; 458 pp.
———. 1950b. *Grundzüge einer Theorie der phylogenetischen Systematik.* Deutscher Zentralverlag; Berlin; [4]+370 pp.
———. 1952. *Die Larvenformen der Dipteren. III. Teil.* Akademie-Verlag; Berlin, Germany; 628 pp.
———. 1953. Kritische Bemerkungen zum phylogenetischen System der Insekten. *Beiträge zur Entomologie* 3: 1–85.
———. 1954. Flügelgeader und System der Dipteren unter Berüchsichtigung der aus dem Mesozoikum beschriebenen Fossilien. *Beiträge zur Entomologie* 4: 245–388.
———. 1958. Die Familien der Diptera Schizophora und ihre phylogenetischen Verwandtschaftsbeziehungen. *Beiträge zur Entomologie* 8: 505–688.
———. 1960. Die Dipteren-Fauna von Neuseeland als Systematisches und Tiergeographisches Problem. *Beiträge zur Entomologie* 10: 221–329.
———. 1965a. Die Acalyptratae des baltischen Bernsteins und ihre Bedeutung für die Erfoschung der phylogenetischen Entwicklung dieser Dipteren-Gruppe. *Stuttgarter Beiträge zur Naturkunde* 145: 1–215.
———. 1965b. Phylogenetic systematics. *Annual Review of Entomology* 10: 97–116.
———. 1966. *Phylogenetic Systematics.* University of Illinois Press; Chicago, Illinois; [2]+263 pp.
———. 1969. *Die Stammesgeschichte der Insekten.* Kramer; Frankfurt am Main, Germany; 436 pp.
———. 1970. Insektenfossilien aus der unteren Kreide. III. Empidiformia ("Microphorinae") aus der unteren Kreide und aus dem baltischen Bernstein; ein Vertreter der Cyclorrhapha aus der unteren Kreide. *Stuttgarter Beiträge zur Naturkunde* 232: 1–28.
———. 1971. Neue Untersuchungen über die Familien der Diptera Schizophora (Diptera: Schizophora). *Stuttgarter Beiträge zur Naturkunde* 226: 1–76.
———. 1973. Ordnung Diptera (Zweiflügler). *Handbuch der Zoologie* 4(2) (Lfg. 20): 1–337.
———. 1981. *Insect Phylogeny.* Wiley; Chichester, UK; xxii+514 pp.
Henriksen, K. L. 1922. Eocene insects from Denmark. *Danmarks Geologiske Undersoegelse (series 2)* 37: 1–36.
Henry, C. S. 1972. Eggs and rapagula of *Ululodes* and *Ascaloptynx* (Neuroptera: Ascalaphidae): A comparative study. *Psyche* 79: 1–22.
———. 1977. The behavior and life histories of two North American ascalaphids. *Annals of the Entomological Society of America* 70: 179–95.
———. 1994. Singing and cryptic speciation in insects. *Trends in Ecology and Evolution* 9: 388–92.
Henry, T. J. 1997. Phylogenetic analysis of family groups within the infraorder Pentatomomorpha (Hemiptera: Heteroptera), with emphasis on the Lygaeoidea. *Annals of the Entomological Society of America* 90: 275–301.
Henson, W. R. 1957. Notes on a collection of *Grylloblatta campodeiformis* Walker. *Canadian Field Naturalist* 71: 149–52.
Henwood, A. 1992a. Exceptional preservation of dipteran flight muscle and the taphonomy of insects in amber. *Palaios* 7: 203–12.
———. 1992b. Soft-part preservation of beetles in Tertiary amber from the Dominican Republic. *Palaeontology* 35: 901–12.
Hepburn, H. R. 1971. Proboscis extension and recoil in Lepidoptera. *Journal of Insect Physiology* 17: 637–56.
———, and S. E. Radloff. 1998. *Honeybees of Africa.* Springer Verlag; Berlin, Germany; xv+370 pp.
———, D. R. Smith, S. E. Radloff, and G. W. Otis. 2001. Infraspecific categories of *Apis cerana*: Morphometric, allozymal and mtDNA diversity. *Apidologie* 32: 3–23.
Heraty, J. M., J. B. Woolley, and D. C. Darling. 1997. Phylogenetic implications of the mesofurca in Chalcidoidea (Hymenoptera), with emphasis on Aphelinidae. *Systematic Entomology* 22: 45–65.
Herendeen, P. S., S. Magallón-Puebla, R. Lupia, P. R. Crane, and J. Kobylinska. 1999. A preliminary conspectus of the Allon Flora from the Late Cretaceous (Late Santonian) of Central Georgia, U.S.A. *Annals of the Missouri Botanical Garden* 86: 407–71.
Hering, E. M. 1951. *Biology of the Leaf Miners.* Dr. W. Junk; the Hague, the Netherlands; iv+420 pp.
Herman, L. H. 1975. Revision and phylogeny of the monogeneric subfamily Pseudopsinae for the world (Staphylinidae: Coleoptera). *Bulletin of the American Museum of Natural History* 155: 245–317.
———. 1986. Revision of *Bledius*, Part. IV. Classification of species groups, phylogeny, natural history, and catalogue (Coleoptera, Staphylinidae, Oxytelinae). *Bulletin of the American Museum of Natural History* 184: 1–368.
———. 2001. Catalog of the Staphylinidae (Insecta: Coleoptera). 1758 to the end of the second millennium. *Bulletin of the American Museum of Natural History* 265: 1–4218.
Herre, E. A., C. A. Machado, E. Bermingham, J. D. Nason, D. M. Windsor *et al.* 1996. Molecular phylogenies of figs and their pollinating wasps. *Journal of Biogeography* 23: 521–30.
Herter, K. 1943. Zur Fortpflanzungsbiologie eines lebendgebärenden Ohrwürmes (*Prolabia arachidis* Yersin). *Zeitschrift für Morphologie und Ökologie der Tiere* 40: 158–80.
Hey, J., and R. M. Kliman. 1993. Population genetics and phylogenetics of DNA sequence variation at multiple loci within the *Drosophila melanogaster* species complex. *Molecular Biology and Evolution* 10: 804–22.
Hickey, L. J., and J. A. Doyle. 1977. Early Cretaceous fossil evidence for angiosperm evolution. *Botanical Review* 43: 1–104.
———, and R. W. Hodges. 1975. Lepidopteran leaf mine from the Early Eocene Wind River Formation of northwestern Wyoming. *Science* 189: 718–20.
Higashi, M., N. Yamamura, and T. Abe. 2000. Theories on the sociality of termites. Pp. 169–87. *In* Abe, T., D. E. Bignell, and M. Higashi (eds.), *Termites: Evolution, Sociality, Symbioses, Ecology.* Kluwer Academic Publishers; Dordrecht, the Netherlands; xxii+466 pp.
Hildebrand, A. R., M. Pilkington, M. Connors, C. Ortiz-Aleman, *et al.* 1995. Size and structure of the Chicxulub crater revealed by horizontal gravity gradients and cenotes. *Nature* 376: 415–17.
Hillis, D. M. 1994. Homology in molecular biology. Pp. 339–68. *In* Hall, B. K. (ed.), *Homology: The Hierarchical Basis of Comparative Biology.* Academic Press; New York, New York; xvi+483 pp.
Hinton, H. E. 1955. On the structure, function, and distribution of the prolegs of the Panorpoidea, with a criticism of the Berlese-Imms theory. *Transactions of the Royal Entomological Society, London* 106: 455–556.
———. 1958. Phylogeny of the panorpoid orders. *Annual Review of Entomology* 3: 181–206.
———. 1971. Some neglected phases in metamorphosis. *Proceedings of the Royal Entomological Society of London (C)* 35: 55–63.
Hinz, I. 1987. The Lower Cambrian microfauna of Comley and Rushton, Shropshire/England. *Palaeontographica A* 198: 41–100.
Hirashima, Y., and K. Yamagishi. 1975. Embolemidae of Japan, with description of a new species of *Embolemus* from Hachijo Island (Hymenoptera, Bethyloidea). *Esakia* 9: 25–30.
Hirsch, F. 1986. Die Mundwerkzeuge von *Phthirus* [sic] *pubis* L. (Anoplura). *Zoologische Jahrbücher, Abteilung für Anatomie und Ontogenie der Tiere* 114: 167–204.
Hirst, S. 1923. On some arachnid remains from the Old Red Sandstone (Rhynie Chert bed, Aberdeenshire). *Annals and Magazine of Natural History* 9: 455–74.
——. 1953. *The Conquest of Plague: A Study of the Evolution of Epidemiology.* Clarendon Press; Oxford, UK; xvi+478 pp.
———, and S. Maulik. 1926. On some arthropod remains from the Rhynie chert (Old Red Sandstone). *Geological Magazine* 63: 69–71.
Hitchcock, E. 1858. *Ichnology of New England: A Report on the Sandstone of the Connecticut Valley, Especially its Fossil Footmarks.* William White; Boston, Massachusetts; xii+220 pp.
Hoare, R. J. B. 1999. The Gelechioidea. Pp. 131–58. *In* Kristensen, N. P. (ed.), *Handbuch der Zoologie: Eine Naturgeschichte der Stämme des Tierreiches: Band IV: Arthropoda: Insecta: Teilband 35: Lepidoptera, Moths and Butterflies: Volume 1: Evolution, Systematics, and Biogeography.* Walter de Gruyter; Berlin, Germany; x+491 pp.
———, and J. S. Dugdale. 2003. Description of the New Zealand incurvarioid *Xanadoses nielseni*, gen. nov., sp. nov. and placement in Cecidosidae (Lepidoptera). *Invertebrate Systematics* 17: 47–58.
Hodgson, C. J. 2002. Preliminary phylogeny of some non-margarodid Coccoidea (Hemiptera) based on adult male characters. *Bollettino di Zoologia Agraria e di Bachicoltura* 33: 129–37.
———, and D. R. Miller. 2002. A new genus and two new species of felt scales (Hemiptera: Coccoidea: Eriococcidae) from Chile, with comments on zoogeographical affinities between eriococcid faunas of southern South

America and New Zealand. *Systematic Entomology* 27: 191–209.

Hodkinson, I. D., and D. Casson. 1991. A lesser predilection for bugs: Hemiptera (Insecta) diversity in tropical rain forests. *Biological Journal of the Linnean Society* 43: 101–9.

Hoelzer, G. A., and D. J. Melnick. 1994. Patterns of speciation and limits to phylogenetic resolution. *Trends in Ecology and Evolution* 9: 104–7.

Holland, J. N., and T. H. Fleming. 1999. Mutualistic interactions between *Upiga virescens* (Pyralidae), a pollinating seed-consumer, and *Lophocereus schottii* (Cactaceae). *Ecology* 80: 2074–84.

Hölldobler, B., and E. O. Wilson. 1990. *The Ants.* Harvard University Press; Cambridge, Massachusetts; xii+732 pp.

Holloway, B. A. 1976. A new bat-fly family from New Zealand (Diptera: Mystacinobiidae). *New Zealand Journal of Zoology* 3: 279–301.

———, and E. S. Nielsen. 1999. Biogeography of the Lepidoptera. Pp. 423–62. *In* Kristensen, N. P. (ed.), *Handbuch der Zoologie: Eine Naturgeschichte der Stämme des Tierreiches: Band IV: Arthropoda: Insecta: Teilband 35: Lepidoptera, Moths and Butterflies: Volume 1: Evolution, Systematics, and Biogeography.* Walter de Gruyter; Berlin, Germany; x+491 pp.

Holmes, N. D. 1954. Food relations of the wheat stem sawfly, *Cephus cinctus* Nort. (Hymenoptera: Cephidae). *Canadian Entomologist* 86: 159–67.

Holser, W. T., and M. Magaritz. 1987. Events near the Permian-Triassic boundary. *Modern Geology* 11: 155–80.

Hong, Y.-C. 1979. On Eocene *Philolimnias* gen. nov. (Ephemeroptera, Insecta) in amber from Fushan Coalfield, Liaoning Province. *Scientia Sinica* 12: 331–9.

———. 1981. *Eocene Fossil Diptera Insecta in Amber of Fushun Coalfield.* Geological Publishing House; Beijing, China; vi+166 pp. [In Chinese]

———. 1983. *Middle Jurassic Fossil Insects in North China.* Geological Publishing House; Beijing, China; 223 pp. [In Chinese]

———. 1998. Establishment of fossil entomofaunas and their evolutionary succession in north China. *Entomologia Sinica* 5: 283–300.

———, and W.-L. Wang. 1987. Miocene Emboptera [sic] and Coleoptera (Insecta) of Shanwang, Shandong Province, China. *Professional Papers of Stratigraphy and Palaeontology* 17: 257–62. [In Chinese, with English summary]

Honigberg, B. M. 1970. Protozoa associated with termites and their role in digestion. Pp. 1–36. *In* Krishna, K., and F. M. Weesner (eds.), *Biology of Termites, Volume II.* Academic Press; New York, New York; vii+643 pp.

Hooker, J. D. 1844–1860. *The Botany of the Antarctic Voyage of H.M. Discovery Ships Erebus and Terror, in the Years 1839–1843, Under the Command of Captain Sir James Clark Ross, Kt., R.N., F.R.S., &c* [3 volumes]. Reeve; London, UK; Vol. I: *Flora Antarctica* [1844–1847, part 1. Botany of Lord Auckland's group and Campbell's Island, [i]+1–208 pp.; part 2. Botany of Fuegia, The Falklands, Kerguelen's land, &c., [iii]+209–574 pp.; +198 pls.], Vol. II: *Flora Novae-Zelandiae* [1852–1855, part 1. Flowering plants, [viii]+xxxix+312 pp.; part 2. Flowerless plants, [iii]+378 pp.; +130 pls.], Vol. III: *Flora Tasmaniae* [1855–1860, part 1. Dicotyledones, viii+vii+cxxviii+18+359 pp.; part 2. Monocotyledones and Acotyledones, [iv]+422 pp.; +200 pls.]

Hopkin, S. P. 1997. *Biology of the Springtails (Insecta: Collembola).* Oxford University Press; Oxford, UK; x+330 pp.

———, and H. J. Read. 1992. *The Biology of Millipedes.* Oxford University Press; Oxford, UK; xii+233 pp.

Hopkins, D. M., J. V. Matthews, J. A. Wolfe, and M. L. Silberman. 1971. A Pliocene flora and insect fauna from the Bering Strait Region. *Palaeogeography, Palaeoclimatology, and Palaeoecology* 9: 211–31.

Horak, M. 1999. The Tortricoidea. Pp. 199–216. *In* Kristensen, N. P. (ed.), *Handbuch der Zoologie: Eine Naturgeschichte der Stämme des Tierreiches: Band IV: Arthropoda: Insecta: Teilband 35: Lepidoptera, Moths and Butterflies: Volume 1: Evolution, Systematics, and Biogeography.* Walter de Gruyter; Berlin, Germany; x+491 pp.

Hormiga, G., M. Arnedo, and R. C. Gillespie. 2003. Speciation on a conveyor belt: Sequential colonization of the Hawaiian Islands by *Orsonwelles* spiders (Araneae, Linyphiidae). *Systematic Biology* 52: 70–88.

Hörnschemeyer, T. 1994. Ein fossiler Tenebrionidae *Ceropria*? *messelense* n. sp. (Coleoptera: Tenebrionidae: Diaperinae) aus dem Mitteleozän der Grube Messel bei Darmstadt. *Courier Forschungsinstitut Senckenberg* 170: 75–83.

———. 1998. Morphologie und Evolution des Flügelgelenks der Coleoptera und Neuropterida. *Bonner Zoologische Monographien* 43: 1–126.

———. 1999. Fossil insects from the Lower Permian of Nierdermoschel (Germany). Pp. 57–59. *In* Vršanský, P. (ed.), *Proceedings of the First Palaeoentomological Conference, Moscow, 1998.* AMBA Projects; Bratislava, Slovakia; 199+[2] pp.

———. 2002. Phylogenetic significance of the wing-base of the Holometabola (Insecta). *Zoologica Scripta* 31: 17–29.

Hou, X.-G., and J. Bergström. 1995. Cambrian lobopodians – ancestors of extant onychophorans? *Zoological Journal of the Linnean Society* 114: 3–19.

Houck, M. A., ed. 1994. *Mites: Ecological and Evolutionary Analyses of Life-History Patterns.* Chapman and Hall; New York, New York; xxii+357 pp.

Houston, T. F. 1984a. Bionomics of a pollen-collecting wasp, *Paragia tricolor* (Hymenoptera: Vespidae: Masarinae), in Western Australia. *Records of the Western Australian Museum* 11: 141–51.

———. 1984b. Biological observations of bees in the genus *Ctenocolletes* (Hymenoptera: Stenotritidae). *Records of the Western Australian Museum* 11: 153–72.

———. 1987. A second contribution to the biology of *Ctenocolletes* bees (Hymenoptera: Apoidea: Stenotritidae). *Records of the Western Australian Museum* 13: 189–201.

Hovmöller, R., T. Pape, and M. Källersjö. 2002. The Palaeoptera problem: Basal pterygote phylogeny inferred from 18S and 28S rDNA sequences. *Cladistics* 18: 313–23.

Howard, L. O. 1886. The excessive veracity of the female mantis. *Science* 8: 326.

Howden, H. F. 1981. Zoogeography of some Australian Coleoptera exemplified by the Scarabaeoidea. Pp. 1009–35. *In* Keast, A. (ed.), *Ecological Biogeography of Australia.* Junk; The Hague; the Netherlands; xix+1434 pp. [3 volumes]

Howland, D. E., and G. M. Hewitt. 1994. DNA analysis of extant and fossil beetles. Pp. 49–51. *In* Eglinton, G., and R. L. F. Kay (eds.), *Biomolecular Palaeontology.* NERC Earth Sciences Directorate Special Publication no. 94/1; UK.

Hoy, R. R., and D. Robert. 1996. Tympanal hearing in insects. *Annual Review of Entomology* 41: 433–50.

Hsiao, T. H., and D. M. Windsor. 1999. Historical and biological relationships from Hispinae inferred from 12S mtDNA sequence data. Pp. 39–50. *In* Cox, M. L. (ed.), *Advances in Chrysomelidae Biology.* Backhuys Publishers; Leiden, the Netherlands; xii+671 pp.

Hsü, J. 1983. Late Cretaceous and Cenozoic vegetation in China, emphasizing their connections with North America. *Annals of the Missouri Botanical Garden* 70: 490–508.

Huang, F.-S. 1980. The order Zoraptera and its geographical distribution. Pp. 1087–90. *In* Liu, D.-S. (ed.), *Geological and Ecological Studies of Qinghai-Xizang Plateau* [2 Volumes]. Science Press; Beijing, China; xxv+2138 pp.

Hubbard, M. D. 1990. Mayflies of the world: A catalog of the family and genus group taxa (Insecta: Ephemeroptera). *Flora and Fauna Handbook* 8: 1–119.

———, and J. Kukalová-Peck. 1980. Permian mayfly nymphs: New taxa and systematic characters. Pp. 19–31. *In* Flanagan, J. F., and K. E. Marshall (eds.), *Advances in Ephemeroptera Biology.* Plenum Press; New York, New York; xiii+552 pp.

———, and W. L. Peters. 1978. *Environmental Requirements and Pollution Tolerance of Ephemeroptera.* Environmental Protection Agency; Springfield, Virginia; vi+461 pp.

———, and E. F. Riek. 1978. New name for a Triassic mayfly from South Africa (Ephemeroptera). *Psyche* 83: 260–1.

Huber, J. T. 1986. Systematics, biology, and hosts of the Mymaridae and Mymarommatidae (Insecta: Hymenoptera): 1758–1984. *Entomography* 4: 185–243.

Huber, P., N. G. McDonald, and P. E. Olsen. 2003. Early Jurassic insects from the Newark Supergroup, northeastern United States. Pp. 206–23. *In* LeTourneau, P. M., and P. E. Olsen (eds.), *The Great Rift Valleys of Pangea in Eastern North America* [Volume 2]. Columbia University Press; New York, New York; xi+384 pp.

Hudson, W. G., J. H. Frank, and J. L. Castner. 1988. Biological control of *Scapteriscus* spp. mole crickets (Orthoptera: Gryllotalpidae) in Florida. *Bulletin of the Entomological Society of America* 34: 192–8.

Hueber, F. M., and J. H. Langenheim. 1986. Dominican amber tree had African ancestors. *Geotimes* 31: 8–10.

Huelsenbeck, J. P. 1997. Is the Felsenstein zone a fly trap? *Systematic Biology* 46: 69–74.

———. 1998. Systematic bias in phylogenetic analysis: Is the Strepsiptera problem solved? *Systematic Biology* 47: 519–37.

———. 2001. A Bayesian perspective on the Strepsiptera problem. *Tijdschrift voor Entomologie* 144: 165–78.

———, B. Larget, and D. Swofford. 2000. A compound Poisson process for relaxing the molecular clock. *Genetics* 154: 1879–92.

———, F. Ronquist, R. Nielsen, and J. P. Bollback. 2001. Bayesian inference of phylogeny and its impact on evolutionary biology. *Science* 294: 2310–14.

Hull, F. M. 1945. A revisional study of the fossil Syrphidae. *Bulletin of the Museum of Comparative Zoology* 95: 251–355.

———. 1949. The morphology and inter-relationships of the genera of syrphid flies, recent and fossil. *Transactions of the Zoological Society of London* 26: 257–408.

———. 1962. Robber flies of the world. The genera of the family Asilidae. *Bulletin of the United States National Museum* 224: 1–907.

Humphries, C. J., and L. Parenti. 1986. *Cladistic Biogeography*. Clarendon Press; Oxford, UK; xii+98 pp.

Hünicken, M. A. 1980. A giant fossil spider (*Megarachne servinei*) from Bajo de Veliz, Upper Carboniferous, Argentina. *Boletin de la Academia Nacional de Ciencias, Cordoba* 53: 317–28.

Hurd, P. D., Jr., R. F. Smith, and J. W. Durham. 1962. The fossiliferous amber of Chiapas, México. *Ciencia* 21: 107–18.

Hwang, U. W., W. Kim, D. Tautz, and M. Friedrich. 1998. Molecular phylogenetics at the Felsenstein zone: Approaching the Strepsiptera problem using 5.8S and 28S rDNA sequences. *Molecular Phylogenetics and Evolution* 9: 470–80.

Hyatt, A., and J. M. Arms. 1890. *Insecta*. Heath; Boston, Massachusetts; xxviii+300 pp.

ICZN [International Commission on Zoological Nomenclature]. 1996. Megalodontidae Konow, 1897 (Insecta, Hymenoptera): Spelling emended to Megalodontesidae, so removing the homonymy with Megalodontidae Morris & Lycett, 1853 (Mollusca, Bivalvia). *Bulletin of Zoological Nomenclature* 53: 66–7.

———. 1999. *International Code of Zoological Nomenclature* [Fourth Edition]. International Trust for Zoological Nomenclature; London, UK; xxix+306 pp.

Ikeda, Y., and R. Machida. 1998. Embryogenesis of the dipluran *Lepidocampa weberi* Oudemans (Hexapoda, Diplura, Campodeidae). *Journal of Morphology* 237: 101–15.

Illies, H. 1941. Das Insekten führende Alttertiär von Havighout. *Zeitschrift für Geschiebeforschung* 17: 16–24.

Illies, J. 1966. Katalog der rezenten Plecoptera. *Das Tierreich* 81: ixxx+1–632.

———. 1967. Megaloptera and Plecoptera (Ins.) aus den jungpliozänen Süßwassermergeln von Willershausen. *Bericht der Naturhistorischen Gesellschaft zu Hannover* 111: 47–55.

Inoue, T., O. Kitade, T. Yoshimura, and I. Yamaoka. 2000. Symbiotic associations with protists. Pp. 275–88. *In* Abe, T., D. E. Bignell, and M. Higashi (eds.), *Termites: Evolution, Sociality, Symbioses, Ecology*. Kluwer Academic Publishers; Dordrecht, the Netherlands; xxii+466 pp.

Iturralde-Vinent, M. A., and R. D. E. MacPhee. 1996. Age and paleogeographical origin of Dominican amber. *Science* 273: 1850–2.

Ivanov, V. D., and I. D. Sukatsheva. 2002. Order Trichoptera Kirby, 1813. The caddisflies. Pp. 199–219. *In* Rasnitsyn, A. P., and D. L. J. Quicke (eds.), *History of Insects*. Kluwer Academic Publishers; Dordrecht, the Netherlands; xii+517 pp.

Iwata, K. 1953. Biology of *Eumenes* in Japan. *Mushi* 25: 25–47.

———. 1972. *Evolution of Instinct: Comparative Ethology of Hymenoptera*. Mano Shoten; Kanagawa Prefecture, Japan; iii+[3]+503 pp. [In Japanese]

———. 1976. *Evolution of Instinct: Comparative Ethology of Hymenoptera*. Amerind Publishing; New Delhi, India; ix+[1]+535 pp.

Izzo, T., S. M. J. Pinent, and L. A. Mound. 2002. *Aulacothrips dictyotus* (Heterothripidae), the first ectoparasitic thrips (Thysanoptera). *Florida Entomologist* 85: 281–3.

Jacobs, B. F., J. D. Kingston, and L. L. Jacobs. 1999. The origin of grass-dominated ecosystems. *Annals of the Missouri Botanical Garden* 86: 590–643.

Jacobson, E. 1911. Biological notes on the hemipteron *Ptilocerus ochraceus*. *Tijdschrift voor Entomologie* 54: 175–9.

Jaenike, J. 1981. Criteria for ascertaining the existence of host races. *American Naturalist* 117: 830–4.

Jamieson, B. G. M. 1987. *The Ultrastructure and Phylogeny of Insect Spermatozoa*. Cambridge University Press; Cambridge, UK; xvi+320 pp.

Janzen, D. H. 1979. How to be a fig. *Annual Review of Ecology and Systematics* 10: 13–51.

———. 1981. Patterns of herbivory in a tropical deciduous forest. *Biotropica* 13: 271–82.

———. 1984. Natural history of *Hylesia lineatea* (Saturniidae, Hemileucinae) in Santa Rosa National Park, Costa Rica. *Journal of the Kansas Entomological Society* 57: 490–514.

Jarzembowski, E. A. 1977. Insect fossils from the Wealden of the Weald. *Proceedings of the Geologists' Association* 87: 443–7.

———. 1980. Fossil insects from the Bembridge Marls, Palaeogene of the Isle of Wight, southern England. *Bulletin of the British Museum (Natural History) (Geology)* 33: 237–93.

———. 1984. Early Cretaceous insects from southern England. *Modern Geology* 9: 71–93.

———. 1987. *Early Cretaceous Insects from Southern England*. Ph.D. dissertation; University of Reading; Reading, UK; 421 pp.

———. 1991. The Weald Clay of the Weald: Report of the 1988/89 field meetings. *Proceedings of the Geologists' Association* 102: 83–92.

———. 1992. Fossil insects from the London Clay (Early Eocene) of southern England. *Tertiary Research* 13: 87–94.

———. 1994. Fossil cockroaches or pinnule insects? *Proceedings of the Geologists' Association* 105: 305–11.

———. 1995a. Early Cretaceous insect faunas and palaeoenvironment. *Cretaceous Research* 16: 681–93.

———. 1995b. Fossil caddisflies (Insecta, Trichoptera) from the Early Cretaceous of southern England. *Cretaceous Research* 16: 695–703.

———. 1999. Arthropods 2: Insects. Pp. 149–60. *In* Swift, A., and D. M. Martill (eds.), *Fossils of the Rhaetian Penarth Group*. The Palaeontological Association; London, UK; 312 pp.

———, and M. B. Mostovski. 1999. A new species of *Sinonemestrius* (Diptera, Brachycera) from the Weald Clay (Lower Cretaceous, southern England), with a discussion of its affinities and stratigraphical implications. *Cretaceous Research* 21: 761–5.

———, and A. J. Ross. 1996. Insect origination and extinction in the Phanerozoic. Pp. 65–78. *In* Hart, M. B. (ed.), *Biotic Recovery from Mass Extinction Events*. [Geological Society Special Publication 102]. Geological Society; Tulsa, Oklahoma; vi+392 pp.

———, X. Martínez-Delclòs, G. Bechly, A. Nel, R. Coram, and F. Escuillie. 1998. The Mesozoic non-calopterygoid Zygoptera: Description of new genera and species from the Lower Cretaceous of England and Brazil and their phylogenetic significance (Odonata, Zygoptera, Coenagrionoidea, Hemiphlebioidea, Lestoidea). *Cretaceous Research* 19: 403–44.

Jaschof, M., and R. K. Didham. 2002. Rangomaramidae fam. nov. from New Zealand and implications for the phylogeny of the Sciaroidea (Diptera: Bibionomorpha). *Studia Dipterologica, supplement* 11: 1–59.

Jeannel, R. 1942. Coléopteres carabiques, 2. *Fauna de France* 40: 573–1173.

———. 1961. La Gondwanie et le peuplement de l'Afrique. *Annales de la Musee Royal de l'Afrique Centrale* 102: 1–161.

Jell, P. A., and P. M. Duncan. 1986. Invertebrates, mainly insects, from the freshwater, Lower Cretaceous, Koonwara Fossil Bed (Korumburra Group), South Gippsland, Victoria. *Memoir of the Association of Australasian Palaeontologists* 3: 111–205.

———, and K. J. Lambkin. 1993. Middle Triassic orthopteroid (Titanoptera) insect from the Esk Formation at Lake Wivenhoe. *Memoirs of the Queensland Museum* 33: 258.

Jennings, J. T., and A. D. Austin. 1994a. Revision of *Pseudofoenus* Kieffer (Hymenoptera: Gasteruptiidae), a hyptigastrine wasp genus endemic to New Zealand. *Invertebrate Taxonomy* 8: 1289–1303.

———, and A. D. Austin. 1994b. Revision of the genus *Crassifoenus* Crosskey (Hymenoptera: Gasteruptiidae), with a description of a new species from Western Australia. *Records of the Western Australian Museum* 16: 575–91.

———, and A. D. Austin. 1997a. Revision of the Australian endemic genus *Hyptiogaster* Kieffer (Hymenoptera: Gasteruptiidae), with descriptions of seven new species. *Journal of Natural History* 31: 1533–62.

———, and A. D. Austin. 1997b. Revision of *Aulacofoenus* Kieffer (Hymenoptera: Gasteruptiidae), hyptiogastrine wasps with a restricted Gondwanic distribution. *Invertebrate Taxonomy* 11: 943–76.

———, and A. D. Austin. 2002. Systematics and distribution of world hyptiogastrine wasps (Hymenoptera: Gasteruptiidae). *Invertebrate Systematics* 16: 735–811.

Jeram, A. J., P. A. Selden, and D. Edwards. 1990. Land animals in the Silurian: Arachnids and myriapods from Shropshire, England. *Science* 250: 658–61.

Jervis, M. A. 1980a. Life history studies on *Aphelopus* species (Hymenoptera, Dryinidae) and *Chalarus* species (Diptera, Pipunculidae), primary parasites of typhlocybine leafhoppers (Homoptera, Cicadellidae). *Journal of Natural History* 14: 769–80.

———. 1980b. Ecological studies on the parasite complex associated with typhlocybine leafhoppers (Homoptera, Cicadellidae). *Ecological Entomology* 5: 123–6.

———, N. A. C. Kidd, M. G. Fitton, T. Huddleston, and H. A. Dawah. 1993. Flower-visiting by hymenopteran parasitoids. *Journal of Natural History* 27: 67–105.

Jin, Y. G., Y. Wang, W. Wang, Q. H. Shang, C. Q. Cao, and D. H. Erwin. 2000. Pattern of marine mass extinction near the Permian-Triassic boundary in South China. *Science* 289: 432–6.

Johannsmeier, M. F. 1983. Experiences with the Cape bee in the Transvaal. *South African Bee Journal* 55: 130–8.

Johnson, C., D. Agosti, J. H. Delabie, K. Dumpert, D. J. Williams *et al.* 2001. *Acropyga* and *Azteca* ants (Hymenoptera: Formicidae) with scale insects (Sternorrhyncha: Coccoidea): 20 million years of intimate symbiosis. *American Museum Novitates* 3335: 1–18.

Johnson, E. W., D. E. G. Briggs, R. J. Suthren, J. L. Wright, and S. P. Tunnicliff. 1994. Non-marine arthropod traces from the subaerial Ordovician Borrowdale Volcanic Group, English Lake District. *Geological Magazine* 131: 395–406.

Johnson, K. P. and E. L. Mockford. 2003. Molecular systematics of Psocomorpha (Psocoptera). *Systematic Entomology* 28: 409–16.

———, and M. F. Whiting. 2002. Multiple genes and the monophyly of Ischnocera (Insecta: Phthiraptera). *Molecular Phylogenetics and Evolution* 22: 101–10.

———, R. H. Cruickshank, R. J. Adams, V. S. Smith, *et al.* 2003. Dramatically elevated rate of mitochondrial substitution in lice (Insecta: Phthiraptera). *Molecular Phylogenetics and Evolution* 26: 231–42.

Johnson, J. B., and K. S. Hagen. 1981. A neuropterous larva uses an allomone to attack termites. *Nature* 289: 506–7.

Johnson, N. F. 1988. Midcoxal articulations and the phylogeny of the order Hymenoptera. *Annals of the Entomological Society of America* 81: 870–81.

———. 1999. The fossil pelecinids *Pelecinopteron* Brues and *Iscopinus* Kozlov (Hymenoptera: Proctotrupoidea: Pelecinidae). *Proceedings of the Entomological Society of Washington* 100: 1–6.

———, and L. Musetti. 1999. Revision of the proctotrupoid genus *Pelecinus* Latreille (Hymenoptera: Pelecinidae). *Journal of Natural History* 33: 1513–43.

———, L. Musetti, and J.-W. Janzen. 2001. A new fossil species of the Australian endemic genus *Peradenia* Naumann & Masner (Hymenoptera: Proctotrupoidea, Peradeniidae) from Baltic amber. *Insect Systematics and Evolution* 32: 191–4.

Johnson, S. D. 1996. Pollination, adaptation and speciation models in the Cape flora of South Africa. *Taxon* 45: 59–66.

———, and K. E. Steiner. 2000. Generalization versus specialization plant pollination systems. *Trends in Ecology and Evolution* 15: 140–3.

Johnston, J. E. 1993. Insects, spiders, and plants from the Tallahatta Formation (Middle Eocene) in Benton County, Mississippi. *Mississippi Geology* 14: 71–82.

———. 1999. Caddisfly cases from the Middle Eocene (Lower Lutetian) of Mississippi, USA. Pp. 61–4. *In* Vršanský, P. (ed.), *Proceedings of the First Palaeoentomological Conference, Moscow, 1998*. AMBA Projects; Bratislava, Slovakia; 199+[2] pp.

Jolivet, P. 1997. *Biologie des Coléoptères Chrysomélides*. Société Nouvelle des Éditions Boubée; Paris, France; 279 pp.

———, and M. L. Cox, eds. 1996. *Chrysomelidae Biology, Volume 1: The Classification, Phylogeny, and Genetics*. SPB Academic Publishing; New York, New York; 444 pp.

———, and T. J. Hawkeswood. 1995. *Host-Plants of Chrysomelidae of the World: An Essay about the Relationships between the Leaf-Beetles and their Food-Plants*. Backhuys Publishers; Leiden, the Netherlands; 281 pp.

Jones, M. B., and M. H. Depledge, eds. 1997. The biology of Crustacea. *Journal of the Marine Biological Association of the United Kingdom* 77: 1–158.

Jordan, S., C. Simon, and D. Polhemus. 2003. Molecular systematics and adaptive radiation of Hawaii's endemic damselfly genus *Megalagrion* (Odonata: Coenagrionidae). *Systematic Biology* 52: 89–109.

Joron, M. 2003. Mimicry. Pp. 714–26. *In* Resh, V. H., and R. T. Cardé (eds.), *Encyclopedia of Insects*. Academic Press; San Diego, California; xxviii+[1]+1266 pp.

———, I. R. Wynne, G. Lamas, and J. Mallet. 2001. Variable selection and the coexistence of multiple mimetic forms of the butterfly *Heliconius numata*. *Evolutionary Ecology* 13: 721–54.

Judson, M. L. I. 2000. *Electrobisium acutum* Cockerell, a cheiridiid pseudoscorpion from Burmese amber, with remarks on the validity of the Cheiridioidea (Arachnida, Chelonethi). *Bulletin of the Natural History Museum, London (Geology)* 56: 79–83.

Kabakov, O. N. 1988. A new lamellicorn beetle of the genus *Copris* (Insecta) from the Pliocene of Georgia. *Paleontologicheskiy Zhurnal* 1988: 110–11. [In Russian]

Kaiho, K., Y. Kajiwara, T. Nakano, Y. Miura, H. Kawahata *et al.* 2001. End-Permian catastrophe by bolide impact: Evidence of a gigantic release of sulfur from the mantle. *Geology* 29: 815–18.

Kaiser, T. 2000. Proposed fossil insect modification to fossil mammalian bone from Plio-Pleistocene hominid-breaing deposits of Laetoli (northern Tanzania). *Annals of the Entomological Society of America* 93: 693–700.

Kaltenbach, A. P. 1978. Mecoptera. *Handbuch der Zoologie, IV.* 2/28: 1–111.

———. 1996. Unterlagen für eine Monographie der Mantodea des südlichen Afrika: 1. Artenbestand, geographische Verbreitung und Ausbreitungsgrenzen (Insecta: Mantodea). *Annales Naturhistorisches Museum Wien, B* 98: 193–346.

Kalugina, N. S. 1991. New Mesozoic Simuliidae and Leptoconopidae and the origin of bloodsucking in the lower dipteran insects. *Paleontologicheskiy Zhurnal* 26: 142–6. [In Russian].

Kambhampati, S. 1995. A phylogeny of cockroaches and related insects based on DNA sequence of mitochondrial ribosomal RNA genes. *Proceedings of the National Academy of Sciences, U.S.A.* 92: 2017–20.

———. 1996. Phylogenetic relationships among cockroach families inferred from mitochondrial 12S rRNA gene sequence. *Systematic Entomology* 21: 89–98.

———, and P. Eggleton. 2000. Taxonomy and phylogeny of termites. Pp. 1–23. *In* Abe, T., D. E. Bignell, and M. Higashi (eds.), *Termites: Evolution, Sociality, Symbioses, Ecology*. Kluwer Academic Publishers; Dordrecht, the Netherlands; xxii+466 pp.

———, K. M. Kjer, and B. L. Thorne. 1996a. Phylogenetic relationships among termite families based on DNA sequence of mitochondrial 16S ribosomal RNA gene. *Insect Molecular Biology* 5: 229–38.

———, P. Luykx, and C. A. Nalepa. 1996b. Evidence for sibling species in *Cryptocercus punctulatus*, the wood roach, from variation in mitochondrial DNA and karyotype. *Heredity* 76: 485–96.

Kamp, J. W. 1973. Numerical classification of the orthopteroids, with special reference to Grylloblattodea. *Canadian Entomologist* 105: 1235–49.

Kaplin, V. G. 1985. On the classification and phylogeny of the Machilidae (Thysanura, Microcoryphia). *Entomological Review* 64: 117–31.

Karny, H. H. 1922. Zorapteren aus Süd-Sumatra. *Treubia* 3: 14–37.

———. 1932. Psocoptera. *Insects of Samoa* 7: 117–29.

Kasparyan, D. R. 1994. A review of ichneumon flies of the Townesitinae subfam. nov. (Hymenoptera, Ichneumonidae) from the Baltic amber. *Paleontologicheskiy Zhurnal* 4: 86–96. [In Russian]

———. 2001. A new genus and species of the subfamily Ghilarovitinae from Baltic amber (Hymenoptera: Paxylommatidae). *Zoosystematica Rossica* 10: 97–9.

Kathirithamby, J. 1989. Review of the order Strepsiptera. *Systematic Entomology* 14: 41–92.

———. 1991. Strepsiptera. Pp. 684–95. *In* Naumann, I. D. (ed.), *The Insects of Australia: A Textbook for Students and Research Workers, Volume 2* [2nd Edition]. Cornell University Press; Ithaca, New York; [2]+543–1137 pp.

———, and D. Grimaldi. 1993. Remarkable stasis in some Lower Tertiary parasitoids: Descriptions, new records, and review of Strepsiptera in the Oligo-Miocene amber of the Dominican Republic. *Entomologica Scandinavica* 24: 31–41.

———, and J. S. Johnston. 2003. The discovery after 94 years of the elusive female of a myrmecolacid (Strepsiptera), and the cryptic species of *Caenocholax fenyesi* Pierce *sensu lato*. *Proceedings of the Royal Society of London, Series B, Biological Sciences, Supplement* 271:S5–S8.

———, L. D. Ross, and J. S. Johnston. 2003. Masquerading as self? Endoparasitic Strepsiptera (Insecta) enclose themselves in host-derived epidermal bag. *Proceedings of the National Academy of Sciences, USA* 100: 7655–9.

Kato, M. 2000. Anthophilous insect community and plant-pollinator interactions on Amami Islands in the Ryukyu Archipelago, Japan. *Contributions from the Biological Laboratory, Kyoto University* 29: 157–252.

———, and T. Inoue. 1994. Origin of insect pollination. *Nature* 368: 195.

———, and T. Inoue. 1995. Pollination biology of *Gnetum* (Gnetaceae) in a lowland mixed dipterocarp forest in Sarawak. *American Journal of Botany* 82: 862–8.

———, A. Takimura, and A. Kawakita. 2003. An obligate pollination mutualism and reciprocal diversification in the tree genus *Glochidion* (Euphorbiaceae). *Proceedings of the National Academy of Sciences, USA* 100: 5264–7.

Kavanaugh, D. H. 1998. Investigations of phylogenetic relationships among some basal grade Carabidae (Coleoptera): A report of work in progress. *Bollettino del Museo Regionale di Scienze Naturali-Torino* 1998: 329–41.

Kay, E. A., ed. 1994. *A Natural History of the Hawaiian Islands, Selected Readings II.* University of Hawaii Press; Honolulu, Hawai; xi+519 pp.

Keller, G., T. Adatte, W. Stinnesbeck, M. Rebolledo-Vieyra, *et al.* 2004. Chicxulub impact predates the K-T boundary mass extinction. *Proceedings of the National Academy of Sciences, USA* 101: 3753–8.

Keltner, J., and W. P. McCafferty. 1986. Functional morphology of burrowing in the mayflies *Hexagenia limbata* and *Pentagenia vittigera*. *Zoological Journal of the Linnean Society* 87: 139–62.

Kemp, A. 2002. Amino acid residues in conodont elements. *Journal of Paleontology* 76: 518–28.

Kemp, T. S. 1999. *Fossils and Evolution.* Oxford University Press; Oxford, UK; 284 pp.

Kennel, J. von, and F. Eggers. 1933. Die abdominalen Tympanalorgane der Lepidopteren. *Zoologische Jahrbücher, Abteilung für Anatomie und Ontogenie der Tiere* 57: 1–104.

Kenrick, P. 1999. Botany: The family tree flowers. *Nature* 402: 358–9.

Kent, D. S., and J. A. Simpson. 1992. Eusociality in the beetle *Australoplatypus incompertus* (Coleoptera: Platypodidae). *Naturwissenschaften* 79: 86–7.

Kerdelhue, C., I. LeClainche, and J.-Y. Rasplus. 1999. Molecular phylogeny of the *Ceratosolen* species pollinating *Ficus* of the subgenus *Sycomorus sensu stricto*: Biogeographical history and origins of the species-specificity breakdown cases. *Molecular Phylogenetics and Evolution* 11: 401–14.

Kessel, E. L. 1955. The mating activities of balloon flies. *Systematic Zoology* 4: 97–104.

———. 1959. Introducing *Hilara wheeleri* Melander as a balloon maker, and notes on other North American balloon flies (Diptera: Empididae). *Wasmann Journal of Biology* 17: 221–30.

Kessler, S. 1962. Courtship rituals and reproductive isolation between the races or incipient species of *Drosophila paulistorum*. *American Naturalist* 96: 117–21.

Kethley, J. B., R. A. Norton, P. M. Bonamo, and W. A. Shear. 1989. A terrestrial alicorhagiid mite (Acari: Acariformes) from the Devonian of New York. *Micropaleontology* 35: 367–73.

Kevan, D. K. McE. 1977. The higher classification of the orthopteroid insects: A general review. *Lyman Entomological Museum and Research Laboratory Memoirs* 4: 1–31.

———. 1982. Phasmatoptera. Pp. 379–83. *In* Parker, S. P. (ed.), *Synopsis and Classification of Living Organisms.* McGraw-Hill; New York, New York; xviii+1166 pp.

———. 1986. A rationale for the classification of orthopteroid insects – the saltatorial orthopteroids or grigs – one order or two? *Proceedings of the 4th Triennial Meeting of the Pan American Acridological Society* 1985: 49–67.

Kevan, P. G., and H. G. Baker. 1983. Insects as flower visitors and pollinators. *Annual Review of Entomology* 28: 407–53.

———, W. G. Chaloner, and D. B. O. Savile. 1975. Interrelationships of early terrestrial arthropods and plants. *Palaeontology* 18: 391–417.

Khalil, M. A. K., R. A. Rasmussen, J. R. J. French, and J. A. Holt. 1990. The influence of termites on atmospheric trace gases: CH_4, CO_2, $CHCl_3$, N_2O, CO, H_2, and light hydrocarbons. *Journal of Geophysical Research* 95: 3619–34.

Kieffer, J. J. 1914. Bethylidae. *Das Tierreich* 41: 11–595.

Kiester, A. R., and E. Strates. 1984. Social behaviour in a thrips from Panama. *Journal of Natural History* 18: 303–14.

Kim, K. C. 1985a. Evolutionary relationships of parasitic arthropods and mammals. Pp. 3–82. *In* Kim, K. C. (ed.), *Coevolution of Parasitic Arthropods and Mammals.* John Wiley and Sons; New York, New York; xiii+800 pp.

———. 1985b. Evolution and host associations of Anoplura. Pp. 197–232. *In* Kim, K. C. (ed.), *Coevolution of Parasitic Arthropods and Mammals.* John Wiley and Sons; New York, New York; xiii+800 pp.

———. 1985c. Evolutionary aspects of the disjunct distribution of lice on Carnivora. Pp. 257–94. *In* Kim, K. C. (ed.), *Coevolution of Parasitic Arthropods and Mammals.* John Wiley and Sons; New York, New York; xiii+800 pp.

———, and P. H. Adler. 1985. Patterns of insect parasitism in mammals. Pp. 157–96. *In* Kim, K. C. (ed.), *Coevolution of Parasitic Arthropods and Mammals.* John Wiley and Sons; New York, New York; xiii+800 pp.

———, and H. W. Ludwig. 1978a. The family classification of the Anoplura. *Systematic Entomology* 3: 249–84.

———, and H. W. Ludwig. 1978b. Phylogenetic relationships of parasitic Psocodea and taxonomic position of the Anoplura. *Annals of the Entomological Society of America* 71: 910–22.

———, and H. W. Ludwig. 1982. Parallel evolution, cladistics, and classification of parasitic Psocodea. *Annals of the Entomological Society of America* 75: 537–48.

Kimsey, L. S. 1991. Relationships among the tiphiid wasp subfamilies (Hymenoptera). *Systematic Entomology* 16: 427–38.

———, and R. M. Bohart. 1990. *The Chrysidid Wasps of the World.* Oxford University Press; Oxford, UK; ix+[1]+652 pp.

Kinchin, I. M. 1994. *The Biology of Tardigrades.* Portland Press; London, UK; xi+186 pp.

King, A. B. S., and J. L. Saunders. 1984. *Las Plagas Invertebradas de Cultivos Anuales Alamenticios en América Central.* ODA; London, UK; 182 pp.

King, D. G. 1991. The origin of an organ: Phylogenetic analysis of evolutionary innovation in the digestive tract of flies (Insecta: Diptera). *Evolution* 45: 568–88.

Kinsey, A. C. 1919. Fossil Cynipidae. *Psyche* 26: 44–9.

———. 1937. Order Hymenoptera, family Cynipidae. Pp. 21–7. *In* Carpenter, F. M., J. W. Folsom, E. O. Essig, A. C. Kinsey, C. T. Brues, *et al.* (eds.), *Insects and Arachnids from Canadian Amber.* University of Toronto Press (University of Toronto Geological Series 40); Toronto, Canada; 62 pp.

Kinzelbach, R. 1971. *Morphologische Befunde an Fächerflüglern und ihre Phylogenetische Bedeutung (Insecta: Strepsiptera).* Schweizerbart'sche-Verlagsbuchhandlung; Stuttgart, Germany; xiii+256 pp.

———. 1978. Fächerflügler (Stepsiptera). Gustav Fischer; Jena, Germany; 166 pp.

———. 1979. Das erste neotropische Fossil der Fächerflügler (Stuttgarter Bernsteinsammlung: Insecta, Strepsiptera). *Stuttgarter Beiträge zur Naturkunde, serie B, Geologie und Paläontologie* 52: 1–14.

———. 1983. Fächerflügler aus dem dominikanischen Bernstein (Insecta: Strepsiptera, Myrmecolacidae). *Verhandlungen des Naturwissenschaftlichen Vereins in Hamburg* 26: 29–36.

———. 1990. The systematic position of Strepsiptera (Insecta). *American Entomologist* 35: 292–303.

———, and H. Lutz. 1985. Stylopid larva from the Eocene – A spotlight on the phylogeny of the stylopids (Strepsiptera). *Annals of the Entomological Society of America* 78: 600–2.

———, and H. Pohl. 1994. The fossil Strepsiptera (Insecta: Strepsiptera). *Annals of the Entomological Society of America* 87: 59–70.

Kirby, W. 1802. *Monographia Apum Angliae, or, An Attempt to Divide into Their Natural Genera and Families, such Species of the Linnean Genus* Apis, *as have been Discovered in England.* White; Ipswich, UK; [vol. 1] xxii+258 pp., [vol. 2] 387 pp.

———, and W. Spence. 1826. *An Introduction to Entomology: Or Elements of the Natural History of Insects* [vol. III]. Longman *et al.*; London, UK; v+[iii]+732 pp.

Kirk, W. D. J. 1984. Pollen-feeding in thrips (Insecta: Thysanoptera). *Journal of Zoology, London* 204: 107–17.

Kirkendall, L. R., D. S. Kent, and K. F. Raffa. 1997. Interactions among males, females and offspring in bark and ambrosia beetles: The significance of living in tunnels for the evolution of social behavior. Pp. 181–215. *In* Choe, J. C., and B. J. Crespi (eds.), *The Evolution of Social Behavior in Insects and Arachnids.* Cambridge University Press; Cambridge, UK; vii+541 pp.

Kishino, H., J. L. Thorne, and W. J. Bruno. 2001. Performance of a divergence time estimation method under a probabilistic model of rate evolution. *Molecular Biology and Evolution* 18: 352–61.

Kitching, I. J. 1984. An historical review of the higher classification of the Noctuidae (Lepidoptera). *Bulletin of the British Museum of Natural History (Entomology)* 49: 153–234.

———, and J.-M. Cadiou. 2000. *Hawkmoths of the World: An Annotated and Illustrated Revisionary Checklist (Lepidoptera: Sphingidae)*. Cornell University Press; Ithaca, New York; viii+226 pp.

———, and J. E. Rawlins. 1999. The Noctuoidea. Pp. 355–402. *In* Kristensen, N. P. (ed.), *Handbuch der Zoologie: Eine Naturgeschichte der Stämme des Tierreiches: Band IV: Arthropoda: Insecta: Teilband 35: Lepidoptera, Moths and Butterflies: Volume 1: Evolution, Systematics, and Biogeography*. Walter de Gruyter; Berlin, Germany; x+491 pp.

Kjer, K. M., R. J. Blahnik, and R. W. Holzenthal. 2001a. Phylogeny and Trichoptera (Caddisflies): Characterization of signal and noise within multiple datasets. *Systematic Biology* 50: 781–816.

———, R. J. Blahnik, and R. W. Holzenthal. 2001b. Phylogeny of caddisflies (Insecta, Trichoptera). *Zoologica Scripta* 31: 83–91.

Klapálek, F. 1904. Ueber die Gonopoden der Insekten und die Bedeutung derselben für die Systematik. *Zoologischer Anzeiger* 27: 449–53.

———. 1905. Noch einige bemerkungen über die Gonopoden der Insekten. *Zoologischer Anzeiger* 28: 255–9.

Klass, K.-D. 1997. The external male genitalia and the phylogeny of Blattaria and Mantodea. *Bonner Zoologische Monographien* 42: 1–341.

———. 1997/8. The ovipositor of Dictyoptera (Insecta): Homology and ground-plan of the main elements. *Zoologischer Anzeiger* 236: 69–101.

———. 1998. The proventriculus of the Dicondylia, with comments on evolution and phylogeny in Dictyoptera and Odonata (Insecta). *Zoologischer Anzeiger* 237: 15–42.

———. 2001a. Morphological evidence on Blattaria phylogeny: "Phylogenetic histories and stories" (Insecta, Dictyoptera). *Deutsche Entomologische Zeitschrift* 48: 223–65.

———. 2001b. The female abdomen of the viviparous earwig *Hemimerus vosseleri* (Insecta: Dermaptera: Hemimeridae), with a discussion of the postgenital abdomen of Insecta. *Zoological Journal of the Linnean Society* 131: 251–307.

———. 2002. [Reponse to Tilgner, 2002]. *Science* 297: 731a.

———, ed. 2003. Proceedings of the 1st Dresden meeting on insect phylogeny: "Phylogenetic Relationships within the insect orders" (Dresden, September 19–21, 2003). *Entomologische Abhandlungen* 61: 119–70.

———, O. Zompro, N. P. Kristensen, and J. Adis. 2002. Mantophasmatodea: A new insect order with extant members in the Afrotropics. *Science* 296: 1456–9.

———, O. Zompro, and J. Adis. 2003a. Ordnung Mantophasmatodea. Pp. 161–6. *In* Dathe, H. H. (ed.), *Lehrbuch der Speziellen Zoologie, Band I: Wirbellose Tiere. 5. Teil: Insecta*. Spektrum Akademischer Verlag; Heidelberg, Germany; xii+[1]+961 pp.

———, M. D. Picker, J. Damgaard, S. van Noort, and K. Tojo. 2003b. The taxonomy, genitalic morphology, and phylogenetic relationships of southern African Mantophasmatodea (Insecta). *Entomologische Abhandlungen* 61: 3–67.

Kleinow, W. 1966. Untersuchungen zum Flügelmechanismus der Dermapteren. *Zeitschrift für Morphologie und Ökologie der Tiere* 56: 363–416.

Kliman, R. M., and J. Hey. 1993. DNA sequence variation at the period locus within and among species of the *Drosophila melanogaster* complex. *Genetics* 133: 375–87.

———, P. Andolfatto, J. A. Coyne, F. Depaulis, M. Kreitman *et al.* 2000. The population genetics of the origin and divergence of the *Drosophila simulans* complex species. *Genetics* 156: 1913–31.

Klimaszewski, J., and D. K. McE. Kevan. 1986. A new lacewing-fly (Neuroptera: Planipennia) from Canadian Cretaceous amber, with an analysis of its fore wing [sic] characters. *Entomological News* 97: 124–32.

Klompen, H., and D. Grimaldi. 2001. First Mesozoic record of a parasitiform mite: A larval argasid tick in Cretaceous amber (Acari: Ixodida: Argasidae). *Annals of the Entomological Society of America* 94: 10–15.

Kluge, N. J. 1989. A question of the homology of the tracheal gills and paranotal processi of the mayflies [sic] larvae and wings of the insects with reference to the taxonomy and phylogeny of the order Ephemeroptera. *Chteniya Pamyati Nikolaya Aleksandrovicha Kholodkovskogo* 41: 48–77. [In Russian]

———. 1993. New data on mayflies (Ephemeroptera) from fossil Mesozoic and Cenozoic resins. *Paleontological Journal* 27: 35–49.

———. 1996. A new suborder of Thysanura for the Carboniferous insect originally described as larva of *Bojophlebia*, with comments on characters of the orders Thysanura and Ephemeroptera. *Zoosystematica Rossica* 4: 71–5.

——. 1997. Classification and phylogeny of the Baetidae (Ephemeroptera) with description of the new species from the Upper Cretaceous resins of Taimyr. Pp. 527–35. *In* Landolt, P., and M. Sartori (eds.), *Ephemeroptera and Plecoptera: Biology – Ecology – Systematics*. Mauron and Tinguely and Lacht SA; Fribourg, Switzerland; xi+569 pp.

———. 1998. Phylogeny and higher classification of Ephemeroptera. *Zoosystematica Rossica* 7: 255–69.

———. 2000. *Modern Systematics of Insects. Part 1. Principles of Systematics of Living Organisms and General System of Insects with Classification of Primary Wingless and Paleopterous Insects*. Saint Petersburg, Russia; 332+[4] pp. [In Russian]

———. 2004. *The Phylogenetic System of Ephemeroptera*. Kluwer Academic; Dordrecht, the Netherlands; xiii+442 pp.

Knight, O. L. 1950. Fossil insects of Belmont, N.S.W. *Records of the Australian Museum* 22: 251–3.

Knisley, C. B., D. L. Reeves, and G. T. Stephens. 1989. Behavior and development of the wasp *Pterombrus rufiventris hyalinatus* Krombein (Hymenoptera: Tiphiidae), a parasite of larval tiger beetles (Coleoptera: Cicindelidae). *Proceedings of the Entomological Society of Washington* 91: 179–84.

Knoll, A. H. 1984. Patterns of extinction in the fossil record of vascular plants. Pp. 21–68. *In* Nitecki, M. H. (ed.), *Extinctions*. University of Chicago Press; Chicago, Illinois; ix+354 pp.

Knoll, F. 1921. *Bombylius fuliginosus* und die Farbe der Blümen. *Abhandlungen der Zoologisch-Botanische Gesellschaft in Wien* 12: 17–119.

———. 1956. *Die Biologie der Blüte*. Springer Verlag; Berlin, Germany; 164 pp.

Koch, M. 1997. Monophyly and phylogenetic position of the Diplura (Hexapoda). *Pedobiologia* 41: 9–12.

Koeniger, G., N. Koeniger, M. Mardan, G. Otis, and S. Wongsiri. 1991. Comparative anatomy of male genital organs in the genus *Apis*. *Apidologie* 22: 539–52.

Koeniger, N., G. Koeniger, M. Gries, S. Tingek, and A. Kelitu. 1996. Reproductive isolation of *Apis nuluensis* Tingek, Koeniger and Koeniger, 1996 by species-specific mating time. *Apidologie* 27: 353–9.

Kohler, R. E. 1994. *Lords of the Fly:* Drosophila *Genetics and the Experimental Life*. University of Chicago Press; Chicago, Illinois; xv+321 pp.

Kohring, R. 1998. REM-Untersuchungen an harzkonservierten Arthropoden. *Entomologia Generalis* 23: 95–106.

———, and T. Schlüter. 1989. Historische und paläontologische Bestandsaufnahme des Simetits, eines fossilen Harzes mutmässlich mio/pliozänens Alters aus Sizilien. *Documenta Naturae* 56: 33–58.

———, and T. Schlüter. 1993. Sciariden (Insecta: Diptera: Nematocera) aus dem Oberpliozän von Willershausen. *Berliner Geowissenschaftlichen Abhandlungen, Reihe E, Paläobiologie* 9: 191–9.

Kokubu, H., and P. Duelli. 1983. Adult food of sponge flies: Observations on the crop and gut contents of *Sisyra terminalis* Curtis (Planipennia, Sisyridae). *Neuroptera International* 2: 157–62.

Kolbe, W. 1968. Der Einfluss der Waldameise auf die Verbreitung von Käfern in der Bodenstreueines Eichen-Birken-Waldes. *Natur und Heimat* 3: 120–4.

———. 1969. Käfer im Wirkungbereich der Roten Waldameise. *Entomologische Zeitschrift* 79: 269–80.

Königsmann, E. 1976. Das phylogenetische System der Hymenoptera. Teil 1: Einführung, Grundplanmerkmale, Schwestergruppe und Fossilfunde. *Deutsche Entomologische Zeitschrift* 23: 253–79.

———. 1977. Das phylogenetische System der Hymenoptera. Teil 2: "Symphyta". *Deutsche Entomologische Zeitschrift* 24: 1–40.

———. 1978a. Das phylogenetische System der Hymenoptera. Teil 3: "Terebrantes" (Unterordnung Apocrita). *Deutsche Entomologische Zeitschrift* 25: 1–55.

———. 1978b. Das phylogenetische System der Hymenoptera. Teil 4: Aculeata (Unterordnung Apocrita). *Deutsche Entomologische Zeitschrift* 25: 365–435.

Kononova, E. L. 1975. A new aphid family (Homoptera, Aphidinea) from the Upper Cretaceous of the Taymyr. *Entomological Review* 54: 60–8.

———. 1976. Extinct aphid families (Homoptera, Aphidinea) of the Late Cretaceous. *Paleontological Journal* 1976: 352–60.

———. 1977. New aphid species (Homoptera, Aphidinea) from Upper Cretaceous deposits of the Taymyr. *Entomological Review* 56: 588–600.

Konow, F. W. 1897. Ueber fossile Blatt-und Halmwespen. *Entomologische Nachrichten* 23: 36–8.

Kosmowska-Ceranowicz, B. 1987. Charakterystuka mineralogiczno-petrograficzna bursztynonośnych osadów Eocenu w okolicach Chłapowo oraz osadów Paleogenu Połnocnej Polski. *Biuletyn Instytutu Geologicznego* 356: 29–50.

———, and C. Müller. 1985. Lithology and calcareous nannoplankton in amberbearing Tertiary sediments from boreholes Chłapowo (northern Poland). *Bulletin of the Polish Academy of Sciences* 33: 119–28.

Koteja, J. 1974. Comparative studies on the labium in the Coccinea (Homoptera). *Zeszyty Naukowe Akademii Rolniczej w Krakowie* 89: 1–162.

———. 1984. The Baltic amber Matsucoccidae (Homoptera, Coccinea). *Annales Zoologici* 37: 437–96.

———. 1985. Essay on the prehistory of the scale insects (Homoptera, Coccinea). *Annales Zoologici* 38: 461–503.

———. 1986. Morphology and taxonomy of male Ortheziidae (Homoptera, Coccinea). *Polskie Pismo Entomologiczne* 56: 323–74.

———. 1990. Paleontology. Pp. 149–63. *In* Rosen, D. (ed.), *Armored Scale Insects: Their Biology, Natural Enemies, and Control.* Elsevier; Amsterdam, the Netherlands; xxvi+688 pp.

———. 1996. The scale insects (Homoptera: Coccinea) a day after. Pp. 65–88. *In* Schaeffer, C. W. (ed.), *Studies on Hemipteran Phylogeny.* Entomological Society of America (Thomas Say Publications); Lanham, Maryland; iii+244 pp.

———. 1999. *Eomatsucoccus andrewi* sp. nov. (Hemiptera: Sternorrhyncha: Coccinea) from the Lower Cretaceous of southern England. *Cretaceous Research* 20: 863–6.

———. 2000a. Advances in the study of fossil coccids (Hemiptera: Coccinea). *Polskie Pismo Entomologiczne* 69: 187–218.

———. 2000b. Scale insects (Homoptera, Coccinea) from Upper Cretaceous New Jersey amber. Pp. 147–229. *In* Grimaldi, D. (ed.), *Studies on Fossils in Amber, with Particular Reference to the Cretaceous of New Jersey.* Backhuys Publishers; Leiden, the Netherlands; viii+498 pp.

———. 2001. Essays on coccids (Hemiptera: Coccinea). Paleontology without fossils? *Prace Muzeum Ziemi, PAN* 46: 41–53.

Koulianous, S., R. Schmid-Hempel, D. W. Roubik, and P. Schmid-Hempel. 1999. Phylogenetic relationships within the corbiculate Apinae (Hymenoptera) and the evolution of eusociality. *Journal of Evolutionary Biology* 12: 380–4.

Kovalev, O. V. 1994. Palaeontological history, phylogeny and the system of brachycleistogastromorphs and cynipomorphs (Hymenoptera, Brachycleistogastromorpha infraorder n., Cynipomorpha infraorder n.) with description of new fossil and recent families, subfamilies and genera. *Entomologicheskoe Obozrenie* 73: 385–426, 495.

———. 1995. New taxa of fossil cynipoids (Hymenoptera, Cynipoidea) from the Cretaceous and Palaeogene. *Amber and Fossils* 1: 9–16.

———. 1996. New higher taxa of cynipoids (Hymenoptera: Cynipoidea): Renaming of a family and description of a new family, new subfamily and genus. *Entomologicheskoe Obozrenie* 75: 408–16.

Kovalev, V. G. 1990. Muscida. *Trudy Paleontologischeskogo Instituta Akademii Nauk, SSSR* 239: 123–77. [In Russian]

Kozlov, M. A. 1968. Jurassic Proctotrupoidea (Hymenoptera). Pp. 237–40. *In* Rohdendorf, B. B. (ed.), *Jurassic Insects of Karatau.* Nauka; Moscow, Russia; 252 pp. [In Russian]

———. 1974. An early Cretaceous ichneumon of the family Pelecinidae (Hymenoptera, Pelecinoidea). *Paleontologicheskiy Zhurnal* 1974: 144–6. [In Russian]

———, and A. P. Rasnitsyn. 1979. On the limits of the family Serphitidae (Hymenoptera, Proctotrupoidea). *Entomologicheskoe Obozrenie* 58: 402–16. [In Russian]

Kozlov, M. V. 1988. Paleontology of the Lepidoptera and problems of phylogeny of the order of the Papilionoidea. Pp. 16–69. *In* Ponomarenko, A. G. (ed.), *Cretaceous Biocenotic Crisis and the Evolution of Insects.* Nauka Press; Moscow, Russia. [In Russian]

———. 1988–90. A brief review and a key to Palearctic species of the genus *Micropterix* (Lepidoptera, Micropterigidae), 1–4. *Vestnik Zoologii* 1988 (4): 8–14; 1989 (6): 26–31; 1990 (2): 21–26; 1990 (3): 28–33. [In Russian]

———. 1989. New Lepidoptera (Papilionida) from Upper Jurassic and Lower Cretaceous. *Paleontologicheskiy Zhurnal* 4: 37–42. [In Russian]

Kral, K. 1999. Binocular vision and distance estimation. Pp. 114–40. *In* Prete, F. R., H. Wells, P. H. Wells, and L. E. Hurd (eds.), *The Praying Mantids.* Johns Hopkins University Press; Baltimore, Maryland; xiv+362 pp.

Krantz, G. W. 1970. *A Manual of Acarology.* Oregon State University; Corvallis, Oregon; 335 pp.

Krassilov, V. A., and F. Bacchia. 2000. Cenomanian florule of Nammoura, Lebanon. *Cretaceous Research* 21: 785–99.

———, and A. P. Rasnitsyn. 1982. A unique find: Pollen in the intestine of Early Cretaceous sawflies. *Paleontological Journal* 4: 80–95.

———, and A. P. Rasnitsyn. 1997. Pollen in the guts of Permian insects: First evidence of pollinivory and its evolutionary significance. *Lethaia* 29: 369–72.

———, and A. P. Rasnitsyn. 1999. Plant remains from the guts of fossil insects: Evolutionary and paleoecological inferences. Pp. 65–72. *In* Vršanský, P. (ed.), *Proceedings of the First Palaeoentomological Conference, Moscow, 1998.* AMBA Projects; Bratislava, Slovakia; 199+[2] pp.

Kraus, O. 1974. On the morphology of Paleozoic diplopods. *Symposium of the Zoological Society of London* 32: 13–22.

———. 1998. Phylogenetic relationships between higher taxa of tracheate arthropods. Pp. 295–303. *In* Fortey, R. A., and R. H. Thomas (eds.), *Arthropod Relationships.* Chapman and Hall; London, UK; xii+383 pp.

———. 2001. "Myriapoda" and the ancestry of the Hexapoda. *Annales de la Société Entomologique de France* 37: 105–27.

———, and C. Brauckmann. 2003. Fossil giants and surviving dwarfs. Arthropleurida and Pselaphognatha (Atelocerata, Diplopoda): Characters, phylogenetic relationships and construction. *Verhandlungen der Naturwissenschaftlichen Vereins in Hamburg* 40: 5–50.

———, and M. Kraus. 1994. Phylogenetic system of the Tracheata (Mandibulata): On "Myriapoda"-Insecta interrelationships, phylogenetic age and primary ecological niches. *Verhandlungen des Naturwissenschaftlichen Vereins in Hamburg* 35: 95–226.

Krausse, A. 1906. Ueber die Systematik der Insekten. *Insektenbörse* 23: 115–16, 120.

———, and M. Wolff. 1919. Eine Übersicht über die bisher aufgestellten fossilen und rezenten Insektenordnungen. *Archiv für Naturgeschichte* 85: 151–71.

Krell, F.-T. 2000. The fossil record of Mesozoic and Tertiary Scarabaeoidea (Coleoptera: Polyphaga). *Invertebrate Taxonomy* 2000: 871–905.

Krenn, H. W. 1990. Functional morphology and movements of the proboscis of Lepidoptera (Insecta). *Zoomorphologie* 110: 105–14.

———, and N. P. Kristensen. 2000. Early evolution of the proboscis of Lepidoptera (Insecta): External morphology of the galea in basal glossatan moth lineages, with remarks on the origin of the pilifers. *Zoologischer Anzeiger* 239: 179–96.

Kress, W. J., and J. H. Beach. 1994. Flowering plant reproductive systems. Pp. 161–82. *In* McDade, L. A., K. S. Bawa, H. A. Hespenheide, and G. S. Hartshorn (eds.), *La Selva: Ecology and Natural History of a Neotropical Rain Forest.* University of Chicago Press; Chicago, Illinois; x+486 pp.

Krishna, K. 1961. A generic revision and phylogenetic study of the family Kalotermitidae (Isoptera). *Bulletin of the American Museum of Natural History* 122: 309–408.

———. 1970. Taxonomy, phylogeny, and distribution of termites. Pp. 127–53. *In* Krishna, K., and F. M. Weesner (eds.), *Biology of Termites, Volume II.* Academic Press; New York, New York; vii+643 pp.

———, and A. E. Emerson. 1983. A new fossil species of termite from Mexican amber, *Mastotermes electromexicus* (Isoptera, Mastotermitidae). *American Museum Novitates* 2767: 1–8.

———, and D. Grimaldi. 1991. A new fossil species from Dominican amber of the living Australian termite genus *Mastotermes* (Isoptera: Mastotermitidae). *American Museum Novitates* 3201: 1–10.

———, and D. Grimaldi. 2003. The first Cretaceous Rhinotermitidae (Isoptera): A new species, genus, and subfamily in Burmese amber. *American Museum Novitates* 3390: 1–10.

Kristensen, N. P. 1975. The phylogeny of hexapod "orders". A critical review of recent accounts. *Zeitschrift für Zoologische Systematik und Evolutionsforschung* 13: 1–44.

———. 1976. Remarks on the family-level phylogeny of butterflies (Insecta, Lepidoptera, Rhopalocera). *Zeitschrift für Zoologische Systematik und Evolutionsforschung* 14: 25–33.

———. 1978a. Phylogenetic methodology in hexapod high-level systematics: Results and perspectives. *Norwegian Journal of Entomology* 25: 84–5.

———. 1978b. A new family of Hepialoidea from South America, with remarks on the phylogeny of the suborder Exoporia (Lepidoptera). *Entomologica Germanica* 4: 272–94.

———. 1981. Phylogeny of insect orders. *Annual Review of Entomology* 26: 135–57.

———. 1984. Studies on the morphology and systematics of primitive Lepidoptera. *Steenstrupia* 10: 141–91.

———. 1989a. Insect phylogeny based on morphological evidence. Pp. 295–306. *In* Fernholm, B., K. Bremer, and H. Jörnvall (eds.), *The Hierarchy of Life: Molecules and Morphology in Phylogenetic Analysis*. Elsevier; Amsterdam, the Netherlands; xiii+499 pp.

———. 1989b. The New Zealand scorpionfly (*Nannochorista philpotti* comb. n.): Wing morphology and its phylogenetic significance. *Zeitschrift für Zoologische Systematik und Evolutionsforschung* 27: 106–14.

———. 1991. Phylogeny of extant hexapods. Pp. 125–40. *In* Naumann, I. D. (ed.), *The Insects of Australia: A Textbook for Students and Research Workers, Volume 1* [2nd Edition]. Cornell University Press; Ithaca, New York; xvi+[1]+542 pp.

———. 1995. Forty years' insect phylogenetic systematics. Hennig's 'Kritische Bemerkungen' . . . and subsequent developments. *Zoologische Beiträge* 36: 83–124.

———. 1997. Early evolution of the Trichoptera + Lepidoptera lineage: Phylogeny and the ecological scenario. *Mémoires Museum National d'Histoire Naturelle* 173: 253–71.

———. 1999a. Phylogeny of endopterygote insects, the most successful lineage of living organisms. *European Journal of Entomology* 96: 237–53.

———. 1999b. The non-glossatan moths. Pp. 41–50. *In* Kristensen, N. P. (ed.), *Handbuch der Zoologie: Eine Naturgeschichte der Stämme des Tierreiches: Band IV: Arthropoda: Insecta: Teilband 35: Lepidoptera, Moths and Butterflies: Volume 1: Evolution, Systematics, and Biogeography*. Walter de Gruyter; Berlin, Germany; x+491 pp.

———, (ed.). 1999c. *Handbuch der Zoologie: Eine Naturgeschichte der Stämme des Tierreiches: Band IV: Arthropoda: Insecta: Teilband 35: Lepidoptera, Moths and Butterflies: Volume 1: Evolution, Systematics, and Biogeography*. Walter de Gruyter; Berlin, Germany; x+491 pp.

———, and E. S. Nielsen. 1979. A new subfamily of micropterigid moths from South America. A contribution to the morphology and phylogeny of the Micropterigidae, with a generic catalogue of the family (Lepidoptera: Zeugloptera). *Steenstrupia* 5: 69–147.

———, and E. S. Nielsen. 1982. South American micropterigid moths: Two new genera of the *Sabatinca*-group (Lepidoptera: Micropterigidae). *Entomologica Scandinavica* 13: 513–29.

———, and A. W. Skalski. 1999. Palaeontology and phylogeny. Pp. 7–25. *In* Kristensen, N. P. (ed.), *Handbuch der Zoologie: Eine Naturgeschichte der Stämme des Tierreiches: Band IV: Arthropoda: Insecta: Teilband 35: Lepidoptera, Moths and Butterflies: Volume 1: Evolution, Systematics, and Biogeography*. Walter de Gruyter; Berlin, Germany; x+491 pp.

Kristensen, S. E., and A. C. Wilson. 1986. A review of the coal and lignite resources of Western Australia. Pp. 87–97. *In* Berkman, D. A. (ed.), *13th Congress of Mining and Metallurgical Institutions: Geology and Exploration* [Volume 2]. Congress of Mining and Metallurgical Institutions and Australasian Institute of Mining and Metallurgy; Victoria, Australia.

Krombein, K. V. 1968. Studies in the Tiphiidae, X. *Hylomesa*, a new genus of myzine wasp parasitic on the larvae of longicorn beetles (Hymenoptera). *Proceedings of the United States National Museum* 124: 1–22.

———. 1978. Biosystematics of Ceylonese wasps. III. Life history, nest, and associates of *Paraleptomenes mephitis* (Cameron) (Hymenoptera: Eumenidae). *Journal of the Kansas Entomological Society* 51: 721–34.

———. 1983. Three new egg parasites of South African walking sticks (Hymenoptera: Chrysididae, Amesiginae). *Journal of the Entomological Society of Southern Africa* 46: 139–46.

———. 1986. Three cuckoo wasps from Siberian and Baltic amber (Hymenoptera: Chrysididae: Amiseginae and Elampinae). *Proceedings of the Entomological Society of Washington* 88: 740–7.

———, and B. B. Norden. 1997a. Nesting behavior of *Krombeinictus nordenae* Leclerq, a sphecid wasp with vegetarian larvae (Hymenoptera: Sphecidae: Crabroninae). *Proceedings of the Entomological Society of Washington* 99: 42–9.

———, and B. B. Norden. 1997b. Bizarre nesting behavior of *Krombeinictus nordenae* LecLerq (Hymenoptera: Sphecidae, Crabroninae). *Journal of South Asian Natural History* 2: 145–54.

———, P. D. Hurd, Jr., D. R. Smith, and B. D. Burks, eds. 1979. *Catalog of Hymenoptera in American North of Mexico*. Smithsonian Institution Press; Washington, D.C.; xvi+2735 pp. [3 volumes]

Krüger, L. 1923. Neuroptera succinica baltica. Die im baltischen Bernstein eingeschlossenen Neuropteren des Westpreußischen Provinzial-Museums (heute Museum für Naturkunde und Vorgeschichte) in Danzig. *Stettiner Entomologische Zeitung* 84: 68–92.

Krzemiński, W. 1992. Triassic and Lower Jurassic stage of Diptera evolution. *Mitteilungen der Schweizerischen Entomologischen Gessellschaft* 65: 39–59.

———, and J. Ansorge. 2000. On *Protobrachyceron* Handlirsch, 1920 (Diptera: Brachycera) from the Lower Jurassic of Germany. *Polskie Pismo Entomologiczne* 69: 231–37.

———, and N. L. Evenhuis. 2000. Review of Diptera palaeontological records. Pp. 536–64. *In* Papp, L., and B. Darvas (eds.), *Contributions to a Manual of Palaearctic Diptera* [Volume 1]. Science Herald; Budapest, Hungary; 978 pp.

———, and E. Jarzembowski. 1999. *Aenne triassica* sp. n., the oldest representative of the family Chironomidae (Insecta: Diptera). *Polskie Pismo Entomologisczne* 68: 445–9.

———, and E. Krzemińska. 2003. Triassic Diptera: Descriptions, revisions, and phylogenetic relations. *Acta Zoologica Cracoviensia* 46 (Supplement): 153–84.

———, and C. Lombardo. 2001. New fossil Ephemeroptera and Coleoptera from the Ladinian (middle Triassic) of Canton Ticino (Switzerland). *Rivista Italiana di Paleontologia e Stratigrafia* 107: 69–78.

———, E. Krzemińska, and F. Papier. 1994. *Grauvogelia arzvilleriana* sp.n. – the oldest Diptera species (Lower/middle Triassic of France). *Acta Zoologica Cracoviensia* 37: 95–9.

Kuhn, O. 1937. Insekten aus dem Buntsandstein von Thüringen. *Beiträge zur Geologie der Thüringen* 4: 190–3.

———. 1961. Die Tier- und Pflanzenwelt der Solnhofener Schiefers, mit vollständigem Arten- und Schriftenwerzeichnis. *Geologica Bavarica* 48: 5–68.

Kukalová, J. 1955. *Permoedischia* n. g. (Protorthoptera) a *Moraviptera* n. g. (Palaeodictyoptera) z moravského permu. *Sbornik Ustredniho ustavu Geologicheko* 21: 541–75.

——. 1958. Paoliidae Handlirsch (Insecta-Protorthoptera) aus dem Oberschlesischen Steinkohlenbecken. *Zeitschrift für Geologische Wissenschaften* 7: 935–59.

——. 1960. New Palaeodictyoptera (Insecta) of the Carboniferous and Permian of Czechoslovakia. *Sbornik Ustredniho ustavu Geologicheko, oddil Geologicky* 25: 239–50.

——. 1963. Permian insects of Moravia. Part I. Miomoptera. *Sbornik Geologickych ved Paleontologie* 1: 7–52.

——. 1964. Permian insects of Moravia. Part II. Liomopteridae. *Sbornik Geologickych ved Paleontologie* 3: 39–118.

——. 1965. Permian Protelytroptera, Coleoptera and Protorthoptera (Insecta) of Moravia. *Sbornik Geologickych ved Paleontologie* 6: 61–98.

———. 1969a. Revisional study of the order Palaeodictyoptera in the Upper Carboniferous shales of Commentry, France, part I. *Psyche* 76: 163–215.

———. 1969b. Revisional study of the order Palaeodictyoptera in the Upper Carboniferous shales of Commentry, France, part II. *Psyche* 76: 439–86.

———. 1969c. On the systematic position of the supposed Permian beetles, Tshecardocoleidae [sic], with a description of a new collection from Moravia. *Sborník Geologických Ved, Paleontologie* 11: 139–62.

———. 1970. Revisional study of the order Palaeodictyoptera in the Upper Carboniferous shales of Commentry, France, part III. *Psyche* 77: 1–44.

Kukalová-Peck, J. 1974. Wing-folding in the Paleozoic insect order Diaphanopterodea (Paleoptera), with a description of new representatives of the family Elmoidae. *Psyche* 81: 315–33.

———. 1975. Megasecoptera from the Lower Permian of Moravia. *Psyche* 82: 1–19.

———. 1978. Origin and evolution of insect wings and their relation to metamorphosis, as documented by the fossil record. *Journal of Morphology* 156: 53–126.

———. 1983. Origin of the insect wing and wing articulation from the arthropodan leg. *Canadian Journal of Zoology* 61: 1618–69.

———. 1985. Ephemeroid wing venation based upon new gigantic Carboniferous mayflies and basic morphology, phylogeny and metamorphosis of pterygote insects (Insecta, Ephemeroptera). *Canadian Journal of Zoology* 63: 933–55.

———. 1987. New Carboniferous Diplura, Monura, and Thysanura, the hexapod ground plan, and the role of thoracic side lobes in the origin of wings (Insecta). *Canadian Journal of Zoology* 65: 2327–45.

———. 1991. Fossil history and the evolution of hexapod structures. Pp. 141–79. *In* Naumann, I. D. (ed.), *The Insects of Australia: A Textbook for Students and Research Workers, Volume 1* [2nd Edition]. Cornell University Press; Ithaca, New York; xvi+[1]+542 pp.

———. 1992. The 'Uniramia' do not exist: The ground plan of the Pterygota as revealed by Permian Diaphanopterodea from Russia (Insecta: Palaeodictyopteroidea). *Canadian Journal of Zoology* 70: 236–55.

———. 1997. Arthropod phylogeny and "basal" morphological structures. Pp. 269–79. *In* Fortey, R. A., and R. H. Thomas (eds.), *Arthropod Relationships*. Chapman and Hall; London, UK; xii+383 pp.

———, and C. Brauckmann. 1990. Wing folding in pterygote insects and the oldest Diaphanopterodea from the early Late Carboniferous of West Germany. *Canadian Journal of Zoology* 68: 1104–111.

———, and J. F. Lawrence. 1993. Evolution of the hind wing in Coleoptera. *Canadian Entomologist* 125: 181–258.

———, and S. B. Peck. 1976. Adult and immature Calvertiellidae (Insecta: Palaeodictyoptera) from the Upper Paleozoic of New Mexico and Czechoslovakia. *Psyche* 83: 79–83.

———, and S. B. Peck. 1993. Zoraptera wing structures: Evidence for new genera and relationship with the blattoid orders (Insecta: Blattoneoptera). *Systematic Entomology* 18: 333–50.

———, and R. Willmann. 1990. Lower Permian mecopteroid-like insects from Central Europe (Insecta: Endopterygota). *Canadian Journal of Earth Sciences* 27: 459–68.

Kulicka, R. 1978. *Mengea tertiaria* (Menge) (Strepsiptera) from the Baltic amber. *Prace Muzeum Ziemi, PAN* 29: 141–5.

———. 1979. *Mengea mengei* sp. n. from the Baltic amber. *Prace Muzeum Ziemi, PAN* 32: 109–12.

———. 2001. New genera and species of Strepsiptera from the Baltic amber. *Prace Muzeum Ziemi, PAN* 46: 3–16.

Kumar, P., and P. Kumar. 2001. Phthirapteran insect and larval Acanthocephala from the Late Triassic sediments of the Satpura Basin, India. *Journal of the Palaeontological Society of India* 46: 141–6.

Kuperus, W. R., and W. Chapco. 1996. Reanalysis of some classic orthopteroid phylogenies. *Journal of Orthoptera Research* 5: 205–11.

Kuschel, G. 1969. Biogeography and ecology of South American Coleoptera. Pp. 709–22. *In* Fittkau, E. J., J. Illies, H. Klinge, G. H. Schwabe, and H. Sioli (eds.), *Biogeography and Ecology in South America* [Volume 2]. Junk; The Hague, the Netherlands; pp. 449–946 + xi pp.

———. 1983. Past and present of the relict family Nemonychidae (Coleoptera, Curculionoidea). *GeoJournal* 7: 499–504.

———. 1992. Reappraisal of the Baltic amber Curculionoidea described by E. Voss. *Mitteilungen aus dem Geologisch-Paläontologischen Institut der Universität Hamburg* 73: 191–215.

———. 1995. A phylogenetic classification of Curculionoidea to families and subfamilies. *Memoirs of the Entomological Society of Washington* 14: 5–33.

Kusnezov, N. J. 1941. A revision of the amber Lepidoptera. *Paleontologicheskiy Institut, Akademiya Nauk SSSR* 136: 1–58. [In Russian]

Kyte, F. T. 1998. A meteorite from the Cretaceous-Tertiary boundary. *Nature* 396: 237–9.

Labandeira, C. C. 1997. Insect mouthparts: Ascertaining the paleobiology of insect feeding strategies. *Annual Review of Ecology and Systematics* 28: 153–93.

———. 1998. Early history of arthropod and vascular plant associations. *Annual Review of Earth and Planetary Sciences* 26: 329–77.

———, 1999. Insects and other hexapods. Pp. 603–24. *In* Singer, R. (ed.), *Encyclopedia of Paleontology* [vol. 1: A–L, vol. 2: M–Z]. Fitzroy Dearborn; London, UK; xix+1435 pp.

———. 2000. The paleobiology of pollination and its precursors. Pp. 233–69. *In* Gastaldo, R. A., and W. A. DiMichele (eds.), *Phanerozoic Terrestrial Ecosystems*. Paleontological Society Papers [Volume 6], Paleontological Society; Pittsburgh, Pennsylvania; 308 pp.

———. 2001. Rise and diversification of insects. Pp. 82–8. *In* Briggs, D. E. G., and P. R. Crowther (eds.), *Paleobiology II*. Blackwell Science; London, UK; 583 pp.

———. 2002. The history of association between plants and animals. Pp. 26–261. *In* Herrera, C. M., (ed.), *Plant-Animal Interactions: An Evolutionary Approach*. Blackwell; London, UK; 320 pp.

———, and B. S. Beall. 1990. Arthropod terrestriality. Pp. 214–56. *In* Mikulic, D. G. (ed.), *Arthropods Paleobiology*. University of Tennessee Press; Knoxville, Tennessee; v+315 pp.

———, and T. L. Phillips. 1996a. A Carboniferous insect gall: Insight into early ecological history of the Holometabola. *Proceedings of the National Academy of Sciences, USA* 93: 8470–8474.

———, and T. L. Phillips. 1996b. Insect fluid-feeding on Upper Pennsylvanian tree ferns (Palaeodictyoptera, Marattiales) and the early history of the piercing-and-sucking functional feeding group. *Annals of the Entomological Society of America* 89: 157–83.

———, and J. J. Sepkoski. 1993. Insect diversity in the fossil record. *Science* 261: 310–15.

———, D. L. Dilcher, D. R. Davis, and D. L. Wagner. 1994. Ninety-seven million years of angiosperm-insect association: Paleobiological insights into the meaning of coevolution. *Proceedings of the National Academy of Sciences, USA* 91: 12278–82.

———, T. L. Phillips, and R. A. Norton. 1997. Oribatid mites and the decomposition of plant tissues in Paleozoic coal-swamp forests. *Palaios* 12: 319–53.

———, B. S. Beall, and F. M. Hueber. 1988. Early insect diversification: Evidence from a Lower Devonian bristletail from Québec. *Science* 242: 913–16.

———, B. A. LePage, and A. H. Johnson. 2001. A *Dendroctonus* bark engraving (Coleoptera: Scolytidae) from a middle Eocene *Larix* (Coniferales: Pinaceae): Early or delayed colonization? *American Journal of Botany* 88: 2026–39.

———, K. R. Johnson, and P. Lang. 2002. Preliminary assessment of insect herbivory across the Cretaceous-Tertiary boundary: Major extinction and minimum rebound. Pp. 297–327. *In* Hartman, J. H., K. R. Johnson, and D. J. Nichols (eds.), *The Hell Creek Formation and the Cretaceous-Tertiary Boundary in the Northern Great Plains: An Integrated Continental Record of the End of the Cretaceous*. Geological Society of America Special Paper 361; Boulder, Colorado; vi+520 pp.

Lacau, S., A. Nel, C. Villemant, J.-J. Menier, M. J. Orliac, and G. de Plöeg. 2000. A fossil Scolebythidae from the lowermost Eocene amber of France (Insecta: Hymenoptera). *Annals of the Entomological Society of America* 93: 701–6.

Lachaise, D., M.-L. Cariou, J. R. David, F. Lemeunier, L. Tsacas, and M. Ashburner. 1988. Historical biogeography of the *Drosophila melanogaster* species subgroup. *Evolutionary Biology* 22: 159–225.

Lack, D. 1968. *Darwin's Finches: An Essay on the General Biological Theory of Evolution*. Peter Smith; Gloucester, Massachusetts; x+204 pp.

Lacy, R. 1980. The evolution of eusociality in termites: A haplodiploid analogy? *American Naturalist* 116: 449–51.

La Greca, M. 1980. Origin and evolution of wings and flight in insects. *Bollettino di Zoologia* 47 (supplement): 65–82.

Lai, P.-Y. 1988. Biological control: A positive point of view. *Proceedings of the Hawaiian Entomological Society* 28: 179–90.

Laicharting, J. N., von. 1781. *Verzeichniss und Beschreibung der Tyroler Insecten* [vol. 1]. Fuessely; Zürich, Switzerland; xii+248 pp.

Lambkin, K. J. 1986. A revision of the Australian Mantispidae (Insecta: Neuroptera) with a contribution to the classification of the family. I. General and Drepanicinae. *Australian Journal of Zoology, Supplementary Series* 116: 1–142.

———. 1988. A re-examination of *Lithosmylidia* Riek from the Triassic of Queensland with notes on Mesozoic 'osmylid-like' fossil Neuroptera (Insecta: Neuroptera). *Memoirs of the Queensland Museum* 25: 445–58.

———. 1992. A record of *Austrosialis* Tillyard from the Queensland Paleocene (Insecta: Megaloptera: Sialidae). *Queensland Naturalist* 31: 84–6.

Lameere, A. 1917. Étude sur l'évolution des éphémères. *Bulletin de la Société Zoologique de France* 42: 41–82.

Landa, V., and T. Soldán. 1985. Phylogeny and higher classification of the order Ephemeroptera: A discussion from the comparative anatomical point of view. *Studie Československá Akademie Věd, Prague* 4: 1–121.

Lane, R. 1997. The species concept in blood-sucking vectors of human diseases. Pp. 273–89. *In* Claridge, M. F., H. A. Dawah, and M. R. Wilson (eds.), *Species: The Units of Biodiversity*. Chapman and Hall; London, UK; xvi+439 pp.

Lane, R. P. 1984. Host specificity of ectoparasitic midges on butterflies. Pp. 105–8. *In* Vane-Wright, R. I., and P. R. Ackery (eds.), *The Biology of Butterflies*. Academic Press; London, UK; xxiv+429 pp.

Langenheim, J. H. 1966. Botanical source for amber from Chiapas, Mexico. *Ciencia* 24: 201–11.

———. 1969. Amber: A botanical inquiry. *Science* 163: 1157–69.

———. 1995. Biology of amber-producing trees: Focus on case studies of *Hymenaea* and *Agathis*. Pp. 1–31. *In* Anderson, K. B., and J. C. Crelling (eds.), *Amber, Resinite, and Fossil Resins*. American Chemical Society; Washington, D.C.; xvii+297 pp.

———. 2003. *Plant Resins: Chemistry, Evolution, Ecology, and Ethnobotany*. Timber Press; Portland, Oregon; 586 pp.

Lankester, E. R. 1904. The structure and classification of the Arthropoda. *Quarterly Journal of the Microscopical Society* 47: 523–82.

LaPolla, J. S., S. P. Cover, and U. G. Mueller. 2002. Natural history of the mealybug-tending ant, *Acropyga epedana*, with descriptions of the male and queen castes. *Transactions of the American Entomological Society* 128: 367–76.

Larew, H. 1986. The fossil gall record: A brief summary. *Proceedings of the Entomological Society of Washington* 88: 385–8.

———. 1987. Two cynipid wasp acorn galls preserved in the La Brea Tar Pits (Early Holocene). *Proceedings of the Entomological Society of Washington* 89: 831–3.

———. 1992. Fossil galls. Pp. 50–9. *In* Shorthouse, J. D., and O. Rohfritsch (eds.), *Biology of Insect-Induced Galls*. Oxford University Press; Oxford, UK; x+285 pp.

Larink, O. 1997a. Apomorphic and plesiomorphic characteristics in Archaeognatha, Monura, and Zygentoma. *Pedobiologia* 41: 3–8.

———. 1997b. Aspekte der Speziellen Zoologie von Archaeognatha und Zygentoma. *Abhandlungen und Berichte des Naturkundemuseums Goerlitz* 69: 119–34.

Larochelle, A., and M. C. Larivière. 2003. *A Natural History of the Ground Beetles (Coleoptera: Carabidae) of America North of Mexico*. Pensoft Publishers; Sofia, Bulgaria; 583 pp.

Larsson, S. G. 1978. Baltic amber – A palaeobiological study. *Entomonograph* 1: 1–192.

LaSalle, J., and I. D. Gauld, eds. 1993. *Hymenoptera and Biodiversity*. CABI Publishing; Wallingford, UK; xi+348 pp.

Latreille, P. A. 1796. *Précis des Caractères Génériques des Insectes Disposés dans un Ordre Naturel, &c.* F. Bourdeaux; Brive, France; xiii+[1]+201+[7] pp.

———. 1802a. *Histoire Naturelle des Fourmis, et Recueil de Mémoires et d'Observations sur les Abeilles, les Araignées, les Faucheurs, et autres Insectes*. Crapelet; Paris, France; xvi+445 pp.

———. 1802b. *Histoire Naturelle Générale et Particulière des Crustacés et des Insectes, Tome 3*. Dufart; Paris, France; xii+467 pp.

Laufer, B. 1927. Insect-musicians and cricket champions of China. *Field Museum of Natural History, Anthropology Leaflet* 22: 1–27.

Laurentiaux, D. 1950. Les insectes houillers do Limbourg hollandais. *Mededelingen van de Geologische Stichting* 4: 13–22.

———. 1952. Découverte d'un rostre chez *Stenodictya lobata* Brgt. (Paléodictyoptère sténodictyide) et le problème des Protohémiptères. *Bulletin de la Société Geologique de France (serie 6)* 2: 233–47.

———. 1953. Classe des Insectes (Insecta Linné 1758). Pp. 397–527. *In* Piveteau, J. (ed.), *Traité de Paléontologie, Tome III*. Masson et Cie; Paris, France; 1063 pp.

———, and F. Laurentiaux-Vieira. 1980. Un type singulier d'insecte protorthoptéroïde du westphalien supérieur du Pas-de-Calais. *Annales de la Société Géologique du Nord* 99: 407–13.

Laurentiaux-Vieira, F., and D. Laurentiaux. 1986. Presence du genre *Zdenekia* Kuk. (protorthopteres paoliides) dans le Westphalien inferieur de Belgique. *Annales de la Société Géologique du Nord* 105: 195–201.

Lauterbach, K.-E. 1972. Schlüsselereignisse in der Evolution der Stammgruppe der Euarthropoda. *Zoologische Beiträge* 19: 251–99.

Law, J. H., and B. J. Crespi. 2002. The evolution of geographic parthenogenesis in *Timema* walking-sticks. *Molecular Ecology* 11: 1471–89.

Lawrence, D. A., and B. P. J. Williams. 1987. Evolution of drainage systems in response to Acadian deformation: The Devonian Battery Point Formation, eastern Canada. *Society of Economic Paleontologists and Mineralogists Special Publication* 39: 287–300.

Lawrence, J. F. 1982. Coleoptera. Pp. 482–553. *In* Parker, S. P. (ed.), *Synopsis and Classification of Living Organisms*. McGraw-Hill; New York, New York; xviii+1166 pp.

———. 1991. Order Coleoptera. Pp. 144–658. *In* Stehr, F. W. (ed.), *Immature Insects* [Volume 2]. Kendall-Hunt; Dubuque, Iow; xvi+975 pp.

———. 1995. *Electribius* Crowson: Alive and well in Mesoamerica, with notes on *Ctesibius* Champion and the classification of the Artematopodidae. Pp. 411–31. *In* Pakaluk, J., and S. A. Slipinski (eds.), *Biology, Phylogeny and Classification of Coleoptera*. Muzeum I Instytut Zoologii PAN; Warsaw; xii+558 pp.

———. 1999. The Australian Ommatidae (Coleoptera: Archostemata): New species, larva and discussion of relationships. *Invertebrate Taxonomy* 13: 369–90.

———, and E. B. Britton. 1991. Coleoptera (beetles). Pp. 543–683. *In* Naumann, I. D. (ed.), *The Insects of Australia: A Textbook for Students and Research Workers, Volume 2* [2nd Edition]. Cornell University Press; Ithaca, New York; [2]+543–1137 pp.

———, and A. F. Newton. 1982. Evolution and classification of beetles. *Annual Review of Ecology and Systematics* 13: 261–90.

———, and A. F. Newton. 1995. Families and subfamilies of Coleoptera (with selected genera, notes, references and data on family-group names). Pp. 779–1092. *In* Pakaluk, J., and S. A. Ślipiński (eds.), *Biology, Phylogeny, and Classification of Coleoptera: Papers Celebrating the 80th Birthday of Roy A. Crowson*. Museum i Instytut Zoologii PAN; Warsaw, Poland; x+1092 pp. [2 volumes]

———, A. M. Hastings, M. J. Dallwitz, T. A. Paine, and E. J. Zurcher. 2000a. *Beetles of the World*. CSIRO Publishing; Victoria, Australia; CD-ROM.

———, A. M. Hastings, M. J. Dallwitz, T. A. Paine, and E. J. Zurcher. 2000b. *Beetle Larvae of the World*. CSIRO Publishing; Victoria, Australia; CD-ROM.

Lawrence, P. N. 1985. Ten species of Collembola from Baltic amber. *Prace Muzeum Ziemi, PAN* 37: 101–4.

Leakey, L. S. B. 1952. Lower Miocene invertebrates from Kenya. *Nature* 169: 624–5.

Lee, B.-H., U.-W. Hwang, W. Kim, K.-H. Park, and J.-T. Kim. 1995. Phylogenetic study of the suborder Arthropleona (Insecta: Collembola) based on morphological characters and 18S rDNA sequence analysis. *Polskie Pismo Entomologiczne* 64: 261–77.

Leestmans, R. 1983. Les lépidoptères fossils trouvés en France (Insecta Lepidoptera). *Linneana Belgica* 9: 64–89.

Lehane, M. J. 1991. *Biology of Blood-Sucking Insects*. HarperCollins Academic; London, UK; 288 pp.

Lelej, A. S. 1996. Males of the genus *Protomutilla* (Hymenoptera, Mutillidae) from Baltic amber. *Paleontologicheskiy Zhurnal* 1986: 104–6. [In Russian]

———, and P. G. Nemkov. 1997. Phylogeny, evolution and classification of Mutillidae (Hymenoptera). *Far Eastern Entomologist* 46: 1–24.

Lemaire, C., and J. Minet. 1999. The Bombycoidea and their relatives. Pp. 321–55. *In* Kristensen, N. P. (ed.), *Handbuch der Zoologie: Eine Naturgeschichte der Stämme des Tierreiches: Band IV: Arthropoda: Insecta: Teilband 35: Lepidoptera, Moths and Butterflies: Volume 1: Evolution, Systematics, and Biogeography*. Walter de Gruyter; Berlin, Germany; x+491 pp.

Lemche, H. 1940. The origin of winged insects. *Videnskabelige Meddelelser fra Dansk Naturhistorisk Forening* 104: 127–68.

Lemeunier, F., and M. Ashburner. 1976. Relationships within the *melanogaster* species subgroup of the genus *Drosophila*. II. Phylogenetic relationships between six species based upon their polytene chromosome banding sequences. *Proceedings of the Royal Society of London, Series B, Biological Sciences* 193: 275–94.

Lent, H., and P. Wygodzinsky. 1979. Revision of the Triatominae (Hemiptera, Reduviidae), and their significance as vectors of Chagas' Disease. *Bulletin of the American Museum of Natural History* 163: 123–520.

Leo, N. P., N. J. H. Campbell, X. Yang, K. Mumcuoglu, and S. C. Barker. 2002. Evidence from mitochondrial DNA that head lice and body lice of humans (Phthiraptera: Pediculidae) are conspecific. *Journal of Medical Entomology* 39: 662–6.

Lepeletier de Saint Fargeau, A. L. M. 1835. *Histoire Naturelle des Insectes-Hyménoptères* [Volume 1]. Roret; Paris, France; 547 pp.

———. 1841. *Histoire Naturelle des Insectes-Hyménoptères* [Volume 2]. Roret; Paris, France; 680 pp.

Lesquereux, L. 1882. On some specimens of Permian fossil plants from Colorado. *Bulletin of the Museum of Comparative Zoology* 7: 243–7.

Leston, D., J. G. Pendergast, and T. R. E. Southwood. 1954. Classification of the terrestrial Heteroptera (Geocorisae). *Nature* 174: 91.

Lever, R. J. A. W. 1933. The greater spike moth (*Tirathaba rufivena*, Walk.) and its parasite *Apanteles tirathabae*, Wilk.). *British Soloman Islands Protectorate Agricultural Gazette* 3: 7–8.

Lewis, D. J. 1974. The biology of Phlebotomidae in relation to *Leishmaniasis*. *Annual Review of Entomology* 19: 363–84.

Lewis, R. E. 1998. Résumé of the Siphonaptera (Insecta) of the world. *Journal of Medical Entomology* 35: 377–89.

———, and D. Grimaldi. 1997. A pulicid flea in Miocene amber from the Dominican Republic (Insecta: Siphonaptera: Pulicidae). *American Museum Novitates* 3205: 1–9.

Lewis, S. E. 1969. Fossil insects from the Latah Formation (Miocene) of eastern Washington and northern Idaho. *Northwest Science* 43: 9–115.

———. 1971. A new species of fossil Diptera (Diopsidae) from the Ruby River Basin (Oligocene) of Montana. *Annals of the Entomological Society of America* 64: 959–60.

———. 1973. Two new species of fossil crane flies (Diptera: Tipulidae) from the Ruby River Basin (Oligocene) of southwestern Montana. *Annals of the Entomological Society of America* 66: 706–7.

———. 1978. An immature fossil Ephemeroptera (Ephemeridae) from the Ruby River Basin (Oligocene) of southwestern Montana. *Annals of the Entomological Society of America* 71: 479–80.

———. 1985. Miocene insects from the Clarkia deposits of northern Idaho. Pp. 245–64. *In* Smiley, C. J. (ed.), *Late Cenozoic History of the Pacific Northwest: Interdisciplinary Studies on the Clarkia Fossil Beds of Northern Idaho*. American Association for the Advancement of Science, Pacific Division; San Francisco, California; 417 pp.

———. 1992. Insects from the Klondike Mountain Formation (middle Eocene) of Republic, Washington. *Washington Geology* 22: 39–40.

———. 1994. Evidence of leaf-cutting bee damage from the Republic sites (middle Eocene) of Washington. *Journal of Paleontology* 68: 172–3.

———, and J. C. Luhman. 1988. Fossil Ichneumonidae (Hymenopterra [sic]) from the Ruby River Basin (Oligocene) of southwestern Montana. *Occasional Papers in Paleobiology, St. Cloud State University* 2: 1–11.

———, P. M. Heikes, and K. L. Lewis. 1990. Entomofauna from Miocene deposits near Juliaetta, Idaho. *Occasional Papers in Paleobiology, St. Cloud State University* 4: 1–22.

Lewis, T. 1973. *Thrips: Their Biology, Ecology, and Economic Importance*. Academic Press; London, UK; xv+349 pp.

Li, W.-H. 1997. *Molecular Evolution*. Sinauer Associates; Sunderland, Massachusetts; xv+487 pp.

———, M. Gouy, K. H. Wolfe, and P. M. Sharp. 1989. Angiosperm origins. *Nature* 342: 131–2.

Lidgard, S., and P. R. Crane. 1988. Quantitative analysis of early angiosperm radiation. *Nature* 331: 344–6.

Liebherr, J. K., and K. W. Will. 1998. Inferring phylogenetic relationships within Carabidae (Insecta, Coleoptera) from characters of the female reproductive tract. *Bollettino del Museo Regionale di Scienze Naturali-Torino* 1998: 107–70.

———, and J. V. McHugh. 2003. Coleoptera (beetles, weevils, fireflies). Pp. 209–30. *In* Resh, V. H., and R. T. Cardé (eds.), *Encyclopedia of Insects*. Academic Press; San Diego, California; xxviii+[1]+1266 pp.

———, and E. C. Zimmerman. 1998. Cladistic analysis, phylogeny and biogeography of the Hawaiian Platynini (Coleoptera: Carabidae). *Systematic Entomology* 23: 137–72.

Liebig, V., and G. Gruber. 2000. The Messel drillings of 1980: Redescription and interpretation (Grube Messel, southern Hesse, Germany). *Terra Nostra* 2000/6: 254–8.

Lienhard, C. 1990. Revision of the western Palaearctic species of *Liposcelis* Motschulsky (Psocoptera: Liposcelidae). *Zoologische Jahrbücher, Abteilung für Systematik* 117: 117–74.

———, and C. N. Smithers. 2002. *Psocoptera (Insecta): World Catalogue and Bibliography*. Muséum d'Histoire Naturelle, Genève; Geneva, Switzerland; xli+745 pp.

Lin, C. P., and T. K. Wood. 2002. Molecular phylogeny of the North American *Enchenopa binotata* (Homoptera: Membracidae) species complex. *Annals of the Entomological Society of America* 95: 162–71.

Lin, Q.-B. 1976. The Jurassic fossil insects from western Liaoning. *Acta Paleontologica Sinica* 15: 97–118. [In Chinese, with English summary].

———. 1978. Upper Permian and Triassic fossil insects of Guizhou. *Acta Palaeontologica Sinica* 17: 313–7. [In Chinese, with English summary]

———. 1980. Fossil insects. Pp. 211–34. *In, Fossils of Mesozoic Deposits of Volcanic Origin in Zhejiang and Anhui*. Science Press; Beijing, China; 234 pp. [In Chinese]

———. 1982. Insecta. Pp. 148–55. *In, Paleontological Atlas of East China. Part 3. Volume of Mesozoic and Caenozoic*. Geological Publishing House; Beijing, China; 405 pp. [In Chinese]

———. 1986. Early Mesozoic fossil insects from South China. *Palaeontologia Sinica* 170: 1–112. [In Chinese]

———. 1998. Cretaceous insects of China. *Cretaceous Research* 15: 305–16.

———, and W.-J. Han. 1985. A new cockroach from the Upper Shihhotse Formation (Upper Permian) in Yu County, Henan Province. *Acta Palaeontologica Sinica* 24: 122–4. [In Chinese, with English summary]

———, and D.-Y. Huang. 2001. Description of *Caenoephemera shangyuanensis*, gen. nov., sp. nov. (Ephemeroptera), from the Yixian Formation. *Canadian Entomologist* 133: 747–54.

———, and X.-Y. Liang. 1988. A Permian cockroach tegmen from Gongxian, Henan, China. *Acta Palaeontologica Sinica* 27: 640–2. [In Chinese, with English summary]

———, Y.-M. Yao, W.-D. Xiang, and Y.-R. Xia. 1988. An Oligocene micropalaeoentomofauna from Gubei District of Shandong and its ecological environment. *Acta Micropalaeontologica Sinica* 5: 331–45.

Lindahl, T. 1993. Instability and decay of the primary structure of DNA. *Nature* 362: 709–15.

Lindroth, C. H. 1948. Interglacial insect fossils from Sweden. *Arsbok Sveriges Geologiska Undersökning, Series C* 42: 1–29.

———. 1969. The ground beetles (Carabidae, exc. Cicindelidae) of Canada and Alaska. *Opuscula Entomologica Supplimenta* 1–846.

———. 1992. *Ground Beetles (Carabidae) of Fennoscandia: A Zoogeographic Study*. Intercept Press; Andover, UK; [vol. 1] xxviii+630 pp., [vol. 2] x+271 pp., [vol. 3] xii+814 pp.

Lindsley, D. L., and G. G. Zimm. 1992. *The Genome of* Drosophila melanogaster. Academic Press; New York, New York; viii+1133 pp. +10 maps.

Linnaeus, C. 1737. *Critica Botanica: In qua Nomina Plantarum Generica, Specifica & Variantia examini Subjiciuntur, Selectiora Confirmantur, Indigna Rejiciuntur: Simulque Doctrina circa Denominationem Plantarum Traditur, seu, Fundamentorum Botanicorum pars IV. Accedit Johannis Browallii de Necessitate Historiae Naturalis Discursus*. Conradum Wishoff; Lugduni Batavorum [Leiden], the Netherlands; [16]+270+[54]+24 pp.

———. 1758. *Systema Naturae per Regna Tria Naturae, Secundum Classes, Ordines, Genera, Species, cum Characteribus, Differentiis, Synonymis, Locis* [vol. 1, ed. 10, Reformata]. Salviae; Holmiae [Stockholm], Sweden; 824 pp.

Liou, L. W., and T. D. Price. 1994. Speciation by reinforcement of premating isolation. *Evolution* 48: 1451–9.

Little, E. L., Jr. 1983. North American trees with relationships in eastern Asia. *Annals of the Missouri Botanical Garden* 70: 605–15.

Liu, F.-G. R., and M. M. Miyamoto. 1999. Phylogenetic assessment of molecular and morphological data for eutherian mammals. *Systematic Biology* 48: 54–64.

Liu, Z. 1998a. Phylogenetic systematics and historical biogeography of macrocynipoids parasitizing woodboring insects. *Acta Universitatis Agriculturae Sueciae, Silvestria* 62: 1–26.

———. 1998b. A new species of *Ibalia* from Borneo, with a revised phylogeny and historical biogeography of Ibaliidae (Hymenoptera: Cynipoidea). *Journal of Hymenoptera Research* 7: 149–56.

———. 2001. Phylogeny, biogeography, and revision of the subfamily Dallatorrellinae (Hymenoptera: Liopteridae). *American Museum Novitates* 3353: 1–23.

Lloyd, D. C. 1952. Biological observations on some thynnids of western Patagonia. *Bulletin of Entomological Research* 42: 707–19.

Lloyd, D. G., and M. S. Wells. 1992. Reproductive biology of a primitive angiosperm, *Pseudowintera colorata* (Winteraceae), and the evolution of pollination systems in the Anthophyta. *Plant Systematics and Evolution* 181: 77–95.

Lloyd, J. E. 1983. Bioluminescence and communication in insects. *Annual Review of Entomology* 28: 131–60.

Lo, N., G. Tokuda, H. Watanabe, H. Rose, M. Slaytor *et al.* 2000. Evidence from multiple gene sequences indicates that termites evolved from wood-feeding cockroaches. *Current Biology* 10: 801–4.

———, C. Bandi, H. Watanabe, C. Nalepa, and T. Beninati. 2003. Evidence for cocladogenesis between diverse dictyopteran lineages and their intracellular endosymbionts. *Molecular Biology and Evolution* 20: 907–13.

Loconte, H., and D. W. Stevenson. 1990. Cladistics of the Spermatophyta. *Brittonia* 42: 197–211.

Lohmann, H. 1996. Das phylogenetische System der Anisoptera (Odonata). *Entomologische Zeitschrift* 106: 209–66.

Lomholdt, O. 1982. On the origin of the bees (Hymenoptera: Apidae, Sphecidae). *Entomologica Scandinavica* 13: 185–90.

Longino, J. T. 1991. *Azteca* ants in *Cecropia* trees: Taxonomy, colony structure, and behavior. Pp. 271–88. *In* Huxley, C. R., and D. F. Cutler (eds.), *Ant-Plant Interactions*. Oxford University Press; Oxford, UK; xviii+601 pp.

Lopez-Vaamonde, C., J. Y. Rasplus, G. D. Weiblen, and J. M. Cook. 2001. Molecular phylogenies of fig wasps: Partial cocladogenesis of pollinators and parasites. *Molecular Phylogenetics and Evolution* 21: 55–71.

Lourenço, W. R. 2001. A remarkable scorpion fossil from the amber of Lebanon. Implications for the phylogeny of Buthoidea. *Comptes Rendus de l'Academie des Sciences, Serie II A, Sciences de la Terre et des Planetes* 332: 641–6.

———. 2002. The first scorpion fossil from the Cretaceous amber of Myanmar (Burma). New implications for the phylogeny of Buthoidea. *Comptes Rendus Palevolution* 2: 97–101.

———, and W. Weitschat. 2001. Description of another fossil scorpion from Baltic amber, with considerations on the evolutionary levels of Cenozoic Buthoidea. *Mitteilungen aus dem Geologisch-Paläontologischen Institut der Universität Hamburg* 85: 277–83.

Lovallo, N., B. A. McPheron, and D. L. Cox-Foster. 2002. Effects of the polydnavirus of *Cotesia congregata* on the immune system and development of non-habitual hosts of the parasitoid. *Journal of Insect Physiology* 48: 517–26.

Lovejoy, T. E. 1985. Strategies for preserving species in the wild. Pp. 97–114. *In* Hoage, R. J. (ed.), *Animal Extinctions: What Everyone Should Know.* Smithsonian Institution Press; Washington, D.C.; xviii+192 pp.

Low, A. B., and A. G. Rebelo. 1998. *Vegetation of South Africa, Lesotho and Swaziland: A Companion to the Vegetation Map of South Africa, Lesotho and Swaziland.* Department of Environmental Affairs and Tourism; Pretoria, South Africa.

Lubkin, S. H., and M. S. Engel. 2005. *Permocoleus*, new genus, the first Permian beetle (Coleoptera) from North America. *Annals of the Entomological Society of America* 98: 73–6.

Lucas, J. 1989. The structure and function of antlion pits: Slope asymmetry and predator-prey interactions. *Animal Behaviour* 38: 318–30.

Lucas, S. G. 1999. The epicontinental Triassic, an overview. *Zentralblatt für Geologie und Paläontologie* 1: 475–96.

Ludwig, R. 1877. Fossile Crocodiliden aus der Tertiärformation der Mainzer Beckens. *Palaeontographica Supplement* 3: 1–52.

Luhman, J. C., R. W. Holzenthal, and J. K. Kjaerandsen. 1999. New host records of a ceraphronid (Hymenoptera) in Trichoptera pupae. *Journal of Hymenoptera Research* 8: 126.

Lukashevich, E. D. 1996. Mesozoic Dixidae (Insecta: Diptera) and systematic position of *Dixamima* Rohdendorf, 1964 and *Rhaetomyia* Rohdendorf, 1962. *Paleontologischeskiy Zhurnal* 1996: 48–53. [In Russian]

———, and M. B. Mostovski. 2003. Hematophagous insects in the fossil record. *Paleontological Journal* 37: 153–61.

———, and D. E. Shcherbakov. 1997. The first find of net-winged midges (Blephariceridae, Diptera) in the Mesozoic. *Neues Jahrbuch für Geologische und Paläontologie, Monatshefte* 1997: 639–46.

———, J. Ansorge, W. Krzemiński, and E. Krzemińska. 1998. Revision of Eoptychopterinae (Diptera: Eoptychopteridae). *Polskie Pismo Entomologiczne* 67: 311–43.

Lupia, R. 1999. Discordant morphological disparity and taxonomic diversity during the Cretaceous angiosperm radiation: North American pollen record. *Paleobiology* 25: 1–28.

Lutz, H. 1990. Systematische und palökologische Untersuchungen an Insekten aus dem Mittel-Eozän der Grube Messel bei Darmstadt. *Courier Forschungsinstitut Senckenberg* 124: 1–165.

———. 1992. Giant ants and other rarities: The insect fauna. Pp. 53–67. *In* Schaal, S., and W. Ziegler (eds.), *Messel – An Insight into the History of Life and of the Earth.* Claredon Press; Oxford, UK; 322 pp.

———, and F.-O. Neuffer. 2001. A climatic archive with hide and hair. *German Research* 2001: 8–11.

———, H. Frankenhäuser, and F.-O. Neuffer. 1998. *Fossilfundstätte Eckfelder Maar – Archiv eines mitteleozänen Lebensraumes in der Eifel.* Landessammlung für Naturkunde Rheinland Pfalz; Mainz, Germany; 51 pp.

———, F.-O. Neuffer, F.-J. Harms, S. Schaal, N. Micklich *et al.* 2000. Tertiary maars as fossil deposits: Eckfeld, Messel, Randeck, Höwenegg, Öhnigen. *Mainzer Naturwissenschaftliches Archiv, Beiheft* 24: 125–60.

Lyal, C. H. C. 1985a. A cladistic analysis and classification of trichodectid mammal lice (Phthiraptera: Ischnocera). *Bulletin of the British Museum (Natural History), Entomology* 51: 187–346.

———. 1985b. Phylogeny and classification of the Psocodea, with particular reference to the lice (Psocodea: Phthiraptera). *Systematic Entomology* 10: 145–65.

———. 1987. Co-evolution of trichodectid lice (Insecta: Phthiraptera) and their mammalian hosts. *Journal of Natural History* 21: 1–28.

Lyell, C. 1830–3. *Principles of Geology; Being an Attempt to Explain the Former Changes of the Earth's Surface, by Reference to Causes now in Operation.* J. Murray; London, UK; [vol. 1, 1830] xv+511 pp., [vol. 2, 1832] xxii+330 pp., [vol. 3, 1833] xxxi+[1]+398+109 pp.

Maa, T. C. 1953. An inquiry into the systematics of the tribus Apidini or honeybees (Hym.). *Treubia* 21: 525–640.

MacGinitie, H. D. 1969. The Eocene Green River flora of northwestern Colorado and northeastern Utah. *University of California Publications in Geological Sciences* 83: 1–140.

Machado, C. A., E. A. Herre, S. McCafferty, and E. Bermingham. 1996. Molecular phylogenies of fig pollinating and non-pollinating wasps and the implications for the origin and evolution of the fig-fig wasp mutualism. *Journal of Biogeography* 23: 531–42.

———, E. Jousselin, F. Kjellberg, S. G. Compton, and E. A. Herre. 2001. Phylogenetic relationships, historical biogeography and character evolution of fig-pollinating wasps. *Proceedings of the Royal Society of London, Series B, Biological Sciences* 268: 685–94.

———, R. M. Kliman, J. A. Markert, and J. Hey. 2002. Inferring the history of speciation from multilocus DNA sequence data: The case of *Drosophila pseudoobscura* and close relatives. *Molecular Biology and Evolution* 19: 472–88.

MacKay, M. R. 1970. Lepidoptera in Cretaceous amber. *Science* 167: 379–80.

Mackay, R. J., and G. B. Wiggins. 1979. Ecological diversity in the Trichoptera. *Annual Review of Entomology* 24: 185–208.

MacLeod, E. G. 1967. Studies on the systematics of the Berothidae, Part I: A redescription of the genus *Sphaeroberotha* Návas, with a critique of the taxonomic characters used in the Berothinae (Neuroptera). *Psyche* 74: 342–52.

———. 1970. The Neuroptera of the Baltic amber. I. Ascalaphidae, Nymphidae, and Psychopsidae. *Psyche* 77: 147–80.

———, and P. A. Adams. 1967. A review of the taxonomy and morphology of the Berothidae, with the description of a new subfamily from Chile (Neuroptera). *Psyche* 74: 237–65.

———, and P. E. Spiegler. 1961. Notes on the larval habitat and developmental peculiarities of *Nallachius americanus* (McLachlan) (Neuroptera: Dilaridae). *Proceedings of the Entomological Society of Washington* 63: 281–6.

MacNaughton, R. B., J. M. Cole, R. W. Dalrymple, S. J. Braddy, D. E. G. Briggs, and T. D. Lukie. 2002. First steps on land: Arthropod trackways in Cambrian-Ordovician eolian sandstone, southeastern Ontario, Canada. *Geology* 30: 391–4.

MacPhee, R. D. E., and D. A. Grimaldi. 1996. Mammal bones in Dominican amber. *Nature* 380: 489–90.

Maddison, D. R., M. D. Baker, and K. A. Ober. 1999. Phylogeny of carabid beetles as inferred from 18S ribosomal DNA (Coleoptera: Carabidae). *Systematic Entomology* 24: 103–38.

Maekawa, K., K. Osamu, and T. Matsumoto. 1999. Molecular phylogeny of orthopteroid insects based on the mitochondrial cytochrome oxidase II gene. *Zoological Science* 16: 175–84.

Mahner, M. 1993. Systema cryptoceratorum phylogeneticum (Insecta, Heteroptera). *Zoologica* 143: 1–302.

Makarkin, V. N. 1990. New lacewings (Neuroptera) from the Upper Cretaceous of Asia. Pp. 63–8. *In* Akimov, I. A. (ed.), *Novosti Faunistiki i Sistematiki: Sbornik Nauchnyhk Trudov [News of Faunistics and Systematics: Collected Scientific Papers].* Naukova Dumka; Kiev, Ukraine; 184+[1] pp. [In Russian]

———. 1991. Miocene Neuroptera from northern Caucasus and Sikhote-Alin'. *Paleontologicheskiy Zhurnal* 1991: 57–68. [In Russian]

———. 1994. Upper Cretaceous Neuroptera from Russia and Kazakhstan. *Annales de la Société Entomologique de France* 30: 283–92.

———. 1996. Fossil Neuroptera of the Lower Cretaceous of Baisa, East Siberia. Part 5. Mantispidae. *Russian Entomological Journal* 5: 91–3.

———. 1998. New Tertiary Neuroptera (Insecta) from the Russian Far East. *Tertiary Research* 18: 77–83.

Malaise, R. 1941. Gattungstabelle der Blattwespen (Hym. Tenth.) der Welt. *Entomologisk Tidskrift* 62: 131–140.

Malicky, H. 1984. Ein Beitrag zur Autökologie und Bionomie der aquatischen Netzflüglergattung *Neurorthus* [sic] (Insecta, Neuroptera, Neurorthidae). *Archiv für Hydrobiologie* 101: 231–46.

Mallet, J. 1993. Speciation, raciation, and color pattern evolution in *Heliconius* butterflies: Evidence from hybrid zones. Pp. 226–60. *In* Harrison, R. G. (ed.), *Hybrid Zones and the Evolutionary Process.* Oxford University Press; Oxford, UK; x+364 pp.

Mallet, J. B. L., and M. Joron. 2000. Evolution of diversity in warning color and mimicry: Polymorphisms, shifting balance, and speciation. *Annual Review of Ecology and Systematics* 30: 201–33.

Mallis, A. 1971. *American Entomologists*. Rutgers University Press; New Brunswick, New Jersey; xvii+549 pp.

Malyshev, S. I. 1935. The nesting habits of solitary bees. *Eos* 11: 201–309.

———. 1964. A comparative study of the life and development of primitive gasteruptiids (Hymenoptera, Gasteruptiidae). *Entomological Review* 43: 267–71.

———. 1966. *Genesis of the Hymenoptera and the Phases of Their Evolution*. Nauka; Moscow, Russia; 328+[1] pp. [In Russian]

———. 1968. *Genesis of the Hymenoptera and the Phases of Their Evolution*. Methuen; London, UK; viii+319 pp.

Mani, M. S. 1964. *Ecology of Plant Galls*. Junk; The Hague; the Netherlands; xii+434 pp.

Manley, D. G., and G. O. Poinar, Jr. 1991. A new species of fossil *Dasymutilla* (Hymenoptera: Mutillidae) from Dominican amber. *Pan-Pacific Entomologist* 67: 200–5.

———, and G. O. Poinar, Jr. 1999. A second species of fossil *Dasymutilla* (Hymenoptera: Mutillidae) from Dominican amber. *Pan-Pacific Entomologist* 75: 48–51.

Manning, J. C., and P. Goldblatt. 1997. The *Moegistorhynchus longirostris* (Diptera: Nemestrinidae) pollination guild: Long-tubed flowers and a specialized long-proboscid fly pollination system in southern Africa. *Plant Systematics and Evolution* 206: 51–69.

Mansell, M. W. 1980. The Crocinae of southern Africa (Neuroptera: Nemopteridae). 1. The genera *Laurhervasia* Navás and *Thysanocroce* Withycombe. *Journal of the Entomological Society of Southern Africa* 43: 341–65.

———. 1981. The Crocinae of southern Africa (Neuroptera: Nemopteridae). 2. The genus *Concroce* Tjeder. *Journal of the Entomological Society of Southern Africa* 44: 91–106.

———. 1986. Biogeography and phylogeny of the Crocinae (Neuroptera: Nemopteridae). Pp. 77–85. *In* Gepp, J., H. Aspöck, and H. Hölzel (eds.), *Recent Research in Neuropterology: Proceedings of the Second International Symposium on Neuropterology*. Druckhaus Thalerhof; Graz, Austria; 176 pp.

———. 1992. The systematic position of Nemopteridae (Insecta: Neuroptera: Myrmeleontidae). Pp. 233–41. *In* Canard, M., H. Aspöck, and M. W. Mansell (eds.), *Current Research in Neuropterology: Proceedings of the Fourth International Symposium on Neuropterology*. SACCO; Toulouse, France; 414 pp.

Mant, J. G., F. P. Schiestl, R. Peakall, and P. H. Weston. 2002. A phylogenetic study of pollinator conservatism among sexually deceptive orchids. *Evolution* 56: 888–98.

Manton, S. M. 1964. Mandibular mechanisms and the evolution of arthropods. *Philosophical Transactions of the Royal Society, London (B)* 247: 1–183.

———. 1966. The evolution of arthropodan locomotory mechanisms. 9. Functional requirements and body design in Symphyla and Pauropoda and the relationships between Myriapoda and Pterygota. *Zoological Journal of the Linnean Society* 46: 103–41.

———. 1972. The evolution of arthropodan locomotory mechanisms. 10. Locomotory habits, morphology and evolution of the hexapod classes. *Zoological Journal of the Linnean Society* 51: 203–400.

———. 1973. Arthropod Phylogeny – A modern synthesis. *Journal of Zoology* 171: 111–30.

———. 1977. *The Arthropoda: Habits, Functional Morphology and Evolution*. Clarendon Press; Oxford, UK; xxii+527 pp.

———. 1979. Functional morphology and the evolution of the hexapod classes. Pp. 387–465. *In* Gupta, A. P. (ed.), *Arthropod Phylogeny*. Van Nostrand Reinhold; New York, New York; xx+762 pp.

Maples, C. G. 1989. *Paolia vetusta* Smith, 1871 (Insecta: Protorthoptera), from the Mansfield Formation (Pennsylvanian), Indiana. *Journal of Paleontology* 63: 886–9.

Marchal-Papier, F. 1998. *Les Insects du Buntsandstein des Vosges (NE de la France). Biodiversité et Contribution aux Modalités de la Crise Biologique du Permo-Trias*. Ph.D. Dissertation, Université Louis Pasteur de Strasbourg; Strasbourg, France; 177+[60] pp.

———, A. Nel, and L. Grauvogel-Stamm. 2000. Nouveaux orthoptères (Ensifera, Insecta) du Trias des Vosges (France). *Acta Geologica Hispanica* 35: 5–18.

Marden, J. H. 2003. The surface-skimming hypothesis for the evolution of insect flight. *Acta Zoologica Cracoviensia* 46 (supplement): 73–84.

———, and M. G. Kramer. 1994. Surface-skimming stoneflies: A possible intermediate stage in insect flight evolution. *Science* 266: 427–30.

———, and M. G. Kramer. 1995. Locomotor performance of insects with rudimentary wings. *Nature* 377: 332–4.

Mardulyn, P., and S. A. Cameron. 1999. The major opsin in bees (Insecta: Hymenoptera): A promising nuclear gene for higher level phylogenetics. *Molecular Phylogenetics and Evolution* 12: 168–76.

Mari-Mutt, J. A. 1983. Collembola in amber from the Dominican Republic. *Proceedings of the Entomological Society of Washington* 85: 575–87.

———, and P. F. Bellinger. 1990. A catalog of the Neotropical Collembola, including Nearctic areas of Mexico. *Flora and Fauna Handbook* 5: 1–237.

Marshall, A. G. 1981. *The Ecology of Ectoparasitic Insects*. Academic Press; London, UK; xvi+459 pp.

Marshall, I. K. 2003. A morphological phylogeny for four families of amblyceran lice (Phthiraptera: Amblycera: Menoponidae, Boopiidae, Laemobothriidae, Ricinidae). *Zoological Journal of the Linnean Society* 138: 39–82.

Martill, D. M. 1988. Preservation of fish in the Cretaceous Santana Formation of Brazil. *Palaeontology* 31: 1–18.

———. 1990. Macromolecular resolution of fossilized muscle from an elopomorph fish. *Nature* 346: 171–2.

Martin, J., V. Guryev, S. S. MacDonald, A. Blinov, and D. H. D. Edward. In press. Phylogenetic relationships of *Archaeochlus* Brundin, *Austrochlus* Cranston, and *Afrochlus* Freeman (Diptera: Chironomidae), basal genera with a Gondwanan connection. *Cimbebasia*.

Martin, J. W., and G. E. Davis. 2001. An updated classification of the Recent Crustacea. *Science Series, Natural History Museum of Los Angeles County* 39: 1–124.

Martin, R. E. 1999. *Taphonomy: A Process Approach*. Cambridge University Press; Cambridge, UK; xvi+508 pp.

Martínez-Delclòs, X. 1989. Insectos del Cretácico inferior de las Hoyas (Cuenca). Pp. 51–82. *In*, *La Fauna del Pasado en Cuenca*. Actas del I Curso de Paleontología; Cuenca, Spain.

———. 1991. *Les Calcàries Litogràfiques del Cretacico Inferior del Montsec. Deu Anys de Campanyes Paleontològiqies*. Instituta d'Estudis Ilerdencs; Lleida, Spain; 160 pp.

———. 1993. Blátidos (Insecta, Blattodea) del Cretácico Inferior de España. Familias Mesoblattinidae, Blattulidae y Poliphagidae. *Boletín Geológico y Minero* 104–5: 516–38.

———, and J. Martinell. 1995. The oldest known record of social insects. *Journal of Paleontology* 69: 594–9.

Martinez-Torres, C. Buades, A. Latorre, and A. Moya. 2001. Molecular systematics of aphids and their primary endosymbionts. *Molecular Systematics and Evolution* 20: 437–49.

———, D. E. G. Briggs, and E. Peñalver. 2004. Taphonomy of insects in carbonates and amber. *Palaeogeography, Palaeoclimatology, Palaeoecology* 203: 19–64.

Martins-Neto, R. G. 1989. Novos insetos terciários do Estado de São Paulo. *Revista Brasiliensis Geociencias* 19: 375–86.

———. 1993. Nova espécie de borboleta (Lepidoptera: Nymphalidae: Satyrinae) da Formação Tremembe, Oligoceno do Estado de São Paulo. *Acta Geologica Leopoldensia* 16: 5–16.

———. 1997. Neurópteros (Insecta, Planipennia) da Formação Santana (Cretáceo Inferior), Bacia do Araripe, nordeste do Brasil. X-descrição de novos taxa (Chrysopidae, Babinskaiidae, Myrmeleontidae, Ascalaphidae e Psychopsidae). *Revista Universidade Guarulhos, série Ciências Exatas e Tecnológicas* 2: 68–83.

———. 1998. A paleoentomofauna de Formação Tremembe (Bacia do Taubaté) Oligoceno do Estado de São Paulo: Descripção de novos lepidópteros (Insecta). *Acta Geologica Leopoldensia* 21: 75–82.

———. 1999. Estado actual del conocimiento de la paleoentomofauna brasileña. *Revista de la Sociedad Entomológica Argentina* 58: 71–85.

———. 2000. Remarks on the neuropterofauna (Insecta, Neuroptera) from the Brazilian Cretaceous, with keys for the identification of the known taxa. *Acta Geologica Hispanica* 35: 97–118.

———, and O. F. Gallego. 1999. The Triassic insect fauna from Argentina. 1. Auchenorrhyncha, Miomoptera and Ensifera. *Revista Española de Paleontología* 14: 191–202.

———, and M. A. Vulcano. 1989. Amphiesmenoptera (Trichoptera + Lepidoptera) na Formação Santana (Cretáceo Inferior) Bacia do Araripe, Nordeste do Brasil. I – Lepidoptera (Insecta). *Anais da Academia Brasileira de Ciencias* 61: 459–66.

———, J. C. Kucera-Santos, F. R. M. Viera, and L. M. C. Fragoso. 1993. Nova espécie de

Borboleta (Lepidoptera: Nymphalidae: Satyrinae) da Formaçao Tremembé, Oligoceno do Estado de São Paulo. *Acta Geologica Leopoldensia* 16: 5–16.

———, O. F. Gallego, and R. N. Melchor. 2003. The Triassic insect fauna from South America (Argentina, Brazil and Chile): A checklist (except Blattoptera and Coleoptera) and descriptions of new taxa. *Acta Zoologica Cracoviensia* 46 (Supplement): 229–56.

Martius, C., R. Wassmann, U. Thein, A. Bandeira, H. Rennenberg *et al.* 1993. Methane emission from wood-feeding termites in Amazonia. *Chemosphere* 26: 623–32.

Martynov, A. V. 1923. On two basic types of insect wings and their significance for the general classification of insects. Pp. 88–9. *In* Deryugina, K. M. (ed.), *Trudy Pervogo Vserossijskogo S'exda Zoologov, Anatomov i Gistologov*. Nauka; Moscow, Russia; 507 pp. [In Russian]

———. 1924. *Caddisflies (Trichoptera)*. Practical entomology, Part 5; Leningrad, Union of Soviet Socialist Republics; iv+388 pp. [In Russian]

———. 1925a. Über zwei Grundtypen der Flügel bei den Insecten und ihre Evolution. *Zeitschrift für Morphologie und Ökologie der Tiere* 4: 465–501.

———. 1925b. To the knowledge of fossil insects from Jurassic beds in Turkestan. 1. Raphidioptera. *Izvestiya Rossiiskoi Akademii Nauk, 6 seriya* 19: 233–46.

———. 1925c. To the knowledge of fossil insects from Jurassic beds in Turkestan. 3. Hymenoptera, Mecoptera. *Izvestiya Rossiiskoi Akademii Nauk, 6 seriya* 19: 753–62.

———. 1928. A new fossil form of Phasmatodea from Galinko (Turkestan) and on Mesozoic phasmids in general. *Annals and Magazine of Natural History* 1: 319–28.

———. 1938. Essays on the geological history and phylogeny of the insect orders. 1. Palaeoptera and Neoptera-Polyneoptera. *Trudy Paleontologischeskogo Instituta Akademii Nauk, SSSR* 7: 1–150. [In Russian]

Martynova, O. M. 1958. New insects from the Permian and Mesozoic deposits of the USSR. *Materialy k Osnovam Palaeontologiy* 2: 69–94. [In Russian]

Marvaldi, A. E., and J. J. Morrone. 2000. Phylogenetic systematics of weevils (Coleoptera: Curculionoidea): A reappraisal based on larval and adult morphology. *Insect Systematics and Evolution* 31: 43–58.

———, A. S. Sequeira, C. W. O'Brien, and B. D. Farrell. 2002. Molecular and morphological phylogenetics of weevils (Coleoptera, Curculionoidea): Do niche shifts accompany diversification? *Systematic Biology* 51: 761–85.

Masner, L. 1969. A scelionid wasp surviving unchanged since Tertiary (Hymenoptera: Proctotrupoidea). *Proceedings of the Entomological Society of Washington* 71: 397–400.

———. 1978. A revision of the New World species of *Leptoteleia* Kieffer (Hymenoptera: Scelionidae), egg parasites of crickets. *Canadian Entomologist* 110: 353–80.

———. 1993a. Superfamily Proctotrupoidea. Pp. 537–57. *In* Goulet, H., and J. T. Huber (eds.), *Hymenoptera of the World: An Identification Guide to Families*. Agriculture Canada; Ottawa, Canada; vii+668 pp.

———. 1993b. Superfamily Platygastroidea. Pp. 558–65. *In* Goulet, H., and J. T. Huber (eds.), *Hymenoptera of the World: An Identification Guide to Families*. Agriculture Canada; Ottawa, Canada; vii+668 pp.

———. 1993c. Superfamily Ceraphronoidea. Pp. 566–9. *In* Goulet, H., and J. T. Huber (eds.), *Hymenoptera of the World: An Identification Guide to Families*. Agriculture Canada; Ottawa, Canada; vii+668 pp.

———, and P. Dessart. 1967. La reclassification des catégories taxonomiques supérieures des Ceraphronoidea (Hymenoptera). *Institut Royale des Science Naturelles de Belgique* 43: 1–33.

Mason, W. R. M. 1981. Paxylommatidae: The correct family-group name for *Hybrizon* Fallén (Hymenoptera: Ichneumonoidea), with figures of unusual antennal sensilla. *Proceedings of the Entomological Society of Washington* 113: 433–9.

———. 1987. Discovery of the female *Apozyx* (Hymenoptera: Apozygidae) and comments on its phylogenetic position. *Proceedings of the Entomological Society of Washington* 89: 226–9.

———. 1993. Superfamilies Evanioidea, Stephanoidea, Megalyroidea, and Trigonalyoidea. Pp. 510–20. *In* Goulet, H., and J. T. Huber (eds.), *Hymenoptera of the World: An Identification Guide to Families*. Agriculture Canada; Ottawa, Canada; vii+668 pp.

Masselot, G., and A. Nel. 1999. *Pseudokageronia thomasi* gen. nov., sp. nov. from the Upper Miocene of Murat (France) (Ephemeroptera: Heptageniidae). *Ephemera* 1: 61–73.

Massoud, Z. 1967. Contribution à l'étude de *Rhyniella praecursor* Hirst *et* Maulik 1926, collembole fossile du Dévonien. *Revue d'Ecologie et de Biologie du Sol* 4: 497–505.

Mathews, S., and M. J. Donoghue. 1999. The root of angiosperm phylogeny inferred from duplicated phytochrome genes. *Science* 286: 947–50.

Matile, L. 1990. Recherches sur la systématique et l'evolution des Keroplatidae (Diptera, Mycetophiloidea). *Mémoires Museum National d'Histoire Naturelle* 148: 1–682.

Matsuda, R. 1965. Morphology and evolution of the insect head. *Memoirs of the American Entomological Institute* 4: 1–334.

———. 1970. Morphology and evolution of the insect thorax. *Memoirs of the Canadian Entomologist* 76: 1–431.

———. 1976. *Morphology and Evolution of the Insect Abdomen, with Special Reference to Developmental Patterns and their Bearings upon Systematics*. Pergamon Press; New York, New York; xiii+534 pp.

———. 1981. The origin of insect wings (Arthropoda: Insecta). *International Journal of Insect Morphology and Embryology* 10: 387–98.

Matsumura, F. 1985. *Toxicology of Insecticides* [2nd Edition]. Plenum Press; New York, New York; xx+598 pp.

Matsuura, M., and S. Yamane. 1990. *Biology of the Vespine Wasps*. Springer Verlag; Berlin, Germany; xx+323 pp.

Matthews, J. V., Jr. 1970. Two new species of *Micropeplus* from the Pliocene of western Alaska with remarks on the evolution of Micropeplinae (Coleoptera: Staphylinidae). *Canadian Journal of Zoology* 48: 779–88.

Maxwell, M. R. 1999. Mating behavior. Pp. 69–92. *In* Prete, F. R., H. Wells, P. H. Wells, and L. E. Hurd (eds.), *The Praying Mantids*. Johns Hopkins University Press; Baltimore, Maryland; xiv+362 pp.

May, R. M., M. Maria, and J. Guimand. 1964. Action différentielle des rayon X et ultraviolets sur le Tardigrade *Macrobiotus areolatus* à l'état actif et desséché. *Bulletin Biologique de la France et de la Belgique* 48: 349–67.

May, R. M. 1988. How many species are there on earth? *Science* 241: 1441–9.

———. 1992. How many species inhabit the earth? *Scientific American* 267: 42–8.

Mayr, E. 1942. *Systematics and the Origin of Species from the Viewpoint of a Zoologist*. Columbia University Press; New York, New York; xiv+334 pp.

———. 1963. *Animal Species and Evolution*. Harvard University Press; Cambridge, Massachusetts; xiv+797 pp.

Mayr, G. 2000. Die Vögel der Grube Messel – Ein Einblick in die Vogelwelt Mitteleuropas vor 49 Millionen Jahren. *Natur und Museum* 130: 365–78.

———. 2001. Comments on the osteology of *Masillapodargus longipes* Mayr 1999 and *Paraprefica major* Mayr 1999, caprimulgiform birds from the middle Eocene of Messel (Hassen, Germany). *Neues Jahrbuch für Geologie und Paläontologie, Monatshefte* 2001(2): 65–76.

———. 2002a. A new specimen of *Salmila robusta* (Aves: Gruiformes: Salmilidae n. fam.) from the middle Eocene of Messel. *Paläontologische Zeitschrift* 76: 305–16.

———. 2002b. A new species of *Plesiocathartes* (Aves: ?Leptosomidae) from the middle Eocene of Messel, Germany. *PaleoBios* 22: 10–20.

Mazzarolo, L. A., and D. S. Amorim. 2000. *Cratomyia macrorrhyncha*, a Lower Cretaceous brachycerean fossil from the Santana Formation, Brazil, representing a new species, genus and family of the Stratiomyomorpha (Diptera). *Insect Systematics and Evolution* 31: 91–102.

Mazzini, M., and V. Scali, eds. 1987. *Stick Insects: Phylogeny and Reproduction*. University of Siena; Siena, Italy; 224 pp.

McAlpine, J. F. 1970. First record of calypterate flies in the Mesozoic era (Diptera: Calliphoridae). *Canadian Entomologist* 102: 342–6.

———. 1989. Phylogeny and classification of the Muscomorpha. Pp. 1397–1518. *In* McAlpine, J. F., and D. M. Wood (eds.), *Manual of Nearctic Diptera, Volume 3*. Agriculture Canada Monograph 32; Ottawa, Canada; vi+1333–1581 pp.

———, and J. E. H. Martin. 1969. Canadian amber: A paleontological treasure chest. *Canadian Entomologist* 101: 819–38.

———, and D. M. Wood, eds. 1989. *Manual of Nearctic Diptera, Volume 3*. Agriculture Canada Monograph 32; Ottawa, Canada; vi+1333–1581 pp.

———, B. V. Peterson, G. E. Shewell, H. J. Teskey, J. R. Vockeroth, and D. M. Wood, eds. 1981. *Manual of Nearctic Diptera, Volume 1*. Agriculture Canada Monograph 27; Ottawa, Canada; vi+674 pp.

———, B. V. Peterson, G. E. Shewell, H. J. Teskey, J. R. Vockeroth, and D. M. Wood, eds. 1987. *Manual of Nearctic Diptera, Volume*

2. Agriculture Canada Monograph 28; Ottawa, Canada; xi+675–1332 pp.

McCafferty, W. P. 1987. New fossil mayfly in amber and its relationships among extant Ephemeridae (Ephemeroptera). *Annals of the Entomological Society of America* 80: 472–4.

———. 1990. Ephemeroptera. *Bulletin of the American Museum of Natural History* 195: 20–50.

———. 1991. Toward a phylogenetic classification of the Ephemeroptera (Insecta): A commentary on systematics. *Annals of the Entomological Society of America* 84: 343–60.

———. 1997a. Ephemeroptera. Pp. 20–50. *In* Poole, R. W., and P. Gentili (eds.), *Nomina Insecta Nearctica, A Checklist of the Insects of North America. Volume 4: Non-Holometabolous Orders*. Entomological Information Services; Rockville, Maryland; 731 pp.

———. 1997b. Discovery and analysis of the oldest mayflies (Insecta, Ephemeroptera) known from amber. *Bulletin de la Société d'Histoire Naturelle de Toulouse* 133: 77–82.

———, and Y. J. Bae. 1992. Filter-feeding habits of the larvae of *Anthopotamus* (Ephemeroptera: Potamanthidae). *Annales de Limnologie* 28: 27–34.

———, and G. F. Edmunds, Jr. 1979. The higher classification of the Ephemeroptera and its evolutionary basis. *Annals of the Entomological Society of America* 72: 5–12.

———, and V. P. Provonsha. 1986. Comparative mouthpart morphology and evolution of the carnivorous Heptageniidae (Ephemeroptera). *Aquatic Insects* 8: 83–9.

———, and N. D. Sinitshenkova. 1983. *Litobrancha* from the Oligocene in eastern Asia (Ephemeroptera: Ephemeridae). *Annals of the Entomological Society of America* 76: 205–8.

———, and T.-Q. Wang. 2000. Phylogenetic systematics of the major lineages of pannote mayflies (Ephemeroptera: Pannota). *Transactions of the American Entomological Society* 126: 9–101.

McCobb, L. M. E., I. J. Duncan, E. A. Jarzembowski, B. A. Stankiewicz, M. A. Wills, and D. E. G. Briggs. 1998. Taphonomy of the insects from the Insect Bed (Bembridge Marls), late Eocene, Isle of Wight, England. *Geological Magazine* 135: 553–63.

McCormick, J. P., and J. E. Carrel. 1997. Cantharidin biosynthesis and function in meloid beetles. Pp. 307–50. *In* Prestwich, G. D., and G. J. Blomquist (eds.), *Pheromone Biochemistry*. Academic Press; London, UK; 367 pp.

McEvoy, M. V., and B. A. Underwood. 1988. The drone and species status of the Himalayan honey bee, *Apis laboriosa* (Hymenoptera: Apidae). *Journal of the Kansas Entomological Society* 61: 246–9.

McEvoy, P. B. 2002. Insect-plant interactions on a planet of weeds. *Entomologia Experimentalis et Applicata* 104: 165–79.

McIver, J. D., and G. Stonedahl. 1993. Myrmecomorphy: Morphological and behavioral mimicry of ants. *Annual Review of Entomology* 38: 351–79.

McKay, I. J., and R. J. Rayner. 1986. Cretaceous fossil insects from Orapa, Botswana. *Journal of the Entomological Society of Southern Africa* 49: 7–17.

McKenna, M. C., and S. K. Bell. 1997. *Classification of Mammals Above the Species Level*. Columbia University Press; New York, New York; xii+631 pp.

McKenzie, H. L. 1967. *Mealybugs of California, with Taxonomy, Biology, and Control of North American Species (Homoptera, Coccoidea, Pseudococcidae)*. University of California Press; Berkeley, California; viii+525 pp.

McKenzie, J. A. 1996. *Ecological and Evolutionary Aspects of Insecticide Resistance*. R. G. Landes; Austin, Texas; 185 pp.

McKittrick, F. A. 1964. Evolutionary studies of cockroaches. *Cornell University Agricultural Experiment Station Memoir* 389: 1–197.

Medvedev, S. G. 1994. Morphological basis of the classification of fleas. *Entomological Review* 73: 30–51

———. 1998. Classification of the flea order and its theoretical foundations. *Entomological Review* 77: 1080–93.

Medway, L. 1958. On the habit of *Arixenia esau* Jordan (Dermaptera). *Proceedings of the Entomological Society of London (A)* 33: 191–5.

Meeuse, A. D. J., A. H. de Meijer, O. W. P. Mohr, and S. M. Wellinga. 1990. Entomophily in the dioecious gymnosperm *Ephedra aphylla* Fork. (= *E. alta* C. A. Mey), with some notes on *Ephedra campylopoda* C. A. Mey. III. Further anthecological studies and relative importance of entomophily. *Israel Journal of Botany* 39: 113–23.

Meeuse, B. J. D. 1961. *The Story of Pollination*. Ronald Press; New York, New York; 243 pp.

Meinander, M. 1972. A revision of the family Coniopterygidae (Planipennia). *Acta Zoologica Fennica* 136: 1–357.

———. 1975. Fossil Coniopterygidae (Neuroptera). *Notulae Entomologicae* 55: 53–7.

———. 1979. The phylogeny and geographical distribution of the Aleuropteryginae (Neuroptera, Coniopterygidae). *Annales Entomologici Fennici* 45: 16–23.

———. 1998. Coniopterygidae (Neuroptera) in amber from the Dominican Republic. *Journal of Neuropterology* 1: 33–6.

Melander, A. L. 1949. A report on some Miocene Diptera from Florissant, Colorado. *American Museum Novitates* 1407: 1–63.

Mellor, P. S., J. Boorman, and M. Baylis. 2000. *Culicoides* biting midges: Their role as arbovirus vectors. *Annual Review of Entomology* 45: 307–40.

Melo, G. A. R. 1999. Phylogenetic relationships and classification of the major lineages of Apoidea (Hymenoptera), with emphasis on the crabronid wasps. *Scientific Papers, Natural History Museum, University of Kansas* 14: 1–55.

———. 2000. Biology of an extant species of the scolebythid genus *Dominibythus* (Hymenoptera: Chrysidoidea: Scolebythidae), with description of its mature larva. Pp. 281–4. *In* Austin, A. D., and M. Dowton (eds.), *Hymenoptera: Evolution, Biodiversity and Biological Control*. CSIRO; Collingwood, Australia; xi+468 pp.

———, and L. A. O. Campos. 1993. Nesting biology of *Microstigmus myersi* Turner, a wasp with long-haired larvae (Hymenoptera: Sphecidae, Pemphredoninae). *Journal of Hymenoptera Research* 2: 183–8.

Melville, R. V. 1995. *Towards Stability in the Names of Animals: A History of the International Commission on Zoological Nomenclature 1895–1995*. International Trust for Zoological Nomenclature; London, U K; viii+ [4]+92 pp.

Mendes, L. F. 1990. An annotated list of generic and specific names of Machilidae (Microcoryphia, Insecta) with identification keys to genera and geographical notes. *Estudios, Ensaios e Documentos (Lisboa)* 155: 1–127.

———. 1991. On the phylogeny of the genera of Lepismatidae (Insecta: Zygentoma). Pp. 1–13. *In* Veeresh, G. K., J. Rajagopal, and C. Viraktamath (eds.), *Advances in Management and Conservation of Soil Fauna*. IBH Publishing; New Delhi, India; xvi+925 pp.

———. 1994. Evolutionary relationships among the Nicoletiidae (Insecta, Zygentoma). *Acta Zoologica Fennnica* 195: 98–103.

———. 1997. First contribution to the study of the Dominican amber Zygentoma (Insecta): Family Ateluridae. *Pedobiologia* 41: 40–3.

———. 1998. Second contribution to the study of the Dominican Republic amber Zygentoma (Insecta): Family Lepismatidae. *Canadian Entomologist* 130: 899–904.

———. 2002. Taxonomy of Zygentoma and Microcoryphia: Historical overview, present status and goals for the new millennium. *Pedobiologia* 46: 225–33.

Menke, A. S., and A. P. Rasnitsyn. 1987. Affinities of the fossil wasp, *Hoplisidea kohliana* Cockerell (Hymenoptera: Sphecidae: Sphecinae). *Psyche* 94: 35–8.

Merian, M. S. 1705. *Metamorphosis Insectorum Surinamensium: In Qua Erucae ac Vermes Surinamenses, cum Omnibus suis Transformationibus, ad Vivum Delineatur & Describuntur, Singulis Eorum in Plantas, Foores & Fructus Collocatis, in Quibus Reperta sunt . . .* Sumptibus Auctoris; Amstelodami [Amsterdam]; the Netherlands; [2]+60 pp., 60 pls.

Merrit, R. W., and K. W. Cummins. 1978. *An Introduction to the Aquatic Insects of North America*. Kendall-Hunt; Dubuque, Iowa; xiii+441 pp.

Mertz, D. F., C. C. Swisher, J. L. Franzen, F.-O. Neuffer, and H. Lutz. 2000. Numerical dating of the Eckfeld maar fossil site, Eifel, Germany: A calibration mark for the Eocene time scale. *Naturwissenschaften* 87: 270–4.

Meunier, M. F. 1917. Un Pompilidae de l'ambre de la Baltique. *Tijdschrift voor Entomologie* 60: 181–4.

Meyer, H. W. 2002. *The Fossils of Florissant*. Smithsonian Books; Washington, D.C.; xiii+[i]+258 pp.

Meyer, J. 1987. *Plant Galls and Gall Inducers*. Gebruder Bornträger; Berlin, Germany; viii+291 pp.

Michelsen, V. 1996. Neodiptera: New insights into the adult morphology and higher level phylogeny of Diptera (Insecta). *Zoological Journal of the Linnean Society* 117: 71–102.

———. 1996/7. A revised interpretation of the mouthparts in adult fleas (Insecta, Siphonaptera). *Zoologischer Anzeiger* 235: 217–23.

———. 1999. Wood gnats of the genus *Sylvicola* (Diptera, Anisopodidae): Taxonomic status, family assignment, and review of nominal species described by J. C. Fabricius. *Tijdschrift voor Entomologie* 142: 69–75.

———. 2000. Oldest authentic record of a fossil calyptrate fly (Diptera): A species of Anthomyiidae from early Coenozoic Baltic amber. *Studia Dipterologica* 7: 11–18.

Michener, C. D. 1944. Comparative external morphology, phylogeny, and a classification of the bees (Hymenoptera). *Bulletin of the American Museum of Natural History* 82: 151–326.

———. 1952. The Saturniidae (Lepidoptera) of the Western Hemisphere: Morphology, phylogeny and classification. *Bulletin of the American Museum of Natural History* 98: 335–501.

———. 1965. A classification of the bees of the Australian and South Pacific regions. *Bulletin of the American Museum of Natural History* 130: 1–362.

———. 1974. *The Social Behavior of the Bees: A Comparative Study.* Harvard University Press; Cambridge, Massachusetts; xii+404 pp.

———. 1979. Biogeography of the bees. *Annals of the Missouri Botanical Garden* 66: 277–347.

———. 1990. Reproduction and castes in social halictine bees. Pp. 77–121. *In* Engels, W. (ed.), *Social Insects: An Evolutionary Approach to Castes and Reproduction.* Springer Verlag; Berlin, Germany; 265 pp.

———. 2000. *The Bees of the World.* Johns Hopkins University Press; Baltimore, Maryland; xiv+[1]+913 pp.

———, and L. Greenberg. 1980. Ctenoplectridae and the origin of long-tongued bees. *Zoological Journal of the Linnean Society* 69: 183–203.

———, and D. Grimaldi. 1988a. A *Trigona* from Late Cretaceous amber of New Jersey (Hymenoptera: Apidae: Meliponinae). *American Museum Novitates* 2917: 1–10.

———, and D. Grimaldi. 1988b. The oldest fossil bee: Apoid history, evolutionary stasis, and antiquity of social behavior. *Proceedings of the National Academy of Sciences, U.S.A.* 85: 6424–6.

———, and R. R. Sokal. 1957. A quantitative approach to a problem in classification. *Evolution* 11: 130–62.

Mickel, C. E. 1928. Biological and taxonomic investigations on the mutillid wasps. *Bulletin of the United States National Museum* 143: 1–351.

———. 1967. A review of the mutillid genus *Chyphotes* Blake (Hymenoptera; Mutillidae; Apterogyninae). *Transactions of the American Entomological Society* 93: 125–234.

Mickoleit, E. 1963. Untersuchungen zur Kopfmorphologie der Thysanopteren. *Zoologische Jahrbücher, Abteilung für Anatomie und Ontogenie der Tiere* 81: 101–50.

Mickoleit, G. 1971. Das Exoskelett von *Notiothauma reedi* MacLachlan, ein Beitrag zur Morphologie und Phylogenie der Mecoptera (Insecta). *Zeitschrift für Morphologie der Tiere* 69: 318–62.

———. 1973. Über den Ovipositor der Neuropteroidea und Coleoptera und seine phylogenetische Bedeutung (Insecta, Holometabola). *Zeitschrift fur Morphologie der Tiere* 74: 37–64.

———. 1975. Die Genital- und Postgenitalsegmente der Mecoptera-Weibchen (Insecta, Holometabola). 1. Das Exoskeletttt. *Zeitschrift für Morphologie der Tiere* 80: 97–135.

———. 1976. Die Genital- und Postgenitalsegmente der Mecoptera-Weibchen (Insecta, Holometabola). Das Dach der Genitalkammer. *Zoomorphologie* 85: 133–56.

———. 1978. Die phylogenetischen Beziehungen der Schnabelfliegen-Familien aufgrund morphologischer Ausprägungen der weiblichen Genital- und Postgenitalsegmente (Mecoptera). *Entomologica Germanica* 4: 258–71.

Middlekauff, W. W. 1974. Larva of wood-boring sawfly *Syntexis libocedrii* Rohwer (Hymenoptera: Syntexidae). *Pan-Pacific Entomologist* 50: 288–90.

Midgley, J. J., and W. J. Bond. 1991. How important is biotic pollination and dispersal to the success of the angiosperms? *Philosophical Transactions of the Royal Society, London (B)* 333: 209–15.

Mierzejewski, P. 1976. On application of scanning electron microscope to the study of organic inclusions from the Baltic amber. *Annals of the Geological Society of Poland* 46: 291–5.

Miller, B. R., M. B. Crabtree, and H. M. Savage. 1997. Phylogenetic relationships of the Culicomorpha inferred from 18S and 5.8S ribosomal DNA sequences (Diptera: Nematocera). *Insect Molecular Biology* 6: 105–14.

Miller, D. R. 1984. Phylogeny and classification of the Margarodidae and related groups (Homoptera: Coccoidea). *Verhandlungen des Zehnten Internationalen Symposiums über Entomofaunistik Mitteleuropas (SIEEC X), 15–20 August, 1983, Budapest* 1984: 321–4.

Miller, H. K., and F. W. Fisk. 1971. Taxonomic implications of the comparative morphology of cockroach proventriculi. *Annals of the Entomological Society of America* 64: 671–87.

Miller, J. S. 1987a. Host-plant relationships in the Papilionidae (Lepidoptera): Parallel cladogenesis or colonization? *Cladistics* 3: 105–20.

———. 1987b. Phylogenetic studies in the Papilioninae (Lepidoptera: Papilionidae). *Bulletin of the American Museum of Natural History* 186: 365–512.

———. 1991. Cladistics and classification of the Notodontidae (Lepidoptera: Noctuoidea) based on larval and adult morphology. *Bulletin of the American Museum of Natural History* 204: 1–230.

———, and J. W. Wenzel. 1995. Ecological characters and phylogeny. *Annual Review of Entomology* 40: 389–415.

Miller, J. Y., and F. M. Brown. 1989. A new Oligocene fossil butterfly, *Vanessa amerindica* (Lepidoptera: Nymphalidae) from the Florissant Formation, Colorado. *Bulletin of the Allyn Museum* 126: 1–9.

Miller, K. B. 2001. On the phylogeny of the Dytiscidae (Insecta: Coleoptera) with emphasis on the morphology of the female reproductive system. *Insect Systematics and Evolution* 32: 45–92.

Miller, N. C. E. 1956. *The Biology of the Heteroptera.* Leonard Hill Books; London, UK; 162 pp.

Miller, S. E. 1983. Late Quaternary insects of Ranco La Brea and McKittrick, California. *Quaternary Research* 20: 90–104.

Miller, T. D. 1986. New species of the genus *Sierolomorpha* from New Mexico (Hymenoptera: Sierolomorphidae). *Pan-Pacific Entomologist* 62: 121–3.

Miller, W. R., J. D. Miller, and H. Heatwole. 1996. Tardigrades of the Australian Antarctic Territories: The Windmill Islands, East Antarctica. *Zoological Journal of the Linnean Society* 116: 175–84.

Mills, H. B., and J. H. Pepper. 1937. Observations on *Grylloblatta campodeiformis* Walker. *Annals of the Entomological Society of America* 30: 269–74.

Mills, J. S., R. White, and L. J. Gough. 1984. The chemical composition of Baltic amber. *Chemical Geology* 47: 15–39.

Milne, L. J., and M. Milne. 1976. The social behavior of burying beetles. *Scientific American* 235: 84–9.

Minet, J. 1983. Etude morphologique et phylogénétique des organs tympaniques des Pyraloidea. I. – Généralités et homologies (Lepidoptera, Glossata). *Annales de la Société d'Entomologique de France* 19: 175–207.

———. 1991. Tentative reconstruction of the ditrysian phylogeny (Lepidoptera: Glossata). *Entomologica Scandinavica* 22: 69–95.

———, and T. Bourgoin. 1986. Phylogenie et classification des Hexapodes (Arthropoda). *Cahiers Liaison, O.P.I.E.* 20: 23–8.

———, and M. J. Scoble. 1999. The drepanoid/geometroid assemblage. Pp. 301–20. *In* Kristensen, N. P. (ed.), *Handbuch der Zoologie: Eine Naturgeschichte der Stämme des Tierreiches: Band IV: Arthropoda: Insecta: Teilband 35: Lepidoptera, Moths and Butterflies: Volume 1: Evolution, Systematics, and Biogeography.* Walter de Gruyter; Berlin, Germany; x+491 pp.

Mingram, J. 1998. Laminated Eocene maar-like sediments from Eckfeld (Eifel region, Germany) and their short-term periodicities. *Palaeogeography, Palaeoclimatology, Palaeoecology* 140: 289–305.

Mirmoaydi, A., and A. Yassayie. 1999. *Dilar golestani* sp. n. (Neuroptera: Dilaridae) from Iran. *Journal of the Entomological Society of Iran* 18: 9–15.

Mitchell, A., C. Mitter, and J. C. Regier. 2000. More taxa or more characters revisited: Combining data from nuclear protein-encoding genes for phylogenetic analyses of Noctuoidea (Insecta: Lepidoptera). *Systematic Biology* 49: 202–24.

———, F. A. H. Sperling, and D. A. Hickey. 2002. Higher-level phylogeny of mosquitoes (Diptera: Culicidae): mtDNA data support a derived placement for *Toxorhynchites. Insect Systematics and Evolution* 33: 163–74.

Mitchell, P., and D. Wighton. 1979. Larval and adult insects from the Paleocene of Alberta, Canada. *Canadian Entomologist* 111: 777–82.

Mitter, C., B. Farrell, and B. Wiegmann. 1998. The phylogenetic study of adaptive zones: Has phytophagy promoted insect diversification? *American Naturalist* 132: 107–28.

———, B. Farrell, and D. J. Futuyma. 1991. Phylogenetic studies of insect-plant interactions: Insights into the genesis of diversity. *Trends in Ecology and Evolution* 6: 271–306.

Miura, T., K. Maekawa, O. Kitade, T. Abe, and T. Matsumoto. 1998. Phylogenetic relationships among subfamilies in higher termites (Isoptera: Termitidae) based on mitochondrial COII gene sequences. *Annals of the Entomological Society of America* 91: 515–23.

Miyamoto, M. M., C. A. Porter, and M. Goodman. 2000. c-Myc gene sequences and the phylogeny of bats and other eutherians. *Systematic Biology* 49: 501–14.

Mockford, E. L. 1967. Some Psocoptera from the plumage of birds. *Proceedings of the Entomological Society of Washington* 69: 307–9.

———. 1971a. Psocoptera from sleeping nests of the dusky-faced wood rat in southern California (Psocoptera: Atropidae; Psoquillidae, Liposcelidae). *Pan-Pacific Entomologist* 47: 127–40.

———. 1971b. Parthenogenesis in psocids (Insecta: Psocoptera). *American Zoologist* 11: 327–39.

———. 1972. New species, records, and synonymy of Florida *Belaphotroctes* (Psocoptera: Liposcelidae). *Florida Entomologist* 55: 153–63.

———. 1987. Order Psocoptera. Pp. 196–214. *In* Stehr, F. W. (ed.), *Immature Insects, Volume 1*. Kendall Hunt; Dubuque, Iowa; xiv+754 pp.

———. 1993. North American Psocoptera. *Flora and Fauna Handbook* 10: xviii+1–455 pp.

Möhn, E. 1960. Eine neue Gallmücke aus der niederrheinischen Braunkohle, *Sequoiomyia krauseli* n. g., n. sp. (Diptera, Itonididae). *Senckenbergiana Lethaea* 41: 513–22.

Mohr, B., and E. M. Friis. 2000. Early angiosperms from the Aptian Crato Formation (Brazil): A preliminary report. *International Journal of Plant Sciences, Supplement* 161: S155–67.

Molumby, A. 1995. Dynamics of parasitism in the organ-pipe wasp, *Trypoxylon politum*: Effects of spatial scale on parasitoid functional response. *Ecological Entomology* 20: 159–68.

Momose, K., T. Yumoto, T. Nagamitsu, M. Kato, *et al.* 1998. Pollination biology in a lowland dipterocarp forest in Sarawak, Malaysia. 1. Characteristics of the plant-pollinator community in a lowland dipterocarp forest. *American Journal of Botany* 85: 1477–1501.

Monsserat, V. 1977. A systematic and alphabetical list of Neurorthidae [sic] and Sisyridae (Neuroptera). *Nouvelle Revue d'Entomologie* 7: 91–6.

Moore, B. P. 1979. Chemical defense in carabids and its bearing on phylogeny. Pp. 193–203. *In* Erwin, T. L., G. E. Ball, and D. R. Whitehead (eds.), *Carabid Beetles: Their Evolution, Natural History, and Classification*. Junk; The Hague, the Netherlands; ix+635 pp.

Moran, N. A., M. A. Munson, P. Baumann, and H. Ishikawa. 1993. A molecular clock in endosymbiotic bacteria is calibrated using the insect hosts. *Proceedings of the Royal Society of London, Series B, Biological Sciences* 253: 167–71.

Moran, V. C., and T. R. E. Southwood. 1982. The guild composition of arthropod communities in trees. *Journal of Animal Ecology* 51: 289–306.

Mori, Y. 2001. Insect fossils found from the Plio-Pleistocene Tokai Group. *Science Report of the Toyohashi Museum of Natural History* 11: 49–53. [In Japanese]

Moritz, G. 1991. Phylogenie der Thysanoptera (Insecta) aus ontogenetischer Sicht. *Mitteilungen aus dem Zoologischen Museum in Berlin* 67: 141–55.

Morris, D. C., and L. Mound. 2003. Thysanoptera phylogeny: The molecular future. *Entomologische Abhandlungen* 61: 153–5.

———, L. A. Mound, M. P. Schwarz, and B. J. Crespi. 1999. Morphological phylogenetics of Australian gall-inducing thrips and their allies: The evolution of host-plant affiliations, domicile use and social behavior. *Systematic Entomology* 24: 289–99.

———, L. A. Mound, and M. P. Schwarz. 2000. *Advenathrips inquilinus*: A new genus and species of social parasites (Thysanoptera: Phlaeothripidae). *Australian Journal of Entomology* 39: 53–7.

Morrissey, R., and J. S. Edwards. 1979. Neural function in an alpine grylloblattid: A comparison with the house cricket, *Acheta domesticus*. *Physiological Entomology* 4: 241–50.

Morse, J. C. 1997. Phylogeny of Trichoptera. *Annual Review of Entomology* 42: 427–50.

Mostovski, M. B. 1995. New taxa of ironomyiid flies (Diptera, Phoromorpha, Ironomyiidae) from Cretaceous deposits of Siberia and Mongolia. *Paleontological Journal* 30: 318–31.

———. 1996. New species of the genus *Mesosolva* Hong, 1983 (Diptera, Archisargidae) from the Jurassic of Kazakhstan and Mongolia. *Bulletin of the Museum of Northern Arizona* 60: 329–32.

———. 1999. A brief review of brachycerous flies (Diptera, Brachycera) in the Mesozoic, with descriptions of some curious taxa. Pp. 103–10. *In* Vršanský, P. (ed.), *Proceedings of the First Palaeoentomological Conference, Moscow, 1998*. AMBA Projects; Bratislava, Slovakia; 199+[2] pp.

———, and E. A. Jarzembowski. 2000. The first brachycerous flies (Diptera: Rhagionidae) from the Lower Jurassic of Gondwana. *Paleontological Journal, Supplement 3* 34: S367–9.

———, and X. Martínez-Delclòs. 2000. New Nemestrinoidea (Diptera: Brachycera) from the Upper Jurassic – Lower Cretaceous of Eurasia, taxonomy and palaeobiology. *Entomological Problems* 31: 137–48.

Moulton, J. K. 2000. Molecular sequence data resolves basal divergences within Simuliidae (Diptera). *Systematic Entomology* 25: 95–113.

Mound, L. A. 1971. The feeding apparatus of thrips. *Bulletin of Entomological Research* 60: 547–8.

———. 2003. Thysanoptera. Pp. 1127–32. *In* Resh, V. H., and R. T. Cardé (eds.), *Encyclopedia of Insects*. Academic Press; San Diego, California; xxviii+[1]+1266 pp.

———, and B. S. Heming. 1991. Thysanoptera (thrips). Pp. 510–15. *In* Naumann, I. D. (ed.), *The Insects of Australia: A Textbook for Students and Research Workers, Volume 1* [2nd Edition]. Cornell University Press; Ithaca, New York; xvi+[1]+542 pp.

———, and R. Marullo. 1998. Two new basal-clade Thysanoptera from California with Old World affinities. *Journal of the New York Entomological Society* 106: 81–94.

———, and D. C. Morris. 1999. Abdominal armature and the systematics of *Xaniothrips* species (Thysanoptera: Phlaeothripidae), kleptoparasites of domicile-producing thrips on Australian *Acacia* trees. *Australian Journal of Entomology* 38: 170–88.

———, and K. O'Neill. 1974. Taxonomy of the Merothripidae, with ecological and phylogenetic considerations (Thysanoptera). *Journal of Natural History* 8: 481–509.

———, B. S. Heming, and J. M. Palmer. 1980. Phylogenetic relationships between the families of recent Thysanoptera (Insecta). *Zoological Journal of the Linnean Society* 69: 111–41.

———, B. J. Crespi, and A. Tucker. 1998. Polymorphism and kleptoparasitism in thrips (Thysanoptera: Phlaeothripidae) from woody galls on *Casuarina* trees. *Australian Journal of Entomology* 37: 8–16.

Muche, H. 1981. Die Cephidae der Erde (Hym., Cephidae). *Deutsche Entomologische Zeitschrift* 28: 234–95.

Muesebeck, C. F. W. 1963. A new ceraphronid from Cretaceous amber (Hymenoptera: Proctotrupoidea). *Journal of Paleontology* 37: 129–30.

Müller, A. H. 1982. Über Hyponome fossiler und rezenter Insekten, erster Beitrag. *Freiberger Forschungsheft C* 366: 7–27.

Müller, H. 1872. Anwendung der Darwinschen Lehre auf Bienen. *Verhandlungen des Naturhistorischen Vereines der Preussischen Rheinlande und Westphalen* 29: 1–96.

Müller, K. J., and I. Hinz. 1992. Cambrogeorginidae fam. nov., soft-integumented problematica from the Middle Cambrian of Australia. *Alcheringa* 16: 333–53.

———, and I. Hinz-Schallreuter. 1993. Palaeoscolecid worms from the Middle Cambrian of Australia. *Palaeontology* 36: 549–92.

———, and D. Walossek. 1986. Arthropod larvae from the Upper Cambrian of Sweden. *Transactions of the Royal Society of Edinburgh (Earth Sciences)* 77: 157–79.

———, and D. Walossek. 1991. Ein Blick durch das "Orsten"-Fenster in die Arthrpodenwelt vor 500 Millionen Jahren. *Verhandlungen der Deutschen Zoologischen Gesellschaft* 84: 281–94.

———, D. Walossek, and A. Zakharov. 1995. 'Orsten' type phosphatized soft-integument preservation and a new record from the Middle Cambrian Kuonamka Formation in Siberia. *Neues Jahrbuch für Geologie und Paläontologie, Abhandlungen* 197: 101–18.

Munroe, D. D. 1974. The systematics, phylogeny, and zoogeography of *Symmerus* Walker and *Australosymmerus* Freeman (Diptera: Mycetophilidae: Ditomyiinae). *Memoirs of the Entomological Society of Canada* 92: 1–183.

Munroe, E., and M. A. Solis. 1999. The Pyraloidea. Pp. 233–56. *In* Kristensen, N. P. (ed.), *Handbuch der Zoologie: Eine Naturgeschichte der Stämme des Tierreiches: Band IV: Arthropoda: Insecta: Teilband 35: Lepidoptera, Moths and Butterflies: Volume 1: Evolution, Systematics, and Biogeography*. Walter de Gruyter; Berlin, Germany; x+491 pp.

Munson, M. A., P. Baumann, M. A. Clark, L. Baumann, N. A. Moran, *et al.* 1991. Evidence for the establishment of aphid-eubacterium endosymbiosis in an ancestor of four aphid families. *Journal of Bacteriology* 173: 6321–4.

Muona, J. 1993. Eucnemidae and Throscidae from Baltic amber (Coleoptera). *Entomologische Blätter* 89: 15–45.

Nagashima, T. 1990. Eye structure of externally eyeless grylloblattids (Insecta, Notoptera). *Bulletin of the Sugadaira Montane Research Center, University of Tsukaba* 11: 89–93.

———, H. Ando, and G. Fukushima. 1982. Life history of *Galloisiana nipponensis* (Caudell *et*

King). Pp. 43–59. *In* Ando, H. (ed.), *Biology of the Notoptera*. Kashiyo-Insatsu Co. Ltd.; Nagano, Japan; vi+194 pp.

Nagatomi, A. 1982. The genera of Rhagionidae (Diptera). *Journal of Natural History* 16: 31–70.

Nagy, C. 1969. Sur la sous-famille Mesitinae [sic] Berland (Hym., Bethylidae). *Lucrările Statiunii der Cercetări Marine "Prof. Ioan Borcea" Agigea* 3: 275–300.

———. 1971. First record of the Old World species of *Sierolomorpha* Ashm. (Hym, Heterogynoidea). *Reichenbachia* 13: 247–9.

———. 1973. Revisionary studies on the family Plumariidae Bischoff (Hymenoptera, Heterogynoidea [sic]). *Folia Entomologica Hungarica* 26 (supplement): 255–67.

———. 1975. A new genus of Scolebythidae (Hymenoptera) from South Africa and Australia. *Journal of the Entomological Society of Southern Africa* 38: 75–8.

Naisbit, R., C. D. Jiggins, and J. Mallet. 2003. Mimicry: Developmental genes that contribute to speciation. *Evolution and Development* 5: 269–80.

Nakahara, W. 1958. The Neurorthinae [sic], a new subfamily of the Sisyridae (Neuroptera). *Mushi* 32: 19–32.

Nakata, S., and T. C. Maa. 1974. A review of the parasitic earwigs (Dermaptera: Arixeniina; Hemimerina). *Oriental Insects* 16: 307–74.

Nalepa, C. A. 1984. Colony composition, protozoan transfer and some life history characteristics of the woodroach *Cryptocercus punctulatus* Scudder (Dictyoptera: Cryptocercidae). *Behavioral Ecology and Sociobiology* 14: 273–9.

———. 1991. Ancestral transfer of symbionts between cockroaches and termites: An unlikely scenario. *Proceedings of the Royal Society of London, Series B, Biological Sciences* 246: 185–9.

———, and C. Bandi. 2000. Characterizing the ancestors: Paedomorphosis and termite evolution. Pp. 53–75. *In* Abe, T., D. E. Bignell, and M. Higashi (eds.), *Termites: Evolution, Sociality, Symbioses, Ecology*. Kluwer Academic Publishers; Dordrecht, the Netherlands; xxii+466 pp.

———, and M. Lenz. 2000. The ootheca of *Mastotermes darwiniensis* Froggatt (Isoptera: Mastotermitidae): Homology with cockroach oothecae. *Proceedings of the Royal Society of London, Series B, Biological Sciences* 267: 1809–13.

———, G. W. Byers, C. Bandi, and M. Sironi. 1997. Description of *Cryptocercus clevelandi* (Dictyoptera: Cryptocercidae) from the northwestern United States, molecular analysis of bacterial symbionts in its fat body, and notes on biology, distribution, and biogeography. *Annals of the Entomological Society of America* 90: 416–24.

Namkung, J. 1982. The taxonomy of the Grylloblattodea and their distribution in Korea. Pp. 29–42. *In* Ando, H. (ed.), *Biology of the Notoptera*. Kashiyo-Insatsu Co.; Nagano, Japan; vi+194 pp.

Naomi, S.-I. 1985. The phylogeny and higher classification of the Staphylinidae and their allied groups (Coleoptera, Staphylinoidea). *Esakia* 23: 1–27.

Naumann, C. M. 1987. On the phylogenetic significance of two Miocene zygaenid moths (Insecta, Lepidoptera). *Paläontologische Zeitschrift* 61: 299–308.

Naumann, I. D. 1983. The biology of mud-nesting Hymenoptera (and their associates) and Isoptera in rock shelters of the Kakadu region, Northern Territory. Pp. 129–89. *In* Gillespie, D. (ed.), *The Rock Art Sites of Kakadu National Park: Some Preliminary Research Findings for Their Conservation and Management*. Australian National Parks and Wildlife Service (Special Publication 10); Canberra, Australia; 216 pp.

———. 1987. A new megalyrid (Hymenoptera: Megalyridae) parasitic on a sphecid wasp in Australia. *Australian Journal of Zoology* 26: 215–22.

———. 1990. The aculeate wasps and bees (Hymenoptera) of Norfolk and Philip Islands. *Australian Entomological Magazine* 17: 17–28.

———, ed. 1991a. *The Insects of Australia: A Textbook for Students and Research Workers* [2 volumes]. Cornell University Press; Ithaca, New York; xvi+[1]+1–542+[2]+543–1137 pp.

———, 1991b. Hymenoptera (wasps, bees, ants, sawflies). Pp. 916–1000. *In* Naumann, I. D. (ed.), *The Insects of Australia: A Textbook for Students and Research Workers, Volume 2* [2nd Edition]. Cornell University Press; Ithaca, New York; [2]+543–1137 pp.

———, and L. Masner. 1985. Parasitic wasps of the proctotrupoid complex: A new family from Australia and a key to world families (Hymenoptera: Proctotrupoidea *sensu lato*). *Australian Journal of Zoology* 33: 761–83.

Needham, J. G. 1930. *A Manual of the Dragonflies of China; A Monographic Study of the Chinese Odonata*. The Fan Memorial Institute of Biology; Peiping, China; 344 pp.

———, and M. H. Anthony. 1903. The skewness of the thorax in Odonata. *Journal of the New York Entomological Society* 11: 117–25.

———, S. W. Frost, and B. H. Tothill. 1928. *Leaf-Mining Insects*. Williams and Wilkins; Baltimore, Maryland; viii+351 pp.

———, J. R. Traver, and Y.-C. Hsu. 1935. *The Biology of Mayflies, with a Systematic Account of North American Species*. Comstock Publishing; Ithaca, New York; xiv+759 pp.

———, M. J. Westfall, Jr., and M. L. May. 2000. *Dragonflies of North America*. Scientific Publishers; Gainesville, Florida; xv+939+[1] pp.

Neff, J. L., B. B. Simpson, N. L. Evenhuis, and G. Dieringer. 2003. Character analysis of adaptations for tarsal pollen collection in the Bombyliidae (Insecta: Diptera): The benefits of putting your foot in your mouth. *Zootaxa* 157: 1–14.

Nekrutenko, Yu. P. 1965. Tertiary nymphalid butterflies and some phylogenetic aspects of systematic lepidopterology. *Journal of Research on the Lepidoptera* 4: 149–58.

Nel, A. 1987. Deux nouveaux termites du tertiaire du sud-est de la France (Dictyoptera, Hodotermitidae, Kalotermitidae). *Revue Francaise d'Entomologie* 9: 17–20.

———. 1988a. Les Sialidae (Megaloptera) fossiles des diatomites de Murat (Cantal, France) et de Bes-Konak (Anatolie, Turque). *Neuroptera International* 5: 39–44.

———. 1988b. Deux nouveaux Mantispidae (Planipennia) fossiles de l'Oligocene du sud-est de la France. *Neuroptera International* 5: 103–9.

———. 1988c. Redescription de *Eosirex ligniticus* Piton, 1940 (Hymenoptera Symphyta Siricidae). *Entomologiste* 44: 287–92.

———. 1991a. Descriptions et révisions de trois "Siricidae" fossiles du Cénozoïque (Hymenoptera). *Bulletin de la Société de Entomologique de France* 96: 247–53.

———. 1991b. Nouveaux insectes neuroptéroïdes fossiles de l'Oligocène de France (Neuroptera et Megaloptera). *Bulletin du Muséum National d'Histoire Naturelle, Série 4, Section C (Sciences de la Terre Paléontologie, Géologie, Minéralogie)* 12: 327–49.

———. 1993. Nouveaux raphidioptères fossiles du Cénozoïque de France et d'Espagne (Raphidioptera, Raphidiidae, Inocelliidae). *EPHE Biologie et Evolution des Insectes* 6: 99–108.

———, and H. Descimon. 1984. Une nouvelle espèce de Lépidoptère fossile du Stampien de Céreste (04) (Lepidoptera, Satyridae). *Géologique Mediterranean* 11: 287–93.

———, and M. Henrotay. 1994. Les Chrysopidae Mésozoïques. État actuel des connaissances. Description d'un nouveau genre et nouvelle espèce dans le Jurassique Inférieur (Lias) (Insecta: Neuroptera). *Annales de la Société Entomologique de France* 30: 293–318.

———, and E. Jarzembowski. 1997. New fossil Sisyridae and Nevrorthidae (Insecta: Neuroptera) from Eocene Baltic amber and Upper Miocene of France. *European Journal of Entomology* 94: 287–94.

———, and J.-C. Paicheler. 1993. Les Isoptera fossiles. État actuel des connaissances, implications paléoécologiques et paléoclimatogiques [Insecta, Dictyoptera]. Pp. 103–79. *In* Nel, A., X. Martínez-Delclòs, and J.-C. Paicheler, *Essai de Révision des Aeschinioidea [Insecta, Odonata, Anisoptera] / Les Isoptera Fossiles [Insecta, Dictyoptera]*. CNRS Editions [Cahiers de Paléontologie]; Paris, France; 179 pp.

———, and R. Roy. 1996. Revision of the fossil "mantid" and "ephemerid" species described by Piton from the Palaeocene of Menat (France) (Mantodea: Chaeteessidae, Mantidae; Ensifera: Tettigonioidea). *European Journal of Entomology* 93: 223–34.

———, J. Nel, and C. Balme. 1993. Un nouveau Lépidoptère Satyrinae fossile de l'Oligocène du Sud-Est de la France (Insecta, Lepidoptera, Nymphalidae). *Linneana Belgica* 14: 20–36.

———, G. Gand, G. Fleck, O. Béthoux, J. Lapeyrie, and J. Garric. 1999a. *Saxonagrion minutus* nov. gen. *et* sp., the oldest damselfly from the Upper Permian of France (Odonatoptera, Panodonata, Saxonagrionidae nov. fam.). *Geobios* 32: 883–8.

———, G. Gand, J. Garric, and J. Lapeyrie. 1999b. The first recorded protozygopteran insects from the Upper Permian of France. *Palaeontology* 42: 83–97.

———, J.-J. Menier, G. de Ploëg, G. Hodebart, and L. Danvin. 2002a. *Eosialis*, a new alderfly genus in French lowermost Eocene amber (Insecta, Megaloptera, Sialidae). *Geobios* 35: 313–19.

———, J.-J. Menier, A. Waller, G. Hodebart, and G. de Ploëg. 2002b. New fossil spongillaflies from the lowermost Eocene amber of France (Insecta, Neuroptera, Sisyridae). *Geodiversitas* 25: 109–17.

———, A. Waller, V. Albouy, J.-J. Menier, and G. de Ploëg. 2002c. New fossil earwigs from the lowermost Eocene amber of Paris basin (France) (Insecta, Dermaptera, family incertae sedis). *Geodiversitas* 25: 119–29.

———, G. Perrault, V. Perrichot, and D. Néraudeau. 2004. The oldest ant in the Lower Cretaceous amber of Charente-Maritime (SW France) (Insecta: Hymenoptera: Formicidae). *Geologica Acta* 2: 23–9.

Nelson, C. H. 1984. Numerical cladistic analysis of phylogenetic relationships in Plecoptera. *Annals of the Entomological Society of America* 77: 466–73.

Nelson, C. R., and W. D. Tidwell. 1987. *Brodioptera stricklani* n. sp. (Megasecoptera: Brodiopteridae), a new fossil insect from the Upper Manning Canyon Shale Formation, Utah (lowermost Namurian B). *Psyche* 94: 309–16.

Nelson, D. R., and R. P. Higgins. 1990. Tardigrada. Pp. 393–419. *In* Dindal, D. L. (ed.), *Soil Biology Guide*. Wiley and Sons; New York, New York; xx+1349 pp.

Nelson, G., and N. Platnick. 1981. *Systematics and Biogeography: Cladistics and Vicariance*. Columbia University Press; New York, New York; xi+567 pp.

Nemkov, P. G. 1988. New genus of digger wasp from Baltic amber. *Paleontologicheskiy Zhurnal* 1988: 119–21. [In Russian]

———. 1990. A new genus from fossorial remains of the tribe Gorytini from Oligocene layers of Primorky Kray. *Paleontologicheskiy Zhurnal* 1990: 123–5. [In Russian]

Néraudeau, D., R. Allain, V. Perrichot, B. Videt, F. L. de Broin *et al.* 2003. Découverte d'un depot paralique à bois fossils, amber insectifère et restes d'Iguanodontidae (Dinosauria, Ornithopoda) dans le Cénomanien inférieur de Fouras (Charente-Maritime, Sud-Ouest de la France). *Comptes Rendus Palevolution* 2: 221–30.

Neuffer, F.-O., G. Gruber, H. Lutz, and H. Frankenhäuser. 1996. *Das Eckfeld Maar – Zeuge tropischen Lebens in der Eifel*. Landessammlung für Naturkunde Rheinland Pfalz; Mainz, Germany; 102 pp.

New, T. R. 1972. Some Brazilian Psocoptera from bird nests. *The Entomologist* 105: 153–60.

———. 1974. Psocoptera. *Handbooks for the Identification of British Insects* 1: 1–102.

———. 1975. The biology of Chrysopidae and Hemerobiidae (Neuroptera) with reference to their usage as biocontrol agents: A review. *Transactions of the Royal Entomological Society, London* 127: 115–40.

———. 1978. Note on the habitat of the presumed larva of *Austroneurorthus* (Neuroptera, Neurorthidae [sic]). *Australian Entomological Magazine* 5: 9.

———. 1984. Intergeneric relationships in recent Nymphidae. Pp. 125–31. *In* Gepp, J., H. Aspöck, and H. Hölzel (eds.), *Progress in World's Neuropterology: Proceedings of the First International Symposium on Neuropterology*. Druckhaus Thalerhof; Graz, Austria; 265 pp.

———. 1986. A review of the biology of Neuroptera Planipennia. *Neuroptera International, Supplemental Series* 1: 1–57.

———. 1987. Biology of the Psocoptera. *Oriental Insects* 21: 1–109.

———. 1989. Planipennia: Lacewings. Pp. 1–132. Fischer, M. (ed.), *Handbuch der Zoologie: Ein Naturgeschichte der Stämme des Tierreiches, Band IV Arthropoda: Insecta, Teilband 30*. Walter de Gruyter; Berlin, Germany; 132 pp.

———. 1995. The order Zoraptera (Insecta) from Christmas Island, Indian Ocean. *Invertebrate Taxonomy* 9: 243–46.

———. 1996. Neuroptera. Pp. 1–104. *In* Wells, A. (ed.), *Zoological Catalogue of Australia, Volume 28: Neuroptera, Strepsiptera, Mecoptera, Siphonaptera*. CSIRO; Melbourne, Australia; xiii+230 pp.

———. 1999. Neuroptera and biological control. *Stapfia* 60: 147–66.

———. 2002. Prospects for extending the use of Australian lacewings in biological control. *Acta Zoologica Academiae Scientiarum Hungaricae (supplement 2)* 48: 209–16.

———, and G. Theischinger. 1993. Megaloptera: Alderflies, Dobsonflies. Pp. 1–97. Fischer, M. (ed.), *Handbuch der Zoologie: Ein Naturgeschichte der Stämme des Tierreiches, Band IV Arthropoda: Insecta, Teilband 33*. Walter de Gruyter; Berlin, Germany; 102 pp.

Newton, A. F., Jr. 1985. South temperate Staphylinoidea (Coleoptera): Their potential for biogeographic analysis of austral disjunctions. Pp. 180–220. *In* Ball, G. E. (ed.), *Taxonomy, Phylogeny, and Zoogeography of Beetles and Ants: A Volume Dedicated to the Memory of Philip Jackson Darlington, Jr. (1904–1983)*. W. Junk-Kluwer; Dordrecht, the Netherlands; xiii+514 pp.

———, and M. K. Thayer. 1988. A critique on Naomi's phylogeny and higher classification of Staphylinidae and allies (Coleoptera). *Entomologia Generalis* 14: 63–72.

Nicholas, C. J., A. A. Henwood, and M. Simpson. 1993. A new discovery of Early Cretaceous (Wealden) amber from the Isle of Wight. *Geological Magazine* 130: 847–50.

Nielsen, C. 2001. *Animal Evolution: Interrelationships of the Living Phyla* [2nd Edition]. Oxford University Press; Oxford, UK; x+563 pp.

Nielsen, E. S. 1985. Primitive (non-ditrysian) Lepidoptera of the Andes: Diversity, distribution, biology and phylogenetic relationships. *Journal of Research on the Lepidoptera, Supplement* 1: 1–16.

———, and D. R. Davis. 1985. The first southern hemisphere prodoxid and the phylogeny of the Incurvarioidea (Lepidoptera). *Systematic Entomology* 10: 307–22.

———, and N. P. Kristensen. 1989. Primitive ghost moths. *Monographs on Australian Lepidoptera* 1: 1–206.

———, and N. P. Kristensen. 1996. The Australian moth family Lophocoronidae and the basal phylogeny of the Lepidoptera: Glossata. *Invertebrate Taxonomy* 10: 1199–1302.

Nijhout, H. F. 1986. Pattern and pattern diversity on lepidopteran wings. *Bioscience* 36: 527–33.

———. 1991. *The Development and Evolution of Butterfly Wing Patterns*. Smithsonian Institution Press; Washington, D.C.; xvi+297 pp.

Niklas, K. J., R. M. Brown, Jr., and R. Santos. 1985. Ultrastructural states of preservation in *Clarkia* angiosperm leaf tissues: Implications on modes of fossilization. Pp. 143–60. *In* Smiley, C. J. (ed.), *Late Cenozoic History of the Pacific Northwest*. Pacific Division of the American Association for the Advancement of Science; San Francisco, California; 417 pp.

Nikolskaya, M. N. 1978. Serphitidae. Pp. 646–7. *In* Medvedev, G. S. (ed.), *Keys to the Insects of the European Part of the USSR, Volume III: Hymenoptera, Part II*. Nauka; Moscow, Russia; xii+1341 pp. [In Russian]

Nitecki, M. H., ed. 1979. *Mazon Creek Fossils*. Academic Press; New York, New York; xv+581 pp.

Nixon, K. C., and W. L. Crepet. 1993. Late Cretaceous fossil flowers of Ericalean affinity. *American Journal of Botany* 80: 616–23.

———, W. L. Crepet, D. Stevenson, and E. M. Friis. 1994. A reevaluation of seed plant phylogeny. *Annals of the Missouri Botanical Garden* 81: 484–533.

Noirot, C. 1970. The nests of termites. Pp. 73–126. *In* Krishna, K., and F. M. Weesner (eds.), *Biology of Termites, Volume II*. Academic Press; New York, New York; vii+643 pp.

———. 1995. The gut of termites (Isoptera). Comparative anatomy, systematics, phylogeny. I. Lower termites. *Annales de la Société Entomologique de France* 31: 197–226.

———. 2001. The gut of termites (Isoptera). Comparative anatomy, systematics, phylogeny. II. Higher termites (Termitidae). *Annales de la Société Entomologique de France* 37: 431–71.

———, and J. P. E. C. Darlington. 2000. Termite nests: Architecture, regulation and defence. Pp. 121–39. *In* Abe, T., D. E. Bignell, and M. Higashi (eds.), *Termites: Evolution, Sociality, Symbioses, Ecology*. Kluwer Academic Publishers; Dordrecht, the Netherlands; xxii+466 pp.

Noll, F. B. 2002. Behavioral phylogeny of corbiculate Apidae (Hymenoptera; Apinae), with special reference to social behavior. *Cladistics* 18: 137–53.

Nordlander, G. 1984. What do we know about parasitic cynipoids (Hymenoptera)? *Entomologisk Tidskrift* 105: 36–40.

———, Z. Liu, and F. Ronquist. 1996. Phylogeny and historical biogeography of the cynipoid wasp family Ibaliidae (Hymenoptera). *Systematic Entomology* 21: 151–66.

Norell, M. A., and M. J. Novacek. 1992. The fossil record and evolution: Comparing cladistic and paleontologic evidence for vertebrate history. *Science* 255: 1690–3.

Normark, B. B. 2003. Evolution of alternative genetic systems in insects. *Annual Review of Entomology* 48: 397–423.

Norstog, K. J. 1987. Cycads and the origin of insect pollination. *American Scientist* 75: 300–6.

———, P. K. S. Fawcett, and A. P. Vovides. 1992. Beetle pollination of two species of *Zamia*: Evolutionary and ecological considerations. *Palaeobotanist* 41: 149–58.

Norton, R. A., P. M. Bonamo, J. D. Grierson, and W. A. Shear. 1988. Oribatid mite fossils from a terrestrial Devonian deposit near Gilboa, New York. *Journal of Paleontology* 62: 259–69.

Novacek, M. J. 1992. Fossils, topologies, missing data, and the higher level phylogeny of eutherian mammals. *Systematic Biology* 41: 58–73.

———. 1999. 100 million years of land vertebrate evolution: The Cretaceous-early Tertiary transition. *Annals of the Missouri Botanical Garden* 86: 230–58.

Novokshonov, V. G. 1997. *Early Evolution of Scorpionflies (Insecta: Panorpida)*. Nauka Press; Moscow, Russia; 140 pp. [In Russian]

———. 1998. Some problems of scorpionfly (Mecoptera) evolution. *Entomological Review* 78: 378–90.

———. 2002. Order Panorpida Latreille, 1802. The scorpionflies. Pp. 194–8. *In* Rasnitsyn, A. P., and D. L. J. Quicke (eds.), *History of Insects*. Kluwer Academic Publishers; Dordrecht, the Netherlands; xii+517 pp.

Noyes, J. S. 1989. The diversity of Hymenoptera in the tropics with special reference to Parasitica in Sulawesi. *Ecological Entomology* 14: 197–207.

———. 1990. The number of described chalcidoid taxa in the world that are currently regarded as valid. *Chalcid Forum* 13: 9–10.

Nur, U. 1980. Evolution of unusual chromosome systems in scale insects (Coccoidea: Homoptera). Pp. 97–117. *In* Blackman, R. L., G. M. Hewitt, and M. Ashburner (eds.), *Insect Cytogenetics*. Blackwell Scientific; Oxford, UK; viii+278 pp.

Nuttall, M. J. 1980. Insect parasites of *Sirex*. *Forest and Timber Insects in New Zealand* 47: [not paginated].

Oatman, E. R. 1985. *Aphanogmus floridanus*, a primary parasite of *Feltiella acarivora*, a predator of spider mites on strawberry (Hymenoptera: Ceraphronidae; Diptera: Cecidomyiidae). *Proceedings of the Entomological Society of Washington* 87: 714–16.

Obordo, C. O., L. L. Burrows, and C. C. Labandeira. 1994. Insect-mediated herbivory from the Late Pennsylvanian and Early Permian red-bed floras of north-central Texas. *Geological Society of America, Abstracts with Programs* 26: 123.

Oeser, R. 1961. Ovipositor der Hymenopteren, IV. Vergleichende Untersuchungen ueber die Teile des Ovipositors. *Mitteilungen aus dem Zoologischen Museum in Berlin* 37: 46–62.

Ogden, T. H., and M. F. Whiting. 2003. The problem with "the Paleoptera Problem": Sense and sensitivity. *Cladistics* 19: 432–42.

Ohgushi, R., S. F. Sakagami, and S. Yamane. 1990. Nest architecture of the stenogastrine wasps: Diversity and evolution (Hymenoptera, Vespidae): A comparative review. Pp. 41–72. *In* Sakagami, S. F., R. Ohgushi, and D. W. Roubik (eds.), *Natural History of Social Wasps and Bees in Equatorial Sumatra*. Hokkaido University Press; Sapporo, Japan; ix+274 pp.

Ohl, M. 1997. Die Evolution von Cleptoparasitismus zwischen nahverwandten Bienen- und Grabwespengruppen (Hymenoptera: Apoidea: "Sphecidae", Apiformes). *Mitteilungen der Deutschen Gesellschaft für Allgemeine und Angewandte Entomologie* 11: 785–8.

———. 1999. A revision of *Stizoides* Guérin-Méneville, 1844: Taxonomy, phylogenetic relationships, biogeography, and evolution (Hymenoptera: Apoidea: "Sphecidae"). *Mitteilungen des Museums für Naturkunde in Berlin, Zoologische Reihe* 75: 63–169.

———, and D. Linde. 2003. Ovaries, ovarioles, and oocytes in apoid wasps, with special reference to cleptoparasitic species (Hymenoptera: Apoidea: "Sphecidae"). *Journal of the Kansas Entomological Society* 76: 147–59.

Oldroyd, H. 1964. *The Natural History of Flies*. W. W. Norton; New York, New York; xiv+324 pp.

Oliver, K. M., J. A. Russell, N. A. Moran, and M. S. Hunter. 2003. Facultative bacterial symbionts in aphids confer resistance to parasitic wasps. *Proceedings of the National Academy of Sciences, USA* 100: 1803–7.

Ollivier, C. 1985. Gisements fossilifères d'Europe occidentale recelant une ichtyofaune cenozoique. *Minéruax et Fossiles* 11: 15–21.

Olmi, M. 1984. A revision of the Dryinidae (Hymenoptera). *Memoirs of the American Entomological Institute* 37: 1–1913. [2 volumes]

———. 1988. New Oriental and Neotropic [sic] Dryinidae (Hymenoptera Chrysidoidea). *Frustula Entomologica* 20: 152–67.

———. 1995. Dryinids and embolemids in amber (Hymenoptera Dryinidae *et* Embolemidae). *Redia* 78: 253–71.

———. 1996. A revision of the world Embolemidae (Hymenoptera Chrysidoidea). *Frustula Entomologica* 18: 85–146.

———. 1999. A new species of *Anteon* Jurine from Venezuela (Hym., Dryinidae). *Entomologist's Monthly Magazine* 135: 213–14.

———. 2002. Descriptions of two new species of Dryinidae from Botswana: *Bocchus martellii* and *Gonatopus martellii* (Hymenoptera Chrysidoidea). *Bollettino di Zoologia Agraria e di Bachicoltura* 34: 273–80.

———, and G. Bechly. 2001. New parasitic wasps from Baltic amber (Insecta: Hymenoptera: Dryinidae). *Stuttgarter Beiträge zur Naturkunde, Serie B, Geologie und Paläontologie* 306: 1–58.

Olsen, P. E. 1986. A 40-million-year lake record of early Mesozoic orbital climatic forcing. *Science* 234: 842–8.

———, C. L. Remington, B. Cornet, and K. S. Thomson. 1978. Cyclic change in Late Triassic lacustrine communities. *Science* 201: 729–33.

———, D. V. Kent, H.-D. Sues, C. Koeberl, H. Huber *et al.* 2002. Ascent of dinosaurs linked to an iridium anomaly at the Triassic-Jurassic boundary. *Science* 296: 1305–7.

O'Neill, K. M. 2001. *Solitary Wasps: Behavior and Natural History*. Cornell University Press; Ithaca, New York; xiii+[1]+406 pp.

Oosterbroek, P., and G. Courtney. 1995. Phylogeny of the nematocerous families of Diptera (Insecta). *Zoological Journal of the Linnean Society* 115: 267–311.

Opler, P. A. 1973. Fossil lepidopterous leaf mines demonstrate the age of some insect-plant relationships. *Science* 179: 1321–3.

Oppenheim, P. 1888. Die Insectenwelt des lithographischen Schiefers in Bayern. *Palaeontographica* 34: 215–47.

Orfila, R. N. 1949. Nuevos "Stephanidae" (Hym., Ichneum.) neotropicales. *Acta Zoologica Lilloana* 7: 339–52.

O'Riain, M. J., J. U. M. Jarvis, R. Alexander, R. Buffenstein, and C. Peeters. 2000. Morphological castes in a vertebrate. *Proceedings of the National Academy of Sciences, USA* 97: 13194–7.

Osorio, D., and J. P. Bacon. 1994. A good eye for arthropod evolution. *BioEssays* 16: 419–24.

———, M. Averof, and J. P. Bacon. 1995. Arthropod evolution: Great brains, beautiful bodies. *Trends in Evolution and Ecology* 10: 449–54.

Ossianilsson, F. 1949. Insect drummers, a study on the morphology and function of the sound-producing organ of Swedish Homoptera Auchenorrhyncha with notes on their sound production. *Opuscula Entomologia, Supplementum* 10: 139 pp.

Oswald, J. D. 1990. Raphidioptera. *Bulletin of the American Museum of Natural History* 195: 154–63.

———. 1993a. Phylogeny, taxonomy, and biogeography of extant silky lacewings (Insecta: Neuroptera: Psychopsidae). *Memoirs of the American Entomological Society* 40: 1–65.

———. 1993b. Revision and cladistic analysis of the world genera of the family Hemerobiidae (Insecta: Neuroptera). *Journal of the New York Entomological Society* 101: 143–299.

———. 1993c. A new genus and species of brown lacewing from Venezuela (Neuroptera: Hemerobiidae), with comments on the evolution of the hemerobiid forewing radial vein. *Systematic Entomology* 18: 363–70.

———. 1994. A new phylogenetically basal subfamily of brown lacewings from Chile (Neuroptera: Hemerobiidae). *Entomologica Scandinavica* 25: 295–302.

———. 1996. A new brachypterous *Nusalala* species from Costa Rica, with comments on the evolution of flightlessness in brown lacewings (Neuroptera: Hemerobiidae). *Systematic Entomology* 21: 343–52.

———. 1998a. Rediscovery of *Polystoechotes gazullai* Navás (Neuroptera: Polystoechotidae). *Proceedings of the Entomological Society of Washington* 100: 389–94.

———. 1998b. Annotated catalogue of the Dilaridae (Insecta: Neuroptera) of the world. *Tijdschrift voor Entomologie* 141: 115–28.

———. 1999. The brown lacewing genus *Notiobella* (Neuroptera: Hemerobiidae) from Dominican amber. *Journal of the New York Entomological Society* 107: 297–303.

———, and N. M. Schiff. 2001. A new species of the genus *Dilar* Rambur (Neuroptera: Dilaridae) from Borneo. *Proceedings of the Entomological Society of Washington* 103: 74–80.

Otis, G. W. 1991. A review of the diversity of species within *Apis*. Pp. 29–49. *In* Smith, D. R. (ed.), *Diversity in the Genus* Apis. Westview Press; Boulder, Colorado; xiv+265 pp.

———. 1996. Distribution of recently recognized species of honey bees (Hymenoptera: Apidae; *Apis*). *Journal of the Kansas Entomological Society, supplement* 69: 311–33.

Otte, D. 1981. *The North American Grasshoppers, Volume I: Acrididae: Gomphocerinae and Acridinae*. Harvard University Press; Cambridge, Massachusetts; ix+[2]+275 pp.

———. 1984. *The North American Grasshoppers, Volume II: Acrididae: Oedipodinae*. Harvard University Press; Cambridge, Massachusetts; ix+[2]+275 pp.

———. 1994. *The Crickets of Hawaii: Origin, Systematics, and Evolution*. The Orthopterists' Society and Academy of Natural Sciences; Philadelphia, Pennsylvania; vi+396 pp.

O'Toole, C., and A. Raw. 1991. *Bees of the World*. Blanford; London, UK; 192 pp.

Owen, R. 1866. *On the Anatomy of Vertebrates, Volume 1, Fishes and Reptiles*. Longmans, Green, and Company; London, UK; xlii+650 pp.

Packard, A. S. 1869. *Guide to the Study of Insects, and a Treatise on Those Injurious and Beneficial to Crops: For the Use of Colleges, Farm Schools, and Agriculturists.* Naturalist's Book Agency; Salem, Massachusetts; viii+702 pp.

———. 1886. A new arrangement of the orders of insects. *American Naturalist* 20: 808.

Paclt, J. 1956. *Biologie der primär flügellosen Insekten.* Fischer; Jena, Germany; 285 pp.

———. 1957. Diplura. *Genera Insectorum* 212: 1–123.

———. 1963. Thysanura. Fam. Nicoletiidae. *Genera Insectorum* 216: 1–58.

———. 1967. Thysanura. Fam. Lepidotrichidae, Maindroniidae, Lepismatidae. *Genera Insectorum* 218: 1–86.

———. 1972. Grundsätzliches zur Chorologie und Systematik der Felsenspringer. *Zoologischer Anzeiger* 188: 422–9.

Page, R. D. M. 1994. Parallel phylogenies: reconstructing the history of host-parasite assemblages. *Cladistics* 10: 155–73.

———, R. D. Price, and R. A. Hellenthal. 1995. Phylogeny of *Geomydoecus* and *Thomomydoecus* pocket gopher lice (Phthiraptera: Trichodectidae) inferred from cladistic analysis of adult and first instar morphology. *Systematic Entomology* 20: 129–43.

———, P. L. M. Lee, S. A. Becher, R. Griffiths, and D. H. Clayton. 1998. A different tempo of mitochondrial DNA evolution in birds and their parasitic lice. *Molecular Phylogenetics and Evolution* 9: 276–93.

Pagés, J. 1997. Notes sur les Diploures Rhabdours (Insectes, Aptérygotes) no. 1 – Diplura Genavensia XXII. *Revue Suisse de Zoologie* 104: 869–96.

Pagliano, G. 2002. Revisione della sottofamiglia Apterogyninae (Hymenoptera: Bradynobaenidae). *Museo Regionale di Scienze Naturali Monografie, Turin* 34: 1–387.

Palmer, A. 1957. Miocene arthropods from the Mojave Desert California. *United States Geological Survey Professional Paper* 294-G: 237–77, 5 pls.

Panfilov, D. V. 1980. New representatives of lacewings (Neuroptera) from the Jurassic of Karatau. Pp. 82–111. *In* Dolin, V. G., D. V. Panfilov, A. G. Ponomarenko, and L. N. Pritykina (eds.), *Fossil Insects of the Mesozoic.* Naukova Dumka; Kiev, Ukraine; 133+[3] pp. [In Russian]

Pantel, P. J. 1915. Notes orthoptérologiques. VI. Le "vomer sous-anal" n'est par le "titillateur", étude des segments abdominaux et principalement du segment terminal des mâles, ches les Phasmides. *Annales del la Société Entomologique de France* 84: 173–243.

Papageorgis, C. 1975. Mimicry in Neotropical butterflies. *American Scientist* 63: 522–32.

Papavero, N., ed. 1966–???. *A Catalogue of the Diptera of the Americas South of the United States.* Departamento de Zoologia, Secretaria da Agricultura; São Paulo, Brazil (numerous fascicles, published infrequently).

———. 1977. *The World Oestridae (Diptera), Mammals, and Continental Drift.* Junk; The Hague, the Netherlands; vii+240 pp.

Pape, T. 1992. Phylogeny of the Tachinidae family-group (Diptera: Calyptratae). *Tijdschrift voor Entomologie* 135: 43–86.

———. 2000. The future of entomological collections. *Entomologica Austriaca* 4: 3–7.

———. 2001. Phylogeny of Oestridae (Insecta: Diptera). *Systematic Entomology* 26: 133–71.

Papier, F. and L. Grauvogel-Stamm. 1995. Les Blattodea du Trias: Le genre *Voltziablatta* n. gen. du Buntsandstein Superieur des Vosges (France). *Palaeontographica* 235: 141–62.

———, and A. Nel. 2001. Les Subioblattidae (Blattodea, Insecta) du Trias d'Asie Centrale. *Paläontologische Zeitschrift* 74: 533–42.

———, L. Grauvogel-Stamm, and A. Nel. 1994. *Subioblatta undulate* n. sp., une nouvelle blatte (Subioblattidae Schneider) du Buntsandstein supérieur (Anisien) des Vosges (France). Morphologie, systématique et affinities. *Neues Jahrbuch für Geologie und Paläontologie, Montashefte* 1994: 277–90.

Pardi, L., and S. Turillazzi. 1985. Biologia delle Stenogastrinae (Hymenoptera, Vespoidea). *Atti della Accademia Nazionale Italiana di Entomologia Rendiconti* 30–32: 25–43.

Parfin, S. I., and A. B. Gurney. 1956. The spongilla-flies, with special reference to those of the Western Hemisphere (Sisyridae, Neuroptera). *Proceedings of the United States National Museum* 105: 421–529.

Park, L. 1995. Geochemical and paleoenvironmental analysis of lacustrine arthropod-bearing concretions of the Barstow Formation, southern California. *Palaios* 10: 44–57.

———, and K. F. Downing. 2001. Paleoecology of an exceptionally preserved arthropod fauna from lake deposits of the Miocene Barstow Formation, southern California, U.S.A. *Palaios* 16: 175–84.

Parkin, E. A. 1942. Symbiosis of siricid woodwasps. *Annals of Applied Biology* 29: 268–74.

Parnell, J. R. 1963. Three gall midges (Diptera: Cecidomyidae) and their parasites found in the pods of broom (*Sarothamnus scoparius* (L.) Wimmer). *Transactions of the Royal Entomological Society, London* 115: 261–75.

Pass, G. 1987. The "cercus heart" in stoneflies – A new type of accessory circulatory organ in insects. *Naturwissenschaften* 74: 440–1.

———. 2000. Accessory pulsatile organs: Evolutionary innovations in insects. *Annual Review of Entomology* 45: 495–518.

Pate, V. S. L. 1947a. A conspectus of the Tiphiidae, with particular reference to the Nearctic forms (Hymenoptera, Aculeata). *Journal of the New York Entomological Society* 55: 115–45.

———. 1947b. Neotropical Sapygidae, with a conspectus of the family (Hymenoptera: Aculeata). *Acta Zoologica Lilloana* 4: 393–426.

Paterson, A. M., G. P. Wallis, L. J. Wallis, and R. D. Gray. 2000. Seabird and louse coevolution: Complex histories revealed by 12S rRNA sequences and reconciliation analyses. *Systematic Biology* 49: 383–99.

Patterson, C. 1981. Significance of fossils in determining evolutionary relationships. *Annual Review of Ecology and Systematics* 12: 195–223.

Patton, J. L., and M. F. Smith. 1994. Paraphyly, polyphyly, and the nature of species boundaries in pocket gophers (genus *Thomomys*). *Systematic Biology* 43: 11–26.

Paulus, H. F. 1979. Eye structure and the monophyly of Arthropoda. Pp. 299–383. *In* Gupta, A. P. (ed.), *Arthropod Phylogeny.* Van Nostrand Reinhold; New York, New York; xx+762 pp.

Pavan, M. 1963. *Ricerche Biologiche e Mediche su Pederina e su Estratti Purificati di* Paederus fuscipes *Curt. (Coleoptera, Staphylinidae).* Industrie Lito-Tipografiche Mario Ponzio; Pavia, Italy; 93 pp.

———. 1982. Summary of the present data on pederin. *Publicazioni dell'Instituto di Entomologia dell'Universita di Pavia* 1982: 1–61.

Pawlowski, J., R. Szadziewski, D. Kmieciak, J. Fahrni, and G. Bittar. 1997. Phylogeny of the infraorder Culicomorpha (Diptera: Nematocera) based on 28S RNA gene sequences. *Systematic Entomology* 21: 167–78.

Peakall, R. 1990. Responses of male *Zaspilothynnus trilobatus* Turner wasps to females and the sexually deceptive orchid it pollinates. *Functional Ecology* 4: 159–67.

Pearman, J. V. 1936. The taxonomy of Psocoptera: Preliminary sketch. *Proceedings of the Royal Entomological Society of London (B)* 5: 58–62.

———. 1960. Some African Psocoptera found on rats. *The Entomologist* 93: 246–50.

Pearson, H. H. W. 1929. *Gnetales.* Cambridge University Press; Cambridge, UK; vi+[1]+194 pp.

Pedroni, G. 1999. Primo contributo allo studio dei cerambicidi del contrafforte pliocenico, valli del Setta e Savena (Appenino Bolognese) (Insecta Coleoptera Cerambycidae). *Quaderno di Studi e Notizie di Storia Naturale della Romagna* 12: 25–36.

———. 2002. Sui colleotteri Curculionoidea del contrafforte Pliocenico, delle gole di Scascoli e della Valle del Savena (Appennino Tosco-Emiliano) e prima segnala zione Italiana di *Simo variegatus* (Boheman, 1843) (Insecta Coleoptera Curculionoidea). *Quaderno di Studi e Notizie di Storia Naturale della Romagna* 17: 43–56.

Peigler, R. S. 1993. Wild silks of the world. *American Entomologist* 39: 151–61.

Pellens, R., P. Grandcolas, and I. Domingos da Silva-Neto. 2002. A new and independently evolved case of xylophagy and the presence of intestinal flagellates in the cockroach *Parasphaeria boleiriana* (Dictyoptera, Blaberidae, Zetoborinae) from the remnants of the Brazilian Atlantic forest. *Canadian Journal of Zoology* 80: 350–9.

Pellisier-Scott, M. 1994. Competition with flies promotes communal breeding in the burying beetle, *Nicrophorus tomentosus. Behavioral Ecology and Sociobiology* 34: 367–73.

———. 1996. Communal breeding in burying beetles. *American Scientist* 84: 376–82.

———. 1997. Reproductive dominance and differential ovicide in the communally breeding burying beetle *Nicrophorus tomentosus. Behavioral Ecology and Sociobiology* 40: 313–20.

Pellmyr, O. 1992. The phylogeny of a mutualism: Evolution and coadaptation between *Trollius* and its seed-parasitic pollinators. *Biological Journal of the Linnean Society* 47: 337–65.

———. 1999. A systematic revision of the yucca moths in the *Tegeticula yuccasella* complex north of Mexico. *Systematic Entomology* 24: 243–71.

———. 2003. Yuccas, yucca moths, and coevolution: A review. *Annals of the Missouri Botanical Garden* 90: 35–55.

———, and C. J. Huth. 1994. Evolutionary stability of mutualism between between yuccas and yucca moths. *Nature* 372: 257–60.

———, and J. Leebens-Mack. 1999. Forty million years of mutualism: Evidence for Eocene origin of the yucca-yucca moth association. *Proceedings of the National Academy of Sciences, U.S.A.* 96: 9178–83.

———, L. B. Thien, G. Bergström, and I. Groth. 1990. Pollination of New Caledonian Winteraceae: Opportunistic shifts or parallel radiation with their pollinators. *Plant Systematics and Evolution* 173: 143–57.

Peñalver Mollá, E. 1998. *Estudio Tafonómica y Paleoecológica de los Insectos del Mioceno de Rubielos de Mora (Teruel).* Instituto de Estudios Turolenses; Teruel, Spain; 177+[1] pp.

———, and X. Martínez-Delclòs. 2000. Insectos del Mioceno Inferior de Ribesalbes (Castellon, España). Hymenoptera. *Treballs del Museu de Geologia de Barcelona* 9: 97–153.

———, X. Martínez-Delclòs, and A. Arillo. 1999. Yacimientos con insectos fósiles en España. *Revista Española de Paleontología* 14: 231–45.

Penney, D. 2000. Anyphaenidae in Miocene Dominican Republic amber (Arachnida, Araneae). *Journal of Arachnology* 28: 223–6.

———. 2001. Advances in the taxonomy of spiders in Miocene amber from the Dominican Republic (Arthropoda: Araneae). *Palaeontology* 44: 987–1009.

———. 2002. Spiders in Upper Cretaceous amber from New Jersey (Arthropoda: Araneae). *Palaeontology* 45: 709–24.

———, C. P. Wheater, and P. A. Selden. 2003. Resistance of spiders to Cretaceous-Tertiary extinction events. *Evolution* 57: 2599–2607.

Penny, N. D. 1977. A systematic study of the Boreidae (Mecoptera). *University of Kansas Science Bulletin* 51: 141–217.

———. 1996. A remarkable new genus and species of Ithonidae from Honduras (Neuroptera). *Journal of the Kansas Entomological Society* 69: 81–6.

———. 1997. *World Checklist of Extant Mecoptera Species.* California Academy of Sciences; San Francisco, California. *http://www. calacademy.org/research/entomology/masecoptera.*

———, P. A. Adams, and L. A. Stange. 1997. Species catalog of the Neuroptera, Megaloptera, and Raphidioptera of American north of Mexico. *Proceedings of the California Academy of Sciences* 50: 39–114.

Penteado-Dias, A. M., and C. van Achterberg. 2002. First record of the genus *Probethylus* Ashmead (Sclerogibbidae: Probethylinae) from Brazil, with description of a new species. *Zoologische Mededelingen* 76: 105–7.

Penz, C. M. 1999. Higher level phylogeny for the passion-vine butterflies (Nymphalidae, Heliconiinae) based on early stage and adult morphology. *Zoological Journal of the Linnean Society* 127: 277–344.

Percival, M. S. 1965. *Floral Biology.* Pergamon Press; Oxford, UK; xv+243 pp.

Peters, W. L., and J. G. Peters. 1970. *Proceedings of the First International Conference on Ephemeroptera.* Brill; Leiden, the Netherlands; viii+344 pp.

———, and J. G. Peters. 2000. Discovery of a new genus of Leptophlebiidae: Leptophlebiinae (Ephemeroptera) in Cretaceous amber from New Jersey. Pp. 127–31. *In* Grimaldi, D. (ed.), *Studies on Fossils in Amber, with Particular Reference to the Cretaceous of New Jersey.* Backhuys Publishers; Leiden, the Netherlands; viii+498 pp.

Petrov, A. V., and V. V. Zherikhin. 2000. Fossil bark beetles of the genus *Scolytus* from the Neogene of France (Coleoptera: Curculionidae: Scolytinae). *Paleontological Journal* 34: 344–6.

Petrulevičius, J. F. 1999. Insectos del Cenozoico de la Argentina. *Revista de la Sociedad Entomológica Argentina* 58: 71–85.

———, and R. G. Martins-Neto. 2000. Checklist of South American Cenozoic insects. *Acta Geologica Hispanica* 35: 135–47.

Petrunkevitch, A. 1913. A monograph of the terrestrial Palaeozoic Arachnida of North America. *Transactions of the Connecticut Academy of Arts and Sciences* 18: 1–137.

———. 1945. *Clacitro fisheri*: A new fossil arachnid. *American Journal of Science* 243: 320–9.

Peus, F. 1968. Über die beiden Bernstein-Flöhe (Insecta, Siphonaptera). *Palaeontological Journal* 42: 62–72.

Pfau, H. K. 1986. Untersuchungen zur Konstruktion, Funktion und Evolution der Flugapparrates der Libellen (Insecta, Odonata). *Tijdschrift voor Entomologie* 129: 35–123.

———. 1971. Struktur und Funktion des sekundären Kopulationsapparates der Odonaten (Insecta, Palaeoptera), ihre Wandlung in der Stammesgeschichte und Bedeutung für die adaptive Entfaltung der Ordnung. *Zeitschrift für Morphologie und Ökologie der Tiere* 70: 281–371.

———. 1991. Contributions of functional morphology to the phylogenetic systematics of Odonata. *Advances in Odonatology* 5: 109–41.

Picker, M. D. 1984. A new genus and species of spoon-winged lacewing (Neuroptera: Nemopteridae) from Namaqualand, South Africa. *Journal of the Entomological Society of Southern Africa* 47: 259–68.

———, J. F. Colville, and S. van Noort. 2002. Mantophasmatodea now in South Africa. *Science* 297: 1475.

Piek, T., ed. 1986. *Venoms of the Hymenoptera.* Academic Press; London, UK; xi+570 pp.

Piel, O. 1933. *Monema flavescens* Wkr. and its parasites (Lepidoptera, Heterogeneidae). *Lignan Science Journal* 12 (supplement): 173–201.

Pielinska, A. 1997. Inclusions of wood in the amber collections of the Museum of the Earth in Warsaw. *Metalla, Bochum* 66: 25–8.

Pierce, N. E., M. F. Braby, A. Heath, D. J. Lohman, J. Mathew, D. B. Rand, and M. A. Travassos. 2002. The ecology and evolution of ant association in the Lycaenidae (Lepidoptera). *Annual Review of Entomology* 47: 733–71.

Pierce, W. D. 1951. Fossil arthropods from Onyx-Marble. *Bulletin of the Southern California Academy of Sciences* 50: 34–49.

———. 1963. Fossil arthropods of California. 25. Silicified leafhoppers from Calico Mountains nodules. *Bulletin of the Southern California Academy of Sciences* 62: 69–82.

———. 1966. Fossil arthropods of California. 29. Silicifed Miocene pupae of ceratopogonid flies. *Bulletin of the Southern California Academy of Sciences* 65: 81–98.

———, and J. Gibron, Sr. 1962. Fossil arthropods of California. 24. Some unusual fossil arthropods from the Calico Mountains nodules. *Bulletin of the Southern California Academy of Sciences* 61: 143–51.

———, and R. A. Kirkby. 1959. Fossil insects from Montana. 1. A new fossil nemopterid (Neuroptera). *Bulletin of the Southern California Academy of Sciences* 58: 47–50.

Pike, E. M. 1994. Historical changes in insect community structure as indicated by hexapods of Upper Cretaceous Alberta (Grassy Lake) amber. *Canadian Entomologist* 126: 695–702.

———. 1995. *Amber Taphonomy and the Grassy Lake, Alberta Amber Fauna.* Ph.D. Dissertation, University of Calgary; Calgary, Canada; 264 pp.

Pijl, L. van der. 1960. Ecological aspects of flower evolution, 1. Phyletic evolution. *Evolution* 14: 256–65.

Pilgrim, R. L. C. 1991. External morphology of flea larvae (Siphonaptera) and its significance in taxonomy. *Florida Entomologist* 74: 386–96.

Pimentel, D., H. Acquay, M. Biltonen, P. Rice, M. Silva *et al.* 1992. Environmental and economic costs of pesticide use. *Bioscience* 42: 750–60.

Pinhey, E. C. G. 1951. The dragonflies of Southern Africa. *Memoir of the Transvaal Museum* 5: xv+1–335.

———. 1961. Dragonflies (Odonata) of Central Africa. *Occasional Papers of the Rhodes-Livingstone Museum* 14: 1–110.

Pinna, M. C. C., de. 1991. Concepts and tests of homology in the cladistic paradigm. *Cladistics* 7: 367–94.

Pinto, I. D. 1956. Artrópodes da Formação Santa Maria (Triássico Superior) do Rio Grande do Sul, com notícias sobre alguns restos vegetais. *Boletim da Sociedade Brasileira de Geologia, São Paulo* 5: 76–87.

———. 1986. Carboniferous insects from Argentina; III, Familia Xenopteridae Pinto, nov., ordo Megasecoptera. *Pesquisas* 18: 23–9.

———. 1994. *Sphecorydaloides lucchesei*, a new Carboniferous megasecopteran Insecta from Argentina. *Pesquisas* 21: 85–9.

———, and L. Ornellas. 1974. A new insect, *Triassoblatta cargnini* Pinto *et* Ornellas, sp. n., a Triassic blattoid from the Santa Maria Formation, southern Brazil. *Anais da Academia Brasileira de Ciências, Rio de Janeiro* 46: 512–21.

———, and L. P. de Ornellas. 1978. New fossil insects from the White Band Formation (Permian), South Africa. *Pesquisas* 10: 96–104.

———, and L. Pinto. 1981. Permian insects from Parana Basin, south Brazil. 3. Homoptera – 1 – Pereboridae. *Anais do Congresso Latino Americano Paleontologia* 2: 209–19.

———, and M. Mendes. 2002. A second Upper Paleozoic blattoid (Insecta) from Betancourt, Chubut Province, Argentina. *Revista Brasileira de Paleontologia* 4: 45–50.

———, and I. Purper. 1978. A new genus and two species of plecopteran insects from the Triassic of Argentina. *Pesquisas* 10: 77–86.

Pirrung, B. M. 1992. Geologische und geophysikalische Untersuchungen am tertiären "Eckfeld Maar", Südwesteifel. *Mainzer Naturwissenschaftliches Archiv* 30: 3–21.

———. 1998. Zur Entstehung isolierter alttertiärer Seesedimente in zentraleuropäischen Vulkanfeldern. *Mainzer Naturwissenschaftliches Archiv, Beiheft* 20: 1–117.

———, G. Büchel, and W. Jacoby. 2001. The Tertiary volcanic basins of Eckfeld, Enspel and Messel (Germany). *Zeitschrift der Deutschen Geologischen Gesellschaft* 152: 27–59.

Pix, W., G. Nalbach, and J. Zeil. 1993. Strepsipteran forewings are haltere-like organs of equilibrium. *Naturwissenschaften* 80: 371–4.

———, J. M. Zanker, and J. Zeil. 2000. The optometer response and spatial resolution of the visual system in male *Xenos vesparum* (Strepsiptera). *Journal of Experimental Biology* 203: 3397–3409.

Platnick, N. 1976. Drifting spiders or continents? Vicariance biogeography of the spider subfamily Laroniinae (Araneae: Gnaphosidae). *Systematic Zoology* 25: 101–9.

Ploëg, G., de, D. Dutheil, E. Gheerbrant, M. Godinot, A. Jossang *et al.* 1998 Un nouveau gisement paleontologique Konservat-Lagerstatte a la base de l'Eocene dans la region de Creil (Oise). *Strata, Serie 1, Communications* 9: 108–10.

Plotnick, R. E. 1983. *Patterns in the Evolution of the Eurypterids.* Ph.D. Dissertation; University of Chicago; Chicago, IL; xiv+411 pp.

Pohl, H. 2002. Phylogeny of the Strepsiptera based on morphological data of the first instar larvae. *Zoological Scripta* 31: 123–34.

———, and R. Kinzelbach. 1995. Neufunde von Fächerflüglern aus dem Baltischen und Dominikanischen Bernstein (Strepsiptera: Bohartillidae & Myrmecolacidae). *Mitteilungen des Geologisch-Paläontologischen Instituts und Museum der Universität Hamburg* 78: 197–209.

———, and R. Kinzelbach. 2001. First record of a female stylopid (Strepsiptera: ? Myrmecolacidae) parasite of a prionomyrmecine ant (Hymenoptera: Formicidae) in Baltic amber. *Insect Systematics and Evolution* 32: 143–6.

———, R. G. Beutel, and R. Kinzelbach. In press. Protoxenidae fam. n. from Baltic amber – A "missing link" in Strepsiptera phylogeny. *Zoologica Scripta.*

Poisson, R. 1951. Ordre de Hétéroptères. Pp. 1657–1803. *In* Grassé, P. P. (ed.), *Traité de Zoologie, Tome 10.* Masson et Cie; Paris, France; pp. 978–1948.

Polaszek, A., and P. Chinwada. 2000. *Dendrocerus rodhaini* (Hymenoptera: Megaspilidae): A hyperparasitoid of the maize stem borer, *Busseola fusca* Fuller (Lepidoptera: Noctuidae), in Zimbabwe. *African Entomology* 8: 299–302.

Polhemus, J. T. 2000. North American Mesozoic aquatic Heteroptera (Insecta, Naucoroidea, Nepoidea) from the Todilto Formation, New Mexico. *New Mexico Museum of Natural History and Science Bulletin* 16: 29–40.

Polis, G. A., ed. 1990. *The Biology of Scorpions.* Stanford University Press; Stanford, California; xxiii+587 pp.

Pollock, D. A., and B. B. Normark. 2002. The life cycle of *Micromalthus debilis* LeConte (1878) (Coleoptera: Archostemata: Micromalthidae): Historical review and evolutionary perspective. *Journal of Zoological Systematics and Evolutionary Research* 40: 105–12.

Ponomarenko, A. G. 1963. Paleozoic beetles of Cupedoidea from the European part of the U.S.S.R. *Paleontologicheskiy Zhurnal* 1963: 70–85. [In Russian]

———. 1969. Historical development of Archostematan beetles. *Trudy Paleontologischeskogo Instituta Akademii Nauk, SSSR* 125: 1–240. [In Russian]

———. 1975. Family Dryinidae. Pp. 104–5. *In* Rasnitsyn, A. P. Hymenoptera-Apocrita of the Mesozoic. *Trudy Paleontologicheskogo Instituta, Akademii Nauk SSSR* 147: 1–132. [In Russian]

———. 1976. A new insect from the Cretaceous of Transbaikalia – A possible parasite of pterosaurians. *Paleontological Journal* 10: 339–43.

———. 1977a. Paleozoic members of Megaloptera (Insecta). *Paleontologischeskiy Zhurnal* 1977: 78–86. [In Russian]

———. 1977b. Suborder Adephaga. Pp. 1–104. *In* Arnol'di, L. V., V. V. Zherikhin, L. M. Nikritin, and A. G. Ponomarenko (eds.), *Mesozoic Coleoptera.* Trudy Paleontologischeskogo Instituta Akademii Nauk, SSSR 161; 204 pp.

———. 1981. New Dryinidae (Hymenoptera) from the Late Cretaceous of the Taymyr and Canada. *Paleontologicheskiy Zhurnal* 1981: 139–43. [In Russian]

———. 1985. Fossil insects from the "Solnhofener Plattenkalke" in the Museum of Natural History, Vienna. *Annalen des Naturhistorische Museum Wien, A* 87: 135–44.

———. 1995. The geological history of beetles. Pp. 155–71. *In* Pakaluk, J., and S. A. Ślipiński (eds.), *Biology, Phylogeny, and Classification of Coleoptera: Papers Celebrating the 80th Birthday of Roy A. Crowson.* Museum i Instytut Zoologii PAN; Warsaw, Poland; x+1092 pp. [2 volumes]

———. 2000. New alderflies (Megaloptera: Parasialidae) and glosselytrodeans (Glosselytrodea: Glosselytridae) from the Permian of Mongolia. *Paleontological Journal, Supplement 3* 34: 309–11.

———. 2002a. Order Corydalida Leach, 1815 (= Megaloptera Latreille, 1802). The alderflies and dobsonflies. Pp. 180–3. *In* Rasnitsyn, A. P., and D. L. J. Quicke (eds.), *History of Insects.* Kluwer Academic Publishers; Dordrecht, the Netherlands; xii+517 pp.

———. 2002b. Superorder Scarabaeidea Laicharting, 1781. Order Coleoptera Linné, 1758. The Beetles. Pp. 164–76. *In* Rasnitsyn, A. P., and D. L .J. Quicke (eds.), *History of Insects.* Kluwer Academic Publishers; Dordrecht, the Netherlands; xii+517 pp.

———. 2003. Ecological evolution of beetles (Insecta: Coleoptera). *Acta Zoologica Cracoviensia* 46 (Supplement): 319–28.

Ponomaryova, G. Yu., V. G. Novokshonov, and S. V. Naugolnykh. 1998. *Tshekarda – A Locality of Permian Fossil Plants and Insects.* Perm University; Perm, Russia; 92 pp. [In Russian]

Popadic, A., D. Rusch, M. Peterson, B. T. Rogers, and T. C. Kaufman. 1996. Origin of the arthropod mandible. *Nature* 380: 395.

———, A. Abzhanov, D. Rusch, and T. C. Kaufman. 1998. Understanding the genetic basis of morphological evolution: The role of homeotic genes in the diversification of the arthropod bauplan. *International Journal of Developmental Biology* 42: 453–61.

Popham, E. J. 1961. On the systematic position of *Hemimerus* Walker, a case for ordinal status. *Proceedings of the Entomological Society of London (B)* 30: 19–25.

———. 1985. The mutual affinities of the major earwig taxa (Insecta, Dermaptera). *Zeitschrift für Zoologische Systematik und Evolutionsforschung* 23: 199–214.

———. 1990. Dermaptera. *Bulletin of the American Museum of Natural History* 195: 69–75.

Popov, Yu. A. 1971. The historical development of bugs of the infraorder Nepomorpha (Heteroptera). *Proceedings of the Paleontological Institute of the Academy of Sciences of the USSR* 129: 1–230. [In Russian]

———. 1980. Superorder Cimicidea Laicharting, 1781. Pp. 58–68. *In* Rohdendorf, B. B., and A. P. Rasnitsyn (eds.), *Historical Development of the Class Insecta.* Trudy Paleontologicheskogo Instituta 175: 1–269.

———, and D. E. Shcherbakov. 1996. Origin and evolution of Coleorrhyncha as shown by the fossil record. Pp. 9–30. *In* Schaefer, C. W. (ed.), *Studies on Hemipteran Phylogeny.* Entomological Society of America; Lanham, Maryland; iii+244 pp.

———, W. R. Dolling, and P. E. S. Whalley. 1994. British Upper Triassic and Lower Jurassic Heteroptera and Coleorrhyncha (Insecta: Hemiptera). *Genus* 5: 307–47.

Porsch, O. 1910. *Ephedra campylopoda*, eine entomophile Gymnosperme. *Berichte der Deutschen Botanischen Gesellschaft* 28: 404–12.

Potgieter, C. J., T. J. Edwards, R. M. Miller, and J. van Staden. 1999. Pollination of seven *Plectranthus* spp. (Lamiaceae) in southern Natal, South Africa. *Plant Systematics and Evolution* 218: 99–112.

Powell, J. A. 1980. Evolution of larval food preferences in microlepidoptera. *Annual Review of Entomology* 25: 133–59.

———. 1984. Biological interrelationships of moths and *Yucca schottii. University of California Publications in Entomology* 100: 1–93.

———. 1992. Interrelationships of yuccas and yucca moths. *Trends in Ecology and Evolution* 7: 10–15.

———, and W. J. Turner. 1975. Observations on oviposition behavior and host selection in *Orussus occidentalis* (Hymenoptera: Siricoidea). *Journal of the Kansas Entomological Society* 48: 299–307.

———, C. Mitter, and B. Farrell. 1999. Evolution of larval food preferences. Pp. 403–22. *In* Kristensen, N. P. (ed.), *Handbuch der Zoologie: Eine Naturgeschichte der Stämme des Tierreiches: Band IV: Arthropoda: Insecta: Teilband 35: Lepidoptera, Moths and Butterflies: Volume 1: Evolution, Systematics, and Biogeography.* Walter de Gruyter; Berlin, Germany; x+491 pp.

Prance, G. T., and J. R. Arias. 1975. A study of the floral biology of *Victoria amazonica* (Poepp.) Sowerby (Nymphaeaceae). *Acta Amazonica* 5: 109–39.

Prell, H. B., von. 1912. Beitrage zur Kenntnis der Proturen. II. Anamerentoma und Holomerentoma, eine Eintheilung der Hexapoden. *Zoologischer Anzeiger* 39: 357–65.

Prendini, L. 2000. Phylogeny and classification of the superfamily Scorpionoidea Latreille 1802 (Chelicerata, Scorpiones): An exemplar approach. *Cladistics* 16: 1–78.

Prentice, M. 1991. Morphological analysis of the tribes of Apidae. Pp. 51–69. *In* Smith, D. R. (ed.), *Diversity in the Genus* Apis. Westview Press; Boulder, Colorado; xiv+265 pp.

———. 1998. *The Comparative Morphology and Phylogeny of Apoid Wasps (Hymenoptera: Apoidea)*. Ph.D. Dissertation, University of California; Berkeley, California; 1439 pp.

———, G. O. Poinar, Jr., and R. Milki. 1996. Fossil scolebythids (Hymenoptera, Scolebythidae) from Lebanese and Dominican amber. *Proceedings of the Entomological Society of Washington* 98: 802–11.

Prete, F. R., H. Wells, P. H. Wells, and L. E. Hurd, eds. 1999. *The Praying Mantids*. Johns Hopkins University Press; Baltimore, Maryland; xiv+362 pp.

Price, R. D. 1975. The *Geomydoecus* (Mallophaga: Trichodectidae) of the southeastern USA pocket gophers (Rodentia: Geomyidae). *Proceedings of the Entomological Society of Washington* 77: 61–5.

———, R. A. Hellenthal, R. L. Palma, K. P. Johnson, and D. H. Clayton. 2003. The chewing lice: World checklist and biological overview. *Illinois Natural History Survey Special Publication* 24: x+1–501.

Priesner, H. 1936. *Aphanogmus steinitzi* spec. nov., ein coniopterygiden-parasit (Hymenoptera Proctotrupoidea). *Bulletin de la Société Royale Entomologique d'Egypte* 20: 248–51.

Prischam, D. A., and C. A. Sheppard. 2002. A world view of insects as aphrodisiacs, with special reference to Spanish fly. *American Entomologist* 48: 208–20.

Pritchard, G., and P. Scholefield. 1978. Observations on the food, feeding behaviour, and associated organs of *Grylloblatta campodeiformis* (Grylloblattodea). *Canadian Entomologist* 110: 205–12.

Pritykina, L. N. 1981. New Triassic dragonflies from central Asia. *Trudy Paleontologicheskogo Instituta* 183: 1–42. [In Russian]

Proctor, M., M. Yeo, and A. Lack. 1996. *The Natural History of Pollination*. Timber Press; Portland, Oregon; 479 pp.

Proctor, W. 1946. *Biological Survey of the Mount Desert Region, Part VII: The Insect Fauna*. Wistar Institute of Anatomy and Biology; Philadelphia, Pennsylvania; 566 pp.

Prunescu, C. C. 1996. Plesiomorphic and apomorphic character states in the class Chilopoda. *Memoires du Museum National d'Histoire Naturelle* 169: 299–306.

Pruvost, P. 1927. Sur une aile d'insectes fossile trouvée au sondage de Gulpen. *Jaarverstag van voor het Nederlandsche Mijngebied te Heerlen* 1926: 76–7.

———. Description d'un insecte fossile des couches de la Lukuga (Kivu). *Memoires de l'Institut geologique de l'Universite de Louvain* 9: 1–8.

Fukowski, E. 1933. Ökologische Untersuchungen an *Necrophorus* F. *Zeitschrift für Morphologie und Ökologie der Tiere* 27: 518–86.

Pulawski, W. J., and A. P. Rasnitsyn. 1980. On the taxonomic position of *Hoplisus sepultus* Cockerell, 1906, from the Lower Oligocene of Colorado (Hymenoptera, Sphecidae). *Polskie Pismo Entomologiczne* 50: 393–6.

Pullin, A. S., ed. 1995. *Ecology and Conservation of Butterflies*. Chapman and Hall; London, UK; xiv+363 pp.

Putman, W. L., and D. H. Herne. 1966. The role of predators and other biotic agents in regulating the population density of phytophagous mites in Ontario peach orchards. *Canadian Entomologist* 98: 808–20.

Qiu, Y. L., J. Lee, F. Bernasconi-Quadroni, D. E. Soltis, P. S. Soltis *et al.* 1999. The earliest angiosperms: Evidence from mitochondrial, plastid and nuclear genomes. *Nature* 402: 404–7.

———, J. Lee, F. Bernasconi-Quadroni, D. E. Soltis, P. S. Soltis, *et al.* 2000. Phylogeny of basal angiosperms: Analyses of five genes from three genomes. *International Journal of Plant Sciences, Supplement* 161: S3–27.

Quartau, J. A. 1986. An overview of the paranotal theory on the origin of the insect wings. *Publicações do Instituto de Zoologia "Dr. Augusto Nobre", Faculdade de Ciências do Porto* 194: 1–42.

Queinnec, E. 2001. Insights into arthropod head evolution. Two heads in one: The end of the "endless dispute"? *Annales de la Société Entomologique de France* 37: 51–69.

Quicke, D. L. J. 1997. *Parasitic Wasps*. Chapman and Hall; London, UK; xvi+[1]+470 pp.

———, M. G. Fitton, J. R. Tunstead, S. N. Ingram, and P. V. Gaitens. 1994. Ovipositor structure and relationships within the Hymenoptera, with special reference to the Ichneumonoidea. *Journal of Natural History* 28: 635–82.

———, H. H. Basibuyuk, M. G. Fitton, and A. P. Rasnitsyn. 1999. Morphological, palaeontological and molecular aspects of ichneumonoid phylogeny (Hymenoptera, Insecta). *Zoologica Scripta* 28: 175–202.

Ragge, D. R. 1977. Classification of the Tettigonioidea. *Lyman Entomological Museum and Research Laboratory Memoir* 4: 44–6.

Raasch, G. O. 1947. *Wellington Formation in Oklahoma*. Ph.D. dissertation, University of Wisconsin; Madison, Wisconsin; 157 pp.

Radchenko, V. G., and Yu. A. Pesenko. 1994. *Biology of the Bees (Hymenoptera, Apoidea)*. Russian Academy of Sciences; Saint-Petersburg, Russia; 350 pp. [In Russian]

Rähle, W. 1970. Untersuchungen an Kopf und Prothorax von *Embia ramburi* Rimsky-Korsakov 1906 (Embioptera, Embiidae). *Zoologische Jahrbücher, Abteilung für Anatomie und Ontogenie der Tiere* 87: 248–330.

Rahm, G. 1921. Biologische und Phsyiologische Beiträge zur Kenntnis der Moosfauna. *Zeitschrift für Allgemeine Physiologie* 20: 1–34.

———. 1924. Weitere physiologische Versuche mit niederen Temperaturen. Ein Beitrag zur Lösung des Kaltenproblems. *Verhandlungen der Deutschen Zoologischen Geselleschaft* 29: 106–11.

———. 1925. Beiträge zur Kenntnis der Moostierwelt der preußischen Rheinlande. *Archiv für Naturgeschichte, Abteilung A* 90: 153–214.

Rajchel, J., and A. Uchman. 1998. Insects boring in Oligocene wood, Kliwa Sandstones, Outer Carpathians, Poland. *Annales Societatis Geologorum Poloniae* 68: 219–24.

Ramazzotti, G., and W. Maucci. 1983. Il phylum Tardigrada [3 Edizione riveduta e aggiornata]. *Memorie dell'Istituto di Idrobiologia Dott Marco de Marchi* 41: 1–1012.

Rampino, M. R., and A. C. Adler. 1998. Evidence for abrupt latest Permian mass extinction of foraminifera: Results of tests for the Signor-Lipps effect. *Geology* 26: 415–8.

Ramsköld, L. and J.-Y. Chen. 1998. Cambrian lobopodians: Morphology and phylogeny. Pp. 107–50. *In* Edgecombe, G. D. (ed.), *Arthropod Fossils and Phylogeny*. Columbia University Press; New York, New York; [2]+347 pp.

———, and G. D. Edgecombe. 1991. Trilobite monophyly revisited. *Historical Biology* 4: 267–83.

———, and X.-G. Hou. 1991. New early Cambrian animal and onychophoran affinities of enigmatic metazoans. *Nature* 351: 225–8.

Ransome, H. M. 1937. *The Sacred Bee in Ancient Times and Folklore*. Houghton Mifflin Company; New York, New York; 308 pp.

Rao, C. N., and S. C. Shah. 1959. Fossil insects from the Gondwana of India. *Indian Minerals* 13: 3–5.

Rapp, W. F. 1961. Corrodentia in cliff swallow nests. *Entomological News* 72: 195.

Rasa, O. A. E. 1998. Biparental investment and reproductive success in a subsocial desert beetle: the role of maternal effort. *Behavioral Ecology and Sociobiology* 43: 105–13.

———. 1999. Division of labour and extended parenting in a desert tenebrionid beetle. *Ethology* 105: 37–56.

Rasnitsyn, A. P. 1964. New Triassic Hymenoptera from Central Asia. *Paleontologicheskiy Zhurnal* 1964: 88–96. [In Russian]

———. 1968. New Mesozoic sawflies (Hymenoptera, Symphyta). Pp. 190–236. *In* Rohdendorf, B. B. (ed.), *Jurassic Insects of Karatau*. Nauka; Moscow, Russia; 252 pp. [In Russian]

———. 1969. The origin and evolution of lower Hymenoptera. *Trudy Paleontologicheskogo Instituta, Akademii Nauk SSSR* 123: 1–196. [In Russian]

———. 1972. Late Jurassic hymenopterous insects (Praeaulacidae) of Karatau. *Paleontologicheskiy Zhurnal* 1972: 70–87. [In Russian]

———. 1975. Hymenoptera-Apocrita of the Mesozoic. *Trudy Paleontologicheskogo Instituta, Akademii Nauk SSSR* 147: 1–132. [In Russian]

———. 1977a. New Hymenoptera from the Jurassic and Cretaceous of Asia. *Paleontologicheskiy Zhurnal* 1977: 98–108. [In Russian]

———. 1977b. A new family of sawflies (Hymenoptera, Tenthredinoidea, Electrotomidae) from the Baltic amber. *Zoologicheskiy Zhurnal* 59: 1304–8. [In Russian]

———. 1977c. New Paleozoic and Mesozoic insects. *Paleontological Journal* 11: 60–72.

———. 1980. Origin and evolution of hymenopterous insects. *Trudy Paleontologischeskogo Instituta Akademii Nauk, SSSR* 174: 1–190. [In Russian]

———. 1983a. Hymenoptera from the Jurassic of East Siberia. *Biologicheskogo Moskovskogo Obshchestva Ispytatelej Prirody, Otdel Geologiya* 58: 85–94. [In Russian]

———. 1983b. Fossil Hymenoptera of the superfamily Pamphiloidea. *Paleontologicheskiy Zhurnal* 1983: 54–68. [In Russian]

———. 1983c. Ichneumonoidea (Hymenoptera) from the Lower Cretaceous of Mongolia. *Contributions of the American Entomological Institute* 20: 259–65.
———. 1983d. First find of a moth from the Jurassic. *Doklady Akademie Nauk, SSSR* 84: 467–71. [In Russian]
———. 1985. Jurassic insects of Siberia and Mongolia. *Trudy Paleontologischeskogo Instituta Akademii Nauk, SSSR* 211: 1–192.
———. 1986. Review of the fossil Tiphiidae, with description of a new species (Hymenoptera). *Psyche* 93: 91–101.
———. 1988a. An outline of evolution of the hymenopterous insects (order Vespida). *Oriental Insects* 22: 115–45.
———. 1988b. Sepulcids and the origin of stem sawflies (Hymenoptera: Cephidae, Sepulcidae). *Trudy Vsesoyuznogo Entomologicheskogo Obshchestva* 70: 68–73. [In Russian]
———. 1990. New representatives of the hymenopterous family Praeaulacidae from Early Cretaceous Buryatia and Mongolia. *Vestnik Zoologii* 1990: 27–31. [In Russian]
———. 1991. Early Cretaceous evaniomorphous hymenopterans. *Vestnik Zoologii* 1991: 128–32. [In Russian]
———. 1992. *Strashila incredibilis,* a new enigmatic mecopteroid insect with possible siphonapteran affinities from the Upper Jurassic of Siberia. *Psyche* 99: 323–33.
———. 1993. Archaeoscoliinae, an extinct subfamily of scoliid wasps (Insecta: Vespida = Hymenoptera: Scoliidae). *Journal of Hymenoptera Research* 2: 85–95.
———. 1996. New Early Cretaceous Embolemidae (Vespida = Hymenoptera: Chrysidoidea). *Memoirs of the Entomological Society of Washington* 17: 183–7.
———. 1998. On the taxonomic position of the insect order Zorotypida = Zoraptera. *Zoologischer Anzeiger* 237: 185–94.
———. 1999. Taxonomy and morphology of *Dasyleptus* Brongniart, 1885, with description of a new species (Insecta: Machilida: Dasyleptidae). *Russian Entomological Journal* 8: 145–54.
———. 2000. An extremely primitive aculeate wasp in the Cretaceous amber from New Jersey (Vespida: ?Sierolomorphidae). Pp. 327–32. *In* Grimaldi, D. (ed.), *Studies on Fossils in Amber, with Particular Reference to the Cretaceous of New Jersey.* Backhuys Publishers; Leiden, the Netherlands; viii+498 pp.
———. 2002. Superorder Vespidea Laicharting, 1781. Order Hymenoptera Linné, 1758 (=Vespida Laicharting, 1781). Pp. 242–54. *In* Rasnitsyn, A. P., and D. L. J. Quicke (eds.), *History of Insects.* Kluwer Academic Publishers; Dordrecht, the Netherlands; xii+517 pp.
———. 2003. On the skimming hypothesis of the origin of insect flight. *Acta Zoologica Cracoviensia* 46 (Supplement): 85–8.
———, and J. Ansorge. 2000a. Two new Lower Cretaceous hymenopterous insects (Insecta: Hymenoptera) from Sierra del Montsec, Spain. *Acta Geologica Hispanica* 35: 59–64.
———, and J. Ansorge. 2000b. New Early Cretaceous hymenopterous insects (Insecta: Hymenoptera) from Sierra del Montsec (Spain). *Paläontologische Zeitschrift* 74: 335–41.
———, and O. V. Kovalev. 1988. The oldest Cynipoidea (Hymenoptera, Archaeocynipidae fam. n.) from the Early Cretaceous Transbaikalia. *Vestnik Zoologii* 1988: 18–21.
———, and V. A. Krassilov. 1996a. First find of pollen grains in the gut of Permian insects. *Paleontological Journal* 30: 484–90.
———, and V. A. Krassilov. 1996b. Pollen in the gut contents of fossil insects as evidence of coevolution. *Paleontological Journal* 30:716–22.
———, and X. Martínez-Delclòs. 1999. New Cretaceous Scoliidae (Vespida = Hymenoptera) from the Lower Cretaceous of Spain and Brazil. *Cretaceous Research* 20: 767–72.
———, and X. Martínez-Delclòs. 2000. Wasps (Insecta: Vespida = Hymenoptera) from the Early Cretaceous of Spain. *Acta Geologica Hispanica* 35: 65–95.
———, and C. D. Michener. 1991. Miocene fossil bumble bee from the Soviet Far East with comments on the chronology and distribution of fossil bees (Hymenoptera: Apidae). *Annals of the Entomological Society of America* 84: 583–9.
———, and D. L. J. Quicke, eds. 2002. *History of Insects.* Kluwer Academic Publishers; Dordrecht, the Netherlands; xii+517 pp.
———, and A. J. Ross. 2000. A preliminary list of arthropod families present in the Burmese amber collection at The Natural History Museum, London. *Bulletin of the Natural History Museum, London (Geology)* 56: 21–4.
———, and M. J. Sharkey. 1988. New Eoichneumonidae from Early Cretaceous of Siberia and Mongolia (Hymenoptera: Ichneumonoidea). Pp. 169–97. *In* Gupta, V. K. (ed.), *Advances in Parasitic Hymenoptera Research.* E. J. Brill; Leiden, the Netherlands; 546 pp.
———, and V. V. Zherikhin. 1999. First fossil chewing louse from the Lower Cretaceous of Baissa, Transbaikalia (Insecta, Pediculida = Phthiraptera, Saurodectidae fam.n.). *Russian Entomological Journal* 8: 253–5.
———, E. A. Jarzembowski, and A. J. Ross. 1998. Wasps (Insecta: Vespida = Hymenoptera) from the Purbeck and Wealden (Lower Cretaceous) of southern England and their biostratigraphical and palaeoenvironmental significance. *Cretaceous Research* 19: 329–91.
———, W. J. Pulawski, and X. Martínez-Delclòs. 1999. Cretaceous digger wasps of the new genus *Bestiola* Pulawski and Rasnitsyn (Hymenoptera: Sphecidae: Angarosphecinae). *Journal of Hymenoptera Research* 8: 23–34.
———, J. Ansorge, and W. Zessin. 2003. New hymenopterous insects (Insecta: Hymenoptera) from the Lower Toarcian (Lower Jurassic) of Germany. *Neues Jahrbuch für Geologie und Paläontologie, Abhandlungen* 227: 321–42.
Rasplus, J.-Y., C. Kerdelhue, I. LeClainche, and G. Mondor. 1998. Molecular phylogeny of fig wasps: Agaonidae are not monophyletic. *Comptes Rendus de l'Academie des Sciences, serie III, Sciences de la Vie* 321: 517–27.
Raup, D. M. 1979. Size of the Permo-Triassic bottleneck and its evolutionary implications. *Science* 206: 217–18.
———, and J. J. Sepkoski. 1984. Periodicity of extinctions in the geological past. *Proceedings of the National Academy of Sciences, U.S.A.* 81: 801–5.
Raven, P. H., and D. Axelrod. 1974. Angiosperm biogeography and past continental movements. *Annals of the Missouri Botanical Garden* 61: 539–673.
Rawlings, G. B. 1957. *Guiglia schauinslandi* (Ashmead), a parasite of *Sirex noctilio* (Fabricius) in New Zealand. *Entomologist* 90: 25–36.
Ray, J. 1710. *Historia Insectorum.* A. & J. Churchill; London, UK; xv+[1]+400 pp.
Raymond, A., P. Cutlip, and M. Sweet. 2000. Rates and processes of terrestrial nutrient cycling in the Paleozoic: The world before beetles, termites, and flies. Pp. 235–83. *In* Allmon, W., and D. J. Bottjer (eds.), *Evolutionary Paleoecology, the Ecological Context of Macroevolutionary Change.* Cambridge University Press; Cambridge, UK; 320 pp.
Rayner, R. J., M. K. Bamford, D. J. Brothers, X. Dippenaar-Schoeman, I. J. McKay, *et al.* 1998. Cretaceous fossils from the Orapa diamond mine. *Paleontologica Africana* 33: 55–65.
Redborg, K. E. 1998. Biology of the Mantispidae. *Annual Review of Entomology* 43: 175–94.
Reddy, D. N. R., and Y. K. Kotikal. 1986. Pit construction and prey capture behaviour of larvae of the ant lion *Myrmeleon tenuipennis* (Neuroptera: Myrmeleontidae). *Journal of Soil Biology and Ecology* 6: 112–16.
Rédei, G. P. 1974. Steps in the evolution of genetic concepts. *Biologisches Zentralblatt* 93: 385–424.
Redtenbacher, J. 1906. [Untitled Sections]. *In* Brunner von Wattenwyl, K., and J. Redtenbacher (eds.), *Die Insektenfamilie der Phasmiden.* Engelmann; Leipzig, Germany; 589 pp., 27 pls. [Issued in three lieferung: 1906, 1907, 1908]
Reed, D. L., and M. S. Hafner. 1997. Host specificity of chewing lice on pocket gophers: A potential mechanism for cospeciation. *Journal of Mammalogy* 78: 655–60.
Rees, M. 1999. Exploring our universe and others. *Scientific American* 1999(Dec.): 80–3.
Regier, J. C., and J. W. Shultz. 2001a. Elongation factor-2: A useful gene for arthropod phylogenetics. *Molecular Phylogenetics and Evolution* 20: 136–48.
———, and J. W. Shultz. 2001b. A phylogenetic analysis of Myriapoda (Arthropoda) using two nuclear protein-encoding genes. *Zoological Journal of the Linnean Society* 132: 469–86.
Rehn, A. C. 2003. Phylogenetic analysis of higher-level relationships of Odonata. *Systematic Entomology* 28: 181–239.
Rehn, J. A. G., and J. W. H. Rehn. 1936. A study of the genus *Hemimerus* (Dermaptera, Hemimerina, Hemimeridae). *Proceedings of the Academy of Natural Sciences, Philadelphia* 87: 457–508.
Reid, C. A. M. 1995. A cladistic analysis of subfamilial relationships in the Chrysomelidae *sensu lato* (Chrysomeloidea). Pp. 559–631. *In* Pakaluk, J., and S. A. Ślipiński (eds.), *Biology, Phylogeny, and Classification of Coleoptera: Papers Celebrating the 80th Birthday of Roy A. Crowson.* Museum i Instytut Zoologii PAN; Warsaw, Poland; x+1092 pp. [2 volumes]
———. 2000. Spilopyrinae Chapuis: A new subfamily in the Chrysomelidae and its systematic placement (Coleoptera). *Invertebrate Taxonomy* 14: 837–62.
Remane, A. 1952. *Die Grundlagen des Natürlichen Systems der Vergleichenden Anatomie*

und der Phylogenetik. Geest and Portig; Leipzig, Germany; vi+400 pp.

REMINGTON, C. L. 1954. The "Apterygota." Pp. 495–505. *In* KESSEL, E. L. (ed.), *A Century of Progress in the Natural Sciences, 1853–1953*. California Academy of Sciences; San Francisco, California; x+[1]+807 pp.

REMPEL, J. G. 1975. The evolution of the insect head: The endless dispute. *Quaestiones Entomologicae* 11: 7–24.

REN, D. 1995. Insects. Pp. 54–197. *In* REN, D., L. LU, Z. GUO, and S. JI (eds.), *Faunae and Stratigraphy of Jurassic-Cretaceous in Beijing and Adjacent Areas*. Geological Publishing House; Beijing, China; vii+[1]+222 pp., 32 pls. [In Chinese, with English summary]

———. 1997. Studies on the late Mesozoic snakeflies of China (Raphidioptera: Baissopteridae, Mesoraphidiidae, Alloraphidiidae). *Acta Zootaxonomica Sinica* 22: 172–88. [In Chinese, with English summary]

———. 1998a. Flower-associated brachycera flies as fossil evidence for Jurassic angiosperm origins. *Science* 280: 85–8.

———. 1998b. Late Jurassic Brachycera from northeastern China (Insecta: Diptera). *Acta Zootaxonomica Sinica* 23: 65–82.

———, and Z.-G. GUO. 1996. On the new fossil genera and species of Neuroptera (Insecta) from the Late Jurassic of northeast China. *Acta Zootaxonomica Sinica* 21: 461–79.

———, M. S. ENGEL, and W. LÜ. 2002. New giant lacewings from the middle Jurassic of Inner Mongolia, China (Neuroptera: Polystoechotidae). *Journal of the Kansas Entomological Society* 75: 188–93.

RENTZ, D. C. F. 1980. A new family of ensiferous Orthoptera from the coastal sands of southeast Queensland. *Memoirs of the Queensland Museum* 20: 49–63.

———, and D. K. MCE. KEVAN. 1991. Dermaptera (earwigs). Pp. 360–8. *In* NAUMANN, I. D. (ed.), *The Insects of Australia: A Textbook for Students and Research Workers, Volume 1* [2nd Edition]. Cornell University Press; Ithaca, New York; xvi+[1]+542 pp.

RESH, V. H., and R. T. CARDÉ, eds. 2003. *Encyclopedia of Insects*. Academic Press; San Diego, California; xxviii+[1]+1266 pp.

RETALLACK, G. J. 1984. Trace fossils of burrowing beetles and bees in an Oligocene paleosol, Badlands National Park, South Dakota. *Journal of Paleontology* 58: 571–92.

———. 1995. Permian-Triassic life crisis on land. *Science* 267: 77–80.

RETTENMEYER, C. W. 1970. Insect mimicry. *Annual Review of Entomology* 15: 43–74.

REX, G. M., and J. GALTIER. 1986. Sur l'évidence d'interactions animal-vegetal dans le Carbonifère inferieur Français. *Compte Revue de Academie des Sciences Paris* 17: 1623–6.

REYNE, A. 1927. Untersuchungen über die Mundteile der Thysanopteren. *Zoologische Jahrbücher, Abteilung für Anatomie und Ontogenie der Tiere* 49: 391–500.

RIBERA, I., J. E. HOGAN, and A. P. VOGLER. 2002a. Phylogeny of hydradephagan water beetles inferred from 18S rRNA sequences. *Molecular Phylogenetics and Evolution* 23: 43–62.

———, R. G. BEUTEL, M. BALKE, and A. P. VOGLER. 2002b. Discovery of Aspidytidae, a new family of aquatic Coleoptera. *Proceedings of the Royal Society of London, Series B, Biological Sciences* 269: 2351–6.

RICE, C. M., W. A. ASHCROFT, D. J. BATTEN, A. J. BOYCE, J. B. D. CAULFIELD *et al.* 1995. A Devonian auriferous hot spring system, Rhynie, Scotland. *Journal of the Geological Society of London* 152: 229–50.

RICE, H. M. A. 1969. An antlion (Neuroptera) and a stonefly (Plecoptera) of Cretaceous age from Labrador, Newfoundland. *Geological Survey of Canada*, Paper 68-65: 1–12.

RICHARDS, A. G. 1933. Comparative skeletal morphology of the noctuid tympanum. *Entomologica Americana* 13: 1–43.

RICHARDS, O. W., and A. H. HAMM. 1939. The biology of the British Pompilidae (Hymenoptera). *Transactions of the Society for British Entomology* 6: 51–114.

RICHARDS, P. A., and A. G. RICHARDS. 1969. Acanthae: A new type of cuticular process in the proventriculus of Mecoptera and Siphonaptera. *Zoologische Jahrbücher, Abteilung für Anatomie und Ontogenie der Tiere* 86: 158–76.

RICHARDS, W. R. 1966. Systematics of fossil aphids from Canadian amber (Homoptera: Aphididae). *Canadian Entomologist* 98: 746–60.

RICHARDSON, E. S., JR. 1956. Pennsylvanian invertebrates of the Mazon Creek area, Illinois. *Fieldiana Geology* 12: 3–76.

RICHARDSON, J. B., and D. C. MACGREGOR. 1986. Silurian and Devonian spore zones of the Old Red Sandstone continent and adjacent regions. *Geological Survey of Canada Bulletin* 364: 1–79.

RICHTER, G., and G. KREBS. 1999. Larvenstadien von Eintagsfliegen (Insecta: Ephemeroptera) aus Sedimenten des eozänen Messelsees. *Natur und Museum* 129: 21–8.

RICKLEFS, R. E. 1979. *Ecology* [2nd edition]. Chiron Press; New York, New York; xii+996 pp.

———, 1990. *Ecology* [3rd Edition]. Freeman; New York, New York; xii+896 pp.

RIDSILL-SMITH, J. 1991. Competition in dung insects. Pp. 111–129. *In* BAILEY, W. J., and J. RIDSDILL-SMITH (eds.), *Reproductive Behaviour of Insects: Individuals and Populations*. Chapman and Hall; London, UK; 339 pp.

RIEGER, C. 1976. Skelett und Muskultur des Kopfes und Prothorax von *Ochterus marginatus* Latreille. *Zoomorphologie* 83: 109–91.

RIEK, E. F. 1952. Fossil insects from the Tertiary sediments at Dinmore, Queensland. *University of Queensland Papers, Department of Geology* 4: 15–22.

———. 1953. Fossil mecopteroid insects from the Upper Permian of New South Wales. *Records of the Australian Museum* 23: 55–87.

———. 1954. Further Triassic insects from Brookvale, N.S.W. (orders Orthoptera, Saltatoria, Protorthoptera, Perlaria). *Records of the Australian Museum* 23: 161–8.

———. 1955. Fossil insects from the Triassic beds at Mt. Crosby, Queensland. *Australian Journal of Zoology* 3: 654–91.

———. 1962. Fossil insects from the Triassic at Hobart, Tasmania. *The Papers and Proceedings of the Royal Society of Tasmania* 96: 39–40.

———. 1968. Undescribed fossil insects from the Upper Permian of Belmont, New South Wales (with an appendix listing the described species). *Records of the Australian Museum* 27: 303–10.

———. 1970a. Hymenoptera (wasps, bees, ants). Pp. 867–959. *In* WATERHOUSE, D. F. (ed.), *The Insects of Australia: A Textbook for Students and Research Workers*. Melbourne University Press; Melbourne, Australia; xiii+1029 pp.

———. 1970b. Lower Cretaceous fleas. *Nature* 227: 746–7.

———. 1971. The presumed heads of Homoptera (Insecta) in the Australian Upper Permian. *Palaeontology* 14: 211–21.

———. 1973. Fossil insects from the Upper Permian of Natal, South Africa. *Annals of the Natal Museum* 21: 513–32.

———. 1974a. Upper Triassic insects from the Molteno Formation, South Africa. *Palaeontologica Africana* 17: 19–31.

———. 1974b. The Australian moth-lacewings (Neuroptera: Ithonidae). *Journal of the Australian Entomological Society* 13: 37–54.

———. 1976a. A new collection of insects from the Upper Triassic of South Africa. *Annals of the Natal Museum* 22: 791–820.

———. 1976b. An unusual mayfly (Insecta: Ephemeroptera) from the Triassic of South Africa. *Paleontologica Africana* 19: 149–91.

———, and J. KUKALOVÁ-PECK. 1984. A new interpretation of dragonfly wing venation based upon Early Carboniferous fossils from Argentina (Insecta: Odonatoidea) and basic character states in pterygote wings. *Canadian Journal of Zoology* 62: 1150–66.

RIETSCHEL, S. 1983. *Aleurochiton petri* n. sp., eine Mottenschildlaus (Homoptera, Aleyrodina) aus dem Pliozän von Neu-Isenburg, Hessen. *Carolinea* 41: 97–100.

RIGHTMYER, M. G. 2004. Phylogeny and classification of the parasitic bee tribe Epeolini (Hymenoptera: Apidae, Nomadinae). *Scientific Papers, Natural History Museum, University of Kansas* 33: 1–51.

RILEY, C. V. 1892. The yucca moth and Yucca pollination. *Missouri Botanical Garden, Third Annual Report* 99–158.

RILEY, J., A. A. BANAJA, and J. L. JAMES. 1978. The phylogenetic relationships of the Pentastomida: The case for their inclusion within the Crustacea. *International Journal of Parasitology* 8: 245–54.

RIOU, B. 1992. Premiere decouverte de *Cimbex* fossiles dans les terrains du Miocene ardechois. *EPHE [Ecole Pratique des Hautes Etudes] Travaux du Laboratoire de Biologie et Evolution des Insectes* 5: 115–20.

———. 1999. Descriptions de quelques insectes fossiles du Miocène supérieur de la Montagne d'Andance. *EPHE [Ecole Pratique des Hautes Etudes] Travaux du Laboratoire de Biologie et Evolution des Insectes* 11/12: 123–33.

RITCHIE, J. M. 1987. Trace fossils of burrowing Hymenoptera from Laetoli. Pp. 433–8. *In* LEAKEY, M. D., and J. M. HARRIS (eds.), *Laetoli: A Pliocene Site in Northern Tanzania*. Oxford University Press; Oxford, UK; xxii+561 pp.

RITCHIE, M. G., and J. M. GLEASON. 1995. Rapid evolution of courtship song pattern in *Drosophila willistoni* sibling species. *Journal of Evolutionary Biology* 8: 463–79.

RITZKOWSKI, S. 1997. K-Ar-Altersbestimmungen der bernsteinführenden Sedimente des Samlandes (Paläogen, Bezirk Kaliningrad). *Metalla, Bochum* 66: 19–23.

Rivers, R. L., Z. B. Mayo, and T. J. Helms. 1979. Biology, behavior and description of *Tiphia berbereti* (Hymenoptera: Tiphiidae) a parasite of *Phyllophaga anxia* (Coleoptera: Scarabaeidae). *Journal of the Kansas Entomological Society* 52: 362–72.

Rix, A. 1999. Fossil insects of the southeast Queensland Tertiary deposits. *Fossil Collector* 57: 21–8.

Robertson, C. 1904. Synopsis of Anthophila. *Canadian Entomologist* 36: 37–43.

———. 1928. *Flowers and Insects; Lists of Visitors of Four Hundred and Fifty-Three Flowers.* Privately printed; Carlinville, Illinois; 221 pp.

Robinson, G. S. 1988. A phylogeny for the Tineoidea (Lepidoptera). *Entomologica Scandinavica* 19: 117–29.

———, and E. S. Nielsen. 1993. *Tineid Genera of Australia.* CSIRO; Canberra, Australia; xvi+244 pp.

Rodd, N. W. 1951. Some observations on the biology of Stephanidae and Megalyridae (Hymenoptera). *Australian Zoologist* 11: 341–6.

Rodriguez-Trelles, F., R. Tarrio, and F. J. Ayala. 2002. A methodological bias toward overestimation of molecular evolutionary time scales. *Proceedings of the National Academy of Sciences, USA* 99: 8112–15.

Roeder, K. D. 1963. *Nerve Cells and Insect Behavior.* Harvard University Press; Cambridge, Massachusetts; viii+188 pp.

———. 1965. Moths and ultrasound. *Scientific American* 212: 94–102.

———. 1967. *Nerve Cells and Insect Behavior.* Harvard University Press; Cambridge, Massachusetts; xiii+238 pp.

———, and A. E. Treat. 1970. An acoustic sense in some hawkmoths (Choerocampinae). *Journal of Insect Physiology* 18: 1249–64.

Roesler, R. 1944. Die Gattungen der Copeognathen. *Stettiner Entomologisches Zeitung* 105: 117–66.

Rogers, A. F. 1938. Fossil termite pellets in opalized wood from Santa Maria, California. *American Journal of Science* 36: 389–92.

Rogers, B. T., and T. C. Kaufman. 1996. Structure of the insect head as revealed by the EN protein pattern in developing embryos. *Development* 122: 3419–32.

Rohdendorf, B. B., ed. 1961. The oldest dipteran infraorders from the Triassic of middle Asia. *Palaeontologicheskiy Zhurnal* 1961: 90–100. [In Russian]

———. 1962. Order Diptera. Pp. 444–502. *In* Rohdendorf, B. B. (ed.), *Fundamentals of Paleontology, Volume 9: Arthropoda: Tracheata and Chelicerata.* Akademii Nauk; Moscow; 561 pp. [In Russian: English translation in Rohdendorf, B. B., ed. 1991.

———. 1964. The historical development of two-winged insects. *Trudy Paleontologischeskogo Instituta Akademii Nauk, SSSR* 100: 1–311. [In Russian; English translation: Hocking, B., H. Oldroyd, and G. E. Ball, eds. 1974. *The Historical Development of Diptera.* University of Alberta Press; Alberta, Canada; xv+360 pp.]

———. 1968. *Jurassic Insects of Karatau.* Nauka; Moscow, Russia; 252 pp. [In Russian]

———, ed. 1991. *Fundamentals of Paleontology, Volume 9: Arthropoda: Tracheata and Chelicerata.* Smithsonian Institution Libraries; Washington, D.C.; xxxi+894 pp.

Rohr, D. M., A. J. Boucot, J. Miller, and M. Abbott. 1986. Oldest termite nest from the Upper Cretaceous of west Texas. *Geology* 14: 87–8.

Rohwer, R. A. 1908a. A fossil larrid wasp. *Bulletin of the American Museum of Natural History* 24: 519–20.

———. 1908b. On the Tenthredinoidea of the Florissant shales. *Bulletin of the American Museum of Natural History* 24: 521–30.

———. 1908c. The Tertiary Tenthredinoidea of the expedition of 1908 to Florissant, Colo. *Bulletin of the American Museum of Natural History* 24: 591–5.

———. 1909. Three new fossil insects from Florissant, Colorado. *American Journal of Science* 18: 533–6.

———, and R. A. Cushman. 1917. Idiogastra, a new suborder of Hymenoptera with notes on the immature stages of *Oryssus*. *Proceedings of the Entomological Society of Washington* 19: 89–98.

Roig-Alsina, A. 1994. A new genus of Plumariidae, with notes on the relationships among the genera of the family (Hymenoptera, Chrysidoidea, Plumariidae). *Mitteilungen der Münchner Entomologischen Gesellschaft* 84: 91–6.

———, and C. D. Michener. 1993. Studies of the phylogeny and classification of long-tongued bees (Hymenoptera: Apoidea). *University of Kansas Science Bulletin* 55: 123–62.

Roisin, Y. 2000. Diversity and the evolution of caste patterns. Pp. 95–119. *In* Abe, T., D. E. Bignell, and M. Higashi (eds.), *Termites: Evolution, Sociality, Symbioses, Ecology.* Kluwer Academic Publishers; Dordrecht, the Netherlands; xxii+466 pp.

Rokas, A., J. Kathirithamby, and P. W. H. Holland. 1999. Intron insertion as a phylogenetic character: The engrailed homeobox of Strepsiptera does not indicate affinity with Diptera. *Insect Molecular Biology* 8: 527–30.

Rolfe, W. D. I., and C. M. Beckett. 1984. Autecology of Silurian Xiphosurida, Scorpionida, and Phyllocarida. *Special Papers in Palaeontology* 32: 27–37.

Ronquist, F. 1994. Evolution of parasitism among closely related species: Phylogenetic relationships and the origin of inquilinism in gall wasps (Hymenoptera: Cynipidae). *Evolution* 48: 241–66.

———. 1995a. Phylogeny and classification of the Liopteridae, an archaic group of cynipoid wasps (Hymenoptera). *Entomologica Scandinavica, Supplement* 46: 1–74.

———. 1995b. Phylogeny and early evolution of the Cynipoidea (Hymenoptera). *Systematic Entomology* 20: 309–35.

———. 1999. Phylogeny, classification and evolution of the Cynipoidea. *Zoologica Scripta* 28: 139–64.

———, and J. L. Nieves-Aldrey. 2001. A new subfamily of Figitidae (Hymenoptera, Cynipoidea). *Zoological Journal of the Linnean Society* 133: 483–94.

———, A. P. Rasnitsyn, A. Roy, K. Eriksson, and M. Lindgren. 1999. Phylogeny of the Hymenoptera: A cladistic reanalysis of Rasnitsyn's (1988) data. *Zoologica Scripta* 28: 13–50.

Rosenthal, G. A., C. G. Hughes, and D. H. Janzen. 1982. L-Canavanine, a dietary nitrogen sources for the seed predator *Caryedes brasiliensis* (Bruchidae). *Science* 217: 353–5.

Rosler, O., R. Rohn, and L. Albamonte. 1981. Libelula Permiana do estado de São Paulo, Brasil (Formação Irati): *Gonvanoptilon brasiliense* gen. et sp. nov. *Anais do Congresso Latino Americano Paleontologia* 2: 221–32.

Ross, A. J. 1997. Insects in amber. *Geology Today* 13: 24–28.

———. 1998. *Amber: The Natural Time Capsule.* Natural History Museum; London, UK; 73 pp.

———. 2001. The cockroaches (Blattodea) of the Purbeck Limestone Group and Wealden Supergroup (Lower Cretaceous) of southern England. Pp. 59–60. *In* Krzemińska, E., and W. Krzemiński (eds.), *Second International Congress on Palaeoentomology: Abstracts Volume.* Polish Academy of Sciences; Kraków, Poland; 94 pp.

———, and E. Cook. 1995. The stratigraphy and palaeontology of the Upper Weald Clay (Barremian) at Smokejacks Brickworks, Ockley, Surrey, England. *Cretaceous Research* 16: 705–16.

———, and E. A. Jarzembowski. 1993. Arthropoda (Hexapoda; Insecta). Pp. 363–426. *In* Benton, M. J. (ed.), *The Fossil Record 2.* Chapman and Hall; London, UK; xvii+845 pp.

———, and E. A. Jarzembowski. 1996. A provisional checklist of fossil insects from the Purbeck Group of Wiltshire. *Wiltshire Archaeological and Natural History Magazine* 89: 106–15.

———, and P. V. York. 2000. A list of type and figured specimens of insects and other inclusions in Burmese amber. *Bulletin of the Natural History Museum, London (Geology)* 56: 11–20.

———, E. A. Jarzembowski, and S. J. Brooks. 2000. The Cretaceous and Cenozoic record of insects (Hexapoda) with regard to global change. Pp. 288–302. *In* Culver, S. J., and P. F. Rawson (eds.), *Biotic Response to Global Change: The Last 145 Million Years.* Cambridge University Press; Cambridge, UK; xiii+501 pp.

Ross, E. S. 1956. A new genus of Embioptera from Baltic amber. *Mitteilungen aus dem Geologischen Staatsinstitut in Hamburg* 25: 76–81.

———. 1970. Biosystematics of the Embioptera. *Annual Review of Entomology* 15: 157–71.

———. 1984. A synopsis of the Embiidina of the United States. *Proceedings of the Entomological Society of Washington* 86: 82–93.

———. 1987. Studies in the insect order Embiidina: A revision of the family Clothodidae. *Proceedings of the California Academy of Sciences* 45: 9–34.

———. 2000a. EMBIA: Contributions to the biosystematics of the insect order Embiidina. Part 1: Origin, relationships and integumental anatomy of the insect order Embiidina. *Occasional Papers of the California Academy of Sciences* 149: 1–53.

———. 2000b. EMBIA: Contributions to the biosystematics of the insect order Embiidina. Part 2: A review of the biology of Embiidina. *Occasional Papers of the California Academy of Sciences* 149: 1–36.

———. 2001. EMBIA: Contributions to the biosystematics of the insect order Embiidina. Part 3: The Embiidae of the Americas (order Embiidina). *Occasional Papers of the California Academy of Sciences* 150: 1–86.

———. 2003a. EMBIA: Contributions to the biosystematics of the insect order Embiidina. Part 4: Andesembiidae, a new Andean family of Embiidina. *Occasional Papers of the California Academy of Sciences* 153: 1–13.

———. 2003b. EMBIA: Contributions to the biosystematics of the insect order Embiidina. Part 5: A review of the family Anisembiidae with descriptions of new taxa. *Occasional Papers of the California Academy of Sciences* 154: 1–123.

Ross, H. H. 1955. Evolution of the insect orders. *Entomological News* 66: 197–208.

———. 1956. *Evolution and Classification of the Mountain Caddisflies.* University of Illinois Press; Urbana, Illinois; vii+213 pp.

———. 1964. Evolution of caddis worm cases and nets. *American Zoologist* 4: 209–20.

Ross, K. G., and R. W. Matthews, eds. 1991. *The Social Biology of Wasps.* Cornell University Press; Ithaca, New York; xvii+678 pp.

Roth, L. M. 1991. Blattodea, Blattaria (cockroaches). Pp. 320–320. *In* Naumann, I. D. (ed.), *The Insects of Australia: A Textbook for Students and Research Workers, Volume 1* [2nd Edition]. Cornell University Press; Ithaca, New York; xvi+[1]+542 pp.

Rothschild, M. 1983. *Dear Lord Rothschild: Birds, Butterflies, and History.* Arrow Press; London, UK; [xxiii]+398 pp.

Rothschild, W. von, and K. Jordan. 1903. A revision of the lepidopterous family Sphingidae. *Novitates Zoologicae* 9 (Supplement): cxxxv + 1–972.

Rothwell, G. W., and A. C. Scott. 1983. Coprolites within the marattiaceous fern stems (*Psaronius magnificus*) from the Upper Pennsylvanian of the Appalachian Basin, U.S.A. *Palaeogeography, Palaeoclimatology, Palaeoecology* 41: 227–32.

Roubik, D. W., and Q. D. Wheeler. 1982. Flightless beetles and stingless bees: Phoresy of scotocryptine beetles (Leiodidae) on their meliponine hosts (Apidae). *Journal of the Kansas Entomological Society* 55: 125–35.

Rougier, G. W., and M. J. Novacek. 1998. Early mammals: Teeth, jaws, and finally. . . a skeleton! *Current Biology* 8: 284–87.

Rouland-Lefevre, C., M. N. Diouf, A. Brauman, and M. Neyra. 2002. Phylogenetic relationships in *Termitomyces* (family Agaricaceae) based on the nucleotide sequence of ITS: A first approach to elucidate the evolutionary history of the symbiosis between fungus-growing termites and their fungi. *Molecular Phylogenetics and Evolution* 22: 423–9.

Rouse, G. W., and K. Fauchauld. 1997. Cladistics and polychaetes. *Zoologica Scripta* 26: 139–204.

Rowell, C. H. F., and P. K. Flook. 1998. Phylogeny of the Caelifera and the Orthoptera as derived from ribosomal gene sequences. *Journal of Orthoptera Research* 7: 147–56.

Rowland, J. M., and W. D. Sissom. 1980. Report on a fossil palpigrade from the Tertiary of Arizona, and a review of the morphology and systematics of the order (Arachnida: Palpigradida). *Journal of Arachnology* 8: 69–86.

Roy, K., and L. Fåhraeus. 1989. Tremadocian (Early Ordovician) nauplius-like larvae from the Middle Arm Point Formation, Bay of Islands, western Newfoundland. *Canadian Journal of Earth Sciences* 26: 1802–6.

Roy, R. 1999. Morphology and taxonomy. Pp. 19–42. *In* Prete, F. R., H. Wells, P. H. Wells, and L. E. Hurd (eds.), *The Praying Mantids.* Johns Hopkins University Press; Baltimore, Maryland; xiv+362 pp.

Rozefelds, A. C. 1988. Insect leaf mines from the Eocene Angelsea locality, Victoria, Australia. *Alcheringia* 12: 51–7.

———, and M. De Baar. 1991. Silicified Kalotermitidae (Isoptera) frass in conifer wood from a mid-Tertiary rainforest in central Queensland, Australia. *Lethaia* 24: 439–42.

———, and I. Sobbe. 1987. Problematic insect leaf mines from the Upper Triassic Ipswich Coal Measures of southeastern Queensland, Australia. *Alcheringia* 11: 51–7.

Rozen, J. G., Jr. 1971. *Micromalthus debilis* LeConte from amber of Chiapas, Mexico (Coleoptera, Micromalthidae). *University of California Publications in Entomology* 63: 75–6.

———. 1984. Nesting biology of diphaglossine bees (Hymenoptera, Colletidae). *American Museum Novitates* 2786: 1–33.

———. 1989. Morphology and systematic significance of first instars of the cleptoparasitic bee tribe Epeolini (Anthophoridae: Nomadinae). *American Museum Novitates* 2957: 1–19.

———. 1991. Evolution of cleptoparasitism in anthophorid bees as revealed by their mode of parasitism and first instars (Hymenoptera: Apoidea). *American Museum Novitates* 3029: 1–36.

———. 1996. A new species of the bee *Heterosarus* from Dominican amber (Hymenoptera: Andrenidae; Panurginae). *Journal of the Kansas Entomological Society, Supplement* 69: 346–52.

———. 2000. Systematic and geographic distributions of Neotropical cleptoparasitic bees, with notes on their modes of parasitism. Pp. 204–10. *In* Bitondi, M. M. G., and K. Hartfelder (eds.), *Anais do IV Encontro sobre Abelhas.* Universidade de São Paulo; Ribeirão Preto, Brazil; xxix+[1]+363 pp.

———. 2003. Eggs, ovariole numbers, and modes of parasitism of cleptoparasitic bees, with emphasis on Neotropical species (Hymenoptera: Apoidea). *American Museum Novitates* 3413: 1–36.

———, and H. Özbek. 2003. Oocytes, eggs, and ovarioles of some long-tongued bees (Hymenoptera: Apoidea). *American Museum Novitates* 3393: 1–35.

Rudolf, D. 1982. Occurrence, properties, and biological implications of the active uptake of water vapour from the atmosphere in Psocoptera. *Journal of Insect Physiology* 28: 111–21.

———. 1983. The water-vapour uptake system of the Phthiraptera. *Journal of Insect Physiology* 29: 15–25.

Rundle, A. J., and J. Cooper. 1971. Occurrence of a fossil insect larva from the London Clay of Herne Bay, Kent. *Proceedings of the Geologists' Association* 82: 293–5.

Rupke, N. A. 1994. *Richard Owen: Victorian Naturalist.* Yale University Press; New Haven, Connecticut; xvii+462 pp.

Russell, L. K. 1979. A new genus and a new species of Boreidae from Oregon (Mecoptera). *Proceedings of the Entomological Society of Washington* 81: 22–31.

———. 1982. The life history of *Caurinus dectes* Russell, with a description of the immature stages (Mecoptera: Boreidae). *Entomologica Scandinavica* 13: 225–35.

Rust, J. 1999. Fossil insects from the Fur and Olst Formations ("mo-clay") of Denmark (Upper Paleocene/lowermost Eocene). Pp. 135–9. *In* Vršanský, P. (ed.), *Proceedings of the First Palaeoentomological Conference, Moscow, 1998.* AMBA Projects; Bratislava, Slovakia; 199+[2] pp.

———. Fossil record of mass moth migration. *Nature* 405: 530–1.

———, and N. M. Andersen. 1999. Giant ants from the Paleogene of Denmark with a discussion of the fossil history and early evolution of ants (Hymenoptera: Formicidae). *Zoological Journal of the Linnean Society* 125: 331–48.

———, A. Stumpner, and J. Gottwald. 1999. Singing and hearing in a Tertiary bushcricket. *Nature* 399: 650.

Rutten, M. G. 1966. Geologic data on atmospheric history. *Palaeogeography, Palaeoclimatology, Palaeoecology* 2: 47–57.

Ruttner, F. 1988. *Biogeography and Taxonomy of Honeybees.* Springer Verlag; Berlin, Germany; xxii+284 pp.

Saether, O. A. 2000. Phylogeny of the Culicomorpha (Diptera). *Systematic Entomology* 25: 223–34.

Sakagami, S. F., and C. D. Michener. 1962. *The Nest Architecture of the Sweat Bees (Halictinae): A Comparative Study of Behavior.* University of Kansas Press; Lawrence, Kansas; iv+135 pp.

———, T. Matsumura, and K. Ito. 1980. *Apis laboriosa* in Himalaya, the little known world largest honeybee (Hymenoptera: Apidae). *Insecta Matsumurana* 19: 47–77.

Samigullin, T. K., W. F. Martin, A. V. Troitsky, and A. S. Antonov. 1999. Molecular data from the chloroplast rpoC1 gene suggest a deep and distinct dichotomy of contemporary spermatophytes into two monophyla: Gymnosperms (including Gnetales) and angiosperms. *Journal of Molecular Evolution* 49: 310–15.

Samways, M. J. 1994. *Insect Conservation Biology.* Chapman and Hall; London, UK; xiii+358 pp.

Sanderson, M. G. 1996. Biomass of termites and their emissions of methane and carbon dioxide: A global database. *Global Biogeochemical Cycles* 10: 543–57.

Sanderson, M. J. 1997. A nonparametric approach to estimating divergence times in the absence of rate constancy. *Molecular Biology and Evolution* 14: 1218–32.

Sandoval, C. P., and V. R. Vickery. 1996. *Timema douglasi* (Phasmatoptera: Timematodea), a new parthenogenetic species from southwestern Oregon and northern California, with notes on other species. *Canadian Entomologist* 128: 79–84.

———, and V. R. Vickery. 1998. *Tinema* [sic] *coffmani* (Phasmatoptera: Timematodea) a new species from Arizona and description of the female of *Timema ritensis*. *Journal of Orthoptera Research* 7: 103–6.

———, D. A. Carmean, and B. J. Crespi. 1998. Molecular phylogenetics of sexual and parthenogenetic *Timema* walking-sticks.

Proceedings of the Royal Society of London, Series B, Biological Sciences 265: 589–95.
Sands, W. A. 1969. The association of termites and fungi. Pp. 495–524. *In* Krishna, K., and F. M. Weesner (eds.), *Biology of Termites, Volume I.* Academic Press; New York, New York; xiii+598 pp.
———. 1987. Ichnocoenoses of probable termite origin from Laetoli. Pp. 409–33. *In* Leakey, M. D., and J. M. Harris (eds.), *Laetoli: A Pliocene Site in Northern Tanzania.* Oxford University Press; Oxford, UK; xxii+561 pp.
Santiago-Blay, J. A. 1994. Paleontology of leaf beetles. Pp. 1–68. *In* Jolivet, P. H., M. L. Cox, and E. Petitpierre (eds.), *Novel Aspects of the Biology of Chrysomelidae.* Kluwer Academic Publishers; Dordrecht, the Netherlands; xxiii+582 pp.
———, C. C. Labandeira, L. Pribyl, C. Hotton, and L. D. Martin. 2001. The Sundance insect fauna (middle Jurassic) of northern Wyoming and southern Montana. *Abstracts with Programs, Geological Society of America* 32: A266.
Sattler, K. 1991. A review of wing reduction in Lepidoptera. *Bulletin of the British Museum of Natural History (Entomology)* 60: 243–88.
Saunders, W. B., R. H. Mapes, F. M. Carpenter, and W. C. Elsik. 1974. Fossiliferous amber from the Eocene (Claiborne) of the Gulf Coastal Plain. *Geological Society of America Bulletin* 85: 979–84.
Savage, N. M., M. A. Lindorfer, and D. A. MacMillan. 1990. Amino acids from Ordovician conodonts. *Courier Forschungsreisen-Institut Senckenbergiana* 118: 267–75.
Savolainen, V., M. W Chase, S. B. Hoot, C. M. Morton, D. E. Soltis *et al.* 2000. Phylogenetics of flower plants based on combined analysis of plastid atpB and rbcL gene sequences. *Systematic Biology* 49: 306–62.
Schaal, S., and W. Ziegler. 1992. *Messel – An Insight into the History of Life and of the Earth.* Clarendon Press; Oxford, UK; 322 pp.
Schaefer, C. W., and P. L. Mitchell. 1983. Food plants of the Coreoidea (Hemiptera: Heteroptera). *Annals of the Entomological Society of America* 76: 591–615.
Schawaller, W. 1979. Erstnachweis der Ordnung Geisselspinnen in Dominikanischem Berstein (Stuttgarter Bersteinsammlung: Arachnida, Amblypygi). *Stuttgarter Beiträge zur Naturkunde, Serie B, Geologie und Paläontologie* 50: 1–12.
———. 1982. Der erte Pseudoskorpion (Chernetidae) aus Mexikanischen Bernstein (Stuttgarter Bersteinsammlung: Arachnida, Pseudoscorpionoidea). *Stuttgarter Beiträge zur Naturkunde, Serie B, Geologie und Paläontologie* 85: 1–9.
———. 1991. The first Mesozoic pseudoscorpion, from Cretaceous Canadian amber. *Palaeontology* 34: 971–6.
———, W. A. Shear, and P. M. Bonamo. 1991. The first Palaeozoic pseudoscorpions (Arachnida, Pseudoscorpionida). *American Museum Novitates* 3009: 1–17.
Schawaroch, V. 2002. Phylogeny of a paradigm lineage: The *Drosophila melanogaster* species group (Diptera: Drosophilidae). *Biological Journal of the Linnean Society* 76: 21–37.
Schedl, W. 1991. Hymenoptera, Unterordnung Symphyta: Pflanzenwespen. *Handbuch der Zoologie* 31: 1–136.
Scheller, U., and J. Wunderlich. 2001. First description of a fossil pauropod, *Eopauropus balticus* n. gen. n. sp. (Pauropoda: Pauropodidae), in Baltic amber. *Mitteilungen aus dem Geologisch-Paläontologischen Institut der Universität Hamburg* 85: 221–7.
Schenck, A. 1861. Die nassauischen Bienen. *Jahrbücher des Vereins für Naturkunde im Herzogthum Nassau* 14: 1–414.
———. 1869. Beschreibung der nassauischen Bienen, Zweiter Nachtrag. *Jahrbücher des Vereins des Nassauischen Vereins für Naturkunde* 21–2: 1–114.
Schidlowski, M. 1971. Probleme der atmosphärischen Evolution im Präkambrium. *Geologische Rundschau* 60: 1351–84.
Schildknecht, H., U. Maschwitz, and H. Winkler. 1968. Zur evolution der Carabiden-Wehrdrüsensekrete. *Naturwissenschaften* 55: 112–17.
Schlee, D. 1969a. Morphologie und Symbiose; ihre Beweiskraft für die Verwandtschaftsbeziehungen der Coleorrhyncha (Insectga, Hemiptera). Phylogenetische Studien an Hemiptera IV: Heteropteroidea (Heteroptera + Coleorrhyncha) als monophyletische Gruppe. *Stuttgarter Beiträge zur Naturkunde* 210: 1–27.
———. 1969b. Sperma-Übertragung (und andere Merkmale in ihrer Bedeutung für das phylogenetische System der Sternorrhyncha. Phylogenetische Studien an Hemiptera I. Psylliformes (Psyllina + Aleyrodina) als monophyletische Gruppe. *Zeitschrift fur Morphologie der Tiere* 64: 95–138.
———. 1969c. Die Verwandtschaftsbeziehungen innerhalb der Sternorrhyncha aufgrund synapomorpher Merkmale. Phylogenetische Studien an Hemiptera II. Aphidiformes (Aphidina + Coccina) als monophyletische Gruppe. *Stuttgarter Beiträge zur Naturkunde* 199: 1–19.
———. 1969d. Der Flügel von *Sphaeraspis* (Coccina), prinzipiell identisch mit Aphidina-Flügeln. Phylogenetische Studien an Hemiptera V: Synapomorphie Flügelmerkmale bei Aphidina und Coccina. *Stuttgarter Beiträge zur Naturkunde* 211: 1–11.
———. 1970. Verwandtschaftsforschung an fossilen und rezenten Aleyrodina (Insecta, Hemiptera). *Stuttgarter Beiträge zur Naturkunde* 213: 1–71.
———. 1978. In memoriam Willi Hennig 1913–1976. Eine biographische Skizze. *Entomologica Germanica* 4: 377–91.
———. 1980. *Bernstein-Raritäten.* Staatliches Museum für Natukunde; Stuttgart, Germany; 88 pp.
———. 1984. Bernstein-Neuigkeiten. *Stuttgarter Beiträge zur Naturkunde, Serie C, Allgemeinverstandliche Aufsätze* 18: 1–100.
———. 1986. *Der Bernsteinwald.* Staatliches Museum für Naturkunde; Stuttgart, Germany; 100 pp.
———. 1990. Das Bernstein-Kabinett. *Stuttgarter Beiträge zur Naturkunde, Serie C, Allgemeinverstandliche Aufsätze* 28: 1–100.
———, and H. G. Dietrich. 1970. Insektenführender Bernstein aus der Unterkreide des Libanon. *Neues Jahrbuch für Geologie und Paläontologie, Montshefte* 1970: 40–50.
Schliephake, G. 1975. Beitrag zur phylogenetischen Systematik bei Thysanoptera (Insecta). *Beiträge zur Entomologie* 25: 5–13.
———. 1990. Beiträge zur Kenntnis fossiler Fransenflügler (Thysanoptera, Insecta) aus dem Bernstein des Tertiär. 1. Beitrag: Stenurothripidae. *Zoology (Journal of Pure and Applied Zoology), Delhi* 2: 163–84.
———. 1993. Beiträge zur Kenntnis fossiler Fransenflügler (Thysanoptera, Insecta) aus dem Bernstein des Tertiär. 2. Beitrag: Aeolothripidae (Melanthripinae) und Thripidae (Dendrothripinae und Thripinae). *Zoologisches Jahrbücher, Abteilung für Systematik, Ökologie und Geographie der Tiere* 120: 215–51.
———. 1997. Beiträge zur Kenntnis fossiler Fransenflügler (Thysanoptera, Insecta) aus dem Bernstein des Tertiär des Bitterfelder Raumes (3. Beitrag: Thripidae, Panchaetothripinae). *Entomologische Nachrichten und Berichte* 41: 66–7.
———. 1999. Fossil thrips (Thysanoptera, Insecta) of the Baltic (North- and Baltic Sea) and Saxonian (Bitterfeldian) Tertiary amber from the collections of Hoffeins. *Deutsche Entomologische Zeitschrift* 46: 83–100.
———. 2000. Neue Fransenflügler aus dem Baltischen und Bitterfelder Bernstein (Insecta: Thysanoptera). *Mitteilungen aus dem Geologisch-Paläontologischen Institut der Universität Hamburg* 84: 219–29.
———. 2001. Thysanoptera (Insecta) of the Tertiary amber of the Museum of the Earth, Warsaw, with keys to the species of the Baltic and Bitterfeld amber. *Prace Muzeum Ziemi* 46: 16–38.
———, and K. Klimt. 1979. Thysanoptera, Fransflügler. Pp. 132–6. *In* Stresemann, E. (ed.), *Exkursionsfauna für die Gebiete der DDR und der BRD. Wirbellose, Band II/1: Insekten – Erster Teil.* Volk und Wissen Volkseigener Verlag; Berlin, Germany; 504 pp.
Schlüter, T. 1978. Zur Systematik und Palökologie harzkonservierter Arthropoda einer Taphozönose aus dem Cenomanium von NW-Frankreich. *Berliner Geowissenschaftliche Abhandlungen, Reihe A, Geologie und Paläontologie* 9: 1–150.
———. 1982. *Cimbrochrysa moleriensis* n. g. n. sp. und *Hypochrysa hercyniensis* n. sp., zwei fossile Chrysopidae-Arten (Insecta: Planipennia) aus dem europäischen Tertiär. *Neues Jahrbuch für Geologie und Paläontologie, Monatshefte* 1982: 257–64.
———. 1983. A fossiliferous resin from the Cenomanium [sic] of the Paris- and the Aquitanian Basins of northwestern France. *Cretaceous Research* 4: 265–9.
———. 1984. Phylogeny of Chrysopidae. Pp. 1–8. *In* Canard, M., Y. Séméria, and T. R. New (eds.), *Biology of Chrysopidae.* Junk Publishers; The Hague, the Netherlands; 294 pp.
———. 1986. The fossil Planipennia – A review. Pp. 103–11. *In* Gepp, J., H. Aspöck, and H. Hölzel (eds.), *Recent Research in Neuropterology: Proceedings of the Second International Symposium on Neuropterology.* Druckhaus Thalerhof; Graz, Austria; 176 pp.
———. 1990. Fossil insect localities in Gondwanaland. *Entomologica Generalis* 15: 61–76.
———. 2000. *Moltenia rieki* n. gen., n. sp. (Hymenoptera: Xyelidae?), a tentative sawfly

from the Molteno Formation (Upper Triassic), South Africa. *Paläontologische Zeitschrift* 74: 75–8.

———. 2003a. Dragonflies preserved in transparent gypsum crystals from the Messinian (Upper Miocene) of Alba, northern Italy. *Acta Zoologica Cracoviensia* 46 (Supplement): 373–9.

———. 2003b. Fossil insects in Gondwana – Localities and palaeodiversity trends. *Acta Zoologica Cracoviensia* 46 (Supplement): 345–71.

———, and R. Kohring. 1990. Die Zwergwespengattung *Palaeomymar* (Hymenoptera: Proctotrupoidea: Serphitidae) aus dem mio/pliozänen Simetit Siziliens. *Berliner Geowissenschaftlische Abhandlungen, Reihe A, Geologie und Paläontologie* 124: 115–21.

———, and R. Kohring. 2001. Dragonflies preserved in gypsum crystals from the Messinian (Upper Miocene) of northern Italy. Pp. 60–1. *In* Krzemińska, E., and W. Krzemiński (eds.), *Second International Congress on Palaeoentomology: Abstracts Volume.* Polish Academy of Sciences; Kraków, Poland; 94 pp.

———, and W. Stürmer. 1984. Die Identifikation einer fossilen Rhachiberothinae-Art (Planipennia: Berothidae oder Mantispidae) aus mittelkretazischem Bernstein NW-Frankreichs mit Hilfe röntgenographischer Methoden. Pp. 49–55. *In* Gepp, J., H. Aspöck, and H. Hölzel (eds.), *Progress in World's Neuropterology: Proceedings of the First International Symposium on Neuropterology.* Druckhaus Thalerhof; Graz, Austria; 265 pp.

Schmidt-Ott, U. 2000. The amnioserosa is an apomorphic character of cyclorrhaphan flies. *Development, Genes, and Evolution* 210: 373–6.

———, M. Gonzalez-Gaitan, H. Jackle, and G. M. Technau. 1994. Number, identity, and sequence of *Drosophila* head segments as revealed by neural elements and their deletion patterns in mutants. *Proceedings of the National Academy of Sciences, USA* 91: 8363–7.

———, M. Gonzalez-Gaitan, and G. M. Technau. 1995. Analysis of neural elements in head-mutant *Drosophila* embryos suggests segmental origin of the optic lobes. *Roux's Archives of Developmental Biology* 205: 31–44.

Schmiedeknecht, H. L. O. 1882. *Apidae Europaeae* [Volume 1]. Gumperdae and Berolini; Berlin, Gemany; xiv+866 pp.

Schmitt, M. 1996. The phylogenetic system of the Chrysomelidae – History of ideas and present state of knowledge. Pp. 57–96. *In* Jolivet, P. H. A., and M. L. Cox (eds.), *Chrysomelidae Biology, Volume 1: The Classification, Phylogeny, and Genetics.* SPB Academic Publishing; Amsterdam, the Netherland.

Schmitz, J., and R. F. A. Moritz. 1998. Molecular phylogeny of Vespidae (Hymenoptera) and the evolution of sociality in wasps. *Molecular Phylogenetics and Evolution* 9: 183–91.

Schneider, J. 1983. Die Blattodea (Insecta) des Paleozoicums, Teil I. Systematik, Ökologie und Biostratigraphie. *Freiberger Forschungshefte, Reihe C* 382: 106–46.

———. 1984. Die Blattodea (Insecta) des Paleozoicums, Teil II. Morphogenese der Flügelstrukturen und Phylogenie. *Freiberger Forschungshefte, Reihe C* 391: 5–34.

Schneider, J. W., and R. Werneburg. 1998. *Arthropleura* und Diplopoda (Arthropoda) aus dem Unter-Rotliegend (Unter-Perm, Assel) des Thüringer Waldes (südwest-Saale-Senke). *Veröffentlichungen Naturhistorisches Museum Schloss Bertholdsburg Schleusingen* 13: 19–36.

Schoch, G., von. 1884. Ueber die Gruppirung der Insekten-Ordnungen. *Mitteilungen der Schweizerischen Entomologischen Gesellschaft* 7: 34–6.

Scholtz, G. 1995. Head segmentation in Crustacea – An immunocytochemical study. *Zoology* 98: 104–14.

———. 2002. The Articulata hypothesis – or what is a segment? *Organisms, Diversity and Evolution* 2: 197–215.

———, B. Mittmann, and M. Gerberding. 1998. The pattern of *Distal*-less expression in the mouthparts of crustaceans, myriapods and insects: New evidence for a gnathobasic mandible and the common origin of Mandibulata. *International Journal of Developmental Biology* 42: 801–10.

Schopf, J. M. 1975. Modes of fossil preservation. *Review of Palaeobotany and Palynology* 20: 27–53.

Schorn, H. E., H. I. Scudder, D. E. Savage, and J. R. Firby. 1989. General stratigraphy and paleontology of the Miocene continental sequence in Stewart Valley, Mineral Valley, Nevada, USA. Pp. 96–115. *In* Liu, G., R. Tsuchi, and Q. Lin (eds.), *Proceedings of the International Symposium on Pacific Neogene Continental and Marine Events.* Nanjing University Press; Nanjing, China.

Schram, F. R., ed. 1983a. Crustacean phylogeny. *Crustacean Issues* 1: xi+1–372.

———. 1983b. Lower Carboniferous biota of Glencartholm, Eskdale, Dumfriesshire. *Scottish Journal of Geology* 19: 1–15.

———. 1986. *Crustacea.* Oxford University Press; Oxford, UK; xiv+606 pp.

———, and C. H. J. Hof. 1998. Fossils and the interrelationships of major crustacean groups. Pp. 233–302. *In* Edgecombe, G. D. (ed.), *Arthropod Fossils and Phylogeny.* Columbia University Press; New York, New York 347 pp.

Schuh, R. T. 1995. *Plant Bugs of the World (Insecta, Heteroptera, Miridae): Systematic Catalogue, Distributions, Host List, and Bibliography.* New York Entomological Society; New York, New York; xii+1329 pp.

———, and J. A. Slater. 1995. *True Bugs of the World (Hemiptera: Heteroptera): Classification and Natural History.* Cornell University Press; Ithaca, New York; xii+337 pp.

———, and P. Štys. 1991. Phylogenetic analysis of cimicomorphan family relationships (Heteroptera). *Journal of the New York Entomological Society* 99: 298–350.

Schulmeister, S. 2003a. Review of morphological evidence on the phylogeny of basal Hymenoptera (Insecta), with a discussion of the ordering of characters. *Biological Journal of the Linnean Society* 79: 209–43.

———. 2003b. Simultaneous analysis of basal Hymenoptera (Insecta): Introducing robust-choice sensitivity analysis. *Biological Journal of the Linnean Society* 79: 245–75.

———, W. C. Wheeler, and J. M. Carpenter. 2002. Simultaneous analysis of the basal lineages of Hymenoptera (Insecta) using sensitivity analysis. *Cladistics* 18: 455–84.

Schultz, T. R., and R. Meier. 1995. A phylogenetic analysis of the fungus-growing ants (Hymenoptera: Formicidae: Attini) based on morphological characters of the larvae. *Systematic Entomology* 20: 337–70.

———, M. S. Engel, and M. Prentice. 1999. Resolving conflict between morphological and molecular evidence for the origin of eusociality in the "corbiculate" bees (Hymenoptera: Apidae): A hypothesis-testing approach. *University of Kansas Natural History Museum Special Publication* 24: 125–38.

———, M. S. Engel, and J. S. Ascher. 2001. Evidence for the origin of eusociality in the corbiculate bees (Hymenoptera: Apidae). *Journal of the Kansas Entomological Society* 74: 10–16.

Schulz, W. A. 1907. Hymenoptera, Fam. Trigonaloidae [sic]. *Genera Insectorum* 61: 1–24.

Schuster, J. C., and L. B. Schuster. 1985. Social behavior in passalid beetles (Coleoptera: Passalidae): Cooperative brood care. *Florida Entomologist* 68: 266–72.

———, and L. B. Schuster. 1997. The evolution of social behavior in Passalidae (Coleoptera). Pp. 260–9. *In* Choe, J. C., and B. J. Crespi (eds.) *The Evolution of Social Behavior in Insects and Arachnids.* Cambridge University Press; Cambridge, UK; vii+541 pp.

Schuster, M., P. Duringer, A. Nel, M. Brunet, P. Vignaud, and H. T. Mackaye. 2000. Decouverte de termiteres fossiles dans les sites a vertebres du Pliocene tchadien: Description, identification et implications paleoecologiques. *Comptes Rendus de l'Academie des Sciences, Serie II A, Sciences de la Terre et des Planetes* 331: 15–20.

Schuster, R., and P. W. Murphy, eds. 1991. *The Acari: Reproduction, Development and Life History Strategies.* Chapman and Hall; London, UK; xxiii+554 pp.

Schweigert, G., and G. Bechly. 2001. Bibliographie zur Geologie und Paläontologie des Randecker Maars (Unter-Miozän, Südwestdeutschland) 1825–2000. *Stuttgarter Beiträge zur Naturkunde, serie B, Geologie und Paläontologie* 302: 1–12.

Scoble, M. J. 1986. The structure and affinities of the Hedyloidea: A new concept of butterflies. *Bulletin of the British Museum (Natural History) (Entomology)* 53: 251–86.

———. 1995. *The Lepidoptera: Form, Function, and Diversity* [2nd Edition]. Oxford University Press; Oxford, UK; xi+404 pp.

———, ed. 1999. *Geometrid Moths of the World: A Catalogue (Lepidoptera, Geometridae)* [Volumes 1 & 2]. The Natural History Museum; London, UK; xxv+1016 pp., CD-ROM.

———, and A. Aiello. 1990. Moth-like butterflies (Hedylidae: Lepidoptera): A summary, with comments on the egg. *Journal of Natural History* 24: 159–64.

Scotese, C. R. 1991. Jurassic and Cretaceous plate tectonic reconstruction. *Palaeogeography, Palaeoclimatology, Palaeoecology* 87: 493–501.

Scott, A. C. 1991. Evidence for plant-arthropod interactions in the fossil record. *Geology Today* 7: 58–61.

———, and T. N. Taylor. 1983. Plant/animal interactions during the Upper Carboniferous. *Botanical Review* 49: 259–307.

———, J. Stephenson, and W. G. Chaloner. 1991. Interaction and coevolution of plants and arthropods during the Palaeozoic and Mesozoic. *Philosophical Transactions of the Royal Society, London (B)* 335: 129–65.

———, J. Stephenson, and M. E. Collinson. 1994. The fossil record of leaves with galls. Pp. 447–70. *In* Williams, M. A. J. (ed.) *Plant Galls: Organisms, Interactions, Populations*. Clarendon Press; Oxford, UK; xiv+487 pp.

Scott, J. A. and D. M. Wright. 1990. Butterfly phylogeny and fossils. Pp. 152–208. *In* Kudrna, O. (ed.), *Butterflies of Europe, II*. AULA-Verlag; Wiesbaden, Germany.

Scourfield, D. J. 1940a. The oldest known fossil insect. *Nature* 145: 799–801.

———. 1940b. The oldest known fossil insect (*Rhyniella praecursor* Hirst & Maulik) – further details from additional specimens. *Proceedings of the Linnean Society of London* 152: 113–31.

Scudder, G. G. E. 1961. The comparative morphology of the insect ovipositor. *Transactions of the Royal Entomological Society, London* 113: 25–40.

Scudder, S. H. 1875. Fossil butterflies. *Memoirs of the American Association for the Advancement of Science* 1: 1–99.

———. 1877. Description of two new species of Carabidae found in the Inter-glacial deposits at Scarborough Hts., in Toronto, Canada. *United States Geological Survey of the Territories* B3: 763–4.

———. 1878a. An account of some insects of unusual interest from the Tertiary rocks of Colorado and Wyoming. *Bulletin of the United States Geological and Geographical Surveys of the Territories* 4: 519–43.

———. 1878. Additions to the insect-fauna of the Tertiary beds at Quesnel, British Columbia. *Geological Survey of Canada, Report of Progress 1876–1877* 1878: 457–64.

———. 1885. Systematische Übersicht der fossilen Myriopoden [sic], Arachnoideen, und Insekten. Pp. 721–831. *In* Zittel, K. A. (ed.), *Handbuch der Palaeontologie, Abteilung I, Zweiter Band, Palaeozoologie*. Oldenbuch; Munich, Germany; 831 pp.

———. 1886. Systematic review of our present knowledge of fossil insects including myriapods and arachnids. *Bulletin of the United States Geological Survey* 5: 1–128.

———. 1889. The fossil butterflies of Florissant. *United States Geological Survey, Annual Report* 8: 439–74.

———. 1890a. The Tertiary insects of North America. *Report of the United States Geological Survey of the Territories* 13: 1–734.

———. 1890b. The insects of the Triassic beds at Fairplay, Colorado. *Memoirs of the Boston Society of Natural History* 4: 457–72.

———. 1892. Some insects of special interest from Florissant, Colorado and other points in the Tertiaries of Colorado and Utah. *United States Geological Survey Bulletin* 93: 1–25.

Seeger, W. 1975. Funktionsmorphologie an Spezialbildungen der Fühlergeissel von Psocoptera und anderen Paraneoptera (Insecta); Psocoptera als monophyletische Gruppe. *Zeitschrift für Morphologie der Tiere* 81: 137–59.

———. 1979. Spezialmerkmale an Eihüllen und Embryonen von Psocoptera im Vergleich zu anderen Paraneoptera (Insecta); Psocoptera als monophyletische Gruppe. *Stuttgarter Beiträge zur Naturkunde, Serie A, Biologie* 329: 1–57.

Seeley, T. D. 1985. *Honeybee Ecology: A Study of Adaptation in Social Life*. Princeton University Press; Princeton, New Jersey; x+201 pp.

———. 1995. *The Wisdom of the Hive: The Social Physiology of Honey Bee Colonies*. Harvard University Press; Cambridge, Massachusetts; xiv+295 pp.

Seelinger, G., and U. Seelinger. 1983. On the social organisation, alarm and fighting in the primitive cockroach *Cryptocercus punctulatus* Scudder. *Zeitschrift für Tierpsychologie* 61: 315–33.

Seeno, T. N., and J. A. Wilcox. 1982. Leaf beetle genera (Coleoptera: Chrysomelidae). *Entomography* 1: 1–221.

Séguy, E. 1951. Ordre des Diptères (Diptera Linné, 1758). Pp. 449–744. *In* Grassé, P. P. (ed.), *Traité de Zoologie, Tome 10*. Masson et Cie; Paris, France; 975 pp.

Sehnal, F., P. Švacha, and J. Zrzavý. 1996. Evolution of insect metamorphosis. Pp. 3–58. *In* Gilbert, L. I., J. R. Tata, and B. G. Atkinson (eds.), *Metamorphosis: Postembryonic Reprogramming of Gene Expression in Amphibian and Insect Cells*. Academic Press; San Diego, California; xvi+687 pp.

Seilacher, A., W.-E. Reif, and F. Westphal. 1985. Sedimentological, ecological and temporal patterns of fossil Lagerstätten. *Philosophical Transactions of the Royal Society, London (B)* 311: 5–23.

Seki, K, and M. Toyoshima. 1998. Preserving tardigrades under pressure. *Nature* 395: 853–4.

Selden, P. A. 1990. Lower Cretaceous spiders from the Sierra de Montsech, northeast Spain. *Palaeontology* 33: 257–85.

———. 1992. Revision of the fossil ricinuleids. *Transactions of the Royal Society of Edinburgh, Earth Sciences* 83: 595–634.

———. 1996. Fossil mesothele spiders. *Nature* 379: 498–9.

———. 2001. Eocene spiders from the Isle of Wight with preserved respiratory structures. *Palaeontology* 44: 695–729.

———. 2002. Missing links between *Argyroneta* and Cybaeidae revealed by fossil spiders. *Journal of Arachnology* 30: 189–200.

———, and J. A. Dunlop. 1998. Fossil taxa and relationships of chelicerates. Pp. 303–31. *In* Edgecombe, G. D. (ed.), *Arthropod Fossils and Phylogeny*. Columbia University Press; New York, New York; 347 pp.

———, and J.-C. Gall. 1992. A Triassic mygalomorph spider from the northern Vosges, France. *Palaeontology* 35: 211–35.

———, and A. J. Jeram. 1989. Palaeophysiology of terrestrialisation in the Chelicerata. *Transactions of the Royal Society of Edinburgh, Earth Sciences* 80: 303–10.

———, and W. A. Shear. 1996. First Mesozoic solpugid (Arachnida), from the Cretaceous of Brazil, and a redescription of the Palaeozoic solpugid. *Palaeontology* 39: 583–604.

———, W. A. Shear, and P. M. Bonamo. 1991. A spider and other arachnids from the Devonian of New York, and reinterpretations of Devonian Araneae. *Palaeontology* 34: 241–81.

Sellards, E. H. 1904. A study of the structure of Paleozoic cockroaches, with descriptions of new forms from the Coal Measures. *American Journal of Science* 18: 113–34, 213–27.

Sellés-Martínez, J. 1996. Concretion morphology, classification, and genesis. *Earth Science Reviews* 41: 177–210.

Sellick, J. T. C. 1995. Phasmida (stick insect) eggs from the Eocene of Oregon. *Palaeontology* 37: 913–21.

———. 1997a. Descriptive terminology of the phasmid egg capsule, with an extended key to the phasmid genera based on egg structure. *Systematic Entomology* 22: 97–122.

———. 1997b. Eggs and classification – The phasmid connection. *Linnean* 13: 29–34.

———. 1997c. The range of egg capsule morphology within the Phasmatodea and its relevance to the taxonomy of the order. *Italian Journal of Zoology* 64: 97–104.

———. 1998. The micropylar plate of the eggs of Phasmida, with a survey of the range of plate form within the order. *Systematic Entomology* 23: 203–28.

Sellnick, M. 1931. Milben im Bernstein. *Bernsteinforschungen* 2: 148–80.

Sendel, N. 1742. *Historia Succinorum, Corpora Aliena Involvientum et Naturae Opera Pictorum et Caelatorum ex Regiis Augustorum cimeliis Dresdae conditis aeri Insculptorum Conscripta*. Gleditschium; Lipsiae [Leipzig], Germany; viii+[2]+328 pp., xiii pls.

Sepkoski, J. J., Jr. 1989. Periodicity in extinction and the problem of catastrophism in the history of life. *Journal of the Geological Society of London* 146: 7–19.

Serrão, J. E. 2001. A comparative study of the proventricular structure in corbiculate Apinae (Hymenoptera, Apidae). *Micron* 32: 379–85.

Shabica, C. W., and A. A. Hay, eds. 1997. *Richardson's Guide to the Fossil Fauna of Mazon Creek*. Northeastern Illinois University; Chicago, Illinois; xvii+308 pp.

Sharkey, M. J., and A. Roy. 2002. Phylogeny of the Hymenoptera: A reanalysis of the Ronquist *et al.* (1999) reanalysis, emphasizing wing venation and apocritan relationships. *Zoologica Scripta* 31: 57–66.

———, and D. B. Wahl. 1992. Cladistics of the Ichneumonoidea (Hymenoptera). *Journal of Hymenoptera Research* 1: 15–24.

Sharov, A. 1948. Triassic Thysanura from the Ural forelands. *Doklady Akademii Nauk, SSSR* 61: 517–19. [In Russian]

———. 1957. Distinctive Paleozoic wingless insects of a new order Monura (Insecta, Apterygota). *Doklady Akademii Nauk, SSSR* 115: 795–8. [In Russian]

———. 1966. *Basic Arthropodan Stock, with Special Reference to Insects*. Pergamon Press; Oxford, UK; xii+271 pp.

———. 1968. Phylogeny of the Orthopteroidea. *Trudy Paleontologischeskogo Instituta Akademii Nauk, SSSR* 118: 1–216. [In Russian]

———. 1973. Morphological features and mode of life of the Palaeodictyoptera. *Chteniya Pamyati Nikolaya Aleksandrovicha Kholodkovskogo* 25: 45–63. [In Russian]

Sharp, D. 1895. Insects. Part 1. Introduction, Aptera, Orthoptera, Neuroptera, and a portion of Hymenoptera (Sessiliventres and Parasitica). Pp. 83–565. *In* Sedgwick, A. (ed.), *Cambridge Natural History* [Volume 5]. Macmillan; London, UK; xi+584.

———. 1898. Some points in the classification of the Insecta Hexapoda. *Proceedings of the 4th International Congress of Zoology* 4: 246–9.

Sharpe, S. C. F. 1932. Eurypterid trails from the Ordovician. *American Journal of Science* 24: 355–61.

Shaw, K. L. 1995. Biogeographic patterns of two independent Hawaiian cricket radiations (*Laupala* and *Prognathogryllus*). Pp. 39–56. *In* Wagner, W. L., and V. A. Funk (eds.), *Hawaiian Biogeography: Evolution on a Hot Spot Archipelago*. Smithsonian Institution Press; Washington, D.C.; xvii+467 pp.

Shaw, M. R. 1981. Delayed inhibition of host development by the non-paralyzing venoms of parasitic wasps. *Journal of Invertebrate Pathology* 37: 215–21.

———. 1983. On evolution of endoparasitism: The biology of some genera of Rogadinae (Braconidae). *Contributions of the American Entomological Institute* 20: 307–28.

Shaw, S. R. 1988. *Carminator*, a new genus of Megalyridae (Hymenoptera) from the Oriental and Australian regions, with a commentary on the definition of the family. *Systematic Entomology* 13: 101–13.

———. 1990. Phylogeny and biogeography of the parasitoid wasp family Megalyridae (Hymenoptera). *Journal of Biogeography* 17: 569–81.

———. 2003. A new *Cryptalyra* species from Colombia (Hymenoptera: Megalyridae). *Zootaxa* 248: 1–4.

Shcherbakov, D. E. 1996. Origin and evolution of the Auchenorrhyncha as shown by the fossil record. Pp. 31–45. *In* Schaefer, C. W. (ed.), *Studies on Hemipteran Phylogeny*. Entomological Society of America; Lanham, Maryland; iii+244 pp.

———, and Yu. A. Popov. 2002. Superorder Cimicidea Laicharting, 1781. Order Hemiptera Linné, 1758. The bugs, cicadas, plantlice, scale insects, etc. (= Cimicida Laicharting, 1781, = Homoptera Leach, 1815 + Heteroptera Latreille, 1810). Pp. 143–56. *In* Rasnitsyn, A. P., and D. L. J. Quicke (eds.), *History of Insects*. Kluwer Academic Publishers; Dordrecht, the Netherlands; xii+517 pp.

———, and P. Wegeriek. 1991. Creaphididae, a new and the oldest aphid family from the Triassic of middle Asia. *Psyche* 98: 81–5.

———, E. D. Lukashevich, and V. A. Blagoderov. 1995. Triassic Diptera and initial radiation of the order. *Dipterological Research* 6: 75–115.

Shear, W. A. 1981. Two fossil millipedes from the Dominican amber (Diplopoda: Chytodesmidae, Siphonophoridae). *Myriapodologica* 1: 51–4.

———, and P. M. Bonamo. 1988. Devonobiomorpha, a new order of centipeds (Chilopoda) from the middle Devonian of Gilboa, New York State, USA, and the phylogeny of centiped orders. *American Museum Novitates* 2927. 1–30.

———, and J. Kukalová-Peck. 1990. The ecology of Paleozoic terrestrial arthropods: The fossil evidence. *Canadian Journal of Zoology* 68: 1807–34.

———, P. M. Bonamo, J. D. Grierson, W. D. I. Rolfe, E. L. Smith, and R. A. Norton. 1984. Early land animals in North America: Evidence from Devonian age arthropods from Gilboa, New York. *Science* 224: 492–4.

———, P. A. Selden, W. D. I. Rolfe, P. M. Bonamo, and J. D. Grierson. 1987. New terrestrial arachnids from the Devonian of Gilboa, New York (Arachnida, Trigonotarbida). *American Museum Novitates* 2901: 1–74.

———, J. M. Palmer, J. A. Coddington, and P. M. Bonamo. 1989. A Devonian spinneret: Early evidence of spiders and silk use. *Science* 246: 479–81.

———, J. T. Hannibal, and J. Kukalová-Peck. 1992. Terrestrial arthropods from Upper Pennsylvanian rocks at the Kinney Brick Quarry, New Mexico. *New Mexico Bureau of Mines and Mineral Resources Bulletin* 138: 135–41.

———, A. J. Jeram, and P. A. Selden. 1998. Centiped legs (Arthropoda, Chilopoda, Scutigeromorpha) from the Silurian and Devonian of Britain and the Devonian of North America. *American Museum Novitates* 3231: 1–16.

Shedrinsky, A. M., D. Grimaldi, T. P. Wampler, and N. S. Baer. 1991. Amber and copal: pyrolysis gas chromatographic (PyGC) studies of provenance. *Wiener berichte über Naturwissenschaft in der Kunst* 6/7/8: 37–66.

Shelford, R. 1910. On a collection of Blattidae preserved in amber, from Prussia. *Journal of Linnean Society, Zoology* 30: 336–54.

Shellman-Reeve, J. S. 1997. The spectrum of sociality in termites. Pp. 52–93. *In* Choe, J. C., and B. J. Crespi (eds.), *The Evolution of Social Behavior in Insects and Arachnids*. Cambridge University Press; Cambridge, UK; vii+541 pp.

Sheppard, P. M., J. R. G. Turner, K. S. Brown, W. W. Benson, and M. C. Singer. 1985. Genetics and evolution of Müllerian mimicry in *Heliconius* butterflies. *Philosophical Transactions of the Royal Society, London (B)* 308: 433–613.

Shimizu, A. 1994. Phylogeny and classification of the family Pompilidae (Hymenoptera). *Tokyo Metropolitan University Bulletin of Natural History* 20: 1–142.

Shivarudrappa, T. V. 1977. A rare fossil diplopod from the intertrappean sediments of Gurmatkal, Gulbarga district, Karnataka State. *Journal of the Mysore University, Section B, Science* 27: 111–12.

Shoshani, J., and M. C. McKenna. 1998. Higher taxonomic relationships among extant mammals based on morphology, with selected comparisons of results from molecular data. *Molecular Phylogenetics and Evolution* 9: 572–84.

Shull, V. L., A. P. Vogler, M. D. Baker, D. R. Maddison, and P. M. Hammond. 2001. Sequence alignment of 18S ribosomal RNA and the basal relationships of adephagan beetles: Evidence for monophyly of aquatic families and the placement of Trachypachidae. *Systematic Biology* 50: 945–69.

Shultz, J. W. 1989. Morphology and locomotor appendages in Arachnida: Evolutionary trends and phylogenetic implications. *Zoological Journal of the Linnean Society* 108: 335–65.

———. 1990. Evolutionary morphology and phylogeny of Arachnida. *Cladistics* 6: 1–38.

———, and J. C. Regier. 1997. Progress toward a molecular phylogeny of the centipede orders (Chilopoda). *Entomologica Scandinavica, Supplement* 51: 25–32.

Sierwald, P., W. A. Shear, R. M. Shelley, and J. E. Bond. 2003. Millipede phylogeny revisited in the light of the enigmatic order Siphoniulida. *Journal of Zoological Systematics and Evolutionary Research* 41: 87–99.

Silberglied, R. E. 1984. Visual communication and sexual selection among butterflies. Pp. 207–24. *In* Vane-Wright, R. I., and P. R. Ackery (eds.), *The Biology of Butterflies*. Academic Press; London, UK; xxiv+429 pp.

Silsby, J. 2001. *Dragonflies of the World*. Natural History Museum; London, UK; vii+216 pp.

Silvestri, F. 1912. Die Thysanuren des baltischen Bernstein. *Schriften der Physikalisch-ökonomischen Gesellschaft* 53: 42–66.

———. 1913. Descrizione di un nuovo ordine di insetti. *Bollettino del Laboratorio di Zoologia Generale e Agraria, Portici* 7: 193–209.

Simiczyjew, B. 2002. Structure of the ovary in *Nannochorista neotropica* Navás (Insecta: Mecoptera: Nannochoristidae) with remarks on mecopteran phylogeny. *Acta Zoologica* 83: 61–6.

Simmons, N. B., and J. H. Geisler. 1998. Phylogenetic relationships of *Icaronycteris*, *Archaeonycteris*, *Hassianycteris*, and *Palaeochiropteryx* to extant bat lineages, with comments on the evolution of echolocation and foraging strategies in Microchiroptera. *Bulletin of the American Museum of Natural History* 235: 1–182.

Simon, C. 1988. Evolution of 13- and 17-year periodical cicadas (Homoptera: Cicadidae: *Magicicada*). *Bulletin of the Entomological Society of America* 1988: 163–76.

Simon-Benito, J. C., V. M. Ortuño, and D. Espantaleon. 2002. Colembolos (Collembola, Insecta) del ambar cretacico de Alava (cuenca vasco-cantabrica, norte de España). *Estudios del Museo de Ciencias Naturales de Alava* 17: 83–91.

Simonsen, T. J., and N. P. Kristensen. 2001. *Agathiphaga* wing vestiture revisited: Evidence for complex early evolution of lepidopteran scales (Lepidoptera: Agathiphagidae). *Insect Systematics and Evolution* 32: 169–75.

Simpson, G. G. 1944. *Tempo and Mode in Evolution*. Columbia University Press; New York, New York; xviii+237 pp.

———. 1945. The principles of classification and a classification of mammals. *Bulletin of the American Museum of Natural History* 85: 1–350.

———. 1948. The beginning of the Age of Mammals in South America. *Bulletin of the American Museum of Natural History* 91: 1–232.

Simutnik, S. A. 2002. A new genus of encyrtid wasps (Hymenoptera, Chalcidoidea, Encyrtidae) from Late Eocene Rovno amber (Ukraine). *Vestnik Zoologii* 36: 99–102.

Sinclair, A. R. E. 1975. The resource limitation of trophic levels in tropical grassland ecosystems. *Journal of Animal Ecology* 44: 497–520.

Sinclair, B. J., J. M. Cumming, and D. M. Wood. 1994. Homology and phylogenetic implications of male genitalia in Diptera: Lower Brachycera. *Entomologica Scandinavica* 24: 407–32.

Sinéty, R., de. 1901. Recherches sur la biologie et l'anatomie de Phasmes. *La Cellule* 19: 118–278.

Sinitshenkova, N. D. 1975. Mayfly larvae of the family Hexagenitidae (Insecta, Ephemeroptera). *Paleontologicheskiy Zhurnal* 1975: 82–7. [In Russian]

———. 1976. New Early Cretaceous mayflies (Insecta, Ephemeroptera) from eatern Transbaikalia. *Paleontologicheskiy Zhurnal* 1976: 85–93. [In Russian]

———. 1986. Mayflies. Ephemerida (=Ephemeroptera). *Trudy Sovmestnaya Sovetsko-Mongolskaya Paleontologicheskaya Ekspeditsiya* 28: 45–56. [In Russian]

———. 1999. A new mayfly species of the extant genus *Neoephemera* from the Eocene of North America (Insecta: Ephemerida = Ephemeroptera: Neoephemeridae). *Paleontologischeskiy Zhurnal* 4: 67–9.

———. 2000a. New Jersey amber mayflies: The first North American Mesozoic members of the order (Insecta: Ephemeroptera). Pp. 111–25. *In* Grimaldi, D. (ed.), *Studies on Fossils in Amber, with Particular Reference to the Cretaceous of New Jersey*. Backhuys Publishers; Leiden, the Netherlands; viii+498 pp.

———. 2000b. The first fossil prosopistomatid mayfly from Burmese amber (Ephemeroptera, Prosopistomatidae). *Bulletin of the Natural History Museum, London (Geology)* 56: 25–8.

———. 2000c. A review of Triassic mayflies, with a description of new species from Western Siberia and Ukraine (Ephemerida = Ephemeroptera). *Paleontological Journal, Supplementary Issue* 34: 275–83.

Skalski, A. W. 1973. Studies on the Lepidoptera from fossil resins, part II. *Epiborkhausenites osbscurotrimaculatus* gen. *et* spec. nov. (Oecophoridae) and tineid-moth discovered in Baltic amber. *Acta Palaeontologica Polonica* 18: 153–60.

———. 1976. Les lépidoptères fossils de l'ambre. Etat actuel de nos connaissances. *Linneana Belgica* 6/7: 154–69; 8: 195–208; 9: 221–33.

———. 1977. Studies on the Lepidoptera from fossil resins, Part I. General remarks and description of new genera and species of the families Tineidae and Oecophoridae from the Baltic amber. *Prace Muzeum Ziemi, PAN* 26: 3–24.

———. 1979. Records of oldest Lepidoptera. *Nota Lepidopterologica* 2: 61–6.

———. 1984. A new lower Cretaceous Lepidoptera (Homoneura). *Bulletin of the Polish Academy of Sciences, Biological Sciences* 32: 389–92.

———. 1988a. A new fossil trichogrammatid from the Sicilian amber (Hymenoptera, Chalcidoidea, Trichogrammatidae). *Fragmenta Entomologica* 21: 111–16.

———. 1988b. *Stauropolia nekrutenkoi* gen. *et* sp. n. (Lepidoptera, Arctiidae) – A new fossil tiger-moth from Miocene of the Caucasus. *Vestnik Zoologii* 4: 21–5.

———. 1990. An annotated review of fossil records of lower Lepidoptera. *Bulletin of the Sugadeira Montane Research Center, Tsukuba University* 11: 125–8.

Skorikov, A. S. 1929. Eine neue Basis für eine Revision der Gattung *Apis* L. *Reports on Applied Entomology, Leningrad* 4: 249–70. [In Russian, with Germany summary]

Sleeper, E. L. 1968. A new fossil weevil from Nevada (Coleoptera: Curculionidae). *Bulletin of the Southern California Academy of Sciences* 67: 196–8.

Smiley, C. J., J. Gray, and L. M. Higgins. 1975. Preservation of Miocene fossils in unoxidized lake deposits, Clarkia, Idaho. *Journal of Paleontology* 49: 833–44.

Smiley, J. 1978. Plant chemistry and the evolution of host specificity: New evidence from *Heliconius* and *Passiflora*. *Science* 201: 745–7.

Smit, F. G. A. M. 1987. *An Illustrated Catalogue of the Rothschild Collection of Fleas (Siphonaptera) in the British Museum (Natural History), vol. 7, Malacopsylloidea (Malacopsyllidae and Rhopalopsyllidae)*. British Museum (Natural History); London, UK; [7]+380 pp.

Smith, A. B. 1994. *Systematics and the Fossil Record: Documenting Evolutionary Patterns*. Blackwell; Oxford, UK; viii+223 pp.

Smith, D. [David] R. 1978. Suborder Symphyta (Xyelidae, Pararchexyelidae, Parapamphiliidae, Xyelydidae, Karatavitidae, Gigasiricidae, Sepulcidae, Pseudosiricidae, Anaxyelidae, Siricidae, Xiphydriidae, Paroryssidae, Xyelotomidae, Blasticotomidae, Pergidae). *Hymenopterorum Catalogus* 14: 1–193.

———. 1988. A synopsis of the sawflies (Hymenoptera: Symphyta) of American south of the United States: Introduction, Xyelidae, Pamphiliidae, Cimbicidae, Diprionidae, Xiphydriidae, Siricidae, Orussidae, Cephidae. *Systematic Entomology* 13: 205–61.

———. 1990. A synopsis of the sawflies (Hymenoptera, Symphyta) of American south of the United States: Pergidae. *Revista Brasileira de Entomologia* 34: 7–200.

———, and G. O. Poinar, Jr. 1992. Sawflies (Hymenoptera: Argidae) from Dominican amber. *Entomological News* 103: 117–24.

———, and N. M. Schiff. 2002. A review of the siricid woodwasps and their ibaliid parasitoids (Hymenoptera: Siricidae, Ibaliidae) in the eastern United States, with emphasis on the mid-Atlantic region. *Proceedings of the Entomological Society of Washington* 104: 174–94.

Smith, D. [Deborah] R., L. Villafuerte, G. Otis, and M. R. Palmer. 2000. Biogeography of *Apis cerana* F. and *A. nigrocincta* Smith: Insights from mtDNA studies. *Apidologie* 31: 265–79.

Smith, F. 1861. Descriptions of new genera and species of exotic Hymenoptera. *Journal of Entomology* 1: 146–55.

———. 1865. On the species and varieties of the honey-bees belonging to the genus *Apis*. *Annals and Magazine of Natural History, series 3* 15: 372–380.

Smith, J. B. 1897a. A classification of the orders of insects. *Science* 2: 671–7.

———. 1897b. A classification of the orders of insects. *Entomologist* 30: 188–93.

Smith, K. V. G. 1986. *A Manual of Forensic Entomology*. British Museum (Natural History); London, UK; 205 pp.

Smith, R. F., ed. 1973. *History of Entomology*. Annual Reviews; Palo Alto, California; vii+[1]+517 pp.

Smith, S. I. 1871. Notice of a fossil insect from the Carboniferous formation of Indiana. *American Journal of Science, Series 3* 1: 44–6.

Smith, V. S. 2000. Basal ischnoceran louse phylogeny (Phthiraptera: Ischnocera: Goniodidae and Heptapsogasteridae). *Systematic Entomology* 25: 73–94.

———. 2001. Avian louse phylogeny (Phthiraptera: Ischnocera): A cladistic study based on morphology. *Zoological Journal of the Linnean Society* 132: 81–144.

Smithers, C. N. 1967. A catalogue of the Psocoptera of the world. *Australian Zoology* 14: 1–145.

———. 1972. The classification and phylogeny of the Psocoptera. *Memoirs of the Australian Museum* 14: 1–349.

———. 1991. Psocoptera (psocids, booklice). Pp. 412–20. *In* Naumann, I. D. (ed.), *The Insects of Australia: A Textbook for Students and Research Workers, Volume 1* [2nd Edition]. Cornell University Press; Ithaca, New York; xvi+[1]+542 pp.

Sneath, P. H. A., and R. R. Sokal. 1973. *Numerical Taxonomy: The Principles and Practice of Numerical Classification*. Freeman; San Francisco, California; xv+573 pp.

Snodgrass, R. E. 1905. A revision of the mouthparts of the Corrodentia and the Mallophaga. *Transactions of the American Entomological Society* 31: 297–307.

———. 1935. *Principles of Insect Morphology*. McGraw-Hill; New York, New York; ix+667 pp.

———. 1937. The male genitalia of orthopteroid insects. *Smithsonian Miscellaneous Collections* 96: 1–107.

———. 1938. Evolution of the Annelida, Onychophora, and Arthopoda. *Smithsonian Miscellaneous Collections* 57: 1–159.

———. 1946. The skeletal anatomy of fleas (Siphonaptera). *Smithsonian Miscellaneous Collections* 104: 1–89.

Sokal, R. R., and P.H.A. Sneath. 1963. *Principles of Numerical Taxonomy*. Freeman; San Francisco, California; xvi+359 pp.

Sokoloff, A. 1972/1974/1977. *The Biology of* Tribolium, *with Special Emphasis on Genetic Aspects*. Clarendon Press; Oxford, UK; 962 pp. [3 volumes]

Soli, G. E. E. 1997. On the morphology and phylogeny of Mycetophilidae, with a revision of *Coelosia* Winnertz (Diptera, Sciaroidea). Entomologica Scandinavica, Supplement 50: 1–139.

Solis, A., and C. Mitter. 1992. Review and preliminary phylogenetic analysis of the subfamilies of the Pyralidae (*sensu stricto*) (Lepidoptera: Pyraloidea). *Systematic Entomology* 17: 79–90.

Soltis, D. E., P. S. Soltis, M. W. Chase, M. E. Mort, D. C. Albach *et al.* 2000. Angiosperm phylogeny inferred from 18S rDNA, rbcL and aptB sequences. *Botanical Journal of the Linnean Society* 133: 381–461.

Soltis, P. S., D. E. Soltis, and C. J. Smiley. 1992. An rbCL sequence from a Miocene *Taxodium* (bald cypress). *Proceedings of the National Academy of Sciences, USA* 89: 449–51.

Soós, A., ed. 1984–1994. *Catalogue of Palaearctic Diptera* [volumes 1–14]. Akadémiai Kiadó; Budapest, Romania; 5540 pp.

Sorenson, J. T., B. C. Campbell, R. J. Gill, and J. D. Steffen-Campbell. 1995. Non-monophyly of Auchenorrhyncha ("Homoptera"), based upon 18S rDNA phylogeny: Eco-evolutionary and cladistic implications within pre-Heteropteroidea Hemiptera (s.l.) and a proposal for new monophyletic suborders. *Pan-Pacific Entomologist* 71: 31–60.

SORG, M. 1986. Grabwespen der Gattung *Passaloecus* aus fossilen Harzen (Hymenoptera, Sphecoidea, Pemphredoninae) *Passaloecus microceras* n. sp., Baltischer Bernstein, oberes Eozän *Passaloecus munax* n. sp., Bitterfelder Bernstein, unteres Miozän. *Paläontologische Zeitschrift* 60: 277–84.

SPAHR, U. 1987. Ergänzungen und Berichtigungen zu R. Keilbach's Bibliographie und Liste der Bernsteinfossilien-Ordnung Hymenoptera. *Stuttgarter Beiträge zur Naturkunde, Serie B, Geologie und Paläontologie* 127: 1–121.

SPANGLER, H. G. 1988. Moth hearing, defense, and communication. *Annual Review of Entomology* 33: 59–81.

SPEED, M. P., and J. R. G. TURNER. 1999. Learning and memory in mimicry: II. Do we understand the mimicry spectrum? *Biological Journal of the Linnean Society* 67: 281–312.

SPEIDEL, W., H. FANGER, and C. M. NAUMANN. 1996. The phylogeny of the Noctuidae (Lepidoptera). *Systematic Entomology* 21: 219–51.

SPEIGHT, M., M. D. HUNTER, and A. D. WATT. 1999. *Ecology of Insects: Concepts and Applications*. Blackwell; Oxford, UK; ix+350 pp.

SPENCE, J. R. 1986. Interactions between the scelionid egg parasitoid *Tiphodytes gerriphagus* (L.) (Hymenoptera) and its gerrid hosts (Heteroptera). *Canadian Journal of Zoology* 64: 2728–38.

SPILMAN, T. J. 1976. A new species of fossil *Ptinus* from fossil wood rat nests in California and Arizona (Coleoptera, Ptinidae), with a postscript on the definition of a fossil. *Coleopterist's Bulletin* 30: 239–44.

SPRADBERRY, J. P. 1973a. A comparative study of the phytotoxic effects of siricid woodwasps on conifers. *Annals of Applied Biology* 75: 309–20.

———. 1973b. *Wasps: An Account of the Biology and Natural History of Solitary and Social Wasps, with Particular Reference to those of the British Isles*. Sidgwick and Jackson; London, UK; xvi+408 pp.

———. 1975. The biology of *Stenogaster concinna* van der Vecht, with comments on the phylogeny of Stenogastrinae (Hymenoptera: Vespidae). *Journal of the Australian Entomological Society* 14: 309–18.

SPRINGATE, N. D. 1994. Megalodontidae Morris & Lycett, 1853 (Mollusca, Bivalvia) and Megalodontidae Konow, 1897 (Insecta, Hymenoptera): Proposed removal of homonymy. *Bulletin of Zoological Nomenclature* 51: 230–1.

SQUIRES, R. L. 1979. Middle Pliocene dragonfly nymphs, Ridge Basin, Transverse Ranges, California. *Journal of Paleontology* 53: 446–52.

SRIVASTAVA, A. K. 1988. An insect wing from the Lower Gondwana of India. *Journal of Paleontology* 62: 827–8.

STÅHLS, G., H. HIPPA, R. ROTHERAY, J. MUONA, and F. GILBERT. 2003. Phylogeny of the Syrphidae (Diptera) inferred from combined analysis of molecular and morphological characters. *Systematic Entomology* 28: 433–50.

STANGE, L. A. 1970. Revision of the ant-lion tribe Brachynemurini of North America (Neuroptera: Myrmeleontidae). *University of California Publications in Entomology* 55: 1–192.

———, and R. B. MILLER. 1990. Classification of the Myrmeleontidae based on larvae (Insecta: Neuroptera). Pp. 151–69. *In* MANSELL, M. W., and H. ASPÖCK (eds.), *Advances in Neuropterology: Proceedings of the Third International Symposium on Neuropterology*. Department of Agriculture Development; Pretoria, South Africa; vi+298 pp.

———, and G. J. WILLINER. 1981. Una nueva especie de *Veurise* Navás de Bolivia con notas sobre Nemopteridae en America del sur (Insecta: Neuroptera). *Physis, Seccion C, los Continentes y los Organismos Terrestres* 39: 35–9.

STANHOPE, M. J., O. MADSEN, V. G. WADDELL, G. C. CLEVEN, *et al.* 1998. Highly congruent molecular support for a diverse superordinal clade of endemic African mammals. *Molecular Phylogenetics and Evolution* 9: 501–8.

STANICZEK, A. H., and G. BECHLY. 2002. First fossil record of the mayfly family Baetischidae from Baltic amber (Insecta: Ephemeroptera). *Stuttgarter Beiträge zur Naturkunde, serie B, Geologie und Paläontologie* 322: 1–11.

STANKIEWICZ, B. A., D. E. G. BRIGGS, R. P. EVERSHED, and I. J. DUNCAN. 1997a. Chemical preservation of insect cuticle from the Pleistocene asphalt deposits of California, USA. *Geochimica et Cosmochimica Acta* 61: 2247–52.

———, D. E. G. BRIGGS, R. P. EVERSHED, M. B. FLANNERY, and M. WUTTKE. 1997b. Preservation of chitin in 25-million-year-old fossils. *Science* 276: 1541–3.

———, D. E. G. BRIGGS, R. P. EVERSHED, R. F. MILLER, and A. BIERSTEDT. 1998a. The fate of chitin in Quaternary and Tertiary strata. Pp. 211–25. *In* STANKIEWICZ, B. A., and P. F. VAN BERGREN (eds.), *Nitrogen-Containing Macromolecules in the Bio- and Geosphere*. American Chemical Society Symposium Series 707. Oxford University Press; Oxford, UK; xiii+370 pp.

———, A. C. SCOTT, M. E. COLLINSON, P. FINCH, B. MÖSLE *et al.* 1998b. Molecular taphonomy of arthropod and plant cuticles from the Carboniferous of North America: Implications for the origin of kerogen. *Journal of the Geological Society, London* 155: 453–62.

———, H. N. POINAR, D. E. G. BRIGGS, R. P. EVERSHED, and G. O. POINAR, JR. 1998c. Chemical preservation of plants and insects in natural resins. *Proceedings of the Royal Society of London, Series B, Biological Sciences* 265: 641–7.

STARR, C. K. 1985. A simple pain scale for field comparison of hymenopteran stings. *Journal of Entomological Science* 20: 225–31.

STATZ, G. 1936a. Ueber neue Funde von Neuropteren, Panorpen und Trichopteren aus den tertiären Schiefern von Rott am Siebengebirge. *Decheniana* 93: 208–55.

———. 1936b. Ueber alte und neue fossile Hymenopterenfunde aus den tertiären Ablagerungen von Rott am Siebengebirge. *Decheniana* 93: 256–312.

———. 1938. Neue Funde parasitischer Hymenopteren aus dem Tertiär von Rott am Siebengebirge. *Decheniana* 98: 71–144.

———. 1940. Neue Dipteren (Brachycera et Cyclorrhapha) aus dem Oberoligocän von Rott. *Palaeontoographica* 91: 120–74.

STEINMANN, H. 1986. Dermaptera: Catadermaptera I. *Das Tierreich* 102: xiv+343 pp.

———. 1989a. Dermaptera: Catadermaptera II. *Das Tierreich* 105: xix+504 pp.

———. 1989b. Dermaptera: Eudermaptera I. *Das Tierreich* 106: xviii+558 pp.

———. 1989c. *World Catalogue of Dermaptera*. Kluwer Academic Publishers; Dordrecht, the Netherlands; 934 pp.

———. 1993. Dermaptera: Eudermaptera II. *Das Tierreich* 108: xxii+1–711 pp.

———. 1997a. World catalogue of Odonata. Volume 1. Zygoptera. *Das Tierreich* 110: xxi+1–500 pp.

———. 1997b. World catalogue of Odonata. Volume 2. Anisoptera. *Das Tierreich* 111: xiv+1–636 pp.

STELZL, M. 1990. Nahrungsanalytische Untersuchungen an Hemerobiiden-imagines (Insecta, Planipennia). *Mitteilungen der Deutschen Gesellschaft für Allgemeine und Angewandte Entomologie* 7: 670–6.

STEPHENSON, J., and A. C. SCOTT. 1992. The geological history of insect-related plant damage. *Terra Review* 4: 542–52.

STERN, D. L., and W. A. FOSTER. 1997. The evolution of sociality in aphids: A clone's-eye view. Pp. 150–65. *In* CHOE, J. C., and B. J. CRESPI (eds.), *The Evolution of Social Behavior in Insects and Arachnids*. Cambridge University Press; Cambridge, UK; vii+541 pp.

STEVENSON, D. W., K. J. NORSTAG, and P. K. S. FAWCETT. 1998. Pollination biology of cycads. Pp. 277–94. *In* OWENS, S. J., and P. J. RUDALL (eds.), *Reproductive Biology in Systematics, Conservation and Economic Botany*. Royal Botanic Gardens, Kew; London, UK; ix+491 pp.

STILWELL, M. A. 1965. Hypopleural organs of the woodwasp larva *Tremex columba* (L.) containing fungus *Daedalea unicolor* Bull. ex Fries. *Canadian Entomologist* 97: 783–4.

———. 1966. Woodwasps (Siricidae) in conifers and the associated fungus, *Stereum chailletii*, in eastern Canada. *Forest Science* 72: 121–7.

STOCKWELL, S. A. 1989. *Revision of the Phylogeny and Higher Classification of Scorpions (Chelicerata)*. Ph.D. Dissertation; University of California; Berkeley, CA; iv+413 pp.

STOFFOLANO, J. G., JR., N. E. WOODLEY, A. BORKENT, and L. R. S. YIN. 1988. Ultrastructural studies of the abdominal plaques of some Diptera. *Annals of the Entomological Society of America* 81: 503–10.

STOLTZ, D. B. 1993. The polydnavirus life cycle. Pp. 167–87. *In* BECKAGE, N. E., S. N. THOMPSON, and B. A. FEDERICI (eds.), *Parasites and Pathogens of Insects. Volume 1: Parasites*. Academic Press; San Diego, California; xviii+364 pp.

STONE, A., C. W. SABROSKY, W. W. WIRTH, R. H. FOOTE, and J. R. COULSON, eds. 1965. *A Catalogue of the Diptera North of Mexico*. Agriculture Handbook 276; Washington, D.C.; iv+1696 pp.

STORCH, V. 1993. Pentastomida. Pp. 115–42. *In* HARRISON, F. W., and M. E. RICE (eds.), *Microscopic Anatomy of Invertebrates, Volume 12: Onychophora, Chilopoda and Lesser Protostomata*. Wiley-Liss; New York, New York; xiv+484 pp.

———, and B. G. M. JAMIESON. 1992. Further spermatological evidence for including the Pentastomida (tongue worms) in the Crustacea. *International Journal of Parasitology* 22: 95–108.

Stork, N. E. 1987. Arthropod faunal similarity of Bornean rain forest trees. *Ecological Entomology* 12: 219–26.

———. 1988. Insect diversity: Facts, fiction, and speculation. *Biological Journal of the Linnean Society* 35: 321–37.

———. 1991. The composition of the arthropod fauna of Bornean lowland rain forest trees. *Journal of Tropical Ecology* 7: 161–80.

———. 1996. Measuring global biodiversity and its decline. Pp. 41–68. *In* Reaka-Kudla, M. L., D. E. Wilson, and E. O. Wilson (eds.), *Biodiversity II: Undertanding and Protecting our Biological Resources.* Joseph Henry Press; Washington, D.C.; v+551 pp.

———, J. Adis, and R. K. Didham, eds. 1997. *Canopy Arthropods.* Chapman and Hall; London, UK; xvi+567 pp.

Størmer, L. 1977. Arthropod invasion of land during Late Silurian and Devonian times. *Science* 197: 1362–4.

Storozhenko, S. Y. 1997. Classification of order Grylloblattida (Insecta), with description of new taxa. *Far Eastern Entomologist* 42: 1–20.

———. 1998. *Systematics, Phylogeny and Evolution of the Grylloblattids (Insecta: Grylloblattida).* Dal'nauka; Vladivostok, Russia; 207 pp. [In Russian]

———, and J. K. Park. 2002. A new genus of the ice crawlers (Grylloblattida: Grylloblattidae) from Korea. *Far Eastern Entomologist* 114: 18–20.

Stoutamire, W. P. 1983. Wasp-pollinated species of *Caladenia* (Orchidaceae) in southwestern Australia. *Australian Journal of Botany* 31: 383–94.

Strand, M. R. 1986. The physiological interactions of parasitoids with their hosts and their influence on reproductive strategies. Pp. 97–136. *In* Waage, J. K., and H. C. J. Greathead (eds.), *Insect Parasitoids.* Academic Press; London, UK; xvii+389 pp.

Straus, A. 1977. Gallen, Minen und andere Frasspuren im Pliokän von Willershausen am Harz. *Verhandlungen des Botanischen Vereins der Provinz Brandenburg* 113: 43–80.

Strong, D. R., J. H. Lawton, and R. Southwood. 1984. *Insects on Plants: Community Patterns and Mechanisms.* Harvard University Press; Cambridge, Massachusetts; vi+313 pp.

Stuart, A. E. 2000. *The Utility of Behaviour in Macroevolutionary Studies: Analyses of Caddisfly (Trichoptera) Case Building Behaviour.* Ph.D. Dissertation, University of Toronto; Toronto, Canada; 268 pp.

———, and D. C. Currie. 2001. Using caddisfly (Trichoptera) case-building behaviour in higher level phylogeny reconstruction. *Canadian Journal of Zoology* 79: 1842–54.

———, and D. C. Currie. 2002. Behaviour is not reliably inferred from end-product structure in caddisflies. *Ethology* 108: 837–56.

Stuckenberg, B. R. 1999. Antennal evolution in the Brachycera (Diptera), with a reassessment of terminology relating to the flagellum. *Studia Dipterologica* 6: 33–48.

———. 2000. *Namamyia*, a new genus of Vermileonidae (Diptera) from Namaqualand, with a conspectus of the southern Africa genera and an account of their adaptations for anthophily. *Annals of the Natal Museum* 41: 181–202.

———. 2001. Pruning the tree: A critical review of classifications of the *Homeodactyla* (Diptera, Brachycera), with new perspectives and an alternative classification. *Studia Dipterologica* 8: 3–41.

Sturm, H. 1952. Die Paarung bei *Machilis* (Felsenspringer). *Naturwissenschaften* 39: 308.

———. 1955. Beiträge zur Ethologie einiger mitteldeutscher Machiliden. *Zeitschrift für Tierpsychologie* 12: 337–63.

———. 1984. Zur Systematik, Biogeographie und Evolution der sudamerikanischen Meinertellidae (Machiloidea, Archaeognatha, Insecta). *Zeitschrift für Zoologische Systematik und Evolutionsforschung* 22: 27–44.

———. 1994a. Diskontinuierlich aktive Gene und Evolution – Eine Diskussion am Beispiel der Archaeognatha. *Journal of Zoological Systematics and Evolutionary Research* 32: 241–63.

———. 1994b. Beziehungen zwischen Paarungsbiologie und Taxonomie bei Felsenspringern (Machiloidea, Archaeognatha, Insecta). *Verhandlungen Westdeutscher Entomologentag* 1993: 131–7.

———. 1995a. Chaetotaxy of tergites in Archaeognatha (Insecta) and its taxonommic significance. *Zoologischer Anzeiger* 234: 85–100.

———. 1995b. Relationships between mating behaviour and taxonomy in bristletails (Machiloidea, Archaeognatha, Insecta). *Polskie Pismo Entomologiczne* 64: 53–61.

———. 1997. Fossilgeschichte und Taxonomie der Felsenspringer und der fischchenartigen Insekten (Archaeognatha, Zygentoma, "Apterygota", Insecta). *Mitteilungen der Deutschen Gesellschaft für Allgemeine und Angewandte Entomologie* 11: 811–16.

———. 1998. Erstnachweis fischchenartiger Insekten (Zygentoma, Insecta) fur das Mesozoikum (Untere Kreide, Brasilien). *Senckenbergiana Lethaea* 78: 135–40.

———. 2003a. Archaeognatha (bristletails). Pp. 57–9. *In* Resh, V. H., and R. T. Cardé (eds.), *Encyclopedia of Insects.* Academic Press; San Diego, California; xxviii+[1]+1266 pp.

———. 2003b. Zygentoma (Thysanura, silverfish). Pp. 1203–5. *In* Resh, V. H., and R. T. Cardé (eds.), *Encyclopedia of Insects.* Academic Press; San Diego, California; xxviii+[1]+1266 pp.

———, and C. Bach de Roca. 1993. On the systematics of the Archaeognatha (Insecta). *Entomologia Generalis* 18: 55–90.

———, and R. Machida. 2001. *Handbuch der Zoologie: Eine Naturgeschichte der Stämme des Tierreiches: Band IV: Arthropoda: Insecta: Teilband 37: Archaeognatha.* Walter de Gruyter; Berlin, Germany; [2]+213 pp.

———, and L. F. Mendes. 1998. Two new species of Nicoletiidae (Zygentoma, "Apterygota", Insecta) in Dominican amber. *American Museum Novitates* 3226: 1–11.

Stürmer, W., and J. Bergström. 1976. The arthropods *Mimetaster* and *Vachonisia* from the Devonian Hunrück Shale. *Paläontologische Zeitschrift* 52: 57–81.

Sturtevant, A. H. 1942. The classification of the genus *Drosophila*, with description of nine new species. *University of Texas Publication* 4213: 6–51.

———. 1965. *A History of Genetics.* Harper and Row; New York, New York; viii+165 pp.

Štys, P. 1969. Revision of fossil and pseudofossil Enicocephalidae (Heteroptera). *Acta Entomologica Bohemoslavika* 66: 352–65.

Subias, L. S., and A. Arillo. 2002. Oribatid mite fossils from the Upper Devonian of South Mountain, New York and the Lower Carboniferous of County Antrim, northern Ireland (Acariformes, Oribatida). *Estudios Museo Ciencias Naturales de Álava* 17: 93–106.

Sugiomoto, A., D. E. Bignell, and J. A. MacDonald. 2000. Global impact of termites on the carbon cycle and atmospheric trace gases. Pp. 409–35. *In* Abe, T., D. E. Bignell, and M. Higashi (eds.), *Termites: Evolution, Sociality, Symbioses, Ecology.* Kluwer Academic Publishers; Dordrecht, the Netherlands; xxii+466 pp.

Sukatsheva, I. D. 1982. Historical development of the order Trichoptera. *Trudy Paleontologischeskogo Instituta Akademii Nauk, SSSR* 192: 1–111 pp. [In Russian]

———. 1985. Jurassic caddisflies of southern Siberia. Pp. 115–19. *In* Rasnitsyn, A. P. (ed.), *Jurassic Insects of Siberia and Mongolia.* Trudy Paleontologischeskogo Instituta Akademii Nauk, SSSR 211: 192 pp. [In Russian]

———. 1994. Cases of the Early Jurassic caddisflies (Insecta, Trichoptera) from Mongolia. *Paleontological Journal* 28: 99–113.

———. 1999. The Lower Cretaceous caddisfly (Trichoptera) case assemblages. Pp. 163–5. *In* Vršanský, P. (ed.), *Proceedings of the First Palaeoentomological Conference, Moscow, 1998.* AMBA Projects; Bratislava, Slovakia; 199+[2] pp.

———, and A. P. Rasnitsyn. 1992. First members of the family Boreidae (Insecta, Panorpida) from the Upper Jurassic of Mongolia and the Lower Cretaceous of Transbaykalia. *Paleontological Journal* 26: 168–172.

Sun, G., D. L. Dilcher, S. Zheng, and Z. Zhou. 1998. In search of the first flower: A Jurassic angiosperm, *Archaefructus*, from northeast China. *Science* 282: 1692–5.

———, W. Ji, D. L. Dilcher, S. Zheng, K. C. Nixon, and X. Wang. 2002. Archaefructaceae, a new basal angiosperm family. *Science* 296: 899–904.

Süss, H. 1980. Fossile Kambium-Minierer der Familie Agromyzidae (Diptera) in tertiären Laub- und Nadelholzresten. *Zeitschrift für Geologisches Wissenschaften* 8: 1217–25.

Sustera, O. 1958. Übersicht des Systems der paläarktischen und mitteleuropäischen Gattung der Superfamilie Apoidea. *Acta Entomologica Musei Nationalis Prague* 32: 443–63.

Suzuki, K. 1996. Higher classification of the family Chrysomelidae (Coleoptera). Pp. 3–54. *In* Jolivet, P., and M. L. Cox (eds.), *Chrysomelidae Biology, Volume 1: The Classification, Phylogeny, and Genetics.* SPB Academic Publishing; New York, New York; 444 pp.

Svensson, S. A. 1972. *Boreus* Latreille, 1825. A synopsis of described species. Studies on some winter-active insects 1. *Entomologica Scandinavica* 3: 26–32.

Swammerdam, J. 1685. *Historia Insectorum Generalis, in qua Quaecunque ad Insecta Eorumque Mutationes Spectant, Dilucide ex Sanioris Philosophiae & Experientiae Principiis Explicantur. Cum Figuris & Indicibus Necessariis.* Jordanum Luchtmans; Lugduni, Batavorum [Leiden, the Netherlands]; 16+212+[17] pp., 13 pls.

Swofford, D. L., G. J. Olsen, P. J. Waddell, and D. M. Hillis. 1996. Phylogenetic inference. Pp. 407–514. *In* Hillis, D. M., C. Moritz, and

B. K. Mable (eds.), *Molecular Systematics* [2nd Edition]. Sinauer Associates; Sunderland, Massachusetts; xvi+655 pp.

Symmons, S. 1952. Comparative anatomy of the mallophagan head. *Transactions of the Zoological Society of London* 27: 349–436.

Szabó, J., and J. Oehlke. 1986. Neue Proctotrupoidea aus dem baltischen Bernstein. *Beiträge zur Entomologie* 36: 99–106.

Szadziewski, R. 1988. Biting midges (Diptera: Ceratopogonidae) from Baltic amber. *Polski Pismo Entomologiczne* 58: 3–283.

———. 1990. Biting midges (Diptera: Ceratopogonidae) from Sakhalin amber. *Prace Muzeum Ziemi, PAN* 41: 77–81.

———. 1993. Biting midges (Diptera, Ceratopogonidae) from Miocene Saxonian amber. *Acta Zoological Cracoviensia* 35: 603–56.

———. 1996. Biting midges from Lower Cretaceous amber of Lebanon and Upper Cretaceous Siberian amber of Taimyr (Diptera, Ceratopogonidae). *Studia Dipterologica* 3: 23–86.

———. 2000. Biting midges (Diptera: Ceratopogonidae) from the Lower Cretaceous amber of Jordan. *Polski Pismo Entomologiczne* 69: 251–6.

———, and A. Arillo. 1994. Biting midges (Diptera: Ceratopogonidae) from the Lower Cretaceous amber from Alava, Spain. *Polski Pismo Entomologiczne* 67: 291–8.

———, and W. L. Grogan. 1998a. Biting midges from Dominican amber. III. Species of the tribes Culicoidini and Ceratopogonini (Diptera: Ceratopogonidae). *Insecta Mundi* 12: 39–52.

———, and W. L. Grogan. 1998b. Biting midges from Dominican amber. IV. Species of the tribes Dasyheleini and Forcipomyiini (Diptera: Ceratopogonidae). *Polski Pismo Entomologiczne* 67: 255–90.

———, and T. Schlüter. 1992. Biting midges (Diptera: Ceratopogonidae) from Upper Cretaceous (Cenomanian) amber of France. *Annales de la Société d'Entomologique de France* 28: 73–81.

Szucsich, N. U., and H. W. Krenn. 2000. Morphology and function of the proboscis in Bombyliidae (Diptera, Brachycera) and implications for proboscis evolution in Brachycera. *Zoomorphology* 120: 79–90.

———, and H. W. Krenn. 2002. Flies and concealed nectar sources: Morphological innovations in the proboscis of Bombyliidae (Diptera). *Acta Zoologica* 83: 183–92.

Szumik, C. 1994. *Oligembia vetusta*, a new fossil teratembiid (Embioptera) from Dominican amber. *Journal of the New York Entomological Society* 102: 67–73.

———. 1996. The higher classification of the order Embioptera: A cladistic analysis. *Cladistics* 12: 41–64.

———. 1998. Uma nueva especie de Anisembiidae (Insecta, Embioptera) en ámbar dominicano *Revista Brasileira de Entomologia* 42: 7–8.

———. 2004. Phylogenetic systematics of Archembiidae (Embiidina, Insecta). *Systematic Entomology* 29: 215–37.

———, J. Edgerly-Rooks, and C. Y. Hayashi. 2003. Phylogenetics of Embioptera (= Embiidina). *Entomologische Abhandlungen* 61: 131.

Tanaka, H., D. W. Roubik, M. Kato, F. Liew, and G. Gunsalam. 2001. Phylogenetic position of *Apis nuluensis* of northern Borneo and phylogeography of *A. cerana* as inferred from mitochondrial DNA sequences. *Insectes Sociaux* 48: 44–51.

Tappan, H. 1974. Molecular oxygen and evolution. Pp. 81–135. *In* Hayashi, O. (ed.), *Molecular Oxygen in Biology: Topics in Molecular Oxygen Research.* North-Holland; Amsterdam, the Netherlands; viii+367 pp.

Tasch, P. 1957. Flora and fauna of the Rhynie chert: A paleoecological reevaluation of published evidence. *University of Wichita Bulletin, University Studies 36* 32: 1–23.

———. 1962. Vertical extension of Mid-continent Leonardian insect occurrences. *Science* 135: 378–9.

———. 1963. Paleolimnology: Part 3 – Marion and Dickinson Counties, Kansas, with additional sections in Harvey and Sedgwick Counties: Stratigraphy and biota. *Journal of Paleontology* 37: 1233–51.

———. 1969. Branchiopoda. Pp. 128–91. *In* Moore, R. C. (ed.), *Treatise on Invertebrate Paleontology, Part R, Arthropoda 4.* University of Kansas Press; Lawrence, Kansas; xxxvi+398 pp.

———. 1973. Antarctic and other Gondwana non-marine deposits. *Antarctic Journal of the United States* 8: 272–3.

———. 1987. Fossil Conchostraca of the Southern Hemisphere and Continental Drift. Paleontology, Biostratigraphy, and Dispersal. *Memoirs of the Geological Society of America* 165: 1–290.

———, and J. R. Zimmerman. 1959. New Permian insects discovered in Kansas and Oklahoma. *Science* 130: 1656.

———, and J. R. Zimmerman. 1962. The *Asthenohymen-Delopterum* bed – A new Leonardian insect horizon in the Wellington of Kansas and Oklahoma. *Journal of Paleontology* 36: 1319–33.

Tauber, C. A., and M. J. Tauber. 1968. *Lomamyia latipennis* (Neuroptera: Berothidae) life history and larval description. *Canadian Entomologist* 100: 623–9.

Tauber, M. J., C. A. Tauber, and S. Masaki. 1986. *Seasonal Adaptations of Insects.* Oxford University Press; Oxford, UK; xv+411 pp.

Taylor, T. N., and A. C. Scott. 1983. Interactions of plants and animals during the Carboniferous. *Bioscience* 33: 488–93.

Templeton, A. R. 1977. Analysis of head shape differences between two interfertile species of Hawaiian *Drosophila*. *Evolution* 31: 630–41.

Terra, P. S. 1995. Revisão sistemática dos géneros de louva-a-deus da Região Neotropical (Mantodea). *Revista Brasiliensis Entomologia* 39: 13–94.

Theischinger, G., and W. W. K. Houston. 1988. Megaloptera. Pp. 23–32. *In* Walton, D. W. (ed.), *Zoological Catalogue of Australia, Volume 6.* Australian Government Publishing Service; Canberra, Australia; 316 pp.

Théobald, N. 1937. *Les Insectes Fossiles des Terrains Oligocènes de France.* George Thomas; Nancy, France; 473 pp.

Thewissen, J. G. M., E. M. Williams, L. J. Roe, and S. T. Hussain. 2001. Skeletons of terrestrial cetaceans and the relationships of whales to artiodactyls. *Nature* 413: 277–81.

Thiele, H.-U. 1977. *Carabid Beetles in Their Environments: A Study on Habitat Selection by Adaptations in Physiology and Behavior.* Springer Verlag; Berlin, Germany; xvii+369 pp.

Thien, L. B. 1980. Patterns of pollination in the primitive angiosperms. *Biotropica* 12: 1–13.

———, H. Azuma, and S. Kawano. 2000. New perspectives on the pollination biology of basal angiosperms. *International Journal of Plant Sciences, Supplement* 161: S225–35.

Thomas, M. A., K. A. Walsh, M. R. Wolf, B. A. McPheron, and J. H. Marden. 2000. Molecular phylogenetic analysis of evolutionary trends in stonefly wing structure and locomotor behavior. *Proceedings of the National Academy of Sciences, USA* 97: 13178–83.

Thompson, G. J., O. Kitade, N. Lo, and R. H. Crozier. 2000. Phylogenetic evidence for a single, ancestral origin of a "true" worker caste in termites. *Journal of Evolutionary Biology* 13: 869–81.

Thompson, I., and D. S. Jones. 1980. A possible onychophoran from the middle Pennsylvanian Mazon Creek Beds of northern Illinois. *Journal of Paleontology* 54: 588–96.

Thomsen, E., and S. A. Schack-Pedersen. 1997. *Geology and Palaeontology of the Mo-clay.* Aarhus Geoscience, Geological Institute; Aarhus, Denmark; 68 pp.

Thomson, C. G. 1872. *Skandinaviens Hymenoptera* [Volume 2]. Lund; Berlin, Germany; 286 pp.

Thorne, B. L. 1990. A case for ancestral transfer of symbionts between cockroaches and termites. *Proceedings of the Royal Society of London, Series B, Biological Sciences* 241: 37–41.

———. 1991. Ancestral transfer of symbionts between cockroaches and termites: An alternative hypothesis. *Proceedings of the Royal Society of London, Series B, Biological Sciences* 246: 191–5.

———, 1997. Evolution of eusociality in termites. *Annual Review of Ecology and Systematics* 28: 27–54.

———, and J. M. Carpenter. 1992. Phylogeny of the Dictyoptera. *Systematic Entomology* 17: 253–68.

———, and J. F. A. Traniello. 2003. Comparative social biology of basal taxa of ants and termites. *Annual Review of Entomology* 48: 283–306.

———, D. A. Grimaldi, and K. Krishna. 2000. Early fossil history of the termites. Pp. 77–93. *In* Abe, T., D. E. Bignell, and M. Higashi (eds.), *Termites: Evolution, Sociality, Symbioses, Ecology.* Kluwer Academic Publishers; Dordrecht, the Netherlands; xxii+466 pp.

Thorne, J. L., and H. Kishino. 2002. Divergence time and evolutionary rate estimation with multilocus data. *Systematic Biology* 51: 689–703.

Thornton, I. 1996. *Krakatau: The Destruction and Reassembly of an Island Ecosystem.* Harvard University Press; Cambridge, Massachusetts; xi+346 pp.

Thorp, R. W. 1979. Structural, behavioral, and physiological adaptations of bees (Apoidea) for collecting pollen. *Annals of the Missouri Botanical Garden* 66: 788–812.

Throckmorton, L. S. 1966. The relationships of the endemic Hawaiian Drosophilidae. *University of Texas Publications in Genetics* 6615: 335–96.

Tichomerova, A. L. 1968. Staphylinid beetles from the Jurassic of Karatau (Coleoptera, Staphylinidae). Pp. 139–54. *In* Rohdendorf, B. B. (ed.), *Jurassic Insects of Karatau.* Nauka; Moscow, Russia; 252 pp. [In Russian]

Tiegs, O. W. 1947. The development and affinities of the Pauropoda, based on a study of *Pauropus sylvaticus*. *Quarterly Journal of Microscopical Science* 88: 165–336.

———, and S. M. Manton. 1958. The evolution of the Arthropoda. *Biological Reviews* 33: 255–337.

Tilgner, E. 2000. The fossil record of Phasmida (Insecta: Neoptera). *Insect Systematics and Evolution* 31: 473–80.

———. 2002. Mantophasmatodea: A new insect order? *Science* 297: 731a.

———, T. G. Kiselyova, and J. V. McHugh. 1999. A morphological study of *Timema cristinae* Vickery with implications for the phylogenetics of Phasmida. *Deutsche Entomologische Zeitschrift* 46: 149–62.

Tillyard, R. J. 1917a. *The Biology of Dragonflies*. Cambridge University Press; Cambridge, UK; xii+396 pp.

———. 1917b. Permian and Triassic insects from New South Wales, in the collection of Mr. John Mitchell. *Proceedings of the Linnean Society of New South Wales* 42: 721–56.

———. 1922a. An insect wing in a crystal of selenite. *Records of the Geological Survey of New South Wales* 10: 205–7.

———. 1922b. Mesozoic insects of Queensland. No. 9. Orthoptera, and additions to the Protorthoptera, Odonata, Hemiptera and Planipennia. *Proceedings of the Linnean Society of New South Wales* 47: 447–70.

———. 1922c. The life-history of the moth-lacewing *Ithone fusca* Newman (Neuroptera, Planipennia). *Bulletin of Entomological Research* 13: 205–23.

———. 1925. A new fossil insect wing from Triassic beds near Deewhy, N. S. W. *Proceedings of the Linnean Society of New South Wales* 50: 374–7.

———. 1928a. The larva of *Hemiphlebia mirabilis* Selys (Odonata). *Proceedings of the Linnean Society of New South Wales* 53: 193–206.

———. 1928b. Some remarks on the Devonian fossil insects from Rhynie chert beds, Old Red Sandstone. *Transactions of the Royal Entomological Society of London* 76: 65–71.

———. 1928c. Kansas Permian insects. Part 12. The family Delopteridae, with a discussion of its ordinal position. *American Journal of Science* 16: 469–84.

———. 1928d. Kansas Permian insects. Part 10. The new order Protoperlaria: A study of the typical genus *Lemmatophora* Sellards. *American Journal of Science* 16: 185–220.

———. 1932. Kansas Permian insects. Part 15. The order Plectoptera. *American Journal of Science* 23: 97–135, 237–72.

———. 1936. Kansas Permian insects. Part 16. The oder Plectoptera (cont'd.): the family Doteridae, with a note on the affinities of the order Protohymenoptera. *American Journal of Science* 32: 435–53.

———. 1937a. Kansas Permian insects. Part 17. The order Megasecoptera and additions to the Palaeodictyoptera, Odonata, Protoperlaria, Copeognatha and Neuroptera. *American Journal of Science* 33: 81–110.

———. 1937b. Kansas Permian insects. Part 18. The order Embiaria. *American Journal of Science* 33: 241–51.

———, and B. Dunstan. 1916. Mesozoic and Tertiary insects of Queensland and New South Wales. *Queensland Geological Survey Publication* 253: xiv+1–60.

Tindale, N. B. 1945. Triassic insects of Queensland. 1. *Eoses*, a probable lepidopterous insect from the Triassic Beds of Mt. Crosby, Queensland. *Proceedings of the Royal Society of Queensland* 56: 37–46.

———. 1985. A butterfly-moth (Lepidoptera: Castaniidae) from the Oligocene shales of Florissant, Colorado. *Journal of Research on the Lepidoptera* 24: 31–40.

Tischlinger, H. 2001. Bemerkungen zur Insekten-Taphonomie der Solnhofener Plattenkalke. *Archaeopteryx* 19: 29–44.

Tjeder, B. 1944. A note on the food of the adult *Sisyra fuscata* F. (Neuroptera, Sisyridae). *Entomologisk Tidskrift* 65: 203–4.

———. 1959. Neuroptera-Planipennia: The lace-wings of southern Africa, 2, family Berothidae. Pp. 256–351. *In* Hanström, B., P. Brinck, and G. Rudbeck (eds.), *South African Animal Life, 6*. Swedish Natural Science Research Council; Stockholm, Sweden.

———. 1974. Taxonomic notes on species of the Crocini (Neuroptera, Nemopteridae). *Entomologica Scandinavica* 5: 300–4.

Tobias, V. J. 1987. New taxa of braconids (Hymenoptera, Braconidae) from Baltic amber. *Entomologicheskoe Obozrenie* 66: 845–59. [In Russian]

Tollerton, V. P., Jr. 1989. Morphology, taxonomy, and classification of the order Eurypterida Burmeister, 1843. *Journal of Paleontology* 63: 642–57.

Tomka, I., and C. Elpers. 1991. Problems in the phylogeny of the Ephemeroptera. Pp. 115–34. *In* Alba-Tercedor, J., and A. Sanchez-Ortega (eds.), *Overview and Strategies of Ephemeroptera and Plecoptera*. Sandhill Crane Press; Gainesville, Florida; xiv+588 pp.

Torchio, P. F. 1972. *Sapyga pumila* Cresson, a parasite of *Megachile rotundata* (F.) (Hymenoptera: Sapygidae; Megachilidae). I: Biology and description of immature stages. *Melanderia* 10: 1–22.

———. 1979. An eight-year field study involving control of *Sapyga pumila* Cresson (Hymenoptera: Sapygidae), a wasp parasite of the alfalfa leafcutter bee, *Megachile pacifica* Panzer. *Journal of the Kansas Entomological Society* 52: 412–19.

Townes, H. K. 1949a. The Nearctic species of the family Stephanidae (Hymenoptera). *Proceedings of the United States National Museum* 99: 361–70.

———. 1949b. The Nearctic species of Evaniidae (Hymenoptera). *Proceedings of the United States National Museum* 99: 525–39.

———. 1950. The Nearctic species of Gasteruptiidae (Hymenoptera). *Proceedings of the United States National Museum* 100: 85–145.

———. 1973. Two ichneumonids (Hymenoptera) from the Early Cretaceous. *Proceedings of the Entomological Society of Washington* 75: 216–19.

———. 1977. A revision of the Rhopalosomatidae (Hymenoptera). *Contributions of the American Entomological Institute* 15: 1–34.

Trant, C. A., and P. G. Gensel. 1985. Branching in *Psilophyton*: A new species from the Lower Devonian of New Brunswick, Canada. *American Journal of Botany* 72: 1256–73.

Trautner, J., and K. Geigenmüller. 1987. *Tiger Beetles, Ground Beetles: Illustrated Key to the Cicindellidae and Carabidae of Europe*. Margraf Press; Aichtal, Germany; 188 pp.

Trewin, N. H. 1989. The Rhynie hot-spring deposit. *Earth Science Conservation* 26: 10–12.

———. 1994. Depositional environment and preservation of biota in the Lower Devonian hot-springs of Rhynie, Aberdeenshire, Scotland. *Transactions of the Royal Society of Edinburgh, Earth Sciences* 84: 433–42.

Tröster, G. 1991. Eine neue Gattung der Elateridae (Insecta: Coleoptera) *Macropunctum* gen. n. aus der Messel-Formation des unteren Mittel-Eozän der Fundstelle Messel. *Courier Forschungsinstitut Senckenberg* 139: 99–117.

———. 1992. Fossile Insekten aus den mitteleozänen Tonsteinen der Grube Messel bei Darmstadt. *Mitteilungen des Internationalen Entomologischen Vereins* 17: 191–208.

———. 1993. Zwei neue mitteleuropäische Arten der Gattung *Tenomerga* Neboiss 1984 aus dem Mitteleozän der Grube Messel und des Eckfelder Maares (Coleoptera: Archostemata: Cupedidae). *Mainzer Naturwissenschaftliches Archiv* 31: 169–76.

———. 1994. Fossile Elateridae (Insecta: Coleoptera) aus dem Unteren Mitteleozän (Lutetium) der Grube Messel bei Darmstadt. *Courier Forschungsinstitut Senckenberg* 170: 11–64.

Trueman, J. W. H. 1996. A preliminary cladistic analysis of odonate wing venation. *Odonatologica* 25: 59–72.

———, B. E. Pfeil, S. A. Kelchner, and D. K. Yeates. 2004. Did stick insects *really* regain their wings? *Systematic Entomology* 29: 138–9.

Truman, J. W., and L. W. Riddiford. 1999. The origins of insect metamorphosis. *Nature* 401: 447–52.

Tschinkel, W. R. 2001. Colonies in space. *Natural History* 4: 64–5.

Tshernova, O. A. 1971. Mayfly from fossil pitch of Polar Siberia (Ephemeroptera: Leptophlebiidae). *Entomologischeskoe Obozrenie* 50: 612–18. [In Russian]

———, and N. D. Sinitshenkova. 1974. A new fossil genus and species of the mayfly family Hexagenitidae (Ephemeroptera) from the South of the European part of the USSR and its connection with Recent mayflies. *Entomologischeskoe Obozrenie* 53: 130–6. [In Russian]

Turbeville, J. M., D. M. Pfeifer, K. G. Field, and R. A. Raff. 1991. The phylogenetic status of arthropods, as inferred from 18S rRNA sequences. *Molecular Biology and Evolution* 8: 669–86.

Turillazzi, S. 1987. Distinguishing features of the social behaviour of Stenogastrinae wasps. Pp. 492–5. *In* Eder, J., and H. Rembold (eds.), *Chemistry and Biology of Social Insects*. J. Peperny Verlag; Munich, Germany; xxxviii+757 pp.

———. 1989. The origin and evolution of social life in the Stenogastrinae (Hymenoptera, Vespidae). *Journal of Insect Behavior* 2: 649–61.

———. 1990a. Social biology of *Liostenogaster vechti* Turillazzi 1988 (Hymenoptera Stenogastrinae). *Tropical Zoology* 3: 69–87.

———. 1990b. Socialita ed architettura del nido nelle vespe Stenogastrinae (Hymenoptera, Vespidae). *Atti della Accademia Nazionale Italiana di Entomologia Rendiconti* 36–38: 183–206.

———. 1991. The Stenogastrinae. Pp. 74–98. *In* Ross, K. G., and R. W. Matthews (eds.), *The Social Biology of Wasps*. Cornell University Press; Ithaca, New York; xvii+678 pp.

———. 1996. *Polistes* in perspective: Comparative social biology and evolution in *Belonogaster* and Stenogastrinae. Pp. 235–47. *In* Turillazzi, S., and M. J. West-Eberhard (eds.), *Natural History and Evolution of Paper-Wasps*. Oxford University Press; Oxford, UK; xiv+400 pp.

——— 1999. New species of *Liostenogaster* van der Vecht 1969, with keys to adults and nests (Hymenoptera Vespidae Stenogastrinae). *Tropical Zoology* 12: 335–58.

Turkin, N. I. 1997. Preliminary results of microscopic research of tangential wood imprints in Baltic amber. *Metalla, Bochum* 66: 55–6.

Turner, J. R. G. 1977. Butterfly mimicry: The genetical evolution of an adaptation. *Evolutionary Biology* 10: 163–206.

———. 1981. Adaptation and evolution in *Heliconius*: A defense of neo-Darwinism. *Annual Review of Ecology and Systematics* 12: 99–121.

———. 1984. Mimicry: The palatability spectrum and its consequences. Pp. 141–61. *In* Vane-Wright, R. I., and P. R. Ackery (eds.), *The Biology of Butterflies*. Academic Press; London, UK; xxi + 427 pp.

Tuxen, S. L. 1958. Relationships of Protura. *Proceedings of the 10th International Congress on Entomology* 1: 493–7.

———. 1959. Phylogenetic significance of entognathy in entognathous apterygotes. *Smithsonian Miscellaneous Contributions* 137: 379–416.

———. 1963. Phylogenetical trends in the Protura, as shown by relationships between recent genera. *Zeitschrift für Zoologische Systematik und Evolutionsforschung* 1: 277–310.

———. 1964. *The Protura: A Revision of the Species of the World, with Keys for Determination*. Hermann; Paris, France; 360 pp.

———. 1967a. The entomologist J. C. Fabricius. *Annual Review of Entomology* 12: 1–14.

———. 1967b. Australian Protura, their phylogeny and zoogeography. *Zeitschrift für Zoologische Systematik und Evolutionsforschung* 5: 1–53.

———, ed. 1970. *Taxonomist's Glossary of Genitalia in Insects*. Munksgaard; Copenhagen, Denmark; 359 pp.

Ulmer, G. 1912. Die Trichoptera des Baltischen Bernsteins. *Schriften der Physikalisch-ökonomischen Gesellschaft* 10: 1–380.

Ulrich, W. 1927. Über das bisher einzige strepsipteron aus dem baltischen Bernstein und über eine Theorie der Mengeinen-biologie. *Zeitschrift für Wissenschaftliche Biologie, A, Zeitschrift für Morphologie und Ökologie der Tiere* 8: 45–62.

———. 1943. Die Mengeiden (Mengenillini) und die Phylogenie der Strepsipteren. *Zeitschrift für Parasitenkunde* 3: 62–101.

Underwood, B. A. 1990. Time of drone flight of *Apis laboriosa* Smith in Nepal. *Apidologie* 21: 501–4.

Usinger, R. L. 1966. *Monograph of Cimicidae (Hemiptera–Heteroptera)*. Entomological Society of America (Thomas Say Publications); College Park, Maryland; 585 pp.

Ussher, J. 1650. *Annales Veteris Testamenti, a Prima Mundi Origine Deducti: Una cum rerum Asiaticarum et AEgyptiacarum Chronico, a Temporis Historici Principio usque ad Maccabaicorum initia Producto*. J. Flesher *et al.*; Londini [London], UK; [10]+554+[10] pp.

Uvarov, B. P. 1928. *Locusts and Grasshoppers: A Handbook for Their Study and Control*. Imperial Bureau of Entomology; London, UK; xiii+352 pp.

———. 1966. *Grasshoppers and Locusts: A Handbook of General Acridology*. Cambridge University Press; Cambridge, UK; 465 [vol. 1] pp., 588 [vol. 2] pp.

Val, F. C. 1977. Genetic analysis of the morphological differences between two interfertile species of Hawaiian *Drosophila*. *Evolution* 31: 611–39.

Valentine, B. D. 1986. Grooming behavior in Embioptera and Zoraptera (Insecta). *Ohio Journal of Science* 86: 150–2.

Valentine, J. W., T. C. Foin, and D. Peart. 1978. A provincial model of Phanerozoic marine diversity. *Paleobiology* 4: 55–66.

van Dijk, D. E. 1997. Insect faunas of South Africa from the Upper Permian and the Permian/Triassic boundary. *Palaeontologia Africana* 34: 43–8.

———, and H. Geertsema. 1999. Permian insects from the Beaufort Group of Natal, South Africa. *Annals of the Natal Museum* 40: 137–71.

van Dijk, M. A. M., O. Madsen, F. Catzelis, M. J. Stanhope, W. W. de Jong, and M. Pagel. 2001. Protein sequence signatures support the African clade of mammals. *Proceedings of the National Academy of Sciences, USA* 98: 188–93.

Vane-Wright, D. 2004. Butterflies at that awkward age. *Nature* 428: 477–80.

Vane-Wright, R. I. 1981. Mimicry and its unknown ecological consequences. Pp. 157–68. *In* Greenwood, P. H., and P. L. Forey (eds.), *Chance, Change, and Challenge: The Evolving Biosphere*. British Museum of Natural History; London, UK; viii+311 pp.

Vanzolini, P. E., and L. R. Guimarães. 1955. Lice and the history of South American land mammals. *Revista Brasileira Entomologia* 3: 13–46.

Vårdal, H., G. Sahlén, and F. Ronquist. 2003. Morphology and evolution of the cynipoid egg (Hymenoptera). *Zoological Journal of the Linnean Society* 139: 247–60.

Veevers, J. J., C. M. Powell, and S. R. Roots. 1991. Review of seafloor spreading around Australia. 1. Synthesis of the patterns of spreading. *Australian Journal of Earth Sciences* 38: 373–89.

Verhoeff, K. W. 1902. Ueber die verwandtschaftliche Stellung von *Hemimerus*. *Sitzungsberichte der Gesellschaft naturforschender Freunde zu Berlin* 1902: 87–9.

Vermeij, G. J. 1987. *Evolution and Escalation: An Ecological History of Life*. Princeton University Press; Princeton, New Jersey; xv+527 pp.

Via, L., and S. Calzada. 1987. Artropodos fosiles triasicos de Alcover-Montral. 1. Insectos. *Cuadernos de Geologia Iberica* 11: 273–80.

Vickery, V. R. 1993. Revision of *Timema* Scudder (Phasmatoptera: Timematodea) including three new species. *Canadian Entomologist* 125: 657–92.

———, and D. K. McE. Kevan. 1983. A monograph of the orthopteroid insects of Canada and adjacent regions. *Lyman Entomological Museum Research Laboratory* 13: 216–37.

———, and D. K. McE. Kevan. 1985. The grasshoppers, crickets and related insects of Canada and adjacent regions. Ulonata: Dermaptera, Cheleutoptera, Notoptera, Dictuoptera, Grylloptera, and Orthoptera. *The Insects and Arachnids of Canada* 14: 1–918.

———, and C. P. Sandoval. 1997. *Timema bartmani* (Phasmatoptera: Timematodea: Timematidae), a new species from southern California. *Canadian Entomologist* 129: 933–6.

———, and C. P. Sandoval. 1998. *Timema monikensis* sp. nov. (Phasmatoptera: Timematodea: Timematidae), a new parthenogenetic species in California. *Lyman Entomological Museum Research Laboratory* 22: 1–3.

———, and C. P. Sandoval. 1999. Two new species of *Timema* (Phasmatoptera: Timematodea: Timematidae), one parthenogenetic, in California. *Journal of Orthoptera Research* 8: 45–7.

———, and C. P. Sandoval. 2001. Descriptions of three new species of *Timema* (Phasmatoptera: Timematodea: Timematidae) and notes on three other species. *Journal of Orthoptera Research* 10: 53–61.

Vilhelmsen, L. 1996. The preoral cavity of lower Hymenoptera (Insecta): Comparative morphology and phylogenetic significance. *Zoologica Scripta* 25: 143–70.

———. 1997. The phylogeny of lower Hymenoptera (Insecta), with a summary of the early evolutionary history of the order. *Journal of Zoological Systematics and Evolutionary Research* 35: 49–70.

———. 2001. Phylogeny and classification of the extant basal lineages of the Hymenoptera (Insecta). *Zoological Journal of the Linnean Society* 131: 393–442.

———. 2003. Phylogeny and classification of the Orussidae (Insecta: Hymenoptera), a basal parasitic wasp taxon. *Zoological Journal of the Linnean Society* 139: 337–418.

———, N. Isidoro, R. Romani, H. H. Basibuyuk, and D. L. J. Quicke. 2001. Host location and oviposition in a basal group of parasitic wasps: The subgenual organ, ovipositor apparatus and associated structures in the Orussidae (Hymenoptera, Insecta). *Zoomorphology* 121: 63–84.

Vishniakova, V. N. 1968. Mesozoic roaches with the external ovipositor and peculiarity of their reproduction (Blattodea). Pp. 55–86. *In* Rohdendorf, B. B. (ed.), *Jurassic Insects of Karatau*. Nauka Press; Moscow, Russia; 252 pp. [In Russian]

———. 1975. Psocoptera in Late-Cretaceous insect-bearing resins from the Taimyr. *Entomological Review* 54: 63–75.

———. 1980. Earwigs (Insecta, Forficulida) from the Upper Jurassic of the Karatau Range. *Paleontologischeskiy Zhurnal* 1980: 78–94. [In Russian]

———. 1981. New Paleozoic and Mesozoic lophioneurids (Thripida, Lophioneurida). *Trudy Paleontologischeskogo Instituta Akademii Nauk, SSSR* 193: 43–63. [In Russian]

Vogel, S. 1954. *Blütenbiologische Typen als Elemente der Sippengliederung*. Fischer; Jena, Germany; x+338 pp.

Voigt, E. 1952. Ein Haareinschluß mit Phthirapteren-Eiern im Bernstein. *Mitteilungen aus dem Geologischen Staatsinstitut in Hamburg* 21: 59–74.

Volk, P. 1995. An entomological study of apartment 4A: My wild kingdom. *The New York Times Magazine* 1995 (5 March): 44–5.

Vollrath, F. 1986. Eusociality and extraordinary sex ratios in the spider *Anelosimus eximius* (Araneae: Theridiidae). *Behavioral Ecology and Sociobiology* 18: 283–7.

Von Dohlan, C. D., and N. A. Moran. 1995. Molecular phylogeny of the Homoptera: A paraphyletic taxon. *Journal of Molecular Evolution* 41: 211–23.

———, and N. A. Moran. 2000. Molecular data support a rapid radiation of aphids in the Cretaceous and multiple origins of host alternation. *Biological Journal of the Linnean Society* 71: 689–717.

Vovides, A. P. 1991. Insect symbionts of some Mexican cycads in their natural habitat. *Biotropica* 23: 102–4.

Vršanský, P. 1997. *Piniblattella* gen. nov. – The most ancient genus of the family Blattellidae (Blattodea) from the Lower Cretaceous of Siberia. *Entomological Problems* 28: 67–79.

———. 1999. Two new species of Blattaria (Insecta) from the Lower Cretaceous of Asia, with comments on the origin and phylogenetic position of the families Polyphagidae and Blattulidae. *Entomological Problems* 30: 85–91.

———, and J. Ansorge. 2001. New Lower Cretaceous polyphagid cockroaches from Spain (Blattaria, Polyphagidae, Vitisminae subfam. nov.). *Cretaceous Research* 22: 157–62.

———, S. Y. Storozhenko, C. C. Labandeira, and P. Ihringova. 2001. *Galloisiana olgae* sp. nov. (Grylloblattodea: Grylloblattidae) and the paleobiology of a relict order of insects. *Annals of the Entomological Society of America* 94: 179–84.

———, V. N. Vishniakova, and A. P. Rasnitsyn. 2002. Order Blattida Latreille, 1810. The cockroaches. Pp. 263–9. *In* Rasnitsyn, A. P., and D. L. J. Quicke (eds.), *History of Insects.* Kluwer Academic Publishers; Dordrecht, the Netherlands; xii+517 pp.

Waage, J. K. 1979. The evolution of insect/vertebrate associations. *Biological Journal of the Linnean Society* 12: 187–224.

———, and H. C. J. Greathead, eds. 1986. *Insect Parasitoids.* Academic Press; London, UK; xvii+389 pp.

———, and G. G. Montgomery. 1976. *Cryptoses choloepi*: A coprophagous moths that lives on a sloth. *Science* 193: 157–8.

Wägele, J.-W., and B. Misof. 2001. On quality of evidence in phylogeny reconstruction: A reply to Zrzavý's defence of the 'Ecdysozoa' hypothesis. *Journal for Zoological Systematics and Evolutionary Research* 39: 165–76.

———, T. Reikson, P. Lockhart, and B. Misof. 1999. The Ecdysozoa: Artifact or monophylum? *Journal for Zoological Systematics and Evolutionary Research* 37: 211–23.

Waggoner, B. J. 1999. Fossil oak leaf galls from the Stinking Water paleoflora of Oregon (middle Miocene). *PaleoBios* 19: 8–14.

———, and M. F. Poteet. 1996. Unusual oak leaf galls from the Middle Miocene of northwestern Nevada. *Journal of Paleontology* 70: 1080–4.

Wagner, R., C. Hoffeins, and H. W. Hoffeins. 2000. A fossil nymphomyiid (Diptera) from the Baltic and Bitterfeld amber. *Systematic Entomology* 25: 115–20.

Wagner, W. E. J. 1996. Convergent song preferences between female field crickets and acoustically orienting parasitoid flies. *Behavioral Ecology* 7: 279–85.

Wagner, W. L., and V. A. Funk, eds. 1995. *Hawaiian Biogeography: Evolution on a Hot Spot Archipelago.* Smithsonian Institution Press; Washington, D.C.; xvii+467 pp.

Walden, K. K. O., and H. M. Robertson. 1997. Ancient DNA from amber fossil bees? *Molecular Biology and Evolution* 14: 1075–7.

Walker, E. M. 1914. A new species of Orthoptera, forming a new genus and family. *Canadian Entomologist* 51: 131–9.

———. 1937. *Grylloblatta*, a living fossil. *Transactions of the Royal Society of Canada, Section V, Series III* 26: 1–10.

Walker, G. P., and P. J. Cameron. 1981. The biology of *Dendrocerus carpenteri* (Hymenoptera: Ceraphronidae), a parasite of *Aphidius* species, and field observations of *Dendrocerus* species as hyperparasites of *Acyrthosiphon* species. *New Zealand Journal of Zoology* 8: 531–8.

Walker, J. A. 2003. Mantophasmatodea – A new order of insects. *Bulletin of the Amateur Entomologists' Society* 62: 72–8.

Walker, M. V. 1938. Evidence of Triassic insects in the Petrified Forest National Monument, Arizona. *Proceedings of the United States National Museum* 85: 137–41.

Wallace, B. 1974. The adaptation of *Drosophila virilis* to life on an artificial crab. *American Naturalist* 112: 971–3.

Wallace, B. 1978. The adaptation of *Drosophila virilis* to life on an artificial crab. *American Naturalist* 112: 971–3.

Walossek, D., and K. J. Müller. 1994. Pentastomid parasites from the Lower Palaeozoic of Sweden. *Transactions of the Royal Society of Edinburgh, Earth Sciences* 85: 1–37.

———, and K. J. Müller. 1998. Early arthropod phylogeny in light of the Cambrian 'Orsten' fossils. Pp. 185–231. *In* Edgecombe, G. D. (ed.), *Arthropod Fossils and Phylogeny.* Columbia University Press; New York, New York; 347 pp.

———, and H. Szaniawski. 1991. *Cambrocaris baltica* n. gen. n. sp, a possible stem-lineage crustacean from the Upper Cambrian of Poland. *Lethaia* 24: 363–78.

———, C. S. Li, and C. Brauckmann. 1990. A scorpion from the Upper Devonian of Hubei Province, China (Arachnida, Scorpionida). *Neues Jahrbuch für Geologie und Paläontologie, Monatshefte* 1990: 169–80.

———, I. Hinz-Schallreuter, J. H. Shergold, and K. J. Müller. 1993. Three-dimensional preservation of arthropod integument from the Middle Cambrian of Australia. *Lethaia* 26: 7–15.

———, J. E. Repetski, and K. J. Müller. 1994. An exceptionally preserved parasitic arthropod, *Heymonsicambria taylori* n. sp. (Arthropoda incertae sedis: Pentastomida), from Cambrian-Ordovician boundary beds of Newfoundland, Canada. *Canadian Journal of Earth Sciences* 31: 1664–71.

Waloszek, D. 2003. The 'Orsten' window: A three-dimensionally preserved Upper Cambrian meiofauna and its contribution to our understanding of the evolution of Arthropoda. *Paleontological Research* 7: 71–88.

———, and J. A. Dunlop. 2002. A larval sea spider (Arthropoda: Pycnogonida) from the Upper Cambrian 'Orsten' of Sweden, and the phylogenetic position of pycnogonids. *Palaeontology* 45: 421–46.

Walter, D. E., and H. C. Proctor. 1999. *Mites: Ecology, Evolution and Behaviour.* CABI Publishing; Wallingford, UK; x+322 pp.

Wang, T. 1993. *Synopsis on the Classification of Mantodea from China.* Scientific and Technical Literature Publishing House; Shanghai, China. [In Chinese]

Wang, T.-Q., and W. P. McCafferty. 1995. Relationships of the Arthropleidae, Heptageniidae, and Pseudironidae (Ephemeroptera: Heptagenioidea). *Entomological News* 106: 251–6.

———, W. P. McCafferty, and Y. J. Bae. 1997. Sister relationships of the Neoephemeridae and Caenidae (Ephemeroptera: Pannota). *Entomological News* 108: 52–6.

Wappler, T., and M. S. Engel. 2003. The middle Eocene bee faunas of Eckfeld and Messel, Germany (Hymenoptera: Apoidea). *Journal of Paleontology* 77: 908–21.

———, T., V. S. Smith, and R. C. Dalgleish. 2004. Scratching an ancient itch: An Eocene bird louse fossil. *Proceedings of the Royal Society of London, Series B, Biological Sciences, (Supplement 5: Biology Letters)* 271: 5255–8.

Ward, J. V. 1992. *Aquatic Insect Ecology. Part 1. Biology and Habitat.* John Wiley and Sons; New York, New York; 438 pp.

Ward, P. S. 1994. *Adetomyrma*, an enigmatic new ant genus from Madagascar (Hymenoptera: Formicidae), and its implications for ant phylogeny. *Systematic Entomology* 19: 159–75.

———, and S. G. Brady. 2003. Phylogeny and biogeography of the ant subfamily Myrmeciinae (Hymenoptera: Formicidae). *Invertebrate Systematics* 17: 361–86.

Warncke, K. 1977. Ideen zum naturlichen System der Bienen. *Mitteilungen der Münchner Entomologischen Gesellschaft* 67: 39–63.

Wasbauer, M. S. 1995. Pompilidae. Pp. 522–539. *In* Hanson, P. E., and I. D. Gauld (eds.), *The Hymenoptera of Costa Rica.* Oxford University Press; Oxford, UK; xx+893 pp.

Waser, N. M. 1998. Pollination, angiosperm speciation, and the nature of species boundaries. *Oikos* 81: 198–201.

Waters, S. B. 1989a. A new hybotine dipteran from the Cretaceous of Botswana. *Palaeontology* 32: 657–67.

———. 1989b. A Cretaceous dance fly (Diptera: Empididae) from Botswana. *Systematic Entomology* 14: 233–41.

———, and A. Arillo. 1999. A new genus of Hybotidae (Diptera, Empidoidea) from Lower Cretaceous amber of Alava (Spain). *Studia Dipterologica* 6: 59–66.

Watson, J. A. L., G. Theischinger, and H. M. Abbey. 1991. *The Australian Dragonflies: A Guide to the Identification, Distributions and Habitats of Australian Odonata.* CSIRO; Canberra, Australia; vii+278 pp.

Way, M. J. 1963. Mutualism between ants and honeydew-producing Homoptera. *Annual Review of Entomology* 8: 307–44.

Wcislo, W. T., and J. H. Cane. 1996. Floral resource utilization by solitary bees (Hymenoptera: Apoidea) and exploitation of their stored foods by natural enemies. *Annual Review of Entomology* 41: 257–80.

———, M. J. West-Eberhard, and W. G. Eberhard. 1988. Natural history and behaviour of a primitively social wasp, *Auplopus semialatus*, and its parasite, *Irenangelus eberhardi* (Hymenoptera: Pompilidae). *Journal of Insect Behavior* 1: 247–60.

Weaver, J. S., III. 1984. The evolution and classification of Trichoptera, part 1: The groundplan of Trichoptera. Pp. 413–19. *In* Morse, J. C. (ed.), *Proceedings of the Fourth International Symposium on Trichoptera*. Junk; The Hague, the Netherlands; xxii+486 pp.

———, and J. C. Morse. 1986. Evolution of feeding and case-making behavior in Trichoptera. *Journal of the North American Benthological Society* 5: 150–8.

Webb, B. A., and L. Cui. 1998. Relationships between polydnavirus genomes and viral gene expression. *Journal of Insect Physiology* 44: 785–93.

———, and M. D. Summers. 1990. Venom and viral expression products of the endoparasitic wasp *Campoletis sonorensis* share epitopes and related sequences. *Proceedings of the National Academy of Sciences, USA* 87: 4961–5.

Weber, H. 1969. Die Elefantenlaus (*Haematomyzus elefantis* Piaget 1869) – Versuch einer konstruktionsmorphologischen Analyse. *Zoologica (Stuttgart)* 116: 1–155.

Weber, W. A. 1965. Theodore Dru Alison Cockerell, 1866–1948. *University of Colorado Studies, Series in Biobibliography* 1: 1–124.

Wedmann, S. 1998. First records of fossil tremecine hymenopterans. *Palaeontology* 41: 929–38.

———. 2000. Die Insekten der oberoligozänen Fossillagerstätte Enspel (Westerwald, Deutschland): Systematik, Biostrationomie und Paläoökologie. *Mainzer Naturwissenschafticħes Archiv* 23: 1–154.

Weele, H. W., vander. 1910. *Collections Zoologiques du Baron E. de Selys-Longchamps: Catalogue Systémtique et Descriptif; Fasc. 5: Megaloptera*. Hayez; Brussels, Belgium; 93 pp.

Wegener, A. 1924. *The Origin of Continents and Oceans* [3rd Edition]. Meuthen and Co.; London, UK; 212 pp.

Wegierek, P. 2000. A new genus and species of aphid (Hemiptera: Aphidinea) from New Jersey amber. Pp. 141–5. *In* Grimaldi, D. (ed.), *Studies on Fossils in Amber, with Particular Reference to the Cretaceous of New Jersey*. Backhuys Publishers; Leiden, the Netherlands; viii+498 pp.

Wehr, W. C. 1998. Middle Eocene insects and plants of the Okanongen Highlands. Pp. 125–135. *In* Martin, J. E. (ed.), *Contributions to the Paleontology and Geology of the West Coast, in Honor of V. Standish Mallory*. Research Report of the Thomas Burke Memorial Washington State Museum [volume 6]; Seattle, Washington.

Weiblen, G. D. 2002. How to be a fig wasp. *Annual Review of Entomology* 47: 299–330.

———. 2001. Phylogenetic relationships of fig wasps pollinating functionally dioecious *Ficus* based on mitochondrial DNA sequences and morphology. *Systematic Biology* 50: 243–67.

———. 2002. How to be a fig wasp. *Annual Review of Entomology* 47: 299–330.

———, and G. L. Bush. 2002. Speciation in fig pollinators and parasites. *Molecular Ecology* 11: 1573–8.

Weidner, H. 1969. Die Ordnung Zoraptera oder Bodenläuse. *Entomologische Zeitschrift* 79: 29–51.

———. 1970. Zoraptera (Bodenläuse). *Handbuch der Zoologie* 4: 1–12.

———. 1972. Copeognatha (Psocodea). *Handbuch der Zoologie* 4: 1–94.

———. 1979. Eine weitere Mitteilung über Termiten as dem Pliozän von Willershausen, Harz (Insecta, Isoptera). *Bericht der Naturhistorischen Gesellschaft zu Hannover* 122: 91–95.

Weinstein, P., and A. D. Austin. 1991. The host relationships of trigonalyid wasps (Hymenoptera: Trigonalyidae), with a review of their biology and catalogue to world species. *Journal of Natural History* 25: 399–433.

Weintraub, J. D., and J. S. Miller. 1987. [Review] The structure and affinities of the Hedyloidea: A new concept of butterflies. *Cladistics* 3: 299–304.

Weissman, D. B. 2001. Communication and reproductive behaviour in North American Jerusalem crickets (*Stenopelmatus*) (Orthoptera: Stenopelmatidae). Pp. 351–75. *In* Field, L. H. (ed.), *The Biology of Wetas, King Crickets and their Allies*. CABI Publishing; Oxon, UK; xx+540 pp.

Weitschat, W. 1997. Bitterfelder Bernstein – Ein eozäner Bernstein auf miozäner Lagerstätte. *Metalla, Bochum* 66: 71–84.

———, and W. Wichard. 1998. *Atlas der Pflanzen und Tiere im Baltischen Bernstein*. Verlag Dr. Friedrich Pfeil; Munich, Germany; 256 pp.

———, and W. Wichard. 2002. *Atlas of Plants and Animals in Baltic Amber*. Verlag Dr. Friedrich Pfeil; Munich, Germany; 256 pp.

Weller, S. J., and D. P. Pashley. 1995. In search of butterfly origins. *Molecular Phylogenetics and Evolution* 4: 235–46.

———, D. P. Pashley, J. A. Martin, and J. L. Constable. 1994. Phylogeny of noctuoid moths and the utility of combining independent nuclear and mitochondrial genes. *Systematic Biology* 43: 194–211.

Wenzel, J. W. 1990. A social wasp's nest from the Cretaceous period, Utah, USA, and its biogeographical significance. *Psyche* 97: 21–9.

———. 1991. Evolution of nest architecture. Pp. 480–519. *In* Ross, K. G., and R. W. Matthews (eds.), *The Social Biology of Wasps*. Cornell University Press; Ithaca, New York; xvii+678 pp.

———. 1992. Behavioral homology and phylogeny. *Annual Review of Ecology and Systematics* 23: 361–81.

———. 1993. Application of the biogenetic law to behavioral ontogeny: A test using next architecture in paper wasps. *Journal of Evolutionary Biology* 6: 229–47.

———. 1998. A generic key to the nests of hornets, yellowjackets, and paper wasps worldwide (Vespidae: Vespinae, Polistinae). *American Museum Novitates* 3224: 1–39.

West-Eberhard, M. J. 1978. Polygyny and the evolution of social behavior in wasps. *Journal of the Kansas Entomological Society* 51: 832–56.

Westerkamp, C. 1996. Pollen in bee-flower relations. *Botanica Acta* 109: 325–32.

Westfall, M. J., Jr., and M. L. May. 1996. *Damselflies of North America*. Scientific Publishers; Gainesville, Florida; vii+[3]+649+[1] pp.

Weygoldt, P. 1969. *The Biology of Pseudoscorpions*. Harvard University Press; Cambridge, MA; xiv+145 pp.

———. 1980. Towards a cladistic classification of the Chelicerata. *Proceedings of the 8th International Congress of Arachnology, Wien* 1980: 331–4.

———. 1986. Arthropod interrelationships – The phylogenetic-systematic approach. *Zeitschrift für Zoologische Systematik und Evolutionsforschung* 24: 19–35.

———. 1996. Evolutionary morphology of whip spiders: Towards a phylogenetic system (Chelicerata: Arachnida: Amblypygi). *Journal of Zoological Systematics and Evolutionary Research* 34: 185–202.

———, and H. F. Paulus. 1979. Untersuchungen zur Morphologie, Taxonomie, und Phylogenie der Chelicerata. *Zeitschrift für Zoologische Systematik und Evolutionsforschung* 17: 85–116.

Whalley, P. E. S. 1978. New taxa of fossil and recent Micropterigidae with a discussion of their evolution and a comment on the evolution of the Lepidoptera (Insecta). *Annals of the Transvaal Museum* 31: 72–86.

———. 1980. Neuroptera (Insecta) in amber from the Lower Cretaceous of Lebanon. *Bulletin of the British Museum of Natural History (Geology)* 33: 157–64.

———. 1983. *Fera venatrix* gen. and sp. n. (Neuroptera, Mantispidae) from amber in Britain. *Neuroptera International* 2: 229–33.

———. 1985. The systematics and palaeogeography of the Lower Jurassic insects of Dorset, England. *Bulletin of the British Museum (Natural History) (Geology)* 39: 107–89.

———. 1986a. A review of the current fossil evidence of Lepidoptera in the Mesozoic. *Biological Journal of the Linnean Society* 28: 253–71.

———. 1986b. Insects from the Italian Upper Triassic. *Rivista del Museo Civico di Scienze Naturali "Enrico Caffi", Bergamo* 10: 51–60.

———. 1988. Insect evolution during the extinction of the Dinosauria. *Entomologica Generalis* 13: 119–24.

———, and E. A. Jarzembowski. 1981. A new assessment of *Rhyniella*, the earliest known insect, from the Devonian of Rhynie, Scotland. *Nature* 291: 317.

———, and E. A. Jarzembowski. 1985. Fossil insects from the Lithographic limestone of Montsech (late Jurassic-early Cretaceous), Lérida Province, Spain. *Bulletin of the British Museum of Natural History (Geology)* 38: 381–412.

Wheeler, A. G., Jr. 2001. *Biology of the Plant Bugs (Hemiptera: Miridae). Pests, Predators, Opportunists*. Cornell University Press; Ithaca, New York; xv+507 pp.

Wharton, R. A. 1989. Final instar larva of the embolemid wasp, *Ampulicomorpha confusa* (Hymenoptera). *Proceedings of the Entomological Society of Washington* 91: 509–12.

Wheeler, D. A., C. P. Kyriacou, M. L. Greenacre, Q. Yu, *et al.* 1991. Molecular transfer of a species-specific behavior from *Drosophila simulans* to *Drosophila melanogaster*. *Science* 251: 1082–5.

WHEELER, E. A., and P. BAAS. 1991. A survey of the fossil record for dicotyledonous wood and its significance for evolutionary and ecological wood anatomy. *IAWA Bulletin* 12: 275–332.

WHEELER, Q. D. 1986. Revision of the genera of Lymexylidae (Coleoptera: Cucujiformia). *Bulletin of the American Museum of Natural History* 183: 115–210.

———, and R. MEIER, eds. 2000. *Species Concepts and Phylogenetic Theory: A Debate*. Columbia University Press; New York, New York; xii+[1]+230 pp.

WHEELER, W. C. 1998. Sampling, groundplans, total evidence and the systematics of arthropods. Pp. 87–96. *In* FORTEY, R. A., and R. H. THOMAS (eds.), *Arthropod Relationships*. Chapman and Hall; London, UK; xii+383 pp.

———, P. CARTWRIGHT, and C. Y. HAYASHI. 1993a. Arthropod phylogeny: A combined approach. *Cladistics* 9: 1–39.

———, R. T. SCHUH, and R. BANG. 1993b. Cladistic relationships among higher groups of Heteroptera: Congruence between morphological and molecular data sets. *Entomologica Scandinavica* 24: 121–37.

———, M. WHITING, Q. D. WHEELER, and J. M. CARPENTER. 2001. The phylogeny of the extant hexapod orders. *Cladistics* 17: 113–69, 403–4.

WHEELER, W. M. 1910. *Ants: Their Structure, Development, and Behavior*. Columbia University Press; New York, New York; xxv+663 pp.

WHEELER, W. M. 1915. The ants of the Baltic amber. *Schriften der Physikalischökonomischen Gesellschaft* 55: 1–142.

———. 1942. Studies on Neotropical ant-plants and their ants. *Bulletin of the Museum of Comparative Zoology* 90: 1–262.

WHITE, D. A., and L. B. THIEN. 1985. The pollination of *Illicium parviflorum* (Illiciaceae). *The Journal of the Elisha Mitchell Scientific Society* 101: 15–18.

WHITFIELD, J. B. 1990. Parasitoids, polydnaviruses and endosymbiosis. *Parasitology Today* 6: 381–4.

———. 1992a. Phylogeny of the non-aculeate Apocrita and the evolution of parasitism in the Hymenoptera. *Journal of Hymenoptera Research* 1: 3–14.

———. 1992b. The polyphyletic origin of endoparasitism in the cyclostome lineages of Braconidae (Hymenoptera). *Systematic Entomology* 17: 273–86.

———. 1997. Molecular and morphological data suggest a single origin of the polydnaviruses among braconid wasps. *Naturwissenschaften* 84: 502–7.

———. 1998. Phylogeny and evolution of host-parasitoid interactions in Hymenoptera. *Annual Review of Entomology* 43: 129–51.

———. 2000. Phylogeny of microgastroid braconid wasps, and what it tells us about polydnavirus evolution. Pp. 3–7. *In* AUSTIN, A. D., and M. DOWTON (eds.), *Hymenoptera: Evolution, Biodiversity and Biological Control*. CSIRO; Collingwood, Australia; xi+468 pp.

———. 2002. Estimating the age of the polydnavirus/braconid wasp symbiosis. *Proceedings of the National Academy of Sciences, USA* 99: 7508–13.

———, N. F. JOHNSON, and M. R. HAMERSKI. 1989. Identity and phylogenetic significance of the metapostnotum in nonaculeate Hymenoptera. *Annals of the Entomological Society of America* 82: 663–73.

WHITING, M. F. 1994. Cladistic analysis of the alderflies of America north of Mexico (Megaloptera: Sialidae). *Systematic Entomology* 19: 77–91.

———. 1998. Phylogenetic position of the Strepsiptera: Review of molecular and morphological evidence. *International Journal of Insect Morphology and Embryology* 27: 53–60.

———. 2001. Mecoptera is paraphyletic: Multiple genes and a phylogeny for Mecoptera and Siphonaptera. *Zoological Scripta* 31: 93–104.

———. 2002. Phylogeny of the holometabolous insect orders: Molecular evidence. *Zoological Scripta* 31: 3–15.

———, and W. C. WHEELER. 1994. Insect homeotic transformation. *Nature* 368: 696.

———, J. C. [sic: M.] CARPENTER, Q. D. WHEELER, and W. C. WHEELER. 1997. The Strepsiptera problem: Phylogeny of the holometabolous insect orders inferred from 18S and 28S ribosomal sequences and morphology. *Systematic Biology* 46: 1–68.

———, S. BRADLER, and T. MAXWELL. 2003. Loss and recovery of wings in stick insects. *Nature* 421: 264–7.

WHITINGTON, P. M., T. MEIER, and P. KING. 1991. Segmentation, neurogenesis and formation of early axonal pathways in the centipede, *Ethmostigmus rubripes* (Brandt). *Roux's Archives of Developmental Biology* 199: 349–63.

WHITTLAKE, E. B. 1981. Fossil plant galls. Pp. 729–31. *In* KAISER, H. E. (ed.) *Neoplasms – Comparative Pathology of Growth in Animals, Plants, and Man*. Williams and Wilkins; Baltimore, Mdarylan; xxxii+908 pp.

WICHARD, W. 1997. Schlammfliegen aus Baltischen Bernstein (Megaloptera, Sialidae). *Mitteilungen aus dem Geologisch-Paläontologischen Institut der Universität Hamburg* 80: 197–211.

———, W. ARENS, and G. EISENBEIS. 2002. *Biological Atlas of Aquatic Insects*. Apollo Books; Stenstrup, Denmark; 339 pp.

WICKLER, W. 1968. *Mimicry in Plants and Animals*. McGraw-Hill; New York, New York;. 153 pp.

WIEBES, J. T. 1979. Co-evolution of figs and their insect pollinators. *Annual Review of Entomology* 10: 1–12.

WIEGERT, R. G., and F. C. EVANS. 1967. Investigations of secondary productivity in grasslands. Pp. 499–518. *In* PETRUSEWICZ, K. (ed.), *Secondary Productivity of Terrestrial Ecosystems: Principles and Methods*. Wydawnictwo Naukowe; Warsaw, Poland; vi+879 pp. [2 volumes]

WIEGMANN, B. M., C. MITTER, and F. C. THOMPSON. 1993. Evolutionary origin of the Cyclorrhapha (Diptera): Tests of alternative morphological hypotheses. *Cladistics* 9: 41–81.

———, C. MITTER, J. C. REGIER, T. P. FRIEDLANDER, D. M. WAGNER, and E. S. NIELSEN. 2000a. Nuclear genes resolve Mesozoic-aged divergences in the insect order Lepidoptera. *Molecular Phylogenetics and Evolution* 15: 242–59.

———, S.-C. TSAUR, D. K. YEATES, and B. K. CASSEL. 2000b. Monophyly and relationships of the Tabanomorpha (Diptera: Brachycera) based on 28S ribosomal gene sequences. *Annals of the Entomological Society of America* 93: 1031–8.

———, J. C. REGIER, and C. MITTER. 2002. Combined molecular and morphological evidence on the phylogeny of the earliest lepidopteran lineages. *Zoologica Scripta* 31: 67–81.

———, D. K. YEATES, J. L. THORNE, and H. KISHINO. 2003. Time flies, a new molecular time-scale for brachyceran fly evolution without a clock. *Systematic Biology* 52: 745–56.

WIELAND, G. R. 1925. Rhaetic crane flies from South America. *American Journal of Science* 9: 21–8.

———. 1926. South American fossil insect discovery. *American Journal of Science* 12: 130–5.

WIER, A., M. DOLAN, D. GRIMALDI, R. GUERRERO, J. WAGENSBERG, and L. MARGULIS. 2002. Spirochete and protist symbionts of a termite (*Mastotermes electrodominicus*) in Miocene amber. *Proceedings of the National Academy of Sciences, USA* 99: 1410–13.

WIGGINS, G. B. 1977. *Larvae of the North American Caddisfly Genera (Trichoptera)*. University of Toronto Press; Toronto, Canada; 401 pp.

———. 1996. *Larvae of the North American Caddisfly Genera (Trichoptera)* [2nd Edition]. University of Toronto Press; Toronto, Canada; xiii+457 pp.

WIGHTON, D. C. 1982. Middle Paleocene insect fossils from south-central Alberta. Pp. 577–8. *In* MAMET, B., and M. J. COPELAND (eds.), *Proceedings of the 3rd North American Paleontological Convention* [Volume 2]. Universite de Montreal; Ottawa, Canada; xxii+599 pp.

WIGNALL, P. B., and A. HALLAM. 1992. Anoxia as a cause of the Permian/Triassic mass extinction: Facies evidence from northern Italy and the western United States. *Palaeogeography, Palaeoclimatology, Palaeoecology* 93: 21–46.

———, and A. HALLAM. 1993. Griesbachian (earliest Triassic) palaeoenvironmental changes in Salt Range, Pakistan and southeast China and their bearing on the Permo-Triassic mass extinction. *Palaeogeography, Palaeoclimatology, Palaeoecology* 102: 215–37.

———, and A. HALLAM. 1996. Facies change and the end-Permian mass extinction in S.E. Sichuan, China. *Palaios* 11: 587–96.

———, and R. J. TWITCHETT. 1996. Oceanic anoxia and the End Permian mass extinction. *Science* 272: 1155–8.

———, H. KOZUR, and A. HALLAM. 1996. On the timing of palaeoenvironmental changes at the Permo-Triassic (P/Tr) boundary using conodont biostratigraphy. *Historical Biology* 12: 39–62.

WIKSTRÖM, N., V. SAVOLAINEN, and M. W. CHASE. 2001. Evolution of the angiosperms: Calibrating the family tree. *Proceedings of the Royal Society of London, Series B, Biological Sciences* 268: 2211–20.

WILBY, P. R., D. E. G. BRIGGS, and B. RIOU. 1996. Mineralization of soft bodied invertebrates in a Jurassic metalliferous deposit. *Geology* 24: 847–50.

WILDE, V., and H. FRANKENHÄUSER. 1998. The Middle Eocene plant taphocoenosis from Eckfeld (Eifel, Germany). *Review of Paleobotany and Palynology* 101: 7–28.

———, H. Frankenhäuser, and H. Lutz. 1993. Algenreste aus den mitteleozänen Sedimenten des Eckfelder Maares bei Manderscheid in der Eifel. *Mainzer Naturwissenschaftliches Archiv* 31: 127–48.

Wiley, E. O. 1978. The evolutionary species concept reconsidered. *Systematic Zoology* 27: 17–26.

Wilf, P., C. C. Labandeira, W. J. Kress, C. L. Staines, D. M. Windsor *et al.* 2000. Timing the radiations of leaf beetles: Hispines on gingers from Latest Cretaceous to Recent. *Science* 289: 291–4.

———, C. C. Labandeira, K. R. Johnson, P. D. Coley, and A. D. Cutter. 2001. Insect herbivory, plant defense, and early Cenozoic climate change. *Proceedings of the National Academy of Sciences, USA* 98: 6221–6.

Wilford, G. E., and P. J. Brown. 1994. Maps of late Mesozoic-Cenozoic Gondwana break-up: Some paleogeographical implications. Pp. 5–13. *In* Hill, R. S. (ed.), *History of the Australian Vegetation: Cretaceous to Recent.* Cambrige University Press; Cambridge, UK; x+433 pp.

Will, K. W. 1995. Plecopteran surface-skimming and insect flight evolution. *Science* 270: 1684–5.

———, A. B. Attygalle, and K. Herath. 2000. New defensive chemical data for ground beetles (Coleoptera: Carabidae): Interpretations in a phylogenetic framework. *Biological Journal of the Linnean Society* 71: 459–81.

Wille, A. 1959. A new stingless bee (Meliponini) from the amber of Chiapas, Mexico. *Journal of Paleontology* 33: 849–52.

———, and L. Chandler. 1964. A new stingless bee from the Tertiary amber of the Dominican Republic (Hymenoptera; Meliponini). *Revista de Biología Tropical* 12: 187–95.

Williams, G. A. 2001. Thrips (Thysanoptera) pollination in Australian subtropical rainforests, with particular reference to pollination of *Wilkiea huegeliana* (Monimiaceae). *Journal of Natural History* 35: 1–21.

Williams, L. H. 1972. Anobiid beetle eggs consumed by a psocid (Psoc., Liposcelidae). *Annals of the Entomological Society of America* 65: 533–6.

Williams, M. A. J., ed. 1994. *Plant Galls: Organisms, Interactions, Populations.* Clarendon Press; Oxford, UK; xiv+487 pp.

Williams, N. H. 1978. A preliminary bibliography on euglossine bees and their relationships with orchids and other plants. *Selbyana* 2: 345–55.

Willmann, R. 1978. Mecoptera (Insecta, Holometabola). *Fossilium Catalogus, Animalia* 124: 1–139.

———. 1981. Das Exoskelett der männlichen Genitalien der Mecoptera (Insecta). 1. Morphologie. II. Die phylogenetischen Beziehungen der Schnabelfliegen-Familien. *Zeitschrift für Zoologische Systematik und Evolutionsforschung* 19: 96–150.

———. 1983. Widersprüchliche Rekonstruktionen der Phylogenese am Beispel der Ordnung Mecoptera (Schabelfliegen; Insecta, Holometabola). *Paläontologische Zeitschrift* 57: 285–308.

———. 1987. The phylogenetic system of the Mecoptera. *Systematic Entomology* 12: 519–24.

———. 1989. Evolution und phylogenetisches System der Mecoptera (Insecta, Holometabola). *Abhandlungen der Senckenbergischen Naturforschenden Gesellschaft* 544: 1–153.

———. 1990a. Die Bedeutung paläontologischer Daten für die zoologische Systematik. *Verhandlungen der Deutschen Zoologischen Gesellschaft* 83: 277–89.

———. 1990b. The phylogenetic position of the Rhachiberothinae and the basal sister-group relationships within Mantispidae (Neuroptera). *Systematic Entomology* 15: 253–65.

———. 1990c. Insekten aus der Fur-Formation von Dänemark (Moler, ob. Paleozän/ unt. Eozän?), 1, Allgemeines. *Meyniana* 42: 1–14.

———. 1994a. Raphidiodea aus dem Lias und die Phylogenie der Kamelhalsfliegen (Insecta: Holometabola). *Paläontologische Zeitschrift* 69: 167–97.

———. 1994b. Die phylogenetischen Position ursprünglicher Mantispidae (Insecta, Planipennia) aus dem Mesozoikum und Alt-Tertiär. *Verhandlungen des Naturwissenschaftlichen Vereins in Hamburg* 34: 177–203.

———. 1997. Advances and problems in insect phylogeny. Pp. 269–79. *In* Fortey, R. A., and R. H. Thomas (eds.), *Arthropod Relationships.* Chapman and Hall; London, UK; xii+383 pp.

———. 1999. The Upper Carboniferous *Lithoneura lameerei* (Insecta, Ephemeroptera?). *Paläontologische Zeitschrift* 73: 289–302.

———. 2003. Die phylogenetischen Beziehungen der Insecta: Offene Fragen und Probleme. *Verhandlungen Westdeutscher Entomologentag* 2001: 1–64.

———, and S. J. Brooks. 1991. Insekten aus der Fur-Formation von Dänemark (Moler, ob. Paleozän / unt. Eozän?). 6. Chrysopidae (Neuroptera). *Meyniana* 43: 125–35.

Willmer, P. 1990. *Invertebrate Relationships: Patterns in Animal Evolution.* Cambridge University Press; Cambridge, UK; xiv+400 pp.

Wills, M. A., D. E. G. Briggs, R. A. Fortey, and M. Wilkinson. 1995. The significance of fossils in understanding arthropod evolution. *Verhandlungen der Deutschen Zoologischen Gesellschaft* 88: 203–15.

———, D. E. G. Briggs, R. A. Fortey, M. Wilkinson, and P. H. A. Sneath. 1998. An arthropod phylogeny based on fossil and recent taxa. Pp. 33–105. *In* Edgecombe, G. D. (ed.), *Arthropod Fossils and Phylogeny.* Columbia University Press; New York, New York; 347 pp.

Wilson, E. O. 1961. The nature of the taxon cycle in the Melanesian ant fauna. *American Naturalist* 95: 169–93.

———. 1971. *The Insect Societies.* Harvard University Press; Cambridge, Massachusetts; x+548 pp.

———. 1985a. Invasion and extinction in the West Indian ant fauna: Evidence from the Dominican amber. *Science* 229: 265–7.

———. 1985b. Ants from the Cretaceous and Eocene amber of North America. *Psyche* 92: 205–16.

———. 1987. The earliest known ants: An analysis of the Cretaceous species and an inference concerning their social organization. *Paleobiology* 13: 44–53.

———. 1992. *The Diversity of Life.* Norton; New York, New York; 424 pp.

———. 2003. Pheidole *in the New World: A Dominant, Hyperdiverse Ant Genus.* Harvard University Press; Cambridge, Massachusetts; [5]+794 pp.

———, and D. J. Farish. 1973. Predatory behaviour in the ant-like wasp *Methoca stygia* (Say) (Hymenoptera: Tiphiidae). *Animal Behaviour* 21: 292–5.

———, and R. W. Taylor. 1964. A fossil ant colony: New evidence of social antiquity. *Psyche* 71: 93–103.

———, W. L. Brown, Jr., and F. M. Carpenter. 1967. The first Mesozoic ants, with the description of a new subfamily. *Psyche* 74: 1–19.

Wilson, H. M., and D. M. Martill. 2001. A new japygid dipluran from the Lower Cretaceous of Brazil. *Palaeontology* 44: 1025–31.

Wilson, M. V. H. 1977. New records of insect families from the freshwater Middle Eocene of British Columbia. *Canadian Journal of Earth Sciences* 14: 1139–55.

———. 1978. Paleogene insect faunas of western North America. *Quaestiones Entomologia* 14: 14–34.

Wimberger, P. H., and A. De Queiroz. 1996. Comparing behavioral and morphological characters as indicators of phylogeny. Pp. 206–33. *In* Martins, E. P. (ed.), *Phylogenies and the Comparative Method in Animal Behavior.* Oxford University Press; Oxford, UK; x+415 pp.

Wing, S. L. 2000. Evolution and expansion of flowering plants. Pp. 209–31. *In* Gastaldo, R. A., and W. A. DiMichele (eds.), *Phanerozoic Terrestrial Ecosystems.* The Paleontological Society; Pittsburgh, Pennsylvania; 308 pp.

Wingstrand, K. G. 1972. Comparative spermatology of a pentastomid, *Raillietiella hemidactyli*, and a branchiuran crustacean, *Argulus foliaceus*, with a discussion of pentastomid relationships. *Biologiske Skrifter, Danske Videnskabernes Selskabs* 19: 1–72.

———. 1973. The spermatozoa of the thysanuran insects *Petrobius brevistylis* Carp. and *Lepisma saccharina* L. *Acta Zoologica* 54: 31–52.

Winston, M. L. 1987. *The Biology of the Honey Bee.* Harvard University Press; Cambridge, Massachusetts; viii+281 pp.

———, and C. D. Michener. 1977. Dual origin of highly social behavior among bees. *Proceedings of the National Academy of Sciences, USA* 74: 1135–7.

Withycombe, C. L. 1924. Some aspects of the biology and morphology of the Neuroptera: with special reference to the immature stages and their possible phylogenetic significance. *Transactions of the Entomological Society of London* 1924: 303–411.

Wlodarczyk, J. 1963. Psocoptera of some bird nests. *Fragmenta Faunistica* 10: 361–6 [in Polish].

Wolcott, C. D. 1912. Middle Cambrian Branchiopoda, Malacostraca, Trilobita, and Merostomata: Cambrian geology and paleontology, part 2. *Smithsonian Miscellaneous Collections* 57: 109–44.

Wolfe, J. A., and W. Wehr. 1987. Middle Eocene dicotyledonous plants from Republic, northeastern Washington. *United States Geological Survey* 1597: iv+1–25, 16 pls.

Wood, D. M. 1987. Tachinidae. Pp. 1193–1269. *In* McAlpine, J. F., B. V. Peterson, G. E. Shewell, H. J. Teskey, J. R. Vockeroth, and D. M. Wood (eds.), *Manual of Nearctic Diptera, Volume 2*. Agriculture Canada Monograph 28; Ottawa, Canada; vi+675–1332 pp.

———, and A. Borkent. 1989. Phylogeny and classification of the Nematocera. Pp. 1333–70. *In* McAlpine, J. F., and D. M. Wood (eds.), *Manual of Nearctic Diptera, Volume 3*. Agriculture Canada Monograph 32; Ottawa, Canada; vi+1333–1581 pp.

Wood, K. V. 1995. The chemical mechanism and evolutionary development of beetle bioluminescence. *Photochemistry and Photobiology* 62: 662–73.

Wood, S. P., A. L. Panchen, and T. R. Smithson. 1985. A terrestrial fauna from the Scottish Lower Carboniferous. *Nature* 314: 355–6.

Wood, T. G. 1978. Food and feeding habits of termites. Pp. 55–80. *In* Brian, M. V. (ed.) *Production Ecology of Ants and Termites*. Cambridge University Press; Cambridge, UK; xvi+409 pp.

———, and W. A. Sands. 1978. The role of termites in ecosystems. Pp. 245–92. *In* Brian, M. V. (ed.) *Production Ecology of Ants and Termites*. Cambridge University Press; Cambridge, UK; xvi+409 pp.

Wood, T. K. 1980. Intraspecific divergence in *Enchenopa binotata* Say (Homoptera: Membracidae) effected by host plant adaptation. *Evolution* 34: 147–60.

———, and S. I. Guttman. 1982. Ecological and behavioural basis for reproductive isolation in the sympatric *Enchenopa binotata* complex (Homoptera: Membracidae). *Evolution* 36: 233–42.

Woodley, N. E. 1989. Phylogeny and classification of the "orthorrhaphous" Brachycera. Pp. 1371–96. *In* McAlpine, J. F., and D. M. Wood (eds.), *Manual of Nearctic Diptera, Volume 3*. Agriculture Canada Monograph 32; Ottawa, Canada; vi+1333–1581 pp.

———, and D. J. Hilburn. 1994. The Diptera of Bermuda. *Contributions of the American Entomological Institute* 28: 1–64.

Woodward, H. 1865. A monograph of the British fossil Crustacea. *Monograph of the Palaeontological Society* 19: 1–43.

Woodworth, C. W. 1906. The wing veins of insects. *University of California Publications, Technical Bulletins, Entomology* 1: 1–152.

———. 1907. The classification of insects. *Entomological News* 18: 243–7.

———. 1930. The arrangement of the major orders of insects. *Psyche* 37: 157–62.

Woolley, T. A. 1988. *Acarology: Mites and Human Welfare*. Wiley and Sons; New York, New York; xxii+484 pp.

Wootton, R. J. 1979. Function, homology, and terminology in insect wings. *Systematic Entomology* 4: 81–93.

———. 1981. Palaeozoic insects. *Annual Review of Entomology* 26: 319–44.

———. 1988. The historical ecology of aquatic insects: An overview. *Palaeogeography, Palaeoclimatology, Palaeoecology* 62: 477–92.

———. 1990. Major insect radiations. Pp. 187–208. *In* Taylor, P. D., and G. P. Larwood (eds.), *Major Evolutionary Radiations*. Clarendon Press; Oxford, UK; xi+437 pp.

———. 1992. Functional morphology of insect wings. *Annual Review of Entomology* 37: 113–40.

———, and J. Kukalová-Peck. 2000. Flight adaptations in Palaeozoic Palaeoptera (Insecta). *Biological Reviews* 75: 129–67.

Worssam, B. C. 1978. *The Stratigraphy of the Weald Clay*. Institute of Geological Sciences Report 78/11, British Geological Survey; London, UK; 23 pp.

Wray, G. A., J. S. Levinton, and L. H. Shapiro. 1996. Molecular evidence for deep pre-Cambrian divergences among metazoan phyla. *Science* 274: 568–73.

Wu, C.-I., and M. F. Palopoli. 1994. Genetics of postmating reproductive isolation in animals. *Annual Review of Genetics* 28: 283–308.

———, D. E. Perez, A. W. Davis, N. A. Johnson, E. L. Cabot *et al.* 1993. Molecular genetic studies of postmating reproductive isolation in *Drosophila*. Pp. 199–212. *In* Takahata, N., and A. G. Clark (eds.), *Mechanisms of Molecular Evolution: Introduction to Molecular Paleopopulation Biology*. Sinauer; Sunderland, Massachusetts; x+250 pp.

Wu, Z. 1983. On the significance of Pacific intercontinental discontinuity. *Annals of the Missouri Botanical Garden* 70: 577–90.

Wunderlich, J. 1986. *Spinnenfauna Gestern und Heute: Fossile Spinnen in Bernstein und ihre heute lebenden Verwandten*. Eric Bauer; Wiesbaden, Germany; 283 pp.

———. 1988. Die fossilen Spinnen im dominikanischen Bernstein. *Beiträge zur Araneologie* 2: 1–378.

———. 2000. Zwei neue Arten der Familie Falltürspinnen (Araneae: Ctenizidae) aus dem Baltischen Bernstein. *Entomologische Zeitschrift* 110: 345–8.

Wygodzinsky, P. 1961. On a surviving representative of the Lepidotrichidae (Thysanura). *Annals of the Entomological Society of America* 54: 621–7.

———. 1966. A monograph of the Emesinae (Reduviidae, Hemiptera). *Bulletin of the American Museum of Natural History* 133: 1–614.

———, and K. Schmidt. 1991. Revision of the New World Enicocephalomorpha (Heteroptera). *Bulletin of the American Museum of Natural History* 200: 1–265.

Yager, D. D. 1992. Ontogeny and phylogeny of the cyclopean mantis ear. *Journal of the Acoustical Society of America* 92: 2421.

———. 1999a. Structure, development, and evolution of insect auditory systems. *Microscopy Research and Techniques* 47: 380–400.

———. 1999b. Hearing. Pp. 93–113. *In* Prete, F. R., H. Wells, P. H. Wells, and L. E. Hurd (eds.), *The Praying Mantids*. Johns Hopkins University Press; Baltimore, Maryland; xiv+362 pp.

———, and R. R. Hoy. 1986. The cyclopean ear: A new sense for the praying mantis. *Science* 231: 727–9.

———, and R. R. Hoy. 1987. The midline metathoracic ear of the praying mantis, *Mantis religiosa*. *Cell and Tissue Research* 250: 531–41.

———, M. L. May, and M. B. Fenton. 1990. Ultrasound-triggered, flight-gated evasive maneuvers in the praying mantis *Parasphendale agrionina*. I. Free flight. *Journal of Experimental Biology* 152: 17–39.

Yamada, Y. 1987. Characteristics of the oviposition of a parasitoid, *Chrysis shanghaiensis* (Hymenoptera: Chrysididae). *Applied Entomology and Zoology* 22: 456–64.

Yamane, S. 1993. Evolution of nest architecture and societies in stenogastrine wasps (Hymenoptera: Vespidae). Pp. 267–327. *In* Inoue, T., and S. Yamane (eds.), *Evolution of Insect Societies: Comparative Sociology of Bees, Wasps and Ants*. Hakuhin-sha Publishing; Tokyo, Japan; vii+532 pp.

Yamin, M. A. 1979. Flagellates of the orders Trichomonadida Kirby, Oxymonadida Grassé, and Hypermastigida Grassi & Foà reported from lower termites (Isoptera families Mastotermitidae, Kalotermitidae, Hodotermitidae, Termopsidae, Rhinotermitidae, and Serritermitidae) and from the wood-feeding roach *Cryptocercus* (Dictyoptera: Cryptocercidae). *Sociobiology* 4: 1–120.

Yeates, D. K. 1994. The cladistics and classification of the Bombyliidae (Diptera: Asiloidea). *Bulletin of the American Museum of Natural History* 219: 1–191.

———. 2001. Relationships of extant lower Brachycerea (Diptera): A quantitative synthesis of morphological characters. *Zoologica Scripta* 31: 105–21.

———, and M. E. Irwin. 1996. Apioceridae (Insecta: Diptera): Cladistic reappraisal and biogeography. *Zoological Journal of the Linnean Society* 116: 247–301.

———, and B. M. Wiegmann. 1999. Congruence and controversy: Toward a higher-level phylogeny of Diptera. *Annual Review of Entomology* 44: 397–428.

———, M. E. Irwin, and B. M. Wiegmann. 2003. Ocoidae, a new family of asiloid flies (Diptera: Brachycera: Asiloidea), based on *Ocoa chilensis* gen. and sp. n. from Chile, South America. *Systematic Entomology* 28: 417–31.

Yeo-Moon, S., and W. Kim. 1996. Phylogenetic position of the Tardigrada based on the 18S ribosomal RNA gene sequences. *Zoological Journal of the Linnean Society* 116: 61–9.

Yin, W.-Y. 1983. Grouping the known genera of Protura under eight families with keys for determination. *Contributions of the Shanghai Institute of Entomology* 3: 151–63. [In Chinese]

———. 1984. A new idea on phylogeny of Protura with approach to its origin and systematic position. *Scientia Sinica, Series B (Chemical, Biological, Agricultural, Medical, and Earth Sciences)* 27: 149–60.

Ying, T.-S. 1983. The floristic relationships of the temperate forest regions of China and the United States. *Annals of the Missouri Botanical Garden* 70: 597–604.

Yokoyama, A., and M. Tsuneyoshi. 1958. Discovery of a hymenopterous ectoparasite of *Oligotoma japonica* Okajima (Embioptera). *Kontyû* 26: 25–8.

Yoshikawa, K., R. Ohgushi, and S. F. Sakagami. 1969. Preliminary report on entomology of the Osaka City University 5th Scientific Expedition to Southeast Asia 1966. With descriptions of two new genera of stenogasterine [sic] wasps by J. van der Vecht. *Nature Life, Southeast Asia* 6: 153–82.

Yoshimoto, C. M. 1975. Cretaceous chalcidoid fossils from Canadian amber. *Canadian Entomologist* 107: 499–528.

Yoshizawa, K. 2002. Phylogeny and higher classification of suborder Psocomorpha (Insecta: Psocodea: 'Psocoptera'). *Zoological Journal of the Linnean Society* 136: 371–400.

———, and K. P. Johnson. 2003a. Phylogenetic position of Phthiraptera (Insecta: Paraneoptera) and elevated rate of evolution of mitochondrial 12S and 16S rDNA. *Molecular Phylogenetics and Evolution* 29: 102–14.

———, and K. P. Johnson. 2003b. Phylogenetic position of Phthiraptera (Insecta: Paraneoptera) and elevated rate of evolution in mitochondrial 12S and 16S rDNA. *Molecular Phylogenetics and Evolution* 29: 102–14.

———, and T. Saigusa. 2001. Phylogenetic analysis of paraneopteran orders (Insecta: Neoptera) based on forewing base structure, with comments on the monophyly of Auchenorrhyncha (Hemiptera). *Systematic Entomology* 26: 1–13.

Zajic, J. 2000. Vertebrate zonation of the nonmarine Upper Carboniferous – Lower Permian basins of the Czech Republic. *Courier Forschungsinstitut Senckenberg* 223: 563–75.

Zalessky, M. D. 1932. Observations sur les nouveaux insectes permiens de l'Europe orientale. *Bulletin de la Société Geologique de France* 2: 183–210.

Zanis, M. J., D. E. Soltis, P. S. Soltis, S. Mathews, and M. J. Donoghue. 2002. The root of angiosperms revisited. *Proceedings of the National Academy of Sciences, USA* 99: 6848–53.

Zatwarnicki, T. 1996. A new reconstruction of the origin of the eremoneuran hypopygium and its implications for classification (Insecta: Diptera). *Genus* 3: 103–75.

Zeigler, P. 1969. *The Black Death*. Collins; London, UK; 319 pp.

Zessin, W. 1983. Revision der mesozoischen Familie Locustopsidae unter Berücksichtigung neuer Funde (Orthopteroida, Caelifera). *Deutsche Entomologische Zeitschrift* 30: 173–237.

———. 1985. Neue oberliassische Apocrita und die Phylogenie der Hymenoptera (Insecta, Hymenoptera). *Deutsche Entomologische Zeitschrift* 32: 129–42.

Zeuner, F. E. 1939. *Fossil Orthoptera Ensifera*. British Museum; London, UK; xiii+321 pp., 80 pls. [plates bound in separate volume]

———. 1955. A fossil blattid from the Permian of Rhodesia. *Annals and Magazine of Natural History* 8: 685–8.

———. 1959. Jurassic beetles from Grahamland, Antarctica. *Palaeontology* 1: 407–9.

———. 1961. A Triassic insect fauna from the Molteno Beds of South Africa. *Verhandlungen, 11th International Kongress Entomologie, Wien* 1960: 304–6.

———, and F. J. Manning. 1976. A monograph on fossil bees (Hymenoptera: Apoidea). *Bulletin of the British Museum of Natural History (Geology)* 27: 149–268.

Zhang, H.-C., and J.-F. Zhang. 2000. A new genus and two new species of Hymenoptera (Insecta) from the Upper Jurassic Yixian Formation of Beipiao, western Liaoning. *Acta Micropalaeontologica Sinica* 17: 286–90.

———, A. P. Rasnitsyn, and J.-F. Zhang. 2002a. Two ephialtitid wasps (Insecta, Hymenoptera, Ephialtitidae) from the Yixian Formation of western Liaoning, China. *Cretaceous Research* 23: 401–7.

———, A. P. Rasnitsyn, and J.-F. Zhang. 2002b. Pelecinid wasps (Insecta: Hymenoptera: Proctotrupoidea) from the Yixian Formation of western Liaoning, China. *Cretaceous Research* 23: 87–98.

———, A. P. Rasnitsyn, and J.-F. Zhang. 2002c. The oldest known scoliid wasps (Insecta, Hymenoptera, Scoliidae) from the Jehol biota of western Liaoning, China. *Cretaceous Research* 23: 77–86.

Zhang, J.-F. 1985. New data on the Mesozoic fossil insects from Laiyang in Shandong. *Geology of Shandong* 1: 23–39. [In Chinese, with English summary]

———. 1989. *Fossil Insects from Shanwang, Shandong, China*. Shandong Science and Technology Publishing House; Jinan, China; 459 pp., 92 pls. [In Chinese]

———. 1993. New Miocene species of Bibionidae (Insecta: Diptera) with discussion on taxonomic position of *Clothonopsis miocenica*. *Acta Palaeontologica Sinica* 32: 141–50. [In Chinese, with English summary]

———, B. Sun, and X.-Y. Zhang. 1994. *Miocene Insects and Spiders from Shanwang, Shandong*. Science Press; Beijing, China; v+298 pp. [In Chinese, with English summary]

Zhelochovtzev, A. N., and A. P. Rasnitsyn. 1972. On some Tertiary sawflies (Hymenoptera, Symphyta) from Colorado. *Psyche* 79: 315–27.

Zherikhin, V. V. 1978. Development and changes of the Cretaceous and Cenozoic faunal assemblages (Tracheata and Chelicerata). *Trudy Paleontologischeskogo Instituta Akademii Nauk, SSSR* 165: 1–198. [In Russian]

———. 2003. Insect trace fossils, their diversity, classification and scientific importance. *Acta Zoologica Cracoviensia* 46 (supplement): 59–66.

———, and K. Yu. Eskov. 1999. Mesozoic and Lower Tertiary resins in former USSR. *Estudios Museo Ciências Naturales de Alava* 14 (Num. Espec. 2): 119–31.

———, and V. G. Gratshev. 1993. Obrieniidae, fam. nov., the oldest Mesozoic weevils (Coleoptera, Curculionidae). *Paleontological Journal* 27: 50–69.

———, and A. J. Ross. 2000. A review of the history, geology and age of Burmese amber (Burmite). *Bulletin of the Natural History Museum, London (Geology)* 56: 3–10.

———, and I. D. Sukatsheva. 1973. On the Cretaceous insect-bearing "ambers" (retinites) from North Siberia. Pp. 3–48. *In* Narchuk, E. P. (ed.), *Problems in Insect Palaeontology: XXIV Annual Lectures in Memory of N. A. Kholodkovskogo, 1–2 April 1971*. Nauka Press; Leningrad, Union of Soviet Socialist Republics.

———, M. B. Mostovski, P. Vršanský, V. A. Blagoderov, and E. D. Lukashevich. 1999. The unique Lower Cretaceous locality Baissa and other contemporaneous fossil insect sites in North and West Transbaikalia. Pp. 185–91. *In* Vršanský, P. (ed.), *Proceedings of the First Palaeoentomological Conference, Moscow, 1998*. AMBA Projects; Bratislava, Slovakia; 199+[2] pp.

Ziegler, P. 1969. *The Black Death*. Collins; London, UK; 319 pp.

Zimmerman, P. R., J. P. Greenberg, S. O. Wandiga, and P. J. Crutzen. 1982. Termites: A potentially large source of atmospheric methane, carbon dioxide, and molecular hydrogen. *Science* 218: 563–5.

Zompro, O. 2001. The Phasmatodea and *Raptophasma* n. gen., Orthoptera incertae sedis, in Baltic amber (Insecta: Orthoptera). *Mitteilungen aus dem Geologisch-Paläontologischen Institut der Universität Hamburg* 85: 229–61.

———, J. Adis, and W. Weitschat. 2002. A review of the order Mantophasmatodea (Insecta). *Zoologischer Anzeiger* 241: 269–79.

———, J. Adis, P. E. Bragg, P. Naskrecki, K. Meakin, *et al.* 2003. A new genus and species of Mantophasmatidae (Insecta: Mantophasmatodea) from the Brandberg Massif, Namibia, with notes on behaviour. *Cimbebasia* 19: 13–24.

Zrzavý, J. 2001. Ecdysozoa versus Articulata: Clades, artifacts, prejudices. *Journal of Zoological Systematics and Evolutionary Research* 39: 159–63.

———. 2003. Gastrotricha and metazoan phylogeny. *Zoologica Scripta* 32: 61–82.

———, and P. Štys. 1997. The basic body plan of arthropods: Insights from evolutionary morphology and developmental biology. *Journal of Evolutionary Biology* 10: 353–67.

———, S. Mihulka, P. Kepka, A. Bezdek, and D. Tietz. 1998a. Phylogeny of the Metazoa based on morphological and 18S ribosomal DNA evidence. *Cladistics* 14: 249–85.

———, V. Hypša, and M. Vlášková. 1998b. Arthropod phylogeny: Taxonomic congruence, total evidence and conditional combination approaches to morphological and molecular data sets. Pp. 97–107. *In* Fortey, R. A., and R. H. Thomas (eds.), *Arthropod Relationships*. Chapman and Hall; London, UK; xii+383 pp.

———, V. Hypša, and D. F. Tietz. 2001. Myzostomida are not annelids: Molecular and morphological support for a clade of animals with anterior sperm flagella. *Cladistics* 17: 170–98.

Zuidema, H. P. 1950. A new fossil insect and plant locality in Montana. *Papers of the Michigan Academy of Science, Arts, and Letters* 34: 119–123.

zur Strassen, R. 1973. Insektenfossilen aus der unteren Kreide – 5. Fossile Franseflügler aus mesozoischem Bernstein des Libanon (Insecta: Thysanoptera). *Stuttgarter Beiträge zur Naturkunde, Serie A, Biologie* 256: 1–51.

Zwick, P. 1967. Beschreibung der aquatischen Larve von *Neurorthus* [sic] *fallax* (Rambur) und Errichtung der neuen Planipennierfamilie Neurorthidae [sic] fam. nov. *Gewässer und Abwässer* 44/45: 65–86.

———. 1973. Insecta Plecoptera: Phylogenetisches System und Katalog. *Das Tierreich* 94: xxxii+1–465.

———. 1980. Plecoptera (Steinfliegen). *Handbuch der Zoologie* 4: 1–115.

———. 2000. Phylogenetic system and zoogeography of the Plecoptera. *Annual Review of Entomology* 45: 709–46.

INDEX

B

D

E

N

O

Q

R

S

T